Laboratory Diagnosis of Infectious Diseases
Principles and Practice

Senior Editors
A. Balows W.J. Hausler, Jr. E.H. Lennette

Laboratory Diagnosis of Infectious Diseases

Principles and Practice

VOLUME I

Bacterial, Mycotic, and Parasitic Diseases

Editors
A. Balows W.J. Hausler, Jr. M. Ohashi A. Turano

With 218 Illustrations, 16 in Full Color

Springer-Verlag New York Berlin Heidelberg
London Paris Tokyo

A. BALOWS, Center for Infectious Diseases, Centers for Disease Control, Atlanta, GA 30333, USA

W. J. HAUSLER, JR., Hygienic Laboratory, Oakdale Hall, University of Iowa, Iowa City, IA 52242, USA

E. H. LENNETTE, Viral and Rickettsial Disease Laboratory (Emeritus Director), California Department of Health Services, and California Public Health Foundation, Berkeley, CA 94704, USA; and Institut Virion, CH-8803 Ruschlikon/Zurich, Switzerland

M. OHASHI, Tokyo Metropolitan Research Laboratory of Public Health, Hyakunin-cho, Shinjuku-ku, Tokyo 160, Japan

A. TURANO, Istituto di Microbiologia dell 'Universitá, Spedali Civili, Brescia, Italy

Library of Congress Cataloging-in-Publication Data
Laboratory diagnosis of infectious diseases: principles and practices.
p. cm.
Includes bibliographies and index.
Contents: v. 1. Bacterial, mycotic, and parasitic diseases / A. Balows . . . [et al.], editors—v. 2. Viral, rickettsial, and chlamydial diseases / E. H. Lennette, P. Halonen, F. A. Murphy, editors.
1. Diagnostic microbiology—Laboratory manuals. I. Balows. Albert.
[DNLM: 1. Communicable Diseases—diagnosis. 2. Diagnosis, Laboratory. 3. Virus Diseases—diagnosis. WC 25 L1226]
QR67.L328 1988
616.9'0475—dc19
DNLM/DLC
Printed on acid-free paper. 88-6581

Softcover reprint of the hardcover 1st edition 1988

Typeset by Bi-Comp, York, Pennsylvania.

9 8 7 6 5 4 3 2 1

ISBN-13:978-1-4612-8393-5 e-ISBN-13:978-1-4612-3898-0
DOI: 10.1007/978-1-4612-3898-0

Foreword

This two volume work stems from the belief of the Editors that infectious diseases are not only very much with us today but, more importantly, that they will continue to play a significant global role in morbidity and mortality in all people. A continuing need for an informed and knowledgeable community of laboratory scientists is fundamental. Data describing the global impact of infectious diseases are difficult to come by. Fortunately, a recent thoughtful and provocative publication by Bennett et al. (1987) provides us with data derived from several consultants that clearly delineate the impact of infectious diseases on the United States today. In the mid 1980s, almost three-quarters of a billion cases of infectious diseases occurred annually in the United States, resulting in mortality approaching 200,000. These cases represent over two million years lost before age 65 and almost two billion days lost from work, school, or other usual daily activities. It would be very helpful if similar data exist or could be obtained from other developed or developing nations.

Infectious diseases or, more specifically, the agents that cause them usually have little regard for age or sex, but nutritional or immune status, underlying or pre-existing disease, and habitat frequently contribute to the establishment of acute and often chronic infections. Without question, the scientific advances in biology and medicine that have occurred during the past 75 years and the successes that will take place between now and the end of this century will be major forces in keeping the above-cited numbers from increasing. In all likelihood, these numbers should continue a worldwide downward trend.

A master plan to diagnose, treat, control, and prevent infectious diseases has evolved over the past several decades. Ingredients of this master plan include in-depth study and research in all aspects of infectious diseases to completely characterize the etiologic agents and fully describe the natural history of each disease. Inclusive aspects, although the list is by no means complete, are host–parasite interactions, immunology, pathology, microbiology, parasitology, virology, and epidemiology. The most important resource in gaining the necessary knowledge in these areas is well-informed people—particularly those who deal with infectious diseases on a daily basis.

There are several excellent textbooks dealing with medical microbiology, and there are equally well-recognized books devoted to infectious diseases. The Editors of this work, on the other hand, were persuaded that there was a need for a publication that would bring together the most pertinent and relevant information on the principles and practice of the laboratory diagnosis of infectious diseases and include clinical relationships. While this two volume text is directed toward the role of the laboratory in infections—after all, that is where the definitive diagnosis of an infectious disease is most often made—ample consideration is given to the disease, its symptomatology, pathogenesis, epidemiology, treatment, prevention, and control.

This work is designed to keep the clinician and the clinical laboratory scientist informed on nearly all the commonly known infectious diseases and their etiologic agents about which there are useful and reliable data. To achieve this goal, international authorities were enlisted as authors. The international approach using a chapter format designed to give the reader access to both a synopsis and a more detailed discussion on each disease or group of related diseases is unique. Dividing the work into two volumes was a convenient and practical way to make the information readily available in a useful manner, and these volumes therefore constitute the complete work. The Editors welcome comments from the users of this set.

Bennett, J. V., S. D. Holmberg, M. F. Rogers, and S. L. Solomon. 1987. Infections and parasitic diseases in closing the gap: the burden of unnecessary illness, p. 102–104. In: R. W. Amler and H. B. Dull (ed.). Oxford University Press, New York.

The Senior Editors

Albert Balows
William J. Hausler, Jr.
Edwin H. Lennette

The Associate Editors

Pekka Halonen
Frederick A. Murphy
Makoto Ohashi
Adolfo Turano

Foreword

Preface

Much too often, we hear that laboratories do not diagnose disease, they only analyze specimens involved in infectious processes. The truth is that laboratories do analyze specimens involved in infectious processes, but if those in the laboratories do not understand the complex interrelationships of etiologic agent, pathology, epidemiology, treatment, and control, their work is vacuous. *Laboratory Diagnosis of Infectious Diseases: Principles and Practice* was conceptualized by the Senior Editors to draw these interrelationships together for most of the known infectious diseases in an easily understood and appropriately referenced format. It is for this specific purpose that the users of this two volume work will find a summary box at the beginning of each disease addressed, which provides the reader with a comprehensive overview of the disease.

Although infectious diseases of humans are universal problems, their global distribution is not uniform. The Editors and Associate Editors designed this work to be international in scope by enlisting authors from around the world who are known for their knowledge and expertise in one or more infectious diseases.

Laboratory Diagnosis of Infectious Diseases: Principles and Practice has been conveniently divided into two volumes. Volume I addresses all those diseases of bacterial, mycotic, and parasitic etiology. Volume II continues with those diseases of viral, rickettsial, and chlamydial etiology. The two volumes are to be considered complementary to each other; serious efforts have been made by the Editors to avoid duplication and promote cross-referencing between volumes. We hope that you, the user, will find these two volumes to be handy reference guides whether you are a laboratory scientist, clinician, epidemiologist, infectious disease researcher, academician, or student.

The introductory chapters are general in nature and provide background for those that follow. Volume I is arranged alphabetically by the disease. As one might expect, there are exceptions to this format when a group of organisms causes no known single disease or group of diseases or when one organism may cause more than a single disease. In these few instances, we have chosen to assign chapter titles and contents to generally recognized infections. Examples are the following: Anaerobic Bacterial Infections, Glucose Nonfermenting Gram-Negative Bacilli Infections, Gram-Negative Aerobic Bacteria—Miscellaneous Infections. Chapter titles chosen for Volume II necessitate a different approach.

Using specific disease names as chapter titles in Volume II for the viral, rickettsial, and chlamydial diseases would cause endless repetition and continuous cross-referencing. The Editors decided that an alphabetized taxonomic-based order of chapters would be the most useful and beneficial. Within each chapter, however, format of presentation remains much the same.

The magnitude of this two volume work with 151 chapters and almost as many senior authors from around the world caused great concern with uniform style and presentation. A fixed set of guidelines with some flexibility was prepared for the authors to follow. In this way, some uniformity is apparent, but did not deny the authors their own style of expression and presentation.

The editors acknowledge with deep gratitude the contributions of the authors, because without devotion to their tasks, this collection of work would not be possible. The names of these prominent and internationally known scientists are given where their work appears. In addition, there is a large group of professional associates, staff of the publisher, friends, and colleagues who contributed much, but whose names are not listed for fear of overlooking even just one. To all these important but frequently unrecognized contributors, we extend our thanks for your patience, help, and forebearance.

The Editors

Albert Balows
William J. Hausler, Jr.
Makoto Ohashi
Adolfo Turano

Contents: Volume I

Section III Mycotic Infections

Section IV Parasitic Infections

Contents: Volume II

Contributors

D. G. AHEARN, Laboratory for Microbial and Biochemical Sciences, Department of Biology, Georgia State University, Atlanta, GA 30303, USA

L. AJELLO, Division of Mycotic Diseases, Center for Infectious Diseases, Centers for Disease Control, Atlanta, GA 30333, USA

WIDAD AL-NAKIB, MRC Common Cold Unit, Harvard Hospital, Salisbury, Wiltshire SP2 8BW, England

W. L. ALBRITTON, Department of Microbiology, University of Saskatchewan, Saskatoon, Canada S7N 0W0

DANIEL AMSTERDAM, Division of Clinical Microbiology and Immunology, Erie County Laboratory and Medical Center, and Departments of Microbiology and Immunology, State University of New York School of Medicine, Buffalo, NY 14215, USA

LARRY J. ANDERSON, Respiratory and Enterovirus Branch, Division of Viral Diseases and Hospital Infections Program, Centers for Disease Control, Atlanta, GA 30333, USA

M. J. ANDERSON, Department of Medical Microbiology, Faculty of Clinical Sciences, University College and Middlesex School of Medicine, London WC1E 6JJ, England

RAY R. ARTHUR, Department of Immunology and Infectious Diseases, The Johns Hopkins University, School of Hygiene and Public Health, Baltimore, MD 21205, USA

PERTTI P. ARSTILA, Department of Virology, University of Turku, Kiinamyllynkatu 13, SF-20520, Turku, Finland

GEORGE M. BAER, Rabies Laboratory, Viral and Rickettsial Zoonoses Branch, Division of Viral Diseases, Center for Infectious Diseases, Centers for Disease Control, Lawrenceville, GA 30246, USA

MARILYN S. BARTLETT, Department of Pathology, Indiana University School of Medicine, Indianapolis, IN 46223, USA

GAURI BAZAZ-MALIK, Department of Pathology, Vallabhbhai Patch Chest Institute, University of Delhi, Delhi 110 007, India

C. M. BECK-SAGUE, Sexually Transmitted Diseases Laboratory Program, Center for Infectious Diseases, Centers for Disease Control, Atlanta, GA 30333, USA

R. W. BENNETT, Division of Microbiology, Food and Drug Administration, Washington, D.C. 20204, USA

JENNIFER M. BEST, Department of Virology, United Medical and Dental Schools of Guys' and St. Thomas' Hospitals, St. Thomas' Campus, London SE1 7EH, England

JEFFREY W. BIER, Division of Microbiology, Center for Food Science and Applied Nutrition, Food and Drug Administration, Washington, D.C. 20204, USA

WALTER W. BOND, Division of Viral Diseases and Hospital Infections Program, Center for Infectious Diseases, Centers for Disease Control, Atlanta, GA 30333, USA

ANN CALI, Department of Biological Sciences, Boyden Hall, Rutgers University, Newark, NJ 07102, USA

CHARLES H. CALISHER, Division of Vector-Borne Viral Diseases, Center for Infectious Diseases, Centers for Disease Control, Fort Collins, CO 80522, USA

MARIO E. CAMARGO, Laboratorio de Immunologia e Soroepidemiologia, Instituto de Medicina Tropical de São Paulo, São Paulo, Brazil

MARIO CAMPA, Institute of Microbiology, University of Pisa, 56100 Pisa, Italy

JOSEPH C. CHAN, Division of Infectious Diseases, Department of Medicine, University of Miami School of Medicine, Miami, FL 33101, USA

TIMOTHY J. CLEARY, Department of Pathology, University of Miami School of Medicine, Miami, FL 33101, USA

JOHN H. CROSS, Division of Tropical Public Health, Department of Preventive Medicine and Biometrics, Uniformed Services University of the Health

Sciences, F. Edward Hebert School of Medicine, Bethesda, MD 20814-4799, USA

William L. Current, Eli Lilly Research Laboratories, Lilly Corporate Center, Indianapolis, IN 46285, USA

Richard F. D'Amato, Division of Microbiology, The Catholic Medical Center of Brooklyn and Queens, Jamaica, NY 11432, USA

A. F. DiSalvo, Bureau of Laboratories, South Carolina Department of Health and Environmental Control, Columbia, SC 29202, USA

Frances W. Doane, Department of Microbiology, Faculty of Medicine, University of Toronto, Toronto, Ontario M5S 1A8, Canada

Gary V. Doern, Department of Clinical Microbiology, University of Massachusetts Medical Center, Worcester, MA 01605, USA

W. Lawrence Drew, Clinical Microbiological Laboratory, Mt. Zion Hospital and Medical Center, San Francisco, CA 94611, USA

Richard W. Emmons, Viral and Rickettsial Disease Laboratory, Division of Laboratories, California Department of Health Services, Berkeley, CA 94704, USA

G. Fadda, Institute of Microbiology, University of Sassari, 07100 Sassari, Italy

S. Faine, Department of Microbiology, Monash University, Clayton, Victoria, Australia 3168

Giuseppe Falcone, Institute of Microbiology, University of Pisa, 56100 Pisa, Italy

W. Edmund Farrar, Infectious Diseases and Immunology Division, Department of Medicine, Medical University of South Carolina, Charleston, SC 29425, USA

Frank Fenner, The John Curtin School of Medical Research, The Australian National University, Canberra 2601, Australia

Bernard N. Fields, Department of Microbiology and Molecular Genetics and the Shipley Institute of Medicine, Harvard Medical School, and Division of Infectious Diseases, Department of Medicine, Brigham & Womens Hospital, Boston, MA 02115, USA

Sydney M. Finegold, Research Service and Infectious Disease Section, West Los Angeles Wadsworth V.A. Medical Center, Los Angeles, CA 90073; and Departments of Medicine and Microbiology and Immunology, UCLA School of Medicine, Los Angeles, CA 90024, USA

Daniel B. Fishbein, Viral and Rickettsial Zoonoses Branch, Division of Viral Diseases, Center for Infectious Diseases, Centers for Disease Control, Atlanta, GA 30333, USA

Susan Fisher-Hoch, Division of Viral Diseases and Hospital Infections Program, Center for Infectious Diseases, Centers for Disease Control, Atlanta, GA 30333, USA

D. O. Freedman, Laboratory of Parasitic Diseases, National Institute of Allergy and Infectious Disease, National Institutes of Health, Bethesda, MD 20892, USA

Morris L. V. French, Department of Pathology, James Whitcomb Riley Hospital for Children, Indiana University Medical Center, Indianapolis, IN 46223, USA

Damien A. Gabis, Silliker Laboratories, Inc., Chicago Heights, IL 60411, USA

H. R. Gamble, Helminthic Diseases Laboratory, Animal Parasitology Institute, Agricultural Research Service, USDA, Beltsville, MD 20705, USA

Lynne S. Garcia, Department of Clinical Microbiology, UCLA Medical Center, Los Angeles, CA 90024, USA

Robert M. Genta, Department of Pathology and Laboratory Medicine, V.A. Hospital, University of Cincinnati Medical Center, Cincinnati, OH 45267, USA

W. Lance George, Medical and Research Services, Wadsworth Division, West Los Angeles V.A. Medical Center, Los Angeles, CA 90073, and Department of Medicine, UCLA School of Medicine, Los Angeles, CA 90024, USA

Michael A. Gerber, Department of Pediatrics, University of Connecticut School of Medicine, Farmington, CT 06032, USA

Mary Ann Gerencser, Department of Microbiology, West Virginia University Medical Center, Morgantown, WV 26506, USA

M. J. R. Gilchrist, Microbiology Laboratory, Veterans Administration Medical Center, Cincinnati, OH 45220, USA

A. González-Mendoza, División de Patológía Experimental de Occidente, Unidad de Investigación Biomédica, Instituto Mexicano del Seguro Social, Centro Medico de Occidente, Guadalajara, Jalisco, Mexico

Robert C. Good, Division of Bacterial Diseases, Center for Infectious Diseases, Centers for Disease Control, Atlanta, GA 30333, USA

Monica Grandien, Department of Virology, National Bacteriological Laboratory, S-105 21 Stockhom, Sweden

I. D. Gust, Virus Laboratory, Fairfield Hospital for Communicable Diseases, Fairfield, Victoria 3078, Australia

Stephen C. Hadler, Division of Viral Diseases and Hospital Infections Program, Center for Infectious Diseases, Centers for Disease Control, Atlanta, GA 30333, USA

Pekka Halonen, Department of Virology, University of Turku, Kiinamyllynkatu 13, SF-20520, Turku, Finland

Maurice W. Harmon, Division of Viral Diseases,

Center for Infectious Diseases, Centers for Disease Control, Atlanta, GA 30333, USA

S. M. HARMON, Division of Microbiology, Food and Drug Administration, Washington, D.C. 20204, USA

CHARLES L. HATHEWAY, Division of Bacterial Diseases, Center for Infectious Diseases, Centers for Disease Control, Atlanta, GA 30333, USA

KENNETH L. HERRMANN, Division of Viral Diseases, Center for Infectious Diseases, Centers for Disease Control, Atlanta, GA 30333, USA

JOHN C. HIERHOLZER, Respiratory and Enteric Viruses Branch, Division of Viral Diseases, Centers for Disease Control, Atlanta, GA 30333, USA

ANTHONY E. HILGER, Department of Allied Health Professions, University of North Carolina, Chapel Hill, NC 27514, USA

GEORGE V. HILLYER, Department of Pathology, University of Puerto Rico, Medical Science Campus, San Juan, Puerto Rico 00936, USA

YORIO HINUMA, Institute for Virus Research, Kyoto University, Kyoto 606, Japan

LARRY A. HOLCOMB, Hygienic Laboratory, Oakdale Hall, University of Iowa, Iowa City, IA 52242, USA

B. HOLMES, Public Health Laboratory Service, National Collection of Type Cultures, Central Public Health Laboratory, London NW9 5HT, England

IAN H. HOLMES, Department of Microbiology, University of Melbourne, Parkville 3052, Victoria, Australia

DONALD R. HOPKINS, Center for Infectious Diseases, Centers for Disease Control, Atlanta, GA 30333, USA

HENRY D. ISENBERG, Department of Microbiology, Long Island Jewish Medical Center, New Hyde Park, NY 11042, and Professor of Clinical Pathology, State University of New York at Stony Brook, Stony Brook, NY 11794, USA

WILLIAM R. JARVIS, Division of Viral Diseases and Hospital Infections Program, Center for Infectious Diseases, Centers for Disease Control, Atlanta, GA 30333, USA

STIG JEANSSON, Department of Clinical Virology, Institute of Medical Microbiology, University of Göteborg, 41346 Göteborg, Sweden

RUSSELL C. JOHNSON, Department of Microbiology, School of Medicine, University of Minnesota, Minneapolis, MN 55455, USA

JERRY E. JONES, Department of Family Practice, College of Medicine, University of Kentucky, Lexington, KY 40536, USA

S. L. JOSEPHSON, Hygienic Laboratory, Oakdale Hall, University of Iowa, Iowa City, IA 52242, USA

H. KAMO, Department of Medical Zoology, Tottori University School of Medicine, Yonago 683, Japan

JACOBA G. KAPSENBERG, Rijksinstituut voor Volksgezondheid en Milieuhygiene, Laboratorium voor Virologie, 3720 BA Bilthoven, The Netherlands

D. S. KELLOGG, Sexually Transmitted Diseases Laboratory Program, Center for Infectious Diseases, Centers for Disease Control, Atlanta, GA 30333, USA

ALAN KENDAL, Influenza Branch, Center for Infectious Diseases, Centers for Disease Control, Atlanta, GA 30333, USA

MICHAEL P. KILEY, Division of Viral Diseases, Center for Infectious Diseases, Centers for Disease Control, Atlanta, GA 30333, USA

MICHAEL M. KLIKS, Division of Comparative Medicine, John A. Burns School of Medicine, University of Hawaii, Honolulu, HA 96822, USA

T. KOYAMA, Department of Parasitology, National Institute of Health, Shinagawa-ku, Tokyo 141, Japan

DONALD J. KROGSTAD, Departments of Pathology and Internal Medicine Division of Laboratory Medicine, Washington University School of Medicine, St. Louis, MO 63110, USA

YASUO KUDOH, The First Division of Bacteriology, Department of Microbiology, Tokyo Metropolitan Research Laboratory of Public Health, Sinjuku-ku Tokyo 160, Japan

MONIQUE LAFON, Rabies Unit, Institut Pasteur, 75724 Paris Cedex 15, France

S. A. LARSEN, Sexually Transmitted Diseases Laboratory Program, Center for Infectious Diseases, Centers for Disease Control, Atlanta, GA 30333, USA

WILLIAM K. K. LAU, Department of Medicine, John A. Burns School of Medicine, University of Hawaii, Honolulu, HA 96822, USA

DIANE S. LELAND, Department of Pathology, James Whitcomb Riley Hospital for Children, Indiana University Medical Center, Indianapolis, IN 46223, USA

DAVID A. LENNETTE, Virolab, Inc., Berkeley, CA 94710, USA

EVELYNE T. LENNETTE, Virolab, Inc., Berkeley, CA 94710, USA

JAY A. LEVY, Department of Medicine and Cancer Research Institute, Universtiy of California School of Medicine, San Francisco, CA 94143, USA

C. C. LINNEMANN, JR., Division of Infectious Diseases, University of Cincinnati College of Medicine, Cincinnati, OH 45229, USA

S. A. LOCARNINI, Virus Laboratory, Fairfield Hospital for Communicable Diseases, Fairfield, Victoria 3078, Australia

E. G. LONG, Division of Parasitic Diseases, Centers for Disease Control, Atlanta, GA 30333, USA

CARLOS LOPEZ, Division of Viral Diseases and Hospital Infections Program, Center for Infectious Diseases, Centers for Disease Control, Atlanta, GA 30333, USA

ERIK LYCKE, Department of Clinical Virology, Institute of Medical Microbiology, University of Göteborg, 41346 Göteborg, Sweden

S. E. MADDISON, Division of Parasitic Diseases, Center for Infectious Diseases, Centers for Disease Control, Atlanta, GA 30333, USA

CHARLES RICHARD MADELEY, Department of Virology University of Newcastle upon Tyne and Royal Victoria Infirmary, Newcastle upon Tyne NE1 4LP, England

EL SHEIKH MAHGOUB, Department of Medical Microbiology and Parasitology, Faculty of Medicine, University of Khartoum, Khartoum, Republic of Sudan

D. D. MARTIN, Life Sciences Division, U.S. Army Dugway Proving Ground, Dugway, UT 84022, USA

HENRY M. MATHEWS, Division of Parasitic Diseases, Center for Infectious Diseases, Centers for Disease Control, Atlanta, GA 30333, USA

JOSEPH B. MCCORMICK, Special Pathogens Branch, Division of Viral Diseases, Center for Infectious Diseases, Centers for Disease Control, Atlanta, GA 30333, USA

JOSEPH E. MCDADE, Viral and Rickettsial Zoonoses Branch, Division of Viral Diseases, Center for Infectious Diseases, Centers for Disease Control, Atlanta, GA 30333, USA

MICHAEL R. MCGINNIS, Department of Pathology, University of Texas Medical Branch, Galveston, TX 77550, USA

PATRICIA A. MERZ, Department of Virology and Pathology, N.Y. State Office of Mental Retardation and Developmental Disabilities, Institute for Basic Research in Developmental Disabilities, Staten Island, NY 10314, USA

F. MEUNIER, Service de Médecine et Laboratoire d'Investigation Clinique, Clinique H. J. Tagnon, Institut Jules Bordet, 1000 Brussels, Belgium

M. MIYAJI, Department of Fungal Infections, Research Center for Pathogenic Fungi and Microbial Toxicoses, Chiba University, Inohana, Chiba 280, Japan

THOMAS P. MONATH, Division of Vector-Borne Viral Diseases, Center for Infectious Diseases, Centers for Disease Control, Fort Collins, CO 80522, USA

NELSON P. MOYER, Hygienic Laboratory, Oakdale Hall, University of Iowa, Iowa City, IA 52242, USA

FREDERICK A. MURPHY, Center for Infectious Diseases, Centers for Disease Control, Atlanta, GA 30333, USA

PATRICK R. MURRAY, Division of Laboratory Medicine, Department of Pathology, Washington University School of Medicine, St. Louis, MO 63110, USA

K. DARWIN MURRELL, Helminthic Diseases Laboratory, Animal Parasitology Institute, Agricultural Research Service, USDA, Beltsville, MD 20705, USA

KAZUNARI NAKAMURA, The First Department, National Institute for Leprosy Research, Higashimurayama, Tokyo 189, Japan

JAMES H. NAKANO, Center for Infectious Diseases, Centers for Disease Control, Atlanta, GA 30333, USA

THOMAS R. NAVIN, Division of Parasitic Diseases, Centers for Disease Control, Atlanta, GA 30333, USA

RICARDO NEGRONI, Centro de Micrología de la Cátedra de Microbiología, Parasitología, e Immunología, Facultad de Medícine, Universidad de Buenos Aires, Buenos Aires, Argentina

ERLING NORRBY, Department of Virus Research, Karolinska Institutet, S-105 21 Stockholm, Sweden

T. B. NUTMAN, Laboratory of Parasitic Diseases, National Institute of Allergy and Infectious Disease, National Institutes of Health, Bethesda, MD 20892, USA

MAKOTO OHASHI, Tokyo Metropolitan Research Laboratory of Public Health, Shinjuku-ku, Tokyo 160, Japan

CLAES ÖRVELL, Department of Virology, National Bacteriological Laboratory and Karolinska Institute, School of Medicine, S-105 21 Stockholm, Sweden

SIOBHAN O'SHEA, Department of Virology, United Medical and Dental Schools of Guys' and St. Thomas' Hospitals, St. Thomas' Campus, London SE1 7EH, England

ROBERT L. OWEN, Departments of Medicine, Epidemiology, and International Health, University of California–San Francisco, and Cell Biology and Aging Section, V.A. Medical Center, San Francisco, CA 94121, USA

A. A. PADHYE, Division of Mycotic Diseases, Center for Infectious Diseases, Centers for Disease Control, Atlanta, GA 30333, USA

NICHOLAS E. PALUMBO, Department of Medicine, John A. Burns School of Medicine, University of Hawaii, Honolulu, HA 96822

DEMOSTHENES PAPPAGIANIS, Department of Medical Microbiology and Immunology, University of California-Davis School of Medicine, Davis, CA 95616, USA

HERBERT PFISTER, Institut für Klinische und Molekulare Virologie, Universität Erlangen-Nürnberg, D-8520 Erlangen, West Germany

PETER PIOT, Department of Microbiology, Institute of Tropical Medicine (Prince Leopold), B-2000 Antwerp, Belgium

F. PIRALI, Institute of Microbiology, Brescia University, 25100 Brescia, Italy

HELEN M. POLLOCK, Department of Pathology, University of South Alabama Hospitals and Clinics, Mobile, AL 36617, USA

THOMAS J. QUAN, Plague Branch, Division of Vector-Borne Viral Diseases, Center for Infectious Diseases, Centers for Disease Control, Ft. Collins, CO 80522, USA

H. S. RANDHAWA, Department of Medical Mycology, Vallabhbhai Patel Chest Institute, University of Delhi, Delhi 110 007, India

M. RANKI, Orion Genetic Engineering Laboratory, Orion Corporation Ltd., SF-00380 Helsinki, Finland

ANNETTE C. REBOLI, Infectious Diseases and Immunology Division, Department of Medicine, Medical University of South Carolina, Charleston, SC 29425, USA

H. B. REES, JR., 3584 Millstream Lane, Salt Lake City, UT 84109, USA

MARCIA L. RHOADS, Helminthic Diseases Laboratory, Animal Parasitology Institute, Agricultural Research Service, USDA, Beltsville, MD 20705, USA

F. O. RICHARDS, JR., Division of Parasitic Diseases, Center for Infectious Diseases, Centers for Disease Control, Atlanta, GA 30333, USA

LEE W. RILEY, Divisions of Infectious Diseases and Geographic Medicine, Stanford University Medical Center, Stanford, CA 94305, USA

HAN-JONG RIM, Department of Parasitology, Institute for Tropical Endemic Diseases, College of Medicine, Korea University, Chongno-ku, Seoul 110, Korea

MICHAEL G. RINALDI, Department of Pathology, University of Texas Health Science Center at San Antonio and Veterans Administration Mycology Reference Laboratory, Laboratory Service, Audie L. Murphy Memorial Veterans' Hospital, San Antonio, TX 78284, USA

ALLAN R. RONALD, Department of Medicine, Faculty of Medicine, University of Manitoba, Winnipeg, Manitoba R3E 0Z3, Canada

TRENTON K. RUEBUSH, II, Division of Parasitic Diseases, Center for Infectious Diseases, Centers for Disease Control, Atlanta, GA 30333, USA

RAYMOND W. RYAN, Department of Laboratory Medicine, Division of Clinical Microbiology, University of Connecticut School of Medicine, Farmington, CT 06032, USA

G. SATTA, Institute of Microbiology, University of Siena Medical School, Siena, Italy

JULIUS SCHACHTER, Department of Laboratory Medicine, University of California-San Francisco, San Francisco, CA 94143, USA

PETER M. SCHANTZ, Helminthic Diseases Branch, Division of Parasitic Diseases, Center for Infectious Diseases, Centers for Disease Control, Atlanta, GA 30333, USA

H. J. SHADOMY, Department of Microbiology and Immunology, Virginia Commonwealth University, Richmond, VA 23298, USA

KEERTI V. SHAH, Department of Immunology and Infectious Diseases, The Johns Hopkins University, School of Hygiene and Public Health, Baltimore, MD 21205, USA

ROBERT E. SHOPE, Yale Arbovirus Research Unit, Department of Epidemiology and Public Health, School of Medicine, Yale University, New Haven, CT 06510, USA

JOHN H. SILLIKER, Silliker Laboratories of California, Inc., Carson, CA 90746, USA

JAMES W. SMITH, Department of Pathology, Indiana University School of Medicine, Indianapolis, IN 46223, USA

M. A. SMITH, Environmental Technology, Utah Technical College, Orem, UT 84058, USA

H. SÖDERLUND, Orion Genetic Engineering Laboratory, Orion Corporation Ltd., SF-00380 Helsinki, Finland

S. J. STEWART, Laboratory of Microbial Structure and Function, Rocky Mountain Laboratory, NIAID, National Institutes of Health, Hamilton, MT 59841, USA

PIERRE SUREAU, Rabies Unit, Institut Pasteur, 75724 Paris Cedex 15, France

NORMAN S. SWACK, Hygienic Laboratory, Oakdale Hall, University of Iowa, Oakdale Hall, Iowa City, IA 52242, USA

A.-C. SYVÄNEN, Orion Genetic Engineering Laboratory, Orion Corporation Ltd., SF-00380 Helsinki, Finland

MICHIAKI TAKAHASHI, Department of Virology, Research Institute for Microbial Diseases, Osaka University, Suita, Osaka, Japan

GREGORY A. TANNOCK, Respiratory and Enteric Viruses Branch, Divsion of Viral Diseases, Center for Infectious Diseases, Centers for Disease Control, Atlanta, GA 30333, USA

A. A. TERRENI, Bureau of Laboratories, South Carolina Department of Health and Environmental Control, Columbia, SC 29202, USA

RICHARD C. TILTON, Division of Microbiology, University of Connecticut School of Medicine, Farmington, CT 06032, USA

M. Tsubokura, Department of Veterinary Microbiology, Faculty of Agriculture, Tottori University, Koyama-cho, Tottori 680, Japan

A. Turano, Institute of Microbiology, Brescia University, 25100 Brescia, Italy

Kenneth L. Tyler, Department of Microbiology and Molecular Genetics, Harvard Medical School, and Department of Neurology, Massachusetts General Hospital, Boston, MA 02115, USA

David A. J. Tyrrell, MRC Common Cold Unit, Harvard Hospital, Salisbury, Wiltshire SP2 8BW, England

P. van der Auwera, Service de Médecine et Laboratoire d'Investigation Clinique, Clinique H. J. Tagnon, Institut Jules Bordet, 1-1000 Brussels, Belgium

M. H. V. Van Regenmortel, Institute de Biologie Moléculaire et Cellulaire de CNRS, 67084 Strasbourg Cedex, France

P. E. Varaldo, Institute of Microbiology, University of Ancona Medical School, 60131 Ancona, Italy

Neylan A. Vedros, Department of Biomedical and Environmental Health Sciences, School of Public Health, Department of Biomedical and Environmental Health Sciences, Earl Warren Hall, University of California, Berkeley, CA 94720, USA

G. S. Visvesvara, Division of Parasitic Diseases, Center for Infectious Diseases, Centers for Disease Control, Atlanta, GA 30333, USA

G. Wadell, Department of Virology, University of Umea, S-901 87, Umea, Sweden

Duard L. Walker, Department of Medical Microbiology, University of Wisconsin School of Medicine, Madison, WI 53706, USA

Kenneth W. Walls, 4006 Northlake Creek Court, Tucker, GA 30084, USA

N. G. Warren, Division of Consolidated Laboratory Services, Microbiological Sciences Bureau, Richmond, VA 23298, USA

Haruo Watanabe, Department of Bacteriology, National Institute of Health, Kamiosaki, Shinagawa-ku, Tokyo 141, Japan

Irene Weitzman, Bureau of Laboratories, Department of Health, City of New York, New York, NY 10016, USA

Hazel W. Wilkinson, Division of Bacterial Diseases, Center for Infectious Diseases, Centers for Disease Control, Atlanta, GA 30333, USA

Henryk M. Wisniewski, New York State Office of Mental Retardation and Developmental Disabilities, Institute for Basic Research in Developmental Disabilities, Staten Island, NY 10314, USA

H. Yamaguchi, Institute of Applied Microbiology, University of Tokyo, Bunkyo-ku, Tokyo 113, Japan

Naoki Yamamoto, Department of Virology and Parasitology, Yamaguchi University School of Medicine, Ube, Yamaguchi 755, Japan

Yutaka Zinnaka, Department of Microbiology, Toho University School of Medicine, Ohta-ku, Tokyo 143, Japan

Arie J. Zuckerman, Department of Medical Microbiology and WHO Collaborating Centre for Reference and Research on Viral Hepatitis, London School of Hygiene and Tropical Medicine, London WC1E 7HT, England

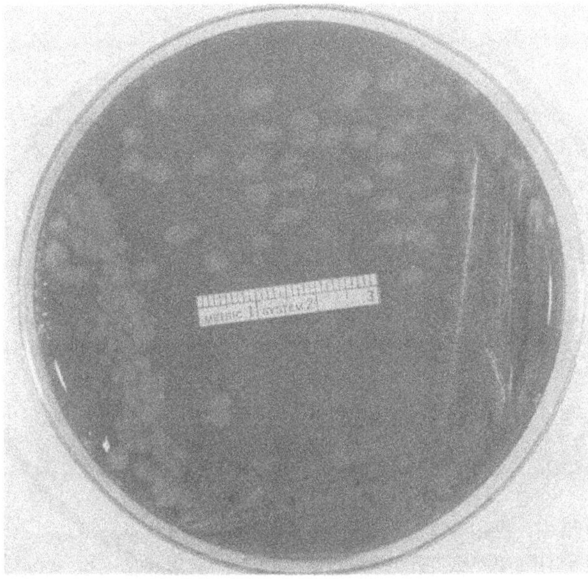

FIG. 1. Colonies of *Bacillus anthracis* grown on 5% blood agar for 18 to 24 h at 37°C. Note colonies with extended and irregular borders described as "medusa head" or "comet tails." (U.S. Army photograph; reproduced with permission.) For further discussion, see page 60.

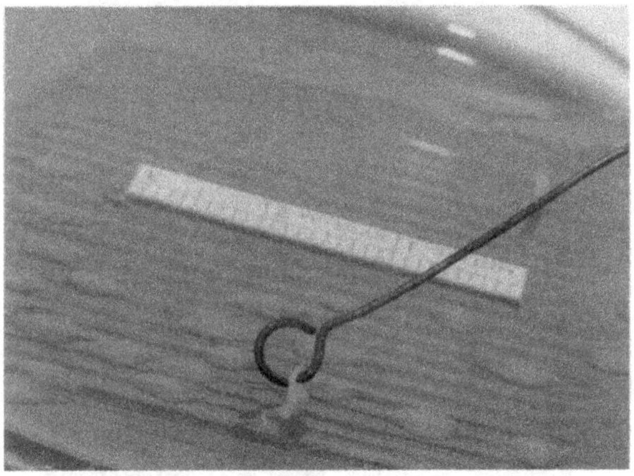

FIG. 2. Photomicrograph of *Bacillus anthracis* stained by the Gram method. Note unstained vacuolated areas or spores within the vegetative cell and little or no evidence of swelling of the cell. (U.S. Army photograph; reproduced with permission.) For further discussion, see page 62.

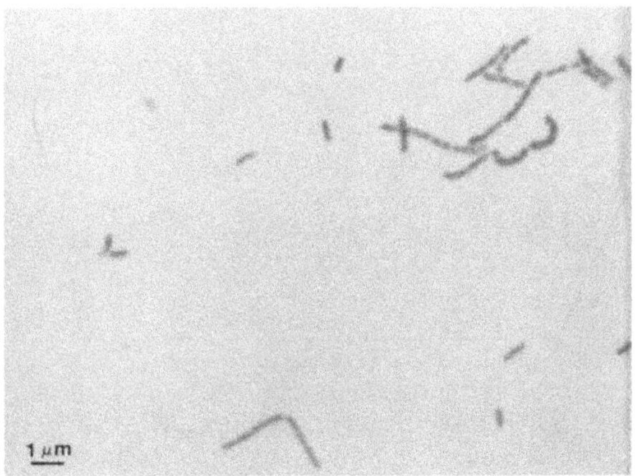

FIG. 3. Colonies of *Bacillus anthracis* grown on blood agar at 37°C. Note malleability or the tendency of the cells to adhere to each other sufficiently to permit sculpting with inoculating needle. For further discussion, see page 63.

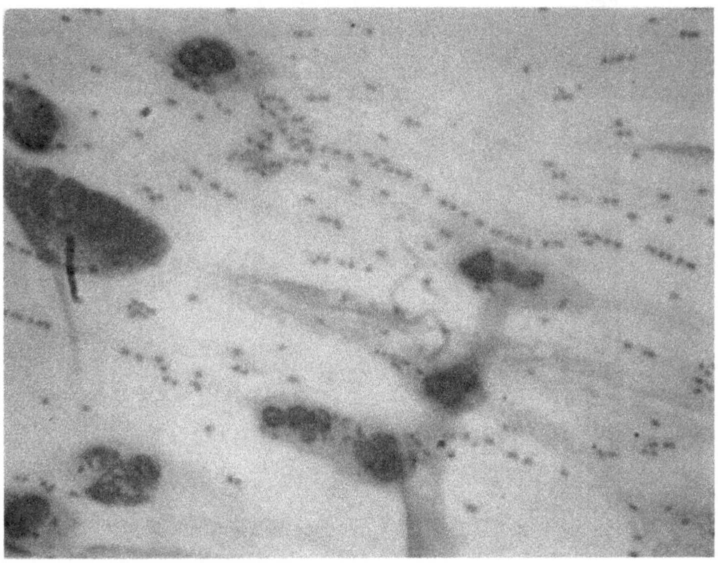

FIG. 1. Gram-stained smear of expectorated sputum from a patient with *Branhamella catarrhalis* pneumonia, revealing numerous polymorphonuclear leukocytes and numerous intracellular and extracellular gram-negative diplococci. For further discussion, see page 136.

Color Plate 2

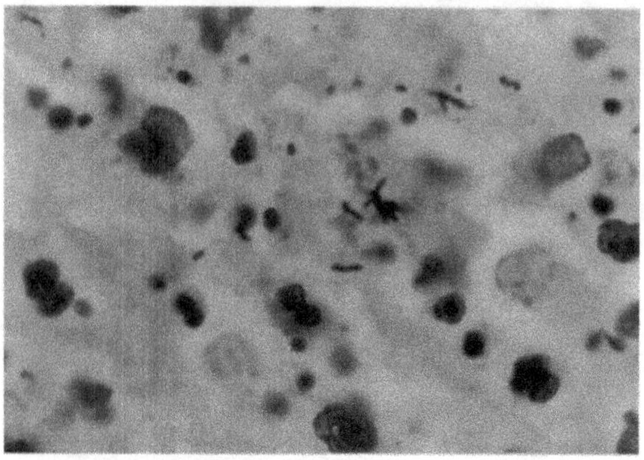

FIG. 1. Warthin–Starry stained section of a lymph node, showing pleomorphic bacilli. For further discussion, see page 169.

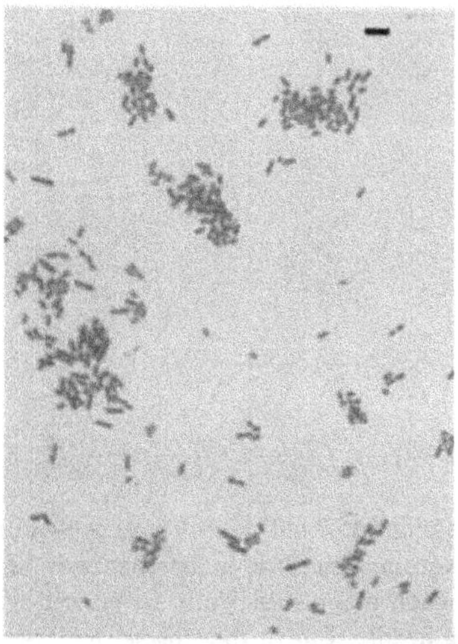

a

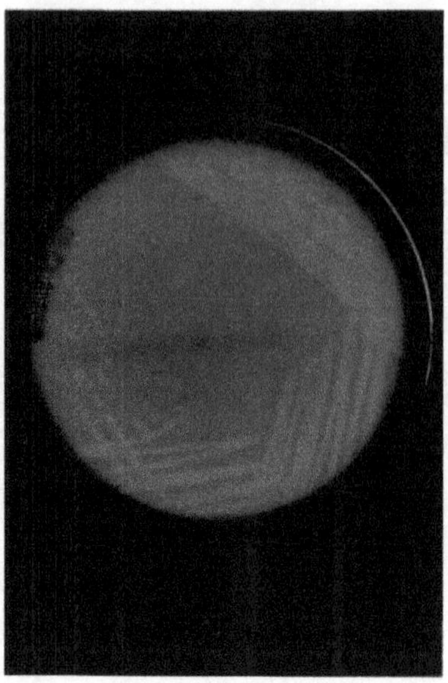

b

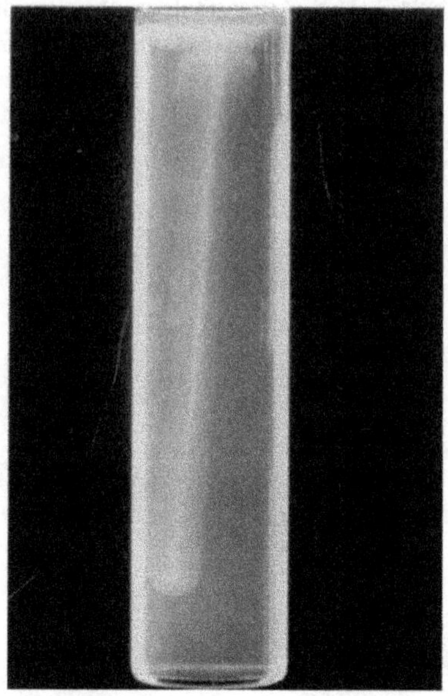

c

d

FIG. 2. (a) Gram stain morphology of *Listeria monocytogenes*. When grown in culture (shown), *Listeria* is increasingly gram-variable with age of the culture. Cell morphology is that of cocci, coccobacilli, or bacilli. Organisms may appear in pairs, short chains, or palisade layers, resembling diphtheroids. Bar = 2 μm. (b) Hemolytic reaction of *Listeria monocytogenes*. When grown on sheep, rabbit, or horse blood agar, the organism is generally beta-hemolytic. On sheep blood agar, the hemolytic zone may be very narrow; in some cases, it may be necessary to scrape off the colony to reveal hemolysis under the colony. The plate shows the faint hemolysis characteristic at 24 h; after 48 h, hemolysis is often more pronounced. (c) Motility of *Listeria monocytogenes* in semisolid agar. When grown at room temperature, the organisms are motile, forming an umbrella shape a few millimeters below the surface of the agar. (d) CAMP reactions of *L. monocytogenes* and group B streptococci. When streaked perpendicularly to *S. aureus*, the zone of hemolysis of the group B streptococci takes the form of an arrowhead at the juncture between the two streaks. For *Listeria monocytogenes*, the zone of the hemolysis at the juncture is more rounded. In the photograph, two group B streptococcal streaks are flanked by three *Listeria* steaks. For further discussion, see pages 356 and 357.

Color Plate 3

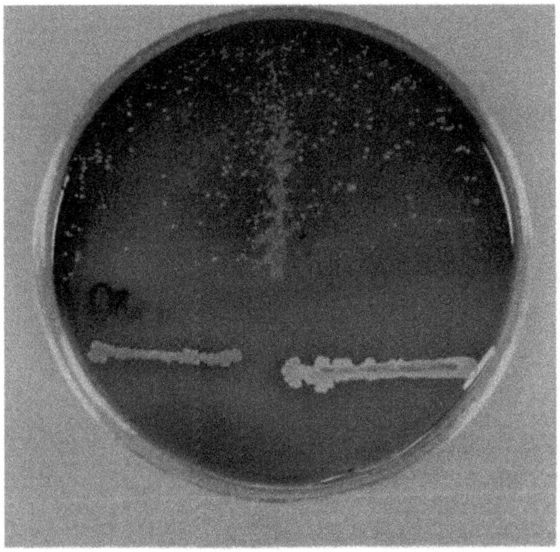

FIG. 1. Inhibition of *Bordetella pertussis* by *Staphylococcus aureus*. The photograph shows colonies of *B. pertussis* growing at a distance from two strains of *S. aureus*. The occurrence of *S. aureus* in a nasopharyngeal specimen is relatively common, and many strains of *S. aureus* are inhibitory to *B. pertussis* growth, with zones of inhibition ranging as high as several centimeters. Thus, it is useful to incorporate antistaphylococcal antibiotics (methicilllin, cephalexin) into the primary isolation medium. For further discussion, see page 406.

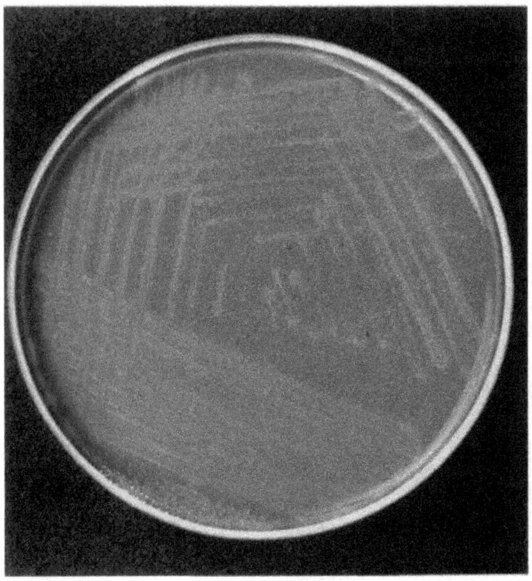

FIG. 2. Growth of *B. pertussis* on Bordet-Gengou medium. The hazy zone of hemolysis is difficult to visualize without magnification unless a large number of colonies occur relatively close to one another. For further discussion, see page 406.

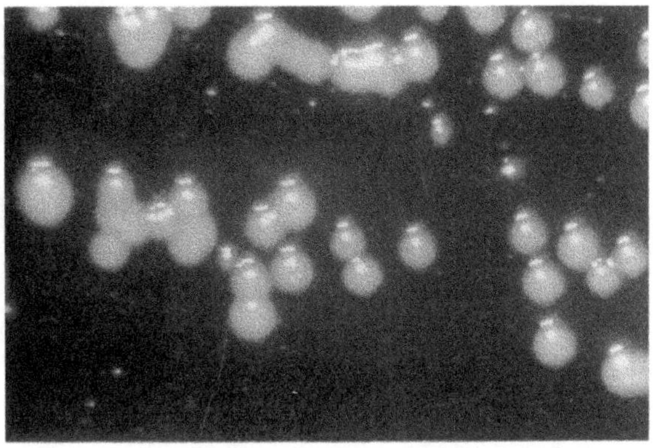

FIG. 3. Growth of *B. pertussis* on charcoal medium. On charcoal medium, colonies of *B. pertussis* are white, opaque, and convex. When observed under the stereomicroscope, they resemble a bisected pearl. For further discussion, see page 407.

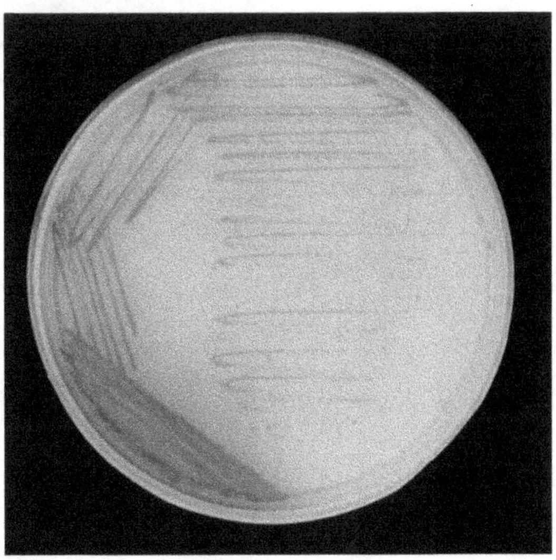

FIG. 4. Biochemical identification of *B. pertussis:* In addition to serologic means, the differentiation of *B. pertussis* from *B. parapertussis* may be accomplished by biochemical testing. The hydrolysis of urea and browning of nutrient agar are characteristics of *B. parapertussis*. The photograph is a strain of *B. parapertussis* growing on Mueller-Hinton agar. Browning on this medium is first associated with the colonies, but after prolonged incubation diffuses throughout the medium. For further discussion, see page 407.

Color Plate 4

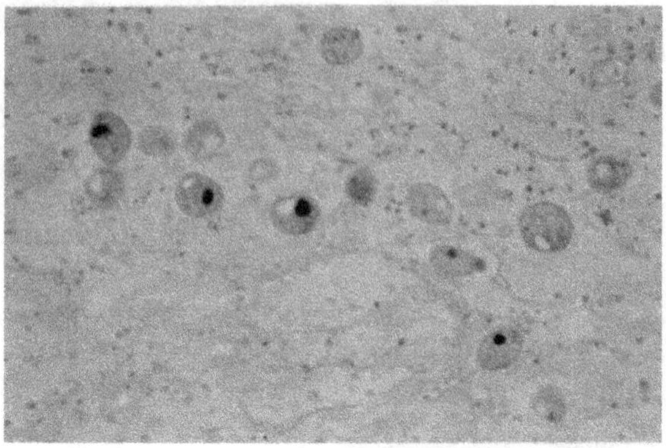

FIG. 1. Many *Balantidium coli* trophozoites in tissue from the large intestine of a pig. Darkly stained macronuclei are found. Hematoxylin and eosin stain. Magnification ×100; reduced by 37%. For further discussion, see page 788.

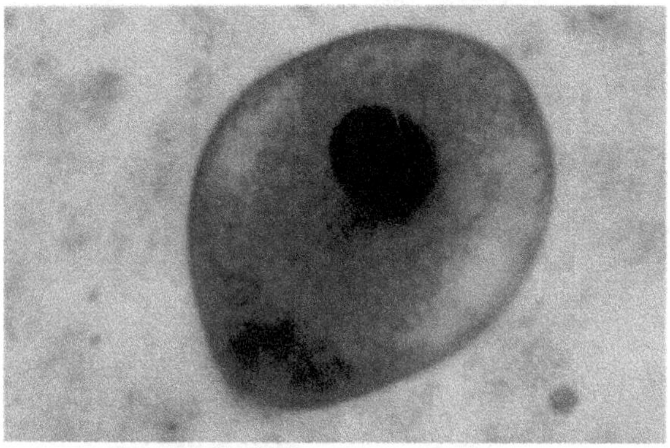

FIG. 2. *Balantidium coli* trophozoite from the feces of a chimpanzee. Hematoxylin and fast green FCF stain. Magnification ×650; reduced by 37%. For further discussion, see page 789.

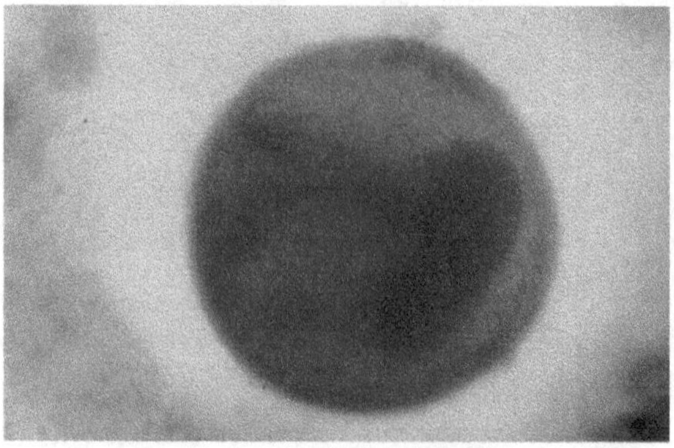

FIG. 3. *Balantidium coli* cyst from the feces of a chimpanzee. Hematoxylin and fast green FCF stain. Magnification ×700; reduced by 37%. For further discussion, see page 789.

SECTION I

General Principles and Practice

CHAPTER 1

Good Laboratory Practices: Scope and Purpose

HENRY D. ISENBERG and RICHARD F. D'AMATO

Definition

The scope and purpose of clinical microbiology are deceptively simple: namely, the accurate provision of pertinent information about the microorganisms in a given specimen while such information is still of significance to the clinician and especially the patient (Isenberg and Berkman, 1962; Steel, 1962). This definition encompasses the entire discipline: both the scientific aspect of microbiology and its application to the health care field. The hybrid term "clinical microbiology," however, does not imply a specialty that reflects purely clinical concerns, as do designations such as clinical hematology, clinical endocrinology, or clinical immunology. Clinical microbiology remains a strict laboratory specialty devoted to the recognition of living organisms and of their products or constituents that may be harmful to a particular or to many human hosts. Nevertheless, both aspects of the designation are manifested in the practices of the discipline and each modifies the other to achieve the goal of the specialty: the recognition of etiological agents of infectious disease and, whenever possible, the designation of appropriate therapeutic agents.

Distinguishing Attributes of Clinical Microbiology

Primarily, clinical microbiology is a confirmatory rather than a diagnostic endeavor, although more than a century has passed since the relationship between the microbial world and infectious disease was demonstrated. Many reasons, by no means excuses, account for this difference between clinical microbiology and other laboratory specialities designed to recognize pathologic departures from the established normal realm of chemical, immunologic, and cellular body constituents. Foremost among several differences between these disciplines of laboratory medicine is the ubiquity of microorganisms on all surfaces of this planet, especially their intimate association with all living forms, including humans. Although it is comparatively simple to set the limits of normalcy in chemical and physiologic terms, such limits are difficult if not impossible to establish for the microbiota of the human biosphere (Isenberg and D'Amato, 1985). Autochthonous microorganisms are present in certain niches of the human anatomy, but their presence is by no means universal. So-called sterile areas of the human body, only occasionally contaminated with microbes, abound. Transient organisms may share a number of areas briefly, for prolonged periods, or they may become resident participants of that particular microbiota. The specific role any microorganism may play in the dynamic first encounter with the skin and mucous membranes of the human host depends on the particular attributes of the microorganism itself, the total effect exerted by the resident microbial consortium colonizing a particular site, the availability of appropriate nutrients and atmospheric conditions in that locale, and the tolerance expressed by the sum of the host factors operative in that location, including epitope receptors (Isenberg and Balows, 1981). Therefore, one can expect to find microorganisms on various body surfaces, but their recovery does not indicate infectious disease. The temptation to call these organisms the normal microbiota is great, but such a designation ignores the circumstances when impaired host factors permit the participation of these "normal" organisms in the complication of underlying disease processes. These members of the resident microbiota must be considered amphibionts (Rosebury, 1961) or "sometime pathogens." Thus, the comfortable normal range that serves our colleagues in other health care laboratories is denied the clinical microbiologist, who must interpret recovery of microorganisms in terms of the history and symptomatology of a particular patient.

A second reason for the confirmatory nature of clinical microbiology is the extant methodology of microbiology, a methodology that unifies the disparate prokaryotic and primitive eukaryotic forms we describe as microorganisms (Stanier, 1978). The absolute identification of any given microorganism still demands that it be separated completely from all of its companions in a specimen, raised in total isolation under completely artificial circumstances, and identified by a number of biochemical, physiologic, and serologic examinations. The dogma of the pure culture technique, this final arbiter of microbiologic taxonomy, has neither completely yielded to the latest molecular manipulations nor been supplanted by advances in immunochemistry. In general, the time-honored, labor- and time-intensive, unnatural study of microorganisms remains in place and prevents the immediate diagnosis of a microbial agent involved in disease production.

The Dynamic Nature of Clinical Microbiology

The determinant role of each human host in the pathologic expression of a microbial presence also plays a significant part in the deliberations of the clinical microbiologist. Traditionally, however, microbiologists and physicians have ignored almost completely the cellular, humoral, and even general impact of the host on the diagnosis of infectious disease. Instead, the recovery of an organism serves as the sole indicator. The growing number of microbes of environmental origin, without readily recognizable attributes of harmfulness to humans—recovered from the patient in medical facilities or from community residents in remission from diseases that compromise immunity—compel the clinical microbiologist to introduce historical patient information for the proper interpretation of all microorganisms isolated from the patient. Of course, this practice should always have been integrated into the evaluation of microbiologic analyses. The more recent incursions of organisms from the large microbial reservoir in nature reflect in part the selective pressures exerted by antimicrobic therapy in medical institutions in concert with the widespread addition of these drugs to foodstuffs and manufactured goods subject to spoilage. This role of microbes, until recently considered commensal organisms or saprophytes, in complicating the recovery of patients must be appreciated from several divergent views. Their presence in specimens from noncompromised patients, personnel or healthy individuals, denotes pathology only rarely, emphasizing the importance of host information in the interpretation of results. Within hospitals resistant microorganisms may colonize the skin and mucous membranes of patients. Their entry into usually sterile, only transiently contaminated organs and tissues is aided and encouraged by various support devices such as catheters, medications, protheses, etc. Their presence in the medical setting may be the consequence of air, water, vegetables, flowers, visitors, and hospital personnel.

Awareness of the different origins of microorganisms isolated from patient specimens affects expectations and interpretations of laboratory results, but can also impact on the analytical process itself. The alert scientist, therefore, should insist on pertinent information when a specimen is received, modify the analysis in accordance with the patient's history, and base the interpretation of results on this information.

An additional aspect of the clinical microbiologist's responsibilities transcends concern for the infected individual and focuses on the risk to other patients, hospital personnel, and the community at large. While epidemiologists have the primary responsibility for this aspect, it is often in the microbiology laboratory that such organisms and viruses are discovered. Of course, one need not curtail this concern to the isolation of etiologic agents of global significance; cluster epidemics with multiply resistant bacteria, outbreaks of diarrhea or pustulosis, catheter-associated urinary tract infections, central-line infections, and the entire control of nosocomial complications are based on such observations. In addition, the laboratory recognition of an appreciable number of etiological agents warrants reporting to appropriate governmental agencies (Isenberg, 1982).

The Impact of Advanced Technology

We must also consider that rapid technological advances fall well within the scope of clinical microbiology, claims of other laboratory disciplines notwithstanding. Although still in their infancy, molecular and immunologic probes directed at specific etiological agents are becoming available with increasing frequency. These approaches hold the promise of providing immediate pertinent information to the clinician. Although these methods represent a rapidly expanding field, their ability to serve all the needs of infectious disease diagnoses and therapeutic guidance will take considerable time, during which classic microbiological technologies must function in tandem with advanced modalities.

Clinical microbiologists must expand their scope and activities to meet the recent advances made in understanding the determinant role played by the individual host in the overt clinical manifestation of infectious disease. Specific, particular, and peculiar attributes of an individual patient's immunity—in the broadest sense—in relation to infection with a specific microorganism should fall within the purview of microbiological analyses. Unfortunately, we lack practically all information and have neglected the methodologies that would permit measurement of a patient's granulocytic activity against a specific bacterium, fungus, or virus; we do not understand the influence of several host metabolic cascades on the success of a particular microorganism in exerting its harmful effect; and we have not yet determined which host epitopes permit attachment and/or entry of any microbe or virus across membranes. Host–parasite interactions are the required corollary to the rapid diagnostic tests of the future if successful prevention and palliation of infectious disease is to be realized.

Voluntary Limits of Clinical Microbiology

The temptation is great to characterize as completely as possible each and every microorganism or virus isolated from a specimen, especially since clinical microbiologists are painfully aware that not all microorganisms from contaminated body sites have been domesticated successfully. Undoubtedly, such pursuits would increase further our astonishment at the variety and diversity of the microbial world in or on our bodies. But no practical goal would be served by such efforts; indeed, these efforts are contraindicated by the major charge to clinical microbiologists to expeditiously communicate to clinicians the presence of potential etiological agents for the observed patient symptoms. The present financial strictures resulting from prospective payment policies makes such actions totally irresponsible. The purpose of analytical clinical microbiology is not to detail the minutiae of the microbiota, but to discern in general and particular terms how the objectives of clinical microbiology can be attained. Of course, the isolation of a bacterium or fungus that resembles morphologically or by biochemical activity well-established pathogens should lead invariably to identification efforts regardless of the source of isolation. Similarly, isolation of microorganisms from usually sterile body sites warrants complete analysis.

The Clinical Specimen

The success of these efforts rests squarely on the awareness of the clinician on what specimen is required for a particular examination, the manner of obtaining that specimen without breaching its microbiological integrity, and the expeditious and proper delivery of that specimen to the laboratory. The skills microbiology personnel bring to the examination of any specimen are of no consequence if procurement and transport are not appropriate. One of the first responsibilities of clinical microbiologists must be the education of the clinical staff. It should not be assumed that all physicians are fully aware of these requirements. The last few years have seen a shift in the education of medical professionals that emphasize advances in molecular biology and directs special attention to the psychosocial needs of patients. Both efforts are laudable, but they do lead to a neglect of teaching and understanding the role of microorganisms in health and disease along with a veritable avoidance of exposing students to the laboratory aspects of microbiology. The clinical microbiologist must jump into the breach and exert every effort to present in the most palatable fashion the requirements for collecting and transporting clinical specimens to the laboratory. This effort requires time, patience, and the recognition that it is an unending process. Each microbiologist must adapt instructions to suit the needs of each group and, if necessary, specific individuals. However, the opportunity to make each encounter as broad an instructive experience as possible should not be missed. A discussion of the usual microbiota encountered in a specific anatomic locus can be expanded to include the role of these microbes in that site, their potential to become involved in disease production, or their beneficial activities in preventing the entry of other organisms. Above all, each practitioner must learn that the lesion, the pathologic process, or the nidus of the infection is the area of interest and that the specimen from a given site must be cultured without contamination by organisms from neighboring areas.

This is a propitious time to indicate why anaerobic cultures are not performed on specimens from sites colonized by these bacteria, and to explain the reasons for rejecting sputum specimens teaming with squamous epithelial cells but without granulocytes, and the reasons for insisting on midstream urine specimens for culture. Obviously, many more important instructions that could be imparted at this juncture can neither be enumerated nor delivered in toto at any one encounter with groups or individuals. However, all encounters should be tailored to invite dialogue and physicians and their associates should be encouraged to ask before acting. Clinical microbiologists should remember that their efforts have no meaning until a clinician can act on the reported result. Therefore, they must create an atmosphere of availability, advice, assistance, and cooperation that encourages clinical colleagues to seek assistance. This process of education or explanation should not be limited to the procurement of specimens; it is a given for all activities in the clinical microbiology laboratory. Clinical microbiologists can no longer be satisfied with reporting their identifications and antibiograms. The interpretation of their findings in light of the clinical situation and the patient's history is, as stated previously, an integral part of the clinical microbiologist's responsibilities and an exercise universally welcomed if provided in a tactful manner (Isenberg, 1982).

Interest in the patient's welfare and cooperation can also be demonstrated by the clinical microbiologist's request that a smear accompany each specimen whenever a lesion or wound specimen is submitted for culture or when aspirates are submitted. If necessary, a smear can be prepared in the laboratory. Preparation of a smear can even be applied to urine specimens, especially if such a specimen plays a significant role in the differential diagnosis. Parenthetically, uncentrifuged urine speicmens should be used for this purpose, and the air-dried specimen fixed with absolute methanol rather than heat to avoid loss of cells and microorganisms dislodged by urinary crystaloids. A stained smear of all such specimens provides immediate, pertinent information. The type(s) of bacteria and fungi may be discerned, the cellular response of the host ascertained, and cursory quantitation of all microbial constituents reported. Important results must be reported immediately to provide the physician of record with corroborating evidence for clinical impressions and a more intelligent approach to therapeutic intervention, if indicated. The course of microbiologic analysis is also helped, as culture media and incubation conditions can be selected more judiciously. Smears also serve as quality control measures for evaluating isolates from cultures.

Communicating findings in emergency situations (cerebrospinal fluid, blood, wounds, and the like) and detecting microorganisms that may require isolation of a patient is imperative; experience will teach the laboratorian that such significant reports should be made to responsible individuals, preferably the clinician or designee who is responsible for the patient. A telephone call to a clerk or secretary is often inadequate; if direct contact cannot be made, messages to have a responsible individual return the call should be left and followed up with additional telephone contacts if inordinate delays in such communications are encountered. A record of these contacts must be maintained in the microbiology laboratory.

Advances in technology have provided the laboratorian with the capacity to identify a few specific organisms directly in certain specimens. For example, the microscopic visualization of certain morphotypes on smears of cerebrospinal fluid may be confirmed with commercial reagents that involve antigen recognition by specific antibodies on latex particles or by coagglutination. Some clinical microbiologists may feel that such confirmation is redundant; however, these examinations are mandatory, especially if no bacteria are detected in these specimens despite the clinical impression, the presence of pus cells, or a history of partial treatment. Results may lead to initiation of appropriate therapy. However, cultural analysis of the cerebrospinal fluid must nonetheless be conducted. Antimicrobic susceptibility testing and the determination of β-lactamase activities in infections caused by *Haemophilus influenzae, Neisseria gonorrhoeae,* or *Bacteroides* species are important adjuncts that must be provided as rapidly as possible. The findings of antigen detection tests and antibiotic susceptibilities deserve immediate communication under emergent circumstances.

Clinical microbiologists must provide the means to ensure safe transport of specimens to the laboratory. For most cultures, a transport medium that prevents disturbing the distribution of microbial components, buffers the minimal metabolic activities during this period, prevents general or selective microbial proliferation, and preserves appropriate moisture and atmosphere should be provided. When volumes in excess of 1 ml or biopsy specimens are procured, sterile tubes or other appropriate containers can be used with the proviso that immediate delivery to the laboratory is desirable. Special arrangements for blood cultures usually reflect the preferences and the degree of automation in the microbiology laboratory. Urine specimens have always been problematic. Their collection and transport require ongoing education of all personnel involved. Messenger and charting services, often not within the jurisdiction of laboratory administration, require special attention. Employees of such services should learn to appreciate the importance of their function in the care of patients, especially that undue delay in the delivery of specimens may have severe consequences for very ill patients. Periodic reminders may prove helpful in curtailing unnecessary delays.

Laboratory Records

The methods for logging and tracking specimens are numerous. Obviously, concern with these activities must be great, must be spelled out in each laboratory, and must be made to conform with the rapid changes that impinge on these important activities. Advice on these matters has been provided in several excellent texts (Finegold and Baron, 1986; Isenberg et al., 1985; Miller and Wentworth, 1985; Washington, 1985). Maintenance of an incident log record is a requirement that serves to record specimens that were not properly received or handled, arrived broken, or arrived with some contents spilling and contaminating the outside of the transport vehicles. Immediate notification of responsible individuals must be practiced; the date, the name of the contact, and the time of contact must be registered and the corrective action noted. Although some individuals may be sufficiently persuasive to convince laboratory personnel to perform the requested analysis, only supervisory staff should have the privilege of waiving the need for repeat specimens in circumstances clearly defined by the director of the service.

Analysis

Clinical microbiologists need not apologize for the many variations they have invented in the analyses of clinical specimens. These different approaches often reflect the desire to accommodate procedures to the demands specific to the patient mix of the facility, the geographic location, local habits, past experience, the mind-set inculcated during training and recent advances. All approaches are intended to achieve rapid and accurate recognition of most medically important organisms. Protocols for routine cultural analyses based on the anatomic source of the specimen should be available to guide the technical staff. These directions should take into account the potentially harmful organisms that may be expected in the specimen and devise the concatenation of selective and enriched media in a fashion aimed as well at the recognition of the normal microbiota that can be expected in some routine specimens. This routine approach must, however, be extremely elastic and accommodate the special situations engendered by the patient's history, the smear results, and the service or the specialty of the responsible clinician.

Microbiologists and Clinicians

The clinical microbiologist must respond to the interests of the clinician (Isenberg, 1982). In particular, the relationship between the infectious disease specialist and the microbiologist should be one of cooperation, complementarity, and equality. The difficulty in prescribing the achievement of such a relationship is obvious but this goal can be accomplished if the professional ego of each is made subservient to the welfare of the patient. Dialogue between the laboratorian and the infectious disease clinician should be frequent, constant, reasonable, and within the capability of the laboratory. When a patient's condition dictates referral of a specimen to a reference laboratory, this referral should be accomplished with dispatch. Clinical microbiologists need not defend the limitations of their ability to isolate and recognize rarely encountered, potentially pathogenic organisms. If suspected, arrangements within the laboratory can be made if possible; otherwise the specimen should be sent to a local, state or federal governmental reference laboratory. The inability to use recently described tests for the rapid recognition or isolation of certain microbes can be explained to clinicians on the basis of the experimental nature of the new approach, the lack of general experience with the method, and the undetermined specificity and sensitivity of the examination. The financial strictures combined with professional standards may slow the introduction of new modalities. The situation deserves very careful explanation to the inquiring specialists.

Most clinicians request microbiologic services that reflect their specialty in the hope of finding clues to the specific pathologic process, mostly in the manner that clinical chemistry results are used. Explanations and interpretations by the clinical microbiologist can guide the expectations, for example, of the general practitioner or pediatrician regarding a throat culture that routinely and especially in community residents will yield information only on the presence of *Streptococcus pyogenes* or, in very special circumstances, on the presence of toxigenic *Corynebacterium diphtheriae*. The oncologist, in contrast, will be interested in fungi and gram-negative rods isolated from this locus. These examples underline the need for laboratory flexibility and constant communication.

Emergencies

Clinical emergency situations demand a sympathetic and immediate response by the staff of the microbiology laboratory. Meningitis, septicemia, pneumonia, extensive wound infections, and gangrene are examples of opportunities to demonstrate the concern of laboratorians for patient care. Although this aspect transcends the technical requirements, appropriate responsiveness in keeping with what might be called the art of clinical microbiology is indicated (Isenberg, 1982). Emergencies or persistent unexplained symptoms suggestive of infection require the involvement of the laboratory staff. Unreasonable demands by clinicians may occur; the concerned physician may be grasping at straws with requests for examinations beyond the capability of the laboratory, such as experimental direct fluorescent antibody tests, gas–liquid chromatographic analyses, toxin detection, use of molecular probes, and the like. The explanations outlined previously regarding the nature of these analyses should be proffered in an atmosphere of good will. Immediate performance of Gram stains on a suitable specimen can signal to the clinician the willingness of the staff to make efforts to help. The clinical microbiologist should discuss with the clinician the potential etiological agents in light of the patient's history and other clinical laboratory findings. This discussion should be followed by a constant reporting of results, no matter how preliminary, as soon as they become available. During incubation, cultures should be inspected at intervals of no greater than 6 hours and the clinician apprised of the observations. Antimicrobial susceptibility should be performed as soon as an adequate inoculum is available, and results should be communicated at shorter than usual intervals with an explanation of their preliminary nature.

Conclusions

Today's clinical microbiologists may be on the horns of a dilemma. A constant barrage of new modalities is reported in the literature that appears helpful in escaping the pure culture dogma (D'Amato et al., 1981). These modalities range from rapid manual multiple-compartmental systems for the identification of medically important microorganisms to automation of many or all steps involved in the process (mostly combined with computer identification schemes or susceptibility interpretations) to monoclonal antibody technology and molecular probes. At the same time, there is the reasonable demand that all analytic approaches be pertinent to the diagnosis and treatment of patients, that the new modalities be cost-effective (i.e., equal in cost or preferably less expensive when their total impact on reagents, labor, and patients' hospital stays is considered), and that their accuracy, sensitivity, specificity, and predictive values be equal to or greater than those of existing methods. These divergent demands must be coupled with the understandable desire of each laboratory director to conduct evaluations of products intended to replace existing methods, especially when data generated by collaborative efforts of several investigators to establish the limits of new procedures have not been published in peer-reviewed journals. Another facet of this problem is the thoughtful considerations expressed by Bartlett (1982, 1985) that warn against obfuscating the diagnosis and therapy of the clinician with the plethora of identifications and antimicrobial profiles provided by clinical microbiology services. In addition, infectious disease practitioners (Louria, 1982; McCabe and Stottmeier, 1982) have introduced serious reservations about the clinical significance of the universal identification of microorganisms to the species level. These experts recognize the necessity of such identification in many situations, but do not see any advantage to performing this step on all isolates. Most clinical microbiologists are in complete agreement with this view and undoubtedly insist that such efforts are applied judiciously in their laboratories. Unfortunately, guidelines for this task that encompass all situations are not available. In general, representatives of the expected normal microbiota in specimens from sites that are usually contaminated need not be identified to the species level. A minimal number of phenotypically significant procedures that imply the identity of frequently encountered etiologic agents (e.g., *Escherichia coli* in a routine urine specimen) may suffice for the care of noncritically ill patients. Unless the laboratory staff is apprised of the clinical impression, however, lingering doubts about the propriety of such shortcuts shall continue. Laboratory directors, therefore, must insist that an absolute minimum of patient information accompany each specimen. Otherwise, the need to maintain cultures on file for weeks or longer to meet a future need for more detailed analysis would defeat the advantages gained. Epidemiologic considerations also impinge on the need to identify certain select microorganisms (Isenberg, 1977; Isenberg and Smith, 1981). The clinician or the microbiologist cannot escape recent advances that, for example, may result in interpreting a strain of *E. coli* isolated from a diarrheic stool specimen, devoid of other etiologic agents, as a potentially toxigenic or invasive representative or as the carrier of colonizing factors inimical to the patient's health. Thus, the clinical microbiologist is constantly on a tightrope with many factors poised to upset a most precarious balance. Only constant attention to these many factors permits one to chart a reasonable course for the conduct of professional activities.

The scope and purpose of clinical microbiology remains very clear despite intellectual and practical obstacles. The performance of microbiological analyses is directed at rendering assistance to the physician treating an ailing patient and providing pertinent information to prevent the spread of harmful agents to other patients, hospital personnel, or the community. These goals must be accompanied by constant efforts to improve the accuracy and speed of providing this information, in addition to searching for a better understanding of the host–parasite equilibrium, a disturbance of which determines the degree of harm etiologic agents can inflict. Such efforts may even lead to an explanation of the role the microbial world plays in maintaining the health of humans and all living creatures.

Literature Cited

Bartlett, R. C. 1982. How far to go—how fast to go in 1982, p. 12–44. *In* V. Lorian (ed.), Significance of medical microbiology in the care of patients, 2nd ed. The Williams & Wilkins Co., Baltimore.

Bartlett, R. C. 1985. Quality control in clinical microbiology, p. 14–23. *In* E. H. Lennette, A. Balows, W. J. Hausler, Jr., and H. J. Shadomy (ed.), Manual of clinical microbiology, 4th ed. American Society for Microbiology, Washington, D.C.

D'Amato, R. F., B. Holmes, and E. J. Bottone. 1981. The systems approach to diagnostic microbiology. Crit. Rev. Microbiol. **9**:1–44.

Finegold, S. M., and E. J. Baron. 1986. Bailey and Scott's diagnostic microbiology, 7th ed. C. V. Mosby Co., St. Louis.

Isenberg, H. D. 1977. Significance of environmental microbiology in nosocomial infections and the care of the hospitalized patient, p. 220–234. *In* V. Lorian (ed.), Significance of medical microbiology in the care of patients. The Williams & Wilkins Co., Baltimore.

Isenberg, H. D. 1982. Microbiology and the ailing patient, p. 1–11. *In* V. Lorian (ed.), Significance of medical microbiology in the care of patients, 2nd ed. The Williams & Wilkins Co., Baltimore.

Isenberg, H. D., and A. Balows. 1981. Bacterial pathogenicity in man and animals, p. 83–122. *In* M. P. Starr, H. Stolp, H. G. Trüper, A. Balows, and H. G. Schlegel (ed.), The prokaryotes—a handbook of habitats, isolation and identification of bacteria. Springer-Verlag KG, Berlin.

Isenberg, H. D., and J. I. Berkman. 1962. Microbial diagnosis in a general hospital. Ann. N.Y. Acad. Sci. **98**:647–669.

Isenberg, H. D., and R. F. D'Amato. 1985. Indigenous and pathogenic microorganisms of humans, p. 73–98. *In* E. H. Lennette, A. Balows, W. J. Hausler, Jr., and H. J. Shadomy (ed.), Manual of clinical microbiology, 4th ed. American Society for Microbiology, Washington, D.C.

Isenberg, H. D., and J. K. Smith. 1981. The epidemiology and control of nosocomial and opportunistic infections, p. 119–139. *In* A. Balows and W. J. Hausler, Jr. (ed.), Diagnostic procedures for bacterial, mycotic and parasitic infections. American Public Health Association, Washington, D.C.

Isenberg, H. D., J. A. Washington II, A. Balows, and A. C. Sonnenwirth. 1985. Collection, handling and processing of specimens, p. 73–98. *In* E. H. Lennette, A. Balows, W. J. Hausler, Jr., and H. J. Shadomy (ed.), Manual of clinical microbiology, 4th ed. American Society for Microbiology, Washington, D.C.

Louria, D. B. 1982. The speciation polemic: an analysis of the debate largely from a physician's point of view, p. 45–52. *In* V. Lorian (ed.), Significance of medical microbiology in the care of patients, 2nd ed. The Williams & Wilkins Co., Baltimore.

McCabe, W. R., and K. B. Stottmeier. 1982. Clinical significance of speciation of aerobic gram negative bacilli, p. 56–63. *In* V. Lorian (ed.), Significance of medical microbiology in the care of patients, 2nd ed. The Williams & Wilkins Co., Baltimore.

Miller, J. M., and B. B. Wentworth (ed.). 1985. Methods of quality control in diagnostic microbiology. American Public Health Association, Washington, D.C.

Rosebury, T. 1961. Microorganisms indigenous to man. McGraw–Hill Book Co., New York.

Stanier, R. Y. 1978. What is microbiology?, p. 1/1–32. *In* J. R. Norris and M. H. Richmond (ed.), Essays in microbiology. John Wiley & Sons, Chichester.

Steel, K. J. 1962. The practice of bacterial identification. Symp. Soc. Gen. Microbiol. **12**:405–432.

Washington, J. A. (ed.). 1985. Laboratory procedures in clinical microbiology, 2nd ed. Springer-Verlag, New York.

CHAPTER 2

Quantification Methods in Microbiology

A. TURANO and F. PIRALI

Introduction

The quantitative evaluation of microorganisms isolated and identified in different biological samples is important in microbiology. Since 1956, the determination of bacterial concentration in urine of patients has been used in diagnosing urinary tract infections (Kass, 1956). In alimentary and industrial microbiology, the evaluation of bacterial concentrations in water and food may give important information about potability and edibility (International Commission on Microbiological Specification for Foods, 1978). Likewise, quantitative evaluations of microorganisms in operating rooms, drug preparation laboratories, intensive care units, and other environments can also be invaluable. In such settings, the comparison of quantitative results acquired from different samples during the same day and in subsequent days allows environmental contamination and in some cases its origin to be documented.

Counts of Viable and Nonviable Microorganisms

Microbial counts give rise to both technical and interpretative problems. First, it is important to distinguish between viable and nonviable microorganisms. Viable microorganisms are commonly present in aggregates. Often when using the usual microbial counting methods, an aggregate of bacteria is counted as a single bacterium. Thus, the term colony-forming unit (CFU) has been used to indicate viable bacteria that are evaluated by counting the number of visible colonies on the agar plate.

Bacterial Count Limits

Only rarely are bacteria uniformly distributed in samples. Thus, quantitative results may not always be reproducible, especially if a small sample is used for the count. Counting methods depend on the specific medium used and its pH and incubation requirements. Therefore, the accuracy of quantitative determinations may be low, depending on the materials and technical methods used. A single determination generally is not reliable; averaging several determinations provides greater accuracy, especially if the counting method is not standardized.

Classification of the Counting Methods

The methods used in bacterial counting are numerous, and may be categorized as physical or biological. Generally, physical methods are used in a total bacterial count and do not depend on viability, whereas biological methods evaluate the viability of microorganisms.

Methods can also be classified as direct, which evaluate the number of bacteria directly, and indirect, which determine bacterial numbers through various metabolites and activities of the microorganism. Most indirect methods are used only in research. In this chapter, we will consider the most common microbial counting methods.

Common Bacterial Counting Methods

Direct Methods

Agar Plates

Using agar plates, a predetermined quantity of sample is seeded in Petri dishes to which a still-liquid medium has been added after careful mixing (Anderson and Stuart, 1935; Reed and Reed, 1948). After solidification, microorganisms enclosed in the gel grow until they develop into visible colonies that are easily counted with a lens or an electronic apparatus. In practice, the sample is diluted and seedings of different dilutions are made. When the number of colonies is between 20 and 200, the colonies are counted. This method is based on the principle that every viable microorganism will reproduce and yield a visible colony. We presume that bacteria are isolated single cells, are uniformly distributed in the sample, and are not present in aggregates. Although counting the bacteria on the plate is easily accomplished, it is time-consuming. However, this method has the advantage of high sensitivity. Theoretically, if only 1 bacterium per ml is present in a sample, one visible colony should develop after seeding 1 ml of the specimen.

Membrane Filtration

If the number of microorganisms in the sample is very low, it is possible to increase the sensitivity of the agar plate method by using molecular or membrane filters with pores that retain bacteria (Dick, 1967; Guthrie and Reeder, 1969; Winn et al., 1966). The membranes are placed in the agar plate where, after growth, the colonies are counted. This method is a variant of the simple count in plates specially designed to collect bacteria from a large volume of air or liquid.

Turbidimetric Methods

Bacterial suspensions can be counted by turbidimetric methods (Wallis et al., 1981). With turbidimetric methods, absorbed light is proportional to the number of bacterial cells reached by a light beam. By use of a spectrophotometer, colorimeter, or nephelometer, it is possible to measure the optical density (absorbance) of the bacterial suspension in a quartz container. The optical density is equal to the relationship between the log of the intensity of the incident light and the intensity of the reflected light.

This method is very rapid and simple, but it is applicable only to bacterial concentrations high enough to cause turbidity in the suspension. Furthermore, colored bacterial suspensions and extraneous material in the suspension change the results. Additionally, this method is not able to distinguish between viable and nonviable bacteria because both cause turbidity in the suspension.

Electronic Counters

Electronic bacterial counters operate on much the same principle as do hematologic counters (Pelczar et al., 1977). Even without modification, such an apparatus can be used to count cells.

A wide variety of bacterial counting equipment is commercially available. A predetermined volume of bacterial suspension is placed in a container and passed through a 5- to 10-μm orifice placed between two electrical poles. When a microorganism passes through the orifice, a change in resistance is produced, amplified, and registered electronically, allowing the microorganism to be counted.

Microscopic Counts

Microscopic counting methods are rapid, simple, and require minimal equipment. Bacteria can be easily counted in various cell count chambers, such as Buerker and Nageotte chambers that are generally suitable for blood or cephalhorachidian liquor cell counts (MacKay-Scollay, 1969). The Helber and Petroff-Hansen chambers are also used for counting bacteria. These are thick glass chambers divided into squares covered by glass; square volume is 1/20,000 mm^3. If, for instance, 5 bacteria for every square of the chamber are counted, the concentration of suspension will be 100,000,000/ml. It may not be possible to count a dense bacterial suspension; such suspensions should be diluted before microscopic evaluation. This method cannot distinguish between viable and nonviable cells, and dilution processing may result in mistakes.

The complete filling of the chamber volume is very important. The position of the coverglass is correct when Newton's ring can be seen between the two glass surfaces. After the chamber is loaded, it is important to wait 4 to 5 min for the bacteria to sediment and to average 2 or 3 counts. Careful cleaning of the chamber and coverglass is necessary to avoid bacterial adherence.

Indirect Methods

Determination of Nitrogen Content

The concentration of a bacterial population can be evaluated by measuring its nitrogen content, which is 14% of the weight of the bacteria (Baker, 1967). The determination of bacterial nitrogen is rather difficult and time-consuming, but it allows for the collection of bacterial cells without a culture medium. However, this process requires very controlled conditions to avoid every source of nonbacterial nitrogen. This method is applicable only to bacteria in high concentrations and generally is reserved for research work.

Determination of Cell Weight

This is the most straightforward method for evaluating the concentration of a bacterial suspension (Baker, 1967). Generally, the method is very reliable and reproducible, but it can be used only for high-density bacterial suspensions, usually in research.

Biochemical Evaluation

Biochemical evaluation of bacterial concentration involves counting metabolites produced by bacteria from substrate in the culture medium (Clarke and Cowan, 1952; Eliot and Pryles, 1964). Its most common use is in determining the amount of lactic acid produced from bacterial glucose fermentation. The concentration of lactic acid is proportional to the bacterial concentration of the suspension. This type of count is generally used only in research.

Bacterial Counts in Clinical Specimens

The methods described below are suitable for counting bacteria in biological material except material such as urine from patients with urinary tract infections, cephalhorachidian liquor from patients with meningitis, or spring water submitted for a potability judgment. Particular consideration should be given to samples obtained by swab (such as pharyngeal, skin, ear, and so on) or solid materials obtained from a biopsy. In these cases, different quantification techniques must be used.

Microflora quantification has a special role in clinical microbiology, where speed in obtaining results that have a diagnostic or prognostic significance for the patient is vital. Some methods described in the scientific literature for bacterial quantification in biological samples are presented next.

Stool Microflora

The quantification of microflora in stool specimens is useful in studying the bacterial ecology of the colon and in diagnosing infectious enterocolitis (Flocks et al., 1968). Five grams of fresh stool are homogenized in 45 ml of sterile water. Several dilutions in sterile water of the sample are then made: 1 : 10, 1 : 100, 1 : 1,000 and so on until 1 : 1,000,000. An inoculum of 0.1 ml for every dilution is seeded in selective media for enterobacteria, anaerobes, and fungi and in nonselective media, and the resulting solutions are incubated at 37°C. The number of counted colonies is then multiplied by the dilution factor (e.g., 10, 100, or 1,000) and by 10 to express the value per milliliter.

Skin Microflora

Swab Method

A swab of cotton or rayon moistened in phosphate buffer plus 0.01 of Triton×100 (Sigma Chemical Co., St. Louis, Mo.) is applied 10 to 12 times to the skin surface. The swab is then put in 5 ml of sterile buffer and centrifuged for 30 min. Progressive dilutions of buffer are made, and from every dilution, 0.1 ml is spread on agar plates and incubated for 48 h at 37°C. The number of colonies grown is then counted and multiplied by the dilution factor.

Cellophane Tape-Stripping Method

Using sterile gloves, the tape is pressed to a marked area of skin (Updegraff, 1964). The tape is then placed with the sticky side up in a Petri dish, and 15 ml of cooled agar is added. The agar is allowed to solidify and is incubated for 2 to 4 days at 37°C. The number of colonies is then counted and divided by the area of the tape to obtain the number of bacteria per square centimeter of skin.

Plastic Bag Technique

This method is primarily used to test the effectiveness of handwashing soaps. A plastic bag is placed in a bucket with its top folded over the rim and fastened securely. Sterile water is poured into the bag, and the technician washes his or her hands in this water for 2 min using the test soap and scrubbing well. The plastic bag is closed and fastened. Within 15 min, a 10-fold serial dilution (up to 1 : 1,000) of the liquid in the bag is made. Next, 1 ml of each dilution is placed in a Petri dish, spread with a sterile microbiologic loop, and incubated at 37°C. The colonies are counted and the number is multiplied by the dilution factor.

Wound and Burn Microflora

Biopsy Culture Method

In this method, the wound surface is first cleaned with alcohol (Brentano and Gravens, 1967). A biopsy specimen of the open granulating wound is obtained, and the tissue is weighed and aseptically homogenized in 9 ml of thioglycollate broth. Serial 10-fold dilutions are made of homogenized tissue in sterile water, and two Petri agar plates are spread with 0.1 ml of different dilutions. One plate each of each dilution is aerobically and anaerobically incubated for 48 h. The number of colonies is counted and then multiplied by the dilution factor. This value is then divided by the weight of the tissue to obtain the number of bacteria per gram of tissue.

Contact Plate Method

Contact plates are microbiologic plates filled to a convex surface with tryptic soy agar or other medium, with the bottom surface divided into predetermined areas (Hall and Harnett, 1964). The lid is removed and the plate is inverted so that the agar touches the surface to be tested. The plate is pressed gently for 2 or 3 sec but is not allowed to skid along the surface. The plate is incubated for 2 days, and the number and types of colonies are recorded.

Sputum Microflora

The quantification of bronchial microflora is useful in studying the bacterial ecology of the respiratory tract in healthy subjects and patients with infectious bronchopneumopathies.

N-acetyl cysteine (Monroe et al., 1969), dithiothreitol (Shah and Dyer, 1966), alpha-amylase (Balows et al., 1969), and sodium ethylhexylsulphate (Kilbourn et al., 1968) can be used to fluidify the samples. In quantitative sputum cultures, sputum is placed in a centrifuge tube containing glass beads. The tubes should not be more than 20% full. An equal amount of fluidifying agent is added and mixed for 1 to 3 min. Serial dilutions up to 1 : 10,000,000 are made in the diluent. Cultures of every dilution in blood agar, salt mannitol agar, chocolate agar, Sabouraud agar, and MacConkey agar are made using a 0.01-ml calibrated loop. The number of colonies is counted and multiplied by the dilution factor.

Urine Microflora

Calibrated Loop Method

In this method, a 0.01-ml calibrated loop, which may be bent at a 45° angle, is inserted into a well-mixed specimen, and the loop is removed without being tipped (Hoeprich, 1960). The loop is streaked across the center of the agar plate, covering the plate's full diameter. With cross-hatching, the entire surface is spread from this streak, and the plate is incubated for 12 to 18 h. The colonies are counted, and the number is multiplied by the dilution factor to determine the concentration of bacteria per milliliter of sample.

Agar Slide Dip Method

With this technique, culture medium can be put on a sterile microscope slide or other commercially prepared apparatus (Naylor and Guttman, 1967). Such a prepared slide is removed from the refrigerator, dipped in urine, and then drained by touching it to absorbent paper and leaving it standing upright for a few minutes. Once dried, the slide is incubated for 12 to 18 h in an unsealed container. The colonies on the slide are then counted. If it is possible to see 50 to 100 colonies on the slide, the bacterial concentration of urine will be nearly 10,000 CFU/ml. If the growth is confluent, the bacterial concentration will be greater than 1,000,000 CFU/ml.

Spread Plate Method

With a pipette or loop, 0.01 ml of a specimen or its dilution is placed on the agar plate. Using a bent wire, the inoculum is spread over the surface of the plate. The plate is incubated for 12 to 18 h, after which the colonies on the plate are counted. To obtain the number of colonies per milliliter of urine, the count is multiplied by the dilution factor.

Modified and Simplified Methods

Numerous simplified methods of bacterial quantification have been developed because most traditional methods are time-consuming and not available for routine work. The sensitivity of these modified methods differs depending on equipment and media used.

A new medium for counting bacteria in water, called R2-A (containing yeast extract, peptone, glucose, soluble starch, and sodium pyruvate) has been proposed by Reasoner and Geldreich (1985). Parallel studies of bacterial counting methods on plates have documented that R2-A allows the determination of a larger number of microorganisms than is possible with traditional media, especially if the incubation period is prolonged. The largest number of bacteria from water was counted after 14 days' incubation at 20°C (Reasoner and Geldreich, 1985).

The reliability of bacterial counting has been evaluated because results obtained from one sample can vary widely. Woolfrey (1981), when counting bacteria of skin eschars divided into two equal parts, found wide variability in the results and a low correlation between bacterial concentration in eschars and the clinical probability of septicemia in patients with infected burn wounds. Such discrepancies in quantitative data in clinical microbiology have been widely debated and have resulted in controversy, especially over determining bacterial counts in anatomic sites where a wide variety of saprophytic flora is present. This is the case in the respiratory tract, which is sterile below the larynx but colonized by microflora above. In such conditions, distinguishing between saprophytic and pathogenic microorganisms may be very difficult. However, interest in quantitative bacteriology derives from the opinion that the quantitative data may be valuable for complete and quick diagnosis of bronchopneumopathies.

Determining the concentration of a single bacterial species isolated from sputum, tracheoaspirate, bronchoaspirate, or other bronchial material could, at least in some cases, exclude the possibility that a bacterial species is an oral contaminant. In fact, by this method, it is easy to detect a species that has a pathogenic role in bronchopneumopathy because a pathogenic species is more numerous than are normal colonizing bacteria of the respiratory tract. Furthermore, the same quantitative examination allows us to study the ratios of different bacterial species, delineating an exact map of the complex bacterial ecology of the respiratory tract. The greatest limits to the use of quantitative methods in respiratory bacteriology are their time-consuming nature and lack of availability for routine laboratory use. In light of this problem, we will present the most important technical aspects of a rapid method for the quantification of respiratory microflora performed at our laboratory.

Rapid Method for Isolating and Counting Bacteria in the Respiratory Tract

During the last 30 years, many investigators have studied the quantification of bronchial microflora with conflicting results in relation to the different diseases studied and different methods used. All investigators, however, have indicated the need for a rapid and easy method of quantification. In our laboratory, we developed a simple method to count the microorganisms in the respiratory tract.

In this method, the clinical specimen is collected in a plastic sterile container and quickly transported to the laboratory. It is put inside a sterile plastic bag with a 0.1% dithiothreitol solution in a 1:10 dilution and homogenized mechanically for 2 min. This material is seeded with a 10-μl calibrated loop, making parallel streaks on special plates of sheep blood agar, salt mannitol agar, MacConkey agar, Sabouraud agar, and chocolate agar plus factor V. The streaks must be made by dipping the loop into the clinical specimen only one time for each agar plate used. The loop will progressively discharge its contents on the agar plate, and the bacteria grow in streaks in proportion to the initial bacterial concentration. The chocolate agar plate is incubated in 7% carbon dioxide; the others are incubated in ambient air. All are incubated at 37°C for 18 to 24 h.

We have compared results from this method with results of the agar dilution methods, thereby fixing the numerical value corresponding to the growth on each streak. If the bronchial secretion has been diluted to 1:10, the values are roughly 100,000/ml on the first streak, 1,000,000/ml on the third streak, 10,000,000/ml on the sixth streak, and 100,000,000/ml on the 13th streak, with intermediate values on the other streaks. The simultaneous use of selective and nonselective media on the same plate allows the comparison of total and partial bacterial concentrations for each species. By making streaks, the progressive dilution of the bronchial material permits bacterial colonies to be isolated in the plates within 12 to 18 h for biochemical assays.

Our data suggest that bacteria are responsible for a lung infection only if they are present in sputum or bronchoaspirates above a certain concentration. However, there are some exceptions, such as in immunodeficient patients in whom lower concentrations of microorganisms are potential pathogens. Special mention must be made about patients affected by chronic bronchopneumopathies or patients undergoing antibiotic therapy; in these patients, strict correlation between quantitative microbiologic data and clinical status is not valid. Naturally, judgment is influenced by available clinical information and previous microbiologic reports. Our data, however, suggest that the simple identification of a bacterium assumed to be a pathogen in sputum is not enough to demonstrate its etiologic role. Only the evaluation of its concentration in bronchial secretions can document its pathogenic role.

Literature Cited

Anderson, E. H. and C. A. Stuart. 1935. A quantitative differential method for counting mixed cultures of bacteria. J. Bacteriol. **30**:207–209.

Baker, F. J. 1967. Handbook of bacteriological technique, 2nd ed. Butterworth, London.

Balows, A., W. Allen, P. E. Greenup, A. M. Walters. 1969. Quantitative sputum cultures: an approach to better clinical microbiology. Bact. Proc. **M**:254.

Brentano, L. and D. L. Gravens. 1967. A method for quantification of bacteria in burn wounds. Appl. Microbiol. **15**:670–671.

Clarke, P. H. and S. T. Cowan. 1952. Biochemical methods for bacteriology. J. Gen. Microbiol. **6**:187–197.

Crone, P. B. 1948. The counting of surface colonies of bacteria. J. Hyg. **46**:426–430.

Dick, S. L. 1967. The enumeration of bacteria in urinary tract infections. Can. J. Med. Techn. **29**:15–22.

Eliot, C. R. and C. V. Pryles. 1964. Observations on the use of triphenyltetrazolium for the detection of bacteriuria. Pediatrics **34**:421–433.

Flocks, M. H., W. Gershengoren, and L. R. Freedman. 1968. Methods of the quantitative study of the aerobic and anaerobic intestinal bacterial flora of man. Yale J. Biol. Med. **41**:50–61.

Guthrie, R. K. and D. J. Reeder. 1969. Membrane filter-fluorescent antibody method for detection and enumeration of bacteria in water. Appl. Microbiol. **17**:399–401.

Hall, L. B. and M. J. Harnett. 1964. Measurement of the bacterial contamination on surfaces in hospitals. Public Health Rep. **79**:1021–1024.

Hoeprich, P. D. 1960. Culture of urine. J. Lab. Clin. Med. **56**:899–907.

International Commission on Microbial Specifications for Food. 1978. Microorganisms in foods. Their significance and methods of enumeration, 2nd ed. University of Toronto Press, Toronto.

Kass, E. H., 1956. Asymptomatic infection of the urinary tract. Trans. Assoc. Am. Phys. **69**:56–64.

Kilbourn, J. P., R. A. Campbell, J. L. Grach, and M. D. Willis. 1968. Quantitative bacteriology of sputum. Am. Rev. Resp. Dis. **98**:810–818.

Litsky, B. Y. and W. Litsky, 1967. The evaluation of single-use bar soaps for surgical scrub. Hosp. Management **103**:74–86.

MacKay-Scollay, E. M. 1969. A simple quantitative and qualitative microbiological screening test for bacteriuria. J. Clin. Pathol. **22**:651–653.

Monroe, P. W., H. G. Muchmore, F. G. Felton, J. K. Pirtle. 1969. Quantitation of microorganisms in sputum. Appl. Mocrobiol. **18**:214–220.

Naylor, J. R. and D. Guttman, 1967. The dip-slide: a modified dip inoculum transport medium for the laboratory diagnosis of infections of the urinary tract. J. Hyg. **65**:367–371.

Pelczar, M. J., R. D. Reid, and E. C. S. Chan. 1977. Microbiology. McGraw-Hill, New York.

Reasoner, D. J. and E. E. Geldreich. 1985. A new medium for enumeration and subculture of bacteria from potable water. Appl. Env. Microb. **49**:1–7.

Reed, R. W. and G. B. Reed. 1948. Drop plate method of counting viable bacteria. Can. J. Res. **E**:316–326.

Shah, R. R. and W. E. Dyer. 1966. Use of dithiothreitol to replace N-acetyl L cysteine for routine sputum digestion-decontamination for the culture of mycobacteria. Am. Rev. Resp. Dis. **94**:454.

Ulrich, J. A. 1964. Technics of skin sampling for microbial contaminants. Health Lab. Sci. **1**:133–136.

Updegraff, D. M. 1964. A culture method of quantitatively studying the microorganisms in the skin. J. Invest. Dermatol. **43**:129–137.

Wallis, C., J. Melnick, and C. Longoria. 1981. Colorimetric method for rapid determination of bacteriuria. J. Clin. Microbiol. **14**:342–346.

Winn, W. R., M. L. White, W. T. Carter, W. T. Miller, and S. M. Finegold. 1966. Rapid diagnosis of bacteremia with quantitative differential membrane filtration culture. J. Am. Med. Assoc. **197**:539–548.

Woolfrey, B. F. 1981. An evaluation of burn wound quantitative microbiology. Am. J. Clin. Pathol. **75**:532–537.

CHAPTER 3

Immunologic Methods for Detection of Microbial Antigens

RICHARD C. TILTON

Early diagnosis and prompt institution of specific antimicrobial therapy are required for the effective treatment of infectious disease. Traditionally, the laboratory has isolated and identified the specific etiologic agent and, when possible, determined its antimicrobial susceptibility profile. If cultural studies were not possible, then a diagnosis could be made by observing the appearance or disappearance of specific antibodies. Although the detection of antibodies remains important in the diagnosis of infectious disease, the process is time consuming, because for optimal results acute and convalescent serum samples should be collected 2 to 3 weeks apart.

A variety of immunologic techniques are available to detect microbial antigens directly in body fluids. They include counterimmunoelectrophoresis, latex agglutination, coagglutination, enzyme immunoassay, fluorescent immunoassay, fluorescence polarization immunoassay, and radioimmunoassay. Counterimmunoelectrophoresis, latex agglutination, and coagglutination are manual techniques that have not been automated. Enzyme immunoassay, fluorescence immunoassay, fluorescence polarization immunoassay, and radioimmunoassay, however, are more aptly suited to semiautomation and several instruments are commercially available for detection of microbial antigens. Not all of these instruments automate the entire process; instead, they are sophisticated instruments that utilize digital circuitry for process and analytic control.

Counterimmunoelectrophoresis

Counterimmunoelectrophoresis was originally described in 1959 by Bussard and was first used clinically for the detection of "Australia antigen" (HbsAg). Radioimmunoassay soon replaced counterimmunoelectrophoresis for this function; however, it then became a valuable immunologic tool for rapid detection of both microbial antibodies and antigens. Counterimmunoelectrophoresis is a rapid precipitin reaction in which the reactants are excited by an electric current. In 1901, Vincent and Bellot first described the use of the tube precipitin reaction for the detection of meningococcal antigen in cerebrospinal fluid. Dochez and Avery identified the capsular polysaccharide of the pneumococcus in patients' urine as early as 1917. The basic principle of antigen detection is more than 60 years old. However, the precipitin reaction is a function of the precipitation of antibody and soluble antigen at the equivalence point, and this reaction may take up to 18 h. Counterimmunoelectrophoresis, on the other hand, combines the advantages of immunodiffusion and electrophoresis. The antigen is placed in a well on the cathodic side of a solid support and the antibody is placed on the anodic side. The antigen, if negatively charged, migrates toward the anode and the antibody, which has a weak negative charge, also migrates toward the anode. Positively charged buffer ions sweep the antibody molecule to the cathode, a phenomenon called *endoosmotic flow*. If conditions of voltage, current, buffer, pH, and antigen–antibody concentration and quality of antisera are optimal, a visible precipitin line appears between the two wells after as little as 30 min of electrophoresis.

Currently, counterimmunoelectrophoresis is not available in kit form. Some laboratories continue to use it for antigen detection tests, as it is more economical of reagents than agglutination procedures. For most, however, counterimmunoelectrophoresis has become a transitional technology, being rapidly replaced by latex agglutination and coagglutination techniques. The prepackaging of these antigen detection kits make them much more convenient, albeit less economical.

Latex Agglutination

Latex polystyrene beads were first used to detect rheumatoid factor in serum. Hence, the general term *latex fixation,* while applying to the original work on rheumatoid factor, has been used clinically to describe a variety of latex agglutination tests. Although incorrect, it points out the broad visibility of latex as a sensitized inert carrier of antigens and antibodies.

Styrene was first discovered almost 200 years ago by the German chemist Neuman. In 1839, Simon, an Englishman, converted styrene to polystyrene. Polystyrene was not widely used until World War II, when its value as synthetic rubber proved a major turning point in the war. Latex, then, is tiny microscopic particles of polystyrene. A comprehensive review of latex agglutination by Hechemy and Michaelson (1984) was published.

Simply stated, antigens or immunoglobulin (Ig)G antibody are nonspecifically absorbed to the surface of these latex polystyrene beads, which are of uniform diameter (usually 0.8 μm). Addition of the homologous antibody or antigen visibly agglutinates the milky-white latex suspension. Although latex agglutination tests can be done in test tubes, they are usually performed on slides for ease of operation.

Depending on the system, the procedure for the detection of antigen or antibody by latex agglutination is quite simple. A drop or two of the latex reagent is mixed with a suspension of the colony, the body fluid, or antigenic extract to be tested. The suspension is mixed and incubated at room temperature with occasional mechanical agitation of the slide. The presence of agglutination constitutes a positive test. One of the biggest drawbacks to latex agglutination is nonspecific reactions with specimens such as urine, sputum, serum, and synovial fluid. False-positive agglutination can sometimes be eliminated by heating the specimen to 60°C for 15 min or adding ethylenediaminetetraacetate.

There are two ways to perform latex agglutination tests: (1) Latex agglutination inhibition, and (2) direct agglutination of sensitized latex particles. Latex agglutination inhibition is used primarily for detection of low molecular weight antigens. Its use in microbiology is limited to the measurement of aminoglycoside antibiotics in body fluids. Direct agglutination, on the other hand, is used for high molecular weight substances and is applicable to either antigen or antibody detection. The pH range of latex agglutination is narrow (7.0 to 8.0). The reaction takes place at room temperature on slides, in tubes, or in microdilution trays. Sensitized latex is provided either as a liquid reagent or dried in capsules. Tube tests are 10 to 100 times more sensitive. Because the tube latex agglutination test is an equilibrium reaction, specificity is increased. The disadvantage of the tube latex agglutination test is that the minimum incubation period is 2 h, compared with 3 to 5 min with the slide agglutination test.

A variety of factors may interfere in the latex agglutination test, including complement fraction C1Q when testing for rheumatoid factor, lipemic serum, rheumatoid factor when testing for either antigens or antibodies in serum, and nonspecific aggregation of latex. C1Q interference may be removed by diluting the serum before testing or heating the serum at 56°C for 30 min. Lipemic serum often can be clarified by high-speed centrifugation. Rheumatoid factor can be removed in several ways, including heat, addition of ethylenediaminetetraacetate or dithiothreitol, or coating the latex with $F(ab')_2$ fragments of IgM. Nonspecific aggregation of latex may occur when the protein concentration of the sample is low and destabilization of the latex reagent occurs. On the other hand, an excessive protein concentration may lead to overstabilization of the latex and reduced aggregation. Prozone phenomena are also seen with latex agglutination, and samples containing high concentrations of antigen may require dilution.

Latex agglutination has been reported to be more sensitive than either counterimmunoelectrophoresis or coagglutination, but less sensitive than enzyme immunoassay, fluorescence immunoassay, or radioimmunoassay.

Coagglutination

The coagglutination technique for the detection of pneumococcal antigens was first introduced by Kronvall (1973). Certain strains of *Staphylococcus aureus,* including Cowan strain I, contain a cell surface protein called protein A. Antibodies (IgG1, 3, and 4) adhere to protein A by their F_c portion, leaving the Fab ends free to react with homologous antigen. The presence of antigen results in visible agglutination of the staphylococcal particles. The agglutination of sensitized protein A-containing staphylococci may be less distinct than with latex. The procedure, like latex agglutination, is subject to nonspecific agglutination in body fluids such as cerebrospinal fluid, serum, and urine. The preparation of coagglutination reagents is technically simple and, once prepared, the sensitized staphylococci are more stable than latex reagents. The reagent is stable under refrigeration and its use is similar to that of a latex reagent.

Enzyme Immunoassay

The search for methodology with the advantages of radioimmunoassay and few of the disadvantages culminated in the discovery of the enzyme immunoassay technique. Engvall and Perlmann (1971) first used an enzyme-linked immunosorbent assay to measure rabbit IgG.

The enzyme-linked immunosorbent assay, now termed *enzyme immunoassay,* has wide application in clinical microbiology. The concept is similar to that of radioimmunoassay. An antibody or antigen is bound to a solid support, a plastic tube, tray, paddle, or polystyrene beads. The complementary substance forms an antigen–antibody complex. In the simplest procedure, an anti-species enzyme-labeled antibody is added to the developing "sandwich."

After separation of the bound and free enzyme-labeled antibody (or antigen), substrate is added and the resulting color formation indicates the presence of either antigen or antibody. The reaction may be halted before reading or kinetic analysis performed. There are two major types of enzyme immunoassays: (1) homogenous assay and (2) heterogenous assay.

The homogenous assay, sometimes called enzyme-mediated immunosorbent test (EMIT) is a competitive assay. The substance to be tested, usually a low molecular weight product such as an aminoglycoside antibiotic (e.g., tobramycin), is enzyme labeled. When antibody to tobramycin is added, the enzymatic activity decreases. In the clinical test, serum that purportedly contains some tobramycin is mixed with the enzyme-labeled tobramycin. Both bound and unbound antibiotic compete for binding sites on the antibody. An excess of unbound tobramycin complexes antibody to the extent that it is unable to inhibit the activity of the labeled tobramycin complex.

In the enzyme immunoassay double-sandwich technique, the solid phase is coated with specific antibody. The antigen-containing body fluid is layered over the sensitized solid phase, to which is added an enzyme-labeled antibody and substrate. For antibody determinations, the antigen is bound to the solid phase. The sera are incubated, and the test antibody binds to the antigen. After washing, an enzyme-labeled antiglobulin and substrate are added.

Although the methodology of enzyme immunoassay is not conceptually difficult, there are many variables to be controlled. In the author's experience, the identification and control of the variables is essential for a successful test. These variables include the enzyme used, the substrate, the reaction termination reagent, the solid phase, incubation times, antigen–antibody concentration, and visual versus spectrophotometric reading of results.

Fluoroimmunoassay

Fluorescence immunoassay systems have not been widely used for detection of microbial antigens, but have become increasingly popular for therapeutic drug monitoring. The major difference between fluorescence immunoassay and either enzyme immunoassay or EMIT is that the indicator system is a fluorescing molecule. A fluorescent response may be achieved by directly determining the extent to which fluorescent labeled antigens compete with non-fluorescent antigens for binding sites on an antibody or by the use of fluorescent substrates for alkaline phosphatase or β-galactosidase. These high-energy substrates do not necessarily increase the sensitivity of an immunoassay procedure. Sensitivity is a function of the antigen–antibody response. The use of fluorescent substrates, however, markedly reduces the time necessary to achieve the sensitivity limits of the test. The 2-h incubations required with enzyme immunoassay can be reduced to minutes with fluorescent detection methods. The high-energy substrates may allow immunoassays to compete in the same "time to detection" frame as counterimmunoelectrophoresis, latex agglutination, and coagglutination.

Fluorescence Polarization Immunoassay

Fluorescence polarization immunoassay, like EMIT and fluorescent immunoassay, is a competitive protein-binding process. For example, a gentamicin molecule is labeled with fluorescein. There is competition for antibody binding sites between fluorescent and unlabeled gentamicin. In fluorescence polarization immunoassay, the binding of fluorescein–gentamicin to antibody is measured by determining its fluorescence polarization. The polarization of the fluorescent signal is directly proportional to the amount of labeled gentamicin bound and inversely proportional to the unlabeled antibiotic.

Amplified Enzyme Immunoassay Systems

Many attempts have been made to maximize the sensitivity of enzyme immunoassays by amplifying the signal. These attempts have included the use of various combinations of antigen–antibody sandwiches, as well as the addition of various labels and the use of unique solid phases. An innovation that combines several features of latex agglutination and enzyme immunoassay was published recently (Miotti et al., 1986). This system is called *centrifugation-augmented solid-phase immunoassay* (CASPIA). Miotti et al. (1986) describe CASPIA as follows:

> The wells of microtiter plates are coated with antibody directed at the antigen to be measured. Control wells are prepared by the binding of nonimmune antibody at the same

concentration. Unbound immunoglobulin is removed by washing, and unbound sites on the solid phase are blocked by the addition of a nonreactive protein such as bovine serum albumin. Particulate antibody to the antigen and the sample to be tested are added. The sample and the particulate antibody can be added simultaneously or sequentially depending upon the nature of the specimen. If there is antigen (Ag) in the sample, it will bind to the solid-phase antibody and to the particulate antibody causing the particles to adhere to the solid phase. The microplate is centrifuged at 1,500 g for 15 min. The presence of antigen in the solid phase is manifested by particles remaining on the solid phase. Specimens not containing antigen or specimens reacted with solid phase coated with control immunoglobulin will result in the particles migrating to the bottom of the wells. The presence of the particles on the side of the wells can be seen macroscopically, recorded by a photocopying device, or measured by a microplate colorimeter.

When CASPIA was evaluated for detection of *Haemophilus influenzae* type B antigen and group A *Streptococcus* polysaccharide antigen, markedly higher sensitivity was observed. CASPIA, as well as similar amplification technologies, can be easily automated and may be applicable to a wide variety of clinical tests.

Radioimmunoassay

Radioimmunoassay is a technique that combines the specificity of immunology and the sensitivity of radiochemistry. The principal use of radioimmunoassay is in endocrinology for the assay of hormones; in most areas of laboratory medicine, however, it is also a useful tool. There are many variations of this competitive protein binding assay (another term for radioimmunoassay). In most applications, however, labeled (usually with ^{125}I) antigen competes with unlabeled antigen for available binding sites on a specified amount of homologous antibody. After equilibrium between the bound and the unbound antigen is reached, the bound compound is separated from the unbound by centrifugation or precipitation. In some applications the antibody is coated on tubes, and separation of the phases is as easy as decanting the fluid in the tube.

Although microbiologic research employs radioimmunoassay widely, it is rarely used routinely in clinical microbiology laboratories except for the determination of aminoglycoside antibiotics in body fluids.

Time-Resolved Fluoroimmunoassay

One of the biggest problems in any fluorescence immunoassay is background fluorescence. The fluorescent signal of the analyte to be detected is usually very low and often masked by interfering signals. Time-resolved fluoroimmunoassay uses chelates of europium and terbium as labels instead of isotopes, enzymes, or other fluorescing compounds (Soini, 1985). The advantage of these lanthamide chelates is their high quantum yield, which enables emission peaks to be easily seen above background. The instrumentation of this method also reduces background by selective detection of long-decay fluorescence; that is, the detection can be programmed to read emission signals between 400 and 800 μs after excitation. This programming cancels out the short-lived background fluorescence (such as that caused by serum) that occurs up to 100 μs after excitation. Time-resolved fluoroimmunoassay has been widely used in Europe for viral antigen detection. The instrument is sold by Wallac (Turku, Finland).

Summary of Clinical Applications

Counterimmunoelectrophoresis has been used primarily for the detection of microbial antigens in body fluids (Tilton, 1978). Virtually any body fluid may be tested, including cerebrospinal fluid, urine, serum, pleural fluid, synovial fluid, peritoneal fluid, abscess drainage, and pericardial fluid. In general, those fluids with less protein (cerebrospinal fluid, urine) are easier to process, as there are fewer spurious precipitin lines or areas of nonspecific precipitation of protein or lipoprotein.

Edwards (1971) first demonstrated the polysaccharide capsule of *Neisseria meningitidis* in the serum of patients with meningococcemia. The same polysaccharide was detected by counterimmunoelectrophoresis in cerebrospinal fluid samples from 47 of 68 patients and cultures from 42 of 68 patients. Since these early reports, much has been published on the rapid diagnosis of bacterial meningitis by counterimmunoelectrophoresis. The most commonly detected antigen components are *H. influenzae, N. meningitidis,* group B streptococci, and *Escherichia coli* K1. The combination of cerebrospinal fluid, serum, and urine counterimmunoelectrophoresis can identify virtually 100% of children with *H. influenzae* type B meningitis and 60% of patients with pneumococcal meningitis, but less than 50% of those with meningococcal meningitis (Feigin et al., 1976). The poor results with meningococcal meningitis are in contrast to those of some earlier investigations and are probably due to the serogroups involved. It is also recognized that group B meningococcal polysaccharide cannot be detected by a routine agarose counterimmunoelectrophoresis system (Tilton, 1978).

The value of counterimmunoelectrophoresis for the prospective diagnosis of early-onset group B streptococcal infection has been reported. Edwards and Baker (1979) found group B streptococcal antigen in the cerebrospinal fluid of an infant with both meningitis and septicemia, in the serum of 3 of 7 septicemic infants, and in the concentrated urine of 6 of 7 patients. In all 7 cases, group B *Streptococcus* was eventually cultured from the blood. Jacobs et al. (1981) report similar utility of counterimmunoelectrophoresis for detecting group B *Streptococcus*.

Anhalt et al. (1978) list the following five applications of counterimmunoelectrophoresis in infectious disease: (1) Detection of antigen in body fluids; (2) determination of antibody titers (poor sensitivity); (3) prognostic assessment; (4) identification and/or typing of clinical isolates; and (5) elucidation of the role of circulating antigens in the pathogenesis of disease.

Latex Agglutination

Ingram et al. (1975) showed that both counterimmunoelectrophoresis and latex agglutination tests were positive in 75% of patients with *H. influenzae* meningitis. Kaldor et al. (1977) studied 95 patients with purulent meningitis and 63 control patients with other diseases. With the exception of one urine specimen, none of the controls had a positive latex agglutination reaction, whereas 62 of 95 cerebrospinal fluid specimens, 10 of 14 sera, and 11 of 17 urine specimens were positive by latex agglutination.

In a similar study, Leinonen and Herva (1977) tested 103 cerebrospinal fluid samples for *H. influenzae* type B and *N. meningitidis* groups A, C, and D by latex agglutination. They reported that latex agglutination was at least as sensitive as counterimmunoelectrophoresis, but had little success detecting *N. meningitidis* group B antigen in cerebrospinal fluid by either method.

Bromberger et al. (1980) reported in a preliminary study that group B streptococci could be detected in body fluids more sensitively by latex agglutination than by counterimmunoelectrophoresis, especially in urine. The ability of latex agglutination to detect cryptococcal polysaccharide in serum or cerebrospinal fluid has been well documented (Muchmore et al., 1978) and is more sensitive than demonstration of the capsule of *Cryptococcus neoformans* by india ink.

A recent and important application of latex agglutination has been for the rapid detection of group A β-hemolytic streptococci directly in throat cultures. The importance of streptococcal pharyngitis was noted in a statewide survey conducted in Rhode Island in 1983. In the state's total population of just under 1 million, 157,000 cultures were performed to aid in the diagnosis of pharyngitis. Recognizing that the incidence of pharyngitis may not be uniform across the country, approximately 30 to 40 million throat cultures are done in the United States each year. Depending on age and time of year, 15 to 40% of these patients will have group A β-hemolytic streptococci isolated from the throat. Although the incidence of rheumatic fever after infection with this organism has declined markedly, the potential for this and other sequelae remains a concern.

Two test kits for group A β-hemolytic streptococci were introduced within months of each other: The first was Directigen (BBL, Cockeysville, Md.), a 70-min latex agglutination test; the second, the Culturette group A Strep test (Marion Scientific, Div. Marion Laboratories, Inc., Kansas City, Mo.), a 10-min latex agglutination test. Although both kits had essentially identical sensitivity and specificity, time (10 min) was an obvious advantage with the Culturette kit. Gerber et al. (1984) published one of the first reports to compare these two kits with the culture method, the data from which are provided in Table 1. In this study, the majority of false-negative rapid tests were from those patients whose culture plates showed less than 10 colonies; that is, they had relatively few organisms in the throat.

At the time of this writing, an additional 40 products have become available for detection of group A β-hemolytic streptococci since the introduction of the two kits. These products are latex agglutination, coagglutination, and solid-phase enzyme immunoassay, all of which have been reviewed by Tilton (1987). The sensitivity of the solid-phase enzyme immunoassay kits is 90 to 95%, whereas that of latex agglutination ranges from 80 to 90%. Specificity for both methods is greater than 95%.

Several advantages are gained by the user of a rapid "strep" kit. It is well known that traditional microbiologic testing in the physician's office laboratory is of modest quality. Several studies have shown that the sensitivity and specificity of group A strepto-

TABLE 1. Comparison of 10-min and 70-min detection tests for group A β-hemolytic streptococci versus culture[a]

	Blood agar culture	
Test	*Positive*	*Negative*
Culturette (10 min)		
Positive	90	3
Negative	18	228
Total	108	231
Directigen (70 min)		
Positive	71	1
Negative	14	177
Total	85	178

[a] Modified from Bass and Tilton (1985).

coccal identification in this setting is markedly reduced compared with that of a professionally directed clinical or private laboratory. These kits provide the physician's office laboratory with an accurate, sensitive, and reproducible technique. Although the kits are not perfect, they offer a distinct advantage over traditional culture methods.

There are significant clinical advantages as well. The ability to determine the etiology of a sore throat while a patient waits in the office enables the physician to prescribe appropriate antibiotics. If the rapid group A β-hemolytic *Streptococcus* test is negative and if signs and symptoms are suggestive of viral pharyngitis, then antibiotics should be withheld. (Antibiotic usage for viral pharyngitis is contraindicated.) If the test is positive, then an appropriate antibiotic can be prescribed, usually penicillin. There are two reasons to treat streptococcal sore throat: (1) To prevent sequelae, in which case early diagnosis and treatment are not quite as important; and (2) to alter the disease course, for which early intervention is critical.

Two studies (Krober et al., 1987; Randolph et al., 1985) have shown that early treatment reduces symptoms and infectivity in 24 to 48 h. This reduction in the length of morbidity (1 to 2 days as opposed to 4 to 5 days) is important for working parents as well as for the day-care center or school.

Bass and Tilton (1985) have summarized the problem as follows:

> Pharyngitis due to group A beta-hemolytic streptococci is a relatively common infection that poses difficulties both in diagnosis and management. Evidence shows that reliable diagnoses cannot be made on the basis of clinical signs and symptoms. Throat cultures are expensive and time consuming and must be performed according to a rigorous technique to be reliable. The great majority of pharyngeal infections in children are not due to group A beta-hemolytic streptococci. Therefore, treatment prescribed before culture results are available will, according to objective evidence, often delay patients' return to school or work by as much as 2 days. A test that identifies [group A beta-hemolytic streptococci] in throat swabs while the patient is still in the physician's office will be an aid in more rapid diagnosis and more efficient treatment.

Enzyme Immunoassay

The clinical applications of enzyme immunoassay are many. Although it has been used in the past primarily to detect antibody for the diagnosis of infectious disease, antigen detection by enzyme immunoassay promises rapid diagnosis of etiologic agents directly from body fluids. Table 2 summarizes antigen detection by enzyme immunoassay.

An example of the clinical utility of enzyme immunoassay for antigen detection is the rapid detection of *Chlamydia trachomatis* in genital specimens. Ryan et al. (1986) recently reported on a commercially available enzyme immunoassay for *C. trachomatis*, which has been shown to be a major cause of sexually transmitted diseases throughout the world. An enzyme immunoassay (Abbott Laboratories, North Chicago, Ill.) has been developed that detects chlamydial antigen directly in urogenital specimens. Specimens from 1,074 patients belonging to one of three risk groups were evaluated. When compared with cell culture, the sensitivity and specificity of the enzyme immunoassay for symptomatic men and women attending a sexually transmitted disease clinic were 82 and 100% and 91.3% and 95.0%, respectively. A moderate-risk group, consisting of female patients seen at either urology or gynecology clinics for genitourinary symptoms, was also evaluated. The sensitivity and specificity of the test on this group were 96 and 96.7%, respectively. A population of women at low risk was also screened for chlamydial infection. In this group, the sensitivity and specificity of the enzyme immunoassay were 89.3 and 93.2%, respectively. This rapid test is a highly specific and sensitive procedure for the detection of chlamydial antigen in genital specimens from high-risk female patients as well as symptomatic male patients. However, the population to be tested must be

TABLE 2. Microbial antigen detection by enzyme immunoassay

Aflatoxin B (*Aspergillus*)
Brucella abortus
Candida albicans
Epstein–Barr viral capsid antigen
Escherichia coli toxin (heat labile, heat stable)
Hepatitis A
Hepatitis B surface antigen
Herpes simplex virus
Rotavirus
Salmonella typhi
Schistosoma mansoni
Staphylococcus aureus (enterotoxin)
Toxoplasma gondii
Vibrio cholera (enterotoxin)
Yersinia enterocolitica
Legionella pneumophila
Clostridium difficile toxin
Cytomegalovirus
Chlamydia trachomatis
Giardia lamblia
Pneumocystis carinii
Influenza virus
Parainfluenza
Adenovirus
Respiratory syncytial virus
Human immunodeficiency virus
Cryptococcus neoformans

chosen carefully. It may make little sense to screen a low-risk population (incidence, 1 to 2%) with this or similar tests.

Sensitivity and Specificity of Antigen Detection Tests

Of the antigen detection methods, counterimmunoelectrophoresis is the least sensitive, followed by coagglutination, latex agglutination, enzyme immunoassay, fluorescence immunoassay, and radioimmunoassay. Much of the published work on the sensitivity of antigen detection systems has dealt with commercial products for detection of microbial antigens in spinal fluid. A wide range of sensitivity is observed as a function of the test modality. Various factors influence detection, including bacterial antigen concentration in the cerebrospinal fluid, a prozone effect caused by antigen excess, and the particular phase of the disease process during which the sample was taken. In the early stages of disease, there may be an undetectable threshold level of antigen or such a high concentration of antigen that the undiluted specimen will be negative by counterimmunoelectrophoresis and agglutination techniques. Dilution of the specimen may be required to demonstrate positivity.

Currently available tests for cryptococcal antigen, herpes simplex virus, respiratory syncytial virus, adenovirus, hepatitis B infection, and rotavirus using a monoclonal antibody are quite specific. Few, if any, known cross-reactions occur.

A number of bacterial cross-reactivities have been encountered in counterimmunoelectrophoresis, coagglutination, and latex agglutination. Although such cross-reactions may occur, they are observed rarely with clinical specimens. Cross-reactivity occurs between *N. meningitidis* group B and *E. coli* K1. An antigen of *E. coli* K1 appears to be identical to one of *N. meningitidis* group B. Even a monoclonal antibody for *N. meningitidis* cross-agglutinates with *E. coli* K1.

There may also be nonspecific cross-reactions, particularly in urine and serum. Agglutination tests often show spontaneous nonspecific agglutination. Treatment of these body fluids with heat, ethylenediaminetetraacetate, proteinase, or a reducing agent such as *n*-acetyl cysteine may eliminate this nonspecificity.

Summary

Antigen detection methods are now used widely in clinical microbiology. They are economical, easy to perform, and have the potential for revealing the etiologic agent of an infectious disease in a time frame that affects the precise and discriminate treatment of a patient.

Literature Cited

Anhalt, J. P., G. E. Kenny, and M. W. Rytel. 1978. Detection of microbial antigens by counterimmunoelectrophoresis, p. 1–11. *In* T. L. Gavan (ed.), Cumitech 8. American Society for Microbiology, Washington, D.C.

Bass, J. W., and R. C. Tilton. 1985. Group A beta hemolytic streptococcal pharyngitis. A monograph. Marion Scientific, Div. Marion Laboratories, Inc., Kansas City, Mo.

Bromberger, P. I., B. Chandler, H. Gezon, J. E. Haddow. 1980. Rapid detection of neonatal group B streptococcal infections by latex agglutination. J. Pediatr. **96:**104–106.

Bussard, A. 1959. Description d'une technique combinant simultanement l'electrophorese et la precipitation immunologique dans un gel: l'electrophorese. Biochim. Biophys. Acta **34:**258.

Dochez, A. R., and D. T. Avery. 1917. The elaboration of specific soluble substance by pneumococcus during growth. J. Exp. Med. **26:**477.

Edwards, E. A. 1971. Immunologic investigations of meningococcal antigens. I. Group-specific *Neisseria meningitidis* antigens present in the serum of patients with fulminant meningococcemia. J. Immunol. **106:**314–317.

Edwards, M. S., and C. J. Baker. 1979. Prospective diagnosis of early-onset Group B streptococcal infection by countercurrent immunoelectrophoresis. J. Pediatr. **94:**286–288.

Engvall, E., and P. Perlmann. 1971. Enzyme-linked immunosorbent assay (ELISA). Quantitative assay of immunoglobulin G. Immunochemistry **8:**871–874.

Feigin, R. D., M. Wong, P. G. Shackelford. 1976. Countercurrent immunoelectrophoresis of urine as well as of CSF and blood for diagnosis of bacterial meningitis. J. Pediatr. **89:**773–775.

Gerber, M. A., L. J. Spadaccini, L. L. Wright, and L. Deutsch. 1984. Latex agglutination tests for rapid identification of group A streptococci directly from throat swabs. J. Pediatr. **105:**702.

Hechemy, K. E., and E. E. Michaelson. 1984. Latex particle assays in laboratory medicine. Parts I and II. Lab. Manage. **June:**27–38, **July:**26–33.

Ingram, D. L., R. J. O'Reilly, and P. J. Pond. 1975. Diagnosis of *Haemophilus influenzae* b and other meningitides: gram stain, latex agglutination and countercurrent immunoelectrophoresis. Pediatr. Res. **91:**341.

Jacobs, R. F., T. Yamauchi, and K. D. Eiseinach. 1981. Detection of streptococcal antigen by counterimmunoelectrophoresis. Am. J. Clin. Pathol. **75:**203–208.

Kaldor, J., R. Asynowicz, and D. C. P. Buist. 1977. Latex agglutination in diagnosis of bacterial infections with special reference to patients with meningitis and septicemia. Am. J. Clin. Pathol. **68:**284–289.

Krober, M. S., J. W. Bass, and G. N. Michels. 1985. Streptococcal pharyngitis: placebo-controlled double-blind evaluation of clinical response to penicillin therapy. J. Am. Med. Assoc. **253:**1271–1274.

Kronvall, G. 1973. A rapid slide agglutination method for typing pneumococci by means of specific antibody absorbed to Protein-A containing staphylococci. J. Med. Microbiol. **6:**187.

Leinonen, M., and E. Herva. 1977. The latex agglutination test for the diagnosis of meningococcal and *Haemophilus influenzae* meningitis. Scand. J. Infect. Dis. **9**:187–191.

Miotti, P. G., R. P. Viscidi, J. Eiden, E. Cerny, and R. H. Yolken. 1986. Centrifugation augmented solid phase immunoassay (CASPIA) for the rapid diagnosis of infectious diseases. J. Infect. Dis. **154**:301–308.

Muchmore, H. G., F. G. Felton, and E. N. Scott. 1978. Rapid presumptive identification of *Cryptococcus neoformans*. J. Clin. Microbiol. **8**:166–170.

Randolph, M. F., M. A. Gerber, and K. K. DeMeo, 1985. Effects of antibiotic therapy on the clinical course of streptococcal pharyngitis. J. Pediatr. **106**:870.

Ryan, R. W., I. Kwasnik, O. Steingrimsson, J. Gudmundsson, H. Thorarinsson, and R. C. Tilton. 1986. Rapid detection of *Chlamydia trachomatis* by an enzyme immunoassay method. Diagn. Microbiol. Infect. Dis. **5**:225–234.

Soini, E. 1985. The principle of time resolved fluorometry, p. 414–421. *In* K. O. Habermehl (ed.), Rapid methods and automation in microbiology and immunology. Springer-Verlag KG, Berlin.

Tilton, R. C. 1978. Counterimmunoelectrophoresis in biology and medicine. Crit. Rev. Lab. Med. **9**:347–365.

Tilton, R. C. 1987. Update on rapid tests for group A streptococci. Med. Lab. Prod. **2**:16–17.

Vincent, M. H., and M. Bellot. 1901. Diagnostic de la meningite cerebrospinale a meningocoques par la "precipitoreaction". Bull. Acad. Natl. Med. (Paris) **61**:326.

CHAPTER 4

Principles of Antibiotic Testing in the Laboratory

DANIEL AMSTERDAM

Laboratory Studies: In vitro evaluations of the interaction of antimicrobial agents, bacterium, and host, as represented in the laboratory by body fluids that incorporate the administered drug to the compartment, can be quantified by a variety of methods. It is necessary that the laboratory adhere to acceptable protocols to execute the determinations because variations in inoculum size, phase of inoculum, medium components, drug size and stability, incubation environment and interval, and volume of the assay system have been shown to modify the outcome. For these reasons, quality control standards and the use of standard reference strains with known reactivity are important and necessary.

Evaluations of Drug-Organism Interaction: Several parameters are used to quantify the outcome of the interaction between organism and the antimicrobial agent. These include the impregnated disk agar diffusion end point, the minimum inhibitory concentration (MIC), the minimum bactericidal concentration (MBC), the minimum antibiotic concentration (MAC), and the postantibiotic effect (PAE). For some antibiotics, resistance markers can be directly and rapidly determined.

Assessment of Bactericidal Activity: The MBC, derived directly from the MIC and killing curves (i.e., the rate of reduction in the number of bacteria with time) are used to assess the bactericidal activity of drug (or patient's serum containing drug). This can be achieved for single or drug combinations.

Evaluations of Host-Drug Interaction: Serum or other body fluid obtained at specified dosing intervals is used to determine the level of antimicrobial agent in that compartment. This determination is frequently accomplished when agents are particularly toxic or when there is need to determine maximum dosing regimens.

Evaluations of Host-Organism Interaction: Serum (or other body fluid) is used to determine the serum bactericidal test or titer (SBT) of serum, which is the concentration of drug in body fluid that inhibits (or kills) the organism offending the patient. An alternative to the SBT is the serum bactericidal rate (SBR) or killing curve, which evaluates the rate at which single or multiple drugs kill bacteria. The SBT and SBR integrate in vitro activity and in vivo pharmacokinetics of antimicrobial agents.

Perspectives

Evolution of Susceptibility Test Methods

While the earliest association of humans and antimicrobial agents may have come about through casual and later purposeful contact with soil laden with antibiotic-producing strains of *Streptomyces* and other *Actinomyces*, the first method for assessing antimicrobic susceptibility was developed by Fleming (1929) after his serendipitous observation of the action of *Penicillium notatum*. Using the first agar diffusion technique, the "ditch plate" (Fig. 1), he excavated a strip of agar in the shape of a ditch from a petri dish and replaced it with medium containing

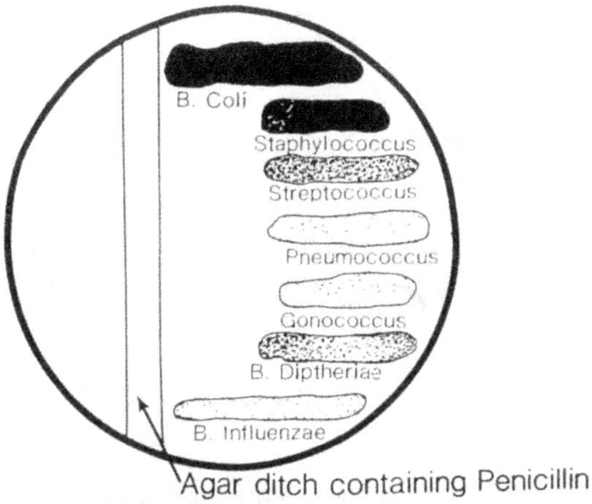

FIG. 1. Schematic representation of Fleming's "ditch plate" antibiotic susceptibility test. A solution of penicillin was placed in the ditch. Of the several species that were streaked perpendicular to the ditch, *"B. coli"* and *"B. influenzae"* were considered "resistant."

the extract of potential antimicrobial activity (for Fleming, the mold extract of penicillin). At right angles, he streaked various strains across the ditch. Fleming did not observe the "radial zones" of inhibition that we recognize today. Instead, he saw lanes of inhibition, and concluded that the lane in the streak of microorganisms that was furthest away from the antimicrobial-laden ditch represented activity against that particular strain (Fleming, 1929). Seemingly, Fleming was more interested in isolating *"B. influenzae"* than inhibiting the other susceptible microorganisms.

The modern need for antimicrobic susceptibility testing was born most likely because of financial considerations. Before World War II, the production of penicillin was limited and extremely expensive. Therefore, it was necessary to develop procedures for predicting when the use of penicillin would be effective in an individual with a particular infectious disease. During the 1940s, several antibiotics were discovered. Although the broth susceptibility method was the first method developed for assessing in vitro efficacy of antimicrobial agents and still serves today as a reference method, it was replaced by the antibiotic-impregnated filter paper strip (Foster and Woodruff, 1943) and later by disks (Vincent and Vincent, 1944).

Toward the end of the 1950s, the status of antimicrobial susceptibility testing was in chaos as a result of the lack of acceptable standard procedures. To rectify this, an international group of experts met in 1960, outlined general guidelines, and reported on the need for standards (World Health Organization, 1961). Afterwards, the Federal Register (1961) recognized the use of antibiotics in the laboratory. In 1969, the Supreme Court of the United States rendered the decision that sustained the right of the Food and Drug Administration (FDA) to rule on the certification of antibiotic susceptibility disks. The court averred that the disk is a "drug" since it is "intended for the use of diagnosis, cure medication, treatment, or prevention of disease in man or other animals."

After the WHO report, an international collaborative group was formed to address the several problems of standardization. The group's report (Ericsson and Sherris, 1971) was later followed by adoption of the procedures of Bauer et al. (1966) in the Federal Register (1972, 1973).

The Laboratory's Role

In clinical microbiology, the practice involves dealing with patients' specimens. It is the specimen that creates the challenge, and the microbiologist must initiate several operations that will determine:

Detection: First, determine if there are any microorganisms present in the specimen. This is best exemplified by specimens from putatively sterile body sites such as blood and spinal fluid.

Recovery: Recover ("capture") from the specimen any microorganisms associated with it. To achieve this, the laboratory must provide the specimen with an appropriate medium (media) and environmental milieu to determine the number and distribution of the microorganisms present in that specimen.

Isolation: From among the one or many colonial types that may appear under the environmental conditions presented to the specimen, the microbiologist must isolate the one or several microorganisms that are presumably linked to the patient's disease state.

Characterization: According to Linnean principles in the classic biological approach, this would mean assigning a genus and species designation. This could come about in several different ways as a result of biochemical fermentation/utilization of various substrates, susceptibility to specific phages, identification of a specific and unique (target) nucleotide sequence, bacteriocin typing, toxicity typing (that is, determining the activity of selected heavy metal compounds or other organic inhibitors), determining the reactive epitopes on the bacterial surface for serotyping, and determining the susceptibility of the isolate to an array antimicrobial agents. When integrated, all of these operations can be referred to as laboratory diagnosis in the microbiology setting. The process of isolating and characterizing the microorganism with a genus and species apellation is one aspect of the process. Another, and probably the characterization that is most closely welded to the final outcome of the patient's disease, is the antibiotic susceptibility test.

The Nature of Laboratory Studies

Medical studies are frequently defined as those procedures performed on patients or on specimens taken from patients that provide clinical data beyond that which can be obtained by history-taking and physical examination. A laboratory examination or study is usually interpreted as a medical study performed in the laboratory by competent personnel. If a laboratory study is conducted on a specimen rather than directly from a patient, it is referred to as a clinical laboratory study.

Laboratory studies that quantify the interraction of bacteria and antimicrobial agents may be of two types. They can be quantitative, with the level of quantification depending on the clinical needs or the sophistication of the method of measurement, or they can be of binary quantification, which characterizes qualitative studies. In this case the results would be reported as either S—susceptible—or R—resistant (or I—intermediate).

No laboratory study, no one susceptibility study, can adequately and satisfactorily answer the question, "Given this specimen containing this organism with a resulting susceptibility to several agents, what will be the clinical outcome for the patient?" This is the province of clinical diagnosis. The laboratory can, however, select appropriate procedures that would more accurately reflect the host's physiology and underlying condition or conditions and determine bacteriologic outcome. There have been recent attempts in clinical laboratory studies to integrate microbiologic and pharmacologic data so as to more accurately mirror the kinetic events in the patient. These are included in this discussion.

General Considerations

Factors that Modify the Outcome of Antibiotic Therapy

It is recognized that effective chemotherapy embodies the idea of selective toxicity and requires that the antimicrobial agent reach the site of infection in sufficient concentration to inhibit and preferably kill the offending microorganism with little or no toxicity to the host. Many factors modulate this interaction and interrelationship of drug, host, and microorganism, and only the clinician with the necessary laboratory data can knowledgeably evaluate and integrate the entire complex before a therapeutic decision can be initiated.

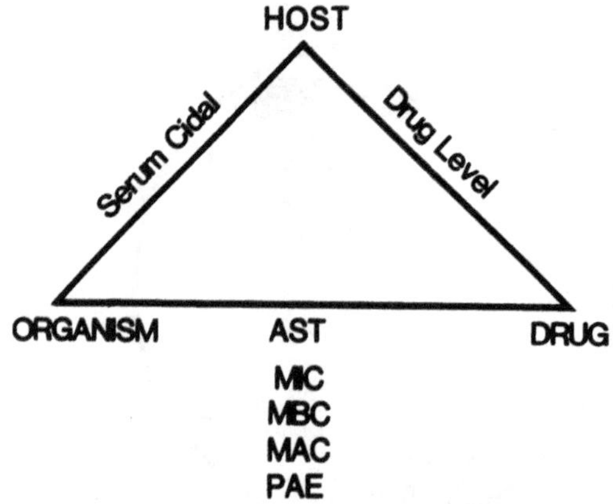

FIG. 2. The triad of host, drug, and organism depicting the tests that can be applied to evaluate interacting components. AST = antimicrobial susceptibility test; MAC = minimum antibacterial concentration; MBC = minimum bactericidal concentration; MIC = minimum inhibitory concentration; PAE = postantibiotic effect.

Many factors contribute to the interaction between organism, drug, and host and modulate the eventual outcome. These include the nature of the microorganism, the class of antimicrobial agent, and the milieu in which these are tested and evaluated. Each component will be examined from the laboratory's perspective.

In the laboratory, the host is represented by the specimen, which contains either the infectious agent or body fluid incorporating the administered antimicrobial compound. The specimen containing offending microorganism or microorganisms or fluid-containing antimicrobial agents is representative of the host rather than the integrated complex physiology and biochemistry of its source. Classic laboratory studies attempt to isolate one component of the complex and evaluate interactions with drugs or bacteria, or both (Fig. 2).

The Organism

The bacterium possesses its own genetic background, structure, and metabolic behavior, which has much to do with the development of resistance and disease-producing potential. Resistance may be acquired directly on host cell division by two bacterial daughter cells, chromosomally, or from resistance plasmids, covalently closed circular DNA molecules. R plasmids can carry resistance to one or several antimicrobial agents. Thus, bacteria develop resistance to antibiotics by acquisition of R plasmids or by mutation. In the latter case, this is accomplished by a genetic event that results in a change in the existing cellular structure, which in turn modifies target sites within the microorganism, making them indifferent to the presence of antibiotic or impermeable to a particular class of antimicrobial agents. When an organism harbors R plasmids, resistance is conferred as a result of the synthesis of distinct proteins that enzymatically inactivate or modify the antimicrobial agent, alter the target sites for the antibiotic, block the transport of the antibiotic to the bacterial cell, or bypass the metabolic steps inhibited by the antimicrobial agent.

Although it is convenient to consider the organism as a single entity, for studies of microorganisms, we think in terms of populations. Thus, it is possible to have more than one potentially susceptible (or resistant member in a population) in a specimen belonging to the same species—as is the case with methicillin-resistant staphylococci—and isolating a single colony may not be representative of the specimen and, in turn, the patient.

The Host

The host has its own genetic background, prior experience (e.g., sensitivity to penicillin), immunologic profile, underlying disease, specific site and pathology of the lesion and its biochemical environment, and the host environment (i.e., within the hospital or the community). One needs to remember that the patterns of antibiotic use and the problems they produce are quite different in hospitals as compared with in office practice. In office practice, the agents used most frequently are the broad to medium spectrum oral compounds, chiefly for respiratory infections. In the hospital environment where gram-negative enteric bacteria have largely replaced the *Staphylococcus* as the leading cause of bacteremia just as the *Staphylococcus* replaced the pneumococcus and the Group A *Streptococcus*, there is a different pattern of antibiotic usage.

The Drug

In the triad, the antimicrobial agent represented by the "drug" possesses its own characteristics that affect solubility, pathway of administration, toxicity, protein binding, distribution, absorption, stability, metabolite modification, and pathway of excretion and elimination—properties, which in one word, can be called the pharmacokinetics of the drug.

In the laboratory, an attempt is made to monitor these associations, usually by isolating two of the three members of the triad. The interraction of organism and drug is the thrust of this chapter, and deals with the assessment of the susceptibility of antimicrobial agents to bacteria isolated from the host (Fig. 2). The host-drug component of the triad is exemplified by the determination of the levels of antimicrobial agent in host body fluid (usually serum) after the agent has been appropriately administered. Lastly, one can consider the closing link of the triangle—the host and organism—as the determination of serumcidal activity or the killing properties of the serum containing none to several antimicrobial agents in it.

Several parameters have been used to evaluate the outcome of the interaction between offending microorganisms and therapeutic intervention, and can readily be measured in the laboratory. These include the minimum inhibitory concentration (MIC), the minimum bactericidal concentration (MBC) determined directly from the outcome of the MIC, the minimum antibiotic concentration (MAC), and post-antibiotic effect (PAE) activity.

In the final analysis, it is necessary to realize that in the laboratory setting, there is a continuous exposure of 10^5 to 10^6 CFU/ml to a static drug concentration for the entire incubation period. This situation differs from that which occurs in vivo, where larger (or smaller) numbers of bacteria at the infected site are exposed to fluctuating gradients of drug concentrations.

Methodologic Factors that Modify the Results of Susceptibility Tests

Inoculum Size

In general, large bacterial populations are less promptly and completely inhibited than smaller ones. Also, the likelihood of the emergence of resistant mutants is greater in a large population. This is exemplified by the recent emergence and recognition of methicillin-resistant *Staphylococcus aureus* and *Staphylococcus epidermidis*. To more accurately determine these particular populations, it is necessary to depress the growth of the more rapidly growing susceptible population and to extend the incubation interval beyond 18 to 24 h so as to readily detect the presence of the slower-growing resistant population (Thornsberry, 1984). There is less chance of observing the resistant population if the initial population size is small.

A therapeutically equivalent inoculum has never been established. Observed "inoculum effects" are attributable to emergence of resistant variants, enzymatic destruction of drug, or the unavailability of sufficient drug.

For microbroth assays, an inoculum density of 5×10^5 CFU/ml is recommended (Jones et al, 1985), resulting in 5×10^4 CFU/ml per well (0.1 ml volume). The inoculum should represent three to five morphologically similar ("pure") colonies.

Phase of Inoculum

The metabolic activity of bacterial populations differs from the inoculum phase from which they are taken. Some organisms are more readily inhibited when they are actively dividing, as is the classic case with penicillin. Today, it is recognized that beta-lactam drugs in general more rapidly and completely inhibit faster-growing bacterial populations (Tuomanen, 1986).

Medium Components

Salt concentrations generally inhibit aminoglycosides and enhance the activity of penicillin against proteus species. Divalent cations, especially Mg^{2+} and Ca^{2+} antagonize aminoglycosides. Currently, several sources (Jones et al., 1985; National Committee for Clinical Laboratory Standards, [NCCLS] 1986a, 1986b) recommend the supplementation of broth medium with 25 mg/liter of magnesium and 50 mg/liter of calcium, after Reller et al. (1974).

The competitive action of para-aminobenzoic acid (PABA) antagonizes the sulfonamides. Mueller-Hinton medium minimizes this effect. Serum proteins can bind to penicillin in varying degrees and thus inactivate them, whereas carbohydrates enhance the inhibition produced by nitrofurantoin. Thymidine antagonizes the activity of sulfamethoxazole/trimethoprim (SXT). The pH can modify drug activity and acid pH enhances the activity of nitrofurantoin, whereas at alkaline pH aminoglycosides are more active. At pH 5.0, the MIC of gentamicin is 100 to 200 times that at pH 8.5.

For these reasons, there has been a recent undertaking in the United States to standardize the production of Mueller-Hinton medium, and in the United States the reference medium is currently Mueller-Hinton. These recommendations are now incorporated into a proposed document distributed by the NCCLS (1986a).

Drug Size

The size of compounds markedly affects their diffusion in agar. It is well recognized that polypeptide antimicrobial agents (e.g., polymyxin) diffuse more slowly than do smaller molecular weight compounds. This can be adjusted for in terms of interpretation of end points by correlations with MIC and regression analysis (see the section Agar Diffusion Tests).

Drug Stability

At incubator environmental temperature (35 to 37°C), certain drugs (like penicillin) are less stable and are more readily inactivated on continued incubation. Some compounds (like the aminoglycosides and chloramphenicol) are much more stable. Thus, extended incubation intervals will lead to drug inactivation.

Incubation Environment

The gaseous environment should mimic that for aerobic organisms. Anaerobic conditions are known to inhibit the action of the aminoglycosides, as oxygen is required to drive gentamicin across the cell membrane and lack of it will alter the outcome of test results.

Incubation Interval

During a short incubation interval (<12 h), bacteria are not killed but only inhibited (see the section PAE). The longer the incubation interval, the greater the chance for resistant mutants to emerge, and this is, in fact, the practice for determining the outgrowth of methicillin-resistant *Staphylococcus aureus* (MRSA) and methicillin-resistant *Staphylococcus epidermidis* (MRSE) (Thornsberry, 1984). Also, the morphologic effects (cell elongation) induced by β-lactam agents in gram-negative bacilli without the concomitant alteration in the cell population will alter the optical density in broth susceptibility systems. Thus, the results obtained after a brief incubation interval will be different from those obtained after an 18-h interval (Eng et al., 1985).

Quality Control and Standardization

For in vitro susceptibility results to be meaningful and of value in selecting appropriate agents and monitoring their use for the treatment of infection, they need to be accurate and reproducible. Because of the potential for variations, emphasis has been focused on strict adherence to methods and reference procedures (NCCLS, 1984, 1985). The contribution by voluntary, consensus advisory groups to the development of quality control parameters has played a significant role in the high level of performance obtained by most laboratories. The key to this performance is the application of standard reference strains with known reactivity in assessment of qualitative (disk diffusion) and quantitative end points.

A major source of potential variations has already been recognized as the medium—Mueller-Hinton for agar disk-diffusion assays. Ericcson and Sherris (1971) had described in their early report the effects of media on susceptibility results. Over the years, manufacturers have made several adjustments. The recently proposed standard (NCCLS, 1986a) will ensure that the reference standard for Mueller-Hinton medium will respond in an expected reproducible manner and that results will be comparable provided that all other factors are constant.

Early on, it was considered necessary to test reference strains on a daily basis. However, the impetus for reducing costs in laboratories and new guidelines from the Health Care Finance Administration obviate the previous requirement for routine daily testing (Miller et al., 1986).

Selection of Antimicrobial Compounds for Evaluation

As the number of approved compounds continues to expand each year—seemingly at almost a logarithmic rate—the challenge for the laboratory is to select rationally a limited number of antimicrobial agents for testing. To test all licensed drugs would be impractical and, in this era of limiting laboratory costs, uneconomical. In the past, this process was aided in part by the U.S. Food and Drug Administration (FDA) as it recognized "class" compounds—or as they apply to agar disk-diffusion testing, "class disks." However, as the number of chemical classes and congeners within each class increases, the unique pharmacokinetic properties frequently dictate separate disks for each newly approved compound.

What each laboratory and clinical setting needs to do is to develop a strategy for testing compounds based on previously accrued information on resistance patterns of class agents. Patient demographics as well as patterns of antibiotic usage (and abuse) will serve to assist in this selection process.

Guidelines to assist the laboratory in this selection process have been proposed by the NCCLS (1986b) and are included here for convenience (Table 1). It should be noted that the represented groups of organisms are those for which it is most difficult to predict, with any degree of certainty, a susceptibility result based on prior experential data. The agents in the primary and secondary groups achieve peak levels in serum, whereas those noted in the urine group will attain maximum concentration in that compartment. Final drug selection should represent a consensus opinion at each site or medical center, and should include input from the microbiologist, pharmacist, and infectious disease physician. As noted, these decisions require information about the antimicrobial nature (toxicity, pharmacokinetics) and the probability of resistance for each "drug-bug" pair previously used. Aquisition and use costs also need to be considered. Frequently, these data are incorporated in a quarterly or annual summary report distributed by the institution.

Although costs and test constraints may limit the number of agents (usually 8 to 12) tested, the advent of hospital and/or laboratory information systems can specify the distribution of reports of susceptibility test results to the patient's chart. In a sense, the problem of selecting agents for testing is allied to the question of, "What information about which agent should be routinely reported?". In this fashion, compounds that are on a "restricted" or "limited" formulary status need not be routinely reported, but the data can be stored for consultation and future consideration of "routine testing and reporting." The laboratory may also wish to consider testing those agents that are soon to be approved by the FDA in the anticipation of acquiring a data base for setting new priorities.

In summary the laboratory should:

1. test organisms recovered from significant sites (e.g., blood, CSF) to provide a guide for rational therapy,
2. evaluate the susceptibility of selected nosocomial agents to determine variations in resistance patterns,
3. determine and monitor antibiotic patterns as epidemiologic markers as evidence for the effective use of antibiotics for coverage, and
4. study the activity of new (and newer) agents.

Evaluations of Drug-Organism Interaction

Assessments of Bacteriostatic Activity

The MIC

The MIC can be determined in liquid (i.e., broth) or in agar using serial two-fold or less dilutions of the drug or gradients of the drug concentration. Standard reference methods in the United States for these techniques have been documented from two sources (Jones et al., 1985; NCCLS, 1985; Washington, 1985) on antimicrobial susceptibility testing. Dilutions tests are used to assess the minimum concentration of an antimicrobial agent that will inhibit (or kill) a microorganism. As can be seen in Fig. 3, the lowest concentration of the graded antibiotic values that will inhibit the growth of the test microorganism is defined as the minimal inhibitory concentration (MIC). We now refer to those assays done in the 1960s in 2.0-ml volumes, macrobroth dilution tests, and assays done in 0.1-ml volumes as a microbroth dilution. For macro- or microbroth procedures, the end point is read macroscopically as turbidity; for agar dilution, the area of the spot inoculum is examined for the presence or absence of colonial growth.

The apparent, precise, quantitative result afforded by dilution testing (MIC or minimum bactericidal concentration [MBC] see below) would be meaningless unless the value reported was related to some parameter measurable in the patient. That parameter is the concentration of antimicrobial agent achievable in the patient's serum, plasma, or other

TABLE 1. Suggested groupings of antimicrobial agents that should be considered for routine testing and reporting by clinical microbiology laboratories[a]

Enterobacteriaceae[b]	*Pseudomonas*[b]	*Staphylococci*	*Enterococci*[c]	*Nonenterococcal streptococci*
Primary				
Amikacin	Amikacin	Cephalothin[d]	Penicillin G[e] *or*	Penicillin G
Ampicillin	Azlocillin *or*	Clindamycin	Ampicillin	
Cefazolin[d]	Piperacillin	Erythromycin	Vancomycin[f]	
Cephalothin[d]	Carbenicillin *or*	Oxacillin *or*		
Chloramphenicol[g]	Mezlocillin *or*	Methicillin[d]		
Gentamicin	Ticarcillin	Penicillin G		
Tetracycline[h]	Gentamicin	Tetracycline[h]		
Tobramycin	Netilmicin	Vancomycin		
	Tobramycin			
Secondary				
Amoxicillin/clavulanic acid	Aztreonam	Amikacin *or*		Cephalothin
Aztreonam	Cefoperazone *or*	Gentamicin *or*		Chloramphenicol[g]
Cefotaxime *or*	Ceftazidime	Kanamycin *or*		Clindamycin
Ceftazidime *or*	Cefotaxime *or*	Netilmicin *or*		Erythromycin
Ceftizoxime *or*	Ceftriaxone *or*	Tobramycin		Tetracycline
Ceftriaxone *or*	Moxalactam	Amoxicillin/clavulanic		Vancomycin
Moxalactam	Chloramphenicol[g,k]	acid[d]		
Carbenicillin *or*	Imipenem	Chloramphenicol[g]		
Ticarcillin	Ticarcillin/clavulanic	Imipenem		
Cefamandole *or*	acid	Trimethoprim/		
Cefonicid *or*	Trimethoprim/	sulfamethoxazole		
Cefuroxime	sulfamethoxazole[k]			
Cefoperazone				
Cefotetan				
Cefoxitin				
Imipenem				
Kanamycin				
Mezlocillin or				
Piperacillin				
Netilmicin				
Ticarcillin/clavulanic acid				
Trimethoprim/				
sulfamethoxazole				
Urine				
Cinoxacin *or*	Ceftizoxime	Nitrofurantoin	Erythromycin	Nitrofurantoin
Nalidixic acid	Norfloxacin	Norfloxacin	Nitrofurantoin	Norfloxacin
Nitrofurantoin	Sulfisoxazole	Sulfisoxazole	Norfloxacin	
Norfloxacin	Tetracycline[h,k]	Trimethoprim	Tetracycline	
Sulfisoxazole	Trimethoprim/	Trimethoprim/		
Trimethoprim	sulfamethoxazole[k]	sulfamethoxazole		
Trimethoprim/				
sulfamethoxazole				

[a] Selection of the most appropriate antimicrobial agents to test is a decision best made by each clinical laboratory in consultation with the medical staff, infectious disease practitioners, and the pharmacy. The lists presented are agents of proven clinical efficacy for that organism group and additionally, those with acceptable in vitro test performance.
[b] Also applies to *Acinetobacter* spp.
[c] Antimicrobial agents not listed in this column, such as the cephalosporins, clindamycin, and the aminoglycosides, should not be tested and/or reported against the enterococci because the reporting of their results can be dangerously misleading.
[d] The cephalothin susceptibility test results *cannot* be relied upon to detect cephalosporin-resistance among methicillin-resistant *Staphylococcus* spp. Amoxicillin/clavulanic acid and ticarcillin/clavulanic acid have a similar and comparable spectrum as the penicillinase-resistant penicillins and the cephalosporins against intrinsically oxacillin-resistant (heteroresistant) staphylococci, and heteroresistant staphylococcal strains resistant to oxacillin or methicillin should currently also be considered refractory to amoxicillin/clavulanic acid and ticarcillin/clavulanic acid. Cefazolin should *not* be used to predict the susceptibility of other first-generation cephalosporins (cephalothin remains preferred) because of a significant number of false-susceptible errors, principally among the *E. coli* isolates.
[e] Penicillin G susceptibility may be used to predict the susceptibility to ampicillin, ampicillin analogs, amoxicillin, and

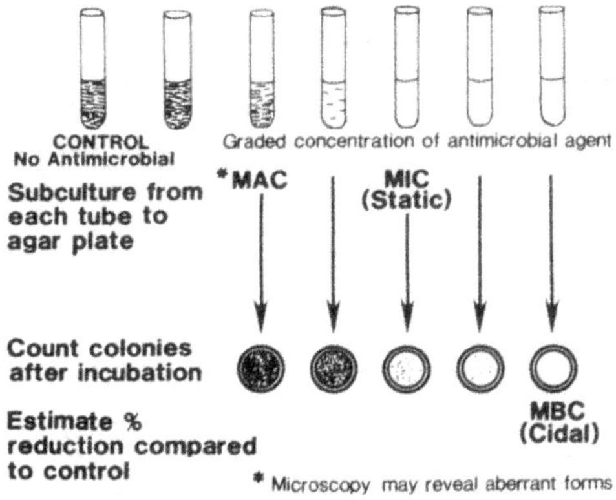

FIG. 3. Procedural derivation and interpretations for MIC, MAC, and MBC. See text for further details.

clinically relevant body fluid. A microorganism may be considered to be susceptible if a peak serum level equivalent to four times the MIC can be safely achieved with the recommended therapeutic regimen for that antimicrobial agent. At minimum, it is presumed that the drug must achieve serum concentrations greater than the MIC to be effective (Kunin, 1981). Thus, the range of antibiotic concentrations tested needs to be related to attainable fluid levels for that drug since there would be small value to extend the assay beyond these limits.

From an examination of the MIC, it follows that a limited number of dilutions could serve to yield semiquantitative data about an organism's susceptibility. Schneierson and Amsterdam (1959) reported on such a two-concentration method. One tube contained antibiotic in relatively high concentration (near the "break point" for "S" versus "R"), and the other contained a lower concentration. A "resistant" organism grew in both tubes; an organism of "intermediate" susceptibility was able to grow only in the tube containing the broth with the lower antibiotic concentration; and a "susceptible" organism did not grow in either of the two antibiotic concentrations.

Currently, this approach is considered to have merit because it combines the quantitative features of broth dilution and the associated automated capability of that method. Several commercial suppliers of prepared microbroth susceptibility tests make this available as the "category" or "break point" method.

AGAR DISK DIFFUSION TESTS

Historically, agar diffusion methods for determining antimicrobial susceptibility were developed after earlier broth dilution methods. As the number of antimicrobial agents began to increase, it soon became apparent that agar methods were more convenient and less time-consuming when several different antibiotic-impregnated paper disks could readily be "dropped" on the agar surface of one petri dish.

Two dynamic events are initiated after agar medium is inoculated and disks are applied. The antibiotic diffuses into the medium, producing a gradual changing gradient of concentration, and the organism, after an initial lag period, multiplies at a logarithmic rate. The kinetics of these events were documented by Cooper (1963).

The appearance of a delimited radial zone free of bacterial growth surrounding the paper disk impregnated with antibiotic indicates susceptibility of the organism to the drug as a function of the solubility of the antibiotic in the liquid phase of the agar medium, the diffusion constant of the drug, and the potency of the antibiotic against the organism. In practice, contrary to theoretical expectations, the edge of the zone of inhibition is frequently not sharp, and one can observe a gradient of growth in response to the decreasing antibiotic gradient (Barry, 1976).

The appearance of a zone of inhibition does not imply clinical susceptibility. To establish interpretative zone criteria, at least 100 strains representative of one to several targeted species for a designated antimicrobial agent need to be tested by MIC and disk diffusion methods. Scattergrams are plotted with the MIC (log) on the ordinate, and the corresponding zone diameter (mm) is plotted on the abscissa (Fig. 4). When the method of least squares is

acylamino-penicillins to which enterococci are also "moderately susceptible." However, combination therapy of penicillin G or ampicillin, plus an aminoglycoside, is usually indicated for serious enterococcal infections such as endocarditis.

[f] Vancomycin is often used for serious enterococcal infections in patients with significant penicillin allergy.

[g] Not routinely tested against organisms isolated from the urinary tract.

[h] Doxycycline or minocycline may be tested in place of or in addition to tetracycline for some isolates of *S. aureus* and nonfermentative, gram-negative bacilli (e.g., *Acinetobacter* spp.).

[j] Represents secondary antimicrobial agents that may require testing as a primary agent in those institutions harboring endemic or epidemic resistance to one or more of the primary drugs, especially in the same family (e.g., β-lactams or aminoglycosides), as an epidemiologic aid or for treatment of patients allergic to primary drugs (see nonenterococcal streptococci). This designation was applied to some drugs in the *Pseudomonas* list due to lack of long-term clinical experience with that drug or in the case of the newer cephalosporins, a monobactam, the carbapenem, and ticarcillin/clavulanic acid, their first clinical application.

[k] May be indicated for testing some *Pseudomonas* spp. other than *P. aeruginosa*, or for *Acinetobacter* spp.

[l] Only tested with organisms recovered from urinary tract infections.

Reprinted with permission of the National Committee for Clinical Laboratory Standards, 1987.

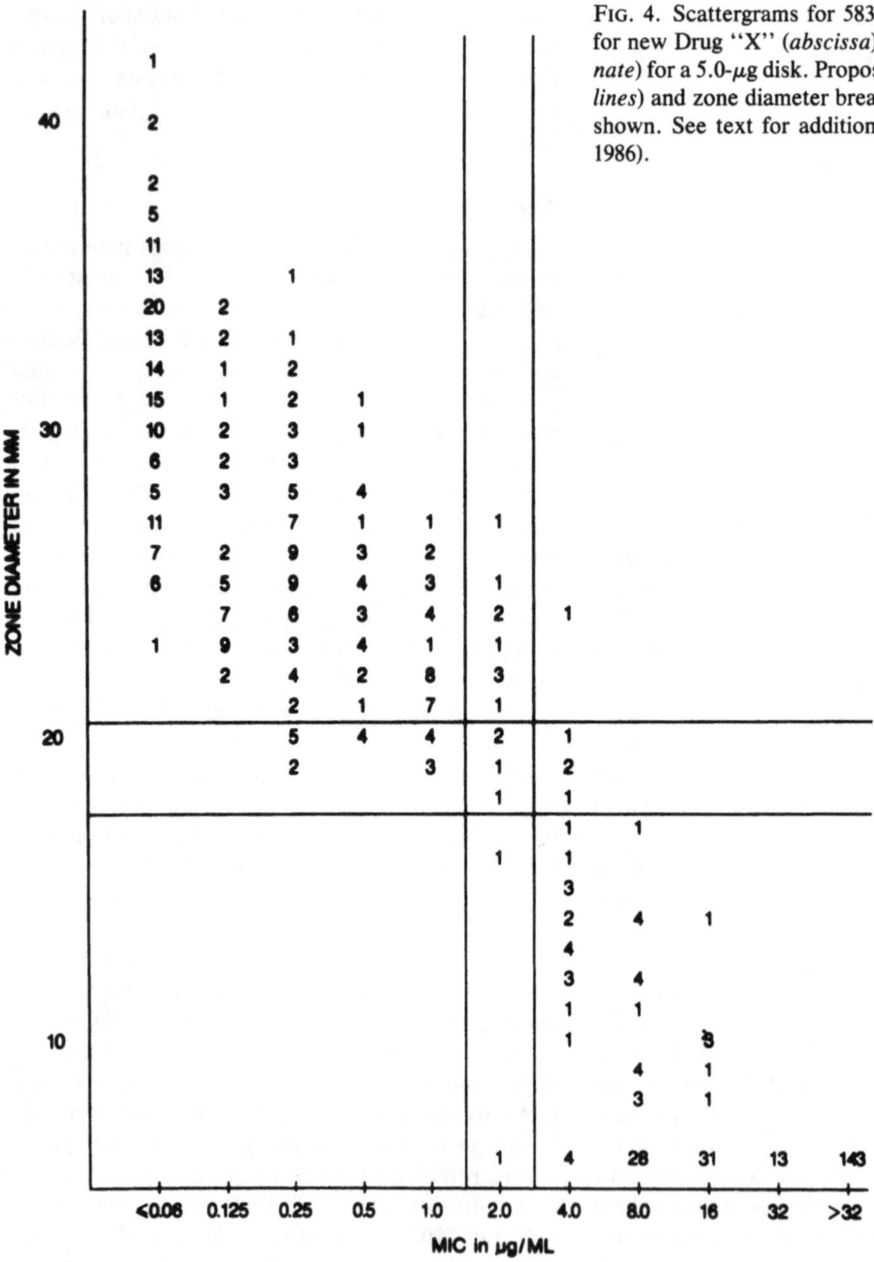

FIG. 4. Scattergrams for 583 bacterial isolates with MICs for new Drug "X" (*abscissa*) versus zone diameters (*ordinate*) for a 5.0-µg disk. Proposed MIC break points (*vertical lines*) and zone diameter break points (*horizontal lines*) are shown. See text for additional details (after Fuchs et al., 1986).

applied, a linear relationship is determined between the MIC and zone diameter, and a straight regression line can be obtained that displays "best fit" (Barry and Thornsberry, 1985). In some cases when the scatter does not permit a uniform distribution and a valid regression analysis, the error rate-bounded method of Metzler and De Haan (1974) is applied.

Comparative Features of MIC and Disk Diffusion Methods

A clear advantage of the agar disk diffusion method has been the relative ease with which several agents can be compared and evaluated. This overriding distinction is not detracted by the apparent loss in a quantitative end point when compared with dilution methods. Although quantitative results derived from MIC assay would seem to be more accurate, it is necessary to consider that there is a definitive relationship between the two parameters. Other comparative features are noted in Table 2.

The MIC_{50} and MIC_{90}

Clearly, one of the reasons for testing the organisms's response to an antimicrobial agent is to evaluate new compounds and compare their efficacy with previously evaluated, therapeutically effective, safe compounds. The MIC_{50} and MIC_{90} (and MIC_{100}) representing the MICs at which 50 and 90% (and

TABLE 2. Comparative features of disk diffusion and broth dilution methods

Feature	Disk diffusion	Broth dilution[a]
Contamination readily observed	Yes	No[b]
Tests for most resistant members of inoculum	No	Yes
Inoculum effect	No[c]	Yes[c]
Quantitative result	No[d]	Yes
MBC result	No	Yes
Direct test from clinical material	Yes	No
Applicable for slow-growing and fastidious organisms	No[e]	Yes
Readily amenable to automation	No	Yes

[a] Micro (0.1 ml) or macro (1.0 to 2.0 ml).
[b] Contaminants can be detected on subculture to appropriate medium.
[c] Slow-growing subpopulations (e.g., MRSA/E) require additional incubation interval and/or medium advantage.
[d] Zone diameters are inversely related to MIC and required regression analysis for development of interpretive guidelines (S,I,R).
[e] Current standards apply to aerobic and anaerobic organisms.
I = intermediate or indeterminant; MRSA = methicillin-resistant *S. aureus*; MRSE = methicillin-resistant *S. epidermidis*; R = resistant; S = susceptible.

100%) of the population of isolates tested respond to a compound are an efficient means of evaluating this type of comparative data. There is some question as to the minimum number of tested isolates that are required (usually 10) to obtain a reliable figure and whether the parameter should be based on raw data or derived from calculations (Schmidt, 1987). Calculations of the MIC_{50} and MIC_{90} are derived from the methods originally outlined by Reed and Muench (1938).

SUB-MIC, THE MAC

Concentrations of antibiotic below the MIC can quantitatively affect the growth of bacteria and structurally alter them. Lorian (1986) has shown that the minimum antibiotic concentration (MAC) is the lowest concentration of drug that will induce a demonstrable morphologic or ultrastructural modification in an organism. The inhibitory MAC is the drug concentration of organism that results in a 90% reduction of growth in 5.5 h as compared with control (Fig. 3). In general, MIC/MAC ratios are higher for aminoglycosides than beta-lactam agents.

Intermittent administration of antibiotics may produce levels that are reflective of sub-MIC (MAC). These levels can modify the bacterium's surface, increase chemotaxis (Forsgren and Gnarpe, 1982), increase opsonization of susceptible bacteria and ingestion by phagocytes, and augment phagocyte bactericidal function (Horne and Tomasz, 1981, Daschner, 1985).

PAE

The postantibiotic effect (PAE) describes the continued inhibition of bacterial replication after a limited exposure to drug. The PAE can be determined by exposing a broth culture in logarithmic phase to an antibiotic concentration above the MIC for at least 1 to several hours. The antibiotic is then removed by enzymatic inactivation, repeated washing, or dilution. The culture is sampled serially to define the rate of regrowth. The PAE was first observed by Eagle et al. (1950) for penicillin, but it has since been described for both gram-positive and gram-negative bacteria and for several classes of antimicrobial agents (Craig and Gudmundsson, 1985). In most cases, the duration of PAE increases with longer exposure times or higher drug concentrations up to a maximum.

Sub-MIC effects (MAC) and PAE activity appear to characterize the time course of antimicrobial activity more accurately than do the MIC and MBC alone. This information can provide a theoretical basis for dosing, depending on the pathogen and the agent administered.

DIRECT TESTS FOR ASSESSMENT OF RESISTANCE

When recognized bacterial pathogens known to be exquisitely susceptible to antimicrobial agents (e.g., *Neisseria gonorrhoeae, Haemophilus influenzae, Staphylococcus, Bacteroides*) develop resistance to a specific agent or class, it is frequently associated with the organism's acquisition of an R plasmid capable of destroying the therapeutic agent (see the preceding section The Organism). For the species just noted that produce beta-lactamase, penicillin or ampicillin would be ineffective. The inactivating enzyme can be readily and rapidly detected by chemical means (within 1 min for some tests), and thus the cost and delay of performing routine susceptibility studies can be avoided. Beta-lactamase can be detected by iodometric, acidometric, and colorimetric methods (Schoenknecht et al., 1985). Perhaps the method most widely used with the greatest application is the chromogenic cephalosporin (nitrocefin, Glaxo) substrate test, which is available commercially as a paper-impregnated disk. Similarly, chloramphenicol-resistant bacteria that elaborate chloramphenicol acetyl-transferase (CAT), which inactivates chloramphenicol can be detected. A rapid (70 min) colorimetric assay for CAT based on earlier described work is commercially available (Azemun et al., 1981).

These directed tests are used as rapid adjuncts to

provisional therapy (*H. influenzae* meningitis) and as epidemiologic markers to monitor changing resistance patterns within a community or medical center.

The advances in recombinant nucleic acid technology hold the promise for additional direct tests that will target specific and unique bacterial nucleotide sequences as indicators of resistance to specific agents (Tenover, 1986; Cooksey and Mayer, 1987). These test results will have a negative decision input when applied to decision-making therapy.

Assessments of Bactericidal Activity

The minimal inhibitory concentration as outlined here is satisfactory for guiding optimal therapy in most situations. However, there does arise specific clinical conditions in which there is a need for complete killing of the offending microorganism at the locus of infection, such as in cases of endocarditis or osteomyelitis. The parameter that most readily supplies this information is the minimum bactericidal concentration (MBC) as derived from the MIC. Some investigators prefer the term "minimum lethal concentration (MLC)" as this applies to both bacteria and fungi. Here, however, we will only address the effect on bacteria. Two other methods attempt to address the killing action of bacteria—the determination of the killing curve and the serumcidal effect.

The MBC

The MBC (see Fig. 3) is founded on the broth dilution method of susceptibility testing utilizing the MIC as the end point. Whereas it is anticipated that for bactericidal drugs (aminoglycosides and beta-lactams) the MIC and MBC will be common, it is realized that for bacteristatic drugs the MBC can be several dilutions greater. A procedural problem associated with this test is the definition of the MBC as it relates to the number of survivors that remain in the population since 100% elimination is an impossibility to achieve (Anhalt and Washington, 1985; Schoenknecht et al., 1985). Several investigators have utilized the definition of a reduction of the number of bacteria present in the inoculum to 99.99% of the original inoculation. This 10^{-3} ($-\log_{10}3$) reduction is somewhat arbitrary since there is no convincing evidence that 99.0% or 98.0% less accurately portrays the outcome of results. Very recently in the United States, the NCCLS (1987b) has undertaken to write a proposed standard for these tests in which the 1,000-fold reduction of original inoculum is used as a conventional standard. When the ratio of MBC to MIC is 32 or greater for a given bacterium-antimicrobial agent combination, the organism is said to be "tolerant" to the action of the antimicrobial agent, and it is questionable whether favorable clinical response or outcome can be achieved. The problems of tolerance and its clinical implication have been extensively reviewed (Handwerger and Tomasz, 1985; Tuomanen, 1986), and the reader can refer to these reviews for additional clarification. Procedural modifications will give rise to varied results (Sherris, 1986). These have been defined earlier in factors that modify the outcome of organism-drug interactions. In addition to those previously noted, variables uncovered include the material in which the test is executed (glass versus plastic) as bacteria adhere variably to the inside and, thus, alter estimates of survivors; the method and length of mixing and the reincubation intervals after mixing; and the volume of liquid (broth) used.

Killing Curve

The time course of antimicrobial activity can be plotted by determining the killing rate or curve of die-off of bacteria when exposed to antimicrobial agents (Figs. 5 and 6). This approach has been used to evaluate and compare new drugs and study differences and changes in the susceptibility of clinically significant bacterial pathogens. Although such determinations may be used as a guide for chemotherapy, they have less meaning in individual cases and apply more generally to classes of drugs. For example, when one compares the beta-lactam antimicrobial agents with the aminoglycosides, the former are characterized by a slower dose-independent initial bactericidal activity. The extent of the cidal action is related to the duration that serum levels exceed the MIC. As levels fall below the MIC, there is immediate regrowth of gram-negative organisms and streptococci. In contrast, aminoglycosides demonstrate rapid dose-de-

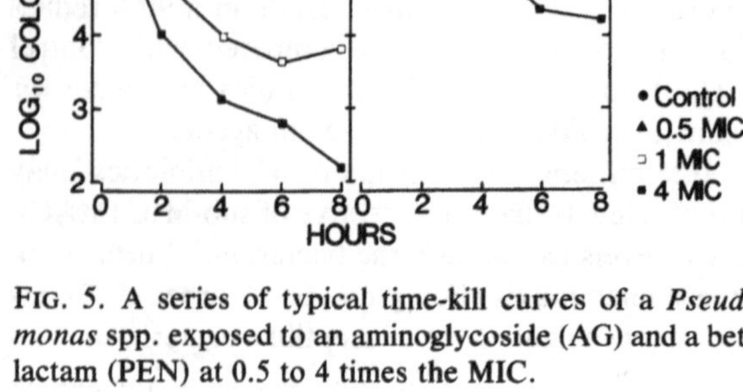

FIG. 5. A series of typical time-kill curves of a *Pseudomonas* spp. exposed to an aminoglycoside (AG) and a beta-lactam (PEN) at 0.5 to 4 times the MIC.

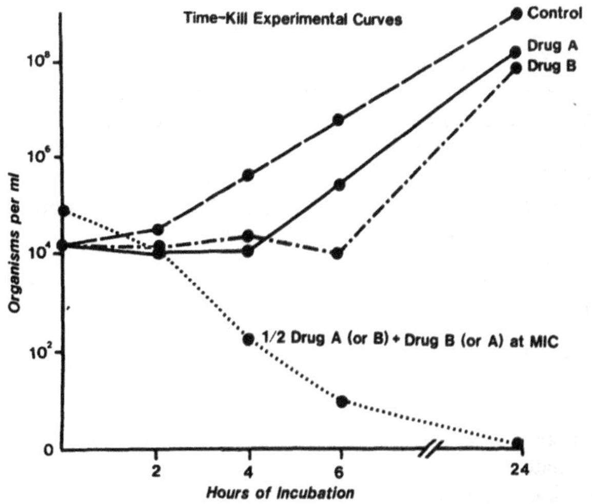

FIG. 6. Rate of kill (time-kill) curves for two antimicrobial drugs ("A" and "B"), indicating kill rates for each, and 0.5A (or B) + drug B (or A) at MIC (*dotted line*), demonstrating a synergistic effect.

pendent initial bactericidal activity, followed by a bacteriostatic phase that can last several hours after serum concentrations fall below the MIC.

The kinetics of antimicrobial activity as described provide a theoretical basis for frequency of dosing and depend on the pathogen and the antibiotic used.

In practice, the killing curve test is done in large volumes (1 to 2 ml); usually in glass tubes. Only one antibiotic concentration representing the average obtainable blood level or a limited number representing multiples or fractions of the blood level is used. At specific intervals from 0 to 24 h incubation, colony counts are performed and plotted with the log survivors on the ordinate and the time on the abscissa (see Fig. 5). Here it can be seen that the aminoglycoside (AG) demonstrates a drug with concentration-dependent bactericidal activity. At increasing drug concentrations, there is an increase in the magnitude and rate of killing. Beta-lactam compounds (PEN in Fig. 5) demonstrate little concentration-dependent bactericidal activity.

ANTIBIOTIC COMBINATIONS

There are situations where the use of multiple (usually two) agents is preferred and advantageous. These are: 1) for the treatment of dual and multiple bacterial infections in which the responsible organisms are not susceptible to a common drug; 2) to kill the infecting population rapidly so as to prevent the development of drug resistance and overcome bacterial tolerance; and 3) to diminish drug toxicity when alternative therapy is unavailable.

Ideally, the multiple use of antimicrobial agents is employed when these therapeutic modalities can lead to more than an additive effect; antagonistic actions are to be avoided. We do not mean here the use of marketed compounds in a fixed ratio (e.g., trimethoprim and sulfamethoxazole [TMP/SMX], the copenicillins Timentin [ticarcillin and clavulanic acid] or Augmentin [amoxacillin and clavulanic acid]).

The basic method for evaluating the combined effect of two antimicrobials is the two-dimensional broth dilution or "checkerboard" (chessboard to some) titration. In this assay, serial two-fold dilutions of two drugs, alone and in combination, are tested against a standardized inoculum. The lowest concentration of each drug in combination that inhibit (MIC) or kill (MBC) the organism is plotted as an isobologram (Fig. 7). Results can also be expressed as a fractional inhibitory concentration (FIC) of each drug by dividing the MIC in the combination by the MIC of drug A alone. Fractional bactericidal concentrations (FBCs) can be similarly calculated (Anhalt and Washington, 1985).

If the rate of bactericidal activity (killing curve [Fig. 6]) is used, a synergistic effect is defined as a decrease (−log 2) in colony-forming units as compared with the activity of a single agent. Alternatively, a synergistic effect can be evidenced by a four-fold decrease in the MIC of each drug in the combination (Rahal, 1978).

The serum bactericidal titer and the serum bactericidal rate have also been used to estimate synergy.

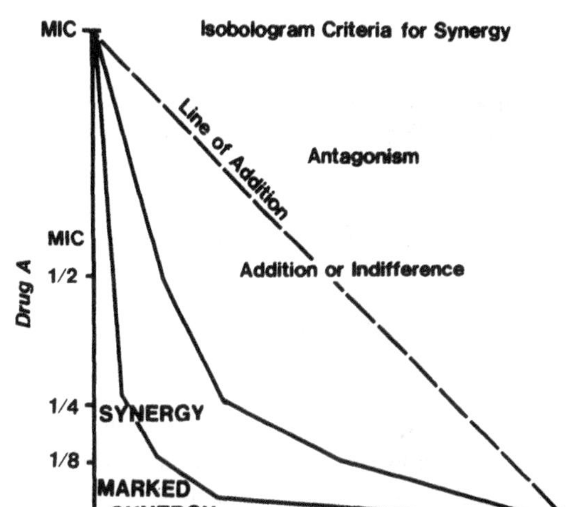

FIG. 7. Isobologram assessing three possible interactions of Drugs A and B tested (MIC or MBC) singly and at varying combinations. The *straight* (*dashed*) *line* represents values for each drug separately and indicates indifference. The concavity of the *curves* toward the coordinate indicates synergism. Antagonism would be indicated by a line that would bow away from the coordinates.

TABLE 3. Assessment of antimicrobial combinations: criteria for determination of synergy

Method	*Procedure*	*Interpretation*	*References*
Serial dilution—"checkerboard" technique	Determine MIC of multiple combinations of each drug	(a) 0.25 Reduction of each antibiotic used in combination to 0.25 MIC of each drug alone	Schoenknecht et al., 1985 Anhalt and Washington, 1985
		(b) Calculate fractional ratio (MIC or MBC) FIC (FBC) $= \frac{MIC_A\ (A + B)}{MIC_A}$ FIC < 1.0 = synergy FIC > 1.0 = antagonism FIC $= 1.0$ = indifference	Anhalt and Washington, 1985
	From MIC plot isobolograms	(c) Synergy = bow toward coordinates Antagonistic = bow outward Indifference = line of addition (see Fig. 7)	Schoenknecht et al., 1985
Killing curve	Killing cidal effect over a 24-h period; plot log survivors over a 24-h period (see Fig. 6).	10 to 100-fold (1 to 2 $\log_{10}$) increased killing in combination as compared with most effective of either drug—synergy	Schoenknecht et al., 1985 Anhalt and Washington, 1985
SBT	Determine SBT and calculate free drug level	Relationship between combined ratios of free drug and respective MBCs to SBT (see text)	Robinson et al., 1985
SBR	Determine SBT and plot vs. time; calculate area under curve (AUBC)	When AUBC for drug combination is significantly greater than sum of two drugs—synergy	Barriere et al., 1985
SBR	Determine SBT using 1 : 2 dilutions of serum and plot at 2, 4, 6, and 8 h	Significant difference in plot of single versus combination drugs	Drake et al., 1983

The methods and criteria for interpreting results are summarized in Table 3.

Evaluations of Host-Drug Interaction

Levels of Antimicrobial Agents in Body Fluids

Attempts to estimate the level of antimicrobial agents are as old as the in vitro assessments of activity. Although it is common to relate drug levels to serum (or plasma), antibacterial activity can be quantitated in cerebrospinal fluid, urine, osteocytes, and the aqueous humor of the eye.

Average achievable drug levels following package-insert directions of oral and parenteral drugs are known prior to market distribution of any compound. Yet, confusion exists as to what concentration constitutes a therapeutic level that will result in a positive clinical outcome. Part of this problem stems from a lack of concordance in relating the antimicrobial level needed to inhibit or kill the infecting organism and the achievable in vivo drug levels (Hewitt and McHenry, 1978). Generally, levels two to four times that of the MIC are considered desirable.

In patients with severe bacterial infections, the widest application of monitoring drug concentrations (levels) in body fluids is not to ensure adequate concentration of drug, but to prevent toxic levels. Serum drug assays are of prime importance for monitoring levels of potential hepatoxic, nephrotoxic, and ototoxic antimicrobials, especially in patients with underlying liver or kidney disease. The aminoglycosides are of major concern because of their narrow therapeutic range and the potential of auditory and renal damage.

Serum samples that best represent the anticipated rapid changes in blood levels of aminoglycosides (and other drugs with similar pharmacokinetic properties) are denoted as peak and trough or "valley" specimens. Depending on the route of administration, the peak level is the specimen collected 30 to 45 min after completion of an intravenous infusion, 60 min after an intramuscular dose, or 90 min after an oral dose. The trough level is represented by the specimen collected 30 min before the following dose. After collection, specimens should be placed on ice and, if not immediately assayed, frozen at −20°C so as to minimize degradation or interaction between compounds in the event multiple agents were administered.

Several methods are currently used (Edberg et al., 1984). The most common are agar diffusion approaches being derived from the early work of Sabath et al. (1971). In these methods, the assay or "test" organism, usually *Bacillus subtilis*, *B. globigii*, *Staphylococcus epidermidis*, *Klebsiella pneumoniae*, or others, of proven susceptibility to the drug in question is used. Ideally, the assay organism should be exquisitely susceptible to the drug to be assayed, resistant to all other antimicrobial agents, and be rapidly growing so that results can be obtained within the same day. When the assay organism is generally susceptible to many antimicrobial compounds, it is necessary to negate one (or more) agents if the patient is on combination therapy. This can readily be achieved with penicillinases and beta lactamases for the respective compound substrates. A standard curve is constructed encompassing several (three to five) duplicate or triplicate drug standards. Using the patient's specimen and standards, the maximal zone of growth inhibition measured around a disk or well is plotted on the standard curve that relates the log of drug concentration (ordinate) to zone of inhibition (abscissa). The value of each unknown patient sample can be interpolated, and the concentration of the agent determined.

Other nonmicrobiological assays are available utilizing radiolabeled enzyme and antibody (Stevens et al., 1975) and immunoassays coupled to enzymatic or fluorescent labels (Buchanan et al., 1983; Edberg et al., 1984). These latter techniques are adaptable to automated equipment, are capable of rapid turnaround time, and results are highly reproducible (Amsterdam et al., 1984).

Evaluations of Host-Organism Interaction

Serum Bactericidal Test/Titer (SBT)

This estimate of antimicrobial activity most dramatically represents the efficacy of the antimicrobial agent in the patient during therapy. Antimicrobial agents singly or in combination can be used in this, what would seem to be, the most logical approach to monitoring chemotherapy. Several procedural problems are extant, and variations in the procedure exist as originally described by Schlichter and MacLean (1947). In its original form, the test was only for bacteriostatic activity, and was later modified to include bactericidal activity. The NCCLS (1987a) has proposed procedural guidelines.

The SBT represents a variation in the broth dilution procedure in that serum from the patient instead of the antimicrobial agent is serially diluted. The test cannot be executed unless the laboratory has previously isolated a significant pathogen presumed to be associated with the patient's condition. Although the cidal activity of nascent patient serum can be evaluated, the test should be applied so as to measure the effect of singly or multiply administered antimicrobial agents.

Evaluation of SBT can be of help in monitoring therapy of bacterial endocarditis, osteomyelitis or septic arthritis, meningitis, and, in cancer patients, bacteremia. Frequently, both peak and trough serum specimens are evaluated since there is no evidence indicating the contrary. The advantage of the SBT is that in addition to considering the pathogen, it evaluates the pharmacokinetic properties of the drug(s) (absorption, elimination, protein binding, and effects of metabolite and parent compound) and the effects of multiple drug interactions (antagonistic to synergistic).

Clinical studies (Wolfson and Swartz, 1985) have attempted to evaluate the usefulness of the SBT, but have been plagued with problems associated with the standardization of the test procedure. The largest error appears to be associated with the use of a suitable physiologic diluent. A review of the SBT in endocarditis indicated that peak titers of 1:64 and greater and trough titers of 1:32 and greater predicted bacteriologic cure but not bacteriologic failure (Wolfson and Swartz, 1985). Results from a multicenter collaborative study utilizing a standardized protocol for the performance of the test agreed with the previous finding, and further indicated that the traditionally accepted SBT of 1:8 had statistically significant predictive accuracy at trough levels only (Weinstein et al., 1985). The SBT is a poor predictor of bacteriologic failure and ultimate clinical outcome (Weinstein et al., 1986).

As shown by Stratton et al. (1982), the SBT is related to the MBC as follows:

$$SBT = \frac{AL}{MBC}$$

where SBT is the measured bactericidal titer, and AL/MBC is the calculated reciprocal serum bactericidal titer (AL = antibiotic level). This would appear to be acceptable in that as the serum level increases (or the MBC decreases), the serum titer will increase.

The SBT has been used as an experimental tool to assess new drugs and drug combinations and for determining antimicrobial activity in body compartments other than serum. Robinson et al. (1985) used the serum bactericidal titer for determining the combined effectiveness of antimicrobials. They demonstrated that when more than one drug is administered, the relationship between the combined ratios of free drug (drug-f) and their respective MBCs in broth (MBC_B) to the SBT is expressed as:

$$SBT = \Sigma \frac{\text{drug-f}}{MBC_B}$$

The assumption is that when the SBT does not equal the sum of the free drug concentration/MBC ratios, it is suggestive that an interaction has occurred between the two antimicrobial agent. Thus, when

$$SBT > \Sigma \frac{\text{drug-f}}{MBC_B}$$

synergy is suggested (see Table 3).

Serum Bactericidal Rate (SBR)

The time-kill curve evaluates the rate at which single or multiple drugs kill bacteria. The SBR is a methodologic approach that represents the rate of serum killing; it integrates in vitro activity and in vivo pharmacokinetics of antimicrobial drugs and, therefore, reflects very accurately what happens in the host.

Drake et al. (1983) proposed the SBR as an adjunct to the SBT. In an animal model for endocarditis, they demonstrated that the SBR can be of value when limited serum dilutions are plotted over an 8-h period. They proposed that these methods are adaptable for the clinical microbiology laboratory (see Table 3).

In a novel approach to the application of the SBR, Barriere et al. (1985) measured the area under the bactericidal curve (AUBC) to determine the total synergistic bactericidal activity. When the calculated value of the AUBC for a combination was greater than the sum of individual drug, the combination was considered synergistic (Table 3).

In the future, laboratory assessments of antimicrobial activity will no doubt attempt to incorporate microbiologic and pharmacologic attributes of antibiotics. As currently conceived, models of in vitro kinetics are too tedious and labor-intensive to execute as a per patient service. They may, however, be incorporated into computer models and serve as an algorithm to assist the clinician in the delivery of rational, safe, and effective therapy.

Literature Cited

Amsterdam, D., C. Kalinka, and W. R. Bartholomew. 1984. Comparative immunoassays for quantitation of tobramycin serum levels. Antimicrob. Newsletter **1**:56–57.

Anhalt, J. P. and J. W. Washington II. 1985. Bactericidal tests. p. 731–745. *In* J. A. Washington (ed.), Laboratory procedures in clinical microbiology, 2nd ed., Springer-Verlag, New York.

Azemun, P., T. Stall, M. Roberts, and A. L. Smith. 1981. Rapid detection of chloramphenicol resistance in *Haemophilus influenzae*. Antimicrob. Agents Chemother. **20**:168–170.

Barriere, S. L., E. Ely, J. E. Kapusnik, and J. G. Gambertoglio. 1985. Analysis of a new method for assessing activity of combinations of antimicrobials: area under the bactericidal activity curve. J. Antimicrob. Chemother. **16**:49–59.

Barry, A. L. 1976. The antimicrobic susceptibility test: principles and practices, p. 163–179. Lea & Febiger, Philadelphia, Pennsylvania.

Barry, A. L. and Thornsberry, C. 1985. Susceptibility tests: Diffusion test procedures, p. 978–987. *In* E. H. Lennette, A. Balows, W. J. Hausler, Jr., and H. J. Shadomy (ed.), Manual of clinical microbiology, 4th ed., American Society for Microbiology, Washington, D.C.

Bauer, A. W., W. M. M. Kirby, J. C. Sherris, and M. Turck. 1966. Antibiotic susceptibility testing by a standardized single disk method. Am. J. Clin. Pathol. **45**:493–496.

Buchanan, A. G., E. Witwicki, and W. L. Albritton. 1983. Serum aminoglycoside monitoring by enzyme immunoassay, biological, and fluorescence immunoassay procedures. Am. J. Med. Tech. **49**:438–441.

Cooksey, R. C. and L. W. Mayer. 1987. Identification of antibacterial resistance mechanisms: Advances in laboratory assays. Antimicrob. Newsletter **4**:57–66.

Cooper, K. E. 1963. The theory of antibiotic inhibition zones. p. 1–86. *In* F. Kavanagh (ed.), Analytical microbiology, Academic, Press, New York.

Craig, W. A. and S. Gudmundsson. 1985. The postantibiotic effect, p. 515–536. *In* V. Lorian (ed.), Antibiotic in laboratory medicine, 2nd ed., William & Wilkins, Baltimore, Maryland.

Daschner, F. D. 1985. Antibiotics and host defense with special reference to phagocytosis by human polymorphonuclear leukocytes. J. Antimicrob. Chemother. **16**:135–141.

Drake, T. A., C. J. Hackbarth, and M. Sande. 1983. Value of serum tests in combined drug therapy of endocarditis. Antimicrob. Agents Chemother. **24**:653–657.

Eagle, H., R. Fleischman, and A. D. Musselman. 1950. Effect of schedule of administration on the therapeutic efficacy of penicillin. Am. J. Med. **9**:280–299.

Edberg, S. C., A. L. Barry, and L. S. Young. 1984. Cumitech 20. Therapeutic drug monitoring: antimicrobial agents. J. A. Morello (Coordinating ed.), American Society for Microbiology, Washington, D.C.

Ericsson, H. M. and J. C. Sherris. 1971. Antibiotic sensitivity testing. Report of an international collaborative study. Acta Pathol. Microbiol. Scand. Sect. [B] **Suppl**:217.

Eng, R. H. K., C. M. Cherubin, S. M. Smith, and F. Buccini. 1985. Inoculum effect of β-lactam antibiotics in Enterobacteriaceae. Antimicrob. Agents Chemother. **28**:601–606.

Federal Register. 1961. Antibiotics intended for use in the laboratory diagnosis of disease. Fed. Register **26**:2596.

Federal Register. 1972. Rules and regulations. Antibiotic susceptibility disks. Fed. Register **37**:20525–20529.

Federal Register. 1973. Rules and regulations. Antibiotic susceptibility disks: correction. Fed. Register **38**:2576.

Fleming A. 1929. On the antibacterial action of cultures of a penicillium with special reference to their use in the isolation of *B. influenzae*. Br. J. Exp. Pathol. **10:** 226–236.

Forsgren, A. and H. Gnarpe. 1982. The effect of antibacterial agents on the association between bacteria and leukocytes. Scand. J. Infect. Dis. **33(suppl.):**115–120.

Foster, J. W. and H. B. Woodruff. 1943. Microbiological aspects of penicillin. J. Bacteriol. **46:**187–202.

Handwerger, S. and A. Tomasz. 1985. Antibiotic tolerance among clinical isolates of bacteria. Rev. Infect. Dis. **7:**368–386.

Hewitt, W. L. and M. C. McHenry. 1978. Blood level determinations of antimicrobial drugs: some clinical considerations. Med. Clin. North Am. **62:**1119–1140.

Horne, D. and A. Tomasz. 1981. Hypersusceptibility of penicillin-treated group B streptococci to bactericidal activity of human polymorphonuclear leukocytes. Antimicrob. Agent Chemother. **19:**745–753.

Jones, R. N., A. L. Barry, T. L. Gavan, and J. A. Washington II. 1985. Susceptibility tests: microdilution and macrodilution broth procedures, p. 972–977. *In* E. H. Lennette, A. Balows, W. J. Hausler, Jr., and H. J. Shadomy (ed.). Manual of clinical microbiology, 4th ed., American Society for Microbiology, Washington, D.C.

Kunin, C. M. 1981. Dosage schedules of antimicrobial agents: a historial review. Rev. Infect. Dis. **3:**4–11.

Lorian, V. 1986. Effect of low antibiotic concentrations on bacteria, p. 596–668. *In* V. Lorian (ed.), Antibiotics in laboratory medicine, 2nd ed., Williams & Wilkins, Baltimore, Maryland.

Metzler, C. M. and R. M. De Haan. 1974. Susceptibility of anaerobic bacteria: statistical and clinical considerations. J. Infect. Dis. **130:**588–594.

Miller, J. M., B. V. Andersen, and S. P. Caudill. 1986. New quality control frequency guidelines for antimicrobic susceptibility testing. Antimicrob. Newsletter **3:**77–79.

National Committee for Clinical Laboratory Standards. 1984. Performance standards for antimicrobial disk susceptibility tests. NCCLS publication M2-A3, 3rd ed., National Committee for Clinical Laboratory Standards, Villanova, Pennsylvania.

National Committee for Clinical Laboratory Standards. 1985. Dilution procedures for susceptibility testing of aerobic bacteria. NCCLS publication M7-A, National Committee for Clinical Laboratory Standards, Villanova, Pennsylvania.

National Committee for Clinical Laboratory Standards. 1986a. Evaluating production lots of dehydrated Mueller-Hinton agar; proposed standard. NCCLS publication M6-P, National Committee for Clinical Laboratory Standard, Villanova, Pennsylvania.

National Committee for Clinical Laboratory Standards. 1986b. Performance standards for antimicrobial susceptibility testing. NCCLS publication M100-S, National Committee for Clinical Laboratory Standards, Villanova, Pennsylvania.

National Committee for Clinical Laboratory Standards. 1987a. Methodology for the serum bactericidal test; proposed guideline. NCCLS document M21-P, National Committee for Clinical Laboratory Standards, Villanova, Pennsylvania.

National Committee for Clinical Laboratory Standards. 1987b. Methods for determining bactericidal activity of antimicrobial activity of antimicrobial agents; proposed guideline. NCCLS document M26-P, National Committee for Clinical Laboratory Standards, Villanova, Pennsylvania.

Rahal, J. R. 1978. Antibiotic combinations: the clinical relevance of synergy and antagonism. Medicine **57:**179–195.

Reed, L. J. and H. Muench. 1938. A simple method of estimating fifty percent endpoints. Am. J. Hyg. **27:**493–497.

Reller, L. B., F. D. Schoenknecht and M. A. Kenny. 1974. Antibiotic testing of *P. aeruginosa*: selection of a control strain and criteria for magnesium and calcium content in media. J. Infect. Dis. **130:**454–463.

Robinson, A., R. C. Bartlett, and M. F. Mazens. 1985. Antimicrobial synergy testing based on antibiotic levels, minimal bactericidal concentration, and serum bactericidal activity. Am. J. Clin. Pathol. **84:**328–333.

Sabath, L. D., J. I. Casey, P. A. Ruch, L. I. Stumpf, and M. Finland. 1971. Rapid microassay of gentamicin, kanamycin, neomycin, streptomycin, and vancomycin in serum or plasma. J. Lab. Clin. Med. **78:**457–463.

Schlichter, J. G. and H. MacLean. 1947. A method of determining the effective therapeutic level in the treatment of subacute bacterial endocarditis with penicillin. Am. Heart J. **34:**209–211.

Schmidt, L. H. 1987. The MIC_{50}/MIC_{90}: assessments of in vitro activities of antimicrobial agents that facilitate comparative agent-agent and agent-species susceptibility comparisons. Antimicrob. Newsletter **4:**1–8.

Schneierson, S. S. and D. Amsterdam. 1959. A simplified tube procedure for the routine determination of bacterial sensitivity to antibiotics. Am. J. Clin. Pathol. **31:**81–86.

Schoenknecht, F. D., L. D. Sabath, and C. Thornsberry. 1985. Susceptibility tests: special tests, p. 1000–1008. *In* E. H. Lennette, A. Balows, W. H. Hausler, Jr., and H. J. Shadomy (ed.), Manual of clinical microbiology, 3rd ed., American Society for Microbiology, Washington, D.C.

Sherris, J. C. 1986. Problems in in vitro determination of antibiotic tolerance in clinical isolates. Antimicrob. Agent Chemother. **30:**633–637.

Stevens, P., L. S. Young, and W. L. Hewitt. 1975. Radioimmunoassay, radioenzymatic assay and microbioassay of gentamicin: a comparative study. J. Lab. Clin. Med. **86:**349–359.

Stratton, C. W., M. P. Weinstein, and L. B. Reller. 1982. Correlation of serum bactericidal activity with antimicrobial agent level and minimal bactericidal concentration. J. Infect. Dis. **145:**160–168.

Tenover, F. C. 1986. Studies of antimicrobial resistance genes using DNA probes. Antimicrob. Agents Chemother. **29:**721–725.

Thornsberry, C. 1984. Methicillin-resistant (heteroresistant) staphylococci. Antimicrob. Newsletter **1:**43–47.

Tuomanen, E. 1986. Phenotype tolerance. The search for β-lactam antibiotics that kill nongrowing bacteria. Rev. Infect. Dis. **8(suppl.):**S279–S291.

Vincent, J. G. and H. W. Vincent. 1944. Filter paper modification of the Oxford cup penicillin determination. Proc. Soc. Exp. Biol. Med. **55:**162–164.

Washington, J. A. II. 1985. Susceptibility tests: agar dilution, p. 967–971. *In* E. H. Lennette, A. Balows, W. J. Hausler, Jr., and H. J. Shadomy (ed.), Manual of clinical microbiology, 4th ed., Americal Society for Microbiology, Washington, D.C.

Weinstein, M. P., C. W. Stratton, A. Ackley, H. B. Hawley, P. A. Robinson, B. D. Fisher, D. V. Alcid, D. S. Stephens, and L. B. Reller. 1985. Multicenter collaborative evaluation of a standardized serum bactericidal test as a prognostic indicator in infective endocarditis. Am. J. Med. **78:**262–269.

Weinstein, M. P., C. W. Stratton, and L. B. Reller. 1986. Current status of the serum bactericidal test as a moni-

tor of therapeutic efficacy in serious infections. Antimicrob. Newsletter **2**:9–14.
Wolfson, J. S. and M. N. Swartz. 1985. Serum bactericidal activity as a monitor of antibiotic therapy. N. Engl. J. Med. **312**:969–974.
World Health Organization. 1961. Standardization of methods for conducting microbic sensitivity tests. Second report of the Expert Committee on Antibiotics, p. 1–24. World Health Organization Technical Reports Series No. 210.

SECTION II

Bacterial Infections

CHAPTER 5

Anaerobic Bacterial Infections (Non-Spore-Forming)

SYDNEY M. FINEGOLD

Disease: Embraces infections of all types in any tissue or organ of the body. Major types of anaerobic infections, in terms of frequency encountered, include pleuropulmonary, intraabdominal, and female genital tract infections of various types. Other infections in which anaerobes are commonly involved are brain abscess, chronic sinusitis and otitis media, dental infections, bite wound infections, soft tissue infections, and bone and joint infections.

Etiologic Agents: Numerous species of *Bacteroides, Fusobacterium, Peptostreptococcus, Actinomyces,* and other species of non-spore-forming anaerobic gram-positive bacilli.

Source: Virtually always the indigenous flora of the host, particularly the oral, intestinal, and female genital tracts.

Clinical Manifestations: Very variable. Common features are gas production, tissue destruction, abscess formation, foul odor, and features typical of the specific infection involved (e.g., pneumonia, brain abscess).

Pathology: Variable. Purulence, abscess formation, tissue invasion with necrosis and gas production, and granuloma formation (in the case of actinomycosis).

Laboratory Diagnosis: Recovery of non-spore-forming anaerobic organisms, alone or together with clostridia and/or aerobic or facultative organisms, from appropriate specimens in patients with compatible clinical illness. Demonstration of such organisms in tissues or specimens by direct means such as fluorescent antibody technique or DNA probes.

Epidemiology: Not pertinent. Even when non-spore-forming anaerobic infection is associated with surgical manipulation and implantation of a foreign body (e.g., ventriculoatrial shunt, implanted intrauterine device), the etiologic agent comes from the host's indigenous flora.

Treatment: Surgical debridement, surgical or percutaneous drainage of abscesses, and antimicrobial agents.

Prevention and Control: Careful surgical technique, maintaining good oxygenation of tissues when possible, avoidance of aspiration of material into the respiratory tract, antimicrobial prophylaxis for dental and selected other procedures, and minimizing the spread of anaerobic bacteria from mucosal surfaces into deeper tissues.

Description of Disease

Non-spore-forming anaerobic bacteria of clinical significance are found primarily in the genera *Bacteroides, Fusobacterium, Peptostreptococcus, Actinomyces, Arachnia, Bifidobacterium,* and *Propionibacterium.* These organisms are prevalent among the normal flora of the human body (Table 1). They take part, in a mixed aerobic/facultative and anaerobic flora, in a variety of infections throughout the body.

Pathogenesis

As with clostridial infections, infection due to non-spore-forming anaerobes occurs chiefly in an anaerobic environment, which usually results from impaired circulation, trauma, surgery, or malignant or other disease. These factors often cause a breach in the mucosal surface of the oral, intestinal, or female genital tract and thus permit endogenous anaerobes to enter deeper tissues to produce the infection. Prior or concurrent infection with nonanaerobes may re-

TABLE 1. Incidence of various anaerobes as normal flora in humans[a]

	Non-spore-forming bacilli						
	Gram-positive			Gram-negative		Cocci	
Site	Clostridium	Actinomyces	Propioni-bacterium	Bacteroides	Fuso-bacterium	Gram positive	Gram negative
Skin	0[b]	0	2	0	0	1	0
Upper respiratory tract[c]	0	1	1	1	1	1	1
Mouth	±	1	±	2	2	2	2
Intestine	2	±	±	2	±	2	±
External genitalia	0	0	U	1	1	1	0
Urethra	±	0	0	1	1	±	U
Vagina	±	0	1	1		2	1

[a] Adapted from Sutter et al., Wadsworth Anaerobic Bacteriology Manual, 4th ed., 1985.
[b] U = Unknown; 0 = not found or rare; ±, irregular; 1 = usually present; 2 = usually present in large numbers.
[c] Includes nasal passages, nasopharynx, oropharynx, and tonsils.

duce the oxygen content or the oxidation-reduction potential at the site, thus facilitating involvement of anaerobes. Other factors involved in synergy, which is common in mixed infections, include production of nutrients essential for the growth of anaerobes and production of toxins or other virulence factors. Putative virulence factors produced by gram-negative anaerobic bacilli (the most commonly encountered in infections of the non-spore-forming anaerobes) include ability to adhere to surfaces such as the peritoneal mesothelium or gingival crevice epithelium, production of capsular material (by *Bacteroides fragilis,* in particular; the capsule inhibits macrophage migration, is antiphagocytic for aerobes and anaerobes, and promotes abscess formation), production of superoxide dismutase and catalase (which confers oxygen tolerance), production of immunoglobulin proteases, production of hyaluronidase, collagenase, chondroitin sulfatase, neuraminidase, fibrinolysin, and heparinase and other coagulation-promoting factors, and production of lipopolysaccharide, "leukotoxin," butyrate, and soluble inhibitors of chemotaxis.

In the case of aspiration pneumonia and other pleuropulmonary infections involving anaerobes, the pathogenesis is usually different, with aspiration of oropharyngeal or gastric contents the major background factor and periodontal disease or gingivitis also important in some patients. Such aspiration, of course, carries oral anaerobes and other organisms from the mouth down into the bronchial tree.

Clinical Manifestations

Non-spore-forming anaerobic bacteria are involved in a wide variety of infections; the clinical manifestations depend on the site and type of infection.

Circulatory Infections

The chief circulatory infections involving non-spore-forming anaerobes are bacteremia and endocarditis, although mycotic aneurysm or other endovascular infection may occur in appropriate circumstances. In most series, anaerobes have been recovered from 10 to 25% of all positive blood cultures. The most common isolate, *Propionibacterium acnes,* usually represents a skin contaminant but may actually cause bacteremia in the setting of an implanted prosthetic or intravascular device. Clinically significant bacteremia accounts for about 5 to 15% of all bacteremias. Among these, the *Bacteroides fragilis* group is by far the most commonly encountered. This finding may reflect, to some extent, the ease with which this group of organisms is cultured and identified, but it is clearly a very important pathogen as well. The second most common blood culture isolate of significance is various *Peptostreptococcus* species. Anaerobes are found with some frequency in polymicrobial bacteremia. For the most part, anaerobic bacteremia will not be distinctive clinically from other bacteremias; suppurative thrombophlebitis, metastatic infection, and hyperbilirubinemia are features that may suggest involvement of anaerobes.

The incidence of anaerobes as causes of endocarditis is quite low. The most common isolates are anaerobic cocci, most of which are actually microaerophilic streptococci and belong in the genus *Streptococcus. Bacteroides* (and perhaps other anaerobic) endocarditis tends to produce large vegetations and embolic phenomena in large vessels.

Central Nervous System Infections

The chief infection of the central nervous system involving anaerobes is brain abscess; indeed, anaerobes are the primary cause of this serious problem. Background factors are often sinus, ear, or mastoid infections, oral or dental infection, pulmonary infection, or hematogenous spread from other foci. The anaerobes most commonly involved are *Bacteroides fragilis* and other β-lactamase-producing *Bacteroides, Fusobacterium, Peptostreptococcus, Actinomyces*, and *Clostridium*. Other important organisms are microaerophilic and viridans group streptococci and, on occasion, other nonanaerobes.

Anaerobes may be involved frequently in subdural empyema and sometimes in epidural abscess; they are quite uncommon in meningitis in the absence of brain abscess or subdural empyema as associated features.

Ophthalmic Infections

Recent work indicates that anaerobes are found in eye infections more frequently than was appreciated previously. Anaerobic infections of the eye are especially likely to occur in the settings of trauma, surgery, use of soft contact lenses, use of topical corticosteroids, and antibacterial therapy. Peptostreptococci are among the most commonly encountered anaerobes in eye infection. *Propionibacterium acnes* has been noted a number of times recently in infection secondary to implantation of artificial lenses. One of the few distinctive clinical syndromes of anaerobic infection of the eye is lacrimal canaliculitis involving *Actinomyces*, which leads to accumulation of concretions in the lacrimal canal. All types of eye infections may involve anaerobes, including the most serious ones such as endophthalmitis and orbital cellulitis.

Head and Neck Infections

Infections of the head and neck in which anaerobes may play a role include acute and, more commonly, chronic sinusitis and otitis media, mastoiditis, membranous tonsillitis (Vincent's angina), peritonsillar abscess, deep neck space infections, odontogenic infections, and infection following radical neck dissection or other head and neck surgery. The usual oral anaerobes are those most often involved: peptostreptococci, *Bacteroides* other than the *B. fragilis* group (although this group may be found in 5 to 20% of certain infections in this area), and fusobacteria. *Fusobacterium necrophorum* is a key pathogen in Vincent's angina and its complications.

Oral and Dental Infections

Root canal infection, periapical abscess, osteomyelitis of the jaw, periodontitis and abscess, stomatitis, noma, and other odontogenic infections are the processes in the oral area that may involve anaerobic bacteria.

Pleuropulmonary Infections

Anaerobes are an important cause of pleuropulmonary infection; however, because of the difficulty in collecting appropriate specimens, the role of anaerobes in this setting is often overlooked. Aspiration pneumonia and lung abscess are primarily anaerobic infections. Spontaneously occurring empyema (i.e., empyema in the absence of previous surgery or trauma) also frequently involves anaerobes. Along with the usual oral anaerobes, the *B. fragilis* group is found in about 7% of such infections, and nonanaerobes such as streptococci of the viridans group are common. In hospitalized patients, colonization of the oropharynx with nosocomial pathogens such as staphylococci and various gram-negative bacilli may occur; such organisms may then be involved in any subsequent aspiration pneumonia, along with anaerobes. The most commonly encountered anaerobes are *Bacteroides, Fusobacterium nucleatum*, and *Peptostreptococcus*.

Other Intrathoracic Infections

Mediastinitis, in the absence of prior cardiac surgery, most often involves anaerobes with preceding processes including perforation of the esophagus, extension of a retropharyngeal abscess, or extension of cellulitis or abscess of dental origin from the neck. Pericarditis and myocardial abscess seldom involve anaerobes.

Intraabdominal Infections

Anaerobes play a prominent role in intraabdominal infections, with the bowel as their point of origin; thus, the *Bacteroides fragilis* group is found in most patients. Other anaerobes encountered include other *Bacteroides, Peptostreptococcus*, and *Clostridium; E. coli* and enterococci are also commonly found, and in patients with prior hospitalization and antimicrobial therapy, more resistant gram-negative bacilli such as *Enterobacter* and *Pseudomonas* may be encountered. Wound and other postoperative infections will have a similar infecting flora. In biliary tract infection, *Clostridium perfringens* and, less commonly, *Bacteroides fragilis* are the anaerobes found, along with coliforms and enterococci. Anaerobes are probably the predominant pathogens in liver abscess. The most common anaerobic isolates are *Peptostreptococcus*, the *Bacteroides fragilis* group, other *Bacteroides, Fusobacterium necrophorum*, other fusobacteria, clostridia, and *Actinomyces*.

Retroperitoneal Infections

Anterior retroperitoneal abscess secondary to appendicitis or diverticulitis will frequently involve anaerobic bacteria. Anaerobes have not been recovered often from psoas abscesses.

Perineal and Perirectal Infections

As would be anticipated from the location, colonic anaerobes (including the *B. fragilis* group) are commonly involved in perineal and perirectal infections.

Female Genital Tract Infections

Essentially all pathologic processes of the female genital tract infection, whether obstetric or gynecologic in nature, involve anaerobes. Peptostreptococci and *Bacteroides* (including the *B. fragilis* group) are the most common anaerobic isolates, but clostridia are important because of the devastating illnesses they may produce.

Urinary Tract Infections

Anaerobes are thought to be rare causes of urinary tract infection, but are undoubtedly much more important than has been appreciated previously. Older reports described various anaerobes in all types of infections of the urinary tract. There are data to indicate that at least 1.3% of cases of bacteriuria involve anaerobes. There have been a number of reports of anaerobic bacteremia following manipulation of the urinary tract, such as urethral dilatation.

Skin, Subcutaneous Tissue, and Muscle Infections

Anaerobes are significant pathogens in a wide variety of minor to serious infections of the skin and underlying soft tissues, including cellulitis, hidradenitis suppurativa, pyoderma, cutaneous and subcutaneous abscesses, "fetid foot" infections, infected decubitus ulcers, bite wound infections, bacterial synergistic gangrene, burn wound infections, necrotizing fasciitis, anaerobic streptococcal myositis, anaerobic muscle abscesses in narcotic addicts, and synergistic nonclostridial anaerobic myonecrosis (synergistic necrotizing cellulitis). Clostridial infections, including gas gangrene, are discussed elsewhere in this volume. Infected decubitus ulcers are an important source of anaerobic bacteremia, particularly with the *B. fragilis* group.

Bone and Joint Infections

The role of anaerobes requires considerably more study, but there is much evidence to indicate that these organisms are of major importance in bone and joint infections. Most anaerobic bone infection is of the lower extremities and pelvis; lower extremity infection is usually associated with diabetes mellitus, other causes of vascular insufficiency or neuropathy, or trauma. The specific anaerobes involved in bone infection vary according to the site of the infection. Overall, *Peptostreptococcus* species are by far the most common; strains of the *B. fragilis* group are found in about 10 to 15% of cases. Purulent anaerobic arthritis is uncommon, but not rare; the *B. fragilis* group is the most common isolate when the source is hematogenous. *Fusobacterium necrophorum* is much less common in both bone and joint infections than was true before the antimicrobial era.

Miscellaneous Anaerobic Infections

Secondary anaerobic infection of luetic chancres and of other venereal diseases occurs on occasion. Anaerobic balanoposthitis, the "sixth venereal disease" previously known as genital fusospirochetosis, is not at all rare. Scrotal abscesses and abscesses of the groin and buttock involve anaerobes with some frequency.

Subareolar and recurrent breast abscesses often involve anaerobic bacteria. These abscesses represent special problems, as they may form chronic sinuses draining foul-smelling pus. There is evidence that the pathogenesis is similar to that of hidradenitis suppurativa, with infection in or around a lactiferous duct lined by squamous epithelium and filled with keratin plugs.

Actinomycosis

Actinomycosis is a chronic destructive process of soft tissue and bone caused by various species of *Actinomyces* and by *Arachnia*. Pathologically, the disease may be suppurative, granulomatous, or both; fibrosis may account for the distinctive woody character of the process on palpation. Indeed, the process may be so firm to palpation that on some occasions thoracic surgeons have resected a lung on the mistaken impression that they were dealing with cancer. Chronic draining sinuses are common. Classically, actinomycosis is of three types: cervicofacial, thoracic, and abdominal. However, any part of the body may be involved. In recent years, a number of cases have been described in connection with the use of intrauterine devices. Some of these latter cases, as well as cervicofacial actinomycosis, have yielded an organism other than *Actinomyces* or *Arachnia: Eubacterium nodatum*, an organism that resembles *Actinomyces* closely.

Pediatric Anaerobic Infections

All of the previously described anaerobic infections may be seen in children as well as in adults, although the incidence may vary because of differences in the incidence of predisposing or underlying diseases. The newborn infant and the neonate are prone to develop somewhat different types of anaerobic infections either as complications of labor and delivery or as complications of congenital malformations.

Pathology

The pathology varies with the specific entity. Characteristic of anaerobic infection is tissue destruction, which may vary from minor degrees of necrosis to extensive gangrenous changes and purulence, often with abscess formation. Gas production in tissues is not uncommon. In certain situations, such as chronic subareolar breast abscess, hidradenitis suppurativa, perirectal abscess, and actinomycosis, there is a tendency for formation of draining sinuses. Actinomycosis is distinctive, as noted previously, by virtue of production of abscesses and granulomata, as well as fibrotic reaction.

Diagnostic Features

A number of clues to anaerobic infection are given in Table 2. Most are nonspecific (e.g., tissue necrosis, gas in tissues) but nonetheless are very useful in suggesting the possibility or likelihood of infection with anaerobic bacteria. Foul or putrid odor is the only clue that is specific; no other microorganisms causing infection in humans produce such odors other than the anaerobes. It must be appreciated, however, that such odors are not invariably present; thus, the absence of such an odor does not rule out the possibility of anaerobic infection. Certain organisms, such as some anaerobic cocci, do not produce the volatile amines or organic acids responsible for the distinctive odor and, in some cases, the examiner will not be able to detect an odor because an infectious process may be totally walled off. There may also be features typical of the specific type of infection, such as pneumonia, appendicitis, and the like; these features will not ordinarily suggest that the process likely involves anaerobic bacteria except for those types of infection that frequently involve anaerobes (e.g., brain abscess, endodontic infection, aspiration pneumonia). Table 3 lists infections commonly involving anaerobes, with data on the incidence of anaerobes in each. Certainly the unique morphology of certain anaerobic organisms (notably the gram-negative anaerobic bacilli) will often suggest involvement of anaerobes and even specific anaerobes.

Laboratory diagnosis of non-spore-forming anaerobic infections is discussed below under Etiologic Agents.

TABLE 2. Clues to anaerobic infection[a]

Certain hints suggest to the microbiologist that a given specimen is likely to contain anaerobic bacteria:

1. Foul odor to specimen
2. Location of infection in proximity to a mucosal surface
3. Infections secondary to human or animal bite
4. Gas in specimen
5. Previous therapy with aminoglycoside antibiotics (such as gentamicin or amikacin) in the absence of concomitant effective coverage vs. anaerobes
6. Black discoloration of blood-containing exudates; these exudates may fluoresce red under ultraviolet light (infections involving pigmented *Bacteroides*)
7. Presence of "sulfur granules" in discharges (actinomycosis)
8. Unique morphology on Gram stain
9. Failure of organisms seen on Gram stain of original exudate to grow aerobically
10. Growth in anaerobic zone of fluid media or of agar deeps
11. Anaerobic growth on media containing 75 to 100 μg/ml kanamycin, neomycin, or paromomycin (or medium also containing 7.5 μg/ml vancomycin, in the case of gram-negative anaerobic bacilli) or on other selective media such as BBE, CCFA and RIF (see Appendix)
12. Characteristic colonies on anaerobic agar plates (e.g., *F. nucleatum* and *C. perfringens*)
13. Young colonies of pigmented *Bacteroides* (growth from anaerobic blood agar plate) may fluoresce red under ultraviolet light

[a] Adapted from Sutter et al., Wadsworth Anaerobic Bacteriology Manual, 4th ed., 1985.

Prognosis

The prognosis varies widely with the nature of the specific type of infection, the age and general condition of the patient, the nature of any accompanying or underlying problems, and the specific flora (both anaerobic and any accompanying flora) and its resistance to antimicrobial agents. In general, however, infection with non-spore-forming anaerobic bacteria tends to have a distinctly better prognosis than is true for the more serious clostridial infections. As with all infections, the sooner the diagnosis is made or suspected and appropriate medical and surgical therapy undertaken, the better the prognosis.

The prognosis of brain abscess, which remained poor (mortality greater than 50%) long after introduction of antimicrobial agents, has improved considerably in recent years. This improvement can be attributed to a great extent to the introduction of metronidazole for therapy of brain abscess; this agent achieves high levels in the brain substance and is highly bactericidal. Another important factor was the introduction of computerized tomography and magnetic resonance imaging, which facilitated earlier diagnosis and permitted easy follow-up of treated patients.

Beyond the example of brain abscess, the prognosis in anaerobic infections generally has improved with the availability of many effective antimicrobial agents, improvements in diagnostic and surgical

TABLE 3. Infections commonly involving anaerobes[a]

Infection	Incidence (%)
Bacteremia	20
Bacteremia secondary to tooth extraction	84
Ocular infections	38
Corneal ulcers	7
Central nervous system	
Brain abscess	89
Extradural or subdural empyema	10
Head and neck	
Chronic sinusitis[b]	52
Acute sinusitis	7
Chronic otitis media[c]	56
	59
	33
Cholesteatoma	92
Neck space infections	100
Wound infection following head and neck surgery	95
Bite wounds	47
Dental and oral	
Orofacial, of dental origin	94
Root canal infection[c]	95
	100
Periodontal abscess	100
Dental abscess, endodontic origin[c]	100
	90
Thoracic	
Aspiration pneumonia[c]	93
	62
	100
Lung abscess[c]	93
	85
Bronchiectasis	76
Empyema (nonsurgical)	62
Abdominal	
Intraabdominal infection (general)[c]	86
	90
	81
	94
Appendicitis with peritonitis	96
Liver abscess	52
Other intraabdominal infection (post-surgery)	93
Wound infection following bowel surgery	
Biliary tract[c]	45
	41
Obstetric-gynecologic	
Miscellaneous types[c]	100
	74
	72
Pelvic abscess	88
Vulvovaginal abscess	75
Vaginal cuff abscess	98
Septic abortion, sepsis[c]	67
	63
Pelvic inflammatory disease[c]	25
	48
Soft tissue and miscellaneous	
Nonclostridial crepitant cellulitis	75
Pilonidal sinus	73+
Diabetic foot ulcers	95
Infected diabetic gangrene (deep tissue culture)	85
Soft tissue abscess	60
Cutaneous abscess	62
Decubitus ulcers with bacteremia	63
Osteomyelitis	40
Gas gangrene (clostridial myonecrosis)	
Breast abscess	
Perirectal abscess	

[a] Adapted from Sutter et al., Wadsworth Anaerobic Bacteriology Manual, 4th ed., 1985.
[b] Of 28 cultures, 23 (82%) yielding heavy growth of one or more organisms had only anaerobes present.
[c] Multiple studies with different incidence figures.

techniques, and introduction of nonsurgical modes of abscess drainage in selected cases.

Epidemiologic Aspects

Epidemiology is not pertinent in the case of the non-spore-forming anaerobic infections. Essentially, the organisms involved invariably come from the indigenous flora of the infected patient. Although some of the hardier, more oxygen-tolerant, non-spore-forming anaerobes may survive in the environment for hours, there is no evidence that this source may serve to infect the patient. Even infection associated with surgical procedures and implantation of foreign material (e.g., *P. acnes* and a ventriculoatrial shunt, *Actinomyces* or *Eubacterium nodatum* and an implanted intrauterine device) involves organisms from the host's indigenous flora.

Etiologic Agents

Characterization

The two most commonly encountered groups of non-spore-forming anaerobes in clinical infection are the gram-negative anaerobic bacilli (chiefly those in the genera *Bacteroides* and *Fusobacterium*) and anaerobic gram-positive cocci (virtually all of which are now in the genus *Peptostreptococcus*). The primary pathogens among the non-spore-forming gram-positive bacilli are the causative agents of actinomycosis (various species of *Actinomyces* and *Arachnia pro-*

pionica). *Propionibacterium acnes* is seen almost exclusively in infections associated with implanted materials, such as shunts, but may also be found in endocarditis. *Bifidobacterium dentium* (previously *B. eriksonii*) is found as a rare cause of serious pulmonary infection. *Eubacterium nodatum* may be found in infections associated with intrauterine devices; some of these are undoubtedly called *Actinomyces* because of their close resemblance. A few obligately anaerobic streptococci in the genus *Streptococcus* are the true "anaerobic streptococci"; the others should be called peptostreptococci.

Specimen Collection and Transport

Anaerobic bacteriology is only as good as the specimen on which one works. The best anaerobic bacteriology is wasted on an improperly collected or transported specimen; indeed, it may even be misleading if the specimen is contaminated with normal flora. Thus, one must take special pains to be certain that specimen collection and transport are done in optimal fashion. Microbiologists should instruct clinicians who are not aware of these problems or who are insufficiently informed regarding specific procedures that may be used effectively.

Details on various techniques for specimen collection and transport may be found in *The Wadsworth Anaerobic Bacteriology Manual* (Sutter et al., 1985) and *The Manual of Clinical Microbiology* (Lennette et al., 1985). Table 4 lists a number of recommended methods of specimen collection. The basis of all of these methods is to avoid contaminating the clinical specimen with indigenous flora of the region from which the specimen was collected. As these anaerobes (and other organisms) from the normal flora may be pathogenic, particularly in situations in which anaerobic culture is deemed desirable, it is clearly of paramount importance to avoid picking up normal flora during the sampling procedure. In terms of transport, the principle is to protect the specimen from exposure to oxygen, as certain anaerobes are relatively oxygen sensitive, particularly in specimens of small volume that are relatively thin. In such specimens, even brief periods of exposure to air may result in significant death of the organisms. Many different transport devices are available commercially; one may readily put together such setups in the laboratory as well.

Isolation and Identification Procedures

The approach to isolation and identification procedures, as well as to preliminary and definitive identification, will be given in somewhat abbreviated form in this and subsequent sections; further details are available in *The Wadsworth Anaerobic Bacteriology Manual* (Sutter et al., 1985).

For cultivation of the vast majority of anaerobes of clinical significance, processing on the bench and the use of an anaerobic jar technique are entirely satisfactory. *Clostridium difficile* is not very aerotolerant; specimens to be set up for this organism are best handled in an anaerobic chamber (glovebox).

A scheme for initial processing of specimens is given in Fig. 1. All specimens should be examined directly by Gram stain and, when indicated, by darkfield or other special technique; one should note the number and type of host cells and the relative numbers of the various morphologic types of bacte-

TABLE 4. Recommended specimen collection methods for anaerobic culture[a]

Source	*Procedure*
Pulmonary	Percutaneous transtracheal aspiration or direct lung puncture (Medi-Tech® plugged double lumen catheter and bronchial brush for bronchoscopic specimens may also be valid)
Pleural	Thoracentesis
Urinary tract	Suprapubic percutaneous bladder aspiration, nephrostomy specimens
Abscesses	Needle and syringe aspiration of closed abscess; use of swabs much less desirable
Female genital tract	Use culdocentesis to obtain specimens, when possible, after decontaminating the vagina with povidone–iodine; Medi-Tech® double catheter and bronchial brush or sterile swab for uterine cavity specimens (not suitable for postpartum endometritis; see note under Pulmonary)
Sinus tracts or draining wounds	Aspiration by syringe and small plastic catheter introduced as deeply as possible through decontaminated skin orifice; specimen obtained at surgery from depths of wound or underlying bone lesion always preferable; curettings and tissue biopsies provide excellent material

[a] Adapted from Sutter et al., Wadsworth Anaerobic Bacteriology Manual, 4th ed., 1985.

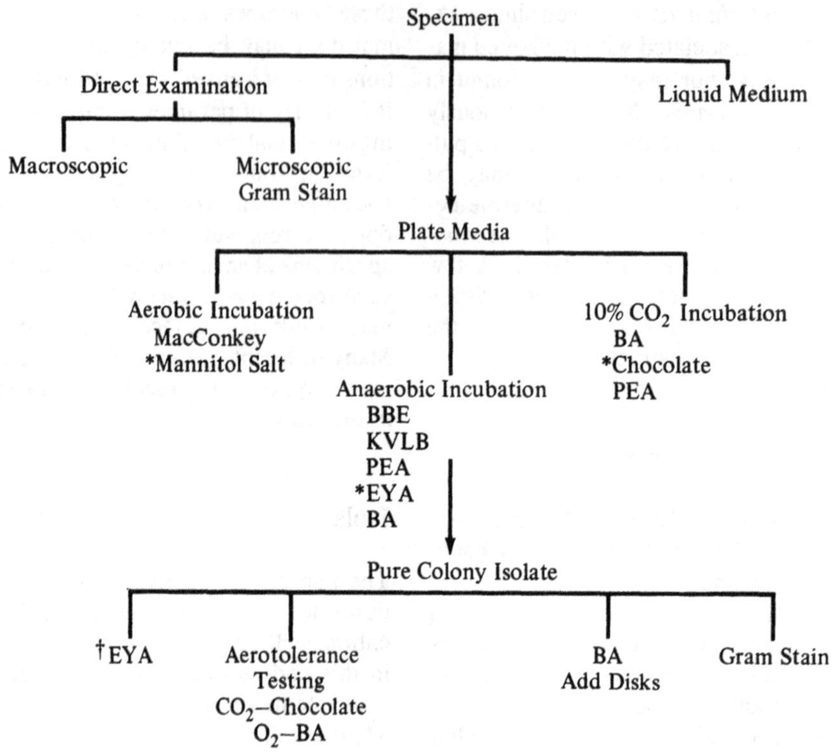

FIG. 1. Initial processing of specimens. *Optional, depending on specimen source or Gram stain. †For suspected lipase producers and all *Clostridium* spp. (Adapted from Sutter et al., Wadsworth Anaerobic Bacteriology Manual, 4th ed., 1985.)

ria present. Macroscopic examination of the specimen is also important; one may be able to detect pus, blood, odors, and the like. One should set up at least a MacConkey or similar agar plate aerobically and a blood agar (BA) plate under 10% carbon dioxide. For anaerobic incubation, the following agar plates are recommended: blood, phenylethyl alcohol (PEA), *Bacteroides* bile esculin (BBE), and kanamycin–vancomycin laked blood (KVLB). An egg yolk agar (EYA) plate may also be helpful. It is also desirable to inoculate a tube of thioglycolate broth, which serves as a backup in the event of jar failure and may permit recovery of organisms inhibited on plate media by bacteriostatic substances in the specimen. After incubation, each colony type on the anaerobic plate(s) is Gram stained and subcultured to blood agar anaerobically and aerobically and to chocolate agar under 10% carbon dioxide for determination of aerotolerance; keep in mind that most medically important bacteria that grow aerobically are facultative and therefore also grow anaerobically. It may also be worthwhile to subculture to an egg yolk agar plate for lipase and lecithinase detection and to rabbit blood (preferably laked) agar for better detection of pigment. Organisms that are obtained in pure culture and that are determined to be obligate anaerobes can then be processed for identification, as outlined in the following section.

Preliminary Identification

Gram-negative anaerobic bacilli that produce colonies greater than 1 mm in diameter on *Bacteroides* bile esculin agar are presumptively identified as belonging to the *B. fragilis* group. Gram-negative coccobacilli or bacilli whose colonies show brown or black pigmentation on blood-containing media (usually best on laked rabbit blood agar) or that fluoresce brick red under long-wave ultraviolet light are in the pigmented *Bacteroides* group (sometimes called the *B. melaninogenicus* group). Gram-negative anaerobic bacilli whose colonies pit the agar are tentatively placed in the *Bacteroides ureolyticus* group. Slender gram-negative anaerobic rods with pointed ends whose colonies show speckling or a breadcrumb appearance are presumptively identified as *Fusobacterium nucleatum*. Anaerobic cocci can be subdivided to some extent by Gram reaction. Gram-positive anaerobic cocci are virtually certain to be in the genus *Peptostreptococcus*.

Additional simple tests can further characterize the isolates to permit more educated presumptive identification or even, in some cases, definitive identification. Included among these tests are motility, spot indole, catalase, nitrate disk, sodium polyanethol sulfonate disk, special antibiotic identification disks (vancomycin, 5 μg; kanamycin, 1000 μg;

and colistin, 10 μg), glutamic acid decarboxylase, 20% bile test (looking for inhibition, no effect or stimulation), urease, spore tests, certain growth stimulation tests, certain kits to detect enzymatic reactions, and kits with miniaturized biochemical tests.

Definitive Identification

For definitive identification, one must often use (in addition to the tests described above) prereduced anaerobically sterilized (PRAS) biochemical testing and gas–liquid chromatography for identification and quantitation of volatile fatty acids and other organic acids, such as lactic and succinic acids, as metabolic end-products or determination of type and quantity of medium-length cellular fatty acid chains.

Table 5, a key to differentiating the genera of anaerobic bacteria, depends heavily on GLC data. Tables 6, 7, and 8 give the characteristics and differentiation between species of bile-resistant *Bacteroides*, nonpigmented bile-sensitive *Bacteroides*, and the

TABLE 5. Differentiation of genera of anaerobes[a]

Gram-negative bacilli	
I. Nonmotile or peritrichous flagella	
A. Produce butyric acid (without isobutyric and isovaleric acid)	*Fusobacterium*
B. Produce major lactic acid	*Leptotrichia*
C. Produce acetic acid and hydrogen sulfide; reduce sulfate	*Desulfomonas*
D. Not as above	*Bacteroides*
II. Polar flagella	
A. Fermentative	
1. Produce butyric acid	
2. Produce succinic acid	
a. Spiral-shaped cells	
b. Ovoid cells	*Succinimonas*
3. Produce propionic and acetic acids	*Anaerovibrio*
B. Nonfermentative; produce succinic acid from fumarate	*Wolinella*
III. Tufts of flagella on concave side of curved cells	
A. Fermentative	*Selenomonas*
B. Nonfermentative	*Mobiluncus*
IV. Bipolar tufts of flagella	*Anaerobiospirillum*
Gram-negative cocci	
I. Produce propionic and acetic acids	*Veillonella*
II. Produce butyric and acetic acids	*Acidaminococcus*
III. Produce isobutyric, butyric, isovaleric, valeric, and caproic acids	*Megasphaera*
Gram-positive cocci	
I. Require a fermentable carbohydrate	
A. Produce butyric (plus other acids)	*Coprococcus*
B. Do not produce butyric acid	*Ruminococcus*
II. Do not require a fermentable carbohydrate	
A. Lactic acid sole major product	*Streptococcus*
B. Not as above	*Peptostreptococcus* or *Peptococcus*
Spore-forming bacilli	*Clostridium*
Gram-positive non-spore-forming bacilli	
I. Produce propionic and acetic acids as major products	
A. Usually catalase positive	*Propionibacterium*
B. Catalase negative	*Arachnia*
II. No propionic acid produced	
A. Produce acetic and lactic acids (acetic > lactic)	*Bifidobacterium*
B. Produce lactic acid as sole major acid	*Lactobacillus*
C. Produce moderate acetic acid plus one of the following:	
1. Major succinic and lactic acids	
2. Major succinic acid	*Actinomyces*
D. Other: butyric and/or others, acetic or no major acids	*Eubacterium*

[a] Adapted from Sutter et al., Wadsworth Anaerobic Bacteriology Manual, 4th ed., 1985.

TABLE 6. Characteristics of bile-resistant *Bacteroides* species[a,b]

Bacteroides *species*	*Growth in 20% bile*	*Indole production*	*Catalase*	*Esculin hydrolysis*	*Fermentation of* Sucrose	Maltose	Rhamnose	Salicin	Trehalose	Arabinose	Xylan	*Fatty acids from PYG*
B. fragilis group												
B. distasonis	+	–	$+^{-}$	+	+		V	+	+	$-^{+}$		A p S (pa) (ib iv l)
B. caccae (3452A homology group)	+	–	–	+	+		$+^{-}$	$+^{-}$	+	+		
B. fragilis	+	–	+	+	+		–	–	–	–		A p S, pa
B. vulgatus	+	–	$-^{+}$	$+^{-}$	+		+	–	–	+		A p S
B. ovatus	+	+	$-^{+}$	+	+		+	+	+	+	+	A p S pa (ib iv l)
B. thetaiotaomicron	+	+	$+^{-}$	+	+		+	$-^{+}$	+	+	–	A p S pa
B. uniformis	W+	+	$-^{+}$	+	+		$-^{+}$	$+^{-}$	–	+		a p l S (ib iv)
Other												
B. eggerthii	+	+	–	+	–	+	$+^{-}$	–	–	+		A p S (ib iv l)
B. splanchnicus	W+	+	–	+	–	–	–	–	–	+		A P ib b iv S (l)

[a] Adapted from Sutter et al., Wadsworth Anaerobic Bacteriology Manual, 4th ed., 1985.
[b] Reactions: – = negative reaction; + = positive reaction for majority of strains and includes weak as well as strong acid production from carbohydrates in saccharolytic organisms; $+^{-}$ = most strains positive, reaction helpful if positive; $+^{W}$ = most strains positive, some weakly positive; w^{-} = most strains weakly positive, some negative; $-^{+}$ = most strains negative, some positive; $-^{w}$ = most strains negative, some weakly positive; V = variable reaction; (+) = delayed reaction; c = clot formed in milk; d = milk digested.
Prereduced anaerobically sterilized carbohydrates: + = pH < 5.5; W = pH 5.5–5.7; – = pH > 5.7.
Gas in PYG agar deep: – = no gas detected; 2 = splits agar horizontally; 3 = agar displaced halfway to top of the tube; 4 = agar displaced to top of the tube.
Fatty acids: A = acetic; P = propionic; IB = isobutyric; B = butyric; IV = isovaleric; V = valeric; IC = isocaproic; C = caproic; C′ = caprylic; L = lactic; S = succinic; PA = phenylacetic.
Note: 1) Capital letters indicate major metabolic products; 2) small letters indicate minor products; 3) parentheses indicate a variable reaction; 4) isoacids are primarily from carbohydrate-free media (such as PY) in the case of saccharolytic organisms.

TABLE 7. Characteristics of nonpigmented bile-sensitive *Bacteroides* species[a,b]

Bacteroides *species*	Fermentation of: *Glucose*	*Sucrose*	*Lactose*	*Arabinose*	*Xylose*	*Salicin*	*Xylan*	*Esculin hydrolysis*	*α-Fucosidase*	*Zoogleal mass*	*Indole production*	*Gelatin hydrolysis*	*Fatty acids from PYG*
Saccharolytic													
Pentose fermenters													
B. oris	+	+	+	+	+	+		+	–	–	–		A S (p ib iv)
B. buccae	+	+	+	+	+	+		+	+	–	–		A S (p ib b iv l)
B. zoogleoformans	+	+	+	V	V	V		+		+	+⁻		A P S (ib iv)
Pentose nonfermenters													
B. oralis	+	+	+	–	–	+		+			–		A S (l)
B. buccalis	+	+	+	–	–	–	–	+			–		a iv S
B. veroralis	+	+	+	–	–	–	+	+			–		a S
Saccharolytic and proteolytic													
B. bivius[c]	+	–	+	–	–	–		–			–	+	A iv S (ib)
B. disiens	+	–	–	–	–	–		–			–	+	A S (p ib iv)
Nonsaccharolytic or weakly saccharolytic													
B. capillosus	W⁻	–	–	–	–	–		+			–	–	a s (p l)
B. praeacutus[d]	–	–	–	–	–	–		–			–	+	A p ib B IV a (l)
B. putredinis	–	–	–	–	–	–		–			+	+	a P ib b IV S (l)

[a] Adapted from Sutter et al., Wadsworth Anaerobic Bacteriology Manual, 4th ed., 1985.
[b] See footnote *b*, Table 6, for explanation of abbreviations and symbols.
[c] Most strains are black pigmented on laked rabbit blood agar after prolonged incubation; see text for differential characteristics that separate *B. bivius* from the other pigmented *Bacteroides* species.
[d] This organism has now been reclassified as *Tissierella praeacuta*.

TABLE 8. Characteristics of pigmented *Bacteroides* species[a,b]

Bacteroides *species*[c]	*Indole production*	*Lipase*	*Fermentation of* Glucose	*Fermentation of* Lactose	*Fermentation of* Cellobiose	*Esculin hydrolysis*	*Phenylacetic acid production*	*Fatty acids from PYG*
B. asaccharolyticus[c]	+	−	−	−	−		−	A p ib B iv S
B. gingivalis	+	−	−	−	−		+	a p ib B IV s pa
B. intermedius	+	+	+	−	−			A iv S (p ib)
B. corporis	−	−	+	−	−			A ib iv S (b)
B. melaninogenicus	−	−	+	+	−	−		A S (ib iv l)
B. denticola	−	−	+	+	−	+		A S (ib iv l)
B. loescheii	−	−+	+	+	+	V		a S (l)

[a] Adapted from Sutter et al., Wadsworth Anaerobic Bacteriology Manual, 4th ed., 1985.
[b] See footnote *b*, Table 6, for explanation of abbreviations and symbols.
[c] *B. endodontalis*, a newly described pigmented *Bacteroides*, is phenotypically indistinguishable from *B. asaccharolyticus*. *B. bivius* forms black pigmented colonies on laked rabbit blood agar after prolonged incubation. See Table 7 for further characteristics and the text for differentiation from other pigmented *Bacteroides* species.

pigmented *Bacteroides*, respectively. The characteristics of *Fusobacterium* species are noted in Table 9, of gram-positive anaerobic cocci in Table 10, and of gram-positive non-spore-forming bacilli in Table 11. Figure 2 is a flow chart for anaerobic cocci.

Treatment, Prevention, and Control

Antimicrobial Susceptibility

Increasing resistance to antimicrobial agents has become a definite problem among the anaerobes, especially the gram-negative bacilli, in recent years. One of the major mechanisms of resistance is β-lactamase production, which renders many (but certainly not all) β-lactam drugs such as penicillins and cephalosporins less active. The list of anaerobes known to produce such enzymes is noted in Table 12. Other mechanisms of resistance known or thought to be operative in anaerobic bacteria include plasmid-mediated transferable resistance, transposons, changes in outer membrane porins, and changes in penicillin-binding proteins.

Techniques for doing susceptibility tests with anaerobic bacteria are detailed in *The Wadsworth Anaerobic Bacteriology Manual* (Sutter et al., 1985). In general, they are comparable to those used for non-anaerobes; however, the disk diffusion test is not generally practical for use with anaerobes. The two most practical tests, particularly for laboratories doing only small numbers of such tests, are the broth disk elution procedure as modified by Kurzynski et al. (using thioglycolate broth) and the commercially available microbroth dilution trays. Instructions for preparation of broth-disk tubes are provided in Table 13.

General patterns of susceptibility of anaerobes (including clostridia, for comparison) are given in Table 14. More detailed information on susceptibility patterns is provided in Table 15 (activity of various drugs against *Bacteroides*) and Table 16 (activity versus other anaerobes, including *Clostridium*).

It should be appreciated that there are potential pitfalls in in vitro susceptibility testing. Variables include technique used, medium, additives, size and age of inoculum, time and type of incubation, and breakpoint used to determine resistance. Testing too few strains may give a false picture, as may failure to test recent isolates rather than old stock strains. Clustering of resistance about the breakpoint may make interpretation difficult. Choosing too many or too few of the different species in the *B. fragilis* group to test may influence the results considerably. There is no standardized method or agreed upon breakpoints that would be acceptable to even the majority of investigators in the field. Our own experience with the *Brucella* blood agar plate dilution technique has been generally favorable over many years. It must be kept in mind that in the patient many factors operate that cannot be simulated in the laboratory. The age and general condition of the patient, whether or not required debridement of necrotic tissue and drainage of abscesses is carried out promptly, the nature and resistance of the non-anaerobes present in a mixed infection, synergistic activity between various organisms in mixed infections, and many other factors will influence the outcome. The normal host defense mechanisms of the body, of course, will lead to recovery of many patients with a variety of infections even in the absence of therapy. *(text continued on page 56)*

TABLE 9. Characteristics of *Fusobacterium* species[a,b]

Fusobacterium species	Distinctive cellular morphology	Indole production	Growth in 20% bile	Lipase	Gas in glucose agar	Fermentation of Glucose	Fermentation of Fructose	Fermentation of Mannose	Esculin hydrolysis	Lactate converted to propionate	Threonine converted to propionate	Fatty acids from PYG
F. nucleatum	Slender, pointed ends	+	−	−	$-^{2}$	$-^{w}$	$-^{w}$	−	−	−	+	a p B (L s)
F. gonidiaformans	Gonidia forms	+	−	−	4^{2}	−	−	−	−	−	+	A p B (l s)
F. necrophorum	None[c]	+	$-^{+}$	+	4^{2}	$-^{w}$	$-^{w}$	−	−	+	+	a p B (l s)
F. naviforme	Boat shape	+	−	−	−	w^{-}	−	−	−	−	−	a B L (p s)
F. varium	None	$+^{-}$	+	−	4	+	+	+	−	−	+	a p B L (s)
F. mortiferum	Bizarre, round bodies	−	+	−	4	$+^{w}$	$+^{w}$	$+^{w}$	+	−	+	a p B (v l a)
F. russii		−	−	−	2	−	−	−	−	−	−	a B L

[a] Adapted from Sutter et al., Wadsworth Anaerobic Bacteriology Manual, 4th ed., 1985.
[b] See footnote *b*, Table 6, for explanation of abbreviations and symbols.
[c] May have same microscopic morphology as *F. mortiferum*.

TABLE 10. Characteristics of anaerobic gram-positive cocci[a,b]

Organism	Indole production	Butyric acid production	Isocaproic acid production	Lactic acid only major end-product	Fermentation of Glucose	Fermentation of Cellobiose	Fermentation of Lactose	Fermentation of Maltose	Fermentation of Sucrose	Urease	Cell size 0.6 μm	Fatty acids from PYG
Peptostreptococcus asaccharolyticus	+	+	−	−	−	−	−	−	−	−		B (A l p)
P. prevotii	−	+	−	−	$-^{w}$	−	−	w^{-}	w^{-}	$-^{+}$		B (A L p)
P. tetradius	−	+	−	−	+	−	−	+	+	+		B L (a p)
P. anaerobius	−	v	+	−	w^{-}	−	−	w^{-}	w^{-}	−		A IC (ib b iv)
P. productus	−	−	−	−	+	+	+	+	+	−		A l s
P. micros	−	−	−	−	−	−	−	−	−	−	−	A (s)
P. magnus	−	−	−	−	−	−	−	−	−	−	+	A
Streptococcus sp.	−	−	−	+	+	V	V	+	+			L (a)

[a] Adapted from Sutter et al., Wadsworth Anaerobic Bacteriology Manual, 4th ed., 1985.
[b] See footnote *b*, Table 6, for explanation of abbreviations and symbols.

TABLE 11. Level III identification of gram-positive non-spore-forming bacilli[a,b]

Organism	Nitrate reduction	Catalase	Indole production	Esculin hydrolysis	Urease	Red colony
Actinomyces sp.	+	$-^{+}$		$+^{-}$	V	
A. israelii	$+^{-}$	−	−	+	−	−
A. odontolyticus	+	−	−	$+^{-}$	−	+
A. naeslundii	+	−	−	$+^{-}$	+	−
A. viscosus	$+^{-}$	+	−	V	+	−
A. meyeri	$-^{+}$	−	−	$-^{+}$	−	−
Propionibacterium sp.	V	V	$-^{+}$	V		
P. acnes	$+^{-}$	$+^{-}$	$+^{-}$	−		
P. granulosum	−	+	−	−		
P. avidum	−	+	−	+		

[a] Adapted from Sutter et al., Wadsworth Anaerobic Bacteriology Manual, 4th ed., 1985.
[b] See footnote *b*, Table 6, for explanation of abbreviations and symbols.

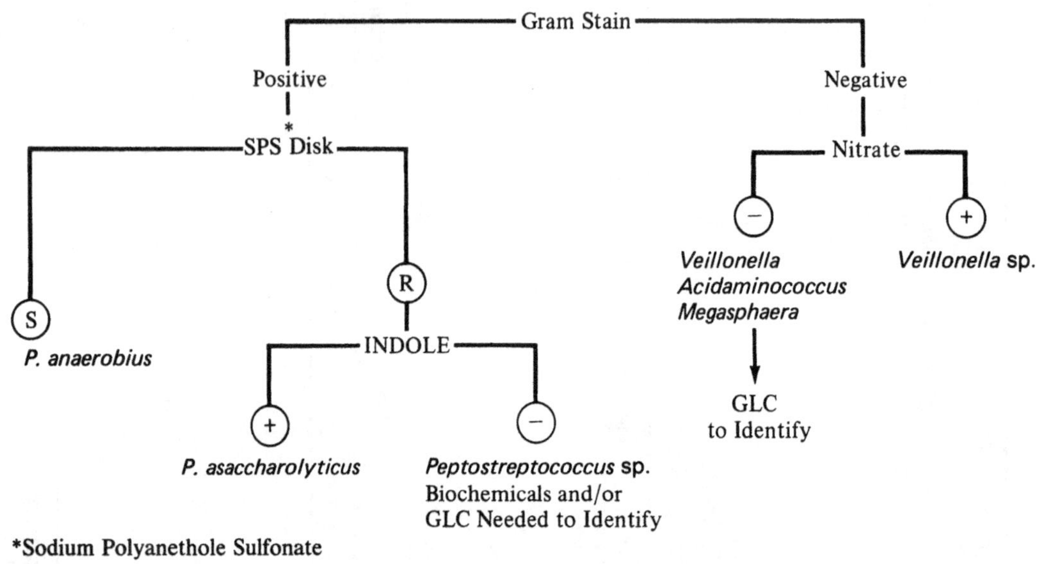

FIG. 2. Flow chart for anaerobic cocci. (Adapted from Sutter et al., Wadsworth Anaerobic Bacteriology Manual, 4th ed., 1985.)

Oxygen toler-ance	*Glucose*	*Mannitol*	*Raffinose*	*Ribose*	*Tre-halose*	*Erythritol*	*Maltose*	*Fatty acids from PYG*
	+							A S (L)
AM	+	+	$+^-$	$+^-$	$+^-$		+	A L S
AM	+	−	−	−	−		V	A S
MF	+	−	+	V	$+^-$		+	A L S
AM	+	−	+	$-^+$	V		+	A L S
A	+	−	−	$+^w$	−		+	A S
AM	+							A P
AM	+					$-^+$	−	A p (iv L s)
	+					−	$+^-$	A p (iv s)
	+					$+^w$	+	A p (iv s)

TABLE 12. β-Lactamase-producing anaerobes[a,b]

Bacteroides fragilis group
Pigmented *Bacteroides*
B. oralis group
B. coagulans
B. bivius
B. disiens
B. hypermegas[c]
B. multiacidus[d]
B. "ruminicola" group (*B. oris, B. buccae*)
B. splanchnicus
Fusobacterium nucleatum
Clostridium ramosum
C. clostridiiforme
C. butyricum

[a] Adapted from Sutter et al., Wadsworth Anaerobic Bacteriology Manual, 4th ed., 1985.
[b] Some, but not all, strains of these species produce β-lactamase.
[c] This organism is now reclassified as *Megamonas hypermegas*.
[d] This organism is now reclassified as *Mitsuokella multiacidus*.

TABLE 13. Preparation of broth-disk tubes[a]

Drugs	*Disk content*	*No. of disks (5 ml medium)*	*Test concentration/ml*[b]
Ampicillin	10 μg	2	4 μg
Carbenicillin	100 μg	3	60 μg
Cefoperazone	75 μg	2	30 μg
Other cephalosporins[c]	30 μg	3	18 μg
Chloramphenicol	30 μg	3	18 μg
	30 μg	2	12 μg
Clindamycin[d]	2 μg	8	3.2 μg
	2 μg	4	1.6 μg
Metronidazole	80 μg	1	16 μg
Mezlocillin	75 μg	4	60 μg
Penicillin G	10 U	1	2 U
Piperacillin	100 μg	3	60 μg
Tetracycline	5 μg	3	3 μg
Ticarcillin	75 μg	4	60 μg

[a] From Sutter et al., Wadsworth Anaerobic Bacteriology Manual, 4th ed., 1985.
[b] Concentrations listed are not always the same as those recommended by Kurzynski and co-workers. Other concentrations can be obtained by altering the numbers of disks or the volume of medium. If more than four disks are used, they should be reduced prior to addition to the medium.
[c] Includes cefamandole, cefotaxime, cefoxitin, and moxalactam.
[d] A 10-μg disk may become available and would be preferable. Two 10-μg disks added to 5 ml of medium will give a concentration of 4 μg/ml. One 10-μg disk added to 5 ml of medium will give a concentration of 2 μg/ml. If oral therapy is to be used, the lower concentration should be tested.

TABLE 14. In vitro susceptibility of anaerobes to antimicrobial agents[a,b]

Bacteria	*Chloram-phenicol*	*Clinda-mycin*	*Erythro-mycin*[c]	*Metro-nidazole*	*Cefoxitin*	*Ureido- and carboxy-penicillins*[d]	*Penicillin G*[e]	*Tetra-cycline*	*Vanco-mycin*[c]
Anaerobic cocci	+++	++ to +++	++ to +++	+++[f,g]	+++	+++	+++	+ to ++	+++
B. fragilis group	+++	++ to +++	+ to ++	+++	++ to +++	++ to +++	+	+ to ++	+
Other *Bacteroides* sp.	+++	+++[f]	++ to +++	+++	+++[f]	+++[f]	++ to +++	++	+
F. varium	+++	+ to ++	+	+++	+++[f]	+++[f]	+++[f]	++	+
Other *Fusobacterium* sp.	+++	+++	+	+++	+++[h]	+++[h]	++++	+++	+
C. perfringens	+++	+++[h]	+++	+++	+++	+++	+++	++	+++
Other *Clostridium* sp.	+++	++	++ to +++	+++	+ to ++	+++	+++	++	++ to +++
Non-spore-forming GPR	+++	++ to +++	+++	+ to ++	+++	+++	++++	++ to +++	++ to +++

[a] Adapted from Sutter et al., Wadsworth Anaerobic Bacteriology Manual, 4th ed., 1985.

[b] + = Poor or inconsistent activity; ++ = moderate activity; +++ = good activity; ++++ = good activity, good pharmacologic characteristics, low toxicity, drug of choice.

[c] Not approved by FDA for anaerobic infections.

[d] Piperacillin, mezlocillin, azlocillin, carbenicillin, and ticarcillin.

[e] Other penicillins and cephalosporins are frequently less active. Ampicillin and amoxicillin are roughly comparable to penicillin G in activity. Addition of beta-lactamase inhibitors, such as clavulanic acid and sulbactam, remarkably increases the activity against beta-lactamase producers, such as the *B. fragilis* group.

[f] A few strains are resistant.

[g] Microaerophilic streptococci (officially in the genus *Streptococcus*) are often resistant to metronidazole.

[h] Rare strains are resistant.

TABLE 15. Activity of various drugs against *Bacteroides*

Drug	Breakpoint	Percentage susceptible at breakpoint			
		B. fragilis *species*	B. fragilis *group*[a]	B. gracilis	*Other*
Chloramphenicol	16	100	100	100	100
Imipenem	8	100	100	100	100
Ampicillin/sulbactam[b]	16	100	100	NT[c]	100
Metronidazole	16	100	100	93	99
Clindamycin	4	93	81	61	100
Cefoxitin	32	92	75	78	99
Piperacillin	128	84	85	78	97
Penicillin G	16	5	6	67	70
Cefotaxime	32	50	50	78	96
Ceftizoxime	32	43	45	89	97
Cefoperazone	32	57	51	67	94
Moxalactam	16	78	55	78	77
Cefotetan	32	85	56	NT	80

[a] Includes *B. fragilis* species.
[b] This particular combination of a β-lactam drug and a β-lactamase inhibitor is not yet, but soon to be marketed. Similar combinations presently available commercially are amoxicillin plus clavulanic acid and ticarcillin plus clavulanic acid.
[c] NT = Not tested.

TABLE 16. Susceptibility of anaerobes other than *Bacteroides*

Drug	Percentage susceptible at breakpoint[a]			
	Fusobacterium	Anaerobic *cocci*	Clostridium	Actinomyces, Arachnia
Chloramphenicol	100	100	100	100
Imipenem	77	100	100	100
Ampicillin/sulbactam	97	100	100	100
Metronidazole	100	98–99	99	25
Piperacillin	99	100	100	100
Clindamycin	92	97	90	100
Cefoxitin	99	100	65	100

[a] See Table 15 for breakpoints.

Treatment

The principles of therapy of non-spore-forming anaerobic infections are 1) modification of the environment so that it is difficult for anaerobes to proliferate, and 2) checking the spread of anaerobes into healthy tissues. Measures to control the local environment include debridement of dead tissue, drainage of pus, elimination of obstruction, decompression of swollen tissues restricted by a tight fascia, release of trapped gas, improvement of circulation to the affected part, and improvement of tissue oxygenation. Antimicrobial agents play an important role in limiting the spread of anaerobes into healthy tissues.

The important role of surgical therapy will be evident from the above discussion. Certain situations represent exceptions, however. Surgery is rarely indicated presently for lung abscess; indeed, it is usually contraindicated because of the danger of spread of infection. In certain cases of intraabdominal infection, drainage of collections of pus may be effected by percutaneous aspiration, with or without catheter drainage, guided by ultrasound or computerized tomography. Conventional oxygen therapy may be useful. Hyperbaric oxygen would seem to be highly desirable from the theoretical point of view. In fact, however, there is no evidence that it does play a role in non-spore-forming anaerobic infection; there is not even clear-cut evidence of a beneficial effect in clostridial myonecrosis. On occasion, one may demonstrate effectiveness of topically applied oxygen-releasing compounds such as hydrogen or zinc peroxide.

Antimicrobial therapy is important in most anaerobic infections, whether or not debridement and drainage have been done. The preceding discussion on antimicrobial susceptibility identifies the compounds that may be useful. In the case of certain agents such as cefotaxime, ceftizoxime and cefotetan, data available at present are not adequate, in my opinion, to justify using them routinely in such anaerobic or mixed infections as intraabdominal infection. In other cases, such as with clindamycin and cefoxitin, the incidence of resistance of the *B. fragilis* group is high enough in certain centers to indicate that these compounds should not be used in seriously ill patients in whom these organisms may be playing a significant role. In general, antimicrobial agents should be used in high dosage and for extended periods.

Prevention and Control

A limited number of useful measures are available. Good surgical technique is certainly important. Good oxygenation of tissues must be maintained when possible. Patients who are not fully conscious must be watched concerning the possibility of aspiration of material into the respiratory tract. Appropriate antimicrobial prophylaxis should be used for dental and selected other procedures such as transurethral resection.

Literature Cited

Finegold, S. M. 1977. Anaerobic bacteria in human disease. Academic Press, Inc., New York.

Finegold, S. M., W. L. George, and M. E. Mulligan. 1986. Anaerobic infections. Year Book Medical Publishers, Inc., Chicago.

Krieg, N. R., and J. G. Holt (ed.). 1984. Bergey's manual of systematic bacteriology, 9th ed., vol. 1. The Williams & Wilkins Co., Baltimore.

Kurzynski, T. A., J. W. Yrios, A. G. Helstad, and C. R. Field. 1976. Aerobically incubated thioglycolate broth disk method for antibiotic susceptibility testing of anaerobes. Antimicrob. Agents Chemother. **10:**727–732.

Lennette, E. H., A. Balows, W. J. Hausler, Jr., and H. J. Shadomy (ed.). 1985. Manual of clinical microbiology, 4th ed. American Society for Microbiology, Washington, D.C.

Sneath, P. H. A., N. S. Mair, M. E. Sharpe, and J. G. Holt (ed.): 1986. Bergey's manual of systematic bacteriology, 9th ed., vol. 2. The Williams & Wilkins, Co., Baltimore.

Sutter, V. L., D. M. Citron, M. A. C. Edelstein, and S. M. Finegold. 1985. Wadsworth anaerobic bacteriology manual, 4th ed. Star Publishing Co., Belmont, Calif.

CHAPTER 6

Anthrax

H. B. REES, JR., D. D. MARTIN, and M. A. SMITH

Disease: Anthrax ("black bain," woolsorter's disease, Siberian fever).

Etiologic Agent: *Bacillus anthracis.*

Source: Contaminated soil, water, animal hides, and by-products of infected domestic livestock and wild herbivores.

Clinical Manifestations: Initial symptoms depend primarily on the portal of entry (namely, a cutaneous lesion, or internally as respiratory or gastrointestinal infections). The cutaneous infection is the most frequently observed form. It appears first as a macular lesion, enlarges to become vesicular, develops a black eschar after vesicular eruption in 5 to 7 days, and then dries and retracts. Symptoms of the respiratory and intestinal forms, which appear within 2 to 7 days of incubation, include sudden onset, mild upper respiratory involvement (without a productive cough), fever, headache, nausea, and discomfort or mild pain localized near the heart and stomach. In untreated human cases where toxemia has developed, cyanosis, dyspnea, shock, and rapid collapse may occur within 2 to 4 days. Death usually follows in hours or a few days.

Pathology: The systemic phase of the disease begins with rapid multiplication of the microorganisms in the mediastinal lymph nodes. A general septicemia with toxemia may follow, with rapid growth of bacilli in the spleen and liver. The potent toxin produces vascular and endothelial cell injury. If thrombosis occurs, circulatory failure, cyanosis, and death usually follow.

Laboratory Diagnosis: Isolation of a gram-positive rod from the blood, tissue, and fomites that is susceptible to γ bacteriophage constitutes confirmatory diagnosis. Characterization of the isolated strain by biochemical procedures remains presumptive.

Epidemiology: Worldwide in its distribution, cycles of multiplication of the anthrax bacillus that occur in contaminated, alkaline soils under ideal climatic conditions maintain the species in endemic areas. Transportation of the contaminated products of domestic animals (e.g., wool, mohair, bone meal) produces unsuspected infections and unknowingly increases the hazard of contaminating areas and herds of livestock previously free of the disease. Infections by ingestion or inhalation occur more frequently in colder regions of the world; cutaneous infections, in the temperate zones.

Treatment: Penicillin or penicillin–streptomycin. Therapy should *not* be delayed until a presumptive laboratory diagnosis is available, because delaying treatment a few hours may drastically decrease a favorable prognosis.

Prevention and Control: Incinerating infected carcasses in situ, immunizing domestic livestock and persons in endemic areas, and educating all those likely to be affected.

Description of Disease

Anthrax (*carbon,* woolsorter's disease, malignant pustule, malignant anthrax, milzbrand, ragpicker's disease, tanner's disease, Siberian or splenic fever) is an infection that begins as a localized lesion and may progress to an acute, lethal toxemia in untreated human and animal cases. Infected sheep, cattle, and goats and their by-products serve as the primary source of human infection. The public health and economic significance of the disease is great due to its worldwide distribution and the loss inflicted whenever outbreaks occur in herds of domestic livestock. A single species of the *Bacillus* group is presently recognized, but several other species of the group are toxigenic and frequently difficult to distinguish from strains of *Bacillus cereus.*

Communicability and Transmissibility

Anthrax infection in humans occurs through contact with infected carcasses or spore-contaminated products. Dust from hides, wool, and mohair constitutes the primary source of infections in industry. New foci of infection arise from feeding meal produced from the bones of infected animals. Person-to-person contact can occur where hygiene and custom sanction the use of communal toilet articles, such as towels, shaving brushes, and the like (Heyworth et al., 1975). These investigators also suggest that vultures may have been responsible for the spread of the disease in animals from village to village in Gambia. Other suggestions for the transmission of anthrax in nature include those of Krinsky (1977), who implicated tabanids, houseflies, and deerflies without stating the role insects may play in transmission, which is probably minor at best.

Clinical Manifestations

The clinical manifestations are addressed according to the portal of entry as the initial site of infection (i.e., external and internal). Untreated cases usually follow a similar course after the initial stage of the disease regardless of the portal of entry. In the later systemic phase of the disease, bacilli that are found in the mediastinal lymph nodes initially produce septicemia, then toxemia before death occurs. Initially the cutaneous or carbuncle form of the disease is external and has been reported in cattle, horses, swine, dogs, and rabbits as well as humans. The lesion first appears macular in nature, then becomes papular within a day or two with a zone of erythema 1 to 2 cm in diameter. Later, a large vesicle develops and discolors the surrounding tissue to a dark bluish-black color. This vesicle ruptures 4 to 6 days later and produces the diagnostically important black eschar. This lesion usually increases in size before drying and retracting within 10 to 14 days. As the infection progresses, the bacilli multiply and migrate via the lymphatic system and tend to concentrate in the cervical lymph nodes and pharyngeal area. Swelling and hemorrhage occur and a reddish, gelatinous exudate is produced. All untreated infections, including cutaneous anthrax, may have fatal outcomes.

The internal forms of the initial infection are pulmonary, gastrointestinal, and oral–oropharyngeal. The pulmonary infection may be septic, fulminant, or apoplectic in nature and presents similar symptoms in humans and animals. No visible form of ulceration is seen for anthrax acquired by inhalation. The disease occurs with a sudden onset and is often fatal unless recognized and treated promptly with appropriate antibiotics. Initially, a mild upper respiratory infection occurs and is characterized by a nonproductive cough. Other symptoms include fever, headache, general discomfort, nausea, and pain that seems generalized around the heart and stomach. The signs of infection intensify 2 to 4 days later, with dyspnea, cyanosis, shock, and rapid collapse. Death may occur within hours or a few days. The gastrointestinal portal of entry may result whenever inadequately cooked or raw meat from an infected animal is consumed. The symptoms produced are fulminating in appearance and indistinguishable from many other diseases. In addition to the rapid onset and fever, abdominal pain with tenderness may occur in the right upper and lower quadrants. Other symptoms may include vomiting and the presence of blood in the stools. Oral–oropharyngeal forms are observed as mucosal lesions within the mouth or the oropharynx that can progress to membranous necrosis, cervical adenopathy, and edema. Twenty-four cases of this nature were reported recently in Thailand (Sirisanthana et al., 1984).

Pathology

Once the infection has been established in the host, several substances, including a toxin produced by the microorganism, produce profound pathologic changes. An unusual capsule produced by *Bacillus anthracis* is composed of polyglutamic acid (Record and Wallis, 1956; Zwartouw and Smith, 1956) rather than the usual polysaccharide found in many bacterial capsules. This material, called an *aggressin* by Zwartouw and Smith (1956), enables the bacteria to resist phagocytosis by the cellular immune system of the host. As in many bacterial diseases, the microorganisms per se appear not to be responsible for the lethal effect in the advanced stage of the disease; rather, it is due to a toxin released by the bacteria under specific growth conditions (Smith and Stoner, 1967). The ability of the microorganism to produce the toxin appears to be controlled by a plasmid within the bacterium (Mikesell et al., 1983).

The holotoxin, or complete toxin, consists of three separate components (Fish and Lincoln, 1968): 1) protective antigen, 2) lethal factor, and 3) edema factor. The name *protective antigen* is derived from the observation that it induced a protective immune response in animals, and it has been used as an immunizing agent (Belton and Strange, 1954). The lethal factor, when combined with protective antigen and injected into laboratory animals, produces death. Leppla (1982, 1984) has demonstrated that the combination of protective antigen and edema factor elevates dramatically the amount of intracellular cyclic adenosine monophosphate in tissue culture cells. Injection of edema factor combined with protective antigen subcutaneously induces edema around the injection site. When any of the three components are injected separately, no biologic activity is observed (Haines et al., 1965).

Animals that have been infected experimentally with *B. anthracis* demonstrate various physiologic effects, including meningoencephalitic (Tahernia and Hashemi, 1972) and kidney involvement (Nalin et al., 1977; Smith and Keppi, 1954). Whether these effects are primary or secondary, however, is not known.

Clinical and Diagnostic Features

Clear-cut diagnostic features for anthrax are limited and suggestive at best. The possibility that a patient may be infected with the anthrax bacillus should be considered 1) if he presents a carbuncle type of lesion (previously described in detail under Clinical Manifestations) or 2) if he has malaise and is employed as a farmer, butcher, veterinarian, or in any other occupation that involves handling or exposure to wool, mohair, leather, or bone products. Another feature of lesser value is a nonproductive cough in a febrile patient.

Prognosis

The fatality rate for anthrax has declined steadily with the introduction of antibiotic therapy. Wilson and Miles (1975) cite a rate of 4.6% (17 deaths) for 369 cases reported in England and Wales for a 25-year period from 1945 to 1969. The fatality rate in untreated, systemic cases approaches 100%; however, if the disease is suspected or recognized early in its course and antibiotic therapy started, few deaths occur.

Epidemiologic Aspects

Occurrence and Transmission

Anthrax occurs throughout the world and continues to create sizable economic losses of livestock and dairy herds. The number of human cases reported and the number of deaths have declined steadily for the last three or four decades in the more developed countries. In Third World countries, however, human cases continue to be reported sporadically. Heyworth et al. (1975) collected data on 448 cases that occurred from 1970 to 1974 in Gambia.

Transmission of anthrax to humans occurs most often by the cutaneous route, where 95 to 98% of cases are observed (Rees and Smith, 1981). A survey of some 2,400 cases reported in Russia for the period 1920 to 1959 (Zaporozhachenko, 1962) found 94% to be cutaneous, 5.2% intestinal, and 0.7% respiratory. Of the 176 cases of human anthrax in the United States surveyed by Brachman (1966) from 1955 to 1966, 95% were cutaneous and 4.5% were respiratory; no cases of intestinal infection were discovered during the period of study. In cases of cutaneous origin, 54% of the lesions occurred on the hands and arms, 31% on the head and neck, 0.6%, on the trunk, and 0.5% on the leg. In Iran, a study of 468 cases by Parvizpour (1979) over a 13-year period showed that 59.8% were in men, 40.2% in women. Based on occupational background, 27.9% were found to be industrial workers. Some 20.7% contracted anthrax through animals, 74.2% had a history of contacts with animal products, and 5% became infected through using the toilet articles of infected persons.

Susceptibility, Resistance, and Carriage

The susceptibility of any given animal species to anthrax infection, as measured by the number of organisms present per milliliter of blood at death, is inversely proportional to its susceptibility to the lethal effects of anthrax toxin (Lincoln et al., 1967). These investigators ranked the susceptibility to infection in descending order, from most to least susceptible species, as follows: mouse, guinea pig, rabbit, rhesus monkey, chimpanzee, goat, sheep, horse, cow, unimmunized human, rat, dog, swine, and immunized human. In contrast, the reverse order by species corresponds to their susceptibility to anthrax toxin. The order of susceptibility to anthrax in nature, particularly animals infected by the gastrointestinal route, may alter the species ranking reported by Lincoln and colleagues and reduce the minimal number of organisms required to initiate infection (Rees et al., 1977).

The hardiness of the anthrax spore and the ability of the organism to undergo a spore–vegetative cell–spore cycle in nature might seem to preclude any need of the bacillus for a carrier state. Heyworth et al. (1975) found that subclinical infection, particularly in sheep and goats, played a role in the maintenance of the disease foci. They also believed that vultures and insects were possible vectors for the spread of anthrax in Gambia. In another study (Parvizpour, 1979), 5% of cases became infected through the use of communal towels ("spidabs").

Etiologic Agent

Characterization

Bacillus anthracis is a large, gram-positive, spore-forming, aerobic–facultatively anaerobic bacillus. The organism produces large, raised, opaque, grayish, granular colonies within 24 h when grown on ordinary laboratory media at 37°C. Borders of colonies are irregular and may resemble a "medusa head" or "comet tails" (Fig. 1). Smears prepared from colonies demonstrate long chains of rods; smears prepared from broth cultures show either single cells or short chains. Capsules may be observed in stained smears from infected animal tissues, but

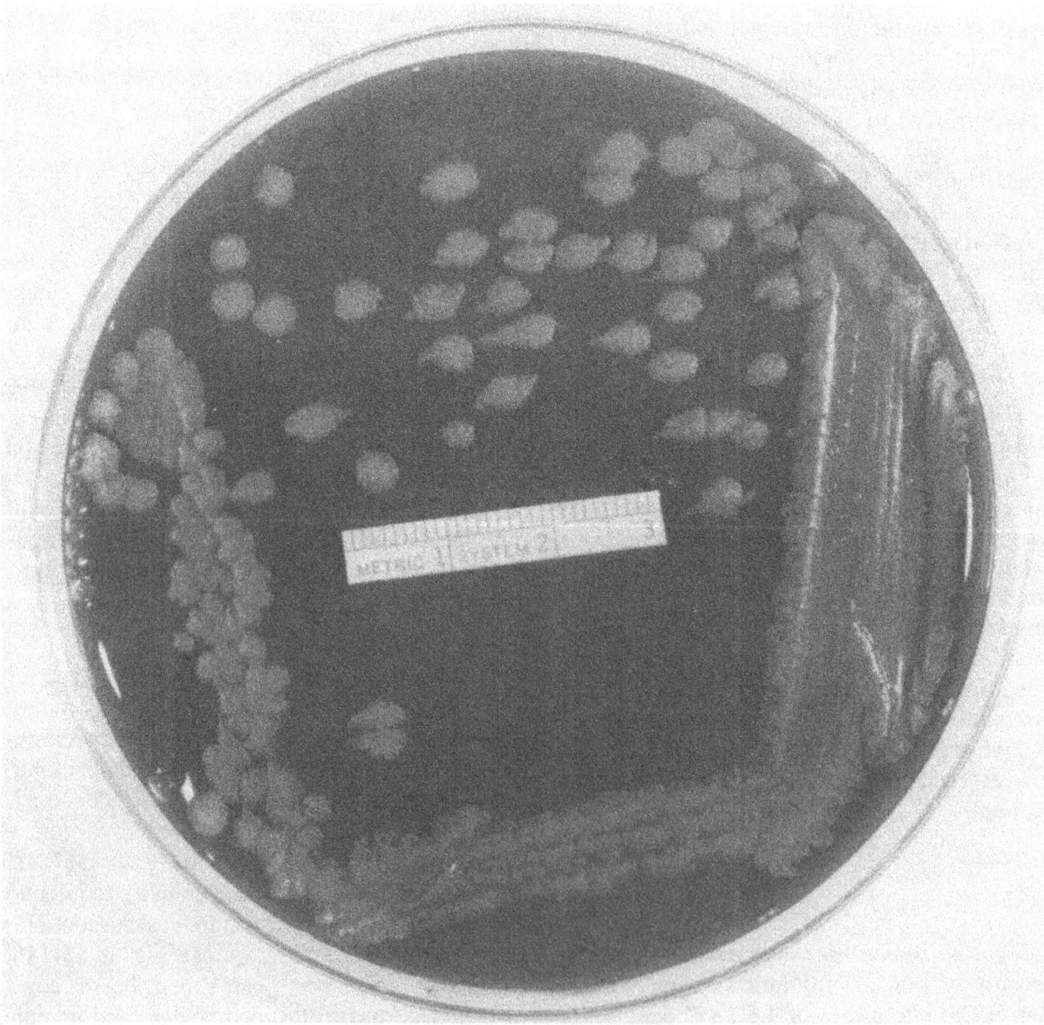

FIG. 1. Colonies of *Bacillus anthracis* grown on 5% blood agar for 18 to 24 h at 37°C. Note colonies with extended and irregular borders described as "medusa head" or "comet tails." A color plate of this figure appears in Color Plate 1 at the beginning of this volume. (U.S. Army photograph; reproduced with permission.)

are lacking when the organism is grown on ordinary media unless the bicarbonate ion and an increased atmosphere of carbon dioxide are present during incubation.

Specimen Collection and Transport

Unusual care is required in the collection and transport of all specimens suspected of containing anthrax bacilli or their extremely resistant spores to reduce possible respiratory hazards for investigators and individuals nearby. Adherence to stringent safety procedures and use of proper microbiologic containment facilities cannot be overemphasized.

Clinical Specimens

Successful isolation of the bacilli from clinical specimens is influenced by the stage of infection and whether or not specimens are collected before antibiotic administration. Gold (1967) found that penicillin given 6 h before collection eliminated the bacilli from fluid of vesicular lesions. Also, the ability to demonstrate the microorganism in the bloodstream is usually limited to a period of hours before the patient dies. In veterinary specimens collected from carcasses, putrefaction often eliminates the vegetative forms of the bacillus. Field specimens should be limited to smears or swabs of exudate only. Autopsies performed in the field substantially increase the possibility of infecting attending personnel and contaminating the soil, thereby establishing a potential foci for future infection of livestock. Authorities in several countries (e.g., Great Britain) have recognized this hazard and enacted legislation to ban individuals from performing autopsies on animals suspected of succumbing to anthrax infection.

The order in which specimens are discussed below represents the descending preference and is based on availability and designed to minimize the risk of infecting personnel and the environment. Note, however, that environmental specimens are addressed separately.

Smears for Microscopy

Make at least two impression smears of vesicular fluids from human or animal lesions and transport the prepared slides in sealable closures to the laboratory for staining and examination.

Swab Samples for Laboratory Examination

Saturate several sterile cotton swabs with vesicular fluid, blood, or the like and place them in sterile, screw-cap tubes for transportation to the laboratory. Unless the samples are small enough to dry on the cotton swab, refrigerate them during transit and during any holding period at the laboratory.

Blood and Tissues

Included in the category of blood and tissue samples are human and animal blood (taken during the terminal stage of illness) and lymph node, spleen, peritoneal exudate, or other tissues obtained by biopsy or at necropsy. As described previously, place these materials in sterile containers, secure closure with tape, and place the original specimen container within a crush-proof outer container for transport to the laboratory. Consider freezing tissue specimens to afford maximal preservation of the bacilli and to minimize the hazard to personnel should the specimen container be broken in transit. Place larger specimens within plastic bags, which should be enclosed in additional bags to prevent any leakage of fluids during transport. Lincoln et al. (1964) suggest that tissue specimens from autopsy can be conveniently packaged by initially encasing them in a surgical glove by turning the contaminated glove inside out when it is removed from the hand. In addition, Lincoln and colleagues caution against the possible loss of viability caused by repeated freezing and thawing of the specimen. Blood specimens should be refrigerated, not frozen, if collected aseptically for determination of the anthrax toxin in the plasma fraction.

Serum samples from convalescent or acutely ill patients may contain antibodies to the protective antigen. Demonstrating these antibodies, however, is usually performed only in a few specially equipped research laboratories. Several serologic tests that are reported include passive hemagglutination (Feeley et al., 1970), measurement of the antigenic combining power (Gruber and Wright, 1967), and the diffusion precipitin reaction (Ray and Kadull, 1964).

Environmental Specimens

All specimens that do not originate from clinical cases, such as hair, bones, hides, bone meal, fertilizers, soil, water, and industrial air samples, may be regarded as environmental specimens. Ideally, sample sizes of approximately 500 g should be collected (Davies and Harvey, 1955). In the case of soil, water, and air, randomness and other statistical considerations must be satisfied to quantify any degree of contamination. Soil samples must be collected to a depth of 15 cm (Fox et al., 1975). Liquid samples require refrigeration during transport.

Isolation and Preliminary Identification

The diminishing frequency with which human cases of anthrax are encountered increases the possibility that clinicians and pathologists may fail to recognize anthrax infections. The microbiologist may also tend to regard the occasional colonies of gram-positive rods as contaminants (e.g., *B. cereus*, which is closely related to *B. anthracis* and ubiquitous in its distribution) rather than the anthrax bacilli.

Bacilli isolated with the procedures detailed in this section are identified definitively, as described subsequently, by 1) their susceptibility to lysis by the γ bacteriophage, 2) their pathogenicity for inoculated laboratory animals, or 3) their serologic characteristics.

Microscopic Examination

Laboratory personnel who receive smears to be stained for the possible presence of anthrax bacilli should fix them for 3 to 5 min in Zenker fluid (2.5% aqueous potassium bichromate and 8% mercuric chloride) or 10% Formalin (37% aqueous formaldehyde) for 10 min to ensure inactivation of any spores present. Smears fixed by passage through a flame as many as 70 times retain spore viability of 20 to 30% (Buravtseva, 1968). One of the smears should be Gram stained or stained by the method of Giemsa or with 1% aqueous methylene blue. Another smear should be stained by any capsule-staining method to determine the presence of encapsulated bacilli. Feeley and Brachman (1974) suggest staining the second smear by a fluorescent-labeled antibody technique rather than with a capsule stain. Staining for *B. anthracis* specific antigens will provide more definitive information. If only one smear is received, staining for the presence of specific antigen by the fluorescent antibody procedure should be chosen. Microscopically, *B. anthracis* appears as a large, square-ended, gram-positive, spore-forming rod 1.0 to 1.3 μm wide and 3 to 10 μm long. Spores, when present, are oval and located centrally to subterminally within the vegetative cell. Spores produce little or no swelling within the cell (Fig. 2).

Finding microorganisms from the environment that meet the preceding description, when such laboratory findings are combined with supporting clinical observations, is sufficient evidence for a tentative diagnosis.

Fig. 2. Photomicrograph of *Bacillus anthracis* stained by the Gram method. Note unstained vacuolated areas or spores within the vegetative cell and little or no evidence of swelling of the cell. A color plate of this figure appears in Color Plate 1 at the beginning of this volume. (U.S. Army photograph; reproduced with permission.)

Culture Examination

Streak samples of tissue, blood, sputum, or spinal fluid directly to the surface of 5% antibiotic-free blood [from sheep, rabbit or human, either defibrinated or citrated (Feeley and Brachman, 1974)], peptone-based agar. Incubate streaked plates for 18 h at 37°C. Feces and any other specimens possibly containing large numbers of contaminating bacteria should be streaked to a selective medium or subjected to growth in an enrichment broth first. The selective PLET* agar (Knisely, 1966) is the recommended medium. It must be determined that each batch of medium will permit known cultures of *B. anthracis* to develop colonies and to inhibit the growth of colonies of *B. cereus* substantially. After incubation, examine plates carefully with good illumination, preferably with the aid of a dissecting microscope. Little or no hemolysis of the blood in agar occurs in 14 to 18 h at 37°C. Some strains of *B. anthracis* may produce questionable, weak hemolysis (Leise et al., 1959) if cultures are incubated for more than 24 h. Characteristics of colony formation include the tenacity of the cells, which allows various shapes to be molded by manual manipulation with an inoculating needle (Fig. 3). Colonies of *B. cereus* usually lack this characteristic.

Isolation from Environmental Specimens

Process samples by the steps recommended by Rees et al. (1977), who showed recovery of as few as seven spores of *B. anthracis* per 10 g of artificially seeded clayey soils. Briefly summarized, procedures include the following steps: Prepare 20% soil suspensions or 6% hair (wt/vol) in 0.03 mM KH_2PO_4 buffer (pH 7.2) containing 0.25% polyethylene-(20)-sorbitan monooleate (Tween 80). Homogenize the soil suspensions in a mechanical blender at high speed

* An acronym for a medium containing *p*olymyxin, *l*ysozyme, disodium *e*thylenediaminetetraacetate (EDTA), and *t*hallous acetate.

FIG. 3. Colonies of *Bacillus anthracis* grown on blood agar at 37°C. Note malleability or tendency of the cells to adhere to each other sufficiently to permit sculpting with inoculating needle. A color plate of this figure appears in Color Plate 1 at the beginning of this volume.

within a bacteriologic safety cabinet for 1.5 min; agitate bone or hair in a mechanical shaker for 30 min. Afterward, allow approximately 5 min for the aerosol hazard within the container to subside. Excess debris can be removed by centrifugation in unbreakable bottles inside a safety enclosure at 400 × *g* for 5 min before proceeding to the final centrifugation. Afterward, place the original suspension (or supernatant fluid from the first centrifugation step) into other centrifuge bottles and centrifuge at 6,000 × *g* for 30 min. Decant the resultant supernatant fluid into 5% phenol or hypochlorite solution (Feeley and Brachman, 1974). Resuspend the sediment in 10 ml of detergent-free buffer. Mix thoroughly for 2 min to resuspend sediments and transfer them to sealable tubes. Submerge the sealed tubes in a 70°C water bath for 10 min to destroy contaminant saprophytes and to activate the spores. Allow the samples to cool to room temperature before inoculating media or animals as detailed below.

Large volumes of water, washings of bone or hair, and air impinged into liquids may be concentrated more easily by filtration through a sterile cellulose or polycarbonate membrane filter (0.4- to 0.7-μm mean pore diameter). Then, culture for isolation by placing the filter directly on solid medium or by subjecting it to the enrichment techniques described below.

Sampling air in processing facilities and factories requires devices that concentrate large volumes of air either directly onto the surface of a bacteriologic medium (Andersen, 1958) or into liquid by impingement (Buchanan et al., 1968; Litton Systems, 1966). Estimates of the size distribution of viable bacterial particles in the air can be obtained with the Andersen sampler. Such estimates provide important information to epidemiologists, because particles measuring 5 μm or less are needed to infect primates with aerosols of anthrax spores (Dalldorf et al., 1971). The disadvantages of using the Andersen sampler include 1) substantial reduction in the volume of air sampled (i.e., 28.3 liter/min versus at least 1,000 liter/min with other sampling devices) and 2) the risk of obscuring *B. anthracis* by overgrowth of saprophytes present unless a highly selective medium is used initially (Biegeleisen et al., 1962). In contrast, other devices that sample greater volumes of air increase the degree of detection markedly and offer more options in enrichment, cultivation, or otherwise demonstrating the presence of these spores.

Enrichment Procedures

Enrichment techniques include the use of one of the selective media recommended previously without the agar. Inoculate the selected enrichment broth heavily and incubate it for 12 to 18 h at 37°C before streaking to a solid medium for isolating colonies. The entire pads on which samples have been concentrated by filtration may be subjected to this procedure as well.

Inoculation of Laboratory Animals

Perform the following steps in a bacteriologic safety cabinet: Emulsify a loopful of an 18- to 24-h growth from an isolated colony in a sterile test tube with a small amount of sterile saline solution. Continue to dilute and mix the suspension until no visible turbidity can be observed. Inoculate several 18- to 25-g mice or 250- to 400-g guinea pigs subcutaneously with 0.2 or 5.0 ml, respectively, of the prepared suspension. House the inoculated animals in a safety cabinet or enclosure and observe them frequently. (Broth cultures suspected to contain anthrax bacilli should not be inoculated because many *Bacillus* species also produce lethal toxins.)

Other Historic Procedures

Other laboratory tests that support a tentative identification of *B. anthracis* include lack of motility for most strains grown either in broth examined by hanging drop mount or in a semisolid motility medium for 12 to 18 h at 37°C. Brown and Cherry (1955) reported finding several motile strains of *B. anthracis* among the strains that they examined. Similarly, Brown et al. (1955) induced motility in 6 of 80 strains treated with ω bacteriophage.

Another unique test is "the string of pearls" method, which is based on the susceptibility of this pathogen to penicillin. In this procedure a drop of young (i.e., 12 to 18 h) broth culture is placed on two plates of a nutrient agar that contain either 0.5 or 10 U of penicillin per ml. After incubating the inoculated media at 37°C for 3 to 6 h, place a cover slip directly on the growth present on the medium. Examine the underlying growth microscopically under low magnification for the presence of large, rounded cells (or a "string of pearls"). Further supporting evidence is little or no growth of the inoculum placed on the medium containing 10 U of penicillin per ml. Li and Kuo (1968) found only 13 (4.8%) of the 269 strains of *B. anthracis* they examined to be atypical in the "string of pearls" test; in contrast, 23 (8.2%) of the 279 "nonanthrax" strains studied showed typical results. The remainder were negative to the "string of pearls" test.

Another method useful in identifying the more virulent strains of the anthrax bacillus exploits their ability to encapsulate when incubated under an increased atmosphere of carbon dioxide on a medium containing bicarbonate ions. A candle jar provides sufficient concentrations of carbon dioxide (i.e., the inoculated medium and a lighted candle are placed in a container and sealed to allow the flame to extinguish before incubation at 37°C). The required medium may be prepared simply by combining 1 part 7% aqueous, filter-sterilized sodium bicarbonate with 9 parts of sterile nutrient agar cooled to 45°C before mixing and pouring the medium into sterile culture plates. Examine inoculated plates after 21 to 24 h of incubation for the presence of mucoid colonies, a characteristic that is very presumptive of *B. anthracis*. Unfortunately, the absence of mucoid growth does not eliminate the possibility of anthrax bacilli, because nonencapsulating strains exist. A similar medium was described by Meynell and Meynell (1964), who found that the presence of glucose and several peptones inhibits capsular formation.

Determining the ability of a given isolate to metabolize various substrates is frequently more confusing than helpful in identifying *B. anthracis*. The principal and general distinguishing biochemical reactions ascribed for *B. anthracis* include 1) lack or slow reduction for litmus milk in 72 h and no reduction of methylene blue within 24 h; 2) inability to liquefy gelatin in 7 days; and 3) no fermentation of salicin in 24 h. Other characteristics that are shared with species of *B. cereus* include the ability to ferment glucose, maltose, or sucrose without producing gas and the frequent inability to ferment arabinose, mannitol, and xylose.

Definitive Identification

Send cultures of all isolates, either suspected to be or identified as *B. anthracis,* for confirmation to appropriate state or other governmental health laboratories. Within the United States, the national laboratory for confirmation is the Centers for Disease Control in Atlanta, where specimens must be submitted by state health authorities for acceptance. Also, cases of anthrax in both humans and animals are legally reportable to the appropriate health and veterinary officials in many countries of the world.

Bacteriophage Susceptibility

Susceptibility of suspected isolates to lysis by γ bacteriophage offers both a rapid and a reliable means of distinguishing *B. anthracis* from other *Bacillus* species. The following steps are recommended: Place approximately 0.1 ml of a 18-h broth culture of suspected anthrax bacilli on the surface of a dry peptone–agar plate and spread the suspension evenly over the surface. Inoculate additional plates with young broth cultures of known isolates of both *B. anthracis* and *B. cereus*. Leave the lids of the plates ajar to permit excess moisture to evaporate. Mark two identifying circles on the bottom of each these streaked plates with a wax or felt-tip pen; identify each marked area as either the γ bacteriophage or a diluent control. Place 1 drop of undiluted γ-bacteriophage suspension on the surfaces of all streaked plates within the appropriately marked areas. Repeat the inoculation of the plates with either sterile, aqueous 0.85% sodium chloride or nutrient broth. Incubate all plates at 37°C. Examine all plates for bacterial growth periodically. Within 4 h of incubation, lysis may be observed as areas of little or no growth within the areas of the plate inoculated with the bacteriophage suspension. Examine all plates for lysis after an additional 18 h of incubation. The presence of lysis within the marked area of the plate with the unknown isolate confirms the presence of anthrax bacilli. In contrast, the corresponding areas of the plate with *B. cereus* should show confluent bacterial growth. The negative control (diluent or broth) areas should also show confluent growth.

If the results of the test for bacteriophage susceptibility are needed more rapidly, Feeley and Brachman (1974) suggest preparing additional sets of test plates as described previously and streaking the unknown test plate with a suspension prepared from the growth of a suspect colony diluted to approximately 10^8 organisms per ml, which is the equivalent of a no. 1 McFarland nephelometric standard. This latter procedure, which skips the step for growing the organisms in broth, may save valuable time in identifying the microorganism; however, it should not replace the original procedure described.

Pathogenicity for Laboratory Animals

Virulent strains of *B. anthracis* usually kill inoculated mice and guinea pigs within 2 to 10 days. The length of time required to cause death is generally inversely proportional to the number of microorganisms inoculated. However, deaths that are observed in the first 24 h after injection are frequently caused by specimens that are grossly contaminated with other microorganisms.

A necropsy should be performed on each animal that dies. The site of injection in animals infected with anthrax shows extensive hemorrhage with a gelatinous exudate. Smears prepared from these exudates and stained should reveal gram-positive encapsulated rods as described previously.

Serologic Procedures

Use of serologic techniques provides investigators with specificity and certainty in identifying *B. anthracis* rapidly. Unfortunately, the requisite equipment and reagents and trained personnel are frequently lacking in all but a few research laboratories and large public health laboratories.

The first of these methods is staining smears of blood, tissue impressions, exudates, and the like with specific *B. anthracis* fluorescent-labeled antibody, which has been discussed previously under Microscopic Examination. Other serologic procedures published, such as those described by Feeley et al. (1970), Gruber and Wright (1967), and Ray and Kadull (1964), employ acute and convalescent serum samples for specific identification. Few of these techniques, however, are employed outside the research laboratory.

Treatment, Prevention, and Control

Treatment

Before the availability of antibiotic therapy, excising the lesions surgically appeared to improve the prognosis of infected individuals (Freeman, 1979), but this practice has been abandoned and replaced by chemotherapy alone. Penicillin is the therapeutic agent of choice. Gold (1967) recommended an initial injection of 10,000 to 50,000 U of crystalline penicillin with subsequent doses of 10,000 to 50,000 U every 3 to 4 h. A dose of 600,000 U of penicillin was recommended as a maintenance regime every 12 to 24 h for a day or two until the edema and erythema began to subside, then reduced to 300,000 to 600,000 u. Gold also preferred penicillin to tetracyclines as a therapeutic agent because the bacilli disappeared from serous fluid of lesions more rapidly when penicillin was administered than when tetracyclines were given. Patients usually experience the Jarisch–Herxheimer reaction—elevation in body temperature—approximately 2 h after the initial injection of antibiotic.

The development of antibiotic resistance would appear to be of little medical importance outside the laboratory. Few, if any, reports of naturally resistant strains have appeared in the literature despite reports of experimentally produced penicillinase for *B. anthracis* by Buravtseva (1968), Horodniceanu and Sasrman (1962), and Prokupek (1969).

In a clinical case that has advanced to toxemia, neutralization of the toxin by administering hyperimmune serum, where available, is essential to alter the otherwise fatal outcome. Fedotova and Ulanova (1970) found that dilutions as great as 10^{-4} of hyperimmune globulins neutralized anthrax toxin when the mixtures were injected into the skin of guinea pigs. In cases where the patient is gravely ill, one may consider administering serum processed from the blood of persons who survived a previous infection or even from individuals currently immunized against anthrax as a therapeutic measure. Administration of cortisone along with antibiotic is reported to have been beneficial for Iranian cases with massive edema (Parvizpour, 1979).

Prevention and Control

Immunization

Persons at greatest risk because of occupation should receive the initial series of immunizations and maintain their immune status with a nonviable protective-antigen booster. An effective vaccine (Wright, 1975) is not widely available, but sources and availability may be obtained by inquiries to local or state government public health officials. In the Soviet Union, administering a vaccine of nonencapsulated spores by the aerosol route has been found effective in immunizing both humans and animals (Gefen et al., 1972).

Immunization of livestock with the nonsporulating Sterne strain of *B. anthracis* provides an excellent means of protecting animals pastured in anthrax endemic areas. It has been successful during outbreaks in preventing additional livestock losses (Rees et al., 1977). In areas that experience outbreaks of the disease infrequently (i.e., every several decades), the expense of maintaining the immune status of the pastured livestock may be prohibitive. Instead, state veterinary authorities may advise vigilance until it becomes necessary to apply quarantine measures and immunize herds that become affected.

Decontamination and Disinfection

Steam under pressure in an autoclave and the disinfectants routinely used in hospitals and laboratories are effective in rendering anthrax spores and vegetative cells nonviable. Elsewhere, however, the resistance of anthrax spores to drying and their tolerance to extreme temperatures makes decontamination difficult if not impossible.

Enclosed facilities such as laboratories and industrial buildings are best decontaminated by aerosolizing gaseous formaldehyde, either by sublimating paraformaldehyde (with or without steam) or by spraying solutions of Formalin after all openings to the enclosure have been sealed. Young et al. (1970) reported that a concentration of 1.38 to 1.62 mg of paraformaldehyde per liter of air is sufficient to reduce the viability of spores 10-fold or more. Formaldehyde condenses and polymerizes on surfaces of many paints and building materials; hence, several days of flushing with clean outside air are required before affected areas are inhabitable again. Adding 10% methanol to the solution before spraying reduces the degree of polymerization.

Delicate instruments small enough to be placed inside chambers that can be pressurized, such as autoclaves, as well as other articles that might be damaged by formaldehyde vapors may be disinfected effectively by subjecting them to 10% concentrations of ethylene oxide for at least 3 h (Uchida et al., 1973). Lower concentrations (1.5% at 20°C or 0.5% at 30°C) have effectively killed spores seeded to filter paper or collagen in 72 h. Articles of clothing heavily contaminated with anthrax spores may also be disinfected by treatment with a 4% solution of Formalin containing a 1% cationic detergent for 4 h (Darlow, 1970).

Disinfection of soil contaminated with anthrax spores is rarely achieved. Stellmacher and Zerbe (1972) found chemical disinfection of livestock runs of sandy soil possible when 5.0% Formalin was applied, but not runs composed of loam or any type of soil covered with vegetation.

Carcasses of animals suspected to have died of anthrax infection should not be moved to other sites for disinfection and disposal. The carcass should be burned in situ by igniting and burning it with whatever fuel may be permitted by local ordinances.

Environmental Awareness and Consideration

Two theories have been advanced to explain the recurrence of anthrax in pasturelands after long periods of absence. The first theory focuses on the persistence of spores that remain static in the soil until ingested by a suscepible host. In support of this mechanism is the report of Novel and Pongrantz (1969), who found anthrax spores to remain viable for as long as 68 years. Evidence not supporting this hypothesis is failure of these spores to infect livestock despite continual grazing of various stock for many years (Rees et al., 1977). The other theory, advanced by Van Ness (1971), proposes that a dynamic state exists in nature in which the anthrax bacilli undergo cycles of propagation whenever environmental factors meet the requirements for their growth. Other investigators have shown that anthrax bacilli do propagate in the soil. Maly (1972) found that the organisms grew exponentially at pH 7.2 in certain moist soil of East Slovakia at 20°C. Likewise, Lomakovskaya (1960) grew the bacilli in a sample of soil from a muddy swamp incubated at 37°C for 24 h. The detailed requirements for spores that reside in the soil to germinate and grow are unclear. Some of the features that seem to be essential for the spore–vegetative cell–spore cycle to occur in any given geographic area and remain endemic are 1) alkaline soils, 2) above-normal rainfall in amounts sufficient to create "incubator" sites of decayed vegetation, 3) excessive moisture and coolness in the subsequent spring, and 4) drought in mid- to late summer (Van Ness, 1971).

To whichever theory one may subscribe, an animal must ingest a minimal number of spores for infection to recur at sites previously endemic for anthrax. At present, it is not known whether livestock become infected by ingesting numerous doses of spores or if a single dose is sufficient. Schlingman et al. (1956) estimated that approximately 10^7 laboratory-grown anthrax spores constituted the minimal oral dose for cattle, which have been shown to be most susceptible to anthrax infection via the alimentary route (Grosso and Arrieta, 1961). The dose might be much smaller for anthrax spores grown or conditioned in the soil (Rees et al., 1977).

Regardless of the resolution of these findings, an awareness of the ecology of anthrax serves to contain outbreaks in pastures endemic for the disease and aids, or complicates, as the case may be, the epidemiology of such outbreaks.

Literature Cited

Andersen, A. A. 1958. New sampler for the collection, size, and enumeration of viable airborne particles. J. Bacteriol. **76:**471–484.

Belton, F. C., and R. E. Strange. 1954. Studies on a protective antigen produced *in vitro* from *B. anthracis,* medium and methods of production. Br. J. Exp. Pathol. **35:**144–152.

Biegeleisen, J. S., Jr., W. B. Cherry, P. Skaliy, and M. C. Moody. 1962. The demonstration of *B. anthracis* in environmental specimens by conventional and fluorescent antibody techniques. Am. J. Hyg. **75:**230–239.

Brachman, P. S. 1966. Human anthrax in the United States, p. 111–114. *In* G. L. Hobby (ed.), Antimicrobial Agents and Chemotherapy–1965. American Society for Microbiology, Washington, D.C.

Brown, E. R., and W. B. Cherry. 1955. Specific identification of *B. anthracis* by means of a variant bacteriophage. J. Infect. Dis. **96**:34–39.

Brown, E. R., W. B. Cherry, M. C. Moody, and M. A. Gordon. 1955. The induction of motility in *B. anthracis* by means of bacteriophage lysates. J. Bacteriol. **69**:590–602.

Buchanan, L. M., H. M. Decker, D. E. Frisque, C. R. Phillips, and C. M. Dahlgren. 1968. Novel multislit large volume air sampler. Appl. Microbiol. **16**:1110–1123.

Buravtseva, N. P. 1968. Effect of fixing smears over a fmale on the viability of anthrax spores. Biol. Abstr. **49**: entry no. 57451. (Translated from Russian; Osobo opasnye infektsii na Kavkaze. Stavropol, 224, 1966.)

Dalldorf, F. G., A. F. Kaufmann, and P. S. Brachman. 1971. Woolsorters' disease: an experimental model. Arch. Pathol. **92**:418–426.

Darlow, H. M. 1970. Decontamination methods for overalls of hide and wool handlers. Proc. R. Soc. Med. **63**:59–60.

Davies, D. G., and R. W. S. Harvey. 1955. The isolation of *B. anthracis* from bones. Lancet **2**:86–87.

Fedotova, Y. M., and A. A. Ulanova. 1970. Study of specific antitoxic activity of therapeutic globulin against anthrax. Biol. Abstr. **51**: entry no. 91162. (Zh. Mikrobiol. Epidemiol. Immunobiol. **47**:47–51, 1970.)

Feeley, J. C., and P. S. Brachman. 1974. *Bacillus anthracis*, p. 143–147. *In* E. H. Lennette, E. H. Spaulding, and J. P. Truant (ed.), Manual of clinical microbiology, 2nd ed. American Society for Microbiology, Washington, D.C.

Feeley, J. C., T. M. Buchanan, P. S. Hayes, and P. S. Brachman. 1970. Anthrax indirect microhemagglutination test. Bacteriol. Proc. **126**:282–284.

Fish, D. C., and R. E. Lincoln. 1968. *In vivo*-produced anthrax toxin. J. Bacteriol. **95**:919–924.

Fox, M. D., J. M. Boyce, J. C. Feeley, B. T. Wood, A. F. Kaufmann, and E. J. Gangarosa. 1975. Animal anthrax, Falls County, Texas. U.S. Public Health Service, EPI no. 75-6-2. Centers for Disease Control, Atlanta.

Freeman, B. A. 1979. Anthrax, p. 631–637. *In* B. A. Freeman (ed.), Burrow's textbook of microbiology, 21st ed. The W. B. Saunders Co., Philadelphia.

Gefen, N. E., V. M. Shustikov, O. A. Rudneva, S. F. Fedyaev, G. T. Tsimlyanskii, E. A. Golubitskii, and E. S. Lazareva. 1972. A study of reactogenicity, harmlessness and immunological efficacy of aerosol method of immunization against anthrax in trial on man. Biol. Abstr. **54**: entry no. 184. (Zh. Microbiol. Epidemiol. Immunobiol. **47**:81–85, 1971.)

Gold, H. 1967. Treatment of anthrax. Fed. Proc. **26**:1563–1568.

Grosso, A. M., and A. P. Arrieta. 1961. Experimental anthrax in cattle. Biol. Abstr. **36**: entry no. 42149. (Rev. Invest. Ganad. **1**:123–128, 1960.)

Gruber, J., and G. G. Wright. 1967. Iodine-131 labeling of purified microbial antigens by microdiffusion. Proc. Soc. Exp. Biol. Med. **126**:282–284.

Haines, B. W., F. Klein, and R. E. Lincoln. 1965. Quantitative assay for crude anthrax toxins. J. Bacteriol. **89**:74–83.

Heyworth, B., M. E. Ropp, U. G. Voos, H. I. Meinel, and H. M. Darlow. 1975. Anthrax in Gambia: an epidemiological study. Br. Med. J. **4**:79–82.

Horodniceanu, T., and A. Sasrman. 1962. Contribution to the study of bacterial penicillinase. III. Relationships between penicillinase production and susceptibility to penicillin in *B. anthracis*. Biol. Abstr. **37**: entry no. 10224. (Translated from Roumanian; Arch. Roum. Pathol. Exp. Microbiol. **20**:257–262, 1961.)

Knisely, R. F. 1966. Selective medium for *B. anthracis*. J. Bacteriol. **92**:784–786.

Krinsky, W. L. 1977. Animal disease agents transmitted by horse flies and deer flies (Dipter; Tabanidae). Biol. Abstr. **63**: entry no. 59748. (J. Med. Entomol. **13**:225–275, 1976.)

Leise, J. M., C. H. Carter, H. Friedlander, and S. W. Freed. 1959. Criteria for the identification of *B. anthracis*. J. Bacteriol. **77**:655–660.

Leppla, S. H. 1982. Anthrax toxin edema factor: a bacterial adenylate cyclase that increase cyclic AMP concentrations in eukaryotic cells. Proc. Natl. Acad. Sci. USA **79**:3162–3166.

Leppla, S. H. 1984. *B. anthracis* calmodulin-dependent adenylate cyclase: chemical and enzymatic properties and interactions with eucaryotic cells. Adv. Cyclic Nucleotide Protein Phosphorylation Res. **17**:189–198.

Li, L., and H. P. Kuo. 1968. A study of string-of-pearls test of *B. anthracis*. Biol. Abstr. **49**: entry no. 40917. (Translated from Chinese; Acta Microbiol. Sin. **11**:80–87, 1965.)

Lincoln, R. E., J. S. Walker, F. Klein, and B. W. Haines. 1964. Anthrax. Adv. Vet. Sci. **9**:327–368.

Lincoln, R. E., J. S. Walker, F. Klein, A. J. Rosenwald, and W. I. Jones, Jr. 1967. Value of field data for extrapolation in anthrax. Fed. Proc. **26**:1558–1562.

Litton Systems. 1966. Model M large-volume air sampler: instruction manual. Applied Science Division report no. 3028. Litton Systems, Inc., Minneapolis.

Lomakovskaya, V. M. 1960. On the propagation of the microbes of Siberian anthrax in soils of various types. Biol. Abstr. **35**: entry no. 45444. (Sb. Nauchn. Tr. L'vovsk Zoovet. Inst. **8**:44–52, 1956.)

Maly, E. 1972. Saprophytic vegetation of *B. anthracis* in humusgley soil. Biol. Abstr. **54**: entry no. 28821. (Folia Fac. Med. Univ. Comenianae Bratisl. **8**(Suppl.):253–263, 1970.)

Meynell, E., and G. G. Meynell. 1964. The roles of serum and carbon dioxide in capsule formation by *B. anthracis*. J. Gen. Microbiol. **34**:153–164.

Mikesell, P., B. E. Ivins, J. D. Ristroph, and T. M. Dreier. 1983. Evidence for plasmid-mediated toxin production in *B. anthracis*. Infect. Immunol. **39**:371–376.

Nalin, D. R., B. Sultana, R. Sahunja, A. K. Islam, M. A. Rahim, M. Islam, B. S. Costa, N. Mawla, and W. B. Greenough. 1977. Survival of a patient with intestinal anthrax. Am. J. Med. **62**:130–132.

Novel, E., and E. Pongrantz. 1969. The survival of spores of *B. anthracis* and two other bacterial species. Pathol. Microbiol. **33**:180–184.

Parvizpour, D. 1979. Human anthrax in Iran. An epidemiological study of 468 cases. Biol. Abstr. **68**: entry no. 18521. (Int. J. Zoonoses **5**:69–74, 1978.)

Prokupek, K. 1969. The penicillinase activity of anthrax and cereus strains. Biol. Abstr. **50**: entry no. 76564. (Translated from Czech; Vet. Med. (Prague) **13**:149–158, 1968.)

Ray, J. G., Jr., and P. J. Kadull. 1964. Agar-gel precipitin technique in anthrax antibody determinations. Appl. Microbiol. **12**:349–354.

Record, B. R., and R. G. Wallis. 1956. Physiochemical examination of polyglutamic acid from *B. anthracis* grown *in vivo*. Biochem. J. **63**:443–447.

Rees, H. B., Jr., and M. A. Smith. 1981. Anthrax, p. 215–233. *In* A. Balows and W. J. Hausler, Jr. (ed.), Diagnostic procedures for bacterial, mycotic and parasitic infections. American Public Health Association, Washington, D.C.

Rees, H. B., Jr., M. A. Smith, J. C. Spendlove, R. S. Fraser, T. Fukushima, and A. G. Barbour, Jr. 1977. Epidemiology and laboratory investigations of bovine anthrax in two Utah countries in 1975. Public Health Rep. **92:**176–186.

Schlingman, A. S., H. B. Devlin, G. G. Wright, R. J. Maine, and M. C. Manning. 1956. Immunizing activity of alum-precipitated protective antigen of *B. anthracis* in cattle, sheep and swine. Am. J. Vet. Res. **17:**256–261.

Sirisanthana, T., N. Navachareon, P. Tharavichitkul, V. Sirisanthana, and A. E. Brown. 1984. Outbreak of oral–oropharyngeal anthrax: an unusual manifestation of human infection with *B. anthracis*. Am. J. Trop. Med. Hyg. **33:**144–150.

Smith, H., and J. Keppie. 1954. Observations on experimental anthrax: demonstration of a specific lethal factor produced *in vivo* by *B. anthracis*. Nature (London) **173:**869–870.

Smith, H., and H. B. Stoner. 1967. Anthrax toxic component. Fed. Proc. **26:**1554–1557.

Stellmacher, W., and J. Zerbe. 1972. Disinfection of livestock outdoor runs, pastures and ground. Biol. Abstr. **54:** entry no. 53719. (Translated from German; source not given, 1971.)

Tahernia, A. C., and G. Hashemi. 1972. Survival in anthrax meningitis. Pediatrics **50:**329–332.

Uchida, K., A. Terada, S. Tanaka, Y. Ichiman, and S. Nagano. 1973. Effect of ethylene oxide sterilization. Biol. Abstr. **56:** entry no. 59913. (Translated from Japanese; Bull. Nippon Vet. Zootech. Coll. **21:**1–7, 1972.)

Van Ness, G. B. 1971. Ecology of anthrax. Science **172:**1303–1307.

Wilson, G. S., and A. Miles. 1975. Anthrax, p. 2208–2224. *In* Topley and Wilson's principles of bacteriology, virology and immunity, 6th ed. The Williams & Wilkins Co., Baltimore.

Wright, G. G. 1975. Anthrax, p. 237–250. *In* W. T. Hubert, W. F. McCulloch, and P. R. Schnurrenberger (ed.), Diseases transmitted from animals to man, 6th ed. Charles C. Thomas, Publisher, Springfield, Ill.

Young, L. S., J. C. Feeley, and P. S. Brachman. 1970. Vaporized formaldehyde treatment of a textile mill contaminated with *B. anthracis*. Arch. Environ. Health **20:**400–403.

Zaporozhachenko, A. Y. 1962. Epidemiological and clinical characteristics of anthrax. Defense Documentation Center, Defense Supply Agency, translation no. 723, A.D. no. 299067. (Translated from Russian; Zh. Microbiol. Epidemiol. Immunobiol. **33:**111–115, 1962.)

Zwartouw, H. T., and H. Smith. 1956. Polyglutamic acid from *B. anthracis* growth *in vivo;* structure and aggressin activity. Biochem. J. **63:**437–453.

CHAPTER 7

Infections Due to Nonanthrax *Bacillus* Species, *Kurthia*, and *Rothia*

ANNETTE C. REBOLI and W. EDMUND FARRAR

Disease: A wide variety of infections: 1) local, usually involving the eye, 2) mixed, and 3) disseminated including endocarditis and meningitis.
Etiologic Agents: Of 34 nonanthrax *Bacillus* species, nine have caused infection in man.
Source: Widespread in nature, especially in soil, water, dust, and decaying organic matter. Also found as contaminants of dialysates, blood for transfusion, and from cultures of heroin and drug paraphernalia.
Clinical Manifestations: For ocular infection: intense pain followed by chemosis, periorbital swelling, and severe proptosis. Corneal ring abscess develops with irreversible loss of vision and loss of the eye. Manifestations of disseminated infections are the same as for other bacterial infections.
Pathology: Pathologic changes in organs infected by *Bacillus* species are the inflammatory reactions caused by bacteria in general. Toxins may be responsible for some pathologic changes.
Laboratory Diagnosis: Bacteriologic isolation of etiologic agent from blood, normally sterile body fluids, or closed spaces of a susceptible host in the appropriate clinical setting. Species identification is based on the results of two morphologic features and 11 biochemical tests.
Epidemiology: Worldwide.
Treatment: Almost all strains are sensitive to the aminoglycosides, tetracycline, chloramphenicol, erythromycin, vancomycin, and clindamycin. Susceptibility to β-lactams is high for *B. subtilis* and low for *B. cereus*. For treatment of severe ocular infections, early vitrectomy and intravitreal antibiotics combined with systemic and periocular antibiotics. Clindamycin with or without an aminoglycoside may be the treatment of choice for endocarditis or sepsis with *B. cereus*. Surgical drainage of closed space infections. Replacement of infected prosthetic heart valves.
Prevention and Control: No effective control measures.

Disease: *Kurthia:* Endocarditis. *Rothia:* Endocarditis, dental caries, periodontal disease, abscess.
Etiologic Agents: Two *Kurthia* species. One *Rothia* species, *R. dentocariosa*.
Source: *Kurthia:* fresh and spoiled meat, water, feces, and abattoirs. *Rothia:* normal flora of the mouth and pharynx.
Clinical Manifestations: Similar to other bacterial infections.
Pathology: No distinctive pathologic changes.
Laboratory Diagnosis: Bacteriologic isolation from blood, a normally sterile body fluid, or a closed space in the appropriate clinical setting.
Epidemiology: Worldwide.
Treatment: *Kurthia:* Penicillin, aminoglycosides, erythromycin, chloramphenicol, or trimethoprim–sulfamethoxazole. Cases of endocarditis may require valve replacement. *Rothia:* Penicillin combined with an aminoglycoside.
Prevention and Control: No effective control measures.

Bacillus Infections

Description of Disease States

Introduction

The genus *Bacillus* contains 34 species of aerobic, sporulating rods. Because of phenotypic and genetic diversity, there is much confusion and uncertainty in taxonomy. Except for *B. anthracis,* these species once were considered nonpathogenic saprophytes and frequent laboratory contaminants. They are now recognized as significant pathogens in man and insects. Species important in human disease include *B. cereus, B. megaterium, B. subtilis, B. sphaericus, B. circulans, B. brevis, B. lentiformis, B. macerans,* and *B. pumilus. Bacillus thuringiensis, B. larvae, B. lentimorbus, B. popilliae,* and some strains of *B. sphaericus* are insect pathogens.

Clinical Manifestations of Disease

Bacillus infections have been classified as 1) local, usually involving the eye or an isolated organ that was previously damaged; 2) mixed, in which a *Bacillus* species is found in the company of other organisms; and 3) disseminated, in which the organism is consistently cultured from the blood or spinal fluid of a seriously ill person (Farrar, 1963). *B. cereus* is the next best known mammalian pathogen after *B. anthracis*. *B. subtilis* has been used as a synonym for aerobic sporeformers other than *B. anthracis,* and many isolates described as *B. subtilis* in the early literature were probably *B. cereus* (Weinstein and Colburn, 1950).

Local Infections

The eye has been the organ most commonly infected by *Bacillus* species, especially *B. cereus*. They can cause conjunctivitis, keratitis, iridocyclitis, dacryocystitis, orbital abscess, and panopthalmitis. *B. cereus* is one of the most destructive organisms to infect the eye (Davey and Tauber, 1987; O'Day et al., 1981). Panopthalmitis usually occurs in the setting of hematogenous dissemination, as in intravenous drug abusers, or following penetrating trauma (Davey and Tauber, 1987; Pearson, 1970; Young et al., 1980). Clinical features are fairly characteristic, with intense pain followed by chemosis, periorbital swelling, and severe proptosis. Within 48 h a corneal ring abscess develops. The end result is irreversible loss of vision and eventual loss of the eye (Davey and Tauber, 1987; O'Day et al., 1981). A characteristic feature of infection with *B. cereus* is its frequent association with fever and leukocytosis, which are not usually seen in panopthalmitis caused by other bacteria (Davey and Tauber, 1987). This organism should be considered whenever ocular infection occurs in the setting of penetrating injury with probable soil contamination. Vitreous aspiration is diagnostic. Empiric therapy should begin before isolation of an organism and the availability of antibiotic susceptibility data.

The first case of human infection with *B. thuringiensis* was reported by Samples (1983) in a healthy farmer who splashed an insecticide containing the organism into his eye. The organism was recovered from a corneal ulcer. The clinical course was much less fulminant than with *B. cereus* infection.

Mixed Infections

Mixed infections have included cellulitis, wound infections, infected burns, necrotizing fasciitis, pyelonephritis, and pneumonia (Bias, 1929; Farrar, 1963; Ihde et al., 1973; Logan et al., 1985; Pearson, 1970; Pennington et al., 1976; Tuazon et al., 1979). Streptococcal organisms are frequently isolated along with the *Bacillus* species. It is frequently unclear what role the *Bacillus* plays in these infections. It may be a copathogen, a secondary invader, or a colonizer. Ihde et al. (1973) found that whether or not the *Bacillus* was treated did not affect outcome in surgical wounds; however, they did note a change in the character of the drainage from sanguinous to serous after the *Bacillus* was no longer recovered in culture.

Disseminated Infections

The category of disseminated infections includes bacteremia, endocarditis, meningitis, and other such infections. Endocarditis due to *Bacillus* organisms is an uncommon disease; only about a dozen cases have been reported in the English-language literature (Block et al., 1978; Craig et al., 1974; Farrar, 1963; Finland and Barnes, 1970; Oster and Kong, 1982; Reller, 1973; Tuazon et al., 1979; Weller et al., 1979). *B. cereus* has been the most common species found in cases of endocarditis (Block et al., 1978; Oster and Kong, 1982; Tuazon et al., 1979; Yeh et al., 1967). In three cases there was infection of a prosthetic valve (Block et al., 1978; Oster and Kong, 1982; Yeh et al., 1967). More than half the cases have occurred in intravenous drug users (Craig et al., 1974; Reller, 1973; Tuazon et al., 1979; Weller et al., 1979), probably relating to the frequent presence of *Bacillus* organisms on injection paraphernalia and in street heroin (Tuazon et al., 1974).

Bacillus endocarditis cannot be distinguished from other bacterial causes of endocarditis on the basis of clinical features. The majority of patients have had low-grade fever and heart murmurs. Some have had hepatosplenomegaly, and a few have had peripheral stigmata.

Meningitis due to *Bacillus* organisms has been reported primarily in situations in which there is either direct access of exogenous organisms to the subarachnoid space (spinal anesthesia, head trauma, neurosurgical procedures), or in individuals whose immune function is depressed (newborn state, alcoholism, hematologic malignancy), although at least one case has been reported in an otherwise healthy individual (Allen and Wilkinson, 1969; Boyette et al., 1951; Colpin et al., 1981; Farrar, 1963; Weinstein and Colburn, 1950). Causative organisms have included *B. subtilis, B. pumilus, B. circulans, B. sphaericus,* and *B. cereus*. Single cases of postoperative ventriculitis due to *B. licheniformis* and *B. cereus* have been reported (Young et al., 1982). The latter infection was in a patient with a ventriculoperitoneal shunt.

Bacteremia has also occurred in association with infections in a variety of body sites. Some of these may actually have been cases of endocarditis, and some were treated as such. A relatively common situation has been pneumonia with bacteremia in individuals with acute leukemia and granulocytopenia (Coonrod et al., 1971; Ihde et al., 1973; Pennington et al., 1976; Sathmary, 1958). The granulocytopenic patient is at risk for life-threatening bacteremia following seemingly minor trauma (Guiot et al., 1986). These organisms have also been implicated in several cases of osteomyelitis (Pearson, 1970; Solny et al., 1977; Tuazon et al., 1979). An outbreak of five cases of bacteremia occurred in a group of hemodialysis patients exposed to contaminated dialysis fluid and equipment (Curtis et al., 1967). Individual cases of bacteremia in association with volvulus and peritonitis (Sugar and McCloskey, 1977), infected ventriculojugular shunt cured only by removal of the shunt (Cox et al., 1957), and sepsis in a newborn infant, perhaps related to a contaminated blood transfusion (Yow et al., 1949), have been reported.

Hypersensitivity Reactions

Allergic reactions have been described in workers exposed to autolysates of *B. subtilis* used as enzymes in laundry products (Dubos, 1971; Flindt, 1969; Greenberg et al., 1970; Pepys et al., 1969). These reactions were mainly dermatitis and asthma. Those involved showed immediate and late reactions to inhalation and skin prick tests (Flindt, 1969). Hypersensitivity pneumonitis following exposure to wood dust contaminated with *B. subtilis* has been described (Johnson et al., 1980).

Pathology

The pathologic changes in organs infected by *Bacillus* species are the inflammatory reactions caused by bacteria in general. There have not been any special or pathognomonic changes described in humans. Subcutaneous injection of *B. cereus* in mice, guinea pigs, and rabbits has caused deep, hemorrhagic, ulcerative skin lesions and death (Burdon and Wende, 1960). Direct cytotoxicity of *B. cereus* culture filtrate for tissue culture cells has been noted (Bonventre, 1965). *B. cereus* filtrates were noted to kill mice and rats within minutes after intravenous injections. Pathologically there were widespread thromboses. The filtrates were noted to be hemolytic and dermonecrotic.

The pathogenesis and pathologic changes in mice with experimental infection with *B. cereus* have been studied (Burdon et al., 1967). Pathologic changes included focal areas of coagulative necrosis and inflammatory cells in the pulmonary arterioles and capillaries. Thrombi were found in the pulmonary vessels. Because no organisms were seen in tissue sections, toxins were believed to be responsible for illness and death. This hypothesis was further supported by the finding that antitoxic serum had a positive protective action.

B. cereus produces distinct extracellular products: hemolysin, phospholipase C, and a lethal toxin (Coolbaugh and Williams, 1978; Turnbull et al., 1979). Two hemolysins have been described. One has a molecular weight of 52,000, causes immediate lysis of red blood cells, and is probably "cereolysin" (Coolbaugh and Williams, 1978). The other has a molecular weight of 31,000. *B. cereus* is not the only species that produces hemolysins. At least 18 *Bacillus* species have the capability (Bernheimer and Grushoff, 1967). The lysins of some species resemble streptolysin O. Paper chromatography has demonstrated that hemolysins are not the same as phospholipases (Stein and Logan, 1965). Phospholipases and lecithinases may be identical, however. Lecithinases may be responsible for much of the ocular pathology, whereas hemolysins may be responsible for changes in quality of drainage from infected wounds (Ihde et al., 1973).

Diagnostic Features

Except for the characteristic signs and symptoms of panopthalmitis, most infections with *Bacillus* species do not have distinctive clinical presentations, and although serious infections are being recognized with increasing frequency, endocarditis, meningitis, and sepsis are still sufficiently rare that the question of contaminant versus pathogen is still valid. Diagnosis of true infection rests on the isolation of the organism from blood, normally sterile body fluids, or closed spaces (such as the pleural space) of a susceptible host in the appropriate clinical setting. A positive Gram stain and growth in pure culture are contributory evidence of the *Bacillus* being a true pathogen, whereas growth only in thioglycolate broth (indicating a small number of organisms) or the presence of *Bacillus* organisms in mixed culture with other species is suggestive of a contaminant. Although infection can occur in healthy persons, it is most often seen in immunocompromised patients such as those with hematologic malignancies, parenteral drug users, or patients with severe trauma (Ihde et al., 1973; Pennington et al., 1976; Tuazon et al., 1979).

Prognosis

Morbidity and mortality for disseminated infections were high prior to the antibiotic era. Except for panopthalmitis, which usually results in loss of vision and of the eye, other infections by *Bacillus* species have a good prognosis unless the underlying disease (such as malignancy) supervenes. *B. cereus* endocarditis in intravenous drug users has a relatively benign course and favorable prognosis. Of patients with meningitis, approximately 40% survived; however, most have had central nervous system sequelae.

Epidemiologic Aspects

Occurrence

Bacillus species are widespread in nature. They are found in soil, water, milk, dust, decaying organic matter, vegetables, air, wool, and feces (Norris, 1981). They are known contaminants of dialysates and dialyzers and occasionally they cause bacteremia, as Curtis et al. (1967) reported in five patients on hemodialysis. They have also been isolated from blood associated with a contaminated transfusion (Yow et al., 1949) and from cultures of heroin and drug paraphernalia (Tuazon et al., 1974). Approximately 50% of these cultures contain *Bacillus* species, making them the most common organism isolated. In one case of endocarditis occurring in an intravenous drug user, the water used as a diluent contained the organism (Reller, 1973). Periodically, outbreaks of positive cultures are linked to contaminated equipment used for diagnostic procedures, such as the pseudooutbreak of positive cultures linked to a contaminated bronchoscope (Goldstein and Abrutyn, 1985).

Etiologic Agents

Characterization

Bacillus species are rod-shaped organisms that range from approximately 0.4 to 2 μm in width and from 3 to 12 μm in length. They are straight or slightly curved, have parallel sides, and have round or square ends. Arrangement is variable. Most organisms occur singly or in pairs, but occasionally groups or long chains are seen. Motility is a common feature, with most strains having 4 to 12 flagella in a peritrichate arrangement.

Organisms of the genus *Bacillus* are usually gram positive but may appear gram negative, especially when colonies are older. Sudan black stain reveals the presence of fat globules in the cells of most species (Burdon, 1956). The globules can be large and diffuse, as in *B. megaterium,* or small and discrete, as in *B. cereus*. *B. subtilis, B. pumilus,* and *B. licheniformis* rarely possess globules.

The cell wall contains meso-diaminopimelic acid. Teichoic acids have been detected in the cell wall of some species. Several species (*B. subtilis, B. licheniformis,* and *B. megaterium*) produce a carbohydrate capsule, the polysaccharides of which cross-react with antisera from other genera such as *Streptococcus pneumoniae, Haemophilus influenzae,* and *Neisseria meningitidis*. Capsule formation is enhanced by aerobic incubation with added carbon dioxide and serum, albumin, or bicarbonate of a specific concentration in the medium (Meynell and Meynell, 1964). Blue rods surrounded by a pink capsule are demonstrated after staining with polychrome methylene blue.

This genus is diverse in both phenotypic and genetic characterization. The base composition (guanine plus cytosine content) of DNA ranges from 32 to 70 mol%. DNA–DNA hybridization studies support this heterogeneity.

All species can form endospores, which vary in shape from spherical to ellipsoidal. They may be in a terminal, subterminal, midline, or lateral position. Some have a diameter greater than that of the vegetative cell, thereby causing the cell to have a swollen appearance. These swollen rods resemble clostridia, but sporulation of *Bacillus* species occurs only in the presence of oxygen. Because of their thick cell wall coats, spores may not be evident on routine Gram staining. They are very resistant to heat and drying and even to some chemical disinfectants. They are best detected in older colonies with a phase-contrast microscope and appear to be very refractile. Unlike vegetative cells, they may be acid-fast. The appearance of spores of 19 species has been looked at under the electron microscope using the carbon replica technique (Bradley and Franklin, 1958). Most species showed distinctive patterns. The pattern is a species characteristic or, at most, common to two related species.

None of the proposed classification schemes provides a way to distinguish clearly the different species or among strains within a species. The spore is important as a taxonomic character and forms the basis of one long-standing classification scheme based on the shape and position of the spore and the presence or absence of swelling of the vegetative cell (Wilson, 1983). Species in group I are gram positive and have ellipsoidal spores that do not cause swelling. This group consists of a large cell subgroup having a rod diameter of 0.9 μm or more. Members include *B. anthracis, B. cereus, B. cereus* variant *mycoides, B. megaterium,* and *B. thuringiensis*. The small cell subgroup has a rod diameter of less than 0.9 μm and consists of *B. licheniformis, B. pumilus, B. subtilus,* and *B. coagulans*. In group II the sporangia are swollen by oval spores. The gram-variable *B. circulans, B. polymyxa,* and *B. stearothermophilus* belong to this group. *B. sphaericus, B. pantothenicus,* and *B. pasteurii* belong to group III. They are also gram variable and possess sporangia swollen by circular spores. Species in group I have simple nutritional requirements, those in group II are more complex, and those in group III are the most fastidious and complex. DNA–DNA hybridization studies support the homology of group I and, in fact, suggest that *B. anthracis, B. cereus,* and *B. thuringiensis* should be a single species (Kaneko et al., 1978; Seki et al., 1978).

Although there is antigenic heterogeneity with many different spore serotypes and cell wall types, there is also a great deal of cross-reactivity between spores from different species (Tomcsik and Baumann-Grace, 1959). Spores have surface antigens that are characteristic of the spore and distinct from those of the parent cell as well as others in common with the parent cell (Doak and Lamanna, 1948). The common antigens are well below the surface. Some small cell species are distinguishable by their spore antigens (Lamanna, 1940). With the exception of *B. cereus* and *B. megaterium*, most large cell species are not.

Motile species have flagellar or H antigens. A serotyping scheme for *B. cereus* has been proposed based on agglutinins against the flagellar (H) antigens (Taylor and Gilbert, 1975). Eighteen strains were used to prepare agglutinating sera, which were used to type 137 isolates. The majority were H serotype I. There were 18 H serotypes in all. Serotyping of *B. cereus* has been useful in epidemiologic studies of food poisoning. H serotyping schemes also exist for *B. thuringiensis* and *B. sphaericus* (Claus and Berkeley, 1986). In general, antigenic composition is of little help in distinguishing other species or strains from one another.

Nutrition, Growth, and Metabolism

Most *Bacillus* species grow well on ordinary nutrient media, with improved growth in the presence of glucose. All are aerobic, and some (*B. cereus, B. anthracis, B. licheniformis,* and *B. coagulans*) can grow anaerobically. Growth under strict anaerobic conditions on nutrient agar distinguishes *B. cereus* from *B. megaterium*. This genus has members that are mesophilic, psychrophilic, and thermophilic. For most species the optimal temperature for growth ranges from 25 to 37°C. Small-cell species such as *B. subtilis* grow better at higher temperatures. Temperature requirement cannot be used to define a species, as the ability to grow at higher temperatures is variable within the same species. Some species are acidophilic, others alkaliphilic, but most grow best at pH 7 with a range from pH 2 to 9. Some species are salt tolerant, whereas others require it for growth. Alkaliphilic species have an obligate sodium requirement.

The morphologic appearance of colonies is influenced by the media on which they are growing. On 5% sheep blood agar, single colonies may be 2 mm or more in diameter at 24 h. Colonies of *B. cereus* are α- or β-hemolytic with a lavender color. They may be small, shiny, and compact or large and feathery. Colonies of other *Bacillus* species are usually large, flat, and dull with a ground-glass appearance. Colonies of *B. circulans* are motile on the surface of nutrient agar. There are two types of motion: migration of the entire colony at a defined speed and rotary motion (Roberts, 1935). Spore formation is often stimulated on esculin agar or an acidified media such as triple sugar iron (TSI). Heat shocking of vegetative cells may be necessary to produce sporulation. In broth, a surface layer may form. There may or may not be turbidity.

Some species form pigment, which may be brown, yellow, orange, pink, or black depending on the media on which they are growing. Pigment formation is increased by carbohydrate in the media. It is a variable, late-appearing feature and of little differential value.

On carbohydrate media most species produce acid only. (*B. polymyxa* and *B. licheniformis* also can produce gas.) Commonly fermented sugars include glucose, maltose, and sucrose. Lactose is rarely utilized. The end products of glucose fermentation vary. Some species produce lactic acid. *B. subtilis, B. licheniformis,* and *B. cereus* produce 2,3-butanediol and glycerol. *B. polymyxa* produces ethanol and hydrogen in addition to 2,3-butanediol. Ethanol, acetone, acetic acid, and formic acid are produced by *B. macerans* (Gibson and Gordon, 1974). *B. sphaericus* does not use carbohydrates effectively. It uses amino acids and organic acids and produces urease, which converts urea to ammonia. During growth the pH of the medium becomes alkaline. It should be kept in mind that fermentation reactions are often weak and unreliable.

Most species are catalase positive. They can be oxidase positive or negative. Some produce hydro-

TABLE 1. Abbreviated scheme for identification of gram-positive bacilli

Genus	Morphology[a]	Test or substrate[a]: Catalase	Grows better anaerobically	Motility	Esculin	H_2S (TSI)	TSI acid slant/butt	Nitrate reduction	Urease	Hemolysis	Comments
Bacillus	Med/large, regular rods	+	–	v	+	–	+/+, v/v	v	v	B, A, –	Endospore forming
Kurthia	Chains, plump rods	+	–	+	–	–	–/–	–	+	–	Strict aerobe, feathery growth
Listeria monocytogenes	Small rods, coccobacilli	+	–	+	+	–	+/+	–	–	B	Umbrella motility
Oerskovia	Coccoid, branching filaments	+	–	+	+	–	+/+	v	v	–, A	Yellow pigment
Rothia dentocariosa	Coccoid, diphtheroid, branching filaments	+	–	–	+	–	+/+	+	–	–	Coccoid in broth
Mycobacterium fortuitum complex	Slender rods	+	–	–	–	–	–/–	+	v	–	Acid fast
Rhodococcus equi	Coccoid, branching filaments	+	–	–	v	–	–/–	v	v	–	Pink, coral pigment
Bacterionema	Regular rods, "whips"	+	–	+	–	–	+/+ –/–	+	–	v	Acid fast
Corynebacterium	Diphtheroid, curved, pleomorphic	+	–	–[b]	–	–	v/v	v	v	v	
Arcanobacterium	Diphtheroid, curved, pleomorphic,	–	–/+	–	v	–	+/+	–	–	B	
Arachnia	Diphtheroid, pleomorphic rods	–	+	–	–	–	+/+	–	–	–	
Actinomyces	Diphtheroid, branching filaments	–[c]	+	–	+	–	+/+	+, v	–	–	
Lactobacillus	Thin rods, thick rods, chains	–	+/v	–	+/v	–	+/+	–	–	–, A	Some grow on tomato juice agar
Erysipelothrix	Pleomorphic chains, rods	–	–	–	–	+	+/+	–	–	–, A	
Propionibacterium	Diphtheroid	+	+	–	v	–	+/+	–	–	B, A, –	
Clostridium	Straight or curved rods, some long filaments	–	+	v	v	+	v/v	+	v	B/A	Endospore forming

[a] Small (<1 mm diameter at 24 h); medium (1–2 mm diameter at 24 h); large (>2 mm diameter at 24 h); + = >90% of strains positive; – = <10% of strains positive; v = variable; B = β-hemolytic; A = α-hemolytic.
[b] Some species are positive.
[c] *A. viscous* is positive.
Modified from Clarridge and Weissfeld (1984).

gen sulfide, reduce nitrate, and give a positive Voges-Proskauer reaction. In general, they are indole negative. Other important biochemical reactions include starch hydrolysis, gelatin liquefaction, methylene blue reduction, peptonization of milk, and reduction of litmus when a clot is formed in litmus milk and digested.

In addition to hemolysins and toxins already discussed, various other extracellular products such as amylases and antibiotics (polymyxin, bacitracin, tyrocidin, gramacidin, and licheniformin) are produced by members of this genus. *B. cereus* is distinguished by the production of lecithinases, which produce an opaque zone around colonies on egg yolk agar and a milky turbidity in egg yolk broth (Costlow, 1958). Other species may exhibit a lower level of lecithinase activity.

Specimen Collection, Transport, and Maintenance

Clinical specimens such as blood, spinal fluid, and wound exudates do not require special handling for *Bacillus* species. Prompt plating to prevent growth of contaminants is advocated. Tissue specimens should be macerated and inoculated into a peptone broth. As spores are viable for long periods, maintenance of cultures after initial isolation is usually not difficult. Preservation in sterile soil has been suggested, but nutrient agar with manganese sulfate is equally good for sporulation and maintenance and is more convenient than soil extract (Norris et al., 1981). *Bacillus* species may be stored on agar slants, frozen at −70°C, or freeze-dried for long-term maintenance. Freeze-drying may alter the character of the organisms as can repeated subculturing at frequent intervals.

Isolation and Identification

An initial Gram stain should be done on any suitable clinical material. It is the most rapid and easiest way to assess the quality of the sample and to have a preliminary categorization of the pathogen. If gram-positive bacilli are seen, further testing will be needed to distinguish *Bacillus* species from other gram-positive rods (Table 1). *Bacillus* species may at times resemble gram-negative bacilli. In contrast to the nonfermenting gram-negative rods, *Bacillus* species do not usually grow on enteric agars. In addition, *Bacillus* species are sensitive to vancomycin, whereas gram-negative nonfermenters are not, and a KOH test can be performed in which gram-negative rods show a viscous thread (Gregersen, 1978). Although these tests are helpful, they are not perfectly reliable.

It is fairly easy to differentiate between *B. anthracis* and other *Bacillus* species with the exception of *B. cereus*. Table 2 shows important differential characteristics between *B. anthracis* and other pseudoanthrax bacilli.

If the clinical situation demands a determined effort to isolate *Bacillus* organisms, specimens can be suspended in distilled water or nonnutrient buffer

Table 2. Differences between *B. anthracis* and the pseudoanthrax bacilli

B. anthracis	*Anthrax-like or pseudoanthrax bacilli*
Nonmotile	Generally motile[a]
Capsulated	Noncapsulated[a]
Grows in long chains	Grows in short chains
No turbidity or pellicle in broth	Often turbidity and pellicle in broth
No growth on penicillin agar (10 μg/ml)	Usually good growth on penicillin agar
Inverted fir-tree growth in gelatin	Fir-tree growth absent or atypical[a]
Methylene blue reduced weakly	Methylene blue usually reduced strongly
Hemolysis of sheep cells weak	Hemolysis of sheep cells often strong
Liquefaction of gelatin slow	Liquefaction of gelatin usually rapid[a]
Lecithinase reaction weakly positive	Lecithinase reaction strongly positive[a]
Ferments salicin slowly	Often ferments salicin rapidly
Polysaccharide precipitin reaction strongly positive	Polysaccharide precipitin reaction weakly positive
Produces toxin, neutralized by *B. anthracis* antitoxin	Any toxic substances produced not neutralized by *B. anthracis* antitoxin
Pathogenic to laboratory animals	Mostly nonpathogenic; if pathogenic, produce disease unlike anthrax
Susceptible to γ phage	Insusceptible to γ phage
Culture filtrates nontoxic to tissue culture cells	Culture filtrates toxic to tissue culture cells[a]

[a] *B. cereus*.
Modified from Wilson (1983).

and pasteurized (heated at 70 to 80°C for approximately 10 min) to kill contaminating vegetative cells. Spores will survive. Air-drying and treatment with ethanol are also effective for selecting out sporulating organisms. Caution is needed, however, as organisms may be present only in the vegetative state and no growth will be achieved. In addition, heat treatment of spores may cause mutations. Treatment with ethanol may therefore be more desirable.

Isolation is relatively easy, as most medically important *Bacillus* species grow well on blood agar or simple nutrient agar, which is composed of a meat extract in peptone with a pH of 6.8 to 7.0. They should be incubated aerobically at 35 to 37°C for 24 h. Some strains require enrichment techniques using specific broth, incubation under specific conditions of pH, temperature, and salt concentration, or plating on specific agar media. Colony morphology and microscopic appearance form the basis of preliminary identification.

Colony morphology should be observed from young (18- to 24-h) cultures grown on nutrient agar under aerobic conditions. The microscopic appearance of cells is best observed on slides freshly coated with a thin layer of 2% water agar and examined with a phase-contrast microscope (Claus and Berkeley, 1986). Colonies suspected of being *B. cereus* should be subcultured and checked for motility, hemolysis, and susceptibility to penicillin. Serotyping with antibody directed against flagellar antigens can be employed.

The common procedure for isolation and presumptive identification of *B. cereus* is based on the egg yolk reaction (Doyle et al., 1985). Material to be cultured is placed onto blood agar containing polymyxin to suppress gram-negative growth. Plates are incubated at 37°C overnight, and suspect colonies are then inoculated on mannitol egg yolk agar by stabbing various sites on the plate and incubated at 35 to 37°C overnight. As *B. cereus* is lecithinase positive and does not ferment mannitol, plates are examined for halos around sites of inoculation and for evidence of fermentation by a change in the background color from violet to yellow. Some strains of *B. cereus* give a weak or negative egg yolk reaction. On PEMBA (polymyxin pyruvate–egg yolk–mannitol–bromothymol blue–agar) media, *B. cereus* has a

TABLE 3. Characteristics useful to differentiate between some species of *Bacillus*

	Characteristic[a]						
Bacillus *species*	*Rods 1.0 μm wide or wider*	*Parasporal bodies*	*Growth in anaerobic agar*	*Growth at 50°C*	*Growth at 65°C*	*Growth in 4% NaCl*	*Growth in 7% NaCl*
anthracis	+	−	+	−	−	+	+
megaterium	+	−	−	−	−	ND	+
cereus	+	0	+	−	−	ND	+
thuringiensis	+	+	+	−	−	+	+
licheniformis	−	−	+	+	−	+	+
subtilis	−	−	−	+	−	+	+
pumilus	−	−	−	+	−	+	+
firmus	−	−	−	−	−	+	+
coagulans	0	−	+	+	−	−	−
polymyxa	−	−	+	−	−	−	−
macerans	−	−	+	+	−	−	−
circulans	−	−	0	+	−	0	0
stearothermophilus	0	−	−	+	+	0	−
alvei	0	−	+	−	−	0	−
laterosporus	−	+	+	+	−	0	−
brevis	−	−	−	+	−	−	−
sphaericus	0	−	−	−	−	0	0
larvae	−	−	+	−	−	−	+
popilliae	−	+	+	−	−	−	+
lentimorbus	−	−	+	−	−	−	−
pasteurii	−	−	+	−	−	+	+
pantothenticus	−	−	+	0	−	+	+

[a] V-P = Voges-Proskauer; + = greater than 85% of strains positive; − = greater than 85% of strains negative; 0 = variable character; ND = no data available.
Modified from Norris (1981) and Wilson (1983).

distinct colonial appearance (Holbrook and Anderson, 1980). After incubation at 37°C for 24 to 48 h, colonies of *B. cereus* are rough, dry, and crenated or rhizoid. They have a distinctive bright blue color. (*B. thuringiensis* has the same appearance.)

The nature of members of this genus is such that strains may not exhibit the important characteristics of the genus, let alone the species. Because of the number of species and variant strains, it may not be possible to speciate quickly. Norris (1981) suggests basing species identification on the results of two morphologic features and 11 tests: width of the rod, parasporal body in the sporangium, catalase production, Voges-Proskauer reaction, pH in Voges-Proskauer broth, growth in anaerobic agar, growth at 50°C, growth at 65°C, growth in 7% sodium chloride, production of acid and gas from glucose, reduction of nitrate to nitrite, hydrolysis of starch, and casein decomposition. Citrate utilization may not be useful because of disagreement in interpretation. Wilson (1983) advocates also testing for gelatin liquefaction, urease production, and growth in 4% sodium chloride. A scheme for identification of *Bacillus* species is provided in Table 3. Details for performance of these tests are given in *Bergey's Manual of Systematic Bacteriology* (Claus and Berkeley, 1986).

Phage typing using bacteriophage γ has been used for typing *B. anthracis, B. thuringiensis, B. sphaericus*, and *B. stearothermophilus*.

Serologic tests (namely, precipitation and complement fixation) have been used to separate *B. anthracis* from other *Bacillus* species, but antisera to *B. anthracis* react to some extent with other bacilli, limiting their usefulness. Agglutination tests have been found to be of little value, as these bacteria have the tendency to agglutinate spontaneously (Sievers and Zetterberg, 1940).

B. anthracis and other *Bacillus* species can be differentiated by lectin agglutination assays (Cole et al., 1984). The agglutination can be carried out on a microscope slide. *B. anthracis* and *B. mycoides* give a positive agglutination with soybean lectin (*Glycine max*). Lectin agglutination with *Helix pomatia* (snail lectin) is positive only for *B. mycoides*. Spores have lectin receptors, and heating at 100°C decreases the agglutination.

A comparison of gas–liquid chromatography, DNA–DNA hybridization, API systems, and classic

Characteristic[a]								
Catalase	*V-P reaction*	*Acid from glucose*	*Acid and gas in glucose*	*NO_3 reduced to NO_2*	*Starch hydrolyzed*	*pH in V-P medium <6.0*	*Hydrolysis of casein*	*Lecithinase*
+	+	+	–	+	+	+	+	+
+	–	+	–	0	+	0	+	–
+	+	+	–	+	+	+	+	+
+	+	+	–	+	+	+	+	+
+	+	+	–	+	+	0	+	–
+	+	+	–	+	+	0	+	–
+	+	+	–	–	–	+	+	–
+	–	+	–	+	+	–	+	–
+	+	+	–	0	+	+	0	–
+	+	+	+	+	+	0	+	–
+	–	+	+	+	+	–	–	–
+	–	+	–	0	+	0	0	–
0	–	+	–	0	+	+	0	–
+	+	+	–	–	+	+	+	–
+	–	+	–	+	–	–	+	+
+	–	+	–	0	–	–	+	–
+	–	–	–	–	–	–	0	–
–	–	+	–	0	–	–	+	ND
–	–	+	–	–	–	–	–	ND
–	–	+	–	–	–	–	–	ND
ND	–	ND	–	+	–	ND	0	–
+	–	+	–	0	+	+	0	–

biochemical tests for speciation of some *Bacillus* species found API systems to be superior (O'Donnell et al., 1980). This impression was confirmed by Logan and Berkeley (1984) in their study of 600 *Bacillus* strains. API systems supplemented with spore examination, motility testing and gas from carbohydrate fermentation were more rapid, accurate, and reproducible than classic biochemical tests.

Treatment

Almost all strains of *Bacillus* species are sensitive to the aminoglycosides, tetracycline, chloramphenicol, erythromycin, vancomycin, and clindamycin (Coonrod et al., 1971; Tuazon et al., 1979). Susceptibility to penicillin, ampicillin, methicillin, and cephalosporins is usually species related and has been shown to be high for *B. subtilis* and low for *B. cereus*. *B. pumilus* is intermediate between *B. subtilis* and *B. cereus*. *B. cereus* is the most resistant species, showing resistance to the penicillins and cephalosporins that may be attributable to its ability to produce a β-lactamase that hydrolyzes these agents (Sabath and Abraham, 1965). For treatment of severe ocular infections, early vitrectomy and intravitreal antibiotics combined with systemic and periocular antibiotics appear to offer the best chance for success (Davey and Tauber, 1987). Clindamycin with or without an aminoglycoside may be the treatment of choice for endocarditis or sepsis with *B. cereus* (Tuazon et al., 1979). Other species can be treated with penicillins if they are found to be susceptible. Surgical drainage may be necessary in closed space and soft tissue infections. Infected prosthetic heart valves may need replacement.

Kurthia Infections

Kurthia are aerobic, nonsporulating, gram-positive rods that have been isolated from fresh and spoiled meat, water, feces, milk, waste water, and air from abattoirs (Wenner and Rettger, 1919). Only two species, *K. zopfii* and *K. gibsonii*, are recognized currently (Keddie and Shaw, 1986). The first and only report of proven infection caused by *Kurthia* was a case of endocarditis in an intravenous drug abuser (Pancoast et al., 1979). The organism caused the destruction of the aortic valve, a fistula from the left ventricular outflow tract to the left atrium, and an abscess of a mitral valve cusp. *Kurthia* species had been isolated previously from the eye, feces, sputum, and a pilonidal cyst, but in none of these cases was there clear association with infection (Elston, 1961). They have also been isolated from the upper intestinal tract of 6 of 25 patients with acute diarrhea and only 1 of 24 control patients, but the significance is unclear (Jarumilinta et al., 1976).

The organism measures approximately 2 to 12 μm in length and 0.8 μm in width. At 37°C it is motile via numerous peritrichate flagella, whereas at 20°C it is nonmotile. *Kurthia* species grow at temperatures ranging from 5 to 44°C, with the best growth occurring at 20 to 30°C (Keddie, 1981). In liquid media they can be filamentous and break up into chains of coccoid bodies. *K. zopfii*, the type species, forms small, gray, rhizoid colonies with a "Medusa head" appearance in agar when viewed under low power.

Members of this genus are catalase positive and oxidase negative. They do not ferment sugars, reduce nitrate, or produce indole or acetoin. Some strains produce hydrogen sulfide weakly (Jones, 1975). A single streak inoculated into a gelatin slope will produce the characteristic featherlike growth after incubation in a vertical position at 22°C for 2 to 3 days. *K. zopfii* does not liquefy gelatin.

There are no selective media for isolation. Methods that utilize their unusual cultural properties (the gelatin streak) can be used for isolation. In yeast nutrient gelatin (YNG), filamentous outgrowths appear in 2 to 3 days. To obtain a pure culture, a piece of the gelatin can be removed and streaked onto yeast extract nutrient agar (YNA) (Keddie, 1981). *Kurthia* may also be isolated by direct plating of clinical material onto YNA. Some *Bacillus* species (*B. cereus* variant *mycoides*) may produce similar filamentous outgrowths in gelatin. Slope cultures on YNA should remain viable for at least 6 months at room temperature. Long-term preservation can be accomplished by freeze-drying.

Identification of *Kurthia* is based on appearance and on the result of biochemical tests. *Kurthia* can be confused with some *Bacillus* species and with *Brochotrix thermosphacta*. *Brochotrix thermosphacta* is similar morphologically, but ferments carbohydrates and is nonmotile. *Bacillus* species are distinguished from *Kurthia* by endospore formation, carbohydrate fermentation, and gelatin liquefaction. *K. zopfii* can be differentiated from *K. gibsonii*. It produces acid from ethanol but not from glycerol, it does not grow at 45°C, and its colonies have no distinctive pigmentation. *K. gibsonii* grows at 45°C, produces acid from glycerol but not from ethanol, and has cream- or yellow-pigmented colonies.

The only experience with treatment of infection caused by *Kurthia* is the case of endocarditis (Pancoast et al., 1979). The organism was susceptible to penicillin, ampicillin, carbenicillin, cephalothin, gentamicin, amikacin, erythromycin, chloramphenicol, and trimethoprim–sulfamethoxazole. It was resistant to oxacillin, clindamycin, and kanamycin. The process was so destructive that the patient required valve replacement.

Rothia Infections

The genus *Rothia* contains a single species, *R. dentocariosa*. It is a pleomorphic, gram-positive, nonsporulating rod with irregular swellings along its body and clubbed ends. It is nonmotile, nonencapsulated, and non-acid-fast. It may have a coccoid, diphtheroid, or filamentous shape or be a mixture of these. Coccoid forms predominate in broth.

R. dentocariosa is part of the normal flora of the mouth and pharynx. It has been isolated from salvia and from dental plaque and is a cause of dental caries and periodontal disease (Barksdale, 1979; Lesher et al., 1977). *Rothia* has been isolated from many sources, including blood, sputum, urine, cerebrospinal fluid, nares, throat, and various types of wounds (Brown et al., 1979). No clinical data are available on these isolates, so it is impossible to know whether *Rothia* was the etiologic agent of infection or a contaminant. It has been isolated in pure culture from a periappendiceal abscess and from a pilonidal abscess along with group F streptococcus (Lutwick and Rockhill, 1978).

Three cases of *Rothia* endocarditis have been reported (Broeren and Peel, 1984; Pape et al., 1979; Schafer et al., 1979). All three patients had preexisting heart murmurs; one had a history of recent dental work (Schafer et al., 1979). In two of the cases, positive blood cultures were not detected by growth visualization of the bottles. Routine blind culture from four nonturbid blood cultures yielded the organism in one instance (Pape et al., 1979); in the other, there was no visible turbidity in broth but the organism was seen on Gram stain (Schafer et al., 1979). In all the cases, the organism grew after a 7- to 8-day incubation period.

A pleomorphic gram-positive rod with many of the characteristics of *Rothia* has been cultured from an excised lymph node of a patient with cat-scratch disease (Gerber et al., 1985). Morphologically similar to the bacterium observed in Warthin-Starry stains of lymph node sections from patients with cat-scratch disease, it has been suggested to be the etiologic agent of cat-scratch disease. Subcutaneous and intraperitoneal injection of the isolated organism into mice produced ulcerating skin lesions that were histologically granulomatous, with microabscesses consistent with the picture of cat-scratch disease. Others have disputed the claim that *Rothia* may be the cause of this disease (Hadfield et al., 1985). The appearance on Gram stain, size, and wall structure of *Rothia* are not characteristic of the cat-scratch disease bacillus found in more than 100 patients. In addition, *Rothia* was tested by indirect fluorescent antibody using sera from patients with known cat-scratch disease, and no reaction was noted.

Rothia is catalase positive. It reduces nitrate and hydrolyzes esculin. It is indole and urease negative. Lactic acid and acetic acid are produced from glucose fermentation, but lactose and mannitol are not fermented.

Rothia grows well on enriched media (brain–heart infusion or trypticase soy broth) and on simpler media such as nutrient broth. The best growth occurs under aerobic conditions at 35 to 37°C in the presence of CO_2. Young colonies (less than 24 h old) grown anaerobically are usually filamentous and resemble *Actinomyces* (Gerencser and Bowden, 1986). Under aerobic conditions young colonies are larger and smooth or granular. Mature colonies (4 to 7 days old) are 2 to 6 mm in diameter and off-white in color. They may be smooth and creamy or rough and wrinkled.

Enriched media are used for primary isolation and maintenance. Cultures should be incubated aerobically at 37°C for 2 to 7 days (Gerencser and Bowden, 1986). Selective media have been described (Beighton and Colman, 1976). Short-term maintenance is accomplished on slants or in deep agar butts of brain–heart infusion or trypticase soy agar kept at room temperature. Under such conditions, cultures remain viable for up to 4 months. Freeze-drying is used for long-term maintenance.

Rothia may appear coccoid or diphtheroidal on Gram stain and can be mistaken for *Streptococcus* or *Corynebacterium*. *Rothia* resembles *Actinomyces* morphologically, but grows better aerobically and differs in the end products of glucose metabolism. It differs from *Nocardia* in its metabolism and cell wall composition. The genus *Rothia* can be distinguished from similar genera serologically using fluorescent antibody. There are few cross-reactions. The major drawback is that not all *Rothia* isolates react with the available antisera. The taxonomy of *Rothia*-like organisms is currently being studied (Gerencser and Bowden, 1986).

The organism is very sensitive to many antibiotics, including penicillins, cephalosporins, erythromycin, tetracycline, and aminoglycosides. Penicillin is the drug of choice in cases of endocarditis (Broeren and Peel, 1984; Pape et al., 1979; Schafer et al., 1979). It is usually combined with an aminoglycoside for the first 2 weeks of treatment. Penicillin- and cephalosporin-resistant strains occasionally occur. A β-lactamase isolated from a penicillin- and cephalexin-resistant strain hydrolyzed cephalosporins more efficiently than penicillin and was encoded by a 47-megadalton plasmid (Minato and Abiko, 1983).

Literature Cited

Agarwala, B., T. Baffes, and A. Grossman. 1975. Bacterial endocarditis following open heart surgery. Illinois Med. J. **148**:38–39.

Allen, B. T., and H. A. Wilkinson III. 1969. A case of meningitis and generalized Schwartzman reaction caused by *Bacillus sphaericus*. Johns Hopkins Med. J. **125:**8–13.

Barksdale, L. 1979. Identifying *Rothia dentocariosa*. Ann. Intern. Med. **91:**786–788.

Beighton, D., and G. Colman. 1976. A medium for the isolation and enumeration of oral *Actinomycetaceae* from dental plaque. J. Dent. Res. **55:**875–878.

Bernheimer, A. W., and P. Grushoff. 1967. Extracellular hemolysins of aerobic sporogenic bacilli. J. Bacteriol. **93:**1541–1543.

Bias, W. J. 1927. A case of pathogenicity of *Bacillus subtilis*. J. Infect. Dis. **41:**313–315.

Block, C. S., M. L. Levy, and V. U. Fritz. 1978. *Bacillus cereus* endocarditis. A case report. S. Afr. Med. J. **53:**556–557.

Bonventre, P. F. 1965. Differential cytotoxicity of *Bacillus anthracis* and *Bacillus cereus* culture filtrates. J. Bacteriol. **90:**284–285.

Boyette, D. P., and F. L. Rights. 1952. Heretofore undescribed aerobic spore forming *Bacillus* in child with meningitis. J. Am. Med. Assoc. **148:**1223–1224.

Bradley, D. E., and J. G. Franklin. 1958. Electron microscope survey of the surface configuration of spores of the genus *Bacillus*. J. Bacteriol. **76:**618–630.

Broeren, S. A., and M. M. Peel. 1984. Endocarditis caused by *Rothia dentocariosa*. J. Clin. Pathol. **37:**1298–1300.

Brown, J. M., L. K. Georg, and L. C. Waters. 1969. Laboratory identification of *Rothia dentocariosa* and its occurrence in human clinical materials. Appl. Microbiol. **17:**150–156.

Burdon, K. L. 1956. Useful criteria for the identification of *Bacillus anthracis* and related species. J. Bacteriol. **71:**25–42.

Burdon, K. L., J. S. Davis, and R. D. Wende. 1967. Experimental infection of mice with *Bacillus cereus:* studies of pathogenesis and pathologic changes. J. Infect. Dis. **117:**307–316.

Burdon, K. L., and R. D. Wende. 1960. On the differentiation of anthrax bacilli from *Bacillus cereus*. J. Infect. Dis. **107:**224–234.

Clarridge, J. E., and A. S. Weissfeld. 1984. Aerobic asporogenous gram-positive bacilli. Clin. Microbiol. Newsl. **6:**115–118.

Claus, D., and R. C. W. Berkeley. 1986. Genus *Bacillus,* p. 1105–1139. *In* P. H. A. Sneath, N. S. Mair, M. E. Sharpe, and J. G. Holt (ed.), Bergey's manual of systematic bacteriology, 1st ed., vol. 2. The William & Wilkins Co., Baltimore.

Cole, H. B., J. W. Ezzell, Jr., K. F. Keller, and R. J. Doyle. 1984. Differentiation of *Bacillus anthracis* and other *Bacillus* species by lectins. J. Clin. Microbiol. **19:**48–53.

Colpin, G. G. D, H. F. L. Guiot, R. F. A. Simonis, and F. E. Zwaan. 1981. *Bacillus cereus* meningitis in a patient under gnotobiotic care. Lancet **2:**694–695.

Coolbaugh, J. C., and R. P. Williams. 1978. Production and characterization of two hemolysins of *Bacillus cereus*. Can. J. Microbiol. **24:**1289–1295.

Coonrod, J. D., P. J. Leadley, and T. C. Eickhoff. 1971a. Antibiotic susceptibility of *Bacillus* species. J. Infect. Dis. **123:**102–105.

Coonrod, J. D., P. J. Leadley, and T. C. Eickhoff. 1971b. *Bacillus cereus* pneumonia and bacteremia. Am. Rev. Respir. Dis. **103:**711–714.

Costlow, R. D. 1958. Lecithinase from *Bacillus anthracis*. J. Bacteriol. **76:**317–325.

Cox, R., G. Sockwell, and B. Landers. 1959. *Bacillus subtilis* septicemia. Report of a case and review of the literature. N. Engl. J. Med. **261:**894–896.

Craig, C. P., W. S. Lee, and M. Ho. 1974. *Bacillus cereus* endocarditis in an addict (letter). Ann. Intern. Med. **80:**418.

Curtis, J. R., A. J. Wing, and J. C. Coleman. 1967. *Bacillus cereus* bacteremia: a complication of intermittent hemodialysis. Lancet **1:**136–138.

Davey, R. T., and W. B. Tauber. 1987. Posttraumatic endophthalmitis: the emerging role of *Bacillus cereus* infection. Rev. Infect. Dis. **9:**110–123.

Doak, B. W., and C. Lamanna. 1948. On the antigenic structure of the bacterial spore. J. Bacteriol. **55:**373–380.

Doyle, R. J., K. F. Keller, and J. W. Ezzell. 1985. *Bacillus,* p. 211–215. *In* E. H. Lennette, A. Balows, W. J. Hausler, and H. J. Shadomy (eds.), Manual of clinical microbiology, 4th ed. American Society for Microbiology, Washington, D.C.

Dubos, R. 1971. Toxic factors in enzymes used in laundry products. Science **173:**259–260.

Elston, H. R. 1961. *Kurthia bessonii* isolated from clinical material. J. Pathol. Bacteriol. **81:**245–247.

Farrar, W. E., Jr. 1963. Serious infections due to "nonpathogenic" organisms of the genus *Bacillus*. Review of their status as pathogens. Am. J. Med. **34:**134–141.

Finland, M., and M. W. Barnes. 1970. Changing etiology of bacterial endocarditis in the antibacterial era. Experiences at Boston City Hospital 1933–1965. Ann. Intern. Med. **72:**341–348.

Flindt, M. L. H. 1969. Pulmonary disease due to inhalation of derivatives of *Bacillus subtilis* containing proteolytic enzyme. Lancet **1:**1177–1181.

Gerber, M. A., A. K. Sedgwick, T. J. MacAlister, K. B. Gustafson, M. Ballow, and R. C. Tilton. 1985. The aetiological agent of cat scratch disease. Lancet **1:**1236–1239.

Gerencser, M. A., and G. H. Bowden. 1986. Genus *Rothia,* p. 1342–1346. *In* P. H. A. Sneath, N. S. Mair, M. E. Sharpe, and J. G. Holt (ed.), Bergey's manual of systematic bacteriology, 1st ed., vol. 2. The Williams & Wilkins Co., Baltimore.

Gibson, T., and R. E. Gordon. 1974. *Bacillus,* p. 529–550. *In* R. E. Buchanan and N. E. Gibbons (ed.), Bergey's manual of determinative bacteriology, 8th ed. The Williams and Wilkins Co., Baltimore.

Goldstein, B., and E. Abrutyn. 1985. Pseudo-outbreak of *Bacillus* species: related to fibreoptic bronchoscopy. J. Hosp. Infect. **6:**194–200.

Greenberg, M., J. F. Milne, and A. Watt. 1970. Survey of workers exposed to dusts containing derivatives of *Bacillus subtilis*. Br. Med. J. **2:**629–633.

Gregersen, T. 1978. Rapid method for distinction of gram-negative from gram-positive bacteria. Eur. J. Appl. Microbiol. Biotechnol. **5:**123–127.

Guiot, H. F. L., M. M. De Planque, D. J. Richel, and J. W. van't Wout. 1986. *Bacillus cereus*: a snake in the grass for granulocytopenic patients. J. Infect. Dis. **153:**1186.

Hadfield, T. L., D. J. Wear, A. M. Margileth, and C. K. English. 1985. Is *Rothia dentocariosa* the cause of cat scratch disease? Lancet **2:**720–721.

Holbrook, R., and J. M. Anderson. 1980. An improved selective and diagnostic medium for the isolation and enumeration of *Bacillus cereus* in foods. Can. J. Microbiol. **26:**753–759.

Ihde, C., and D. Armstrong. 1973. Clinical spectrum of infection due to *Bacillus* species. Am. J. Med. **55:**839–845.

Jarumilinta, R., M. Miranda, and V. M. Villarejos. 1976. A bacteriological study of the intestinal mucosa and luminal fluid of adults with acute diarrhoea. Ann. Trop. Med. Parasitol. **70:**165–179.

Johnson, C. L., I. L. Bernstein, J. S. Gallagher, P. F. Bonventre, and S. M. Brooks. 1980. Familial hypersensitivity pneumonitis induced by *Bacillus subtilis*. Am. Rev. Respir. Dis. **122:**339–348.

Jones, D. 1975. A numerical taxonomic study of coryneform and related bacteria. J. Gen. Microbiol. **87:**52–96.

Kaneko, T., R. Nozaki, and K. Aizawa. 1978. Deoxyribonucleic acid relatedness between *Bacillus anthracis, Bacillus cereus* and *Bacillus thuringiensis*. Microbiol. Immunol. **22:**639–641.

Keddie, R. M. 1981. The genus *Kurthia,* p. 1888–1893. *In* M. Starr, N. Stolp, H. Truper, A. Balows, and H. Schlegel (ed.), The prokaryotes. A handbook on habitats, isolation, and identification of bacteria. Springer-Verlag, New York.

Keddie, R. M., and S. Shaw. 1986. Genus *Kurthia,* p. 1255–1258. *In* P. H. A. Sneath, N. S. Mair, M. E. Sharpe, and J. G. Holt (ed.). Bergey's manual of systematic bacteriology. The Williams & Wilkins Co., Baltimore.

Lamanna, C. 1940. The taxonomy of the genus *Bacillus*. J. Bacteriol. **40:**347–360.

Lesher, R. J., V. F. Gerencser, and D. J. Morrison. 1977. Presence of *Rothia dentocariosa* strain 477 serotype 2 in gingiva of patients with inflammatory periodontal disease. J. Dent. Res. **56:**189.

Logan, N. A., and R. C. W. Berkeley. 1984. Identification of *Bacillus* strains using the API system. J. Gen. Microbiol. **130:**1871–1882.

Logan, N. A., D. C. Old, and H. M. Dick. 1985. Isolation of *Bacillus circulans* from a wound infection. J. Clin. Pathol. **38:**838–839.

Lutwick, L. I., and R. C. Rockhill. 1978. Abscess associated with *Rothia dentocariosa*. J. Clin. Microbiol. **8:**612–613.

Meynell, E., and G. G. Meynell. 1964. The roles of serum and carbon dioxide in capsule formation by *Bacillus anthracis*. J. Gen. Microbiol. **34:**153–164.

Minato, K., and Y. Abiko. 1984. β-Lactam antibiotics resistant *Rothia dentocariosa* from infected postoperative maxillary cyst: studies on R-plasmid and β-lactamase. Gen. Pharmacol. **15:**287–292.

Norris, J. R., R. C. W. Berkeley, N. A. Logan, and A. G. O'Donnell. 1981. The genera *Bacillus* and *Sporolactobacillus,* p. 1711–1742. *In* M. Starr, N. Stolp, H. Truper, A. Balows, and H. Schlegel (ed.), The prokaryotes. A handbook on habitats, isolation, and identification, vol. 2. Springer-Verlag, New York.

O'Day, D. M., R. S. Smith, C. R. Gregg, P. C. B. Turnbull, W. S. Head, J. A. Ives, and P. C. Ho. 1981. The problem of *Bacillus* species infection with special emphasis on the virulence of *Bacillus cereus*. Ophthalmology (Rochester) **88:**833–838.

O'Donnell, A. G., J. R. Norris, R. C. W. Berkeley, D. Claus, T. Kaneko, N. A. Logan, and R. Nozaki. 1980. Characterization of *Bacillus subtilis, Bacillus pumilus, Bacillus licheniformis,* and *Bacillus amyloliquefaciens* by pyrolysis gas-liquid chromatography, deoxyribonucleic acid–deoxyribonucleic acid hybridization, biochemical tests and API systems. Int. J. Syst. Bacteriol. **30:**448–459.

Oster, H. A., and T. Q. Kong. 1982. *Bacillus cereus* endocarditis involving a prosthetic valve. South. Med. J. **75:**508–509.

Pancoast, S. J., P. D. Ellner, J. A. Jahre, and H. C. Neu. 1979. Endocarditis due to *Kurthia bessonii*. 1979. Ann. Intern. Med. **90:**936–937.

Pape, J., C. Singer, T. E. Kiehn, B. J. Lee, and D. Armstrong. 1979. Infective endocarditis caused by *Rothia dentocariosa*. Ann. Intern. Med. **91:**746–748.

Pearson, H. E. 1970. Human infections caused by organisms of the *Bacillus* species. Am. J. Clin. Pathol. **53:**506–515.

Pennington, J. E., N. D. Gibbons, J. E. Strobeck, G. L. Simpson, and R. L. Myerowitz. 1976. Bacillus species infection in patients with hematologic neoplasia. J. Am. Med. Assoc. **235:**1473–1474.

Pepys, J., F. E. Hargreave, J. L. Longbottom, and J. Faux. 1969. Allergic reactions of the lungs to enzymes of *Bacillus subtilis*. Lancet **1:**1181–1184.

Reller, L. B. 1973. Endocarditis caused by *Bacillus subtilis*. Am. J. Clin. Pathol. **60:**714–718.

Roberts, J. L. 1935. A new species of the genus *Bacillus* exhibiting mobile colonies on the surface of nutrient agar. J. Bacteriol. **29:**229–236.

Sabath, L. D., and E. P. Abraham. 1965. Cephalosporinase and penicillinase activity of *Bacillus cereus*. Antimicrob. Agents Chemother. p. 392–397.

Samples, J. R., and H. Buettner. 1983. Ocular infection caused by a biological insecticide. J. Infect. Dis. **148:**614.

Sathmary, M. N. 1958. *Bacillus subtilis* septicemia and generalized aspergillosis in a patient with acute myeloblastic leukemia. N.Y. State J. Med. **58:**1870–1876.

Schafer, F. J., E. J. Wing, and C. W. Norden. 1979. Infectious endocarditis caused by *Rothia dentocariosa*. Ann. Intern. Med. **91:**747–748.

Seki, T., C. K. Chung, H. Mikami, and Y. Oshima. 1978. Deoxyribonucleic acid homology and taxonomy of the genus *Bacillus*. Int. J. Syst. Bacteriol. **28:**182–189.

Sievers, O., and B. Zetterberg. 1940. A preliminary investigation into the antigenic characters of spore-forming, aerobic bacteria. *J. Bacteriol.* **40:**45–56.

Stein, M. W., and G. F. Logan. 1965. Characterization of the phospholipases of *Bacillus cereus* and their effects on erythrocytes, bone and kidney cells. J. Bacteriol. **90:**69–81.

Solny, M. N., G. R. Failing, and J. S. Borges. 1977. *Bacillus cereus* osteomyelitis (letter). Arch. Intern. Med. **137:**401.

Sugar, A. M., and R. V. McCloskey. 1977. *Bacillus licheniformis* sepsis. J. Am. Med. Assoc. **238:**1180–1181.

Taylor, A. J., and R. J. Gilbert. 1975. *Bacillus cereus* food poisoning: a provisional serotyping scheme. J. Med. Microbiol. **8:**543–550.

Tomcsik, J., and J. B. Baumann-Grace. 1959. Specific exosporium reaction of *Bacillus megaterium*. J. Gen. Microbiol. **21:**666–675.

Tuazon, C. U., R. Hill, and J. N. Sheagren. 1974. Microbiologic study of street heroin and injection paraphernalia. J. Infect. Dis. **129:**327–329.

Tuazon, C. U., H. W. Murray, C. Levy, M. N. Solny, J. A. Curtin, and J. N. Sheagren. 1979. Serious infections from *Bacillus* species. J. Am. Med. Assoc. **241:**1137–1140.

Turnbull, P. C. B., K. Jorgensen, J. M. Kramer, R. J. Gilbert, and J. M. Parry. 1979. Severe clinical conditions associated with *Bacillus cereus* and the apparent involvement of exotoxins. J. Clin. Pathol. **32:**289–293.

Weinstein, L., and C. G. Colburn. 1950. *Bacillus subtilis* meningitis and bacteremia. Arch. Intern. Med. **86:**585–594.

Weller, P. F., A. Nicholson, and N. Braslow. 1979. The spectrum of *Bacillus* bacteremias in heroin addicts. Arch. Intern. Med. **139:**293–294.

Wenner, J. J., and L. F. Rettger. 1919. A systematic study of the *Proteus* group of bacteria. J. Bacteriol. **4:**331–353.

Wilson, G. 1983. *Bacillus*: the aerobic spore-bearing bacilli, p. 422–441. *In* G. Wilson, A. Miles, and M. T. Parker (ed.), Topley and Wilson's principles of bacteriology, virology and immunity. The Williams & Wilkins Co., Baltimore.

Yeh, T. J., I. N. Anabtawi, V. E. Cornett, A. White, W. H. Stern, and R. G. Ellison. 1967. Bacterial endocarditis following open-heart surgery. Ann. Thorac. Surg. **3**:29–36.

Young, E. J., R. J. Wallace, Jr., C. D. Ericsson, R. A. Harris, and J. Clarridge. 1980. Panophthalmitis due to *Bacillus cereus*. Arch. Intern. Med. **140**:559–560.

Young, R. F., R. N. Yoshimori, D. L. Murray, and P. J. Chou. 1982. Postoperative neurosurgical infections due to *Bacillus* species. Surg. Neurol. **18**:271–273.

Yow, M. D., J. B. Reinhart, and L. J. Butler. 1949. *Bacillus subtilis* septicemia treated with penicillin. J. Pediatr. **35**:237–239.

CHAPTER 8

Bacillus cereus Food Poisoning

R. W. BENNETT and S. M. HARMON

Disease: *Bacillus cereus* food poisoning.
Etiologic Agents: *Bacillus cereus* and possibly *Bacillus thuringiensis* bacteria or toxic substances.
Source: Soil; widely distributed in the environment; a frequent contaminant in many foods.
Clinical Manifestations: Acute abdominal cramps and diarrhea (diarrheal type), which occur within 4 to 16 h, or nausea and vomiting (emetic type), which occur within 1 to 5 h after consumption of contaminated foods. Some patients experience symptoms of both types.
Laboratory Diagnosis: Quantitative bacteriologic examination of implicated foods to demonstrate presence of *B. cereus* strains with enterotoxigenic activity; serologic or biological assays to confirm enterotoxigenic activity of isolates from food, vomitus, or feces.
Epidemiology: Associated with consumption of foods containing excessive populations of *B. cereus* organisms or preformed material with enterotoxic activity; symptoms are similar to intoxication caused by enterotoxins of other bacteria such as *Clostridium perfringens* and *Staphylococcus aureus*.
Treatment: Symptomatic; no specific treatment is presently available.
Prevention and Control: Limiting preventable contamination by good food handling practices and proper holding of prepared foods (especially adequate refrigeration) to minimize proliferation of the organism.

Description of Disease

Foodborne Intoxications

Although *Bacillus cereus* was long considered to be a harmless saprophite, its role as a food poisoning organism has been known since the 1950s (Goepfert et al., 1972; Hauge, 1955). Consumption of foods containing millions of *B. cereus* cells per gram frequently results in food poisoning. Two types of illness have been attributed to consumption of foods contaminated with *B. cereus*. The first and best known type is characterized by abdominal pain and diarrhea, not unlike the symptoms produced by consumption of foods containing large numbers of viable enterotoxigenic cells of *Clostridium perfringens* (Hobbs et al., 1953; McClung, 1945). The diarrheal illness caused by *B. cereus* has an incubation period of 4 to 16 h; symptoms usually last for 12 to 24 h.

The second type of illness is characterized by an acute attack of vomiting, which commences 1 to 5 h after a meal. Diarrhea is not a common feature, but it is known to occur (Melling et al., 1976). Most investigators consider *B. cereus* food poisoning to be a true intoxication rather than a foodborne infection.

Diarrheal Activity

The diarrheal strains produce an extracellular protein that elicits diarrhea in monkeys and can be detected in culture fluids by using serological assays (Bennett, 1985). Although it has not been completely characterized, this toxin is reported to be a protein with a molecular weight of approximately 55,000 to 60,000 (Ezepcuk and Fluer, 1973). For use in diagnostic tests, specific antibodies can be raised against the protein in rabbits. The diarrheal component is heat-labile and is inactivated by heating for 5 min at 56°C. Recent studies have shown that it is also produced by some strains of *Bacillus thuringiensis* (Bennett and Harmon, 1986).

Emetic Activity

Less is known about the emetic component than the diarrheal entity, but apparently it is not produced by all strains of *B. cereus*. When fed to monkeys with rice, large populations of strains implicated in outbreaks with symptoms of vomiting elicited a vomiting response (Melling et al., 1976). The emetic activity appears to be associated with a peptide of low molecular weight, which is extremely resistant to heat (withstanding 120°C for more than 1 h) and is produced as a cell-free component in broth cultures. Besides the emetic response in monkeys, heat-treated culture fluids of emetic strains injected intravenously induced vomiting in cats (Bennett, unpublished observations). A similar response in cats is produced by injection of heated culture fluid of certain *Bacillus subtilis* and *Bacillus licheniformis* cultures (Bennett, 1987) that have been implicated epidemiologically in food poisoning (Gilbert et al., 1981; Kramer et al., 1982). It is not known whether the toxin (or toxins) produced by these species is the same as that produced by emetic *B. cereus* strains.

Diagnostic Features

Traditionally, examination of implicated foods for the presence of large numbers (10^6 per g) of *B. cereus* has indicated involvement of the organism in foodborne disease. The culture techniques are simple and can be performed in most food hygiene laboratories; however, until recently there were no reliable means of demonstrating the enterotoxigenicity of isolates. A variety of biological assays, including the rabbit ligated ileal loop (Spira and Goepfert, 1972) and the vascular permeability skin test (Glatz and Goepfert, 1973), have been used by different investigators to imply the enterotoxigenicity of strains, but correlation of these results with diarrheagenic activity in monkeys has not been demonstrated. At present, there are two reliable assays for these foodborne *B. cereus* metabolites: the microslide gel diffusion test for the diarrheal metabolite (Bennett, 1985) and the kitten test for demonstrating emetic activity of culture fluids heated for 1.5 h at 100°C (Bennett, unpublished observations). The monkey feeding test is a possible but impractical alternative (Goepfert, 1974; Melling et al., 1976).

Epidemiologic Aspects

Foods implicated in past outbreaks include vanilla puddings, cooked meat and vegetable dishes, boiled and fried rice, certain dairy products, and vegetable sprouts. In most instances, the causative strain was found to be present in large numbers in the implicated food. One or sometimes both categories of symptoms (i.e., vomiting and diarrhea) are experienced by the victims. Because symptoms of these intoxications mimic those caused by enterotoxins of other bacteria such as *C. perfringens* (diarrheal type) and *Staphylococcus aureus* (emetic type), laboratory confirmation is essential before a definite diagnosis can be made.

Etiologic Agent

Taxonomic Position of *B. cereus*

Within the genus *Bacillus*, a great diversity of species and strains exists. The work of Gordon (1973), Gordon et al. (1973), and Smith et al. (1946) has brought a measure of order to the classification of bacilli and greatly facilitated the identification of the more common species. *B. cereus* is classified as a large-celled member of Group 1 (i.e., those species that have a cell width greater than 0.9 μm and whose spores do not appreciably swell the sporangium). There has been considerable argument about whether *Bacillus anthracis*, *B. thuringiensis*, and *Bacillus mycoides* should be accorded species status or considered as varieties of the parent species *B. cereus*. Certainly, they are closely related and, for all practical purposes, the other members of this group differ from *B. cereus* by only a single characteristic, namely, 1) the pathogenicity of *B. anthracis* for animals, 2) the production of endotoxin crystals by *B. thuringiensis*, and 3) rhizoid growth by *B. mycoides*. Loss of the specific property of each variety has been reported. Thus, absolute separation of this group into separate species may not be possible in all instances. Nevertheless, the typical characteristics of *B. cereus* strains appear to be quite stable, and the other biotypes can be readily distinguished from them when the variant properties are evident (Harmon, 1982). Therefore, for the purposes of this chapter, each will be considered as a separate species (except the rhizoid strains, which have been accepted as merely a variant of *B. cereus*). In addition, procedures will be described for differentiating the other biotypes from typical *B. cereus* strains.

Characteristics of *B. cereus* and Culturally Similar Species

The most salient features of the four biotypes of the *B. cereus* group are summarized in Table 1. Some caution is necessary in applying and interpreting tests for identification purposes, since even within the species, variability and strain heterogeneity are relatively common. Some tests are more valuable or more easily performed than others and, for that reason, are advocated for identification of *B. cereus*. The term egg yolk reaction is used to describe the turbidity developed in egg yolk or agar that contains egg yolk. The responsible agent is an extracellular substance (or substances) referred to as egg yolk turbidity factor, lecithinase, or phospholipase.

Although it has been established that *B. cereus* produces phospholipase C, there is some evidence that the turbidity developed in egg yolk may be due to a more complex series of events than the action of a single enzyme (Kushner, 1957). Quantitation of *B. cereus* in a given specimen is usually performed by a simple surface-plating technique. In the United States and many parts of Europe, use is made of 1) the ability of *B. cereus* organisms to produce turbidity surrounding colonies growing on agar containing egg yolk, and 2) the resistance of *B. cereus* to the antibiotic polymyxin B to create a selective and differential plating medium. Four such media have been described and, in three of them, mannitol is incorporated to enhance differentiation (Holbrook and Anderson, 1980; Kim and Goepfert, 1971; Mossell et al., 1967; Szabo et al. 1984). In the United Kingdom, it is not uncommon for blood agar to be used as the primary plating medium. Some use of blood agar has also been made elsewhere (Harmon et al., 1984; Schiemann, 1978). Colonies of appropriate morphology that show the characteristic alteration of the surrounding medium are considered to be presumptive *B. cereus*. Confirmatory tests include microscopic examination of sporulating cultures and a variety of other determinations (as described later in this chapter).

Specimen Collection, Preparation, and Dilution

Specimens should be collected using aseptic techniques, transported to the laboratory without freezing, and kept at or near 5°C until examined. Insulated containers and gel-type refrigerants are suitable for this purpose. Specimens should be stored refrigerated for up to 48 h, but frozen for storage for longer periods to discourage overgrowth by psychrophiles. Data are not available to suggest that refrigeration or freezing will cause an unacceptable decrease in viable *B. cereus* cells. Homogenization and dilution of samples are essentially the same as recommended for standard plate counts. A 25-g portion of food is blended with 225 ml of Butterfield's buffer for 2 min in a Waring blender and diluted appropriately before testing.

TABLE 1. Characteristics of *Bacillus cereus* and culturally similar species

Feature	B. cereus	B. cereus *var.* mycoides	B. thuringiensis	B. anthracis
Egg yolk reaction	+[a]	+	+	(+)[b]
Acid from mannitol	–	–	–	–
Catalase	+	+	+	+
Gram reaction	+	+	+	+
Motility	±[c]	–[d]	±	–
Hemolysis (sheep RBC)	+	(+)	(+)	–
Rhizoid growth	–	+	–	–
Toxin crystals produced	–	–	+	–
Anaerobic utilization of glucose	+	+	+	+
Reduction of nitrate	+	+	+	+
VP reaction	+	+	+	+
Tyrosine decomposition	+	(+)	+	(+)
Resistance to lysozyme	+	+	+	+

[a] + = 90 to 100% of strains are positive.
[b] (+) = usually weakly positive.
[c] ± = 50 to 90% of strains are positive.
[d] – = Less than 10% of strains are positive.
Reprinted from Compendium of Methods for the Microbiological Examination of Foods, Marvin L. Speck, ed., APHA, Washington, D.C., p. 459, with permission of the American Public Health Association, Washington, D.C.

Methods of Enumeration

A surface-plating technique is usually employed to determine viable *B. cereus* populations in outbreak foods. The mannitol-egg yolk-polymyxin (MYP) medium specified for use in the official method of the Association of Official Analytical Chemists (AOAC) for enumerating *B. cereus* in foods (Lancette and Harmon, 1980) is preferred. Other plating media of similar efficacy such as polymyxin-pyruvate-egg yolk-mannitol-bromothymol blue agar (PEMBA) (Holbrook and Anderson, 1980), KG agar (Kim and Goepfert, 1971), or blood agar (Schiemann, 1978) are used in some laboratories. In those instances where *B. cereus* counts are expected to be less than 1,000 per g, the alternative AOAC most probable number (MPN) method using tryptic-soy-polymyxin broth is recommended (Lancette and Harmon, 1980).

Isolation and Identification Procedures

Presumptive colonies from plating media should be transferred to nutrient agar slants and tested to confirm their identity. The differential and confirmatory tests specified in the official AOAC methods (Lancette and Harmon 1980; Harmon 1982) are preferred. Lecithinase-positive, mannitol-negative colonies on MYP agar are highly indicative of *B. cereus;* however, because of the similarity of reactions by other members of the *B. cereus* group, including *B. anthracis, B. thuringiensis,* and rhizoid strains classified as *B. cereus* var. *mycoides,* isolates must be tested to differentiate them from other *Bacillus* species before an identification can be made. The method of Harmon (1982) seems preferable for this because it is relatively simple and will markedly increase confidence in the classification of strains.

The distinguishing characters are: 1) lack of motility—weak or no hemolysis on tryptic-soy-sheep blood agar = *B. anthracis*; 2) presence of stainable toxin crystals in cultures grown on nutrient agar = *B. thuringiensis*; 3) rhizoid growth on nutrient agar plates = *B. cereus* var. *mycoides*; and 4) motile, hemolytic nonrhizoid strains without delta-endotoxin crystals = *B. cereus*. The biochemical tests specified for use by Lancette and Harmon (1980) are also useful for confirming isolates as members of the *B. cereus* group. These are ability to: 1) grow anaerobically and produce acid from glucose; 2) reduce nitrate to nitrite; 3) produce acetylmethylcarbinol [Voges-Proskauer (VP) test]; 4) decompose L-tyrosine; and 5) grow in the presence of 0.001% lysozyme. Some details of the differential and biochemical tests are presented later in this chapter.

Differentiating Members of the *B. cereus* Group

Motility Tests

Semisolid *B. cereus* (BC) motility medium (Harmon, 1982) or direct microscopic examination may be used to determine motility. The BC motility medium is inoculated by stabbing down the center with a 3-mm loopful of culture, and then incubated for 18 to 24 h at 30°C. Motile strains produce diffuse growth out into the medium away from the stab. Nonmotile strains (except *B. cereus* var. *mycoides*) grow only in and along the stab. Rhizoid strains of *B. cereus* var. *mycoides* usually produce characteristic fuzzy growth in semisolid media because of expansion of the filamentous growth, but are not motile by means of flagella. Doubtful results should be confirmed by the alternative microscopic motility test, which is performed by adding 0.2 ml of sterile distilled water to a nutrient agar slant and then inoculating the slant with a 3-mm loopful of culture. After incubation at 30°C for 6 to 8 h, a loopful of liquid culture from the base of the slant is suspended in a drop of water on a clean slide, covered with a cover slip, and examined immediately for motility. *B. cereus* and *B. thuringiensis* are usually actively motile, whereas *B. anthracis* and the typically rhizoid strains of *B. cereus* var. *mycoides* are nonmotile.

Rhizoid Growth

To test for rhizoid growth, a predried nutrient agar plate is inoculated near the center with a 2-mm loopful of culture and incubated for 24 to 48 h at 30°C. If the culture is rhizoid, root or hairlike structures will develop up to several centimeters from the point of inoculation. This property is characteristic only of strains that are classified as *B. cereus* var. *mycoides*.

Hemolytic Activity

To test for hemolytic activity, the bottom of a standard tryptic-soy-sheep blood agar plate is marked into 6 to 8 equal segments. Each segment is labeled and inoculated near its center by gently touching the agar surface with a 2-mm loopful of culture. The plate is then incubated at 30 to 32°C for 24 h and checked for hemolytic activity as indicated by a zone of complete hemolysis surrounding the growth. *B. cereus* is usually strongly hemolytic, whereas *B. thuringiensis* and *B. cereus* var. *mycoides* are often weakly hemolytic or produce hemolysis only under the growth. *B. anthracis* is usually nonhemolytic. One must take caution, however, because nonmotile, nonhemolytic cultures could be *B. anthracis* and should be handled with special care.

Detection of Toxin Crystals

The endotoxin crystals of *B. thuringiensis* may be detected by phase-contrast microscopy, but are probably most conveniently detected by staining. To perform this test, a nutrient agar slant is inoculated with a loopful of culture and incubated at 30°C for 24 h, followed by holding at room temperature for 2 to 3 days to permit sporangiolysis. A smear is made on a clean slide, air-dried, and lightly heat-fixed. The smear is further fixed by flooding the slide with methanol. After 30 s, the methanol is poured off and the slide is dried thoroughly by passing through a flame. The smear is stained by flooding the slide with a 0.5% aqueous basic fuchsin or TB carbol fuchsin ZN (Difco Laboratories, Detroit, MI) and gently heating it from below until steam is seen. After 1 to 2 min, the slide is heated again until steam is seen, held for 30 s, and the stain is poured off. The slide is rinsed thoroughly in tap water, dried without blotting, and examined microscopically under oil immersion for the presence of free spores and darkly stained tetragonal toxin crystals. Free toxin crystals are usually abundant within 3 days, but will not be detectable by staining until the sporangia have lysed. *B. thuringiensis* produces endotoxin crystals that usually can be detected by staining, but are not produced by other members of the *B. cereus* group.

Interpreting Results

On the basis of test results, isolates that are actively motile, strongly hemolytic, and do not produce rhizoid growth or endotoxin crystals should be tentatively identified as *B. cereus*. Nonmotile or weakly hemolytic strains of *B. cereus* that may occasionally be encountered can be differentiated from *B. anthracis* by their resistance to penicillin and gamma-bacteriophage (Leise et al., 1959). Acrystalliferous variants of *B. thuringiensis* and nonrhizoid strains derived from *B. cereus* var. *mycoides* cannot be differentiated from *B. cereus* by the tests described.

Biochemical Tests

In some instances, greater certainty in the identification of isolates may be desired. The following tests confirm the most salient characteristics of the *B. cereus* group and virtually eliminate the possibility of confusing strains of this group with any other *Bacillus* species.

Reactions on Mannitol-Egg Yolk-Polymyxin (MYP) Agar

This test can usually be omitted if the isolates are picked from MYP agar, and the reactions of all isolates are typical. However, the test should be included with isolates from PEMBA, KG, or blood agar to check for mannitol fermentation. The bottom of each plate should be marked into 6 or 8 equal segments. Each segment is labeled and inoculated near its center by touching the agar with a 2-mm loopful of culture. The plate is incubated at 30 to 35°C for 24 h and checked for lecithinase production as indicated by a zone of precipitation surrounding the growth. Mannitol is not fermented if the growth and surrounding medium are eosin pink. *B. cereus* and other members of the *B. cereus* group usually produce lecithinase, but do not ferment mannitol.

Anaerobic Glucose Fermentation

The ability of isolates to metabolize glucose anaerobically is determined by inoculating phenol red dextrose broth with a loopful of culture and incubating the tube in an anaerobic jar. A color change from red to yellow indicates that acid was produced. Acid is produced from glucose anaerobically by *B. cereus* and other members of the *B. cereus* group.

Nitrate Reduction

Nitrates are usually reduced to nitrite by *B. cereus* and other members of the *B. cereus* group. Nitrate broth is inoculated and incubated at 35°C for 24 h. Nitrite is detected by adding 0.25 ml each of sulfanilic acid and alpha-naphthol solutions. An orange color that develops within 10 min indicates the presence of nitrite.

VP Test

Acetylmethylcarbinol is produced from glucose by *B. cereus* and other members of the *B. cereus* group. Modified VP medium is inoculated and incubated at 35°C for 48 h. The presence of acetylmethylcarbinol is determined by adding 0.2 ml of 40% potassium hydroxide and 0.6 ml of 5% alcoholic alpha-naphthol solution to 1 ml of culture in a test tube. Addition of a few crystals of creatine speeds the reaction.

Tyrosine Decomposition

The ability of *B. cereus* and culturally similar species to decompose tryosine is determined by inoculating a slant of nutrient agar containing 0.5% tyrosine and incubating for 48 to 72 h at 35°C. Decomposition of tyrosine is indicated by clearing of the medium immediately under the growth due to dissolution of tyrosine crystals. This progresses to a depth of 3 to 4 mm after 3 days of incubation. *B. cereus* and other members of the *B. cereus* group (except *B. anthracis*) readily decompose tyrosine.

Lysozyme Resistance

The ability of isolates to grow in the presence of lysozyme is determined by inoculating nutrient broth containing a final concentration of 0.001% lysozyme. Growth that develops during 48 h of incubation at 35°C indicates that the organism is resistant to lysozyme. *B. cereus* as well as other members of the *B. cereus* group are resistant to lysozyme.

Production and Identification of the Diarrheal Entity

Preparation of Materials and Media

BHIG, 0.1%

Dissolve by stirring and adjust brain heart infusion (BHI) broth containing 0.1% glucose to pH 7.4. Distribute medium in 30-ml portions in 125-ml flasks and autoclave at 121°C for 10 min.

No. 1 McFarland Standard

Prepare turbidity standard No. 1 of McFarland nephelometer scale. Mix 1 part 1% $BaCl_2$ with 99 parts 1% H_2SO_4 in distilled water.

1.5% Gel Diffusion Agar for Gel Diffusion Slides

Prepare fluid base for agar in distilled water as follows: sodium chloride 0.85%; sodium barbital 0.88%; merthiolate (crystalline) 1 : 10,000, available from Eli Lilly and Co., Terre Haute, Indiana. Adjust the pH to 7.4. Prepare agar by adding 1.5% Noble special agar (Difco 0142-01). Melt agar mixture in Arnold sterilizer (steamer) and filter while hot, in a steamer, through two layers of analytical grade filter paper; dispense in small (15 to 25 ml) portions in 4 oz prescription bottles. (Remelting more than twice may break down purified agar.)

Thiazine Red R Stain

Prepare 0.1% solution of thiazine red R stain in 1.0% acetic acid. One stain source is Matheson, Coleman and Bell, Norwood, Ohio; color index No. 14780. A Wheaton horizontal staining dish with removable slide rack or a Coplin jar may be used for staining the microslides.

Preparation of Slides

Wrap double layers of electrician's plastic insulating tape (¾-in. wide, Scotch Brand Vinyl plastic No. 33, 3M Co. or comparable) around both sides of glass slide, leaving a 2.0-cm space in the center. Apply tape as follows: Start a piece of tape about 9.5 to 10-cm long approximately 0.5 cm from the edge of the under surface of the slide and wrap tightly around the slide twice. Wipe the area between tapes with cheesecloth soaked with 95% ethanol, and then with dry cheesecloth. Coat the upper surface area between tapes with 0.2% agar (bacteriological grade) in distilled water as follows: Melt 0.2% bacteriologic grade agar and maintain at 55°C or higher in screw-capped bottle. Hold the slide over a beaker placed on a hot plate adjusted to 65 to 85°C and pour or brush 0.2% agar over the slide between two pieces of tape. Let any excess agar drain into the beaker. Return the agar collected in the beaker to the original container for reuse. Wipe the under surface of the slide. Place the slide on a tray and dry in a dust-free atmosphere (e.g., an incubator). Note that if a slide is not clean, agar will roll off the slide without coating it uniformly.

Preparation of Slide Assembly

Prepare plastic templates as described by Bennett and McClure (1975) (see Fig. 1 for specifications). Spread a thin film of silicone grease (Dow Corning Co., Midland, Michigan) on the side of the template that will be placed next to agar (i.e., the side with the smaller holes). Place approximately 0.4 ml of melted and cooled (55 to 60°C) 1.2% gel diffusion agar between tapes. Lay the silicone-coated template immediately on the melted agar and edges of bordering tapes. Place one edge of the template on one of the tapes and bring the opposite edge to rest gently on the other tape. Place the slide in a prepared Petri dish (see the following discussion) soon after the agar solidifies, and label the slide with a number, date, or other information. The templates can be obtained from the Division of Microbiology, U.S. Food and Drug Administration, Washington, D.C. 20204.

Preparation of Petri Dishes for Slide Assemblies

Maintain the necessary high humidity by saturating two strips of synthetic sponge (about ½-in. wide × ½-in. deep, 2½-in. long) with distilled water and placing them in 20 × 150-mm Petri dishes. From 2 to 4 slide assemblies can be placed in each dish.

Capillary Pipettes

Pasteur capillary pipettes should be prepared by drawing out approximately 7 mm (diameter) of glass tubing in a flame; however, 1-ml tuberculin syringes or 25-μl precision pipettes (Medical Laboratory Automation, Inc., Mt. Vernon, New York, or equivalent) may be used for dispensing the reagents and materials (e.g., culture fluids) under test.

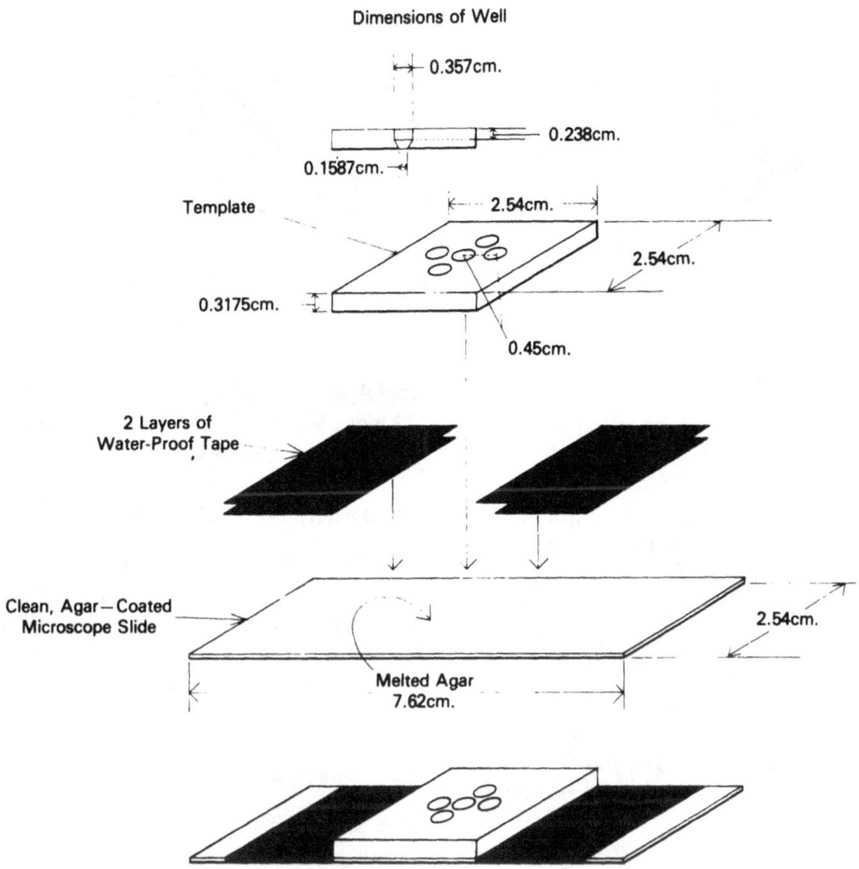

FIG. 1. Microslide diffusion assembly with specifications for Plexiglas template. (Reproduced from Compendium of Methods for the Microbiological Examination of Foods, Marvin L. Speck, ed., APHA, Washington, D.C., p. 441, with permission of the American Public Health Association.)

DEBUBBLERS

Prepare fine glass rods for removing trapped air bubbles in the template after dispensing reagents by pulling glass tubing very fine as in making capillary pipettes. Break glass rods into approximately 6-cm lengths and seal ends by flaming.

PREPARATION OF RECORD SHEET

A record sheet of the assays is prepared by drawing the "hole" pattern of the template or alternatively using an ink stamp pattern of the template arrangement. The materials that are placed in wells of the template should be indicated on the record sheet. A recommended arrangement for antiserum, enterotoxic reference preparation, and test material is shown in Fig. 2 for determining the enterotoxigenic activity of *B. cereus* isolates.

PREPARATION OF STANDARDS

The reagent standards (enterotoxic reference preparation and its respective homologous antiserum) should be diluted or rehydrated from the lyophilized state in accordance with the specific instructions. The reagents will, generally, have been standardized at concentrations to produce a line of precipitate approximately halfway between the antigen reference and antiserum wells in the microslide test. These preparations should produce faint but distinct reference lines. The precipitate lines should be enhanced by staining (see preceding discussion on thiazine red R stain). Reference preparation and homologous antiserum are not readily available; however, they can be obtained from the Division of Microbiology, U.S. Food and Drug Administration, Washington, D.C. 20204.

Production of Diarrheal Antigen

Of the various methods described for the production of the diarrheal metabolite, cultivation of *B. cereus* in BHI broth supplemented with 0.1% glucose, pH 7.4 (Spira and Goepfert, 1972) is simple and does not require any special apparatus other than a shaker. A loopful of 18 to 24-h growth from a nutrient agar slant is added to 3 to 5 ml of sterile distilled water. This

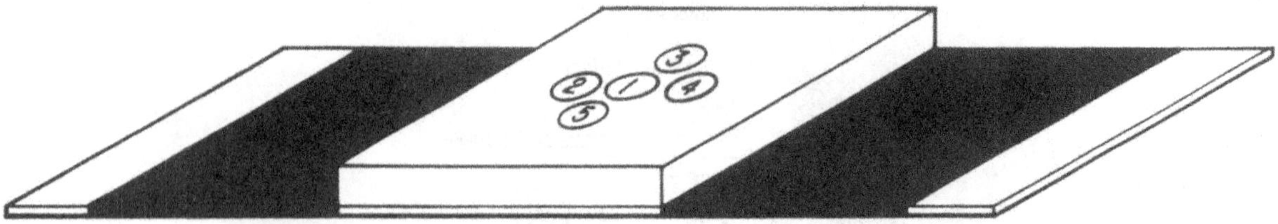

FIG. 2. Reagent arrangement for the serologic identification of *B. cereus* diarrheal antigen: (1) antiserum to *B. cereus* antigen; (2) preparation under test; (3) *B. cereus* toxic reference; (4 and 5) preparation under test.

aqueous suspension should contain approximately 300 million organisms per ml, which is a turbidity equivalent of No. 1 on the McFarland nephelometer scale (McFarland, 1907). A 0.5-ml amount of the aqueous suspension of the test culture is added to the BHIG (see preceding discussion on BHIG, 0.1%) delivered from a 1.0-ml pipette. The BHIG shake flask is then rotated at 84 to 125 cycles/min for 12 h at 31°C. Good growth is usually obtained after 12 h of incubation. A suitable quantity of culture is transferred to a 50-ml centrifuge tube, and organisms are removed by high speed centrifugation (e.g., 10 min at 32,800 × *g*). The culture fluid is decanted into a test tube and examined serologically for the presence of the diarrheal antigen.

Serologic Assay for Diarrheal Component

The microslide gel double diffusion method (Crowle, 1958), which has been used successfully for the serotype identification of the staphylococcal enterotoxins (Casman et al., 1969), can be used with some modification for the identification of *B. cereus* diarrheal antigen.

Addition of Reagents and Test Materials to Slides

After preparing a record sheet (described previously) and following the reagents arrangement shown in Fig. 2, a suitable dilution of antiserum is placed in the center well, and the homologous reference preparation is placed in the upper peripheral well. The material(s) under examination is placed in wells adjacent to the well containing reference preparation. Control slides are prepared using only the reference preparation and antiserum. The wells are filled to convexity with reagents and test materials using pipettes. Bubbles are removed from all wells by probing with fine glass rods. It is best to fill wells and remove bubbles against a dark background. Glass rods are inserted into all wells to remove trapped air bubbles that may not always be visible. Slides are allowed to remain at room temperature in covered Petri dishes containing moist sponge strips (described previously) for 48 to 72 h at 25°C or 24 h at 37°C before examination of the precipitate reactions.

Staining the Slide Reaction

The template is removed by sliding to one side. If necessary, slides are cleaned by dipping momentarily in water and wiping the bottom of the slide. Lines of precipitation are enhanced by immersing slide(s) in thiazine red R stain for 5 to 10 min before examination.

Interpretation of Slide Reaction

The slide is examined for lines of precipitation by holding it over a light source and against a dark background. Presence of antigen in the test material is indicated by a line of precipitation that coalesces with the reference line of precipitation (Fig. 3). If the

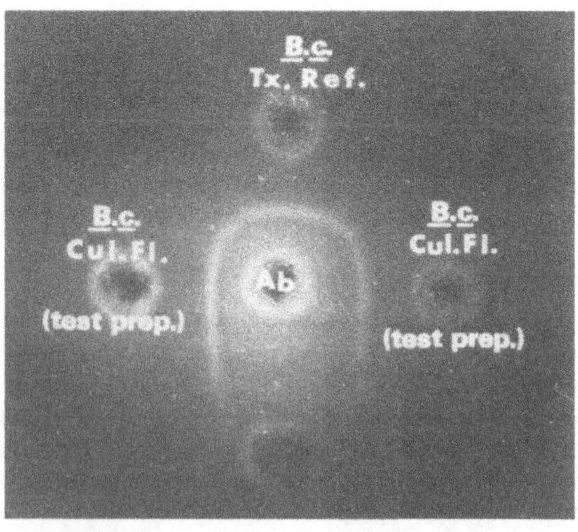

FIG. 3. Diarrheal antigen-antibody line of precipitation with the microslide test. Interpretation of reaction: *B. cereus* (*B. c.*) culture fluids (Cul. Fl.) (right, left, and adjacent to the reference toxin [Tx. Ref.]) contain the diarrheal component as indicated by their lines of precipitation coalescing with the reference line.

concentration of the antigen in the test material is excessive, the formation of a reference line will be inhibited, and the test material must be diluted and retested. A typical line formation of a diluted preparation is shown in Fig. 4. Occasionally, atypical precipitate patterns form and may be difficult for inexperienced operators to interpret; however, these lines intersect the reference line. One of the most common atypical reactions is the formation of lines that are not related to toxin but are caused by other antigens in the test material. The resulting precipitate pattern should be transposed to the record sheets.

Preservation of Slide Reactions

It may be desirable to preserve the slide reaction as a permanent record of assay results. The preserved slide may be useful in litigation cases or for other purposes. If the slide precipitate reactions are to be preserved, the reactant liquids remaining on the slide should be rinsed away by momentarily dipping the slide in distilled water and then immersing it for 10 min in each of the following baths: 0.1% thiazine red R in 1% acetic acid, 1% acetic acid (2X), and 1% acetic acid containing 1% glycerol. The excess fluid should be drained from the slide, and the slide should be dried at 35 to 37°C. After prolonged storage, lines of precipitation may not be visible until the slide is immersed in water.

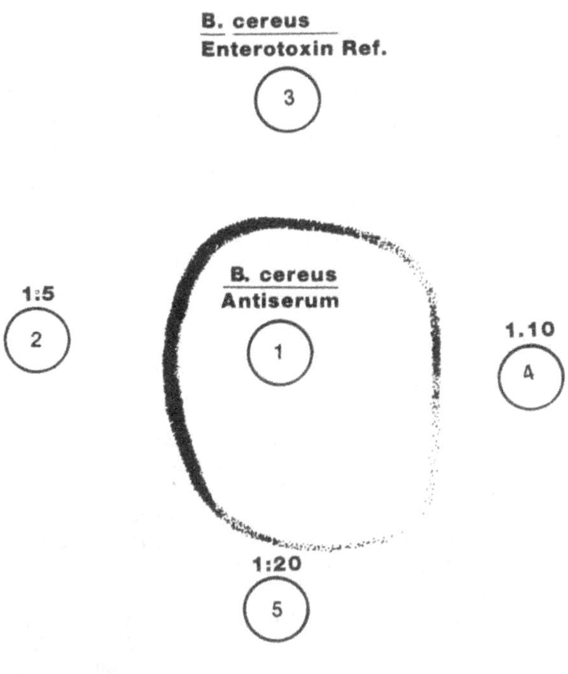

FIG. 4. Typical lines of precipitation of a diluted *B. cereus* culture fluid using the microslide test.

Recovery of Used Slides and Templates

Slides should be cleaned without removing the tape. They are rinsed with tap water, brushed to remove the agar gel, boiled in detergent solution for 15 to 20 min, rinsed for about 5 min in hot running water, and boiled in distilled water. Slides can then be dried by being placed on end in a test tube rack (or equivalent) and put in an incubator. If the slides cannot be uniformly coated with hot 0.2% bacteriological grade agar, they are not clean enough and should be rewashed. When being cleaned, templates should not be exposed to excessive heat or plastic solvents. Templates should be placed in a container, and a hot detergent solution should be poured over them. They are soaked for 10 to 15 min and brushed with a soft nylon brush to remove residual silicone grease. They are then rinsed sequentially in tap water, distilled water, and 95% ethanol, and are air-dried on a towel.

Production and Identification of the Emetic Component

The fact that conditions for production of the emetic substance are not as well defined as those for the diarrheal entity has resulted in limited methodology for both its production and identification. Serologic identification of the emetic component has been prevented thus far by its low molecular weight (about 5,000 to 10,000), which suggests that it may be a peptide with poor antigenicity.

Production of Emetic Substance

Because rice or rice-based foods have been associated more often with the emetic type of food poisoning, most of the laboratory media preparations investigated have contained rice. Melling et al. (1976) recommended a medium containing rice for demonstrating emetic activity. These authors found that autoclaved soaked rice (1 part rice: 4 parts saline [w/v]) was a good growth medium. Liquefaction of an overnight rice culture (30°C) with diastase produces a liquor, which after dialysis for 24 h at 4°C against 10% polyethylene glycol, can be used in biological assays. Small porosity (3,500 to 5,000 or not higher than 6,000 to 8,000 MW exclusion) dialyzer tubing can be used (Bennett, unpublished observations). Other media components that stimulate growth of *B. cereus* include the addition of beef, chicken, or egg proteins or Casamino acids (with B-complex vitamins) to rice (Morita and Woodburn, 1977).

Biological Assay

Reasonable success (50% demonstrating emesis) has been achieved with rice slurries fed to monkeys by intragastric tube (Melling et al., 1976). More recently, emesis has been demonstrated by intravenous injection of kittens with concentrated heated (1.5 h, 100°C) culture fluids produced by culturing *B. cereus* in BHI broth supplemented with 0.1% glucose, pH 7.4 (Bennett, unpublished observations).

Treatment, Prevention, and Control

Treatment

Because the food poisoning illnesses caused by *B. cereus* are intoxications rather than infections, the medical management involves treating symptoms much as other foodborne intoxications. *B. cereus* poisoning is usually self-limiting without recognized post-illness complications. However, certain categories of patients, notably young children, the elderly, and immunosuppressed persons, may be more seriously affected and may require a more thorough follow-up. Exposure does not confer significant protection (immunity) to a person subsequently exposed. As with all foodborne intoxications and infections, prevention is preferable to treatment.

Prevention and Control

Because the sequence of events leading to *B. cereus* foodborne illness is like that of other mesophilic enterotoxigenic pathogens such as *C. perfringens* and *S. aureus,* measures that are effective in preventing outbreaks with the latter organisms are also likely to be effective for controlling *B. cereus*. They include limiting background contamination wherever possible and maintaining good food handling practices with special emphasis on adequate refrigeration. *B. cereus* spores are widely distributed in the environment (particularly in soil), and, thus, are a ubiquitous threat; in addition, spores can resist heat (20 min or more at 95°C), drying, and other adverse conditions, and can germinate and proliferate in susceptible foods at ordinary room temperatures (Harmon et al., 1987; Johnson et al., 1982, 1983). Although not all *B. cereus* strains are enterotoxigenic, the majority apparently are; therefore, it is prudent to avoid preparation and storage practices that allow survival and growth of the organism. Proliferation of *B. cereus* in foods to a level sufficient to cause food poisoning is not accompanied by visible signs of spoilage, and organoleptic changes are usually too subtle to provide a warning. Thus, prevention rests on two major principles: avoidance of preventable contamination through good food handling practices, and "without-fail" refrigeration of susceptible cooked and raw foods. The only exceptions are those foods where limitations of pH, water activity, and so forth are adequate to prevent bacterial growth (e.g., dehydrated, fermented, and pickled foods). These parameters are ill-defined for most *Bacillus* species, including *B. cereus*.

Although it has not been specifically linked with food poisoning, the practice of applying viable preparations of *B. thuringiensis* to food crops as an insecticide by both home gardeners and commercial growers may present a special hazard. The ability of *B. thuringiensis* cultures to induce fluid accumulation in the ligated rabbit ileum (Spira and Goepfert, 1972) and to produce diarrhea in monkeys (Bennett and Harmon, 1986), plus the easily demonstrable presence of the diarrheal antigen in culture fluids of most *B. thuringiensis* strains with the microslide gel diffusion test (Bennett and Harmon, 1986) raise doubt about the safety of this practice. Certainly, thorough rinsing of leafy vegetables that have been treated with *B. thuringiensis* is a minimal precaution; there is no published information to suggest that *B. thuringiensis* is less capable of proliferating in foods than are the enterotoxigenic *B. cereus* strains. To the contrary, the two species behave similarly in susceptible foods at ordinary foodholding temperatures.

Literature Cited

Bennett, R. W. 1985. Serological and biological activities of *Bacillus cereus* diarrheal toxin. Seminar on *Bacillus cereus* enterotoxins: current concepts and developments. Presented at the 85th Annual Meeting of the American Society for Microbiology, Las Vegas, Nevada, March 30, 1985.

Bennett, R. W. 1987. Production of an emetic toxin by *Bacillus subtilis* and *B. licheniformis*. IFT Abstracts, Paper 75. Presented at the 1987 IFT Annual Meeting, Las Vegas, NV, June 16–19, 1987.

Bennett, R. W. and S. M. Harmon. 1986. Production and detection of a diarrheal toxin by *Bacillus thuringiensis*. IFT Abstracts 1986:255.

Bennett, R. W. and F. McClure. 1975. Collaborative study of the serological identification of staphylococcal enterotoxins by the microslide gel double diffusion test. J. Assoc. Off. Anal. Chem. **59**:594–601.

Casman, E. P., R. W. Bennett, A. E. Dorsey, and J. E. Stone. 1969. The microslide gel double diffusion test for the detection and assay of staphylococcal enterotoxins. Health Lab. Sci. **6**:185–198.

Crowle, A. J. 1958. A simplified microdouble-diffusion agar precipitin technique. J. Lab. Clin. Med. **52**:784–787.

Ezepcuk, Y. V. and F. S. Fluer. 1973. The enterotoxin effect. Mod. Medizin **3**:20–25.

Gilbert, R. J., P. C. B. Turnbull, J. M. Parry, and J. M. Kramer. 1981. *Bacillus cereus* and other *Bacillus species*: their part in food poisoning and other clinical infections, p. 298–314. *In* R. C. W. Berkely and M. Goodfellow (ed.), The aerobic endospore-forming bacteria: classification and identification. Society for General Microbiology Special Publication No. 4. Academic Press, London and New York.

Glatz B. A. and J. M. Goepfert. 1973. Extracellular factor synthesized by *Bacillus cereus* which evokes a dermal reaction in guinea pigs. Infect. Immunol. **8**:25–29.

Goepfert, J. M. 1974. Monkey feeding trials in the investigation of the nature of *Bacillus cereus* food poisoning. Proc. IV Int. Congr. Food Sci. Technol. **3**:178–181.

Goepfert, J. M., W. M. Spira, and H. U. Kim. 1972. *Bacillus cereus:* food poisoning organism: a review. J. Milk Food Technol. **35**:213–227.

Gordon, R. E. 1973. The genus *Bacillus,* p. 71–88. *In* Handbook of microbiology, vol. 1. The Chemical Rubber Co., Cleveland, Ohio.

Gordon, R. E., W. C. Haynes, and C. Hor-Nay Pang. 1973. The genus *Bacillus*. USDA Agriculture Handbook, No. 427, Washington, D.C.

Harmon, S. M. 1982. New methods for differentiating members of the *Bacillus cereus* group: collaborative study. J. Assoc. Off. Anal. Chem. **65**:1133–1139.

Harmon, S. M., D. A. Kautter, and F. D. McClure. 1984. Comparison of selective plating media for enumeration of *Bacillus cereus* in foods. J. Food Prot. **47**:65–67.

Harmon, S. M., D. A. Kautter, and H. M. Solomon. 1987. *Bacillus cereus* contamination of seeds and vegetable sprouts grown in a home sprouting kit. J. Food Prot. **50**:62–65.

Hauge, S. 1955. Food poisoning caused by aerobic sporeforming bacilli. J. Appl. Bacteriol. **18**:591–595.

Hobbs, B. C., M. E. Smith, C. L. Oakley, G. H. Warrack, and J. C. Cruickshank. 1953. *Clostridium welchii* food poisoning. J. Hygiene **51**:75–101.

Holbrook, R. and J. M. Anderson. 1980. An improved selective diagnostic medium for the isolation and enumeration of *Bacillus cereus* in foods. Can. J. Microbiol. **26**:753–759.

Johnson, K. M., C. L. Nelson, and F. F. Busta. 1982. Germination and heat resistance of *Bacillus cereus* spores from strains associated with diarrheal and emetic foodborne illness. J. Food Sci. **47**:1268–1271.

Johnson, K. M., C. L. Nelson, and F. F. Busta. 1983. Influence of temperature on germination and growth of spores of emetic and diarrheal strains of *Bacillus cereus* in a broth medium and in rice. J. Food. Sci. **48**:286–287.

Kim, H. U. and J. M. Goepfert. 1971. Enumeration and identification of *Bacillus cereus* in foods: 24-hour presumptive test medium. Appl. Microbiol. **22**:581–587.

Kramer, J. M., P. C. B. Turnbull, G. Munshi, and R. J. Gilbert. 1982. Identification and characterization of *Bacillus cereus* and other *Bacillus* species associated with foods and food poisoning, p. 261–286. *In* J. E. L. Corry, F. Roberts, and F. A. Skinner (ed.), Isolation and identification methods for food poisoning organisms. Society for Applied Bacteriology, Technical Series No. 17, Academic Press, London and New York.

Kushner, D. J. 1957. An evaluation of the egg yolk reaction as a test for lecithinase activity. J. Bacteriol. **73**:297–302.

Lancette, G. A. and S. M. Harmon. 1980. Enumeration and confirmation of *Bacillus cereus* in foods: collaborative study. J. Assoc. Off. Anal. Chem. **63**:581–586.

Leise, J. M., C. H. Carter, H. Friedlander, and S. W. Freed. 1959. Criteria for the identification of *Bacillus anthracis*. J. Bacteriol. **77**:655–659.

McClung, L. S. 1945. Human food poisoning due to growth of *Clostridium perfringens* in freshly cooked chicken. J. Bacteriol. **50**:229–231.

McFarland, J. 1907. The nephelometer: an instrument for estimating the number of bacteria in suspensions used for calculating the opsonic index and for vaccines. J. Am. Med. Assoc. **49**:1176.

Melling, J., B. J. Capel, P. C. B. Turnbull, and R. J. Gilbert. 1976. Identification of a novel enterotoxigenic activity associated with *Bacillus cereus*. J. Clin. Pathol. **29**:938–940.

Morita, T. N. and M. J. Woodburn. 1977. Stimulation of *Bacillus cereus* growth in cooked rice combinations. J. Food Sci. **42**:1232–1235.

Mossell, D. A. A., M. J. Koopman, and E. Jongerius. 1967. Enumeration of *Bacillus cereus* in foods. Appl. Microbiol. **15**:650–653.

Schiemann, D. A. 1978. Occurrence of *Bacillus cereus* and bacteriological quality of Chinese take-out food. J. Food Prot. **41**:450–454.

Smith, N. R., R. E. Gordon, and F. E. Clark. 1946. Aerobic mesophilic sporeforming bacteria. USDA Misc. Publ. No. 559, Washington, D.C.

Spira, W. M. and J. M. Goepfert. 1972. *Bacillus cereus* induced fluid accumulation in rabbit ileal loops. Appl. Microbiol. **24**:341–348.

Szabo, R. A., E. C. D. Todd, and M. K. Rayman. 1984. Twenty-four hour isolation and confirmation of *Bacillus cereus* in foods. J. Food Prot. **47**:856–860.

CHAPTER 9

Bacterial Vaginosis

PETER PIOT

Disease: Vaginosis (bacterial vaginosis, anaerobic vaginosis, nonspecific vaginitis, *Gardnerella*-associated vaginitis).
Etiologic Agents: Not well defined; anaerobic bacteria (including *Bacteroides* spp., *Mobiluncus* spp., and anaerobic cocci), *Gardnerella vaginalis,* and *Mycoplasma hominis* may play a role.
Source: Human genital tract.
Clinical Manifestations: Homogeneous and thin abnormal vaginal discharge and vaginal malodor; postpartum and postabortal fever; wound infection; balanoposthitis.
Pathology: Overgrowth of vaginal bacteria; possibly synergistic interaction between *G. vaginalis* and various anaerobes; increased production of succinate and amines; epithelial cell shedding; no inflammatory reaction.
Laboratory Diagnosis: Three of the following four: 1) thin/homogeneous discharge; 2) pH greater than 4.5; 3) clue cells; and 4) fishy amine odor. Isolation of *G. vaginalis* is not recommended.
Epidemiology: Most common cause of vaginal discharge; worldwide; risk factors include use of intrauterine device and sexual promiscuity.
Treatment: Oral metronidazole.

Description of Disease

Bacterial vaginosis (''nonspecific vaginitis'') is a disease characterized by a malodorous vaginal discharge and an overgrowth of the normal bacterial flora of the vagina with various anaerobic bacterial species, *Gardnerella vaginalis,* and *Mycoplasma hominis*. Although the disease entity is now well defined, the etiology and pathogenesis of bacterial vaginosis are still unclear. It is the most common cause of vaginal discharge, and has been implicated as a risk factor for premature delivery.

Communicability and Transmissibility

Although sexual intercourse is a risk factor for bacterial vaginosis, it has not been clearly established that it is, indeed, a sexually transmitted disease. After experimental inoculation of grivet monkeys with *G. vaginalis* and *Mobiluncus* sp., profuse vaginal discharge appeared 5 days after challenge (Mårdh et al., 1984), but the incubation period in humans is not well defined.

Clinical Manifestations

Women with bacterial vaginosis typically complain of excessive vaginal discharge and malodor, the latter often reported as more intense after sexual intercourse. The discharge is thin, homogeneous, gray, and uniformly adherent to the vaginal wall. There are no signs of gross inflammation and, on colposcopy, no abnormally dilated vessels or increased density of vessels can be seen on the vaginal wall. The pH of the vaginal discharge is increased above 4.5 (usually 5.0 to 5.5).

In one study, bacterial vaginosis in pregnant women was significantly associated with preterm labor, premature rupture of membranes, and amniotic fluid infection (Gravett et al., 1986).

Extragenital infections with microorganisms implicated in bacterial vaginosis do occur. Thus, *G. vaginalis* is a common blood isolate in women with postpartum and postaboral fever and postpartum endometritis (Eschenbach et al., 1984; Venkataramani and Rathbun, 1976). Bacteremia with *G. vaginalis* may be transient in most cases. it can also be recovered from infected episiotomy wounds, vaginal abscesses, and endometrial aspirates in women with postpartum endometritis, often in combination with anaerobes associated with vaginosis (Eschenbach et al., 1986). *G. vaginalis* can cause fatal and nonfatal neonatal infection, including septicemia and soft tissue infection (Leighton et al., 1982; Platt, 1971). Isolation of this agent was also reported from urine of pregnant women and of immunocompromised patients with urinary tract infections (McFadyen and Eykyn, 1968; McDowall et al., 1981; Petit and Mouton, 1985).

Balanoposthitis associated with *G. vaginalis* and *Bacteroides* spp. has also been reported (Kinghorn et al., 1982). *Mobiluncus* spp. can also be isolated from extragenital infections in women (Glupczynski et al., 1984).

Pathogenesis

The pathogenesis and etiology of bacterial vaginosis are still controversial issues. Both *G. vaginalis* and anaerobic bacteria have been isolated from the vagina of nearly all women with bacterial vaginosis, as compared with approximately 50% of women without this condition. In addition, the concentrations of *G. vaginalis* and anaerobes are drastically increased (10^7 to 10^8 colony-forming units per milliliter of vaginal fluid) in bacterial vaginosis. Anaerobes implicated in vaginosis include mainly *Bacteroides* spp., peptococci, peptostreptococci, *Mobiluncus curtisii*, and *Mobiluncus mulieris*. *M. hominis* is also associated with bacterial vaginosis. This mixed bacterial flora replaces the normal vaginal flora, which is dominated by facultative lactobacilli (Pheiffer et al., 1978; Piot et al., 1982a; Spiegel et al., 1980, 1983; Thomason et al., 1984a).

Analysis of vaginal fluid from women with vaginosis by thin layer chromatography shows an increased concentration of amines such as putrescine and cadaverine, probably as a result of decarboxylation of ornithine and lysine (Chen et al., 1979). Gas-liquid chromatography shows a distinct pattern of organic fatty acids in vaginal secretions, with an increase in the succinate/lactate ratio (>0.4). Thus, whereas the concentration of lactate is reduced in bacterial vaginosis, succinate, acetate, propionate, and butyrate are present in increased concentrations (Spiegel et al., 1980). The vaginal redox potential, which is normally +170 mV (+137 to +322 mV), is reduced to a mean of −92 mV (+71 to −257 mV) (Holmes et al., 1985).

It is not clear how to fit these data together in a model for the pathogenesis of bacterial vaginosis. One hypothesis favors some as yet unidentified factor leading to a synergistic interaction between the different microorganisms associated with the condition. Vaginal epithelial cell shedding and the malodorous discharge may be a result of adhesion of *G. vaginalis* to vaginal epithelial cells, as well as to the effect of amines mentioned above. In any event, histopathology and the low number of leukocytes in vaginal fluid indicate that bacterial vaginosis is a superficial event with limited local inflammatory response. Protective immunity has not been reported.

Diagnostic Features

Bacterial vaginosis should be considered in women complaining of excessive vaginal discharge or vaginal malodor. The diagnosis of bacterial vaginosis is based on the presence of at least three of the four following signs: 1) a characteristic homogeneous gray discharge; 2) a vaginal pH greater than 4.5; 3) the presence of clue cells; and 4) release of a fishy amine odor from vaginal secretions mixed with 10% potassium hydroxide (KOH) (Amsel et al., 1983).

Differential diagnosis includes other causes of vaginal or cervical discharge, such as trichomoniasis, candidiasis, gonorrhea, and chlamydial infection.

Epidemiologic Aspects

Bacterial vaginosis is the most common cause of vaginal discharge in women seen for primary health care and by gynecologists. It occurs worldwide. The disease seems to be limited to sexually active women, but is also commonly found after sexual abuse in children (Hammerschlag et al., 1985). Sexual promiscuity and a recent new sex partner are risk factors, and the use of intrauterine contraceptive devices predisposes to the development of bacterial vaginosis (Amsel et al., 1983; Piot et al., 1986). However, sexual transmission of the syndrome has not been firmly proven, though a high proportion of male sex partners of infected women carry *G. vaginalis* in their urethra. Factors implicated in the acquisition of vaginosis remain to be defined. Up to 50% of sexually active women are infected with *G. vaginalis*, but only a small proportion of these are symptomatic.

Etiologic Agents and Laboratory Diagnosis

Characterization of *G. vaginalis* and *Mobiluncus* spp.

Although neither *G. vaginalis* nor *Mobiluncus* spp. are considered to be the sole etiologic agent of vaginosis, their microbiological characteristics will be discussed here since these species are associated with the disease, and are not described elsewhere in this volume.

G. vaginalis

G. vaginalis was previously classified as *Haemophilus vaginalis* or *Corynebacterium vaginale,* but has been transferred to the single species genus *Gardnerella* on the basis of phenotypic and genetic features (Greenwood and Pickett, 1980; Piot et al., 1980). The type strain is 594 Gardner and Dukes (ATCC 14018; NCTC 10287). This genus is not closely related to any other taxon, as based on DNA hybridization and on the unique trilayered laminar cell wall. The guanine-plus-cytosine content is 42 to 44 mol %.

Even the Gram reaction of the cell wall is a controversial matter. Whereas the cell wall morphologically resembles that of a gram-negative bacterium, its chemical composition is more typical of that of a gram-positive organism (Greenwood and Pickett, 1980; Harper and Davis, 1982; O'Donnell et al., 1984). The cells are small (0.5 × 1.5 μm), nonencapsulated, pleomorphic, and nonmotile.

Catalase, oxidase, nitrate reduction, urease, and indole tests are negative. *G. vaginalis* is facultatively anaerobic and fermentative, with acetic acid as the major end product of fermentation, and lactic and formic acid as minor end products. Most strains have fastidious growth requirements, and obligately anaerobic strains have been described (Malone et al., 1975). Optimal growth temperature and pH are 35°C, and 6 to 7, respectively. After incubation at 35°C for 48 h, colonies on human bilayer agar are 0.3 to 0.5 mm in diameter. A characteristic beta-hemolysis is produced on human and rabbit, but not on sheep, blood agar media.

Mobiluncus curtisii AND *Mobiluncus mulieris*

Organisms of the genus *Mobiluncus* are curved, anaerobic, motile rods with multiple subpolar flagella. Their guanine-plus-cytosine contents range from 49 to 52 mol %. The cell wall is gram-variable multilayered. Succinate and acetate are produced, and growth is stimulated by rabbit serum. Strains grow at 35 to 42°C, with an optimal pH of 6.0 to 7.2. Colonies are translucent, smooth, convex, and small (2 to 3 mm diameter after 5 days of incubation) (Spiegel and Roberts, 1984).

Mobiluncus curtisii (type strain: ATCC 35241) includes small (±1.7 μm) gram-variable rods, which are stimulated by arginine. They produce ornithine, citrulline, and ammonia from arginine. Nitrate reduction varies between strains. Two subspecies have been proposed by Spiegel and Roberts (1984): *M. curtisii* subsp. *curtisii,* and *M. curtisii* subsp. *holmesii* (Table 1).

Mobiluncus mulieris (type strain: ATCC 35243) consists of larger (±2.9 μm) gram-negative rods, which produce acid from glycogen and are CAMP test positive.

TABLE 1. Characteristics of *Mobiluncus* spp.[a]

Characteristic	M. curtisii *subsp.* curtisii	M. curtisii *subsp.* holmesii	M. mulieris
Length of cells (μm)	1.7	1.7	2.9
Gram reaction	Variable	Variable	Negative
Stimulated by arginine	+	+	−
NH_4^+ from arginine	+	+	−
Hippurate hydrolyzed	+	+	−
CAMP reaction	Weak	Weak	Strong
Produces >1 meq of acetate per 100 ml	−	−	+
Acid (pH < 5.5) from glycogen	−	−	+
Acid (pH ≤ 6.0) from melibiose	+(−)	−(+)	−
Acid (pH < 5.5) from trehalose	−	−	V
Nitrate reduced	−	+	−(+)
Migrates through soft agar	+	−	V

[a] From Spiegal and Roberts (1982); with permission.

Specimen Collection and Transport

Vaginal fluids are collected with cotton-tipped or calcium alginate swabs. Both a wet mount and a Gram-stained smear should be prepared. If culture is to be performed, a swab can be placed into a transport medium such as Amies or Stuart. Isolation media for *G. vaginalis* should be inoculated within 24 h when such transport media are used. For the isolation of *Mobiluncus* spp., specimens should be handled as for strictly anaerobic bacteria. The vaginal pH can be measured using pH paper strips. Since the pH of cervical mucus is different from that of the vagina, the exocervix should not be touched. When 10% KOH is mixed with vaginal fluid from women with vaginosis, a fishy odor is instantly released.

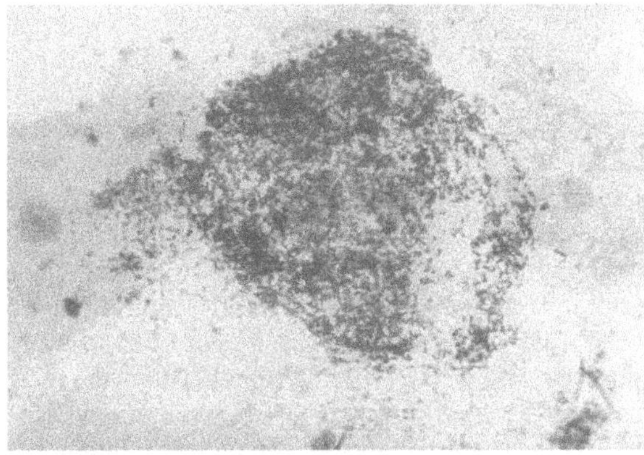

FIG. 1. Clue cell (Gram stain, 1,000× magnification).

Direct Examination

Microscopic examination of vaginal fluid is the most valid and practical microbiological test for the diagnosis of bacterial vaginosis, and is more relevant for this purpose than culture of *G. vaginalis*, *Mobiluncus* spp., or anaerobes. The Gram stain is preferable to a wet-mount preparation because it allows for a better evaluation of the vaginal flora and because a preparation may be available for later examination. Microscopic changes associated with vaginosis are characteristic and include the presence of clue cells (usually representing 20% or more of vaginal epithelial cells), accompanied by a mixed flora consisting of large numbers of small gram-negative and gram-variable rods and coccobacilli in the absence of lactobacilli (longer gram-positive rods). Gram-negative curved rods (*Mobiluncus* spp.) may be present as well. There is a strong inverse correlation between the numbers of *G. vaginalis* morphotypes and *Lactobacillus* spp. morphotypes (Spiegel et al., 1983a). Clue cells are vaginal epithelial cells covered by numerous coccobacilli (Fig. 1).

Isolation

Isolation of bacteria associated with vaginosis is not recommended as a routine procedure.

G. vaginalis

Human blood bilayer agar (HBT agar) (Totten et al., 1982), a semiselective medium, is recommended for culture of genital and extragenital specimens. Alternative solid media include V-agar (Greenwood et al., 1977) and Columbia agar with colistin and nalidixic acid (CCNA agar). Plates should be incubated at 35 to 37°C in a humid atmosphere containing 5% CO_2 or in a candle jar, and examined for 2 days. Strictly anaerobic strains of *G. vaginalis* have been described, but are rare, and routine anaerobic incubation is not recommended (Malone et al., 1975). After 48 h of incubation, colonies are 0.3 to 0.5 mm in diameter and exhibit a diffisue beta-hemoloysis. Strains should be subcultured every 3 days, or preserved by freezing at −70°C or by lyophilization.

Mobiluncus SPP.

Mobiluncus spp. can be isolated on enriched media such as human blood Columbia agar (Sprott et al., 1983) or a semiselective fetal calf serum/rabbit blood bilayer agar (Thomason et al., 1984b). Plates should be incubated anaerobically at 35 to 37°C for 3 to 5 days, and examined for minute translucent colonies that may be weakly beta-hemolytic.

Preliminary Identification

G. vaginalis

Catalase-negative colonies that produce a beta-hemolysis with diffuse edges on human blood bilayer agar should be Gram stained and examined for the presence of typical gram-negative to gram-variable coccobacilli. Such isolates can be identified as *G. vaginalis* with a high degree of confidence (Greenwood and Pickett, 1979; Piot et al., 1982b; Taylor and Phillips, 1983). The growth of *G. vaginalis* should be reported semiquantitatively.

Mobiluncus SPP.

Small translucent convex colonies that grow strictly anaerobically should be Gram stained for curved rods. Small gram-variable curved rods are presumably *M. curtisii*, whereas larger gram-negative curved rods are presumably *M. mulieris*.

Definitive Identification

G. vaginalis

Selected characteristics of *G. vaginalis* are shown in Table 2. Tests yielding a maximal discriminative value include hippurate and starch hydrolysis, and alpha- and beta-glucosidase (Piot et al., 1982b).

Hippurate hydrolysis is detected by suspending a loopful of bacteria from an overnight culture in 0.5 ml of a 1% solution of sodium hippurate in 0.067 M Sörensen phosphate buffer (pH 6.4). Tubes are incubated for 2 h at 35°C, and 0.2 ml of a solution of 3.5% ninhydrin in equal parts of acetone and butanol is added. A positive test is indicated by the development of a deep purple color within 5 minutes.

Starch hydrolysis is determined by flooding Lugol iodine over a inoculated Mueller-Hinton agar plate enriched with 5% horse serum.

For the demonstration of alpha- and beta-glucosidase, a loopful of bacteria from an overnight culture is inoculated in 0.5 ml of a substrate solution consisting of 0.1% 4-nitrophenyl-alpha-D-glycopyranoside or 4-nitrophenyl-beta-D-glycopyranoside, respectively, in 0.067 M Sörensen phosphate buffer (pH 8.0). After 4 h incubation at 35°C in a water bath, tubes are examined for the development of yellow color.

Acid production from various sugars is not recommended for the identification of genital isolates of *G. vaginalis* since the test results may be inconsistent and other vaginal organisms frequently give the same reactions. However, extragenital isolates should always be fully characterized, and acid production from the sugars listed in Table 1 may be determined by the rapid method of Greenwood et al. (1977). Tests should be performed with large inocula from overnight cultures since reactions with older cultures may yield inconsistent results.

Inhibition tests with the compounds listed in Table 1 also provide useful differential reactions with *G. vaginalis*, vaginal lactobacilli, and unclassified catalase-negative coryneform bacteria. The latter two groups are resistant to metronidazole, trimethoprim, and bile, and give variable reactions with sulfonamide. Routine antimicrobial susceptibility testing of genital isolates of *G. vaginalis* is not recommended.

Most *G. vaginalis* strains produce acetic and lactic acids, which are also produced by bifidobacteria and some unclassified vaginal coryneforms.

Isolates can be subdivided into at least 25 serovars (Ison et al., 1986) and into 6 biovars on the basis of reaction patterns for lipase, beta-galactosidase, and hippurate hydrolysis (Piot et al., 1984).

Mobiluncus spp.

Anaerobic curved rods should be differentiated into *M. curtisii* and *M. mulieris* using anaerobic techniques (see Table 1).

TABLE 2. Selected features of *Gardnerella vaginalis* (results for 175 strains)[a]

Test	*Test result*	*% Positive*
Beta-hemolysis (human blood-bilayer agar)	+	99
Hippurate hydrolysis	+	90
Starch hydrolysis	+	100
Alpha-glucosidase	+	100
Beta-glucosidase	−	0
Beta-galactosidase	V[a]	45
Lipase	V	64
Acid from		
Arabinose	V	38
Glucose	+	100
Maltose	+	100
Mannitol	−	0
Starch	+	100
Sucrose	+	85
Xylose	V	44
Zone of inhibition with		
Metronidazole (50 μg)	+	92
Trimethoprim (5 μg)	+	100
Sulfonamide (1 mg)	−	0

[a] From Piot et al. (1982); with permission.
[b] Variability between strains.

Treatment

Oral metronidazole is the treatment of choice of bacterial vaginosis, with oral amoxicillin as a less effective alternative drug. Sulfonamide-containing vaginal creams are ineffective (Piot, 1985). The need for treating male sex partners is controversial.

Literature Cited

Amsel, R., P. A. Totten, C. A. Spiegel, K. C. S. Chen, D. Eschenbach, and K. K. Holmes. 1983. Nonspecific vaginitis: diagnostic criteria and microbial and epidemiologic associations. Am. J. Med. **74**:14–22.

Chen, K. C. S., P. S. Forsyth, T. M. Buchanan, and K. K. Holmes. 1979. Amine content of vaginal fluid from untreated and treated patients with nonspecific vaginitis. J. Clin. Invest. **63**:828–835.

Eschenbach, D. A., M. G. Gravett, K. C. S. Chen, U. B. Hoyme, and K. K. Holmes. 1984. Bacterial vaginosis during pregnancy. p. 213–222. *In* P. A. Mårdh and D. Taylor-Robinson (ed.), Bacterial vaginosis. Almqvist & Wiksell, Stockholm.

Eschenbach, D. A., K. Rosene, L. S. Tompkins, H. Watkins, and M. G. Gravett. 1986. Endometrial cultures obtained by a triple lumen method among afebrile and febrile postpartum women. J. Infect. Dis. **153:**1038–1045.

Glupczynski, Y., M. Labbé, F. Crokaert, F. Pepersack, P. Van Der Auwera, and E. Yourassowsky. 1984. Isolation of *Mobiluncus* in four cases of extragenital infections in adult women. Eur. J. Clin. Microbiol. **5:**433–435.

Gravett, M. G., M. P. Nelson, T. De Rouen, C. Critchlow, D. A. Eschenbach, and K. K. Holmes. 1986. Independent associations of bacterial vaginosis and *Chlamydia trachomatis* infection with adverse pregnancy outcome. J. Am. Med. Assoc. **256:**1899–1903.

Greenwood, J. R. and M. J. Pickett. 1979. Salient features of *Haemophilus vaginalis*. J.Clin. Microbiol. **9:**200–204.

Greenwood, J. R. and M. J. Pickett. 1980. Transfer of *Haemophilus vaginalis* Gardnere and Dukes to a new genus. *Gardnerella, G. vaginalis* (Gardner and Dukes) comb. nov. Int. J. Syst. Bacteriol. **30:**170–178.

Greenwood, J. R., M. J. Pickett, W. J. Martin, and E. G. Mack. 1977. *Haemophilus vaginalis* (*Corynebacterium vaginale*): method for isolation and rapid biochemical identification. Health Lab. Sci. **14:**102–106.

Hammerschlag, M. R., M. Cummings, B. Doraiswamy, P. Cox, and W. M. McCormack. 1985. Nonspecific vaginitis following sexual abuse in children. Pediatrics **75:**1028–1031.

Harper, J. J. and G. H. G. Davis. 1982. Cell wall analysis of *Gardnerella vaginalis* (*Haemophilus vaginalis*). Int. J. Syst. Bacteriol. **32:**48–50.

Holmes, K. K., K. C. S. Chen, C. M. Lipinski, and D. A. Eschenbach. 1985. Vaginal redox potential in bacterial vaginosis (nonspecific vaginitis). J. Infect. Dis. **152:**379–382.

Ison, C. A., D. G. Harvey, A. Tanna, and C. S. F. Easmon. 1986. Serotyping of *Gardnerella vaginalis*. Abstract 242, 26th International Conference on Antimicrobiol Agents and Chemotherapy, New Orleans, American Society for Microbiology, Washington, D.C.

Kinghorn, G. R., B. M. Jones, F. H. Chowdhury, and I. Geary. 1982. Balanoposthitis associated with *Gardnerella vaginalis* infection in men. Br. J. Vener. Dis. **58:**127–129.

Leighton, P. M., B. Bulleid, and R. Taylor. 1982. Neonatal cellulitis due to *Gardnerella vaginalis*. Pediatr. Infect. Dis. **1:**339–340.

Malone, B. H., M. Schreiber, N. J. Schneider, and L. V. Holdeman. 1975. Obligately anaerobic strains of *Corynebacterium vaginale* (*Haemophilus vaginalis*). J. Clin. Microbiol. **2:**272–275.

Mårdh, P. A., E. Holst, and B. R. Møller. 1984. The grivet monkey as a model for study of vaginitis. Challenge with anaerobic curved rods and *Gardnerella vaginalis*, p. 201–206. *In* P. A. Mårdh and D. Taylor-Robinson (ed.). Bacterial vaginosis. Almqvist & Wiksell, Stockholm.

McDowall, D. R. M., J. D. Buchanan, K. F. Fairley, and G. L. Gilbert. 1981. Anaerobic and other fastidious microorganisms in asymptomatic bacteriuria in pregnant women. J. Infect. Dis. **144:**114–122.

McFadyn, I. D., M. B. Glasy, and S. J. Eykin. 1968. Suprapubic aspiration of urine in pregnancy. Lancet **1:**1112–1114.

O'Donnell, A. G., D. E. Minnikin, M. Goodfellow, and P. Piot. 1984. Fatty acid, polar lipid and wall amino acid composition of *Gardnerella vaginalis*. Arch. Microbiol. **138:**68–71.

Petit, P. L. C. and R. P. Mouton. 1985. *Gardnerella vaginalis* in urinary tract infections of immunocompromised patients. Eur. J. Clin. Microbiol. **4:**357.

Pfeifer, T. A., P. S. Forsyth, M. A. Durfee, H. M. Pollock, and K. K. Holmes. 1978. Nonspecific vaginitis: role of *Haemophilus vaginalis* and treatment with metronidazole. N. Engl. J. Med. **298:**1429–1434.

Piot, P., E. Van Dyck, M. Goodfellow, and S. Falkow. 1980. A taxonomic study of *Gardnerella vaginalis* (*Haemophilus vaginalis*) Gardner and Dukes, 1955. J. Gen. Microbiol. **119:**373–396.

Piot, P., E. Van Dyck, P. A. Totten, and K. K. Holmes. 1982a. Identification of *Gardnerella* (*Haemophilus*) *vaginalis*. J. Clin. Microbiol. **15:**19–24.

Piot, P., E. Van Dyck, P. Godts, and J. Vanderheyden. 1982b. The vaginal microbial flora in non-specific vaginitis. Eur. J. Clin. Microbiol. **1:**301–306.

Piot, P. 1985. Bacterial vaginosis. An evaluation of treatment. Scand. J. Urol. Nephrol. **S86:**229–235.

Piot, P., E. Van Dyck, M. Peeters, J. Hale, P. A. Totten, and K. K. Holmes. 1984. Biotypes of *Gardnerella vaginalis*. J. Clin. Microbiol. **20:**677–679.

Piot, P., D. Avonts, M. Sercu, P. Heyerick, I. Vandermeeren, and A. Meheus. 1986. Increased risk for genital infection with *Chlamydia trachomatis* in a cohort of women using oral contraception, p. 217–221. *In* J. D. Oriel, G. Ridgway, J. Schachter, D. Taylor-Robinson, M. Ward (ed.). Chlamydial infections. Cambridge University Press, Cambridge.

Platt, M. S. 1971. Neonatal *Haemophilus vaginalis* (*Corynebacterium vaginale*) infection. Clin. Pediatr. **10:**513–516.

Spiegel, C. A., R. Amsel, D. Eschenbach, F. Schoenknecht, and K. K. Holmes. 1980. Anaerobic bacteria in nonspecific vaginitis. N. Engl. J. Med. **303:**601–607.

Spiegel, C. A., R. Amsel, and K. K. Holmes. 1983a. Diagnosis of bacterial vaginosis by direct Gram stain of vaginal fluid. J. Clin. Microbiol. **18:**170–177.

Spiegel, C. A., D. A. Eschenbach, R. Amsel, and K. K. Holmes. 1983b. Curved anaerobic bacteria in bacterial (nonspecific) vaginosis and their response to antimicrobial therapy. J. Infect. Dis. **148:**817–822.

Spiegel, C. A. and M. Roberts. 1984. *Mobiluncus* gen. nov., *Mobiluncus curtisii* susp. *curtisii* sp. nov., *Mobiluncus curtisii* subsp. *holmesii* subsp. nov., and *Mobiluncus mulieris* sp. nov., curved rods from the human vagina. Int. J. Syst. Bacteriol. **34:**177–184.

Sprott, M. S., H. R. Ingham, R. S. Pattman, R. L. Eisenstadt, G. R. Short, H. K. Narang, P. R. Sisson, and J. B. Selkon. 1983. Characteristics of motile curved rods in vaginal secretions. J. Med. Microbiol. **16:**175–182.

Taylor, E. and I. Phillips. 1983. The identification of *Gardnerella vaginalis*. J. Med. Microbiol. **16:**83–92.

Thomason, J. L., P. C. Schreckenberger, W. N. Spellacy, L. J. Riff, and L. J. LeBeau. 1984a. Clinical and microbiological characterization of patients with nonspecific vaginosis associated with motile, curved anaerobic rods. J. Infect. Dis. **149:**801–809.

Thomason, J. L., P. C. Schreckenberger, L. J. LeBeau, L. M. Wilcoski, and W. N. Spellacy. 1984b. A selective and differential agar for anaerobic comma-shaped bacteria recovered from patients having motile rods and nonspecific vaginitis, p. 125–128. *In* P. A. Mårdh and D. Taylor-Robinson (ed.). Bacterial vaginosis. Almqvist & Wiksell International, Stockholm.

Totten, P. A., R. Amsel, J. Hale, P. Piot, and K. K. Holmes. 1982. Selective differential human blood bilayer media for isolation of *Gardnerella* (*Haemophilus*) *vaginalis*. J. Clin. Microbiol. **15:**141–147.

Venkataramani, T. K. and H. K. Rathbun. 1976. *Corynebacterium vaginale* (*Haemophilus vaginalis*) bacteremia: clinical study of 29 cases. Johns Hopkins Med. J. **139:**93–97.

CHAPTER 10

Bartonellosis

TIMOTHY J. CLEARY

Disease: Bartonellosis (Carrión disease; Oroya fever and verruga peruana).
Etiologic Agent: *Bartonella bacilliformis*.
Source: Human reservoir; transmitted to humans by a sandfly vector, *Lutzomyia* species.
Clinical Manifestations: Oroya fever is the systemic form of disease; it presents as an acute febrile illness and may have an associated hemolytic anemia. Fever, myalgias, and arthralgias are common manifestations. Verruga peruana, the cutaneous form of disease, is characterized by the appearance of wartlike eruptions after the primary infection.
Pathology: Organisms multiply in vascular endothelial cells during the initial phase of disease. The bacteria are then released into the peripheral circulation, where they adhere to and invade erythrocytes. Phagocytosis of infected erythrocytes occurs in the liver, spleen, and lymphatic system. Lesions in the cutaneous form of disease are highly vascular. Organisms are found in the cytoplasm of endothelial cells and in the extracellular milieu.
Laboratory Diagnosis: Stained blood smears for observation of the parasitized erythrocytes. Isolation of the bacteria from blood specimens and cutaneous lesions.
Epidemiology: Restricted to the habitat of the vector in the Andes Mountain area of Peru, Ecuador, and Colombia.
Treatment: Chloramphenicol is the drug of choice.
Prevention and Control: No immunization is available. Methods to control the vector with insecticides and to limit its access to humans are successful.

Description of Disease

Bartonellosis is a human disease with two distinct clinical presentations: a bacteremic form, which may result in a severe anemia, and a cutaneous form with skin eruptions. The organism is transmitted to humans via the bite of the phlebotomine sandfly vector (Hertig, 1942). The disease is restricted geographically to the Andes Mountain regions of South America. Peru is the principal country in which infections occur; however, infections also occur in Ecuador and Colombia (Weinman, 1944a,b, 1981). *Bartonella bacilliformis* is the only species that causes this disease, and humans are the only known natural reservoir for the organism in nature.

Communicability and Transmissibility

Bartonellosis is not communicable from person to person. After inoculation by the vector, the organism multiplies in the vascular endothelial cells, and eventually a bacteremia will result. The incubation period is approximately 3 weeks. If untreated, the cutaneous phase follows the primary episode in 1 to 3 weeks. Infected individuals act as reservoirs for the organism in nature (Weinman and Pinkerton, 1937).

Inoculation of individuals with infected blood or skin tissue may result in an infection.

Clinical Manifestations

The clinical manifestations of the disease after the initial infection varies within the susceptible population. Asymptomatic infections do occur; however, most infections are accompanied by fever, malaise, myalgias, and arthralgias (Ricketts, 1949; Weinman, 1944a,b). The value of blood cultures to establish the diagnosis of this disease has been demonstrated by many investigators (Herrer, 1953; Ricketts, 1949; Weinman, 1944a,b). Some of these individuals develop a progressive hemolytic anemia, which is characteristic of Oroya fever. Factors determining severity of infection are unknown. However, it is known that epidemics with a resultant high mortality are more likely to occur in a population with no prior exposure to the organism than to those living in the endemic area (Weinman, 1944a,b). Verruga peruana, with its characteristic skin lesions, is the subsequent cutaneous form of disease. These lesions may occur after a primary infection regardless of the previous clinical manifestations.

Oroya Fever

Oroya fever occurs as a febrile hemolytic anemia with organisms parasitizing the erythrocytes. The anemia that develops is proportional to the number of erythrocytes that are parasitized and eventually phagocytized by the reticuloendothelial system (Reynafarje and Ramos, 1961). In severe cases 60 to 90% of the erythrocytes may contain organisms. Multiple organisms are usually visualized within the erythrocytes. Red blood cell counts may fall from normal to less than 1 million per cubic millimeter within a few days. Physical findings also associated with this illness include extreme pallor, weakness, tachycardia, and generalized lymphadenopathy.

Verruga Peruana

Verruga peruana is characterized by red, wartlike skin eruptions on any part of the body surface. These manifestations occur 1 or more weeks after resolution of the bacteremic phase of the illness. Initially, a small red papule develops that progressively enlarges to 1 to 2 cm in the following days to weeks. Verruga may appear as multiple lesions over the body, as nodular forms of various sizes and shapes, or as ulcerating lesions that represent infection in the deeper tissues (Ricketts, 1949). Successive eruptions may occur over 1 to 3 months, and the lesions may be or tend to be in all stages of development. The cutaneous lesions are nontender, but arthralgias and myalgias may accompany the eruptions. These lesions may become secondarily infected with other microorganisms.

Pathology

After inoculation of organisms from the arthropod vector, there is an initial multiplication of organisms within the vascular endothelial cells. Once released into the peripheral circulation, the organisms parasitize the red blood cells. Electron microscopy studies have demonstrated that the organisms adhere to and penetrate erythrocytes (Cuadra and Takano, 1969). In vitro studies show that the organisms adhere to a glycolipid moiety on the erythrocyte (Walker and Winkler, 1981). *Bartonella* does not have the capacity to lyse erythrocytes (Reynafarje and Ramos, 1961). Phagocytosis of parasitized erythrocytes by the reticuloendothelial system, particularly by cells in the liver and spleen, accounts for the destruction of red blood cells. Immunologic functions may also serve to modulate the illness. Early in the infection, organisms associated with erythrocytes are predominantly rod shaped. Concomitant with the decrease in the number of infected erythrocytes, the organism changes to a coccoid morphology within the red blood cell (Cuadra, 1986a,b). Eventually the organisms localize in the cutaneous tissue, where they continue to proliferate. The verruga is highly vascular. Organisms may be intracellular in the endothelial cells or extracellular in the fibrous interstitium associated with the verruga (Recavarren and Lumbreras, 1972).

Diagnostic Features

The classic form of Oroya fever is easily recognized. Diagnosis is difficult, however, in patients who present with a nonspecific febrile illness. Physicians in the endemic area are more familiar with this illness and would be likely to make the correct assessment in their patients. For health care workers outside the endemic area, a history of travel to the endemic area would aid in diagnosis. Regardless, blood cultures obtained during the acute form of disease should yield the organism. The diagnosis of verruga peruana should be suggested by the characteristic lesions found on patients from the endemic area. Biopsy specimens from representative lesions should be submitted for culture and histologic examination. Blood cultures from patients with verruga peruana should also be obtained.

Prognosis

The overall mortality in untreated Oroya fever is reported to be as high as 40 to 70%. The high mortality is not necessarily related to the severity of the anemia associated with the illness, but rather to complicating infections by other organisms, particularly *Salmonella* species, that occur in these compromised hosts (Ricketts, 1948). Mortality is markedly reduced if patients are treated with the appropriate antibiotics. The infection can be self-limiting in those individuals who recover from the initial bacteremic episode without treatment. Little, if any, mortality is associated with the cutaneous form of the illness. Most of the lesions will resolve within a few months. Secondary infections may occur in these lesions and lead to delays in healing and even more serious complications. Immunity usually develops after recovery.

Epidemiology

The cutaneous form of the disease was recognized long before its association with the bacteremic phase (Schultz, 1968; Weinman, 1944a,b). Many names were applied to the cutaneous phase that reflected its epidemiologic location. It was called the warts of the Andes because of its prevalence in the mountains and also verruca peruviana (Peruvian warts) because infections were thought only to occur in Peru. In 1870, while a railroad was being constructed from Lima to Oroya, Peru, many workers developed fever and anemia. Mortality from this epidemic was high. The cause of the illness was unknown, but it acquired the name *Oroya fever*. After the outbreak, it was noted that cutaneous lesions developed in several individuals who recovered from the acute infection. This observation suggested that the two features were probably different manifestations of the same illness. Several years later a Peruvian medical student, Daniel Carrión, was intentionally inoculated with a blood sample from a patient with skin lesions. Within 21 days he developed the classic manifestations of Oroya fever and eventually died.

Alberto Barton, in 1909, described the organisms in erythrocytes from an infected individual. Noguchi and Battistini (1926) were the first to isolate the organism in culture from the blood of a patient with anemia. The organism was isolated from a cutaneous lesion on a monkey that was experimentally infected the following year (Noguchi, 1927).

Because this organism is transmitted to humans via a vector, the geographic distribution of the disease is limited to areas where the vector survives in nature. The *Lutzomyia* sandfly inhabits certain valleys in the Andes Mountains region of Peru, Ecuador, and Colombia. The vector is not found at higher elevations (above 8,000 feet [approximately 24.4 km]) where the temperature is less than 10°C or in arid regions (below 2,500 feet [approximately 7.6 km]) where the annual rainfall is low. No other insect vector has been shown to transmit this organism (Hertig, 1942; Weinman and Pinkerton, 1937).

Serologic data from individuals who live in the endemic area indicate that 64% of this population has been infected at some time in the past (Knobloch et al., 1985). Individuals who lived outside the endemic area were negative for antibodies to *Bartonella*.

Etiologic Agent

Characterization

B. bacilliformis is classified in the order *Rickettsiales* in the family *Bartonellaceae* (Ristic and Kreier, 1984). The genus contains only one species. It is a motile, aerobic, gram-negative rod measuring 0.2 to 0.5 μm by 0.3 to 3 μm. Flagella are only found in culture and appear as a tuft of 1 to 10 unipolar filaments. Nucleic acid hybridization studies have not been performed to compare this species with other members of the *Rickettsiales*.

B. bacilliformis can be isolated on a variety of media supplemented with animal sera and/or blood. Walker and Winkler (1981) demonstrated two colony types when they cultured ATCC 35686 and ATCC 35685 isolates on heart infusion agar supplemented with 10% rabbit serum, 0.1% glucose, and 0.5% rabbit hemoglobin. The distinct colony types were designed T1 and T2. Subcultures of each type contained both colony types after 7 days of incubation at 29°C. T1 colonies were small, dark, and friable with an entire edge. T2 colonies were larger, lighter in color, and not friable with an irregular edge.

Specimen Collection, Transport, and Processing

Multiple blood cultures should be obtained during the bacteremic illness. Blood smears should also be prepared to examine for parasitized erythrocytes. Biopsy samples of infected tissue should be obtained aseptically. Tissue samples may be placed between moistened, sterile gauze pads for transport to the laboratory. Blood cultures obtained from patients with the cutaneous form of disease have yielded the organism.

Microscopic Identification

Blood smears may be stained with Wright or Giemsa stain for visualization of the organism. Several of the quick staining methods are useful for work in the field. Infected red blood cells may carry more than one organism. During the acute stage of the infection, the organism has a rod-shaped appearance. When multiple organisms infect the cell, they may appear in groups or overlaying one another. As the infection proceeds, the number of infected erythrocytes decreases and the organism appears more coccoid. In stained tissue sections, organisms are found in the cytoplasm of infected cells or extracellularly in the fibrous interstitium of the lesion.

Isolation and Identification

Bartonella will grow on solid or semisolid media containing rabbit, horse, or human blood. No growth occurs on unsupplemented nutrient media. Growth will occur in the properly supplemented broth cultures, but the culture may not become turbid. The preferred blood culture isolation medium is *Leptospira* enriched medium. Inoculated tubes are incubated at 28 to 37°C for 2 to 4 weeks. The greater the number of infected erythrocytes, the faster the isolation time. Growth occurs only in the oxygen-rich portion of the tube, and the morphology of the organism is best visualized using darkfield microscopy or Giemsa-stained preparations. The morphology is highly variable. Rod-shaped organisms are easily visible; however, they also tend to aggregate and mask the individual appearance of the organism.

There are no specific biochemical tests for identification of the organism. No hemolysis occurs on blood agar and no carbohydrates are fermented.

Immunologic Aspects

A diagnostic change in antibody titer can be found when acute and convalescent sera are tested. However, these tests are performed in only a few reference laboratories. A recent publication (Knobloch et al., 1985) outlined methods for the detection of antibodies using an indirect fluorescent-antibody test, an indirect hemagglutination test, and an enzyme-linked immunosorbent assay. Approximately 64% of 187 individuals tested from a *Bartonella* endemic area in Peru were found positive for antibodies when a combination of these tests was used. Sera from individuals outside the endemic area and other healthy controls were below the cutoff for a positive test. No cross-reactivity between *B. bacilliformis* and other bacteria was detected. The enzyme-linked immunosorbent assay was more sensitive than the other assays. Immunoglobulin IgM antibody was detected in 10 sera using the indirect fluorescent-antibody test. Five individuals had recently recovered from an acute infection; three had Oroya fever within the previous 3 years and two were without any prior symptoms. These results again demonstrate that the disease is endemic within certain localities and that asymptomatic infections can occur.

Treatment, Prevention, and Control

Treatment

Bartonella is susceptible to a variety of antibiotics, including penicillin, tetracycline, streptomycin, and chloramphenicol. Within 24 h of initiating appropriate therapy in a symptomatic patient, there is a rapid resolution of fever and a marked decrease in the number of parasitized erythrocytes. Chloramphenicol is the drug most frequently employed, because it is effective against *Salmonella* species that may cause a secondary illness (Urteaga and Payne, 1955). Treatment of individuals during this initial bacteremic phase of infection will prevent or at least markedly reduce the cutaneous form of disease. Treatment of individuals who present with the cutaneous form of disease does not appear to alter the course of infection.

Prevention and Control

As no vaccine is available for immunization, all methods of prevention and control are directed toward measures that reduce the likelihood of vector contact (Weinman, 1944a,b). Insecticides have proved effective in the elimination of the sandfly, but their use over wide areas has not been practical. Methods that reduce access of the vector to humans have received the most attention. Because the sandfly feeds during the evening, one effective measure is to leave the endemic area during this time. Other measures include screening of windows and securing of floors, walls, and ceilings to prevent access to the house, use of insect repellent, and wearing clothing that protects the body from bites.

Literature Cited

Cuadra, M. 1986a. *Bartonella bacilliformis*, p. 444–449. *In* A. I. Braude, C. E. Davis, and J. Fierer (ed.), Infectious diseases and medical microbiology, 2nd ed. The W. B. Saunders Company, Philadelphia.

Cuadra, M. 1986b. Bartonellosis, p. 1219–1225. *In* A. I. Braude, C. E. Davis, and J. Fierer (ed.), Infectious diseases and medical microbiology, 2nd ed. The W. B. Saunders Company, Philadelphia.

Cuadra, M., and J. Takano. 1969. The relationship of *Bartonella bacilliformis* to the red blood cell as revealed by the electron microscope. Blood **33**:708–716.

Herrer, A. 1953. Carrión's disease II. Presence of *Bartonella bacilliformis* in the peripheral blood of patients with the benign form. Am. J. Trop. Med. Hyg. **2**:645–649.

Hertig, M. 1942. Phlebotomus and Carrión's disease. Am. J. Trop. Med. (Suppl.) **22**:1–80.

Knobloch, J., L. Solano M., O. Alverez G., and E. Delgado A. 1985. Antibodies to *Bartonella bacilliformis* as determined by fluorescence antibody test, indirect haemagglutination and ELISA. Trop. Med. Parasitol. **36**:183–185.

Noguchi, H. 1927. Etiology of Oroya fever. VI. Pathological changes observed in animals experimentally infected with *Bartonella bacilliformis*. J. Exp. Med. **45**:437–455.

Noguchi, H., and T. S. Battistini. 1926. Etiology of Oroya fever. I. Cultivation of *Bartonella bacilliformis*. J. Exp. Med. **43**:851–864.

Recavarren, S., and H. Lumbreras. 1972. Pathogenesis of the verruga of Carrión's disease. Am. J. Pathol. **66**:461–464.

Reynafarje, C., and J. Ramos. 1961. The hemolytic anemia of human bartonellosis. Blood **17**:562–578.

Ricketts, W. E. 1948. *Bartonella bacilliformis* anemia (Oroya fever). A study of thirty cases. Blood **3**:1025–1049.

Ricketts, W. E. 1949. Clinical manifestations of Carrión's disease. Arch. Intern. Med. **84**:751–781.

Ristic, M., and J. P. Kreier. 1984. *Bartonellaceae* Gieszczykiewicz 1939, 25al, p. 717–718. *In* N. R. Krieg and J. G. Holt (ed.), Bergey's manual of systematic bacteriology, vol. 1. The Williams & Wilkins Co., Baltimore.

Schultz, M. G. 1968. A history of bartonellosis (Carrión's disease). Am. J. Trop. Med. Hyg. **17**:503–515.

Urteaga, O. B., and E. H. Payne. 1955. Treatment of the acute febrile phase of Carrión's disease with chloramphenicol. Am. J. Trop. Med. **4**:507–511.

Walker, T. S., and H. H. Winkler. 1981. *Bartonella bacilliformis:* colonial types and erythrocyte adherence. Infect. Immun. **31**:480–486.

Weinman, D. 1944a. Bartonella and human bartonellosis. Trans. Am. Philos. Soc. **33**:246–287.

Weinman, D. 1944b. Public health aspects of bartonellosis. Trans. Am. Philos. Soc. **33**:339–345.

Weinman, D. 1981. Bartonellosis and anemias associated with *Bartonella*-like structures, p. 235–248. *In* A. Balows and W. J. Hausler, Jr. (ed.), Diagnostic procedures for bacterial, mycotic and parasitic infections, 6th ed. American Public Health Association, Washington, D.C.

Weinman, D., and H. Pinkerton. 1937. Carrión's disease. IV. Natural sources of bartonella in the endemic zone. Proc. Soc. Exp. Biol. Med. **37**:596–598.

CHAPTER 11

Borelliosis (Relapsing Fever)

RUSSELL C. JOHNSON

Disease: Relapsing fever (louseborne [epidemic] relapsing fever, tickborne [endemic] relapsing fever).

Etiologic Agents: Numerous species of *Borrelia* including *Borrelia recurrentis, Borrelia hermsii, Borrelia duttoni,* and *Borrelia crocidurae.*

Clinical Manifestations: An acute recurrent illness characterized by febrile episodes that wane spontaneously but usually reappear with decreasing intensity and duration over a period of days to weeks. Shaking chills, fever, headache, and fatigue accompany febrile periods.

Pathology: Miliary and necrotic lesions in spleen. Hepatic, renal, and cerebral congestion occurs, and myocarditis is probably the most common cause of death.

Laboratory Diagnosis: Examination of blood for spirochetes by microscopy and animal inoculation. Serologic tests are not available.

Epidemiology: Worldwide in tropical and temperate climates with the exception of a few areas in the Southwest Pacific. Disease transmitted by body lice and the soft ticks, *Ornithodoros.*

Treatment: The drug of choice is tetracycline. Erythromycin may be used as an alternative agent.

Prevention and Control: Avoidance of the vectors by personal hygiene, delousing, and wearing protective clothing.

Description of Disease

Relapsing fever (louseborne [epidemic] relapsing fever, tickborne [endemic] relapsing fever) is an acute febrile, septicemic illness of humans caused by spirochetes of the genus *Borrelia. Borrelia recurrentis* is the etiologic agent of louseborne relapsing fever and is transmitted by the body louse. The vectors of tickborne relapsing fever are the soft ticks, and a number of *Borrelia* species can be transmitted by these arthropods. The relapsing fevers are widely distributed in both tropical and temperate climates. Large epidemics of louseborne relapsing fever have occurred throughout Europe, Africa, Asia, and South America, and cases continue to be reported from African countries. Little is known about the worldwide incidence of tickborne relapsing fevers because of their sporadic occurrence.

Communicability and Transmissibility

Humans are the only known reservoir for *Borrelia recurrentis,* the spirochete responsible for louseborne relapsing fever. Transmission of louseborne relapsing fever takes place via the infected body louse, *Pediculus humanus humanus.* In contrast tickborne relapsing fever is a zoonosis with rodents serving as the major reservoir of the spirochetes. The tickborne relapsing fever borreliae are transmitted by a large variety of soft-shelled ticks of the genus *Ornithodoros* (Burgdorfer, 1976b; Felsenfeld, 1971). Person-to-person transmission of the relapsing fever with the exception of maternal-fetal transmission (Fuchs and Oyama, 1969) has not been reported.

Clinical Manifestations

The clinical manifestations of louseborne and tickborne relapsing fevers are similar. After an incubation period of 4 to 18 days during which the spirochetes have multiplied to a concentration of 10^6 to 10^8 cells per milliliter of blood, there is a sudden onset of shaking chills, fever, headache, and fatigue. Myalgia, arthralgia, anorexia, dry cough, and abdominal pain often increase in intensity for several days. Because of the large number of spirochetes present in the blood, they are readily visible on blood smears during this early phase of the disease. These symptoms may be continuous or intermittent and frequently increase in intensity over several days. After 3 to 7 days, an afebrile period of several days to several weeks occurs followed by another febrile symptomatic episode. Untreated tickborne relapsing fever is characterized by additional relapses, whereas they are unusual in louseborne relapsing fever. Generally, the relapses become sequentially milder and shorter as the afebrile intervals are extended (Bryceson et al., 1970; Southern and Sandford, 1969). Because the symptoms of relapsing fever are not pathognomonic, it is often a misdiagnosed illness (Fihn and Larson, 1980).

Pathology

Subsequent to exposure to an infected arthropod vector, the relapsing *Borreliae* gain access to the blood and lymphatic circulation where they multiply and a septicemia ensues. The borreliae do not appear to proliferate in extravascular sites and are rapidly killed by phagocytic cells in the presence of specific antibody. Relapses are due to spontaneous antigenic changes in the spirochetes (Barbour et al., 1982; Stoenner et al., 1982). These spirochetes are not known to possess specific toxins. However, a heat-stable pyrogenic factor resides in the spirochetal cell that stimulates mononuclear phagocytes to elaborate leukocyte pyrogen (Butler et al., 1979). The thrombocytopenia of relapsing fever is caused by sequestration of platelets and disseminated intravascular coagulation. In fatal cases of louseborne relapsing fever, miliary and necrotic lesions and microabscesses are present in the spleen. In addition, hepatic, renal tubular and cerebral congestion, and pulmonary alveolar hemorrhages are observed. Myocarditis is probably the most common cause of death. The Jarish-Herxheimer-like reaction occurs commonly after treatment of louseborne relapsing fever and may be responsible for some of these lesions. Tickborne relapsing fever in the United States is usually a milder self-limiting disease without complications. However, in Africa, tickborne relapsing fever can be more severe and prolonged than the louseborne form (Bryceson et al., 1970; Southern and Sandford, 1969).

Abortion frequently occurs if relapsing fever is acquired during pregnancy.

Diagnostic Features

The symptoms of relapsing fever are nonspecific, making a diagnosis extremely difficult in nonendemic areas (Fihn and Larson, 1980). Patients seen in these areas should be questioned for a history of travel and potential exposure in an endemic area. In contrast, diagnosing relapsing fever on clinical grounds is relatively simple in endemic areas of louseborne disease.

Specific diagnosis of relapsing fever depends on (1) the demonstration of spirochetes in blood smears stained by Giemsa's or Wright's method, or (2) inoculation of young rats or mice with blood or other tissues and examining tail blood for spirochetes for 1 to 10 days. The last method is the most sensitive for detecting the relapsing fever borreliae. Serologic tests for relapsing fever are not generally available (Burgdorfer, 1976a).

Malaria, leptospirosis, typhoid fever, hepatitis, rat bite fever, rickettsial infections, Colorado tick fever, and dengue are among the disease to be considered in the differential diagnosis of isolated cases of relapsing fever during the initial febrile episode. These diagnoses can be excluded on the basis of epidemiologic considerations, occurrence of relapses, and the demonstration of *Borrelia* in the blood (Bryceson et al., 1970; Southern and Sandford, 1969).

Prognosis

The mortality rate of untreated cases of louseborne relapsing fever may be as high as 40%. With current treatment the mortality rate is 5% or less in both forms of relapsing fever. Permanent sequelae are infrequent, but some residual visual defects may occur. Central nervous system abnormalities, nephritis, and endocarditis are uncommon complications (Bryceson et al., 1970; Southern and Sandford, 1969).

Epidemiologic Aspects

Occurrence and Transmission

Relapsing fever occurs worldwide in tropical and temperate climates with the exception of a few areas in the Southwest Pacific, such as New Zealand and Australia.

Louseborne relapsing fever is primarily a disease of persons compelled to live under unfavorable hygienic conditions. It is stated that 50 million persons suffered from this form of relapsing fever during the first half of this century and experienced a mortality rate of 10%. Approximately 1 million reported infections, and 9 million unreported cases occurred during World War II and its aftermath with a fatality rate of 5%. More recently, louseborne relapsing fever has been reported primarily from African countries, particularly Ethiopia and Sudan and is endemic in the South American Andes. Louseborne relapsing fever has disappeared from the United States except for imported cases (Felsenfeld, 1971).

The human body louse, *Pediculus humanus humanus* ingests *Borrelia recurrentis* while feeding on a person who is borrelemic. Multiplication of the spirochete is restricted to the hemolymph and central ganglion of the louse. Because of this, transovarial transmission of the borreliae does not occur. Also, *B. recurrentis* is only released from the louse when it is injured by host activities such as scratching or when lice are "popped" between the teeth. The louse remains infective for its entire life (approximately 1 month). Humans are the only known reservoir of this disease. Louseborne relapsing fever is called epidemic relapsing fever because it can be rapidly disseminated under conditions of overcrowding and poor personal hygiene.

Less is known about the incidence of tickborne relapsing fever because of its sporadic occurrence. The vectors of this disease are species of the soft-shelled tick *Ornithodoros*. A number of different species of *Borrelia* are transmitted by the bite of these ticks (Table 1). In contrast to louseborne relapsing fever, tickborne relapsing fever is a zoonosis. Because it is not transmitted from human to human and because its incidence is sporadic, tickborne relapsing fever is called endemic relapsing fever. When humans invade the habitats of the tick, the opportunity for the transmission of the relapsing fever borreliae exists. *Ornithodoros* species become infected by feeding on borrelemic animals, primarily rodents, and all tissues of the tick are invaded. Transovarial

TABLE 1. Characteristics and distribution of louseborne and tickborne borreliae*

Borrelia *spp.*	*Arthropod vector*	*Animal reservoir*	*Distribution*	*Disease*
B. recurrentis (syn. *B. obermeyeri, B. novyi*)	*Pediculus h. humanus*	Humans	Worldwide	Louseborne, epidemic relapsing fever
B. duttonii	*Ornithodoros moubata*	Humans	Central, Eastern, Southern Africa	East African tickborne, endemic relapsing fever
B. hispanica	*O. eraticus* (large variety)	Rodents	Spain, Portugal, Morocco, Algeria, Tunisia	Hispano-African tickborne relapsing fever
B. crocidurae, B. merionesi, B. microti, B. dipodilli	*O. erraticus* (small variety)	Rodents	Morocco, Libya, Egypt, Iran, Turkey, Senegal, Kenya	North African tickborne relapsing fever
B. persica	*O. tholozani* (syn. *O. papillipes, O. crossi?*)	Rodents	From West China and Kashmir to Iraq and Egypt, U.S.S.R., India	Asiatic-African tickborne relapsing fever
B. caucasica	*O. verrucosus*	Rodents	Caucasus to Iraq	Caucasian tickborne relapsing fever
B. latyschewii	*O. tartakovskyi*	Rodents	Iran, Central Asia	Caucasian tickborne relapsing fever
B. hermsii	*O. hermsi*	Rodents, chipmunks, and tree squirrels	Western United States	American tickborne relapsing fever
B. turicatae	*O. turicata*	Rodents	Southwestern United States	American tickborne relapsing fever
B. parkeri	*O. parkeri*	Rodents	Western United States	American tickborne relapsing fever
B. mazzottii	*O. talaje* (*O. dugesi?*)	Rodents	Southern United States, Mexico, Central and South America	American tickborne relapsing fever
B. venezuelensis	*O. rudis* (syn. *O. venezuelensis*)	Rodents	Central and South America	American tickborne relapsing fever

* From Burgdorfer, 1976b.

transmission occurs and spirochetes are present in salivary and coxal (basal segment of appendage) secretions. The *Borrelia* present in the saliva and coxal secretions enter the host through the bite wound. Because *Ornithodoros* ticks are mostly nocturnal fast feeders (5 to 30 min) patients are usually not aware of this painless tick bite. The tick vectors most important in the Eastern Hemisphere are *O. moubata*, *O. erraticus*, *O. tholozani*, and *O. verrucosus*, and in the Western hemisphere *O. hermsi*, *O. turicata*, and *O. rudis* are important vectors (Burgdorfer, 1976b; Felsenfeld, 1971). The largest outbreak of tickborne relapsing fever in the Western Hemisphere took place on the North Rim of the Grand Canyon in Arizona in 1973. Sixty-two campers who spent several nights in log cabins contracted the disease, which was transmitted by *O. hermsi* (Boyer et al., 1977).

Etiologic Agents

Characterization

The relapsing fever borreliae are helically shaped bacteria 0.2 to 0.5 × 5 to 20 μm and have 3 to 10 loose coils. They possess the general structural characteristics that distinguish them from other bacteria. A multilayered outer membrane or envelope is the outermost structure of the cell. The flexible, helical-shaped peptidoglycan layer and the cytoplasmic membrane are located beneath the outer membrane and encompass the cytoplasmic contents of the cell. The structures surrounded by the outer membrane are collectively referred to as the protoplasmic cylinder. A unique feature of the spirochetes is the location of the flagella, which lie beneath the outer membrane and are wound around the protoplasmic cylinder. They are called periplasmic flagella because of their location in the periplasmic space. One end of each periplasmic flagellum is inserted into the protoplasmic cylinder at a subterminal position and the free end extends toward the opposite cell end. The relapsing fever borreliae have 15 to 20 periplasmic flagella per cell end. *Borrelia* are considered to be gram-negative but are visualized most effectively with Giemsa's stain (Hovind-Hougen, 1976; Johnson, 1981; Kelly, 1984).

Borrelia are slow-growing microaerophilic spirochetes that grow well in the temperature range of 30 to 35°C. The generation time of *B. hermsii* in artificial medium at 35°C is 12 h. *Borrelia* have not been grown on solid medium as isolated colonies (Kelly, 1971, 1976).

Borreliae are catalase-negative chemoorganotropic bacteria. Glucose is fermented to lactic acid. Metabolic characteristics are not useful for species identification (Kelly, 1984).

Specimen Collection

Borreliae are circulating in the blood during the febrile periods and it is the tissue examined for spirochetes. The largest number of spirochetes are present in blood during the initial febrile period. As the relapses become milder and shorter, the number of circulating borreliae decreases.

Identification and Isolation Procedures

For rapid diagnosis, unstained wet preparations of the patients blood can be examined for spirochetes by darkfield or phase-contrast microscopy. The spirochetes are not as readily visible by brightfield microscopy. A drop of blood is placed on a microscope slide and mixed with a drop of saline or citrate solution and covered with a coverslip. Spirochetes are identified by their characteristic shape and motility. Care must be exercised not to mistakenly identify fibrils and extrusions from cells as spirochetes.

Borrelia can also be detected by examining stained blood smears by brightfield microscopy. Relapsing fever borreliae are readily stained with aniline dyes. The Giemsa stain is most commonly used for the demonstration of spirochetes in blood smears. A thick blood smear is prepared by placing 1 to 2 drops of blood on a slide and stirring the blood with a toothpick in a circular motion over an area of approximately 1 cm in diameter. The preparation is air dried for at least 30 min before it is stained by the Giemsa method, which requires dehemoglobinization. The spirochetes will be stained blue.

Suckling swiss mice or rats are the animals of choice for detecting borreliae by animal inoculation. Animals are injected intraperitoneally with 0.1 to 0.2 ml of a 1 : 1 dilution of patient blood specimen, they will be detected in the test animal blood 3 to 6 days postinoculation. Blood samples should be examined daily for 14 days before the animal is considered noninfected. The animal inoculation test is a more sensitive technique for detecting borreliae than direct examination of patients blood (Burgdorfer, 1976a; Felsenfeld, 1971).

Isolation of the relapsing fever borreliae by in vitro cultivation is not recommended at this time.

Preliminary Identification

Identification of the relapsing fever borreliae is based on their characteristic shape and motility and on epidemiologic considerations. Phenotypic characteristics for distinguishing the relapsing fever borreliae from one another are not known. Accordingly for most cases of relapsing fever, identification of the spirochetes is presumptive and is based largely on vector specificity (Davis, 1956).

Immunologic Aspects

Serodiagnosis of relapsing fever is complicated by the antigenic variability manifested by these spirochetes. Serologic procedures used previously including agglutination, immune adhesion, immobilization, and borreliolysin have not been effective for detecting antibodies. With the recent advances in culture media, the use of the indirect immunofluorescent test with cultured borreliae as antigens has promise as a useful serologic test (Burgdorfer, 1976a; Felsenfeld, 1971).

Treatment, Prevention, and Control

Treatment

The relapsing fevers respond favorably to treatment with tetracycline and erythromycin. Tetracycline is the antimicrobial agent of choice except in pregnant women and young children (Bryceson et al., 1970; Southern and Sandford, 1969). Recent studies on the treatment of *B. recurrentis* infections in Ethiopia indicate that a single 500-mg oral dose of tetracycline or erythromycin is as effective in clearing spirochetemia and preventing relapse as longer courses. Penicillin G therapy results in slow clearance of spirochetes and frequent relapses (Butler et al., 1978). Similar studies have not been conducted for the treatment of the tickborne relapsing fevers.

Antimicrobial treatment provokes a Jarisch-Herxheimer reaction in most patients with louseborne relapsing fever and some cases of tickborne relapsing fever. The reaction can be life threatening. Antipyretic or antiinflammatory drugs have not been effective for controlling this reaction. The Jarish-Herxheimer reaction is best managed by anticipating the reaction and providing intensive nursing care and intravenous fluid support during the first day of therapy (Butler et al., 1978).

Prevention and Control

Both forms of relapsing fever are prevented by avoiding their vectors. The most important measure for controlling louseborne relapsing fever is delousing of clothing and bodies with insecticides and instructing people to bathe and wash clothing. The detection and treatment of cases of louseborne relapsing fever reduces the reservoir of infection resulting in decreased transmission of the disease. With the tickborne relapsing fevers, the habits of *Ornithodoros* and humans determine the personal and environmental control measures needed. Human dwelling should be rodent proof to prevent accessibility to animal hosts of vector ticks. Partial protection can be achieved in unsatisfactory buildings by treating the interior with insecticides. Personal precautions in potential habitats of vector ticks consist of wearing clothing that covers as much of the skin as possible. This can be accomplished by tucking pants into boots or socks and sleeve cuffs and collars should be close fitting. Application of repellents such as diethyl-m-toluamide to the skin may be helpful. Because of the short feeding time of *Ornithodoros* ticks, periodic skin inspection and removal of ticks may not prevent relapsing fever. Vaccines are not available for the relapsing fevers (Burgdorfer, 1976b; Felsenfeld, 1971).

Literature Cited

Barbour, A. G., S. L. Tessier, and H. G. Stoenner. 1982. Variable major proteins of *Borrelia hermsii*. J. Exp. Med. **156:**1312–1324.

Boyer, K. M., R. S. Munford, G. O. Maupin, C. P. Pattison, M. D. Fox, A. M. Barnes, W. L. Jones, and J. E. Maynard. 1977. Tick-borne relapsing fever: an interstate outbreak originating at Grand Canyon National Park. Am. J. Epidemiol. **105:**469–479.

Bryceson, A. D. M., E. H. O. Parry, P. L. Perine, D. A. Warrell, D. Vukotich, and C. S. Leithead. 1970. Louse-borne relapsing fever. A clinical and laboratory study of 62 cases in Ethiopia and a reconsideration of the literature. Quart. J. Med. **39:**129–170.

Burgdorfer, W. 1976a. The diagnosis of the relapsing fevers, p. 225–234. *In* R. C. Johnson (ed.), The biology of parasitic spirochetes. Academic Press, Inc., New York.

Burgdorfer, W. 1976b. The epidemiology of the relapsing fevers, p. 191–200. *In* R. C. Johnson (ed.), The biology of parasitic spirochetes. Academic Press, Inc., New York.

Butler, T., P. K. Jones, and C. K. Wallace. 1978. *Borrelia recurrentis* infection: single dose antibiotic regimens and management of Jarisch-Herxheimer reaction. J. Infect. Dis. **137:**573–577.

Butler, T., P. Hazen, C. K. Wallace, S. Awoke, and A. Habte-Michael. 1979. Infection with *Borrelia recurrentis:* pathogenesis of fever and petechiae. J. Infect. Dis. **140:**665–675.

Davis, G. E. 1956. The identification of spirochetes from human cases of relapsing fever by xenodiagnosis with comments on local specificity of tick vectors. Exp. Parasitol. **5:**271–275.

Felsenfeld, O. 1971. Borrelia. Strains, vectors, human and animal borreliosis. Warren H. Green, Inc., St. Louis.

Fihn, S., and E. B. Larson. 1980. Tick-borne relapsing fever in the Pacific Northwest: an underdiagnosed illness? West. J. Med. **133:**203–209.

Fuchs, P. C., and A. A. Oyama. 1969. Neonatal relapsing fever due to transplacental transmission of borrelia. J. Am. Med. Assoc. **208:**690–692.

Hyde, F. W., and R. C. Johnson. 1984. Genetic relationship of Lyme disease spirochetes to *Borrelia, Treponema,* and *Leptospira.* J. Clin. Microbiol. **20:**151–154.

Hovind-Hougen, K. 1976. *Treponema* and *Borrelia* morphology, p. 7–28. *In* R. C. Johnson (ed.), The biology of parasitic spirochetes. Academic Press, Inc., New York.

Johnson, R. C. 1981. Introduction to the spirochetes, p. 533–537. *In* M. P. Starr, H. Stolp, H. G. Truper, A. Balows, and H. G. Schlegel (eds.), The prokaryotes, a handbook on habitats, isolation and identification of bacteria. Springer-Verlag, New York.

Kelly, R. 1971. Cultivation of *Borrelia hermsi.* Science **173:**443.

Kelly, R. T. 1976. Cultivation and physiology of relapsing fever borreliae, p. 87–94. *In* R. C. Johnson (ed.), The biology of parasitic spirochetes. Academic Press, Inc., New York.

Kelly, R. T. 1984. *Borrelia,* p. 57–62. *In* N. R. Krieg and J. G. Hott (eds.), Bergey's manual of systematic bacteriology. The Williams & Wilkins Co., Baltimore.

Southern, P. M., and J. P. Sanford. 1969. Relapsing fever. A clinical and microbiological review. Medicine **48:**129–149.

Stoenner, H. G., T. Dodd and C. Larsen. 1982. Antigenic variation of *Borrelia hermsii.* J. Exp. Med. **156:**1312–1324.

CHAPTER 12

Botulism

CHARLES L. HATHEWAY

Disease: Botulism (foodborne botulism, wound botulism, infant botulism, and toxicoinfectious botulism).

Etiologic Agents: *Clostridium botulinum,* toxin types A,B,C,D,E,F,G; botulinogenic strains of other clostridial species.

Source: Toxin—produced in preserved or stored foods by the organism; Organism—soils and sediments in lakes, lagoons, streams, seas, and oceans.

Clinical Manifestations: Descending symmetric flaccid paralysis beginning with muscles innervated by cranial nerves, progressing to weakness of extremities and respiratory involvement; occular signs and symptoms (dilation of pupils, ptosis, blurred and double vision); dryness of mouth, difficulty in speaking, fatigue, and dyspnea; gastrointestinal disturbances may accompany neurologic problems.

Pathology: Prevention of transmission of nerve impulses to muscles by blocking acetylcholine release from the nerve endings; toxin in the circulation is bound by receptors on the nerve endings; toxin in the intestine enters the circulation by absorption through the lymphatic vessels; toxin may be produced by organisms growing in an infected wound or colonizing the intestines, and disseminated systemically by the circulation.

Laboratory Diagnosis: Identification of botulinal neurotoxin in the patient's serum or feces or in food consumed by the patient prior to onset of illness, using mouse bioassay; isolation of *Clostridium botulinum* from the patient's feces or from an infected wound; identification of toxin requires neutralization of toxicity by one (and only one) of the monospecific botulinal antitoxin reagents; observation of signs of botulism in mice injected with unneutralized samples is desirable.

Epidemiology: Foodborne botulism recognized worldwide; incidence predominates in countries with food preservation customs conducive to spore survival, outgrowth, and toxin production in foods that are not subsequently heated before consumption. Wound botulism very rare, with less than 50 cases recognized, mostly in the United States. Infant botulism recognized in only a few countries; it has become the most frequent form of botulism in the United States.

Treatment: Intensive supportive care often including mechanical ventillatory assistance and tracheostomy or endotracheal tube insertion to guarantee airway patency; therapeutic antitoxin often advised for foodborne and wound cases, but not for infant cases.

Prevention and Control: Strict adherence to safe food-processing practices by food industry; education of public on safe home-preserving methods for foods; avoidance of feeding honey to infants younger than 1 year of age.

Botulism is a neuroparalytic illness of humans and various animal species that occurs worldwide. This chapter attempts to present an understanding of the laboratory, clinical, and epidemiologic aspects of the disease based on the experience of the Centers for Disease Control (CDC) from investigating botulism in the United States. Although some features differ in different parts of the world, this experience should provide insights that are helpful regardless of geographic location.

Description of Disease

Botulism has long been perceived as a type of food poisoning caused by a potent neurotoxin resulting from the growth in preserved foods by an anaerobic, spore-forming bacillus known as *Clostridium botulinum*. This foodborne illness has long been recognized in Europe, associated with blood and meat sausages (thus, the name *botulism* from *botulus*, the Latin word for sausage). The bacterial and toxigenic etiology of the illness was discovered and reported by van Ermengem (1897) after a comprehensive investigation of an incident in Belgium in 1895. The outbreak was caused by consumption of raw, salted ham, involving 13 serious cases, 3 of them fatal, and 10 milder cases. The isolated toxigenic organism was named *Bacillus botulinus* by van Ermengem. Subsequent investigations of incidents of human botulism and of outbreaks of botulism in various mammalian and avian species over the next 40 years revealed at least five different types of organisms, the toxins of which could be distinguished serologically. In addition, some significant cultural and physiologic differences were found among the toxigenic strains of organisms isolated from various sources. By 1970, seven toxin types (A, B, C, D, E, F, and G) were identified, and four physiologically different groups of organisms (I, II, III, and IV) were described, which were capable of producing botulinal neurotoxin (Smith, 1977).

It was also discovered that botulism could occur in humans without the consumption of spoiled food. It became evident that toxin produced by *C. botulinum* in an infected wound could cause paralytic botulism, which was clinically indistinguishable from the foodborne illness (Merson and Dowell, 1973). Some observers had speculated that *C. botulinum* might be able to colonize the gut and cause botulism because of absorption and dissemination of the toxin produced in the intestine. This speculation met with considerable skepticism because of the widespread distribution of *C. botulinum* in the environment and failure to document cases of this nature. This form of the illness was finally confirmed when six cases of extreme weakness and muscular impairment in infants in California were recognized as botulism and documented by toxicologic and cultural analysis of fecal specimens (Midura and Arnon, 1976; Pickett et al. 1976). No toxic foods and no infected wounds were associated with any of those cases, nor with the more than 500 cases of infant botulism confirmed since then. All cases involved infants younger than 1 year of age. Colonization of the infants has been considered to be due to lack of a stable normal flora, which exists in more mature humans and inhibits the germination and multiplication of *C. botulinum*.

Occasionally, rare, single case incidents of adult botulism in which no food could be incriminated have been observed and suggest that botulism in children and adults might occasionally be due to intestinal colonization (McCroskey and Hatheway, 1986). For this reason, in 1979, the CDC established a fourth category of botulism, namely, *undetermined classification* (Centers for Disease Control, 1979). Cases of human botulism involving individuals older than 12 months for which no food or wound source has been implicated are placed in this category.

Thus, at present four categories of botulism have been established: foodborne, wound, infant, and undetermined classification.

Foodborne Botulism

In contrast to the European experience with the association of botulism with sausages and other cured meat products, most of the foodborne botulism in the United States has been caused by processed vegetable products. Typically, an outbreak will be found to be due to consumption of home-canned green beans, red beets, peppers, mushrooms, or asparagus. Rarely are commercially canned foods incriminated. Strict adherence to processing methods that assure 1) destruction of all bacterial spores, or 2) prevention of germination and outgrowth of spores in the product has resulted in a commendable safety record for commercial products. Some cases of botulism have resulted from mishandling of frozen commercial foods. Consumption of frozen dinners that have been heated in an oven and subsequently left at ambient temperatures for as long as several days has been responsible for a few cases (Centers for Disease Control, unpublished data).

The largest outbreaks of botulism in the United States in recent years have been associated with restaurants. The largest, involving 59 persons, was caused by a hot sauce containing jalapeno peppers prepared and preserved by an improper home-canning process (Terranova et al., 1978). Two restaurant outbreaks were caused by potato salad (Center for Diseases, 1978, Seals et al., 1981; 34 and 7 cases, respectively), and one involving 28 cases was due to sauteed onions (MacDonald et al., 1985).

Since 1950, 906 cases of foodborne botulism in the United States have been documented by clinical, epidemiologic, and laboratory evidence. The distribution and fatality rates by toxin type are shown in Fig. 1. Most of the foodborne botulism in the continental United States involved either toxin type A or type B. In contrast, 32 of the 44 botulism outbreaks in Alaska of known toxin type since 1947 have been toxin type E (Wainwright et al., 1986). The latter is usually associated with fish or other marine-source foods. Alaskan botulism is associated with foods prepared by traditional methods of the native Eskimo and Indian populations. These include dried, smoked, fermented or salted fish; fermented fish eggs, fermented whale, seal, and walrus blubber; seal meat stored under seal oil; and fermented beaver tails.

Foodborne botulism is confirmed by identifying botulinal toxin in samples of the epidemiologically incriminated food; in the serum of patients in the early phase of their illness (usually within 1 or 2 days of the onset of symptoms) or in the stools, vomitus, or gastric contents of the patients; or by isolating *C. botulinum* from patients' stools.

Wound Botulism

Wound botulism is the rarest form of this rare disease. Through 1985, 33 cases were reported in the United States. The first case was recognized in 1943 (Davis et al., 1951). It involved a 15-year-old girl who had experienced multiple fractures of both legs in a fall from a building. Infection subsequently developed in both legs, leading to paralytic signs consistent with botulism. She died 16 days after injury. Cultural examination yielded a mixture of organisms including *C. botulinum* type A. Of the 33 investigated cases, 25 yielded confirmatory evidence (either toxin in serum or *C. botulinum* from a wound specimen, or both); 17 cases were type A, 7 were type B, and 1 was caused by a mixture of type A and B (Centers for Disease Control, unpublished data). Five of the cases were fatal. In a published report of cases for which laboratory data were available, 7 (35%) of 20 had positive serum tests and 14 (64%) of 22 had positive cultures of wound specimens (Hatheway and McCroskey, 1981). Most cases involved infections in traumatic wounds contaminated with soil. At least one case occurred in a clinical setting due to a necrotic bowel (Centers for Disease Control, unpublished data, 1984). MacDonald et al. (1985) described two cases of botulism (laboratory confirmed) and four of botulismlike illness (unconfirmed) involving chronic drug abusers. When diagnosing cases of botulismlike paralysis, physicians must be aware of the possibility of an infected wound being the source of the paralytic toxin.

Infant Botulism

Clinical, Epidemiologic, and Microbiological Aspects

A syndrome of reversible cranial nerve dysfunction in infants had been observed for some time without its etiology having been established (Clay et al., 1977; Grover et al., 1974). Among the possible causes of the illness under consideration in such cases was the involvement of toxins affecting the neuromuscular junction. Botulism seemed rather unlikely because there were no likely food vehicles for botulinal toxin and, when tested, no toxin was found in the patients' serum. In six cases studied in 1976, the disease was established as botulism by the detection of botulinal toxin in the feces and recovery of *C.*

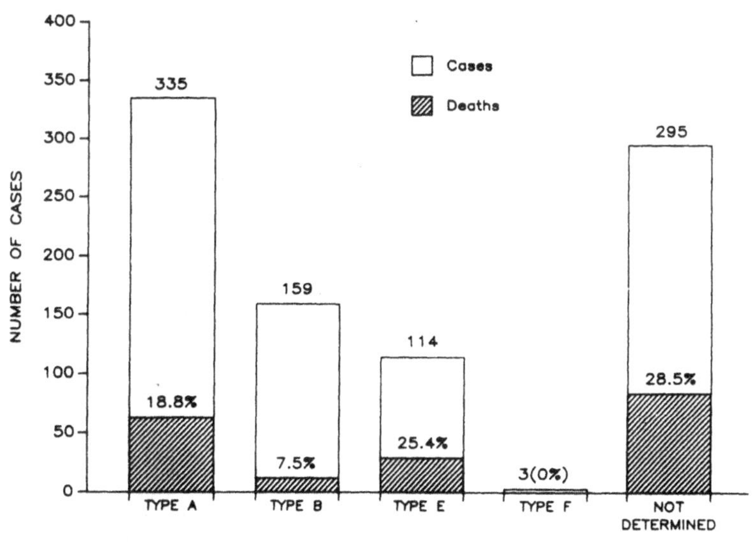

FIG. 1. Cases, deaths, and case-fatality ratios for foodborne botulism, by toxin type, United States, 1950–1985.

botulinum (Midura and Arnon, 1976; Pickett et al., 1976). Coproexamination, which had just become established as part of the routine procedures for investigation of botulism (Dowell et al., 1977), provided the means of establishing the etiology in the absence of toxic food and serum specimens. The conclusion was that the illness was due to absorption of toxin produced in the gut by the organisms that germinated from ingested spores and colonized the infants. This is supported by the persistence of both organism and toxin (Arnon et al., 1977) and by animal model studies (Sugiyama and Mills, 1978). One question that remains is: how do the babies recover from the paralysis at a time when high levels of the toxin often still are present in the feces?

Honey has been the only food implicated as the source of spores for cases of infant botulism (Arnon et al., 1979). Samples of house dust and soils obtained from around the parents' homes have yielded the same *C. botulinum* toxin types as the organism isolated from the infants (Chin et al., 1979; Istre et al., 1986). Spores have also been detected in corn syrup (California State Department of Health Services, 1984; Kautter et al., 1982), although culture-positive samples have not been linked to any cases.

Since the recognition of infant botulism in 1976, the number of confirmed cases has increased, making it the most common form of the disease in the United States. The number of cases and deaths, the sex distribution, and the toxin types of confirmed cases from 1975 through 1985 are listed in Table 1. The ages range from 1 to 48 weeks; 94% were 26-weeks-old or younger, and 47% occurred in California. The case fatality rate for infant botulism has been about 2%. Cases are distributed equally between males and females and between toxin types A and B. The latter is unexpected since in foodborne botulism in the United States, type A cases outnumber type B by more than 2:1 (Fig. 1). Outside the United States, cases of infant botulism have been confirmed in Australia, Canada, Czechoslovakia, England (Arnon 1983), Argentina (California State Department of Health Services, 1983), Sweden (Jansson et al., 1985), Italy (Aureli et al., 1986), and Japan (Dr. G. Sakaguchi, personal communication, 1987).

Among the unusual cases in the United States was one in which a mixture of two *C. botulinum* strains (one type A and the other type B) appeared to have been responsible. In two cases, the strain of the organism produced an atypical toxin, which required a mixture of type B and type F antitoxin for neutralization; toxin was produced by a single strain, but appeared to be a mixture of about 90% type B and 10% type F (McCroskey and Hatheway, 1984). In addition, one case of type F infant botulism (Hoffman et al., 1982) was reported in which the toxigenic organism was quite different from any other strain isolated from a patient with botulism. This organism resembled *C. barati* in its cultural and physiologic features (Hall et al., 1985). The two cases of type E infant botulism confirmed in Italy were caused by organisms which appear to be toxigenic strains of *C. butyricum* (Aureli et al., 1986; McCroskey et al., 1986).

Toxin is usually detected in relatively high titers in the feces during the acute phase of the illness. The fecal extracts typically will be lethal for mice at a dilution of 1:100, although the titers of positive extracts have ranged from less than 1:10 to 1:100,000. Because in the cases confirmed between 1975 and 1978, toxin was readily detected and *C. botulinum* was easily recovered from the feces, while no toxin was detected in serum, the latter appeared to be of little use and was usually not requested. In subsequent experience at the CDC, however, 9 of 67 infants with botulism whose sera were tested had de-

TABLE 1. Cases, sex distribution, and toxin types of confirmed infant botulism in the United States 1975–1985

Year	*Cases*	*Deaths*	*Male*	*Female*	*Toxin type A*	*Toxin type B*	*Other*
1975	1	0	1	0	0	1	0
1976	14	0	11	3	5	9	0
1977	43	2	19	24	27	16	0
1978	39	1	21	18	21	18	0
1979	24	0	8	16	13	10	1(F)
1980	66	1	36	30	31	34	1(BF)
1981	71	2	42	29	40	30	1(BF)
1982	60	2	33	27	29	31	0
1983	79	4	36	43	33	46	0
1984	99	0	43	56	42	56	1(A+B)
1985	71	0	32	39	36	35	0
Total	567	12	282	285	277	286	4

monstrable levels of botulinal toxin; 8 (36%) of 22 type A cases and 1 (2.3%) of 43, type B cases were positive (Hatheway and McCroskey, 1987).

The CDC has found complete concordance of positive toxin and culture results with cases clinically consistent with infant botulism (Hatheway and McCroskey, 1987). The California Department of Health Services obtained negative results on specimens from 395 healthy infants, with but one exception (Arnon, 1985). Thompson et al. (1980), however, reported the isolation of *C. botulinum* from the feces of three normal control infants and nine infants with neurologic diseases that were "clearly not infant botulism." They reported that the asymptomatic carriers as well as the infants in the Salt Lake City area who acquired botulism "characteristically resided in neighborhoods where conditions were favorable for soil growth of *C. botulinum*."

Infant Botulism as a Possible Cause of Sudden Death in Infants

The similarity in age of infants developing botulism and that of those with Sudden Infant Death Syndrome (SIDS) led Arnon et al. (1978) to undertake a study on whether some SIDS cases were due to botulism. They examined fecal samples obtained from autopsies of 212 cases, from which positive cultures were observed in 10 (4.7%). Two of the 10 were also positive for toxin. The organisms were type A in nine cases and type B in one. More recently, a study in Switzerland reported that botulism appeared to be the cause of death in 9 (12.8%) of 70 cases of SIDS (Sonnabend et al., 1985). The Swiss results are difficult to accept without question because: 1) seven of the nine involved toxin types C, F, and G, which are very rare causes, or a rather doubtful cause in the case of type G, of human botulism; and 2) no *hospitalized* cases of infant botulism of any type have yet been documented in Switzerland. Other studies involving fewer numbers of cases have failed to implicate botulism as a cause. *C. botulinum* was not found in cultures of intestinal specimens from 88 cases investigated at the CDC (Centers for Disease Control, unpublished data). Similarly, negative results were obtained in a German study of 41 cases (Heinzel et al., 1985) and in a Canadian study of 33 cases (Gurwith et al., 1981).

Botulism: Undetermined Classification

With the recognition of the wound and infant forms of botulism, it became apparent that when botulism is diagnosed but there is no plausible food vehicle for the toxin, one cannot conclusively presume that the case is due to ingestion of toxin in an unknown or overlooked food item. The source of the toxin might be organisms colonizing an undetected wound or the intestinal tract. Intestinal colonization had been considered as a cause of botulism since the 1920s. However, before the recognition of infant botulism and understanding of its mechanism, the notion was generally dismissed. Since *C. botulinum* spores are widely distributed in the environment and are present even in commonly consumed fresh foods, one would expect to see a high incidence of botulism due to these organism if they were capable of colonization. One can hypothesize, however, that under certain rare circumstances, the conditions in the intestinal tract of adults (or children older than 1 year of age) might be conducive to germination and outgrowth of *C. botulinum* spores. A few recent cases that well support this hypothesis have been documented (Chia et al., 1986; McCroskey and Hatheway, 1986). In the cases in which intestinal colonization was indicated, 1) typical signs and symptoms of botulism were apparent, 2) botulism was confirmed by positive laboratory evidence in one or more clinical specimens, and 3) no food vehicle was incriminated by the detection of preformed toxin. In some of the cases, botulinal toxin or *C. botulinum* was detected in serum or stool for long periods (in one case, the stool was culture-positive 119 days after onset of illness). In some of them, the gastrointestinal conditions had been altered by surgery or treatment with antibiotics or antacids.

From 1979 to 1986, 15 cases of botulism in the United States were recorded as undetermined classification (Centers for Disease Control, unpublished data). It is probable that if they were properly defined, some would have belonged in each of the following categories: foodborne, wound, and intestinal toxicoinfectious botulism.

Botulism in Nonhuman Species

Although the veterinary aspects of botulism are beyond the scope of this chapter, one should point out that botulism in animals and birds is a much more extensive problem than in humans. Smith provides an excellent overview of the occurrence of animal botulism and susceptibility of various animal species to the different types of botulinal toxin (Smith, 1977). A recent monograph on avian botulism is now available (Eklund and Dowell, 1987). Types C and D botulism are common in domestic and wild birds and animals, but are extremely rare in humans. Studies of animal botulism provide some valuable insights about human botulism. Although types C and D botulism are not expected in humans, the investigator must be aware of their existence and be prepared to recognize these types and isolate the organisms if they are encountered. A neuroparalytic condition in

TABLE 2. Symptoms and signs of illness observed in foodborne botulism outbreaks; number of outbreaks in which one or more patients experienced symptom or exhibited sign[a]

	Type A	Type B	Type E	Type F	Und[b]	Total	% with sign or symptom
Outbreaks	*34*	*15*	*10*	*1*	*44*	*104*	
Cases	*97*	*46*	*36*	*3*	*90*	*272*	
Symptoms							
Blurred vision, diplopia photophobia	31	13	9	1	40	94	90.4
Dysphagia	27	14	3		35	79	76.0
Generalized weakness	22	12	4		22	60	57.7
Nausea and/or vomiting	15	13	10	1	19	58	55.8
Dysphonia	25	8	5		19	57	54.8
Dizziness or vertigo	8	4	5		15	32	30.8
Abdominal pain, cramps, fullness	5	6	3		7	21	20.2
Diarrhea	5	6			5	16	15.4
Urinary retention or incontinence	2	2	1		2	7	6.7
Sore throat	4	2	1			7	6.7
Constipation		2		1	3	6	5.8
Paresthesias	1					1	1.0
Signs							
Respiratory impairment	32	7	7		30	76	73.1
Specific muscle weakness or paralysis	23	9	3		13	48	46.2
Eye muscle involvement, including ptosis	16	9	3	1	17	46	44.2
Dry throat, mouth, or tongue	7	6	2		7	22	21.2
Dilated, fixed pupils	3	4	2		8	17	16.3
Ataxia	3	1		1	4	9	8.7
Postural hypotension			1		2	3	2.9
Nystagmus	1		1		1	3	2.9
Somolence			1			1	1.0

[a] Table from Centers for Disease Control (1979A). The outbreaks occurred 1953–1973.
[b] Toxin type undetermined or unspecified.

foals known as shaker foal syndrome (Rooney and Pricket, 1967) appears to be a botulism paralysis analogous to infant botulism in humans (Whitlock and Messick, 1986). Assistance in diagnosis of botulism in avian and wildlife species can be obtained through the National Wildlife Health Laboratories in Madison, Wisconsin, and in domestic animals, through the University of Pennsylvania School of Veterinary Medicine, New Bolton Center, Kennett Square, Pennsylvannia.

Clinical Aspects of Human Botulism

Diagnosis of Botulism

While laboratory investigation, as described below, must be initiated as soon as possible, the initial diagnosis for the purpose of patient treatment must be made using clinical features of the illness, food history, and other background information. The frequencies with which various symptoms and signs of patients involved in foodborne outbreaks are observed are listed in Table 2 by toxin type. Most of the features with the highest frequencies are neurologic and are to be emphasized over the less specific features such as the gastrointestinal symptoms. In cases such as wound botulism in which there is no food vehicle for the toxin, the latter features would probably be absent. The key features observed in 44 cases of infant botulism are listed in Table 3 (Long et al.,

TABLE 3. Observations on 44 cases of infant botulism in Pennsylvania[a]

Physical findings or complications	*%*
Weakness	100
Hypotonia	95
Constipation	95
Failure of oral feeding	93
Diminished gag or sucking reflex	91
Respiratory failure	89
Facial diparesis or ptosis	84
Decreased spontaneous movement	82
Decreased deep tendon reflexes	57

[a] From Long et al. (1985); features observed in >50% of cases.

1985). Electrophysiologic studies can aid in the diagnosis. Certain features are characteristic of both adult (Cherington, 1982) and infant (Cornblath et al., 1983) botulism—an incremental response (facilitation) usually occurs in the evoked muscle action potential to high rates (20 to 50 Hz) of repetitive nerve stimulation, and electromyography shows low-amplitude evoked muscle action potentials in response to a single supramaximal nerve stimulation.

When considering a diagnosis of botulism, the physician should contact the state or local health department immediately. Expert consultation on the case can be made available, and arrangements can be made for obtaining therapeutic antitoxin and having specimens tested. Alerting health authorities to the existence of an individual case could result in recognition of a more serious common-source outbreak.

Patient Care

Treatment for patients with botulism consists primarily of intensive supportive care, including monitoring of respiratory function, mechanical ventilatory assistance when necessary, nasogastric or parenteral nutritional support, and antitoxin treatment. Early and improved delivery of ventilatory support in cases leading to respiratory failure seems to have been the most significant intervention responsible for lowering the botulism fatality rate in recent years (Morris, 1981). Recovery from botulism requires regeneration of the nerve endings that have been poisoned with toxin. If the patient can be kept alive through the paralytic period long enough for this regeneration to take place, recovery should be somewhat assured.

Botulinal antitoxin will effectively neutralize unbound toxin in the circulation, but it will not reverse binding or prevent the subsequent action of bound toxin molecules. Therefore, antitoxin needs to be administered as early as possible. The decision to give antitoxin must be based on clinical and epidemiologic features and should not be delayed until after laboratory results are available. In the United States, the request for antitoxin should be made through the state health department when first considering the diagnosis. The antitoxin will be released by the CDC from the nearest quarantine station at the request of the state health department. Since the antitoxin is expensive, is in limited supply, and has a brief shelf-life, it cannot be stocked in hospitals.

The therapeutic antitoxin currently available in the United States is a trivalent type A, B, and E product of equine origin, with a stated potency of 7,500, 5,500, and 8,500 International Units (IU) per vial for the three antitoxin types, respectively. These are minimal values, and assayed values should exceed the stated unitage. In one study, assays on patients treated with 2 to 4 vials of the product showed substantial levels of circulating antitoxins within 24 h after injection, with persistent half-lives of 5 to 7 days. Peak levels were sufficient to overwhelm any serum toxin level anticipated in a patient with botulism (Hatheway et al., 1984). Two vials should be sufficient, and may be given either i.m. or i.v., although the latter would seem to be the most efficient means of obtaining an immediate high titer. Even if paralysis progresses after the treatment, no additional antitoxin should be given. Progression may be due to toxin bound before treatment. Precautions to minimize the risk of adverse reactions are described in the package insert. They include attempting to ascertain the patient's history of allergies or previous exposure to horse serum, skin testing with 0.1 ml of a 1 : 100 dilution of the serum, and having epinephrine immediately available in case of an anaphylactic reaction.

Generally, noninfant cases in the United States are treated with antitoxin shortly after diagnosis, whereas infant cases are not; the failure to detect toxin in the serum in the initially observed infant cases led to the conclusion that antitoxin was unnecessary. In a typical period (1979 to 1980), 29 of 35 adults and none of 90 infants with botulism were treated with antitoxin (Morris, 1981).

In an 11-year period, adverse reactions to the antitoxin were reported in 9% of 268 recipients and included serum sickness (3.7%), urticaria (2.6%), and anaphylaxis (1.9%) (Black and Gunn, 1980). Serum sickness occurred more frequently in persons who received more than 4 vials.

Guanidine is now seldom used to treat botulism, and reports of benefits are anecdotal. In view of adverse side effects, it cannot be recommended for general use (Morris, 1981). Antibiotics that are shown to inhibit *C. botulinum* in *vitro* do not appear to prevent the growth of the organism in wound botulism, nor to aid in eliminating the toxin and organism from the gut of infants with botulism.

Etiologic Agent

Toxin Types of the Organisms

Shortly after the original isolation of *Bacillus botulinus* by van Ermengem (1897), it became apparent that characteristics of the strains responsible for different outbreaks of botulism varied. The organism responsible for an outbreak of botulism in Darmstadt, Germany, due to consumption of bean salad (Landman, 1904) was found to be more proteolytic and produced a toxin that was not neutralized by antitoxin to the van Ermengem strain (Leuchs, 1910). These two strains were lost before they could be assigned toxin types corresponding to the present designations. The type A and type B designations were established in 1919 with strains isolated from outbreaks of human botulism in the United States (Burke, 1919). The generic name *Clostridium* was adopted in preference to the name *Bacillus*, as recommended by the Committee on Classification of the Society of American Bacteriologists (Bengston, 1924). A third toxin type (type C) was isolated almost simultaneously from fly larvae that caused an outbreak of "limberneck" in chickens in the United States (Bengtson, 1922) and from specimens from an outbreak of "bulbar paralysis" in cattle in Australia (Seddon, 1922). Noting that the antitoxin prepared against the Bengtson strain neutralized toxin from both strains, whereas antitoxin against the Seddon strain neutralized only the homologous toxin, Gunnison and Meyer (1929) suggested that the types be designated as C-alpha and C-beta, respectively. An outbreak of botulism in cattle in South Africa was found to be caused by an organism similar to the type C strains, but the toxin was serologically distinct; it was designated as type D (Meyer and Gunnison, 1928).

Outbreaks of human botulism due to consumption of fish contaminated with a type of toxin not previously identified occurred in New York State (Hazen, 1937) and the Ukraine (Gunnison et al., 1936–1937; Kushnir et al., 1937) in 1934. Toxin from those strains was designated as type E. *C. botulinum* type F was recognized in 1958 when it caused an outbreak in Denmark, involving four of five people who had consumed liver paste (Moller and Scheibel, 1960).

The most recently recognized type of botulinal toxin, type G, was discovered in a culture of a soil sample in Argentina in 1969 (Gimenez and Ciccarelli, 1970), unrelated to any outbreak of botulism. The toxigenic organism, designated as *C. botulinum* type G, has characteristics quite different from any previously identified with the capability of producing botulinal neurotoxin; however, the toxin produced the same clinical picture of botulism paralysis in various animal species in experimental studies (Ciccarelli et al., 1977). Although one laboratory has reported the recovery of *C. botulinum* type G from autopsy specimens (Sonnabend et al., 1981, 1985), no clear evidence exists that it has caused botulism in humans or animals except in experimental studies. The toxin titers in ordinary cultures are much lower than those obtained with the other types, although high titers can be obtained in dialysis cultures (Ciccarelli et al., 1977).

Thus, seven serologically distinguishable types of botulinal toxin (types A to G) have so far been identified, although some minor variations exist among toxins of a given type. Human foodborne botulism is almost entirely restricted to types A, B, and E; two outbreaks of type F botulism have been documented (Midura et al., 1972; Moller and Scheibel, 1960). Although there are rare reports of human incidents of types C and D in the European literature, no cases have been adequately documented in the United States. Wound botulism has been either type A or type B (Centers for Disease Control, 1979; unpublished data). Infant botulism has been entirely type A or type B, except for one case of type F (Hoffman et al., 1982), two cases of type E in Italy (Aureli et al., 1986), and some toxin variants that are primarily type B (Hatheway and McCroskey, 1987; McCroskey and Hatheway, 1984). All of the cases of unclassified botulism in the United States have been either type A or B, except for one case of type F (Green et al., 1983). In summary, human botulism is caused by toxin types A, B, E, and, rarely, F; types C and D are associated with animal botulism. Type G botulism has yet to be adequately documented.

Physiologic Characteristics of the Organisms

Regardless of the type of toxin produced, all strains of organisms included under the name *Clostridium botulinum* are anaerobic, gram-positive, spore-forming rods. Photomicrographs of some representative strains are shown in Fig. 2. The key feature of the organisms currently included in this species is the ability to produce the botulism neurotoxin. This toxin causes the typical flaccid paralysis in animals, involving the skeletal musculature and paralyzing the respiratory system; the latter is usually responsible for death in fatal cases. The Gram reaction may appear weak or negative in cultures more than 24 hours after inoculation, and the subterminal spores may be difficult to visualize in some strains. The lipase reaction, which is visualized as a pearly luster film around the colonies on egg yolk agar (EYA) (McClung and Toabe, 1947), is a very important distinguishing characteristic of all strains, except unfortunately, for those that produce type G toxin. This reaction serves as an effective marker in isolating the organism from clinical and environmental specimens.

Striking differences exist among strains that share the ability to produce the neurotoxin. Some of the differential characteristics are proteolytic activity, heat stability of spores, optimal growth temperature, and metabolic end products. Smith (1977) divides the organisms currently known as *C. botulinum* into four culturally distinct groups, each designated by a Roman numeral (Groups I to IV).

Group I

All strains of *Clostridium botulinum* type A that have been characterized to date belong to this group of strongly proteolytic organisms. Almost all of the

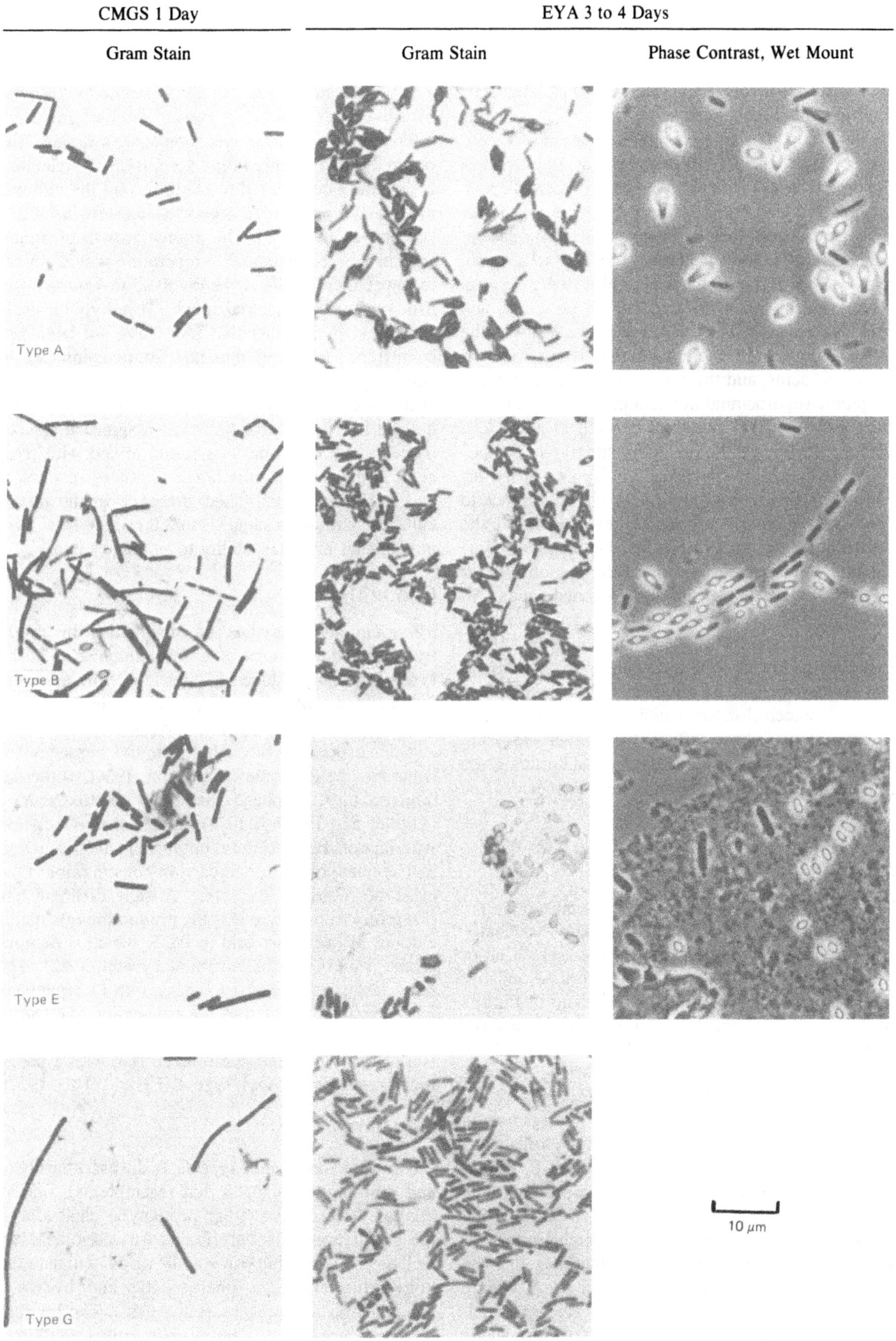

FIG. 2. Photomicrographs of *Clostridium botulinum*, toxin types A, B, E, and G. Row 1: *C. botulinum* type A, CDC 5356. Row 2: *C. botulinum* type B (proteolytic), CDC 5448. Row 3: *C. botulinum* type E, CDC 5545. Row 4: *C. botulinum* type G, CDC 1353. Photography by Don Howard.

type B strains involved in botulism and isolated from environmental specimens in the United States belong to this group. The first encountered strain of type F (Moller and Scheibel, 1960) as well as about half of the type F strains isolated from environmental specimens in the United States belong to Group I. These organisms digest various proteins including casein, meat particles, and coagulated egg and serum proteins. They ferment glucose and usually several other carbohydrates, produce lipase on EYA, and hydrolyze esculin. They produce a variety of volatile fatty acid end products including isobutyric and isovaleric acids, and the nonvolatile product phenylpropionic (hydrocinnamic) acid in peptone yeast extract glucose (PYG) medium (Dowell et al., 1977), detectable by gas-liquid chromatography (GLC) (Lombard and Dowell, 1982). The spores of the organisms in this group have a very high resistance to heat. Because of this property, they can survive the heat-processing temperatures used in the older home canning methods and are, therefore, responsible for most of the botulism from home canned foods. Although optimal growth temperature is approximately 35°C, at least some of these strains will grow at temperatures as low as 15°C (Schmidt, 1964). An organism that appears to be identical to *C. botulinum* in Group I, except for toxigenicity, is *C. sporogenes*. Genetic studies show sufficient DNA homology between some strains of the toxigenic and nontoxigenic organisms to indicate they are the same species (Lee and Riemann, 1970).

Group II

The organisms in this group have been commonly referred to as the "nonproteolytic" or "saccharolytic" strains of *C. botulinum*. These include all of the type E strains that have been described, most of the strains implicated in type B botulism in Europe (usually involving meat products such as sausages and salted, uncooked hams), and about half of the type F strains encountered from various sources. The one documented case of type F foodborne botulism in the United States was due to an organism in this group (Midura et al., 1972). These organisms liquefy gelatin (as do the other groups of *C. botulinum*), but do not digest other complex proteins such as casein, egg white, or heat-coagulated serum. They ferment a variety of sugars including glucose, mannose, fructose, trehalose, and sometimes maltose and sucrose. They produce lipase on EYA, but do not hydrolyze esculin. The metabolic acid end products produced in PYG detectable by GLC are acetic and butyric acids. The spores of these organisms are much less heat resistant. It is probably because of this that they rarely cause botulism associated with foods that are heated in the process of home canning. The type E strains are usually associated with botulism due to fish or other marine products, and are often found in marine environmental samples. The optimal growth temperature for Group II organisms is generally considered to be 30°C, and the minimal growth temperature is reported as low as 3.3°C (Eklund et al., 1967). The poorer growth of organisms in this group at body temperature would appear to lower their ability to be involved in wound botulism or intestinal colonization. This hypothesis is supported by the fact that they have not been isolated from such cases thus far. Nontoxigenic organisms that resemble Group II *C. botulinum* in all respects except for toxigenicity are encountered in nature, but these have not been assigned a species name. Sometimes they are found mixed with toxigenic organisms in enrichment cultures of clinical and food specimens. Their presence in the mixed cultures sometimes suggests that they may be *C. botulinum* that lost their ability to produce toxin.

Group III

The main characteristics that distinguish this group from the other groups of *C. botulinum* are: 1) the types of toxin produced (C and D); 2) the apparent fastidiousness of growth requirements in laboratory media; 3) the instability of toxigenicity of most strains, especially through the initial passages for isolating single colonies (Bengtson, 1924); 4) the demonstrability of phage mediation of toxigenicity (Eklund and Poysky, 1974); and 5) the association with (almost restriction to) botulism in avian and animal species, often due to carrion consumption. One clear physiologic characteristic that distinguishes this group from Group II is the production of a major amount of propionic acid in PYG, which is demonstrable by GLC (Lombard and Dowell, 1982). The relationship of *C. novyi* to this group of organisms has been demonstrated by the conversion of *C. botulinum* type C into *C. noyi* type A by curing the organism of its own phage, then infecting it with a phage recovered from *C. novyi* type A (Eklund et al., 1974).

Group IV

Clostridium botulinum type G is an asaccharolytic and proteolytic organism that resembles *C. subterminale*. It resembles other proteolytic clostridia in the production of isobutyric and isovaleric acids as well as acetic and butyric volatile acids, and also the nonvolatile products, phenylacetic and hydroxyphenylacetic acids (Moss et al., 1980). The level of toxin produced by this organism in ordinary cultures is rather low, especially in comparison with type A and B strains, but high titers can be obtained in dialysis cultures (Ciccarelli et al., 1977). Two strains of this organism were isolated from soils in different provinces in Argentina (Ciccarelli et al., 1976; Gimenez and Ciccarelli, 1970), and other isolates were reported from autopsy specimens in Switzerland (Sonnabend et al., 1981, 1985).

Spores are usually rare or not seen by microscopic examination (Fig. 2), but Lynt et al. (1984) succeeded in obtaining substantial spore crops for study. They found that the major portion of spores had a low heat resistance, being destroyed at 80°C within 10 min. The spores surviving that treatment (less than 1% of the total), however, had a heat resistance similar to that of the Group I strains.

DNA hybridization studies have shown a distinction between *C. botulinum* type G and the type strains of *C. subterminale* and *C. hastiforme* (Suen et al., 1986). All strains of *C. botulinum* type G were very closely related to each other and also quite closely related to 2 strains previously identified as *C. subterminale* and 1 strain identified as *C. hastiforme*. Therefore, *C. botulinum* type G is genetically related to some strains of nontoxigenic organisms. Those strains had been isolated from clinical specimens in the United States. *C. botulinum* type G and the related nontoxigenic strains can be distinguished from *C. subterminale* and *C. hastiforme* by the production of an indole derivative (Lombard and Dowell, 1983; Suen et al., 1986).

Other Neurotoxigenic Clostridia Causing Botulism

In 1979, a case of infant botulism was confirmed by identifying type F toxin in the feces, but isolating the causative organism was difficult because it lacked the lipase characteristic, which had served as a perfectly reliable marker in all previous cases. The problem was compounded because the fecal specimen contained nontoxigenic organisms that possessed all the expected phenotypic characteristics. That organism was identified as *C. sporogenes*. The lipase-negative organism that produced type F toxin was identified as *C. barati* on the basis of its phenotypic characteristics (Hall et al., 1985), and this was confirmed later by genetic analysis (Suen, 1986).

In 1984 a case of type E infant botulism that presented the same problems was confirmed in Italy. The toxigenic organism was also lipase-negative, and after isolation was found to correspond phenotypically with *C. butyricum* (McCroskey et al., 1986). The organism was found to be closely related to the type strain of *C. butyricum* by DNA homology studies (Suen, 1986). The stool specimen from this baby also contained *C. sporogenes,* which caused some confusion similar to that in the type F case. A second case of type E infant botulism caused by an identical organism was confirmed in Italy 7–1/2 months after the first case (Aureli et al., 1986).

Taxonomic Problems with Clostridia Capable of Producing Botulinal Neurotoxin

The characteristics of the neurotoxigenic clostridia and their nontoxigenic phenotypic analogs described in this section are listed in Table 4. Although no nontoxigenic species is listed for Group II, nontoxigenic organisms are encountered in nature, sometimes in specimens examined in botulism investigations (Centers for Disease Control, unpublished data). If we are consistent with the scheme of referring to all clostridia that produce botulism neurotoxin as *C. botulinum*, strain 2990, which is related to *C. barati*, would be identified as *C. botulinum* type F, Group V and strain 5262 (related to *C. butyricum*) would be identified as *C. botulinum* type E, Group VI (Table 4). It would seem more appropriate to assign a separate species name for each group and include both toxigenic and nontoxigenic organisms within each. It would then be understood that botulinal neurotoxin can be produced by some strains of at least six different *Clostridium* species.

Botulinal Neurotoxin

Botulinal neurotoxin is regarded as the most poisonous substance known, on the basis of the mass of weight of animals that could be killed by a unit of weight of the toxin. One mg of purified type A toxin contains about 1.5×10^8 mouse intraperitoneal (i.p.) LD_{50} (Sugiyama, 1980). That amount would kill (theoretically) 7.5×10^7 20-g mice. One gram of the toxin would thus kill 1.5×10^9 kg or 1.65×10^6 tons of mice. Toxins of types B through F have similar to slightly lower specific toxicities.

The nature of all of the types of botulinal neurotoxin that have been studied (types A through F) is similar (Sugiyama, 1980). The toxins are proteins with molecular weights of approximately 150,000 daltons. The active form consists of two peptide chains, approximately 100,000 and 50,000 daltons, joined by a disulfide bridge. It is postulated that the neurotoxin is synthesized as a single-chain polypeptide with a molecular weight of 150,000 daltons, with at least one intrachain disulfide bond. In this state, it has relatively low biological activity. Upon exposure to endogenous enzymes of the proteolytic strains, or exogenous proteolytic enzymes in the case of nonproteolytic strains, the single peptide chain is cleaved at a site between the disulfide link, resulting in a "nicked" molecule with the dichain structure (Simpson, 1986). This is the activated form of the toxin, which has full potency. Toxins from nonproteolytic strains may be activated by enzymes from other organisms in a mixed culture or by adding trypsin to a pure culture.

TABLE 4. Differential characteristics of clostridia capable of producing botulinal neurotoxin and of analogous nontoxigenic clostridia

	Type of toxin	*Heat resistance of spores*	*Egg yolk agar reactions*		*Glucose fermentation*	*Esculin hydrolysis*
		Temp./D value	*Lecithinase*	*Lipase*		
Group I						
Toxigenic organisms						
C. botulinum type A	A		−	+	+	+
C. botulinum type B	B		−	+	+	+
C. botulinum type F	F	112°C/1.23 (high)	−	+	+	+
Nontoxigenic analog						
C. sporogenes	−		−	+	+	+
Group II						
Toxigenic organisms						
C. botulinum type B	B		−	+	+	−
C. botulinum type E	E	80°C/0.6–1.25	−	+	+	−
C. botulinum type F	F	(low)	−	+	+	−
Group III						
Toxigenic organisms						
C. botulinum type C	C		−	+	+	−
C. botulinum type D	D	104°C/0.12–0.9	−	+	+	−
Nontoxigenic analog		(intermediate)				
C. novyi type A	−		+	+	+	−
Group IV						
Toxigenic organisms						
C. botulinum type G	G		−	−	−	−
Nontoxigenic analogs						
C. subterminale	−		−	−	−	−
C. hastiforme	−		−	−	−	−
C. subterminale	−		−	−	−	−
"Group V"						
Strain 2990	F		+	−	+	+
Nontoxigenic analog						
C. barati	−		+	−	+	+
"Group VI"						
Strain 5262	E		−	−	+	+
Nontoxigenic analog						
C. butyricum	−		−	−	+	+

[a] PYG = Peptone-yeast extract-glucose; A = acetic; P = propionic; IB = isobutyric; B = butyric; IV = isovaleric; PA = phenylacetic; PP = phenylpropionic (hydrocinnanmic).
[b] Indole derivative, as described by Lombard and Dowell (1983).

The paralytic action of botulinal neurotoxin is caused by blockage of the release of the neurotransmitter acetylcholine at the neuromuscular junction. The larger chain of the toxin molecule has a binding site with an affinity for receptors on the nerve endings of the specific muscle fibers that are affected. After the binding step, a portion of the molecule, which includes the smaller chain or a part of it, enters into the nerve cell and prevents the release of acetylcholine (Simpson, 1986; Sugiyama, 1980). Thus, a flaccid paralysis results from loss of nerve function. Recovery of function entails regeneration of new nerve endings. The mechanism of the blocking action inside the nerve cell has not been elucidated.

C. botulinum type C produces more than one kind of lethal toxin. This is believed to be responsible for the differences between C-alpha and C-beta strains, although the literature is a little confusing on the matter. Nevertheless, two toxins designated as C_1 and C_2 have been purified and characterized. Toxin C_1 appears to be structurally and pharmacologically

			Acid products in PYG[a]							
			Volatile					Nonvolatile		
Gelatin liquefaction	*Casein digestion*	*Indole derivative*[b]	*A*	*P*	*IB*	*B*	*IV*	*PA*	*PP*	*Representative strain*
+	+	+	+	−	+	+	+	−	+	ATCC 25763
+	+	+	+	−	+	+	+	−	+	ATCC 7949
+	+	+	+	−	+	+	+	−	+	ATCC 35415
+	+	+	+	−	+	+	+	−	+	ATCC 3584
+	−	−	+	−	−	+	−	−	−	ATCC 25765
+	−	−	+	−	−	+	−	−	−	ATCC 9564
+	−	−	+	−	−	+	−	−	−	ATCC 27321
+	−	−	+	+	−	+	−	−	−	ATCC 17849
+	−	−	+	+	−	+	−	−	−	ATCC 27517
+	−	−	+	+	−	+	−	−	−	ATCC 17861
+	+	+	+	−	+	+	+	+	−	ATCC 27322
+	+	−	+	−	+	+	+	+	−	ATCC 25774
+	+	−	+	−	+	+	+	+	−	ATCC 33268
+	+	+	+	−	+	+	+	+	−	CDC 16525
−	−	−	+	−	−	+	−	−	−	ATCC 43756
−	−	−	+	−	−	+	−	−	−	ATCC 27638
−	−	−	+	−	−	+	−	−	−	ATCC 43181
−	−	−	+	−	−	+	−	−	−	ATCC 19398

similar to all of the other botulinal neurotoxins, whereas toxin C_2 is different. Toxin C_2 consists of two separate molecules (55,000 and 105,000 daltons, respectively), which are not covalently linked. It is thus referred to as a *binary* toxin. It does not block neuromuscular transmission, but causes hypotension, hemorrhaging in the lungs, and collection of fluids around the lungs and in the trachea (Simpson, 1982). The in vivo effects of toxin C_2 appear to be due to changes in membrane permeability or cellular secretion.

Laboratory Aspects

Investigation and Confirmation of Botulism

First, since botulism is so rare, it is recommended that diagnostic investigation be carried out in a limited number of reference laboratories. Many laboratories, even at the state or regional level in some areas, may go for years without encountering a single positive specimen. Proper interpretation of results requires an ongoing experience with both positive and negative specimens. One of the problems is recognition of nonspecific reactions. Nonspecific toxic substances are sometimes found in food and fecal samples and, rarely in serum. An experienced laboratorian can recognize signs of botulism in mice, and can usually distinguish between those and nonbotulinal toxic effects. When using specific antitoxin reagents with marginally lethal, nonspecifically toxic samples, a fortuitous death/survival pattern may suggest a specific neutralization. Experience will dictate that a guarded interpretation be made, that early reports are provisional, that tests be repeated when possible, and that results are consistent. Furthermore, personnel working with *C. botulinum* and its toxin should be properly protected and trained, and this is not warranted for the rare occasional need, when such work could be easily and satisfactorily handled in a reference laboratory.

Botulism is confirmed by one or more of the following:

1. Detecting and identifying botulinal neurotoxin in the serum, stool, gastric contents, or vomitus of the patient.
2. Detecting and identifying botulinal neurotoxin in a food consumed by the patient prior to onset of illness.
3. Recovery of *C. botulinum* from the patient's stool, other gastrointestinal specimen, or wound specimen.

Safety Precautions

The following safety measures for work with *C. botulinum* and botulinal toxin are recommended by the Centers for Disease Control:

1) Education of each employee: The worker should be thoroughly instructed as to the nature of the hazard of working with *C. botulinum* and its toxin. He or she should know how to handle transfers, where to place contaminated materials and equipment, and how to decontaminate spills. The worker must be adept at all required manipulations.

2) Protection of mouth and skin from contact with toxin: No mouth pipetting, smoking, eating, or drinking should be done in the laboratory. Wear disposable rubber gloves when handling toxic materials to prevent contamination of hands and subsequent transfer of toxin to mouth or face. Remove gloves before handling "clean" items such as the telephone.

3) Prevention of aerosols: Centrifuge cultures or materials containing toxin in closed containers or a sealed rotor. The outside of the containers or rotor must not be contaminated with toxic material. Pressurized containers containing toxic materials should be opened under a hood or inside a plastic bag.

4) Spills: Botulinal toxin is rapidly inactivated with alkali. Flood the area of the spill with 5 volumes or more of 0.1 N NaOH and cover with paper towels. After a few minutes, wipe up with paper towels. Autoclave towels before discarding. Work surfaces may also be cleansed with NaOH.

5) Decontamination of equipment and waste materials: Botulinal toxins are destroyed by exposure to 100°C for 10 min. Because of the lethal nature of the toxins, a wide margin of safety is necessary. Autoclaving of toxic materials should be done for 60 min at 121°C. This assures that sufficient heat has penetrated to inactivate all toxin and that all spores are killed. If volumes greater than 1 liter are involved, autoclaving should be done for 2 h.

6) Immunization of personnel: All persons working directly with known cultures of *C. botulinum* or botulinal toxins should be immunized with pentavalent botulism toxoid, which is available from CDC.

7) Restricted work area: Work with *C. botulinum* or botulinal toxin should be done only in certain laboratory rooms and at certain bench areas. Unrelated work should not be permitted in these areas. Visitors and other workers should not be permitted to sit down at these sites nor to handle the equipment or items used in the work with the organism or toxin.

Collection of Specimens

Materials suitable for laboratory investigation of botulism are listed in Table 5. Ten ml or more of serum obtained as close to onset of illness and before anti-

TABLE 5. Suitable specimens for laboratory confirmation of botulism[a]

Suspected form of botulism	*Suitable materials*
Foodborne	Clinical materials: serum, gastric contents, stool, return from sterile water enema, vomitus
	Autopsy samples: serum, gastric and intestinal contents
	Food samples
Infant	Feces
	Return from sterile water enema
	Serum
	Autopsy samples: intestinal contents from different levels of small and large intestine
	Food and environmental samples as appropriate for the investigation
Wound	Serum
	Exudate, tissue or swab samples of wound
	Feces or return from sterile water enema (wound may not be source)

[a] Adapted from Dowell (1979).

toxin administration, stool specimens (about 25 g), and gastric content and vomitus samples (when available) are suitable patient specimens in suspected foodborne botulism incidents. When wound botulism is suspected, a wound specimen in addition to serum and stool samples should be obtained for culturing. A series of stool specimens (at least 2 specimens) is sufficient for confirming or ruling out infant botulism, but serum (if available) may be tested for toxin. The latter are often not obtained to avoid unnecessary trauma to the infants. Foods temporally associated with the onset of illness of noninfant cases must be considered as possible vehicles for ingested toxin; home canned foods of low acidity (pH > 4.6), low salt or sugar (Aw ≥ 0.94), held at temperatures at which the organism can grow, and that are consumed without prior heating are likely vehicles. Only on rare occasions are commercially processed foods incriminated. Representative samples of suspected foods should be transferred aseptically into sterile specimen containers, or the food may be submitted in the original containers. Washings from empty containers with only remnants of foods adhering to the interior surfaces can be tested for toxin and cultured. Specimens should be refrigerated during transport to the laboratory (e.g., shipped in insulated packages containing several cool-packs) and while holding before laboratory examination. Freezing has been shown to inactivate purified toxin, so it is usually not recommended for botulism specimens, although it may not have such a deleterious effect on toxin in specimens. Specimens that have already been frozen should remain frozen until they are ready to be tested. Arrangements should be made with the reference laboratory before shipment of specimens. This is usually initiated by contacting the state health department for investigations within the United States. In some states, the decision will be made to test the specimens in the state or local health department laboratory; in others, arrangements will be made to send the specimens to the CDC. When commercial products are suspected, much of the testing of foods is done in U.S. Food and Drug Administration laboratories. The immediate contact with the state health department will be helpful for consultation on the diagnosis, making available the therapeutic antitoxin if needed, arranging for laboratory testing, and assuring that emergency action will be taken if necessary to limit the extent of the outbreak.

Laboratory Procedures

The procedures presented in this section are a detailed description of laboratory methods partially published elsewhere (Centers for Disease Control, 1979a; Hatheway, 1979).

Detection and Identification of Botulinal Toxin

The mouse bioassay is the method of choice in testing for botulinal neurotoxins in investigative specimens. Mice are extremely sensitive to the toxins. They react to biological activity of the toxins, not to antigenic structures, and permit the observation of signs of botulism. They will react similarly to new or atypical, unexpected antigenic variants of the neurotoxin, even if they have not been recognized previously and for which no serological reagents are available. Caution and experience are required for recognizing and overcoming occasional problems of nonspecific mouse deaths (somewhat common with fecal specimens). Specificity of toxin test results is established by use of a sufficient variety of antitoxins (usually types A, B, and E monovalent and polyvalent) and observation of mice for signs of botulism before death. Failure to achieve a specific neutralization may be due: 1) entirely to a lethal nonbotulinal substance; 2) to a mixture of botulinal toxin, which is actually neutralized by one of the antitoxins, and a nonbotulinal substance which is not affected by any of the antitoxins; or 3) an excessive amount of botulinal toxin that cannot be entirely neutralized by the antitoxin. In the latter case, titration of the toxicity using 10-fold dilutions should indicate a suitable dilution at which the neutralization test may successfully be repeated. When there is a mixture of toxic substances, the botulinal toxin often has the greater titer, and repetition of the neutralization test at a dilution closer to the lethal end point will succeed in identifying the botulinal toxin.

Mice. ICR strain, male white mice, weighing 18 to 25 g, are used routinely at the CDC laboratory, but little difference in suitability was noted with several laboratory strains, (ICR, CFW, or Balb/c) either sex, or weights ranging from 12 to 33 g (Hatheway et al., 1985).

Diagnostic Antitoxins. Lyophilized monovalent A, B, C, D, E, F, and G, and polyvalent (mixture of types A, B, C, D, E, and F) antitoxins can be purchased by qualified laboratories from CDC. When reconstituted with 50% glycerol in the volume indicated on the label, the monovalent reagents contain from 2 to 10 IU per ml. The indicated unitage is only approximate and, thus, the reagents cannot serve as accurate quantitative standards. (International Antitoxin Standards are available from the World Health Organization for that purpose.) It must be kept in mind that mice can tolerate only a limited amount of glycerol. The CDC reagents are reconstituted on the premise that only 0.1 ml will be included in the mixture injected into mice. Some published laboratory methods for identifying botulinal toxins are performed by i.p. injection of 0.5 ml (1.0 IU) of antitoxin into mice 30 min before injection of the test sample (Kautter and Lynt, 1978). In that case, the reconstitution cannot be made with 50% glycerol because 0.5 ml would be lethal for the mice.

The unitage designations for the reconstituted antitoxins are helpful for estimating the maximal amount of toxin in the injected dose that can be neutralized by the added antitoxin. It is generally assumed that 1 IU of antitoxin will neutralize 10,000 mouse LD_{50}, except for type E for which 1 IU will neutralize 1,000 LD_{50}. The actual quantities vary for toxins from different strains and for different preparations of toxin from a given strain (Hatheway et al., 1985).

Test Materials. Liquid specimens can be tested by direct i.p. inoculation into mice. Solid and semisolid specimens such as foods and stools must be extracted, with the extracts then tested. Specimens are extracted by suspending them in gelatin phosphate buffer (0.2% gelatin, 0.4% Na_2HPO_4, adjusted to pH 6.2 with HCl, sterilized by autoclaving) and recovering the supernatant after centrifugation. Satisfactory suspension of some specimens may require cutting them into small pieces and grinding with mortar and pestle. Sterile utensils and containers must be used to assure that organisms are not being introduced into materials that are going to be cultured. The ratio of buffer to solid material may vary from 1 ml/g to 5 ml/g, depending on the dryness of the specimen or the consistency of the suspension. The suspensions are held for several hours or overnight in the refrigerator before centrifugation to assure solubilization of the toxin. Incubation in the refrigerator may be omitted initially, and will not be needed if the test of the extract is positive without that step. If it is negative, however, another portion of the suspension should be retested after the holding period.

Setting Up the Test. Portions of liquid samples and extracts are neutralized before mouse injection by treatment of aliquots with the several antitoxins to be used; with the exception of serum, aliquots of samples are also heat-treated (100°C, 10 min) to demonstrate heat lability, and trypsinized (1 part 1.5% trypsin solution [Difco 1:250] plus 4 parts of test sample) to test for enhancement of activity of low concentrations of toxin. The test material is distributed into a series of tubes, as illustrated in Table 6. To 1 ml of sample is added 0.25 ml of antitoxin or trypsin in the designated tubes; the mixtures are incubated for 30 min at 37°C. The sample for heat treatment is plugged with cotton (to reduce evaporation) and placed in a boiling water bath for 10 min. After treatment, a sufficient volume of material from each tube for injecting 2 mice is drawn into a separate syringe. The mice are injected i.p. with 0.4 ml of sample or 0.5 ml of sample mixed with antitoxin or trypsin. Each set of mice is identified by a distinctive dye marking.

Serum is not heat-treated because heat coagulates the sample, making it unsuitable for injection. Trypsin treatment of serum has not been recommended because it is generally assumed that toxin reaching the circulation would be fully activated by exposure to digestive enzymes. Results in some recent investigations, however, indicate that trypsinization of serum samples may sometimes enhance the demonstrability of the toxin (Centers for Disease Control, unpublished observations; A. H. W. Hauschild, personal communication).

Observing and Recording Test Results. In each investigation, the mice in the first tests should be observed at short intervals (e.g., 2, 6, 10, and 18 h after injection) in an attempt to detect signs of botulism. Sometimes an early confirmation of botulism can be

TABLE 6. Protocol for mouse toxicity and neutralization test for liquid samples and extracts of solid specimens

Tube no.	*Treatment*[a]	*Test material (ml)*	*Reagent (ml)*	*Amount to syringe (ml)*	*Amount injected/mouse (ml)*
1	Untreated	1.0	—	0.8	0.4
2	Heated[b]	1.0	—	0.8	0.4
3	Trypsin[c]	1.0	0.25	1.0	0.5
4	Anti A	1.0	0.25	1.0	0.5
5	Anti B	1.0	0.25	1.0	0.5
6	Anti E	1.0	0.25	1.0	0.5
7	Polyvalent[d]	1.0	0.25	1.0	0.5

[a] After adding trypsin or antitoxin, the mixtures are incubated for 30 min at 37°C; antitoxin reagents contain approximately 1 IU/0.1 ml.
[b] Heat-treated samples are processed by placing in a boiling water bath for 10 min.
[c] 0.5% Aqueous solution of trypsin (Difco 1:250) is used as the reagent (column 4).
[d] Polyvalent reagent contains type A, B, C, D, E, and F antitoxins.

made provisionally from the appearance of the distinctive signs and the protective effect of one specific reagent and the polyvalent reagent. Mouse deaths due to botulinal toxin usually occur within 24 h and rarely after 48 h. Signs of botulism in mice are generalized weakness, especially pronounced in the rear quarters, usually accompanied by loss of abdominal tone and an indrawing of the abdominal and loin areas (wasp-shaped body) and development of respiratory difficulty, progressing to gasping and finally respiratory cessation. Figure 3 shows the abnormal body conformation resulting from botulism paralysis in mice. Ptosis may be apparent in the later stages, and is recognized by a slit-eyed rather than the normal round-eyed appearance. At the time of death, a moment of convulsion may be observed. Mice receiving neutralized samples will show none of these signs, but may show some effects of other deleterious substances. A record should be made of the observed signs and of death or survival of the mice in each test. Protected mice must survive until the test is terminated at the end of 4 days. Sometimes nonspecific deaths occur on the third and fourth day after specific neutralization has clearly been observed through the second day. These deaths are due to substances other than botulinal toxin or other pathogens included in the injected sample. Nonspecific deaths sometimes make conclusive judgment difficult. Such problems often can be resolved by repeating the tests at a higher dilution of the sample.

It may be noted that on rare occasions, sera from patients ultimately diagnosed with Guillain-Barré syndrome have caused signs in mice suggestive of botulism, but that these signs were not eliminated by any of the antitoxins. Furthermore, the signs were more transitory than those caused by botulinal toxin, with surviving mice improving during the 4-day observation period; with botulism, the signs become more pronounced over the 4-day period. High levels of medication may result in serum levels that may have an effect on mice. Residues of medication in feces have interfered with mouse tests, but these substances may be eliminated by dialysis (Horwitz et al., 1976). High salt content in food extracts may also be injurious to mice, but may likewise be removed by

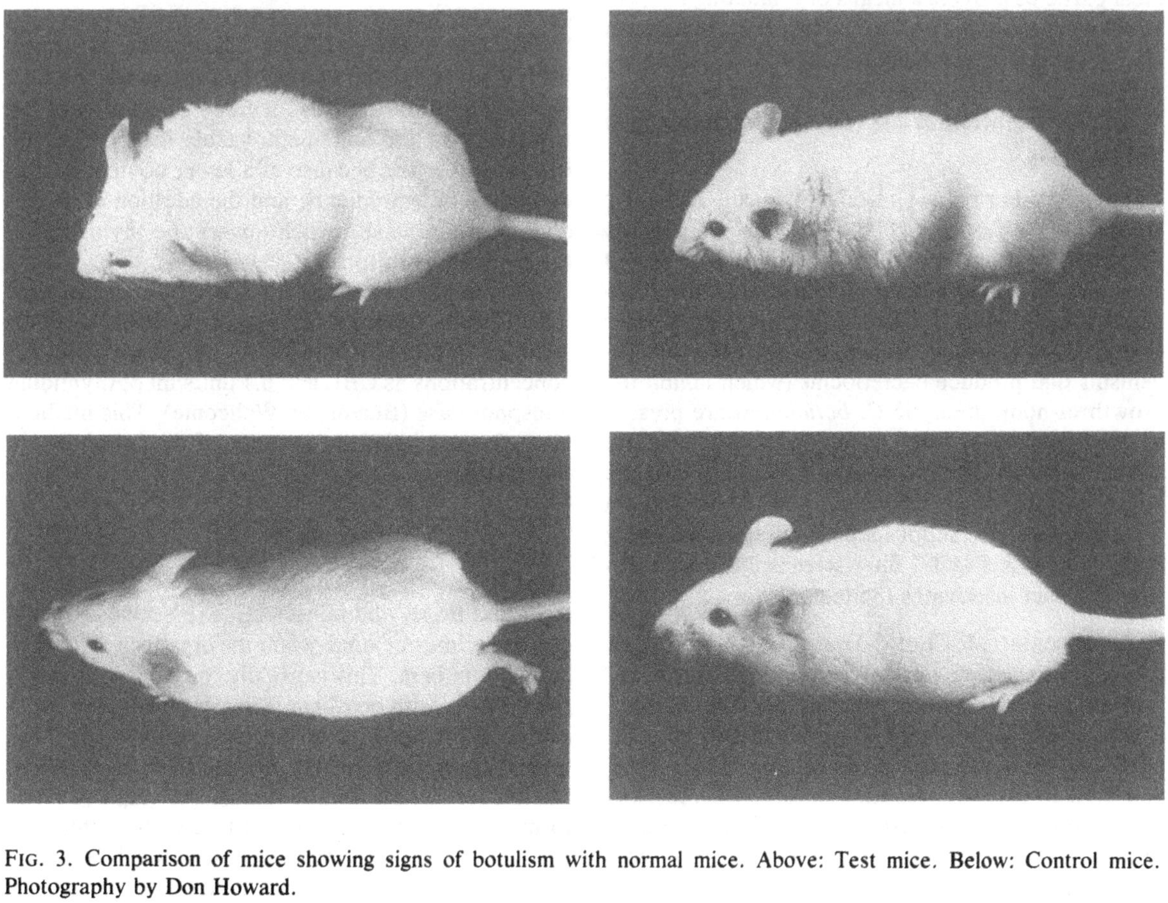

FIG. 3. Comparison of mice showing signs of botulism with normal mice. Above: Test mice. Below: Control mice. Photography by Don Howard.

dialysis to permit satisfactory toxin testing (Centers for Disease Control, unpublished observations).

Culturing Specimens for Recovery of Clostridium botulinum

Recovery of *C. botulinum* from specimens is very important in botulism investigations. Positive stool cultures from a suspected foodborne case, especially when specimens have been taken late after onset, may be the only confirmatory results obtained. Recovery of an organism that produces the same type of toxin from specimens of patients whose other specimens tested toxin-positive affirms the validity of the toxicity tests. Examination of the causative organism from some cases of human botulism has revealed some very interesting and unexpected facts, especially the wide variety of clostridia that can cause botulism.

Media. CHOPPED MEAT-GLUCOSE-STARCH (CMGS): This is prepared as described by Dowell et al. (1977). It is satisfactory for enrichment cultures of food, clinical and environmental specimens, and for culturing of pure isolates of most *Clostridium* species. The starch in the medium promotes germination of spores by binding certain inhibiting substances, and the meat particles provide a means of evaluating the proteolytic capability of isolated strains. This medium is also satisfactory for holding cultures for extended periods. Clostridia usually remain viable for months at ambient temperature and for years when frozen in this medium.

TRYPTICASE-PEPTONE-GLUCOSE-YEAST EXTRACT BROTH WITH TRYPSIN (TPGYT): This prepared as described by Lilly et al. (1971) and by Kautter and Lynt (1978). This medium is sometimes more successful than CMGS for isolating type E (and other Group II) organisms. Sometimes nontoxigenic organisms that produce bacteriocins (which inhibit the growth of nonproteolytic *C. botulinum*) are present in the specimens being examined. The medium, which contains 2% trypticase, 0.5% peptone, 2% yeast extract, 0.4% glucose, and 0.1% sodium thioglycollate, is supplemented after autoclaving with 1/10 volume of 1% filter-sterilized trypsin solution. Trypsin inactivates the bacteriocins.

CDC MODIFIED MCCLUNG-TOABE EGG YOLK AGAR (EYA): This is a nonselective solid medium for direct isolation or isolation from enrichment cultures. This medium, as prepared according to Dowell et al. (1977), is a modification of the medium of McClung and Toabe (1947) by the addition of 0.5% yeast extract and use of trypticase instead of proteose peptone. Egg yolk in the medium shows the lipase reaction in the vicinity of the colonies. This is a key characteristic for recognition of *C. botulinum*.

SELECTIVE EGG YOLK AGAR NO. 1—CLOSTRIDIUM BOTULINUM ISOLATION (CBI) MEDIUM: This medium, described by Dezfulian et al. (1981), incorporates three antibiotics to which Group I *C. botulinum* is resistant into the EYA medium (just described). Since Group II organisms are sensitive to trimethoprim, it cannot be recommended for all investigations. This medium is prepared according to the formulation for EYA, but after adding the egg yolk suspension aseptically to the autoclaved constituents, the filter-sterilized antibiotic solutions are added to 1 liter of medium as follows:

1. 25 ml 1% cycloserine (Vega Biochemicals, Tucson, Arizona),
2. 4 ml 1.9% sulfamethoxazole (Burroughs-Wellcome Co., Research Triangle Park, North Carolina), and
3. 4 ml 0.1% trimethoprim (Burroughs-Wellcome).

The antibiotics are dissolved in distilled water. For solubilization, the sulfamethoxazole requires careful alkalinization with 1N NaOH, while the trimethoprim requires slight acidification with 0.1N HCl. The plates are stored in a refrigerator and have a limited shelf-life of 2 weeks, since the cylcoserine loses activity after longer periods.

SELECTIVE EGG YOLK AGAR NO. 2—BOTULINUM-SELECTIVE MEDIUM (BSM): This was developed by Mills et al. (1985) to improve the suppression of fecal flora over that of the CBI medium. The action of the trimethoprim and sulfamethoxazole in BSM should be more effective because of a lower content of thymidine in the ingredients, and the addition of thymidine phosphorylase, which lowers the thymidine in the medium even further.

The medium contains 2.5% dehydrated heart-infusion broth, 2% agar, 3% egg yolk suspension (instead of 10% in CBI and EYA), the same antibiotic concentrations as CBI, and 0.1 units/ml of thymidine phosphorylase (Burroughs-Wellcome). This medium has a 2 week shelf-life because of the instability of the cycloserine.

Direct Isolation of C. botulinum from Specimens. Isolation of *C. botulinum* from stool and food specimens by direct streaking of samples onto selective (CBI and BSM) and nonselective (EYA) agar media can sometimes be done when the organism is present in high numbers. This is usually the case with stools from infants with botulism. Selective media are superior to EYA because of the high numbers of other organisms in the feces (Glasby and Hatheway, 1985). Studies for comparison of media have employed uniform inoculation of the agar surface with diluted suspensions of the specimens (Glasby and Hatheway, 1985; Mills et al., 1985), but dilution streaking of a few drops over the surface of the plate offers a good likelihood of obtaining isolated colonies using a single plate. Inoculated plates are incubated at 35°C in an anaerobic chamber or an anaerobic jar. The plates may be observed periodically up to 48 h for colonies with features suggestive of *C. botulinum*. Recognition of colonies depends on the lipase reaction, which is positive for the causative organism in all but the rarest of cases (Hall et al., 1985; McCroskey et al., 1986). Microscopic observation should show the expected cellular morphology (Fig. 2). Toxigenicity of the isolated organism must be established by use of the mouse test on broth cultures. Isolation of the toxigenic organism from directly streaked plates allows one to discontinue further work on enrichment cultures, which may be in progress.

Enrichment Cultures. Samples of suspensions of stools or foods, wound tissue specimens, rectal swabs, swabs from wounds, and environmental specimens are cultured in CMGS broth for at least 4 days in an anaerobic chamber or jar, after which culture supernatants are assayed for botulinal toxin by mouse tests. Samples of the enrichment cultures are also streaked on selective (CBI and BSM) and/or nonselective egg yolk agar (EYA). Usually, cultures containing demonstrable toxin will readily yield toxigenic organisms recognizable by their lipase reaction on the agar plates. Isolation of toxigenic organisms may be difficult, however, if nontoxigenic lipase-positive organisms are present in the culture. In some cases, toxigenic organisms are isolated from enrichment cultures when the toxicity tests on the enrichment culture supernatants are negative.

When inoculating enrichment cultures, two tubes of CMGS are usually inoculated, with one receiving a heat treatment (80°C water bath, 10 min) to eliminate competing vegetative organisms, leaving the heat-stable spores to germinate in the culture. Some specimens seem to lack heat-stable spores and toxigenic organisms grow only in the unheated culture. Certain strains, usually Group II, produce spores with low heat stability, and sometimes recovery is not successful from either the untreated or the heated enrichment cultures. In these cases, spore selection by alcohol treatment may be attempted; add an equal volume of ethanol to the sample and incubate 1 h at room temperature before inoculating into the medium.

Growth of *C. botulinum* type E has sometimes been inhibited by bacteriocins produced by accompanying organisms in the samples (Lilly et al., 1971). This problem may be overcome by inoculating the specimens into TPGYT medium for enrichment. The trypsin in the medium inactivates the bacteriocins. Thus, when investigating outbreaks caused by type E or other Group II strains, one must consider alcohol treatment and TPGYT if other methods fail.

Characterization of Isolated Organisms. In the investigation of human botulism, one expects the causative organism to belong to either Group I or Group II (Table 2). For minimum characterization, the following determinations should be done:

1. lipase reaction,
2. toxigenicity and specific neutralization,
3. Gram stain or phase-contrast observation for presence and location of spores (Fig. 2),
4. growth under aerobic (candle jar) and anaerobic conditions,
5. digestion of milk proteins, and
6. volatile fatty acid end products (GLC procedures according to Lombard and Dowell, 1982).

This information is sufficient to identify the isolate as *C. botulinum* and to establish its toxin type and physiologic group. If any unusual features are noted, it would be of interest to do a complete characterization as described by Dezfulian and Dowell (1980) or Dowell and Hawkins (1974).

Experimental Laboratory Approaches

In Vitro Methods for Detection of Toxin

For several reasons, in vitro methods have been sought for detecting and identifying botulinal toxins. Mice are expensive, and special facilities are needed to house them. Tests are maintained for 4 days, and this is often misunderstood as requiring 4 days before any results can be obtained. In addition, bioassays seem obsolete, subjective, and less precise than methods that employ measurement of physical or chemical changes using sophisticated instrumentation. In recent times, there has been a general movement away from the use of laboratory animals. Among methods described in the literature are immunodiffusion (ID) and reversed passive hemagglutination (RPHA) (Sakaguchi et al., 1974), radioimmunoassay (RIA) (Boroff and Shu-Chen, 1973), and enzyme-linked immunosorbent assay (ELISA) (Notermans and Kozaki, 1981). There are several problems which make it doubtful that any of the methods will be able to completely replace the conventional mouse bioassay. Sensitivity, specificity, and the serological variability of the toxins are responsible for technical problems, while the low incidence of botulism may make it impractical to have test kits or reagents available for investigators.

Mouse tests can detect as low as 5 to 10 pg (1 to 2 LD_{50}) of type A toxin protein in the injected sample. Shone et al. (1985) described a monoclonal antibody-based immunoassay for type A botulinal toxin, which uses an enzyme amplification system with a sensitivity "comparable to the mouse bioassay." The sensitivity of serologic tests is usually determined for the antigen from the homologous strain or, at most, for a rather limited number of other strains. Serological properties of toxins produced by strains designated as the same toxin type vary. Monoclonal reagents, in particular, will have to be evaluated against such variants. An adequate test system for investigating human botulism must be able to detect and identify all types of botulinal toxin, including toxins with serological properties never before encountered. Mouse tests seem to provide a universal detection system.

A major obstacle to a practical in vitro test for botulism diagnosis is the low incidence of botulism. Demand will not be sufficient to warrant commercial development and marketing. Whether laboratory confirmation is done by conventional or serological tests, it will continue to be done, for the most part, in special reference laboratories.

In vitro tests can be used for detecting toxin in specimens and cultures in investigations in which the toxin type has already been established by conventional methods, and with experimental samples, especially when large numbers must be assayed. Such methods can also be used with confidence in measuring antibodies to toxins of known types and from known strains. The ELISA tests could be used, for example, in determining the immune status of persons immunized with botulinal toxoids and assaying therapeutic antitoxins and diagnostic reagents. They no doubt will have considerable usefulness in many experimental studies.

Fluorescent Antibody Techniques

Fluorescent antibody techniques (FA) have been proposed for use in the laboratory confirmation of botulism. An evaluation of reagents prepared against type A, B, E, and F strains (both Groups I and II) of *C. botulinum* showed that the reagent made against each group and toxin type reacted with 100% of the evaluation strains of the homologous toxin type within the group, with considerable cross-reaction with other toxin types, especially within each group, but with only limited cross-reaction with nontoxigenic organisms (Glasby and Hatheway, 1983). Using a trivalent combination of reagents against organisms of Group I toxin types A, B, and F, tests were performed on 77 fecal samples from hospitalized infants for whom botulism tests were requested. Of the 34 specimens that were culture-positive for *C. botulinum* at the time of the FA study, 29 showed FA-positive bacilli on the direct smear of the specimen, and enrichment cultures were positive after 24 to 72 h of incubation for all 34 specimens (Glasby and Hatheway, 1984). All of the 43 fecal samples that were culture-negative were also FA-negative on both the direct smears and the 72 h enrichment cultures. All of the specimens that were toxin-positive in the mouse test of the specimen extract were FA-positive on the direct smear. Thus, it appears that the FA test on stools is a good method for presumptive confirmation of infant botulism. Since the reagents occasionally react with some nontoxigenic organisms and since the toxin type could not be determined by FA results, confirmation of toxicity and type determination had to be done by mouse tests. Although the FA approach may provide presumptive confirmation a few hours earlier, it really offers little to the investigation in a laboratory that routinely performs mouse tests. At the present time, FA reagents for *C. botulinum* are not commercially available.

Immunity to Botulism

Although botulinal neurotoxins are antigenic, neutralizing antibodies have not been detected in the serum of persons who have recovered from botulism (CDC, unpublished data; Dickson, 1926–1927). Persons who recovered from type E botulism showed no evidence of an antigenic stimulus; they failed to show an anamnestic response when injected with type E botulinal toxoid (Koenig et al., 1964). Because of such observations, a conclusion has been expressed by the cliche, "the lethal dose is less than the antigenic dose." It has not been possible to confirm botulism retrospectively by demonstrating antitoxin in the serum.

Active immunization of humans and animals using toxoids is done quite successfully. Pentavalent (ABCDE) toxoids have been developed for human use (Cardella, 1964). They are available for immunization of laboratory workers who are at risk because of work with *C. botulinum* and its toxins. The toxoid is distributed as an Investigational New Drug (IND) by the CDC Drug and Immunobiological Service, Atlanta, Georgia. The toxoid may be requested by a licensed physician who will agree to serve as an "investigator." The agreement will require the physician to follow the IND protocol for this product. Immunization involves 3 initial series injections at 0, 2, and 12 weeks, and a first booster at 52 weeks. Each recipient is evaluated subsequently at 2-year intervals to determine if boosters are needed.

Literature Cited

Arnon, S. S. 1983. Infant botulism: a pan-Pacific perspective. Med. J. Aust. 1:4–5.

Arnon, S. S. 1985. Infant botulism, p. 39–57. *In* S. P. Borriello (ed.), Clostridia in gastrointestinal disease. CRC Press, Boca Raton, Florida.

Arnon, S. S., T. F. Midura, S. A. Clay, R. M. Wood, and J.

Chin. 1977. Infant botulism: epidemiological, clinical and laboratory aspects. J. Am. Med. Assoc. **237**:1946–1951.

Arnon, S. S., T. F. Midura, K. Damus, B. Thompson, R. M. Wood, and J. Chin. 1979. Honey and other environmental risk factors for infant botulism. J. Pediatr. **94**:331–336.

Arnon, S. S., T. F. Midura, K. Damus, R. M. Wood, and J. Chin. 1978. Intraintestinal infection and toxin production by *Clostridium botulinum* as one cause of sudden infant death syndrome. Lancet **1**:1273–1277.

Aureli, P., L. Fenicia, B. Pasolini, M. Gianfranceschi, L. M. McCroskey, and C. L. Hatheway. 1986. Two cases of type E infant botulism caused by neurotoxigenic *Clostridium butyricum* in Italy. J. Infect. Dis. **154**:207–211.

Bengtson, I. A. 1922. Preliminary note on a toxin-producing anaerobe isolated from the larvae of *Lucilia caesar*. Public Health Rep. **37**:164–170.

Bengtson, I. A. 1924. Studies on organisms concerned as causative factors in botulism. Hygiene Laboratory Bulletin No. 136, U.S. Public Health Service, Government Printing Office, Washington, D.C.

Black, R. E., and R. A. Gunn. 1980. Hypersensitivity reactions associated with botulinal antitoxin. Am. J. Med. **69**:567–570.

Boroff, D. A., and G. Shu-Chen. 1973. Radioimmunoassay for type A toxin of *Clostridium botulinum*. Appl. Microbiol. **25**:545–549.

Burke, G. S. 1919. Notes on *Bacillus botulinus*. J. Bacteriol. **4**:555–565.

California State Department of Health Services. 1983. Infant botulism in California and elsewhere. Calif. Morbid. Wkly. Rep. No. 15.

California State Department of Health Services. 1984. Honey, corn syrups and infant botulism. Calif. Morbid. Wkly. Rep. No. 14.

Cardella, M. A. 1964. Botulinum toxoids. p. 113–130, In: K. H. Lewis and K. Cassell, Jr. (ed.), Botulism, proceedings of a symposium. U.S. Public Health Service Publication No. 999-FP1, Washington, D.C.

Centers for Disease Control. 1978. Botulism in New Mexico. Morbid. Mortal. Weekly Rep. **27**:138–145.

Centers for Disease Control. 1979. Botulism—United States, 1978. Morbid. Mortal. Weekly Rep. **28**:73–75.

Centers for Disease Control. 1979a. Botulism in the United States 1899–1977. Handbook for epidemiologists, clinicians and laboratory workers. U.S. Department of Health and Human Services, U.S. Public Health Service, Atlanta, Georgia.

Cherington, M. 1982. Electrophysiologic methods as an aid in diagnosis of botulism: a review. Muscle Nerve **5**:S28–S29.

Chia, J. K., J. B. Clark, C. A. Ryan, and M. Pollack. 1986. Botulism in an adult associated with foodborne intestinal infection with *Clostridium botulinum*. N. Engl. J. Med. **315**:239–241.

Chin, J., S. S. Arnon, and T. F. Midura. 1979. Food and environmental aspects of infant botulism in California. Rev. Infect. Dis. **1**:693–696.

Ciccarelli, A. S., A. M. Giulietti, and D. F. Gimenez. 1976. Segunda cepa de *Clostridium botulinum* tipo G. IV. Jornadas Argentinas de Microbiologia, Buenos Aires.

Ciccarelli, A. S., D. N. Whaley, L. M. McCroskey, D. F. Gimenez, V. R. Dowell, Jr., and C. L. Hatheway. 1977. Cultural and physiological characteristics of *Clostridium botulinum* type G and susceptibility of certain animals to its toxin. Appl. Environ. Microbiol. **34**:843–848.

Clay, S. A., J. C. Ramseyer, L. S. Fishman, and R. P. Sedgwick. 1977. Acute motor unit disorder: infant botulism? Arch. Neurol. **34**:236–243.

Cornblath, D. R., J. T. Sladky, and A. J. Summer. 1983. Clinical electrophysiology of infant botulism. Muscle Nerve **8**:448–452.

Davis, J. B., L. H. Mattman, and M. Wiley. 1951. *Clostridium botulinum* in a fatal wound infection. J. Am. Med. Assoc. **146**:646–648.

Dezfulian, M., and V. R. Dowell, Jr. 1980. Cultural and physiological characteristics and antimicrobial susceptibility of *Clostridium botulinum* isolates from foodborne and infant botulism cases. J. Clin. Microbiol. **11**:604–609.

Dezfulian, M., L. M. McCroskey, C. L. Hatheway, and V. R. Dowell, Jr. 1981. Selective medium for isolation of *Clostridium botulinum* from human feces. J. Clin. Microbiol. **13**:526–531.

Dickson, E. C. 1926–1927. The present status of the botulism problem. Pathogenicity toxicology and immunology. J. Prevent. Med. **1**:71–83.

Dowell, V. R., Jr. 1979. Collection, handling and shipment of specimens for laboratory confirmation of anaerobic bacterial diseases. U.S. Dept. HHS, PHS, Centers for Disease Control, Atlanta, Georgia.

Dowell, V. R., Jr., and T. M. Hawkins. 1974. Laboratory methods in anaerobic bacteriology. Center for Disease Control Laboratory Manual. U.S. Dept. HHS, PHS, Centers for Disease Control, Atlanta, GA.

Dowell, V. R., Jr., G. L. Lombard, F. S. Thompson, and A. Y. Armfield. 1977. Media for isolation, characterization and identification of obligately anaerobic bacteria. U.S. Dept. HHS, PHS, Centers for Disease Control, Atlanta, GA.

Dowell, V. R., Jr., L. M. McCroskey, C. L. Hatheway, G. L. Lombard, J. M. Hughes, and M. H. Merson. 1977. Coproexamination for botulinal toxin and *Clostridium botulinum:* a new procedure for laboratory diagnosis of botulism. J. Am. Med. Assoc. **238**:1829–1832.

Eklund, M. W., and V. R. Dowell, Jr. (ed.). 1987. Avian botulism: an international perspective. Charles C Thomas, Springfield, Illinois.

Eklund, M. W., and F. T. Poysky. 1974. Interconversion of type C and D strains of *Clostridium botulinum* by specific bacteriophages. Appl. Microbiol. **27**:251–258.

Eklund, M. W., F. T. Poysky, J. A. Meyers, and G. A. Pelroy. 1974. Interspecies conversion of *Clostridium botulinum* type C to *Clostridium novyi* type A by bacteriophage. Science **186**:456–458.

Eklund, M. W., D. I. Wieler, and F. T. Poysky. 1967. Outgrowth and toxin production of non-proteolytic type B *Clostridium botulinum* at 3.3–5.6°C. J. Bacteriol. **93**:1461–1462.

Gimenez, D. F., and A. S. Ciccarelli. 1970. Another type of *Clostridium botulinum*. Zentralbl. Bakteriol., Abt. 1 Orig. **215**:221–224.

Glasby, C., and C. L. Hatheway. 1983. Fluorescent antibody reagents for the identification of *Clostridium botulinum*. J. Clin. Microbiol. **18**:1378–1383.

Glasby, C., and C. L. Hatheway. 1984. Evaluation of fluorescent-antibody tests as a means of confirming infant botulism. J. Clin. Microbiol. **20**:1209–1212.

Glasby, C., and C. L. Hatheway. 1985. Isolation and enumeration of *Clostridium botulinum* by direct inoculation of infant fecal specimens on egg yolk agar and *Clostridium botulinum* isolation media. J. Clin. Microbiol. **21**:264–266.

Green, J., H. Spear, and R. R. Brinson. 1983. Human botulism type F: a rare type. Am. J. Med. **75**:893–895.

Grover, W. D., G. J. Peckham, and P. H. Berman. 1974. Recovery following cranial nerve dysfunction and muscle weakness in infancy. Dev. Med. Child Neurol. **16:**163–171.

Gunnison, J. B., J. R. Cummings, and K. F. Meyer. 1936–1937. *Clostridium botulinum* type E. Proc. Soc. Exp. Biol. Med. **35:**278–280.

Gunnison, J. B., and K. F. Meyer. 1929. Cultural study of an international collection of *Clostridium botulinum* and *parabotulinum*. J. Infect. Dis. **45:**119–134.

Gurwith, M. J., C. Langston, and D. M. Citron. 1981. Toxin-producing bacteria in infants. Am. J. Dis. Child. **135:**1104–1106.

Hall, J. D., L. M. McCroskey, B. J. Pincomb, and C. L. Hatheway. 1985. Isolation of an organism resembling *Clostridium barati* which produces type F botulinal toxin from an infant with botulism. J. Clin. Microbiol. **21:**654–655.

Hatheway, C. L. 1979. Laboratory procedures for cases of suspected infant botulism. Rev. Infect. Dis. **1:**647–651.

Hatheway, C. L., M. C. Ferreira, and L. M. McCroskey. 1985. Evaluation of various factors in the mouse toxicity test for identification of botulinal toxin (abstr.) Presented at the 85th Annual Meeting of the American Society for Microbiology, Las Vegas, Nevada.

Hatheway, C. L., and L. M. McCroskey. 1981. Laboratory investigation of human and animal botulism, p. 165–180. *In* G. E. Lewis, Jr. (ed.), Biomedical aspects of botulism. Academic Press, New York.

Hatheway, C. L., and L. M. McCroskey. 1987. Examination of feces and serum for diagnosis of infant botulism in 336 patients. J. Clin. Microbiol. **25:**2334–2338.

Hatheway, C. L., J. D. Snyder, J. E. Seals, T. A. Edell, and G. E. Lewis. 1984. Antitoxin levels in botulism patients treated with trivalent equine botulism antitoxin to toxin types A, B, and E. J. Infect. Dis. **150:** 407–412.

Hazen, E. L. 1937. A strain of *B. botulinus* not classified as type A, B, or C. J. Infect. Dis. **60:**260–264.

Heinzel, G., V. Lenk, and V. Schneider. 1985. Ergbnisse bakteriologischer Untersuchungen zum plotzlichen unerwarteten Sauglingstod unter besonderer Berucksichtigung des Sauglingsbotulismus. Beitr. Gerichtl. Med. **43:**378–380.

Hoffman, R. E., B. J. Pincomb, M. R. Skeels, and M. J. Burkhart. 1982. Type F infant botulism. Am. J. Dis. Child. **136:**270–271.

Horwitz, M. A., C. L. Hatheway, and V. R. Dowell, Jr. 1976. Laboratory diagnosis of botulism complicated by pyridostigmine treatment of the patient. Am. J. Clin. Pathol. **66:**737–742.

Istre, G. R., R. Compton, T. Novotny, J. E. Young, C. L. Hatheway, and R. S. Hopkins. 1986. Infant botulism: three cases in a small town. Am. J. Dis. Child. **140:**1013–1014.

Jansson, L., I. Bjerre, C. Schalen, and P. A. Mardh. 1985. Forsta falled av infantil botulism i sverige? Lakartidningen **82:**955–956.

Kautter, D. A., T. Lilly, H. M. Solomon, and R. K. Lynt. 1982. *Clostridium botulinum* spores in infant foods: a survey. J. Food Protect. **45:**1028–1029.

Kautter, D. A., and R. K. Lynt. 1978. *Clostridium botulinum*. p. xiv-1–xiv-15, *In* Bacterial analytic manual, 5th ed. U.S. Food and Drug Administration, Washington, D.C.

Koenig, M. G., A. Spikard, M. A. Cardella, and D. E. Rogers. 1964. Clinical observations on type E botulism in man. Medicine **43:**517–544.

Kushnir, E. D., T. M. Brun, and S. S. Paikina. 1937. Istochniki infitsirovaniya krasnoy ribi mikrobom botulizma. Zh. Mikrobiol. Epidemiol. Immunobiol. **19:**80–85.

Landman, G. 1904. Ueber die Ursache der Darmstadter Bohnen Vergiftung. Hyg. Rundschr. **14:**449–452.

Lee, W. H., and H. Riemann. 1970. The genetic relatedness of proteolytic *Clostridium botulinum* strains. J. Gen. Microbiol. **64:**85–90.

Leuchs, J. 1910. Beitrage zur Kenntnis des Toxins und Antitoxins des *Bacillus botulinus*. Z. Hyg. Infektionskr. **65:**55–84.

Lilly, T., Jr., S. M. Harmon, and D. A. Kautter. 1971. An improved method for detection of *Clostridium botulinum* type E. J. Milk Food Technol. **34:**492–497.

Lombard, G. L., and V. R. Dowell, Jr. 1982. Gas-liquid chromatography analysis of the acid products of bacteria. U.S. Department of Health and Human Services, U.S. Public Health Service, Centers for Disease Control, Atlanta, Georgia.

Lombard, G. L., and V. R. Dowell, Jr. 1983. Comparison of three reagents for detecting indole production by anaerobic bacteria in microtest systems. J. Clin. Microbiol. **18:**609–613.

Long, S. S., J. L. Gajewski, L. W. Brown, and P. H. Gilligan. 1985. Clinical, laboratory, and environmental features of infant botulism in southeastern Pennsylvania. Pediatrics **75:**935–941.

Lynt, R. K., H. M. Solomon, and D. H. Kautter. 1984. Heat resistance of *Clostridium botulinum* type G in phosphate buffer. J. Food Protect. **47:**463–466.

MacDonald, K. L., G. W. Rutherford, S. M. Friedman, J. R. Dietz, B. R. Kaye, G. F. McKinley, J. H. Tenny, and M. L. Cohen. 1985. Botulism and botulism-like illness in chronic drug abusers. Ann. Intern. Med. **102:**616–618.

MacDonald, K. L., R. F. Spengler, C. L. Hatheway, N. T. Hargrett, and M. L. Cohen. 1985. Type A botulism from sauteed onions. J. Am. Med. Assoc. **253:**1275–1278.

McClung, L. S., and R. Toabe. 1947. The egg yolk plate reaction for the presumptive diagnosis of *Clostridium sporogenes* and certain species of the gangrene and botulinum groups. J. Bacteriol. **53:**139–147.

McCroskey, L. M., and C. L. Hatheway. 1984. Atypical strains of *Clostridium botulinum* isolated from specimens in infant botulism cases. (abstr.). Presented at the 84th Annual Meeting of the American Society for Microbiology, St. Louis, Mo.

McCroskey, L. M., and C. L. Hatheway. 1986. Laboratory findings in adult botulism cases that suggest colonization of the intestinal tract (abstr.). Presented at the 86th Annual Meeting of the American Society for Microbiology, Washington, D.C.

McCroskey, L. M., C. L. Hatheway, L. Fenicia, B. Pasolini, and P. Aureli. 1986. Characterization of an organism that produces type E botulinal toxin but which resembles *Clostridium butyricum* from the feces of an infant with type E botulism. J. Clin. Microbiol. **23:**201–202.

Merson, M. H., and V. R. Dowell, Jr. 1973. Epidemiologic, clinical and laboratory aspects of wound botulism. N. Engl. J. Med. **289:**1005–1010.

Meyer, K. F., and J. B. Gunnison. 1928. *Cl. botulinum* type D sp. Nov. Proc. Soc. Exp. Biol. Med. **26:**88–89.

Midura, T. F., and S. S. Arnon. 1976. Infant botulism: identification of *Clostridium botulinum* and its toxin in faeces. Lancet **2:**934–936.

Midura, T. F., G. S. Nygaard, R. M. Wood, and H. L.

Bodily. 1972. *Clostridium botulinum* type F: isolation from venison jerky. Appl. Microbiol. **24:**165–167.

Mills, D. C., T. F. Midura, and S. S. Arnon. 1985. Improved selective medium for the isolation of lipase-positive *Clostridium botulinum* from feces of human infants. J. Clin. Microbiol. **21:**947–950.

Moller, V., and I. Scheibel. 1960. Preliminary report on the isolation of an apparently new type of *Cl. botulinum.* Acta Pathol. Microbiol. Scand. **48:**80.

Morris, J. G., Jr. 1981. Current trends in therapy of botulism in the United States, p. 317–326. *In* G. E. Lewis, Jr. (ed.), Biomedical aspects of botulism. Academic Press, New York.

Moss, C. W., C. L. Hatheway, M. A. Lambert, and L. M. McCroskey. 1980. Production of phenylacetic and hydroxyphenylacetic acids by *Clostridium botulinum* type G. J. Clin. Microbiol. **11:**743–745.

Notermans, S., and S. Kozaki. 1981. Isolation and identification of botulinum toxins using the ELISA, p. 181–189. *In* G. E. Lewis, Jr. (ed.), Biomedical Aspects of Botulism. Academic Press, New York.

Pickett, J., B. Berg, E. Chaplin, and M. A. Brunstetter-Shafer. 1976. Syndrome.of botulism in infancy: clinical and electrophysiologic study. N. Engl. J. Med. **295:**770–772.

Rooney, J. R., and M. E. Prickett. 1967. Shaker foal syndrome. Mod. Vet. Pract. **48:**44–45.

Sakagucki, G., S. Sakaguchi, S. Kozaki, S. Sugii, and I. Ohishi. 1974. Cross-reaction in reversed phase passive hemagglutination between *C. botulinum* type A and B toxins and its avoidance by use of antitoxic component immunoglobulin isolated by affinity chromatography. Jpn. J. Med. Sci. Biol. **27:**161–172.

Schmidt, C. F. 1964. Spores of *Clostridium botulinum:* Formation, resistance, germination, p. 69–82. *In* K. H. Lewis and K. Cassel, Jr. (ed.). Botulism. Proceedings of a symposium. U.S. Public Health Service Publication. No. 999-FP1, Washington, D.C.

Seals, J. E., J. D. Snyder, T. A. Edell, C. L. Hatheway, C. J. Johnson, R. C. Swanson, and J. M. Hughes. 1981. Restaurant-associated type A botulism: transmission by potato salad. Am. J. Epidemiol. **113:**436–444.

Seddon, H. R. 1922. Bulbar paralysis in cattle due to the action of a toxicogenic bacillus with a discusison of the relationship of the condition to the forage poisoning (botulism). J. Comp. Pathol. Ther. **35:**147–190.

Shone, C., P. Wilton-Smith, N. Appleton, P. Hambleton, N. Modi, S. Gatley, and J. Melling. 1985. Monoclonal antibody-based immunoassay for type A *Clostridium botulinum* is comparable to the mouse bioassay. Appl. Environ. Microbiol. **50:**63–67.

Simpson, L. L. 1982. A comparison of the pharmacological properties of *Clostridium botulinum* type C_1 and C_2 toxins. J. Pharmacol. Exp. Ther. **223:**695–701.

Simpson, L. L. 1986. Molecular pharmacology of botulinum toxin and tetanus toxin. Ann. Rev. Pharmacol. Toxicol. **26:**427–453.

Smith, L. DS. 1977. Botulism: the organism, its toxins, the disease. Charles C Thomas, Springfield, Illinois.

Sonnabend, O. A., W. F. F. Sonnabend, R. Heinzle, T. Sigrist, R. Dirnhofer, and U. Krech. 1981. Isolation of *Clostridium botulinum* type G and identification of type G botulinal toxin in humans; report of five sudden unexpected deaths. J. Infect. Dis. **143:**22–27.

Sonnabend, O. A. R., W. F. F. Sonnabend, U. Krech, G. Molz, and T. Sigrist. 1985. Continuous microbiological and pathological study of 70 sudden and unexpected infant deaths. Toxigenic intestinal *Clostridium botulinum* infection in 9 cases of sudden infant death syndrome. Lancet **1:**237–241.

Suen, J. C. 1986. DNA homology studies on *Clostridium botulinum* and related clostridial species (Dissertation, D.P.H.), University of North Carolina, Chapel Hill, North Carolina.

Suen, J. C., C. L. Hatheway, A. G. Steigerwalt, and D. J. Brenner. 1986. Use of DNA relatedness to study taxonomic relationships among *Clostridium botulinum* type G and similar clostridial species (abstr.). Presented at the 86th Annual Meeting of the American Society for Microbiology, Washington, D.C.

Sugiyama, H. 1980. *Clostridium botulinum* toxin. Microbiol. Rev. **44:**419–448.

Sugiyama, H., and D. C. Mills. 1978. Intraintestinal toxin in infant mice challenged intragastrically with *Clostridium botulinum* spores. Infect. Immun. **21:**59–63.

Terranova, W., J. G. Breman, R. P. Locey, and S. Speck. 1978. Botulism type B: epidemiologic aspects of an extensive outbreak. Am. J. Epidemiol. **108:**150–156.

Thompson, J. A., L. A. Glascow, and J. R. Warpinski. 1980. Infant botulism: clinical spectrum and epidemiology. Pediatrics **66:**936–942.

van Ermengem, E. 1897. Ueber einen neuen anaeroben Bacillus und seine Beziehungen zum Botulismus. Z. Hyg. Infektionskr. **26:**1–56.

Wainwright, R. B., W. L. Heyward, J. P. Middaugh, C. L. Hatheway, A. P. Harpster, and T. R. Bender. 1986. Botulism in Alaska: epidemiology and clinical findings, 1947–1985. (abstr.) Presented at the 26th Annual Meeting of the Interscience Conference on Antimicrobial Agents and Chemotherapy, New Orleans, La.

Whitlock, R. H., and J. B. Messick, 1986. Botulism in cattle and horses. p. 57–62, Proc. 4th Annual Forum, American College of Veterinary Internal Medicine, Washington, D.C.

CHAPTER 13

Branhamella catarrhalis and *Neisseria* Species

GARY V. DOERN

Etiologic Agents: *Branhamella catarrhalis, Neisseria lactamica, N sicca, N. subflava, N. mucosa, N. flavescens, N. cinerea, N. elongata,* and *N. canis.*
Source: Commensal bacterial flora of the oropharynx.
Clinical Manifestations: *B. catarrhalis*—acute otitis media, maxillary sinusitis, and bronchopulmonary infections; *Neisseria* spp.—human disease uncommon.
Laboratory Diagnosis: *B. catarrhalis* infections—Gram-stained smear and culture; *Neisseria* spp.—culture.
Treatment: *B. catarrhalis*—third generation cephalosporins, chloramphenicol, erythromycin, tetracycline, and the aminoglycosides. Penicillin and ampicillin ineffective because of high prevalence of beta-lactamase production; *Neisseria* spp.—usually penicillin, but resistance has been documented.

Taxonomy and Nomenclature

During the past decade, the taxonomy of *Branhamella* and *Neisseria* species has been extensively revised and expanded. Inturn, this has led to the renaming of certain species, as well as the addition of new ones. As the tools for defining bacterial genus and species relatedness become more refined, bacterial nomenclature necessarily becomes more complex. Unfortunately, as the taxonomist achieves higher degrees of precision, organism nomenclature often becomes more confusing. Among bacteria, with the possible exception of the *Enterobacteriaceae*, nowhere is this paradox more evident than with the *Branhamella* and *Neisseria* species.

Branhamella

Among the four currently recognized species of *Branhamella*, *B. catarrhalis* is the sole human pathogen. This organism, previously referred to as *Micrococcus* and then *Neisseria catarrhalis*, received its present genus-species designation in 1970 when, based on DNA homology studies, genetic transformation experiments, and analysis of DNA base composition, Catlin (1970) proposed that *B. catarrhalis* warranted a genus designation separate from the *Neisseria* species. Subsequent analyses of cell envelope protein and lipopolysaccharide composition supported this separation (Johnston et al., 1976).

More recently, it has been suggested, based on observations of genetic and physiologic relatedness, that *Branhamella* be designated as one of two subgenera within the genus *Moraxella*, the other subgenus being *Moraxella* (Bøvre, 1979). Indeed, in the latest edition of *Bergey's Manual of Systematic Bacteriology B. catarrhalis* is referred to as *Moraxella (Branhamella) catarrhalis* (Bøvre, 1984a). Although this designation may be taxomonically correct, it serves only to promulgate the confusion that already exists regarding the name of this organism. Furthermore, it has not received widespread use. For these reasons, the terminology *B. catarrhalis* will be used herein.

Three other species within the genus *Branhamella* have been recognized: *B. caviae*, *B. ovis*, and *B. cuniculi* (Morello et al., 1985). These organisms have been found to be associated with various animals (i.e., guinea pigs, sheep, and rabbits, respectively) but not with humans and, therefore, will not be discussed further.

Neisseria

In addition to the two commonly recognized pathogens (*N. gonorrhoeae* and *N. meningitidis*), a variety of saprophytic species are now classified in the *Neisseria* genus. These include *N. lactamica*, *N. sicca*, *N. mucosa*, *N. subflava*, *N. flavescens*, *N. cinerea*, *N. elongata*, *N. canis*, and *N. denitrificans* (Bøvre, 1984b). The last organism, *N. denitrificans*, is found in the upper respiratory tract of guinea pigs and has not been recovered from humans. It will not be discussed further.

Branhamella catarrhalis Infections

Disease Description

Brahamella catarrhalis, previously considered a harmless commensal of the upper respiratory tract of healthy humans, is now recognized as an important human pathogen capable of causing a wide variety of infectious diseases (Doern, 1986). These include meningitis, endocarditis, septic arthritis, conjunctivitis in both neonates and children, keratitis, and sepsis without a focus of infection in patients with underlying lymphoproliferative disorders. Acute urethritis may also be caused by *B. catarrhalis* (Doern and Gantz, 1982). These diseases, however, are uncommon. In contrast, *B. catarrhalis* is the second or third most common bacterial cause of acute otitis media (Shurin et al., 1983) and maxillary sinusitis (Wald et al., 1981). Only *Streptococcus pneumoniae* and perhaps *Haemophilus influenzae* are more common. *Branhamella catarrhalis* is also frequently associated with bronchopulmonary infections (Doern et al., 1981, McLeod et al., 1983, Slevin et al., 1984). Indeed, at least with respect to the etiology of bacterial lower respiratory tract infections in adults, *B. catarrhalis* is exceeded only by *S. pneumoniae* and *H. influenzae*.

The subject of *Branhamella* bronchopulmonary infections has recently been extensively reviewed (Doern, 1986). However, selected features of this disease warrant brief discussion. *Branhamella catarrhalis* can be associated with a diverse spectrum of lower respiratory tract clinical manifestations. These range from acute exacerbation of chronic bronchitis to frank bacterial pneumonia. Nearly all patients with *Branhamella* bronchopulmonary infections have some form of antecedent functional or structural abnormality of the lungs. The most common is chronic obstructive pulmonary disease.

Patients with acute exacerbation of chronic bronchitis usually manifest with dyspnea, increased cough, and purulent sputum production. Constitutional findings such as fever are usually absent. Chest X rays are usually normal. In contrast, patients with *Branhamella* pneumonia, in addition to productive cough and dyspnea, often have fever as high as 103°F, significantly increased respirations, tachycardia, and in some cases, hypotension. Pleuritic chest pain is noted in many patients. Auscultation may reveal localized signs of consolidation. Chest roentgenograms may demonstrate diffuse peribronchial interstital infiltrates or, more commonly, dense alveolar lobar consolidation. Pleural effusions with empyema or concommitant bacteremia, or both, have been observed, but are extremely uncommon.

Epidemiology and Pathogenesis

Branhamella catarrhalis is a normal inhabitant of the upper respiratory tract of healthy humans (Brorson and Malmvall, 1981). Acute otitis media and maxillary sinusitis probably occur following contiguous spread of the organism from the upper respiratory tract into the middle ear or sinuses. *Branhamella* bronchopulmonary infections arise after endobronchial seeding of the lungs. Hematogenous seeding of the lungs is extremely uncommon; indeed, it may not occur. Similarly, direct human to human tranmission via airborne droplets is thought not to occur.

Branhamella catarrhalis otitis media and maxillary sinusitis are largely diseases of children. In contrast, *Branhamella* bronchopulmonary infections invariably occur in adults, particularly in the elderly. This is probably related to the opportunistic nature of *B. catarrhalis* as a lower respiratory tract pathogen. As mentioned previously, the large majority of patients who develop *Branhamella* bronchopulmonary infections have some antecedent predisposing condition. Examples include chronic bronchitis, chronic obstructive pulmonary disease, emphysema, alcoholism, hematogenous malignancy, tumor, and humoral immune deficits. These conditions are most common in adults, particularly elderly adults. Furthermore, the normal age-related depression of nonspecific host defense mechanisms (such as mucociliary clearance and the cough reflex) may also explain why *Branhamella* bronchopulmonary infections become more common with advancing age.

Although there is some anecdotal evidence that suggests that lower respiratory tract infections due to *B. catarrhalis* are more common among women than men, data from published surveys indicate that the sexes are affected equally (Doern, 1986; McLeod et al., 1983). A seasonal pattern of human infections has, however, been clearly documented. The vast majority of *B. catarrhalis* bronchopulmonary infections occur during the winter months.

Diagnosis

The diagnosis of *Branhamella* infections is best made by Gram-stained smear and culture of appropriate clinical specimens. The specimens of choice from patients with otitis media or maxillary sinusitis are middle ear fluid (Luotonen et al., 1981) and sinus aspirates (Wald et al., 1981), respectively. Gram-stained smears typically reveal large numbers of polymorphonuclear leukocytes and numerous intracellular and extracellular gram-negative diplococci, morphologically indistinguishable from *Neisseria* species (Fig. 1). As will be discussed, *B. catarrhalis* is readily propagated on standard nonselective solid media such as 5% agar sheep blood agar and enriched chocolate agar. Most patients with *Branhamella* otitis media and sinusitis yield the organism in pure culture. In some patients, however, *B. catarrhalis* is found to cooccur with *Streptococcus pneumoniae* or *Haemophilus influenzae*, or both (Luotonen et al., 1981; Shurin et al., 1983).

The optimal specimen from patients suspected of having *Branhamella* bronchopulmonary infections is expectorated sputum. Two extensive studies have shown equivalent Gram-stained smear and culture findings when expectorated sputa were compared with transtracheal aspirates (Aitken and Thornley, 1983, Thornley et al., 1982). In other words, there is usually no need to perform invasive procedures in an attempt to obtain representative clinical specimens from patients with *B. catarrhalis* lower respiratory tract infections; expectorated sputum is nearly always sufficient. Gram-stained smears are identical to those described for *Branhamella* otitis media and sinusitis (Fig. 1). Similarly, predominant growth of *B. catarrhalis* is usually observed in culture. At the present time, there exists no serologic procedure of diagnostic utility for *Branhamella* infections.

The importance of expectorated sputum Gram stain smear as a simple, inexpensive, and rapid means for making the diagnosis of *B. catarrhalis* bronchopulmonary infections cannot be overstated. First of all, nearly all patients with acute exacerbation of chronic bronchitis or frank bacterial pneumonia due to *Branhamella* produce expectorated sputa with Gram-stained smears such as that described previously. Furthermore, since *Neisseria* infections of the lower respiratory tract, including infections caused by *N. meningitidis*, are extremely uncommon, expectorated sputum Gram stains that reveal numerous polymorphonuclear leukocytes and gram-negative diplococci with little evidence of contamination can be considered diagnostic for *B. catarrhalis* infection.

Therapy

Until recently, penicillin G and ampicillin, together with their oral congeners, have been considered the drugs of choice for treatment of *B. catarrhalis* infections in patients who are not allergic to penicillin. Currently, these agents are of diminished clinical utility, since it is now recognized that many clinically significant isolates of *B. catarrhalis* procedure beta-lactamase and are, thus, resistant to the penicillin antimicrobials (Doern, 1986). In the author's institution, 100% of 24 sputum isolates of *B. catarrhalis* examined during 1985 produced beta-lactamase (unpublished observations).

The beta-lactamase of *B. catarrhalis* is constitutively produced in very low concentrations (Labia

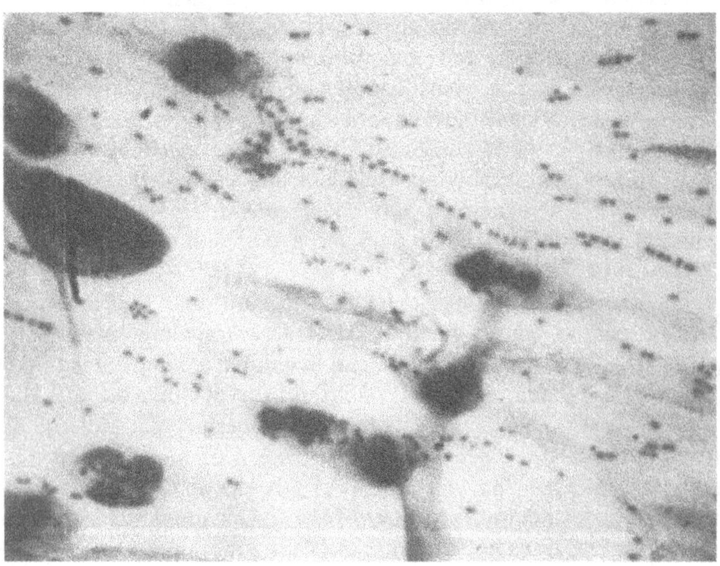

FIG. 1. Gram-stained smear of expectorated sputum from a patient with *Branhamella catarrhalis* pneumonia, revealing numerous polymorphonuclear leukocytes and numerous intracellular and extracellular gram-negative diplococci. A color plate of this figure appears in Color Plate 1 at the beginning of this volume.

et al., 1986). There is circumstantial evidence that, in at least some strains, it is plasmid-mediated (Kamme et al., 1986). Most of the enzyme remains tightly cell-associated (Kamme et al., 1986). Finally, it is highly active against the chromogenic cephalosporin nitrocephin (Stobberingh et al., 1986), but is inhibited by beta-lactamase inhibitors such as clavulanic acid and sulbactam (Farmer and Reading, 1986). These characteristics impact on the laboratory detection of beta-lactamase production by clinical isolates of *B. catarrhalis* (discussed next).

Branhamella catarrhalis beta-lactamase can be distinguished into several different types on the basis of isoelectric patterns and substrate affinity profiles. The two most common types have been designated Ravasio and 1908 (Farmer and Reading, 1986). They are mutually exclusive, that is, strains produce one type of beta-lactamase or the other, but not both. In the United States, strains of *B. catarrhalis* that produce the Ravasio-type beta-lactamase are perhaps 10-fold more common than those that produce the 1908-type beta-lactamase (Nash et al., 1986). In addition, Ravisio-type beta-lactamase-producing strains usually have higher penicillin and ampicillin minimum inhibitory concentrations (MICs) than do strains that produce the 1908-type beta-lactamase. This is consistent with the observation that isolates of *B. catarrhalis* with the Ravsio-type beta-lactamase produce quantitatively greater amounts of enzyme. This is also consistent with the results of numerous published surveys of the in vitro activity of penicillin and ampicillin versus clinical isolates of *B. catarrhalis* (Ahmad et al., 1984; Alvarez et al., 1985; Brorson et al., 1981; Doern et al., 1980). Invariably, at least some beta-lactamase-producing strains are found to have penicillin and ampicillin MICs as low as 0.1 to 0.5 μg/ml. Based on MICs, these strains would be characterized as being penicillin and ampicillin susceptible. It is likely that strains with low penicillin and ampicillin MICs produce the 1908-type beta-lactamase. Irrespective of low MICs, all strains of *B. catarrhalis* that are determined to produce beta-lactamase should be considered resistant to penicillin and ampicillin because therapeutic failures have been noted when these antimicrobials have been used to treat patients with beta-lacatmase-positive *Branhamella* infections.

Oral antimicrobials that are of clinical utility in treating localized, noninvasive *Branhamella* infections include cefaclor, trimethoprimsulfamethoxazole, erythromycin, tetracycline, and the combination of amoxicillin-clavulanate (Augmentin) (Renkonen, 1983). Invasive infections such as frank bacterial pneumoniae, which warrant parenteral therapy, can be treated with agents such as third generation cephalosporins, the aminoglycosides, antipseudomonas penicillins, and chloramphenicol (Alvarez et al., 1985). In the United States, resistance of clinical isolates of *B. catarrhalis* to these antimicrobials is extremely uncommon.

Infections Caused by *Neisseria* Species

Disease Description

With the exception of *N. canis*, each of the *Neisseria* species listed in Table 1 may be found as part of the oropharyngeal commensal flora of healthy humans. Despite its name, *Neisseria canis* is found in the oral cavity of cats (Hoke and Vedros, 1982). These organisms, often referred to collectively as the "nonpathogenic" or "saprophytic" *Neisseria*, are clearly opportunistic with respect to their ability to cause infection. Although infections caused by these organisms are extremely uncommon, they do occur. Examples of systemic infections caused by *Neisseria* species include endocarditis, usually associated with

TABLE 1. Identifying characteristics of *Branhamella catarrhalis* and *Neisseria* species

Organism	*Growth on* Neisseria-*selective media*	*Fermentation or utilization of*					*Reduction of*		*DNAse production*
		Glucose	*Maltose*	*Lactose*	*Sucrose*	*Fructose*	NO_3	NO_2	
B. catarrhalis	V	0	0	0	0	0	+	+	+
N. lactamica	+	+	+	+	0	0	0	+	0
N. sicca	0	+	+	0	+	+	0	+	0
N. subflava	0	+	+	0	V	V	0	+	0
N. mucosa	0	+	+	0	+	+	+	+	0
N. flavescens	0	0	0	0	0	0	0	+	0
N. cinerea	V	0	0	0	0	0	0	+	0
N. elongata	0	0	0	0	0	0	0	+	0
N. canis	0	0	0	0	0	0	+	+	0

V = variable; 0 = none; + = positive.

previous valvular abnormalities, meningitis, osteomyelitis, and sepsis without an evident focus of infection (Herbert and Ruskin, 1981). *Neisseria canis* has been recovered from an arm wound of a child who was bitten by a cat (Hoke and Vedros, 1982).

Diagnosis

The diagnosis of systemic infections caused by *Neisseria* species is dependent on the recovery of these organisms from representative clinical specimens. Growth of *Neisseria* species is readily accomplished using common plated media such as 5% sheep blood agar and chocolate agar (Morello et al., 1985). *Neisseria* species also grow well in standard blood culture broth (Feder and Garibaldi, 1984). Because all of the *Neisseria* species except *N. canis* constitute part of the commensal microbial flora of healthy humans, these organisms are frequently observed as culture contaminants. Assessing the clinical significance of an isolate of *Neisseria* species, therefore, requires consideration of the quality of the clinical specimen, the frequency with which an individual isolate is recovered, and the nature of the infectious disease process.

Therapy

The *Neisseria* species are nearly always susceptible to penicillin. For this reason, in the patient who is not allergic to penicillin, penicillin is the drug of choice for empiric therapy (Feder and Garibaldi, 1984). In the patient who is allergic to penicillin, chloramphenicol may be used. Resistance to these agents, although extremely uncommon, has been reported. It is appropriate, therefore, for the clinical laboratory to perform susceptibility testing on clinically significant isolates of *Neisseria* species as a means of guiding antimicrobial therapy.

Organism Characterization

General Description

Branhamella catarrhalis and the *Neisseria* species listed in Table 1 are gram-negative aerobic organisms. With the exception of *N. elongata*, these organisms exist as cocci, 0.5 to 1.5 μm in diameter, usually appearing in pairs with their adjacent sides retracted (i.e., kidney bean-shaped gram-negative diplococci) (Doern and Morse, 1980; Morello et al., 1985). *Neisseria elongata* is a small coccobacillary organism, 0.5 by 1.0 to 1.5 μm in size (Bøvre, 1984b). *Branhamella catarrhalis* and all of the *Neisseria* species are nonmotile and produce cytochrome oxidase. With the exception of *N. elongata*, they all possess catalase activity.

Growth Characteristics

B. catarrhalis and the human *Neisseria* species listed in Table 1 grow readily on standard plated media such as 5% sheep blood agar and enriched chocolate agar. In addition, they can be propagated on nonenriched media such as trypticase soy agar and nutrient agar. Optimal growth is obtained at 35 to 37°C in 5 to 7% CO_2; however, growth does occur in ambient air and at room temperature. *Neisseria lactamica* and some strains of *B. catarrhalis* and *N. cinerea* also grow on *Neisseria*-selective media such as modified Thayer-Martin medium, Martin-Lewis medium, or New York City medium (Doern and Morse, 1980; Knapp et al., 1984; Morello et al., 1985). Growth on these media is due to colistin resistance. This observation is relevant to the diagnosis of gonococcal infections of the genitourinary and upper respiratory tracts. Since *N. lactamica*, *B. catarrhalis*, and *N. cinerea* are commonly found as part of the normal bacterial flora of the upper respiratory tract and have been recovered, al beit infrequently, from genital tract specimens (Doern and Gantz, 1982; Knapp et al., 1984), growth of an oxidase-positive, gram-negative diplococcus on *Neisseria*-selective media inoculated with specimens from these sites does not, a priore, mean that the isolate is *N. gonorrhoeae*.

It has been suggested that colony pigmentation is a useful means of distinguishing among the *Neisseria* species and *B. catarrhalis* (Morello et al., 1985). *Neisseria flavescens* typically produces colonies with a distinct, deep-yellow pigmentation. With this exception, however, pigmentation of other *Neisseria* species and *B. catarrhalis* is variable and often indistinct, and should not be relied on as primary identification criterion.

Biochemical Characterization

Neisseria species and *B. catarrhalis* should be examined for their ability to utilize glucose, maltose, lactose, sucrose, and fructose (Table 1). This may be accomplished using 1 to 2% concentrations of carbohydrates in any of a variety of fermentation basal media containing a pH indicator. These include cysteine-trypticase agar (Brown, 1975), New York City fermentation medium (Faur et al., 1973), cysteine proteose peptone agar (Robinson and Oberhofer, 1983), or proteose peptone agar containing NaCl and K_2HPO_4 (Knapp and Holmes, 1983). Tubes containing the five different carbohydrates are inoculated heavily with growth from solid media, incubated for up to 72 h at 35 to 37°C in ambient air, and examined daily for a color change indicative of acid production due to carbohydrate fermentation. Care should be taken when inoculating tubes to avoid contamination. A tube containing only fermentation basal medium and no carbohydrate should also be inoculated. This serves as a negative control. Determination of beta-galactosidase activity utilizing O-nitrophenyl-B-D-galactopyranoside (ONPG) as a substrate may be substituted for the lactose fermentation test.

Same-day determination of carbohydrate utilization patterns can be accomplished using several different methods. The traditional method involves preparing a heavy suspension of growth from solid medium in 2.0% solutions of carbohydrates in potassium phosphate buffer containing potassium chloride and phenol red indicator (Brown, 1974; Kellog and Turner, 1973). Suspensions are incubated for up to 4 h in 37°C in ambient air, and examined periodically for the presence of yellow coloration as evidence of carbohydrate utililization. Two prepackaged, miniaturized kit versions of this procedure, the RIM-Neisseria kit (Austin Biologicals Laboratories, Inc., Austin, Texas) and Quad-Ferm + (Analytab Products, Inc., Plainview, New York) are now commercially available. Both procedures require a heavy inoculum. Results are available within 30 to 60 minutes with the RIM-Neisseria test (Robinson et al., 1985) and within 2 h with the Quad-Ferm + procedure. Alternatively, filter paper disks impregnated with carbohydrates (Minitek, BBL, Baltimore, Maryland) can be used for same-day determination of carbohydrate utilization patterns (Morse and Bartenstein, 1976). Commercially prepared paper disks containing the desired carbohydrates are placed in wells in a plastic dish. A heavy suspension of test organism is prepared in a solution containing proteose peptone and NaHCO. Aliquots of this suspension are then placed in the disk-containing wells, and the plastic dish is incubated at 37°C in ambient air for up to 6 h. The wells are examined periodically for the presence of a color change indicative of carbohydrate degredation. Each of these same-day procedures detects the presence of preformed enzymes and, thus, requires a heavy inoculum. Because of this, although these same-day procedures can be considered rapid, at least in comparison to traditional overnight techniques, they invariably require subcultures from primary isolation plates to achieve sufficient inoculum to perform the test.

Among the organisms listed in Table 1, *B. catarrhalis*, *N. flavescens*, *N. cinerea*, *N. elongata*, and *N. canis* are asaccharolytic. As a result, procedures in addition to carbohydate fermentation or utilization tests are necessary for their identification. *Branhamella catarrhalis* and *N. canis* are the only organisms in this group that reduce nitrate. *Branhamella catarrhalis*, however, produces extracellular DNAse, *Neisseria canis* does not. *Neisseria elongata* is uniquely catalase-negative and is coccobacillary in shape. *Neisseria flavescens* and *N. cinerea* may be distinguished on the basis of the former's ability to synthesize polysaccharide from 5% sucrose.

Alternative Approaches to Identification

A variety of commercially available methods has been developed for the identification of *Neisseria* species which are not predicated on carbohydrate fermentation or utilization patterns. Two of these, Gonocheck II (DuPont de Nemours, Co., Wilmington, Delaware) and Identicult-Neisseria (Scott Laboratories, Inc., Fiskeville, Rhode Island) are based on the detection of three enzymes: beta-galactosidase, proline aminopeptidase, and gamma-glutamayl-aminopeptidase. Both of these tests are limited to the characterization of organisms that grow on *Neisseria*-selective media and are primarily aimed at identifying *N. gonorrhoeae* (produces proline aminopeptidase) and *N. meningitidis* (produces gamma-glutamyl-aminopeptidase). Two organisms (*B. catarrhalis* and *N. lactamica*) discussed in this chapter, however, may be recovered on *Neisseria*-selective media and can be identified with these procedures. *Branhamella catarrhalis* does not produce any of the three enzymes; *N. lactamica* produces beta-galactosidase.

The Gonocheck II procedure is a 30-min test in which the activity of all three enzymes may be detected on the basis of specific color changes in a single tube (Welborn et al., 1984). An inoculum of 10 colonies is required. With the Identicult-Neisseria test, the activities of the three enzymes are detected colorimetrically on 3 reagent-impregnated filter paper disks attached to a cardboard strip. Results are available within approximately 15 min; an inoculum of 2 to 3 colonies per disk is required.

Another commerically available system, the RapID NH System (Innovative Diagnostics Systems, Inc., Decatur, Georgia) is a chromogenic substrate, conventional biochemical micromethod, which permits species-level identification of *B. catarrhalis* and *Neisseria* species within 4 h (Robinson and Oberhofer, 1983). The following characteristics are measured: phosphatase activity, nitrate reduction, beta-galactosidase production, proline and gamma-glutamyl-aminopeptidase activity, resazurin reduction glucose and sucrose utilization, indole production, ornithine decarboxylation, and urea hydrolysis. The first eight characteristics are used in the identification of *B. catarrhalis* and *Neisseria* species. The individual reactants are contained in a plastic cartridge. The cartridge is inoculated with a heavy suspension of test organism prepared in an inoculation fluid provided by the manufacturer. The cartridges are incubated for 4 h at 35°C in ambient air, and individual reactants are examined for color changes representative of positive reactions. Identifications are achieved using an octal code index.

Finally Vitek Systems, Inc. (Hazelwood, Missouri) has recently introduced an automated method for identifying *Neisseria* species and *B. catarrhalis*. A molded plastic card containing 16 reactants is inoculated with a heavy suspension of test organism. After 4 h of incubation at 35°C, the card is examined by the AutoMicrobic System (AMS) instrument, and an identification result is generated. The inoculation, incubation, and examination steps are all performed automatically by the AMS machine. The user, however, has to perform a spot oxidase and spot indole test manually. The results of these two determinations plus the ability of the test organism to grow on *Neisseria*-selective medium must be noted in the instrument to achieve an identification.

Antimicrobial Susceptibility Testing

All clinically significant isolates of *Branhamella catarrhalis* should be examined for beta-lactamase production. The procedure of choice for detecting *B. catarrhalis* beta-lactamase is a test employing the chromogenic cephalosporin, nitrocephin, as a substrate. The test may be performed according to the conventional method (O'Callaghan et al., 1972) or utilizing filter paper disks impregnated with nitrocephin (BBL, Cockeysville, MD). In both instances, a small inoculum is required, and results are available within 5 min. Acidometric procedures (Escamilla, 1976), the iodometric method (Catlin, 1975), and chromogenic cephalosporin tests that use pyridine-2-azo-dimethylaniline cephalosporin (PADAC) as a substrate are not suitable for detecting *B. catarrhalis* beta-lactamase. False-negative results may be obtained with these assays, perhaps because of the unusual properties of the *B. catarrhalis* beta-lactamase (i.e., it is produced in low concentrations and it is tightly cell-bound). The fact that nitrocephin-based tests seem to work is probably related to the high affinity of the *B. catarrhalis* beta-lactamase for nitrocephin. Beta-lactamase-producing isolates of *B. catarrhalis* should be considered resistant to penicillin, ampicillin, and their analogs.

Direct in vitro tests for the activity of penicillin and ampicillin are problematic. As stated previously, quantitative susceptibility tests often reveal MICs to penicillin and ampicillin among beta-lactamase-producing strains of *B. catarrhalis* that would be considered susceptible (i.e., MICs of 0.1 to 0.5 μg/ml). This is probably related to the low concentrations of beta-lactamase produced by *B. catarrhalis*, especially 1908-type strains. Disk diffusion susceptibility testing has not been standardized for *B. catarrhalis*, that is, insufficient information exists upon which to base recommendations pertaining to test methods or interpretive categories when testing penicillin or ampicillin. It should be noted, however, that there is at least some evidence that suggests that the currently recommended method for disk diffusion susceptibility testing of nonfastidious bacteria may be suitable for testing *B. catarrhalis* (Oberhofer and Towle, 1982). A zone of inhibition of 29 mm or more surrounding disks containing 10 units of penicillin or 10 μg of ampicillin best defines strains as being susceptible to these antimicrobials (R. J. Wallace, personal communication).

Because in vitro activity of other antimicrobials versus *B. catarrhalis* is highly predictable in the United States (see the preceeding discussion), there is little need to perform susceptibility testing with other agents. In summary, in vitro susceptibility testing of *B. catarrhalis* is quite reasonably restricted to performance of a nitrocephin beta-lactamase test.

Antimicrobial susceptibility testing of the *Neisseria* species discussed in this chapter should be restricted only to those isolates that have been determined to be of clinical significance. When such testing is considered to be warranted, a quantitative susceptibility test procedure should be employed, utilizing Mueller-Hinton broth as a medium. Disk diffusion susceptibility testing should not be performed because it has not been evaluated and proven to apply to *Neisseria* species.

Literature Cited

Ahmad, F., D. T. McLeod, M. J., Croughan, and M. A. Calder. 1984. Antimicrobial susceptibility of *Branhamella* isolates from bronchopulmonary infections. Antimicrob. Agents Chemother. **26**:424–430.

Aitken, J. M. and P. E. Thornley. 1983. Isolation of *Branhamella catarrhalis* from sputum and tracheal aspirates. J. Clin. Microbiol. **18**:1262–1266.

Alvarez S., M. Jones, S. Holtsclaw-Berk, J. Guarderas, and S. L. Berk. 1985. In vitro susceptibilities and beta-lactamase production of 53 clinical isolates of *Branhamella catarrhalis*. Antimicrob. Agents Chemother. **27**:646–651.

Bøvre, K. 1979. Proposal to divide the genus *Moraxella* Lwoff 1939 emend. Henriksen and Bovre 1968 into two subgenera, subgenus *Moraxella* (Lwoff 1939) and subgenus *Branhamella* (Catlin 1970) Bovre 1979. Int. J. Syst. Bacteriol. **29**:403–406.

Bøvre, K. 1984a. *Moraxella*, p. 296–303. *In* N. R. Krieg (ed.), Bergey's manual of systematic bacteriology, vol. 1. Williams and Wilkins, Co., Baltimore, Maryland.

Bøvre, K. 1984b. *Neisseriaceae*, p. 288–290. *In* N. R. Krieg (ed.), Bergey's manual of systematic bacteriology, vol. 1. Williams and Wilkins, Baltimore, Maryland.

Brorson, J-E. and B-E. Malmvall. 1981. *Branhamella (Neisseria) catarrhalis* and other bacteria in the nasopharynx of children with longstanding cough. Scand. J. Infect. Dis. **13**:111–116.

Brorson, J-E., J. Martinell, and H. Wilske. 1981. *Branhamella catarrhalis*: antibiotic susceptibility and beta-lactamase production. J. Antimicrob. Chemother. **7**:208–214.

Brown, W. J. 1974. Modification of the rapid fermentation test for *Neisseria gonorrhoeae*. Appl. Microbiol. **27**:1027–1030.

Brown, W. J. 1975. A comparison of fermentation methods for the confirmation of *Neisseria gonorrhoeae*. Health Lab. Sci. **13**:54–59.

Catlin, B. W. 1970. Transfer of the organism named *Neisseria* catarrhalis to *Branhamella* gen. nov. Int. J. Syst. Bacteriol. **20**:155–159.

Catlin, B. W. 1975. Iodometric detection of *Haemophilus influenzae* beta-lactamase: rapid presumptive test for ampicillin resistance. Antimicrob. Agents Chemother. **7**:265–270.

Doern, G. V. 1986. *Branhamella catarrhalis*—an emerging human pathogen. Diagn. Microbiol. Infect. Dis. **4**:191–201.

Doern, G. V. and N. M. Gantz. 1982. Isolation of *Branhamella (Neisseria) catarrhalis* from men with urethritis. Sex. Transm. Dis. **9**:202–205.

Doern, G. V., M. J. Miller, and R. E. Winn. 1981. *Branhamella (Neisseria) catarrhalis* systemic disease in humans. Arch. Intern. Med. **141**:1690–1694.

Doern, G. V. and S. A. Morse. 1980. *Branhamella (Neisseria) catarrhalis*: criteria for laboratory identification. J. Clin. Microbiol. **11**:193–199.

Doern, G. V., K. G. Siebers, L. M. Hallick and S. A. Morse. 1980. Antibiotic susceptibility of beta-lactamase-producing strains of *Branhamella (Neisseria) catarrhalis*. Antimicrob. Agents Chemother. **17**:24–29.

Escamilla, J. 1976. Susceptibility of *Haemophilus influenzae* to ampicillin as determined by use of a modified one-minute beta-lactamase test. Antimicrob. Agents Chemother. **9**:196–198.

Farmer, T. and C. Reading. 1986. Inhibition of the beta-lactamase of *Branhamella catarrhalis* by clavulanic acid and other inhibitors. Drugs **31(suppl. 3)**:70–78.

Faur, V. C., M. H. Weisburd, M. E. Wilson, and P. S. May. 1973. A new medium for the isolation of pathogenic *Neisseria* (NYC medium). I. Formulation and comparisons with standard media. Health Lab. Sci. **10**:44–52.

Feder, H. M. and R. A. Garibaldi. 1984. The significance of nongonococcal, nonmeningococcal *Neisseria* isoaltes from blood cutures. Rev. Infect. Dis. **6**:181–187.

Herbert, D. A. and J. Ruskin. 1981. Are the "nonpathogenic" *Neisseria* pathogenic? Am. J. Clin. Pathol. **75**:739–742.

Hoke, C. and N. A. Vedros. 1982. Characterization of atypical aerobic gram negative cocci isolated from humans. J. Clin. Microbiol. **15**:906–914.

Johnson, K. G., J. J. McDonald, and M. A. Perry. 1976. Studies on the cellular and free lipopolysaccharides from *Branhamella catarrhalis*. Can. J. Microbiol. **22**:460–467.

Kamme, C., I. Eliasson, B. K. Knutson and M. Vang. 1986. Plasmid-mediated beta-lactamase in *Branhamella catarrhalis*. Drugs **31(suppl. 3)**:55–63.

Kellogg, D. S. and E. M. Turner. 1973. Rapid fermentation confirmation of *Neisseria gonorrhoeae*. Appl. Microbiol. **25**:550–552.

Knapp, J. S. and K. K. Holmes. 1983. Modified oxidation-fermentation medium for detection of acid production from carbohydrates by *Neisseria* spp. and *Branhamella catarrhalis*. J. Clin. Microbiol. **18**:56–62.

Knapp, J. S., P. A. Totter, M. H. Mulks, and B. H. Minshew. 1984. Characterization of *Neisseria cinerea*, a nonpathogenic species isolated on Martin-Lewis medium selective for pathogenic *Neisseria* species. J. Clin. Microbiol. **19**:63–67.

Labia, R., M. Barthelemy, C. B. LeBouguennec, and A. B. Hoi-Dang Van. 1986. Classification of beta-lactamase from *Branhamella catarrhalis* in relation to penicillinases produced by other bacterial species. Drugs **31(suppl. 3)**:40–47.

Luotonen, J., E. Herva, P. Karma, M. Timonen, M. Leinonen, and P. H. Makela. 1981. The bacteriology of acute otitis media in children with special reference to *Streptococcus pneumoniae* as studied by bacteriologic and antigen detection methods. Scand. J. Infect. Dis. **13**:177–183.

McLeod, D. T., J. T. Power, M. A. Calder and A. Seaton. 1983. Bronchopulmonary infection due to *Branhamella catarrhalis*. Br. Med. J. **287**:1446–1453.

Morello, J. A., W. M. Janda, and M. Bohnhoff. 1985. *Neisseria* and *Branhamella*, p. 176–192, *In* E. H. Lennette, A. Balows, W. J. Hausler, and H. J. Shadomy (eds.). Manual of clinical microbiology 4th ed. American Society for Microbiology, Washington, D.C.

Morse, S. A. and L. Bartenstein. 1976. Adaptation of the Minitek system for the rapid identification of *Neisseria gonorrhoeae*. J. Clin. Microbiol. **2**:8–13.

Nash D. R., R. J. Wallace, Jr., V.A. Streingrub, and P. A. Shurin. 1986. Isoelectric focusing of beta-lactamase from sputum and middle ear isolates of *Branhamella catarrhalis* recovered in the United States. Drugs **31(suppl. 3)**:48–54.

O'Callaghan, C. H., A. Morris, S. Kirby, and A. H. Shingler. 1972. Novel method for detection of beta-lactamase by using a chromogenic cephalosporin substrate. Antimicrob. Agents Chemother. **1**:283–288.

Oberhofer, T. R. and D. W. Towle. 1982. Evaluation of the rapid penicillinase paper strip test for detecton of beta-lactamase. J. Clin. Microbiol. **15**:196–199.

Renkonen, O. V. 1983. Antibacterial activity of nine oral antibiotics against *Streptococcus pneumoniae*, *Haemophilus influenzae* and *Branhamella catarrhalis*. Scand. J. Infect. Dis. **39**:106–110.

Robinson, A., S. B. Griffith, D. G. Moore, and J. R. Carlson. 1985. Evaluation of the RIM system and Gono Gen test for identification of *Neisseria gonorrhoeae* from clinical specimens. Diagn. Microbiol. Infect. Dis. **3**:125–130.

Robinson, M. J. and T. R. Oberhofer. 1983. Identification of pathogenic *Neisseria* species with the Rapid NH system. J. Clin. Microbiol. **17**:400–407.

Shurin, P. A., C. D. Marchant, C. H. Kim, G. F. Hare, C. E. Johnson, M. A. Tutihasi, and L. J. Knapp. 1983. Emergence of beta-lactamase-producing strains of *Branhamella catarrhalis* as important agents of otitis media. Ped. Infect. Dis. **2**:34–39.

Slevin, N. J., J. Aitken, and P. E. Thornley. 1984. Clinical and microbiologic features of *Branhamella catarrhalis* bronchopulmonary infections. Lancet **1**:782–784.

Stobberingh, E. E., H. J. van Eck, A. W. Houben, and C. P. A. van Boven. 1986. Analysis of the relations between ampicillin resistance and beta-lacatamase production in *Branhamella catarrhalis*. Drugs **31(suppl. 3)**:23–27.

Thornley, P. E., J. Aitken, C. J. Drennan, J. MacVicar, and N. J. Slevin. 1982. *Branhamella catarrhalis* infection of the lower respiratory tract: reliable diagnosis by sputum examination. Br. Med. J. **285**:1537–1542.

Wald, E. R., G. J. Milmoe, A. Bowen, J. Ledesma-Medina, N. Salamone, and C. D. Bluestone. 1981. Acute maxillary sinusitis in children. N. Engl. J. Med. **304**:749–756.

Welborn, P. P., C. T. Uyeda, and N. Ellison-Birang. 1984. Evaluation of Gonocheck II as a rapid identification system for pathogenic *Neisseria* species, J. Clin. Microbiol. **20**:680–683.

CHAPTER 14

Brucellosis

NELSON P. MOYER and LARRY A. HOLCOMB

Disease: Brucellosis (undulant fever, Malta fever, Mediterranean fever).

Etiologic Agents: *Brucella melitensis* biotypes 1 through 3, *Brucella abortus* biotypes 1 through 7 and 9, *Brucella suis* biotypes 1 through 4 and *Brucella canis*.

Source: Domestic livestock and some wild mammals or their products by direct contact, ingestion, or inhalation.

Clinical Manifestations: Acute or insidious onset of intermittent fever, headache, malaise, chills, sweating, arthralgia, myalgia, weight loss, and depression. May be acute, chronic, or subclinical.

Pathology: Reticuloendothelial disease with granulomatous foci in liver, spleen, lung, bone, kidney, gonads, heart, or brain.

Laboratory Diagnosis: Bacteriologic isolation of etiologic agent from blood, bone marrow, or tissue. Tube agglutination test titer of 1 : 160 or greater, with a fourfold rise considered diagnostic. 2-Mercaptoethanol test useful in diagnosis of chronic cases. Homologous antigen required for serodiagnosis of *B. canis*.

Epidemiology: Worldwide, especially in Mediterranean countries of Europe and in Africa, Central Asia, and Central and South America. Occupational disease of livestock producers, veterinarians, and abattoir workers. Sporadic cases or outbreaks from consumption of contaminated food or dairy products.

Treatment: Tetracycline plus streptomycin, aminoglycosides, cotrimoxazole, and rifampicin. Surgical drainage of abscess or valve replacement in endocarditis.

Prevention and Control: Eradication in domestic animals through test and slaughter and immunization, mandatory pasteurization, protective equipment, and education.

Description of Disease

Brucellosis is an infectious disease of the genitourinary tract of sheep, goats, pigs, cattle, dogs, and other animals and of the reticuloendothelial system of humans. The damage sustained from brucellosis includes abortion of livestock, loss of protein resources through contamination of meat, milk, and milk products, and loss of human productivity through prolonged illness. The four species of brucellae important in human medicine include *Brucella melitensis*, found in sheep and goats; *Brucella suis*, which infects pigs; *Brucella abortus*, a cattle pathogen; and *Brucells canis*, which causes abortion in dogs. *Brucella ovis*, a pathogen of sheep, and *Brucella neotomae*, which occurs in the desert rat, have not been shown to cause human disease.

Communicability

Brucellosis is communicable to humans from animals or their products by direct contact, ingestion, inhalation, and conjunctival inoculation. Travelers to endemic areas risk infection if they drink raw milk or eat fresh cheese. Person-to-person transfer has been reported but remains extremely rare (Goossens et al., 1983), as are cases resulting from blood transfusions and bone marrow transplants (Naparstek et al., 1982). *B. canis* infections occur in dog handlers, laboratory workers, and children with infected pet dogs (Roth and Gleckman, 1985).

Clinical Manifestations

The incubation period of brucellosis varies from 5 to 35 days or more, with an average of 10 to 14 days in acute cases. Onset is characteristically gradual, with weakness, malaise, and moderate increases in temperature. Fever may present in irregular cycles that repeat over several months. The acute temperatures may vary from 38 to 40°C. Drenching sweats may occur in the late afternoon or evening. Other symptoms include myalgias, headache, loss of appetite, and mental depression. The clinical manifestations of brucellosis were reviewed by Young (1983).

Asymptomatic Brucellosis

Veterinarians and abattoir workers exposed to brucellae over time may develop high titers of agglutinating antibody without manifesting symptoms of disease.

Acute Brucellosis

Acute brucellosis normally follows exposure to infected animals or animal products. Bacteremia develops without localizing signs. Symptoms include malaise, irregular fever, chills, sweating, aches, weight loss, and dry cough. Apathy, insomnia, depression, and confusion are significant complaints.

Relapsing Brucellosis

Relapse can occur within the first 3 months after treatment. Symptoms are similar to those of acute attacks, but may be exacerbated by sensitivity to the toxins of brucellae.

Localized Disease

Localized brucellosis may result after subclinical or mild illness. Organisms are sequestered in bone, joints, cerebrospinal fluid, liver, kidney, spleen, or skin but are not found in blood. Serologic titers are low or absent, and diagnosis is difficult.

Chronic Brucellosis

Chronic brucellosis is defined as persistence of disease, with clinical symptoms and serologic evidence of infection lasting more than 1 year. Characteristics of chronic brucellosis include anxiety and severe depression in addition to the malaise and fever of the acute attack.

Hypersensitivity

Previous exposure to brucellae may induce a hypersensitivity to toxins produced by the organism. Subsequent contact evokes an allergic response characterized by local redness, edema, and pain. Severe reactions may include fever and chills, but cultures are negative.

Strain 19 Disease

Vaccine-associated brucellosis occurs in veterinarians as a result of accidental contact with strain 19. Disease caused by vaccine exposure is generally milder than natural disease, although the usual symptoms of acute brucellosis are produced.

Pathology

Brucellae gain entrance through abrasions of the skin and directly through the gastrointestinal mucosa and the conjunctiva. They pass through the lymph to the regional nodes, where they multiply unless destroyed by the cellular immune response. If infection is established, organisms appear in peripheral blood within a few hours, where they are phagocytized by polymorphonuclear leukocytes and carried to the liver, spleen, bone marrow, and other lymph nodes, in which they may establish focal infections. Brucellae multiply in the phagosomes, which burst, releasing organisms that are ingested by phagocytic reticuloendothelial cells lining sinusoids. Intracellular brucellae are protected from serum antibody and from some antibiotics. This aspect of their life cycle results in a latent phase of disease with gradual release of endotoxin into the blood, producing the relapsing fever and eventually resulting in the hypersensitivity of chronic disease. Abscesses may persist for years, but small granulomas usually resolve or calcify. Chronic brucellosis may result in abscesses of subcutaneous tissue, testis, epididymis, ovary, gall bladder, kidney, and brain.

Brucellae localize in the placenta and seminal vesicles of susceptible animals, causing abortion and infertility. These tissues are high in erythritol, which stimulates growth of the organism. Erythritol is not found in the human placenta, and abortion is unusual; however, ovarian and pelvic abscesses, chronic salpingitis, and cervicitis may be caused by brucellae. Brucellae contain endotoxin, and treatment of acute cases frequently induces a Herxheimer-like reaction, suggesting that the manifestations of chronic brucellosis are largely attributable to increasing sensitivity to brucellar antigen.

Diagnostic Features

The diagnosis of brucellosis is difficult because of the insidious onset, the generalized nature of patient complaints, and the variability of clinical findings on physical examination. Brucellosis should be considered in every patient with a history of occupational exposure who has been febrile for more than 1 week and who has generalized myalgia and arthralgia. Cervical and axillary lymphadenopathy is common, and leukopenia is highly suggestive. Musculoskeletal complaints include arthritis of the hip, knee, and lumbar spine. Crosby et al. (1984) studied 38 patients infected with *B. melitensis* and found fever (98%), hepatomegaly (87%), anemia (74%), malaise (74%), anorexia (68%), sweating (68%), headache (63%), splenomegaly (61%), arthralgia (55%), myalgia (52%), and adenopathy (50%) to be the primary symptoms. Hematologic abnormalities included neutropenia (21%), lymphopenia (63%), leukopenia (45%), and thrombocytopenia (39.5%).

Acute brucellosis clinically mimics influenza, hepatitis, malaria, toxoplasmosis, typhoid, typhus, tularemia, and miliary tuberculosis. Brucellosis may resemble mononucleosis, except that heterophile titers are low and antibody to cytomegalovirus and Epstein–Barr virus is not found. Chronic brucellosis may be confused with lymphoma, and it may be difficult to differentiate renal and spinal brucellosis from tuberculosis.

Complications

Retinal detachment, cutaneous lesions, cystitis, cholecystitis, prostatitis, and pyelonephritis occur. Ear damage includes hearing loss, tinnitus, and dizziness. Meningitis and cardiac involvement are rare in acute brucellosis, but myocarditis, pericarditis, and endocarditis have been reported. Bone and joint involvement is a moderately frequent complication of brucellosis. Hepatitis and hepatic granulomas are complications of chronic brucellosis.

Prognosis

The mortality in untreated brucellosis is less than 3%, with recovery the usual outcome. With antimicrobial therapy, death is rare, but relapse may occur in 2 to 10% of patients. Complicated disease involving meningoencephalitis, osteomyelitis, or endocarditis is associated with mortality. Patients recovering from disease exhibit protective immunity against reinfection.

Epidemiologic Aspects

The worldwide distribution of brucellosis in animals and humans was reported by Thimm (1982) and by Matyas and Fujikura (1984). Although domestic animals such as sheep, goats, cattle, pigs, and dogs remain the primary sources of human infection, contact with semidomesticated animals and wild animals has been associated with cases of brucellosis. Thimm (1982) reported brucellosis in reindeer, buffalo, yaks, camels, and a wide variety of wild animals. Caribou in the United States, the Soviet Union, and Canada are infected with *B. suis* biotype 4.

Occurrence

Brucellosis has been eradicated from herds in northern Europe except Scotland and Northern Ireland. The prevalence of disease is low in central Europe and high in Spain, Malta, Greece, and Italy. In Africa, 37 of 49 countries report human cases. Eradication programs in North America have been highly successful in Canada and have virtually eliminated *B. suis* in swine. *B. abortus* persists in the southeastern and south central United States. *B. melitensis* is not found in the United States but is common in Mexico and Central America. South American countries report moderate levels of *B. abortus* infection in cattle and *B. suis* infection in swine and low to moderate levels of *B. melitensis* infection in goats. In Asia, 17 of 38 countries reported brucellosis, with the highest incidence in Iran, India, Pakistan, Bangladesh, Sri Lanka, Laos, Mongolia, and Hong Kong. The incidence of brucellosis in Australia is low except for *B. ovis* in sheep.

Transmission

In developed countries, intermittent occupational exposure of livestock producers, abattoir workers, butchers, veterinarians, and laboratory workers constitutes the main source of infection, whereas nomadic herdsman in developing countries are continually exposed to brucellosis through their daily contact with animals. In cold to temperate climates, acute cases occur in the spring and summer related to abortion, parturition, and postpartum care of animals. Desert societies and arctic dwellers are at increased risk of infection because of the lack of water for personal hygiene and lack of fuel except dung, which is also a source of infection. Tropical and subtropical climates experience acute cases year round. The incidence of infection in children is low in developed countries, but high in countries where children tend goats and sheep. Raw milk and soft cheeses from sheep cause human infections in southern France, Italy, Greece, eastern Mediterranean countries, and central Asia. Goat milk and cheese is associated with cases in Mediterranean countries, Africa, Asia, and Central and South America. Consumption of imported goat cheese is associated with cases of brucellosis caused by *B. melitensis* in the United Kingdom and the United States.

Etiologic Agents

Characterization

Brucellae are gram-negative coccobacilli or short rods 0.5 to 0.7 μm by 0.6 to 1.5 μm arranged singly or in pairs, short chains, or groups. True capsules, endospores, and flagella are not produced. Growth occurs aerobically and many strains require carbon dioxide, especially on primary isolation; no growth occurs under anaerobic conditions. The optimal growth temperature is 37°C, although growth occurs between 20 and 40°C. Optimal pH range is 6.6 to 7.4.

Thiamine, niacin, and biotin are required growth factors, and some strains grow better when serum, blood, or hemin (factor X) and nicotinamide adenine dinucleotide (factor V) are supplied.

Lipopolysaccharide antigens A and M are useful for differentiating biovars of *B. melitensis*, *B. abortus*, and *B. suis*. Smooth species of *Brucella* cross-react in agglutination tests with unabsorbed antiserum. Cross-reactions occur with *Francisella tularensis*, *Vibrio cholerae*, *Yersinia enterocolitica* 0:9 and 0:16, *Escherichia coli* 0:157, and *Salmonella* 0:30. Nonsmooth strains of *B. canis* cross-react with *Pasteurella multocida*, mucoid *Pseudomonas aeruginosa*, *Pseudomonas maltophilia*, and *Actinobacillus equuli* (Corbel et al., 1984). *Brucella* species are not closely related to any other genus but are homologous within the genus based upon DNA hybridization. A recent study of 51 strains led investigators to propose that only one species, *B. melitensis*, should be recognized and that former species should be designated as biovars, for example, *Brucella melitensis* biovar *abortus 1* (Verger et al., 1985).

Specimen Collection

Acute serum should be obtained as soon as possible after onset of the disease, followed by convalescent serum collected 14 to 21 days after onset. Additional sera may be required to establish a diagnosis when immunoglobulin (Ig)G and IgA studies are indicated.

Blood cultures may be positive after the second day of fever, and multiple specimens should be drawn during the febrile period. Cultures should not be discarded as negative until after 4 weeks. Tryptose broth, brain–heart infusion broth, and *Brucella* broth have been used successfully in biphasic bottles. Other specimens important in the diagnosis of human brucellosis include bone marrow, lymph node aspirates, urine, synovial fluid, spinal fluid, abscess aspirates, and liver or spleen biopsy.

Isolation and Preliminary Identification

Herr and Roux (1981) studied several media for primary recovery of brucellosis from milk and tissue specimens and concluded that no single medium detected all strains. Optimal recovery may be expected using tryptose, brain–heart infusion, or *Brucella* agar and from selective and nonselective serum dextrose agar.

Blood cultures should be incubated with 5 to 10% carbon dioxide at 37°C for up to 30 days. Etemadi et al. (1984) obtained improved recovery of brucellae from blood, bone marrow, and spinal fluid using lysis and centrifugation of specimens before inoculation into primary culture media.

Gotuzzo et al. (1986) compared recovery of *B. melitensis* from bone marrow and blood culture in 50 patients with a clinical diagnosis of brucellosis. Bone marrow cultures were positive in 92%, whereas blood culture was positive in 70% of these patients. The average time required for growth was 4.32 days for bone marrow and 6.65 days for blood cultures. Bone marrow cultures were positive in acute, subacute, and chronic brucellosis, whereas blood cultures were positive only in acute cases.

Fluid media have been used for enrichment and primary culture (Brodie and Sinton, 1975). Recovery is enhanced and recognition time is reduced, but growth in static culture is poor and there is a tendency toward dissociation, resulting in an inability to confirm isolates serologically with the agglutination test.

Fibrous clots, exudates, and tissue are aseptically ground and inoculated onto 5% sheep blood agar, *Brucella* agar containing 5% serum, or serum dextrose agar (Oxoid Ltd., London, England). When significant contamination is likely, a selective medium should be inoculated in addition to the basal medium. Overgrowth of cultures can be controlled by adding 1.4 ml of 0.1% aqueous crystal violet (certified) per liter of medium before sterilization; however, this concentration of crystal violet may inhibit small numbers of *B. suis* and *B. melitensis*. The medium developed by Kuzdas and Morse (1953) containing bacitracin, polymyxin B, cycloheximide, and circulin has proved effective in clinical studies (Weed, 1957). Other selective media used successfully include a medium containing bacitracin, polymyxin B, nalidixic acid, vancomycin, cycloheximide, and nystatin (Farrell, 1974) and Thayer–Martin medium (Brown et al., 1970).

Inoculated plates are incubated at 35 to 37°C in an atmosphere of 5 to 10% carbon dioxide for 10 days. The high humidity that develops in sealed chambers or jars under these conditions promotes rapid growth of molds, which may contaminate and render cultures valueless. A layer or tray of dry calcium chloride placed in the bottom of the chamber or jar will help to control humidity.

Plates should be examined for signs of growth. After 4 to 5 days, *Brucella* colonies are 2 to 7 mm in diameter, spheroid, moist, slightly opalescent, translucent, and bluish white in reflected light. These characteristics may vary somewhat with pH and available moisture. Isolated colonies should be transferred to several tryptose agar slants or slants of similar media and incubated under 5 to 10% carbon dioxide tension at 37°C for 48 h. If sufficient growth is not obtained, reincubation for another 24 h is required. Clinical laboratories must rely upon growth characteristics, carbon dioxide requirement, hydrogen sulfide production, and urease activity combined with knowledge of the patient history and perhaps the results of serologic tests for presumptive identification of brucellae.

Definitive Identification and Biotyping

Common criteria for the definitive identification of the species and biotypes of the genus *Brucella* as shown in Table 1, are (1) agglutination in monospecific sera, (2) requirement of increased carbon dioxide for growth, (3) urease production, (4) production of hydrogen sulfide, and (5) growth in the presence of basic fuchsin and thionin in solid media (Hausler et al., 1985). Gram stains are prepared and examined for gram-negative pleomorphic coccobacilli. Nonhemolytic colonies that are gram-negative coccobacilli, do not ferment lactose or glucose, are obligate aerobes, and are oxidase and urease positive should be tested for agglutination to anti-smooth *Brucella* serum.

Agglutination in Monospecific Serum

A dense suspension of the organism to be tested is prepared in 0.5% phenolized saline and heated at 60°C for 1 h. To perform a slide agglutination test, a drop of the suspension is added to a drop of each monospecific antiserum and mixed. Agglutination should occur within 1 min with one of the antisera. Control cultures of *B. abortus* biotype 1, *B. melitensis* biotype 1, and *B. ovis* or *B. canis* should be used for this test. To perform the tube agglutination test, a cell suspension from a fresh slant is prepared in 0.5% phenolized saline. The suspension is then heated at 60°C for 1 h and diluted to obtain 78% light transmission at a wavelength of 650 nm in a spectrophotometer. The monospecific tube test is performed in duplicate (i.e., one set of tubes for each of the monospecific sera). The serum is diluted just beyond its known titer by the double-dilution method, starting at 1:5. An equal amount of the antigen suspension, prepared as described, is then added to each tube plus the saline control. The tubes are incubated for 24 h at 37°C and read. Usually the strain being studied will be agglutinated by one of the sera to its known titer, but not at all by the other serum. Strains of *B. melitensis* biotype 3 and *B. abortus* biotype 7 are agglutinated by both monospecific sera at varying titers.

Requirements for Carbon Dioxide

Requirements for additional carbon dioxide are determined by inoculation of cultures onto duplicate tryptose agar slants; one slant is incubated in air, the other in an atmosphere of 5 to 10% carbon dioxide. As mutants occur that are no longer dependent on additional carbon dioxide, this test should be conducted immediately after initial isolation of the organism. Good growth in carbon dioxide and scanty or no growth in air indicates a requirement for carbon dioxide.

Urease Test

A semiquantitative urease test can be performed by preparing a dense suspension of the organisms and placing a loopful of the suspension on a slant of Christensen urea agar. *B. suis*, *B. canis*, and *B. neotomae* will turn the indicator pink within 10 min, whereas strains of *B. abortus* and *B. melitensis* are usually positive in 1 to 12 h. Most strains of *B. ovis* are urease negative (Corbel and Hendry, 1985).

Production of Hydrogen Sulfide

Production of hydrogen sulfide is detected by suspending a lead acetate paper strip directly over but not touching the inoculated surface of a tryptose agar slant. The strip is examined daily for 4 days and replaced with a fresh strip each day. Production of hydrogen sulfide blackens the strip.

Dye Sensitivity Tests

Growth in the presence of thionin and basic fuchsin is determined by the procedure of Corbel and Brinley-Morgan (1984). The concentration of basic fuchsin and thionin (Allied Chemical and Dye Co., New

TABLE 1. Differential characteristics of the species and biotypes in the genus *Brucella*

Species	Biotype	CO_2 required	H_2S production	Urease activity	Growth on dyes[a]: Basic fuchsin II	Growth on dyes[a]: Thionin I	Thionin II	Thionin III	Thionin blue (1:500,000)	Erythritol (1 mg/ml)	Penicillin (5 U)	Agglutination in sera[b]: A	M	R	Lysis by phage Tbillsi: RTD[c]	$10^4 \times$ RTD	Most common host reservoir
B. melitensis	1	−	−	Variable	+	−	+	+	+	+	+	−	+	−	−	−	Sheep, goats
	2	−	−	Variable	+	−	+	+	+	+	+	+	−	−	−	−	Sheep, goats
	3	−	−	Variable	+	−	+	+	+	+	+	+	+	−	−	−	Sheep, goats
B. abortus	1	±	+	1–2 h	+	−	−	−	+	+	+	+	−	−	+	+	Cattle
	2	+	+	1–2 h	−	−	−	−	−	±	−	+	−	−	+	+	Cattle
	3	±	+	1–2 h	+	+	+	+	+	+	+	+	−	−	+	+	Cattle
	4	±	+	1–2 h	+	−	−	−	+	+	+	−	+	−	+	+	Cattle
	5	−	−	1–2 h	+	−	+	+	+	+	+	−	+	−	+	+	Cattle
	6	−	±	1–2 h	+	−	+	+	+	+	+	+	−	−	+	+	Cattle
	7	−	±	1–2 h	+	−	+	+	+	+	+	+	+	−	+	+	Cattle
	9	±	+	1–2 h	+	−	+	+	+	+	+	−	+	−	+	+	Cattle
Strain 19	1	−	+	1–2 h	+	−	−	−	−	−	−	+	−	−	+	+	Vaccine
B. suis	1	−	+	0–30 min	−	+	+	+	±	+	−	+	−	−	−	+	Pigs
	2	−	−	0–30 min	−	+	+	+	−	+	−	+	−	−	−	+	Pigs, horses
	3	−	−	0–30 min	+	+	+	+	+	+	±	+	−	−	−	+	Pigs
	4	−	−	0–30 min	+	+	+	+	−	+	−	+	+	−	−	+	Reindeer
B. canis		−	−	0–30 min	−	+	+	+	±	+	−	−	−	+	−	−	Dogs
B. neotomae		−	+	0–30 min	−	−	−	+	+	+	−	+	−	−	−	+	Wood rat
B. ovis		+	−	Negative	+	+	+	+	−	−	−	−	−	+	−	−	Sheep (rams)

[a] Species differentiation is obtained on tryptic soy or tryptose agar with the following graded concentrations of dyes: 1:25,000 (I), 1:50,000 (II), 1:100,000 (III). Other concentrations may be preferable with other growth media. Interpretation of results should be controlled with the reference strains of each species. Tests should be conducted in CO_2 for those strains requiring CO_2.

[b] A, Monospecific *B. abortus* antiserum; M, monospecific *B. melitensis* antiserum; R, anti-rough serum.

[c] RTD, Routine test dilution.

York, N.Y.) used in these tests is between 10 and 40 μg of dye per ml of medium (1:25,000 and 1:100,000). The actual concentration of dye is that which will differentiate among control cultures of *B. melitensis*, *B. abortus*, and *B. suis* biotypes. This differentiation will depend on the basal medium used, as well as the bacteriostatic action and purity of the dye, and must be determined by using control cultures. Plates are incubated at 37°C in carbon dioxide for 4 days. Growth on three of four quadrants of a plate is regarded as resistance to the dye. Other growth inhibition tests include thionin blue, erythritol, and penicillin (Morgan, 1961).

Bacteriophage Typing

The usefulness of bacteriophage lytic patterns for classification of brucellae has been reported by several investigators. Lytic patterns with Tbilisi (Tb), Weybridge (Wb), Berkeley (Bk_2), Firenze (Fi), and R bacteriophages are useful for taxonomic and epidemiologic studies. Tbilisi bacteriophages are useful in differentiating *B. abortus* biotypes 3 and 9 from *B. melitensis* (Table 1). Firenze (Fi) bacteriophage lysis is recommended as an additional characteristic for identification of *B. suis* biotype 4 (Thomas and Corbel, 1982) and bacteriophages R/O and R/C are useful for identification of *B. ovis* and *B. canis* that appear only in a nonsmooth state (Corbel et al., 1979). Procedures for bacteriophage typing are presented by Alton et al. (1975) and by Corbel and Brinley-Morgan (1984).

Immunologic Aspects of Brucellosis

Serodiagnosis of brucellosis is especially difficult because no one test or combination of tests has proved effective for recognition of all forms of the disease. The selection and interpretation of serologic tests for diagnosis of brucellosis must include consideration of the forms of the disease and the mechanism of exposure, as these factors influence the classes of immunoglobulins present and the relative amounts of each class. The initial response to antigenic stimulation is production of IgM, followed by IgG and IgA (Parratt et al., 1977). Production of IgA after skin sensitization to *Brucella* antigen has been implicated in the localized hypersensitivity reactions suffered by veterinarians (McAllister, 1976).

The finding of IgM in a healthy individual indicates recent or intermittent exposure to *Brucella* antigen or active infection, which may be subclinical or acute (Elberg, 1973). In patients with a history of brucellosis, the presence of IgM indicates recrudescence of an underlying infection (Young, 1983). In acute brucellosis, IgM titers rise rapidly to levels of 1:640 or greater and fall to undetectable levels in months. IgM titers decline rapidly with treatment and are a good indicator of response to therapy. IgM contributes to the combined titers in the tube agglutination, complement fixation, microagglutination, slide agglutination, and rose bengal agglutination tests. The indirect immunofluorescent-antibody test, microimmunodiffusion, radioimmunoassay and enzyme-linked immunoassay may be made immunoglobulin class specific for IgM, IgG, and IgA.

Because the onset of brucellosis is usually insidious and patients frequently delay seeking medical attention until incapacitated, IgG predominates in high titer when serologic specimens are first examined. A single serum titer of 1:160 or greater or a fourfold rise in titer on the tube agglutination test is considered significant in populations not commonly exposed to brucellae. In abattoir workers, veterinarians, herdsmen, and others occupationally or avocationally exposed to antigen, a titer of 1:320 or more or 1:640 or more is considered significant, as these individuals frequently have background IgG titers of 1:40 to 1:160 in the absence of clinical symptoms (Jamieson and Rich, 1981). Each laboratory must establish the diagnostic cutoffs for tests used and educate physicians in interpretation of the local laboratories' test results. Among the considerations in establishment of significant titers are tests employed, reagents used, batch variation between antigens, and nature of the patient population served. Ascribing significance to titers of 1:80 may result in overdiagnosis of brucellosis in populations occupationally exposed to antigen.

Presence of IgG without IgM indicates convalescence, chronic infection, or continuous exposure over a long period, as is common in veterinarians and abattoir workers. The presence of IgG has traditionally been regarded as the best indicator for diagnosis of brucellosis, as it agglutinates, fixes complement, and can be revealed in its nonagglutinating form by the anti-human globulin (Coombs) agglutination test. IgG rises to titers of 1:640 or greater during the first weeks of disease and persists for months to years in untreated patients. IgG titers fall rapidly in treated cases. The higher titers of late acute brucellosis may be missed because of the prozone phenomenon unless agglutination tests include dilutions of 1:80 and 1:320. The role of "blocking antibody" in serodiagnosis of brucellosis has been attributed to nonagglutinating IgG and IgA, and the anti-human globulin (Coombs) agglutination test has been recommended to measure IgG in chronic brucellosis (Kerr et al., 1968). However, the criteria for diagnosis of chronic brucellosis are found in patients recovering from acute disease, so there is no clear serologic distinction between laboratory results from convalescent and chronic patients (Farrell et al., 1975b). Since the anti-human globulin agglutination test responds to nonspecific antibody, it is no longer recommended when the enzyme immunoassay and radioimmunoassay tests for IgG and IgA are available. The presence of IgA indicates active infection. IgA may or may not agglutinate and does not fix complement.

Smooth strains of brucellae are used for the tube agglutination test, 2-mercaptoethanol agglutination test, and anti-human globulin agglutination test. *B. abortus* strain 1119-3 (U.S. Department of Agriculture, National Animal Disease Laboratory, Ames, Iowa), strain 99 (World Health Organization/Food and Agricultural Organization Collaborating Center for Reference and Research on Brucellosis, Central Veterinary Laboratory, Weybridge, United Kingdom), or other national reference strains that have been prepared and standardized against the International Standard for *B. abortus* must be used to ensure reproducibility and reliability of diagnostic tests. U.S. Department of Agriculture or equivalent antigens are adequate for serodiagnosis of infections caused by *B. abortus*, *B. suis*, and *B. melitensis*. For serodiagnosis of disease caused by *B. canis*, a homologous antigen must be prepared using the method of Alton et al. (1975).

Tube Agglutination Test

The tube agglutination test remains the classic test for serodiagnosis of brucellosis (Meyer, 1986). In low-incidence areas or for populations infrequently exposed to brucellae, a single serum titer of 1 : 160 or greater is considered significant and a titer of 1 : 80 or greater is considered suggestive. In endemic areas and for patients occupationally exposed to brucellae, titers of 1 : 320 or greater or even 1 : 640 or greater have been recommended as a cutoff for significance. A fourfold increase in titer is considered significant in all populations.

2-Mercaptoethanol Agglutination Test

The addition of 2-mercaptoethanol to the tube agglutination test disrupts IgM so that only IgG is measured (Meyer, 1986). Buchanan and Faber (1980) reported that the 2-mercaptoethanol test was superior to the tube agglutination test in determining response to therapy and that a negative 2-mercaptoethanol test is strong evidence against a diagnosis of chronic brucellosis. Klein and Behan (1981) substituted dithiothreitol for 2-mercaptoethanol and found the tests equivalent. Titers of 1 : 40 to 1 : 80 in either test indicate active disease in low-incidence populations, whereas titers of 1 : 80 or greater are significant in endemic areas as for the occupationally exposed patient.

Coombs Test

The anti-human globulin agglutination (Coombs) test is performed by the method of Meyer (1986). Test results are variable and difficult to interpret apart from clinical symptoms. Generally, a titer of 1 : 80 or greater is present in cases of chronic brucellosis. Otero et al. (1982) adapted the Coombs test for microtiter equipment, thereby increasing the ease of performance and enabling batch testing of large numbers of specimens. Ansorg et al. (1984) described a coagglutination test for diagnosis of brucellosis. The procedure is essentially the same as the Coombs test, except that anti-human globulin was replaced by staphylococcal protein A and sera were treated with magnesium chloride–heparin solution. The sensitivity and specificity of the coagglutination test for detection of nonagglutinating antibody surpassed that of the Coombs test.

Complement Fixation Test

MacDonald and Elmslie (1967) reported the usefulness of the complement fixation test for diagnosis of chronic brucellosis. Farrell et al. (1975a) introduced the conglutinating complement fixation test to improve diagnosis of early acute brucellosis. A method for detection of complement-fixing antibodies by an enzyme-linked immunosorbent assay was described by Hinchliffe and Robertson (1983).

Indirect Immunofluorescent-Antibody Test

Henderson et al. (1976) compared the indirect immunofluorescent-antibody test to the tube agglutination test, the 2-mercaptoethanol tube agglutination test, the complement fixation test, and the Coombs test using sera from veterinarians. They recommended tube agglutination and complement fixation as the most consistent tests for diagnosis in a high-risk population. Although the indirect immunofluorescent-antibody test is immunoglobulin class specific, it has not been widely used for diagnosis of brucellosis.

Counterimmunoelectrophoresis Test

Counterimmunoelectrophoresis has been used to study the antigenic lipopolysaccharides and proteins that cross-react with *Yersinia enterocolitica* 0:9. Diaz et al. (1976) reported that the counterimmunoelectrophoresis test detected a protein antigen in 94% of sera for *B. canis* and that counterimmunoelectrophoresis offered promise for clinical diagnosis as a rapid, sensitive, and easily performed confirmatory test. To date, the method remains primarily a research tool.

Microimmunodiffusion Test

The microimmunodiffusion test has been used to elucidate the antigenic nature of various cellular extracts of *Brucella* species, but has not been widely used in human diagnosis (Diaz et al., 1976). McMahon et al. (1979) found that the microimmunodiffusion test detected 97% of cases when the tube agglutination titer was 1 : 160 or greater. The test has not been adequately studied for diagnosis of human brucellosis.

Radioimmunoassay Test

Since its introduction by Wilson et al. (1976), radioimmunoassay has been advocated as a sensitive immunoglobulin class-specific test for *Brucella* antibody. Hewitt and Payne (1984) used radioimmunoassay to measure antibody response in acute, relapsing, chronic, and asymptomatic brucellosis and affirmed its usefulness and concordance with conventional tests. Changes in relative proportions of immunoglobulin class were detectable only by radioimmunoassay. Radioimmunoassay was used to confirm early IgM responses in acute brucellosis.

Rose Bengal Test

Morgan and colleagues (1969) used the rose bengal acidified card test antigen in a plate agglutination test that has been widely used in Europe and the Soviet Union for clinical diagnosis, epidemiologic investigations, and blood donor screens (Caces et al., 1979; Cernyseva et al., 1977; Diaz et al., 1978). Altwegg and Bohl (1985) reported 100% sensitivity of the rose bengal test for patients with culturally proven brucellosis or serologic evidence of active infection. They prefer the rose bengal test for initial screening and the tube agglutination test for retesting of positive sera. The rose bengal test is marketed in Europe as the Brucelloslide-Test (Bio Merieux, Charbonnieres-les-Bains, France).

Enzyme Immunoassay Test

Carlsson et al. (1976) reported detection of *Brucella* antibody by enzyme immunoassay and found the test to be 10- to 100-fold more sensitive than the tube agglutination test. Magee (1980) demonstrated that enzyme immunoassay had sensitivity comparable to that of radioimmunoassay and could detect class-specific antibody. Marmonier et al. (1980) compared enzyme immunoassay to the tube agglutination, rose bengal, complement fixation, and indirect immunofluorescent-antibody tests and found excellent sensitivity and specificity in enzyme immunoassay. Sippel et al. (1982) reported that enzyme immunoassay was an excellent method for screening large populations for brucellar antibody and for differentiation between acute and chronic brucellosis. DeKlerk and Anderson (1985) found that enzyme immunoassay was the most sensitive serodiagnostic test for detection of IgM in acute brucellosis and was particularly useful for diagnosis of brucellosis in occupationally exposed workers.

Slide Agglutination Test

The slide agglutination test is frequently performed by hospital laboratories as part of their febrile agglutination test battery. Although commercial antigen remains available, the slide agglutination test is insensitive and therefore not recommended as a screening test for brucellosis (Hall and Manion, 1953).

Card Agglutination Test

The card agglutination test is widely used in veterinary screening; however, Russell et al. (1978) found it insensitive for diagnosis of human brucellosis and discouraged its use.

Microagglutination Test

Brown et al. (1981) described the microagglutination test using safranine-O-stained antigen to screen human sera for brucellosis. This test was more convenient to perform than the slide or tube agglutination tests, compared favorably with the latter. The microagglutination test is used in the authors' laboratory to screen sera, and all reactive sera with titers of 1 : 80 or greater are retested using the tube agglutination and 2-mercaptoethanol tests. Polt and Schaefer (1982) adapted the microagglutination test for detection of antibody to *B. canis* in human sera. Bettelheim et al. (1984) found a microagglutination test using hematoxylin-stained antigen useful for screening human populations for brucellosis.

Passive Hemagglutination Test

The passive hemagglutination test, employing alkaline-hydrolyzed smooth lipopolysaccharide antigen and sheep erythrocytes, is used in the Soviet Union (Versilova et al., 1974). Field evaluations of the test showed it to have diagnostic value in acute and subacute brucellosis, where 91% of cases were detected using passive hemagglutination alone. The test was less promising in chronic disease, where only 63% of cases gave positive results. A titer of 1 : 100 or greater is considered significant in humans. Coombs et al. (1978) developed the mixed reverse passive anti-globulin hemagglutination test to measure the presence and amounts of specific immunoglobulin classes in brucellosis. Although the test may provide insight into the immunologic mechanism of the disease, it offers little promise as a routine diagnostic test for brucellosis. Renoux (1980) recommended the chromic chloride modification of the passive hemagglutination test for diagnosis of both human and animal brucellosis after finding positive tests at titers of 1 : 200 or greater only in proven cases of disease. He recommended the modified passive hemagglutination test as a simple, sensitive screening test that could replace the conventional battery of tests commonly used. Sera examined by passive hemagglutination did not cross-react with *Yersinia* 0 : 9.

Treatment, Control, and Prevention

Treatment

Treatment of brucellosis with antimicrobial chemotherapeutic agents was reviewed by Havas (1980). Pratt et al. (1978) successfully treated a patient with endocarditis caused by *B. melitensis* using a combination of rifampicin, streptomycin, and tetracycline. Llorens-Terol and Busquets (1980) used rifampicin alone and with cotrimoxazole for successful treatment of children. Ariza et al. (1985) reported a relapse rate of 38.8% in patients treated with rifampicin at 15 mg/kg per day in a single morning dose plus 100 mg of doxycycline every 12 h for 30 days. Patients treated with tetracycline or doxycycline plus streptomycin according to the recommendations of the World Health Organization experienced a relapse rate of 7.1%. These authors also reported a 40% relapse rate in patients treated with cotrimoxazole, 240 to 1,200 mg every 12 h orally for 45 days. Doxycycline and rifampicin show synergy and have been used to treat patients in whom the organism was resistant to cotrimoxazole and tetracycline was undesirable (Mortensen et al., 1986; Nunan et al., 1984). Montanari et al. (1985) suggested that treatment with rifampicin and doxycycline offered the best alternative to streptomycin and doxyclore in acute and complicated brucellosis. Tosi and Nelson (1982) reported successful treatment of a 17-month-old child with *B. canis* infection using moxalactam. No single drug regimen is effective in chronic brucellosis, where repeated courses of therapy are common. Analgesics, vitamin B complex, and vitamin C are sometimes added to combat the despondency and malnutrition that accompany chronic disease. Surgical drainage of abscesses is recommended and valve replacement is frequent in endocarditis, although newer drug combinations offer the promise of successful antimicrobial treatment. When given therapeutically for 6 months to patients with chronic brucellosis, levamisole restores normal monocyte phagocytosis and T-lymphocyte numbers and patients become free of symptoms (Boura et al., 1984).

The World Health Organization recommendation for treatment of acute brucellosis includes oral tetracycline, 250 to 500 mg four times daily, or doxycycline, 200 to 300 mg once or twice daily, plus intramuscular inoculation of streptomycin, 1,000 mg daily for 3 weeks. Intramuscular gentamicin, 360 mg daily for 4 weeks, may be more effective than streptomycin. Oral cotrimoxazole, three tablets (trimethoprim 80 mg, sulfamethoxazole 400 mg) twice daily for 2 weeks followed by two tablets twice daily for a minimum of 6 weeks is effective. Oral rifampicin, 20 mg/kg per day, may be used with tetracycline or cotrimoxazole.

Control

Brucellosis in animals persists wherever control measures are lacking because of insufficient administrative organization or fiscal resources to undertake an eradication program. The success of eradication programs in the Scandinavian countries, Northern and Central Europe, Canada, and the United States should serve as models in demonstrating the importance and feasibility of eliminating this ravaging disease. Education of legislators is vital for enactment of laws to protect the public health while providing indemnity for losses suffered by producers in an eradication program.

Prevention

Prevention of human infection in endemic areas must address protection from environmental contamination and education in the rudiments of personal and food hygiene. Pasteurization of milk and dairy products and thorough cooking of meats virtually eliminates the risk of brucellosis for populations not occupationally exposed to animals. Prevention of brucellosis in a high-risk occupation such as livestock producer, veterinarian, and abattoir worker depends on prevention of contact through use of protective equipment. Despite all precautions, occasional abattoir outbreaks continue to occur (Davos et al., 1981). Education programs should be designed to inform workers about the disease, its symptoms, mode of transmission, need for early diagnosis, and proper use of protective equipment. In the laboratory, prevention of infection depends upon strict adherence to laboratory safety guidelines, use of a biologic safety cabinet for specimen and cultural manipulations producing aerosols, and avoidance of direct skin contamination or accidental inoculation.

Literature Cited

Alton, G. G., L. M. Jones, and D. E. Pietz. 1975. Laboratory techniques in brucellosis. W.H.O. Monogr. Ser. 55, p. 47–52, 149–152.

Altwegg, M., and E. Bohl. 1985. Evaluation of a rapid, reliable, and inexpensive screening test for the serological diagnosis of human brucellosis. Zentralbl. Bakteriol. Abt. I Orig. B. Hyg. **260**:65–70.

Ansorg, R. A., S. Heine, and C. J. Kraus. 1984. Substitution of anti-human globulin by protein A-bearing staphylococci in the detection of *Brucella* antibodies. Med. Microbiol. Immunol. **173**:233–240.

Ariza, J., F. Gudiol. R. Pallares, G. Rufi, and P. Fernandez-Viladrich. 1985. Comparative trial of rifampicin–doxycycline versus tetracycline–streptomycin in the therapy of human brucellosis. Antimicrob. Agents Chemother. **28**:548–551.

Bettelheim, K. A., W. J. Maskill, R. V. Metcalfe, and J. L. Pearce. 1984. The use of the microagglutination technique to determine the antibody status of healthy New Zealanders to *Brucella abortus*. J. Hyg. **92**:401–410.

Boura, P., M. Raptopoulou-Gigi, E. Acriviadis, and G. Goulis. 1984. Re-evaluation of the effect of levamisole in chronic brucellosis: *in vitro* and *in vivo* effect on monocyte phagocytosis. J. Immunopharmacol. **6**:135–146.

Brodie, J., and G. P. Sinton. 1975. Fluid and solid media for isolation of *Brucella abortus*. J. Hyg. **74**:359–367.

Brown, G. M., C. R. Ranger, and D. J. Kelley. 1970. Selective medium for isolation of *Brucella ovis*. Cornell Vet. **61**:265–280.

Brown, S. L., G. C. Klein, F. T. McKinney, and W. L. Jones. 1981. Safranin-O-stained antigen microagglutination test for detection of *Brucella* antibodies. J. Clin. Microbiol. **13**:398–400.

Buchanan, T. M., and L. C. Faber. 1980. 2-Mercaptoethanol *Brucella* agglutination test: usefulness for predicting recovery from brucellosis. J. Clin. Microbiol. **11**:691–693.

Caces, E., H. De Lauture, S. Vol, J. Tichet, and P. Boulard. 1979. The systematic detection of human brucellosis by the rose bengal test on agricultural people after a study in the middle-west of France on 89,000 workers. Comp. Immunol. Microbiol. Infect. Dis. **1**:107–114.

Carlsson, H. E., B. Hurvell, and A. A. Lindberg. 1976. Enzyme-linked immunosorbent assay (ELISA) for titration of antibodies against *Brucella abortus* and *Yersinia enterocolitica*. Acta Pathol. Microbiol. Scand. **84**:168–176.

Cernyseva, M. I., E. N. Knjazeva, and L. S. Egorova. 1977. Study of the plate agglutination test with rose bengal antigen for diagnosis of human brucellosis. Bull. W.H.O. **55**:669–674.

Coombs, R. R. A., L. Edebo, A. Feinstein, and B. W. Gurner. 1978. The class of antibodies sensitizing bacteria measured by mixed reverse passive antiglobulin haemagglutination (MRPAH). Immunology **34**:1037–1044.

Corbel, M. J., C. D. Bracewell, E. L. Thomas, and K. P. W. Gill. 1979. Techniques in the identification and classification of *Brucella*, p. 71–122. *In* F. A. Skinner and D. W. Lovelock (ed.), Identification methods for microbiologists, Society for Applied Bacteriology, technical series no. 14, 2nd ed. Academic Press, Inc., New York.

Corbel, M. J., and W. J. Brinley-Morgan. 1984. *Brucella*, p. 377–388. *In* J. G. Holt and N. R. Krieg (ed.), Bergey's manual of systematic bacteriology, vol. 1. The Williams & Wilkins Co., Baltimore.

Corbel, M. J., and D. M. F. D. Hendry. 1985. Urease activity of *Brucella* species, Res. Vet. Sci. **38**:252–253.

Crosby, E., L. Llosa, M. M. Quesada, C. Carrillo P., and E. Gotuzzo. 1984. Hematologic changes in brucellosis. J. Infect. Dis. **150**:419–424.

Davos, D. E., C. F. Cargill, M. R. Kyrkou, J. A. Jamieson, and G. E. Rich. 1981. Outbreak of brucellosis at a South-Australian abattoir. 2. Epidemiological investigations. Vet. Med. J. Aust. **2**:657–660.

DeKlerk, E., and R. Anderson. 1985. Comparative evaluation of the enzyme-linked immunosorbent assay in the laboratory diagnosis of brucellosis. J. Clin. Microbiol. **21**:381–386.

Diaz, R., E. Maravi-Poma, G. Delgado, and A. Rivero. 1978. Rose bengal plate agglutination and counterimmunoelectrophoresis tests on spinal fluid in the diagnosis of *Brucella* meningitis. J. Clin. Microbiol. **7**:236–237.

Diaz, R., E. Maravi-Poma, and A. Rivero. 1976. Comparison of counterimmunoelectrophoresis with other serological tests in the diagnosis of human brucellosis. Bull. W.H.O. **53**:417–424.

Elberg, S. S. 1973. Immunity to *Brucella* infection. Medicine (Baltimore) **52**:339–356.

Etemadi, H., A. Raissadat, M. J. Pickett, Y. Zafari, and P. Vahedifar. 1984. Isolation of *Brucella* spp. from clinical specimens. J. Clin. Microbiol. **20**:586.

Farrell, I. D. 1974. The development of a new selective medium for the isolation of *Brucella abortus* from contaminated sources. Res. Vet. Sci. **16**:280–286.

Farrell, I. D., P. M. Hinchliffe, and L. Robertson. 1975a. The use of the conglutinating complement fixation test in the diagnosis of human brucellosis. J. Hyg. **74**:29–33.

Farrell, I. D., L. Robertson, and P. M. Hinchliffe. 1975b. Serum antibody response on acute brucellosis. J. Hyg. **74**:23–28.

Goossens, H., L. Marcelis, P. DeKeyser, and J. P. Butzler. 1983. *Brucella melitensis:* person to person transmission? Lancet **1**:773.

Gotuzzo, E., C. Carrillo, J. Guerra, and L. Llosa. 1986. An evaluation of diagnostic methods for brucellosis—the value of bone marrow culture. J. Infect. Dis. **153**:122–125.

Hall, W. H., and R. E. Manion. 1953. Comparison of the Coombs test with other methods for *Brucella* agglutinins in human serum. J. Clin. Invest. **32**:96–106.

Hausler, W. J., Jr., N. P. Moyer, and L. A. Holcomb. 1985. *Brucella*, p. 382–386, *In* E. H. Lennette, A. Balows, W. J. Hausler, Jr., and H. J. Shadomy (ed.), Manual of clinical microbiology. American Society for Microbiology, Washington, D.C.

Havas, L. 1980. Problems and new developments in the treatment of acute and chronic brucellosis in man. Acta Trop. **37**:281–286.

Henderson, R. J., D. M. Hill, A. A. Vickers, J. M. Edwards, and H. Tillett. 1976. Correlation between serological and immunofluorescence results in the investigation of brucellosis in veterinary surgeons. J. Clin. Pathol. **29**:35–38.

Herr, S., and D. Roux. 1981. The efficacy of bacteriological procedures for the isolation of *Brucella abortus* from abattoir material. Onderstepoort J. Vet. Res. **48**:7–12.

Hewitt, W. G., and D. J. H. Payne. 1984. Estimation of IgG and IgM *Brucella* antibodies in infected and non-infected persons by a radioimmune technique. J. Clin. Pathol. **37**:692–696.

Hinchliffe, P., and L. Robertson. 1983. Detection of complement fixation by enzyme immunosorbent assay (COMPELISA). J. Clin. Pathol. **36**:100–103.

Jamieson, A., and G. Rich. 1981. Brucellosis and serological titers (correspondence). Aust. Fam. Physician **10**:319.

Kerr, W. R., W. J. McCaughey, J. D. Coghlan, D. J. H. Payne, R. A. Quaife, L. Robertson, and I. D. Farrell. 1968. Techniques and interpretations in the serological diagnosis of brucellosis in man. J. Med. Microbiol. **1**:181–193.

Klein, G. C., and K. A. Behan. 1981. Determination of *Brucella* immunoglobulin G agglutinating antibody titer with dithiothreitol. J. Clin. Microbiol. **14**:24–25.

Kuzdas, C. D., and E. V. Morse. 1953. A selective medium for the isolation of brucellae from contaminated materials. J. Bacteriol. **66**:502–503.

Llorens-Terol, J. and R. M. Busquets. 1980. Brucellosis treated with rifampicin. Arch. Dis. Child. **55**:486–488.

MacDonald, A., and W. H. Elmslie. 1967. Serological investigations in suspected brucellosis. Lancet **1**:380–382.

Magee, J. T. 1980. An enzyme-labelled immunosorbent assay for *Brucella abortus* antibodies. J. Med. Microbiol. **13**:167–172.

Marmonier, A., J. P. Sthal, M. Micoud, and P. Le Noc. 1980. Application de la technique immunoenzymatique ELISA (enzyme-linked-immunosorbent-assay) au diagnostic serologique des brucelloses humaines. II. Comparaison aux autres techniques serologiques existantes. Pathol. Biol. **19**:83–87.

Matyas, Z., and T. Fujikura. 1984. Brucellosis as a world problem. Third International Symposium on Brucellosis, Algiers, Algeria, 1983. Dev. Biol. Stand. **56**:3–20.

McAllister, T. A. 1976. Laboratory diagnosis of human brucellosis. Scott. Med. J. **21**:129–131.

McMahon, K. J., E. D. Renner, G. W. Allmaras, and D. K. Anderson. 1979. An agar-gel immunodiffusion test for detection of *Brucella* antibodies in human serum. Can. J. Microbiol. **25**:850–854.

Meyer, M. E. 1986. Immune response to brucellae. *In* N. R. Rose, H. Friedman, and J. L. Fahey (ed.), Manual of clinical immunology. American Society for Microbiology, Washington, D.C.

Montanari, M., T. Gandini, D. Torre, and A. Dietz. 1985. La terapia della brucellosi: esperienza clinica con l'associazione rifampicina-doxiciclina. Minn. Med. **76**:1407–1410.

Morgan, W. J. B. 1961. The use of the thionin blue sensitivity test in the examination of *Brucella*. J. Gen. Microbiol. **25**:135–139.

Morgan, W. J. B., D. J. MacKinnon, J. R. Lawson, and G. A. Cullen. 1969. The rose bengal plate agglutination test in the diagnosis of brucellosis. Vet. Rec. **85**:636–641.

Mortensen, J. E., D. G. Moore, J. E. Clarridge, and E. J. Young, 1986. Antimicrobial susceptibility of clinical isolates of *Brucella*. Diagn. Microbiol. Infect. Dis. **5**:163–169.

Naparstek, E., C. S. Block, and S. Slavin. 1982. Transmission of brucellosis by bone marrow transplantation. Lancet **1**:574–575.

Nunan, T. O., S. J. Eykyn, and N. F. Jones. 1984. Brucellosis with mesangial IgA nephropathy: successful treatment with doxycycline and rifampicin. Br. Med. J. **288**:1802.

Otero, J. R., A. Fuertes, E. Palenque, and A. R. Noriega. 1982. Microtiter-adapted method that facilitates the Coombs test for brucellosis. J. Clin. Microbiol. **16**:737–738.

Parratt, D., K. H. Nielsen, and R. C. White. 1977. Radioimmunoassay of IgM, IgG and IgA *Brucella* antibodies. Lancet **1**:1075–1078.

Polt, S. S., and J. Schaefer. 1982. A microagglutination test for human *Brucella canis* antibodies. Am. J. Clin. Pathol. **77**:740–744.

Pratt, D. S., J. H. Tenney, C. M. Bjork, and L. B. Reller. 1978. Successful treatment of *Brucella melitensis* endocarditis. Am. J. Med. **64**:897–900.

Renoux, M. 1980. A passive hemagglutination test for the detection of *Brucella* infection. J. Immunol. Methods **32**:349–355.

Roth, R. M., and R. A. Gleckman. 1985. Human infections derived from dogs. Postgrad. Med. **77**:169–180.

Russell, A. O., C. M. Patton, and A. K. Kaufmann. 1978. Evaluation of the card test for diagnosis of human brucellosis. J. Clin. Microbiol. **7**:454–458.

Sippel, J. E., N. A. El-Masry, and Z. Farid. 1982. Diagnosis of human brucellosis with ELISA. Lancet **2**:19–21.

Thimm, B. M. 1982. Brucellosis: distribution in man, domestic and wild animals. Springer-Verlag, New York.

Thomas, E. L., and M. J. Corbel. 1982. Differentiation of *Brucella suis* biotype 4 from other biotypes by Firenze phage. Br. Vet. J. **138**:11–17.

Tosi, M. F., and T. J. Nelson. 1982. *Brucella canis* infection in a 17-month-old child successfully treated with moxalactam. J. Pediatr. **101**:725–727.

Verger, J.-M., F. Grimont, P. A. D. Grimont, and M. Grayon. 1985. *Brucella*, a monospecific genus as shown by deoxyribonucleic acid hybridization. Int. J. Syst. Bacteriol. **35**:292–295.

Versilova, P. A., M. I. Cernyseva, R. G. Aslanjan, and E. N. Knjazeva. 1974. Diagnosis of human and animal brucellosis by the indirect hemagglutination test. Bull. W.H.O. **51**:191–197.

Weed, L. A. 1957. Use of a selective medium for isolation of *Brucella* from contaminated surgical specimens. Am. J. Clin. Pathol. **27**:482–485.

Wilson, D. V., M. J. Thornley, and R. R. A. Coombs. 1976. A solid phase assay with radioactively-labeled antibody for the detection of *Brucella abortus*. J. Med. Microbiol. **10**:281–292.

Young, E. J. 1983. Human brucellosis. Rev. Infect. Dis. **5**:821–842.

CHAPTER 15

Campylobacteriosis

NELSON P. MOYER and LARRY A. HOLCOMB

Disease: Gastroenteritis, colitis, proctitis, septicemia, gastritis, and ulcers of the stomach and duodenum. Abortion and reproductive disorders of livestock.

Etiologic Agents: *Campylobacter jejuni, C. coli, C. hyointestinalis, C. laridis, C. cinaedi, C. fennelliae, C. fetus* subspecies *fetus*, and *C. pylori* in humans.

Source: Domestic livestock and poultry, wild and caged birds, direct contact, contaminated meat, raw milk, and water.

Clinical Manifestations: Bloody diarrhea, abdominal pain, fever, nausea. Disseminated focal infections and ulcers of gastric and duodenal mucosa.

Pathology: Invasive or secretory diarrhea. Cytotoxins with and without hemolytic activity and enterotoxin with choleralike mechanism. Attachment and penetration of intestinal and gastric mucosa with ulceration. Septicemia in compromised hosts.

Laboratory Diagnosis: Presumptive identification from characteristic tumbling, darting motility, and corkscrew morphology in wet mounts examined by phase-contrast or darkfield microscopy. Culture on selective and differential media for definitive identification.

Epidemiology: Zoonosis with worldwide distribution. Sporadic disease of children in developing countries. Outbreaks associated with contaminated food and water.

Treatment: Antibiotic therapy not recommended except for severe or disseminated cases, in which erythromycin, chloramphenicol, and aminoglycosides are effective. Bismuth compounds are active against *C. pylori*.

Prevention and Control: Personal hygiene and food precautions, including pasteurization of milk.

Description of Disease

Although *Campylobacter* species (formerly *Vibrio fetus* and "related vibrio") have been recognized as zoonotic pathogens since 1909, their significance in human medicine was not appreciated until 1947 when the first culture of *V. fetus* from human blood was reported. Rare isolations over the next decade established the organism as an opportunistic pathogen of debilitated patients. Elizabeth King first observed the thermophilic nature of organisms associated with gastroenteritis, but characterization of these organisms was postponed until selective media were developed for their isolation from fecal specimens. The introduction of selective media developed by Skirrow (1977), Butzler and Skirrow (1979), and Blaser et al. (1979) initiated an intensive investigation of members of the genus *Campylobacter* that established their clinical importance as human enteric pathogens with worldwide distribution. The taxonomy and nomenclature of the genus is changing rapidly as biotyping, serotyping, and hybridization studies of human and veterinary isolates reveal previously unrecognized species. The spectrum of disease produced by *Campylobacter* species includes septicemia, gastroenteritis, gastritis, gastric and duodenal ulcer, and various disseminated opportunistic infections in humans and reproductive disorders and intestinal disease in animals.

Campylobacter jejuni has emerged as a major human enteric pathogen that is isolated from fecal specimens with a frequency equal to or greater than that of salmonellae and shigellae in most parts of the world. Disease is characterized by invasive and sometimes secretory gastroenteritis (Walker et al., 1986). *Campylobacter coli* is found in healthy pigs, but causes gastroenteritis in humans. Because most laboratories do not differentiate *C. coli* and *C. jejuni*, the true incidence of disease caused by either of these enteric pathogens is unknown (Finch and Riley, 1984). *Campylobacter laridis* is found in the intestinal contents of seagulls and other avian hosts and mammals, including humans, where it is an infrequent cause of gastroenteritis (Benjamin et al., 1983).

Campylobacter fetus subspecies *fetus* causes disseminated disease in humans and abortion in sheep and cattle (Francioli et al., 1985). *Campylobacter fetus* subspecies *venerealis* has not been implicated in human disease, but causes abortion and infertility in cattle. *Campylobacter sputorum* subspecies *bubulus* may be isolated from the reproductive tracts of cattle and sheep, but has not been linked to animal or human disease. *Campylobacter sputorum* subspecies *mucosalis* causes intestinal disease in pigs, but human disease has not been reported. *Campylobacter concisus* has been isolated from the gingival crevices of humans with peridontal disease, but its etiologic role is unknown. *Campylobacter nitrofigilis* is associated with marine vegetation and does not cause disease in animals or humans. *Campylobacter fecalis* is found in feces of sheep and cattle, where it may cause enteric disease, but is not found in human feces. *Campylobacter hyointestinalis* has been isolated from swine with proliferative ileitis (Gebhart et al., 1983) and humans with gastroenteritis (Edmonds et al., 1987). *Campylobacter fennelliae* has been isolated from the rectum of homosexual men with proctitis (Totten et al., 1985). *Campylobacter cinaedi* has been isolated from blood and rectal swabs of homosexual men with intestinal disease (Totten et al., 1985). *Campylobacter pylori* (formerly *C. pyloridis*) is associated with gastritis and ulcers of the stomach and duodenum (Marshall, 1986; Marshall and Goodwin, 1987). Aerotolerant *Campylobacter* species are associated with abortion in cattle and pigs and mastitis in cattle. Growth occurs aerobically at 30°C in Ellinghausen–McCullough–Johnson–Harris medium after primary isolation in a microaerophilic atmosphere. No human diseases are attributed to these organisms.

Communicability

Environmental contamination and contact with infected animals or their products are primary means of transmission for campylobacteriosis. Disease occurs as sporadic cases, in clusters of cases, and in widespread outbreaks. Fecal–oral transmission occurs directly by person-to-person contact or indirectly through meat, poultry, eggs, raw milk, and untreated water. Flies may serve as a vector for fecal contamination of raw milk, where as few as 500 organisms have been shown to be infective (Rosef and Kapperud, 1983). Contamination of other foods by arthropod vectors is unlikely to cause foodborne outbreaks, as these organisms do not grow in foods and a relatively high infective dose of 10^5 to 10^6 organisms is usually required to establish infection. Rates of contamination for potential sources have been reported for surface water (53%), cattle (43%), poultry (91%), pigs (88%), dogs (49%), cats (53%), and infected humans (1.6%).

Clinical Manifestations

Infections caused by *Campylobacter* may be classified into three distinctive syndromes: intestinal disease, bacteremia with disseminated disease, and ulcers of the gastric or duodenal mucosa.

Intestinal disease caused by the thermophilic *Campylobacter* may vary from a brief, relatively asymptomatic, and self-limiting enteritis to a fulminating enterocolitis. Onset occurs 1 to 7 days after exposure in active cases, and disease is characterized by diarrhea, abdominal pain, fever, malaise, headache, dizziness, myalgias, backache, and arthralgias. Nausea may occur, but vomiting is uncommon. Stools are mucoid to liquid, foul smelling, and bile stained, progressing to profusely watery diarrhea containing blood and pus. Patients experience 8 to 15 or more stools per day for a period of 2 to 7 days. Symptoms usually subside within 1 week. Dehydration is uncommon in adults. Prolonged shedding of organisms (2 weeks to 3 months) occasionally occurs in immunocompromised and debilitated patients. Relapse occurs in one-fourth of patients, although recurrent illness is generally milder than the initial episode. Diarrhea is usually self-limiting and antibiotics do not reduce the severity of symptoms, although they do decrease the number of relapses and shorten the duration of shedding. Stool culture is required for definitive diagnosis of infections caused by thermophilic *Campylobacter*. Blaser et al. (1986) reported mild transient bacteremia in normal hosts infected with *C. jejuni* and *C. coli*, but blood cultures are usually negative. Serum-resistant strains of these organisms were capable of causing sustained bacteremia in normal hosts, and both serum-sensitive and serum-resistant strains caused disseminated infections in immunocompromised patients. *C. laridis* causes both enteritis and septicemia in immunodeficient patients (Tauxe et al., 1985). *C. cinaedi*, *C. fennelliae* and *C. hyointestinalis* have been isolated from stools and rectal swabs from 17.6% of homosexual men with proctitis, enteritis, or proctocolitis, and *C. cinaedi* has been isolated from blood samples from homosexual men with intestinal symptoms (Fennell et al., 1984; Totten et al., 1985; Edmonds et al., 1987). These organisms were isolated from 8% of homosexual male volunteers.

Extraintestinal infections are typically caused by *C. fetus* subspecies *fetus*. Infections have been associated with immunodeficiency caused by long-term steroid use, splenectomy, hypogammaglobulinemia, agammaglobulinemia, malignancy, and acquired immune deficiency. Stools of bacteremic patients are rarely culture positive (Francioli et al., 1985). *C. fetus* subspecies *fetus* has also been reported to cause premature labor and neonatal sepsis, cholecystitis, sepsis and meningitis in an alcoholic patient, sepsis following "nutritional therapy" in Mexico, infection of an abdominal aneurysm, and gastroenteritis. *C. pylori* and gastric *Campylobacter*-like organisms have been strongly associated with active chronic gastritis, gastric ulcers, and duodenal ulcers, but their causal effect remains controversial.

Pathology

The mechanisms by which *Campylobacter* causes diarrhea have just begun to be elucidated. The finding of dysenteric stools suggests mucosal invasion by a process analogous to shigellosis. In vitro studies using cell culture and chick and cattle embryos have shown attachment and penetration of epithelium; however, the Sérény test is negative (Klipstein et al., 1985). The occurrence of profuse watery diarrhea in some cases suggests that some strains produce enterotoxin. Cytotoxin has been demonstrated in tissue culture. Kazmi et al. (1984) found evidence for both invasive and secretory mechanisms using a neonatal mouse model, and studies in chickens, rabbits, calves, dogs, and hamsters support these findings. Johnson and Lior (1986) reported a heat-labile cytotoxin with hemolytic activity and a heat-stable enterotoxin with GM-1 binding capability in most human isolates of *C. jejuni* and *C. coli*. A heat-stable cytotoxin with hemagglutination activity but not hemolytic action was present in some strains. Cytotoxic activity has been demonstrated in Vero, HeLa, Chinese hamster ovary, MRC-5, and HEp-2 cell cultures. Environmental strains exhibit less invasive and cytotoxic activity than clinical strains, and strains from India and Mexico generally produce choleralike toxin.

Endoscopic examination reveals bloody exudate and inflamed, granular, friable mucosal surfaces with fine ulcerations. Colon X ray after barium enema reveals pancolitis or segmental colitis having small aphthoid ulcers with or without halos and presenting a stippled appearance (Tielbeek et al., 1985). Lymphoid hyperplasia is characteristic. Histologic findings may be unremarkable in mild cases to characteristic of acute infectious colitis in severe cases. Chronic cases are typical of ulcerative colitis. Early epithelial changes during the first week of active infection include neutrophilic infiltrates and development of superficial ulcers with accumulation of exudate. Cryptitis progresses to crypt abscess. The lamina propria become edematous with diffuse infiltrates of granulocytes, histiocytes, plasma cells, and lymphocytes. Between 7 and 14 days, inflammation resolves with epithelial regeneration. Edema of the lamina propria persists, and lymphoid hyperplasia is common. Few inflammatory effects are visible after 14 days, although lymphocytes and plasma cells persist in the lamina propria and crypts. Hematologic findings are characterized by a left shift suggestive of shigellosis. Peripheral leukocytosis is common, but function tests are usually normal (Blaser et al., 1979).

Pathologic examination of gastric biopsy is necessary to distinguish epithelial cell changes of bacterial from nonbacterial gastritis and gastric ulcer, as *C. pylori* has been cultured from gastric mucosa in the absence of disease. Isolation of *C. pylori* is usually associated with inflammation of the antrum and the presence of polymorphonuclear leukocytes. Histologic examination of gastric biopsy reveals patchy distribution of organisms beneath the gastric mucus. Antral epithelial cells are irregular with bulging surfaces. Depletion of microvilli and marked reduction in numbers of cytoplasmic mucous secretory glands are typical. Bacterium–cell interaction is postulated to combine with gastric acid erosion to cause ulcers. The Warthin–Stary silver stain, acridine orange stain, Gram stain, and phase-contrast microscopy have been used to demonstrate organisms in gastric mucosal biopsy, and these rapid microscopic methods provide useful presumptive information. Culture requires 3 to 5 days for isolation of organisms and another 2 to 4 days for definitive identification.

Diagnosis

Intestinal campylobacteriosis presents a broad spectrum of disease from proctitis in homosexual men to toxic megacolon. Acute gastroenteritis and enterocolitis are the most frequent manifestations. Complications include lymphadenitis, mesenteriales, cystitis, cholecystitis, pancreatitis, hepatitis, hemolytic–uremic syndrome, peritonitis, erythema nodosum, septic arthritis, osteomyelitis, and septic abortion.

Circulating antibody to *Campylobacter* is found in 10 to 24% of patients with appendicitis, and *C. jejuni* has been isolated from the appendix. Antibody to *C. jejuni* is also present in 38% of patients with Guillain–Barré syndrome. Acute enterocolitis caused by *Campylobacter* must be differentiated from shigellosis, salmonellosis, amebiasis, viral enteritis, ulcerative colitis, appendicitis, and Crohn's disease (Tielbeek et al., 1985). Thorson et al. (1985) examined stools directly by several methods and determined the predictive value of fecal leukocytes at 77%, darkfield microscopy at 83%, and Gram stain at 75%. Blood cultures for *C. jejuni* and *C. coli* are positive in immunocompromised hosts and patients with disseminated infections.

C. fetus subspecies *fetus* causes infections of pregnant women, neonates, children, and immunocompromised or debilitated patients. Complications of bacteremia include endocarditis, pericarditis, meningitis, arthritis, thrombophlebitis, lung abscess, hepatitis, and pneumonia (Torphy and Bond, 1979). Bacteremia caused by *C. fetus* subspecies *fetus* must be differentiated from brucellosis and malaria. Bacteremia occurs as a secondary infection to cirrhosis, malignancy, diabetes mellitus, cardiovascular disease, and alcoholism.

C. pylori is the only *Campylobacter* species that produces urease, and this feature has been exploited diagnostically. Marshall and Langton (1986) reported that gastric acid levels of less than 1 mmol/liter are indicative of *C. pylori* infection. Other investigators have developed a biopsy urease test with a reported sensitivity of 88% and specificity of 100%. The test involves direct inoculation of a crushed biopsy specimen into Christiansen urea broth by the endoscopist. Development of a pink color indicates the presence of *C. pylori*. The *Campylobacter*-like organism (CLO) test (commercially available in Europe) is similar to the broth test, except that an agar-containing medium is employed. The rapid tests for enzymatic activity offer little advantage over direct microscopic observation of the organism, and culture is required for definitive identification.

Prognosis

Intestinal campylobacteriosis is rarely fatal except in children and in elderly, debilitated, or immunocompromised patients. Relapses and chronic infection are more common in children, where death may occur as a result of dehydration and malnutrition, especially in developing countries. Extraintestinal infection caused by *C. fetus* subspecies *fetus* may be serious to life threatening and must be treated with antibiotics. The finding of *C. pylori* in gastric and duodenal ulcers and gastritis is a recent development, and the long-term relationship between the organism and disease has not been examined.

Epidemiologic Aspects

Intestinal campylobacteriosis is a zoonotic disease that is incidentally transmitted to humans (Blaser et al., 1983). The incidence is higher in rural than in urban areas, and cases in developed countries exhibit a marked seasonal variation with peak incidence in the summer and fall months, although cases occur throughout the year. Cases occur most frequently in small children (1 to 2 years old) and young adults (20 to 29 years old), presumably from direct transmission and exposure to raw milk in children and occupational or food exposure in adults. National surveillance programs have only recently been established in developed countries, and little epidemiologic information is available. When laboratories include a search for *Campylobacter* in the routine enteric culture procedure, the number of *Campylobacter* isolations frequently exceeds those of *Salmonella* and *Shigella* isolations. Thompson et al. (1986) reported monthly isolation rates for *Campylobacter* between 1.1 and 7.4% in Canada.

Veterinarians, livestock producers, dog handlers, poultry processors, packing plant workers, and butchers are at risk of occupationally acquired infections. Food handlers may also acquire the disease from raw animal products, as the carcasses of poultry, cattle, sheep, and pigs show high isolation rates at slaughter and organisms are detectable in meat at the retail level. Oosterom and co-workers (1984) reported an association between consumption of home-barbecued chicken and incidence of campylobacteriosis, but only a slight association with caged birds and no association with household pets. Hopkins and colleagues (1984) identified risk factors for the general population of developed countries as ingestion of untreated water, raw milk, and undercooked chicken and contact with household pets. Outbreaks have occurred among military personnel, in nurseries, and in day-care centers from consumption of food or raw milk and from contaminated public water supplies. Although the incidence of campylobacteriosis is low in developed countries, where asymptomatic carriage is of minimal significance, the incidence of infection in developing countries is high and asymptomatic carriage and acute disease are significant public health problems in children.

Intestinal campylobacteriosis has been reported

from countries in Africa, Asia, and Latin America, but incomplete epidemiologic information impedes complete understanding of its global significance. Population studies in Africa and Asia show an incidence of disease between 18.7 and 38% and asymptomatic carriage rates of 8.6 to 40% in children. No sex-specific prevalence has been documented, and the role of protective immunity from chronic exposure is not understood.

Campylobacter has been found in the feces of most warm-blooded animals examined, and the thermophilic species are especially common in wild birds, which may serve as a natural reservoir for the organism. Counts of organisms in avian feces may reach 10^4 to 10^7 CFU/g. Significant carriage is found in migratory waterfowl, pigeons, gulls, and crows. The high optimal growth temperature of the thermophilic *Campylobacter* species may be a result of adaptation to the high body temperature of birds. Manser and Dalziel (1985) reported intestinal carriage of *C. coli* in pigs at 66%, and *C. jejuni* was isolated from 24% of cattle and 22% of sheep examined.

Campylobacter species differ from common agents of foodborne illness because they neither grow to infective numbers nor produce a toxin in food, as they require a microaerophilic atmosphere and an elevated temperature for optimal growth. Between 1980 and 1982, however, 23 foodborne outbreaks were reportedly caused by *Campylobacter* (Finch and Blake, 1985). Of these outbreaks, 14 were traced to raw milk, 5 were traced to infected beef, poultry, or eggs, and 4 resulted from food-handling errors (Klein et al., 1986). The incubation periods ranged from 18 to 168 h, and symptoms were typical of intestinal campylobacteriosis.

Serotyping

Serotyping of *Campylobacter* is useful in epidemiologic investigations of transmission from animals to humans, person-to-person transmission, and food, milk, and waterborne outbreaks of disease. Two systems are widely used for serotyping *C. jejuni*. The method of Penner and Hennessy (1980) detects heat-stable soluble antigens by passive hemagglutination. The method of Lior et al. (1982) is based upon detection of heat-labile antigenic factors by slide agglutination of live bacteria with whole cell antisera absorbed with homologous heated and heterologous unheated cross-reactive antigens. To date, 21 serogroups are recognized, and most strains isolated from human cases are typable.

Patton et al. (1985) compared the serotyping methods of Penner and Lior using 1,405 isolates of *Campylobacter* collected from human, animal, and environmental sources during epidemiologic studies and found 96.1% of isolates were typable by the Penner method and 92.1% were typable by the Lior method. The Lior method was easier to perform and produced rapid results. Cultures frequently reacted in multiple antisera with the Penner method, but multiple reactions were rare with the Lior method, making it easier to interpret. Neither serotyping method typed all strains from outbreaks. Where one serotype was associated with an outbreak, either method was acceptable, but when multiple serotypes were encountered, the two methods complemented each other. The most common serotypes of *C. jejuni* in the United States are Lior 1, 4, 6, 16, and 36 and Penner 1, 2, 4, 5−, and 16 (Patton et al., 1985). Africa, Europe, and North America show similar serotype distribution by the Penner scheme (Lastovica et al., 1986).

As serotyping is performed in few reference laboratories in the United States, antibiograms and plasmid profiles may provide useful information and be more readily available. Other recently reported methods for typing *Campylobacter* include use of lectins (Gill and Corbel, 1986) and coagglutination (Fricker et al., 1986). Grajewski et al. (1985) introduced a bacteriophage typing system for *C. jejuni* and *C. coli* using phage isolated from poultry feces. Fourteen bacteriophages were selected for epidemiologic investigations as an adjunct or alternative to serotyping.

Etiologic Agents

Members of the genus *Campylobacter* are slender, curved rods 0.2 to 0.5 μm wide and 0.5 to 5 μm long. Organisms may have one or more spirals and appear as comma, S, or gull-wing forms when cells occur in chains. Cells in older cultures may form spherical or coccoid bodies. *Campylobacter* species are non-spore-forming, oxidase positive, Gram negative, and motile by means of a single polar flagellum at one or both ends of the cell. Motile cells observed in wet mounts have a characteristic corkscrewlike motion. The genus is microaerophilic, requiring 3 to 6% O_2 for growth. Carbohydrates are neither oxidized or fermented.

Collection, Transport, and Storage of Specimens

Feces or rectal swabs collected from patients with diarrhea should be inoculated onto selective media within 2 h of collection. If inoculation of media must be delayed, the rectal swab or a swab prepared from a fresh stool should be placed in modified Cary–Blair transport medium (Wang et al., 1983). Cary–Blair transport medium can also be used to transport feces for culture of *Salmonella* and *Shigella*. Fecal specimens in Cary–Blair medium may be held at 4°C for 1 to 2 weeks before the number of *Campylobacter* are greatly reduced. Gastric biopsies for the culture of *C. pylori* should be transported in 0.5 ml of 20% glucose. The biopsies can be kept at 4°C for up to 5 h before inoculation to media with little loss of organism viability (Goodwin et al., 1985).

Media

Isolation of *Campylobacter* species requires special selective techniques because of their microaerophilic nature and the contaminating microorganisms found in specimens containing them. *Campylobacter* can be cultured using differential filtration followed by selective enrichment in broths, direct plating on selective agar media, or a combination of these methods. Filtration as described by Goossens et al. (1986) is relatively easy and inexpensive to perform, but suffers from low sensitivity. Modifications of Skirrow medium (Skirrow, 1977), Butzler medium (Butzler and Skirrow, 1979) and Blaser medium (Blaser et al., 1979) are available commercially for the isolation of *Campylobacter* from human fecal specimens. Other selective media utilizing different basal media and antibiotic supplements have been described (Bolton and Robertson, 1982; Goossens et al., 1983). Bolton et al. (1984), Goossens et al. (1986), and Karmali et al. (1986) have described charcoal-based, blood-free selective *Campylobacter* media, which may be of particular value to developing countries where supplies of sterile blood may be limited.

Ng et al. (1985a) studied the inhibitory effects of antibiotics used in selective *Campylobacter* media on strains of *C. jejuni* and *C. coli* and found colistin and polymyxin B to be inhibitory to *C. coli*. Comparative studies by Goossens et al. (1986), Karmali et al. (1986), and Merino et al. (1986) found that selective *Campylobacter* media with cefoperazone had higher isolation rates for *Campylobacter* species and lower fecal flora contamination than did selective media containing colistin or polymyxin B.

Ng et al. (1985b) compared the various basal media used in selective *Campylobacter* media and concluded that Mueller–Hinton agar (Difco Laboratories, Detroit, Mich.; Oxoid Ltd., London, England), *Campylobacter* agar base (Difco), and *Brucella* agar (BBL Microbiology Systems, Cockeysville, Md.) gave the best growth for *Campylobacter* species. Selective *Campylobacter* media containing cefazolin or cephalothin will inhibit the growth of *C. fetus* subspecies *fetus* and some strains of *C. coli;* therefore, filtration procedures or selective media not containing these antibiotics should be used when culturing these organisms from nonsterile specimens (Brooks et al., 1986). Specimens from normally sterile sites may be inoculated onto nonselective media such as blood agar.

A number of selective enrichment broths have been developed to increase the isolation rate of *Campylobacter* species from clinical, food, and environmental specimens. Using a selective enrichment broth in combination with a selective plate medium or using two different selective plating media produced a 10% increase in *Campylobacter* isolation rates (Wells et al., 1981). At present, routine use of selective enrichment broths for clinical specimens is not recommended. Enrichment procedures are most useful for improperly handled stool specimens or where the specimens may contain low numbers of *Campylobacter*.

A selective medium for the isolation of *C. pylori* as recommended by Goodwin et al. (1985) consists of brain–heart infusion base (Oxoid) containing 7% horse blood and 1% Isovitalex (BBL) with vancomycin (3 mg/liter), nalidixic acid 10 mg/liter), and amphotericin B (2 mg/liter). Media supplemented with 0.025% ferrous sulfate, 0.025% sodium metabisulfite, and 0.025% sodium pyruvate (FBP) according to the method of George et al. (1978) may be inhibitory to some strains of *C. pylori* that are sensitive to sodium metabisulfite.

Inoculation of Media

Fecal specimens or rectal swabs can be inoculated directly onto selective media or by emulsifying the feces in broth or phosphate-buffered saline and inoculating 1 to 2 drops onto each plate. If blood or mucus is present, it should be cultured. The plates are then streaked for isolation. Blood cultures should be inoculated into a liquid blood culture medium and a nonselective agar medium such as blood agar. Specimens from other normally sterile sites may be inoculated onto a nonselective medium.

Gastric biopsies should be crushed in sterile saline and 1 to 2 drops inoculated onto a selective agar medium. Also, 1 drop should be put on each of two glass slides. One slide should be Gram stained and the other slide examined by phase-contrast microscopy.

Incubation of Media

Media inoculated with fecal specimens should be incubated at 42 to 43°C in a microaerophilic atmosphere containing 5% O_2, 10% CO_2, and 85% N_2. The higher incubation temperature suppresses growth of the normal fecal flora while permitting the thermophilic *Campylobacter* species to grow. Media inoculated with blood or other nonfecal specimens, or when a nonthermophilic *Campylobacter* species is suspected in feces, should be incubated at 35 to 37°C in a microaerophilic atmosphere.

Several systems are available for producing a microaerophilic atmosphere. The evacuation-replacement system uses an anaerobic container without a catalyst. The container is evacuated twice to 38 cm (15 mm Hg [1 mm Hg = 133.3 Pa]) and refilled each time with a 10% H_2–90% N_2 gas mixture. A second method uses a commercially available disposable gas-generating envelope in an anaerobic jar. A new envelope must be used each time the jar is opened. A third system, also commercially available, is a plastic bag incubation system (BioBag). A simple and inexpensive system involves use of the Fortner principle (Karmali and Fleming, 1979). A rapidly growing facultative anaerobe is incubated with the *Campylobacter* cultures in a jar or a plastic bag. As the facultative anaerobe grows, the oxygen content is lowered and the CO_2 content increased. A disadvantage of this system is the longer incubation time usually required for growth of *Campylobacter*. Pennie et al. (1984) reported a system using steel wool, copper sulfate, sodium bicarbonate, and water to produce an adequate atmosphere for *Campylobacter* growth. A candle jar may be used if the medium has been supplemented with FBP, which is commercially available and enhances tolerance for oxygen by inactivating superoxides. Disadvantages of this system include longer incubation times and inhibition of growth of some oxygen-sensitive *Campylobacter*. Media inoculated with gastric biopsy specimens for *C. pylori* should be incubated at 36°C in high humidity (92 to 95%) in a microaerophilic environment for up to 6 days.

Examination of Plates

Plates should be examined after 24, 48, and 72 h of incubation. *Campylobacter* colonies are gray or colorless and nonhemolytic. They can be flat and watery with irregular edges, or round and convex with entire edges. The colonies may be pinpoint in size or spread over large areas of the plate. Suspicious colonies should be screened with the oxidase test, with the Gram stain, and by observing motility and morphology using either darkfield or phase-contrast microscopy. Use of 0.3% carbol fuchsin is recommended as a counterstain because *Campylobacter* species stain poorly with safranine. A presumptive identification for *Campylobacter* species may be made if the organism presents a tumbling or darting motility, is oxidase positive, and appears as Gram-negative spiral or curved rods.

Identification Methods

At present, few tests are available in clinical laboratories for definitive identification of *Campylobacter* species and subspecies. Presumptively identified *Campylobacter* isolates can be definitively identified with the tests described below. To prepare the inoculum for these tests, suspend 24- to 48-hour growth from a blood agar plate in heart infusion or *Brucella* broth to a turbidity matching a McFarland No. 1 standard. Test media are inoculated with 2 drops of cell suspension from a Pasteur pipette. Media are incubated in a microaerophilic atmosphere at 37°C for 3 days.

Catalase

To test for catalase, 3% hydrogen peroxide reagent is added to the growth on a *Brucella* agar or heart infusion agar slant. Any bubble formation is considered positive.

Hydrogen Sulfide Production

To test for hydrogen sulfide production, inoculate a triple sugar iron agar slant or sulfide indole motility medium with a needle. Also, suspend a lead acetate strip over *Brucella* broth containing 0.02% cysteine. Blackening of the agar or strip constitutes a positive reaction.

Hippurate Test

The hippurate hydrolysis test is performed as described by Morris and Patton (1985). A loopful of an overnight growth from a blood agar plate is emulsified in 0.4 ml of 1% aqueous sodium hippurate to produce a dense suspension. Incubate for 2 h in a 37°C water bath. After incubation, slowly add 0.2 ml of ninhydrin solution (3.5% ninhydrin in a 1 : 1 mixture of acetone and butanol) down the side of the tube to form an overlay. Incubate for 10 min in a 37°C water bath. A deep purple color is considered positive. Weak color reactions are considered negative. Store the ninhydrin solution in a dark bottle at room temperature.

Sensitivity to Antimicrobial Agents

Inoculate a blood agar plate with broth inoculum to produce confluent growth. Place a 30-μg nalidixic acid disk and a 30-μg cephalothin disk on the agar surface. Incubate at 37°C for 24 to 48 h. Any zone of inhibition is considered a susceptible reaction.

Nitrate Reduction

Inoculate a nitrate broth and incubate for 48 to 72 h. Add 2 to 3 drops each of sulfanilic acid (0.85% in 5 N acetic acid) and *N,N*-dimethyl-1-naphthylamine (0.6% in 5 N acetic acid). Development of a red color indicates a positive reaction.

Growth Conditions

Growth in aerobic and anaerobic atmospheres can be determined on either heart infusion agar slants or blood agar plates. Growth at 25, 37, and 42°C can be determined on heart infusion agar slants, blood agar plates, or thioglycolate broth. Growth in 1% glycine and 3.5% NaCl is determined in *Brucella* broth containing 0.16% agar.

Differentiation

Campylobacter species are usually divided into two groups, the catalase-positive and the catalase-negative *Campylobacter*. The *Campylobacter* species most often associated with human disease are catalase positive. The catalase-positive *Campylobacter* species can be differentiated on the basis of growth temperatures. The thermophilic *Campylobacter* species, which grow at 42°C, but not at 25°C, and which cause disease in humans, are *C. jejuni*, *C. coli*, and *C. laridis*. Resistance to nalidixic acid differentiates *C. jejuni* and *C. coli* from *C. laridis*. *C. jejuni* can hydrolyze hippurate to glycine and benzoic acid, whereas *C. coli* and the other *Campylobacter* species do not hydrolyze hippurate.

The catalase-positive *Campylobacter* species that grow at 25°C but usually not at 42°C are *C. fetus* and *C. hyointestinalis*. *C. fetus* subspecies *fetus* can be differentiated from *C. fetus* subspecies *venerealis* by the 1% glycine test and from *C. hyointestinalis* by the failure to produce H_2S in triple sugar iron agar.

C. fennelliae, *C. cinaedi*, and *C. pylori* grow at 37°C but not at 25 or 42°C. *C. pylori* will hydrolyze urea, whereas none of the other species have this capacity. Table 1 lists additional tests for identification of *Campylobacter* species.

Environmental Methods

The methods for detection of *Campylobacter* species in foods, milk, and water must address the fastidious physiologic characteristics of the organism and the relatively low numbers of viable cells present in these samples. Unlike isolation of *Campylobacter* from feces, culture of environmental samples includes an enrichment step to resuscitate injured organisms before isolation on selective media. The methods of Doyle and Roman (1982) are recommended for isolation of *Campylobacter* from milk and food. Direct plating on modified Butzler agar, blood-free *Campylobacter* medium, or Campy blood agar plates is superior to any most probable number method yet devised. Enrichment cultures should be subcultured to Campy blood agar plates for isolation of *Campylobacter*. Doyle and Roman (1982) have estimated the sensitivity of their method as capable of detecting 1 cell per 20 g of chicken. Lovett et al. (1983) used a modified Park medium to detect 1 cell per ml of milk. The method involved centrifugation of milk samples followed by enrichment in *Brucella* broth containing FBP, vancomycin (15 mg/liter), trimethoprim (7.5 mg/liter), and polymyxin B (5000 U/liter), filtration through a 0.65-μm membrane filter, and culture of the filter on Skirrow agar at 42°C in a microaerophilic atmosphere. Blaser and Cody (1986) developed a membrane filter method for isolation of *C. jejuni* from low-turbidity water as an adjunct in investigation of waterborne outbreaks and as a potential screen for finished water.

Immunology

An understanding of the immune response in campylobacteriosis is useful for retrospective diagnosis in the absence of culture evidence, epidemiologic study of populations at risk, and study of the pathogenesis of the disease syndromes. Several epidemiologic studies have demonstrated that immunoglobulin (Ig)M, IgG, and IgA antibodies are produced in campylobacteriosis. Studies in developing countries show that children progressively acquire a tolerance toward *C. jejuni* infection with chronic exposure, whereas adults are frequently asymptomatic carriers (Glass et al., 1983).

A classic immune response occurs in acute intestinal disease. IgM appears first, followed by IgG and IgA during the second week of infection. Both IgM and IgG persist for 3 months or more, whereas IgA remains elevated for 1 month or less (Spreeuwel et al., 1985). Secretory IgA was found in the gut of seven of eight patients with naturally occurring infections in a study by Winsor et al. (1986). IgA titers were 1 : 16 to 1 : 64 in fecal extracts, whereas control specimens gave titers of less than 1 : 4. Western blot analysis was attempted to identify the antigens responsible for the IgA response; that investigation remains incomplete. Healthy individuals with occupational exposure to *C. jejuni* produce IgM without overt disease (Svedhem et al., 1983). Blaser and Duncan (1984) showed that chronic raw milk consumption produced elevated IgG titers, but not IgM or IgA antibody.

Several laboratory tests show promise for further elucidation of the immune response and for diagnostic testing. Blaser and co-workers (1979) reported IgG titers of more than 1 : 32 by the immunofluorescence test in convalescent sera from 17 of 20 patients with positive stool cultures, whereas control sera showed titers of 1 : 2 to 1 : 16. The enzyme immu-

TABLE 1. Differential characteristics of *Campylobacter* species[a]

Organism	Growth					Biochemical reaction								Other test responses				
	25°C	36°C	42°C	1% Glycine	3.5% NaCl	Oxidase	Catalase	Glucose[b]	Nitrate reduction	Nitrite reduction	H_2S in TSI	H_2S on lead acetate paper	Hippurate hydrolysis[c]	Nalidixic acid[d]	Cephalothin[d]	Hydrogen[e]	C-19:0 fatty acid	Urease
C. jejuni	−	+	+	+	−	+	+	−	+	−	−	+	+	S	R	−	+	
C. coli	−	+	+	+	−	+	+	−	+	−	−	+	−	S	R	−	d	
C. laridis	−	+	+	+	−	+	+	−	+		−	+	−	R	R	−	−	
"*C. fecalis*"	−	+	+	+	+	+	+	−	+		+	+	−	d	S	−	−	
C. fetus subsp. *fetus*	+	+	d	+	−	+	+	−	+	−	−	d	−	R	S	−	−	
C. fetus subsp. *venerealis*	+	+	−	−	−	+	+	−	+	−	−	d	−	d	S	−	−	
C. hyointestinalis	d	+	d	+	−	+	+	−	+	−	+	+	−	R	S			
C. sputorum subsp. *sputorum*	−	+	+	+	−	+	−	−	+	+	+	+	−	R	S	−	−	
C. sputorum subsp. *bubulus*	d	+	−	+	+	+	−	−	+	+	+	+	−	d	S	−	−	
C. sputorum subsp. *mucosalis*	−	+	+	−	−	+	−	−	+	+	+	+	−	d	S	+	−	
C. concisus	−	+	−	−	−	+	−	−	+	+	+	+		R		+		
C. cinaedi	−	+	−	+	−	+	+	−	+		−	+	−	S	S			
C. fennelliae	−	+	−	+	−	+	+	−	−		−	+	−	S	S			
C. pylori		+				+	+	−[f]	−			+		R				+
"Aerotolerant *Campylobacter*"	+	+	−	d	−	+	+	−	−		−	−	−	S	R			

[a] + = 90% or more of the strains are positive; − = 90% or more of the strains are negative; d = 11 to 89% of the strains are positive; R = resistant; S = susceptible; TSI = triple sugar iron. Table modified from Morris and Patton (1985).
[b] Ferments or oxidizes glucose.
[c] Ninhydrin test.
[d] 30-μg disk.
[e] Hydrogen required for microaerophilic growth.
[f] Oxidation of glucose not indicated.

noassay procedure detects total immunoglobulins or may be made class specific for IgM, IgG, and IgA (Blaser et al., 1984). Svedhem et al. (1983) developed a gel diffusion enzyme immunoassay for quantitative detection of IgM and IgG antibody to *C. jejuni*, which has research advantages but little practical diagnostic utility. Mascart-Lemone et al. (1985) used radioimmunoassay to study IgA response to *C. jejuni* and found an age-related gradient effect in healthy subjects, with the lowest IgA titers in children and the highest in the elderly. Polymeric IgA1 was associated with acute cases of enteritis, and the form of IgA1 synthesized in the lamina propria predominated.

The direct fluorescent antibody test has been used by veterinarians for many years, and Hebert et al. (1983) developed conjugates with serogroup specificity for *C. jejuni*, *C. coli*,and *C. fetus* subspecies *fetus;* however, these reagents have not been tested on human clinical or environmental specimens. Hodge et al. (1986) reported diagnostic use of direct fluorescent antibody reagents for detection of *C. jejuni* and *C. coli* in human fecal specimens and concluded that the sensitivity of the method was too low for routine diagnostic use with conjugates presently available.

Oosterom et al. (1985) compared enzyme immunoassay to complement fixation and found a high degree of correlation; however, no test has emerged with sufficient reliability for the routine serodiagnosis of campylobacteriosis.

Seventy-two percent or more of patients with severe gastritis and gastric or duodenal ulcer where *C. pylori* was isolated developed complement fixation titers of 1 : 5 or more (Von Wulffen et al., 1986). The response to *C. pylori* infection is predominantly IgG and IgA; IgM appears to play no significant role, which is a characteristic response to chronic infection. Goodwin et al. (1987) measured antibody to *C. pylori* by enzyme immunoassay (ELISA) in sera of 160 patients from whom gastric biopsy specimens were obtained. A titer equal to or greater than 300 ELISA units was characteristic of patients harboring *C. pylori* in the stomach, whereas patients with titers of less than 150 ELISA units were highly unlikely to have *C. pylori* in the stomach.

Investigations are under way to determine the specific antigens and products that contribute to pathogenesis. For *C. jejuni*, flagellin, outer membrane proteins, lipopolysaccharide, and enterotoxins are all potentially immunogenic (Nachamkin and Hart, 1985).

Treatment

Campylobacter gastroenteritis is usually self-limiting, and antibiotics are reserved for severe cases. Supportive treatment is recommended in mild cases. No antibiotic has been shown to reduce the duration of symptoms, but treatment decreases the period of shedding and may prevent relapse or establishment of a carrier state (Robins-Browne et al., 1983). Erythromycin is the drug of choice, but chloramphenicol, gentamicin, tobramycin, amikacin, and tetracyclines have been used. Plasmid-mediated tetracycline resistance has been reported in up to 30% of *Campylobacter* strains examined and may result from widespread use of antibiotics in poultry and hog feed. Resistance to metronidazole, cefazolin, ampicillin, and carbenicillin has been reported. Antibiotics with good in vitro activity include clindamycin, thiamphenicol, niridazole, and three derivatives of nitrothiazole (Hof et al., 1984; Vanhoof et al., 1980).

Fliegelman et al. (1985) evaluated ciprofloxacin and found it very active in vitro against *Campylobacter, Salmonella, Shigella, Yersinia, Vibrio,* and *Eschrichia,* but clinical trials are needed before it can be acclaimed as an effective drug for patients with bacterial gastroenteritis. Elharrif et al. (1985) studied the action of macrolides and related compounds on *C. jejuni* and *C. coli* and found that human isolates of both species had similar susceptibilities, whereas porcine isolates of *C. coli* were resistant to erythromycin. Wang et al. (1984) reported differences in clindamycin susceptibility between human and porcine strains. The relationship between drugs in hog feeds and antimicrobial resistance patterns in *C. coli* has not been prospectively investigated; however, Bradbury and Munroe (1985) found plasmids in 53% of *Campylobacter* strains examined from cattle, pigs, and chicks. These *C. jejuni* and *C. coli* strains were resistant to bacitracin, ampicillin, streptomycin, and penicillin G. Resistance to tetracycline and gentamicin was plasmid mediated when detected. Tenover et al. (1985) also reported plasmid-mediated tetracycline resistance in 32% of 688 strains that had plasmids. Taylor (1986) cloned the plasmid responsible for tetracycline resistance in *C. jejuni* and showed hemology between it and *Streptococcus* class M determinant. Bopp et al. (1985) examined 31 *Campylobacter* strains from 11 outbreaks. Nineteen contained plasmids and were resistant to tetracycline, ampicillin, metronidazole, and carbenicillin. These strains were susceptible to erythromycin, clindamycin, chloramphenicol, tobramycin, gentamicin, streptomycin, and kanamycin. Antibiograms, plasmid profiles, and serotypes have epidemiologic usefulness during investigations of outbreaks. Multiple resistance has been reported in isolates from patients with acquired immune deficiency syndrome (Dworkin et al., 1986).

Flores et al. (1985) published antibiograms of *Campylobacter*-like organisms isolated from homosexual males with and without intestinal disease and found that, unlike *C. jejuni* and *C. coli,* these strains were susceptible to rifampicin, nalidixic acid, and cefoxolin. As most isolation media contain one or more drugs inhibitory for *Campylobacter*-like organisms, they are likely to be missed in most clinical laboratories. *Campylobacter*-like organisms are resistant to erythromycin and clindamycin; therefore, tetracycline or doxycycline may be the therapeutic drugs of choice.

The antimicrobial susceptibility of *C. pylori* differs from that of *C. jejuni* or *C. fetus* subspecies *fetus*. *C. pylori* is sensitive to penicillin, ampicillin, erythromycin, panomycin, metronidazole, cephalothin, rifampicin, naladixic acid, trimethoprim, and vancomycin and resistant to colistin, polymyxin B, and sulfonamides (Goodwin et al., 1986). Colloidal bismuth has antibacterial activity and should be tried before surgery. Type B (Whitehead) gastritis is reversible, and both active and chronic inflammation resolves after eradication of *C. pylori* (Marshall, 1986).

Control

Because campylobacteriosis is primarily a zoonotic disease, control depends upon the adequacy of veterinary public health measures. Education and personal hygiene are more important than antibiotics in control of the disease. Cleaning and disinfection of stables and proper disposal of animal wastes reduces the risk of food and milk contamination, but prevention of infection in those occupationally exposed to animals remains a problem. Sanitation and personal hygiene in the slaughterhouse, dairy barn, and poultry processing plant minimize opportunities for transmission to humans. Cleaning surfaces with detergent and hot water or steam followed by drying effectively reduces the number of viable *Campylobacter*. Hypochlorite, 70% ethanol, iodophores, phenolic disinfectants, and quaternary ammonium compounds are equally effective. Refrigeration or freezing of foods rapidly reduces the number of viable cells; however, *Campylobacter* have been cultured from meats stored at −20°C for several months. Thermophilic *Campylobacter* are sensitive to low pH and are not found in acidic foods. They survive in milk at 4°C for up to 3 weeks, but die off in 3 days at 25°C. Cheese making reduces the number of viable organisms. Pasteurization of milk and thorough cooking of meats and poultry are effective controls, as are handwashing and avoiding recontamination of cooked foods.

Literature Cited

Benjamin, J., S. Leaper, R. J. Owen, and M. B. Skirrow. 1983. Description of *Campylobacter laridis,* a new species comprising the nalidixic acid resistant thermophilic *Campylobacter* (NARTC) group. Curr. Microbiol. **8:**231–238.

Blaser, M. J., I. D. Berkowitz, F. M. LaForce, J. Cravens, L. B. Reller, and W.-L. L. Wang. 1979. *Campylobacter* enteritis: clinical and epidemiologic features. Ann. Intern. Med. **91:**179–185.

Blaser, M. J., and H. J. Cody. 1986. Methods for isolating *Campylobacter jejuni* from low-turbidity water. Appl. Environ. Microbiol. **51:**312–315.

Blaser, M. J., and D. J. Duncan. 1984. Human serum antibody response to *Campylobacter jejuni* infection as measured in an enzyme-linked immunosorbent assay. Infect. Immun. **44:**292–298.

Blaser, M. J., D. Hovorson, I. G. Ely, D. J. Duncan, W.-L. L. Wang, and W. R. Brown. 1984. Studies of *Campylobacter jejuni* in patients with inflammatory bowel disease. Gastroenterology **86:**33–38.

Blaser, M. J., G. P. Perez, P. F. Smith, C. Patton, F. C. Tenover, A. J. Lastovica, and W.-L. L. Wang. 1986. Extraintestinal *Campylobacter jejuni* and *Campylobacter coli* infections: host factors and strain characteristics. J. Infect. Dis. **153:**552–559.

Blaser, M. J., D. N. Taylor, and R. Feldman. 1983. Epidemiology of *Campylobacter jejuni* infections. Epidemiol. Rev. **5:**157–176.

Bolton, F. J., D. N. Hutchinson, and D. Coates. 1984. Blood-free selective medium for isolation of *Campylobacter jejuni* from feces. J. Clin. Microbiol. **19:**169–171.

Bolton, F. J., and L. Robertson. 1982. A selective medium for isolating *Campylobacter jejuni/coli*. J. Clin. Pathol. **35:**462–467.

Bopp, C. A., K. A. Birkness, I. K. Wachsmuth, and T. J. Barrett. 1985. *In vitro* antimicrobial susceptibility, plasmid analysis and serotyping of epidemic-associated *Campylobacter jejuni*. J. Clin. Microbiol. **21:**4–7.

Bradbury, W. C. and D. L. G. Munroe. 1985. Occurrence of plasmids and antibiotic resistance among *Campylobacter jejuni* and *Campylobacter coli* isolated from healthy and diarrheic animals. J. Clin. Microbiol. **22:**339–346.

Brooks, B. W., M. M. Garcia, A. D. E. Fraser, H. Lior, R. B. Stewart, and A. M. Lammerding. 1986. Isolation and characterization of cephalothin-susceptible *Campylobacter coli* from slaughter cattle. J. Clin. Microbiol. **24:**591–595.

Butzler, J. P., and M. B. Skirrow. 1979. *Campylobacter* enteritis. Clin. Gastroenterol. **8:**737–765.

Doyle, M. P., and D. J. Roman. 1982. Recovery of *Campylobacter jejuni* and *Campylobacter coli* from inoculated foods by selective enrichment. Appl. Environ. Microbiol. **43:**1343–1353.

Dworkin, B., G. P. Wormser, R. A. Abdoo, F. Cabello, M. E. Aguero, and S. L. Sivak. 1986. Persistence of multiple antibiotic-resistant *Campylobacter jejuni* in patients with the acquired immune deficiency syndrome. Am. J. Med. **80:**965–970.

Edmonds, P., C. M. Patton, P. M. Griffin, T. J. Barrett, G. P. Schmid, C. N. Baker, M. A. Lambert, and D. J. Brenner. 1987. *Campylobacter hyointestinalis* associated with human gastrointestinal disease in the United States. J. Clin. Microbiol. **25:**685–691.

Elharrif, Z., F. Megraud, and A.-M. Marchand. 1985. Susceptibility of *Campylobacter jejuni* and *Campylobacter coli* to macrolides and related compounds. Antimicrob. Agents Chemother. **28:**695–697.

Fennell, C. L., P. A. Totten, T. C. Quinn, D. L. Patton, K. K. Holmes, and W. E. Stamm. 1984. Characterization of *Campylobacter*-like organisms isolated from homosexual men. J. Infect. Dis. **149:**58–66.

Finch, M. J., and P. A. Blake. 1985. Foodborne outbreaks

of campylobacteriosis: the United States experience 1980–1982. Am. J. Epidemiol. **122**:262–268.

Finch, M. J., and L. W. Riley. 1984. *Campylobacter* infections in the United States. Arch. Intern. Med. **144**:1610–1612.

Fliegelman, R. M., R. M. Petrak, L. J. Goodman, J. Segreti, G. M. Trenholme, and R. L. Kaplan. 1985. Comparative *in vitro* activities of twelve antimicrobial agents against *Campylobacter* species. Antimicrob. Agents Chemother. **27**:429–430.

Flores, B. M., C. L. Fennell, K. K. Holmes, and W. E. Stamm. 1985. *In vitro* susceptibilities of *Campylobacter*-like organisms to twenty antimicrobiol agents. Antimicrob. Agents Chemother. **28**:188–191.

Francioli, P., J. Herzstein, J.-P. Grob, J.-J. Vallotton, G. Mombelli, and M. P. Glauser. 1985. *Campylobacter fetus* subspecies *fetus* bacteremia. Arch. Intern. Med. **145**:289–292.

Fricker, C. R., J. Uradzinski, M. M. Alemohammad, R. W. A. Park, C. Whelan, and R. W. A. Girdwood. 1986. Serotyping of campylobacters by co-agglutination on the basis of heat-stable antigens. J. Med. Microbiol. **21**:83–86.

Gebhart, C. J., G. E. Ward, K. Chang, and H. J. Kurtz. 1983. *Campylobacter hyointestinalis* (new species) isolated from swine with lesions of proliferative ileitis. Am. J. Vet. Res. **44**:361–367.

George, H. A., P. S. Hoffman, R. M. Smibert, and N. R. Krieg. 1978. Improved media for growth and aerotolerance of *Campylobacter fetus*. J. Clin. Microbiol. **8**:36–41.

Gill, K. P. W., and M. J. Corbel. 1986. Lectin agglutination for the differentiation of *Campylobacter fetus* serotypes. Vet. Rec. **118**:158.

Glass, R. I., B. J. Stoll, M. I. Huq, M. J. Struelens, M. Blaser, and A. K. M. G. Kibriya. 1983. Epidemiologic and clinical features of endemic *Campylobacter jejuni* infection in Bangladesh. J. Infect. Dis. **148**:292–296.

Goodwin, C. S., P. Blake, and E. Blincow. 1986. The minimum inhibitory and bactericidal concentrations of antibiotics and anti-ulcer agents against *Campylobacter pyloridis*. J. Antimicrob. Chemother. **17**:307–314.

Goodwin, C. S., E. D. Blincow, J. R. Warren, T. E. Waters, C. R. Sanderson, and L. Easton. 1985. Evaluation of cultural techniques for isolating *Campylobacter pyloridis* from endoscopic biopsies of gastric mucosa. J. Clin. Pathol. **38**:1127–1131.

Goodwin, C. S., E. Blincow, G. Peterson, C. Sanderson, W. Cheng. B. Marshall, J. R. Warren, and R. McCulloch. 1987. Enzyme-linked immunoassay for *Campylobacter pyloridis:* correlation with presence of *C. pyloridis* in the gastric mucosa. J. Infect. Dis. **155**:488–494.

Goossens, H., M. De Boeck, and J.-P. Butzler. 1983. A new selective medium for the isolation of *Campylobacter jejuni* from human faeces. Eur. J. Clin. Microbiol. **2**:389–394.

Goossens, H., M. De Boeck, H. Coignau, L. Vlaes, C. Van Den Borre, and J.-P. Butzler. 1986. Modified selective media for isolation of *Campylobacter* spp. from feces: comparison with Preston medium, a blood-free medium and a filtration system. J. Clin. Microbiol. **24**: 840–843.

Grajewski, B. A., J. W. Kusek, and H. M. Gelfand. 1985. Development of a bacteriophage typing system for *Campylobacter jejuni* and *Campylobacter coli*. J. Clin. Microbiol. **22**:13–18.

Hebert, C. A., D. G. Hollis, R. E. Weaver, A. G. Steigerwalt, R. M. McKinney, and D. J. Brenner. 1983. Serogroups of *Campylobacter jejuni, Campylobacter coli,* and *Campylobacter fetus* as defined by direct immunofluorescence. J. Clin. Microbiol. **17**:529–538.

Hodge, D. S., J. F. Prescott, and P. E. Shewen. 1986. Direct immunofluorescence microscopy for rapid screening of *Campylobacter* enteritis. J. Clin. Microbiol. **24**:863–865.

Hof, H., M. Rieffert, V. Sticht-Groh, O. Zak, and E. H. Schweizer. 1984. Antibacterial activities of nitrothiazole against *Campylobacter jejuni* and *Campylobacter coli*. Antimicrob. Agents Chemother. **26**:498–500.

Hopkins, R. S., R. Olmsted, and G. R. Istre. 1984. Endemic *Campylobacter jejuni* infection in Colorado: identified risk factors. Am. J. Public Health **74**:249–250.

Johnson, W. M., and H. Lior. 1986. Cytotoxic and cytotonic factors produced by *Campylobacter jejuni, Campylobacter coli,* and *Campylobacter laridis*. J. Clin. Microbiol. **24**:275–281.

Karmali, M. A., and P. C. Fleming. 1979. Application of the Fortner principle to isolation of *Campylobacter* from stools. J. Clin. Microbiol. **10**:245–247.

Karmali, M. A., A. E. Simor, M. Roscoe, P. C. Fleming, S. S. Smith, and J. Lane. 1986. Evaluation of a blood-free, charcoal-based, selective medium for the isolation of *Campylobacter* organisms from feces. J. Clin. Microbiol. **23**:456–459.

Kazmi, S. U., B. S. Roberson, and N. J. Stern. 1984. Animal-passed, virulence-enhanced *Campylobacter jejuni* causes enteritis in neonatal mice. Curr. Microbiol. **11**:159–164.

Klein, B. S., J. M. Vergeront, M. J. Blaser, P. Edmonds, D. J. Brenner, D. Janssen, and J. P. Davis. 1986. *Campylobacter* infection associated with raw milk: an outbreak of gastroenteritis due to *Campylobacter jejuni* and thermotolerant *Campylobacter fetus* subsp. *fetus*. J. Am. Med. Assoc. **255**:361–364.

Klipstein, F. A., R. F. Engert, H. Short, and E. A. Schenk. 1985. Pathogenic properties of *Campylobacter jejuni:* assay and correlation with clinical manifestations. Infect. Immun. **50**:43–49.

Lastovica, A. J., E. Le Roux, R. V. Congi, and J. L. Penner. 1986. Distribution of sero-biotypes of *Campylobacter jejuni* and *C. coli* isolated from paediatric patients. J. Med. Microbiol. **21**:1–5.

Lior, H., D. L. Woodward, J. A. Edgar, L. J. Laroche, and P. Gill. 1982. Serotyping of *Campylobacter jejuni* by slide agglutination based on heat-labile antigenic factors. J. Clin. Microbiol. **15**:761–768.

Lovett, J., D. W. Francis, and J. M. Hunt. 1983. Isolation of *Campylobacter jejuni* from raw milk. Appl. Environ. Microbiol. **46**:459–462.

Manser, P. A., and R. W. Dalziel. 1985. A survey of *Campylobacter* in animals. J. Hyg. **95**:15–21.

Marshall, B. J. 1986. *Campylobacter pyloridis* and gastritis. J. Infect. Dis. **153**:650–657.

Marshall, B. J., and S. R. Langton. 1986. Urea hydrolysis in patients with *Campylobacter pyloridis* infection. Lancet **1**:965–966.

Marshall, B. J., and C. S. Goodwin. 1987. Revised nomenclature of *Campylobacter pyloridis*. Int. J. Syst. Bacteriol. **37**:68.

Mascart-Lemone, F., J. Duchateau, E. Rummens, J. Oosterom, J.-P. Butzler, and D. L. Delacroix. 1985. Characterization of serum IgA anti-*Campylobacter jejuni* antibodies in healthy subjects and in patients convalescent from an acute *C. jejuni* enteritis, p. 80. *In* A. D. Pearson, M. B. Skirrow, H. Lior, and B. Rowe (eds.) Campylobacter III. Public Health Laboratory Service, London.

Merino, F. J., A. Agulla, P. A. Villasante, A. Diaz, J. V. Saz, and A. C. Velasco. 1986. Comparative efficacy of seven selective media for isolating *Campylobacter jejuni*. J. Clin. Microbiol. **24**:451–452.

Morris, G. K., and C. M. Patton. 1985. *Campylobacter*, p. 302–308. *In* E. H. Lennette, A. Ballows, W. J. Hausler, Jr., and H. J. Shadomy (eds.), Manual of clinical microbiology. American Society for Microbiology, Washington, D.C.

Nachamkin, I., and A. M. Hart. 1985. Western blot analysis of the human antibody response to *Campylobacter jejuni* cellular antigens during gastrointestinal infection. J. Clin. Microbiol. **21**:33–38.

Ng, L.-K., M. E. Stiles, and D. E. Taylor. 1985a. Inhibition of *Campylobacter coli* and *Campylobacter jejuni* by antibiotics used in selective growth media. J. Clin. Microbiol. **22**:510–514.

Ng, L.-K., M. E. Stiles, and D. E. Taylor. 1985b. Comparison of basal media for culturing *Campylobacter jejuni* and *Campylobacter coli*. J. Clin. Microbiol. **21**: 226–230.

Oosterom, J., C. H. Den Uyl, J. R. J. Banffer, and J. Huisman. 1984. Epidemiological investigations on *Campylobacter jejuni* in households with a primary infection. J. Hyg. **92**:325–332.

Oosterom, J., C. H. Den Uyl, J. R. J. Banffer, S. Lauwers, J. Huisman, A. E. Busschbach, F. G. J. Poelma, and R. Bellemans. 1985. Evaluation of an enzyme-linked immunosorbent assay (ELISA) for the detection of *Campylobacter jejuni* antibodies, and comparison with a complement fixation test (CFT). J. Microbiol. **51**:321–331.

Patton, C. M., T. J. Barrett, and G. K. Morris. 1985. Comparison of the Penner and Lior methods for serotyping *Campylobacter* spp. J. Clin. Microbiol. **22**:558–565.

Penner, J. L., and J. N. Hennessy. 1980. Passive haemagglutination technique for serotyping *Campylobacter fetus* subsp. *jejuni* on the basis of soluble heat-stable antigens. J. Clin. Microbiol. **12**:732–737.

Pennie, R. A., J. N. Zunino, C. E. Rose, Jr., and R. L. Guerrant. 1984. Economical, simple method for production of the gaseous environment required for cultivation of *Campylobacter jejuni*. J. Clin. Microbiol. **20**: 320–322.

Robins-Browne, R. M., M. K. R. Mackenjee, M. N. Bodasing, and H. M. Coovadia. 1983. Treatment of *Campylobacter*-associated enteritis with erythromycin. Am. J. Dis. Child. **137**:282–285.

Rosef, O., and G. Kapperud. 1983. House flies (*Musca domestica*) as possible vectors of *Campylobacter fetus* subsp. *jejuni*. Appl. Environ. Microbiol. **45**:381–383.

Skirrow, M. B. 1977. *Campylobacter* enteritis: a "new" disease. Br. Med. J. **2**:9–11.

Spreeuwel, J., P. Van, G. C. Duursma, C. J. L. M. Meijer, R. Bax, P. C. M. Rosekrans, and J. Lindeman. 1985. *Campylobacter* colitis: histological immunohistochemical and ultrastructural findings. Gut **26**:945–951.

Svedhem, A., H. Gunnarsson, and B. Kaijser. 1983. Diffusion-in-gel enzyme-linked immunosorbent assay for routine detection of IgG and IgM antibodies to *Campylobacter jejuni*. J. Infect. Dis. **148**:82–92.

Tauxe, R. V., C. M. Patton, P. Edmonds, T. J. Barrett, D. J. Brenner, and P. A. Blake. 1985. Illness associated with *Campylobacter laridis*, a newly recognized *Campylobacter* species. J. Clin. Microbiol. **21**:222–225.

Taylor, D. E. 1986. Plasmid-mediated tetracycline resistance in *Campylobacter jejuni:* expression in *Escherichia coli* and identification of homology with streptococcal class M determinant. J. Bacteriol. **165**:1037–1039.

Tenover, F. C., S. Williams, K. P. Gordon, C. Nolan, and J. J. Plorde. 1985. Survey of plasmids and resistance factors in *Campylobacter jejuni* and *Campylobacter coli*. Antimicrob. Agents Chemother. **27**:37–41.

Thompson, J. S., F. E. Cahoon, and D. S. Hodge. 1986. Rate of *Campylobacter* spp. isolation in three regions of Ontario, Canada, from 1978 to 1985. J. Clin. Microbiol. **24**:876–878.

Thorson, S. M., J. A. Lohr, S. Dudley, and R. L. Guerrant. 1985. Value of methylene blue examination, darkfield microscopy and carbol–fuchsin Gram stain in the detection of *Campylobacter* enteritis. J. Pediatr. **106**:941–943.

Tielbeek, A. V., G. Rosenbusch, H. L. Muytjens, S. H. Yap, S. P. Strijk, and C. Boetes. 1985. Roentgenologic changes of the colon in *Campylobacter* infection. Gastrointest. Radiol. **10**:358–361.

Torphy, D. E., and W. W. Bond. 1979. *Campylobacter fetus* infections in children. Pediatrics **64**:898–903.

Totten, P. A., C. L. Fennell, F. C. Tenover, J. M. Wezenberg, P. L. Perine, W. E. Stamm, and K. K. Holmes. 1985. *Campylobacter cinaedi* (sp.nov.) and *Campylobacter fennelliae* (sp.nov.): two new *Campylobacter* species associated with enteric disease in homosexual men. J. Infect. Dis. **151**:131–139.

Vanhoof, R., B. Gordts, R. Dierickx, H. Coignau, and J.-P. Butzler. 1980. Bacteriostatic and bactericidal activities of 24 antimicrobial agents against *Campylobacter fetus* subsp. *jejuni*. Antimicrob. Agents Chemother. **18**:118–121.

Von Wulffen, H., J. Heesemann, G. H. Butzow, T. Loning, and R. Laufs. 1986. Detection of *Campylobacter pyloridis* in patients with antrum gastritis and peptic ulcers by culture, complement fixation test, and immunoblot. J. Clin. Microbiol. **24**:716–720.

Walker, R. I., M. B. Caldwell, E. C. Lee, P. Guerry, T. J. Trust, and G. M. Ruiz-Palacios. 1986. Pathophysiology of *Campylobacter* enteritis. Microbiol. Rev. **50**:81–94.

Wang, W.-L. L., L. B. Reller, and M. J. Blaser. 1984. Comparison of antimicrobial susceptibility patterns of *Campylobacter jejuni* and *Campylobacter coli*. Antimicrob. Agents Chemother. **26**:351–353.

Wang, W.-L. L., L. B. Reller, B. Smallwood, N. W. Luechtefeld, and M. J. Blaser. 1983. Evaluation of transport media for *Campylobacter jejuni* in human fecal specimens. J. Clin. Microbiol. **18**:803–807.

Wells, J. G., C. A. Bopp, and M. J. Blaser. 1981. Evaluation of selective media for the isolation of *Campylobacter jejuni*, p. 80–84. *In* D. G. Newell (ed.), *Campylobacter*. MTD Press Unlimited, Boston.

Winsor, D. K., Jr., J. J. Mathewson, and H. L. DuPont. 1986. Western blot analysis of intestinal secretory immunoglobulin. A response to *Campylobacter jejuni* antigens in patients with naturally acquired *Campylobacter* enteritis. Gastroenterology **90**:1217–1222.

CHAPTER 16

Cat Scratch Disease

RICHARD C. TILTON

Disease: Cat scratch disease (cat scratch fever).
Etiologic Agents: Bacterial (genus/species unidentified).
Source: Presumably cats, dogs, and other animals implicated.
Clinical Manifestations: Regional lymphadenitis, encephalitis, and Parinaud's syndrome (oculoglandular syndrome).
Pathology: Lymphocytic infiltrate with epithelioid granuloma formation; necrosis of the granuloma.
Laboratory Diagnosis: Warthin-Starry silver stain of the lymph node to reveal characteristic microorganisms.
Epidemiology: Cat scratch disease is seasonal, most cases occurring during the summer; not easily acquired and no person-to-person transmission has been shown.
Treatment: None (at present); surgery is not recommended.
Prevention and Control: Avoid cats.

Description of Disease

Cat scratch disease (CSD) was first recognized at the University of Paris by Debre and co-workers (1950). They described CSD in a 10-year-old boy who had many scratches on his arm as well as regional lymphadenitis. These investigators did not isolate the organism or speculate on its etiology. Cat scratch disease is now recognized as a subacute regional lymphadenitis that occurs after inoculation of organisms through the skin. Although complications of CSD have been reported, it usually resolves in 2 to 3 months without treatment.

Several case reports as well as clinical reviews have appeared in the literature over the past half-century (Carithers, 1985; Spaulding and Hennessy, 1960; Warwick, 1967). These reports have established CSD as a distinct clinical syndrome. However, the exact identification of the etiologic agent, now thought to be a bacterium, is elusive.

Etiologic Agent

Numerous efforts have been made to isolate bacteria, viruses, mycoplasma, chlamydia, and fungi from excised lymph nodes, drainage, and the eyes of patients with Parinaud syndrome (a form of CSD). With the exception of one recent study (Gerber et al., 1985), there has been little success in defining an etiologic agent even when multiple artificial and cell culture systems have been employed. Cultural and microscopic studies have implicated chlamydia, viruses, and acid-fast bacilli in CSD, but none has withstood scrutiny (Hadfield et al., 1985b).

In the past, *Chlamydia* was the leading candidate for the etiology of CSD due to some early descriptions of inclusion bodies in lymph node materials (Boyer and Cherry, 1981) and the similarity of CSD to lymphogranuloma venereum. Some serologic studies have also shown an increase in specific antibodies to certain chlamydial antigens in patients with

CSD (Spaulding and Hennessy, 1960). However, several lines of investigation have shown that *Chlamydia* is probably not the etiologic agent. Present techniques of chlamydial isolation permit most chlamydiae to be readily cultured in cell lines or visualized with fluorescent antibodies, but they cannot be isolated from suspect specimens. In addition, patients with CSD do not respond to skin testing for lymphogranuloma venereum. Antibiotics used to treat chlamydial infection, such as tetracycline and erythromycin, seem to have no activity against CSD.

Atypical mycobacteria were cultured from the lymph nodes of eight patients who presumably had CSD (Boyd and Craig, 1961). However, other investigations using atypical mycobacterial skin tests showed no difference in skin test reactivity between children with or devoid of a history of CSD.

Because no bacterial agents had been isolated, many early investigators thought that a viral etiology was possible. Turner et al. (1960) isolated a herpeslike organism from chick embryos that were inoculated with pus from lymph nodes of patients with CSD, and Kalter et al. (1969) saw herpeslike particles by electron microscopy from such patients. None of these findings, however, has been confirmed immunologically.

Although some of the newer techniques for viral diagnosis have not been applied to patients with CSD, there seems to be little evidence that viruses are involved. Recently, Wear et al. (1983) described the presence of pleomorphic bacilli in lymph node sections from 29 of 34 patients with CSD. The organisms were within capillary walls in areas of follicular hyperplasia and within microabscesses. At present, the best way to visualize these organisms in lymph nodes is by using the Warthin–Starry silver impregnation stain. These pleomorphic bacteria have been described not only from lymph nodes, but also from primary skin lesions of patients with CSD and conjunctival biopsy specimens of patients with Parinaud oculoglandular syndrome.

Hadfield et al. (1985a) indicated that the causative agent is a small, pleomorphic, gram-negative rod, which can be seen clearly in lymph nodes that have not yet become suppurative. In such specimens, the alleged bacteria are much more difficult to observe. These bacteria are not acid fast and are gram negative when stained with the Brown–Hopp tissue Gram stain. They are approximately 0.2 μm in diameter and 0.5 to 1 μm in length, but appear larger when stained with a silver impregnation stain. Warthin–Starry stained sections of lymph node showing the bacteria can be seen in Fig. 1.

Microscopic details have been confirmed by many laboratories, including our own. Hadfield et al. (1985a) again reported the presence of bacilli in lymph nodes of patients with CSD and showed electron micrographs of the etiologic agent. These lymph nodes showed focal necrosis. The silver-stained slide revealed numerous bacteria, and the tissue Gram stain suggested that the bacteria were small, pleomorphic, and gram negative. The electron micrographs were of poor quality, and the ultrastructure of the host cells and the microorganisms were very difficult to see. According to the authors, the ultrastructure of the cell wall consisted of an outer glycocalyx and a thin peptidoglycanlike layer surrounding the protoplasm.

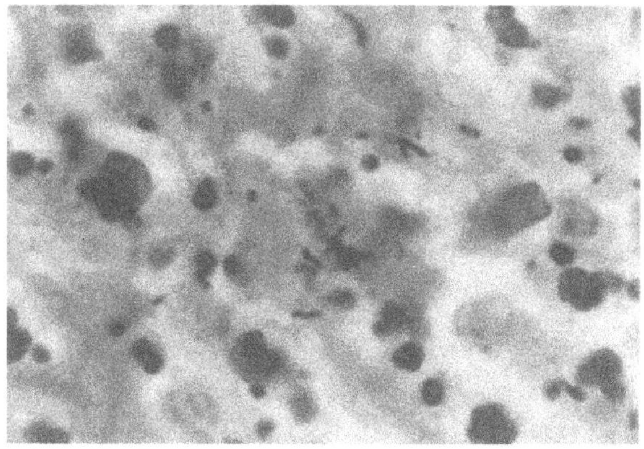

FIG. 1. Warthin–Starry stained section of a lymph node showing pleomorphic bacilli. A color plate of this figure appears in Color Plate 2 at the beginning of this volume.

Margileth et al. (1984) and Hadfield et al. (1985a) have attempted to grow the organism, with little success to date. They have inoculated many artificial media incubated under a wide variety of environmental conditions. Animal inoculations and experiments have also been unsuccessful. These investigators reported injecting mice, guinea pigs, monkeys, and the embryos of chickens, quails, and duck. Various organisms, including rickettsiae and legionellae, have been eliminated by failure to demonstrate serologic reactivity in the host.

Gerber et al. (1985) reported on the isolation of a suspected etiologic agent of CSD. They indicated that a highly pleomorphic gram-positive bacterium was cultured from an excised lymph node of a patient with CSD. The organism was morphologically identical to bacteria observed in Warthin–Starry stains of lymph node sections from many patients with CSD. Extensive biochemical and physiologic analysis of this isolate suggested that it was a member of the genus *Rothia*. This organism resembles a bacterium that was identified as the etiologic agent of Parinaud oculoglandular syndrome more than 70 years ago. In 1913, Verhoeff observed microorganisms in conjunctival scrapings from patients with Parinaud syndrome. He described these organisms as irregularly curved filaments with intensely gram-positive granules and called them *Leptothrix*. Years later, Verhoff and King (1933) isolated a similar gram-negative bacterium. Subsequently, however, Verhoeff (1940) reported an improved method of staining this bacterium and described it again as gram positive. Unfortunately, this controversy has not yet been resolved. The report of Gerber et al. (1985) is in conflict with those of Hadfield et al. (1985a), Margileth et al. (1984), and Wear et al. (1983), who claim to observe a gram-negative, not a gram-positive bacterium. Transmission electron micrographs of a broth culture of the isolated organism showed all of the morphologic forms identified by light microscopy. These organisms were 0.5 to 1 μm in diameter and 0.5 to 2.5 μm in length. A lymph node specimen from another patient with CSD in which many organisms were demonstrated on Warthin–Starry stains was also examined by transmission electron microscopy. The organisms in the lymph node lacked an attached cell wall, so it was impossible to determine whether they were gram positive or gram negative.

There is a risk in attempting to determine the Gram reaction of a bacterium from a tissue stain. Bacteria that are dead or devitalized usually stain gram negative regardless of their original Gram reaction. The Gram stain is a variable reaction. Many bacteria stain gram positive in one part of their life cycle and gram negative in another. It is recognized that highly pleomorphic branching structures are more consistent with gram-positive than with gram-negative bacteria. The observation, then, of gram-negative bacteria in a Brown–Hopp stain and electron micrographs of degenerate organisms must be interpreted with caution. Young cells of *Rothia* are gram positive, but as the culture ages, the stain concentrates in granules that form along the filament, whereas the filament itself becomes gram negative. The etiologic agent of CSD has not been definitively identified.

Source

It is assumed that the primary source of the organism is the cat, as most patients report contact with cats. Carithers (1985) indicates that cats that transmit CSD are not ill, have no distinctive features, and are usually young. In his series of 1,200 patients, only 12 patients had no established contact with cats.

Many observations suggest contact with other mammals, but are unsubstantiated. A scratch is not necessary, as any break in the skin such as a bite or other injury can cause inoculation of the organism. Margileth et al. (1984) reported an animal handler who developed CSD following the bite of a monkey; however, cats were also in the laboratory. In addition, inoculations of the CSD organism through insect bites have been reported, and there have been anecdotal reports of disease transmission by wood splinters, rabbit claws, porcupine quills, thorns, fish hooks, and the like. Carithers (1985) suggests that repetition of this "overused list" should be abandoned, but this view is not universally accepted.

Clinical Manifestations

Cat scratch disease is the most common cause of unilateral regional lymphadenitis in children. The incubation period after the scratch or injury is approximately 7 to 12 days, at which time small erythematous papules may appear at the inoculation site. These papules persist for 1 to 4 weeks, then disappear spontaneously as the lymphadenitis appears. The area of the lymphadenitis is usually close to the part of the body that was scratched. In approximately one-half of cases, only a single node is involved, but in the other half there is multiple node involvement. These nodes are tender, red, and warm, and in approximately one-third of cases will suppurate spontaneously. The most common nodes involved are axillary, followed by cervical, submandibular, and inguinal. The lymphadenitis usually resolves in about 4 to 6 weeks. Most patients with CSD have no symptoms, except for swollen lymph nodes. In those who do, symptoms consist of low-grade fever, fatigue, and sometimes headaches. Complications of CSD occur very infrequently. The most common and dramatic complication of CSD is encephalitis. However, children with CSD encephalitis usually recover without sequelae. Other complications include erythema nodosum, osteolytic lesions, arthralgia, mesenteric adenitis, urethritis, thrombocytopenia purpura, iritis, lymphedema, thyroiditis, and hepatitis.

Parinaud oculoglandular syndrome is a distinctive form of CSD in which the site of inoculation is the eyelid or conjunctiva. A lesion develops, followed by conjunctivitis and preauricular lymphadenitis.

Pathology

The histologic findings in the lymph node sections are consistent with but not pathognomonic of CSD; other diseases may also cause similar symptoms. The early histologic changes consist of a lymphocytic infiltrate with epithelioid granuloma formation similar to those seen in lymphomas or sarcoid.

Later changes consist of infiltration by polymorphonuclear leukocytes with additional granuloma formation. These granulomas fuse and eventually form necrotic areas. Large sinuses develop that eventually rupture through the node capsule into the surrounding tissue. In these later stages, the changes most resemble tularemia, lymphogranuloma venereum, brucellosis, or mycobacterial infection.

Because of the pathology of the process, it is not surprising that patients with CSD do not respond to antibiotics to which the suspected etiologic agent is susceptible. This finding should not be used as a criterion for suspecting or ruling out the microorganism.

Laboratory Diagnosis

Margileth et al. (1984) have suggested five criteria for diagnosing CSD, any three of which can be highly suggestive: (1) A history of animal contact, particularly with a cat, as well as a scratch, primary inoculation of a site, or an eye lesion; (2) a positive skin test; (3) negative studies for other causes of infectious lymphadenitis; (4) characteristic histopathology of a lymph node biopsy specimen; and (5) the presence of "typical" silver-staining microorganisms with aberrant morphology in sections of the lymph node, in the primary skin lesion, or in the eye. Laboratory diagnosis is difficult because the etiologic agent has not been isolated with any regularity from lymph node tissues. At the present time, a Warthin–Starry stain of the lymph node tissue or the pus can be done. The slide should be examined microscopically for characteristic organisms. In the early stage of the lymph node lesion, many bacteria can be seen. In the late stages of the disease, however, particularly in necrotic lymph nodes, it is very difficult to find these organisms. A person who has been trained to examine smears of lymph node lesions expected to contain the CSD bacillus is necessary, as the silver-stained bacteria are very difficult to differentiate from stain debris.

A skin test known as the Hanger Rose skin test (McGovern et al., 1955) has been widely used for diagnosis. The skin test antigen, however, may not be readily available to clinicians. It has been produced from aspirated pus from patients with CSD. The skin test is not standardized, and there is variability in how and individual responds to the different skin test antigens (Bradstreet and Dishero, 1977). Many do not recommend skin testing for CSD, as there is no assurance that the skin test material is safe. The Committee on Infectious diseases of the American Academy of Pediatrics strongly discourages the use of the CSD skin test. One of the biggest problems in the diagnosis of CSD is the possibility of lymphadenitis being of neoplastic process, hence the necessity of a precise diagnosis.

Epidemiology

Cat scratch disease is seasonal. Most cases occur during the summer months through October to early November. Although unusual, there have been multiple cases within families, and veterinarians appear to have an occupational risk for CSD. The attack rate of CSD is difficult to determine. Because there is no adequate diagnostic test, the disease may be underreported. It does not appear to be easily acquired, person-to-person transmission has not been shown, and there is no evidence that reinfection occurs. The disease has been reported throughout the United States and from most other countries.

Treatment, Prevention, and Control

Treatment

Cat scratch disease is most likely caused by a bacterium. It should hold, then, that progression of the disease should be altered by administration of antibiotics. Such is not the case, however, no doubt because of the isolated nature of the microorganisms within the necrotic lymph node. A number of antibiotics have been used for treatment of CSD, including penicillin, tetracycline, erythromycin, first-generation cephalosporins, cephaclor, chloramphenicol, and trimethoprim–sulfamethoxazole. There is no evidence that antibiotics affect the natural history of the disease.

If the lymph nodes have suppurated, needle aspiration may be indicated. Surgical excision of these nodes is contraindicated. Most clinicians suggest that incision and drainage should not be done, as such treatment may prolong the suppuration and cause scar formation. Most patients with CSD have a benign course.

Prevention and Control

There appears to be little one can do to prevent CSD except to avoid cats, particularly young kittens who have a tendency to scratch. If a particular cat is known to have transmitted CSD, the owners might consider declawing the cat, although it is unusual for more than one case to be attributed to a single animal. There are more than 10 million cats in the

United States, and the benefits of cat ownership seem to outweigh the risks of possible disease.

Literature Cited

Boyd, G. L., and G. Craig. 1961. Etiology of cat-scratch fever. J. Pediatr. **59:**313–317.

Boyer, K. M., and J. D. Cherry. 1981. Cat scratch disease, p. 1649–1654. *In* R. O. Feigin and J. D. Cherry (ed.), Textbook of pediatric infectious diseases, vol. II. The W. B. Saunders Co., Philadelphia.

Bradstreet, C. M. R., and M. W. Dishero. 1977. Cat scratch fever skin test antigen. Lancet **1:**913.

Carithers, H. A. 1985. Cat scratch disease: an overview based on a study of 1200 patients. Am. J. Dis. Child. **139:**1124–1133.

Debre, R., M. Lamy, M. Jammet. 1950. La maladie des griffes de chat. Bull. Mem. Soc. Med. Hop. Paris **66:**76–79.

Gerber, M. A., A. K. Sedgwick, T. J. MacAlister, K. B. Gustafson, H. Ballow, R. C. Titton. 1985. The aetiological agent of cat scratch disease. Lancet **1:**1236–1239.

Hadfield, T. L., R. H. Malaty, A. Van Dellen, D. J. Wear, A. Margileth. 1985a. Electron microscopy of the bacillus causing cat-scratch disease. J. Infect. Dis. **152:**643–645.

Hadfield, T. L., C. Schlagel, and A. Margileth. 1985b. Stalking the cause of cat-scratch disease. Diagn. Med. **8:**23–27.

Kalter, S. S., C. S. Kim, and R. L. Heberling. 1969. Herpes-like virus particles associated with cat-scratch disease. Nature (London) **224:**190.

Margileth, A. M., D. J. Wear, T. L. Hadfield, C. J. Schlagel, G. T. Spigel, J. E. Muhlbauer. 1984. Cat-scratch disease: bacteria in skin at the primary inoculation site. J. Am. Med. Assoc. **252:**928–931.

McGovern, J. J., L. J. Kunz, and F. M. Blodgett. 1955. Non-bacterial regional lymphadenitis ("cat-scratch fever"). An evaluation of the diagnostic intradermal test. N. Engl. J. Med. **252:**166–171.

Spaulding, W. B., and J. N. Hennessy. 1960. Cat-scratch disease. A study of eighty-three cases. Am. J. Med. **28:**504–509.

Turner, W., N. J. Bigley, M. C. Dodd, G. Anderson. 1960. Hemagglutinating virus isolated for cat scratch disease. J. Bacteriol. **80:**430–435.

Verhoeff, F. H. 1913. Parinaud's conjunctivitis: a mycotic disease due to a hitherto undescribed filamentous organism. Arch. Ophthalmol. **42:**345–351.

Verhoeff, F. H. 1940. Improved method of staining within tissues: leptotriches of Parinaud's conjunctivitis and gram-positive microorganisms. J. Am. Med. Assoc. **115:**1546–1547.

Verhoeff, F. H., and M. J. King. 1933. Leptotrichosis conjunctivae (Parinaud's conjunctivitis). Arch. Ophthalmol. **9:**701–714.

Warwick, W. J. 1967. The cat scratch syndrome, many diseases or one disease? Prog. Med. Virol. **9:**256–301.

Wear, D. J., A. M. Margileth, T. L. Hadfield, G. W. Fischer, C. J. Schlagel, F. M. King. 1983. Cat scratch disease: a bacterial infection. Science **221:**1403–1404.

CHAPTER 17

Chancroid

ALLAN R. RONALD

Disease: Chancroid (soft chancre, ulcus molle).
Etiologic Agent: *Haemophilus ducreyi.*
Source: Humans with ulcers; spread by sexual intercourse.
Clinical Manifestations: Acute ulcer on the external genitalia with inguinal lymphadenopathy and formation of buboes.
Pathology: Ulcer with a dense inflammatory exudate of both neutrophils and lymphocytes. Regional lymphadenitis progressing to suppuration.
Laboratory Diagnosis: Bacteriologic isolation of *H. ducreyi* from ulcer exudate.
Epidemiology: Worldwide, occurring in the industrialized world in discrete outbreaks, often at major ports; clusters can occur anywhere. In developing countries, the disease is frequently endemic.
Treatment: Erythromycin, ceftriaxone, trimethoprim–sulfamethoxazole.
Prevention and Control: Established methods to control sexually transmitted diseases; disease control programs in prostitutes; contact tracing; condoms.

Description of Disease

Chancroid (soft chancre, ulcus molle) is a sexually transmitted disease that produces ulcers on the external genitalia and suppurative inguinal lymphadenitis. *Haemophilus ducreyi* is the etiologic agent.

Communicability and Transmissibility

Haemophilus ducreyi is spread by sexual contact. Acquisition by other routes, including maternal–infant spread, has not been reported. Although autoinoculation has occurred, it is rare. Fomites have not been shown to play any role in the transmission of *H. ducreyi.* The incubation period for chancroid varies from 4 to 30 days, with a median duration of 7 days (Hammond et al., 1980).

Clinical Manifestations

The chancroid ulcer begins as a tender papule that ulcerates within 12 to 24 h. The ulcer is painful, irregular, and usually not indurated. It is sharply demarcated from the surrounding skin, which is usually not inflamed. The ulcer base is covered with dirty, necrotic, gray-yellow exudate and bleeds easily when scraped. Ulcers can range in size from a diameter of 1 mm to 4 to 5 cm (Figs. 1 and 2).

The common sites of the chancroid ulcer in men, in descending importance, are the prepuce, the coronal sulcus, the skin of the shaft of the penis, the glans, the urethral meatus, and the scrotum and surrounding perineal skin. Single ulcers are found in about 50% of men. Ulcers in women occur frequently on the labia majora, with the forchette, the labia minora, the perineum, the medial aspects of the thigh, and the perianal region less commonly involved. Ulcers on the vaginal wall are rare. Cervical ulcers occur in about 10% of women with chancroid and seem to be associated with accompanying cervicitis. The mean number of ulcers in women is 4.5. Ulcers in women are usually less painful, and as many as one-third of women with chancroid are unaware of their infection (Plummer et al., 1985).

Atypical presentations of chancroid are common.

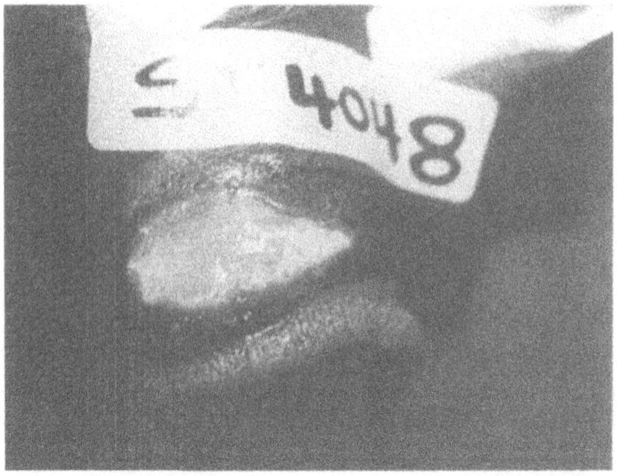

FIG. 1. Irregular painful 1.5 × 0.8 cm ulcer. Culture positive for *H. ducreyi.*

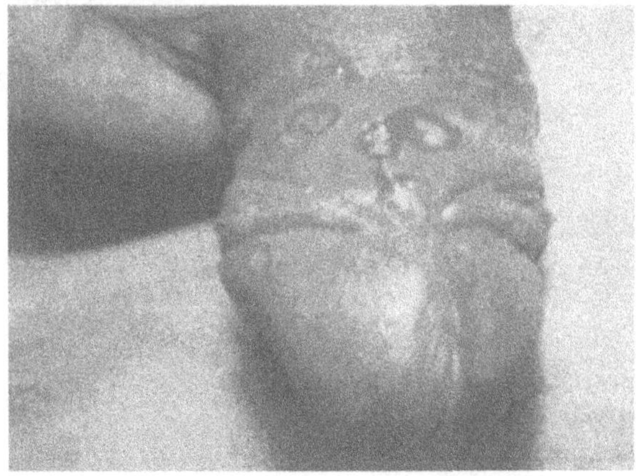

FIG. 3. Herpeslike superficial ulcers on ventral surface of the penile shaft. Culture positive for *H. ducreyi.*

These presentations include herpetiform features (Fig. 3), a rash that resembles a folliculitis, and a "septic sore presentation" with a large abscess in the foreskin. About 20% of ulcers are indurated and mimic the classic chancre of primary syphilis (Fig. 4).

Inguinal lymphadenitis occurs in about 50% of patients, with the onset usually within 5 to 10 days of initial ulceration. Suppuration of lymph nodes with the formation of an inguinal abscess occurs in 15 to 30% of patients. The abscess may rupture and lead to large, deep inguinal ulcers.

Pathology

The histology of the chancroid ulcer has not been shown to be specific for *H. ducreyi* infection. Interstitial edema, polymorphonuclear leukocyte infiltration, round cell infiltration, and disruption of normal subcutaneous cellular elements have been described following *H. ducreyi* cutaneous infection. Few biopsies, however, have been conducted. Regional inguinal lymph nodes are involved, with a pyogenic suppurative process with microabscesses that coalesce to become large pyogenic collections.

Healing occurs with minimal scar formation on the skin of the external genitalia and in the regional lymph nodes. Except for occasional patients who have residual complications such as urethral fistulas or phimosis, no chronic structural damage occurs to either the genitalia or the lymphatic system.

Diagnostic Features

The diagnosis of chancroid should be considered in any patient with genital ulcer disease. Classic clinical features are present in only 50 to 70% of patients who are infected with *H. ducreyi,* and chancroid can be

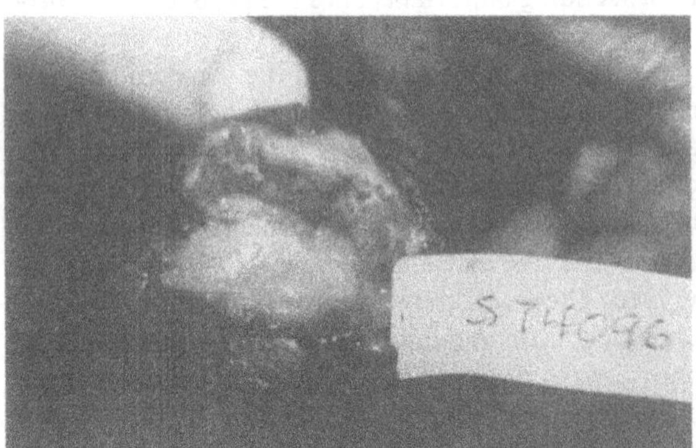

FIG. 2. Deep painful necrotic ulcer on coronal sulcus. Culture positive for *H. ducreyi.*

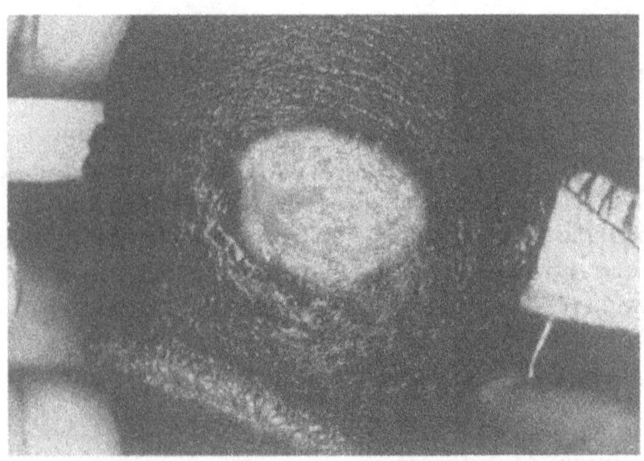

FIG. 4. A 1.5 cm indurated ulcer mimicking the chancre of primary syphilis. Darkfield negative and culture positive for *H. ducreyi.*

readily confused with primary syphilis, ulcerative herpetic lesions, and granuloma inguinale (Chapel et al., 1977; Kraus et al., 1982; Salzman et al., 1984). In industrialized societies, cultures for *H. ducreyi* should be available for patients with 1) a clinical syndrome compatible with chancroid; 2) genital ulcers that are dark field negative and culture negative for *Herpes simplex* virus; 3) an epidemiologic history of contact with an individual from an endemic region; 4) sexual contact with a patient with chancroid; and 5) ulcers on the genitalia during an epidemic of chancroid.

In developing countries, laboratory resources are not routinely available for the diagnosis of *H. ducreyi*. However, reference laboratories should be able to undertake periodic surveys of genital ulcers to confirm the proportion due to *H. ducreyi* and to collect strains for determination of antimicrobial susceptibility.

Darkfield microscopy is essential for the investigation of genital ulcer disease and should be repeated at least twice. Serology is also mandatory to exclude syphilis. *Treponema pallidum* occurs concomitantly with *H. ducreyi* in 3 to 5% of patients with genital ulcer disease in many developing countries. In the absence of laboratory facilities to exclude syphilis, all patients with genital ulcer disease should be treated routinely for syphilis with 2.4 million units of benzathine penicillin, regardless of the presumed diagnosis.

Chancroid rarely causes systemic disease. Patients occasionally may present with an inguinal bubo, and the ulcer only becomes apparent when careful examination of the genitalia is performed. However, most patients with a genital ulcer and an acute suppurative bubo are infected with *H. ducreyi*. Although *Herpes simplex* and *T. pallidum* are associated with acute inguinal lymphadenopathy, these infections rarely, if ever, suppurate.

A positive culture for *H. ducreyi*, as for Neisseria gonorrhoeae, is abnormal. Studies from Sheffield, England, that suggest that *H. ducreyi* commonly colonizes the genital tract have not been confirmed by other workers in England, Winnipeg, or Kenya (Kinghorn et al., 1982). Although asymptomatic infection of the smegma was described by Brahms in 1924, this finding has not been confirmed by current technology.

Currently, no serologic or other laboratory procedure is available to diagnose *H. ducreyi* infection. A positive culture for *H. ducreyi* remains a sine qua non for the definitive diagnosis of chancroid.

Prognosis

Chancroid is self-limited, and ulcers resolve in 4 to 12 weeks with no specific treatment. Soldiers admitted to hospitals and treated with local antiseptics, before sulfonamides became available, were hospitalized for an average of 36 days. In women, ulcers have been known to persist for up to 6 months in the absence of treatment. Ongoing draining inguinal abscesses are a feature of untreated chancroid. Metastatic lesions have not been reported outside the genital tract, however, and chancroid has not been found to cause pelvic inflammatory disease or neonatal infection. Urethritis in men may be a manifestation of *H. ducreyi* infection (Kunimoto et al., 1987).

Patients recovering from chancroid have no known natural immunity, and reinfection can occur immediately on reexposure. Relapse, due to the ulceration at the original site, has been reported to occur in 5 to 10% of patients after recovery. It occurs in the absence of reexposure both in patients who have been given specific treatment and in those who have had no treatment.

Relationship to Human Immunodeficiency Virus

Recent studies in Africa provide some evidence to suggest that chancroid may be a risk factor for the heterosexual spread of human immunodeficiency virus (HIV-1). In a setting of sexually transmitted disease, among men who are HIV-1 seropositive, 66% have a history of recent genital ulcer disease compared to 31% of those who are seronegative (Greenblatt et al., 1987b). In addition, in a prospective study of men who acquired chancroid from sexual contacts who were also infected with HIV-1, the risk of seroconversion in the 12 weeks after acquisition of chancroid was about 10% (Cameron et al., 1987). This risk was significantly increased compared to the risk of concomitant acquisition of gonococcal infection and HIV-1. Finally, among prostitutes, genital ulcers appear to be a significant risk factor for acquisition of HIV-1 (Plummer et al., 1987). Further studies to determine the specific risk factors associated with chancroid that result in acquisition of HIV-1 are urgently required.

Epidemiologic Aspects

Occurrence and Transmission

Chancroid is endemic in many developing countries. In Western societies, chancroid occurs among individuals who have traveled to endemic areas and occurs in clusters, particularly at major ports of entry. During the past 12 years, five discreet outbreaks have been described in the United States and Canada. In these infection clusters, as well as during earlier studies among military populations, prostitutes have been implicated as the reservoir for chancroid. In Africa, 80% of men give a history of a recent contact with a prostitute. In the outbreak in Winnipeg in 1975 to 1977, prostitutes were the continuing source of chancroid. In both the California and Winnipeg epidemics, epidemiologic investigation and treatment of female sexual contacts eradicated the disease from the community (Blackmore et al., 1985; Hammond et al., 1980). No ongoing evidence of a continuing reservoir of *H. ducreyi* was present in these communities. Studies from Sheffield, England suggested that *H. ducreyi* commonly coinfects herpetic ulcers and can be frequently isolated from genital secretions and the oropharynx of otherwise healthy individuals (Kinghorn et al., 1982). These results have not been reproduced by other investigators (Piot et al., 1983; Plummer et al., 1983a).

In an epidemiologic study carried out in East Africa, the source contacts of 10 men with chancroid were investigated and all 10 women had genital ulcers (Plummer et al., 1983a). An investigation of 29 secondary contacts of men with chancroid showed 17 to have ulcers. Three additional women had positive cultures for *H. ducreyi* from the cervix in the absence of ulcers.

Circumcision decreases susceptibility to infection. During the Vietnam conflict, the odds ratio for infection for uncircumcised American troops was 3 : 1 compared with circumcised men. The risk of transmission from an infected person to his or her partner during intercourse has not been established. Presumably, individuals remain infectious as long as they have untreated ulcers. There is no good evidence for a sustained carrier state.

Chancroid has been presumed to be a disease more prevalent among impoverished individuals and those with low standards of cleanliness. This prevalence, however, is probably a reflection of sexual outlets available to these individuals and the lack of medical care provided to prostitutes. Large populations of prostitutes with no health care have a very high prevalence of genital ulcer disease (Plummer et al., 1985). In Western societies, prostitutes make a limited contribution to the epidemiology of sexually transmitted diseases. Some such diseases, particularly gonococcal and chlamydial infections, are widely distributed, are often asymptomatic in the population, and remain common and endemic. Chancroid, with a reservoir usually confined to prostitutes with ulcers, disappears.

Disease Susceptibility and Resistance

Haemophilus ducreyi presumably gains access through direct invasion through healthy or possibly abraded genital skin. There is no other known route of entry. With the appropriate infectious challenge, it is presumed that all individuals are susceptible to infection. However, the inoculum of organisms required to cause infection or factors that alter susceptibility are unknown.

Etiologic Agent

Characterization and Conditions for Culture

Haemophilus ducreyi is a poorly staining gram-negative nonmotile rod. Kilian (1976) confirmed that the cell envelope is a trilaminar structure, made up of two electron-dense layers separated by a translucent layer. He also showed that the guanosine-plus-cytosine content was 0.38 mole fraction and demonstrated the absence of heme synthethase, the enzyme necessary to catalyze the insertion of iron into the protoporphyrin nucleus. The porphyrin test is positive because of the inability of *H. ducreyi* to synthesize porphyrin from δ-aminoleavulinic acid. *Haemophilus ducreyi* is the only human pathogen in the genus *Haemophilus* with a requirement for hemin as demonstrated by the porphyrin test.

Albritton et al. (1984) investigated the DNA-relatedness within the genus *Haemophilus* and showed that although the organism shares some antigenic relationships with other members of the genus, DNA–DNA hybridization indicates a very distant relationship to *H. influenzae,* with a DNA homology of only 0.18. This observation was confirmed by the inefficient heterospecific transformation of streptomycin resistance markers from *H. influenzae* to *H. ducreyi.*

Haemophilus ducreyi has few biochemical characteristics that permit specific identification and, to date, there is no proven immunologic procedure to identify isolates as *H. ducreyi.* All strains of *H. ducreyi* are hydrogen sulfide and indole negative and require hemin for growth. The latter, however, cannot be demonstrated by growth around a disk impregnated with hemin on trypicase agar, owing to the failure of this medium to support the growth of *H. ducreyi.* The requirement for hemin can be shown on a gonococcal agar base supplemented with glucose (0.1%), glutamine (0.01%), and cysteine (0.05%). Strains of *H. ducreyi* grow optimally at hemin levels of 25 to 50 mg/liter; most other strains of Haemophilus that require hemin grow well at hemin concentrations of 1 to 10 mg/liter (Hammond et al., 1978a).

Optimal growth occurs with added serum. Sottnek et al. (1979) demonstrated that fetal bovine serum was superior to other sera and that Mueller–Hinton agar and heart infusion agar were superior to Columbia agar, *Protease* peptone no. 3 agar, and gonococcal base agar. Hafiz et al. (1984) described the Sheffield medium, which consists of *Protease* peptone no. 3, gelatin, rice starch, and agar with an added supplement consisting of glutamine, glucose, and ferric nitrate combined with equine hemin (0.02%) and 5% defibrinated horse blood. This medium was reported to result in more luxuriant growth than did previously described media; we have not been able to confirm this observation, however, and on primary isolation most strains of *H. ducreyi* do not grow on the Sheffield medium (McDonald et al., 1987; Piot et al., 1983). In addition, Sturm and Zanen (1984) reported that "Beiling agar," a mixture of two parts of Columbia agar base and one part of hemolyzed horse blood, supplemented with 2.5% yeast dialysate gave excellent results. In a comparative study, we are unable to show that this medium was superior to gonococcal agar or Mueller–Hinton agar (Kunimoto et al., 1986).

Table 1 summarizes the results of primary *H. ducreyi* isolation in eight studies. Unfortunately, many studies do not compare one medium with another. From our prospective studies, we have concluded that Mueller–Hinton agar and gonococcal base agar each provide nutrients uniquely required by some strains of *H. ducreyi* (Dylewski et al., 1986; Kunimoto et al., 1986; Nsanze et al. 1984). In each of three prospective studies, between 15 and 25% of isolates did not grow on primary isolation on Mueller–Hinton agar, and 10 to 15% did not grow on gonococcal base agar. On subsequent cultures, isolates could rapidly adapt to the alternate medium, and this trait was lost. However, until further nutritional studies identify the specific nutritional requirements of this organism, we advocate the use of both agar media for the primary isolation of *H. ducreyi.*

Growth of *H. ducreyi* occurs both aerobically and anaerobically and many isolates, particularly on primary isolation, are carbon dioxide dependent. Growth is almost invariably enhanced by carbon dioxide. Strains of *H. ducreyi* require an environment with 100% humidity and grow best at a neutral pH (Sturm and Zanen, 1984).

The temperature of incubation is very important. Strains of *H. ducreyi* do not grow if the temperature is 37°C or higher, particularly on primary isolation, and most grow best between 30 and 33°C. Many isolates grow at temperatures as low as 22°C.

Specimen Collection and Transport

Both cotton and calcium alginate have been shown to be suitable for specimen collection. Swabs should be taken from the purulent ulcer base. However, we have not been able to show that the ulcer requires extensive cleaning and recommend that, after one swab is used to remove superficial debris and scab, a second swab be used to collect the exudate from the ulcer. Subsequent material from the ulcer base can then be used for darkfield examination and for viral culture for *H. simplex.*

TABLE 1. Solid media for primary isolation of *Haemophilus ducreyi*

Results	*Agar*	*Supplements*	*Primary culture results*
Dylewski et al. (1986)	1) Meuller–Hinton[a]	5% chocolated horse blood, vancomycin	53 of 111
	2) Gonococcus[b]	2% bovine hemoglobin, 5% fetal calf serum, 1% CVA,[b] (cofactors, vitamins, amino acids) vancomycin	74 of 111
Kunimoto et al. (1986)	1) Bieling medium	See Kunimoto et al. (1986)	24 of 31
	2) Meuller–Hinton[a]	5% chocolated horse blood, vancomycin	20 of 31
	3) Gonococcus[b]	2% bovine hemoglobin, 5% fetal calf serum, 1% CVA,[b] vancomycin	26 of 31
McDonald et al. (1987)	1) Sheffield medium[c]		1 of 87
	2) Meuller–Hinton[a]	5% chocolated horse blood, vancomycin	41 of 87
	3) Gonococcus[b]	2% bovine hemoglobin, 5% fetal calf serum, 1% CVA,[b] vancomycin	30 of 87
Hammond et al. (1978b)	Gonococcus	Bovine hemoglobin,[d] 1% IsoVitaleX, 1%,[a] vancomycin	7 of 16
Sottnek et al. (1979)	Heart infusion	5% defibrinated rabbit blood, 10% fetal bovine serum, vancomycin	16 of 17
Hafiz et al. (1984)	Proteos peptone no. 3	Rice starch, gelatin, glutamine, ferric nitrate, 5% equine blood, vancomycin	9 of 10
Hannah and Greenwood (1982)	Chocolate agar (clinical standards)	Vancomycin	51 of 62
Nzanse et al. (1984)	1) Meuller–Hinton[a]	5% chocolated horse blood, vancomycin	122 of 163
	2) Gonococcus[b]	2% bovine hemoglobin, 5% fetal calf serum, 1% CVA,[b] vancomycin	143 of 163

[a] BBL Microbiology Systems, Cockeysville, Md.; [b] GIBCO Laboratories, Grand Island, N.Y.; [c] Oxoid Ltd., London, England; [d] Difco Laboratories, Detroit, Mich.

A satisfactory transport system has not been developed. The organisms will only survive for 2 to 4 h on a swab, and survival is improved if the inoculated swab is kept refrigerated. For optimal results, however, the specimen should be inoculated immediately onto the agar surface after collection. Blood cultures and serologic studies for *H. ducreyi* are not useful.

Isolation and Identification Procedures

A Gram stain should be done from the ulcer exudate. Unfortunately, the exact role of the Gram stain in the diagnosis of *H. ducreyi* infection in patients with genital ulcer disease continues to be controversial. Although experts in Thailand found the Gram stain to be specific and sensitive, this observation could not be confirmed in Kenya (Nsanze et al., 1981; Taylor et al., 1984). *Haemophilus ducreyi* stains faintly with safranine and crystal violet. The classic spatial description of *H. ducreyi* with "school of fish" and "railroad tracking" is seen occasionally, but is the exception rather than the rule. Alternate techniques for immediate identification of *H. ducreyi* in ulcer or bubonic exudate are urgently needed. Although a monoclonal antibody has been described to a 47,000-megadalton outer membrane protein, which detected *H. ducreyi* in lesions in rabbits in a field study, it failed to detect *H. ducreyi* in clinical material (Greenblatt et al., 1987a; Hansen and Loftus, 1984).

For reasons discussed previously, we prefer to use two media for routine primary isolation of *H. ducreyi*. Gonococcal base agar, with added bovine hemoglobin, fetal calf serum, and vitamin supplement, and Mueller–Hinton agar, with 5% horse blood and vitamin supplement, are both satisfactory. Vancomycin, 3 mg/liter, is added to both media. We no longer encourage the use of human or rabbit liquid blood culture for initial enrichment of *H. ducreyi* from the clinical specimen.

The sensitivity of a single culture in a study in which two cultures were obtained on two different days was 88 and 85%, respectively (Dylewski et al., 1986). Overall, the sensitivity of a positive culture for *H. ducreyi* in patients with clinical chancroid is about 80%.

The numbers of *H. ducreyi* in ulcer exudates appear to be substantial and probably are in the range of 10^7 to 10^8/ml of pus. Frequently, hundreds of gram-negative rods per oil-immersion field are seen on the initial Gram stain. As no other gram-negative rods are recovered in either aerobic or anaerobic cultures in many instances, we presume that these organisms are all *H. ducreyi*. On the other hand, in bubonic pus, the Gram stain usually shows no organisms unless the bubo has ruptured and an inguinal ulcer is present. Cultures of bubonic pus are negative for *H. ducreyi* in more than 90% of patients.

A candle extinction jar with a moistened paper towel is an excellent environment for culture of *H. ducreyi*. The culture should be incubated at 31 to 33°C. Conditions favorable for the culture of *Campylobacter*, except for incubation temperature, are also suitable for culture of *H. ducreyi*. However, growth is scanty or nonapparent at 24 h, and typical colonies are usually visible only at 48 to 72 h. Culture plates should not be discarded as negative until the fifth day.

Identification

Although initially colonies of *H. ducreyi* are difficult to distinguish from other organisms, certain useful features permit recognition of *H. ducreyi*. Characteristically, colonies are yellow-gray, dome shaped, and difficult to remove with an inoculating needle. They vary in size from 0.2 to 1.5 mm, giving a pleomorphic appearance and the impression of a mixed culture. The colonies are extremely cohesive and can be pushed intact across the agar surface. Colonies clump firmly together when dispersed in a saline suspension. Slight α-hemolysis occurs on rabbit or sheep blood agar, but has not been noted with blood from other species. Although Hafiz et al. (1982) noted that colonies of *H. ducreyi* aggregate starch, this characteristic has not been incorporated in routine identification.

Individual colonies are selected for Gram staining and for subculture. The Gram stain smear seldom shows classic morphology.

There are few biochemical characteristics and no immunologic reagents for specific identification of *H. ducreyi*. We proceed by demonstrating a requirement for hemin and a positive oxidase test with *N,N,N,N*-tetramethyl-*p*-phenylinedyamine. No characteristic biotypes of *H. ducreyi* have been described. Our group has identified differences in proteins on agar gel electrophoresis and in plasmid profiles (McNicol and Ronald, 1984; Odumeru et al., 1983). However, the usefulness of these characteristics to establish typing schema for *H. ducreyi* requires further investigation.

Immunologic Aspects

Presumably, both a humoral and a cell-mediated immune response occurs to infection of *H. ducreyi*. However, studies to demonstrate an immunologic response have been limited to date. Schalla et al. (1986) have shown that dot immunobinding to *H. ducreyi* antigen occurs in patients with clinical chancroid. Unfortunately, in most instances, *H. ducreyi* was not cultured from these patients. Other immunologic studies are required to determine the kinetics of the immune response to *H. ducreyi* infection. Although earlier experimental inoculation studies in humans suggest that there is no acquired immunity to this pathogen, this finding needs to be confirmed. At one time, immunization with crude *H. ducreyi* antigen was used to treat chancroid. At present, no immunologic studies are available to diagnose *H. ducreyi* in patients with genital ulcer disease.

Treatment, Prevention, and Control

Treatment

Chancroid ulcers and buboes ultimately heal with scarring without any specific antimicrobial treatment. Many patients have persistent ulcers and buboes for months before diagnosis. Patients may have progressive destruction of genital tissue and extensive scarring if untreated. Circumcision, heat application, and topical antibacterial solutions have all been used with variable benefit. Since the classic article by Hanschell in 1938, chancroid has been successfully treated with sulfonamides. The organism is rapidly eradicated from the ulcer. Cultures taken 24 h after initiation of therapy are almost always negative. The ulcer base becomes granulomatous and epithelialization quickly proceeds. The average duration after successful therapy depends on the diameter of the ulcer, with an average time to complete healing of 9 days in men and 5 days in women. The tetracyclines, streptomycin, kanamycin, and chloramphenicol have all been investigated and found to be satisfactory treatment alternatives to the sulfonamides for chancroid.

During the 1970s, resistant strains of *H. ducreyi* began to emerge and treatment failures were common after treatment with both sulfonamides and tetracycline (Fast et al., 1983). During the past decade, 11 different plasmids have been found in *H. ducreyi;* these plasmids have now been shown to mediate resistance to the sulfonamides, the penicillins, kanamycin, streptomycin, tetracycline, chloramphenicol, and trimethoprim (McNicol and Ronald, 1984). Due to the rapid emergence of resistance, treatment regimens have been undergoing continuous change. In most regions of the world, trimethoprim in combination with sulfonamides is an excellent treatment choice. In Thailand and perhaps in other parts of Southeast Asia, however, the prevalence of resistance to sulfonamides and trimethoprim make this regimen ineffective (Taylor et al., 1985b).

The duration of treatment regimens has also been investigated prospectively in several studies. Although it has been stated that treatment should be continued until complete reepithelization of the ulcer occurs, this has been shown to be unnecessary. *Haemophilus ducreyi* is eradicated within 48 h by antimicrobial activity above the minimum inhibitory concentration of the isolate. As a result, short-term and single-dose regimens are as effective as longer courses in prospective clinical studies conducted in Kenya (Plummer et al., 1983b).

The Centers for Disease Control presently recommend 500 mg of erythromycin four times per day for 7 days as a standard treatment regimen. The long-acting third-generation cephalosporin ceftriaxone has been shown in two separate studies to be very effective if prescribed as an intramuscular dose of 250 mg (Bowmer et al., 1987; Taylor et al., 1985b). Amoxicillin combined with clavulanic acid is also effective when prescribed for 3 days (Ndinya-Achola et al., 1986). Clavulanic acid blocks the β-lactamase, permitting amoxicillin to inhibit bacterial growth. The quinolones are also very effective in vitro and have been shown in preliminary clinical studies to cure chancroid. Ciprofloxacin in a dose of 500 mg twice daily for 3 days cured all of 40 patients treated in a study in Kenya (Naamara et al., 1987).

Antimicrobial susceptibility is determined with the agar dilution technique (Hammond et al., 1978b). Gonococcal base agar enriched with 1% hemoglobin and 1% IsoVitalex (consisting of cofactors, vitamins, and amino acids) is prepared, and antimicrobial agents are incorporated at 50°C to yield final concentrations increasing in a $\log_2$ dilution series from 0.05 to 128 mg/liter. Susceptibilities to sulfonamide and trimethoprim are determined on the same media enriched with 0.1% glucose, 0.01% glutamine, and 0.025% hemin with added 5% lysed horse blood. The agar is poured to a depth of 4 mm and can be stored for up to 1 week at 4°C. Following inoculation with a Steers replicator, the plates are incubated in 5% carbon dioxide at 33°C in an atmosphere saturated with water vapor and read after 48 h of incubation.

Antimicrobial susceptibility studies indicated during outbreaks of chancroid and in specific instances in which patients with culture-positive ulcers fail to respond to treatment. As antimicrobial susceptibility varies markedly from one geographic area to another, periodic surveys of susceptibility on a country-wide basis are necessary to make recommendations for therapeutic regimens. Table 2 summarizes the antimicrobial susceptibility of 1,000 strains of *H. ducreyi* collected from eight countries on four continents during the past 6 years.

Relapse after complete ulcer healing occurs at the site of the original ulcer in about 5% of patients. This recurrence has been noted as part of the natural history of chancroid and also following otherwise effective treatment. Chancroid relapses respond to the

TABLE 2. Antimicrobial susceptibility of 1,000 isolates of *Haemophilus ducreyi*

Antimicrobial agent	*Range of MIC[a] (mg/liter)*	*Resistance*	*Mechanism of resistance*
Sulfonamides	20–>1024	High level >30%	Plasmid mediated
		Moderate level >30%	Unknown
Trimethoprim	0.125–32	10%	Transposon on chromosome
Erythromycin	0.0005–0.06	None known	—
Tetracycline	0.25–32	>50%	Plasmid mediated
Streptomycin	0.5–>64	10%	Plasmid mediated
Kanamycin	1.0–>128	10%	Plasmid mediated
Chloramphenicol	0.3–>64	10%	Plasmid mediated
Spectinomycin	1–16	None known	—
Cefotaxime	0.001–0.06	None known	—
Ceftriaxone	0.001–0.03	None known	—
Rifampin	0.0001–1.0	Only in laboratory strains	Unknown
Enoxacin	0.008–0.25	None known	—
Ciprofloxacin	0.004–0.06	None known	—
Rosoxacin	0.001–0.015	None known	—
Fleroxacin	0.001–0.03	None known	—
Ampicillin	0.05–>128	95%	Plasmid mediated

[a] MIC = Minimum inhibitory concentration.

original therapeutic agent with rapid resolution. Patients may also reinfect if the sexual partner is not investigated and treated. We recommend that all sexual contacts of patients with clinical chancroid be treated with the same regimen found to be effective in the patient.

Prevention and Control

Strategies to investigate the control or eradication of chancroid have not been a priority. In Winnipeg and in California, widespread treatment of prostitutes with ulcers and prostitutes named as sexual contacts of men with genital ulcer disease effectively eradicated the disease in both populations. In Kenya, careful prospective studies have shown that all source contacts and most secondary contacts have ulcers. Unfortunately, in women many ulcers are relatively subclinical, and the symptoms do not lead to treatment referral.

Recent studies have shown an association between acquisition of chancroid and HIV. In retrospective and prospective studies, this association appears to account for a considerable proportion of the heterosexual infectiousness of HIV in East Africa. As a result, chancroid eradication programs to impede the spread of HIV may soon become a global priority.

Condoms can prevent acquisition of *H. ducreyi.* Control strategies should screen individuals at risk for chancroid, particularly women engaged in prostitution. The efficacy of single-dose treatment regimens in this population requires further investigation.

Literature Cited

Albritton, W. L., J. K. Setlow, M. Thomas, F. Sottnek, and A. G. Steigerwalt. 1984. Heterospecific transformation in the genus *Haemophilus*. Mol. Genet. **193:**358–363.

Blackmore, C. A., K. Limpakarnjanart, J. G. Rigau-Perez, W. L. Albritton, and J. R. Greenwood. 1985. An outbreak of chancroid in Orange County, California: descriptive epidemiology and disease-control measures. J. Infect. Dis. **151:**840–844.

Bowmer, M. I., H. Nsanze, L. J. D'Costa, J. Dylewski, L. Fransen, P. Piot, and A. R. Ronald. 1987. Single dose ceftriaxone for chancroid. Antimicrob. Agents Chemother. **31:**67–69.

Brahms, J. 1924. Isolation of ducrey bacillus from the smegma of thirty men. J. Am. Med. Assoc. **82:**1166–1167.

Cameron, D. W., F. A. Plummer, J. N. Simonsen, J. O. Ndinya-Achola, L. J. D'Costa, P. Piot, and A. R. Ronald. 1987. Female to male heterosexual transmission of HIV infection in Nairobi. Presented at the 3rd International Conference on AIDS, Washington, D.C.

Chapel, T. A., W. J. Brown, C. Jeffries, and J. A. Stewart. 1977. How reliable is the morphological diagnosis of penile ulcerations? Sex. Transm. Dis. **4:**150–152.

Dylewski, J., H. Nsanze, G. Maitha, and A. R. Ronald. 1986. A laboratory diagnosis of *Haemophilus ducreyi:* sensitivity of culture media. Diagn. Microbiol. Infect. Dis. **4:**241–245.

Fast, M. V., H. Nsanze, F. A. Plummer, L. J. D'Costa, I. W. Maclean, and A. R. Ronald. 1983. Treatment of chancroid: a comparison of sulfamethoxazole and

trimethoprim–sulfamethoxazole. Br. J. Vener. Dis. **59**:320–324.
Greenblatt, R. M., S. L. Lukehart, F. A. Plummer, L. J. D'Costa, W. Naamara, A. R. Ronald, and K. K. Holmes. 1987a. Diagnosis of genital ulcer disease. (In preparation.)
Greenblatt, R. M., S. L. Lukehart, F. A. Plummer, T. C. Quinn, C. W. Critchlow, L. J. D'Costa, W. Naamar, A. R. Ronald, and K. K. Holmes. 1987b. Genital ulceration as a risk factor for human immunodeficiency virus infection in Kenya. Presented at the 3rd International Conference on Aids, Washington, D.C.
Hafiz, S., G. R. Kinghorn, and M. G. McEntegart. 1982. Starch aggregation as a presumptive test for *Haemophilus ducreyi*. Lancet **2**:872.
Hafiz, S., M. G. McEntegart, and G. R. Kinghorn. 1984. Sheffield medium for cultivation of *Haemophilus ducreyi*. Br. J. Vener. Dis. **60**:196–198.
Hammond, G. W., C. J. Lian, J. C. Wilt, W. L. Albritton, and A. R. Ronald. 1978a. Determination of the hemin requirement of *Haemophilus ducreyi:* evaluation of the porphyrin test and media used in the satellite growth test. J. Clin. Microbiol. **7**:243–246.
Hammond, G. W., C. J. Lian, J. C. Wilt, and A. R. Ronald. 1978b. Comparison of specimen collection and laboratory techniques for isolation of *Haemophilus ducreyi*. J. Clin. Microbiol. **7**:39–43.
Hammond, G. W., M. Slutchuk, J. Scatliff, E. Sherman, J. C. Wilt, and A. R. Ronald. 1980. Epidemiologic, clinical, laboratory and therapeutic features of an urban outbreak of chancroid in North America. Rev. Infect. Dis. **2**:867–879.
Hannah, H., and J. R. Greenwood. 1982. Isolation and rapid identification of *Haemophilus ducreyi*. J. Clin. Microbiol. **16**:861–864.
Hanschell, H. M. 1938. Sulphanilamide in the treatment of chancroid. Lancet **1**:886–888.
Hansen, E. J., and T. A. Loftus. 1984. Monoclonal antibodies reactive with all strains of *Haemophilus ducreyi*. Infect. Immun. **44**:196–198.
Kilian, M. 1976. A taxonomic study of the genus *Haemophilus*, with the proposal of a new species. J. Gen. Microbiol. **93**:9–62.
Kinghorn, G. R., S. Hafiz, and M. G. McEntegart. 1982. Pathogenic microbial flora of genital ulcers in Sheffield with particular reference to herpes simplex and *Haemophilus ducreyi*. Br. J. Vener. Dis. **58**:377–380.
Kraus, S. J., B. S. Werman, J. W. Biddle, F. O. Sottnek, and E. P. Ewing, Jr., 1982. Pseudogranuloma inguinale caused by *Haemophilus ducreyi*. Arch. Dermatol. **118**:494–497.
Kunimoto, D. Y., F. A. Plummer, W. Naamara, L. J. D'Costa, J. O. Ndinya-Achola, and A. R. Ronald. 1987. Urethral infection with *Haemophilus ducreyi* in men. Sex. Transm. Dis. (In press.)
Kunimoto, D. Y., L. Slaney, J. Koss, L. J. D'Costa, F. A. Plummer, J. O. Ndinya-Achola, and A. R. Ronald. 1986. Field testing of modified Bieling media for the isolation of *Haemophilus ducreyi* in Kenya. Eur. J. Clin. Microbiol. **5**:673–675.
McDonald, K., D. Cameron, G. Irungu, L. J. D'Costa, F. A. Plummer, J. O. Ndinya-Achola, and A. R. Ronald. 1987. Comparison of Sheffield media with standard media in the isolation of *Haemophilus ducreyi*. Presented at the 7th Annual Meeting of the International Society for S.T.D. Research, Atlanta.
McNicol, P. J., and A. R. Ronald. 1984. The plasmids of *Haemophilus ducreyi*. J. Antimicrob. Chemother. **14**:561–573.
Naamara, W., F. A. Plummer, R. Greenblatt, L. J. D'Costa, J. O. Ndinya-Achola, and A. R. Ronald. 1987. Treatment of chancroid with ciprofloxacin: a prospective randomized clinical trial. Am. J. Med. **82(suppl. 4A)**:317–320.
Ndinya-Achola, J. O., H. Nsanze, P. Karasira, L. Fransen, L. J. D'Costa, A. R. Ronald, and P. Piot. 1986. Three day oral course: augmentin in the treatment of chancroid. Genitourin. Med. **62**:202–204.
Nsanze, H., M. V. Fast, L. J. D'Costa, P. Tukei, J. Curran, and A. Ronald. 1981. Genital ulcers in Kenya. Clinical and laboratory study. Br. J. Vener. Dis. **57**:378–381.
Nsanze, H., M. V. Fast, L. J. D'Costa, P. Tukei, J. Curran, and A. R. Ronald. 1984. Comparison of media for the primary isolation of *Haemophilus ducreyi*. Sex. Transm. Dis. **11**:6–9.
Odumeru, J. A., A. R. Ronald, and W. L. Albritton. 1983. Characterization of cell proteins of *Haemophilus ducreyi* by polyacyralamide gel electrophoresis. J. Infect. Dis. **148**:710–714.
Piot, P., H. Nsanze, and A. R. Ronald. 1983. Isolating *Haemophilus ducreyi*. Lancet **1**:909.
Plummer, F. A., L. J. D'Costa, H. Nsanze, J. Dylewski, P. Karasira, and A. R. Ronald. 1983a. Epidemiology of chancroid and *Haemophilus ducreyi* in Nairobi. Lancet **2**:1293–1295.
Plummer, F. A., L. J. D'Costa, H. Nsanze, P. Karasira, I. W. Maclean, and A. R. Ronald. 1985. Clinical and microbiologic studies of genital ulcers in Kenyan women. Sex. Transm. Dis. **12**:193–197.
Plummer, F. A., H. Nsanze, L. J. D'Costa, P. Karasira, I. W. Maclean, R. H. Ellison, and A. R. Ronald, 1983b. Single dose therapy of chancroid with trimethoprim–sulfametrole. N. Engl. J. Med. **309**:67–71.
Plummer, F. A., J. N. Simonsen, E. N. Ngugi, D. W. Cameron, P. Piot, A. R. Ronald, and J. O. Ndinya-Achola. 1987. Incidence of human immunodeficiency virus (HIV) infection and related disease in a cohort of Nairobi prostitutes. Presented at the 3rd International Conference on AIDS, Washington, D.C.
Salzman, R. S., S. J. Kraus, R. G. Miller, F. O. Sottnek, and G. S. Kleris. 1984. Chancroidal ulcers that are not chancroid. Cause and epidemiology. Arch. Dermatol. **120**:636–639.
Schalla, W. O., L. L. Sanders, G. P. Schmid, M. R. Tam, and S. A. Morse. 1986. Use of dot-immunobinding and immunofluorescence assays to investigate clinically suspected cases of chancroid. J. Infect. Dis. **153**:879–887.
Sottnek, F. O., J. W. Biddle, S. J. Draus, R. E. Weaver, and J. A. Stewart. 1979. Isolation and identification of *Haemophilus ducreyi* in a clinical study. J. Clin. Microbiol. **12**:170–174.
Sturm, A. W., and H. C. Zanen. 1984. Characteristics of *Haemophilus ducreyi* in culture. J. Clin. Microbiol. **19**:672–674.
Taylor, D. N., C. Duangmani, C. Suvongse, R. O'Connor, C. Pitarangsi, K. Panikabutra, and P. Echeverriap. 1984. The role of *Haemophilus ducreyi* in penile ulcers in Bangkok, Thailand. Sex. Transm. Dis. **11**:148–151.
Taylor, D. N., P. Echeverria, S. Hanchalay, C. Pitarangsi, L. Slootmans, and P. Piot. 1985a. Antimicrobial susceptibility and characterization of outer membrane proteins of *Haemophilus ducreyi* isolated in Thailand. J. Clin. Microbiol. **21**:442–444.
Taylor, D. N., C. Pitarangsi, P. Echeverria, K. Panikabutra, and C. Suvongse. 1985b. Comparative study of ceftriaxone and trimethoprim–sulfamethoxazole for the treatment of chancroid in Thailand. J. Infect. Dis. **152**:1002–1006.

CHAPTER 18

Cholera

YUTAKA ZINNAKA

Disease: Cholera.

Etiologic Agents: *Vibrio cholerae* serovar (serogroup) O1, subtypes Ogawa, Inaba, and Hikojima. Biovars (biotypes) *cholerae* and *eltor* are recognized. The former is also called the "classical" biotype.

Source: Water and foods contaminated by feces of patients and carriers. Shellfish and plankton may be the natural reservoir.

Clinical Manifestations: Acute onset with diarrhea and vomiting. Diarrhea is usually watery and colorless (rice-water stool). No fever or abdominal pain in typical cases. Dehydration, acidosis, and circulatory collapse are the main symptoms. Renal failure is often associated with inadequate treatment.

Pathology: Hypersecretion of small intestine. Destruction of mucosa is minimal. Some inflammatory signs in lamina propria.

Laboratory Diagnosis: Bacteriologic isolation of etiologic agent from the stool. Direct microscopic examination of watery stool is helpful. Serologic examination may be used for retrospective diagnosis.

Epidemiology: Originally an epidemic disease in Indian subcontinent, with worldwide expansion several times since the beginning of the 19th century. Etiologic agent of the present (seventh) pandemic is mainly of the *eltor* biotype, which is known to produce more asymptomatic carriers and mild cases than the classical biotype. Tropical countries in Asia and Africa are the main endemic foci, although sporadic cases in industrialized countries or in other continents are found among travelers returning from such countries or people who have ingested contaminated (mainly imported) food.

Treatment: Rehydration is essential. Intravenous electrolyte solution with alkali is used for initial stage, followed by oral solution containing glucose. Oral rehydration only may be used for mild cases. Tetracyclines should be administered until the vibrio disappears from the stool.

Prevention and Control: Killed whole cell vaccine has only a limited capacity for prevention. Trials have been conducted to create a more effective vaccine. Improved sanitation and education to improve personal hygiene are the most important preventive measures. Hospitalization with enteric precautions and early treatment of patients and carriers are required. No strict isolation is necessary.

Description of Disease

Cholera is an epidemic diarrheal disease caused by *Vibrio cholerae* serovar O1. Other serovars of *V. cholerae* may or may not cause similar symptoms. According to past records, cholera was known as a disease mainly endemic in the Indian subcontinent and its vicinity. Since the beginning of the 19th century, however, the disease has spread beyond this border several times to cause worldwide pandemics. The present (seventh) pandemic, which started in 1961, has been a serious international health problem. Natural infection by *V. cholerae* O1 in animals is not known.

Communicability and Transmissibility

Cholera infection usually occurs through the ingestion of water or food contaminated with the feces of patients or carriers. Parenteral infection with *V. cholerae* O1 has been reported rarely, although such infection has occurred with non-O1 vibrios. Laboratory-acquired infection occurs rarely, if at all. The incubation period in cholera ranges from a few hours to 5 days (average, 2 to 3 days). After recovery, vibrios are excreted in the stool for a few days only. This carrier state may persist for several weeks or more in some cases. A chronic carrier who harbors the vibrio in the gallbladder has been reported.

Clinical Manifestations

Cholera (cholera asiatica) is a human disease caused exclusively by *V. cholerae* serovar O1 irrespective of subtype or biovar. Infections with other serovars of *V. cholerae* (non-O1 *V. cholerae*) sometimes produce a disease indistinguishable from cholera, but the symptoms are mostly mild or atypical. Infection occurs perorally, and the vibrios that survive killing by stomach acidity multiply rapidly in the small intestine. The onset of the disease is characterized by an uncertain but unpleasant feeling in the stomach and bowel, followed by a large amount of effortless, watery diarrhea. In typical cases, the stool is colorless and rice water-like, but may be yellowish or brownish. Vomiting, which is also effortless and watery, ensues. The amount of diarrhea may reach 10 liters per day, causing severe dehydration, with abnormality of plasma electrolytes (acidosis), circulatory collapse, and cyanosis. The skin becomes dry and cold; its turgor disappears, resulting in "washerwoman's hand." Other symptoms include hypotension, unpalpable pulse, tachycardia, tachypnea, anuria, and cramps of the skeletal muscles (most common in the gastrocnemius). Although patients (except children) usually remain conscious, the voice becomes weak and sometimes almost inaudible. Increases in body temperature and abdominal pain are rare symptoms. Loss of isotonic body fluid results in an increased specific gravity of the plasma, an elevation of plasma protein levels, a decreased level of bicarbonate, and a low arterial pH. Loss of body fluid is estimated to be 5% for mild, 6 to 9% for moderate, and 10% or more for severe dehydration. Plasma levels of sodium remain normal, but slight and moderate increases in levels of chloride and potassium, respectively, are observed. The illness may last from 12 h to several days, depending on the adequacy of treatment. In severe cases, death may occur within a few hours of onset, and mortality in such cases may exceed 50%. The major causes of death are hypovolemic shock, uncompensated acidosis, and uremia. Mortality may fall to 1% or less with proper treatment.

In addition to typical cases, those with mild diarrhea only are also common during outbreaks. Such mild cases, as well as asymptomatic carriers, are more frequent in outbreaks with the *eltor* biovar than in those with the classical biovar.

Pathology

Vibrio cholerae multiplies in the small intestine to 10^8/ml or more. The organism produces an enterotoxin (cholera toxin, choleragen, or cholera enterotoxin) that causes the watery diarrhea through alteration of the ion transport mechanism of mucosal cells. The presence of toxic substances other than the cholera toxin has been also reported, but their pathogenic significance remains unclear. As *V. cholerae* does not invade or destroy the mucosal cells, the epithelium remains almost intact (Gangarosa et al., 1960). Inflammatory responses are also seen in the lamina propria. In a case of infection with the *eltor* biovar, a destructive process on the mucosal surface was described (Asakura et al., 1973). In severe cases with renal failure, acute tubular necrosis may be observed.

It is difficult to produce the symptoms of human cholera in animals. However, oral feeding with *V. cholerae* produces choleralike diarrhea in the dog (Sack and Carpenter, 1969) and fluid accumulation in the small intestine and diarrhea in the suckling rabbit (Dutta and Habbu, 1955). Fluid accumulation is also produced in the intestinal loop of the rabbit (De and Chatterjee, 1953) and in other laboratory animals by direct inoculation of *V. cholerae*. The mechanism of this fluid accumulation is thought to be the same as in human cholera.

Cholera toxin is a protein with a molecular weight of about 84,000 (Finkelstein and LoSpalluto, 1969), 0.2 μg of which causes a positive loop test in the rabbit and 0.25 to 1 μg a positive Dutta test. Permeability factor, which increases the permeability of capillaries in the skin, was found by Craig (1965) in choleraic stool and culture filtrates of *V. cholerae* and later determined to be identical to cholera toxin. The cholera toxin molecule consists of two kinds of subunits, 1 A and 5 or 6 B, linked noncovalently. After the toxin molecule adsorbs onto the receptor (GM_1 ganglioside on the surface of the target cell) with subunit B, subunit A enters the cell and increases the intracellular level of cyclic AMP by activating adenylate cyclase, which in turn produces hypersecretion of electrolytes and water from the mucosal surface (Holmgren, 1978). This mechanism to increase intracellular cyclic AMP takes place in any kind of cell, with various biologic responses according to the cell type, as long as cholera toxin can adsorb onto it.

Diagnostic Features

Definitive diagnosis requires isolation of *V. cholerae* O1 from the feces or vomitus of patients. Diarrhea due to enterotoxigenic *Escherichia coli* and some strains of the non-O1 group of *V. cholerae* are difficult to distinguish from true cholera unless the bacteriologic examination is performed properly.

Prognosis

In the past, the fatality rate of cholera was as high as 50% or more. As a result of improved treatment, however, the rate has dramatically decreased, and fatality at present is less than 1%. After recovery, patients exhibit some degree of immunity against both somatic antigen and cholera toxin, but reinfection has been reported.

Epidemiologic Aspects

The main sources of cholera infection are patients and carriers. There are two types of carriers: convalescent and contact (healthy). It is believed that strains of the *eltor* biovar produce more carriers than the classical biovar, which in turn produces more severe infection (Gangarosa and Mosley, 1974). The duration of positive stool cultures after recovery is usually less than several days; in some cases, however, positive cultures continue for 3 weeks or longer, and sometimes the vibrios are harbored in the gallbladder for years with intermittent excretion into the stool (Azurin et al., 1967). Infection usually occurs through ingestion of water or food contaminated with *V. cholerae*. A review by Felsenfeld (1965) details the survival of *V. cholerae* in foods, beverages, and fomites.

Recent observations provide evidence that the vibrios survive interepidemic periods in environmental water on the surface of chitinous (crustacean) plankton (Huq et al., 1983; Nalin, 1976) and shellfish (Blake et al., 1980; Davis and Sizemore, 1982). Concentration of *V. cholerae* on the surfaces of water plants has also been reported (Spira et al., 1981). These observations may explain the mechanism of multicentric outbreaks of cholera in the endemic area and sporadic outbreaks in otherwise nonendemic areas.

In the endemic area of Bangladesh and West Bengal, attack rates are 10 times greater in children 1 to 5 years old than in adults. However, in areas where cholera has been newly introduced, attack rates are as high in adults as in children. Patients who have undergone gastrectomy or are in a hypochlorohydrial state are especially susceptible (Gitelson, 1971; Sack et al., 1972).

Etiologic Agent

Characterization

General Characteristics

Vibrio cholerae is a gram-negative rod of medium size (0.3 to 0.6 by 1.0 to 5.0 μm), usually with a commalike curvature. The organism is vigorously motile, with a monotrichous flagellum, and grows well on ordinary culture media between 30 and 40°C (optimum, 37°C). The optimal pH range for growth is 7.6 to 9.0, with limits of approximately pH 6 and 9.6. Unlike other *Vibrio* species, *V. cholerae* multiplies readily without NaCl, but cannot grow in 8% NaCl. Media with this unusual NaCl concentration are used for biologic characterization. On nutrient agar colonies are 2 to 3 mm in diameter, low convex, circular with entire edges, smooth, colorless, and slightly translucent, but become slightly brownish after prolonged growth. By means of the Lankford oblique light technique, a characteristic homogeneous structure is observed (Finkelstein and Gomez, 1963). Growth is rapid in alkaline peptone water (pH 8.2 to 9.0) and usually shows homogeneous turbidity, with abundant growth near the surface. The organism is facultatively anaerobic and catalase and indophenol oxidase positive. Acid is produced from glucose, but no gas. Lysine and ornithine are usually decarboxylized. Indole is produced from tryptophan. Since *V. cholerae* reduces nitrate to nitrite, nitrosoindole is produced from indole by adding drops of concentrated H_2SO_4 to peptone water culture; because of its color, the reaction is called *cholera red reaction,* although other bacterial species may also exhibit a positive reaction. The important biologic characteristics of *V. cholerae* and those of other pathogenic *Vibrio* species are summarized in Table 1. Refer to Bauman et al. (1984) for further descriptions.

Serologic Properties

The O1 group of *V. cholerae* ("cholera vibrio") is divided into three serotypes (subtypes): Ogawa, Inaba, and Hikojima. Each serotype has an antigenic formula: AB(c), AC, and ABC, respectively. Most authors describe Ogawa type as AB, but the pres-

TABLE 1. Characteristics of pathogenic vibrios used for differentiation[a]

Characteristic	V. cholerae *O1*	V. cholerae *non-O1*	V. mimicus	V. para-haemolyticus	V. fluvialis	V. furnissii	V. hollisae	V. algino-lyticus	V. vulnificus	V. metsch-nikovii	V. damsela
V. cholerae O1 polyvalent serum	+	−	−	−	−	−	−	−	−	−	−
Growth in NaCl											
0%	+	+	+	−	−	−	−	−	−	−	−
3%	+	+	+	+	+	+	+	+	+	+	+
8%	−	−	−	+	d	d	−	+	−	d	−
Oxidase	+	+	+	+	+	+	+	+	+	−	+
Decarboxylation											
Lysine	+	+	+	+	−	−	−	+	+	d	d
Ornithine	+	+	+	+	−	−	−	−	d	−	−
Arginine dihydrolase	−	−	−	−	+	+	−	−	−	d	+
Voges–Proskauer reaction	d	d	−	−	−	−	−	+	−	+	+
Gas from glucose	−	−	−	−	−	+	−	−	−	−	−
Acid from											
Sucrose	+	+	−	−	+	+	−	+	d	+	−
L-Arabinose	−	−	−	d	+	+	+	−	−	−	−
D-Mannitol	+	+	+	+	+	+	−	+	d	+	−
ONPG	+	+	+	−	d	d	−	−	d	d	−
Entero-pathogenicity	+	+	+	+	+	+	+	−	−	−	−

[a] + (or −) = Reaction present (or absent) in ≥90% of strains; d = reaction differs from strain to strain; ONPG = *o*-nitrophenyl-β-D-galactopyranoside.

ence of a small amount of C antigen has been suggested (Sakazaki and Donovan, 1984). Change of serotype from Ogawa to Inaba has been reported both in vitro and in vivo. There is only one type of H antigen in *V. cholerae*.

Biovars

Vibrio cholerae is divided into two biovars (biotypes), *cholerae* (classical) and *eltor*, according to biologic characteristics. Originally the strains that differed from "classical" *V. cholerae* only in their hemolysin production against goat or sheep red blood cells were called *V. eltor* (El Tor vibrio, after the place of its first isolation) and were generally considered less pathogenic or nonpathogenic. However, a cholera epidemic caused by this novel type of vibrio was recognized in Indonesia in the late 1930s, and after the outbreak of the seventh cholera pandemic since 1961 a general concensus that both belong to the same species, *V. cholerae*, was established. Differences between the two biovars are shown in Table 2. From the taxonomic point of view, such differences are regarded as insignificant (Desmarchelier and Reichelt, 1981), and some "atypical" strains have also been reported. For instance, most recent isolates of the *eltor* biovar are nonhemolytic by the original method, but hemolytic in broth with added glycerol. Furthermore, in some instances, contradictory results would be obtained in sensitivity tests with Mukerjee bacteriophage IV and to polymyxin B, which are thought to be reliable markers for differentiation (W.H.O. Scientific Working Group, 1980). Nevertheless, such terminology survives because of its usefulness for clinical or epidemiologic purposes.

Table 2. Differentiation of biovars of *Vibrio cholerae*[a]

Test	*Biovar* cholerae (*classical*)	eltor
Hemolysis (sheep red blood cells)	−	+[b]
Hemagglutination (chicken red blood cells)	−[b]	+[b]
Voges–Proskauer reaction	−[b]	+[b]
Polymyxin B (50 μg)	S	R
Bacteriophage IV	S	R

[a] +, Positive; −, negative; S, sensitive; R, resistant.
[b] Unusual results may be observed.

Specimen Collection and Transport

Fresh stool samples taken before administration of any antibiotic drug are the most suitable material. Cotton swabs are generally used for collection of specimens from acute cases. Glass rods with flattened or ground ends may be also applicable. The swabs should be inserted beyond the anus, and contamination from the perineum should be avoided. Some investigators recommend use of a sterile rectal catheter lubricated with liquid paraffin. As a cotton swab can adsorb only 0.1 to 0.2 ml of the fluid, use of two or more swabs per case is preferable. When the specimen is taken from a convalescent patient, the cotton swab should be moistened with transport medium or alkaline peptone water before application. A plastic screw-cap container with an attached spoon is also useful and especially convenient for transportation. Use of buckets or pans that could contain residual disinfectant should be avoided. If the specimen cannot be processed immediately after collection, transport medium should be used for preservation. Cary–Blair medium is recommended for this purpose, because *V. cholerae* remains viable for 5 days or more in this medium. A rectal swab or a spatula used for collecting material should be inserted into the medium. Glycerol–salt solution may also be used, but in this case *V. cholerae* cannot be expected to remain viable for more than 1 day after collection. Specimens may also be inoculated directly into alkaline peptone water (1 g/10 ml of medium) if the cultures can be incubated within 8 to 15 h after collection (see Enrichment Culture).

Microscopic Examination of Stool

Watery choleraic stool can be examined directly under the microscope. For this purpose, one drop of fresh stool is placed on a clean glass slide, covered with a cover slip, and examined with a darkfield or phase-contrast microscope at a magnification of 200× to 400×. *Vibrio cholerae* can be observed as comma- or spiral-shaped corpuscles with vigorous movement (Benenson et al., 1964). Such motility would be inhibited by adding type-specific preservative-free antiserum; thus, the serotype of the vibrio is predicted before the isolation culture is completed. The dilution of the antiserum should be determined beforehand with a fresh peptone water culture of *V. cholerae*.

The fluorescent-antibody technique is also applicable for detection of *V. cholerae* directly from the stool or after enrichment culture (Finkelstein and Gomez, 1963; Sack and Barua, 1965; Zinnaka et al., 1965). However, the need for a fluorescent microscope and good-quality conjugated sera may restrict its application. Unlike shigellosis and other inflammatory diarrheas, no leukocytes are seen in smear specimens of choleric stool.

Enrichment Culture

One gram (or 1 ml) of fresh specimen or a rectal swab is placed in 10 ml of alkaline peptone water (pH 8.6) and incubated at 37°C for 8 to 15 h (preferably 8 h). A swab preserved in Cary–Blair medium is taken out and squeezed in 0.5 to 1 ml of fresh peptone water, the remainder of which is transferred to a tube containing fresh alkaline peptone water. The cotton swab can be used after squeezing for streaking an isolation plate. The second enrichment culture may improve the isolation rate. For this purpose, several loopfuls to 0.5 ml of the primary culture is taken from the surface and transferred to the fresh alkaline peptone water. Monsur tellurite–bile salt broth is more selective than alkaline peptone water; however, longer incubation (up to 12 h) is necessary. When the specimen is directly inoculated in alkaline peptone water, the time needed for transportation should be calculated in the total incubation time, namely, incubation for 1 h at 37°C is equivalent to 2 h of transportation at 25 to 30°C or 3 h at 20°C.

Isolation and Identification Procedures

A generalized scheme for isolation of *V. cholerae* is shown in Fig. 1. Bacteriologic examination should be started as soon as possible after receiving the specimens. As seen in this scheme, for isolation of *V. cholerae*, direct plating and plating after enrichment are usually performed in parallel. In most acute cases, the vibrio can be isolated by direct plating, and preliminary identification is possible 1 day after

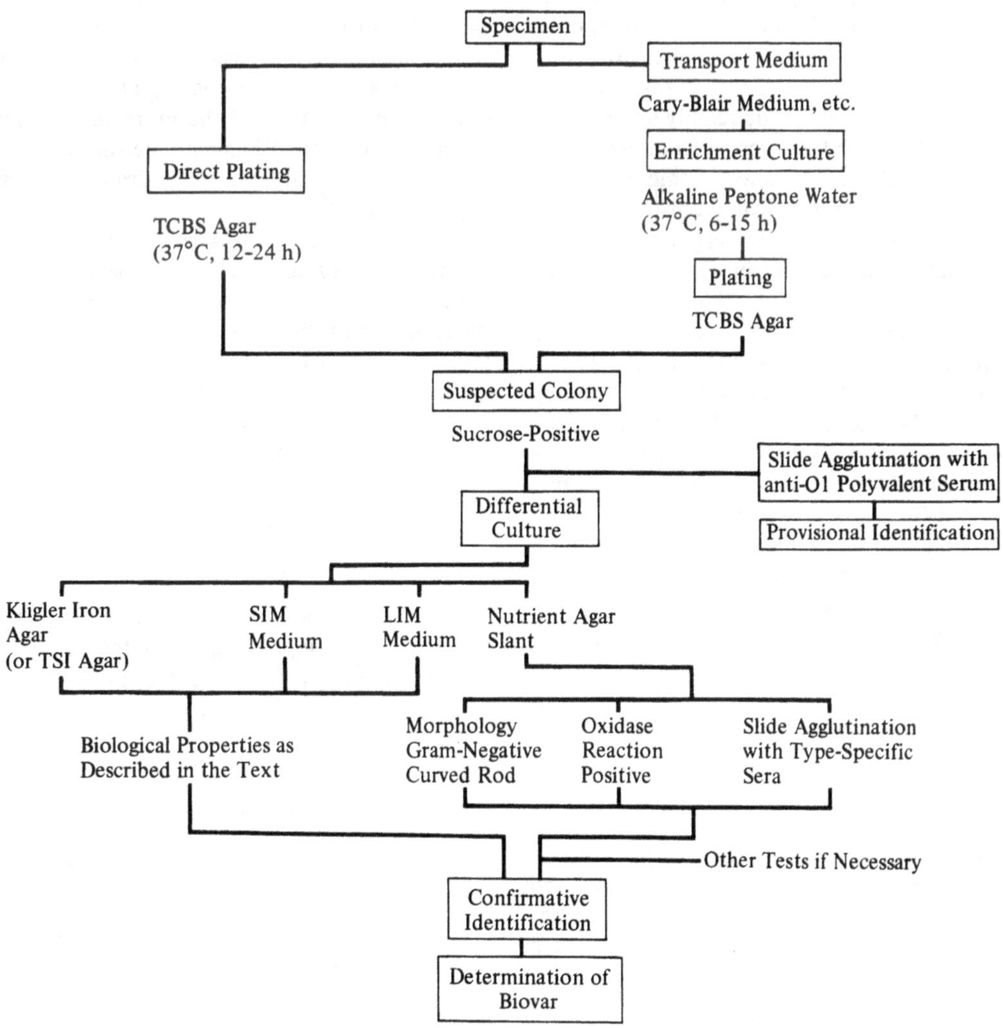

FIG. 1. Scheme for the isolation and identification of *Vibrio cholerae* O1.

arrival of the specimens. On the other hand, in specimens obtained after antibiotic treatment or from convalescent patients or carriers, the number of vibrios is usually small, and the vibrio can be isolated only after enrichment culture.

Isolation Culture

Alkaline nutrient agar (pH 8.2 to 8.6) may be used for isolation of *V. cholerae,* but selective media are much preferable. Thiosulfate–citrate–bile salt–sucrose (TCBS) agar, which is highly selective for *V. cholerae* and other pathogenic vibrios, is commercially available from several manufacturers. Most gram-negative organisms other than *Vibrio* species and gram-positive bacteria are inhibited, but because of the high selectivity of the medium, dense inoculation of the specimen is necessary. On TCBS agar, *V. cholerae* forms smooth, yellowish (sucrose-fermenting), medium-sized (2 to 3 mm in diameter) colonies after overnight incubation. Colonies of non-sucrose-fermenting vibrios (e.g., *V. parahaemolyticus* and *V. mimicus*) remain green. With the *eltor* biovar, colonies often turn from yellow to greenish after prolonged incubation. It has been said that colonies grown on TCBS agar may not be readily agglutinable with specific antiserum; however, such difficulty can be avoided by careful examination. Some laboratory strains of the classical biovar will not grow on TCBS agar because of its high selective capacity.

Taurocholate–tellurite–gelatin (TTG) agar may be used for isolation (Monsur, 1961), although its selectivity is inferior to that of TCBS agar. *Vibrio cholerae* forms small (diameter, 1 to 2 mm at 24 h, 3 to 4 mm at 48 h) translucent colonies with a cloudy zone in the medium surrounding the colonies (gelatinase positive). After 24 h, colonies exhibit dark centers due to tellurite reduction. Because the medium does not contain sucrose, the color of the colonies remains unchanged. Modified TTG medium, which contains 4-methylumbelliferyl-beta-D-galactoside (150 μg/ml), is reported to allow easy differentiation of *V. cholerae* from *V. parahaemolyticus* by detecting the β-galactosidase activity of the former through its bright blue fluorescence under ultraviolet light (O'Brien and Colwell, 1985).

Biologic Tests for Identification

Suspected colonies on selective agar plates are collected and transferred to nutrient agar slants and other media for further investigation. With cultures on nutrient agar, oxidase testing (positive), gram staining (gram-negative curved rod), and serologic testing (see below) must be performed. The media necessary for identification include Kligler iron agar (lactose negative, glucose positive, no gas production, hydrogen sulfide negative), SIM medium (hydrogen sulfide negative, indole pyruvic acid (IPA) negative, indole positive motility) and Moeller or Falkow medium containing lysine for decarboxylation (lysine positive). The latter two may be replaced by LIM medium (lysine positive, indole positive, motility). Triple sugar iron medium can be used instead of Kligler iron agar; however, keep in mind that with this medium the color of the slant changes to yellow by fermentation not only of sucrose, but of lactose. If necessary, the other characteristics listed in Table 1 should be examined.

Determination of Serotypes

For identification of *V. cholerae,* an agglutination test with polyvalent and type-specific antisera is necessary. Suspected colonies on the isolation plate or the pure culture on the slant should be subjected to slide agglutination with polyvalent O1 serum (containing antibodies to A, B, and C antigens) and type-specific Ogawa (anti-B) and Inaba (anti-C) sera successively. Agglutination of Ogawa strain is clear and rapid; that of Inaba strain is usually weak and slow. Interpretation of the results is shown in Table 3. Some Ogawa strains may be agglutinable with Inaba monospecific serum and therefore misdiagnosed as Inaba or Hikojima. Diagnostic serum should be examined with typical laboratory strains to confirm its agglutinin titer.

If the isolate is agglutinated by O1 polyvalent antiserum and has typical biologic properties, it is identified as *V. cholerae* O1. If agglutination is negative, the isolate may be a non-O1 strain. Agglutination with antisera specific to O antigens other than O1 should be performed before a final decision is made. Support from reference laboratories should be requested if necessary.

Determination of Biovar

Hemolysin Production

The result of hemolysin production by *V. cholerae* may be variable depending on the technique utilized. Therefore, the method used for the test should always be recorded with the result. The simplest tech-

Table 3. Serotypes of *Vibrio cholerae* O1[a]

Type	O antigen	Polyvalent antiserum	Factor sera: Anti-Ogawa (B)	Factor sera: Anti-Inaba (C)
Ogawa	AB(c)	+	+	−
Inaba	AC	+	−	+
Hikojima	ABC	+	+	+

[a] +, Positive; −, negative.

nique is the modified Greig test (Feeley and Pittman, 1963). The isolate is inoculated in 2 ml of heart infusion broth and incubated at 37°C for 24 h. A 5% suspension of sheep red blood cells (0.2 ml) in saline is added to the culture and reincubated at 37°C for 2 h, then kept at 4°C overnight before final reading. Most recent isolates of the *eltor* biovar do not exhibit hemolysis in this test. Therefore, addition of 1% glycerol to heart infusion broth, by which most *eltor* strains exhibit hemolysis, has been recommended (Sakazaki et al., 1971). As some strains of the classical biovar may exhibit hemolysis in the presence of glycerol, differentiation of biovar by means of hemolysin production may be of little value.

Hemagglutination Test

Chicken red blood cell suspension (2.5% in saline) is prepared by repeated centrifugation and washing of fresh chicken blood (Finkelstein and Mukerjee, 1963). One drop of the suspension is placed on a clean glass slide and mixed with a pure culture of the test strain grown on nutrient agar. Positive agglutination occurs with strains of the *eltor* biovar. Newly isolated strains of the classical biovar are usually negative in hemagglutination; however, it must be kept in mind that laboratory strains of the classical biovar maintained for long periods may not always be negative in this reaction (Zinnaka et al., 1964).

Polymyxin B Sensitivity

A loopful of overnight peptone water culture of the isolate is spread on a nutrient agar (or Mueller–Hinton agar) plate to a diameter of about 2 to 3 cm. After the liquid is absorbed into the agar layer, a sensitivity disk containing 50 units of polymyxin B is placed on it, and the plate is incubated overnight at 37°C. Classical strains are sensitive to polymyxin B, and a clear inhibitory zone around the disk is observed; *eltor* strains are resistant, however, and no inhibitory zone is formed.

Sensitivity to Bacteriophage IV

An overnight culture of the isolated strain is spread on a nutrient agar plate as described previously. One loopful of Mukerjee bacteriophage IV at a routine test dilution is placed on it and incubated overnight at 37°C. Classical strains are sensitive to this bacteriophage, and a clear lytic spot is produced. In contrast, *eltor* strains are resistant. The routine test dilution should be predetermined by spotting each of 10-fold dilutions of bacteriophage IV on the culture of the standard host strain (classical biovar).

Bacteriophage Typing

It is evident that bacteriophage typing is a very helpful tool for epidemiologic analysis of communicable diseases, and that of *V. cholerae* O1 may not be an exception. Several typing schemes have been reported; however, the practical value of these techniques may be limited because of their unstability, and the development of an improved scheme has been awaited. Among these typing schemes, that proposed recently by the Scientific Research Antiplague Institute (Rostov, Don, Soviet Union) appears to be more stable and more likely to be applicable as an epidemiologic tool than those reported previously (Drozhevkina and Artyunov, 1979). Another scheme proposed by the Public Health Laboratory (Maidstone, England) (Lee and Furniss, 1981) may not be as stable and reproducible as the Rostov scheme. The *eltor* strains of *V. cholerae* are divided into three types by the "prophage" typing proposed by Takeya and colleagues (Mukerjee and Takeya, 1974), but application of this typing system is also limited because of its small number of types.

Detection of Cholera Toxin

When the O1 group of *V. cholerae* is isolated in newly infected or nonendemic areas, determination of toxigenicity may be required to determine its epidemiologic significance. In ordinary laboratories, however, traditional methods to detect cholera toxin such as the permeability factor test (Craig, 1965), the Y1 adrenal cell or Chinese hamster ovary cell assay (Turnbull et al., 1985), the enzyme-linked immunosorbent assay (Honda et al., 1984), and the like would not be appropriate. Reverse passive agglutination with antibody-bound latex particles may be the most suitable technique for routine use (VET-RPLA detection kit, Denka-Seiken, Tokyo, distributed by Oxoid Ltd., London, England; SERO Toxi-LT, Eiken, Tokyo, Japan). For this purpose, supernatant or filtrate from an overnight shake culture of the test strain with peptone water–yeast extract broth (3% peptone, 0.5% yeast extract, 0.5% NaCl, pH 7.2) is serially diluted and incubated with antibody-bound latex suspension for 16 to 18 h at room temperature (Ito et al., 1983; Sugiyama, 1983). As the heat-labile toxin of enterotoxigenic *E. coli* shares the common antigenic determinant of cholera toxin, a positive reaction may be observed with such cultures.

Serodiagnosis

Serodiagnosis of cholera is not used routinely for diagnosis of acute cholera, but usually for epidemiologic purposes or retrospective diagnosis. A tube agglutination test with both standard Ogawa and Inaba strains of *V. cholerae* (either live or killed) is used most frequently (Barua, 1974). Microtechnique (Benenson et al., 1968b) is more convenient for a large number of samples, although it is less sensitive than the standard tube method. The complement-dependent vibriocidal test, which uses live vibrio as antigen, is more sensitive than the agglutination test. The plate method (Finkelstein, 1962) has great sensitivity, but the technique is rather complicated and time consuming for routine use. Among the modifications, microtechnique (Benenson et al., 1968a) is less sensitive but preferred because of its simplicity and ability to handle large numbers of samples. An enzyme-linked immunosorbent assay developed recently for detection of antitoxin (Levine et al., 1985) is expected to attain routine use.

Treatment, Prevention, and Control

Treatment

Prompt fluid therapy is essential for treatment of cholera. At admission of the patient, an intravenous fluid containing 5 g NaCl, 4 g sodium bicarbonate, and 1 g KCl in 1 liter of water (Dacca solution) or Ringer lactate should be infused rapidly (50 to 100 ml/min) until a strong radial pulse is restored. Subsequent infusion of this solution equal to gastrointestinal loss should be continued to maintain fluid and electrolyte balance. Overhydration should be avoided.

Peroral solution can replace the intravenous solution, especially in mild cases. In severe cases, this is also applicable, but only after the patient has recovered from hypovolemic shock. This solution, which consists of 3.5 g NaCl, 1.5 g KCl, 2.9 g trisodium citrate dihydrate or 2.5 g sodium bicarbonate, and 20 g glucose or 40 g sucrose in 1 liter of water, should be given perorally in a volume matching the fluid loss of the patient (Programme for Control of Diarrhoeal Diseases, 1984).

Although antibiotic therapy cannot improve the physiologic impairment of the patient, it dramatically reduces the duration and volume of diarrhea, as well as the duration of vibrio excretion. Oral administration of tetracyclines (40 to 50 mg/kg per day) is recommended.

Prevention and Control

Active immunization with commercially available killed whole-cell vaccine (1×10^{10}/ml) has no value in epidemic control. Two 0.5-ml injections of the vaccine, given 1 week apart, provide only partial protection (approximately 50%) for short durations (3 to 6 months) in the endemic area. Trials with other types of vaccines, including toxoid, toxin subunit, mutagenized or genetically engineered live *V. cholerae,* and combinations of different types of antigens, have also been conducted (Levine et al., 1983). Some have proved effective; side effects have also been reported, however, and none of these vaccines is commercially available to date.

As it is difficult to protect individuals by active immunization, the only measure to control cholera currently available is to block the route of infection. Isolation (hospitalization) of patients with enteric precautions is important. Disinfection of feces and vomitus as well as of linens and articles used by patients, fly control, effective handwashing by staff and attendants, and maintaining clean wards are the basic principles of patient care. Transmission of infection from patients to hospital staff has not been reported conclusively.

For management of contacts, surveillance for 5 days from last exposure should be conducted. Chemoprophylaxis is used for household contacts. Tetracycline (1 g/day for adults and 50 mg/kg per day for children for 5 days) or a single dose of doxycycline (200 mg for adults or 4 to 6 mg/kg for children) is recommended. However, mass chemoprophylaxis has never succeeded in limiting the spread of cholera.

Improvement of environmental sanitary conditions (e.g., water supplies, lavatories, and sewage systems) is very important for control of cholera. Health education should be enforced. Further information on the topics discussed in this chapter may be found in the literature (Benenson, 1985; Carpenter et al., 1977; Programme for Control of Diarrhoeal Disease, 1986).

Literature Cited

Asakura, H., A. Morita, T. Morishita, M. Tsuchiya, H. Fukumi, M. Ohashi, C. Uylangco, and A. Castro. 1973. Pathologic findings from intestinal biopsy specimens in human cholera. Am. J. Dig. Dis. **18:**271–278.

Azurin, J. C., K. Kobari, D. Barua, M. Alvero, C. Z. Gomez, J. J. Dizon, E. Nakano, R. Suplido, and L. Ledesma. 1967. A long-term carrier of cholera: Cholera Dolores. Bull. W.H.O. **37:**745–749.

Barua, D. 1974. Laboratory diagnosis of cholera, p. 85–126. *In* D. Barua and W. Burrows (ed.), Cholera. The W. B. Saunders Co., Philadelphia.

Baumann, P., A. L. Furniss, and J. V. Lee. 1984. Vibrio, p. 518–538. *In* N. R. Krieg and J. G. Holt (ed.), Bergey's manual of systematic bacteriology, vol. 1. The Williams & Wilkins Co., Baltimore.

Benenson, A. S. (ed.). 1985. Control of communicable diseases in man, 14th ed. The American Public Health Association, Washington, D.C.

Benenson, A. S., M. R. Islam, and W. B. Greenough. 1964. Rapid identification of *Vibrio cholerae* by darkfield microscopy. Bull. W.H.O. **30:**827–831.

Benenson, A. S., A. Saad, and W. H. Mosley. 1968a. Serological studies in cholera. 2. The vibriocidal antibody response of cholera patients determined by a microtechnique. Bull. W.H.O. **38:**277–285.

Benenson, A. S., A. Saad, and M. Paul. 1968b. Serological studies in cholera. 1. Vibrio agglutinin response of cholera patients determined by a microtechnique. Bull. W.H.O. **38:**267–276.

Blake, P. A., D. T. Allegra, J. D. Snyder, T. J. Barett, L. McFarland, C. T. Caraway, J. C. Feeley, J. P. Craig, J. V. Lee, N. D. Puhr, and R. A. Feldman. 1980. Cholera—a possible endemic focus in the United States. N. Engl. J. Med. **302**:305–309.

Carpenter, C. C. J., A. A. F. Mahmoud, and K. S. Warren. 1977. Algorithms in the diagnosis and management of exotic diseases. XXVI. Cholera. J. Infect. Dis. **136**:461–464.

Craig, J. P. 1965. A permeability factor (toxin) found in cholera stools and culture filtrates and its neutralization by convalescent cholera sera. Nature (London) **207**:614–616.

Davis, J. W., and R. K. Sizemore. 1982. Incidence of *Vibrio* species associated with blue crabs (*Callinectus sapidus*) collected from Galveston Bay, Texas. Appl. Environ. Microbiol. **43**:1092–1097.

De, S. N., and D. N. Chatterjee. 1953. An experimental study of the mechanism of action of *Vibrio cholerae* on the intestinal mucous membrane. J. Pathol. Bacteriol. **66**:559–562.

Desmarchelier, P. M., and J. L. Reichelt. 1981. Phenotypic characterization of clinical and environmental isolates of *Vibrio cholerae* from Australia. Curr. Microbiol. **5**:123–127.

Drozhevkina, M. S., and Y. I. Artyunov. 1979. Phage-typing of *Vibrio cholerae* using a new collection of phages. J. Hyg. Epidemiol. Microbiol. Immunol. (Prague) **23**:340–347.

Dutta, N. K., and M. K. Habbu. 1955. Experimental cholera in infant rabbits: a method of chemotherapeutic investigation. Br. J. Pharmacol. **10**:153–159.

Feeley, J. C., and M. Pittman. 1963. Studies on the hemolytic activity of El Tor vibrios. Bull. W.H.O. **28**:347–356.

Felsenfeld, O. 1965. Notes on food, beverages and fomites contaminated with *V. cholerae*. Bull. W.H.O. **33**:725–734.

Finkelstein, R. A. 1962. Vibriocidal antibody inhibition (VAI) analysis: a technique for the identification of the predominant vibriocidal antibodies in serum and for the detection and identification of *V. cholerae* antigens. J. Immunol. **89**:264–271.

Finkelstein, R. A., and C. Z. Gomez. 1963. Comparison of methods for the rapid recognition of cholera vibrios. Bull. W.H.O. **28**:327–332.

Finkelstein, R. A., and J. J. LoSpalluto. 1969. Pathogenesis of experimental cholera. Preparation and isolation of choleragen and choleragenoid. J. Exp. Med. **130**:185–202.

Finkelstein, R. A., and S. Mukerjee. 1963. Hemagglutination: a rapid method for differentiating *V. cholerae* and El Tor vibrios. Proc. Soc. Exp. Biol. Med. **112**:355–359.

Gangarosa, E. J., W. R. Beisel, C. Benyajati, H. Sprinz, and P. Piyaratn. 1960. The nature of the gastrointestinal lesion in asiatic cholera and its relation to pathogenesis: a biopsy study. Am. J. Trop. Med. Hyg. **9**:125–135.

Gangarosa, E. J., and W. H. Mosley. 1974. Epidemiology and surveillance of cholera, p. 381–403. *In* D. Barua and W. Burrows (ed.), Cholera. The W. B. Saunders Co., Philadelphia.

Gitelson, S. 1971. Gastrectomy, achlorohydria and cholera. Isr. J. Med. Sci. **7**:663–667.

Holmgren, J. 1978. Cholera toxin and the cell membrane, p. 333–366. *In* J. Jeljaszwicz and T. Waldstroem (ed.), Bacterial toxins and cell membrane. Academic Press, Inc. (London), Ltd., London.

Honda, T., M. Sato, and T. Miwatani. 1984. Differential detection of cholera enterotoxin and *Escherichia coli* heat-labile enterotoxin by enzyme-linked immunosorbent assays with antibodies specific to the two toxins. J. Clin. Microbiol. **20**:664–667.

Huq, A., E. B. Small, P. A. West, M. I. Huq, R. Rahman, and R. Colwell. 1983. Ecological relationships between *V. cholerae* and planktonic crustacean copepods. Appl. Environ. Microbiol. **45**:275–283.

Ito, T., S. Kuwahara, and T. Yokota. 1983. Automatic and manual latex agglutination tests for measurement of cholera toxin and heat-labile enterotoxin of *Escherichia coli*. J. Clin. Microbiol. **17**:7–12.

Lee, J. V., and A. L. Furniss. 1981. Discussion 1: the phage-typing of *Vibrio cholerae* serovar O1, p. 119–122. *In* T. Holme, J. Holmgren, M. H. Merson, and R. Moellby (ed.), Acute enteric infections in children. New prospects for treatment and prevention. Elsevier North-Holland Publishing Co., Amsterdam.

Levine, M. M., J. B. Kaper, R. E. Black, and M. L. Clements. 1983. New knowledge on pathogenesis of bacterial enteric infections as applied to vaccine development. Microbiol. Rev. **47**:510–550.

Levine, M. M., C. R. Young, R. E. Black, Y. Takeda, and R. A. Finkelstein. 1985. Enzyme-linked immunosorbent assay to measure antibodies to purified heat-labile enterotoxins from human and porcine strains of *Escherichia coli* and to cholera toxin: application in serodiagnosis and seroepidemiology. J. Clin. Microbiol. **21**:174–179.

Monsur, K. A. 1961. A highly selective gelatin taurocholate-tellurite medium for isolation of *Vibrio cholerae*. Trans. R. Soc. Trop. Med. Hyg. **55**:440–442.

Mukerjee, S., and K. Takeya. 1974. Vibriophages and vibriocin, p. 61–83. *In* D. Barua and W. Burrows (eds.), Cholera. The W. B. Saunders Co., Philadelphia.

Nalin, D. R. 1976. Cholera, copepods, and chitinase. Lancet **2**:958.

O'Brien, M., and R. Colwell. 1985. Modified taurocholate-tellurite-gelatin agar for improved differentiation of *Vibrio* species. J. Clin. Microbiol. **22**:1011–1013.

Programme for Control of Diarrhoeal Diseases. 1984. A manual for treatment of acute diarrhoea. WHO/CDD/SER/80.2, rev.1. World Health Organization, Geneva.

Programme for Control of Diarrhoeal Diseases. 1986. Guideline for cholera control. WHO/CDD/SER/80.4, rev. 1. World Health Organization, Geneva.

Sack, G. H., N. F. Pierce, K. N. Hennessey, R. C. Mitra, R. B. Sack, and D. N. Guha Mazumuder. 1972. Gastric acidity in cholera and noncholera diarrhea. Bull. W.H.O. **47**:31–36.

Sack, R. B., and D. Barua. 1965. The fluorescent antibody technique in the direct examination of cholera stool, p. 50–56. *In* O. A. Bushnell and C. S. Brookhyser (eds.), Proceedings of the Cholera Research Symposium, Honolulu, January 24–29, 1965. U.S. Public Health Service Publication No. 1328. U.S. Government Printing Office, Washington, D.C.

Sack, R. B., and C. C. J. Carpenter. 1969. Experimental canine cholera. 1. Development of the model. J. Infect. Dis. **119**:138–149.

Sakazaki, R., and T. J. Donovan. 1984. Serology and epidemiology of *Vibrio cholerae* and *Vibrio mimicus*, p. 271–289. *In* T. Bergan (ed.), Methods in microbiology, vol. 16. Academic Press, Inc. (London), Ltd., London.

Sakazaki, R., K. Tamura, and M. Murase. 1971. Determination of the hemolytic activity of *V. cholerae*. Jpn. J. Med. Sci. Biol. **24**:83–91.

Spira, W. M., A. Huq, Q. S. Ahmed, and Y. A. Saad. 1981. Uptake of *Vibrio cholerae* biotype *eltor* from contaminated water by water hyacinth (*Eichornia crassipes*). Appl. Environ. Microbiol. **42**:550–553.

Sugiyama, J. 1983. Studies on the reversed passive latex agglutination for the detection of enterotoxin. Part 1. Detection of cholera toxin and heat-labile enterotoxin of *Escherichia coli*. Yamaguchi Igaku **32**:259–267.

Turnbull, P. C. B., J. V. Lee, M. D. Miliotis, C. S. Still, M. Isacson, and Q. S. Ahmed. 1985. In vitro and in vivo cholera toxin production by classical and El Tor isolates of *Vibrio cholerae*. J. Clin. Microbiol. **21**:884–890.

W.H.O. Scientific Working Group. 1980. Cholera and other vibrio-associated diarrhoeas. Bull. W.H.O. **58**:353–374.

Zinnaka, Y., S. Shimodori, and K. Takeya. 1964. Hemagglutinating activity of *Vibrio comma*. Jpn. J. Microbiol. **8**:97–103.

Zinnaka, Y., S. Shimodori, and K. Takeya. 1965. Application of fluorescent antibody technique to the detection of cholera vibrio. Jpn. J. Infect. Dis. **39**:51–58.

CHAPTER 19

Miscellaneous Clostridial Infections

SYDNEY M. FINEGOLD and W. LANCE GEORGE

Disease: Tetanus, botulism, gangrenous cholecystitis, postabortal sepsis, clostridial infection associated with malignancy, common or miscellaneous infections, and enteric infections, including pseudomembranous colitis, enteritis necroticans, and food poisoning.
Etiologic Agents: Numerous species of *Clostridium*.
Source: Indigenous flora of host (chiefly that of the intestinal and female genital tracts) and, in special circumstances, food and the hospital and other environments such as soil.
Clinical Manifestations: Very variable. Classic findings of tetanus or botulism; sepsis with or without shock and intravascular hemolysis; wound infection, pneumonia, and the like; tissue necrosis or gangrene; gas in tissues; severe diarrhea; mild food poisoning.
Pathology: Variable. Purulence, abscess formation, tissue invasion with necrosis and gas production, pseudomembrane or plaque formation in bowel mucosa.
Laboratory Diagnosis: Recovery of various clostridial species, alone or together with other organisms, from appropriate specimens in patients with compatible illness.
Epidemiology: Pertinent in cases of tetanus, botulism, and enteric diseases involving clostridia. Implicated may be soil, food, the hospital environment, and antimicrobial agents.
Treatment: Surgical debridement and drainage, antimicrobial agents, and antitoxin.
Prevention and Control: Immunization, proper preparation and storage of food, aseptic surgical technique, and good infection control procedures.

Description of Diseases

Clostridia are anaerobic gram-positive spore-forming bacilli. However, some species (*C. carnis*, *C. histolyticum*, and *C. tertium*) grow aerobically and others become aerotolerant on subculture. Of the 85 species recognized in the 9th edition of *Bergey's Manual of Systematic Bacteriology*, 20 are important pathogens in humans. The organisms are widely distributed in nature, but the most commonly encountered clostridial diseases in humans are of endogenous origin. Clostridia are involved in serious intoxications such as tetanus and botulism and also in a variety of infections that are variable in their manifestations and severity.

Pathogenesis

Clostridial infection of soft tissue, like other anaerobic infections, occurs primarily in an anaerobic environment due to impaired circulation, trauma, surgery, or malignancy or other disease. A breach in the mucosal surface of the intestinal or female genital tract permits endogenous clostridial flora to enter the tissues to set up the infection. The majority of clostridial infections are mixed infections involving other anaerobes as well as aerobic and/or facultative bacteria.

Capsules may be important in protecting clostridia, especially *C. perfringens*, from the body's host defenses. The major virulence factors, however, are exotoxins produced by many clostridial species. The toxins produced by *C. tetani* and *C. botulinum* are among the most potent poisons known to humans. Tetanus toxin (tetanospasmin) binds to gangliosides in the central nervous system and suppresses the central inhibitory balancing influences on motor neuron activity; this leads to spasticity and convulsions and intensified reflex responses to stimuli. This toxin also acts on the sympathetic nervous system and on the neurocirculatory and neuroendocrine systems. Botulinal toxin attaches to individual motor nerve terminals to prevent acetylcholine release at nerve endings. Other species of clostridia produce numerous toxins. For example, *C. perfringens* produces an α-toxin (a lecithinase, phospholipase C); this toxin is a hemolysin, is also active against white blood cells and platelets, and has necrotizing activity. This organism produces many other toxins—a proteolytic enzyme, an RNAse, another hemolysin, a collagenase, hyaluronidase, neuraminidase, an enterotoxin, and toxins producing increased capillary permeability. The β-toxin of type C *C. perfringens* is the key factor in necrotizing enterocolitis. *C. difficile* produces at least three toxins—a cytotoxin, an enterotoxin, and a motility-altering toxin. A number of other clostridia produce important toxins.

The normal bowel flora is certainly an important defense mechanism against infant (and probably some cases of adult) botulism and against pseudomembranous colitis due to *C. difficile* and certain other enteric infections or intoxications due to clostridia. Infant botulism is not seen after the age of 8 months, by which time the normal intestinal flora is well established. Antitoxic immunity is important in preventing tetanus, botulism, and enteritis necroticans and may be important in other situations as well.

Clinical Manifestations

Clostridia are involved in a wide variety of disease processes; the clinical manifestations vary widely according to the nature of the specific infection or intoxication.

Tetanus

Early in the course of tetanus there is tension or cramping and twitching in muscles about a wound, tightness of jaw muscles, and mild pain in facial muscles. With more advanced disease, there is trismus (lockjaw), an inability to open the mouth because of spasm of the masseter muscles. Facial muscles may go into spasm producing the characteristic grotesque grinning expression known as risus sardonicus. Spasms or contractions of muscles of the trunk and extremities may be widespread resulting in board-like rigidity, painful intermittent tonic convulsions, and opisthotonus (backward bowing of the back so that the heels and the back of the head approach each other). Extended painful tonic convulsions may be precipitated by even minor external stimuli. There may be spasm of pharyngeal and laryngeal muscles leading to difficulty in swallowing, cyanosis, and even respiratory arrest and death.

Botulism

As in the cases of tetanus, patients have no fever and have clear mental processes. There is a distinctive descending symmetrical motor paralysis first involving muscles supplied by the cranial nerves. Common symptoms include diplopia, dysarthria, and dysphagia. Pupils are often dilated and fixed. Mucous membranes of the mouth, tongue, and pharynx may be extremely dry and painful. Gastrointestinal symptoms are variable and are seen in about one-third of patients.

Gangrenous Cholecystitis

Clostridia may be involved in acute cholecystitis indistinguishable from that involving other organisms such as *E. coli*, but may lead to a gangrenous and/or gas-forming process (emphysematous cholecystitis) or widespread visceral gas gangrene.

Postabortal Sepsis

Clostridia are involved in only a small percentage of postabortal infections, but may produce the most dramatic and severe illness of this type. With clostridial sepsis due to *C. perfringens*, there may be a very dramatic clinical picture consisting of hemolytic anemia, hemoglobinemia, hemoglobinuria, disseminated intravascular coagulation, hyperbilirubinemia, shock, and anuria.

Clostridial Infection and Malignancy

There is a distinct association between "spontaneous" myonecrosis or lesser infection with *C. septicum* or less frequently, *C. perfringens* and malignancy or other disease of the cecum.

Role in Common or Miscellaneous Infections

Although most of the clinical pictures described are dramatic and even life threatening, *C. perfringens* and a number of other clostridia are more commonly involved in infections of diverse types throughout the body, of the type caused by nonspore-forming anaerobes and other bacteria. Included among these are such conditions as brain abscess, subdural empyema, aspiration pneumonia, thoracic empyema, intraabdominal infection, infection related to gynecologic disease or surgery, wound infection after other types of surgery, and soft tissue infections. Some of these types of infections may be complicated by bacteremia.

Pseudomembranous Colitis

C. difficile is now known to be the primary cause of pseudomembranous colitis related to antimicrobial or antineoplastic therapy. Rarely, cases of this disease are caused by type C *C. perfringens* and other clostridia, as well as by other organisms. Nonspecific colitis after antimicrobial therapy is probably also due to *C. difficile* commonly and diarrhea without colitis may develop on occasion. There is also evidence to implicate type A *C. perfringens* in this latter entity.

Enteritis Necroticans

This condition, known as "pig-bel" in New Guinea, is a severe enteric infection due to type C *C. perfringens*. The disease is often fulminant with abdominal pain, bloody diarrhea, vomiting, peritonitis, and shock. There may be gangrene of the bowel.

Food Poisoning

This common entity involves certain serotypes of type A *C. perfringens*. It is a benign, self-limited process usually lasting only 24 hours. Symptoms include abdominal pain, nausea, and acute watery diarrhea without fever.

Pathology

The pathology varies with the specific entity. The most dramatic changes are seen in clostridial myonecrosis (gas gangrene); this is discussed in another chapter. In other kinds of infection, there are the typical changes seen in other anaerobic infections—invasion of polymorphonuclear leukocytes as part of the inflammatory response, abscess formation, varying degrees of tissue necrosis or gangrene, and gas in the tissues at times. In the case of enteritis necroticans, there may be frank gangrene of the bowel, usually segmental but complete. In pseudomembranous colitis, the hallmark of the disease is the yellow elevated plaque, with or without frank pseudomembrane. In so-called nonspecific colitis secondary to antimicrobial therapy, findings vary from mild inflammatory changes to hemorrhagic or severe inflammatory changes simulating chronic ulcerative colitis. With pseudomembranous colitis, complications include toxic megacolon and perforation.

Diagnostic Features

The diagnosis of tetanus and botulism are made on clinical grounds; this is not difficult in the case of well-developed and classical findings, but may be a problem early in the course of illness. Botulism may be confirmed by demonstration of the toxin in the serum, gastric contents, feces, and/or material from underlying wounds or in implicated food. Gangrenous cholecystitis may be suspected clinically and confirmed at surgery. Emphysematous cholecystitis may be diagnosed by radiographic, ultrasound, or computed tomographic studies before surgery. The dramatic picture of postabortal sepsis with intravascular hemolysis described above is diagnostic. Blood cultures may be used to confirm the diagnosis; Gram stain of the buffy coat may be positive. The other infections, aside from those involving the bowel, are diagnosed in the same way that brain abscess, thoracic empyema, and so on are ordinarily diagnosed. Gram stain may clearly establish the presence of clostridia in appropriate specimens. Culture of blood and of other body fluids or exudates will provide definitive diagnosis.

There are various tests available for detection of the cytopathic toxin or enterotoxin of *C. difficile,* and the organism itself may be cultured in cases of pseudomembranous colitis. However, the organism is commonly present in patients on antibiotic therapy, even in the absence of gastrointestinal symptoms. Even the toxins may be found in the absence of disease in patients on antibiotics or in infants. The definitive way to make the diagnosis is to demonstrate the pathognomonic yellow plaques and/or pseudomembrane on colonoscopy or proctoscopy. Enteritis necroticans is readily diagnosed because of the distinctive clinical picture and because it virtually always occurs in the setting of massive pork feasting. The organism can be isolated from the bowel and from the pigs feasted upon and antibody to the β-toxin can be demonstrated. *C. perfringens* food poisoning is diagnosed on epidemiologic grounds and can be confirmed by demonstrating high counts of the offending organism in the implicated food as well as in the stools of patients.

Prognosis

The prognosis is relatively poor in many clostridial disease processes. In particular, the intoxications have a guarded prognosis. In tetanus, the mortality remains high—often more than 50%, especially in the very young and very old and in developing nations. The prognosis is better in botulism, but patients must be managed carefully in an intensive care unit. Obviously, the sooner the diagnosis is suspected and appropriate therapy undertaken, the better the prognosis.

Prognosis is also poor in patients with gangrenous or emphysematous cholecystitis, in postabortal sepsis with intravascular hemolysis, and in *C. septicum* or *C. perfringens* infections in patients with cecal malignancy—again varying with the speed with which the diagnosis is suspected and appropriate therapy given. In the other nonenteric infections, the prognosis varies with the nature, severity, and duration of the infection and the age and condition of the patient. Death is now uncommon in pseudomembranous colitis, but 20 to 25% of patients suffer relapses after therapy is stopped; there may be multiple relapses. Enteritis necroticans is a severe illness, but the prognosis has improved with the advent of better therapy and early surgery, when indicated. Clostridial food poisoning is a very benign disease.

Epidemiologic Aspects

There are seven antigenically different botulinal toxins designated by the letters A through G; types A, B, and E are the major causes of human botulism. Food products responsible for outbreaks of botulism include vegetables, fish, fruits, and condiments. Peppers are involved commonly and mushrooms have been common among commercially prepared foods. Few outbreaks have been related to beef, milk products, pork, poultry, and other vehicles. Most of the fish-related outbreaks have been due to type E. In adults, most disease has been thought to be due to ingestion of preformed toxin. However, local production of toxin in a wound infected by *C. botulinum* may lead to botulism, and there is increasing evidence to suggest that, even in adults, the organism may be ingested and produce toxin in the gastrointestinal tract. The latter mechanism is the one that operates in infant botulism; implicated foods are honey and corn syrup.

The spores of *C. tetani* are widespread and extremely resistant, even to autoclaving. The organism is found in the gastrointestinal tract of humans and various animals. The bacterium enters the body through wounds ranging from severe to trivial and unnoticed. No wound is found in 10 to 20% of cases. Decubitus or other cutaneous ulcers may serve as portals of entry. Tetanus may occur postoperatively, after childbirth, or in relation to infection such as otitis media. Two-thirds of predisposing injuries occur in the home and 20% in farms or gardens.

The epidemiology of *C. difficile*-induced colitis is not yet fully understood. It is certainly clear now, however, that this may be a hospital-acquired disease and that patients with it should be isolated. Why only some of the many patients receiving antimicrobial agents develop this disease remains unknown. In any case, clusters or outbreaks are relatively common in addition to sporadic cases. A number of typing systems for *C. difficile* have been developed that should ultimately lead to improved knowledge of the associated disease. These involve agglutination tests, polyacrylamide gel electrophoresis, Western blots, and so on. Enteritis necroticans is seen only sporadically in Western countries but is still endemic in New Guinea. Susceptible natives have been on a low-protein diet previously; this predisposes to low levels of digestive proteases so that they cannot degrade the β-toxin of the clostridia involved. The proteases are also blocked by ingestion of trypsin inhibitors in sweet potatoes, ingested as part of the pork feast. Clostridial food poisoning often occurs in outbreaks related to ingestion of contaminated meat, poultry, or gravy. Common features of outbreaks are reheating of food that has been previously cooked and then cooled slowly or left unrefrigerated; this permits germination of spores and rapid multiplication of bacterial cells. The vegetative cells sporulate within the gastrointestinal tract and release enterotoxin.

With the other types of clostridial infection under consideration in this chapter, the organism is found in the indigenous flora and escapes to set up infection in the manner outlined in the section on pathogenesis.

Etiologic Agents

Characterization

As noted earlier, some species of clostridia are capable of aerobic growth and others become aerotolerant on subculture. Differentiation between aerotolerant clostridia and *Bacillus* species is based on the findings that *Clostridium* species sporulate anaerobically only, grow much better anaerobically, and are almost always catalase-negative, whereas *Bacillus* species sporulate aerobically only, usually grow better aerobically, and are usually catalase-positive. A few species of clostridia are gram-negative, and older cultures may appear gram-negative. A further problem is that sporulation may be difficult to demonstrate in some species.

On the basis of the disease, syndromes they produce or are involved in, clostridia may be placed in five groups:

1. The gas gangrene group. This includes type A *C. perfringens*, *C. novyi*, *C. septicum*, *C. bifermentans*, *C. histolyticum*, *C. sordellii*, *C. sporogenes*, and others.
2. *C. tetani*, the etiologic agent of tetanus.

3. The botulism group. This includes various types of *C. botulinum* as well as strains of such organisms as *C. barati* and *C. butyricum* that produce an identical toxin.
4. The enteric group. This includes *C. difficile* and other clostridia that much less commonly produce antimicrobial-induced pseudomembranous colitis and types A and C *C. perfringens*.
5. The miscellaneous infection group (wound infection, abscesses, bacteremia, and the like). This group includes *C. perfringens, C. ramosum, C. bifermentans, C. sphenoides, C. sporogenes*, and a number of others.

Specimen Collection and Transport

The importance of and techniques for optimum specimen collection and transport are covered in the chapter on Anaerobic Bacterioses.

Isolation and Identification Procedures

Isolation and identification procedures for the clostridia are covered in the chapter on Anaerobic Bacterioses as part of a discussion on bacteriologic procedures for anaerobes in general.

Preliminary Identification

The colonial and microscopic morphology of clostridia vary widely. As noted earlier, it may be difficult to determine that they are anaerobes, that they are gram-positive, and that they are sporeformers. However, in some cases the morphology is distinctive enough to suggest that a clostridium is involved and even to indicate a particular species. Examples of such unique morphology include medusa-head colonies, the double zone of hemolysis around colonies of *C. perfringens* on sheep blood agar, lipase- and lecithinase-positive colonies on egg yolk agar, the large colonies that fluoresce chartreuse on cycloserine-cefoxitin-fructose agar (a highly selective medium for *C. difficile*), the box car-like shape of *C. perfringens* (especially in a clinical Gram stain in which the white blood cells have been severely damaged or destroyed by the organism's toxin), and gram-positive rods whose cells clearly contain spores. There are tests available for detection of toxins or other products of *C. difficile*, such as the cytotoxicity test in tissue culture and a latex agglutination test. The use of the Nagler plate, the reverse cAMP test, the ethanol and heating tests for spores, and the use of antibiotic identification disks are discussed in the *Wadsworth Anaerobic Bacteriology Manual* (Sutter et al., 1985).

Definitive Identification and Biotyping

Cultural and biochemical reactions, along with shape and location of spores, as indicated in Tables 1 and 2 will be useful for the diagnosis and identification in the vast majority of commonly encountered clostridial infections. There is a mouse lethality test and also an ELISA procedure for detection and confirmation of botulinal toxin in gastrointestinal contents and in implicated food specimens. In enteritis necroticans, β-toxin or antibody to it may be sought. As noted earlier, in clostridial food poisoning, epidemiologic studies of outbreaks would demonstrate high counts of a certain type of *C. perfringens* in the implicated food and in the feces of the patients.

Detection of *C. difficile* and its toxins or products has already been mentioned. Biotyping of *C. difficile* strains has been used on an investigational basis for outbreaks and for other epidemiologic studies. Typing systems include unique biochemical characteristics or antibiograms, agglutination tests, agarose electrophoresis for detection of plasmid DNA, crossed immunoelectrophoresis for detection of extracellular antigens and toxins, polyacrylamide gel electrophoresis of soluble proteins, bacteriophages, bacteriocins, electroblot transfer, immunoblotting, and combinations of various of the aforementioned systems.

TABLE 1. Characteristics of Nagler-positive *Clostridium* species

	Indole production	*Gelatin hydrolysis*	*Urease*	*Growth enhanced by 1% mannose*	*Phenylacetic acid produced*	*Glutamic acid decarboxylase*
C. perfringens	–	+	–			
C. barati	–	–	–			
C. bifermentans	+	+	–	–	–	–[a]
C. sordellii	+	+	+–	+	V	+

[a] Only a few strains have been tested.
From Sutter et al., Wadsworth anaerobic bacteriology manual, 4th ed., 1985.
See Table 2 for key.

TABLE 2. Characteristics of gram-positive spore-forming bacilli[a]

	Gelatin hydrolysis	Glucose fermentation	Lecithinase	Lipase	Indole	Butyric acid produced in PYG	Isoacids produced in PYG	Aerobic growth	Urease
Saccharolytic—proteolytic									
C. bifermentans	+	+	+	−	+	(+)	(+)	−	−
C. sordellii	+	+	+	−	+	(+)	(+)	−	$+^{-}$
C. perfringens	+	+	+	−	−	+	−	−	
C. novyi type A	+	+	+	+	−	+	−	−	
C. sporogenes	+	+	−	+	−	+	+	−	
C. cadaveris	+	+	−	−	+	+	−	−	
C. septicum	+	+	−	−	−	+	−	−	
C. difficile	+	+	−	−	−	+	+	−	
C. putrificum	+	+	−	−	−	+	+	−	
Saccharolytic—nonproteolytic									
C. barati	−	+	+	−	−	+	−	−	
C. tertium	−	+	−	−	−	+	−	+	
C. butyricum	−	+	−	−	−	+	−	−	
C. innocuum	−	+	−	−	−	+	−	−	
C. ramosum	−	+	−	−	−	−	−	−	
C. clostridiiforme	−	+	−	−	$-^{+}$	−	−	−	
Asaccharolytic—proteolytic									
C. tetani	+	−	−	−	V	+	−	−	
C. hastiforme	+	−	−	−	−	+	+	−	
C. subterminale	+	−	$-^{+}$	−	−	+	+	−	
C. histolyticum	+	−	−	−	−	−	−	$+^{-}$	
C. limosum	+	−	+	−	−	−	−	−	

[a] *C. botulinum* types vary in proteolytic, saccharolytic, and lipase reactions. Send suspected isolates or suspected *C. Botulinum*-containing material to the appropriate local or state agency.
From Sutter et al., Wadsworth anaerobic bacteriology manual, 4th ed., 1985.
Reactions: − = negative reaction; + = positive reaction for majority of strains and includes weak as well as strong acid production from carbohydrates in saccharolytic organisms; $+^{-}$ = most strains positive, reaction helpful if positive; $+^{w}$ = most strains positive, some weakly positive; w^{-} = most strains weakly positive, some negative; $-^{+}$ = most strains negative, some positive; $-^{w}$ = most strains negative, some weakly positive; V = variable reaction; (+) = delayed reaction; c = clot formed in milk; d = milk digested.

Treatment, Prevention, and Control

Antimicrobial Susceptibility

The usual patterns of susceptibility of clostridia to some of the more commonly used antimicrobial agents are noted in Table 3. In general, the most resistant clostridial species are *C. difficile* and *C. ramosum*. Penicillin G is generally excellent against clostridia, but some strains (including some of *C. perfringens*) are resistant, and others, such as some strains of *C. ramosum*, are relatively resistant (MICs as high as 8 units/ml). *Clostridium ramosum*, often overlooked or misidentified, is commonly encountered in infection. About 15% of strains of this species are highly resistant to clindamycin and many strains are resistant to tetracyclines and erythromycin. Some 20 to 30% of strains of a number of clostridial species other than *C. perfringens* are resistant to clindamycin. One-third of clostridia other than *C. perfringens* are resistant to cefoxitin. Metronidazole and chloramphenicol are essentially always active against clostridia. Techniques for antimicrobial susceptibility studies with anaerobes are discussed in the chapter on Anaerobic Bacterioses.

Treatment

There are several approaches to therapy of clostridial infections and intoxications. In infection, surgical debridement of necrotic tissue, drainage of abscesses, and other important surgical principles are generally the most important measures to be used. In lesser infections, surgical intervention may not be required. In certain situations, such as intraabdominal abscess, it may be feasible to drain collections of pus percutaneously under guidance of ultrasound or computed tomographic scans; a catheter may be left in place in the abscess, if indicated. Antimicrobial therapy is essentially always important.

Milk reaction	Lactose	Maltose	Fructose	Cellobiose	Arabinose	Mannose	Xylose	Nitrate reduction	Spore location	End products from PYG
d	–	W^-	V	–	–	$-^w$	–	–	OS	A (p ib b iv ic l s)
d	–	W^+	V	–	–	$-^w$	–	–	OS	A (p ib b iv IC l)
d^c	+	+	+	$-^+$	–	+	–	$+^-$	—	A B L (p s)
c	–	V	$-^w$	–	–	–	–	–	OS	A P B
d	–	$-^w$	$-^w$	–	–	–	–	–	OS	A B ib iv (p v ic l s)
cd	–	–	V	–	–	$-^w$	–	–	OT	A B (l s)
cd	+	+	+	$+^w$	–	+	–	V	OS	A B (p l)
–	–	–	+	$+^w$	–	$+^-$	$-^w$	–	OS	A ib B iv ic (v l)
d	–	$-^w$	$-^w$	–	–	–	–			A ib B iv (p v ic l s)
c	$+^w$	$+^w$	+	+	–	+	–	V		A B L (s)
c	+	+	+	+	–	+	$+^-$	$+^-$	OT	A B L (s)
c	+	+	+	+	$+^-$	+	+	–	OS	A B F (l s)
–	–	–	+	+	–	+	$-^w$	–	OT	a B L (s)
c	+	+	+	+	–	+	$-^w$	–	R/OT	A L (s)
c	$+^-$	$+^w$	+	$+^-$	V	$+^w$	+	$+^-$	OS	A (l s)
d^-	–	–	–	–	–	–	–	–	RT	A P B (p l s)
$-^d$	–	–	–	–	–	–	–	$-^+$	S	A B iv ib (p ic)
d^c	–	–	–	–	–	–	–	–	OS	A B ib IV (p ic l s)
d	–	–	–	–	–	–	–	–	OS	A (l s)
d	–	–	–	–	–	–	–	–	S	A (l s)

PRAS carbohydrates: + pH <5.5; W pH 5.5–5.7; – pH >5.7.
Gas in PYG agar deep: –, no gas detected; 2, splits agar horizontally; 3, agar displaced halfway to top of the tube; 4, agar displaced to top of the tube.
Fatty acids: A = acetic, P = propionic, IB = isobutyric, B = butyric, IV = isovaleric, V = valeric, IC = isocaproic, C = caproic, C′ = caprylic, L = lactic, S = succinic, PA = phenylacetic.
Note: 1) capital letters indicate major metabolic products; 2) lower case letters indicate minor products; 3) parentheses indicate a variable reaction; 4) isoacids are primarily from carbohydrate-free media (such as PY) in the case of saccharolytic organisms.

Effective agents have been discussed. In general, it is important to use high-dose therapy and to treat for extended periods to avoid relapse. In the case of the intoxications, measures to eliminate toxin still present and to prevent additional elaboration of toxin are important along with the use of appropriate antitoxin and measures to combat the effects of the toxins. Patients with tetanus will require special nursing care in a quiet room, preferably within a respiratory or other intensive care unit, to minimize external stimuli that will precipitate convulsions and drugs to tranquilize, sedate, and relieve muscle spasm. Botulism patients should also be in an intensive care unit under close observation. For pseudomembranous colitis due to *C. difficile* or other clostridia, oral metronidazole, vancomycin, or bacitracin are the antimicrobial agents of choice. Some workers feel that bile acid-binding resins are helpful, but this never has been documented in a controlled study. Corticosteroids and antiperistaltic agents should be avoided. In enteritis necroticans, surgery to remove necrotic bowel is mandatory. Antitoxin to the β-toxin of type C *C. perfringens* should be administered.

Prevention and Control

Highly effective and safe immunization is available for prevention of tetanus. Adequate cleansing and debridement of wounds should also be practiced. With regard to botulism, proper procedures for home canning of foods is crucial. In the case of identification of a patient with botulism, all others who may have been exposed to the same toxin source must be sought immediately and observed closely; if the exposure to the toxin is definite, they should receive antitoxin. Rational use of antimicrobial agents and selection of agents with relatively little impact on the indigenous colonic flora, when feasible, should minimize the risk of patients developing pseudomembranous colitis. Isolation of cases of this disease and

TABLE 3. In vitro susceptibility of clostridia to antimicrobial agents

Bacteria	*Chloramphenicol*	*Clindamycin*	*Erythromycin*[c]	*Metronidazole*	*Cefoxitin*	*Ureido- and carboxypenicillins*[a]	*Penicillin G*[d]	*Tetracycline*	*Vancomycin*[c]
C. perfringens	+++	+++[b]	+++	+++	+++	+++	+++	++	+++
Other *Clostridium* spp.	+++	++	++ to +++	+++	+ to ++	+++	+++	++	++ to +++

Key: + = Poor or inconsistent activity; ++ = moderate activity; +++ = good activity; ++++ = good activity, good pharmacologic characteristics, low toxicity, drug of choice.

[a] Piperacillin, mezlocillin, azlocillin, carbenicillin, ticarcillin.

[b] Rare strains are resistant.

[c] Not approved by FDA for anaerobic infections.

[d] Other penicillins and cephalosporins are frequently less active. Ampicillin and amoxicillin are roughly comparable to penicillin G in activity. Addition of β-lactamase inhibitors such as clavulanic acid and sulbactam remarkably increases the activity against β-lactamase producers.

good infection control practices should minimize spread within the hospital setting. There is effective active immunization available for those at risk of exposure to enteritis necroticans. Good practices of food handling should minimize the risk of clostridial food poisoning.

Literature Cited

Finegold, S. M. 1977. Anaerobic bacteria in human disease. Academic Press, Inc., New York.

Finegold, S. M., W. L. George, and M. E. Mulligan. 1986. Anaerobic infections. Year Book Medical Publishers, Inc., Chicago.

Rolfe, R. D., and S. M. Finegold (eds.). 1987. Clostridium difficile: its role in intestinal disease. Academic Press, Inc., New York. (In press.)

Sneath, P. H. A., N. S. Mair, M. E. Sharpe, and J. G. Holt. 1986. Bergey's manual of systematic bacteriology, vol. 2, 9th ed. The Williams & Wilkins Co., Baltimore.

Sutter, V. L., D. M. Citron, M. A. C. Edelstein, and S. M. Finegold. 1985. Wadsworth anaerobic bacteriology manual, 4th ed. Star Publishing Co., Belmont, Ca.

Willis, A. T. 1969. Clostridia of wound infection. Butterworths, London.

CHAPTER 20

Clostridial Myonecrosis (Gas Gangrene)

W. LANCE GEORGE and SYDNEY M. FINEGOLD

Disease: Clostridial myonecrosis (gas gangrene).
Etiologic Agents: Histotoxic clostridia, particularly *Clostridium perfringens*, *C. novyi*, and *C. septicum*.
Source: Indigenous flora of the host; soil.
Clinical Manifestations: Severe pain and local findings, including edema, serous or serosanguineous drainage, bulla formation, and muscle necrosis.
Pathology: Progressive muscle necrosis due to effects of α toxin.
Laboratory Diagnosis: Recovery of appropriate species of clostridia in the presence of a compatible illness.
Epidemiology: Contamination of injured tissues with soil containing histotoxic clostridia or bowel flora.
Treatment: Extensive surgical debridement and antimicrobial therapy.
Prevention and Control: Debridement of contaminated traumatic wounds plus antimicrobial prophylaxis and prompt treatment of all wound infections.

Description of Disease

Certain species of *Clostridium*, sometimes termed the *histotoxic clostridia*, produce a necrotizing infection termed *gas gangrene* or *clostridial myonecrosis* (MacLennan, 1962). This disease has the potential to be rapidly lethal; prompt diagnosis and treatment can be lifesaving.

Pathogenesis

Clostridial myonecrosis, as the name implies, is a destructive process of muscle that most commonly develops at a site of tissue ischemia produced by accidental trauma or surgery (Finegold, 1977). Organisms may be introduced into the wound either from soil or other exogenous contaminated material or from the patient's own colonic flora. Histotoxic clostridia introduced into ischemic tissue can proliferate and produce a variety of putative toxins, including α toxin, a phospholipase C that hydrolyzes lecithin (Sneath et al., 1986). This toxin, apparently as a consequence of hydrolyzing membrane lecithin, disrupts cell membranes and produces cell death; it appears to the principal virulence factor in clostridial myonecrosis. The disease can progress quite rapidly. The α toxin produces injury and death of muscle at the primary site of infection; adjacent muscle becomes ischemic (due to local processes that are poorly understood) and provides a milieu that will support the growth of, and toxin production by, the infecting *Clostridium*. An important aspect of clostridial myonecrosis is the ability of the process, once established, to invade healthy adjacent tissues.

Other forms of clostridial myonecrosis include so-called metastatic or spontaneous gas gangrene (myonecrosis) and uterine myonecrosis, which is essentially gas gangrene of the uterine musculature; the latter infection almost always follows instrumentation of the gravid uterus and has been seen most commonly in association with criminal abortion (Finegold, 1977; Swartz, 1985). An unusual aspect of uterine myonecrosis is the high incidence of associated bacteremia and intravascular hemolysis.

Clinical Manifestations

The incubation period of clostridial myonecrosis is variable, but probably averages 3 to 4 days; it may, however, be as short as 8 to 12 h (Finegold, 1977; Finegold et al., 1986). Typically, the first sign of myonecrosis is the sudden onset of severe, unremitting pain at the site of the infection. The pain steadily increases in severity and spreads outward as the infection spreads; prominent local swelling and blanching of the overlying skin occur; and a thin, but sometimes copious, serous, serosanguinous, or hemorrhagic exudate develops. The wound usually has a sweet odor; however, a foul, offensive odor has been noted on occasion. The skin may then develop a purple or blue discoloration, and bullae filled with serosanguinous fluid develop. The sensorium may be typically clear, with the patient being apprehensive and having a sense of impending doom; toxic delirium usually occurs late in the course of disease. Later in the illness (usually hours), the patient may become hypotensive and delirious and develop renal failure. Moderate fever is typically present; although gas may be present in the tissues that surround the wound, it may be difficult to detect early in the illness and may become abundant only in the terminal stages of disease.

Pathology

The involved muscle undergoes a series of rapidly evolving changes, which often can be detected only by inspection at surgery (Finegold, 1977; Finegold et al., 1986; Swartz, 1985). Early in the illness, the muscle is pale and edematous; subsequently, contractility disappears, the blood supply is lost, and gas may be detectable. There is progressive reddening and purple mottling of the muscle, then dark discoloration as the muscle becomes necrotic. All of these findings may be present simultaneously in various areas of the same muscle, because the changes occur rapidly and the process spreads outward from the initial site of infection.

Microscopic examination of the wound exudate or muscle fragments may reveal large gram-positive bacilli with square ends, but not spores (*Clostridium perfringens*), or, less frequently, thinner gram-positive bacilli in which occasional spores may be seen (species of *Clostridium* other than *C. perfringens*); few polymorphonuclear leukocytes are in the exudate, and those present usually appear "smudged" because of the lytic effects of the α toxin (Finegold, 1977). The histologic appearance of the muscle is dependent on the stage of disease.

Diagnostic Features

The diagnosis must be made largely on clinical grounds. The disease tends to have a rather characteristic (but often unrecognized) presentation that should prompt immediate inspection of involved muscles at surgery. The presence of clostridia-like organisms in exudate (or fragments of tissue) and of smudged leukocytes is suggestive of this disease. It is important to realize that many wounds may become simply colonized with clostridia; hence, the results of examination of wound exudate must be interpreted in the context of the patient's illness. Because the process may progress rapidly in 2 to 4 h, valuable time should not be lost awaiting the results of diagnostic tests. Confirmation of the role of clostridia in the disease process, of course, requires recovery of the organism; rapid methods for detection of *C. perfringens*, the most common cause of clostridial myonecrosis, are described in another chapter of this text. Initial attempts to establish the diagnosis should never be based on culture methods because of the inherent delay (Finegold et al., 1986). Clostridial myonecrosis is a bacterial disease in which the microbiologist must urge the clinician to establish the diagnosis by immediate surgical exploration.

Prognosis

The prognosis is relatively poor compared with that of other types of bacterial infections and is dependent on the site of myonecrosis and the speed with which a definitive diagnosis is made and therapy instituted. The prognosis is best when an extremity is involved and worst when there is truncal involvement, particularly if complete debridement cannot be effected.

Epidemiologic Aspects

Histotoxic clostridia can be recovered from soil, which is undoubtedly the source of infection in some grossly contaminated wounds. Another important source of infection is the patient's endogenous microflora. An extremely high percentage of individuals have *C. perfringens* as part of their normal colonic flora; occasionally, other histotoxic clostridia may be present in the gut. Surgery involving the colon and surgical or accidental wounds that become contaminated with fecal flora are the most common antecedent conditions. A key element in the development of clostridial myonecrosis is a prior insult, such as trauma to the muscle that results in ischemia or devitalization of tissue. Spontaneous myonecrosis is usually associated with malignancies of the cecum or colon with secondary clostridial bacteremia, leading to seeding of remote muscle groups (Swartz 1985).

Etiologic Agents

The organisms most commonly involved in clostridial myonecrosis are *C. perfringens* (approximately 80 to 95% of cases), *Clostridium novyi* (10 to 40% of cases), and *Clostridium septicum* (5 to 20% of cases). *Clostridium bifermentans*, *Clostridium fallax*, and *Clostridium histolyticum* are documented but infrequent causes of myonecrosis (Caplan and Kluge, 1976; Finegold, 1977; Swartz, 1985). Criteria for identification of clostridia to the species level are given in another chapter of this text.

Treatment and Prevention

The most important treatment of clostridial myonecrosis is immediate and extensive surgical exploration of the involved site to remove all injured and necrotic tissue; reoperation is often necessary to remove additional tissues that have become devitalized (Altemeier and Fullen, 1971; Finegold, 1977). If complete surgical excision (or amputation) of involved tissue is not possible, the prognosis is bleak and survival unlikely. High-dose penicillin G has been the antimicrobial agent of choice since the 1940s; however, more recent in vitro data and limited animal studies have suggested that penicillin G may be less active than was believed previously (Swartz, 1985). Alternative agents would include chloramphenicol and possibly metronidazole, although clinical data for the latter agent are quite limited (Swartz, 1985). The wound should be cultured for both anaerobes and nonanaerobes and, if the primary infection is thought to be polymicrobial, appropriate broad-spectrum antimicrobial therapy should be given empirically.

The benefit of hyperbaric oxygen therapy remains controversial. Its use should be considered when the involved tissues cannot be completely removed surgically, such as when myonecrosis involves the paravertebral muscles. The efficacy of equine polyvalent gas gangrene antitoxin has never been established; because of the potential side effects of this preparation, it should not be used (Finegold, 1977; Swartz, 1985).

Prevention of clostridial myonecrosis requires prompt surgical debridement of contaminated traumatic wounds and of infected postoperative wounds. Antimicrobial treatment of infected wounds is always important; although good data are lacking, antimicrobial prophylaxis of grossly contaminated traumatic wounds is probably of value.

Literature Cited

Altemeier, W. A., and W. D. Fullen. 1971. Prevention and treatment of gas gangrene. J. Am. Med. Assoc. **217**:806.

Caplan, E. S., and R. M. Kluge. 1976. Gas gangrene: review of 34 cases. Arch. Intern. Med. **136**:788.

Finegold, S. M. 1977. Anaerobic bacteria in human disease. Academic Press, Inc., New York.

Finegold, S. M., W. L. George, and M. E. Mulligan. 1986. Anaerobic infections. Year Book Medical Publishers, Inc., Chicago.

MacLennan, J. D. 1962. The histotoxic clostridial infections of man. Bacteriol. Rev. **26**:177.

Sneath, P. H. A., N. S. Mair, M. E. Sharpe, and J. G. Holt (ed.). 1986. Bergey's manual of systematic bacteriology, 9th ed., vol. 2. The Williams & Wilkins Co., Baltimore.

Swartz, M. N. 1985. Myositis, p. 613. *In* G. L. Mandell, R. G. Douglas, Jr., and J. E. Bennett (ed.), Principles and practice of infectious diseases, 2nd ed. John Wiley & Sons, Inc., New York.

CHAPTER 21

Other Corynebacterioses

HELEN M. POLLOCK

Disease: Different diseases depending on the species, including cutaneous and respiratory diseases, bacteremia, and lymphadenitis.
Etiologic Agents: *Corynebacterium ulcerans, Corynebacterium pseudotuberculosis, Corynebacterium* JK, and other corynebacteria.
Source: Domestic livestock and their products for some; endogenous human colonization for others.
Clinical Manifestations: Multiple presentations compatible with pharyngitis, pneumonitis, lymphadenitis, endocarditis, and bacteremia.
Pathology: Ranging from acute inflammatory to chronic granulomatous lesions.
Laboratory Diagnosis: Bacteriologic isolation.
Epidemiology: Zoonoses or nonsocomial depending on the organism.
Treatment: Erythromycin, tetracycline, and vancomycin for group JK.
Prevention and Control: Devised according to the epidemiology of the organism.

Description of Disease

The genus, *Corynebacterium,* consists of a variety of species, many of which are ill defined. It has only been in recent years that sufficient characterization of human pathogens has occurred to delineate their epidemiologic and infectious roles. Only the more commonly described species with relatively clear taxonomic descriptions will be included.

Communicability and Transmissibility

The natural habitats of each of these organisms is distinct (Table 1) (Lipsky et al., 1982). In many cases not a great deal is known about their transmission to humans. Others, such as *Corynebacterium* JK and the lipophilic cutaneous corynebacteria have recently been well delineated. Infections with *Corynebacterium* JK are endogenously acquired as restriction endonuclease patterns were the same for a given patient, but differed between patients (Khabbaz et al., 1986). The spread of colonization between patients may be associated with adherence to handwashing (Quinn et al., 1984).

Clinical Manifestations

The clinical manifestations are also varied. Whereas some species, such as *C. ulcerans* and *C. pseudotuberculosis,* are known human pathogens, others are generally nonpathogenic and primarily infect compromised hosts (Table 1).

Respiratory Tract

C. ulcerans causes a mild to severe pharyngitis, which often is not associated with other findings. This species has been known to carry the phage responsible for diphtheria toxin and has, on occasion, caused a diphtheria-like illness (Buck et al., 1985).

Arcanobacterium hemolyticum (*C. hemolyticum*) has been associated with pharyngitis in young adults. About one-half of the patients exhibit a skin rash that is scarlatiniform or rubellaform in character. It tends to be distributed peripherally as an erythematous macular papular eruption.

Pneumonitis is the usual presentation of human infections with *C. equi.* It causes a suppurative pneumonia in young fowls, and humans acquire it by inhalation. Patients who present with clinical symptoms are usually immunocompromised.

TABLE 1. Some nondiphtheria corynebacterioses

Species	*Reservoir*	*Acquisition*	*Human infections*	*Host immunocompromised*	*Therapy*	*Comment*
C. ulcerans	Raw milk, cattle	Exposure to cattle and raw milk	Pharyngitis	No	Erythromycin	Some produce diphtheria toxin
C. pseudotuberculosis	Domestic and wild mammals, sheep most common lymphadenitis	Extensive contact with rural animals	Lymphadenitis, necrotizing/ supprative granulomatis inflammation	No	Pencillin, erythromycin, surgical drainage	Dermonecrotic toxin
C. hemolyticum	Human Pharynx	Presumed endogenous	Cutaneous chronic ulcers, pharyngitis often with skin rash	No	Erythromycin	
C. pyogenes	Domestic farm animals commensal, pathogen	Contact with animals	Few cases Varied	No	?	
C. pseudodiphtheriticum	Human, throats	Endogenous	Endocarditis	Prosthesic valve, other organisms in natural valves	?	Soon after prosthesis
C. equi	Soil of animal runs	Exposure to soils where animals are kept	Pneumonitis	Steroids	?	
C. bovis	Bovine udder	Contact with bovine products	Varied	Trauma, prosthesis	Erythromycin and rifampin	
C. xerosis	Human conjunctiva	Endogenous	Varied	Prothesis, malignancy, surgery	?	
Corynebacterium JK	Human skin; broad spectrum antibiotics encourage high counts	Endogenous	Bacteremia Pneumonia	Severely	Vancomycin	

Endocarditis

C. pseudodiphtheriticum has been reported as the causative agent in 8% of cases reported due to coryneform endocarditis. Other species include *C. xerosis* and others also have been reported. These infections usually follow the placement of a prosthetic valve.

Lymphadenitis

C. pseudotuberculosis is a cause of lymphadenitis in humans and animals. Nine of 10 human cases were subacute and chronic disease, with a relapsing pattern. Extensive contact with sheep has been associated with this infection.

Immunocompromised Settings

Oncology wards have been associated with infections caused by organisms, which up to now have not been a great problem. *Corynebacterium* JK, resistant to many antibiotics, has been associated with bacteremic episodes in these patients. Intravenous catheters and other procedures, as well as the heavy use of broad-spectrum antibiotics, have contributed to this situation. This resistance is long-lasting and is not plasmid mediated (Khabbaz et al., 1986).

Pathology

The pathology of these infections is variable and mimics the more typical pathology caused by organisms more commonly associated with each clinical syndrome.

The lymphadenitis caused by *C. diphtheriticum* shows the typical pathology of a necrotizing and/or suppurative inflammatory granuloma. It is morphologically similar to that produced by other organisms.

Diagnostic Features

For most of these organisms the diagnostic features are not particularly unique. Culture confirmation is required to implicate any of these coryneforms in the etiology of infection.

Epidemiologic Aspects

The reservoirs of these organisms are varied (Table 1), and transmission usually depends on more than casual contact in most cases. Colonization usually precedes infection and, with the exception of *C. ulcerans* and *C. pseudotuberculosis,* the organisms may not have a high level of virulence. Immunocompromised hosts are required for overt infection to manifest in many cases.

Etiologic Agents

The speciation of these organisms has been difficult due to differences in definitions and methodologies. Isolation and identification procedures requires the addition of serum to accommodate the lipophilic corynebacteria. *Corynebacterium* JK has been readily separated from the rest of the diphtheroids by its multiple resistance.

In general, most species fit the description for the genus of being catalase positive, short gram-positive nonsporing rods. Biochemical patterns have been compiled with attention to distinguishing characteristics (Coyle et al., 1985). Recently, identification profiles using two commercial identification systems, Minitek and AP120S systems (Kelly et al., 1984; Slifkin et al., 1986), have been shown to be useful.

Treatment, Prevention, and Control

The treatment is variable depending on the organism (Table 1). Prevention and control involve the general principles of epidemiologic control as they pertain to the particular situation.

Literature Cited

Buck, G. A., R. E. Cross, T. P. Wong, J. Loera, and N. Groman. 1985. DNA relationships among some tox-bearing corynebacteriophages. Infect. Immun. **49**:679–684.

Coyle, M. B., D. G. Hollis, and N. B. Groman. 1985. *Corynebacterium* spp. and other coryneform organisms, p. 193–204. *In* E. H. Lennette, A. Balows, W. J. Hausler, Jr., and H. J. Shadomy (ed.), Manual of clinical microbiology 4th ed. American Society for Microbiology, Washington, D.C.

Kelly, M. C., I. D. Smith, R. J. Ansley, J. H. Thornley, and R. P. Rennie. 1984. Rapid identification of antibiotic-resistant corynebacteria with the AP120S System. J. Clin. Microbiol. **19**:245–247.

Khabbaz, R. R., J. B. Kaper, M. R. Moody, S. C. Schimpff, and J. H. Tenney. 1986. Molecular epidemiology of group JK *Corynebacterium* on a cancer ward: lack of evidence for patient-to patient transmission. J. Infect. Dis. **154**:95–99.

Lipsky, B. A., A. C. Goldberger, L. S. Thompkins, and J. J. Plorde. 1982. Infections caused by nondiphtheria corynebacteria. Rev. Infect. Dis. **4**:1220–1235.

Quinn, J. P., P. M. Arnow, D. Weil, and J. Rosenbluth. 1984. Outbreak of JK diphtheroid infections associated with environmental contamination. J. Clin. Microbiol. **19**:668–671.

Slifkin, M., G. M. Gil, and C. Engwall. 1986. Rapid identificaiton of group JK and other corynebacteria with the Minitek System. J. Clin. Microbiol. **24**:177–180.

CHAPTER 22

Diphtheria

HELEN M. POLLOCK

Disease: Diphtheria.
Etiologic Agent: *Corynebacterium diphtheriae*.
Source: Human upper respiratory tract and skin infections acquired by inhalation or direct contact.
Clinical Manifestations: Range from mild upper respiratory symptoms to fatal cardiac and neurologic malfunction.
Pathology: Localized leathery fibrous membrane in the upper airway, with or without cardiac and neurologic symptoms, resulting from interruption of protein synthesis by a protein toxin.
Laboratory Diagnosis: Bacteriologic isolation from skin or upper airway; test for toxigenicity to evaluate virulence.
Epidemiology: Worldwide; controlled in nations with vaccine programs. Sporadic outbreaks occur in individuals with low antitoxin levels.
Treatment: Penicillin and erythromycin are the antibiotics of choice. Antitoxin should be given to prevent serious complications.
Prevention and Control: Mass vaccine programs and prophylaxis of case contacts.

Description of Disease

Diphtheria is a disease of worldwide distribution, which reflects one of the victories of control of the communicable organism, *Corynebacterium diphtheriae*. Epidemics are sporadic in developed countries in recent years, primarily resulting from vaccination programs. Classic diphtheria manifests systemic signs, particularly cardiac and neurologic, resulting from the activity of an exotoxin produced by the organism at a remote site, usually the oropharynx. Humans are the only host for this pathogen.

Communicability and Transmissibility

The patterns of this disease have changed over the years. Previously, diphtheria was considered a disease of rural settings; in more recent years, it has been more commonly found in urban settings. Although children have been considered the group at risk, three recent epidemics have been described among adults with alcoholic problems (Hellstrom, 1986; Malmvall, 1986; Pederson et al., 1977). Transmission from oropharyngeal cases is by aerosol droplets and inhalation. As cutaneous infections contaminate the environment to a greater extent, more contacts of cutaneous cases harbor the organism in their oropharynx than do contacts of oropharyngeal cases (Koopman and Campbell, 1975). Fomite transmission and skin-to-skin transmission have also been suggested (Belsey et al., 1969). Efforts to control transmission by disinfection of the environment were unsuccessful in the Seattle Skid Road epidemic (Pederson et al., 1977). Because the organism remained viable in dust for long periods, cutaneous infections outnumbered pharyngeal infections in this outbreak. Routine screening of skin lesions of the affected population, seen at a downtown free clinic during this period, probably contributed to the large number of cutaneous cases identified. General observations suggest that poor hygiene contributes to the spread of the organism (Koopman and Campbell, 1975) as well as closeness and duration of contact.

Clinical Manifestations

The clinical range of this disease spans the total spectrum, including lack of symptoms to hypertoxicity resulting in fatalities.

Asymptomatic Diphtheria

Corynebacterium diphtheriae may be rarely a part of the normal upper respiratory tract flora of humans. Although a survey for pharyngeal infection was conducted in the King County Jail from 1959 to 1972, only five cases were reported (Pederson et al., 1977). The number of asymptomatic infections is not very high in vaccinated populations (Wilson and Miles, 1964).

Upper Respiratory Tract Diphtheria

Symptoms generally appear 2 to 6 days following exposure and colonization. Patients with nasal diphtheria rarely develop intoxicating manifestations. A yellowish membrane and mucopurulent discharge usually occur.

Early diphtheritic membranes may be thin mucilagenous structures, which may have areas exhibiting the more characteristic grayish-green coloration by the time the patient seeks medical attention. Some bleeding may occur with attempts to remove it. Symptoms include tender nodes in the neck, sore throat, pain on swallowing, and nausea and vomiting.

The membrane may spread to the uvula, soft palate, and pharynx and proceed to the lower airway, where it has been known to produce a cast of the main portion of the bronchial tree. Death resulting from hypoxia is a usual result, and removal of the membrane to open the airway is required.

Cutaneous Diphtheria

Skin lesions occur more commonly in the Southeastern United States. The Seattle epidemic, however, had a high rate of cutaneous isolations, usually concomitantly with *Streptococcus pyogenes*. Skin lesions due to *C. diphtheriae* are indistinguishable from those caused by other bacteria. The spectrum includes ecthyma, impetigo, infected insect bites, nearly healed lesions, and pyoderma. The severity of the lesion is not related to the type or toxigenicity of the isolate.

Myocarditis

Cardiac manifestations occur late in the course of the disease; patients may even be recovering from pharyngeal infection. Heart failure is the most common cause of death following diphtheria. Cardiac involvement is manifest by delay in peripheral nerve conduction velocity, which precedes the cardiac rhythmic changes. Acute circulatory failure may occur in 9% of pharyngeal cases. The mortality is higher in patients demonstrating electrocardiographic changes and elevated serum glutamic-oxaloacetic transaminase. The condition does not usually respond well to digitalis glycosides because of the physiologic and structural damage to myocardial fibers. Salt restriction and substances used to control arrhythmias are used as supportive therapy.

Neuritis

Peripheral nerve involvement manifests as a motor defect ranging from mild weakness to paralysis. Cranial nerves appear to be involved first, possibly resulting from retrograde movement of the toxin along the nerve fibers. Demyelinization of nerve fibers contributes to the paralysis (Youmans, 1985). The onset of these complications occurs 4 to 6 weeks after the onset of infection.

Pathology

Diphtheria is initially an upper respiratory tract disease. Pseudomembranes formed by strains that do not produce exotoxin are indistinguishable from typical membranes. The membrane consists of bacteria, fibrin, and cellular infiltrates consisting of epithelial cells and polymorphonuclear and mononuclear leukocytes.

While the respiratory symptoms are primarily due to membrane formation, the systemic cardiac and neurologic signs result from transportation of an exotoxin to remote locations. Experimental data has suggested that the toxin may be carried via the bloodstream (Coyle et al., 1985), but evidence of toxemia in active cases appears to be lacking. Although the diphtheria bacillus does not invade into deeper tissues, the toxin acts to increase necrosis at the local site as well as to produce cardiac symptoms. The mechanism of transport of the toxin in natural disease remains to be elucidated.

On an experimental cellular level, a great deal is known about the mode of action of diphtheria toxin. The toxin first binds to specific glycoprotein receptors (Hranitzky et al., 1985; Proia et al., 1979). Through a series of complex steps a few molecules reach the cytoplasm (Morris et al., 1985). The toxin is a single polypeptide chain that may be divided into two fragments, A and B. Fragment B is responsible for attachment to sensitive cells, whereas fragment A exhibits enzymatic activity. Fragment A is translocated into the cytosol after endocytosis. It catalyzes the covalent addition to elongation factor 2 of the ADP-ribosyl portion of NAD^+. Elongation factor is thus inactivated and protein synthesis arrested (Collier, 1975). Cell death ensues, causing reduced function of the involved organ system.

The interesting aspect of this toxin is its genetic regulation. A number of bacteriophages carry (tox^+), a structural gene for diphtheria toxin (DT). These phages have the ability to convert nontoxigenic strains of *C. diphtheriae* to produce toxin. A comparison of these phages (α, β, δ, L, and P) demonstrated similar or identical restriction enzyme patterns of their DNAs, and all their *BamHI* restriction fragments exhibited homology with β phage DNA (Buck et al., 1985).

Diagnostic Features

Because the disease is presently rare, the diagnosis and indices of suspicion are very low. Fatality rates at the beginning of epidemics are frequently high for this reason.

The features representing upper respiratory tract symptoms are not unique. The membrane is similar to those produced by streptococci and the mixed anaerobes contributing to Vincent's angina. Direct alkaline methylene blue-stained smears of the exudate may assist separation of these etiologies, but this technique is not recommended. Negative direct smears do not rule out the diagnosis, and smears are generally unreliable. Streptococcal pharyngitis may present with a fiery red throat, whereas grayish membrane development involving the uvula is highly suggestive of diphtheria. Treatment cannot await bacteriologic confirmation, which may take 1 to 4 days; thus, suspicion based on signs and symptoms is crucial.

Prognosis

Cardiac and neurologic involvement are serious prognostic signs. Early identification of myocarditis will reduce morbidity and mortality. Bleeding from the upper airway is also associated with a grave prognosis.

Epidemiologic Aspects

Diphtheria is a classic example of the impact of immunization programs. The epidemiology differs markedly in nations with mass vaccination versus those without such programs.

Occurrence and Transmission

Diphtheria had a death rate of 300 per 10^6 population in 1901 in England and Wales. After initiation of a mass immunization program, this rate dropped to 73 per 10^6 population in 1951 (Wilson and Miles, 1964). By 1959, no deaths were recorded. Outbreaks of diphtheria have often been reported in immunized populations, but rarely with any fatalities. Mass immunizations have also succeeded in reducing the carrier rate in many populations, thus, reducing the risk of disease.

Recent outbreaks have been in the adult populations as opposed to earlier ones, which were primarily in children. Serosurveys of the adult population indicate that many do not have protective levels of antitoxin (MMWR, 1986). Only five cases were reported in the United States from 1980 to 1984.

Susceptibility, Resistance, and Carrier States

Resistance to classic diphtheria with toxemia is in practical terms determined by the presence of circulating antitoxin. During the early phases of vaccination programs, pharyngeal diphtheria occurred in children with adequate levels of antitoxin as determined by negative Schick tests (Wilson and Miles, 1964). As the carrier rate is extremely low, clusters of cases occur when a carrier comes in contact with a susceptible population. Apparently the vaccine affects the ability of the organism to colonize the oropharynx. The prevalence of carriage appears to be related to the occurrence of active disease in susceptible individuals.

Etiologic Agent

The agent associated with and clearly the major cause of diphtheria is *Corynebacterium diphtheriae*. The genus *Corynebacterium*, originally designated for this organism, consists of non-spore-forming, gram-positive, non-acid-fast rods occurring in angular arrangements. The taxonomy of this genus has yet to be clarified, but new tools of chemotaxonomy suggest that only organisms with meso-diaminopimelic acid, arabinose, galactose, and shorter-chained mycolic acids (28 to 36 carbons) should be included in the genus. *C. diphtheriae* is only found in humans.

Characterization

The term diphtheroid, commonly used in medicine, includes the usual nonpathogenic skin coryneforms, denoting their morphologic resemblance to *C. diphtheriae*. Strains of *C. diphtheriae* are catalase positive, nonhemolytic, and utilize glucose and maltose, but not usually sucrose, mannitol, or xylose. Nitrates are reduced by most strains. It may be separated from other pathogenic members of the genus *C. ulcerans* and *C. pseudotuberculosis* by the failure to hydrolyze urea. Strains may produce toxin.

Specimen Collection and Transport

Lyman and Youngstrom (1956) reported that collection of specimens from both nasopharynx and throat are essential in cases and contacts. As many as 20% of positive cultures may be missed with a single sample. Nasopharyngeal wire swabs, which may be used to sample deep into the posterior nares, and cotton swabs rubbed firmly on areas of inflammation and membrane formation in the throat are used for collection in symptomatic cases. Samples should be taken from the tonsillar fossae, posterior pharynx, and uvula in asymptomatic individuals. Skin lesions are washed with sterile saline, the crust removed, and a cotton swab applied firmly to the wound base. Specimens transported in less than 24 h may be placed on a Loeffler slant for shipment to the laboratory. Longer shipping times require the use of silica gel transporters.

Isolation

Specimens are inoculated onto a selective tellurite medium and a Loeffler serum slant as well as 5% sheep blood agar for primary isolation. The tellurite medium inhibits the growth of most contaminating flora and allows selection of candidate colonies. *C. diphtheriae* reduces tellurite to tellurium, producing a brown to black colony. Not all tellurite media are equally good for this purpose. A comparison of Tinsdale (TM) (Tinsdale 1942) and Frobisher cystine–tellurite agar (CT) (Frobisher, 1937) for the isolation of *C. diphtheriae* from skin lesions and nasopharynges yielded a positive diagnosis using CT on 17 of 207 specimens. Only 10 of these isolates grew on Tinsdale medium. Cystine–tellurite also inhibited *Staphylococcus aureus* more effectively (Pollock and Stevens, unpublished data). A critical balance between cystine and tellurite is required to achieve maximum selectivity without inhibiting the isolation of diphtheria bacilli. Media are incubated at 35 to 37°C and examined at 24 and 48 h.

Identification

After 24 h, growth on Loeffler slants and suspicious colonies from plated media are examined.

Presumptive

Smears from Loeffler slants are stained with alkaline methylene blue. The typical pleomorphic, club-shaped rods with metachromatic granules are presumptive evidence of the presence of *C. diphtheriae*. This starvation medium enhances granule formation, which represent storage of phosphates.

Colonies on tellurite are blackened and show gram-positive rods with angular alignment. Metachromatic granules are not generally produced on these or blood agar media.

Confirmatory

Species confirmation requires a biochemical battery to separate *C. diphtheriae* from other members of the genus. Biochemical tests are performed on standard fermentation and nitrate media with added serum. Tests may be positive in 24 h, but negative tests should be held 72 h before discarding.

Toxin Assay

Although biochemical tests distinguish the species, determination of toxin production requires specialized methodology.

For the in vivo test for toxigenicity, a single guinea pig may be used. Broth cultures of the test strain and control strain are injected intracutaneously. Five hours later, 500 units of diphtheria antitoxin are given intraperitoneally. A second intracutaneous injection of the previous cultures is placed below the first as a control. Readings at 24 and 48 h will reveal a central necrotic area surrounded by erythema with toxigenic strains.

The in vitro test, known as the Elek test, is performed in agar gels. Sterile filter paper soaked in diphtheria antitoxin is placed in melted agar in a petri dish and allowed to sink below the surface. Test strains are streaked in a line perpendicular to the strip. Toxin-producing strains form a precipitin line at a 45° angle to the filter paper and culture streak. Plates should be incubated up to 72 h. This test is not as sensitive as the in vivo test, and low level toxin producers may give false-negative results.

Epidemiology Typing Methods

The three colony types of *C. diphtheriae*, known as *mitis*, *intermedius*, and *gravis*, are used as epidemiologic markers. *Mitis* colonies are medium, smooth, and black on tellurite-containing media. *Intermedius* colonies are small and very black. The *gravis* colonies are grayish, flat, and have irregular edges. The subtype *mitis* includes a few sucrose-positive and nitrate-negative strains. Fermentation of starch and glycogen distinguishes *gravis* from *mitis* and *intermedius* (Coyle et al., 1985; Wilson and Miles, 1964). Although all types may produce severe disease, these characteristics are useful epidemiologic markers. It is of interest that the types may change as an epidemic runs its course. Often, strains isolated toward the end of an outbreak have lost the phage required for toxin production.

More definitive markers using bacteriophage and agglutination techniques are available for tracking outbreaks (Youmans, 1985). At least 35 types have been delineated using bacteriophage lysis patterns.

Different types appear to predominate in different countries (McClosky, 1985).

Treatment, Prevention, and Control

Control and treatment of diphtheria requires a double approach. One is to control growth and transmission of the organism; the other is protection against the effects of the toxin.

Treatment

As bacteriologic confirmation requires 1 to 4 days, the decision concerning therapy must be made strictly on clinical grounds. Administration of diphtheria antitoxin as soon as possible is of primary consideration. The dose is adjusted to the severity of disease as indicated by the extent of progression of the membrane. In addition, absolute bed rest and penicillin or erythromycin for elimination of the organism are required. Patients must be put in isolation for 10 to 14 days. Any airway supportive therapy to maintain the opening is instituted as required (McClosky, 1985).

Prevention and Control

The major effort contributing to the dramatic reduction in cases has been the routine vaccination of children. It appears that the use of toxoid alone has also interrupted the transmission of the organism in the developed world. The unavailability of vaccine has made the problem more difficult in the developing world. Recent outbreaks in adult populations raise concerns about the waning levels of antitoxin in this population and their susceptibility to the respiratory symptoms of disease. Control of an epidemic is primarily aimed at prophylactic antibiotic therapy and booster vaccination of contacts of cases.

Literature Cited

Belsey, M. A., T. M. M. Sinclair, M. R. Roder, and D. R. LeBlanc. 1969. *Corynebacterium diphtheriae* skin infections in Alabama and Louisiana. N. Engl. J. Med. **280**:135–141.

Buck, G. A., R. E. Cross, T. P. Wong, J. Loera, and N. Groman. 1985. DNA relationships among some tox-bearing corynebacteriophages. Infect. Immun. **49**:679–684.

Collier, J. R. 1975. Diphtheria toxin: mode of action and structure. Bacteriol. Rev. **39**:54–85.

Coyle, M. B., D. G. Hollis, and N. B. Groman. 1985. *Corynebacterium* spp. and other coryneform organisms, p. 193–204. *In* E. H. Lennette, A. Balows, W. A. Hausler, Jr., and H. J. Shadomy (ed.), Manual of clinical microbiology, 4th ed. American Society for Microbiology, Washington, D.C.

Frobisher, M. 1937. Cystine–tellurite agar for *C. diphtheriae*. J. Infect. Dis. **60**:99–105.

Hellstrom, L. 1986. Diphtheria in Stockholm December 1984–Jan. 1985. Communicable Disease Rep. 86/32.

Hranitzky, K. W., D. L. Durham, D. A. Hart, and L. Eidels. 1985. Role of glycosylation in expression of functional diphtheria toxin receptors. Infect. Immun. **49**:336–343.

Koopman, J. S., and J. Campbell. 1975. The role of cutaneous diphtheria infections in a diphtheria epidemic. J. Infect. Dis. **131**:239–244.

Lyman, E. D., and J. A. Youngstrom. 1956. Diphtheria cases and contacts: is it necessary to take cultures from both nose and throat? Nebr. State Med. J. **41**:361–362.

Malmvall, B. 1986. Diphtheria in Gothenberg, Aug. 1984–Jan. 1985. Communicable Dis. Rep. 86/32.

McClosky, R. V. 1985. *Corynebacterium diphtheriae* (diphtheria), p. 1171–1174. *In* G. L. Mandell, R. G. Douglas, Jr., and J. E. Bennett (ed.), Principles and practice of infectious diseases, 2nd ed. John Wiley & Sons, Inc., New York.

Morbidity and Mortality Weekly Report. 1986. Annual Summary 1984. Morbid. Mortal. Weekly Rep. **33(54)**:23. HHS Pub. No. (CDC) 86-8241.

Morris, R. F., A. S. Gerstein, P. F. Bonventre, and C. B. Sallinger. 1985. Receptor-mediated entry of diphtheria toxin into monkey kidney (Vero) cells: electron microscopic evaluation. Infect. Immun. **50**:721–727.

Pederson, A. H. B., J. Spearmen, E. Tronca, M. Bader, and J. Harnisch. 1977. Diphtheria on Skid Road, Seattle, Wash 1972–1978. Public Health Rep. **92**:336–342.

Proia, R. L., L. Eidels, and D. A. Hart. 1979. Diphtheria toxin-binding glycoproteins on hamster cells: candidates for diphtheria toxin receptors. Infect. Immun. **25**:786–791.

Tinsdale, G. F. W. 1947. A new medium for the isolation and identification of *C. diphtheriae* based on the production of hydrogen sulfide. J. Pathol. Bacteriol. **59**:461–466.

Wilson, G., and A. A. Miles. 1964. Diphtheria and other diseases due to corynebacteria, p. 1668–1720. *In* Topley and Wilson's principles of bacteriology and immunity, 5th ed., vol. II. The Williams & Wilkins Co., Baltimore.

Youmans, G. P. 1985. Diphtheria, *In* Youmans, Paterson, and Sommers (ed.), The biologic and clinical basis of infectious diseases. The W. B. Saunders Co., Philadelphia.

CHAPTER 23

Donovanosis

D. S. KELLOGG

Disease: Donovanosis (granuloma inguinale, granuloma venereum).
Etiologic Agent: *Calymmatobacterium granulomatis.*
Source: Human sexual contact primarily, but other modes are likely.
Clinical Manifestations: Chronic, specific infection manifested clinically by ulcerating granulomatous lesions involving, in most cases, the skin and subcutaneous tissues. Secondary involvement of the bones, joints, and cervix has been reported. Associated infections with pyogenic organisms and other agents of sexually transmitted disease are common.
Pathology: Luxuriant granulation tissue infiltrated with mononuclear cells and histiocytes.
Laboratory Diagnosis: Demonstration of the causative microorganism in a properly prepared and stained tissue smear is currently the only effective procedure. Electron microscopic examination of tissue sections can be effective. Cultural and serologic procedures are not useful.
Epidemiology: Primarily a disease of tropical and subtropical regions, but not restricted to them. Usually affects persons in the second and third decades of life. There is a male bias (2.5 : 1) and a greater prevalence among individuals of low socioeconomic status and poor personal hygiene.
Treatment: Tetracycline, chloramphenicol, gentamicin, sulfoxazole, and cotrimoxazole are effective. A combination of lincomycin and erythromycin has been satisfactory for pregnant patients.
Prevention and Control: Improved personal hygiene and availability of primary care have resulted in decreased incidences.

Description of Disease

Donovanosis (granuloma inguinale, granuloma venereum) is a chronic specific infection involving principally the skin and subcutaneous tissues of humans and manifested clinically by ulcerating granulomatous lesions. There is only one species of the causative organism, *Calymmatobacterium granulomatis,* which is currently classified as an unassigned genus associated with the family *Enterobactericeae* (Krieg, 1984).

Communicability and Transmissibility

Donovanosis is transmitted between humans principally by sexual contact (Lal and Nicholas, 1970). There is substantial evidence of a close relationship between *C. granulomatis* and the fecal organism *Klebsiella rhinosclermatis,* suggesting rectal intercourse as a source of this infection (Goldberg, 1959). Other modes of infection are likely (Spagnola et al., 1984). The infection can be transmitted between humans by subcutaneous inoculation of exudate (Dienst et al., 1938). Laboratory infections have never been documented. The incubation period has been estimated to vary between 8 days and 12 weeks.

Clinical Manifestations of Disease

Onset begins with a small papule that erodes the skin surface and progresses into a granulomatous, typically hypertrophic, velvety granulation tissue that is somewhat indurated and beefy red with a well-defined rolled border (Kuberski, 1980). These lesions bleed easily with trauma, but are not usually painful unless they are secondarily infected. Generally the lesions occur on the genitalia, but in 50% of both men and women they are also found in the anal or inguinal area. Growth of lesions is usually through extension to adjacent skin, especially along groin folds. Differences in descriptions of lesions and photographic data indicate some variation in lesion characteristics between tropical and subtropical environments.

Pathology

The affected dermis is usually infiltrated by a dense, mixed cellular infiltrate composed predominantly of mononuclear cells, with occasional foci of neutrophils and histiocytes. The epithelial border of the lesions frequently shows acanthosis. Caseation, suppuration, and epithelioid giant cells are not typical. There is a striking intracellular association of *C. granulomatis* within histiocytes, resulting in the diagnostically characteristic "pathognomonic cells" (Pund and Greenblatt, 1937). These enlarged cells (20 to 90 μm) have an eccentric nucleus and individual or groups of cells in cystlike cytoplasmic compartments. Polymorphonuclear leukocytes are seen in appreciable numbers only in the surface aspects of the lesion. Pathognomonic cells are scattered in variable numbers in the upper half of the granulation tissue, particularly near the margins of the ulcer. They are very abundant in exuberant granulation tissue.

Diagnostic Features

Donovanosis should always be considered when the lesions have a history of considerable duration and slow progression. The diagnosis of this disease is most difficult in its earliest stages, when it is readily confused with chancroid (Verdich, 1984). In the later stages of donovanosis, genital carcinomas, lymphogranuloma venereum, syphilis, and anogenital cutaneous amebiasis must be considered in the differential diagnosis (Kuberski et al., 1980). The only effective laboratory procedure in the diagnosis of donovanosis is a properly prepared and stained tissue smear (Cannefax, 1948).

Prognosis

Untreated, the lesions caused by donovanosis will mutilate and destroy the genital tissues and spread to other areas of the body by either secondary or systemic dissemination. Treatment with effective antibiotics for adequate periods will effect a cure. Mortality is rare with donovanosis, but its infrequent occurrence follows dissemination and death due to secondary causes such as pneumonia, cardiac failure, and hemorrhage.

Epidemiologic Aspects

Occurrence and Transmission

Donovanosis occurs principally in, but is not restricted to tropical and subtropical areas; Western New Guinea, Southeast India, and Guyana are considered endemic areas. Case reporting is incomplete in some areas and nonexistent in others. In the United States fewer than 100 cases are reported annually (Sexually Transmitted Diseases Summary, 1985). Donovanosis is considered a venereal disease; however, a variable percentage (1 to 50%) of marital partners of infected patients contract the disease, and its frequency is low in the very young and the elderly. The male: female ratio is 2.5 : 1, and there is a preponderance of cases in those aged 20 and 30 years (Maddocks et al., 1976).

Susceptibility, Resistance, and Carriage

Susceptibility may be greater in darker-skinned races, although this finding may be more a matter of socioeconomic and environmental circumstances. No resistance has been documented. There is no spontaneous regression without therapy even when high levels of circulating antibodies are present (Maddocks et al., 1976).

Etiologic Agent

Characterization

Calymmatobacterium granulomatis is a gram-negative, nonmotile, asporogenic, encapsulated coccobacillus measuring 0.5 to 1.5 by 1 to 2 μm. It is classified as an unassigned genus associated with the family *Enterobacteriaceae*. Morphologically and antigenically, *C. granulomatis* resembles members of the genus *Klebsiella*, particularly *K. rhinoscleromatis*. It has been cultured only under microaerophilic to anaerobic conditions and will not grow on the surface of ordinary simple or complex laboratory media. The organism has been grown in special liquid media, egg yolk slant fluid, and yolk sacs of 5-day chicken embryos. The optimal temperature is 37°C. There are no cultures of this microorganism in existence.

Specimen Collection

An area of granulomatous tissue near the periphery of the lesion is carefully cleaned with several saline-soaked gauze squares and similarly dried (Cannefax, 1948). The appearance of blood does not affect specimen characteristics. Using a punch biopsy forceps, secure a small piece of tissue (about the size of a match head) and smear and underside of the specimen over the surface of a glass slide (Greenblatt and Barfield, 1952). Do not respread any area, and cease spreading when the specimen begins to smear dry. The dry areas will contain rubbed, broken cells of no diagnostic value.

Smear Staining and Examination

Stain the smear with Wright, Giemsa, or pinacyanole stain according to usual procedures, except with the Wright stain. With the latter, allow the dye to remain on the smear for 1.5 min before diluting with phosphate buffer (pH 6.4). A positive specimen will exhibit large mononuclear cells with cytoplasm containing small, straight or curved dumbbell-shaped (safety-pin-shaped) rods that are blue to deep purple in color and surrounded by pink capsules. Broken tissue cells are not satisfactory for examination, as there are many other extracellular objects of similar size and characteristics in the specimen. The large "pathognomonic cells" can be detected under low magnification (10×) with some experience and confirmed at a higher magnification (45×) (Pund and Greenblatt, 1937).

Culture Specimen Examination

A specimen obtained in a manner similar to that used for smears is placed between two sterile glass slides and mashed with a twisting motion. Separate the slides and add 0.2 to 0.5 ml of sterile saline to each slide surface. Emulsify the tissue debris with a sterile toothpick or loop.

Culture Characterization

For isolation in chick embryos, inoculate 0.2 ml of the tissue debris suspension with the yolk sac of a 5-day-old chicken embryo and incubate at 37°C for 72 h. Growth will occur in the yolk sac fluid and will demonstrate the characteristic morphology of *C. granulomatis*.

For isolation on "Dulaney slants," inoculate 0.2 ml of the tissue debris suspension onto coagulated egg yolk slants, then cover three-quarters of the slant with Lockes salt solution. Close the tubes and incubate upright at 37°C for 48 to 72 h. Slants made of yolks from range-fed chickens (or ducks, geese, and turkeys) were effective, whereas slants of yolks from diet-fed birds were not (Goldberg et al., 1953).

Immunologic Aspects

No serologic procedure is currently available either for detecting antibodies to *C. granulomatis* in human sera or for identifying the organism as an isolate; however, antibodies have been detected in human sera by complement fixation procedures. Significant serum titers were found only in sera from patients whose lesions persisted for more than 3 months (Goldberg et al., 1953). The lack of protection despite high levels of circulating antibodies is similar to that found in other diseases in which the organism is located intracellularly (e.g., lepromatous leprosy, tuberculosis, and chronic mucocutaneous candidiasis).

Treatment, Prevention, and Control

Treatment

The treatment of choice is tetracycline or sulfasoxazole for a minimum of 2 to 3 weeks until the lesions have healed completely. This disease tends to relapse after successful treatment and healing, so prolonged follow-up is necessary. Chloramphenicol, gentamicin, streptomycin, aureomycin, and cotrimoxazole are also effective. The combination of lincomycin and erythromycin has been satisfactory for the treatment of pregnant patients. Penicillin is not efficacious, and ampicillin or erythromycin alone has given inconsistent results (Kuberski, 1980).

Prevention and Control

A general improvement in living conditions and a broader understanding of the epidemiology of donovanosis should reduce the incidence of this disease. Increased diagnostic effectiveness and the application of effective treatment regimens should minimize the incidence of donovanosis in the absence of an available vaccine.

Literature Cited

Cannefax, G. R. 1948. The technic of the tissue spread method for demonstrating Donovan bodies. J. Vener. Dis. Info. **19**:201–204.

Dienst, R. B., R. B. Greenblatt, and E. S. Sanderson. 1938. Cultural studies on the Donovan bodies of granuloma inguinale. J. Infect. Dis. **62**:112–114.

Goldberg, J. 1959. Studies on granuloma inguinale IV. Growth requirements of *Donovania granulomatis* and its relationship to the natural habitat of the organism. Br. J. Vener. Dis. **35**:266–268.

Goldberg, J. 1964. Studies on granuloma inguinale VII.

Some epidemiological considerations of the disease. Br. J. Vener. Dis. **40:**140–145.

Goldberg, J., R. H. Weaver, and H. Packer. 1953. Studies on granuloma inguinale II. The complement fixation test in the diagnosis of granuloma inguinale. Am. J. Syphilis Gonorrhea Vener. Dis. **37:**71–76.

Greenblatt, R. B., and W. E. Barfield. 1952. Newer methods in the diagnosis and treatment of granuloma inguinale. Br. J. Vener. Dis. **28:**123–128.

Krieg, N. R. 1984. Facultatively anaerobic gram-negative rods, family 1. Enterobacteriaceae, p. 585–587. *In* Bergeys manual of systemic bacteriology, vol. 1. American Society for Microbiology, Washington, D.C.

Kuberski, T. 1980. Granuloma inguinale (donovanosis). A review. Sex. Transm. Dis. **7:**29–36.

Kuberski, T., J. M. Papadimitriou, and P. Phillips. 1980. Ultrastructure of *Calymmatobacterium granulomatis* in lesions of granuloma inguinale. J. Infect. Dis. **142:**744–749.

Lal, S., and C. Nicholas. 1970. Epidemiological and clinical features in 165 cases of granuloma inguinale. Br. J. Vener. Dis. **46:**461–463.

Maddocks, I., E. M. Anders, and E. Dennis. 1976. Donovanosis in Papua New Guinea. Br. J. Vener. Dis. **52:**190–196.

Pund, E. R., and R. B. Greenblatt. 1937. Specific histology of granuloma inguinale. Arch. Pathol. **23:**224–229.

Sexually Transmitted Diseases Summary. 1985. U.S. Department of Health and Human Services, Technical Information Services, Atlanta.

Spagnola, D. V., P. R. Coburn, J. J. Cream, and B. S. Azadian. 1984. Extragenital granuloma inguinale (donovanosis) diagnosed in the United Kingdom: a clinical, histological and electron microscopical study. J. Clin. Pathol. **37:**945–949.

Verdich, J. 1984. *Haemophilus ducreyi* infection resembling granuloma inguinale. Acta. Dermato-Venereol. **64:**452–455.

CHAPTER 24

Enteric Bacteriosis

RICHARD F. D'AMATO and HENRY D. ISENBERG

Disease: Enteric bacteriosis (hospital-acquired or nosocomial infections; infections with various gram-negative rods acquired by patients during hospitalization).

Etiologic Agents: Most commonly involved genera: *Aeromonas*, *Citrobacter*, *Edwardsiella*, *Enterobacter*, *Escherichia*, *Ewingella*, *Hafnia*, *Klebsiella*, *Kluyvera*, *Morganella*, *Plesiomonas*, *Proteus*, *Providencia*, and *Serratia*.

Clinical Manifestations: These bacteria complicate the recovery of hospitalized patients, most of whom have disease- or therapy-impaired immunity and require catheters, central lines, mediports, or prostheses. These patients usually have received or are being treated with antimicrobial agents; many are on immunosuppressive or antineoplastic therapy.

Pathology: Bacteremia, pneumonia, urinary tract infections, wound infections, central nervous system infections, abscess formation in various organs, and colonization and infection of implants, prostheses, and catheters.

Laboratory Diagnosis: Bacteriologic isolation of etiologic agents from blood, other body fluids, wounds, and other anatomic loci and exudates. Antimicrobial susceptibilities of isolates characteristically reveal resistance to commonly used therapeutic modalities. Molecular epidemiologic analysis may identify responsible resistance plasmids.

Epidemiology: These bacteria, ubiquitous in nature, are involved in complicating the recovery of patients in hospitals. Very few individuals are admitted to institutions with community-acquired disease involving these organisms except *Klebsiella* and *Proteus*. The latter two, when present on admission, usually display greater antimicrobial susceptibility than the institutional representatives.

Treatment: Must be based on antimicrobial susceptibilities of the isolated bacteria. Incision and drainage of abscesses and removal of catheters or prostheses may be indicated, especially when antimicrobial therapy regimens fail repeatedly.

Prevention and Control: Isolation or cohorting of patients infected or colonized with multiply resistant representatives of this group of bacteria. Strict adherence to advocated personnel practices to prevent spread of bacteria must be enforced. Continuing education of all personnel is imperative. Institutionwide restriction of select antimicrobial agents may be useful in controlling further emergence of multiply resistant variants.

Introduction

Molecular biology has provided the taxonomist occupied with prokaryotic nomenclature with the tools to classify bacteria in a fashion denied earlier investigators, who based their analysis of prokaryotes on biochemical and immunochemical variables. The family *Enterobacteriaceae* encompasses a large collection of bacteria that share a number of characteristics and form large population centers within the family grouped together by identical biochemical and physiologic attributes. Species and intraspecies differences are discerned by biochemical activities and epitope variations. However, representatives outside of these dense population centers defy traditional classification attempts. Molecular biology and its most stringent DNA hybridization qualifications were required to place these outliers appropriately (Brenner, 1985). As stated by Farmer et al. (1985), "In 1972 there were only 11 genera and 26 species in the family *Enterobacteriaceae*. Today there are 22 genera, 69 species and 29 biogroups or enteric groups." Since the publication of this benchmark paper in classifying *Enterobacteriaceae*, additional species and genera have emerged. We seem to be in a logorithmic growth phase of *Enterobacteriaceae* taxonomy, with the end hardly in sight. It must be emphasized, to the credit of the bacterial taxonomist, that new species and genus designations have been withheld until adequate laboratory tests could be identified that allowed the clinical microbiologist to recognize these newly designated entities. An important caveat that one must appreciate is that molecular biology is not invariably the end-all of classification. As an example, shigellae may be considered members of the genus *Escherichia* on the basis of DNA reactions. To designate a *Shigella* as an *Escherichia* would be clinically and epidemiologically unacceptable, but should serve to highlight the spectrum of potentially host-harmful activities that reside within the realm of DNA homologies. Molecular biology permits us to recognize a species, but it does not as yet provide the tools to identify important variants within the species (Isenberg and Isenberg, 1985). Thus, we are able to recognize the bacterial "dog" but we cannot differentiate between the St. Bernard and the dachshund.

The vastness of the family *Enterobacteriaceae* makes it a given that many species are not necessarily encountered in the intimate human biosphere or in the intestinal tract of humans and animals. Widely distributed in nature, it is the adaptability of members of the family that bring rarely encountered or unknown species or subspecies within the confines of humans, possibly as a result of selective pressures extant in the communities and medical facilities. The dynamic nature and selectivity brought about by many influences have resulted in the recognition of underlying disease complications by a number of newly identified species almost exclusively in the hospital setting. This new role of formerly neglected *Enterobacteriaceae* representatives does not obviate the place of many established and recognized genera and species of the family. *Escherichia coli* is a member of the enteric bacteriosis group; it is discussed in Chapter 26 (*Escherichi coli* Infections) of this volume. This chapter addresses the more recently recognized members of the family that affect human health. Included are bacteria that were considered mostly as commensals in the past, but have now been shown to participate in complicating the recovery of patients in medical institutions. This rather large collection of *Enterobacteriaceae* will be supplemented by the inclusion of the genera *Aeromonas* and *Plesiomonas*.

These bacteria of various habitats in nature and wide spectra of activity share one characteristic with regard to the human host. They may be occasional or accidental residents of skin or mucous membranes, but are of little significance in disease production unless the human host has been compromised in some fashion. Thus, the isolation of the majority of these organisms is rare in patients in the community setting. Patients with an infection by one of these bacteria admitted to medical facilities are most often patients in remission from an underlying disease. In the institutional setting, where a variety of procedures and devices promote colonization of the human host, these bacteria are usually encountered as infectious agents. The institutional use of antimicrobial drugs is quite probably the most important factor in concentrating these bacteria in the modern hospital. The ability of these microorganisms to resist the commonly used antibiotial agents provides the advantage to replace the normal microbiota and, thus, to complicate the recovery of the affected patients. Although molecular epidemiology has not as yet become an established procedure in clinical microbiology laboratories, such studies often indicate that resistance plasmids enable these bacteria to play their role as nosocomial challengers and that their encroachment on the established microbiota may in part reflect the ease of information exchange between the newcomers to the human environment. Therefore, plasmid epidemiology may have to supplement bacterial taxonomy to cope with this challenge to the recovery of patients.

In keeping with the family name, enteric bacteriosis encompasses these less commonly encountered representatives of *Enterobacteriaceae* as well as the aeromonads. This chapter does not include all of the recently designated genera and species of *Enterobacteriaceae*.

Bacterial Sources

Environmental Occurrence and Habitats

The representatives of enteric bacteriosis are widely distributed in nature. The origins for entry into the human environment are listed in Table 1 along with the clinical specimens from which they were isolated. Table 1 is not exhaustive, but renders the sources presently listed (Krieg and Holt, 1984; Lennette et al., 1985; Starr et al., 1981); however, for each genus other sources may exist. One must also keep in mind that isolation does not necessarily signify habitat. Thus, a bacterial animal resident may be recovered from inanimate or plant environments frequented by the particular animal host. The genus list-

TABLE 1. Sources of the agents of enteric bacteriosis

	Environmental sources						Clinical sources							
Genus	*Animals*	*Water*	*Sewage*	*Soil*	*Food*	*Plants*	*Blood*	*Urine*	*Central nervous system*	*Wounds*	*Respiratory tract*	*Feces*	*Autopsy specimens*	*Medical devices*
Citrobacter	+	+	+	+	+		+	+	+	+	+	+	+	+
Klebsiella	+	+	+	+	+	+	+	+	+	+	+	+	+	+
Enterobacter	+	+	+	+	+	+	+	+	+	+	+	+	+	+
Serratia	+	+				+	+	+	+	+	+	+	+	+
Hafnia	+	+	+	+			+	+		+	+	+	+	+
Edwardsiella	+						+	+	+	+		+		+
Proteus	+	+		+			+	+	+	+	+	+	+	+
Providencia	+	+		+			+	+	+	+	+	+	+	+
Morganella	+						+	+		+	+	+	+	+
Kluyvera		+	+	+	+		+	+		+	+	+	+	
Ewingella					+		+	+		+	+	+		
Aeromonas	+	+	+	+	+		+	+	+	+		+		
Plesiomonas	+	+					+		+	+		+		

ing may exclude a specific species with an environmental predilection not shared by other species in the genus.

The table emphasizes the role of animals as vectors for the entrance of enteric bacteriosis organisms into the human biosphere. Although urban environments decrease this interaction, pets may serve this function in addition to the bacteria introduced to humans in foods containing bacteria that reflect the normal microbiota of the animal host or products of an environment in contact with this host or its wastes. The zoonotic origin of many bacteria, including the etiologic agents of enteric bacteriosis, has been neglected and deserves greater emphasis in the search for the source of microorganisms that intrude into the environment of hospitalized patients.

It is not surprising that most genera can be detected in water. Most potable water in developed countries contains a very low number of bacteria and allegedly never any representatives of that ill-defined group, the coliforms. Although all representatives can be isolated from patients' feces, their role in nature does not include the intestinal tract as a preferred niche. Bacteria such as the aeromonads are common in aqueous habitats; the representatives of *Enterobacteriaceae* are probably only transient residents of this environment; *Edwardsiella tarda* has not been recovered from this source. Sewage can also harbor many of the agents of enteric bacteriosis; those not included may occur in numbers so low that they have escaped detection. They may eventually be recovered from this source. This group of bacteria is well represented among isolates from soil; similarly, omission from this category in the table does not preclude future isolation from this source. Food generally has yielded *Citrobacter*, *Klebsiella*, *Enterobacter*, *Kluyvera*, *Ewingella*, and *Aeromonas* spp. The role of these bacteria in food has not been clarified; it is not clear if food serves merely as an accidental vehicle for the dissemination of these bacteria or if their presence indicates a degree of spoilage. To date, the bacteria listed have not been implicated in community outbreaks of food poisoning or diarrhea, although *Aeromonas* has been suspected as a possible etiologic agent of the latter.

A variety of plants harbor *Klebsiella*, *Enterobacter*, and *Serratia* spp. The first two especially display enhanced growth responses in the presence of plant-derived carbohydrates. This connection may also explain the presence of *Klebsiella* in industrial effluents.

The remaining sources listed indicate merely the ubiquity of these bacteria in nature; absence from these sources cannot be implied for the bacteria not listed. No systematic effort to define habitats has been undertaken, making definite statements difficult.

Clinical Occurrence

Table 1 also lists the clinical specimens that yielded representatives of the enteric bacteriosis. The listing should be interpreted as confirming the isolation of these bacteria in these specimens without signifying involvement in disease production. Although most of the reports in the literature imply the presence of overt infections in compromised patients, the exact role of the bacteria remains somewhat vague. Certainly, repeated isolations from blood, spinal fluids, and wounds leave little doubt about involvement in the disease picture; the bacterial presence in urine, sputum, and autopsy specimens may reflect colonization in response to antimicrobial therapy or prolonged hospitalization. Interpretation of the laboratory isolation of all agents of enteric bacteriosis cannot be accomplished without a patient history that details the underlying disease, recent therapy, degree of immune compromise, length of hospital stay, the presence of any prosthesis including catheters, requirements for respiratory assistance, and hematologic and pathologic reports. Involvement in disease production or establishing the degree of pathogenicity cannot be based on the mere laboratory isolation of the bacterium. We must acknowledge the determinant role played by the host in the overt clinical manifestations of disease. These caveats notwithstanding, this group of bacteria can be present in a variety of clinical specimens; blood, urine, and wounds especially have been the sources. The rather recent genus designations such as *Ewingella* and *Kluyvera* and the lack of interest in the laboratory recognition of aeromonadlike bacteria undoubtedly have influenced the representation of these bacteria in specimen categories. The ability to suggest that certain tissues or organs may present a selective advantage for colonization or the expression of disease involvement is hampered by the paucity of studies aimed at such interactions. At this stage of our understanding of enteric bacteriosis, one can only state that serendipity seems to govern the intrusion of a specific bacterium into the human biosphere and that the consequences of this chance encounter are controlled by a particular host. Several members of the group are known constituents of the normal microbiota and, thus, are amphibionts or sometime pathogens (Rosebury, 1961); other representatives may share this property, but have escaped laboratory detection. It is surmised that most members of the group are introduced at the time of underlying disease from extraneous sources (Isenberg and Smith, 1981).

The agents of enteric bacteriosis share especially the ability to resist the commonly used antimicrobial agents. Many species of the group apparently do not respond constitutively to a considerable array of drugs; others, especially the *Enterobacter* spp., synthesize inducible β-lactamases, whereas all seem to harbor, expand, and mobilize resistance plasmids with ease. It is this latter property that requires special attention. Different species responsible for nosocomial infections can share the same plasmids. Therefore, therapy should be directed against the plasmid rather than the specific bacterium. The need for molecular epidemiology to control hospital-acquired infections is obvious. One can only hope that simplification and increasing awareness of techniques to deal with this problem shall become part of the clinical microbiologist's approach to the problem (Lennette et al., 1985). In addition, microbiologists must recognize the dominant genus or species of the group resident in each particular institution, constantly monitor the antimicrobial profiles of these nosocomiants, and apprise the clinicians, especially the infectious disease specialists and epidemiologists, of shifts in the microbial responses to anti-infective therapy. Simultaneously, and in seeming contradiction, the isolation of a potentially nosocomiant bacterium warrants a careful study of its response to antimicrobial agents and surveillance for possible changes in that response during therapy. Thus, although a common plasmid endowing several species with common multiply resistant drug responses may be present in any given institution, isolates from newly admitted or recently infected patients deserve a thorough analysis of their antimicrobial response patterns.

Laboratory Diagnosis

Isolation

All members of this group can be isolated readily on the media presently employed in most United States laboratories for the detection of members of the family *Enterobacteriaceae*. All grow well on MacConkey, eosin methylene blue, and blood agars. On selective media, colonial morphology and dye uptake, pigmentation and tests for oxidase, catalase along with motility, established by one of several methods, can help immediately to direct, at least, initial grouping of the bacteria that constitute the enteric bacteriosis. Tne usual caveats about properly collected and transported specimens, the distribution and number of the suspect colonies in relation to the specimen source and the need for patient information of course pertain to these isolations.

Approach to Genus Recognition

Many approaches are available to identify these bacteria to the species level; these methods range from testing for individual biochemical activities to manual system approaches to automated and computer-enhanced analyses (Lennette et al., 1985). Although the dichotomous keys to the identification of bacteria are no longer the vogue, it seems appropriate to outline a guide to the recognition of the genera involved (Isenberg and MacLowry, 1976). This approach is used here (Fig. 1) as a suspicion index to help laboratorians without the more sophisticated modalities to arrive at the appropriate genus with a minimal number of individual tests. Genera with similar biochemical reactions are included in the table to avoid the obfuscation created by a few select species. It is apparent that motile fermentative bacteria should be screened for oxidase to recognize quickly the presence of aeromonad like organisms. Nonmotile bacteria can be separated readily with adonitol, serologic testing for *Shigella* spp., and acetoin production. The motile, oxidase-negative, fermentative gram-negative rods require examination for DNase, urease, phenylalanine deaminase, and acetoin production; xylose, melibiose, cellobiose, and arabinose fermentation; O-nitrophenyl-β-D-galactoside and esculin hydrolysis; citrate utilization; and ornithine and lysine decarboxylases, as well as H_2S production. The judicious use of the reactions listed will limit this first attempt at grouping the bacteria to a few tests.

Once genus identification has been established, the test to identify the isolate to the species level may be undertaken. It is at this juncture that clinical microbiologists should establish if such further efforts are helpful for the diagnosis, and therapeutic regimens and any epidemiologic implications generated by this group of bacteria in a particular hospital setting. Knowledge of the antibiotic profile of the isolate and appropriate patient information are required to reach a sensible decision. The inherent difficulty in designating a bacterium as a colonizing organism in contrast to one involved in complicating the recovery of a patient remains a challenge, especially with specimens such as sputum, urine, throat swabs, feces, surface wounds, and the proximal areas of indwelling catheters. Repeated isolation of the same bacterium in large or dominating numbers from properly collected and transported specimens improves the chance of clinical significance for such an isolate. When these bacteria have been recovered from blood, cerebrospinal fluid, deep wounds, transtracheal aspirates, or biopsy specimens, the likelihood of disease involvement is considerable. Under the circumstances enumerated, further analyses are indicated. The finding of a few colonies of the agents of enteric bacteriosis in a polymicrobial specimen should not lead to additional efforts under ordinary circumstances. Although clinical microbiologists have always been tempted to identify as many of the

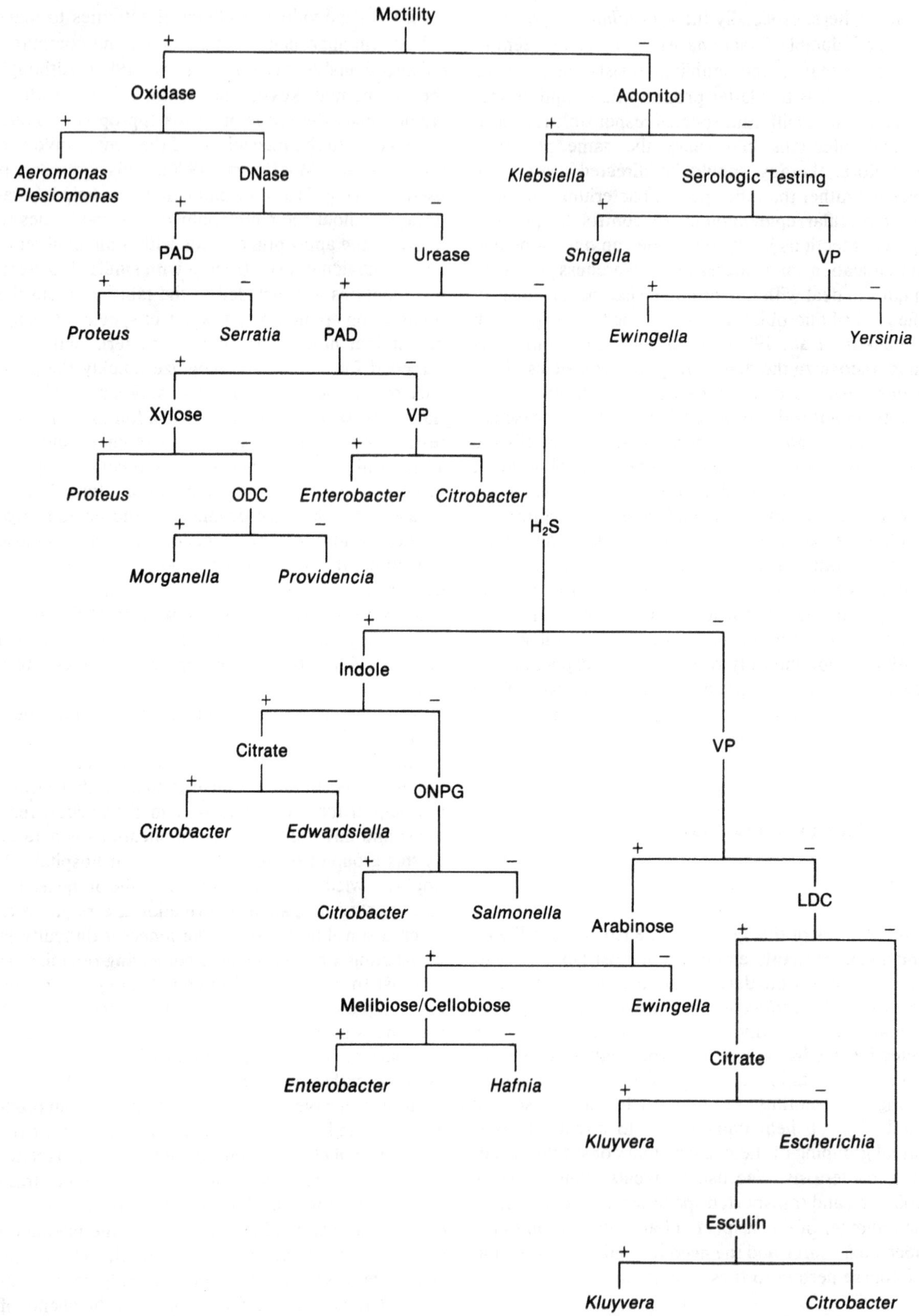

FIG. 1. Guide to identification of the genera involved in enteric bacteriosis. PAD = Phenylalanine deaminase; VP = Voges–Proskauer reaction; ODC = ornithine decarboxylase; ONPG = *O*-*N*itrophenyl-β-D-galactoside; LDC = lysine decarboxylase.

TABLE 2. Differentiation of *Klebsiella* species

Test	Klebsiella *species*[a]			
	K. pneumoniae	K. oxytoca	K. planticola	K. terrigena
Indole	0	+	d	0
Gas from lactose at 44.5°C	+	0	0	0
Growth at 10°C	0	+	+	+
Fermentation of inulin	0	+	d	d
Fermentation of D-melizitose	0	d	0	+
Utilization of m-hydroxy-benzoate	0	+	0	+
Methyl red	0	0	d	+

[a] + = ≥90% positive; 0 = ≤10% positive; d = 11 to 89% positive.

bacteria in a specimen as possible, the clinical significance of such information may be of little consequence (Bartlett, 1985). In fact, in the hands of clinicians not experienced in infectious disease, such information may be damaging, as it may lead to therapeutic measures not required. The recent fiscal constraints on health care demand that the clinical microbiologist assist clinician colleagues by reporting only those isolates with obvious or highly suspect roles in the disease manifested in the patient; dialogue between all parties concerned is imperative.

Several genera and species included and classified as members of the enteric bacteriosis group make such a decision very difficult. Farmer et al. (1985) stated that three species of *Enterobacteriaceae* comprised 80 to 95% of all clinical isolates from the family. These three species are *Escherichia coli*, *Klebsiella pneumoniae*, and *Proteus mirabilis*. *Klebsiella pneumoniae* and *Proteus mirabilis* are included in this discussion because multiply-resistant strains are isolated from nosocomial infections with increasing frequency while antibiotic susceptible strains represent the large majority of community isolates. Such strict interpretation may not find universal acceptance, as these bacteria are involved in disease production in patients with diminished immunity or other underlying diseases not sufficiently severe to warrant hospitalization initially. The observations of Farmer and colleagues (1985) direct analytical efforts in the clinical microbiology setting to the rapid, accurate recognition of *K. pneumoniae* and *P. mirabilis*.

Agents of Enteric Bacteriosis

Klebsiella

The various *Klebsiella* spp. can be distinguished by the seven tests listed in Table 2. *Klebsiella oxytoca* has been isolated with some frequency from clinical specimens; although feces seem to be the major specimen yielding these bacteria, blood, urine, and respiratory tract specimens, for example, have also harbored them. As the name implies, *Klebsiella planticola* has as its habitat vegetables, seeds, leaves, and the like, but has been found in drinking water and a few clinical specimens. While it may appear redundant, we must mention once more that these and most other species mentioned may never become frequent in the general clinical setting, but they do have the potential to do so. The laboratorians' ability to recognize them at such times must be considered important, as only such identification can lead to their containment. *Klebsiella terrigena* has not as yet been isolated from clinical specimens. The species *Klebsiella pneumoniae* consists of three subspecies listed in Table 3. *Klebsiella ozaenae* and *Klebsiella rhinoscleromatis* are considered biochemically inactive variants of *K. pneumoniae* that have adapted to produce specific disease manifestations (Krieg and Holt, 1984). However, the two subspecies *K. ozaenae* and *K. rhinoscleromatis* are seen very rarely in clinical specimens in the United States and have not been encountered in patients at risk for enteric bacteriosis.

TABLE 3. Differentiation of *Klebsiella pneumoniae* subspecies

Test	K. pneumoniae *subspecies*[a]		
	K. pneumoniae	K. ozaena	K. rhino-scleromatis
Acid from lactose	+	+	0
Acid from dulcitol	d	0	0
Voges–Proskauer	+	0	0
Urease	+	d	0
Citrate	d	d	0
Mucate	+	d	0
Lysine decarboxylase	+	d	0

[a] + = ≥90% positive; 0 = ≤10% positive; d = 11 to 89% positive.

TABLE 4. Differentiation of *Proteus* species

Test	Proteus *species*[a]		
	P. mirabilis	P. vulgaris	P. penneri
Indole production	0	+	0
Ornithine decarboxylase	+	0	0
Acid from maltose	0	+	+
Esculin hydrolysis	0	d	0
Chloramphenicol susceptibility[b]	+	d	0

[a] + = ≥90% positive; 0 = ≤10% positive; d = 11 to 89% positive.
[b] Usual clinical isolates; exceptions can be found.

The *Proteus* Group

Proteus mirabilis, the other most frequent clinical isolate, belongs to a group of bacteria that includes the genera *Providencia* and *Morganella.* These bacteria elaborate phenylalanine deaminase and urease (Fig. 1) and may be differentiated from one another by DNase production (some *Proteus* spp.), swarming, xylose fermentation by DNase-negative *Proteus* representatives that separate them from *Morganella* and *Providencia.* These two genera may be distinguished from one another by ornithine decarboxylase, an enzyme elaborated by *Morganella* but not by *Providencia* spp. Table 4 lists the select tests to separate the clinically significant species of *Proteus. Proteus mirabilis* and *Proteus vulgaris* are well-known organisms long identified without difficulties in the clinical microbiology laboratory. Variations within the species *P. vulgaris* are discernible phenotypically, especially salicin fermentation and esculin hydrolysis (Farmer et al., 1985). Although these differences may reflect the need for taxonomists to establish new species, the difference is of little consequence to the clinical circumstances. Clinical microbiologists should keep this versatility in mind, especially if they pursue the epidemiology of cluster epidemic. It is this type of phenotypic difference that led to the establishment of *Proteus penneri* (Hickman et al., 1982), a bacterium recognized with increasing frequency in the institutional setting, where its characteristic great ability to resist antimicrobial agents undoubtedly provides the opportunity to complicate the recovery of agranulocytic patients. These three members of the genus are widely distributed in the human environment and many are endogenous minority constituents of the normal human microbiota. As has been mentioned previously, the division of bacteria into colonizers and those involved in disease must be based on the clinical history and laboratory information. Another species, *Proteus myxofaciens* (Brenner et al., 1978), has not as yet intruded into the human biosphere.

The genus *Providencia* has played a significant role in nosocomial outbreaks of urinary tract infections, especially *Providencia rettgeri* and *Providencia stuartii.* These two bacteria, differentiated as shown in Table 5, are rarely found in feces. Their natural habitat and mode of entry into the compromised patient is not clear. *Providencia alcalifaciens* has been isolated from diarrheal stools of infants and children, but no definite etiologic role can be assigned. *Providencia rustigianii,* a new species proposed in 1983 by Hickman-Brenner et al., has been isolated occasionally from diarrheic stools, but even less is known about its causal relationship to the observed symptoms. This genus provides an excellent illustration of the need to identify bacteria to the species level when the clinical circumstances for such an analysis are appropriate.

Morganella morganii, the third member of the *Proteus* group, has been suspected as an etiologic agent of so-called summer diarrhea (Penner, 1981). The role of this bacterium as the causal agent of diarrhea is quite controversial and has not been defined satisfactorily. This bacterium is being recovered with

TABLE 5. Differentiation of *Providencia* species

Test	Providencia *species*[a]			
	P. alcalifaciens	P. rettgeri	P. rustigianii	P. stuartii
Urease	0	+	0	d
Acid production from				
myo-Inositol	0	+	0	+
Adonitol	+	+	0	0
Arabitol	0	+	0	0
Trehalose	0	0	0	+
Galactose	0	+	+	+

[a] + = ≥90% positive; 0 = ≤10% positive; d = 11 to 89% positive.

increasing frequency from wounds, urine, respiratory tract specimens, and the like as a member of the enteric bacteriosis group.

Serratia

The several species of the genus *Serratia* are widely distributed in nature. They may be found in fresh water; those recovered from a marine environment are salt tolerant, but do not require Na^+, the hallmark of truly marine bacteria. *Serratia* spp. are ubiquitous in nature and are found on leaves, shrubs, fruits, vegetables, herbs, mushrooms, and mosses (Krieg and Holt, 1984). Plants may serve as the vector to introduce these bacteria into the human biosphere. Many insects are colonized by members of the genus however, and some may be pathogenic for the insect species from which they have been isolated. Several species have become established in hospitals, where they complicate the recovery of compromised patients. The frequency with which these bacteria are involved in nosocomial disease varies from hospital to hospital within each geographic area. This difference may be a reflection of the antimicrobial agents used in each institution as well as in the communities served by the institutions. The ingestion of raw vegetables in salads accompanying patients' meals and the presence of flowers in patients' rooms may play a role in the colonization of patients and disease manifestations caused by representatives of *Serratia*.

The genus *Serratia* is distinguished from other genera of the family *Enterobacteriaceae* by several biochemical activities (Fig. 1). *Serratia* spp. produce extracellular deoxyribonuclease, gelatinase at 22°C, and lipase and are resistant to colistin and cephalothin. The combination of these biochemical activities and the responses to the two antimicrobial agents is not encountered in other *Enterobacteriaceae*. DNase, lipase, and gelatinase may be produced by *Proteus* spp., but not as constantly as by *Serratia*. The response to the two antibiotics is helpful in distinguishing lipase-, gelatinase-, and DNase-positive strains of *Proteus* from *Serratia*. DNase is produced by members of the CDC-designated enteric group 68. These microorganisms fail to elaborate lipase and gelatinase. Most *Cedecea* spp. produce lipase, but are negative for DNase and gelatin hydrolysis. There are presently six species in the genus, which may be differentiated as shown in Table 6.

Several *Serratia* spp. included in Table 6 have not as yet impinged on the human environment. Of these, *Serratia ficaria* in particular has only recently been introduced into the human biosphere (Brouillard et al., 1984; Gill et al., 1981; Pien and Farmer, 1983).

The most common species is *Serratia marcescens*, which is the only species that has been isolated with some frequency from hospitalized patients. It is, as shown in Table 6, distinguished from the other serratiae by its inability to ferment L-arabinose. This inability is an important biochemical differentiation, as not all *S. marcescens* produce the typical red pigment prodigiosin (2-methyl-3-amyl-6-methoxyprodigiosene). However, this pigment is not an exclusive characteristic of *S. marcescens*. Strains of *Serratia rubidaea* and *Serratia plymuthica* can elaborate the same or a very similar water-insoluble compound. There are 19 biotypes or biovars of S. *marcescens*, which are important for the epidemiologic differentiation of the species. Biotyping is based mostly on carbon sources, but this procedure is of little practical use in the average clinical microbiology labora-

TABLE 6. Differentiation of *Serratia* species

Test	Serratia *species*[a]					
	S. marcescens	S. liquefaciens	S. rubidaea	S. plymuthica	S. odorifera	S. ficaria
DNase (25°C)	+	+	+	+	+	+
Lipase	+	+	+	d	d	+
Gelatinase (22°C)	+	+	+	d	+	+
Lysine decarboxylase	+	+	d	0	+	0
Ornithine decarboxylase	+	+	0	0	d	0
Distinct odor	0	0	0	0	+	+
Prodigiosin production	d	0	0	d	0	d
Fermentation of						
L-Arabinose	0	+	+	+	+	+
D-Arabitol	0	0	+	0	0	+
D-Sorbitol	+	+	0	d	+	+
Adonitol	d	0	+	0	d	0

[a] + = ≥90% positive; 0 = ≤10% positive; d = 10 to 90% positive.

tory. However, only six *S. marcescens* biotypes produce the pigment. The nonpigmented representatives are most frequently involved in nosocomial disease; certain of the nonpigmented biogroups are almost exclusively recovered from cases of nosocomial infections, namely, biotypes A5, A8abc, and TCT. The Centers for Disease Control recognize another variant group, *S. marcescens* biogroup 1, which is isolated almost exclusively from urine specimens (Farmer et al., 1985). These organisms belong phenotypically to the genus, do not ferment L-arabinose, and are able to express only some of the other typical attributes of S. *marcescens*, such as motility, a positive Voges–Proskauer reaction, citrate utilization, gelatinase, lysine, and ornithine decarboxylase production, KCN tolerance, and gas production on glucose. *Serratia marcescens* displays a very complex antigenic spectrum. There are 20 somatic and 20 flagellar antigens recognized, but serologic analyses of these organisms is complicated by crossreactions as well as the lack of commercially available reagents. There seems to be a relationship between biotypes and antigenic composition. Several members of other genera in *Enterobacteriaceae* may cross-react with antisera to *S. marcescens* somatic antigens. A potentially useful distinction can be achieved with bacteriocin and bacteriophage typing, as well as the demonstration of distinct proteinases.

The most important attribute of *S. marcescens* is its ability to resist antimicrobial agents and become a reservoir for the transmission of resistance to other bacteria, especially members of the family *Enterobacteriaceae*, in the intimate biosphere of hospitalized patients. The distribution of *S. marcescens* in nature appears to prepare this bacterium to respond effectively to the ecologic challenges of other microorganisms or to express this attribute when other members of the human microbiota are curbed or eliminated by agents related to those, *S. marcescens* may have encountered in its natural environment. It is rare that patients are admitted with community-acquired *S. marcescens* infections; such individuals are frequently granulocytopenic or immunosuppressed by disease or therapy. *Serratia marcescens* is an opportunist in respiratory tract and urinary tract colonizations and infections, at times invading the bloodstream. It may contaminate indwelling catheters in various anatomic sites, colonize or complicate surgical and traumatic wounds, and contaminate equipment used for respiratory assistance and the entire inanimate environment of the patient. Therefore, the dissemination of a particular strain to other patients can take place quite readily, especially to patients compromised by disease or therapy and receiving antimicrobial prophylaxis. The role of hospital personnel in the spread of this organism is assumed to be considerable.

Serratia liquefaciens (*Serratia proteamaculans* and *Enterobacter liquefaciens*) differs from *S. marcescens* in its ability to ferment L-arabinose and failure of all members of the species to produce pigment. It is a common inhabitant of environmental sources, including water, plants, insects, and foods. It is encountered less frequently than *S. marcescens* in human specimens, but can become established as a nosocomial bacterium with antimicrobial resistance patterns similar to those of *S. marcescens*. This organism has been reported more frequently in England than in many other countries, where it has been especially recovered from blood cultures. Molecular biologic analysis indicates that this species may consist of two distinct types. We might anticipate that the taxonomists may divide the present species. Clinical microbiologists should, therefore, be aware that *S. liquefaciens* may represent a group of phenotypically related bacteria. The clinical significance of *S. liquefaciens* is its potential complicating ability in compromised patients. If the organism is frequently isolated from different clinical specimens or patients, the clinical microbiologist should be alert to the possibility of a cluster epidemic.

The other species of *Serratia* have been encountered comparatively rarely in clinical specimens. *Serratia rubidaea* (*S. marinorubra*) is a pigment-producing organism that has been isolated from water, including marine sources, plants, and insects. Clinically it has been found in specimens of the respiratory tract, wounds, feces, and on rare occasions from bacteremia. Further investigation of its role in human disease is required. *Serratia plymuthica* (Clark and Janda, 1985) is found primarily in water and on plant surfaces and is a rare complicator of respiratory tract infections. The exact role of this organism in complicating the recovery of these patients has not been established. An infrequently encountered species, *S. odorifera*, can be identified by a pungent vegetablelike odor. This bacterium may eventually be shown to be involved in nosocomial infections.

It is important for clinical microbiologists to remember that the antibiotic susceptibility of all *Serratia* spp. varies and that each isolate may display a spectrum of resistance. The bacteria are usually resistant to the penicillins, the tetracyclines, and the macrolid antimicrobial agents. The new cephalosporins may be effective against certain strains. The aminoglycosidic aminocyclitols often successfully inhibit many isolates. During therapy however, *Serratia* may very quickly manifest resistance to a drug to which it was originally susceptible.

Citrobacter

The genus *Citrobacter* (Farmer, 1981) is most easily differentiated from other members of the family *En-*

terobacteriaceae by its ability to utilize citrate as a sole carbon source and by not producing acetylmethylcarbinol or elaborating lysine decarboxylase (Fig. 1). The organism is considered a normal inhabitant of the intestinal tracts of humans and animals, including mammals, birds, reptiles, and some insects. It is also found in soil, water, sewage, and foods, as well as in industrial wastewater. Its role in disease production is that of an opportunistic bacterium taking advantage of the suppressed immunity of the human host. However, these organisms, especially *Citrobacter freundii* and *Citrobacter diversus*, have been suspected of causing diarrhea in humans.

Table 7 lists the differentiating characteristics of the three species in this genus. *Citrobacter freundii* remains the most commonly isolated representative. This bacterium is usually indole negative, whereas *C. diversus* and *C. amalonaticus* are positive. *Citrobacter amalonaticus* is unable to utilize malonate as a sole carbon source. This bacterium joins *C. freundii* in the ability to grow in the presence of KCN, a characteristic not shared by *C. diversus*.

It is of interest that several antigenic investigations of *C. freundii* indicate 42 somatic and more than 70 flagellar antigens. Antigens of *C. freundii* are related to those of the salmonellae and escherichiae. *Citrobacter diversus*, on the other hand, has from 6 to 17 O antigens and at least 7 flagellar antigens. *Citrobacter amalonaticus* has been reported to possess 13 somatic antigens, which seem to be related to those of *Shigella dysenteriae* and *Shigella boydii*.

Citrobacter freundii, a not uncommon organism in clinical specimens, usually displays susceptibility to the aminoglycosides in addition to chloramphenicol and the polymyxins. Its susceptibility to the remaining antimicrobial agents, especially the penicillins and cephalosporins, remains extremely individual.

Citrobacter freundii has been isolated from clinical specimens such as urine, sputum, blood, spinal fluid, otitis media, surgical and traumatic wounds, and abscesses and from post-mortem specimens. Occasionally, throat cultures, especially those of patients on antimicrobial therapy, may yield the organism. This variety of involvement reflects the opportunistic propinquities of this bacterium. The species *C. diversus* and *C. amalonaticus* have been isolated from similar specimens. Although both are less frequently isolated, this finding may not pertain to specific localities.

The lack of lysine decarboxylase production helps in the differentiation of the citrobacters from the salmonellae, escherichiae, and most of the enterobacters. The ability of these organisms to utilize citrate as the sole carbon source aids in their differentiation from escherichiae. The inability of *Citrobacter* spp. to produce acetylmethylcarbinol differentiates the members of this genus from *Enterobacter* spp. As *C. freundii* and *C. amalonaticus* can grow in the presence of KCN, this test may help differentiate them from the salmonellae and escherichiae. The inability of *C. freundii* to produce indole helps in the separation from *E. coli*.

TABLE 7. Differentiation of *Citrobacter* species

	Citrobacter *species*[a]		
Test	C. freundii	C. diversus	C. amalonaticus
Indole	0	+	+
Citrate	+	+	d
H_2S	d	0	0
KCN	+	0	+
Malonate	d	+	0
Adonitol	0	+	0
Raffinose	d	0	d
Melibiose	d	0	d
Arabitol	0	+	0

[a] + = ≥90% positive; 0 = ≤10% positive; d = 11 to 89% positive.

Enterobacter

Enterobacter spp. (Brenner, 1981) are widely distributed in nature. They may be recovered from human and animal feces, where they seem to be transient rather than permanent residents. They are also found in soils, sewage, milk and dairy products, animal hides, and meat. They have been isolated from fish found in contaminated water, in plants such as grasses, wheat, corn, sugar cane, and bananas, and in fresh water. However, *Enterobacter sakazakii* and *Enterobacter gergoviae*, the most recently described members of the genus, have only been encountered in humans. It will be necessary to study their environmental sources to designate their particular natural niches. The various species of this genus can be identified readily in the clinical microbiology laboratory as shown in Table 8.

Enterobacter spp. are involved most frequently in complicating the recovery of hospitalized patients. As with the other etiologic agents of enteric bacteriosis, patients are rarely admitted with overt *Enterobacter* infections unless underlying disease or therapy administered while they were ambulatory compromised their immune mechanisms. The members of the genus constitute approximately 5% of all nosocomial infections; they have been isolated from blood, urinary tract and surgical wound infections, pneumonia, and burn wounds. The frequency with which these bacteria are encountered may reflect their ability to resist an appreciable number of antimicrobial agents. They do so by a variety of mechanisms, such as constitutive chromosomal, inducible

TABLE 8. Differentiation of *Enterobacter* species

Test	Enterobacter *species*[a]				
	E. aerogenes	E. cloacae	E. agglomerans	E. gergoviae	E. sakazakii
Lysine decarboxylase	+	0	0	+	0
Arginine dihydrolase	0	+	0	0	+
Ornithine decarboxylase	+	+	0	+	+
KCN	+	+	d	0	+
Sucrose	+	+	d	+	+
Adonitol	+	d	0	0	0
Sorbitol	+	+	d	0	0
Raffinose	+	+	d	+	+
Pigment	0	0	d	0	+

[a] + = ≥90% positive; 0 = ≤10% positive; V = 11 to 89% positive.

chromosomal, or plasmid-mediated resistance. Their lack of response to many antimicrobial drugs enhances their chances of survival, colonization, and disease complication. These bacteria can contribute to the resistance plasmid pool in an institution by their ability to transfer the extrachromosomal resistance factors to susceptible bacteria.

In the analysis of clinical specimens, it is not unusual to ignore the members of the genus if they are encountered in small numbers. These bacteria are recognized easily on MacConkey agar or eosin methylene blue agar, as the most frequently seen members ferment lactose readily. *Enterobacter agglomerans* and *E. gergoviae* do not attack lactose and, thus, may appear as nonfermenting colonies. Even these non-lactose-fermenting organisms can readily be recognized as *Enterobacter* spp. because they hydrolyze O-nitrophenol-β-D-galactoside. They share biochemical activities typical of the genus (Fig. 1). The biochemical activities of *E. agglomerans* challenge the clinical microbiologist. This species may be recognized by its failure to decarboxylate the usual three amino acids included in identification schemes, the late production of a yellow pigment (72%), motility (89.4%), arabinose fermentation (98%), rhamnose fermentation (approximately 96% with an additional 4% producing acid after prolonged incubation), and indole production (20%). The other *Enterobacter* spp. can be easily distinguished from one another and from the members of the family *Enterobacteriaceae* by their biochemical activities.

The various somatic and capsular antigens of the genus have not been evaluated sufficiently to be of help in their detection or classification. However, these antigens, bacteriocins, plasmids, and bacteriophages are now under intensive investigation for epidemiologic purposes and for molecular and immunologic probes designed to detect their presence in clinical specimens.

The members of the genus are quite versatile in their response to antimicrobial agents. It is important, therefore, that each clinically significant isolate be tested for its antibiotic susceptibility. Although the laboratory tests may indicate susceptibility to ampicillin and the first-generation cephalosporins, clinical experience has shown that these drugs have little effect in the treatment of patients with *Enterobacter* infections. It must also be remembered that laboratory susceptibility studies, especially microdilution and rapid methods, fail to detect inducible β-lactamase resistance of *Enterobacter* spp. to second- and third-generation cephalosporins and the broad-spectrum penicillins. These drugs are potent inducers of various *Enterobacter* β-lactamases found in the periplasmic space. Here, these enzymes can combine even with drugs they do not hydrolyze and possibly present a barrier for the diffusion of these antimicrobials into the bacterial cytoplasm or prevent their combination with appropriate penicillin-binding proteins (Sanders and Sanders, 1979).

Edwardsiella

Edwardsiella tarda, the only species in this genus, is an organism widely distributed in the animal kingdom (Farmer, 1981b). It is found in pets, domestic animals, and rodents and in aquatic environments may be recovered from birds, fish, frogs, turtles, and marine species. The bacterium is also present in environments inhabited by these animals; at least it is assumed that the environmental inanimate sources are contaminated by the animals carrying *E. tarda*. These organisms are found in the stools and other sources of healthy animals, although they are suspected of causing red disease in pond-cultured eels and emphysematous putrefactive disease of channel catfish. At the present time, the ecologic niche of this bacterium is the animal intestine. It is characterized by growing more slowly than other *Enterobacteriaceae* on ordinary media. The organism, along with representatives of *Serratia*, *Proteus*, *Providencia*, and *Morganella*, is resistant to colistin, and it does not ferment mannitol, xylose, trehalose, or adonitol. As shown in Fig. 1, the organism can be separated from the other representatives of the family by the following characteristics: it is motile, oxidase negative, DNase negative, and urease negative, and it produces H_2S and indole, but does not utilize citrate as the sole carbon source.

Edwardsiella tarda has been suspected of causing bloody diarrhea, possibly in conjunction with *Entamoeba histolytica*. It has rarely been isolated from the feces of healthy people. The demonstration of involvement in diarrhea apparently depends on the methods of processing the specimen, the geographic locale, and the season. Suspicion of involvement in diarrhea has been established mostly in developing countries. The organism is not readily found in industrialized areas. It is important that increases in antibody titer be demonstrated in conjunction with the diarrheal disease.

As a member of the enteric bacteriosis group, this organism is most frequently encountered complicating surgical and other wounds. Usually this bacterium is accompanied by other microorganisms, making it extremely difficult to define its role as a nosocomially significant etiologic agent. It has rarely been found in meningitis, endocarditis, bacteremia or urinary tract infections. *Edwardsiella tarda*, like the other agents of enteric bacteriosis, readily displays resistance to antimicrobial agents. Therefore, its isolation makes it imperative that an antimicrobial profile be obtained. If prolonged therapy is required for the patient, during-treatment-cultures yielding *E. tarda* should be reexamined for their susceptibilities.

Hafnia

Hafnia alvei is an organism found in animals, birds, soil, sewage, and water (Sakazaki, 1981). Its precise habitat and niche have not yet been determined. It has been implicated as a possible agent of diarrhea and gastroenteritis, although no conclusive causal role has been assigned. *Hafnia* has been recovered from normal stools. Its opportunistic role has involved feces, sputum, urine, wounds, abscesses, autopsy material, and abdominal cavity as well as upper respiratory tract specimens. Usually it has been found in association with other bacteria. Isolation, in pure culture, has invariably been from patients with severe underlying diseases. Most of these patients were on antibiotic therapy before the emergence of *Hafnia*. Thus, *Hafnia* is another agent of enteric bacteriosis that is selected primarily because of antimicrobial therapy in patients who have immune deficiencies as a result of disease or therapy.

The organism, as shown in Fig. 1, is readily distinguished from other members of the family *Enterobacteriaceae*. It resembles the genus *Enterobacter* in most of its biochemical activities. It can be separated from members of that genus by its inability to ferment melibiose and cellobiose.

Ewingella

Ewingella americana is a new species in a new, recently described genus (Farmer, 1985). To date, very little about its normal habitat is known. The organism seems quite resistant to a number of antimicrobial agents and resembles the other agents of enteric bacteriosis in this regard. As shown in Table 1, the bacterium has been recovered from a variety of clinical sources such as blood, wounds, respiratory tract specimens, and stools. Environmentally, only food has been implicated as a potential source. It is of interest that this bacterium seems to cause small cluster epidemics in institutions where it becomes established.

As shown in Fig. 1, the organism is a nonmotile member of the family *Enterobacteriaceae* that does not ferment adonitol and produces acetylmethylcarbinol. It is, thus, an organism that could readily be confused with *Yersinia* spp.

Kluyvera

As indicated in Table 1, representatives of the genus *Kluyvera* (Farmer, 1985) are found in water, sewage, soil, and food. The exact habitat of these organisms has not yet been established. Members of the genus have come to the attention of clinical microbiologists by fulfilling the typical role of representatives of the enteric bacteriosis, namely, complicating the recovery of hospitalized patients. The bacterium has been encountered in blood, urine, wounds, respiratory tract specimens, feces, and autopsy samples. The organisms are characterized by their readily acquired resistance to antimicrobial agents. The patients whose recovery is complicated are usually immune compromised individuals. The two species in this genus, *Kluyvera ascorbata* and *Kluyvera cryocrescens*, can be differentiated as shown in Table 9. Although

TABLE 9. Differentiation of *Kluyvera* species

	Kluyvera *species*[a]	
Test	K. ascorbata	K. cryocrescens
D-Ascorbate	+	0
Glucose growth and fermentation at 5°C for 21 days	0	+

[a] + = ≥90% positive; 0 = ≤10% positive.

differentiation into species may not be imperative in the clinical microbiology laboratory, the genus itself can be differentiated as shown in Fig. 1. *Kluyvera* is a motile organism that is DNase negative, urease negative, and does not produce hydrogen sulfide or acetylmethylcarbinol; however, it may or may not elaborate lysine decarboxylase. The lysine decarboxylase-positive organisms can be separated from members of the genus *Escherichia* by their ability to utilize citrate. The lysine decarboxylase-negative representatives all attack esculin, which differentiates them from *Citrobacter*. Isolation of the organism from clinical specimens, especially if in pure culture or as the dominant isolate, should invariably lead to an evaluation of its antimicrobial profile. It is also advisable that during-treatment-cultures yielding the bacterium should be reexamined for antimicrobial susceptibility.

Aeromonas and *Plesiomonas*

Aeromonas spp. (von Graevenitz, 1985) are widely distributed in aquatic environments. They are found primarily in stagnating water, although flowing water and brackish marine water have yielded these organisms. Soil and food materials are implicated as sources for *Aeromonas* spp. Certain animals, especially frogs, reptiles, and fish, may display infections by *Aeromonas* spp.

The genus *Plesiomonas* (von Graevenitz, 1985), represented primarily by *Plesiomonas shigelloides*, is also a waterborne bacterium, although it does not tolerate seawater as well as *Aeromonas*. It has been isolated from surface water, from the intestines of freshwater fish, and from animals, including snakes, monkeys, dogs, cats, goats, cattle, and toads.

These bacteria are included among the agents of enteric bacteriosis without the implication of a relationship to *Enterobacteriaceae*. Both genera belong to the family *Vibrionaceae* and possess the characteristics of that family. These straight, rod-shaped, round-ended, gram-negative bacteria are usually motile by a single polar flagellum, although peritrichous flagella may be encountered in very young cultures growing on solid media. Their carbohydrate metabolism is both oxidative and fermentative. The most distinguishing characteristic of these bacteria is their ability to elaborate oxidase in addition to catalase.

These bacteria share with other members of the enteric bacteriosis group the ability to complicate the recovery of hospitalized patients. *Aeromonas* has been isolated from blood, urine, spinal fluid, the ear, wounds, and feces. *Plesiomonas*, on the other hand, has been isolated only from blood, spinal fluid, wounds, and feces. Human infections with *Aeromonas* involve four categories. As described by von Graevenitz (1985), cellulitis or wound infections may be related to human exposure to water or soil. Acute diarrheal diseases of short duration, sometimes bloody or choleralike, occur on a worldwide basis and affect any age. This phase of *Aeromonas* infection resembles traveler's diarrhea and occurs most frequently in children less than 5 years of age. With malignancies, especially acute leukemia or hepatic, biliary, or pancreatic disease, septicemia due to *Aeromonas* may be encountered. Finally, various soft tissue infections and rare cases of urinary tract infections, meningitis, peritonitis, otitis and endocarditis have been reported. *Aeromonas* spp. may be differentiated as indicated in Table 10. All three species have been isolated from clinical specimens, especially those from compromised patients. The appearance of the aeromonads as nosocomially significant agents may depend very much on the geographic distribution of individual institutions. *Plesiomonas shigelloides*, on the other hand, has been isolated from the stools of diarrheic patients living in tropical or subtropical areas of the world. It is assumed that these infections may be waterborne, as carriers are very rare except in endemic areas. Cellulitis, septicemia, and neonatal meningitis have been reported with *Plesiomonas*, although its role in nosocomial infection has not been completely established.

The organisms are separated from the other members of the enteric bacteriosis group by their ability to produce oxidase (Fig. 1). *Plesiomonas* may be separated from *Aeromonas* by its much lesser fermentation of carbohydrates and lack of gas formation. *Plesiomonas* does not elaborate DNase, caseinase, or amylase, liquefy gelatin, or produce lipase. It does, however, elaborate ornithine decarboxylase.

As one would expect, *Aeromonas* spp. are resistant to penicillin, ampicillin, carbenicillin, and ticarcillin, but may be susceptible to the ureido-penicillins as well as some of the cephalosporins. As with other agents of enteric bacteriosis, however, they can readily acquire resistance to these drugs. At this

TABLE 10. Differentiation of *Aeromonas* species

	Aeromonas species[a]		
Test	A. hydrophila	A. caviae	A. sobria
Esculin hydrolysis	+	+	0
Growth in KCN	+	+	0
Acid from salicin	+	+	0
Gas from glucose	+	0	+
Voges-Proskauer	+	0	d

[a] + = ≥90% positive; 0 = ≤10% positive; d = 11 to 89% positive.

juncture, most of the strains respond to the aminoglycosidic aminocyclitols, chloramphenicol, tetracycline, and trimethoprim–sulfamethoxazole. Antibiotic susceptibility studies of *P. shigelloides* have not been conducted in sufficient numbers to permit a summary of this organism's reaction to antimicrobial agents. For both genera, the previous suggestion pertains, namely, antimicrobial susceptibility testing is indicated when these bacteria are isolated in large numbers or as the sole agents from clinical specimens submitted for microbiologic analysis.

The various, seemingly different bacteria that constitute the agents of enteric bacteriosis share the ability to resist antimicrobial agents intrinsically or acquire this resistance by various mechanisms. This property distinguishes all organisms that cause enteric bacteriosis, in addition to their predominance in hospital-acquired infectious complications of agranulocytic patients. Control of enteric bacteriosis entails erecting effective barriers between the environment and personnel in institutions and educating all personnel regarding the threats posed by these environmental bacteria to the welfare of the patients.

Literature Cited

Bartlett, R. C. 1985. Quality control in clinical microbiology, p. 14–23. *In* E. H. Lennette, A. Balows, W. J. Hausler, Jr. and H. J. Shadomy (ed.), Manual of clinical microbiology, 4th ed. American Society for Microbiology, Washington, D.C.

Brenner, D. J. 1978. Characterization and clinical identification of *Enterobacteriaceae* by DNA hybridization. Prog. Clin. Pathol. **7**:71–117.

Brenner, D. J. 1981. The genus *Enterobacter*, p. 1173–1180. *In* M. P. Starr, H. Stolp, H. G. Trüper, A. Balows, and H. G. Schlegel (ed.), The prokaryotes—a handbook of habitats, isolation and identification of bacteria. Springer-Verlag KG, Berlin.

Brenner, D. J. 1985. Taxonomy, classification and nomenclature of bacteria, p. 1–7. *In* E. H. Lennette, A. Balows, W. J. Hausler, Jr., and H. J. Shadomy (ed.), Manual of clinical microbiology, 4th ed. American Society for Microbiology, Washington, D.C.

Brenner, D. J., J. J. Farmer III, G. R. Fanning, A. G. Steigerwalt, P. Klykken, H. G. Wathen, F. W. Hickman, and W. H. Ewing. 1978. Deoxyribonucleic acid relatedness of *Proteus* and *Providencia* species. Int. J. Syst. Bacteriol. **28**:269–282.

Brouillard, J. A., W. Hanson, and A. Compere. 1984. Isolation of *Serratia ficaria* from human clinical specimens. J. Clin. Microbiol. **19**:902–904.

Clark, R. B., and J. M. Janda. 1985. Isolation of *Serratia plymuthica* from a human skin site. J. Clin. Microbiol. **21**:656–657.

Farmer, J. J. III. 1981a. The genus *Citrobacter*, p. 1140–1147. *In* M. P. Starr, H. Stolp, H. G. Trüper, A. Balows, and H. G. Schlegel (ed.), The prokaryotes—a handbook on habitats, isolation and identification of bacteria. Springer-Verlag KG, Berlin.

Farmer, J. J. III. 1981b. The genus *Edwardsiella*, p. 1135–1139. *In* M. P. Starr, H. Stolp, H. G. Trüper, A. Balows, and H. G. Schlegel (ed.), The prokaryotes—a handbook on habitats, isolation and identification of bacteria. Springer-Verlag KG, Berlin.

Farmer, J. J. III, B. R. Davis, F. W. Hickman-Brenner, A. McWhorter, G. P. Huntley-Carter, M. A. Asbury, C. Riddle, H. G. Wathen-Grady, C. Elias, G. R. Fanning, A. G. Steigerwalt, C. M. O'Hara, G. K. Morris, P. B. Smith, and D. J. Brenner. 1985. Biochemical identification of new species and biogroups of *Enterobacteriaceae* isolated from clinical specimens. J. Clin. Microbiol. **21**:46–76.

Gill, V. J., J. J. Farmer III, P. A. Grimont, M. A. Asbury, and C. L. McIntosh. 1981. *Serratia ficaria* isolated from a human clinical specimen. J. Clin. Microbiol. **14**:234–236.

Hickman, F. W., A. G. Steigerwalt, J. J. Farmer III, and D. J. Brenner. 1982. Identification of *Proteus penneri* sp. nov., formerly known as *Proteus vulgaris* indole negative or as *Proteus vulgaris* biogroup 1. J. Clin. Microbiol. **15**:1097–1102.

Hickman-Brenner, F. W., J. J. Farmer III, A. G. Steigerwalt, and D. J. Brenner. 1983. *Providencia rustigianii:* a new species in the family *Enterobacteriaceae*, formerly known as *Providencia alcalifaciens* biogroup 3. J. Clin. Microbiol. **17**:1057–1060.

Isenberg, H. D., and G. A. Isenberg. 1985. Evolving bacterial nomenclature. Infect. Med. **2**:141–147.

Isenberg, H. D., and J. D. MacLowry. 1976. Automated methods and data handling in bacteriology. Annu. Rev. Microbiol. **30**:483–505.

Isenberg, H. D., and J. K. Smith. 1981. The epidemiology and control of nosocomial and opportunistic infections, p. 119–139. *In* A. Balows and W. J. Hausler, Jr. (ed.), Diagnostic procedures for bacterial, mycotic and parasitic infections. American Public Health Association, Washington, D.C.

Krieg, N. R., and J. G. Holt (ed.). 1984. Bergey's manual of systematic bacteriology, vol. 1. The Williams & Wilkins Co., Baltimore.

Lennette, E. H., A. Balows, W. J. Hausler, Jr., and H. J. Shadomy (ed.). 1985. Manual of clinical microbiology, 4th ed. American Society for Microbiology, Washington, D.C.

Penner, J. L. 1981. The tribe *Proteae*, p. 1204–1224. *In* M. P. Starr, H. Stolp, H. G. Trüper, A. Balows, and H. G. Schlegel (ed.), The prokaryotes—a handbook on habitats, isolation and identification of bacteria. Springer-Verlag KG, Berlin.

Pien, F. D., and J. J. Farmer III. 1983. *Serratia ficaria* isolated from a leg ulcer. South. Med. J. **76**:1591–1592.

Rosebury, T. 1961. Microorganisms indigenous to man. McGraw–Hill Book Co., New York.

Sakazàki, R. 1981. The genus *Hafnia*, p. 1181–1186. *In* M. P. Starr, H. Stolp, H. G. Trüper, A. Balows, and H. G. Schlegel (ed.), The prokaryotes—a handbook on habitats, isolation and identification of bacteria. Springer-Verlag KG, Berlin.

Sanders, C. C., and W. E. Sanders. 1979. Emergence of resistance to cefamandole: possible role of cefoxitin-induced beta-lactamases. Antimicrob. Agents Chemother. **15**:792–797.

Starr, M. P., H. Stolp, H. G. Trüper, A. Balows, and H. G. Schlegel (ed.). 1981. The prokaryotes—a handbook on habitats, isolation and identification of bacteria. Springer-Verlag KG, Berlin.

von Graevenitz, A. 1985. *Aeromonas* and *Plesiomonas*, p. 278–281. *In* E. H. Lennette, A. Balows, W. J. Hausler, Jr., and H. J. Shadomy (ed.), Manual of clinical microbiology, 4th ed. American Society for Microbiology, Washington, D.C.

CHAPTER 25

Erysipelothrix Infections

JOSEPH C. CHAN

Disease: Erysipeloid (cutaneous infection, septicemia, endocarditis), erysipelas in animals.
Etiologic Agent: *Erysipelothrix rhusiopathiae* serovars 1-22.
Source: Domestic livestock especially pigs, and other animals such as birds and fish by direct contact.
Clinical Manifestations: A localized cutaneous infection (erysipeloid), diffuse cutaneous infection, septicemia and endocarditis. May be acute or subacute.
Pathology: Epidermal edema with polymorphonuclear leukocyte infiltration, infiltration of lymphocytes, and mast cells in the corium.
Laboratory Diagnosis: Isolation of etiologic agent from tissue or blood. Tube agglutination test has been used in swine erysipelas. There is no information regarding the value of this test in humans.
Epidemiology: Worldwide in distribution. The majority of reported infections have involved direct contact of organic matter such as swine carcasses, fish, clams, crabs, and poultry. Occupational disease of veterinarians, abattoir workers, fish handlers, and others exposed to meat and fish products in industry and the home.
Treatment: Penicillin G, ampicillin, cephalosporins; second choices include tetracycline and erythromycin.
Prevention and Control: Control in domestic animals by vaccination and improvement of their environment; protective gloves and apron should be worn by individuals handling animals and their products.

Description of Disease

The history of swine erysipelas is probably as old as the domestication of the pig. The causative organism *Erysipelothrix rhusiopathiae* was first isolated from a laboratory mouse in 1878 by Robert Koch and from swine in 1882 by Louis Pasteur. *Erysipelothrix rhusiopathieae,* from the Greek erysipelas (a disease), thrix (a hair or thread), rhusius (reddish), and pathus (a disease), means a reddish disease on the hair and skin. *Erysipelothrix* infection in humans is an infectious disease involving skin, blood, and heart valves. The infection in pigs is manifested in three forms: an acute septicemic form, a subacute cutaneous form with well-defined rhomboidal hemorrhagic patches or wheals (''diamonds''), and a chronic form characterized by polyarthritis and occasionally endocarditis. Humans acquire the disease by accidental inoculation through handling of infected animal products. The disease has immense economic impact throughout the major swine-producing areas in the United States and other countries as well.

Communicability and Transmissibility

Erysipeloid is communicable from animals to humans by direct cutaneous contact. There have been two reported cases, one septicemia and one endocarditis, after ingestion of undercooked pork (Grieco and Sheldon, 1970). It is still not certain whether infection can occur after contacting soil previously exposed to animals. Many investigators believe that *E. rhusiopathiae* can lead a saprophytic existence in soil. Human-to-human transfer of infection has not been reported.

Clinical Manifestations

Most human infections by *E. rhusiopathiae* fall into three clinical categories: (1) a mild, localized, cutaneous form; (2) a severe generalized cutaneous form; and (3) a septicemic form often associated with endocarditis (Grieco and Sheldon, 1970; Woodbine, 1950).

Localized Cutaneous Form

The infection usually develops after an injury while handling infected organic matter. The lesion of erysipeloid (resembles streptococcal erysipelas), which usually occurs on a finger, is painful. It manifests in 2 to 7 days after the inoculation. It is characterized by a dark purple-red zone, sharply defined, and somewhat raised. About 3 to 5 days later, the lesion extends peripherally, while the central part of the lesion fades and desquamates. It is common for the spread of erythema to reach neighboring fingers or the palm of the hand.

General symptoms are few and there is usually no fever. Lymphangitis and regional lymphadenopathy may occur in less than 10% of cases. Arthralgia of the neighboring interphalangeal joints can occur and may lead to chronic articular stiffness.

The average duration of erysipeloid with topical treatment before penicillin was 17.4 days (14 to 28 days) (King, 1946). Regression of the lesion can occur without treatment. The disease is self-limiting in most cases; it can run a variable course of 1 to 32 weeks.

Severe Generalized Cutaneous Form

Patients with this form of the disease continue to have new lesions that develop centrifugally from the original site of inoculation. New lesions can also appear at remote areas. With increasing number of lesions, bullous vesiculation has been described (Ehrlich, 1946). Of a total of 715 cases reported in three series, only a single case of this form was described (Klauder, 1938; King, 1946; Nelson, 1955).

Septicemic Form With or Without Endocarditis

Systemic infections can occur with or without any detectable cutaneous site of entry. Clinical presentations are similar to other forms of bacteremia: fever, rigor, malaise, headache, lethargy, and musculoskeletal pain. However, the course of the illness is highly variable. The patient can present with septic shock within hours of onset of symptoms or as recurrent fever for a 1 month duration (Normann and Kihlstrom, 1985; Ognibene et al., 1985). The longest symptom interval before diagnosis was reported to be 1 year (Grieco and Sheldon, 1970). As the majority of the reported cases of *E. rhusiopathiae* bacteremia had evidence of endocarditis, all patients should be treated as having endocarditis even if they have no underlying valvular disease.

Occupational or recreational contact of infected material such as pork, fish, or crabs can be elicited from more than half of the bacteremic cases. About 40% of the cases also have cutaneous erysipeloid lesions.

Favorable outcome is usually achieved with 4 to 6 weeks of high-dose intravenous penicillin therapy. Morbidity and mortality has been quite low and these have been associated with embolic complications or valvular dysfunction.

Pathology

Erysipeloid is characterized by marked inflammation and edema of the skin. The absence of suppuration and pitting on pressure help to differentiate this infection from other pyogenic skin infections. Microscopical examination of stained tissue, reveals the epidermis being edematous, infiltrated with polymorphonuclear leukocytes, and necrotic. The deeper corium is infiltrated with lymphocytes and mast cells. The organism is concentrated in the corium around the capillaries (Grieco and Sheldon, 1970). The rhomboidal wheals ("diamond skin") described in pigs and not in human infections, are apparently secondary to thrombotic vasculitis of end arterioles (Julian, 1968).

The common findings in swine arthritis are pannus formation, destruction of cartilage at the site of pannus attachment, subchondral cellular reaction, and intraarticular fibrous adhesions. These latter histopathologic findings are similar to human rheumatoid arthritis (Hadler, 1976). With bacteremia and endocarditis, the basic pathologic processes are capillary hyperemia with plugging of vessels by mononuclear cells and bacteria, and vegetations on involved heart valves.

Diagnostic Features

Erysipeloid should be considered in every patient with cutaneous lesions (especially those occurring on the hand) and a recent history of direct contact with domestic animals, meat products especially pork, fish, and crabs. As the name implies, streptococcal erysipelas is the most common diagnosis to be excluded. Central fading with slow centrifugal spreading of the skin lesions and the lack of constitutional disturbance are main diagnostic features of erysipeloid.

All patients with fever and constitutional symptoms, especially those with 2 to 4 weeks duration, and a positive history of occupational or recreational exposure as mentioned above should be considered as having the bacteremic form of erysipeloid. This diagnosis should be considered highly probable if the findings of suspicious cutaneous lesions are also present.

Definitive diagnosis requires isolation of *E. rhusiopathiae* from soft tissue biopsy specimens or blood. So far, no serologic test is reliable for the diagnosis of infection in humans.

Prognosis

Resolution of all signs and symptoms of the cutaneous infection is the rule even without antibiotic therapy. Relapses may occur after apparent cure. The mortality rate of the septicemic form is low, and the patient usually recovers after 4 weeks of penicillin therapy. There is no lasting immunity, and the patient can presumably have repeated infections.

Epidemiologic Aspects

Occurrence and Transmission

Erysipeloid occurs worldwide, but the incidence of this infection is not known. In an earlier report, 115 cases were discovered from records of a hand clinic in a period of 54 months (King, 1946). During 30 years of practice as an industrial physician in Colorado, Nelson (1955) encountered about 500 cases. There has not been any recent report of large series of erysipeloid in the English literature. Only sporadic case reports of *E. rhusiopathiae* septicemia with or without endocarditis are found in the recent medical literature (Kramer et al., 1982; Morris, 1980; Muirhead and Reid, 1980; Simerkoff and Rahal, 1973). Although the true incidence of human infection may have decreased due to increased protection of individuals at risk, the apparent decline also may be attributable to selective reporting of severe cases only. On the contrary, the prevalence of acute erysipelas in swine has increased since the late 1960s. Furthermore, the prevalence of *E. rhusiopathiae* isolation rate from swine in the United States is still 29% (Wood, 1984).

Most *E. rhusiopathiae* infection in humans can invariably be traced to contact with animals, fish, shell fish, and products derived from animals. The organism probably gains access through injuries of the skin. Direct transmission to humans from swine appears to be uncommon.

Susceptibility, Resistance, and Carrier States

E. rhusiopathiae is resistant to salting and putrefaction and can probably exist saprophytically in favorable surroundings. Epizootic outbreaks may be linked to indirect spread through contamination of soil, food, and water. The organism has been recovered from tonsils, intestinal mucosa, and other internal organs of apparently normal swine; therefore, a carrier state can exist in swine (Woodbine, 1950).

Etiologic Agent

Characterization

Erysipelothrix rhusiopathiae is a straight, or slightly curved, slender, gram-positive, nonsporing rod. Rods usually 0.2 to 0.4 μm in diameter and 0.8 to 2.5 μm in length occur singly in short chains, in pairs at an angle to give V-forms, or in groups with no particular arrangement. The organism decolorizes readily and may appear to be gram-negative. It is nonmotile.

The organism is microaerophilic. Optimum temperature of growth is between 35 to 37°C, but growth occurs between 5 and 42°C. It is catalase-negative, oxidase-negative, and citrate- and methyl red-negative. Fermentation of glucose results in lactic acid and is best demonstrated in nutrient broth with phenol red as indicator. It exists in a rough (R) and a smooth (S) form. S-form colonies are convex, circular, and transparent. R-form colonies are larger, flatter, more opaque, and have an irregular edge. It causes a greenish discoloration on blood agar, but a slight clearing can occur after 48 hours incubation. In triple sugar iron agar (TSI) slants, it produces an acid slant and butt. Production of H_2S in the butt is a characteristic, which readily distinguishes this organism from other morphologically similar gram-positive bacilli.

Specimen Collection and Transport

Full-thickness biopsy specimens of skin from the edge of the erysipeloid lesions is necessary to recover the organism. Positive cultures are rarely obtained from swabs of a lesion. Specimens should be placed in an infusion broth containing 1% glucose after collection. In patients with septicemia and/or endocarditis, routine blood culture techniques are adequate.

Isolation

Commercially available blood culture media are satisfactory for the primary isolation from patients with septicemia or endocarditis. Any growth in the broth is subcultured to a blood-agar plate, which is then incubated at 35 to 37°C for 24 to 48 hours.

Isolation from skin biopsy specimens may require selective media if the sample is contaminated with other bacteria. The specimen should be placed in 10 ml of tryptose broth supplemented with 5% horse serum by volume, 400 μg/ml of kanamycin, 500 μg/ml of neomycin, and 25 μg/ml of vancomycin (Wood, 1965). After overnight incubation, some of the liquid portion is removed, placed into a sterile test tube, and centrifuged for 20 minutes at approximately 1400 $\times$ *g*. The supernatant is discarded, and the sediment is resuspended in 1 to 2 ml physiologic saline (0.85%). A portion of this liquid is plated on modified blood-azide (MBA) medium (Harrington and Hulse, 1971). The plate is examined for colonies after incubation.

Identification

Colonies on blood-agar plates appear α-hemolytic, and their morphologies were described in an earlier section. These colonies should be Gram stained and examined for gram-positive bacilli. *E. rhusiopathiae* is catalase and oxidase negative. It ferments glucose and lactose, but not xylose, mannitol, or sucrose. A few strains may produce a weakly acidic reaction in maltose by 6 to 7 days. In TSI slants, it produces both an acid slant and an acid butt, which is blackened by H_2S production. H_2S production is a distinguishing characteristic of *E. rhusiopathiae*.

Bacteria that are likely candidates for differentiation in the clinical laboratory include the genera *Corynebacterium* and *Listeria*. Both of these groups of organisms produce a β-like hemolysis on blood agar and do not produce H_2S. Furthermore, *Listeria monocytogenes* is motile and catalase-positive.

Rarely, the mouse protection test is considered necessary to identify new isolates of *E. rhusiopathiae* with certainty. Commercially available horse antierysipelothrix serum is satisfactory for the test. The mice that receive the unknown culture (if it is *E. rhusiopathiae*) and antiserum should survive; those receiving the culture alone should die within 6 days (Kandler and Weiss, 1986).

Immunologic Aspects

There is no immunoserologic procedure that can be routinely used in a clinical laboratory. In veterinary research, *E. rhusiopathiae* may be identified serologically. According to the Kucsera system of designating the serovars using arabic numbers, there are 22 serovars (Kandler and Weiss, 1986). Serovars 1 and 2 are most common, and the other 20 are relatively rare. The double agar gel diffusion precipitation test is the recommended method for determining serotypes of isolates. Identification of serotypes is confirmed by reactions of identity with reference antigen-antibody systems (Wood et al., 1981; Wood and Harrington, 1978).

Treatment, Prevention, and Control

Treatment

The organism is highly susceptible to penicillin and the minimal inhibitory concentration ranges between 0.002 to 0.02 μg/ml (Simerkoff and Rahal, 1973). Other antibiotics that are also effective include ampicillin, oxacillin, cephalosporins, and clindamycin. Erythromycin and tetracycline can only be used if the isolated organism is tested to be susceptible to these antibiotics. A recent Japanese survey showed 43% of their strains to be resistant to erythromycin and tetracycline (Takahashi et al., 1984).

Cutaneous infection is self-limiting. During the prepenicillin era, patients treated with various topical regimens and/or serum had a slightly abbreviated course. Intramuscular penicillin had been shown to shorten the course of erysipeloid from an average of 17.4 to 2 days (King, 1946). The optimal dosage and duration of penicillin treatment has not been defined; oral penicillin for 5 to 7 days is probably adequate for most uncomplicated cases.

Bacteremia and/or endocarditis should be treated with at least 12 million units daily of penicillin intravenously. As the majority of patients with bacteremia also have endocarditis, all patients with positive blood cultures should be treated for at least 4 weeks.

Prevention and Control

Prevention of erysipeloid for individuals in a high-risk occupation such as veterinarians, poultry workers, fishers, meat packers, and abattoir workers depends on protective equipment. These individuals should avoid unprotected direct contact with animals and their by-products. Controlling swine erysipelas has been successful in this country due to management practices such as improved waste disposal, improved animal environment, vaccination, and use of antibiotics in feed. However, human infection can only be controlled by avoiding direct contact with infected materials because this organism can be found ubiquitously in many animals.

Literature Cited

Ehrlich, J. C. 1946. *Erysipelothrix rhusiopathiae* infection in man. Arch. Intern. Med. **78:**565–577.

Grieco, M. H., and C. Sheldon. 1970. *Erysipelothrix rhusiopathiae*. Ann. N.Y. Acad. Sci. **174:**523–532.

Hadler, N. M. 1976. A pathogenetic model for erosive synovitis. Arthr. Rheum. **19:**256–266.

Harrington, R., and D. C. Hulse. 1971. Comparison of two plating media for the isolation of *Erysipelothrix rhusiopathiae* from enrichment broth culture. Appl. Microbiol. **22:**141–142.

Julian, R. J. 1968. Shapes of infectious disease lesions. Can. Med. Assoc. J. **98:**1016.

Kandler, O., and N. Weiss. 1986. Regular, nonsporing Gram-positive rods, p. 1208–1249. *In* Holt, J. G., et al. (ed.), Bergey's manual of systematic bacteriology, vol. 2, Williams & Wilkins, Baltimore.

King, P. F. 1946. Erysipeloid, survey of 115 cases. Lancet **2:**196–198.

Klauder, J. V. 1938. Erysipeloid as an occupational disease. J. Am. Med. Assoc. **111:**1345–1348.

Kramer, M. R., M. E. Gombert, and M. L. Corrado, 1982. *Erysipilothrix rhusiopathiae* endocarditis. South. Med. J. **75:**892.

Morris, C. A. 1980. *Erysipelothrix* endocarditis in the immunocompromised host. J. Infect. **2:**283.

Muirhead, N., and T. M. S. Reid. 1980. *Erysipelothrix rhusiopathiae* endocarditis. J. Infect. **2:**83–85.

Normann, B., and E. Kihlstrom. 1985. *Erysipelothrix rhusiopathiae* septicemia. Scand. J. Infect. Dis. **17:**123–124.

Ognibene, F. P., R. E. Cunnion, and V. Gill. 1985. *Erysipelothrix rhusiopathiae* bacteremia presenting as septic shock. Am. J. Med. **78:**861–864.

Simerkoff, M. S., and J. J. Rahal, Jr. 1973. Acute and subacute endocarditis due to *Erysipelothrix rhusiopathiae*. Am. J. Med. Sci. **266:**53–57.

Takahashi, T., T. Sawada, and K. Ohmae. 1984. Antibiotic resistance of *Erysipelothrix rhusiopathiae* isolated from pigs with chronic swine erysipelas. Antimicrob. Agents Chemother. **25:**385–386.

Wood, R. L. 1965. A selective liquid medium utilizing antibiotics for isolation of *Erysipelothrix insidiosa*. Am. J. Vet. Res. **26:**1303–1308.

Wood, R. L., and R. Harrington. 1978. Serotypes of *Erysipelothrix rhusiopathiae* isolated from swine and from soil and manure of swine pens in the United States. Am. J. Vet. Res. **39:**1833–1840.

Wood, R. L., R. Harrington, and D. R. Haubrich. 1981. Serotypes of previously unclassified isolates of *Erysipelothrix rhusiopathiae* from swine in the United States and Puerto Rico. Am. J. Vet. Res. **42:**1248–1250.

Wood, R. L. 1984. Swine erysipelas—a review of prevalence and research. J. Am. Vet. Med. Assoc. **184:**944–949.

Woodbine, M. 1950. *Erysipelothrix rhusiopathiae*. Bacteriology and chemotherapy. Bacteriol. Rev. **14:**161–178.

CHAPTER 26

Infectious Diseases Associated with *Escherichia Coli*

LEE W. RILEY

Diseases: *Intestinal infections* (diarrhea, dysentery, hemorrhagic colitis). *Urinary tract infections* (asymptomatic bacteriuria, cystitis, pyelonephritis, prostatitis). *Bacteremia. Neonatal meningitis.*

Etiologic Agents: *Intestinal infections*—enterotoxigenic *Escherichia coli* (ETEC), enteroinvasive *E. coli* (EIEC), enteropathogenic *E. coli* (EPEC), and enterohemorrhagic *E. coli* (EHEC). *Urinary tract infections—E. coli,* usually strains carrying uropathogenic determinants. *Bacteremia—E. coli,* usually strains belonging to the same serogroups associated with urinary tract infections. *Neonatal meningitis—E. coli,* usually strains carrying K1 acidic polysaccharide capsular antigen.

Source: *Intestinal infections*—possibly human intestinal tract, probably cattle (EHEC). *Urinary tract infections*—human intestinal tract. *Bacteremia*—human intestinal tract. *Neonatal meningitis*—human intestinal tract.

Clinical Manifestations: *Intestinal infections*—acute, watery diarrhea sometimes leading to severe dehydration (ETEC); dysenterylike diarrhea (EIEC); protracted, relapsing watery diarrhea (EPEC); grossly bloody diarrhea with little or no fever (EHEC). *Urinary tract infections*—dysuria, urgency, lower abdominal pain (cystitis); fever rigor, nausea, vomiting, flank pain (pyelonephritis). *Bacteremia*—poor feeding, jaundice (in neonates), fever or hypothermia, respiratory distress, bleeding, shock (in adults and neonates). *Neonatal meningitis*—signs of bacteremia, convulsions; sequelae such as seizure disorders, hearing deficits, and mental and motor disabilities may occur.

Pathogenesis: *Intestinal infections*—interaction of the intestinal mucosa with colonization factors, heat-labile (LT) and heat-stable (ST) enterotoxins (ETEC); enterocyte invasiveness factors (EIEC); enteroadhesiveness factors (EPEC); and unknown factor(s). *Urinary tract infections*—interaction of urinary tract mucosa with adhesins (type 1 fimbriae, P-fimbriae, X-adhesins); possibly hemolysin, capsular, and O antigens. *Bacteremia*—possible serum resistance effect of the O polysaccharides; immature host immunity; shock associated with endotoxin or lipopolysaccharide. *Neonatal meningitis*—acidic polysaccharide capsule K1.

Laboratory Diagnosis: *Intestinal infections*—isolation of *E. coli* from feces and determining its virulence factors. *Urinary tract infections*—isolation and quantitation of *E. coli* by urine culture. *Bacteremia*—isolation of *E. coli* from blood. *Neonatal meningitis*—isolation of *E. coli* from cerebrospinal fluid.

Epidemiology: *Intestinal infections*—worldwide occurrence with predominance of ETEC, EIEC, and EPEC in developing countries, and EHEC in North America; foodborne, waterborne, and person to person transmissions occur. *Urinary tract infections*—community and nosocomial occurrence; risk factors include anatomic and functional urologic defects, pregnancy, P blood group, sexual intercourse in some women, and urologic procedures. *Bacteremia*—worldwide, community, and nosocomial occurrence; neonates, especially premature and low birth weight infants, and adults with hematologic malignancies are at increased risk. *Neonatal meningitis*—worldwide,

continued

continued
community, and nosocomial occurrence; premature and low birth weight infants are at increased risk.

Treatment: *Intestinal infections*—oral rehydration therapy in most cases; intravenous rehydration in advanced dehydration; antibiotics usually not beneficial, except for EIEC infection. *Urinary tract infections*—single dose treatment with a sulfonamide, amoxicillin, or trimethoprim-sulfamethoxazole in most uncomplicated acute infections in young females; longer duration of treatment for recurrent and complicated infections in the female and for any infection in the male. *Bacteremia*—parenteral antibiotics, with the initial conventional regimen of a penicillin with an aminoglycoside or a cephalosporin until monotherapy can be offered; supportive care for shock is important. *Neonatal meningitis*—intravenous ampicillin and an aminoglycoside; new beta-lactam antibiotics are an appropriate substitute for an aminoglycoside, but comparative clinical trials not completed.

Prevention and Control: *Intestinal infections*—prevention of mortality and morbidity by oral rehydration therapy, breast milk, and maternal education. *Urinary tract infections*—low dose, long-term antimicrobial prophylaxis in high risk groups; screening for bacteriuria in pregnant women; strict adherence to aseptic techniques in catheter care. *Bacteremia*—prenatal care; prevention of urinary tract infections; passive immunotherapy using antibodies raised against bacterial surface products under evaluation. *Neonatal meningitis*—prevention of bacteremia.

Escherichia coli first described by Theodore Escherich in 1885, was initially regarded as a nonpathogenic commensal present in feces. Over time, however, the pathogenic potential of some *E. coli* strains became recognized and today, diseases associated with *E. coli* span a wide spectrum of clinical manifestations, epidemiologic patterns, pathogenesis, diagnostic considerations, and management. Hence, each disease associated with *E. coli* must be addressed separately.

In developing countries, *E. coli* is a major cause of intestinal infections, which are associated with high childhood mortality and morbidity. In contrast, in the developed countries, *E. coli* is more often associated with community-acquired and nosocomial extraintestinal infections. A satisfactory diagnosis of extraintestinal *E. coli* infections requires the isolation of the organism from a sterile site, whereas the diagnosis of intestinal *E. coli* infections relies not only on the primary isolation, but also on the determination of the characteristic virulence property of the organism. Therefore, infections associated with *E. coli* may be broadly categorized into intestinal and extraintestinal infections.

Currently, intestinal infections are associated with four distinct groups of *E. coli*—enterotoxigenic *E. coli* (ETEC), enteroinvasive *E. coli* (EIEC), enteropathogenic *E. coli* (EPEC), and enterohemorrhagic *E. coli* (EHEC) (Levine, 1987). They are distinguished on the basis of characteristic clinical manifestations, epidemiologic features, virulence determinants, and serotypic differences associated with each group. More recently, a fifth category of *E. coli* associated with diarrhea—referred to as enteroadherent *E. coli* (Mathewson et al., 1985) or autoagglutinating *E. coli* (Nataro et al., 1987)—has been proposed. Its clinical, epidemiologic, and virulence characteristics as well as additional evidence for its role in diarrhea pathogenesis are under investigation.

Among extraintestinal infections, *E. coli* is the predominant causative agent of urinary tract infections (UTI), bacteremia, and neonatal meningitis. These infections, therefore, will be discussed separately.

Etiologic Agents: General Features

Characterization and Identification

Escherichia coli is a facultative anaerobe belonging to the genus *Escherichia* in the family *Enterobacteriaceae*. It accounts for more than 99% of the four recognized species of *Escherichia* isolated from clinical specimens (others are *E. fergusonii*, *E. hermannii*, and *E. vulneris*) (Farmer et al., 1985).

E. coli can be screened for by a characteristic morphology and a spot indole test. Indole-positive, flat, lactose-positive colonies on MacConkey agar can be tentatively identified as *E. coli*. Definitive identification involves a battery of biochemical tests: *E. coli* typically produces gas upon glucose fermentation, and gives a positive reaction with indole, lysine, arabinose, mannitol, O-nitrophenyl-β-D-galactopyranoside (ONPG), trehalose, and xylose; it gives a negative reaction with DNase, H_2S, phenylalanine deaminase, urease, Voges-Proskauer, inositol, and potassium cyanide (KCN). Many rapid detection kits from commercial sources are available, and are reviewed elsewhere (D'Amato et al., 1985).

Serologic Examination

The species *E. coli* is further differentiated into serotypes. The scheme, first proposed by Kauffman in the 1940s (Kauffman, 1947), is based on the somatic O, flagellar H, and envelope or capsular K antigens. The O antigen is the repeating polysaccharide side chains of the lipopolysaccharide. There are currently over 170 O and 55 H antigenic groups recognized. The K antigen describes a cell envelope or capsule that surrounds the O antigen, which prevents bacterial agglutination by the antiserum raised against the O antigen. They were previously divided into three types (L, A, and B) according to the effect of heat on their agglutinability by the O antiserum. The L-type antigen is removed by heating for 1 h at 100°C, and the A-type antigen must be heated for 2 h at 121°C before the bacteria can be agglutinated with the O-specific antiserum. The B-type antigen, found frequently in EPEC group, is not considered a true K antigen, but instead is thought to be part of the O antigen (Ørskov et al., 1977). The L, A, and B designations are no longer used for *E. coli* serotyping descriptions, and the K designation is not used for EPEC.

The standard procedure for serotyping *E. coli* has been described in detail by Edwards and Ewing (1972). The basic procedure for the O antigen determination is the tube agglutination. The strains are initially tested with a pooled O antisera, and the agglutinated strains are then tested against an individual O or OK antiserum, which must be diluted out to an end point titer. To complete the serotyping, an agglutination test against the H antiserum must be performed.

The procedures for specimen collection, transport, and isolation vary according to the *E. coli* infection, and hence are discussed under each respective section.

Intestinal Infections

Description of Disease

Clinical Manifestations

Enterotoxigenic *E. coli* (ETEC) produces watery diarrhea with nausea and abdominal cramping, which in children may lead to severe dehydration. Fever is mild to moderate, and mucus and bloody discharges are usually absent. The site of infection is the proximal small intestine.

Enteroinvasive *E. coli* (EIEC) on the other hand produces dysenterylike symptoms with crampy abdominal pain, malaise, and diarrhea containing mucus and often blood. White blood cells are abundant in stool, and fever may be high. The colon is the predominant site of infection with EIEC.

Enterohemorrhagic *E. coli* (EHEC) produces a distinct enteric illness called hemorrhagic colitis (Riley et al., 1983). After an incubation period of 3 to 4 days, the typical illness begins with severe abdominal cramps, followed within hours by watery diarrhea and later by grossly bloody diarrhea. Fever is characteristically absent or low grade, and mucus production is scanty. The illness lasts about 8 days, and is usually self-limited.

The clinical manifestations associated with EPEC cover a wider spectrum. A characteristic feature of an illness associated with the so-called classic EPEC is the prolonged, relapsing episodes of watery diarrhea. Fever rarely exceeds 39°C in such cases. Cases of EPEC-associated bloody diarrhea have also been reported (McKay and Wahle, 1955). The reports of EPEC-associated diarrheal outbreaks in infant nurseries in Europe and North America in the 1940s and 1950s often described a watery diarrheal illness with a protracted course and relapses that led to high case/fatality rates (Bray, 1945; Giles et al., 1949). Other EPEC outbreaks feature milder illness. Sporadic EPEC infections in developing countries may cause illness more severe than that caused by other enteric pathogens.

The observation of the heterogeneous clinical features of EPEC infections relates partly to terminology. Until recently, the term EPEC has been operationally used to include all *E. coli* strains associated with diarrhea that lacked the recognized virulence properties of ETEC or EIEC. The *E. coli* serotypes frequently found to be associated with infantile diarrhea through the 1950s (Ewing et al., 1963), which could not be subsequently classified into the ETEC or EIEC group, came to be known as the "classic" EPEC serotypes. Later, other serotypes of *E. coli* associated with diarrhea were identified and classified as EPEC. Therefore, EPEC may include a variety of *E. coli* strains with distinct virulence properties associated with different clinical manifestations.

Complications and Prognosis

The extraintestinal spread of these diarrheagenic *E. coli* is rare. Diarrhea associated with ETEC exerts a significant negative effect on the bimonthly weight gain of children (Black et al., 1984), and the prolonged diarrhea associated with EPEC may also lead to nutritional complications. An unusual transient paresis of the lower extremities has been reported with EIEC (Neubauer et al., 1982). An infection with EHEC may be complicated by hemolytic-uremic syndrome (HUS) (Carter et al., 1987; Ryan et al., 1986; Spika et al., 1986) and thrombotic thrombocytopenic purpura (TTP) (Morrison et al., 1985; Ryan et al., 1986).

Deaths occur with diarrhea produced by all four groups. Mortality rates are most reliably obtained from investigations of outbreaks, especially those in developed countries. When EPEC outbreaks were more common in Europe and North America, case/fatality rates reaching 50% were reported (Bray, 1945; Giles et al., 1949). In an outbreak of hemorrhagic colitis caused by *E. coli* O157:H7 in a nursing home in Canada, a case/fatality rate of 35% was reported (Carter et al., 1987). Deaths confirmed to be caused by ETEC or EIEC are difficult to obtain in developing countries, and do not occur among travelers from developed countries who acquire the infection abroad.

Pathogenesis

The understanding of the pathogenic features of diarrheagenic *E. coli* is crucial to the diagnosis of these pathogens. The mechanism of diarrhea production by diarrheagenic *E. coli* is a multifactorial process that has several common features. Strains of each group have 1) distinct surface antigens or adhesins that enable the bacteria to initiate the pathogenic process by attaching to and colonizing the intestinal mucosal surface, 2) plasmids that encode major virulence determinants, and 3) distinct lipopolysaccharide O antigenic polysaccharides that may serve as virulence determinants or cofactors of virulence.

Enterotoxigenic *E. coli* attaches and colonizes the surface of the intestinal mucosa via fimbrial or fibrillar adhesins. A family of antigenically distinct adhesins has been identified among different ETEC serogroups, including CFA/I (Evans et al., 1975; Evans et al., 1978), CFA/II (Evans and Evans, 1978), and at least three additional colonization factors (McConnell et al., 1985; Tacket et al., 1987; Thomas et al., 1982). All of these colonization factors are encoded by plasmids.

Two distinct classes of enterotoxins (as defined originally by the rabbit ligated-ileal loop dilatation) are produced by *E. coli* belonging to the ETEC group—heat-labile (LT) and heat-stable (ST) toxins. They are both encoded by plasmids—almost always by the same plasmids that encode the colonization factors. The heat-labile toxin (inactivated by heating at 60°C for 30 min) is structurally, functionally, and antigenically related to cholera toxin (Field, 1979). After binding to the GM_1 ganglioside receptor on the enterocytes, the toxin activates the adenylate cyclase system, which leads to the intracellular accumulation of the cyclic AMP. The increase in intracellular cyclic AMP alters the membrane permeability to the electrolytes, which results in net secretion of fluid and electrolytes into the gut lumen. This fluid secretion produces the characteristic watery diarrhea of ETEC infections.

Two forms of ST are recognized—STa (ST I) and STb (ST II). The STa form, which is soluble in methanol, produces fluid accumulation in the suckling mouse; two distinct structural genes, which share 69% DNA homology, are recognized for STa (sometimes referred to as STh for ST human, and STp for ST porcine) (Moseley et al., 1982). A component of solubilized rat intestinal brush border membrane has been recently shown to bind specifically to a synthetic peptide analogue of STa (Gariepy and Schoolnik, 1986), but the structure of the binding site of STa in the enterocyte is unknown. The STa form causes intracellular accumulation of cyclic GMP by activating the guanylate cyclase system, which alters the membrane permeability, thus causing net fluid secretion. The STb form is insoluble in methanol, does not produce fluid accumulation in the suckling mouse, and causes diarrhea in pigs, but not in humans. The mode of action of STb is unknown.

Enterotoxigenic *E. coli* strains that produce these enterotoxins, especially those that produce both LT and ST, fall into a restricted number of O serogroups (Merson et al., 1976). The role of the ETEC-specific O antigens in pathogenesis, if any, is unknown.

Enteroinvasive *E. coli* share many of the biochemical, antigenic, and pathogenic features of *Shigella*. The mechanism of mucosal attachment by EIEC is unknown, but like *Shigella,* EIEC preferentially colonizes the colon. *E. coli* belonging to this group causes diarrhea by the following series of events: 1) mucosal penetration; 2) intracellular proliferation and spread within colonic enterocytes; 3) local mucosal destruction; and 4) elicitation of inflammatory response involving polymorphonuclear leukocytes, leading to mucosal ulceration and production of mucus, blood, and fluid. Despite its invasiveness, EIEC rarely penetrates the lamina propria to enter the lymphatic system. Hence, bacteremia is extremely unusual with serotypes belonging to this group.

A large plasmid 120 to 140 megadaltons (Mdal) is associated with invasiveness of both EIEC and *Shigella* (Harris et al., 1982; Sansonetti et al., 1981). In *Shigella,* in addition to the plasmid, at least four distinct loci in the chromosome have been shown to encode virulence determinants (Sansonetti et al., 1983). One of these determinants (which is sometimes encoded by a plasmid in some strains of *Shigella*) has been shown to be the O polysaccharide

(Timmis et al., 1986). A mutant strain devoid of the O antigen fails to elicit conjunctivitis in guinea pigs—the standard test for invasiveness (see Sereny test below)—although the ability to invade HeLa cells is preserved. Hence, the O antigens of *Shigella*-like *E. coli* play a role as a virulence factor, but the nature of this role is unknown.

Ultrastructural studies of intestinal biopsy specimens from children with EPEC-induced diarrhea show that the bacteria attach to the intestinal epithelium in multiple microcolonies and cause effacement of the microvilli (Rothbaum et al., 1982), features that may be critical in the pathogenesis of diarrhea caused by EPEC. Several investigators have recently demonstrated that some *E. coli* strains belonging to the classic EPEC group (but not ETEC, EIEC, or EHEC) attach to HEp-2 or HeLa tissue culture cells in clusters—a pattern that has been referred to as localized adherence (LA) (Cravioto et al., 1979; Scaletsky et al., 1984). LA is usually plasmid-mediated, and *E. coli* strains that produce this characteristic have been shown to be associated with diarrhea by both epidemiologic and human volunteer studies (Levine et al., 1985; Nataro et al., 1985). An EPEC adhesiveness factor (EAF) has been proposed to be the putative adhesin responsible for LA, but it has not yet been characterized.

The ability to differentiate EPEC by tissue culture adherence pattern is one example of how EPEC group may be subgrouped according to newly identified virulence determinants. It was found that among the so-called EPEC serogroups, LA is demonstrated only by strains belonging to the most commonly isolated serogroups—O55, O111, O119, O127, O128, and O142. This subgroup has been proposed to be called Class I EPEC serogroup (Nataro et al., 1985). The Class II serogroup, which is clearly associated with diarrhea, includes strains that either do not attach or attach to HEp-2 cells in diffuse pattern, and includes serogroups O44, O86, and O114 (Nataro et al., 1985).

Another study showed that in an O111:NM EPEC strain, a 54-Mdal plasmid that encodes LA also encodes the expression of the organism's O polysaccharide (Riley et al., 1987). A plasmid-cured derivative strain, lacking the polysaccharide side-chains, interacted with HeLa cells in a pattern that was distinct from LA—it was internalized by HeLa cells. Additional studies suggested that the polysaccharide side-chain may block the phagocytosis-like response by the eukaryotic cell, and is a possible cofactor in the interaction of EPEC with mucosal cells.

The pathogenesis of diarrhea and hemorrhagic colitis by EHEC is unknown. Strains of *E. coli* O157:H7 associated with hemorrhagic colitis produce a family of cytotoxins called Shiga-like toxins which are active against Vero cells, but the role of these cytotoxins in the pathogenesis of hemorrhagic colitis remains unestablished. One of these cytotoxins—Shiga-like toxin 1 (SLT-1) or Verotoxin-1 (VT-1)—is antigenically, structurally, and genetically related to the Shiga toxin produced by *Shigella dysenteriae* type 1 (O'Brien and Holmes, 1987). Shiga-like toxin-2 (SLT-2) or Verotoxin-2 (VT-2) is antigenically distinct from the Shiga toxin; both SLT-1 and SLT-2 are specified by distinct bacteriophages (O'Brien et al., 1984; Strockbine et al., 1986). *E. coli* serotypes other than O157:H7 that produce Vero cell active cytotoxins have been isolated from sporadic cases of a clinical syndrome consistent with hemorrhagic colitis, but none of these strains have yet been epidemiologically associated with the illness (Bopp et al., 1987).

E. coli O157:H7 has been shown to cause diarrhea and characteristic mucosal lesions in gnotobiotic pigs (Francis et al., 1986). In perhaps the best designed animal study to date, SLT was shown not to be essential for the illness seen in gnotobiotic pigs (Tzipori et al., 1987). The illness and intestinal mucosal pathology detected in piglets fed strains of EHEC that produced SLT and piglets fed derivative strains that did not produce SLT were indistinguishable; no diarrhea was elicited in piglets fed strains of *E. coli* K12 and piglets fed *E. coli* K12 strains transduced with the phage encoding SLT.

The strain *E. coli* O157:H7 also carries a 60-Mdal plasmid that reportedly encodes fimbriae that mediate attachment to Henle 407 intestinal cells (Karch et al., 1987). In the same gnotobiotic pig study, this plasmid was found to be unimportant for the pathogenesis of disease observed in this animal model (Tzipori et al., 1987). Additional studies are necessary to establish the virulence determinants of EHEC.

Diagnostic Features

Because of the heterogeneous clinical manifestations of intestinal infections associated with *E. coli*, diagnosis by signs and symptoms is difficult. In areas endemic for cholera, the disease caused by ETEC is indistinguishable from that caused by *Vibrio cholerae*. Diarrhea produced by EIEC may be difficult to distinguish clinically from an intestinal infection with *Shigella* sp., *Salmonella* sp., *Campylobacter jejuni*, *Yersinia enterocolitica*, or *Entamoeba histolytica*. The typical illness caused by EHEC may be distinguished clinically from bloody diarrhea or dysentery seen in shigellosis, *Campylobacter* and EIEC enteritis, amebiasis, necrotizing enterocolitis, or pseudomembranous colitis by the lack of prominent fever. The first appearance of an inflammatory bowel disease could be mistaken for hemorrhagic colitis, but early colonoscopic and microbiologic examinations should help to differentiate the two diseases. Hence, the presence in stool of other conventional enteropathogens must be ruled out before *E. coli* can be considered as an etiologic agent. Then the definitive diagnosis involves the isolation of *E. coli* from stool and its further characterization, as discussed below.

Epidemiologic Aspects

Occurrence and Transmission

Although true incidence estimates of diarrhea caused by the four *E. coli* groups are unavailable, survey studies suggest that ETEC and EPEC are the leading bacterial causes of endemic and probably epidemic forms of diarrhea in many developing countries. Enterotoxigenic *E. coli* has been isolated from 11% of children with diarrhea in one study in the Philippines (Echeverria et al., 1978) to 35% of patients with diarrhea in Dhaka, Bangladesh (Sack et al., 1977). Enterotoxigenic *E. coli* infections are seen in both urban and rural areas of developing countries. In contrast, EPEC infections appear to be a disease of infants in large urban centers of developing countries, such as Sao Paulo, Brazil (Toledo et al., 1983), Mexico City, Mexico (Donta et al., 1977), Djakarta, Indonesia (Joe et al., 1966), and Lagos, Nigeria (Agbonlahor and Odugbemi, 1982). In Brazil, EPEC accounted for 30% of infants with diarrhea in Sao Paulo, the largest city in South America (Toledo et al., 1983), whereas it was responsible for only 4.6% of the diarrhea cases in the rural northeast (Guerrant et al., 1983). When sought, EIEC does not appear to be a major cause of diarrhea in developing countries. Information about EHEC from countries outside of North America, Europe, and Japan does not currently exist.

In developed countries, all four groups of *E. coli* diarrhea are usually recognized as part of the outbreaks. However, in the United States and Canada, EPEC and ETEC are important causes of endemic diarrhea among residents of some Indian reservations (Gurwith and Williams, 1977; Sack et al., 1975). Enteropathogenic *E. coli* was a common cause of infantile or "summer diarrhea" outbreaks in both institutional and community settings through the 1960s, but by the beginning of the mid-1970s, for an unexplained reason these outbreaks began to decline. In the United States, the most recent outbreaks of EPEC diarrhea have occurred in daycare centers (Paulozzi et al., 1986). Diarrhea outbreaks due to ETEC, EIEC, and EHEC can also occur in institutional and community settings. The first recognized outbreak of hemorrhagic colitis associated with EHEC occurred in 1982, and at least seven outbreaks associated with EHEC serotype O157:H7 were reported in North America through 1985—more than any other *E. coli* diarrhea outbreaks (Riley, 1987). Enterohemorrhagic *E. coli* also causes sporadic infections (Pai et al., 1984; Remis et al., 1984), and in western Canada and the United States Northwest, EHEC is now recognized as the most frequent cause of bloody diarrhea. In fact, EHEC is now probably the most common cause of *E. coli*-associated diarrhea in the United States.

A study of diarrhea among British troops in South Arabia (Rowe et al., 1970) demonstrated that ETEC is an important cause of diarrhea among travelers from developed countries to the developing countries. Enterotoxigenic *E. coli* was shown to be responsible for 45 to 72% of cases of "Turista" among travelers to Mexico (Gorbach et al., 1975; Merson et al., 1976).

Foodborne transmissions can occur with all four groups of diarrheagenic *E. coli*, and waterborne transmission of ETEC, EIEC, and EPEC have been reported. Contaminated milk was often implicated in hospital and nursery outbreaks of EPEC diarrhea in developed countries through the 1960s. Dairy products have also caused large outbreaks of EIEC and ETEC diarrhea in the United States, and raw milk may occasionally transmit EHEC (Martin et al., 1986). Transmission from person to person is an important mode of transmission in institutional settings, and can occur with all four groups of *E. coli*.

In the tropics, the peak occurrence of ETEC and EPEC diarrhea is the rainy season. Enteropathogenic *E. coli* diarrhea was frequently referred to as "summer diarrhea" in the developed countries. In Canada, EHEC-induced diarrhea is most frequently seen during the summer months. The number of recognized infections is not sufficient to determine the seasonal variation of EIEC-induced diarrhea.

The intestinal tracts of humans, mammals, and birds are the reservoir of *E. coli*, but the sources of pathogenic *E. coli* are unestablished. Strains of ETEC that cause diarrhea in animals do not cause disease in humans, and human ETEC strains are rarely isolated from farm animals. A recent isolation of *E. coli* O157:H7 from cattle (Martin et al., 1986) and the frequent association of the ingestion of beef with hemorrhagic colitis suggest that cattle may be a reservoir of EHEC.

Susceptibility

In developing countries, symptomatic disease can occur in all age groups after an infection with ETEC or EIEC, but the diarrhea caused by EPEC is generally confined to infants, especially those less than 6 months of age. However, volunteer studies in developed countries have shown that adults can develop symptomatic disease with EPEC (Levine et al., 1978). The reason for this age specificity and apparent life-long protection of EPEC-induced diarrhea in developing countries is unexplained. Diarrhea associated with EHEC, which has only been reported in developed countries, affects all age groups, but children and institutionalized elderly persons appear to be particularly susceptible to the severe manifestations of the disease (Riley, 1987).

Breast-feeding appears to be protective for ETEC- and EPEC-induced diarrhea. The recognized high frequency of EPEC infections among infants in urban centers of developing countries may be related to the lower frequency of breast-feeding practices among the urban mothers. Hospitals in these large urban centers may also be an important nidus for the transmission of EPEC. These patterns are similar to those that existed in developed countries in the 1940s to the 1960s.

Etiologic Agent

Collection and Transport

Stool specimens should be collected early in the course of the disease. Evacuated stool specimens are preferable to rectal swab specimens, and multiple specimens should be obtained. If the specimen cannot be plated immediately, it should be placed in an appropriate transport media or a stool preservative such as buffered glycerol. In addition to culture, stool specimens should be examined for the presence of leukocytes and blood.

Isolation and Definitive Identification

Because pathogenic *E. coli* strains could be overgrown by the normal gut flora or be present in small numbers if the specimen is collected late in the course of the illness, appropriate enrichment and plating media must be used to isolate the organism. As it is beyond the scope of this discussion to detail the primary isolation and identification procedures, the reader is referred to the methodologies described by Edwards and Ewing (1972).

Because most *E. coli* strains found in feces are nonpathogenic, the definitive identification of diarrheagenic *E. coli* depends not only on the primary isolation of the organism, but also on further differentiation of these organisms. These methods have included serotyping and determination of recognized virulence markers. The antigenic differentiation of *E. coli* by serotypes is still the most widely used procedure for typing *E. coli* by the reference laboratories. Serotyping is particularly useful in: 1) the preliminary determination of the etiologic role of an *E. coli* strain in the absence of any knowledge about its virulence factors, and 2) studying modes of transmission, geographic distribution, and incidence.

Despite the utility of serotyping, however, the identification of enteric *E. coli* pathogens by this method has been labor-intensive and cumbersome. Furthermore, the procedure is generally confined to the reference laboratories. Hence, simpler and more rapid techniques to identify diarrheagenic *E. coli* were necessary. With the recent discoveries of the various *E. coli* virulence factors, the approach to the diagnosis of diarrheagenic *E. coli* no longer needs to rely completely on the serotyping procedure, and can be performed by most modestly equipped laboratories. These new methods include those that use: 1) tissue culture, 2) immunologic assays, and 3) nucleic acid hybridization. The validity of these methods is based on the association of the positive test results with diarrhea in epidemiologic investigations, human volunteer studies, and animal models.

It should be noted that two animal models are still widely used to assess virulence determinants of EIEC and ST-producing ETEC. Sereny test (or guinea pig conjunctivitis test) is used to test the invasiveness of EIEC (Day et al., 1981; Sereny, 1955). The infant mouse assay is used to detect STa (Dean et al., 1972).

Tissue Culture Assays

A variety of tissue culture cells are available to test for LT, enteroinvasiveness, and enteroadherence. Some of these tests are detailed next.

Y-1 mouse adrenal cell assay for LT. The Y-1 adrenal cells maintained by standard procedures are resuspended and diluted in growth medium to about 2.5×10^5 cells/ml (determined by hemocytometer). A 0.2-ml aliquot of this suspension is distributed into each well of a 96-well tissue culture plate, and the plate is sealed and incubated, usually for 3 days at 37°C until the cells are confluent. The wells are replaced with fresh medium, and 0.025 ml of heated and unheated bacterial filtrate are added to the wells. The plate is resealed and incubated for 24 h at 37°C. The cells are then fixed with methanol and stained with Giemsa stain (5 to 10%, vol/vol), washed, dried, and examined by light microscopy. A positive test for the presence of LT is indicated by the rounding of over 90% of the cells with the unheated filtrate and absence of change within the heated sample (Donta et al., 1974). This test can be performed with bacterial cultures instead of culture filtrates (Sack and Sack, 1975).

HEp-2 cell invasion assay. The following procedure is adapted from the method of Small et al. (1987). Bacteria are grown overnight at 37°C in Trypticase soy agar plates, and the next day, they are suspended in 3 ml of brain-heart infusion broth and grown to log phase with shaking for 2 h. They are then resuspended in growth medium to a density of about 3 × 10^7 organisms/ml, and 1 ml of this suspension is distributed into a well of a 24-well tissue culture plate to achieve a ratio of about 100 bacteria/HEp-2 cell. At this point, the plate may be centrifuged at 800 × *g* for 10 min to increase the efficiency of invasion, although this step is not vital. The infected monolayer is then incubated for 2 h in 5% CO_2 at 37°C, washed three times with phosphate-buffered saline (PBS) (pH 7.4), covered with fresh growth medium containing gentamicin (100 μg/ml), and incubated for 1 h at 37°C. (Since gentamicin cannot penetrate the eukaryotic cell membrane, bacteria that remain outside the tissue culture cells will be killed.) The monolayer is washed six times with PBS, and treated with 200 μl of 1% Triton X-100 for 5 min to release the intracellular bacteria. Trypticase soy broth (800 μl) is added, and the plate is incubated for 30 min at 37°C. Dilutions of this suspension are plated on Tryticase soy agar plates, and colony-forming units are determined. The test should be done in triplicate. Noninvasive and known invasive *E. coli* strains should be used as controls in each test. The efficiency of invasion is determined by the number of bacteria that invade the tissue culture cells (final CFU) divided by the number of bacteria in the original inoculum per well. This test can also be performed with HeLa cells.

HEp-2 or HeLa cell adhesion assay. Several methods to examine patterns of attachment by *E. coli* to tissue culture cells exist (Nataro et al., 1987; Riley et al., 1987; Scaletsky et al., 1984). The following procedure is similar to that used for bacterial invasion, except that the gentamicin step is omitted. Briefly, the tissue culture cells are suspended and diluted in growth medium to 1 to 2 × 10^5 cells/ml, and 1.5 ml of this suspension is distributed to each well of a 24-well polystyrene tissue culture plate. The plate is incubated at 37°C in a CO_2 incubator until cells cover about 50% of the bottom of the well. The wells are then replaced with fresh growth medium containing 0.5% D-mannose (which is needed to inhibit nonspecific binding by *E. coli* that express type 1 fimbriae) and no antibiotics. To each well, 0.1 ml of overnight bacterial growth in Trypticase soy broth is added, and the plate is incubated for 3 h at 37°C. The medium is then removed and the infected monolayer is washed six times with PBS (pH 7.4), fixed with methanol for 5 min, stained with Giemsa stain (10%, vol/vol), washed with distilled water, and dried. The wells are examined under an inverted light microscope for the characteristic patterns of bacterial attachment.

Immunologic Assays

Two immunologic methods to detect LT are now in wide use. They include the GM_1-based enzyme-linked immunosorbent assay (GM_1-ELISA) (Sack et al., Svennerholm and Holmgren, 1978) and the Biken test (Honda et al., 1981).

Because of the poor antigenicity of STa, the development of simple immunologic detection methods for STa has lagged behind that for LT. A radioimmunoassay for STa, a competitive ELISA based on glutaraldehyde-coupled STa-human serum albumin conjugate (Lockwood and Robertson, 1984), and an ELISA that uses monoclonal antibodies raised against BSA-conjugated STa purified from a human isolate (Thompson et al., 1984) have been developed. These new tests are undergoing evaluation.

An ELISA based on the detection of an antigen associated with the invasiveness of *Shigella* and EIEC has been developed in Hungary (Pai et al., 1985) and is undergoing evaluation. No immunologic methods, other than serotyping, are currently available for the detection of EPEC or EHEC.

GM_1-ELISA for LT. The following procedure is currently used at the Centers for Disease Control (Atlanta, Georgia) and is adapted from the procedure of Sack et al. (1980). The inner 60 wells of a 96-well polyvinyl microtiter plate are filled with 0.1 ml of GM_1 ganglioside diluted to 1 μg/ml in PBS, and the plate is incubated overnight at 25°C in a moist chamber. The plates are washed three times with 0.2 ml PBS/Tween per well and stored until use. The bacterial culture supernatant (0.05 ml) and 0.05 ml of diluent (PBS/Tween, 1% rabbit serum) are added to the wells, and the plate is again incubated overnight at 25°C. The plates are washed three times with 0.2 ml per well of PBS/Tween, and to each well, 0.1 ml of goat anti-LT diluted 1 : 5,000 in the above diluent is added. The plate is reincubated at 37°C and washed three times. To each well is added 0.1 ml of alkaline phosphatase-conjugated rabbit anti-goat antibody, diluted 1 : 600 in PBS/Tween, 1% rabbit serum, and the plate is incubated for 1 h at 37°C. The wells are washed three times, and 0.1 ml of enzyme substrate (p-nitrophenyl phosphate, dissolved in diethanolamine buffer—5 mg tablet per 5 ml of buffer) is added to each well. The substrate is also added to well A1 for the plate reader. The reaction is stopped after 30 min by adding 0.05 ml of 3 M NaOH to each well, including A1. The yellow color produced by this reaction is determined visually and by an ELISA plate reader at 405 nm. Each plate should include three negative and one positive control. The optical density of the sample is divided by the mean of the optical density of the three negative controls, and a resulting value of 2.0 or greater indicates the presence of LT.

Biken test for LT. The antiserum raised against LT must first be purified by affinity chromatography. The bacterial strains to be tested are inoculated on an agar plate to form four discrete growth zones surrounding a central site at which a well will be punched out. The inner edge of these growths should be about 4 mm from the edge of the central well. The plate is incubated for 48 h at 37°C, and a polymixin B (500 units) disk is then placed on top of each bacterial growth. The plate is reincubated for another 5 to 6 h and then 20 μl of antiserum is placed in the central well. The plate is then reexamined after 24 h for the development of precipitation line between the bacterial growth and the central well.

DNA Hybridization Probes

The detection of an infectious agent by a DNA gene probe is based on the principle that a single-strand DNA molecule can hybridize with another DNA (or RNA) strand that is complementary or nearly complementary to it. With the possible exception of EHEC, the recognized major virulence genes of diarrheagenic *E. coli* reside almost always on a plasmid. The gene probes used to detect ETEC, EIEC, and EPEC are, therefore, based on fragments of DNA cloned from such virulence plasmids. Moseley et al. (1980) used a radiolabeled restriction endonuclease-cleaved DNA fragments that encode LT and ST to hybridize homologous DNA sequences of *E. coli* colonies grown on nitrocellulose paper overlaid on a solid growth medium. Others have used radiolabeled 2.5 kilobase (kb)- and biotinylated 17-kb DNA probes cloned from the large plasmid associated with enteroinvasion to detect EIEC as well as *Shigella* (Boileau et al., 1984; Gomes et al., 1987; Wood et al., 1986). A radiolabeled 1-kb DNA probe cleaved from the enteroadhesiveness factor (EAF) gene specifically detects EPEC strains that exhibit localized adherence with HEp-2 or HeLa cells (Nataro et al., 1985).

Except to answer specific epidemiologic questions, the routine use of radiolabeled, cloned DNA probes, especially in places where diarrhea caused by *E. coli* is endemic, is often viewed to be impractical, expensive, and technically unfeasible. However, with attempts to improve and simplify the techniques, the practical application of this technology in such areas is rapidly becoming a possibility. These attempts have included the following: 1) the construction of synthetic oligonucleotide gene probes based on the known nucleotide sequences of fragments from virulence plasmids; 2) the use of nonradioactive label to tag the probe; 3) the processing of the specimen containing the target gene to increase sensitivity, reduce background signal, and shorten specimen preparation time; and 4) the amplification of the target DNA sequences.

The problem in the use of radioisotopes in developing countries, and the occasional problem of the vector DNA sequences of cloned probes hybridizing with a complementary sequence in the target DNA, have motivated the construction of synthetic oligonucleotide probes tagged with nonradioactive labels such as biotin or alkaline phosphatase. Oligonucleotides conjugated to alkaline phosphatase can be lyophilized and stored at room temperature. Hence, such products can be mailed anywhere in the world, and with minimal equipment and reagents, clinical specimens can be tested. So far, such gene probes have been constructed to detect ST- and LT-producing ETEC (Seriwatana et al., 1987). The DNA sequence data of EIEC invasiveness probe and EPEC EAF probe will soon become available to construct specific oligonucleotide probes.

The specimens containing the target genes are usually prepared on a matrix such as the nitrocellulose paper on Whatman 541 filter paper (Whatman Ltd., England). The isolated bacteria are spotted on the filter paper overlaid onto a growth medium and allowed to form separate colonies ("colony hybridization"), or the stool specimen itself can be directly inoculated onto the filter paper and allowed to proliferate in a small zone ("stool blot," "stool growth," "macrocolony"). To improve sensitivity and reduce background signals, additional processing of the colonies or stool growths, such as lysing and deproteinizing the cells, may be necessary.

The target DNA sequence amplification method is already applied in the study of human genetic disorders (Saiki et al., 1986; Scharf et al., 1986). The polymerase chain reaction (PCR) procedure is used to enzymatically amplify a gene segment of interest in a specimen in vitro. The procedure involves repeated cycles of DNA denaturation, oligonucleotide primer annealing, and primer extension by a DNA polymerase. After 20 cycles, which can be completed in less than 2 h, this procedure can yield greater than 10^5-fold increase in a 110-base pair target gene fragment. Such a procedure will facilitate epidemiologic studies in which sensitivity of a probe is critical, such as studies of persistent enteric infections or asymptomatic carriage, as well as the detection of a pathogen in stool before it is cultured.

Treatment, Prevention, and Control

Treatment

The mainstay of therapy for most cases of diarrhea caused by *E. coli* is fluid and electrolyte replacement. Administration of oral rehydration solutions (ORS) is an adequate treatment for most cases, and will prevent severe dehydration and death, especially among children infected with ETEC.

The use of antimicrobial agents, especially in developing countries and hospital settings, must be weighed against the possibility of establishing resistant strains. Pockets of multiple drug-resistant strains of ETEC, EPEC, and probably EIEC exist in South and Central America and South and Southeast Asia. There is some evidence that antibiotics may reduce the duration of diarrhea and bacterial excretion due to ETEC and EPEC, and the latter may be important in breaking the chain of transmission within institutions or families. However, persistent diarrhea associated with EPEC occurs, even with the elimination of the *E. coli* in stool. As with *Shigella* sp., antimicrobial agents are probably effective in reducing the duration of illness caused by EIEC. As dysentery may not be as responsive to oral rehydration solutions as is watery, dehydrating diarrhea, appropriate antimicrobial agents must be considered in severe manifestations of EIEC diarrhea. The duration of illness and hospitalization in EHEC infection is not affected by antimicrobial agents (Riley et al., 1983).

The complications of EHEC infection such as hemolytic-uremic syndrome and thrombotic thrombocytopenic purpura must be anticipated, especially in children and elderly patients, and appropriate supportive measures be provided.

Prevention and Control

The provision of clean water and foods, adequate nutrition, and improving the socioeconomic standards are, of course, the solution to the problem of diarrhea in general. Given the current realities, however, alternative methods must be considered. Encouraging use of ORS and breastmilk and maternal education are some of the most effective, already available ways to prevent morbidity and mortality due to diarrhea.

Several vaccine candidates based on the recognized major virulence determinants of diarrheagenic *E. coli* have been developed. A recent volunteer study showed that oral administration of live, nontoxigenic, fimbriated strain of ETEC induced secretory IgA response in the intestines and was protective against diarrhea (Levine, 1987). Vaccines based on the virulence determinants encoded by the 120 to 140-Mdal plasmids of *Shigella* spp. are under investigation, and may be applied to the protection against EIEC infections (Timmis et al., 1986). The virulence determinants of EPEC and EHEC must be better characterized before new vaccines can be considered.

Urinary Tract Infections

Description of Disease

Escherich was also the first to show that *E. coli* was present in the urine of young girls suffering from urinary tract infections. Today, *E. coli* is recognized as the most common cause of both community-acquired and nosocomial urinary tract infections (Rubin, 1987).

Clinical Features

Urinary tract infections (UTI) may present with asymptomatic bacteriuria, cystitis, or pyelonephritis. Cystitis is an infection limited to the bladder, and symptoms may include dysuria, urgency, frequency, and lower abdominal pain. Pyelonephritis develops from the bacterial invasion of the kidney parenchyma, and symptoms may include fever, rigor, nausea, vomiting, and lower back or flank pain. However, the clinical differentiation of the sites of infection is difficult and unreliable. Ureteral catheterization and bladder washout studies suggest that up to 50% of patients with lower urinary tract symptoms may have infection in the kidney also (Sanford, 1975). Furthermore, urethritis associated with gonorrhea or chlamydial infection shares many of the symptoms of cystitis. Renal infarction, calculi, and obstructive nephropathy without infection can produce the symptoms of bacterial pyelonephritis.

Asymptomatic bacteriuria may be thought of as a phase of UTI interrupted by symptomatic episodes. Asymptomatic bacteriuria early in pregnancy is known to predispose to overt pyelonephritis later in pregnancy. Prospective follow-up surveys for bacteriuria in young school girls showed that although bacteriuria itself is generally benign, greater morbidity was associated with girls found to be bacteriuric (Gillenwater et al., 1979). The morbidity included recurrent infections, hospitalization for pyelonephritis, urologic surgery, and nephrectomy.

Complications and Prognosis

Complications of acute UTI may include chronic pyelonephritis that may lead to renal failure or hypertension. However, chronic pyelonephritis rarely occurs after early childhood in the urinary tract that has no anatomic or functional defects. Surgical procedures or instrumentations of the urinary tract may trigger *E. coli* sepsis and endotoxemia. In the male, UTI may be complicated by prostatitis or epididymitis. More than 80% of the cases of prostatitis is caused by *E. coli* (Rubin, 1987), and *E. coli* is the most frequently isolated organism in epididymitis of the older men.

Pyelonephritis may develop in 2 to 4% of pregnant women during the third trimester. If the infection is treated appropriately, major complications are rare. However, in the preantibiotic era, it was often accompanied by sepsis, premature birth, and renal stone disease after delivery. More recent studies suggest that symptomatic UTI during pregnancy increases the risk of delivering low birth weight babies (McGrady et al., 1985; Sever et al., 1978).

There is no evidence that bacteriuria or recurrent infections lead to increased mortality or major morbidity, such as end-stage renal disease, in the absence of any underlying anatomic or functional defects. Mortality rates from pyelonephritis do not differ by sex or race. However, increased mortality does occur in patients with nosocomial UTI, and a large proportion of this mortality includes elderly patients who develop complications of prostatitis or who undergo urologic instrumentations.

Pathogenesis

The pathogenesis of *E. coli* UTI involves the interaction of *E. coli* virulence factors with host factors. More than 95% of the cases of UTI results from infection through the urethra up into the bladder, and the source of the offending organism is usually the fecal flora of the infected patient. It was previously thought that *E. coli* strains responsible for UTI were those that predominate in the feces, but recent evidence suggests that certain *E. coli* strains possess properties that render them uropathogenic. Postulated virulence properties of uropathogenic *E. coli* have included somatic (O) antigens (Lindberg et al., 1975), capsular (K) antigens (Glynn and Howard, 1970; McCabe et al., 1975a), hemolysins, and adhesins (Svanborg Edén et al., 1976; Svanborg Edén and Hansson, 1978).

As is recognized for diarrheagenic *E. coli*, a restricted set of serogroups of *E. coli* appears to be associated with UTI. The serogroups identified in European and North American studies are shown in Table 1. Urinary tract infection-associated O serogroups are more prevalent in the symptomatic than in asymptomatic infections (Lindberg et al., 1975). The O antigens of uropathogenic strains may determine the organisms' immunogenicity and serum sensitivity, which in turn may influence their infectivity, but their role in the pathogenesis of UTI is not known.

The evidence for the pathogenic role of capsular antigens and hemolysins follows similar observations—that strains exhibiting these phenotypic characteristics are more frequently identified in the urine than in the feces. However, there is good clinical and experimental evidence for the pathogenic role of adhesins. At least three classes of adhesins appear to be involved in the colonization of the urinary tract during infection. Most strains of *E. coli*, including uropathogenic *E. coli*, produce type 1 fimbriae which bind specifically to D-mannose-containing receptors

TABLE 1. *Escherichia coli* serogroups associated with human intestinal and extraintestinal infections

Intestinal infections[a]				*Urinary tract infections*[f]	*Bacteremia*[f]	*Meningitis*[f]
ETEC[b]	*EPEC*[c]	*EIEC*[d]	*EHEC*[e]			
06	018	028	0157	01	01	01
08	026	029		02	02	06
015	044	0112		04	04	07
020	055	0124		06	06	016
025	086	0136		07	07	018
027	0114	0143		08	08	083
063	0111	0144		09	09	
078	0119	0152		011	011	
080	0125	0164		018	018	
085	0126	0167		022	022	
0114	0127			025	025	
0115	0128ab			062	075	
0128ac	0142			075		
0148	0158					
0153						
0159						
0167						

[a] Listed in this table if the serogroup is included in these diarrheagenic *E. coli* groups by two or more of the following references: Ørskov et al., 1977; Levine, 1987; Rowe, 1979.
[b] Enterotoxigenic *E. coli*.
[c] Enteropathogenic *E. coli*.
[d] Enteroinvasive *E. coli*.
[e] Enterohemorrhagic *E. coli*.
[f] From Ørskov et al., 1977.

on a variety of eukaryotic cells. Its binding can be reversed by mannose, and hence these fimbriae have also been termed mannose-sensitive fimbriae (MS fimbriae). Type 1 fimbriae can bind to the Tamm-Horsfall glycoproteins in the urinary mucus (Ørskov et al., 1980). Hence, urinary mucus may serve as a barrier against mucosal colonization by *E. coli* strains that exhibit type 1 fimbriae, and may explain why the introduction of most fecal strains into the urinary tract does not lead to sustained colonization. By the same token, changes in host factors that lead to decreased clearance, such as obstruction to flow or vesicoureteral reflux, may predispose such hosts to develop UTI with strains that may otherwise be nonuropathogenic.

Some strains of uropathogenic *E. coli* produce mannose-resistant (MR) fimbriae that bind to the carbohydrate moiety (α-D-Gal*p*-(1 $\rightarrow$ 4)-β-D-Gal*p*-) of the P blood group antigen (Kallenius et al., 1980; Svanborg Edén et al., 1981). In the uroepithelium, the globoseries glycosphingolipids are the receptors for these fimbriae (Leffler and Svanborg Edén, 1981), which have been variously termed P fimbriae, gal-gal pili, and pap-pili (pyelonephritis-associated pili). A mouse model experiment showed that *E. coli* K12 clone expressing both MS and gal-gal pili was able to colonize and invade the mouse kidneys with a dose of 10^6 organisms; a clone expressing only the gal-gal pili colonized but did not invade, and a clone that expressed only the MS fimbriae neither colonized nor invaded the kidneys with the same doses (O'Hanley et al., 1985).

Finally, uropathogenic *E. coli* adhesins that are mannose-resistant and lack the activity of the P fimbriae have been termed X adhesins. These may include fimbrial as well as nonfimbrial adhesins (Walz et al., 1985). Their significance in the pathogenesis of UTI remains to be established.

Diagnostic Features

One unique feature to be considered in the diagnosis of UTI is the notion of "significant bacteriuria," first proposed by Kass (1956). Since sterile urine in the bladder can become contaminated during passage through the distal urethra, when diagnosing UTI by urine culture, one must distinguish a true infection from contaminated urine. Significant bacteriuria is defined as bacterial colony counts greater than 10^5 cfu/ml in the voided urine specimen. It should be noted that up to 30% of patients with UTI may have counts less than 10^5 cfu/ml (Rubin, 1987), especially patients who may have been taking bacteriostatic antimicrobial agents, when urine pH is low, when urine flow from a focal site of infection is partially or completely obstructed, or when the specimen is collected early in the incubation period after inoculation of the bladder.

Methods to localize UTI include both invasive and noninvasive procedures. In the bladder washout method (Fairley et al., 1967), a neomycin solution is introduced into the bladder through a catheter, followed by 2 liters of sterile water to wash out the bladder. Then urine is collected at regular intervals and cultured. Patients with cystitis only will have sterile urine in all of the postwashout samples, whereas patients with upper tract infection will have bacteria in these samples. This method does not localize the infection to one kidney, which requires ureteral catheterization.

Response to therapy, especially to single antibiotic therapy, may help to localize the infection. Early relapse after therapy suggests upper urinary tract infection. Nonspecific serologic tests such as the determination of the rise in C-reactive protein are sometimes used to diagnose acute pyelonephritis. The antibody-coated bacteria (ACB) test claims to be able to distinguish pyelonephritis from cystitis. Detection of *E. coli* in urine coated with human immunoglobulins was found to correlate with the results of the bladder washout test in distinguishing pyelonephritis from cystitis (Jones et al., 1979; Thomas et al., 1979). However, false-positive results have been reported in patients with prostatitis, nephrotic syndrome, bladder neoplasm, and indwelling catheter in place. High rates of false-negative results occur in children with acute pyelonephritis (Rubin, 1987).

Epidemiologic Aspects

Occurrence and Transmission

Much of the data on the incidence of UTI comes from studies conducted in developed countries. In the United States, *E. coli* accounts for over 90% of the community-acquired UTI (Rubin, 1987). The National Nosocomial Infections Study estimates that the nosocomial UTI rate is about 13/1000 discharges, and *E. coli* accounts for over 30% of all causes of nosocomial UTI (Centers for Disease Control, 1986). Urologic instrumentation plays a major role in the transmission of *E. coli* infections in hospitals.

The incidence of UTI varies according to sex and age. Among infants with UTI, males predominate in the first 3 months of life (Ginsburg and McCracken, 1982). *E. coli* accounts for over 80% of the cases of UTI in infants. After infancy, the rates of infection become substantially higher in females, with the overall incidence (per year) ranging from 1.5 to 3% among preschool children, to 1.2% among school age children, to 2.5% among adults of reproductive age, and 10 to 30% among the geriatric population (Rubin, 1987).

Susceptibility

Both host and external factors influence susceptibility to UTI. Well documented host factors include: 1) demographic characteristics such as sex and age; 2) anatomic or functional defects that lead to obstruction or reflux; 3) pregnancy; 4) sexual intercourse in some female subjects; and 5) P_1 blood group (Lomberg et al., 1983). The most important external factor influencing increased susceptibility to UTI is instrumentation.

Although there is no convincing evidence that diabetics have a greater chance of developing bacteriuria than do nondiabetics, other medical problems associated with diabetes (for example, neurogenic bladder and other problems requiring hospitalization) may create conditions (for example, prolonged catheterization) that increase their susceptibility to UTI.

Etiologic Agent

Collection and Transport

Several methods for the collection of urine specimens are recommended, and the procedure varies with the type of patient—ambulatory female, male, infants, and bedridden patients (Kunin, 1987). In the ambulatory female and male, the midstream collection of urine is recommended by some clinicians. For infants, although suprapubic aspiration is the method of choice, a strap-on plastic device, after proper cleansing, may be used. If the urine culture result is negative by this method, it is unnecessary to perform a suprapubic aspiration. For the bedridden patient, if possible, every attempt to collect clean-voided urine should be made before urethral catheterization is considered. If catheterization is required, aseptic technique must be strictly followed.

In the female, the clean-voided collection method is about 80% accurate with one specimen, 90% with two specimens, and virtually 100% with three specimens containing the same organism. In the circumsised male, one specimen is sufficient for diagnosis.

A specimen that cannot be examined and cultured within 1 h of collection should be refrigerated. Also, to avoid delay in specimen processing, the urine can be immediately plated on an agar surface or inoculated into a transport medium.

Isolation and Definitive Identification

Urine specimens are examined microscopically, chemically, and by quantitative culture methods. The microscopic examination includes Gram stain of unspun urine or methylene blue-stained or unstained centrifuged urine sediment. The presence of one organism per immersion oil field of a gram-stained specimen correlates with significant bacteriuria. The presence of more than 20 bacteria in the urinary sediment is considered positive for infection.

Chemical tests are used for screening for bacteriuria. Most of these tests are nonspecific, and must be followed up with cultures (Kunin, 1987). A colorimetric method to detect bacteriuria in 2 min has been developed (Wallis et al., 1981), but high false-negative test results have been reported (Wu et al., 1985).

The standard urine culture methods include the pour plate and the streak plate methods. For the detection of *E. coli*, the streak plate method is adequate. One loopful each of the urine specimen is streaked onto an eosin-methylene blue (EMB) plate and a blood plate. One hundred colonies per plate represents 10^5 bacteria per ml of urine. Other simple culture techniques adaptable for office use or epidemiologic studies are available and reviewed by Kunin (1987). Automated, rapid detection systems that provide quantification, speciation, and antimicrobial susceptibility information are also becoming available in diagnostic laboratories and large hospitals.

Treatment, Prevention, and Control

Treatment

Unlike the treatment for most enteric *E. coli* infections, antimicrobial agents are the mainstay of therapy for *E. coli* UTI. The decisions concerning diagnosis, selection of appropriate antimicrobial agents, duration of therapy, and follow-up depend on host characteristics such as age, sex, pregnancy, underlying disease, and anatomic or functional deficits. Although the standard duration of treatment has been 10 to 14 days, in a young female patient with uncomplicated UTI—the most common presentation of UTI—one dose of an antimicrobial agent such as sulfonamide, amoxicillin, or trimethoprim/sulfamethoxazole is adequate, especially if treatment is initiated early in the course of the infection (Rubin, 1987). Urine should become sterile within 24 to 48 h of treatment. Single dose regimen is not recommended for patients with complicated infections, such as those with diabetes and anatomic or functional defects, clinical manifestation suggestive of bacterial tissue invasion, frequent recurrent infection (usually defined as more than two episodes per year), or in males. The cure rate appears to be considerably reduced in the presence of antibody-coated bacteria.

A common problem in uncomplicated UTI is recurrent infection. Most recurrent infections are due to reinfections with different strains of *E. coli*, rather than relapse with the same organism. In such infections, drug susceptibility tests should guide the choice of drugs, and because *E. coli* resistance to trimethoprim/sulfamethoxazole remains low in recurrent *E. coli* infection, this drug is recommended (Kunin, 1987). Courses varying from 3 days to 2 weeks have been shown to be effective in eliminating the cycle of recurrent infection in many women. In pregnancy, ampicillin or amoxicillin are safe drugs to use for uncomplicated *E. coli* UTI.

Patients with acute pyelonephritis may be too nauseated to take oral drugs. Such patients should be hospitalized and started on parenteral drugs. The empirical combination drug therapy of ampicillin and an aminoglycoside is frequently used until a single parenteral or oral drug can be substituted. The new antimicrobial agents such as the broad spectrum beta-lactam antibiotics are just as effective as the aminoglycosides for *E. coli* infections, and they lack the possible nephrotoxicity or ototoxicity of the aminoglycosides. However, their bioavailability, inducibility of drug resistance, side effects, and cost should be considered in comparison with the proven drugs of choice before use.

Longer duration of therapy is recommended for both uncomplicated (2 weeks) and recurrent or complicated (6 to 12 weeks) infection in male patients (Kunin, 1987). Complicated infections may include acute prostatitis. Severe forms of acute prostatitis should be treated parenterally with a combination that includes an aminoglycoside. Milder forms of acute prostatitis usually respond to the drugs given for uncomplicated bacteriuria. Chronic prostatitis is difficult to treat and requires long-term therapy.

Of course, all of these regimens requiring antimicrobial intervention should be accompanied by close monitoring of clinical signs, urine bacteriology, and appropriate follow-up care.

Prevention and Control

Prevention of UTI is directed at persons who are recognized to be at increased risk for such infection. They include: 1) women or girls with frequent recurrences; 2) pregnant women; 3) hospitalized patients who undergo repeated catheterization; 4) persons with neurogenic bladder requiring intermittent or chronic catheterization; 5) men with chronic prostatitis; and 6) men undergoing urologic procedures. In women with recurrent infection, short or intermediate duration of treatment may fail to eradicate recurrent bacteriuria. In such patients, low dose, long-term prophylaxis is recommended (Kunin, 1987). A postcoital dose of an antibiotic appears to reduce the occurrence of UTI in women who develop UTI after sexual intercourse (Vosti, 1975).

Although screening is no longer recommended for women with recurrent infection, it is an important part of prevention in pregnant women. All pregnant women are recommended to be screened for bacteriuria during the first trimester and again during the third trimester. Screening and treatment of bacteriuria during the first trimester will reduce the expected incidence of pyelonephritis by 10-fold in pregnant women (Stamm, 1984).

Prevention of recurrent infection in catheterized persons is difficult. Of course, the most effective prevention is avoiding the use of the indwelling catheter when it is not necessary. Aseptic techniques and catheter care are the only currently effective measures that can be taken to prevent UTI in these patients (Kunin, 1984).

New knowledge on the virulence determinants of uropathogenic *E. coli* has provided an opportunity to adopt novel approaches to the prevention and control of UTI. Immunization with gal-gal pili protected against renal colonization and invasion of an intraurethral challenge of uropathogenic *E. coli* in mice (O'Hanley et al., 1985). Another study showed that a eukaryotic cell surface receptor analogue that binds the gal-gal fimbriae competitively blocked attachment to the bladder mucosa by the uropathogenic *E. coli* in mice (Svanborg Edén et al., 1982). However, human clinical trials to show efficacy of protection with such reagents are still needed.

Bacteremia

Description of Disease

With the advent and changes in the use of antimicrobial agents, intensive care units, and new invasive procedures, the spectrum of pathogens that cause bacteremia has fluctuated both in time and place. However, despite these changes, *E. coli* has persisted as one of the most important causes of bacteremia in both neonates and adults.

Clinical Features

In neonates (infants 28 days old or younger), bacteremia occurs as "early onset" and "late onset" infections (Klein and Marcy, 1976). The early onset disease, defined as an illness occurring during the first week of life, tends to be more severe than the late onset disease, and the mortality rates are higher. *E. coli* may cause either of these diseases. Prominent signs of bacteremia in neonates may include poor feeding, fever or hypothermia, respiratory distress, apnea, vomiting, diarrhea, jaundice, hepatomegaly, and lethargy. Focal signs indicative of primary infections include pneumonia, otitis, omphalitis, urinary tract infection, septic arthritis, peritonitis, and osteomyelitis.

Since blood cultures are not routinely obtained from asymptomatic neonates, the incidence of asymptomatic *E. coli* bacteremia is unknown. Transient bacteremia in a mildly ill young child (usually between the ages of 6 and 24 months) who presents with fever as an outpatient is not an uncommon problem for the clinician, but the pathogens responsible for such cases most often include *Streptococcus pneumoniae, Haemophilus influenzae, Salmonella* sp., group B streptococcus, and *Neisseria meningitidis*. *E. coli* is rarely isolated in such cases.

As in neonates, the clinical findings in bacteremia in adults caused by *E. coli* cannot be distinguished from those associated with other gram-negative or gram-positive pathogens. The spectrum of clinical manifestations ranges from fever and rigor with changes in mental status such as lethargy or increased agitation, to a fulminant course leading to shock.

Complications and Prognosis

Approximately 30% of neonates with bacteremia will progress to develop meningitis. Most of the sequelae of bacteremia are actually those of the meningitis, which are discussed below. Shock occurs in 30 to 40% of cases of gram-negative bacteremia, and develops 2 to 8 h after the initial manifestation of bacteremia. In both adults and neonates, septicemia leading to shock may be complicated by organ failures, bleeding, or acute disseminated intravascular coagulation (DIC).

The mortality rate in neonatal septicemia was nearly 100% in the preantibiotic era. With the introduction of sulfonamides and penicillins, it declined to about 60%, and more recent studies show overall rates to have declined further to around 20% (Freedman et al., 1981). However, several studies suggest that much of the decrease can be attributed to the decrease in mortality from septicemia due to specific pathogens, especially group B streptococcus (Freedman et al., 1981), and that the decline in mortality from *E. coli* septicemia has not been so dramatic. In a 15-year study (1969 to 1983) of septicemia in a hospital in Sweden, the overall mortality rate decreased from 27% in 1969 to 1973 to 12% in 1979 to 1983, and cases of septicemia caused by *E. coli* also decreased from about 25 to 11%, respectively (Bennett et al., 1985). However, during the same period, neonates with *E. coli* septicemia had the highest mortality rates—50% between 1969 and 1973 and over 30% between 1979 and 1983 (Bennet et al., 1985).

In adults, the prognosis of *E. coli* bacteremia depends mostly on host factors. A poor prognosis in bacteremic patients is associated with: 1) hematologic malignancies; 2) absence of fever; 3) shock or DIC; 4) pulmonary infection; 5) persistent bacteremia despite appropriate antibiotic administration; and 6) low or decreasing absolute granulocyte count ($<1000/mm^2$) during the infection (Bodey et al., 1986; Weinstein et al., 1983). The genitourinary tract is the main primary site of infection in adults who develop *E. coli* bacteremia. The mortality rate is high in such patients because they often are elderly or diabetic patients who undergo urologic instrumentation or surgery.

Pathogenesis

The pathogenic process of bacteremia begins with the entry of *E. coli* into the bloodstream. In neonates, a primary source, especially for early onset bacteremia, is usually not apparent. Transplacental infection by *E. coli* does not occur, and the initial colonization of the newborn most likely occurs during passage through the birth canal. Amniotic fluid is normally inhibitory to *E. coli*, but if it is meconium-stained, it may favor its growth (Florman and Teubner, 1969). Bacteremia may follow urinary tract infection or manipulation in about 20% of cases of neonatal bacteremia (DuPont and Spink, 1969).

In adults, 40 to 60% of cases of *E. coli* bacteremia can be traced to the genitourinary tract as the primary source, and another 25 to 30% may have a gastrointestinal source (bowel, biliary) (DuPont and Spink, 1969; Weinstein et al., 1983). However, the source may be inapparent in about 25% of adult *E. coli* bacteremia cases (DuPont and Spink, 1969).

Most of the *E. coli* serogroups associated with bacteremia are also associated with UTI (Table 1). Only serogroups O6, O8, and O25 are found in the diarrheagenic *E. coli* group. Some studies suggest that certain O polysaccharide antigens may promote survival of the bacteria in the blood stream by rendering the organism serum resistant (McCabe et al., 1978). The evidence for the pathogenic role of *E. coli* acidic polysaccharide capsular antigen in bacteremia is not as strong as it is for neonatal meningitis (Ørskov and Ørskov, 1985).

One well studied *E. coli* bacterial product associated with many of the clinical features of bacteremia, especially shock, is the endotoxin, or the lipopolysaccharide (LPS). Shock is manifested by hypotension and is accompanied by peripheral vasoconstriction, tissue hypoperfusion, and oliguria or anuria. Vasodilatation and increased vascular permeability may occur in some parts of the circulatory system. In the lungs, increased capillary and alveolar permeability may produce adult respiratory distress syndrome (ARDS).

The LPS of gram-negative bacteria is composed of three distinct regions: 1) the hydrophobic region called lipid A, which is linked via the 2-keto-3-deoxy-octonate (KDO) to 2) the inner region, called the core glycolipid, and 3) the outer hydrophilic region, which is the O antigenic polysaccharide. The lipid A is thought to be the primary effector of many of the manifestations of the endotoxic shock (Young et al., 1977). A variety of endogenous mediators have also been proposed to trigger the different pathophysiologic phases of shock. One intriguing recent proposal is that the endotoxin exerts its action via the activation of a macrophage hormone called cachectin or tumor necrosis factor (TNF) (Beutler and Cerami, 1987). The evidence for the association of cachectin with endotoxic shock is as follows: 1) in rat and rabbits, cachectin can induce responses similar to those produced by LPS—shock, metabolic acidosis, electrolyte disturbances, interstitial pneumonitis, acute tubular necrosis, pulmonary artery thrombosis, and gastrointestinal ischemia and hemorrhage; 2) cachetin may induce production of leukotrienes and platelet-activating factor, which have been suggested to be involved in endotoxic shock; 3) cachetin may alter the permeability of the endothelial surface and allow leakage of water and electrolytes into the extravascular space; 4) cachectin can release interleukin-1 by monocytes and endothelial cells; interleukin-1 may induce fever, hypotension, and thrombocytopenia; 5) cachectin itself can directly affect the hypothalamus to induce fever; and 6) cachectin activates polymorphonuclear leukocytes to attach to endothelial cells, which may explain the peripheral neutropenia seen in septic shock (Beutler and Cerami, 1987). Hence, it appears that this macrophage hormone can mediate most of the pathophysiologic events of endotoxic shock. The mechanism of stimulation of the macrophage by LPS to induce this endogenous mediator is under investigation.

Diagnostic Features

Early and aggressive initiation of therapy is critical in the management of *E. coli* bacteremia and, hence, its diagnosis must rely not only on rapid confirmatory procedures, but on the best judgment of clinicians and microbiologists and on the understanding of the local epidemiologic circumstances of the pathogens. In neonates, the nonfocal manifestations may be difficult to distinguish clinically from infections such as the congenital infections of toxoplasmosis, rubella, cytomegalovirus, and herpes simplex virus. Noninfectious diseases such as the hemolytic disease of the newborn can mimic the signs of *E. coli* bacteremia. Definitive diagnosis involves isolation of *E. coli* from blood.

Epidemiologic Aspects

Occurrence and Transmission

The incidence of *E. coli* bacteremia varies according to time and reporting institutions. In the United States and Europe, *E. coli* was an unusual cause of bacteremia until the 1950s, when it replaced the gram-positive bacteria as the most common cause of bacteremia in both neonates and adults (Bennet et al., 1985; Dupont and Spink, 1969; Freedman et al., 1981). In a 50-year study of neonatal sepsis at Yale-New Haven Hospital, the incidence of early onset and late onset *E. coli* sepsis averaged 0.85 and 1.27/1000 live births, respectively (Freedman et al., 1981). Studies from most parts of the world show that *E. coli* is often the most predominant blood isolate from neonates. *E. coli* is the most frequently-isolated pathogen in bacteremia among cancer patients with granulocytopenia in hospitals participating in the study of the International Antimicrobial Therapy Cooperative Group of the European Organization of Research and Treatment (EORTC) (EORTC, 1987). However, more recently, especially in large teaching hospitals of the United States, gram-positive organisms such as Group B streptococcus, *Staphylococcus aureus,* and coagulase-negative staphylococci are beginning to reemerge and surpass *E. coli* as the most common agents of bacteremia. In the most recent report of the National Nosocomial Infections Surveillance System (NNIS) that considered 51 hospitals in the United States, the overall nosocomial rate of bacteremia was 2.5/1000 discharges per year, and *E. coli* accounted for 10.1% of all blood isolates after coagulase-negative staphylococci (14.9%) and *Staphylococcus aureus* (12.3%) (Centers for Disease Control, 1986). Compared with the university hospitals, the incidence of gram-negative bacteremia appears to be less (1.8 to 2.0/1000 admissions) in community hospitals (McCabe et al., 1975b; Scheckler, 1977; Wolff and Bennett, 1974). Several studies suggest that *E. coli* is more frequently isolated from community-acquired than from hospital-acquired cases of bacteremia (DuPont and Spink, 1969; Scheckler, 1977).

Hospitals play an important role as places of transmission of *E. coli* bacteremia. The different rates of isolation by hospital services suggest that transmission is influenced by the types of patients seen and procedures practiced in these services. *E. coli* bacteremia from transfusion of contaminated blood products has been reported (Arnow et al., 1986).

Susceptibility

Well documented risk factors for neonatal bacteremia include: 1) premature delivery or low birth weight, 2) premature rupture of the membrane, 3) maternal peripartum infection, 4) septic or traumatic delivery, and 5) hypoxia (Klein and Marcy, 1976). The immature immune response and phagocytotic function of low birth weight newborns may increase their risk for bacteremia. These newborns also have decreased levels of lactoferrin and transferrin. These iron-binding proteins, if present in the serum at sufficient levels, may provide protection against bacteremia since iron is known to promote growth of *E. coli* in serum.

In adults, the risk of developing bacteremia increases substantially with the presence of an underlying disease, especially hematologic malignancies. Other factors such as urologic instrumentation and surgery contribute to increased incidence of bacteremia among patients with genitourinary disease.

Etiologic Agents

Collection and Transport

E. coli bacteremia is usually intermittent. Hence, three separate blood cultures should be obtained during a 24 to 48-h period. Although the likelihood of recovering a pathogen after three blood cultures is low, if an antimicrobial therapy had already been initiated, an additional one to three cultures may increase the yield. Also, the use of culture media containing reagents to absorb (for example, adsorbent resins) or inactivate (for example, penicillinase) the antimicrobial agents in the blood specimen should be considered. However, it is recommended that such systems be used in conjunction with the conventional systems.

In neonates, blood, urine, and cerebrospinal fluid (CSF) should be obtained for culture as soon as there is any clinical suspicion of bacteremia. Usually, 0.2 ml of blood is sufficient to detect *E. coli* bacteremia in neonates (Klein and Marcy, 1976). In adults, collection of 10 to 20 ml is recommended. The standard types of culture and transport medium, isolation and subculturing procedures, and identification of the organism from blood specimens are detailed elsewhere (Washington, 1975).

Isolation and Definitive Identification

Since the specimen (blood) examined in the diagnosis of bacteremia comes from a sterile site, unlike the intestinal infections and UTI, any *E. coli* isolated from blood must be regarded as a causative agent until shown otherwise. The rapid diagnosis of bacteremia itself can be facilitated by an automated detection of $^{14}CO_2$ generation by organisms in blood cultured in medium containing ^{14}C-labeled substrates (Damato et al., 1983). Further identification of the organism and its antimicrobial susceptibility patterns may be rapidly determined by a number of newly available automated or manual procedures, which are reviewed elsewhere (D'Amato et al., 1985).

Bacterial quantification of blood cultures may be useful. Meningitis has been reported to develop only in neonates who have greater than 1,000 bacterial colonies/ml of blood (Dietzman et al., 1974). The methods of obtaining CSF and urine from neonates are discussed in other sections of this chapter.

Treatment, Prevention, and Control

Treatment

Early administration of antimicrobial agents is the mainstay of therapy of *E. coli* bacteremia. Since *E. coli* bacteremia is clinically indistinguishable from bacteremia caused by other pathogens, and more than 50% of patients with untreated bacteremia will die within 72 h, the empirical administration of antimicrobial agent is critical. The local institutional knowledge of the prevalent organisms and their antimicrobial susceptibility, as well as the clinical characteristics of the patient, should be taken into consideration in selecting the drugs. The most recent National Nosocomial Infections Study in the United States found that aminoglycoside resistance among hospital isolates of *E. coli* is relatively low (1.4% for tobramycin in nonteaching hospitals to 2.6% for gentamicin in small teaching hospitals and 2.6% for tobramycin in large teaching hospitals) (Centers for Disease Control, 1986). Hence, in the United States, an empirical regimen that includes an aminoglycoside, together with a penicillin (including penicillinase-resistant or antipseudomonal penicillins), or a cephalosporin may be appropriate. Once the *E. coli* is identified, a single drug regimen may be used. A recent study of cancer patients found that a single appropriate antibiotic was as effective as a combination in such patients (Bodey et al., 1986).

Of course, in patients who develop shock, adjunct measures such as blood pressure support with vasoactive amines, ventilation, and fluid and electrolytes should be provided. In contrast to earlier studies (Schumer, 1976; Sprung et al., 1984), more recent studies of the use of corticosteroids in septic shock suggest that they do not prevent or reverse shock or decrease mortality, and in patients with initially elevated serum creatinine levels (>2 mg/deciliter [dl]), the use of methylprednisolone may increase mortality (Bone et al., 1987). The use of methylprednisolone may also prolong the duration of secondary infections (Veterans Administration Systemic Sepsis Cooperative Study Group, 1987).

Another potential adjunctive therapy in gram-negative bacteremia that has shown some success in improving the outcome is the granulocyte transfusion (Steeper and McCullough, 1985). However, the recovery of the patient's own granulocyte count to greater than 1000/μl is the most important determinant of favorable outcome, even in transfused patients.

Prevention and Control

Prevention of bacteremia is directed at reducing the recognized risk factors for the infection. In neonates, prenatal care may decrease the incidence of prematurity and provide early recognition of low birth weight infants, maternal peripartum infection, or premature rupture of the membrane, which are all factors associated with bacteremia. Neonatal intensive care facilities may have contributed to the increase and change in the pattern of neonatal bacteremia (Bennet et al., 1985; LaGamma et al., 1983; Townsend and Wentzel, 1981), but may also have prevented deaths of infected premature and low birth weight infants, who otherwise may not have survived.

In adults, proper catheter care and prompt treatment of patients with urinary tract disease will decrease the chances of bacteremia. In patients with hematologic malignancies, monitoring clinical signs and laboratory parameters such as the granulocyte count may provide early recognition of bacteremia.

The efficacy of prophylactic administration of antimicrobial agents to prevent neonatal or adult bacteremia in patients recognized to have predisposing factors is equivocal. Such an intervention must be weighed against the establishment of resistant pathogens in the hospital environment. The efficacy of prophylactic granulocyte transfusion is also not established (Quie, 1987).

Another recent approach to prevention involves immunotherapy, based on the antibody raised against the core glycolipid of the gram-negative bacterial LPS. For many years, it has been recognized that gram-negative bacteria share core glycolipids that are immunologically cross-reacting. When the serum from healthy volunteers immunized with heat-killed *E. coli* rough mutant lacking the O polysaccharide (*E. coli* J5) was administered in a randomized controlled trial to patients with culture-proven gram-negative bacteremia, it significantly improved the survival rate (Ziegler et al., 1982). In another study (Baumgartner et al., 1985), when anti-J5 plasma was administered to patients who underwent abdominal surgery, it did not prevent the occurrence of gram-negative bacteremia, but did significantly reduce the occurrence of shock and death. The mechanism of protection by the J5 antibody is under investigation.

Neonatal Meningitis

Description of Disease

Clinical Features

E. coli is the predominant etiologic agent of neonatal meningitis. Many of the early signs of neonatal sepsis are shared by neonatal meningitis, including fever, lethargy, vomiting, diarrhea, respiratory distress, and jaundice. Other early signs of meningitis may include change in consciousness and hyperactivity. Convulsions were reported in 40% of neonates with meningitis in a multihospital review (Klein and Marcy, 1976). Bulging fontanelle or nuchal rigidity occur, but are less commonly seen in neonates than in older children.

Complications and Prognosis

About 20 to 50% of neonates who survive meningitis develop long-term complications ranging from mild perceptual difficulties to seizure disorders, mental and motor disabilities, hearing deficits, and hydrocephalus (Fitzhardinge et al., 1974). Relapse, even after an appropriate antimicrobial treatment, can occur in a minority of cases.

Mortality rates from all causes of neonatal meningitis exceed 30% in most series. As is discussed next, mortality rates of *E. coli* neonatal meningitis depend both on strain and host characteristics. A mortality rate in excess of 30% has been reported among neonates infected with *E. coli* strains that carry K1 capsular polysaccharide (McCracken et al., 1974). In a recent study from the Netherlands, *E. coli* meningitis case/fatality rates exceeding 35% were associated with gestational age of less than 35 weeks, birth weight less than 2,500 g, and the presence of congenital defects, compared with the rate of about 12% among neonates without any known risk factors (Mulder et al., 1984).

Pathogenesis

In neonates, *E. coli* meningitis is almost always a complication of bacteremia. The proportion of neonates with *E. coli* bacteremia who develop meningitis varies from about 10% in the 50-year series at Yale-New Haven Hospital (Freedman et al., 1981) to 40% in a 20-year study in West Germany (Speer et al., 1985).

The association of neonatal meningitis with strains of *E. coli* that possess an acidic polysaccharide capsular antigen K1 has been long recognized (Robbins et al., 1974). Nearly 80% of *E. coli* isolates from cases of neonatal meningitis in the United States and Europe carry this K1 antigen. In addition, the K1 strains associated with meningitis fall into restricted O antigen groups (Table 1).

The K1 antigen, which is a (α-2-8)-linked sialic acid polymer, is chemically and immunologically related to the capsular polysaccharide of *Neisseria meningitidis* type B, another common cause of meningitis. The pathogenic relationship of this polysaccharide to meningitis is unknown. However, the neonatal brain contains a glycopeptide with (α-2-8)-linked sialic acid stretches that cross-react immunologically with the K1 antigen (Finne et al., 1983), and this cross-reacting product disappears within the first few weeks of life. The age-specific susceptibility of *E. coli* meningitis may be related to the presence of such material in the neonatal brain.

Diagnostic Features

The diagnostic considerations applied to bacteremia also apply to meningitis. In neonates, *E. coli* meningitis cannot be distinguished clinically from other frequent causes of neonatal meningitis such as group B streptococcus, other gram-negative bacteria, and *Listeria monocytogenes*. Noninfectious conditions that produce similar signs include kernicterus, intracranial hemorrhage, congenital central nervous system defects, and withdrawal signs in neonates born from drug-addicted mothers. The CSF should be examined early for definitive diagnosis.

Epidemiologic Aspects

Occurrence and Transmission

As with *E. coli* bacteremia, the incidence of *E. coli* neonatal meningitis varies with time and place. From the data on overall incidence of neonatal meningitis and the proportion of cases attributed to *E. coli*, the incidence of *E. coli* meningitis can be estimated. In studies that provide such data, the incidence (per 1,000 live births) ranges from about 0.1 in the years 1976 to 1982 in the Netherlands (Mulder et al., 1984) to about 0.4 between 1962 and 1982 in Gottingen, West Germany (Speer et al., 1985), and to 0.5 between 1969 and 1983 in Dallas, Texas (McCracken, 1984).

The modes of transmission of *E. coli* in neonatal meningitis are the same as those for neonatal bacteremia. In adults, *E. coli* meningitis is not common. If it occurs, it results almost always from intracranial surgical procedures or head trauma.

Susceptibility

In most studies, a higher proportion of meningitis is observed in male infants. Although premature rupture of the membrane is an important risk factor for neonatal bacteremia, the most important predisposing factor for *E. coli* meningitis appears to be the low birth weight (< 2,500 g) (Mulder et al., 1984). Other identified risk factors include congenital defects and multiple deliveries (Mulder et al., 1984).

Etiologic Agents

Collection and Transport

Cerebrospinal fluid (CSF) should be obtained early in the course of the disease, cultured, and its profiles examined immediately. A caveat is that the normal CSF profile of the neonate is different from that of older children and adults. Polymorphonuclear leukocytes can be present, protein concentration may be higher, and glucose concentration may be lower in the CSF of normal term or premature infants compared with that in older infants or children (Klein and Marcy, 1976).

Isolation and Definitive Identification

The isolation of *E. coli* from the CSF confirms the diagnosis of *E. coli* meningitis. In addition to culture, the Gram stain of the CSF should be obtained. *E. coli* may remain in the CSF for several days, even after the initiation of appropriate antimicrobial treatment. A rapid assay, based on a gelation reaction of the limulus (horseshoe crab) amebocyte lysate with the bacterial endotoxin (LPS) in the CSF, has been utilized to diagnose meningitis, but the test has had equivocal results and is not regularly used (Ross et al., 19975). The DNA hybridization methods may soon play a role in the rapid diagnosis of *E. coli* meningitis.

Treatment, Prevention, and Control

Treatment

Many neonates with meningitis, especially those with low birth weight, require supportive and intensive care. But, as in bacteremia, the rapid elimination of the bacteria is the main goal of treatment of meningitis. An intravenous administration of the combination of ampicillin and an aminoglycoside has been the conventional initial therapy for gram-negative meningitis (McCracken, 1984). However, McCracken (1984) found that the case/fatality rates among treated neonates remained unchanged at about 18 to 19% over a 15-year period. In addition, intrathecal or intraventricular administration of gentamicin to increase its CSF bactericidal concentration did not improve outcome from gram-negative meningitis (McCracken and Mize, 1976, McCracken et al., 1980). These observations led to the reexamination of antimicrobial therapy for gram-negative meningitis.

Many of the new beta-lactam antibiotics have high in vitro activities against gram-negative bacteria that cause meningitis, and achieve CSF bactericidal titers superior to that of aminoglycosides. However, animal studies indicate that most of these new antibiotics do not appear to be superior to the conventional regimens in improving outcome (Kim, 1985). To date, only moxalactam has been examined in comparison with ampicillin-amikacin in a large clinical trial (McCracken et al., 1984), and no significant difference in short-term outcome between the ampicillin-moxalactam treatment group and the ampicillin-amikacin treatment group was demonstrated. Although one can argue for the substitution of a beta-lactam antibiotic with less toxicity and better CSF penetration for an aminoglycoside, other factors such as the aminoglycoside resistance of the *E. coli*, renal insufficiency, and poor initial response to the conventional therapy should be considered in the decision to select the newer antimicrobial agents.

The adjunctive administration of corticosteroids with the rationale to reduce inflammation and cerebral edema in neonatal meningitis remains controversial. Animal studies indicate that corticosteroids can decrease the CSF penetration of the antibiotic, but the decrease is usually not sufficient to affect the bactericidal concentrations (Scheld and Brodeur, 1983). Their adjunctive therapeutic role in *E. coli* neonatal meningitis has yet to be established.

Prevention and Control

Measures to prevent bacteremia will greatly reduce the occurrence of *E. coli* meningitis because the latter is a complication of the former. These measures were discussed above.

Miscellaneous Infections

E. coli is associated with infections of most human organ systems. It is a leading cause of peritonitis in patients with liver cirrhosis and hepatic abscess, and is recovered from infections of the skin and soft tissue, especially in diabetics. The pathogenic mechanisms vary according to the organ system and host factors, but the virulence determinants of the *E. coli* in these miscellaneous infections have not yet been well characterized. The definitive diagnosis of these infections involves aspiration or biopsy to recover the *E. coli* from fluid, abscess, or tissue specimens, following standard sterile techniques. Management includes appropriate antimicrobial therapy, as well as drainage of abscesses. Prevention and control vary according to the sites of infection.

In addition to the direct invasion of organ tissues, some strains of *E. coli* are associated with a systemic syndrome in which the organism itself does not appear to participate directly in tissue injury. This is the hemolytic uremic syndrome (HUS), which is a disease, usually of small children, that begins with a prodrome of diarrhea followed by a triad of microangiopathic hemolytic anemia, thrombocytopenia, and acute renal failure (Gasser et al., 1955). Karmali et al. (1983) demonstrated that sporadic cases of HUS were significantly associated with recovery from stool of *E. coli* strains that produce Vero cell-active cytotoxins (*Shiga*-like toxins). The bacteria is not recovered from extraintestinal sources. The pathogenic mechanism and the relationship of the cytotoxins to the disease are under investigation (see discussion on Hemorrhagic Colitis). The cytotoxins can be assayed for by tissue culture (Vero cells or HeLa cells) (Marques et al., 1986).

Hence, after over 100 years since its discovery, *E. coli* is not only recognized to be a major cause of many infectious diseases, but continues to be discovered as the etiologic agent of previously undiagnosed illnesses and provides much challenge for improved diagnostic procedures.

Literature Cited

Agbonlahor, D. E. and T. O. Odugbemi. 1982. Enteropathogenic, enterotoxigenic, and enteroinvasive *Escherichia coli* isolated from acute gastroenteritis patients in Lagos, Nigeria. Trans. R. S. Trop. Med. Hyg. **76:**265–67.

Arnow, P. M., L. M. Weiss, D. Weil, and N. R. Rosen. 1986. *Escherichia coli* sepsis from contaminated platelet transfusion. Arch. Intern. Med. **146:**321–324.

Baumgartner, J-D., M. P. Glauser, J. A. McCutchan, E. J. Ziegler, G. van Melle, M. R. Klauber, M. Vogt, E. Muehlen, R. Luethy, R. Chiolero, and S. Geroulanos. 1985. Prevention of gram-negative shock and death in surgical patients by antibody to endotoxin core glycolipid. Lancet **ii:**59–63.

Bennet, R., M. Eriksson, B. Melen, and R. Zetterstrom. 1985. Changes in the incidence and spectrum of neonatal septicemia during a fifteen-year period. Acta Paediatr. Scand. **74:**687–690.

Beutler, B. and A. Cerami. 1987. Cachectin: More than a tumor necrosis factor. N. Engl. J. Med. **316:**379–385.

Black, R. E., K. H. Brown, and S. Becker. 1984. Effects of diarrhea associated with specific enteropathogens on the growth of children in rural Bangladesh. Pediatrics **6:**799–805.

Bodey, G. P., L. Elting, H. Kassamali, and B. Lim. 1986. *Escherichia coli* bacteremia in cancer patients. Am. J. Med. **81:**85–95.

Boileau, C. R., H. M. D'Hauteville, and P. J. Sansonetti. 1984. DNA hybridization technique to detect *Shigella* species and enteroinvasive *Escherichia coli*. J. Clin. Microbiol. **20:**959–961.

Bone, R. C., C. J. Fisher, T. P. Clemmer, G. J. Slotman, C. A. Metz, R. A. Balk, and the Methylprednisolone Severe Sepsis Study Group. 1987. A controlled clinical trial of high-dose methylprednisolone in the treatment of severe sepsis and septic shock. N. Engl. J. Med. **317**:653–658.

Bopp, C. A., K. D. Greene, F. P. Downes, E. G. Sowers, J. G. Wells, and I. K. Wachsmuth. 1987. Unusual verotoxin-producing *Escherichia coli* associated with hemorrhagic colitis. J. Clin. Microbiol. **25**:1486–1489.

Bray, J. 1945. Isolation of antigenically homogeneous strains of *Bact. coli neopolitanum* from summer diarrhoea of infants. J. Pathol. Bacteriol. **57**:239–247.

Carter, A. O., A. A. Borczyk, J. A. K. Carlson, B. Harvey, J. C. Hockin, M. A. Karmali, C. Krishnan, D. A. Korn, and H. Lior. 1987. A severe outbreak of *Escherichia coli* O157:H7-associated hemorrhagic colitis in a nursing home. N. Engl. J. Med. **317**:1496–1500.

Centers for Disease Control. 1986. Surveillance summaries 1986. **35**:17ss–29ss.

Cravioto, A., R. J. Gross, S. M. Scotland, and B. Rowe. 1979. An adhesive factor found in strains of *Escherichia coli* belonging to the traditional infantile enteropathogenic serotypes. Curr. Microbiol. **3**:95–99.

Damato, J. J., M. T. Collins, M. V. Rothlauf, and J. K. McClatchy. 1983. Detection of mycobacteria by radiometric and standard plate procedures. J. Clin. Microbiol. **17**:1066–1073.

D'Amato, R. F., J. C. McLaughlin, and M. J. Ferraro. 1985. Rapid and mechanized/automated methods for the detection and identification of bacteria and yeasts, p. 52–65. *In:* E. H. Lennette, A. Balows, W. J. Hausler Jr., and H. J. Shadomy (ed.). Manual of clinical microbiology. American Society for Microbiology, Washington, D.C.

Day, N. P., S. M. Scotland, and B. Rowe. 1981. Comparison of an HEp-2 tissue culture test with the Sereny test for detection of enteroinvasiveness in *Shigella* spp. and *Escherichia coli*. J. Clin. Microbiol. **13**:596–597.

Dean, A. G., Y. C. Ching, R. G. Williams, and L. B. Harden. 1972. Test for *Escherichia coli* enterotoxin using infant mice: application in a study of diarrhea in children in Honolulu. J. Infect. Dis. **125**:407–411.

Dietzman, D. E., G. W. Fischer, and F. D. Schoenknecht. 1974. Neonatal *Escherichia coli* septicemia—bacterial counts in blood. J. Pediatr. **85**:128–130.

Donta, S. T., H. W. Moon, and S. C. Whipp. 1974. Detection of heat-labile *Escherichia coli* enterotoxin with the use of adrenal cells in tissue culture. Science **183**:334–336.

Donta, S. T., R. B. Wallace, S. C. Whipp, and J. Olarte. 1977. Enterotoxigenic *Escherichia coli* and diarrheal disease in Mexican children. J. Infect. Dis. **135**:482–485.

DuPont, H. L. and W. W. Spink. 1969. Infections due to gram-negative organisms: an analysis of 860 patients with bacteremia at the University of Minnesota Medical Center, 1958–1966. Medicine **48**:307–332.

Echeverria, P., N. R. Blacklow, J. L. Vollet, C. V. Ulyangco, G. Cukor, V. B. Soriano, H. C. Dupont, J. H. Cross, F. Ørskov, and I. Ørskov. 1978. Reoviruslike agent and enterotoxigenic *Escherichia coli* infections in paediatric diarrhea in the Philippines. J. Infect. Dis. **138**:326–332.

Edwards, P. R. and W. H. Ewing. 1972. Identification of enterobacteriaceae, 3rd ed. Burgess Publishing Co., Minneapolis, Minnesota.

European Organization for Research and Treatment of Cancer (EORTC). International Antimicrobial Therapy Cooperative Group. 1987. Ceftazidime combined with a short or long course of amikacin for empirical therapy of gram-negative bacteremia in cancer patients with granulocytopenia. N. Engl. J. Med. **317**:1692–1698.

Evans, D. G. and D. J. Evans Jr. 1978. New surface-associated heat-labile colonization factor antigen (CFA/II) produced by enterotoxigenic *Escherichia coli* of serogroups O6 and O8. Infect. Immun. **21**:638–647.

Evans, D. G., D. J. Evans Jr, W. S. Tjoa, and H. L. DuPont. 1978. Detection and characterization of colonization factor of enterotoxigenic *Escherichia coli* isolated from adults with diarrhea. Infect. Immun. **19**:727–736.

Evans, D. G., R. P. Silver, D. J. Evans Jr., D. G. Chase, and S. L. Gorbach. 1975. Plasmid-controlled colonization factor associated with virulence in *Escherichia coli* enterotoxigenic for humans. Infect. Immun. **12**:656–667.

Ewing, W. H., B. R. Davis, and T. S. Montague. 1963. Studies on the occurrence of *Escherichia coli* serotypes associated with diarrheal disease. Atlanta, Ga, U.S. Department of Health Service, Communicable Disease Center.

Fairley, K. F., A. A. G. Bond, A. B. Brown, and P. Habersberger. 1967. Simple test to determine the site of urinary tract infections. Lancet **ii**:7513–7514.

Farmer, J. J., III, B. R. Davis, F. W. Hickman-Brenner, A. McWhorter, G. P. Huntley-Carter, M. A. Asbury, C. Riddle, H. G. Wathen-Grady, C. Elias, G. R. Fanning, A. G. Steigerwalt, C. M. O'Hara, G. K. Morris, P. B. Smith, and D. J. Brenner. 1985. Biochemical identification of new species and biogroups of Enterobacteriaceae isolated from clinical specimens. J. Clin. Microbiol. **21**:46–76.

Field, M. 1979. Modes of action of enterotoxins from *Vibrio cholerae* and *Escherichia coli*. Rev. Infect. Dis. **1**:918–925.

Finne, J., M. Leinonen, and P. H. Makela. 1983. Antigenic similarities between brain components and bacteria causing meningitis. Lancet **ii**:354–357.

Fitzhardinge, P. M., M. Kazemi, M. Ramsay, and L. Stern. 1974. Long-term sequelae of neonatal meningitis. Dev. Med. Child. Neurol. **16**:3–10.

Florman, A. L. and D. Teubner. 1969. Enhancement of bacterial growth in amniotic fluid by meconium. J. Pediatr. **74**:111–114.

Francis, D. H., J. E. Collins, and J. R. Duimstra. 1986. Infection of gnotobiotic pigs with an *Escherichia coli* O157:H7 strain associated with an outbreak of hemorrhagic colitis. Infect. Immun. **51**:953–956.

Freedman, R. M., D. L. Ingram, I. Gross, R. A. Ehrenkranz, J. B. Warshaw, and R. S. Baltimore. 1981. A half century of neonatal sepsis at Yale. Am. J. Dis. Child. **135**:140–144.

Gariepy, J. and G. K. Schoolnik. 1986. Design of a photoreactive analogue of the *Escherichia coli* heat-stable enterotoxin STIb: use in identifying its receptor on rat brush border membranes. Proc. Natl. Acad. Sci. USA **83**:483–487.

Gasser, C., E. Gautier, A. Stgech, R. E. Siebenmann, and R. Dechslin. 1955. Hamolytischuramische syndrome: Bilatorale Nierenrindennekrosen bei akuten erworbenen hamolytischen Anamien. Schweiz. Med. Wschr. **85**:905–909.

Giles, C., G. Sangster, and J. Smith. 1949. Epidemic gastroenteritis of infants in Aberdeen during 1947. Arch. Dis. Child. **24**:45–53.

Gillenwater, J. Y., R. B. Harrison, and C. N. Kunin. 1979. Natural history of bacteriuria in schoolgirls. N. Engl. J. Med. **301**:396–399.

Ginsburg, C. M., and G. H. McCracken. 1982. Urinary tract infections in young infants. Pediatrics **69**:409–412.

Glynn, A. A. and C. J. Howard. 1970. The sensitivity to complement of strains of *Escherichia coli* related to their K antigens. Immunology **18**:331–346.

Gomes, T. A., M. R. F. Toledo, L. R. Trabulsi, P. K. Wood, and J. G. Morris Jr. 1987. DNA probes for identification of enteroinvasive *Escherichia coli*. J. Clin. Microbiol. **25**:2025–2027.

Gorbach, S. L., B. H. Kean, D. G. Evans, D. J. Evans, and D. Bessudo. 1975. Travelers' diarrhea and toxigenic *Escherichia coli*. N. Engl. J. Med. **292**:933–936.

Guerrant, R. L., L. V. Kirchhoff, D. S. Shields, M. K. Nations, J. Leslie, M. A. de Sousa, J. G. Araujo, L. L. Correia, K. T. Sauer, K. E. McClelland, F. L. Trowbridge, and J. M. Hughes. 1983. Prospective study of diarrheal illnesses in Northeastern Brazil: patterns of disease, nutritional impact, etiologies, and risk factors. J. Infect. Dis. **148**:986–997.

Gurwith, M. J. and T. W. Williams. 1977. Gastroenteritis in children: a two-year review in Manitoba. I. Etiology. J. Infect. Dis. **136**:239–247.

Harris, J. R., I. K. Wachsmuth, B. R. Davis, and M. L. Cohen. 1982. High molecular weight plasmid correlates with *Escherichia coli* enteroinvasiveness. Infect. Immun. **37**:1295–1298.

Honda, T., S. Taga, Y. Takeda, and T. Miwatani. 1981. Modified Elek test for detection of heat-labile enterotoxin of enterotoxigenic *Escherichia coli*. J. Clin. Microbiol. **13**:1–5.

Joe, L. K., B. Rukmono, S. Oemijati, K. Sahab, K. W. Newell, S. T. Hway, and R. W. Talogo. 1966. Diarrhoea among infants in a crowded area of Djakarta, Indonesia. Bull. Wld. Hlth. Org. **34**:197–210.

Jones, S. R., J. W. Smith, and J. P. Sanford. 1979. Localization of urinary tract infections by detection of antibody-coated bacteria in urine sediment. N. Engl. J. Med. **290**:591–593.

Källenius, G., R. Mollby, S. B. Svenson, J. Winberg, and H. Hultberg. 1980. Identification of a carbohydrate receptor recognized by uropathogenic *Escherichia coli*. Infection Suppl. **3**:S288–S293.

Karch, H., J. Heesemann, R. Laufs, A. D. O'Brien, C. O. Tacket, and M. M. Levine. 1987. A plasmid of enterohemorrhagic *Escherichia coli* O157:H7 is required for expression of a new fimbrial antigen and for adhesion to epithelial cells. Infect. Immun. **55**:455–461.

Karmali, M. A., B. T. Steele, M. Petric, and C. Lim. 1983. Sporadic cases of haemolytic-uraemic syndrome associated with faecal cytotoxin and cytotoxin-producing *Escherichia coli* in stools. Lancet **i**:619–620.

Kass, E. H. 1956. Asymptomatic infections of the urinary tract. Trans. Assoc. Am. Phys. **69**:56–63.

Kauffman, F. 1947. The serology of the coli group. J. Immunol. **57**:71–100.

Kim, K. S. 1985. Comparison of cefotaxime, imipenem-cilastatin, ampicillin-gentamicin, and ampicillin-chloramphenicol in the treatment of experimental *Escherichia coli* bacteremia and meningitis. Antimicrob. Agents Chemother. **28**:433–436.

Klein, J. O. and S. M. Marcy. 1976. Bacterial infections, p. 747–802. *In:* J. S. Remington and J. O. Klein (ed), Infectious diseases of the fetus and newborn infant. W. B. Saunders Co., Philadelphia.

Kunin, C. M. 1984. Genitourinary infections in the patient at risk: Extrinsic risk factors. Am. J. Med. **76(suppl)**:131–139.

Kunin, C. M. 1987. Detection, prevention, and management of urinary tract infections. Lea & Febiger, Philadelphia. 447 p.

LaGamma, E. F., L. M. Drusin, A. W. Mackles, S. Machalek, and P. A. M. Auld. 1983. Neonatal infections. Am. J. Dis. Child. **137**:838–841.

Leffler, H. and C. Svanborg Edén. 1981. Chemical identification of a glycosphingolipid receptor for *Escherichia coli* attaching to human urinary epithelial cells and agglutinating human erythrocytes. FEMS Microbiol. Lett. **8**:127–134.

Levine, M. M. 1987. *Escherichia coli* that cause diarrhea: Enterotoxigenic, enteropathogenic, enteroinvasive, enterohemorrhagic, and enteroadherent. J. Infect. Dis. **155**:377–389.

Levine, M. M., E. J. Bergquist, D. R. Nalin, D. H., Waterman, R. B. Hornick, C. R. Young, S. Sotman, and B. Rowe. 1978. *Escherichia coli* strains that cause diarrhoea but do not produce heat-labile or heat-stable enterotoxins and are non-invasive. Lancet **i**:1119–1122.

Levine, M. M., J. P. Nataro, H. Karch, M. M. Baldini, J. B. Kaper, R. E. Black, M. L. Clements, and A. D. O'Brien. 1985. The diarrheal response of man to some classical enteropathogenic *Escherichia coli* is dependent upon a plasmid encoding an enteroadhesiveness factor. J. Infect. Dis. **152**:550–559.

Lindberg, U., U. Jodal, L. A. Hanson, and B. Kaijser. 1975. Asymptomatic bacteriuria in schoolgirls. IV. Difficulties of level diagnosis and the possible relation to the character of infecting bacteria. Acta Paediatr. Scand. **64**:574–580.

Lockwood, D. E. and D. C. Robertson. 1984. Development of a competitive enzyme-linked immunosorbent assay (ELISA) for *Escherichia coli* heat-stable enterotoxin (STa). J. Immunol. Methods **75**:295–307.

Lomberg, H., L. Å. Hanson, B. Jacobsson, U. Jodal, H. Leffler, and C. Svanborg Edén. 1983. Correlation of P blood group, vesicoureteral reflux, and bacterial attachment in patients with recurrent pyelonephritis. N. Engl. J. Med. **308**:1189–1192.

Marques, L. R. M., M. A. Moore, J. G. Wells, I. K. Wachsmuth, and A. D. O'Brien. 1986. Production of Shiga-like toxin by *Escherichia coli*. J. Infect. Dis. **154**:338–341.

Martin, M. L., L. D. Shipman, J. G. Wells, M. E. Potter, and K. Hedberg. 1986. Isolation of *Escherichia coli* O157:H7 from dairy cattle associated with two cases of haemolytic-uremic syndrome. Lancet **ii**:1043.

Mathewson, J. J., P. C. Johnson, H. L. DuPont, D. R. Morgan, S. A. Thornton, L. V. Wood, and C. D. Ericsson. 1985. A newly recognized cause of travelers' diarrhea: enteroadherent *Escherichia coli*. J. Infect. Dis. **151**:471–475.

McCabe, W. R., P. C. Carling, S. Bruins, and A. Greely. 1975a. The relation of K antigen to virulence of *Escherichia coli*. J. Infect. Dis. **131**:6–10.

McCabe, W. R., B. Kaijser, J. Ollings, M. Uwaydah, and L. A. Hanson. 1978. *Escherichia coli* in bacteremia: K and O antigens and serum sensitivity of strains from adults and neonates. J. Infect. Dis. **138**:33–41.

McCabe, W. R., S. M. Wolff, and J. V. Bennett. 1975b. Incidence of gram-negative bacteremia. N. Engl. J. Med. **292**:111.

McConnell, M. M., L. V. Thomas, N. P. Day, and B. Rowe. 1985. Enzyme-linked immunosorbent assays for the detection of adhesion factor antigens of enterotoxigenic *Escherichia coli*. J. Infect. Dis. **152**:1120–1127.

McCracken, G. H. Jr. 1984. Management of neonatal meningitis, 1984. J. Antimicrob. Chemother. Suppl. B **14**:23–31.

McCracken, G. H. Jr. and S. G. Mize. 1976. A controlled study of intrathecal antibiotic therapy in gram-negative enteric meningitis in infancy. J. Pediatr. **89**:66–72.

McCracken, G. H. Jr., S. G. Mize, and N. Threlkeld. 1980. Intra-ventricular gentamicin therapy in gram-negative bacillary meningitis of infancy. Report of the Second

Neonatal Meningitis Cooperative Study Group. Lancet **i**:787–791.
McCracken, G. H. Jr., L. D. Sarff, M. P. Glode, S. G. Mize, M. S. Schiffer, J. B. Robbins, E. C. Gotchlich, I. Orskov, and F. Orskov. 1974. Relation between *Escherichia coli* K1 capsular polysaccharide antigen and clinical outcome in neonatal meningitis. Lancet **ii**:246–250.
McCracken, G. H. Jr., N. Threlkeld, S. G. Mize, C. J. Baker, S. L. Kaplan, I. Faingezicht, W. E. Feldman, U. Schaad, and the Neonatal Meningitis Cooperative Study Group. 1984. Moxalactam therapy for neonatal meningitis due to gram-negative bacilli. J. Am. Med. Assoc. **252**:1427–1432.
McGrady, G. A., J. R. Daling, and D. R. Peterson. 1985. Maternal urinary tract infection and adverse fetal outcomes. Am. J. Epidemiol. **121**:377–381.
McKay, D. G. and G. H. Wahle. 1955. Epidemic gastroenteritis due to *Escherichia coli* O111:B4. Arch. Pathol. **60**:679–693.
Merson, M. H., F. Ørskov, I. Ørskov, R. B. Sack, I. Huq, and F. T. Koster. 1979. Relationship between enterotoxin production and serotypes in enterotoxigenic *Escherichia coli*. Infect. Immun. **23**:325–329.
Merson, M. H., G. K. Morris, D. A. Sack, J. G. Wells, J. C. Feeley, R. B. Sack, W. B. Creech, A. Z. Kapikian, and E. J. Gangarosa. 1976. Travelers' diarrhea in Mexico. N. Engl. J. Med. **294**:1299–1305.
Morrison, D. M., D. L. J. Tyrrell, and L. D. Jewel. 1985. Colonic biopsy in verotoxin-induced hemorrhagic colitis and thrombotic thrombocytopenic purpura (TTP). Am. J. Clin. Pathol. **86**:108–112.
Moseley, S. L., P. Echeverria, J. Seriwatana, C. Tiparat, W. Chaicumpa, T. Sakuldaipeara, and S. Falkow. 1982. Identification of enterotoxigenic *Escherichia coli* by colony hybridization using three enterotoxin gene probes. J. Infect. Dis. **145**:863–869.
Moseley, S. L., I. Huq, A. R. M. A. Alim, M. So, M. Samadpour-Motalebi, and S. Falkow. 1980. Detection of enterotoxigenic *Escherichia coli* by DNA hybridization. J. Infect. Dis. **142**:8792–8798.
Mulder, C. J. J., L. van Alphen, and H. C. Zanen. 1984. Neonatal meningitis caused by *Escherichia coli* in the Netherlands. J. Infect. **9**:177–184.
Nataro, J. P., M. M. Baldini, J. P. Kaper, R. E. Black, N. Bravo, and M. M. Levine. 1985. Detection of an adherence factor in enteropathogenic *Escherichia coli* with a DNA probe. J. Infect. Dis. **152**:560–565.
Nataro, J. P., J. B. Kaper, R. Robins-Browne, V. Prado, P. Vial, and M. M. Levine. 1987. Patterns of adherence of diarrheagenic *Escherichia coli* to HEp-2 cells. Pediatr. Infect. Dis. J. **6**:829–831.
Neubauer, M., E. Aldova, and J. Duben. 1982. Outbreak of enterocolitis caused by an enteroinvasive *E. coli*. Zbl. Bakt. Hyg. **252**:507–513.
O'Brien, A. D. and R. K. Holmes. 1987. Shiga and shiga-like toxins. Microbiol. Rev. **51**:206–220.
O'Brien, A. D., J. W. Newland, S. F. Miller, R. K. Holmes, H. W. Smith, and S. B. Formal. 1984. Shiga-like toxin-converting phases from *Escherichia coli* strains that cause hemorrhagic colitis or infantile diarrhea. Science **226**:694–696.
O'Hanley, P., D. Lark, S. Falkow, and G. K. Schoolnik. 1985. Molecular basis of *Escherichia coli* colonization of the upper urinary tract in BALB/c mice: gal-gal pili immunization prevents *E. coli* pyelonephritis. J. Clin. Invest. **75**:347–360.
Ørskov, I., A. Ferencz, and F. Ørskov. 1980. Tamm-Horsfall protein or uromucoid is the normal urinary slime that traps type 1 fimbriated *Escherichia coli*. Lancet **i**:887.
Ørskov, I. and F. Ørskov. 1985. *Escherichia coli* in extra-intestinal infections. J. Hyg. Camb. **95**:551–575.
Ørskov, I., F. Ørskov, B. Jann, and K. Jann. 1977. Serology, chemistry, and genetics of O and K antigens of *Escherichia coli*. Bacteriol. Rev. **41**:667–710.
Pai, C. H., R. Gordon, H. V. Sims, and L. E. Bryan. 1984. Sporadic cases of hemorrhagic colitis associated with *Escherichia coli* O157:H7. Ann. Intern. Med. **101**:738–742.
Pal, T., A. S. Pacsa, L. Emody, S. Voros, and E. Selley. 1985. Modified enzyme-linked immunosorbent assay for detecting enteroinvasive *Escherichia coli* and virulent *Shigella* strains. J. Clin. Microbiol. **21**:415–418.
Paulozzi, L. J., K. E. Johnson, L. M. Kamahele, C. R. Clausen, L. W. Riley, and S. D. Helgerson. 1986. Diarrhea associated with adherent enteropathogenic *Escherichia coli* in an infant and toddler center, Seattle, Washington. Pediatrics **77**:296–300.
Quie, P. G. 1987. The white cells: Use of granulocyte transfusions. Rev. Infect. Dis. **9**:189–193.
Remis, R. S., K. L. MacDonald, L. W. Riley, N. D. Puhr, J. G. Wells, B. R. Davis, P. A. Blake, and M. L. Cohen. 1984. Sporadic cases of hemorrhagic colitis associated with *Escherichia coli* O157:H7. Ann. Intern. Med. **101**:624–626.
Riley, L. W. 1987. The epidemiologic, clinical, and microbiologic features of hemorrhagic colitis. Ann. Rev. Microbiol. **41**:383–407.
Riley, L. W., L. N. Junio, L. B. Libaek, and G. K. Schoolnik. 1987. Plasmid-encoded expression of lipopolysaccharide O-antigenic polysaccharide in enteropathogenic *Escherichia coli*. Infect. Immun. **55**:2052–2056.
Riley, L. W., R. S. Remis, S. D. Helgerson, H. B. McGee, J. G. Wells, B. R. Davis, R. J. Hebert, E. S. Olcott, L. M. Johnson, N. T. Hargrett, P. A. Blake, and M. L. Cohen. 1983. Hemorrhagic colitis associated with a rare *Escherichia coli* serotype. N. Engl. J. Med. **308**:681–685.
Robbins, J. B., G. H. McCracken Jr., E. C. Gotschlich, F. Ørskov, I. Ørskov, and L. Å. Hanson. 1974. *Escherichia coli* K1 capsular polysaccharide associated with neonatal meningitis. N. Engl. J. Med. **290**:1216–1220.
Ross, S., W. Rodriguez, G. Controni, G. Korengold, S. Watson, and W. Khan. 1975. Limulus lysate test for gram-negative bacterial meningitis: bedside application. J. Am. Med. Assoc. **233**:1366–1369.
Rothbaum, R. A., J. McAdams, R. Gianella, and J. C. Partin. 1982. A clinicopathologic study of enterocyte-adherent *Escherichia coli:* a cause of protracted diarrhea in infants. Gastroenterology **83**:441–454.
Rowe, B. 1979. The role of *Escherichia coli* in gastroenteritis. Clin. Gastroenterol. **8**:625–644.
Rowe, B., J. Taylor, and K. A. Bettelheim. 1970. An investigation of travellers' diarrhea. Lancet **i**:1–5.
Rubin, R. H. 1987. Infections of the urinary tract. Inf. Dis. Sci. Am. **7**:1–10.
Ryan, C. A., R. V. Tauxe, G. W. Hosek, J. G. Wells, P. A. Stoez, H. W. McFadden, P. W. Smith, G. F. Wright, and P. A. Blake. 1986. *Escherichia coli* O157:H7 diarrhea in a nursing home: clinical, epidemiological, and pathological findings. J. Infect. Dis. **154**:631–638.
Sack, R. B., N. Hirschhorn, I. Brownlee, R. A. Cash, W. E. Woodward, and D. A. Sack. 1975. Enterotoxigenic *E. coli* associated diarrhoeal disease in Apache children. N. Engl. J. Med. **292**:1041–1045.
Sack, D. A., S. Huda, P. K. B. Neogi, R. R. Daniel, and

W. M. Spira. 1980. Microtiter ganglioside enzyme-linked immunosorbent assay for *Vibrio* and *Escherichia coli* heat-labile enterotoxins and antitoxin. J. Clin. Microbiol. **11**:35–40.

Sack, D. A., J. C. McLaughlin, R. B. Sack, F. Ørskov, and I. Ørskov. 1977. Enterotoxigenic *Escherichia coli* isolated from patients at a hospital in Dacca. J. Infect. Dis. **135**:275–280.

Sack, D. A. and R. B. Sack. 1975. Test for enterotoxigenic *Escherichia coli* using Y1 adrenal cells in miniculture. Infect. Immun. **11**:334–336.

Saiki, R. K., T. L. Bugawan, G. T. Horn, K. B. Mullis, and H. A. Erlich. 1986. Analysis of enzymatically amplified -globin and HLA-DQ DNA with allele-specific oligonucleotide probes. Nature **324**:163–666.

Sanford, J. P. 1975. Urinary tract symptoms and infections. Ann. Rev. Med. **26**:485.

Sansonetti, P. J., T. L. Hale, G. J. Dammin, C. Kapfer, H. H. Collins Jr., and S. B. Formal. 1983. Alterations in the pathogenicity of *Escherichia coli* K-12 after transfer of plasmid and chromosomal genes from *Shigella flexneri*. Infect. Immun. **39**:1392–1402.

Sansonetti, P. J., D. J. Kopecko, and S. B. Formal. 1981. *Shigella sonnei* plasmids: evidence that a large plasmid is necessary for virulence. Infect. Immun. **34**:75–83.

Scaletsky, I. C. A., M. L. M. Silva, and L. R. Trabulsi. 1984. Distinctive patterns of adherence of enteropathogenic *Escherichia coli* to HeLa cells. Infect. Immun. **45**:534–536.

Scharf, S. J., G. T. Horn, and H. A. Erlich. 1986. Direct cloning and sequence analysis of enzymatically amplified genomic sequences. Science **233**:1076–1078.

Scheckler, W. E. 1977. Septicemia in a community hospital. 1970 through 1973. J. Am. Med. Assoc. **237**:1938–1941.

Scheld, W. M. and J. P. Brodeur. 1983. Effect of methylprednisolone on entry of ampicillin and gentamicin into cerebrospinal fluid in experimental pneumococcal and *Escherichia coli* meningitis. Antimicrob. Agents Chemother. **23**:108–112.

Schumer, W. 1976. Steroids in the treatment of clinical septic shock. Ann. Surg. **184**:333–341.

Sereny, B. 1955. Experimental *Shigella* keratoconjunctivitis. Acta Microb. Acad. Sci. Hung. **2**:293–296.

Seriwatana, J., P. Echeverria, D. N. Taylor, T. Sakuldaipeara, S. Changchawalit, and O. Chivoratanond. 1987. Identification of enterotoxigenic *Escherichia coli* with synthetic alkaline phosphatase-conjugated oligonucleotide DNA probes. J. Clin. Microbiol. **25**:1438–1441.

Sever, J. L., J. H. Ellenbert, and D. Edmonds. 1978. Urinary tract infections during pregnancy: maternal and pediatric findings, p. 19–21. *In:* E. H. Kass, W. Brumfitt. (ed.). Infections of the urinary tract. University of Chicago Press, Chicago.

Small, P. L. C., R. R. Isberg, and S. Falkow. 1987. Comparison of the ability of enteroinvasive *Escherichia coli, Salmonella typhimurium, Yersinia pseudotuberculosis,* and *Yersinia enterocolitica* to enter and replicate within HEp-2 cells. Infect. Immun. **55**:1674–1679.

Speer, C. P., D. Hauptmann, P. Stubbe, and M. Gahr. 1985. Neonatal septicemia and meningitis in Gottingen, West Germany. Ped. Infect. Dis. **4**:36–41.

Spika, J. S., J. E. Parsons, D. Nordenberg, J. G. Wells, R. A. Gunn, and P. A. Blake. 1986. Hemolytic-uremic syndrome and diarrhea associated with *Escherichia coli* O157:H7 in a day care center. J. Pediatr. **109**:287–291.

Sprung, C. L., P. V. Caralis, E. H. Marcial, M. Pierce, M. A. Gelbard, W. M. Long, R. C. Duncan, M. D. Tendler, and M. Karpf. 1984. The effects of high-dose corticosteroids in patients with septic shock: a prospective controlled study. N. Engl. J. Med. **311**:1137–1143.

Stamm, W. E. 1984. Prevention of urinary tract infections. Am. J. Med. 76(suppl):148–154.

Steeper, T. A. and J. McCullough. 1985. Platelet and granulocyte transfusion, p. 247–279. *In:* P. H. Wiernik, (ed). Contemporary issues in clinical oncology. Vol. 4. Leukemias and lymphomas. Churchill Livingstone, New York.

Strockbine, N. A., L. R. Marques, J. W. Newland, H. W. Smith, R. K. Holmes, and A. D. O'Brien. 1986. Two toxin-converting phages from *Escherichia coli* O157:H7 strain 933 encode antigenically distinct toxins with similar biologic activities. Infect. Immun. **53**:135–140.

Svanborg Edén, C., R. Freter, L. Hagberg, R. Hull, H. Leffler, and G. K. Schoolnik. 1982. Inhibition of experimental ascending urinary tract infection by an epithelial cell-surface receptor analogue. Nature **298**:560–562.

Svanborg Edén, C., L. Hagberg, L. Å. Hanson, T. Korhonen, H. Leffler, and S. Ollig. 1981. Adhesion of *Escherichia coli* in urinary tract infection, p. 161–187. *In:* K. Elliot, M. O'Connor, and J. Whelan (ed), Adhesion and microorganism pathogenicity. Ciba Foundation Symposium 80: Pitman Medical, Tunbridge Wells.

Svanborg Edén, C. and H. A. Hansson. 1978. *Escherichia coli* pili as possible mediators of attachment to human urinary tract epithelial cells. Infect. Immun. **21**:229–237.

Svanborg Edén, C., L. Å. Hanson, U. Jodal, U. Lindberg, and A. S. Akerlund. 1976. Variable adherence to normal human urinary-tract epithelial cells of *Escherichia coli* strains associated with various forms of urinary tract infection. Lancet **ii**:490–492.

Svennerholm, A. M. and J. Holmgren. 1978. Identification of *Escherichia coli* heat-labile enterotoxin by means of a ganglioside immunosorbent assay (GM1 ELISA) procedure. Curr. Microbiol. **1**:19–23.

Tacket, C. O., D. R. Maneval, M. M. Levine. 1987. Purification, morphology, and genetics of a new fimbrial purtative colonization factor of enterotoxigenic *Escherichia coli*. Infect. Immun. **55**:1063–1069.

Thomas, L. V., A. Cravioto, S. M. Scotland, and B. Rowe. 1982. New fimbrial antigenic type (E8775) that may represent a colonization factor in enterotoxigenic *Escherichia coli* in humans. Infect. Immun. **35**:1119–1124.

Thomas, V., A. Shelokov, M. Forland. 1979. Antibody-coated bacteria in the urine and the site of urinary-tract infection. N. Engl. J. Med. **290**:588–590.

Thompson, M. R., H. Brandwein, M. LaBine-Racke, and R. A. Gianella. 1984. Simple and reliable enzyme-linked immunosorbent assay with monoclonal antibodies for detection of *Escherichia coli* heat-stable enterotoxins. J. Clin. Microbiol. **20**:59–64.

Timmis, K. N., S. Sturm, and H. Watanabe. 1986. Genetic dissection of pathogenesis determinants of *Shigella* and enteroinvasive *Escherichia coli,* p. 107–126. *In:* J. Holmgren, A. Lindberg, R. Mollby (eds.). Development of vaccines and drugs against diarrhea. 11th Nobel Conference. 1985. Studentlitterature, Lund, Sweden.

Toledo, M. R. F., M. C. B. Alvariza, J. Murahovschi, S. R. T. S. Ramos, and L. R. Trabulsi. 1983. Enteropathogenic *Escherichia coli* serotypes and endemic diarrhea in infants. Infect. Immun. **39**:586–589.

Townsend, T. R. and R. P. Wentzel. 1981. Nosocomial bloodstream infections in a newborn intensive care unit. A case-matched control study of morbidity, mortality, and risk. Am. J. Epidemiol. **114**:73–80.

Tzipori, S., H. Karch I. K. Wachsmuth, R. M. Robins-Browne, A. D. O'Brien, H. Lior, M. L. Cohen, J. Smithers, and M. M. Levine. 1987. Role of a 60-megadalton plasmid and shiga-like toxins in the pathogenesis of infection caused by enterohemorrhagic *Escherichia coli* O157:H7 in gnotobiotic pigs. Infect. Immun. **55:**3117–3125.

Veterans Administration Systemic Septis Cooperative Study Group. 1987. Effect of high-dose glucocorticoid therapy on mortality in patients with clinical signs of systemic sepsis. N. Engl. J. Med. **317:**659–665.

Vosti, K. L. 1975. Recurrent urinary tract infections: prevention by prophylactic antibiotics after sexual intercourse. J. Am. Med. Assoc. **231:**934–940.

Wallis, C., J. L. Melnick, and C. J. Longoria. 1981. Colorimetric method for rapid determination of bacteriuria. **14:**342–446.

Walz, W., M. A. Schmidt, A. F. Labigne-Roussel, S. Falkow, and G. K. Schoolnik. 1985. AFA-I, a cloned afibrial X-type adhesin from a human pyelonephritic *Escherichia coli* strain. Eur. J. Biochem. **152:**315–321.

Washington, J. A. 1975. Blood cultures. Principles and techniques. Mayo Clin. Proc. **50:**91–98.

Weinstein, M. P., J. R. Murphy, L. B. Reller, and K. A. Lichtenstein. 1983. The clinical significance of positive blood cultures: a comprehensive analysis of 500 episodes of bacteremia and fungemia in adults. II. Clinical observations, with special reference to factors influencing prognosis. Rev. Infect. Dis. **5:**54–70.

Wolff, S. M. and J. V. Bennett. 1974. Gram-negative rod bacteremia. N. Engl. J. Med. **291:**733–734.

Wood, P. K., J. G. Morris Jr., P. L. C. Small, O. Sethabutr, M. R. F. Toledo, L. R. Trabulsi, and J. B. Kaper. 1986. Comparison of DNA probes and the Sereny test for identification of invasive *Shigella* and *Escherichia coli* strains. J. Clin. Microbiol. **24:**498–500.

Wu, T. C., E. C. Williams, S. Y. Koo, and J. D. MacLowry. 1985. Evaluation of three bacteriuria screening methods in a clinical hospital. J. Clin. Microbiol. **21:**796–799.

Young, L. S., W. J. Martin, R. D. Meyer, R. J. Weinstein, and E. T. Anderson. 1977. Gram-negative rod bacteremia: microbiologic, immunologic, and therapeutic considerations. Ann. Intern. Med. **86:**456–471.

Ziegler, E. J., J. A. McCutchan, J. Fierer, M. P. Glauser, J. C. Sadoff, H. Douglas, and A. I. Braude. 1982. Treatment of gram-negative bacteremia and shock with human antiserum to a mutant *Escherichia coli*. N. Engl. J. Med. **307:**1225–1230.

CHAPTER 27

Glucose Nonfermenting Gram-Negative Rods

B. HOLMES

Disease: Few specific. The organisms occur in a wide variety of human clinical specimens where their pathogenic role is often uncertain.

Etiologic Agents: The gram-negative nonfermenters of clinical interest comprise the following genera: *Acinetobacter, Agrobacterium, Alcaligenes* (including *Achromobacter xylosoxidans*), *Bordetella, Branhamella, Brucella* (these last three genera all covered in other chapters), *Chryseomonas, Eikenella* (actually a facultative anaerobe), *Flavimonas, Flavobacterium, Janthinobacterium, Kingella* (also a facultative anaerobe), *Moraxella, Neisseria* (covered in another chapter), *Oligella, Pseudomonas* (again, covered in another chapter), *Shewanella,* and *Weeksella.*

Source: Organisms in this group are frequently free-living in soil and water; they may, therefore, colonize moist sites and enter the body by various routes. Others are normal inhabitants of mucous membranes in man and other warm-blooded animals.

Clinical Manifestations: Patient generally has underlying condition predisposing to infection. Symptoms vary with focus of this infection.

Pathology: Varies according to site of isolation of the organism.

Laboratory Diagnosis: Isolation of bacterium from almost any specimen of body fluid; infrequently recovered from feces.

Epidemiology: Worldwide. Can be opportunistic pathogens of compromised patients.

Treatment: Varies with organism. Some highly resistant, some susceptible even to penicillin. Restoration of defense mechanisms and removal of contaminated material from patient. Surgical drainage of abscesses; valve replacement in endocarditis.

Prevention and Control: Sterilization of moist sites by adequate autoclaving or disinfection.

Description of Disease

Some gram-negative nonfermenters are associated with specific diseases—*Bordetella pertussis* with whooping cough, *Brucella* with brucellosis, *Neisseria gonorrhoeae* with gonorrhea, *Neisseria meningitidis* with meningitis, *Pseudomonas mallei* with glanders, and *Pseudomonas pseudomallei* with melioidosis—but these are dealt with in other chapters. The nonfermenters covered in this chapter are rarely associated with specific diseases. Those organisms that are free-living in soil and water can colonize moist sites in a hospital, such as the nebulizer reservoirs of respiratory equipment, or disinfectant solutions. Other organisms are normal inhabitants of mucous membranes. Strains of both types of organism may cause localized infections in otherwise healthy hosts, but more widespread infections in the debilitated individual. Isolates of the group will frequently be of doubtful clinical significance, but they are important as opportunistic pathogens because several are resistant to a wide range of antimicrobial agents.

The free-living organisms important in human medicine include *Acinetobacter*, *Alcaligenes*, and *Flavobacterium* (*Pseudomonas* is the subject of a separate chapter), which may be recovered from almost any specimen of body fluid. Those important organisms living on mucous membranes include *Eikenella* and *Moraxella* (*Bordetella* and *Neisseria* are dealt with elsewhere). From the mouth, *Eikenella* strains may cause serious infections related to the mouth, upper and lower respiratory tract sinuses, lips, face and brain abscesses; from the intestine, *Eikenella* may cause abdominal infection, wound infection, liver and other abscesses, and peritonitis. *Moraxella* strains may occur in similar sites, although infections are generally less serious; *M. lacunata* may be associated with conjunctivitis and keratitis. Other genera are of less importance.

Communicability and Transmissibility

Developments in modern surgery, intensive care, and immunosuppressive drugs enable clinically compromised patients to exist in suboptimal physiologic conditions which leave the patient open to infection. In addition, broad-spectrum antimicrobial agents remove the normal flora and thereby permit colonization by bacteria that may not be indigenous to the host, who has no real defense to this challenge. The Gram-negative nonfermenters are more prevalent in this situation than other bacterial groups, for two reasons. First, many show resistance to a wide range of antimicrobial agents. Second, many free-living organisms are able to grow in water with minimal organic requirements; *Pseudomonas cepacia*, for example, will grow in distilled water (see Holmes, 1986). Such organisms thus have a great potential for cross-infection. Living in aqueous environments, they are able to colonize moist sites and so form a focus of infection. They may grow in disinfectants and therefore be conveyed via the skin to nonautoclavable, patient-monitoring apparatus so that they may colonize ventilation humidifiers, suction apparatus, and so forth. Such organisms may infect the susceptible patient by various routes. The bacteria do not need to grow actively in a patient to cause infection; some do not grow at 37°C (such as certain strains of *Pseudomonas fluorescens*). Such bacteria can, however, colonize burns and surface wounds. The only species of *Janthinobacterium*, *J. lividum*, is a psycrophile (growth range, 2° to 32°C) and hence is unlikely to cause human infection. This species will, however, grow in stored refrigerated products such as blood, and other organisms may be found in intravenous solutions. Such contaminated products, if given to a patient, may result in gram-negative shock. Those nonfermenters that are normal inhabitants of mucous membranes are less likely to be involved in cross-infection. *Moraxella* species may be recovered from eyes or blood cultures. Among the general population, malnutrition, physiologic stress, and drug addiction are factors likely to predispose to such infections.

Clinical Manifestations of Disease

The organisms covered in this chapter are predominantly associated with hospital-acquired infections and predominantly with the following conditions: necrotizing pneumonia, septicemia (with and without endocarditis), infections of the urinary tract, and wounds. Only these conditions (and the well-defined condition of neonatal meningitis due to *Flavobacterium meningosepticum*) will therefore be considered in the following sections. The gram-negative nonfermenters are not associated with enteric diseases and are encountered only rarely in fecal specimens (those reported from this source include *Alcaligenes faecalis*, Group Vd, and *Shewanella putrefaciens*). It must be appreciated that these organisms, when isolated, are more often than not of no etiologic significance, and may only be colonizers rather than playing a true pathogenic role. It is, however, often difficult to distinguish between colonization and infection, even with knowledge of the patient's clinical condition. Often, few clinical data are available on these organisms, so that the material presented below is sometimes given in general terms, with more information on specific gram-negative nonfermenters where such information is available. Diseases are often described as they appear in "normal" patients with such infections. However, in the immunocompromised patient, who is more likely to become infected with the gram-negative nonfermenters, the clinical presentation may well be atypical.

Other conditions with which these organisms may be associated include pancreatic abscess (Group Vd), cerebral ventriculitis or meningitis (*Acinetobacter* spp., *Alcaligenes xylosoxidans* subspecies *xylosoxidans* [synonym: *Achromobacter xylosoxidans*], *Eikenella corrodens, Flavobacterium meningosepticum, Moraxella* spp., and *Weeksella zoohelcum*), external ear infections (*Alcaligenes faecalis, A. xylosoxidans* subspecies *xylosoxidans, Shewanella putrefaciens*), cervical lymphadenopathy (with septicemia, *A. xylosoxidans* subspecies *xylosoxidans*), osteomyelitis (*Alcaligenes faecalis, E. corrodens*), peritonitis (*Acinetobacter* spp., *E. corrodens*), periodontal disease with bone destruction (*E. corrodens*), and brain, liver, and other abscesses (*E. corrodens*).

Pathology

Descending the air passages in a healthy individual, the number of potentially pathogenic microorganisms should decline progressively, and none should be found in the peripheral portion of the lung. Foreign particles caught in the normal mucoid secretions are continuously ejected by ciliary activity and cough. If the occasional microbe passes this barrier and enters the alveoli, then protection is afforded by alveolar macrophages and IgG antibodies. When the defences are impaired, conditions may permit the entrance, growth, and invasion of bacteria and the development of inflammation. The route of infection may thus be by endogenous aspiration from the pharynx, but also occasionally by bacteremia from an extrapulmonary source, or, most likely for the gram-negative nonfermenters, exogenous aspiration (e.g., nosocomial infection from contaminated reservoir nebulizers). There may be infection of the alveolar septae and/or alveoli. The basement membrane of the alveolar septum may be destroyed to leave a pulmonary scar on healing.

The blood is sterile in normal individuals, but hospital patients often have organisms circulating in the bloodstream. This may be a harmless bacteremia associated with localized sepsis, but can be associated with more serious infections. In septicemia, the bacterial load in the bloodstream is considerably heavier than is the case with bacteremia. The bacteria may actually be multiplying as they spread. Metastatic abscess formation may occur. Septicemic shock is due to the damaging action of endotoxins and exotoxins (produced respectively by gram-negative and gram-positive bacteria) at a cellular level and the body's response to this damage. Most organs may be affected, usually with microthrombus formation, leading to tissue necrosis and hemorrhage. The organs most at risk owing to their large blood flow and susceptibility to hypotension are the kidneys and lungs. If bacteria invade the endothelial lining of the cardiac valves, then the resulting inflammatory changes may lead to serious damage of the valves. Endocarditis is most frequently seen in those whose hearts are already abnormal, through congenital or degenerative heart disease, heart surgery, or, less commonly now, previous rheumatic heart disease. As a result of the inflammatory changes, vegetations of fibrin and blood platelets may form on the surface of the valve cusps. The vegetations can form on artificial heart valves and become foci of infective endocarditis. Fragments of vegetations may detach, form emboli in the bloodstream, and then lodge in such sites as the brain (causing meningitis, cerebral abscess, or mycotic aneurysm), kidneys, skin, and spleen. Local spread of infection into surrounding tissues can lead to myocardial abscess, infarction, conduction abnormalities and arrhythmias, and pericarditis.

The normal urinary tract is sterile above the level of the distal urethra. In most cases infection is by the ascending route, but may occasionally arise by hematogenous spread. Most urinary tract infections due to the gram-negative nonfermenters covered in this chapter occur in hospitalized patients with indwelling urinary catheters. Insertion of a urethral catheter will cause urinary tract infection in 1 to 6% of previously uninfected patients; hence it is no longer acceptable to pass a catheter for the sole purpose of collecting diagnostic samples of urine. Passing a catheter into an infected urinary tract may also cause the organism to enter the bloodstream. Problems frequently arise with catheters left in position for lengthy periods. Despite all precautions, these often give rise to an ascending urinary tract infection, the organism reaching the bladder along the layer of mucus between the urethra and the outside of the catheter. Symptoms such as dysuria and frequency probably result from an acute inflammatory response in the bladder or urethra. In more serious infections in the upper urinary tract involving the kidneys, pyelonephritic scars form as a result of inflammatory changes, beginning in the calyces and extending to the cortex in a segmental fashion. Infiltration with mononuclear cells follows the original changes of acute inflammation, and, still later, areas of fibrosis lead to kidney distortion, scar formation, and loss of function in the areas involved.

Wound infections result in inflammation, which is the local reaction of the body to any form of damage to its cells. Inflammation localizes the effects of the infection and tends to prevent its spread to other organs and tissues. Inflammation results in the production of an inflammatory exudate in the tissue spaces in the affected area. The exudate contains white blood cells and antibodies, which leak through openings in the walls of capillaries. Polymorphonuclear leukocytes are found in large numbers in the exudate of acute inflammation, and, when the reaction starts to subside, increasing numbers of macrophages appear at the site of the reaction; both types of cells take up infecting microorganisms by phagocytosis. Where tissue damage is pronounced, pus may form. Initially the pus will lie in a cavity or abscess in the tissue formed as a result of tissue destruction. Without surgical incision, abscess cavities may enlarge and then rupture, either to the body surface or into some hollow cavity or organ.

Diagnostic Features

The cardinal signs of all pneumonias are chills, fever, marked perspiration, pleuritic chest pain over the affected area, cough, and sputum. Occasional elderly patients, or those with myxoedema or those in shock, are almost afebrile. Cyanosis may be present, and tachycardia proportional to the temperature. An abnormal chest X-ray is necessary to confirm the clinical diagnosis; in the case of *Acinetobacter,* at least, bronchopneumonia is usually seen, but cavitation of the lung is rarely observed. Most pneumonias caused by *Acinetobacter* spp. are nosocomial, but when the occasional community-acquired pneumonia due to this organism is seen, it is usually bacteremic, and up to about one-third of the patients may be without serious underlying disease. Leukopenia is common in such cases, probably a reflection of the overwhelming nature of the infection in these patients.

Specimens of expectorated sputum are generally purulent, may be tenacious and yellow-green to bloody and, effectively, should provide as much information as a biopsy of the infected alveolus. Such specimens should, therefore, be collected by experienced personnel. Microscopic examination of the sputum should reveal many polymorphonuclear leukocytes, few epithelial cells, and numerous gram-negative bacterial cells. The same predominant gram-negative organism should be cultured from two or more consecutive sputum specimens if the diagnosis is to be confirmed. In some cases the same organism as from the sputum will also be recovered from the blood of the patient. Bacteremia arising from the urinary tract or pelvis may cause pneumonias due to gram-negative nonfermenters. The diagnosis of pneumonia will also be confirmed if the infecting organism is recovered from pleural fluid. The clinical course of the patient and the response to antimicrobial therapy must be compatible with the diagnosis.

Septicemia is generally secondary to some local lesion and often follows an operation or invasive diagnostic instrumentation, such as cystoscopy. It is evidenced by high fever, rigors, or even shock, and there may be vomiting and diarrhea. There is a marked rise in body temperature. In the elderly, however, septicemia may present more subtly: general malaise, confusion, altered levels of consciousness, falls, and immobility. Other features include hyperventilation (possibly mimicking pulmonary emboli), hypotension, jaundice, and nausea. Fever may be mild or absent, and rigors are less common. Bacterial endocarditis is evidenced by abnormal cardiac murmurs, anemia, enlargement of the heart and spleen, fever, signs and symptoms due to embolism, and wasting. Presentation of endocarditis in the elderly may also be subtle, and this may delay diagnosis if symptoms and signs are attributed to other diseases. Such delays may lead to heart failure and valve rupture with a consequent increase in mortality.

In infants, urinary tract infection does not present a characteristic clinical picture, and bacteriuria may be present in asymptomatic adult patients. Acute pyelonephritis in adults may be recognizable from loin pain, kidney tenderness, fever, occasionally rigors due to bacteremia, dysuria, frequency, and often hematuria. However, apart from such patients, in only about half of patients with dysuria and frequency can bacteriuria be demonstrated. Urinary tract infection may thus be difficult to diagnose clinically, and bacteriologic confirmation of the diagnosis is generally essential. A count of 10^5 colonies/ml or greater of a single species in a mid-stream urine specimen is almost invariably indicative of bladder bacteriuria. However, 15 to 20% of urine specimens from patients with urinary tract infection with aerobic organisms yield counts below 10^5 colonies/ml. Purity of the culture (on a noninhibitory medium) may be a better guide to its significance than rigid adherence to a particular count value.

The area of acute inflammation around a wound appears red and hot owing to a temporarily increased blood supply. The area will be swollen by the presence of exudate in the tissue spaces and painful and tender to the touch, resulting from the increased tension within these spaces and pressure on nearby sensory nerve endings. However, these signs may be absent in the immunocompromised patient.

In the well-known, albeit rare, condition of neonatal meningitis due to *Flavobacterium meningosepticum,* symptoms are typical of neonatal meningitis: refusal to feed, respiratory difficulties, unstable temperature, jaundice, convulsions or lethargy, failure to gain weight, and abnormal tension or bulging of the fontanelle. The spinal fluid displays an elevated cell count with polymorphonuclear predominance, elevation of the protein level (120 to 1,260 mg/100 ml), and decrease in the glucose levels (often below 20 mg/100 ml). A direct smear of the sediment should be positive; blood cultures are also often positive.

Prognosis

Pneumonia is very often fatal in patients debilitated by some other severe or prolonged disease. In later life, chronic diseases are often terminated by pneumonia, and the prognosis then is accordingly bad. The presence of bacteremia is not necessarily serious, the depth of cyanosis and the rapidity and weakness of the pulse may more clearly indicate the severity of the infection. The more widespread the involvement, generally the worse are the chances for recovery. Infective endocarditis is uncommon, as is pericarditis which may result from direct extension of the inflammatory process from the lungs through the pleurae. Small, multiple lung abscesses may occur, especially in prolonged cases. Other complications include bronchial obstruction, atelectatic collapse, and empyema. Bacteremias secondary to the pneumonia may cause distant foci of infection, such as septic arthritis or meningitis.

The gram-negative nonfermenters covered in this chapter are more likely to be associated with nosocomial pneumonias than with those community-acquired. Nosocomial pneumonias arising in intensive care units are likely to be more severe, but the mortality rate in occasional cases of true pneumonia, is likely to be less than that caused by other nonfermenters, such as *Pseudomonas aeruginosa.* Airway colonization by gram-negative nonfermenters is, however, much more likely than infection. These organisms often colonize the upper or lower respiratory tract of patients who have undergone tracheostomy or who are in intensive care; true infection is difficult to diagnose (and may involve invasive techniques, such as special cuffed tracheostomy or percutaneous biopsy). Although the overall mortality rate is around 50% for true pneumonias due to gram-negative bacteria in general, the rate will be lower for the nonfermentative species considered in this chapter.

Mortality in septicemia remains high on the order of 15 to 20%, despite the intensive treatment now available. Mortality is higher for patients in shock, on the order of 50%. The mode of death is cardiac arrest due to severe metabolic disturbance and hypotension. Death may follow at a later stage owing to secondary organ damage, such as severe lung damage or renal failure. The mortality rate for infective endocarditis in general is on the order of 30 to 40%, and most deaths are due to secondary factors such as heart failure. These rates are likely to be lower for the gram-negative nonfermenters covered in this chapter.

The absence or presence of renal involvement is the major deciding factor in the outcome of urinary tract infection. Where infection is confined to the bladder or below, there is no risk of the serious consequences of bacterial infection of the kidneys. Such patients may continue to have episodes of bacteriuria, with or without symptoms, and the frequency of such attacks will probably be determined by the persistence of predisposing factors. Even when appropriate antimicrobial chemotherapy is given, the cure rate in nosocomial urinary tract infection will be approximately only 60 to 70% compared with higher cure rates of 80 to 85% expected with short-course treatment of community-acquired urinary tract infection. This is a reflection of the more complex problems encountered in a hospital, where there is an adverse age structure among the patients treated and where there are more people with recurrent urinary tract infection, many of whom will have functional and structural abnormalities of the urinary tract (renal stones, tumors, prostatic enlargement, bladder diverticulum), which complicate treatment and affect its outcome. Also those bacteria causing nosocomial urinary tract infections tend to be less susceptible to the usual antimicrobial agents than are the flora encountered in community-acquired infections. Infection of the upper urinary tract is potentially far more serious. Renal scarring may occur, resulting in loss of function. If both kidneys are involved, then renal failure may occur and dialysis or transplantation may be required.

Wounds have generally localized reactions, but if the invading bacteria are not destroyed by the inflammatory response or antimicrobial therapy where appropriate, the infection may spread into the draining lymphatics to cause lymphadenitis and lymphangitis. The invading bacteria may also spread by the bloodstream to distant tissues to cause secondary (metastatic) infections.

In neonatal meningitis due to *Flavobacterium meningosepticum,* the case fatality rate is about 55%, and the complication of a hydrocephalus develops in most survivors. Only rarely are there no sequelae.

Epidemiologic Aspects

Occurrence and Transmission

Although a few nonfermenters appear to have a local distribution (e.g., *Pseudomonas pseudomallei* appears to be largely confined to the soil and surface waters of South East Asia, although there have been some cases of melioidosis in Mexico), some have been largely eradicated from certain parts of the world (e.g., *Pseudomonas mallei* is now hardly ever seen in developed countries), and still more have yet to be described outside specific areas (e.g., all strains of *Flavobacterium thalpophilum* so far described have been from the United States), most of the nonfermenters described here probably occur worldwide. Compared with developed countries, the underdeveloped ones have not experienced such substantial improvements to diet, immunization programs, public health, and standard of living, leading to a reduction in the incidence of infectious diseases such as diphtheria and tuberculosis. Also, such countries have enjoyed less benefit of developments in intensive care, immunosuppressive drugs, and modern surgery, which enable clinically compromised patients to exist in suboptimal physiologic conditions. Most of the nonfermenters are thus of less importance and interest in underdeveloped countries, where they are also less likely to be encountered.

Many of the nonfermenters occur so infrequently in human clinical material that when one is isolated and has apparent clinical significance, an individual report of the case appears in the literature (there are, for example, only two reports of infective endocarditis due to *Kingella denitrificans* [Goldman et al., 1980; Swann and Holmes, 1984]). Surveys show the rare occurrence of certain organisms, for example, in a review of 280 cases of neonatal meningitis in the Netherlands, only one case involved *Flavobacterium meningosepticum* (Mulder and Zanen, 1984). More isolates are seen in outbreak situations, for example, 46 isolates of *Achromobacter xylosoxidans* were encountered over an 18-month period by Reverdy et al. (1984). The common source of the organism was the deionized water of the hemodialysis system; two patients in the intensive care unit died, while 37 became colonized. The most common nonfermenter seen, of those covered in this chapter, also most likely to cause outbreaks of infection, is *Acinetobacter baumannii* (*sensu stricto; Acinetobacter calcoaceticus sensu lato*).

Since these organisms are mainly opportunists in the debilitated individual suffering from such conditions as cirrhosis of the liver, renal failure, acute leukemia, rheumatoid arthritis, and systemic lupus erythematosus, they will be more prevalent in neonates and the elderly, occurring in other age groups following accident, surgery, or ill health due to drug addiction, malnutrition, or physiologic stress.

Susceptibility, Resistance, and Carrier States

Those nonfermenters that are parasites of mucous membranes are naturally already present in the host, in the upper respiratory tract or oral cavity (or intestines, at least in the case of *Eikenella*) and are ready to cause infection when the normal defense mechanisms of the host are seriously impaired. Free-living forms enter the human body when the host is debilitated in some way, through skin wounds, through the eye, by inhalation, and by ingestion. *Acinetobacter* spp. inhabit moist areas of the human body (antecubital fossa, axilla, groin, toe webs), but are not a frequent cause of wound infections. Since, however, most nonfermenters are infrequently reported from feces or in association with gastrointestinal disease (except *Eikenella*), it seems likely that most ingested nonfermenters will be killed by normal stomach acidity. Transmission of free-living nonfermenters to man will generally be via a contaminated aqueous source, often thought unlikely to support bacterial growth (e.g., a disinfectant or distilled water), but in which the organism may survive for several years unless proper sterilization procedures are undertaken. Such organisms thrive in manure and soil, their natural environment. Hand carriage of *Acinetobacter* spp. appears to be an important mode in which the organism is spread in nosocomial epidemics.

Etiologic Agents

Characterization

The glucose nonfermenting gram-negative rods produce an oxidative reaction, alkaline reaction, or effect no change to the medium in the glucose oxidation-fermentation (OF) test of Hugh and Leifson (1953); in general, they have oxidative pathways or none at all. The term nonfermenting is one of convenience (some indeed are, strictly speaking, fermenters) and is used to group together a number of bacterial genera that share certain properties. Since the grouping is one of convenience, it does not necessarily have any taxonomic significance. Medical microbiologists group these bacteria together, first, because they are mostly inactive in the tests used to characterize other bacteria (e.g., tests used for the enterobacteria). Second, they are grouped together because many of these bacteria are encountered in similar clinical situations, sometimes being responsible for cross-infection and sometimes acting as opportunistic pathogens.

The characteristics of the gram-negative nonfermenters are given in specialized books on the subject (e.g., Gilardi, 1985; Oberhofer, 1985) and also in the relevant sections of *Manual of Clinical Microbiology* (1985, 4th ed., American Society for Microbiology, Washington D.C.) and *Bergey's Manual of Systematic Bacteriology* (1984, Volume 1, Williams & Wilkins, Baltimore). Their characteristics are also given in the probability matrix of Holmes et al. (1986a).

The majority of the nonfermenters described below have cells in the shape of rods; *Acinetobacter* is an exception, since on initial isolation the cells take the form of cocci and thus can be confused with *Branhamella* and *Neisseria* (both covered in another chapter). Almost all are strict aerobes, will grow aerobically on nutrient agar, and produce catalase; *Eikenella* and *Kingella* are exceptions, as they are facultative anaerobes which do not produce catalase, and some strains of *Eikenella* may need blood agar plus an atmosphere with increased CO_2 in order to grow. Most produce oxidase, but members of the genera *Acinetobacter, Chryseomonas,* and *Flavimonas* are oxidase negative. Some are motile by peritrichous flagella, some by polar flagella, and some are nonmotile. They can be conveniently discussed in the following categories.

The Strictly Aerobic Non- or Weakly Saccharolytic Generally Nonmotile Genera

This section includes the genera *Acinetobacter, Moraxella, Oligella,* and *Weeksella*. Of these, only strains of *Acinetobacter* are oxidase-negative, and all are nonmotile except for one species of *Oligella*. *Bordetella* and *Brucella* would also come into this section, but they are dealt with in other chapters.

Juni (1984) recognized only one species of *Acinetobacter, A. calcoaceticus;* his definition of the species included *A. lwoffii*. However, it is a simple matter to distinguish these two species by biochemical tests. Strains of *A. calcoaceticus* give an oxidative reaction in glucose OF medium and produce acid from glucose in peptone–water medium and acid from 10% (wt/vol) lactose. Strains of *A. lwoffii* yield an alkaline reaction in glucose OF medium and negative results in the other two tests. All strains of *A. calcoaceticus* and some of *A. lwoffii* will also produce acid from arabinose and xylose in suitable media such as ammonium salt sugars (Holmes et al., 1975) or OF sugars. The genus was known for some time to be heterogeneous in its nutritional characters (Baumann et al., 1968), as well as genetically (Johnson et al., 1970). Further DNA–DNA hybridization studies (Bouvet and Grimont, 1986) show that *Acinetobacter* comprises at least 12 genospecies. Two of the latter are *A. calcoaceticus* and *A. lwoffii,* and four others have been named *A. baumanii, A. haemolyticus, A. johnsonii,* and *A. junii;* the remaining six are as yet unnamed. The phenotypic distinction of the genospecies, however, depends primarily on nutritional characters, which are unlikely to be performed routinely in the clinical laboratory.

All species of *Moraxella* are nonsaccharolytic. Although most strains will grow on nutrient agar, some will not grow at 37°C unless in a moist atmosphere, and growth is said to be best when the culture is incubated in a closed jar (Henriksen, 1952). Growth is generally improved by the addition of blood or serum, although no specific growth factors are known. The *Moraxella* species that may be encountered in human clinical specimens are *M. atlantae, M. lacunata, M. nonliquefaciens, M. osloensis,* and *M. phenylpyruvica*. Of these, only strains of *M. lacunata* produce gelatinase and liquefy serum, only strains of *M. osloensis* accumulate lipid inclusions in the cells following growth on β-hydroxybutyrate (some strains also produce acid from ethanol), only in strains of *M. atlantae* and *M. phenylpyruvica* is growth stimulated by bile, but most strains of the latter species, in contrast to *M. atlantae,* reduce nitrate and grow both at 5°C and in the presence of 6% NaCl (some strains also produce urease). Strains of *M. phenylpyruvica* deaminate phenylalanine and tryptophan in suitable media, but if too sensitive a method is used, strains of several other *Moraxella* species will also give positive results in these tests.

Oligella contains two species. *O. urethralis* was

formerly included in *Moraxella* as *M. urethralis,* but this species was always somewhat atypical of that genus. The cells, although rods, were never plump like those of typical moraxellas, and genetic studies showed no close relationship to the "true" moraxellas. The latter studies (see Bøvre, 1984; Rossau et al., 1986) indicated that *M. urethalis* could form a new genus in the family *Neisseriaceae*. An extension of the genetic studies (Rossau et al., 1987) surprisingly revealed that the new genus should include the peritrichously flagellated *Alcaligenes*-like organism previously known as Group IVe (Clark et al., 1984). For the latter organism, the name *Oligella ureolytica* has been proposed (Rossau et al., 1987). The two species are easily separated from each other on the basis of motility and urease production (both positive in *O. ureolytica*). The latter character also differentiates *O. ureolytica* from *Alcaligenes*. *O. urethralis* strains can be distinguished from *Moraxella* strains principally by being able to reduce nitrite; strains also fail to reduce nitrate, are able to grow at 42°C, and (like *M. osloensis* only) accumulate lipid inclusions after growth on β-hydroxybutyrate.

Weeksella contains the two species *W. virosa* (formerly Group IIf) and *W. zoohelcum* (formerly Group IIj; see Clark et al., 1984; Holmes et al., 1986b,c). As well as being nonmotile, strains of both species also produce indole (if a sensitive method is used such as xylene extraction, followed by the addition of Ehrlich reagent) and produce gelatinase, and were hence considered *Flavobacterium*-like (they also fail to reduce nitrate). Unlike *Flavobacterium,* however, strains of *Weeksella* are nonsaccharolytic (only one *Flavobacterium* species is nonsaccharolytic), they are not yellow-pigmented, they are not highly resistant to antimicrobial agents, and they are not free-living. *Weeksella zoohelcum* strains differ from those of *W. virosa* in producing urease and in giving negative results in the following tests: growth at 42°C, growth on MacConkey agar (without crystal violet), and growth on β-hydroxybutyrate.

The Strictly Aerobic Non- or Weakly Saccharolytic Motile Genera

This category covers *Alcaligenes* (including *Achromobacter*), *Shewanella,* and the alkali-producing pseudomonads (such as *Pseudomonas pseudoalcaligenes*); the latter are, however, dealt with in a separate chapter.

In the past, clinical microbiologists thought of *Achromobacter* as comprising peritrichously flagellated, saccharolytic bacteria, and *Alcaligenes* as similarly flagellated but nonsaccharolytic bacteria. Although there is at least one unnamed taxon (Group Vd; see Clark et al., 1984; also see below) that conforms to that old view of *Achromobacter,* the only named species in the genus is *A. xylosoxidans*. The latter species produces acid from only a few carbohydrates: principally glucose and xylose, but occasional strains also produce acid from ethanol, fructose, and mannitol. Strains of *A. xylosoxidans* are thus easily distinguished from those of *Alcaligenes,* where strains do not produce acid from carbohydrates (some strains will, however, attack ethanol). Despite this ease of phenotypic distinction, DNA-relatedness studies show that *Achromobacter xylosoxidans* and *Alcaligenes denitrificans* are so closely related that they should be considered subspecies of a single species. Accordingly, it has been proposed (Kiredjian et al., 1986) that *Achromobacter xylosoxidans* be transferred to the genus *Alcaligenes,* where its specific epithet has priority over that in *Alcaligenes denitrificans*. The effect is to leave no species in the genus *Achromobacter,* and in *Alcaligenes* we have *Alcaligenes xylosoxidans* subspecies *xylosoxidans* and *Alcaligenes xylosoxidans* subspecies *denitrificans*. The other species of *Alcaligenes* are *A. faecalis* and *A. piechaudii*. Details of their differentiation are given in Kiredjian et al. (1986), but the differentiation can be done most easily by tests for reduction of both nitrate and nitrite: strains of both subspecies of *Alcaligens xylosoxidans* reduce both, strains of *A. faecalis* reduce nitrite but not nitrate, and strains of *A. piechaudii* reduce nitrate but not nitrite. For the nitrite reduction test, however, it is important to use the precise method of Kiredjian et al. (1986) since, by different methods, other authors find that different strains of *Alcaligenes xylosoxidans* subspecies *xylosoxidans* give different results in this test (Clark et al. 1984; Holmes et al. 1986a). Although long recognized as a separate species, "*A. odorans*" has been shown in several studies and by various means to be synonymous with *A. faecalis;* the name "*A. odorans*" no longer has standing in nomenclature.

The organism long known as "*Pseudomonas putrefaciens*" became known as *Alteromonas putrefaciens,* and the latest proposal is to transfer the species to the new genus *Shewanella* (MacDonell and Colwell, 1985). Although there is some heterogeneity within the species, there have as yet been no proposals for its subdivision. Strains of the species are easily recognizable because they produce a salmon-pink pigment, they produce copious H_2S (they are the only nonfermenter to give a positive result for this character on triple sugar–iron–agar), and they decarboxylate ornithine.

The Strictly Aerobic Saccharolytic Genera

This category includes the motile genera *Achromobacter, Agrobacterium, Chryseomonas, Flavimonas, Janthinobacterium,* and the oxidizing pseudomonads (which are covered in a separate chapter). There is a single genus whose strains are nonmotile: *Flavobacterium* (including *Sphingobacterium*). It contains one species whose strains are nonsaccharolytic, *F. odoratum*.

Achromobacter xylosoxidans, which produces acid from only a few carbohydrates, is dealt with above under *Alcaligenes*. Although no named species are now left in *Achromobacter,* there are some biochemically similar organisms known as Group Vd (Clark et al., 1984) which deserve mention here. These organisms show a fairly close phylogenetic relationship (from rRNA/DNA in vitro hybridization) to *Agrobacterium* (De Ley et al., 1987), and strains of both genera produce acid from a very wide range of carbohydrates. Strains of Group Vd do not produce acid from lactose or raffinose, nor do they hydrolyze esculin or produce 3-ketolactose, whereas *Agrobacterium tumefaciens* (synonym: *A. radiobacter*) strains do. Although not true members of the genus *Agrobacterium,* certain yellow-pigmented strains also able to produce 3-ketolactose have been ascribed to the so-called "*Agrobacterium* yellow group" (Holmes and Roberts, 1981; Swann et al., 1985). Strains of the latter taxon are phylogenetically closely related to *Pseudomonas paucimobilis* (formerly Group IIk-1; dealt with in a separate chapter), and performance of the 3-ketolactose test may be necessary to distinguish the two organisms.

Proposals of names for two hitherto unnamed groups of bacteria—Groups Ve-1 and Ve-2—were made at about the same time by two different groups of workers. This caused some confusion, but was rectified by Holmes et al. (1987); the latest proposal is of *Chryseomonas luteola* (synonym: *Chryseomonas polytricha*) for Group Ve-1, and *Flavimonas oryzihabitans* for Group Ve-2. Although strains of both taxa are yellow-pigmented and do not produce oxidase, there is a difference of at least 6% in mol% G+C content, so they are clearly not closely related phylogenetically. Strains of *C. luteola* produce a dihydrolase for arginine, hydrolyze esculin, grow at 42°C, hydrolyze tyrosine, and produce β-galactosidase, whereas strains of *F. oryzihabitans* give negative results in these tests.

Janthinobacterium contains the single species *J. lividum*. Strains resemble the fermentative organism *Chromobacterium violaceum* in frequently producing a purple pigment. They differ, however, not only in their mode of metabolism, but also in their growth temperature range, *J. lividum* will not grow above 32°C.

As well as being nonmotile, strains of *Flavobacterium* are generally yellow-pigmented and highly resistant to antimicrobial agents; they can, therefore, be easily recognized in the clinical laboratory. Some subdivision is, however, possible within the genus. Strains of *F. breve, F. meningosepticum,* and *Flavobacterium* species Group IIb are proteolytic, saccharolytic, and produce indole. Strains of *F. odoratum* are proteolytic, nonsaccharolytic, and do not produce indole (strains also reduce nitrite but not nitrate). Strains of *F. multivorum, F. spiritivorum,* and *F. thalpophilum* (also *Sphingobacterium mizutae,* see below) are nonproteolytic, saccharolytic, and do not produce indole. Some species, at least in the latter group, also possess novel sphingolipids in their cell walls, and on this evidence it has been proposed that they be transferred to a new genus *Sphingobacterium* (Yabuuchi et al., 1983). Since, however, this character is not one that can be tested for routinely, we prefer to leave these species in *Flavobacterium* pending the availability of data from other techniques such as rRNA–DNA hybridization. Within the sphingobacterium group, only *F. thalpophilum* produces acid from adonitol, grows at 42°C, and reduces nitrate. Only strains of *F. spiritivorum* produce acid from ethanol and mannitol. Strains of *F. multivorum* give negative results in all these tests. *Flavobacterium* species Group IIb has been shown, by DNA-relatedness studies, to comprise several genospecies. The name *F. indoltheticum* has been proposed for the whole of Group IIb (Yabuuchi et al., 1983) and the name *F. gleum* for certain strains within Group IIb that show a high degree of DNA-relatedness to one another (Holmes et al., 1984). Until the nomenclatural problems are resolved, we shall continue to use the name *Flavobacterium* species Group IIb. It is not always easy to clearly distinguish strains of *F. breve, F. meningosepticum,* and *Flavobacterium* species Group IIb. Strains of *F. breve* differ from those of the other two species in producing acid only from glucose and maltose and in not hydrolyzing esculin. As well as hydrolyzing esculin, strains of both *F. meningosepticum* and Group IIb may produce acid from several other carbohydrates (in addition to glucose and maltose), principally from glycerol and trehalose. Strains of *F. meningosepticum* are generally only weakly yellow-pigmented, may produce acid from lactose, and do not hydrolyze starch; strains of *Flavobacterium* species Group IIb are almost always brightly yellow-pigmented, do not produce acid from lactose, and may hydrolyze starch. *Flavobacterium*-like strains formerly known as Group IIf and Group IIj are now placed in the genus *Weeksella* (see above). The taxonomic position of other unnamed *Flavobacterium*-like taxa—Groups IIe, IIh, and IIi (Clark et al., 1984)—remains undetermined.

The Facultatively Anaerobic Genera

This section includes *Eikenella* and *Kingella*. Although strains of both genera are nonmotile and produce oxidase but not catalase, strains of the former genus are nonsaccharolytic and those of the latter saccharolytic, in suitable media.

Eikenella contains the single species *E. corrodens*. Although some strains will grow poorly on nutrient agar, most will grow best (in an atmosphere with increased CO_2) on blood agar, where the colonies frequently appear to corrode the agar surface. Strains generally also decarboxylate lysine and ornithine, so these features, together with the lack of catalase activity and corroding of the agar surface, make the organism relatively easy to identify.

There are three species of *Kingella;* they all ferment glucose. Strains of *K. kingae* (formerly *Moraxella kingii*) also ferment maltose. Only strains of *K. denitrificans* reduce nitrate and generally nitrite; they do not ferment maltose, but some will ferment dextrin. Only strains of *K. indologenes* produce indole and ferment maltose; they differ from the other two species in being able to ferment fructose, mannose, and sucrose (they also ferment dextrin). A close phylogenetic relationship has been demonstrated by rRNA–DNA hybridizations (Rossau et al., 1986) between the "true neisserias," *K. kingae* and *K. denitrificans,* but *K. indologenes* is not closely related to them.

Specimen Collection and Transport

The nonfermenters covered in this chapter may occasionally be recovered from feces, but they are much more likely to be recovered from body fluids: blood, cerebrospinal fluid, peritoneal fluid, pleural fluid, pus, and urine. As with any specimen, they should be transported to the laboratory without delay, and if a large volume of pus is drained, it is better to submit the full amount rather than a proportion or a swab of it. It is unlikely that sera will need to be taken except in occasional, serious infections when it is necessary to determine that adequate levels of antimicrobial agent are being maintained.

In the specific diagnosis of pneumonia, biopsy of an involved alveolus and isolation of a bacterium are ideal. Diagnosis may be facilitated by culture of expectorated sputum, aspiration of the contents of an infected bronchus through a fiberoptic bronchoscope, transtracheal aspiration, or occasionally open lung biopsy. Diagnosis of pneumonia can be complicated by contamination of sputum by normal flora during passage through the oropharynx, a difficulty that can be circumvented by transtracheal aspiration or open lung biopsy.

Isolation and Identification Procedures

For the isolation from blood of the organisms described in this chapter, two blood culture bottles may be necessary (although single bottle systems, for the recovery of both aerobic and anaerobic organisms, are now commercially available). One is vented to provide the conditions necessary for growth of the strictly aerobic organisms (all the free-living forms and some of those that inhabit mucous membranes) and the facultatively anaerobic organisms (some of those inhabiting mucous membranes). The other is unvented in order to suppress growth of the strictly aerobic organisms, so that the more fastidious, facultatively anaerobic organisms, *Eikenella* and *Kingella,* have a chance to appear. Blood cultures should be incubated at 35° to 37°C and examined regularly for growth. Only one set is necessary to diagnose septicemia, but if infective endocarditis is suspected, then sets of blood cultures should be taken on three separate occasions, when the temperature of the patient is raised, or over a period of 3 to 4 h, whereupon up to 90% of cases should yield a positive result.

Subcultures from blood culture bottles and specimens of other body fluids can be plated on blood agar. This medium will support the growth of even the more fastidious organisms that normally inhabit mucous membranes. The free-living forms will grow on MacConkey agar, but most of the more fastidious organisms will not. *Kingella denitrificans* may be recovered on Thayer-Martin medium from pharyngeal swabs taken for screening for *Neisseria meningitidis*.

Inoculated plates are incubated at 35° to 37°C, one aerobically (on which all the organisms considered here should grow), and one anaerobically to recover the more fastidious facultative anaerobes that might otherwise be overgrown. Anaerobic culture of sputum specimens is generally not performed, since the results are difficult to interpret. Selective media are rarely used to recover the organisms described here.

Although plates should be examined daily for growth, the nonfastidious organisms should all have grown well after overnight incubation. Colonial appearance will vary from species to species [and sometimes even from strain to strain within a species (e.g., Holmes et al., 1977b)], but they will generally have an entire edge, be low convex, circular, smooth, and shining, and some may be pigmented: yellow in *Chryseomonas, Flavimonas,* and *Flavobacterium;* salmon-pink in *Shewanella putrefaciens;* and violet in *Janthinobacterium.* Isolated colonies should be transferred to nutrient agar slants and nutrient broth, and incubated for a further 24 h. Growth from the agar slant should be removed to perform the Gram stain, the oxidase test, the glucose oxidation-fermentation (OF) test, and for making a stock culture (e.g., nutrient agar or Dorset egg slope in metal screw-cap bijou bottles). Pigment production should be recorded; then the catalase test may be performed. Additional biochemical tests can be done by inoculation with the broth culture, or, alternatively, isolated colonies from the primary isolation plate can be removed and emulsified in distilled water for the inoculation of the tests.

Preliminary Identification

The Gram stain should confirm that the test organism is gram-negative, but the cell morphology should also be recorded, since it may be useful to know whether one is dealing with cocci, cocco-bacilli, or rods. The growth conditions for the isolation plates will have indicated whether the test organism will grow at 37°C and whether it is a strict aerobe or facultative anaerobe. This will be confirmed by the results of the OF test, which will also indicate whether the organism is an oxidizer or nonoxidizer. With the results available also for production of catalase, oxidase, and pigment, the performance of the test for motility (in a hanging-drop preparation) should permit identification to genus level in most cases without further testing (see above, under Characterization). Serologic and fluorescent antibody tests are not available for these organisms, and further characterization is by means of biochemical or carbon-source utilization tests.

Definitive Identification and Typing

Common criteria for the definitive identification of the species of nonfermenters covered here have been given above and in the standard texts mentioned there. The methods for the tests are given above, under characterization, and are too numerous to repeat here. Since different methods may produce different results, it is important to use the methods relevant to the identification scheme being followed.

Fermentative organisms use the Embden-Meyerhoff pathway, and in nonaerated cultures large amounts of organic acids are accumulated, and a low pH is obtained. Most nonfermenters use various oxidative pathways, and less acid is produced. Many nonfermenters also produce large amounts of alkali from peptone, so that the small amounts of acid produced from carbohydrates may be neutralized. In peptone–water–sugars, because of the large amount of peptone present, acid production by nonfermenters is rarely detected. Low-peptone–Hugh and Leifson sugar medium is better, but best of all for nonfermenters are ammonium salt sugars, which contain no peptone and which contain instead ammonium salts as a nitrogen source. Some of the more fastidious organisms may fail to grow in this medium, but the problem may be overcome by the addition of growth factors. Bromocresol purple or bromothymol blue may be used as the indicator.

In the decarboxylase media of Møller (1955), fermentative organisms will initially produce acid from the glucose in the medium (yellow), but if decarboxylation occurs, then the reaction will return to neutral (gray) and proceed to alkalinity (purple). Nonfermenters do not produce acid from glucose in the decarboxylase media, and therefore the control tube remains unchanged in color (gray). As with fermentative organisms, however, if decarboxylation does occur, then an alkaline reaction is observed (purple).

Several commercial systems are also now available for the identification of nonfermenters.

Most of the nonfermenters covered in this chapter are so uncommon that if two or three strains of a species were encountered in a single hospital, then one could reasonably suspect that they were multiple isolates of a single strain. There is increased interest in typing of *Acinetobacter* isolates, because outbreaks of infection due to this organism do occur, and recent studies have been made of bacteriocin typing (Andrews, 1986), polyacrylamide-gel electrophoresis of whole cell proteins (Alexander et al., 1984), and serotyping (Das and Ayliffe, 1984). None of these schemes has, however, been fully evaluated for its wider application. Several serotypes have been determined in *F. meningosepticum* over the years, and the series now runs from A to O.

Immunologic Aspects

None of the organisms described here is likely to produce a disease for which serodiagnosis would provide useful information either for the diagnosis of acute infections or the management of chronic cases.

Treatment, Prevention, and Control

Treatment

Most of the organisms described in this chapter will be encountered rarely in clinical specimens, and when they are, they will frequently be of doubtful clinical significance, and no treatment will be indicated. Sometimes they will be clearly contaminants, such as a series of isolates of *Alcaligenes xylosoxidans* subspecies *xylosoxidans* (*Achromobacter xylosoxidans*) grown from blood culture bottles and found to originate from a chlorlexidine gluconate-based product used to disinfect the skin prior to withdrawing blood (Holmes et al., 1977a). Sometimes they will colonize a site following antimicrobial therapy to eradicate an infecting organism. For example, Holmes et al. (1979) reported the repeated isolation of a strain of *Flavobacterium odoratum* (a highly drug-resistant organism) from urine following a course of carfecillin and tobramycin to clear an infection caused by *Pseudomonas aeruginosa*. In such a case, the colonizing organism will probably disappear of its own accord upon discontinuance of the antimicrobial therapy.

Occasional strains of several species described here will be considered clinically significant. When they are infecting wound abscesses, these should be surgically drained, since antimicrobial therapy will not help; in some situations, antimicrobial agents may be deemed useful to prevent the spread of bacteria locally through the tissues or distantly via the bloodstream. Assessing appropriate antimicrobial chemotherapy will obviously depend on the susceptibility of the individual strain, especially as susceptibility will be found to vary both from strain to strain and species to species. Variation may also be seen from country to country, from hospital to hospital, and even within one hospital; also levels of resistance can increase during epidemic situations. In vitro susceptibility tests may need to include a wide range of antimicrobial agents, including those undergoing clinical trials and probably combinations of several agents. When an organism is only partially susceptible to the antimicrobial agents tested, a combination of agents may be needed. Pairs of antimicrobial agents may then be tested against the organism in the laboratory in vitro to look for evidence of bacterial synergy. With so diverse a bacterial group, it is possible to make only broad statements. The nonfermenters described in this chapter fall into three main categories with regard to antimicrobial susceptibility.

All the free-living organisms are penicillin-resistant, but one group is also resistant to most other antimicrobial agents. Strains of *Achromobacter* Group Vd show susceptibility to aminoglycosides, carbenicillin, nalidixic acid, tetracycline, and trimethoprim-sulfamethoxazole. Strains of *Alcaligenes xylosoxidans* subspecies *xylosoxidans* (*Achromobacter xylosoxidans*) are susceptible to carbencillin, cefoperazone, ceftazidime, doxycycline, minocycline, polymyxin B, trimethoprim-sulfamethoxazole, and ureidopenicillins (but not tetracycline, in contrast to *Achromobacter* Group Vd). Both species of *Acinetobacter* are susceptible to minocycline and N-formimidoyl thienamycin (imipenem), and *A. lwoffii* strains are also often susceptible to ampicillin and chloramphenicol. Piperacillin may also be effective and marginally more so than mezlocillin and ticarcillin. Determination of antimicrobial susceptibilities of *Flavobacterium* strains by the disk diffusion method may indicate false susceptibility to several antimicrobial agents, and the tube dilution technique is recommended for the accurate assessment of susceptibilities (von Graevenitz and Grehn, 1977). Strains of *F. meningosepticum* and of *Flavobacterium* species Group IIb show susceptibility to nalidixic acid, rifampin, and trimethoprim-sulfamethoxazole and varying susceptibility to clindamycin and erythromycin; only rifampin shows a consistently bactericidal effect. *Flavobacterium breve, F. spiritivorum,* and *F. thalpophilum* are similar to *F. meningosepticum,* although *F. breve* strains are also susceptible to chloramphenicol, erythromycin, and tetracycline; *F. odoratum* is also susceptible to clindamycin, erythromycin, and thienamycin; *F. multivorum* resembles *F. breve* and is also susceptible to cefotaxime, ceftazidime, and ceftriaxone.

The second susceptibility group comprises free-living, penicillin-resistant organisms, but ones that are susceptible to most antimicrobial agents. *Agrobacterium* isolates display a unique susceptibility to carbenicillin, gentamicin, polymyxin B, tetracycline, and trimethoprim-sulfamethoxazole, but resistance to amikacin and tobramycin. Strains of *Alcaligenes faecalis* (including "*Alcaligenes odorans*") exhibiting resistance to cefuroxime and chloramphenicol are frequently encountered. *Alcaligenes xylosoxidans* subspecies *denitrificans* (*Alcaligenes denitrificans*) strains resemble those of *A. faecalis* except that susceptibility is more often shown to carbenicillin, cefamandole, cefoperazone, and the ureidopenicillins, but less often to the aminoglycosides. Strains of *Chryseomonas luteola* and *Flavimonas oryzihabitans* show resistance only to first- and second-generation cephalosporins, although susceptibility to ampicillin, chloramphenicol, and trimethoprim-sulfamethoxazole varies from strain to strain. Isolates of *Shewanella putrefaciens* show resistance to cephalothin and occasionally to ampicillin, carbenicillin, and cefamandole.

The third group comprises those organisms that are not free-living. Strains are susceptible to penicillin, highly so in *Kingella* and *Moraxella*. They are also susceptible to most other antimicrobial agents. *Eikenella corrodens* strains are usually resistant to clindamycin, lincomycin, and metronidazole, while occasional strains of *Moraxella osloensis* may be penicillin-resistant. Strains of *Kingella* and *Moraxella* show variable susceptibility to aminoglycosides; although amikacin has been most effective, some strains can inactivate this agent. Strains of *Oligella ureolytica* are resistant to ampicillin and carbenicillin, and many are also resistant to tetracycline. Strains of both *Weeksella* species are resistant to tobramycin and show varying susceptibility to gentamicin; *W. zoohelcum* strains are also resistant to polymyxin B. Fuller details of antimicrobial susceptibility are given by Lyons (1985) and von Graevenitz (1985).

Susceptibility of a strain to a particular antimicrobial agent does not necessarily mean that treatment with that agent will be successful. Conversely, in *Acinetobacter* strains the minimal inhibitory concentrations (MICs) for amikacin, gentamicin, netilmicin, sissomicin, and tobramycin are usually well below achievable serum levels (Daschner and Nopper, 1980; Garcia et al., 1983). Nevertheless, treatment of serious *Acinetobacter* infections should include an aminoglycoside and piperacillin or ticarcillin until susceptibility results are available (Lyons, 1985). Such combinations are synergistic against some strains showing moderate resistance to an aminoglycoside alone (Glew et al., 1977). Although *Acinetobacter* strains show varying susceptibility to trimethoprim-sulfamethoxazole, this drug has been used successfully as oral therapy in a case of endocarditis (Noble et al., 1981). Cefuroxime has been used successfully to treat cases of *Acinetobacter* pneumonia (Guérisse, 1981), but as yet little clinical experience has been reported of treating such infections with third-generation cephalosporins. In the treatment of neonatal meningitis due to *Flavobacterium meningosepticum*, the clinician will need to beware of rapidly acquired resistance, that intrathecal as well as intravenous administration may be necessary, and that frequent monitoring of levels in the spinal fluid (which have to exceed the MICs of the isolates) is also necessary (Holmes, 1987).

Although patient isolation is generally unnecessary, care should be taken not to spread the infecting organism from patient to patient or to infect further patients from the same source of contamination. Surgery may be required to replace heart valves in cases of endocarditis or to drain abscesses and excise devitalized tissue.

Prevention and Control

For those organisms that are normal inhabitants of mucous membranes and which may occasionally cause infection in the compromised individual, no control measures are possible. For the free-living forms, however, prevention of disease is dependent on adequate disinfection or sterilization of moist sites. Such sites may occur in hemodialysis systems, intravenous solutions, respirators, and even in disinfectants themselves.

Manufacturers need to be encouraged to produce equipment in which disinfection or sterilization is made easy, perhaps with parts particularly at risk of harboring microorganisms that can be removed for autoclaving. In the specific case of inhalation therapy equipment, any water used should be discarded and replaced every 8 h. The equipment should be decontaminated every 24 h with ethylene oxide gas (if available) by nebulizing with 0.25% peracetic acid in water for 5 min (hydrogen peroxide may be used) or by soaking in 2% glutaraldehyde. There are, however, problems associated with the use of ethylene oxide, glutaraldehyde, and peracetic acid, and many would consider a hot wash followed by adequate drying to be sufficient. Tubing, medication, mouthpieces, and nebulizers should be replaced every 24 h. Effluent air mists produced by the equipment should be regularly monitored with blood agar plates, open for 5 min, for bacterial contamination; alternately, the water may be sampled directly. Sterile disposable gloves and catheters, together with meticulous attention to aseptic technique, should be used when suctioning patients with endotracheal tubes and tracheostomies. Skin carriage by hospital personnel has also been implicated as the source of nosocomial outbreaks of pneumonia due to *Acinetobacter* spp. Nursing staff may need to be reminded to fill fresh containers with disinfectant creams instead of topping up existing containers as their contents become depleted. Perhaps most important, the appointment of a control-of-infection nurse is highly recommended.

Laboratory-acquired infection is unlikely under normal microbiology laboratory safe working conditions. None of the organisms covered here requires working in a biological safety cabinet. Although aerosol production, direct skin contamination, or accidental inoculation should be avoided, serious infections are unlikely from such accidents.

Literature Cited

Alexander, M., F. Ismail, P. J. H. Jackman, and W. C. Noble. 1984. Fingerprinting *Acinetobacter* strains from clinical sources by numerical analysis of electrophoretic protein patterns. J. Med. Microbiol. **18**:55–64.

Andrews, H. J. 1986. *Acinetobacter* bacteriocin typing. J. Hosp. Infect. **7**:169–175.

Baumann, P., M. Doudoroff, and R. Y. Stanier. 1968. A study of the *Moraxella* group. II. Oxidative-negative species (genus *Acinetobacter*). J. Bacteriol. **95**:1520–1541.

Bouvet, P. J. M., and P. A. D. Grimont. 1986. Taxonomy of the genus *Acinetobacter* with the recognition of *Acinetobacter baumannii* sp. nov., *Acinetobacter haemolyticus* sp. nov., *Acinetobacter johnsonii* sp. nov., and *Acinetobacter junii* sp. nov. and emended descriptions of *Acinetobacter calcoaceticus* and *Acinetobacter lwoffii*. Int. J. Syst. Bacteriol. **36**:228–240.

Bøvre, K. 1984. Genus II. *Moraxella* Lwoff 1939, 173 emend. Henriksen and Bøvre 1968, 391, p. 296–303. *In* N. R. Krieg and J. G. Holt (ed.), Bergey's manual of systematic bacteriology, vol. 1. The Williams & Wilkins Co., Baltimore.

Clark, W. A., D. G. Hollis, R. E. Weaver, and P. Riley. 1984. Identification of unusual pathogenic gram-negative aerobic and facultatively anaerobic bacteria. U.S. Department of Health and Human Services, Centers for Disease Control, Atlanta.

Das, B. C., and G. A. J. Ayliffe. 1984. Serotyping of *Acinetobacter calcoaceticus*. J. Clin. Pathol. **37**:1388–1391.

Daschner, F., and S. Nopper. 1980. Susceptibility of nosocomial *Acinetobacter anitratus* strains to 14 antibiotics. J Antimicrob. Agents Chemother. **6**:415–416.

De Ley, J., W. Mannheim, P. Segers, A. Lievens, M. Denijn, M. Vanhoucke, and M. Gillis. 1987. Ribosomal ribonucleic acid cistron similarities and taxonomic neighborhood of *Brucella* and CDC group Vd. Int. J. Syst. Bacteriol. **37**:35–42.

Garcia, I., V. Fainstein, B. LeBlanc, and G. P. Bodey. 1983. In vitro activities of new β-lactam antibiotics against *Acinetobacter* spp. Antimicrob. Agents Chemother. **24**:297–299.

Gilardi, G. L. (ed.) 1985. Nonfermentative gram-negative rods: laboratory identification and clinical aspects. (Microbiology series vol. 16). Marcel Dekker, Inc., New York.

Glew, R. H., R. C. Moellering, Jr., and L. J. Kunz. 1977. Infections with *Acinetobacter calcoaceticus* (*Herellea vaginicola*): clinical and laboratory studies. Medicine **56**:79–97.

Goldman, I. S., P. D. Ellner, E. L. Francke, G. J. Garvey, H. C. Neu, and N. Squilla. 1980. Infective endocarditis due to *Kingella denitrificans*. Ann. Intern. Med. **93**:152–153.

Guérisse, P. 1981. Cefuroxime for *Acinetobacter* infection. Lancet **2**:96.

Henriksen, S. D. 1952. *Moraxella:* classification and taxonomy. J. Gen. Microbiol. **6**:318–328.

Holmes, B. 1986. The identification of *Pseudomonas cepacia* and its occurrence in clinical material. J. Appl. Bacteriol. **61**:299–314.

Holmes, B. 1987. Identification and distribution of *Flavobacterium meningosepticum* in clinical material. J. Appl. Bacteriol. **62**:29–41.

Holmes, B., S. P. Lapage, and H. Malnick. 1975. Strains of *Pseudomonas putrefaciens* from clinical material. J. Clin. Pathol. **28**:149–155.

Holmes, B., R. J. Owen, A. G. Steigerwalt, and D. J. Brenner. 1984. *Flavobacterium gleum*, a new species found in human clinical specimens. Int. J. Syst. Bacteriol. **34**:21–25.

Holmes, B., C. A. Pinning, and C. A. Dawson. 1986a. A probability matrix for the identification of gram-negative, aerobic, non-fermentative bacteria that grow on nutrient agar. J. Gen. Microbiol. **132**:1827–1842.

Holmes, B., and P. Roberts. 1981. The classification, identification and nomenclature of agrobacteria: Incorporating revised descriptions for each of *Agrobacterium tumefaciens* (Smith & Townsend) Conn 1942, *Agrobacterium rhizogenes* (Riker et al.) Conn 1942, and *Agrobacterium rubi* (Hildebrand) Starr & Weiss 1943. J. Appl. Bacteriol. **50**:443–467.

Holmes, B., J. J. S. Snell, and S. P. Lapage. 1977a. Strains of *Achromobacter xylosoxidans* from clinical material. J. Clin. Pathol. **30**:595–601.

Holmes, B., J. J. S. Snell, and S. P. Lapage. 1977b. Revised description, from clinical isolates, of *Flavobacterium odoratum* Stutzer and Kwaschnina 1929, and designation of the neotype strain. Int. J. Syst. Bacteriol. **27**:330–336.

Holmes, B., J. J. S. Snell, and S. P. Lapage. 1979. *Flavobacterium odoratum:* a species resistant to a wide range of antimicrobial agents. J. Clin. Pathol. **32**:73–77.

Holmes, B., A. G. Steigerwalt, R. E. Weaver, and D. J. Brenner. 1986b. *Weeksella virosa* gen. nov., sp. nov. (formerly group IIf), found in human clinical specimens. Syst. Appl. Microbiol. **8**:185–190.

Holmes, B., A. G. Steigerwalt, R. E. Weaver, and D. J. Brenner. 1986c. *Weeksella zoohelcum* sp. nov. (formerly group IIj), from human clinical specimens. Syst. Appl. Microbiol. **8**:191–196.

Holmes, B., A. G. Steigerwalt, R. E. Weaver, and D. J. Brenner. 1987. *Chryseomonas luteola* comb. nov. and *Flavimonas oryzihabitans* gen. nov., comb. nov., *Pseudomonas*-like species from human clinical specimens and formerly known, respectively, as groups Ve-1 and Ve-2. Int. J. Syst. Bacteriol. **37**:245–250.

Hugh, R., and E. Leifson. 1953. The taxonomic significance of fermentative versus oxidative metabolism of carbohydrates by various gram negative bacteria. J. Bacteriol. **66**:24–26.

Johnson, J. L., R. S. Anderson, and E. J. Ordal. 1970. Nucleic acid homologies among oxidase-negative *Moraxella* species. J. Bacteriol. **101**:568–573.

Juni, E. 1984. Genus III. *Acinetobacter* Brisou and Prévot 1954, 727^{AL}, p. 303-307. *In* N. R. Krieg and J. G. Holt (ed.), Bergey's manual of systematic bacteriology, vol. 1. The Williams & Wilkins Co., Baltimore.

Kiredjian, M., B. Holmes, K. Kersters, I. Guilvout, and J. De Ley. 1986. *Alcaligenes piechaudii,* a new species from human clinical specimens and the environment. Int. J. Syst. Bacteriol. **36**:282–287.

Lyons, R. W. 1985. Ecology, clinical significance, and antimicrobial susceptibility of *Acinetobacter* and *Moraxella*, p. 159–179. *In* G. L. Gilardi (ed.), Nonfermentative gram-negative rods—laboratory identification and clinical aspects. Marcel Dekker, Inc., New York.

MacDonell, M. T., and R. R. Colwell. 1985. Phylogeny of the Vibrionaceae, and recommendation for two new genera, *Listonella* and *Shewanella*. Syst. Appl. Microbiol. **6**:171–182.

Møller, V. 1955. Simplified tests for some amino acid decarboxylases and for the arginine dihydrolase system. Acta Pathol. Microbiol. Scand. **36:**158–172.

Mulder, C. J. J., and H. C. Zanen. 1984. A study of 280 cases of neonatal meningitis in The Netherlands. J. Infect. **9:**177–184.

Noble, R. C., R. M. Cooper, A. L. Jarvis, P. L. Caples, and E. P. Todd. 1981. Trimethoprim-sulfamethoxazole therapy for infective endocarditis. South. Med. J. **74:**1299–1303.

Oberhofer, T. R. 1985. Manual of nonfermenting gram-negative bacteria. John Wiley & Sons, Inc., New York.

Reverdy, M. E., J. Freney, J. Fleurette, M. Coulet, M. Surgot, D. Marmet, and C. Ploton. 1984. Nosocomial colonization and infection by *Achromobacter xylosoxidans*. J. Clin. Microbiol. **19:**140–143.

Rossau, R., K. Kersters, E. Falsen, E. Jantzen, P. Segers, A. Union, L. Nehls, and J. De Ley. 1987. *Oligella,* a new genus including *Oligella urethralis* comb. nov. (formerly *Moraxella urethralis*) and *Oligella ureolytica* sp. nov. (formerly CDC group IVe): relationship to *Taylorella equigenitalis* and related taxa. Int. J. Syst. Bacteriol. **37:**198–210.

Rossau, R., A. Van Landschoot, W. Mannheim, and J. De Ley. 1986. Inter- and intrageneric similarities of ribosomal ribonucleic acid cistrons of the *Neisseriaceae*. Int. J. Syst. Bacteriol. **36:**323–332.

Swann, R. A., S. J. Foulkes, B. Holmes, J. B. Young, R. G. Mitchell, and S. T. Reeders. 1985. '*Agrobacterium* yellow group' and *Pseudomonas paucimobilis* causing peritonitis in patients receiving continuous ambulatory peritoneal dialysis. J. Clin. Pathol. **38:**1293–1299.

Swann, R. A., and B. Holmes. 1984. Infective endocarditis caused by *Kingella denitrificans*. J. Clin. Pathol. **37:**1384–1387.

von Graevenitz, A. 1985. Ecology, clinical significance, and antimicrobial susceptibility of infrequently encountered glucose-nonfermenting gram-negative rods, p. 181–232. *In* G. L. Gilardi (ed.), Nonfermentative gram-negative rods—laboratory identification and clinical aspects. Marcel Dekker, Inc., New York.

von Graevenitz, A., and M. Grehn. 1977. Susceptibility studies on *Flavobacterium* II-b. FEMS Microbiol. Lett. **2:**289–292.

Yabuuchi, E., T. Kaneko, I. Yano, C. W. Moss, and N. Miyoshi. 1983. *Sphingobacterium* gen. nov., *Sphingobacterium spiritivorum* comb. nov., *Sphingobacterium multivorum* comb. nov., *Sphingobacterium mizutae* sp. nov., and *Flavobacterium indologenes* sp. nov.: glucose-nonfermenting gram-negative rods in CDC groups IIK-2 and IIb. Int. J. Syst. Bacteriol. **33:**580–598.

CHAPTER 28

Gonorrhea

PETER PIOT

Disease: Gonorrhea.
Etiologic Agent: *Neisseria gonorrhoeae.*
Source: Human genital tract by sexual contact or at birth.
Clinical Manifestations: Urethritis, cervicitis, proctitis, pharyngitis, bartholinitis, conjunctivitis, epididymitis, endometritis, salpingitis, disseminated gonococcal infection (arthritis/dermatitis), chorioamnionitis. May be symptomatic or asymptomatic.
Pathology: Infections of mucous membranes lined by nonsquamous epithelium. Local epithelial cell destruction with polymorphonuclear leukocytes.
Laboratory Diagnosis: Isolation of etiologic agent. Gram stain for selected specimens. Antigen detection by enzyme immunoassay.
Epidemiology: Sexually transmitted disease. Worldwide.
Treatment: Many antimicrobials are effective, depending on local antimicrobial susceptibility patterns. In many areas, a high proportion of strains are penicillinase-producing and/or chromosomally resistant to penicillin. Amoxicillin-probenecid, tetracycline, procaine penicillin-probenecid, spectinomycin, norfloxacin, ceftriaxone.
Prevention and Control: Early diagnosis and treatment of infected individuals. Eye prophylaxis at birth. Contact tracing. Health education. Condom use.

Description of Disease

Gonorrhea is a common sexually transmitted disease in humans caused by *Neisseria gonorrhoeae*. Gonococci primarily colonize mucous membranes lined by columnar or transitional epithelium including the male and female urethra, endocervix, rectum, throat, and conjunctiva. Both local and distant complications occur such as salpingitis, epididymitis, and dermatitis/arthritis. When not properly managed, serious sequelae including ectopic pregnancy and infertility may occur. Gonococcal infection during labor is a risk factor for premature delivery and is transmitted to the neonate, thus causing opthalmia neonatorum. Asymptomatic infection occurs in both men and women.

Communicability and Transmissibility

Gonorrhea is transmitted sexually, and even in infants transmission by fomites or by nonsexual contact is probably very rare (Gutman and Wilfert, 1984). The rate of transmission of gonococcal infection varies with the type of sexual practice. Following a single exposure to vaginal intercourse with an infected woman, the risk of acquiring a urethral infection for men is approximately 20% (Hooper et al., 1978). Though the transmission rate from the male urethra to the uterine cervix after one exposure is not precisely known, it is thought to be higher than the other way around. About 90% of secondary sex contacts of men with gonococcal urethritis are infected. Both the rectum and the pharynx can be a source of infection, but the rate of transmission is not well defined (Tice and Rodriguez, 1981). The rate of transmission at birth from an infected mother to her neonate is about 40% (Laga et al., 1986b).

The incubation period in gonorrhea is on the average 2 to 5 days but may vary between 1 and 14 days.

Clinical Manifestations of Disease

The clinical expression of gonococcal infection is very diverse and includes symptomatic, asymptomatic, and complicated infection of different anatomic sites.

Uncomplicated Infection in Adults

The most common manifestation of infection in men is acute anterior urethritis, which is usually characterized by purulent urethral discharge and dysuria. In approximately one-fourth of cases, urethral discharge is mucoid, and a small but unknown proportion of men with a urethral gonococcal infection never develop signs or symptoms of urethritis (Handsfield et al., 1974).

The endocervix is the primary site of urogenital infection in women, but urethral infection is found in 70 to 90% of infected females as well. Vulvovaginitis occurs in prepubertal girls. Clinical manifestations may include vaginal discharge, dysuria, intermenstrual bleeding, purulent or mucopurulent endocervical discharge, erythema, friability, and cervical ectopy. A majority of women with uncomplicated gonorrhea are asymptomatic or have mild symptoms.

Anorectal infection is present in up to half of women with gonorrhea and is a common site of infection in homosexual men. It is thought that the majority of patients with anorectal infection are asymptomatic. In less than 5% of patients, the pharynx is the sole site of infection. Pharyngeal infection occurs in 10 to 20% of women with gonorrhea and in 10 to 25% of infected homosexual men (Wiesner et al., 1973). It is not clear whether this is associated with pharyngeal symptoms nor what its epidemiological role is. The spontaneous-cure rate of pharyngeal gonorrhea approaches 100% within 12 weeks (Wallin and Siegel, 1979). Gonococcal conjunctivitis is rare in adults and is probably due to autoinnoculation.

Complicated Infection

Local complications in males include tysonitis, cowperitis, periurethral abscess, seminal vesiculitis, epididymitis, and prostatitis. Urethral stricture is still common in developing countries.

Approximately 10% of women with gonococcal cervicitis develop salpingitis (pelvic inflammatory disease or PID) as a result of ascending infection. Sequelae of salpingitis include chronic menstrual abnormalities, infertility, and ectopic pregnancy, due to tubal scarring and blockage. Perihepatitis (Fitz-Hugh-Curtis syndrome) may develop by spread of *N. gonorrhoeae* through the peritoneal cavity. Other, less severe complications are bartholinitis and skenitis.

Disseminated gonococcal infection (DGI) develops in a small proportion of infected individuals, and is characterized by the typically arthritis-dermatitis syndrome (Barr and Danielsson, 1971; Holmes et al., 1971). Disseminated gonococcal infection has been associated with particular strains of *N. gonorrhoeae,* requiring arginine, hypoxanthine, and uracil (AHU) for growth, and which are resistant to the bactericidal action of human serum (Eisenstein et al., 1977; Morello et al., 1976).

Infection in Pregnancy and in Infants

Gonorrhea during pregnancy is associated with chorioamnionitis, premature rupture of membranes and premature delivery (Brunham et al., 1984). Gonococcal ophthalmia neonatorum is characterized by a profuse purulent conjunctival exudate. If untreated, corneal ulceration and compromised vision, including blindness, may be the result. Pharyngeal infection, DGI, and scalp abscess may also occur.

Pathology

Gonococci normally infect mucous membranes lined by columnar, transitional, or stratified squamous epithelium. Critical steps in the pathogenesis of gonococcal infection include attachment to the mucosa mediated by pili, IgA1 protease, and possibly other surface proteins and extracellular products, followed by mucosal invasion resulting in the presence of gonococci between and within mucosal cells (Buchanan et al., 1978; Swanson and Mayer, 1984). The infection is characterized by a strong polymorphonuclear leukocytic response, submucosal microabscesses, and exudation into the lumen of the infected organ. Both oxidative and nonoxidative mechanisms may be responsible for the leukocyte bactericidal activity against *N. gonorrhoeae.*

In an experimental model using organ cultures of human fallopian tubes, gonococci initially attach to nonciliated cells, resulting in tissue damage by sloughing of ciliated cells. This is followed by invasion of the nonciliated epithelial cells by gonococci which then enter subepithelial tissues (McGee et al., 1981). Gonococcal lipopolysaccharide may be critical in damaging mucosal cells (Gregg et al., 1981). Gonococcal peptidoglycan plays a role in the pathogenesis of arthritis (Rosenthal and Fleming, 1985).

Diagnostic Features

Gonorrhea should be considered in patients with clinical syndromes and manifestations listed above as well as in patients who have had sexual contact with persons with gonorrhea and in men and women who may have been exposed to infection with *N. gonorrhoeae* during sexual intercourse.

Whereas symptomatic gonococcal urethritis and conjunctivitis may be presumptively diagnosed by microscopy, diagnosis requires isolation of *N. gonorrhoeae* from the pertinent clinical specimen. Clinical differential diagnosis is unreliable. Most syndromes caused by *N. gonorrhoeae* can also be due to *Chlamydia trachomatis*. The etiology of pelvic inflammatory disease is complex, including *N. gonorrhoeae, C. trachomatis, Mycoplasma hominis, Enterobacteriaceae,* and anaerobes.

Prognosis

If untreated, noncomplicated gonorrhea heals spontaneously in most cases. However, approximately 10 to 15% of women and a smaller proportion of men with gonorrhea develop severe complications and sequelae such as epididymitis, salpingitis, infertility, and ectopic pregnancy. Gonococcal ophthalmia neonatorum may lead to blindness. There is an apparent lack of protective immunity following natural infection.

Epidemiologic Aspects

Gonorrhea is a common, worldwide, sexually transmitted disease. It is the most frequently reported bacterial infection in the United States, where over 2 million cases are diagnosed annually. In the United Kingdom in 1984, 53,802 cases of gonorrhea were reported (Communicable Disease Surveillance Centre, 1986). For the past decade, the incidence of gonorrhea has been declining in Europe and North America. The incidence is much higher in many developing countries, where up to 10% of pregnant women may be infected with *N. gonorrhoeae* and 3% of newborns develop gonococcal ophthalmia neonatorum (Laga et al., 1986). Incidence rates of complications and sequelae of gonococcal infection follow the same time and geographical trends.

N. gonorrhoeae is a fragile microorganism with complex growth requirements and a narrow growth temperature range of 35° to 37°C. Its only natural habitat is the human mucous membranes, and this organism dies off very rapidly outside the human body. Thus transmission of the agent requires contact between mucous membranes.

Risk factors for gonorrhea are those for sexually transmitted diseases in general and include number of sex partners, age, socioeconomic level, and history of previous gonorrhea. Conflicting data exist on the impact of hormonal contraception in women on the risk of acquiring gonorrhea. Asymptomatically infected men and women, together with a promiscuous core group of frequent transmitters, probably play a key role in the spread of gonococcal infection.

Etiologic Agent

Characterization

N. gonorrhoeae belongs to the family *Neisseriaceae*, along with *Branhamella, Kingella, Moraxella,* and *Acinetobacter* species. Neisseriae are small (±0.6 to 1 μm diameter), encapsulated, nonmotile, gram-negative diplococci with kidney or coffee bean appearance. They have complex growth requirements, with an optimal growth temperature of 35° to 37°C, and they produce cytochrome oxidase and catalase.

The guanine plus cytosine content of *N. gonorrhoeae* is 49 to 50 mol%. Growth is enhanced by CO_2. After 24 h of incubation, colonies are 0.5 to 1 mm in diameter, smooth, gray to white, and opaque. At least five different colony types (T1 to T5) can be identified on primary culture. After nonselective subculture, T1 and T2 colonies rapidly dissociate to the larger types, T3 to T5.

N. gonorrhoeae produces acid from glucose by oxydation but not from lactose, sucrose, maltose, or fructose. Glucose is metabolized by a combination of Entner-Doudoroff and the pentose phosphate pathways. Nitrate is not reduced, and about half of the strains reduce nitrite.

N. gonorrhoeae strains can be subdivided into at least 35 auxotypes and into at least 46 serovars (Catlin, 1978; Knapp et al., 1985; Sandström et al., 1985).

Specimen Collection

Specimens should be collected from appropriate anatomic sites, depending on the clinical manifestations and on the sex, sexual preference, and sexual practices of the individual. In several countries, collection of anorectal specimens is recommended as a routine practice in women. For screening (in contrast to diagnosing) women, only endocervical specimens are recommended, because this site provides the highest yield.

As a rule, disinfectants should not be used to lubricate instruments (which may be lubricated with warm water), and specimens should only be collected using materials that are nontoxic to the gonococcus. These include spur-free platinum-iridium bacteriological loops (for immediate inoculation) and calcium alginate, Dacron, or treated, nontoxic cotton swabs. Specimens should be inoculated immediately onto isolation medium or nutrient transport medium, or swabs placed into nonnutrient transport medium. Duplicate specimens are recommended for culture and for Gram stain.

Urethral samples can be collected using a swab with a narrow diameter or a sterile bacteriological loop inserted 3 to 4 cm into the urethra. After gentle rotation, the swab is withdrawn. Purulent discharge may be collected directly on a swab or bacteriological loop, eventually after milking the urethra. Patients should not have urinated for at least 2 h before sampling. Culture of prostatic fluid does not increase the rate of isolation of *N. gonorrhoeae* in cases of urethritis and is not recommended.

Endocervical specimens are collected during pelvic examination using a speculum. After wiping off cervical mucus with a cotton ball, a swab is introduced into the endocervix and is then rotated with a ringing motion for at least 10 s before withdrawal.

Anorectal specimens are obtained by inserting a swab 4 to 5 cm into the anal canal and swabbing the crypts. Mucopurulent exudate may be collected directly under anoscopy.

For oropharyngeal specimens, the posterior pharynx and the tonsillar crypts are swabbed.

Conjunctival exudate is collected on swabs for microscopy and culture.

In cases of bartholinitis, draining pus is collected on a swab. Material from closed abscesses is aspirated with a syringe and needle.

Ideally, specimens in cases of salpingitis should be collected from the fallopian tubes during laparascopy or laparotomy. Alternatively, culdocentesis to obtain a cul-de-sac specimen may be obtained. However, such samples may be contaminated by the vaginal flora or from puncturing the intestine. Cervical, urethral, and anorectal specimens should be obtained in all cases of pelvic inflammatory disease.

Blood cultures for *N. gonorrhoeae* should be performed as soon as DGI is suspected. Although conventional blood culture techniques can be used, one should assure that the medium used supports the growth of gonococci. Sodium polyanetholsulfonate (SPS) is toxic for some strains of gonococci, particularly the AHU auxotype.

Skin lesions in cases of DGI are rarely culture-positive for *N. gonorrhoeae*. A small punch biopsy should be taken. Some investigators have obtained reasonable results with heat-fixed material from the bottom of the pustule using fluorescein-conjugated gonococcal antiserum.

Joint fluid should be inoculated into blood culture bottles and also seeded immediately onto gonococcal isolation media.

Transport

Ideally, specimens for the isolation of *N. gonorrhoeae* should be inoculated directly onto the culture medium and promptly incubated. When such conditions are not available, an appropriate transport medium should be used.

Amie's modification of Stuart's holding medium (Phillips and Nash, 1985) is frequently used and contains activated charcoal to absorb toxic substances that may be present in swabs or sticks. Although some loss is inevitable, gonococci survive in this medium for at least 6 h at room temperature and may be used for transportation times of up to 12 h.

Alternatively, a nutritive transport system can be used comprising a gonococcal selective isolation medium and a CO_2-generating tablet or ampul in an individual plastic bag. Examples of such systems include JEMBEC (Ames Co., Elkart, IN), Bio-Bag (Marion Scientific, Kansas City, MO), and Gono-Pak System (Nasco, Modesto, CA). The Transgrow system consists of a screw-capped bottle containing gonococcal isolation medium and filled with CO_2. In all such nutritive system, preincubation before transport is recommended.

Direct Examination

Smears that are sufficiently thin should be Gram-stained and examined for intracellular gram-negative diplococci. Gram-stained smears from patients with urethral discharge and from cases of mucopurulent conjunctivitis are very accurate for the diagnosis of gonococcal infection and should always be performed in such cases. However, the sensitivity and specificity of a Gram-stained smear for the diagnosis of cervical gonorrhea vary between 50 and 80% and 60 and 100%, respectively. Thus, its positive predictive value in a population with a low incidence of gonorrhea may be unacceptably low.

Isolation Procedures

Genital, rectal, and oropharyngeal specimens should be inoculated on selective isolation media such as modified Thayer-Martin agar (MTM), New York City medium (NYC), or Martin-Lewis agar (ML) (Phillips and Nash, 1985). These media are selective for *N. gonorrhoeae* because of the addition of combinations of antibiotics including vancomycin (3 $\mu g/$

ml), colistin (7.5 μg/ml), nystatin (12.5 μg/ml), and trimethoprim (5 μg/ml) (MTM); vancomycin (4 μg/ml), colistin (7.5 μg/ml), anisomycin (20 μg/ml), and trimethoprim (5 μg/ml) (ML); and vacomycin (2 μg/ml), colistin (5.5 μg/ml), amphotericin B (1.2 μg/ml), and trimethoprim (3 μg/ml) (NYC). For specimens from normally sterile sites or conjunctival swabs, a nonselective chocolate agar should be used. Since up to 10% of gonococci in a given geographical area may be susceptible to 3 to 4 μg/ml of vancomycin, biplates containing a selective medium and chocolate agar are recommended, particularly for genital specimens (Mirrett et al., 1981). All batches of media should be checked for their ability to support the growth of small numbers of recently isolated gonococci.

Common blood culture media without SPS or with added gelatin (in a 1% final concentration) to neutralize SPS (Staneck and Vincent, 1981) can be used for the isolation of *N. gonorrhoeae* from blood or joint fluid.

Plates and bottles are incubated in a CO_2 incubator with a 5% CO_2 level, a candle extinction jar, or an anaerobic jar containing a CO_2-generating envelope. High humidity (>70%) is important for optimal recovery of gonococci. Media are incubated at 35°C for 2 to 3 days.

Examine plates after overnight incubation and again after 48 h of incubation for gray to white, opaque, convex, and glistening colonies. The colonies have a diameter of 0.5 to 1 mm after 24 h and of up to 3 mm after 48 h of incubation. Nutritionally fastidious isolates such as AHU strains grow more slowly and product smaller colonies. At least five colony types of different size and morphology can be observed under a stereoscopic microscope (Kellogg et al., 1963).

Presumptive Identification

Suspicious colonies should be tested for oxidase by rubbing a portion of a colony on a filter paper impregnated with a 1% solution of dimethyl- or tetramethyl-p-phenylenediamine dihydrochloride. Oxidase-positive colonies produce a dark purple color, usually within 10 s. Bacteriologic loops made of nichrome may generate false-positive reactions. Alternatively, the reagent solution can be poured over suspected colonies for rapid development of a dark blue to purple color. However, when this technique is used, oxidase-positive colonies should be subcultured immediately after the test, since the reagent is lethal for gonococci.

A Gram stain should be performed on all suspicious colonies. Oxidase-positive, gram-negative diplococci growing on one of the selective media listed above and isolated from urogenital specimens can be presumably identified as *N. gonorrhoeae*.

Identification

All presumptive gonococcal isolates should be fully identified. Fresh overnight subcultures on nonselective media should be used in all identification procedures. Since gonococci autolyze as they age, isolates should be subcultured every 2 days.

Standard criteria for the definitive identification of *N. gonorrhoeae* include acid production from different carbohydrates, as well as additional tests listed in Table 1, in case the carbohydrate reaction pattern is deviant.

Acid production from glucose, maltose, lactose, sucrose, and fructose can be determined using cystine-tryptic digest agar (CTA) containing filter-sterilized carbohydrate. Only reagent-grade carbohydrates should be used. A carbohydrate-free control tube should always be included. Either semisolid CTA media containing 1% carbohydrate or CTA agar slants with 2% carbohydrate may be used. In the latter method, which is the more sensitive, the surface of the slant is inoculated with a loopful of colonies. When a heavy inoculum of an overnight culture is used to inoculate CTA media, results can be read after 24 h of incubation at 35°C. The fastidious AHU strains may require an additional 24 h of incubation. If no acid production is observed after 48 h, fresh CTA media should be inoculated with a larger inoculum of an overnight culture. False-positive maltose reactions may occur because of contamination of the maltose with glucose. The ONPG test (o-nitrophenyl-beta-D-galactopyranoside) for the detection of beta galactosidase may be substituted for the lactose reaction.

Rapid carbohydrate degradation tests are now preferred by many laboratories when identifying gonococci. Small tubes containing 0.4 ml of 4% carbohydrate solutions in phosphate buffer with phenol red indicator (K_2HPO_4 0.04 g, KH_2PO_4 0.01 g, KCl 0.8 g, phenol red 5% 1 ml, aqd 100 ml, pH 7.2) are inoculated heavily with colonies of an overnight culture on blood agar and incubated at 35°C. Clear-cut results are available within 1 to 4 h.

Rapid carbohydrate tests are also commercially available, with carbohydrate reagents presented as solutions or in impregnated paper disks (RIM-Neisseria kit, Austin Biological Laboratories, Austin, TX; Minitek, BBL Microbiology Systems, Cockeysville, MD).

Chromogenic enzyme substrate systems, eventually in combination with conventional tests, can also

TABLE 1. Characteristics of *Neisseria* species and *Branhamella catarrhalis*

	N. gonorrhoeae	N. meningitidis	N. lactamica	N. sicca	N. subflava	N. mucosa	N. flavescens	N. cinerea	N. catarrhalis
Acid production from									
Glucose	+	+	+	+	+	+	−	−	−
Maltose	−	+	+	+	+	+	−	−	−
Lactose	−	−	+	−	−	−	−	−	−
Sucrose	−	−	−	+	V[a]	+	−	−	−
Fructose	−	−	−	+	V	+	−	−	−
Beta galactosidase	−	−	+	−	−	−	−	−	−
DNase	−	−	−	−	−	−	−	−	+
Reduction of									
Nitrate	−	−	−	−	−	+	−	−	+
Nitrite	−	V	+	+	+	+	+	+	+
Polysaccharide synthesis	−	−	−	−	V	+	+	−	−
Growth on									
MTM, ML, or NYC agar	+	+	+	−	−	−	−	V	V
22°C	−	−	V	V	V	+	+	V	+
Nutrient agar	−	−	+	+	V	+	+	+	+

[a] Variable reactions.

be used for the rapid identification of *N. gonorrhoeae*. They usually include prolyl and gamma glutamyl aminopeptidase substrate tests and should be heavily inoculated with a MacFarland no. 3 suspension of bacteria. The reactions are read after 30 min to 4 h of incubation at 35°C (GO Check, Roche Diagnostica, Switzerland; Diagnostic Tablets, Rosco, Taastrup, Denmark).

N. gonorrhoeae can also be identified in a coagglutination test using a *Staphylococcus aureus* strain coated with polyclonal or monoclonal gonococcal antibody (Phadebact Gonococcus test, Pharmacia Diagnostics, Piscataway, NJ; Gono Gen Test, Micro-Media Systems, Potomac, MD). Results are available within 30 s to 2 min of mixing a boiled suspension of organisms with the coagglutination reagent.

Gonococci can be *subtyped* by auxotyping and serotyping (Catlin, 1978; Knapp et al., 1985). Monoclonal antibody reagents for coagglutination are now available (Knapp et al., 1985; Sandström et al., 1985).

Antimicrobial Susceptibility Testing

Gonococcal isolates should be routinely tested for beta-lactamase production using a chromogenic cephalosporin assay (Cefinose, BBL; Nitrocefine, Oxoid) or an acidometric assay (Marion Laboratories). Colonies are rubbed on disks impregnated with the test substrate and observed for color change according to the test used. Tests for beta lactamase should be performed directly from the primary isolation plate.

In areas where chromosomally mediated penicillin resistance to penicillin or spectinomycin resistance occurs, gonococcal strains may be tested for these agents in a disk agar diffusion susceptibility test. This should also be the case for strains from patients who fail to respond to therapy. A no. 1 MacFarland suspension of gonococci is swabbed over a chocolate plate containing the Diagnostic Sensitivity Test (Oxoid, Basingstoke, U.K.). Disks with 10 U penicillin and 100 μg spectinomycin are then placed on the agar. Plates are incubated at 35°C for 24 h in a 5% CO_2 atmosphere. Zone diameters that suggest resistance are <20 mm for penicillin and no zone for spectinomycine. Whenever possible, any resistance other than that due to beta-lactamase production should be confirmed in an agar dilution test.

Gonococcal Antigen Detection

N. gonorrhoeae antigen can be detected in urethral and endocervical specimens in an enzyme immunoassay using beads coated with absorbed rabbit antibody against the major outer membrane protein of *N. gonorrhoeae* (Gonozyme, Abbott Laboratories, North Chicago, IL). Results are available in 3 h. The sensitivity and specificity of this assay for the diagnosis of gonococcal urethritis were 87 to 100% and 94 to 100%, respectively. For endocervical specimens, the respective figures were 78 to 91% and 95 to 100%

(Aardoom et al., 1982; Danielsson et al., 1983; Schachter et al., 1984; Stamm et al., 1984).

Disadvantages of this assay include the nonavailability of gonococci for susceptibility testing and the poor sensitivity in asymptomatic urethral infection. Because of its low predictive value in low-prevalence populations, it is not recommended for screening patients belonging to low-risk groups.

Treatment, Prevention, and Control

Treatment

N. gonorrhoeae strains are susceptible to numerous antimicrobial agents belonging to different groups of antibiotics. However, treatment of patients with gonorrhea in a given area (country) should be based on the results of programs monitoring the in vitro susceptibility of gonococcal isolates to recommended antibiotics. Thus, in several parts of the world, penicillinase-producing strains of *N. gonorrhoeae* (PPNG) or strains that are chromosomally resistant to penicillin or tetracycline are frequent.

In general, single-dose treatment is preferred. Since in 25 to 40% of cases of gonorrhea, a chlamydial infection is also present, a combined treatment may be preferred, including single-dose therapy against gonorrhea and a multiple-day regimen with tetracycline or erythromycine against *C. trachomatis* infection. Recommended antimicrobials for gonorrhea may include procaine penicillin plus probenicid or amoxicillin plus probenicid, spectinomycin, cefoxitin, cefuroxime, ceftriaxone, tetracycline, norfloxacin, ciprofloxacin, and ofloxacin. Gonococcal ophthalmia neonatorum may be treated with single-dose ceftriaxone (Laga et al., 1986a).

All patients should be cultured for *N. gonorrhoeae* after termination of therapy.

Prevention and Control

Early diagnosis and treatment, case holding, and contact tracing, together with identification of asymptomatic men and women in high-risk groups through screening, are the cornerstones of gonorrhea-control programs.

Primary prevention relies on health education and the use of barrier contraception, in particular condoms. Gonococcal ophthalmia neonatorum can be prevented by screening pregnant women for gonococcal infection or by eye prophylaxis at birth. Both silver nitrate and tetracycline ointment are effective for the prevention of gonococcal neonatal conjunctivitis, including infections due to PPNG and tetracycline-resistant strains (Laga et al., 1987).

Literature Cited

Aardoom, H. A., D. de Hoop, C. O. A. Iserief, M. F. Michel, and E. Stolz. 1982. Detection of *Neisseria gonorrhoeae* antigen by a solid-phase enzyme immunoassay. Br. J. Vener. Dis. **58**:359–362.

Barr, J., and D. Danielson. 1971. Septic gonococcal dermatitis. Br. Med. J. **1**:482–485.

Brunham, R. C., K. K. Holmes, and D. Eschenbach. 1984. Sexually transmitted diseases in pregnancy, p. 782–816. *In* K. K. Holmes, P.-A. Mårdh, P. F. Sparling, and P. J. Wiesner (eds.), Sexually transmitted diseases. McGraw-Hill, New York.

Buchanan, T. M., K. C. S. Chen, R. B. Jones, J. F. Hildebrandt, W. A. Pearce, M. A. Hermodson, J. C. Newland, and D. L. Luchtel. 1978. Pili and principal outer membrane protein of *Neisseria gonorrhoeae:* immunochemical, structural, and pathogenic aspects, p. 145–154. *In* G. F. Brooks, E. F. Gotschlich, K. K. Holmes, W. D. Sawyer, and F. E. Young (ed.), Immunobiology of *Neisseria gonorrhoeae*. American Society for Microbiology, Washington, D.C.

Catlin, B. W. 1978. Characteristics and auxotyping of *Neisseria gonorrhoeae*, p. 345–380. *In* T. Bergan and J. R. Norris (ed.), Methods in microbiology, Vol. 10. Academic Press, London.

Communicable Disease Surveillance Centre. 1986. Sexually transmitted disease surveillance in Britain—1984. Br. Med. J. **293**:942–943.

Danielsson, D., H. Moi, and L. Forslin. 1983. Diagnosis of urogenital gonorrhoea by detecting gonococcal antigen with a solid phase enzyme immunoassay (gonozyme). J. Clin. Pathol. **36**:647–649.

Eisenstein, B. I., T. J. Lee, and P. F. Sparling. 1977. Penicillin sensitivity and serum resistance are independent attributes of strains of *Neisseria gonorrhoeae* causing disseminated gonococcal infection. Infect. Immun. **15**:834–841.

Fransen, L., H. Nsanze, V. Klaus, P. Van der Stuyft, L. D'Costa, R. C. Brunham, and P. Piot. 1986. Ophthalmia neonatorum in Nairobi, Kenya: the roles of *Neisseria gonorrhoeae* and *Chlamydia trachomatis*. J. Infect. Dis. **153**:862–869.

Gregg, C. R., M. A. Melly, C. G. Hellerqvist, J. G. Coniglio, and Z. A. McGee. 1981. Toxic activity of purified lipopolysaccharide of *Neisseria gonorrhoeae* for human fallopian tube mucosa. J. Infect. Dis. **143**: 432–439.

Gutman, L. T., and C. M. Wilfert. 1984. Gonococcal diseases in infants and children, p. 238–243. *In* K. K. Holmes, P.-A. Måardh, P. F. Sparling, and P. J. Wiesner (ed.), Sexually transmitted diseases. McGraw-Hill, New York.

Handsfield, H. H., T. O. Lipman, J. P. Harnisch, E. Tronca, and K. K. Holmes. 1974. Asymptomatic gonorrhea in men: diagnosis, natural course, prevalence and significance. N. Engl. J. Med. **290**:117–123.

Hendrickson, D. A. 1985. Reagents and stains, p. 1093–1107. *In* E. H. Lennette, A. Balows, W. J. Hausler, Jr., and H. J. Shadomy (ed.), Manual of clinical microbiology, 4th ed. American Society for Microbiology, Washington, D.C.

Holmes, K. K., P. J. Wiesner, and A. H. B. Pedersen.

1971. The gonococcal arthritis-dermatitis syndrome. Ann. Intern. Med. **75**:470–471.

Hooper, R. R., G. H. Reynolds, O. G. Jones, A. Zaidi, P. J. Wiesner, K. P. Latimer, A. Lester, A. F. Campbell, W. O. Harrison, W. W. Karney, and K. K. Holmes. 1978. Cohort study of venereal disease. 1. The risk of gonorrhea transmission from infected women to men. Am. J. Epidemiol. **108**:136–144.

Kellogg, D. S. Jr., W. L. Peacock, Jr., W. E. Deacon, L. Brown, and C. I. Pirkle. 1963. *Neisseria gonorrhoeae*. I. Virulence genetically linked to clonal variation. J. Bacteriol. **85**:1274–1279.

Knapp, J. S., E. G. Sandström, and K. K. Holmes. 1985. Overview of epidemiological and clinical applications of auxotype/serovar classification of *Neisseria gonorrhoeae*. p. 6–12. *In* G. K. Schoolnik (ed.), The pathogenic *Neisseriae*. American Society for Microbiology, Washington, D.C.

Laga, M., W. Namaara, R. C. Brunham, L. J. D'Costa, H. Nsanze, P. Piot, D. Kunimoto, J. O. Ndinya-Achola, L. Slaney, A. R. Ronald, and F. A. Plummer. 1986a. Single dose therapy of gonococcal ophthalmia neonatorum with ceftriaxone. N. Engl. J. Med. **315**:1382–1385.

Laga, M., F. A. Plummer, H. Nsanze, W. Namaara, R. C. Brunham, J. O. Ndinya-Achola, G. Maitha, A. R. Ronald, J. D'Costa, V. Bhullar, J. Y. Mati, M. Cheang, and P. Piot. 1986b. Epidemiology of ophthalmia neonatorum in Kenya. Lancet **2**:1145–1148.

Laga, M., F. A. Plummer, P. Datta, W. Namaara, J. O. Ndinya-Achola, H. Nsanze, R. C. Brunham, and P. Piot. 1986. Comparison of tetracycline and silver nitrate for the prophylaxis of chlamydial and gonococcal ophthalmia neonatorum, p. 301–304. *In* D. Oriel, G. Ridgway, J. Schachter, D. Taylor-Robinson, and M. Ward (ed.), Chlamydial infections. Cambridge University Press, Cambridge.

McGee, Z. A., P. P. Johnson, and D. Taylor-Robinson. 1981. Pathogenic mechanisms of *Neisseria gonorrhoeae:* observations on damage to human fallopian tubes in organ culture by gonococci of colony type 1 or type 4. J. Infect. Dis. **143**:413–422.

Mirrett, S. L., B. Reller, and J. S. Knapp. 1981. *Neisseria gonorrhoeae* strains inhibited by vancomycin in selective media and correlation with auxotype. J. Clin. Microbiol. **14**:94–99.

Morello, J. A., S. A. Lerner, and M. Bohnhoff. 1976. Characteristics of atypical *Neisseria gonorrhoeae* from disseminated and localized infections. Infect. Immun. **13**:1510–1516.

Phillips, E., and P. Nash. 1985. Culture media, p. 1051–1092. *In* E. H. Lennette, A. Balows, W. J. Hausler, Jr., and H. J. Shadomy (ed.), Manual of clinical microbiology, 4th ed. American Society for Microbiology, Washington, D.C.

Rosenthal, R. S., and T. J. Fleming. 1985. Arthropathic properties of gonococcal peptidoglycan, p. 352–359. *In* G. K. Schoolnik (ed.), The pathogenic *Neisseriae*. American Society for Microbiology, Washington, D.C.

Sandström, E., P. Lindell, B. Härfast, F. Blomberg, A.-C. Ryden, and S. Bygdeman. 1985. Evaluation of a new set of *Neisseria gonorrhoeae* serogroup w-specific monoclonal antibodies for serovar determination, p. 6–12. *In* G. K. Schoolnik (ed.), The pathogenic *Neisseriae*. American Society for Microbiology. Washington, D.C.

Schachter, J., W. M. McCormack, R. F. Smith, R. M. Parks, R. Bailey, and A. C. Ohlin. 1984. Enzyme immunoassay for diagnosis of gonorrhoea. J. Clin. Microbiol. **19**:57–59.

Stamm, W. E., B. Cole, C. Fennell, P. Bonin, A. S. Armstrong, J. E. Hermann, and K. K. Holmes. 1984. Antigen detection for the diagnosis of gonorrhea. J. Clin. Microbiol. **19**:399–362.

Staneck, J. L., and S. Vincent. 1981. Inhibition of *Neisseria gonorrhoeae* by sodium polyanetholsulfonate. J. Clin. Microbiol. **13**:463–467.

Swanson, J., and L. W. Mayer. 1984. Biology of *Neisseria gonorrhoeae*, p. 187–204. *In* K. K. Holmes, P.-A. Mårdh, P. F. Sparling, and P. J. Wiesner (ed.), Sexually transmitted diseases. McGraw-Hill, New York.

Tice, A. W., and V. L. Rodriguez. 1981. Pharyngeal gonorrhea. J. Am. Med. Assoc. **246**:2717–2719.

Wallin, J., and M. S. Siegel. 1979. Pharyngeal *Neisseria gonorrhoeae:* coloniser or pathogen? Br. Med. J. **1**:1462–1463.

Wiesner, P. J., E. Tronca, P. Bonin, A. H. B. Pedersen, and K. K. Holmes. 1973. Clinical spectrum of pharyngeal gonococcal infection. N. Engl. J. Med. **288**:181–185.

CHAPTER 29

Infections Caused by Miscellaneous Gram-Negative Aerobic Bacteria

PATRICK R. MURRAY

Etiologic Agent and Most Common Diseases: *Cardiobacterium hominis*—endocarditis; *Actinobacillus actinomycetemcomitans*—juvenile periodontitis, endocarditis: *Kingella* spp.—endocarditis; *Capnocytophaga*—juvenile periodontitis, septicemia: CDC group DF-2—septicemia, wound infections: *Pasteurella multocida*—wound infection, chronic respiratory disease, septicemia.

Source: Human (*Cardiobacterium, Actinobacillus, Kingella, Capnocytophaga*), dog (DF-2, *Pasteurella*), or cat (*Pasteurella*) upper respiratory tract.

Predisposing Conditions: Valvular heart disease (*Cardiobacterium, Actinobacillus, Kingella*); dental manipulations or oral cavity disease (*Cardiobacterium, Actinobacillus, Kingella, Capnocytophaga*); hematologic malignancy (*Capnocytophagia*); animal bite or contact (DF-2, *Pasteurella*); splenectomy (DF-2); liver disease (DF-2; *Pasteurella*).

Treatment: Penicillin drug of choice, although resistance observed with *Actinobacillus* and *Kingella*; all susceptible to ampicillin except *Actinobacillus*; all susceptible to tetracycline and chloramphenicol; all resistant to penicillinase-resistant penicillins (oxacillin, nafcillin, methicillin) except *Kingella*; erythromycin resistance observed with *Cardiobacterium*, *Actinobacillus*, and *Pasteurella*; clindamycin resistance observed with all except *Capnocytophaga* and DF-2; aminoglycoside resistance observed with all except *Cardiobacterium* and *Kingella*; all resistant to vancomycin.

Laboratory Diagnosis: *Cardiobacterium hominis*—slow-growing, pleomorphic, gram-negative bacillus, oxidase-positive, catalase-negative, indole-positive (weak), ferments glucose, maltose, mannitol, and sucrose; *Actinobacillus actinomycetemcomitans*—slow-growing, small, gram-negative bacillus, requires CO_2 for growth, oxidase-negative, catalase-positive, ferments glucose, maltose, and mannitol; *Kingella*—slow-growing coccobacillus in pairs and chains, oxidase-positive, catalase-negative, ferments glucose and maltose; *Capnocytophaga*—slow-growing, slender, fusiform, gram-negative bacillus that requires CO_2 for growth, yellow-pigmented colonies, oxidase-negative, catalase-negative, ferments glucose, lactose, maltose, and sucrose; CDC group DF-2—slow-growing, long, thin, gram-negative bacillus, oxidase-positive, catalase-positive, ferments glucose, lactose, and mannitol; *Pasteurella multocida*—small, gram-negative, coccobacillus, musty odor, oxidase-positive, catalase-positive, ferments glucose, mannitol, and sucrose, reduces nitrate, indole-positive.

Cardiobacterium hominis

Epidemiology

Natural Habitat

Cardiobacterium hominis, formerly called CDC group IId, is a rare cause of subacute bacterial endocarditis. The organism inhabits the upper respiratory tract in man and has been detected in 68% of healthy individuals (Slotnick et al., 1964). Carriage appears to be unrelated to the age or sex of the individual, and antibody against *C. hominis* is not detected in carriers.

Method of Transmission

Disease is associated with spread of *C. hominis* from the oropharynx. Dental manipulations or oral cavity disease is reported in almost one-half of patients with *C. hominis* disease (Wormser and Bottone, 1983).

Description of Disease

Spectrum of Infection

As the name implies, *Cardiobacterium hominis* has a predilection for the human heart. Although the number of reported cases is small (approximately 40), infectious endocarditis is the only disease caused by *C. hominis* in humans. *C. hominis* endocarditis has been reported from Great Britain, France, Germany, Belgium, and Scandinavia, as well as from the United States.

Clinical Manifestations

The median age for infected patients is 50 years, with disease more common in males than females (2.3:1). The organism has a relatively low virulence potential with disease unable to be established in experimental animals (Wormser and Bottone, 1983) and human infections primarily in individuals with a history of valvular heart disease. The usual course of disease is insidious. The patients are usually afebrile and present with vague flulike symptoms, fatigue, and malaise. The duration of symptoms before diagnosis can be as long as 18 months with the average of 5 months, which is significantly longer than with endocarditis caused by related gram-negative bacilli (Ellner et al., 1979). This reflects both the low virulence of *C. hominis* and the difficulties in culturing the organism. The left side of the heart is involved almost exclusively (Robison and Vitelli, 1985).

Treatment and Prognosis

Treatment

C. hominis is susceptible to penicillin, ampicillin, cephalothin, tetracycline, aminoglycosides, and chloramphenicol. Resistance has been reported to clindamycin, erythromycin, oxacillin, and vancomycin. Treatment for 3 weeks with penicillin G or ampicillin in a 12-g daily dose has been recommended for patients with endocarditis of natural valves (Geraci et al., 1978). Susceptibility tests are sometimes difficult to perform because of the fastidious growth requirements of the organism.

Prognosis

Blood cultures usually become negative immediately after therapy is initiated. A bacteriologic cure is achieved in approximately 90% of patients, although valve replacement may be required for patients with refractory congestive heart failure.

Etiologic Agent

Characteristics

Pale-staining, gram-negative bacillus (0.5 × 2 μm), which occasionally appears as pleomorphic teardrop-shaped forms when grown on yeast extract–deficient media (Savage et al., 1977). Slow-growing in liquid media with blood cultures requiring incubation for 5 to 14 days before growth is detected. Discrete grayish floccules on the bottom of the blood culture broth typically develop in the absence of turbidity. Growth on solid media usually requires incubation in an atmosphere of 5% CO_2 and increased humidity. Blood is not required for growth. *C. hominis* will grow on Mueller Hinton agar but not on MacConkey, eosine methylene blue, Salmonella-Shigella, or Hektoen enteric agars. Pinpoint colonies on blood or chocolate agar plates developed within 2 or 3 days. The isolates are nonhemolytic on blood agar, and surface translocation (gliding) is observed (Wormser and Bottone, 1983).

Identification

C. hominis is oxidase-positive, catalase-negative, nonmotile, and fermentative, although serum should be added to the carbohydrate reactions to ensure adequate growth. Production of indole differentiates *C. hominis* from similar organisms (e.g., *Actinobacillus*, *Haemophilus aphrophilus*, and *Eikenella*), although indole production is weak, and it must be extracted with xylene for detection. The typical biochemical

TABLE 1. Biochemical properties of miscellaneous aerobic gram-negative bacilli

	Percent positive reactions[a]					
Reactions	Cardiobacterium hominis	Actinobacillus actinomycetem-comitans	Kingella kingae	Capnocytophaga	*CDC group DF-2*	Pasteurella multocida
Catalase	3	99	0	7	100	98
Oxidase	100	19 (2)[b]	100	7	96	97
Motility	0	0	0	5 (40)	0	0
Hemolysis (beta)	0	0	84	0	0	0
Gas from glucose	0	28	0	0	0	0
Acid from:						
Glucose	78 (22)	83 (16)	42 (48)	90 (10)	85	100
Lactose	0	0	0	75 (11)	100	8
Maltose	72 (28)	81 (15)	76 (24)	86 (14)	100	2
Mannitol	50 (50)	66 (16)	0	0	0	78
Sucrose	61 (39)	0 (15)	0	90 (9)	0	100
Xylose	0	33 (9)	0	0	0	67
TSI slant, acid	96	100	28 (4)	73	17	98 (1)
TSI butt, acid	83	100	12	55	15	99
Growth on MAC agar	0	4 (1)	33 (3)	0	0	2 (1)
Simmons citrate	0	0	0	0	0	0
Urease	0	0	0	0	0	0
Nitrate reductase	0	100	3 (3)	63	0	99
Gas from nitrate	0	0	0	0	0	0
Indole production	97	0	0	0	0	99
Gelatin hydrolysis	0	0	0	0	0	0
Esculin hydrolysis	4	0	0	81 (2)	77	0

[a] Adapted from Weaver et al. (1985).
[b] Percent of isolates with delayed reactivity.

reactions observed with isolates are summarized in Table 1.

IMMUNOLOGY

Whereas antibodies have not been detected in healthy individuals colonized with *C. hominis*, agglutinating antibodies have been reported in nine patients with endocarditis. The antibody titers ranged from 1:128 to 1:3,000 (Wormser and Bottone, 1983). Additional data will have to be collected to determine the usefulness of this test as a diagnostic or prognostic tool. *C. hominis* is not antigenically related to other fastidious gram-negative bacilli associated with endocarditis.

Actinobacillus

Epidemiology

NATURAL HABITAT

Five species of *Actinobacillus* have been described: *A. actinomycetemcomitans* and four species that are primarily animal pathogens—*A. lignieresii*, *A. equuli*, *A. suis*, and *A. capsulatus*. *Actinobacillus actinomycetemcomitans* is a normal resident of the human oral cavity. The organism was isolated from gingival sites in 17% of healthy individuals compared with 20% of patients with adult periodontitis and 96% of patients with localized juvenile periodontitis (Zombon et al., 1983).

METHOD OF TRANSMISSION

Endocarditis is associated with spread of *Actinobacillus* from the oral cavity following dental manipulations or in patients with periodontitis. Underlying cardiac disease is reported for virtually all patients.

Description of Disease

SPECTRUM OF INFECTION

Actinobacillus is commonly associated with juvenile periodontitis. Additionally, approximately 60 cases of endocarditis caused by this organism have been reported in the medical literature, as well as rare reports of pericarditis, coronary arteritis, osteomyelitis, brain abscess, meningitis, thyroid abscess, urinary tract infection, and respiratory tract disease. *Actinobacillus* infections have a worldwide geographic distribution with documented reports from Europe, Israel, Saudi Arabia, Canada, and the United States.

Clinical Manifestations

Endocarditis caused by *Actinobacillus* usually presents in a subacute form with a prolonged interval (average 3.5 months) between the initial signs and symptoms and the etiologic diagnosis. The symptoms most frequently reported include fever, fatigue, weight loss, and night sweats. Splenomegaly, hematuria, anemia, and an elevated erythrocyte sedimentation rate have also been reported. The course of endocarditis can be complicated by multiple emboli and congestive heart failure. *Actinobacillus* endocarditis is primarily an adult disease (average age 40 years), although the range is 7 to 68 years. Disease is more common in males than females (3:1).

Treatment and Prognosis

Treatment

Strains are generally susceptible to chloramphenicol, tetracycline, carbenicillin, and streptomycin. Penicillin, ampicillin, and erythromycin are less effective, and resistance has been reported with clindamycin, vancomycin, and methicillin. In contrast with *Cardiobacterium*, only about 50% of isolates are inhibited by 4 μg/ml of penicillin and ampicillin, and approximately 10% of strains have high level resistance (Slots et al., 1980; Yogev et al., 1986). If the isolate is susceptible in vitro, patients with infections of natural valves can be successfully treated with ampicillin alone in a daily dosage of 12 g administered intravenously for 3 weeks. For patients with artificial heart valves or infections with resistant organisms, prolonged therapy with a penicillin combined with an aminoglycoside has been recommended (Geraci et al., 1980). If the response is unsatisfactory, a third antibiotic such as tetracycline or chloramphenicol should be added (Siegman-Igra et al., 1984).

Prognosis

Actinobacillus endocarditis has a fatality rate of approximately 25%. The most frequent complications leading to death are congestive heart failure and embolism (Siegman-Igra et al., 1984).

Etiologic Agent

Characteristics

Actinobacillus is a small, facultative, anaerobic, gram-negative bacillus with a strict requirement for growth in a CO_2-enriched atmosphere. Growth after 2 or 3 days on chocolate and blood agar plates appears as small, adherent, dry colonies with mixed morphology. A starlike opaque center will be seen on some colonies. The organism fails to grow on MacConkey agar. Growth in broth media is granular. A selective medium for the isolation of *Actinobacillus* in mixed cultures has been reported (Slots, 1982), but for infections other than periodontitis the organism is usually isolated in pure culture.

Identification

Actinobacillus closely resembles *Haemophilus aphrophilus* but is differentiated by a positive catalase reaction (can be weak), absolute requirement for growth in CO_2, and an inability to ferment lactose and sucrose. Other closely related organisms (e.g., *Cardiobacterium*, *Kingella*, and *Streptobacillus*) are generally catalase-negative. Definitive identification tests for *A. actinomycetemcomitans* are summarized in Table 1.

Immunology

Specific antibodies against *Actinobacillus* have been measured in a patient with endocarditis. Although the antibody titers in this patient decreased during convalescence, the diagnostic or prognostic value of this test has not been defined (Pierce et al., 1984).

Kingella

Epidemiology

Natural Habitat

The genus of *Kingella* consists of three species: *K. kingae*, *K. denitrificans*, and *K. indologenes*. *K. kingae*, initially named *Moraxella* new species 1 by King, is a commensal of the human respiratory tract. One study reported isolating the organism in nasopharyngeal cultures of 1% of healthy individuals (Henriksen, 1969). *K. denitrificans* and *K. indologenes* have been isolated less frequently in upper respiratory tract specimens.

Method of Transmission

Infection of tissue with *Kingella* results from bacteremia following dental manipulations or oral disease.

Description of Disease

Spectrum of Infection

Kingella isolates have been associated with endocarditis, bacteremia, osteomyelitis, septic arthritis, and intervertebral disk infections. Endocarditis is usually associated with underlying valvular heart disease.

Clinical Manifestations

In contrast with *Cardiobacterium* and *Actinobacillus*, endocarditis caused by *K. kingae* occurs primarily in children and young adults (median age 21 years; half between 4 and 30) and has a rapid onset (Odum et al., 1984). Other infections with *K. kingae* are also reported commonly in children with 70% of the strains isolated at the CDC from patients <16 years old. This predilection for disease in the young is not observed with the other *Kingella* species. Four cases of endocarditis caused by *K. denitrificans* have been reported—all in adult patients, and three of the four patients with prosthetic heart valves. *K. indologenes* has been isolated from patients with infections of the eye.

Treatment and Prognosis

Treatment

K. kingae is highly susceptible to penicillin, ampicillin, carbenicillin, cloxacillin, oxacillin, cephalothin, aminoglycosides, erythromycin, tetracycline, chloramphenicol, and sulfonamides (Bosworth, 1983). Resistance to clindamycin and vancomycin has been reported. Treatment of endocarditis for up to 8 weeks with penicillin alone or in combination with an aminoglycoside has been recommended with successful therapy reported for as short a period as 1 week. *K. denitrificans* is reported to be moderately resistant to penicillin (Khan et al., 1986). Empiric therapy with ampicillin (12 g/day, i.v.) is recommended until susceptibility results are available. Susceptibility data for *K. indologenes* are not available.

Prognosis

Virtually all patients survive with the prompt initiation of therapy and surgical intervention when indicated.

Etiologic Agents

Characteristics

Gram-negative coccobacillus arranged in pairs and short chains. Bacteria have a tendency to retain the primary stain, particularly from young cultures. Colonies grow in the presence or absence of 5% CO_2 but grow poorly in an anaerobic atmosphere. Colony size after incubation for 48 h is 0.5 to 1 mm. Two types of colonial morphology have been described: a spreading, pitting type, and a smooth, convex form. *Kingella* fails to grow on unsupplemented peptone media, and growth is only marginally improved on nutrient agar by the addition of blood or serum. X and V factors are not required. Most isolates fail to grow on MacConkey agar, but *K. denitrificans*, originally named TM-1, will grow on Thayer-Martin medium and can be mistakenly identified as *Neisseria gonorrhoeae*.

Identification

The genus *Kingella* is oxidase-positive, catalase-negative, and nonmotile. *K. kingae* is differentiated from *K. denitrificans* by its beta-hemolysis on sheep or horse blood agar, fermentation of maltose, and absence of nitrate reductase activity. *K. indologenes* is separated from the other two species by its production of indole. The biochemical tests important for the identification of *K. kingae* are summarized in Table 1.

Capnocytophaga

Epidemiology

Natural Habitat

Three species of *Capnocytophaga* have been described: *C. ochracea* (formerly *Bacteroides ochraceus*), *C. gingivalis*, and *C. sputigena*. *Capnocytophaga* species colonize the subgingival sulcus and oral cavity of healthy adults (Moore et al., 1983).

Method of Transmission

Infection occurs following spread of the organism from the oral cavity into the bloodstream.

Description of Disease

Spectrum of Infection

Capnocytophaga has been associated with juvenile periodontitis, bacteremia, septic arthritis, osteomyelitis, empyema, mediastinal abscess, lung abscess, subphrenic abscess, cervical lymphadenitis, sinusitis, conjunctivitis, and keratitis.

Clinical Manifestations

Most infections occur in patients with hematologic malignancy and profound neutropenia (Hawkey et al., 1984). Oral mucosal ulceration is usually present and is presumably the portal of entry into the bloodstream. Nonimmunocompromised patients are generally older and have an infection that has originated from either the oropharynx or the gastrointestinal tract (Parenti and Snydman, 1985).

Treatment and Prognosis

Treatment

Capnocytophaga is susceptible to penicillin, ampicillin, carbenicillin. chloramphenicol, erythromycin, tetracycline, clindamycin, and metronidazole. Isolates are uniformly resistant to oxacillin, vancomycin, aminoglycosides, and polymyxin (Hawkey et al., 1984).

Prognosis

Survival is excellent with appropriate therapy. Although the incidence of death is higher in infected immunocompromised patients, death is usually caused by the patient's underlying disease.

Etiologic Agent

Characteristics

Capnocytophagia are slender, fusiform, gram-negative bacilli that grow in an anaerobic or CO_2-enriched atmosphere but not in air. Some strains will not grow initially in a 5% CO_2 atmosphere (the group was originally classified as an anaerobe). Growth is usually slow with 1- to 2-mm colonies observed on chocolate agar after incubation for 48 h. Pitting of the agar may be seen. Smaller colonies are seen on blood agar. Colonies are flat and thin and have entire edges. Yellow pigmentation is usually noted, as is a sourish, bitter-almond odor. The genus belongs to the class *Flexibacteriae*, which is characterized by their gliding motility on solid media (best seen on tryptic soy agar with blood).

Identification

Negative oxidase and catalase tests separate *Capnocytophaga* from *Actinobacillus*, CDC group DF-2, *Eikenella*, and *H. aphrophilus*. Fermentation of carbohydrates is also helpful for separating *Capnocytophaga* from *Cardiobacterium* and CDC group DF-2; however, reactivity with carbohydrates is generally weak. Separation of the individual species is difficult and not generally performed (Kristiansen et al., 1984). Biochemical reactivity of *Capnocytophaga* is summarized in Table 1.

CDC Group DF-2

Epidemiology

Natural Habitat

DF-2 is part of the oropharyngeal flora of healthy dogs.

Method of Transmission

Disease is associated with dog bites or other exposure to canine secretions (Rubin, 1984).

Description of Disease

Spectrum of Infection

Diseases caused by DF-2 include septicemia, necrotizing wound infections, endocarditis, cellulitis, meningitis, brain abscess, glomerulonephritis, and blepharitis.

Clinical Manifestations

DF-2 infections have been reported in 33 patients between the age of 14 and 77 years (median 50 years) with a preponderance of infections in males (3:1) and in splenectomized patients (47%). Alcoholism was a significant contributing factor in one-third of the patients described in the literature. Infections primarily occur during the summer and fall months. Two patterns of disease have been described: a fulminant form of septic shock with disseminated intravascular coagulation in splenectomized patients, and a milder form of disease in patients with an intact spleen (Kalb et al., 1985).

Treatment and Prognosis

Treatment

Susceptibility tests are difficult to perform because the organism grows poorly. Limited studies indicate that isolates are uniformly susceptible to penicillin, ampicillin, cephalothin, tetracycline, chloramphenicol, erythromycin, clindamycin, and carbenicillin. Resistance is noted to all aminoglycosides, colistin, vancomycin, and methicillin, and varible susceptibility has been observed to trimethoprim-sulfamethoxazole (Rubin, 1984). Penicillin is considered the therapeutic drug of choice (Butler et al., 1977).

Prognosis

Survival has been reported for 25 (76%) of 33 infected patients.

Etiologic Agent

Characteristics

Fermentative bacteria that grow poorly on agar media (dysgonic fermenters) have been classified into biochemically related groups until additional information is available to permit proper genus classification. The first group of organisms, CDC group DF-1, has now been classified in the genus *Capnocytophaga*; while a second group is currently named CDC group DF-2. DF-2 are long, thin, gram-negative bacilli that grow slowly on chocolate agar supplemented with cysteine or on heart infusion agar with 5% rabbit or sheep blood. The organism grows poorly on tryptic soy agar with blood and fails to grow on MacConkey agar. Initial growth is characterized by pinpoint purplish colonies with a convex, smooth, circular morphology when the media are incubated in a humid environment supplemented with 5% CO_2. Growth is poor in an anaerobic atmosphere. Growth of DF-2 in blood culture broths is influenced by the magnitude of bacteremia. The organism will be isolated within 1 to 3 days from patients with fulminant infections, incubation for more than a week is required for less severe infections.

Identification

DF-2 is an oxidase-positive (weak), catalase-positive, nonmotile, fermentative organism. Definitive identification tests are summarized in Table 1. Growth in the biochemical tests may be poor even with supplementation with serum. Direct measurement of constitutive enzyme activity by use of a heavy inoculum has been used successfully to circumvent this problem (Heltberg et al., 1984).

Pasteurella

Epidemiology

Natural Habitat

Six species of *Pasteurella* are currently listed in the *Bergey's Manual of Systematic Bacteriology*: *P. multocida*, *P. haemolytica*, *P. pneumotropica*, *P. ureae*, *P. aerogenes*, and *P. gallinarum* (Holt and Krieg, 1984). *P. multocida* is the most common human pathogen; the other species are rarely associated with human disease. The natural reservoir of *P. multocida* is the upper respiratory tract of healthy animals such as dogs, cats, rabbits, swine, and birds. Carriage is more common in cats (52 to 70%) than in dogs (12 to 54%) (Stein et al., 1983). *P. multocida* is primarily an animal pathogen with death due to septicemia common—hence, the source of the name.

Method of Transmission

Infection is associated with animal contact such as a cat scratch or animal bite. When an animal exposure can be documented, approximately 75% of infections are caused by cats (Francis et al., 1975). However, no history of animal exposure is documented in 15% of infected patients.

Description of Disease

Spectrum of Infection

P. multocida is associated with three general types of disease: (1) localized infection following an animal bite or scratch, with cellulitis, lymphadenitis, and purulence usually present; (2) chronic respiratory disease in elderly patients with underlying pulmonary disease; and (3) systemic infection in patients with impaired resistance. The latter category of infections includes meningitis, brain abscess, sinusitis, otitis, endocarditis, and renal abscess. These frequently occur in patients with underlying hepatic disease (e.g., cirrhosis). Sixty percent of nontraumatic *P. multocida* disease is restricted to the pulmonary tract, and two-thirds of these infections occur in patients with chronic pulmonary disease (Hubbert and Rosen, 1970).

Clinical Manifestations

The median time from injury to localized manifestations is 1 day, with a presentation of inflammation, pain, swelling at the wound site, drainage, and regional lymphadenopathy. The clinical presentation of respiratory disease or systemic infection reflects the site of tissue involvement. Infections are equally common for females and males, and no seasonal incidence has been noted.

Treatment and Prognosis

Treatment

P. multocida is susceptible to penicillin, ampicillin, carbenicillin, cephalothin, chloramphenicol, and tetracycline. Oxacillin, clindamycin, erythromycin, and aminoglycosides have relatively low activity. Penicillin is the antibiotic of choice, with tetracycline or a cephalosporin an acceptable alternative (Stevens et al., 1979).

Prognosis

Rapid improvement is observed with the institution of appropriate antibiotic therapy and surgical debridement when indicated.

Etiologic Agent

Characteristics

Pasteurella isolates are small, nonmotile, gram-negative coccobacilli that frequently exhibit bipolar staining. *P. multocida* is a facultative anaerobe and will grow on many nonselective media including sheep blood and chocolate agars, but not on MacConkey agar. Colonies are smooth, convex, butyrous, and 1 to 2 mm after 24 h of incubation. A green discoloration but no hemolysis will be seen on blood agar media, and a characteristic musty odor due to indole production is usually detected.

Identification

P. multocida has positive reactions with catalase, oxidase, and indole production. Other biochemical reactions that can be used to separate this species from other related bacteria are summarized in Table 1.

Literature Cited

Bosworth, D. E. 1983. *Kingella* (*Moraxella*) *kingae* infections in children. Am. J. Dis. Child. **137**:650–653.

Butler, T., R. E. Weaver, T. K. Venkata Ramani, C. T. Uyeda, R. A. Bobo, J. S. Ryu, and R. B. Kohler. 1977. Unidentified gram-negative rod infection: a new disease of man. Ann. Intern. Med. **86**:1–5.

Ellner, J. J., M. S. Rosenthal, P. I. Lerner, and M. C. McHenry. 1979. Infective endocarditis caused by slow-growing, fastidious, gram-negative bacteria. Medicine (Baltimore) **58**:145–158.

Francis, D. P., M. A. Holmes, and G. Brandon. 1975. *Pasteurella multocida:* infections after domestic animal bites and scratches. J. Am. Med. Assoc. **233**:42–45.

Geraci, J. E., P. R. Greipp, C. J. Wilkowske, W. R. Wilson, and J. A. Washington II. 1978. *Cardiobacterium hominis* endocarditis: four cases with clinical and laboratory observations. Mayo Clin. Proc. **53**:49–53.

Geraci, J. E., W. R. Wilson, and J. A. Washington II. 1980. Infective endocarditis caused by *Actinobacillus actinomycetemcomitans:* report of four cases. Mayo Clin. Proc. **55**:415–419.

Hawkey, P. M., H. Malnick, S. Glover, N. Cook, and J. A. Watts. 1984. *Capnocytophaga ochracea* infections: two cases and a review of the published work. J. Clin. Pathol. **37**:1066–1070.

Heltberg, O., H. E. Busk, A. Bremmelgaard, J. E. Kristiansen, and W. Frederiksen. 1984. The cultivation and rapid enzyme identification of DF-2. Eur. J. Clin. Microbiol. **3**:241–243.

Henriksen, S. D. 1969. Corroding bacteria from the respiratory tract. *Moraxella kingii.* Acta Pathol. Microbiol. Scand. **75**:85–90.

Holt, J. G., and N. R. Krieg. 1984. Bergey's manual of systematic bacteriology, vol. 1. The Williams and Wilkins Co., Baltimore.

Hubbert, W. T., and M. N. Rosen. 1970. *Pasteurella multocida* infections in man unrelated to animal bites. Am. J. Public Health **60**:1109–1117.

Kalb, R., M. H. Kaplan, M. J. Tenebaum, G. R. Joachim, and S. Samuels. 1985. Cutaneous infection at dog bite wounds associated with fulminant DF-2 septicemia. Am. J. Med. **78**:687–690.

Khan, J. A., S. Sharp, K. R. Mann, and J. Brewer. 1986. Case report. *Kingella denitrificans* prosthetic endocarditis. Am. J. Med. Sci. **291**:187–189.

Kristiansen, J. E., A. Bremmelgaard, H. E. Busk, O. Heltberg, W. Frederiksen, and T. Justesen. 1984. Rapid identification of *Capnocytophaga* isolated from septicemia patients. Eur. J. Clin. Microbiol. **3**:236–240.

Moore, W. E. C., L. V. Holdeman, E. P. Cato, R. M. Smibert, J. A. Burmeister, and R. R. Ranney. 1983. Bacteriology of moderate (chronic) periodontitis in mature adult humans. Infect. Immun. **42**:510–515.

Odum, L., K. T. Jensen, and T. D. Slotsbjerg. 1984. Endocarditis due to *Kingella kingae*. Eur. J. Clin. Microbiol. **3**:263–266.

Parenti, D. M., and D. R. Snydman. 1985. *Capnocytophaga* species: infections in nonimmunocompromised and immunocompromised hosts. J. Infect. Dis. **151**:140–147.

Pierce, C. S., W. R. Bartholomew, D. Armsterdam, E. Neter, and J. J. Zambon. 1984. Endocarditis due to *Actinobacillus actinomycetemcomitans* serotype c and patient immune response. J. Infect. Dis. **149**:479.

Robison, W. J., and A. S. Vitelli. 1985. Infectious endocarditis caused by *Cardiobacterium hominis*. South Med. J. **78**:1020–1021.

Rubin, S. J. 1984. DF-2: a fastidious fermentative gram-negative rod. Eur. J. Clin. Microbiol. **3**:253–257.

Savage, D. D., R. L. Kagan, N. A. Young, and A. E. Horvath. 1977. *Cardiobacterium hominis* endocarditis: description of two patients and characterization of the organism. J. Clin. Microbiol. **5**:75–80.

Siegman-Igra, Y., D. Schwartz, O. Ophir, and N. Konforti. 1984. Endocarditis caused by *Actinobacillus actinomycetemcomitans*. Eur. J. Clin. Microbiol. **3**:556–559.

Slotnick, I. J., J. A. Mertz, and M. Dougherty. 1964. Fluorescent antibody detection of human occurrence of an unclassified bacterial group causing endocarditis. J. Infect. Dis. **114**:503–505.

Slots, J. 1982. Selective medium for isolation of *Actinobacillus actinomycetemcomitans*. J. Clin. Microbiol. **15**:606–609.

Slots, J., R. T. Evans, P. M. Lobbins, and R. J. Genco. 1980. In vitro antimicrobial susceptibility of *Actinobacillus actinomycetemcomitans*. Antimicrob. Agents Chemother. **18**:9–12.

Stein, A. A., M. A. Fialk, A. Blevins, and D. Armstrong. 1983. *Pasteurella multocida* septicemia: experience at a cancer hospital. J. Am. Med. Assoc. **249**:508–509.

Stevens, D. L., J. W. Higbee, T. R. Oberhofer, and E. D. Everett. 1979. Antibiotic susceptibilities of human isolates of *Pasteurella multocida*. Antimicrob. Agents Chemother. **16**:322–324.

Weaver, R. E., D. G. Hollis, and E. J. Bottone. 1985. Gram-negative fermentative bacteria and *Francisella tularensis*. *In* E. H. Lennette, A. Balows, W. J. Hausler, Jr., and H. J. Shadomy (ed.), Manual of clinical microbiology 4th ed. American Society for Microbiology, Washington, D.C.

Wormser, G. P., and E. J. Bottone. 1983. *Cardiobacterium*

hominis: review of microbiologic and clinical features. Rev. Infect. Dis. **5:**680–691.

Yogev, R., D. Shulman, S. T. Shulman, and W. G. Glogowski. 1986. In vitro activity of antibiotics alone and in combination against *Actinobacillus actinomycetemcomitans*. Antimicrob. Agents Chemother. **29:**179–181.

Zombon, J. J., L. A. Christersson, and J. Slots. 1983. *Actinobacillus actinomycetemcomitans* in human periodonatal disease: prevalence in patient groups and distribution of biotypes and serotypes within families. J. Periodontol. **54:**707–711.

CHAPTER 30

Group A Streptococcal Infections

MICHAEL A. GERBER

Diseases: Pharyngitis, impetigo.
Etiologic Agents: Group A streptococci.
Source: Humans, occasionally food or milk. Direct contact, inhalation, or ingestion.
Clinical Manifestations: Pharyngitis: acute onset of fever, sore throat, headache, nausea, vomiting, and abdominal pain associated with pharyngeal injection and exudate, cervical lymphadenitis, palatal petechiae, and rash. Impetigo: discrete skin lesions with a thick, purulent, golden crust on top of an erythematous base associated with pruritis and regional adenopathy without systemic symptoms.
Laboratory Diagnosis: Bacteriologic isolation of etiologic agent from throat or skin lesion or identification of group A carbohydrate antigen with antigen detection test. Significant antibody titer rise to extracellular enzymes.
Epidemiology: Pharyngitis primarily a disease of young, school-age children; more common in temperate climates; peaks in late winter and early spring. Impetigo primarily a disease of pre-school-age children; more common in warm, moist environments; peaks late summer and early fall. Both occur primarily as endemic disease, but can occur in epidemics.
Treatment: Benzathine penicillin G, oral penicillin V; erythromycin for penicillin-allergic.
Prevention and Control: Identify and treat patients with disease as quickly as possible to reduce the risk of transmission to others. Prophylactic benzathine penicillin G during epidemic of impetigo.

Description of Disease

The most common group A streptococcal infections are acute pharyngitis and impetigo. These infections are significant not only because of the acute morbidity associated with them but also because they may be followed by nonsuppurative sequelae.

Clinical Manifestations

Acute Pharyngitis

Group A streptococci are an important cause of acute pharyngitis, particularly in young children. Streptococcal pharyngitis is characterized by the sudden onset of fever, sore throat, headache, nausea, vomiting, and abdominal pain. On examination, the pharynx and tonsillar areas are erythematous, and a whitish exudate may be present on the tonsils. The cervical lymph nodes are usually enlarged and tender. Petechiae of the soft palate may also be present. In constrast to other types of pharyngitis, streptococcal pharyngitis is not usually associated with cough, hoarseness, coryza, conjunctivitis, or diarrhea. If the strain of group A streptococci that is producing the pharyngitis is lysogenized by a bacteriophage capable of inducing the production of erythrogenic toxin, scarlet fever will occur. The rash of scarlet fever typically appears on the second day of the illness and consists of a diffuse, erythematous, punctiform eruption that gives the skin a sandpaperlike texture. The distribution of the rash is variable, but the trunk and inner aspects of the arms and thighs are often most involved. The face is flushed and red, and there is a pale area around the mouth. The erythema usually disappears by the 6th to 9th day of the illness and is followed by desquamation of the face and body and later by desquamation of the palms and soles. The enanthem of scarlet fever includes a tongue that may be bright red with enlarged papillae or a tongue that has red papillae protruding from beneath a white coating.

Streptoocccal pharyngitis, either with or without a rash, is usually a benign, self-limited disease. Untreated, the patient will be afebrile and asymptomatic in 4 or 5 days; however, treatment with appropriate antibiotics, particularly early in the course of the illness, can dramatically shorten the clinical course (Randolph et al., 1985). The suppurative complications of streptococcal pharyngitis, which include cervical lymphadenitis, otitis media, sinusitis, retropharyngeal abscess, and peritonsillar abscess, are seen almost exclusively in untreated patients.

Impetigo

Impetigo, also known as streptococcal pyoderma, is a primary, superficial infection of the skin. Impetigo begins as tiny papules that rapidly develop into vesicles and then pustules with little or no surrounding erythema. These vesicles and pustules are short-lived and are infrequently seen. They quickly rupture, leaving a thin, amber crust associated with slight pruritis and regional adenopathy without systemic symptoms. In several days, these lesions develop a thick, purulent, golden crust on top of an erythematous base. The lesions gradually extend in size and increase in number, although central clearing and healing of individual lesions are sometimes seen. Impetigo occurs most frequently on the legs with involvement of the arms, face, and trunk, seen with decreasing frequency. Except in epidemic situations, survey cultures of normal skin infrequently yield group A streptococci, and experimental inoculation of normal skin with this organism does not usually result in clinical disease or invasion of the skin. During outbreaks of impetigo, group A streptococci appear on normal skin several days before the lesions of impetigo develop. After an insult to the integrity of the skin, such as an insect bite, an abrasion, or even undetectable trauma, skin lesions may appear. Colonization of the respiratory tract with group A streptococci almost always occurs after the development of skin lesions. However, this respiratory tract colonization is transient and does not usually result in clinical disease (Ferrieri et al., 1972).

Untreated impetigo is a mild, usually chronic but occasionally self-limited disease. However, untreated impetigo may lead to local extension of the lesions, spread to distant areas of the body, and transmission of the group A streptococci to others. Antibiotic treatment can minimize the spread to others and prevent local extension and spread to other areas of the skin.

Nonsuppurative Sequelae

Although the pathogenesis of rheumatic fever and poststreptococcal glomerulonephritis is still poorly understood, both are well-recognized, nonsuppurative sequelae of group A streptococcal infections. Glomerulonephritis can occur following either streptococcal pharyngitis or impetigo, usually after a latent period of approximately 1 to 2 weeks for the former and 3 weeks for the latter. Rheumatic fever can occur approximately 2 to 4 weeks after an episode of streptococcal pharyngitis, but it is not a complication of impetigo. Appropriate antibiotic therapy of streptococcal pharyngitis can prevent the subsequent occurrence of rheumatic fever. However, antibiotic therapy of either streptococcal pharyngitis or impetigo has not been shown to be effective in preventing the subsequent occurrence of glomerulonephritis. The attack rate of acute rheumatic fever in the United States following an untreated episode of streptococcal pharyngitis had, in the past, been reported to be between 0.3 and 3%. Although the current attack rate in the United States is unknown, the rapidly declining incidence of rheumatic fever in this country would suggest that the current attack rate is much lower. The attack rate of glomerulonephritis varies with the prevalence of nephritogenic strains in the community. In outbreaks of pharyngitis or impetigo caused by nephritogenic strains of group A streptococci, the incidence of glomerulonephritis may be as high as 10 to 15%.

Other Infections

Group A streptococci can cause other skin and soft-tissue infections besides impetigo, including cellulitis, erysipelas, lymphangitis, abscesses, paronychia, and blistering distal dactylitis. Vaginitis in prepubertal girls and perianal cellulitis have also been reported to be caused by this organism. In addition, group A streptococci can cause pneumonia, sepsis, meningitis, osteomyelitis, and acute necrotizing fasciitis (Breese and Hall, 1978).

Immunologic Aspects

A true streptococcal infection is defined as a positive culture associated with evidence of an immunologic response to this organism; streptococcal carriage is defined as a positive culture with no evidence of an immunologic response (Kaplan, 1980). An immunologic response to the group A streptococci is usually demonstrated by measurement of the antibodies directed against the extracellular antigens: [antistreptolysin O (ASO), antideoxyribonuclease B (ADB), and antihyaluronidase (AH)]. Three weeks after the onset of streptococcal pharyngitis, approximately 80% of untreated children will have a significant increase in either ASO or ADB titer. If antibodies to multiple antigens are measured, over 95% of children with streptococcal pharyngitis will demonstrate a serologic response. Following an episode of impetigo, there is often only a feeble ASO response, and other antibody tests, such as ADB or AH, may be more sensitive in this situation (Peter and Smith, 1977).

Antibodies to group A carbohydrate can be detected approximately 3 weeks after uncomplicated streptococcal pharyngitis. Antigenic cross-reactivity between the group A carbohydrate antigen and a structural glycoprotein in human heart valves has been demonstrated; however, the implications of this finding with respect to the pathogenesis of rheumatic heart disease have yet to be determined. Antibodies to M proteins can be detected approximately 30 to 60 days after the onset of infection, and their presence correlates with type-specific immunity. Antigenic cross-reactivity between an M protein–related moiety and human myocardial cells has also been demonstrated; however, the implications of this finding with respect to the pathogenesis of rheumatic heart disease have yet to be determined. Since there are more than 70 different M types and serologic cross-reactivity between them is minimal, multiple streptococcal infections can occur in the same individual. Antibiotic therapy, particularly if given early in the illness, aborts the anti-M response and suppresses the magnitude of the ASO, ADB, and AH response.

Diagnostic Features

Group A streptococcal pharyngitis cannot be distinguished from pharyngitis caused by other agents on the basis of clinical findings alone. Therefore, an accurate diagnosis of streptococcal pharyngitis depends on some form of bacteriologic confirmation, either with a throat culture or with an antigen detection test. However, there are certain signs and symptoms that, if present, make a streptococcal etiology more likely, whereas the presence of other signs and symptoms, makes a nonstreptococcal etiology more likely (Wannamaker, 1984). There are also certain epidemiologic factors, such as age, season, and prevalence of agents in the community, that when taken into consideration, are also suggestive of either a streptococcal or nonstreptococcal etiology for the pharyngitis. For example, streptococcal pharyngitis is relatively unusual in children less than 3 or more than 16 years of age, and in temperate climates it is primarily a disease of the winter and early spring. Group C and group G streptococci have been shown to be responsible for outbreaks of acute pharyngitis that are clinically indistinguishable from pharyngitis caused by group A streptococci. The role of group C and group G streptococci in sporatic cases of acute pharyngitis has yet to be determined. However, the nonsuppurative sequelae of acute rheumatic fever and acute glomerulonephritis do not occur after either group C or group G streptococcal infections.

It has been demonstrated that only about half of the patients with symptoms of pharyngitis and a positive throat culture for group A streptococci have an immunologic response to this organism and are truly infected. The other half are merely streptococcal carriers with a concurrent nonstreptococcal pharyngitis (Kaplan et al., 1971). There are currently no reliable clinical or laboratory methods for distinguishing patients with streptococcal infections from patients who are streptococcal carriers, except to review serologic data retrospectively.

Although bacteriologic investigations of children with impetigo have demonstrated that many of these lesions will grow *Staphylococcus aureus* in addition to group A streptococci and that some of these lesions will grow *S. aureus* alone, observations suggest that the group A streptococcus is actually causing the impetigo and that the *S. aureus* is merely a secondary invader (Dajani et al., 1972). Bullous impetigo is a particular type of impetigo, characterized by large, evanescent bullae that, when ruptured, leave a red base with a varnishlike, thin crust. Bullous impetigo, in contrast to impetigo, is caused by *S. aureus*.

Epidemiologic Aspects

Some strains of group A streptococci preferentially produce pharyngitis, some impetigo, some both, and a few neither. The group A streptococci that cause pharyngitis are spread from one individual to another by large airborne droplets. Therefore, patients with streptococcal pharyngitis frequently have a history of close contact with a well-documented case of streptococcal infection. Significant extrahuman or animal reservoirs of these organisms do not exist, except for occasional outbreaks due to the contamination of food or milk. Group A streptococci dissociate rapidly outside the human host. Organisms recovered from clothing, bedding, or house dust can be identified as group A but are noninfectious when inoculated into the throats of volunteers. Almost 97% of untreated patients will carry group A streptococci in their pharynx for 4 months or more. In addition, as many as 20% of asymptomatic school-age children may carry group A streptococci in their throats, and during epidemics, the carriage rate can be considerably higher. In temperate climates, streptococcal pharyngitis is a more common problem and demonstrates more of a seasonal variation in incidence than in warmer climates. In temperate climates, the incidence of streptococcal pharyngitis and asymptomatic streptococcal carriage peak in the late winter and early spring. Group A streptococcal pharyngitis is primarily a disease of young, school-age children and is relatively unusual in patients either <3 years or >16 years old.

Impetigo is primarily an endemic disease of preschool-age children, but it can occur in epidemics. Impetigo is more common in warm, moist environments than in cool, dry ones. In temperate climates, epidemics of impetigo tend to occur in the late summer and early fall. In warmer climates, seasonal epidemics, similar to those seen in temperate climates, do occur; but they are superimposed upon a continuous endemic rate. The spread of skin streptococci is usually by direct contact and is enhanced by crowding and poor living conditions. Therefore, patients with impetigo frequently have a history of close contact with a well-documented case of streptococcal infection. Household pets and indigenous biting insects rarely carry group A streptococci. People living in equivalent conditions who practice good hygiene have a lower incidence of impetigo than those whose hygiene is poor. Trauma in the form of an insect bite, abrasion, or minor burn also plays an important role in the pathogenesis of impetigo.

Etiologic Agents

Characterization

Streptococci are gram-positive, facultative anaerobic cocci that grow in chains and survive relative extremes of temperature and humidity. They can be classified using two different systems. The first is the Lancefield system, which identifies streptococci as belonging to group A through group L on the basis of a serologic identification of the carbohydrate antigen in the cell wall. Most of the streptococci that cause acute pharyngitis and impetigo belong to group A. The second classification system is based on the type of hemolysis produced by streptococci on blood agar plates. Alpha hemolysis, or partial hemolysis, leaves a zone of green or brownish-green color around the colony on the blood agar plate; beta hemolysis, or complete hemolysis, leaves a clear zone; and gamma hemolysis, or no hemolysis, leaves no zone. About 96% of group A streptococci are beta-hemolytic, but streptococci from groups B, G, F, and D can also produce beta hemolysis.

The group A streptococcus is surrounded by a nonantigenic hyaluronic acid capsule that is responsible for the smooth, shiny appearance of the colonies on blood agar plates. This capsule appears to be a virulence factor that helps the organism resist phagocytosis. The cell wall of the group A streptococci consists of three major components: protein, carbohydrate, and peptidoglycan. The major protein fraction consists of two antigenically distinct proteins, termed M and T, which are used to serotype group A streptococci. The M proteins, which appear as roughened fimbriae, extending from the cell wall as seen on electron microscopy, are a virulence factor that helps the organism resist phagocytosis. The T proteins do not appear to be a virulence factor. The middle layer of the cell wall consists of the group-specific carbohydrate of group A streptococci, which consists of a polymer of n-acetylglucosamine and rhamnose. The inner layer of the cell wall, or peptidoglycan, gives the organism its shape and rigidity and has the same basic structure as in all gram-positive cocci.

The group A streptococci secretes at least 20 distinct extracellular products, including enzymes (streptolysins, hyaluronidase, streptokinase, DNAses, NADase) and toxins (erythrogenic toxins). The streptolysins are responsible for the beta hemolysis observed on blood agar plates. Streptolysin O is the oxygen-labile enzyme that produces subsurface hemolysis; streptolysin S is the oxygen-stabile enzyme that produces surface hemolysis. It has been suggested that streptolysin O has a direct cardiotoxic effect and is involved in the pathogensis of rheumatic heart disease; however, this has never been proved. A specific role for hyaluronidase in the pathogenesis of tissue lesions has not been defined, but it is believed to assist in the spread of streptococci through connective tissue rich in hyaluronic acid or in producing local damage to the ground substance of such tissues. Streptokinase promotes fibrinolytic activity by converting plasminogen to plasmin and probably enhances the spread of streptococci through connective tissue. Four antigenically distinct deoxyribonucleases are produced by group A streptococci, with DNAse B being produced in the largest amounts by most of the strains causing human infections. The pathologic significance of these deoxyribonucleases, as well as that of NADase, is as yet unknown. Three immunologically distinct, rash-producing erythrogenic toxins have also been identified. Erythrogenic toxin production is mediated by lysogeny, but the manner in which the toxin operates at the cellular level is unclear.

Specimen Collection and Transport

The composition (cotton, Dacron, rayon) of the swab used to obtain a culture for group A streptococci is not nearly as important as the manner in which that culture is obtained. For a throat culture, the tongue should be depressed and the throat adequately illuminated while the swab is vigorously passed over the posterior pharynx and both tonsils (or tonsillar fossae). If any exudate is present, it should be swabbed. The swab should not come in contact with the tongue, buccal mucosa, or lips. For a skin culture, the swab should be moistened if the lesion itself is dry. If vesicles are present, they should be punctured and the fluid collected on the swab. If scabs are present, they should be lifted and the underlying surfaces swabbed. The throat or skin swab can be put directly onto a blood agar plate or held for up to 2 h in a paper envelope or dry, sterile test tube. If the delay before plating is going to be longer than 2 h, the swab should be put into a holding media (Stuart's or Amie's). If the delay before plating is going to be longer than a day, the swab should be put into an aluminum foil packet with a drying agent such as silica gel. Group A streptococci survive well for even weeks on a dried swab, and, in fact, this procedure may facilitate the isolation of streptococci, because other organisms in the pharynx survive drying less well.

Questions have been raised about the reliability of a single throat culture. However, review of existing data suggests that only about one patient in 10 with group A streptococci in the upper respiratory tract would be missed if a single throat culture were performed. In addition, most of the patients shown to have group A streptococci on duplicate throat cultures that would have been missed if only a single throat culture had been performed have weakly positive cultures, suggesting that most of them may be carriers and are not truly infected (Gerber, 1984).

Preliminary Identification

The streptococcal cultures should be performed on enriched infusion agar with defibrinated blood (Facklam and Carey, 1985). Some believe that the isolation rate of group A streptococci can be increased with the use of selective media. However, the clinical relevance of this increased yield has not been determined and could merely reflect an uncovering of streptococcal carriers for whom treatment is not required. The agar should be free of reducing sugars, as these may interfere with the expression of hemolysis. Sheep blood is preferred, because it will inhibit the growth of *Haemophilus haemolyticus* which can be confused with group A streptococci. The agar should be about 4 to 6 mm thick, and the concentration of blood should be approximately 5%. If the agar is either too thick or too thin or the concentration of blood too high or too low, it may be difficult to distinguish the different types of hemolysis.

The swab should be repeatedly streaked with rotation across approximately one-third of the blood agar plate. Secondary and tertiary streaking should be performed with a sterile inoculating loop in order to obtain well-isolated colonies. Several surface stabs should be made at the intersection of the primary and secondary streak with the loop tip and a bacitracin disk placed in this area. A 0.02- to 0.04-unit bacitracin disk, not the 2- to 10-unit disk (for antibiotic sensitivity testing), should be used. The plate should then be incubated at 35 to 37°C for 18 to 24 h. Anaerobic incubation can increase the isolation rate of group A, group C and group G beta-hemolytic streptococci (Schwartz et al., 1985). However, the significance of group C or group G streptococci as a cause of endemic pharyngitis has yet to be determined. In addition, it has been suggested that many of the additional patients with group A streptococci that are identified with anaerobic incubation may be streptococcal carriers and not truly infected. Therefore, it may be difficult to justify the additional cost and effort of anaerobic incubation. Some people prefer to incubate the blood agar plates in a candle extinction jar as a way of enhancing the hemolysis.

After 18 to 24 h, plates should be examined for the presence of smooth, domed, transparent or translucent colonies, approximately 0.5 mm in diameter, surrounded by a zone of beta-hemolysis. Group A streptococci that produce primarily streptolysin O may demonstrate beta hemolysis only in the subsurface stabs where the oxygen tension is reduced. Negative plates should be reincubated for an additional 18 to 24 h and then reexamined. *Staphylococcus aureus* may be present and may also be beta-hemolytic; however, these colonies are larger and are opaque or pigmented, and they are catalase-positive whereas the streptococci are catalase-negative.

Enterobacteriaceae may be present and may be beta-hemolytic; however, these colonies resemble *S. aureus*, are catalase-positive, and are gram-negative on Gram staining. The appearance of beta-hemolytic group C or group G colonies does not differ enough from that of group A colonies to be of any value in identification. However, a presumptive identification of group A streptococci may be provided by the selective inhibition of such strains by bacitracin. Approximately 96% of group A streptococci will be sensitive to bacitracin, whereas about 80% of non–group A streptococci will be resistant to bacitracin. If the number of beta-hemolytic streptococcus colonies in the area of the bacitracin disk is small, or if there are other beta-hemolytic organisms in the area of the bacitracin disk, it may be difficult to determine bacitracin sensitivity on the primary plate. One can then test for bacitracin sensitivity on a secondary plate inoculated with a single colony of beta-hemolytic streptococci.

Estimation of the number of colonies of group A

streptococci cultured from the throat has been advocated as a means of differentiating the carrier state from true infection. A number of studies suggest that patients with true streptococcal pharyngitis are more likely to have strongly positive cultures than are patients who are merely streptococcal carriers. However, there is so much overlap that differentiation of patients with streptococcal infections from patients who are streptococcal carriers cannot be accurately made on the basis of degree of positivity of the throat culture alone (Gerber, 1984).

Definitive Identification—Grouping, Serotyping

Definitive grouping of a beta-hemolytic streptococci can be performed in a number of different ways, all of which ultimately depend on identification of the group-specific carbohydrate in the cell wall. Isolated colonies on a blood agar plate, or a sample from a 4- to 6-h broth culture, can be identified as group A by immunofluorescence staining using a fluorescein-conjugated antibody to the group A carbohydrate. There are also a number of grouping procedures which consist of the extraction of the group A carbohydrate antigen with either hot hydrochloric acid, hot formamide, autoclaving, nitrous acid, *Streptomyces albus* enzyme, pronase B enzyme, or *S. albus* lysozyme followed by a capillary precipitin test with group-specific streotococcal antisera. Recently, several commercial slide agglutination tests, which use group-specific streptococcal antisera, conjugated to either latex particles or protein A–rich staphylococcal cells, to identify group A carbohydrate antigen extracted from colonies on blood agar plates with a micronitrous acid procedure. Commercial enzyme immunoassay procedures for the identification of group A carbohydrate antigen extracted from colonies on blood agar plates with a micronitrous acid procedure have also been developed.

Group A streptococci can be serotyped with respect to the M and T proteins by means of a precipitin reaction between M extract and an agglutination reaction between T extract with type-specific antiserum. The ability to produce opacity in horse serum is a characteristic of certain M types of group A streptococci. All members of a particular M type are either serum opacity reaction (SOR)-positive or SOR-negative. The SOR can, therefore, be used in conjunction with the M and T typing in studying epidemiologic patterns of human streptococcal disease (Maxted and Widdowson, 1972).

Antigen Detection Tests

Several commercial antigen detection tests for the identification of group A streptococcal antigen directly from throat swabs are now available (Gerber, 1986). After obtaining a throat swab as one would for the performance of a blood agar culture, group A carbohydrate antigen is either enzymatically or chemically extracted using the materials in the kit. The presence of the group A carbohydrate antigen is then detected using either an enzyme immunoassay, latex agglutination, or coagglutination system, depending on the particular kit being used. Various procedures require from 10 to 60 min to perform. Some of these procedures are simple enough to be used in a physician's office; others are relatively complex and could only be used in a diagnostic microbiology laboratory. In general, the specificity of these antigen detection tests has been very good when compared with blood agar cultures—usually 95% or higher. In contrast. the sensitivity of these tests has varied from 50 to 95% and has depended on the setting in which the test is being performed, on the experience of the person performing the test, and on the exact culture procedure that is used as the standard of comparison. It is conceivable that as more sensitive procedures are developed, these antigen detection tests may replace blood agar plate cultures. Currently, however, because of their relatively low sensitivity, a negative antigen detection test should be confirmed with a blood agar culture. On the other hand, because the specificity of these tests is quite high, one need not perform a blood agar plate culture if the antigen detection test is positive.

Treatment

There are several reasons for attempting to diagnose and treat streptococcal pharyngitis (Gerber and Markowitz, 1985): (1) It is important for the prevention of nonsuppurative sequelae. It has been established that eradication of the streptococci from the upper respiratory tract of a patient with acute pharyngitis will prevent that patient from developing acute rheumatic fever. Although acute rheumatic fever has become a relatively rare disease in this country, there are data that suggest that the dramatic decline in the incidence of this disease is due, at least in part, to efforts to diagnose and treat streptococcal pharyngitis. Antibiotic therapy has not, however, been shown to be effective in preventing the other nonsuppurative sequela of streptococcal pharyngitis—poststreptococcal glomerulonephritis. (2) Antibiotic therapy can prevent the patient from developing suppurative sequelae of streptococcal pharyngitis, such as peritonsilliar abscess, retropharyngeal abscess, and cervical lymphadenitis. These complications were relatively common in the preantibiotic era, but their incidence has declined markedly with the use of antibiotic therapy. (3) Eradication of the infecting streptococci from the patient's upper respiratory tract with antibiotic therapy prevents spread of that organism to other individuals. (4) Antibiotic therapy, particularly if begun early in the disease, can shorten the clinical course of the illness.

The most effective treatment for streptococcal pharyngitis is a single intramuscular injection of benzathine penicillin G. Preparations containing procaine penicillin, in addition to benzathine penicillin G, are associated with decreased pain at the injection site. Alternatively, a 10-day course of oral pencillin V therapy may be given. Antibiotics other than penicillin offer no advantage, and their use should be limited to patients who are allergic to penicillin. Erythromycin is the drug of choice for patients who are allergic to penicillin. Although erythromycin-resistant strains of group A streptococci are prevalent in some areas of the world (Maruyama et al., 1979), they are still rare in the United States. An oral cephalosporin is an acceptable alternative for the patient allergic to pencillin. Tetracyclines and trimethroprim-sulfamethoxazole are ineffective in the treatment of group A streptococcal pharyngitis and should not be used.

Although impetigo is usually a mild and occasionally self-limited disease, antibiotic treatment is indicated in order to prevent local extension of the lesions, spread to other areas of the body, and transmission of the infection to others. However, there is no evidence that antibiotics can prevent the only nonsuppurative sequela of impetigo—acute glomerulonephritis. Intramuscular benzathine penicillin G is the most effective form of therapy (Baltimore, 1985). If compliance is assured, a 7-day course of oral penicillin V (or erythromycin in the patient allergic to pencillin) is an effective alternative. Penicillinase-producing staphylococci that may be present in the streptococcal impetigo are usually commensals and do not interfere with the efficacy of penicillin therapy. Therefore, penicillinase-resistant penicillins offer no advantage over penicillin V. Topical antibiotic therapy of impetigo is not as effective as systemic therapy, and the use of topical antibiotics in a patient who is receiving systemic antibiotics is unnecessary. However, in a patient with a solitary lesion or a few small lesions, topical antibiotics alone may be sufficient therapy. Bacitracin is the topical antibiotic of choice. Although cleanliness and good hygiene are important in preventing streptococcal impetigo, scrubbing with hexachlorophene and removing crusts do not appear to be beneficial and may actually delay healing of the lesions.

Prevention and Control

The most important way to control group A streptococcal infections and their sequelae is to identify patients with this disease as quickly as possible. The prompt initiation of appropriate antibiotic therapy in these patients will then reduce the risk of their transmitting the group A streptococci to others. During epidemics of impetigo, especially in association with poststreptococcal glomerulonephritis, prophylactic benzathine penicillin G should be considered as a means of reducing the number of new cases. Although a considerable amount of effort has gone into the development of a safe, effective group A streptococcal vaccine, there is currently no such vaccine available. Recent identification of epitopes on the M protein molecule that share antigenic determinants with human heart tissue, and the separation of these heartreactive antigens from the peptides of the M protein molecule that induce protective antibody production, may eliminate concerns about the possible deleterious effect of a vaccine (Dale and Beachey, 1985).

Literature Cited

Baltimore, R. S. 1985. Treatment of impetigo: a review. Pediatr. Infect. Dis. **4**:597–601.

Breese, B. B., and C. B. Hall. 1978. Beta hemolytic streptococcal disease, p. 176–220. Houghton-Mifflin, Boston.

Dajani, A. S., P. Ferrieri, and L. W. Wannamaker. 1972. Natural history of impetigo. II. Etiologic agents and bacterial interaction. J. Clin. Invest. **51**:2863–2871.

Dale, J. B., and E. H. Beachey. 1985. Multiple heart cross-reactive epitopes of streptococcal M proteins. J. Exp. Med. **161**:113–122.

Facklam, R. R., and R. B. Carey. 1985. Streptococci and aerococci, p. 154–175. *In* E. H. Lennette, A. Balows, W. J. Hausler, Jr., and H. J. Shadomy (ed.), Manual of clinical microbiology, 4th ed. American Society for Microbiology, Washington, D.C.

Ferrieri, P., A. S. Dajani, L. W. Wannamaker, and S. S. Chapman. 1972. Natural history of impetigo. I. Site sequence of acquisition and familial patterns of spread of cutaneous streptococci. J. Clin. Invest. **51**:2851–2862.

Gerber, M. A. 1984. Diagnosisi of pharyngitis: methodology of throat culture, p. 61–72. *In* S. T. Shulman (ed.), Pharyngitis: management in an era of declining rheumatic fever. Praeger, New York.

Gerber, M. A. 1986. Diagnosis of group A beta-hemolytic streptococcal pharyngitis: use of antigen detection tests. Diagn. Microbiol. Infect. Dis. **4**:5S–15S.

Gerber, M. A., and M. Markowitz. 1985. Management of streptococcal pharyngitis reconsidered. Pediatr. Infect. Dis. **4**:518–526.

Kaplan, E. L. 1980. The group A streptococcal upper respiratory tract carrier state: an enigma. J. Pediatr. **97**:337–345.

Kaplan, E. L., F. H. Top, Jr., B. A. Dudding, and L. W. Wannamaker. 1971. Diagnosis of streptococcal pharyngitis: differentiation of active infection from the carrier state in the symptomatic child. J. Infect. Dis. **123**:490–501.

Maruyama, S., H. Yoshioka, K. Fujita, M. Takimoto, and Y. Satake. 1979. Sensitivity of group A streptococci to antibiotics: prevalence of resistance to erythromycin in Japan. Am. J. Dis. Child. **133**:1143–1145.

Maxted, W. R., and J. P. Widdowson. 1972. The protein antigens of group A streptococci, p. 251–266. *In* L. W. Wannamaker and J. M. Matsen (ed.), Streptococci and streptococcal diseases: recognition, understanding, and management. Academic Press. New York.

Peter, G., and A. L. Smith. 1977. Group A streptococcal infections of the skin and pharynx. N. Engl. J. Med. **297:**311–317.

Randolph, M. F., M. A. Gerber, K. K. DeMeo, and L. Wright. 1985. Effect of antibiotic therapy on the clinical course of streptococcal pharyngitis. J. Pediatr. **106:**870–875.

Schwartz, R. H., M. A. Gerber, and P. McCoy. 1985. Effect of atmosphere of incubation on the isolation of group A streptococci from throat cultures. J. Lab Clin. Med. **106:**88–92.

Wannamaker, L. W. 1984. Diagnosis of pharyngitis: clinical and epidemiologic features, p. 33–46. *In* S. T. Shulman (ed.), Pharyngitis: management in an era of declining rheumatic fever. Praeger, New York.

CHAPTER 31

Haemophilus influenzae Infections

W. L. ALBRITTON

Disease: Pneumonia, meningitis, epiglottitis, septic arthritis, cellulitis, otitis media, conjunctivitis, neonatal infection, and other clinical syndromes.
Etiologic Agents: *Haemophilus influenzae* serotype b; *H. influenzae* serotype a, c, d, e, f; nonserotypable *H. influenzae*.
Source: Human respiratory and genital tract by direct contact or inhalation.
Clinical Manifestations: Usually, acute onset of fever and other symptoms dependent on organ system localization.
Pathology: Acute pyogenic infection following primary bacteremia from respiratory source for serotypable strains; isolated mucosal infection with direct extension for nonserotypable strains with occasional secondary bacteremia.
Laboratory Diagnosis: Culture isolation of etiologic agent from blood, CSF, joint fluid, biopsy material, tissue aspirate, sputum, or mucosal swabs. Presumptive nonculture diagnosis can be made by detection of type b capsular antigen in normally sterile body fluids such as CSF or joint fluid.
Epidemiology: Predominantly a disease of children, with highest asymptomatic carriage rate for serotype b strains in childhood. Colonization with both typable and nontypable strains occurs. Maternal genital tract is usual source of neonatal infections. Otitis media due to nonserotypable strains is one of most common infections in childhood, and systemic infections with serotype b strains are most important bacterial infections in children.
Treatment: Ampicillin, chloramphenicol, and certain third-generation cephalosporins.
Prevention and Control: Eradication of carriage in close contacts of cases with rifampin. Immunization with polyribose phosphate (PRP) provides some protection from infection with serotype b strains.

Description of Disease

Since its original description by Pfeiffer in 1892 in association with respiratory disease, *Haemophilus influenzae* has been found in a number of clinical syndromes involving several organ systems. Encapsulated strains were recognized by Pittman (1931) in association with invasive disease and six serotypes (a–f) have been described based on capsular antigen. Over 95% of all invasive disease has been associated with strains producing the serotype b capsule, although invasive disease due to other capsular serotypes and nonencapsulated strains has been described. Nonencapsulated strains are largely responsible for local mucosal disease of the respiratory tract and contiguous structures such as otitis media, conjunctivitis, genitourinary tract infection, and invasive neonatal disease acquired from the maternal genital tract. Infections due to *H. influenzae* are perhaps the most important bacterial infections of childhood and are associated with significant morbidity and mortality.

Communicability and Transmissibility

Person-to-person transfer by respiratory secretions is by far the most common route of transmission.

Although *H. influenzae*–like strains are isolated from animal sources, there are no data to support zoonotic sources for strains associated with human disease. The maternal genital tract appears to be the source for organisms associated with neonatal infections.

Clinical Manifestations of Disease

The clinical manifestations of infection with *H. influenzae* depend primarily on the presence or absence of organ system localization. Infections due to capsular serotype b strains are more often systemic, and the spectrum of disease is broad with an age-related organ system localization (Dajani et al., 1979; Hirschmann and Everett, 1979; Turk and May, 1967). Strains of capsular serotypes other than b and nonencapsulated strains occasionally cause disease that is indistinguishable clinically from disease due to serotype b strains (Berk et al., 1982; Eng et al., 1980; Greene, 1978). Nonencapsulated strains are most often found in neonatal infections and infections contiguous to mucosal surfaces such as otitis media, conjunctivitis, and genitourinary tract infections (Liston and Foshee, 1982).

Pneumonia

The association of *H. influenzae* with respiratory disease was first made by Pfeiffer (1892) during the 1889–1892 pandemic of influenza with the original description of the organism. Since the isolation of the influenza virus in 1933, *H. influenzae* has been associated with acute lower respiratory tract infections in children (Asmar et al., 1978; Ginsburg et al., 1979; Honig et al., 1973; Jacobs and Harris, 1979) and adults (Berk et al., 1982; Levin et al., 1977; Quintiliani, 1971; Wallace et al., 1978). There are no unique features that would differentiate infections due to *H. influenzae* from other pyogenic lower respiratory tract infections, and radiological features vary from minimal bronchopneumonia to lobar consolidation with or without effusion, abscess, and empyema.

Meningitis

With the decline of meningococcal meningitis since 1950, *H. influenzae* is the most common cause of bacterial meningitis in childhood (Michaels, 1971) and is infrequently a cause of bacterial meningitis in adults (Spagunolo et al., 1982). The onset may be acute or subacute when associated with infection outside the CNS. Fever and seizures are common early in the disease, and acute complications include subdural effusions, inappropriate secretion of antidiuretic hormone, cranial nerve palsies, and anemia (Kaplan and Oski, 1980; Lindberg et al., 1977). Early mortality is reported between 5 and 10%. Persistent or recurrent fever is common (Rutman and Wald, 1981), and empyema and abscess have been reported (Feldman and Schwartz, 1983). Late sequelae include hydrocephalus, neurosensory hearing loss, and intellectual and motor deficits (Jadavji et al., 1986; Lindberg et al., 1977; Taylor et al., 1984).

Epiglottitis

H. influenzae is by far the most common cause of epiglottitis. Sporadic reports have found *Streptococcus pneumoniae* and very rarely other organisms in association with epiglottitis. Disease usually occurs in older children (mean age 3 years compared to 1 to 2 years for meningitis) and is less common to rare in certain genetic groups such as Native Americans and Inuit. Onset is usually sudden, and the child refuses to swallow or stay in the recumbent position. Diagnosis is usually established clinically with support of a lateral roentgenogram of the neck or direct visualization of the epiglottis at the time of intubation (Molteni, 1978; Weber et al., 1976).

Septic Arthritis

H. influenzae is the most common cause of septic arthritis in children, particularly in infants under the age of 2 years. Monoarticular disease is most common with involvement of the large joints. Concurrent infection may be present (Rotbart and Glode, 1985). Osteomyelitis due to *H. influenzae* is much less common. A case of subacute osteomyelitis or Brodie's abscess has also been reported (Kurlandsky et al., 1979). Diagnosis is usually established by examination of synovial fluid and radiological findings or radionucleotide scan.

Cellulitis

Although cellulitis due to *H. influenzae* has been reported in adults (Drapkin et al., 1977; Shaw and Plouffe 1979) and in older children at uncommon sites (Minnefor et al., 1972), the clinical manifestations are usually associated with periorbital or buccal infections in infants and young children.

Preseptal cellulitis presents as periorbital swelling without evidence of trauma. Bacteremia is common, and associated ethmoiditis has been reported. Pupillary response to light, ocular motility, and visual acuity are usually intact but difficult to evaluate because of eyelid edema (Gellady et al., 1978; Goldberg et al., 1978; Smith et al., 1978). For this reason, computed tomography and sinus roentgenograms may be used to support the clinical diagnosis.

In the absence of trauma, buccal cellilitis in infants less than 1 year of age is most often due to *H. influenzae*. Typically, onset is acute with fever and irritability. Unilateral involvement is most common, and bacteremia or concomitant infection at other sites may be present (Bada and Wright, 1974; Weimert and Kwyer, 1978).

Miscellaneous Infections

Serotype b strains of *H. influenzae* have been reported in isolated case reports or small series in association with almost any infection, including peritonitis, CSF shunt infection, pancreatic abscess, endocarditis, pericarditis, uvulitis, and glossitis and should be considered potential pathogens in almost all pyogenic infections in childhood (Rogers et al., 1960).

Otitis Media

Acute otitis media is one of the most common infections of childhood, with over 25% of all children having at least one episode before the age of 5 years. *H. influenzae* and *S. pneumoniae* are the most common bacterial agents associated with otitis media outside the newborn period. Unlike in systemic infections discussed so far, strains of *H. influenzae* associated with otitis media are usually nonencapsulated. Organisms reach the middle ear by direct extension from the nasopharynx, and middle ear disease is usually associated with eustachian tube dysfunction. Onset may be acute, subacute, or chronic, and recurrent infection is common. Clinical findings of fever, ear pain, irritability, and pulling at the ear are variable and unreliable. Examination of the tympanic membrane should be a routine part of the physical examination of both well and ill children. Hearing loss and its impact on language development and learning may be significant in children with recurrent or chronic disease (Paradise, 1980; Rowe, 1975).

Conjunctivitis

Strains of *H. influenzae* isolated from the conjunctivae were originally described by Koch in 1883 and isolated by Weeks in 1886 in association with epidemic conjunctivitis. Pittman and Davis (1956) characterized some strains associated with severe conjunctivitis and called them *H. aegyptius*. Kilian (1976) has suggested that strains of *H. aegyptius* belong to *H. influenzae* biotype III, and the current literature does not distinguish between these species. *H. influenzae* (*H. aegyptius*) is the most common cause of endemic bacterial conjunctivitis in children (Gigliotti et al., 1981), and differential clinical features include bilateral purulent conjunctivitis with a higher association of otitis media, the conjunctivitis-otitis media syndrome (Bodor et al., 1985).

Neonatal infections

As with otitis media and conjunctivitis, strains of *H. influenzae* associated with neonatal infection are usually nonencapsulated. Although not common, the clinical spectrum resembles that of other bacterial pathogens in this age group, particularly the group B streptococcus. Premature infants born to mothers with obstetrical complications appear to be at highest risk, and the organism is almost invariably found in association with maternal genitourinary tract infection or colonization. Strains of biotype IV *H. influenzae* are more frequently isolated in this clinical setting, when compared to infections with nontypable strains associated with other sites (Campognone and Singer, 1986; Friesen and Cho, 1986; Wallace et al., 1983).

Pathology

Pathological findings are consistent with pyogenic infection of the particular organ system involved. A specific role for *H. influenzae* toxins has not been demonstrated. A possible role for surface appendages such as fimbriae in the initial colonization of the respiratory or genital mucosa has been suggested, but the data are inconclusive (Mason et al., 1985). Pathogenic strains of *H. influenzae* have been reported to produce an IgA protease, but its role in maintaining colonization of the mucosa or in invasion is not known (Farley et al., 1986; Kilian et al., 1979; Male, 1979). Following initial colonization, the serotype b capsule has unquestionably been associated with invasive disease. Organ system localization follows an initial bacteremia, and factors leading to particular organ system tropism have not been identified. It is clear that certain host/organism factors may be involved however as noted by the relative distribution of clinical syndromes particularly epiglottitis, among population groups. Strains of *H. aegyptius*, *H. influenzae* biotype III, and *H. influenzae* biotype IV clearly show tissue tropism with their association with conjunctivitis and genitourinary/neonatal infections, respectively, but the basis for such an association is not known.

Diagnostic Features

H. influenzae should be considered in almost any febrile clinical syndrome in infancy and childhood that is not a clearly recognizable viral syndrome. The major clinical syndromes, however, have been described above. There are insufficient distinguishing characteristics to establish a clinical diagnosis, and laboratory support for the diagnosis is usually needed.

Definitive diagnosis requires the isolation of *H. influenzae*, and care must be exercised in interpretation of a positive culture from mucosal surfaces where *H. influenzae* may be part of the normal commensal flora. Isolation of *H. influenzae* from a normally sterile body fluid such as CSF, synovial fluid, or blood is sufficient to establish a definitive diagnosis.

A presumptive diagnosis may be made by direct examination of Gram-stained specimens of normally sterile fluid or direct antigen detection by CIE, coagglutination, or latex agglutination.

Prognosis

The prognosis for *H. influenzae* infections depends very much on the associated clinical syndrome, presence or absence of underlying disease, age of the patient, and time interval between onset and institution of appropriate therapy and management. Morbidity and mortality are dependent on the particular clinical syndrome. Mortality varies from 0% to over 15% for young infants with meningitis and premature infants with neonatal infections. Long-term sequelae of certain clinical syndromes such as meningitis and otitis media may be substantial.

Epidemiologic Aspects

Occurrence and Transmission

Geographic distribution of *H. influenzae* infections is worldwide, although regional differences in reported incidence are known. Several epidemiologic studies have demonstrated increased risk factors in certain North American population groups such as blacks, Native Americans, and Inuit (Cochi et al., 1986; Lerman, 1982; Tarr and Peter, 1978). There is a clear age-related risk, with highest incidence of disease reported in the group 6 months to 5 years old. Seasonal variation in certain clinical syndromes has been reported and may be related to increased respiratory droplet transmission. Miniepidemics have been described in certain institutional and day-care settings, and household crowding has been associated with increased risk of disease.

Susceptibility, Resistance, and Carrier States

As noted earlier, premature infants born to mothers with obstetrical complications are at increased risk of infection in the presence of genitourinary tract infection or carriage. Genital carriage of *H. influenzae* has been reported to be 0.5 to 5% (Albritton et al., 1982; Sturm, 1986).

Genetically determined susceptibility to infection has been reported but has been difficult to establish (Cochi et al., 1986). A genetic variant of UMP kinase has been positively associated with invasive *H. influenzae* disease due to serotype b strains in Alaskan Eskimos (Petersen et al., 1985). Differences in incidence of specific clinical syndromes, such as epiglottitis, have also been suggested to have a genetic basis but require more definitive studies.

Levels of antibody to the serotype b capsule are inversely correlated with incidence of disease, accounting for the age-related risk increase during the latter half of the first year of life with declining transplacentally acquired antibody from the mother. Selective IgG subclass deficiency may be associated with increased risk of recurrent infections (Umetsu et al., 1985).

Respiratory carriage of nonencapsulated strains is common (30 to 50%), with carriage of serotype b strains in the general population being much lower (3 to 5%). Carriage rates in closed institutional settings may be much higher (Moxon, 1986). Household contacts of patients with invasive disease demonstrate an age-dependent increased risk of disease (Granoff and Daum, 1980; Ward, 1979).

Etiologic Agents

Characterization

Haemophilus influenzae is the type species of the genus *Haemophilus* as established by the American Committee on Classification and Nomenclature in 1920. The genus originally contained fastidious gram-negative coccobacilli and included such organisms as *Bordetella pertussis*, *Moraxella*, *H. ducreyi*, and *H. aegyptius* (Winslow et al., 1920). The ninth edition of Bergey's *Manual of Systematic Bacteriology* includes *H. influenzae* as one of 16 species of *Haemophilus*, all of which are gram-negative, facultatively anaerobic, rod-shaped bacteria requiring hemin or certain other porphyrins (X factor) and/or nicotinamide adenine dinucleotide (NAD) or certain other definable coenzymelike substances (V factor). Strains of *H. influenzae* are genetically related to other members of the genus *Haemophilus* and the closely related genera *Actinobacillus* and *Pasteurella*.

GC content of *H. influenzae* DNA is 39%, and DNA-DNA hybridization results suggest that strains of *H. aegyptius* and *H. influenzae* are highly related and that certain strains may warrant subspecies status, being only about 60% related to the type strain of the species *H. influenzae* NCTC 8143.

H. influenzae is transformable under growth conditions known to promote competence, and transfer of both plasmid and chromosomal genes by conjugation has been shown. Bacteriophages have been described in *H. influenzae* as well, but transfer of genetic information by this route has not been demonstrated. Six serovars (a–f) and eight biovars

(I–VIII) have been described. High-frequency loss of capsular type has been demonstrated in the laboratory (Hoiseth et al., 1985), but nonencapsulated strains in nature do not appear to be genetically derived from encapsulated strains (Musser et al., 1985).

Growth occurs aerobically or anaerobically and is enhanced by CO_2 and moisture. Desmethylmenaquinone has been shown to be the sole respiratory quinone in *H. influenzae* (Hollander and Mannheim, 1975). Acid is produced from glucose and certain other carbohydrates, but results are dependent on strain and media.

Hemolytic strains would ordinarily be identified as *H. haemolyticus* and hemagglutinating strains as *H. aegyptius*, but recent studies may force reevaluation of the current taxonomy and nomenclature of the genus *Haemophilus*.

Specimen Collection and Transport

Appropriate specimens for culture include blood, CSF, synovial fluid, tissue aspirates, biopsy material, and mucosal swabs. Prompt transportation to the laboratory is essential, or specimens should be placed in transport media or inoculated and incubated before transportation. Several transport media have been evaluated, but where culture diagnosis requires transportation of the specimen to a remote laboratory, nonculture antigen detection methods should be used in conjunction with isolation.

Isolation and Identification

Media and Growth Conditions

Most commercially available blood culture media and media used in automatic growth detection systems have been modified to support growth of *H. influenzae*, but internal batch quality control is essential to assure reliable performance (Schell et al., 1979). Selective liquid broth cultures have been described but are rarely used in routine diagnostic laboratories.

After Grassberger's (1897) original description of satellitism, many laboratories used blood agar medium with a V factor excreting "feeder" strain, such as *Staphylococcus aureus*, as a cross streak. Colonies of *H. influenzae* can be found adjacent to the feeder strain. Because of the limited surface area of the plate for growth with this method, most laboratories routinely use a chocolatized blood agar base supplemented with V factor and other nutritional supplements. Defined media have also been described but are rarely used in diagnostic laboratories (Herriott et al., 1970). Again, internal batch quality control is essential to ensure media performance.

Preliminary Identification

Inoculated plates are incubated at 35 to 37°C in a moist CO_2-enriched atmosphere, either in candle jars or in environmentally controlled incubators. Growth is usually visible at 24 to 48 h as 1 to 2-mm grayish, translucent colonies. Colonies of gram-negative coccobacillary or pleomorphic rods are picked for subculture to factor plates, commercial identification media, and conversion of δ-aminolevulinic acid to porphyrins. Preliminary identification as *H. influenzae* is usually made for nonhemolytic, gram-negative coccobacilli, requiring hemin and NAD for growth.

Misidentification of organisms as *H. influenzae* is common. Many laboratories do not routinely check for hemolysis, resulting in misidentification of *H. haemolyticus* as *H. influenzae*. Nutritionally deficient medium used for factor determination may contain trace amounts of hemin allowing growth of *H. influenzae* around the V factor strip, resulting in misidentification of *H. influenzae* as *H. parainfluenzae*. Many algorithms developed for identification of *H. influenzae* in commercial systems by biochemical tests would misidentify the uncommon biotypes. Such problems in misidentification are rarely of clinical significance when the organisms are isolated from sterile body fluids, because most clinicians would manage the patient as if *H. influenzae* were identified, and direct antigen detection systems and serotyping reagents would correctly identify serotype b organisms. Definitive identification (Table 1) should always be made, however, when *Haemophilus* species other than *H. influenzae* and organisms other than serotype b are isolated from sterile body fluids, when the etiology of the clinical condition is unclear, and for epidemiological studies.

Definitive Identification

Factor requirements should be unambiguously determined. Because of the possibility of trace amounts of hemin in the medium or carryover from the inoculum interfering with factor requirements by growth around factor disks or strips on nutritionally deficient media, determination of the presence of enzymes necessary for the conversion of δ-aminolevulinic acid to porphyrins is recommended (Kilian 1974). When carefully controlled, the porphyrin test is a reliable indicator of hemin requirement.

Nonhemolytic, gram-negative, coccobacilli requiring V factor for growth that are negative in the porphyrin test can be further identified as to serotype by agglutination with commercially available sera (Himmelreich et al., 1985). Several additional commercial reagents, such as latex particles, are available for identification of serotype b strains (Ingram et al., 1979). Cross-reaction with organisms carrying chemically similar capsular antigen and the presence

TABLE 1. Differential identification characteristics of the genus *Haemophilus*[a]

Species	*X-factor requirement*	*V-factor requirement*	*Indole*	*Urease*	*Ornithine decarboxylase*	*Oxidase*	*Catalase*	*ONPG*	*Glucose, acid*	*Sucrose, acid*	*Lactose, acid*
H. influenzae	+	+	V[b]	V	V	+	+	−	+	−	−
H. haemolyticus	+	+	V	−	−	+	+	−	+	−	−
H. aegyptius	+	+	−	+	−	+	−	−	+	−	−
H. parainfluenzae	−	+	V	V	V	+	V	V	+	+	−
H. segnis	−	−	−	−	−	−	V	V	W[c]	W	−
H. parahaemolyticus	−	+	−	+	V	+	+	V	+	+	−
H. paraphrophilus	−	+	−	−	−	+	−	+	+	+	+
H. aphrophilus	−[d]	−	−	−	−	−	−	+	+	+	+
A. actino-mycetemcomitans	−	−	−	−	−	+	+	−	+	−	−

[a] All members of the genus *Haemophilus* are gram-negative, facultatively anaerobic, rod-shaped bacilli; have G + C content of 37 to 44%; and reduce nitrate to nitrite.
[b] Variable.
[c] Weak.
[d] *H. aphrophilus* is reported to require hemin on initial isolation, but requirement is lost on subculture.

TABLE 2. Biotype identification of *H. influenzae*

Biotype	*Indole*	*Urease*	*Ornithine decarboxylase*
I	+	+	+
II	+	+	−
III	−	+	−
IV	−	+	+
V	+	−	+
VI	−	−	+
VII	+	−	−
VIII	−	−	−

of antibodies to other than capsular antigen in commercially available polyclonal reagents is a problem but rarely causes difficulty in the clinical laboratory. The development of monoclonal reagents to capsular and noncapsular antigens may improve the specificity of these reagents (Belmaaza et al., 1986).

Kilian (1978) described a system for biotyping *H. influenzae* based on the presence of urease and ornithine decarboxylase activities and the production of indole. Eight biotypes (Table 2) have been described (Gratten, 1983; Kilian, 1976; Oberhofer and Back, 1979; Sottnek and Albritton, 1984). Although biotyping has been useful in epidemiological studies, it is rarely if ever required in the routine identification of *H. influenzae*.

Nonculture Diagnostic Methods

Since serotype b strains of *H. influenzae* are associated with > 95% of invasive disease, several nonculture methods have been developed for direct detection of serotype b capsular antigen in body fluids such as CSF, urine, serum, and sputum (Daum et al., 1982; Ingram et al., 1972; Ward et al., 1978). Latex agglutination is more sensitive than counterimmunoelectrophoresis and is widely used in clinical laboratories for establishing a rapid presumptive diagnosis. Latex agglutination may also be useful in establishing a diagnosis after instituting antibiotics or when transportation of specimens to remote laboratories results in unacceptably low rates of culture isolation. Care must be taken, however, since the specificity of nonculture diagnostic methods may be unacceptably low in some clinical situations.

Treatment, Prevention, and Control

Treatment

The introduction of sulfonamides and type-specific rabbit immune serum resulted in reasonable therapeutic results in the treatment of *H. influenzae* infections (Alexander, 1943). The introduction of chloramphenicol and then ampicillin with efficacy in bacterial meningitis due to *H. influenzae*, *S. pneumoniae*, and *N. meningitidis* led to single-drug therapy with ampicillin (Smith, 1976). The emergence of ampicillin-resistant strains carrying plasmids coding for a TEM-1 β-lactamase (Elwell et al., 1975) resulted in a return to combination initial therapy with chloramphenicol and ampicillin. The subsequent isolation of chloramphenicol and ampicillin-resistant strains has led some to suggest third-generation cephalosporins with activity against *H. influenzae* as the initial therapy of choice in serious infections. The emergence of β-lactamase-negative, ampicillin-resistant strains due to alteration of binding affinity of the penicillin-binding proteins (Parr and Bryan, 1984) and tolerant strains (Bergeron and Lavoie, 1985) has made it necessary to routinely determine antimicrobial susceptibility in clinical isolates.

Intravenous administration of appropriate dosage is essential, and 7 to 10 days is the usual duration for most infections. The usual recommended duration of therapy for bone and joint infections is 4 to 6 weeks.

Prevention and Control

Chemoprophylaxis with rifampin is currently recommended for household contacts of patients with invasive disease due to serotype b strains of *H. influenzae* if children under the age of 6 years are in the home. Some authors recommend chemoprophylaxis in certain institutional and day-care settings (Brunell et al., 1984) and for patients themselves, because the usual therapeutic regimens may fail to eradicate carriage (Li and Wald, 1986). Timing of chemoprophylaxis in patients treated with chloramphenicol is important because of induction by rifampin of hepatic enzymes involved in acetylating chloramphenicol resulting in lower serum levels (Prober, 1985).

Immunization offers the best potential for prevention. Capsular polysaccharide vaccines are currently available for prevention of invasive disease due to serotype b strains of *H. influenzae* in children over 18 months of age (Cochi et al., 1985; Peltola et al., 1984). Such vaccines are poorly immunogenic in infants under 12 months of age, and efforts are currently directed to increasing immunogenicity in the young infant by coupling the polysaccharide to carrier protein such as tetanus or diphtheria toxoid or outer membrane protein. Such vaccines should allow expanded recommendations for prevention of invasive disease due to serotype b strains (Brunell et al., 1985).

No preventive or control measures have been evaluated for serotypable strains other than serotype b or for nonserotypable strains. The high incidence and substantial morbidity associated with otitis media and the low incidence but high mortality associated with neonatal infections suggests that alternative control strategies for nonserotypable strains of *H. influenzae* are needed.

Literature Cited

Albritton, W. L., J. L. Brunton, M. Meier, M. N. Bowman, and L. A. Slaney. 1982. *Haemophilus influenzae:* comparison of respiratory tract isolates with genitourinary tract isolates. J. Clin. Microbiol. **16:**826–831.

Alexander, H. E. 1943. Treatment of *Haemophilus influenzae* infections and of meningococci and pneumococcic meningitis. Am. J. Dis. Child. **66:**172–187.

Asmar, B. I., T. L. Slovis, J. O. Reed, and A. S. Dajani. 1978. *Haemophilus influenzae* type b pneumonia in 43 children. J. Pediatr. **93:**389–393.

Bada, H., and S. P. Wright. 1974. *Hemophilus influenzae* cellulitis. Clin. Pediatr. **13:**658–660.

Belmaaza, A., J. Hamel, S. Mousseau, S. Montplaisir, and B. R. Brodeur. 1986. Rapid diagnosis of severe *Haemophilus influenzae* serotype b infections by monoclonal antibody enzyme immunoassay for outer membrane proteins. J. Clin. Microbiol. **24:**440–445.

Bergeron, M. G., and G. Y. Lavoie. 1985. Tolerance of *Haemophilus influenzae* to β-lactam antibiotics. Antimicrob. Agents Chemother. **28:**320–325.

Berk, S. L., S. A. Holtsclaw, S. L. Wiener, and J. K. Smith. 1982. Nontypeable *Haemophilus influenzae* in the elderly. Arch. Intern. Med. **142:**537–539.

Bodor, F. F., C. D. Marchant, P. A. Shurin, and S. J. Barenkamp. 1985. Bacterial etiology of conjunctivitis–otitis media syndrome. Pediatrics **76:**26–28.

Brunell, P. A., et al., Committee on Infectious Diseases, American Academy of Pediatrics. 1984. Revision of recommendation for use of rifampin prophylaxis of contacts of patients with *Haemophilus influenzae* infection. Pediatrics **74:**301–302.

Brunell, P. A., et al., Committee on Infectious Diseases, Academy of Pediatrics. 1985. *Hemophilus* type b polysaccharide vaccine. Pediatrics **76:**322–324.

Campognone, P., and D. B. Singer. 1986. Neonatal sepsis due to nontypable *Haemophilus influenzae*. Am. J. Dis. Child. **140:**117–121.

Cochi, S. L., C. V. Broome, and A. W. Hightower. 1985. Immunization of US children with *Hemophilus influenzae* type b polysaccharide vaccine: a cost-effectiveness model of strategy assessment. J. Am. Med. Assoc. **253:**521–529.

Cochi, S. L., D. W. Fleming, A. W. Hightower, K. Limpakarnjanarat, R. R. Facklam, J. D. Smith, R. K. Sikes, and C. V. Broome. 1986. Primary invasive *Haemophilus influenzae* type b disease: a population-based assessment of risk factors. J. Pediatr. **108:**887–896.

Dajani, A. S., B. I. Asmar, and M. C. Thirumoorthi. 1979. Systemic *Haemophilus influenzae* disease: an overview. J. Pediatr. **94:**355–364.

Daum, R. S., G. R. Siber, J. S. Kawou, and R. R. Russell. 1982. Evaluation of a commercial latex particle agglutination test for rapid diagnosis of *Haemophilus influenzae* type b infection. Pediatrics **69:**466–471.

Drapkin, M. S., M. E. Wilson, S. M. Shrager, and R. H. Rubin. 1977. Bacteremic *Hemophilus influenzae* type b cellulitis in the adult. Am. J. Med. **63:**449–452.

Elwell, L. P., J. D. DeGraaff, D. Seibert, and S. Falkow. 1975. Plasmid-linked ampicillin resistance in *Haemophilus influenzae* type b. Infect. Immun. **12:**404–410.

Eng, R. H. K., M. L. Corrado, D. Cleri, and M. R. Sierra. 1980. Non-type b *Haemophilus influenzae* infections in adults with reference to biotype. J. Clin. Microbiol. **11:**669–671.

Farley, M. M., D. S. Stephens, M. H. Mulks, M. D. Cooper, J. V. Bricker, S. S. Mira, and A. Wright. 1986. Pathogenesis of IgA1 protease-producing and -nonproducing *Haemophilus influenzae* in human nasopharyngeal organ cultures. J. Infect. Dis. **154:**752–759.

Feldman, W. E., and J. Schwartz. 1983. *Haemophilus influenzae* type b brain abscess complicating meningitis: case report. Pediatrics **72:**473–475.

Friesen, C. A., and C. T. Cho. 1986. Characteristic features of neonatal sepsis due to *Haemophilus influenzae*. Rev. Infect. Dis. **8:**777–780.

Gellady, A. M., S. T. Shulman, and E. M. Ayoub. 1978. Periorbital and orbital cellulitis in children. Pediatrics **61:**272–277.

Gigliotti, F., W. T. Williams, F. G. Hayden, and J. O. Hendley. 1981. Etiology of acute conjunctivitis in children. J. Pediatr. **98:**531–536.

Ginsburg, C. M., J. B. Howard, and J. D. Nelson. 1979. Report of 65 cases of *Haemophilus influenzae* b pneumonia. Pediatrics **64:**283–286.

Goldberg, F., A. S. Berne, and F. A. Oski. 1978. Differentiation of orbital cellulitis from prespeptal cellulitis by computed tomography. Pediatrics **62:**1000–1005.

Granoff, D. M., and R. S. Daum. 1980. Spread of *Haemophilus influenzae* type b: recent epidemiologic and therapeutic considerations. J. Pediatr. **97:**854–860.

Grassberger, R. 1897. Beiträge zür Bakteriologie der Influenca. Z. Hyg. Infektionskr. **25:**453–475.

Gratten, M. 1983. *Haemophilus influenzae* biotype VII. J. Clin. Microbiol. **18:**1015–1016.

Greene, G. R. 1978. Meningitis due to *Haemophilus influenzae* other than type b: case report and review. Pediatrics **62:**1021–1025.

Herriott, R. M., E. Y. Meyer, M. Vogt, and M. Modan. 1970. Defined medium for growth of *Haemophilus influenzae*. J. Bacteriol. **101:**513–516.

Himmelreich, C. A., S. J. Barenkamp, and G. A. Storch. 1985. Comparison of methods for serotyping isolates of *Haemophilus influenzae*. J. Clin. Microbol. **21:**158–160.

Hirschmann, J. V., and E. D. Everett. 1979. *Haemophilus influenzae* infectious in adults: report of nine cases and a review of the literature. Medicine **58:**80–93.

Hoiseth, S. K., C. J. Connelly, and E. R. Moxon. 1985. Genetics of spontaneous, high-frequency loss of b capsule expression in *Haemophilus influenzae*. Infect. Immun. **49:**389–395.

Hollander, R., and W. Mannheim. 1975. Characterization of hemophilic and related bacteria by their respiratory quinones and cytochromes. Int. J. Syst. Bacteriol. **25:**102–107.

Honig, P. J., P. S. Pasquariello, and S. E. Stool. 1973. *H. influenzae* pneumonia in infants and children. J. Pediatr. **83:**215–219.

Ingram, D. L., P. Anderson, and D. H. Smith. 1972. Countercurrent immunoelectrophoresis in the diagnosis of systemic diseases caused by *Hemophilus influenzae* type b. J. Pediatr. **81:**1156–1159.

Ingram, D. L., A. M. Collier, E. Pendergrass, and S. H. King. 1979. Methods for serotyping nasopharyngeal isolates of *Haemophilus influenzae:* slide agglutination, Quellung reaction, countercurrent immunoelectrophoresis, latex agglutination, and antiserum agar. J. Clin. Microbiol. **9:**570–574.

Jacobs, N. M., and V. J. Harris. 1979. Acute *Haemophilus* pneumonia in childhood. Am. J. Dis. Child. **133:**603–605.

Jadavji, T., W. D. Biggar, R. Gold, and C. G. Prober. 1986.

Sequelae of acute bacterial meningitis in children treated for seven days. Pediatrics **78:**21–25.
Kaplan, K. M., and F. A. Oski. 1980. Anemia with *Haemophilus influenzae* meningitis. Pediatrics **65:**1101–1104.
Kilian, M. 1974. A rapid method for the differentiation of *Haemophilus* strains. The porphyrin test. Acta Pathol. Microbiol. Scand. **82B:**835–842.
Kilian, M. 1976. A taxonomic study of the genus *Haemophilus* with the proposal of a new species. J. Gen. Microbiol. **93:**9–62.
Kilian, M., J. Mestecky, and R. E. Schrohenloher. 1979. Pathogenic species of the genus *Haemophilus* and *Streptococcus pneumoniae* produce immunoglobulin A1 protease. Infect. Immun. **26:**143–149.
Kurlandsky, L. E., P. H. Quinn, and E. M. Sills. 1979. *Haemophilus influenzae* as a cause of Brodie's abscess in an infant. Johns Hopkins Med. J. **144:**15–17.
Lerman, S. J. 1982. Systemic *Haemophilus influenzae* infection: a study of risk factors. Clin. Pediatr. **21:**360–364.
Levin, D. C., M. I. Schwarz, R. A. Matthay, and F. M. LaForce. 1977. Bacteremic *Haemophilus influenzae* pneumonia in adults: a report of 24 cases and a review of the literature. Am. J. Med. **62:**219–224.
Li, K. I., and E. R. Wald. 1986. Use of rifampin in *Haemophilus influenzae* type b infections. Am. J. Dis. Child. **140:**381–385.
Lindberg, J., U. Rosenhall, O. Nylen, and A. Ringner. 1977. Long-term outcome of *Haemophilus influenzae* meningitis related to antibiotic treatment. Pediatrics **60:**1–6.
Liston, T. E., and W. S. Foshee. 1982. Invasive disease due to nontypable *Haemophilus influenzae* in children. South. Med. J. **75:**753–754.
Male, C. 1979. Immunoglobulin A1 protease production by *Haemophilus influenzae* and *Streptococcus pneumoniae*. Infect. Immun. **26:**254–261.
Mason, E. O., S. L. Kaplan, B. L. Wiedermann, E. P. Norrod, and W. A. Stenback. 1985. Frequency and properties of naturally occurring adherent piliated strains of *Haemophilus influenzae*. Infect. Immun. **49:**98–103.
Michaels, R. H. 1971. Increase in influenzal meningitis. N. Engl. J. Med. **285:**666–667.
Minnefor, A. B., J. J. Murray, and P. H. Davis. 1972. *Hemophilus influenzae* cellulitis of the lower extremity. Am. J. Dis. Child. **124:**920–921.
Molteni, R. A. 1976. Epiglottitis: incidence of extraepiglottic infection: report of 72 cases and review of the literature. Pediatrics **58:**526–531.
Moxon, E. R. 1986. The carrier state: *Haemophilus influenzae*. J. Antimicrob. Chemother. **18 (Suppl. A):**17–24.
Musser, J. M., S. J. Barenkamp, D. M. Granoff, and R. K. Selander. 1986. Genetic relationships of serologically nontypable and serotype b strain of *Haemophilus influenzae*. Infect. Immun. **52:**183–191.
Oberhofer, T. R., and A. E. Back. 1979. Biotypes of *Haemophilus* encountered in clinical laboratories. J. Clin. Microbiol. **10:**168–174.
Paradise, J. L. 1980. Otitis media in infants and children. 1980. Pediatrics **65:**917–943.
Parr, T. R., and L. E. Bryan. 1984. Mechanism of resistance of an ampicillin-resistant, β-lactamase-negative clinical isolate of *Haemophilus influenzae* type b to β-lactam antibiotics. Antimicrob. Agents Chemother. **25:**747–753.
Peltola, H., H. Käyhty, M. Virtanen, and P. H. Mäkelä, 1984. Prevention of *Hemophilus influenzae* type b bacteremic infections with the capsular polysaccharide vaccine. N. Engl. J. Med. **310:**1566–1569.
Petersen, G. M., E. M. Scott, J. I. Rotter, D. R. Silimperi, D. B. Hall, and J. I. Ward. 1985. Uridine monophosphate kinase 3: a genetic marker for susceptibility to *Haemophilus influenzae* type b disease. Lancet **2:**417–418.
Pfeiffer, R. 1892. Vorlaufige Mitteilungen uber die Evreger der Influenza. Dtsch. Med. Wochenschr. **18:**28.
Pittman, M. 1931. Variation and type specificity in the bacterial species *Haemophilus influenzae*. J. Exp. Med. **53:**471–492.
Pittman, M., and D. J. Davis. 1950. Identification of the Koch-Weeks bacillus (*Hemophilus aegyptius*). J. Bacteriol. **59:**413–426.
Prober, C. G., T. Jadavji, and S. J. Soldin. 1985. Effect of rifampin on chloramphenicol levels. N. Engl. J. Med. **312:**788–789.
Quintiliani, R., and P. J. Hymans. 1971. The association of bacteremic *Haemophilus influenzae* pneumonia in adults with typable strains. Am. J. Med. **50:**781–786.
Rogers, K. B., K. Zinnemann, and W. P. Foster. 1960. The isolation and identification of *Haemophilus* spp. from unusual lesions in children. J. Clin. Pathol. **13:**519–524.
Rotbart, H. A., and M. P. Glode. 1985. *Haemophilus influenzae* type b septic arthritis in children: report of 23 cases. Pediatrics **75:**254–259.
Rowe, D. S. 1975. Acute suppurative otitis media. Pediatrics **56:**285–294.
Rutman, O. L., and E. R. Wald. 1981. Fever in *Haemophilus influenzae* type b meningitis. Clin. Pediatr. **20:**192–195.
Schell, R. F., J. L. LeFrock, J. P. Babu, and D. B. Robinson. 1979. Recovery of *Haemophilus influenzae* from twenty-three blood culture media. J. Clin. Microbiol. **9:**84–87.
Shaw, R. A., and J. F. Plouffe. 1979. *Haemophilus influenzae* cellulitis in an adult. Arch. Intern. Med. **139:**368–369.
Smith, A. L. 1976. Antibiotics and invasive *Haemophilus influenzae*. N. Engl. J. Med. **294:**1329–1331.
Smith, T. F., D. O'Day, and P. F. Wright. 1978. Clinical implications of preseptal (periorbital) cellulitis in childhood. Pediatrics **62:**1006–1009.
Sottnek, F. O., and W. L. Albritton. 1984. *Haemophilus influenzae* biotype VIII. J. Clin. Microbiol. **20:**815–816.
Spagunolo, P. J., J. J. Ellner, P. I. Lerner, M. C. McHenry, F. Flatauer, P. Rosenberg, and M. S. Rosenthal. 1982. *Haemophilus influenzae* meningitis: the spectrum of disease in adults. Medicine **61:**74–85.
Sturm, A. W. 1986. Isolation of *Haemophilus influenzae* and *Haemophilus parainfluenzae* from genital tract specimens with a selective culture medium. J. Med. Microbiol. **21:**349–352.
Tarr, P. I., and G. Peter. 1978. Demographic factors in the epidemiology of *Haemophilus influenzae* meningitis in young children. J. Pediatr. **92:**884–888.
Taylor, H. G., R. H. Michaels, P. M. Mazur, R. E. Bauer, and C. B. Liden. 1984. Intellectural, neuropsychological, and achievement outcomes in children six to eight years after recovery from *Haemophilus influenzae* meningitis. Pediatrics **74:**198–205.
Turk, D. C., and J. R. May. 1967. *Haemophilus influenzae:* its clinical importance. English Universities Press, London.
Umetsu, D. T., D. M. Ambrosino, I. Quinti, G. R. Siber, and R. S. Geha. 1985. Recurrent sinopulmonary infection and impaired *Haemophilus influenzae* antibody response to bacterial capsular polysaccharide antigen in children with selective IgG-subclass deficiency. N. Engl. J. Med. **313:**1247–1251.

Wallace, R. J., D. M. Musher, and R. R. Martin. 1978. *Hemophilus influenzae* pneumonia in adults. Am. J. Med. **64**:87–93.

Wallace, R. J., C. J. Baker, F. J. Quinones, D. G. Hollis, R. E. Weaver, and K. Wiss. 1983. Nontypable *Haemophilus influenzae* (biotype 4) as a neonatal, maternal, and genital pathogen. Rev. Infect. Dis. **5**:123–136.

Ward, J. I., G. R. Siber, D. W. Scheifele, and D. H. Smith. 1978. Rapid diagnosis of *Hemophilus influenzae* type b infections by latex particle agglutination and counterimmunoelectrophoresis. J. Pediatr. **93**:37–42.

Ward, J. I., D. W. Fraser, L. J. Baraff, and B. D. Plikaytis. 1979. *Haemophilus influenzae* meningitis: a national study of secondary spread in household contacts. N. Engl. J. Med. **301**:122–126.

Weber, M. L., R. Desjardins, G. Perreault, G. Rivard, and Y. Turmel. 1976. Acute epiglottitis in children—treatment with nasotracheal intubation: report of 14 consecutive cases. Pediatrics **57**:152–155.

Weimert, T. A., and T. A. Kwyer. 1978. *Haemophilus influenzae* cellulitis. Arch. Otolaryngol. **104**:544–545.

Winslow, C. E. A., J. Broadhurst, R. E. Buchanan, C. Krumwiede, L. A. Rogers, and G. H. Smith. 1920. The families and genera of the bacteria. Final report of the Society of American Bacteriologoists on characterization and classification of bacterial types. J. Bacteriol. **5**:191–229.

CHAPTER 32

Infections Due to *Haemophilus* Species Other than *Haemophilus influenzae*

W. L. ALBRITTON

Diseases: Conjunctivitis, genital ulcer (chancroid), endocarditis, brain abscess, pneumonia, and other syndromes.

Etiologic Agents: *Haemophilus aegyptius, H. ducreyi, H. parainfluenzae, H. aphrophilus, H. paraphrophilus, H. haemolyticus, H. parahaemolyticus, H. paraphrohaemolyticus, H. segnis.*

Source: Human respiratory and genital tract by direct contact or inhalation.

Clinical Manifestations: Acute or subacute onset of fever and other symptoms dependent on organ system localization.

Pathology: Acute pyogenic infection. Isolated mucosal infection with or without local lymphadenopathy or secondary organ system localization following primary bacteremia.

Laboratory Diagnosis: Culture isolation of etiologic agent from blood, sputum, tissue, or mucosal swabs. Careful clinical correlation may be required to establish diagnosis when isolation is from mucosal surface contaminated by commensal flora.

Epidemiology: *Haemophilus aegyptius* is primarily associated with conjunctivitis in children. *Haemophilus ducreyi* is only found as a sexually transmitted genital ulcer disease. *H. haemolyticus* has not been associated with invasive disease but may be isolated as part of the normal respiratory flora. All other species are found in association with clinical syndromes commonly caused by *H. influenzae*, especially nonencapsulated strains, but are more commonly found in the syndromes of endocarditis and brain abscess, probably as a result of bacteremic localization from their origin as part of the normal oral flora.

Treatment: Ampicillin, chloramphenicol, and certain third-generation cephalosporins. *Haemophilus ducreyi* is usually susceptible to erythromycin and certain third-generation cephalosporins.

Prevention and Control: *Haemophilus ducreyi* infections are best prevented by avoidance of exposure and expectant treatment of sexual contacts of known cases of chancroid. Effective prevention and control strategies for *H. aegpytius* have not been described. All other species may be part of the normal oral or respiratory tract flora. For this reason, appropriate use of prophylactic antibiotics during dental procedures in high-risk patients with underlying cardiac disease may be beneficial.

Haemophilus aegyptius

Description of Disease

Robert Koch's microscopic observations on material from acute catarrhal conjunctivitis during his study of cholera in Egypt in 1883 and the subsequent isolation of the bacillus by Weeks in 1886 represent the first clinical descriptions of disease due to members of the genus *Haemophilus*. The Koch-Weeks bacillus was included in the original description of the genus *Haemophilus* by the American Committee on Classification and Nomenclature in 1920 and has been designated by a number of names. Pittman and Davis (1950) characterized isolates during a field investigation of acute conjunctivitis and proposed the currently accepted nomenclature. Kilian et al. (1976) suggested that *H. aegyptius* is a hemagglutinating variant of *H. influenzae* biotype III, and this has led to some confusion regarding the clinical description of the disease. Several investigators have described *H. influenzae* in association with acute conjunctivitis (Bodor et al., 1985; Gigliotti et al., 1981), which tends to be purulent, bilateral, and often associated with otitis media. The syndrome of acute catarrhal conjunctivitis due to *H. aegyptius* has historically been a more severe disease with typical onset in one eye with watering, vascular injection, and pain followed by purulent discharge and bilateral involvement with significant edema of the eyelids (Davis and Pittman, 1950; Dawson, 1960). Until recently, *H. aegyptius* had not been described in association with systemic disease. Marraro et al. (1977) reported a case of pneumonia due to *H. aegyptius* (*H. influenzae*, biotype III), and bacteremia complicating purulent conjunctivitis, Brazilian purpuric fever, has recently been described in association with *H. aegyptius* infection (CDC, 1986).

Epidemiologic Aspects

The failure to differentiate *H. influenzae* from *H. aegyptius* has led to much confusion in recent years regarding the epidemiology of infections due to these organisms. The conjunctivitis–otitis media syndrome is probably the most common clinical syndrome and is due to nontypable strains of *H. influenzae*, predominantly biotypes II and III. As with other *H. influenzae* infections, it is seen most often in infants and young children and has shown seasonal variation, being most common in winter months. Epidemic acute catarrhal conjunctivitis due to *H. aegyptius* has been described infrequently in North America in recent years but occurred in the southern United States and California in summer months and occurred predominantly in preschool and school-age children. The bacteriology and epidemiology of Brazilian purpuric fever has not been adequately described.

Etiologic Agent

Characterization

H. aegyptius is a long, slender, gram-negative bacillus with G + C content of 38 to 40 mol% and has 80 to 100% DNA homology with the type strain of *H. influenzae*. For this reason, *H. aegyptius* has been considered a biotype III strain of *H. influenzae* (Kilian et al., 1976). Hemagglutination of human red cells, lack of xylose fermentation or indole production, and serological typing were used by Pittman and Davis (1950) to identify strains of *H. aegyptius*. Mazloum et al. (1982) differentiated strains of *H. aegyptius* from strains of *H. influenzae* by their failure to grow on factor-supplemented trypticase soy agar and their sensitivity to trooleandomycin. A comparison of outer-membrane protein profiles has also suggested differences between *H. aegyptius* and *H. influenzae* (Carlone et al., 1985).

Isolation and Identification

Most media suitable for isolation of *H. influenzae* will support the growth of *H. aegyptius*. One study comparing media for isolation of *Haemophilus* species from seasonal conjunctivitis in Tunisia gave best isolation rates with IsoVitaleX-supplemented chocolate agar (Vastine et al., 1974). At this time, no unequivocal tests are available for identification of *H. aegyptius* (Table 1). Strains of *H. aegyptius* appear as long, slender, gram-negative rods on smear of conjunctival scrapings and as approximately 1-mm grayish, translucent colonies after 24 h incubation in a moist CO_2-enriched atmosphere. Biotyping demonstrates the absence of indole production and ornithine decarboxylase activity and the presence of urease activity (biotype III *H. influenzae*). Most strains do not produce acid from xylose, but Tiller (1982) has shown variable results with this characteristic.

Presence of mannose-resistant hemagglutinins (Scott and Old, 1981) has been shown by several investigations to be a variable characteristic (Davis et al., 1950; Ivler et al., 1963), and Martel et al. (1986) demonstrated trooleandomycin susceptibility in strains of *H. influenzae*. Thus no phenetic traits are sufficiently distinguishing to identify *H. aegyptius*, and additional studies are needed to determine the clinical relevance of distinguishing *H. aegyptius* from *H. influenzae* and other X and V factor–requiring strains of *Haemophilus* associated with acute conjunctivitis.

Treatment, Prevention, and Control

Strains of *H. aegyptius* show the same general pattern of antimicrobial susceptibility as do strains of *H. influenzae*, although ocular isolates of *Haemophilus* have a higher incidence of plasmids than do other respiratory isolates (Roberts et al., 1986). Ampicillin, chloramphenicol, and certain third-generation cephalosporins are the antimicrobials of choice. Optic preparations of sulfonamides and chloramphenicol are widely used as well, although well-designed clinical trials demonstrating their efficacy have not been published. Eye-to-eye transmission may be reduced by careful handwashing and improved general hy-

TABLE 1. Differential identification characteristics of *Haemophilus* and related species[a]

Species	X-factor[b] requirement	V-factor requirement	Hemolysis	Lysine decarboxylase	Glucose, acid	Sucrose, acid	Lactose, acid	Xylose, acid	Alkaline phosphatase	Nitrate reduction
H. influenzae	+	+	−	V	+	−	−	V	+	+
H. aegyptius	+	+	−	−	+	−	−	V	+	+
H. haemolyticus	+	+	+	−	+	−	−	V	+	+
H. parainfluenzae	−	+	−	V	+	+	−	−	+	+
H. parahaemolyticus	−	+	+	−	+	+	−	−	+	+
H. paraphrohaemolyticus	−	+	+	−	+	+	−	−	+	+
H. aphrophilus	−[c]	−	−	−	+	+	+	−	+	+
H. paraphrophilus	−	+	−	−	+	+	+	−	+	+
H. segnis	−	+	−	−	+	+	−	−	+	+
H. ducreyi	+	−	−	−	V	−	−	−	+	+
A. actinomycetemcomitans	−	−	−	−	+	−	−	V	+	+
E. corrodens	−	−	−	+	−	−	−	−	−	+
C. hominis	−	−	−	−	+	+	−	−	ND[d]	−

[a] + = More than 90% of strains positive; − = more than 90% of strains negative; V = 10–90% of strains positive.
[b] X-factor requirement determined by porphyrin test.
[c] Reported to require hemin on initial isolation, but requirement lost on subculture.
[d] Not determined.

giene. No specific control strategies have been developed for *H. aegyptius* infections.

Haemophilus ducreyi

Description of Disease

Chancroid, or soft chancre (ulcus molle), was first differentiated from syphilis by Bassereau in 1852. *Haemophilus ducreyi* was first described by Auguste Ducrey (1889), although he was unable to grow the organism in vitro. *H. ducreyi* appears to be associated only with genital ulcer disease or direct inoculation infections. The chancre usually begins as a tender papule after a 3- to 5-day incubation period with subsequent erosion and ulceration over 24 to 48 h. Secondary ulcers ("kissing ulcers") may occur on contiguous skin owing to autoinoculation. Unlike syphilis, the ulcer is painful, and unilateral inguinal adenitis occurs in about 30% of patients and progresses to fluctuant bubo formation in about one-third. Women more often have minor asymptomatic ulcers in the vaginal fourchette without inguinal adenopathy. The differential diagnosis of genital ulcer disease includes a number of conditions but especially syphilis, herpes, lymphogranuloma venereum, granuloma inguinale, and trauma.

Epidemiologic Aspects

In tropical areas, the incidence of chancroid may be second only to gonorrhea as a sexually transmitted disease. In North America since 1960, disease due to H. *ducreyi* has more commonly been associated with urban epidemics (Blackmore et al., 1985, Hammond et al., 1980). Epidemics have frequently been associated with prostitutes but a role for asymptomatic carriage or asymptomatic ulcers in transmission has not been clearly established. *H. ducreyi* has been isolated only from humans, and asymptomatic carriage has only occasionally been described. The reported incidence of clinical disease due to *H. ducreyi* is about 10 times higher in men than in women.

Etiologic Agents

Characterization

H. ducreyi was originally described by Ducrey as a nonmotile, pleomorphic, gram-negative bacillus occurring in streptobacillary chains and parallel chains known as "schools of fish." Because of its fastidious nature, it was originally included in the genus *Haemophilus* as described by Winslow et al. (1920). It was not until 1937, however, that Lwoff and Pirosky (1937) described hemin (X factor) requirement, and *H. ducreyi* remains the only hemin-only-requiring species of *Haemophilus* associated with human disease. Kilian (1976) determined the G + C content to be 38 mol%, and these criteria should be sufficient for inclusion of the organism in the genus *Haemophilus*. However, recent work suggests that *H. ducreyi* is genetically unrelated to *H. influenzae*, the type species of the genus (Casin et al., 1985). Strains of *H. ducreyi* are notably deficient in carbohydratase activity but show a wide spectrum of peptidase activity (Casin et al., 1982).

Isolation and Identification

Direct inoculation of material from the base of the ulcer onto selective solid medium is the recommended procedure for isolation. Various base formulations have been used, but most include a blood source and vancomycin or other selective antibiotics (Hammond et al., 1978; Hannah and Greenwood, 1982;; Oberhofer and Back, 1982; Sottnek et al., 1980). Inoculated media are incubated in a water-saturated atmosphere containing 5% CO_2 such as a candle extinction jar. An incubation temperature of 33 to 35°C is important, and reduced isolation rates are found with an incubation temperature of 37°C. After 24 to 48 h of incubation, small, nonmucoid, yellowish, semiopaque, nonpitting colonies may be seen which can be pushed intact across the surface of the agar. Recovery of viable organisms declines after 48 h, so early subculture is recommended.

Gram stain of colonies usually reveals gram-negative coccobacilli in parallel chains (railroad-track or school-of-fish patterns) or clumps and whorls (fingerprint patterns). Streptobacillary chains or pleomorphic bacilli are more commonly seen in liquid culture or suspension.

Although nitrate reduction, alkaline phosphatase, and oxidase activity are frequently used in the identification of *H. ducreyi* (Table 1), demonstration of hemin requirement in the absence of NAD requirement is the major identification determinant. Since many clinical isolates fail to grow on nutritionally deficient media used for satellitism around factor disks, the preferred test for hemin requirement is the porphyrin test of Kilian (1974). *H. ducreyi* may be differentiated from morphologically similar organisms isolated from genital ulcers on the basis of hemin requirement and the presence of alkaline phosphatase activity (Ursi et al., 1982).

Treatment, Prevention, and Control

Strains of *H. ducreyi* show the same general pattern of antimicrobial susceptibility as do other *Haemophilus* species. Plasmid-mediated antibiotic resistance is commonly due to transposon-associated resistance determinants and plasmids shared by other *Haemophilus* species and the *Enterobacteriaceae* (McNicol and Ronald, 1984). Recent recommendations for therapy include erythromycin, trimethoprim, and certain third-generation cephalosporins (Kraus et al., 1982).

Few studies are available on control of endemic chancroid. Urban outbreaks of chancroid in North America have been controlled by aggressive contact tracing and empirical treatment of genital ulcer disease and sexual contacts with appropriate antibiotics.

Haemophilus parainfluenzae

Description of Disease

Russell and Fildes (1928) first reported disease in man due to *H. parainfluenzae* in association with endocarditis. A recent review of the literature (Albritton, 1982) found reports of infections due to *H. parainfluenzae* in association with the clinical syndromes of endocarditis, meningitis, pneumonia, septic arthritis, epiglottitis, brain abscess, and urinary tract infection, with endocarditis and meningitis accounting for the majority. There are no sufficiently distinguishing clinical features to differentiate infection due to *H. parainfluenzae* from infection due to *H. influenzae* and certain other species of *Haemophilus*, although non-*H. influenzae* infections account for less than 3% of all *Haemophilus* infections (Trollfors et al., 1985).

Epidemiologic Aspects

The epidemiology of *H. parainfluenzae* infections is poorly described. Most reported cases have insufficient microbiological data to confirm identification. Many laboratories report all V-factor-only-requiring strains as *H. parainfluenzae*, and even factor requirement has been shown to be erroneous in nearly 20% of tests when using various satellite phenomena (Kilian, 1980; Klein and Blazevic, 1969). Thus, many reported cases of infection due to *H. parainfluenzae* may represent infection due to *H. influenzae* or other V-factor-only-requiring species such as *H. paraphrophilus* or *H. segnis*.

Etiologic Agent

Characterization

H. parainfluenzae was originally described by Rivers (1922) during studies on the influenza bacillus following the 1918 pandemic, and he showed that certain influenzalike bacilli required only one of the two accessory growth factors present in blood—the heat-labile V factor, or NAD. Strains are gram-negative bacilli and may show less pleomorphism than *H. influenzae*. As with other non-hemin-requiring species of *Haemophilus*, the G + C content is somewhat higher than hemin-requiring species (40 to 44 mol% compared to 38 to 40 mol% for hemin-requiring species). Although surface appendages and mannose-resistant hemagglutinins have been described (Kahn and Gromkova, 1981), no special pathogenic characteristics have been described (Tuyau and Sims, 1983). Based on the presence of indole production, urease activity, and ornithine decarboxylase activity, eight biotypes of *H. parainfluenzae* have been described by Kilian (1976) and others, as for *H. influenzae*. Most strains of H. *parainfluenzae* isolated from clinical material, however, are negative for indole production. In addition, most V-factor-only-requiring strains produce acid from sucrose (Table 1).

Isolation and Identification

H. parainfluenzae can usually be isolated on any medium supporting the growth of *H. influenzae*, and most laboratories use supplemented chocolate agar. After incubation at 35 to 37°C for 18 to 24 h, colonies of *H. parainfluenzae* appear as 1 mm, grayish, translucent colonies which may show pleomorphic gram-negative bacilli on Gram stain. V-factor requirement should be established by satellite growth around NAD-containing disks on nutritionally deficient medium. The absence of X-factor requirement should be established by the porphyrin test using δ-aminolevulinic acid as substrate, as described by Kilian (1974). Most clinical laboratories do not require further identification, but in certain cases and for reported or epidemiological studies, confirmation in a reference laboratory or additional testing is required for definitive identification (Table 1). As a minimum, isolates should be tested for hemolysis on blood agar using one or more species of red blood cells with batch quality assurance testing to ensure that no inhibiting antibodies are present in the source of blood and tested for acid production from lactose. CO_2 enhancement of growth can be assessed under carefully controlled conditions, but constant humidity and temperature are essential for test interpretation.

Treatment, Prevention, and Control

Strains of *H. parainfluenzae* show the same general patterns of susceptiblity as strains of *H. influenzae*. Ampicillin, chloramphenicol ,and certain third-generation cephalosporins are the antimicrobials of choice. Antibiotic-resistant strains have been described (Scheifele et al., 1982), and Brunton et al. (1986) have recently demonstrated a shared pool of antibiotic resistance determinants and plasmids among certain species of *Haemophilus*. No specific prevention methods or control strategies have been proposed for *H. parainfluenzae* infections.

Other Species

Description of Disease

Hemolytic X- and V-factor-requiring strains, *H. haemolyticus*, have not been reported in clearly documented cases of systemic disease since the original description of hemolytic strains by Pritchett and Stillman (1919) was modified by Pittman (1953) to separate *H. haemolyticus* from the V-factor-only-requiring strains, *H. parahaemolyticus*.

H. parahaemolyticus, *H. segnis* and the CO_2-requiring species *H. aphrophilus*, *H. paraphrophilus*, and *H. paraphrohaemolyticus* have all been associated with systemic disease in man, especially endocarditis and brain abscess. A variety of isolated case reports have appeared, however, and the less commonly reported species may from time to time be reported in almost any pyogenic infection (Albritton, 1982). The clinically related species *Actinobacillus actinomycetemcomitans*, *Eikenella corrodens*, and *Cardiobacterium hominis* share the spectrum of infections seen with these *Haemophilus* species (Savage et al., 1977; Suwanagool et al., 1983; Zambon, 1985).

Epidemiologic Aspects

This group of organisms share a common ecological niche, the oral mucosa, and are part of the normal flora of man. As such, they should be considered in any pyogenic infection associated with human or animal bites or with dental manipulation and damage to the oral mucosa. Systemic infections are presumed to be preceded by bacteremic seeding of the involved organs. As with other oral mucosal organisms, the reasons for selective involvement of the endocardium and brain are not understood.

Etiologic Agents

Characterization

All of the organisms included in this section are fastidious gram-negative bacilli and are members of the family *Pasteurellacae*. *Actinobacillus actinomycetemcomitans* requires neither hemin nor NAD for growth but has a G + C content of 44 to 48 mol% and is related to *H. aphrophilus* and *H. paraphrophilus* (Coykendall et al., 1983). Hemin requirement for *H. aphrophilus* is seen only on initial isolation, as has been described for *Eikenella corrodens* (Rubin et al., 1985). CO_2-enhanced growth is seen with *E. corrodens* and *Cardiobacterium hominis* (Savage, 1977). Hemolysis provides no clinically or biologically useful characterization. Thus the major characteristics of hemin requirement, hemolysis and CO_2-enhanced growth, used to define the species of *Haemophilus* are not consistent with genetic characterization of the organisms. Work is under way to redefine the characteristics of the family *Pasteurellaceae* and the species of the genus *Haemophilus*. Until such work is completed, useful characterization of this group of organisms is not available.

Isolation and Identification

The "other" species of *Haemophilus* and related bacteria can be isolated on media suitable for the isolation of *Haemophilus influenzae*. Since the species discussed in this section are part of the normal oral flora, care must be taken in interpreting the isolation of these organisms from other than normally sterile body fluids such as blood, synovial fluid, CSF, and uncontaminated tissue. The failure to properly identify isolates has made the literature regarding infections with these organisms confusing if not uninterpretable. Because of the lack of clinical significance to differentiation of members of this group, many clinical laboratories report all V-factor-only requiring strains as *H. parainfluenzae*. For practical management, if one knows the antimicrobial susceptibility, this should be sufficient. If, however, one is interested in publishing unusual cases or conducting epidemiological studies, then additional tests for species identification are required (Table 1). In addition to V-factor requirement and the porphyrin test, nitrate reduction, hemolysis, lysine decarboxylase activity, acid production from glucose, sucrose, lactose, and xylose should be determined with alkaline phosphatase added where *H. ducreyi* may be a consideration. Inclusion of differential test results in the description of the organism or case definition would be useful for future meta-analyses. Reports from reference laboratories confirming identification of the organism should also include the differential test results used.

Treatment, Prevention, and Control

Antimicrobial susceptibility of this group of organisms parallels that of other *Haemophilus* species, including *H. influenzae*. Strains are usually sensitive to aminopenicillins, but resistant strains have been isolated, and definitive treatment should be supported by in vitro susceptibility tests. No specific prevention or control strategies have been considered. The regimens usually recommended for dental manipulations in patients at risk for endocarditis should be effective for this group of organisms.

Literature Cited

Albritton, W. L. 1982. Infections due to *Haemophilus* species other than *H. influenzae*. Annu. Rev. Microbiol. **36:**199–216.

Blackmore, C. A., K. Limpakarnjanarat, J. G. Rigau-Perez, W. L. Albritton, and J. R. Greenwood. 1985. An outbreak of chancroid in Orange County, California: descriptive epidemiology and disease-control measures. J. Infect. Dis. **151:**840–844.

Bodor, F. F., C. D. Marchant, P. A. Shurin, and S. J. Barenkamp. 1985. Bacterial etiology of conjunctivitis–otitis media syndrome. Pediatrics **76:**26–28.

Brunton, J., M. Meier, N. Erhman, D. Clare, and R. Almawy. 1986. Origin of small β-lactamase-specifying plasmids in *Haemophilus* species and *Neisseria gonorrhoeae*. J. Bacteriol. **168:**374–379.

Carlone, G. M., F. O. Sottnek, and B. D. Plikaytis. 1985. Comparison of outer membrane protein and biochemical profiles of *Haemophilus aegyptius* and *Haemophilus influenzae* biotype III. J. Clin. Microbiol. **22:**708–713.

Casin, I., F. Grimont, P. A. D. Grimont, and M. J. Sanson-LePors. 1985. Lack of deoxyribonucleic acid relatedness between *Haemophilus ducreyi* and other *Haemophilus* species. Int. J. Syst. Bacteriol. **35:**23–25.

Casin, I. M., M. J. Sanson-LePors, M. F. Gorce, M. Ortenberg, and Y. Perol. 1982. The enzymatic profile of *Haemophilus ducreyi*. Ann. Microbiol. **133:**379–388.

Centers for Disease Control. 1986. Brazilian purpuric fever: *Haemophilus aegyptius* bacteremia complicating purulent conjunctivitis. Morbid. Mortal. Weekly Rep. **35:**553–554.

Coykendall, A. L., J. Setterfield, and J. Slot. 1983. Deoxyribonucleic acid relatedness among *Actinobacillus actinomycetemcomitans, Haemophilus aphrophilus,* and other *Actinobacillus* species. Int. J. Syst. Bacteriol. **33:**422–424.

Davis, D. J., and M. Pittman. 1950. Acute conjunctivitis caused by *Haemophilus*. Am. J. Dis. Child. **79:** 211–222.

Davis, D. J., M. Pittman, and J. J. Griffitts. 1950. Hemagglutination by the Koch-Weeks bacillus (*Haemophilus aegyptius*). J. Bacteriol. **59:**427–431.

Dawson, C. R. 1960. Epidemic Koch-Weeks conjunctivitis and trachoma. Am. J. Ophthalmol. **49:**801–808.

Ducrey, A. 1889. Experimentelle Untersuchungen ulcer den Ansteckungsstoff des weichen Schankers and uber die Bubonen. Montasch. Prakt. Dermatol. **9:**387–405.

Gigliotti, F., W. T. Williams, F. G. Hayden, and J. O. Hendley. 1981. Etiology of acute conjunctivitis in children. J. Pediatr. **98:**531–536.

Hammond, G. W., C. J. Lian, J. C. Wilt, and A. R. Ronald. 1978. Comparison of specimen collection and laboratory techniques for isolation of *Haemophilus ducreyi*. J. Clin. Microbiol. **7:**39–43.

Hammond, G. W., M. Slutchuk, J. Scatliff, E. Sherman, J. C. Wilt, and A. R. Ronald. 1980. Epidemiologic, clinical, laboratory, and therapeutic features of an urban outbreak of chancroid in North America. Rev. Infect. Dis. **2:**867–879.

Hannah, P., and J. R. Greenwood. 1982. Isolation and rapid identification of *Haemophilus ducreyi*. J. Clin. Microbiol. **16:**861–864.

Ivler, D., H. M. Preston, and B. Portnoy. 1963. Direct hemagglutination by *Hemophilus influenzae*. Proc. Soc. Exp. Biol. Med. **114:**232–234.

Kahn, M. E., and R. Gromkova. 1981. Occurrence of pili on and adhesive properties of *Haemophilus parainfluenzae*. J. Bacteriol. **145:**1075–1078.

Kilian, M. 1974. A rapid method for the differentiation of *Haemophilus* strains. The porphyrin test. Acta Pathol. Microbiol. Scand. Sect. B **82:**835–842.

Kilian, M. 1976. A taxonomic study of the genus *Haemophilus* with the proposal of a new species. J. Gen. Microbiol. **93:**9–62.

Kilian, M. 1985. *Haemophilus*. p. 387–393. *In* E. H. Lennette, A. Balows, W. J. Hausler, and H. J. Shadomy (ed.), Manual of clinical microbiology, 4th ed. American Society for Microbiology, Washington, D.C.

Kilian, M., C. H. Mordhorst, C. R. Dawson, and H. Lautrop. 1976. The taxonomy of haemophili isolated from conjunctivae. Acta Pathol. Microbiol. Scand. Sect. B **84:**132–138.

Klein, M., and D. J. Blazevic. 1969. Differentiation of *Haemophilus* species using satellite methods. Am. J. Med. Technol. **35:**695–701.

Kraus, S. J., H. W. Kaufman, W. L. Albritton, C. Thornsberry, and J. W. Biddle. 1982. Chancroid therapy: a review of cases confirmed by culture. Rev. Infect. Dis. Suppl. **4:**848–856.

Lwoff, A., and I. Pirosky. 1937. Determination du facteur de croissance pour *Haemophilus ducreyi*. C. R. Seances Soc. Biol. **124:**1169–1171.

Marraro, R. V., F. K. McCleskey, and J. L. Mitchell. 1977. Pneumonia due to *Haemophilus influenzae* (*H. aegyptius*) biotype 3. J. Clin. Microbiol. **6:**172–173.

Martel, A. Y., F. O. Sottnek, M. L. Thomas, and W. L. Albritton. 1986. Susceptibility of *Haemophilus aegyptius* to trooleandomycin: lack of taxonomic value. Can. J. Microbiol. **32:**289–293.

Mazloum, H. A., M. Kilian, Z. M. Mohamed, and M. D. Said. 1982. Differentiation of *Haemophilus aegyptius* and *Haemophilus influenzae*. Acta Pathol. Microbiol. Immunol. Scand. Sect. B **90:**109–112.

McNicol, P. J., and A. R. Ronald. 1984. The plasmids of *Haemophilus ducreyi*. J. Antimicrob. Agents Chemother. **14:**561–564.

Oberhofer, T. R., and A. E. Back. 1982. Isolation and cultivation of *Haemophilus ducreyi*. J. Clin. Microbiol. **15:**625–629.

Pittman, M. 1953. A classification of the hemolytic bacteria of the genus *Haemophilus: Haemphilus haemolyticus* Bergey et al. and *Haemophilus parahaemolyticus* nov. spec. J. Bacteriol. **65:**750–751.

Pittman, M., and D. J. Davis. 1950. Identification of the Koch-Weeks bacillus (*Haemophilus aegyptius*). J. Bacteriol. **59:**413–426.

Pritchett, I. W., and E. G. Stillman. 1919. The occurrence of *Bacillus influenzae* in throats and saliva. J. Exp. Med. **29:**259–266.

Rivers, T. M. 1922. Influenza-like bacilli: growth of influenza-like bacilli on media containing only an autoclave-labile substance as an accessory food factor. Bull. Johns Hopkins Hosp. **33:**429–431.

Roberts, M. C., T. A. Bell, K. I. Sandstrom, A. L. Smith, and K. K. Holmes. 1986. Characterization of *Haemophilus* spp. isolated from infant conjunctivitis. J. Med. Microbiol. **21:**219–224.

Rubin, S. J., P. A. Granato, and B. L. Wasilauskas. 1985.

Glucose-nonfermenting gram-negative bacteria. p. 330–349. *In* E. H. Lennette, A. Balows, W. J. Hausler, and H. J. Shadomy (ed.), Manual of clinical microbiology, 4th ed. American Society for Microbiology, Washington, D.C.

Russell, D. S., and P. Fildes. 1928. A case of endocarditis in man associated with *Bacillus parainfluenzae* Rivers, 1922. J. Pathol. Bacterol. **31:**651–656.

Savage, D. D., R. L. Kagan, N. A. Young, and A. E. Horvath. 1977. *Cardiobacterium hominis* endocarditis: description of two patients and characterization of the organism. J. Clin. Microbiol. **5:**75–80.

Scheifele, D. W., S. J. Fussell, and M. C. Roberts. 1982. Characterization of ampicillin-resistant *Haemophilus parainfluenzae*. J. Antimicrob. Agents Chemother. **21:**734–739.

Scott, S. S., and D. C. Old. 1981. Mannose-resistant and eluting (MRE) hemagglutinins, fimbriae and surface structure in strains of *Haemophilus*. FEMS Microbiol. Lett. **10:**235–240.

Sottnek, F. O., J. W. Biddle, S. J. Kraus, R. E. Weaver, and J. A. Stewart. 1980. Isolation and identification of *Haemophilus ducreyi* in a clinical study. J. Clin. Microbiol. **12:**170–174.

Suwanagool, S., M. M. Rothkopf, S. M. Smith, D. LeBlanc, and R. Eng. 1983. Pathogenicity of *Eikenella corrodens* in humans. Arch. Intern. Med. **143:**2265–2268.

Tiller, F. W. 1982. Biochemical differentiation of *Haemophilus influenzae:* additional characterization of biotypes by carbohydrate fermentation patterns. Zentralbl. Bakteriol. Microbiol. Hyg. 1 Abt. Orig. A **253:**236–246.

Trollfors, B., J. E. Brorson, B. Claesson, and T. Sandberg. 1985. Invasive infections caused by *Haemophilus* species other than *Haemophilus influenzae*. Infection **13:**12–14.

Tuyau, J. E., and W. Sims. 1983. Aspects of the pathogenicity of some oral and other haemophili. J. Med. Microbiol. **16:**467–475.

Ursi, J. P., E. VanDyck, R. C. Ballard, W. Jacob, P. Piot, and A. Z. Meheus. 1982. Characterization of an unusual bacterium isolated from genital ulcers. J. Med. Microbiol. **15:**97–103.

Vastine, D. W., C. R. Dawson, I. Hoshiwara, C. Yoneda, T. Daghfous, and M. Messadi. 1974. Comparison of media for the isolation of *Haemophilus* species from cases of seasonal conjunctivitis associated with severe endemic trachoma. Appl. Microbiol. **28:**688–690.

Winslow, C. E. A. J., Broadhurst, R. E. Buchanan, C. Krumwiede, L. A. Rogers, and G. H. Smith. 1920. The families and genera of the bacteria: final report of the Committee of the Society of American Bacteriologists on characterization and classification of bacterial types. J. Bacteriol. **5:**191–229.

Zambon, J. J. 1985. *Actinobacillus actinomycetemcomitans* in human periodontal disease. J. Clin. Periodontol. **12:**1–20.

CHAPTER 33

Legionellosis

HAZEL W. WILKINSON

Disease: Legionellosis (*Legionella* pneumonia, Legionnaires disease, and hypothesized but unproved, Pontiac fever).

Etiologic Agents: *Legionella pneumophila* serogroups 1 to 14, *Legionella bozemanii* serogroups 1 and 2, *Legionella dumoffii, Legionella gormanii, Legionella micdadei, Legionella longbeachae* serogroups 1 and 2, *Legionella jordanis, Legionella wadsworthii, Legionella sainthelensi, Legionella feeleii* serogroups 1 and 2, *Legionella hackeliae* serogroups 1 and 2, *Legionella maceachernii, Legionella birminghamensis, Legionella cincinnatiensis,* and presumptively, *Legionella oakridgensis.*

Source: Natural and man-made water reservoirs. Inhalation, ingestion, and aspiration hypothesized as modes of acquisition.

Clinical Manifestations: Acute onset of high fever, recurrent chills, productive or nonproductive cough, myalgia, malaise, relative bradycardia, radiographic evidence of pneumonia, and commonly, evidence of multisystem involvement, including diarrhea, headache, confusion, elevated liver enzymes, hyponatremia, hypophosphatemia, hematuria, and proteinuria.

Pathology: Pneumonia with dense alveolar exudate consisting of neutrophils, macrophages, fibrin, and antigenic material. Hematogenous, lymphatic, endobronchial, and interstitial spread, with legionellae demonstrated occasionally in bone marrow, heart, brain, spleen, liver, kidneys, ileum, and peritoneum. Pericarditis and endocarditis in the absence of pneumonia.

Laboratory Diagnosis: Presumptive identification of isolated culture by Gram stain and growth characteristics, or of *Legionella* cells in tissue specimens or respiratory secretions by direct immunofluorescence assay (DFA). Definitive identification of culture isolates showing typical growth and Gram-staining characteristics by slide agglutination test, DFA, or gene probe (the latter, only to genus level). Gas-liquid chromatographic and DNA hybridization analyses required for atypical strains. Fourfold rise in serum indirect immunofluorescence assay titer to ≥ 128 against *L. pneumophila* serogroup 1 antigen, considered diagnostic, and against all other *Legionella* antigens, considered presumptive.

Epidemiology: Worldwide distribution. Sporadic cases and epidemics, including nosocomial outbreaks. Immunosuppressed patients and those with debilitating, underlying disease at high risk.

Treatment: Erythromycin and, for recalcitrant cases, rifampin.

Prevention and Control: Additional data required to prove effectiveness of various suggested procedures, such as hyperchlorination or heat treatment of hospital water supplies during outbreaks, sterilization of fluids used in lavage procedures, and, for immunocompromised patients, erythromycin prophylaxis.

Description of Disease

Legionellosis (*Legionella* pneumonia, Legionnaires disease) is an infectious disease of humans caused by the gram-negative bacterium *Legionella*. Although two distinct syndromes have been described, *Legionella* species have been isolated only from patients with *Legionella* pneumonia (McDade et al., 1977). A flulike illness, Pontiac fever, was attributed to infection with *L. pneumophila* (Glick et al., 1978) or *L. feeleii* (Herwaldt et al., 1984); however, supportive evidence was based solely on serologic tests with cross-reactive antigens and on demonstration of the organisms in the environment and therefore remains unconfirmed.

Legionella pneumonia is caused by at least 14 of 25 characterized species and additional *Legionella*-like organisms (LLO) not yet assigned to species or serogroup (Table 1). *L. pneumophila* serogroup 1 predominates in both epidemic and sporadic cases (Reingold et al., 1984), and most published data are derived from studying this serogroup. Information on other species and serogroups is noted parenthetically in this chapter.

Communicability and Transmissibility

Legionella species are ubiquitous in freshwater habitats, both natural and man-made. Hypothesized modes of transmission include inhalation of aerosolized bacteria, such as those generated by cooling towers (Fraser, 1980), showers (Dennis et al., 1984), dust (Thacker et al., 1978), or respiratory equipment (Arnow et al., 1982); ingestion of legionellae; and aspiration of contaminated water (Dournon et al., 1982a; Muder et al., 1986). Laboratory-acquired infections are rare, and person-to-person transmission has not been documented.

TABLE 1. *Legionella* species characterized as of August 1, 1987

Human pathogen documented by culture	*Presumptive pathogen by DFA*	*Not yet isolated from human disease*
L. pneumophila (14)[a]	*L. oakridgensis*	*L. anisa*
		L. santicrucis
L. bozemanii (2)		*L. steigerwaltii*
L. dumoffii		*L. parisiensis*
L. gormanii		*L. spiritensis*
L. micdadei		*L. jamestowniensis*
L. longbeachae (2)		*L. cherrii*
L. jordanis		*L. rubrilucens*
L. wadsworthii		*L. erythra*
L. feeleii (2)		*L. israelensis*
L. sainthelensi		
L. hackeliae (2)		
L. maceachernii		
L. birminghamensis		
L. cincinnatiensis		

[a] Number of serogroups.

Clinical Manifestations of Disease

Legionella pneumonia, Pontiac fever, and asymptomatic infection have been attributed to *Legionella* species, because the bacteria were isolated from clinical specimens (for pneumonia only) or because antibody titer increases to *Legionella* whole cells were demonstrated in sera from individuals exposed to environmental sources of the bacteria (pneumonia, Pontiac fever, asymptomatic infection). Therefore, the etiologic agent was proved directly only in pneumonia cases.

LEGIONELLA PNEUMONIA

The illness cannot be definitively diagnosed by clinical signs and symptoms, although suspicion is heightened by evidence of multisystem involvement, negative routine cultures, and prompt response of the patient to erythromycin therapy. *L. pneumophila* serogroup 1 pneumonia (Legionnaires disease) predominates in published case reports. The small number of cases reported for pneumonia due to other legionellae suggest that the clinical picture is similar for all illness caused by *Legionella* species and serogroups. The incubation period is 2 to 10 days (Fraser et al., 1977).

Typically, the patient presents with acute onset of high fever (>39.4°C), recurrent chills, nonproductive cough, myalgia, malaise, relative bradycardia, and, often, evidence of multisystem involvement: diarrhea, headache, confusion, elevated liver enzymes, hyponatremia, hypophosphatemia, hematuria, and proteinuria (Kirby et al., 1980; Swartz, 1979; Tsai, 1979). Moderate leukocytosis with a left shift and elevated erythrocyte sedimentation rate are also characteristic, although leukopenia has also been reported (Miller, 1979). Within several days, the cough may become productive, with occasional hemoptysis. Pneumonia progresses from patchy bronchopneumonia to lobar and multilobar consolidation. Empyema (*L. bozemanii*), pulmonary cavitation, and abscesses have been reported (Brettman et al., 1986; Dowling et al., 1983; Edelstein et al., 1981; Lewin et al., 1979).

Reported extrapulmonary manifestations include myocarditis, endocarditis, and pericarditis, the latter two also in the absence of pneumonia (Nelson et al., 1984); acute renal failure (Fenves, 1985; Tsai et al., 1979); myoglobinemia and rhabdomyolysis (Posner et al., 1980); brainstem and cerebellar dysfunction

(Baker et al., 1981); disseminated intravascular coagulation and thrombocytopenia (Gregory, 1979); and peritonitis (Dournon et al., 1982a). Also reported, but apparently rare, are wound infection (*L. pneumophila* serogroup 4) (Brabender et al., 1983), cutaneous abscess (*L. micdadei*) (Ampel et al., 1985), and sinusitis associated with AIDS (Schlanger, 1984).

Clinical improvement generally precedes radiographic resolution, which can take several months. Relapse has been reported even after appropriate therapy (Sanders et al., 1980). Fatalities occur most often in patients with underlying illness or immunosuppression.

Pontiac Fever

Pontiac fever is an epidemiologically defined, flulike illness with incubation period of 5 to 66 h and short duration of illness (Glick et al., 1978). It is indistinguishable clinically from various other respiratory syndromes caused by viruses. Patients have fever, chills, myalgia, headache, and sometimes a dry cough. The illness resolves without treatment, usually within 1 week. No deaths have been reported. The illness has been described only in epidemic reports in which the only evidence linking Pontiac fever to *Legionella* infection was the isolation of *Legionella* species from an environmental source to which the patients were exposed (or from an exposed sentinel guinea pig) and the patients' antibody responses to *Legionella* antigens. However, legionellae are commonly found in environmental water specimens, and the serologic data require culture confirmation.

Asymptomatic Infection

The observation that some sera from apparently well individuals tested during outbreak investigations show seroconversions to *Legionella* antigens has led to the hypothesis that exposure to environmental sources of *Legionella* can cause asymptomatic infection (Herwaldt et al., 1984). However, several equally plausible, alternative explanations include exposure and subsequent serologic responses to cross-reactive antigens in the environment or in the commensal flora and polyclonal stimulation of the immune system by unrelated mitogens.

Pathology

The most widely hypothesized route of infection is inhalation of aerosols containing *Legionella* cells. In animal models, small droplets (2 to 5 μm) were necessary for infection, and this is consistent with observations of human lung tissue in which the organisms were demonstrated primarily in pulmonary bronchioles, alveolar ducts, and alveolar sacs (Hicklin et al., 1980), and only rarely in larger bronchioles. The primary lesion becomes confluent, frequently lobar, and often multilobar, with no preferential distribution (Baskerville, 1984; Winn and Myerowitz, 1981).

The histopathology has been described as an acute fibrinopurulent pneumonia with dense, alveolar exudate consisting of neutrophils, macrophages, and abundant fibrin and antigenic material (Carrington, 1979; Hicklin et al., 1980). Consolidation is evenly distributed in all lobes. Abscesses have been observed in some cases. There is only slight necrosis. Larger airways can become partially plugged by debris containing phagocytes. Individual *Legionella* cells can be seen in and around phagocytic cells by Dieterle silver impregnation staining (Chandler et al., 1977) and, more specifically, by DFA (Hicklin et al., 1980). Hematogenous, lymphatic, endobronchial, and interstitial spread of the organism probably occurs by phagocyte migration, but whether extrapulmonary manifestations are caused by direct invasion of the legionellae or by their toxins is unknown. *Legionella* toxins include hemolysin (Baine et al., 1979), cytotoxins (Baine, 1985; Friedman, 1982), endotoxin (Wong et al., 1979), and extracellular proteases (Conlan et al., 1986; Dreyfus and Iglewski, 1986; Muller, 1980). *Legionella* species have been cultured from or demonstrated by DFA in blood, lymph nodes, bone marrow, heart, brain, spleen, liver, kidneys, ileum, and the peritoneum (Dournon et al., 1982a; Edelstein et al., 1979; Evans and Winn, 1981; Humayun et al., 1981; Schurmann et al., 1983; Watts et al., 1980).

Untreated or inadequately treated patients who survive may have long-term pulmonary changes, including fibrosis and reduced diffusion capacity (Baskerville, 1984). Although intact bacteria are usually cleared from the lung within 3 weeks (Hicklin et al., 1980), soluble antigen has been detected in the urine of some patients for approximately 1 year after the acute illness (Kohler and Sathapatayavongs, 1983).

Diagnostic Features

Legionella pneumonia cannot be distinguished clinically or radiographically from pneumonias with other etiologies. However, it should be suspected whenever routine stains and cultures of respiratory tissues or secretions are negative for other pathogens. Suspicion is further heightened for patients with high fever ($\geq$39.4°C), relative bradycardia, encephalopathy, diarrhea, hyponatremia, and abnormal liver function tests, and for those who respond promptly to erythromycin therapy. Patients undergoing immunosuppressive therapy for renal or cardiac transplants and those with underlying malignancies or other debilitating disease are especially susceptible to legionellosis (Cohen et al., 1979; Dowling et al., 1984; Fuller et al., 1985; Glazier et al., 1983; Gump et al., 1979; Johnson et al., 1985; Sarovolatz et al., 1979), which should be suspected immediately in these patients if lower respiratory tract symptoms develop. If lung tissue is available for histopathologic staining, the organisms can be seen in sections with the Dieterle silver impregnation stain (Chandler et al., 1977).

The diagnosis can be confirmed only by isolation and subsequent identification of the bacteria or, for *L. pneumophila* serogroup 1 infection, by detection of soluble antigen in urine, demonstration of the organism in lung tissue or respiratory secretions, or detection of at least a fourfold rise in antibody levels to an IFA titer of at least 128. For *Legionella* species or serogroups other than *L. pneumophila* serogroup 1, culture identification is the only method shown to be sufficiently specific, and all other positive test results should be considered presumptive.

Prognosis

The case-fatality ratio for *L. pneumophila* pneumonia is related to immunosuppression and to prompt, appropriate therapy. In one report (Kirby et al., 1980), 25% of untreated, nonimmunosuppressed patients died, whereas only 7% of those treated with erythromycin died. Of the immunosuppressed patients, 80% of the untreated patients died as opposed to 24% of those treated. Similar outcomes were reported in another investigation (Broome et al., 1979), in which 4% of patients treated with erythromycin died compared with 17% who were treated with penicillins, cephalosporins, or aminoglycosides. Relapse or prolonged convalescence in patients treated for 3 weeks (Sanders et al., 1980) or less (Kirby et al., 1980; Saravolatz et al., 1979) has been reported.

Epidemiologic Aspects

Occurrence and Transmission

Legionellosis occurs worldwide, but the true incidence of disease is unknown, because it is both underdiagnosed and underreported. The number of reported cases probably reflects the degree of awareness and laboratory capability of a given region. Freshwater sources, both natural and man-made, are reservoirs of the bacteria (Fliermans, 1984) which grow in association with other organisms, including amoebae (Tyndall and Domingue, 1982), cyanobacteria (Tison et al., 1980), algae (Pope et al., 1982), protozoa (Fields et al., 1984), and other bacteria (Wadowsky and Yee, 1985). These parasitic, commensal, or symbiotic relationships may provide legionellae with both environmental protection and nutrients. *Legionella* species have been demonstrated in lakes, rivers, streams, riparian soil, hot-water tanks, humidifiers, tap water, rubber washers and sediment in plumbing systems, shower heads and their aerosols, ice machines, respiratory therapy equipment, hot-water taps, and cooling tower sediments (Bollin et al., 1985; Ciesielski et al., 1984; Cordes et al., 1981; Fliermans et al., 1981; Howland and Pope, 1983; Jones et al., 1984; Kaan et al., 1985; Politi et al., 1979; Schofield et al., 1985; Stout et al., 1985). In epidemics, cooling towers have been the most often suspected sources of *Legionella* infections.

The largest number of legionellae in cooling towers are found in the summer and fall months (Howland and Pope, 1983). This coincides with the seasonality of legionellosis which occurs in large common-source outbreaks, sporadic cases, and as a series of cases in areas of high endemicity (Baine, 1979). The bacterium was first identified in human specimens during the investigation of the 1976 epidemic in Philadelphia (Fraser et al., 1977). Retrospective serologic studies suggested that two previous epidemics of pneumonia were caused by *L. pneumophila* strains (Osterholm, et al., 1983; Thacker et al., 1978), but the serologic data could not be confirmed by culture. However, the organism was cultured subsequently from human specimens that had been stored since 1947 (McDade et al., 1979), which showed that legionellosis is not a new disease but only a recently recognized disease.

The mode of acquisition of sporadic and epidemic legionellosis can only be speculated on with available data. Hypothesized sources include aerosolized spray from cooling towers, shower heads, hot-water taps, and hospital respiratory therapy equipment. Inhalation, ingestion, and aspiration are suggested modes of acquisition (Muder et al., 1986). Patients with debilitating underlying disease and, particularly, those who are immunosuppressed are at high risk. Nosocomial outbreaks of *Legionella* pneumonia have occurred in bone marrow, cardiac, and renal transplant recipients. Other risk factors include cigarette smoking, excessive alcohol consumption, and advanced age (Baine, 1979). In epidemic cases (Eickhoff, 1979), males have predominated.

Susceptibility, Resistance, and Carrier State

The heightened susceptibility of patients receiving immunosuppressive drugs to *Legionella* pneumonia and the finding of high antibody titers in some patients with a fatal outcome suggest that cellular immunity is more important than humoral immunity in host defense. Animal experiments and histopathologic observations support this hypothesis. In guinea pigs, serum antibody titers did not correlate with protection against aerosol challenge: immunized animals died earlier than controls (Baskerville et al., 1983). In the lung, *Legionella* cells are phagocytosed by macrophages, monocytes, and neutrophils. Opsonization is enhanced by antibody and complement, either by the classical pathway (*L. pneumophila;* Verbrugh et al., 1985) or alternative pathway (*L. micdadei*; Steffensen et al., 1985). Host factors determined the ultimate outcome of ingestion of the legionellae by phagocytes. Factors that do not prevent or that enhance the multiplication of the bacteria with eventual lysis of the phagocytes include immunosuppressive drugs, such as hydrocortisone (Nash et al., 1984) and cyclophosphamide (Widen et al., 1984), and *Legionella* toxins that inhibit or decrease acidification of phagosomes (Horwitz and Maxfield, 1984), phagosome-lysosome fusion (Horwitz, 1983b), hexose monophosphate shunt enzymes, oxygen consumption, iodination (Friedman et al., 1982), and superoxide anion production (*L. pneumophila*—Lochner et al., 1985; *L. micdadei*—Saha et al., 1985). Factors that inhibit the growth of or enhance phagocyte killing of legionellae include lymphokines (Horwitz, 1983a), including interferon (*L. pneumophila*—Bhardwaj et al., 1986; *L. micdadei*—Whitaker-Dowling et al., 1986), and erythromycin (Miller et al., 1984).

Legionella isolates show differing degrees of virulence for various animal, tissue, and protozoan models, but the significance of this finding in human disease is speculative. Carriage has not been found to be significant. Normal throat flora inhibits the growth of *L. pneumophila* cells (Flesher et al., 1980), and pharyngeal colonization is infrequent (Bridge and Edelstein, 1983). However, symbiotic relationships between *L. pneumophila* and *L. micdadei* and *Haemophilus influenzae* and *Neisseria meningitidis* have been noted (Stout et al., 1986a).

Etiologic Agents

Characterization

There are 25 species in the genus *Legionella,* family *Legionellaceae*. Legionellae are aerobic, gram-negative bacilli, 0.5 μm wide × 1–2 μm long. Under adverse growth conditions, they can be pleomorphic with lengths of up to 20 μm. They counterstain poorly with safranin and somewhat better with carbol fuchsin. The cellular fatty acids of *Legionella* species are predominantly branch chained, which is a distinguishing characteristic of the genus along with the presence of ubiquinones containing 10 or more isoprene units (Moss, 1981; Moss et al., 1983). Dihydroxy fatty acids of chain length > 10 carbons distinguish several *Legionella* species from all other procaryotes (Mayberry, 1981).

Legionella strains do not ferment or oxidize carbohydrates but, instead, obtain their carbon and energy from the metabolism of amino acids (George et al., 1980). They produce catalase but not nitrate reductase or urease (Table 2) (Bercovier et al., 1986; Brenner et al., 1985; Weaver and Feeley, 1979; Wilkinson, 1987). Most strains have single, polar flagella and produce a brown pigment when they are grown on tyrosine-supplemented yeast extract agar. Many strains produce a pigment that fluoresces yellow-green, blue-white, or red-pink under long-wave ultraviolet light. Most strains also liquefy gelatin and produce beta-lactamase, and some species hydrolyze hippurate and produce oxidase.

The ultrastructure of *Legionella* cells is typical of a gram-negative bacterium, with a double-membrane outer envelope and pinching-type cell division (Chandler et al., 1979). Immunochemical analyses did not reveal diaminopimelic acid (Keel et al., 1979) but did show that 2-keto-3-deoxy octonate was present (Wong et al., 1979). The lipopolysaccharide (LPS) partitioned in the phenol instead of aqueous phase and therefore differed from the LPS of other gram-negative bacteria (Conlan and Ashworth, 1986; Otten et al., 1986). Also atypical was the weak pyrogenicity of *Legionella* endotoxin in rabbits, although it did give a strong reaction in the Limulus amoebocyte assay (Wong et al., 1979). Serogroup specificity is probably determined primarily by LPS structure (Ciesielski et al., 1986; Conlan and Ashworth, 1986; Otten et al., 1986). However, both species-specific and genuswide proteins in legionellae were identified by SDS-PAGE profiles and immunoblots (Sampson et al., 1986). The genuswide 60-kilodalton (kDa) (range, 57- to 62-kDa) protein may provide improved diagnostic tests if its *Legionella*-specific epitopes are shown to be sufficiently sensitive (Plikaytis et al., 1987).

In the environment, legionellae survive extreme temperature ranges and grow in water, perhaps only in association with other organisms that provide them with nutrients and protection (Fliermans, 1984). Isolated in the laboratory, *Legionella* strains show more restrictive growth requirements. Best growth has been obtained so far on media that contain a detoxicant, such as charcoal, a source of iron and other growth factors; cysteine and other amino acids; and alpha-ketoglutarate (Edelstein, 1981; Feeley et al., 1979; Keathley and Winn, 1985). The

TABLE 2. *Legionella* reactions on biochemical tests[a]

Species	Nitrate reduction	Urease	Glucose fermentation	Gelatin liquefaction	Hippurate hydrolysis	Catalase	Oxidase	Beta-lactamase	Auto fluorescence
L. anisa	−[b]	−	−	+	−	+	+	+	+/− (BW)
L. birminghamensis	−	−	−	+	−	+	+/−	+	+ (YG)
L. bozemanii	−	−	−	+	−	+	+/−	+/−	+ (BW)
L. cherrii	−	−	−	+	−	+	−	+	+ (BW)
L. cincinnatiensis	−	−	−	+	−	+	−	+	−
L. dumoffii	−	−	−	+	−	+	−	+	+ (BW)
L. erythra	−	−	−	+	−	+	+	+	+ (RP)
L. feeleii	−	−	−	−	+/−	+	−	−	−
L. gormanii	−	−	−	+	−	+	−	+	+ (BW)
L. hackeliae	−	−	−	+	−	+ (W)	+	+	−
L. israelensis	−	−	−	+ (W)	−	+	−	+	−
L. jamestowniensis	−	−	−	+	−	+	−	+	−
L. jordanis	−	−	−	+	−	+	+/−	+	−
L. longbeachae	−	−	−	+	−	+	+	+/−	−
L. maceachernii	−	−	−	+	−	+	+	−	−
L. micdadei	−	−	−	−	−	+	+	−	−
L. oakridgensis	−	−	−	+	−	+	−	+ (W)	−
L. parisiensis	−	−	−	+	−	+	+	+	+ (BW)
L. pneumophila	−	−	−	+	+	+	+/−	+	+ (YG)
L. rubrilucens	−	−	−	+	−	+	−	+	+ (RP)
L. sainthelensi	−	−	−	+	−	+	+	+	−
L. santicrucis	−	−	−	+	−	+	+	+	−
L. spiritensis	−	−	−	+	+ (W)	+	+	+	−
L. steigerwaltii	−	−	−	+	−	+	−	+	+ (BW)
L. wadsworthii	−	−	−	+	−	+	−	+	−

[a] All strains required cysteine to grow except for *L. oakridgensis*, which required it only when first isolated. Reprinted with modifications from Wilkinson (1987), with permission.
[b] + = positive; − = negative; +/− = not always positive; W = weakly positive; BW = blue-white; RP = red-pink; YG = yellow-green (not usually observed on charcoal-containing agar for *L. pneumophila*).

requirement for cysteine helps to presumptively identify legionellae. Some species are capnophilic, but most grow without added CO_2 (Edelstein, 1983). They grow best on media maintained at pH 6.9 and when incubated at 35°C.

Colonies are round, convex, and glistening gray with an entire edge. Growth may be mucoidal to the extent that dilution of the inoculum is necessary to isolate colonies. Strains are identified by growth requirements and by serogroup-specific antigens. Isolates showing multiple or no serologic reactivity are identified to the genus level by gas-liquid chromatography (GLC) and to the species level by DNA hybridization.

Specimen Collection and Transport

The kind of specimen obtained usually depends on the patient's condition, which determines whether invasive techniques are used for specimen collection. Lung tissue, bronchial washings, transtracheal aspirates, pleural fluid, chest drainage, and sputum should be cultured, and, if DFA conjugates are available, they can be stained before culture results are available. Blood specimens should be cultured. Acute serum should be obtained within 1 week pre- or postonset and should be stored and tested with a convalescent serum specimen drawn 3 to 6 weeks after onset. Specimens should be refrigerated to reduce overgrowth by contaminants or, for sera, should be frozen at −20 or −70°C.

Unfixed speicmens can be shipped cold with ice packs if they are in transit no longer than 3 days, but should be shipped frozen if they had been stored frozen before shipment. Formalin-fixed and paraffin-embedded specimens can be shipped at room temperature.

Isolation and Identification Procedures (Cherry and McKinney, 1979; Wilkinson, 1987)

Inoculate lung tissue and respiratory secretions on buffered charcoal-yeast extract agar (BCYE). First prepare the inoculum as follows.

With lung tissue, make a fresh cut with a sterile

scalpel and press the fresh cut on and push it across the surface of a BCYE plate. Dilute the tissue 1 : 10 (v/v) in sterile distilled water, and either grind it with a mortar and pestle with alundum or use a homogenizer. Dilute the suspension 1 : 10 in sterile distilled water and inoculate both the 1 : 10 and 1 : 100 dilutions onto BCYE plates.

Concentrate respiratory secretions, such as bronchial washings, pleural fluid, and chest drainage, by centrifuging them at 2,000 × *g* for 15 min. Discard the supernatant fluid, and resuspend the sediment to one-tenth the original volume in sterile distilled water. Inoculate each specimen onto duplicate BCYE plates, one with a heavy inoculum (large drop) and the other with a light inoculum (small drop).

Do not concentrate or dilute sputum. Inoculate sputum specimens onto BCYE plates (heavy and light inocula) and also onto an antimicrobial-supplemented BCYE plate (available commercially for *Legionella* isolation). If subsequent examination shows that the sputum was too contaminated to allow isolation of *Legionella* colonies, acid treatment of the specimen should be attempted (Buesching et al., 1983; Wilkinson, 1987). Mix equal volumes of sputum and 0.2 M HCl-KCl, pH 2.0, and allow the suspension to sit at room temperature for 15 min. Then add 0.1 N KOH to the mixture until the pH is 6.9. Heat can also be used to enrich the specimen for *Legionella* species by placing the sputum in a 60°C water bath for 3 min. Inoculate BCYE plates as above.

Place all BCYE plates in bags or cans to maintain adequate moisture, and incubate them at 35°C for up to 2 weeks. Inspect the plates daily for *Legionella*-like colonies, which usually appear within 2 to 4 days. Transfer isolated colonies to fresh BCYE plates and to BCYE-without-cysteine plates. Incubate and examine as before. *Legionella* colonies are distinguished by their failure to grow on BCYE-without-cysteine.

Inoculate blood specimens into BACTEC (Chester et al., 1983; Rihs et al., 1985) or biphasic BCYE media (Feeley et al., 1979). Incubate at 35°C for 1 week, and subculture daily onto BCYE plates. Alternatively, blood specimens can be plated directly onto BCYE plates (Reinhardt et al., 1987).

Preliminary Identification (Wilkinson, 1987)

Legionella strains can be presumptively identified by their growth characteristics: slow-growing colonies on BCYE agar that are gray, glistening, convex, and round with entire edges, and absence of growth on BCYE-without-cysteine. *Legionella* strains also either fail to grow on or grow poorly on blood agar plates, which can be used instead of the preferred BCYE-without-cysteine when the latter is unavailable. Other organisms with similar growth characteristics include *Capnocytophaga sputigena, Bordetella pertussis*, and *Francisella tularensis*.

Gram stain suspected *Legionella* colonies, preferably with carbol fuchsin instead of safranin counterstain. *Legionella* cells are small, gram-negative bacilli; 0.5 μm wide × 1 to 2 μm long. Depending on their physiologic state, some cells may be pleomorphic, with lengths of up to 20 μm.

Definitive Identification and Serogrouping

No biochemical scheme has been found to definitively identify *Legionella* species. They are identified instead by their reactivity with species- and group-specific antisera, either with slide agglutination tests (SAT) or with DFA. Strains that do not react with antisera to known species and serogroups and those that react with multiple antisera must be identified as belonging to the *Legionella* genus by their GLC profiles and are speciated by DNA relatedness to known species. A commercially available ^{125}I-labeled gene probe can also be used for identification of *Legionella* cultures to the genus level (Wilkinson, et al., 1986).

Slide Agglutination Test (Thacker et al., 1985; Wilkinson, 1987)

Prepare a heavy suspension of the suspected *Legionella* cells in 10% neutral Formalin, and heat it in a boiling water bath for 15 min. Mix a drop of the suspension with a drop of *Legionella* antiserum on a glass slide, and rock the slide for 30 s. Examine the slide for agglutination with indirect lighting over a dark background. A ≥ 3+ reaction (75 to 100% of cells agglutinated) is considered positive. Use known positive and negative *Legionella* strains as controls for each antiserum.

Direct Immunofluorescence Assay (Cherry and McKinney, 1979; Wilkinson, 1987)

Make colony suspensions in a small amount of 1% Formalin in 0.01 M phosphate-buffered saline (PBS), pH 7.6, and make smears on a glass FA slide. Allow the smear to air-dry and gently heat-fix it. Add a drop of conjugate to the smear, and incubate the slide in a moist chamber at room temperature for 20 min. Tap the bottom edge of the slide on a towel to remove excess conjugate. Rinse the slide briefly with a running stream of PBS, and then place it in a PBS bath for 5 min. Rinse the slide briefly with a running stream of distilled water. Allow it to air-dry, apply buffered glycerol, pH 9, mounting fluid, and a cover slip. Examine the slide on an epiilluminated FA microscope equipped with fluorescein filters with the 40X dry objective. Confirm fluorescent cells with 50× and 100× oil objectives. For controls, include a normal conjugate, PBS instead of conjugate, and known positive and negative *Legionella* strains. Bacilli that stain ≥3+ and that are morphologically typical of *Legionella* cells are considered DFA-positive.

The DFA can also be used to stain *Legionella* cells directly in tissue or respiratory secretions. For tissue, make smears from a 1 : 10 suspension in sterile distilled water. Make direct smears of sputum. Centrifuge other respiratory secretions, and make smears from the sediment resuspended to one-tenth volume. Air-dry, gently heat-fix, and, for tissues not already formalinized, cover the smear with 10% Formalin for 10 min, drain, rinse with distilled water, and allow to air-dry. Deparaffinized sections can be used without fixation, because they are already fixed to the slide. Stain with conjugates as described above. Morphologically typical bacilli that stain ≥3+ with a *Legionella* conjugate are considered presumptively positive. Reported cross-reactions include some strains of *Pseudomonas, Flavobacterium-Xanthomonas, Bacteroides,* and *Bordetella* species. Confirm presumptively positive DFA results with culture, if possible.

Gene Probe

The probe can be obtained from Gen-Probe, Inc., San Diego, CA, and consists of ^{125}I-labeled DNA complementary to *Legionella* ribosomal RNA. It identifies strains to the genus level only. Suspend the suspected *Legionella* culture in distilled water to a density of approximately MacFarland no. 2 standard. Add 100 μl of probe solution to 100 μl of culture suspension in a watertight test tube. Incubate it completely submerged in a 72°C water bath for 2 h. Add 5 ml of hydroxyapatite solution to the tube, and mix well by inversion. Incubate in a 72°C water bath for 5 min. Mix by inversion, centrifuge, and discard the supernatant fluid. Add 5 ml of wash buffer to the pellet, and mix by inversion. Centrifuge, discard the supernatant fluid, and count radioactivity in a gamma counter. Calculate the percentage of probe bound by subtracting the CPM of a negative control from the mean CPM of duplicate test samples, dividing by the total CPM in 100 μl of probe solution, and multiplying by 100. Calculate the ratio of this value to the percentage bound by an *L. pneumophila* serogroup 1 control strain. A ratio of ≥0.2 was considered positive for *Legionella* strains in two independent evaluations (Edelstein, 1986; Wilkinson, et al., 1986). The protocol given in the package insert should be examined for any changes recommended by the manufacturer. The probe is being marketed for detection of legionellae in primary clinical specimens as well as in isolated cultures. However, formalinized specimens cannot be tested with the probe, and other primary specimens may give false-negative (Edelstein et al., 1987) or false-positive (Laussucq et al., 1987) test results. Therefore, the probe can be used as a screening test on *Legionella* isolates, and can be used to identify *Legionella*-like organisms for which serologic reagents are unavailable. It should not be used to replace standard methods.

Immunologic Aspects

Most patients with Legionnaires disease excrete soluble *Legionella* antigens in their urine (Kohler, 1986) and produce serum IgG, IgM, and IgA that react with *Legionella* antigens (Wilkinson, 1986). Antigenuria can be detected with radio- or enzyme immunoassays or with latex agglutination reagents (Kohler, 1986), none of which are available commercially. The antigen detection tests are generally available only in research laboratories.

Serum immunoglobulins can be measured with an indirect immunofluorescence assay (IFA) (Wilkinson et al., 1981) and with an enzyme immunoassay (EIA) (Sampson et al., 1983). Paired acute- and convalescent-phase sera must be tested concomitantly.

For IFA, prepare slides by adding a drop of antigen suspension (heat-killed *Legionella* cells in water containing 0.5% normal hen yolk sacs) to each well of a multiwell FA slide. Remove excess liquid with the same pipette. Allow the smears to air-dry, and place the slide in an acetone bath for 15 min. Air-dry and then place a drop of each serum dilution from 1 : 32 to 1 : 256 on an antigen-bearing well. The serum should be diluted 1 : 16 in 3% normal yolk sac suspension and subsequently twofold in PBS. Incubate the slide in a moist chamber at 37°C for 30 min. Rinse the slide in a running stream of PBS, and then place it in a PBS bath for 10 min. Air-dry or gently blot dry, and place a drop of fluorescein isothiocyanate–labeled serum conjugate in each well. Incubate, rinse, and dry as before. Add buffered glycerol mounting fluid, pH 9, and a cover slip. Read reactions with the 40× dry objective on an epiilluminated FA microscope equipped with fluorescein filters. The serum titer is the reciprocal of the highest dilution giving 1+ fluorescence intensity (barely visible yellow-green fluorescence). Controls include known positive and negative sera, PBS instead of conjugate, and PBS instead of serum and conjugate. With *L. pneumophila* serogroup 1 antigens, a ≥ fourfold increase in titer to ≥128 from the acute-phase (±1 week pre- or postonset) to the convalescent-phase (3 to 6 weeks postonset) of illness is considered positive in a patient with pneumonia. Similar titer increases against other *Legionella* species or serogroup antigens can only be considered presumptively positive, because the diagnostic relevance of such titers is unknown. Single or standing titers of ≥256 may indicate exposure to the organism but are not diagnostic, because titers can persist for several years after onset and because 20% of some normal population groups have *Legionella* titers ≥128. The identity of the infecting *Legionella* strain cannot be determined by IFA titers.

Treatment, Prevention, and Control

Treatment

Antimicrobial efficacy may differ from the results expected from susceptibility tests, probably because of the intracellular location of most *Legionella* cells. Erythromycin is currently the drug of choice; 2 to 4 g should be given intravenously or by mouth daily for 3 weeks (Kirby et al., 1980). Rifampin is added to the regimen in recalcitrant cases (*L. bozemanii;* Parry et al., 1985), but it should not be given alone because of the risk of inducing one-step multiple resistance. In vitro erythromycin resistance has been induced in *L. pneumophila* and *L. micdadei* (Dowling et al., 1985). Good clinical responses to imipenem (Farrell et al., 1985) and to combination erythromycin–fusidic acid therapy (Friis-Moller et al., 1985) have also been noted. Isolation of patients is not recommended.

Prevention and Control

Legionellae are ubiquitous in the environment and cannot be permanently eradicated from potential environmental disseminators, such as cooling towers. Instead, routine maintenance, including cleaning and biocide application, has been recommended to reduce numbers of *Legionella* cells and the sediments they inhabit. Routine culturing of cooling towers, as advocated by several companies that sell such services, is costly and provides no useful information.

Alternative strategies include protection of high-risk individuals from exposure to and infection with *Legionella* strains. Hyperchlorination of hospital water supplies (Fischer-Hoch et al., 1981) and heating the water to ≥60°C (Best et al., 1984; Stout et al., 1986b) have reduced numbers of viable legionellae and coincided with cessation of nosocomial outbreaks. However, this approach is controversial because of cost, potential hazards, and damage to plumbing (Parry et al., 1985). *Legionella* cells are susceptible to ultraviolet light (Antopol and Ellner, 1979) but can subsequently photoreactivate (Knudson, 1985), thereby decreasing the efficacy of this method. Special precautions may be required for immunosuppressed patients, such as sterilization of lavage fluids (Dournon et al., 1982b) and erythromycin prophylaxis (Vereerstraeten et al., 1986). Additional data are needed to determine the efficacy of these theoretically plausible control strategies.

Literature Cited

Ampel, N. M., F. L. Ruben, and C. W. Norden. 1985. Cutaneous abscess caused by *Legionella micdadei* in an immunosuppressed patient. Ann. Intern. Med. **102:**630–631.

Antopol, S. C., and P. D. Ellner. 1979. Susceptibility of *Legionella pneumophila* to ultraviolet radiation. Appl. Environ. Microbiol. **38:**347–348.

Arnow, P. M., T. Chou, D. Weil, E. N. Shapiro, and C. Kretzschmar. 1982. Nosocomial Legionnaires' disease caused by aerosolized tap water from respiratory devices. J. Infect. Dis. **146:**460–467.

Baine, W. B. 1979. Clinical manifestations of Legionnaires' disease and recommended therapy. p. 3–8. *In* G. L. Jones and G. A. Hebert (eds.), Legionnaires': the disease, the bacterium and methodology. Centers for Disease Control, Atlanta.

Baine, W. B. 1985. Cytolytic and phospholipase C activity in *Legionella* species. J. Gen. Microbiol. **131:**1383–1391.

Baine, W. B., J. K. Rasheed, D. C. Mackel, C. A. Bopp, J. G. Wells, and A. F. Kaufmann. 1979. Exotoxin activity associated with the Legionnaires' disease bacterium. J. Clin. Microbiol. **9:**453–456.

Baker, P. C., T. R. Price, and C. D. Allen. 1981. Brain stem and cerebellar dysfunction with Legionnaires' disease. J. Neurol. Neurosurg. Psychiatry **44:**1054–1056.

Baskerville, A. 1984. Pathology and Pathophysiology. p. 136–141. *In* C. Thornsberry, A. Balows, J. C. Feeley, and W. Jakubowski (eds.), *Legionella:* Proceedings of the 2nd International Symposium, American Society for Microbiology, Washington, D.C.

Baskerville, A., R. B. Fitzgeorge, J. W. Conlan, L. A. E. Ashworth, D. H. Gibson, and C. P. Morgan. 1983. Studies on protective immunity to aerosol challenge with *Legionella pneumophila.* Zbl. Bakt. Hyg. I Abt. Orig. A **255:**150–155.

Bercovier, H., A. G. Steigerwalt, M. Derhi-Cochin, C. W. Moss, H. W. Wilkinson, R. F. Benson, and D. J. Brenner. 1986. Isolation of legionellae from oxidation ponds and fishponds in Israel and description of *Legionella israelensis* sp. nov. Int. J. Syst. Bacteriol. **36:**368–371.

Best, M., A. Goetz, and V. L. Yu. 1984. Heat eradication measures for control of nosocomial Legionnaires' disease. Am. J. Infect. Control **12:**26–30.

Bhardwaj, N., T. W. Nash, and M. A. Horwitz. 1986. Interferon-activated human monocytes inhibit the intracellular multiplication of *Legionella pneumophila.* J. Immunol. **137:**2662–2669.

Bollin, G. E., J. F. Plouffe, M. F. Para, and B. Hackman. 1985. Aerosols containing *Legionella pneumophila* generated by shower heads and hot-water faucets. Appl. Environ. Microbiol. **50:**1128–1131.

Brabender, W., D. R. Hinthorn, M. Asher, N. J. Lindsey, and C. Liu. 1983. *Legionella pneumophila* wound infection. J. Am. Med. Assoc. **250:**3091–3092.

Brenner, D. J., A. G. Steigerwalt, G. W. Gorman, H. W. Wilkinson, W. F. Bibb, M. Hackel, R. L. Tyndall, J. Campbell, J. C. Feeley, W. L. Thacker, P. Skaliy, W. T. Martin, B. J. Brake, B. S. Fields, H. V. McEachern, and L. K. Corcoran. 1985. Ten new species of *Legionella*. Int. J. Syst. Bacteriol. **35**:50–59.

Brettman, L. R., D. DeHertogh, E. L. Rank, and M. A. Mandour. 1986. *Legionella bozemanii* and empyema. Ann. Intern. Med. **105**:146–147.

Bridge, J. A., and P. H. Edelstein. 1983. Oropharyngeal colonization with *Legionella pneumophila*. J. Clin. Microbiol. **18**:1108–1112.

Broome, C. V., S. A. J. Goings, S. B. Thacker, R. L. Vogt, H. N. Beaty, D. W. Fraser, and Field Investigation Team. 1979. The Vermont epidemic of Legionnaires' disease. Ann. Intern. Med. **90**:573–577.

Bueschi ng, W. J., R. A. Brust, and L. W. Ayers. 1983. Enhanced primary isolation of *Legionella pneumophila* from clinical specimens by low-pH treatment. J. Clin. Microbiol. **17**:1153–1155.

Carrington, C. B. 1979. Pathology of Legionnaires' disease. Ann. Intern. Med. **90**:496–499.

Chandler, F. W., M. D. Hicklin, and J. A. Blackmon. 1977. Demonstration of the agent of Legionnaires' disease in tissue. N. Engl. J. Med. **297**:1218–1220.

Chandler, F. W., R. M. Cole, M. D. Hicklin, J. A. Blackman, and C. S. Callaway. 1979. Ultrastructure of the Legionnaires' disease bacterium. Ann. Intern. Med. **90**:642–647.

Cherry, W. B., and R. M. McKinney. 1979. Detection of Legionnaires' disease bacteria in clinical specimens by direct immunofluorescence. p. 91–103. *In* G. L. Jones and G. A. Hebert (eds.), Legionnaires': the disease, the bacterium and methodology. Centers for Disease Control, Atlanta.

Chester, B., E. G. Poulos, M. J. Demaray, E. Albin, and T. Prilucik. 1983. Isolation of *Legionella pneumophila* serogroup 1 from blood with nonsupplemented blood culture bottles. J. Clin. Microbiol. **17**:195–197.

Ciesielski, C. A., M. J. Blaser, and W.-L. L. Wang. 1984. Role of stagnation and obstruction of water flow in isolation of *Legionella pneumophila* from hospital plumbing. 1984. Appl. Environ. Microbiol. **48**:984–987.

Ciesielski, C. A., M. J. Blaser, and W.-L. L. Wang. 1986. Serogroup specificity of *Legionella pneumophila* is related to lipopolysaccharide characteristics. Infect. Immun. **51**:397–404.

Cohen, M. L., C. V. Broome, A. L. Paris, W. T. Martin, and J. R. Allen. 1979. Fatal nosocomial Legionnaires' disease: clinical and epidemiologic characteristics. Ann. Intern. Med. **90**:611–613.

Conlan, J. W., and L. A. E. Ashworth. 1986. The relationship between the serogroup antigen and lipopolysaccharide of *Legionella pneumophila*. J. Hyg. (Lond.) **96**:39–48.

Conlan, J. W., A. Baskerville, and L. A. E. Ashworth. 1986. Separation of *Legionella pneumophila* proteases and purification of a protease which produces lesions like those of Legionnaires' disease in guinea pig lung. J. Gen. Microbiol. **132**:1565–1574.

Cordes, L. G., A. M. Wiesenthal, G. W. Gorman, J. P. Phair, H. M. Sommers, A. Brown, V. L. Yu, M. H. Magnussen, R. D. Meyer, J. S. Wolf, K. N. Shands, and D. W. Fraser. 1981. Isolation of *Legionella pneumophila* from hospital shower heads. Ann. Intern. Med. **94**:195–197.

Dennis, P. J. L., A. E. Wright, D. A. Rutter, J. E. Death, and B. P. C. Jones. 1984. *Legionella pneumophila* in aerosols from shower baths. J. Hyg. (Lond.) **93**:349–353.

Dournon, E., A. Bure, J. L. Kemeny, J. L. Pourriat, and D. Valeyre. 1982a. *Legionella pneumophila* peritonitis. Lancet **1**:1363.

Dournon, E., A. Bure, N. Desplaces, M. F. Carette, and C. Mayaud. 1982b. Legionnaires' disease related to gastric lavage with tap water. Lancet **1**:797–798.

Dowling, J. N., F. J. Kroboth, M. Karpf, R. B. Yee, and A. W. Pasculle. 1983. Pneumonia and multiple lung abscesses caused by dual infection with *Legionella micdadei* and *Legionella pneumophila*. Am. Rev. Respir. Dis. **127**:121–125.

Dowling, J. N., A. W. Pasculle, F. N. Frola, M. K. Zaphyr, and R. B. Yee. 1984. Infections caused by *Legionella micdadei* and *Legionella pneumophila* among renal transplant recipients. J. Infect. Dis. **149**:703–713.

Dowling, J. N., D. A. McDevitt, and A. W. Pasculle. 1985. Isolation and preliminary characterization of erythromycin-resistant variants of *Legionella micdadei* and *Legionella pneumophila*. Antimicrob. Agents Chemother. **27**:272–274.

Dreyfus, L. A., and B. H. Iglewski. 1986. Purification and characterization of an extracellular protease of *Legionella pneumophila*. Infect. Immun. **51**:736–743.

Edelstein, P. H. 1981. Improved semiselective medium for isolation of *Legionella pneumophila* from contaminated clinical and environmental specimens. J. Clin. Microbiol. **14**:298–303.

Edelstein, P. H. 1983. Culture diagnosis of *Legionella* infections. Zbl. Bakt. Hyg. I Abt. Orig. A **255**:96–101.

Edelstein, P. H. 1986. Evaluation of the Gen-probe DNA probe for the detection of legionellae in culture. J. Clin. Microbiol. **23**:481–484.

Edelstein, P. H., R. N. Bryan, R. K. Enns, D. E. Kohne, and D. L. Kacien. 1987. Retrospective study of Gen-Probe rapid diagnostic system for detection of legionellae in frozen clinical respiratory tract samples. J. Clin. Microbiol. **25**:1022–1026.

Edelstein, P. H., R. D. Meyer, and S. M. Finegold. 1979. Isolation of *Legionella pneumophila* from blood. Lancet **1**:750–751.

Edelstein, P. H., R. D. Meyer, and S. M. Finegold. 1981. Long-term followup of two patients with pulmonary cavitation caused by *Legionella pneumophila*. Am. Rev. Respir. Dis. **124**:90–93.

Eickhoff, T. C. 1979. Epidemiology of Legionnaires' disease. Ann. Intern. Med. **90**:499–502.

Evans, C. P., and W. C. Winn, Jr. 1981. Extrathoracic localization of *Legionella pneumophila* in Legionnaires' pneumonia. Am. J. Clin. Pathol. **76**:813–815.

Farrell, I. D., J. Barker, P. L. Chiodini, J. G. P. Hutchinson, and A. M. Geddes. 1985. The activity of imipenem of *Legionella pneumophila*, with a note on the treatment of two cases. J. Antimicrob. Chemother. **16**:61–65.

Feeley, J. C., R. J. Gibson, G. W. Gorman, N. C. Langford, J. K. Rasheed, D. C. Mackel, and W. B. Baine. 1979. Charcoal-yeast extract agar: primary isolation medium for *Legionella pneumophila*. J. Clin. Microbiol. **10**:437–441.

Fenves, A. Z. 1985. Legionnaires' disease associated with acute renal failure: a report of two cases and review of the literature. Clin. Nephrol. **23**:96–100.

Fields, B. S., E. B. Shotts, Jr., J. C. Feeley, G. W. Gorman, and W. T. Martin. 1984. Proliferation of *Legionella pneumophila* as an intracellular parasite of the ciliated protozoan *Tetrahymena pyriformis*. Appl. Environ. Microbiol. **47**:467–471.

Fischer-Hoch, S. P., C. L. R. Bartlett, J. O. Tobin, M. B. Gillett, A. M. Nelson, J. E. Pritchard, M. G. Smith, R. A. Swann, J. M. Talbot, and J. A. Thomas. 1981. Investigation and control of an outbreak of Legionnaires' disease in a district general hospital. Lancet **1**:932–936.

Flesher, A. R., D. L. Kasper, P. A. Modern, and E. O. Mason, Jr. 1980. *Legionella pneumophila:* growth inhibition by human pharyngeal flora. J. Infect. Dis. **142**:313–317.

Fliermans, C. B. 1984. Philosophical ecology: *Legionella* in historical perspective. p. 285–289. *In* C. Thornsberry, A. Balows, J. C. Feeley, and W. Jakubowski (eds.), *Legionella:* Proceedings of the 2nd International Symposium, American Society for Microbiology, Washington, D.C.

Fliermans, C. B., W. B. Cherry, L. H. Orrison, S. J. Smith, D. L. Tison, and D. H. Pope. 1981. Ecological distribution of *Legionella pneumophila*. Appl. Environ. Microbiol. **41**:9–16.

Fraser, D. W. 1980. Legionellosis: evidence of airborne transmission. Ann. N.Y. Acad. Sci. **353**:61–66.

Fraser, D. W., T. R. Tsai, W. Orenstein, W. E. Parkin, H. J. Beecham, R. G. Sharrar, J. Harris, G. F. Mallison, S. M. Martin, J. E. McDade, C. C. Shepard, P. S. Brachman, and Field Investigation Team. 1977. Legionnaires' disease: description of an epidemic of pneumonia. N. Engl. J. Med. **297**:1189–1197.

Friedman, R. L., J. E. Lochner, R. H. Bigley, and B. H. Iglewski. 1982. The effects of *Legionella pneumophila* toxin on oxidative processes and bacterial killing of human polymorphonuclear leukocytes. J. Infect. Dis. **146**:328–334.

Friis-Moller, A., C. Rechnitzer, L. Nielsen, and S. Madsen. 1985. Treatment of *Legionella* lung abscess in a renal transplant recipient with erythromycin and fusidic acid. Eur. J. Clin. Microbiol. **4**:513–515.

Fuller, J., M. M. Levinson, J. R. Kline, and J. Copeland. 1985. Legionnaires' disease after heart transplantation. Ann. Thorac. Surg. **39**:308–311.

George, J. R., L. Pine, M. W. Reeves, and W. K. Harrell. 1980. Amino acid requirements of *Legionella pneumophila*. J. Clin. Microbiol. **11**:286–291.

Glazier, M. C., R. B. Kohler, and R. L. Campbell. 1983. Legionnaires' disease in postoperative neurosurgical patients. J. Neurosurg. **59**:596–600.

Glick, T. H., M. B. Gregg, B. Berman, G. Mallison, W. W. Rhodes, Jr., and I. Kassanoff. 1978. Pontiac fever: an epidemic of unknown etiology in a health department. I. Clinical and epidemiologic aspects. Am. J. Epidemiol. **107**:149–160.

Gregory, D. W., W. Schaffner, R. H. Alford, A. B. Kaiser, and Z. A. McGee. 1979. Sporadic cases of Legionnaires' disease: the expanding clinical spectrum. Ann. Intern. Med. **90**:518–521.

Gump, D. W., R. O. Frank, W. C. Winn, Jr., R. S. Foster, Jr., C. V. Broome, and W. B. Cherry. 1979. Legionnaires' disease in patients with associated serious disease. Ann. Intern. Med. **90**:538–542.

Herwaldt, L. A., G. W. Gorman, T. McGrath, S. Toma, B. Brake, A. W. Hightower, J. Jones, A. L. Reingold, P. A. Boxer, P. W. Tang, C. W. Moss, H. Wilkinson, D. J. Brenner, A. G. Steigerwalt, and C. V. Broome. 1984. A new *Legionella* species, *Legionella feeleii* species nova, causes Pontiac fever in an automobile plant. Ann. Intern. Med. **100**:333–338.

Hicklin, M. D., B. M. Thomason, F. W. Chandler, and J. A. Blackmon. 1980. Pathogenesis of acute Legionnaires' disease pneumonia. Am. J. Clin. Pathol. **73**:480–487.

Horwitz, M. A. 1983a. Symbiotic interactions between *Legionella pneumophila* and human leukocytes. Int. Rev. Cytol. **14**:307–328.

Horwitz, M. A. 1983b. The Legionnaires' disease bacterium (*Legionella pneumophila*) inhibits phagosome-lysosome fusion in human monocytes. J. Exp. Med. **158**:2108–2126.

Horwitz, M. A., and F. R. Maxfield. 1984. *Legionella pneumophila* inhibits acidification of its phagosome in human monocytes. J. Cell Biol. **99**:1936–1943.

Howland, E. B., and D. H. Pope. 1983. Distribution and seasonality of *Legionella pneumophila* in cooling towers. Curr. Microbiol. **9**:319–323.

Humayun, H. M., T. J. Bird, J. T. Daugirdas, R. C. Fruin, M. M. Shawky, and T. S. Ing. 1981. Legionnaires' disease bacillus in bone marrow. Can. Med. Assoc. J. **125**:1084–1085.

Johnson, J. T., V. L. Yu, M. G. Best, R. M. Vickers, A. Goetz, R. Wagner, H. Wicker, and A. Woo. 1985. Nosocomial legionellosis in surgical patients with head-and-neck cancer: implications for epidemiological reservoir and mode of transmission. Lancet **2**:298–300.

Jones, E., P. Checko, A. Dalton, J. Cope, J. Barbaree, G. Klein, W. Martin, and C. Broome. 1984. Nosocomial Legionnaires disease associated with exposure to respiratory therapy equipment, Connecticut. p. 225–227. *In* C. Thornsberry, A. Balows, J. C. Feeley, and W. Jakubowski (eds.), *Legionella:* proceedings of the 2nd International Symposium, American Society for Microbiology, Washington, D.C.

Kaan, J. A., A. M. Simoons-Smit, and D. M. MacLaren. 1985. Another source of aerosol causing nosocomial Legionnaires' disease. J. Infect. **11**:145–148.

Keathley, J. D., and W. C. Winn, Jr. 1985. Comparison of media for recovery of clinical isolates of *Legionella pneumophila*. J. Clin. Pathol. **83**:498–499.

Keel, J. A., W. R. Finnerty, and J. C. Feeley. 1979. Fine structure of the Legionnaires' disease bacterium. Ann. Intern. Med. **90**:652–655.

Kirby, B. D., K. M. Snyder, R. D. Meyer, and S. M. Finegold. 1980. Legionnaires' disease: report of sixty-five nosocomially acquired cases and review of the literature. Medicine **59**:188–205.

Knudson, G. B. 1985. Photoreactivation of UV-irradiated *Legionella pneumophila* and other *Legionella* species. 1985. Appl. Environ. Microbiol. **49**:975–980.

Kohler, R. B. 1986. Antigen detection for the rapid diagnosis of *Mycoplasma* and *Legionella* pneumonia. Diagn. Microbiol. Infect. Dis. **4**:47S–59S.

Kohler, R. B., and B. Sathapatayavongs. 1983. Recent advances in the diagnosis of serogroup 1 *L. pneumophila* pneumonia by detection of urinary antigen. Zbl. Bakt. Hyg. I Abt. Orig. A **255**:102–107.

Laussucq, S., D. Schuster, W. J. Alexander, H. Wilkinson, and J. Spika. 1987. Cluster of respiratory illness associated with positive DNA probe for *Legionella*. Abstr. Annu. Meet. ICAAC, New York.

Lewin, S., L. R. Brettman, E. J. C. Goldstein, R. S. Holzman, H. Devila, F. Taubman, M. F. Sierra, and P. H. Edelstein. 1979. Legionnaires' disease: a cause of severe abscess-forming pneumonia. Am. J. Med. **67**:339–342.

Lochner, J. E., R. H. Bigley, and B. H. Iglewski. 1985. Defective triggering of polymorphonuclear leukocyte oxidative metabolism by *Legionella pneumophila* toxin. J. Infect. Dis. **151**:42–46.

Mayberry, W. R. 1981. Dihydroxy and monohydroxy fatty acids in *Legionella pneumophila*. J. Bacteriol. **147**:373–381.

McDade, J. E., C. C. Shepard, D. W. Fraser, T. R. Tsai, M. A. Redus, W. R. Dowdle, and Laboratory Investigation Team. 1977. Legionnaires' disease: isolation of a bacterium and demonstration of its role in other respiratory disease. N. Engl. J. Med. **297**:1197–1203.

McDade, J. E., D. J. Brenner, and F. M. Bozeman. 1979. Legionnaires' disease bacterium isolated in 1947. Ann. Intern. Med. **90**:659–661.

Miller, A. C. 1979. Early clinical differentiation between

Legionnaires' disease and other sporadic pneumonias. Ann. Intern. Med. **90**:526–528.

Miller, M. F., J. R. Martin, P. Johnson, J. T. Ulrich, E. J. Rdzok, and P. Billing. 1984. Erythromycin uptake and accumulation by human polymorphonuclear leukocytes and efficacy of erythromycin in killing ingested *Legionella pneumophila*. J. Infect. Dis. **149**:714–718.

Moss, C. W. 1981. Gas-liquid chromatography as an analytical tool in microbiology. J. Chromatogr. **203**:337–347.

Moss, C. W., W. F. Bibb, D. E. Karr, and G. O. Guerrant. 1983. Chemical analysis of the genus *Legionella:* fatty acids and isoprenoid quinones. INSERM Colloq. **114**:375–381.

Muder, R. R., V. L. Yu, and A. H. Woo. 1986. Mode of transmission of *Legionella pneumophila*. Arch. Intern. Med. **146**:1607–1612.

Muller, H. E. 1980. The action of proteolytic enzymes of *Legionella pneumophila* on human serum proteins. Zbl. Bakt. I. Abt. Orig. A **246**:74–79.

Nash, T. W., D. M. Libby, and M. A. Horwitz. 1984. Interaction between the Legionnaires' disease bacterium (*Legionella pneumophila*) and human alveolar macrophages. J. Clin. Invest. **74**:771–782.

Nelson, D. P., E. R. Rensimer, C. M. Burke, and T. A. Raffin. 1984. Cardiac legionellosis. Chest **86**:807–808.

Osterholm, M. T., T. D. Y. Chin, D. O. Osborne, H. B. Dull, A. G. Dean, D. W. Fraser, P. S. Hayes, and W. N. Hall. 1983. A 1957 outbreak of Legionnaires' disease associated with a meat packing plant. Am. J. Epidemiol. **117**:60–67.

Otten, S., S. Iyer, W. Johnson, and R. Montgomery. 1986. Serospecific antigens of *Legionella pneumophila*. J. Bacteriol. **167**:893–904.

Parry, M. F., L. Stampleman, J. H. Hutchinson, D. Folta, M. G. Steinberg, and L. J. Krasnogor. 1985. Waterborne *Legionella bozemanii* and nosocomial pneumonia in immunosuppressed patients. Ann. Intern. Med. **103**:205–210.

Plikaytis, B. B., G. M. Carlone, C.-P. Pau, and H. W. Wilkinson. 1987. Purified 60-kilodalton *Legionella* protein antigen with *Legionella*-specific and nonspecific epitopes. J. Clin. Microbiol. **25**:2080–2084.

Politi, B. D., D. W. Fraser, G. F. Mallison, J. V. Mohatt, G. K. Morris, C. M. Patton, J. C. Feeley, R. D. Telle, and J. V. Bennett. 1979. A major focus of Legionnaires' disease in Bloomington, Indiana. Ann. Intern. Med. **90**:587–591.

Pope, D. H., R. J. Soracco, H. K. Gill, and C. B. Fliermans, 1982. Growth of *Legionella pneumophila* in two-membered cultures with green algae and cyanobacteria. **7**:319–321.

Posner, M. R., M. A. Caudill, R. Brass, and E. Ellis. 1980. Legionnaires' disease associated with rhabdomyolysis and myoglobinuria. Arch. Intern. Med. **140**:848–850.

Reingold, A. L., B. M. Thomason, B. J. Brake, L. Thacker, H. W. Wilkinson, and J. N. Kuritsky. 1984. *Legionella* pneumonia in the United States: the distribution of serogroups and species causing human illness. J. Infect. Dis. **148**:819.

Reinhardt, J. F., C. Nakahama, and P. H. Edelstein. 1987. Comparison of blood culture methods for recovery of *Legionella pneumophila* from the blood of guinea pigs with experimental infection. J. Clin. Microbiol. **25**:719–721.

Rihs, J. D., V. L. Yu, J. J. Zuravleff, A. Goetz, and R. R. Muder. 1985. Isolation of *Legionella pneumophila* from blood with the BACTEC system: a prospective study yielding positive results. J. Clin. Microbiol. **22**:422–424.

Saha, A. K., J. N. Dowling, K. L. LaMarco, S. Das, A. T. Remaley, N. Olomu, M. T. Pope, and R. H. Glew. 1985. Properties of an acid phosphatase from *Legionella micdadei* which blocks superoxide anion production by human neutrophils. Arch. Biochem. Biophys. **243**:150–160.

Sampson, J. S., B. B. Plikaytis, and H. W. Wilkinson. 1986. Immunologic response of patients with legionellosis against major protein-containing antigens of *Legionella pneumophila* serogroup 1 as shown by immunoblot analysis. J. Clin. Microbiol. **23**:92–99.

Sampson, J. S., H. W. Wilkinson, V. C. W. Tsang, and B. J. Brake. 1983. Kinetic-dependent enzyme-linked immunosorbent assay for detection of antibodies to *Legionella pneumophila*. J. Clin. Microbiol. **18**:1340–1344.

Sanders, K. L., D. H. Walker, and T. J. Lee. 1980. Relapse of Legionnaires' disease in a renal transplant recipient. Arch. Intern. Med. **140**:833–834.

Saravolatz, L. D., K. H. Burch, E. Fisher, T. Madhavan, D. Kiani, T. Neblett, and E. L. Quinn. 1979. The compromised host and Legionnaires' disease. Ann. Intern. Med. **90**:533–537.

Schlanger, G., L. I. Lutwick, M. Kurzman, B. Hoch, and F. W. Chandler. 1984. Sinusitis caused by *Legionella pneumophila* in a patient with the acquired immune deficiency syndrome. Am. J. Med. **77**:957–960.

Schofield, G. M., and R. Locci. 1985. The persistence of *Legionella pneumophila* in non-sterile, sterile and artificial hard waters and their growth pattern on tap washer fittings. J. Appl. Bacteriol. **59**:519–527.

Schurmann, D., G. Grosse, I. Horbach, and F. J. Fehrenbach. 1983. Pulmonary and extrapulmonary manifestations of *L. pneumophila*. Zbl. Bakt. Hyg. I Abt. Orig. A **255**:120–126.

Steffensen, D. O., D. L. Weinbaum, and J. N. Dowling. 1985. Opsonic requirements for phagocytosis of *Legionella micdadei* by polymorphonuclear leukocytes. Infect. Immun. **49**:695–699.

Stout, J. E., V. L. Yu, and P. Muraca. 1985. Isolation of *Legionella pneumophila* from the cold water of hospital ice machines: implications for origin and transmission of the organism. Infect. Control **6**:141–146.

Stout, J. E., M. G. Best, V. L. Yu, and J. D. Rihs. 1986a. A note on symbiosis of *Legionella pneumophila* and *Tatlockia micdadei* with human respiratory flora. J. Appl. Bacteriol. **60**:297–299.

Stout, J. E., M. G. Best, and V. L. Yu. 1986b. Susceptibility of members of the family *Legionellaceae* to thermal stress: implications for heat eradication methods in water distribution systems. Appl. Environ. Microbiol. **52**:396–399.

Swartz, M. N. 1979. Clinical aspects of Legionnaires' disease. Ann. Intern. Med. **90**:492–495.

Thacker, S. B., J. V. Bennett, T. F. Tsai, D. W. Fraser, J. E. McDade, C. C. Shepard, K. H. Williams, Jr., W. H. Stuart, H. B. Dull, and T. C. Eickhoff. 1978. An outbreak in 1965 of severe respiratory illness caused by the Legionnaires' disease bacterium. J. Infect. Dis. **138**:512–519.

Thacker, W. L., B. B. Plikaytis, and H. W. Wilkinson. 1985. Identification of 22 *Legionella* species and 33 serogroups with the slide agglutination test. J. Clin. Microbiol. **21**:779–782.

Tison, D. L., D. H. Pope, W. B. Cherry, and C. B. Fliermans. 1980. Growth of *Legionella pneumophila* in association with blue-green algae (cyanobacteria). Appl. Environ. Microbiol. **39**:456–459.

Tsai, T. F., D. R. Finn, B. D. Plikaytis, W. McCauley, S. M. Martin, and D. W. Fraser. 1979. Legionnaires' disease: clinical features of the epidemic in Philadelphia. Ann. Intern. Med. **90**:509–517.

Tyndall, R. L., and E. L. Domingue. 1982. Cocultivation of

Legionella pneumophila and free-living amoebae. Appl. Environ. Microbiol. **44**:954–959.

Verbrugh, H. A., D. A. Lee, G. R. Elliott, W. F. Keane, J. R. Hoidal, and P. K. Peterson. 1985. Opsonization of *Legionella pneumophila* in human serum: key roles for specific antibodies and the classical complement pathway. Immunology **54**:643–653.

Vereerstraeten, P., J.-C. Stolear, E. Schoutens-Serruys, N. Maes, J.-P. Thys, C. Liesnard, F. Rost, P. Kinnaert, and C. Toussaint. 1986. Erythromycin prophylaxis for Legionnaire's disease in immunosuppressed patients in a contaminated hospital environment. Transplantation **41**:52–54.

Wadowsky, R. M., and R. B. Yee. 1985. Effect of non-*Legionellaceae* bacteria on the multiplication of *Legionella pneumophila* in potable water. Appl. Environ. Microbiol. **49**:1206–1210.

Watts, J. C., M. D. Hicklin, B. M. Thomason, C. S. Callaway, and A. J. Levine. 1980. Fatal pneumonia caused by *Legionella pneumophila*, serogroup 3: demonstration of the bacilli in extrathoracic organs. Ann. Intern. Med. **92**:186–188.

Weaver, R. E., and J. C. Feeley. 1979. Cultural and biochemical characterization of the Legionnaires' disease bacterium. p. 19–25. *In* G. L. Jones and G. A. Hebert (eds.), Legionnaires': the disease, the bacterium and methodology. Centers for Disease Control, Atlanta.

Whitaker-Dowling, P., J. N. Dowling, L. Liu, and J. S. Youngner. 1986. Interferon inhibits the growth of *Legionella micdadei* in mouse L cells. J. Interferon Res. **6**:107–114.

Widen, R., T. Klein, and H. Friedman. 1984. Enhanced susceptibility of cyclophosphamide-treated mice to infection with *Legionella pneumophila*. J. Infect. Dis. **149**:1023–1024.

Wilkinson, H. W. 1986. Serodiagnosis of *Legionella pneumophila* disease. p. 395–398. *In* N. R. Rose, H. Friedman, and J. L. Fahey (eds.), Manual of clinical laboratory immunology, 3rd ed. American Society for Microbiology, Washington, D.C.

Wilkinson, H. W. 1987. Hospital-laboratory diagnosis of *Legionella* infections. Centers for Disease Control, Atlanta, 42 p.

Wilkinson, H. W., D. D. Cruce, and C. V. Broome. 1981. Validation of *Legionella pneumophila* indirect immunofluorescence assay with epidemic sera. J. Clin. Microbiol. **13**:139–146.

Wilkinson, H. W., J. S. Sampson, and B. B. Plikaytis. 1986. Evaluation of a commercial gene probe for identification of *Legionella* cultures. J. Clin. Microbiol. **23**:217–220.

Winn, W. C., Jr. and R. L. Myerwitz. 1981. The pathology of the *Legionella* pneumonias. Hum. Pathol. **12**:401–422.

Wong, K. H., C. W. Moss, D. H. Hochstein, R. J. Arko, and W. O. Schalla. 1979. "Endotoxicity" of the Legionnaires' disease bacterium. Ann. Intern. Med. **90**:624–627.

CHAPTER 34

Leprosy

KAZUNARI NAKAMURA

Disease: Leprosy (Hansen's disease).

Etiologic Agent: *Mycobacterium leprae* (leprosy bacillus, Hansen's bacillus).

Source: The only reservoir of proved significance is man. In wild animals, armadillos, mangabey monkeys, and chimpanzees, however, naturally acquired leprosy has been observed.

Clinical Manifestations: A slowly progressive disease with long remissions and recurrent exacerbations affecting the skin, mucous membrane, and certain other tissues with peripheral nerve involvement. Manifestations vary in a broad spectrum between two distinct major types, namely, lepromatous and tuberculoid. Patients with lepromatous leprosy have marked diffuse infiltration and multiple nodular skin lesions distributed bilaterally or symmetrically. Skin lesions may ulcerate. Conjunctivitis, iridocyclitis, and keratitis lead to blindness. Deformities of the nose, face (leontiasis), fingers, and toes are considerable. Invasion of mucous membranes of the upper respiratory tract causes chronic nasal discharge and voice change. In tuberculoid leprosy, skin lesions are few and usually localized, sharply demarcated, and anesthetic or hypesthetic; relatively early involvement of peripheral nerves is diagnostic. Borderline infections are labile, with a tendency to shift toward the lepromatous type in untreated cases and toward the tuberculoid type in treated cases.

Pathology: Tissue changes are characterized by an infectious granulomatous process. In the lepromatous type, granulomas are composed of large lipid-containing macrophages (lepra cells or foaming cells) in which there are numerous organisms, forming small or giant globi and represent the multibacillary status. Foci of epithelioid cells frequently coalesce, forming Langhan's giant cells. In the tuberculoid type, histopathologic features are characterized by localized epithelioid cell granulomas surrounded by lymphocytes, plasma cells, and scattered giant cells. The bacilli are scanty in this type of lesion, which indicates the paucibacillary status.

Laboratory Diagnosis: Because the leprosy bacillus cannot be cultured, diagnosis is supported by the demonstration of the organism in biopsy specimens of suspected lesions or by scrapings or washings from ulcerated areas by acid-fast staining. In tuberculoid lesions, the organisms are sparse, and diagnosis is confirmed by histopathologic evaluation of the biopsy specimens. Skin reaction known as lepromin or the Mitsuda test gives prognostic information, but is of no diagnostic value. Serodiagnostic methods, such as fluorescent leprosy antibody absorption test and enzyme-linked immunosorbent assay using phenolic glycolipid 1 extracted from *M. leprae* as antigen, have recently been developed.

continued

continued

Epidemiology: Leprosy occurs worldwide, especially in the rural tropics and subtropics where socioeconomic conditions are poor. Areas of high prevalence include Southeast Asia, Papua New Guinea, some Pacific islands, India, tropical Africa, and some parts of Central and South America. Untreated multibacillary leprosy cases are the main source of infection. The mode of transmission is not clearly understood, but household and prolonged close contact are important. Susceptibility is universal. The incubation period is 6 months to 20 years, the average probably being 3 to 5 years.

Treatment: Combined chemotherapy with rifampicin, clofazimine, and dapsone (4,4-diaminodiphenyl sulfone) together with ethionamine or protionamide is advisable. Surgical treatment is used for deformities and wrinkling skin. Physiotherapy is undertaken for rehabilitation.

Prevention and Control: In many countries, case reporting is obligatory. Early diagnosis, regularly administered chemotherapy, and contact isolation for active cases of lepromatous leprosy are essential for controlling the disease. Periodic examination of household and other close contacts deserves high priority. A policy of rigid segregation of patients from society leads to concealment of the disease. Health education of patients and their family (emphasizing the availability of effective therapy) and the suppressed infectivity of patients under treatment are important. In controlled studies carried out in Uganda and New Guinea, prophylactic Bacillus Calmette-Guérin (BCG) inoculation was shown to be effective in reducing the incidence of tuberculoid leprosy among contacts, but little protection was observed in a study in Burma. Protection due to chemoprophylaxis is of a temporary nature only.

Leprosy is a chronic communicable disease in humans caused by leprosy bacilli (*Mycobacterium leprae*), which affect peripheral nerves, skin, eye, and mucous membranes. Naturally acquired leprosy has also been observed in some animals such as the nine-banded armadillo, mangabey monkey, and chimpanzee (Walsh et al., 1981).

Clinical Manifestations

Manifestations of this disease vary in a continuous wide spectrum. They are classified according to the following five forms: tuberculoid, borderline tuberculoid, mid-borderline, borderline lepromatous, and lepromatous (Ridley and Jopling, 1966). Tuberculoid and lepromatous forms are further subdivided into polar and subpolar types, respectively (polar tuberculoid, subpolar tuberculoid, polar lepromatous, and subpolar lepromatous) (Jopling, 1978). These various manifestations are probably a result of the cell-mediated immunologic status of the host. Patients with an uncertain prognosis are classified as having indeterminate leprosy.

Indeterminate Leprosy

Skin lesions in this group present as a single flat macule or a few flat macules without plaques or nodules. The lesions are usually slightly hypopigmented or slightly erythematous, and are often found on the exposed surface of the skin. Such lesions are indicative of an early and transitory stage of the disease.

Tuberculoid Leprosy

The polar tuberculoid lesion is generally stable. In contrast, the subpolar tuberculoid lesion is unstable. Lesions are solitary or few in number. They are reddish, brownish, or hypopigmented with slight atrophy. Usually, only peripheral nerves in the area are involved.

Borderline Leprosy

A broad spectrum of manifestations exists between the extremes of tuberculoid and lepromatous disease. Lesions of patients in this group present varied histologic features, ranging from a tuberculoid appearance (borderline tuberculoid), to a mixed histology or a mid-borderline appearance, and to a more lepromatous or a borderline lepromatous appearance. These forms are labile, representing a tendency to shift toward the lepromatous state in untreated patients and toward the tuberculoid state in treated patients.

Skin lesions in borderline tuberculoid patients

may be multiple. They are erythematous or hypochromic and are associated with thickened peripheral and cutaneous nerves. Skin lesions in patients with mid-borderline leprosy are also multiple and brownish, circular, or oval in shape. Erythematous lesions are characteristic of this type. The lesions are without sensation (anaesthetic) in varying degrees.

The skin lesions in patients with borderline leprosy are multiform, varying in size. They are thickened or infiltrated and red or brown in color. They are numerous and widespread, often appearing bilaterally. In patients with this type of leprosy, the bacilli are found in both skin lesions and other tissues such as lymph node, liver, and skeletal muscle. Clinical signs or symptoms due to such involvement are observed.

Lepromatous Leprosy

Immunocompromised patients develop subpolar to polar lepromatous disease. Subpolar disease is labile, having a tendency to shift toward the polar type. Skin lesions of the subpolar type closely resemble to those of polar type. However, a small number of subpolar asymmetric lesions may present with asymmetric nerve damage or thickened nerves.

Skin lesions of the polar type in lepromatous patients are usually symmetric and extensive, presenting as papules, macules, nodules, and diffuse infiltrations. Some lesions may lead to ulceration. Peripheral nerve involvement is severe in the later stage of the disease. Involvement of the nasal mucosa leads to crusting, obstructed breathing, and epistaxis. Leprous keratoconjunctivitis and iridocyclitis ultimately lead to blindness. Deformities such as pendulous and nodular ear lobes, saddle-nose, claw hand, and claw or hammer toes are common. Body hair, eyelashes, and eyebrows are lost. The patient's voice becomes hoarse, and the upper incisor teeth fall out; nerve involvement in symmetric (especially in the latter stage of the disease).

Histoid leprosy (leproma) is a special manifestation of lepromatous leprosy, and is characterized by firm, erythematous, round or oval, shiny, and glistening papules or nodules. They may appear in skin or subcutaneous tissue in which infiltration has healed or in other previously uninfected sites. Relapse usually develops when treatment is discontinued or when the causative organisms develop resistance to promine, although it may appear without any prior chemotherapy.

Lepra Reactions

Clinical manifestations include acute adverse episodes called "lepra reactions," type 1 and type 2. Type 1 reactions result from a rapid improvement or deterioration in the cell-mediated immune response of patients to the antigens of *M. leprae*. Improved reactions, which are also termed "reversal reactions," occur in tuberculoid, borderline, and subpolar lepromatous patients, producing fever, neuritis, edema, and erythema in preexisting lesions and sometimes progressing to new lesions. The deteriorating reaction, which may result from the increased number of leprosy bacilli and macrophages in the lesion, produces relapse-like manifestations, and the lepromin test may become negative.

The type 2 reaction differs from the type 1 (improved) reaction in that it involves humoral antibody responses and presents an immune complex syndrome. It usually occurs in polar or subpolar lepromatous patients, occasionally developing in those with borderline leprosy after treatment. The reaction has been called erythema nodosum leprosum (ENL), which is preferable to describing it as a "type 2 reaction" because the reaction should include a much broader range of symptoms besides ENL. The main clinical signs include erythematous, slightly raised, nodular skin lesions, neuritis, bone pain, joint pain, painful dactylitis, lymphadenitis, rhinitis, epistaxis, iritis, and epididymo-orchitis, or a combination thereof. The skin lesions may be present as a pustulation or ulceration, appearing on the surface of the face, arms, and thighs. Several lesions of ENL may become vesicular or bullous and break down with erythma necroticans possessing an Arthus reaction (with other evidence of humoral antibody production to have autoantibodies such as rheumatoid factor and antinuclear factor) and increased IgG, IgM, and CIq levels.

Lucio leprosy is a peculiar and severe form of type 2 reaction, appearing in some cases of the diffuse type of lepromatous disease. Painful and tender reddish skin becomes necrotic and ulcerated, and falls off leaving an atrophic scar.

Pathology

Histologic findings may not help to distinguish the indeterminate group of patients from those with dermatitis due to other causes. Lymphocytes and macrophages are moderately localized in the skin, infiltrating around blood vessels and dermal appendages such as sweat glands and nerves. Mild perineural infiltration of macrophages is typically seen in patients with indeterminate leprosy. Leprosy bacilli in skin lesions are absent or very scanty. In some patients, however, an increasing number of bacilli is discernible by using Harada's stain method.

Tuberculoid foci are characterized by the presence of epithelioid cell granulomas (no caseation), with or without Langhan's giant cells. Epithelioid cell granulomas with a large number of lymphocytes develop in the basal layer of epidermis and sweat glands and in nerves undergoing destruction. Acid-fast bacilli are also absent or very scanty in the lesions.

Borderline tuberculoid foci are characterized by the presence of a clear narrow subepidermal zone, which are distinguishable from tuberculoid foci that do not have the clear zone. Langhan's giant cells may be numerous, with a few bacilli within the swollen nerves resembling sarcoidosis. However, borderline tuberculoid foci have no endoplasmic reticulum in the epithelioid cells.

Mid-borderline foci are characterized by the demonstration of diffusely spread epithelioid cell granulomas, not focalized by lymphocytes, with a clear subepidermal zone. Langhan's giant cells are absent. A nerve may be surrounded and partially infiltrated by epithelioid cell granuloma, which will ultimately result in the destruction of sweat glands.

Borderline lepromatous foci are characterized by the presence of macrophages (histiocytic cell) granulomas, some having an epithelioid-like pattern. Sometimes the foci show a tendency to undergo a foamy change. Numerous acid-fast bacilli are observed (multibacillary status), but they do not form large globi. This granuloma is differentiated from that of lepromatous lesions by the area of dense lymphocyte infiltration. A clear subepidermal zone is present. Nerves show lamination and infiltration of the perineurium.

Histologically, subpolar lepromatous foci show the presence of histiocytic cell granuloma with a clear subepidermal zone. Histiocytes may resemble epithelioid cells that produce a slight foamy change in the cytoplasm. Sometimes the lesions include multinucleated Touton giant cells.

Polar lepromatous foci are characterized by the presence of histiocytic cell granuloma with a clear subepidermal zone. Histiocytes in an active stage of the disease are large nonfoamy cells, but in an inactive stage are large foamy cells. Lymphocytes are usually scanty. Nerves may be normal or show some structural damage with hyalinization or fibrosis. A large number of globi or giant globi is present. In patients with this stage of the disease, granulomatous lesions are seen in the liver, spleen, testicle, epididymis, lymph node, bone marrow, adrenal gland, gum, tongue, pharynx, and larynx.

Early ENL lesions show a marked edema and acute inflammatory reactions that are very rich in polymorphonuclear leukocytes, possessing microabscesses in patients with severe type 2 lepra reactions. The histology in such cases is similar to that in Arthus reaction, usually involving the lower half or the dermis and subcutaneous tissues. Blood vessels show edema and infiltration with acute inflammatory cells, followed by endothelial cell hyperplasia narrowing the lumen. Leprosy bacilli in reactive sites are usually few and degenerated.

Histoid lesions consist of the localized foci of intertwining polyhedral and spindle-shaped histiocytes. They look like those of neoplasm, having packets of a large number of acid-fast bacilli with a clear subepidermal zone.

Diagnostic Features

Leprosy bacillus cannot be artificially cultured. Therefore, a laboratory diagnosis is possible through histopathologic findings and is based on the presence of acid-fast bacteria in skin smears or secretions. When the presence of acid-fast bacteria cannot be confirmed, histopathologic findings such as thickening or enlargement of peripheral nerves and local infiltration of epithelioid cells, giant cells and lymphocytes are useful in diagnosis. The presence of the acid-fast bacteria is ascertained by staining methods such as Ziehl-Neelsen's staining, Fite-Faraco's staining, and Harada's staining. Confirmation of the clump of acid-fast bacilli, particularly the characteristic lepra globi, is important.

The lepromin reaction reflects the cell-mediated immune response and is used to determine the disease type. A strong positive lepromin reaction is generally identified with the tuberculoid type in which the organisms are sparse, and a negative reaction with the lepromatous type in which leprosy bacilli abound. In vitro serodiagnostic methods such as the fluorescent leprosy antibody absorption test and enzyme-linked immunosorbent assay using phenolic glycolipid 1 (a specific antigen of *M. leprae*) as the antigen have recently been developed. They have not yet, however, been fully proved satisfactory in the diagnosis of the disease.

Diseases that must be differentiated from leprosy are: lupus vulgaris, pityriasis vulgaris, annulus granuloma, mycosis fungoides, tinea vesicolor psoriasis, necrobiosis lipoidica diabeticorum, atypical mycobacteriosis, erythema nodosum induratum, Kaposi's sarcoma, pachydermoperiostosis, and sarcoidosis. When there are skin eruptions but no evidence of the leprosy bacillus, sensory disorders will play an important role in the diagnosis of leprosy. Sarcoidosis is histopathologically similar to borderline tuberculoid leprosy and may also reveal mild hypalgesia at the site of the lesion. However, it can be identified by bilateral hilar lymph node swelling, which does not occur in leprosy. Sarcoidosis can also be distinguished from leprosy by the degree of endoplasmic reticulum in the epithelial cell at electronmicroscopic examination.

Both diseases can be immunologically differentiated. The Kveim reaction (which is used for diagnosis of sarcoidosis) is highly specific, and if an antigen with a high specificity is used, 80% of patients with sarcoidosis will test positive, whereas the patients with tuberculosis and leprosy will test negative. Although patients with the lepromatous type of the disease have weak T-lymphocyte function, patients with sarcoidosis have no immunodeficiency and the T-lymphocyte function of the local lesion is accelerated. Furthermore, a lowered tuberculin response is frequently observed in sarcoidosis cases at the onset of the disease, whereas the response in leprosy cases is positive in 95% in the tuberculoid cases, 85% in the mid-borderline cases, and 75% in the lepromatous cases. This is another aspect that differentiates leprosy from sarcoidosis. In the case of other diseases, however, it is extremely difficult to make an accurate diagnosis when mononeuritis occurs without any accompanying skin lesions.

Prognosis

Leprosy is a disease with an extremely low mortality rate and a relatively good prognosis. It is said that death in patients with leprosy occurs as a result of asphyxia due to glottic edema, but occasionally there are mortalities due to toxemia or necrotic lesions caused by severe leprosy reactions. A positive lepromin response means a good prognosis, and a negative response means an unfavorable prognosis.

Epidemiologic Aspects

Occurrence

The numbers of registered cases of leprosy in the respective WHO regions for the years 1966, 1976, and 1985 are shown in Table 1 (Noordeen, 1985). The number of leprosy cases worldwide in 1985 is estimated to be approximately 12 million. Leprosy is still a serious disease that requires urgent countermeasures in developing countries in tropical and subtropical areas such as Africa, Central and South America, and Southeast Asia (especially in the rural areas). The major factor related to prevalence, however, is socioeconomic conditions, not climate.

Mode of Transmission

The primary mode of transmission is prolonged close contact with a patient who has ulcerating lepromatous skin lesions or bacilli secretions from the nasal mucosa. Invasion of the respiratory tract or of skin wounds by bacilli is considered to be the major route of transmission, but the possibility of infection from the digestive tract or mediation by insect bites cannot be negated. Bacilli excreted in nasal secretions remain viable for at least 7 days. The majority of patients are believed to lose their infectiousness with DDS (4,4′-diamino-diphenylsulfone). Patients with leprosy were formerly believed to be the sole infectious agents of leprosy, but the occurrence of naturally acquired leprosy has recently been observed among nine-banded armadillos (Thomas et al., 1985; Walsh et al., 1975), mangabey monkeys (Meyers et al., 1980), and chimpanzees (Donham and

TABLE 1. Registered leprosy cases by WHO regions for the years 1966, 1976, and 1985

WHO regions	*No. of registered cases (no. of countries)* 1966	1976	1985[a]
Africa	1,685,526 (38)	1,398,220 (38)	1,008,857 (43)
Americas	177,813 (47)	241,248 (40)	305,999 (41)
Eastern Mediterranean	40,963 (10)	63,236 (20)	60,513 (20)
Europe	19,589 (20)	20,452 (20)	17,091 (11)
Southeast Asia	790,851 (6)	1,748,468 (8)	3,737,157 (9)
Western Pacific[b]	117,003 (30)	128,325 (22)	145,823 (24)
Total	2,831,775 (151)	3,599,949 (148)	5,275,440 (149)

[a] Or the most recent year for which data are available.

[b] Does not include the People's Republic of China for which the number of registered cases is not available; however, the number of estimated cases currently is 200,000 (cited from Noordeen, 1985).

Leininger, 1977), demonstrating that there exists the rare possibility of infection from these animals.

Susceptibility and Resistance

Racial differences in susceptibility have not been proven. There are, however, reports of the involvement of HLA, but these have not yet been substantiated. The average incubation period is 4 years for tuberculoid leprosy, and approximately twice as long for lepromatous leprosy. Patients younger than 3 years of age are rare. There are more male than female patients. The persistence and spectrum of manifestations in this disease are dependent on the development of effective cell-mediated immunity. The Mitsuda reaction, which involves reading the response 28 days after intracutaneous injection of autoclaved *M. leprae,* will be positive for tuberculoid leprosy, but will also yield positive results in a proportion of normal adults. Judging from the fact that the rate of positive results among the general population increases with age and the fact that specific lymphocyte transformation and specific *M. leprae* antibodies are highly prevalent among close contacts of such patients, it is known that only a proportion of the infected persons become patients.

Etiologic Agent

Leprosy bacillus was identified as the pathogen of leprosy in 1873 by the Norwegian researcher Dr. Henrik Armauer Hannsen (1955). In spite of the efforts of many researchers, the bacillus has not yet been cultivated successfully in artificial media or cell cultures.

Leprosy bacillus is acid-resistant and is stained red by acid-fast staining. It is a nonsporulating, gram-positive bacillus or coccus-like bacillus, 0.3 to 0.5 μm in width, and 0.6 to 7.0 μm in length. Bacillus in the infected tissue usually appears in the form of clumps, (i.e., globi containing >100 bacilli). The globi are usually more prevalent in the lepromatous type of the disease, but are also observed in the borderline lepromatous type. In the tuberculoid or indeterminate types, only a few are observed microscopically in several fields, and generally they cannot be observed as globi.

Leprosy bacilli have the following distinguishing characteristics:

1. acid-fast bacillus that cannot be cultured in artificial media or cell cultures;
2. ability to oxidize D-dopa (3,4-dihydroxyphenylalanine) as compared with other mycobacteria (Prabhakaran and Kirchheimer, 1966);
3. predilection towards the peripheral nerves;
4. glycin in the cell wall components, which differs from the other acid-fast bacteria (Draper, 1976);
5. G + C contents of DNA are 55.8%, which is lower than that of other acid-fast bacteria (Imaeda et al., 1982);
6. leprosy bacillus has a specific antigen, namely, phenolic glycolipid 1 (Brennan and Barrow, 1980);
7. treatment of the bacillus with pyridine causes the rapid loss of acid-fastness (Convit and Pinardi, 1972);
8. micolic acid is identified as C_{20} in *M. leprae* (Kusaka et al., 1981);
9. cytochrome c, cytochrome a, and catalase activity cannot be found in the whole cells, particle fraction, or the cell-free extract of *M. leprae* (Mori et al., 1985); and
10. induces localized proliferation at the inoculation site but does not cause systemic disease when injected into the foot-pad of the normal mouse (Shepard, 1960).

Diagnostic Procedures

Demonstration of Leprosy Bacillus in Skin Smears

Collection of Materials

Wipe the lesion with ether, incise it with a scalpel while holding the lesion steady with the fingers, scratch the wound to collect the tissue fluid, and spread it over a glass slide. After drying at room temperature, fix by heating with a Bunsen burner. Make sure that the depth of the incision is sufficiently deep to collect more than 3 mm of dermal tissue. Otherwise, the test may yield faulty results.

Staining

The Ziehl-Neelsen method is the primary technique used for staining leprosy bacillus, but special caution is required because the acid-fastness of this bacillus is weaker than that of other mycobacteria. Harada's periodic-acid carbol pararosaniline stain (PACP) method is sometimes used in conjunction with the Ziehl-Neelsen method (Harada and Kasai, 1978).

Ziehl-Neelsen Method

Procedure

1. Cover with carbol fuchsin on the fixed smear slide, and allow to stand for 20 min at room temperature.

2. Tip the slide to remove the stain, and wash in a gentle stream of tap water.
3. Pour a 1% hydrochloric acid-alcohol mixture onto the slide, and allow to stand for 30 s to 1 min.
4. Wash the acid-alcohol mixture away with running water. If the smear is deep pink, treat it again with acid-alcohol and wash with running water.
5. Cover the slide with methylene blue for about 30 s to 1 min for counter-staining.
6. Rinse the slide in tap water and allow to dry.

In this staining method, the leprosy bacilli appear red and the background and nonacid-fast organisms appear blue.

Stains

Carbol Fuchsin Tween 80 Solution

Dissolve 3.5 g of basic fuchsin in 12.5 g of pure phenol and heat the solution to 80°C in a water bath. Add 25 ml of 95% alcohol when the solution is cool. Mix well and make up to 300 ml with distilled water, and then add 30 drops of Tween 80.

Acid-Alcohol Mixture

One percent hydrochloric acid in 70% ethanol.

Counter-Staining Solution

Two grams of methylene blue in 1% HCl added 100 ml of 70% ethanol, or 10-fold dilution of Loeffler alkaline (0.01% KOH) methylene blue solution.

Evaluation of the Results

The number of bacilli in smears is microscopically observed by using an oil immersion objective and can be recorded by either the Ridley (1959) or WHO system (Table 2). For a more comprehensive description, the former is recommended.

Bacteriologic Index

The bacteriologic index is the average obtained by dividing the total bacterial count of the respective test sites (as evaluated by the Ridley method) by the number of test sites. To obtain the bacteriologic index, it is desirable to collect materials from as many locations as possible, from at least 4, and preferably 6 to 8 lesions. The bacteria in the nasal secretions, however, are not included in the bacteriologic index evaluation. To ascertain the changes in this index in the patient, it is recommended to check the bacteriologic index three or four times a year.

Morphologic Index

Leprosy bacilli that live in favorable conditions have sufficient protoplasm so that the whole bacillus is stained well by fuchsin when stained by the Ziehl-Neelsen method. This is called the solid type. When conditions are not favorable for the bacilli, such as during chemotherapy or in the resorbing stage, the stainability of the acid-fastness is partially or totally decreased, and the morphology of the mycobacterium becomes fragmented and inconsistent. Such types are called nonsolid. Besides these two types, there is a third type called the granular form, but this will be excluded from calculations of the bacterial count. The morphologic index (MI), therefore, implies the percentage of solid bacteria among the total number of bacteria (solid + nonsolid).

$$MI = \frac{\text{Number of solid}}{\text{Number of solid} + \text{Number of nonsolid}} \times 100\%$$

The morphologic index of leprosy bacilli in untreated lepromatous skin lesions is generally more than 30%, but this figure is reduced by chemother-

TABLE 2. Criteria on bacterial calculation

	No. of bacteria
Criteria by Ridley	
6+	Over 100 bacilli per field
5+	100 to 1,000 bacilli per field on average
4+	10 to 100 bacilli per field on average
3+	1 to 10 bacilli per field on average
2+	1 to 10 bacilli per 10 fields
1+	1 to 10 bacilli per 100 fields
0	No bacillus in the any field
Criteria by WHO	
(+)4	Many clumps of bacilli in an average field
(+)3	Large number of bacilli in every field
(+)2	1 to 10 bacilli in 100 fields
(+)1	Less than 1 bacillus in each field on average
(−)	No bacillus in the 100 fields

apy. Both the morphologic and bacteriologic indices are used as markers of the efficacy of chemotherapy or changes in the clinical status of the disease.

Histopathologic Examination of the Lesion

Skin or nerve biopsy is used for histopathologic examination. To take a skin biopsy, clean the lesion with ethanol, then, under anesthesia, excise the skin. The section should include both affected and healthy skin and be deep enough to include subdermal fat tissue. The excised section is fixed in 10% formalin. In the case of a nerve biopsy, a thickened sensory nerve is suitable for examination. The formalin-fixed tissues are dehydrated, blocked in paraffin, and sectioned in 5 μm-thick samples in the usual manner. Albumin-glycerin is used as adhesive.

Some sections from the lesions are stained by hematoxylin-eosin for routine histopathology. The Ziehl-Neelsen stain method, Fite-Faraco stain method, or Harada's periodic-acid methenamine silver stain method (Harada, 1976) are used for staining acid-fast bacilli.

In the preparation of tissue specimens for detection of the leprosy bacillus, the method of deparaffinization differs from that used for other purposes. If the leprosy bacillus specimen is exposed to solvents such as alcohol, chloroform, and xylol, the lipid of the bacillus may be dissolved and staining may not be very clear. The specimen must, therefore, be deparaffinized in a solution of xylol and peanut oil at a concentration of 2 : 1 or 1 : 1, then stained with phenol fuchsin without going through an alcohol stage (Fite-Faraco staining method).

Because the leprosy bacillus has an affinity for the nerves, it is important to examine the peripheral nerves carefully for the presence of the bacteria. It is also necessary to observe the histologic reactions specific to each disease type as a reference for classification of the disease as tuberculoid or lepromatous (see Pathology).

Animal Models

Mouse Footpad Inoculation Method (Shepard, 1960)

1. Normal mice, about 5 weeks old, are used.
2. Bacterial suspension is prepared in Hanks solution with 0.1% bovine serum albumin added, and 0.03 ml is administered subcutaneously either in the foot-pad or dorsum of the hind foot.
3. A ⅕-sized or two step needle is recommended to ensure accuracy of the injection.
4. The mice are kept at 20 to 25°C.
5. The mice are sacrificed every month starting from the 3rd month after inoculation, and observed for the presence or absence of bacterial proliferation (small globi).
6. To calculate the number of bacteria, soft tissue is removed from the inoculated foot, scraped off from the bone area, then emulsified.
7. In the pathologic examination, the leprosy bacilli should be visible, localized in the fibers of the cross-striated muscle.

Five to 10 months after inoculation, the bacterial counts should be 10^5 to 10^7/foot. Bacterial proliferation can be observed by the subcutaneous footpad inoculation method in the golden hamster or the rat, but systemic results cannot be observed as is the case in the mouse. No proliferation can be observed by using the footpad method in the guinea pig, rabbit, or chicken.

Nude Mouse, Lasat Mouse, and Nude Rat Models

The combined infection method can be used to obtain many bacilli quickly. The feet, upper lip, external ear, hip, tail, and genitals are inoculated simultaneously. The inoculated sites will show lepromatous histopathologic lesions when the leprosy bacillus is injected into these animals.

Nine-Banded Armadillo Model (Kirchheimer and Storrs, 1971)

This animal will produce systemic lepromatous lesions when inoculated intravenously in the vena saphena with about 10^9 bacilli. An animal model for the tuberculoid type has not yet been developed.

Serologic Examination

Lepromin Test

The lepromin test is a test for cellular immunity against the leprosy bacillus and is used for the classification of the disease type. However, it is not a diagnostic test because it is positive in many people who have never been exposed to leprosy. It is a skin test employing an antigen of heat-treated lepromatous tissue (Lepromin or Mitsuda antigen).

Lepromin (Mitsuda antigen) is prepared in the following manner. Heat-treated leprosy tissue with a high bacterial count is emulsified with the use of physiologic saline solution, and the large particles are removed by natural sedimentation. The number of bacilli in the suspension is adjusted to 1.5 to 1.6 × 10^8/ml to produce the standard preparation. Recently, however, researchers recommend four-fold dilution (Abe et al., 1963).

The Lepromin test reaction in the tuberculoid

type of disease is strongly positive, but is negative in the lepromatous and borderline types of disease. Positive results indicate that cellular immunity exists, and negative results mean that cellular immunity has been impaired or lost. This test is, therefore, useful in the evaluation of prognosis.

Lepromin fluid (0.1 ml) is injected intracutaneously in the flexural side of the forearm. The erythematous reaction obtained after 48 h is called the Fernandez reaction or early reaction. This response to the protein antigen of the leprosy bacillus is a delayed allergic reaction similar to the tuberculin reaction in tuberculosis. The erythematous papulonodule that appears 3 to 4 weeks after the inoculation is called the Mitsuda reaction or late reaction.

In the late (Mitsuda) reaction, the reaction of the phagocytes to the insoluble components of the bacillus is added to the early reaction.

The histologic picture will reveal epithelioid cells and occasionally tuberculoid granuloma composed of giant cell lymphocytes. This is due to the antigen-antibody reaction between the antibody for the bacilli and the residual antigen in the inoculation site, and is visible to the naked eye as an erythematous papulonodule. Sometimes, the late reaction might be positive in healthy subjects. This is believed to occur because the subject is sensitized by the injected antigen. Therefore, repeated examinations within short spans of time should be avoided. To avoid this kind of reaction, partially purified Dharmendra antigen or soluble antigen (leprosin, leprolin) is sometimes used. The evaluation criteria for the lepromin reaction are shown in Table 3.

Serologic Method

One in vitro method for the detection of antibody to the specific antigen of the leprosy bacillus is the fluorescent leprosy antibody absorption test (FLA-ABS) (Abe et al., 1980). There is also the ELISA method using phenolic glycolipid 1, which is a specific antigen of the leprosy bacillus. The antigen that is used in these methods is a lipid that is extracted from the bacilli or lepromatous lesions that has potent antigenicity in the sugar chain moiety. This antigen has a drawback in that it is insoluble and difficult to handle. Recently, however, Fujiwara et al. (1984) developed a simpler serodiagnostic method by using synthesized epitope of phenolic glycolipid 1. The epitope is conjugated with bovine serum albumin after treatment with octyl alcohol or phenylpropionate, and used as an artificial antigen in this method.

Another way of detecting the bacillary antigen in the serum is the serum antibody competition test (SACT). The ultrasonicated leprosy bacillus is reacted with the test serum, then reacted with the ^{125}I-labeled monoclonal antibody to the specific bacillary antigen, and the antibody titer of the test serum can be determined by the degree of inhibition of binding to the labeled antibody (Sinha et al., 1985).

Immunopathologic Aspects

Leprosy is a disease that exerts different immune responses depending on the host's resistance to infection. The humoral immunity and cellular immunity of the host interact to produce a variety of disease types. The various types of leprosy that have been mentioned previously may be explained immunologically as follows. The lepromatous type is characterized by acceleration of humoral immunity and lack of cellular immunity, which allow the leprosy bacilli to proliferate systemically. In the tuberculoid type, the proliferation of the mycobacteria is suppressed by acceleration of cellular immunity and the lesion is localized. The borderline group is in be-

Table 3. Criteria on lepromin reaction

Early reaction (Fernandez reaction), readings after 48 h	
—	Nothing to see, erythema without induration, or erythema and induration less than 5 mm in diameter
+1	Erythema and induration 5 to 10 mm in diameter
+2	Erythema and induration 11 to 15 mm in diameter
+3	Erythema and induration 16 mm or more in diameter
Late reaction (Mitsuda reaction), readings after 4 weeks	
−	Nothing to see
±	A papule less than 3 mm in diameter
+1	An erythematous papule 3 to 5 mm in diameter
+2	An erythematous papule 6 to 10 mm in diameter
+3	An erythematous nodule larger than 10 mm or one of any size that ulcerates

tween these two types and shows unstable immune responses. Premature immunologic response in the early stage of the disease is classified as the indeterminate group.

Immunogenetic research on leprosy has made remarkable progress in recent years and yielded many reports on the relationship between leprosy and human leukocyte antigen (HLA). These studies have suggested the existence of a disease-sensitive gene related to HLA. This gene is believed to be an immunosuppressor gene (Is gene), and may be acting to inhibit the development of the lepromatous type of leprosy through the action of the suppressor T cells. The immune response to *M. leprae*-derived antigen is high in patients with tuberculoid and borderline tuberculoid type disease and low in patients with lepromatous type disease. The following two mechanisms may be responsible for the low response in the lepromatous type: 1) the presence of antigen-specific inhibitory T cells in the peripheral blood lymphocytes (PBL); and 2) the absence or scarcity of T-cell clones that respond to the leprosy antigen.

According to Bloom and Mehra (1984), the lymphocytes of patients with the lepromatous and borderline lepromatous type of disease reveal inhibition of the immune response in the presence of lepromin or PGL-I. This phenomenon is due to the activity of the OKT8 cell (suppressor T cell) and is not observed in tuberculoid patients or healthy subjects.

The T4 : T8 ratio in lymphocytes is relatively similar among the different disease types as far as the peripheral lymphocytes are concerned. In the lymphocytes of the skin lesion, however, the ratio is remarkably different: 2 : 1 in the tuberculoid type and 0.7 : 1 in the lepromatous type. Furthermore, an immunohistopathologic examination of the distribution of T4 and T8 cells shows that T8 cells are prevalent in the vicinity of the granuloma in the tuberculoid type, and that both types of cells can be seen inside the granuloma in the lepromatous type. This finding is believed to explain the mechanism of decline of cellular immune response in the lepromatous types of disease.

Treatment

The first definitive chemotherapy for leprosy was introduced by Faget et al. (1943), who used intravenous injection of Promin in 1943. Subsequently, the oral administration of Dapson (DDS; 4,4-diaminodiphenylsulfon), the basic form of promin, was attempted with extremely good results. Since then, effective drugs such as thiourea compounds, long-acting sulphonamides, clofazimine, and rifampicin have been developed.

With the appearance of Dapson-resistant strains in 1963, and following the report in 1977 of a case of infection and onset caused by a resistant strain, the World Health Organization has recommended multidrug therapy for the chemotherapy of leprosy (World Health Organization, 1982). According to the recommendations, patients with the paucibacillary form—the tuberculoid type and some borderline tuberculoid types—are administered rifampicin (600 mg once a month) and Dapson (50 to 100 mg/day, every day) for 6 months.

Patients with the multibacillary form—borderline tuberculoid, mid-borderline, borderline lepromatous, and lepromatous—are administered rifampicin (600 mg once a month), Dapson (100 mg every day), and clofazimine (Lamprene, B663) (300 mg once a month or 50 mg every day) for 2 consecutive years. Ethionamid or protionamid (250 to 375 mg) may be given every day in place of clofazimine.

The duration of treatment is 2 to 3 years for the tuberculoid type and indeterminate group, 5 to 6 years for the borderline tuberculoid type, 8 to 9 years for mid-borderline type, and 10 to 12 years for the borderline lepromatous and lepromatous types. Lifelong therapy is required for some lepromatous type patients (namely, until the lepromin reaction turns positive).

Prevention and Control

To control leprosy, we must undertake measures to control the sources of infection. The conventional method of preventing the spread of leprosy has been to isolate patients from society. With the advent of efficacious drugs for the treatment of leprosy, however, prevention and control measures have been radically changed. Patients with leprosy can now be controlled by chemotherapy without hospitalization. The World Health Organization is presently adopting the policy of early detection as well as proper and regular treatment of these patients. For regions in which leprosy is prevalent, WHO recommends educating and training general hospital, clinic, and public health staff about leprosy and strengthening systems of persistent surveillance and guidance by experts in the field of leprosy.

The following four conditions are necessary for the control of leprosy:

1. strengthening the efforts to discover patients, encouraging activities of specialized leprosy task forces;
2. establishment of a primary health care system as part of the total leprosy control system;
3. widespread use of multidrug therapy; and
4. preventive measures using vaccination.

In previous times, BCG inoculation was expected to prevent leprosy infection, but it did not produce satisfactory results. Today, a combination vaccine using BCG and inactivated *M. leprae* is used for prevention. The vaccine is made by mixing 6×10^8 *M. leprae* cells with 0.2 or 0.02 mg of BCG, and is given every 3 months. This immunotherapy method was tested in 577 subjects and resulted in a positive conversion of the Mitsuda reaction in 84% of the subjects after one or two administrations; it is now being attempted in many regions. This is a promising vaccine that can be used for both immunoprophylaxis and immunotherapy (Convit, 1980). A live vaccine using recombinant BCG and vaccinia virus is also being developed.

Literature Cited

Abe, M., M. Maeda, T. Nakayama, K. Yanagisawa, N. Asami, N. Tachikawa, K. Okamura, S. Kobayashi, and K. Hayashi. 1963. Studies on the preparation, standardization and preservation of lepromin. II. Standardization by bacillary counting. La Lepro. **32:**129–132 (in Japanese).

Abe, M., F. Minagawa, Y. Yoshino, T. Ozawa, K. Saikawa, and T. Saito. 1980. Fluorescent leprosy antibody absorption (FLA-ABS) test for detecting subclinical infection with *Mycobacterium leprae*. Int. J. Lepr. **48:**109–119.

Bloom, B. R. and V. Mehra. 1984. Immunological unresponsiveness in leprosy. Immunol. Rev. **80:**5–28.

Brennan, P. J. and W. W. Barrow. 1980. Evidence for species-specific lipid antigens in *M. leprae*. Int. J. Lepr. **48:**382–387.

Convit, J. 1980. Planning of a vaccine trial in Venezuela, p. 24–36. TDR/IMMLEP (Sub-Trials). World Health Organization.

Convit, J. and M. E. Pinardi. 1972. A simple method for the differentiation of *Mycobacterium leprae* from other mycobacteria through routine staining technics. Int. J. Lepr. **40:**130–132.

Donham, K. J. and J. R. Leininger. 1977. Spontaneous leprosy-like disease in a chimpanzee. J. Inf. Dis. **136:**132–136.

Draper, P. 1976. Cell walls of *Mycobacterium leprae*. Int. J. Lepr. **44:**95–98.

Faget, G. H., R. C. Pogge, F. A. Johansen, J. F. Dinan, B. M. Prejean, and C. G. Eccles. 1943. The promin treatment of leprosy. Publi. Health Rep. **58:**1729–1741.

Fujiwara, T., S. W. Hunter, S. N. Cho, G. O. Aspinall, and P. J. Brennan. 1984. Chemical synthesis and serology of the di- and tri-saccharides of the phenolic glycolipid antigens from the leprosy bacillus and preparation of a disaccharide protein conjugate for serodiagnosis of leprosy. Infect. Immun. **43:**245–252.

Hansen, G. A. 1873. Undersogelser angaende spedalskhedens arsager (causes of leprosy), reprinted in translation from Norsk Magazine for Laegevidenskaben 4:1 (1874). Int. J. Lepr. **23:**307–309 (1955).

Harada, K. 1976. Periodic acid-methenamine silver stain for mycobacteria in tissue sections. Stain Technol. **51:**278–280.

Harada, K. and T. Kasai. 1978. Two methods of demonstrating leprosy bacilli in smears. Int. J. Lepr. **46:**167–171.

Imaeda, T., W. F. Kirchheimer and L. Barksdale. 1982. DNA isolated from *Mycobacterium leprae:* genome size, base ratio, and homology with other related bacteria as determined by optical DNA-DNA reassociation. J. Bacteriol. **150:**414–417.

Jopling, W. H. 1978. Handbook of leprosy, 2nd ed., p. 39. William Heinemann Medical Books LTD, London.

Kirchheimer, W. F. and E. E. Storrs. 1971. Attempts to establish the armadillo (*Dasypus novemcinctus* Linn) as a model for the study of leprosy 1. Report of lepromatoid leprosy in an experimentally infected armadillo. Int. J. Lepr. **39:**693–702.

Kusaka, T., K. Kohsaka, Y. Fukunishi, and H. Akimori. 1981. Isolation and identification of mycolic acid in *M. leprae* and *M. lepraemurium*. Int. J. Lepr. **49:**406–416.

Meyers, W. M., G. P. Walsh, H. L. Brown, Y. Fukunishi, C. H. Binford, P. J. Gerone, and R. H. Wolf. 1980. Naturally-acquired leprosy in a mangabey monkey (*Cercocebus sp*). Int. J. Lepr. **48:**495–496.

Mori, T., Y. Miyata, K. Kohsaka, and M. Makino. 1985. Respiration *Mycobacterium leprae*. Int. J. Lepr. **53:**600–609.

Noordeen, S. K. 1985. Current global strategy for leprosy control, p. 18–23. *In:* Proceedings of the first international leprosy symposium in China, The China Leprosy Association, Beijing, P.R.C.

Prabhakaran, K. and W. F. Kirchheimer. 1966. Use of 3,4-dihydroxyphenylalanine oxidation in the identification of *M. leprae*. J. Bacteriol. **92:**1267–1268.

Ridley, D. S. 1959. Bacterial indices, p. 371. *In:* Cochrane (ed.), Leprosy in theory and practice. London, John Wright & Son's Ltd.

Ridley, D. S. and W. H. Jopling. 1966. Classification of leprosy according to immunity. A five-group system. Int. J. Lepr. **34:**255–273.

Shepard, C. C. 1960. The experimental disease that follows the injection of human leprosy bacilli into foot-pads of mice. J. Exp. Med. **112:**445–454.

Sinha, S., U. Sengupta, G. Ramu, and J. Ivanys. 1985. Serological survey of leprosy and control subjects by a monoclonal antibody-based immunoassay. Int. J. Lepr. **53:**33–38.

Thomas, D. A., J. S. Mines, T. M. Mack, D. C. Thomas, and T. H. Rea. 1985. Armadillo exposure among Mexican-born patients with lepromatous leprosy. Int. J. Lepr. **53:**724–725.

Walsh, G. P., E. E. Storrs, H. P. Burchfield, E. H. Cottrell, M. F. Vidrine, and C. H. Binford. 1975. Leprosy-like disease occurring naturally in armadillos. J. Reticuloendothel. Soc. **18:**347–351.

Walsh, G. P., W. M. Meyers, C. H. Binford, P. H. Geron, R. H. Wolf, and J. R. Leininger. 1981. Leprosy—a zoonosis. Lep. Rev. **52(suppl 1):**77–83.

Wheeler, P. R. and D. Gregory. 1980. Superoxide dismutase peroxidatic activity and catalase in *M. leprae* purified from armadillo liver. J. Gen. Microbiol. **121:**457–464.

World Health Organization (WHO). 1982. Technical Report Series, No. 675.

CHAPTER 35

Leptospirosis

S. FAINE

Disease: Leptospirosis (Weil's disease in its severe form; canecutter's, swineherd's, or harvest disease or 7-day or autumnal fever in its mild form).
Etiologic Agent: One of the serovars of *Leptospira interrogans*.
Source: Rodents, livestock, and wild mammals or marsupials by direct contact with infected urine or tissues; indirect contact through contaminated surface waters, mud, or earth.
Clinical Manifestations: Acute onset of severe headache, muscle pains, fever, and rash, followed in severe cases by meningitis, bleeding, renal failure, and hepatitis. Complete recovery in nonfatal cases, with occasional prolonged mental changes.
Pathology: Generalized toxic damage to blood vascular endothelium leading to hemorrhage, ischemia and tissue necrosis, myositis, interstitial nephritis, hepatitis, and meningitis. Carrier state persists in animal renal tubules after clinical recovery.
Laboratory Diagnosis: Growth of infecting serovar of *Leptospira* from blood, urine, or cerebrospinal fluid in the febrile period. Immunostaining of leptospires in tissues. Serologic presumptive diagnosis by fourfold rise in agglutinating titer or single high titer (1 : 800 or greater) to the presumed infecting serovar of *Leptospira* in a clinically suspect patient.
Epidemiology: Worldwide. Many carrier animal species in tropics. Mainly domestic livestock and rodents in temperate climates, where there is a strong occupational association in animal handlers, abattoir workers, and water-sport participants. Usually sporadic, seasonal, and geographic.
Treatment: Penicillin or erythromycin. Supportive measures for renal or other organ failure.
Prevention and Control: Surveillance; immunization of livestock and dogs; protective clothing; education and awareness of public, physicians, and veterinarians. Chemoprophylaxis with doxycycline.

Description of Disease

Leptospirosis in humans is an acute, febrile zoonosis occurring in all parts of the world. Its reservoir is in the chronically infected proximal renal tubules of carrier animals. The causal organisms, *Leptospira interrogans,* are excreted in urine and contaminate soil, mud, ground waters, streams, and rivers. Fresh infections of animals or humans occur from direct inoculation with urine or indirectly from contaminated waters. The carriers may be wild or domestic animals, especially rodents and small marsupials, cattle, pigs, and dogs. Almost every mammal (including aquatic) and marsupial has been known to be a carrier of leptospires. Humans are almost never chronic carriers, but suffer acute infections, sometimes with longer-term sequelae.

The disease varies in severity according to the

infecting serovar of *Leptospira* and the age, health, and nutritional status of the patient. It may range from a mild influenzalike illness to severe renal and hepatic failure, myocarditis, hemorrhage, and death (Weil's disease). Congenital infection occurs in humans and in animals, where it causes loss of productivity in livestock through abortion, stillbirth, loss of milk production, or failure to thrive. There is a strong worldwide occupational association with animal production (dairy farming, pig breeding, slaughter) and a universal risk from rodent-carrier-mediated infection is seen especially in tropical countries, where many serovars may be present locally.

All pathogenic leptospiras are classified as *Leptospira interrogans*, of which there are about 200 serologic variants (serovars) recognized by cross-agglutination. Common antigens allow the arrangement of serovars into serogroups, within each of which there is substantial cross-agglutination. Classification is significant clinically and epidemiologically because each serovar may be associated with a characteristic degree of severity of illness, a narrow natural host range (sometimes with restricted geographic distribution), or both. The main associations recognized throughout the world are serovars *hardjo* with cattle; *pomona* with pigs, cattle, and rodents; and *icterohaemorrhagiae, copenhageni, bataviae, autumnalis, australis,* and *javanica* with rats and small rodents.

Communicability and Transmissibility

Humans may be infected indirectly from animals by contact with contaminated water, soil, or mud in a moist environment, or by direct infection from urine, fresh carcasses, or organs. Person-to-person transmission has been reported (Spinu et al., 1963) but is extremely rare and insignificant because human patients and convalescents do not generally excrete leptospires for long, and elementary hygiene prevents infection from infected human urine or blood. The usual incubation period varies from 3 to 14 days. Second infections with the same serovar occur rarely, if ever; however, infection with an unrelated serovar may occur in people who have recovered from leptospirosis. Laboratory infections have been reported very rarely.

Clinical Manifestations of Disease

Leptospirosis commences suddenly with headache, fever (typically to 102°F [39°C]), malaise, muscle pains, conjunctival suffusion (red eyes), and a transient rash on the palate and skin. Thereafter, the illness may be mild and self-limiting or severe and possibly fatal. The mild type may be serious and incapacitating, but seldom leads to kidney or liver failure, hemorrhage, or death (Faine, 1982).

Mild Type

The description "mild type" is applicable worldwide to illness characteristically seen in infection with serovars *hardjo, grippotyphosa,* or *pomona,* but with others as well. The initial symptoms may be followed by a transient remission in 3 to 5 days, which may infrequently proceed to an exacerbation and include "aseptic" meningitis, renal failure, and abdominal or chest pains, reflecting the generalized pathology. The clinical differential diagnosis includes severe influenza, viral meningitis, acute abdominal conditions, and glomerulonephritis. Recovery is usually complete, but weakness, fatigue, depression, and even psychosis may prevent a return to normal living for weeks or months.

Severe Type

Characteristically, the severe type is seen in severe infections with serovars *icterohaemorrhagiae, copenhageni, bataviae, javanica,* and others. Illness worsens, usually rapidly after onset, so that renal failure appears within 7 to 10 days, sometimes accompanied or followed by skin and mucosal hemorrhages, jaundice, hemoptysis, myocarditis, or liver failure, leading to death if untreated. Peritoneal or extracorporeal dialysis is frequently lifesaving, and subsequent recovery of renal function is usually complete; the patient eventually recovers to normal activity provided that other organs (e.g., heart) have survived serious damage. Relapses and chronic infection have been reported, but are poorly documented. Leptospirosis in pregnancy carries the risks of intrauterine infection and fetal death (Faine et al., 1984).

Pathology

Leptospires enter through small cuts or abrasions, possibly through sodden skin, or by inhalation of aerosols of urine. They spread immediately and circulate in the bloodstream, although they are phagocytosed by fixed macrophages of the liver and lung. This diagnostically significant bacteremic phase lasts approximately 1 to 7 days. After the number of leptospires in the blood and tissues reaches a critical level of toxicity, due to leptospiral endotoxin, lesions and consequent symptoms appear. The primary lesion is damage to the endothelium of small blood vessels, leading to localized ischemia resulting in renal tubular necrosis, hepatocellular damage, meningitis, myositis, and placentitis. Hemorrhage occurs in severe cases, as do jaundice and frequently platelet deficiency. There is usually a mild granulocytosis and splenomegaly. Once immunity develops, leptospires are removed from the circulation and from tissues and organs by phagocytosis, after opsonization, as seen by the appearance of circulating antibodies that can agglutinate and opsonize leptospires. Tissue damage, although severe, may be reversible and followed by complete repair (e.g., kidney, liver); however, long-lasting damage (e.g., myocarditis) may be a complication and lead to scarring, which is well recognized in the kidneys of pigs and dogs (Faine, 1980, 1982).

Immunity is apparently solely humoral and can be passively transferred via the placenta and colostrum or by serum. Intracellular sequestration of leptospires and cellular immunity are not important in resistance to reinfection. However, leptospires may persist in the anterior chamber of the eye in humans and animals, leading to hypersensitive (allergic) inflammation (uveitis) on further contact with leptospires and eventual panophthalmitis and blindness in horses.

Diagnostic Features

Clinical diagnosis of leptospirosis requires the confidence of current experience supported by laboratory confirmation. The symptoms, although characteristic, are not pathognomonic. The epidemiologic environment, clinical history, symptoms, and findings on physical examination may allow a clinical diagnosis, but laboratory diagnosis is usually required. The main clinically useful features are the sudden onset, severity of headache and fever, almost invariable accompanying proteinuria, initially mild or transient nitrogen retention, palatal and skin rash, conjunctival suffusion, and extreme muscle tenderness, especially in the calf and back muscles (Faine, 1982). Presenting symptoms may mimic or indeed result from an abdominal emergency (e.g., acute cholecystitis), hemorrhage (hemoptysis), renal failure, or liver failure (jaundice).

Prognosis

Patients almost invariably recover completely from the mild type of leptospirosis. Antibiotic treatment hastens recovery and prevents complications and progression to secondary pathology. Long-term sequelae (see Clinical Manifestations of Disease) are uncommon and usually temporary. If leptospirosis progresses to the severe form with renal failure, hemorrhages in vital organs, and liver failure, renal dialysis and supportive treatment are almost always lifesaving; however, severe myocarditis may still result in death. Recovery from renal failure appears to be complete and leaves no lasting renal or other organ impairment.

Epidemiologic Aspects

Occurrence and Transmission

Leptospirosis occurs worldwide wherever there is a risk of direct or indirect contact with the urine or kidneys of infected animals or their infected products of conception. The organisms die when dry or in acid environments (pH less than 7.0); therefore, transmission is confined to wet alkaline environments. The most frequent sources of infection are urine, kidneys, surface waters, mud, and soil. The disease is not transmitted by ingestion of food or by inhalation of airborne particles.

The main sources of nonoccupational leptospirosis affecting city and rural dwellers in temperate climates are rodents, dogs, leisure activities, and travel (e.g., hobby farming, rafting, camping, boating, swimming). City dwellers may also be occupationally exposed (e.g., sewer workers, drainers, abbatoir workers, cattle-yarders, butchers, meat inspectors, veterinarians). Rural dwellers are always at risk, especially in tropical countries where they are in close contact with potentially infected field or domestic rodents and with domestic animals in a warm, wet climate, usually without protective clothing or footwear. Their habitation is often subjected to flooding. In temperate climates carriers among pigs, cattle, and dogs are the most important sources of leptospirosis in humans.

In principle, avoidance of direct or indirect contact with infected animals or ground water or soil would be a useful preventive measure. Avoidance should include adequate rodent control in the human habitat. Where such contacts are unavoidable, prevention of leptospirosis in the domestic animal population may be controlled by immunization of animals. Vaccines for dogs, pigs, and cattle are used widely and successfully. Vaccines for humans are available only in certain countries in Asia.

The roles of nonmammalian animals in the epidemiology of leptospirosis are unclear. Pathogenic leptospires have been found in coldblooded animals, which also develop antileptospiral antibodies, but pathology of leptospiral origin has not been described in them. These animals may be passive, uninfected carriers (Minette, 1983). Biting arthropods (ticks, lice) have transmitted experimental leptospirosis, but they are not usually regarded as significant vectors. Birds can be infected experimentally in the egg or soon after hatching; although they may be runted, they appear not to become carriers if they survive.

Susceptibility, Resistance, and Carrier States

Almost all young animals appear to be susceptible, either before or after birth, provided they are not protected by maternal specific immunoglobulins passively transferred through the placenta or by colostrum, in the case of mammals or marsupials. Infection may be overt and serious or subclinical; however, either form may result in the renal carrier state, which cannot be established without previous systemic infection. It is not known whether the general and overlapping association of host species and serovar of *Leptospira* is based on biologic and immunologic factors, ecologic relationships, or a combination of factors. To some extent this association, where it is known, may be an epidemiologic clue to the source of epidemic or endemic leptospirosis. On the other hand, practically nothing is known regarding the basis of the apparent complete resistance to infection by certain, or even most, leptospire serovars seen in some animal species, or about the increasing resistance with age of adult animals, except that the rate of development of specific immunoglobulins plays a part.

The renal carrier state is central to the persistence and epidemiology of leptospirosis. Leptospires colonize the luminar surface of the renal proximal tubular cells, closely interdigitating with the rough tubular epithelial surface; there they may persist for weeks to years, often for the lifetime of the animal in some leptospiral serovar–host associations. Specific adhesion factors have not been described. Excretion in urine may be intermittent or continuous and is associated with elevated immunoglobulin (Ig)M serovar-specific antileptospiral antibodies in the blood serum as detected by enzyme immunoassay. Antileptospiral immunoglobulin is present in the tubular and bladder urine, but does not kill the leptospires. Leptospires cannot survive in acidic bladder urine, but remain viable in alkaline urine. Consequently, herbivores and animals whose diet produces an alkaline urine are relatively more important as shedders than are producers of acid urine.

Etiologic Agent

The etiologic agent of leptospirosis is a serovar of *Leptospira interrogans*. Although serovars of *Leptospira biflexa* and *Leptonema* have reportedly been isolated from animals or patients with leptospirosis, evidence that they cause disease is inadequate.

Characterization and Properties

Leptospira interrogans, a thin, helical spirochete 10 to 20 μm long and 0.2 μm in cross section, has been well described recently (Faine and Stallman, 1982; Johnson and Faine, 1984). Even when stained intensely, the organism is poorly visualized by light microscopy because it is very thin. It is best seen by darkfield microscopy. Phase-contrast microscopy is less effective in practice because the motility of the leptospire moves it in and out of optic range.

The fine helical winding, characteristic motility, and usually hooked ends characterize the leptospire at first observation. Threads of fibrin and proteins in laboratory specimens and in blood cultures can be confused with leptospires even by experienced workers. Immunostaining with fluorescein- or peroxidase-labeled antibodies provides a specific stain for leptospires in tissues or specimens. Staining with traditional silver deposition methods either on microscope slides or in tissue sections is tedious, painstaking, and sometimes difficult to interpret.

Morphologic and Cultural Properties

Leptospires in specimens are recognized by their characteristic shape and motility on darkfield microscopy. Electron microscopy is not used routinely, but demonstrates a helical cell body (cytoplasmic cylinder) enclosed in a cell wall surrounded by an outer envelope or membrane.

Flagella ("axial filaments"), one inserted subterminally at each end, pass through the coils of the helix to the middle of the leptospire, where they may meet or overlap slightly.

Morphologically, *L. interrogans* and *L. biflexa* appear so similar that they cannot be differentiated by light or electron microscopy. One or both ends are charactericaliy hooked; this shape, however, is variable. Straight-ended variants, which are less translationally motile and frequently less virulent than the hooked variants, also occur. Leptospires can pass through 0.2-μm membrane filters.

The chemical composition of leptospires is characterized by a high lipid content, which is not clearly associated with serovar or virulence. Fatty acids unique to leptospires have been described.

Various enzymes (catalase, lipases, phospholipases), toxins (including hemolysins and lipases), and lipopolysaccharide have been described, the clinical, pathologic, and diagnostic significance of which has either not been clearly established or not exploited and widely used in diagnostic laboratory practice.

Cultivation

Leptospires grow relatively easily under aerobic conditions in a variety of fluid media, although fresh isolates are sometimes slow to adapt to culture in the laboratory. Growth is usually enhanced by aeration. Maximum growth rates are about 6 to 8 h doubling time at an optimum temperature of 28 to 30°C, reaching yields of about 5×10^8 to 10^9 per ml; the culture is turbid and birefringent, but not opaque. Cultures are commonly incubated for 3 to 10 days to reach maximum growth. Growth on solid media (0.7 to 1.0% agar) usually produces hemispheric, translucent, spreading, subsurface colonies with a gray edge; some strains will grow on the surface of 2% agar medium. In tubes of solid or semisolid media, leptospires grow in one or more disks 1 to 20 mm below the surface, as well as in discrete colonies below the disks.

Essential growth requirements are 1) a nutritional source of long chain fatty acids (Tween 20, 40, or 80 or rabbit serum is commonly used); 2) a means of detoxifying the fatty acids, usually bovine or other serum albumin or rabbit or other mammalian serum; 3) a simple source of nitrogen, usually supplied as an ammonium salt; 4) pH between 7.0 and 7.8, (optimal, 7.4); and 5) essential additional factors, including thiamine, cyanocobalamin, iron (as ferrous salts or hemoglobin), and other trace elements. These essentials are supplied in media variously modified from but based on the formulations of Ellinghausen–Macullough–Johnson–Harris (EMJH) or Korthof. Media may be solidified by the addition of 0.7 to 1.0% agar; semisolid (0.1 to 0.2% agar) media are used for maintenance of stock cultures. Leptospires can be adapted with care and difficulty to grow in protein-free media (Bey and Johnson, 1978) or low-protein (0.1 to 0.2% bovine serum albumin) medium, in both of which continuous frequent subculture is required to maintain viability.

The heat-sensitive nature of the detoxicants requires that media undergo final sterilization through a filter. However, all solutions and water for solutions to be filtered should be heat sterilized first to avoid contamination of media with leptospires that pass through the filter.

In fluid media, growth appears first as a faint birefringent turbidity; confirmation by darkfield microscopy is necessary at this early stage.

Genetic and Molecular Classification

The approved classification of *Leptospira* is by serology; however, an additional system of classification by DNA relatedness (DNA homology) shows four major groups. Also, comparisons of restriction endonucleases and DNA on electrophoretic gels and characteristic profiles in polyacrylamide gel electrophoresis have been described and used for identification (Marshall et al., 1981). These methods and their systems of classification are not in routine use at present.

Differentiation from Nonpathogenic Leptospires

Strains of *L. biflexa* grow at less than 13°C, and are sensitive to serum, 8-azaguanine, copper sulfate, and 1.0% NaCl. There is no reliable difference in the microscopic appearance in cultures (Faine, 1982; Johnson and Faine, 1984).

Biosafety Considerations

Freshly isolated leptospires are potentially highly virulent. Small numbers can cause infection in experimental animals and, presumably, in people. The need for direct entry into tissues means that cultures are not highly infectious in the laboratory. Laboratory infections have occurred from needle-stick or cut injuries rather than accidental inhalation or ingestion. Spillage on laboratory benches or floors should be treated with a suitable disinfectant (halogen, phenolic, detergent) before mopping up. Dried splashes and spillages should not pose a threat, but leptospires can survive freezing in tissues. Precautions in handling infected animal tissues and carcasses have been described (Faine, 1982).

Etiologic Diagnosis

Cultural, microscopic, and serologic methods are available. Leptospires have been cultivated, albeit sometimes with difficulty, from blood, tissues, urine, cerebrospinal fluid and anterior chamber fluid, and exudates in the acute stages (up to about 10 to 14 days). Culture is the only means of establishing the etiologic agent with certainty. Blood culture is the most successful method, being positive in most selected cases that are recognized and tested early in infection. Darkfield microscopy of unstained fluids or tissue specimens is relatively insensitive and also unproductive, as even experts may confuse fibrin and other protein threads with leptospires. Immunostaining with antileptospiral serum followed by fluorescein- or peroxidase-labeled conjugates is helpful because it is serospecific as well as precise, but it must be technically well controlled (Faine, 1982). Small numbers of leptospires can be missed, however, and at least a good guess of the appropriate serogroup must be made to choose the correct primary antileptospiral serum.

Serologic tests for presumptive diagnosis are used most frequently. They cannot identify the infecting serovar. Experience is required for their interpretation, especially in endemic areas or in constantly exposed populations (see Serologic Diagnosis).

Specimen Collection and Transport

Blood for culture should be taken during the febrile period. One-half milliliter is inoculated into 10 to 50 ml of EMJH culture medium, which is more "sensitive" and produces better results than Korthof-type media for primary isolation. Several cultures may be taken at intervals.

Urine should be alkaline or the patient given an alkalinizing regime before testing. Freshly voided urine, taken with aseptic precautions if possible, may be cultivated or examined by darkfield.

Leptospires do not survive long in urine that is acidic or contaminated with other microorganisms. The addition of 5-fluorouracil to culture media facilitates primary isolation from slightly contaminated urine (Faine, 1982; Johnson and Rogers, 1964). Neither blood nor urine should be refrigerated during transport.

Standard procedures (see Specimens Required) may be followed to obtain serum for serologic tests. The use of paired serum specimens taken 5 to 10 days apart is recommended; however, the first specimen should be tested without delay, because a "positive" reaction can be very useful diagnostically.

Isolation and Identification of Etiologic Agent

Culture Media and Methods and Darkfield Microscopy

Best results have been obtained using EMJH, tween albumin (TA), or Ellis medium (Ellis et al., 1985; Faine, 1982). Cultures should be incubated for up to 28 days at 28 to 30°C and examined every 2 to 3 days for macroscopic and, if necessary, microscopic evidence of growth. Subculturing when a small amount of growth has occurred may assist in early adaptation of the leptospires to laboratory cultivation. Cultures are examined by darkfield microscopy by placing a loopful on a slide without a cover slip. The use of a low-power dry darkfield condenser (e.g., numerical aperture 0.8) and a 20 to 25× objective is strongly recommended to assist in rapid scanning of the droplet. Positive cultures are subcultured by taking up about 0.5 to 1.0 ml in a sterile Pasteur pipette and inoculating 0.1 to 0.5 ml into each of several screw-cap tubes or bottles containing approximately 5 to 20 ml of medium that has been brought to room temperature. Either screw caps or close-fitting closures are required because of the prolonged incubation time. Transfer volumes of about 2 to 10% are recommended for routine subculturing with a pipette. Subcultures should be incubated at 28 to 30°C for 3 to 10 days (up to 30 days if necessary to achieve maximum growth). Isolates may be stored at room temperature or in liquid nitrogen. Only relatively sparse growth and faint turbidity (up to 10^8 per ml) may be obtained initially, or even regularly, with some strains.

Preliminary identification of an isolate is made serologically by initial slide agglutination of a heavy culture with a loopful of standard antiserum (Faine, 1982), confirmed by darkfield microscopy for evidence of agglutination. Indirect immunofluorescence with specific or monoclonal serospecific sera, if available, is also helpful. Further presumptive identification is made by agglutination to titer using a range of standard antisera and reading the endpoint by darkfield microscopy as described previously. Final identification can be made only by agglutination after cross absorption of homologous rabbit antisera raised from the isolate and compared with standard reference antisera and standard reference strains. It is tedious, expensive, and usually performed only in reference or specialist laboratories (Dikken and Kmety, 1978; Faine, 1982; Johnson and Faine, 1984).

An indication of the serovar can be obtained from comparative banding patterns on restriction-endonuclease-treated DNA electrophoresis (Marshall et al., 1981), but the DNA patterns, although characteristic, do not always correlate with serologic tests, which to date remain the only accepted criteria. No growth characteristics on solid media or enzyme or other cultural characteristics have been accepted as criteria for diagnosis or identification.

Noncultural Methods for Identification in Tissues or Exudates

Immunostaining with specific antisera or monoclonal antibodies can provide a presumptive serogroup diagnosis if the leptospires can be visualized in tissues. Enzyme immunoassay and radioimmunoassay (Chappel et al., 1985; Faine, 1982) may be used for detection of leptospiral antigen in contaminated tissues; these methods have not yet been refined for routine use. DNA probes have been used to detect leptospires in blood and urine (Terpstra et al., 1987).

Maintenance of Laboratory Strains and Stock Cultures and Production of Antisera

The requirements for maintaining laboratory strains and stock cultures and producing antisera are mainly for a specialist or reference laboratory (Faine, 1982) and are not recommended for routine laboratories.

Serologic Diagnosis

Of the numerous tests described (Faine, 1982), only the microscopic agglutination test (MAT) is used routinely at present, although enzyme immunoassays are appropriate for special purposes.

Specimens Required

Blood specimens, obtained by venipuncture and allowed to clot to provide serum samples, should be taken at the earliest opportunity during the illness and forwarded immediately for examination. They should be followed by a second sample and later samples, if required, at intervals of about 5 to 7 days. In remote areas finger- or ear-prick (capillary) blood may be soaked onto a standard-sized sterilized filter paper calibrated to hold a known volume (e.g., 0.1 ml), allowed to dry, and labeled for identification. The specimen is then sent to a laboratory, where the serum is eluted and used for titration.

Principles and Techniques

Dilutions and Observations

The serum sample is diluted in the laboratory in tubes or in 0.25-ml volumes in microtitration trays. Dilutions are usually started in the range of 1 : 10 to 1 : 50 and then doubled, although other dilution systems are in use. Suitable diluents are culture medium base or buffered physiologic saline, pH 7.2. An equal volume of well-grown cultures of leptospires, preferably in log phase and standardized to approximately 10^8 per ml, is added to each tube or well. A "positive" control of known titer and a "negative" control of diluent alone are always included. The tests are incubated at room temperature or up to 30°C for 1 to 4 h.

A loopful from each well or tube is placed on a microscope slide and examined without a cover slip for agglutination. A low-power darkfield microscope, using a long working distance 25× objective, should be used. Modifications include the use in some high-volume specialist laboratories of flat-bottomed microtitration trays and direct observation in the well with a special long working distance objective or an inverted microscope with a suitable long working distance darkfield condenser. The disadvantage is that scratched plates cause optic distortion and interference, preventing direct reading of the tests.

Microscopic Reading of Endpoint

Reading the endpoint microscopically in the MAT is facilitated by using an agglutination endpoint of 50%. The culture used for the test is diluted 1 : 2 in diluent, then diluted an additional 50% (i.e., 1 : 4 altogether) to provide a density equivalent to 50% free leptospires. The culture being tested is matched microscopically for density to provide an approximate comparison with a 50% agglutination endpoint.

Choice of Cultures for Agglutination

The MAT is highly sensitive and relatively specific. There are many possible serovars, but one or more from each serogroup known to react with patient sera after infection with any member of the serogroup is sufficient; that is, the strain is chosen for its cross-reactivity within the serogroup when reacting with human (or animal) sera. However, experts find that local and often recent isolates of a serovar react more clearly, specifically, and frequently than standard strains of the same serovar from reference laboratory collections (Dikken and Kmety, 1978; Faine, 1982).

Enzyme Immunoassay

Enzyme immunoassay using sonicated (or boiled) antigens are gaining increasing use, especially with specific anti-IgG or anti-IgM conjugates (Adler et al., 1980; Terpstra et al., 1980). Detection of reactive IgM in enzyme immunoassays correlates with urinary excretion in cattle and pigs (Ballard et al., 1984). Other tests such as latex agglutination, complement fixation, hemagglutination, sensitized erythrocyte lysis and hemolysin tests (Faine, 1982), never widely used, are being discarded in favor of MAT or enzyme immunoassay. New forms of enzyme immunoassay and other immunologic tests (immunoblotting, dot blot) are under development.

Interpretation

In endemic areas, most people can be expected to have a titer against locally prevalent leptospires as a result of past clinical or subclinical infection. Sera may react with one or more serovars, but titers are usually low (less than 1 : 100). A new infection will cause an increase in antibodies specific for the infecting serovar; in the first 2 to 3 weeks, however, there may be cross-reacting antibodies, usually to a lower titer. Hence, the infecting serovar cannot be ascertained from the highest reacting serovar in MAT.

A rising titer in sera collected 5 to 10 days apart, or a single titer of 1 : 800 or greater in the presence of compatible clinical findings, is indicative of leptospirosis. In nonendemic areas almost any titer of 1 : 50 or greater to a pathogenic serovar of *Leptospira* may be suspicious in patients with indicative clinical features. A fourfold or greater increase in titer or a high single titer of 1 : 800 or greater is highly suggestive. A checklist of key clinical observations in acute illness and interpretations of serologic tests appears on pages 50 and 51 of the World Health Organization Guidelines (Faine, 1982).

If clinical suspicion and serologic testing occur relatively late, such as more than 14 days after the date of infection (*not* the date of first symptoms), the titer may rise little if at all beyond its currently high level. Such static high titers usually indicate recent infection.

Treatment, Prevention, and Control

Treatment

Antibiotic treatment is effective within the first 7 to 10 days of infection. It should be given immediately on diagnosis or suspicion. Rapid, effective antibiotic treatment can reduce the progression of symptoms and the likelihood of sequelae and thus prevent the severe manifestations of illness. Antibiotics will eliminate leptospires from accessible tissues, but will not reverse established pathologic changes; the latter must run their course and be treated symptomatically or by appropriate nonantibiotic therapy.

The antibiotic of choice is benzylpenicillin by injection in doses of 5 MU per day for 5 days. A variation is to give half the dose in procaine penicillin for the first 2 days of the regime. Patients genuinely hypersensitive to penicillin may be given erythromycin, 250 mg four times daily for 5 days. Tetracyclines have been used, but are contraindicated in patients with renal insufficiency (which includes almost all cases of leptospirosis), in children, or in pregnant women. Doxycycline has been used as a chemoprophylactic for short-term exposure (Takafuji et al., 1984), but cannot be recommended for routine continuous use or for long-term occupational exposure.

Prevention and Control

Prevention is difficult. Surveillance can establish the extent of risk and likely sources of infection (Faine, 1982). Measures for occupational hygiene, such as protective clothing and avoidance of splash from urine or water, are often useful but hard to implement because they impede work or are unacceptable to both workers and employers on grounds of cost and perceived low effectiveness. It is not practicable to advise dwellers in tropical villages, for example, to avoid walking or working in wet conditions or contact with pigs, other livestock, or dogs, as well as soil or water contaminated by the animals' urine.

Human vaccines have been used successfully in China after floods that exposed large populations to the risk of leptospirosis, as well as in Japan and Vietnam. Repeated annual revaccination is recommended to retain immunity. The duration of immunity following either natural infection or vaccination is not well documented or established. Immunity against the same serovar is believed to be long lasting, even lifelong, but reinfections with the same or a different serovar have been recorded. Significant debility has been reported to result from vaccine reactions following single or repeated immunization.

Vaccination of livestock can reduce urinary shedding and risk to human handlers, especially (and perhaps only) when associated with a high level of education, awareness, cooperation, and hygiene in the community and administration. Prenatal vaccination of sows protects young piglets from acute leptospirosis, but not from being carriers as adults, and thus fails to reduce the risks to handlers, abattoir workers, butchers, and the like. Chemotherapy of livestock with dihydrostreptomycin has been used, but is relatively expensive and impractical for widespread control.

Literature Cited

Adler, B., A. M. Murphy, S. A. Locarnini, and S. Faine. 1980. Detection of specific antileptospiral immunoglobulins M and G in human serum by solid-phase enzyme-linked immunoadsorbent assay. J. Clin. Microbiol. **11**:452–457.

Ballard, S. A., B. Adler, B. D. Millar, R. N. Chappel, R. T. Jones, and S. Faine. 1984. The immunoglobulin response of swine following experimental infection with *Leptospira interrogans* serovar *pomona*. Zentralbl. Bakteriol. Mikrobiol. Hyg. I Abt. Orig. A Med. Mikrobiol. Infektionskr. Parasitol. **256**:510–517.

Bey, R. F., and R. C. Johnson. 1978. Protein-free and low-protein media for cultivation of *Leptospira*. Infect. Immun. **19**:562–569.

Chappel, R. J., B. Adler, S. A. Ballard, S. Faine, R. T. Jones, B. D. Millar, and J. A. Swainger. 1985. Enzymatic radioimmunoassay for detecting *Leptospira interrogans* serovar *pomona* in the urine of experimentally-infected pigs. Vet. Microbiol. **10**:279–286.

Dikken, H., and E. Kmety. 1978. Serological typing methods of leptospires, p. 260–295. *In* T. Bergan and J. R. Norris (ed.), Methods in microbiology, vol. 11. Academic Press, Inc., New York.

Ellis, W. A., J. Montgomery, and J. A. Cassels. 1985. Dihydrostreptomycin treatment of bovine carriers of *Leptospira interrogans* serovar *hardjo*. Res. Vet. Sci. **39**:292–295.

Faine, S. 1980. Leptospirosis—here, now. Pathology **13**: 1–5.

Faine, S. 1982. Guidelines for the control of leptospirosis. Offset publication no. 67. World Health Organization, Geneva.

Faine, S., B. Adler, W. Christopher, and R. Valentine. 1984. Fatal congenital human leptospirosis. Zentralbl. Bakteriol. Mikrobiol. Hyg. I Abt. Orig. A Med. Mikrobiol. Infektionskr. Parasitol. **257**:548.

Faine, S., and N. J. Stallman. 1982. Amended description of the genus *Leptospira* Noguchi 1917 and the species *L. interrogans* (Stimson 1907) Wenyon 1926 and *L. biflexa* (Wolbach and Binger 1914) Noguchi 1918. Int. J. Syst. Bacteriol. **32**:461–463.

Johnson, R. C., and S. Faine. 1984. Genus I. *Leptospira* Noguchi 1917, 755, p. 62–67. *In* N. R. Krieg and J. G. Holt (ed.), Bergey's manual of systematic bacteriology, vol. 1. The Williams & Wilkins Co., Baltimore.

Johnson, R. C., and P. Rogers. 1964. 5-Fluorouracil as a selective agent for the growth of leptospirae. J. Bacteriol. **88:**422–426.

Marshall, R. B., B. Wilton, and A. J. Robinson. 1981. Identification of *Leptospira* serovars by restriction-endonuclease analysis. J. Med. Microbiol. **14:**163–166.

Minette, H. P. 1983. Leptospirosis in poikilothermic vertebrates—a review. Int. J. Zoonoses **10:**111–121.

Spinu, I., V. Topciu, Trinh Thi Hang Quy, and Vo Van Hung. 1963. L'homme comme réservoir de virus dans une épidémie de leptospirose survenue dans la jungle. Arch. Roum. Pathol. Exp. Microbiol. **22:**1081–1100.

Takafuji, E. T., J. W. Kirkpatrick, R. N. Miller, and J. J. Karwacki. 1984. An efficacy trial of doxycycline chemoprophylaxis against leptospirosis. N. Engl. J. Med. **310:**497–500.

Terpstra, W. J., G. S. Ligthart, and G. J. Schoone. 1980. Serodiagnosis of human leptospirosis by enzyme-linked-immunosorbent-assay (ELISA). Zentralbl. Bakteriol. Mikrobiol. Hyg. I Abt. Orig. A Med. Mikrobiol. Infektionskr. Parasitol. **247:**400–405.

Terpstra, W. J., G. J. Schoone, G. S. Ligthart and J. Ter Schegget. 1987. Detection of *Leptospira interrogans* in clinical specimens by *in situ* hybridization using biotin-labelled DNA probes. J. Gen. Microbiol. **133:**911–914.

CHAPTER 36

Listeriosis

M. J. R. GILCHRIST

Disease: Listeriosis.
Etiologic Agent: *Listeria monocytogenes.*
Source: Reservoirs in the inanimate environment and domestic and feral animals serve as sources of human colonization and/or infection by direct or indirect means, especially contamination of food.
Clinical Manifestations: Disease may be asymptomatic (detected retrospectively in epidemic situations), a mild, flulike illness (often reported in pregnant females after delivery of affected infants), or severe, characterized by bacteremia, meningitis, or localizing infections (most commonly found in the immunocompromised—e.g., infants, those undergoing immunosuppressive therapy, and the aged). Primary cutaneous infections occur in those who handle animals.
Pathophysiology: A primary bacteremia with subsequent seeding of sites such as meninges, placenta, endocardium, and viscera. Infection or colonization of maternal genital tract with extension to placenta.
Laboratory Diagnosis: Specific culture of *L. monocytogenes* from body fluids (blood, cerebrospinal fluid, pleural, peritoneal, vitreous, bile), tissues (liver, spleen, lung, brain), or cutaneous wounds.
Epidemiology: Documented incidence is primarily in developed countries where specific diagnosis is available and where large populations of immunosuppressed patients exist. True incidence is presumably worldwide. Occupational disease of veterinarians and abattoir workers.
Treatment: Combined therapy with ampicillin and gentamicin for prolonged time period (3 weeks).
Prevention and Control: Minimize the contamination of food products by careful control of food production techniques.

Description of Disease

Man and animals are infected. Clinical manifestations depend on the species and immune status of the animal. A primary bacteremia may be followed by a localizing infection. Neurotropic in many animals, the organism produces a preponderance of meningitis in humans and encephalitis in ruminants (Gray and Killinger, 1966). The organism also exhibits a tropism for the products of conception in humans and many animals, resulting in abortion, stillbirth, or neonatal infections.

Communicability and Transmissibility

The specific details of acquisition of listeriosis by humans are currently under investigation. A hypothetical model for the epidemiology of human listeriosis is shown in Fig. 1. *Listeria monocytogenes* is found in soil, water, vegetable matter, and in a vari-

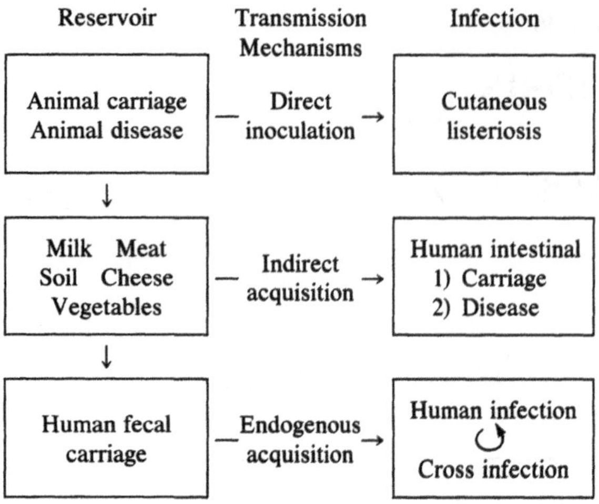

FIG. 1. Hypothetical model for epidemiology of human listeriosis.

ety of domestic and feral animals, and human listeriosis may be acquired from any of these sources. Animal carriage or disease produces human infection via direct inoculation or contamination of food products. Human intestinal colonization is thought to be a mechanism of endogenous acquisition and may also be a mechanism for human-to-human transmission via the fecal-oral route. Hospital-acquired cases of listeriosis are documented (Green and Macaulay, 1978; Stamm et al., 1982), but the mechanism of transmission is not understood. Although sporadic cases of listeriosis are known to occur, they are primarily in neonates and the immunocompromised (Albritton et al., 1976) and may signal a food-borne epidemic (Schlech et al., 1983) that is undetected, because *Listeria* infection is not a reportable disease or because the affected population is a small proportion of the the community.

Clinical Manifestations in Humans

Where human listeriosis is concerned, the species epithet "monocytogenes" is misleading. Monocytosis, first described in laboratory animals infected with *L. monocytogenes,* is an atypical feature of most types of human infection. A polymorphonuclear response in peripheral blood, CSF, and tissue is commonly observed (Shackelford and Feigin, 1977). Protean in its manifestations, human listeriosis can encompass one or a combination of the following forms: cutaneous, septicemia, meningitis/encephalitis, oculoglandular, cervicoglandular, granulomatosis septica, typhoid-pneumonic (Seeliger and Finger, 1983), or localizing infections of other sterile sites.

Immunocompromised Adults

In adults, patients undergoing immunosuppressive therapy for renal transplantation or cancer treatment experience the majority of cases of listeriosis. The most common form of listeriosis in adult humans is that of meningitis (Nieman and Lorber, 1980), but bacteremia, empyema, endocarditis, encephalitis, endophthalmitis, peritonitis (Myers et al., 1983), hepatitis (Yu et al., 1982), and cholecystitis (Gordon and Singer, 1986) are reported. Listeric and tuberculous meningitis have many features in common and may be mistaken for one another.

Perinatal Disease

Whereas disease in the pregnant female is often limited to a mild, flulike illness, *Listeria* is known to produce placental infections that result in abortion, stillbirth, or neonatal infections of considerable morbidity and mortality. Neonates experience bacteremia as a common feature, but some experience hyperthermia and others hypothermia. Clinical signs and symptoms are variable and not specific for listeriosis.

Cutaneous Disease

Not uncommonly, wound infections of animal caretakers and abattoir workers are due to *L. monocytogenes*. No characteristic manifestations are known; diagnosis is by specific culture. The specific etiology of many such infections may be missed if *Listeria* is overlooked in the laboratory as a contaminating "diphtheroid."

Pathophysiology

Systemic disease due to *L. monocytogenes* is generally acquired following ingestion of foods contaminated with the organism. The portal of entry of the organism is the bloodstream via the gastrointestinal tract, with the possible intermediary role of the lymphatic system. Bacteremia due to *L. monocytogenes* would then result in the seeding of localized sites, including the meninges or the placenta. Colonization or infection of the female genital tract from organisms in the gastrointestinal tract may produce antepartum seeding of the membranes, fetus, and placenta or intrapartum seeding of the infant. In one study of perinatal infections, maternal vaginitis was a common feature (Filice et al., 1978).

Diagnostic Features

Culture is the usual specific means of diagnosis. In postmortem samples, histopathologic techniques that reveal short, non-acid-fast rods in tissue sections suggest listeriosis in the appropriate patient population. Liver, spleen, adrenals, brain, lungs, esophagus, tonsils, lymph nodes, thymus, bone marrow, myocardium, testes, skeletal muscle, and the intestinal tract have been found to contain focal necroses or miliary granulomas. In the meninges, conjunctiva, and epithelial lining of the middle ear and nasal sinuses, a suppurative reaction is found (Seeliger and Finger, 1983).

Perinatal Infections

In five infants with listeric infection, culture-documented bacteremia and white blood cell counts in the normal range with 58 to 65% polymorphonuclear leukocytes were common features. Thrombocytopenia, a relatively low hematocrit, and positive cultures of CSF, skin, endotracheal aspirates, urine, stool, amniotic fluid, and maternal cervix were inconstant features (Ahlfors et al., 1977).

Immunocompromised Adults

In 102 renal transplant recipients, diagnostic features included meningitis (50%) or parenchymal CNS disease (10%), primary bacteremia (30%), and urinary tract infection (1%). In some cases, pneumonia (7%) accompanied the bacteremia or central nervous system infection (Stamm et al., 1982). Of several patients with meningitis, neurologic signs were minimal, suggesting that immunosuppressed patients with listeric bacteremia should be treated for meningitis (Gantz et al., 1975).

Prognosis

Infant mortality rates of 40 to 54% have been cited (Ahlfors et al., 1977). Aggressive ventilatory and pharmacologic therapy for pneumonia and early institution of antibiotics may reduce this mortality rate. In renal transplant recipients, the overall mortality rate was 26% (Stamm et al., 1982).

Epidemiologic Aspects

Occurrence and Transmission

In the past decade there have occurred a number of outbreaks of human listeriosis that are helping to elucidate the nature of disease transmission. In all cases a food product was implicated epidemiologically, and in several cases the etiologic agent was isolated from the implicated food and found to be the same strain (serovar and phage type) as that of the isolates from the epidemic. Table 1 enumerates the major features of three well-documented common-source outbreaks.

Susceptibility, Resistance, and Carrier States

The most susceptible population in which disease is manifest is the immunocompromised patient. Although most cases of listeriosis are in the immunocompromised, recent epidemics in North America have revealed infection among some normal hosts (Table 1). It is not known whether such infections of normal hosts result from higher inocula, relatively virulent organisms, or other unknown factors. Data on the carrier state in humans are often in conflict, perhaps an artifact of different study design parameters or inefficiencies in culture methods.

Etiologic Agent

Characterization

Although characteristically a small, non-acid-fast, nonspore-forming, nonbranching, gram-positive rod

Table 1. Features of common-source outbreaks of *L. monocytogenes* infections

Year	Location	*Cases (deaths[a])*		*Epidemiologically implicated source*	*Environmental isolate*
		Perinatal	*Adults*		
1981	Nova Scotia (Schlech et al., 1983)	34 (16)	7 (2)	Cole slaw[b]	Cole slaw
1983	Massachusetts (Fleming et al., 1985)	7 (2)	42 (12)	Pasteurized milk	None
1985	Los Angeles (CDC, 1985)	93 (29)	49[c] (18)	Mexican-style soft cheese	Cheese plant

[a] Deaths and stillbirths.
[b] Made from cabbage grown on soil fertilized with sheep manure.
[c] Forty-seven were immunocompromised.

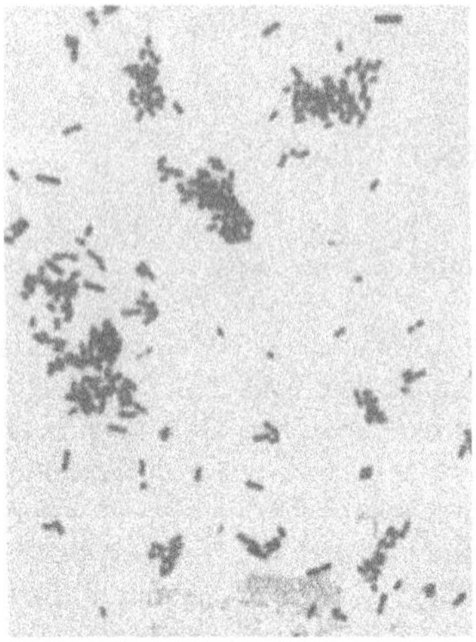

a

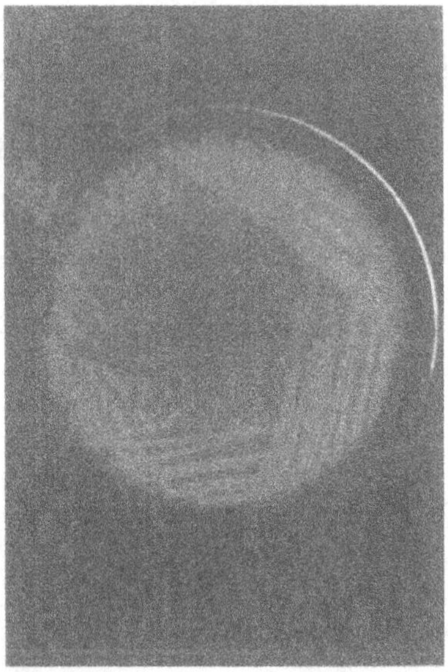

b

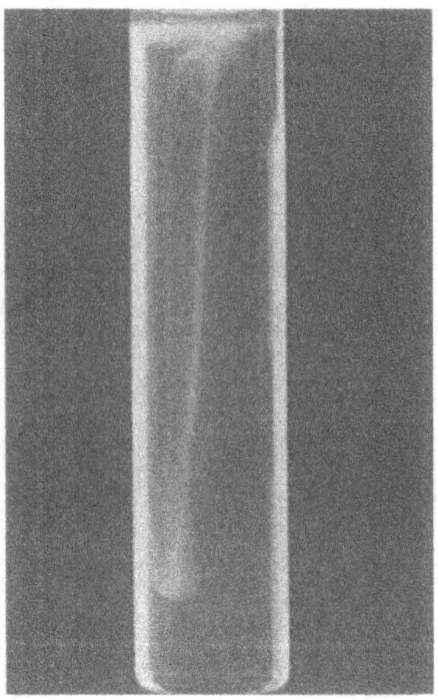

c

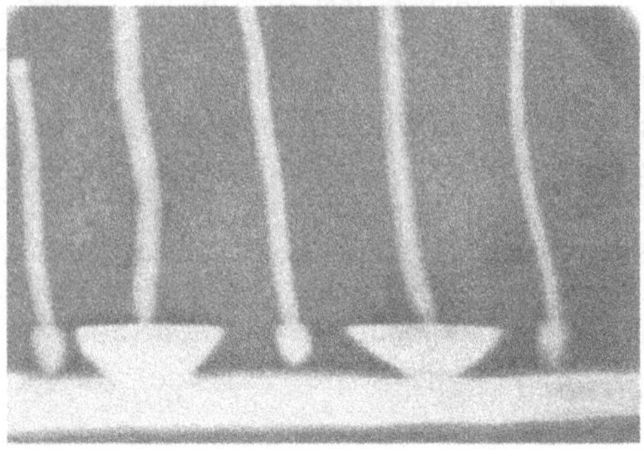

d

FIG. 2. (a) Gram stain morphology of *Listeria monocytogenes*. When grown in culture (shown), *Listeria* is increasingly gram-variable with age of the culture. Cell morphology is that of cocci, coccobacilli, or bacilli. Organisms may appear in pairs, short chains, or palisade layers, resembling diphtheroids. Bar = 2 μm. (b) Hemolytic reaction of *Listeria monocytogenes*. When grown on sheep, rabbit, or horse blood agar, the organism is generally beta-hemolytic. On sheep blood agar, the hemolytic zone may be very narrow; in some cases, it may be necessary to scrape off the colony to reveal hemolysis under the colony. The plate shows the faint hemolysis characteristic at 24 h; after 48 h, hemolysis is often more pronounced. (c) Motility of *Listeria monocytogenes* in semisolid agar. When grown at room temperature, the organisms are motile, forming an umbrella shape a few millimeters below the surface of the agar. (d) CAMP reactions of *L. monocytogenes* and group B streptococci. When streaked perpendicularly to *S. aureus,* the zone of hemolysis of the group B streptococci takes the form of an arrowhead at the juncture between the two streaks. For *Listeria monocytogenes,* the zone of the hemolysis at the juncture is more rounded. In the photograph, two group B streptococcal streaks are flanked by three *Listeria* streaks. The color plates of these figures appear in Color Plate 2 at the beginning of this volume.

(Fig. 2a) measuring 0.5 × 1 to 2 μm (Seeliger and Jones, 1986), *L. monocytogenes* may be pleomorphic and gram-variable and resemble cocci or coccobacilli, particularly in clinical material. When occurring in palisade formation as a gram-positive rod, *Listeria* organisms have been mistaken for "diphtheroids," and when resembling gram-positive cocci or gram-negative coccobacilli, they have been misidentified as streptococci and *Haemophilus,* respectively.

A facultative anaerobe, *L. monocytogenes,* will grow under either aerobic or anaerobic conditions. Preferring temperatures in the moderate range (optimal: 30 to 37°C), *L. monocytogenes* will grow at a wide range of temperatures (limits: 1 to 45°C) (Seeliger and Jones, 1986). On the surface of blood agar, the organism produces colonies that are small, entire, convex, bluish-gray, and translucent. Although *L. monocytogenes* is characterized as beta-hemolytic, some laboratory strains are nonhemolytic, and many fresh clinical isolates are so weakly hemolytic that the colony must be scraped off the surface of the blood agar to visualize the hemolysis underneath. At best there is a narrow zone of beta hemolysis (Fig. 2b) that may be more easily visualized on rabbit blood agar than on sheep blood agar and that is more easily visualized on sheep blood agar if incubation is extended for several days.

Temperature-dependent motility is a characteristic of *L. monocytogenes*. When growth occurs at 35 to 37°C, flagella are absent or functionally compromised, and motility is limited or absent. Growth at room temperature allows for production of functional flagella. Organisms grown in broth at room temperature and observed in wet-mount preparations exhibit tumbling motility, best characterized as end-over-end movement. With organisms grown in semisolid agar at room temperature, motility is expressed as an umbrella-shaped turbid area a few millimeters below the surface of the agar (Fig. 2c).

Specimen Collection and Transport

Specimens from affected sites (blood, CSF, fluids, and tissues) are collected and transported by the usual methods employed by the laboratory. Birth byproducts, feces, and environmental samples should be stored on ice or at 4°C until they can be delivered to the laboratory.

Isolation and Identification Procedures

Routine laboratory methods are applicable for culture of normally sterile body fluids. *L. monocytogenes* is readily cultivable by inoculation of blood agar and incubation at 35°C for 24 to 72 hr. For culture of contaminated food, environmental samples and by-products of birth, there is some evidence that cold enrichment may enhance recovery. Specimens are incubated in broth at 4°C for 7 to 30 days (with some authors citing several months), removed to a 35°C incubator for 18 h, and then plated and incubated for 24 to 72 h (Welshimer, 1981). Plating on a clear medium facilitates detection of the organism if there are a variety of other microorganisms in the sample. Observation of the colonies using light transmitted through the agar at an oblique angle produces a differential effect. *Listeria* colonies are blue-green, but considerable experience is required to distinguish these colonies from those of organisms that produce a similar effect. A variety of selective media and culture protocols are currently under trial to determine the best modes of recovery of the organism from foods.

Presumptive Identification

A simple presumptive identification is based on the following characteristics of *L. monocytogenes:* motile at room temperature, beta-hemolytic, catalase-positive, bile esculin–positive, hydrogen sulfide–negative in triple-sugar iron agar or Kligler iron agar, acid from glucose, and facultative oxygen requirements. A simple, dichotomously branching decision tree to distinguish *Listeria* from other nonsporing, non-acid-fast, nonbranching, gram-positive bacilli cannot be constructed because of the diverse characteristics of many of these organisms. If one constructs a table, however, requiring two or more characteristics *simultaneously*, then the individual members of other genera can be eliminated (Table 2).

Problems in Identification

The misidentification of *Listeria* as a streptococcus is a common occurrence in laboratories. Beta-hemolytic and morphologically similar to elongated cocci, *L. monocytogenes* may be misidentified as a group B streptococcus (Kontnick et al., 1977), because both are hippurate-positive and CAMP-positive (Fig. 2d), or it may be misidentified as an enterococcus (Wetzler et al., 1968), because both are bile esculin–positive and tolerant to 6.5% NaCl. Laboratories that routinely employ the catalase reaction (streptococci, negative; *Listeria*, positive) in all identification schemes for streptococci will alleviate this potential for error.

Definitive Identification and Typing

In most laboratories, the identification scheme outlined above is sufficient for the presumptive identification of human isolates of *L. monocytogenes*. Isolates should be referred to a public health laboratory for confirmatory testing and typing. Since environmental sources and foods may contain several species of *Listeria*, specific species determination must be performed. *L. monocytogenes* may be distinguished from other species of *Listeria* by the follow-

TABLE 2. Differentiation among *Listeria* and other nonspore-forming, non-acid-fast, nonbranching, gram-positive bacilli

Characteristics required	*Genera excluded*
Motile *and* catalase-positive *and* beta-hemolytic	*Corynebacterium* spp., *Bacterionema*, *Rothia*, and *Oerskovia*
Catalase-positive *and* motile *and* hydrogen sulfide–negative	*Erysipelothrix*
Motile *and* catalase-positive	*Lactobacillus*
Facultative oxygen requirement *and* acid from glucose	*Kurthia*

TABLE 3. Differentiation of the species in the genus *Listeria*[a]

	Beta hemolysis	CAMP reaction		Acid produced from			Nitrate reduced
		S. aureus	R. equi[b]	Xylose	Mannitol	Rhamnose	
L. monocytogenes	+[c]	+	−	−	−	+	−
L. innocua	−	−	−	−	−	V[d]	−
L. seeligeri	+	+	−	+	−	−	
L. welshimeri	−	−	−	+	−	V	
L. ivanovii[e]	+[f]	−	+	+	−	−	−
L. grayi[g]	−	−	−	−	+	−	−
L. murrayi[g]	−	−	−	+	−	V	+
L. denitrificans[g]	−	−	−	+	−	−	+

[a] Adapted from Seeliger and Jones (1986).
[b] *Rhodococcus equi* NCTC 1621.
[c] Not all strains positive; e.g., type strain ATCC 15313 is negative.
[d] Variable.
[e] Formerly *L. monocytogenes* serotype 5.
[f] Wide-zone or multiple-zone hemolysis.
[g] Taxonomic affiliation with *Listeria* uncertain or controversial.

ing characteristics: beta hemolysis, production of acid from rhamnose but not xylose or mannitol, a CAMP reaction with *S. aureus* but not with *R. equi*, and failure to reduce nitrate (Seeliger and Jones, 1986). Table 3 illustrates these features.

Within the species *L. monocytogenes*, there exist a number of serovars. These are designated 1/2a, 1/2b, 1/2c, 3a, 3b, 3c, 4a, 4ab, 4b, 4c, 4d, 4e, 5, 7,. Once identified as an *L. monocytogenes*, the isolate may be further characterized by determination of the serovar. Since there is cross-reactivity within the genus *Listeria* and with other genera, serologic characteristics should not be used to identify the isolate. Isolates of different serovars are presumed not to belong to the same epidemic. Since most epidemics are caused by one of a few serovars, isolates of the same serovar cannot, a priori, be presumed to be from the same epidemic. Further characterization of the epidemiologically implicated organisms may require phage typing (Taylor et al., 1981) or isoenzyme analysis (Bibb et al., 1986).

Serologic Diagnosis

Because of cross-reactivity of *Listeria* with other organisms or because of prior exposure to *Listeria*, many humans have low titers to *Listeria* without any history of disease. Thus a fourfold rise in titer is required for definitive diagnosis. Unfortunately, in sporadic cases, acute sera are rarely available by the time the diagnosis of listeriosis is entertained. For epidemiologic evaluation of ongoing epidemics, it is often useful to employ serologic methods, but serology has had little application in the diagnosis of sporadic listeric disease (Issekutz et al., 1986).

Treatment, Prevention, and Control

Treatment

Since the organism is an intracellular parasite with a propensity to invade sequestered sites (meninges, placenta), therapy may not be based solely upon an evaluation of in vitro susceptibility results but must respect the requirement for the antibiotic to enter mammalian cells and to cross the blood-brain or placental barrier. Not complemented by the assistance of an intact immune system, therapy with bacteriostatic agents is generally unsatisfactory. Penicillin or ampicillin therapy is bacteriostatic for *L. monocytogenes*, but synergistic action with aminoglycosides has been demonstrated (Moellering et al., 1972) and found to occur at the low concentrations of aminoglycoside in CSF (Trautman et al., 1985). The recommended therapy for listeriosis is primarily based on case reports. Practice has shown that a prolonged course (>3 weeks) of combination therapy with an aminoglycoside and a beta lactam antibiotic (e.g., penicillin or ampicillin plus gentamicin) minimizes relapses (Stamm et al., 1982). Trimethoprim-sulfamethoxazole has recently been proposed as a candidate for a clinical trial, since it is bactericidal with *L. monocytogenes*, and therapeutic levels are achieved in the sites of interest (Spitzer et al., 1986).

Prevention and Control

Definitive methods for prevention of *L. monocytogenes* infections will require a more thorough understanding of the means by which *Listeria* has come to contaminate food products. New evidence suggests that in many cases of contaminated dairy products, the organism has been introduced after pasteurization. Dairy plants need to be reevaluated to make the milk processing safer.

Literature Cited

Ahlfors, C. E., B. W. Goetzman, C. C. Halsted, M. P. Sherman, and R. P. Wennberg. 1977. Neonatal listeriosis. Am. J. Dis. Child. **131**:405–408.

Albritton, W. L., G. L. Wiggins, and J. C. Feely. 1976. Neonatal listeriosis: distribution of serotypes in relation to age at onset of disease. J. Pediatr. **88**:481–483.

Bibb, W. F., T. A. Kuffner, and R. E. Weaver. 1986. Typing of *Listeria monocytogenes* by isoenzyme analysis. Abstracts, Annual Meeting of the American Society for Microbiology, Washington, D.C.

Bortolussi, R., W. F. Schlech, III, and W. L. Albritton. 1985. Listeria. p. 205–208. *In* E. H. Lennette, A. Balows, W. J. Hausler, Jr., and H. J. Shadomy (ed.), Manual of clinical microbiology, 4th ed. American Society for Microbiology, Washington, D.C.

Centers for Disease Control. 1985. Listeriosis outbreak associated with Mexican-style cheese—California. Morbidity Mortality Weekly Rep. **34**:357–359.

Filice, G. A., H. F. Cantrell, A. B. Smith, P. S. Hayes, J. C. Feeley, and D. W. Fraser. 1978. *Listeria monocytogenes* infection in neonates: investigation of an epidemic. J. Infect. Dis. **138**:17–23.

Fleming, D. W., S. L. Cochi, K. L. Macdonald, J. Brondum, P. S. Hayes, B. D. Plikaytis, M. B. Holmes, A. Audurier, C. V. Broome, and A. L. Reingold. 1985. Pasteurized milk as a vehicle of infection in an outbreak of listeriosis. N. Engl. J. Med. **312**:404–407.

Gantz, N. M., R. L. Myerowitz, A. A. Medeiros, G. F. Carrera, R. E. Wilson, and T. F. O'Brien. 1975. Listeriosis in immunosuppressed patients: a cluster of eight cases. Am. J. Med. **58**:637–643.

Green, H. T., and M. B. Macauley. 1978. Hospital outbreak of *Listeria* septicaemia: a problem of cross infection? Lancet **2**:1039–1040.

Gordon, S., and C. Singer. 1986. *Listeria monocytogenes* cholecystitis. J. Infect. Dis. **154**:918–919.

Gray, M. L., and A. H. Killinger. 1966. *Listeria monocytogenes* and listeric infections. Bacteriol. Rev. **30**:309–382.

Issekutz, T. B., S. H. S. Lee, and R. Bortolussi. 1986. Immune response to *Listeria monocytogenes*. p. 402–407. *In* N. R. Rose, H. Friedman, and J. L. Fahey (eds.), Manual of clinical immunology, 3rd ed. American Society for Microbiology, Washington, D.C.

Kontnick, C., A. von Graevenitz, and V. Piscitelli. 1977. Differential diagnosis between *Streptococcus agalactiae* and *Listeria monocytogenes* in the clinical laboratory. Ann. Clin. Lab. Sci. **7**:269–276.

Moellering, R. C., G. Medoff, I. Leech. C. Wennersten, and L. J. Kunz. 1972. Antibiotic synergism against *Listeria monocytogenes*. Antimicrob. Agents Chemother. **1**:30–34.

Myers, J. P., G. Peterson, and A. Rashid. 1983. Peritonitis due to *Listeria monocytogenes* complicating continuous ambulatory peritoneal dialysis. J. Infect. Dis. **148**:1130.

Nieman, R. E., and B. Lorber. 1980. Listeriosis in adults: a changing pattern. Report of eight cases and review of the literature, 1968–1978. Rev. Infect. Dis. **2**:207–227.

Schlech, W. F. III, P. M. Lavigne, R. A. Bortolussi, A. C. Allen, E. V. Haldane, A. J. Wort, A. W. Hightower, S. E. Johnson, S. H. King, E. S. Nicholls, and C. V. Broome. 1983. Epidemic listeriosis—evidence for transmission by food. N. Engl. J. Med. **308**:203–206.

Seeliger, H. P. R., and H. Finger. 1983. Listeriosis. p. 264–289. *In* R. S. Remington and J. O. Klein (ed.), Infectious diseases of the fetus and newborn infant, 2nd ed. W. B. Saunders, Philadelphia.

Seeliger, H. P. R., and D. Jones. 1986. Listeria. p. 1235–1245. *In* P. H. A. Sneath, N. S. Mair, and E. Sharpe (ed.), Bergey's manual of systematic bacteriology, Vol. 2. Williams and Wilkins, Baltimore.

Shackelford, P. G., and R. D. Feigin. 1977. *Listeria* revisited. Am. J. Dis. Child. **131**:391–392.

Spitzer, P. G., S. M. Hammer, and A. W. Karchmer. 1986. Treatment of *Listeria monocytogenes* infection with trimethoprimsulfamethoxazole: case report and review of the literature. Rev. Infect. Dis. **8**:427–430.

Stamm, A. M., W. E. Dismukes, B. P. Simmons, C. G. Cobbs, A. Elliott, P. Budrich, and J. Harmon. 1982. Listeriosis in renal transplant recipients: report of an outbreak and review of 102 cases. Rev. Infect. Dis. **4**:665–682.

Taylor, A. G., J. McLauchlin, H. T. Green, M. B. Macaulay, and A. Audurier. 1981. Hospital cross-infection with *Listeria monocytogenes* confirmed by phage-typing. Lancet **1**:1106.

Trautmann, M., J. Wagner, M. Chahin, and T. Weinke. 1985. *Listeria* meningitis: report of ten recent cases and review of current therapeutic recommendations. J. Infect. Dis. **10**:107–114.

Welshimer, H. J. 1981. The genus *Listeria* and related organisms. p. 1680–1687. *In* M. P. Starr (ed.), The prokaryotes: a handbook on habitats, isolation and identification of bacteria, Vol. 2. Springer-Verlag, New York.

Wetzler, T. F., N. R. Freeman, M. L. French, L. A. Renkowski, W. C. Eveland, and B. J. Carver. 1968. Biological characterization of *Listeria monocytogenes*. Health Lab. Sci. **5**:46–62.

Yu, V. L., W. P. Miller, E. J. Wing, J. M. Romano, C. A. Ruiz, and F. J. Bruns. 1982. Disseminated listeriosis presenting as acute hepatitis. Case reports and review of hepatic involvement in listeriosis. Am. J. Med. **73**:773–777.

CHAPTER 37

Lyme Disease and Related Disorders

RUSSELL C. JOHNSON

Disease: Lyme disease and related disorders (Lyme borreliosis, erythema migrans disease, Bannwarth's syndrome).

Etiological Agent: *Borrelia burgdorferi*

Clinical Manifestations: A multisystemic illness that often begins with a characteristic rash (erythema migrans chronicum) accompanied by flu-like or meningitis-like symptoms. Weeks to months later some patients develop meningitis, cranial or peripheral neuropathies, or myocarditis. Weeks to years later, many patients develop arthritis, chronic skin lesions, and neurologic abnormalities. Maternal-fetal transmission occurs.

Pathology: Synovial hyperplasia, erosion of joint cartilage and bone, nerve demyelation, and atrophy of skin in chronic disease.

Laboratory Diagnosis: Bacteriologic isolation not recommended. Indirect immunofluorescence assay or enzyme-linked immunosorbant assay of serum, cerebral spinal fluid, or synovial fluid for anti-*B. burgdorferi* antibodies. Immunofluorescence assay titers of 1:256 or greater are considered to be evidence of Lyme disease in patients with compatible clinical symptoms.

Epidemiology: Disease widespread. Present in North America, Europe, Russia, Asia, Africa, and Australia. Arthropod-borne disease, with *Ixodes* species of tick serving as the primary vector of *B. burgdorferi.*

Treatment: For early Lyme disease the drug of choice is tetracycline, then penicillin followed by erythromycin. For late Lyme disease high-dose intravenous penicillin G or ceftriaxone (Rocephin) is recommended.

Prevention and Control: Avoidance of vectors. Prompt removal of ticks, wearing protective clothing, and use of insect repellants helpful.

Description of Disease

Lyme disease (Lyme borreliosis, erythema migrans disease, Bannwarth's syndrome) is a multisystem illness caused by a newly described spirochete, *Borrelia burgdorferi* (Burgdorfer et al., 1982; Johnson et al., 1984). *Ixodes* species of ticks are the primary vectors of this disease (Anderson et al., 1983; Bosler et al., 1983; Burgdorfer et al., 1982). Lyme disease was first identified in the United States in 1975, after a mysterious outbreak of arthritis among the residents of Lyme, Connecticut (Steere et al., 1977). The disease is also present in 13 countries from Europe to Australia. The first illness caused by this spirochete was described in Europe in 1909 (Afzelius, 1921). The disease is characterized by the development of an expanding annular red rash-like skin lesion, erythema chronicum migrans (ECM), at the site of the tick bite. The ECM may be followed by cardiac, joint, and neurologic abnormalities (Steere et al., 1983). The arthritis and neurologic abnormalities may be progressive and chronic in some patients (Reik et al., 1986). It is also becoming a disease of veterinary importance, especially for dogs (Kornblatt et al., 1985; Lissman et al., 1984; Magnarelli et al., 1985).

Communicability and Transmissibility

Lyme disease and related disorders is a tick-borne zoonosis, with rodents serving as the major reservoir for *Borrelia burgdorferi*. With the exception of maternal-fetal transmission (Schlesinger, 1985), person-to-person transmission of Lyme disease has not been reported. However, nonarthropod transmission of *Borrelia burgdorferi* has been reported in rodents (Burgess et al., 1986).

Clinical Manifestations

Lyme disease can be divided into three stages (Steere et al., 1985). However, there is considerable overlap and not all patients exhibit all stages. Approximately 1 week after exposure to an infected *Ixodes* tick, the first stage of the disease ensues. It is characterized by the development of an erythematous expanding, annular skin lesion, erythema chronicum migrans (ECM), at the site of the tick bite. Approximately 50% of the cases of Lyme disease in the United States have multiple ECM, whereas a solitary ECM is typical of European cases of the illness. The average ECM is 15 cm in diameter with a bright red expanding edge and a central area of clearing. Constitutional symptoms associated with stage 1, which usually lasts for 3 weeks, include malaise, fatigue, headache, fever and chills, stiff neck, arthralgias, and myalgias. About 25% of cases of Lyme disease in adults and 50% of cases in children do not experience the ECM. Weeks to months later the disease enters its second stage. Approximately 10% of the patients will experience transient cardiac dysfunction (atrioventricular block, myopericarditis, and left ventricular dysfunction). Neurologic abnormalities including meningitis, cranial neuropathies (especially facial palsies), and sensory or motor radiculopathies occur in about 37% of the patients (Steere et al., 1983). Months to 2 years after the tick bite, stage 3 or the chronic form of Lyme disease may occur. In the United States approximately 53% of untreated cases have recurrent attacks of oligoarthritis primarily affecting the large joints. The knee is the most common site. Episodic bouts of arthritis may occur for 3 to 7 years, and 10% of these patients will develop chronic arthritis (Steere et al., 1977, 1983). Neurologic dysfunction experienced during this stage ranges from fatigue to progressive nerve damage resulting in dementia and paralysis (Reik et al., 1986). A chronic skin lesion, acrodermatitis chronica atrophicans (ACA), is also a feature of stage 3 (Weber et al., 1984). Neurologic abnormalities and ACA occur more frequently and arthritis to a lesser extent in Europe as compared with the United States (Stanek et al., 1985).

When ECM is present, Lyme disease is easily diagnosed because it is pathognomonic. In the absence of ECM, Lyme disease may mimic many other disorders.

Pathology

After the bite of an infected tick, *Borrelia burgdorferi* enters the blood and lymphatic circulation where they multiply and are distributed to various organs of the host. In contrast to the relapsing fever borreliae, *Borrelia burgdorferi* is only present in low numbers in the blood during the first week of illness. Spirochetes are present in the synovial tissue of patients with established arthritis. Associated with their presence is synovial hyperplasia, villous hypertrophy, lymphoplasma cellular infiltration, deposition of fibrin, and damage to small blood vessels. In chronic cases of Lyme arthritis, erosion of cartilage and bone may occur (Johnston et al., 1985). Demyelination may be associated with progressive neurologic abnormalities. Atrophy of skin and dilated small blood vessels are present in ACA. These spirochetes are not known to possess specific toxins. One patient with coexistent Lyme disease and babesiosis died due to pancarditis with diffuse lymphoplasmacytic infiltrate. The presence of spirochetes in the myocardium suggested that the patient died from Lyme disease (Marcus et al., 1985).

Adverse outcomes including in utero and neonatal death have been associated with contracting Lyme disease during pregnancy (Markowitz et al., 1986; Schlesinger et al., 1985).

Diagnostic Features

During early Lyme disease, stage 1, an expanding, erythematous skin lesion, ECM, develops at the site of the tick bite in approximately 70% of the cases (Steere et al., 1983). The presence of ECM is pathognomonic for Lyme disease. If the ECM does not occur or is not recognized and a history of a tick bite is lacking, diagnosis of the disease becomes more difficult. Depending on the stage of Lyme disease, it can mimic many other disorders such as rheumatoid arthritis, Reiters syndrome, acute rheumatic fever, multiple sclerosis, and aseptic meningitis.

A definite diagnoses of Lyme disease can be made by isolation of *Borrelia burgdorferi* from patients. However, this is presently a low-yield procedure. Because spirochetes are only present in low numbers, they are rarely seen by direct examination of blood, skin transudates of ECM, cerebral spinal fluid, or synovial fluid. Lyme serology is currently the most important diagnostic test available. Determination of antibody titers is accomplished by indirect immunofluorescent assay (IFA) or enzyme-linked immunosorbent assay (ELISA) (Craft et al., 1984; Magnarelli et al., 1984; Russell et al., 1984). Maximum specific IgM titers are reached between the third and sixth week of the illness, and IgG titers peak months later. An IFA titer of $\geq 1:256$ is considered by a number of laboratories to be evidence of Lyme disease in a patient with compatible clinical symptoms. Because of the slow antibody response, serology is only positive in about 50% of the cases of early Lyme disease. Some cross-reactivity occurs between *Treponema pallidum* and *Borrelia burgdorferi* both in IFA and ELISA assays. The lack of reactivity of antibodies to *B. burgdorferi* in the rapid plasma reagin and the microhemagglutination assay for *Treponema pallidum* can be used to differentiate cases of Lyme disease from those of syphilis. The relapsing fever borreliae are closely related to *B. burgdorferi*, and there is strong serologic cross-reactivity between these spirochetes. This is not a diagnostic problem except in geographic areas where both relapsing fever and Lyme disease occur, such as the Western United States. A major complication of Lyme disease serology is the lack of a significant antibody response in a number of patients with the illness and the lack of a standardized assay. The observation that patients with Lyme disease may excrete antigens of *Borrelia burgdorferi* in their urine provides the possibility of an additional diagnostic test (Hyde et al., 1987).

Proposed criteria for Lyme disease are: confirmed case: ECM, or in the absence of ECM, involvement of at least one of three organ systems (cardiovascular system, nervous system, or joint involvement) and a positive serologic test; probable case: in the absence of both ECM and positive serology, involvement of at least two organ systems or involvement of at least one organ system and a history of tick bite.

During early Lyme disease patients often have a moderate elevation of the sedimentation rate. Tests for antinuclear antibodies and rheumatoid factor are usually negative (Steere et al., 1983).

During inflammatory arthritis, synovial fluid contains 500 to 100,000 cells per cubic millimeter, predominately polymorphonuclear leukocytes. When neurologic abnormalities occur, a mild lymphocyte pleocytosis is found in the cerebrospinal fluid. Anti-*B. burgdorferi* antibodies are usually present in the synovial fluid and the cerebrospinal fluid of patients with arthritis and meningitis, respectively (Stiernstedt et al., 1985).

Prognosis

Lyme disease is rarely a fatal illness. Early antimicrobial therapy will prevent the majority of patients from developing the late sequelae of Lyme disease. However, about 50% of the treated patients will still experience mild fatigue, recurrent headaches, and migratory arthralgias. Untreated patients may experience neurologic abnormalities, arthritis, and chronic skin lesions for years. The damage may be progressive in some cases resulting in erosion of joint bone and hypertrophy of synovial tissue and permanent neurologic damage resulting in some cases of dementia and paralysis. Treatment of late or complicated Lyme disease is less effective than the early stage of the disease.

Epidemiologic Aspects

Occurrence and Transmission

Cases of Lyme disease have been reported from the United States, Europe, Russia, Asia, Africa, and Australia.

Lyme disease is primarily a tick-borne disease and the major vectors are the hard-shelled *Ixodes* ticks. In the United States, from the New England coast through the northern midwest, the vector of *B. burgdorferi* is *I. dammini*, whereas in the western states the disease is transmitted by *I. pacificus* (Burgdorfer et al., 1985). *Ixodes ricinus* is responsible for the transmission of the disease in Europe. Some cases of Lyme disease in the United States have been associated with the bite of the ticks *Ambyloma americanum* and *Ixodes scapularis* (Burgdorfer and Kierans, 1983). Mosquitoes, horse flies, and deer flies contain spirochetes after an infected blood meal. However, it is not known whether they can mechanically transmit this agent (Magnarelli et al., 1986). Long-range dispersion of ticks occurs by birds, and short-range spread occurs by mammals. In contrast to the common wood tick, *Dermacentor* species, which feeds on humans when in the adult form, *Ixodes* ticks bite humans in the larval, nymphal, and adult forms. The number of Lyme disease cases in the United States peaks in June and July when the nymphal form is active. Only 30% of patients were aware of a preceding tick bite because of the small size of the nymph: 1 to 2 mm. In contrast to the immature forms, the adult tick will only feed on large mammals, with the deer being the preferred host. Because deer play an important role in the life cycle of the tick, most cases of Lyme disease occur in areas where deer are present. *Borrelia burgdorferi* is transmitted while the tick is feeding. The mechanism of spirochete transmission, for example, via saliva or feces, has not been elucidated as yet. The *Ixodes* ticks are slow feeders requiring several days to become fully engorged. Early removal of ticks, within 24 hours, should reduce the possibility of infection.

Studies with dogs and mice indicate that the spirochete can be passed among these animals without the intervention of the tick. Spirochetes have been seen in the urine of mice and this may be a means of transmission among these animals. As there is a high incidence of infected dogs in endemic areas, they may carry infected ticks into homes. The possibility that their urine may be infectious requires investigation. Maternal-fetal transmission of *B. burgdorferi* does occur.

Etiologic Agent

Characterization

Borrelia burgdorferi are helically shaped bacteria with dimensions of 0.18 to 0.25 μm by 4 to 30 μm. Seven or eleven periplasmic flagella are located at each cell end. The other structural features and staining and physiologic characteristics are the same as those described for the relapsing fever borrelia. The antigenic variation present in the relapsing fever borreliae has not been observed in *B. burgdorferi* (Johnson et al., 1984).

Specimen Collection

Isolation of *B. burgdorferi* from patients with Lyme disease has been a low-yield procedure. Spirochetes have been isolated from the blood, ECM skin lesions, and cerebral spinal fluid. Serum should be obtained for serologic testing. During arthritis and neurologic disease, synovial fluid and cerebral spinal fluid, respectively, can be assayed for antibodies.

Isolation and Identification Procedures

Because of the paucity of *B. burgdorferi* in tissues and body fluids, direct examination of specimens by darkfield or phase-contrast microscopy or by cultural procedures is not recommended. Spirochetes have been visualized in tissues when stained by silver-impregnation methods. Spirochetes can be detected in ticks by darkfield examination of the midgut removed by dissection. *Borrelia burgdorferi* can be isolated from infected ticks and animals. The isolation medium consists of Barbour-Stoenner-Kelly (BSK) medium (Barbour, 1984) plus 0.15% agarose. The midgut of the tick and the blood, liver, kidneys, and spleen of animals are usually cultured. Samples of 1 to 2 drops of blood are added to 6 to 10 ml of media. A 10% (wt/vol) homogenate of organs in BSK medium without agarose is prepared. After allowing the larger tissue debris to settle, duplicate 1 : 10 dilutions of the supernatant are cultured in the isolation medium at 30 to 34°C. Cultures are examined by darkfield microscopy for at least 6 weeks for the presence of spirochetes (Johnson et al., 1984).

Identification

The preliminary identification of isolated spirochetes is established by the indirect immunofluorescence (IFA) test using rabbit antisera to *B. burgdorferi*. Because of possible cross-reactivity of the antisera with other spirochetes, definitive identification is accomplished by the use of monoclonal antibodies directed to *B. burgdorferi* specific surface antigens (Barbour et al., 1985).

Immunologic Aspects

Because definitive diagnosis of Lyme disease by isolation of the spirochetes is currently a low-yield procedure, determination of antibody titers is presently the most useful diagnostic test. Two serologic tests used to detect total immunoglobulins or class-specific IgM and IgG antibodies are the IFA and the ELISA assays. Both tests are of similar sensitivity and specificity. The ELISA is preferred by some laboratories because the procedures are more easily analyzed and automated and because absorbance values can be statistically analyzed. Specific IgM antibodies usually are present within 1 to 3 weeks after disease onset, and specific IgG antibodies take 3 weeks or longer to develop. The IgM response is not necessarily limited to early Lyme disease. It may persist in patients with prolonged illness, and a new IgM response may appear late in the illness (Craft et al., 1986). Cross-reactivity is largely associated with the IgM antibodies. For routine confirmation of Lyme disease, an IFA or ELISA for total immunoglobulins should be adequate. An assay for specific IgG antibodies may be useful for differentiating later stages of Lyme disease from other neurologic or arthritic disorders. These serologic tests will not discriminate between Lyme disease and relapsing fevers, although homologous titers are sometimes higher. Cross-reactivity occurs between *B. burgdorferi* and the pathogenic *Treponema*. The rapid plasma reagin test and the microhemagglutination test can be used to differentiate cases of Lyme disease from those of syphilis, yaws, and pinta. Sera of Lyme disease patients are nonreactive in these tests (Magnarelli et al., 1987).

The IFA or ELISA for class-specific or total immunoglobulins are usually negative during early Lyme disease (within 2 weeks after onset of ECM). Serologic testing of the cerebral spinal fluid of patients with central nervous system disorders may facilitate diagnosis because reactivity of antibody in the CSF may be higher than that of serum antibodies (Stiernstedt et al., 1985).

Treatment, Prevention, and Control

Treatment

Most cases of early Lyme disease respond favorably to treatment with oral tetracycline and penicillin. Erythromycin has a lower level of therapeutic activity. Tetracycline, 250 mg four times daily for 10 days, is the drug of choice except in pregnant women and young children. Adequate treatment of early Lyme disease prevents major late sequelae such as arthritis, myocarditis, and neurologic abnormalities. However, regardless of the antimicrobial administered, approximately 50% of the patients experienced minor late complications, which include recurrent episodes of headache or joint and muscle pain and lethargy (Steere et al., 1983). The Jarish-Herxheimer reaction is experienced by approximately 10% of the patients. High-dose intravenous penicillin is necessary for the treatment of the late sequelae of Lyme disease. Patients who develop meningitis and cranial or peripheral neuropathies respond favorably to intravenous penicillin G, 20 million units per day for 10 days. The same treatment for established arthritis resulted in a cure for 55% of the patients (Steere et al., 1985). Patients who failed to respond to penicillin or tetracycline therapy have been effectively treated with ceftriaxone (Rocephin) 2 to 4 g per day for 10 days (Dattwyler et al. 1987). Ceftriaxone has the potential of being a highly effective antibiotic for the treatment of Lyme disease and related disorders (Johnson et al., 1987).

Prevention and Control

Lyme disease and related disorders are prevented by avoiding the arthropod vectors. When in potential habitats of the *Ixodes* ticks, such as woodland inhabited by deer, wear clothing that covers as much of the skin as possible. Tuck pants into socks or boots. Wear light-colored clothing to better see the ticks and use insect repellants. Periodic skin inspection and tick removal may prevent Lyme disease. The risk of transmission of *B. burgdorferi* by nymphal *I. dammini* to rodents increased with duration of tick attachment (Piesman et al., 1987). Although a vaccine for Lyme disease is not currently available, the potential for a vaccine in the near future exists (Johnson et al., 1986).

Literature Cited

Afzelius, A. 1921. Erythema chronicum migrans. Acta Derm. Venereol. **2:**120–125.

Anderson, J. F., R. C. Johnson, L. A. Magnarelli, and F. W. Hyde. 1986. Involvement of birds in the epidemiology of the Lyme disease agent *Borrelia burgdorferi.* J. Infect. Immunol. **51:**394–396.

Anderson, J. F., L. A. Magnarelli, W. Burgdorfer, and A. G. Barbour. 1983. Spirochetes in *Ixodes dammini* and mammals from Connecticut. Am. J. Trop. Med. Hyg. **32:**818–824.

Barbour, A. G. 1984. Isolation and cultivation of Lyme disease spirochetes. Yale J. Biol. Med. **57:**521–525.

Barbour, A. G., R. A. Heiland, and T. R. Howe. 1985. Heterogeneity of major proteins in Lyme disease borreliae: a molecular analysis of North American and European isolates. J. Infect. Dis. **152:**478–484.

Bosler, E. M., J. L. Coleman, J. L. Benach, D. A. Massey, J. P. Hanrahan, W. Burgdorfer, and A. G. Barbour. 1983. Natural distribution of the *Ixodes dammini* spirochete. Science **220:**321–322.

Burgdorfer, W., A. G. Barbour, S. F. Hayes, J. L. Benach, E. Grunwaldt, and J. P. Davis. 1982. Lyme disease: a tick-borne spirochetosis? Science **216:**1317–1319.

Burgdorfer, W., and J. E. Kierans. 1983. Ticks and Lyme disease in the United States. Ann. Intern. Med. **99:**121.

Burgdorfer, W., R. S. Lane, A. G. Barbour, R. A. Gresbrink, and J. R. Anderson. 1985. The western black-legged tick, *Ixodes pacificus:* a vector of *Borrelia burgdorferi.* Am. J. Trop. Med. Hyg. **34:**925–930.

Burgess, E. C., T. E. Amundson, J. P. Davis, R. A. Kaslow, and R. Edelman. 1986. Experimental inoculation of *Peromyscus* spp. with *Borrelia burgdorferi:* evidence of contact transmission. Am. J. Trop. Med. Hyg. **35:**355–359.

Craft, J. E., D. K. Fischer, G. T. Shimamoto, and A. C. Steere. 1986. Antigens of *Borrelia burgdorferi* recognized during Lyme disease: appearance of a new immunoglobulin M response and expansion of immunoglobulin G response late in illness. J. Clin. Invest. **78:**934–939.

Craft, J. E., R. L. Grodzicki, and A. C. Steere. 1984. Antibody response in Lyme disease: evaluation of diagnostic tests. J. Infect. Dis. **149:**789–795.

Dattwyler, R. J., J. J. Halperin, H. Pass, and J. Luft. 1987. Ceftriaxone-effective therapy in refactory Lyme disease. J. Infect. Dis. **155:**1322–1325.

Hyde, F. W., C. Shelborne, and R. C. Johnson, 1987. Detection of *Borrelia burgdorferi* antigens in the urine of experimentally infected dogs and mice and humans with Lyme disease. J. Clin. Microbiol (submitted).

Johnson, R. C., C. Kodner, and M. Russell. 1986. Active immunization of hamsters against experimental infection with *Borrelia burgdorferi.* Infect. Immunol. **54:**713–714.

Johnson, R. C., C. Kodner, and M. Russell. 1987. In vitro and in vivo susceptibility of the Lyme disease spirochete, *Borrelia burgdorferi,* to four antimicrobial agents. Antimicrob. Agents Chemother. **31:**164–167.

Johnson, R. C., N. Marek, and C. Kodner. 1984. Infection of Syrian hamsters with Lyme disease spirochetes. J. Clin. Microbiol. **20:**1099–1101.

Johnson, R. C., G. P. Schmid, F. W. Hyde, A. G. Steigerwalt, and D. J. Brenner. 1984. *Borrelia burgdorferi* sp. nov.: etiologic agent of Lyme disease. Int. J. Syst. Bacteriol. **34:**496–497.

Johnston, Y., P. H. Duray, A. C. Steere, M. Kashgarian, S. Malawista, and P. Askenase. 1985. Lyme arthritis:

spirochetes found in microangiopathic lesions. Am. J. Pathol. **118**:26–34.

Kornblatt, A. N., P. H. Urband, and A. C. Steere. 1985. Arthritis caused by *Borrelia burgdorferi* in dogs. J. Am. Vet. Med. Assoc. **186**:960–964.

Lissman, B. A., E. M. Bosler, H. Camay, B. G. Ormiston, and J. L. Benach. 1984. Spirochete-associated arthritis (Lyme disease) in a dog. J. Am. Vet. Med. Assoc. **185**:219–220.

Magnarelli, L. A., J. F. Anderson, and A. G. Barbour. 1986. The etiological agent of Lyme disease in deer flies, horse flies and mosquitoes. J. Infect. Dis. **154**:355–358.

Magnarelli, L. A., J. F. Anderson, and R. C. Johnson. 1987. Cross-reactivity in serological tests for Lyme disease and other spirochetal infections. J. Infect. Dis. **156**:183–188.

Magnarelli, L. A., J. F. Anderson, A. F. Kaufmann, L. L. Lieberman, and G. D. Whitney. 1985. Borreliosis in dogs from southern Connecticut. J. Am. Vet. Med. Assoc. **186**:955–959.

Magnarelli, L. A., J. M. Meegan, J. F. Anderson, and W. A. Chappell. 1984. Comparison of an indirect fluorescent-antibody test with an enzyme-linked immunosorbent assay for serological studies of Lyme disease. J. Clin. Microbiol. **20**:181–184.

Marcus, L. C., A. C. Steere, P. H. Duray, A. E. Anderson, and E. B. Mahoney. 1985. Fatal pancarditis in a patient with coexistent Lyme disease and babesiosis. Ann. Intern. Med. **103**:374–376.

Markowitz, L. E., A. C. Steere, J. L. Benach, J. D. Slade, and C. V. Broome. 1986. Lyme disease during pregnancy. J. Am. Med. Assoc. **255**:3394–3396.

Piesman, J., T. N. Mather, R. J. Sinsky, and A. Spielman. 1987. Duration of tick attachment and *Borrelia burgdorferi* transmission. J. Clin. Microbiol. **25**:557–558.

Reik, L., W. Burgdorfer, and J. O. Donaldson. 1986. Neurological abnormalities in Lyme disease without erythemia chronicum migrans. Am. J. Med. **81**:75–78.

Russell, H., J. S. Sampson, G. P. Schmid, H. W. Wilkinson, and B. Plikaytis. 1984. Enzyme-linked immunosorbent assay and indirect immunofluorescence assay for Lyme disease. J. Infect. Dis. **149**:465–470.

Schlesinger, P. A., P. H. Duray, B. A. Burke, A. C. Steere, and M. T. Stillman, 1985. Maternal-fetal transmission of the Lyme disease spirochete, *Borrelia burgdorferi*. Ann. Intern. Med. **103**:67–68.

Stanek, G., G. Wewalka, V. Groh, R. Neumann, and W. Kristoferitsch. 1985. Differences between Lyme disease and European arthropod-borne borrelia infections. Lancet **i**:401.

Steere, A. C., N. H. Bartenhagen, J. E. Craft, G. J. Hutchinson, J. H. Newman, D. W. Rahn, L. H. Sigal, P. N. Sigal, P. N. Spieler, K. S. Stenn, and S. E. Malawista. 1983. The early clinical manifestations of Lyme disease. Ann. Intern. Med. **99**:76–82.

Steere, A. C., J. Green, R. T. Schoen, E. Taylor, G. J. Hutchinson, D. W. Rahn, and S. E. Malawista. 1985. Successful parenteral penicillin therapy of established Lyme arthritis. N. Engl. J. Med. **312**:869–874.

Steere, A. C., R. L. Grodzicki, A. N. Kornblatt, J. E. Craft, A. G. Barbour, W. Burgdorfer, G. P. Schmid, E. Johnson, and S. E. Malawista. 1983. The spirochetal etiology of Lyme disease. N. Engl. J. Med. **308**:733–740.

Steere, A. C., G. J. Hutchinson, D. W. Rahn, L. H. Sigal, J. E. Craft, E. T. DeSanna, and S. E. Malawista. 1983. Treatment of early manifestations of Lyme disease. Ann. Intern. Med. **99**:22–26.

Steere, A. C., S. E. Malawista, J. A. Hardin, S. Ruddy, P. W. Askenase, and W. A. Andiman. 1977. Erythema chronicum migrans and Lyme arthritis—the enlarging clinical spectrum. Ann. Intern. Med. **86**:685–698.

Stiernstedt, G. T., M. Granstrom, B. Hederstedt, and B. Skoldenberg. 1985. Diagnosis of spirochetal meningitis by enzyme-linked immunosorbent assay and indirect immunofluorescence assay in serum and cerebrospinal fluid. J. Clin. Microbiol. **21**:819–825.

Weber, K., G. Schlerz, B. Wilske, and V. Preac-Mursic. 1984. Zur Klinik und Atiologie der Acrodermatitis chronica atrophicans. Hautarzt **35**:571–577.

CHAPTER 38

Melioidosis and Glanders

NEYLAN A. VEDROS

Diseases: Melioidosis and glanders (farcy).

Etiologic Agents: *Pseudomonas pseudomallei* (ATCC 23343, WRAIR 286, and NBL 121); *Pseudomonas mallei* (ATCC 23344; NBL 7).

Source: *P. pseudomallei:* soil and water; *P. mallei:* infectious discharges from equines.

Clinical Manifestations: *Melioidosis:* Subclinical, subacute, acute, and chronic. Pulmonary features predominate with productive cough, fever, blood-tinged mucoid sputum. Abdominal pain, diarrhea, and muscle tenderness are frequent. *Glanders:* Skin and musculoskeletal system with nodules and granulomatous lesions which can lead to septicemia (lymphadenopathy, jaundice, and splenomegaly).

Pathology: *Melioidosis:* Abscesses and foci in lungs, liver, spleen, and kidneys. Characteristic "collar" of congested capillaries demarcate the granulomatous lesion. *Glanders:* Nodules and abscesses that calcify with age and can ulcerate.

Laboratory Diagnosis: Isolation of etiologic agent on glycerol-nutrient agar with inhibitors. *Melioidosis:* CF titer of greater than 64 and indirect hemagglutination titer of greater than 40 are diagnostic. Immunofluorescent antibody assay using IgM-tagged sera is specific. *Glanders:* Agglutination titer of greater than 320 is diagnostic.

Epidemiology: *Melioidosis:* Endemic in Southeast Asia, Guam, Madagascar, and Australia. Majority of cases worldwide occur between latitude 20N and latitude 20S. *Glanders:* Only found in countries where equine glanders still exists. In developed countries, found only in laboratory personnel.

Treatment: *Melioidosis:* Tetracycline combined with chloramphenicol or cotrimoxazole alone. *Glanders:* Sulfadiazine.

Prevention and Control: *Melioidosis:* Avoidance of contact with contaminated soil and water. *Glanders:* Avoidance of contact with discharges of infected equines.

Description of Diseases

Although melioidosis and glanders differ in many aspects, such as in the ecology of their causative agents, host susceptibility, mode of transmission, and epidemiology, there are equally many similarities. The two diseases are therefore discussed together in this chapter.

Melioidosis is a glanders-like disease of humans and other mammals such as wild rodents. The disease was first recognized by Whitmore and Kirshnaswami (1912) and given the name melioidosis by Stanton and Fletcher (1925) to reflect its similarity both clinically and pathologically to glanders, a chronic and debilitating disease of equines. The disease is relatively uncommon in humans but when contracted can be severe and fatal.

Glanders (farcy) is a contagious disease of horses, asses, and mules and is one of the oldest diseases with well-defined description in medical history. The contagious character of the disease via nasal discharge and draining ulcers among horses was described as early as the time of Constantine the Great. The disease is characterized in equines by the formation of tumor-like swellings or nodules in the nasal passages, trachea, lungs, and other internal organs and in the skin. The term "farcy" is usually limited to the visible appearance of external lesions of the skin.

Etiology

Melioidosis: The causative organism is *Pseudomonas pseudomallei,* first isolated and described by Whitmore in 1913 and formerly known as Whitmore's bacillus. The taxonomic position of this organism, like the agent of glanders, has evolved through a series of classifications, for example, *Pfeiferella whitmori, Pfeiferella pseudomallei, Bacillus whitmori, Flavobacterium pseudomallei, Actinobacillus pseudomallei, Loefferella whitmori,* and *Mallomyces pseudomallei* (Redfern et al., 1966). Because of its resemblance in some cultural and physiologic characteristics to the aerobic pseudomonads, Haynes (1957) formerly included this organism in the genus *Pseudomonas.*

Glanders: The causative organism is *Pseudomonas mallei,* first named *Bacillus mallei* by Zopf (1885). The taxonomy of this microorganism has evolved similarly to that described for *P. pseudomallei* but in a much more ill-defined and vague manner, for example, as recently as 1964 it was proposed to transfer *P. mallei* to the genus *Acinetobacter* (Steel and Cowan, 1964). The close similarity to *P. pseudomallei* in biochemical activities, DNA homology, and the like (Mandel, 1966; Redfern et al., 1966) and its specific susceptibility to temperate bacteriophages from *P. pseudomallei* resulted in *mallei* being assigned to the genus *Pseudomonas.* It has been hypothesized that *P. mallei* evolved from the free-living *P. pseudomallei* when the glanders bacillus adapted to a strict parasitic existence (Fournier, 1967).

Transmission

Melioidosis: P. pseudomallei can be isolated from a wide variety of animals including rats, rabbits, guinea pigs, sheep, goats, pigs, and horses. It is generally accepted that zoonosis plays no role in human meliodosis but rather that humans acquire the infection from contaminated soil and water through skin abrasions or by inhalation (editorial, 1975).

Glanders: Among equines the disease is transmitted by discharges from the nasopharynx, lungs, or skin to healthy animals via food, feed boxes, mangers, clothing, and other stable equipment. Carnivorous animals feeding on infected carcasses can get the disease. Humans acquire the infection in a similar manner by contact with infectious discharges that enter via aerosols or skin abrasions (Mendelsohn, 1950). Aerosols are particularly dangerous for laboratory workers (Howe and Miller, 1947; Jennings, 1963), and in fact *P. mallei* is considered to be one of the most dangerous microorganisms with which to work. Human-to-human transmission has been reported.

Clinical Manifestations

Melioidosis: The incubation period of melioidosis is not known with certainty, but any patient with a mild, self-limiting fever and living in or having visited an endemic area should be suspected of suffering from melioidosis. Manifestations in other patients with overt melioidosis varies from the acute to the chronically ill with the unique feature of clinical manifestations appearing months to years after the initial infection.

Subacute

Pulmonary features predominate in subacute melioidosis. Fever, and productive cough often with blood-streaked mucoid sputum, lung abcesses, pleural involvement, and joint pains may occur. Many subacute cases, identified by serology, may occur in the absence of developing acute infections, resemble tuberculosis, and predominate in endemic areas.

Acute

Acutely ill patients have high fever, chills, a productive cough with blood-tinged mucoid sputum, diarrhea, and abdominal pain. The latter two manifestations were cardinal signs in a patient from Denmark returning from a visit to Kenya (Bremmelgaard et al., 1982). The acutely ill patient may recover and then dramatically worsen with an overwhelming septicemia, increased confusion and tremors, profuse diarrhea that resembles cholera, pustular rash, and become comatose and die of septicemic shock. Generally the patient presents with sudden onset of abdominal pain, muscle tenderness, fever, lethargy, and shaking chills (Lee et al., 1985). Physical examination may indicate pneumonia, empyema, lung abcesses, mild jaundice, hepatomegaly, splenomegaly, acute or subacute arthritis, and septic pustules.

Chronic

Chronic cases of melioidosis vary in their clinical manifestations. The disease may persist for years and show evidence of lung abcesses, osteomyelitis, liver or spleen abcesses, and lymphadenopathy. Fistulas may often appear on the facial area, beginning as early as 1 year of age (Patamasucon et al., 1982).

Glanders: The clinical manifestations of human glanders depends upon the route of transmission and the virulence of the particular strain of *P. mallei*. The skin and musculoskeletal system are the sites of the most common manifestations followed by the lymphatics and nose (Robins, 1906). Once the organism enters the skin, within a week a small nodule appears with surrounding lymphangitis. This lesion develops the typical farcy ("pipes" and "buds"), which leads to a granulomatous lesion and potential disfigurement if located on mucous membranes of the facial area.

Similar to other facultative intracellular bacterial pathogens that cause localized infections (e.g., bubonic plague), septicemia can develop. This is the most serious form of glanders and is manifested by anorexia, nausea, fever, malaise, and myalgia. Generalized pustular eruptions with ulceration occurs and the symptoms depend upon the body location. In the lungs necrotizing pneumonia causes pleuritic pain and severe cough, whereas if the ulceration occurs in the nose or on the lips, the patient may have problems in eating, breathing, and swallowing. Cervical lymphadenopathy, jaundice, and spenomegaly are common, and invariably the patient dies within 1 to 2 weeks if not treated. The metastasis of developing glanders, particularly in the viscera and musculoskeletal system, can lead to a fatal episode of septicemia years after the initial infection. In the developing countries of Asia, Africa, and the Middle East, glanders must be differentiated from other granulomatous diseases, particularly cutaneous leismaniasis.

Pathology

Melioidosis: Except for the case in which the introduction of *P. pseudomallei* occurs via traumatic wounds or burns, the majority of infections are initiated via the pulmonary tract (even by inspiration of contaminated water by drowning; Achana et al., 1985; Lee et al., 1985; or naturally in captive cetaceans in Asia, Leong et al., 1985). Foci are noted in the lungs, liver, spleen, and kidney postmortem. These foci coalesce to form abcesses that vary in size with the severity of the illness. The lesions have a characteristic histology (Piggot and Hochholzer, 1970), that is, a collection of polymorphonuclear neutrophils surrounded by zones of congestion. The larger abcesses have a center of necrotic casseous material containing nuclear debris. Congested capillaries at the periphery form a "collar," which demarcates the lesion from surrounding tissue. In acute cases in humans (and in cetaceans) the microorganisms can be isolated from the blood, heart, brain, and multiple abcesses (Thin et al., 1970).

Glanders: The tuberculoid nodules in acute pulmonary glanders are hemorrhagic, dark in color, and range in size from 1 to 4 mm. The nodules become firm and may be calcified as they age (Frothingham, 1901). In other parts of the body, the nodules often ulcerate with a thick, purulent discharge that is highly infectious. In the acute nodules the polymorphonuclear leukocytes predominate, but in the older or chronic nodules giant cells and epitheliod cells are seen at the periphery of the necrotic center. Discharge from ulcerative nodules in the subcutaneous tissues (farcy) are drained by the local lymphatics, which become firm and resemble cords ("farcy-pipes") or the local nodes enlarge and become firm ("farcy-buds"). The pathology of glanders therefore is very similar to other bacterial pathogens that have a facultative intracellular cycle that results in tuberculoid and other types of granulomatous lesions.

Diagnostic Features

In melioidosis and glanders a clinical diagnosis is possible but must be confirmed by isolation of the causative agent or by serologic demonstration of specific antibodies.

The protean clinical manifestations of melioidosis have been well documented in French (Duroux, 1965) and U.S. military personnel in Vietnam (Clayton et al., 1973; Kishimoto et al., 1971). Infections confined to the lungs must be differentiated from tuberculosis, staphylococcal pneumonia, or pulmonary mycoses (Spotnitz et al., 1967). The septicemic form mimics several diseases, such as typhoid, malaria, leptospirosis, typhus, and plague. The disease may also mimic hepatitis A and B, mononucleosis, and cytomegalovirus infections. There is no suitable culture filtrate (tuberculin-like) material available to aid in the diagnosis. A detailed patient history is critical when a compatible syndrome exists.

Glanders, like melioidosis, has no clearly defined clinical diagnostic features. Examination of pus from skin nodules, nasal discharge, or sputum will show many polymorphonuclear leukocytes and small gram-negative bacteria that stain irregularly, but this is not diagnostic. The causative agent must be isolated and serologic evaluations conducted (Howe and Miller, 1947). A history of association with equines aids in the diagnosis, but acute glanders can be confused with brucellosis, melioidosis, and typhoid. Chronic glanders must be differentiated from sporotrichosis and other mycotic skin infections. In certain circumstances the use of "mallein," a cultural filtrate, can be used to detect a state of delayed hypersensitivity specific for glanders as used in equines (Howe et al., 1971).

Prognosis

In the early literature mortality due to melioidosis was estimated to be approximately 95% (Nigg, 1963).

The rarity of the disease, association of the disease with predisposing debilitating diseases (Rimington, 1962), and the possibility that many infections are subclinical (Nigg, 1963) makes any general statement on prognosis meaningless. Early diagnosis and vigorous long-term appropriate treatment are essential for successful management of the disease (Cooper, 1967; Flemma et al., 1969), but death can occur despite therapy (Flemma et al., 1969).

Human cases of glanders are very rare, but in the few reports available in the literature, treatment was successful in all cases (Ansasi and Minou, 1951; Howe and Miller, 1947). In the less developed countries where glanders still exists, most patients with untreated acute septicemic glanders die (Davis, 1981).

Epidemiologic Aspects

Occurrence and Transmission

Melioidosis: Since the original description of melioidosis from autopsy of victims in Rangoon (Whitmore and Krishnaswami, 1912), the majority of cases have occurred in Southeast Asia and around the world between latitude 20N to latitude 20S (Howe et al., 1971); Redfern et al., 1966). Sporadic cases have been reported from Korea, the Phillippines, Central and South America, the West Indies, Turkey, Taiwan, and a Danish citizen who visited Kenya (Biegelsisen et al., 1964; Bremmelgaard et al., 1982; Howe et al., 1971; Lee et al., 1985). The disease is not only endemic in Southeast Asia but also in Australia, Guam, and Madagascar (Johnson, 1967; Rimington, 1962). In the Western hemisphere cases usually occur in those who have visited endemic areas, but in the United States and Canada human-to-human transmission has been noted by venereal exchange (McCormack et al., 1975) and possibly by transplacental transfer (Osteraas et al., 1971). One case has been acquired from the environment (Nussbaum et al., 1980) and two well-defined cases of laboratory infections (Green and Tuffnell, 1968; Scheck et al., 1981). On the basis of serologic surveys, it is estimated that from 2,500 to 250,000 U.S. military personnel were infected in Vietnam (Clayton et al., 1973; Kishimoto et al., 1971). Active cases with death have been described in these military personnel (Brundage et al., 1968; Rubin et al., 1963), but the real danger is the fear of recrudescence as these veterans get older and become compromised medically for a variety of reasons (Sanford, 1971).

The causative agent, *P. pseudomallei,* is consistently being isolated from animals and the environment in areas where it has been newly introduced, for example, in horses and the soil in the Paris Zoo (Dodin, 1976; Dodin and Galimand, 1976), in animals and rice fields in Iran (Pourtaghva et al., 1975), and in pigs and swamp muds of Upper Volta and Niger (Dodin and Ferry, 1975). Whenever serologic surveys are conducted in these areas, as was done in Southeast Asia (Nigg, 1963), the results indicate that humans have been exposed to *P. pseudomallei* but have subclinical infections (Dodin and Ferry, 1975). Because zoonotic infections are not considered important in human melioidosis, the presence of this free-living pathogen in contaminated soil and water poses a continuous threat. The low incidence of overt clinical disease is probably due to the fact that *P. pseudomallei* has not evolved into a very successful parasite, has little invasive power, and has a restricted host and tissue specificity (Redfern et al., 1966).

Glanders: Human cases of glanders have essentially disappeared from developed countries due to stringent laws that have eliminated the disease in equines. This confirms the fact that unlike *P. pseudomallei, P. mallei* has no reservoir in soil and water but rather has adapted to a specific host-parasite relationship in equines (see review by Howe, 1950). The geographic range of the disease is therefore coterminus with that of the natural host, that is, equines. The exception are those cases that occur in laboratory workers through inhalation of accidentally created aerosols (Howe and Miller, 1947). This experience may indicate that humans are readily susceptible to infection by *P. mallei* at relatively small dosages. Contact with concentrated cultures in the laboratory or with discharges from infected equines and their contaminated food, bedding, and the like can result in the acquisition of glanders. Studies in experimental animals, however, indicate that strains of *P. mallei* from natural cases in equines vary in their virulence and must be considered in the epidemiology of glanders (Minett, 1959).

Etiologic Agents

Characterization

Pseudomonas mallei and *P. pseudomallei* belong to family *Pseudomonadaceae,* of which the type genus is *Pseudomonas,* one of the most complex groups of gram-negative bacteria (Palleroni, 1984). Both *P. mallei* and *P. pseudomallei* have been assigned to rRNA group II and, based on rRNA and DNA homologies, have been clustered with *P. cepacia, P. gladioli,* and *P. caryophylli* (Palleroni, 1973).

P. mallei and *P. pseudomallei* share many similar characteristics between themselves and with the genus *Pseudomonas.* They were therefore earlier called the pseudomallei group of the aerobic pseudomonads (Redfern et al., 1966; Stanier et al., 1966).

Both are oxidase-positive, gram-negative rods, aerobic with a strictly respiratory type of metabolism with oxygen as the terminal electron acceptor, and they can grow anaerobically when nitrate is the terminal electron acceptor, hydrolyse arginine, do not produce a soluble pigment, accumulate poly-B-hydroxybutyrate, liquefy gelatin, cleave catechol and protocatechuate at the ortho bond, grow at 41°C but not 4°C, reduce nitrates, and have a mol% G+C of 69 to 69.5.

A considerable number of substrates (42) can be used by *P. pseudomallei* but not by *P. mallei*, whereas 21 substrates can be used by *P. mallei* but not by *P. pseudomallei* (see Palleroni, 1984 and Redfern et al., 1966 for details).

The major differences between *P. mallei* and *P. pseudomallei* are: *P. mallei* is nonmotile (no flagella), uses xylose as a substrate but not D-ribose, levulinate, or erythritol; *P. pseudomallei* is motile with polar tufts of three or more flagella and has the opposite reactions to *P. mallei* with the substrates noted above.

Specimen Collection and Transport

The source of tissue specimens for the isolation of *P. pseudomallei* depends on the clinical syndrome. Reports in the literature indicate that blood was the best source for successful isolation in septicemic cases followed by sputum and urine but never feces. When an obvious lesion/abcess is present, this material was generally best for isolation. The specimens should be cultured as soon as possible and if transported to the laboratory should be kept at ambient temperatures. Standard procedures for water sampling by filtration should be followed, and suspensions of soil in sterile saline that are plated shortly after collection have been successful.

Pseudomonas mallei should be cultured as soon as possible from nasal drainage, pus from skin or nodules, or sputum depending on the clinical syndrome. Such procedures have been routinely successful in infected equines, but data on human infections are not readily available. No isolates were obtained by a well-equipped microbiology laboratory with trained personnel in six cases of laboratory-acquired infections (Howe and Miller, 1950).

Isolation

Melioidosis: P. pseudomallei can be isolated readily from blood and other tissues on a variety of common laboratory media under aerobic conditions at 37°C. The author has routinely used for clinical specimens brain heart infusion (BHI) agar or trypticase soy agar (TSA) supplemented with 1:200,000 crystal violet and 1% glycerol. Culture plates should be examined daily and kept in the incubator for at least 5 days before discarding. Within 24 to 48 h small, dewdrop, round, entire colonies appear. On BHI-glycerol or nutrient agar-glycerol with no crystal violet, the colonies appear glistening, mucoid, and creamy in color, becoming rugged, wrinkled, and "piled up" upon longer incubation (Nigg et al., 1956; Stanton et al., 1924). No colonial form is particularly associated within virulence and can vary from cream to bright orange. The pigment is insoluble (Brygoo and Richard, 1952), and the culture has a musty, slightly aromatic odor typical of the pseudomonads. A chemically defined salts medium containing ammonia as the sole source of nitrogen and supplemented with glucose and threonine has been used successfully for isolation from infected animals and from soil and water (Dodin et al., 1976; Dodin and Galimand, 1976).

Glanders: P. mallei can be isolated on regular laboratory media similar to *P. pseudomallei*. The growth is slower and the colonies show less variation than *P. pseudomallei* (Redfern et al., 1966). Small, dewdrop, glistening colonies appear in 48 to 72 h and vary in color from a smooth white/cream color which darkens slightly with age (Minett, 1959). A variety of inhibitors for contaminating bacteria has been used such as crystal violet as noted, proflavine, malachite green-fuchsin, and thionin (Miller et al., 1948; Pini, 1957). All 15 strains used in one study grew slowly on minimal salts medium containing ammonia as the sole nitrogen source (Redfern et al., 1966).

Identification

Both *P. mallei* and *P. pseudomallei* must be distinguished from each other, from *P. aeruginosa*, and from other species assigned to rRNA group II (*P. carophylli, P. gladioli, P. picketti, P. soleancearum*, and *P. cepacia*). The latter is particularly important as *P. pseudomallei* was mistakenly identified as *P. cepacia* in a serious laboratory-acquired infection (Schlech et al., 1981).

Pseudomonas mallei, P. pseudomallei, and *P. caryophylli* produce arginine dihydrolase, whereas the other four species do not. *P. caryophylli* is a plant pathogen but can be distinguished from *P. mallei* and *P. pseudomallei* by its inability to hydrolyze gelatin.

A variety of methods are available to distinguish *P. mallei* and *P. pseudomallei* from contaminating gram-negative bacteria and other pseudomonads (Faskas-Himsley, 1968; Fournier et al., 1953), but the most important differentiation must be made with *P. aeruginosa*. The latter will grow on salmonella-shigella agar, whereas *P. mallei* and *P. pseudomallei* will not and this is a simple procedure to include in the isolation (Wetmore and Gochennour, 1956). Further, *P. aeruginosa* produces hemolysis on blood agar and produces the characteristic blue-green pigment. The most useful technique, however, is fluorescent antibody (Moody et al., 1956; Thomason et al., 1956).

Pseudomonas mallei is the only nonmotile pseudomonad and can rapidly be distinguished from *P. pseudomallei* by either the hanging-drop method or by flagella stain. Other tests include: *P. mallei* uses D-xylose as a carbon source, whereas *P. pseudomallei* is negative and the latter uses D-ribose, levulinate, and erythritol as carbon sources, and *P. mallei* is negative (Palleroni, 1984). In the author's opinion, the final distinction between the two species is with the use of highly specific antisera in the agglutination test. The antisera to *P. pseudomallei* should include antibodies to both types 1 and 2. This procedure was developed by Dodin and Galimand (1976) and has proved extremely useful.

Immunologic Aspects

As stated earlier, serology plays a key role in the diagnosis of melioidosis. The two most commonly used tests are the complement fixation (CF) developed by Nigg and Johnston (1961) and the indirect hemagglutination (IH) by Ileri (1965). A number of different antigens have been used in these tests: IH: the Boivin antigen (Fournier et al., 1953), a carbohydrate antigen (Nigg, 1963), and culture filtrates (Ileri, 1965; Laws, 1967); CF: a sonicated cell extract (Nigg, 1963), and aqueous extracts (Cravitz and Miller, 1950; Laws, 1967a, 1967b). In the early studies on healthy Thai men and women subclinical meliodosis was detected using a crude aqueous extract in the CF test and both a protein and polysaccharide antigen in the IH test. This exhaustive study indicated that the polysaccharide antigen was specific for *P. pseudomallei* but that the protein antigen, which had been shown to be effective in monitoring the immune response in experimental animals (Dannenberg and Scott, 1958), cross-reacted with other pseudomonads (notably *P. aeruginosa*) in humans.

The difficulty in diagnosis of melioidosis clinically, the increase in infections in U.S. Armed Forces personnel who served in Vietnam (Cooper, 1967; Flemma et al., 1969; Weber et al., 1969), and the dormant or inapparent infections that may appear or recrudesce months or years later in these personnel (Sanford, 1970) has prompted a refinement and reevaluation of the IH and CF tests and a search for alternative procedures. In an extensive study of 212 patients (119 culturally proved cases), the IH test using as an antigen the supernatant of a heat-killed 2-week-old protein-free broth culture (as developed by Rice et al., 1951) was very specific for *P. pseudomallei*. There were only minimal cross-reactions with *P. aeruginosa* (2 of 60) at low titers (1:20 to 1:40). The CF test using a sonicated cell extract (as developed by Nigg, 1963) was less specific and cross-reacted with *P. aeruginosa* at titers of 1:4 to 1:64. Both tests detected antibodies to *P. pseudomallei* in patients 1 week after the onset of disease, with high titers by 3 weeks, and maximum titers in 4 to 5 months. Some transient seronegative reactions (19%) occurred during the course of the disease when tested by CF and IH (but not both) and antibodies could be detected for as long as 2 years after infection (Alexander et al., 1970). In a later study (Klein, 1980) using the same IH test as described, noted that patients with elevated titers to *P. pseudomallei* had antibody titers to *Legionella pneumophilia*. If the clinical syndromes are compatible it appears that *L. pneumophilia* should be included as a control in the IH test for melioidosis.

The immunofluorescent antibody (IFA) assay has been shown to be a reliable and rapid method for screening of antibodies to *P. pseudomallei* in human infections (Ashdown, 1981a). The method used IgG and IgM and was both sensitive and specific, for example, negative for 8 antisera to *P. cepacia*, 13 antisera to *P. aeruginosa*, and 5 antisera to *P. stutzeri*, at dilutions above 1:4. The procedure used washed cells of *P. pseudomallei*. The importance of the IFA and the use of IgM-tagged antibody for the detection of acute and subclinical infections in 36 hospitalized patients was further confirmed in a study that also included the IH and CF tests. The IFA-IgM test also was shown to be a valuable aid in monitoring the activity and treatment of patients with melioidosis (Ashdown, 1981b).

Because of the danger in working with *P. pseudomallei*, there is a need for a readily available antigen that is stable and can be used in the automated IH titrators. This would be extremely useful at many diagnostic centers. Pyruvic-aldehyde stabilized erythrocytes sensitized with the antigen described by Alexander et al. (1970) was tested in an automated IH titration system and was found to be specific and practical (Hambie et al., 1977).

Recently Haberberger (1985) reported on the use of the kinetic-ELISA test and imunoelectroblotting for detection of specific antibodies to *P. pseudomallei* in human and animal sera. The antigen was an LPS-derived polysaccharide isolated from both types 1 and 2. The antigen was further characterized in the ELISA using monoclonal antibodies to whole cells of *P. pseudomallei*. These studies offer the potential of an antigen-trapping test as a serologic aid in the diagnosis of acute melioidosis.

Beginning in the early part of this century the majority of studies on glanders has been in equines. Both the agglutination and complement fixation test have been used successfully. In the precipitation test using both a protein and alcohol precipitated polysaccharide from a glycerin broth culture, the polysaccharide was found to be the most specific (Sakamoto, 1930). In human cases of glanders, both the CF and agglutination test were evaluated with several antigens. A phenolized broth culture was gently shaken with glass beads for 2 to 4 days, filtered, and approximately 10^9/ml of the organisms used in the agglutination reaction. The supernatant of a 4-day broth culture, commercial mallein, and sonicated cell supernatant were evaluated by CF. The CF tests cross-reacted with other pseudomonads, whereas the agglutination reaction was specific for *P. mallei*. The patients all reacted with a positive mallein skin reaction and it appeared that an agglutinin titer of greater than 1:320 was diagnostic (Cravitz and Miller, 1950; Howe and Miller, 1947).

Treatment

Melioidosis: The sensitivity of *P. pseudomallei* to chemotherapeutic agents has been extensively studied, but no specific regimen has as yet been established. The microorganism often varies in sensitivity within the same culture and from patient to patient. In one study (Malizia et al., 1969) *P. pseudomallei* was sensitive in vitro to tetracycline, chloramphenicaol, and novobiocin at levels that could be attained in vivo. There is some controversy as to the efficacy of novobiocin in vivo. In other studies *P. pseudomallei* was found to be sensitive to sulfadiazine but resistant to penicillin, streptomycin, polymyxin B, and variable in its sensitivity to kannamycin. Various combinations of chemotherapeutic agents were antagonistic in vivo and need to be further assessed clinically (Eickoff et al., 1970). Similar sensitivity patterns were noted in control of infection in mice (Hobby et al., 1969).

Tetracycline appears to be the most effective of the regimens tested, given at 3 g daily until cultures are negative and the lung lesions are cleared or stabilized is essential. In very toxic patients the dosage of tetracycline (3 g) is combined with 3 g daily of chloramphenicol. In severe septicemia, the 3 g of tetracycline plus 4 to 6 g of chloramphenicol are given intravenously (Thin, 1981). Prolonged treatment during a 3 to 6-month period is essential for effective management of melioidosis as relapse can occur if treatment is interrupted (Howe et al., 1971). Since tetracycline resistance has emerged (Bauman and Morita, 1970), the combination of tetracycline and chloramphenicol with appropriate supportive measures has been emphasized (Thin et al., 1970). *P. pseudomallei* is also sensitive in vitro to cotrimoxazole and has been successful in treatment (John, 1976). This chemotherapeutic agent has some theoretical advantages over tetracycline and chloramphenicol because it is bactericidal and the others are bacteriostatic.

There are so few well-documented cases of human glanders in the postantibiotic era that no optimum treatment has been established. *P. mallei* is resistant in vitro to penicillin and sensitive to sulfonamides, streptomycin, tetracycline, and chloramphenicol. There is some controversy over the effectiveness of streptomycin and tetracycline in the treatment of experimental glanders (Davis, 1981), but sulfadiazine given in dosages that maintained the blood level at 10 to 11 mg% for 20 days was effective in the treatment of 6 patients (Howe and Miller, 1947).

Prevention and Control

There is no available vaccine for immunization of humans against melioidosis. Avoiding undue exposure to contaminated soil and water, particularly after heavy rains in endemic areas, is still the best preventive measure. The microorganisms can readily be isolated from surface, run-off waters. Traumatic injuries (car accidents, etc.) and infections from contaminated waters entering the body will still occur, but every effort should be made to improve drainage in those areas where it is possible. In captive marine mammals and hoofstock in Asia, where melioidosis has been intensively studied since 1977, contaminated waters only during the rainy season (May through October) were the source of infections (Vedros et al., 1983).

Glanders in equines has been virtually eliminated from the Western world since the British Farcy Order of 1907. Animals with clinical signs of the disease or positive with the mallein skin test should be slaughtered, carcasses burned, and the entire area including bedding, stable accessories, and the like disinfected. Contacts with positive mallein reactions also should be destroyed. The limited host range of *P. mallei* therefore indicates that prevention centers on the avoidance of contact with the discharges of infected equines and their food, blankets, and so on. Contact and infection through accidental cuts and abrasions is particularly important for those conducting necropsies of infected animals (Howe and Miller, 1947).

Literature Cited

Achana, V., K. S. Silpapojkul, W. Thininta, and S. Kalanowakul. 1985. Acute *Pseudomonas pseudomallei* pneumonia and septicemia following aspiration of contaminated water: a case report. SE Asian J. Trop. Med. Pub. Hlth. **16**:500–504.

Alexander, A. D., D. X. Huxsoll, A. B. Warner, Jr., V. Shepler, and A. Dorsey. 1970. Serological diagnosis of human melioidosis with indirect hemagglutination and complement fixation tests. Appl. Microbiol. **20:**825–833.

Ansasi, M. and M. Minou. 1951. Deux cas de morve chronique humain traites par les sulfamides. Ann. Inst. Pasteur. **81:**98–107.

Ashdown, L. R. 1981a. Demonstration of human antibodies to *Pseudomonas pseudomallei* by indirect fluorescent antibody staining. Pathology **13:**597–601.

Ashdown, L. R. 1981b. Relationship and significance of specific immunoglobulin M antibody response in clinical and subclinical melioidosis. J. Clin. Microbiol. **14:**361–364.

Bauman, B. B. and E. T. Morita. 1967. Systemic melioidosis presenting as a myocardial infarct. Ann. Intern. Med. **67:**836–842.

Biegelsisen, J. Z., R. Mosquerai, and W. B. Cherry. 1964. A case of human melioidosis: clinical, epidemiological and laboratory findings. Am. J. Trop. Med. **13:**89–99.

Bremmelgaard, A., I. B. Bygbjerg, and N. Hoiby. 1982. Microbiological and immunological studies in a case of human melioidosis diagnosed in Denmark. Scand. J. Infect. Dis. **14:**271–275.

Brundage, W. G., C. J. Thuss, and D. C. Walden. 1968. Four cases of melioidosis in U.S. soldiers in Vietnam. J. Trop. Med. **17:**183–191.

Brygoo, E. R. and C. Richard. 1952. Pouvoir chromogene de *Malleomyces pseudomallei*. Ann. Inst. Pasteur. (Paris) **83:**822–825.

Clayton, A. J., R. S. Lisella, and D. G. Martin. 1973. Melioidosis: a serological survey in military personnel. Milit. Med. **138:**24–26.

Cooper, E. B. 1967. Melioidosis. J. Am. Med. Assoc. **200:**452–553.

Cravitz, L. and W. R. Miller. 1950. Immunologic studies with *Malleomyces mallei* and *Malleomyces pseudomallei*. I. Serological relationships between *M. mallei* and *M. pseudomallei*. J. Infect. Dis. **86:**46–51.

Dannenber, A. M. and E. M. Scott. 1958. Melioidosis: pathogenesis and immunity in mice and hamsters. I. Studies with virulent strains of *Malleomyces pseudomallei*. J. Exp. Med. **107:**153–166.

Davis, C. E. 1984. Glanders (farcy), p. 1812–1815. *In* A. I. Braude (ed.), Medical microbiology and infectious diseases. W. B. Saunders, Philadelphia.

Dodin, A., M. Galimand, M. A. Chove, and R. Sanson. 1976. Le bacille de Witmore. Germe actualite. Med. et Maladies Infectieuses **10:**395–398.

Dodin, A. 1976. Actualite de la Melioidose. Rec. Med. Vet. **152:**285–288.

Dodin, A. and R. Ferry. 1974. Recherche epidemiologique du bacille de Whitmore en Afrique. Bull. Soc. Pathol. Exot. **67:**121–126.

Dodin, A. and M. Galimand. 1976. Le bacille de-Whitmore. Rec. Med. Vet. **152:**323–325.

Duroux, A. 1965. La meliodose pulmonaire. Poumon. Coeur. **21:**355–372.

Editorial. 1975. Melioidosis. Lancet **2:**962.

Eickoff, T. C., J. V. Bennett, P. S. Hayesand, and J. Feeley. 1970. *Pseudomonas pseudomallei:* susceptibility to chemotherapeutic agents. J. Infect. Dis. **121:**95–102.

Faskas-Himsley, H. 1968. Selection and rapid identification of *Pseudomonas pseudomallei* from other gram negative bacteria. Am. J. Clin. Pathol. **49:**850—856.

Flemma, R. J., F. C. DiVicenti, L. N. Notin, and B. A. Pruitt. 1969. Hospital Tribune, X, No. 20. March 10, p. 1.

Flemma, R. J., F. C. DiVicenti, L. N. Dotin, and B. A. Pruitt. 1969. Pulmonary melioidosis. Ann. Thorac. Surg. **7:**491–499.

Fournier, J., P. de Lajudie, and L. Chambon. 1953. Recherches sdur les procedes de laboratoire applicables au diagnostic de la melioidose. Med. Trop. **13:**683–697.

Fournier, J. 1967. The thermostatic antigens of *Pseudomonas pseudomallei* and of *Malleomyces mallei* and their communities. Ann. Inst. Pasteur **112:**93–98.

Frothingham, L. 1901. The diagnosis of glanders by the Strauss method. J. Med. Res. **6:**331–335.

Green, R. N., and P. G. Tuffnell. 1968. Laboratory acquired melioidosis. Am. J. Med. **44:**599–605.

Haberberger, R. 1985. K-ELISA detection of antibodies to *Pseudomonas pseudomallei* in animals and humans. Proc. 85th Ann. Meeting of ASM **14:**391.

Hambie, E. A., S. A. Larsen, M. Feller, W. L. Jones, and J. C. Feeley. 1977. Use of stable, sensitized cells in an indirect microhemagglutination test for melioidosis. J. Clin. Microbiol. **5:**167–171.

Haynes, W. C. 1957. Order I. Pseudomonadales, p. 100. *In* Bergeys manual of determinative bacteriology, 7th ed. William and Wilkins, Baltimore.

Hobby, G. L., T. F. Lenert, J. Maier-Engallena, and E. DeNoia-Cicenia. 1969. The control of experimental infection in mice produced by *Pseudomonas pseudomallei*. Am. Rev. Resp. Dis. **99:**952–954.

Howe, C. and W. R. Miller. 1947. Human glanders: report of six cases. Ann. Intern. Med. **26:**93–115.

Howe, C. 1950. Glanders, p. 185–202. *In* Oxford system of medicine, vol. 5. Oxford University Press, Oxford.

Howe, C., M. Sampath, and N. Spotnitz. 1971. The pseudomallei group. A review. J. Infect. Dis. **124:**598–606.

Ileri, S. Z. 1965. The indirect hemagglutination test in the diagnosis of melioidosis in goats. Br. Vet. J. **121:**164–170.

Jennings, W. E. 1963. Glanders, p. 264. *In* T. G. Hull (ed.), Diseases transmitted from animal to man, 5th ed. C. C. Thomas, Springfield, IL.

John, J. F. Jr. 1976. Trimethoprin-sulfamethoxazole therapy of pulmonary melioidosis. Am. Rev. Respir. Dis. **114:**1021–1025.

Johnson, D. W. 1967. Melioidosis: a report of four cases from Torres Strait. Med. J. Austr. **2:**587–588.

Kishimoto, R. A., G. L. Brown, E. B. Blair, and D. Wenkheimer. 1971. Melioidosis: serological studies of U.S. Army personnel returning from Southeast Asia. Milit. Med. **136:**694–698.

Klein, G. C. 1980. Cross-reaction to *Legionella pneumophilia* antigen in sera with elevated titers to *Pseudomonas pseudomallei*. J. Clin. Microbiol. **11:**27–29.

Laws, L. 1967a. Melioidosis in animals in North Queensland. V. Serological methods of diagnosis in sheep. Queensland J. Agr. Anim. Sci. **24:**207–221.

Laws, L. 1967b. Melioidosis in animals in North Queensland. VI. Serological methods of diagnosis in goats, pigs, cattle, horses and man. Queensland J. Agr. Anim. Sci. **24:**223–234.

Lee, N., J-L. Wu, C-H. Lee, and W-C. Tsai. 1985. *Pseudomonas pseudomallei* infection from drowning. The first reported case in Yaiwan. J. Clin. Microbiol. **23:**352–354.

Leong, E., D. D. Hammond, and N. A. Vedros. 1985. *Pseudomonas pseudomallei* infection in a dolphin (*Tursiops qilli*): a case study. Aquatic Mammals **1:**20–22.

Malizia, W. F., G. A. Weast, W. G. Brundage, and D. C. Walden. 1969. Melioidosis-laboratory studies. Health Lab. Sci. **6:**27–39.

Mandel, M. 1966. Deoxyribonucleic acid base composition in the genus *Pseudomonas*. J. Gen. Microbiol. **43:**273–292.

McCormack, J. B., D. J. Sexton, J. G. McMurray, E. Carey, P. Hayes, and R. A. Feldman. 1975. Human to human transmission of *Pseudomonas pseudomallei*. Ann. Intern. Med. **83:**512–513.

Mendelsohn, R. W. 1950. Glanders. U.S. Armed Forces Med. **1:**781–784.

Miller, W. R., L. Pannel, L. Cravitz, W. A. Tanner, and M. Ingalls. 1948. Studies on certain biological characteristics of *Malleomyces pseudomallei*. I. Morphology, cultivation, viability and isolation from contaminated specimens. J. Bacteriol. **55:**115–126.

Minett, F. C. 1959. Glanders (and melioidosis), p. 296–318. *In* A. W. Stableforth and I. A. Galloway (eds.), Infectious diseases of animals. Diseases due to bacteria, vol. 1. Butterworth Science Publishers, London.

Moody, M. D., M. Goldman, and B. Thomason. 1956. Staining bacterial smears with fluorescent antibody. I. General methods for *Malleomyces pseudomallei*. J. Bacteriol. **72:**357–361.

Nigg, C. and M. M. Johnston. 1961. Complement fixation test in experimental clinical and subclinical melioidosis. J. Bacteriol. **82:**159–168.

Nigg, C. 1963. Serologic studies on subclinical melioidosis. J. Immunol. **91:**18–28.

Nussbaum, J. J., D. S. Hull, and M. J. Carter. 1980. *Pseudomonas pseudomallei* in an anophthalmic orbit. Arch. Ophthamol. **98:**1224–1225.

Osteraas, G. R., J. M. Hardman, J. W. Bass, and C. Wilson. 1971. Neonatal melioidosis. Am. J. Dis. Child. **122:**446–448.

Palleroni, N. J. 1984. Bergey's manual of systematic bacteriology, p. 141, vol. 1. William and Wilkins, Baltimore.

Palleroni, N. J., R. Kunisawa, R. Contupolou, and M. Douderoff. 1973. Nucleic acid homologies in the genus *Pseudomonas*. Int. J. Syst. Bacteriol. **22:**333–339.

Patamasucon, P., U. B. Schard, and J. D. Nelson. 1982. Melioidosis. J. Pediatr. **100:**175–182.

Piggot, J. A. and L. Hochholzer. 1970. Human melioidosis. A histopathologic study of acute and chronic melioidosis. Arch. Pathol. (Chicago) **90:**101–111.

Pini, A. 1957. Isolation of *Pfeifferella mallei* on malachite green-fuchsin agar and on thionin agar. Vet. Ital. **8:**103–110.

Pourtaghva, M., M. Machoun, and A. Dodin. 1975. Mise en evidence de *Pseudomonas pseudomallei* (Bacille de Whitmore) dans la boue des rizieres iranians. Soc. Path. Exot. **68:**367–370.

Redfern, M. S., N. J. Palleroni, and R. Y. Stanier. 1966. A comparative study of *Pseudomonas pseudomallei* and *Bacillus mallei*. J. Gen. Microbiol. **43:**293–313.

Rice, C. E., H. Koust, and R. C. Duthie. 1951. Studies by complement fixation methods of malleins produced in broth and synthetic media. I. Relative immunizing activities in horses and rabbits. Can. J. Compar. Med. **15:**284–291.

Rimington, R. A. 1962. Melioidosis in North Queensland. Med. J. Austr. **1:**50–53.

Robins, G. D. 1906. Chronic glanders in man. Studies from the Royal Victoria Hospital, Montreal. ii, no. 1.

Rubin, H. L., A. D. Alexander, and R. H. Yager. 1963. Melioidosis—a military medical problem? Milit. Med. **128:**538–542.

Sakamoto, K. 1930. Glanders. J. Immunol. **18:**331–339.

Sanford, J. P. 1971. Recrudescent melioidosis: a Southeast Asian legacy. Am. Rev. Respir. Dis. **104:**452–453.

Schlech, W. F., J. B. Turchick, R. E. Westlake, G. C. Klein, J. D. Bond, and R. E. Weaver. 1981. Laboratory-acquired infection with *Pseudomonas pseudomallei* (melioidosis). N. Engl. J. Med. **305:**1133–1135.

Spotnitz, M., J. Rudnitzky, and J. J. Rambaud. 1967. Melioidosis pneumonitis. J. Am. Med. Assoc. **202:**950–954.

Stanier, R. Y., J. J. Palleroni, and M. Douderoff. 1966. The aerobic pseudomonads: a taxonomic study. J. Gen. Microbiol. **43:**159–271.

Stanton, A. T., W. Fletcher, and K. Kanagarayer. 1924. Two cases of melioidosis. J. Hyg. (Camb.) **23:**268–276.

Stanton, A. T. and W. Fletcher. 1925. Melioidosis and its relation to glanders. J. Hyg. (Camb.) **23:**347–360.

Steel, K. J. and S. T. Cowan. 1964. La rattachment de bacterium *Anitratum, Moraxella lwoffi, Bacillus mallei,* et *Haemophilus parapertussis* au genre *Acinetobacter* Breson and Prevot. Ann. Inst. Pasteur. **106:**479–481.

Thin, R. N. T., M. Brown, J. B. Stewart, and C. J. Garrett. 1970. Melioidosis: a report of ten cases. Quart. J. Med. **39:**115–127.

Thin, R. N. T. 1981. Melioidosis, p. 921–924. *In* A. I. Braude (ed.), Medical microbiology and infectious diseases. W. B. Saunders, Philadelphia.

Thomason, B. M., M. D. Moody, and M. Goldman. 1956. Staining bacterial smears with fluorescent antibody. II. Rapid detection of varying numbers of *Malleomyces pseudomallei* in contaminated materials and infected animals. J. Bacteriol. **72:**362–367.

Vedros, N. A., E. Liong, and D. D. Hammond. 1983. A potential polysaccharide vaccine for melioidosis in marine mammals. Proc. 14th Ann. Conf. Intl. Assoc. Aquatic An. Med., p. 17.

Weber, D. R., L. E. Douglas, W. G. Brundage, and T. C. Stallkamp. 1969. Acute varieties of melioidosis occurring in U.S. soldiers in Vietnam. Am. J. Med. **46:**234–244.

Wetmore, P. W. and W. S. Gochennour, Jr. 1956. Comparative studies of the genus *Malleomyces* and selected *Pseudomonas* species. I. Morphological and cultural characteristics. J. Bacteriol. **72:**79–89.

Whitmore, A. and C. S. Krishnaswami. 1912. An account of the discovery of a hitherto undescribed infective disease occurring among the population of Rangoon. Indian Med. Gaz. **47:**262–267.

Whitmore, A. 1913. An account of a glanders-like disease occurring in Rangoon. J. Hyg. (Camb.) **13:**1–34.

Zopf, W. 1885. Die Spaltspilze, 3rd ed. Edward Trewendt., Breslau, p. 1–127.

CHAPTER 39

Meningococcal Infections

HELEN M. POLLOCK

Disease: Meningococcal disease (epidemic cerebrospinal fever).
Etiologic Agent: *Neisseria meningitidis* groups A, B, C, D, X, Y, Z, W135.
Source: Human respiratory aerosol transmission acquired by inhalation.
Clinical Manifestations: May be subacute with transient fever to fulminant in onset leading to rapid death. Pneumonia, septicemia, and meningitis are the common forms.
Pathology: Central nervous system responses reflect acute inflammatory changes. Septic manifestations include disseminated intravascular coagulation and intravascular damage leading to shock.
Laboratory Diagnosis: Bacteriologic isolation from blood, spinal fluid, or transtracheal aspirate. Immunoantigenic techniques give presumptive diagnoses. Serodiagnosis is not indicated.
Epidemiology: Worldwide distribution with increased incidence in the meningitis belt of Central Africa. Epidemics occur in closed populations, with sporadic cases occurring in endemic areas.
Treatment: Penicillin or chloramphenicol are the antibiotics of choice. Supportive therapy required in cases of shock.
Prevention and Control: Prophylactic therapy of case contacts with sulfadiazine or rifampin. Vaccination of populations showing risk of epidemic spread.

Description of Disease

Meningococcal disease is an infection of humans, which manifests itself in a broad spectrum of clinical symptoms, including respiratory, meningeal, and septicemic. It is a worldwide public health problem with epidemics emerging in 30 countries as recently as the 1970s (Peltola, 1983). Despite many developments and investigations, it remains a major public health problem that is still not well understood.

Communicability and Transmission

Meningococcal disease is transmitted from person to person by the respiratory route through the nasopharynx. Transmission has been studied using military recruits and following them from their initial entry to a base. Data from monitoring the nasopharyngeal carriage rate indicates that they fluctuate widely. Although many hypotheses have been evaluated through the years, there is little in the way of concrete evidence to implicate the facilitation of person-to-person transmission by any particular factors (Griffiss, 1982). As this agent is transmitted by aerosols dispersed from the respiratory tract, its parallel course with viral upper respiratory tract infections may reflect the role of coughing and sneezing in facilitating transmission.

Clinical Manifestations

The onset of disease is varied. It may manifest as a transient fever, bacteremia, or meningitis, with or without bacteremia or pneumonia. Patients may progress from one of these forms to another during the course of their illness. At times the progression is fulminating, leading to death.

Meningococcemia

Upper respiratory tract symptoms may precede the onset of meningococcemia. Fever, rash, and arthralgia may also accompany this state along with the development of complicating sepsis. Malaise, fever, and generalized weakness accompany these phenomena, and the release of endotoxin by the organisms may cause the patient to go into shock. Often, these shock signs dominate the clinical picture and lead to disseminated intravascular coagulopathies (DIC) and widespread intravascular damage. The cutaneous lesions, which are often a hallmark of meningococcal bacteremia, are definitely petechial and may be purpuras, resembling lesions of Rocky Mountain Spotted fever, rubella and rubeola, as well as various cutaneous eruptions associated with immediate allergic reactions to penicillin and other agents. The lesions are initially peripheral involving wrists, palms, ankles, and soles. They may, however, be concentrated on the trunk later in the course of disease (Apicella, 1985).

Meningitis and Meningoencephalopathy

Neurologic problems in meningococcal meningitis differ from those of other types of purulent meningitis. The responses may range from total alertness to lack of response. Although meningeal irritation is usually present, the other focal signs seen in other types of purulent meningitis tend to be absent.

There may be a progression of the disease to the accumulation of pus around the brain. Patients in this situation are profoundly obtunded with alteration of reflexes. This finding occurs also in pneumococcal disease.

Pneumonia

Primary meningococcal pneumonia is not clinically distinguishable from other types of bacterial pneumonias. It often appears to be preceded by viral upper respiratory tract infections and has been associated with influenza epidemics. There is evidence that meningococcal pneumonia is not a rare phenomenon and may occur without preceding blood stream invasion (Irwin et al., 1975).

Pathology

Commensal colonization of the upper respiratory mucosal surfaces by *N. meningitidis* is common. Meningococci attach to nonciliated epithelial cells of the nasopharynx at the primary site of entry. These cells have a greater number of receptors for pili of *N. meningitidis* than do epithelial cells from other anatomic sites (Stephens et al., 1983). This colonization results in loss of ciliary activity by ciliated cells and a sloughing of these cells, even though meningococci are not directly attached to them. The affect of the organism is apparently not attributable to piliation, outer membrane protein, or lipopolysaccharide (Stephens et al., 1986). Other as yet unidentified surface components may be involved. Subsequent to attachment, meningococci may be found in phagocytic vacuoles of the apical portion of some of these cells. There is little evidence that the organisms are destroyed at this stage. In experimental organ culture, organisms were found in the subepithelial space 18 to 24 h after infection. The factors determining which epithelial cells are invaded are not known.

In susceptible individuals, the organisms penetrate into deeper tissues, resulting in bacteremia and/or meningitis. The exact progression of these events remains unclear.

Hemorrhagic Lesions

The vascular changes are widespread, involving intravascular damage and coagulation (DeVoe, 1982). Damage to endothelial cells with accompanying inflammatory changes in the vascular walls, thrombosis, and necrosis of vessels are all contributing factors in the presentation of disease. Biopsy material from early cutaneous lesions reveals capillary dilatation and the conspicuous absence of cellular exudate. Polymorphonuclear leukocyte infiltration occurs later in the development of these lesions. Systemic lesions may be produced by mechanisms that are similar to those occurring in skin. Deposition of fibrin in small vessels is an important contributor to the clinical manifestations.

Whereas myocarditis has been a common finding, the Waterhouse-Frederichsen syndrome has been seen in only 3 to 4% of the cases. Death in fulminant meningococcemia may be attributed to the effects of acute endotoxic shock and various cardiopathies.

Meningitis

An acute inflammatory exudate, consisting primarily of PMNs, is in the leptomeninges. Full development of the disease results in pus in the subarachnoid space. Although the pus does not invade the tissue per se, there are changes of congestion and edema in the underlying tissue. Severe hyperemia and swelling with petechial hemorrhages suggest endotoxic involvement.

Diagnostic Features

Distinctive diagnostic features are not easily identified as these infections have such a wide variety of clinical presentations. A high index of suspicion and a careful search for specific signs are necessary for diagnosing the sporadic cases. The usual signs of meningitis such as headache and stiff neck may be present only 50% of the time (Apicella, 1985).

Petechial lesions, often occurring in clusters where clothing may cause pressure on the skin, are a hallmark of meningococcemia. General muscle tenderness also may be a sign.

The neurologic symptoms seen in meningitis involve meningeal irritation. Focal signs and seizures seen in other types of meningitis appear to be less common. Infants and small children may not be brought for examination except in advanced disease because fever and vomiting are the only signs they often manifest. Geiseler et al. (1980) studied a series of cases of meningitis during a 22-year period. There were 396 cases of meningococcal meningitis in this series. Positive Gram stains with negative culture were found in 57 of these cases. The overall sensitivity of cerebrospinal fluid (CSF) smears was 75%, with a 98.4% specificity. The numbers of organisms present in the CSF have an effect on the percent of positive smears with $\leq 10^3$ CFU/ml (25%), $>10^3$ to 10^5 CFU/ml (60%), and $>10^5$ CFU/ml (97%) (LaScolea and Dryja, 1984). Although this study was small, it was an indication of the effect of numbers of organisms on smear results. The sensitivities of cultures and Gram stains are also reduced by prior antimicrobial therapy. The most dramatic impact on these tests is caused by chloramphenicol. Patients on this drug, however, had slightly higher CSF protein and leukocyte levels than their untreated counterparts (Geiseler et al., 1980).

The advent of immunologic tests for meningococcal antigens has partially solved the problem of diagnosing previously treated cases in which these tests may be positive in the absence of a positive culture. A positive antigen test would not be expected in a patient who did not have an elevated CSF leukocyte count because an insufficient number of organisms are usually present. At times, a tap repeated several hours after the initial sample, shows the necessary changes in cellular and chemical formulae, as well as positive antigen tests to make the diagnosis. Many sporadic cases of meningococcal meningitis are caused by group B, which gives very poor results with immunologic methods and is generally not included in most commercial kits. A negative test for meningococcal antigen does not, therefore, rule out meningococcal meningitis caused by this group.

Pneumonia due to this organism has few distinct features. Diagnosis, using sputum culture, is of no value because of the high nasopharyngeal carriage rate in many populations. The incidence of bacteremia in this disease is very low, making blood cultures of limited value. Definitive bacteriologic diagnosis requires lower airway specimens collected by transtracheal aspirate or protected catheter bronchoscopy (Pollock et al., 1983).

Prognosis

The petechial rash may be a valuable indication of potential bleeding problems due to intravascular coagulation in meningococcemia. In meningitis, there appears to be some relationship between numbers of organisms present in untreated cases and clinical outcome (LaScolea and Dryja, 1984).

Epidemiologic Aspects

The epidemiology of meninogoccal disease has been a complex mystery, which has demonstrated many different patterns. The multiplicity of these patterns has intrigued many investigators for some time, with some interesting relationships shedding light on the seemingly inconsistent patterns previously observed.

Occurrence and Transmission

Meningococcal disease has a worldwide occurrence with the epidemiologic patterns differing in developed countries compared with underdeveloped nations. Before 1940, epidemic patterns were similar, whereas distinctive differences are predominant in more recent outbreaks of the 1970s (Peltola, 1983).

The incidence shows a seasonal distribution in fall and spring and appears to predominate in children under school age in industrialized nations. Young school-aged children appear to represent the peak age in the meningitis belt. Often, age distribution shifts to older populations during the early stages of an epidemic.

Classical patterns of disease are seen in a tropical region of Africa defined by the southern edge of the Sahara extending almost to the equator and from Mali in the east to Guinea in the west (J. Kennedy, meteorologist, KWWL-TV, Waterloo, Iowa; Peltola, 1983). The average incidence of disease in this "meningitis belt" is about 70 cases per 100,000 population as opposed to an approximate high of 2 per 100,000 population in the United States. The factors responsible for these differences are not clear.

The organisms are transmitted from person to person via the respiratory route. *N. meningitidis* has a particular preference for nasopharyngeal mucosal surfaces, and humans are the only natural host. Although the acquisition of these strains has been studied prospectively in various closed military populations, the incidence of disease cannot be attributed simply to acquisition of any strain. It appears that there are "epidemic" strains, which enter some groups and not others, causing disease. The carriage rate among various groups is often very low for the epidemic strain even though carriage rates to other strains of *N. meningitidis* may be 30 to 70%. Multiple factors appear to converge on the dynamics of the situation to determine whether a single case or epidemic will occur.

Susceptibility, Resistance, and Carrier State

Griffiss (1982) has proposed an immunoepidemiologic model that fits most of the data relating to the relationships between the carrier state and the acquisition of disease by susceptible individuals. Susceptible individuals have been found to have major differences in their immune responses to the specific disease-producing strain when compared with individuals who remained asymptomatic. The noncarrier seems to be the individual at risk. Factors that influence this relationship are the rate of transmission and the particular subcapsular protein and lipopolysaccharide antigens determining the subtype of the serogroup.

Nasopharyngeal colonization with meningococci stimulates serum antibody production to the strain as well as antibodies representing a broad spectrum of specificities. Fewer susceptible individuals with disease have bactericidal activity to the infecting strain (56%) compared with controls (82.2%) (Goldschneider et al., 1969). In a prospective study by these investigators, susceptibility correlated with the absence of serotype-specific bactericidal activity. It also has been demonstrated that serum IgA blocks the bacteriolytic activity of IgM and IgG (Griffiss, 1975). Absorption of the IgA from serum drawn at the time of dissemination restored the bacteriolytic activity of recruits who had acquired the disease (Griffiss and Bertram, 1977). This phenomenon is common and explains the variable behavior of meningococcal disease over space and time better than models proposing a genetic defect which produces a deficiency in the components of complement responsible for the lytic pathway. Whereas normal levels of serum IgA exhibit protective opsonic activity, excessive amounts create a prozone effect reducing antibacterial activity. There is evidence to suggest that cross-reacting respiratory and gastrointestinal normal flora may stimulate circulating IgA, which can react with meningococci. The levels of circulating IgA do not remain constant, and an individual may have higher levels at one point compared with another. In addition, a greater number of individuals who did not have disease in the group A meningococcal outbreak in the Pacific Northwest had normal oropharyngeal flora that inhibited the growth of meningococci, than did their case counterparts (Filice et al., 1985). It is apparent that multiple factors affect carriage, colonization, and dissemination of the organism.

The majority of carriers have strains representing groups B, C, and Y. Serotype 2, responsible for most of the cases due to groups B and C, is not commonly found in the carrier state. Group A meningococci are found in the carrier state associated with epidemics, but are not common in endemic cycles with sporadic cases. Although 20 to 40% of a given population may be meningococcal carriers, the sporadic emergence of cases has little to do with this fact.

Much is yet to be learned about the complex relationships that determine the carrier and the disease states. It appears that a complex relationship of inhibitory effects and stimulation of cross-reacting serum IgA blocking antibody by commensal flora may play a major role in the epidemiologic patterns observed.

Etiologic Agent

Neisseria meningitidis is one of two major pathogens in a genus which contains many members that make up the commensal flora of the oropharynx. They compose a closely related generic group as defined by DNA homology.

Characterization

N. meningitidis is a member of the family *Neisseriaceae*. It is a gram-negative coccus that occurs singly and in pairs with flattened adjacent sides. It often retains some crystal violet on Gram stain, and may appear to be almost gram-positive due to the purplish cast added to the usual gram-negative reaction. This species requires enriched media for growth and is enhanced by CO_2 and a relative humidity of 40%. *N. meningitidis* is catalase positive and produces cytochrome oxidase as do other members of the genus *Neisseria*.

It may be distinguished from nonpathogenic members of the genus by failure to grow on nutrient agar at 35°C. Like other members of the genus *Neisseria,* it degrades carbohydrates producing acid by-products. It may be distinguished from *N. gonorrheae* by its production of acid from maltose and from *N. lactamica* by its failure to produce acid from lactose. Many studies in the older literature included strains of *N. lactamica* among those counted as meningococci.

Specimen Collection and Transport

Specimens should be collected before antibiotic therapy, if possible. Because meningococci are fragile, transport to the laboratory should be rapid with measures taken to keep the specimen moist and protected from temperature extremes.

Cerebrospinal Fluid

The second tube from a lumbar puncture is less likely to have extraneous blood and bacterial contamination. Specimens less than 2 ml in volume may reduce chances for isolating the organism as the numbers may be small.

Blood

Two sets of blood cultures, 10 to 20 ml each drawn at separate times from separate sites, usually provide adequate opportunity for isolation. Blood should be inoculated directly into blood culture broth and transported immediately to the laboratory. There is evidence that some strains of meningococci are inhibited by the concentration of sodium polyanetholsulfonate in most commercial media. These effects may be moderated by increasing the amount of blood inoculated to 10% or more (Rintala and Pollock, 1978), adding 1 to 2% gelatin (Pai and Sorger, 1981) or by using a DuPont Isolater System (Scribner and Welch, 1984).

Skin Lesions

Petechial lesions may be sampled by taking a biopsy or injecting saline into the lesion and aspirating material into a syringe, which is transported immediately.

Joint Fluid

This may be transported in a sterile tube, preferably heparinized to avoid clotting.

Conjunctiva

Conjunctival swabs should be placed in Amies' transport medium to keep them moist. Stuart's transport is less desirable, as the glycerol phosphate present in this formula may allow the overgrowth of other organisms.

Nasopharyngeal

Nasopharyngeal specimens are often taken to screen immediate contacts of known cases. Due to the high carrier rate of *N. meningitidis*, they have no value for diagnosing disease.

Lower Respiratory Tract

Isolation of the organism from expectorated sputum may not be properly interpreted due to the frequent carriage of meningococci in many populations. Culture diagnoses of pneumonia requires specially collected lower airway specimens, such as those collected by transtracheal aspiration or protected-catheter bronchoscopy brush.

Isolation Procedures

N. meningitidis grows well on 5% blood agar at 35 to 37°C in 5 to 10% CO_2. Because humidification enhances growth, the incubator should have a minimum of 50% relative humidity. Although neisseriae do not grow well in broth media, most specimens should be inoculated into tryptic soy or similar all purpose broth for isolation of other organisms. Blood culture bottles must be blindly subcultured because gross turbidity is not always seen in positive cultures.

Specimens that are likely to contain oral flora should be plated onto blood agar and a selective medium for pathogenic neisseriae such as Modified Thayer Martin or New York City agar. These specimens may contain organisms that inhibit the growth of *N. meningitidis*.

Identification Procedures

Preliminary identification shows isolated colonies that are smooth, round, and convex, which are often gray and transparent and 1 mm or more in diameter after 24 h.

Presumptive Identification

Suspicious colonies are screened using a Gram stain and oxidase test. The Gram stain reveals typical kidney bean-shaped gram-negative cocci. The oxidase test uses tetramethyl-p-phenylenediamine dihydrochloride to demonstrate the characteristic cytochrome oxidase produced by the genus *Neisseria*. Oxidase-positive colonies, fitting the appropriate Gram stain morphology, are likely to be *Neisseria* spp. There are a number of genera such as *Kingella* and *Moraxella*, however, that also may fit these criteria.

Definitive Identification

Degradation of specific carbohydrates is the generally accepted confirmatory criterion for separating *N. meningitidis* from other oxidase-positive gram-negative cocci. Many formulations of standard and rapid test media have been described. A panel of carbohydrates including glucose, maltose, sucrose, and lactose is generally recommended. These carbohydrates are added to a basal medium, usually containing phenol red as the indicator of acid production. Phenol red requires a lowering of the medium pH to 6.8 to exhibit a color change, whereas other indicators require more acid production and are less sensitive indicators. Strains of meningococci that fail to degrade maltose are, for the most part, sulfonamide resistant (Morello et al., 1985).

Epidemiologic Markers

Strains of *N. meningitidis* have several chemical moieties that define antigenic schema useful in strain delineation (Frasch et al., 1985). Serogroups A, B, C, D, X, Y, Z, 28E, and W135 are defined by a group-specific capsular polysaccharide. Fifteen serotypes defined by subcapsular outer membrane protein and lipopolysaccharide antigens have been described in groups B and C. Subsequent investigations have shown that the outer membrane proteins can be divided into five classes based on peptide mapping. Classes 2 and 3 of these are found in all meningococcal strains and show antigenic variations that are epidemiologically relevant. Subtyping schema have been shown to be essential for specifically separating epidemic from the more common carrier strains.

Another tool, which has recently been used in Norway, was the addition of DNA fingerprinting to serotyping and sulfonamide susceptibility. Dialyzed DNA was cleaved with Hind III restriction endonuclease. The grouping of isolates by this method may not exactly correlate with serotyping results (Kristiansen et al., 1986) as restriction patterns showed variations of up to four bands among the 50 delineated. Further studies will clarify the possible divergence of clones of genetic material as organisms are transmitted through a population.

Immunologic Aspects

During the past 10 years, commercial reagents have become readily available for the detection of antigens in clinical specimens and has become one of the most useful tools in the rapid diagnosis of acute bacterial meningitis. These reagents are able to detect as little as 10 ng/ml of polysaccharide depending on the particular antigen being tested (Leinonen and Käyhty, 1978). Available methods include coagglutination, counterimmunoelectrophoresis, and reverse latex agglutination. In general, the agglutination methods are at least fourfold more sensitive than CIE and are more easily made available on a stat basis.

Serologic diagnosis by testing for rise in antibody titer is too slow to be clinically useful and is mainly used in prevalence surveys because nasopharyngeal carriage stimulates bactericidal antibody production. The background rate of persons having meningococcal antibody in most populations is high.

Treatment, Prevention, and Control

With the emergence of sulfonamide-resistant strains, other regimens have been used for prophylaxis and control.

Therapy

Penicillin or ampicillin are the usual antibiotics of choice, and resistance to these agents has not as yet been described. Chloramphenicol is the usual choice for those individuals who have a penicillin allergy. Ten to 14 days of therapy is usually sufficient (Apicella, 1985).

The greatest problem is addressing the septic complications of this disease. Supportive therapy dealing with cardiovascular complications is essential. In addition, DIC is a major part of the life-threatening aspects and must be dealt with carefully. The use of heparin in this setting is open to question.

Prevention

In settings where sporadic cases occur, household contacts have been shown to be at 500 to 800 times greater risk than the general population. Other closed populations such as dormitories and military groups are at similar risk. Most secondary cases occur within 10 days. Sulfadiazine may be used for susceptible isolates and rifampin for sulfonamide-resistant strains as chemoprophylactic agents to prevent disease in these contacts.

Control

Vaccines exist for all of the major groups except group B as the group B antigen is insufficiently immunogenic in humans to elicit a protective antibody response. Vaccines have been useful in the protection of various populations and reduction of cases. In the Seattle epidemic, however, cases appeared during the vaccination program (Counts et al., 1984). The role of circulating blocking IgA in this setting remains to be elucidated.

Literature Cited

Apicella, M. A. 1985. *Neisseria meningitidis*. *In* Mandell, Douglas, and Bennett (eds.), Principles and practice of infectious diseases, 2nd ed. John Wiley & Sons, New York

Counts, G. W., D. F. Gregory, J. G. Spearman, B. A. Lee, G. A. Filice, K. K. Holmes, J. M. Griffiss, and the Pacific Northwest Meningococcal Study Group. 1984. Group A meningococcal disease in the US pacific northwest: epidemiology, clinical features and effect of a vaccination control program. Rev. Infect. Dis. **6**:640–648.

De Voe I. W. 1982. The meningococcus and mechanisms of pathogenicity. Microbiol. Rev. **46**:162–190.

Filice, G. A., P. S. Hayes, G. W. Counts, J. M. Griffiss, and D. W. Fraser. 1985. Risk of group A meningococcal disease: bacterial interference and cross-reactive bacteria among mucosal flora. J. Clin. Microbiol. **22**:152–156.

Frasch, C. E., W. D. Zollinger, J. T. Poolman. 1985. Serotype antigens of *N. meningitidis* and a proposed scheme for designation of serotypes. Rev. Infect. Dis. **7**:504–510.

Geiseler, P. J., K. E. Nelson, S. Levin, K. T. Reddi, and V. K. Moses. 1980. Community-acquired purulent meningitis: a review of 1316 cases during the antibiotic era, 1954–1976. Rev. Infect. Dis. **2**:725–745.

Goldschneider, I., E. C. Gotschlick, M. S. Artenstein. 1969. Human immunity to meningococcus. I. The role of humoral antibodies. J. Exp. Med. **129**:1307–1326.

Griffiss, J. M. 1975. Bactericidal activity of meningococcal antisera: blocking by IgA of lytic antibody in human convalescent sera. J. Immunol. **114**:1779–1784.

Griffiss, J. M., and M. A. Bertram. 1977. Immunoepidemiology of meningococcal disease in military recruits. II. Blocking of serum bactericidal activity by circulating IgA early in the course of invasive disease. J. Infect. Dis. **136**:733–739.

Griffiss, J. M. 1982. Epidemic meningococcal disease: synthesis of a hypothetical immunoepidemiologic model. Rev. Infect. Dis. **4**:159–172.

Irwin, R. S., W. K. Woelk, W. L. Coudon, III. 1975. Primary meningococcal pneumonia. Ann. Intern. Med. **82**:493–498.

Kristiansen, B., B. Sorensen, B. Bjorvatn, E. S. Falk, E. Fosse, K. Bryn, L. O. Frholm, P. Gaustad, K. Bvre. 1986. An outbreak of group B meningococcal disease: tracing the causative strain of *Neisseria meningitidis* by DNA fingerprinting. J. Clin. Microbiol. **23**:764–767.

LaScolea, Jr. J., and D. Dryja. 1984. Quantitation of bacteria in cerebrospinal fluid and blood of children with meningitis and its diagnostic significance. J. Clin. Microbiol. **19**:187–190.

Leinonen, M., and H. Käyhty. 1978. Comparison of countercurrent immunoelectrophoresis, latex agglutination and radioimmunoassay in detection of soluble capsular polysaccharide antigens of *Hemophilus influenzae* type b and *Neisseria meningitidis* of groups A or C. J. Clin. Pathol. **31**:1172–1176.

Morello, J. A., W. M. Janda, and M. Bohnhoff. 1985. Neisseria and Branhamella. *In* E. H. Lennette, A. Balows, W. J. Hausler, Jr., H. J. Shadomy (eds.), Manual of clinical microbiology, 4th ed. American Society for Microbiology, Washington, D.C.

Pai, C., and S. Sorger. 1981. Enhancement of recovery of *Neisseria meningitidis* by gelatin in blood culture media. J. Clin. Microbiol. **14**:20–23.

Peltola, H. 1983. Meningococcal disease: still with us. Rev. Infect. Dis. **5**:71–91.

Pollock, H. M., E. L. Hawkins, J. R. Bonner, T. Sparkman, J. B. Bass, Jr. 1983. Diagnosis of bacterial pulmonary infections with quantitative protected catheter cultures obtained during bronchoscopy. J. Clin. Microbiol. **17**:255–259.

Rintala, L., and H. M. Pollock. 1978. Effects of two blood culture anticoagulants on the growth of *Neisseria meningitidis*. J. Clin. Microbiol. **7**:332–336.

Scribner, R. K., and D. F. Welch. 1984. Neutralization of the inhibitory effect of sodium polyanetholesulfonate on *Neisseria meningitidis* in blood cultures processed with the Du Pont isolator system. J. Clin. Microbiol. **20**:40–42.

Stephens, D. S., L. H. Hoffman, and Z. A. McGee. 1983. Interaction of *Neisseria meningitidis* with human nasopharyngeal mucosa: attachment and entry into columnar epithelial cells. J. Infect. Dis. **148**:369–376.

Stephens, D. S., A. M. Whitney, M. A. Melly, L. H. Hoffman, M. M. Farley, and C. E. Frasch. 1986. Analysis of damage to human ciliated nasopharyngeal epithelium by *Neisseria meningitidis*. Infect. Immunol. **51**:579–585.

CHAPTER 40

Mycobacterioses Excluding Tuberculosis

G. FADDA

Disease: Nontuberculous mycobacterial disease.
Etiologic Agents: *Mycobacterium avium complex, M. kansasii, M. fortuitum complex, M. scrofulaceum, M. marinum,* and *M. ulcerans* are among the most common species encountered.
Source: Environmental sources (water, soil, dust, dairy products), by direct inoculation, ingestion, or inhalation.
Clinical Manifestations: Chronic pulmonary disease in adults; local lymphadenitis in children; local lesions of the skin and subcutaneous tissues; disseminated disease.
Pathology: Granulomatous lesions indistinguishable from lesions due to *M. tuberculosis.*
Laboratory Diagnosis: Bacteriologic recovery of mycobacteria from any source suspected of being the site of a mycobacterial infection.
Epidemiology: Worldwide. Often an opportunist infection.
Treatment: Based on antituberculosis drugs (isoniazid, rifampin, ethambutol) and amikacin, erythromycin, and cotrimoxazole. Surgical excision in lymphadenopathies and localized pulmonary disease due to multiple drug-resistant nontuberculous *Mycobacterium.*
Prevention and Control: No need to report cases, but clusters should be reported to the public health officials, and do not need to isolate patients. Eradication measures are not indicated.

Description of Disease

Mycobacteria other than tubercle bacilli (*Mycobacterium tuberculosis, M. bovis, M. africanum*) and *M. leprae* are now recognized as important human pathogens. Some of these species can cause serious, even fatal, diseases in animals and humans. These microorganisms, once considered only environmental and saprophytic species, have been called by many names: anonymous, opportunistic, atypical, and paratubercle bacilli. However, the terms "non-tuberculous mycobacteria" (NTM) and "mycobacteria other than tubercle bacilli" (MOTT) are preferable (Sommers and Good, 1985).

A list of major types of infections caused by NTM include: (1) chronic pulmonary infections resembling tuberculosis in adults; (2) local lymphadenitis in children; (3) local lesion of the skin and deeper tissues; and (4) disseminated disease (Grange and Yates, 1986).

From the clinical point of view, mycobacteria other than tubercle bacilli are divisible into two groups: (1) environmental species that are recognized as etiologic agents of recognizable disease, such as *M. ulcerans, M. marinum, M. lepraemurium,* and *M. paratuberculosis*; and (2) free-living mycobacteria that may be associated with disease, such as *M. kansasii, M. scrofulaceum, M. avium, M. intracellulare* (*M. avium complex*), and *M. fortuitum complex* (Good, 1985). However, other species or complexes of NTM also may cause disease (Table 1) (Wolinski, 1979).

TABLE 1. Nontuberculous mycobacteria that can be isolated from human clinical specimens

Group	*Clinically significant species*	*Usually clinically insignificant*
Photochromogens	*M. kansasii*	None
	M. marinum	
	M. simiae	
	M. asiaticum	
Schotochromogens	*M. scrofulaceum*	*M. gordonae*
	M. szulgai	*M. flavescens*
Nonphoto-chromogens	*M. avium*	*M. gastri*
	M. intracellulare	*M. nonchromogenicum*
	M. ulcerans	
	M. xenopi	*M. terrae*
	M. malmoense	*M. triviale*
	M. haemophilium	
Rapid growers	*M. fortuitum*	*M. phlei*
	M. chelonae	*M. smegmatis*
		M. vaccae

Communicability and Transmissibility

Person-to-person transmission usually does not occur, but family cases and nosocomial outbreaks in hospitals have been reported (Wolinsky, 1979).

Many of the potentially pathogenic species of nontuberculous mycobacteria may be found in the environment, and occur especially in soil and water. The human population is constantly exposed to them through ingestion, inhalation of aerosols, and inoculation of traumatized skin. How the NTM colonize humans is usually obvious in localized soft tissue disease, but it remains obscure for most cases of pulmonary disease. Studies of epidemiology of infection by NTM suggest that the natural water-to-air pathway and its associated bubble-burst and jet-droplet mechanism are a potentially significant source of human respiratory pathogens. Nevertheless, other sources such as soil and dust may be equally or more important (Wicher, 1984). Predisposing conditions such as pneumoconiosis, chronic obstructive pulmonary disease, healed tuberculosis, immunosuppression, malignant diseases, and surgery are important in the pathogenesis of infections due to NTM (Grange and Yates, 1986).

Clinical Manifestations

A list of diseases associated with NTM and their causative agents is reported in Table 2.

Pulmonary Disease

Chronic pulmonary disease in adults indistinguishable from tuberculosis represents the most important infection due to mycobacteria other than tubercle bacilli. The most common pulmonary pathogens are *M.*

TABLE 2. Nontuberculous mycobacteria associated with disease in humans

Disease	*Most common species*	*Other reported species*
Chronic pulmonary disease (adults)	*M. avium complex, M. kansasii, M. fortuitum complex*	*M. xenopi, M. szulgai, M. simiae, M. scrofulaceum, M. malmoense, M. asiaticum, M. terrae complex*
Lymphadenitis (children)	*M. scrofulaceum, M. avium complex*	*M. fortuitum complex, M. kansasii*
Skin and soft tissue		
Swimming pool granuloma	*M. marinum*	
Sporotrichoid	*M. marinum, M. kansasii*	*M. haemophilium*
Abscess	*M. fortuitum complex*	
Buruli ulcers	*M. ulcerans*	
Skeletal	*M. kansasii*	*M. fortuitum complex*
Bone, tendon, joints	*M. avium complex*	*M. marinum, M. gordonae*
Postsurgical infection (mammaplasty)	*M. fortuitum complex*	
Laparatomy, other type of surgery	*M. fortuitum complex*	
Disseminated	*M. avium complex, M. fortuitum complex, M. kansasii*	*M. scrofulaceum*, others
Keratitis	*M. fortuitum complex*	
Endocarditis	*M. fortuitum complex*	
Lymphadenitis	*M. fortuitum complex*	

kansasii, *M. avium complex*, and *M. fortuitum complex*. Other species such as *M. xenopi*, *M. szulgai*, *M. simiae*, *M. scrofulaceum*, *M. malmoense*, *M. terrae complex*, *M. asiaticum*, and *M. gordonae* also are reported as causative agents of respiratory infection. According to Wolinski (1979), the average case of pulmonary disease associated with nontuberculous mycobacteria would be a middle-aged man suffering from a long-standing lung disease with a few months history of productive cough, night sweats, a low-grade fever, and moderate weight loss. The true cause of infection is uncovered first by the microbiology laboratory; the sputum is positive for acid-fast bacilli (AFB) on smear. The chest radiogram shows a thin-walled cavity with fibrosis. Usually there is no family history or obvious source of tuberculosis; however, preexisting pulmonary lesions such as pneumoconiosis, previous tuberculosis, chronic bronchitis, chronic obstructive lung disease, bronchiectasis, and chronic aspiration from esophageal disease are associated in most cases of NTM pulmonary infections.

Local Lymphadenitis

Infected lymph nodes are usually high in the neck; however, preauricular, parotid, and postauricular location and femoral, epitrochlear, and axillary areas also may be involved.

These infections occur almost exclusively in children 1.5 to 5 years old, who appear to be healthy with little or no associated pain or tenderness. The infected nodes progress to softening, rupture, sinus formation, and prolonged drainage. The usual outcome is complete healing. The etiologic agent that is most often involved is *M. scrofulaceum*. *M. avium complex* is rarely isolated; *M. kansasii* and *M. fortuitum complex* are reported in a small proportion of patients.

Soft Tissue Infections

Cutaneous Granulomas

The most common mycobacterial soft tissue infections are "swimming pool" and "fish tank" granulomas due to *M. marinum* (Collins et al., 1985) and "sporotrichoid" caused by *M. marinum* and *M. kansasii*. The lesions usually start as groups of papules on an extremity, with subsequent progression to ulceration and scab formation. These lesions usually heal spontaneously, but they may persist for a long time. The infectious microorganism may be introduced into the skin in a variety of ways such as abrasions, scratches, punctures from fish and shrimp fins, oyster shells, and so on in swimming pools and fish tanks.

Ulcers

Solitary lesions on an arm or leg, most often in children 5 to 14 years of age, starting as an erythematous nodule and progressing gradually to a large indolent ulcer with a necrotic base and undermined edges of skin and underlying tissue are now known as "Bairnsdale ulcer" in Australia or "Buruli ulcer" in Africa (Wolinsky, 1979).

The causative agents, identified and named *Mycobacterium ulcerans*, can be seen in large clusters in the necrotic areas. The reservoir in nature of this organism has not been conclusively determined, and the manner by which humans become infected is unknown. Trauma to the skin or microabrasions probably play an important role.

Localized Abscesses

Infection site abscesses and abscesses after trauma or surgery have been frequently reported. The most common species associated with these infections are *M. fortuitum* or *M. chelonae*.

Site abscesses after injection of one of the many contaminated substances such as insulin, penicillin, iron dextran, BCG, histamine, and diphtheria-pertussis-tetanus-polio vaccine have been documented. War wounds, penetrating iron spikes, surgical incisions, lawn mower blades, and thorns have also served to inoculate NTM into the tissue.

Disseminated Disease

Nonpulmonary disease resembling cryptic disseminated tuberculosis with multiorgan involvement usually occurs in association with hematologic malignancies, congenital or acquired immunodeficiency, therapeutic immunosuppression after transplant surgery, and acquired immunodeficiency syndrome (AIDS). Most cases of disseminated disease are due to *M. avium complex* (Good, 1985), *M. chelonae* (Grange and Yates, 1986), or *M. kansasii* (Wolinsky, 1979).

Other Sites of Disease

Lesions at sites other than those described are uncommon. However, rare infections of bones, joints, tendon sheaths, genitourinary tract, and meninges may be seen. The etiologic agents more frequently reported are *M. kansasii*, *M. fortuitum complex*, and *M. avium complex* (Wolinsky, 1979).

Mycobacterial Infections in Patients with Acquired Immunodeficiency Syndrome

In AIDS patients, multiple infections with otherwise nonpathogenic organisms are commonly seen. Among these, infection with *Mycobacterium avium complex* is one of the most commonly observed processes.

In patients with AIDS, *M. avium complex* has been cultured from different organs and blood. The *Mycobacterium* is often present in impressive masses and clumps. In these patients, there is a poor inflammatory response to the microorganism due to a profoundly deficient cellular response. In patients with AIDS coming from a high tuberculosis incidence area, *M. tuberculosis* seems to predominate.

M. asiaticum, M. flavescens, M. fortuitum, M. gordonae, and *M. kansasii* rarely have been found. *M. avium complex* strains recovered from AIDS patients were mostly serotype 4, 8, 1, and 6 (Good, 1985).

In AIDS patients, treatment of *M. avium complex* infection with conventional antituberculosis drugs is usually unsuccessful. The new rifabutine (spiropiperydil rifamycin) and clofazimine appear effective for the in vitro inhibition of mycobacteria. However, until therapeutic studies conducted under current protocols are completed, the actual benefits of this treatment will not be known.

Pathology

The majority of nontuberculous mycobacteria live freely in the environment and occasionally can cause serious, even fatal, disease in animals and humans. Cutaneous ulcers can be caused by direct implantation of the *Mycobacterium* into skin cuts or abrasions.

Close contact with soil and the tendency that children have to put everything in their mouths could be the possible sources of infection in acquired lymphadenitis. Other types of infection are likely to be caused by inhalation of infected dust and ingestion of contaminated water, dairy products, or raw vegetables. Thus, the organisms penetrate into the body through skin, mucous membranes, or respiratory tract. Usually a granulomatous lesion develops without becoming large enough to be noticeable. Long-latent infections can become active and/or disseminated if the host's resistance is depressed. However, all these statements are based more on presumptions than on facts.

Diagnostic Features

Mycobacterioses excluding tuberculosis are clinically, radiologically, and histologically indistinguishable from tuberculosis (Good, 1985). The diagnosis of opportunistic mycobacterial disease is made bacteriologically; only culture procedures can unequivocally determine the species responsible (Grange and Yates, 1986). Isolation of a nontuberculous *Mycobacterium* from a biopsy specimen, together with histologic evidence of granulomatous disease, usually poses no interpretational problems. However, because of the occurrence of NTM as commensal in clinical specimens such as sputum, a decision as to its clinical relevance must be made.

According to Wolinsky (1979) the following information is required to decide whether an isolate represents a diseased state: the species of *Mycobacterium,* the quantity of growth on primary culture, repeated isolations of the same organism from the specimen taken at different times, and the host's risk factors.

Even greater care must be taken with *Mycobacterium* isolated from the urinary tract. This specimen is frequently contaminated with acid-fast bacilli; contrary to a popular myth, *M. smegmatis* is rarely found (Grange and Yates, 1986). Experience with immunologic tests, such as differential skin testing or detection of specific antibodies, is limited (Comstock, 1982).

Prognosis

The mortality rate in untreated nontuberculous mycobacterial disease such as chronic pulmonary and disseminated infections depends on the host-predisponing conditions. Life-threatening mycobacteriaemia usually occurs in immunocompromised patients, and rarely in previously healthy people. The usual outcome of lymph nodes, skin, and subcutaneous infections is complete healing.

Epidemiologic Aspects

Occurrence and Transmission

Opportunistic mycobacterial disease occurs worldwide. Studies from various parts of the world indicate the different species predominating as a cause of disease in different geographic regions.

The relative frequency of infections is determined by the distribution of the various species of nontuberculous mycobacteria in the environment. It also depends on the opportunities that they have for contact with the human population and the susceptibility of the potential host. However, nontuberculous mycobacterioses are not reported in many countries; therefore, only a rough estimate of their relative frequency may be obtained from the numbers of mycobacteria isolated from human specimens and from the environment by large reference laboratories.

The most common NTM isolated from the environment are *M. fortuitum complex, M. gordonae, M.*

TABLE 3. Distinctive properties of *Mycobacterium* species encountered in clinical specimens

Species	Growth	Colony type	Pigment	Niacin	Susceptible to T2H	Susceptible to PNB	Susceptible to TCZ	Nitrate reduction
M. ulcerans	S	R	N	−	−	+	−	−
M. tuberculosis	S	R	N	+	−	+	+	+
M. bovis	S	RT	N	−	+		+	−
M. marinum	M	SR	P	+/−	−	−	−	−
M. kansasii	S	SR	P	−	−	−	+	+
M. simiae	S	S	P	+	−	−	−	−
M. asiaticum	S	S		−	−	−	−	−
M. scrofulaceum	S	S	S	−	−	−	−	−
M. szulgai	S	SR	S	−	−	−	−	+
M. gordonae	S	S	S	−	−	−	−	−
M. flavescens	M	S	S	−	−	−	−	+
M. xenopi	S	SF	S	−	−	−	−	−
M. avium	S	ST/R	N	−	−		−	−
M. intracellulare	S	ST/R	N	−	−	−	−	−
M. gastri	S	SR	N	−	−	−	+	−
M. malmoense	S	S	N	−	−			−
M. haemophilium	S	R	N	−	−		−	−
M. nonchromogenicum	S	SR	N	−	−		−	−
M. terrae	S	SR	N	−	−		−	+
M. triviale	M	R	N	−	−			+
M. fortuitum	R	SF	N	−	−			+
M. chelonae	R	SR	N	V	−			−
M. phlei	R	R	S	−	−			+
M. smegmatis	R	RS	N	−				+
M. vaccae	R	S	S	−				+

Growth: M = moderate; R = rapid; S = slow.
Colony type: F = filamentous; R = rough; S = smooth; SR = intermediate in roughness; T = thin or transparent.
Pigment: N = nonphotochromogenic; P = photochromogenic; S = schotochromogenic; V = variable.

avium complex, M. scrofulaceum, and *M. kansasii.* Mycobacteria have been found in soil, sewage, well water, and house dust specimens, as well as in domestic and wild animals and birds. Isolation rates for such organisms appear to be on the increase mostly in regions where infections due to *M. tuberculosis* are on the decline. In the United States, Great Britain, and other western countries, the incidence of such mycobacterioses is increasing in absolute terms and in relation to tuberculosis.

In high-incidence areas of tuberculosis, patients susceptible to mycobacterioses will develop tuberculosis rather than one of the other mycobacterioses. However, even in these regions, infections after successfully treated tuberculosis appear to be caused by NTM (Good, 1985).

Studies of the frequency of nontuberculous mycobacterioses and the etiologic agents were carried out in several countries. In the United States, in Dallas and New Orleans, higher frequencies of *M. kansasii* pulmonary disease than in other parts of the country were reported.

In Cleveland, Houston, and St. Louis, an approximately equal number of pulmonary infections associated with *M. kansasii, M. scrofulaceum,* and the members of the *M. avium complex* was seen. In the southern United States, Japan, and Western Australia, the most frequently isolated organisms were the members of *M. avium complex.* England has a predominance in NTM lung disease of *M. kansasii* and *M. xenopi.* In France, Great Britain, and Ontario, *M. xenopi* is endemic (Wicher, 1984).

Skin-testing surveys with two or more antigens have been widely used. However, because of the cross-reactivity within the members of the genus *Mycobacterium,* results indicate only the extent of infections, not the organism responsible for the skin reactivity (Comstock, 1982).

Susceptibility, Resistance, and Carrier State

In contrast to tuberculosis, such diseases are seldom or never transmitted directly from one individual to

Catalase semiquantitative									
>45 mm	*68°C*	*Tween hydrolysis*	*Tellurite reduction*	*Tolerance to 5% NaCl*	*Iron uptake*	*Arylsulfatase*	*Growth on MacConkey agar*	*Urease*	*Pyrazinamidase*
−	+	−	−/+	−		−		−	−
−	−	−	−/+	−	−	−	−	+	+
−	−	−	−/+	−	−	−	−	+	−
−	+	+	−/+	−	−	−	−	+	+
+	+	+	+	−	−	−	−	+	−
+	+	−	−	−	−	−		+	+
+	+	+	−/+	−	−	−		−	−
+	+	−	+/−	−	−	−	−	+	+/−
+	+	−/+	−	−	−	−	−	+	+
+	+	+	−/+	−	−	−	−	−	−/+
+	+	+	−/+	+	−	−	−	+	+
−	+	−	+	−	−	−	−/+	−	V
−	+	−	+	−	−	−	−/+	−	+
−	+/−	−	−/+	−	−	−	−	−	+
−	−	+	+	−	−	−		+	−
−	−	+	−	−	−	−	−	V	+
−	−	−	−	−	−	−	V	−	+
+	+	+	−	−	−	−	V	−	V
+	+	+	−	−	−	−	−	−	V
+	+	+	+	+	−	−/+	+	−	V
+	+	V	+	+	+	+	+	+	+
+	V	V	+	V	−	+	+	+	+
+	+	+	+	+	+	−	−		
+	+	+	+	+	+	−	−		
+	+	+	+	V	+	−	−		

another. Almost all cases follow acquisition of nontuberculous mycobacteria from their immediate environment. Direct implantation into skin cuts and abrasions, inhalation of infected dust, and ingestion of contaminated materials provide possible sources of infection. However, there are several reports indicating possible person-to-person transmission, nosocomial outbreaks, and postsurgery infections (Grange and Yates, 1986).

Etiologic Agents

Characterization of Clinically Significant Mycobacteria

Mycobacteria have been classified in the order *Actinomycetales*, family *Mycobacteriaceae*, genus *Mycobacterium*. They are nonmotile, nonsporulating, obligative aerobe, gram-positive, pleiomorphic long rods (2.0 to 4.0 × 0.3 to 0.5 nm). Nontuberculous mycobacterium have been divided into four Runyon provisional groups (American Thoracic Society, 1981) on the basis of the relative speed of growth and according to the capacity of the cultures to become yellow on exposure to light (photochromogens, group 1) or to produce an orange color when they are grown in the dark (scotochromogens, group 2), to remain colorless (nonphotochromogens, group 3), or to grow in less than 1 week (rapid growers, group 4). Although Runyon classifications are still widely used in the subgrouping of NTM, it is necessary, from the clinical view, to identify an isolate as close as possible to select the proper management.

Several of the 54 species presently recognized (Wayne and Kubica, 1986) in the genus *Mycobacterium,* arranged on the basis of pigment production, growth rate, and clinical significance, are listed in Table 3.

Mycobacteria with similar biochemical characteristics, which are difficult to differentiate by means of common laboratory tests, are usually grouped in complexes (Table 4) (Wayne and Kubica, 1985).

TABLE 4. Nontuberculous mycobacteria complexes

Avium	Fortuitum
M. avium	*M. fortuitum*
M. intracellulare	*M. chelonae*
MAIS	Terrae
M. avium complex	*M. terrae*
M. scrofulaceum	*M. nonchromogenicum*
	M. triviale

From the clinical view, it is sufficient to identify an NTM isolate to a complex level, but for epidemiologic studies, the microorganism can be further differentiated by means of lipid analysis using thin-layer chromatography (TLC) (Jenkins, 1981) and by serologic means (McClatchy, 1981). Bacteriophage typing has not been applied to the opportunistic pathogens and saprophytes (Good, 1985).

The following describes the more clinically significant or frequently isolated species of mycobacteria other than tubercle bacilli. Others species such as *M. gordonae, M. asiaticum,* and *M. terrae complex* that can be isolated from human clinical specimens, but that are not ordinarily associated with disease, are detailed elsewhere (Kent and Kubica, 1985).

MYCOBACTERIUM KANSASII

This slow-growing organism (14 to 21 days) grows in a wide range of temperatures (22 to 39°C). The morphology of the yellow photochromogenic colonies varies from smooth to rough; rough colonies are readily emulsified. Exposure to light induces the appearance of red crystals of β-carotene, which are controlled by an active oxygen-dependent photoinducible enzyme. *M. kansasii* must be distinguished from others photochromogenic mycobacteria such as *M. marinum* and *M. simiae* (weakly photochromogenic), *M. szulgai* (variably photochromogenic at low temperatures), and from the rapidly growing nonpathogenic *M. vaccae*.

On an acid-fast-stained smear, cells of *M. kansasii* are cross-barred. In cultures, the catalase is strongly positive, whereas nitrates are not reduced.

M. kansasii is pathogenic for mice, hamsters, and silicotic guinea pigs. In nature, the organism reservoir is unknown; they are occasionally found in deer, swine, cattle, and in water. They have been reported in humans with chronic pulmonary disease resembling tuberculosis, mycobacteremia, and infections of the skeletal system.

Most strains are susceptible to rifampin and slightly resistant to other agents, including ethambutol and streptomycin.

M. AVIUM COMPLEX

This complex consists of *M. avium* and *M. intracellulare*. These show such similar properties and biochemical reactions that the distinction cannot be made with common laboratory examinations. However, DNA-relatedness studies have demonstrated that they are two separate species.

The organisms are slow growers (14 to 21 days) in a wide range of temperatures: *M. avium* (32 to 43°C) and *M. intracellulare* (22 to 43°C.)

The colonies are smooth, but rough strains are encountered. They are nonphotochromogenic; aging colonies may become yellow.

Seroagglutinations afford the most useful means of separation in the clinical laboratory. *M. avium* consists of 10 types (1–6, 8–11) and *M. intracellulare* of 18 types (7, 12–28). Serotypes 1 and 2 are usually isolated from diseased birds, and are virulent for rabbits and chickens. They can also be isolated from other animals. Strains of types 1, 2, and 3 usually demonstrate full virulence (classic *M. avium*); types 4, 5, 6, 8, 9, 10, and 11 share variable degrees of virulence (intermediate group). As a rule, strains of the others groups are not virulent (intracellulare group). Serotype 8 is the most common isolate from humans and animals. *M. avium complex* are the most pathogenic for fowls and other birds and can be isolated from various domestic and exotic animals, but have rarely been found in soil. They produce infections indistinguishable from tuberculosis (chronic pulmonary and skeletal diseases and disseminated infection, especially in the immunocompromised host).

M. SCROFULACEUM

This organism was isolated by Prissick and Masson (1956) from children with cervical lymphadenitis. Short to long rods or filaments are slow growers in a wide range of temperatures (22 to 39°C). The colonies grow optimally at 35°C, are orange in color (scotochromogens), are buttery in consistency and are usually smooth, but rough strains are encountered. Colonies darken on prolonged exposure to light. By seroagglutination, three serovars (41–43) within this species were identified. Type 41 dominates the European strains. *M. scrofulaceum* are rather nonpathogenic to laboratory animals. At times, they are found in soil, raw milk, oysters, and water. They are most commonly associated with cervical lymphadenitis in children, but have been occasionally found in human sputum, gastric lavage specimens as a casual resident, and occasionally from patients suffering from chronic pulmonary disease. *M. scrofulaceum* in vitro is resistant to antituberculosis drugs; some strains are susceptible to rifampin, ethambutol, and streptomycin.

Because several strains presented some characteristics close enough to those of the members of *M. avium complex,* Hawkins (1977) suggested classifying them as intermediate *M. avium-intracellulare-scrofulaceum complex* (MAIS complex) until their taxonomic standing could be classified. Some authors misinterpreted such a suggestion, referring to a "MAIS complex," which includes the three species. However, recent studies have confirmed that *M. scrofulaceum* is a species distinct from both *M. avium* and *M. intracellulare*.

M. fortuitum complex

This group consists of *M. fortuitum* and *M. chelonae*. Among the rapid growers, they are most frequently associated with disease in humans and are usually isolated on routine culture media in the clinical laboratory. They may be confused with diphtheroides if an acid-fast stain is not performed (Sommers and McClatchy, 1983). They grow at 22 to 40°C. The colonies can be either smooth or rough; usually they are not pigmented. *M. fortuitum* and *M. chelonae* can grow on MacConkey agar prepared without crystal violet and have strongly positive arylsulfatase. The nitrate reduction test is positive in cultures of *M. fortuitum* and negative in cultures of *M. chelonae*.

M. fortuitum consists of two serotypes, whereas in *M. chelonae* it is possible to recognize a uniform antigenic structure.

Both *M. fortuitum* and *M. chelonae* strains are pathogenic for mice and less so for guinea pigs and rabbits.

The *M. fortuitum* reservoirs of infection in nature are represented by water, soil, and dust. They are rarely isolated from cattle and cold-blooded animals. *M. chelonae* may be found in the soil.

The members of this complex have been associated with soft tissue abscesses and skin or corneal infections. Diffuse pulmonary infections in immunosuppressed patients as well as contamination of porcine heart valves and breast implants have been reported (Sommers and McClatchy, 1983).

Most strains of *M. fortuitum* and *M. chelonae* are highly resistant to the antituberculosis drugs. Some are susceptible to ethionamide (35%), viomycin (24%), kanamycin (21%), and capreomycin (21%).

M. marinum

M. marinum cultures are photochromogenic, with an optimal growth temperature of 31 to 33°C. This organism has been isolated from fresh and brackish water, contaminated swimming pools, and a tropical fish aquarium. In humans, it causes skin lesions and soft tissue infections ("swimming pool granuloma"). When injected into laboratory animals such as mice, guinea pigs, rats, and rabbits, it produces disease in the superficial and cooler parts of the body.

Two serotypes have been recognized. *M. marinum* strains are resistant to isoniazid, paraminosalicylic acid (PAS), and streptomycin, but they are susceptible to rifampin and ethambutol.

M. simiae

M. simiae is a photochromogenic niacin-positive *Mycobacterium,* growing at an optimal temperature of 37°C. The organism, isolated originally in Hungary from monkeys imported from India and later described in Cuba where it was called *M. habana,* is rarely associated with human pulmonary disease. *M. simiae* strains are virulent to mice, which develop progressive fatal infections.

This organism can be found as a contaminant of water supplies (Wolinsky, 1979) and is resistant to all antituberculosis drugs except cycloserine and ethionamide. Antigenic cross-reactivity between *M. simiae* and *M. avium complex* types 18 and 19 has been observed (Good, 1985).

M. szulgai

Many of the strains are scotochromogenic at 35°C and photochromogenic at 24°C. This organism has been associated with chronic pulmonary disease resembling tuberculosis and very rarely is reported to cause infection in the palmar tendon and the olecranonal bursa (Davidson, 1976).

M. ulcerans

This very slowly growing (more than 4 weeks' incubation) organism is nonphotochromogenic and has a strict optimal growth temperature of 33°C. Colonies on inspissated egg medium are minute, rough, and yellow. In humans, *M. ulcerans* causes indolent extensive necrosis of the skin and subcutaneous tissue. Lesions are associated with an extracellular heat-labile toxin of molecular weight of 10^5, which produces severe cytopathogenic effects in tissue culture.

The organism can experimentally infect rats and mice, inducing progressive disease only in cooler parts of the body.

M. ulcerans has been isolated in Australia, Mexico, New Guinea, Malaya, and Africa, and rarely in the United States from patients traveling or working in areas where the disease is endemic (Tsang and Faber, 1973).

Most of the strains are susceptible in vitro to rifampin, streptomycin, and cycloserine, and are resistant to isoniazide, ethambutol, PAS, and ethionamide.

M. xenopi

These organisms were first isolated from skin granulomas of the toad *Xenopus laevi* (Marks and Schwabacher, 1965).

After 2 or more weeks' incubation on inspissated egg media, they produce smooth and nonpigmented colonies that can become yellow on aging. *M. xenopi* grow best at 42 to 45°C and have occasionally been associated with chronic pulmonary disease (Koizumi and Sommers, 1980). More frequently, they seem to be implicated in waterborne nosocomial outbreaks (Gross et al., 1976).

By seroagglutination it was possible to recognize only one serotype. Most of the strains examined are susceptible to antituberculosis drugs.

M. haemophilium

This recently found species was isolated from subcutaneous granulomas in Israel (Sompolinsky et al., 1978) and from skin lesions in immunosuppressed hosts in Australia (Dawson and Jennis, 1980).

M. haemophilium can grow on inspissated egg media supplemented with hemin, hemoglobin, or ferric ammonium citrate. Its uniqueness lies in the fact that it requires ferric ions for good growth (Dawson and Jennis, 1980). In these media, nonpigmented, rough to smooth colonies appear after 2 to 4 weeks' incubation at 32°C. These organisms, which are virulent for guinea pigs and mice, show a uniform antigenic structure.

M. malmoense

This species, isolated in Sweden (Schröder and Juhlin, 1977), is usually associated with pulmonary disease and is considered as the etiologic agent of disease. Smooth and nonpigmented colonies appear on inspissated egg media in 1 week or more at a temperature range of 32 to 37°C. Only one serovar has been demonstrated.

Specimen Collection

Mycobacteria can be isolated from a variety of clinical specimens including sputum, urine, pus, tissue biopsy specimens, and any source suspected of being the site of a mycobacterial infection. Recovery and identification of *M. tuberculosis* from a clinical specimen is a clear indication that the patient has tuberculosis.

This is not the case with mycobacteria other than tubercle bacilli. In fact, such organisms are present and abound in the environment and colonize water pipes, taps, and distilled water supplies.

They can also be isolated from contaminated laboratory reagents. As a consequence, great care must be taken in the collection, transport, and laboratory examination of possibly contaminated clinical specimens. However, collection, transport, and isolation protocols are similar to those used in the diagnosis of tuberculosis, which are discussed in more detail Chapter 52 (Tuberculosis) of this book.

Isolation and Identification Procedures

Digestion and Decontamination

The high cell wall lipid content of mycobacterial cells makes them more resistant to killing by chemical agents than other types of bacteria that contaminate sputum, urine, and other clinical specimens and can overgrow the slowly growing mycobacteria. To reduce undesirable bacteria and to liquefy mucus, almost all specimens must be treated with a decontaminating agent that kills microorganisms other than mycobacteria. Methods commonly used for digestion and decontamination of clinical specimens are reported by Sommers and McClatchy (1983).

After treatment with the decontaminating solutions, the acid and alkali used are neutralized. Mixtures are then centrifuged at high speed (3,000 to 3,800 *g*) to concentrate mycobacteria (Sommers and McClatchy, 1983).

Because of the high toxicity of strong acid and alkaline solutions, many mycobacteria are also killed. Therefore, carefully controlled periods of decontamination and proper neutralization are mandatory.

Alternatively, the chloride or bromide salt of cetylpyridinium appears to be a milder decontaminating agent. This reagent may be added to the clinical specimen before shipment from outpatient clinics to the reference laboratory. Concentration and inoculation can begin upon arrival, without any further treatment of the clinical specimen (Smithwick et al., 1975).

Sterile body fluids and tissue biopsy specimens can be homogenized and inoculated into culture media without any decontamination.

Microscopic Examination

The high cell wall lipid content of most mycobacteria makes them able to bind the dye carbolfuchsin and to resist destaining with acid alcohol. All clinical specimens submitted for diagnosis of a mycobacterial infection should be examined for acid-fast bacilli. Direct microscopy is much less sensitive than culture.

A positive acid-fast smear provides only a presumptive diagnosis because the actual species of *Mycobacterium* cannot be determined by a morphologic observation. However, the acid-fastness of mycobacteria together with their characteristic size and shape are useful for the early detection of mycobacterial infection and can help to observe the patient's response to chemotherapy; it also confirms the acid-fast characteristics of isolated cultures.

Two types of acid-fast stains are commonly used: (1) carbolfuchsin stains by Ziehl-Neelsen (hot stain) or Kinyoun (cold stain) procedures; and (2) fluorochrome stain and auramine O, with or without rhodamine.

Smears stained with carbolfuchsin procedures must be scanned with an oil immersion objective. Smears stained with the fluorochrome technique can be scanned with the 25× objective, which will increase the field of view, permitting inspection of a larger area in a shorter time (Sommers and McClatchy, 1983).

Recovery of Mycobacteria

The media commonly used for the isolation of NTM from clinical specimens are similar to those used for the recovery of *M. tuberculosis*. They are: (1) egg-base media (Lowenstein-Jensen, Petragnani American Thoracic Society); (2) agar-base media (Middlebrook and Cohn, 7H10-7H11); (3) liquid media (7H9 broth, Dubos broth); and (4) selective media (the Gruft modification of Lowenstein-Jensen, Mycobactosel, Selective 7H11) (American Thoracic Society, 1981; Kent and Kubica, 1985; Sommers and McClatchy, 1983; Sommers and Good, 1985).

An increased amount of CO_2 (5 to 10%) in the incubation atmosphere has been shown to improve the growth of mycobacteria on primary isolation.

According to Sommers and Good (1985), culture media selected for the recovery of mycobacteria from a clinical specimen should include an egg medium (Lowenstein-Jensen), a nonselective agar medium (Middlebrook 7H11), and one selective medium.

The culture should be examined after 3 and 7 days and weekly thereafter for 6 to 8 weeks. Cultures that show grossly visible colonies are removed for acid-fast stain and further identification.

Identification of Nontuberculous Mycobacteria

Details of the techniques used to identify *Mycobacterium* species can be found in several publications (Jenkins et al., 1985; Kent and Kubica, 1985; Sommers and Good, 1985; Sommers and McClatchy, 1983; Tsukamura, 1981). Disease-associated nontuberculous mycobacteria must be identified to species. For the other mycobacteria, the group is sufficient, except when the organisms are isolated from patients who have a natural or induced immunodeficiency. Nontuberculous mycobacteria may be identified by their rate of growth, optimal temperature of growth, colony photoreactivity, biochemical properties, and growth inhibition tests (Tsukamura, 1981). In problem cases, more specialized laboratories use other tests in addition to those described (Table 4).

Culture Identification

When mycobacteria are recovered from any clinical specimen on one of the egg-inspissated or agar-media inoculations, the laboratory must differentiate them from the tubercle bacilli. The role of the colony morphology in the identification of NTM is limited. Freshly isolated strains may show smooth colonies, whereas strains stored in the laboratory may show rough colonies (Tsukamura, 1981).

Pigmentation and Photoreactivity

A suspension in broth culture of the NTM is inoculated into several tubes of Lowenstein-Jensen medium. One of them is incubated in continuous light, the others are wrapped in aluminum foil so that they are not exposed to light. Pigment production is detected several days after growth is first noted in the light-exposed control cultures.

Photochromogenic species—*M. kansasii, M. marinum, M. simiae,* and *M. asiaticum*—produce a yellow or orange pigment only on the slopes exposed to light. Scotochromogenic species—*M. scrofulaceum, M. szulgai, M. xenopi, M. gordonae,* and *M. flavescens*—produce a yellow or orange pigment on media either incubated in the dark or exposed to light. Nonphotochromogenic species—*M. avium, M. haemophilium,* and *M. intracellulare*—produce nonpigmented colonies.

Growth Rate and Temperature

Slopes of Lowenstein-Jensen medium inoculated with a mycobacterial broth culture are incubated at each of these temperatures: 24°, 32°, 37°, and 42°C, in incubators equipped with accurate thermostats.

Rapidly growing organisms—*M. chelonae, M. fortuitum,* and many other saprophyte species—will commonly produce visible colonies in less than 1 week. Slow-growing colonies will appear after 3 weeks.

The majority of *Mycobacterium* species can be grown at 35 to 37°C. However, *M. ulcerans, M. marinum,* and some variants of *M. chelonae* and *M. fortuitum* grow best at skin temperature (30 to 32°C).

Therefore, all specimens from suspect skin infections or subcutaneous infections should be incubated at both 35 to 37°C and 32°C. *M. avium,* the major cause of tuberculosis in birds, as well as *M. xenopi* (a species not commonly found as a cause of infection in humans) grow best at 42°C.

Arylsulphatase

Tubes of Middlebrook 7H9 medium prepared as recommended (Kent and Kubica, 1985; Sommers and Good, 1985) and inoculated with several drops of mycobacterial suspension are incubated at 37°C for 3 days (rapid-growing organism) or 14 to 21 days (slow-growing organism). When good growth is obtained, a few drops of ammonia are added. A pink color indicates the presence of arylsulphatase (Jenkins et al., 1985).

Only *M. fortuitum* and *M. chelonae* split phenolphtalein from tripotassium phenolphtalein within 3 days. Slowly growing mycobacteria, such as *M. marinum, M. szulgai, M. xenopi, M. triviale,* and *M. flavescens,* are identified in 14 to 21 days.

Catalase (Semiquantitative Test)

A Lowenstein-Jensen butt is inoculated with 0.2 ml of bacterial suspension. When good growth is observed, 2 ml of a mixture of equal parts of hydrogen peroxide (100 volumes) and Tween 80 (0.5%) are added. The height in millimeters of the column of bubbles is recorded after 5 minutes in an upright position at room temperature.

This semiquantitative test is useful for the differentiation of *M. xenopi, M. avium,* and *M. intracellulare* (<45 mm foam) from *M. scrofulaceum* and *M. gordonae* (>45 mm foam) and to distinguish *M. kansasii* (>45 mm foam) from *M. marinum* and *M. gastri* (<45 mm foam).

Tellurite Reduction

A few drops of 0.4% potassium tellurite are added to a 14 to 18 day mycobacterial culture tube in Middlebrook 7H9. The appearance of a jet-black precipitate after a further 4 days or more indicates a positive test.

M. avium complex strains and rapid-growing mycobacteria reduce tellurite.

Tween 80 Hydrolysis

Heavy inoculum, prepared from actively growing colonies, is added to a tube containing Tween medium (0.6 mol/liter phosphate buffer; Tween 80 0.5%; 2% of a 0.1% solution of neutral red). A change in color from amber to red indicates that the mycobacteria possess a lipase that splits Tween 80 into oleic acid and polyoxilated sorbitol.

With rare exceptions, strains that hydrolyze Tween 80 are not clinically significant (*M. gastri, M. terrae complex, M. triviale*).

Nitrate Reduction

Nitrate is reduced to nitrite by mycobacteria that produce nitroreductase. To a suspension, obtained from solid culture that is 3 to 4 weeks old, are added the following: (1) 2 ml of $NaNO_3$ (0.01 M in 0.022 M phosphate buffer), shake mixture and incubate for 2 h at 37°C; (2) one drop of 1 : 2 dilution of concentrated HCl, two drops of aqueous solution of sulfanilamide (0.2%) and two drops of aqueous solution of N-(1-naphtyl) etylendiamine dihydrochloride (0.1%).

A red color indicates the presence of nitrites and a positive test. *M. kansasii, M. szulgai,* and *M. fortuitum* are nitrate reductase positive; *M. bovis, M. marinum, M. simiae,* the *M. avium complex, M. xenopi, M. gastri, M. malmoense,* and *M. chelonae* are negative.

NaCl 5%

M. flavescens, M. triviale, and *M. chelonae* grow on egg-base culture medium containing 5% NaCl. Other NTM do not tolerate such a concentration of salt.

MacConkey Agar

Among the rapidly growing mycobacteria, only the *M. fortuitum-chelonae complex* can grow on MacConkey agar medium without crystal violet.

Growth Inhibition Tests

Three Lowenstein-Jensen slopes containing, respectively, thiacetazone (TCZ) at 10 mg/liter, thiofene-2-carboxilic acid hidrazide (T2H) at 5 mg/liter, and 4-(p)-nitrobenzoic acid (PNB) at 500 mg/liter inoculated with a few drops of a mycobacterial suspension obtained from solid culture 3 to 4 weeks old are incubated for 3 weeks. Only *M. bovis* is sensitive to T2H. *M. kansasii, M. scrofulaceum,* and *M. malmoense,* as well as most strains of *M. tuberculosis,* are sensitive to TCZ. PNB inhibits the growth of the tubercle bacilli and *M. ulcerans.*

Lipid Identification and Serologic Tests

Identification of NTM by lipid patterns from thin-layer chromatography (Jenkins, 1980) is useful in epidemiologic studies. The technique identifies members within a complex and among serotypes not distinguishable by other means.

The seroagglutination test is a useful tool in epidemiologic studies of NTM, particularly of the complex *M. avium* and *M. scrofulaceum* types.

Rapid Methods

Only in the last few years have rapid and automated methods been used for culture of microorganisms and for antimicrobial susceptibility tests in bacteriology. Among these procedures, one of the most widely used is the BACTEC radiometric system (Johnson Laboratories, Inc., Cockeysville, Md). This method uses media containing ^{14}C-labeled substrates which, when metabolized by bacteria, yield detectable levels of $^{14}CO_2$, which are read as growth index (GI) measurements. The same principle has been used to detect growth in pure cultures and for drug susceptibility testing of mycobacteria by using a Middlebrook 7H12 liquid medium, which contains ^{14}C-palmitic acid as a labeled substrate (Fadda et al., 1984, 1986; Middlebrook et al., 1977; Siddiqi et al., 1981). This method was further developed to provide a rapid differentiation of tubercle bacilli from nontuberculous mycobacteria by selective susceptibility to p-nitro-α-acetylamino-β-hydroxypropiophenone (NAP). Reported data show that a concentration of 5 mg/liter of NAP can effectively separate the *M. tuberculosis complex* from NTM species in an average time of 6 days (Lazlo and Siddiqi, 1984).

Fluorimetric assay of enzymatic activity, thin layer chromatography of mycobactins, and serologic procedures measuring the immunologic relatedness of the mycobacterial catalase are under evaluation.

Treatment, Prevention, and Control

Treatment

Because the nontuberculous mycobacterial diseases are not contagious and cannot be transmitted from person to person, there is no need to report single cases or isolate the patients. At present, no universal protocols for the management of such infection have been developed. Proper management of diseases due to opportunistic pathogenic mycobacteria requires clinical experience to decide on treatment and therapy. These will depend on the species of NTM isolated, the site and extension of the lesions, and the host's predisponing conditions (Bailey et al., 1983).

Total surgical excision of the involved nodes is the preferred treatment for lymphadenopathy in young patients as well as in postinjection abscesses. Incision and drainage are likely to cause a chronically draining sinus (Comstock, 1982).

Surgery also appears to be indicated in the treatment of localized pulmonary disease due to multiple drug-resistant bacteria such as the members of *M. avium complex, M. scrofulaceum, M. fortuitum complex,* and *M. szulgai* (Moran et al., 1983).

Wide excision and skin grafting, as well as the use of a local increase in temperature, are indicated in the Buruli ulcer caused by the caloriphobic *M. ulcerans* (Wolinsky, 1979).

Antimicrobial drug treatment has been recommended for the therapy of most deep or systemic infections. The choice of drug regimes depends mainly on the drug susceptibility of the causal agent. However, the combination of drugs to which the nontuberculous mycobacteria is resistant in vitro often proves effective in vivo. For these reasons, chemotherapy of nontuberculous mycobacterial disease is usually empirical and based on retrospective surveys. Therapeutic regimens are usually based on the antituberculosis drugs.

Infections due to the relatively drug-susceptible *M. kansasii, M. xenopi, M. marinum,* and *M. malmoense* will usually respond well to a combination of isoniazid, rifampin, and ethambutol given for 18 to 24 months (Banks et al., 1985) or for 12 months in *M. kansasii* infections (Schraufnagel et al., 1984).

Standard triple antituberculosis therapy prolonged for 18 to 24 months seems to be effective in vivo for the members of *M. avium complex, M. scrofulaceum, M. fortuitum complex,* and *M. szulgai,* which are in vitro multiple drug-resistant bacteria (Hunter et al., 1981).

In recent years, some clinicians have treated patients with erythromycin, amikacin, deoxycycline, and cotrimoxazole, which are not normally used in antituberculosis therapy. There is some evidence that these nonconventional drugs are very effective for infections due to *M. chelonae, M. kansasii, M. scrofulaceum,* and *M. avium complex.*

The combination of the antileprosy drug, clofazimine, with the new spiro-piperidyl-rifamycin rifabutine (Farmitalia Carlo Erba, Milan) has been recommended for the therapy of disseminated *M. avium complex* infections in patients with AIDS.

Prevention and Control

Because nontuberculous mycobacterial diseases are not contagious and cannot be transmitted from person to person, there is no need to report single cases or isolate the patient. Clusters of cases should be reported to public health officials. As it is not possible to isolate humans from the mycobacteria in their environment, measures to prevent infection from environmental sources are not indicated. Chemotherapeutic prophylaxis is rarely indicated; immunization procedures are neither used nor available.

Literature Cited

American Thoracic Society. 1981. Ad hoc committee of the scientific assembly on tuberculosis, diagnostic standards and classification of tuberculosis and other mycobacterial disease, 14th ed. Am. Rev. Respir. Dis. **123**:343–358.

Bailey, W. C., R. K. Albert, P. T. Davidson, L. S. Farer, J. Glassroth, E. Kendig, R. G. London, and L. S. Inselman. 1983. Treatment of tuberculosis and other mycobacterial diseases. Am. Rev. Respir. Dis. **12**:790–796.

Banks, J., P. A. Jenkins, and P. A. Smith. 1985. Pulmonary infection with *Mycobacterium malmoense*—a review of treatment and response. Tubercle **66**:197–203.

Boisvert, H. 1977. L'ulcere cutane a *Mycobacterium ulcerans* au Camroun. II. Etude bacteriologique. Bull. Soc. Pathol. Exot. **70**:125–131.

Collins, C. M., J. M. Grange, W. C. Noble, and M. D. Yates. 1985. *Mycobacterium marinum* infections in man. J. Hyg. (London) **94**:135–149.

Comstock, G. W. 1982. Mycobacterioses other than tuberculosis and leprosy, p. 633–640. *In* A. S. Evans and H. A. Feldman (ed.), Bacterial infections of humans. Epidemiology and control. Plenum Medical Book Company, New York and London.

Davidson, P. T. 1976. *Mycobacterium szulgai.* A new pathogen causing infection of the lung. Chest **68**:799–801.

Dawson, D. J., and F. Jennis. 1980. Mycobacteria with a growth requirement for ferric ammonium citrate, identified as *Mycobacterium haemophilium*. J. Clin. Microbiol. **11**:190–192.

Fadda, G., and S. L. Roe. 1984. Recovery and susceptibility testing of *Mycobacterium tuberculosis* from extrapulmonary specimens by the BACTEC radiometric method. J. Clin. Microbiol. **19**:720–721.

Fadda, G., S. Cossellu, S. L. Roe, L. Rubattu, G. Satta, S. Zanetti. 1984. Evaluation of a rapid radiometric method in the clinical bacteriology laboratory. Clin. Chem. Newsl. **4**:72–77.

Fadda, G., G. Foddai, L. Mureddu, F. Sanciu, S. Zanetti. 1986. Radiometric investigation of susceptibility of *Mycobacterium tuberculosis. In* M. Casel (ed.), Mycobacteria of clinical interest. Elsevier Science Publishing Co., New York.

Good, R. C. 1985. Opportunistic pathogens in the genus *Mycobacterium*. Ann. Rev. Microbiol. **39**:347–369.

Grange, J. M., and M. D. Yates. 1986. Infections caused by opportunistic mycobacteria: a review. J. R. Soc. Med. **79**:226–229.

Gross, W., J. E. Hawkins, and R. Murphy. 1976. *M. xenopi* in clinical specimens I. Water as a source of contamination. Am. Rev. Respir. Dis. **113**:78.

Hawkins, J. E. 1977. Scotochromogenic mycobacteria which appear intermediate between *M. avium/intracellulare* and *M. scrofulaceum*. Am. Rev. Respir. Dis. **116**:963–964.

Hunter, A. M., I. A. Campbell, P. A. Jenkins, and A. P. Smith. 1981. Treatment of pulmonary infections caused by mycobacteria of the *Mycobacterium avium-intracellulare complex*. Thorax **36**:326–329.

Jenkins, P. A. 1981. Lipid analysis for identification of mycobacteria. An appraisal. Rev. Infect. Dis. **3**:862–866.

Jenkins, P. A., L. R. Duddrige, C. M. Collins, and M. D. Yates. 1985. Mycobacteria, p. 275–296. *In* C. H. Collins and J. M. Grange (ed.), Isolation and identification of microorganisms of medical and veterinary importance. The Society for Applied Bacteriology, Technical Series No. 21. Academic Press, London.

Kent, P. T., and G. P. Kubica. 1985. Public health mycobacteriology. A guide for the level III laboratory. U.S. Department of Health and Human Services, Public Health Service, Centers for Disease Control, Atlanta, Georgia.

Koizumi, J. M., and H. M. Sommers. 1980. *Mycobacterium xenopi* and pulmonary disease. Am. J. Clin. Pathol. **73**:826–830.

Lazlo, A. S., S. H. Siddiqi. 1984. Evaluation of a rapid radiometric differentiation test for the *Mycobacterium tuberculosis complex* by selective inhibition with p-nitro-α-acetylamino-β-hydroxypropiphenone. J. Clin. Microbiol. **19**:694–698.

Marks, J., and H. Schwabacher. 1965. Infection due to *Mycobacterium xenopi*. Br. Med. J. **1**:32–33.

McClatchy, J. K. 1981. The seroagglutination test in the study of nontuberculous mycobacteria. Rev. Infect. Dis. **3**:867–870.

Middlebrook, C., Z. Reggiardo, and W. D. Tigertt. 1977. Automable radiometric detection of growth of *Mycobacterium tuberculosis* in selective media. Am. Rev. Respir. Dis. **115**:1067–1069.

Moran, J. F., L. G. Alexander, E. W. Staub, W. G. Young, and W. C. Sealy. 1983. Long-term results of pulmonary resection for atypical mycobacterial disease. Am. Thorac. Surg. **35**:597–604.

Prissick, F. M., and A. M. Masson. 1956. Cervical lymphoadenitis in children caused by chromogenic mycobacteria. Can. Med. Assoc. J. **3**:91–95.

Schraufnagel, D. E., J. A. Leech, M. N. Schraufnagel, and B. Poliak. 1984. Short course chemotherapy for mycobacterioses kansasii. Can. Med. Assoc. J. 130:34–38.

Schröder, K. H., and I. Juhlin. 1977. *Mycobacterium malmoense* sp.nov. Int. J. Syst. Bacteriol. **27**:241–246.

Siddiqi, S. H., J. P. Libonati, G. Middlebrook. 1981. Evaluation of a rapid method for drug susceptibility testing of *Mycobacterium tuberculosis*. J. Clin. Microbiol. **13**:908–912.

Smithwick, R. W., C. B. Stratigos, and M. L. Davvid. 1975. Use of cetylpiridinium chloride and sodium chloride for the decontamination of sputum specimens that are transported to the laboratory for the isolation of *Mycobacterium tuberculosis*. J. Clin. Microbiol. **1**:411–413.

Sommers, H. M., and J. K. McClatchy. 1983. Laboratory diagnosis of the mycobacterioses. Cumulative techniques and procedures in clinical microbiology, n. 16. American Society for Microbiology, Washington, D.C.

Sommers, H. M., and R. C. Good. 1985. *Mycobacterium*, p. 216–248. *In* E. M. Lennette, A. Balows, W. J. Hansler Jr., and H. J. Shadomy (ed.), Manual of clinical microbiology. American Society for Microbiology, Washington, D.C.

Sompolinsky, D., A. Lagziel, D. Naveh, and T. Yankilevitz. 1978. *Mycobacterium haemophilum* sp.nov., a new pathogen of humans. Int. J. Syst. Bacteriol. **28**:67–75.

Tsang, A. Y., and E. R. Faber. 1973. The primary isolation of *Mycobacterium ulcerans*. Am. J. Clin. Pathol. **59**:688–692.

Tsukamura, M. 1981. A review of the methods of identification and differentiation of mycobacteria. Rev. Infect. Dis. **3**:841–861.

Wayne, L. G., and G. P. Kubica. 1986. Genus *Mycobacterium*, p. 1436–1457. *In* P. H. A. Smeath, N. S. Mair, and E. Sharpe (ed.), Bergey's manual of systematic bacteriology, vol. 2. Williams & Wilkins Co., Baltimore.

Wicher, K. 1984. Nontuberculous mycobacteria. An overview, p. 56–77. *In* F. Ginesu, and G. Fadda (ed.), La tubercolosi: un problema risolto o ancora attuale? Lotta contro la tubercolosi e le malattie polmonari sociali, vol. 54 (suppl).

Wolinsky, E. 1979. Nontuberculous mycobacteria and associated diseases. Am. Rev. Respir. Dis. **119**:107–159.

CHAPTER 41

Mycoplasmoses

NORMAN S. SWACK

Disease: Respiratory: Primary atypical pneumonia. Urogenital: Nongonococcal urethritis, acute pelvic inflammatory disease, nonspecific vaginitis.

Etiologic Agents: Respiratory: *Mycoplasma pneumoniae*. Urogenital: *Mycoplasma hominis, Ureaplasma urealyticum*.

Source: Respiratory: Direct contact with infected persons or articles containing nasopharyngeal discharges from an acutely ill patient. Urogenital: Sexual contact, newborn during passage through the birth canal.

Clinical Manifestations: Respiratory: Fever, nonproductive cough that may be severe, headache, and malaise. Onset is gradual and may progress to pneumonia. Urogenital: Ubiquitous in asymptomatic men and women and has been isolated from the urethra and the vagina. Isolated from patients with urethral discharge with or without dysuria.

Pathology: Respiratory: Upper respiratory tract infections, tracheitis, or bronchitis account for most clinical cases. Pneumonias are characterized by bronchiolitis and interstitial-mononuclear and granulocytic infiltration. Peribronchial infiltrate of lymphocytes and plasma cells. Urogenital: Urethritis and acute nonspecific urethroprostatitis. Strongly implicated in salpingitis.

Laboratory Diagnosis: Respiratory: Isolation of etiologic agent from throat washings, sputum, throat and/or nasopharyngeal swabs, or tracheal aspirates in selective media. Serodiagnosis is made by demonstration of a fourfold or greater rise in antibody. Urogenital: In females by isolation of etiologic agent from vaginal, cervical, urethral swab, urine, blood, fallopian tube, endometrial washings or biopsy, placenta or fetus. In males by isolation from urethral swabs or urine.

Epidemiology: Worldwide. *Mycoplasma pneumoniae* causes sporadic or endemic infections. Epidemic-like spread may occur in institutions or in the military. Respiratory disease observed mostly in school-age and young adult populations mostly during the fall and winter. Genital *Mycoplasma* species may be isolated from asymptomatic persons, particularly in the sexually active population.

Treatment: The tetracyclines and erythromycins are effective.

Prevention and Control: No effective methods for control have been proposed. Isolation of patients with respiratory infection appears to be effective in preventing the spread in some instances. Second attacks may occur.

Description of Diseases

Clinical Manifestations

Respiratory Infections

Human infections may range from asymptomatic to serious bronchial pneumonia. The incubation period varies from 1 to 3 weeks, with most cases occurring 2 weeks after exposure (Foy et al., 1966). The onset of disease is gradual and subtle with symptoms of headache, fever, malaise, sore throat, and cough.

Primary atypical pneumonia is the best described syndrome caused by *Mycoplasma pneumoniae*. After increased severity of symptoms of headache, malaise, and fever during a 2 to 4-day period, coughing becomes prominent (Foy et al., 1970). Although, the cough is commonly nonproductive, sputum may later become purulent or mucoid and contain small amounts of blood. Chest pain is also common. Evidence of pulmonary consolidation is often detected by x-ray only. Whereas multiple lobe involvement is not unusual, unilateral segmental bronchopneumonia of a lower lobe is commonly observed (Brolin and Wernstedt, 1978; Finnegan et al., 1981). Untreated, fever may persist for days or weeks and improvement is progressive but slow. Death is rare.

The most commonly observed syndrome due to infection with *M. pneumoniae* is tracheobronchitis (Foy et al., 1966). Bronchitis and frequent coughing are most frequently observed, often with coughing and substernal discomfort. Wheezing may occur in infants and older children. Onset and clinical signs are similar to those of pneumonia, but symptoms are fewer and milder, and improvement is more rapid.

Pharyngitis may occur as the major clinical aspect of disease or may be present as a symptom of a lower respiratory tract infection (Foy et al., 1966). Rhinitis due to *M. pneumoniae* has been reported in infants less than 1 year of age (Mok et al., 1979).

Other infections for which *M. pneumoniae* has been implicated include bullous myringitis, aseptic meningitis, myocarditis, and pericarditis (Ponka, 1979; Skoldenberg, 1972).

Mycoplasma hominis also has been shown to be capable of colonizing the respiratory tract and may cause serious disease, particularly in the newborn (Mufson, 1983).

Urogenital Infections

Mycoplasmal agents, particularly *Ureaplasma urealyticum* and *Mycoplasma hominis*, have been associated with a number of different clinical conditions, but a causal relationship has been established for relatively few (McCormack et al., 1973, Taylor-Robinson and McCormack 1980; Zlatnik et al., 1986). The best evidence for a mycoplasmal etiology has been shown for nongonococcal urethritis, pyelonephritis, pelvic inflammatory disease, postabortal fever, and postpartum fever. Although mycoplasmal agents have been implicated in many other diseases, proof of causal relationship is still lacking.

Nongonococcal urethritis has been the subject of numerous studies for which a mycoplasma species, specifically *Ureaplasma urealyticum*, has been established. Although not a major cause, it no doubt is responsible for some cases of urethritis (Bowie et al., 1977; Coufalik et al., 1979; Oriel, 1983; Stimson et al., 1981; Taylor-Robinson et al., 1977).

Pyelonephritis in some cases is caused by *Mycoplasma hominis*, and in some patients the presence of antibody has been demonstrated. Studies emphasize that *M. hominis* is responsible for a small percentage of these infections but does not play a major role in this disease (Thomsen, 1982).

Pelvic inflammatory disease is another infection for which there is more than a single etiologic agent. This disease includes inflammation of the fallopian tubes, parametria, pelvic peritoneum, and uterine endometrium, and it usually includes fever and abdominal pain. *Mycoplasma hominis* may be isolated from the cervix of healthy women but is recovered from the genital tract of 50 to 70% of women with pelvic inflammatory disease (Eisenbach et al., 1975; Mardh and Westrom, 1970; Moller et al., 1981; Moller, 1983).

Postabortal fever due to *M. hominis* has been reported in 10% of women with this disease. Such patients characteristically experience a low-grade fever for a few days after delivery (Taylor-Robinson and McCormack, 1980).

Postpartum fever is similar to postabortal fever clinically. *Mycoplasma hominis* has been isolated from the blood of such patients accompanied by an antibody response (Platt et al., 1980; Wallace et al., 1978).

Communicability and Transmissibility

Respiratory Infections

Infections generally appear first in children of school age who subsequently may be a source of infections in families (Foy et al., 1966). As close contact is required for infection, spread is slow (Foy and Alexander, 1969, Foy et al., 1971; Steinberg et al., 1969). Spread in families is particularly extensive, especially in children. Similar spreading may occur in populations involving close contact, such as the military. The period of communicability is usually about a week to 10 days.

Urogenital Infections

Mycoplasma infections of the genital tract occur as a result of sexual contact and may be found in both females and males (McCormack et al., 1972, 1973). Infants may be infected during the birth process, and mycoplasmal agents have been isolated from both infant males and females (Klein et al., 1969). Such infections do not generally persist.

Pathology

Respiratory Infections

Infection with mycoplasma is acquired by contact with respiratory secretions from an infected person. The organism has a predilection for the lower respiratory tract where it causes injury of the mucosal cells and ciliostasis (Denny, 1981). The presence of circulating antigen-antibody complexes during acute infection suggests that an immunologic basis may be a significant factor in severe infections (Biberfeld and Norbeg, 1974). Patchy areas of lung infiltration, bronchitis, bronchiolitis, and alveolar pneumonitis have been observed in fatal cases. Leukocytosis may occur after the first week.

Genital Infections

Mycoplasmal agents multiply on the surface of mammalian cells. Although *U. urealyticum* and *M. hominis* are implicated in nongonoccal urethritis and other genital infections, the lack of specific pathognomonic characters for these two agents make it difficult to distinguish colonization from infection. The best evidence exists for urethritis caused by *U. urealyticum,* whereas the role of *M. hominis* still requires considerable clarification (Taylor-Robinson and McCormack, 1980).

Diagnostic Features

Because of the similarity of clinical manifestations to those caused by bacteria and viruses, diagnosis may be difficult. Isolation of *M. pneumoniae* is relatively slow because of the periods of incubation required. Serologic testing is generally the procedure of choice, especially the complement-fixation test. A four-fold or greater rise in antibody during the infection is diagnostic. As complement fixing antibody is present for a short period, a high single titer, usually 1 : 128 or greater, is generally indicative of a recent or current infection. Most patients develop high titers of cold agglutinins. These agglutinins are not specific and have been observed in other infections.

Genital mycoplasma species are similar to *M. pneumoniae* in that they are implicated in infections for which there are a number of etiologic agents. Serologic testing has not been helpful, and isolation in culture is the laboratory procedure used most often.

Prognosis

Convalesence to infection with *M. pneumoniae* may be prolonged, but recovery is generally the rule. Reinfection may occur. Recovery is hastened with antibiotics such as the tetracyclines or erythromycin. Resolution of infection with genital mycoplasma species is also hastened with antibiotics, usually the tetracyclines.

Epidemiologic Aspects

Occurrence and Transmission

Mycoplasmal pneumonia occurs worldwide as a sporadic, endemic, or occasionally epidemic infection of humans mostly in temperate climates. Infection may be present continuously throughout the year, particularly in large populations. Infections during epidemic periods may be three to five times that observed during other periods and may last for as long as 2 years. These outbreaks characteristically have a slow onset, and infections are found mostly in older children and young adults (Foy et al., 1970, 1971, 1979). The contribution of *M. pneumoniae* to the total number of cases of pneumonia may vary considerably. Most cases have been studied at the community level (Foy et al., 1979) or in defined populations such as the military (Mogabgab, 1968). *Mycoplasma hominis* infrequently inhabits the respiratory tract and evidence for association with disease is weak. Transmission for respiratory infections is by infected droplets from an infected patient. The agent attaches to the respiratory epithelium and interferes with the action of the cilia. The organism may be shed in the respiratory secretions for a week before symptoms appear and may continue for weeks thereafter.

Genital infections with *U. urealyticum* and *M. hominis* are a result of sexual contact. Colonization in human populations is rarely observed before puberty and increases as the number of sexual partners increases. Infants may be infected during birth, but such infections do not generally persist. In those cases where persistence has been observed, infections occurred in females (Hammerschlag et al., 1978).

Susceptibility, Resistance, and Carrier States

Risk of respiratory infection due to *M. pneumoniae* is highest in those 5 to 20 years of age. However, the most severe infections are found in those in the middle-age and elderly groups, and this occurrence has led to the hypothesis that severe infection is the result of an immunologic response to reinfection. Reinfection including recurrence of pneumonia occurs (Foy et al., 1977). No carrier state has been described, but the agent may be shed for as long as 14 weeks (Grayston et al., 1967).

Genital infections occur after puberty and increase in relation to the number of different sexual partners. No carrier state has been described, but mycoplasmal species have been isolated from asymptomatic healthy males and females.

Etiologic Agents

Characterization

Mycoplasma and *Ureaplasma* species comprise the two genera of the family *Mycoplasmataceae*. They are the smallest free-living organisms, having diameters of 0.2 to 0.3 μm. Unlike the true bacteria, they do not possess a cell wall. They are extremely pleomorphic, are enclosed in a single membrane, divide by binary fission, and are gram-negative. Most are facultative anaerobes and have been cultured on artificial media. The genus *Ureaplasma* has only one species, and differs from the genus *Mycoplasma* in that it metabolizes urea, forms small, rough colonies (10 to 50 μm) on solid media, is inhibited by thallium acetate, and prefers an acid pH. A number of *Mycoplasma* species have been isolated from humans. Many are isolated only occasionally and only a relatively few are associated with disease. The more commonly encountered species are listed in Table 1.

Specimen Collection and Transport

Because mycoplasma infection may extend for weeks, the time of specimen collection is not critical. *Mycoplasma* species have been isolated from a number of different locations in humans, but most specimens for study are for diagnosis of respiratory or urogenital infections. Because *Mycoplasma* have fastidious growth requirements, specimens should be inoculated onto growth media as quickly as possible. A number of different transport media have been used, such as trypticase soy or tryptose phosphate broth supplemented with 0.5% bovine albumin. Mycoplasma growth medium also may be used. Penicillins, polymyxins, or amphotericin B may be added to transport or selective media to retard bacterial growth but have no effect on *Mycoplasma* species. Specimens should be kept at 4°C or on wet ice until inoculation. If inoculation must be delayed past 48 hours, specimens should be stored at −70°C.

Isolation and Identification Procedure

Mycoplasma pneumoniae is the organism of prime concern for respiratory infections caused by this group of agents. Media for isolation generally require supplements of horse serum and yeast extract, which must be tested for inhibitors and growth-promoting properties before use. The most frequently used medium is Edward-Hayflick agar (Hayflick, 1965) or more recently SP-4 medium (Tully et al., 1979).

TABLE 1. Properties of selected human *Mycoplasma* species

Organism	*Frequency of isolation*	*Arginine hydrolysis*	*Glucose catabolism*	*Urease*	*Common site of colonization*	*Associated disease*
M. buccale	Rare	+	−	−	Oropharynx	None
M. faucium	Rare	+	−	−	Oropharynx	None
M. fermentans	Rare	+	+	−	Urogenital tract joint fluid	Questionable
M. genitalium	Rare	−	+	−	Urogenital tract	None
M. hominis	Common	+	−	−	Urogenital tract	Postpartum fever, pelvic inflammatory disease
M. lipophilium	Rare	+	−	−	Oropharynx	None
M. orale	Common	+	−	−	Oropharynx	None
M. pneumoniae	Rare (except during outbreaks)	−	+	−	Upper and lower respiratory tract	Primary atypical pneumonia, bronchitis, tracheitis
M. salivarium	Common	+	−	−	Oral cavity	None
U. urealyticum	Common	−	−	+	Urogenital tract	Nongonoccal urethritis

Various techniques and media exist for the isolation of human mycoplasmal species. This is also true for the three species most commonly considered to be responsible for human diseases. Comparative studies of these procedures have not been thoroughly investigated. Techniques for isolation may use solid and/or liquid medium, or more often a combination of both (Valleca et al., 1975). Broth media offer greater sensitivity than those containing agar, but it is necessary to subculture to a solid medium to obtain a pure culture and test for species identification. It is the usual laboratory practice to inoculate both liquid and solid media with the clinical specimen simultaneously. Because the mycoplasmas generally grow more slowly than bacteria and fungi, selective media are prepared to inhibit overgrowth.

Procedures for isolation generally involve the inoculation of 0.1 to 0.2 ml of specimen in transport medium or diluted specimen into the liquid phase of diaphasic medium-containing phenol red. Turbidity or a color change within the first 5 days suggests contamination when testing for *M. pneumoniae* and within the first 2 days for genital mycoplasma. Subcultures to a solid medium should be made when a definite color change is observed or blindly at weekly intervals. Cultures are considered negative for *M. pneumoniae* after 4 weeks and after 2 weeks for other *Mycoplasma* species. Solid media are examined microscopically for the presence of colonies typical of *Mycoplasma* species. Colony morphology varies according to species, number of subcultures, and conditions of crowding on agar medium. Colonies vary in appearance to those with dense centers, where the agar has been penetrated, surrounded a less dense area where growth is on the surface of the medium ("fried-egg" appearance) to circular colonies with a homogeneous appearance to colonies with irregular borders. Diameters vary from 50 to 500 μm. *Ureaplasma* colonies are generally considerably smaller, measuring from 10 to 50 μm and thus require purified agar to avoid confusion with particulate impurities.

Methylene-blue glucose diphasic medium is selective for *M. pneumoniae*. Methylene blue inhibits other *Mycoplasma* species and some bacteria under aerobic conditions. Change in the indicator before a week of incubation is generally due to contamination.

Inoculation of glucose and arginine diphasic media will provide presumptive evidence for glucose or arginine metabolizers. *M. pneumoniae, M. fermentans,* and *M. hominis* are detected rapidly using these media, but other species may require three or more passages before a color change is apparent.

Arginine and urea media, broth and solid, are used for isolation of genital *Mycoplasma* species. Use of a candle jar or anaerobic conditions will favor growth of *Ureaplasma*. *Mycoplasma hominis* colonies will be detectable within 5 days. Observations and urease staining for *Ureaplasma* colonies should be done at 48 hours.

Identification of Isolates

Mycoplasma pneumoniae has biologic properties significantly different from other respiratory *Mycoplasma* species to provide a basis for species identification. Flooding the agar plate with a 0.2% suspension of chick or guinea pig erythrocytes and incubation at 37°C followed by gentle washing will demonstrate hemadsorption. Colonies are observed microscopically for adsorbed erythrocytes. This test should not be performed on cultures older than 1 week.

The production of hydrogen peroxide also may be used to differentiate *M. pneumoniae* from other *Mycoplasma* species. This test is performed by incubation of cultures after overlaying with blood agar and observing for hemolysis (Clyde, 1963). Serologic procedures also have been used for identification such as immunofluorescence (Del Giudice et al., 1967) or growth inhibition (Clyde, 1964). Serologic reagents for these tests are not readily available.

For the genital *Mycoplasma* species, organisms that are readily subcultured and test positive for the hydrolysis of urea are identified as *U. urealyticum*. Colonial morphology is also of value for recognition of *Ureaplasma* species.

Large colony isolates on arginine media obtained from urethral or cervical specimens are generally *M. hominis*. Although growth inhibition tests for identification are possible serologically, reagents are not readily available.

Identification of Other *Mycoplasma* Species

Whereas all *Mycoplasma* species may be identified serologically, the lack of a supply of standardized serologic reagents makes this type of testing impractical for routine studies. Isolation and culture on arginine, glucose, and urea media allows for the division into three groups. Further subdivision is often possible from a knowledge of source of specimens and biologic properties.

Immunologic Aspects

Serologic testing for the diagnosis of infections due to *M. pneumoniae* constitutes the major laboratory activity for this organism. This is due in part to the slow growth of the organism on laboratory media and the lack of a capability for culture in many clinical laboratories. The complement-fixation procedure is most generally used for serodiagnosis. Antibodies develop in the second or third week of illness and may persist for up to 12 months (Fernald et al., 1967). A fourfold or greater increase in antibody titer between acute and convalescent serum specimens is diagnostic. Single-serum titers of 1 : 128 or greater, or the demonstration of IgM, are suggestive of an acute stage of infection.

False-positive reactions have been observed in persons with bacterial meningitis probably due to the glycolipid component of the antigen (Kleemola and Kaghty, 1983). Cross-reactions also have been reported with antibody to *Streptococcus pneumoniae* (Allen and Prescott, 1978) and in Wasserman-positive sera (Plackett et al., 1969).

No diagnostic serologic procedures are available for measurement of antibodies to *U. urealyticum* or *M. hominis*. Recent studies indicate that these genital species are serologically heterogeneous and there exist seven *M. hominis* serotypes and at least 16 serotypes of *U. urealyticum* (Lin, 1985).

Cold hemagglutinins may be demonstrated in half of those infected with *M. pneumoniae*. Serum dilutions are mixed with human erythrocytes and incubated overnight at 4°C. Positive reactions are indicated if the hemagglutination disappears upon warming to 37°C. This phenomenon has been observed in other diseases, and is therefore not specific for *M. pneumoniae*.

Mycoplasma and L-Phase Bacteria

Some bacteria give rise to cell wall-defective variants. These variants may arise spontaneously or may be induced by impairing cell wall synthesis such as with penicillin or high salt concentration (Kleineberger, 1935, Pierce, 1942, Sharp, 1954). The cells of L-forms closely resemble mycoplasma, and they produce "fried-egg" colonies on soft, serum-containing agar. *Mycoplasma* species have not been shown to have any relationship to bacteria.

Treatment, Prevention, and Control

Treatment

Infections due to *M. pneumoniae* are effectively treated with tetracycline and its derivatives and erythromycin. Because of the time required for laboratory assistance in the establishment of a diagnosis, decisions concerning antimicrobial therapy, particularly for pneumonia, are usually made on clinical evaluation. This type of evaluation is of little value in cases of pharyngitis or tracheobronchitis because of the resemblance to virus-caused diseases.

The tetracyclines are also effective in the treatment of urogenital mycoplasmal infections. The existence of tetracycline-resistant *Ureaplasma* strains have been observed in some cases of nongonococcal urethritis. In these cases erythromycin is often substituted.

Prevention and Control

No satisfactory methods for the prevention of mycoplasmal diseases have been established. No vaccines for human use are currently available. Immunity from *M. pneumoniae* infection is not complete, and reinfections are common. Methods effective for treatment appear to have no effect on the rate of transmission.

Literature Cited

Allen, P. Z., and B. Prescott. 1978. Immunochemical studies on a *Mycoplasma pneumoniae* polysaccharide fraction: cross-reactions with type 23 and 32 antipneumococcal rabbit sera. Infect. Immun. **20:**421–429.

Biberfeld, G., and R. Norbeg. 1974. Circulating immune complexes in *Mycoplasma pneumoniae* infection. J. Immunol. **112:**413–415.

Bowie, W. R., S-P. Wang, E. R. Alexander, J. Floyd, P. S. Forsyth, H. M. Pollack, J. L. Lin, T. M. Buchanan, and K. K. Holmes. 1977. Etiology of nongonococcal urethritis: evidence for *Chlamydia trachomatis* and *Ureaplasma urealyticum*. J. Clin. Invest. **59:**735–742.

Brolin, I., and L. Wernstedt. 1978. Radiographic appearance of mycoplasmal pneumonia. Scand. J. Respir. Dis. **59:**179–189.

Clyde, W. A., Jr. 1964. Mycoplasma species identification based upon growth inhibition by specific antisera. J. Immunol. **92:**958–965.

Clyde, W. A., Jr. 1963. Hemolysis in identifying Eaton's pleuropneumonia-like organism. Science **139:**55.

Coufalik, E. D., D. Taylor-Robinson, G. W. Csonka. 1979. Treatment of nongonococcal urethritis with rifampicin as a means of defining the role of *Ureaplasma urealyticum*. Br. J. Vener. Dis. **55:**36–43.

Del Giudice, R. A., N. F. Robillard, and T. R. Carski. 1967. Immunofluorescence identification of Mycoplasma on agar by use of incident illumination. J. Bacteriol. **93:**1205–1209.

Denny, F. W. 1981. Atypical pneumonia and the armed forces epidemiological board. J. Infect. Dis. **143:**305–316.

Eisenbach, D. A., T. M. Buchanan, H. M. Pollock, P. S. Forsyth, E. R. Alexander, J. L. Lin, S-P. Wang, B. B. Wentworth, W. M. McCormack, and K. K. Holmes. 1975. Polymicrobial etiology of acute pelvic inflammatory disease. N. Engl. J. Med. **293:**166–171.

Fernald, G. W., W. A. Clyde, Jr., and F. W. Denny. 1967. Nature of the immune response to *Mycoplasma pneumoniae*. J. Immunol. **98:**1028–1038.

Finnegan, O. C., S. J. Fowles, and R. J. White. 1981. Radiographic appearances of mycoplasma pneumonia. Thorax **36:**469–472.

Foy, H. M., and E. R. Alexander. 1969. *Mycoplasma pneumoniae* infections in childhood. Adv. Pediatr. **16:**301–323.

Foy, H. M., G. E. Kenny, R. McMahan, A. M. Mansy, and J. T. Grayston. 1970. *Mycoplasma pneumoniae* pneumoniae in an urban area. Five years of surveillance. J. Am. Med. Assoc. **214:**1666–1672.

Foy, H. M., G. E. Kenny, R. McMahan, G. Kaiser, and G. Grayston. 1971. *Mycoplasma pneumoniae* in the community. Am. J. Epidemiol. **93:**55–67.

Foy, H. M., G. E. Kenny, M. K. Cooney, and I. D. Allan. 1979. Long-term epidemiology of infections with *Mycoplasma pneumoniae*. J. Infect. Dis. **139:**681–687.

Foy, H. M., G. E. Kenny, R. Sefi, H. D. Ochs, and I. D. Allan. 1977. Second attacks of pneumonia due to *Mycoplasma pneumoniae*. J. Infect. Dis. **135:**673–677.

Foy, H. M., J. T. Grayston, G. E. Kenny, E. R. Alexander, and R. McMahan. 1966. Epidemiology of *Mycoplasma pneumoniae* infecton in families. J. Am. Med. Assoc. **197:**859–866.

Grayston, J. T., G. E. Kenny, H. M. Foy. 1967. Epidemiological studies of *Mycoplasma pneumoniae* in civilians. Ann. NY Acad. Sci. **143:**436–446.

Hammerschlag, M. R., S. Alpert, I. Rosner, P. Thurston, D. Semine, D. McComb, and W. M. McCormack. 1978. Microbiology of the vagina in children: normal and potentially pathogenic organisms. Pediatrics **62:**57–62.

Hayflick, L. 1965. Tissue cultures and mycoplasmas. Tex. Rep. Biol. Med. **23:**285–303.

Kleemola, M., and H. Kayhty. 1983. Increase in titers of antibodies to *Mycoplasma pneumoniae* in patients with purulent meningitis. J. Infect. Dis. **146:**284–288.

Klein, J. O., D. Buckland, and M. Finland. 1969. Colonization of newborn infants by mycoplasmas. N. Engl. J. Med. **280:**1025–1030.

Kleineberger, E. 1935. The natural occurrence of pleuropneumonia-like organisms in apparent symbiosis with *Streptobacillus moniliformis* and other bacteria. J. Pathol. Bacteriol. **40:**93–105.

Lin, J. L. 1985. Human mycoplasmal infections: serologic observations. Rev. Infect. Dis. **7:**216–231.

Mardh, P-A., and L. Westrom. 1970. Antibodies to *Mycoplasma hominis* in patients with genital infections and in healthy controls. Br. J. Vener. Dis. **46:**390–397.

McCormack, W. M., P. C. Almeida, P. E. Bailey, E. M. Grady, and Y-H. Lee. 1972. Sexual activity and vaginal colonization with genital mycoplasmas. J. Am. Med. Assoc. **221:**1375–1377.

McCormack, W. M., P. Braun, Y-H. Lee, J. O. Klein, and E. H. Kass. 1973. The genital mycoplasmas. N. Engl. J. Med. **288:**78–89.

McCormack, W. M., Y-H. Lee, and S. H. Zinner. 1973. Sexual experience and urethral colonization with genital mycoplasmas. A study in normal men. Ann. Intern. Med. **78:**696–698.

Mogabgab, W. J. 1968. *Mycoplasma pneumoniae* and adenovirus respiratory illnesses in military and university personnel. 1959–1966. Am. Rev. Resp. Dis. **97:**345–358.

Mok, J. Y. Q., J. M. Inglis, and H. Simpson. 1979. *Mycoplasma pneumoniae* infection. Acta Paediatr. Scand. **68:**833–839.

Moller, B. R. 1983. The role of mycoplasmas in the upper genital tract of women. Sex. Transm. Dis. **10:**281–284.

Moller, B. R., P-A. Mardh, S. Ahrons, and E. Nussler. 1981. *Chlamydia trachomatis, Mycoplasma hominis,* and *Neisseria gonorrhoeae* infections in patients with signs of pelvic inflammatory disease. Sex. Transm. Dis. **8:**198–203.

Mufson, M. A. 1983. *Mycoplasma hominis:* a review of its role as a respiratory tract pathogen of humans. Sex. Transm. Dis. **10:**335–340.

Oriel, J. D. 1983. Role of genital mycoplasmas in nongonococcal urethritis and prostatitis. Sex. Transm. Dis. **10:**263–270.

Pierce, C. H. 1942. *Streptobacillus moniliformis,* its associated L_1 form, and other pleuropneumonia organisms. J. Bacteriol. **43:**780.

Plackett, P., B. P. Marmion, E. J. Shaw, and R. M. Lemcke. 1969. Immunochemical analysis of *Mycoplasma pneumoniae*. 3. Separation and chemical identification of serologically active lipids. Aust. J. Exp. Med. Sci. **47:**171–195.

Platt, R., J. L. Lin, J. W. Warren, B. Rosner, K. C. Edelin, and W. M. McCormack. 1980. Infection with *Mycoplasma hominis* in postpartum fever. Lancet **2:**1217–1221.

Ponka, A. 1979. The occurrence and clinical picture of serologically verified *Mycoplasma pneumoniae* infections with emphasis on central nervous system, cardiac, and joint manifestations. Ann. Clin. Res. **24:**1–60.

Sharp, J. T. 1954. L-colonies from hemolytic-steptococci: new technique in the study of L-forms of bacteria. Proc. Soc. Exp. Biol. Med. **87:**94–97.

Skoldenberg, B. 1972. On the role of viruses in acute infectious diseases of the central nervous system. Clinical and laboratory studies on hospitalized patients. Scand. J. Infect. Dis. **3(suppl):**1–96.

Steinberg, P., R. J. White, S. L. Fuld, R. R. Gutekunst, R. M. Chanock, and L. B. Senterfit. 1969. Ecology of *Mycoplasma pneumoniae* infections in marine recruits at Parris Island, South Carolina. Am. J. Epidemiol. **82:**62–73.

Stimson, J. B., J. Hale, W. R. Bowie, and K. K. Holmes. 1981. Tetracycline-resistant *Ureaplasma urealyticum:* a cause of persistent nongonococcal urethritis. Ann. Intern. Med. **94:**192–194.

Taylor-Robinson, D., and W. M. McCormack. 1980. Medical progress: the genital mycoplasmas. N. Engl. J. Med. **302:**1003–1010, 1063–1067.

Taylor-Robinson, D., G. W. Csonka, and M. J. Prentice. 1977. Human intraurethral inoculation of ureaplasmas. Quart. J. Med. **46:**309–326.

Thomsen, A. C. 1982. Mycoplasma infections in the upper urinary tract. Dan. Med. Bull. **29:**309–322.

Tully, J. G., D. L. Rose, R. F. Witcomb and R. P. Wenzel. 1979. Enhanced isolation of *Mycoplasma pneumoniae* from throat washings with a newly modified culture medium. J. Infect. Dis. **139:**478–482.

Valleca, W. M., B. R. Bird, and F. T. Forrester. 1975. Laboratory diagnosis of mycoplasma infections. *In* Course 8226-C, U.S. Dept. of Health, Education, and Welfare, Public Health Service, Centers for Disease Control, Atlanta, GA.

Wallace, R. J., Jr., S. Alpert, K. Browne, J. L. Lin, and W. M. McCormack. 1978. Isolation of *Mycoplasma hominis* from blood cultures in patients with postpartum fever. Obstet. Gynecol. **51:**181–185.

Zlatnik, F. J., L. F. Burmeister, and N. S. Swack. 1986. Chorionic mycoplasmas and prematurity. J. Reprod. Med. **31:**1106–1108.

CHAPTER 42

Pertussis

M. J. R. GILCHRIST and C. C. LINNEMANN, JR.

Disease: Pertussis (whooping cough).
Etiologic Agent: *Bordetella pertussis.*
Source: Man, usually with symptomatic infection; droplet transmission.
Clinical Manifestations: Catarrhal phase with nonspecific respiratory symptoms followed by coughing phase. Coughing may be paroxysmal and punctuated with an inspiratory whoop. Vomiting or apneic episodes are not uncommon in infants. Lymphocytosis is the usual case. Mild coughing illness is often manifest where infection is modified by prior vaccination or disease.
Pathophysiology: Organism resides on the ciliated cells of the respiratory epithelium and mediates disease by the production of toxin(s).
Laboratory Diagnosis: Culture of the etiologic agent from respiratory specimens. The direct fluorescent antibody procedure should be used only as an adjunct to culture, as its sensitivity and specificity are extremely dependent on the experience and skill of the staff. Occasionally, serology may be the only diagnostic tool available whenever specific antimicrobial therapy already has been applied or wherever adequate culture media cannot be obtained.
Epidemiology: Disease is worldwide in distribution. Prevalence of symptomatic disease is greatest where vaccination is less commonly practiced. In vaccinated populations, the reservoir of infection may be the vaccinated individual who is asymptomatic or mildly ill.
Treatment: Erythromycin is the drug of choice to reduce the duration of shedding, but will not affect duration or severity of symptoms unless therapy is initiated very early in the course of the disease.
Prevention and Control: Vaccine available to protect susceptible children under 7 years of age. Erythromycin for prophylaxis of contacts.

Description of Disease

The term "pertussis" refers to a clinical syndrome that begins with a catarrhal phase of 1 to 2 weeks duration, proceeds to a paroxysmal coughing phase of several weeks duration, and terminates with a convalescent phase. This term was substituted for the common terminology, whooping cough, when it was widely recognized that a whoop is not present in all cases of pertussis. With the more recent recognition that not all infections are associated with intensive cough, it is apparent that the term "pertussis" does not adequately describe the spectrum of clinical illness caused by the bacterium, *Bordetella pertussis*. After a number of etiologic agents (adenovirus, chlamydia, respiratory syncytial virus, and the like) were isolated from patients with the characteristic constellation of symptoms of pertussis, the term "pertussis syndrome" was coined to suggest the possibility of other etiologies. More recent studies indicate that viruses do not cause pertussis (Keller et al., 1980). The scope of this chapter encompasses the spectrum of human diseases due to the *Bordetellae,* to the exclusion of other agents that might cause the pertussis syndrome.

Clinical Manifestations

After an incubation period of 7 to 10 days, the illness begins with a catarrhal phase that is indistinguishable from a viral respiratory infection. The ensuing staccato cough that occurs in paroxysms, often punctuated with an inspiratory whoop, is pathognomonic for clinical pertussis. Lymphocytosis is relatively common and often quite pronounced. Young infants may experience atypical pertussis; of a population of documented cases of pertussis, more than half were given other admission diagnoses: acute bronchitis, bronchiolitis, pneumonia, respiratory distress, fever, or Reye's syndrome/vomiting (Sotomayor et al., 1985). Retrospective chart review revealed that, in contrast to patients with bronchiolitis, acute bronchitis, or pneumonia, children with pertussis were unlikely to have a temperature higher than 38.4°C or to wheeze. Nevertheless, pertussis was not entertained as a diagnosis until the patient developed posttussive vomiting and paroxysmal coughing.

During an investigation of a case of pertussis in a family, it is not unusual to discover that one or more adult family members has respiratory symptoms or a prolonged cough, perhaps dismissed as "smoker's cough," that can be culturally documented as pertussis. Individuals who received a full course of vaccine or who experienced pertussis as a child, may suffer from pertussis as adolescents or adults, presumably due to waning immunity (Trollfors and Rabo, 1981). In many cases these bouts of pertussis are less severe, perhaps modified by the immunologic history of the patient. However, in some cases in adults, symptoms are similar to those in children (Linnemann and Nasenbeny, 1977), with the exception that a whoop is rare and some complain of persistent sore throat (Maclean, 1982).

Pathology

The organism colonizes the ciliated cells in the respiratory tract but does not invade the host (Pittman, 1979). Where bacilli adhere to cilia, there is impaired ciliary activity and eventual extrusion of ciliated cells from the mucosal surface. The action of bacterial toxins is thought to elicit the disease manifestations. Several biologically active components of *B. pertussis,* including pertussis toxin, adenylate cyclase, and hemolysin, appear to contribute to the pathogenicity of the organism (Weiss et al., 1984). Coughing persists for several weeks after viable bacteria are cleared; the mechanism of prolonged toxic effects remains to be elucidated.

Diagnostic Features

When the characteristic cough and subsequent whoop are present, the diagnosis is easy. However, early in the illness or in young infants or adults who may have atypical illness, the diagnosis requires a high index of suspicion. For instance, the infant may present with infrequent paroxysms of coughing, exaggerated at night but not associated with a whoop. Vomiting, cyanosis, or apneic episodes may be the reason that the parents bring the child to the physician. The physician should suspect *Bordetella* infections in any patient with respiratory symptoms and a history of contact with another patient with pertussis or any patient with persistent coughing with or without whooping.

Prognosis

In the prevaccine period, the mortality rate was dependent upon the age (10% for those 2 to 3 months old and decreasing to 0.1% for those at 5 years of age; Mims, 1977). Surveillance conducted in the United States during 1979 to 1981 revealed an overall mortality rate of 0.5%, with mortality limited to the unvaccinated patient under 1 year of age (CDC, 1982). To improve their prognosis, young children with pertussis should be hospitalized and monitored very closely. Complications include apnea, pneumonia, seizures, atelectasis, hypoxia, intracerebral bleeding, hypoglycemia, and persistent neurologic abnormalities (Geller, 1984). In adults with pertussis, 65% missed ≥2 weeks of work and 2% were hospitalized (Trollfors and Rabo, 1981).

Epidemiology

The epidemiology of pertussis has changed since the development of pertussis vaccine. Widespread pertussis vaccination, which began in the United States in the 1950s and in many other parts of the world in the 1950s and 1960s, resulted in a dramatic decrease in reported cases of pertussis. In developed countries, it is unlikely for most children to experience pertussis and many young physicians have not seen a case of pertussis during their training. To have experienced clinical pertussis as a child is common only in older adults. With the low prevalence of pertussis in young children, neither young nor older adult is likely to have their immunity boosted by frequent reexposure to the agent and waning immunity may be responsible for the outbreaks that have occurred in adolescent and young adult populations. This population, many of whom experience mild symptoms, may serve as a reservoir for *B. pertussis* (Linnemann et al., 1974).

Prevalence

As a result of widespread administration of pertussis vaccine in the United States in the 1950s, the reported cases per 100,000 population were reduced from 20 in the late 1950s to less than 1 in the mid-1980s (CDC, 1984).

Communicability

B. pertussis is transmitted from human to human by droplet transmission and is a highly contagious disease with attack rates within the home ranging from 19% for vaccinated contacts to 72% for unvaccinated contacts (Grob et al., 1981a). In a study in which serologic follow-up of families with pertussis was conducted, 83% of family members seroconverted but only 43% of the secondary cases were symptomatic (Mertsola et al., 1983a).

Carrier State

Chronic carriers of *B. pertussis* do not exist (Linnemann et al., 1968), but recent studies suggest that transient asymptomatic infections occur in some patients (Broome et al., 1981; Mertsola et al., 1983a). These patients are less likely to be a reservoir of infection than the mildly symptomatic individual with coughing.

Etiologic Agent

Bordetella pertussis is the etiologic agent. Named after Bordet, who with Gengou first isolated the organism in 1906, this agent causes intensive coughing and thus carries the species epithet, *"pertussis." B. parapertussis,* markedly different in its growth characteristics, has been implicated as a cause of the pertussis syndrome. Apparently nontoxigenic (Pittman, 1979), *B. parapertussis* may be a phenotypic or genotypic (Granstrom and Askelof, 1982) variant of *B. pertussis*. If some trigger (e.g., loss of a lysogenic bacteriophage; Kumazawa and Yoshikawa, 1978) causes *B. pertussis* to revert to *B. parapertussis*, this would explain the reports of dual or sequential infections due to *B. pertussis* and *B. parapertussis* and the occurrence of lymphocytosis in patients infected with *B. parapertussis* (Linnemann and Perry, 1977), an organism that produces no lymphocytosis-promoting factor. *B. bronchiseptica,* a respiratory pathogen of animals, may colonize the human nasopharynx and rarely cause symptomatic infections in humans (Goodnow, 1980).

Characterization

Tiny coccobacilli (0.2 to 0.5 × 0.5 to 2.0 μm), these organisms occur singly or in pairs, are gram negative, and often stain in a bipolar pattern (Pittman, 1984). A strict aerobe, *B. pertussis* grows best at 35 to 37°C. In the family *Bordetellae*, only *B. bronchiseptica* is motile.

Laboratory Diagnosis

Specific laboratory diagnosis is important because undiagnosed cases of mild pertussis may serve as the source of outbreaks in the community and in hospitals. Once a diagnosis has been made, the patient can be treated and made noninfectious and exposed contacts can be treated prophylactically.

Specimen Collection and Transport

Exquisitely sensitive to environmental insult, *B. pertussis* is killed by the fatty acids present in cotton swabs or adsorbed to glass test tubes. The organism must be collected on Dacron (Ross and Cumming, 1981) or calcium alginate swabs and transported in acid-cleaned glassware or in disposable plastic containers. Charcoal or other agents that adsorb fatty acids may facilitate the protection of *B. pertussis* from exposure to fatty acids during transport and cultivation.

Due to its presumed susceptibility to drying, *B. pertussis* was originally cultured by directing the patient to cough on the surface of a culture plate (cough plate) to ensure that secretions did not dry out. However, immediate inoculation of culture plates or immersion of swabs in transport media obviates the need for cough plates. A swab of the posterior nasopharynx transported in 1% casamino acid solution or a nasopharyngeal aspirate (Wort, 1983) transported in a closed container provides satisfactory preservation of the viability of the organism for periods not exceeding 2 h. For longer periods the specimen should be transported on the charcoal transport medium of Jones and Kendrick (1969) or Regan and Lowe (1977).

Isolation

To isolate *B. pertussis* in culture it is necessary to use specially formulated media. Bordet-Gengou medium, containing potatoes, glycerol, and 20 to 50% sheep or horse blood, is successful in growing clinical isolates if freshly prepared just before inoculation or if wrapped in Parafilm and stored in the cold room for a maximum of 7 days. More recent formulations for the primary isolation of *B. pertussis* consist of media that contain charcoal (Jones and Kendrick, 1969; Regan and Lowe, 1977) or cyclodextrin (Aoyama et al., 1986).

During passage on laboratory media, *B. pertussis* undergoes a series of transformations in which it loses toxigenic properties and gains the ability to grow on media that are not permissive for the growth of fresh clinical isolates. Thus, some commercially prepared media, though capable of growing laboratory control strains, will not support the growth of fresh clinical isolates of *B. pertussis*. To optimize for the recovery of *B. pertussis* it is necessary to prepare fresh media in the laboratory. Regan-Lowe medium has been used for routine isolation (Gilligan and Fisher, 1984). Use of two plating media in parallel, Bordet-Gengou medium and a charcoal medium, for primary inoculation has been found to optimize the recovery of fresh clinical isolates (Ahmad and Calder, 1984).

Formerly, penicillin was incorporated into isolation media to reduce the numbers of nasopharyngeal flora present on the culture medium. Now it is known that some staphylococci produce diffusable substances that are inhibitory to the growth of *B. pertussis* and act at a distance of several centimeters from the staphylococcal colonies (Fig. 1). Thus, it is useful to include at least one medium with an antistaphylococcal antibiotic; methicillin at 2.5 μg/ml (Broome et al., 1979) or cephalexin at 40 μg/ml (Sutcliffe and Abbott, 1972) are the two antibiotics most commonly used.

Once inoculated, culture plates should be incubated at 35°C for 5 to 7 days in a moist chamber. Growth of *Bordetella pertussis* appears in 3 to 5 days, whereas the other *Bordetellae* often appear within 1 to 3 days. On Bordet-Gengou medium, *B. pertussis* colonies are small, white, and convex, resembling bisected pearls and have a very narrow hazy zone of hemolysis (Fig. 2). Unless there are numerous colonies closely spaced so that hemolysis appears as the aggregate of many colonies, the inexperienced eye may not readily observe hemolysis. This is particularly true because Bordet-Gengou medium contains relatively large amounts of blood and is generally poured rather thickly to enhance the recovery of organisms. On charcoal media, the organisms appear as small colonies resembling mercury droplets or bisected pearls (Fig. 3).

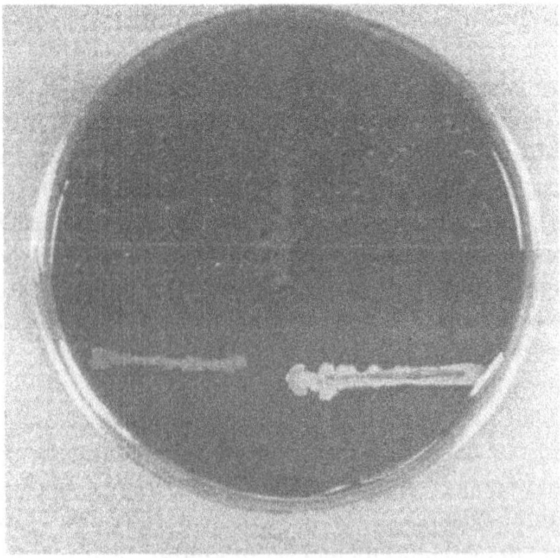

FIG. 1. Inhibition of *Bordetella pertussis* by *Staphylococcus aureus*. The photograph shows colonies of *B. pertussis* growing at a distance from two strains of *S. aureus*. The occurrence of *S. aureus* in a nasopharyngeal specimen is relatively common, and many strains of *S. aureus* are inhibitory to *B. pertussis* growth, with zones of inhibition ranging as high as several centimeters. Thus, it is useful to incorporate antistaphylococcal antibiotics (methicillin, cephalexin) into the primary isolation medium. A color plate of this figure appears in Color Plate 3 at the beginning of this volume.

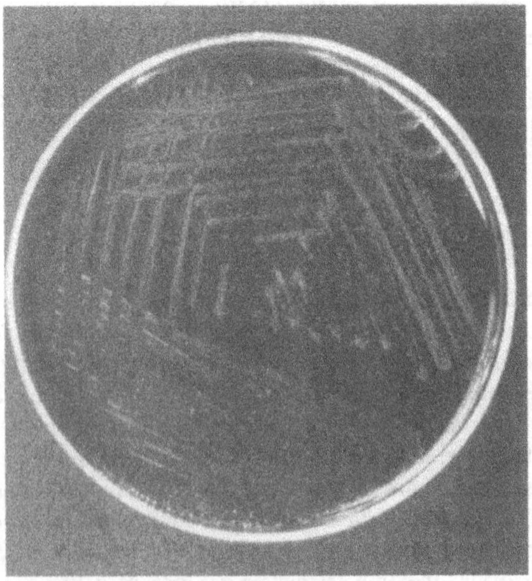

FIG. 2. Growth of *B. pertussis* on Bordet-Gengou medium. The hazy zone of hemolysis is difficult to visualize without magnification unless a large number of colonies occur relatively close to one another. A color plate of this figure appears in Color Plate 3 at the beginning of this volume.

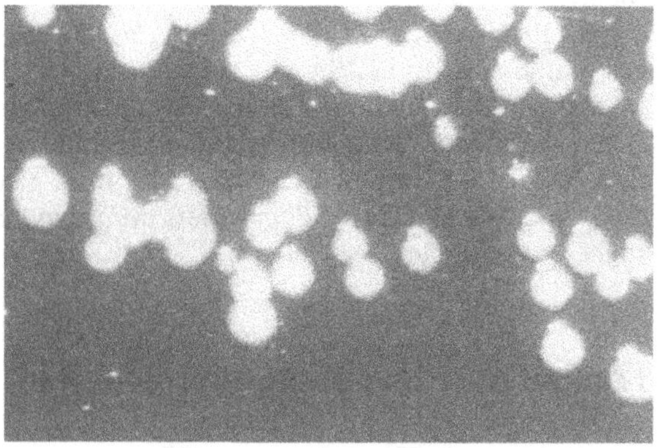

FIG. 3. Growth of *B. pertussis* on charcoal medium. On charcoal medium, colonies of *B. pertussis* are white, opaque, and convex. When observed under the stereomicroscope, they resemble a bisected pearl. A color plate of this figure appears in Color Plate 3 at the beginning of this volume.

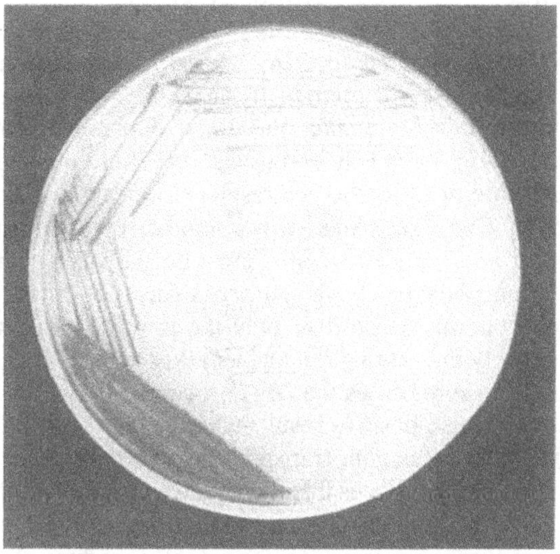

FIG. 4. Biochemical identification of *B. pertussis*. In addition to serologic means, the differentiation of *B. pertussis* from *B. parapertussis* may be accomplished by biochemical testing. The hydrolysis of urea and browning of nutrient agar are characteristics of *B. parapertussis*. The photograph is a strain of *B. parapertussis* growing on Mueller-Hinton agar. Browning on this medium is first associated with the colonies, but after prolonged incubation diffuses throughout the medium. A color plate of this figure appears in Color Plate 3 at the beginning of this volume.

Identification

PRESUMPTIVE IDENTIFICATION

Colonies characteristic of *Bordetella* should be Gram stained and evaluated for the presence of small, gram-negative rods. When present, the organisms can be presumptively identified by performance of the direct fluorescent antibody (DFA) procedure or by agglutination with specific antisera (Parker and Payne, 1985). Requiring a smaller volume of cell mass, the DFA procedure is preferable for rapid identification if only a small number of colonies are present. Careful refinement of the DFA is necessary as an excessive ratio of organisms/FA reagent will quench the reaction and the organisms will be falsely negative. Because haemophili will grow on some isolation media for *Bordetella* and some may cross-react with *B. pertussis* antisera, it is often prudent to rule out *Haemophilus* by demonstrating no X/V factor dependence.

CONFIRMATORY IDENTIFICATION

Distinguishing among the *Bordetellae* requires a knowledge of growth rate, urease activity, browning on peptone (nutrient) agar (Fig. 4), and growth on MacConkey agar. For the expected reactions of the respective species, refer to Table 1. Satisfactory confirmation of *Bordetella pertussis* and *B. parapertussis* may be accomplished in this fashion. For complete confirmation of *B. bronchiseptica,* it is necessary to rule out organisms of the genus *Alcaligenes,* by appropriate biochemical means.

TABLE 1. Differentiation among the *Bordetellae*

	Bordetella pertussis			
Characteristic	*Fresh isolate*	*Laboratory strain*	*Bordetella parapertussis*	*Bordetella bronchiseptica*
Time to appearance of visible colonies on Bordet-Gengou medium	3–5 days	1–3 days	1–2 days	1 day
Growth on blood agar medium	−	+	+	+
Growth on MacConkey agar	−	−	−	+
Hydrolysis of urea	−	−	+	+++
Browning on nutrient agar	−	−	+	−

Direct Fluorescent Antibody Tests

Although available to many laboratories that do not have facilities to prepare fresh media and perform cultures for *Bordetella,* the direct fluorescent antibody (DFA) test may be misleading. There may be a high rate of false-positive results (Linnemann et al., 1974). Cross-reactions with a number of organisms occur and the microscopist must be skilled in ruling out morphologically similar organisms that cross-react, counting as positive only the tiny rods with an intensely fluorescing rim and a darker center (doughnuts). In some cases the DFA procedure may *appear* to yield false-positive results when, in fact, it is poor specimen collection, transportation, or culture techniques that produces falsely negative culture results. Careful standardization and monitoring of the DFA procedure is necessary to consistently produce clinically applicable DFA results. Using a well standardized DFA procedure during a year in which there were 45 culture-positive specimens, the Children's Hospital Medical Center laboratory produced the following results: 31 of the 45 culture-positive specimens were DFA positive (69% sensitivity); and 31 of 32 DFA positive specimens were culture positive (specificity, 99.9%; positive predictive value, 97%). If a laboratory does not simultaneously culture its DFA specimens, it cannot ensure that the DFA procedure retains its optimal performance parameters. A detailed description of a DFA procedure is published (Parker and Payne, 1985).

Serology

Early serologic methods required that acute and convalescent sera be evaluated simultaneously (Macaulay, 1979). The recent development of ELISA technology for the detection of IgM, IgA, and IgG allows for serodiagnosis during the acute phase of the illness. As many patients with the pertussis syndrome have had recent immunizations for pertussis, the test also must distinguish seroconversion due to active disease from seroconversion due to immunization. A number of reports (Viljanen et al., 1982; Granstrom et al., 1982; Mertsola et al., 1983b; Hakansson et al., 1984; Nagel et al., 1984) suggest the value of the ELISA techniques in the diagnosis of *B. pertussis* infection. An ELISA kit for detection of *B. pertussis* antibodies is commercially available.

Other Techniques for Diagnosis

There are a variety of techniques which have been evaluated or suggested recently as possible alternatives in the diagnosis of pertussis. These include counterimmunoelectrophoresis (Boreland and Gillespie, 1984), detection of adenyl cyclase (Confer and Eaton, 1985), ELISA for detection of IgA in nasal secretions (Goodman et al., 1981), and refinements in antigen detection through the use of monoclonal antibodies (Frank and Parker, 1984). A gene probe that was specific for *B. pertussis/parapertussis* might be useful diagnostically.

Treatment, Prevention, and Control

Treatment

Therapy with erythromycin, chloramphenicol, ampicillin, or oxytetracycline has no effect on the course of pertussis infection when therapy is begun during the paroxysmal phase (Bass et al., 1969). In adults with pertussis, erythromycin given within 4 days of onset of the catarrhal phase modifies duration of symptoms and reduces the number who suffer from whooping attacks (Trollfors and Rabo, 1981), but early institution of therapy is seldom possible. Routine therapy with erythromycin, which eliminates shedding of the organism within 3 days (Bass et al., 1969) and thus renders the patient noninfectious, should be continued for 14 days because bacteriologic relapse has been documented after 10 days of therapy (Linnemann et al., 1974).

Prevention

Vaccine

Because widespread distribution of vaccine was instituted in the 1950s, the prevalence of pertussis has been dramatically reduced. Vaccine administration cannot be completed early enough to protect the very young from disease, but even a partial course of immunization will moderate the disease (Grob et al., 1981a). Infants and children who had received two to four immunizations had fewer complications such as abnormal chest roentgenograms, apnea, vomiting, and requirement for intubation or supplemental oxygen, than those who had received zero or one immunization (Sotomayer et al., 1985). Vaccine does not provide lifelong immunity and the vaccinated, mildly ill adult may serve as a major reservoir of infection of the infant (Nelson, 1978). Adverse reactions to the vaccine have been widely publicized and have resulted in reduced acceptance of the vaccine with concomitant increases in the prevalence of disease in some countries. Acellular vaccines without the adverse reactions of the existing vaccine are currently in trial.

Control

Erythromycin Prophylaxis

Studies of erythromycin prophylaxis for individuals exposed to pertussis have shown conflicting results (Bass, 1986). Some have suggested that prophylaxis is effective, and others, including a placebo-controlled study, have suggested that prophylaxis is ineffective. Despite the conflicting experiences, it is interesting that all studies using erythromycin estolate, which gives the highest serum levels, have suggested that erythromycin prophylaxis is useful (Bass, 1986).

Literature Cited

Ahmad, F., and M. A. Calder. 1984. Isolation of *Bordetella pertussis:* benefits of using both Bordet-Gengou and charcoal media. J. Clin. Pathol. **37:**1071–1077.

Aoyama, T., Y. Muraee, T. Iwata, A. Imaizumi, Y. Suzuki, and Y. Sato. 1986. Comparison of blood-free medium (cyclodextrin solid medium) with Bordet-Gengou medium for clinical isolation of *Bordetella pertussis.* J. Clin. Microbiol. **23:**1046–1048.

Bass, J. W. 1986. Erythromycin for treatment and prevention of pertussis. Pediatr. Infect. Dis. **5:**154–157.

Bass, J. W., E. L. Klenk, J. B. Kotheimer, C. C. Linnemann, and M. H. D. Smith. 1969. Antimicrobial treatment of pertussis. J. Pediatr. **75:**768–781.

Boreland, P. C., and S. H. Gillespie. 1984. Counterimmunoelectrophoresis in the diagnosis of whooping cough. J. Clin. Pathol. **37:**950–955.

Broome, C. V., S. R. Preblud, B. Bruner, J. E. McGowan, P. S. Hayes, P. P. Harris, W. Elsea, D. W. Fraser. 1981. Epidemiology of pertussis, Atlanta, 1977. J. Pediatr. **98:**362–367.

Broome, C. V., D. W. Fraser, J. W. English. 1979. Pertussis—diagnostic methods and surveillance, p. 19–22. *In* C. Manclark and J. Hill (eds.), International symposium on pertussis. DHEW Pub # (NIH) 79-1830, U.S. Government Printing Office, Washington, D. C.

Centers for Disease Control. 1982. Pertussis surveillance, 1979–1981. Morbid. Mortal. Weekly Rep. **31:**333–336.

Centers for Disease Control Annual Summary 1984. Reported morbidity and mortality in the U.S. Morbidity Mortality Weekly Rep., 1986; **32:**42–44.

Confer, D. L., and J. W. Eaton. 1985. *Bordetella* adenylate cyclase: host toxicity and diagnostic utility. Proc. 4th. Int. Symp. on Pertussis. Dev. Biol. Std. Skarger. Basel. **61:**3–10.

Frank, D. W., and C. D. Parker. 1984. Isolation and characterization of monoclonal antibodies to *Bordetella pertussis.* J. Biol. Stand. **12:**353–365.

Geller, R. J. 1984. The pertussis syndrome: a persistent problem. Pediatr. Infect. Dis. **3:**182–186.

Gilligan, P. H., and M. C. Fisher. 1984. Importance of culture in laboratory diagnosis of *Bordetella pertussis* infections. J. Clin. Microbiol. **20:**891–893.

Goodman, Y. E., A. J. Wort, F. L. Jackson. 1981. Enzyme-linked immunosorbent assay for detection of pertussis immunoglobulin A in nasopharyngeal secretions as an indicator of recent infection. J. Clin. Microbiol. **13:**286–292.

Goodnow, R. A. 1980. Biology of *Bordetella bronchiseptica.* Microbiol. Rev. **44:**722–738.

Granstrom, M., and P. Askelof. 1982. Parapertussis: an abortive pertussis infection? Lancet **ii:**1249–1250.

Granstrom, M., G. Granstrom, A. Lindfors, and P. Askelof. 1982. Serologic diagnosis of whooping cough by an enzyme-linked immunosorbent assay using fimbrial hemagglutinin as antigen. J. Infect. Dis. **146:**741–745.

Grob, P. R., M. J. Crowder, and J. F. Robbins. 1981a. Effect of vaccination on severity and dissemination of whooping cough. Br. Med. J. **282:**1925–1928.

Grob, P. R., M. Spencely, and H. P. Lambert. 1981b. Whooping cough in infants: antimicrobial prophylaxis. Lancet **i:**772–773.

Hakansson, S., C. G. Sundin, M. Granstrom, and B. Gastrin. 1984. Diagnosis of whooping cough—a comparison of culture, immunofluorescence and serology with ELISA. Scand. J. Infect. Dis. **16:**281–284.

Jones, G. L., and P. L. Kendrick. 1969. Study of a blood free medium for transport and growth of *Bordetella pertussis.* Health Lab. Sci. **6:**40–45.

Keller, M. A., R. Aftandelians, and J. D. Connor. 1980. Etiology of pertussis syndrome. Pediatr. **66:**50–55.

Kumazawa, H. H., and M. Yoshikawa. 1978. Conversion of *Bordetella pertussis* to *Bordetella parapertussis.* J. Hyg. **81:**15–23.

Linnemann, C. C., J. W. Bass, and M. H. D. Smith. 1968. The carrier state in pertussis. Am. J. Epidemiol. **89:**422–427.

Linnemann, C. C., and J. Nasenbeny. 1977. Pertussis in the adult. Ann. Rev. Med. **28:**179–185.

Linnemann, C. C., and E. B. Perry. 1977. *Bordetella parapertussis:* recent experience and a review of the literature. Am. J. Dis. Child. **131:**560–563.

Linnemann, C. C., J. C. Partin, P. H. Perlstein, and G. S. Englender. 1974. Pertussis: persistent problem. J. Pediatr. **85:**589–591.

Macauley, M. E. 1979. The serological diagnosis of whooping cough. J. Hyg. Camb. **83:**95–102.

Maclean, D. W. 1982. Adults with pertussis. J. R. Coll. Gen. Pract. **238:**298–300.

Mertsola, J., O. Ruuskanen, E. Eeroa, and M. K. Viljanen. 1983a. Intrafamilial spread of pertussis. J. Pediatr. **103:**359–363.

Mertsola, J., O. Ruuskanen, T. Kuronen, and M. K. Viljanen. 1983b. Serologic diagnosis of pertussis: comparison of enzyme-linked immunosorbent assay and bacterial agglutination. J. Infect. Dis. **147:**252–257.

Mims, C. A. 1977. The pathogenesis of infectious disease. Academic Press, New York.

Nagel, J., S. de Graaf, and D. Schijf-Evers. 1984. Improved serodiagnosis of whooping cough caused by *Bordetella pertussis* by determination of IgG anti-LPF antibody levels. Dev. Biol. Stand. **61:**325–330.

Nelson, J. D. 1978. The changing epidemiology of pertussis in young infants: the role of adults as reservoirs of infection. Am. J. Dis. Child. **32:**371–373.

Parker, C. D., and B. J. Payne. 1985. *Bordetella,* p. 394–399. *In* E. Lennette, A. Balows, W. Hausler, and H. Shadomy (eds.), Manual of clinical microbiology, 4th ed., American Society for Microbiology, Washington, D. C.

Pittman, M. 1984. Genus *Bordetella* Morena-Lopez 1952, 178, p. 388–393. J. G. Holt (ed.), Bergey's manual of systematic bacteriology, Vol. 1, Williams & Wilkins, Baltimore.

Pittman, M. 1979. Pertussis toxin: the cause of the harmful effects and prolonged immunity of whooping cough. A hypothesis. Rev. Infect. Dis. **1:**401–412.

Regan, J., and F. Lowe. 1977. Enrichment medium for the isolation of *Bordetella*. J. Clin. Microbiol. **6**:303–309.

Ross, P. W., and C. G. Cumming. 1981. Isolation of *Bordetella pertussis* from swabs. Br. Med. J. **283**:403–404.

Sotomayor, J., L. B. Weiner, and J. A. McMillan. 1985. Inaccurate diagnosis in infants with pertussis. Am. J. Dis. Child. **139**:724–727.

Sutcliffe, E. M., and J. D. Abbott. 1972. Selective medium for the isolation of *Bordetella pertussis* and *parapertussis*. Br. Med. J. **6**:732–733.

Trollfors, B., and E. Rabo. 1981. Whooping cough in adults. Br. Med. J. **283**:696–697.

Viljanen, M. K., O. Ruuskanen, C. Granberg, and T. T. Salmi. 1982. Serological diagnosis of pertussis: IgM, IgA and IgG antibodies against *Bordetella pertussis* measured by enzyme-linked immunosorbent assay (ELISA). Scand. J. Infect. Dis. **14**:117–122.

Weiss, A. A., E. L. Hewlett, G. A. Myers, and S. Falkow. 1984. Pertussis toxin and extracytoplasmic adenylate cyclase as virulence factors of *Bordetella pertussis*. J. Infect. Dis. **150**:219–222.

Wort, A. J. 1983. Bacteriological diagnosis of pertussis. Lancet **i**:766.

CHAPTER 43

Plague

THOMAS J. QUAN

Disease: Plague, the "Black Death," pest, or peste.
Etiologic Agent: *Yersinia pestis.*
Source: Bite(s) by infective fleas of rodents or other mammals, direct contact with infected mammals or their products, or inhalation of contaminated airborne droplets.
Clinical Manifestations: Acute onset of fever, myalgia, malaise, shaking chills, and lymphadenopathy marked by pain with or without swelling in groin, axilla, neck, or other lymph node sites.
Pathology: Reticuloendothelial disease with foci in regional lymph nodes, spleen, liver, lung, or bone marrow.
Laboratory Diagnosis: Bacteriologic isolation of the etiologic agent from blood, bubo aspirate, sputum, or tissue. Passive hemagglutination titer >16 or fourfold change in titer is considered diagnostic in persons with no known vaccination or prior plague infection.
Epidemiology: Parts of Africa, Asia Minor, Asia, North America, and South America.
Treatment: Streptomycin, tetracycline, chloramphenicol, sulfa (especially sulfadiazine).
Prevention and Control: Ectoparasite and rodent control; isolation of human patients; prophylaxis or abortive therapy of exposed persons; vaccination of persons at high risk.

Description of Disease

Transmission

Plague is an infectious disease of rodents that affects other animals, including humans, only incidentally. Humans acquire the disease usually by the bite of infective rodent fleas, by direct contact with infected animals or their products, or, less commonly, via inhalation of contaminated airborne droplets expelled by patients or animals with plague pneumonia. Other than pneumonic transmission, little or no person-to-person transfer of the agent has been documented. Human-to-flea-to-human transfer has been suggested in some Andean plague outbreaks in recent years (Rust, 1984). Infection after ingestion of undercooked or raw meat from infected animals has been documented (Christie et al., 1980; Fedorov, 1960).

Clinical Manifestations

Signs and symptoms of plague vary according to the clinical form of the disease.

Bubonic Plague

Typically, plague follows a bubonic course, which is manifested by acute onset 2 to 7 days after exposure and is marked by fever, headache, malaise, myalgia, shaking chills, and pain in the groin, axilla, or neck. Other symptoms may include nausea, vomiting, and diarrhea. At the time of onset, or 1 to 3 days later, a swelling (bubo, thus "bubonic plague") is noted in the affected groin, axilla, or neck; the swelling may become very large, is usually excruciatingly painful, and may interfere with leg, arm, or head movement, depending on location. The bubo is simply the enlarged, inflamed lymph node draining the area of the body that served as the point of entry of the organisms. This point of entry may or may not be marked by the development of an ulcerated lesion. Untreated, the disease may be rapidly progressive and fulminant. Bacteria are disseminated hematogenously to virtually all organ systems, including the lungs.

Septicemic Plague

Approximately 10% of the human plague cases in the United States do not develop external lymphadenopathy. Infections without buboes are termed primary septicemic plague and may be difficult to diagnose. This clinical form is a progressive gram-negative sepsis that, without rapid initiation of specific antibiotic therapy, is often fatal. Septicemic plague may occur in any patient, but most cases have involved either elderly (>55 years) or young (<10 years) patients.

Plague Pneumonia

During either bubonic or septicemic courses, as the bacteria progressively invade and destroy the integrity of the alveolar membranes of the lung, organisms mixed with blood, mucus, and sputum may be expelled from the patient via coughing or choking and transmitted to new hosts via airborne droplets, obviating the intermediary services of arthropod vectors. Pneumonia subsequent to bubonic or septicemic plague is termed secondary plague pneumonia. Plague acquired by inhalation of infective droplets is termed primary plague pneumonia. Progressive and rapidly fatal primary plague pneumonia is highly infectious.

Other Manifestations

Other clinical presentations of plague have been described. Meningitis occurs in approximately 5% of patients and seems to be associated with cervical or axillary lymphadenopathy (Poland and Barnes, 1979). Factors predisposing toward development of meningitic courses may include youth (more common in young people <20 years) and inadequate, inappropriate, or delayed antibiotic therapy. Endophthalmitis and gastroenteritis also have been reported as primary manifestations of illnesses later confirmed as plague (Centers for Disease Control (CDC), 1981, 1985 [in prep.]; Welty et al., 1985). Pestis minor or ambulatory plague are terms for those cases of plague that involve only mild manifestations of symptoms, and often need no chemotherapy for recovery.

Pathogenesis

After plague organisms enter the body via a flea bite, they are ingested by phagocytes. Those cells ingested by polymorphonuclear leukocytes are destroyed, but those ingested by monocytes survive and multiply, acquiring resistance to further phagocytic action. It is these phagocytosis-resistant plague organisms that are acquired through lesions or breaks in the skin by direct contact with infected animal tissues. The bacterial cells are carried by the monocytes from the point of entry, at which a lesion may be formed, to regional lymph nodes where they continue to multiply. Buboes may form in any lymph node, but usually occur in the groin (inguinal or femoral) after flea bites on the legs, or in the axilla after direct contact exposure. Less commonly affected are nodes in cervical, epitrochlear, popliteal, intrathoracic, or abdominal areas. In an untreated patient, the organisms spread from the bubo via blood and/or lymphatic circulatory routes throughout the body, causing secondary plague septicemia, secondary plague pneumonia, and/or plague meningitis.

Progression of bubonic plague to secondary pneumonia is a common, serious complication. Progressive lung tissue destruction, induced by plague pneumonitis, permits expulsion of bacteria-laden airborne droplets during coughing, sneezing, or labored respiration. Primary plague pneumonia, which follows the direct entry of plague organisms into the lungs, is a fulminant, rapidly fatal form of the disease that is fortunately rare in this country. Close-to-intimate contact between a patient with plague pneumonia and the new host is apparently necessary for transmission because no secondary cases of primary plague pneumonia have been associated with any of the 31 patients who developed plague pneumonia during 1975 to 1985 in the United States (Centers for Disease Control, 1975, 1976, 1977, 1978, 1980, 1981, 1982, 1983–1984, 1985 [in prep.]). Three of four human cases of primary plague pneumonia during this period were associated with confirmed secondary plague pneumonias in domestic pets (two cats, one dog). The source of infection for the fourth case was not determined, but the patient, who died, had been a veterinarian specializing in small animal medicine (Centers for Disease Control, 1975, 1977, 1980, 1982).

In addition to, or instead of, pulmonary complications, progression of bubonic or septicemic courses may involve other organs or systems, for example, the central nervous system producing plague meningitis; and intravascular coagulation disorders (DIC) and petechial hemorrhaging, due perhaps to the action of plague endotoxin(s). If these events involve the lungs, respiration is compromised and suffocation occurs (Poland and Barnes, 1979).

Diagnostic Features

Bubonic plague should be considered in the differential diagnoses of all patients with sudden onset of fever, headache, severe malaise, shaking chills, and lymphadenopathy or pain in the area of affected nodes which may or may not be swollen, *and* a compatible medical history. A compatible history would include residing in or having visited plague endemic areas, having evidence of insect bites, or having known or suspected contact with rodents or associated animals in endemic areas. In primary septicemia, characteristic buboes in superficial, palpable lymph nodes are lacking, preventing or delaying a correct diagnosis, especially in those patients whose history is incompletely known or who seek medical attention in nonendemic areas.

Primary plague pneumonia may be difficult to separate correctly from pneumonias of other etiologies unless plague is seriously suspected on the basis of known exposure to plague-infected patients or animals.

Once a diagnosis of plague is seriously considered, specimens should be taken and appropriate antibiotic therapy should be initiated without awaiting even preliminary laboratory results. Although person-to-person transmission of *Y. pestis* occurs only in pneumonic cases, all plague patients should be placed in strict isolation for at least 2 days after initiation of specific antibiotic therapy to minimize possibility of transmission should pneumonia occur before any antimicrobial effect and to minimize and control the numbers of potential contacts.

Prognosis

The case-fatality rate of human plague is variable. Factors that may increase the case-fatality rate include the clinical course, the time between onset of symptoms and seeking medical attention, the time required to achieve a correct diagnosis and initiation of specific therapy, and the age and general health of the patient. Primary plague pneumonia is invariably fatal unless specific treatment is instituted within about 18 h of onset. The case-fatality rate for primary septicemic plague hovers at about 50%, usually because of delay in achieving a correct diagnosis caused by the lack of specific signs and symptoms. In simple, uncomplicated bubonic plague, the described mortality of 50 to 60% in untreated cases should near zero with prompt diagnosis and treatment.

Sequelae are rare but may include cavitation of the lung, abscessation, suppuration, and spontaneous drainage of the bubo.

Epidemiologic Aspects

Occurrence

Plague currently (1976 to 1985) is enzootic in rodent populations, and human cases have been reported from areas of 19 countries in North America, South America, Africa, the Middle East, and Asia. In the United States, all naturally acquired human plague cases since 1924 have been directly associated with plague in wild rodent or associated mammals and their fleas. In other regions, plague in commensal rat populations is the predominant source of human infections. Strong, but circumstantial, evidence of human-to-flea-to-human transmission of *Y. pestis* was reported in plague outbreaks in Andean foci of Bolivia, Ecuador, and Peru (Rust, 1984).

Plague cycles in the various endemic foci vary substantially according to the mammalian hosts and the flea species involved, macro- and microclimatic factors such as temperature and humidity, and socioeconomic, cultural, or other life style factors of the human populations (diet, nature of housing, attitude toward rodent or ectoparasite infestations, and the like).

Selected epidemiologic features of human plague cases in the United States are presented in Table 1. The data presented indicate that persons of both sexes, all ages, and all races are susceptible to plague infection. There is an apparent seasonal distribution, but the infection may be acquired at any time; summer cases are usually those acquired by flea bite and winter cases are commonly associated with direct contact with infected animals. In the United States, acquisition of plague infection by human patients has occurred only in western states where suitable hosts and vectors occur and favorable climatic, habitat, and other environmental factors prevail. Laboratory workers in all areas, however, must maintain competence in isolating and identifying the agent, since travel by infected persons from exposure sites in endemic areas to nonendemic areas is not uncommon.

There is an increasing awareness of the role of domestic cats in the plague cycle in the United States (Kaufmann et al., 1981; Rollag et al., 1981). Since 1975 cats have been confirmed as the sources of two human cases of primary pneumonia, and in a third

TABLE 1. Selected epidemiologic features of human plague cases in the United States, 1956–1985

A. Case-fatality rate by age and sex

Age (yr)	No. died/total (case fatality rate, %) Male	Female	Total
Juvenile (<20)	15/78 (20.0)	10/61 (15.8)	25/139 (18.0)
Adult (>20–69)	19/80 (23.8)	3/50 (6.0)	22/130 (16.9)
Elderly (>70)	1/4 (25.0)	3/6 (50.0)	4/10 (40.0)
Total	35/163 (21.5)	16/117 (13.7)	51/279 (18.3)

B. Racial distribution

Race (% of total)	Total	No. died	Case-fatality rate (%)
Anglo (45.2%)	141	27	19.1
Hispanic (21.7%)	55	8	14.5
American Indian (32.2%)	81	15	18.5
Other[a] (0.9%)	2	1	50.0
Total	279	51	18.3

C. Case distribution by month of onset

Month	No. of cases	No. died	Case-fatality rate (%)
January	3	1	33.3
February	9	1	11.1
March	3	0	0.0
April	12	4	33.3
May	35	8	22.8
June	52	9	17.3
July	65	10	15.4
August	43	9	20.9
September	28	2	7.1
October	14	3	21.4
November	7	2	28.6
December	7	2	28.6
Unknown	1	0	0.0
Total	279	51	18.3

D. Geographic distribution

State	No. of cases (no. deaths)	Percentage of total cases	Percentage of total fatalities
Arizona	41 (5)	14.7	9.8
California	27 (7)	9.6	13.7
Colorado	23 (5)	8.2	9.8
Idaho	1 (1)	0.3	2.0
Nevada	3 (2)	0.1	3.9
New Mexico	163 (25)	58.4	49.0
Oregon	9 (3)	3.2	5.9
Texas	2 (1)	0.7	2.0
Utah	7 (2)	2.5	3.9
Washington	1 (0)	0.3	0.0
Wyoming	2 (0)	0.7	0.0
Total	279 (51)	100.0	100.0

[a] 1 Iranian survived; 1 Japanese died.

case a cat was the suspected source. Additionally, cats, through bites and scratches or by passive carriage of infected mammals or fleas, have served as sources in at least 13 cases. (Centers for Disease Control, 1977, 1978, 1980, 1981, 1982, 1984; Weniger et al., 1984).

Susceptibility and Resistance

A spectrum of susceptibility to plague infection, ranging from highly resistant to highly susceptible, exists not only among various rodent species, but also among the individuals within a species (Quan et al., 1985; Shepherd et al., 1986; Williams et al., 1979). Moderately resistant rodents (for example, *Microtus* sp., *Peromyscus* sp.) represent the natural plague reservoir, harboring the plague organism with little ill effect. During periodic low-level bacteremias, the organisms are acquired during feeding by fleas, which after an extrinsic incubation period may transmit the organisms to new hosts of the same or more susceptible species.

Resistant rodents permit long-term maintenance of plague organisms at low, often undetected levels in enzootic areas, but when introduced into populations of susceptible rodents, amplification of the disease through rapid transmission of bacteria among individual hosts results in large-scale, decimating epizootics.

In the flea gut, bacteria acquired during an infective blood meal multiply and grow into a mass which eventually blockades the stomach. The mass prevents the entry of fresh blood meals and causes the regurgitation of the blood, which now contains plague organisms, back into the donor host. This constitutes biologic transmission of the agent; mechanical transmission, for example, by contaminated mouth parts is possible, but only for a short time after the flea feeds on the infected host. Plague organisms are fragile and do not survive well in unprotected environments.

Etiologic Agent

Plague, in all its clinical forms, is caused by the bacterium, *Yersinia pestis*, a member of the family *Enterobacteriaceae*. Criteria for suspect, presumptive, and confirmed identifications of *Y. pestis* are given in Table 2. The bacteria are slow growing, but nonfastidious, facultatively anaerobic, nonsporing, non-acid-fast, gram-negative rods or coccobacilli. They appear as fat bipolar rods ("closed safety pins" or barrel shaped) when stained by Wayson's (or Giemsa's or Wright's) methods. Typically, after Gram staining, the bacteria measure 0,5 to 0.8 × 1.0 to 2.0

TABLE 2. Criteria for the identification of *Yersinia pestis*

A. 1. Microscopic morphology and size:
Gram-negative rods/coccobacilli
Bipolar rods after Wayson (or Giemsa) staining $0.5–0.8 \times 1.0–2.0\ \mu m$
2. Colony size and morphology:
Invisible pinpoint on blood agar at 24 h, 28–37°C
1.0–2.0 mm with hammered metal surface at 48 h, 28–37°C
3. Growth in stationary broth:
Pellicle formation or stalactite type growth; never uniformly turbid

B. 4. Positive qualitative isocitrate lyase test
5. Positive slide or tube agglutination of formalin-killed suspension of 37°C 24 h culture by known antiplague serum
6. Positive specific fluorescent staining with *Y. pestis;* fraction 1 antiserum conjugated with fluorescein isothiocyanate (or other suitable FA reagent for *Y. pestis*) with 37°C culture or in tissues of laboratory animals inoculated with the culture

C. 7. Sensitivity to lysis by specific *Y. pestis* bacteriophage at both 20–25°C and 37°C.

Cultures meeting criteria in Section A are suspect *Y. pestis;* suspect *Y. pestis* meeting one or both criteria in Section B are presumptive *Y. pestis;* presumptive *Y. pestis* meeting the criterion in Section C are confirmed *Y. pestis*.

μm. On even highly enriched media, the bacteria produce invisible-to-tiny colonies after 24 h at either 28°C or 37°C; by 48 h the colonies reach 1.0 to 2.0 mm in diameter and have a characteristic hammered-metal appearance. *Y. pestis* will grow on many of the selective or inhibitory agar media used for enteric bacteria, but growth is slow, taking perhaps as long as 10 days to be recognized on the highly inhibitory varieties. Pure cultures in stationary broths do not produce uniform turbidity; a pellicle of bacterial growth is seen only at the bottom and along one side of the culture tube. Yield of bacterial cells in a 24 h broth culture is seldom more than about 10^8/ml.

Specimen Collection and Transport

Specimens of human or mammalian origin should be collected before initiation of antimicrobial therapy. For recovery of bacteria, the specimens of choice include blood, bubo aspirate, swabs/scrapings of skin lesions, sputum, and cerebrospinal fluid. Tissues, collected by surgical excision, biopsy, or excised at necropsy that may yield cultures of the agent include spleen, liver, lung, lymph node, and bone marrow. Stool and urine cultures have not been useful in the laboratory diagnosis of plague.

Specimens should be transmitted to the laboratory by the speediest method possible. If delays of more than a day are expected the specimens should be frozen and transported on dry ice if possible. Alternatively, specimens may be triturated and inoculated into Cary Blair Transport Medium and shipped to the laboratory under ambient temperature conditions.

Isolation and Identification Procedures

Bacteria present in the tissues may be presumptively identified by fluorescent antibody staining techniques of heat-fixed impression smears; visualization of organisms consistent with *Y. pestis* may be achieved by Gram and Wayson (or Giemsa or Wright) staining techniques. For isolation of *Y. pestis*, portions of the specimen are streaked on blood agar plates (5% sheep blood in blood agar base), MacConkey agar plates, or other inhibitory enteric media and inoculated into a nutrient or enriched broth medium. Additionally, a cystine heart agar medium supplemented with 8% sheep blood should be inoculated with a portion of the specimen for recovery of *Francisella tularensis*, the agent of tularemia, which, in acute stages, is a disease clinically similar to plague. Media inoculated for the recovery of *Y. pestis* should be incubated at 37°C for at least 10 days. Growth is usually evident by 48 to 72 h, although growth for initial samplings of specimens may be delayed.

Isolation of *Y. pestis* from highly contaminated specimens, such as sputum, or tissues excised at postmortem examination of humans or animals dead more than 24 h, may be accomplished by inoculating a triturated saline suspension of the specimen subcutaneously (or percutaneously) into susceptible laboratory animals. The inoculated animals are held with adequate housing, food, and water and observed daily for at least 10 days. Tissues are excised from animals dying or killed when moribund during the observation period and processed as above.

Procedures for Confirmatory Identification

The qualitative isocitrate lyase test (Hillier and Charnetzky, 1981; Quan et al., 1982) is performed by suspending a loopful of organisms grown 24 h at 28 to 37°C in 0.1 ml of MOPS-Triton buffer (0.1 M morpholinepropanesulfonic acid [pH 7.7], 5 mM $MgCl_2$, 1 mM $Na_2CaEDTA$, 0.2% Triton X-100) in a 13×70 mm tube. Add 0.1 ml DL isocitric acid (8 μM) in MOPS-Triton. Incubate 15 min at ambient temperature (25 to 28°C). Add 0.05 ml 0.2% phenylhydrazine in 0.4 mM oxalic acid. Heat to boiling and cool immediately by plunging the tube into an ice water bath. Add 0.1 ml 12 N HCl and mix. Add 0.05 ml 5% aqueous potassium ferricyanide. Isocitrate lyase activity is indicated by the formation of a red precipitate.

For bacterial slide agglutination tests for *Y. pestis,* formalin (final concentration 0.5 to 1.0%) is added to a suspension of the suspect culture. The suspension is allowed to stand at least 1 h, preferably overnight, at room temperature. The formolized suspension is centrifuged, and the bacterial pellet is washed twice in physiologic saline and resuspended to a McFarland #4 turbidity standard in physiologic saline. One drop of the bacterial suspension is added to one drop of a 1 : 150 dilution of high titered antiplague serum on a Boerner type test slide, mixed gently, and read for agglutination in 10 to 15 min with the aid of a stereoscopic microscope. Known positive and negative controls must be tested concurrently to ensure reliable results.

Fluorescent antibody (FA) tests may be performed on smears of cultures grown at 37°C or on impression smears of tissues (lymph nodes [bubo aspirates], spleen, liver, lung) obtained from patients or animals suspected of plague infection. Specimens are smeared in duplicate on FA quality microscope slides, air dried, and gently heat-fixed. To the first smear is added a drop of fluorescein isothiocyanate conjugated anti-*Y. pestis* (Fraction 1) serum, and to the second smear is added a drop of the same conjugate to which unlabeled anti-*Y pestis* serum has been added (inhibition conjugate). The slide is incubated at room temperature for 30 min and washed twice (10 min each) in phosphate buffered saline, pH 7.2. The slide is air dried and examined by fluorescence microscopy using a BG-12 primary filter and a Wratten 15 secondary filter. *Y. pestis* demonstrates bright apple-green staining in the first smear and a two- to threefold reduction in brightness in the second (inhibition) smear. Fluorescent antibody conjugates and protocols for use are available from plague reference laboratories.

Y. pestis is susceptible to lysis by a specific bacteriophage at temperatures from 20 to 37°C. Other organisms may be lysed by this bacteriophage, but only at temperatures in excess of 30°C. The test may be performed by streaking the suspect culture(s) and known positive and negative cultures in parallel lines on each of two blood agar plates. Perpendicular to these lines is placed a paper strip that has been impregnated with the bacteriophage suspension and lyophilized. One plate is incubated at 37°C and the other at ambient temperature (not in excess of 28°C) for 24 h. Cultures showing inhibition of growth adjacent to the paper strip at both temperatures are *Y. pestis;* those showing no inhibition or inhibition only at the higher temperatures are not *Y. pestis*. Bacteriophage-impregnated paper strips (or suspensions) are available, as are control cultures, from plague reference laboratories.

Biochemical characterization tests are performed by standard methods such as described by Clark et al. (1984). Results of the more common tests for *Y. pestis* are given in Table 3. Other tests of identification confirmation, strain characterization, and serodiagnostic methods (such as passive hemagglutination, ELISAs, radioimmunoassays, in vitro virulence marker determinations, and so on) require reagents that are not commercially available. Protocols and limited supplies of the required reagents or arrangements for performance of the specific tests may be made available on an individual basis by plague reference laboratories.

Identification of bacterial isolates can be initiated at the hospital laboratory using standard biochemical reactions (see Tables 2 and 3) or commercial identification kits, but it is strongly recommended that subcultures of suspect isolates be sent by the most rapid means to the nearest laboratory having competence in confirmation of *Y. pestis* (state laboratory or Cen-

TABLE 3. Characterization test results for *Yersinia pestis*

Oxidase	–	Anaerogenic fermentation	
Catalase	+	of:	
Growth in/on:		Adonitol	–
d-Tartrate	+	Amygdalin	(+)
MacConkey Agar	+	Arabinose	+
Citrate	–	Cellobiose	–
KCN	–	Dulcitol	–
Urea hydrolysis	–	Fructose	+
Esculin hydrolysis	+	Galactose	+
Starch hydrolysis	w	Glycerol	d[a]
Phenylalanine	–	Glucose	+
deamination		Lactose	–
Arginine dihydrolase	–	Maltose	d
Arginine decarboxylase	–	Mannitol	+
Lysine decarboxylase	–	Mannose	+
Ornithine decarboxylase	–	Melibiose	d
Indol	–	Raffinose	–
Methyl red	+	Rhamnose	d
Voges-Proskauer	–	Salicin	d
Nitrate reduction	d*	Sorbitol	+
Gluconate	–	Sucrose	–
Malonate	–	Trehalose	d
DNAse	+	Xylose	d
Hydrogen sulfide	–		
Motility 22C/37C	–/–		
ONPG	+		

+ = Positive; – = negative; w = weak positive; (+) = delayed positive; d = strains differ.

[a] U. S. isolates of *Y. pestis* do not ferment glycerol, but do reduce nitrate

ters for Disease Control through the state laboratory).

When appropriately applied and interpreted, commercial biochemical identification kits are very useful in the laboratory diagnosis of *Y. pestis*. Because of the slow growth of *Y. pestis*, some of these kits may require inocula prepared from suspensions of more than one colony as recommended for most enteric organisms, and/or longer incubation periods than recommended by the manufacturer. Misidentifications made when using these kits have usually involved identifying other agents (*Enterobacter agglomerans*, *Klebsiella* sp., *Shigella* sp.) as *Y. pestis*, often because other characteristics, such as colony morphology and size, mucoid nature, growth pattern in broth, and odor, have not been taken into account. Less frequently *Y. pestis* has been identified as another organism (usually *Enterobacter* sp.); however, this mistake can have severe, often fatal, consequences for the patients involved.

Serologic Diagnosis

Antibodies to *Y. pestis* usually may be demonstrated in sera from patients recovering from plague by about 7 or 8 days after onset. Confirmation of plague infection, in addition to or in lieu of isolating the organism, may be based on finding a fourfold change in specific antibody titer in paired sera drawn 10 to 14 days apart, or by finding a titer of >16 in a single serum from a patient who has no history of prior vaccination or previous plague infection.

Antibodies may be detected using (in increasing order of sensitivity) bacterial agglutination slide or tube tests (Bahmanyar and Cavanaugh, 1976), passive hemagglutination tests (World Health Organization, 1970), radioimmunoassay (Hudson et al., 1980; Soergel et al., 1982), and enzyme immunosorbent assays (Cavanaugh et al., 1979, Williams et al., 1976, 1979, 1982). However, since reagents for these tests are not commercially available, specimens should be submitted to reference or state health department laboratories for processing.

Measurable levels of antibody to one or more *Y. pestis* antigens are found in most patients and animals that recover from the disease. Recovery from active infection probably confers a long-term immunity to recurrent infections which, if they do occur, should produce much milder disease.

Antimicrobial Sensitivity

Routine in vitro antibiotic sensitivity testing of isolates of *Y. pestis* is not recommended because in vitro results do not correlate well with in vivo efficacy. Resistance of *Y. pestis* isolates to the drugs of choice has not been documented in the United States. For drug therapy, therefore, streptomycin, tetracycline, or chloramphenicol may be recommended empirically.

Disinfection

Outside the mammalian body or the flea gut, *Y. pestis* is fragile and readily destroyed by dessication or by the action of various commonly used disinfectants, for example, phenols, hypochlorous acid, iodine compounds, quaternary ammonium compounds, and so on. Disinfection of materials contaminated with plague organisms is best accomplished by autoclaving at 121°C for 15 min or by immersion in one of the above fluid disinfectants for 60 min or more.

Treatment, Prevention, and Control

Treatment

Streptomycin is the drug of choice for the treatment of plague, but patients with simple, uncomplicated bubonic plague diagnosed early respond well to treatment with tetracycline. For treatment of patients with advanced bubonic plague, or those with pneumonia or septicemia, streptomycin is recommended. For those patients with complications such as meningitis or endophthalmitis, chloramphenicol should be used. Other drugs may also be effective, but have not been fully evaluated. Ampicillin and amoxicillin have suboptimal antiplague activities and cannot be recommended. Antimicrobial therapy should continue for 10 days or for 3 or 4 days after clinical recovery (Poland and Barnes, 1979; Welty, 1984).

For prophylaxis or abortive therapy, tetracycline is suggested; alternatively, trimethoprim-sulfamethoxazole may be used for persons with known or strongly suspected exposures. Two cautions need to be stressed regarding the use of antibiotics. First, selection of an antimicrobial regimen must be carefully made so that potential benefit to the patient exceeds the potential risk associated with the specific antibiotic (ototoxicity and renal damage by streptomycin; aplastic anemia by chloramphenicol; potential dental damage; or fetal risks by tetracycline), and second, administration of doses must be such that the patient is not overwhelmed by the release of large amounts of endotoxin caused by lysis of plague organisms.

Prevention and Control

Management of Patients

Patients with diagnoses of suspect plague should be kept in strict isolation for 48 h after initiation of specific therapy. Person-to-person transmission is highly improbable in the absence of pneumonia and/or draining lesions, but isolation procedures (1) limit potential transmission should pneumonia or drainage ensue after hospitalization but before therapy is effective, and (2) provide a means of identifying and minimizing the number of persons potentially exposed to patients who develop pneumonia (Poland and Barnes, 1979; Welty, 1984).

Case Contact Management

Casual contacts of plague patients and those who have had similar exposure to probable sources of infection generally do not require antibiotic prophylaxis, but should be placed under disease surveillance for 6 to 10 days. Those developing fever or other symptoms during the observation period should undergo medical evaluation and, if needed, begin tetracycline or other appropriate therapy.

Close family members or others having intimate contact with patients who display signs of pneumonia should be immediately placed on abortive therapy and monitored closely for development of plague-like illness.

Public Health Considerations

Plague is an internationally reportable disease (World Health Organization, 1974). When a human plague case is diagnosed or is suspected, local, state, and federal public health officials should be notified immediately and given specific information regarding the patient. Preliminary investigations to define family and other contacts who might have been exposed to the patient or common sources of infection and determine the potential for other human cases from the same or related host-vector populations. The investigations may be extended or intensified when the diagnosis of plague is verified or aborted if it is invalidated.

Environmental Control Measures

If the identified zootic source of infection is considered to pose a significant threat to the public health, (e.g., involves areas of high human use such as residential developments or recreational areas), rodent and vector control measures should be implemented using effective, approved pesticides appropriately and specifically applied. Ectoparasite control must precede or be concomitant with rodent control measures. The best plague prevention programs include aggressive, ongoing rodent population management. A thorough discussion of surveillance and control techniques is presented by Barnes (1982).

Safety Considerations

For routine clinical and diagnostic microbiologic procedures *Y. pestis* recently was downgraded to a class 2 bacterial agent (Centers for Disease Control, 1984). The organism should be handled as carefully in the laboratory as are other pathogens, and strict adherence to standard aseptic microbiologic techniques should prevent laboratory-acquired infections. The organism must be treated as a class 3 agent for all procedures involving production of aerosols, or necessitating production of large volumes of culture.

Vaccination

A vaccine (USP Plague Vaccine, Cutter Laboratories, Berkeley, California, USA) is available and recommended for all persons whose occupations require repeated contact with the agent (e.g., laboratory workers in plague reference laboratories) or the host and/or vectors in endemic areas. It is not suited nor recommended for the public at large or for those with only casual potential exposures.

Literature Cited

Bahmanyar, M., and D. C. Cavanaugh. 1976. Plague manual. World Health Organization, Geneva, Switzerland.

Barnes, A. M. 1982. Plague surveillance and control in the United States. Symp. Zool. Soc. London **50:**237–270.

Cavanaugh, D. C., M. K. Fortier, D. M. Robinson, J. E. Williams, and J. H. Rust, Jr. 1979. Application of the ELISA technique to problems in the serological diagnosis of plague. Bull. PAHO **13:**399–402.

Centers for Disease Control. 1975. Vector Borne Diseases Division 1975 Report. Part II. Plague Activities. Centers for Disease Control, Fort Collins, CO.

Centers for Disease Control. 1976. Vector Borne Diseases Division 1976 Report. Part II. Plague Activities. Centers for Disease Control, Fort Collins, CO.

Centers for Disease Control. 1977. Vector Borne Diseases Division 1977 Report. Part II. Plague Activities. Centers for Disease Control, Fort Collins, CO.

Centers for Disease Control. 1978. Vector Borne Diseases Division 1978 Report. Part II. Plague Activities. Centers for Disease Control, Fort Collins, CO.

Centers for Disease Control. 1979. Vector Borne Diseases Division 1979 Report. Part II. Plague Activities. Centers for Disease Control, Fort Collins, CO.

Centers for Disease Control. 1980. Plague Surveillance, Reference and Research 1980 Report. Centers for Disease Control, Fort Collins, CO.

Centers for Disease Control. 1981. Plague Surveillance, Reference and Research 1981 Report. Centers for Disease Control, Fort Collins, CO.

Centers for Disease Control. 1982. Plague Surveillance, Reference and Research 1982 Report. Centers for Disease Control, Fort Collins, CO.

Centers for Disease Control. 1983–1984. Plague Surveillance, Reference and Research 1983–1984 Report. Centers for Disease Control, Fort Collins, CO.

Centers for Disease Control. 1984. Biosafety in Microbiological and Biomedical Laboratories. HHS Publication No. (CDC)84-8395. US Government Printing Office, Washington, D.C.

Centers for Disease Control. 1985 (in prep). Plague Surveillance, Reference and Research 1985 Report. Centers for Disease Control, Fort Collins, CO.

Christie, A. B., T. H. Chen, and S. S. Elberg. 1980. Plague in camels and goats: their role in human epidemics. J. Infect. Dis. **141:**724–726.

Clark, W. A., D. G. Hollis, R. E. Weaver, and P. Riley. 1984. Identification of unusual pathogenic gram-negative and facultatively anaerobic bacteria. Centers for Disease Control, Atlanta, GA.

Fedorov, V. N. 1960. Plague in camels and its prevention in the USSR. Bull. W.H.O. **23:**275–281.

Hillier, S. L., and W. T. Charnetzky. 1981. Rapid diagnostic test that uses isocitrate lyase activity for identification of *Yersinia pestis*. J. Clin. Microbiol. **13:**661–665.

Hudson, B. W., K. Wolff, and T. Butler. 1978. The use of solid phase radioimmunoassay techniques for serodiagnosis of human plague infection. Bull. PAHO **14:**244–250.

Kaufmann, A. F., J. M. Mann, T. M. Gardiner, F. Heaton, J. D. Poland, A. M. Barnes, and G. O. Maupin. 1981. Public health implications of plague in domestic cats. J. Am. Vet. Med. Assoc. **179:**875–878.

Poland, J. D., and A. M. Barnes. 1979. Plague, p. 517–597. *In* J. H. Steele (ed.), CRC handbook series in zoonoses. Section A: bacterial, rickettsial, and mycotic diseases. CRC Press, Boca Raton, FL.

Quan, T. J., J. J. VanderLinden, and K. R. Tsuchiya. 1982. Evaluation of a qualitative isocitrate lyase assay for rapid presumptive identification of *Yersinia pestis* cultures. J. Clin. Microbiol. **15:**1178–1179.

Quan, T. J., A. M. Barnes, L. G. Carter, and K. R. Tsuchiya. 1985. Experimental plague in rock squirrels, *Spermophilus variegatus* (Erxleben). J. Wildl. Dis. **21:**205–210.

Rollag, O. J., M. R. Skeels, L. J. Nims, J. P. Thilsted, and J. M. Mann. 1981. Feline plague in New Mexico: report of five cases. J. Am. Vet. Med. Assoc. **179:**1381–1383.

Rust, J. H. 1984. Plague: epidemiology and control. Epidemiol. Bull. PAHO **5:**1–5.

Shepherd, A. J., P. A. Leman, and D. E. Hummitzsch. 1986. Experimental plague infection in South African wild rodents. J. Hyg. Camb. **96:**171–183.

Soergel, M. E., F. L. Schaffer, and H. F. Blank. 1982. Solid phase radioimmunoassay for detection of plague antigen in animal tissue. J. Clin. Microbiol. **16:**953–956.

Welty, T. K. 1984. Plague, p. 44–45. *In* H. F. Conn (ed.), Current therapy. W. B. Saunders, Co., Philadelphia.

Welty, T. K., J. Grabman, E. Kompare, G. Wood, E. Welty, J. VanDuzen, P. Rudd, and J. Poland. 1985. Nineteen cases of plague in Arizona: a spectrum including ecthyma gangrenosum due to plague and plague in pregnancy. West. J. Med. **142:**641–646.

Weniger, B. G., A. J. Warren, V. Forseth, G. W. Shipps, T. Creelman, J. Gorton, and A. M. Barnes. 1984. Human bubonic plague transmitted by a domestic cat scratch. J. Am. Med. Assoc. **251:**927–928.

Williams, J. E., E. M. Atas, and D. C. Cavanaugh. 1976. A comparison of serological tests for detecting antibodies to plague. Bull. W.H.O. **54:**3513.

Williams, J. E., M. A. Moussa, and D. C. Cavanaugh. 1979. Experimental plague in the California ground squirrel. J. Infect. Dis. **140:**618–621.

Williams, J. E., L. Arntzen, D. M. Robinson, D. C. Cavanaugh, and M. Isaacson. 1982. Comparison of passive hemagglutination and enzyme linked immunosorbent assay for serodiagnosis of plague. Bull. W.H.O. **60:**777–781.

World Health Organization. 1974. International Health Regulations (1969). W. H. O. Geneva, Switzerland.

World Health Organization Committee on Plague. 1970. Passive hemagglutination test. W. H. O. Tech. Rep. Ser. **447:**23–25.

CHAPTER 44

Pneumococcal Diseases

F. MEUNIER and P. VAN DER AUWERA

Disease: Pneumococcal diseases: pulmonary infections, primary sepsis, meningitis, otitis media, sinusitis, other localized infections, and recurrent invasive disease.
Etiologic Agent: *Streptococcus pneumoniae,* 84 known capsular serotypes.
Source: Healthy carrier state; acquisition by inhalation.
Clinical Manifestations: Usually acute and severe, occasionally with sepsis, which confers a poor prognosis.
Pathology: Pyogenic infection with polymorphonuclear leukocyte reaction.
Laboratory Diagnosis: Bacteriologic isolation of *S. pneumoniae* from blood, purulent material, cerebrospinal fluid, or any relevant clinical specimen. Detection of soluble antigens by various techniques in serum, urine, cerebrospinal fluid, sputum, or pleural or pericardial fluids.
Epidemiology: Worldwide; a leading cause of acute pneumonia, meningitis, otitis media, and sinusitis. Rate of healthy carriage in rhinopharynx is 40 to 60%.
Treatment: Penicillin G, erythromycin; multiresistance, including resistance to penicillin, increasing worldwide. Surgical drainage in pleural empyema and pericarditis.
Prevention and Control: Vaccination recommended in selected populations. Antibiotic prophylaxis after splenectomy in the presence of posttraumatic leakage of cerebrospinal fluid, in children less than 3 years old with sickle cell anemia and thalassemia, and in patients with bone marrow transplantation.

Description of the Disease

Clinical Manifestations

Pulmonary Infections

Lobar Pneumonia

Streptococcus pneumoniae is responsible for 26 to 78% of acute pneumonia in outpatients and 10 to 67% in hospitalized patients (Austrian, 1981; Austrian and Gold, 1964; Bartlett et al., 1986; Mufson, 1981; Rytel, 1979; Woodhead et al., 1985). Of patients with pneumococcal pneumonia, 80 to 90% have respiratory symptoms during the preceding 2 to 3 weeks (Austrian, 1981; Heffron, 1979). Typically, the presentation of acute pneumococcal pneumonia is an abrupt onset of fever (up to 40°C) with shaking chills, cough, slight expectoration, and intense pleural pain. Cough may produce bloody, "rusty," or greenish material. Herpetic oral lesions may be present concomitantly. The right lung, particularly the upper or lower lobe, is more frequently affected than the left. Involvement of both lungs occurs in less than 10% of cases.

Numerous complications of pneumococcal pneumonia have been described (Table 1). Empyema is a severe complication most frequently documented in older patients. It is generally unilateral and massive, but may be sacculated or interlobar, bilocular, or even bilateral. The common accompanying symptoms are persistence of fever, cough, chills, sweats, and dyspnea; pain is uncommon. Sterile pleural effusion with serous fluid, which is often associated with slow resolution of temperature, may occur during treatment (Austrian and Gold, 1964; Gransden et al.,

TABLE 1. Complications on pneumococcal pneumonia[a]

Complication	*Incidence (%)*
Bacteremia	22–40
Empyema	7–19
Pericarditis or endocarditis	0.3–4
Otitis media or mastoiditis	2–13
Meningitis	0.4–1.9
Arthritis	0.4–0.9
Parotiditis	0.1
Peritonitis	0.16

[a] Modified from Heffron (1979).

1985). Other very rare complications have been described, such as soft tissue abscesses, cellulitis, mediastinitis, lung abscess, and gangrene (Heffron, 1979).

Bronchopneumonia

Pneumococcal lung infection may present as acute bronchopneumonia (9 to 15% of patients) instead of the classic lobar pattern (Heffron, 1979). This finding seems especially true in hospitalized patients, 61% of whom show a diffuse bronchopneumonic pattern. Such patients are usually older and have several concomitant predisposing factors. This presentation is more difficult to diagnose, as bacteremia is much less frequent than during lobar pneumonia, and the interpretation of sputum culture remains difficult. In the study by Ort et al. (1983), the clinical signs and symptoms of these two forms did not differ except in the incidence of bacteremia, which occurs in 54% of patients with lobar pneumonia but in only 9% of those with bronchopneumonia.

PRIMARY SEPSIS

Occult Bacteremia

Positive blood cultures for *S. pneumoniae* have been reported in specific circumstances, such as in febrile children treated at an outpatient clinic. In fact, unsuspected bacteremia has been observed in children aged 6 to 24 months who had high temperatures (39.4°C), elevated white blood cell counts (greater than 15,000/mm^3) and no apparent signs of infection. A small proportion of these patients may develop severe complications and require an extensive follow-up (Bratton et al., 1977; Myers et al., 1974). *Streptococcus pneumoniae* is responsible for 61 to 78% of cases of occult bacteremia in pediatric outpatients. Occult bacteremia has also been observed in a small number of adults and usually has a benign clinical course (Spitalny et al., 1982).

Postsplenectomy Sepsis

The increased risk of fatal septicemia after splenectomy has been well defined in the literature (Wara, 1981; Zarrabi and Rosner, 1984). *Streptococcus pneumoniae* is the causative pathogen in 50% of these patients, often causing septic shock and occasionally purpura fulminans. Other pathogens commonly identified after splenectomy are other encapsulated species such as *Haemophilus influenzae, Neisseria meningitidis,* and *Staphylococcus aureus.* The reasons for the high prevalence of *S. pneumoniae* are the presence of an antiphagocytic polysaccharidic capsule requiring specific immune opsonization, the rapid multiplication of the organism, and the lack of specific antibody of the immunoglobulin G2 subclass in children less than 3 years old. Overwhelming *S. pneumoniae* septicemia occurs in patients with a nonfunctional spleen as well as in those with anatomic asplenia (as a result of surgery for trauma, Hodgkin's disease, or lymphoma). It has been demonstrated that ectopic splenic tissue in humans does not restore the altered antibody response observed after splenectomy.

Sepsis in Immunocompromised Patients in the Absence of a Recognizable Focus

In addition to pneumococcal sepsis in patients with splenectomy or occult bacteremia, a third syndrome of pneumococcal sepsis occurs, mainly 25 to 32% of cases of bacteremia in immunocompromised hosts without any recognizable focus (Chou et al., 1983; Folland et al., 1974). Granulocytopenia does not appear to create a major risk for acquiring pneumococcal septicemia. During three consecutive multicenter trials of the European Organization for Research on Treatment of Cancer (EORTC International Antimicrobial Therapy Cooperative Group, 1978, 1983, 1986), in which 1,626 febrile episodes in patients with less than 1,000 polymorphonuclear leukocytes per μl were evaluated, 408 had documented septicemia, including 15 with *S. pneumoniae,* corresponding to a prevalence of 0.9%. Although most cancer centers have reported an increased incidence of bacteremia attributable to gram-positive cocci in neutropenic patients, *S. pneumoniae* seems to be isolated more frequently in such clinical settings. In a review of *S. pneumoniae* sepsis in patients with cancer, one-third of the patients presented with septic shock (Bernard et al., 1981).

Neutropenia, by itself, has not been associated with increased mortality from pneumococcal sepsis (Bernard et al., 1981; Folland et al., 1974). In one study (Chou et al., 1983), however, a relationship between the neutrophil count during bacteremia and the outcome was reported. Polymicrobial sepsis, which may be present in 10 to 15% of these patients, may worsen the prognosis.

Meningitis

Streptococcus pneumoniae has been identified as a leading cause of bacterial meningitis in 15 to 40% of all cases in adults and 10% of those in children. It is also the most common cause of bacterial meningitis in patients more than 60 years of age (Gorse et al., 1984; Schlech et al., 1985). Signs and symptoms of acute pneumococcal meningitis are comparable to those of meningitis caused by other pyogenic bacteria, such as *H. influenzae* or *N. meningitidis*. These signs and symptoms include fever and nuchal rigidity in almost all patients; in addition, change in mental status, paresis, positive Babinski sign, cerebellar symptoms, and cranial nerve paralysis are considered typical of pneumococcal meningitis (Tugwell et al., 1976).

Children less than 5 years of age exhibit the highest attack rates (3 to 11 per 10^5 individuals per year). Meningitis in the elderly is associated with pneumonia, neurologic deficit, seizures, and hydrocephalus more frequently than in younger patients (Gorse et al., 1984).

Otitis Media and Sinusitis

Streptococcus pneumoniae is responsible for 1 million cases of otitis media each year in United States. Usually, the first episode of pneumococcal otitis media occurs in children 3 to 24 months old. Otitis media is extremely common: at least two-thirds of children less than 6 years old have experienced at least one episode. *Streptococcus pneumoniae* is responsible for 25 to 50% of these episodes. Acute otitis media caused by *S. pneumoniae* may become chronic, resulting in learning difficulties because of hearing loss. Predominant capsular serotypes (6, 14, 19, and 23) have been recognized in approximately 60% of all episodes (Austrian et al., 1977).

Streptococcus pneumoniae is responsible for 20 to 36% of cases of acute sinusitis in both children and adults. Acute sinusitis is usually preceded by a common cold of viral etiology. Signs and symptoms include facial pain, purulent nasal discharge, headache, and nasal obstruction. Fever may be present in one-half of patients. Specific microbial etiology can be determined only by culture of the pus obtained by sinus puncture to avoid contamination by the commensal oropharyngeal flora.

Peritonitis

Primary pneumococcal peritonitis occurs in children between 4 and 10 years old, mainly girls without underlying disease (Golden and Shaw, 1972). However, it has also been described in association with nephrotic syndrome (Pahmer, 1940), cirrhosis, systemic lupus erythematosus, and immune deficiency. Occasionally, *S. pneumoniae* has been recognized as the causative agent of superinfection in ascites due to hepatic cirrhosis. The source of peritoneal contamination may be either hematogeneous or vaginal colonization (Legrand et al., 1985).

Other Localized Diseases

A few cases of endocarditis have been described, usually associated with pneumonia or meningitis. The endocarditis may have caused the meningitis by septic embolization, or both localizations may be the result of a disseminated infection. Pericarditis remains a severe complication of pneumococcal infections despite antibiotic use (Kaufman et al., 1973). Bone and joint infections represent less than 1% of invasive pneumococcal infections. A few cases of young children with facial cellulitis have also been reported. Adults with malignancy may develop pneumococcal epiglottitis (Lederman et al., 1982).

Recurrent Invasive Pneumococcal Diseases

Heffron (1979) reported a true relapse rate in 0.06% of patients with lobar pneumonia. In South African gold miners, however, recurrent pneumonia was observed in 14%; all cases were caused by a different serotype of *S. pneumoniae*. The recurrence occurred within 30 days after resolution of the first episode (Heffron, 1979). However, Burman et al. (1985) reported 10 patients with multiple episodes of invasive pneumococcal infections who had an interval of at least 3 months between episodes. Underlying diseases in this study included alcoholism, IgG2 deficiency, multiple myeloma, splenectomy, chronic otitis and chronic leakage of cerebrospinal fluid (CSF) after cranial fracture.

Pathology

Pneumococcal lobar pneumonia is characterized by several stages (Heffron, 1979). However, different parts of the lung may present different stages of the disease.

During the first (engorgement) stage, there is initial edema with hyperemia. During the second stage, polymorphonuclear leukocytes and red cells migrate, and the lung appears reddish gray (red hepatization). During the third stage, the lung tissue is ischemic, with coagulation of fibrinous exudates (gray hepatization). The fourth stage (resolution) is characterized by progressive disappearance of fibrin.

Consolidation is usually present within 24 h of the

onset of disease. The red and gray stages of hepatization last 2 to 3 days each. Virulent pneumococci are present during red hepatization; during gray hepatization, however, the organisms degenerate and lose their virulence. In the first three stages, polymorphonuclear leukocytes are the predominant cell type and pneumococci are numerous; during the resolution stage, macrophages become the predominant cell in the exudate, and their functions result in complete disappearance of pneumococci.

Pneumococcal meningitis is characterized by a thick purulent exudate, which mainly affects the basal cistern. During resolution, this exudate becomes organized and may result in obstructed flow of CSF and hydrocephalia. The mechanism by which the infection and inflammation in the subarachnoidal space lead to brain damage is unknown. It has been atttributed to increased intracranial pressure and brain edema. Anaerobiosis during meningitis leads to an increased lactate concentration in the CSF. During experimental meningitis in the rabbit, penicillin and dexamethasone, but not methyl prednisolone, have been shown capable of reducing the elevated CSF pressure and lactate level (Täuber et al., 1985).

Glomerulonephritis, which may follow pneumococcal infection, has been described at autopsy in 9.4% of patients who died of pneumococcal infections and in 0.7% of those with lobar pneumonia. None of the patients seemed to have developed hypertension; however, albuminuria, hematuria, retention of nitrogen, anemia, and slight edema were reported (Heffron, 1979; Rytel, 1979). Later studies have described acute proliferative glomerulonephritis complicating pneumococcal infections. In a few cases, this glomerulonephritis was associated with deposits of type-specific capsular polysaccharide, C1q, C3, properdin, C4, and IgM or IgG. Immune complexes have been demonstrated in a few patients, both in serum and in renal tissue (Preheim et al., 1978).

Predisposing Factors

Table 2 shows the incidence of pneumococcal infections in the general population. Heffron (1979) has shown that several factors increase the risk of pneumococcal infections: youth (less than 2 years old); old age (more than 60 years old); alcoholism; low socioeconomic status, especially vagrancy (Gransden et al., 1985); black race; and occupation (e.g., gold mining in South Africa). Neonates have a low risk, probably due to the presence of maternal antibodies.

TABLE 2. Risk and associated mortality of *Streptococcus pneumoniae* infections

Underlying disease	*Risk*	*Increased severity*	*Mortality (%)*
Irrespective to underlying disease			
Pneumonia	15–100/10^5/yr		10–20
Bacteremia	0.6–5/10^5/yr		15–43
Meningitis	0.03–0.49/10^5/yr		13–65
No underlying disease			
Bacteremia			2–9
With underlying disease			
Neoplastic disease			
Hodgkin's disease	5–15/1,000	Yes	
Hodgkin's disease with splenectomy	15.6/1,000	Yes	30
Multiple myeloma	9–12.5/1,000/yr	Yes	30
Chronic lymphocytic leukemia	1.5/1,000/yr	Yes	30
Bone marrow transplant	Increased	Yes	
Nonneoplastic disease			
Traumatic asplenia	5–22/1,000/yr	Yes	50–83
Sickle cell anemia	70–100/1,000/yr	Yes	35
Thalassemia with splenectomy	200/1,000	Yes	50
Nephrotic syndrome	Increased	Yes	
Chronic cardiovascular disease	Increased	Yes	40–≥50
Liver cirrhosis	Increased	Yes	70–80
Chronic pulmonary disease	Increased	Yes	
Elderly	Increased	Yes	
Renal transplantation	10–24/1,000/yr	Yes	
Diabetes mellitus	Not increased	No	11–28.5
Systemic lupus erythematosus	Not increased	Yes	
Chronic renal failure	Not increased	No	

The following underlying diseases increase the risk of developing pneumococcal lobar pneumonia: influenza virus infection; chronic obstructive bronchial disease; chronic cardiac failure; chronic alcoholism; and cancer, including multiple myeloma, chronic lymphocytic leukemia, and bronchial carcinoma (Bernard et al., 1981; Chou et al., 1983; Folland et al., 1974). Whether diabetes increases the risk for pneumococcal infection remains controversial.

Bone marrow transplantation is also a major risk factor for pneumococcal infection. An incidence of 27% in long-term survivors of bone marrow transplantation was reported by Winston et al. (1979). The spectrum of diseases includes pneumonia, sinusitis, otitis media, and meningitis, with eventual bacteremia. These episodes occurred within 3 years after transplantation.

Chronic lung disease has generally been recognized as a major risk factor for pneumococcal pneumonia and bronchopneumonia and is associated with increased mortality. However, a recent review concluded that this association has never been conclusively substantiated (Williams and Moser, 1986). Fedson and Baldwin (1982) have shown that a recent episode of hospital care was associated with a very high risk of pneumococcal pneumonia, and that most of the deaths from pneumococcal pneumonia occurred among the very small proportion of the population previously discharged with high-risk conditions.

The anatomic structure of the spleen allows prolonged contact between phagocytic cells and pneumococci, resulting in phagocytosis of nonimmune particles. Sequestration by the liver requires increased efficiency of phagocytosis through immune opsonization to compensate for the reduced time of contact. Pneumococci opsonized by specific antibodies that activate the complement pathway are preferentially sequestrated into the liver. Therefore, in patients with no or few specific antibodies, the liver does not clear *S. pneumoniae* optimally, and splenectomy removes the most effective phagocytic system for the clearance of unopsonized pneumococci. In addition, splenectomized patients have an abnormal response to pneumococcal polysaccharide antigen. Defective opsonization through the alternative complement pathway in some splenectomized patients has also been postulated. Splenectomy has also been associated with a decrease in antibody response to intravenous antigen, decreased IgM synthesis, an occasional decrease in conversion from the production of IgM to that of IgG, and a lack of the phagocytosis-stimulating peptide tuftsin.

In sickle cell disease, the spleen becomes nonfunctional before the patient reaches the age of 3 years. Susceptibility to overwhelming sepsis after splenectomy is related to the age at which the surgery is performed. The risk of *S. pneumoniae* infection after splenectomy for trauma (1.4 to 2.1%) diminishes with increasing age. Children less than 2 years old have a 200- to 600-fold higher risk than those more than 5 years old. The risk of pneumococcal sepsis after splenectomy is 2% in idiopathic thrombocytopenia, 3.5% in congenital spherocytosis, 7.5% in acquired hemolytic anemia, 8.2% in portal hypertension, 8.5% in primary anemia, 11.5% in reticuloendothelial disease, 15.2% in sickle cell disease, and 24.7% in thalassemia (Wara, 1981). Hyposplenism associated with sarcoidosis, amyloidosis, and ulcerative colitis may increase the risk of overwhelming sepsis. Regardless of age, the greatest risk occurs in the first few weeks to months following splenectomy. Defects of the humoral system such as hypogammaglobulinemia or IgG2 deficiency and C2, C3, or C3b inactivator deficiency all create a predisposition to severe infection with *S. pneumoniae*.

Traumatic or postoperative head injury and chronic otitis are major predisposing factors for meningitis in adult patients and are recognized in 30% of them. Leakage of CSF poses a major risk for the development of pneumococcal meningitis (Hyslop and Montgomery, 1980). *Streptococcus pneumoniae* is the most frequent pathogen (>80%) responsible for meningitis associated with this condition. The risk of meningitis is 3 to 14%. At least 50% of cases occur within 2 weeks after trauma, but the delay may range from 1 day to several years. Recurrent meningitis has been associated with persistent leakage of CSF.

Prognosis

The prognosis of pneumococcal bacteremia has been significantly correlated with age, occurrence of complications, severity of underlying disease, and presence or absence of a functional spleen. Of the patients who die of pneumococcal infection, 50 to 70% do so within 24 h after admission as a result of overwhelming infection. The death rate from pneumococcal infections in adults remains high despite antibiotic treatment. Early treatment may prevent complications, but does not prevent irreversible disorders induced by pneumococcal virulence during the first 5 days of illness. The death rate from pneumococcal bacteremia has been correlated with the serotype of the offending pathogen. Serotype 3 has been associated with a fatality rate of up to 50% (Gransden et al., 1985) and correlated with fulminant infection requiring admission to the intensive care unit. In the study of Shlaes et al. (1982), it was suggested that disease caused by nonvaccine serotypes was associated with a higher mortality than infection caused by vaccine serotypes in an area where pneumococcal vaccine was frequently used.

The fatality rate of pneumococcal pneumonia has been correlated with bacteremia as well as with the inoculum (Heffron, 1979). In the absence of bacteremia, the mortality from *S. pneumoniae* infection is 11%; for sepsis, the fatality rate is 32% with 1 to 2 CFU/ml of blood, 33 to 86% for 2 to 50 CFU/ml, and 80 to 100% for more than 50 CFU/ml. Similarly, the detection of capsular polysaccharide in the serum and urine has been correlated with mortality. The occurrence of empyema, pericarditis, or other severe complications worsens the prognosis. Severe community-acquired pneumonia may have a death rate of 75%, especially in patients more than 60 years old (Woodhead et al., 1985). Mufson et al. (1974) reported that mortality with bacteremic pneumococcal disease was 6% in patients 12 to 29 years old, 19% in those 30 to 49 years old, 41% in those 50 to 69 years old, and 61% in those more than 70 years old.

Several series studying pneumococcal sepsis in cancer patients have stressed the severity of this complication in this population. In addition, infection with pneumococci having a minimum inhibitory concentration (MIC) of penicillin greater than 1 μg/ml carries a very high mortality. The overall mortality from pneumococcal meningitis is between 13 and 65%, which is higher than that associated with meningitis caused by *H. influenzae, N. meningitidis,* and group B streptococci (Schlech et al., 1985). Up to 35% of survivors of pneumococcal meningitis may present significant residual nervous system damage. Other survivors may present more subtle changes in behavior and intelligence, which may result in learning difficulties for affected children.

Epidemiologic Aspects

Occurrence and Transmission

The incidence of nonbacteremic pneumococcal pneumonia is difficult to assess and certainly underestimated, as sputum culture is not always performed and reliable data for outpatients is lacking. Furthermore, blood cultures are not routinely performed in all patients. However, the incidence of pneumococcal bacteremia has increased over the last decade, probably because of the increased number of patients in whom blood cultures are performed. The incidence of pneumococcal meningitis is reliably estimated in Northern America and Western Europe, as most patients are hospitalized and adequate samples routinely obtained.

The prevalence of pneumococcal pneumonia is highest during the winter months: 44% of cases occur from January to March, 26% from April to June, 7% from July to September, and 22% from October to December. Epidemic situations may exist, such as that among black South African gold miners. In this population, the risk of developing acute lobar pneumonia appeared maximal during the first days after arrival in this epidemiologic setting: from 0 to 4 days the attack rate was 1.16%, whereas from 15 to 19 days it was 0.6% (Heffron, 1979). In most studies a male : female predominance was observed, with a ratio of 3 : 2; the carriage rate, however, is equally distributed.

Transmission of *S. pneumoniae* probably occurs through pharyngeal carriers. Heffron (1979) reported that contacts had a 12% higher carriage rate than noncontacts. Furthermore, more virulent types were found in contacts. In 5 to 20% of the contacts, the strain recovered from the pharynx was homologous with that causing the primary case. The homologous carriage rate was higher among family contacts (22%) than among medical staff contacts (3%). Of patients with chronic obstructive pulmonary disease or congestive heart failure, 30 to 50% have *S. pneumoniae* in the trachea (Austrian, 1981). This finding may explain the high risk of pneumococcal pneumonia in these patients. The association between carriage of pneumococci in the pharynx and the development of otitis media remains uncertain.

Streptococcus pneumoniae is the third to fifth most common isolate from all blood cultures, causing 4 to 5% of all bacteremias; 80% of cases of pneumococcal bacteremia are community acquired (Ruben et al., 1984). In developed countries, one-third of cases of pneumococcal bacteremia occur in patients older than 60 years, and two-thirds in patients older than 40 years (Gransden et al., 1985; Ruben et al., 1984). Serotypes of pneumococci vary with time (Finland and Barnes, 1977), geographic location, underlying disease (Weisholtz et al., 1983), age, and probably use of pneumococcal vaccine (Schlaes et al., 1982).

Carriage

Streptococcus pneumoniae is a commensal organism of the nasopharynx. The reported prevalence of pneumococcal carriage has varied greatly because of differences in the sampling techniques, culture media, and selective media used. Incubation, animal inoculation, and identification methods also interfere with the results (Austrian, 1986). Recovery is highest (up to 70%) when several methods are used simultaneously. The use of a single blood agar plate is the least sensitive method. If studied over time, 100% of humans seem to be colonized by *S. pneumoniae*. Concomitant colonization with more than one pneumococcal capsular type is not uncommon, and as many as six different types may be recovered during the period of observation. Colonization occurs in the first 2 weeks after birth. Symptomatic colonization with a given type is followed by the development of homologous anticapsular antibodies. These antibodies may explain the decreasing risk of pneumococcal bacteremia in splenectomized patients with increasing age at time of splenectomy.

Vaccination has no effect on the carriage of homotypic organisms in adults if colonization precedes immunization, but reduces acquisition in a previously uncolonized individual. The carriage rate is higher in children than in adolescents or adults, depends on socioeconomical status and race, and is higher during the winter months than during the summer. It is interesting that serotypes 1 and 14 are rarely found in the nasopharynx of healthy individuals, probably because these types are highly invasive.

Streptococcus pneumoniae has a protein adhesin, which is distinct from the capsular polysaccharide and possesses a specific receptor. Strains with high adherence properties have been shown more likely to produce otitis media, although less adherent strains are associated with bacteremia and meningitis. Human milk exerts an antiadhesive activity against attachment of *S. pneumoniae* to human pharyngeal or buccal epithelial cells. The organism can produce IgA protease.

Susceptibility

The susceptibility of *S. pneumoniae* to penicillin has progressively decreased since the introduction of this antibiotic. The MICs reported before 1950 were less than 0.01 μg/ml. Strains relatively resistant to penicillin (MICs less than 1 μg/ml) have been reported since 1967. A few strains with MICs of 2 μg/ml have been reported from South Africa, the United States, and England. Resistance is not due to β-lactamase production; it is chromosomally mediated and involves several additive steps, including a decreased affinity of penicillin-binding proteins (PBP) 1 and 2 and a change in the number of copies per cell. For highly resistant strains (MICs greater than 1 μg/ml), a dramatic change in the PBP profile with nonexpression of PBP 1a and 1b and expression of an unusual PBP 1c, which has a very low affinity for penicillin, has been described (Tomasz et al., 1984). These resistant pneumococci seem to be as virulent as susceptible strains in experimental meningitis. Multiresistant strains may be acquired nosocomially and seem more likely to colonize patients who have received multiple antibiotics (Jacobs et al., 1978; Ward, 1981). It is now estimated that 1 to 2.4% of pneumococci have MICs for penicillin G ranging from 0.12 to 1 μg/ml in the United States, Great Britain, France, Belgium (Vanhoof et al., 1980), Australia, Canada, New Guinea (Feldman et al., 1985), and Japan, but in endemic areas such as South Africa or Australia (Hansman et al., 1985) this estimate may be as high as 15 to 25%. Efficient recombination mechanisms occurring in vivo in *S. pneumoniae* probably contribute to the spread of intrinsic resistance to penicillin. The carriage rate of penicillin-resistant pneumococci is very high in children (up to 19%) in endemic areas such as South Africa. In these areas of high endemicity, multiresistant strains are restricted to a few serotypes, notably 6A, 14, and 19A, and some are tolerant (minimum bactericidal concentration/MIC greater than 32 μg/ml) to the bactericidal activity of penicillin. The clinical correlation of higher MICs and therapeutic failures has been substantiated. Resistance to chloramphenicol may or may not be associated with resistance to penicillin and is due to inactivation of chloramphenicol. Resistance to alternative antimicrobial agents, such as sulfonamides, tetracycline, erythromycin, and lincomycin, has also been reported more frequently.

The sensitivity of *S. pneumoniae* to penicillin should always be tested. Oxacillin disks (1 μg) must be used instead of penicillin to screen for resistance. The recommended criterion is disk zone inhibition: less than 12 mm, resistant (MIC greater than 1 μg/ml) or relatively resistant (MIC 0.12 to 1 μg/ml); 13 to 19 mm, intermediate; and 20 mm, susceptible (Jones et al., 1982). Commercial rapid methods are not reliable for the detection of pneumococcal resistance to penicillin.

Etiologic Agent

Characterization

As a member of the genus *Streptococcus, S. pneumoniae* has characteristics typical of other streptococci. This gram-positive coccus grows in short chains in broth cultures, has no catalase, ferments glucose by the hexose monophosphate pathway to produce lactic acid, and contains a group-specific carbohydrate and M protein. The peptidoglycan structure is similar to that of group A *Streptococcus*. DNA transformation between *S. pneumoniae* and other streptococci has been demonstrated, as well as DNA–RNA homology between *S. pneumoniae* and group A and H streptococci. Unlike most streptococci, *S. pneumoniae* can hydrolyze inulin. It is a facultative anaerobe; however, 10 to 20% of the pneumococci may appear as strict anaerobes or first grow anaerobically in primary culture, especially when isolated from blood cultures. In clinical specimens, *S. pneumoniae* appears as lancet-shaped gram-positive diplococci; a capsule may be visible on Gram staining. *Streptococcus pneumoniae* does not produce catalase and therefore depends on the presence of exogenous catalase to destroy H_2O_2 produced during growth. Growth on blood agar plates produces relatively large (0.5 to 1.5 mm), glistening, semitranslucent colonies, with α-hemolysis of horse, human, rabbit, and sheep red cells. Mucoid colonies are usually produced by types 3 and 37. Occasionally, colonies may be rough. After prolonged incubation, *S. pneumoniae* may become gram negative. It is nonmotile and does not sporulate.

Under anaerobic conditions, β-hemolysis is caused by oxygen-labile hemolysin O. This hemolysis may be observed in 39% of strains grown anaerobically on horse blood agar at 37°C for 24 h and refrigerated for 48 h.

Isolation and Identification Procedures

Streptococcus pneumoniae should be regarded as a fastidious organism, and the viable count is reduced if the specimen is allowed to dry at room temperature. The clinical specimen should be transported in nutrient broth if storage is required, or sent to the clinical laboratory as soon as possible (within 1 h). An enriched culture medium (sheep blood tryptic soy, Columbia, or brain–heart infusion agar) is necessary for isolation. Subculture in broth medium is optimal in brain–heart infusion broth, Todd–Hewitt broth, or broth supplemented with serum or ascites (10%) and glucose. For optimum growth, a pH of 7.8 (7 to 8.3) and a temperature of 37°C (25 to 41°C) are required. An increased CO_2 atmosphere is required by 5 to 10% of isolates. During prolonged incubation, *S. pneumoniae* undergoes autolysis due to the release of *N*-acetyl muramyl 1-alanine amidase. This autolysis occurs in broth culture at 48 h, is pH dependent, and is more likely to occur in the presence of β-lactam antibiotics. Autolysis may result in false-negative blood cultures. Mutants with defective triggering of autolysis are tolerant to the β-lactam antibiotics.

Sputum is not an ideal specimen for the diagnosis of *S. pneumoniae* because the carriage rate is very high (Austrian, 1986). A critical review by Mufson (1981) reported that the rate of isolation was 26 to 78% in cases of pneumonia. Better methods for obtaining a good respiratory specimen include transtracheal or protected fiberoptic bronchial aspiration. Addition of 5 μg/ml gentamicin or 40 μg/ml nalidixic acid to sheep blood agar plates improves recovery in the laboratory of pneumococci from sputum samples by inhibiting the growth of contaminating bacteria.

The following tests are useful in the identification of *S. pneumoniae*.

Tube Bile Solubility

A few drops of 10% sodium deoxycholate are added to a turbid suspension of *S. pneumoniae* in saline. The clearing of the suspension compared to the control tube indicates solubility of the organism in bile. This test may be performed on blood agar plates to which 2 to 3 drops of 2% sodium deoxycholate have been added, followed by incubation at 35°C for 30 min. This test is now used less frequently than the following tests.

Optochin (Ethylhydroxycuprein Hydrochloride) Susceptibility

Growth of *S. pneumoniae* on blood agar plates is inhibited by a disk containing 5 μg of optochin. A growth inhibition zone with a diameter greater than 16 mm in air (greater than 13 mm in CO_2) after 24 h of incubation at 37°C is obtained with *S. pneumoniae*. Other α-hemolytic streptococci have zone diameters of less than 10 mm.

Neufeld Quellung Reaction

The clinical specimen or a loopful from an isolated colony is mixed with an equal quantity of type-specific antiserum or Omniserum, which recognizes 84 serotypes with or without methylene blue. After the antiserum is added, the mixture is examined microscopically using a 100× immersion objective. When the reaction is positive, the capsule will appear prominent compared with the control preparation in saline. This capsular "swelling" is caused by a change in the refractive index due to complexation with the antiserum. The antisera, which are available from the Statens Seruminstitut (Copenhagen, Denmark), consist of Omniserum against the 84 known serotypes, nine pooled sera (A through I), and individual antisera.

Agglutination Tests

Latex particles coated with Omniserum (Pneumoslide, BBL Microbiology Systems, Cockeysville, Md.; Slidex Pneumokit, Bio-Mérieux Charbonnières les Bains, France; Wellcogen *S. pneumoniae*, Wellcome Research Laboratories, Beckenham, England) are used to identify isolated colonies. Agglutination is obtained with *S. pneumoniae*. This test is more than 99% accurate compared with the reference optochin method. Coagglutination (Phadebact *Pneumococcus* test, Pharmacia, Uppsala, Sweden) has been used successfully; sensitivity is 100% with *S. pneumoniae*, but false-positive results have been observed with other optochin-resistant viridans streptococci.

The presence of a positive test (bile solubility or optochin susceptibility) and the correct colony,

Gram staining, and hemolytic results are considered adequate for the speciation of *S. pneumoniae* and for its differentiation from other streptococci (Jones et al., 1982): *Streptococcus pneumoniae* has been isolated from virtually all body sites, even bile. The diagnosis of acute pneumococcal meningitis is rarely difficult: Gram staining usually reveals numerous gram-positive diplococci, and pleocytosis (0.1 to 13 × 10^6 cells/liter) of the CSF with more than 90% polymorphonuclear leukocytes, high protein levels, low glucose levels, and high lactate concentrations (more than 40 mg/dl) is typically seen.

Surface Antigens

Capsular polysaccharides are complex structures that form a gel around the cell. Of the 84 serotypes, the structure of some is known. These capsular polysaccharides are antigenic, and antibodies against them are protective when they belong to the IgG2 class. The type and amount of capsular material is related to virulence, although some highly capsulated serotypes (e.g., serotype 37) are not virulent. Certain serotypes (6, 14, 18, 19, and 23) are responsible for most cases of bacteremia and meningitis in children and serotypes 3 and 19 are a leading cause of otitis media in children. Some serotypes responsible for invasive infections in South Africa (serotypes 2, 25, 45, and 46) and in Papua New Guinea (types 2 and 25, included in the vaccine) are seldom isolated in the United States. Moreover, the spectrum of serotypes causing pneumococcal disease varies significantly from year to year (Finland and Barnes, 1977). Serotyping should be performed in reference centers for all strains isolated from normally sterile body sites using sera from the Statens Seruminstitut to determine which formulation of the polyvalent vaccines available (14-, 17-, or 23-valent) is adequate. Capsular polysaccharides may be demonstrated and typed by the reference quellung reaction, by agglutination, by counterimmunoelectrophoresis, or more recently by immunofluorescence; the three latter methods, however, may provide misleading results. Cross-reactions exist between pneumococcal types 2 and 5, 3 and 8, 7 and 18, and 13 and 30 and with polysaccharides of *Escherichia coli, Klebsiella pneumoniae, Salmonella* spp., *H. influenzae* type b, viridans streptococci, group B *Streptococcus*, or human blood group antigens. The capsular polysaccharide leaks from the capsule into the surrounding body fluid and may be eliminated through the serum in urine after partial degradation; however, the fragments isolated in the urine usually retain their antigenic properties.

C polysaccharide, a group-specific teichoic acid polymer containing choline and galactosamine 6-phosphate, is a major constituent of the cell wall. It is precipitated by a human β-globulin C-reactive protein. The binding of C-reactive protein to C polysaccharide can activate complement and mediate phagocytosis of some nonvirulent serotypes of pneumococci.

M-protein antigen is a type-specific protein analogous to the M protein of group A *Streptococcus*. However, it has no antiphagocytic properties. *R-protein antigen* has been isolated from rough strains of pneumococci.

Toxins

Oxygen-labile O hemolysin (pneumolysin) is related to that of β-hemolytic streptococci and produces β-hemolysis under anaerobic conditions. Pneumolysin is a sulfhydril-activated cytolysin that acts by binding cholesterol to the host cell membrane. It may cause hemolytic anemia in pneumococcal bacteremia in rabbits. Pneumolysin may also directly inhibit phagocytic clearance of pneumococci by inhibiting the respiratory burst, bactericidal activity, and migration of human polymorphonuclear leukocytes. It is an inhibitor of mitogen-induced lymphocyte proliferation and can activate the classic complement pathway and deplete serum opsonic activity. The pneumolysin gene has been cloned recently in *E. coli* (Paton et al., 1986).

Purpurigenic protein induces a purpuric condition in experimental animals. Purpura is an occasional observation in pneumococcal infections. Other substances (virulin, leukocidin, necrotin, and neuraminidase) have been described, but their contribution to virulence requires further clarification. Pneumococcus-induced thrombocytopenia in the rabbit may be explained by the direct action on circulating platelets of a heat-stable, trypsin-resistant pneumococcal substance with a high molecular weight (Goldblum et al., 1985, 1986). Experimental evidence and the clinical presentation of severe pneumococcal infections suggest that one or several toxins may be produced by *S. pneumoniae* and contribute to its virulence.

Detection of Soluble Antigen in Body Fluids

Two substances may be detected in various body fluids during pneumococcal infection: C polysaccharide and type-specific capsular polysaccharide. An enzyme-linked immunosorbent assay with a sensitivity of 82% and a specificity of 94% has been used to detect C polysaccharide in the sputum of patients with pneumonia (Holmberg et al., 1985a). Type-specific capsular polysaccharide has been detected using tube precipitation, counterimmunoelectrophoresis,

latex agglutination, coagglutination, and enzyme-linked immunosorbent assay.

Studies of pneumonia in adults have shown that 40 to 80% of cases are positive for pneumococcal antigen in the urine, and half are also positive in the serum (Bukantz et al., 1942; Cerosaletti et al., 1985; Heffron, 1979; Van der Auwera et al., 1983). Detection of antigen in urine specimens has also proved useful in determining the etiologic agent of lower respiratory infections in children; in one study, pneumococcal antigenuria was found in 24% of children 2 months to 18 years of age (Ramsey et al., 1986). In the same study, antigenuria was found in 16% of children with acute otitis media. Detection in the sputum may be misleading, as *S. pneumoniae* is a normal inhabitant of the pharynx and cross-reacts with several other organisms (notably *H. influenzae*) that are saprophytes of the pharynx and may cause pneumonia.

The sensitivity of these tests will probably not be improved, as the capsular polysaccharide is metabolized and may lose its antigenicity; furthermore, the use of a polyvalent antibody (84 serotypes), which results unavoidably in a low titer against each individual antigen, limits the sensitivity of antigen detection. Several pathogenic strains may not produce enough capsular material to be detected. On the other hand, it has been demonstrated that very sensitive methods can detect pneumococcal polysaccharide in the urine of healthy carriers or convalescent patients several months after the infectious episode, which contributes to a lack of specificity. Capsular polysaccharide has been detected for several months in the urine of infected adults with slowly resolving infections. Types 7F, 14, and 33F, which cannot be detected by counterimmunoelectrophoresis, constitute 5 to 20% of the strains that cause disease. The only truly rapid methods are latex agglutination and coagglutination, which require less than 10 min to be performed. The sample should be treated for 10 min in a boiling water bath to destroy interfering substances, and an aliquot of the treated sample should be tested with uncoated particles to provide a negative control.

These tests have been used in the etiologic diagnosis of acute meningitis with almost 100% sensitivity and specificity (Denis et al., 1981a,b; Tugwell et al., 1976). Furthermore, they may detect the specific antigen even when the patient has been partially treated with effective antibiotics. The amount of antigen present in the serum in bacteremia (Bukantz et al., 1942; Heffron, 1979), or in the CSF in meningitis (Denis et al., 1981a,b) has been correlated with prognosis. Detection of antigen in pleural fluid has been used successfully in the diagnosis of purulent empyema.

Immunologic Aspects

After resolution of pneumonia, antibodies against type-specific capsular polysaccharide and cell wall C polysaccharide have been detected by indirect hemagglutination, radioimmunoassay, and enzyme-linked immunosorbent assay. The determination of these antibodies in the serum probably has no diagnostic role in routine clinical microbiology because of the delay in optimal response, the technical problems caused by the antigenic diversity of the type-specific capsular polysaccharide (Musher et al., 1986), and its low sensitivity for the species-specific cell wall polysaccharide (Holmberg et al., 1985a,b).

Immune opsonization by specific anti-capsular polysaccharide antibodies is required for phagocytosis of *S. pneumoniae*. C3b is not deposited in the absence of specific anti-capsular antibodies in virulent strains; avirulent strains may activate the complement cascade through the alternate pathway. Nascent C3b activated via the alternative pathway is fixed to the pneumococcal cell wall, deep within the capsule; therefore, the capsule may act as a mechanical barrier to the recognition of C3b by the phagocytic cell. Pneumococcal serotypes that differ in their capsular polysaccharides also differ in the amount and site of bound C3b and degradative processing to iC3d (Hotstetter, 1986). These differences may explain the well-known patterns of serotypic variance in resistance to phagocytosis and in stimulation of antibody production. However, it remains unclear which chemical structure of the cell wall of virulent pneumococci protects them against the alternative pathway convertase. Antibodies against C polysaccharide and C-reactive protein are unable to opsonize virulent strains. The immune response against purified polyvalent capsular polysaccharide vaccine is restricted to the IgG2 subclass. This system is immature in children less than 2 years old and is dependent on the spleen function. Immunologic paralysis to pneumococcal polysaccharide has been suggested in a 9-month-old infant with severe pneumococcal meningitis (Pichichero, 1985). During severe pneumococcal infection complement is depleted and the availability of alternate pathway factors is reduced in acute sera; these findings are more pronounced in complicated illness and are associated with high serum concentrations of C-reactive protein (Rabinovitch et al., 1986).

The capsule of virulent pneumococci has antiphagocytic properties. However, patients with defects of phagocytic bactericidal activity, as in chronic granulomatous disease or abnormalities of granules, are not at increased risk of pneumococcal infection. The increased risk of agammaglobulinemia is due to the absence of effective immune opsonization through the classic complement activation pathway, which is a prerequisite for effective phagocytosis by neutrophils. The absence of mucosal antibodies, other than IgA, may play a role in the increased risk for agammaglobulinemic patients. Patients with IgA deficiency are not predisposed to pneumococcal infection. Cell-mediated immunity does not appear to play an essential role in the defense against *S. pneumoniae*. Pneumococci are easily killed once ingested by neutrophils or monocytes, and macrophage activation by T lymphocytes is probably unnecessary. It remains unclear, however, why pneumococci are killed so effectively once they are ingested by phagocytes, even in absence of functional oxidative systems. Cirrhotic patients have serum and ascitic antibodies to pneumococcal capsular polysaccharide (Simberkoff et al., 1983). Nonetheless, *S. pneumoniae* remains a common pathogen isolated from spontaneous peritonitis in cirrhosis.

Treatment, Prevention, and Control

Treatment

Penicillin remains the therapeutic agent of choice for all pneumococcal diseases in nonallergic patients. The treatment of pneumococcal pneumonia consists of penicillin G or V. The study by Brevin et al. (1974) of the relationship between dose and outcome revealed no significant difference in outcome between patients who received 20 million units of penicillin per day and those who received 1.2 million units of penicillin G per day, providing that the strain was highly susceptible (MIC less than 0.1 μg/ml).

The prognosis depends on the severity of the underlying disease as well as the type of infection caused by *S. pneumoniae* (Table 2). The mortality secondary to pneumococcal sepsis remains high despite the administration of adequate antibiotic regimens.

In penicillin-allergic patients, erythromycin is an adequate alternative. New antimicrobial agents have also been investigated recently, but their role in the treatment of *S. pneumoniae* should be extremely limited.

The relationship between the concentration of β-lactam antibiotics in the CSF and bactericidal activity has been studied in a rabbit model of pneumococcal meningitis (Täuber et al., 1984). Maximal bactericidal activity was achieved with a CSF concentration of 10 to 30 × the minimum bactericidal concentration or the organism. This activity can be obtained with high doses of penicillin G (20×10^6 U/day in adults) or with well-penetrating antibiotics such as ceftriaxone, which has been shown as effective as ampicillin plus chloramphenicol in the treatment of meningitis in children, although ceftriaxone sterilized the CSF more rapidly (Del Rio et al., 1983). Other new β-lactam antibiotics such as cefuroxime and cefotaxime have been also used successfully in the treatment of acute pneumococcal meningitis. Chloramphenicol and chloramphenicol plus penicillin were equally effective for the treatment of pneumococcal pneumonia or meningitis in children (Shann et al., 1985a,b).

Therapy for disease caused by resistant pneumococci depends on the level of penicillin resistance, the pattern of resistance to alternative antimicrobial agents, and the site of infection. Except for meningitis, high-dose penicillin remains the treatment of choice for these infections. However, meningitis caused by strains with MICs greater than 1 μg/ml does not respond to penicillin. Chloramphenicol is the second-choice antibiotic for these strains, provided that they are still susceptible.

Gransden et al. (1985) reported that one-half of survivors with pneumococcal bacteremia were treated for 10 days or less and became afebrile within 48 h. Longer treatment was required in patients with pleural effusion (mean, 16 days) and in those infected with serotype 3 (mean, 15 days). Prolonged treatment may be inappropriate when associated with sterile pleural effusion. Persistent fever has been described during effective treatment for meningitis.

Prevention and Control

Vaccination

Three capsular polysaccharide vaccines have been developed: a 14-valent (50 μg of each antigen, Pneumovax; Merck Sharp & Dohme, West Point, Pa.), a 17-valent (50 μg of each antigen, Moniarix; Smith Kline & French Laboratories, Philadelphia, Pa.), and a 23-valent (25 μg of each antigen, Pneumovax 2; Merck Sharp & Dohme). Of the virulent serotypes, 55 to 83% are included in the 14-valent vaccine, 75 to 85% in the 17-valent vaccine, and 80 to 94% in the 23-valent vaccine, depending on the country considered. Weisholtz et al. (1983) have shown that significantly fewer strains isolated from high-risk patients with immune deficiency belong to the distribution of serotypes included in the vaccine than are the strains isolated from moderate- and low-risk patients.

Although it remains controversial, vaccination is recommended for the following patients (Centers for Disease Control, 1984; Health and Public Policy Committee, American College of Physicians, 1986): adults with chronic cardiovascular or pulmonary diseases, splenic dysfunction or asplenia, Hodgkin's disease, multiple myeloma, cirrhosis, alcoholism, renal failure, leakage of CSF, or immunosuppression; those more than 65 years old, even if healthy; and children aged 2 years and older with asplenia (functional or anatomic) such as in sickle cell disease or before splenectomy, nephrotic syndrome, leakage of CSF, or immunosuppression. Recurrent otitis media and sinusitis do not seem to be good indications for vaccine use. The vaccine should probably be administered only once to adults to prevent an Arthus reaction. In children less than 2 years of age, revaccination is not beneficial even after two doses.

Adverse reactions, observed in less than 1% of patients, consist of fever, myalgias, and severe local reactions. It has generally been accepted that a level of 250 ng of antibody nitrogen per ml against a specific type was protective; however, this concept has recently been contested (Musher et al., 1986).

Several serologic studies have suggested that the vaccine may not evoke a protective antibody response in patients with high-risk conditions for whom the vaccine is recommended (Landesman and Schiffman, 1981). Patients with insulin-dependent diabetes have a normal response to polyvalent pneumococcal vaccine (Schiffman, 1980). Patients with an underlying disease of the immune system (such as Hodgkin's disease) or who are splenectomized (functionally or anatomically) do not respond to the pneumococcal vaccine, especially if administered after splenectomy. Patients with sickle cell disease, Hodgkin's disease, or renal or bone marrow transplant respond poorly. Healthy patients who undergo splenectomy have a nearly normal antibody response. Serotypes 6A and 23F have a very low immunogenicity, especially in children. When possible, patients who must be splenectomized or receive immunosuppressive therapy should be immunized within the preceding 2 weeks. As elderly patients are at major risk, combined vaccination against influenza virus and pneumococcal polysaccharides has been tested and proved successful (Grabenstein et al., 1986).

The protective efficacy of vaccination is 60 to 92%. The highest rates (76 to 92%) have been found in healthy young adults, such as South African gold miners in epidemic conditions. Vaccination is also highly effective (60 to 75%) and probably cost-effective in patients with a moderately increased risk of pneumococcal infection, such as the elderly and those with chronic diseases (Bolan et al., 1986; Sisk and Riegelman, 1986); it is less effective (0 to 37%), however, in high-risk groups such as severely immunosuppressed patients, including those with Hodgkin's disease and multiple myeloma (Broome et al., 1980; Shapiro and Clemens, 1984). The efficacy of the vaccine has also been questioned in patients with bronchogenic carcinoma (Klastersky et al., 1986). A new approach, using covalent conjugates between capsular polysaccharides and tetanus toxoid, has been proposed (Schneerson et al., 1986).

Antimicrobial Prophylaxis

The risk of septicemia has been estimated at 7 to 10 per 100 persons per year of observation in children with sickle cell anemia, especially those less than 3 years of age. This septicemia is often fulminant, with fatalities in 35% of patients within 12 h. Risk and mortality have remained unchanged for the past 20 years. A multicenter, randomized, double-blind, placebo-controlled study demonstrated that despite administration of a pneumococcal vaccine to all children studied, those treated with 125 mg of penicillin V potassium twice daily showed an 84% reduction in the incidence of infection compared with those receiving placebo (Gaston et al., 1986). No death from pneumococcal septicemia occurred in the treated group. Prophylactic penicillin therapy should be started no later than 4 months after birth and continued beyond the age of 3 years.

It has been suggested in a controlled study that prophylactic therapy with penicillin (600,000 U of penicillin G benzathine monthly) may provide protection against pneumococcal septicemia in children with splenic dysfunction (John et al., 1984). The serum level of penicillin achieved with this mode of administration is rather low, however, particularly for *S. pneumoniae,* because MICs to penicillin are increasing. Therefore, whether this approach should be recommended remains to be demonstrated.

Similarly, patients with thalassemia should also receive penicillin prophylaxis. Prophylactic penicillin should probably be administered to young splenectomized patients with Hodgkin's disease, especially those who are receiving intensive radiotherapy or antineoplastic chemotherapy. In fact, despite the absence of controlled studies on penicillin prophylaxis in splenectomized patients, the rarity of treatment failure suggests that it may be recommended for all children and adults undergoing splenectomy. The duration of prophylaxis remains controversial; however, because sepsis may occur several years after splenectomy, penicillin prophylaxis may be required indefinitely (Zarrabi and Rosner, 1984, 1986). Meanwhile, the following recommendations seem to be reasonable: vaccination as soon as possible and penicillin prophylaxis for at least 1 month after splenectomy. In any case, the patient should be carefully informed of the risk of fulminant bacteremia and that any febrile episode requires prompt self-administered oral antibiotic therapy (preferably a β-lactam) and medical consult. Bone marrow transplantation may also be considered an indication for long-term penicillin prophylaxis. Chronic liquorrhea, despite the absence of large studies in the literature, constitutes another indication for penicillin prophylaxis.

Control Measures Against Penicillin-Resistant Pneumococci

In the presence of sporadic cases or in the absence of secondary spread, the following control measures may be useful: 1) Restriction of the routine and often injudicious use of broad-spectrum antibiotics; 2) isolation of infected patients and compliance with respiratory precautions; and 3) surveillance of contacts to detect carriers of resistant pneumococci. In addition to these measures, in the presence of an outbreak isolation of susceptible patients from those carrying resistant strains (cohorting) should be accomplished; surveillance of contacts and patients at risk of pneumococcal infection should be reinforced; and, in selected situations, eradication of the carrier state should be attempted using effective antimicrobial agents. In eradication of pharyngeal carriage of erythromycin-susceptible strains, a 7-day course of erythromycin was 41% effective, erythromycin plus rifampin was 96% effective, and minocycline was 88% effective, whereas fusidic acid plus rifampin was only 65% effective. Ineffective agents for this purpose include ampicillin, trimethoprim–sulfamethoxazole, and novobiocin (Ward, 1981). Vaccination of patients more than 2 years old is theoretically recommendable if the resistant strain has a serotype included in the vaccine; however, patients who are at high risk are less likely to be protected.

Literature Cited

Austrian, R. 1981. Some observations on the pneumococcus and on the current status of pneumococcal disease and its prevention. Rev. Infect. Dis. (Suppl.) **3:**1–17.

Austrian, R. 1986. Some aspects of the pneumococcal carrier state. J. Antimicrob. Chemother. Suppl. A **18:**35–45.

Austrian, R., and J. Gold. 1964. Pneumococcal bacteremia with especial reference to bacteremic pneumococcal pneumonia. Ann. Intern. Med. **60:**759–776.

Austrian, R., V. M. Howie, and J. M. Ploussard. 1977. The bacteriology of pneumococcal otitis media. Johns Hopkins Med. J. **141:**104–111.

Bartlett, J. G., P. O'Keefe, F. P. Tally, T. J. Lovie, and S. L. Gorbach. 1986. Bacteriology of hospital-acquired pneumonia. Arch. Intern. Med. **146:**868–871.

Bernard, C., G. Monbelli, and J. Klastersky. 1981. Pneumococcal bacteremia in patients with neoplastic diseases. Eur. J. Cancer Clin. Oncol. **17:**1041–1046.

Bolan, G., C. V. Broome, R. R. Facklam, B. D. Plikaytis, D. W. Fraser, and W. F. Schlech 3rd. 1986. Pneumococcal vaccine efficacy in selected populations in the United States. Ann. Intern. Med. **104:**1–6.

Bratton, L., D. W. Teel, and J. O. Klein. 1977. Outcome of unsuspected pneumococcemia in children not initially admitted to the hospital. J. Pediatr. **90:**703–706.

Brevin, A., L. Arango, K. Hadley, and J. F. Murray. 1974. High-dose penicillin and pneumococcal pneumonia. J. Am. Med. Assoc. **230:**409–413.

Broome, C. V., R. R. Facklam, and D. W. Fraser. 1980. Pneumococcal disease after pneumococcal vaccination: an alternative method to estimate the efficacy of pneumococcal vaccine. N. Engl. J. Med. **303:**549–552.

Bukantz, S. C., P. F. De Gara, and J. G. M. Bullowa. 1942. Capsular polysaccharide in the blood of patients with pneumococcic pneumonia. Arch. Intern. Med. **69:**191–212.

Burman, L. A., R. Norrby, and B. Trollfors. 1985. Invasive pneumococcal infections: incidence, predisposing factors, and prognosis. Rev. Infect. Dis. **7:**133–142.

Centers for Disease Control, Department of Health and Human Services. 1984. Update: pneumococcal polysaccharide vaccine usage. United States. Recommendations of the Immunization Practice Advisory Committee. Ann. Intern. Med. **101:**348–350.

Cerosaletti, K. M., M. C. Roghmann, and D. W. Bentley. 1985. Comparison of latex agglutination and counterimmunoelectrophoresis for the detection of pneumococcal antigen in elderly pneumonia patients. J. Clin. Microbiol. **22:**553–557.

Chou, M. Y., A. E. Brown, A. Blevins, and D. Armstrong. 1983. Severe pneumococcal infection in patients with neoplastic disease. Cancer **51:**1546–1550.

Del Rio, M., A. de Los, D. Chrane, S. Shelton, G. H. McCracken, Jr., and J. D. Nelson. 1983. Ceftriaxone versus ampicillin and chloramphenicol for the treatment of bacterial meningitis in children. Lancet **1:**1241–1244.

Denis, F., M. Cadoz, I. Diop Mar, and M. Saulnier. 1981a. Bacterial antigen concentrations in cerebrospinal fluid and prognosis of purulent meningitis (letter). Lancet **1:**1361–1362.

Denis, F., M. Saulnier, M. Cadoz, Y. Roger d'Albert, J. P. Chiron, S. Mboup, M. Prince-David, and I. Diop Mar. 1981b. Diagnostic etiologique rapide des méningites purulentes grace à un kit méningite au latex. Résultats comparatifs avec les techniques classiques portant sur plus de 1300 méningites purulentes. Med. Malad. Infect. **11:**617–621.

EORTC International Antimicrobial Therapy Project Group. 1978. Three antibiotic regimens in the treatment of infection in febrile granulocytopenic patients with cancer. J. Infect. Dis. **137:**14–27.

EORTC International Antimicrobial Therapy Project Group. 1983. Combination of amikacin and carbenicillin with or without cefazolin as empirical treatment of febrile neutropenic patients. J. Clin. Oncol. **1:**597–603.

EORTC International Antimicrobial Therapy Project Group. 1986. Prospective randomized comparison of three antibiotic regimens for empirical therapy of suspected bacteremic infection in febrile granulocytopenic patients. Antimicrob. Agents Chemother. **29:**263–270.

Fedson, D. S., and J. A. Baldwin. 1982. Previous hospital care as a risk factor for pneumonia. Implications for immunization with pneumococcal vaccine. J. Am. Med. Assoc. **248:**1989–1985.

Feldman, C., J. M. Kallenbach, S. D. Miller, J. R. Thorburn, and H. J. Koornhof. 1985. Community-acquired pneumonia due to penicillin-resistant pneumococci. N. Engl. J. Med. **313:**615–617.

Finland, M., and M. W. Barnes. 1977. Changes in occurrence of capsular serotypes of *Streptococcus pneumoniae* at Boston City Hospital during selected years between 1935 and 1974. J. Clin. Microbiol. **5:**154–166.

Folland, D., D. Armstrong, S. Seides, and A. Blevins. 1974. Pneumococcal bacteremia in patients with neoplastic disease. Cancer **33:**845–849.

Gaston, M. H., J. I. Verter, G. Woods, C. Pegelow, J. Kelleher, G. Presbury, H. Zarkowsky, E. Vichinsky, R. Iyer, J. S. Lobel, S. Diamond, C. T. Holbrook, F. M. Gill, K. Ritchey, and J. M. Falletta for the Prophylactic Penicillin Study Group. 1986. Prophylaxis with oral penicillin in children with sickle cell anemia. A randomized trial. N. Engl. J. Med. **314**:1593–1599.

Goldblum, S. E., T. L. Simon, J. P. Thilsted, and T. L. Badcock. 1985. Pneumococcus-induced thrombocytopenia in rabbits. J. Lab. Clin. Med. **106**:298–307.

Goldblum, S. E., K. M. Wu, and H. H. Tai. 1986. *Streptococcus pneumoniae* induced alteration in levels of circulating thromboxane and prostacyclin: dissociation from granulocytopenia, thrombocytopenia and pulmonary leukostasis. J. Infect. Dis. **153**:71–77.

Golden, G. T., and A. Shaw. 1972. Primary peritonitis. Surg. Gynecol. Obstet. **135**:513–516.

Gorse, G. J., L. D. Thrupp, K. L. Nudleman, F. A. Wyle, B. Hawkins, and T. C. Cesario. 1984. Bacterial meningitis in the elderly. Arch. Intern. Med. **144**:1603–1607.

Grabenstein, J. D., L. J. Smith, D. W. Carter, R. J. Engler, R. Evans III, and R. J. Summers. 1986. Comprehensive immunization delivery in conjunction with influenza vaccination. Arch. Intern. Med. **146**:1189–1192.

Gransden, W. R., S. J. Eykyn, and I. Phillips. 1985. Pneumococcal bacteraemia: 325 episodes diagnosed at St-Thomas's Hospital. Br. Med. J. **290**:505–508.

Hansman, D., S. Morris, M. Gregory, and B. McDonald. 1985. Pneumococcal carriage amongst Australian aborigines in Alice Springs, Northern Territory. J. Hyg. **95**:677–684.

Health and Public Policy Committee, American College of Physicians. 1986. Pneumococcal vaccine. Ann. Intern. Med. **104**:79–86.

Heffron, R. 1979. Pneumonia, with special reference to pneumococcus lobar pneumonia, p. 1–1086. The Commonwealth Fund, New York. Harvard University Press, Cambridge, Mass.

Holmberg, H., T. Holme, A. Krook, T. Olsson, L. Stöjberg, and A. M. Sjögren. 1985a. Detection of C. polysaccharide in *Streptococcus pneumoniae* in the sputa of pneumonia patients by an enzyme-linked immunosorbent assay. J. Clin. Microbiol. **22**:111–115.

Holmberg, H., A. Krook, and A. M. Sjögren. 1985b. Determination of antibodies to pneumococcal C. polysaccharide in patients with community acquired pneumonia. J. Clin. Microbiol. **22**:808–814.

Hotstetter, M. K. 1986. Serotypic variations among virulent pneumococci in deposition and degradation of covalently bound C3b: implications for phagocytosis and antibody production. J. Infect. Dis. **153**:682–693.

Hyslop, N. E., Jr., and W. W. Montgomery. 1980. Diagnosis and management of meningitis associated with cerebrospinal fluid leaks, p. 254–285. *In* J. S. Remington and M. N. Swartz (ed.), Current clinical topics in infectious diseases, vol. 3. McGraw-Hill Book Co., New York.

Jacobs, M. R., H. J. Koornhof, R. M. Robins-Browne, C. M. Stevenson, Z. A. Vermaak, I. Freiman, G. B. Miller, M. A. Witcomb, M. Isaäcson, J. I. Ward, and R. Austrian. 1978. Emergence of multiply resistant pneumococci. N. Engl. J. Med. **299**:735–740.

John, R. B., A. Ranmal, H. Jackson, G. H. Maude, A. Waight Sharma, and G. R. Serjeant. 1984. Prevention of pneumococcal infection with homozygous sickle cell disease. Br. Med. J. **288**:1567–1570.

Jones, R. N., D. C. Edson, and the CAP Microbiology Resource Committee. 1982. The ability of participant laboratories to detect penicillin-resistant pneumococci. A report from the microbiology portion of the College of American Pathologists (CAP) surveys. Am. J. Clin. Pathol. (Suppl.) **78**:659–663.

Kaufmann, C. A., C. Watanakunakorn, and J. P. Phair. 1973. Purulent pneumococcal pericarditis: a continuing problem in the antibiotic era. Am. J. Med. **54**:743–750.

Klastersky, J., P. Mommen, F. Cantraine, and A. Safary. 1986. Placebo controlled pneumococcal immunization in patients with bronchogenic carcinoma. Eur. J. Cancer Clin. Oncol. **22**:807–813.

Landesman, S. H., and G. Schiffman. 1981. Assessment of the antibody response to pneumococcal vaccine in high-risk populations. Rev. Infect. Dis. (Suppl.) **3**:184–196.

Lederman, M. M., J. Lowder, and P. I. Lerner. 1982. Bacteremic pneumococcal epiglottitis in adults with malignancy. Am. Rev. Respir. Dis. **125**:117–118.

Legrand, J. C., E. Jeuniaux, P. Van der Auwera, M. Demay, and G. Pollaert. 1985. Péritonite à pneumocoque. Acta Clin. Belg. **40**:194–197.

Mufson, M. A. 1981. Pneumococcal infections. J. Am. Med. Assoc. **246**:1942–1948.

Mufson, M. A., D. M. Kruss, and R. E. Wasil. 1974. Capsular types and outcome of bacteremic pneumococcal disease in the antibiotic era. Arch. Intern. Med. **134**:505–510.

Musher, D. M., A. J. Chapman, S. Jonsson, D. Briles, and R. E. Baughn. 1986. Natural and vaccine-related immunity to *Streptococcus pneumoniae*. J. Infect. Dis. **154**:245–256.

Myers, M. G., P. F. Wright, A. L. Smith, and D. H. Smith. 1974. Complications of occult pneumococcal bacteremia in children. J. Pediatr. **84**:656–660.

Ort, S., J. L. Ryan, G. Barden, and N. D'Esopo. 1983. Pneumococcal pneumonia in hospitalized patients: clinical and radiological presentation. J. Am. Med. Assoc. **249**:214–218.

Pahmer, M. 1940. Pneumococcus peritonitis in nephrotic and non-nephrotic children. J. Pediatr. **17**:90–105.

Paton, J. C., A. M. Berry, R. A. Lock, D. Hansman, and P. A. Manning. 1986. Cloning and expression in *Escherichia coli* of the *Streptococcus pneumoniae* gene encoding pneumolysin. Infect. Immun. **54**:50–55.

Pichichero, M. E. 1985. Immunological paralysis to pneumococcal polysaccharide in man. Lancet **2**:468–471.

Preheim, L., G. Schiffman, R. Komorowski, S. Koethe, and M. W. Rytel. 1978. Immune complex nephritis associated with pneumococcal infection. Clin. Res. **26**:404a.

Rabinovitch, R. A., S. M. Koethe, J. H. Kalbfleisch, L. C. Preheim, and M. W. Rytel. 1986. Relationships between alternative complement pathway activation, C-reactive protein, and pneumococcal infection. J. Clin. Microbiol. **23**:56–61.

Ramsey, B. W., E. K. Marcuse, H. M. Foy, M. K. Cooney, I. Allan, D. Brewer, and A. L. Smith. 1986. Use of bacterial antigen detection in the diagnosis of pediatric lower respiratory tract infections. Pediatrics **78**:1–9.

Ruben, F. L., C. W. Norden, and Y. Korica. 1984. Pneumococcal bacteremia at a medical/surgical hospital for adults between 1975 and 1980. Am. J. Med. **77**:1091–1094.

Rytel, M. W. (ed.). 1979. Rapid diagnosis in infectious disease, p. 91–103. CRC Press, Inc., Boca Raton, Fla.

Schiffman, G. 1980. Antibody response to polyvalent pneumococcal polysaccharide vaccine in diabetics. J. Am. Med. Assoc. **244**:2621–2624.

Schlech, W. F. III, J. I. Ward, J. D. Band, A. Hightower, D. W. Fraser, and C. V. Broome. 1985. Bacterial meningitis in the United States, 1978 through 1981. The National Bacterial Meningitis Surveillance Study. J. Am. Med. Assoc. **253**:1749–1754.

Schneerson, R., J. B. Robbins, J. C. Parke, Jr., C. Bell, J. J. Schlesselman, A. Sutton, Z. Wang, G. Schiffman, A. Karpas, and J. Shiloach. 1986. Quantitative and qualitative analyses of serum antibodies elicited in adults by

Haemophilus influenzae type b and pneumococcus type 6A capsular polysaccharide-tetanus toxoid conjugates. Infect. Immun. **52**:519–528.

Shann, F., J. Barker, and P. Poore. 1985a. Chloramphenicol alone versus chloramphenicol plus penicillin for bacterial meningitis in children. Lancet **2**:681–683.

Shann, R., J. Barker, and P. Poore. 1985b. Chloramphenicol alone versus chloramphenicol plus penicillin for severe pneumonia in children. Lancet **2**:684–685.

Shapiro, E. D., and J. D. Clemens. 1984. A controlled evaluation of the protective efficacy of pneumococcal vaccine for patients at high risk of serious pneumococcal infections. Ann. Intern. Med. **101**:325–330.

Shlaes, D. M., R. Mandell, S. Bass, and P. J. Spagnuolo. 1982. Bacteremia caused by *Streptococcus pneumoniae* of nonvaccine serotypes. Am. Rev. Respir. Dis. **126**:712–713.

Simberkoff, M. S., G. S. Schiffman, J. R. Spicehandler, N. H. Moldover, and J. J. Rahal, Jr. 1983. Radioimmunoassay and opsonic antibody responses to pneumococcal capsular polysaccharide vaccine in serum and ascitic fluid of cirrhotic patients. J. Clin. Microbiol. **18**:154–159.

Sisk, J. E., and R. K. Riegelman. 1986. Cost effectiveness of vaccination against pneumococcal pneumonia: an update. Ann. Intern. Med. **104**:79–86.

Spitalny, K. C., K. Bromberg, and M. B. Ginsberg. 1982. *Streptococcus pneumoniae* bacteremia without an identifiable focus in adults. Johns Hopkins Med. J. **150**:35–37.

Täuber, M. G., C. A. Doroshow, C. J. Hackbarth, M. G. Rusnak, T. A. Drake, and M. A. Sande. 1984. Antibacterial activity of β-lactam antibiotics in experimental meningitis due to *Streptococcus pneumoniae*. J. Infect. Dis. **149**:568–574.

Täuber, M. G., H. Khayam-Bashi, and M. A. Sande. 1985. Effects of ampicillin and corticosteroids on brain water content, cerebrospinal fluid pressure, and cerebrospinal fluid lactate levels in experimental pneumococcal meningitis. J. Infect. Dis. **151**:528–534.

Tomasz, A., S. Zighelboim-Daum, S. Handwerger, H. Liu, and H. Qian. 1984. Physiology and genetics of intrinsic beta-lactam resistance in pneumococci, p. 393–397. *In* L. Leive and D. Schlessinger (ed.), Microbiology 1984. American Society for Microbiology, Washington, D.C.

Tugwell, P., B. M. Greenwood, and D. A. Warrell. 1976. Pneumococcal meningitis: a clinical and laboratory study. Q. J. Med. **45**:583–601.

Van der Auwera, P., A. André, G. Bulliard, J. C. Legrand, B. Gordts, H. Van Landuyt, and F. Schuyteneer. 1983. Comparison of latex agglutination and counterimmunoelectrophoresis in the diagnosis of acute *Streptococcus pneumoniae* infections. Eur. J. Clin. Microbiol. **2**:534–540.

Vanhoof, R., R. Christiaens, R. Dierickx, H. Coignau, and J. P. Butzler. 1980. In vitro evaluation of the antibiotic susceptibility of *Streptococcus pneumoniae:* pneumococci relatively resistant to penicillin in Brussels. Acta Clin. Belg. **35**:343–348.

Wara, D. W. 1981. Host defense against *Streptococcus pneumoniae:* the role of the spleen. Rev. Infect. Dis. **3**:299–309.

Ward, J. 1981. Antibiotic-resistant *Streptococcus pneumoniae:* clinical and epidemiological aspects. Rev. Infect. Dis. **3**:254–266.

Weisholtz, S. J., B. J. Hartman, and R. B. Roberts. 1983. Effect of underlying disease and age on pneumococcal serotype distribution. Am. J. Med. **75**:199–205.

Williams, J. M., and K. M. Moser. 1986. Pneumococcal vaccine and patients with chronic lung disease. Ann. Intern. Med. **104**:106–109.

Winston, D. J., G. Schiffman, D. C. Wang, S. A. Feig, C. M. Lin, E. L. Marso, W. G. Ho, L. S. Young, and R. P. Gale. 1979. Pneumococcal infections after human bone-marrow transplantation. Ann. Intern. Med. **91**:835–841.

Woodhead, M. A. J. T. Macfarlane, R. G. Rodgers, A. Laverick, R. Pilkington, and A. D. Macrae. 1985. Aetiology and outcome of severe community-acquired pneumonia. J. Infect. **10**:204–210.

Zarrabi, M. H., and F. Rosner. 1984. Serious infections in adults following splenectomy for trauma. Arch. Intern. Med. **144**:1421–1424.

Zarrabi, M. H., and F. Rosner. 1986. Rarity of failure of penicillin prophylaxis to prevent postsplenectomy sepsis. Arch. Intern. Med. **146**:1207–1208.

CHAPTER 45

Diseases Caused by *Pseudomonas*

GIUSEPPE FALCONE and MARIO CAMPA

Disease: Infections caused by *Pseudomonas* spp.

Etiologic Agents: *Pseudomonas aeruginosa, Pseudomonas putida, Pseudomonas pseudomallei,* and *Pseudomonas mallei.* They are active agents of mineralization of organic matter and part of the microflora responsible for food spoilage.

Source: Found in soil and water; colonize plants, animals, and humans.

Clinical Manifestations: Bronchopneumonia, septicemia, meningitis, and urinary tract infections caused by *P. aeruginosa* may be acute or chronic in relation to the host immune reactivity. *P. aeruginosa* is also a cause of severe epidemic diarrhęa in infants, ocular and burn infections, osteomyelitis, and malignant external otitis. *P. pseudomallei* and *P. mallei* are causative agents of melioidosis and glanders, respectively.

Pathology: Necrosis and hemorrhagic lesions, infiltration of mononuclear cells, cytotoxic effects on macrophages and polymorphonuclear leukocytes, and widespread granulomatous lesions.

Laboratory Diagnosis: Bacteriologic isolation from blood, cerebrospinal fluid, or other clinical specimens.

Epidemiology: *P. aeruginosa* is an ubiquitous microorganism and its wide diffusion is of great importance in hospital wards, where patients may acquire the bacteria from common environmental sources by contact with human or inanimate vectors. Hospital-acquired infections especially occur in immunocompromised patients or in subjects who have had prior instrumentation or manipulative procedures. *P. pseudomallei* infections have been observed in an appreciable number of persons returning from Southeast Asia. In particular, *P. mallei* infection is a disease of veterinarians, stable workers, and teamsters.

Treatment: *P. aeruginosa* is resistant to disinfectants and several antimicrobial agents. The antibiogram is necessary.

Prevention and Control: When an outbreak of *P. aeruginosa* infection occurs in a hospital, the collection of epidemiologic data is of major importance to determine the probable site of its origin. Procedures such as biotyping, serotyping, and bacteriocin and bacteriophage typing are necessary.

The Genus *Pseudomonas*

The genus *Pseudomonas* belongs to the family *Pseudomonadaceae,* together with the genera *Xanthomonas, Frateuria,* and *Zoogloea.* These asporogenous microorganisms are straight or slightly curved, gram-negative rods, usually motile by one or several polar phlagella, oxidase- and catalase-positive, aerobic with oxygen as the terminal electron acceptor; several species, however, use also nitrate as the terminal electron acceptor. Pseudomonads are active agents of mineralization of organic matter and part of the microflora responsible for food spoilage. They are found in soil and water and colonize plants, animals, and man. Among the several species known, *P. aeruginosa* and *P. putida* are of medical importance, whereas *P. mallei* and *P. pseudomallei* represent the only true pathogens for animals and humans.

On the basis of rRNA/DNA hybridization studies, *Pseudomonas* species have been assigned to five main rRNA homology groups. The groups are as distant from one another in rRNA homology as any of them is to *E. coli*. DNA/DNA hybridization studies have allowed further subdivision into DNA homology groups (Fig. 1). Since strains with aberrant properties are not infrequent, simple dichotomic determinative keys are of limited value. Tables including multiple characters may be of great practical importance for identification.

Pseudomonas aeruginosa

General Properties

P. aeruginosa is usually identified on the basis of the aspect of the colonies, the characteristic grapelike odor of aminoacetophenone, and the production of

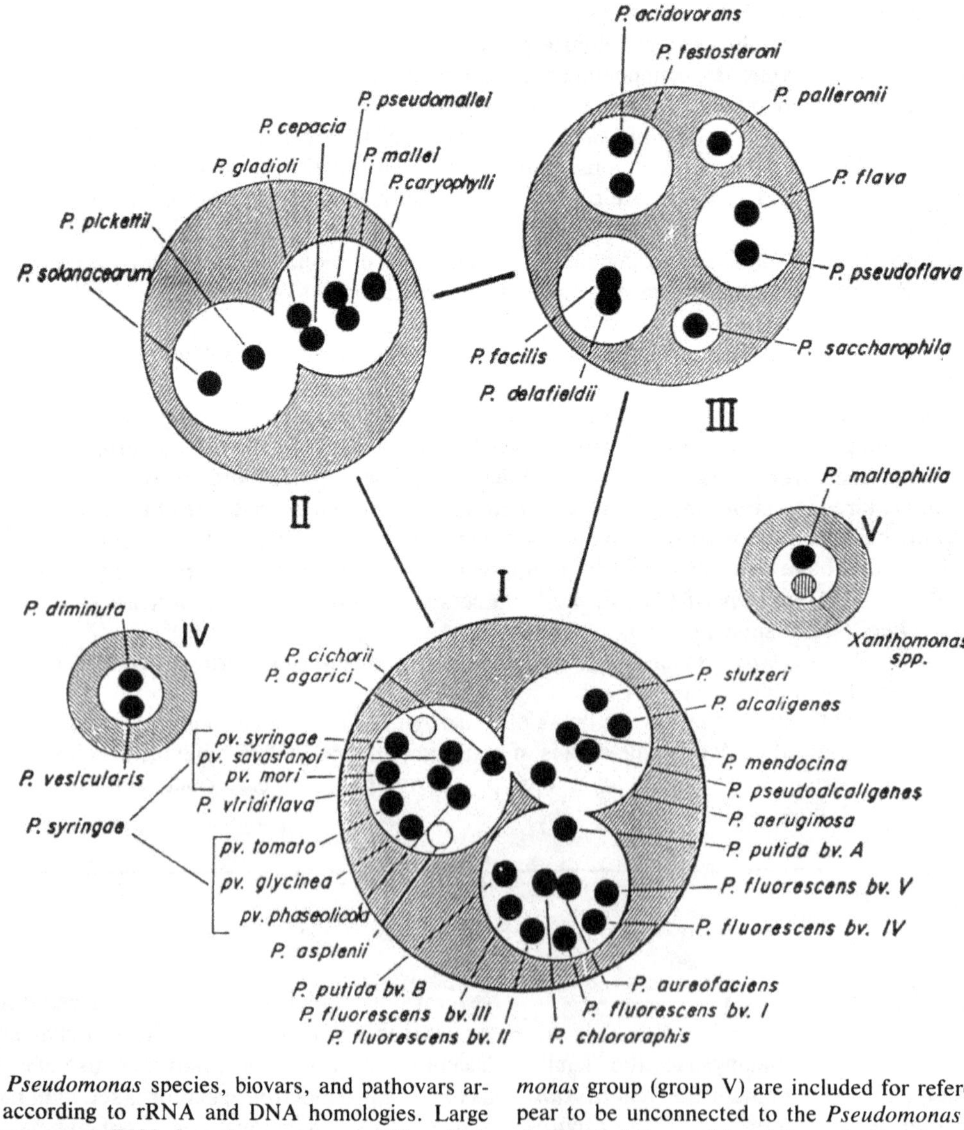

FIG. 1. *Pseudomonas* species, biovars, and pathovars arranged according to rRNA and DNA homologies. Large circles represent rRNA homology groups, within which DNA homology groups are indicated by white circles. The diminuta group (group IV) and the *P. maltophilia–Xanthomonas* group (group V) are included for reference but appear to be unconnected to the *Pseudomonas* groups. Assignment of *P. agarici* and *P. asplenii* to the "fluorescent" DNA homology group within RNA group I is only tentative. (Reprinted with permission from Palleroni, 1984.)

pyocyanin, a blue-colored phenazine pigment, which is soluble in water and chloroform and freely diffuses in the culture medium. Among *Pseudomonas* species, *P. aeruginosa* is the only one known to excrete such a pigment. Since apyocyanogenic strains may show aberrant biochemical patterns, their identification can be difficult. Characteristics of different *Pseudomonas* species have recently been reviewed by Gilardi (1985) and the most important are reported in Table 1. Because the nutritional demand of pseudomonads is modest, they grow easily on common bacteriological media. The isolation of the fluorescent pseudomonads, which are resistant to disinfectants and antimicrobial agents, can usually be accomplished from heavily contaminated specimens by using selective media containing 0.03% cetrimide.

P. aeruginosa, an ubiquitous microorganism, is able to multiply in different environmental conditions owing to its nutritional flexibility and versatility and is capable of surviving for long periods in moist environments, such as tap and distilled water, sink drains, floors, baths, puddles, nebulizers, some pharmaceuticals, shaving brushes, disinfectants, cosmetics, ornamental plants, foodstuffs, etc. Nearly any type of hospital equipment or utensil may be involved in diffusion of infection in medical wards: infusion fluids, ophthalmic solutions, dilute phenolic and benzalkonium chloride solutions, creams and lotions, water decanters, forceps, oral thermometers, suction apparatus, respiratory ventilators, humidifiers.

When an outbreak of *P. aeruginosa* infection occurs in a hospital, the collection of epidemiologic data is of major importance to determine the probable site of its origin. In such conditions, it may become necessary to identify individual strains of *P. aeruginosa* to trace the transmission chain. This can be accomplished by procedures such as biotyping, antibiograms, serotyping, bacteriocin and bacteriophage typing (Bergan, 1978; Govan, 1978; Lanyi and Bergan, 1978). *P. aeruginosa* tends to persist in the hospital environment because of its intrinsic resistance to many antibiotics and antiseptics, owing to the poor permeability of the outer membrane of the microorganism. In evolutionary terms, such a resistance has allowed the survival of *P. aeruginosa* in its major natural habitat, the soil, in which many antibiotic-secreting microorganisms compete.

Ubiquity characterizes *P. aeruginosa* as a saprophyte without a natural reservoir, which makes it difficult to control infection sources, while nutritional versatility makes a commensal relationship of the microorganism with humans possible, even though occasionally and temporarily. Since *P. aeruginosa* behaves as an opportunistic pathogen also for plants, its pathogenicity and its success as an opportunistic human pathogen can be considered the result of an evolutive process not exclusively linked to the human host. Consequently, the diffusion of *P. aeruginosa* to humans does not seem to be the final result of long reciprocal interactions but rather a fortuitous consequence of a selection for other purposes, which has been nevertheless adequate to insure the success of the microorganism as a pathogen.

The evolutive adaptation of *P. aeruginosa* to saprophytism, rather than to commensalism, is emphasized by the low ability of the microorganism to persist in the host. Intentional ingestion of *P. aeruginosa* demonstrated that the microorganism was not detectable in the feces after ingestion of 10^4 bacterial cells, whereas it was regularly cultured only after ingestion of 10^6 bacterial cells; the microorganisms gradually increased in number and finally disappeared after 6 days. Unlike infections caused by commensal microorganisms, *P. aeruginosa* infection is usually exogenous, and the wide environmental diffusion of the microorganism implies an unavoidable contagion. Nonetheless, the carrier rate in human population is low and possibly unstable, but the percentage highly increases during hospitalization. Both healthy carriers and patients infected with *P. aeruginosa* may be a source of cross-infection in the hospital, especially in burn and neonatal units and in wards with immunocompromised patients.

Pathogenic Mechanisms

As a putative pathogen, *P. aeruginosa* must have all the attributes of a successful saprophyte so that it can attach to the tissue, compete with autochthonous flora, persist and multiply, and finally build up a sufficiently large population to initiate its specific pathologic process. It is well known that *P. aeruginosa* in its natural ecosystem is always surrounded by glycocalyx, which encloses microorganisms in microcolonies adherent to solid surfaces. Glycocalyx protects the microorganism from noxious host factors such as professional phagocytes, antibodies, complement, and antibiotics as well as surfactants, which represent the first line of the host defense against bacterial colonization of the respiratory tract. In addition, growth as adherent microcolonies impairs normal function of sloughing of epithelial cells and mucociliary escalator of the respiratory mucosa (Costerton, 1984).

It is generally accepted that virulence is a multifactorial property and is the product of many interacting variables involving both the microorganism and the host. Virulence is therefore a relative concept as immunity is a relative term. As a consequence, minor alterations of the host defenses may create major differences in the apparent virulence of a microorganism. Indeed, *P. aeruginosa* can become

TABLE 1. Characteristics of medical *Pseudomonas* species[a]

Characteristic	P. aeruginosa *(64)*[b]	P. fluorescens *(96)*	P. putida *(111)*	P. acidovorans *(40)*	P. cepacia *(85)*	P. mallei *(8)*	P. maltophilia *(296)*	P. mendocina *(6)*
Acid production								
Fructose	92	98	100	100	100	100	99	100
Galactose	84	99	100	0	100		32	100
Glucose	97	100	100	98	100	100	100	100
Inositol	–	+	–		+	+	–	
Lactose	0	22	23	0	98	0	90	0
Maltose	0	52	32	0	98	0	100	0
Mannitol	67	92	24	98	100	75	0	0
Mannose	83	99	100	0	100		97	100
Rhamnose	27	71	49	0	0	0	0	0
Saccharose	0	56	13	0	80	38	94	0
Xylose	81	98	97	0	100	0	56	100
2-Ketoglutarate	63	71	68	0	12		0	0
ONPG	0	2	3	0	82	0	96	0
Starch hydrolysis	0	0	0	0	0	+/–	0	0
Acylamidase	38	6	1	100	72		0	0
Amino acid attack								
Arginine	98	99	97	0	0	83	0	100
Lysine	0	0	0	0	89	0	100	0
Ornithine	0	0	0	0	42	0	0	0
Phenylalanine	2	2	0	0	0		0	50
Citrate	100	90	90	90	90	0–38	90	
Deoxyribonuclease	13	0	0	0	0		100	0
Esculin hydrolysis	0	0	0	0	72		100	0
Gelatinase, caseinase	58	100	0	3	69	+	100	0
Growth on/at								
1% cetrimide	95	95	86	8	73		0	0
6.5% NaCl	5	3	16	0	0		0	100
1% TTC	75	79	75	3	35		0	0
SS-agar	88	98	97	63	11		0	83
MacConkey	98	100	100	98	89	0	100	100
5°C	0	100	100				0	
42°C	100	0	0				0	
Hemolysis	44	16	1	0	5		0	0
Hydrogen sulfide	0	0	0	0	0	0–25	0	0
Lecithinase	11	92	0	0	28	0	0	0
Lipase	67	42	2	20	100		100	83
Motility	97	100	100	100	99	0	100	100
No. of flagella	1	>1	>1	>1	>1	0	>1	1
Nitrate reductase	36	13	0	93	24	100	46	100
N_2 gas production	59	4	0	0	0	100	0	100
Oxidase	100	100	100	100	84	weak	1	100
PBHB	0	0	0	0	+	+	+	–
Pigments								
Pyocyanin	95	0	0	0	0	0	0	0
Pyoverdin	99	99	99	0	0	0	0	0
Brown	3	0	0					
Orange								
Red	3							
Yellow								
Urease	77	40	50	0	40	50	0	50

[a] References: Doudoroff and Palleroni (1974); Gilardi (1971a,b, 1976); Holmes et al. (1975); Snell (1973); Stanier et al. (1966). All species are catalase-positive and indole-negative. (Reprinted with permission from Bergan, 1981.)
[b] Numbers in parentheses indicate the number of strains examined. The figures in the body of the table indicate percentage positive reactions. When exact figure is unknown, + indicates presence, and – absence of a character. Carbohydrate acidification refers to concentrations of 1% in OF medium. Motility is examined by microscopy after growth in liquid media at 18–20°C.

P. pickettii (*12*)	P. pseudo-alcaligenes (*24*)	P. pseudomallei (*7*)	"P." putrefaciens (*38*)	P. stutzeri (*69*)	P. vescicularis (*3*)	P. alcaligenes (*28*)	P. diminuta (*11*)	P. testosteroni (*7*)
100	100	100	18	100	33	0	0	0
100	4	100	3	97	0	0	0	0
100	96	100	61	100	100	0	0	0
		–	–	+		–		
0	0	100	8	0	0	0	0	0
0	21	100	13	100	100	0	0	0
0	0	100	0	78	0	0	0	0
100	4	100	5	93	0	0	0	0
0	0	71	8	49	0	0	0	0
0	0	83	16	0	0	0	0	0
100	17	100	3	100	33	0	0	0
0	0	0	0	4	0	0	0	0
0	0	0	0	0	0	0	0	0
0	0	0	0	91	0	0	0	0
0	8	0	74	0	0	0	0	0
0	21	100	0	1	0	0	0	0
0	0	0	0	0	0	0	0	0
0	0	0	100	0	0	0	0	0
67	25	0	0	36	0	18	0	0
100		96	80	90	0		0	
0	0	0	100	0	0	0	55	0
0	0	57	13	0	100	0	0	0
0	0	100	97	0	30	4	73	0
0	46	0	0	0	0	7	0	0
0	4	0	90	99	0	0	0	0
0	8	0	0	0	0	8	0	0
0	0	0	60	84	0	0	0	0
100	92	100	100	100	0	100	100	71
			33					
		100	50	+		71	64	14
0	0	43	0	0	0	14	0	0
0	0	0	100	0	0	0	0	0
0	0	100	0	1	0	0	0	0
100	13	83	100	93	0	39	9	43
100	100	100	100	100	100	100	100	100
1	1	>1	1	1	1	1	>1	1
100	96	83	95	67	0	54	0	86
100	0	100	0	100	0	0	0	0
100	100	100	100	100	100	100	100	100
0	+	0	+	–	+	–	+(–)	+
0	0	0	0	0	0	0	0	0
0	0	0	0	0	0	0	0	0
100	0	43	18	22	0	29	0	0

Abbreviations: ONPG = o-nitrophenyl-β-D-galactopyranoside; Arginine = arginine dehydrolase; Lysine = lysine decarboxylase; Ornithine = ornithine decarboxylase; Phenylalanine = phenylalanine deaminase; SS-agar = salmonella-shigella agar; TTC = triphenyltetrazolium chloride; 1 = polar monotrichous; >1 = polar multitrichous; acid production may be weak or delayed from *P. acidovorans*, *P. pickettii*, *P. pseudoalcaligenes*, *P. putrefaciens*, and *P. vesicularis*.

a pathogen in patients with low general or local resistance against infection rather than because of its intrinsic disease-causing capability. Certain populations of patients are especially susceptible to *P. aeruginosa* infection. A common factor among many such patients is an interruption of the natural barriers that protect against bacterial invasion. For example, disruption of the epithelial surface permits the denuded area to be colonized by *P. aeruginosa;* subsequent tissue invasion may lead to devastating infections.

The cell envelopes are of paramount importance for the persistence and growth of *P. aeruginosa* in the host. Superficial components of the microorganism control adhesion and microcolony formation, the passage of nutrients, and the exclusion of humoral and synthetic antibacterial agents from the bacterial cell. On approaching mucosal surfaces, *P. aeruginosa* must first compete with the autochthonous microflora to establish infection. Pyocyanin appears to play a determinative role in this process, owing to its inhibitory effects on the growth of other bacteria. To persist and multiply within the autochthonous adherent bacterial population, *P. aeruginosa* must have access to, and interact with, receptors on the epithelial cells through pili which protrude out of the glycocalyx. In this context, it might be worth noticing that the fibronectin coating, which protects epithelial cells of the respiratory tract from bacterial attachment, may be lost or reduced by the action of salivary proteases. High levels of salivary proteases, together with a concomitant reduction of fibronectin coating of the respiratory cells, may account for the increased colonization of the respiratory tract of patients with cystic fibrosis by *P. aeruginosa*.

It is generally recognized that professional phagocytes, primarily polymorphonuclear leukocytes, represent the first line of the host defense against tissue invasion by *P. aeruginosa*. It has been shown that in healthy mice given an intranasal infection with this microorganism, bacteria were limited to the surface of the nasal mucosa and were cleared rapidly by granulocytes. By contrast, in leukopenic mice, *P. aeruginosa* penetrated interepithelial junctions, and the nasal epithelium underwent necrosis at the site of invasion.

Once local immunologic barriers have been breached, *P. aeruginosa* may become invasive. Several soluble factors produced by the microorganism contribute to its invasiveness.

Extracellular proteases, which are able to liquefy gelatin, dissolve elastin and fibrin, destroy collagen, affect the host physical barriers, and make nutrients available to the microorganism at the site of infection, thus facilitating bacterial proliferation and invasion. Proteases play a significant role in the pathogenesis of the lesions. Elastase exerts a destroying effect on the elastic lamina of blood vessels, and both elastase and alkaline protease produce necrosis. When injected into the skin of animals, proteases induce necrosis and hemorrhagic lesions similar to the ecthyma gangrenosum noted in human infection. In the lungs of experimental animals, proteases cause intraalveolar hemorrhage, necrosis of alveolar septal cells, and infiltration of mononuclear cells, a pathological picture similar to that observed in patients with *P. aeruginosa* pneumonia. In addition, proteases may inhibit host defenses by cleaving IgG and by inactivating complement factors.

The heat-labile hemolysin of *P. aeruginosa,* which is also called phospholipase C, and the heat-stable hemolysin, which is a glycolipid, promote tissue invasion on the basis of their necrotizing properties and their ability to break down lipids and lecithins, in conjunction with phosphatase. In particular, the heat-labile hemolysin seems to have great importance in the pathogenesis of *P. aeruginosa* pneumonia by causing tissue necrosis and destruction of pulmonary surfactants, which leads to atelectasis and other pathologic alterations.

Exotoxin A, an extracellular enzyme produced by most clinical *P. aeruginosa* isolates, is antigenic and inhibits protein synthesis by a mechanism identical to that of diphtheria toxin, although the two toxins have different structures and cellular specificities. The biological activities of exotoxin A in its purified form are various, and its potency on a weight basis exceeds that of other extracellular products of *P. aeruginosa* and of *P. aeruginosa* lipopolysaccharide as well.

Exotoxin A is lethal for a variety of animals including primates, cytotoxic in vitro, and necrotizing for certain tissues in vivo. When injected into the skin of animals, exotoxin A produces bland necrosis and reactive erythema, but not hemorrhage, and severe ocular damage occurs in mice after topical application. Corneal lesions produced by purified exotoxin A are similar to those noted in experimental eye infections. In addition, the inhibitory or cytotoxic effects of exotoxin A on macrophages, polymorphonuclear leukocytes, and bone marrow granulocyte-macrophage progenitor cells suggest that it is an immunotoxin. Recent experiments in mice have also shown that this toxin can induce suppression of the response to T-dependent and T-independent antigens both in vitro and in vivo. There is evidence that exotoxin A is a virulent factor in the human host and that conventional polyclonal antibodies to exotoxin A play a protective role. Monoclonal antibodies to bacterial constituents, such as endotoxin and exotoxin A, which have also been tested for their therapeutic potential, will certainly provide the means to further investigate the pathogenic role of these bacterial products.

Some of the same cell-associated and extracellular products of *P. aeruginosa* responsible for its local invasion also contribute to its dissemination. In addition to the antiphagocytic properties of glycocalyx and lipopolysaccharide, extracellular enzymes affect immune defense mechanisms by inactivating complement factors and by cleaving immunoglobulins. However, endotoxin and exotoxin A are the bacterial components that appear to be primarily responsible for systemic disease. While exotoxin A promotes bacterial dissemination essentially owing to its immunosuppressive effects, the pathological alterations mediated by the endotoxin are roughly comparable to those exerted by endotoxins of other gram-negative bacteria, even if *P. aeruginosa* endotoxin is thought to be less active.

Antibody-mediated opsonophagocytosis has long been considered the major immune protective mechanism against the diffusion of *P. aeruginosa* infection. The reduction in the phagocyte population together with low levels of antibodies is a primary factor in determining *P. aeruginosa* septicemia in patients with severe burns. Also, patients with acute leukemia or cancer, who are frequently neutropenic as a consequence of their disease and related therapy, are especially susceptible to infections with this microorganism. In one study it was found that *P. aeruginosa* septicemia occurred in 42% of patients who had neutropenia, but in none of those who did not. In leukemic dogs, leukocyte transfusions resulted to be highly effective in the control of *P. aeruginosa* septicemia, and the administration of antibodies against bacterial surface antigens or exotoxin A greatly reduced susceptibility to infection.

The role of cell-mediated immunity (CMI) has generally been thought to be limited to modulating the antibody response rather than to play a crucial role in initial resistance to infection. In fact, *P. aeruginosa* persists despite the high levels of specific antibodies in circumstances such as cystic fibrosis. Furthermore, infections due to this microorganism are more frequent and severe in patients with Hodgkin's disease or sarcoidosis characterized by a seriously compromised CMI. It has recently been shown that T lymphocytes from immune animals can adoptively transfer protection against infection with *P. aeruginosa* to nonimmune recipients. These cells turned out to be able to kill the microorganism in vitro by releasing a bactericidal lymphokine; moreover, macrophages enhanced the production of this lymphokine by the secretion of interleukin 1. CMI to *P. aeruginosa* is strictly controlled by a population of suppressor T lymphocytes. Once suppression is removed, however, CMI is capable of providing virtually complete protection against lethal *P. aeruginosa* challenge. This protection was striking in a leukopenic mouse model, in which as few as 5×10^6 T cells from immunized mice protected unimmunized leukopenic recipients, even in the absence of circulating granulocytes or anti–*P. aeruginosa* antibodies.

A decade or more of investigation has proved that one of the prominent pathogenic mechanisms underlying *P. aeruginosa* systemic infection may be related to the ability of the microorganism to inhibit the host humoral and CMI, through the induction of suppressor cells. These cells have been characterized as macrophages by certain investigators and as T lymphocytes by others. B lymphocytes able to suppress contact sensitivity to oxazolone (Ox), a well-defined cell-mediated immune response, have also been found in mice infected intravenously with *P. aeruginosa*. Since *P. aeruginosa* is a polyclonal B-cell activator (PBA), the possibility exists that this microorganism inhibits contact sensitivity with a cellular mechanism identical to that operating in the suppression exerted by other PBAs on contact sensitivity to Ox. It was found, in fact, that such a suppression is mediated by the cooperative action of idiotype-positive anti-Ox B lymphocytes, anti-idiotype B lymphocytes and T lymphocytes, which appear at different times after immunization and interfere with different phases of the immune response. The anti-Ox B lymphocytes can be detected very early after sensitization and induce antiidiotype B lymphocytes. These cells, in turn, activate T suppressor cells that affect the efferent phase of the immune response through the release of soluble factors. Such a complex immunosuppressive circuit is due to the ability of the inducing stimulus to determine a polyclonal activation of B lymphocytes, be it a microorganism or a bacterial constituent.

There is a suggestion that a disturbance in the circulation of the antigen-reactive lymphocytes, as a consequence of their trapping within the foci of infection and/or the draining lymphoid tissues, represents an additional mechanism whereby *P. aeruginosa* may drive the derangement of the host immune reactivity.

The prominent role of the host immune reactivity in determining the onset and course of *P. aeruginosa* infection and the consequent higher and higher diffusion in immunocompromised hosts may indicate that *P. aeruginosa* opportunism, although accidental, will affect the evolution of the microorganism and its relationship with the human host. The contagion mediated by instrumentation and manipulative procedures, such as uretheral catheterization, tracheotomy, lumbar punctures, and intravenous infusion, brings the microorganism in districts that would never be reached in natural conditions. On the other hand, infection in immunocompromised hosts has in the past represented an occasional event without any evolutionary implication for the microorganism. Nowadays, however, the number of immune-deficient patients who can survive owing to the very sophisticated procedures of modern medicine is growing steadily. Such a widening of the susceptible population represents a new condition in the biology of the microorganism, allowing the development of a new relationship between *P. aeruginosa* and the human host. Time may select new variants with higher ability to adapt to humans—that is to say, "humanized variants." Since these variants will be selected on the basis of their higher capability to survive and multiply in the host, they will retain all the pathogenic properties attributed to *P. aeruginosa* intact. The question may arise whether this situation will not lead to a stronger association between *P. aeruginosa* and human pathology in the future.

Pseudomonas pseudomallei and *Pseudomonas mallei* Infections

Unlike the opportunistic pathogen *P. aeruginosa, P. pseudomallei* is a primary pathogen, and melioidosis often occurs in healthy young people. The disease has drawn much attention, because it has been observed in an appreciable number of persons returning from Southeast Asia. It is now quite clear that no natural animal reservoir exists, and the microorganism can be recovered directly from soil and water in endemic areas, which are confined to Southeast Asia, Central and South America, the West Indies, Australia, and Madagascar. The organism, therefore, can be acquired directly from the environment through wound or minor lesions of the skin and mucous membranes. Clinically active cases of human melioidosis are relatively rare, but the mortality is very high. Most episodes are clinically inactive or of a subclinical type and can be recognized only by antibody response. The disease frequently involves the lungs, but it may also take other forms, ranging from an acute septicemia to chronic infections with widespread granulomatous lesions.

P. mallei, a nonmotile pseudomonad found only in animal hosts and isolated by Loeffler and Schütz in 1882, is the causative agent of glanders, one of the oldest diseases known to man and described since the time of Hippocrates. Once an important occupational disease of veterinarians, stable workers, and teamsters, it has disappeared from many regions in the last 50 years. Man is not highly susceptible but is occasionally infected through ingestion, inhalation, or abrasions of the skin or mucous membranes.

Literature Cited

Bergan T. 1978. Phage typing of *Pseudomonas aeruginosa,* p. 167–199. *In* T. Bergan and J. R. Norris (ed.), Methods in microbiology, vol. 10. Academic Press, London, New York, San Francisco.

Bergan T. 1981. Human- and animal-pathogenic members of the genus *Pseudomonas,* p. 666–700. *In:* M. P. Starr, H. Stolp, H. G. Trüper, A. Balows, and H. G. Schegel (ed.), The prokaryotes. Springer-Verlag, Berlin, Heidelberger, New York.

Costerton J. W. 1984. The etiology and persistence of cryptic bacterial infections: a hypothesis. Rev. Infect. Dis. **6(Suppl. 3):**S608–S616.

Doudoroff M., and N. J. Palleroni. 1974. *Pseudomonas,* p. 141–199. *In* R. E. Buchanan and N. E. Gibbons (ed.), Bergey's manual of determinative bacteriology, 8th ed. Williams and Wilkins, Baltimore.

Gilardi G. L. 1971a. Characterization of *Pseudomonas* species isolated from clinical specimens. Appl. Microbiol. **21:**414–419.

Gilardi G. L. 1971b. Antimicrobial susceptibility as a diagnostic aid in the identification of nonfermenting gram-negative bacteria. Appl. Microbiol. **22:**821–823.

Gilardi G. L. 1976. *Pseudomonas* species in clinical microbiology. Mount Sinai J. Med. **43:**710–726.

Gilardi G. L. 1985. *Pseudomonas,* p. 350–372. *In* E. H. Lennette, A. Balows, W. J. Hausler Jr., and H. J. Shadomy, (ed.), Manual of clinical microbiology, 4th ed. American Society for Microbiology, Washington, D.C.

Govan J. R. W. 1978. Pyocin typing of *Pseudomonas aeruginosa,* p. 61–91. *In* T. Bergan and J. R. Norris (ed.), Methods in microbiology, vol. 10. Academic Press, London, New York, San Francisco.

Holmes B., S. P. Lapage, and H. Malnick. 1975. Strains of *Pseudomonas putrefaciens* from clinical material. J. Clin. Pathol. **28:**149–155.

Lanyi B., and T. Bergan. 1978. Serological characterization of *Pseudomonas aeruginosa,* p. 93–168. *In* T. Bergan and J. R. Norris (ed.), Methods in microbiology, vol. 10. Academic Press, London, New York, San Francisco.

Palleroni N. J. 1984. Introduction to the family *Pseudomonadaceae,* p. 655–665. *In* N. R. Krieg and J. G. Holt (ed.), Bergey's manual of systemic bacteriology, vol. 1. Williams and Wilkins, Baltimore.

Snell J. J. S. 1973. The distribution and identification of nonfermenting bacteria. Public Health Laboratory Service Monograph Series. No. 4. Her Majesty's Stationery Office, London.

Stanier, R. Y., N. J. Palleroni, and M. Doudoroff. 1966. The aerobic pseudomonads: a taxonomic study. J. Gen. Microbiol. **43:**159–271.

CHAPTER 46

Rat-Bite Fever

S. L. JOSEPHSON

Disease: Rat-bite fever (Haverhill fever).
Etiologic Agents: *Streptobacillus moniliformis* and *Spirillum minus*
Source: Wild and domestic (laboratory) rodents and animals that ingest rodents.
Clinical Manifestations: Abrupt onset of relapsing fever, chills, vomiting, headache, and rash; occasionally sore throat, lymphadenopathy, lymphangitis, and anemia. Diarrhea and weight loss in children. Complications can include migratory arthritis, bronchitis, pneumonia, hepatosplenomegaly, endocarditis, myocarditis, abscess formation, chorioamnionitis, meningitis, and nephritis.
Pathology: Damage associated with complications, particularly endocarditis and persistent severe arthritis.
Laboratory Diagnosis: Isolation of *S. moniliformis* from blood, joint fluid, cutaneous eruptions, abscesses, or affected tissues. Identification based on cellular morphology, growth characteristics, biochemical profile, and fatty acid profile. Fourfold rise in specific *S. moniliformis* agglutinins is diagnostic. *Sp. minus* identified in blood, lymph node aspirates, or biopsy specimens by darkfield microscopy. Animal inoculations (mice, guinea pigs) to recover *Sp. minus*.
Epidemiology: Worldwide, with *Sp. minus* more common in the Far East and *S. moniliformis* more common in North America. Source of infection is usually a rat bite; occasionally from ingestion of contaminated food, especially milk (Haverhill fever).
Treatment: Penicillin, with tetracycline or streptomycin as alternative choices.
Prevention and Control: Improve sanitation, control wild rat populations, and educate individuals in the proper handling of rodents.

Description of Disease

Rat-bite fever is a systemic disease associated with rat bites and direct contact with rodents or their excreta. Adults occupationally exposed to rats and children exposed to wild rodents account for the majority of cases. Two unrelated bacterial species, *Streptobacillus moniliformis* and *Spirillum minus,* are recognized as the etiologic agents of rat-bite fever.

Communicability and Transmissibility

Both etiologic agents of rat-bite fever are communicable from animals to humans. Although most of the reported cases have involved rat bites, *S. moniliformis* and *Sp. minus* can be transmitted to humans from rodents or rodent-ingesting animals (e.g., cats, dogs, weasels) by way of a bite, scratch, or direct contact. A foodborne version of rat-bite fever, Haverhill fever, can result from ingesting food, usually milk, contaminated with *S. moniliformis*. Human-to-human transmission has not been reported.

Clinical Manifestations

Patients with rat-bite fever typically present with relapsing fever, chills, vomiting, headache, and rash. Furthermore, infants and young children often suffer from diarrhea and weight loss (Raffin and Freemark,

1979). Leukocytosis, sore throat, lymphadenopathy, lymphangitis, myalgia, anemia, and migratory arthralgia and arthritis also may occur. Among the more serious complications of rat-bite fever are bronchitis, pneumonia, hepatosplenomegaly, endocarditis, myocarditis, paricardial effusion, abscess formation, chorioamnionitis, hepatitis, meningitis, meningoencephalopathic involvement, and nephritis (Brown and Nunemaker, 1942; Carbeck et al., 1967; Faro et al., 1980; Kiefer et al., 1981; Roughgarden, 1965).

Pathology

Pathology associated with rat-bite fever is unpredictable. However, complicated cases may produce significant pathologic changes (McHugh et al., 1985). This is particularly evident in cases involving endocarditis (Carbeck et al., 1967; McCormack et al., 1967) and persistent, severe arthritis (Mandel, 1985).

Diagnostic Features

Streptobacillary Rat-Bite Fever

Streptobacillary (*S. moniliformis*) rat-bite fever has a normal incubation period of less than 10 days (range, 1 to 22 days). A patient usually experiences an abrupt onset of fever, chills, headache, and vomiting. Intermittent or relapsing fever occurs in 30% of cases (Roughgarden, 1965). Within several days approximately 50% of patients develop migratory polyarthritis, and 75% develop a maculopapular or petechial rash that often involves the palms and soles (Raffin and Freemark, 1979; Roughgarden 1965). The inoculation site heals rapidly with minimal regional lymphadenitis. Complications can occur especially if the infection is not treated. In the presence of appropriate signs and symptoms, a definitive diagnosis requires the isolation and identification of *S. moniliformis* from fluids or tissues (e.g., blood, joint fluid, cutaneous eruptions, abscesses) and/or the demonstration of specific agglutinins.

Haverhill Fever (Erythema Arthriticum Epidemicum)

Cases of Haverhill fever usually result from the ingestion of milk or milk products contaminated with *S. moniliformis* from rat excrement (Place and Sutton, 1934; Shanson et al., 1983). As with streptobacillary rat-bite fever, chills, vomiting, rash, fever, and arthritis (arthralgia) are common (Shanson et al., 1983). However, gastrointestinal tract and upper respiratory tract symptoms also are prominent (Place and Sutton, 1934). An outbreak of Haverhill fever often occurs in a small-scale epidemic similar to other foodborne outbreaks. A definitive diagnosis of Haverhill fever follows the identification of food as the source of infection and the identification of the etiologic agent, *S. moniliformis,* in body fluids or tissues and/or the detection of specific agglutinins.

Spirillary Rat-Bite Fever (Sodoku)

Spirillary (*Sp. minus*) rat-bite fever, also known as Sodoku (Japanese name), has an incubation period of approximately 14 days (range 1 to 36 days). A patient typically develops a regularly relapsing fever and chills. Several days after the onset of fever, 50% or more of patients will develop macular rashes on the extremities that initially appear dark red or purple and gradually coalesce (Brown and Nunemaker, 1942; Roughgarden, 1965). The inoculation site heals quickly. However, inflammation, ulceration, and induration appear at the inoculation site with the onset of symptoms. This is accompanied by regional lymphadenopathy and lymphangitis. Complications can occur but rarely include arthritis or endocarditis. A preliminary diagnosis of spirillary rat-bite fever can be confirmed by identifying *Sp. minus* directly in blood, lymph node aspirates, or biopsy specimens by use of dark-field microscopy and animal inoculations. *Sp. minus* cannot be cultivated in vitro at this time.

Differential Diagnosis

A clinical presentation of fever, headache, arthritis, arthralgia, and rash on the palms and soles is suggestive of several infectious diseases other than rat-bite fever. A differential diagnosis could include viral infection (e.g., Coxsackie virus), Epstein-Barr virus, Rocky Mountain Spotted fever (Portnoy et al., 1979), meningococcal septicemia, streptococcal disease, and leptospirosis. Serologic tests for syphilis are falsely positive in 25% of patients with streptobacillary rat-bite fever and in 50% of patients with spirillary rat-bite fever (Roughgarden, 1965).

Prognosis

Untreated rat-bite fever may resolve spontaneously within several weeks or persist for months. The mortality rate for untreated streptobacillary rat-bite fever has been calculated to be 12.7% compared with 6.5% for spirillary rat-bite fever (Roughgarden, 1965).

Epidemiologic Aspects

Distribution and Incidence

Rat-bite fever is a rarely reported disease with a worldwide distribution (Bayne-Jones, 1931). Streptobacillary rat-bite fever is more common in North

America (Anderson, et al., 1983), whereas the spirillary form occurs more frequently in the Far East (Krieg, 1984). The true incidence of rat-bite fever is unknown, because the etiologic agents are difficult to isolate and/or identify. Also, similarities in the clinical presentation between rat-bite fever and several other infectious diseases can lead to misdiagnoses, especially in the absence of a rat bite. In the past, more than 50% of cases in the United States have involved children under the age of 12 (Roughgarden, 1965); however, few cases have been reported in children since 1958 (McHugh et al. 1985; Raffin and Freemark, 1979). Instead, nearly half of the recent cases have occurred among adults handling laboratory rodents (Anderson et al., 1983).

Risk Factors

Rat-bite fever traditionally has been associated with low socioeconomic status, overcrowding, and inadequate sanitation. More recently, occupational exposure to laboratory rats and mice has become a prominent risk factor.

Source of Infection

More than 50% of wild and laboratory rats carry *S. moniliformis* in their nasopharynx (Strangeways, 1933). Moreover, 25% of rats screened for *Sp. minus* carried the organism in conjunctival secretions and blood (McDermott, 1928). Even though rats are considered to be the primary reservoir for human infection, the disease can be acquired from other infected rodents, including mice, as well as from dogs, cats, and other carnivores who feed on rodents.

Etiologic Agent

Characterization

S. moniliformis is a pleomorphic, facultatively anaerobic, gram-negative rod, 0.1 to 0.7 μm in diameter and 1 to 5 μm in length. It is nonmotile, nonbranching, nonencapsulated, and non-acid fast. L-phase variants occur spontaneously. The cellular morphology ranges from single cells, some with a central swelling, to long chains or filaments 10 to 150 μm in length with beaded swellings ("string of beads") along the filament (Savage, 1984). Media, growth conditions, and age of culture all affect morphology. For example, cells grown on tryptose–phosphate agar containing 20% horse serum will appear as rods, chains, or filaments, some with fusiform or bulbar swellings. However, cells grown in tryptose–phosphate broth containing 20% horse serum will appear as clumps of rods and filaments with fewer fusiform or bulbar swellings (Savage, 1984). An examination of young cultures often reveals pleomorphic rods and some filaments. Alternatively, older cultures tend to have filaments, sometimes interwoven, with elongated medial swellings and terminal yeast-like projections (Portnoy et al., 1979). The filaments may fragment into chains of bacillary or coccobacillary elements (Roughgarden, 1965). Older cultures do not Gram stain well. Better results may be obtained with Giemsa or Wayson stains (Rogosa, 1985).

Sp. minus is a motile, aerobic, gram-negative spiral bacterium, 0.2 to 0.5 μm in diameter and 3 to 5 μm in length with at least one flagellum at each pole. The spiral characteristically has two to three turns with a wavelength of 0.8 to 1.0 μm (Krieg, 1984). *Sp. minus* characteristically has very angular windings and pointed ends (Kowal, 1961).

Isolation

Cultivation of *S. moniliformis* requires the use of medium supplemented with ascitic fluid, serum, or blood. In the case of blood cultures, a patient's own blood will act as the supplement. *S. moniliformis* has been recovered from clinical specimens with a variety of agar and broth media including sheep blood agar, chocolate agar, Sheadler's agar, and thyoglycolate broth containing blood, 5% ascitic fluid, or 20% calf serum (Faro et al., 1980; Portnoy et al., 1979). Meat infusion and tryptose-based media supplemented with 20% horse serum also are recommended (Savage, 1984). For culturing *S. moniliformis* from blood, Panmede (papain digest of ox liver) can be used as a supplement in brain–heart infusion cystein broth to support early growth of small inocula (Shanson et al., 1985). Liquoid (sodium polyanethol sulfonate), often used in blood culture media to neutralize the antibacterial effect of fresh human blood, will inhibit *S. moniliformis* at concentrations above 0.025% (Lambe et al., 1973). Therefore, only blood culture media without Liquoid should be used when rat-bite fever is suspected. L-phase variants can be grown on heart infusion agar supplemented with horse serum and yeast extract (Rogosa, 1985).

Blood and joint fluid should be citrated to prevent clotting before direct microscopic examination and cultivation (Rogosa, 1985). All clinical specimens should be inoculated into supplemented broth and onto supplemented agar. A substantial quantity of specimen should be used to facilitate the recovery of *S. moniliformis*. Cultures are incubated at 37°C in a humid environment in 7 to 8% CO_2.

S. moniliformis usually forms characteristic "puff balls" in broth culture and produces 1 to 2-mm round, grayish, smooth, and glistening colonies on agar 2 to 6 days after inoculation (Savage, 1984).

L-phase colonies are embedded in agar and have a fried egg appearance. For continued viability, *S. moniliformis* should be subcultured as soon as it is detected in the primary cultures.

At the present time, *Sp. minus* cannot be cultivated in vitro; however, it can be grown in vivo by animal inoculation.

Identification of *S. moniliformis*

Biochemical Characteristics

Bacterial isolates with appropriate cellular morphology and growth characteristics are identified by their biochemical profiles. *S. moniliformis* characteristically does not produce indole; liquefy gelatin; reduce nitrate; or demonstrate oxidase, catalase, or urease activity. Fermentation patterns can be determined by use of 1% carbohydrate in cysteine tryptic agar base (Lambe et al., 1973). Gas is not produced from fermentation. *S. moniliformis* characteristically produces acid from glucose, maltose, fructose, galactose, and salicin but not from xylose, mannitol, lactose, sucrose, glycerol, arabinose, and sorbitol.

The use of gas–liquid chromatography provides rapid identification of *S. moniliformis* grown in broth culture (Rowbotham, 1983). The typical fatty acid profile of *S. moniliformis* consists of palmitic, linoleic, oleic, and stearic acid peaks.

Immunologic Identification

When *S. moniliformis* cannot be isolated from clinical specimens, a patient's serum can be tested for the presence of specific streptobacillus agglutinis. Agglutinins appear as early as 10 days after inoculation (e.g., bite) but may not appear for several months (Raffin and Freemark, 1979). A fourfold rise in titer or a single titer of 1:80 is considered to be diagnostic.

Fluorescent antibody techniques have been used to confirm the identity of *S. moniliformis* (Carbeck et al., 1967).

Identification of *Sp. minus*

Microscopic

Sp. minus can be identified in wet mounts of specimens using darkfield microscopy. Blood specimens also should be stained with Giemsa or Wright's stain. *Sp. minus* is identified on the basis of characteristic cellular morphology described previously.

Animal Inoculation

When *Sp. minus* cannot be detected directly in clinical specimens, four mice and two guinea pigs should be injected intraperitoneally with the patient's blood and wound and lymph node exudates (Brown and Nunemaker, 1942). Peritoneal fluid from inoculated animals should be examined weekly by darkfield microscopy and/or stained smears. Microscopic examination of the animals' blood also may reveal *Sp. minus* 2 or 3 weeks after inoculation (Witzberger and Cohen, 1944).

Treatment, Prevention, and Control

Treatment

Penicillin is the drug of choice for treating rat-bite fever. The spirillary form of the disease is more sensitive then the streptobacillary form to penicillin (Roughgarden, 1965). On the basis of this difference in sensitivity and the tendency of *S. moniliformis* to produce penicillin-resistant L-phase variants, a daily dose of not less than 400,000 to 600,000 units of penicillin is recommended for treating rat-bite fever (Roughgarden, 1965). When an infection does not respond to therapy within 2 days, the dose should be raised to 1.2 million units daily. Even higher doses of penicillin (4.8 to 20 million units/day for 4 to 6 weeks) are recommended in cases involving endocarditis (McCormack et al., 1967). In cases of penicillin resistance or allergy, tetracycline or streptomycin can be substituted for penicillin.

In addition to antimicrobial therapy, entrance wounds should be cleaned and irrigated and tetanus toxoid administered in appropriate instances.

Prevention and Control

Several measures can aid in the prevention of rat-bite fever. These include improving sanitation, controlling wild rat populations, and training individuals in the proper handling of rodents.

Literature Cited

Anderson, L. C., S. L. Leary, and P. J. Manning. 1983. Rat-bite fever in animal research laboratory personnel. Lab. Animal Sci. **33**:292–294.

Bayne-Jones, S. 1931. Rat-bite fever in the United States. Intern. Clin. (41st series) **3**:235–253.

Brown, T. M., and J. C. Nunemaker. 1942. Rat-bite fever: review of American cases with re-evaluation of its etiology and report of cases. Bull. Johns Hopkins Hosp. **70**:201–327.

Carbeck, R. B., J. F. Murphy, and E. M. Britt. 1967. Streptobacillary rat-bite fever with massive pericardial effusion. J. Am. Med. Assoc. **201**:703–704.

Faro, S., C. Walker, and R. L. Pierson. 1980. Amnionitis

with intact amniotic membranes involving *Streptobacillus moniliformis*. Obstet. Gynecol. **55(suppl)**:9S–11S.

Kiefer, H., W. Froscher, and H. P. Mohr. 1981. Rattenbiβkrankheit mit meningoenzephaler Beteiligung. Med. Klin. **76**:653–655.

Kowal, J. 1961. Spirillum fever. Report of a case and review of the literature. N. Engl. J. Med. **264**:123–128.

Krieg, N. R. 1984. Genus *Aquaspirillum* Hylemon, Wells, Krieg, and Jannasch 1973, 361[AL], p. 72–90. *In* N. R. Krieg and J. G. Holt (ed.), Bergey's manual of systematic bacteriology, vol. 1, Williams & Wilkins Co., Baltimore.

Lambe, D. W., A. M. McPhedran, J. A. Mertz, and P. Stewart. 1973. *Streptobacillus moniliformis* isolated from a case of Haverhill fever: biochemical characterization and inhibitory effect of sodium polyanethol sulfonate. Am. J. Clin. Pathol. **60**:854–860.

Mandel, D. R. 1985. Streptobacillary fever. Cleve. Clin. Q. **52**:203–205.

McCormack, R. C., D. Kaye, and E. W. Hook. 1967. Endocarditis due to *Streptobacillus moniliformis*. J. Am. Med. Assoc. **200**:77–79.

McDermott, E. W. 1928. Rat-bite fever: study of experimental disease with critical review of the literature. Quart. J. Med. **21**:433–458.

McHugh, T. P., R. L. Bartlett, and J. I. Raymond. 1985. Rat-bite fever: report of a fatal case. Ann. Emerg. Med. **14**:1116–1118.

Place, E. H., and L. E. Sutton. 1934. Erythema arthriticum epidemicum (Haverhill fever). Arch. Intern. Med. **54**:659–684.

Portnoy, B. L., T. K. Satterwhite, and J. D. Dyckman. 1979. Rat-bite fever misdiagnosed as rocky mountain spotted fever. South. Med. J. **72**:607–609.

Raffin, B. J., and M. Freemark. 1979. Streptobacillary ratbite fever: A pediatric problem. Pediatrics **64**:214–217.

Rogosa, M. 1985. *Streptobacillus moniliformis* and *Spirillum minus*, p. 400–406. *In* E. H. Lennette, A. Balows, W. J. Hausler and H. J. Shadomy (ed.), Manual of clinical microbiology, American Society for Microbiology, Washington, D.C.

Roughgarden, J. W. 1965. Antimicrobial therapy of rat-bite fever. Arch. Intern. Med. **116**:39–54.

Rowbotham, T. J. 1983. Rapid identification of *Streptobacillus moniliformis*. Lancet **2**:567.

Savage, N. 1984. Genus *Streptobacillus* Levaditi, Nicolau and Poincloux 1925, 1188[AL], p. 598–600. *In* N. R. Krieg and J. G. Holt (ed.), Bergey's manual of systematic bacteriology, vol. 1, Williams & Wilkins Co., Baltimore.

Shanson, D. C., B. G. Gazzard, J. Midgley, J. Dixey, G. L. Gibson, J. Stevenson, R. G. Finch, and J. Cheesbrough. 1983. *Streptobacillus moniliformis* isolated from blood in four cases of Haverhill fever. Lancet **2**:92–94.

Shanson, D. C., J. Pratt, and P. Greene. 1985. Comparison of media with and without 'Panmede' for the isolation of *Streptobacillus moniliformis* from blood cultures and observations on the inhibitory effect of sodium polyanethol sulphonate. J. Med. Microbiol. **19**:181–186.

Strangeways, W. I. 1933. Rats as carriers of *Streptobacillus moniliformis*. J. Pathol. Bacteriol. **37**:45–51.

Witzberger, C. M., and H. G. Cohen. 1944. Rat-bite fever: comparison of the spirochetal (Sodoku) and bacillary (Haverhill fever) forms. Arch. Pediatr. **61**:123–133.

CHAPTER 47

Salmonellosis

JOHN H. SILLIKER and DAMIEN A. GABIS

Disease: Salmonellosis (enteric fever, *Salmonella* "food poisoning").

Etiologic Agents: Nearly 2,000 serologic types of the genus *Salmonella*.

Source: The feces of animals, warm- and cold-blooded, wild and domestic.

Clinical Manifestations: 1. Enteric fever: headache, malaise, anorexia, and congestion of mucous membranes. Symptoms generally not referable to the intestinal tract, although a short episode of vomiting and diarrhea sometimes occurs in the first day or two of enteric fever. 2. *Salmonella* food poisoning: diarrhea, abdominal cramps, vomiting, elevated temperature.

Pathology: *Salmonellae* reach the small bowel and invade the lumen where they multiply. Subsequently they penetrate the ileum and, to a lesser extent, the colon, where an inflammatory reaction occurs. Lymphoid follicles can become enlarged and may ulcerate; mesenteric nodes often become swollen. *Salmonellae* sometimes overwhelm mucosal and lymphatic barriers, reach the bloodstream, and give rise to bacteremia and abscesses. This occurs in the enteric fever syndrome, whereas in the "food poisoning" syndrome, the disease is localized in the intestinal tract.

Laboratory Diagnosis: Based on isolation of the organism from clinical specimens. In *Salmonella* food poisoning, diagnosis is generally accomplished by examination of feces, whereas in enteric fever, the organism may be demonstrated in the feces and also in specimens from extraintestinal sites, for example, blood and urine. Postmortem examination may yield positive findings from a variety of sites, including the intestine, spleen, mesenteric glands, bone marrow, and gallbladder. Isolation from feces requires the use of selective and differential media, whereas this may not be necessary with blood or urine specimens. Ultimate identification is based on biochemical and serologic reactions of the isolates.

Epidemiology: Worldwide distribution. Contaminated food and water are the most common source of infection in humans. The epidemiology of human salmonellosis is a reflection of humans' close association with animals.

Treatment: Antibiotic therapy is contraindicated in the treatment of the food poisoning syndrome, in which salmonellosis is a self-limiting infection of the intestine. Enteric fever is effectively treated with chloramphenicol.

Prevention and Control: Prevention requires measures that block contamination of food and water with fecal material. Raw foods of animal origin must be assumed to be contaminated with *Salmonellae* and, therefore, must be treated by the processor or the consumer to destroy the organisms. Water supplies should be tested to ensure freedom from fecal contamination. Public water supplies should be disinfected. Raw milk should never be consumed.

Description of Disease

Salmonellosis is an infectious disease of humans and a wide variety of other vertebrates. Infection with salmonellae may result in an acute, self-limiting gastroenteritis or a systemic infection characterized by septicemia and ultimate localization in extraintestinal sites. The gastrointestinal form is often referred to as the "food poisoning syndrome," although this is a misnomer, as the disease is an infection rather than an intoxication. The systemic disease is referred to as enteric fever, and typhoid fever is the classical form, although other salmonellae may manifest the febrile disease. It must be assumed that any member of the genus *Salmonella* is a human pathogen.

The classification of *Salmonella* has been controversial for many years. Historically, the salmonellae have received species rank based on their antigenic structure, more specifically on their flagellar (H) and somatic (O) antigen components. On this basis, Kelterborn (1967) gives a comprehensive account of the first isolation and geographic distribution of nearly 2,000 *Salmonella* serotypes. This compilation was based on the terminology introduced by Kauffmann and White (Report, 1934). But the Kauffmann-White scheme is not a record of the complete antigenic structure of each organism, as only the antigens of differential importance are recorded in the scheme. If small antigenic differences are taken into account, the "number of serotypes that can be recognized is almost unlimited" (Parker, 1984b).

Ewing (1963) proposed a system of classification of the *Enterobacteriaceae* based on comparative studies of the biochemical reactions given by relatively large numbers of cultures of each genus and species within the family. On this basis the genera *Salmonella, Arizona,* and *Citrobacter* were placed in the tribe *Salmonellae*. Three biochemically distinct species of *Salmonella* were recognized, namely *S. typhi, S. choleraesuis*, and *S. enteritidis*, along with a separate genus for "*Arizona*." *S. typhi* and *choleraesuis* are serologically homogeneous. Within the third species, *S. enteritidis*, serotypes were recognized as varieties.

The recent work of Fitts et al. (1983) confirms at the gene level that the salmonellae constitute a distinct group. Using one strain of *S. typhimurium*, 10 unique sequences of DNA were isolated that appeared to specifically hybridize with all the salmonellae examined by the DNA-DNA hybridization technique (Klausner and Wilson, 1983). Further work (Fitts, 1985) using the Centers for Disease Control reference strain collection of more than 350 *Salmonella* isolates representing all somatic and flagellar antigens indicated that 2 of the 10 DNA sequences reacted with all salmonellae, including *S. arizoniae*. Accordingly, the terminology in this chapter adheres to the more traditional reporting of serotypes. This is consistent with the current approach of the Centers for Disease Control of the U. S. Department of Health and Human Services (Kelley et al., 1985). Ewing's classification system is based primarily on biochemical similarities among the salmonellae. The more traditional approach is based on the antigenic structure of the isolates. The work of Fitts indicates a genetic homogeneity among the salmonellae, and the traditional approach, identifying serotypes, is appropriate.

Communicability and Transmissibility

Salmonellae are introduced into the body by the oral route. Food and water are the most common vehicles, although person-to-person contact may occur (Parker, 1984b). The organisms are in the feces of infected animals, both those suffering disease and healthy carriers. When livestock are slaughtered and dressed, fecal contamination of the carcasses occurs. Since records have been kept, red meat and poultry have been the most important sources of human salmonellosis, and currently there is no acceptable technology to assure that these raw foods can be produced free of salmonellae (Fig. 1; Gangarosa, 1978). The importance of red meat and poultry as vectors of salmonellosis reflects not only their native

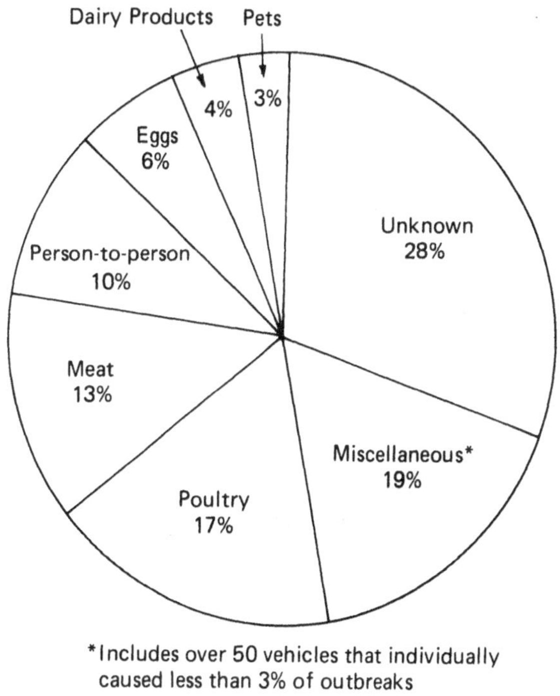

FIG. 1. Modes of transmission in 500 human salmonellosis outbreaks, 1966 to 1975 (Gangarosa, 1978).

contamination with salmonellae, but also the failure of the consumer to handle the raw products so to prevent infection. In Table 1 Bryan (1978) summarizes factors contributing to 61 outbreaks of salmonellosis during a 3-year period in the United States. Though not all the outbreaks involved meat and poultry, the factors listed illustrate the many ways that consumers fail to control salmonellae in foods. Bryan (1980), in a study of 49 meat- and poultry-associated outbreaks reported that inadequate cooking was involved in 25 (51%) of the incidents. In 17 (35%) of the outbreaks, cross-contamination between raw and cooked products occurred. More importantly, in only six outbreaks was the source not traced to inadequate cooking, cross-contamination, or consumption of raw meat. From experience, one or more of these three food handling practice deficiencies is involved in every outbreak.

In the past, egg products were a significant source of human salmonellosis in the United States, but mandatory pasteurization of egg products has been responsible for greatly reducing eggs as an important cause of salmonellosis.

Consumption of contaminated raw shellfish harvested from sewage-polluted waters has caused salmonellosis. The National Shellfish Sanitation Program (U.S. Public Health Service, 1975) monitors shellfish growing areas for sewage pollution and prohibits harvesting when pollution is found.

Raw milk is a relatively frequent cause of human salmonellosis. For example, the consumption of commercially distributed raw milk in California has been a continuing source of *S. dublin*-caused outbreaks (Chin, 1982). Pasteurization of milk destroys *Salmonella*, but the consumption of raw milk carries the hazard not only of salmonellosis but also the risk of infection with other enteric pathogens. There can be no assurance that certified raw milk is free of pathogens, specifically *Salmonella*. The sale of raw milk should be banned in all states.

Sewage-polluted water is always a hazard, as illustrated by a massive outbreak of *S. tyhpimurium* infections in Riverside, California (Report, 1971). As an important safeguard against salmonellosis, municipal water supplies are chlorinated and checked for evidence of fecal contamination.

Person-to-person spread of salmonellosis can occur, particularly in hospitals. The organisms may become established in hospitals and remain for months and even years. The organisms may be isolated from a number of inanimate sites, such as dust, wash basins, and water baths used to warm bottles of milk (see Parker, 1984b for a more detailed discussion of hospital-acquired salmonellosis).

Clinical Manifestations

The food poisoning syndrome is characterized by a self-limiting acute gastroenteritis. Contaminated

TABLE 1. Factors that contributed to the occurrence of 61 outbreaks of foodborne salmonellosis, 1973–1976

	Number				
Contributing factors	*Foodservice (28)*	*Homes (20)*	*Food processing and markets (13)*	*Total (61)*	*Percent*
Factors affecting growth					
Inadequate cooling	20	6		26	43
Preparing foods a day or more before service[a]	7	6		13	16
Inadequate hot storage	10	1		11	13
Use of leftovers[a]	3			3	5
Faulty fermentations			2	2	3
Process failure					
Inadequate reheating	14			14	23
Inadequate cooking	1	3	3	7	11
Factors affecting contamination					
Contaminated raw ingredients	2	11	7	20	33
Cross-contamination[b]	9	3		12	20
Inadequate cleaning	9	1	1	11	18
Infected persons[c]	6	1	1	8	13

[a] Indirect factor.
[b] Data not available for contamination from incoming animals to be processed or raw foods of animal orign.
[c] Unknown whether or not infected persons were initial sources of organism or ate implicated foods.
(Reprinted with permission from Bryan, 1978.)

food or water is the usual but not the only vehicle. The incubation period varies from 6 to 48 h and generally falls within a 12- to 36-h range. Variation in the incubation time may be attributed to the size of the infecting dose, the virulence of the organism, the susceptibility of the host, and the physicochemical composition of the transmitting food. Symptoms include diarrhea, abdominal cramps, vomiting, and fever which generally last from 1 to 7 days. However, the organisms may be excreted in the feces for many weeks after symptoms subside. Salmonellosis may be confused clinically with staphylococcal intoxication, but there are important distinctions. The incubation period in staphylococcal food poisoning is shorter; though the range is 1 to 6 h, most patients show symptoms in 2 to 4 h. Fever is almost always present in cases of salmonellosis, but is absent in staphylococcal intoxication. The acute symptoms of staphylococcal food poisoning normally disappear within 24 h. For distinctions between salmonellosis and other common foodborne diseases see Gilbert et al. (1984).

The symptoms of enteric fever are generally not referable to the intestinal tract, although the route of entry into the body is primarily oral. However, a short episode of vomiting and diarrhea sometimes occurs in the first day or two in typhoid fever. The onset times vary considerably between typhoid and paratyphoid fevers. In typhoid it is usually 8 to 15 days, seldom as short as 5 days, and sometimes as long as 30 or even 35 days; in paratyphoid fever it tends to be shorter and may be so short as to suggest food poisoning. The usual onset symptoms are headache, malaise, anorexia, and congestion of the mucous membranes, especially of the upper respiratory tract (Parker, 1984b). Bacteremia generally occurs in the first week of illness. Coleman and Buxton (1907), in a study of 1,138 cases, reported positive blood cultures in 89% during the first week, 73% in the second week, 60% in the third, 38% in the fourth, and 26% after the fourth week.

Though the reported frequency of isolation from the blood varies (see Parker, 1984b), it is clear that the frequency of isolation from the blood decreases after the first week of disease. On the other hand the number of organisms increases in the feces during the first 2 to 3 weeks of illness. This suggests that the main line of invasion during this period is from the bloodstream to the intestine and not the reverse (Parker, 1984b).

In some cases, the food poisoning gastroenteritis syndrome may be followed by extraintestinal invasion resulting in enteric fever. Extraintestinal infection may occur with a wide variety of serotypes, but is most likely to occur in the very young, the aged, and in debilitated patients initially suffering gastrointestinal illness.

Pathology

Bryan et al. (1979) have described the pathogenesis of *Salmonella* infection as follows: "The current explanation of the pathogenesis of *Salmonella* infection is that a sufficient number of salmonellae reach the small bowel and invade the lumen where they multiply. Subsequently, they penetrate the ileum and, to a lesser extent, the colon, where an inflammatory reaction occurs. Lymphoid follicles can become enlarged and may ulcerate; mesenteric nodes often become swollen. Salmonellae sometimes overwhelm mucosal and lymphatic barriers, reach the bloodstream, and give rise to bacteremia or abscesses." With the food poisoning syndrome the infection is self-limiting and the symptoms are restricted to those of gastroenteritis.

Diagnostic Features

Salmonellosis must be considered in individuals with vomiting, diarrhea, abdominal cramps, and fever. The likelihood of foodborne illness is strengthened if family members or other associates eating the same foods exhibit similar illness. In the food poisoning syndrome definitive diagnosis is based on isolation of *Salmonella* from the patient's feces. The source of infection can be confirmed by isolation of the same serotype from suspect food, water, or other vehicles. However, in food-associated outbreaks multiple foods may be contaminated with the epidemic serotype because of cross-contamination in the food preparation/service environment.

Enteric fever must be suspected in individuals suffering a febrile illness with or without a history of gastrointestinal symptoms. The state of health of family members or other associates may point to a common source of infection. Definitive diagnosis requires isolation of *Salmonella* from feces, blood, urine, or other body site. Postmortem examination may yield positive findings from a variety of sites, including the intestine, spleen, mesenteric glands, bone marrow, and gallbladder (Parker, 1984b). Agglutinins appear in the blood in cases of enteric fever; whereas in the food poisoning syndrome this does not occur.

Prognosis

The case-fatality rate from *Salmonella* food poisoning is low. In the United States during 1962 to 1971, a death rate of 0.41% was reported (Brachman et al., 1973). Most deaths occur in the very young, the old, and in debilitated immunologically compromised individuals. It is probable that the majority of deaths from *Salmonella* food poisoning may be attributed to a generalized infection after the gastroenteritis. In relation to the food poisoning syndrome, mortality rates from enteric fever are high.

Antibiotic treatment is contraindicated in cases of gastroenteritis. Even though the infecting *Salmonella* may show in vitro sensitivity to a given antibiotic, its administration does not shorten the course of illness. Indeed, it may prolong the carrier state (Aserkoff and Bennett, 1969; Dixon, 1965). Chemotherapy is indicated in cases of enteric fever, with chloramphenicol being the antibiotic of choice, although it apparently does not affect the rate of disappearance of the organisms from the intestine or on the number of patients who will become chronic carriers (Parker, 1984b).

Epidemiologic Aspects

Salmonellosis occurs worldwide, but even in industrialized nations, the disease is grossly underreported because it is generally a self-limiting gastroenteritis which may be misdiagnosed as intestinal influenza by the patient or the physician. As a consequence, estimates of the true incidence of disease are based on assumptions derived from epidemiologic evidence. The Committee on Salmonella of the National Research Council conservatively estimated human morbidity at 2,000,000 cases per year (National Research Council, 1969). Using somewhat different assumptions, Hauschild and Bryan (1980) estimated that 740,000 cases per year occur. More recently, Archer and Kvenberg (1985) reasoned that the number of cases of diarrheal disease of foodborne origin, including subsequent person-to-person transfer, is at least 24 million and perhaps as high as 81 million or more cases per year in the United States. No doubt, salmonellosis is a significant portion of these cases.

The incidence of enteric fever in Western Europe, North America, and Australia has declined dramatically since 1880. Data from England and Wales indicate a 200-fold decrease in mortality rate between 1950 and 1952 as compared with 1901. This is a reflection of two facets of the epidemiology of typhoid and paratyphoid fevers. The only reservoir of the typhoid organism and the major source of paratyphoid bacteria is humans. The organisms are excreted in the feces and urine of active cases or carriers and infect other individuals generally through contamination of water or food. The decrease in the incidence of enteric fever can be traced to improved environmental sanitation, primarily the proper disposal of sewage and the provision of pure water. Since the early 1950s, the added effect of chemotherapy has reduced the incidence of enteric fever almost to the vanishing point. For example, in the United States the reported number of cases of typhoid fever was approximately 1.0 per 100,000 in 1955; since 1965 it ranged from 0.2 to 0.3 per 100,000 population. It is estimated that about half of the cases in the United States are acquired during travel to underdeveloped regions of the world (United States Public Health Service, 1983). The situation is similar in the developed countries of Western Europe. Typhoid fever is still common in Asia, Africa, and Central and South America (Parker, 1984b).

In contrast to the epidemiology of typhoid and paratyphoid fevers, most foodborne salmonellosis is caused by nonhost adapted serotypes. Though sewage-polluted water has occasionally caused serious epidemics, the root problem is caused by foods of animal origin. Certain serotypes are adapted to animals, for example, *S. pullorum* in poultry, *S. abortusequi* in horses, *S. abortusovis* in sheep, and *S. cholerasuis* and *S. typhisuis* in pigs. Although these serotypes are animal host adapted, human pyretic illness from them has occurred, sometimes with serious consequences.

Susceptibility, Resistance, and Carrier States

Salmonellae gain access to the human body primarily by the oral route, most commonly through the consumption of food or water, and occasionally by drugs manufactured from animal products. These vehicles usually become contaminated directly or indirectly, with fecal material from animals (including humans) excreting salmonellae, but other routes of transmission do occur. These have been reviewed in *An Evaluation of the Salmonella Problem* (National Research Council, 1969). Infection of the upper respiratory tract before the onset of respiratory symptoms has been reported and suggests airborne transmission. Fomites, such as toys, towels, toilet seats, and other inanimate objects may become contaminated and lead to transmission in the household, between families, in restaurants, and in other institutions including hospitals.

Salmonellae enter the body via the oral route and must pass through the gastric juices before passing into the intestine, which is the site of infection. Though the normal stomach acidity will kill salmonellae, it obviously does not offer complete protection. It was long thought that large numbers of salmonellae are required to produce illness. However, enumeration of *Salmonella* in foods implicated in certain outbreaks indicate that foods containing small numbers of salmonellae have caused infection. It is apparent that food protects salmonellae from the destructive effects of gastric juices. Most human feeding studies leading to the conclusion that infection requires large numbers of organisms involved feeding dilutions of broth cultures, not foods, to the subjects.

Salmonellae are heat sensitive and so are destroyed by pasteurization of milk and proper cooking of foods. They are destroyed in water by chlorination. They are able to survive in environments such as dust and dried and frozen foods. Although they eventually die, viable organisms may be recovered after many weeks, months, or years of storage. The lower the moisture level, the longer the survival. High fat content foods such as milk chocolate tend to promote survival during storage. The chemicophysical composition of the vehicle and storage environment, in addition to water activity, affects the length of survival.

The carrier plays the prime role in the perpetuation of *S. typhi* in humans. Parker (1984b) has distinguished between "temporary" or "transient" carriers who excrete the organism for not more than a year and "chronic" carriers who excrete *S. typhi* for more than a year. Both fecal and urinary carriers occur, but urinary carriers are relatively less common. Parker (1984b) cites papers reporting that 2 to 9% of typhoid patients become chronic fecal carriers The rate of disappearance of *S. paratyphi* B from the feces is more rapid than *S. typhi*. The incidence of chronic carriers is significantly lower with *S. paratyphi* B than with *S. typhi*. Control of carriers is of utmost importance because only humans are the reservoir for *S. typhi* and the paratyphoid organisms. Parker (1984b) indicates that most of those excreting *S. typhi* 3 months after the onset of illness will become chronic carriers and that it is reasonable to encourage them to submit to chemotherapy at this point, even though success of treatment cannot be assured. Further, he indicates that persons who have excreted the organism for over a year very seldom cease spontaneously to do so. In this case cholecystectomy must be considered. Some 70 to 80% of fecal carriers of typhoid and paratyphoid organisms are permanently cured by this procedure.

Human carriers of unadapted serotypes of *Salmonella* also occur, although permanent (chronic) carriers are rare. Generally, in patients with acute gastroenteritis the bacteria are excreted fecally in large numbers, but as recovery progresses the numbers diminish so that after 3 or 4 weeks the bacteria usually cannot be found in stools. A few patients, however, continue to excrete salmonellae much longer, that is, for weeks, months, and occasionally for several years (National Research Council, 1969). Many times analysis of fecal specimens obtained from food handlers, symptomatic or asymptomatic, may reveal the presence of the epidemic serotype, but more often than not the food handler is a victim rather than the cause of the outbreak. Thus, caution should be exercised in attributing the cause of the outbreak to a carrier.

Etiologic Agents

Characterization

The genus, *Salmonella*, is in the the family *Enterobacteriaceae*. In *Bergey's Manual of Systematic Bacteriology, Volume I* (1984), the genus is described as follows:

Straight rods, 0.7 to 1.5 × 2.0 to 5.0 μm, conforming to the general definition of the family *Enterobacteriaceae*. Gram-negative. Usually motile (peritrichous flagella). Facultatively anaerobic. Colonies are generally 2 to 4 mm in diameter. Nitrates are reduced to nitrites. Gas is usually produced from glucose. Hydrogen sulfide is usually produced on triple-sugar iron agar. Indole-negative. Citrate is usually utilized as a sole carbon source. Lysine and ornithine decarboxylase (Müller's) reactions are usually positive. Urease-negative. Phenylalanine and tryptophan are not oxidatively deaminated. Sucrose, salicin, inositol, and amygdalin are usually not fermented. Lipase and deoxyribonuclease are not produced. Pathogenic for humans, causing enteric fevers, gastroenteritis, and septicemia; also may infect many animal species besides humans. Some serotypes are strictly host-adapted. The mol% G + C of the DNA is 50 to 53%.

Five "subgenera" of the genus are recognized in *Bergey's Manual* (1984). This classification is based on the biochemical characteristics shown in Table 2. Clinically, "subgenus I" is of primary concern. Subgenera II and III are frequently isolated from intestinal contents of cold-blooded animals and only rarely from warm-blooded animals. Strains from subgenera IV and V are isolated chiefly from the environment and are rarely pathogenic for humans. The "*Arizona*" group comprises subgenus III. Though members of subgenera II to V are here described as "rarely" pathogenic for humans, the fact is that they have on occasion been linked to human illness and thus must be regarded as potential pathogens of humans (see Parker, 1984a).

Traditionally, biochemical tests have been used extensively to classify members of the family *Enterobacteriaceae*. The identification system resulted from comparative studies of the biochemical reactions given by relatively large numbers of cultures of each genus and species (Ewing, 1963). Broadly based biochemical test results permit establishment of close taxonomic similarities and differences between genera and species within the *Enterobacteriaceae*.

TABLE 2. Differential characteristics of the "subgenera" of the genus *Salmonella*

	"Subgenus"				
	I	*II*	*III*	*IV*	*V*
β-Galactosidase (ONPG test)	−	− or x	+	−	+
Acid production from					
Lactose	−	−	+ or x	−	−
Dulcitol	+	+	−	−	+
Mucate	+	+	d	−	+
Galacturonate	−	+	d	+	+
Utilization of					
Malonate	−	+	+	−	−
d-Tartrate	+	− or x	− or x	− or x	−
Gelatin hydrolysis (film method)	−	+	+	+	−
Growth in presence of KCN	−	−	−	+	+
Habitat of the majority of strains					
Warm-blooded animals	+	−	−	−	−
Cold-blooded animals and environment	−	+	+	+	+

+ = Positive for 90% or more of strains in 1–2 days; d = positive for 11–89% of strains in 1–2 days; − = positive for 0–10% of strains in 1–2 days; x = late and irregularly positive (3–7 days). The temperature for all reactions is 37°C.
(Reprinted with permission from *Bergey's Manual of Systematic Bacteriology,* Volume 1, 1984.)

But, from a practical viewpoint, biochemical testing of cultures for differentiation and identification of the *Enterobacteriaceae* has shortcomings in routine diagnostic work, particularly when rapid detection and identification of salmonellae are desired. This disadvantage is emphasized by Table 3 which shows that the subgenera of salmonellae differ considerably in important biochemical properties. If consideration is given only to subgenus I in which the most clinically significant salmonellae are to be found, the picture is the same. For example, members of subgenus I are listed as lactose-negative; yet lactose-positive strains are encountered and have been implicated in human illness. It is important to note that the commonly used differential media for the isolation and identification of salmonellae contain lactose and that lactose-positive colonies are normally ignored when the media are examined after inoculation and incubation.

TABLE 3. Biochemical differentiation of the four subgenera of *Salmonella*

	Subgenus			
	I	*II*	*III*	*IV*
Acid in dulcitol	+	+	−	−
Acid in lactose	−	−	+/x	−
β-galactosidase	−	−/x	+	−
Acid in salicin	−	−	−	+
Fermentation of D-tartrate	+	−/x	−/x	−/x
Fermentation of mucate	+	+	v	−
Fermentation of malonate	−	+	+	−
Liquefaction of gelatin	−[a]	+	+	+
Growth in KCN medium	−	−	−	+

[a] A few exceptions.
x = Late or irregular results; v = some strains positive, others negative.
(Reprinted with permission from Parker, 1984a)

Salmonellae are also defined according to their antigenic composition. The O antigens are heat-stable polysaccharides located in the cell wall. The H antigens are heat-labile proteins located in the flagella. In addition to the O and H antigens, some salmonellae possess surface polysaccharides that inhibit O agglutination. Heat treatment renders such salmonellae agglutinable by homologous O antisera. The Vi antigen of the typhoid organism is the most important example of these heat-sensitive surface polysaccharide antigens.

Approximately 2,000 serotypes of *Salmonella* are recognized. These have been arranged in the Kauffmann-White diagnostic scheme. The primary division is into groups, with the members of each group sharing a common somatic antigen. An abbreviated antigenic schema is shown in Table 4 (Ewing and Martin, 1974). The 50 serotypes listed account for 96% of the salmonellae isolated from all sources in the United States.

With reference to Table 4 , the common antigen in group B is 4; in group C1 it is 7, in group C2 it is 8, and so on. Most salmonellae are diphasic with respect to H antigens, but only one phase can be initially detected. The antigenic formula for *S. paratyphi B* is 1,4,5,12:b:1,2. The interpretation of the formula indicates the presence of the 1,4,5 and 12 O

TABLE 4. Abbreviated antigenic schema for *Salmonella*

Serotype	*Group*	*Antigenic formula*		
			H antigens	
		O antigens	*Phase 1*	*Phase 2*
S. paratyphi-A	A	1, 2, 12	a	—
S. paratyphi-B	B	1, 4, 5, 12	b	1, 2
S. paratyphi-B, odense	B	1, 4, 12	b	1, 2
S. java	B	1, 4, 5, 12	b	[1, 2]
S. stanley	B	4, 5, 12	d	1, 2
S. schwarzengrund	B	1, 4, 12, 27	d	1, 7
S. saint-paul	B	1, 4, 5, 12	e, h	1, 2
S. reading	B	4, 5, 12	e, h	1, 5
S. chester	B	4, 5, 12	e, h	e, n, x
S. sandiego	B	4, [5], 12	e, h	e, n, z_{15}
S. derby	B	1, 4, 5, 12	f, g	—
S. california	B	4, 5, 12	m, t	—
S. typhimurium	B	1, 4, 5, 12	i	1, 2
S. typhimurium copenhagen	B	1, 4, 12	i	1, 2
S. bredeney	B	1, 4, 12	l, v	1, 7
S. heidelberg	B	1, 4, 5, 12	r	1, 2
S. choleraesuis	C_1	6, 7	c	1, 5
S. choleraesuis kunzendorf	C_1	6, 7	[c]	1, 5
S. braenderup	C_1	6, 7	e, h	e, n, z_{15}
S. montevideo	C_1	6, 7	g, m, s	—
S. oranienburg	C_1	6, 7	m, t	—
S. thompson	C_1	6, 7	k	1, 5
S. infantis	C_1	6, 7	r	1, 5
S. bareilly	C_1	6, 7	y	1, 5
S. tennessee	C_1	6, 7	z_{29}	—
S. muenchen	C_2	6, 8	d	1, 2
S. manhattan	C_2	6, 8	d	1, 5
S. newport	C_2	6, 8	e, h	1, 2
S. blockley	C_2	6, 8	k	1, 5
S. litchfield	C_2	6, 8	l, v	1, 2
S. tallahassee	C_2	6, 8	z_4, z_{32}	—
S. kentucky	C_2	(8), 20	i	z_6
S. miami	D	1, 9, 12	a	1, 5
S. typhi	D	9, 12, Vi	d	—
S. berta	D	9, 12	f, g, t	—
S. enteritidis	D	1, 9, 12	g, m	—
S. dublin	D	1, 9, 12	g, p	—
S. panama	D	1, 9, 12	l, v	1, 5
S. javiana	D	1, 9, 12	l, z_{28}	1, 5
S. pullorum	D	9, 12	—	—
S. anatum	E_1	3, 10	e, h	1, 6
S. meleagridis	E_1	3, 10	e, h	l, w
S. give	E_1	3, 10	l, v	1, 7
S. newington	E_2	3, 15	e, h	1, 6
S. illinois	E_3	(3), (15), 34	z_{10}	1, 5
S. senftenberg	E_4	1, 3, 19	g, s, t	—
S. simsbury	E_4	1, 3, 19	z_{27}	—
S. rubislaw	F	11	r	e, n, x
S. poona	G	13, 22	z	1, 6
S. worthington	G	1, 13, 23	z	l, w
S. cubana	G	1, 13, 23	z_{29}	—
S. florida	H	1, 6, 14, 25	d	1, 7
S. madelia	H	1, 6, 14, 25	y	1, 7
S. cerro	18	18	z_4, z_{23}	—
S. siegburg	18	6, 14, 18	z_4, z_{23}	—
S. minnesota	21	21	b	e, n, x
S. urbana	30	30	b	e, n, x

Adapted from Ewing and Martin, 1974.

antigens; the H antigen in phase 1 is b and the 1 and 2 H antigens are present in phase 2. In most cases initial agglutinations would show either the b antigen (phase 1) or the 1 and 2 antigens (phase 2). Phase reversal can be performed to cause the antigens in the missing phase to be manifested. For example, if a culture of *S. paratyphi B* were found to agglutinate in b antiserum (phase 1) but not in sera containing the 1 and 2 antibodies, phase reversal would be accomplished by inoculating semisolid medium to which b antiserum is added. The H agglutinins in the antiserum would immobilize the homologous flagella (b antigens), and the spreading growth would be in the phase not immobilized (1,2 antigens). H agglutinations carried out on the spreading growth would result in detection of the antigens that originally were not detected. Methodologically, this can be accomplished in a number of ways. One useful approach is the use of a U-tube containing the semisolid medium. The culture is inoculated into one arm and the spreading, motile growth is removed from the opposite arm of the tube.

The Kauffmann-White scheme is not a complete antigenic record, and the antigenic constitution of most organisms is far more complex than is suggested by the Kauffmann-White formulae, in which only antigens of differential importance are recorded. If small antigenic differences are taken into account, the number of serotypes that can be recognized is almost unlimited (Parker, 1984a).

A complete listing of all serotypes of the genus *Salmonella* is to be found in *Bergey's Manual of Systematic Bacteriology,* Volume 1 (1984). Further, this listing also designates the subgenus of each serotype.

The primary approach to the cultural detection of salmonellae focuses on the similarities and contrasts in biochemical properties with other *Enterobacteriaceae,* whereas their ultimate identification and classification is based on antigenic structure. Slavish preoccupation with detailed biochemical testing can not only needlessly waste valuable time, but also, in the hands of the unskilled, can lead to false-negative conclusions. After all testing is complete, the conclusion that a culture is a *Salmonella* rests on the results of antigenic analysis. The sooner antigenic analysis is done in the isolation and identification procedure the better.

Specimen Collection and Transport

Refer to Ewing and Martin (1974) for detailed information regarding the collection and handling of specimens to be analyzed for *Enterobacteriaceae*. Specimens submitted for diagnosis of salmonellosis may be of intestinal or extraintestinal origin. When symptoms suggest enteric fever, blood, urine, feces, purulent material from wounds or abscesses, cerebrospinal fluid, and nasopharyngeal swabs may be submitted. The gastroenteritis associated with the food poisoning syndrome primarily calls for examination of fecal specimens.

As in any type of analytic work, it is axiomatic that the result can be no better than the sample submitted for examination. Specimens must be properly collected and handled from collection to the initiation of analysis. Generally, specimens submitted for examination are analyzed simultaneously for other enteric pathogens.

Isolation and Identification Procedures

Primary Culture/Enrichment

Generally, specimens, whether of extraintestinal or intestinal origin, are inoculated into a broth medium. A nonselective broth may be used for blood (5 ml blood per 100 ml of medium) because blood has no competing flora. Fecal specimens must be enriched in selective broths to increase the proportion of salmonellae in the bacterial population. *Salmonella,* if present, may be greatly outnumbered by other microorganisms. The two media most commonly used are tetrathionate (Müller, 1923) and selenite F (Leifson, 1936) broths. GN broth (Hajna, 1955) also has proven useful for selective enrichment for salmonellae. It is advisable to use at least two selective enrichment broths because certain salmonellae may not be detected if only one broth is used. The selective broths should be inoculated with 1 g of specimen to 10 ml of medium. With respect to extraintestinal specimens other than blood, for example, purulent material from wounds or abscesses, spinal fluid, or nasopharyngeal swabs, the choice between selective and nonselective enrichment depends on the expected flora of the sample. In most cases, nonselective enrichment is indicated for extraintestinal specimens.

Isolation

After primary culture in a broth medium, specimens are inoculated (streaking for isolation) onto agar media. Subculture of primary enrichment broths should be done only on differential and selective agars. Blood broth cultures should be streaked after 18 to 24 h of incubation at 37°C and, if negative, the broths should be reincubated for at least 3 weeks with intermediate subculturings being done at 5-day intervals. Selective enrichment broths should be streaked onto selective/differential agars after 24 h of incubation at 37°C. It is important to streak the culture so that one area of the plate contains large numbers of well isolated colonies. The differential character of the plating media is lost in the crowded areas of the agar, and thus *Salmonella* colonies may be obscured by the commensal flora.

Specimens from extraintestinal sources may be inoculated onto nonselective (blood or infusion agar) and selective/ differential media. There are three categories of selective/differential isolation media (Ewing and Martin, 1974): (1) differential media with little selectivity as regards *Enterobacteriaceae*—MacConkey's, eosin methylene blue (EMB), and deoxycholate agars; (2) differential, moderately selective media—*Shigella-Salmonella* (SS), deoxycholate citrate, Hektoen enteric (HE), and xylose lysine deoxycholate (XLD) agars; and (3) highly selective media—bismuth sulfite (BS) and brilliant green (BG) agars. As with selective enrichment broths, more than one selective/differential medium should be used, as certain salmonellae may be inhibited by one or another agar medium. For example, *S. typhi* does not grow on brilliant green agar. It should be appreciated, further, that with the exception of BS agar, the differential characteristics of these media depend, in part, on carbohydrate use or lack of it. For example, lactose-fermenting salmonellae give the appearance of coliforms on EMB, SS, or BG agars. Such colonies would be given no further consideration as "suspect" *Salmonella*. However, on BS agar these lactose fermenters appear as typical salmonellae and would be subjected to further study. For this reason, BS agar should always be used for direct inoculation or for subculture of primary enrichment cultures.

In our laboratories we used BG, SS, and BS as selective/ differential plating media for years, but recently have replaced BG and SS agars with HE and XLD media. Not to be overlooked is the importance of the analyst's familiarity with the selective/differential medium used. No written descriptions of differential colonial characteristics can substitute for experience in looking at colonies on the medium. The analyst should be encouraged to study further (via biochemical and serologic tests) any colony that appears suspicious. This will, initially, result in "over-picking" of colonies, but with studied experience, skill in the use of a particular medium will develop so that suspect colonies selected for further study will be chosen with great accuracy and confidence.

After incubation, the differential/selective agars must be examined for the presence of suspect *Salmonella* colonies. Bismuth sulfite plates are examined after 24 and 48 h incubation at 37°C. Other differential agars should be examined after 18 to 24 h. Indeed, with most selective/differential media, incubation of the plates beyond 24 h may result in obscured and unreliable readings of the colonial characteristics because pH indicator changes occur with extended incubation. We have found it useful to examine the plates for suspect colonies under the magnifying glass on the Quebec colony counter. Two to three suspect colonies from each medium should be selected for further study.

Identification

Selected colonies should be transferred with a sterile wire needle to slants of triple-sugar iron agar (TSI) and lysine-iron agar (LIA). The surface of the slant is streaked and then the butt of the tube stabbed with the needle. It is very important that the subculture to LIA and TSI slants be made from a well-isolated colony. It must be appreciated that LIA and TSI are used to biochemically screen suspect *Salmonella* colonies. If a mixed culture is introduced into these media, the results will be meaningless. If suspected colonies are found only in crowded areas of the selective/differential isolation agar, it is prudent to restreak suspect colonies to obtain pure cultures before inoculation into TSI and LIA. These two differential media should be examined after 18 to 24 h of incubation at 37°C. This is most important with TSI, as an acid slant at 24 hours often reverts to alkaline with further incubation because the protein components of the medium are used after exhaustion of carbohydrate.

The reactions of *Enterobacteriaceae* in TSI and LIA are shown in Tables 5 and 6, respectively (Ewing and Martin, 1974). Though the notations with respect to gas and H_2S are useful, attention should be

TABLE 5. Reactions of *Enterobacteriaceae* in triple sugar-iron (TSI) agar medium

Genus and species	*Slant*	*Butt*	*Gas*	*H_2S*
Escherichia	A (K)	A	+(−)	−
Shigella	K	A	−	−
Salmonella typhi	K	A	−	+(−)
Other *Salmonella* spp.	K	A	+	+++(−)
Arizona	K (A)	A	+	+++
Citrobacter	K (A)	A	+	+++(−)
Edwardsiella	K	A	+	+++
Klebsiella	A	A	++	−
Enterobacter	A	A	++	−
E. hafniae	K	A	+	−
Serratia	K or A	A	−	−
Proteus vulgaris	A (K)	A	+	+++
P. mirabilis	K (A)	A	+	+++
P. morganii	K	A	−(+)	−
P. rettgeri	K	A	−	−
Providencia	K	A	+ or −	−

A = acid; K = alkaline.
Symbols in parentheses indicate occasional reactions.
(Reprinted with permission from Ewing and Martin, 1974.)

TABLE 6. Reactions of *Enterobacteriaceae* in lysine-iron agar (LIA)

Genus and species	*Slant*	*Butt*	*Gas*	*H_2S*
Escherichia	K	K or N	− or +	−
Shigella	K	A	−	−
Salmonella	K	K or N	−	+(−)
S. typhi	K	K[a]	−	+ or −
S. paratyphi A	K	A	+ or −	− or +
Arizona	K	K or N	−	+(−)
Citrobacter	K	A	− or +	+ or −
Edwardsiella	K	K	− or +	+
Klebsiella	K or N	K or N	+ or −	−
Enterobacter cloacae	K or N	A	+ or −	−
E. aerogenes	K	K or N	+(−)	−
E. hafniae	K	K or N	− or +	−
Serratia	K or N	K or N	−	−
Proteus vulgaris	R	A	−	−(+)
P. mirabilis	R	A	−	−(+)
P. morganii	K or R	A	−	−
P. rettgeri	R	A	−	−
Providencia	R	A	−	−

A = acid; K = alkaline; R = red (oxidative deamination). An alkaline reaction in the *butt* of this medium indicates decarboxylation. Symbols in parentheses indicate occasional reactions.

[a] A rare culture of *S. typhi* may fail to decarboxylate lysine. (Reprinted with permission from Ewing and Martin, 1974.)

focused on the reactions in the butt and on the slant of the cultures, regardless of whether H_2S or gas production is manifested. The reactions of *Pseudomonas* spp. on these two media should be noted. In TSI the organisms produce an alkaline slant and a neutral butt without gas or H_2S. On LIA the slant is alkaline and the butt neutral without gas or H_2S. Thus, the TSI reaction presents no problem, but at times the results in LIA may suggest *Salmonella*. If any doubt exists, however, an oxidase test can be done on the growth on the LIA slant. Pseudomonads are oxidase-positive, salmonellae oxidase-negative. The test can be done in less than a minute, and any oxidase-positive culture can be eliminated from further consideration as a *Salmonella*.

Most *Proteus* and *Providencia* spp. produce a red reaction on LIA slants, due to oxidative deamination. Cultures showing this reaction can be eliminated from further consideration as salmonellae.

The focus of this chapter is on the detection of salmonellae. However, clinical laboratory specimens are submitted for the detection of several enteric pathogens, not only *Salmonella*, and colonies from selective/differential agars may be picked for identification to other media in addition to LIA and TSI (Ewing and Martin, 1974). Two media groups most often suggested for inoculation along with TSI and LIA include (1) motility indole ornithine (MIO) medium, or (2) Christensen's urea agar, Simmon's citrate, ornithine motility, and peptone broth. The authors state that either of these systems permits the laboratory to issue a preliminary report on most isolates at 18 to 24 h.

With respect specifically to identification of salmonellae, any culture which after 24 h incubation at 37°C shows an alkaline slant and acid butt on TSI or an alkaline slant and alkaline butt on LIA should be subjected to further study. The results with the two media must be interpreted in tandem. For example, a culture showing an acid butt and acid slant on TSI is likely not *Salmonella;* but typical reactions on LIA (alkaline slant and alkaline butt) may indicate a lactose-positive or sucrose-positive *Salmonella*. A typical reaction on TSI (alkaline slant and acid butt) but an atypical reaction on LIA (alkaline slant and acid butt) may suggest a lysine-negative *Salmonella*, for example, *S. paratyphi* A. The objective should be to definitively identify an organism as a *Salmonella* as quickly as possible. We have found it useful to conduct serologic tests before further biochemical tests after the suspect cultures have grown on TSI and LIA.

For serologic testing, a subculture from either the LIA or TSI slant is made in an appropriate broth medium, for example, 3 ml of brain-heart infusion (BHI) broth. After 3 to 4 h at 37°C the broth is sufficiently turbid to be used as an antigen for performance of H agglutinations. An equal volume of 0.6% formalinized saline is added to the broth culture. Equal volumes of polyvalent H antiserum and the formalinized culture are incubated for 1 h at 50°C, with appropriate saline and serum controls. A positive reaction, agglutination, is *highly* indicative that the suspect culture is *Salmonella*. It should be appreciated that the H antigens of salmonellae are quite specific to the genus, whereas the O antigens are shared with other members of the *Enterobacteriaceae*. Thus, a positive reaction with polyvalent H antiserum is sufficient to presume that the organism is a *Salmonella, in the absence of confirmatory biochemical tests*. Further serologic tests can be conducted using the forminalized broth culture for H agglutinations and a saline suspension from either the TSI or LIA slants for O agglutinations.

It is feasible to conduct serologic confirmation even at the selective/differential agar isolation stage. One of us (J.H.S.) has advocated subculture of suspect colonies directly from differential/selective media into BHI broth simultaneously with inoculation of TSI and LIA. Within 3 to 4 h the broth cultures are sufficiently turbid to be used for H antigen agglutination studies. A portion of the broth culture may be saved for O agglutinations in case the culture proves positive in H agglutination tests. Thus, it is possible to arrive at a definitive diagnosis, including probable serotype of salmonellosis within 24 h of initiating testing of a given specimen. This time requirement assumes direct plating of the specimen on selective agars with subculture into infusion broth at 18 h (Silliker et al., 1965).

In looking for *Salmonella*, it is important to conduct the serologic tests as early as possible for two reasons: (1) ultimately, the conclusion that an organism is a *Salmonella* rests on antigenic analysis; and (2) biochemical testing that leads to serologic studies is time consuming and carries the risk of erroneous interpretations. Furthermore, proper conclusions from biochemical tests can be drawn only if the cultures used are pure. As pointed out by Parker (1984a): "Failure to appreciate that organisms subcultured from colonies on selective media are frequently impure is one of the commonest pitfalls of enteric bacteriology." If TSI or LIA cultures used for agglutination tests are not pure, the results of such tests are usually not compromised because the tests search for stable *Salmonella* antigens, not potentially variable biochemical characteristics. If the same nonaxenic cultures were to be used in biochemical tests, spurious results could easily lead to the false conclusion that the organism is not *Salmonella*.

Biochemical Tests

Ewing and Martin (1974) advocate relatively extensive biochemical testing to classify organisms in the family *Enterobacteriaceae*. We prefer to use serologic tests as early as possible to confirm the presence of *Salmonella*. However, when serologic studies fail to confirm LIA- and TSI-suspect cultures as *Salmonella*, our philosophy is that biochemical tests are required at this point in the analysis. If biochemical test results are consistent with the genus *Salmonella*, then the possibility exists that the organism possesses *Salmonella* antigens not contained in the antisera used by the laboratory. A *Salmonella* isolate lacking all the antigens represented by the antigenic formulae in Table 4 is a relatively rare occurrence because these 50 serotypes and variants account for 96% of the salmonellae from all sources. A check should be made to determine whether the isolate is motile; if it is nonmotile H antigens will not be detected. If the organism shows *Salmonella* O antigens, is biochemically consistent with the genus, yet negative in H agglutination studies, it is appropriate to conclude preliminarily that the isolate is a *Salmonella* without demonstrable H antigens. If the organism agglutinates in *Salmonella* O antisera, but gives biochemical reactions not consistent with the genus, it is likely that the isolate is not *Salmonella* because the O antigens are shared with a number of other *Enterobacteriaceae*. When conclusive serologic and/or biochemical results are not obtainable, it is wise to submit the culture to an appropriate reference laboratory for final identification.

Table 7 (Kelley et al., 1985) lists 28 biochemical tests to differentiate all currently recognized, clinically significant *Enterobacteriaceae*. A primary battery of 14 tests is claimed to be sufficient to identify most isolates to genus and many isolates to the species level. Fourteen secondary tests may be used to identify more difficult isolates. It is absolutely essential that these tests be applied to pure cultures. Use of mixed cultures carries a high risk of misinterpretation of results which could lead to discarding of *Salmonella*-positive cultures.

Results from biochemical tests in Table 7 are given as percent positive reactions. This table permits a clear understanding of the "usual" reactions to be expected of a given organism, but exceptions to the norm exist in specific cultures that prevent exclusion of a biochemically deviant isolate from consideration as a *Salmonella*. To reduce the risk of misidentifying biochemically deviant cultures, we advocate serologic testing as early as possible.

D'Amato et al. (1985) list many prominent commercially available systems and devices for biochemical characterization of isolates. We do not presume to evaluate these systems, but wish to emphasize that the risks and shortcomings are identical to those inherent to classical biochemical testing methods.

Differentiation of *S. typhi*, *S. partyphi* A, *S. paratyphi* B, and *S. choleraesuis* from other salmonellae is important, as the first three of these invade the blood stream and cause enteric fever. *S. choleraesuis* may produce enteric fever, particularly in children, but may also only cause gastroenteritis. Isolation of these four serotypes should be immediately reported to the attending physician, so that appropriate treatment can be initiated. Tables 8 and 9 show that *S. choleraesuis*, *S. typhi*, *S. enteritidis*, and *S. paratyphi* A can be distinguished from other members of subgenus I on the basis of biochemical reactions. Serologic tests can clearly separate these four organisms. Ewing and Martin (1974) emphasize the importance of the clinical laboratory recognizing these particular serotypes, even though the laboratory may not have the ability to identify other serotypes.

Serologic Tests

Three types of antigens are used in the diagnostic definition of serotypes of salmonellae: the heat-stable O antigens that form a part of the cell wall, the heat-labile H antigens contained in the flagella and surface polysaccharide antigens. The H antigens are highly specific to *Salmonella*, whereas *Salmonella* O antigens are found in many other genera. The surface

TABLE 7. Biochemical characteristics of clinically significant *Enterobacteriaceae*

	% Positive reactions																					
Characteristic	*Cedecea davisae*	*C. neteri*	*Citrobacter amalonaticus*	*C. diversus*	*C. freundii*	*Edwardsiella tarda*	*Enterobacter aerogenes*	*E. agglomerans*	*E. cloacae*	*E. gergoviae*	*E. sakazakii*	*Escherichia coli*	*E. fergusonii*	*Ewingella americana*	*Hafnia alvei*	*Klebsiella oxytoca*	*K. ozaenae*	*K. planticola*	*K. pneumoniae*	*K. rhinoscleromatis*	*Kluyvera ascorbata*	*K. cryocrescens*
Primary test battery																						
D-Adonitol	0	0	0	98	0	0	100	7	25	0	0	5	96	0	0	100	97	100	90	100	0	0
Arginine dihydrolase	50	100	85	65	65	0	0	0	97	0	100	17	4	0	6	0	6	0	0	0	0	0
Citrate utilization	95	100	85	100	95	0	95	50	100	98	100	0	19	97	10	95	30	100	98	0	96	76
DNase	0	0	0	0	0	0	0	0	0	0	0	0	0	0	0	0	0	0	0	0	0	0
D-Glucose, gas production	70	100	97	98	95	100	100	20	100	98	98	90	92	0	98	97	50	100	97	0	93	95
Hydrogen sulfide (TSI)	0	0	0	0	80	100	0	0	0	0	0	0	0	0	0	0	0	0	0	0	0	0
Indole production	0	0	100	100	5	100	0	20	0	0	11	95	96	0	0	100	0	20	0	0	89	86
Lysine decarboxylase	0	0	0	0	0	100	98	0	0	90	0	90	92	0	100	98	40	100	98	0	97	35
Motility	95	100	98	95	95	98	97	85	95	90	91	80	92	60	90	0	0	0	0	0	98	86
Ornithine decarboxylase	95	0	95	100	20	100	98	0	96	100	91	65	100	0	98	2	3	0	0	0	100	100
Phenylalanine deaminase	0	0	0	0	0	0	0	20	0	0	50	0	0	0	0	2	0	0	0	0	0	0
Sucrose	100	100	15	20	30	0	100	75	97	98	100	50	0	0	10	100	20	100	100	75	98	81
Urease	0	0	80	75	70	0	2	20	65	93	0	0	0	0	4	90	10	98	95	0	0	0
Voges-Proskauer	50	50	0	0	0	0	98	70	100	100	100	0	0	97	85	90	0	98	95	0	0	0
Secondary test battery																						
L-Arabinose	0	0	100	100	100	9	100	95	100	98	100	100	96	0	95	98	98	100	100	100	100	100
myo-Inositol	0	0	0	0	3	0	100	15	15	0	75	0	0	0	0	98	55	100	95	95	0	0
KCN, growth in	86	65	95	0	96	0	98	35	98	0	100	3	0	7	95	97	88	100	98	90	92	86
Lactose	19	35	50	35	50	0	95	40	93	55	100	90	0	70	5	100	30	100	98	0	98	95
Malonate	91	100	0	90	15	0	85	65	75	96	18	0	30	0	70	98	3	100	93	95	96	86
D-Mannitol	100	100	100	100	100	0	100	100	100	100	100	98	100	100	100	100	100	100	100	100	100	95
Melibiose	0	0	5	0	50	0	100	50	90	97	100	75	0	0	0	100	97	100	100	100	100	100
ONPG (β-galactosidase)	90	100	100	96	95	0	100	90	100	97	100	95	83	96	90	100	80	100	100	0	100	100
Raffinose	10	0	5	0	30	0	96	30	97	97	100	50	0	0	2	100	90	100	100	75	98	100
L-Rhamnose	0	0	100	100	100	0	100	85	92	98	100	80	92	23	97	100	55	100	100	96	100	100
Salicin	100	100	40	20	5	0	100	65	75	98	100	40	54	90	13	100	97	100	100	98	100	100
D-Sorbitol	0	100	100	98	98	0	100	30	95	0	0	94	0	0	0	100	75	100	100	100	40	45
Trehalose	100	100	100	100	100	0	100	97	100	100	100	98	96	100	100	100	98	100	100	100	100	100
D-Xylose	100	100	100	100	100	0	100	93	100	98	100	95	96	13	98	100	95	100	100	100	100	91

TSI = Triple sugar-iron agar.
(Reprinted with permission from Kelley et al., 1985.)

polysaccharide antigens, for example the Vi antigen of *S. typhi*, also occurs in other genera such as *Citrobacter*. The O antigens are detected by slide agglutination tests using O antisera on saline suspensions of live organisms. H antigens are detected through tube agglutination tests using broth cultures diluted with an equal volume of 0.6% formalinized saline. O agglutinations are carried out at room temperature and H agglutinations are done at 50°C in a water bath. The surface polysaccharide antigens (Vi) are detected by slide agglutination tests as in the detection of O antigens. If such surface antigens are present, the organism may be inagglutinable in homologous O antisera. Heat treatment of the saline suspension (30 min in boiling water) will render the suspension agglutinable in O antisera.

As indicated in the abbreviated antigenic schema for *Salmonella* shown in Table 4 (Ewing and Martin, 1974), the salmonellae are assigned to groups based on predominating O antigens. In this abbreviated schema, serotypes in groups A to H are listed. Three additional groups are also included, namely those containing O factors 18, 21, and 30. Not all known serotypes in the O groups are listed in this table, many others have been identified. Furthermore, higher O groups are recognized, witness to the fact

% Positive reactions

Morganella morganii	*Proteus mirabilis*	*P. penneri*	*P. vulgaris*	*Providencia alcalifaciens*	*P. rettgeri*	*P. stuartii*	*Salmonella choleraesuis*	*S. paratyphi A*	*S. typhi*	*Salmonella, other serotypes*	*Serratia liquefaciens group*	*S. marcescens*	*S. odorifera*	*S. rubidaea*	*Shigella boydii*	*S. dysenteriae*	*S. flexneri*	*S. sonnei*	*Tatumella ptyseos*	*Yersinia enterocolitica*	*Y. frederiksenii*	*Y. kristensenii*	*Y. pestis*	*Y. pseudotuberculosis*	*Y. intermedia*
0	0	0	0	100	100	5	0	0	0	0	5	40	55	98	0	0	0	0	0	0	0	0	0	0	0
0	0	0	0	0	0	0	55	15	3	70	0	0	0	0	18	3	2	2	0	0	0	0	0	0	1
0	65	0	15	98	95	93	25	0	0	95	90	98	98	95	0	0	0	0	2	0	15	0	0	0	0
0	0	25	60	0	0	2	0	0	0	2	85	98	100	98	0	0	0	0	0	9	0	0	0	0	0
87	96	45	85	85	10	0	95	100	0	96	75	55	8	30	0	0	3	0	0	5	40	23	0	0	18
2	95	10	95	0	0	0	50	10	97	95	0	0	0	0	0	0	0	0	0	0	0	0	0	0	0
98	2	0	98	100	100	98	0	0	0	0	2	2	55	2	25	45	50	0	0	50	100	30	0	0	99
0	0	0	0	0	0	0	95	0	98	98	90	100	96	55	0	0	0	0	0	0	0	0	0	0	0
95	95	85	95	96	94	85	95	95	97	95	95	97	100	85	0	0	0	0	0	2	0	8	0	0	6
98	100	0	0	0	0	0	100	95	0	97	100	100	40	0	2	0	0	98	0	95	92	92	0	0	100
95	98	100	100	98	98	95	0	0	0	0	0	0	0	0	0	0	0	0	90	0	0	0	0	0	0
0	15	100	97	15	15	50	0	0	0	2	98	100	40	98	0	0	2	2	98	95	100	0	0	0	100
98	90	100	95	0	98	30	0	0	0	0	3	15	2	2	0	0	0	0	0	75	60	77	5	95	77
0	25	0	0	0	0	0	0	0	0	0	93	98	81	100	0	0	0	0	5	2	0	0	0	0	12
0	0	0	0	0	0	0	0	100	2	100	98	0	100	100	94	45	60	95	0	98	100	77	100	50	100
0	0	0	0	0	90	95	0	0	0	35	60	75	100	20	0	0	0	0	0	30	7	15	0	0	18
100	98	100	100	100	97	100	0	0	0	0	90	95	35	25	0	0	0	0	0	2	0	0	0	0	6
0	2	0	2	0	5	2	0	0	2	2	10	2	87	100	2	0	2	2	0	5	40	8	0	0	35
2	2	0	0	0	0	0	0	0	0	0	2	3	0	94	0	0	0	0	0	0	0	0	0	0	6
0	0	0	0	2	100	10	98	100	100	100	100	100	98	100	97	0	95	100	0	98	100	100	97	100	100
0	0	0	0	0	5	0	45	95	100	95	75	0	98	98	15	0	55	25	25	0	0	0	50	70	80
5	0	0	10	0	5	10	0	0	0	2	93	95	100	100	10	40	2	90	0	95	100	70	50	70	93
0	0	0	5	0	5	7	0	0	0	2	85	2	45	98	2	0	40	3	11	5	30	0	0	15	45
0	2	0	5	0	70	0	100	100	0	95	15	0	94	2	0	30	5	75	0	0	100	0	2	70	100
0	0	0	75	0	50	2	0	0	0	0	97	95	65	98	0	0	0	0	55	20	92	15	70	25	100
0	0	0	0	0	0	0	90	95	98	95	95	100	100	2	45	30	30	2	0	98	100	100	50	0	100
13	98	55	30	0	0	98	0	100	100	100	100	100	100	100	85	85	65	100	93	98	100	100	100	100	100
0	98	100	95	0	10	7	98	0	82	97	100	7	100	98	11	4	2	2	9	70	100	85	90	100	100

that approximately 2,000 *Salmonella* serotypes are recognized, and only 50 are listed in this table. Most of the serotypes have diphasic H antigens which, in most cases, exist in only one of the two phases; and it is necessary to carry out phase reversal to detect the antigens in the second phase. A current listing of the antigenic formulae of the genus *Salmonella* is in *Bergey's Manual of Systematic Bacteriology,* Volume 1 (1984).

The ability to detect *Salmonella* antigens, whether H or O, is limited by the antibody content of the diagnostic antisera available to the laboratory. Furthermore, identification of specific serotypes is limited by the availabilty of monospecific antisera. The antisera required for serotyping virtually all the serotypes are commercially available, but the requirements and capabilities for in depth antigenic analysis vary with the laboratory.

Ewing and Martin (1974) suggested three degrees of effort which can be applied to the serologic identification of salmonellae. The first stage is the complete identification of *S. typhi*, *S. partyphi* A, *S. paratyphi* B, and *S. choleraesuis*. Only a few antisera are required to identify these four serotypes and *S. typhimurium*, by far the most common serotype causing gastroenteritis. Required are a polyvalent O antise-

Table 8. Differentiation of species of *Salmonella*[a]

	S. cholerasuis		S. typhi		S. enteritidis	
Test or substrate	*Sign*[b]	*Percent positive*[c]	*Sign*[b]	*Percent positive*[c]	*Sign*[b]	*Percent positive*[c]
Hydrogen sulfide (TSI)	+ or −	60	+^{W}	94.4	+	98
Citrate (Simmons)	(+)	0 (90)	−	0	+	99.3 (0.7)
Ornithine	+	100	−	0	+	99
Gas from glucose	+	100	−	0	+	97.7
Dulcitol	d	5 (15)	d	6.2 (31.3)	+	98.3
Inositol	−	0	−	0	d	42.8 (1)
Trehalose	−	0	+	100	+	100
Arabinose	−	0	−	(6.3)	+	99.3
Rhamnose	+	100	−	0	+	95
Cellobiose	−	0	d	6^{W} (31.3)	(+)	5 (92.8)
Erythritol	(+^{W}) or −	0 (85)	−	0	−	0.6
Sodium acetate	− or (+^{W})	0 (20)	−	0	+	92.4 (2.2)
Mucate	−	0	−	0	+ or −	88.3
Stern's glycerol fuchsin	−	0	−	0	+	98.2

[a] Adapted from Ewing and Martin, 1974. In this context, species *S. enteritidis* includes all *Salmonellae* other than *S. choleraesuis* and *S. typhi.*
[b] Sign: + = 90% or more positive within 1 or 2 days; (+) = positive reaction after 3 or more days (decarboxylase tests: 3 or 4 days); − = no reaction (90% or more); + or − = most cultures positive, some strains negative; − or + = most strains negative, some cultures positive; + or (+) = most reactions occur within 1 or 2 days, some are delayed; d = different reactions, +, (+), −; ^{W} = weakly positive reaction.
[c] Figures in parentheses indicate percentages of delayed reactions (3 or more days).

Table 9. Differentiation of Salmonella *paratyphi A* from other strains of *S. enteritidis*[a]

	Bioserotype paratyphi *A*		S. enteritidis	
Substrate or test	*Sign*[b]	*Percent positive*[c]	*Sign*[b]	*Percent positive*[c]
Hydrogen sulfide (TSI)	− or +^{W}	12.5	+	98
Citrate (Simmons)	− or (+)	0 (25)	+	99.3 (0.7)
Lysine decarboxylase	−	0	+	99.7
Jordan's tartrate	−	0	+	92.5
Inositol	−	0	d	42.8 (1)
Xylose	−	0	+	99
Cellobiose	d	12.5 (6.2)	(+)	5 (92.8)
Glycerol	(+)	0 (100)	d	5.7 (7.2)
Stern's glycerol fuchsin medium	−	0	+	98.2 (0.6)
Sodium acetate	−	(6.2)	+	92.4 (2.2)
Mucate	−	0	+ or −	88.3
Organic acids				
Citrate	−	0	+	96 (4)
D-Tartrate	−	0	+	92.3 (4)
i-Tartrate	−	0	d	4.7 (57.5)
l-Tartrate	−	0	d	11.8 (75.2)

[a] Adapted from Ewing and Martin, 1974. In this context, species *S. enteritidis* includes all *Salmonellae* other than *S. choleraesuis* and *S. typhi.*
[b] Sign: + = 90% or more positive within 1 or 2 days; (+) = positive reaction after 3 or more days (decarboxylase tests: 3 or 4 days); − = no reaction (90% or more); + or − = most cultures positive, some strains negative; − or + = most strains negative, some cultures positive; + or (+) = most reactions occur within 1 or 2 days, some are delayed; d = different reactions, +, (+), −; ^{W} = weakly positive reaction.
[c] Figures in parentheses indicate percentages of delayed reactions (3 or more days).

rum; O antisera for serogroups A to E inclusive (a mixed antiserum that agglutinates all members of E1, E2, E3 and E4); antisera for flagellar antigens a, b, c, d, i, and the 1 complex (mixed 1,2; 1,5, 1,6; and 1,7 antisera, single factor H antisera for factors 2 and 5, and a Vi antiserum produced with *Citrobacter* 029:Vi; Ewing and Martin, 1974).

The second stage is designated by Ewing and Martin (1974) as "intermediate." In this stage, the 50 or 60 most common serotypes of *S. enteritidis* (Table 4) are identified completely. Fifty-one unabsorbed and 32 absorbed antisera are required to serotype 96% of the most common salmonellae, as well as a considerable number of others. Use of a modified second stage is workable because the use of 83 different antisera is burdensome. Antiserum kits permitting determination of O groups A through I are commercially available. By adding single factor antisera for factors 21 and 30, also commercially available, it becomes possible to place 96% of *Salmonella* isolates into a specific O group by using 14 O antisera, albeit the O antigens are not specifically identified. The modified second stage procedure can be done in 10 min.

The Spicer-Edwards agglutination technique can be used to rapidly detect H antigens. Spicer (1956) developed a simplified, rapid method for screening and identifying the most commonly encountered H antigens in salmonellae. He used four main polyvalent H antisera, including two adjunctive antisera for identifying more than 17 of the commonly occurring H antigens. Edwards modified the Spicer antisera by omitting the L complex from the main antisera to minimize cross-reactions and suggested that it be used independently. The reactions of the Spicer-Edwards H antisera are shown in Table 10 (DIFCO, 1984). With the seven antisera in the Spicer-Edwards system all H antigens present in the organisms listed in Table 4, as well as those in a large number of other serotypes, can be detected. The Spicer-Edwards H agglutination procedure can be performed in a little more than an hour. With 14 O and 7 H antisera, all commercially available, the antigens of the serotypes in Table 4 can be detected but not necessarily completely identified. Practically, complete serologic identification of serotypes responsible for food poisoning is not necessary for the physician because the four most important serotypes would have been completely identified in the first stage of serologic testing. As Ewing and Martin (1974) pointed out, complete serologic identification should be carried out. But if the four most serious serotypes are identified at an early point, complete serologic identification of other isolates can be performed by a reference laboratory without compromising treatment of the patient. The serologic data from the modified phase 2 procedure is sufficient for initial epidemiologic studies. The procedures of the modified phase 2 should be within the scope of most clinical laboratories.

If an organism fails to agglutinate in the modified phase 2 antisera, but possesses typical biochemical characteristics, the isolate may be a rare serotype and antibodies to its antigens may not be present in the antisera being used. Polyvalent and monospecific

TABLE 10. Reactions of the *Salmonella* H antisera Spicer-Edwards with H antigens

	Salmonella *H antisera Spicer-Edwards*					Salmonella *H antisera Spicer-Edwards*			
H antigens	*1*	*2*	*3*	*4*	*H antigens*	*1*	*2*	*3*	*4*
a	+	+	+	−	k	−	+	+	+
b	+	+	−	+	r	−	+	−	+
c	+	+	−	−	y	−	+	−	−
d	+	−	+	+	z	−	−	+	+
e,h	+	−	+	−	Z_4 complex[b]	−	−	+	−
G complex[a]	+	−	−	+	z_{10}	−	−	−	+
i	+	−	−	−	z_{29}	−	+	+	−

H antigens	Salmonella *H antisera*
e, n, x; e, n, z_{15};	EN Complex
l, v; l, w; l, z_{13}; l, z_{29};	L Complex
1, 2; 1, 5; 1, 6; 1, 7	1 Complex

[a] The G complex component of *Salmonella* H antisera Spicer-Edwards 1 and 4 reacts with antigens f,g; g,g,s; f,g,t; g,m; g,m,q; g,m,s; g,m,s,t; g,m,t; g,p; q,p,s; g,p,u; g,q; g,s,t; g,t; m,p,t,u, and m,t.
[b] The Z_4 complex component reacts with z_4, z_{23}; z_4, z_{24} and z_4, z_{32}.
(Reprinted with permission from DIFCO Manual, 1984.)

antisera for antigens in "higher groups" are commercially available for a laboratory to use with isolates not agglutinating in modified phase 2 antisera.

Alternatively, it may be assumed that the culture is a *Salmonella* without demonstrating the antigens. Such a culture should be completely serologically identified by a reference laboratory. If the isolate is biochemically characterized as a *Salmonella*, the physician should assume that it is, pending the report from the reference laboratory.

The third phase suggested by Ewing and Martin (1974) is complete serologic identification, which should be done "regardless of the stage that may be adopted." They indicate that services for complete serologic identification should be restricted to a relatively small number of reference laboratories. The expense and labor required for complete serologic identification of all serotypes is not justified by the benefit. It is our belief that phases 1 and 2 should be used by each clinical laboratory that tests stool specimens. With this system, the four most dangerous serotypes will be specifically identified, and the antigens of most other serotypes will be detected, even if not completely identified.

Bacteriophage Typing

Craigie and Yen (1938) were able to classify nearly all of 592 strains of *S. typhi* with series of Vi bacteriophages. There was great correlation between the bacteriophage type and epidemic source. The value of bacteriophage typing is limited because of the untypability of *Salmonella* strains. For example, about a quarter of the strains isolated in Great Britain are untypable. Bacteriophage typing systems also have been developed for S, *paratyphi* B, *S. typhimurium*, and several other serotypes (see Parker, 1984a). If bacteriophage typing systems are available they can be valuable epidemiologic tools, but this technique is beyond the scope of most clinical laboratories.

Impact of Future Developments

Within the past year collaborative studies have led to "first action approval" for the detection of salmonellae in foods using the DNA hybridization technique (Flowers et al., 1986a) and monoclonal antibodies (Flowers et.al. 1986b). Neither of these screening systems requires conventional isolation and characterization of salmonellae; they are rapid and based on the results in foods are quite specific for salmonellae. It may be anticipated that in the near future these techniques may find acceptance in the detection of salmonellae in clinical specimens. If these techniques prove efficacious, the whole approach to the detection of salmonellae in clinical specimens will change.

Treatment, Prevention, and Control

Treatment

Effective chemotherapy for enteric fever first became possible with chloramphenicol. Administration of adequate dosage of this antibiotic early in the disease leads, in most patients, to a rapid clinical cure. There is no evidence that chloramphenicol has any effect on the rate of disappearance of salmonellae from the intestine or on the number of patients who will become chronic carriers. Other drugs have been used with varying success in cases involving chloramphenical-resistant organisms (see Parker, 1984b, for a review of chemotherapy)

Prevention and Control

Salmonellae enter the body through contaminated food and water, although direct contact transmission through fomites occurs. Prevention requires measures that block contamination of food and water with fecal material. Water supplies, especially private systems with wells, should be tested to assure freedom from fecal contamination, and public water supplies should be disinfected. Raw foods of animal origin must be assumed to be contaminated with salmonellae and, therefore, must be treated by the processor or the consumer to destroy the organisms. The consumer must prevent cross-contamination between raw and cooked foods in the kitchen environment. The processor must use manufacturing methods and adequate process controls to ensure destruction of salmonellae during the making of foods and drugs. Human salmonellosis will continue to be a serious public health problem if these control measures are not applied. The incidence of salmonellosis will continue to be proportional to the effort given to enforcement of effective public health control measures.

Literature Cited

Archer, D. L., and J. E. Kvenberg. 1985. Incidence and cost of foodborne disease in the United States. J. Food Prot. **48:**887–894.

Aserkoff, B., and J. V. Bennett. 1969. Effect of antibiotic therapy in acute salmonellosis on the fecal excretion of salmonellae. N. Engl. J. Med. **281:**636–640.

Bergey's Manual of Systematic Bacteriology, Vol. I. 1984. p. 427–458. N. R. Krieg, J. G. Holt (eds.). Academic Press, Baltimore.

Brachman, P. S., A. Taylor, E. J. Gangarosa, M. H. Merson, and W. H. Barker. 1973. Food poisoning in the U.S.A. p. 143–151. *In* B. C. Hobbs and J. H. B. Christian (ed.), The microbiological safety of food. Academic Press, London.

Bryan, F. L. 1978. Factors that contribute to outbreaks of foodborne disease. J. Food Prod. **41:**816–827.

Bryan, F. L. 1980. Foodborne diseases in the United States associated with meat and poultry. J. Food Protect. **43:**140–150.

Bryan, F. L., M. J. Fanelli, and H. Riemann. 1979. *Salmonella* infections, pp. 74–130. *In* H. Riemann and F. L. Bryan (ed.), Foodborne infections and intoxications. Academic Press, New York.

Chin, J. 1982. Raw milk—an editorial. J. Infect. Dis. **146:**440–441.

Coleman, W., and B. H. Buxton. 1907. The bacteriology of the blood in typhoid fever: an analysis of 1,602 cases. Am. J. Med. Sci. **133:**896–903.

Craigie, J., and C. H. Yen. 1938. The demonstration of types of *B. typhosus* by means of preparation of type II Vi phage. Can. J. Publ. Health **29:**448–463, 484–496.

D'Amato, R. F., J. C. McLaughlin, and M. J. Ferraro. 1985. Rapid manual and mechanized/automated methods for the detection and identification of bacteria and yeasts, p. 52–65. *In* E. H. Lennette, A. Balows, W. J. Hausler, and H. F. Shadomy (ed.), Manual of clinical microbiology (4th ed.). American Society for Microbiology, Washington, D.C.

DIFCO. 1984. *Salmonella* serology, p. 797. *In* DIFCO Manual. Difco, Detroit, MI.

Dixon, J. M. S. 1965. Effect of antibiotic treatment on duration of excretion of *Salmonella typhimurium* by children. Br. Med. J. **2:**1343–1345.

Ewing, W. H. 1963. An outline of nomenclature of the family *Enterobacteriaceae*. Int. Bull. Bacteriol. Nomencl. Taxon. **13:**95–110.

Ewing, W. H., and W. J. Martin. 1974. *Enterobacteriaceae*, p. 189–221. *In* Lennette, Spaulding and Truant, (ed.), Manual of clinical microbiology (2nd ed). American Society for Microbiology, Washington, D.C.

Fitts, R. 1985. Development of DNA-DNA hybridization test for the presence of *Salmonella* in foods. Food Technol. **39:**95–102.

Fitts, R., M. Diamond, C. Hamilton, and M. Neri. 1983. DNA-DNA hybridization assay for detection of *Salmonella* spp. in foods. Appl. Environ. Microbiol. **46:**1146–1151.

Flowers, R. S., K. Eckner, D. A. Gabis, B. J. Robison, J. A. Mattingly, and J. H. Silliker. 1986a. Enzyme immunoassay for detection of *Salmonella* in foods: collaborative study. J. Assoc. Off. Anal. Chem. **69:**786–798.

Flowers, R. S., M. J. Klatt, M. A. Mozola, M. S. Curiale, D. A. Gabis, and J. H. Silliker. 1986b. A DNA hybridization assay for the detection of *Salmonella* in foods: collaborative study. J. Assoc. Off. Anal. Chem. **70(3):**521–529.

Gangarosa, E. J. 1978. History and surveillance programs for salmonellosis. Proceedings. National Salmonellosis Seminar. Washington, D.C.

Gilbert, R. J., D. Roberts, and G. Smith. 1984. Foodborne diseases and botulism, p. 477–514. *In* G. R. Smith, (ed.), Topley and Wilson's principles of bacteriology, virology and immunity. Williams & Wilkins, Baltimore.

Hajna, A. A. 1955. A new enrichment broth medium for gram-negative organisms of the intestinal group. Pub. Health Lab. 13:83–89.

Hauschild, A. H. W., and F. L. Bryan. 1980. Estimate of cases of food- and waterborne illness in Canada and the United States. J. Food Prot. **43:**435–440.

Kelley, M. T., D. J. Brenner, and J. J. Farmer III. 1985. *Enterobacteriaceae*, p. 263–277. *In* E. H. Lennette, A. Balows, W. J. Hausler, and H. J. Shadomy, (ed.), Manual of clinical microbiology, (4th ed.). American Society of Microbiology, Washington, D.C.

Kelterborn, E. 1967. *Salmonella* species. First isolations, names and occurrence. Dr. W. Junk, publ. The Hague; 1979 Zbl. Bakt. I Abt. Orig. A 243, 289.

Klausner, A., and T. Wilson. 1983. Gene detection technology opens doors for many industries. Biotechnol. **1:**471–478.

Leifson, E. 1936. New selenite enrichment media for the isolation of typhoid and paratyphoid (*Salmonella*) bacilli. Am. J. Hyg. **24:**423–432.

Müller, L. 1923. Un nouveau milieu d'enrichissement pour la recherche du bacille typhique et des paratyphiques. C.R. Soc. Biol. **89:**434–437.

National Research Council. 1969. An evaluation of the *Salmonella* problem. Committee on *Salmonella*. National Academy of Sciences, Washington, D.C.

Parker, M. T. 1984a. *Salmonella*, p. 332–355. *In* G. R. Smith (ed.), Topley and Wilson's principles of bacteriology, virology and immunity. Williams & Wilkins, Baltimore.

Parker, M. T. 1984b. Enteric infection: typhoid and paratyphoid fever, p. 407–433. *In* G. R. Smith, (ed.), Principles of bacteriology, virology and immunity. Williamd & Wilkins, Baltimore.

Report. 1934. *Salmonella* Sub-Committee of Nomenclature Comm. Int. Soc. Microbiol., J. Hyg. Camb. **34:**333, 1978 W. H. O. Techn. Ser. No. 624.

Report. 1971. A waterborne epidemic of salmonellosis in Riverside, California, 1965: epidemiologic aspects. A collaborative report. Am. J. Epidemiol. **93:**33–48.

Silliker, J. H., P. T. Fagan, J. Y. Chiu, and A. Williams. 1965. Polyvalent H agglutination as a rapid means of screening non-lactose-fermenting colonies for *Salmonella* organisms. Am. J. Clin. Pathol. **43:**548–554.

Spicer, C. C. 1956. A quick method for identifying *Salmonella* "H" antigens. J. Clin. Pathol. **9:**378–379.

U. S. Department of Health, Education and Welfare. 1985. National Shellfish Safety Program. Fed. Reg. **40:**25916–25935.

U. S. Public Health Service. 1983. Typhoid fever, p. 67. *In* Morbid. Mortal. Weekly Rep. Annual Report, 1983.

CHAPTER 48

Shigellosis

HARUO WATANABE

Disease: Shigellosis, bacillary dysentery.
Etiologic Agents: *Shigella* species, subgroups A, B, C, and D.
Source: Eating and drinking contaminated food or water.
Clinical Manifestations: Acute onset of abdominal pain; cramps; fever; diarrhea with blood, pus, and mucus; and tenesmus.
Pathology: Bacterial invasion, cellular filtration, and necrosis confined to the gut wall of the lower part of the colon.
Laboratory Diagnosis: Bacteriologic isolation of causative agent from the feces.
Epidemiology: Epidemic as well as endemic in most parts of the world. A major public health problem in Central and South America, Africa, Asia, and the Western Pacific in particular.
Treatment: Supportive care for relieving dehydration and improving malnourishment. Antibiotics: ampicillin, trimethoprim–sulfamethoxazole, nalidixic acid, and tetracycline. However, drug resistance is now prevalent worldwide.
Prevention and Control: Supply of safe water, effective sewage systems, and stressing good personal hygiene, including handwashing.

Description of Disease

Shigellosis (bacillary dysentery) is an acute infectious disease mainly of the large bowel of humans and higher apes. Its characteristic symptoms are abdominal pain and diarrhea, which often contains blood, mucus, and pus. The causative agents of this disease are bacteria of the genus *Shigella,* which is divided into four subgroups or species: subgroup A (*S. dysenteriae*), subgroup B (*S. flexneri*), subgroup C (*S. boydii*), and subgroup D (*S. sonnei*). All of these organisms cause bacillary dysentery. It is known that certain strains of *Escherichia coli* (enteroinvasive *E. coli*) cause a disease clinically indistinguishable from bacillary dysentery.

Communicability and Transmissibility

The patient excretes the organisms in the feces throughout the acute phase of the disease. Healthy convalescents, symptomless carriers (usually temporary), and chronic carriers (probably for less often) are a potential source of infections to others.

The eating of food or drinking of water contaminated by infected intestinal discharges of the patient or carrier is the principal cause of *Shigella* infection. The organisms may be transferred to food through the imperfectly cleaned fingers of the patient or carrier, or to water through the cross-connection of sewage with a water supply. The role of insects, especially flies, is probably an important factor in the spread of bacillary dysentery when sanitary facilities are primitive or under military field conditions (Kuhns and Anderson, 1944). Transmission of *Shigella* infection from person to person has also been demonstrated (Mosley et al., 1962).

Clinical Manifestations

The incubation period is from 1 to 7 days, usually 1 to 3 days. In the typical case, there is a sudden onset of abdominal pain, cramps, fever, and watery diarrhea due to jejunal involvement. This phase may last 1 to 3 days and be followed by a second diarrhea characterized by frequent, scant stools containing blood, mucus, and pus. Tenesmus (rectal spasms) is quite characteristic of the illness at this point. The severity of the disease varies widely, from simple watery diarrhea to the dysenteric stools clinically described as containing blood, mucus, and pus, and depends upon the condition of the host (age and state of nutrition), the size of the infecting dose, and the serotype of the organism. Infection with *S. dysenteriae* serotype 1 often causes the most severe symptoms and rapid appearance of dysentery after onset of diarrhea. The other three *Shigella* species and other serotypes of *S. dysenteriae* cause a wider spectrum of disease. *Shigella* bacteremia is rare in the immunocompetent host.

Shigellosis is generally self-limited to 7 to 10 days. In the malnourished, it may cause a chronic relapsing disease. In children, central nervous system manifestations such as convulsions, headache, and delirium are often seen. In severe infection a complication of shigellosis, hemolytic–uremic syndrome, can develop (Koster et al., 1978) that is manifested as acute hemolytic anemia, thrombocytopenia, and oliguria. Death from bacillary dysentery may be observed in malnourished children, especially if untreated or treated inappropriately.

Pathology

The lesion attacked by the causative organism is usually confined to the gut wall of the lower part of the colon. Virulent strains of *Shigella* penetrate the mucosa, multiply within, spread to adjacent cells, and further invade lamina propria. The invasive phenotype of *Shigella* is associated with the presence of a large, 120- to 140-megadalton plasmid (Sansonetti et al., 1981, 1982; Watanabe and Nakamura, 1985). There are cellular filtration in the area of lamina propria and submucosa and exudation. Epithelial cells die from the invasion of bacteria and probably from the action of the Shiga toxin, which is cytotoxic in tissue culture (Hale and Formal, 1981) and inhibits intracellular protein synthesis (Olsnes and Eiklid, 1980). The resulting necrosis creates a false membrane; separation of this necrotic membrane causes an ulcer. Necrosis does not usually extend below the muscularis mucosae. Epithelial cells regenerate from remnants of Lieberkühn follicles, and healing generally proceeds from the periphery to the center.

Diagnostic Features

Shigellosis should be suspected when bloody, mucoid diarrhea or the dysentery syndrome occurs. A history of travel to developing countries where the disease is endemic may be helpful information in diagnosing suspected shigellosis. Young children may manifest high fever and/or convulsions before the intestinal disease develops. This presentation can be misleading, particularly because an accompanying meningismus is often noted.

When intestinal symptoms are present, bacteriologic diagnosis should be done by the isolation of the organism from the feces and identification. It is not difficult to isolate the organism during the acute phase of the disease, because numerous organisms are excreted in the mucus or feces. The rectal swab is often valuable at the bedside in acute cases. As *Shigella* bacilli tend to die out rapidly, the stool should be cultured as soon as possible. If delay is unavoidable, it is of some use to suspend the fecal material into a holding medium such as Cary–Blair transport medium. Detection of agglutinins in patient serum is now used infrequently in the diagnosis of dysentery.

The differential diagnosis includes infectious illness with watery and bloody diarrhea, such as salmonellosis; enteritis associated with enteroinvasive *E. coli*, *Vibrio parahaemolyticus*, or *Campylobacter fetus* subspecies *jejuni;* or dysentery such as amebiasis. Microscopic examination of the mucus in the early stage of acute bacillary dysentery may be one way to differentiate it from amebic dysentery. A predominance of polymorphonuclear cells suggests bacillary dysentery, whereas a predominance of mononuclear cells favors a diagnosis of *Entamoeba histolytica* colitis.

Prognosis

Clinical manifestations vary from asymptomatic to mild and to typical, according to the infecting dose of organisms and to various host factors. In well-nourished subjects, shigellosis is generally self-limited to 7 to 10 days. In malnourished children or compromised hosts, death may occur when better therapy is not employed. Mortality is negligible in most Western and industrialized nations, but more common in developing countries.

Chronic Carriage

Shigella bacilli are generally excreted in the stool of recovering patients for only a few weeks. Prolonged carriage of *Shigella* is not the usual pattern. In a study in England, for example, only 2 to 4% of patients excreted the organism as long as 4 months and none beyond 7 months (Thomas and Tillett, 1973).

Epidemiologic Aspects

Occurrence

Bacillary dysentery occurs epidemically as well as endemically in most parts of the world (Mata et al., 1969; Rahaman et al., 1975), but with greatest frequency in Asia, Africa, and Central America. In most developed countries, the incidence of shigellosis decreased dramatically as a result of improved sanitation. Shigellosis is usually found among travelers to areas such as Asia, Africa, and Central America (Gross et al., 1979), where it is endemic or occasionally epidemic. Outbreaks commonly occur under conditions of crowding and poor sanitation, such as in jails, day-care centers, institutions for children, mental institutions, and camps and aboard ships.

Usually, more than one serotype is present in a community. In industrialized countries, *S. sonnei* is now the predominant organism isolated and *S. dysenteriae* is uncommon (Black et al., 1978), whereas in developing countries outbreaks of *S. dysenteriae* serotype 1 are observed (Macaden et al., 1980; Rahaman et al., 1975). Multi-antibiotic-resistant *Shigella* (including *S. dysenteriae* serotype 1) have appeared in many areas of the world, possibly owing to widespread use of antimicrobial agents (Murray, 1986).

Transmission

The natural hosts of *Shigella* are humans and a few nonhuman primates. The route of transmission is inevitably traced to an infected human or, on occasion, to a simian contact. Large numbers of *Shigella* (10^5 to 10^8/g) are present in the stools of clinical cases (Dale and Mata, 1968). Infection may occur after the ingestion of very few (10 to 100) organisms.

The disease is spread mainly by the transfer of the organisms from the infected stools of the patient to the alimentary tract of a noninfected person by person-to-person contact or by contaminated food and water. Dairy foods, such as milk and ice cream, and raw or ready-cooked foods may be contaminated with the feces of patients or carriers. Flies carrying the bacilli on their legs may also contaminate foods. Direct fecal–oral transmission may occur, especially in children and in patients in mental hospitals. Polluted water also plays a part in some outbreaks (Hollister et al., 1955).

Susceptibility and Resistance

There is no evidence that race affects susceptibility to the disease. The disease is more severe in children than in adults. Malnourished persons are particularly susceptible to severe disease and death.

Epidemiologic data suggest acquired immunity to *Shigella* infection. Studies with experimental live oral vaccines have shown short-term serotype-specific protection against infection (DuPont et al., 1972).

Etiologic Agents

Characterization

Shigellae are nonmotile, nonsporulating, gram-negative rods, but are frequently so short on primary isolation that they resemble coccobacilli.

Shigella bacilli, like other enteric bacilli, are aerobes and facultative anaerobes. Optimal temperature for growth is approximately 37°C. Their nutritional requirements are relatively simple, as they will grow on ordinary nutrient media. Most strains grow when glucose and nicotinic acid are added to a medium containing only salts, but some require one or more amino acids. They are relatively resistant to the bacteriostatic action of dyes and various bile salts. Characteristic of the Shiga bacillus is its inability to form catalase. All members of the genus *Shigella* are oxidase negative.

Shigellae ferment glucose, producing acid but no gas, with the exception of a few isolates of *S. flexneri* serotype 6 (Table 1). Lactose is usually not fermented, except by *S. sonnei*. Fermentation of lactose by *S. sonnei* is nearly always delayed. Shigellae do not fermentate mucate, liquefy gelatin, produce H_2S, hydrolyze urea, grow in KCN medium, utilize sodium acetate, or form lysine decarboxylase. They give a positive methyl red and a negative Voges-Proskauer reaction, do not grow on Simmon citrate medium, and do not change the pH of Christensen citrate medium. Their cultural and biochemical char-

TABLE 1. Biochemical reactions of *Shigella*

Test or substrate	Reactions in[a] TSIA	LIM	Christensen urea
Glucose, acid	+		
Glucose, gas	–[b]	–	
Lactose, acid	–[c]		
Sucrose, acid	–[c]		
Hydrogen sulfide	–		
Motility		–	
Indole		–/+	
Lysine decarboxylase		–	
Urease			–

[a] TSIA = Triple sugar-iron agar; LIM = lysine–indole motility medium.
[b] Manchester and Newcastle varieties of *S. flexneri* serotype 6 produce gas from glucose.
[c] *S. sonnei* usually ferments lactose and sucrose slowly.

acteristics indicate that they are most closely related to *E. coli*. The guanine-plus-cytosine content of their DNA is 50 to 52 mol%.

The genus *Shigella* is divided into four main subgroups or species by antigenic and biochemical characteristics: subgroup A (*S. dysenteriae*) typically does not ferment manitol; subgroup B (*S. flexneri*), a group of antigenically interrelated organisms, usually ferments mannitol; subgroup C (*S. boydii*) ferments mannitol but differs from *S. flexneri* antigenically; and subgroup D (*S. sonnei*) ferments mannitol and lactose.

S. dysenteriae serotype 1 forms a toxic product called *Shiga toxin,* which has enterotoxic, cytotoxic, and neurotoxic abilities (Olsnes and Eiklid, 1980). This agent is also produced by some strains of *S. flexneri* and *S. sonnei* (Keusch and Jacewicz, 1977). Certain strains of *E. coli* produce a toxin, "Shigalike toxin," which is antigenically related to Shiga toxin (O'Brien et al., 1982).

S. sonnei may undergo an antigenic variation from S (form I) to R (form II) that affects the somatic antigens. This change is associated with the presence of a 120-megadalton plasmid (Kopecko et al., 1980). *S. dysenteriae* serotype 1 also has a 6-megadalton plasmid that is involved in the formation of the O antigen (Watanabe and Timmis, 1984). In *S. flexneri*, antigenic variations are related to the presence of a specific-temperature bacteriophage (Iseki and Hamano, 1959).

Specimen Collection and Transport

When shigellosis is suspected, the *Shigella* bacilli should be isolated from the feces before antibiotic treatment is initiated. Stools should be hand-delivered to the laboratory and plated without delay. If this approach is impractical, the specimens may be collected into transport medium such as Cary–Blair transport medium, Stewart transport medium, or buffered glycerol solution containing 1% NaCl and be sent to the laboratory. Blood and urine cultures are usually sterile.

Identification

ISOLATION

At least two kinds of solid media with different degrees of selectivity should be used. The most successful in current use are MacConkey agar, xylose–lysine–deoxycholate agar, and tergitol-7-triphenyl tetrazolium chloride agar, which are less selective media. These media permit reasonably good growth of all *Shigella* species. *Salmonella–Shigella* agar, Hektoen enteric agar, and deoxycholate citrate agar are more selective media. *S. sonnei* is suppressed more than other species by *Salmonella–Shigella* agar and deoxycholate citrate agar. *S. dysenteriae* serotype 1 is inhibited on *Salmonella–Shigella* agar.

When few organisms are in the sample, Hajna gram-negative broth can be used for enrichment of fecal specimens. After the sample is incubated for 6 to 8 h in this broth, the resulting culture is streaked onto the differential media as described previously.

Shigellae form characteristic circular, low, convex, translucent colonies on differential media. On primary isolation, *S. sonnei* usually forms small, circular, convex colonies with a smooth surface and an entire edge; some colonies, however, may show a small, tangled-hair-like projection at one or more points on the periphery.

BIOCHEMICAL EXAMINATION

Colonies suspected to be *Shigella* should be examined for biochemical reactions by conventional tube tests or commercially prepared kits. For conventional tube tests, the use of triple sugar iron agar or Kligler iron agar, lysine–indole motility medium or a combination of semisolid indole motility medium and lysine decarboxylase broth, and Christensen urea medium is desirable to identify shigellae. As shown in Table 1, shigellae ferment glucose with the production of acid and no gas (except for a few isolates of *S. flexneri* serotype 6), do not ferment lactose and sucrose (except for *S. sonnei*), and do not produce hydrogen sulfide in triple sugar iron agar. They are not motile, do not form lysine decarboxylase in lysine indole motility medium or a combination of semisolid indole motility medium and lysine decarboxylase broth, and do not produce urease in Christensen urea medium.

Serologic Examination

Isolates suspected to be shigellae should be examined for agglutination in polyvalent antisera for each of the four species. Dense suspensions of bacteria, which are prepared from triple sugar iron agar slants, should be used for the slide agglutination tests. Distinct agglutination within 30 s should be considered a positive reaction. If a suspension fails to agglutinate despite a correlation between the biologic reactions and the characteristics of *Shigella,* it should be boiled for 30 min, cooled, and retested in the same antisera. Many shigellae, especially among subgroups of A and C, possess heat-labile envelope or capsular antigens that inhibit agglutination of unheated bacteria in O antisera. Suspensions that agglutinated in one of the polyvalent antisera should be subjected to the agglutination tests in type-specific and group factor antisera. Classification of *Shigella* serotypes is shown in Table 2.

It is known that the O antigen of most serotypes of *Shigella* is either identical or related to those of the *E. coli* O-antigen groups (Ewing, 1986). In Table 2, *E. coli* O-antigen groups that are identical to or reciprocally cross-react serologically with O antigens of *Shigella* are described. When identification is based solely on agglutination in one of the *Shigella* antisera, misidentification may occur. *E. coli* strains of the Alkalescens–Dispar bioserogroup or enteroinvasive *E. coli* may be identified as *Shigella* because most of these *E. coli* strains are cross-related serologically to *Shigella.* In such cases, biochemical reactions and serologic examination should be considered as a whole to classify the strain as a member of the genus *Shigella.* Table 3 lists the distinguishing characteristics of *Shigella* and *E. coli.* In practice, differentiation of *Shigella* from *E. coli* usually relies on the decarboxylation of lysine, the fermentation of citrate in Christensen medium, and the fermentation of acetate and mucate. Strains that do not have lysine decarboxylase, ferment mucate, or display an alkaline reaction on acetate and Christensen citrate are much more likely to be *Shigella* than *E. coli.*

Table 2. Antigenic formulas of *Shigella* and relationships between O antigens of *Shigella* and *Escherichia coli*

Shigella subgroup and serotype		Relationship with E. coli O-antigen groups: Identical	Reciprocal cross-reactivity
A (*S. dysenteriae*)	1		1
	2	112ac	
	3	124	
	4	*E. coli* 3588-51	159
	5	58	
	6		130
	7		121
	8		38, 23
	9		
	10		144
	11		29
	12	152	3
B (*S. flexneri*)	1a		
	1b		
	2a		13
	2b	147ab	13
	3		
	3b		
	4a		
	4b	135	
	5	129	
	6		
Variant X			
Variant Y			
C (*S. boydii*)	1	149	2
	2	87ab	96
	3		85
	4	53	
	5	79	
	6		76
	7	Coliform 4838-69	
	8	143	
	9		102
	10		105ac
	11	105ab	
	12		7
	13		28ac, 98
	14	32	83
	15	112ab	
	16		41
	17		124
	18		3, 25, 115, 152
D (*S. sonnei*)			

Detection of Virulence Factors

Virulent strains are usually positive in the Serény test (Serény, 1957). Multiple gene products are required for shigellae to be fully virulent (Sansonetti et al., 1983). Large plasmids of *Shigella* species and enteroinvasive *E. coli* are necessary for epithelial cell penetration (Sansonetti et al., 1981, 1982). A DNA fragment of the plasmid can be used as a DNA probe to detect *Shigella* species and enteroinvasive *E. coli* in isolated colonies (Boileau et al., 1984). Detection of a possibly unique virulence antigen by an enzyme-linked immunosorbent assay procedure may

TABLE 3. Differentiation of *Shigella* from *Escherichia coli*

Test	*Reaction in*[a]	
	Shigella	E. coli
Motility	−	v
Gas from D-glucose	−[b]	+[b]
Acid from lactose	−[c]	+[b]
Indole	v	+
Lysine decarboxylase	−	v
Christensen citrate	−	v
Sodium acetate	−[d]	+[b]
Mucate	−[e]	+[b]

[a] + = positive reaction; − = negative reaction; v = some strains positive, others negative.
[b] Some exceptions.
[c] *S. sonnei* usually ferment lactose slowly.
[d] Most cultures of the mannitol-negative bioserotype of *S. flexneri* 4a slowly utilize sodium acetate.
[e] Some strains of *S. sonnei* utilize mucate slowly.

be also useful to identify virulent *Shigella* and enteroinvasive *E. coli* (Pál et al., 1985).

Treatment, Prevention, and Control

Treatment

Patients, especially small children, with shigellosis should receive supportive care for relieving dehydration, improving malnourishment, and limiting the effects of bacterial toxemia. Antibiotics such as ampicillin, trimethoprim–sulfamethoxazole, nalidixic acid, and tetracycline are also of value. However, resistance to these drugs is now prevalent worldwide (Frost et al., 1981; Watanabe, 1963). Treatment with antibiotics, when required, should ideally be based on the results of sensitivity tests on the causative shigella. When these tests are not readily available, it is desirable in severe infections to begin treatment immediately based on the known frequency of resistance to these drugs in the area where the infection was contracted.

Prevention and Control

As shigella infection occurs by eating food contaminated by the uncleaned fingers of a patient or carrier, persons who have had dysentery should not cook or serve food until fecal samples have proved negative for dysentery bacilli. In hospitals, schools, and nurseries, shigella infection is sometimes transmitted by indirect contact via hands and other contaminated objects (lavatory seats, door handles, and water-flushing devices). The best way to prevent such infection is to wash hands after visiting the lavatory and before eating. To prevent an infection from polluted water, ensuring safe water supplies and effective sewage systems is also important.

Oral vaccination has been investigated. A streptomycin-dependent mutant, a spontaneous attenuated mutant, and a spontaneous nonpathogenic mutant have been used as oral vaccines. However, the value of vaccination in the prophylaxis of dysentery remains undecided, and no commercial vaccines are yet available.

Literature Cited

Black, R. E., G. F. Graun, and P. A. Blake. 1978. Epidemiology of common source outbreaks of shigellosis in the United States, 1961–1975. Am. J. Epidemiol. **108**:47–52.

Boileau, C. R., H. M. d'Haureville, and P. J. Sansonnetti. 1984. DNA hybridization technique to detect *Shigella* species and enteroinvasive *Escherichia coli*. J. Clin. Microbiol. **20**:959–961.

Dale, D. C., and L. T. Mata. 1968. Studies of diarrhea disease in Central America. XI. Intestinal bacterial flora in malnourished children with shigellosis. Am. J. Trop. Med. **17**:397–403.

DuPont, H. L., R. B. Hornick, M. J. Snyder, J. P. Libonati, S. B. Formal, and E. J. Gangarosa. 1972. Immunity in shigellosis. II. Protection induced by oral live vaccine or primary injection. J. Infect. Dis. **125**:12–16.

Ewing, W. H. 1986. Identification of Enterobacteriaceae, 4th ed. Elsevier Science Publishing, Inc., New York.

Frost, J. A., B. Rowe, J. Vandepitte, and E. J. Threlfall. 1981. Plasmid characterization in investigation of an epidemic caused by multiple resistant *Shigella dysenteriae* type 1 in central Africa. Lancet **2**:1074–1075.

Gross, R. J., L. V. Thomas, and B. Rowe, 1979. *Shigella dysenteriae, S. flexneri* and *S. boydii* infections in England and Wales: the importance of foreign travel. Br. Med. J. **280**:744–750.

Hale, T. L., and S. B. Formal. 1981. Protein synthesis in HeLa or Henle 407 cells infected with *Shigella dysenteriae, Shigella flexneri* 2a or *Salmonella typhimurium* W118. Infect. Immun. **32**:137–144.

Hollister, A. C., Jr., M. D. Beck, A. M. Gittelsohn, and E. C. Hemphill. 1955. Influence of water availability on *Shigella* prevalence in children of farm labor families. Am. J. Public Health **45**:354–362.

Iseki, S., and Hamano, S. 1959. Conversion of type antigen IV in *Shigella flexneri* by bacteriophage. Proc. Jpn. Acad. Sci. **35**:407–408.

Keusch, G. T., and M. Jacewicz. 1977. The pathogenesis of *Shigella* diarrhea. VI. Toxin and antitoxin in *Shigella flexneri* and *Shigella sonnei* infections in humans. J. Infect. Dis. **135**:552–556.

Kopecko, D. J., O. Washington, and S. B. Formal. 1980. Genetic and physical evidence for plasmid control of *Shigella sonnei* form I surface antigen. Infect. Immun. **29**:207–214.

Koster, F., J. Levin, L. Walker, K. S. K. Tung, R. H. Gilman, M. M. Rahaman, M. A. Majid, S. Islam, and R. C. Williams. 1978. Haemolytic–uremic syndrome after shigellosis. N. Engl. J. Med. **298:**927–933.

Kuhns, D. M., and T. G. Anderson. 1944. A fly-borne bacillary dysentery epidemic in a large military organization. Am. J. Public Health **34:**750–754.

Macaden, R., B. N. Gokul, P. Pereira, and P. Bhat. 1980. Bacillary dysentery due to multidrug resistant *Shigella dysenteriae* 1. Indian J. Med. Res. **71:**178–185.

Mata, L. J., E. J. Gangarosa, A. Cáceres, D. R. Perera, and M. L. Mejicanos. 1969. Epidemic Shiga bacillus in Central America. 1. Etiologic investigations in Guatemala. J. Infect. Dis. **122:**170–180.

Mosley, W. H., B. Adams, and E. D. Lyman. 1962. Epidemiologic and sociologic features of a large urban outbreak of shigellosis. J. Am. Med. Assoc. **182:**1307–1312.

Murray, B. E. 1986. Resistance of *Shigella, Salmonella* and other selected enteric pathogens to antimicrobial agents. Rev. Infect. Dis. **8:**S172–S181.

O'Brien, A. D., G. D. LaVeck, M. R. Thompson, and S. B. Formal. 1982. Production of *Shigella dysenteriae* type 1-like cytotoxin by *Escherichia coli*. J. Infect. Dis. **146:**763–769.

Olsnes, S., and K. Eiklid. 1980. Isolation and characterization of *Shigella* Shiga cytotoxin. J. Biol. Chem. **255:**284–289.

Pál, T., A. S. Pácsa, L. Emödy, S. Vörös, and E. Sélley. 1985. Modified enzyme-linked immunosorbent assay for detecting enteroinvasive *Escherichia coli* and virulent *Shigella* strains. J. Clin. Microbiol. **21:**415–418.

Rahaman, M. M., M. M. Khan, K. M. S. Aziz, M. S. Islam, and A. K. M. G. Kibriya. 1975. An outbreak of dysentery caused by *Shigella dysenteriae* type 1 on a coral island in the Bay of Bengal. J. Infect. Dis. **132:**15–19.

Sansonetti, P. J., T. L. Hale, G. J. Dammin, C. Kapfer, H. H. Collins, Jr., and S. B. Formal. 1983. Alterations in the pathogenicity of *Escherichia coli* K-12 after transfer of plasmid and chromosomal genes from *Shigella flexneri*. Infect. Immun. **39:**1392–1402.

Sansonetti, P. J., D. J. Kopecko, and S. B. Formal. 1981. *Shigella sonnei* plasmids: evidence that a large plasmid is necessary for virulence. Infect. Immun. **34:**75–83.

Sansonetti, P. J., D. J. Kopecko, and S. B. Formal. 1982. Involvement of a plasmid in the invasive ability of *Shigella flexneri*. Infect. Immun. **35:**852–860.

Serény, B. 1957. Experimental keratoconjunctivitis shigellosa. Acta Microbiol. Acad. Sci. Hung. **4:**367–376.

Thomas, M. E. M., and H. E. Tillett. 1973. Dysentery in general practice: a study of cases and their contacts in Enfield and an epidemiological comparison with salmonellosis. J. Hyg. **71:**373–389.

Watanabe, H., and A. Nakamura. 1985. Large plasmids associated with virulence in *Shigella* species have a common function necessary for epithelial cell penetration. Infect. Immun. **48:**391–396.

Watanabe, H., and K. N. Timmis. 1984. A small plasmid in *Shigella dysenteriae* 1 specifies one or more functions essential for O-antigen production and bacterial virulence. Infect. Immun. **43:**391–396.

Watanabe, T. 1963. Infective heredity of multiple drug resistance in bacteria. Bacteriol. Rev. **27:**87–115.

CHAPTER 49

Staphylococcal Diseases

P. E. VARALDO and G. SATTA

Diseases: Staphylococcal infections, staphylococcal food poisoning, scalded-skin syndrome, toxic shock syndrome.

Etiologic Agents: *Staphylococcus aureus, Staphylococcus epidermidis,* and *Staphylococcus saprophyticus;* other human species less documented.

Source: Infected patients, healthy carriers, normal flora of the skin, contaminated foods in food poisoning.

Clinical Manifestations: Symptoms of particular infections involving skin, soft tissues, bloodstream, urinary tract, internal organs and artificial devices; gastroenteritis in food poisoning; high fever, hypotension, diarrhea, erythematous rash, and other symptoms in toxic shock syndrome.

Pathogenesis: Pathogenicity determinants represented by exocellular toxins and enyzmes and structural cell components. Importance of underlying diseases and impaired host defenses.

Laboratory Diagnosis: Isolation of etiologic agent from clinical material (from foods in case of food poisoning). Identification based on morphologic, biochemical, metabolic, and physiologic characters; also with the aid of identification schemes or commercial kits. Determination of susceptibility to methicillin.

Epidemiology: Major natural reservoir in the skin of mammals. Typing performed using appropriate set of bacteriophages or other methods. Emergence of methicillin-resistant staphylococci as nosocomial pathogens.

Treatment: Strictly dependent on susceptibility data. Major problems with methicillin-resistant organisms, for which vancomycin may be the drug of choice in case of severe infections.

Prevention and Control: Especially aimed at interrupting the mode of transmission (*S. aureus*) or at limiting the reservoir (*S. epidermidis*) to prevent spread within hospital wards. Avoidance of food contamination and control of personnel involved in food preparation and distribution to prevent food poisoning.

Description of Diseases

Staphylococcal diseases include a variety of infectious conditions that may differ greatly in severity, sites involved, and pathogenetic mechanisms, also depending on the *Staphylococcus* species considered.

Twelve of the 24 currently recognized *Staphylococcus* species can be found in humans, that is, *S. aureus, S. epidermidis, S. saprophyticus, S. cohnii, S. haemolyticus, S. xylosus, S. warneri, S. capitis, S. hominis, S. simulans* (Kloos and Schleifer, 1975a; Schleifer and Kloos, 1975), and the more recently described *S. saccharolyticus* (Kilpper-Bälz and Schleifer, 1981) and *S. auricularis* (Kloos and Schleifer, 1983).

Clinical Significance

S. aureus and its more common clinical implications have been known for more than a century. Quite recently, however, other morbid conditions of newer nosographic and pathogenetic definition have been associated with this species.

The other human *Staphylococcus* species are often referred to as a whole as coagulase-negative species owing to their lacking in coagulase production, that is the classical distinguishing feature of *S. aureus*. The clinical significance of the coagulase-negative species has become somewhat clearer in recent years, also thanks to the improved knowledge about their taxonomic differentiation. *S. epidermidis* and *S. saprophyticus* are the two coagulase-negative *Staphylococcus* species whose clinical implications are better documented.

S. aureus INFECTIONS

Infections due to *S. aureus* (both community- and hospital-acquired) are still a major source of morbidity and mortality.

The most common diseases caused by this organism are skin and soft tissue infections, including pustular lesions centered on hair follicles (folliculitis, boils, carbuncles), impetigo, cellulitis, and lymphangitis. Both local factors impairing the integrity of the skin (burns, ulcers, surgical and nonsurgical wounds, and so on) and underlying diseases (diabetes mellitus, disorders of neutrophil function immunosuppressive and corticosteroid therapy, and the like) may play important roles in predisposing to these lesions. Deep seated infections and abscesses caused by *S. aureus* can occur at a variety of organs and sites (respiratory tract, bones and joints, cardiovascular system, and so on), either as a direct extension of a superficial infection or through bacteremic diffusion and seeding. In particular, *S. aureus* remains the commonest cause of primary osteomyelitis.

Another frequent and well-known disease caused by *S. aureus* is the gastroenteritis typical of staphylococcal food poisoning. This situation completely differs from the above conditions because of its tight pathogenetic dependence on a heat-stable enterotoxin released in foods where the responsible strain has grown.

S. aureus is also involved in other newly described toxin-mediated diseases originating from an often asymptomatic focus of infection. This is the case of the scalded-skin syndrome, which is caused by an exfoliative toxin and mostly affects infants, and of the toxic shock syndrome, an acute severe illness mostly affecting women during or just after menstruation and associated with the use of tampons.

S. epidermidis INFECTIONS

S. epidermidis is the most prevalent and persistent *Staphylococcus* species on human skin. Usually disregarded in the past in diagnostic laboratories and dismissed as a harmless contaminant, its importance as a pathogen, especially in a hospital-based practice, has been recognized over the past 2 decades and has been better defined with the more precise understanding of its taxonomic position within coagulase-negative staphylococci. *S. epidermidis* is currently regarded as a major cause of infection of indwelling foreign devices such as prosthetic heart valves, vascular and joint prostheses, cerebrospinal fluid shunts, intravenous and peritoneal catheters, and so on. It also may be implicated in native valve endocarditis and occasionally in a variety of infections at other sites, and is an increasingly common cause of bacteremia in immunosuppressed patients.

S. saprophyticus INFECTIONS

The importance of *S. saprophyticus* as a urinary pathogen has been established since the early 1960s, when its taxonomic position was still uncertain and these organisms were usually reported as micrococci or novobiocin-resistant staphylococci. *S. saprophyticus* is currently considered a major cause of acute, often recurrent urinary tract infections mostly affecting young female outpatients. Infection usually occurs as a cystitis, but kidneys may sometimes be involved. *S. saprophyticus* also has been suspected as a possible agent of sexually transmitted nongonococcal urethritis in man.

INFECTIONS CAUSED BY OTHER HUMAN SPECIES

The other human *Staphylococcus* species are being extensively investigated for their clinical relevance and a number of possible implications have been documented. Further studies seem however to be needed to better understand possible specific differences in such implications.

Pathogenesis

Impairment of specific or nonspecific host defenses may have crucial importance to favor staphylococcal infection. A number of exocellular toxins and enzymes as well as structural components of staphylococcal cells may have pathogenetic relevance.

A clear relationship of a particular pathogenicity determinant to a given morbid condition has been established for some *S. aureus* exotoxins: this is the case of enterotoxins in staphylococcal food poisoning, of the exfoliative toxin in the scalded-skin syndrome, and of the toxin(s) involved in the toxic shock syndrome. The role of pathogenicity determinants in the other and more common *S. aureus* infections is probably more complex and is much less understood.

The importance of *S. epidermidis* as an infecting agent in compromised hosts and its special involvement in bacteremias and infections of indwelling foreign devices also may reflect its prevalence in the commensal flora of human skin, but it seems likely that other less known and more specific mechanisms of pathogenicity are implicated.

The grounds for the involvement of *S. saprophyticus* in urinary tract infections seem to be more defined, given the special affinity for urothelial cells demonstrated by this organism in adherence studies.

Epidemiologic Aspects

Natural Habitats

Mammals (particularly their skin) are the major natural reservoir of staphylococci. Some species also have been isolated from birds, and some from alimentary products of animal origin. *S. epidermidis* and *S. hominis* are the major resident species living on humans. For these two species, as well as for *S. capitis* and *S. saccharolyticus*, humans seem to be the sole natural host. The other human species (or certain subspecies or biotypes of them) also can be found on nonhuman primates, other mammalian hosts, and poultry (Kloos, 1980). Some of the human species demonstrate clear preferences for certain sites and niches of the human skin and appendages.

Hospital- and Community-Acquired Infections

Staphylococci are important causes of morbidity both in the community and in the hospital.

In the antibiotic era, two major epidemics of nosocomial infections due to *S. aureus* have occurred. The former, caused by strains secreting high levels of penicillinase, started in the early 1950s spread rapidly worldwide, and lasted about 1 decade. The latter, still in course, has in the intrinsic resistance to methicillin of the responsible strains its outstanding characteristic. This epidemic spread slowly and irregularly in the 1960s and 1970s, and was recognized as an international medical problem of growing importance in the late 1970s to early 1980s. Both infected patients and healthy carriers in the staff may act as the reservoir or the source for intra- and interhospital dissemination of methicillin-resistant strains. Community-acquired outbreaks of infections due to methicillin-resistant *S. aureus* recently have been described, and may in turn be a source of further spreading in hospitals (Saravolatz et al., 1982).

Among coagulase-negative staphylococci, *S. epidermidis* is emerging as an increasingly important nosocomial pathogen, also as a result of the expanding use of therapeutic practices such as immunosuppression or the application of indwelling artificial devices. On the contrary, *S. saprophyticus* proves a typical community pathogen, second only to *Escherichia coli* as a cause of symptomatic urinary tract infections in young female outpatients. In studies predominantly including inpatients, *S. epidermidis* is the most common staphylococcal species isolated from infected urine (Wallmark et al., 1978).

Typing

As a result of the ubiquity of staphylococci, several typing systems have been applied to these organisms to detect their sources and ways of spreading in epidemic outbreaks (nosocomial infections, food poisonings, and so on). Extensive typing studies have focused especially on *S. aureus*. At present, however, there is a growing interest in developing effective typing methods also for coagulase-negative staphylococci.

Phage Typing

Phage typing is the most established method for epidemiologic typing of *S. aureus*. Both the basic set of bacteriophages used and the typing procedures are standardized and updated under the control of an international subcommittee active since the early 1950s. Not very encouraging results have been yielded so far by attempts at investigating coagulase-negative staphylococci by this same epidemiologic approach.

Serologic Typing

Although extensively investigated (at least with *S. aureus*) and shown to be of value in epidemiologic studies, serotyping is generally much less used than phage typing. This mainly depends on some difficulties dealing with the standardization of techniques and the preparation of specific antisera.

Biotyping

A certain number of biotyping schemes were used up to the early 1970s with coagulase-negative human staphylococci. The subsequent recognition of a variety of distinct species among these organisms took away most of its original interest from this approach.

Susceptibility Patterns

Antibiograms and antibiotypes may provide important and easy-to-obtain epidemiologic information about staphylococcal isolates. However, susceptibility patterns are usually markedly influenced by local factors, so that they may be of value within a confined area or community but are not fit for epidemiologic studies of wider proportions.

Plasmid Profiles

Determination of plasmid profiles recently has been attempted for typing purposes particularly with *S. epidermidis* (Parisi and Hecht, 1980). Despite its still limited use, it seems to be a promising approach, even though more suitable for research laboratories than for clinical laboratories.

Other Approaches

Bacteriocin production in staphylococci has long been considered for typing purposes, but with substantially limited results. Very recently, preliminary accounts have been published about new approaches to the problem of staphylococcal typing. This is the case of molecular analysis of plasmids by restriction enzyme digestion and DNA-DNA hybridization (Parisi, 1985), or of exocellular protein analysis by Western Blot using human plasma as a source of antibodies (Krikler et al., 1986).

Etiologic Agents

Characterization

Staphylococci are nonmotile, gram-positive cocci, 0.5 to 1.5 μm in diameter, characteristically dividing in more than one plane to form irregular clusters. The guanine plus cytosine content of their DNA ranges from 30 to 39 mol %.

The staphylococcal cell wall contains peptidoglycan and teichoic acid (Schleifer, 1986). A special protein (protein A), known for its reaction with normal sera through the FC fragment of IgG, is a wall constituent of most *S. aureus* strains. Cell wall teichoic acids contain either glycerol or ribitol residues or both (depending on the species) linked together through phosphodiester bridges, and sugar or *N*-acetylamino sugars also may be present in the polymers usually as glycosyl substituents. Peptidoglycan contains L-lysine as the diamino acid in the third position of the peptide subunit. Depending on the species, interpeptide bridges may be formed by glycine residues only or contain L-serine or L-alanine instead of a glycine residue. Staphylococcal cells are lysed by lysostaphin, more promptly if their interpeptide bridges contain glycine only, and are resistant to lysozyme. The latter may however potentiate the lytic activity of lysostaphin (Cisani et al., 1982).

Different *Staphylococcus* species contain various numbers of penicillin-binding proteins in their membranes (Canepari et al., 1985) and produce specific bacteriolytic enzymes (Valisena et al., 1983).

Metabolism is mainly respiratory or mainly fermentative depending on the species. Most staphylococci are facultative anaerobes, but usually grow more rapidly and abundantly under aerobic conditions. Characteristically anaerobic are the strains of *S. saccharolyticus* (Kilpper-Bälz and Schleifer, 1981) and of *S. aureus* subsp. *anaerobius* (De la Fuente et al., 1985).

At present, the genus *Staphylococcus* is still included in the family *Micrococcaceae*, together with the genera *Micrococcus*, *Stomatococcus*, and *Planococcus* (Schleifer, 1986). It seems however to be a provisional systematic arrangement, since the four genera are not closely related as documented by DNA base composition and cell wall composition studies, nucleic acid hybridization data, and comparative analysis of 16S ribosomal RNA sequence. Based on the latter approach, it has been suggested that the genus *Staphylococcus* is rather phylogenetically related to the genus *Bacillus* (Stackebrandt and Woese, 1979).

Specimen Collection and Transport

No special precautions are usually necessary for collection and transport of clinical specimens. In fact, staphylococci are easily obtained from most infected sites and resist relatively well to drying and to changes of temperature. Also anaerobic strains usually stand normal transport conditions.

Cultivation and Isolation

Staphylococci have little growth requirements and can be cultivated on virtually any nutrient medium at 37°C. Most strains grow between 18 and 40°C and tolerate the presence of 10% NaCl in the growth medium.

Human staphylococci usually yield abundant growth on blood agar within 18 to 24 h (more prolonged incubation may be required by *S. auricularis*. which has a lower growth rate). By this time colonies are circular, 1 to 3 mm in diameter, raised, smooth, glistening, butyrous, and nontransparent.

The use of selective media is advisable for specimens from heavily contaminated sources. Media selective for the presence of 7 to 8% NaCl such as mannitol salt agar have been the most used for years in clinical laboratories. The same selecting principle has been exploited in more recently developed media for direct identification of *S. aureus*, which incorporate in the plate a lysostaphin-containing paper strip and a phosphatase substrate (Grazi et al., 1981). Other selective media such as the one containing potassium tellurite proposed by Baird-Parker (1962) are more suitable to recover organisms damaged by heating, drying, and irradiation, thus being preferred for isolation of staphylococci in food microbiology laboratories.

Identification

In the past, identification of staphylococci in the clinical laboratory was substantially limited to the recognition of *S. aureus*, the other staphylococcal isolates being disregarded as harmless commensals or contaminants. At present, improved knowledge both about taxonomic differentiation within coagulase-negative *Staphylococcus* species and about specific clinical involvements of some of these species makes it advisable to proceed to a more careful identification at species level of clinical *Staphylococcus* isolates.

The major distinguishing characters relevant to the laboratory identification of *Staphylococcus* species are reported in Table 1.

Preliminary Tests

The colony appearance in nonselective media may provide some information. In the clinical laboratory practice, however, the only colonial character that is object of interest is the occurrence of a yellow-orange pigment.

Gram staining may be useful both to check isolated colonies and to perform direct examination of clinical specimens. The latter testing is mainly important in the case of normally sterile fluids, but it may also be of value with other specimens, for example to reveal inflammatory response.

The demonstration of catalase production is a classical test to rule out other common gram-positive cocci such as streptococci. The reaction can be performed by spotting a drop of 3% hydrogen peroxide onto a colony on an agar plate (blood-containing media should be avoided as they may cause false-positive results) and examining for the formation of gas bubbles. Negative catalase reactions have recently been reported to be normal in certain anaerobic staphylococci that may be encountered in veterinary microbiology laboratories (De la Fuente et al., 1985).

Differentiation between staphylococci and micrococci has long been regarded as an important and difficult problem in diagnostic microbiology. Given the practical impossibility to examine DNA base composition in the clinical laboratory, a number of simple and variably reliable methods have been proposed for routine use. They are based on modifications of the classical test of anaerobic acid production from glucose, lysostaphin susceptibility, growth on selective media, production of lytic activity, phage adsorption tests, serologic tests, or modifications of oxidase and benzidine tests. However, it seems likely that the problem of separating staphylococci from micrococci has been somewhat overconsidered, at least in clinical microbiology. In fact, given their much lower growth rate compared with staphylococci, only in the case of specimens requiring particularly long incubations (e.g., blood cultures) can micrococci occasionally become a real problem in the clinical laboratory.

The opportunity of laboratory differentiation between staphylococci and stomatococci can also be considered in light of reports denoting possible involvements of stomatococci in human infections. Stomatococci are variable as to the catalase reaction and are not lysed by lysostaphin.

Coagulase

The tube test for the production of coagulase (staphylocoagulase or free coagulase) is still the most widely used method for routine identification of *S. aureus*. Commercially available citrate- or EDTA-containing rabbit plasma is the most used substrate for the reaction, although plasmas from other sources are also clotted by staphylocoagulase. The tube test may be performed from either broth or agar cultures, by adding 0.1 ml to or resuspending a loopful of bacterial growth into 0.5 ml of reconstituted plasma, respectively. The tube is then incubated at 37°C and examined for clot formation after 4 and 24 h. Any degree of clotting should be regarded as a positive reaction. The test has more limitations than usually believed. False-positive results may occur either due to proteolytic enzymes (occasionally produced by certain staphylococci) capable of activating prothrombin and initiating the clotting process (pseudocoagulase activity), or in the case of contamination by citrate-using organisms when a citrated plasma has been used. False-negative results, however, are more important in laboratory practice, and may depend upon inoculation procedures and inoculum size; source, and lot of plasma used; insufficient time of incubation; or lysis of the clot by staphylokinase (particularly when incubations up to 24 h are needed). But it is important to realize that frankly coagulase-negative *S. aureus* isolates can be found. Coagulase-positive staphylococci other than *S. aureus* may be isolated from animal specimens in veterinary microbiology laboratories.

The slide coagulase test has a different meaning because it does not detect staphylocoagulase but clumping factor, or bound coagulase. The test is performed by suspending a loopful of bacterial growth in a drop of distilled water, then a drop of plasma is

TABLE 1. Differentiation of human *Staphylococcus* species[a]

Character	S. aureus	S. simulans	S. capitis	S. saprophyticus	S. cohnii	S. xylosus	S. epidermidis	S. hominis	S. haemolyticus	S. warneri	S. saccharolyticus	S. auricularis
Colony pigment	(+)	−	−	v	−	(+)	−	v	(−)	(−)	−	−
Anaerobic growth	+	+	(+)	v	v*	(+)	+	−	v*	+	+	−
Nitrate reduction	+	+	(+)	−	−	(+)	(+)	(+)	(+)	(−)	+	v*
Novobiocin resistance	−	−	−	+	+	+	−	−	−	−	−	−
Lysostaphin susceptibility	+	+	−	−	v	v	−	−	−	−	−	−
Lyogroup	I	II	III	IV	IV	IV	V	VI	VI	VI	nd	nd
Production of												
Staphylocoagulase	+	−	−	−	−	−	−	−	−	−	−	−
Clumping factor	(+)	−	−	−	−	−	−	−	−	−	−	−
Phosphatase	+	−	−	−	−	(+)	(+)	−	−	−	−	−
Urease	(+)*	+	(−)	+	−	+	+	(+)	−	+	nd	−
DNase	+	(+)*	(+)*	−	−	−	−	−	−	−	nd	v*
Thermostable DNase	+	(+)*	−	−	−	−	−	−	−	−	nd	nd
Acetoin	+	−	v	+	v	v*	+	v	(+)	+	nd	v*
Hemolytic activity	+	v*	(−)*	−	v*	−	(−)*	(−)*	+	v*	−	−
Aerobic acid from												
Xylose	−	−	−	−	−	+	−	−	−	−	−	−
Maltose	+	−	−	+	(+)*	+	+	+	+	+	−	(+)*
Lactose	+	+	−	(+)	v	v	(+)	v	v	(−)	−	−
Sucrose	+	+	+	+	−	+	+	+	+	+	−	v
Trehalose	+	(+)	−	+	+	+	−	+	+	+	−	+*
Xylitol	−	−	−	(+)	v	(−)*	−	−	−	−	−	−
Mannitol	+	+	+	+	(+)	+	−	−	v	v	−	−
Anaerobic acid from												
Mannitol	+	+	−	−	−	−	−	−	−	−	−	−

[a] + or − = Positive or negative reaction in more than 90% of isolates; (+) or (−) = positive or negative reaction in 70 to 90% of isolates; v = variable; nd = not determined; the symbol * indicates that the reaction, when positive, is usually weak or delayed.

added, and the mixture is observed for clumping within 10 s. False-positive reactions may occur if colonies for testing have been picked up from media with high salt concentrations. Although easy, rapid, and inexpensive, the slide test cannot be recommended for routine diagnosis as an alternative to the tube test as it correlates only to a certain extent with staphylocoagulase (at least 10 to 20% of *S. aureus* strains may give a negative reaction).

Phosphatase

Phosphatase production is one of the most important characteristics for the differentiation of *Staphylococcus* species. This property is present in all *S. aureus* strains and in about 90% of strains of *S. epidermidis*; among human staphylococci, also the majority of strains of *S. xylosus* show a clearly positive reaction.

A variety of techniques can be applied to detect phosphatase activity. In the most traditional method, phenolphthalein diphosphate is incorporated as a substrate (200 μg/ml) in a solid nutrient medium, and phosphatase-positive cultures are indicated by the deep pink color they assume after exposure to ammonia vapors. This last step may be avoided using a more recent method, according to which methyl green (25 μg/ml) is added to the test medium together with phenolphthalein diphosphate: in this medium phosphatase-positive strains grow deep green-stained colonies, whereas phosphatase-negative strains do not (Satta et al., 1979). The technique of Pennock and Huddy (1967) and its modifications use phenolphthalein monophosphate as a substrate and provide a response within 4 h; however, given the particularly high sensitivity of such methods, weak reactions yielded by certain isolates may be of difficult interpretation. Another phosphatase substrate largely used in commercial test systems is *p*-nitrophenylphosphate, which releases yellow-colored *p*-nitrophenol in consequence of hydrolysis by phosphatase.

Novobiocin Resistance

Within human staphylococci, novobiocin resistance is a conventional distinguishing feature of those organisms currently classified as *S. saprophyticus*, *S. cohnii*, and *S. xylosus*, which are a triplet of particularly closely related *Staphylococcus* species. In this context, an MIC of novobiocin of ≥1.6 μg/ml is generally considered as resistance, which may thus be simply inferred from growth in a medium containing this antibiotic concentration. Alternatively, a common agar diffusion method using disks with a potency of 5 μg may be suitable to evaluate this property (resistance corresponds to a zone diameter of ≤16 mm).

Susceptibility to Lysostaphin

Lysostaphin susceptibility can be cheaply determined by an agar diffusion procedure, using previously prepared paper disks (100 sterile disks, 6 mm in diameter, imbued with 1 ml of a sterile 40-μg/ml solution of lysostaphin). Susceptible strains (i.e., showing a zone of growth inhibition around the disk) are mostly from species containing only glycine in their cell wall interpeptide bridges.

Carbohydrate Reactions

Acid production from carbohydrates has been extensively used in the conventional classification and identification of staphylococci. Molten purple agar base (a commercial medium containing bromocresol purple) is supplemented with a solution of filter-sterilized carbohydrate to give a final concentration of 1%. Four to eight cultures may be streaked on the surface of one plate. Plates are incubated at 37°C and examined daily up to 3 to 5 days. A positive reaction is indicated by the development of yellow indicator color in the streak, which may extend into the surrounding medium in case of strong acid production (partitioned plates are more convenient for use when sugars more often involved in strong reactions are tested).

Other Tests

Other conventional tests that can be performed in the diagnostic laboratory and provide useful information about the identity of a staphylococcal isolate include hemolysis; anaerobic growth; nitrate reduction; and production of acetoin, urease, and DNase.

The hemolytic activity can be checked in common blood agar (bovine blood is particularly suitable) incubated at 37°C up to 72 h.

Anaerobic growth can be shown using a semisolid thioglycolate medium incubated without shaking at 37°C after inoculation, and checked daily (up to 3 to 5 days) for growth in the lower portion of the tube.

Production of DNase is best demonstrated in plates of toluidine blue-DNA agar (Lachica et al., 1971) by the appearance of a pink zone of decolorization around the test material. The latter can be applied to the agar either using a paper disk dipped in an overnight broth culture or by filling with the broth a well cut in the agar. If the broth culture used by either procedure has previously been heated for 15 min at 100°C, thermostable DNase will be detected.

Nitrate reduction and production of acetoin and urease can be tested with staphylococcal isolates using the same procedures commonly used in diagnostic laboratories for the identification of other organisms (e.g., enterobacteria).

Identification Schemes

Soon after the development by Schleifer and Kloos of the taxonomic scheme that is the base for the current classification of human staphylococci, the same authors proposed a simplified identification system for routine use (Kloos an Schleifer, 1975b). This method, through a 13-test dichotomic frame, leads to distinguishing 11 positions, 10 of which virtually correspond to as many species. The 13 key characters include coagulase, hemolysis, nitrate reduction, and acid production from 10 different carbohydrates. Despite its rather large use, this abbreviated scheme is perhaps still too complicated for many clinical laboratories. In addition, its dichotomic structure makes nonidentifications virtually impossible, but also makes the method prone to misidentifications.

A simple but less ambitious scheme is the one recommended by the Subcomittee on the Taxonomy of Staphylococci and Micrococci (1976), which only aims at identifying the three species regarded as particularly important in human pathology (i.e., *S. aureus*, *S. epidermidis*, and *S. saprophyticus*). The method is based on six key characters including coagulase, phosphatase, novobiocin resistance, and acid production from sucrose, trehalose, and mannitol. Some nonidentifications of strains of the three species considered are not exceptional given the essentiality of the method. Its a priori disregarding the other *Staphylococcus* species may further discourage its use in many laboratories.

An original approach to staphylococcal classification and identification consists in the analysis of bacteriolytic activity patterns. A simplified version of this method, designed for routine use, is based on the evaluation of lytic activity in five different test media and the determination of phosphatase activity (Varaldo et al., 1980). By these tests staphylococci can be subdivided into six groups (lyogroups) closely related to species. Lyogroups I, II, III, and V equate with *S. aureus*, *S. simulans*, *S. capitis*, and *S. epidermidis*, respectively. The two remaining lyogroups equate each with a cluster of three closely related species: lyogroup IV with *S. saprophyticus*, *S. cohnii*, and *S. xylosus*, and lyogroup VI with *S. hominis* *S. haemolyticus*, and *S. warneri*.

Commercial Identification Kits

Commercially available biochemical test systems are increasingly used for laboratory identification of *Staphylococcus* isolates. Some commercial kits are aimed at only identifying *S. aureus*, for example, by means of a hemagglutination slide test for clumping factor (BBL) or by latex agglutination detecting both clumping factor and protein A (Scott Laboratories). Other more complex miniaturized biochemical test systems are marketed by several manufacturers for routine identification of a certain number of *Staphylococcus* species. Such kits include APl Staph (Api System S.A.), APl Staph-Ident and API 20 GP (Analytab), the AutoMicrobic System Gram-Positive Card (Vitek), Minitek Gram-Positive System (BBL), DMS Staph-Trac (DMS Laboratories), and the MicroScan positive combination panel (American MicroScan). Additional tests may be necessary for more accurate identification of certain species. However, the actual reliability of these methods is often difficult to evaluate. In fact, their availability varies considerably from country to country (most of the above-mentioned kits are not marketed in most European countries). In addition, these methods are frequently modified, improved, and updated by the manufacturers, and need to be used in strict accordance to their respective instructions.

Susceptibility to Methicillin

By now, susceptibility to methicillin should be considered as an integral part of routine identification of any clinical *Staphylococcus* isolate rather than a mere susceptibility datum. The fact that a staphylococcus responsible for an infectious process is susceptible or resistant to methicillin may actually have very important reflections on prognosis and therapy as well as on the control measures to be adopted, as pointed out in a later section of this chapter.

Laboratory detection of methicillin resistance poses special problems primarily resulting from the frequent heterogeneity of its expression (heteroresistance). In other words, within a given heteroresistant strain only a certain, usually small, proportion of cells are able to express the resistance trait under normal conditions. This is the reason for the need to resort to special test conditions aimed at favoring an increased expression of resistance. Of these conditions, the two most widely used are addition of 5% NaCl to the test medium and plate incubation at 30 rather than 35 or 37°C. Except for such precautions, susceptibility to methicillin can be tested by a common agar diffusion procedure. Oxacillin disks are currently preferred to methicillin disks in such test, mainly because of their better stability to refrigeration. Automated systems for susceptibility testing are often unsuited for an accurate detection of methicillin resistance.

Treatment, Prevention, and Control

Treatment

Although staphylococcal infections may respond to various treatments, it should be considered that staphylococci have demonstrated special ability to develop antibiotic resistances, and progressively increasing resistances of clinical isolates to many drugs have recently been noted in some countries (Richardson and Marples, 1982; Varaldo et al., 1984). Careful laboratory testing of antimicrobial susceptibilities are thus essential for a successful therapy.

Penicillinase-mediated penicillin resistance was developed in the early 1950s and has since become a stable property of the great majority of staphylococcal isolates. Thus, the introduction of methicillin and other penicillinase-resistant penicillins represented a significant step forward in antistaphylococcal therapy. However, staphylococci proved able to become resistant also to these drugs, and the present reemergence of staphylococci as nosocomial pathogens is substantially the expression of the current problem of intrinsic resistance to methicillin. In fact, not only methicillin-resistant staphylococci are often resistant in vitro to multiple antibiotics, but also, according to many authors, they should be considered potentially resistant in vivo to penicillins and cephalosporins as a whole, independently of their in vitro susceptibilities to individual drugs. Another important drug to which clinical *Staphylococcus* isolates (particularly those resistant to methicillin) have developed increasing resistance in recent years is gentamicin.

The management of severe infections due to methicillin-resistant staphylccocci is currently a major therapeutic problem. The treatment of choice for such infections is mostly considered to be vancomycin, to which (as well as to the newer related compound teicoplanin) methicillin-resistant staphylococci prove uniformly susceptible in vitro. Another antibiotic which is active against most methicillin-resistant staphylococci is minocycline. Netilmicin, fusidic acid, and rifampin also may be valuable particularly in combination regimens.

Prevention and Control

As well as for therapy, also for prevention and control of staphylococcal infections a major concern is currently dealing with methicillin-resistant organisms. Application of the general principles of infection control (especially effective handwashing) is of primary importance to control the spread of methicillin-resistant *S. aureus* within a hospital. Recognition of carriers among hospital staff and eradication of carriage are also extremely important. Patient isolations, limitations of admissions, and other stringent control measures are of special value in institutions that have experienced extensive outbreaks of methicilli-resistant *S. aureus* infections. Efforts should rather be aimed at preventing spread within wards particularly prone to such outbreaks (e.g., burn, intensive care, and surgical units) than at eliminating methicillin-resistant *S. aureus* from the whole institution. Unlike *S. aureus*, with methicillin-resistant *S. epidermidis* multiple strains are probably involved in nosocomial infections. Therefore, it is likely that methicillin-resistant *S. epidermidis* will be better controlled by limiting its reservoir than by interrupting its mode of transmission (Wenzel, 1986).

For staphylococcal food poisoning, control measures are mainly directed to avoidance of contamination of foods during preparation and distribution, and to the health, hygiene, and education of personnel involved, with emphasis on the screening and treatment of carriers.

Literature Cited

Baird-Parker, A. C. 1962. An improved diagnostic and selective medium for isolating coagulase-positive staphylococci. J. Appl. Bacteriol. **25**:12–19.

Canepari, P., P. E. Varaldo, R. Fontana, and G. Satta. 1985. Different staphylococcal species contain various numbers of penicillin-binding proteins ranging from four (*Staphylococcus aureus*) to only one (*Staphylococcus hyicus*). J. Bacteriol. **163**:796–798.

Cisani, G., P. E. Varaldo, G. Grazi, and O. Soro. 1982. High-level potentiation of lysostaphin anti-staphylococcal activity by lysozyme. Antimicrob. Agents Chemother. **21**:531–535.

De la Fuente, R., G. Suarez, and K. H. Schleifer. 1985. *Staphylococcus aureus* subsp. *anaerobius* subsp. nov., the causal agent of abscess disease of sheep. Int. J. Syst. Bacteriol. **35**:99–102.

Grazi, G., O. Soro, P. E. Varaldo, G. Cisani, and G. Satta. 1981. Selective media for isolation and rapid identification of *Staphylococcus aureus*, p. 109–112. *In* J. Jeljaszewicz (ed.), Staphylococci and staphylococcal infections, Gustav Fischer Verlag, Stuttgart.

Kilpper-Bälz, R., and K. H. Schleifer. 1981. Transfer of *Peptococcus saccharolyticus* Foubert and Douglas to the genus *Staphylococcus: Staphylococcus saccharolyticus* (Foubert and Douglas) comb. nov. Zentralbl. Bakteriol. Parasitenk. Infektionskr. Hyg. Abt 1 Orig. Reihe C **2**:324–331.

Kloos, W. E. 1980. Natural populations of the genus *Staphylococcus*. Ann. Rev. Microbiol. **34**:559–592.

Kloos, W. E., and K. H. Schleifer. 1975a. Isolation and characterization of staphylococci from human skin. II. Descriptions of four new species: *Staphylococcus warneri*, *Staphylococcus capitis*, *Staphylococcus hominis*, and *Staphylococcus simulans*. Int. J. Syst. Bacteriol. **25**:62–79.

Kloos, W. E., and K. H. Schleifer. 1975b. Simplified scheme for routine identification of *Staphylococcus* species. J. Clin. Microbiol. **1**:82–88.

Kloos, W. E., and K. H. Schleifer. 1983. *Staphylococcus auricularis* sp. nov.: an inhabitant of the human external ear. Int. J. Syst. Bacteriol. **33**:9–14.

Krikler, S. J., T. H. Pennington, and D. Petrie. 1986. Typing of strains of *Staphylococcus aureus* by Western Blot analysis of culture supernates. J. Med. Microbiol. **21**:169–171.

Lachica, R. V. F., C. Genigeorgis, and P. D. Hoeprich. 1971. Metachromatic agar-diffusion methods for detecting staphylococcal nuclease activity. Appl. Microbiol. **21**:585–587.

Parisi, J. T. 1985. Coagulase-negative staphylococci and the epidemiological typing of *Staphylococcus epidermidis*. Microbiol. Rev. **49:**126–139.

Parisi, J. T., and D. W. Hecht. 1980. Plasmid profiles in epidemiologic studies of infections by *S. epidermidis*. J. Infect. Dis. **141:**637–643.

Pennock, C. A., and R. B. Huddy. 1967. Phosphatase reaction of coagulase-negative staphylococci and micrococci. J. Pathol. Bacteriol. **93:**685–688.

Richardson, J. F., and R. R. Marples. 1982. Changing resistance to antimicrobial drugs, and resistance typing in clinically significant strains of *Staphylococcus epidermidis*. J. Med. Microbiol. **15:**475–484.

Saravolatz, L. D., D. J. Pohlod, and L. M. Arking. 1982. Community-acquired methicillin-resistant *Staphylococcus aureus* infections: a new source for nosocomial outbreaks. Ann. Intern. Med. **97:**325–329.

Satta, G., G. Grazi, P. E. Varaldo, and R. Fontana. 1979. Detection of bacterial phosphatase activity by means of an original and simple test. J. Clin. Pathol. **32:**391–395.

Schleifer, K. H. 1986. Family I. *Micrococcaceae* Prévot 1961, 31, p. 1003–1035. *In* J. G. Holt (ed.), Bergey's manual of systematic bacteriology, vol. 2, Williams & Wilkins, Baltimore.

Schleifer, K. H., and W. E. Kloos. 1975. Isolation and characterization of staphylococci from human skin. I. Amended descriptions of *Staphylococcus epidermidis* and *Staphylococcus saprophyticus* and descriptions of three new species: *Staphylococcus cohnii, Staphylococcus haemolyticus,* and *Staphylococcus xylosus*. Int. J. Syst. Bacteriol. **25:**50–61.

Stackebrandt, E., and C. R. Woese. 1979. A phylogenetic dissection of the family Micrococcaceae. Curr. Microbiol. **2:**317–322.

Subcommittee on the Taxonomy of Staphylococci and Micrococci. Minutes of the meeting, 9 September 1975. Appendix 1, identification of staphylococci. Int. J. Syst. Bacteriol. **26:**332–334.

Valisena, S., P. E. Varaldo, and G. Satta. 1983. Purification and characterization of three separate bacteriolytic enzymes excreted by *Staphylococcus aureus, Staphylococcus simulans,* and *Staphylococcus saprophyticus*. J. Bacteriol. **151:**636–647.

Varaldo, P. E., P. Cipriani, A. Focà, C. Geraci, A. Giordano, M. A. Madeddu, A. Orsi, R. Pompei, M. Prenna, A. Repetto, S. Ripa, P. Rosselli, G. Russo, F. Scazzocchio, and G. Stassi. Identification, clinical distribution, and susceptibility to methicillin and 18 additional antibiotics of clinical *Staphylococcus* isolates: a nationwide investigation in Italy. J. Clin. Microbiol. **19:**838–843.

Varaldo, P. E., G. Grazi, O. Soro, G. Cisani, and G. Satta. 1980. Simplified lyogroup system, a new method for routine identification of staphylococci: description and comparison with three other methods. J. Clin. Microbiol. **12:**63–68.

Wallmark, G., I. Arremark, and B. Telander. 1978. *Staphylococcus saprophyticus:* a frequent cause of acute urinary tract infection among female outpatients. J. Infect. Dis. **138:**791–797.

Wenzel, R. P. 1986. Methicillin-resistant *S. aureus* and *S. epidermidis* strains: modern hospital pathogens. Infect. Control **7:**118–119.

CHAPTER 50

Streptococcal Infections: Alpha-Hemolytic Streptococci

RAYMOND W. RYAN

Alpha Hemolytic Streptococci: Group D streptococci and viridans streptococci.
Habitat: Group D streptococci usually found as normal flora in the gastrointestinal tract. Viridans streptococci comprise part of the normal oral flora as well as the urogenital and gastrointestinal tract.
Disease: *Enterococcus faecalis* and *Streptococcus bovis* are responsible for the majority of human infections caused by Group D streptococci. They are most frequently found causing endocarditis, urinary tract infection, and soft tissue infection. The viridans streptococci are responsible for 30 to 50% of bacterial endocarditis, whereas *S. mutans* is the major etiologic agent of dental caries.
Antimicrobial Susceptibility: *E. faecalis* infections are usually treated with a synergistic combination of antibiotics, that is, gentamicin and penicillin. Most strains of viridans streptococci are susceptible to penicillin, which is usually the antibiotic of choice.

Group D Streptococci: General Characteristics

Organisms that possess the glycerol teichoic acid group D antigen can be separated into two groups based on their ability to grow in broth containing 6.5% NaCl. The enterococcal species (Table 1) can grow in the presence of 6.5% NaCl, whereas the nonenterococcal species cannot. Until recently, it was considered clinically important to distinguish between these two groups as it was always assumed that the nonenterococcal strains, that is, *S. bovis*, was universally susceptible to penicillin. However, Savitch et al. (1978) have now reported on isolates of *S. bovis* resistant to penicillin. Successful treatment of infections due to enterococci usually require both an aminoglycoside and a penicillin.

Members of the group D streptococci are found as normal flora in the human gastrointestinal tract as well as other animals and birds. *E. faecalis* also can be found as normal flora in the mouth of normal adults. *E. faecalis* and *S. bovis* are responsible for the majority of human infections caused by group D streptococci. They have most frequently been implicated in endocarditis, urinary tract infection, and soft tissue infection.

Pathogenesis

E. faecalis along with *S. bovis* are responsible for 10 to 20% of all cases of bacterial endocarditis (Wilson and Washington, 1977). Individuals developing endocarditis with these organisms usually have certain predisposing factors such as underlying valve disease or prosthetic heart valves. The source of the bacteremia is usually secondary to infection at another site, that is, urinary tract, wound infections, or contaminated intravenous lines (Mandell et al., 1970). Gould et al. (1975) have shown that *E. faecalis* attaches more readily to heart valve tissue than other organisms known to cause endocarditis. It has been suggested that the production of dextran by *E. faecalis* may be an important virulence factor in the establishment of disease.

Klein and co-workers (1977) have described an association between *S. bovis* endocarditis and ade-

TABLE 1. Group D streptococci

Enterococcal species	
E. faecalis	Human pathogen
E. faecium	Rare human pathogen
E. durans	Do not usually cause infection in humans
E. gallinarum	
E. avium	
E. casseliflavus	
E. malodoratus	
Nonenterococcal species	
S. bovis	Human pathogen
S. equinus	Does not usually cause infection in humans

nocarcinoma of the colon. They found patients with this malignancy to have a much higher carriage rate of *S. bovis* in their feces. Only 5 to 10% of normal adults are colonized with this organism. Isolation of this organism from the blood may be the initial indication of underlying gastrointestinal disease.

E. faecalis is commonly encountered in urinary tract infections (UTIs). Approximately 8 to 10% of all UTIs can be attributed to this organism. However, hospitalized patients with indwelling urinary catheters, usually have a higher incidence of infection with *E. faecalis* (Weinstein and Moellering, 1973). Harding and Ronald (1974) have reported that women taking antimicrobial prophylaxis to prevent recurrent UTI are prone to develop infection with *E. faecalis* because of their increased resistance to many of these agents. The role of enterococci in both soft tissue and intraabdominal abscesses is not clear. It is the third most common organism isolated from intraabdominal abscesses, after *E. coli* and *Bacteroides fragilis*. Specific virulence factors have not yet been characterized for these organisms.

Antimicrobial Susceptibility

The necessity to differentiate between enterococcal and nonenterococcal group D streptococci becomes extremely important when antimicrobial therapy is considered. Whereas *S. bovis* is susceptible to penicillin (MBC, 0.2 μg/ml), the enterococci are uniquely antibiotic resistant. They are usually resistant to penicillin, aminoglycosides, cephalosporins, as well as several other antibiotics. Investigators have demonstrated a synergistic effect of penicillin in combination with an aminoglycoside in the treatment of enterococcal infections (Moellering et al., 1971b). Penicillins, including ampicillin, carbenicillin, and penicillin G can be successfully combined with aminoglycosides such as streptomycin, kanamycin, gentamicin, tobramycin, netilmicin, and amikacin to treat serious enterococcal disease. It soon became apparent that not all enterococcal strains were killed by these synergistic combinations of antibiotics. Moellering et al. (1971a) demonstrated high-level resistance to aminoglycosides among clinical isolates of enterococci. They found that isolates with high-level streptomycin resistance (MIC $>$ 2000 μg/ml) were resistant to penicillin/streptomycin synergism, whereas strains without high-level resistance (MIC $<$ 2000 μg/ml) were susceptible to the combination. Further studies with these high level aminoglycoside-resistant isolates revealed that their ribosomes were insensitive or resistant to the effect of aminoglycosides on protein synthesis. Other mechanisms for this resistance include inability of the antibiotic to penetrate the bacterial cell and the production of enzymes that inactivate the aminoglycoside by adenylation, phosphorylation, or acetylation (Kaye, 1982). At present, gentamicin is considered the drug of choice to achieve a synergistic effect, as it has been shown to be effective for the treatment of patients with endocarditis due to strains with high-level streptomycin resistance (Weinstein and Moellering, 1973). The susceptibility of the isolate to 2000 μg/ml gentamicin should be determined before initiating therapy.

Viridans Streptococci: General Characteristics

The viridans streptococci are a heterogenous group of α- and γ-hemolytic organisms that comprise part of the normal oral flora in humans. They are important pathogens in both human and animal infections (Table 2). Viridans streptococci are nutritionally fastidious organisms that require media rich in protein and a CO_2 environment for optimal growth. They appear on sheep blood agar as tiny colonies, usually surrounded by a narrow zone of α-hemolysis (green). They are opportunistic pathogens of low virulence. Several species produce extracellular enzymes that synthesize large polysaccharide capsules of levan or dextran from sucrose. Although they produce no known virulence factors, certain members such as *S. sanguis* produce extracellular proteolytic enzymes (IgA protease, neuraminidase; Straus, 1982). The role of these extracellular products in the pathogenicity of viridans streptococci is not well understood.

TABLE 2. Normal habitat of human viridans streptococci

S. mitior	Dental surface (plaque), pharynx
S. sanguis	Dental surface (plaque)
S. mutans	Dental surface
S. salivarius	Dorsum of tongue, hard palate, pharynx
S. milleri	Dental surface, gingival crevice, pharynx

Many characteristics of the viridans streptococci are variable. Their cell wall structure is similar to that found in the β-hemolytic streptococci in that they possess peptidoglycan, teichoic acid, and membrane-associated lipoteichoic acid; however, they do not have a defined group carbohydrate antigen. No correlation exists between serogroups and physiologic characteristics of these strains. Investigators in both the United States (Centers for Disease Control, Facklam et al.) and Great Britain (Colman et al.) have independently attempted to define and identify the known members of the viridans streptococci.

Pathogenesis: Viridans Streptococci

Although the viridans streptococci are considered to be opportunistic pathogens of low virulence, they account for 30 to 50% of bacterial endocarditis in the United States. Those individuals who develop endocarditis with one of the viridans streptococci usually possess a previously damaged heart valve. *S. mitior, S. mutans,* and *S. sanguis* are the agents primarily for streptococcal endocarditis. There is a relationship between the organism's ability to produce dextran and its ability to adhere to damaged heart valves (Finegold and Baron, 1986). The increased ability of these streptococci to adhere to epithelial cells may be the single most important factor that allows them to establish infection. Predisposing factors such as gingival disease or routine dental manipulation may allow these organisms to enter the bloodstream.

The relationship of viridans streptococci to the formation of dental caries is well understood. *S. mutans* is undoubtedly considered the major etiologic agent of dental caries. The cariogenic property of this organism is demonstrated by its ability to produce dextran and adhere to tooth surfaces where it multiplies and produces lactic acid from carbohydrate fermentation. Gibbons and Barghart (1967) have shown that these acid end products contribute to the development of carious lesions. Several studies have demonstrated that subjects with active caries harbor an increased number of *S. mutans*. The association of *S. sanguis* and *S. mitior* and human caries has been suggested based on their acid production, acid tolerance, and by their presence in dental plaque. The viridans streptococci also have been implicated in other infectious processes such as brain abscesses, aspiration pneumonia, central nervous system infection and liver abscess formation. *S. milleri* (British classification) and *S. intermedius* (CDC classification) are more commonly associated with extracardiac infections.

Nutritionally variant streptococci also have been known to be significant human pathogens. They are referred to as vitamin B_6-dependent streptococci, thiol-requiring streptococci, or satelliting streptococci. These vitamin B_6-dependent streptococci exist as part of the normal oral flora. Routine laboratory media does not usually support the growth of these organisms. To achieve optimal growth, media must be supplemented with either pyridoxal hydrochloride or pyridoxamine dihydrochloride. A report from New York Hospital (Roberts, 1980) indicates that approximately 6% of all cases of bacterial endocarditis are due to these nutritionally variant streptococci.

Major Human Pathogens of Clinical Significance

S. sanguis adheres to the tooth surface through the production of extracellular insoluble dextran. It is found as a common constituent of dental plaque, and produces extracellular proteolytic enzymes (neuramidase, IgA protease). The role of these enzymes in the pathogenesis of disease is unknown. *S. sanguis* is an important cause of bacterial endocarditis.

S. mitior is a common constituent of dental plaque. Some strains produce insoluble dextran from glucose. It is an important cause of bacterial endocarditis.

S. mutans produces insoluble dextran, which allows the organism to adhere to tooth surfaces. *S. mutans* is thought to cause the initiation of carious lesions. It contains glycerol teichoic acid in the cell wall, which allows cross-reaction with antisera produced against group D streptococci, and it is an important cause of bacterial endocarditis.

S. salivarius is associated mainly with epithelial surfaces in the mouth, that is, the dorsal surface of the tongue and the cheeks. It produces levan from sucrose and is easily detected when grown on mitis-salivarius agar.

S. milleri is different from other viridans streptococci in that it does not produce any extracellular carbohydrate polymers. It is not usually localized in a particular site, that is, tooth surfaces. *S. milleri* is the most common viridans streptococci associated with suppurative infections (brain abscesses). Causes less than 10% of all bacterial endocarditis due to viridans streptococci.

Antimicrobial Susceptibility

Most strains of viridans streptococci are susceptible to penicillin, which is usually the antibiotic of choice. However, approximately 17% of the viridans streptococci are moderately resistant to penicillin (MIC $>$ 0.1 μg/ml; Roberts et al. 1979). The laboratory plays a significant role in assisting the physician in diagnosing and treating patients with bacterial endocarditis. Determination of both the minimum inhibitory concentration (MIC) and minimum bactericidal concentration (MBC) should be performed. Some strains of viridans streptococci can exhibit tolerance, that is, when the MBC exceeds the MIC by 32-fold or greater, in which case high-dose intravenous penicillin or combination therapy may be required to eradicate the organism. Tolerance has to be demonstrated with some strains of all species of viridans streptococci except *S. salivarius* (Holloway et al., 1980). For those patients with endocarditis who are unable to receive penicillin, cephalosporins and vancomycin can be used.

Isolation and Identification Procedures

The viridans streptococci and group D streptococci may be isolated from a variety of sources as described. These organisms, excluding the nutritionally variant streptococci, grow readily on 5% sheep blood agar, which also shows their hemolytic patterns. The nutritionally variant streptococci require plates supplemented with 0.001% vitamin B_6 or pyridoxal.

Commercially prepared blood culture bottles containing tryptic soy broth, brain–heart infusion, supplemented peptone, or thioglycollate broth will support the growth of viridans streptococci and group D streptococci. These media will usually support the growth of nutritionally variant streptococci because the human red blood cells provide the necessary growth factors; however, subculture must be made to pyridoxal containing plates or a blood agar plate cross-streaked with *Staphylococcus aureus*. The streptococci will satellite around the *Staphylococcus* colony. Inoculated plates must be incubated at 35 to 37°C in an atmosphere of reduced oxygen tension, either 5% CO_2 or anaerobically. These organisms are facultative and many strains are microaerophilic.

Plates may be examined for growth after 24 h. Colonies of group D streptococci are 0.5 to 1.0 mm in diameter, less opaque than group A streptococci, and are either α-hemolytic or nonhemolytic. The viridans streptococci are smaller than group D streptococci ranging in size from pinpoint to 0.5 mm. They may be α-hemolytic or nonhemolytic, and the colonies can be mucoidal and translucent or glossy and nontranslucent.

Preliminary Identification

The viridans streptococci, including nutritionally variant streptococci and group D streptococci, which include enterococcal and nonenterococcal strains are gram-positive, nonmotile cocci. They produce neither catalase or oxidase, which differentiates them from the *Micrococcus* and *Neisseria*. When Gram stained, they occur in pairs and chains of varying lengths and may appear rod-like as well as coccoidal.

Differentiation of streptococci is based on hemolytic patterns, physiologic characteristics, and antigenic reactions. Different types of hemolysis were described in 1919 by Brown based on the appearance of deep colonies in pour plates made with blood (Hardie, 1986): β-, α-, and α'-hemolysis. The factors affecting hemolysis are the source and concentration of animal blood, and the atmosphere of incubation which also affects colony size. Viridans streptococci as well as the group D streptococci may exhibit either α- or γ-hemolysis. Some viridans streptococci may exhibit α'-hemolysis. Many viridans streptococci, when incubated anaerobically, appear nonhemolytic. Alpha-hemolysis is a result of partial degradation of red blood cells, causing a greenish or brownish discoloration of the medium. Alpha-prime-hemolysis has a small zone of intact or partially lysed erythrocytes next to the colony and a zone of complete hemolysis surrounding this zone.

The group D streptococci, both enterococcal and nonenterococcal, possess a specific carbohydrate antigen (Lancefield's group D) and therefore react with group D-specific antiserum. A new genus, *Enterococcus* recently has been adopted that includes the enterococcal strains; this is based on results of DNA-rRNA hybridization studies. The group D antigen, which is an intracellular glycerol teichoic acid associated with the cytoplasmic membrane, may be demonstrated by a variety of methods including the conventional capillary precipitin test and slide agglutination tests (latex). The latter is commercially available and has the advantage of shortened identification time. In addition to reaction with specific antiserum, two physiologic tests are used to differentiate the enterococcal from nonenterococcal strains: bile-esculin (BE) and tolerance to 6.5% NaCl broth. Presumptive identification of *S. bovis* can be made by performing an additional test, the PYR test (Facklam and Carey, 1985). Aerococci may be bile-esculin positive as well as grow in the presence of 6.5% salt. Reaction with group D antiserum avoids misidentification of aerococci as group D streptococci, although a Gram stain showing tetrad formation is characteristic.

The viridans streptococci, as a rule, do not react with group B- or D-specific antisera, but they may carry antigens on their surface that cross-react with Lancefield groups A, C, F, and G antisera. Several investigators have attempted to differentiate viridans streptococci serologically without much success. The viridans streptococci can be distinguished from *S. pneumoniae* because they are insoluble in bile and resistant to optochin.

TABLE 3. Reactions for medically important strains

	Arabinose	*Sucrose*	*Sorbitol*
E. faecium	+	V	V
E. faecalis	–	+	+
E. durans	–	–	–

Definitive Identification

Strains of the genus *Enterococcus*, *E. faecium* and *E. durans*, and the nonenterococcal strains, *S. equinus* and *S. bovis*, may be identified on the basis of physiologic reactions. Sugar fermentation patterns (L-arabinose, arbutin, melezitose, melibiose, sorbitol, sucrose, and sorbose), decarboxylation of tyrosine, ammonia production from arginine, and reduction of potassium tellurite aid the differentiation of the *Enterococcus* strains. The reactions for the medically important strains are listed in Table 3 (Finegold and Baron, 1986; Mundt, 1986).

The viridans streptococci are primarily identified by physiologic tests. Two major systems exist for the differentiation of these organisms: the British scheme developed by Colman and co-workers and the CDC scheme by Facklam. The British scheme designates eight species including *S. pneumoniae*, and the CDC scheme designates 10 species excluding *S. pneumoniae*. Table 4 lists which species are recognized by both (Facklam and Carey, 1985).

CDCs (Facklam) scheme is based on 30 physiologic characteristics listed in Table 5 (Facklam, 1977). The British scheme is based on many of the same characteristics but also includes cell wall analysis for the presence of rhamnose and ribitol, susceptibility to bacitracin and nitrofurazone, and production of acetoin from glucose (Facklam, 1977). Several modifications of both schemes (Finegold and Baron, 1986; Roberts et al., 1979) have been used to identify the viridans streptococci and also *S. bovis*. Facklam had proposed a guide for the clinical microbiologist to use in the differentiation of viridans streptococci using the following tests: acid formation in mannitol, lactose, inulin and raffinose broth; hydrolysis of hippurate, arginine, and esculin; growth in litmus milk, tolerance to 40% bile; glucan production on 5% sucrose agar and broth; and α- or γ-hemolysis (Facklam, 1977). Additional tests that are helpful but not necessary are acid reactions in carbohydrate broths such as trehalose, salicin, and melibiose. Media and reagents must be prepared according to CDC formulations. For carbohydrate use, bromcresol purple broth must be used. When different broth bases and pH indicators are used, results may not correlate and different speciation may occur.

TABLE 4. Recognized species of viridans streptococci: British and Centers for Disease Control

British	*CDC*
S. salivarius	*S. salivarius*
S. sanguis I	*S. sanguis* I
S. mutans	*S. mutans*
S. acidominimus	*S. acidominimus*
S. uberis	*S. uberis*
S. mitior	*S. mitis*
S. mitior (dextran positive)	*S. sanguis* II
S. milleri (lactose positive)	*S. intermedius* (formerly *S. MG-intermedius*)
S. milleri (lactose negative)	*S. constellatus*
No comparable strain	*S. morbillorum*

TABLE 5. Centers for Disease Control scheme based on 30 physiologic characteristics

1. Hemolysis (hemolytic reaction in poured plates with aid of a microscope).[a]
2. Tolerance to
   - BE[a]
   - Methylene blue milk
   - 10°C[a]
   - 45°C
   - 6.5%,[a] 4.0%, 2.0% NaCl
   - 10% and 40% bile
   - 0.04% tellurite
   - 0.1% tetrazolium
   - Reduction of litmus milk[a]
3. Hydrolysis of
   - Arginine
   - Esculin[a]
   - Hippurate[a]
   - Starch
4. Acid from broths of
   - Mannitol,[a] sorbitol, inulin,[a] lactose,[a] raffinose,[a] sucrose, trehalose, salicin, melibiose, arabinose, and glycerol
5. Reaction in 5% sucrose broth[a]
6. Reaction in 5% sucrose agar[a]

[a] Denotes most useful reactions.

Commercial Systems

Several commercial identification systems are available based on both biochemical and enzymatic reactions. The Minitek System (BBL) uses reagent-impregnated filter paper disks. A small amount of inoculum (McFarland 1 turbidity) is pipetted in wells containing the disks and incubated 48 h. Then an indicator is added. Conventional media must be used for hydrolysis of esculin and arginine and the fermentation of inulin. Good correlation between the Minitek and the conventional method for the viridans streptococci has been shown (Setterstrom et al., 1979).

The Rapid Strep System (API) is a 4- or 24-h system, using 20 microcupules containing dehydrated substrates for the demonstration of enzymatic activity or the fermentation of sugars. The reactions are easily interpretable and a numerical code is generated. Several studies have shown excellent identification of group D streptococci and viridans streptococci without supplemental testing (Appelbaum et al., 1984b; Facklam et al., 1984). Additional tests that may be required to aid in identification include production of levan or dextran, acid production of additional sugars, growth at 10°C, BE, urease hydrolysis, optochin sensitivity, serology, and growth in 6.5% NaCl. A disadvantage of this system is the need for a dense inoculum (equal to or greater than a MacFarland 4 turbidity).

Vitek systems has a gram-positive identification card for its AutoMicrobic System. Suspensions of the organisms equal to a MacFarland 1 turbidity are used. Results are based on 27 tests and are available 4 to 13 h after incubation. Without additional testing, identification rates of the viridans streptococci varies from strain to strain, with an overall low correlation (Appelbaum et al., 1984a; Facklam et al., 1985). In contrast, 97.2% of the group D streptococci strains were identified correctly to species level (Appelbaum et al., 1984a).

RapID STR system from Innovative Diagnostic Systems is a system using 14 chromogenic substances with results generated in 4 h. Inocula equal to a MacFarland 1 turbidity are used and the panels are incubated aerobically at 35°C. Inclusion of optochin testing greatly increased the accuracy of identification of the viridans streptococci (Appelbaum et al., 1986). Additional testing may be required to accurately differentiate between the group D streptococcal species. Current CDC nomenclature is used in the database. This method is rapid, convenient, and reproducible.

Literature Cited

Appelbaum P. C., M. R. Jacobs, W. M. Palko, E. E. Frauenhoffer, and A. Duffett. 1986. Accuracy and reproducibility of the IDS RapID STR system for species identification of streptococci. J. Clin. Microbiol. **23:**843–846.

Appelbaum, P. C., M. R. Jacobs, J. I. Heald, W. M. Palko, A. Duffett, R. Crist, and P. A. Naugle. 1984a. Comparative evaluation of the API 20S system and the AutoMicrobic system gram-positive identification card for species identification of streptococci. J. Clin. Microbiol. **19:**164–168.

Appelbaum, P. C., P. S. Charushiya, M. R. Jacobs, and A. Duffett. 1984b. Evaluation of the Rapid Strep system for species identification of streptococci. J. Clin. Microbiol. **19:**588–591.

Facklam, R. R., G. S. Bosley, D. Rhoden, A. R. Franklin, N. Weaver, and R. Schulman. 1985. Comparative evaluation of the API 20S and AutoMicrobic gram positive identification systems for non-beta-hemolytic streptococci and aerococci. J. Clin. Microbiol. **21:**535–541.

Facklam, R. R., D. L. Rhoden, and P. B. Smith. 1984. Evaluation of the Rapid Strep system for the identification of *Streptococcus* species. J. Clin. Microbiol. **20:**894–898.

Facklam, R. R., and R. B. Carey. 1985. Streptococci and aerococci. *In* E. H. Lennette, A. Balows, W. J. Hausler, Jr., and H. J. Shadomy (ed.), Manual of clinical microbiology (4th ed.), American Society for Microbiology, Washington, D. C.

Facklam, R. R. 1977. Physiological differentiation of viridans streptococci. J. Clin. Microbiol. **5:**184–201.

Finegold, S. M., and E. J. Baron. 1986. Streptococci, including enterococci and pneumococci and aerococcus. *In* Diagnostic microbiology (7th ed.). C. V. Mosby Co., St. Louis, MO.

Gibbons, R. J., and S. Barghart. 1967. Synthesis of extracellular dextran by cariogenic bacteria and its presence in human dental plaque. Arch. Oral Biol. **12:**11–23.

Gould, K., C. H. Ramirez-Ronda, R. K. Holmes. 1975. Adherence of bacteria to heart values in vitro. J. Clin. Invest. **56:**1364–1370.

Hardie, J. M. 1986. Genus streptococcus Rosenbach 1884, 22. *In* J. G. Holt, P. H. A. Sneath, N. S. Mair, M. E. Sharpe (ed.), Bergey's manual of systematic bacteriology, Vol. 2, Williams & Wilkins, Baltimore, MD.

Harding, G. K. M., and A. R. Ronald. 1974. A controlled study of antimicrobial prophylaxis of recurrent urinary infection in women. N. Engl. J. Med. **291:**597–601.

Holloway, Y., J. Dankert, and J. Hess. 1980. Penicillin-tolerance and bacterial endocarditis. Lancet **i:**598.

Kaye, D. 1982. Enterococci: biologic and epidemiologic characteristics and in vitro susceptibility. Arch. Intern. Med. **142:**2006–2009.

Klein, R. S., R. A. Recco, M. T. Catalano, S. C. Edberg, J. I. Casey, and N. H. Steigbigel. 1977. Association of *Streptococcus bovis* with carcinoma of the colon. N. Engl. J. Med. **296:**800–802.

Mandell, G. L., D. Kaye, M. E. Levison. 1970. Enterococcal endocarditis: an analysis of 38 patients observed at the New York Hospital-Cornell Medical Center. Arch. Intern. Med. **125:**258–272.

Moellering, R. C., Jr., C. B. G. Wennersten, T. Medrek, and A. N. Weinberg. 1971a. Prevalence of high-level resistance to aminoglycosides in clinical isolates of enterococci. Antimicrob. Agents Chemother. **170:**335–340.

Moellering, R. C., Jr., C. B. G. Wennersten, and A. N. Weinberg. 1971b. Synergy of penicillin and gentamicin against enterococci. J. Infect. Dis. **124(suppl):**S207–S209.

Mundt, J. O. 1986. Enterococci. *In* J. G. Holt, P. H. A. Sneath, N. S. Mair, and M. E. Sharpe (ed.), Bergey's manual of systematic bacteriology, vol. 2, Williams & Wilkins, Baltimore, MD.

Roberts, R. B. 1985. Viridans and beta hemolytic streptococci. *In* G. L. Mandell, R. G. Douglas, Jr., and J. E. Bennett (ed.), Principles and practice of infectious diseases (2nd ed.), John Wiley and Sons, New York.

Roberts, R. B., A. G. Krieger, and N. L. Schiller. 1979.

Viridans streptococcal endocarditis: the role of various species including pyridoxal-dependent streptococci. Rev. Infect. Dis. **1**:955–966.

Savitch, C. B., A. L. Barry, and P. D. Hoeprich. 1978. Infective endocarditis caused by *Streptococcus bovis* resistant to the lethal effect of penicillin G. Arch. Intern. Med. **138**:931–934.

Setterstrom, J. A., A. Gross, and R. S. Stanko. 1979. Comparison of Minitek and conventional methods for the biochemical characterization of oral streptococci. J. Clin. Microbiol. **10**:409–414.

Straus, D. C. 1982. Protease production by *Streptococcus sanguis* associated with subacute bacterial endocarditis. Infect. Immunol. **38**:1037–1045.

Warren, J. W., J. H. Tenney, J. M. Hoopes. 1982. A prospective microbiologic study of bacteriuria in patients with chronic indwelling urethral catheters. J. Infect. Dis. **146**:719–723.

Weinstein, A. J., and R. C. Moellering, Jr. 1973. Penicillin and gentamicin therapy for enterococcal infections. J. Am. Med. Assoc. **223**:1030–1032.

Wilson, W. R., J. A. Washington, II. 1977. Infective endocarditis: a changing spectrum? Mayo Clin. Proc. **52**:254–255.

CHAPTER 51

Syphilis

S. A. LARSEN and C. M. BECK-SAGUE

Disease: Syphilis (Lues, Pox, Bad Blood).
Etiologic Agent: *Treponema pallidum* subspecies *pallidum.*
Source: Lesions or blood of infected persons. Disease is acquired by sexual contact with infected persons or congenitally by the transplacental transmissions from pregnant women to fetus.
Clinical Manifestations: Solitary, painless skin lesion at site of entry, usually genital, which resolves spontaneously (primary) or generalized rash accompanied by mild malaise or fever and occasionally more severe systematic symptoms, which resolve spontaneously (secondary). Late sequelae of untreated disease include thoracic aneurysm, aortic insufficiency, paresis, psychosis, and long tract signs (teritary).
Pathology: Pathology varies with disease stage. However, the vascular component, that is, endarteritis and periarteritis, is demonstrable in virtually all stages.
Laboratory Diagnosis: Darkfield microscopic identification of typical spirochetes from non-oral mucous membrane lesions, skin lesions, autopsy materials, or direct immunofluorescent examination from these sources as well as from oral mucous membrane lesions. Venereal Disease Research Laboratory (VDRL) or comparable nontreponemal test $\geq 1:1$ with a fourfold rise and confirmation with a treponemal test (fluorescent treponemal antibody or comparable test) considered diagnostic. In cases more than 5 years after initial infection, only treponemal test may persist.
Epidemiology: Worldwide distribution with concentration in urban areas and economically deprived populations. High prevalence related to prostitution and other circumstances where sexually active persons have contact with large numbers of sexual partners.
Treatment: Long-acting injectable penicillin (benzathine penicillin) preparations. In central nervous system involvement, aqueous penicillin G in high dose during a 3 to 4 week period results in a satisfactory clinical response.
Prevention and Control: Treatment of infected persons and their sexual partners. Screening of all sexually active persons in high prevalence areas, particularly pregnant women.

Description of Disease

Syphilis (Lues, Pox, Bad Blood) is a sexually transmitted multisystem disease of humans. Human infection is acquired by sexual contact with infected persons, and rarely, by blood from persons with spirochetemia, or congenitally by transplacental infection from infected mother to fetus. Untreated disease in the adult can result in severe late sequelae of tertiary syphilis, which involves the cardiovascular system, central nervous system, bones and joints, and possibly death. In the congenital form, syphilis causes death in the form of stillbirth or neonatal death in more than 20% of affected fetuses. In untreated persons with congenital syphilis, sequelae include mental retardation, blindness, deafness, and facial, dental, and long-bone stigmata.

The disease has public health significance because of the importance of the congenital form in infant morbidity and mortality in high prevalence areas and the severity of its symptomatology in adult patients who progress to the tertiary stage of acquired syphilis. The prevention of sexual and vertical transmission is extremely cost-effective, even in low prevalence areas. *Treponema pallidum* subspecies *pallidum* is the spirochete that causes syphilis in humans.

Communicability and Transmissibility

Syphilis is transmissible transplacentally and by sexual contact throughout the course of untreated infection, from incubation to the tertiary stage. However, the risk of infection is greatest during the primary and secondary stages, when spirochetemia is greatest and transmission is most effective. During these stages, syphilis is highly infectious to the sexual partner and fetus of the infected person. Only an insignificant percentage of infections are acquired by nonsexual contact with lesions and blood of infected persons. The incubation period of syphilis may vary from 10 to 90 days, median 21 days.

Clinical Manifestations of Disease

Onset of disease is heralded by a solitary painless lesion, often accompanied by regional lymphadenopathy, within 2 to 6 weeks of exposure. The clinical manifestations of syphilis are classified by stage of infection as primary, secondary, latent, benign late, and tertiary syphilis.

Primary Stage

The primary lesion, or chancre, is usually a genital, painless indurated ulcer which appears within 2 to 6 weeks of infection. Because these can be vulvar, anal, or cervical, they may not be noticed. These resolve without treatment within 2 months. Often, the patient will be seronegative early in this stage, but direct microscopic examination of intact lesions can confirm the diagnosis.

Secondary Stage

The secondary stage, which occurs within 6 weeks of healing of the primary lesion is characterized by a generalized hyperpigmented papular rash with palmar and plantar predilection. Virtually any type of rash except vesicular rashes are seen during the secondary stage. Low-grade fever, malaise, and generalized lymphadenopathy are often seen. Rarely, clinically severe hepatitis, and meningitis are seen as well. Multiple flat moist lesions, condylomata lata, are seen in some patients. High titers of nontreponemal tests are seen in virtually 100% of patients in this stage. The diagnosis should be suspected if nontreponemal tests are negative and a prozone phenomenon can be excluded. Lesions of secondary syphilis usually resolve in 2 to 10 weeks, with or without treatment.

Latency

The period after the symptoms of secondary syphilis resolve in the untreated patient is divided into early latency, less than 1 year after initial infection, and late latency, more than 1 year after infection. The late latency stage is without symptoms for the patient and is characterized by gradual reduction in spirochetemia. The infectiousness to the sexual partner and fetus, which is over 80% in early latency, falls to less than 50% in late latency.

Benign Late

The gumma (lesions) of benign late syphilis are rarely seen today. The lesions can occur irom 2 to 40+ years after the onset of initial infection. The gummas resemble the granuloma of tuberculosis and may occur in the skin, bones, mucosae, viscera, muscles, and ocular structure. Depending on the location of the gumma, for example, brain, spinal cord, or trachea, benign late syphilis may not be benign. Over 90% of the lesions can be successfully treated with penicillin (Kampmeier, 1984).

Tertiary Stage

However, in the majority of untreated patients with syphilis, no symptoms referable to syphilis are noted; cardiovascular morbidity and mortality is significantly greater in these persons than in the uninfected person. Most of the symptoms and mortality of tertiary syphilis are referably to syphilitic aortitis, aortic valve insufficiency, and thoracic aneurysm and rupture. Central nervous system involvement causes insanity, often accompanied by delusional and paranoid states and general paresis. Involvement of the posterior white columns causes tabes dorsalis which is characterized by lightning pains, incontinence, ataxia, and joint destruction secondary to insensibility to injury.

Pathology

After sexual intercourse with an infected individual, a person has approximately a 30% chance of becoming infected. The skin and mucous membranes provide effective barriers to infection. Infection probably occurs through microscopic abrasions produced during intercourse. At the site of entry, the organism may be initially ingested but not killed by polymorphonuclear leukocytes (PMNs). Some treponemes may escape ingestion; this mechanism is unclear.

Primary Stage

The primary stage begins with a generalized bacteremia within approximately 30 h after infection; however, preferential multiplication occurs at the site of entry. The primary chancre forms usually within 3 weeks of infection. This lesion contains a mucoid center of hyaluronic acid and chondroitin sulfate surrounded by a cellular infiltrate of primarily mononuclear cells. Both treponemal and nontreponemal antibodies usually appear 1 to 4 weeks after the lesion has formed.

Secondary Stage

The secondary stage's disseminated lesions are attributable to systemic infection. The lesions take many forms due in part to systemic immunity to some degree. Almost every organ and tissue of the body is affected during the secondary stage including virtually all body fluids. Circulating immune complexes, detected in approximately 80% of the secondary cases, may lead to immune complex deposition in the kidneys and subsequent renal damage (Engel and Diezel, 1980).

Latency

The early latent stage, represents a conversion from an acute to a chronic infection. During this period at least one relapse occurs with cutaneous eruptions. After the first year, the infection is suppressed by the host's immune response to the point that no lesions are clinically apparent, and the disease becomes asymptomatic. Yet, treponemes can be isolated from infected tissues. Delayed-type hypersensitivity to pathogenic treponemes is characteristic of this stage.

Benign Late

The pathogenesis of this stage is poorly understood. The gumma is characteristic of an inflammatory process, perhaps due to an allergic response to a few remaining organisms. The center of the granuloma shows coagulative necrosis surrounded by lymphocytes and other mononuclear cells. Vascularized connective tissue extends out from the necrotic center, while proliferating connective tissue encapsulates the lesion. Deep scarring is common in the healing of gummas.

Tertiary Stage

An immunologic basis for the lesions of late cardiovascular syphilis has been proposed. Apparently treponemes spread via the lymphatics and lodge preferentially in the proximal aorta producing an endarteritis, which may also involve the coronary arteries near the ostia. This inflammatory process continues for years. Eventually all three layers of the aortic wall are affected, resulting in patchy necrosis and focal scarring of the aortic media, aortic dilatation, aneurysm formation, fibrous thickening of the adventitia, and atherosclerotic changes of the aortic intima, that is, "tree barking" (Bulkley, 1984).

Neurosyphilis takes a variety of forms; however, all have in common a chronic meningitis, which may produce vascular and parenchymatous sequelae in the cerebrum and spinal cord (Swartz, 1984). In most cases the CSF shows >5 lymphocytes per cubic millimeter and an elevated protein of ≥45 mg/dl. In acute syphilitic meningitis, granular ependymitia and endarteritis may develop. These may further lead to thrombosis, vascular occlusion, and cerebral infarction. As a result of obstruction of CSF flow, several forms of hydrocephalus result. In cerebrovascular neurosyphilis, the vasa vasorum is occluded resulting in destruction of the smooth muscle and elastic tissue of the media and finally infarction. In general paresis, parechymatous neurosyphilis, computed tomography (CT) scan patterns indicate demyelinization of cerebral white matter. On gross examination of the brain, the meninges show varying degrees of thickening consistent with chronic meningitis. Cerebral atrophy particularly of the frontal pole and tips of the temporal lobes is prominent. Whorls of subependynal astrocytes forming granular ependymitis are a characteristic finding. Treponemes can be demonstrated in brain tissue by direct immunofluorescence or Dieterle silver strain.

Diagnostic Features

Primary syphilis should be considered in any person presenting with a solitary genital or anal lesion, particularly if it is painless indurated, and/or ulcerated. Definitive diagnosis in these persons requires visualization of spirochetes on direct microscopic examination of material from the lesions. Although early in this stage at least 15% of patients are seronegative, a positive nontreponemal test in the presence of such a lesion is considered diagnostic.

Secondary syphilis should be considered in persons with multiple characteristic genital lesions or generalized rashes, particularly palmar and plantar rashes in the presence of a positive nontreponemal test such as the VDRL. The diagnosis should be questioned in the presence of a negative nontreponemal serologic test, if a prozone phenomenon can be excluded. This phenomenon is typically seen in persons with extremely high titers and is a false-negative result. Secondary syphilis genital lesions can resemble herpes, and the differential diagnosis of secondary syphilis includes generalized lymphadenopathy.

Early congenital syphilis is characterized by the onset of a nasal discharge in the first weeks of life, often bloody, called snuffles. Other common features of early congenital infection are prematurity and stillbirth, condylomata lata, vesiculobullous lesions of the skin, jaundice, hemolytic anemia, nephrotic syndrome with generalized edema and hepatosplenomegaly. Bone involvement, including osteochonditis and perostitis, can cause pseudoparalysis due to pain in the affected limb. Bone involvement is radiologically demonstrable in more than 90% of affected infants, although it may not be clinically apparent. A definitive diagnosis can be made in such infants by direct microscopic visualization of spirochetes from skin lesion or nasal discharge. In autopsy material, it can be made with Warthin-Starry or Dieterle silver staining or with direct fluorescent antibody staining. Tertiary syphilis should be suspected in patients with congestive heart failure associated with isolated aortic insufficiency or thoracic aneurysms. The diagnosis is suspected in the presence of other valve involvement or in the presence of abdominal aneurysm. In autopsy series anatomic evidence of syphilitic aortitis is found in over half of patients with untreated syphilis.

Asymptomatic neurosyphilis should be suspected when CSF lymphocyte count and proteins are elevated, the serum serology is reactive, and treatment history is unknown or inadequate. Meningeal syphilis should be considered when a patient presents with a reactive nontreponemal test, headache, fever, photophobia, and stiff neck with or without a secondary rash. The diagnosis of meningovascular syphilis should be considered when a cerebrovascular accident occurs in a young person. Again symptoms are headache, dizziness, memory loss, hemiplegic, aphasia, hemiparesis, and seizures, General paresis should be suspected when typical pupillary abnormalities present, which may be accompanied by dementia or seizures.

Prognosis

The mortality rate in untreated acquired syphilis in persons 25 to 50 years of age is significantly higher than in controls, although only one-third of patients develop clinically apparent tertiary syphilis (Termini and Music, 1972). Cardiovascular or central nervous system syphilis is the primary cause of death. With penicillin therapy during the early stages, death is rare, although occasional treatment failures have been reported. After treatment patients can be readily reinfected.

Untreated congenital syphilis is fatal in at least 20% of affected infants. In untreated survivors, long-term sequalae include interstitial keratitis with resultant blindness, hearing loss, mental retardation, anterior bowing of the shins called "sabre shins," scarring of the face ("rhagades"), abnormalities of the molars, and saddle nose deformity secondary to destruction of the septum. With penicillin therapy at birth, such complications are rare.

Epidemiologic Aspects

Occurrence and Transmission

Syphilis occurs worldwide. Since the introduction of penicillin, the number of cases of late manifestations has fallen considerably. Case reporting is incomplete, but thousands of cases are reported on all continents each year.

In the United States, syphilis is the third most common reportable disease, exceeded only by chicken pox and gonorrhea. Critical changes in the distribution and characteristics of reported early syphilis occurred between 1967 and 1982 (Fichtner et al., 1983). These changes included a twofold increase in the ratio of male to female early syphilis cases. This trend was reversed between 1982 and 1986, and the absolute number of reported cases of primary and secondary syphilis as well as the rate per 100,000 males and the percentage of males with same sex partners declined. However, in 1987, an increase in primary and secondary syphilis occurred. Now case rates for primary and secondary syphilis are 13 per 100,000.

Congenital syphilis incidence tends to reflect primary and secondary syphilis incidence in women of childbearing age, and the number of reported cases of early congenital syphilis (less than 1 year of age) has been gradually increasing since 1978, with more than 200 cases reported each year in the years 1984 to 1986.

Primary and secondary syphilis rates per 100,000 population are higher in large urban areas than in less populated ones. Incidence is highest in the southern states.

Most cases of congenital syphilis are the result of failure to obtain prenatal care, where serologic screening and treatment can occur. Errors in management with failure to screen or treat account for almost a third of cases, and occasional failures of prenatal treatment to arrest fetal disease account for a small percentage of cases.

Epidemics of syphilis are often traced to circumstances favoring anonymous sexual encounters with large numbers of partners by infected persons. Decreased accessibility to screening and treatment and difficulties identifying sexual partners of infected persons can be associated with rapid propagation of an outbreak. Such outbreaks have occurred in prison populations and have been linked with prostitution (Kinsie, 1962). Although in the last 2 decades, men with same sex partners have been important in the transmission of syphilis, they have become less so since 1982 (CDC, 1984).

Susceptibility, Resistance, and Carrier States

Treponema pallidum gains access to the human body readily through intact mucous membrane or abraded skin and within a few hours enters the lymphatics and blood to produce systemic infection and metastatic foci long before the appearance of a primary lesion. In humans there is no natural resistance to infection by pathogenic treponemes. The relative importance of innoculum size, sexual practices, and body size remain undefined.

In animal models, resistance to reinfection, or superinfection is demonstrable about 3 months after the primary infection. The majority of human volunteers with untreated latent disease do not develop either a primary lesion or serologic response after innoculation.

Serum antibody does not appear to be protective. However, delayed hypersensitivity to *T. pallidum* has been demonstrated by skin test in late syphilis, Passive transfer experiments in animals suggest that cellular immunity contributes to resolution of early lesion and resistance to infections. Cell-mediated immunity may also be implicated in the pathogenesis of late syphilis lesions.

Etiologic Agent

Characterization

Treponemes are delicate, corkscrew-shaped spirochetes of 6 to 14 μm in length, a width of 0.25 to 0.30 μm, a spiral wavelength of 1.0 to 1.5 μm, and a spiral depth of 0.5 to 1.0 μm.

T. pallidum subspecies *pallidum* is morphologically, immunologically, and chemically indistinguishable from *T. pallidum* subspecies *pertenue,* the causative agent of yaws. These subspecies are also homogeneous based on DNA hybridization. Other members of the genus *Treponema* are not known to cause human disease with the exception of *T. carateum* the causative agent of pinta. The nonpathogenic treponemes are free-living and widely distributed in nature.

Growth outside an animal host is limited. A tissue culture method using rabbit epithelial cells (Fieldsteel et al., 1982) has been the most successful in vitro method developed to date. Growth appears to be best in a mixture of 3 to 5% O_2 and 5% CO_2 in N_2 at 33 to 35°C. Studies indicate that *T. pallidum* metabolizes glucose and pyruvate aerobically and has catalase activity, enzyme superoxide dimutase, and a cytochrome system. Replication time in in vitro is approximately 30 h.

Specimen Collection and Transport

To collect specimens for direct microscopic examination from external lesions, one should clean the lesion with 0.9% saline, dry, and gently abrade to provoke exudate, and the exudate is collected directly on the microscope slide or transferred by bacteriologic loop to the slide. Specimens for the direct immunofluorescent test (DFA-TP) may be collected either on the microscope slide or in capillary tubes. Paraffin-embedded tissues also may be examined by DFA-TP. Transport of specimens for darkfield examination is impossible. Lesion material for DFA-TP collected in capillary tubes sealed with plastic closure or putty should be stored at 2 to 8°C until shipment, but may be sent without refrigerant. Smears prepared on microscope slides should be air-dried and mailed to the laboratory without fixation.

Specimens for serologic tests should be collected by venipuncture, without anticoagulant. The serum should be aseptically separated from the clot and mailed or taken to the laboratory in leakproof tubes. No refrigerant or antimicrobial agent is necessary if transport time is less than 2 days. If transport time is greater than 2 days, then gentamycin sulfate at a final concentration of 0.01% should be added to the serum.

At least 1 ml of spinal fluid should be collected by lumbar puncture into a clean dry sterile tube. A bacteriostatic agent such as thimersol may be added if testing is to be delayed.

Direct Identification Procedures

Definitive diagnosis of primary, secondary, and congenital syphilis is based on visualization of *T. pallidum* on direct microscopic examination. The darkfield examination must be performed immediately after specimens are collected, since identification is based on characteristic motility of *T. pallidum*. The DFA-TP has certain advantages over darkfield microscopy: (1) motile organism are not required, (2) pathogenic treponemes can be differentiated from nonpathogenic treponemes in oral lesions, and (3) tissue sections from biopsy and autopsy material can be examined (Hunter et al., 1984). Only specimens collected from genital and anal lesions and lymph nodes are suitable for darkfield examination.

Direct microscopy is useful in establishing a diagnosis of reinfection, as well as to rule out syphilis as the cause of lesions associated with other sexually transmitted diseases.

When specimens are properly collected, the direct methods are at least 95% sensitive. However, negative darkfield or DFA-TP results do not exclude a diagnosis of syphilis, but may mean that organisms were not present in sufficient numbers in the specimen to be detected, antitreponemal drugs were applied locally or taken systemically, the lesion has approached natural resolution, or was a lesion of late syphilis.

Darkfield Microscopy (Creighton, 1980)

The microscope assembly consists of 10×, 40 to 45×, and 90 to 100× (oil immersion objectives), 10× oculars, a darkfield condensor, and a 6.0 to 6.5 v illuminator. Gingival scrapings containing *T. denticola* biotype microdentium in a drop of saline are used to verify microscope adjustment before specimen collection. In any specimen, organisms are located with the high-dry objective. The suspect organism is then examined with the oil immersion objective. A darkfield-positive slide has delicate, corkscrew-shaped organisms with rigid, uniform, tightly wound, deep spirals, and unique motility. The characteristic motility of *T. pallidum* is a deliberate forward and backward movement with rotation about the longitudinal axis. Rotation maybe accompanied by a soft bending, twisting of the organism or undulation from side to side. When attached to or obstructed by heavier particles the organism may contort, convolute, or bend and thereby distort the coils, but snap back to the organism's original form in a coil-like manner. Organisms that are easily confused with *T. pallidum* are *T. refringens* and *T. denticola*.

Direct Fluorescent Antibody Test for *T. Pallidum* (DFA-TP) (Yobs et al., 1964)

The fluorescent microscope assembly consists of a microscope equipped with ultraviolet and tungsten light source and proper filters. When using an HBO-200 mercury bulb, a BG-12 exciter filter or a KP490 filter may be used in combination with an OG-1 barrier filter (K510 or K530). A 45× high-dry objective is used for scanning the area and a 100×/1.25 oil objective for verification. (For FA with transmitted illumination the oil objective must be fitted with a funnel stop or equipped with a built-in iris diaphragm.) Reagents for the test include phosphate-buffered saline (PBS) pH 7.2 with 2% Tween 80, reagent grade acetone or reagent grade methanol, mounting media (1 part PBS 7.2 plus 9 parts glycerine), fluorescein isothiocyanate (FITC)-labeled anti-*T pallidum* globulin conjugate, and controls. The conjugate should have been absorbed with the nonpathogenic *T. phagedenis* biotype Reiter treponeme, or be a monoclonal antibody to *T. pallidum* and stain only *T. pallidum*. Lyophilized conjugate should be rehydrated with sterile distilled water and diluted with PBS with 2% Tween 80 according to the manufacturer's directions. Appropriate controls must be included in each test run. The positive control may be either smears made from the fluorescent treponemal antibody-absorption (FTA-ABS) test antigen or testicular impression smears made from *T. pallidum*-infected rabbits. The negative control may be smears made from washed nonpathogenic Reiter treponeme cultures, freshly obtained human mouth treponemes, or washed *T. denticola biotype* microdentium from culture.

To perform the test, the specimen sample is mixed to ensure even distribution and approximately 10 μl of test material is placed within each circle or 1-cm area of a clean, grease-free slide and allowed to air dry. If possible, four smears should be prepared for each specimen. Fix slides in acetone for 10 min or in 10% methanol for 20 min, or air dry and gently heat fix. Fix all individual patient slides separately. Cover each circle with approximately 30 μl of diluted conjugate. Place slides in a moist chamber at 35 to 37°C for 30 min. Rinse slides individually with PBS and cover with PBS for a total of 10 min. Follow with a final rinse in distilled water. Reactive controls and nonreactive controls may be rinsed in the same staining dish. Process all unknown smears separately to prevent reactive treponemes from washing onto nonreactive smears. Drain and blot slides. Place a small drop of mounting medium on each circle and apply a cover slip.

Locate treponemes on positive control using tungsten light and 45× objective. Focus scope and switch to 100× and ultraviolet (UV) light to verify 4+ staining intensity with positive control. Scan negative control slide and test slides with 45× objective and UV light switching to 100× when treponemes are located. If FITC-stained treponemes are observed, report: "Treponemes, immunologically specific for *T. pallidum,* were observed by direct immunofluorescence." If no treponemes were observed, report: "No treponemes were observed by direct immunofluorescence."

During fixation and rinsing of smears, treponemes may be washed off the slides, particularly from the control slides, and can adhere to other smears. Therefore, smears from patients must be handled individually to prevent *T. pallidum* organisms from adhering to nonreactive smears, resulting in a false-positive report. If staining dishes are used for controls, they must be cleaned after each use.

Preliminary Serologic Tests

The preliminary or nontreponemal tests for syphilis are used for screening for antibodies formed in response to initial infection or reinfection and to follow the efficacy of treatment. The nontreponemal antigens measure both IgG and IgM antilipid antibodies formed by the host in response to lipoidal material released from damaged host cells early in infection as well as to lipid from the treponeme itself. These antigens are biologically nonspecific and are basically modifications of the Venereal Disease Research Laboratory (VDRL) antigen that contains cardiolipin, cholesterol, and lecithin.

There are currently four nontreponemal tests that are considered standard tests for syphilis and include the microscopic (1) Venereal Disease Research Laboratory (VDRL) slide, and (2) unheated serum reagin (USR); and the macroscopic (3) rapid plasma reagin (RPR) 18-mm circle-card test, and (4) reagin screen test (RST). The VDRL antigen is the only one of the four antigens that has not been stabilized by the addition of the disodium salt of ethylenediamine tetracetic acid (EDTA) and that does not contain choline chloride to eliminate the need to heat inactivate the serum. The RPR card antigen contains sized charcoal particles to aid in the visualization of the antigen-antibody reaction (Portnoy et al., 1962), whereas the RST uses a lipid-soluble dye (March and Stiles, 1980). Although the RPR card test is the most widely used of the nontreponemal tests, the VDRL CSF slide test is still the only recognized procedure for the serodiagnosis of neurosyphilis.

Whereas the overall sensitivity of the nontreponemal tests is approximately 90%, in early primary syphilis up to 28% of patients will have nonreactive nontreponemal test results on initial visit. In addition, in approximately 30% of the cases of late untreated syphilis the patient will present with nonreactive nontreponemal tests. Specificity for the nontreponemal tests is 98%. However, the specificity of the test is greatly influenced by the population tested.

In the primary stage patients with suspect lesions and nonreactive nontreponemal tests should have repeat tests at 1 week, 1 month, and 3 months. Nonreactive tests during the 3-month period exclude the diagnosis of syphilis. The nontreponemal tests are reactive in secondary syphilis, almost without exception. Normally patients with secondary syphilis will have nontreponemal test titers of 1 : 16 or greater, regardless of the test method used. Less than 2% of sera from these patients will exhibit a prozone (Spangler et al., 1964). Nontreponemal test titers in early latent syphilis are similar to those in secondary syphilis. However, as the length of the latent stage increases the titer decreases. After treatment for syphilis, titers should be obtained at 3-month intervals for at least 1 year. After adequate treatment for primary and secondary syphilis, there should be at least a fourfold decline in titer by 3 months (Brown et al., 1985). For most patients, the titer continues to decline until little or no reaction is detected (Fiumara, 1980). Patients treated in the latent stage may show a more gradual decline in titer. Low titers will persist in approximately 50% of patients treated for latent syphilis even after 2 years (Fiumara, 1979). The same nontreponemal test method should be used for all quantitative testing because up to fourfold (2 dils) and occasionally greater difference in titers may be obtained with certain sera using various tests.

A new diagnosis for syphilis, may be established by a fourfold increase in the quantitative nontreponemal test titer.

In congenital syphilis the role of the nontreponemal tests is to monitor the antibody titer. A rising titer in monthly serial bleedings from an infant during a 6-month period is diagnostic for congenital syphilis. If the infant was not infected in utero, by 3 months passively transferred antibodies should no longer be detected in the nontreponemal tests.

All of the nontreponemal tests will occasionally give false-positive results. In general populations, the nonspecificity ranges between 1 and 2%. Acute false-positive reactions lasting less than 6 months usually occur after febrile diseases, immunizations, or during pregnancy. However, rates of false-positive reactions during pregnancy are no greater than seen in the general population, that is, 1 to 2%. Chronic false-positive reactions are more often associated with autoimmune diseases, such as arthritis and lupus or with chronic infections. False-positive rates may exceed 10% in populations where the prevalence of intravenous drug use is high. When false-positive results occur, the titers are usually less than 1 : 8. Just as not all low titers are false-positive, because titers are low in latent syphilis, not all high titers are true-positive. In populations where intravenous drug use is prevalent, approximately 12% of the false-positive titers are 1 : 8 or greater.

VDRL Slide Test (Serum) (Manual, 1969)

VDRL antigen and VDRL-buffered saline are available from a number of commercial sources and should be stored according to the manufacturer's directions. The antigen suspension must be prepared fresh daily. Repeated use may result in antigen deterioration; therefore antigen should be periodically checked with control sera during use. Reagents and glassware should be at room temperature. The antigen suspension is prepared by pipetting 0.4 ml buffered saline to the bottom of a 30-ml-round, glass-stoppered bottle. Next 0.5 ml of antigen is added drop by drop but within 6 s directly into the saline while the bottle is continuously but gently rotated on a flat surface. Continue rotation of bottle for another 10 s, then add 4.1 ml of buffered saline. Place top on bottle and shake from bottom to top approximately 30 times in 10 s. The antigen may be used during 1 (8-h) day. To deliver the antigen use an 18-gauge needle without bevel that delivers 60 ± 2 drops of antigen suspension/ml when syringe and needle are held vertically. The antigen suspension must be checked with control sera of graded reactivity that reproduce established results of reactive, weakly reactive, and nonreactive. For the qualitative and quantitative tests, the sera should be heated in a 56°C water bath for 30 min before testing. Sera and reagents should be at room temperature for testing. Correct room temperature is critical to the reliability and reproducibility of results. Tests should be performed within a temperature range of 23 to 29°C (73 to 85°F). To perform the test place 50 μl of heated serum into 1 ring of slide, then add 1 drop (1/60 ml) of antigen suspension. Slides are rotated for 4 min at 180 ± 2 rpm on a mechanical rotator that circumscribes a 3/4 inch diameter. Tests are read microscopically with 10× ocular and 10× objective immediately after rotation. Specimens exhibiting medium and large flocculation particles are reported as reactive (R), those with small particles are reported as weakly reactive (W), and those with complete dispersion of antigen particles or slight roughness are reported as nonreactive (N). A prozone reaction is occasionally encountered which may be so pronounced that only a weakly reactive or rough nonreactive result occurs. All sera producing such results should be quantitated before a report of the VDRL slide test is submitted.

For the quantitative test 0.9% saline is used as diluent. Twofold serial dilutions are made directly on the slide using 50 μl serum and 50 μl of saline, and the test is completed as described. Titers are reported as the last dilution producing a reactive, not weakly reactive, result.

VDRL Slide Test (Spinal Fluid)

Do not heat spinal fluids. Spinal fluids should be centrifuged before testing (Manual, 1969). The initial antigen suspension for the VDRL spinal fluid test is prepared as described for the VDRL slide test (serum). The antigen is then diluted 1:2 in 10% saline and mixed by gently rotating the bottle or inverting the tube. The diluted antigen should stand at least 5 min but no longer than 2 h before use.

To dispense the antigen, a 21- or 22-gauge needle without bevel that will deliver 100 ± 2 drops when needle and syringe are held vertically is used. For spinal fluid the antigen should be checked with control sera, available commercially. The slide for the VDRL spinal fluid test is 2 × 3 inch, with concavities measuring 16 mm in diameter and 1.75 mm in depth.

For testing, all reagents and spinal fluid should be at room temperature, 23 to 29°C (73 to 85°F). To perform the test, deliver 50 μl of spinal fluid to one concavity of the slide. Add 1 drop of antigen with a 21- or 22-gauge needle and rotate for 8 min at 180 rpm. Tests are read with a 10× ocular and 10× objective. Results are reported as reactive (definite flocculation) or nonreactive (smooth dispersion of antigen particles). Reactive spinal fluids should be quantitated by preparing twofold dilutions in 0.9% saline as described for the VDRL slide test (serum), but using the slide with concavities. Quantitative results are reported as the highest titer that produces a reactive result.

RPR 18-mm Circle Card Test (Manual, 1969)

The RPR card test is available in kit form from a number of manufacturers. Additional reagents for the quantitative procedure are 0.9% saline and a 1:50 dilution of serum nonreactive for syphilis prepared in 0.9% saline. This diluent is used for making dilutions 1:32 or greater.

Before beginning the actual test, the antigen should be transferred to the dispensing bottle according to the manufacturer's directions and the accuracy of the needle checked. The needle is considered to be satisfactory if 30 ± 1 drops are delivered from 0.5 ml. The needle need not be checked daily. The antigen must be tested daily with control sera of graded reactivities of reactive, minimally reactive, and nonreactive. Reagents and sera should be at room temperature, 23 to 29°C (73 to 8°F). To perform the test, place 50 μl of unheated serum into one of the circles on the test card, using a dispensing or automatic pipetting device. Spread the specimen to fill the entire circle and add a drop of antigen suspension to each test area containing serum. The card is rotated for 8 min at 100 ± 2 rpm under a humidity cover. The test reactions are read in the wet state under a high-intensity light immediately after the card is removed from the rotator. Results are reported as reactive or nonreactive. The reactive report includes specimens demonstrating slight but definite flocculation (minimally reactive [RM]) as well as marked and intense flocculation (reactive [R]). The nonreactive (N) report includes specimens that show no flocculation or only a slight "roughness." Specimens giving any degree of reactivity as well as rough nonreactives should be quantitated. To quantitate, 50 μl of 0.9% saline is used as the diluent for 50 μl of serum and twofold dilutions are prepared. The dilution is continued through the 1 : 16 dilution. Starting at the highest dilution spread the specimen within each circle. Add 1 drop of antigen to each circle, and rotate the card as described. Quantitative results are reported as the highest titer that produces a reactive (R,Rm) result. If the end point titer is not reached at the highest dilution (1 : 16), then a 1 : 16 dilution of test serum is prepared in 0.9% saline. Place 100 μl of the 1 : 16 dilution on the card and prepare twofold dilutions in the 1 : 50 dilution of the nonreactive serum.

Confirmatory Serologic Tests

In contrast to the nontreponemal tests, the treponemal tests are reserved for confirmatory testing when the clinical signs and/or history disagree with the reactive nontreponemal test results. The treponemal tests are based on the detection of antibodies formed specifically to the antigenic determinants of the treponemes. They are qualitative procedures and therefore cannot be used to monitor the efficacy of treatment. When treatment is begun early in primary syphilis, however, about 10% of these patients will be nonreactive within 2 years after treatment. Like the nontreponemal tests, the treponemal tests are almost always reactive in secondary and latent syphilis. For the majority of cases, once the treponemal tests are reactive they remain reactive for one's lifetime. In fact, in some patients with late syphilis, a reactive treponemal test may be the only means of confirming the suspected diagnosis.

In congenital syphilis, as in sexually acquired syphilis, the treponemal tests will be reactive. The reactivity in the treponemal tests may be either passively transferred, as is the reactivity of the nontreponemal tests, to the fetus during pregnancy or actually due to infection. Passively transferred treponemal antibodies should be catabolized and become undetectable in the noninfected infant by the age of 6 months.

The greatest value of the treponemal tests is to differentiate true-positive nontreponemal results from false-positive results. However, false-positive treponemal test results occur with about the same frequency (1%) as false-positive nontreponemal test results. Although some false-positive results in the treponemal tests are transient and their cause unknown, a definite association has been made between false-positive treponemal tests and the connective tissue diseases (Kraus et al., 1970). When unexplained reactive results occur in elderly patients, attempts should be made to rule out acquired or congenital syphilis or infections with other treponemes before a diagnosis of false-positive serologic test is made.

There are currently four treponemal tests considered as standard tests for syphilis: the fluorescent antibody absorption (FTA-ABS), the FTA-ABS double staining (DS), the microhemagglutination assay for antibodies to *T. pallidum* (MHA-TP), and the hemagglutination treponemal test (HATTS).

The overall sensitivity of the FTA-ABS test at approximately 98% is greater than that of the other three treponemal tests with sensitivities of approximately 95%. The major difference in sensitivities is found in the primary stage. The specificities of the four treponemal tests range from 98% with the FTA-ABS test to 99% with the other three treponemal tests (Larsen et al., 1984). False-positive results can occur in the treponemal test, but only rarely do false-positive results occur in both the nontreponemal and treponemal tests for the same patient.

FTA-ABS (Hunter et al., 1964) and FTA-ABS DS Tests (Hunter et al., 1979)

Both immunofluorescent tests are based on the same indirect antibody detection principle. The FTA-ABS DS test, which was designed specifically for microscopes with incident illumination, uses a fluorescein isothiocyanate (FITC)-labeled conjugate as the direct stain or counterstain component and a tetramethyl-rhodamine isothiocyanate-TMRITC-labeled antihuman immunoglobulin conjugate for the indirect component of the test. In contrast, the FTA-ABS uses a FITC-labeled antihuman immunoglobulin conjugate as its indirect component. The use of the counterstain in the FTA-ABS DS test eliminates the need for darkfield location of treponemes.

For a microscope with transmitted illumination, the lamp may be an HBO 200, Xenon CS, or HBO 150 plus a 6V 5A tungsten light. The oculars are 10×, the objectives 10×, and 40× dry (fluorite), the condenser is a darkfield D 1.20 to 1.40, and the filters are BG12 or KP490 with K510 or K530.

For a microscope with incident illumination the lamp may be an HBO 50, HBO 100, or HBO 200, the oculars 6× or 8×, and the objectives 40×/1.30 oil, or 63× oil, or 100×/1.25 oil. Filters for FITC are BG38, K480, 2KP490, TK510, and K525, and for TMRITC are BG38, BG36, KP560, K530, TK580, and K590.

Reagents may be purchased in kit form from a variety of commercial manufacturers and include *T. pallidum* subspecies *pallidum* antigen, sorbent, conjugates, and control sera. The FTA-ABS test FITC-labeled antihuman conjugate principally recognizes heavy-chain IgG, but is also capable of detecting IgA and IgM. Therefore, the conjugate should be properly titrated to reduce IgA and IgM activity. The

FTA-ABS DS test TMRITC-labeled antihuman conjugate should recognize heavy-chain IgG only. The FITC-labeled antitreponemal conjugate (counterstain) should stain the treponemes at a 3 to 4+ intensity. The control sera consist of a reactive 4+ control, a minimally reactive 1+ control which may be a dilution of the 4+ control, and a nonspecific serum control. Other reagents are phosphate-buffered saline (PBS), pH 7.2 ± 0.1, 2.0% Tween 80 (polysorbate 80), mounting medium, and acetone.

To prepare antigen slides, resuspend and mix treponemes until adequately dispersed. Completeness of dispersion should be determined by darkfield examination before slides are prepared. On thoroughly cleaned slides, make a very thin antigen smear within each circle using a standard 2-mm, 26-gauge platinum wire loop. Generally, 1/2 loop is sufficient antigen for 2 1-cm id. circles. Allow slides to air dry for at least 15 min, then fix smears in acetone for 10 min and again allow to thoroughly air dry. Slides fixed in this manner may be stored frozen indefinitely.

There are 7 controls for the FTA-ABS tests. Prepare the 2 reactive 4+ controls by adding 50 μl of the reactive control serum to a tube containing 200 μl of PBS and mixing well, and by adding 50 μl of the reactive control serum to a tube containing 200 μl of sorbent, again mixing well. Place 30 μl of each dilution on separate antigen circles. The reactive 4+ control should demonstrate 4+ staining with the 1:5 dilution in PBS and 3+ to 4+ staining with the 1:5 dilution in sorbent. The minimally reactive (1+) control is a dilution of a reactive serum in PBS to a point where the minimal degree of fluorescence that is reported as reactive is observed. This control is *not* further diluted in sorbent. Place 30 μl of the 1+ control on an antigen circle. This control is used as the reading standard for the day's test run. Prepare the two nonspecific serum controls from a serum specimen of a nonsyphilitic individual. The serum should give 2+ staining when diluted 1:5 in PBS and essentially no staining when diluted 1:5 in sorbent. Place 30 μl of each of the two 1:5 dilutions on separate antigen circles. Prepare the two reagent specificity controls by simply adding 30 μl of PBS to 1 antigen smear and 30 μl of sorbent to another antigen smear.

Heat the sera to be tested for 30 min at 56°C. Then, prepare a 1:5 dilution by adding 50 μl of the heated serum to 200 μl of sorbent and mixing at least eight times. Place 30 μl of the diluted serum on the antigen smear, and the slide, along with the control slides, and incubate within a moist chamber at 35 to 37°C for 30 min. After incubation rinse slides with running PBS for approximately 5 s, then place in a staining dish containing PBS. While in the PBS, agitate the slides by dipping in and out at least 30 times. Change the PBS after 5 min and again dip the slides in and out of the fresh PBS at least 30 times over a 5-min period. Next, rinse slides in running distilled water for approximately 5 s, and then gently blot to remove excess water droplets.

If the FTA-ABS DS test is being performed at this point place 30 μl TMRITC-labeled antihuman IgG globulin that has been diluted in PBS containing 2% Tween 80 to its working titer on each smear. The slides are incubated, rinsed, and blotted as above. Then place 30 μl of FITC-labeled antitreponemal globulin that has been diluted in PBS containing 2% Tween 80 to the working titer on each smear, and incubate the slides for 20 min as above and rinse and blot as described.

For the FTA-ABS test after the initial serum incubation, rinse, and blot slides, place 30 μl of FITC-labeled antihuman immunoglobulin properly diluted in PBS containing 2% Tween 80 on each antigen smear. The slides are incubated, rinsed, and blotted after the initial serum incubation.

In both FTA-ABS test techniques mount the slides by placing a small drop of mounting medium on each smear and apply a cover slip. The slides should be examined as soon as possible. However, if a delay in reading is necessary place slides in a darkened room and read within 4 h.

For the FTA-ABS test, the microscope is focused by adjusting the darkfield condenser height using tungsten light and the 10× objective, then switching to the parfocal 40× high-dry objective and the UV light source for reading. The tungsten light is also used to verify the presence of treponemes on nonreactive slides.

For the FTA-ABS DS test, treponemes are located and the microscope focused using the FITC filter system, then switched to the TMRITC filters to read specific fluorescence.

The reading of both tests is the same. The intensity of staining is recorded as ± or <1+ when there is visible staining but it is less than the 1+ reading standard, 1+ when the intensity is equivalent to the minimally reactive 1+ control, and 2+ to 4+ when the intensity is moderate to strong. If there is no staining or treponemes are vaguely visible but without distinct fluorescence the reading is recorded as nonreactive. Results are reported as:

Initial Test Reading	Repeat Test Reading	
Report		
2+ to 4+		Reactive
1+	<1+	Reactive
1+	= 1+	Reactive minimal[a]
<1+		Nonreactive
−		Nonreactive

[a] In the absence of historical or clinical evidence of treponemal infection, this test result should be considered equivocal. A second specimen should be submitted for serologic testing.

MHA-TP (Cox, 1969) and HATTS (Wentworth et al., 1978)

The microhemagglutination tests are virtually identical. The major difference between the two tests is in the erythrocyte carrier used for the *T. pallidum* subspecies *pallidum* antigen. In the HATTS, turkey cells are used as the sensitized and unsensitized cells, whereas sheep cells are used in the MHA-TP for this purpose. The HATTS and MHA-TP are available only in kit form. Contained in the kits are diluent, sensitized erythrocytes (antigen), unsensitized erythrocytes (control), and control sera. For the MHA-TP, rehydrate the lyophilized reagents and control sera with the rehydrating fluid supplied, and store according to the manufacturer's directions. The rehydrated reagents should stand 1 h before using. Dilute unheated test and control sera 1:20 by adding 0.02 ml of serum to 0.38 ml of absorbing diluent in 12 × 75 mm test tubes. Mix the diluted sera and incubate at room temperature (20 to 30°C) for 30 min. In a microtitration tray, prepare further twofold dilutions of the reactive control serum by placing 50 μl of the absorbed reactive control serum in the first well of the tray and serially transferring 25 μl to the 2nd through the 10th wells each of which contain 25 μl of absorbing diluent. In addition, place 25 μl of the 1:20 dilution of the reactive control serum in another well to be used as a control for the unsensitized erythrocytes. For the nonreactive control, place 25 μl of absorbed nonreactive control serum in each of 2 wells. For the day's testing prepare working dilutions of the reconstituted suspensions of both sensitized and unsensitized erythrocytes by adding 1 part of the erythrocyte suspension to 5.5 parts of absorbing diluent (1:6.5 dilution). The quantity of working dilution needed is 75 μl for each serum and control tested plus a slight excess. For reagent controls, place 25 μl of absorbing diluent in each of 2 wells. For testing of patient's serum, place 25 μl of the absorbed test serum (1:20 dilution) into each of 2 wells.

Sensitized erythrocytes (75 μl) placed in each of the 10 wells of the diluted reactive control serum, in 1 well of the reagent control absorbing diluent, in 1 well of the nonreactive control serum, and in 1 well of each of the test sera. Place 75 μl of unsensitized erythrocytes in the second 1:20 dilution of the reactive control, the other reagent control well, the other nonreactive control serum well, and the remaining test sera wells. The final dilution of the test serum is 1:80, as is the first dilution of the reactive control. The trays are gently shaken, stacked, covered with an empty tray, and incubated at room temperature for at least 4 h or overnight. Read the settling patterns of the red blood cells using a tray viewer to visualize the patterns from below. Readings are scored on a scale of – to 4+ hemagglutination.

For the HATTS, rehydrate the lyophilized reagents with the rehydrating water supplied, according to the manufacturer's directions, and store as recommended. The rehydrated reagents should stand 30 min before using. Prepare dilutions of test and control sera which have been heated to 56°C and held for 30 min by adding 20 μl of sera to 300 μl of test diluent pink in 12 × 75 mm test tubes to yield a 1:16 dilution. In the microtitration tray further prepare twofold dilutions of the reactive control serum, prepared as described for the MHA-TP, with test diluent pink used as the diluent. In addition place a second 25 μl aliquot of 1:16 dilution of the reactive control serum in another well to be used as a control for the unsensitized erythrocytes. For the nonreactive control, place 25 μl of the absorbed nonreactive control serum in each of 2 wells. For the day's testing, prepare working dilutions of the sensitized and unsensitized erythrocytes according to the manufacturer's instructions using test diluent and allow to stand 30 min at room temperature before use. The quantity of working dilution needed is 100 μl for each serum and control tested. For reagent controls add 25 μl of test diluent pink to each of 2 wells. For test sera, place 25 μl of the absorbed test serum (1:16 dilution) into each 2 wells. Place 100 μl of sensitized erythrocytes in each of the 10 wells of the diluted reactive control serum, in 1 well of the reagent control, in 1 well of the nonreactive control serum, and in 1 well of each of the test sera. Add 100 μl of unsensitized cells to each of the remaining wells. The final dilution of serum is 1:80. The trays are gently shaken, stacked, covered with an empty tray, and incubated for 1 h at room temperature in an area free from vibration and draft. The incubation may be extended to overnight under these conditions. Read the hemagglutination patterns using a tray viewer. Score readings on a scale of – to 4+.

Reading and Reporting MHA-TP and HATTS

Results are determined and reported according to the following criteria:

Degree of Hemagglutination	Reading	Report
Smooth mat of cells covering entire bottom of well, edges sometimes folded	4+	Reactive
Smooth mat of cells covering less area of well	3+	Reactive
Smooth mat of cells surrounded by red circles	2+	Reactive

Smooth mat of cells surrounded by smaller red circle with hemagglutination outside circle	1+	Reactive
Button of cells having small "hole" in center	±	Nonreactive
Infinite compact button in center of well with or without a very small "hole" in the center	−	Nonreactive

Note: Specimens giving ± hemagglutination should be repeated before results are reported. Specimens repeating at ± or − should be reported as nonreactive, whereas specimens repeating as 1+ should be reported as reactive.

Control sera should give the expected results for the day's testing to be considered valid. The reactive control serum's titer should not vary more than ±1 doubling dilution from the end point titer established by the manufacturer. Nonreactive results should be obtained with the nonreactive control serum and both reagent controls. Each test serum should be nonreactive with the unsensitized erythrocytes. For the MHA-TP, the sera showing nonspecific hemagglutination with the unsensitized erythrocytes may be retested by quantitation of the absorbed sera. Prepare dilutions as for the reactive control serum; however, do the titration in duplicate. To the first row add sensitized cells and add unsensitized erythrocytes to the second row. Report the results as reactive if the hemagglutination with sensitized erythrocytes is at least two doubling dilutions (four times) greater than with the unsensitized cells and the first dilution showing no hemagglutination with unsensitized cells has a 3+ or 4+ reaction with sensitized cells. If the test results do not meet these specifications, then either the results are reported as inconclusive or a second specimen is requested. Sera giving nonspecific hemagglutination with unsensitized erythrocytes in the HATTS are reported as inconclusive, and a second specimen is requested. No provision for quantitative comparisons of results between sensitized and unsensitized cells is made for the HATTS.

Treatment, Prevention, and Control

Treatment

Primary, secondary, and latent syphilis of less than 1 year's duration should be treated with benzathine penicillin G 2.4 million limits total intramuscularly (IM) at a single session (CDC, 1985). Nonpregnant patients who are allergic to penicillin should be treated with tetracycline HCE 500 mg by mouth four times a day for 15 days. Tetracycline appears to be effective, but patient compliance with this regimen may be difficult. For penicillin- and tetracycline-allergic patients, erythromycin 50 mg by mouth four times a day for 15 days is recommended if compliance and serologic follow-up can be assured. Syphilis of more than 1 year's duration, excluding neurosyphilis, should be treated with benzathine penicillin G 2.4 million U IM once a week for 3 successive weeks (7.2 million U total). Although antimicrobials may not affect the pathology of cardiovascular syphilis, the 7.2 million-U regimen is recommended for this presentation as well as late benign syphilis and latent syphilis of more than 1 year's duration or of indeterminate duration.

Within a few hours after the treatment of early syphilis or general paresis a Jarish-Herxheimer reaction may occur. Manifestations of fever, chills, malaise, arthralgia, myalgia, nausea, and headache usually subside within 24 h. Primary lesions may intensify and developing secondary lesions first appear at this time. The Jarish-Herxheimer reaction is not related to a particular type or dose of antimicrobial, but may be mistaken for drug allergy.

Neurosyphilis can be treated with aqueous crystalline penicillin G 12 to 24 million U intravenously (IV) per day (2 to 4 million U every 4 h) for 10 days, followed by benzathine penicillin G 2.4 million U once a week for three doses or aqueus procaine penicillin G 2.4 million U IM daily plus probenecid 500 mg by mouth four times daily both for 10 days, followed by benzathine penicillin G 2.4 million U IM once a week for 3 doses. For patients at all stages of pregnancy who are not allergic to penicillin, this drug should be used in dosage schedules appropriate for the stage of syphilis as advised for the treatment of nonpregnant persons. Pregnant patients who have documented allergy to penicillin can be treated with erythromycin regimens appropriate for the stage of syphilis if compliance and serologic follow-up can be assured. Erythromycin regimens have resulted in failures to arrest fetal disease even in mothers with adequate compliance and satisfactory serologic response, presumably because of erratic transplacental passage. Oral desensitization for penicillin-allergic patients with syphilis and subsequent successful treatment with penicillin regimens have been documented in pregnant women (Wendel, et al., 1985).

Prevention and Control

Primary prevention of syphilis relies on preventing transmission from the infected person to his or her sexual partner. The risk of infection by *T. pallidum* appears to be reduced by application following exposure to genital areas of 30% calomel ointment or sufactant-mepharsen preparations (Stone, et al. 1986).

Use of condoms during intercourse reduces substantially the risk of infection to the partner (Stone, et al. 1986) and some spermicidal gels may have treponemacidal activity.

In the past, control of syphilis has largely depended on secondary prevention. This consists of treatment of sexual partners identified by interview of the infected person during the partner's incubation period or early in the course of his or her disease. Prevention of the morbidity and mortality of congenital syphilis by prenatal treatment of the mother has been the highest goal of the syphilis control program. Infectious syphilis cases in the community are identified by serologic screening of sexually active persons, and evaluation of symptomatic persons with characteristic lesions and identification of sexual partners. For pregnant women prenatal serologic testing at the first visit, with rescreening during the third trimester in high prevalence areas, is vital to the control effort.

After treatment, serologic follow-up to document adequate response to treatment and to detect reinfection guarantees the efficacy of treatment. Particularly in treatment of pregnant women, serologic testing ensures that the infection is arrested.

Retreatment is indicated when clinical signs of symptoms of syphilis persist or recur, when there is a fourfold increase in the titer of a nontreponemal test, or an initially high-titer nontreponemal test fails to decrease fourfold within 3 months. Some patients may maintain stable low titers in nontreponemal tests or may have irreversible anatomic damage.

Cerebrospinal fluid examination is indicated in all persons exhibiting signs of neurosyphilis and is recommended in persons with syphilis of more than 1 year's duration. Cerebrospinal fluid examination before treatment is essential to treatment.

Early congenital syphilis (less than 1 year) should be treated with a regimen effective for neurosyphilis, regardless of the cerebrospinal fluid findings. Infected infants are often asymptomatic at birth and may even be seronegative if maternal infection occurred shortly before birth. Infants should be treated at birth if maternal treatment failed to result in an adequate serologic response or was unknown, if it did not include penicillin in adequate doses, of if adequate follow-up of the infant through the first 15 months of life cannot be ensured.

Recommended regimens for congenital syphilis are aqueous crystalline penicillin G 50,000 U/kg IM or IV daily in 2 divided doses for a minimum of 10 days or aqueus procaine penicillin G 50,000 U/kg IM daily for a minimum of 10 days. Benzathine penicillin 50,000 U/kg in a single dose is used by some consultants in asymptomatic infants whose mothers were adequately treated and are unlikely to be infected, when follow-up cannot be ensured. Penicillin therapy in congenital syphilis after infancy should not exceed the dose used in adult patients with syphilis of more than 1 year's duration.

In the laboratory and clinical setting, prevention of infection depends on gloving before touching infected material, including lesion secretions and blood from patients with primary, secondary, or early congenital syphilis. These precautions need only be applied for 24 h after start of effective therapy on the patient.

Literature Cited

Brown, S. T., A. Zaidi, S. A. Larsen, and G. H. Reynolds. 1985. Serological response to syphilis treatment: a new analysis of old data. J. Am. Med. Assoc. **2593:**1296–1299.

Bulkley, B. H. 1984. Cardiovascular syphilis, p. 334–338. *In* K. Holmes, P. A. Mardh, P. F. Sparling, P. Wiesner (ed.), Sexually transmitted diseases, McGraw-Hill, Inc., New York.

Centers for Disease Control. 1984. Syphilis–United States, 1983. Morbid. Mortal. Weekly Rep. **33:**433–441.

Centers for Disease Control. 1985. STD treatment guidelines. Morbid. Mortal. Weekly Rep. **34(suppl):**94S–99S.

Cox, P. M., L. C. Logan, and L. C. Norins. 1969. Automated, quantitative microhemagglutination assay for *Treponema pallidum* antibodies. Appl. Microbiol. **18:**485–489.

Creighton, E. T. 1980. Darkfield microscopy procedure for detection and identification of *Treponema pallidum*. Centers for Disease Control, Atlanta.

Engel, S., and W. Diezel. 1980. Persistent serum immune complexes in syphilis. Br. J. Vener. Dis. **56:**221–222.

Fieldsteel, A. H., D. L. Cos, and R. A. Moechli. 1982. Further studies on replication of virulent *Treponema pallidum* in tissue cultures of Sf1Ep cells. Infect. Immunol. **35:**449–455.

Fichtner R., S. O. Aral, J. H. Blount, A. A. Zaidi, G. H. Reynolds, and W. W. Darrow. 1983. Syphilis in the United States: 1967–1979. Sex. Transm. Dis. **10:**77–80.

Fiumara, N. J. 1979. Serologic responses to treatment of 128 patients with late latent syphilis. Sex. Transm. Dis. **6:**243–246.

Fuimara, N. J. 1980. Treatment of primary and secondary syphilis: serological response. J. Am. Med. Assoc. **243:**2500–2502.

Hunter, E. F., W. E. Deacon, and P. E. Meyer. 1964. An improved test for syphilis—the absorption procedure (FTA-ABS). Public Health Rep. **17:**410–412.

Hunter, E. F., P. W. Greer, B. L. Swisher, A. R. Simmons, C. E. Farshy, J. A. Crawford, and K. R. Sulzer. 1984. Immunofluorescent staining of *Treponema* in tissues fixed with formalin. Arch. Pathol. Lab. Med. **108:**878–880.

Hunter, E. F., R. M. McKinney, S. E. Maddison, and D. D. Cruce. 1979. Double-staining procedure for the fluorescent treponemal antibody-absorption (FTA-ABS) test. Br. J. Vener. Dis. **55:**105–108.

Kampmeier, R. H. 1984. Late benign syphilis, p. 338–351. *In* K. Holmes, P-A. Mardh, P. F. Sparling, and P. Wiesner (ed.), Sexually transmitted diseases. McGraw-Hill, Inc., New York.

Kinsie, P. M. 1962. Impact of prostitution on syphilis control, p. 149–152. Proceedings of the World Forum on

Syphilis and Other Treponematoses. Centers for Disease Control, Atlanta.

Kraus, S. J., J. R. Haserick, and M. A. Lantz. 1970. Fluorescent treponemal antibody absorption test reactions in lupus erythematosus. N. Engl. J. Med. **282**:1287–1290.

Larsen, S. A., E. F. Hunter, and B. E. McGrew. 1984. Syphilis, p. 1–42. *In* B. B. Wentworth, F. N. Judson (ed.), Laboratory methods for the diagnosis of sexually transmitted diseases. APHA, Washington, D.C.

Manual of Tests for Syphilis. 1969. National Communicable Disease Center. DHEW publication no. (PHS) 411. Atlanta.

March, R. W., and G. E. Stiles. 1980. The reagin screen test: a new reagin card test for syphilis. Sex. Transm. Dis. **7**:66–70.

Portnoy, J., J. H. Brewer, and A. Harris. 1962. Rapid plasma reagin card test for syphilis and other treponematoses. Public Health Rep. **77**:645–652.

Spangler, A. S., J. H. Jackson, N. J. Fiumara, and T. A. Wartin. 1964. Syphilis with a negative blood test reaction. J. Am. Med. Assoc. **189**:87–90.

Stone, K. M., D. A. Grimes, and L. S. Magder. 1986. Personal protection against sexually transmitted diseases. Am. J. Obstet. Gynecol. **155**:180–188.

Swartz, M. N. 1984. Neurosyphilis, p. 318–334. *In* K. Holmes, P-A Mardh, P. F. Sparling, and P. Wiesner (ed.), Sexually transmitted diseases. McGraw-Hill, Inc., New York.

Termini, B. A., and S. I. Music. 1972. The natural history of syphilis. South Med. J. **65**:241–245.

Wendel, G. D., B. J. Strark, R. B. Jamison, R. D. Molina, and T. J. Sullivan. 1985. Penicillin allergy and desensitization in serious infection during pregnancy. N. Engl. J. Med. **312**:1229–1232.

Wentworth, B. B., M. A. Thompson, C. R. Peter, R. E. Bawdon, and D. L. Wilson. 1978. Comparison of a hemagglutination treponemal test for syphilis (HATTS) with other serologic methods for the diagnosis of syphilis. Sex. Transm. Dis. **5**:103–111.

Yobs, A. R., L. Brown, and E. F. Hunter. 1964. Fluorescent antibody technique in early syphilis. Arch. Pathol. **77**:220–225.

CHAPTER 52

Tuberculosis

ROBERT C. GOOD

Disease: Tuberculosis.

Etiologic Agents: *Mycobacterium tuberculosis* phage types 1 through 8, *Mycobacterium bovis,* and *Mycobacterium africanum.*

Source: Strict pathogens that spread from person to person via droplet nuclei or from susceptible animal species and their products, principally milk, by inhalation or ingestion.

Clinical Manifestations: An insidious onset is common, with the first symptoms usually being low-grade fever, night sweats, and chronic weight loss. As the disease progresses, a cough develops with the production of sputum and hemoptysis.

Pathology: Primarily a pulmonary disease, but any organ may be attacked. Granuloma formation usually accompanied by giant cells, the presence of acid-fast bacilli, central necrosis, and eventually cavity formation in the lungs.

Laboratory Diagnosis: Definitive diagnosis of disease is dependent on the isolation and identification of the tubercle bacillus. Presumptive diagnosis is based on detection of acid-fast bacilli in tissue sections or sputum smears.

Epidemiology: Worldwide, especially in developing countries where there is crowding and general health care may not be optimal. Cases are usually sporadic, but may be epidemic from exposure to infectious aerosol.

Treatment: Isoniazid, rifampin, pyrazinamide, ethambutol, and streptomycin. Treatment regimens always include two or more drugs for up to 6 months or more.

Prevention and Control: Case finding followed rapidly by specific therapy to interrupt infectious spread. Infected individuals (tuberculin positive) who do not have symptoms of disease are given one drug (isoniazid or rifampin) as preventive therapy. In many countries, individuals, especially infants, are vaccinated with bacille Calmette–Guérin to prevent complications, such as meningitis in primary infection.

Description of Disease

Tuberculosis is an infectious disease affecting primarily the lungs, but any organ system may be affected. It is primarily a disease of humans, but may occur in lower animals, particularly in cattle. In the latter case, the etiologic agent is usually transmitted to humans through the consumption of raw milk; however, airborne spread from cattle may occur. Tuberculosis is of major public health significance on a worldwide basis. Diagnosis of disease is followed by months of treatment if the drugs are available. When treatment is not available, a long period of progressive decline leads to death. Although the disease has been romanticized by novelists and composers, many prominent members of these and other professions have succumbed (Leff and Leff, 1986). Tuberculosis ranks with malnutrition, malaria, and diarrheal disease as a leading cause of morbidity and mortality. It is estimated that, as we approach the year 2000, tuberculosis will produce the highest mortality of any bacterial disease (Cook, 1985).

Three *Mycobacterium* species produce the classic disease known as tuberculosis: *M. tuberculosis, M. bovis,* and *M. africanum*. They, along with *M. microti,* which produces progressive disease only in certain rodents, are collectively called the *M. tuberculosis* complex. Species in the complex cannot be differentiated in DNA homology studies (Baess, 1979) or by skin test reactivity to culture filtrate antigens. Some confusion exists as to whether *M. africanum* can be separated from *M. bovis* or *M. tuberculosis* with biochemical reactions, as investigators have reported both failure (David et al., 1978) and success (Collins and Yates, 1984). Whereas *M. tuberculosis* and *M. bovis* are found throughout the world, the distribution of *M. africanum* is limited almost exclusively to the central part of Africa. *Mycobacterium bovis* is primarily a pathogen of cattle. An extensive review of *M. bovis,* its occurrence, significance, characteristics of the disease produced, and spread was prepared by Collins and Grange (1983) and should be consulted for a more complete discussion. In countries where extensive tuberculin testing and slaughtering programs have been conducted, tuberculosis in cattle has decreased (Thoen et al., 1984) and *M. bovis* infections in the human population have been controlled.

Communicability and Transmissibility

Tuberculosis is a communicable disease that is primarily transmitted from one person to another by droplet nuclei. These respirable particles measure 1 to 5 μm and are formed by sneezing, coughing, singing, or other forceful expiratory activities. Many bacilli in the aerosol die because of drying or exposure to sunlight, but some viable bacilli remain suspended in air and may infect other individuals, especially in crowded or poorly ventilated areas. The tuberculin reaction will convert to positive 2 to 10 weeks after infection; however, disease may not occur for years, perhaps not at all. Factors associated with the development of disease are incompletely understood, but the ultimate response to infection is related to the natural or acquired resistance of the host, the characteristics of the infecting strain, the size of the inoculum, and the like.

Clinical Manifestations of Disease

The onset of tuberculosis is usually insidious. The first symptoms are nonspecific and often include fatigue, anorexia, weight loss, fever, and night sweats. As the disease progresses it is accompanied by chronic cough, sputum production, and, occasionally, hemoptysis. At this stage the person may have chest pain and will usually seek medical attention.

Pulmonary Tuberculosis

Primary tuberculosis is initiated when one or more tubercle bacilli deposited in an alveolus are ingested by macrophages and segmented neutrophils. Without being inactivated, the bacilli are transported via the lymphatic system; thus, the hilar lymph nodes are infected early. Eventually the bacilli spill over into the blood-stream and spread throughout the body. The primary lesion in one of the lower lobes of the lung will then be replaced with a secondary lesion, usually in the posterior apical area of an upper lobe. Bacilli disseminated to other organs are taken up largely by mononuclear phagocytes. Some of the bacilli that reseed into the lungs will multiply to produce a progressive lesion. This sequence of events, which is only briefly described here, led Youmans (1979) to conclude that "primary pulmonary tuberculosis, or primary tuberculosis, regardless of the portal of entry of the tubercle bacillus, is actually a disease of the reticuloendothelial system." As immunity develops over several days or weeks, multiplication of the bacilli slows and the lesions in the lungs and other organs begin to resolve. This resolution can continue until signs of the disease disappear completely or the only remnant is the *Ghon complex,* a small calcified nodule in the lung, and enlarged hilar lymph nodes. Stead and Dutt (1983) report a 5% risk of developing disease in the first year after infection that rapidly declines as the period since infection increases.

Reinfection tuberculosis, whether the inoculum is endogenous or exogenous, results in a localized, circumscribed lesion that becomes necrotic. Goodwin and Des Prez (1983) suggest that localization of lesions in the posterior apical sites of the upper lobes is due to the deficiency of lymph production in that area and the lack of lung motion needed to make lymph flow. When lesions erode into bronchi, loss of necrotic material leaves an open cavity. Erosion into a blood vessel results in blood becoming mixed with the necrotic exudate (caseum) and producing hemoptysis.

Extrapulmonary Tuberculosis

As described previously, tubercle bacilli first establish a primary lesion in the lung and then spread through the lymphatic vessels to the bloodstream. Disease may develop at any of the sites where the organisms lodge. Tuberculosis in nonpulmonary sites has been seen for centuries, but its relationship to the pulmonary disease was not always recognized. Therefore, the spinal disease was called *Pott's disease,* disease of the skin was called *lupus vulgaris,* and disease of the cervical lymph nodes was called *scrofula.* The extrapulmonary form of the disease is not as much of a public health concern, as bacilli are not spread from the extrapulmonary lesions by aerosol.

The number of cases of pulmonary tuberculosis reported in the United States declined at a steady rate until 1985; the number of extrapulmonary cases remained fairly constant, however, so that they now represent a greater percentage of the total cases of tuberculosis. The Centers for Disease Control (1986c) reported that cases of pulmonary tuberculosis decreased 36.6% from 1975 (29,400 cases) through 1984 (18,642 cases), whereas extrapulmonary cases decreased only 21.3% from 4,589 to 3,613 during the same period. Of the 3,613 cases of extrapulmonary disease reported in 1984, 23% were pleural, 28% were lymphatic, 13% were genitourinary, 10% were bone and/or joint, 10% were miliary, 5% were meningeal, 4% were peritoneal, and 9% were at other sites.

Hypersensitivity

Koch reported in 1891 that if viable tubercle bacilli are inoculated into the skin of normal guinea pigs, a nodule appears after 10 to 14 days, then proceeds to open and drain until the animal dies. If the same dose of bacilli is inoculated in the same way into tuberculous guinea pigs, the site becomes indurated in 1 to 2 days with necrosis and eventual sloughing; however, the ulcer quickly heals. This reaction in tuberculous guinea pigs has been called the *Koch Phenomenon* and represents a delayed-type hypersensitivity (DTH) reaction, which could also be elicited by killed bacilli or culture filtrate. Tuberculin is culture filtrate of *M. tuberculosis* concentrated to about 50% of volume and preserved by adding glycerine. This reagent is known as *Koch's Old Tuberculin.* In 1934, Florence Seibert reported isolation of a purified protein derivative (PPD) of Old Tuberculin. The production medium was chemically defined, but the isolation procedure resulted in a mixture of protein antigens. This tuberculin, PPD has been used for many years to identify infected individuals. The reagent standardized in 1941 by Siebert and her colleague, J. T. Glenn, is known as PPD-S, and five units (0.1 μm of the protein) will produce a mean 12-mm DTH reaction in sensitized guinea pigs and a 16-mm (mean) reaction in humans with active tuberculosis. This reagent is the standard that all commercial preparations of tuberculin, PPD must match in reactivity.

The DTH reaction has been equated to cell-mediated immunity by some, but Youmans (1979) presented evidence that two separate reactions are involved. This separation is not surprising because of the complex antigenic structure of mycobacteria and the possible responses to the number of protein antigens encountered. Janicki et al. (1971) described the 11 major antigens in unheated culture filtrate of *M. tuberculosis* and even greater complexity has been demonstrated by fused-rocket immunoelectrophoresis (IEP) (Chaparas, 1981). As many of the antigens are shared among the various species of *Mycobacterium,* DTH reactions will be elicited with tuberculin, PPD in patients with other mycobacterioses; however, the test remains of diagnostic significance (see the review by Chaparas [1982]). Investigators recognize the need for a more specific skin test reagent (Daniel, 1980), and they are studying selected antigens produced in *Escherichia coli* mutants after insertion of DNA from *M. tuberculosis* (Cohen et al., 1987; Thole et al., 1987; Young et al., 1985).

The tuberculin test that elicits DTH reactions in infected persons is performed by either the multiple puncture test or the Mantoux test. In the former procedure, prongs of a multiple puncture device are coated with either tuberculin, PPD or Old Tuberculin, and the device is then pressed into the skin of the forearm so that the tines penetrate. The reaction size around the points of penetration is of little significance because the dose of antigen delivered is not standardized (Reichman, 1979). The preferred method, the Mantoux test, is the intracutaneous injection of a standard dose of antigen into the volar aspect of the forearm (Snider, 1982). The DTH reaction elicited is read 48 to 72 h later using a good light. The zone of erythema is disregarded, but the area of induration, identified by touch, is measured in millimeters (American Thoracic Society, 1981b; Reichman, 1979). If sizes of reactions in culture-positive patients are graphically illustrated, a symmetric curve results with a mean value at 16 mm. In tests of tuberculin in U.S. Navy recruits, reactions fell into a bimodal curve. After studying these reactions closely, the American Thoracic Society (1981b) established guidelines for interpretation. A person whose tuberculin reaction changes from less than 10 mm in diameter to 10 mm or more and increases by more than 6 mm within a 2-year period is considered newly infected. However, persons exposed to a tuberculous patient are considered positive if the reaction is 5 mm or greater. On the other hand, patients who live in geographic areas where infection with other *Mycobacterium* species is common should not be considered positive unless the reaction exceeds 15 mm.

The DTH response may vary in an uncontrolled way, even when the same product is administered by the same trained individuals. Classification of the patient as newly infected may change because of this variation (Chaparas et al., 1985). Vaccination with the attenuated bacille Calmette–Guérin (BCG) strain of *M. bovis,* the only currently accepted vaccine for tuberculosis, will convert the tuberculin reaction to positive, but often the reaction is less than 10 mm and tends to wane. In general the response to tuberculin will wane in all individuals unless the antigen is presented to maintain the response. After waning, an enhancing or boosting of the reaction may be seen, particularly in serial tuberculin testing (Thompson et al., 1979). Boosting may be seen if the repeat test is made as early as 1 week after the primary test and may be demonstrated for a year or longer. This reaction can cause concern in populations such as hospital employees who are tested at yearly intervals, as a new employee can be boosted by the preemployment test and appear as a new converter on the following test. Therefore, the American Thoracic Society recommends that, in routine periodic testing of adults, a two-stage test be used to prevent interpreting the boosted reaction as conversion (American Thoracic Society, 1981a,b). That is, if the reaction to the first test is less than 10 mm induration and a second test 7 days later is also less than 10 mm, the reaction is interpreted as negative; however, if the second test is greater than 10 mm with at least a 6-mm increase from the first test, the response is interpreted as boosted.

Hypersensitivity to tuberculin may be suppressed by some viral infections, such as measles. The degree of suppression among primary-school children vaccinated with BCG before entry into school was found to be greatest within 20 days of onset of rubella, and gradual restoration of response occurred 30 days after onset (Mori and Shiozawa, 1985). Suppression was not noted in children with subclinical disease or in those who had previously had the disease and were again exposed. Other conditions, such as irradiation, drugs, or disease conditions that lead to immunosuppression, may also suppress the response to tuberculin. This finding is particularly true in patients with acquired immunodeficiency syndrome (AIDS). Anergy, a complete failure to respond to antigen, may be a late manifestation in tuberculosis possibly as a result of antigen excess.

Results of the tuberculin test are not diagnostic of tuberculosis, but they help to indicate a possible disease condition (Sbarbaro, 1985). In view of the variability described for this test, clinical evidence should be used to decide whether or not to treat. Definitive diagnosis of tuberculosis will depend on the culture results.

Pathology

The tissue response to infection with tubercle bacilli is primarily the body's attempt to eradicate the organisms. When the bacilli are first deposited in an alveolus of the naive individual who has not encountered them previously, they are ingested by alveolar macrophages. There is little or no inhibition of growth until the development of hypersensitivity and cellular immunity. Lymphocytes encountering the antigen are activated and release lymphokines that attract and activate macrophages. The activated macrophages produce high concentrations of enzymes that act on the mycobacterial cell wall. The soluble antigens released are responsible for a DTH reaction and the associated tissue necrosis. A tubercle is formed by the organization of local tissue cells, including macrophages and lymphocytes. Caseum, an acellular material with a cheesy consistency, accumulates as a result of the necrotic reaction. As the caseum liquefies in the lungs, it is expelled to leave a rough-walled cavity. The caseous material contains few tubercle bacilli, but many bacilli are found in liquefied caseum (detailed descriptions of the histology and the relationship of the tubercle bacillus to specific pathologic lesions can be found in the text by Canetti [1955]). In rapidly progressive disease, areas of consolidation occur around cavities as a result of a DTH reaction to the soluble antigens. Therefore, fulminating disease can occur in addition to the circumscribed lesion with clearly defined limits.

Lesions may develop in any organ as a result of hematogenous spread of tubercle bacilli. Miliary disease results from the development of lesions in several organs at one time.

Many pathologic changes that occur in tuberculosis are a result of the development of DTH and cell-mediated immunity. Reactions leading to consolidation and necrosis result from DTH, whereas organization into tubercles and localized foci is due to the development of cell-mediated immunity that promotes destruction of the bacilli and their products.

Diagnostic Characteristics

Tuberculosis should be suspected in patients with chronic cough, weight loss, persistent fevers, or night sweats (American Thoracic Society, 1981a). In some instances patients may have an acute febrile illness with chills or a generalized influenzalike illness. In other instances nonspecific symptoms associated with a fever of unknown origin may be the only evidence of disease. This presentation is more difficult to resolve and usually requires repeated chest roentgenograms and cultures of biopsy specimens from a variety of sites, such as lung, pleura, pericardium, liver, peritoneum, and bone marrow. A positive tuberculin reaction, general malaise, and anorexia are part of the total diagnostic complex. Enlarged mediastinal lymph nodes and the appearance of a pulmonary infiltrate, with or without cavity formation, add to the clinical diagnostic features. Characteristics of roentgenograms are usually typical and are pathognomonic for tuberculosis (Farman and Speir, 1986). Tuberculosis should be considered in such population groups as immigrants from geographic areas with a high prevalence of disease, homeless persons, contacts of active tuberculosis cases, and persons residing in areas with a high incidence of disease.

Isolation and identification of *M. tuberculosis* are required for the definitive diagnosis of tuberculosis; however, detection of acid-fast bacilli in smears of sputum is highly suggestive.

Hematogenous tuberculosis is termed *miliary tuberculosis,* as the small nodular lesions in the organs are the size of a millet seed. This form of the disease can occur in patients of any age and often presents as an acute febrile illness accompanied by dyspnea and cyanosis (the acute respiratory distress syndrome), although it can also occur as a more slowly progressive, chronic disease with systemic symptoms. In either expression of the disease, a miliary pattern may be recognized by the diffuse, finely nodular, evenly distributed infiltrate visible on the chest roentgenogram. Meningitis may be a complication, and enlarged liver, spleen, and lymph nodes may be present.

Signs and symptoms of tuberculosis in extrapulmonary organs may be dramatic or cryptic. Clinical manifestations vary with the site of the lesions (Schlossberg, 1983).

Prognosis

Of the noteworthy diseases, morbidity from tuberculosis was ranked sixth in the United States in 1983 after gonorrhea, chickenpox, syphilis, hepatitis, and salmonellosis; however, tuberculosis was the leading cause of death (Comstock, 1986). Case and death rates for tuberculosis declined at a steady rate during the first half of this century; the death rate in particular declined at a more rapid rate after adequate chemotherapy became available in the early 1950s.

In 1983, 23,846 new cases (10.2 per 100,000 population) were reported. The number of reported deaths with tuberculosis was 1,779 (death rate of 0.76 per 100,000); 60% of the patients were 65 years or older and 1.5% were less than 25 years old (Centers for Disease Control, 1986c). The death rate in men of all races is approximately 2.3 times that in women, and the death rate in nonwhites is more than 3 times the rate in whites. The highest death rate, 18.24 per 100,000 population, was in nonwhite men 65 years of age or older. These data suggest that deaths occurred most frequently in persons with reactivation of old infections, which occurred after immunity in the host had declined. However, if the disease is diagnosed early and treated with recommended therapeutic regimens, recovery can be expected.

Epidemiologic Aspects

Occurrence and Transmission

Youmans (1979) reviewed literature documenting epidemic "waves" of tuberculosis that have occurred in recorded history. The hypothesis that the decline in epidemic tuberculosis is due to selection of a more resistant population of human beings is rejected as the major reason for the steady decline in tuberculosis in the last 100 years.

The World Health Organization estimates that 8 to 10 million new cases of tuberculosis occur each year and that 2 to 3 million deaths annually result from the disease. The greatest morbidity and mortality occur in developing countries, but the disease has not yet been eradicated in developed countries. Certainly, countries with "low" rates of disease have certain areas or population groups with extremely high rates. These groups can be native populations, such as on the North American continent, or they can be in previously infected immigrant groups such as those in many European cities (Hershfield, 1979).

In the United States, *M. tuberculosis* is the most commonly isolated *Mycobacterium* species, and *M. bovis* is rarely isolated (Good, 1980; Good and Snider, 1982); *M. tuberculosis* and *M. bovis* are strict pathogens that are not found free in the environment. However, the bacilli will remain viable for a limited time in droplet nuclei generated by coughing, and the like. Sputum will protect the bacilli from the adverse effects of drying and sunlight.

Mycobacterium bovis has been virtually eliminated from cattle, particularly dairy herds, in the United States by a program of tuberculin testing with slaughter of all positive reactors. Raw milk was the primary mode of transmission of this species, and required pasteurization was a major step in preventing transmission. The times and temperatures selected for the pasteurization process are based on laboratory results with *M. tuberculosis* and *M. bovis*. These two major control steps have virtually eliminated disease due to *M. bovis,* but such dramatic measures to control the spread of *M. tuberculosis* have not been discovered.

Comstock (1986) examined tuberculosis as an example of chronic disease epidemiology and evaluated the factors that affect the pools of uninfected and infected persons and the development of disease. Disease may develop either within 1 to 2 months after exposure to tubercle bacilli or many years later

(Glassroth et al., 1980). For the first 2 years, the newly infected individual has a risk of approximately 4% per year of developing disease. As the time after infection increases this risk lessens, but still exists.

Phage Types

The ability to phage type strains of *M. tuberculosis* has led to population studies (Jones et al., 1982; Jones and Woodley, 1983) and to delineation of specific relationships between strains involved in disease (Jones and Good, 1982). Strains can be divided into eight major types with four phages (one phage is a control that lyses all *M. tuberculosis* strains), and each type can be subdivided into 64 subtypes using six additional phages. Phage types of strains isolated during outbreak investigations in communities, factories, nursing homes, and similar institutions (Snider et al., 1984) are fascinating to follow with this new procedure. As this typing becomes more widely available, better epidemiologic studies will be possible. Gruft et al. (1984) have linked drug resistance to phage typing to improve strain identification further.

Host Factors Affecting Infection or Disease

Certain populations emerge as groups at higher risk of developing tuberculosis when data for all cases of the disease are reviewed (Centers for Disease Control, 1986c). Thus, those in the higher risk groups are more than 65 years old, male, and nonwhite. Cases continue to occur in the 0 to 4 year age group, indicating that the disease is still being transmitted. Those infected in this young age group who do not develop disease immediately are at risk for the rest of their lives and may infect others in years to come.

Whether certain populations are more resistant to tuberculosis because of innate characteristics has been addressed by Kushigemachi et al. (1984). Because of factors such as sex, age, and nutrition (body weight) of the host, and virulence and size of the inoculum of the infecting organism, the effect on a particular population could not be evaluated. However, others agree that nutrition and specific dietary substances play a role in susceptibility to disease (Davies, 1985; McMurray et al., 1985; Rook et al., 1986).

Three control procedures were proposed in 1980 (Glassroth et al., 1980): persons ill with tuberculosis must be identified and treated so that they are noninfectious; persons who are infected but not ill must be prevented from becoming infectious; and persons not yet infected must be prevented from becoming infected. Seven years later, the goals are still the same, but we also face a more challenging problem. The number of reported cases of tuberculosis has now leveled off and is actually increasing (Centers for Disease Control, 1987). This alteration in pattern could result from factors that affect host defenses.

Tuberculosis and the Acquired Immunodeficiency Syndrome

The current AIDS epidemic is changing the pattern of many diseases, and it may affect the epidemiologic associations of tuberculosis. Tuberculosis associated with AIDS was found to occur in Florida in a refugee group that was known to have high reactor rates to tuberculin as well as endemic AIDS (Pitchenik et al., 1984; Pitchenik and Rubinson, 1985). Other investigators also have found that tuberculosis is a complication of AIDS in populations associated with a high incidence and prevalence of tuberculosis and tuberculin hypersensitivity (Sunderam et al., 1986). AIDS seems to be partially responsible for the failure to maintain or accelerate the decline in the number of cases of tuberculosis (Centers for Disease Control, 1986a). Further control measures will be needed to correct this problem, but meanwhile all AIDS patients should be examined for tuberculosis and all tuberculosis patients should be tested for the presence of human immunodeficiency virus (HIV) antibody titer (Centers for Disease Control, 1986b). Preventive treatment of tuberculin reactors and treatment of individuals with tuberculosis is recommended in those with titers to HIV.

Etiologic Agents

The association of an infectious agent with tuberculosis has been known since 1868, when Villemin reported transmission of the disease to animals by direct inoculation. Koch's classic studies reported in 1882 clearly defined the etiologic agent of tuberculosis and established criteria for exacting bacteriologic technique and basic infectious disease studies. Koch showed that the tubercle bacillus could be stained using special techniques, and the bacilli were demonstrated in diseased tissue, could be grown on inspissated blood serum, could produce the disease in animals by inoculation with cultures transferred many times on artificial medium, and could transmit the disease from animal to animal. The unique characteristics of the tubercle bacillus have been studied extensively since these first significant experiments.

Precautions recommended by the U.S. Department of Health and Human Services (1984) include Biosafety Level 2 practices, containment equipment, and facilities for preparing acid-fast smears and for culturing sputa or other clinical specimens. Biosafety Level 3 practices, containment equipment, and facilities are recommended for activities involving growing, transferring, and working with cultures. The number of organisms needed to establish infection in at least 50% of exposed individuals is less than 10 viable units. Therefore, care should be taken in handling culture material.

Safety precautions are essential in the laboratory. The ideal mycobacteriology laboratory consists of an isolation room equipped with a biologic safety cabinet, an autoclave in an adjacent room, and work areas in space with adequate air balance and control (Kent and Kubica, 1985). These areas are separated from other laboratory functions. All procedures are planned to prevent production of aerosols. The safe environment will prevent inadvertent infection of laboratory personnel.

Levels of Laboratory Service

The American Thoracic Society and the Centers for Disease Control support the concept of Levels of Laboratory Service for handling and processing mycobacterial specimens in clinical laboratories. Some proficiency test programs are keyed to this concept and rate laboratory performance by level. Development of the levels and the history of the need for the program were reported by Kubica et al. (1975), and the need was reaffirmed in 1983 by the American Thoracic Society (1983a,b). The levels are now generally endorsed (Good et al., 1981; Kent and Kubica, 1985; Sommers and Good, 1985), and may be described briefly as follows.

Level I laboratories collect adequate clinical specimens with appropriate background and identifiers, prepare specimens for transport to a higher-level laboratory, and may prepare and examine smears; 10 to 15 specimens must be examined each week to maintain proficiency. If this number is not examined, specimens should be referred to a Level II laboratory.

Level II laboratories perform the same functions as Level I laboratories and, in addition, process specimens for culture, identify *M. tuberculosis,* may determine drug susceptibility of tubercle bacilli, and retain cultures for at least 6 months for additional or repeat tests. Proficiency in culture and identification of *M. tuberculosis* can be maintained by processing 20 specimens per week.

Level III laboratories are able to perform all activities of Level I and Level II laboratories, identify all *Mycobacterium* species from clinical specimens, may perform drug susceptibility tests, and may conduct research and provide training.

If the Levels of Service concept is properly applied, full expertise and proficiency in mycobacteriology will reside in only a few laboratories, the expense of supporting a full-scale mycobacteriology unit in all clinical laboratories will be avoided, and the work performed by proficient personnel will be of excellent quality.

Characterization

Tubercle bacilli are nonmotile, non-spore-forming, slightly curved or straight rods 0.2 to 0.6 μm wide by 1.0 to 10.0 μm long. The bacilli resist staining with basic dyes because of the high lipid content in the cell wall. The cells usually appear weakly positive on Gram staining. If special methods are used to stain the organisms, the stain is not easily removed, even with acidified organic solvents. Therefore, the mycobacteria are called *acid-fast bacilli.*

Growth occurs aerobically and is enhanced in primary culture by 5 to 10% additional carbon dioxide in the atmosphere. The optimal growth temperature is 35 to 37°C. Temperatures higher than 40°C will kill tubercle bacilli, and the bacilli do not grow well at temperatures below 34°C. Colonies have a characteristic buff color on egg-based media and are rough and corded. Tubercle bacilli grow in pellicles on liquid medium unless a surfactant such as polysorbate 80 is used.

Cultures of *M. tuberculosis* produce niacin and reduce nitrate, and their catalase activity is inactivated by heating the cells to 68°C for 20 min. Cultures of *M. bovis* do not produce niacin or reduce nitrate, and their catalase activity is heat sensitive. *Mycobacterium bovis* is inhibited by thiophen-2-carboxylic acid hydrazide (TCH), whereas *M. tuberculosis* is not.

Specimen Collection and Transport

Specimens examined for tubercle bacilli are usually pulmonary secretions, the most common being sputum. This specimen must be collected carefully; saliva and nasopharyngeal secretions are not representative, although they often may contain bacilli if the sputum is also positive. Sputum, the material brought up from the lungs by a deep cough, is the most adequate specimen. Specimens should be collected on at least 3 consecutive days and processed separately rather than pooled. A specimen of 5 to 10 ml collected in a 50-ml centrifuge tube is preferred. On occasion sputum must be induced by inhaling an aerosol of 5 to 10% saline. These specimens appear watery and similar to saliva; hence, they must be clearly identified as induced so that they will not be discarded as an inappropriate specimen.

Specimens should be sent to the processing laboratory as quickly as possible. If 1 to 2 days are required, cetylpyridinium chloride (CPC) should be added to the specimen (Smithwick et al., 1975). This agent will decontaminate the specimen during shipment, and even exposure for prolonged periods does not affect the recovery of either tubercle bacilli or other mycobacteria.

Young children and some adults cannot provide a sputum specimen. Gastric lavage is one way to obtain a specimen in these cases. The specimen should be taken in the early morning at least 8 h after eating and before the patient gets out of bed. Sterile water (20 to 50 ml) is instilled through a nasogastric tube that has been passed into the stomach. The specimen is withdrawn into a 50-ml sterile syringe and transferred to a sterile container. If the specimen cannot be processed promptly, it should be preserved by adding 100 mg of powdered sodium carbonate.

Bronchial washings, transtrachial aspirates, brushings, and transbronchial biopsies may also be used for the recovery of tubercle bacilli. Urine specimens should be collected on several separate days in the early morning and processed individually rather than as a pooled specimen. The entire voided amount, rather than just a midstream specimen, can be processed. External genitalia should be washed with soap and water before urine is collected. Urine specimens should be stored at 4°C until processed.

Tissue specimens must be collected aseptically and delivered directly to the laboratory for processing. Selected tissue specimens should be placed in sterile distilled water or neutral buffered saline for transport. If the specimens cannot be processed immediately, they should be frozen at −20°C and transported to the laboratory without thawing.

Other specimens, such as normally sterile body fluids, exudates, pleural fluids, and laryngeal swabs, should be collected aseptically in sterile containers and delivered to the laboratory immediately.

To meet the requirements of the Code of Federal Regulations, Title 42, Section 72.25, paragraph (c), subparagraph (1), all specimens for mailing must be placed in a watertight container (screw-cap tube), which is then placed in a second watertight container, usually of metal construction and containing enough nonparticulate absorbent material to take up all liquid in case the container breaks or leaks. One or more of these containers may then be packed into a container of paperboard, wood, or other material to a total volume of 50 ml.

Isolation and Identification Procedures

Details of isolation and identification procedures have been given in several recent publications (Good et al., 1981; Kubica and Good, 1981; Kent and Kubica, 1985; Sommers and Good, 1985; Strong and Kubica, 1981) and in the excellent reviews of differential reactions for identification of mycobacteria by Kubica (1973) and Wayne (1982). A probability matrix for identifying slowly growing mycobacteria has been developed using 23 characteristics (features) (Wayne et al., 1984). A computer program was developed for identification of species on the basis of an overall identity score obtained by matching characteristics of an isolate to those of known species with calculated feature frequencies. This approach to identification of *Mycobacterium* species in the clinical laboratory will increase accuracy and allow reports of additional tests needed for confirmation.

Specimen Digestion and Decontamination

Sputum is the specimen most often examined for detection of tubercle bacilli. This specimen tends to be very mucoid and must be treated with a mucolytic agent. It is highly contaminated and is routinely treated with a caustic agent to kill contaminating organisms. Unfortunately, treatment with the regular caustic agents often kills many tubercle bacilli as well as contaminants. Agents most commonly used are *N*-acetyl-L-cysteine (NALC)–sodium hydroxide, benzalkonium chloride–trisodium phosphate, sodium hydroxide alone, or oxalic acid. Only the NALC–NaOH procedure will be described here.

All glassware should be clean, preferably acid washed, to assure that acid-fast organisms are not carried over into materials used in preparing slides for microscopy.

The following solutions are mixed for the final correct concentrations of NALC–NaOH reagent: 25 ml of 4% aqueous solution of NaOH, 25 ml of 2.9% aqueous solution of sodium citrate (dihydrate). Immediately before use, add 0.25 g of NALC. Larger volumes of the solution can be prepared as long as the ratio of reagents remains the same. NaOH–sodium citrate solutions can be mixed ahead of time and held, but NALC must be added just before use because it will deteriorate on standing.

Add a volume of the complete NALC–NaOH reagent to an equal volume of sputum in a screw-cap, conical, 50-ml centrifuge tube. Mix on a test-tube mixer until the specimen is liquefied, but do not allow the solution to froth, which will inactivate the NALC. After 15 min at room temperature, fill tube to 50 ml with 0.067 M phosphate buffer, pH 6.8 (prepare buffer by dissolving 9.47 g of anhydrous Na_2HPO_4 in 1,000 ml of distilled water [Solution A]; dissolve 9.07 g of KH_2PO_4 in 1,000 ml of distilled water [Solution B]; mix 50 ml of Solution A with 50 ml of Solution B to prepare final solution; confirm pH

6.8 on a pH meter and adjust with appropriate buffer solution). Enclose tubes in sealed centrifuge cups and centrifuge at a relative centrifugal force of 3,000 × *g* for 15 min. The relative centrifugal force is very important to efficient recovery of tubercle bacilli (Rickman and Moyer, 1980) and must be at least 2,500 × *g*.

After centrifuging, carefully decant the supernatant fluid into a discard container containing a disinfectant solution. Remove some of the sediment with a sterile inoculating loop or applicator stick to prepare a 1- by 2-cm smear on a new glass slide. Resuspend the remaining sediment in sterile 0.2% solution of fraction V bovine albumin (dissolve 2 g of bovine albumin fraction V in 100 ml of 0.85% saline by stirring, adjust to pH 6.8 with 4% NaOH and filter sterilize; aseptically dilute 1 part of this stock solution with 9 parts of sterile saline for use).

Process other specimens in a way similar to that described for sputum. For example, laryngeal swabs should be placed in a sterile centrifuge tube with 2 ml of saline or sterile distilled water and 2 ml of NALC–NaOH. Other specimens should be treated in volumes of 2 to 5 ml; even feces can be decontaminated effectively by this procedure and mycobacteria isolated if placed on a selective medium.

If a gastric lavage specimen is mucoid, add 100 mg of NALC powder directly to liquefy, then centrifuge at 3,000 × *g* for 30 min. Discard the supernatant fluid, resuspend the sediment in 2 to 5 ml of sterile water or saline, and process as a sputum specimen. Urine is also precentrifuged so that the sediment can be handled conveniently.

Tissue specimens (approximately 1 g) are homogenized in 3 ml of sterile 0.2% bovine albumin with a glass or Teflon tissue homogenizer and inoculated to media directly or after dilution.

Cetylpyridinium chloride, a quaternary ammonium compound, may inhibit the growth of tubercle bacilli if it is not neutralized. Therefore, all specimens processed with this reagent should be inoculated onto an egg-based medium.

Staining and Preliminary Identification

Smears of sediments are stained with either the Ziehl–Neelsen or Kinyoun carbol fuchsin procedures for bright-field microscopy or the auramine O or Smithwick procedure for fluorescence microscopy (Kent and Kubica, 1985; Smithwick, 1976). Fluorescence microscopy is preferred, as slides can be screened more rapidly than with bright-field microscopy. The number of acid-fast bacilli should be estimated and reported as outlined previously (American Thoracic Society, 1981a; Kent and Kubica, 1985; Sommers and Good, 1985). Detection of acid-fast bacilli by microscopy is presumptive evidence of a mycobacteriosis, probably tuberculosis if the clinical signs are compatible. However, detection of acid-fast bacilli in some specimens, such as spinal fluid, is diagnostic for tuberculous meningitis.

Isolation Media

Inoculate 0.1 ml of the decontaminated specimen onto slants of Lowenstein–Jensen medium or American Trudeau Society (ATS) medium (both egg based) or plates of Middlebrook 7H-10 or Middlebrook 7H-11 medium (both agar based). The inoculum should be spread evenly over the surface of the medium. A vial of BACTEC 12B medium (Johnston Laboratories, Inc., Towson, Md.) made selective by the addition of polymyxin B, amphotericin B, nalidixic acid, trimethoprim, and azlocillin may also be inoculated with 0.5 ml of the decontaminated specimen (recommendations of the manufacturer should be followed for other medium formulations used in radiometric procedures). The BACTEC procedure is based on the ability of mycobacteria to metabolize a radiolabeled substrate to release $^{14}CO_2$ that is collected and measured in an ion chamber. The results, expressed as a growth index per 24-h period, are indicative of mycobacterial growth in the special medium. When the growth index is 100 or greater, part of the broth culture can be removed for an acid-fast stain. When the growth index is 300 or greater, additional tests such as drug susceptibility can be performed. Dubos liquid medium could also be inoculated, but contamination may be a problem unless the medium is made selective.

Specimens obtained from patients with AIDS (blood, bone marrow, feces, and the like) are usually different than those routinely processed in the mycobacteriology laboratory, and methods must be changed to process them effectively (Kiehn and Cammarata, 1986). Specimens may be processed with the Isolator lysis–centrifugation system (Du Pont Co., Wilmington, Del.) or originate in a stool specimen. They may be inoculated into BACTEC 12A broth, onto Lowenstein–Jensen medium slants, and onto 7H10 medium that have been made selective by the addition of antimicrobial agents. As more than one *Mycobacterium* species may be isolated from patients with AIDS, the mycobacteriologists must interpret results carefully.

Colonies of *M. tuberculosis* are rough, heaped up, and buff colored on egg-based media. On agar media they are less pigmented, usually flat (and may spread), and have a rough, dry appearance. The size of the colony varies, and some strains may produce colonies of 0.5 to 1 mm in 21 days, whereas others produce colonies that may be as large as 10 mm. The species is classed as nonphotochromogenic. Cording on agar media can be observed with 20× to 50×

magnification. Colonies of *M. bovis* on an egg-based medium are commonly smooth, pyramidal, and transparent or translucent. Growth can be detected in cultures of *M. tuberculosis* in about 14 days, whereas *M. bovis* cultures grow more slowly. Cultures of *M. bovis* will grow more luxuriantly in medium prepared with sodium pyruvate rather than with glycerol (Collins and Grange, 1983).

The amount of growth on primary culture should be quantitated and recorded using the following plan: if no colonies observed, negative; less than 50 colonies, record actual number; 50 to 100 colonies, 1+; 100 to 200 colonies, 2+; 200 to 500 colonies, 3+ (almost confluent); more than 500 colonies, 4+ (confluent).

Definitive Identification and Biotyping

Standard Differential Tests

The definitive identification of *M. tuberculosis* can be accomplished with comparatively few tests (Kent and Kubica, 1985). These tests include failure of the strains to grow on Lowenstein–Jensen medium within 7 days, production of niacin, reduction of nitrate, and failure of the catalase to withstand 68°C for 20 min. The biochemical tests must be performed as described or aberrant results may be obtained.

Niacin Test. The niacin test is performed using the originally described method or with commercially available paper strips. With either procedure include a positive control (*M. tuberculosis*), a negative control (*M. bovis* or an *M. avium* complex), and a reagent control without organisms. The solutions required for the test are a 4% solution of colorless aniline in 95% ethanol, a 10% solution of cyanogen bromide (a toxic lacrimator that should only be handled in a hood exhausting to the outside) in distilled water, and sterile 0.85% saline. Both the aniline and cyanogen bromide solutions should be stored at 4°C in brown bottles for short periods. Add 1 ml of sterile saline to a well-grown culture on egg-based medium and tilt the slant to make sure the entire surface is contacted. After extraction for 15 min, remove and transfer 0.5 ml to a 16- by 125-mm screw-cap tube and add 0.5 ml of the aniline solution and 0.5 ml of the cyanogen bromide solution. If niacin was extracted from the culture, an immediate yellow color should develop. The color can often be seen best in good light against a white background. Discard opened tubes in containers of disinfectant made alkaline with NaOH to prevent release of toxic fumes of hydrocyanic acid on autoclaving.

Niacin strips are available from Difco Laboratories (Detroit, Mich.) and Remel (Lenexa, Kans.), and manufacturers' instructions should be followed. In brief, extract the niacin from a well-grown culture as described previously and add 0.6 ml of the water extract to a sterile tube that can be sealed (see manufacturers' recommendations for correct size), then insert the strip according to the instructions, ensuring that the middle of the strip does not become wet. After 15 to 20 min with occasional light shaking, observe the liquid in the tubes for development of a yellow color.

Nitrate Reduction. The test for the presence of nitrate reductase can be performed by using solutions, dry crystalline chemicals, or a paper strip. The Virtanen procedure uses four solutions: 1) Substrate: 0.01 M sodium nitrate in 0.022 M phosphate buffer at pH 7.0; add 0.085 g of $NaNO_3$, 0.117 g of KH_2PO_4, and 0.485 g of $Na_2HPO_4 \cdot 12H_2O$ to 100 ml of distilled water and dissolve. Sterilize by autoclaving. 2) Add 50 ml of concentrated HCl to 50 ml of distilled water very carefully. If water is added to the acid, splattering could result in personal injury. 3) Add 0.2 g of sulfanilamide to 100 ml of distilled water. 4) Dissolve 0.1 g of *N*-(1-naphthyl)-ethylenediamine dihydrochloride in 100 ml of distilled water.

To perform the test, emulsify a heavy loop of growth from a 3- to 4-week-old culture in 3 to 4 drops of sterile distilled water in a 16- by 125-mm screw-cap tube. Add 2 ml of nitrate reductase substrate and incubate in a 37°C water bath for 2 h. Add 1 drop of diluted HCl and mix gently. Then add 2 drops of sulfanilamide reagent and 2 drops of *N*-(1-naphthyl)-ethylenediamine dihydrochloride reagent. Look for the solution to become red immediately as a positive sign of nitrate reduction. Compare with a reagent control and a positive control prepared with *M. tuberculosis* or *M. kansasii*. As this test detects the presence of nitrite, a pinch of zinc dust must be added to all negative tubes (i.e., those without color) to see if the red develops. If color does not develop after zinc dust is added, then the reaction has proceeded beyond nitrite and should be recorded as positive. A color standard should be prepared (Kent and Kubica, 1985) to assure that the red reaction is positive (i.e., equivalent to at least the 3+ level).

To perform the test with crystalline reagents (Warren et al., 1983), mix 1 part sulfanilic acid, 1 part *N*-(1-naphthyl)-ethylenediamine dihydrochloride, and 10 parts 1-tartaric acid with a mortar and pestle and store in a dark bottle at room temperature. Prepare a suspension of bacilli in substrate as described previously and incubate for 2 h. Add a small amount of the crystalline mixture to the suspension and interpret the color developed.

Often a combined niacin–nitrate test is more convenient, uses the same reagent formulas already given, and gives more consistent and intense colors in both tests. To perform the test, add 2.5 ml of the nitrate reductase substrate to a well-grown, 4-week-old culture on an egg-based medium, suspend some of the growth in the medium, and incubate upright in a 37°C water bath for 2 h. Transfer 0.6 ml of the substrate to a clean tube and test for niacin. Perform the nitrate test on the fluid remaining in the slant, but ensure that reagents do not run over the surface of the slant. Interpret the results of both tests using the same criteria described previously.

Heat-Stable Catalase. To test for inactivation of the constitutive catalase by heat, suspend several loops or spadesful of growth from a well-grown culture on egg-based medium in 0.5 ml of 0.067 M phosphate buffer, pH 7. Buffer component solutions should be prepared as described under Specimen Digestion and Decontamination and mixed as 61.1 ml of solution A to 38.9 ml of solution B and checked to assure a pH of 7.0. Incubate the suspension in a water bath at 68°C for 20 min, then cool to room temperature. Add 0.5 ml of a reagent solution containing equal portions of 30% hydrogen peroxide (Superoxol; Merck & Co., Inc., Rahway, N.J.) and a 10% aqueous solution of polysorbate 80. The appearance of bubbles within 20 min indicates a positive test. A known culture of *M. tuberculosis* should be included as the negative control and a culture of *M. gordonae* as the positive control.

Mycobacterium bovis and the attenuated BCG strain of *M. bovis* do not produce niacin, do not reduce nitrates to nitrites, and do not have a heat-stable catalase. These reactions are similar to those of many other species. However, strains of *M. bovis* are susceptible to low concentrations of thiophen-2-carboxylic acid hydrazide (TCH), whereas *M. tuberculosis* and other nonphotochromogens are not. Incorporate a filter-sterilized solution of TCH (Aldrich Chemical Co., Inc., Milwaukee, Wis.) into liquefied Middlebrook 7H-10 agar medium to a concentration of 1 to 2 μg/ml and pour into plates as for drug susceptibility tests. Prepare comparable plates of control medium without drug. Inoculate with 0.1 ml of 1 : 1,000 and 1 : 100,000 dilutions of a 7-day-old culture in liquid medium. The inoculum can either be spread evenly over the agar surface with a glass rod or placed on the agar in 3 individual drops from a capillary pipette. Enclose plates in plastic bags to prevent evaporation and incubate for 21 days. Compare growth on drug-containing and drug-free plates. The culture is considered resistant if growth on the drug plate is 1% of that on the control plate.

Strains of *M. tuberculosis* complex are inhibited by *p*-nitro-α-acetylamino-β-hydroxy-propiophenone (NAP), which is an intermediate in the synthesis of chloramphenicol. When this compound is added to a vial of BACTEC 12B medium, members of the *M. tuberculosis* complex can be differentiated from other *Mycobacterium species*. The test should be performed according to the manufacturer's directions (Johnston Laboratories, Inc., Cockeysville, Md.) using as inoculum a vial that has reached a daily growth index output of at least 100. Susceptibility to NAP can also be determined in 7H-9 broth, Dubos broth, or other liquid media and observed for growth after 7 days.

Other systems of identifying tubercle bacilli have been proposed, and the most exact is one that recognizes five variants (Collins et al., 1982, 1985). This division is made on the basis of whether or not the isolate is aerophilic or microaerophilic, susceptible to TCH and pyrazinamide (PZA), and reduces nitrate. The reactions vary from *M. tuberculosis*, which is aerophilic, resistant to TCH, susceptible to PZA, and reduces nitrate to *M. bovis*, which is microaerophilic, susceptible to TCH, resistant to PZA, and does not reduce nitrate. The total identification procedure first calls for the inoculation of three tubes of media to measure growth at 25 and 37°C, pigment production and whether it is influenced by light, and inhibition on Lowenstein–Jensen medium containing 500 mg of *p*-nitrobenzoic acid per liter. Further reactions are then set up to confirm the identification of tubercle bacilli.

Genetic Probes

Mycobacterial DNA is no longer the mystery it once was. One of the immediate benefits of this molecular technology was the development of a DNA probe that is compatible with the ribosomal RNA of *Mycobacterium* species. As the cell contains 1,000 to 10,000 times more RNA than DNA molecules, the test is more sensitive when targeted to RNA. Sensitivity and specificity of the probes against pure cultures were presented at the 1986 and 1987 annual meetings of the American Society for Microbiology. Four probes are currently available from the manufacturer (GenProbe, Inc., San Diego, Calif.), one to identify members of the genus *Mycobacterium*, and one each for *M. tuberculosis* complex, *M. avium*, and *M. intracellulare*. The specificity and sensitivity of these tests to identify cultures are reported to be greater than 95%.

Every laboratory must continue to perform standard procedures while testing new techniques to be able to interpret the results of the new test properly. Even the exact technologies with probes require comparison to ensure proper performance, including sensitivity and specificity, in each laboratory.

Tests for Mycobacterial Products

Mycobacteria produce tuberculostearic acid in quantities that can be diagnostic if found in certain tissues or body fluids. Mårdh et al. (1983) suggested that detection of the acid in spinal fluid by gas chromatography–mass spectrometry could be diagnostic of tuberculous meningitis. Brooks et al. (1987) described a frequency-pulsed electron capture gas–liquid chromatographic method to detect femtomole quantities of this acid in cerebrospinal fluid and serum and correlated the results with culture reports. Selected negative ion monitoring gas chromatography–mass spectrometry has now been proposed as more sensitive than the previously used positive ion monitoring for the detection of tuberculostearic acid in sputum (Larsson et al., 1987). Tuberculostearic acid is produced by all *Mycobacterium* species except *M. gordonae*. Detection of the acid in cerebrospinal fluid is diagnostic of *M. tuberculosis* infection, as only on extremely rare occasions is another *Mycobacterium* species found as the cause of meningitis. The same cannot be said for pulmonary disease; therefore, finding tuberculostearic acid in sputum is diagnostic of a mycobacteriosis, but not necessarily tuberculosis.

Other studies have indicated that fatty acid profiles and pyrolysis/gas–liquid chromatography of cultures can be used to identify species. Recently, high-performance liquid chromatography has been shown to be a potential tool for identification of mycobacteria (Butler et al., 1986).

Immunologic Probes

Because *Mycobacterium* species are so antigenically complex, immunologic probes for identifying species have been difficult to develop. Many of the antigens are cross-reactive (see section on Hypersensitivity); however, enzyme-linked immunosorbent assay procedures have been developed using several rather crude antigens, so the assays suffer from a lack of specificity and predictability.

Kadival et al. (1987) developed a biotin-avidin radioimmunoassay for detecting antigens of *M. tuberculosis* in cerebrospinal fluid. The assay depends on sandwiching mycobacterial antigens from sonicates of cells, either in culture or cerebrospinal fluid, between two antibodies. The assay is sensitive to 20 ng of mycobacterial protein per ml, but only specimens of cerebrospinal fluid with 80 ng/ml or more were considered positive.

Drug Susceptibility Tests

Details for the methods to determine the susceptibility of tubercle bacilli to antituberculosis drugs are given in recent publications (Good et al., 1981; Kent and Kubica, 1985; Sommers and Good, 1985; Strong and Kubica, 1981). Procedures have also been developed for the BACTEC, but only streptomycin, isoniazid, rifampin, and ethambutol can be tested by this radiometric procedure. As the drug concentrations for use in the BACTEC have not been firmly established, the manufacturer must be consulted for specific details of the test. At the 1984 Annual Meeting of the American Society for Microbiology, a report was given showing that laboratories using the BACTEC procedure measured true susceptibility and resistance as reliably as laboratories using conventional procedures.

A new method of measuring the drug susceptibility of tubercle bacilli is unique because the test involves cultured human macrophages and measures drug activity on bacilli inside the cell (Crowle et al., 1984, 1986). This new method probably gives a more accurate prediction of therapeutic response than other in vitro tests and could be used even with the patient's own macrophages to measure the combined effects of drugs and cell defense mechanisms. This method may be used in research studies, but is not applicable to routine drug susceptibility tests.

Treatment, Prevention, and Control

Treatment

Chemotherapy of tuberculosis requires using multiple drugs for an extended period to prevent the emergence of drug-resistant bacilli and to eliminate viable persisters in tissues.

Treatment recommended in the United States (American Thoracic Society, 1986) is a 6-month regimen that includes isoniazid (300 mg/kg maximal daily dose [MDD]), rifampin (600 mg/kg MDD), and pyrazinamide (20 to 30 mg/kg, but not to exceed 2 g MDD) given for 2 months followed by a maintenance dose of isoniazid and rifampin for 4 months. If primary resistance to isoniazid is a possibility, ethambutol is included in the regimen during the first 2 months. Therapy should be changed if drug suceptibility test results indicate resistance to any of the drugs.

Preventive treatment is recommended for many infected (tuberculin-positive) individuals without signs of disease and especially for contacts of persons with infectious (sputum-positive) tuberculosis (American Thoracic Society, 1986). The treatment consists of the MDD of 300 mg of isoniazid for 6 to 12 months. During the course of this treatment, individuals must be monitored for adverse effects and those 35 years old or older should be monitored periodically for abnormal hepatic enzyme function. Rifampin is given if the case contact has disease resistant to isoniazid.

Prevention and Control

Vaccination

A completely effective vaccine against tuberculosis is not available. The regularly used vaccine, the BCG strain of *M. bovis*, is attenuated so that it has little potential to produce progressive disease, yet complications are occasionally reported (Lotte et al., 1984). The BCG vaccine is effective in protecting nonhuman primates from tuberculosis, although this protection is not always complete (Good, 1984; Good and McCarroll, 1978). Studies in nonhuman primates show excellent humoral and hypersensitivity responses to BCG vaccination and virulent *M. tuberculosis* infection (Janicki et al., 1973). Immune mechanisms in human recipients of the vaccine are similar (Daniel, 1980; Edwards and Kirkpatrick, 1986; Grange, 1984; Ivanyi, 1986), but studies with the different routes of vaccine administration, particularly inhalation and intravenous routes, have not been possible. Because of variable results obtained with the vaccine, a major trial was set up in southern India. In this trial BCG was not effective; subsequently, experience with vaccination practice has been used to justify programs to vaccinate children. Ten Dam (1984) has reviewed the study in southern India and summarized conclusions of the World Health Organization study group, which recommended that BCG vaccination, particularly of young children, should be continued but that the effectiveness of the practice should be evaluated without delay. New antigenic substances, such as those used by Oftung et al. (1987), are being produced by the application of molecular technology. These products have the best probability of being specific as vaccines and diagnostic reagents.

Monoclonal antibodies made by one group (Anderson et al., 1986) illustrate potential problems. Of five different monoclonal antibodies prepared against *M. tuberculosis,* one reacted only with *M. tuberculosis* and BCG, two reacted with all mycobacteria examined, and two were more limited in their reactivity patterns but were still cross-reactive. Because epitopes found in *Mycobacterium* species are so similar, additional work is under way in several laboratories to identify the sequence of amino acids in the active peptide and to synthesize these active peptides for further study.

Select Populations

The incidence of tuberculosis in certain populations in the United States is very high, as discussed previously. This finding is particularly true for homeless individuals in large cities. The magnitude of the problem is not well known, because these populations are difficult to screen and to track. Tuberculosis control officers are now evaluating the problem, designing treatment schemes to assure compliance, and applying specific procedures for case finding. Others at high risk of developing tuberculosis include persons who are foreign born, in a racial minority group, elderly, institutionalized, HIV infected, or work in a health care setting.

Literature Cited

American Thoracic Society. 1981a. Diagnostic standards and classification of tuberculosis and other mycobacterial diseases (14th edition). Am. Rev. Respir. Dis. **123:**343–358.

American Thoracic Society. 1981b. The tuberculin skin test. Am. Rev. Respir. Dis. **124:**356–363.

American Thoracic Society. 1983a. Levels of laboratory services for mycobacterial diseases. Am. Rev. Respir. Dis. **128:**1.

American Thoracic Society. 1983b. The levels of service concept in mycobacteriology. Am. Thorac. Soc. News, Summer 1983, pp 19–25.

American Thoracic Society. 1986. Treatment of tuberculosis and tuberculosis infection in adults and children. Am. Rev. Respir. Dis. **134:**355–363.

Anderson, A. B., Z.-L. Yuan, K. Haslov, B. Vergmann, and J. Bennedsen. 1986. Interspecies reactivity of five monoclonal antibodies to *Mycobacterium tuberculosis* as examined by immunoblotting and enzyme-linked immunosorbent assay. J. Clin. Microbiol. **23:**446–451.

Baess, I. 1979. Deoxyribonucleic acid relatedness among species of slowly-growing mycobacteria. Acta Pathol. Microbiol. Scand. (B) **87:**221–226.

Brooks, J. B., M. I. Daneshvar, D. M. Fast, and R. C. Good. 1987. Selective procedures for detecting femtomole quantities of tuberculostearic acid in serum and cerebrospinal fluid by frequency-pulsed electron capture gas–liquid chromatography. J. Clin. Microbiol. **25:**1201–1206.

Butler, W. R., D. G. Ahearn, and J. O. Kilburn. 1986. High-performance liquid chromatography of mycolic acids as a tool in the identification of *Corynebacterium, Nocardia, Rhodococcus,* and *Mycobacterium* species. J. Clin. Microbiol. **23:**182–185.

Canetti, G. 1955. The tubercle bacillus in the pulmonary lesion of man: histobacteriology and its bearing on the therapy of pulmonary tuberculosis. Springer-Verlag, New York.

Centers for Disease Control. 1986a. Tuberculosis and the acquired immunodeficiency syndrome—Florida. Morbid. Mortal. Weekly Rep. **35:**587–590.

Centers for Disease Control. 1986b. Diagnosis and management of mycobacterial infection and disease in persons with human T-lymphotropic virus type III/lymphadenopathy-associated virus infection. Morbid. Mortal. Weekly Rep. **35:**448–452.

Centers for Disease Control. 1986c. Tuberculosis in the United States 1981–1984, publication no. (CDC) 86-8322, Centers for Disease Control, U.S. Department of Health and Human Services, Atlanta.

Centers for Disease Control. 1987. Tuberculosis provisional data—United States, 1986. Morbid. Mortal. Weekly Rep. **36:**254–255.

Chaparas, S. D. 1981. Antigenic relationships among mycobacterial species studied by modified-rocket and crossed immunoelectrophoresis. Rev. Infect. Dis. **3:**934–943.

Chaparas, S. D. 1982. The immunology of mycobacterial infections, p. 139–197. *In* H. D. Isenberg (ed.), Critical reviews in microbiology. CRC Press, Boca Raton, Fla.

Chaparas, S. D., H. M. Vandiviere, I. Melvin, G. Koch, and C. Becker. 1985. Tuberculin test: variability with the Mantoux procedure. Am. Rev. Respir. Dis. **132:**175–177.

Cohen, M. L., L. W. Mayer, H. S. Rumschlag, M. A. Yakrus, W. D. Jones, Jr., and R. C. Good. 1987. Expression of proteins of *Mycobacterium tuberculosis* in *Escherichia coli* and potential of recombinant genes and proteins for development of diagnostic reagents. J. Clin. Microbiol. **25:**1176–1180.

Collins, C. H., and J. M. Grange. 1983. The bovine tubercle bacillus. J. Appl. Microbiol. **55:**13–29.

Collins, C. H., J. M. Grange, and M. D. Yates. 1985. Organization and practice in tuberculosis bacteriology. Butterworth & Co. (Publishers) Ltd., London.

Collins, C. H., and M. D. Yates. 1984. *Mycobacterium africanum* and the 'African' tubercle bacilli. Med. Lab. Sci. **41:**410–413.

Collins, C. H., M. D. Yates, and J. M. Grange. 1982. Subdivision of *Mycobacterium tuberculosis* into five variants for epidemiological purposes: methods and nomenclature. J. Hyg. **89:**235–242.

Comstock, G. W. 1986. Tuberculosis—a bridge to chronic disease epidemiology. Am. J. Epidemiol. **124:**1–16.

Cook, G. C. 1985. Tuberculosis—certainly not a disease of the past! Q. J. Med. **56:**519–521.

Crowle, A. J., J. A. Sbarbaro, F. N. Judson, G. S. Douvas, and M. H. May. 1984. Inhibition by streptomycin of tubercle bacilli within cultured human macrophages. Am. Rev. Respir. Dis. **130:**839–844.

Crowle, A. J., J. A. Sbarbaro, and M. H. May. 1986. Inhibition by pyrazinamide of tubercle bacilli within cultured human macrophages. Am. Rev. Respir. Dis. **134:**1052–1055.

Daniel, T. M. 1980. The immunology of tuberculosis. Clin. Chest Med. **1:**189–201.

David, H. L., M.-T. Jahan, A. Jumin, J. Grandry, and E. H. Lehman. 1978. Numerical taxonomy analysis of *Mycobacterium africanum*. Int. J. Syst. Bacteriol. **28:**467–472.

Davies, P. D. O. 1985. A possible link between vitamin D deficiency and impaired host defence to *Mycobacterium tuberculosis*. Tubercle **66:**301–306.

Edwards, D., and C. H. Kirkpatrick. 1986. The immunology of mycobacterial disease. Am. Rev. Respir. Dis. **134:**1062–1071.

Farman, D. P., and W. A. Speir, Jr. 1986. Initial roentgenographic manifestations of bacteriologically proven *Mycobacterium tuberculosis*. Chest **89:**75–77.

Glassroth, J., A. G. Robins, and D. E. Snider, Jr. 1980. Tuberculosis in the 1980s. N. Engl. J. Med. **302:**1441–1450.

Good, R. C. 1980. Isolation of nontuberculosus mycobacteria in the United States. 1979. J. Infect. Dis. **142:**779–783.

Good, R. C. 1984. Diseases in nonhuman primates, p. 903–924. *In* G. P. Kubica and L. G. Wayne (ed.), The mycobacteria: a sourcebook. Marcel Dekker, Inc., New York.

Good, R. C., and N. E. McCarroll. 1978. BCG vaccination in rhesus monkeys: study of skin hypersensitivity and duration of protective immunity, p. 115–121. *In* R. J. Montali (ed.), Mycobacterial infections of zoo animals. Smithsonian Institution Press, Washington, D.C.

Good, R. C., V. Silcox, and J. O. Kilburn. 1981. Tuberculosis and other mycobacterioses, p. 675–703. *In* A. Balows and W. J. Hausler, Jr. (ed.), Bacterial, mycotic and parasitic infections, 6th ed. American Public Health Association, Washington, D.C.

Good, R. C., and D. E. Snider, Jr. 1982. Isolation of nontuberculous mycobacteria in the United States, 1980. J. Infect. Dis. **146:**829–833.

Goodwin, R. A., and R. M. Des Prez. 1983. Apical localization of pulmonary tuberculosis, chronic pulmonary histoplasmosis, and progressive massive fibrosis of the lung. Chest **83:**801.

Grange, J. M. 1984. The humoral immune response in tuberculosis: its nature, biological role and diagnostic usefulness. Adv. Tuberc. Res. **21:**1–78.

Gruft, H., R. Johnson, R. Claflin, and A. Loder. 1984. Phage-typing and drug-resistance patterns as tools in mycobacterial epidemiology. Am. Rev. Respir. Dis. **130:**96–97.

Hershfield, E. S. 1979. Tuberculosis in the world. Chest **76**(Suppl.)**:**805–811.

Ivanyi, J. 1986. Pathogenic and protective interactions in mycobacterial infections. Clin. Immunol. Allergy **6:**127–157.

Janicki, B. W., S. D. Chaparas, T. M. Daniel, G. P. Kubica, G. L. Wright, Jr., and G. S. Yee. 1971. A reference system for antigens of *Mycobacterium tuberculosis*. Am. Rev. Respir. Dis. **104:**602–604.

Janicki, B. W., R. C. Good, P. Minden, L. F. Affronti, and W. F. Hymes. 1973. Immune responses in rhesus monkeys after bacillus Calmette–Guérin vaccination and aerosol challenge with *Mycobacterium tuberculosis*. Am. Rev. Respir. Dis. **107:**359–366.

Jones, W. D., Jr., and R. C. Good. 1982. Hazel elephant redux. Am. Rev. Respir. Dis. **125:**270.

Jones, W. D., Jr., R. C. Good, N. J. Thompson, and G. D. Kelly. 1982. Bacteriophage types of *Mycobacterium tuberculosis* in the United States. Am. Rev. Respir. Dis. **125:**640–643.

Jones, W. D., Jr., and C. L. Woodley. 1983. Phage-type patterns of *Mycobacterium tuberculosis* from Southeast Asian immigrants. Am. Rev. Respir. Dis. **127:**348–349.

Kadival, G. V., A. M. Samukel, T. B. M. S. Mazarelo, and S. D. Chaparas. 1987. Radioimmunoassay for detecting *Mycobacterium tuberculosis* antigen in cerebrospinal fluids of patients with tuberculosis meningitis. J. Infect. Dis. **155:**608–611.

Kent, P. T., and G. P. Kubica. 1985. Public health mycobacteriology: a guide for the level III laboratory. U.S. Department of Health and Human Services, Centers for Disease Control, Atlanta.

Kiehn, T. E., and R. Cammarata. 1986. Laboratory diagnosis of mycobacterial infections in patients with acquired immunodeficiency syndrome. J. Clin. Microbiol. **24:**708–711.

Kubica, G. P. 1973. Differential identification of mycobacteria: VII. Key features for identification of clinically significant mycobacteria. Am. Rev. Respir. Dis. **107:**9–21.

Kubica, G. P., and R. C. Good. 1981. The genus *Mycobacteria* (except *M. leprae*), p. 1962–1984. *In* M. P. Starr, H. Stolp, H. G. Truper, A. Balows, and H. G. Schlegel (ed.), The prokaryotes: a handbook on habitats, isolation and identification of bacteria. Springer-Verlag KG, Berlin.

Kubica, G. P., W. M. Gross, J. E. Hawkins, H. M. Sommers, A. L. Vestal, and L. G. Wayne. 1975. Laboratory services for mycobacterial diseases. Am. Rev. Respir. Dis. **112:**773–787.

Kushigemachi, M., L. J. Schneiderman, and E. Barrett-Connor. 1984. Racial differences in susceptibility to tuberculosis: risk of disease after infection. J. Chronic Dis. **37:**853–862.

Larsson, L., G. Odham, G. Westerdahl, and B. Olsson. 1987. Diagnosis of pulmonary tuberculosis by selected-ion monitoring: improved analysis of tuberculostearate in sputum using negative-ion mass spectrometry. J. Clin. Microbiol. **25:**893–896.

Leff, A. R., and D. R. Leff. 1986. Tuberculosis: a reflection of political institutions and social concerns in the United States. Perspect. Biol. Med. **30:**27–39.

Lotte, A., O. Wasz-Hockert, N. Poisson, N. Dumitrescu,

M. Verron, and E. Couvet. 1984. BCG complications. Adv. Tuberc. Res. **21**:107–193.

Mårdh, P. A., L. Larsson, N. K. Hoiby, H. C. Engbaek, and G. Odham. 1983. Tuberculostearic acid as a diagnostic marker in tuberculous menintigis. Lancet **1**:367.

McMurray, D. N., M. A. Carlomagno, and C. L. Mintzer. 1985. Influence of the timing of nutritional insult on alterations of cell-mediated immunity in guinea pigs infected with *Mycobacterium bovis* BCG by the respiratory route. Nutr. Res. **5**:285–298.

Mori, T.. and K. Shiozawa. 1985. Suppression of tuberculin hypersensitivity caused by rubella infection. Am. Rev. Respir. Dis. **131**:886–888.

Oftung, F., A. A. Mustafa, R. Husson, R. A. Young, and T. Godal. 1987. Human T cell clones recognize two abundant *Mycobacterium tuberculosis* protein antigens expressed in *Escherichi coli*. J. Immunol. **138**:927–931.

Pitchenik, A. E., C. Cole, B. W. Russell, M. A. Fischl, T. J. Spira, and D. E. Snider. 1984. Tuberculosis, atypical mycobacteriosis, and the acquired immunodeficiency syndrome among Haitian and non-Haitian patients in south Florida. Ann. Intern. Med. **101**:641–645.

Pitchenik, A. E., and H. A. Rubinson. 1985. The radiographic appearance of tuberculosis in patients with the acquired immune deficiency syndrome (AIDS) and pre-AIDS. Am. Rev. Respir. Dis. **131**:393–396.

Reichman, L. B. 1979. Tuberculin skin testing: the state of the art. Chest **76**(Suppl.):764–770.

Rickman, T. W., and N. P. Moyer. 1980. Increased sensitivity of acid-fast smears. J. Clin. Microbiol. **11**:618–620.

Rook, G. A. W., J. Steele, L. Fraher, S. Barker, R. Karmali, J. O'Riordan, and J. Stanford. 1986. Vitamin D_3, gamma interferon, and control of proliferation of *Mycobacterium tuberculosis* by human monocytes. Immunology **57**:159–163.

Sbarbaro, J. A. 1985. Tuberculin test: a re-emphasis on clinical judgment. Am. Res. Respir. Dis. **132**:177–178.

Schlossberg, D. 1983. Tuberculosis. Praeger Publishers, New York.

Seibert, F. B. 1934. The isolation and properties of the purified protein derivative of tuberculin. Am. Rev. Tuberc. **30(Suppl)**:713–720.

Seibert, F. B., and J. T. Glenn. 1941. Tuberculin purified protein derivative: preparation and analyses of a large quantity for standard. Am. Rev. Tuberc. **44**:9–25.

Smithwick, R. W. 1976. Laboratory manual for acid-fast microscopy, 2nd ed. U.S. Department of Health, Education, and Welfare, Center for Disease Control, Atlanta.

Smithwick, R. W., C. B. Stratigos, and H. L. David. 1975. Use of cetylpyridinium chloride and sodium chloride for the decontamination of sputum specimens that are transported to the laboratory for the isolation of *Mycobacterium tuberculosis*. J. Clin. Microbiol. **1**: 411–413.

Snider, D. E., Jr. 1982. The tuberculin skin test. Am. Rev. Respir. Dis. **125**:108–118.

Snider, D. E., Jr., W. D. Jones, and R. C. Good. 1984. The usefulness of phage typing *Mycobacterium tuberculosis* isolates. Am. Rev. Respir. Dis. **130**:1095–1099.

Sommers, H. M., and R. C. Good. 1985. *Mycobacterium*, p. 216–248. *In* E. H. Lennette, A. Balows, W. J. Hausler, Jr., and H. J. Shadomy (ed.), Manual of clinical microbiology, 4th ed. American Society for Microbiology, Washington, D.C.

Stead, W. W., and A. K. Dutt. 1983. Epidemiology and host factors, p. 3–20. *In* D. Schlossberg (ed.), Tuberculosis. Praeger Publishers, New York.

Strong, B. E., and G. P. Kubica. 1981. Isolation and identification of *Mycobacterium tuberculosis:* a guide for the level II laboratory. HHS publication no. (CDC) 81-8390. U.S. Department of Health and Human Services, Centers for Disease Control, Atlanta.

Sunderam, G., R. J. McDonald, T. Maniatis, J. Oleske, R. Kapila, and L. B. Reichman. 1986. Tuberculosis as a manifestation of the acquired immunodeficiency syndrome (AIDS). J. Am. Med. Assoc. **256**:362–366.

ten Dam, H. G. 1984. Research on BCG vaccination. Adv. Tuberc. Res. **21**:79–106.

Thoen, C. O., A. G. Karlson, and E. M. Himes. 1984. *Mycobacterium tuberculosis* complex, p. 1209–1235. *In* G. P. Kubica and L. G. Wayne (ed.), The mycobacteria: a sourcebook. Marcel Dekker, Inc., New York.

Thole, J. R., W. J. Keulen, A. H. J. Kolk, D. G. Groothuis, L. G. Verwald, R. H. Tiesjema, and J. D. A. van Embden. 1987. Characterization, sequence determination, and immunogenicity of a 64-kilodalton protein of *Mycobacterium bovis* BCG expressed in *Escherichia coli* K-12. Infect. Immun. **55**:1466–1475.

Thompson, N. J., J. L. Glassroth, D. E. Snider, Jr., and L. S. Farer. 1979. The booster phenomenon in serial tuberculin testing. Am. Rev. Respir. Dis. **119**:587–597.

U.S. Department of Health and Human Services. 1984. Biosafety in microbiological and biomedical laboratories, 1st ed. HHS publication no. (CDC) 84-8395. U.S. Government Printing Office, Washington, D.C.

Warren, N. G., B. A. Body, and H. P. Dalton. 1983. An improved reagent for mycobacterial nitrate reductase tests. J. Clin. Microbiol. **18**:546–549.

Wayne, L. G. 1982. Microbiology of tubercle bacilli. Am. Rev. Respir. Dis. **125**:31–41.

Wayne, L. G., M. I. Krichevsky, D. Portyrata, and C. K. Jackson. 1984. Diagnostic probability matrix for identification of slowly growing mycobacteria in clinical laboratories. J. Clin. Microbiol. **20**:722–729.

Youmans, G. P. 1979. Tuberculosis. The W. B. Saunders Co., Philadelphia.

Young, R. A., B. R. Bloom, C. M. Grosskensky, J. Ivanyi, D. Thomas, and R. W. Davis. 1985. Dissection of *Mycobacterium tuberculosis* antigens using recombinant DNA. Proc. Natl. Acad. Sci. USA **82**:2583–2587.

CHAPTER 53

Tularemia

S. J. STEWART

Disease: Tularemia (rabbit fever, market men's disease, deerfly fever).
Etiologic Agent: *Francisella tularensis* types A and B.
Source: Wild and domestic mammals or their products, bloodsucking arthropods, or untreated water by direct contact, ingestion, or inhalation.
Clinical Manifestations: Flulike symptoms with many variations. Characteristic ulcer with regional lymphadenopathy.
Pathology: Reticuloendothelial disease with frequent involvement of liver, lungs, heart, spleen, and kidneys.
Laboratory Diagnosis: Bacterial isolation of etiologic agent. Tube agglutination test titer of 1 : 160 or a fourfold increase reaching at least 1 : 320.
Epidemiology: Almost exclusively in the temperate zone of the northern hemisphere. Sporadic cases or outbreaks from exposure to wild mammals or bloodsucking arthropods.
Treatment: Streptomycin, tetracycline, chloramphenicol, and aminoglycoside antibiotics.
Prevention and Control: Public education, protective clothing, and vector control.

Description of Disease

Tularemia ia an infectious disease produced by *Francisella tularensis*. The disease is primarily one of wild mammals, perpetuated in nature by ectoparasites, contaminated environment, cannibalism, and chronic carriers. It gives rise to a wide range of clinical manifestations in man depending on the route of the infection, the virulence of the strain, and the dose. The organism can be readily isolated from mud and water in some endemic areas. Natural infections have been reported in over 100 species of wild and domestic mammals, 25 species of birds, and 70 species of bloodsucking arthropods (Yow, 1981). Human infection is incidental, and human-to-human transmission is rare. The organism is highly infectious for man with as few as 10 viable organisms needed to produce infection (Saslaw et al., 1961). The organism can infect man through inhalation and ingestion, but the primary route is through the skin.

There is a wide range of virulence among isolates. The virulence designations types A and B (Jellison et al., 1961) have been widely accepted and are as follows: *F. tularensis* type A is naturally found only in North America. It produces a more serious disease in man than does type B and is usually associated with ticks, rabbits, or sheep. The LD_{50} for rabbits is $<10^1$ organisms. Unlike type B, it is able to ferment glycerol and produce citrulline ureidase. *F. tularensis* type B can be found throughout the temperate zone of the northern hemisphere (including North America). It produces a mild, often subclinical disease in man and is usually associated with water, muskrats, and beavers. The LD_{50} for type B in rabbits is $>10^7$ organisms.

Communicability and Transmissibility

There are few organisms that will infect man in as many ways as *F. tularensis*. Tularemia in the United States is most frequently acquired through the bite of an infected tick or by exposure to an infected wild rabbit—hence the early name "rabbit fever." Infections among trappers and hunters frequently implicate rabbits, beavers, and muskrats as their sources. Big game animals such as deer and black bears have been responsible for human disease, and game birds such as grouse and pheasants have been found naturally infected. Herdsmen have been infected by their stock; even dogs and cats have transmitted tularemia to their owners. Aerosols generated by shearing infected sheep and handling hay infested with infected rodents have produced local epidemics. Contaminated drinking water has been responsible for sporadic infections.

Clinical Manifestations

Most commonly described as a sudden onset of flu-like symptoms, tularemia is characterized by fever in >80% and chills, headache, and generalized aching in 50% of the reported cases. Other accompanying symptoms are listed with their respective clinical forms below.

Ulceroglandular

The ulceroglandular form of the disease constitutes about 80% of reported cases. The ulcer, considered to be the entry site of the organism through the skin, is one of induration and erythema with focal necrosis that increases with time. It is usually described as having a "punched-out" appearance at 1 to 3 weeks. Ulcers on the upper extremities are usually associated with mammalian vectors—e.g., hunters or trappers handling rabbits and beavers. Lesions located on the lower extremities, back, and head usually result from the bite of a bloodsucking arthropod (tick, deerfly, or mosquito). The regional lymph nodes become enlarged and tender and are frequently followed by necroses. Dissemination of the organism through the blood and its colonization of internal organs is a common complication.

Oculoglandular

Oculoglandular tularemia is fairly uncommon and usually the result of rubbing the eye with contaminated fingers. The conjunctiva is quickly and painfully inflamed with numerous sharply defined, yellowish nodules and pinpoint-size ulcers. Because of the pain and debilitating effect of this infection, the patient is usually presented to the physician before the regional lymphadenopathy is well pronounced (Nohinek and Marr, 1983).

Oropharyngeal

This is usually presented as a sore throat painful beyond what the clinician would expect by the degree of inflammation present, and a dry, nonproductive cough. A primary lesion may be found in the oropharyngeal area. The tonsils quickly become involved and if untreated will enlarge and suppurate. They become covered with a white-yellowish membrane (Ohara, H., 1934). Described as penicillin-resistant pseudomembranous tonsillitis (Fulginiti and Hoyle, 1966), oropharyngeal tularemia has often eluded diagnosticians. Within a few days after initial infection, the cervical lymph nodes become involved. Throughout its course, it mimics the symptoms of infectious mononucleosis (Luotonen et al., 1986). Surgical removal of the tonsils is often required if respiration is impaired.

Glandular

In about 10% of otherwise typical cases of ulceroglandular tularemia, the ulcer is not found or is disregarded, and these are termed "glandular." On occasion, patients with isolated regional lymphadenopathy have undergone lymph node biopsy for suspected malignancy (Evans et al., 1985). Surgical manipulations of otherwise untreated tularemic lymph nodes have initiated severe infections due to the release of large numbers of organisms from within the node (Saslaw et al., 1961).

Typhoidal

Typhoidal tularemia is presented as acute septicemia, the patients are seriously ill, and prostration and shock may develop. An external portal of entry is not evident, and lymphadenopathy is uncommon. Signs and symptoms include toxemia, severe headache, weight loss, and continuous high fever. The patient may be delirious or exhibit meningismus. Such infections are usually thought to result from inhalation or ingestion of the organisms. It can be a rapidly fulminating infection due in part to the possible large number of infecting organisms involved.

Specific antibiotic therapy must be started immediately if typhoidal tularemia is suspected. Mortality rates of 30 to 60% have been reported with this form of the infection (Olsen, 1975). Laboratory-acquired infections frequently take this course and have been traced to conditions compatible with infectious aerosols. Pleuropulmonary and gastrointestinal tularemia often are typhoidal but are associated with their own additional unique symptoms.

Pleuropulmonary

Pleuropulmonary tularemia sometimes results directly from inhalation but more frequently from systemic dissemination of the organisms and their colonization of the pleural cavity (Rubin, 1978). In conjunction with either ulceroglandular or typhoidal symptoms, it is often associated with one or more of the following sequelae: tightening of the chest, difficulty in breathing, rales, pericardial friction rub indicating heart involvement, and a roentgenographic picture of atypical pneumonia. In the past, pulmonary involvement was considered uncommon and indicated a poor prognosis. However, with improved X-ray techniques, it is now being found quite frequently as a complication of ulceroglandular tularemia (Rubin, 1978). Roentgenographic descriptions of atypical pneumonia have often been more misleading than helpful to the diagnostician. Confusion with myocardial infarction (Loftin, 1982), lymphoma (Yow, 1981), tuberculosis (Schmidt et al., 1983), legionellosis (Westerman and McDonald 1983), mycoplasmosis, Q fever and psittacosis (Cunha and Quintilianl, 1979) has been reported.

Gastrointestinal

Gastrointestinal tularemia is almost always traced to the ingestion of contaminated food or water. In addition to the typhoidal symptoms, it is usually accompanied by low-back or abdominal pain and persistent diarrhea. When fatal cases have been autopsied, extensive ulceration throughout the bowel has been found. Whether it is the result of a massive inoculum or dissemination from a single primary lesion has not been determined.

Pathology

Following penetration of the skin, the organism produces a local lesion which progresses into a centrally necrotizing ulcer. Early in the infection, usually before symptoms appear, the organism disseminates through the blood for a short period of time, often colonizing internal organs. It disseminates through the lymph to the regional nodes which become enlarged, tender, and often necrotic. Some nodes have reached 10 cm in diameter, but 2 to 4 cm is more typical. Examination of biopsy material usually reveals abscesses and caseous-type necrosis with infiltration of numerous polymorphonuclear leukocytes. In later stages the inflammation is more granulomatous, resembling that of tuberculosis or sarcoidosis (Luotonen et al., 1986).

The spleen, liver, lungs, heart, and kidneys are frequently involved, with varying degrees of severity. Splenomegaly is common, and splenic infarction has been seen by CT scan (Vento and Oung, 1984). Hepatomegaly sometimes occurs, and tularemic ulcers on the liver have been detected by ultrasound (Gourdeau et al., 1983). Lung involvement produces a confusing roentgenographic picture of atypical pneumonia. In rare instances, involvement of the central nervous system (Hutton and Everett, 1985) and skeletal muscles (Kaiser et al., 1985) have been reported.

Diagnostic Features

Tularemia should be considered whenever a patient experiences a rapid onset of flulike symptoms, toxemia, nonproductive cough, atypical pneumonia, or unexplained persistent diarrhea. The physician's strong index of suspicion is usually given credit for early presumptive diagnosis and treatment long before laboratory identification is completed. Skin involvement, common in Europe, is rare in the United States. Erythema nodosum and multiform or maculopapular skin rash, if present, may mimic Rocky Mountain spotted fever (Syrgala et al., 1984). Rhabdomylosis with myoglobinuria and impaired renal function can be complications of illnesses of long duration (Kaiser et al., 1985). The presence of fever, local lymphadenopathy, and an ulcer should be sufficient to initiate treatment, especially in endemic areas where the patient's history indicates possible exposure. Recent evidence of a tick bite or exposure to wild animals, especially rabbits, should be considered significant. Ingestion of infected wild meat or drinking untreated contaminated water has resulted in infection (Amoss and Sprunt, 1936; Jellison et al., 1950).

A commercially available fluorescent antibody test may be used to demonstrate the presence of the organism in pus, ulcer scrapings, and biopsy material. Conventional microscopy is of little value, because the gram-negative organism is difficult to discern from the tissue background, and polychromatic stains only show that it is very small, occurring singularly and in groups, both intra- and extracellularly.

Tularemia is most frequently diagnosed by serology. Tube agglutination utilizing commercially available antigen is the preferred method. A single titer of 1 : 160 or a fourfold increase from acute to convalescent serum reaching at least 1 : 320 is considered diagnostic. Cross-reactions are rare except for *Brucella*, where titers of 1 : 20 are commonly encountered. The *Brucella* titer, however, does not increase in the convalescent serum.

Prognosis

The prognosis for tularemia once treatment has begun is quite good. Morbidity is difficult to assess because of the myriad of disease types and the unknown percentage of cases that are undiagnosed and cured by serendipity. Prior to the widespread use of streptomycin, the mortality rate was 6.9% (Olsen, 1975). Fatalities are now estimated to be near 1% and are usually the result of typhoidal infections associated with pulmonary involvement, renal failure, massive inocula of type A organisms, delay in initiating specific antibiotic therapy, and unrelated complicating factors.

Epidemiologic Aspects

Occurrence

Tularemia occurs exclusively in the temperate zone of the northern hemisphere, except for a few poorly documented cases in South America (Olsen, 1975). It has been reported in every continental state of the United States with the midcentral states of Arkansas, Missouri, Oklahoma, and Texas reporting about half the nation's total cases (Cox and Everett, 1981).

It has been speculated that > 80% of the reported cases and almost all of the fatalities are from type A infections. Type B is known to have produced isolated epidemics and is suspected to be responsible for the small but significant percentage of people in endemic areas that exhibit positive serology with no compatible history of disease.

In local endemic areas, seasonality is usually sharply defined. Tick-borne tularemia occurs in the spring and summer, rabbit-borne in the fall and winter, and waterborne (usually attributed to beavers or muskrats) in the winter. The reported incidence in man in the United States has declined steadily from a few thousand cases per year in the 1930s to a few hundred per year in the 1980s.

Susceptibility

Man appears to be universally susceptible; age, sex, and race pose no barriers. Local epidemics frequently involve specific age or sex groups, but this selection can usually be shown to be a function of exposure. Recovery from infection is generally considered to result in lifelong immunity, because patients' titers remain elevated for many years, but reinfection, although rare, has been well documented (Burke, 1977).

Etiologic Agent

Characterization

F. tularensis is coccoid to ellipsoid in form and very small (0.2 μm × 0.2 to 0.7 μm). It is gram-negative and nonmotile and occurs singly. Capsules are rare or absent. Being a strict aerobe, it also requires an enriched medium for propagation. All isolates are microscopically and serologically identical, although there are differences in colonial appearance, growth rates, enzymatic activity, and virulence.

The organism is extremely infectious; tularemia has been a leading laboratory-acquired infection since its general recognition in the 1920s (Overholt et al., 1961). It is currently the third most commonly reported laboratory-associated bacterial infection (NIH/USPHS, 1984). It can remain viable for months at low temperatures in substrates as diverse as mud and decaying animal carcasses. It is, however, quite susceptible to common disinfectants such as soap, detergents, and chlorine. Biosafety level 2 is recommended for clinical laboratory work with suspected material, and biosafety level 3 for manipulations of cultures and animal studies (NIH/USPHS, 1984).

Isolation and Identification

During infection, direct isolation of *F. tularensis* from the ulcer, pharynx, conjunctiva, lymph node biopsy, and gastric washing is frequently achieved. Circulating blood seldom reveals the organism; urine and feces almost never do. Cystine glucose blood agar (Stewart, 1981) is the historical medium of choice. Good growth is also achieved with modified Mueller-Hinton broth (Baker et al., 1985), chocolate agar supplemented with IsoVitale X, Thayer-Martin agar (Clark et al., 1984), and charcoal yeast extract agar (Westerman and McDonald, 1983). A blue-gray growth mass will appear overnight from a large inoculum; individual colonies usually take from 3 to 5 days to appear. Overgrowth by contaminating organisms can be reduced by incorporating 100 to 500 units of penicillin per milliliter into the medium. Biochemical reactions are of little value except for determining the virulence type. Cultured organisms can be identified by simple agglutination utilizing commercially available antisera.

Typing

There are two laboratory methods for determining the virulence types A and B. Both glycerol fermentation and citrulline ureidase production correlate reasonably well with the LD_{50} in rabbits.

The glycerol fermentation medium is made by combining 1% glycerol with cystine heart agar and enough bromthymol blue to give a distinct green color. When inoculated, type A ferments the glycerol, producing acid, and turns the medium yellow. Type B, being unable to ferment glycerol, makes the medium more basic (blue).

Citrulline ureidase activity is determined by the enzymatic conversion of citrulline to ornithine. A 48-h culture is washed twice by centrifugation, resuspended in distilled water and lysed by sonication. A reaction mixture of the lysate containing 1 mg/ml of bacterial nitrogen, 53.8 μM DL-citrulline in phosphate buffer, pH 6.5, is incubated overnight at 30°C.

A 1- to 10-μl sample is applied to Whatman No. 1 paper and allowed to migrate at room temperature for 3 to 4 hours using n-butanol-pyridine-water (1 : 1 : 1) as the solvent. The paper is then dried at 65°C for 20 min, sprayed with 0.5% alkaline ninhydrin in 71% ethanol, and developed at 65°C for 22 min. Production of ornithine would indicate a type A organism, because type B organisms do not possess the enzyme. The reaction can be quantitated by elution of the ornithine spot with 71% ethanol and measuring the optical density at 575 μm. Commonly found values range from 5 to 25 μM of ornithine per milligram of bacterial nitrogen for type A organisms (Marchette and Nicholes, 1961).

Treatment and Prevention

Treatment

Streptomycin has long been the drug of choice with defervescence common within 24 h and a history of <5% relapses. The organism is also susceptible to tetracyclines, chloramphenicols, and aminoglycoside antibiotics, but relapse rates with these can exceed 50% (Evans et al., 1985). All isolates tested have been found universally resistant to penicillin, cephalosporin C, and polypeptide antibiotics (Ohara, S. et al., 1978). After recovery, even with specific antibiotic therapy, enlarged necrotic lymph nodes frequently require surgical draining.

Prevention

Education is probably the best prevention: wearing of protective clothing and awareness of ticks when in endemic areas; avoidance of small wild mammals—i.e., rodents, especially those appearing ill; avoid drinking untreated surface water. A live vaccine is available (from the Centers for Disease Control, Atlanta) for people whose occupations place them at unusually high risk. The vaccine reduces disease severity but does not prevent it (Saslaw, 1961).

Tularemia has historically been a disease of animal and meat handlers, market hunters, and fur trappers. But with the demise of the latter occupations and the employment of vaccinations and meat inspection, the incidence in man has shifted. The preschooler/student and retired citizen now constitute about 40% of the reported cases (McChesney and Narain, 1983). This may be a reflection of exposure due to these individuals' increased outdoor leisure activities that place them in close association with the animal hosts, vectors, and contaminated environments.

Literature Cited

Amoss, H. L., and D. H. Sprunt. 1936. Tularemia. Review of literature of cases contracted by ingestion of rabbit. J. Am. Med. Assoc. **106**:1078–1080.

Baker, C. N., D. G. Hollis, and C. Thornsberry. 1985. Antimicrobial susceptibility testing of *Francisella tularensis* with a modified Mueller-Hinton broth. J. Clin. Microbiol. **22**:212–215.

Biosafety in Microbiological and Biomedical Laboratories. 1984. U.S. Department of Health and Human Services and National Institutes of Health No. 84-8395. U.S. Government Printing Office, Washington, D.C.

Burke, D. S. 1977. Immunization against tularemia: analysis of the effectiveness of live *Francisella tularensis* vaccine in prevention of laboratory-acquired tularemia. J. Infect. Dis. **135**:55–60.

Clark, W. A., D. G. Hollis, R. E. Weaver, and P. Riley. 1984. Identification of unusual pathogenic gram-negative aerobic and facultatively anaerobic bacteria. U.S. Department of Health and Human Services, No. 017-023-00149. U.S. Government Printing Office, Washington, D.C., p. 164.

Cox, S. K., and E. D. Everett. 1981. Tularemia: an analysis of 25 cases. Missouri Med. **78**:70–74.

Cunha, B. A., and R. Quintiliani. 1979. The atypical pneumonias. Postgrad. Med. **66**:95–102.

Evans, M. E., D. W. Gregory, W. Schaffner, and Z. A. McGee. 1985. Tularemia: a 30 year experience with 88 cases. Medicine **64**:251–269.

Fulginiti, V. A., and C. Hoyle. 1966. Oropharyngeal tularemia. Rocky Mountain Med. J. **63**:41–42, 67.

Gourdeau, M., F. Lamonthe, M. Ishak, J. Cote, G. Breton, J. P. Villeneuve, and P. D'Amico. 1983. Hepatic abscess complicating ulceroglandular tularemia. Can. Med. Assoc. J. **129**:1286–1288.

Hutton, J. P., and E. D. Everett. 1985. Response of tularemic meningitis to antimicrobial therapy. South. Med. J. **78**:189–190.

Jellison, W. L., D. C. Epler, E. Kuhns, and G. M. Kohls. 1950. Tularemia in man from a domestic rural water supply. Public Health Rep. **65**:1219–1226.

Jellison, W. L., C. R. Owen, J. F. Bell, and G. M. Kohls. 1961. Tularemia and animal populations. Wildlife disease #17. Wildlife Disease Association, American Institute of Biological Sciences. Washington.

Kaiser, A. B., D. Rieves, A. H. Price, M. R. Gelfand, R. E. Parrish, M. D. Decker, and M. E. Evans. 1985. Tularemia and rhabdomyolysis. J. Am. Med. Assoc. **253**:241–243.

Loftin, E. B. 1981. Tularemia simulating acute myocardial infarction. J. Am. Med. Assoc. **246**:2429–2430.

Luotonen, J., J. H. Syrjala, K. Jokinen, S. Sautinen, and A. Salminen. 1986. Tularemia in otolaryngologic practice. Arch. Otolaryngol. Head Neck Surg. **112**:77–80.

Marchette, N. J., and P. S. Nicholes. 1961. Virulence and citrulline ureidase activity of *Pasteurella tularensis*. J. Bacteriol. **82**:26–32.

McChesney, T. C., and J. Narain. 1983. Public health at a glance: a five-year evaluation of tularemia in Arkansas. J. Ark. Med. Soc. **80**:257–262.

Nohinek, B., and J. J. Marr. 1983. Tularemia: two manifestations in urban man. Mo. Med. **80**:687–689, 698.

Ohara, H. 1934. Oto-rhino-laryngologia. Fukuoka **7**:432–437.

Ohara, S., T. Sato, and M. Homma. 1978. Geographical differences in susceptibility among naturally occurring strains of *Francisella tularensis*. Jpn. J. Bacteriol. **33**:415–420.

Olsen, P. F. 1975. Tularemia. *In* W. T. Hubert, W. F. McCulloch, and P. R. Schnurrenberger (ed.), Diseases transmitted from animals to man. Charles C. Thomas, Springfield, IL.

Overholt, E. L., W. D. Tigertt, P. J. Kadull, M. K. Ward, N. D. Charkes, R. E. Rene, T. E. Salzman, and M. Stephens. 1961. An analysis of forty-two cases of laboratory-acquired tularemia. Am. J. Med. **30:** 785–806.

Rubin, S. A. 1978. Radiographic spectrum of pleuropulmonary tularemia. AJR **131:**277–281.

Saslaw, S., H. T. Eiglesbach, H. E. Wilson, J. A. Prior, and S. Carhart. 1961. Tularemia vaccine study. Arch. Intern. Med. **107:**689–701.

Schmid, G. P., D. Catino, S. Siffin, W. Martone, and A. Kaufmann. 1983. Granulomatous pleuritis caused by *Francisella tularensis*. Possible confusion with tuberculosis pleuritis. Am. Rev. Respir. Dis. **128:**314–316.

Stewart, S. J. 1981. Tularemia. *In* A. Balows and W. J. Hausler, Jr. (eds.), Diagnostic procedures for bacterial, mycotic and parasitic infections, 6th ed. American Public Health Association, Washington, D.C.

Syrgala, H., J. Karuonen, and A. Salminen. 1984. Skin manifestations of tularemia. A study of 88 cases in northern Finland during 16 years (1967–1983). Acta Dermatol. Venereol. **64:**513–516.

Vento, A. O., and B. Y. Oung. 1984. Tularemia: atypical presentation. Md. State Med. J. **33:**395–396.

Westerman, E. L., and J. McDonald. 1983. Tularemia pneumonia mimicking legionnaires' disease: isolation of organism on CYE agar and successful treatment with erythromycin. South. Med. J. **76:**1169–1170.

Yow, M. 1981. Tularemia. *In* R. P. Feigin and R. P. Cherry (ed.), Textbook of pediatric infectious diseases. W. B. Saunders, Philadelphia.

CHAPTER 54

Typhoid Fever

MAKOTO OHASHI

Disease: Typhoid fever (*Typhus abdominalis,* enteric fever).

Etiologic Agents: *Salmonella typhi* (*Salmonella* subspecies I, serovar *typhi*), the typhoid bacillus.

Source: Man—both patients and chronic carriers.

Clinical Manifestations: Insidious onset of sustained fever associated with headache, myalgia, arthralgia, nonproductive cough, malaise, anorexia, mental dullness, constipation, rose spots on trunk, a relative bradycardia, splenomegaly, and leukopenia. Intestinal hemorrhage or perforation is a fatal complication.

Pathology: Reticuloendothelial disease with lesions in Peyer's patches in the ileum. Onset of symptoms is associated with bacteremia.

Laboratory Diagnosis: Bacteriologic detection of etiologic agent from blood, feces, or urine. Demonstration of serum antibodies is of limited diagnostic value.

Epidemiology: A widespread communicable disease. It remains a major public health problem in Central and South America, Africa, Asia, and the western Pacific in particular.

Treatment: Chloramphenicol is the drug of choice for acute infection. For cases due to chloramphenicol-resistant strains, ampicillin, amoxicillin, and cotrimoxazole are of proven value. For chronic carriers, prolonged treatment with large doses of ampicillin or amoxicillin with probenecid or cotrimoxazole may be effective.

Prevention and Control: Safe water supplies, effective sewage systems, and surveillance for chronic carriers, including control of their occupational activities as food handlers. Immunization with conventional killed cell vaccine is protective against only small infecting inocula.

Description of Disease

Typhoid fever (*Typhus abdominalis,* enteric fever) is an acute systemic infection caused by *Salmonella typhi* (*Salmonella* subspecies I, serovar *typhi*). The term "enteric fever" includes paratyphoid fever, which is the disease clinically similar to typhoid fever, but the fatality rate is much lower. It is caused by serovar *S. paratyphi A, S. paratyphi B, S. paratyphi C,* or *S. sendai.* Other serovars belonging to the same subspecies, such as *S. choleraesuis, S. panama, S. dublin,* etc., are known to have a high potentiality to cause septicemia and to give rise to pyemic lesions in the internal organs in man. The distinction between paratyphoid fever, except that caused by *S. paratyphi A,* and salmonella septicemia due to these serovars is by no means absolute. Infections due to *S. typhi* and *S. paratyphi A,* which are primarily adapted to man, are characterized by a long incubation period, 10 to 14 days, and those due to the others by a shorter incubation period, 12 to 36 h, and by their clinical manifestation as gastroenteritis.

Because of its communicability and severity in a typical case of infection, enteric fever has an important public health significance, especially in developing countries.

This chapter deals with the disease caused by *S. typhi* and *S. paratyphi A*. With regard to the disease due to other salmonellae, see the chapter on salmonellosis.

Communicability and Transmissibility

Reservoir is man. The causative organisms enter the body by the oral route; i.e., the disease is introduced by food or water contaminated by feces or urine of patients, temporary excreters in convalescence, or chronic carriers. Person-to-person transmission may also occur.

The patients will be communicable as long as they discharge the typhoid bacilli in their excreta, usually from the first week throughout convalescence. About 10% of untreated cases will excrete the bacilli for 3 months after onset of the disease. Chronic carriers excrete the bacilli over a year's time (see below).

Clinical Manifestations

Incubation period in typhoid fever varies from 1 to 3 weeks or longer, depending on size of infecting dose (Hornick et al., 1970). Mostly it is 8 to 15 days.

The typical symptoms of the disease may be preceded by acute gastroenteritis which comes on shortly after consumption of the contaminated water or food, but it is usually very mild and easily overlooked. The onset of major symptoms and signs is associated with bacteremia, and they are relatively nonspecific. The initial symptoms are commonly fever, generalized headache, arthralgia, myalgia, malaise, and anorexia. The body temperature increases in a stepwise fashon over the course of several days, reaching a peak of 39 to 40°C (103–104°F). A short episode of diarrhea may occur, but only in the early stage of illness. Constipation associated with abdominal pain is rather characteristic, especially in the later stages of the disease. A few rose spots in the anterior chest or abdominal areas are found late in the first week of the onset. The spots are slightly raised and erythematous but not tender. Initially they will blanch with pressure, but later they remain fixed as small hemorrhagic lesions.

A relative bradycardia, splenomegaly, and leukopenia are characteristic. Mental cloudiness is seen not infrequently in a typical manifestation. In untreated cases, fever lasts 2 weeks or longer and then resolves by lysis.

In about 1% of patients, intestinal hemorrhage or perforation occurs, representing an acute abdominal crisis, usually during the second or third week of illness. Relapse is seen during convalescence in 5 to 10% of untreated cases.

When the patients receive an appropriate therapy with effective antibiotics, they will respond with a gradual reduction in the symptoms over a period of several days, but relapse may occur more frequently than in untreated cases.

Mild and asymptomatic infections are common. In general, paratyphoid fever including infection due to *S. paratyphi A* results in a mild and short clinical course.

Pathology

Having passed through the stomach, the typhoid bacilli enter the mucosa of the intestine without significant multiplication in the intestinal lumen. They reach the mesenteric glands, further to the bloodstream via the thoracic duct, causing a transitory bacteremia. Active proliferation of the organisms in the reticuloendothelial cells in the liver and the spleen, in particular, proceeds, and secondary bacteremia associated with a generalization of the organisms follows. The incubation period corresponds to the phases of the infection that precede the development of a sustained bacteremia. The frequency with which *S. typhi* can be recovered from blood is highest in the first week of the illness and then it falls gradually; after the fourth week it become quite rare. On the other hand, the organisms are isolated from the feces even in an early stage and until later stages of the disease, because a large number of the organisms reenter the intestinal lumen through the biliary tract or from the intestinal lesions in Peyer's patches in the ileum (Parker, 1984). Ulceration of the intestinal lesions in the Peyer's patches may produce intestinal hemorrhage or perforation.

The gallbladder is a common site of persistent infection. In such cases, the typhoid bacilli are excreted in the feces over long periods of time. Renal abscess and periostitis are sequelae that may develop long after convalescence from the original illness.

Diagnostic Features

Typhoid fever should be suspected in every patient who has a high sustained fever, malaise, anorexia, and constipation associated with abdominal pain. Relative bradycardia, splenomegaly, rose spots, and leukopenia are useful clinical signs of diagnostic value. History of travel to developing countries where the disease is endemic should be followed up by appropriate microbiological examinations.

Definitive diagnosis requires detection of the causative organism from blood, feces, and urine. Positive blood cultures will be given in the onset of fever, especially during the first and second week of the illness, but they become infrequent after that. Stool cultures are positive most frequently after the second week of illness, though they will be so even in the first week if sensitive methods are used. In the patients who have already received antimicrobial agents, the isolation of typhoid bacilli may be suppressed.

The agglutination reaction (Widal test) against somatic (O) and flagellar (H) antigens becomes positive during the course of the disease, reaching peak by the end of the third week of illness. A significant rise in titer in paired sera is diagnostic when the patient has not received vaccine recently. A single Widal test offers no diagnostic assistance, especially in adults in endemic areas (Levine et al., 1978). It should also be recalled that as many as a third of patients confirmed bacteriologically do not develop significant rises in O and H titers. Early treatment with effective antibiotics has a suppressive effect on the antibody response (El-Rooby and Gohar, 1956). Response to Vi antigen is slow and low in titer. However, presence of Vi agglutinins in blood suggests the carrier state (Nolan et al., 1980).

The differential diagnosis may include febrile infectious illness such as brucellosis, tularemia, salmonellosis, miliary tuberculosis, leprospirosis, typhus fever, infectious mononucleosis, malaria, etc.

Prognosis

There is a wide spectrum of clinical manifestations, from asymptomatic to mild to typical, according to infecting dose of the organism and to various host factors. Although the case fatality rate of typhoid fever was formerly more than 10%, it has been reduced to less than 1% since the introduction of chemotherapy. It is well known, however, that a high case fatality rate comparable to that experienced prior to the era of antibiotics was observed in the beginning of the epidemic due to chloramphenical-resistant *S. typhi*, which occurred in Mexico City in the early 1970s (WHO, 1978). The complications, intestinal hemorrhage and perforation, are responsible for almost all the fatalities, especially when their occurrence is unrecognized.

Chronic Carriers

About 2 to 5% of infected cases, whether clinical or subclinical, become chronic or long-term carriers, mostly having persistent infection in the gallbladder. Without any specific symptoms, they excrete enormous numbers of typhoid bacilli in their feces, for a long period of time, continuously or intermittently. These carriers are more commonly found in females than in males, especially of middle age (Ames and Robins, 1943).

Urinary excreters are those who have a chronic lesion of infection in the kidney. These are much less common than the fecal carriers. Many urinary carriers, however, are found in countries in which urinary schistosomiasis is prevalent (Hathout et al., 1966). These carriers play an important role as infectious sources.

Epidemiologic Aspects

Occurrence

Typhoid fever occurs worldwide. Paratyphoid fever due to *S. paratyphi A*, however, is less common, though it is relatively frequent in Asia and Africa.

In most developed countries, the dramatic fall in the incidence of typhoid fever took place as a result of improved sanitation. In western Europe, the United States, and Japan, it has been at a level of less than 0.5 per 100,000 population during the past two decades, and the mortality for the disease had also been reduced almost to the vanishing point. In these areas, the proportion of cases that are travel-associated or imported tends to increase (Taylor et al., 1983).

On the other hand, the disease is still very common in some countries in Central and South America, Africa, Asia, and the South Pacific, where sanitary conditions are poor, and the morbidity rate seems to remain at a level of several tens to several hundreds per 100,000 population, though real figures are difficult to estimate (Edelman and Levine, 1986). In these countries, the disease is prevalent in urban slums and in rural communities, showing an indigenous trend in the same areas.

In endemic areas, there is no seasonal preponderance, and it appears sporadically during the year. Cases acquired by tourists who visit endemic areas usually occur in the summer, coinciding with the peak of travel time.

Associated with increasing urbanization and provision of common water supplies which may sometimes become contaminated, outbreaks begin to occur, and a trend of increase in diagnosed cases in individuals of all ages is observed.

The risk of laboratory-acquired typhoid fever is relatively high (Holmes et al., 1980).

Susceptibility and Resistance

Infection is nearly universal. There are no data to support racial differences in susceptibility to the disease. The two sexes are equally susceptible. However, the incidence of chronic fecal carrier in females is 2 to 3 times greater than in males. The infection of infants is often clinically atypical and may be easily overlooked. A high susceptibility in the elderly and in individuals with gastric achlorhydria and malnutrition is similar to that for other enteric infections. Attack rates decline with age in endemic areas.

Although a high degree of resistance usually follows recovery, second attacks are not uncommon (Marmion et al., 1953).

Mode of Transmission

Outbreaks result from ingestion of contaminated water or food with feces of patients or carriers. The most frequent causes of contamination of water are cross-connection of piped or shallow-well water with seepage of surface water or sewage (Hornick, 1982). In many food-borne epidemics, the source of the contamination is also polluted water. Foods are contaminated by hands of carriers or missed cases. Flies may also contaminate foods, carrying the typhoid bacilli on their legs. Direct fecal-oral transmission may occur, especially in children and inpatients of mental hospitals.

Shellfish, particularly oysters bedded and fattened in sewage-polluted tidal estuaries, are important vehicles of infection. Numerous kinds of foods including milk, dairy products, salads, and soybean curd contaminated through handling by carriers or polluted water have also been documented as sources of typhoid infection.

Etiologic Agents

Characterization

Members of genus *Salmonella* subspecies I, including *S. typhi* and *S. paratyphi A*, are gram-negative, facultatively anaerobic, nonsporulating, oxidase-negative, straight rods. They are also characterized by negative reactions for indole, urease, phenylalanine deaminase, Voges-Proskauer, ONPG (β-galactosidase), DNAse, lipase, growth in KCN, and fermentation of adonitol, sucrose, lactose, raffinose, and salicin and by positive reactions for motility and fermentation of glucose, mannitol and sorbitol.

S. typhi and *S. paratyphi A*, however, have the following exceptional characteristics: (1) Isolates of *S. typhi* produce H_2S in triple sugar iron (TSI) agar as most clinical salmonellae do, but slightly, giving a typical reaction with black band of H_2S formation at the slant-butt interface. Those of *S. paratyphi A* do not form H_2S. (2) Most salmonellae are aerogenic, but *S. typhi* never produces gas from glucose. (3) Citrate is generally utilized by salmonellae, but neither *S. typhi* nor *S. paratyphi A* uses it. (4) Most salmonellae are positive for ornithine- and lysine-decarboxylase, but *S. typhi* is negative for the former and *S. paratyphi A* is negative for the latter. (5) Isolates of *S. typhi* are also characteristically negative for fermentation of arabinose and rhamnose (Le Minor, 1984). These differential characteristics of *S. typhi, S. paratyphi A,* and other salmonellae are summarized in Table 1.

Specimen Collection and Transport

Blood Specimens

It is advisable to obtain the blood specimens early in the course of the disease or in a relapse, preferably before initiation of chemotherapy. Sterile precaution including desinfection of skin should be taken in withdrawing the specimens. Five to ten milliliters of blood should preferably be taken directly into a large volume (150 to 200 ml) of enrichment medium. Alternatively, the specimens should be transported to the laboratory in a sterile tube containing anticoagulant, liquoid (polyanetholsulfonate), which may inhibit the effect of bactericidal substances that might be present in the blood.

Bone marrow aspirate is also a useful specimen, when blood culture fails, in hospitalized patients who are on antibiotic treatment, in particular.

Fecal Specimens

The specimen of choice is a freshly passed stool. The feces should be examined as fresh as possible. If

TABLE 1. Biochemical differentiation between *S. typhi, S. paratyphi A*, and other salmonellae of subspecies I

	Reaction of		
Test	S. typhi	S. paratyphi A	*Others*
Hydrogen sulfide (in TSI)	+[w]	−	+
Gas production from glucose	−	+	+
Lysine decarboxylase	+	−	+
Ornithine decarboxylase	−	+	+
Citrate utilization	−	−	+[a]
Arabinose fermentation	−	+	+[a]
Rhamnose fermentation	−	+	+
Trehalose fermentation	+	+	+[a]

[a] Strains of serovar *S. choleraesuis* give variable reactions for citrate utilization and negative for fermentation of arabinose and trehalose.
+ = more than 90% of strains positive; +[w] = weakly positive reaction; − = more than 90% of strains negative.

immediate examination is impracticable, the specimens may be collected into transport medium such as Cary-Blair's transport medium, Stewart's transport medium, or buffered glycerol solution containing 1% NaCl and sent to the laboratory. Alternatively, several drops of liquid feces or fecal emulsion are placed on a sheet of filter paper and allowed to dry in air, and then the paper is placed in a waterproof envelope and sent to the laboratory (Bailey and Bynoe, 1953).

Isolation Procedure

Blood Culture

For enrichment, thioglycollate medium, tryptic or trypticase soy broth, or heart infusion broth is recommended. These are advantageous, having as broad a microbial coverage as possible. As blood culture media for the typhoid and paratyphoid bacilli, 0.5% sodium taurocholate broth and bile broth are recommended (Kaye et al., 1966). If only the blood clot is available for culture, streptokinase treatment may be advantageous (Watson, 1978). Cultures are incubated at 37°C and plated out daily for at least 10 days. Nonselective media are used for plating after the enrichment.

Fecal Culture

Cultures of the feces shoud be carried out at the same time as the blood culture. Parallel performance of direct plating culture and enrichment with subsequent plating is recommended.

Plating Culture

Use of at least two kinds of solid media of different degrees of selectivity is recommended. Among highly selective ones, bismuth sulfite brilliant green agar of Wilson and Blair (1931) is the medium of choice. Salmonella-shigella (SS) agar may also be used. Less selective media include desoxycholate citrate agar, desoxycholate hydrogen sulfide lactose agar, MacConkey agar, XLD agar, etc. The plates inoculated are incubated aerobically at 37°C for 18 to 24 h.

On bismuth sulfite brilliant green agar, *S. typhi*, and most other salmonellae, round jet black colonies will develop surrounded by a black or brownish-black zone which exhibits a distinctly characteristic metallic sheen when they are well isolated. In the area where the organisms grow congestedly, however, they appear as small, light-green colonies. *S. paratyphi A* yields flat or only slightly raised green colonies.

Enrichment

For enrichment of *S. typhi*, sodium selenite broth (Leifson, 1936) or strontium selenite broth is recommended. Approximately 1 g of feces is inoculated and incubated at 37°C for 12 to 16 h. A more prolonged incubation is disadvantageous, because it allows the overgrowth of coliforms. Rappaport's medium may also be used. The plating culture mentioned above follows the enrichment.

Identification

Biochemical Examinations

For the performance of biochemical tests, several methods are available, including conventional test systems in tubes and commercial kits for identifying *Enterobacteriaceae*. Of the many conventional systems, the following is sufficient to identify *S. typhi* and *S. paratyphi A* and to differentiate them from other salmonellae.

Typical salmonella colonies grown on the plating media are transferred simultaneously to triple sugar iron (TSI) agar or Kligler iron agar and other differential media that allow reading results of lysine decarboxylase, motility, and indole reactions. For the latter, lysine indole motility (LIM) medium or a combination of motility-indole medium such as SIM medium and lysine decarboxylase broth is used.

If the isolates give any of the patterns of reactions in these media after the usual overnight incubation, which are indicated in Table 2, they are suggestive of *S. typhi, S. paratyphi A*, or other salmonellae. Then, they are subjected to the following serological examination.

Serologic Examinations

Isolates that gave biochemical characteristics mentioned above are serologically identified by slide agglutination techniques with grouping or specific antisomatic (O), O 9 and O 2, and anti-Vi sera. Commercial products of the antisera are available. Slide agglutination tests are performed by making a dense suspension of the organism grown on a TSI agar slant. The expected reactions in the agglutination tests of *S. typhi* and *S. paratyphi A* are shown in Table 2 together with their antigenic formulae.

Vi antigen, a heat-labile surface antigen, is present in most of the virulent strains of *S. typhi* newly isolated from the patient but is sometimes defected in isolates from the chronic carrier. The presence of this antigen blocks the agglutinability of the organisms with O antisera. Therefore, the serological reactions of the organisms against anti-Vi and anti-O 9 sera may differ according to the amount of Vi anti-

TABLE 2. Biochemical and serologic reactions of *S. typhi*, *S. paratyphi A*, and other salmonellae of subspecies I

	Reaction of		
Test	S. typhi	S. paratyphi A	*Others*
TSI agar			
Slant/butt	K/A	K/A	K/A
Gas	−	+	+
H_2S	$+^w$	−	+
LIM agar			
Lysine decarboxylase	+	−	+
Indole	−	−	−
Motility	+	+	+
Slide agglutination with			
Anti-Vi	+ or − (−)[a]	−	d
Anti-O 9	− or + (+)	−	d
Anti-O 2	− (−)	+	d
Tube agglutination with			
Anti-H d	+	−	d
Anti-H a	−	+	d
Antigenic formulae	9,12[Vi] : d : −[b]	1,2,12 : a : [1,5]	d

[a] Reactions in parentheses indicate those obtained by using heat-treated antigens.
[b] May possess an R phase H antigen, j or z_{66}.
K = Alkaline (red); A = acid (yellow or blackening), indicating fermentation of glucose but not lactose and sucrose; d = different reactions; + = positive; $+^w$ = weakly positive; − = negative. In LIM agar, strains positive for lysine decarboxylase show violet reaction, and negative strains yellow.

gen. Usually, unheated *S. typhi* typically agglutinate with anti-Vi serum but not with anti-O 9 serum, whereas the reactions are reversed after heating the antigen at 100°C for 30 min.

S. paratyphi A gives positive reactions with anti-O 2 serum.

In this stage of examinations, obtaining the positive results in the agglutination tests with biochemical reactions mentioned above, *S. typhi* and *S. paratyphi A* must be reported almost unmistakably.

Further Confirmation

Additional biochemical examinations for confirmation of the characters of isolates as *Salmonella* are encouraged. With regard to examination items to be covered, see Characterization (above).

Identification of isolates that have atypical characteristics as well as determination of their antigenic formulae based on specific O factor and flagellar (H) antigen testing can be referred to public health laboratories.

For determination of H antigens, a test tube agglutination method is adopted. When the test is needed at clinical laboratories, suspicious colonies grown on the plating media or TSI agar slant are seeded into a trypticase soy broth and incubated overnight. The broth cultures are diluted with an equal volume of saline containing 0.65% Formalin, and they are used as antigen. Results of the H agglutination reaction are read after incubation of tubes containing antigen-antiserum mixture in a 50°C water bath for 2 h. Anti-H sera are also available commercially.

In the H agglutination test, *S. typhi* and *S. paratyphi A* give positive reactions to anti-d and anti-a sera, respectively. However, it has been known that some strains of *S. typhi*, though they are not so commonly encountered, may not react with anti-d serum and possess an R-phase H antigen, j or z_{66} (Guinée et al., 1981; Kauffman, 1936).

Phage Typing

Phage typing method provides a highly discriminatory system of strain identification for epidemiological use, classifying strains of *S. typhi* into 106 phage types and those of *S. paratyphi A* into six. The test is carried out only in experienced national or regional laboratories (International Federation for Enteric Phage-Typing, 1982). The WHO Collaborating International Centre at the Division of Enteric Pathogens, Central Public Health Laboratory, Colindale, United Kingdom is responsible for training and distributing typing phages.

Treatment, Prevention, and Control

Treatment

Chloramphenicol remains the drug of choice; 50 mg/kg/day i.v. is given in four divided doses until the body temperature become normal and then is reduced to 30 mg/kg/day for 10 to 14 days. This drug is also effective in the treatment of paratyphoid fever due to *S. paratyphi A* and others.

Ampicillin, amoxycillin, and trimethoprim-sulfamethoxazole are also of proven value in treatment of acute infection of the disease. In recent years, drug resistance carried by R factors has been a problem in some part of the world (Anderson 1975; Jegathesan and Ohashi, 1985). *S. typhi* may be resistant not only to chloramphenicol but also to ampicillin and, more recently, to trimethoprim-sulfamethoxazole. Therefore, first choice must be made on the basis of the known frequency of resistance to these drugs in the area where the infection was contracted or of results of antibiotic sensitivity testing of the isolates.

Prolonged administration of ampicillin or amoxicillin (6 to 10 g/day for 6 to 9 weeks) with probenecid will often terminate the chronic carrier state if the carrier has no biliary dysfunction and no gallstones. Chloramphenical is not effective in this respect. If there is evidence of gallstones or biliary dysfunction, cholecystectomy associated with ampicillin administration is the treatment of choice; cure is achieved in about 80% of the cases.

Prevention and Control

The most important measures for controling typhoid fever are safe water and effective sewage systems (Cvjetanović et al., 1978). Of course, a nationwide surveillance system for the disease including control of chronic carriers is indispensable. In most countries, typhoid fever is a notifiable disease, and all patients should be reported to the local health department. Notification data, preferably those based on bacteriological confirmation, are indispensable for implementing countermeasures against the disease.

Hospital care and isolation with enteric precautions are required for acute illness, and release from supervision by the local health authority should be based on not less than three consecutive negative cultures of feces and urine taken at least 24 h apart starting the second or third day after the completion of chemotherapy.

In spite of an adequate and thorough treatment of acute illness, it is almost impossible to prevent the development of the carrier state. Therefore, every effort should be made to recognize such cases in convalescents and follow with treatment. All known carriers must be under surveillance; they should be recorded at local health departments together with Vi-phage types of the typhoid bacilli carried. They should be prevented from working in any position that may result in contamination of food or drinking water. They must be required to submit stool specimens once or twice a year to ascertain the continued presence of the organism. Then, they could be ensured as possible infectious sources of infection, and appropriate instruction in personal hygiene could be made.

Once an outbreak occurs, its cause should be investigated by search for unreported cases, carriers, and contaminated food and water and be effectively removed. Identification of phage types of isolates may provide useful information for tracing the transmission route and the infectious source. Sewage culture using Moore's swabs may help in locating carriers (Moore et al., 1952).

Acetone-killed and heat-killed phenolized typhoid vaccines have proven to be effective in reducing the attack rate of the disease in an exposed population (Cvjetanović and Uemura, 1965). However, the protection due to these parenteral vaccines can be overcome by large infecting inocula. In addition, the vaccine not uncommonly causes tenderness at the site of injection, febrile reactions, and general malaise. Mass immunization using the parenteral vaccines is not routinely recommended in developed areas. Recently, oral live attenuated vaccines have been developed, and their effective use in endemic areas is expected (Wahdan et al., 1982). Neverthless, improvements in sanitary conditions, including the provision of safe water supplies and excreta disposal facilities, which must be effective for controling most enteric infections, are more appropriate than mass vaccination.

Literature Cited

Ames, W. R., and M. Robins. 1943. Age and sex as factors in the development of the typhoid carrier state, and a method for estimating carrier prevalence. Am. J. Public Health **33**:221–230.

Anderson, E. S. 1975. The problem and implications of chloramphenicol resistance in the typhoid bacillus. J. Hyg. (Lond.) **74**:289–299.

Bailey, W. R., and E. T. Bynoe. 1953. The "filter paper" method for collecting and transporting stools to the laboratory for enteric bacteriological examination. Can. J. Public Health **44**:468–475.

Cvjetanović, B., and K. Uemura. 1965. The present status of field and laboratory studies of typhoid and paratyphoid vaccines. Bull. W.H.O. **32**:29–36.

Cvjetanović, B., B. Grab, and K. Uemura. 1978. Typhoid fever: an endemic disease with interhuman transmission. Bull. W.H.O. **56(Suppl. 1)**:45–63.

Edelman, R., and M. M. Levine. 1986. Summary of an

international workshop on typhoid fever. Rev. Infect. Dis. **8**:329–349.

El-Rooby, A., and M. A. Gohar. 1956. The effect of chloramphenicol on the agglutinin titre in enteric fevers. J. Trop. Med. Hyg. **59**:47–51.

Gilman, R. H., M. Terminel., M. M. Levine, P. Hernandez-Mendosa, E. Calderone, V. Vasquez, E. Martinez, M. J. Snyder, and R. B. Hornick. 1975. Comparison of trimethoprim-sulfamethoxazole and amoxicillin in therapy of chloramphenicol-resistant and chloramphenicol-sensitive typhoid fever. J. Infect. Dis. **132**:630–636.

Guinée, P. A. M., W. H. Jansen, H. M. E. Maas, L. Le Minor, and R. Beaud. 1981. An unusual H antigen (z_{66}) in strains of *Salmonella typhi*. Ann. Microbiol. (Paris) **132A**:331–334.

Hathout, S. E. D., Y. A. El-Ghaffar, A. Y. Awny, and K. Hassan. 1966. Relation between urinary schistosomiasis and chronic enteric urinary carrier state among Egyptians. Am. J. Trop. Med. Hyg. **15**:156–161.

Holmes, M. B., D. L. Johnson, N. J. Fiumara, and W. M. McCormack. 1980. Acquisition of typhoid fever from proficiency-testing specimens. N. Engl. J. Med. **303**:519–521.

Hornick, R. B. 1982. Typhoid fever. p. 659–676. *In* A. S. Evans and H. A. Feldman (ed.). Bacterial infections of humans. Plenum, New York.

Hornick, R. B., S. E. Greisman, T. E. Woodward, H. L. DuPont, A. T. Dawkins, and M. J. Snyder. 1970. Typhoid fever: pathogenesis and immunologic control. N. Engl. J. Med. **283**:686–691.

International Federation for Enteric Phage-Typing (IFEPT). 1982. The geographical distribution of *Salmonella typhi* and *Salmonella paratyphi A* and *B* phage types during the period 1 January 1970 to 31 December 1973. J. Hyg. (Lond.) **88**:231–254.

Jegathesan, M., and M. Ohashi. 1985. Symposium on problems of antibiotic-resistant enteropathogens in Asia p. 115–142. *In* J. Ishigami, (ed.); Recent advances in chemotherapy, antimicrobial section. University of Tokyo, Press, Tokyo.

Kauffmann, F. 1936. Uber die diphasische Natur der Typhusbacillen. Z. Hyg. **119**:103–118.

Kaye D., M. Palmieri and H. Rocha. 1966. Effect of bile on the action of blood against *Salmonella*. J. Bacteriol. **91**:945–952.

Leifson, E. 1936. New selenite enrichment media for the isolation of typhoid and paratyphoid (*Salmonella*) bacilli. Am. J. Hyg. **24**:423–432.

Le Minor, L. 1984. Genus III, *Salmonella* Lignieres 1900, 389. p. 427–458. *In* N. R. Krieg and J. T. Holt (eds.); Bergey's manual of systematic bacteriology, Vol. 1. Williams and Wilkins, Baltimore.

Levine, M. M., O. Grados, R. H. Gilman, W. E. Woodward, R. Solis-Plaza, and W. Waldman. 1978. Diagnostic value of the Widal test in areas endemic for typhoid fever. Am. J. Trop. Med. Hyg. **27**:795–800.

Marmion, D. E., G. R. E. Naylor, and I. O. Stewart. 1953. Second attacks of typhoid fever. J. Hyg. (Lond.) **51**:260–267.

Moore, B., E. L. Perry, and S. T. Chard. 1952. A survey by the sewage swab method of latent enteric infection in an urban area. J. Hyg. (Lond.) **50**:137–156.

Nolan, C. M., J. C. Feeley, P. C. White, Jr., E. A. Hambie, S. L. Brown, and K. Wong. 1980. Evaluation of a new assay for Vi antibody in chronic carriers of *Salmonella typhi*. J. Clin. Microbiol. **12**:22–26.

Parker, M. T. 1984. Enteric infections: typhoid and paratyphoid fever, p. 407–433. *In* G. R. Smith, (ed.), Topley and Wilson's principles of bacteriology, virology and immunity, Vol. 3. Bacterial diseases. Edward Arnold, London.

Taylor, D. N., R. A. Pollard, and P. A. Blake. 1983. Typhoid in the United States and the risk to the international traveler. J. Infect. Dis. **148**:599–602.

Wahdan, M. H., C. Serie, Y. Cerisier, S. Sallam, and R. Germanier. 1982. A controlled field trial of live *Salmonella typhi* strain Ty21a oral vaccine against typhoid: three-year results. J. Infect. Dis. **145**:292–295.

Watson, K. C. 1978. Laboratory and clinical investigation of recovery of *Salmonella typhi* from blood. J. Clin. Microbiol. **7**:122–126.

Wilson, W. J., and E. M. M. Blair. 1931. Further experience of the bismuth sulfite media in the isolation of *Bacillus typhosus* and *B. paratyphosus B* from faeces, sewage and water. J. Hyg. (Lond.) **31**:138–161.

World Health Organization. 1978. Surveillance for the prevention and control of health hazards due to antibiotic-resistant enterobacteria. Technical Report Series 624. W.H.O., Geneva.

CHAPTER 55

Vibrioses Other than Cholera

YASUO KUDOH

Diseases: Vibrioses other than cholera.

Etiologic Agents: *Vibrio cholerae* non-O 1, *V. mimicus, V. parahaemolyticus, V. fluvialis, V. furnissii, V. hollisae, V. vulnificus, V. alginolyticus, V. damsela,* and possibly *V. metschnikovii.*

Source: Aquatic environment; seafood; coastal and brackish waters by ingestion or direct exposure. Man or animals rarely involved.

Clinical Manifestations: Gastroenteritis of acute onset with diarrhea, abdominal cramps, and sometimes with nausea, vomiting, fever, chills, and headache. A dysentery-like syndrome occasionally involved. Wound infections and primary or secondary septicemias, especially in persons with underlying chronic disease.

Pathology: Secretory diarrhea by cholera-like enterotoxin or enteritis by other unidentified toxic factor(s). Invasiveness may be involved in pathogenesis.

Laboratory Diagnosis: Bacteriologic isolation of causal vibrios from clinical specimens: stool, wound, blood, etc.

Epidemiology: Worldwide, particularly in Japan, Southeast and Western Asia, and the United States. Sporadic cases and common source outbreaks mostly by consumption of contaminated raw or inadequately cooked seafood, or exposure to seawater or brackish water.

Treatment: Fluid replacement and/or antibiotics such as tetracycline or aminoglycosides in gastroenteritis. Good supportive care with antibiotic therapy, surgical debridement, etc., in wound infections and septicemia.

Prevention and Control: Proper handling of raw and cooked seafood.

Description of Disease

Vibrio species cause a variety of human infections which can usually be classified as gastrointestinal or extraintestinal infections (Table 1). At present, 10 *Vibrio* species are known to be pathogenic or potentially pathogenic for humans (Blake et al., 1980; Morris and Black, 1985; Tison and Kelly, 1984).

Species most commonly associated with diarrhea or gastroenteritis include *V. cholerae* (O 1 and non-O 1) and *V. parahaemolyticus*. Recently, *V. fluvialis* (previously known as group F or EF-6 vibrios), V. *furnissii* (formerly classified as biovar II of *V. fluvialis*), *V. mimicus* (formerly classified as sucrose-negative non-O 1 *V. cholerae*), and *V. hollisae* have also been shown to cause or be significantly associated with gastroenteritis. Interestingly, some of these *Vibrio* species may also cause extraintestinal infections such as wound infections and secondary septicemia.

V. vulnificus (previously referred to as L + or Lac + vibrio) is the well-known cause of primary and/or secondary septicemias and wound infections with high mortality. *V. alginolyticus* and *V. damsela* also cause several type of soft-tissue infections and wound infections. *V. metschnikovii* has been impli-

TABLE 1. *Vibrio* species associated with human infections

Organism	*Disease*
V. cholerae O 1	Cholera, (wound infection)[a]
V. cholerae non-O 1	Gastroenteritis, wound infection, ear infection, (septicemia)
V. mimicus	Gastroenteritis, ear infection
V. parahaemolyticus	Gastroenteritis, wound infection, (ear infection), (septicemia)
V. fluvialis	Gastroenteritis
V. furnissii	Gastroenteritis
V. hollisae	Gastroenteritis, (septicemia)
V. vulnificus	Wound infection, septicemia
V. alginolyticus	Wound infection, ear infection
V. damsela	Wound infection
V. metschnikovii	Extraintestinal infection(?)

[a] Infections with rare occurrence in parentheses.

cated as a cause of opportunistic infections, but its pathogenic significance in the illness needs further investigations.

Most, but not all, vibrio infections are associated with consumption of raw or inadequately cooked seafood or direct exposure to seawater.

Communicability and Transmissibility

Vibrioses other than cholera are usually noncommunicable. Most of vibrio gastroenteritis are foodborne, and person-to-person transmission or waterborne infections rarely occur since ingestion of relatively large number of the organism (possibly more than 10^5) is generally required for manifestation of clinical illness. The duration of excretion of vibrios in stools of patients is usually short, mostly within a week. A chronic carrier state rarely exists.

No person-to-person spread has been described in extraintestinal infections.

Clinical Manifestations

V. cholerae O 1 is well known as the cause of epidemic cholera (see chapter on cholera). Other diarrheagenic *Vibrio* species may cause symptoms ranging from severe cholera-like secretory diarrhea to acute gastroenteritis.

The incubation period usually ranges from a few hours to several days, but most commonly 12 to 24 h. The typical symptoms of gastroenteritis include watery diarrhea and abdominal cramps and sometimes nausea, vomiting, low-grade fever, headache, and chills. In severe cases, dehydration, hypotension, or acidosis resulting in death may occur. A dysentery-like illness with bloody or mucoid stools and high temperature has also been recognized in *V. parahaemolyticus* and *V. cholerae* non-O 1 infections.

The disease is generally mild and self-limiting, and the duration is usually less than 3 days. Asymtomatic infections may also occur.

The most serious extraintestinal infections among vibrios is a primary septicemia due to *V. vulnificus*. The disease usually begins within several days after consumption of seafood. Most patients develop malaise which progresses rapidly to fever, chills, prostration, and often severe hypotension, resulting in death (fatality rate nearly 50%). The illness often progresses rapidly from initial symptoms to death within 24 h. At autopsy, there are signs of gram-negative shock, hemorrhage, and necrosis of involved muscles and cutaneous lesions. In most cases, the organism is recovered from both blood and skin lesions of these patients.

Wound infections due to *V. vulnificus* are also very severe and usually progress rapidly. Initial signs of the diseases are swelling and erythema, often followed by the development of vesicles or bullae leading to tissue necrosis. Some patients also have fever and chills.

Extraintestinal infections by the other pathogenic vibrios are generally mild in their clinical course, although there is a tendency to lead to secondary septicemia, particularly in patients with underlying chronic disease.

Pathology

In *V. cholerae* O 1, cholera enterotoxin (CT) elaborated by the organisms is the well-known cause of profuse watery diarrhea (cholera). Among the other known diarrheagenic *Vibrio* species, some strains of non-O 1 *V. cholerae* and *V. mimicus* have been shown to produce an enterotoxin indistinguishable from that produced by *V. cholerae* O 1 (Kudoh et al., 1981; Ohashi et al., 1972; Spira and Fedorka-Cray, 1984), but most do not. These toxigenic strains are known to cause relatively severe, choleralike illness (Kamal and Zinnaka, 1971; Kelly et al., 1982). whereas the other, CT-negative strains may cause mild diarrhea or gastroenteritis by different pathogenic mechanisms. Their virulence factors may consist of heatstable enterotoxin similar to that of *Escherichia coli* heat-stable toxin (Arita et al., 1986), hemolysin/cytotoxin (Yamamoto et al., 1986; Yoh et al., 1986), and invasive or colonizing ability, which needs further investigation.

In *V. parahaemolyticus*, enteropathogenesis is known to correlate well with hemolytic reaction referred to as "Kanagawa phenomenon"; over 95% of clinical isolates are Kanagawa-positive, whereas only about 1% of environmental isolates are positive (Sakazaki et al., 1968). Hemolysin responsible for Kanagawa phenomenon is a heat-stable protein and is shown to have some biological activities such as shown by a positive rabbit ileal loop test (Zen-Yoji et al., 1976), suggesting its possible role in clinical manifestation of the illness. The clinical observation that the organism can cause a dysentery-like syndrome may also suggest the involvement of degenerative changes in mucosal surface of the small intestine as its pathogenesis.

V. fluvialis and *V. furnissii* have been demonstrated to produce several virulence factors such as hemolysin/cytotoxin and some other extracellular enterotoxic substance(s) which were found in infant mice (Kudoh et al., 1983; Nishibuchi et al., 1983), but their nature and role in pathogenicity are still not fully understood.

In *V. vulnificus* extraintestinal infections, increased iron availability, which appears to enhance virulence of the organism, has been demonstrated in an animal experimental model (Simpson and Oliver, 1983). Several proteolytic enzymes and a hemolysin/cytotoxin produced by the organism may also contribute to the illness, although their role in the pathogenesis remains to be established.

Diagnostic Features

Gastroenteritis due to *Vibrio* species generally cannot be differentiated clinically from other enteric infections, even if the patients manifest typical clinical symptoms, although histories of patients with consumption of seafood within few days before the onset of the illness are suggestive of the infection. Mild and asymptomatic infections are common in the disease. Thus, definitive diagnosis should be made by direct isolation of pathogenic vibrios from fecal specimens taken in the acute phase of the illness.

In extraintestinal infections such as wound infection and septicemia, particularly in *V. vulnificus* infections, the recognition of history of recent consumption of seafood or exposure to seawater and underlying disease in patients may strongly suggest the diagnosis. Appearance of characteristic bullous skin lesion is another sign of *V. vulnificus* infections. However, specific diagnosis requires isolation of causal vibrios from clinical specimens taken properly.

In cholera infections, demonstration of a significant rise in titer of vibriocidal, agglutinating or antitoxic antibodies may help in presumptive diagnosis, but in the other vibrioses such serologic tests may be not useful.

Prognosis

The prognosis of vibrioses varies considerably with type of infection, causal organism, and host factors, but the mortality rate of vibrioses other than cholera is generally very low. Gastroenteritis due to diarrheagenic vibrios other than *V. cholerae* O 1 is usually self-limited and requires no therapy. In severe untreated cases, however, death may occur by prompt dehydration, hypotension, or acidosis. Among extraintestinal infections, those by *V. vulnificus* are known to be very serious, especially in patients with underlying chronic disease, and the reported fatality rates have been about 50% in primary septicemia and about 7% in wound infections (Tison and Kelly, 1984).

Occurrence and Transmission

Occurrence

Gastroenteritis due to *V. parahaemolyticus* appears to be most common and occurs worldwide, particularly in Japan, Southeast Asia, and to a lesser extent the United States. The high prevalence in Japan of the infections results exclusively from the national habit of eating raw fish and shellfish. Non-O 1 *V. cholerae* and probably *V. mimicus* gastroenteritis may also be common in areas where cholera is endemic, although the real incidence or prevalence of the disease is still not clear. Other gastroenteritis infections have not so frequently been reported, but they may also occur worldwide. Vibrio gastroenteritis is also important as a travelers' diarrhea (Kudoh and Sakai, 1985).

The diseases occur both sporadically and in common-source outbreaks, most of which are foodborne. In Japan, *V. parahaemolyticus* is the most important cause of food-borne gastroenteritis outbreaks; it is responsible for about 50% of reported bacterial food poisoning (Kudoh and Sakai, 1985). Vehicles of transmission in these oubreaks include not only raw fish and shellfish, but also cooked seafood and other food. Waterborne infections are extremely rare in vibrio gastroenteritis, but polluted water may play a part in some *V. cholera* non-O 1 and *V. mimicus* infections (Kamal and Zinnaka, 1971). Extraintestinal infections may also occur worldwide, although most cases have been documented in limited countries, especially in the United States.

In temperate climates, vibrioses tend to be predominant in the warm months of the year; in tropical areas they may occur at any season of the year. This seasonal tendency may reflect both enhanced opportunity of vibrios to proliferate in seafood and increased prevalence of vibrios in the marine environment during the summer.

Susceptibility and Resistance

Most persons are probably sensitive to vibrios, which affect all age groups. Susceptibility may be increased by host factors such as gastric achlorohydria, antacid therapy, gastrointestinal surgery, or other debilitating condition including malnutrition. Severity of the infections can be related to the virulence of the causal organism, the number of organism ingested, as well as host factors.

In extraintestinal infections, especially in those by *V. vulnificus,* preexisting chronic diseases including liver cirrhosis, diabetes, cholecystitis, alcoholism, leukemia, and immune suppression would increase the risk of the disease. Involvement of immune mechanisms in vibrio infections is not well understood.

Mode of Transmission

In most vibrio gastrointestinal infections, the major route of transmission is ingestion of the organism in raw or inadequately cooked seafood, any food cross-contaminated by kitchen utensils, or seawater. This is mainly related to the fact that vibrios naturally inhabit marine environments. During the warm season, vibrios can be found free in coastal water and estuarine water and to a lesser extent in brackish water, resulting in natural contamination of fish and shellfish. Vibrios can multiply easily in these foods to an infection level when adequate temperature for growth is given.

In *V. cholerae* non-O 1 and *V. mimicus* gastroenteritis, transmission through water contaminated with feces of infected persons may occur in areas where the environmental sanitation is poor. Person-to-person transmission, if any, will be negligible.

In wound infections, direct exposure to seawater or lacerations by shells of marine animals is the primary mode of transmission. Primary septicemia due to *V. vulnificus* is significantly associated with consumption of raw seafood, particularly oysters in the United States.

Etiologic Agents

Characterization

Vibrio species are a facultatively anaerobic gram-negative asporogenous rod (0.5 to 0.8 μm × 1.5 to 3.0 μm) with varying curvature. All species are motile with a sheathed polar monotrichous flagellum. Some species are known to produce short peritrichous flagella (or lateral flagella) when grown on solid media. Growth of *Vibrio* species is stimulated by Na^+, and this is an absolute requirement of most. They are usually sensitive to vibriostatic compound 0/129 (2,4-diamino-6,7-diisopropylpteridine), and some of them are known to be bioluminescent. Except for *V. metschnikovii, Vibrio* species are oxidase-positive and reduce nitrate to nitrite. They ferment D-glucose without gas production (except *V. furnissii* and some strains of *V. damsela*). They do not produce H_2S, phenylalanine deaminase, or urease.

Characters for growth in nutrient broth without NaCl and with 1% NaCl, lysine decarboxylase, ornithine decarboxylase, arginine dihydrolase, Voges-Proskauer, ONPG(O-nitrophenyl-β-D-galactopyranoside) test, gas production from glucose, and acid production from L-arabinose, lactose, sucrose, and salicin are variable among vibrios and prove to be useful for species differentiation (Farmer et al., 1985). The guanine-plus-cytosine content of the vibrio DNA ranges from 38 to 51 mol%.

Specimen Collection and Transport

In gastrointestinal infections, the specimen of choice is a freshly passed stool. Rectal swabs may also be used, but they are probably less effective for convalescent patients or asymptomatic cases. Stool specimens should be collected as early as possible after the onset of the illness, preferably within 24 h, and before any antimicrobial therapy is begun. If the specimen cannot be cultured immediately, it should be placed in a transport medium such as Cary-Blair medium or buffered glycerol saline containing 1% NaCl and shipped to the laboratory without refrigeration.

Specimens taken from materials such as wound, blood, ear, and eye should be collected and processed by routine procedures. There is no special procedure for *Vibrio* species (Farmer et al., 1985).

Isolation

Vibrio species of medical importance can be readily isolated from clinical specimens on media routinely used in the clinical microbiology laboratory. Direct microscopic examination of stool specimens is not recommended for screening *Vibrio* species, because vibrios cannot be easily distinguished from other related, motile, rod-shaped bacteria.

Direct Plating

Use of TCBS (thiosulfate–citrate–bile salts–sucrose) agar, which is a highly selective medium for *Vibrio* species, is the most preferable. It has been pointed out, however, that there will be a lot-to-lot or brand-to-brand variation in selectivity of the medium, and some strains of vibrios, especially those of *V. hollisae,* do not grow well on the medium. In areas where vibrio infections are not common, moderately selective or nonselective enteric medium such as MacConkey agar and blood agar can be used. Blood agar is also preferably used for *V. hollisae*. Inoculated isolation agar plates are incubated aerobically at 37°C for 18 to 24 h.

On TCBS agar, sucrose-fermenting *Vibrio* species, which include *V. cholerae, V. fluvialis, V. furnissii, V. alginolyticus,* and *V. metschnikovii* produce yellow colonies, whereas sucrose nonfermenters such as *V. parahaemolyticus, V. mimicus, V. damsela,* and most of *V. vulnificus* produce blue to green colonies. Colonies of the most of *Vibrio* species on MacConkey agar are colorless to pale pink (lactose-negative or late fermenters), and those on sheep blood agar are variable from beta- or alpha-hemolytic to nonhemolytic.

Enrichment

Alkaline peptone water with 1% NaCl is recommended for enrichment of fecal specimens, especially these obtained from convalescent patients or asymptomatic cases. Approximately 0.5 g of stool specimen is inoculated in the broth and incubated at 37°C for 8 to 15 h, and it is then carefully subcultured from the surface of the broth to isolation agar plates.

Identification

Colonies suspected to be vibrios should be identified to the species level, since pathogenicity and clinical significance may vary among *Vibrio* species. Biochemical identification of *Vibrio* species can be done by using ordinary media and tests commonly used for identifying the family *Enterobacteriaceae*. It is recommended, however, that media used for vibrios should be supplemented to contain 1% NaCl, because most of pathogenic vibrios are halophilic marine bacteria and require Na^{+} for good growth and expression of typical reactions.

Suspicious colonies are preferably tested at first for oxidase by Kovacs reagent, nitrate reduction to nitrite, growth in nutrient broth without NaCl and with 1% NaCl, lysine decarboxylase, ornithine decarboxylase, and arginine dihydrolase. By these seven tests, pathogenic *Vibrio* species can be easily divided into five groups as described by Farmer et al., (1985) (Table 2).

V. cholerae (O 1 and non-O 1) and V. *mimicus* can be differentiated from the other halophilic species by Na^{+} requirement for growth in nutrient broth. Among halophilic vibrios (eight species), *V. metsch-*

TABLE 2. Biochemical reactions useful for differentiation of pathogenic *Vibrio* species

Group/species	*Growth in 0% NaCl*	*Growth in 1% NaCl*	*Oxidase*	*Nitrate to nitrite*	*Lysine*	*Ornithine*	*Arginine*	*Voges-Proskauer*	*Growth in 8% NaCl*	*Growth in 10% NaCl*	*Glucose: gas*	*Sucrose*	*L-Arabinose*	*D-Mannitol*	*Lactose*	*Salicin*
I. *V. cholerae*	+	+	+	+	+	+	–	d	–	–	–	+	–	+	–	–
V. mimicus	+	+	+	+	+	+	–	–	–	–	–	–	–	+	d	–
II. *V. parahaemolyticus*	–	+	+	+	+	+	–	–	+	–	–	–	+	+	–	–
V. vulnificus	–	+	+	+	+	d	–	–	–	–	–	–	–	d	(+)	+
V. alginolyticus	–	+	+	+	+	+	–	+	+	+	–	+	–	+	–	–
III. *V. fluvialis*	–	+	+	+	–	–	+	–	+	–	–	+	+	+	–	–
V. furnissii	–	+	+	+	–	–	+	–	+	–	+	+	+	+	–	–
V. damsela	–	+	+	+	d	–	+	+	–	–	d	–	–	–	–	–
IV. *V. hollisae*	–	+	+	+	–	–	–	–	–	–	–	–	+	–	–	–
V. *V. metschnikovii*	–	+	–	–	d	–	d	+	+	–	–	+	–	+	d	–

+ = ≥90% of strains positive; – = ≤9% of strains positive; d = 10 to 89% strains positive; (+) = delayed reaction.

nikovii is distinguished easily from others by negative reactions in oxidase and nitrate reduction tests. The remaining seven species can be further subdivided into three groups by lysine, ornithine, and arginine reactions—i.e., the group that has positive lysine but negative arginine reactions (*V. parahaemolyticus, V. alginolyticus,* and *V. vulnificus*); the group that is positive arginine but negative lysine and ornithine (*V. fluvialis, V. furnissii,* and *V. damsela*); and the group that is negative lysine, ornithine, and arginine (*V. hollisae*).

Tests useful for further differentiation of the species in these groups include Voges-Proskauer, indole, growth in nutrient broth with 8 and 10% NaCl, gas production from glucose, and acid production from sucrose, lactose, L-arabinose, D-mannitol, and salicin (Table 2). Commercially available kits developed for identification of the family *Enterobacteriaceae* can also be adopted for *Vibrio* species with slight modification in methodology (MacDonald et al., 1982).

Typing

V. cholerae non-O 1, *V. parahaemolyticus, V. mimicus, V. fluvialis, V. furnissii,* and *V. vulnificus* can be subdivided into several types based on O and/or K antigens (Brenner et al., 1982; Kudoh et al., 1983; Shimada and Sakazaki, 1983, 1984). These serotyping systems have proved to be useful for epidemiologic purposes, but usually they are adaptable only for reference laboratories. Biotyping described for non-O 1 *V. cholerae* (Heiberg, 1936), which now include eight types based on fermentation patterns of sucrose, L-arabinose, and D-mannose, is of little practical value, since strains belonging to the species in a strict sense could be divided into two types, I and II.

The Kanagawa hemolysin test is not a typing, but it will be done to learn the pathogenicity of *V. parahaemolyticus* isolates, particularly from outbreak cases. The test is usually done using Wagatsuma blood agar with 5% human or rabbit erythrocytes (Sakazaki and Balows, 1981), immunoassay (Honda et al., 1985; Ohta et al. 1979) or genetic assay (Nishibuchi et al., 1986).

Treatment, Prevention, and Control

Treatment

In patients with mild to moderate vibrio gastroenteritis, therapy with oral electrolyte solution only may be required. However, in severe cases with dehydration, intravenous fluid replacement is highly recommended. Antimicrobial therapy with tetracycline or aminoglycoside may shorten the duration of the illness and the excretion of vibrios in stool and this would apply to patients in whom rehydration therapy alone is not adequate. Plasmid-mediated drug-resistant strains have recently been found in some pathogenic vibrios (Matsushita et al., 1984).

Wound infections may usually be treated with antibiotic therapy or surgical debridement. In *V. vulnificus* primary septicemia, patients should receive good supportive care such as administraion of antibiotics (tetracycline or aminoglycoside) and an appropriate fluid for shock in the early stage of illness, since progression of the disease will be very rapid.

Prevention and Control

As most vibrioses are food-borne infections, their prevention and control will be the same in general principle as those of food-borne diseases. Reservoirs of vibrios are mostly in the marine environment, and natural contamination of fish and shellfish by the organisms is unavoidable, particularly in the warm season. Thus prevention measures should be stressed to assure that seafood is safe for consumption.

The measures include thorough cooking at a high enough temperature to kill contaminated vibrios, proper handling of cooked seafood to prevent cross-contamination from raw seafood or from seawater, and the keeping of seafood both raw and cooked, at a temperature that prevents multiplication of organisms. General measures for prevention of fecal-oral spreading of infections would be not effective.

Literature Cited

Arita, M., T. Takeda, T. Honda, and T. Miwatani. 1986. Purification and characterization of *Vibrio cholerae* non O1 heat-stable enterotoxin. Infect. Immun. **52**:45–49.

Blake, P. A., R. E. Weaver, and D. G. Hollis. 1980. Disease of humans (other than cholera) caused by vibrios. Annu. Rev. Microbiol. **34**:341–367.

Brenner, D. J., B. R. Davis, Y. Kudoh, M. Ohashi, R. Sakazaki, T. Shimada, and H. L. Smith, Jr. 1982. Serological comparison of two collection of *Vibrio cholerae* non O 1. J. Clin. Microbiol. **16**:319–323.

Farmer, J. J. III, F. W. Hickman-Brenner, and M. T. Kelly. 1985. Vibrio, p. 282–301. *In* E. H. Lennette, A. Balows, W. J. Hausler, Jr., and H. J. Shadomy (ed.), Manual of clinical microbiology, 4th ed. American Society for Microbiology, Washington, D.C.

Heiberg, R. 1936. The biochemical reaction of vibrios. J. Hyg. **36**:114–117.

Honda, T., M. Yoh, U. Kongmuang, and T. Miwatani. 1985. Enzyme-linked immunosorbent assays for detection of thermostable direct hemolysin of *Vibrio parahaemolyticus*. J. Clin. Microbiol. **22**:383–386.

Kamal, A. K., and Y. Zinnaka. 1971. Outbreaks of gastroenteritis by non-agglutinable (NAG) vibrios in the republic of the Sudan, October-November 1968. J. Egypt. Pub. Health Assoc. **46:**125–174.

Kelly, M. T., J. W. Peterson, H. E. Sarles, M. Romanko, D. Martin, and B. Hafkin. 1982. Cholera on the Texas gulf coast. J. Am. Med. Assoc. **247:**1598–1599.

Kudoh, Y., and S. Sakai. 1985. Current status of bacterial diarrheal diseases in Japan, p. 83–93. *In* Y. Takeda and T. Miwatani (ed.), Bacterial diarrheal diseases. KTK Science Publishers, Tokyo.

Kudoh, Y., S. Matsushiata, Y. Yamada, M. Tsuno, K. Ohata, S. Sakai, and M. Ohashi. 1981. Enterotoxin producibility and some biological features of O1 and non O1 *Vibrio cholerae* isolates, p. 214–224. *In* S. Kuwahara and Y. Zinnaka (ed.), Proc. 16th Joint Conf. U.S.-Jpn. Coop. Med. Sci. Prog. Gifu. Toho Univ. Tokyo.

Kudoh, Y., M. Tsuno, S. Matsushita, S. Yamada, K. Ohta, S. Sakai, and M. Ohashi. 1983. Enteropathogenicity and some biological features of Group F (EF-6) vibrio isolates, p. 75–86. *In* S. Kuwahara and N. F. Pierce (ed.), Advances in research on cholera and related diarrheas. KTK Science Publishers, Tokyo.

MacDonald, M. T., F. L. Sinleton, and M. A. Hood. 1982. Diluent composition for use of API2OE in characterizing marine and estuarine bacteria. Appl. Environ. Microbiol. **44:**423–427.

Matsushita, S., K. Ohta, M. Tsuno, S. Yamada, Y. Kudoh, S. Sakai, and M. Ohashi. 1985. Antibiotic resistance of *Vibrio cholerae* recently isolated from human sources in Japan and other countries, p. 173–178. *In* S. Kuwahara and N. F. Pierce (ed.), Advances in research on cholera and related diarrheas. KTK Science Publishers, Tokyo.

Morris, J. G. Jr., and R. E. Black. 1985. Cholera and other vibrioses in the United States. N. Engl. J. Med. **312:**343–350.

Nishibuchi, M., R. J. Seidler, D. M. Rollins, and S. W. Joseph. 1983. *Vibrio* factors cause rapid fluid accumulation in suckling mice. Infect. Immun. **40:**1083–1091.

Nishibuchi, M., W. E. Hill, G. Zon, W. L. Payne, and J. B. Kaper. 1986. Synthetic oligodeoxyribonucleotide probed to detect Kanagawa phenomenon-positive *Vibrio parahaemolyticus*. J. Clin. Microbiol. **23:**1091–1095.

Ohashi, M., T. Shimada, and H. Fukumi. 1972. In vitro production of enterotoxin and hemorrhagic principle by *Vibrio cholerae*, NAG. Jpn. J. Med. Sci. Biol. **25:**179–194.

Ohta, K. Y. Kudoh, M. Tsuno, S. Sakai, T. Maruyama, T. Ito, and M. Ohashi. 1979. Development of a sensitive serological assay based on reversed passive hemagglutination for detection of enteropathogenic toxin (Kanagawa hemolysin) of *Vibrio parahaemolyticus* and reevaluation of the toxin producibility of isolates from various sources. Jpn. J. Bacteriol. **34:**837–846 (in Japanese with English summary).

Sakazaki, R., and A. Balows. 1981. The genera *Vibrio, Plesiomonas* and *Aeromonas*, p. 1272–1301. *In* M. P. Starr, H. Stolp, H. G. Trüper, A. Balows, and H. G. Schlegel (ed.), The prokaryotes. Springer-Verlag, Berlin.

Sakazaki, R., K. Tamura, T. Kato, Y. Obara, S. Yamai, and K. Hobo. 1968. Studies on the enteropathogenic, facultatively halophilic bacteria, *Vibrio parahaemolyticus*. III. Enteropathogenicity. Jpn. J. Med. Sci. Biol. **21:**325–331.

Shimada, T., and R. Sakazaki. 1983. Serological studies on *Vibrio fluvialis*. Jpn. J. Med. Sci. Biol. **36:**315–323.

Shimada, T., and R. Sakazaki. 1984. On the serology of *Vibrio vulnificus*. Jpn. J. Med. Sci. Biol. **37:**241–246.

Simpson, L. N., and J. P. Oliver. 1983. Sidrophore production by *Vibrio vulnificus*. Infect. Immun. **41:**644–649.

Spira, W. M., and D. J. Fedorka-Cray. 1984. Purification of enterotoxin from *Vibrio mimicus* that appear to be identical to cholera toxin. Infect. Immun. **45:**679–684.

Tison, D. L., and M. T. Kelly. 1984. *Vibrio* species of medical importance. Diagn. Microbiol. Infect. Dis. **2:**263–276.

Yamamoto, K., Y. Ichinose, N. Nakasone, M. Tanabe, M. Nagahara, J. Sakurai, and M. Iwanaga. 1986. Identity of hemolysin produced by *Vibrio cholerae* non 01 and *V. cholerae* 01 biotype eltor. Infect. Immun. **51:**927–931.

Yoh, M., T. Honda, and T. Miwatani. 1986. Purification and partial characterization of a non-01 *Vibrio cholerae* hemolysin that cross-reacts with thermostable direct hemolysin of *Vibrio parahaemolyticus*. Infect. Immun. **52:**319–322.

Zen-Yoji, H., Y. Kudoh, H. Igarashi, K. Ohta, K. Fukai, and T. Hoshino. 1976. Further studies on characterization and biological activities of an enteropathogenic toxin of *Vibrio parahaemolyticus*, p. 479–498. *In* A. Ohsaka, K. Hayashi, and Y. Sawai (ed.), Animal, plant and microbial toxins, Vol. 1. Plenum, New York.

CHAPTER 56

Yersinioses Other than Plague

M. TSUBOKURA

Disease: Yersinioses (*Yersinia pseudotuberculosis* infection; pseudotuberculosis, *Yersinia enterocolitica* infection).

Etiologic Agents: *Y. pseudotuberculosis* serotypes 1 to 13; *Y. enterocolitica* biotypes 1 to 5 and serogroups 1 to 17.

Source: *Y. pseudotuberculosis* is detectable from healthy animals such as pig, dog, cat, hare, chinchilla, and rat; also meat and water. *Y. enterocolitica* has been recovered from many kinds of healthy animals, foodstuffs, vegetables, and water.

Clinical Manifestations: Fever, abdominal pain, diarrhea, erythema arthritis, appendicitis, nausea, and vomiting are found in both *Y. pseudotuberculosis* and *Y. entrocolitica* infections. Strawberry tongue, desquamation of digits, and myalgia are found occasionally in *Y. pseudotuberculosis* infection.

Pathology: *Y. pseudotuberculosis* and *Y. enterocolitica* infections cause a variety of manifestations in humans including acute mesenteric lymphadenitis, gastroenteritis, appendicitis, septicemia, hepatosplenic abscesses, and terminal ileitis (only *Y. enterocolitica*). Rabbit, hare, guinea pig, monkey, and birds are susceptible to *Y. pseudotuberculosis*. Infected animals show enlarged spleen, mesenteric lymph nodes, and many various-size grayish-white foci in the spleen and liver. Catarrhal and diphtheric enteritis.

Laboratory Diagnosis: Bacteriologic isolation of etiologic agent from blood, mesenteric lymph node, appendix, and stool of patient. Cecal content, mesenteric lymph node, visceral organs, and stool of diseased and healthy animals; tongue of healthy pig. Tube agglutination test titer ≥1 : 160, or rise or decline in titers of fourfold or more.

Epidemiology: Worldwide distribution, in different part of temperate zones, such as European countries, North America, Japan, New Zealand, and South Africa. Rare in tropical areas.

Treatment: Tetracyclines, aminoglycosides, chloramphenicol, and trimethopirim-sulfamethoxazole are active against *Y. pseudotuberculosis* and *Y. enterocolitica* by drug sensitivity tests. The new third-generation drugs and beta-lactam antibiotics are active against *Y. pseudotuberculosis* but not *Y. enterocolitica*. However, therapeutic efficacy has not yet been evaluated.

Prevention and Control: Sanitary management of food and water. Uncooked pork and tongue of pig, unwashed vegetables, and nonchlorinated drinking water should be avoided, and care is necessary for contact with domestic pets.

Description of Disease

There are three species of pathogenic *Yersinia* organisms. These are *Yersinia pestis, Y. pseudotuberculosis* and *Y. enterocolitica*. The latter two species, *Y. pseudotuberculosis* and *Y. enterocolitica* cause enteric disorders in human and animals through oral infection.

In 1910, Albrechit isolated *Y. pseudotuberculosis* from ileocecal lymph nodes of a boy who for suspected appendicitis. Since then, over a period of about 40 years, fewer than 20 cases of *Y. pseudotuberculosis* infection in humans have been reported, and this infection has not aroused much interest. In 1953, Masshoff and Dölle observed pathologic changes in the mesenteric lymph nodes of children suffering from acute appendicitis. They attributed this to possible viral infection, because the pathologic lesion resembled cat scratch fever and lymphogranuloma inguinale. In the following year, Knapp and Masshoff (1954) concluded that the causative agent was *Y. pseudotuberculosis*. Since then, many reports have appeared on human infections with this organism in European countries, North and South America, India, and Japan (Butler, 1983).

Human infection with *Y. enterocolitica* was first reported by Schleifstein and Coleman (1939). Since the second case of human infection was described by Hässig et al. in 1949, the incidence of human infection with *Y. enterocolitica* has increased in recent years, and many cases have been reported from European countries, North and South America, South Africa, and Japan (Bottone, 1981; Butler, 1983; Winblad, 1973).

Y. pseudotuberculosis and *Y. enterocolitica* have become important agents of zoonotic infection.

Communicability

Human infection with both *Yersinia* organisms occurs by the oral route. Direct infection through contact with infected animals or their excreta occurs frequently, although the most likely mode of infection is ingestion of contaminated meat or other infected foodstuffs or by drinking contaminated water or milk. A few cases of person-to-person spread of infection of *Y. enterocolitica* have been reported, but no evidence exists for a similar route of *Y. pseudotuberculosis* infection. The role of transmission by insect vectors remains uncertain. *Y. pseudotuberculosis* has been isolated from ixodic ticks collected from cows. *Y. pseudotuberculosis* is pathogenic to mice, and there is a relationship between serologic types of the organisms and virulence. *Y. enterocolitica* O3 and many other serogroups of strains have been isolated from flies collected in piggeries.

Clinical Manifestations

The clinical symptoms and pathology of *Y. pseudotuberculosis* and *Y. enterocolitica* infections are similar. Common clinical symptoms of *Y. pseudotuberculosis* and *Y. enterocolitica* infections are fever, abdominal pain, diarrhea, nausea, and vomiting. Secondary immunologic complications such as erythema nodosum, arthritis, and renal insufficiency have also been observed. In *Y. pseudotuberculosis* infection, strawberry tongue, desquamation of digits, and myalgia are occasionally found. The clinical symptoms observed in Japan are shown in Table 1.

The organisms cause a variety of manifestations in humans, including acute mesenteric lymphadenitis (chiefly *Y. pseudotuberculosis*), gastroenteritis (chiefly *Y. enterocolitica*), terminal ileitis (only *Y. enterocolitica*), appendicitis, and septicemia.

TABLE 1. Clinical features of outbreaks of *Y. pseudotuberculosis* and *Y. enterocolitica* infections

	Y. pseudotuberculosis[a]			Y. enterocolitica[b]									
	1	*2*	*3*	*1*	*2*	*3*	*4*	*5*	*6*	*7*	*8*	*9*	*10*
No. of patients	188	20	198	188	544	198	296	145	6	184	1,051	641	102
Fever	89.4	86.1	77.1	75.5	49.6	60.6	90.0	68.4	16.7	88.5	26.2	52.6	88.3
Abdominal pain	71.8	79.1	60.0	87.5	63.9	76.3	70.7	84.2	100.0	61.5	77.4	89.7	77.8
Diarrhea	51.6	65.1	32.9	26.5	32.4	35.9	72.7	21.1	50.0	36.5	36.9	55.5	77.8
Erythema	66.5	55.8	48.6										
Nausea / Vomiting	51.1	55.8	48.6										
Nausea				23.5	24.8	7.1					72.6	30.1	
Vomiting				4.0	10.6	12.1	38.2	26.3		3.8	67.9	10.3	22.3
Common cold–like symptoms	50.0	20.0	30.0										
Strawberry tongue	40.4	9.3	20.0										
Desquamation	39.4	41.9	11.4										
Myalgia	9.6	23.3	5.7										
Headache	43.4	53.5	34.3	62.2		60.1	78.2			5.8	58.3		22.3
Malaise				30.6		33.3	30.9					8.0	11.1
Serogroup	5a	5b	4b	O3	O3	O3	O3	O3	O3	O3	O3	O3	O3

[a] Inoue et al. (1985).
[b] Maruyama (1982, 1986).

Epidemiologic Aspects

Geographic Distribution

Yersinia organisms appear to have a worldwide distribution. A large number of human infections with *Y. pseudotuberculosis* and *Y. enterocolitica* have been reported from European countries, North America, Japan, New Zealand, and South Africa, although there is little information on the incidence of this infection in tropical areas (Bottone, 1981; Butler, 1983; Carter et al., 1977; Regamey et al., 1968; Winblad, 1973).

Normal Carriers Among Humans

Normal carriers of *Y. pseudotuberculosis* have seldom been observed. Maruyama (1982), in Japan, reported that *Y. enterocolitica* were isolated from 17 (0.004%) of the 456,044 healthy humans.

Age and Sex

Children appear to be preferentially affected by both *Y. pseudotuberculosis* and *Y. enterocolitica* infections. In European countries, *Y. pseudotuberculosis* infection appears more frequently in the 8-to-15-year age group, and *Y. enterocolitica* infection occurs between 1 and 3 years of age. *Y. pseudotuberculosis* infection in males outnumbers that in females by about 3 to 1.

In Japan, *Y. pseudotuberculosis* infection appears to be concentrated in the 1-to-16 year age group, with a peak at 1 to 3 years of age. Infection in males slightly outnumbers that in females.

Seasonal Incidence

Yersiniosis shows a definite seasonal variation. In European countries and Japan, the monthly prevalence of human infection with *Y. pseudotuberculosis* is higher in colder months. However, human infection with *Y. enterocolitica* appears in warmer or summer seasons. No clear seasonal patterns of *Y. enterocolitica* infection are apparent in the United States.

Isolation of *Y. pseudotuberculosis* and *Y. enterocolitica* from animals occurs more frequently in winter and early spring than in summer.

Reservoirs and Environmental Sources

The asymptomatic carriers of *Y. pseudotuberculosis* are pig, dog, cat, hare, and rat. Isolation of the organism from food and environmental sources is rare (Tsubokura, 1985).

Y. enterocolitica has been isolated from many species of healthy animals and environmental sources (Bottone, 1981; Butler, 1983). However, serotypes associated with human disease have been isolated infrequently from animals. The main, frequently detectable, asymptomatic carriers of the

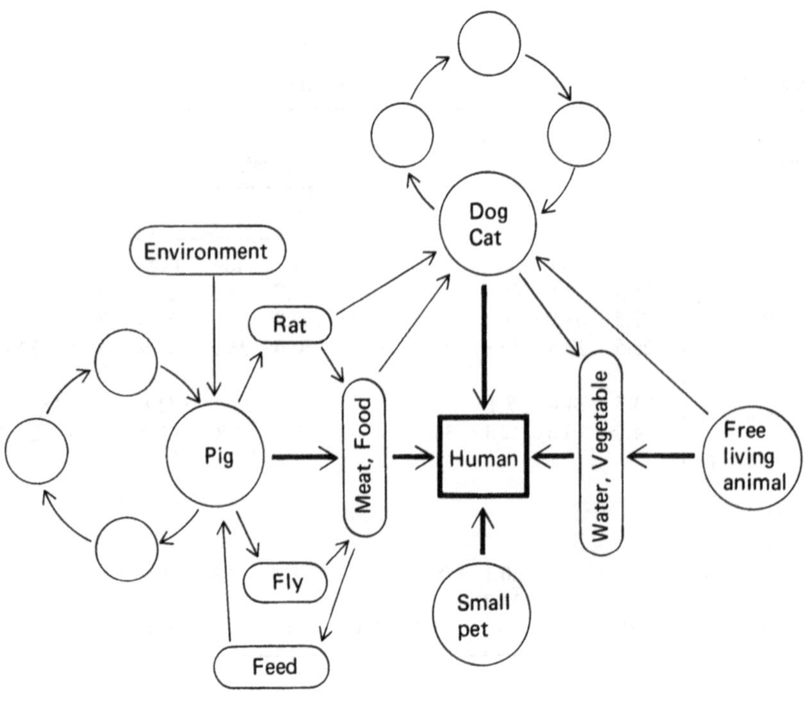

FIG. 1. Yersiniosis cycle.

pathogenic strains of *Y. enterocolitica* are pig, dog, cat, rat, and chinchilla.

Community Outbreaks

Community outbreaks of *Y. pseudotuberculosis* in Hungary, Finland, and Japan (Inoue et al., 1985; Regamey et al., 1968; Tertti et al., 1984) and of *Y. enterocolitica* in the United States, Canada, Czechoslovakia, the Netherlands, and Japan (Bottone, 1981; Butler, 1983; Maruyama, 1982, 1986) have been recognized.

Yersiniosis Cycle

Four major groups—pig, dog/cat, small pet (rodents and birds), and free-living animals (small rodents and birds)—are important in the yersiniosis cycle and transmission to humans (Fig. 1).

Etiologic Agents

Characterization and Classification

Y. pseudotuberculosis and *Y. enterocolitica* are classified into the genus *Yersinia*. They belong to the family *Enterobacteriaceae*.

Morphologic and Biochemical Properties

Y. pseudotuberculosis and *Y. enterocolitica* have similar morphologic and biochemical properties. They are gram-negative coccoid to rod-shaped with a rounded end (0.5 to 1.0 × 1 to 2 μ). Occasionally, long curved filaments may occur. They are facultative anaerobes. When grown at temperatures below 30°C, they are motile by peritrichous flagella, but this is not the case at 37°C (*Y. pseudotuberculosis* is occasionally nonmotile at temperatures below 30°C). They are not capsulated and nonspore-forming. They grow easily on meat-extracted media. Optimum temperature is 30°C. Generation time is about 40 min. Certain strains of *Y. pseudotuberculosis* of serotype 3 produce exotoxin (Thal, 1954). The main characteristics of the two species are shown in Table 2.

TABLE 2. Characteristics of *Y. pseudotuberculosis* and *Y. enterocolitica*

Reaction	Y. pseudo-tuberculosis	Y. entero-colitica
Indole (30°C)	−	d
Methyl red	+	+
Voges-Proskauer (25°C)	−	+
Simmons citrate	−	−
Nitrate reduction	+	+
H_2S (TSI)	−	−
Urease	+	+
Gelatinase (22°C)	−	−
Phenylalanine deaminase	−	−
Lysine decarboxylase	−	−
Arginine dehydrolase	−	−
Ornithine decarboxylase	−	+
KCN	−	−
Motility 25°C	d	+
37°C	−	−
Gas from glucose	−	−
Glucose	+	+
Arabinose	+	+
Cellobiose	−	+
Lactose	−	d
Raffinose	d	−
Rhamnose	+	−
Sucrose	−	+
Treharose	+	+
Xylose	+	d
Melibiose	+	−
Adonitol	−	−
Dulcitol	−	−
Mannitol	+	+
Sorbitol	−	+
Glycerin	+	+
Salicin	+	d
α-Methyl glucosid	−	−
Esculin hydrolysis	+	−
Lipase (Tween 80) (25°C)	−	d
β-Galactosidase (25°C)	+	+
Pigment	−	−
Growth on MacConkey agar (25°C)	+	+
Growth on SS agar (25°C)	− or +	+
G + C %	46.5	48.5 ± 1.5

Tests were done at 37°C unless otherwise specified.

Biotype

No biotype is recognized in *Y. pseudotuberculosis*. However, melibiose nonfermenting strains exist in strains of serotype 3. These strains are avirulent and atoxic.

Biotyping of *Y. enterocolitica* has been demonstrated by several workers. Table 3 shows the different characteristics of biotypes proposed by Wauters (1970). Bercovier et al. (1978) reported that the biotype 3 is divisible to biotypes 3A and 3B. In Japan, biotype 3B has recently been isolated predominantly in patients (Maruyama and Kanzaki, 1985) and healthy pigs (Fukushima and Tsubokura, 1985). It is well known that biotypes are related to serotypes and their sources (Table 4).

TABLE 3. Biotypes of *Y. enterocolitica* (Wauters, 1970)

	Biotype						
Reaction	*1*	*2*	*3*	*4*	*5*	*3A*[a]	*3B*[a]
Lecithinase	+	−	−	−	−	−	−
Indole (29°C)	+	+	−	−	−	−	−
Lactose (oxyd.) 48 h	+	+	+	−	−	+	+
Xylose 48 h	+	+	+	−	−	+	+
Nitrate reduction	+	+	+	+	−	+	+
Treharose	+	+	+	+	−	+	+
Ornithine decarboxylase	+	+	+	+	−	+	+
β-galactosidase	+	+	+	+	−	+	+
Voges-Proskauer[a]	+	+	+	+	+	−	−
Sorbose[a]	+	+	+	+	d	+	−
Inositol[a]	+	+	+	+	+	+	−

[a] Biotypes and reactions of Bercovier et al. 1978.

TABLE 5. The antigenic scheme of *Y. pseudotuberculosis* (Tsubokura et al., 1984)

Group	*Subgroup*	*O antigens*	*H antigens*
1	a	2, 3	a, c
	b	2, 4	a, c
2	a	5, 6, 16	a, d
	b	5, 7, 16, 17	a, d
	c	5, 7, 11, 18	ND
3		8	a
4	a	9, 11	b; a, b
	b	9, 12	a, b, d
5	a	10, 14	a; a, e, (b)
	b	10, 15	a
6		13	a
7		19	ND
8		20	ND

ND = Not determined.

Serologic Properties

Y. pseudotuberculosis possesses 20 somatic (O) and 5 flagella (H) antigens, and 13 serotypes have been classified (Tsubokura et al., 1984) (Table 5). Although *Y. pseudotuberculosis* has been known to possesses O1 antigen, this antigen is excluded from the antigenic scheme because the O1 antigen is thought to be an R antigen.

Y. pseudotuberculosis possesses various common antigens with other enterobacteria. Relationships exist between *Y. pseudotuberculosis* serotype 2 and O4 and O27 of the *Salmonella* B group; *Y. pseudotuberculosis* serotype 3 and *Y. enterocolitica* O20; *Y. pseudotuberculosis* serotype 4 and O9 and O46 of the *Salmonella* D group and O14 of the *Salmonella* H group, and *Escherichia coli* O groups 17 and 77; and between *Y. pseudotuberculosis* serotype 6 and *E. coli* O group 55.

Y. enterocolitica possesses 57 somatic (O), 1 envelope (K), and 19 flagella (H) antigens, and at least 17 serogroups have been classified (Bottone, 1981; Wauters et al., 1971, 1972) (Table 6). There is an antigenic relationship between *Y. enterocolitica* and *Brucella* species and other enterobacteria (Bottone, 1981). In particular, cross-reacting antigens between *Y. enterocolitica* O9 and *Brucella* species are of importance in the serologic diagnosis of brucellosis in domestic animals.

TABLE 4. Biotypes, serotypes, and their related sources

Biotype	*Serotype*	*Source (major)*
1	8 and many types	Man, animals, and environmental materials
2	5B and 9	Man and pig
3	1 and 3[a]	Chinchilla, man[a] and pig[a]
4	3	Man and pig
5	1	Hare

[a] Detected in Japan.

In European countries, 90% or more of strains of *Y. pseudotuberculosis* isolated from humans fell into serotypes 1 and 2 (the majority of the strains being 1a). In Japan, serotypes 1b, 4b, 5a, and 5b are frequently detectable, and serotypes 2a, 2b, 2c, 3, and 4a have also been observed as human pathogens (Tsubokura et al., 1984).

Most *Y. enterocolitica* infections in Canada, Europe, and Japan are due to serotype O3. In contrast, serotype O9 is recorded almost exclusively in Scandinavian countries, while serotype O8 is recorded in the United States (Butler, 1983; Winblad, 1973).

Pathogenicity and Pathogenic Factors

All serotypes of *Y. pseudotuberculosis* are pathogenic to mice, and serotypes 1 to 5 have been isolated from patients. *Y. enterocolitica* serotypes O3, O5B, O8, and O9 are human pathogens.

Both *Y. pseudotuberculosis* and *Y. enterocolitica* have been recognized to possess temperature-dependent virulence factors. The V and W antigens, which are associated with calcium dependency (growth inhibition when cultures are incubated on oxalate medium (Higuchi and Smith, 1961), are synthesized at 37°C. Most marked synthesis of the V and W antigens occurs in the absence of calcium and in the presence of magnesium ions. These conditions approximate the intracellular environment of the host cells. The organism is resistant to phagocytosis in this condition.

Y. pseudotuberculosis and pathogenic *Y. enterocolitica* are sensitive to the killing action of 10% nor-

TABLE 6. The antigenic scheme of *Y. enterocolitica* (Wauters et al., 1971, 1972)

Group	*O, K antigens*[a]	*H antigens*
1	1, 2a, 3	a, b, c
2	2a, 2b, 3	b, c
3	3	a, b, c
	3	c
4	4, 32	b, e, f, i
	4, 33	
5A	5	b, c, d, e, i
	5	a, b, c, d, g
	5, 27	
5B	5	a, b, c
6	6	d, e, f, g, i
	6, 30	a, b, d, g, i
	6, 31	
	6	b, d, e, g, i
7	7, 8	d, e, f, g, h
	7, 8	b, d, e, g, i
	7, 8	b, c, d, k
8	8	b, e, f, i
	8, 19	
9	9	a, b
10	10: K1	b, f, k
	10: K1	b, c, d, e, i
	10	b, c, e, f, i
	10	o
11	11, 23	l
	11, 24	r
12	12, 25	o
	12, 26	l
	12, 26	t
13	13, 7	n
14	14	m
15	15	c
16	16	
	16, 29	p
17	18	q
	18	
	20	
	21	
	22	
	28	s
	34	

[a] In addition, the new O factors, O35 to O57, were recognized (Bottone, 1981; Wauters, 1981).

mal human serum when the organisms are grown at 26°C, whereas they show resistance when grown at 37°C (serum resistant).

Y. pseudotuberculosis and pathogenic *Y. enterocolitica* produce outer membrane proteins (YOPs) in vivo and in vitro at 37°C. These consist of at least five major outer membrane polypeptides. Of them, the V antigen is a protein with a molecular weight of 90,000 and the W antigen is a lipoprotein with a molecular weight of 145,000. Pathogenic *Y. enterocolitica* have a mouse virulence YOP that differs from that of the V antigen. Outer membrane proteins are also associated with autoagglutination and serum resistance.

Pathogenic *Y. enterocolitica* and newly isolated strains of *Y. pseudotuberculosis* bind hemin or Congo red (pigmentation). When they are cultured on Congo red agar plates at 25°C, pigmented colonies appear.

Strains of *Y. enterocolitica* produce a heat-stable enterotoxin when cultured at a low incubation temperature such as 25°C, but not at 37°C. Because the enterotoxin is produced not only by the pathogenic strains but also by the nonpathogenic *Y. enterocolitica* and other nonpathogenic *Yersinia* organisms, the pathogenesis of diarrhea has not been explained. Most virulence factors are plasmid-mediated. Table 7 shows plasmids of *Y. pseudotuberculosis* and *Y. enterocolitica* controlling virulence.

Y. Pseudotuberculosis and *Y. Enterocolitica* Infections in Animals

Y. pseudotuberculosis causes pseudotuberculosis in many kinds of mammals and birds. Hare, guinea pig, rabbit, and monkey are probably the most susceptible to *Yersinia* organisms. Necropsy of animals with systemic infection shows enlarged spleen and mesenteric lymph nodes and many various-size grayish-white foci in the spleen and liver. Catarhal and diphtheric enteritis are often observed.

Hare and chinchilla appear to be highly susceptible to *Y. enterocolitica,* and sporadic infection occurs in other animals such as pig, dog, monkey sheep, guinea pig, rabbit, and cattle (Mair, 1973). The incidence of *Y. enterocolitica* infection in these animals is lower than for *Y. pseudotuberculosis* infection. At necropsy, visible lesions are similar to those of *Y. pseudotuberculosis* infection.

TABLE 7. Plasmids of *Y. pseudotuberculosis* and *Y. enterocolitica* controlling virulence

Virulence factor	*Plasmid (megadalton)*
VW antigens and calcium dependency	42–48
Serum resistance	41–45 (*Y. pseudotuberculosis* is independent)
YOPs	42–46/82
Pigmentation	40–50
Tissue invasive	42–44

Diagnostic Features

Clinical Diagnosis

A clinical diagnosis of yersinioses should be suspected in every patient with fever, abdominal pain, diarrhea, nausea, and vomiting. Symptoms of exanthema, strawberry tongue, and desquamation of the skin in the digits are suspected of being due to *Y. pseudotuberculosis* infection. Secondary immunologic complications such as erythema nodosum, arthritis, and renal insufficiency should also be noted.

Blood and urinary examination may be an aid to the diagnosis of *Yersinia* infection, including positive CRP (≥3mm), elevated WBC count (≥15,000/mm^3), and occasional shift to the left of neutrophils and urinary protein.

Bacteriologic Diagnosis

Because *Yersinia* organisms have the characteristic of low optimum temperature (22 to 30°C) and have a long generation time, care should be taken in their isolation.

Materials for Isolation

The following materials are suitable for the isolation of *Yersinia:* blood, lymph nodes, visceral organs, stool, intestinal contents, foodstuffs, and drinking water. A long time is required for isolation, and it is desirable that the materials be cooled. Carry-Blair medium is suitable for transport of the materials.

Isolation Procedure

Direct Culture

For isolation of *Yersinia* organisms from noncontaminated samples such as blood, lymph nodes, and visceral organs, blood agar or nutrient agar is used. Inoculated plates are incubated at 22 to 28°C for 48 h. Isolation of the organism from contaminated samples such as stools, intestinal contents, and foodstuffs should use selective media such as MacConkey agar, SS (Salmonella and Shigella) agar, and CIN (cefsulodin, irgasan, and novobiocin) agar plates. Incubation is made at 22 to 28°C for 48 h. Isolation from water samples is performed as follows: 1) add 2 ml of 10% $FeCl_2$ solution to 2 liters of water sample and mix well. Leave this at room temperature for 2 to 3 h, after which about 50 ml of precipitated material is collected and centrifuged at 3,000 rpm for 10 min. Approximately 2 ml of centrifuged material is obtained and used for isolation (Fukushima et al., 1984). 2) One liter or more of the water sample is filtered with a membrane filter (0.45 μm pore size). Then the surface of the membrane is touched on the selective media, and the entire membrane is used for enrichment culture.

Enrichment Culture

Isolation of *Yersinia* organisms from blood is done in the culture bottle. The sample is cultivated at 22 to 28°C for 2 to 3 days, and isolation is made on blood agar or MacConkey agar plates at 22 to 28°C for 2 to 3 days. The cold enrichment technique is useful for isolation of *Yersinia* organisms from contaminated samples. About 1 g of sample is suspended in 10 ml of phosphate buffer solution (0.067 *M*, pH 7.6) and the mixture is held at 4 to 7°C for 3 to 4 weeks (Paterson and Cook, 1963). The isolation is made on selective media at weekly intervals. This PBS cold enrichment technique is not suitable for the rapid diagnosis of *Yersinia* infection. As other enrichment media, peptone-mannitol-PBS medium (Fukushima et al., 1984) and modified Rappaport broth (Winblad, 1973) are useful.

Alkali Treatment

This is a new culture method for isolation of *Yersinia* organisms from an enrichment sample (Aulisio et al., 1980). The alkali treatment kills a larger number of contaminating non-*Yersinia* bacteria and thus facilitates the isolation of *Yersinia*. About 0.5 ml of the enriched sample is added to 4.5 ml of 0.5% KOH solution, agitating for 30 seconds with a Vortex mixer, and immediately streaking with a large loop onto selective media.

Character of Colony

Colonies of *Yersinia* organisms are smaller than colonies of other enteropathogenic bacteria. At temperatures below 30°C, the surfaces of colonies are smooth, glistening, and convex, but at 37°C, colonies are flat with an undulate edge and often granular-appearing surface. At 25°C for 48 h, colonies on blood agar or nutrient agar plates are 1.5 to 3 mm in diameter, grayish-white, opaque, circular, and convex. They are nonhemolytic. On SS agar plates, colonies of *Y. enterocolitica* are circular and 1.5 to 2 mm in diameter. They have a pale pink center and a transparent periphery. *Y. pseudotuberculosis* shows scanty growth on SS agar plate. On CIN agar plate, colonies of *Yersinia* organisms are circular, 1 to 3 mm in diameter, and dark pink with a translucent edge. They may be surrounded by a zone of precipitated bile. Certain strains of *Y. pseudotuberculosis* inhibit growth by adding supplement.

Confirmatory Tests

Suspicious colonies are transferred to confirmatory media [TSI (triple-sugar-iron) agar, SIM (sulfide-indole-motility) medium, LIM (lysine-indole-motility) medium, and urease]. Voges-Proskauer tests are performed. Furthermore, some characteristics—lipase, ornithine decarboxylase, rhamnose, cellobiose, sorbitol, and esculin—should be tested, and then serologic and phage typing are performed.

The main characteristics that distinguish the *Yersinia* species are shown in Table 8.

Serologic Diagnosis

Serologic diagnosis is the most reliable and frequently used procedure applicable to human infection with *Yersinia* organisms. The O agglutination test for *Yersinia* organisms appears to be specific. Antibody response is rapid, occurring by day 5 to day 10 of illness; therefore, paired serum samples taken at intervals of 7 to 10 days should be examined. Agglutinin titers elevate, with peak titers at 3 weeks, then decrease gradually (Bottone, 1981).

Agglutination Test

The technique most widely used is the tube agglutination and microtiter method. Sera must be inactivated at 56°C for 30 min, and a serial 2 dilution is made. O antigens are used for the test. The saline suspensions of organisms grown at 25°C are heated at 100°C for 30 min or at 121°C for 2.5 h, and the density is adjusted in a No. 1 MacFarland tube. Because the distribution of serotypes of *Yersinia* organisms differs from one country to another, the most suitable serotype for diagnostic antigens in each country should be selected.

A mixture of antigen and serum is incubated at 50 to 52°C for 18 to 20 h or at 50°C for 2 h and then kept at 5°C overnight. Agglutinin titers of ≥1 : 160 or an increase or decline in their titers of fourfold or more can be judged positive.

Other Serologic Tests

Among the serologic tests other than agglutination for detection of antibodies in *Yersinia* infections, hemagglutination, the complement fixation test (CF), and enzyme-linked immunosorbent assay are applicable.

Relationship Between *Y. pseudotuberculosis* Infection, Kawasaki Syndrome (Mucocutaneous Lymph Node Syndrome), and Izumi Fever

Kawasaki syndrome is an important illness in children in Japan. The clinical features of this syndrome closely resemble those of *Y. pseudotuberculosis* infection. Many cases of *Y. pseudotuberculosis* infection show five or more diagnostic criteria of Kawasaki syndrome (i.e., fever, erythema, rash, desquamation of the skin in digits, strawberry tongue, and swelling of the lymph nodes). Sato et al. (1983) and Chiba et al. (1983) revealed that the anti-

TABLE 8. Characteristics differentiating the species of the genus *Yersinia*[a]

Characteristics	Y. pest.	Y. pseu.	Y. ent.	Y. interm.	Y. fred.	Y. kristen.	Y. ruck.	Y. aldo.[b]
Indole	–	–	d	+	+	d	–	–
Voges-Proskauer (25°C)	–	–	+	+	+	–	d	–
H_2S	–	–	–	–	–	–	–	–
Urease	–	+	+	+	+	+	–	+
Lysine decarboxylase	–	–	–	–	–	–	+	–
Ornithine decarboxylase	–	–	+	+	+	+	+	+
Motility 25°C	–	d	+	+	+	+	d	+
35°C	–	–	–	–	–	–	–	–
Glucose	+	+	+	+	+	+	+	+
Cellobiose	–	–	+	+	+	+	–	–
Lactose	–	–	d	–	d	d	–	·
Rhamnose	–	+	–	+	+	–	–	+
Sucrose	–	–	+	+	+	–	–	–
Melibiose	d	+	–	+	–	–	–	–
Sorbitol	–	–	+	+	+	+	–	+
Esculin hydrolysis	+	+	d	+	+	d	–	·
Lipase (Tween 80)	–	–	d	d	d	d	·	d

[a] Bergey's Manual of Systematic Bacteriology, Vol. 1, 1984.
[b] Bercovier et al., 1984.

Y. pest. = Y. pestis; *Y. pseu.* = Y. pseudotuberculosis; *Y. ent.* = Y. enterocolitica; *Y. interm.* = Y. intermedia; *Y. fred.* = Y. frederiksenii; *Y. kristen.* = Y. kristensenii; *Y. ruck.* = Y. ruckeri; *Y. aldo.* = Y. aldovae.

body against *Y. pseudotuberculosis* is detectable in children with Kawasaki syndrome. The causal agent of Kawasaki syndrome is unknown. Because it is considered that Kawasaki syndrome is not attributed to *Y. pseudotuberculosis* infection, the patient with *Y. pseudotuberculosis* infection should be distinguished from one with Kawasaki syndrome.

Izumi fever is described as an illness in children in Japan. Clinical features of this illness include fever, biphasic rash, erythema nodosum, and abdominal pain. The causal agent is unknown. Sato et al. (1983) have emphasized that *Y. pseudotuberculosis* is the cause of Izumi fever by serologic study.

Treatment and Control

Treatment

Y. pseudotuberculosis and *Y. enterocolitica* are susceptible to many antibiotics by drug sensitivity tests in vitro. Most *Y. pseudotuberculosis* and *Y. enterocolitica* strains are susceptible to tetracyclines, aminoglycosides, chloramphenicol, and trimethoprim-sulfamethoxazole. The new third-generation drugs and beta-lactam antibiotics are active against *Y. pseudotuberculosis,* but not *Y. enterocolitica.* However, efficacy has not been evaluated in a controlled fashion by antibiotic treatment for yersinioses.

Control

Because human infection with *Y. pseudotuberculosis* and *Y. enterocolitica* occurs through ingestion of contaminated food or water and because *Yersinia* organisms are able to grow in refrigerators, sanitary management is needed for foodstuffs and water. The most important reservoir animals of *Y. pseudotuberculosis* and the pathogenic strains of *Y. enterocolitica* appear to be pig, dog, cat, hare, chinchilla, and rat. Thus, uncooked pork and tongue of pig, unwashed vegetables, and nonchlorinated drinking water should be avoided. Care is also necessary for contact with pets.

Literature Cited

Aulisio, C. C. G., I. J. Mehlman, and A. C. Sanders. 1980. Alkali method for rapid recovery of *Yersinia enterocolitica* and *Yersinia pseudotuberculosis* from foods. Appl. Environ. Microbiol. **39**:135–140.

Bercovier, H., J. Brault, N. Barré, M. Treignier, J. M. Alonso, and H. H. Mollaret. 1978. Biochemical, serological, and phage typing characteristics of 459 *Yersinia* strains isolated from a terrestrial ecosystem. Curr. Microbiol. **1**:353–357.

Bercovier, H., A. G. Steigerwalt, A. Guiyoule, G. Huntley-Carter, and D. J. Brenner. 1984. *Yersinia aldovae* (formerly *Yersinia enterocolitica*-like group X2): a new species of Enterobacteriaceae isolated from aquatic ecosystems. Intern. J. Syst. Bacteriol. **34**:166–172.

Bottone, E. J. (ed.). 1981. *Yersinia enterocolitica.* CRC Press, Boca Raton, FL.

Butler, T. 1983. Plague and other *Yersinia* infections. Plenum Press, New York.

Carter, P. B., L. Lafleur, and S. Toma (ed.). 1977. *Yersinia enterocolitica:* biology, epidemiology and pathology. Contrib. Microbiol. Immunol. Vol. 5. Karger, Basel.

Chiba, S., K. Kaneko, N. Hashimoto, and T. Nakao. 1983. *Yersinia pseudotuberculosis* and Kawasaki disease. Pediatr. Infect. Dis. **2**:494.

Fukushima, H., and M. Tsubokura. 1985. *Yersinia enterocolitica* biotype 3B serotype O3 phage type II infection in pigs. Jpn. J. Vet. Sci. **47**:1011–1015.

Fukushima, H., K. Saito, M. Tsubokura, and K. Otsuki. 1984. *Yersinia* spp. in surface water in Matsue, Japan. Zbl. Bakt. Hyg., I. Abt. Orig. B **179**:235–247.

Hässig, A., J. Karrer, and F. Pusterla. 1949. Uber Pseudotuberkulose beim Menschen. Schweiz. Med. Wschr. **79**:971–973.

Higuchi, K., and J. L. Smith. 1961. Studies on the nutrition and physiology of *Pasteurella pestis.* VI. A differential plating medium for the estimation of the mutation rate to avirulence. J. Bacteriol. **81**:605–608.

Inoue, M., H. Nakajima, O. Ueba, K. Tanimoto, T. Ishida, M. Tanaka, and M. Tsubokura. 1985. Community outbreaks with *Yersinia pseudotuberculosis* in Okayama Prefecture, Japan. Media Circle **30**:367–373 (in Japanese).

Knapp, W., and W. Masshoff. 1954. Zur Atiologie der Abszedierenden Retikulozytären Lymphadenitis. Dtsch. Med. Wochenschr. **79**:1266–1271.

Mair, N. S. 1973. Yersiniosis in wildlife and its public health implications. J. Wildlife Dis. **9**:64–70.

Maruyama, T. 1982. Yersiniosis as zoonoses. J. Jpn. Vet. Med. Assoc. **35**:2–8 (in Japanese).

Maruyama, T. 1986. *Yersinia enterocolitica* infection in humans and isolation of the organism from pigs in Japan. 4th Meeting Int. Symp. Yersinia, Melbourne.

Maruyama, T., and M. Kanzaki. 1985. Chronological changes in detection ratio of *Yersinia enterocolitica* serovar O3: biotype 3. Ann. Rep. Tokyo Metro. Res. Lab. P.H. **36**:164–167 (in Japanese with English summary).

Masshoff, W., and W. Dölle. 1953. Über eine besondere Form der sog. mesenterialen Lymphadenopathie: Die abscedierende reticulocytäre Lymphadenitis. Virchows Arch. **323**:664–684.

Paterson, J. S., and R. Cook. 1963. A method for recovery of *Pasteurella pseudotuberculosis* from faeces. J. Pathol. Bacteriol. **85**:241–242.

Regamey, R. H., A. de Barbieri, W. Hennessen, D. Ikić, F. T. Perkins, and J. Ungar (ed.). 1968. International symposium on pseudotuberculos. Symp. Series Immunobiol. Standard, Vol. 9. Karger, Basel.

Sato, K., K. Ouchi, and M. Taki. 1983. *Yersinia pseudotuberculosis* infection in children, resembling Izumi fever and Kawasaki syndrome. Pediatr. Infect. Dis. **2**:123–126.

Schleifstein, J. I., and M. B. Coleman. 1939. An unidentified microorganism resembling *B. lignieri* and *Past. pseudotuberculosis,* and pathogenic for man. N.Y. State J. Med. **39**:1749–1753.

Tertti, R., K. Granfors, O.-P. Lehtinen, J. Mertsola, A.-L.

Mäkelä, I. Välimäki, P. Hänninen, and A. Toivanen. 1984. An outbreak of *Yersinia pseudotuberculosis* infection. J. Infect. Dis. **149**:245–250.
Thal, E. 1954. Untersuchungen über *Pasteurella pseudotuberculosis* unter besonderer Berücksichtigung ihres immunologischen Verhaltens. Nord. Vet. Med. **6**: 829–832.
Tsubokura, M. 1985. The progress of studies on *Yersinia pseudotuberculosis* in Japan. Media Circle **30**:362–366 (in Japanese).
Tsubokura, M., K. Otsuki, Y. Kawaoka, H. Fukushima, K. Ikemura, and Y. Kanazawa. 1984. Addition of new serogroups and improvement of the antigenic designs of *Yersinia pseudotuberculosis*. Curr. Microbiol. **11**: 89–92.
Wauters, G. 1970. Contribution à l'étude de *Yersinia enterocolitica*. Thèse d'Argrég. Vander, Louvain.
Wauters, G., L. Le Minor, and A. M. Chalon. 1971. Antigènes somatiques et flagellaires des *Yersinia enterocolitica*. Ann. Inst. Pasteur **120**:631–642.
Wauters, G., L. Le Minor, A. M. Chalon, and J. Lassen. 1972. Supplément au schéma antigénique de (*Yersinia enterocolitica*). Ann. Inst. Pasteur **122**:951–956.
Winblad, S. (ed.). 1973. *Yersinia, Pasteurella* and *Francisella*. Contrib. Microbiol. Immunol. Vol. 2, Karger, Basel.

SECTION III

Mycotic Infections

CHAPTER 57

Actinomycosis

MARY ANN GERENCSER

Disease: Actinomycosis, lumpy jaw in cattle.
Etiologic Agents: *Actinomyces israelii* and less frequently *Actinomyces naeslundii, Actinomyces viscosus, Actinomyces odontolyticus, Actinomyces meyeri,* and *Actinomyces pyogenes.*
Source: Human oral cavity.
Clinical Manifestations: Persistent swelling, suppuration, abscess formation, draining sinuses; chronic.
Pathology: Subacute or chronic inflammatory reaction with suppuration, necrosis, and fibrosis.
Laboratory Diagnosis: Isolation and identification of the etiologic agent from pus, tissue, or body fluids. Direct examination of clinical material by immunofluorescence when specific antisera are available.
Epidemiology: Worldwide, not related to age, race, season, or occupation. More common in men than in women. Source of the organism is endogenous.
Treatment: Penicillin combined with surgical drainage of lesions.

Description of Disease

Actinomycosis is a chronic, granulomatous, and suppurative disease that may affect any organ of the body. It is generally characterized by the formation of abscesses and draining sinuses. The disease is usually caused by species of *Actinomyces,* but *Arachnia propionica* can cause a clinically identical disease.

All species of *Actinomyces* are normal inhabitants of the oral cavity of humans and/or animals. Disease results when these indigenous bacteria are introduced into the tissues. The disease is not spread from human to human, animal to animal, or animal to human except in the rare case of an infected bite wound.

The genus *Actinomyces* currently contains 11 species. The most important human pathogen is *Actinomyces israelii,* but *A. naeslundii, A. viscosus, A. odontolyticus, A. meyeri,* and *A. pyogenes* also cause human infections. *A. bovis,* the type species of the genus, is found only in animals. To date, the newly described species *A. denticolens, A. howellii, A. hordeovulneris,* and *A. slackii* have been found only in animals. *A. viscosus, A. pyogenes, A. bovis, A. hordeovulneris,* and occasionally *A. israelii* cause infections in cows and other animals.

Actinomycotic infections are almost always mixed infections. A variety of other, usually oral, bacteria can be found in the lesions with the *Actinomyces.* Nevertheless, the *Actinomyces* is considered to be the causative organism because some cases of actinomycosis yield *Actinomyces* in pure culture and typical lesions can be produced in animals by pure cultures. The exact role of the other bacteria found in most lesions is not known. They undoubtedly contribute to the disease process and may add to the difficulty of treating actinomycosis.

Clinical Manifestations

The most common clinical types of actinomycosis are (1) cervicofacial, (2) thoracic, and (3) abdominal infections. Infection may spread to or occur in almost any organ of the body.

Cervicofacial Actinomycosis

Cervicofacial infection may involve the face, neck, jaw, or tongue and usually occurs following trauma. The trauma is often a dental extraction or other dental manipulation, but even minor unrecognized injury to the gums may provide conditions favorable for the entry of the organisms into the tissues. The disease begins with pain and firm swellings along the jaw line. The infection is slowly progressive, but if left untreated draining sinuses will eventually be produced. The infection may spread to the neck, salivary glands, thorax, or sinuses.

Thoracic Actinomycosis

Pulmonary infections usually follow aspiration of bits of dental plaque, calculus, or other oral debris and may involve the lungs, the chest wall, or both. Less commonly, thoracic disease results from direct extension of cervicofacial or abdominal lesions.

Symptoms of actinomycosis are similar to those of other chronic pulmonary infections. Bronchitis, emphysema, bronchiectasis, or other infectious diseases such as tuberculosis are frequently underlying features of thoracic actinomycosis. Thoracic disease may spread extensively to adjacent tissues or organs, as do other forms of the disease. In addition, pulmonary actinomycosis frequently disseminates through the bloodstream. Such dissemination can result in abscesses in the brain and other organs.

Abdominal Actinomycosis

Abdominal infections follow disruption of the intestinal mucosa and are often associated with appendicitis. Persistent purulent drainage after surgery for appendicitis may be the first sign of an *Actinomyces* infection. In other cases, abdominal masses resembling tumors are the first signs of infection.

Other Infections

Actinomycosis may affect the brain, bone, kidneys, heart, subcutaneous tissue, pelvic organs, and eye. Female pelvic infections associated with wearing an intrauterine contraceptive device have been reported with increasing frequency in the past few years.

A lacrimal canaliculitis with persistent discharge from the eye and the formation of concretions in the caniliculi can be caused by *Actinomyces* and *Arachnia*. More than one species of *Actinomyces* or both *Actinomyces* and *Arachnia* are sometimes involved in these infections.

There is some evidence that *A. viscosus,* in particular, is involved in gingivitis and mild forms of periodontitis. It has also been implicated in root surface caries.

Pathology

Multiplication of the bacteria in the tissues produces a subacute or chronic inflammatory reaction followed by suppuration, necrosis, and fibrosis. Lesions consist of abscesses or indurated masses with fibrous walls surrounding a soft central area containing pus. The organisms, particularly in early stages of the disease, may be seen free in the tissue or in pus from draining sinuses. More often, they are found in characteristic yellowish grains or granules called *sulfur granules.* Sulfur granules consist of tangled masses of the bacteria held together by an amorphous material derived partly from the microorganism and partly from the host. When crushed and examined microscopically the granules contain a central mass of filaments with radiating, often clubbed, filaments at the periphery.

The infection usually spreads directly to contiguous tissue through muscle and fascia, although hematogenous spread may sometimes occur. The slow spread of the infection through the tissue eventually leads to the development of chronic, draining sinuses or fistulas.

The mechanism(s) by which *Actinomyces* produces disease is not clear, but may involve the ability of the organisms to suppress some of the immune functions of the host. *Actinomyces* species are known to activate lymphocyte blastogenesis, to stimulate the production of lysosomal enzymes from polymorphonuclear leukocytes, and to be chemotactic.

Pathogenic factors that would aid in tissue invasion and abscess formation have not been demonstrated except for *A. pyogenes,* which produces a soluble toxin and a hemolysin that can be neutralized by antiserum.

Epidemiology and Ecology

Occurrence of Actinomycosis

Actinomycosis is found throughout the world. It is not related to age, race, season, or occupation. Historically, actinomycosis occurs about twice as frequently in men as in women. Since the introduction of antibiotics, the incidence of most forms of the disease has decreased.

Transmission of Actinomycosis

Actinomycosis is of endogenous origin. The bacteria are only found in the mouths of humans and animals and are not found on grasses, in soil, or elsewhere in the environment. In the early literature, actinomycosis was often associated with the habit of chewing on straws. If such an association exists, it is not related to the introduction of exogenous actinomycetes into the mouth on the straws, but to repeated minor trauma to the gums.

In a few cases, actinomycosis might be said to be of exogenous origin. These are cases following a human bite wound or injury in a fist fight (punch actinomycosis). Although the organisms are transmitted from person to person in these cases, this evidence actually supports the idea that actinomycosis is caused by bacteria resident in the human mouth.

Ecology of *Actinomyces*

Actinomyces species occur as normal inhabitants of the oral cavity and tonsillar crypts in humans and other animals. *A. israelii, A. naeslundii, A. viscosus, A. odontolyticus, A. pyogenes,* and *A. meyeri* occur in the human mouth. *Actinomyces* may be isolated from saliva, the mucous membranes, and the tongue, but they are more numerous in dental plaque and in the gingival sulcus. Site distribution within the oral cavity has not been studied for all species, but Ellen (1982) has investigated the oral colonization of *A. viscosus* and *A. naeslundii* in detail. It was shown that *A. naeslundii* colonizes the infant at an early age and is found in high proportions on the oral mucosa, on the tongue, and in saliva. On the other hand, *A. viscosus* only colonizes after teeth erupt and is found mainly on the teeth. These site differences are relative and probably reflect different attachment mechanisms of the two species.

Individuals vary in the numbers and specific species of *Actinomyces* present in a given oral site at any given time, but these bacteria are always prominent members of the oral flora. Of the cultivable bacteria from dental plaque, 15 to 50% are *Actinomyces.*

Some of the *Actinomyces* species isolated from humans are also found in and cause infections in animals. *A. viscosus* was originally isolated from spontaneous periodontal disease in hamsters and has been found in infections in dogs, cats, pigs, and goats. *A. pyogenes* is probably best known for animal infections causing pyogenic disease in a number of domestic animals, particularly mastitis in cows.

The other species of *Actinomyces* have so far only been found in animals. *Actinomyces bovis,* the agent of bovine actinomycosis (lumpy jaw), has only been reported from infections in cattle. Three recently described species, *A. denticolens, A. howellii,* and *A. slackii,* have only been isolated from the dental plaque of cattle. Infections attributable to these species have not been reported. Another new species, *A. hordeovulneris,* was isolated from infections in dogs. Detailed studies of the normal oral flora of most animals have not been done, although Dent (1979) isolated *Actinomyces* from a variety of zoo animals.

Etiologic Agents

Characterization

The genus *Actinomyces* appears in Section 15 (Irregular, Nonsporing, Gram-Positive Rods) in the 1986 *Bergey's Manual of Systematic Bacteriology,* Volume 2 (Schaal, 1986). Ten species are described in the Manual and one additional species, *A. slackii* (Dent and Williams, 1986), that is not included in the Manual has been described.

Actinomyces generally grow well in most rich laboratory media, such as brain–heart infusion broth or agar, trypticase soy broth or agar, or fluid thioglycollate medium with 0.5% yeast extract. *A. hordeovulneris* requires the addition of at least 0.5% fetal calf serum, and other species are stimulated by added serum. Most species also grow better when the medium is supplemented with 0.1% Tween 80. *A. meyeri* requires vitamin K_1 and is stimulated by Tween 80 and a fermentable carbohydrate (Cato et al., 1984). Schaal (1986) recommends the CC medium of Henrich and Korth for growing all *Actinomyces* species.

Oxygen requirements vary among the species, ranging from that of *A. viscosus,* which grows best aerobically with added carbon dioxide, to *A. israelii,* which grows poorly if at all under those conditions. All grow anaerobically in the presence of CO_2. *A. viscosus, A. pyogenes, A. howellii, A. hordeovulneris, A. slackii* and many strains of *A. naeslundii, A. odontolyticus and A. denticolens* will grow on the surface of agar media in air; however, all grow much better in the presence of added carbon dioxide. The genus as a whole should be considered as consisting of facultative or oxygen-tolerant anaerobes with a requirement for carbon dioxide.

Morphology

Actinomyces are Gram-positive, non-acid-fast, nonspore-forming, nonmotile rods, often with diphtheroidal morphology. In Gram-stained smears the staining is often irregular, causing the organisms to have a beaded or barred appearance. Long rods, filaments, and branching filaments (Fig. 1) or rods with clubbed ends are seen frequently, but short to medium-sized rods may be the only morphologic forms seen in any species. Coccobacillary forms are also sometimes seen notably in *A. pyogenes*.

Electron micrographs show the presence of long, thin fibrils on the surface of most strains of *A. naeslundii* and *A. viscosus*. These fimbriae are composed of protein and are involved in the attachment of these species to surfaces and in their coaggregation reactions with other bacteria. Fibrils have been seen on some strains of *A. israelli* and *A. odontolyticus*, but species other than *A. viscosus* and *A. naeslundii* have not been well studied for fibrils.

After 18 to 24 h, *Actinomyces* colonies are quite characteristic. *A. israelii* and *A. hordeovulneris* produce filamentous or "spider" microcolonies consisting of thin, branching filaments that appear to arise from a single point. Most strains of *A. naeslundii* and *A. viscosus* also produce filamentous colonies that have a dense center with medium to long projecting filaments at the edge. Some strains of these species, especially *A. viscosus*, have very short projecting filaments at the colony edge or produce colonies with an entire or irregular edge. *A. bovis*, *A. odontolyticus*, *A. pyogenes*, *A. denticolens*, and *A. meyeri* produce smooth microcolonies. These colonies have a smooth or granular surface and an entire or irregular edge. When observed under the microscope, an optically dark spot may be seen in the center of such colonies. Rare isolates of *A. bovis* and *A. odontolyticus* that produce typical "spider" colonies have been reported. Microcolonies of *A. howellii* and *A. slackii* have not been described.

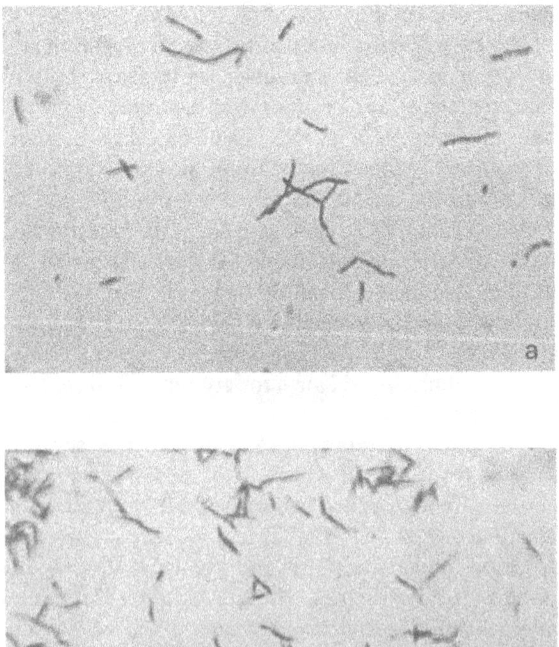

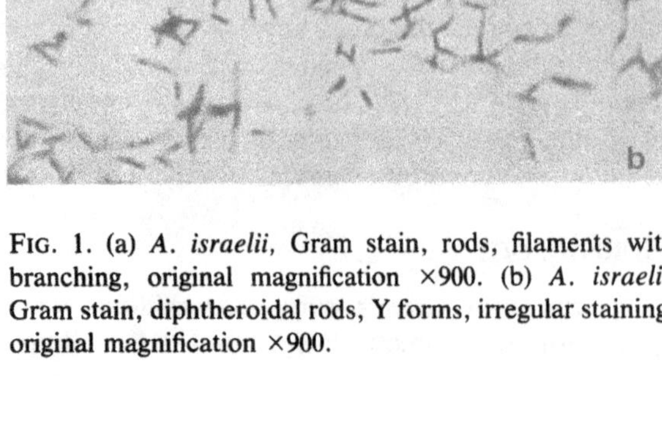

FIG. 1. (a) *A. israelii*, Gram stain, rods, filaments with branching, original magnification ×900. (b) *A. israelii*, Gram stain, diphtheroidal rods, Y forms, irregular staining, original magnification ×900.

Mature colonies of *A. israelii*, *A. hordeovulneris*, and some strains of *A. naeslundii* are rough, with undulate, lobate, or erose edges and very convoluted surfaces. These colonies have been described as molar-tooth, bread-crumb, or raspberrylike colonies.

Those species that typically produce smooth microcolonies also produce smooth mature colonies. These colonies are low convex to umbonate with entire edges and a soft texture.

Colonies on blood agar are usually similar to those on media without blood, except in the case of *A. odontolyticus*. This species produces red colonies on blood agar after 2 to 10 days of incubation. The pigment may only form after anaerobically grown cultures are allowed to stand aerobically for 2 to 3 days. *A. denticolens* may develop a slightly pink colony on blood agar when incubated anaerobically. *A. pyogenes* is β-hemolytic on blood agar, producing tiny colonies with a zone of hemolysis two to three times the diameter of the colony. None of the other species is regularly β-hemolytic, but some species such as *A. meyeri* are sometimes α-hemolytic.

Cell Wall Composition

A. bovis cell wall peptidoglycan contains lysine, alanine, glutamic acid, and aspartic acid. Structurally, it is of the Lys-Lys-D-Asp type, with L-lysine and D-aspartic acid in the interpeptide bridge (Schaal, 1986). *A. pyogenes* is similar but lacks aspartic acid and is representative of a second structural peptidoglycan type in the genus. *A. pyogenes* belongs to the Lys-Ala-Lys-D-Glu type with L-alanine, L-lysine, and D-glutamic acid in the cross-link. The peptidoglycans of all other *Actinomyces* species contain glutamic acid, alanine, ornithine, and lysine in an approximate ratio of 2 : 2 : 1 : 1. Structurally this is a third type of peptidoglycan, and studies indicate that the ornithine is in position 3 in the tetrapeptide and that lysine and glutamic acid are in the interpeptide bridge. Neither diaminopimelic acid nor glycine is found in *Actinomyces* cell walls.

With the exception of *A. israelii* and *A. denticolens*, which contain only galactose, the sugar composition of *Actinomyces* cell walls is complex. Glucose, galactose, rhamnose, 6-deoxytalose, fucose, and mannose may be present in various combinations. Arabinose is not found in *Actinomyces*.

Biochemical Reactions

Despite variation among strains and the method and medium dependency of some tests, the pattern of biochemical reactions of *Actinomyces* species is useful in identification. Biochemical reactions of all 11 species are shown in Table 1. All of the species ferment glucose (not shown in Table 1). The characteristic end products of glucose fermentation as determined by gas chromatography are formic, acetic, lactic, and succinic acids. Propionic acid is not produced.

A. pyogenes is the only proteolytic *Actinomyces* species. *A. pyogenes* liquefies gelatin and Loeffler's serum, hydrolyses casein, and digests clots of litmus milk (Reddy et al., 1982).

Serology

The serologic reactions of *A. bovis, A. israelii, A. odontolyticus, A. naeslundii, A. viscosus,* and *A. meyeri* have been studied using a variety of serologic techniques (Slack and Gerencser, 1975; Schaal and Gatzer, 1985). The newly described species *A. denticolens, A. howellii, A. slackii,* and *A. hordeovulneris* have not been studied serologically, except cross-reactions between *A. hordeovulneris* and antisera to other species have been determined. Based primarily on immunofluorescence studies, each species of *Actinomyces* studied falls into a serologic group that correlates with the species. All of the species contain at least two serovars within the group.

A. bovis does not cross-react serologically with any of the other species, and the two serovars in the species do not react with each other. To date, all serovar 2 strains produce filamentous microcolonies. *A. meyeri* is also serologically distinct from other species. Some strains do not react with antiserum to strain ATCC 33972 (Cato et al., 1984), suggesting more than one serovar in the species.

A. odontolyticus antiserum cross-reacts with two other *Actinomyces* species. Serovar 2 strains may cross-react with *A. viscosus* serovar 1. In addition, cross-reactions are seen between *A. odontolyticus* and *A. pyogenes*.

A. israelii, A. naeslundii, and *A. viscosus* are serologically heterogenous and show varying degrees of serologic relatedness. Two serovars have been described in *A. israelii* that react with each other, usually at low titer. Serovar 1 accounts for the majority of isolates, but some strains do not react with either serovar, suggesting additional types in this species. *A. israelii* has low-titer cross-reactions with some strains of both *A. naeslundii* and *A. viscosus*.

A. naeslundii has at least 3 serovars. Serovar 1 seems to be the most commonly isolated. The three serovars have strain-variable cross-reactions with each other and with *A. viscosus* serovar 2. *A. naeslundii* serovars 2 and 3 appear to be more closely related to each other than to serovar 1, and both of these serovars show a strong relationship to *A. viscosus*.

The two recognized serovars of *A. viscosus* can be correlated with the source of the isolate. Rodent strains of *A. viscosus* belong to serovar 1, whereas human strains belong to serovar 2. Isolates from other animals, such as dogs, seem to belong to or be more closely related to serovar 2. The two serovars have only a one-way cross-reaction. Serovar 2 strains show extensive cross-reactions with *A. naeslundii*, especially with serovars 2 and 3.

The reactions of *A. hordeovulneris* strains with other *Actinomyces* antisera include cross-reactions with *A. israelii* serovar 1 and *A. viscosus* serovar 1 at relatively low titers. The reaction with *A. israelii* was stronger than that with *A. viscosus*. These studies suggest that this species will also be identifiable serologically (Buchanan et al., 1984).

Either direct or indirect immunoflourescence is a useful tool for identifying *Actinomyces* isolates, as interspecies cross-reactions are limited and can generally be removed by dilution or adsorption of the antiserum.

Specimen Collection and Transport

Specimens that should be examined for *Actinomyces* include pus from closed lesions, material collected from draining sinuses, tissue, and body fluids other than urine. Because *Actinomyces* are normal flora of the mouth and are present in saliva, routine sputum specimens should not be cultured for *Actinomyces* in suspected cases of thoracic actinomycosis. Pus should be obtained from a closed, subcutaneous lesion after carefully cleaning the skin. When a draining sinus is present, aspiration of pus with a syringe and an intravenous catheter placed as deeply in the lesion as possible is the preferred collection method. Alternatively, a premoistened gauze pad placed over the lesion may be used to collect pus and granules over an extended period of time. Swabs are the least desirable means of specimen collection and should only be used when necessary.

Specimens should be taken to the laboratory and cultured as soon as possible. If a delay is unavoidable, the specimen should be placed in a suitable anaerobe collecting system container. Several such

TABLE 1. Biochemical characteristics of *Actinomyces* species[a]

	Actinomyces *species*										
Characteristic	A. bovis	A. odonto-lyticus	A. meyeri	A. israelii	A. naeslundii	A. viscosus	A. hordeo-vulneris	A. howellii	A. denti-colens	A. slackii	A. pyogenes
Catalase[b]	−	−	−	−	−	+	+	+	−	+	−
Indole	−	−	−	−	−	−	−	−	−	−	−
Nitrate reduction	−	+	−	v	+	v	−		+		−
Nitrite reduction	−	−		−	v	−[c]			−		−
Urease	−	−	v	−	+	v	−				−
Esculin hydrolysis	+	v	−	+	+	+	+		+	−	−
Starch hydrolysis	+	−	−	v	v	v					v
Ammonia from arginine	−	−	−	+	v	v					−
Valine arylamidase[d]	−	−	+	−	−	−			−		−
N-acetyl-*β*-glucosaminidase[d]	+	−	−	−	−	−			−		+
β-Glucosidase[d]	−	v	−	+	v	−			v		−
β-Galactosidase[d]	−	−	−	+	v	v			+		v
β-Xylosidase	−	−		+	v	v					
Phosphoamidase	−	−	−	v	v	v			+		−
Acid from											
Arabinose	−	−	v	v[e]	−	−	−	v	−	−	v
Glycerol	−	v	−	−	+	+	−	−	−	−	−
Mannitol	−	−	−	v	−	−	−	−	v	−	−
Raffinose	−	−	−	+	+	+	+	+	+	+	−
Trehalose	−	v	−	+	v	v	+	v	−	v	v
Xylose	v	v	+	+	−	−	+	v	−	−	v

[a] Data from Slack and Gerencser (1975), Schaal (1986), Schofield and Schaal (1981), Dent and Williams (1984a,b, 1986), Allen (1985), Buchanan et al. (1984), Cato et al. (1984), and Reddy et al. (1982).
[b] + = ≥90% positive; − = ≤10% positive; v = 11–89% positive.
[c] Method of Schofield and Schall (1980); v = method of Slack and Gerencser (1975).
[d] API enzyme tests.
[e] Serovar-1 strains usually positive.

collection systems are available commercially. Swabs should not be allowed to dry.

Microscopic Examination of Specimens

The pus, body fluid, or other specimen is placed in a sterile petri dish and examined for granules. For pus, filtering the specimen through a piece of sterile guaze may help in locating the granules that cling to the cloth. Granules are white to yellowish, 0.5 to 5.0 mm in diameter, and firm in consistency. For microscopic examination, place a granule in a drop of water on a slide and press gently under a coverslip. Under low power and reduced light, granules have an irregular edge and a ray appearance at the periphery, whereas under high magnification filaments surrounded by hyaline clubs can be seen.

In addition to examining the unstained granules, granules should be crushed and used for making stained smears. If no granules are present, smears should be made from well-mixed pus or the centrifuged sediment of body fluids. Smears are stained with Gram stain and with a modified Kinyoun or other suitable acid-fast stain. The smears are examined for gram-positive, non-acid-fast, diphtheroidal rods and filaments.

Isolation and Cultivation

For primary isolation, crushed granules, well-mixed pus, or other specimens should be inoculated into fluid thioglycollate medium containing 0.5% yeast extract and onto the surface of blood agar plates. One set of plates and the thioglycollate broth should be incubated anaerobically with added carbon dioxide. Cultures may be examined for typical colonies after 2 to 3 days at 35°C, but should be held for at least 14 days before being discarded as negative. Gram-stained smears from the broth culture should be examined for *Actinomyces*. If Gram-positive rods are present in the broth, it should be inoculated onto blood agar plates and treated as were the primary cultures.

A variety of nonselective media are available for the primary isolation of *Actinomyces*. These include brain–heart infusion blood agar and blood agar made with a base of Schaedler agar or trypticase soy agar. CDC Anaerobe blood agar (Dowell et al., 1977) is also an excellent medium for *Actinomyces*. Fluid thioglycollate medium with 0.5% yeast extract or cooked meat medium may be used for broth enrichment cultures.

Actinomyces also grow well on phenylethyl alcohol (PEA) blood agar, a semi-selective medium that inhibits the growth of facultative Gram-negative bacteria.

Selective media for the isolation of *Actinomyces* from pus have not been described; however, Traynor et al. (1981) reported a method for the selective isolation of *Actinomyces* from cervical specimens. They emulsified the material from swabs or beads of pus from an intrauterine device in thioglycollate broth, made serial dilutions in the broth, and spread these dilutions on 10% horse blood agar (Columbia agar base) with and without 2.5 mg/liter metronidazole. They found that the metronidazole medium prevented overgrowth by other bacteria and gave good isolation results.

Several selective media have been devised for the isolation of *Actinomyces* from dental plaque. One of these, described by Beighton and Colman (1976), is intended for the isolation of all *Actinomycetaceae,* whereas the others (Ellen and Balcerzak-Raczkowski, 1975; Korman and Loesche, 1978; Zylber and Jordan, 1982) are intended for the isolation of *A. viscosus* and *A. naeslundii.* Cadmium and fluoride used either singly or together are present in all of these media, often with other inhibitors. The CFAT medium of Zylber and Jordan (1982) contains cadmium sulfate, sodium fluoride, neutral acriflavin, potassium tellurite, and basic fuchsin in a trypticase soy blood agar base.

Identification

Identification of *Actinomyces* to the genus level can usually be done using morphology, analysis of the end products of glucose fermentation, and perhaps a few biochemical tests. If antiserum is available, rapid identification to genus and species can often be done by immunofluorescence. Some isolates fail to react with any of the available antisera. Serologic identification of the organisms can also be done directly on the specimen if organisms have been seen in a Gram stain.

Where serologic identification is not possible, identification to the species level requires a battery of biochemical tests such as those given in Table 1.

Treatment

A. israelii, A. naeslundii, A. odontolyticus, A. viscosus, and *A. bovis* are susceptible to the penicillins and cephalosporins, tetracyclines, chloramphenicol, macrolides, lincomycins, rifamycins, fusidic acid, and vancomycin. They are resistant or relatively resistant to aminoglycosides, metronidazole, peptide antibiotics, and antitubuculosis drugs (Schaal, 1986). Other species have been tested less extensively. *A. pyogenes* is susceptible to penicillins, cephalosporins, tetracyclines, macrolides, and vancomycin.

A. meyeri is susceptible to penicillin G, tetracycline, chloramphenicol, clindamycin, and erythromycin, and *A. hordeovulneris* is susceptible to penicillin G, cephalothin, chloramphenicol, and trimethoprim. Most strains are susceptible to tetracycline. The antibiotic susceptibilities of three new cattle species are not known.

Penicillin in high doses and for prolonged periods of time is the antibiotic of choice for the treatment of actinomycosis. Tetracycline has been recommended for patients allergic to penicillin. In addition to antibiotic treatment, surgical removal of debris and necrotic tissue and drainage of sinus tracts and abscesses are an essential part of successful treatment.

Literature Cited

Allen, S. D. 1985. Gram-positive, nonsporeforming anaerobic bacilli, p. 461–472. *In* E. H. Lennette, A. Balows, W. J. Hausler, Jr., and H. J. Shadomy (ed.), Manual of clinical microbiology. American Society for Microbiology, Washington, D.C.

Beighton, D., and G. Colman. 1976. A medium for the isolation and enumeration of oral *Actinomycetaceae* from dental plaque. J. Dent. Res. **55**:875–878.

Buchanan, A. M., J. L. Scott, M. A. Gerencser, B. L. Beaman, S. Jang, and E. L. Biberstein. 1984. *Actinomyces hordeovulneris* sp. nov., an agent of canine actinomycosis. Int. J. Syst. Bacteriol. **34**:439–443.

Cato, E. P., W. E. C. Moore, G. Nygaard, and L. V. Holdeman. 1984. *Actinomyces meyeri* sp. nov., specific epithet rev. Int. J. Syst. Bacteriol. **34**:487–489.

Dent, V. E. 1979. The bacteriology of dental plaque from a variety of zoo-maintained mammalian species. Arch. Oral Biol. **24**:277–282.

Dent, V. E., and R. A. D. Williams. 1984a. *Actinomyces denticolens* Dent and Williams sp. nov.: a new species from the dental plaque of cattle. J. Appl. Bacteriol. **56**:183–192.

Dent, V. E., and R. A. D. Williams. 1984b. *Actinomyces howellii*, a new species from the dental plaque of dairy cattle. Int. J. Syst. Bacteriol. **34**:316–320.

Dent, V. E., and R. A. D. Williams. 1986. *Actinomyces slackii* sp. nov. from dental plaque of dairy cattle. Int. J. Syst. Bacteriol. **36**:392–395.

Dowell, V. R., Jr., G. L. Lombard, F. S. Thompson, and A. Y. Armfield. 1977. Media for isolation, characterization and identification of obligately anaerobic bacteria. Centers for Disease Control, Atlanta.

Ellen, R. P. 1982. Oral colonization by gram-positive bacteria significant to periodontal disease, p. 98–111. In R. J. Genco and S. E. Mergenhagen (ed.), Host–parasite interactions in periodontal disease. American Society for Microbiology, Washington, D.C.

Ellen, R. P., and I. B. Balcerzak-Raczkowski. 1975. Differential medium for detecting dental plaque bacteria resembling *Actinomyces viscosus* and *Actinomyces naeslundii*. J. Clin. Microbiol. **2**:305–310.

Kornman, K. S., and W. J. Loesche. 1978. New medium for isolation of *Actinomyces viscosus* and *Actinomyces naeslundii* from dental plaque. J. Clin. Microbiol. **7**:514–518.

Reddy, C. A., C. P. Cornell, and A. M. Fraga. 1982. Transfer of *Corynebacterium pyogenes* (Glage) Eberson to the genus *Actinomyces* as *Actinomyces pyogenes* (Glage) comb. nov. Int. J. Syst. Bacteriol. **32**:419–429.

Schaal, K. P. 1986. *Actinomyces*, p. 1383–1418. In P. H. A. Sneath, N. S. Mair, M. E. Sharpe, and J. G. Holt (ed.), Bergey's manual of systematic bacteriology, vol. 2. The Williams & Wilkins Co., Baltimore.

Schaal, K. P., and R. Gatzer. 1985. Serological and numerical phenetic classification of clinically significant fermentative actinomycetes, p. 85–109. In A. Tadashi (ed.), Filamentous microorganisms: biomedical aspects. Japan Scientific Societies Press, Tokyo.

Schofield, G. M., and K. P. Schaal. 1981. A numerical taxonomic study of members of the Actinomycetaceae and related taxa. J. Gen. Microbiol. **127**:237–259.

Slack, J. M., and M. A. Gerencser. 1975. *Actinomyces*, filamentous bacteria: biology and pathogenicity. Burgess Publishing Co., Minneapolis.

Traynor, R. M., D. Parrat, H. L. D. Dugard, and I. D. Duncan. 1981. Isolation of actinomycetes from cervical specimens. J. Clin. Pathol. **34**:914–916.

Zylber, L. J., and H. V. Jordan. 1982. Development of a selective medium for the detection and enumeration of *Actinomyces viscosus* and *Actinomyces naeslundii* in dental plaque. J. Clin. Microbiol. **15**:253–259.

CHAPTER 58

Aspergillosis

MICHAEL G. RINALDI

Disease: Aspergillosis, including allergic, colonizing, mycotoxic, and invasive manifestations.

Etiologic Agents: *Aspergillus fumigatus, Aspergillus flavus, Aspergillus niger,* and *Aspergillus terreus* are the major species pathogenic for humans.

Source: Omnipresent in nature, wherever organic debris are found.

Clinical Manifestations: Allergic conditions (allergic bronchopulmonary aspergillosis, extrinsic allergic alveolitis, hypersensitivity pneumonitis); fungus ball formation (usually in preexisting lung cavities); mycotoxicoses following ingestion/inhalation of toxic metabolites; and tissue invasive aspergillosis in virtually any organ system or location, with the lung as the most frequent site of primary disease. May be acute or chronic in nature.

Pathology: Immunologic damage to lung tissue, invasion of tissue with predilection for vessel thrombosis and resultant infarction and necrosis.

Laboratory Diagnosis: Isolation of etiologic agent from diseased tissue/fluid coupled with microscopic/histopathologic presence of regular, septate hyphae in vivo. Serologic detection of antibody to *Aspergillus* antigens and detection of aspergillal antigen are adjunctive diagnostic tests.

Epidemiology: Worldwide; inhalation of fungal conidia from nature is the primary route of inoculation. The majority of patients with invasive disease have underlying primary disorders which themselves, or through their treatment, compromise cell-mediated immunity.

Treatment and Control: Removal of the inoculum source from the patient in allergic disease; surgery, amphotericin B (and perhaps newer antifungals), and elimination of the myelosuppressive state in invasive disease; pimaricin for ocular infections.

Description of Disease

Aspergillosis comprises a spectrum of diseases caused by species of the fungal genus *Aspergillus,* including allergy (allergic bronchopulmonary aspergillosis, extrinsic allergic alveolitis), colonization (growth of aspergilli occurring in a preexisting cavity in the lung parenchyma = fungus ball; often incorrectly termed "aspergilloma" [Matsumoto and Ajello, 1986; Rinaldi, 1981]), toxicoses (ingestion/inhalation/exposure to mycotoxins or other metabolites produced by aspergilli, e.g., aflatoxin B), and tissue invasion (with potential for dissemination to any body site; particularly noteworthy in patients with abrogated immunity due to a variety of accompanying/underlying disorders).

The various manifestations of aspergillosis may intergrade and/or overlap, that is, aspergilli causing fungus balls may become invasive, or toxins produced by aspergilli in vivo may be responsible for some features of invasive disease. The focus of this chapter will be directed toward *Aspergillus*-caused invasive infections. The major pathogenic species causing human mycoses are *A. fumigatus, A. flavus, A. niger,* and *A. terreus.*

Clinical Manifestations

The variety of clinical entities attributed to species of aspergilli appears to be related to genetic factors and the immune status of the host, as well as to the amount and time of exposure to the fungus.

Ear Infection

Ear infections (otomycoses) are common and most often caused by *A. niger*. The fungus grows in the cerumen and debris of the external auditory canal; most infections are benign (Goodman, 1982). Individuals who swim frequently and retain moisture in the ear canal often complain of otomycosis.

Sinusitis

Species of aspergilli may incite sinusitis, in which a fungus ball may be lying free in or attached to the sinus mucosa (Goodman, 1982). If maxillary sinuses are invaded, further erosion into surrounding structures may occur; hence, this condition merits careful medical observation. Aspergillosis is the most common mycosis of the nose and paranasal sinuses (Zimmerman, 1955).

Keratomycosis

Corneal infections caused by aspergilli are inevitably a result of trauma to the eye with objects containing the fungus. Subsequent invasion may ensue. A high index of suspicion for mycotic keratitis should exist following any eye trauma caused by sources potentially harboring fungi (Rippon, 1982). Additionally, endophthalmitis may result from the hematogenous spread of fungus from extraocular foci (McGinnis, 1980).

Cutaneous Aspergillosis

Skin lesions caused by aspergilli are most often secondary to disseminated disease, but may be primary. When primary, the skin is thick, edematous, and has a purplish appearance. An additional cutaneous manifestation in the immunocompromised host is a necrotic skin ulcer typically covered by a black eschar at sites of minor trauma (Cahill et al., 1967). The skin lesions in disseminated disease start as small, red, discrete papules that become pustular (Prystowsky et al., 1976).

Pulmonary Aspergillosis

As mentioned, there are a variety of pulmonary manifestations of aspergillosis. Some are allergic in nature (allergic bronchopulmonary disease, extrinsic allergic alveolitis), with the formation of fungus balls in preexisting lung cavities and a resultant vigorous immune response. This manifestation may result in severe hemoptysis and may progress to invasive disease. Frankly invasive aspergillosis of the lung is the most common form of invasive disease and is encountered most often in the neutropenic, immunosuppressed host, particularly in those experiencing leukemia. Invasive pulmonary aspergillosis is a grave disease that is acute, progressive, and life threatening. There may be additional, less common, forms of pulmonary aspergillosis, such as chronic necrotizing pulmonary aspergillosis (often described in middle-aged patients with impaired host defenses), empyema, and suppurating granulomatous bronchopneumonia in patients with chronic granulomatous disease (Rinaldi, 1983). Additionally, there are cases of invasive pulmonary aspergillosis in nonimmunocompromised, nonneutropenic hosts (Karam and Griffin, 1986).

Bone Disease

While rarely reported, aspergillal osteomyelitis, particularly sino-orbital aspergillosis, is probably quite common (Tack et al., 1982). Radiographic examination is the major means of recognition (Zinneman, 1982).

Endocarditis

Aspergillus-caused endocarditis clinically resembles bacterial endocarditis except for the marked failure to recover aspergilli from blood. When aspergilli (or moulds in general) are recovered from blood, irrespective of the form of invasive disease, the patient is often moribund. *Aspergillus*-induced endocarditis has most often involved prosthetic valves (Bardana, 1981). Despite chemotherapy and/or surgical intervention, the prognosis of aspergillal endocarditis remains bleak.

Central Nervous System Infection

Expression of aspergillosis as a central nervous system infection has occurred following injection of contaminated drugs in addicts, following iatrogenic procedures in patients, and sometimes without a known portal of entry or focus. The most common manifestation of central nervous system disease is brain abscess and, rarely, meningitis. Central nervous system involvement is rapidly fatal and is also frequently seen in disseminated aspergillosis (Rinaldi, 1983).

Disseminated Aspergillosis

Disseminated aspergillosis is active infection in two or more noncontiguous organs (Young et al., 1970). This disease is relatively "new" and reflects the era of antimicrobics, glucosteroids, and cytotoxins. Most patients experiencing disseminated aspergillosis have received antimicrobial agents, glucosteroids, or both and have underlying disorders, associated diseases, or factors predisposing them to this mycosis: pneumonia, diarrhea, leukemia, hepatitis, transplants, and the like (Tan et al., 1966). Disseminated disease has, however, also been reported in otherwise apparently healthy individuals. Most patients have widespread organ involvement with pulmonary disease occurring most prominently, followed by the central nervous system, kidney, liver, and thyroid gland (Young et al., 1970). The disease is acute and often leads to rapid demise. The most common cause of death is cerebral hemorrhage; other causes frequently seen are bilateral pneumonia, pulmonary infarction with vessel invasion, and brain abscess (Rinaldi, 1983).

Miscellaneous Aspergillosis

Since aspergilli are among the most ubiquitous moulds in the human environment, they are often present in the hospital setting and may incite iatrogenic-caused disease. These fungi may inadvertently gain entrance to susceptible patients by many portals: dialysis bags, air-conditioning ducts, and airborne particles often associated with construction. Such potential inocula are of particular concern with the severely neutropenic patient (e.g., bone-marrow transplant recipients). Other isolated cases of aspergillosis have included genitourinary and gynecologic infections involving tubo-ovarian abscess, vaginal hemorrhage, and cervical and bladder infections (Rinaldi, 1983).

Pathology

Invasive aspergillosis appears to result from 1) suppression of the immune system through underlying disease and/or cytotoxic drug therapy, 2) a local portal of entry for the mould, and 3) a diminished inflammatory response due to glucosteroids and, perhaps, disruption of normal microbial flora by antimicrobial agents. Local or regional invasive disease is most commonly seen in individuals with metabolic disorders, cyanotic heart disease, and heart prostheses and, most importantly, in immunoincompetent hosts secondary to drug therapy, irradiation, congenital immunodeficiency, or malignancy.

Invasive disease may arise de novo or can develop from any of the forms of allergic or colonizing aspergillosis. The major predisposing factors include 1) concomitant glucosteroid therapy, 2) cytotoxic chemotherapy, 3) recent or concurrent therapy with broad-spectrum antimicrobics, 4) leukopenia (less than 1,000 cells/mm^3), and 5) acute leukemia in relapse or acute rejection of an organ transplant. As noted above, there are confirmed reports of invasive aspergillosis leading to death in apparently normal individuals (Karam and Griffin, 1986).

There appears to be little doubt that patients with leukemia, particularly acute lymphocytic and acute myelogenous forms, are especially prone to develop invasive aspergillosis. Patients with lymphoma are also very susceptible to invasive disease (Bodey et al., 1982; Mirsky and Cuttner, 1972).

The term invasive aspergillosis implies histopathologically demonstrated invasion into tissue of aspergillal hyphae, preferably confirmed by culture and/or fluorescent-antibody tests. The most common manifestation of aspergillosis is pulmonary, following inhalation of asexual spores (conidia) of the causative species. It has been demonstrated both in vitro with murine and human cells and in vivo with a murine model that two specific mechanisms of defense against aspergilli must be breached before these moulds can cause progressive disease. Polymorphonuclear leukocytes are the main effector cells instrumental in protection against the mycelial form of the fungus, while macrophages form an effective defense system by rapidly killing conidia. It appears the first line of defense is the conidia-killing macrophages, with the polymorphonuclear leukocytes serving as a second line of defense by killing the hyphae (Schaffner et al., 1982). Natural resistance is impaired when both lines of defense are damaged (e.g., large doses of cortisone).

In invasive pulmonary aspergillosis, hyphae invade bronchial walls and contiguous parenchyma, resulting in acute, necrotizing, purulent pneumonitis. Aspergilli have a marked predilection to invade blood vessels with resultant thrombosis followed by infarction and necrosis. The invasion of blood vessels may then lead to spread of disease to other body sites. Histopathologically, the hyphae of aspergilli are 3 to 4 μm in diameter, typically septate, and characteristically branch repeatedly in a dichotomous fashion with branches oriented in the same direction (Emmons et al., 1977). The hyphae often radiate from a focal point, and the branches exhibit an angle of approximately 45°. As discussed below, other fungi may exhibit a tissue appearance similar, or identical, to aspergilli.

Some strains of *A. terreus* and *A. flavipes* may develop large, spherical (6 to 7 μm in diameter) structures in tissue and culture that are termed aleurioconidia, or aleuriospores (Rippon, 1982). With the previous exception, conidia are not seen in tissue. Because of the acute nature of the disease, the tissue response is most often an acute, pyogenic reaction accompanied by necrosis. However, there have been observations of granulomatous tissue in more chronic presentations of aspergillosis (e.g., paranasal sinuses and/or orbit). Sometimes, in chronic cases, the Splendore–Hoeppli phenomenon of eosinophilic halos surrounding hyphae is observed (Rippon, 1982).

Diagnosis

The diagnosis of aspergillosis, particularly invasive aspergillosis, may present a genuine challenge in contemporary medicine. A definitive diagnosis can be made only by demonstrating fungal elements in pathologic material and by repeated isolation of the fungus in cultures (Bennett, 1985; Rippon, 1982). Tissue identification generally rests on five points: 1) uniform size of hyphae, 3 to 4 μm in width; 2) characteristic dichotomous branching of approximately 45°; 3) staining of hyphae with special stains such as Gomori methenamine silver or periodic acid–Schiff in addition to hematoxylin and eosin; 4) while not routine, demonstration of the organism by fluorescent-antibody techniques; and 5) presence of septa in the hyphae.

If one uses tissue morphology alone to diagnose aspergillosis, it is doubtful that such a diagnosis can be defended without finding the characteristic fruiting heads of the organism. The diagnosis should not be based only on the presence of dichotomously branching hyphae and never on the presence of septate hyphae. Because of the increase in and significance of "opportunistic" mycoses, several nonaspergillal fungal organisms are being documented as etiologic agents. These fungi may cause disease that clinically resembles aspergillosis, and the organism in tissue is indistinguishable from aspergilli. Hence, a point has been reached in medical mycology where use of the morphology of the organism in tissue as the sole criterion for diagnosing aspergillosis is not valid except when distinctive fruiting heads are present in air sacs. Examples of fungi that appear identical to *Aspergillus* in tissue are species of the genera *Fusarium* and *Scedosporium,* well-documented agents of human mycoses in similar patient populations. Without the use of specific immunologic/serologic techniques (i.e., fluorescent-antibody procedures, which are not widely available), there is no way to distinguish such fungi in tissue based only on hyphal appearance (Rinaldi, 1983).

Aspergillus-specific immunofluorescent-antibody testing is available in some reference centers and may be applied to viable and nonviable fungi in culture as well as in clinical materials or tissue sections (Swatek et al., 1985). The immunofluorescent-antibody technique is not entirely specific (Kaufman, 1981).

Many tactics have been employed in attempts to provide a rapid, sensitive, and specific diagnosis of invasive aspergillosis. These methods have included transtracheal aspiration, percutaneous needle biopsy, bronchoscopic procedures involving transbronchial/bronchial biopsy, bronchial washing/brushing/bronchoalveolar lavage, analysis of expectorated sputum, nasal/sputum/respiratory surveillance cultures, and serologic detection of aspergillal antibody or antigen (Albelda et al., 1984; Kahn et al., 1986). All of the above methodologies have restrictions as to sensitivity, specificity, availability, and contraindications. There are proponents of each technique, and all concur that a rapid, specific, and sensitive means of diagnosis is critical in this mycosis. Often, when clinical suspicion is high, empiric antifungal therapy is initiated and open lung biopsy must be considered.

Even with tissue, invasive aspergillosis may be difficult to diagnose as indicated above. Many patients cannot tolerate invasive procedures and some may experience drug toxicity when started on empiric antifungal chemotherapy. For these reasons, investigators continue to search for safe and rapid means of reliable diagnosis.

A variety of serologic tests have been developed for detection of antibodies to aspergilli in patients with aspergillosis. A number of immunoassay formats have been devised and evaluated but, unfortunately, the predictive value of these tests has been unsatisfactory in profoundly immunosuppressed patients with the greatest risk of invasive disease (de Repentigny and Reiss, 1984). A different approach involves detection of antigenemia caused by aspergilli invading tissue. The antigen being sought is a polysaccharide, most likely a cell wall galactomannan (Reiss and Lehmann, 1979). Detection of antigen appears promising and merits further investigation/evaluation for specifically diagnosing invasive disease early enough to permit effective therapy (de Repentigny et al., 1987).

Other avenues of diagnostic activity include further use/development of fluorescent- and/or monoclonal-antibody techniques for detection/recognition of aspergilli in tissue and other specimens (Jones, 1985). An exciting and potential future tool for diagnosis of invasive aspergillosis, as well as other mycoses, is the use of RNA/DNA gene probes. At present, it appears that most important in the diagnosis of invasive aspergillosis is an awareness and high index of suspicion by clinicians and microbiologists of the kind of patient and clinical setting in which this mycosis most commonly occurs.

Prognosis

The outcome of aspergillosis is dependent on the type of disease and the immunologic state of the host. With invasive disease, the prognosis is directly correlated with early diagnosis and therapy, appropriate management of the underlying disorder, and reversal of bone marrow suppression (Herbert and Bayer, 1981). While invasive disease has been successfully treated in both immunosuppressed and immunocompetent hosts (Rinaldi, 1983), the overall mortality still remains greater than 80% (Pennington, 1980). The prognosis of invasive disease in leukemic patients is especially grim (approximately 70% mortality), as contrasted with that of patients who have undergone renal or cardiac transplantation (approximately 25% mortality) (Herbert and Bayer, 1981). In short, invasive aspergillosis is often rapidly fatal and difficult to treat.

Etiologic Agents

Mycology

There are more than 200 species of aspergilli, of which approximately 19 have been authenticated as agents of human disease (Rinaldi, 1983). It seems safe to speculate that the diversity of etiologic species will increase (Opal et al., 1986). Aspergilli are among the most omnipresent moulds in the human environment and grow ubiquitously on or in any location containing organic debris. Many species, including the major pathogens of humans, form prodigious numbers of asexual propagules. Such conidia often become airborne in huge numbers, and no area open to air is free of aspergilli (Swatek et al., 1985). Moulds of the genus *Aspergillus* most often present in an asexual form properly termed the anamorph (Jorgensen and Rinaldi, 1986). Some strains of some species have the ability to undergo sexual reproduction, with a resultant sexual form correctly called the teleomorph. Periodically a teleomorph may be seen in laboratory culture; however, the anamorphic expression of aspergilli is overwhelmingly the morphologic entity seen in the clinical laboratory and whose structures are used in species or group identification.

The anamorphic expression typically observed consists of septate, moniliaceous (light-colored), branching hyphae, which may or may not produce a specialized cell merging with the vegetative hypha called a "foot cell" because of its appearance. If present, the foot cell gives rise to a stalklike, erect hyphal extension, usually nonseptate, termed the conidiophore (phore = a structure that bears something; in this case, conidia). If a foot cell is not present, the conidiophore arises directly from a vegetative hyphal cell. The conidiophore is usually hyaline, but may be colored and varies considerably in length and surface texture depending on the species. The tip of the conidiophore expands and develops into a swollen vesicle, which, again, assumes a shape distinctive for the species. The vesicle then produces on its surface, either entirely or partially, a row of conidiogenous cells (the cells that give rise to the conidia) called phialides. The phialides are generally flask shaped and produce chains of conidia by a budding (blowing-out) process. The conidial chains are produced in a basipetal fashion, with the youngest conidium at the base and the most mature at the tip. Such conidia are descriptively called phialoconidia to reflect their mode of formation. When phialides arise directly from the vesicle, the species is termed uniseriate. Some *Aspergillus* species first produce a row of sterile cells called metulae (singular, metula) on the vesicle surface. From the metulae then arise phialides producing phialoconidia; such species are termed biseriate. Some species, such as *A. flavus*, may demonstrate both conditions in the same head. Phialides are sometimes incorrectly referred to as sterigmata, a term that should be restricted to structures formed by fungi classified in an entirely different taxon, the Basidiomycota (e.g., the mushrooms).

Phialoconidia are generally globose to subglobose in shape and exhibit species-dependent size, color, wall texture, and thickness. The conidial chains are dry, that is, not held together by slime or matrix materials. Metulae and phialides may be arranged on the vesicle in such a manner that the conidial chains radiate from the conidiophore or develop an upward columnar appearance. Conidial chains are easily dislodged and become airborne.

While most laboratory scientists have little trouble assigning an unknown mould to the genus *Aspergillus*, identification to the species level (with the exception of the four most common species pathogenic for humans) is much more difficult. This is a reflection of the large number of species in the genus, the time and care required for detailed morphologic examination including size measurements of various structures, and the relative unfamilarity of most clinical microbiologists with the techniques of taxonomic mycology.

Several helpful sources are available to aid in the identification of aspergilli (Christensen, 1981; McGinnis, 1980; Raper and Fennell, 1965; Swatek et al., 1985). Most of these aids contain keys that allow aspergilli to be placed into one of the many distinctive "groups," the members of each "group" sharing basic common characteristics. With the exception of the Glaucus group (there is no species named *A. glaucus*), each group is named after a representative species within it. Further perusal of the keys

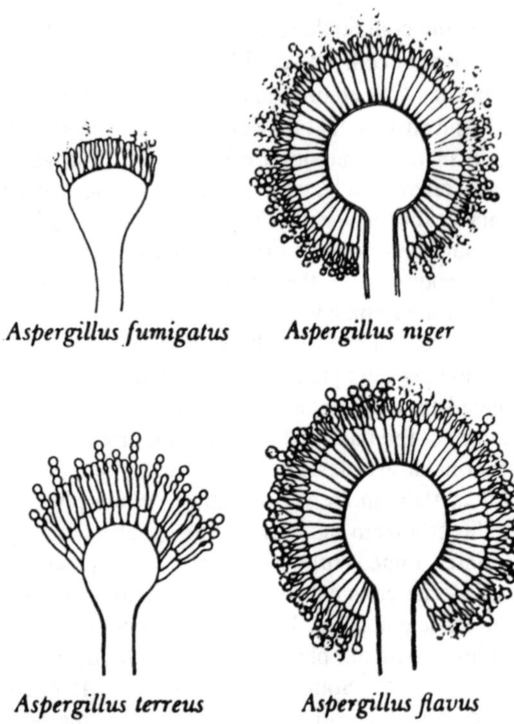

FIG. 1. Microscopic morphology of conidial heads of the four major species of *Aspergillus* causing human disease. Reproduced with permission from Rippon, J. W. 1974. Medical mycology. The pathogenic fungi and the pathogenic actinomycetes. W. B. Saunders Company, Philadelphia, PA.

may allow species identification. Most such keys are based on typical morphologic structures found when the moulds are grown at 25°C on two particular media: Czapek–Dox agar and 2% malt agar. These formulations are available from commercial sources and are detailed in the monograph by Raper and Fennell (1965).

The major pathogenic species, *A. fumigatus,* is characterized by smoky, gray–green colonies (indeed, the specific epithet "*fumigatus*" means smoky, gray–green), the ability to grow up to 50°C, and a microscopic appearance of uniseriate phialides covering only the upper one-third to one-half of the vesicle, with distinctive, tight columnar conidial chains. The clinical significance and characteristic microscopic features of this species mandate its recognition by health-care professionals.

In recent years, *A. flavus* has assumed an increasingly significant role as an etiologic agent of invasive aspergillosis. The colonial color (lime green) and microscopic appearance (radial conidial heads that exhibit both uni- and biseriation), and poor to no growth at 45°C or higher, allow its separation from *A. fumigatus,* from which it differs considerably. The remaining two major pathogenic species, *A. niger* and *A. terreus,* also offer distinctive features useful in identification. Table 1 presents some characteristics used for species identification of the four major pathogens, and Fig. 1 indicates the typical morphology of their conidial heads. Reference laboratories and/or mycologists possessing expertise with members of this genus may be needed for isolates of aspergilli that seem to defy group or species identification.

Nomenclature, Classification, and Taxonomy

While the teleomorphs of some species have been described, it is inevitably the anamorphs that incite disease and are seen in tissue and culture. The clinical microbiologist generally finds fungal taxonomy, classification, and nomenclature to be unfamiliar topics and often confusing as well. The spelling and pronunciation of fungal names are often troublesome, as are the frequent name changes wrought by contemporary fungal taxonomists (Jorgensen and Rinaldi, 1986). Additionally, training in mycology for health-care professionals is often given short shrift and may be taught, of necessity, by individuals who have little background or interest in taxonomic mycology. Often, subject matter pertaining to mycology is presented toward, or at, the conclusion of courses in medical microbiology, and is alloted a comparatively short amount of time.

All too often, on laboratory proficiency tests, an unknown mould sits under the microscope while all mycology books and laboratory guides are laid out and scoured for a picture or drawing that most closely resembles the mystery fungus. With the exception of parasitology, the identification of mould fungi is one of the few remaining areas in clinical microbiology where the eyes, cerebral power, and the microscope are necessary. There are no rapid identification kits, automated machines, or, as yet, gene probes/monoclonal antibodies available for recognition of most moulds. One particularly exciting contemporary methodology for mould identification involves the use of exoantigens (Kaufman, 1981). This technique has proved of tremendous utility in identification of some of the agents of systemic mycoses. Recent studies have shown that the exoantigen technique may be useful in the identification of nonconidiating isolates of *A. fumigatus* and in the correct assignment of aspergilli to appropriate groups (Kaufman et al., 1983; Sekon et al., 1986). These studies indicated correct assignment in 51 of 51 (100%) of aspergilli with no fungi in genera other than *Aspergillus* producing antigens identical to the selected reference antigens. Unfortunately, the necessary reagents are not available commercially, and these techniques remain essentially limited to a very few reference laboratories. In any case, such investi-

TABLE 1. Characteristics of the four major human pathogenic species of *Aspergilli* grown on Czapek–Dox agar at 25°C

Group and species	*Seriation* Uni	Bi	*Colony color*	*Conidiophore*	*Vesicle*	*Conidial head*	*Conidia*	*Comment*
Fumigatus *A. fumigatus*	+		White to green to gray–green; slate gray with age, reverse variable in color	Length up to 300 μm (rarely to 500 μm); width, 5–8 μm; smooth walled; greenish, especially in upper portion	Dome shaped; conidiophore gradually enlarges into vesicle; 20–30 μm in diameter; phialides on upper half only, with axes parallel to that of conidiophore	Strongly columnar, compact	Globose to subglobose; elliptic in some isolates, echinulate to rarely smooth; mostly 2.5–3.0 μm in diameter	One of the most commonly occurring moulds in nature; most common human pathogen; grows well at 50°C
Flavus *A. flavus*	+	+	Yellow to yellowish green to green with age	Length, 400–850 μm (rarely to 2.5 mm); colorless; thick walled; coarsely roughened	Elongate, becoming subglobose to globose; 10–65 (commonly 25–45) μm in diameter	Radiate, splitting into columns with age	Globose, smooth to echinulate, 3–6 (most 3.5–4.5) μm in diameter	Common human pathogen; some strains produce toxins; growth usually enhanced at 37°C; small vesicles uniseriate; dark brown to black sclerotia
Niger *A. niger*		+	White, becoming black; reverse occasionally pale yellow	Length, 1.5–3.0 mm, width, 15–20 μm; smooth walls; colorless with brownish shade on upper half	Globose, most 45–75 μm in diameter	Globose, then radiate, splitting into columns with age	Globose, thick walls, brown, irregularly roughened; most 4–5 μm in diameter	Frequent cause of otomycosis; sometimes associated with cavitary colonization
Terreus *A. terreus*		+	Cinnamon buff to brown, rarely orange–brown	Length, 100–250 μm; width, 4.5–6 μm; flexous, smooth walled, colorless	Hemispheric, dome-like, merging into conidiophore; 10–16 μm in diameter	Long, columnar; uniform in diameter throughout length	Globose to slightly elliptic, smooth, 1.8–2.4 μm in diameter	May form single globose conidia on submerged hyphae (aleurioconidia, 6–7 μm)

gations exemplify the exciting possibilities for future means of aspergillal identification.

Because of the reasons mentioned above, mould identification may be considered a real chore, explaining the sometimes heard microbiologist's lament of having to spend their "turn" in the mycology laboratory. This is unfortunate, as medical mycology has never been as important as in contemporary medicine—the fungi have assumed a role of critical significance in the incitation of infectious diseases. With increased understanding of the prevalence and importance of mycotic disease has come better mycology and resultant changes in nomenclature, taxonomy, and classification.

The aspergilli are classified in the form–division Fungi Imperfecti of the kindgom Fungi. A contemporary classification of the aspergilli is shown in Table 2.

The study of medical mycology, including systematics and nomenclature, is not stagnant, but rather dynamic and in flux. With the powerful tools of molecular biology, electron microscopy, immunology, and contemporary mycologic principles being merged and applied to medical fungi, further change and advance are inevitable (Samson and Pitt, 1985).

Collection, Transport, and Storage of Specimens

Because aspergilli are so common in the human environment, caution must be observed in assigning an etiologic role to isolates obtained from patients. Contamination of respiratory and cutaneous materials or of any step during the processing of clinical specimens may occur and increase the difficulty of judging whether an isolate is clinically significant (Swatek et al., 1985). The collection of material from patients that is suitable for diagnosis is dependent on the site and type of lesion. In many cases, lesions of aspergillosis may be removed in their entirety and be placed in sterile screw-cap bottles containing a drop of sterile saline. Biopsy material is not likely to be contaminated if collected and handled carefully. Swabs are unsatisfactory specimens, but the sediment from centrifuged washings can be used for direct examination and culture. Early-morning sputum samples may be useful in diagnosis, but must be collected carefully to avoid airborne contamination and should be processed rapidly (Roberts et al., 1985). Skin lesions and nails should be washed with soap and water prior to specimen collection. While many specimens may retain their fungal content unchanged over a period of several days if kept at 5°C, rapid examination is preferable because of the potential for bacterial or saprobic fungal overgrowth. As with all specimens collected for microbiologic studies, the success of demonstrating/growing the putative agent is considerably enhanced by proper collection methods and rapid processing by laboratory personnel.

As discussed above, blood rarely yields species of *Aspergillus*, even in far-advanced, disseminated disease. The same may be said for spinal fluid and urine. If the culture of blood is desired, the lysis–centrifugation technique is the preferred method (Swatek et al., 1985).

Specimens such as sputum and exudates should be collected with careful attention to aseptic technique and processed immediately. Multiple early-morning sputa should be collected on consecutive days following thorough cleansing of the patient's mouth (Goodman, 1982). Sputum may require a digestion procedure without NaOH, followed by centrifugation (Rinaldi, 1983).

A common specimen submitted from patients with aspergillosis or suspected aspergillosis is tissue obtained following biopsy or surgery. Tissue specimens may be examined directly following homogenization by mounting in 10 to 20% KOH, placing a coverslip, and gently warming until the tissue softens. As an aside, if zygomycetous fungi such as *Rhizopus* are suspected (zygomycosis may occur in the same patient population as does aspergillosis), tissue should be minced, not homogenized, as the latter procedure will render Zygomycetes nonviable.

Biopsy specimens should be dissected aseptically in sterile saline, which may then be examined microscopically and plated out. A pepsin digestion method has proved useful when dealing with samples from chronic lesions containing caseating and fibrous material (Austwick and Longbottom, 1980). Following digestion, a sediment is obtained that can then be

TABLE 2. Taxonomic status of *Aspergilli* of medical significance

Anamorphic Classification (Imperfect, Asexual)	
Superkingdom:	Eukaryota
Kingdom:	Fungi (Mycota)
Form–division:	Fungi Imperfecti
Form–class:	Hyphomycetes
Form–order:	Hyphomycetales
Form–family:	Moniliaceae
Form–genus:	*Aspergillus* Micheli ex Link: Fries
Major species pathogenic for humans:	
	Aspergillus fumigatus Fresenius, 1863
	Aspergillus flavus Link, 1809
	Aspergillus niger van Tieghem, 1867
	Aspergillus terreus Thom, 1916

directly mounted or centrifuged before clearing in KOH.

All specimens should be covered in sterile containers during transport and, as noted, be processed immediately. While refrigeration of specimens may prevent growth of contaminating bacteria and fungi for 1 to 2 days, the cooler environment may also be detrimental to aspergilli. The same is true for freezing and addition of antibacterial agents to specimens. The most expeditious policy is rapid laboratory processing of all specimens sent for microbiology.

Direct Microscopic Examination

A freshly collected specimen subjected to direct microscopy may be rewarding in making a diagnosis. Both a KOH mount and Gram staining should be performed on small portions of the specimen. An alternative to the above approach involves placing the fresh or Formalin-fixed specimens directly in a solution of equal parts of 40% KOH and Parker Blue-Black Quink (The Parker Pen Co., Janesville, Wis.) or drawing the ink beneath the coverslip after mounting in KOH (Balogh, 1964). This procedure stains the aspergillal hyphae very effectively, and even deeper staining and differentiation occurs if the preparation is kept in a damp chamber overnight (Austwick and Longbottom, 1980). It is important to note that hyphae of other fungi also stain.

Rapid periodic acid–Schiff and silver stains have been devised and are very useful for detection of small hyphal fragments, but these methods are expensive both cost- and timewise. When aspergillal hyphae are present in a specimen, they are usually closely septate with constrictions, may range from 2.5 to 8.0 μm (average, 3.0 μm) wide. There may be only a few hyphal elements present or an entire colony. As noted previously, many fungi other than aspergilli may present with an identical appearance.

While usually parallel and smooth walled, the hyphae may appear pleomorphic and have enlarged, rounded segments. Cytoplasm in hyphal tips is refractile and differs from that of older cells, which are filled with a central vacuole. Occasionally there may be hyphae with no septa, which may lead to confusion with zygomycetous fungi (Swatek et al., 1985). It is common for hyphae to fragment, leaving septate fragments scattered throughout the specimen (Goodman, 1982). If the location of a lesion is in contact with air (e.g., cavity in the lung, ear canal), the fungi may form their typical conidial heads in the specimen.

Tissue submitted for histopathology should be stained by periodic acid–Schiff and Gomori methenamine silver methods in addition to the routine hematoxylin and eosin preparations. Aspergilli in tissue will show the same hyphal characteristics as mentioned above (i.e., septate with acute angle branching), but tissue always should be cultured because other fungi may appear identical histopathologically. *Aspergillus* hyphae in tissue appears identical regardless of the causative species; hence, culture again remains the major means for specific identification of the etiologic agent.

Recent reports have directed attention to a phenomenon known since 1973 (Nime and Hutchins, 1973), in which the occurrence in tissue specimens of birefringent crystals of calcium oxalate are observed in lung lesions from patients with aspergillal fungus balls or invasive pulmonary disease. Such patients consist of both the nonimmunocompromised and those with abrogated immunity typically underlying invasive aspergillosis. The crystals are birefringent and needlelike in rosette or wheat-sheaf-like arrangements and are most often seen in pulmonary cytology specimens. The pathogenesis of such crystal formation is not completely understood; however, the crystals are thought to be formed when oxalic acid produced by the fungus combines with calcium in the patient's tissues.

The crystals are related to the presence of aspergilli, most notably *A. niger*. However, oxalate crystals have also been found in association with *A. flavus* and *A. fumigatus*. While their absence does not rule out aspergillosis, the finding of birefringent, needlelike crystals with rosette or wheat-sheaf-like arrangements in pulmonary cytology specimens should be considered as an important diagnostic marker/aid for the presence of aspergillosis (Farley et al., 1985; Lee et al., 1986).

Culture and Identification

The aspergilli will grow on almost any substrate containing organic material. If the specimen is likely to be contaminated with bacteria, isolation is accomplished on a medium containing antibacterial agents such as chloramphenicol, gentamicin, or penicillin plus streptomycin. It is important to note that traditional mycologic isolation media used in most clinical laboratories, such as Mycosel agar (BBL Microbiology Systems, Cockeysville, Md.) and Mycobiotic agar (Difco Laboratories, Detroit, Mich.), should not be used to isolate aspergilli. Both formulations are essentially Sabouraud dextrose agar (SDA) containing chloramphenicol (to suppress unwanted bacterial growth) and cycloheximide (to suppress unwanted saprobic fungal growth). Cycloheximide (Actidione; The Upjohn Co., Kalamazoo, Mich.) is a selective antieukaryotic agent that is inhibitory to aspergilli.

While there is little doubt that the major medium used for fungal isolation in most clinical laboratories is SDA or a modification thereof, this author believes that SDA is only useful to determine dermatophyte colonial morphology or to maintain fungal stock cultures. Most aspergilli will grow luxuriantly on SDA, but not characteristically. Often, to identify the causative species, the organism must be subcultured to other media and/or set up in slide culture. Sabouraud agar can certainly be used for isolation; however, it is suggested that alternative media be considered that allow for both isolation and identification.

Media that can be recommended for primary isolation are brain–heart infusion agar, 2% malt extract agar, and potato-flake agar (Rinaldi, 1982). These media should be used with and without antibacterial agents, for example, 20 IU of penicillin plus 40 IU of streptomycin per ml, 0.05 mg of chloramphenicol per ml, or 0.05 mg of chlortetracycline per ml. Cultured materials should be incubated at both 25°C and 35 to 37°C and not discarded for at least 4 weeks.

Some aspergilli, documented but rare as human pathogens, will only grow well on media with high osmotic pressure, such as malt salt or Czapek–Dox agar containing 20% sucrose. Examples of such fungi are members of the Glaucus group and *A. restrictus* (Rinaldi, 1983).

As described previously, species identification may require use of Czapek–Dox and 2% malt extract agars in order to effectively utilize the keys available in taxonomic monographs (Raper and Fennell, 1965). These keys include criteria essential in species identification, including colony texture and color; rate of growth; conidiophore morphology; seriation; size, shape, and arrangement of conidia; and the morphologic characteristics of any teleomorph present.

A screening medium useful in presumptive identification of species in the Flavus group has been devised and evaluated (Bothast and Fennell, 1974; Salkin and Gordon, 1975). This medium is called *Aspergillus* differential medium; some members of the Flavus group (*A. flavus* and *A. oryzae*) produce bright, yellow–orange pigment on this substrate, while other medically important *Aspergillus* species do not. This medium should not, however, be employed as the sole method for identifying isolates as members of the Flavus group.

Finally, as discussed earlier, exoantigen techniques offer tremendous potential for rapid aspergillal identification (Sekon et al., 1986). Perhaps the most exciting development for the rapid, sensitive, and specific identification of aspergilli (as well as many other mycotic pathogens) is the use of RNA/DNA gene probes. This powerful molecular technique has been used successfully for identification of mycobacteria, Legionellae, and some viruses (Tompkins, 1985). There is every reason to believe this technology will be fruitfully applied to medical mycology, including the aspergilli.

Serologic Diagnosis

Serologic tests for the detection of antibodies to aspergillal antigens in patients with aspergillosis or suspected aspergillosis have been useful, as such antibodies are generally uncommon in sera from healthy persons. The most reliable tests are immunodiffusion (ID) and counterimmunoelectrophoresis. As noted, these tests have limitations in that immunocompromised patients often have little ability to elaborate antibodies, and positive serologic results alone are not diagnostic while negative serologic tests do not rule out disease (Goodman, 1982).

A majority of individuals with allergic bronchopulmonary aspergillosis and at least 80% of nonimmunocompromised hosts with aspergillal fungus balls respond positively to ID tests (Jones, 1985; Kaufman and Reiss, 1985). Such precipitins are formed less frequently in patients with invasive disease. The greatest number of cases of aspergillosis may be detected by use of *A. fumigatus, A. flavus, A. niger,* and *A. terreus* precipitinogens in separate ID tests performed at the same time (Kaufman and Reiss, 1985, 1986). While the ID test may be helpful in the diagnosis of systemic aspergillosis (Kaufman, 1983) when serial specimens are obtained, its sensitivity in the diagnosis of invasive aspergillosis in patients with acute leukemia does not appear to exceed 60% (Jones, 1985).

Complement fixation (CF) tests have also been described, but their value is open to question. Some investigators feel the CF test exhibits a degree of sensitivity and specificity comparable to the ID test, while others have reported the CF test is less specific or less sensitive (Kaufman, 1980; Kaufman and Reiss, 1985). One approach may be to screen with the ID procedure and submit serum to a reference laboratory for quantitation of antibody by CF testing.

Other means for detection of serum precipitins have been devised and include counterimmunoelectrophoresis, enzyme-linked immunosorbent assay, passive hemagglutination, biotin–avidin-linked immunosorbent assay, and radioimmunoassay (Rinaldi, 1983; Sepulveda et al., 1979; Shamsuddin and Harris, 1983; Wang et al., 1978). Each of these methods has its proponents and limitations. While some of these assays have demonstrated positivity in 70 to 80% of patients with invasive pulmonary aspergillosis (Herbert and Bayer, 1981), the specificity and predictive value of single antibody titers is low. Prospective serial antibody testing in immunosuppressed patients with seroconversion, hence prompting early treatment, offers promise (Fisher et al., 1981; Herbert and Bayer, 1981). At present, many investigators believe that an early and accurate diagnosis of invasive diagnosis can be accomplished by monitoring with ID, counterimmunoelectrophoresis, and other tests for demonstrating lines of identity, critical titers, and seroconversion and/or titer or band changes (Kaufman, 1981).

As mentioned, the patient with invasive aspergillosis may not be capable of mounting an immunologic response, and recent efforts for serodiagnosis have sought detection of *Aspergillus* antigen rather than antibody. Galactomannan antigenemia in aspergillosis has been detected by counterimmunoelectrophoresis, radioimmunoassay, and enzyme immunoassay (Rinaldi, 1983). Recent progress in this area is encouraging and indicates that these tests are highly specific, detect as little as 10 ng of antigen per ml of serum, need not involve radioactivity, are rapid (4 h), utilize well-characterized cell wall galactomannan markers, and offer further promise of even more sensitive testing refinements, for example, double antibody sandwich format (de Repentigny et al., 1987).

All of these tests merit further evaluation with larger numbers of patients to elucidate interactions with other parameters such as antifungal chemotherapy (e.g., therapy may decrease the total number of organisms present—and, therefore, the amount of antigen released—and compromise assay sensitivity [Jones, 1985]) and to define their utility for specifically diagnosing invasive disease early enough to permit efficacious therapy (Rinaldi, 1983).

Antifungal Drugs, Treatment, and the Role of the Laboratory

Antifungal Drugs and Treatment

The treatment of invasive disease is difficult. Early and aggressive therapy may result in cure in a small but definite number of patients with invasive pulmonary aspergillosis (Pennington, 1986). In one series, a 64% mortality rate was associated with early therapy; however, mortality generally exceeds 80% (Albelda et al., 1985; Fisher et al., 1981).

Amphotericin B is the antimicrobic of choice for invasive aspergillosis. It should be given at full therapeutic doses in this mycosis; for example, an adult should receive 0.6 mg/kg intravenously per day within 2 to 3 days of commencement of therapy (Pennington, 1986). The duration of therapy is unknown, and successful doses of amphotericin B have ranged from 100 to 3,100 mg. Often, empiric therapy with amphotericin B is initiated for high-risk patients, especially in the oncologic setting. This antifungal agent demonstrates considerable toxicity, particularly nephrotoxicity; however, it may be the only life-saving measure available. Thus far, the development of resistance to amphotericin B has not been reported.

5-Fluorcytosine, an orally administrable antifungal agent, exhibits strain-dependent activity against aspergilli. There is, however, little reason to use 5-fluorocytosine as monotherapy in aspergillosis. Fungi frequently develop resistance to 5-fluorocytosine in addition to the substantial number of pathogenic aspergilli natively resistant (Rinaldi, 1983). Additionally, the myelosuppressive effects of 5-fluorocytosine preclude its use in most immunosuppressed patients. Some reports have shown in vitro synergism with amphotericin B plus 5-fluorocytosine, as well as enhanced clinical efficacy when the combination has been used for therapy of invasive disease (Pennington, 1986; Rinaldi, 1983). Likewise, reports have suggested amphotericin B plus rifampin may be effective in invasive aspergillosis based on in vitro, animal, and human anecdotal experience (Rinaldi, 1983). Further experience and information is required for assessment of these combination therapies in aspergillosis.

The imidazoles miconazole and ketoconazole, although demonstrating in vitro activity against some isolates of aspergilli, have not proved useful in therapy. The more recent imidazoles itraconazole and fluconazole also demonstrate activity against aspergilli greater than that of the earlier imidazoles (Espinel-Ingroff et al., 1984; Kobayashi et al., 1986; Van Cutsem et al., 1984). The role of these imidazoles in the therapy of aspergillosis remains to be determined.

Because of the increasing significance and prevalence of mycotic diseases, pharmaceutical firms are demonstrating renewed, and in some cases initial, interest in antifungal drugs. At present, there are available, or impending, more antimicrobics to treat mycoses then ever before. It seems likely this trend will continue. There is a definite need for safe, nontoxic, cidal alternatives to amphotericin B for the therapy of invasive aspergillosis and many other fungal diseases.

The polyene pimaricin (natamycin) has been used successfully to treat *Aspergillus*-caused ocular mycoses. This antimicrobic is used strictly for that purpose, as it is too toxic for systemic use (Rippon, 1982).

Antifungal Susceptibility Testing

Laboratory evaluation of antifungal activity has not been widely accomplished. This is due to a variety of reasons: a small number of available drugs to test, lack of standardization, a seeming lack of resistance or its development by medically important fungi, the previously perceived unimportance of invasive mycoses (as contrasted with the uncontested prevalence/significance of bacterial infections), and the lack of correlation of in vitro results with patient therapeutic response. These concerns remain valid; however, recent developments mandate reconsideration of this area of laboratory medicine. Some reasons for antifungal testing are 1) an increased number of broad-spectrum antifungal drugs, with more under development; 2) the documented resistance of medical fungi to amphotericin B, 5-fluorocytosine, and some imidazoles, as well as the likelihood of an increasing prevalence of resistance because of increased drug use; 3) differing efficacy of imidazoles against various fungal species; and 4) efforts currently under way to achieve in vitro standardization (Rinaldi and Howell, 1988).

The current state of antifungal testing is perhaps equivalent to the situation with antibacterial drugs some 30 years ago. Collaborative efforts by research and reference laboratories are under way. There are difficulties, to be sure; however, there appears to be no inherent reason why standardization of antifungal susceptibility testing cannot be achieved (Calhoun et al., 1986; Galgiani, 1986). At present, antifungal susceptibility testing is best accomplished at reference laboratories.

There are detailed considerations of the performance of antifungal testing (as well as the assay of levels of antimycotics in biologic fluids) elsewhere (McGinnis and Rinaldi, 1985; Rinaldi and Howell, 1988; Shadomy et al., 1985). In published reports involving aspergilli, there has been wide variation in susceptibility from strain to strain, with the exception of amphotericin B, to which aspergilli are always susceptible. There is little information available that correlates the results of in vitro tests with outcome of therapy. Correlative problems are amplified in immunocompromised patients with invasive aspergillosis (Rinaldi, 1983; Swatek et al., 1985).

Conclusion and Comments

With limited, toxic, therapeutic agents as the mainstay of treatment, the severe and complicated underlying disorders that cause predisposition to infection, the difficulties in rapid and definitive diagnosis, and the grave prognosis, there is no doubt that invasive aspergillosis is a disease of protean significance. Because the lung is the most frequent target organ, pulmonary aspergillosis must be considered a major clinical problem in contemporary medicine (Pennington, 1986). Aspergillosis was the third most common systemic fungal infection requiring hospitalization in the United States in 1976 and was increasing in frequency more rapidly than any other fungal disease, with a 158% increase in incidence reported between 1970 and 1976 (Fraser et al., 1979). The most recent survey shows that between 1976 and 1980 to 1982 the incidence of aspergillosis requiring hospitalization in the United States almost doubled (Reingold et al., 1986). Such disconcerting figures underscore that mycoses due to species of *Aspergillus* remain a major challenge to clinicians and laboratory workers.

Literature Cited

Albelda, S. M., G. H. Talbot, S. L. Gerson, W. T. Miller, and P. A. Cassileth. 1984. Role of fiberoptic bronchoscopy in the diagnosis of invasive pulmonary aspergillosis in patients with acute leukemia. Am. J. Med. **76:**1027–1034.

Albelda, S. M., G. H. Talbot, S. I. Gerson, W. T. Miller, and P. A. Cassileth. 1985. Pulmonary cavitation and massive hemoptysis in invasive pulmonary aspergillosis: influence of bone marrow recovery in patients with acute leukemia. Am. Rev. Respir. Dis. **131:**115–120.

Austwick, P. K. C., and J. L. Longbottom. 1980. Medically important *Aspergillus* species, p. 620–627. *In* E. H. Lennette, A. Balows, W. J. Hausler, Jr., and J. P. Truant (ed.), Manual of clinical microbiology, 3rd ed. American Society for Microbiology, Washington, D.C.

Balogh, N. 1964. Diagnostik Pilzkrankheiten bei Tieren mit modifizierte Parker Blue-Black Superchrome Tintensarbung. Dtsch. Tieraerztl. Wochenschr. **71:**327–330.

Bardana, E. J. Jr. 1981. The clinical spectrum of aspergillosis. Part 2: classification and description of saprophytic, allergic, and invasive variants of human disease. Crit. Rev. Clin. Lab. Sci. **13:**85–159.

Bennett, J. E. 1985. *Aspergillus* species, p. 1447–1451. *In* G. L. Mandell, R. G. Douglas, Jr., and J. E. Bennett (ed.), Principles and practice of infectious diseases, 2nd ed. John Wiley & Sons, Inc., New York.

Bodey, G. P., R. Bolivar, and V. Fainstein. 1982. Infectious complications in leukemic patients. Semin. Hematol. **19:**193–226.

Bothast, R. J., and D. I. Fennell. 1974. A medium for rapid identification and enumeration of *Aspergillus flavus* and related organisms. Mycologia **66:**365–369.

Cahill, K. M., A. M. El Mofty, and T. P. Kawaguchi. 1967. Primary cutaneous aspergillosis. Arch. Dermatol. **96:**545–547.

Calhoun, D. L., G. D. Roberts, J. N. Galgiani, J. E. Bennett, D. S. Feingold, J. Jorgensen, G. S. Kobayashi, and S. Shadomy. 1986. Results of a survey of antifungal susceptibility tests in the United States and interlaboratory comparison of broth dilution testing of flucytosine and amphotericin B. J. Clin Microbiol. **23:**298–301.

Christensen, M. A. 1981. A synoptic key and evaluation of species in the *Aspergillus flavus* group. Mycologia **72:**1056–1084.

de Repentigny, L., M. Boushira, L. Ste-Marie, and G. Bosisio. 1987. Detection of galactomannan antigenemia by enzyme immunoassay in experimental invasive aspergillosis. J. Clin. Microbiol. **25:**863–867.

de Repentigny, L., and E. Reiss. 1984. Current needs in immunodiagnosis of candidiasis and aspergillosis. Rev. Infect. Dis. **6:**301–312.

Emmons, C. W., C. H. Binford, J. P. Utz, and K. J. Kwon-Chung. 1977. Medical mycology, 3rd ed. Lea & Febiger, Philadelphia.

Espinel-Ingroff, A., S. Shadomy, R. J. Gebhart. 1984. *In vitro* studies with R 51,211 (itraconazole). Antimicrob. Agents Chemother. **26:**5–9.

Farley, M. L., L. Mabry, L. A. Munoz, and H. W. Diserens. 1985. Crystals occurring in pulmonary cytology specimens. Association with *Aspergillus* infection. Acta Cytol. **29:**737–744.

Fisher, B. D., D. Armstrong, B. Yu, and J. W. M. Gold. 1981. Invasive aspergillosis. Progress in early diagnosis and treatment. Am. J. Med. **71:**571–577.

Fraser, D. W., J. I. Ward, L. Ajello, an B. D. Plikaytis. 1979. Aspergillosis and other systemic mycoses—the growing problem. J. Am. Med. Assoc. **242:**1631–1635.

Galgiani, J. N. 1986. Antifungal susceptibility testing. Recent findings and experience. Antimicrob. Newsl. **3:**17–22.

Goodman, N. L. 1982. Aspergillosis. Clin. Microbiol. Newsl. **4:**9–11.

Herbert, P. A., and A. S. Bayer. 1981. Fungal pneumonia. Part 4. Invasive pulmonary aspergillosis. Chest **80:**220–225.

Jones, J. M. 1985. The immunobiology of candidiasis and aspergillosis. Lab. Manage. **23:**53–57.

Jorgensen, J. H., and M. G. Rinaldi. 1986. A clinician's dictionary of bacteria and fungi. Eli Lilly & Co., Indianapolis, Ind.

Kahn, F. W., J. M. Jones, and D. M. England. 1986. The role of bronchoalveolar lavage in the diagnosis of invasive pulmonary aspergillosis. Am. J. Clin. Pathol. **86:**518–523.

Karam, G. H., and F. M. Griffin, Jr. 1986. Invasive pulmonary aspergillosis in nonimmunocompromised, nonneutropenic hosts. Rev. Infect. Dis. **8:**357–363.

Kaufman, L. 1980. Serodiagnosis of fungal diseases, p. 553–572. *In* N. R. Rose and H. Friedman (ed.), Manual of clinical immunology, 2nd ed. American Society for Microbiology, Washington, D.C.

Kaufman, L. 1981. Current methods for serodiagnosing systemic fungus infections and identifying their etiologic agents. Estratto Ig. Mod. **76:**422–442.

Kaufman, L. 1983. Mycoserology: its vital role in diagnosing systemic mycotic infections. Jpn. J. Med. Mycol. **24:**1–8.

Kaufman, L., and E. Reiss. 1985. Serodiagnosis of fungal diseases, p. 924–944. *In* E. H. Lennette, A. Balows, W. J. Hausler, Jr., and H. J. Shadomy (ed.), Manual of clinical microbiology, 4th ed. American Society for Microbiology, Washington, D.C.

Kaufman, L., and E. Reiss. 1986. Serodiagnosis of fungal diseases, p. 446–466. *In* N. R. Rose, H. Friedman, and J. L. Fahey (ed.), Manual of clinical laboratory immunology, 3rd ed. American Society for Microbiology, Washington, D.C.

Kaufman, L., P. G. Standard, and A. A. Padhye. 1983. Exoantigen tests for the immunoidentification of fungal cultures. Mycopathologia **82:**3–12.

Kobayashi, G. S., S. Travis, and G. Medoff. 1986. Comparison of the *in vitro* and *in vivo* activity of the bis-triazole derivative UK 49,858 with that of amphotericin B against *Histoplasma capsulatum*. Antimicrob. Agents Chemother. **29:**660–662.

Lee, S. H., W. G. Barnes, and W. P. Schaetzel. 1986. Pulmonary aspergillosis and the importance of oxalate crystal recognition in cytology specimens. Arch. Pathol. Lab. Med. **110:**1176–1179.

Matsumoto, T., and L. Ajello. 1986. No granules, no mycetomas [letter]. Chest **90:**151–152.

McGinnis, M. R. 1980. Laboratory handbook of medical mycology. Academic Press, Inc., New York.

McGinnis, M. R., and M. G. Rinaldi. 1985. Antifungal drugs: mechanisms of action, drug resistance, susceptibility testing, and assays of activity in biological fluids, p. 223–287. *In* V. Lorian (ed.), Antibiotics in laboratory medicine, 2nd ed. Academic Press, Inc., New York.

Mirsky, H. S., and J. Cuttner. 1972. Fungal infection in acute leukemia. Cancer **30:**348–352.

Nime, F. A., and G. M. Hutchins. 1973. Oxalosis caused by *Aspergillus* infection. Johns Hopkins Med. J. **133:**183–194.

Opal, S. M., L. B. Reller, G. Harrington, and P. Cannady, Jr. 1986. *Aspergillus clavatus* endocarditis involving a normal aortic valve following coronary artery surgery. Rev. Infect. Dis. **8:**781–785.

Pennington, J. E. 1980. *Aspergillus* lung disease. Med. Clin. North Am. **64:**475–490.

Pennington, J. E. 1986. *Aspergillus,* p. 175–189. *In* G. A. Sarosi and S. F. Davies (ed.), Fungal diseases of the lung. Grune & Stratton, Orlando, Fla.

Prystowsky, S. D., B. Vogelstein, D. S. Ettinger, W. G. Merz, H. Kaizer, V. I. Sulica, and W. H. Zinkham. 1976. Invasive aspergillosis. N. Engl. J. Med. **295:**655–658.

Raper, K. B., and D. I. Fennell. 1965. The genus *Aspergillus*. The Williams & Wilkins Co., Baltimore.

Reingold, A. L., X. Dong Lu, B. D. Plikaytis, and L. Ajello. 1986. Systemic mycoses in the United States, 1980–1982. J. Med. Vet. Mycol. **24:**433–436.

Reiss, E., and P. F. Lehmann. 1979. Galactomannan antigenemia in invasive aspergillosis. Infect. Immun. **25:**357–365.

Rinaldi, M. G. 1981. Spontaneous lysis of mycetomas after acute cavitating lung disease [letter]. Am. Rev. Respir. Dis. **124:**342–343.

Rinaldi, M. G. 1982. Use of potato flakes agar in clinical mycology. J. Clin. Microbiol. **15:**1159–1160.

Rinaldi, M. G. 1983. Invasive aspergillosis. Rev. Infect. Dis. **5:**1061–1077.

Rinaldi, M. G., and A. W. Howell. 1988. Antifungal antimicrobics: laboratory evaluation. *In* I. F. Salkin (ed.), Diagnostic procedures for mycotic and parasitic diseases, 7th ed. American Public Health Association, Washington, D.C. (in press).

Rippon, J. W. 1982. Medical mycology. The pathogenic fungi and the pathogenic Actinomycetes, 2nd ed. The W. B. Saunders Co., Philadelphia.

Roberts, G. D., N. L. Goodman, G. A. Land, H. W. Larsh, and M. R. McGinnis. 1985. Detection and recovery of fungi in clinical specimens, p. 500–513. *In* E. H. Lennette, A. Balows, W. J. Hausler, Jr., and H. J. Shadomy (ed.), Manual of clinical microbiology, 4th ed. American Society for Microbiology, Washington, D.C.

Salkin, I. F., and M. A. Gordon. 1975. Evaluation of *Aspergillus* differential medium. J. Clin. Microbiol. **2:**74–75.

Samson, R. A., and J. I. Pitt (eds.). 1985. Advances in *Penicillium* and *Aspergillus* systematics. Plenum Press, New York.

Schaffner, A., H. Douglas, and A. Braude. 1982. Selective protection against conidia by mononuclear and against mycelia by polymorphonuclear phagocytes in resistance to *Aspergillus*. Observations on these two kinds of defense *in vivo* and *in vitro* with human and mouse phagocytes. J. Clin. Invest. **69:**617–631.

Sekon, A. S., P. G. Standard, L. Kaufman, A. K. Garg, and

P. Cifuentes. 1986. Grouping of *Aspergillus* species with exoantigens. Diagn. Immunol. **4**:112–116.

Sepulveda, R., J. L. Longbottom, and J. Pepys. 1979. Enzyme-linked immunosorbent assay (ELISA) for IgG and IgE antibodies to protein and polysaccharide antigens of *Aspergillus fumigatus*. Clin. Allergy **9**:359–371.

Shadomy, S., A. Espinel-Ingroff, and R. Y. Cartwright. 1985. Laboratory studies with antifungal agents: susceptibility tests and bioassays, p. 991–999. *In* E. H. Lennette, A. Balows, W. J. Hausler, Jr., and H. J. Shadomy (ed.), Manual of clinical microbiology, 4th ed. American Society for Microbiology, Washington, D.C.

Shamsuddin, A. M., and C. C. Harris. 1983. Improved enzyme immunoassays using biotin–avidin enzyme complex. Arch. Pathol. Lab. Med. **107**:514–517.

Swatek, F., C. Halde, M. G. Rinaldi, and H. J. Shadomy. 1985. *Aspergillus* species and other opportunistic saprophytic hyaline Hyphomycetes, p. 584–594. *In* E. H. Lennette, A. Balows, W. J. Hausler, Jr., and H. J. Shadomy (ed.), Manual of clinical microbiology, 4th ed. American Society for Microbiology, Washington, D.C.

Tack, K. J., F. S. Rhame, B. Brown, and R. C. Thompson, Jr. 1982. *Aspergillus* osteomyelitis. Report of four cases and review of the literature. Am. J. Med. **73**:295–300.

Tan, K. K., K. Sugai, and K. L. Tan. 1966. Disseminated aspergillosis. Case report and review of the world literature. Am. J. Clin. Pathol. **45**:697–703.

Tompkins, L. S. 1985. DNA methods in clinical microbiology, p. 1023–1028. *In* E. H. Lennette, A. Balows, W. J. Hausler, Jr., and H. J. Shadomy (ed.), Manual of clinical microbiology, 4th ed. American Society for Microbiology, Washington, D.C.

Van Cutsem, J., F. Van Gerven, M-A. Van De Ven, M. Borgers, and P. A. J. Janssen. 1984. Itraconazole, a new triazole that is orally active in aspergillosis. Antimicrob. Agents Chemother. **26**:527–534.

Wang, J. L., R. Patterson, M. Rosenberg, M. Roberts, and B. J. Cooper. 1978. Serum IgE and IgG antibody activity against *Aspergillus fumigatus* as a diagnostic aid in allergic bronchopulmonary aspergillosis. Am. Rev. Respir. Dis. **117**:917–927.

Young, R. C., J. E. Bennett, C. L. Vogel, P. P. Carbone, and V. T. DeVita. 1970. Aspergillosis: the spectrum of disease in 98 patients. Medicine (Baltimore) **49**:147–173.

Zimmerman, L. E. 1955. Fatal fungus infections complicating other diseases. Am. J. Clin. Pathol. **25**:46–65.

Zinneman, H. H. 1982. Sino-orbital aspergillosis. Report of a case and review of the literature. Minn. Med. **55**:661–664.

CHAPTER 59

Blastomycosis

A. F. DiSALVO and A. A. TERRENI

Disease: Blastomycosis.
Etiologic Agent: *Blastomyces dermatitidis.*
Source: Soil enriched with decaying organic material.
Clinical Manifestations: Cutaneous lesions (pustules, verrucous or ulcerated lesions) or subcutaneous abscesses. Acute pulmonary disease with cough, pleuritic pain, chills, and low-grade fever. Chronic pulmonary disease with productive cough, fever, weight loss, dyspnea, and fatigue. Dissemination to other organs may follow, primarily to skin, bone, and genitourinary tract.
Pathology: Tissue responses suppurative and granulomatous.
Laboratory Diagnosis: Isolation of the etiologic agent. Observation of the organism in tissue by fluorescent antibody and/or special stains. Serologic evidence of antibody as demonstrated by a line of identity with positive control serum in an immunodiffusion test supplemented by clinical or pathologic supporting data.
Epidemiology: Primarily North America (Canada, United States, and Mexico) with sporadic cases reported from India, Africa, and the Middle East.
Treatment: Amphotericin B or ketoconazole. Spontaneous remissions occur with primary cutaneous blastomycosis and acute primary pulmonary blastomycosis.
Prevention and Control: Unknown at this time.

Description of Disease

Blastomycosis, a disease of man and animals, may present in humans as a chronic cutaneous lesion, an acute or chronic pneumonia, or a systemic infection. The disseminated disease can affect almost any tissue or organ, but most frequently affects the skin, bones, or genitourinary tract. Exposure to the causative agent, *Blastomyces dermatitidis,* usually occurs during recreational activities, and disease results from the inhalation of infective particles from the environment.

Communicability and Transmissibility

It is generally accepted that blastomycosis is not transmissible from man to man, animal to man, or man to animal. Human beings and animals contract blastomycosis by exposure to the etiologic agent, *B dermatitidis,* in its natural habitat. Outbreaks occur when a group of individuals are exposed to the same point source. However, there have been some reports highly suggestive of intrauterine (Watts et al., 1983) and conjugal (Craig et al., 1970; Farber et al., 1986) transmission, and three well-substantiated cases that support transmission to human beings by a dog bite (Gnann et al., 1983; Jaspers, 1974; Scott, 1955).

The incubation period for blastomycosis is 4 to 6 weeks. The date of exposure or the source of the organism is frequently unknown, but based on several common source outbreaks of the disease, the range of time for incubation can be reliably estimated (DiSalvo, 1983; Klein et al., 1986, 1987).

Clinical Manifestations

With the exception of a "textbook" cutaneous lesion, there are no clinically distinctive signs or symptoms of blastomycosis. Many infections are asymptomatic, mild, or self-limited; therefore, the actual incidence of this disease is unknown. Attention is drawn to blastomycosis only when there is an outbreak or an individual has persistent cutaneous lesions, progressive pulmonary disease, or dissemination. Patients with blastomycosis seek medical assistance most frequently for respiratory symptoms or skin lesions.

Blastomycosis may present with a wide spectrum of illness. In contrast, there have been outbreaks of disease where some of the patients were completely asymptomatic. Evidence of infection was derived from temporal and physical association with the environmental source and from serologic or radiologic changes compatible with blastomycosis and commensurate with those of the remainder of the cohort.

Primary Cutaneous Blastomycosis

The manifestation of primary cutaneous blastomycosis is different from that of secondary blastomycosis, which results from disseminated pulmonary disease. This type of infection occurs infrequently and is essentially an occupational hazard of laboratory workers, anatomic pathologists, and veterinarians (DiSalvo, 1983).

Reported cases have resulted from skin penetration with a hypodermic needle, with a scalpel, and from dog bites. The wounds generally heal initially. After 1 to 3 weeks, a papule, pustule, or nodule develops. The nodules are frequently tender. Traumatic or spontaneous drainage occurs. Lymphadenitis and lymphangitis may occur, but no constitutional signs or symptoms. The disease is usually self-limited and painless and resolves without therapy.

Cutaneous Blastomycosis

Although cutaneous blastomycosis results from dissemination, it is often the presenting symptom. The dermal lesions, which may be single or multiple, range from a few millimeters to several centimeters in diameter. They usually start as papules that become pustules. A typical lesion has a raised border with small pustules at the periphery. Hemorrhagic dots 1 to 2 mm in diameter may be present on the surface of the lesion. The central portion of the lesion may heal, ulcerate, or scar. The lesions may also appear crusted or verrucous or as a papilloma. There is usually little associated pain or discomfort. A potassium hydroxide wet mount of the pustular material will reveal the cells typical of *B. dermatitidis*. There may be periods of remission and exacerbation as the lesions resolve spontaneously, only to reappear elsewhere.

Acute Pulmonary Blastomycosis

The portal of entry of *B. dermatitidis* is the lungs. Acute pulmonary blastomycosis, when detected, is frequently an influenza-like illness with an abrupt onset. Mild to severe pleuritic chest pain of varying duration is common. There may be fever, myalgia, and arthralgia. A productive or nonproductive cough, headache, anorexia, or night sweats may also be present. Erythema nodosum, particularly on the lower extremities, is not uncommon. Radiologic findings are primarily in the upper lung fields (Rabinowitz et al., 1976), and pleural effusion may occur (Jay et al., 1977). The symptoms usually resolve in 2 to 8 weeks without treatment.

Chronic Pulmonary and Systemic Blastomycosis

The symptoms of acute blastomycosis may persist with progressive lung disease, as evidenced by changes in symptoms and radiologic findings. Malaise, anorexia, and fever continue, accompanied by a cough that produces a purulent sputum. Hemoptysis and weight loss are common. Radiologically, there may be evidence of disease ranging from small patchy areas to extensive consolidation, with masslike opacities and cavities in chronic disease (Halvorsen et al., 1984; Rabinowitz et al., 1976). Clinically, the symptoms may resemble tuberculosis, carcinoma, or another fungal disease.

There may be dissemination to other organs, followed by recovery or death. Almost any organ of the body may be affected. Lesions are most common in the skin, bone, urogenital tract, liver, and kidney. Symptoms appropriate for the diseased organ system may be present. Reactivation of pulmonary or extrapulmonary dissemination may occur as long as 3 years after the initial episode (Laskey and Sarosi, 1978).

Pathology

The host reaction in tissue infected with *B. dermatitidis* consists of both acute and chronic inflammatory responses. The mixture of inflammation varies with the duration of the lesion and the tissue infected. The acute reaction is suppurative, with neutrophils and mononuclear leukocytes forming microabscesses.

Yeast cells may be seen within the lesions. The suppurative area is frequently surrounded by chronic inflammation, which consists of epithelioid cells and giant cells, and the yeast forms of *B. dermatitidis* are frequently seen within the giant cells.

Histologically, in cutaneous blastomycosis one may see hyperplastic epidermis, multiple microabscesses, and infiltrates of polymorphonuclear leukocytes, monocytes, and multinucleated giant cells. There is hyperkeratosis and downward proliferation of epidermal cells, and fibrosis is usually present in prolonged infection.

The black dots, which may be seen scattered over the surface of the lesion, appear microscopically to be accumulations of red blood cells in the dermis and epidermis (Leavell, 1965). Degeneration of the walls of papillary blood vessels leads to an extravasation of red blood cells into this tissue.

In tissue sections, *B. dermatitidis* appears as a thick-walled (usually termed *double-walled*) yeast having single buds with a broad base of attachment. The protoplasm usually shrinks, leaving a clear space between the protoplasm and the cell wall.

The organisms vary in size. Typically, *B. dermatitidis* yeast cells are 8 to 15 μm in diameter. Rarely, small yeast forms (2 to 5 μm) may be seen in certain patients (Manwaring, 1949). Large yeast forms (20 to 25 and up to 30 μm, respectively) were reported from environmental isolates by Denton and DiSalvo (1979) and from human tissue by Schwarz and Salfelder (1977). Filamentous forms in tissue are equally rare (Hardin and Scott, 1974). If only nonbudding, round yeasts 8 to 15 μm in diameter are seen, the organisms must be distinguished from *Cryptococcus neoformans, Paracoccidioides brasiliensis, Sporothrix schenckii,* large forms of *Histoplasma capsulatum,* and endospores or immature spherules of *Coccidioides immitis*. Small forms of *B. dermatitidis* may resemble *H. capsulatum* and other yeasts.

The pathognomonic form of *B. dermatitidis* seen in tissue sections is a single bud in which the mother and daughter cells are of equal size, resembling a figure 8. The cells are 8 to 15 μm in diameter, with the broad base of attachment of the bud and the double-walled appearance of the cells. Microscopic observation of *B. dermatitidis* in histologic sections stained with hematoxylin and eosin can sometimes be difficult for the microbiologist with limited experience. The protoplasm stains with the hematoxylin, and multiple nuclei may be visible. The number of yeast cells in the tissue varies considerably. Although fungi are more difficult to see in preparations stained with hematoxylin and eosin, this stain has the benefit of demonstrating the inflammatory response. The peculiar mixed host reaction (suppurative and granulomatous) is characteristic of blastomycosis. When organisms are scarce, as is common in cutaneous blastomycosis, the special stains offer a distinct advantage.

Special Stains

The stain that renders fungi most visible is the Gomori methenamine silver (GMS) stain. The organisms appear black and the background stains light green. The ability to discern the tissue reaction is lost, but this may be of less concern to the microbiologist who is primarily interested in identifying the causative agent. The GMS stains nonviable as well as viable cells. Overstaining (too black) will obliterate the internal structures of *B. dermatitidis* and may also stain artifacts such as red blood cells and blood vessel walls. The combined GMS–hematoxylin and eosin stain is recommended to reveal the tissue reaction as well as to stain the fungi black (Chandler et al., 1980).

The periodic acid–Schiff (PAS) stain and the Gridley stain are additional special stains that are useful for *B. dermatitidis*. Meyer mucicarmine stain is specific for the capsular material of *C. neoformans*. The capsule of this yeast stains pink with a yellow background. Lightly encapsulated or "dry variants" of *C. neoformans* may stain faintly or not at all. *B. dermatitidis* may also be mucicarmine positive; only a few cells will stain, however, and these cells will be lightly colored. In the absence of the typical forms and without benefit of culture, fluorescent-antibody staining may be used to greater advantage.

Fluorescent Antibody Staining

The direct fluorescent antibody technique is used to visualize an antigen–antibody reaction in which the cells of *B. dermatitidis* are the antigen. Antibody to *B. dermatitidis* that is coupled with a fluorescent stain is called a *conjugate*. When the *B. dermatitidis* conjugate is applied to deparaffinized tissue sections or to dried smears of yeast cultures, an antigen–antibody reaction takes place; when observed with an ultraviolet microscope, the yeast of *B. dermatitidis* will fluoresce. *H. capsulatum* may cross-react with *B. dermatitidis*, but these organisms can usually be distinguished by their morphology. The advantages of fluorescent antibody staining are ease of morphologic observation, detection of small numbers of organisms, staining of nonviable organisms, and rapid completion (3 to 4 h).

Specific techniques will not be presented because commercial reagents are not available and because this test must be requested from state public health laboratories or the Centers for Disease Control.

Epidemiologic Aspects

Geographic Occurrence and Transmission

Blastomycosis was described by Gilchrist in 1894. Many of the early instances of this infection were in patients from the Chicago area, and the malady was known as *Chicago disease*. Subsequent recognition of a wider distribution resulted in the term *North American blastomycosis*. This name recognized the distribution of blastomycosis in Mexico, Canada, and the United States and served to distinguish this disease from paracoccidioidomycosis, which occurs almost exclusively in South America and was known as *South American blastomycosis*.

In recent years, autochthonous cases of blastomycosis have been reported from Africa, India, and the Middle East. With few exceptions, patients in North America appear to contract blastomycosis east of the 100th meridian. This area appears to be a natural nidus for *B. dermatitidis* and conforms to Pavlovsky's theory of the natural nidality of disease (Pavlovsky, 1966).

Mycologists, epidemiologists, and clinicians generally agree that human or animal infection occurs by inhalation of the microconidia and fragments of the mycelial form of *B. dermatitidis* as it grows in nature.

Reservoir

The reservoir of *B. dermatitidis* appears to be soil enriched with a high organic content. The specific habitat of *B. dermatitidis* has long baffled mycologists. The proximity of known cases of blastomycosis, in humans and animals, to waterways or alluvium has been described by several investigators. Borelli (1969) was even more specific in associating the incidence of blastomycosis in the United States and Africa with an aquatic ecosystem and the distribution of the salamander. The first isolation of *B. dermatitidis* from a natural source was reported in 1961 (Denton et al., 1961) and confirmed in 1964 (Denton and DiSalvo, 1964) and 1979 (Denton and DiSalvo, 1979). Recent isolations of *B. dermatitidis* from the environment in association with human disease support this description of the reservoir (Bakerspigel et al., 1986; Klein et al., 1986, 1987).

Etiologic Agent

Characterization

The etiologic agent of blastomycosis, the dimorphic fungus *B. dermatitidis* (Gilchrist and Stokes, 1898), is the only species of the genus. At temperatures below 35°C, in the presence of oxygen, this organism grows as a white, cottony mold. The thallus is composed of hyaline, regularly septate mycelium. Small, spherical to subglobose conidia with one cell are produced terminally on slender simple conidiophores. At 35°C, the growth is dry, wrinkled, and soft and consists of large yeastlike cells, which are round to subglobose, thick walled, and 8 to 15 μm in diameter. These cells multiply by budding, with the daughter cell having a broad-based attachment to the parent. There is usually only a single bud, which often becomes as large as the parent cell.

Specimen Collection, Transport, and Processing

All specimens should be collected using aseptic technique and placed in sterile containers with lids that close tightly to prevent leakage. A container that is not or cannot be properly closed may cause contamination of the specimen with organisms from the ambience, pose a hazard for personnel who handle the container, or result in loss of the clinical material. All specimens should be delivered to the laboratory as soon after collection as possible to allow prompt processing, which will reduce the chance of overgrowth of bacteria and saprophytic fungi that may be present. Refrigeration at 4°C may be necessary if immediate delivery is not possible.

The methods for collecting various clinical material are described below.

Sputum

Extensive oral hygiene should be performed before the specimen is collected. Dentures should be removed, teeth brushed thoroughly, and the oral cavity and throat rinsed, followed by gargling with a mouthwash or warm water. Collection of one early-morning specimen provides better recovery of organisms than a pooled specimen. The sputum should be collected from a deep cough. Saliva and mucus are generally useless specimens. The specimen should be collected in a sterile widemouthed jar or other container.

Exudate

Subcutaneous nodules, abscesses, and draining sinus tracts may form, and material from these lesions should be obtained for culture. After the skin is thoroughly disinfected, material may be aspirated with a sterile syringe and needle, placed in a sterile tube, sealed, and sent immediately to the laboratory.

Abscesses in the periphery of an ulcer should be cleaned and opened with a sterile scalpel or hypodermic needle. The small amount of pus obtained should be planted immediately on culture medium and a wet mount prepared at the bedside for a direct microscopic examination in the laboratory. This approach prevents drying of the small amount of exudate on the slide.

Urine

For either a clean catch or a catheterized specimen, the urinary orifice should be cleansed thoroughly with mechanical washing using a disinfectant solution. For suprapubic aspiration, the skin is disinfected as for collection of other body fluids. The urine is placed in a sterile container with a tight lid and immediately delivered to the laboratory or refrigerated at 4°C.

Tissue

Biopsy specimens are minced with sterile scissors and ground with a small amount of broth in a mortar and pestle or a homogenizer. The homogenate is then inoculated on the surface of each plate or slant.

Impression smears may be made by touching the cut surface of a small piece of tissue to a glass slide. Several slides should be made so that they are available for staining and fluorescent antibody studies. A potassium hydroxide preparation of the homogenate (Haley and Callaway, 1978) may also be made.

Isolation and Identification Procedures

All specimens submitted to the laboratory should be examined microscopically and cultured. Penicillin and streptomycin are inexpensive and generally effective against contaminating bacteria. They can be easily incorporated into any medium during preparation.

Cycloheximide is an antifungal drug that inhibits many fungal saprophytes, as well as some pathogens. However, it does not inhibit *B. dermatitidis* and enhances the possibility of its isolation.

It is not worthwhile to culture for primary isolation of *B. dermatitidis* at 35°C, because the fungus grows well at 25°C. Colonies grow slowly at this temperature and generally appear in 10 days to 2 weeks, although the time may vary widely from one isolate to another. At least two different isolation media should be used, one of which should be enriched with blood and one of which should contain cycloheximide. The cultures should be examined every 2 or 3 days for early signs of growth.

Direct Examination of Clinical Material

Each specimen should be examined microscopically in a wet mount. Most clinical material is best examined unstained. Clear specimens may be examined by placing a small amount of material on a glass slide. A coverglass is placed on the slide and it is examined under the low- and high-power objectives of the microscope for the presence of the yeast cells of *B. dermatitidis*. If the specimen contains large amounts of cellular material (i.e., tissue cells, blood cells, debris, and the like), a drop of specimen may be added to a drop of 10% potassium hydroxide on a slide and mounted with a coverglass. It may be necessary to heat the preparation gently. This specimen is then examined as described previously.

The experienced technologist can detect these yeast cells using the low-power objective with reduced light. Under low power, the very black rim of the cell is characteristic. Once detected, the mycologist may then switch to the high-power objective for a more detailed observation. If organisms are not seen under low power, the preparation should still be examined under high power. The typical morphology is apparent under greater magnification. Multiple nuclei may be seen within the cells.

The addition of calcofluor white to potassium hydroxide has been used successfully in the detection of fungi in clinical specimens (Elder and Roberts, 1986; Monheit et al., 1984). Calcofluor white is an industrial brightner that causes fungal elements to fluoresce when observed with a fluorescent microscope (Darken, 1961). This fluorescence renders them much more visible than with potassium hydroxide alone and reduces the scanning time.

Stains

The PAS stain may be used directly on smears of clinical material as well as on histologic sections, although there is a slight difference in technique. The PAS method stains *B. dermatitidis* a bright magenta. The cytoplasm may retract from the cell wall, leaving an unstained area. This procedure also stains some bacteria and certain elements of white blood cells. However, these elements may be differentiated from *B. dermatitidis* as well as other fungi on the basis of size and morphology. Some technologists can detect fungi more easily with this stain than with the potassium hydroxide examination.

Digestion and Concentration

Each laboratory should establish a policy on the concentration of sputa or other viscous specimens. The advantages of concentration are controversial. Some mycologists find a higher yield of pathogenic fungi from concentrated specimens, whereas others claim that this procedure also concentrates contaminating organisms. Most mycologists agree that the problem is more acute with specimens received by mail.

The procedure for concentration is as follows: Place 5 to 10 ml of specimen in a sterile 50-ml, conical, graduated, screw-cap tube. Add an equal volume of digestant (*N*-acetyl-l-cysteine or dithiothreitol without sodium hydroxide). Mix on a mechanical mixer for 15 s, or longer if necessary to liquify the specimen. Add 0.067 M phosphate buffer solution (pH 6.8 to 7.1) to bring the volume to 50 ml. Centrifuge at 2,100 rpm (950 g) for 15 min. Decant the supernatant fluid, leaving 3 to 5 ml in bottom of tube. Resuspend and inoculate 0.1 to 0.2 ml of sediment to each medium used, streaking the material over the entire surface (Haley and Callaway, 1978).

Make a wet preparation of the sediment, unstained or with a drop of lactophenol cotton blue to determine the general character of the specimen (e.g., squamous epithelial cells, white blood cells, red blood cells, phagocytes). Make a potassium hydroxide preparation and a smear for the PAS stain. Incubate cultures at 24 to 30°C for at least 4 weeks and, if possible, for 6 to 7 weeks.

Bronchial washings, urine, and other body fluids may be centrifuged to concentrate the organisms in the sediment. The sediment then should be examined and cultured.

Culture

Petri dishes allow better recovery of the organism, because spreading the inoculum over the larger surface dilutes any inhibitory substance or contaminating organisms and spreads the inoculum for better isolation. Picking colonies for examination and transfer is also easier from petri dishes.

Using plates poses a greater potential hazard, however, particularly for the less experienced technical personnel. Taping them on opposite sides will help to keep the lids in place and filling the plates with 30 ml of agar will retard drying. Placing a pan of water in the incubator will also help to prevent drying.

Many laboratories use tubes of media because they are less expensive, require less space, and they are safer to handle. In addition, there is less drying of the medium. Several tubes should be used for each specimen. Large culture tubes or culture bottles will also increase the surface area.

The choice of media depends, to some extent, on the individual laboratory. As mycotic diseases tend to mimic each other, a battery of media to cover all needs is generally used.

The following is a list of media suitable for the isolation of *B. dermatitidis:*

1. Sabouraud agar (Emmons modification): This medium is more satisfactory than Sabouraud dextrose agar (with 4% dextrose) for the isolation of the *B. dermatitidis*. The addition of chloramphenicol and cycloheximide appears to favor the isolation of *B. dermatitidis*.
2. Brain–heart infusion agar alone or with 5 to 10% sheep blood: Addition of chloramphenicol and cycloheximide is useful with contaminated specimens.
3. Sabhi agar (Gorman, 1967) alone or with 5 to 10% sheep blood: Antibiotics may be added.
4. Yeast extract phosphate agar is recommended for the isolation of some systemic pathogens from contaminated sputum specimens. A drop of concentrated ammonium hydroxide is placed on one side of a plate of yeast extract agar after inoculation. The ammonium hydroxide diffuses into the medium and inhibits development of *Candida* species and saprophytic molds (Smith and Goodman, 1975).
5. Inhibitory mold agar is recommended for the isolation of systemic mold pathogens (Ulrich, 1956). It inhibits many saprophytic molds.

On modified Sabouraud agar with chloramphenicol and cycloheximide incubated at 25°C, *B. dermatitidis* will first appear as a white, cottony growth or as a glabrous colony with a glossy, waxy appearance. If the colony is glabrous at first, after about a week it will generally produce tiny bundles of hyphae, or coremia, often termed *spicules,* which eventually become cottony.

Microscopically, in a teased preparation, the mycelium appears thin and septate, bearing tiny globose to pyriform conidia directly off of the hyphae or on short, slender conidiophores. Occasionally there will be a dumbell-shaped conidium, and sometimes the conidiophore swells just below the conidium as though to give it support. Chlamydospores may be present, as in all fungi. When isolated at 35°C, *B. dermatitidis* grows as a yeast and forms colonies that are cream colored and wrinkled to cerebriform. Microscopically these colonies are seen to be composed of the characteristic single-budding, thick-walled cells 8 to 15 μm in diameter. When *B. dermatitidis* is isolated on a medium containing blood, it should be transferred to a medium without blood to obtain sporulation.

To confirm identification, the mold form must be converted to the yeast form or vice versa. In this case, the mold is transferred to brain–heart infusion blood agar or Kelley agar and incubated at 35°C (Kelley, 1939).

Preliminary Identification

Preliminary identification of an isolate is possible if (1) the yeast form is cultured at 35°C or (2) a fluorescent-antibody technique gives a positive reaction (see Fluorescent-Antibody Staining).

Definitive Identification

Criteria for definitive identification of *B. dermatitidis* are (1) conversion from the typical mold form at 25°C to the typical yeast form at 35°C or vice versa or (2) a positive exoantigen test.

Conversion

When mycelium from a mold colony of *B. dermatitidis* is inoculated to a suitable medium and incubated at 35°C, conversion to the yeast form may occur as early as 2 to 4 days later. In rare instances, it may take many months. However, *B. dermatitidis* is usually one of the easiest of the dimorphic fungi to convert. Often, conversion may be detected first by scraping gently beneath the inoculum. A minute amount of material collected from this location can be examined in a wet mount. The yeastlike cells can be seen microscopically. As the yeast form develops, cream-colored colonies that are raised and cerebriform will appear. It is not unusual for some fragments of hyphae to remain in the 35°C cultures. If conversion is not obtained, identification may be confirmed by the exoantigen test.

In the same manner, if a suspect yeast is isolated at 35°C it should be transferred and incubated at 25°C. In a few days hyphae will form and the white, cottony growth typical of *B. dermatitidis* will develop.

Exoantigen Test

The exoantigen test is an immunodiffusion procedure utilizing reverse serology in which the extract of the organism is the unknown antigen and the antiserum is a known antibody. Exoantigens are soluble antigens that can be extracted from the mycelium with an aqueous solution. The cell-free extracts of the mycelium precipitate with homologous antiserum in agar to form a sharp, clear band of identity with that formed by a control or reference antigen and antiserum.

The conventional method of identifying *B. dermatitidis* involves gross and microscopic observation of morphologic structures and growth rate and demonstration of temperature-dependent changes in form as described previously. This time-consuming procedure requires appropriate education, training, experience, and technique by the laboratorian, and the laboratory facilities must be commensurate with proper microbiologic containment.

Conversion of *B. dermatitidis* has been essential to definitive identification, because the microconidia are not diagnostic. Some isolates may be totally devoid of any characteristic structures and might be discarded as a nonpathogenic species. A mold suggestive of *B. dermatitidis* at 25°C in the presence of cycloheximide and which grows at 35°C without converting to the yeast form should still be suspect.

The immunodiffusion test is performed according to the protocol of the Centers for Disease Control (Kaufman, 1986) or the instructions of the reagent manufacturer. The reference antigen is placed in the central well of a hexagonal pattern. The antiserum is allowed to diffuse for 1 h at room temperature before addition of the antigens. The reference antigen is placed in the upper and lower wells; the unknown extracts and extracts from a second unknown are placed on each side, in adjacent wells.

The charged immunodiffusion plates are placed in a moist chamber and incubated at 25°C. After 24 h, the precipitin reaction is read on a view box with a dark background and oblique light.

Only lines from an unknown extract that form a distinct junction with the reference antigen–antibody precipitation line can be considered positive. Lines of nonidentity or lines of partial identity must not be considered a homologous reaction.

The exoantigen technique has sufficient sensitivity and specificity to detect fungal pathogens in mixed cultures and can be used to identify nonviable cultures. The exoantigen test neither replaces morphologic techniques, practices, or the knowledge of a competent mycologist nor eliminates the need for scrupulous observation of safety practices. Specific instructions for performing the exoantigen test have been published (DiSalvo, 1985), and the test is readily available from most state public health laboratories.

Exfoliative Cytology

B. dermatitidis may be detected by cytologic examination of sputum (Sutliff and Cruthirds, 1973; Trumbull and Chesney, 1981), pleural fluid (Covell et al., 1982), synovial fluid (George et al., 1985), cervical smear (Dyer et al., 1983), and adrenal aspirate (Halvorsen et al., 1982). Stained with the Papanicolaou technique, the tissue form of *B. dermatitidis* appears as it does in other stains. The color of the cytoplasm may be blue or pink. Because this disease is uncommon, the cytotechnologist must be aware of the fungus and be familiar with its morphology, or it will not be recognized.

Immunologic Aspects

Antibodies to *B. dermatitidis* are determined by two methods: complement fixation and immunodiffusion. Because the symptoms of blastomycosis may be suggestive of fungal disease, but not specifically blastomycosis, a battery of serologic tests is indicated. A positive test for blastomycosis is supporting evidence for a clinical impression. Serology assists in determining the etiology and prognosis of the disease (Kaufman, 1986).

Complement Fixation

The complement fixation test has been the "gold standard" of fungal serology for many years, particularly the standardized Laboratory Branch Complement Fixation test developed at the Centers for Disease Control (U.S. Public Health Service, 1965). Because of cross-reactions with other fungal diseases, the complement fixation test is usually performed in a battery with antigens to *C. immitis* and to both yeast and mycelial antigens of *H. capsulatum*. The *B. dermatitidis* antigen is prepared from disrupted yeasts.

A titer of 1 : 8 is considered suspicious, but may be a cross-reaction with other fungal diseases even if it is the only positive antigen. A titer of 1 : 32 is usually considered positive for blastomycosis; however, it must be recognized that the highest titer in a complement fixation battery is not necessarily the homologous reaction. In some patients with blastomycosis, the complement fixation titer to one of the *H. capsulatum* antigens may be higher than that of the *B. dermatitidis* antigen.

In addition to a lack of specificity, other disadvantages of the complement fixation test for blastomycosis are a 50% sensitivity and a delay of 2 to 3 months after the onset of symptoms before the test is positive. The advantage of the complement fixation test is the quantification of results, which can be an aid in establishing a specific diagnosis and in evaluating the patients' response to therapy. In some patients a positive complement fixation test may be the only serologic response. A purified A antigen has been used in the complement fixation test (Green et al., 1980). Although the evaluation was based on a small number of banked samples, it was found to be specific and increased the sensitivity to 70%. The complement fixation test is readily available from most state public health laboratories.

Immunodiffusion

In the immunodiffusion test, antigen and antibody (the patient's serum) are placed in adjacent wells in an agar matrix. The antigens and antibodies diffuse into the agar and form a precipitin band along the point of optimal concentration. The A antigen of *B. dermatitidis* forms the precipitin.

The reagents to perform the immunodiffusion test for blastomycosis have been available commercially for several years, and this test has many advantages. Reagent costs are moderate and the equipment required is minimal. The antigen is a culture filtrate that is also prepared from the yeast form of *B. dermatitidis*. A positive precipitin band (A band) is specific for blastomycosis, and the sensitivity is 80%. Frequently, immunodiffusion antibody will be present within 2 to 3 weeks after the onset of symptoms. The test is not quantitative, but disappearance of the A band usually indicates a favorable prognosis for the patient. This test is available at state public health laboratories and in many larger hospitals.

Enzyme-Linked Immunosorbent Assay

Recently, an enzyme-linked immunosorbent assay for blastomycosis using the purified A antigen was described (Green et al., 1980) and evaluated (Turner et al., 1986). In studying 65 patients with blastomycosis, the enzyme-linked immunosorbent assay was found to be moderately sensitive (80%) and highly specific (98%). A significant drop in titer as the clinical condition of the patient improved was also demonstrated. The investigators even proposed that a reduction in titer might serve as an indicator to discontinue therapy. Many more samples from patients in which the stage of disease or recovery is known will be required to fully evaluate this new procedure.

Interpretation

If the immunodiffusion serology is positive, one specimen can be useful. If the immunodiffusion and complement fixation tests are negative, it is essential to collect additional serum specimens at 2- to 3-week intervals. Ideally, even with cross-reactions to other fungal antigens, a fourfold rise in complement fixation titer is strong evidence of blastomycosis.

Details for performing these serologic tests have not been described. However, they have been published in *Serodiagnosis of Mycotic Diseases* (Palmer et al., 1977), which should be in every laboratory performing fungal serology.

Although the serologic tests for *B. dermatitidis* have room for improvement in sensitivity and early detection of disease, when positive, they provide a useful adjunct to diagnosis in the absence of culture. The complement fixation and immunodiffusion tests furnish different information. The complement fixation test presents quantitative data, which may have prognostic significance and may be positive in the absence of precipitin bands. A positive immunodiffusion test is specific, has a greater sensitivity, and becomes positive earlier in the course of illness. Serologic tests to detect antigen to *B. dermatitidis* would be ideal; however, such tests are not available.

Skin Test

A skin test for blastomycosis is not available commercially.

Treatment, Prevention, and Control

Treatment

It is generally agreed that not all patients with blastomycosis require therapy. If the patient is not in acute distress, most clinicians will evaluate the course of illness for 2 to 3 weeks after the onset of symptoms before initiating therapy.

Amphotericin B is the drug of choice for the treatment of blastomycosis, although there has been considerable success with ketoconazole during the last 5 years. Both agents will be discussed briefly.

Amphotericin B

Amphotericin B exerts its chemotherapeutic effect by damaging the cell membrane of the yeast. This agent binds to sterols in cell membranes, but has a greater affinity for the sterols in the cells of *B. dermatitidis* than for the sterols of the human or animal host cells. This action causes an increased permeability of the membrane and allows intracellular contents to leak out of the yeast cell, resulting in the death of the fungus. Because of this mechanism of action, the growth phase of the organism has little to do with the efficacy of the drug, and amphotericin B may be fungicidal or fungistatic depending on the concentration in the host and the susceptibility of the organism.

This antifungal agent is given intravenously in 5% dextrose, beginning with an initial dose of 5 to 10 mg to evaluate patient response. The dose of amphotericin B is gradually increased until the patient is receiving 50 mg every other day. The objective is to administer a total of at least 2 g during several weeks or months, depending on patient tolerance and clinical response to therapy. The significant drawback with amphotericin B is the frequency of side effects, which may include phlebitis, fever, anorexia, nausea, vomiting, chills, anemia, hypokalemia, and nephrotoxicity. The most important pathology is the renal damage, which is usually reversible if the patient has received less than 4 g of amphotericin B.

It is absolutely essential to monitor the patient frequently for creatinine clearance and serum electrolytes. To ameliorate the side effects of amphotericin B, a combination of heparin, antipyretics, antiemetics, hydrocortisone, or diphenylhydramine hydrochloride are selectively added to the infusion to mitigate the adverse reactions of that patient. Administration of potassium, a high fluid intake, and mannitol to increase diuresis should be considered as adjuncts to amphotericin B therapy.

Amphotericin B has been in use for more than 30 years and has proved to be the most successful treatment for blastomycosis, despite its toxicity. Although therapy with this agent is not without risk and relapses do occur, it has withstood the test of time.

Ketoconazole

Ketoconazole is an imidizole that interferes with the synthesis of ergosterol in the fungal cell membrane. It is fungistatic. The primary advantage of this medication is that it can be administered orally. Reports in the literature since 1982 have been mostly favorable, although relapses have also been reported. Several more years are required before the efficacy of this drug to cure blastomycosis can be determined.

At present, the usual dose of ketoconazole is 400 to 800 mg per day, and treatment may last for 6 months to a year. Experience has been limited, and administration of this agent may be modified as more experience is gained.

The side effects of ketoconazole are nausea, vomiting, anorexia, gynecomastia, hepatotoxicity, anaphylaxis, and drug interactions. However, they occur less frequently than with amphotericin B and are not as persistent or distressful. The adverse effects occur more frequently with the higher dosage. A multi-institution study showed that when daily doses were altered in response to patient tolerance, the side effects of ketoconazole were diminished and clinical response was more effective (National Institute of Allergy and Infectious Disease Mycoses Study Group, 1985). Patients receiving ketoconazole should be monitored for hepatotoxicity and the effects of concomitant medications. Ketonconazole appears to be a reasonable alternative to amphotericin B in patients who have had a relapse with this therapeutic agent or are intolerant of amphotericin B therapy.

Prevention and Control

Knowledge of the ecology and epidemiology of *B. dermatitidis* is just emerging. Until the ecologic niche is more firmly established, it is impossible to outline preventive or control measures for the general public. Good laboratory practices as described subsequently will protect the laboratorian. Observation of general recommendations of the Occupational

Safety and Health Act for industrial hygiene will protect workers from the hazards that may be occupationally related (DiSalvo, 1983).

Safety

The biologic safety cabinet is the single most effective safety device, second only to proper training of laboratory personnel. All diagnostic mycology procedures must be performed under a certified Class II biologic safety cabinet.

It would appear that most clinical material can be handled with minimal risk of aerosol creation, but such is not the case. At least one laboratory infection with *B. dermatitidis* occurred without any awareness of the exposure. This technician removed cellophane tape from plates containing the mycelial form of *B. dermatitidis* and, without opening the plates, placed them in a covered metal container for autoclaving. Her first clinical symptoms led to a diagnosis of osteomyelitis caused by *B. dermatitidis* (Denton et al., 1967).

A similar incident occurred with another laboratory technician who worked with *B. dermatitidis* for 1 day only. He developed symptoms 2 months later, and a diagnosis of pulmonary infection caused by *B. dermatitidis* was made (Baum and Lerner, 1970).

If a biologic safety cabinet had been available at that time, both of these accidents could have been avoided. It is unknown how many similar accidents have occurred and have not been reported in the scientific literature.

Most laboratory accidents occur through carelessness by skin puncture with an inoculating needle or trauma from self-inflicted scalpel wounds (Schwarz, 1983). Safety devices could not have prevented these accidents.

Literature Cited

Bakerspigel, A., J. Kane, and D. Schaus. 1986. Isolation of *Blastomyces dermatitidis* from an earthen floor in southwestern Ontario, Canada. J. Clin. Microbiol. **24:**890–891.

Baum, G. L., and P. I. Lerner. 1970. Primary pulmonary blastomycosis: a laboratory acquired infection. Ann. Intern. Med. **73:**263–265.

Borelli, D. 1969. Reservareas de algunos agentes de micosis. Med. Cutanea **4:**367–370.

Chandler, F. W., W. Kaplan, and L. Ajello. 1980. Blastomycosis, p. 49–31. *In* Color atlas and text of the histopathology of mycotic diseases, Year Book Medical Publishers, Inc., Chicago.

Covell, B. S., E. H. Lowry, Jr., and P. S. Feldman. 1982. Cytologic diagnosis of blastomycosis in pleural fluid. Acta Cytol. **26:**833–836.

Craig, M. W., M. D. Winthrop, and R. A. Green. 1970. Conjugal blastomycosis. Am. Rev. Respir. Dis. **102:**86–90.

Darken, M. 1961. Application of fluorescent brightners in biological techniques. Science **133:**1704–1705.

Denton, J. F., and A. F. DiSalvo. 1964. Isolation of *B. dermatitidis* from natural sites at Augusta, GA. Am. J. Trop. Med. **13:**716–722.

Denton, J. F., and A. F. DiSalvo. 1979. Additional isolations of *Blastomyces dermatitidis* from natural sites. Am. J. Trop. Med. Hyg. **28:**697.

Denton, J. F., A. F. DiSalvo, and M. L. Hirsch. 1967. Laboratory-acquired North American blastomycosis. J. Am. Med. Assoc. **199:**935–936.

Denton, J. F., E. S. McDonough, L. Ajello, and R. J. Ausherman. 1961. Isolation of *Blastomyces dermatitidis* from soil. Science **133:**1126.

DiSalvo, A. F. Blastomycosis. 1983. *In* A. F. DiSalvo (ed.), Occupational mycoses, p. 79–94. Lea & Febiger, Philadelphia.

DiSalvo, A. F. 1985. Identification of mycotic pathogens using antisera to exoantigens. Clin. Immunol. Newsl. **6:**24–29.

Dyer, M. L., T. L. Young, A. A. Kattine, and D. D. Wilson. 1983. Blastomycosis in a Papanicolauo smear. Report of a case with possible venereal transmission. Acta Cytol. **27:**285–287.

Elder, B. L., and G. D. Roberts. 1986. Rapid methods for the diagnosis of fungal infections. Lab. Med. **17:**591–596.

Farber, E. R., M. S. Leary, and T. R. Meadows. 1986. Endometrial blastomycosis acquired by sexual contact. Obstet. Gynecol. **32:**195–199.

George, A. L., Jr., J. T. Hays, and B. S. Graham. 1985. Blastomycosis presenting as monarticular arthritis. Arthritis Rheum. **28:**516–521.

Gilchrist, T. C. 1894. Protozoan dermatitis. J. Cutan. Dis. **12:**496.

Gilchrist, T. C. and W. R. Stokes. 1898. A case of pseudolupus vulgaris caused by a blastomyces. J. Exp. Ed. **3:**53–83.

Gnann, J. W., G. S. Bressler, A. A. Bodet, and C. K. Avent. 1983. Human blastomycosis after a dog bite. Ann. Intern. Med. **98:**48–49.

Gorman, J. W. 1967. Sabhi, a new culture medium for pathogenic fungi. Am. J. Med. Technol. **33:**151–157.

Green, J. H., W. K. Harrell, J. E. Johnson, and R. Benson. 1980. Isolation of an antigen from *Blastomyces dermatitidis* that is specific for the diagnosis of blastomycosis. Curr. Microbiol. **4:**293–296.

Haley, L. D., and C. S. Callaway. 1978. Laboratory methods in medical mycology. U.S. Department of Health, Education, and Welfare publication no. (CDC) 78-8361, p. 33–45. Centers for Disease Control, Atlanta.

Halvorsen, R. A., J. D. Duncan, D. F. Merten, H. S. Gallis, and C. E. Putman. 1984. Pulmonary blastomycosis: radiological manifestations. Radiology **150:**1–5.

Halvorsen, R. A., Jr., D. K. Heaston, W. W. Johnson, P. R. Ashton, and G. M. Burton. 1982. CT guided thin needle aspiration of adrenal blastomycosis. J. Comput. Assist. Tomogr. **6:**389–391.

Hardin, H. F., and D. I. Scott. 1974. Blastomycosis. Occurrence of filamentous forms *in vivo*. Am. J. Clin. Pathol. **62:**104–106.

Jaspers, R. H. 1974. Transmission of blastomycosis from animals to man. J. Am. Vet. Med. Assoc. **164:**8.

Jay, S. J., R. P. O'Neill, N. Goodman, and R. Penman. 1977. Pleural effusion: a rare manifestation of acute pulmonary blastomycosis. Am. J. Med. Sci. **274:**325–328.

Kaufman, L. 1986. Serodiagnosis of fungal diseases, p. 446–466. *In* N. R. Rose, H. Friedman, and J. L. Fahey (ed.), Manual of clinical laboratory immunology, 3rd ed. American Society for Microbiology, Washington, D.C.

Kelley, W. H. 1939. A study of the cell and colony variations of *Blastomyces dermatitidis*. J. Infect. Dis. **64**:293–296.

Klein, B. S., J. M. Vergeront, A. F. DiSalvo, L. Kaufman, and J. P. Davis. 1987. Two outbreaks of blastomycosis along rivers in Wisconsin: isolation of *Blastomyces dermatitidis* from riverbank soil and evidence of its transmission along waterways. Am. Rev. Resp. Dis. **136**:1333–1338.

Klein, B. S., J. M. Vergeront, R. J. Weeks, U. N. Kumar, G. Mathal, B. Varkey, L. Kaufman, R. W. Bradsher, J. F. Stoebig, and J. P. Davis. 1986. Isolation of *Blastomyces dermatitidis* in soil associated with a large outbreak of blastomycosis in Wisconsin. N. Engl. J. Med. **314**:529–534.

Laskey, W., and G. A. Sarosi. 1978. Endogenous activation in blastomycosis. Ann. Intern. Med. **88**:50–52.

Leavell, U. W., Jr. 1965. Cutaneous North American blastomycosis and black dots. Arch. Dermatol. **92**:155–156.

Manwaring, J. H. 1949. Unusual forms of *Blastomyces dermatitidis* in human tissue. Arch. Pathol. **48**:421–425.

Monheit, J. E., D. F. Cowan, and D. G. Moore. 1984. Rapid detection of fungi in tissues using calcofluor white and fluorescence microscopy. Arch. Pathol. Lab. Med. **108**:616–618.

National Institute of Allergy and Infectious Disease Mycoses Study Group. 1985. Treatment of blastomycosis and histoplasmosis with ketoconazole. Ann. Intern. Med. **103**:861–872.

Palmer, D. F., L. Kaufman, W. Kaplan, and J. J. Cavallaro. 1977. Serodiagnosis of mycotic diseases. Charles C. Thomas, Springfield, Ill.

Pavlovsky, E. N. 1966. Natural nidality of transmissible diseases. University of Illinois Press, Urbana.

Rabinowitz, J. G., J. Busch, and W. R. Buttram. 1976. Pulmonary manifestations of blastomycosis: radiological support for a new concept. Radiology **120**:25–32.

Scott, M. J. 1955. Cutaneous blastomycosis: report of case following dog bite. Northwest Med. **54**:225–257.

Schwarz, J. 1983. Laboratory infections with fungi, p. 215–227. A. F. DiSalvo (ed.), Occupational mycoses. Lea & Febiger, Philadelphia.

Schwarz, J., and K. Salfelder. 1977. Blastomycosis. Current topics in pathology, volume 65. Springer-Verlag, New York.

Smith, C. D., and N. L. Goodman. 1975. Improved culture method for the isolation of *Histoplasma capsulatum* and *Blastomyces dermatitidis* from contaminated specimens. Am. J. Clin. Pathol. **63**:276–280.

Sutliff, W. D., and T. P. Cruthirds. 1973. *Blastomyces dermatitidis* in cytologic preparations. Am. Rev. Respir. Dis. **108**:149–151.

Trumbull, M. L., and T. M. Chesney. 1981. The cytological diagnosis of pulmonary blastomycosis. J. Am. Med. Assoc. **245**:836–838.

Turner, S., L. Kaufman, and M. Jalbert. 1986. Diagnostic assessment of an enzyme-linked immunosorbent assay for human and canine blastomycosis. J. Clin. Microbiol. **23**:294–297.

Ulrich, J. A. 1956. Bacteriol. Proc. American Society for Microbiology, Washington, D.C. No 87.

U.S. Public Health Service. 1965. Standardized diagnostic complement fixation method: an adaptation to micro tests. U.S. Public Health Service publication no. 1228 (Public Health monograph no. 74). U.S. Public Health Service, Washington, D.C.

Watts, E. A., P. D. Gard, Jr., and S. W. Tuthill. 1983. First reported case of intrauterine transmission of blastomycosis. Pediatr. Infect. Dis. **2**:308–310.

CHAPTER 60

Candidiasis

D. G. AHEARN

Disease: Candidiasis (yeast infection, candidosis, monoliasis).

Etiologic Agents: *Candida albicans,* less commonly *Candida tropicalis, Candida* (*Torulopsis*) *glabrata, Candida parapsilosis,* or rarely any of approximately 20 other species of the genus.

Source: Principal etiologic agents are endogenous in oral, gastrointestinal, or urogenital tracts of humans and other warm-blooded animals; others in soil and water.

Clinical Manifestations: Diversified, including dermatitis, vulvovaginitis, cystitis, fever, malaise, myositis, hepatic dysfunction, nausea, weight loss, and mental confusion depending on the superficial, locally invasive, or deep nature of the infection.

Pathology: Diversified, including erythematous patches on the skin, mucosal ulceration, macronodular skin lesions, and abscess formation in almost all deep organs.

Laboratory Diagnosis: Repeated isolation of organisms from specimen in conjunction with compatible syndrome. Demonstration of organisms in tissue; serodiagnosis to detect antibody or yeast products such as mannan or arabinitol may be useful.

Epidemiology: Worldwide; vulvovaginitis is the most common syndrome. Deep or systemic candidiasis, mostly of endogenous origin, occurs typically in compromised hosts, frequently immunosuppressed patients on antibacterial therapy or hyperalimentation.

Treatment: Varied with the syndrome; most often the polyene antibiotic nystatin or the imidazole clotrimazole is used for topical infections, whereas oral ketoconazole or intravenous amphotericin B is given for deep infections. Surgical intervention in combination with chemotherapy may be indicated for certain syndromes such as endocarditis.

Prevention and Control: Cutaneous infections (e.g., diaper rash) may be prevented and controlled by drying of skin. Intravenous amphotericin B is sometimes administered to prevent development of disseminated candidiasis in the febrile neutropenic patient.

Description of Disease

Candidiasis (candidosis, monoliasis; the latter is an archaic term) is an infection with varied syndromes and severity of humans and other warm-blooded animals. In humans, the disease may be superficial, locally invasive, or a deep infection. The extent and severity of infection is dependent on the stressed physiologic state or immunocompromise of the host. Infections of the glabrous mucosa, such as thrush and vulvovaginitis, may be periodic, acute or subacute, or chronic and debilitating. Immunosuppressed and postoperative patients are among those most susceptible to deep infection and the prognosis, particularly for the neutropenic patient, is grave.

The species of *Candida* associated with human disease, their sources, and general disease syndromes are listed in Table 1. Superficial infections by other than *Candida albicans* are rare, and *C. albicans* probably is involved in more than 90% of the

TABLE 1. Syndromes and habitats of etiologic agents of candidiasis

	Representative type of infection		*Habitat*	
Species	*Superficial*	*Deep*	*Human association*	*Host free*
Candida albicans	Cutaneous or mucosal, vaginitis (common)	All organs (common)	Mucous membranes (common)	Soil, air, water, plants (rare)
Candida tropicalis	Vaginitis (rare)	Peritonitis, endocarditis	Sputum, urine, feces (common)	Organically enriched soil and aquatic habitats (common)
Candida parapsilosis	Onchomycosis (rare)	Fungemia, endocarditis, endophthalmitis	Skin, sputum (common)	Soil, water, plants, (common)
Candida glabrata	Vaginitis (rare)	Urinary tract, fungemia	Urine, sputum, feces (common)	Soil, water (rare)
Candida guilliermondii (teleomorph, *Pichia guilliermondii*)	Otitis (rare)	Endocarditis, joint infections	Skin, sputum (common)	Soil, water (rare)
Candida krusei	Unknown	Fungemia, endocarditis (rare)	Sputum (occasional)	Soil, water, plants, (common)
Candida kefyr (telemorph, *Kluyveromyces marxianus*)	Unknown	Disseminated candidiasis, pulmonary (rare)	Sputum, feces (occasional)	Dairy products
Candida lusitaniae (teleomorph, *Clavispora lusitaniae*)	Unknown	Fungemia (rare)	Sputum, feces (occasional)	Animal manure, water
Candida famata	Unknown	Fungemia (rare)	Skin, feces	Soil, water
Candida haemulonii	Unknown	Fungemia, peritonitis (rare)	Sputum, urine	Water, fish

deep infections. Strains of this species have been associated with all syndromes of infection and, with the possible exception of invasions by *Candida tropicalis* of the gastrointestinal tract (Ahearn and Lawrence, 1984; Merz et al., 1986), are the most virulent representatives of the genus.

Communicability and Transmissibility

Candidiasis caused by *C. albicans* and *Candida glabrata* are usually of endogenous origin, although both species may be transmitted between sexual partners and from mother to infant at birth. Infections with other species, particularly deep infections, often may be categorized as iatrogenic. The infecting yeast may be of endogenous or exogenous source.

Clinical Manifestations and Pathology

Cutaneous and Mucosal Infections

More than 95% of superficial yeast infections are estimated to be caused by *C. albicans*. Dermatitis, typified by primary vesicopustules that coalesce and develop into sharply demarcated erythematous patches, is produced within folds of the skin, particularly in the obese and elderly. In the latter, candida perleche, characterized by fissures in the corners of the mouth, may be associated with lack of or ill-fitting dentures. Paronychial candidiasis, an acute swollen infection of the nail bed, is seen occasionally in dishwashers. Diaper rash is among the most common cutaneous infections.

Vulvovaginitis, an often chronic disease and the most common *Candida* infection, is typified by an itching and burning sensation accompanied by pruritis. The vulva is inflamed, and ulcerations may spread to the adjacent mucosa. The discharge may be thick and curdlike. The incidence of thrush, represented by white plaques of the oral mucosa, has been reported to be increasing in association with the acquired immunodeficiency syndrome (Belton and Eversole, 1986). In the past, thrush was more often associated with diabetes or old age, particularly in alcoholics. Mucocutaneous candidiasis is a chronic superficial infection of the skin, nail beds, and mucosa, frequently with onset of symptoms during infancy. A variety of immunologic abnormalities are associated with this chronic disease (Cleary, 1985). The disease may be quite disfiguring, but seldom progresses to deep infection.

Localized Infections

Invasive lesions, sometimes severe, of the mucosa of the gastrointestinal, respiratory, and genitourinary tracts may be manifested as diseases such esophagitis, pneumonia, cystitis, and pyelonephritis. Frequently the isolated lesions are found in immunocompromised patients suffering from some other pathology. These infections may eventuate deep infections of most organs.

Systemic or Deep Infections

Deep infections are characterized by multiple parenchymal abscesses typically involving two or more organs. Disseminated candidiasis involves candidemia in addition to deep organ involvement. There is no characteristic clinical symptom that unequivocally indicates a deep candidiasis. Often the only indication is a persistent fever that is unresponsive to antibacterial antibiotics. Overt symptoms may include those of pneumonia, endocarditis, arthritis, peritonitis, meningitis, or endophthalmitis. Hematogenous macronodular skin lesions often associated with myositis are suggestive of disseminated candidiasis.

Diagnostic Features

The demonstration of yeast cells and pseudohyphae in smears and tissue specimens is diagnostic. Repeated isolation of *Candida* from laboratory specimens may be suggestive of infection, but does not eliminate colonization. Cohen et al. (1969) demonstrated that the gastrointestinal tracts of 50 to 60% of asymptomatic normal subjects were colonized with *C. albicans* and occasionally other species of *Candida*. Any isolation of *Candida* from the blood should be considered potentially indicative of candidiasis and evaluated in relation to the clinical syndrome.

Serology and Blood Biochemistry

Because of immunosuppression and colonization without invasion, detection of candidiasis by determination of antibodies to *Candida* antigen has proved difficult. Antigen detection, particularly the presence of mannan, or of a yeast metabolite such as arabinitol in blood and body fluids has offered more promise; however, sensitive and practical procedures for their detection in the routine laboratory are not yet available (Eng et al., 1981; Repentigay et al., 1984).

Prognosis

Superficial candidiasis may be chronic or recurrent, with each episode requiring treatment. In instances where no additional pathology is present, the individual episodes typically are resolved. Localized infections are also responsive to therapy. The prognosis for disseminated candidiasis is grave, with various reports indicating 20 to 70% mortality. The disseminated disease is detected primarily in seriously debilitated patients.

Epidemiologic Aspects

Occurrence and Transmission

The incidences of the various forms of candidiasis can only be estimated, because the disease is not governed by a mandatory registration system. Vulvovaginitis in the United States has been estimated (partially on the basis of unpublished data of prescriptions sold by pharmaceutical houses) to exceed several million cases per year (Ahearn, 1978). The incidence of deep candidiasis for the years 1980 to 1982 was estimated on the basis of hospital discharge data to be 2.6 per million patients, an increase of about 50% over that estimated for 1976 (Reingold et al., 1986).

Susceptibility and Resistance

The etiologic agents of candidiasis are considered adventitious pathogens (i.e., the predisposition of the host is more important than the virulence of the yeast). In superficial infections, maceration of the skin, dampness, or burns create a predisposition to invasion. In addition, immunodeficiencies or immunosuppression, hormonal imbalance, and other underlying disorders facilitate infection. Broad-spectrum antibacterial therapy predisposes patients to infection and is often associated with yeast colonization of the gut or acute vulvovaginitis. The localized and disseminated candidiases are associated particularly with immunosuppression, acute leukemia, organ transplantation, and extensive hyperalimentation. Cytotoxic chemotherapy, such as that administered for leukemia, appears to mediate invasion by *Candida* initially colonizing the gut.

Etiologic Agents

Candida albicans is an imperfect (anamorphic), diploid, dimorphic fungus that grows as a unicellular budding stage in most enriched broths and agars at room temperature. The ultrastructure of the cell wall is representative of ascomycetous yeasts. On cornmeal agar, most strains readily produce pseudohyphae to true hyphae. *Mycotorula*-type pseudohyphae are characteristic of the species, but many strains produce only *Candida* or *Mycotoruloides* types. Terminal and lateral chlamydospores are produced frequently. At 37°C in serum and serum substitutes most strains produce germ tubes, hyphal projections without a constriction at their point of origin. Two serotypes, A (occurring at an incidence of about 75%) and B, are commonly isolated. About 20% of the isolates readily demonstrate an extracellular proteinase. These proteolytic isolates generally demonstrate greater virulence for mice (Rüchel et al., 1986). Yeast and hyphal stages, both occasionally in giant form, are found in tissues of infected animals. The species has a maximal temperature for sustained growth of about 40°C.

Candida tropicalis is the second most virulent species of *Candida*. Some strains demonstrate an inducible extracellular proteinase. This species does not produce germ tubes or chlamydospores, but elongated pseudohyphal cells may develop in serum at 37°C. With current procedures *C. tropicalis* is antigenically identical with *C. albicans* serotype A. The DNA of the species has a guanine-plus-cytosine content of about 35%, similar to that of *C. albicans;* the DNAs of both species, however, show little homology. *Candida tropicalis* is also a diploid, anamorphic yeast with an ascomycetous-type cell wall. In tissues its morphology is often indistinguishable from that of *C. albicans;* however, Walsh and Merz (1986) found that *C. tropicalis,* in contrast to *C. albicans,* produced a band of advancing necrosis in the absence of inflammatory cells in the neutropenic patient. The species has a maximal temperature for sustained growth of about 42 to 44°C.

Candida glabrata reproduces only by unicellular budding. Cells are 3 to 5 μm in their greatest diameter, about half the size of *C. albicans* and *C. tropicalis*. Among other properties, this species differs biochemically from *C. albicans* and *C. tropicalis* by having coenzyme Q_6 rather than Q_9. Maximal temperature for growth is 40°C. *Candida glabrata* is a relatively avirulent yeast; its frequency in humans is high, however, and therefore it is not an uncommon pathogen of the debilitated host (Thurston and Gillespie, 1981). It is most often isolated from the female genitourinary tract as a commensal organism. Its occurrence in high numbers in men, even if colonization only is evident, is suggestive of serious debilitation (Ahearn et al., 1966).

Candida parapsilosis is typically less invasive than *C. albicans* or *C. tropicalis;* like the latter species and *C. glabrata,* however, it is commonly associated with humans. In contrast with the other three species, *C. parapsilosis* is often isolated from human skin (hands) and is more broadly distributed in the host-free environment. This species is frequently isolated from the blood of patients with indwelling intravenous catheters. It will contaminate vacuum pump oil (as will *C. tropicalis*) and may occur in various intravenous fluids (Plouffe et al., 1977; Solomon et al., 1984). This species has been recognized as an agent of endocarditis in drug users and as a contaminant in a commercial eye-irrigating solution related to an oubreak of endophthalmitis (McCray et al., 1986).

Candida krusei is the anamorphic stage of *Issatchenkia orientalis*. This species is common in sewage, on fruits, and on vegetables, and it is not unusual in sputum or feces. Infections, although rare, are apparently increasing in patients undergoing extended chemotherapy for leukemia (Merz et al., 1986).

Candia kefyr (synonym, *C. pseudotropicalis*) is the anamorphic stage of *Kluyveromyces marxianus*. Pseudohyphal production, as with most species of *Candida,* is variable among isolates of this species. The teleomorphic stage is not unusual in clinical specimens, but is probably often missed because of lack of thorough examination for ascospores. This yeast is associated with dairy products and occasionally encountered in sputum, feces, and other habitats (Lutwick et al., 1980). The species is rarely involved in human infections.

Candida guilliermondii is the anamorphic stage of *Pichia guilliermondii*. This species is of widespread occurrence in the environment and may colonize the gut. It is an occasional agent of disseminated candidasis in seriously debilitated patients (Hopfer et al., 1981).

Uncommon species as agents of infection include *Candida famata* (St.-Germain and Laverdiere, 1986) and *Candida haemulonii* (Lavarde et al., 1986), which have recently been implicated as agents of fungemia. *Candida lusitaniae* is the anamorphic stage of *Clavispora lusitaniae*. This species and the anamorph of *Clavispora opuntiae* are occasionally isolated from urine and feces. Both species phenotypically resemble *C. tropicalis*. *Candida lustaniae* was first implicated as an opportunistic pathogen by Holzschu et al. (1979). Iatrogenic infections with such uncommon pathogens can be expected to occur increasingly.

TABLE 2. Physiologic characteristics of yeasts associated with candidiasis[a]

Candida species	Dextrose		Cellobiose	Dulcitol	Erythritol	Galactose		Inositol	Lactose		Maltose		Melibiose	Raffinose		Rhamnose	Sucrose		Trehalose		Xylose	KNO_3	Urease test	37°C	Cycloheximide
	A	F	A	A	A	A	F	A	A	F	A	F	A	A	F	A	A	F	A	F	A	A			
C. albicans	+	+	−	−	−	+	V	−	−	−	+	+	−	−	−	−	+[b]	−[b]	+	V	+[b]	−	−	+	+
C. famata	+	V	+	V	V	+	−	−	V	−	+	−	V	+	V	V	+	V	+	V	+	−	−	V	V
C. glabrata	+	+	−	−	−	−	−	−	−	−	−	−	−	−	−	−	−	−	+	+	−	−	−	+	−
C. guilliermondii	+	+	+	+	−	+	V	−	−	−	+	−	+	+	+	V	+	+	+	+[b]	+	−	−	+	V
C. haemulonii	+	+	−	+	−	+	−	−	−	−	+	−	−	+	V	+	+	+	+	+	+	−	−	+	−
C. kefyr	+	+	+[b]	−	−	+	+	−	+	+	−	−	−	+	+	−	+	+	−	−	V	−	−	+	V
C. krusei	+	+	−	−	−	−	−	−	−	−	−	−	−	−	−	−	−	−	−	−	−	−	+[b]	+	−
C. lusitaniae	+	+	+	−	−	+	+	−	−	−	+	V	−	−	−	−	+	V	+	V	+	−	−	+	V
C. parapsilosis	+	V	−	−	−	+	V	−	−	−	+	−	−	−	−	−	+	−[b]	+	V	+	−	−	+	−[b]
C. tropicalis	+	+	V	−	−	+	+	−	−	−	+	+	−	−	−	−	+[b]	+[b]	+	+	+	−	−	+	V

[a] Reactions determined by Wickerham procedure with agitation used for assimilations: A = assimilation; F = fermentation; V = common strain variation; + = growth; − = negative.
[b] Occasional strain variation or character genetically unstable.

Specimen Collection and Yeast Isolation

Routine methods for the collection and transportation of specimens for bacterial isolation may also be applied to yeasts. Selective isolation media, however, are necessary. Sabouraud dextrose agar fortified with chloramphenicol and with and without cycloheximide is most often recommended (Ahearn and Schlitzer, 1981; Hopfer, 1985). No one medium or incubation temperature is optimal for the growth of all yeasts; therefore, additional media (e.g., chocolate agar) with incubation temperatures of about 30°C as well as 37°C are often used. Some yeasts that cause fungemia from foci such as catheter tips grow poorly at 37°C. Isolation of yeasts from blood samples may be critical. Vented biphasic medium or the use of a lysis–centrifugation blood culture tube offers enhanced results over simple plating procedures (Bille et al., 1983). Unfortunately, the density of yeast cells circulating in the blood during fungemia fluctuates, and the various species have different optimal growth conditions and growth rates. Further information on determining optimal conditions for the isolation of emerging yeast pathogens would be helpful.

Identification

Commercially available systems are available for the presumptive and often specific identification of most yeasts associated with humans (Bowman and Ahearn, 1976; Shinoda et al., 1981). General obser-vations on the presence or absence of the teleomorphic stage, hyphal or pseudohypal production, type of cell wall, and type of budding are necessary. The general physiologic properties of selected species of *Candida* are presented in Table 2. Finite taxonomic details are presented in the monographs of Barnett et al. (1983) and Kreger-van Rij (1984).

Treatment, Prevention, and Control

Superficial infections by *Candida* may respond to proper hygienic practices, such as drying of the infected area and use of absorbent medicated powders. In certain instances, oral thrush or vulvovaginitis may resolve after the administration of antibacterial antibiotics is stopped. Control of an underlying disease such as diabetes also may be beneficial. Meticulous hygiene in the use of catheters or simple removal of a catheter may prevent or resolve fungemia. Surgical intervention may be efficacious in burn infections, mycetoma, or endocarditis, sometimes without specific antifungal chemotherapy. These general procedures seem to be most effective against the less virulent *Candida* (i.e., all but *C. albicans* and *C. tropicalis*). Certain strains of the latter two species also demonstrate low virulence for mice and probably for humans.

Typically, candidiasis is treated with the polyene antibiotic nystatin or the imidazole clotrimazole for topical and mucosal infections and with the polyene amphotericin B or the imidazole ketoconazole for localized and deep infections. Even with chemotherapy the prognosis for deep infections such as endocarditis or disseminated candidiasis, particularly of the neutropenic patient, is grave. Therefore, although early results have been equivocal, prophylactic chemotherapy is being used increasingly. Ketoconazole has been used often in this regard because of the relatively high toxicity of intravenous amphotericin B. The specifics of chemotherapy of candidiasis have been reviewed by Utz and Drouhet (1985).

Literature Cited

Ahearn, D. G. 1978. Medically important yeasts. Annu. Rev. Microbiol. **32:**59–68.

Ahearn, D. G., J. R. Jannach, and F. V. Roth, Jr. 1966. Speciation and densities of yeast in human urine specimens. Sabouraudia **5:**110–119.

Ahearn, D. G., and J. B. Lawrence. 1984. Disseminated candidiasis caused by a sucrose-negative variant of *Candida tropicalis*. J. Clin. Microbiol. **20:**187–190.

Ahearn, D. G., and R. L. Schlitzer. 1981. Yeast infections, p. 991–1012. *In* A. Balows (ed.), Diagnostic procedures in bacterial, mycotic and parasitic infections, 6th ed. American Public Health Association, Washington, D.C.

Barnett, J. A., R. W. Payne, and D. Yarrow. 1983. Yeasts: characteristics and identification. Cambridge University Press, Cambridge.

Belton, C. M., and L. R. Eversole. 1986. Oral hairy leukoplakia: ultrastructural features. Oral Pathol. **15:**493–499.

Bille, J., L. Stockman, G. D. Roberts, C. D. Horstmeier, and D. M. Ilstrup. 1983. Evaluation of a lysis-centrifugation system for recovery of yeasts and filamentous fungi from blood. J. Clin. Microbiol. **18:**469–471.

Bowman, P. I., and D. G. Ahearn. 1976. Evaluation of commercial systems for the identification of clinical yeast isolates. J. Clin. Microbiol. **4:**49–53.

Cleary, T. G. 1985. Chronic mucocutaneous candidiasis, p. 241–252. *In* G. P. Bodey and V. Fainstein (ed.), Candidiasis. Raven Press, New York.

Cohen, R., F. J. Roth, E. Delgado, D. G. Ahearn, and M. H. Kalser. 1969. Fungal flora of the normal human small and large intestine. N. Engl. J. Med. **280:**638–641.

de Repentigay, L., R. J. Kuykendall, F. W. Chandler, J. R. Broderson, and E. Reiss. 1984. Comparison of serum mannan, arabinitol, and mannose in experimental disseminated cndidiasis. J. Clin. Microbiol. **19:**804–812.

Eng, R. H. K., H. Chrael, and M. Buse. 1981. Serum levels of arabinitol in animals and humans. J. Clin. Microbiol. **143:**677–683.

Holzschu, D. L., H. L. Presley, M. Miranda, and H. J. Phaff. 1979. Identification of *Candida lusitaniae* as an opportunistic yeast in humans. J. Clin. Microbiol. **10:**202–205.

Hopfer, R. L. 1985. Mycology of *Candida* infections, p. 1–12. *In* G. P. Bodey and V. Fainstein (ed.), Candidiasis. Raven Press, New York.

Hopfer, R. L., V. Fainstein, M. P. Luna, and G. P. Bodey. 1981. Disseminated candidiasis caused by four different *Candida* species. Arch. Pathol. Lab. Med. **105:**454–455.

Kreger-van Rij, N. J. W. (ed.), 1984. The yeasts. A taxonomic study. Elsevier Science Publishing, Inc., Amsterdam.

Lavarde, V., F. Daniel, H. Saez, M. Arnold, and B. Faguer. 1986. Peritanite mycosique a *Torulopsis haemulonii*. Bull. Soc. Fr. Mycol. Med. **13:**173–176.

Lutwick, L. I., H. J. Phaff, and D. A. Stevens. 1980. *Kluyveromyces fragilis* as an opportunistic fungal pathogen in man. Sabouraudia **18:**69–73.

McCray, E., N. Rampell, S. L. Solomon, W. W. Bond, W. J. Martone, and D. O'Day. 1986. Outbreak of *Candida parapsilosis* endophthalmitis after cataract extraction and intraocular lens implantation. J. Clin. Microbiol. **24:**625–628.

Merz, W. G., J. E. Karp, D. Schron, and R. Saral. 1986. Increased incidence of fungemia caused by *Candida krusel*. J. Clin. Microbiol. **24:**581–584.

Plouffe, J. F., D. G. Brown, J. Silva, Jr., T. Eck, R. L. Stricof, and F. R. Fekety. 1977. Nosocomial outbreak of *Candida parapsilosis* fungemia related to intravenous infusions. Arch. Intern. Med. **137:**1686–1689.

Reingold, A. L., X. Lu, B. D. Plikaytis, and L. Ajello. 1986. Systemic mycosis in the United States, 1980–1982. J. Med. Vet. Mycol. **24:**433–436.

Rüchel, R., B. Boning, and M. Borg. 1986. Characterization of a secretory proteinase of *Candida parapsilosis* and evidence for the absence of the enzyme during infection in vitro. Infect. Immun. **53:**411–419.

Shinoda, T., L. Kaufman, and A. Padhye. 1981. Comparative evaluation of the Iatron serological *Candida* check kit and the API 20C kit for identification of medically important *Candida* species. J. Clin. Microbiol. **13:**513–518.

Solomon, S. L., R. F. Khabbaz, R. H. Parker, R. L. Anderson, M. A. Geraghty, R. M. Furman, and W. J. Martone. 1984. An outbreak of *Candida parapsilosis* bloodstream infections in patients receiving parenteral nutrition. J. Infect. Dis. **149:**98–102.

St.-Germain, G., and M. Laverdiere. 1986. *Torulopsis candida*, a new opportunistic pathogen. J. Clin. Microbiol. **24:**884–885.

Thurston, A. J., and W. J. Gillespie. 1981. *Torulopsis glabrata* osteomyelitis of the spine: a case report and review of the literature. Aust. N.Z. J. Surg. **51:**374–376.

Utz, J. P., and E. Drouhet. 1985. Treatment of *Candida* infections, p. 253–269. *In* G. P. Bodey and V. Fainstein (ed.), Candidiasis. Raven Press, New York.

Walsh, T. J., and W. G. Merz. 1986. Pathologic features in the human alimentary tract associated with invasiveness of *Candida tropicalis*. Am. J. Clin. Pathol. **85:**498–502.

CHAPTER 61

Chromoblastomycosis

A. GONZÁLEZ-MENDOZA

Disease: Chromoblastomycosis (chromomycosis, dermatitis verrucosa, cladosporiosis).

Etiologic Agents: *Cladosporium carrioni, Fonsecaea pedrosoi, Fonsecaea compacta, Phialophora verrucosa, Rhinocladiella aquaspersa.*

Source: Soils, wood, decayed vegetation.

Clinical Manifestations: Slow-growing, painless, verrucous, and nodular lesions limited to the skin, mainly of the legs or arms. Elephantiasis of the affected extremity can be observed in long-term cases. Occasionally, dissemination to internal organs occurs.

Pathology: Chronic inflammatory reaction with formation of granulomata, in which the fungus can be easily observed as thick-walled, chestnut-brown cells with planate divisions (Medlar bodies).

Laboratory Diagnosis: Direct microscopic examination of clinical material in wet mounts, where the above-mentioned cells are easily observed. All of the etiologic agents present the same parasitic morphology. Species are differentiated by culture of clinical material in dextrose agar with added cycloheximide and chloramphenicol. Slide cultures are necessary to identify the diverse sporulation structures on which classification is based.

Epidemiology: Worldwide, but considerably more frequent in underdeveloped countries in tropical and subtropical areas of the world.

Treatment: Early cases: surgical excision with skin graft. Amphotericin B, 5-fluorocytosine, thiabendazole, ketoconazole, and itraconazole, alone or in various combinations.

Prevention and Control: Use of shoes and adequate protective clothing by people in rural areas, where the disease is observed more frequently; opportune medical care; education.

Description of Disease

Chromoblastomycosis is a mycotic disease of the skin and subcutaneous tissue produced by dematiaceous fungi, which present in tissues as thick-walled, brown-pigmented cells that exhibit planate division. It is characterized clinically by the formation of verrucous and nodular slow-growing lesions, which finally produce papillary vegetations that may or may not ulcerate. Lesions are present mainly on the feet and legs, but can occur on any part of the cutaneous surface, such as the face, ears, neck, arms, chest, shoulders, and buttocks. Satellite lesions can be produced in adjacent areas by lymphatic spread and occasionally by hematogenous dissemination to internal organs, including the brain. Extracutaneous mycoses produced by dematiaceous fungi in which microorganisms are present as mycelia or filamentous branching elements in internal organs (mainly the brain and the lungs in the immunodeficient host), known formerly as chromoblastomycoses of the internal organs, are currently regarded as pheohyphomycoses (Ajello, 1975) and are not considered in this chapter.

Synonymy

Chromoblastomycosis has also been known as chromomycosis, verrucous dermatitis, dermatitis verrucosa chromoparasitaria, cladosporiosis, black blastomycosis, figueira, and Pedroso disease. Indeed, chromoblastomycosis is an erroneous term, as it incorrectly presupposes that the fungus is a yeast; however, the term has been retained because its use is firmly established. Moreover, other proposed names are equally inadequate.

Historic Background

Although the disease is seen more frequently in tropical and subtropical areas of the world, the first reported case occurred in Boston and was described by Lane (1915) and Medlar (1915). The patient came from New England and had verrucous lesions on the foot; the fungus was isolated and named *Phialophora verrucosa*. However, some Brazilian investigators, among them Pedroso and Gomes (1920), had observed the disease and identified the etiologic agent as early as 1911, but their observations were not published until 1920. Mycologic studies conducted by the French parasitologist Brumpt (1922) on the strains isolated by the Brazilian investigators showed that, based on their form of sporulation, these dematiaceous fungi were different from those studied by Medlar. For this new species he coined the term *Hormodendrum pedrosoi*. Several decades later, Trejos (1954) reported that at least one species of the genus *Cladosporium* (*C. carrioni*) also could be an etiologic agent of chromoblastomycosis, thus rendering the nomenclature of these fungi difficult to establish. At present, five species are accepted as etiologic agents of chromoblastomycosis (McGinnis, 1983): *Fonsecaea pedrosoi* (Brumpt) Negroni; *Fonsecaea compacta* (Carrión) Carrión; *Phialophora verrucosa* Medlar; *Cladosporium carrioni* (Trejos) Honbo, Padhye, and Ajello; and *Rhinocladiella aquaspersa* (Borelli) Schell, McGinnis, and Borelli.

Clinical Manifestations

The lesions of chromoblastomycosis are limited most frequently to the skin and subcutaneous tissue. In fact, the localization of lesions is determined by the route of fungal inoculation through the skin consequent to a traumatic wound.

The disease is seen more frequently in rural areas among laborers who work without the protection of shoes. The patient may describe the onset of the disease as a small wound that gradually became a warty lesion and remained restricted for months; it is very infrequent that the physician has the opportunity to observe an early lesion. The initial ulcerous wound is replaced by dry, crusted, or verrucoid lesions with an elevated border that extends from one to several centimeters above the surrounding normal skin surface (Fig. 1). The dissemination of lesions takes years; as the disease progresses some lesions become pedunculated, with hard red or grayish nodules in the center that give it a cauliflowerlike appearance (Fig. 2). The lesions are painless, and at the onset of disease discomfort is negligible. For these reasons most patients do not seek early and opportune medical care. Repeated trauma can produce ulceration and secondary infection with pyogenic bacteria, and the lesions acquire a fetid odor. When chromoblastomycosis is well established, extensive fibrosis develops in deeper tissues, lymphatic vessels become blocked, and elephantiasis of the extremity is produced (Fig. 3).

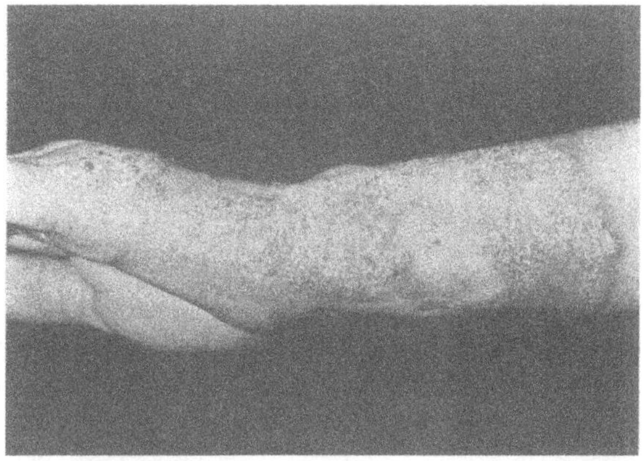

FIG. 1. Erythematous verrucous lesions of hand, wrist, and forearm. The morphology of the lesions is identical to that of tuberculosis verrucosa.

Differential Diagnosis

Initial lesions can be confused with blastomycosis, sporotrichosis, and mycetoma. In well-established disease, the verrucous lesions cannot be differentiated clinically from those of verrucous tuberculosis of the skin, leishmaniasis, tertiary syphilis, and some hypertrophic fibrotic forms of sporotrichosis and mycetoma. For these reasons, chromoblastomycosis can be correctly diagnosed only by laboratory methods.

Pathology

Chromoblastomycosis is confined to the surface of the body. Grossly, the lesions are warty and

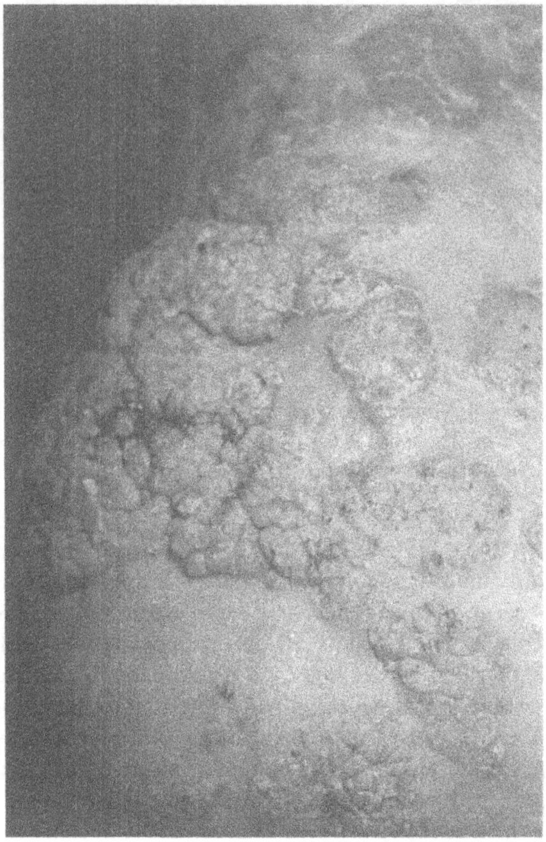

FIG. 2. Nodular, pedunculated, verrucous lesions of ankle and heel showing cauliflowerlike appearance.

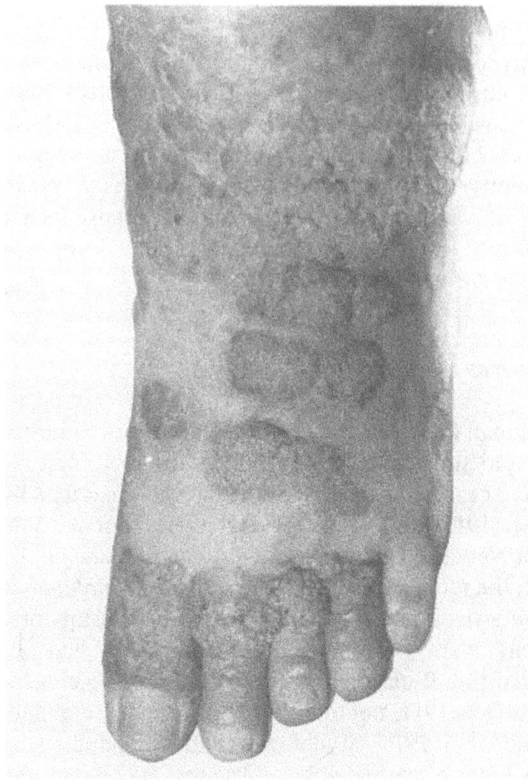

FIG. 3. Chronic nodular verrucous and fibrotic lesions of the foot. Deep-seated fibrosis has produced a moderate degree of elephantiasis.

hyperkeratotic. Microscopically, the epidermis exhibits marked acanthosis with hyperkeratosis (pseudoepitheliomatous hyperplasia). The dermis shows a chronic fibrotic inflammatory reaction interspersed with microabscesses; the histologic picture shows lymphocytes, plasma cells, epithelioid cells, and foreign body giant cells; and numerous neutrophils usually cluster at the center of the granuloma. It is in these granulomas that the fungus is visible as chestnut or light-brown bodies, which appear as rounded yeasts 5 to 12 μm in diameter with thick planate, septal walls (Fig. 4). These sclerotic cells of the fungus are known as *Medlar bodies*. Medlar bodies are also found in the inflammatory reaction in groups or in chains, free or phagocytized by giant cells, mainly in the dermis but sometimes in the hyperplastic epidermal layer. As the microorganism is readily seen in sections stained with hematoxylin and eosin because of its natural chestnut-brown color, special stains are of little or no value. The identification of sclerotic cells within the inflammatory exudate in tissue sections firmly establishes the diagnosis of chromoblastomycosis; nevertheless, it does not permit recognition of the species, as the various fungal etiologic agents of the disease present the same morphology in the parasitic phase.

Epidemiologic Aspects

The disease has been recognized in all parts of the world, but most cases are observed in tropical and subtropical areas. There are several reports of the disease in countries with temperate climates, and it must be remembered that the first reported case in the medical literature came from New England. The etiologic agents of chromoblastomycosis are saprophytic fungi, which are very common and abundant in soils. They belong to the family *Dematiaceae* and can be isolated from wood, pulp, forest litter, and decayed vegetation. The disease is not transmitted from person to person, but by accidental inoculation through the skin and subcutaneous tissue of mycelia and spores of the saprophytic phase. The disease occurs mainly in middle-aged people (i.e., 30 to 50 years of age); it is not common in children. Men are more frequently infected than women, possibly

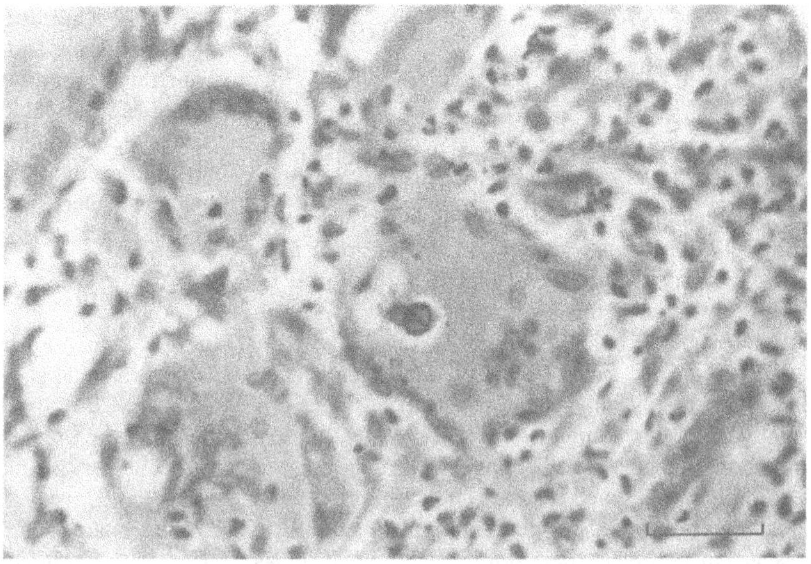

FIG. 4. Sclerotic body with planate septal wall inside a giant cell. Hematoxylin and eosin. Bar = 50 μm.

as a result of occupational activities that place them more readily in contact with the source of infection.

Laboratory Diagnosis

Direct Examination

Laboratory diagnosis of chromoblastomycosis by direct examination of the material obtained from lesions is very easy to perform. Suitable specimens are pus, skin scrapings, and crusts, or fragments of tissue obtained for biopsy. These clinical materials are mounted on microscope slides in a wet preparation of a 10% solution of potassium hydroxide. Typical cells of the fungi consisting of round or ovoid, thick-walled, chestnut bodies with planate divisions are observed. Their size varies between 4 and 12 μm, and occasionally fragments or mycelium originating from such cells also can be observed (Fig. 5). Chromoblastomycosis can be diagnosed with certainty, but the different species cannot be identified.

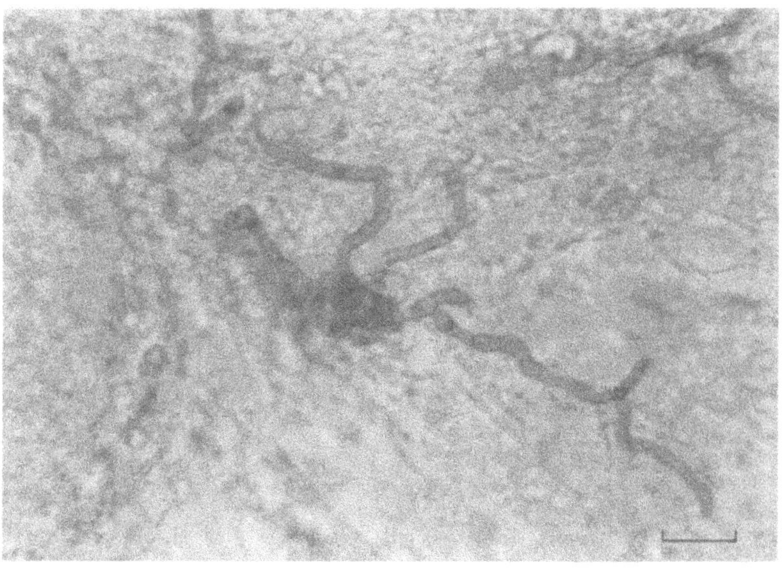

FIG. 5. Skin scrapings with sclerotic rounded body; hyphae originate in the sclerotic cell. Wet mount in KOH. Bar = 20 μm.

Culture Methods

Fungi grow very easily on Sabouraud dextrose agar. Because they are not inhibited by cycloheximide or chloramphenicol, selective media including these antibiotics are preferred. Cultures should be kept at 25 to 28°C for at least 4 weeks before they are discarded as negative. In contrast to rapidly growing saprophytic dematiaceous species, the etiologic agents of chromoblastomycosis grow slowly. The gross morphology of all species is very similar: they grow as molds producing black-brown, olive-gray, gray-black, or black colonies; their texture is velvety to fluffy; and the surface varies from flat to heaped and folded (Fig. 6). Identification of species is complicated and depends on the type of conidia and conidiogenous cells present. According to McGinnis and Schell (1980) and Rippon (1982), four general types of conidiation are found in this group of fungi; their principal features are as follows.

Phialophora Type

In the *Phialophora* type a distinct conidiogenous cell, called the *phialide,* occurs terminally or along the mycelium. The conidiogenous cell is flask shaped and has a rounded, oval, or elongate base, a constricted neck, and an opening that may have a flaring collarette or lip, which measures 3 to 4 μm wide and 4 to 8 μm long. Oval conidia 2 to 3 μm in size are formed at the end of the flask (semiendogenous conidiation) and are extruded through the neck. They may accumulate around the neck, giving the appearance of "flowers in a vase" (Fig. 7A,B).

Rhinocladiella (*Acrotheca*) Type

The conidiophores of the *Rhinocladiella* (*Acrotheca*) type are simple and sometimes poorly differentiated from the vegetative hyphae. They are erect, thick walled, nonbranching, and cylindric; the apical portion is denticulate; and blastic, acropleurogenous, unicellular, oval conidia measuring 2 by 5 μm are produced. Occasionally chains of three to four blastoconidia are formed (Fig. 8).

Cladosporium Type

In the *Cladosporium* type of conidiation the conidiophore is a small stalk, usually slightly enlarged at the distal end. Two or more elliptic conidia are formed at the tip, which bud and form secondary conidia, 2 by 4 μm in size, at their distal poles (acropetalous conidiation). Conidiation continues with the formation of long chains (Fig. 9). The youngest conidium is distal to the conidiophore. Detached conidia show a darkly pigmented scar, called a *disjunctor,* where they were connected to each other.

Fonsecaea Type

Conidiophores of the *Fonsecaea* type are formed by series of ramoconidia functioning as sympodial conidiogenous cells. They have irregularly swollen apices and denticles upon which the conidia are found (Fig. 10A,B).

Many studies utilizing diverse sources of carbon and nitrogen have been conducted to establish physiologic differences among the various species (Fuentes and Bosch, 1960; Honbo et al., 1984). Nevertheless, these studies have not been conveniently standardized and have no practical application.

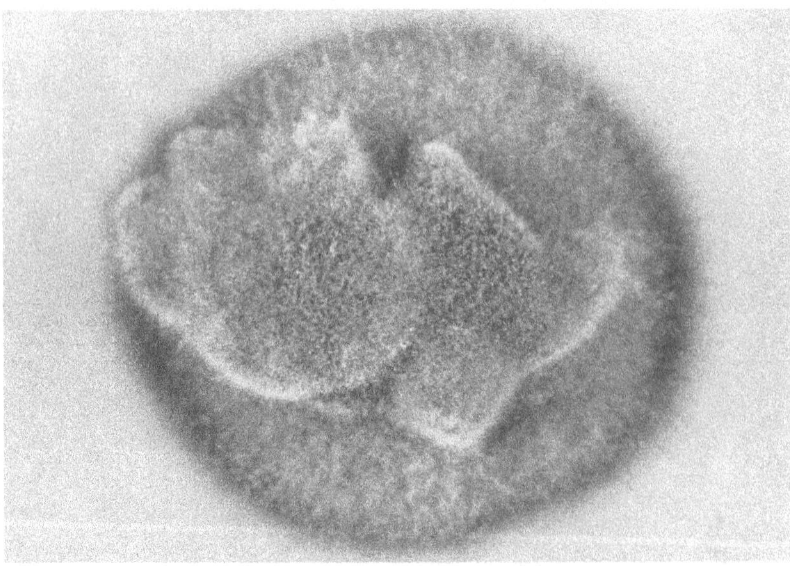

FIG. 6. *Fonsecaea pedrosoi.* Velvety and fluffy black-gray colony growing on potato dextrose agar.

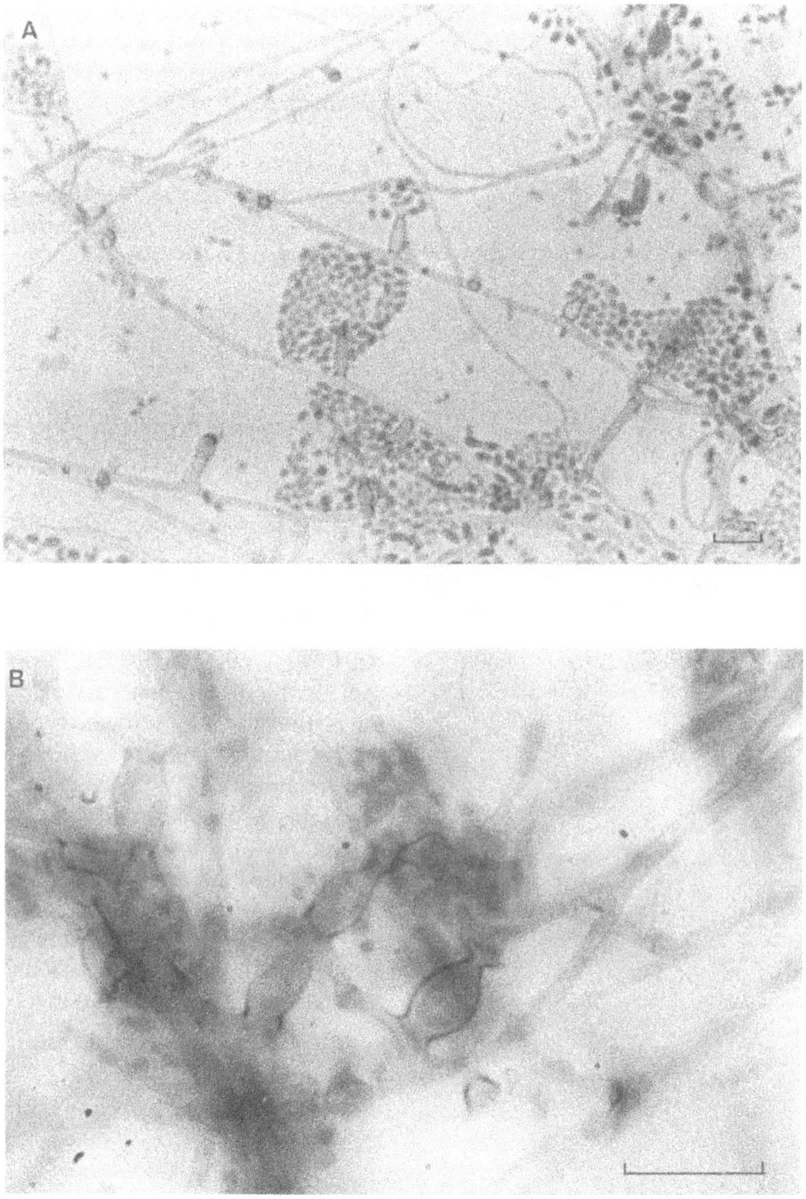

FIG. 7. *Phialophora verrucosa*. (A) Series of phialides with slight variations in morphology. Numerous conidia can be seen around the flaring collarette. Bar = 10 μm. (B) Detail of phialides showing the flask-shaped body, constricted neck, and collarette. Bar = 10 μm.

Description of Species of Chromoblastomycotic Fungi

Cladosporium carrioni (TREJOS) HONBO, PADHYE, AND AJELLO (1984); SYN., *Cladophialophora ajelloi* BORELLI (1980)

Colony Morphology

The organism grows slowly, producing a small, smooth or folded, dark olive-black, compact colony. The edges are entire and marginated by black submerged hyphae.

Microscopic Morphology

Recently, it has been shown (Padhye, 1986) that *C. carrioni* is a polymorphic fungus that produces two anamorphs. The most stable and conspicuous anamorph is the *Cladosporium* type, which is characterized by the production of erect lateral or terminal conidiophores. They are septate, olivaceous brown

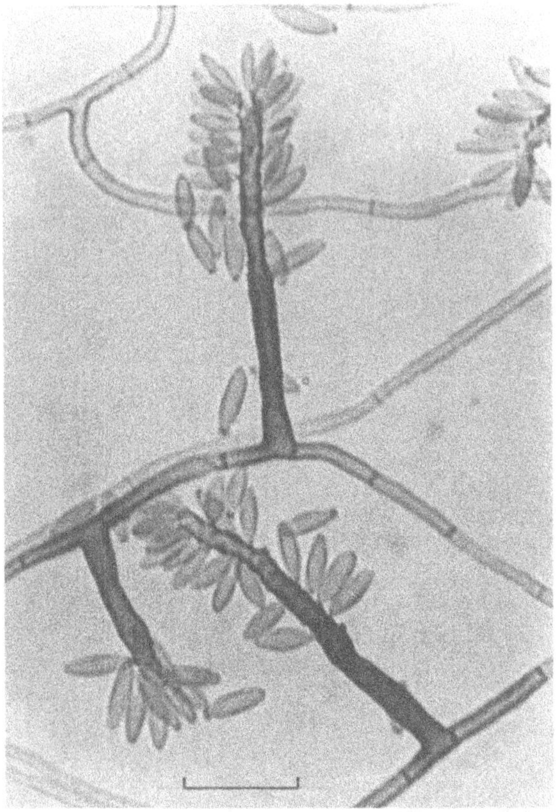

FIG. 8. *Rhinocladiella aquaspersa*. Sympodial conidiophores and one-celled conidia. Bar = 10 μm. (Reproduced courtesy of Dr. M. R. McGinnis.)

in color, and produce long, acropetalous, branched chains of conidia. The conidia are one-celled, subhyaline to brown, and elliptic in shape. Some strains exhibit phialides with collarettes when grown on special media.

Fonsecaea pedrosoi (BRUMPT) NEGRONI (1936); SYN., *Hormodendrum pedrosoi* BRUMPT (1922), *Acrotheca pedrosoi* FONSECA AND ARÊA LEÃO (1923)

Colony Morphology

In culture *F. pedrosoi* grows very slowly, producing a black-brown, gray-black, olive-gray, or black colony. The texture is velvety to fluffy, and the surface varies from flat to heaped and folded. The characteristics of the colony may differ considerably between isolates.

Microscopic Morphology

All three types of conidiation may coexist in this species, the proportion varying with the strain and the medium used for growth. The *Cladosporium* type sometimes predominates, with frequent admixture of the *Rhinocladiella* type and occasional phialides. McGinnis and Schell (1980) studied more than 52 isolates of *F. pedrosoi* and 17 isolates of *F. compacta* and redefined the genus *Fonsecaea* on the basis of its most distinctive, stable conidiophore. This diagnostic form has sympodial conidiogenous cells, often swollen at their apexes, from which primary holoblastic conidia develop on denticles. These conidia,

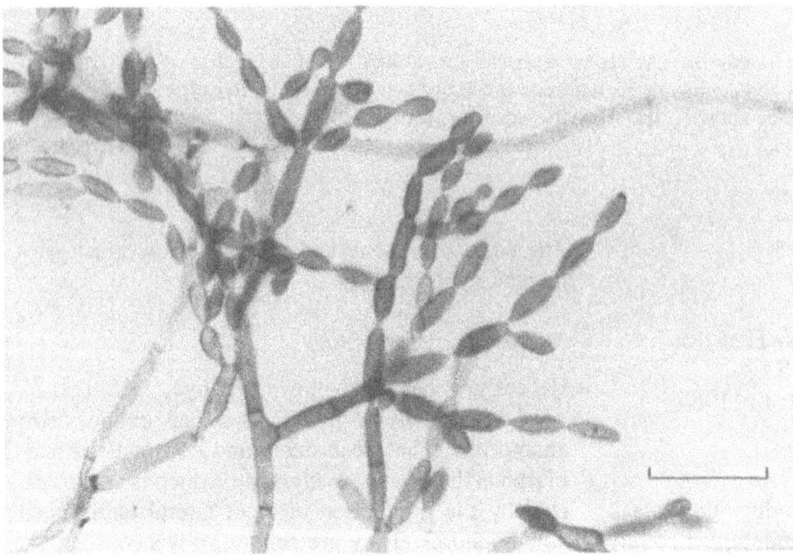

FIG. 9. *Cladosporium carrioni*. Conidiophore is a small stalk with long chains of elliptic conidia. Bar = 10 μm.

in turn, function as sympodial conidiogenous cells, producing secondary conidia in a similar manner. This process may be repeated, resulting in a series of conidia (Fig. 10A,B).

Fonsecaea compacta (CARRIÓN) CARRIÓN (1940); SYN., *Hormodendrum compactum* CARRIÓN (1935)

Colony Morphology

Fonsecaea compacta is a very slow-growing fungus that produces a folded, heaped, brittle colony. Colonies are dark olive-black and develop a brownish-black fuzz with age.

Microscopic Morphology

The conidiophores are similar to those of *F. pedrosoi,* except that conidial heads are more compact. The conidia are subspherical to ovoid, whereas the spores of *F. pedrosoi* are elliptic and large. The other types of conidiation are identical to those seen in *F. pedrosoi;* Borelli (1980) considers this species to be a dysplastic variant of *F. pedrosoi.*

Phialophora verrucosa MEDLAR (1915); SYN., *Fonsecaea pedrosoi* VAR. *phialophorica* CARRIÓN (1940)

Colony Morphology

Phialophora verrucosa grows slowly as a dark olive-gray to black colony, which is initially dome shaped and later becomes flattened. A gray aerial mycelium eventually covers the colony.

Microscopic Morphology

Unlike the other etiologic agents of chromoblastomycosis, *P. verrucosa* has only one type of conidiogenous cell. This cell appears in the form of flask-shaped phialides with distinct cuplike collarettes. Conidia are produced endogenously inside the collarette and are one celled, globose, ellipsoid to ovoid, and hyaline to pale brown. The conidia generally accumulate in clusters at the tip of the phialides.

Rhinocladiella aquaspersa (BORELLI) SCHELL, MCGINNIS, AND BORELLI (1983); SYN., *Acrotheca aquaspersa* BORELLI (1972)

Colony Morphology

This moderately rapid-growing species, initially identified as *Acrotheca aquaspersa* (Borelli, 1972), produces wooly, dark olivaceous gray colonies.

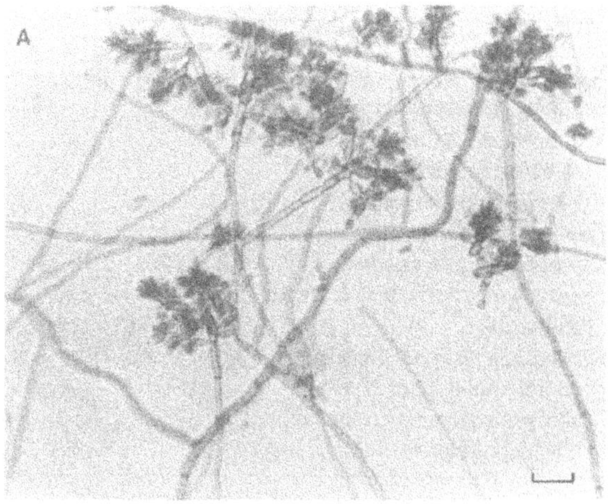

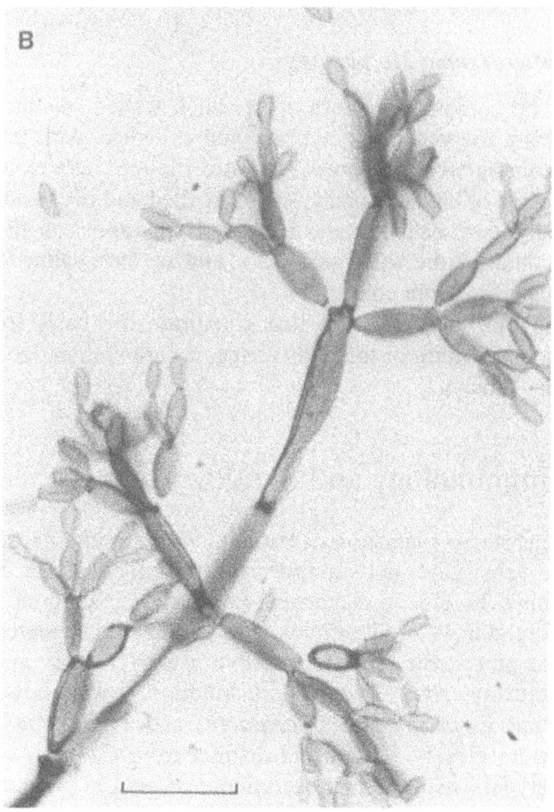

FIG. 10. *Fonsecaea pedrosoi.* (A) *Rhinocladiella* and *Fonsecaea* synamorphs, the latter showing primary conidia functioning as sympodial conidiogenous cells that produce secondary conidia. Bar = 10 μm. (B) *Fonsecaea* and *Cladosporium* synamorphs. Bar = 10 μm.

TABLE 1. Criteria for identification of the etiologic agents of chromoblastomycosis[a]

Conidiation	C. carrioni	F. compacta	F. pedrosoi	P. verrucosa	R. aquaspersa
Cladosporium type (erect conidiophores with branching chains of blastoconidia)	+[b]	+	+	−	−
Exophiala type (annelides)	−	−	−	−	+
Fonsecaea type (sympodial proliferating ramoconidia)	−	+[b]	+[b]	−	−
Phialophora type (phialides with collarettes)	+	+	+	+[b]	−
Rhinocladiella type (erect sympodial conidiophores with acropleurogenous arranged blastoconidia)	−	+	+	−	+[b]

[a] Adapted from McGinnis and Schell (1980).
[b] Basis for identification.

Microscopic Morphology

The conidiophores are erect, thick walled, distinct from the vegetative hyphae, and cylindric, with the conidium-bearing portion toward the top. The elongated or clavate conidia are one celled and occasionally two celled, appear around the apex of the conidiophore with basal scars, and are subhyaline to pale brown in color.

The main features that constitute the basis for identification of the diverse species are summarized in Table 1.

Immunology and Serology

Serologic evaluation of patients with chromoblastomycosis does not yet appear to be a practical procedure to aid in diagnosis. Precipitating antibodies found in 12 of 13 patients with the disease appeared to be specific for the causative agent (Buckley and Murray, 1966). By cross-precipitation, it was shown that *F. pedrosoi, F. compacta,* and *P. verrucosa* were closely related and distinct from *C. carrioni.* Hypersensitivity has been demonstrated in patients with verrucous chromoblastomycosis infected with *F. pedrosoi* (Iwatsu et al., 1979). Although direct microscopic demonstration of fungus is sufficient for diagnosis, the rise and fall of precipitin titers would be of prognostic importance following the institution of therapy.

Several studies have been made of the antigenic relationships between the various etiologic agents of chromoblastomycosis. So far, no consistent pattern has been found that firmly delineates the species involved. Natural resistance to infection is high in normal individuals, and disease occurs only in circumstances of repeated exposure and trauma.

Prognosis and Therapy

Chromoblastomycosis usually remains localized and will not debilitate patients if left untreated. Secondary infection resulting in lymphostasis and elephantiasis may interfere with normal activity. There is also the potential hazard of systemic spread, especially to the brain, depending on the species involved.

In the early stages of the disease, the best therapy probably is wide and deep surgical excision followed by skin grafting. Cryosurgery has also been employed (Ramirez, 1973). Most cases seen in clinical practice are well advanced and require medical management. At present, the drugs employed most frequently in treatment are amphotericin B, thiabendazole, 5-fluorocytosine, ketoconazole, and itraconazole. The use of amphotericin B intralesionally or intravenously has apparently been successful in some cases (Whiting, 1967). Thiabendazole has also produced good results in some instances (Bayles, 1971).

At present, the treatment of choice for chromoblastomycosis is 5-fluorocytosine (González-Ochoa, 1970). When this drug is administered in a dosage of 100 to 500 mg/kg of body weight per day, the lesions rapidly regress and the Medlar bodies disappear from the granulomas. However, resistance and relapse are common and, as a consequence, only a partial cure is achieved. Ketoconazole and itraconazole, used in recent clinical evaluations (Heyl, 1985; Silber et al., 1983), have produced improvement in some cases, but additional clinical trials are necessary before the value of either drug can be assessed in the treatment of chromoblastomycosis.

The combination of 5-fluorocytosine with amphotericin B or thiabendazole has also been used in clinical trials by Lopes et al. (1979) and Solano et al. (1983), who claim to have obtained better results than those produced with the use of single drugs. Despite the preceding evidence, however, at present all drugs and medical procedures proposed to cure the disease are of limited value and chromoblastomycosis, especially in advanced cases, must be regarded as a disease that is difficult to manage and as a medical problem that remains unresolved.

Literature Cited

Ajello, L. 1975. Phaeohyphomycosis: definition and etiology. Pan Am. Health Organ. Sci. Publ. **304:**126–130.

Bayles, M. A. H. 1971. Chromomycosis treatment with thiabendazole. Arch. Dermatol. **104:**476–485.

Borelli, D. 1972. *Acrotheca aquaspersa* nova species de cromomicosis. Acta Cient. Venez. **23:**193–196.

Borelli, D. 1980. Causal agents of chromoblastomycosis (chromomycetes). Pan Am. Health Organ. Sci. Publ. **396:**334–335.

Brumpt, E. 1922. Precis de parasitologie, 3rd ed., p. 1105. Masson et Cie, Paris.

Buckley, H., and I. G. Murray. 1966. Precipitating antibodies in chromomycosis. Sabouraudia **5:**78–80.

Carrión, A. L. 1935. Chromoblastomycosis: preliminary report on a new clinical type of the disease caused by *Hormodendrum compactum* nov sp. Puerto Rico J. Public Health Trop. Med. **10:**543–545.

Carrión, A. L. 1940. The specific fungi of chromoblastomycosis. Puerto Rico J. Public Health Trop. Med. **15:**340–361.

Fonseca, O., and A. E. de Arêa-Leão. 1923. Sur la systématique des champignons produisant des chromoblastomycoses. Comp. Rend. Soc. Biol. (Paris) **89:**762–763.

Fuentes, C. A., and Z. E. Bosch. 1960. Biochemical differentiation of the etiologic agents at chromoblastomycosis from non-pathogenic *Cladosporium* species. J. Invest. Dermatol. **34:**419–422.

González-Ochoa, A. 1970. Curación de la criptococosis y de la cromoblastomicosis con 5-fluorocitosina. Rev. Invest. Salud Publica **30:**63–76.

Heyl, T. 1985. Treatment of chromomycosis with itraconazole. Br. J. Dermatol. **112:**728–729.

Honbo, S., A. A. Padhye, and L. Ajello. 1984. The relationship of *Cladosporium carrionii* to *Cladophialophora ajelloi*. Sabouraudia **22:**209–218.

Iwatsu, T., M. Miyaji, H. Taguchi, S. Okamoto, and N. Kurita. 1979. Skin test active substance prepared from culture filtrate of *Fonsecaea pedrosoi*. Mycopathologia **67:**101–105.

Lane, C. G. 1915. A cutaneous lesion caused by a new fungus (*Phialophora verrucosa*). J. Cutaneous Dis. **33:**840–846.

Lopes, C. F., M. A. Resende, R. J. Alvarenga, and Y. K. Moreira. 1979. Associacao de 5-fluorocitosina e anfotericina B no tratamento da cromomicose. Med. Cutanea Iber. Latin. Am. **7:**1–8.

McGinnis, M. R. 1983. Chromoblastomycosis and phaeohyphomycosis. New concepts, diagnosis and mycology. J. Am. Acad. Dermatol. **8:**1–16.

McGinnis, M. R., and W. A. Schell. 1980. The genus *Fonsecaea* and its relationship to the genera *Cladosporium, Phialophora, Ramichloridium,* and *Rhinocladiella*. Pan Am. Health Organ. Sci. Publ. **396:**215–234.

Medlar, E. M. 1915. Cutaneous infection caused by a new fungus, *Phialophora verrucosa,* with a study of the fungus. J. Med. Res. **32:**507–521.

Negroni, P. 1936. Estudio micológico del primer caso argentino de cromomicosis *Fonsecaea* (n.g.) *pedrosoi* (Brumpt. 1921). Rev. Inst. Bacteriol. (Buenos Aires) **7:**419–426.

Padhye, A. A. 1986. Identification of the etiologic agents of chromoblastomycosis. Pan Am. Health Organ. Sci. Publ. **479:**87–88.

Pedroso, A., and J. M. Gomes. 1920. Sobre quatro casos de dermatite verrucosa produzida pela *Phialophora verrucosa*. Ann. Paulist. Med. Cir. **11:**53–61.

Ramírez, M. 1973. Treatment of chromomycosis with liquid nitrogen. Int. J. Dermatol. **12:**250–254.

Rippon, J. W. 1982. Medical mycology, 2nd ed., p. 263. The W. B. Saunders Co., Philadelphia.

Schell, W. A., M. R. McGinnis, and D. Borelli. 1983. *Rhinocladiella aquaspersa,* new combination for *Acrotheca aquaspersa*. Mycotaxon **17:**341–348.

Silber, J. G., M. E. Gombert, K. N. Green, and A. R. Shalita. 1983. Treatment of chromomycosis with ketoconazole and 5-fluorocytosine. J. Am. Acad. Dermatol. **8:**236–238.

Solano, E., H. Hidalgo, A. Castro, and F. Montero-Gei. 1983. Cromomicosis. Tratamiento con la asociación thiabendazole y 5-fluorocitosina, seis años de seguimiento. Med. Cutanea Iber. Latin. Am. **11:**413–418.

Trejos, A. 1954. *Cladosporium carrionii* n. sp. Rev. Biol. Trop. **2:**75–112.

Whiting, D. A. 1967. Treatment of chromoblastomycosis with high local concentrations of amphotericin B. Br. J. Dermatol. **79:**345–351.

CHAPTER 62

Coccidioidomycosis

DEMOSTHENES PAPPAGIANIS

Disease: Coccidioidomycosis (Coccidiomycosis, San Joaquin Valley fever, desert rheumatism).

Etiologic Agent: *Coccidioides immitis,* a diphasic, multimorphic fungus producing endosporulating spherules in vivo, hyphae and arthroconidia in culture and in nature.

Source: Soil of certain arid and semiarid areas of the Western Hemisphere.

Clinical Manifestations: Acute—onset of malaise, fever, chest pain, cough, anorexia and weight loss, night sweats, arthralgia, myalgia, cephalalgia. Chronic—may begin as acute, followed in weeks or months by appearance of pain or lesion at site of metapulmonary dissemination, or occasionally without recognized antecedent acute stage.

Pathology: Pyogranulomatous disease affecting lungs, thoracic lymph nodes and, when disseminated, all organs and tissues (rare endocardial or gastrointestinal involvement).

Laboratory Diagnosis: Histopathologic demonstration of endosporulating spherules or cultural recovery followed by exoantigen identification, or in vitro or in vivo (intraperitoneal infection of mice) conversion to spherules. Serologic demonstration of "precipitin" (IgM) or complement-fixing (IgG) antibody.

Epidemiology: Endemic to certain areas of the Western Hemisphere: southwestern United States, Mexico, limited sites in Central America, and South American countries (Venezuela, Colombia, Argentina, Paraguay). Disease may occur after one has been in and left an endemic area or, rarely, from items exported from such an area.

Treatment: Amphotericin B and ketoconazole (some other azoles being studied). Surgical removal of diseased pulmonary, cutaneous, joint, or osseous tissue.

Prevention and Control: Reducing exposure to dust (soil) in endemic areas. No effective vaccine for humans.

Description of Disease

Coccidioidomycosis is an infectious disease, usually acquired by inhalation of arthroconidia or, rarely, by their traumatic introduction through the skin.

Communicability and Transmissibility

The disease is not communicable between individuals.* Humans and nonhuman species are infected by arthroconidia that become airborne through disturbance of soil that harbors *Coccidioides immitis.*

* A unique episode involved hospital employees who were infected by exposure to aerosolized hyphal phase from a plaster-of-Paris cast contaminated by exudate from a coccidioidal lesion (Eckmann et al., 1964).

Clinical Manifestations

Asymptomatic Coccidioidomycosis

On the basis of skin test results with coccidioidin, some 60% of individuals infected with *C. immitis* are said to be asymptomatic. [This fraction of asymptomatic skin test converters may be even greater in view of the apparently greater proportion of skin test reactions detected by spherulin than by coccidioidin (Levine et al., 1973).] Some individuals are discovered by means of abnormal chest X-ray and other confirmatory tests to have undergone coccidioidal infections without a *recognizable* prior illness, although the initial illness may have been passed off as a benign respiratory infection.

Acute Primary Coccidioidomycosis

Acute primary coccidioidomycosis ("primary" in analogy with the initial infection in tuberculosis, and contrasted with "secondary" or disseminated coccidioidomycosis) occurs 7 to 28 days after exposure. Most of the 40% or so of infected individuals with acute coccidioidomycosis recover following an illness that may be evident from the following signs, symptoms, and findings (Lundergan et al., 1985; Willett and Weiss, 1945) (Fig. 1): fever, sometimes as high as 40°C; chest pain, often pleuritic and increased with deep breathing (chest pain and malaise may persist for many months after apparent recovery from primary coccidioidomycosis); cough, often nonproductive; malaise; anorexia; weight loss; headache; night sweats; arthralgia, especially of knees and ankles, accompanied by erythema nodosum; myalgia; and various rashes. The chest X-ray shows varied patterns from a small area of pneumonitis to multiple infiltrates, and hilar lymphadenopathy. A transient pulmonary cavity may develop in the original area of pneumonitis. Pleural effusion of varied extent occurs in some patients. Resolution usually occurs within a few weeks (Smith, 1955).

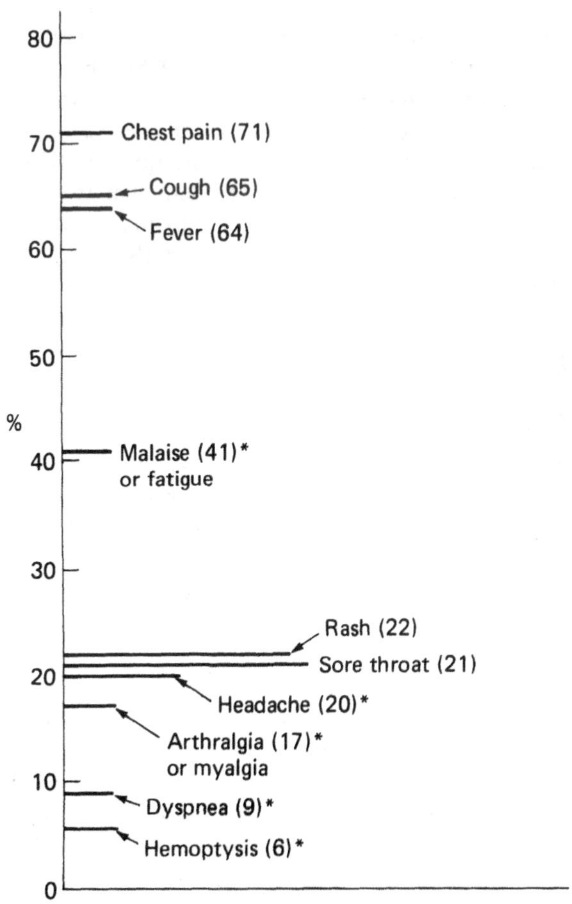

FIG. 1. Frequency (percentage) of signs and symptoms in 494 patients with primary coccidioidomycosis.

Disseminated Coccidioidomycosis

Disseminated coccidioidomycosis usually refers to evidence of spread of coccidioidomycosis outside the thorax. Such extension occurs lymphogenously and hematogenously. Evidence of enlargement of the paratracheal and other mediastinal lymph nodes may precede this, but spread of *C. immitis* to the supraclavicular nodes does not necessarily reflect serious metapulmonary dissemination (Coburn, 1962), as would appearance of lesions in the skin, bones, or meninges, the three most commonly evident sites of dissemination. Osseous coccidoidomycosis can be studied by a combination of radionuclide scan and radiography (Armbruster et al., 1977). Extrapulmonary lesions can be found in the peritoneal cavity, thyroid, adrenal, and prostate glands and virtually all other tissues, although descriptions of endocarditis and involvement of the gastrointestinal tract have been rare (Weisman et al., 1986). Whereas splenic involvement has been frequent in fatal disseminated coccidioidomycosis, splenomegaly has not been a conspicuous antemortem feature. Dissemination may be widespread, that is, with evidence of lesions in many sites, or there may be single sites, for example in the skin or a bone. Lethal coccidioidomycosis may be associated with multiple sites of involvement, but involvement of the meninges alone suffices to kill. Some cases of intrapulmonary coccidioidomycosis without dissemination have been lethal. These have included fulminating infections with acute respiratory failure (Larsen et al., 1985) and a chronic, progressive, fibrocavitary disease (Catanzaro, 1986). Coccidioidal meningitis may ensue directly from a recognized respiratory, febrile, primary, coccidioidal illness, but often it is initially manifested by an aberration of mentation or behavior (Bouza et al., 1981; Caudill et al., 1970).

Factors Influencing Dissemination

Certain factors are associated with increased risk of dissemination. These include racial background, such as Filipino and other Asian, African (American); late pregnancy, when acute primary coccidioidomycosis supervenes; immunosuppressive conditions, as for renal transplantation (Pappagianis, 1980a) or acquired immune deficiency syndrome (AIDS) (Wolf et al., 1986).

Pulmonary Residua

Following acute primary coccidioidomycosis, some 5 to 10% of patients develop pulmonary cavities, cavitating abscesses, or solid lesions (coccidioidomas) (Winn, 1968). A few develop bronchiectasis. These are generally not life threatening and in many instances require no specific therapy, although a cavity may in rare instances bleed severely, rupture into the pleural space, or show accumulation of fluid and possible secondary colonization (infection). A very small proportion of patients develop bilateral, apical, fibrocavitary coccidioidomycosis that may resemble tuberculosis or histoplasmosis and has a poor prognosis. Although *C. immitis* may be identified in the sputum or bronchial washings, biopsy or excision of a coccidioidoma may be needed to establish the presence of *C. immitis* and to distinguish the lesion from malignancy or from those of other microbial etiology.

Viable *C. immitis* probably persists for life in patients with pulmonary cavities, although recovery of the organism from the sputum may be intermittent. Coccidioidomas contain *C. immitis* demonstrable by histopathologic staining, but recovery of the organism on culture has often failed with the presently applied methods (Forseth et al., 1986).

Pathology

Entry of *C. immitis* leads to inflammation that may be evident first in the bronchioles or alveoli or both. The inhaled or inoculated arthroconidia (2 to 4 × 5 to 6 μm) undergo enlargement and change into round, immature spherules, which then mature into endosporulating spherules (20 to 100 μm). This may occur despite phagocytosis by polymorphonuclear leukocytes (PMN) or macrophages. Ingestion by murine PMN may result in the killing of only 20 to 30% of arthroconidia or endospores (Drutz and Huppert, 1983), although immunologically activated PMN (from mice previously stimulated by *Blastomyces dermatitidis*) were said to kill 77.5 ± 7% of endospores but were ineffective versus arthroconidia (Brummer et al., 1985). Peripheral blood monocytes from humans, immune or nonimmune, by themselves showed no phagosome–lysosome fusion and no killing effect on phagocytosed arthroconidia; but in the presence of lymphocytes from immune individuals, monocytes showed phagosome–lysosome fusion and significant killing of arthroconidia (Beaman and Pappagianis, 1985). Mature spherules appear to evoke a macrophage response, the liberated endospores a PMN response. Consequently mixed (pyogranulomatous) responses can be observed, that is, granulomata and microabscesses adjacent to each other. Caseation, fibrosis, hyalinization, and calcification can be seen in coccidioidal lesions. Some of these persist for many years.

The organism often enters the lymphatics to reach the hilar lymph nodes, from which they may extend up the tracheobronchial chain to the supraclavicular nodes. Hematogenous spread may lead to manifest metapulmonary dissemination, although blood-borne dissemination can occur without obvious clinical manifestations, for example, as shown by the presence of *C. immitis* in the urine (Petersen et al., 1976) or by chorioretinal lesions of the eye (Rodenbiker et al., 1981). Primary coccidioidal infection of the skin produces a "chancriform complex," that is, ulcerative lesion and regional lymphadenopathy (Wilson, 1968).

The persistence of recognizable *C. immitis* is evident in the fibrocaseocalcific pulmonary granuloma, in the walls of pulmonary cavities, and in such chronic lesions as fistulae or sinus tracts. Indeed, the spherule or hyphal form may be found in the fibrous wall, mural exudate, or necrotic amorphous or purulent contents of a pulmonary cavity. Virtually all organs and tissues can be affected in disseminated disease, but until recently the gastrointestinal mucosa escaped infection (Weisman et al., 1986). The endocardium is infrequently involved. While the meninges are quite often involved, focal disease of the brain proper has been noted infrequently (Huntington, 1971), although it is evident that the substance of the brain may be affected by miliary spread or by extension from the meninges (Sobel et al., 1984).

Diagnostic Features

Clinical

The varied clinical patterns of coccidioidomycosis should prompt its inclusion in the differential diagnosis of febrile, respiratory infections, with pleuritic chest pain, headache, myalgia, arthralgia, malaise, and anorexia occurring in patients within 4 weeks of their presence in the areas endemic for coccidioidomycosis. It may resemble acute bacterial pneumonias, tuberculosis, other fungal diseases, malignancy (e.g., lymphoma with eosinophilia), or lung cancer. The metapulmonary dissemination usually becomes evident within several weeks. For example, during a well-defined epidemic exposure to airborne *C. immitis* (Pappagianis and Einstein, 1978), clinically apparent meningitis occurred in 22 individuals in a mean period of 5 weeks after onset of the primary pulmonary illness (see Fig. 2). However, occasional cases of dissemination occur many months or, rarely, years after the primary infection.

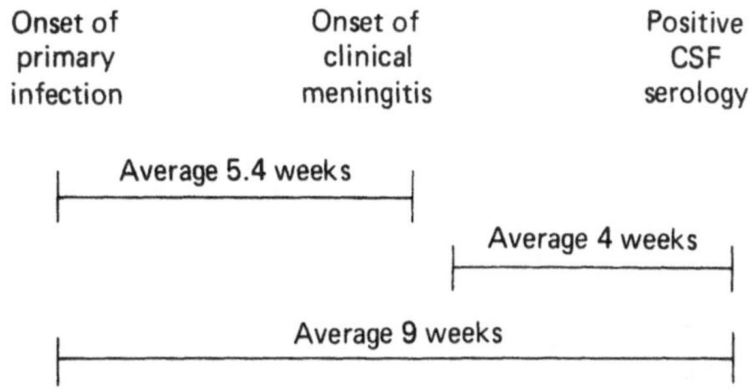

FIG. 2. Coccidioidal meningitis in 22 cases associated with a dust storm in December 1977.

LABORATORY FINDINGS

Nonspecific

The peripheral blood may show an elevation of alkaline phosphatase and eosinophilia. Higher levels of eosinophilia, for instance, more than 15%, are associated with more severe disease (Echols et al., 1982). In coccidioidal meningitis, the cerebrospinal fluid (CSF) glucose is usually depressed (e.g., 20 to 30 mg/dl, but can be lower), the cell count is elevated and is usually predominantly mononuclear, but in early meningitis polymorphonuclear leukocytes may predominate, and there may be eosinophilia. The protein is usually elevated, and the colloidal gold test shows a "first zone" or paretic curve (Bouza et al., 1981; Einstein, 1974; Smith, 1955).

The erythrocyte sedimentation rate is usually increased and serves as a rough guide to activity of the infection.

Specific

The presence of endosporulating spherules is diagnostic (Fig. 3). These can be demonstrated, for example in sputum, with KOH or the cytologic Papanicolaou stain, the latter appearing to be the more sensitive (Warlick et al., 1983). However, immature spherules may be seen that cannot readily be distinguished from *Blastomyces dermatitidis* and perhaps some forms of *Paracoccidioides brasiliensis*, large-cell forms of *Histoplasma capsulatum*, especially var. *duboisii*, and the alga *Prototheca*.

The sputum has on several occasions erroneously been thought to contain *C. immitis* spherules because of the resemblance of artifacts, some of which may be pollen grains. An innate "autofluorescence" of spherules may provide some distinction from artifacts (Kaplan and Clifford, 1964; Kaplan and Kraft, 1969). Alkaline digestion may succeed in destroying nonfungal particles, but one may also incubate sputum or exudate to which is added some glucose–yeast extract or Sabouraud's broth in a hanging-drop microslide. During the next several hours, hyphae can be seen to sprout from the spherule/endospore (Pappagianis et al., 1956).

The chitin of the wall of *C. immitis* can be made visible through the use of calcofluor (Hageage and Harrington, 1984; Roberts and Goodman, 1986). The formulation and application of calcofluor white are described in Appendix I. Treatment with KOH prior to use of the calcofluor can reduce effects of artifacts, although KOH can be mixed with the calcofluor. Some confusion with *C. immitis* spherules has been created by "myospherulosis," which represents an artifact, apparently following the postoperative packing of the nose with petrolatum-coated gauze (Kyriakos, 1977).

Hyphae may be observed in the wall (or even lumen) of pulmonary cavities and occasionally in nonpulmonary tissues (summarized by Meyer et al., 1982; Wolf et al., 1986). These hyphae may resemble those of other organisms.

Epidemiologic Aspects

Occurrence and Transmission

Coccidioidomycosis occurs primarily in the Western Hemisphere, particularly in the southwestern United States (California, Arizona, New Mexico, and Texas); Mexico; rare cases in Guatemala, Honduras and Nicaragua; Venezuela, Colombia, Argentina, and Paraguay. Individuals who have traveled in these areas may not become ill until returning to their homes in nonendemic areas. Rarely, infections have occurred in humans far from the endemic areas handling some item contaminated with *C. immitis*.

Approximately 500 cases and 50 deaths per year are recorded in the United States, most of them coming from relatively few counties. The cases are usually sporadic, but occasional epidemics occur, for example, in archaeologists exposed at a common site. Infection is more likely to occur among those who disturb the soil occupationally. The number of cases increases when the weather is dry and dusty, and diminishes when the rain wets the soil and decreases the airborne infective arthroconidia.

Susceptibility and Resistance

While certain occupations lead to a higher rate of infection, the occurrence of serious forms of the disease is related to certain risk factors. Thus, Blacks, Filipinos, and other Asians are more likely than other ethnic groups to develop severe disease. The adult white female, who is often the most resistant to *Coccidiodes immitis*, becomes highly vulnerable if she acquires a primary acute coccidioidal infection

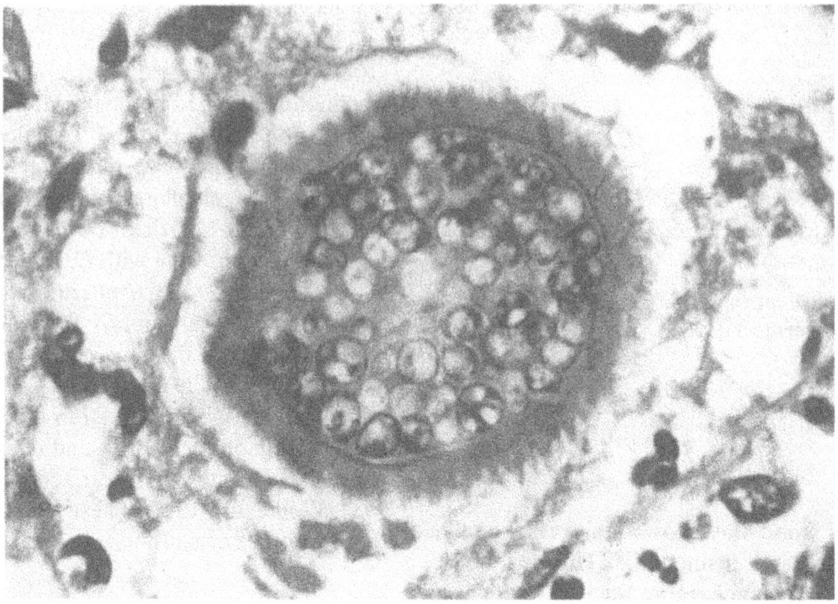

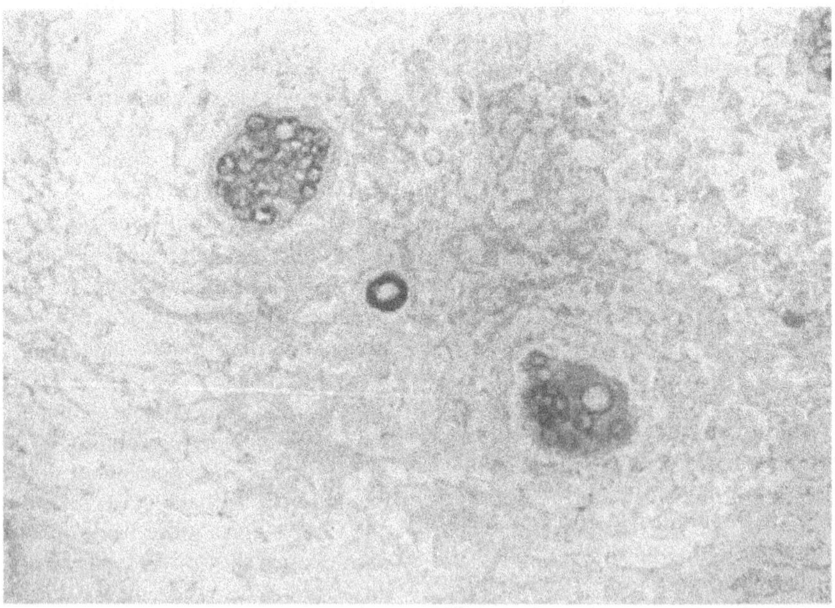

FIG. 3. *Coccidioides immitis* in vivo. Upper left: endospore-containing spherule with shaggy eosinophilic material at periphery (hematoxylin and eosin stain). Upper right: endospore-containing spherule within a multinucleated giant cell (hematoxylin and eosin stain). Lower left: endospores in spherule (wall of spherule is inconspicuous) (periodic acid-Schiff stain). Lower right: ruptured endospore-containing spherule and immature spherules (periodic acid-Schiff plus hematoxylin nuclear stain). All figures reduced by 35%.

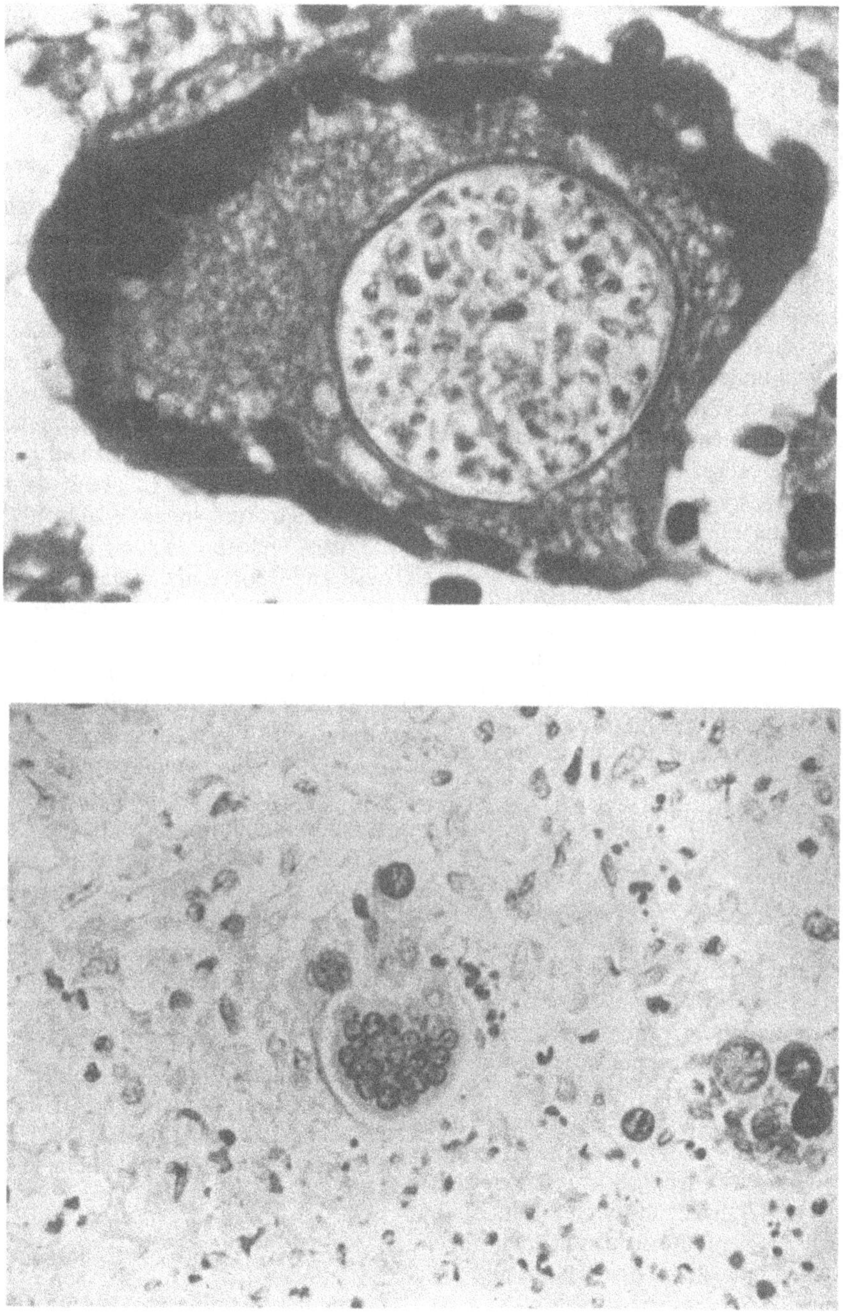

FIG. 3. (*Continued*)

during the latter half of pregnancy. Similarly, immunosuppressive states or medications set the stage for the disease spreading from the initial respiratory infection. Resistance to reinfection usually follows recovery from symptomatic (or nonsymptomatic) infection.

Although humans discharge *C. immitis* in sputum or from infection sites, transmission from human to human does not occur. *C. immitis* may be carried for years or perhaps life in healed or persistent pulmonary lesions, usually posing no threat to the individual or to others. Reactivation may occur in immunosuppressed individuals long after presumed healing has taken place.

Etiologic Agent

Characterization

Coccidioides immitis Rixford and Gilchrist, 1896 is a diphasic, multimorphic fungus that exists in a hyphal-arthroconidal or spherule-endospore (sporangial) form. The former is often referred to as the "saprophytic" or "saprobic" phase, being found in nature and in cultures, infrequently in the human host. The spherule-endospore form(s) is referred to as the "parasitic" phase, being found in the infected mammalian host; but under appropriate conditions it can also be produced in vitro.

The taxonomic placement of *C. immitis* is uncertain, since no sexual stage has yet been demonstrated. The septate hyphae have Woronin-like bodies near the septal pores and, with their enteroarthric arthroconidia that alternate with empty, degenerate cells, resemble certain other organisms in the *Ascomycotina*. The percentage of guanine–cytosine (49.41 to 49.61%) was similar among three strains of *C. immitis* and in the spherule and hyphal forms (Pappagianis et al., 1985b). This compares with the 47.3% and 48% found in *Histoplasma capsulatum* and *Blastomyces dermatitidis*, respectively (Bawdon et al., 1972). Currah (1985) has proposed that *C. immitis* should be in the Family *Onygenaceae* of the Order *Onygenales*, Division *Ascomycotina*.

Variations in colonial and microscopic morphology of the hyphal forms have been noted, including some with considerable range in coloration and texture (Baker et al., 1943; Friedman et al., 1953; Huppert et al. 1967b). Though infrequent, such variability should prompt care on the part of laboratorians in rejecting cultures that do not fit the usual description (see below, Preliminary Identification). There is no compelling indication that there is more than a single species. Foley et al. (1960), using X-irradiation, and Walch and Walch (1967), using ultraviolet irradiation or a chemical mutagen, produced auxotrophic mutants, some of which had altered phenotypes and diminished virulence for laboratory animals.

In a suitable medium (1% glucose–0.5% yeast extract–2% agar), arthroconidia are formed in 5 to 7 days at 34° to 37°C. These arthroconidia, usually 2 to 4 × 5 to 6 μm, barrel-shaped, cylindrical or oval, are produced in apparently specialized aerial hyphae that are sometimes borne on short hyphae 1 to 2 μm in diameter, narrower than the vegetative hyphae (2 to 4 μm). The arthroconidia, usually multinucleate and thick walled, develop with alternating, degenerate, brittle, nonviable cells with thinner walls, which autolyze and readily fracture. This leads to disarticulation and ready dispersion of individual or small chains of the readily airborne arthroconidia. It is these arthroconidia that represent the *highly infectious* form of *C. immitis* that usually initiates infection following inhalation, rarely following percutaneous (traumatic) penetration. [*One must be constantly mindful of the hazard posed by these arthroconidia in the laboratory, particularly in cultures more than 3 days old.*]

In the presence of the host's inflammatory cells and elevated CO_2 (Klotz et al., 1984), the arthroconidia apparently shed an outer wall layer [which contains approximately 50% peptides, 12% neutral carbohydrate, and 12% lipid (Cole et al., 1985)] and all but one nucleus, then round up and enlarge to give a sporangium-like spherule. Nuclear division is followed by formation of cleavage planes that originate with the peripheral cell wall and progressively segment the cytoplasm. In about 48 h in vivo, this results in a mature spherule 20 to 150 μm in diameter with many endospores 2 to 4 μm in diameter (Fig. 3). These endospores are released and are capable of enlarging to produce another generation of spherules with endospores, etc. Such a formation of endosporulating spherules from arthroconidia or endospores can be achieved in vitro in Converse synthetic medium (1957), as modified by Levine et al. (1960) at 35° to 39°C. This medium contains a surface-active agent, Tamol, a neutral sodium salt of a condensed aryl sulfonic acid (Rohm and Haas Company). Lones and Peacock (1960a) showed that maintenance of the spherule-endospore phase is enhanced by increased CO_2. The time from inoculation to formation of mature endosporulating spherules is somewhat variable among different laboratories and may occur at from approximately 30 to 72 h.

The mechanism of causation of disease has not been elucidated. Toxic (Gale et al., 1967), enzymatic [e.g., proteases (Lupan and Nziramasanga, 1986; Nziramasanga and Lupan, 1985; Resnick et al., 1986)], and immunologic/hypersensitivity factors may be involved.

Physiology and Metabolism

Coccidioides immitis grows only aerobically and is catalase positive in both hyphal and spherule phases (Zimmer and Pappagianis, 1986b). The arthroconidia (after a lag of more than 5 h) and 48-h-old hyphae carry on respiration for many hours in the absence of an added carbon source (Pappagianis and Kobayashi, 1960). The respiratory quotient was 0.75 to 0.85, but the nature of the endogenous substrate has not been determined. The endogenous respiration of unstarved hyphae was greater than that of spherules (Lones and Peacock, 1960b).

Enzymes of the Krebs cycle, with the exception of α-ketoglutaric dehydrogenase, were found in extracts of both spherules and hyphae (Lones, 1967).

Enzymes of the pentose phosphate and Embden-Meyerhof pathways were also present in both morphologic forms. Isocitric lyase and malate synthase were present in the hyphal form indicative of a (glyoxylate) bypass for the absent α-ketoglutaric dehydrogenase (Agy and Paznokas, 1985). Although neither acid nor gas is produced from carbohydrate by conventional fermentation tests, Lones and Peacock (1960b) demonstrated that spherules produced more CO_2 from glucose under anaerobic conditions than did the hyphae, and the spherules produced equimolar amounts of ethanol and CO_2 by fermentation of glucose. In the presence of glucose with NH_4Cl as the source if nitrogen, *C. immitis* brought about a drop in the pH of the medium because of the utilization of NH_4^+, not because of the production of organic acids (Pappagianis and Kobayashi, 1958). Baker and Smith (1942) demonstrated that the mycelial phase utilized a variety of carbon sources, extending from 2-carbon (ethanol, acetate) to polysaccharides, and nitrogen sources from simple inorganic (NH_4^+, NO_3^-) to complex organic (peptone). The spherule-endospore (SE) phase can also make use of a broad range of compounds, but appears to differ from the mycelial phase in that SE does not utilize starch or xylose, and endospores but not spherules utilize malate as carbon source (Sippel and Levine, 1969) (Table 1).

None of these physiologic characteristics has been particularly useful for the identification of *C. immitis*. However, *C. immitis* shows tolerance of pH extremes from 3.5 to 9.0. The latter led to development of "K" medium (Appendix II) containing K_3PO_4 and a high pH with or without 0.04% $CuSO_4$ to select for *C. immitis* (Smith, 1943). In this medium, the organism grew sparsely, but its microbial competitors even less.

Specimen Collection and Transport

Selection of the appropriate specimen depends on the form of coccidioidomycosis suspected and the accessibility of clinical specimens. Diagnosis can make use of histopathologic or cytologic, cultural, and serologic testing. Flexible fiberoptic bronchoscopy yields *C. immitis* significantly more often than expectorated sputum (Wallace et al., 1981). For example, biopsy of bronchial tissue, skin, or other site may yield the most secure diagnosis, that is, direct demonstration of *C. immitis*. The biopsy sample must be cultured as well as examined following histologic staining, particularly if no typical endosporulating spherules are seen. In some instances, for example, when there is an old pulmonary lesion, it may be important to know whether viable *C. immitis* is present.

Cultures should be prepared by inoculation onto selective medium, for example, Sabouraud's agar, to which have been added cycloheximide and some antibacterial antibiotic(s), for instance, chloramphenicol (Georg et al., 1951).

Aseptically obtained specimens of body fluids to be used for serologic tests should be preserved by freezing, or by addition of aqueous merthiolate or sodium azide at a final concentration 0.1 mg/ml.

Isolation and Identification Procedures

Recovery of *C. immitis* should be attempted whenever possible to provide confirmation of infection even though serodiagnostic methods (see below) are highly reliable. Whereas sputum may be the most frequent source studied, exudates, effusions (pleural, peritoneal, joint), gastric, bronchial and transtracheal washings, cerebrospinal fluid, peripheral blood (Land et al., 1977), urine, biopsy specimens (lung, lymph node, pleura, synovial membrane, bone marrow) all may be sampled at some time (Howard and Smith, 1983, Huntington, 1957). Cerebrospinal, joint, or pleural fluids may be negative on culture despite other evidence of the presence of *C. immitis*. For example, biopsy of the pleura yields *C. immitis* in many instances when the pleural *fluid* is negative (Lonkey et al., 1976), and synovial membrane may yield *C. immitis* despite negative results with synovia.

Although *C. immitis* is relatively hardy, care is needed to maximize the chance of recovering it. For example, treatment ("digestion") of *sputum* with alkali or hypochlorite for the recovery of *Mycobacterium tuberculosis* may render *C. immitis* nonviable. Therefore, culture for the fungus should be attempted prior to such treatment (Ajello et al., 1951), although deliberate use of mucolytic agents (N-acetyl-l-cysteine and dithiothreitol) and homogenization with glass beads may enhance the recovery of *C. immitis* from sputum (Reep and Kaplan, 1972). The effect of competing organisms, in sputum or other sources, which can overgrow the *C. immitis* can be minimized by use of medium containing cycloheximide and chloramphenicol or the alkaline "K" medium described above. Pretreatment of the specimens for 1 to 4 h with penicillin-G and streptomycin (1,000 units of each per ml of specimen) or with copper sulfate at a final concentration of 0.04% also provide an advantage to *C. immitis*. Yeast extract–phosphate agar (Smith and Goodman, 1975) with a drop of NH_4OH has also served as a satisfactory selective medium for *C. immitis* (Thompson et al., 1977). The antibiotic-treated material can be centrifuged and the sediment inoculated into culture media or test ani-

TABLE 1. Utilization of carbon and nitrogen compounds by *Coccidioides immitis*

Carbon sources	*Hyphae*[a]	*Endospores*[b]	*Spherules*[b]
Pentoses			
Arabinose	−	−	−
Rhamnose	ND	−	−
Ribose	ND	−	−
Xylose	+	−	−
Hexoses			
Glucose	+	+	+
Fructose	+	+	+
Mannose	+	+	+
Galactose	+	+	< +
Disaccharides			
Cellobiose	+		
Lactose	−	−	−
Maltose	+	+(low)	< +
Sucrose	−	−	−
Trehalose	+		
Trisaccharides			
Melezitose	−		
Raffinose	+(low)		
Polysaccharides			
Cellulose	−		
Dextrin	+		
Inulin	+		
Starch	+	−	−
Glucosides			
α-Methylglucoside	+		
Amygdalin	+		
Salicin	+		
Alcohols			
Erythritol	+		
Ethanol	+		
Glycerol	+		
I-Inositol	−		
Mannitol	+		−
Sorbitiol	+		
Organic acids			
Formate	−		
Acetate	+	+	> +(low)
Propionate	+		
Butyrate	−		
Caproate	+		
Lactate	+		
Pyruvate	+	+(low)	> +very low
Oxlate	−		
Tartrate	−		
Succinate	+		
Fumarate	+		
Malate	+	+	−
Gluconate	−		
Citrate	−	−	−
Nitrogen source			
Ammonium	+	+	+
Nitrate	+		
Nitrite	−		
Urea	+		
Acetamide	+		
Alanine	+	+	+
Arginine		+	+

TABLE 1. (*Continued*)

Carbon sources	*Hyphae*[a]	*Endospores*[b]	*Spherules*[b]
Asparagine	+	+	+
Aspartic acid		+	?
Cysteine (cystine)	+		
Glutamic acid	+	+	+
Glycine	+	+	+
Histidine		+(low)	+(low)
Isoleucine		+	+
Leucine		+	+
Lysine		+(low)	+(low)
Methionine		+	?
Phenylalanine		+	+
Proline		+	+
Serine		+	+
Threonine		+	+
Tyrosine	+		
Valine		+	+
Peptone	+		

[a] Baker, E. E., and C. E. Smith. 1942. J. Infect. Dis. **70:**51–53.
[b] Sippel, J. E., and H. B. Levine. 1969. Appl. Microbiol. **18:**522–524.

mals (intraperitoneally in mice, intratesticularly in guinea pigs).

Soil samples from which *C. immitis* is sought can be mixed with water, allowed to settle, and the supernatant containing arthroconidia centrifuged. The centrifuged sediment can be cultured on the above-indicated selective media and inoculated into mice after pretreatment with antibiotics as above.

With a pour-plate technique (sputum placed in the bottom of a petri dish and yeast extract agar poured over it), *C. immitis* was readily recovered despite the presence of abundant *Candida albicans,* whereas *surface* streaking permitted *Candida albicans* to overwhelm the *C. immitis* (Omieczynski et al., 1965).

If it is not possible to culture the sample immediately, freezing in dry ice for up to 72 h may permit recovery of *C. immitis,* although such freezing caused loss of viability of *Histoplasma capsulatum, Blastomyces dermatitidis,* and *Cryptococcus neoformans* (Thompson et al., 1977).

Preliminary Identification

Handling cultures of C. immitis *is extremely hazardous. All manipulations must be carried out in a bacteriologic hood ("glove box") that is secure for such airborne pathogens.* Slide cultures should *not* be made from specimens that might include *C. immitis.* Although it seems unlikely that *C. immitis* would be present in specimens being cultured for dermatophytes, one should be aware that *C. immitis* can cause the Dermatophyte Test Medium to become red, as do dermatophytes, within 7 days of inoculation (Jacobs and Russell, 1973). Cultures customarily produce the infectious arthroconidia in 5 days, so younger cultures pose less threat.

In the usual bacteriologic culture media (blood agar, etc.) and under conditions of free circulation of gases and humidity (i.e., in a Petri dish, cotton-plugged slant, or very loose screw-cap tube), *C. immitis* can produce faintly visible greyish-white growth in 48 to 72 h and clearly visible aerial growth appearing cottony to weblike in 96 to 120 h, although under some circumstances it is advisable to hold cultures 2 to 3 weeks before regarding them as negative (Table 2). (If the culture is enclosed in a screw-cap container that does not permit free circulation of humidity and gases with the outside air, the growth is leathery and closely adherent to the agar.) In uncontaminated cultures, the colonies may attain a diameter 15 to 20 mm with concentric rings of sparse or "loose" growth alternating with dense aerial growth. The colonies typically become granular as abundant

TABLE 2. Size of colonies of *C. immitis* on various media after inoculation with patient specimen

	Diameter (mm) at indicated time				
Medium	*20 h*	*48 h*	*72 h*	*96 h*	*120 h*
Blood agar	0–1	1–4	2–9	4–16	5–18
Brain–heart diffusion	0–1	1–4	3–9	4–16	6–20
Trypticase soy agar	0–1	1–4	2–6	4–16	4–17
Sabouraud's glucose agar	0	1–3	1–6	2–12	3–15
Cornmeal agar	0	0–1	1–3	2–4	2–6

After Creitz and Puckett, 1954.

arthroconidia form, the color becoming faintly cream colored and the underside tan to brown. However, the colonial morphology among different isolates can be sufficiently variable that no clearcut identification should be attempted from the gross appearance. While Huppert et al. (1967b) described colonies of pink, lavender, yellow, buff, as well as the tan-brown of "typical" colonies, the former three are unusual. Under suitable conditions, the formation of the typical chains of arthroconidia can assist in providing a presumptive diagnosis, but the resemblance of other arthroconidium-producing fungi (e.g., *Arachniotus, Auxarthron, Malbranchea, Pseudoarachniotus*) requires that definitive identification be carried out. Mistaken identity has occurred with contaminants found in cultures of CSF as well as pulmonary specimens. This could lead to unnecessary and hazardous therapy. *Coccidioides immitis* grows in the presence of cycloheximide (0.4 mg/ml), which inhibits many saprophytes, so growth in medium containing this antibiotic should be considered as possibly *C. immitis* until the identity is established.

Definitive Identification

Animal Inoculation

If laboratory mice or guinea pigs are available, they may be inoculated with a water or saline suspension prepared from a culture that has been allowed to grow for more than 5 days to permit formation of arthroconidia. (The yield of conidia is greater on 1% glucose–0.5% yeast extract–2% agar than on Sabouraud's or cycloheximide-containing medium). A suspension of the arthroconidia that has just visible turbidity contains some 10^6 cells per ml. Inoculation of 0.5 to 1.0 ml intraperitoneally into three or more mice or intratesticularly in male guinea pigs should permit formation of diagnostic endosporulating spherules within a week. One mouse can be sacrificed at 7 days, the others at 14 and 21 days respectively or examined at the time of spontaneous death. The abdominal viscera and lungs can be examined for the presence of nodular lesions. The guinea pig testis can be aspirated at these times. Microscopic examination of lesions or pus can be carried out in KOH-treated specimens or "squash" preparations of lesions in lactophenol–cotton blue.

In Vitro Conversion to Spherules

This provides a means of obviating the use of animals. It is important to have a culture with arthroconidia present. The arthroconidia can be inoculated into a synthetic medium [that of Converse (1957) as modified by Levine et al. (1960)] either in liquid form static at 40°C (Roberts et al., 1970), or solidified with 1% Ionaagar no. 2 (Brosbe, 1967; Sun et al., 1976). Incubation should be at 40°C in an atmosphere of 20% CO_2–80% air or that supplied by a candle jar. Endosporulating spherules usually are produced in 3 to 5 days by recently isolated cultures of *C. immitis*.

The method appears very useful in the hands of those with considerable practice with the method. Roberts et al. (1970) reported that all of 201 isolates of *C. immitis* produced endosporulating spherules.

Exoantigen Identification

A culture suspected of being *C. immitis* can be grown on Sabouraud–glucose–agar slant until colonial growth reaches at least 15 × 30 mm (this may take 7 to 10 days). The slant is then covered (*very cautiously*) with 8 to 10 ml of 1:5,000 merthiolate in phosphate-buffered saline, pH 7.2. The merthiolate solution serves as a fungicide as well as solvent (Standard and Kaufman, 1977). The culture is allowed to stand in the presence of the merthiolate for 24 h at room temperature. The liquid (5 ml) is aspirated and concentrated 25× in an Amicon Minicon B 15 concentrator. The concentrate is then tested by immunodiffusion with a known reference antigen and anti-*C. immitis* antibody, particularly one directed at the heat-stable HS antigen (Kaufman et al., 1985). While it is desirable to have a pure culture for this test, it was possible to identify *C. immitis* in the presence of contaminants (DiSalvo et al., 1981). The exoantigen test should be able to supplant animal inoculation as well as in vitro spherule formation in the identification of *C. immitis*.

Immunoserologic Aspects

Nature of Antigens

Coccidioidin represents a culture filtrate or autolysate of hyphal phase *C. immitis*. As customarily prepared for use in skin tests, the coccidioidin has been produced by growth of many strains of the organism for 8 to 16 weeks in static culture in a synthetic, nonantigenic asparagine medium. During this long incubation, autolysis takes place. The same type of coccidioidin also can be used as a serologic antigen. In recent years consistently satisfactory coccidioidins for serologic tests have been produced by toluene-induced lysis of young (3-day) hyphae grown in glucose–yeast extract broth (Pappagianis et al., 1957, 1961), or from the culture medium itself (Huppert and Bailey, 1965a).

Spherulin is produced by permitting washed spherules and endospores produced in vitro in synthetic medium by a single strain to undergo spontaneous lysis in distilled water over 40 days (Levine et al., 1969; Scalarone et al., 1973). Serologic antigens can be released in 3 days by toluene-induced lysis of spherules and endospores. Spherulin appears to offer no advantage over coccidioidin in serologic tests; indeed, it shows greater cross reactivity (Huppert et al., 1977).

Both hyphal and spherule/endospore antigens are complex mixtures of proteins and polysaccharides. In recent years progress has been made in distinguishing some of the components responsible for specific serologic reactivity (Calhoun et al., 1986; Cox et al., 1984; Huppert et al., 1978, 1979; Zimmer and Pappagianis, 1986a).

Temporal Sequence of Host Response

Three kinds of response are noted in patients with nondisseminating, acute coccidioidomycosis: development of delayed cutaneous hypersensitivity (DTH) to coccidioidin and/or spherulin, production of immunoglobulin M (IgM) antibody, and, later, of IgG antibody(s). In the careful studies of Charles E. Smith and colleagues, DTH appeared to develop before the IgM antibody [detected by tube precipitin (TP) test] (see Fig. 4). However, for various reasons—perhaps different skin testing techniques or improved sensitivity of serologic methods—the skin test is noted in some patients to be negative at a time when the early antibody is detectable.

Cutaneous Hypersensitivity

Skin tests may be used to assist in the diagnosis, prognosis, or epidemiologic aspects of coccidioidomycosis.

The test is conducted by the intradermal injection of 0.1 ml of coccidioidin or spherulin through a 27-gauge needle, usually in the volar aspect of the forearm. Initially 1 : 100 coccidioidin or "usual-strength" spherulin can be used, with readings made at 24 and 48 h or, if a single reading must be made, at 30 h. Induration greater than 5 mm is regarded as positive, but if the reading is less than 10 mm it is advisable to repeat the test with 1 : 10 coccidioidin or "high-strength" spherulin. Although this has not been systemically studied, it would be advisable to test also with diluent (saline) containing merthiolate 1 : 10,000, which is used as a preservative for the antigens, since some individuals manifest allergy to the merthiolate. A positive skin test provides evidence of prior or current coccidioidal infection, but only demonstration of conversion from negative to positive may provide evidence of a recent coccidioidal infection. A reaction to coccidioidal antigen provides presumptive diagnostic assistance in the case of a pulmonary nodule in a young nonsmoker who has been in an endemic area and who does not react to purified protein derivatives (PPD). Failure to react may result from the skin test being administered too soon after onset of infection or from anergy, which occurs in many patients with disseminated coccidioidomycosis. However, failure to react may occur in nondisseminated disease. For example, a substantial portion (15 to 20%) of patients with a single pulmonary cavity or nodular lesion fail to react (Winn, 1968), but this does not have the connotation of immune deficiency and threat of dissemination. Return of reactivity in the anergic patient with disseminated disease represents a favorable sign.

If a patient previously (e.g., 1 year earlier) shown

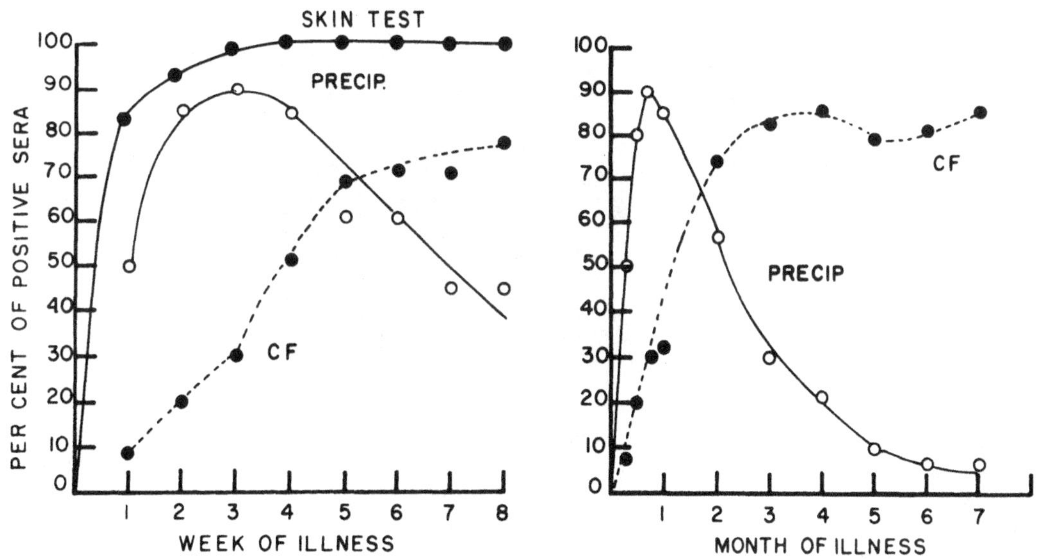

FIG. 4. Temporal sequence of immunoserologic response in primary nondisseminating coccidioidomycosis. (Reprinted from Pappagianis, D. 1980, with the permission of Plenum Medical Book Company, New York, NY.)

to be reactive by coccidioidal skin test now is ill but with negative serology (see below), it is unlikely that the recent illness is coccidioidal. (Exception: the patient could have a symptomatic pulmonary cavity or immunosuppressed state with serology negative.)

Despite the limited usefulness of skin testing in clinical situations involving individual patients, skin tests for epidemiologic studies yield significant data on the distribution and incidence of coccidioidomycosis. *The usual clinical application of coccidioidin or spherulin does not induce skin test reactivity* nor antibody. However, in a study that involved multiple tests with both antigens simultaneously, some individuals developed dermal reactivity, but it is not clear whether this was reactivity to merthiolate and whether it was durable reactivity (Galgiani et al., 1986).

In vitro lymphocyte stimulation by coccidioidin or spherulin is a correlate of skin test responsiveness (Hicks et al., 1986); however, it does not yet appear to represent a practical replacement of skin tests.

Serologic Response

Although there were earlier efforts to utilize serologic tests, not until the 1930's were these brought to consistent usefulness by Charles E. Smith. He demonstrated that patients' sera early in their disease contained an antibody, precipitin, that yielded a precipitate when mixed with coccidioidin in a tube (TP, tube precipitin test) (Smith et al., 1956). Subsequently a corresponding reaction could be demonstrated with precipitin-containing serum and antigen in agar gel double diffusion (ID for immunodiffusion). The responsible antibody has been shown to be IgM (Pappagianis et al., 1965; Sawaki et al., 1966). Latex particles coated with autolysate–coccidioidin agglutinated in the presence of precipitins. Smith also demonstrated that some patients' sera contained antibody that fixed complement in the presence of coccidioidin. This fixation permitted quantitation of antibody. Immunoglobulin G, (IgG) corresponding to the complement-fixing (CF) antibody was found to yield a band of precipitate in the double-diffusion test. However, the historical use of the term "precipitin" should prompt care in its use so that it is clear to the clinician which antibody is meant when the results of the immunodiffusion test are given (Fig. 5) (Pappagianis, 1980b).

Detection of Precipitin IgM

Tube Precipitin. As indicated above, the precipitin antibody(s) can be detected by several methods. The pertinent antigen(s) retain activity after being heated at 60°C for 30 min.

The TP test as described by Smith et al. (1956) entails placing 0.2 ml of serum in each of four 7 × 75

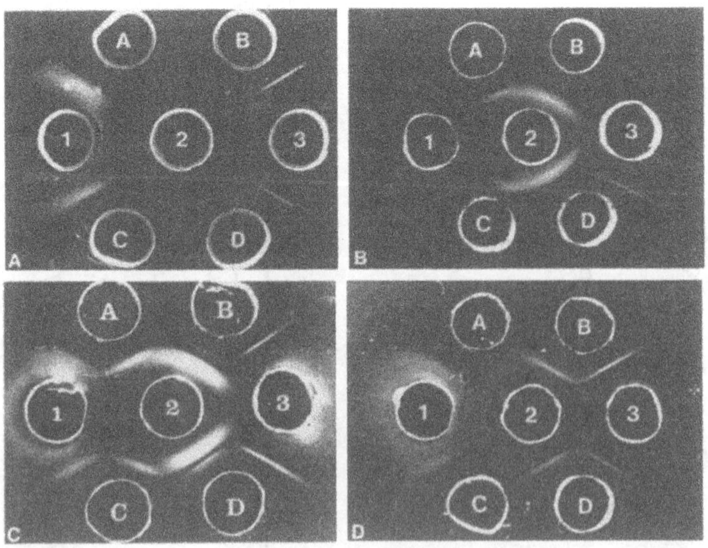

FIG. 5. Sequence of serologic reactivity as indicated by immunodiffusion. The top and bottom left-hand wells (A,C) contain previously heated (60°C) coccidioidin; top and bottom right-hand wells (B,D) contain unheated coccidioidin. Left-hand well contains control precipitin (IgM)-positive human serum to detect IDTP reaction. Right-hand well contains control complement fixation (IgG)-positive serum to detect IDCF reaction. The center well contains patient's test serum, previously concentrated. (A) Negative serum in center well. (B) Center well contains serum from early acute coccidioidomycosis that is IDTP-positive, forming line of identity with precipitin (IgM) control. (C) Center well contains serum from later stage of disease that is both IDTP- and IDCF-positive. (D) Center well contains serum from even later stage of coccidioidomycosis and shows only IDCF reactivity.

mm test tubes. To one (control) tube is added 0.2 ml uninoculated culture medium in which coccidioidin is prepared; to the three other tubes are added respectively 0.2 ml of undiluted, 1 : 10, and 1 : 40 antigen dilutions of coccidioidin selected by comparison with a reference coccidioidin. Smith et al. (1956) further recommended that only the undiluted and 1 : 10 dilutions sufficed to provide optimal ratios of antigen to antibody with most sera. The antigen–antibody mixtures are incubated at 35° to 37°C and examined daily for 4 days, although most positive sera will yield a reaction in 24 h, rarely requiring more than 2 days. A positive reaction is indicated by the presence of a translucent, almost gelatinous button dislodged by flicking the tube sharply. The TP test is a *qualitative* test without significance as a quantitative test, although the diminution and disappearance of the TP antibody may accompany healing of the primary coccidioidal infection.

Latex Agglutination. The development of the latex particle agglutination (LA) provided another *qualitative* test for the detection of TP antibody. Originally, polystyrene latex particles 0.15 to 0.35 μm in diameter were used (Mulder et al., 1966); more recently these have been 0.8-μm particles. The particles are exposed to antigen in culture filtrate or toluene-autolysates derived from multiple strains of *C. immitis* grown in glucose–yeast extract broth. The culture filtrate is heated at 60°C to inactivate CF antigen. The test also calls for heating the serum to inactivate complement. Agglutination by antibody is read at 4 min.

The LA is more sensitive than the TP test, but owing to occurrence of "false-positive" reactions in 6 to 10% of sera, a confirmatory test for the presence of coccidioidal antibody is desirable (Huppert et al., 1968). Because of unacceptably high rates of false-positive reactions, the LA test should not be used with CSF or sera that have been diluted (Pappagianis et al., 1976).

Immunodiffusion. Immunodiffusion was found to yield a band of precipitation when coccidioidin was diffused toward serum from a patient with coccidioidomycosis (Heiner, 1958). With various fractions derived from coccidioidin and human patients' sera, it was found that multiple bands could be detected (Pappagianis et al., 1961). Refinement of the test by Huppert and Bailey (1965a,b) showed that the antibody corresponding to TP antibody could be detected in addition to that corresponding to the CF antibody. As was recognized earlier for TP and CF antigens, heating at 60°C for 30 min eliminated the latter so that one could have antigen only for the detection of antibody involved in TP.

The method applied by Huppert and Bailey (1965a) made use of 5 ml agar in 50-mm diameter plastic petri dishes, which yielded a gel layer of agar 2.8 to 3 mm in depth. Wells 4.5 mm in diameter were cut in the agar with a wall-to-wall distance between wells of 5 mm. The wells could accept 40 to 50 μl of antigen solution or serum (in our experience, the volume is more nearly 40 μl). Various adaptations have been devised that permit use of smaller volumes of reagents, more specimens, and different shapes of dishes. The optimal concentration of antigens is determined by testing serial, twofold dilutions of antigen solution against reference positive serum reactive by TP (that would be included as reference antigen in the ID test). This would permit detection of the most intense band of precipitation, but also one that was placed approximately midway between antigen and serum wells for ease of reading. In our recent experience, antigen suitable for the IDTP test has a protein concentration of about 50 μg/ml and carbohydrate (as mannose) of about 1,250 μg/ml (or 2 and 50 μg, respectively, per well).

We continue to follow the recommendation of Huppert and Bailey (1965a) in prediffusing serum or other fluids for 2 h prior to addition of antigen solution to the apposed well. This is particularly helpful with the IgM antibody to permit this more slowly diffusing immunoglobulin to reach a midpoint reaction site with the more rapidly diffusing antigen(s).

We have made the following modifications that we believe enhance the sensitivity and readability of IDTP (and IDCF) reactions: 1) Serum, CSF, or other fluids are concentrated eight- to ten-fold by evaporation. CSF can readily be concentrated ten-fold, but serum is too viscous to achieve that degree of concentration. Concentration is carried out in a Savant Speed-Vac®, which makes use of a vacuum pump attached to a warmed centrifuge containing the sera, through a cold trap (though other means of concentration can also be used). It is readily demonstrable that such concentration enhances the sensitivity of the ID test. 2) We have substituted 1% Gel-Rite, a heteropolysaccharide gel material from *Pseudomonas* (Shungu et al., 1983), which provides a clearer gel and is less expensive than agarose at the appropriate concentration (Pappagianis et al., 1985a). 3) The same concentrated body fluid is tested simultaneously in the same ID plate for both IDTP (IgM) and IDCF (IgG) reactivity (Fig. 5). This entails the use of positive control IDTP and IDCF sera in independent wells, and heated and unheated coccidioidin respectively, to allow detection of lines of identity corresponding to the two antibodies. Huppert and Bailey (1965b) had proposed a blend of optimal dilutions of IDTP and IDCF into a single preparation. However, we have found that with some sera containing CF antibody at a low titer, for example, 1 : 2 to 1 : 8, the IDTP reaction obscures the IDCF band—this is demonstrable by treating with a solution of 0.05 *M* dithiothreitol (in 2 *M* Tris–HCl buffer, pH 8.6), which damages the IgM, leading to dissolution of the IDTP reaction band and revealing an IDCF band. In the above-described ID method, the earliest detectable reaction is evident in 6 h, but usually fully developed bands of precipitate are evident at 24 h, infrequently only after incubation for 48 or 72 h.

A variation of ID, counterimmunoelectrophoresis (CIE) has provided a more rapid reaction evident in 30 to 90 min. Although this has yielded reactions comparable to IDTP, its application has been more thoroughly studied for the IDCF reaction (Aguilar et al., 1976; Shifrine et al., 1975).

Detection of IgG (CF, IDCF, CIE)

Complement Fixation. The demonstration by C. E. Smith that (some) patients' sera fixed complement in the presence of coccidioidin was extensively evaluated (Smith et al., 1956). The method was similar to the Kolmer method in use for syphilis serology. It involves the mixing of 0.25 ml of various dilutions (starting with 1:2)* of serum or other body fluid previously inactivated at 56°C with 0.25 ml of antigen and 2 units of complement in 0.5 ml. This mixture is incubated for 2 h at 37°C followed by addition of the hemolytic indicator (2% vol/vol sheep erythrocytes "sensitized" with 2 units of anti-sheep cell hemolysin) and incubation at 37°C for an additional hour. The batch and dilution of antigen are selected to be free of anticomplementary activity; however, an anticomplementary control, that is, serum plus complement but without antigen, is essential to evaluate the CF activity of the serum. The highest dilution of serum giving complete "4+" (100%) inhibition of hemolysis gives the titer. The reaction may be read 3+, 2+, or 1+ depending on the extent of hemolysis.

It is important to freeze a positive serum so that it may later be compared *simultaneously* with a subsequent serum from the same patient, inasmuch as some slight variations can be expected in view of the several components involved in the CF test. If a serum yields reactions of 4+ at 1:2, 1:4, 1:8 and 3+ at 1:16, and if a later, simultaneously tested serum from the same patient yields 4+ at 1:2, 1:4, 1:8 and 2+ at 1:16, this does not represent a significant change. A fourfold ("two-tube") or greater difference in titer of simultaneously tested sera is considered a significant change. A 3+ reaction at a 1:2 dilution, but no fixation of complement at 1:4 or higher dilutions, represents an equivocal reaction; on the other hand, if the ID test shows that the serum contains IDCF, the 3+ reaction at the 1:2 dilution may well represent a low level of CF antibody. The importance of concomitant qualitative ID and quantitative CF results has been emphasized by Kaufman and Clark (1974).

Most of the thousands of clinical correlations with CF results were made by Smith et al. (1956), using the 2-h binding of complement at 37°C, and for this reason we have continued to make use of this method for testing sera. However, Smith et al. (1956) demonstrated that overnight (18-h) binding of complement at 4°C provided a more sensitive test, that is, one yielding higher titers than when the same sample was subjected to complement binding for 2 h at 37°C; and this prompted us to adopt this binding of complement in the cold in routine testing of cerebrospinal fluid. It yielded an increase in percentage of positive specimens in cases of coccidioidal meningitis from 75% (Smith et al., 1956) to 95% (Pappagianis and Crane, 1977).

Several other CF methods have been applied, including the microtiter method that makes use of less serum and reagents than the above-described Kolmer method. Some of these, such as the Centers for Disease Control Laboratory Branch Complement Fixation test (LBCF), have had extensive use and appear reliable, but these too should be carried out with simultaneous testing of paired sequential sera from the same patient.

Some sera show "zoning," that is, incomplete fixation of complement in lower (and sometimes all) dilutions. This is usually associated with infections that are severe. While the endpoint is sometimes not sharp, comparison of sequential sera can permit judgment as to changing titers. Some sera are anticomplementary (AC), that is, they appear to interact with and activate or damage complement even in the absence of antigen and therefore cannot yield interpretable CF results. There is no single or clear explanation. However, some specimens, for example, canine sera, frequently show AC activity. In such instances confirmation of the presence of CF antibody can be carried out with immunodiffusion.

IDCF. Antigen satisfactory for this procedure can be prepared by toluene lysis of young hyphae or from culture filtrate (see Nature of Antigens, above). Because the CF-reactive antigen is heat labile, unheated antigen must be used, and the presently available crude preparations also include antigen reactive with precipitin (IgM). However, by appropriate titration of the antigen with known CF-positive sera of different titers, preferably but not obligatorily lacking the IgM antibody, one can select an antigen that

* For the past several years we and others have tested CSF starting with undiluted rather than the 1:2 dilution. This yields more positives, but its significance as an indicator of meningitis has not been adequately assessed, particularly in view of recognition that concentrated CSF may yield positive reactions by ID even in the absence of meningitis (Pappagianis et al., 1972).

will yield a line of identity with control CF antibody in the patient's serum. Prediffusion of sera for 2 h as described above for the IDTP is useful in the IDCF, although others have found the prediffusion unnecessary for the IDCF reaction. We have retained the prediffusion step to allow for the optimal detection of IgM as well as the CF (IgG) antibody in a single, gel-diffusion plate.

The IDCF test with *concentrated* serum or other body fluids has yielded positive reactions with many sera nonreactive by CF (Pappagianis et al., 1986). The reverse—that is, positive by CF, negative by IDCF—has occurred so rarely and in instances not confirmed to be due to coccidioidomycosis, that we feel a negative IDCF virtually excludes the possibility of positive CF. The IDCF has been proposed (along with the LA) as a screening test. We feel that with concentrated serum the ID test alone serves as a sensitive indicator of the presence of CF and precipitin (IgM) antibodies. Antigen preparations for ID are available from several commercial sources. Johnson et al. (1985) have shown wide variation in sensitivity and caution that such commercial kits should be evaluated against a known reference system.

Quantitative IDCF. The good correlation between CF and qualitative IDCF led to efforts by several groups to adapt the IDCF for quantitative use, prompted by the perceived complexities of the CF test. Huppert et al. (1967a) made use of wells 4.5 mm in diameter, separated from one another by 5 mm. Sera were prediffused for 2 h prior to addition of antigen to wells. Their results indicated 80% correlation of titers with the CF titers; 20% differed from the CF titer by more than one tube (twofold dilution). Hampson and Larwood (1967) found a 93% agreement within one-serial dilution between CF and quantitative IDCF. An example of a quantitative immunodiffusion test is shown in Fig. 6.

Wieden et al. (1983) carried out the quantitative IDCF by a method similar to that of Huppert et al. (1967a) and concluded that they could increase the agreement within a twofold dilution of the CF titer (determined by the Kolmer test with 2-h binding of complement at 37°C) by refilling the serum and antigen wells at 24 and 48 h. Readings were made at the customary 72 h. Johnson et al. (1985) found 82% and 87% correlation within one serial dilution of the titer obtained by LBCF.

The quantitative IDCF using appropriate simultaneous testing of successive "paired" sera may provide satisfactory titers. Although the quantitative IDCF has been described as less expensive than the CF (Wood et al., 1982), the refilling of serum and antigen wells used to enhance the correlation of CF and quantitative IDCF titer (Wieden et al., 1983) adds labor, cost, and time in view of a recommended final reading at 72 h. This contrasts with the 4 to 6 h to complete the Kolmer-style CF test. The quantitative IDCF, however, may be practical in a laboratory that carries out few serologic tests and does not wish to await CF results from a reference laboratory. Even the laboratory that routinely carries out the CF test may resort to quantitative IDCF when sera show AC activity or in the rare case when the serum that shows a strong qualitative IDCF reaction fails to fix complement.

Counterimmunoelectrophoresis. A few studies have been carried out to examine the usefulness of counterimmunoelectrophoresis in detecting coccidioidal antibody. Spherulin and coccidioidin have been used (Aguilar et al., 1976; Shifrine et al., 1975). In two studies, the nature of the antigens tested is not clear. Kozub et al. (1977) referred to a "toluene extract of culture filtrates," whereas Graham and Ryan (1980) presumably compared spherulin with coccidioidin but cited as descriptive of their coccidioidin a paper about "spherule coccidioidin," that is, spherulin. Aguilar et al. (1976) demonstrated 98% correlation between CIE and CF and 100% correlation with ID in detection of antibody. Mean titers by CIE appeared to be one serial dilution lower than CF titers.

Kozub et al. (1977) found that when CF titers were 1 : 16 or higher, CIE was 100% sensitive, but at

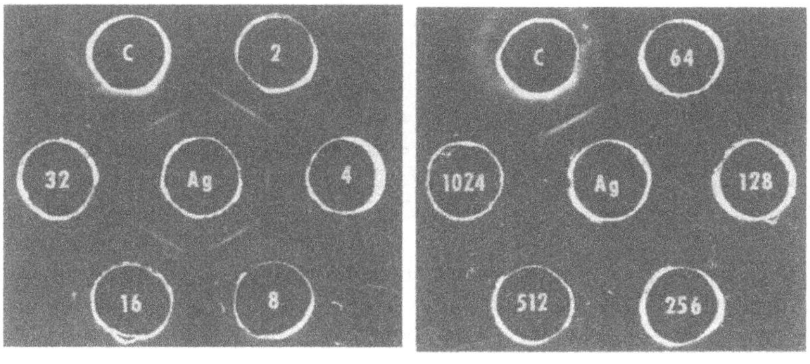

FIG. 6. Quantitative immunodiffusion test. Well C contains control CF (IgG)-positive serum. Ag is unheated coccidioidin. Wells 2 through 1024 represent dilutions of a patient's serum (e.g., 1 : 2, 1 : 4 through 1 : 1024). A bending line of identity with the control shows that the IDCF titer is 1 : 64. This patient's serum also contained IDTP reactivity evident (reaction closest to serum wells) at dilutions of 1 : 2 through 1 : 8.

lower titers the correlation of positive CIE with CF was, respectively, CF 1 : 2, CIE 64% positive; CF 1 : 4, CIE 76% positive; CF 1 : 8, CIE 91% positive. Unfortunately, no additional information was provided as to the validity of *the CF results* in those specimens that yielded no reaction by CIE. In testing canine sera, all that were positive by CF (titers from 1 : 2 to 1 : 128) were also positive by CIE (Shifrine et al., 1975).

Additional Serologic Methods

The *direct fluorescent antibody* has been used to demonstrate *C. immitis* in infected tissue (Kaplan and Clifford, 1964). The reactive antibody was prepared by injecting live *C. immitis* or killed arthroconidia into rabbits. The globulin fraction was precipitated from the rabbit sera and tagged with fluorescein isothiocyanate. The conjugate produced fluorescent staining of endospores and of the contents of spherules in infected tissues, but not of the walls of the spherules. The walls of spherules produced in vitro did take up fluorescent antibody. This appears to represent a sensitive histologic method.

Fluorescent antibody inhibition was used to detect antibody in human sera. Thus, serum from the patient was mixed with fluorescent-labeled rabbit anti-human globulin, then allowed to react with endospores from infected mouse tissue. The presence of antibody in the patient's serum would lead to fixation of fluorescent stain to endospores, and the intensity of this fluorescence would be compared with that of endospores exposed to normal human serum.

An indirect test for detection of antibody is the *agar-gel precipitin inhibition* test. Serial dilutions of the patient's serum are mixed with an appropriate dilution of antigen to permit antigen–antibody reaction, and this mixture then is diffused against a standard antibody-containing serum. The presence of antibody in the patient's serum would bind antigen, thus precluding reaction with the standard serum proportional to the titer of antibody in the original serum. The endpoint is read at 48 h (Ray, 1967).

Application of Serologic Tests to Fluids Other Than Serum

The *serum* should be tested in *all* cases of suspected or confirmed coccidioidomycosis. However, the ID and CF tests can also be applied to other body fluids. For example, CSF, pleural, peritoneal, and joint fluids can be tested when infection of the corresponding compartment is suspected. Some caution is needed in the application of various tests. The LA test proved to be unsuitable for CSF (and has not been evaluated for the other fluids) (Pappagianis et al., 1976). The ID test is useful for detecting antibody in these fluids. In the case of CSF, it has been shown that the IDCF reaction can be obtained in some samples of CSF that have been concentrated even in the absence of coccidioidal meningitis, probably as a result of small amounts of antibody passing from the serum into the CSF (Pappagianis et al., 1972). (In the absence of a serum specimen, CSF reactive by ID indicates the presence of a coccidioidal infection.) The TP and IDTP reactions are infrequently seen with CSF, but when present indicate recent coccidioidal meningitis. Fixation of complement by CSF at the 1 : 2 dilution or higher *usually* indicates the presence of coccidioidal meningitis (fixation of complement by *undiluted* CSF may also indicate meningitis but that has not been established with certainty). While the initial CSF may be negative by CF in some 25% of cases, later specimens are positive in 95% of cases (Pappagianis and Crane, 1977). Fixation of complement by the CSF may also be seen in cases of juxtadural lesions, for example, vertebral abscess. However, in meningitis, the CSF glucose is depressed, and the cells and protein are elevated, whereas the juxtadural lesion usually is accompanied by elevated CSF protein but normal glucose and cells. Contamination of the CSF with blood having a high CF titer could possibly yield a CF titer in the absence of meningitis. This adds to the importance of the laboratory worker knowing the CSF glucose, protein, and cell values.

The CSF may be obtained from the lumbar or cervical subarachnoid space, cisterna magna, or lateral ventricles. As the CSF protein, glucose, and cell levels may differ from one compartment to another, so can the antibody levels vary. For that reason, when one is comparing changes in CF titers, successive specimens from the same anatomic compartment must be compared. Indeed, the ventricular fluid may have normal glucose, protein and cells and be negative by CF despite active basilar meningitis, reflected by abnormalities of the cisternal or lumbar fluid (Goldstein et al., 1972). Fixation of complement by the ventricular fluid indicates either a block to the outflow of CSF or a ventriculitis–ependymitis. A decrease in the CF titer indicates improvement of coccidioidal meningitis. Persisting fixation of complement by CSF indicates persisting meningitis even if glucose, protein, and cells have become normal.

Ordinarily the presence of antibody in the fluid of the pleural, joint, or peritoneal space indicates infection involving those areas. The CF titer is usually lower than that of the serum, but occasionally may be higher. For example, if the meninges represent the only site of metapulmonary dissemination, the CF titer of the CSF may be greater than that of the serum.

It is important to test the *serum* periodically as well as the other body fluids. For example, in a meningitic patient with a ventriculo-peritoneal or ventriculo-atrial shunt, a rising CF titer (with or without reappearance of TP or IDTP reactivity) in the serum indicates systemic reinfection from the meninges.

With the modified Kolmer method utilized by C. E. Smith and continued by us, the critical titer *above which* one has to be concerned with possible metapulmonary dissemination is 1 : 16 (this can vary with the CF method). However, some patients, for example, with *limited* extrapulmonary spread may have low serum CF titers (e.g., 1 : 2 to 1 : 8). Even patients with potentially lethal coccidioidal meningitis may have low *serum* CF titers. CF titers may be low in early acute coccidioidal infection or with chronic pulmonary coccidioidomas or cavities. Any CF titer is significant, particularly if coccidioidal antibody is confirmed by ID. Changes in CF titer generally reflect changes in the clinical condition of patients, decreasing titer indicating improvement, increasing titer worsening. Low CF titers may persist for years in patients who have residual pulmonary cavities.

Clinical Significance and Application of Serologic Findings

Detection of precipitin (IgM) usually precedes that of the CF (IgG)-type antibody. Indeed, the IgM may be the only antibody detectable that can confirm an early, acute, primary coccidioidal infection. Therefore, the appropriate test must be the LA, TP, or ID test, the latter making use of heated antigen and concentrated serum for appropriate sensitivity. Some commercial ID kits are designed and standardized primarily to detect coccidioidal CF (IgG)-type antibody by use of unheated antigen and often will not reveal the presence of the IgM. Thus, they may well be negative in early coccidioidomycosis and mislead the physician. The IgM antibody is infrequently detected in the CSF, but its presence is associated with recent meningitis.

The IgM antibody may persist or seemingly reappears in the serum in some circumstances: in association with pulmonary cavities or bronchopleural fistulae resulting from such cavities, occasional cases of disseminated coccidioidomycosis, and when systemic reinfection occurs from the meninges by way of a ventriculo-peritoneal or -atrial shunt. The latter may provide a clue that there is need for resumption of systemic therapy (in addition to intrathecal). The presence of IgG indicates prior or current infection. In the absence of IgM, the IgG indicates the likelihood that the coccidioidal infection was acquired at least several weeks earlier; however, IgG detected in concentrated serum by ID has been detected several years after the primary coccidioidal infection has resolved.

Serologic tests may be negative with residual coccidioidal pulmonary lesions (coccidioidomas, cavities) that have been present for prolonged periods. Whereas Smith et al. (1956) reported the apparent absence of antibody (titer less than 1 : 2) by the 2-h CF test in 40% of patients with cavities and in 70% of patients with solitary nodules, the greater sensitivity of the IDCF test with concentrated sera than of the CF test (Pappagianis et al., 1986) indicates that there are proportionately fewer negative sera with these types of chronic, coccidioidal, pulmonary lesions than was previously thought. Though few histologically confirmed coccidioidal lesions were accompanied by negative (concentrated) ID, the precise proportion has not yet been established.

The titers of CF (and quantitative IDCF) antibody appear roughly proportional to extent (severity), and CF titers above 1 : 16 should alert the physician to the possibility of disseminated coccidioidomycosis. However, in limited metapulmonary involvement, the serum CF titer may be low, e.g., less than 1 : 8, even with potentially lethal coccidioidal meningitis (Fig. 7) (see above, for testing of fluids other than serum).

Once the presence of coccidioidal antibody has been established, repetition of the *qualitative* IDTP or IDCF is not likely to yield additional, clinically useful information. The IDCF may remain positive for months or even years after recovery; but reemergence of a positive ID in a serum that had become negative can have clinical significance, for example, relapse of coccidioidomycosis in an immunocompromised individual. Once the presence of antibody is established, sequential sera should be tested primarily for CF titer and changes therein. Sequential sera should be tested at 3- to 4-week intervals, since it is unlikely that significant changes in CF titer will occur in less time. When the clinical status of the patient appears stable, the interval between serologic tests can be extended. Serum (or CSF, etc.) *should be saved frozen* so that it can be tested with a later specimen from the same patient. A fourfold increase or decrease in titer when successive sera *are tested simultaneously* is indicative of worsening or improvement respectively.

Although the foregoing guidelines have largely been derived from experience with human coccidioidomycosis, the principles are applicable to non-human patients with coccidioidomycosis, for example, primates, dogs, or horses. Serologic experience has been limited with other species, but the tests appear useful for several species, including sea otters and llamas. In general, human IgM and IgG coccidioidal antibodies serve as satisfactory positive controls in ID tests, forming lines of identity with antibodies produced in other species. The llama poses an exception in that some immunodiffusion bands do not correspond precisely to the lines of reaction of human

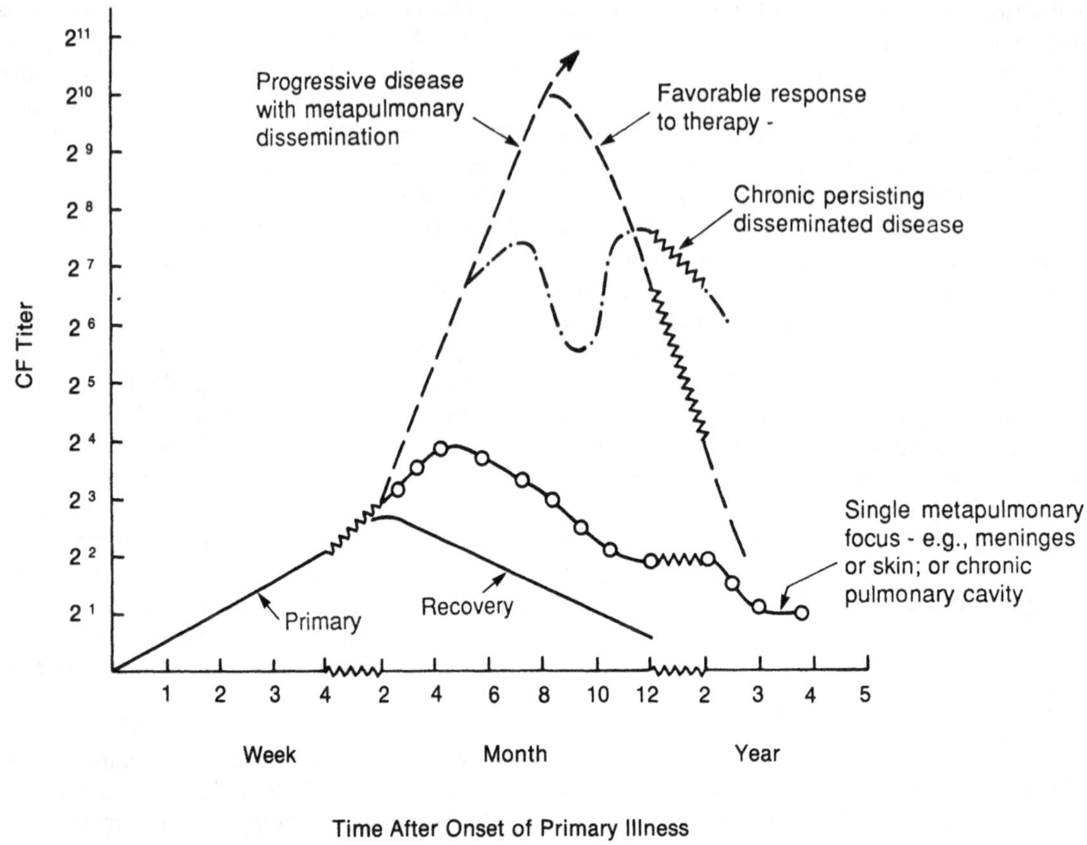

FIG. 7. Patterns of serum complement fixation (CF) titers with various clinical phases of coccidioidomycosis, showing the relationship of CF titer to extent of disease. (Adapted from Campbell 1960; with the permission of the author and the New York Academy of Sciences.)

sera. Latex agglutination, TP, and CIE have been reactive with canine sera in patterns similar to those observed with human sera (Shifrine et al., 1975; Yturraspe, 1971).

The sensitivity and specificity of coccidioidal serology make it virtually indispensable in the diagnosis and management of patients, but isolation or histologic demonstration of *C. immitis* should be sought where practicable both for diagnostic confirmation and progress during therapy. Rarely, reactions that resemble those produced by coccidioidal TP antibody are encountered with sera from patients in whom *C. immitis* or known cross-reacting fungus is not found. Such reactions, usually more or less atypical, necessitate thorough histopathologic and microbiologic review of the patient.

Caution is advised when serologic testing is carried out on the immunocompromised patient. For example, of 22 patients with Acquired Immune Deficiency Syndrome and coccidioidomycosis, 4 were seronegative. Patients immunosuppressed for kidney transplantation and certain non-Hodgkin's lymphomas, sometimes fail to give a detectable serologic response to *C. immitis*. In such instances antigen detection may prove to be useful (Galgiani et al., 1985).

Treatment, Prevention, and Control

Treatment

The marked variability of coccidioidomycosis necessitates consideration and application of varied therapeutic modalities.

Surgical Treatment

Excision can serve for both diagnostic and therapeutic purposes. Thus, a solid pulmonary lesion (coccidioidoma) that may be indistinguishable from a neoplasm on roentgenogram and that is accompanied by negative coccidioidal skin test and/or serologic findings may need to be removed for diagnosis, and this may be curative (though such a lesion in all likelihood poses no threat of further coccidioidal disease). Lobectomy or segmental resection may be required for the removal of a coccidioidal pulmonary cavity if there has been enlargement of the cavity with encroachment on the pleura, persistent and troubling, profuse hemoptysis, or secondary infection not responding to antibacterial antibiotics (Nelson, 1974). However, such a resection may justifiably be deferred in a diabetic patient with a cavity owing to the risk of recurrence of cavitary disease in another, for example, contralateral part of the lung in such patients; repeated resections of the lung would be undesirable in view of the potential loss of functional pulmonary tissue.

Removal of a cutaneous pustule may be curative if, as nearly as can be determined, that represents the single metapulmonary site of involvement. Drainage of abscesses, curettage of infected bone, synovectomy in coccidioidal joint disease, and arthrodesis may be required. In many instances accompanying chemotherapy may be indicated, for instance, perioperatively with synovectomy or resection of a pulmonary cavity, though some experienced thoracic surgeons have felt that amphotericin B has provided no particular benefit as an accompaniment to pulmonary resection.

Medical Treatment

Amphotericin B (AMB) and certain azoles are available. AMB still represents the most secure form of therapy, although its nephrotoxicity and occasional other toxicities (hematopoietic, hepatic) necessitate great care in its administration. It should be used for patients with rapidly progressing, disseminated infection, those very ill with pulmonary disease, and those with primary pulmonary coccidioidomycosis who are among the groups at high risk of dissemination (see above). AMB may deserve a trial in patients with bilateral fibrocavitary coccidioidomycosis, but AMB has not been significantly helpful in bringing about resolution of a coccidioidal pulmonary cavity.

AMB is still the drug of choice in the treatment of coccidioidal meningitis, for which it can be administered by intrathecal (lumbar, cisternal, lateral cervical) or ventricular routes. Initially this should be accompanied by intravenous AMB, although the latter can be discontinued if only the meninges appear to be involved.

Ketoconazole and miconazole have been used in a number of patients with coccidioidomycosis. The latter compound, which is not water soluble and must be given parenterally, has limited usefulness—essentially now only for the intrathecal or intraventricular treatment of coccidioidal meningitis when AMB cannot be given. Ketoconazole can be readily absorbed after oral administration provided gastric acidity is normal, and it lacks the major toxic effects of AMB, although it is not completely devoid of side effects. It can be used particularly in coccidioidomycosis involving soft tissues (Stevens et al., 1983). High oral doses do lead to detectable levels in the CSF, but ketoconazole as yet has not been regarded as sufficiently potent to replace AMB in treating meningitis. Ketoconazole has a suppressive effect on *C. immitis,* but even after it has been administered for several months, discontinuation of therapy is followed by relapse in a substantial proportion of patients. Itraconazole (Ganer et al., 1987) and fluconazole have had limited studies in humans and appear to be at least as effective as ketoconazole but with fewer side effects.

Other Modes of Therapy

Immunotherapy with leukocyte transfer factor has not appeared to represent an effective approach to treating coccidioidomycosis (Catanzaro et al., 1985), although the frequency of anergy in disseminated infections deserves further study of immune reconstitution. On the basis of possible adverse immunological effects of high CF antibody levels or high levels of antigen or antigen–antibody complexes (Cox et al., 1982), plasmapheresis has been proposed but not tested as a therapeutic measure (Pappagianis, 1981).

Prevention and Control

Under some circumstances, reduction of exposure to soil dust (and therefore arthroconidia) has reduced the incidence of coccidioidomycosis. However, in many circumstances, for example, in agricultural or construction work, it is unrealistic to contemplate significant reduction of dust, although machinery with enclosed air-conditioned cabs may provide a lower risk of exposure. Wetting of soil may suppress airborne dust, thereby lessening exposure, for instance, for archaeologists or anthropologists whose digging and sifting may involve limited areas and quantities of soil at a given time. Cleansing soil-contaminated artefacts such as human skeletal remains with water, heating at 60°C for 30 min, or treating with 5% hypochlorite solution for 30 min will assist in decontamination.

As indicated above, recovery from coccidioidomycosis usually provides lasting resistance to reinfection (endogenous reactivation may occur in a presumed immune individual if an immunosuppressive state supervenes). This formed the basis of development of a vaccine consisting of whole, formaldehyde-killed spherules produced in vitro. This vaccine provided protection against lethal respiratory infection with *C. immitis* in mice and monkeys (Levine et al., 1960, 1962). However, at doses considered to be tolerable for humans, the vaccine did not prevent occurrence of clinically apparent coccidioidmycosis (Pappagianis et al., 1986).

Appendix I. Calcofluor Method for Visualization of Fungi (Hageage and Harrington, 1984)

Prepare a 1% wt/vol solution of calcofluor white* by warming in distilled water. This can be stored for at least 1 year at room temperature.

For examination of body fluids, prepare a KOH–calcofluor solution by mixing 5.0 ml of the stock calcofluor solution (0.05 g) with 95 ml of distilled water, to which 10 g of KOH has been added. Add equal volumes of this solution and specimen; examine under a coverslip.

For examination of tissue, prepare an 0.1% solution of calcofluor by 1 : 10 dilution of the stock solution. To 100 ml, add 0.4 g of Evans blue. This can be applied directly to deparaffinized and rehydrated tissue sections.

* Can be obtained as Calcofluor white M2R from Polysciences, Inc., Warrington, Pennsylvania, or as Fluorescent Brightner 28 from Sigma Chemical Co., St. Louis, Missouri.

Appendix II. Coccidioides Synthetic Isolation Agar*

Compound	*Amount*
NH_4Cl	10 g
$NaOOC\text{-}CH_3$ (acetate)	10 g
K_3PO_4	8 g
$CuSO_4$ (10% solution)	4 ml
Agar	20 g
H_2O	1,000 ml

Suspend dry ingredients in water and dissolve with heat. Autoclave at 121°C (15 lb pressure for 15 min); then add cupric sulfate and pour plates with melted mixture.

* From Smith, C. E., 1943. Coccidioidomycosis. Med. Clin. North Am. **27**:790–807.

Literature Cited

Aguilar-Torres, F. G., L. J. Jackson, J. E. Ferstenfeld, D. Pappagianis, and M. Rytel. 1976. Counterimmunoelectrophoresis in the detection of antibodies against *Coccidioides immitis.* Ann. Intern. Med. **85**:740–744.

Agy, M. B., and J. L. Paznokas. 1985. Isocitrate lyase and malate synthase activities of *Coccidioides immitis*. Exp. Mycol. **9**:318–325.

Ajello, L., V. Q. Grant, and M. A. Gutske. 1951. The effect of tubercle bacillus concentration procedures on fungi causing pulmonary mycoses. J. Lab. Clin. Med. **38**:486–491.

Armbruster, T. G., T. G. Goergen, D. Resnick, and A. Catanzaro. 1977. Utility of bone scanning in disseminated coccidioidomycosis. J. Nucl. Med. **18**:450–454.

Baker, E. E., and C. E. Smith. 1942. Utilization of carbon and nitrogen compounds by *Coccidioides immitis* (Rixford and Gilchrist, 1896). J. Infect. Dis. **70**:51–53.

Baker, E. E., E. M. Mrak, and C. E. Smith. 1943. The morphology, taxonomy, and distribution of *Coccidioides immitis* Rixford and Gilchrist 1896. Farlowia **1**:199–244.

Bawdon, R. E., R. G. Garrison, and L. R. Fina. 1972. Deoxyribonucleic acid base composition of the yeast-like and mycelial phases of *Histoplasma capsulatum* and *Blastomyces dermatitidis*. J. Bacteriol. **111**:593–596.

Beaman, L., and D. Pappagianis. 1985. Fate of *Coccidioides immitis* arthroconidia in human peripheral blood monocyte cultures in vitro, p. 170–180. *In* H. E. Einstein and A. Catanzaro (ed.), Coccidioidomycosis. National Foundation for Infectious Diseases, Washington, D.C.

Bouza, E., J. S. Dreyer, W. L. Hewitt, and R. D. Meyer. 1981. Coccidioidal meningitis. An analysis of thirty-one cases and review of the literature. Medicine (Baltimore) **60**:139–172.

Brosbe, E. A. 1967. Use of refined agar for the in vitro propagation of the spherule phase of *Coccidioides immitis*. J. Bacteriol. **93**:497–498.

Brummer, E., L. Beaman, and D. A. Stevens. 1985. Killing of endospores, but not arthroconidia, of *Coccidioides immitis* by immunologically activated polymorphonuclear neutrophils, p. 201–213. *In* H. E. Einstein, and A. Catanzaro (ed.), Coccidioidomycosis. National Foundation for Infectious Diseases, Washington, D.C.

Calhoun, D. L., E. O. Oser, K. O. Dugger, J. N. Galgiani, and J. H. Law. 1986. Humoral antibody responses to specific antigens of *Coccidioides immitis*. J. Infect. Dis. **154**:265–272.

Campbell, C. C. 1960. The accuracy of serologic methods in diagnoses. Ann. NY Acad. Sci. **89**:163–177.

Catanzaro, A. 1986. Coccidioidomycosis, p. 103–123. *In* G. A. Sarosi and S. F. Davies (ed.), Fungal diseases of the lung. Grune and Stratton, Inc., Orlando, Florida.

Catanzaro, A., P. J. Friedman, R. Schillaci, R. Graybill, R. Olshen, H. Einstein, and L. Spitler. 1985. Double-blind study of transfer factor treatment of coccidioidomycosis, p. 330–338. *In* H. E. Einstein and A. Catanzaro (ed.), Coccidioidomycosis. National Foundation for Infectious Diseases, Washington, D.C.

Caudill, R. G., C. E. Smith, and J. A. Reinarz. 1970. Coccidioidal meningitis: a diagnostic challenge. Am. J. Med. **49**:360–365.

Coburn, J. W. 1962. Scalene lymph node involvement in primary and disseminated coccidioidomycosis. Evidence of extrapulmonary spread in primary infection. Ann. Intern. Med. **56**:911–924.

Cole, G. T., L. M. Pope, M. Huppert, S. H. Sun, and P. Starr. 1985. Wall composition of different cell types of *Coccidioides immitis*, p. 112–129. *In* H. E. Einstein and A. Catanzaro (ed.), Coccidioidomycosis. National Foundation for Infectious Diseases, Washington, D.C.

Creitz, J. R., and T. F. Puckett. 1954. A method for cultural identification of *Coccidioides immitis*. Am. J. Clin. Pathol. **24:**1318–1323.

Currah, R. S. 1985. Taxonomy of the Onygenales: Arthrodermataceae, Gymnoascaceae, Myxotrichaceae and Onygenaceae. Mycotaxon **24:**1–216.

DiSalvo, A. F., A. A. Terreni, and A. K. Wooten. 1981. The use of the exoantigen test to identify *Blastomyces dermatitidis, Coccidioides immitis* and *Histoplasma capsulatum* in mixed cultures. Am. J. Clin. Pathol. **75:**825–826.

Drutz, D., and M. Huppert. 1983. Coccidioidomycosis: factors affecting the host–parasite interaction. J. Infect. Dis. **147:**372–390.

Echols, R. M., D. L. Palmer, and G. W. Long. 1982. Tissue eosinophilia in human coccidioidomycosis. Rev. Infect. Dis. **4:**656–664.

Eckmann, B. H., G. L. Schaefer, and M. Huppert. 1964. Bedside interhuman transmission of coccidioidomycosis via growth on fomites. Am. Rev. Respir. Dis. **89:**175–185.

Einstein, H. E. 1974. Coccidioidomycosis of the central nervous system. Adv Neurol **6:**101–106.

Foley, J. M., R. L. Berman, and C. E. Smith. 1960. X-ray irradiation of *Coccidioides immitis* arthrospores: survival curves and avirulent mutants isolated. J. Bacteriol. **79:**480–487.

Forseth, J., J. J. Rohwedder, B. E. Levine, and M. A. Saubolle. 1986. Experience with needle biopsy for coccidioidal lung nodules. Arch. Intern. Med. **146:**319–320.

Friedman, L., D. Pappagianis, R. J. Berman, and C. E. Smith. 1953. Studies on *Coccidioides immitis;* morphology and sporulation capacity of forty-seven strains. J. Lab. Clin. Med. **42:**438–444.

Gale, D., E. A. Lockhart, and E. Kimbell. 1967. Studies of *Coccidioides immitis:* III. Further studies of toxic soluble components of *Coccidioides immitis*, p. 355–372. *In* L. Ajello (ed.) Coccidioidomycosis. University of Arizona Press, Tucson.

Galgiani, J. N., J. I. Ito, and K. O. Dugger. 1985. Enzyme-linked immunosorbent assay detection of coccidioidal antigens in human serum, p. 239–249. *In* H. E. Einstein, and A. Catanzaro (ed.) Coccidioidomycosis. National Foundation for Infectious Diseases, Washington, D.C.

Galgiani, J. N. and the Valley Fever Vaccine Study Group. 1986. Development of dermal hypersensitivity to coccidioidal antigens associated with repeated skin testing. Am. Rev. Respir. Dis. **134:**1045–1047.

Ganer, A., E. Arathoon, and D. A. Stevens. 1987. Initial experience in therapy for progressive mycoses with itraconazole, the first clinically studied triazole. Rev. Infect. Dis. **9**(Suppl. 1):S77–S86.

Georg, L. K., L. Ajello, and M. A. Gordon. 1951. A selective medium for the isolation of *Coccidioides immitis*. Science **114:**387–389.

Goldstein, E., M. J. Winship, and D. Pappagianis. 1972. Ventricular fluid and the management of coccidioidal meningitis. Ann. Intern. Med. **77:**243–246.

Graham, A. R., and K. J. Ryan. 1980. Counterimmunoelectrophoresis employing coccidioidin in serologic testing for coccidioidomycosis. Am. J. Clin. Pathol. **73:**574–577.

Hageage, G. J., and B. J. Harrington. 1984. Use of calcofluor white in clinical mycology. Lab. Med. **15:**109–112.

Hampson, C. R., and T. R. Larwood. 1967. Observations on coccidioidomycosis immunodiffusion test with clinical correlations, p. 211–214. *In* L. Ajello (ed.), Coccidioidomycosis. University of Arizona Press, Tucson.

Heiner, D. C. 1958. Diagnosis of histoplasmosis using precipitin reactions in agar gel. Pediatrics **22:**616–627.

Hicks, M. J., R. M. Hagaman, and R. A. Barbee. 1986. The prevalence of cellular immunity to coccidioidomycosis in a highly endemic area. [Clinical Investigation]. West. J. Med. **144:**425–428.

Howard, P., and J. W. Smith. 1983. Diagnosis of disseminated coccidioidomycosis by liver biopsy. Arch. Intern. Med. **143:**1335–1338.

Huntington, R. W. 1971. Coccidioidomycosis, p. 147–210. *In* R. Baker (ed.), Human infection with fungi, actinomycetes and algae. Springer-Verlag, Berlin.

Huntington, R. W., Jr. 1957. Diagnostic and biologic implications of the histopathology of coccidioidomycosis, p. 36–46. Proceedings of Symposium on Coccidioidomycosis, Public Health Service Publication No. 575, Washington, D.C.

Huppert, M., J. P. Adler, E. H. Rice, and S. H. Sun. 1979. Common antigens among systemic disease fungi analyzed by two-dimensional immunoelectrophoresis. Infect. Immun. **23:**479–485.

Huppert, M., and J. W. Bailey. 1965a. The use of immunodiffusion tests in coccidioidomycosis. I. The accuracy and reproducibility of the immunodiffusion test which correlates with complement fixation. Am. J. Clin. Pathol. **44:**364–368.

Huppert, M., and J. W. Bailey. 1965b. The use of immunodiffusion tests in coccidioidomycosis. II. An immunodiffusion test as a substitute for the tube precipitin test. Am. J. Clin. Pathol. **44:**369–373.

Huppert, M., J. W. Bailey, and P. Chitjian. 1967a. Immunodiffusion as a substitute for complement fixation and tube precipitins in coccidioidomycosis, p. 221–225. *In* L. Ajello (ed.), Coccidioidomycosis. University of Arizona Press, Tucson.

Huppert, M., I. Krasnow, K. R. Vukovich, S. H. Sun, E. H. Rice, and L. J. Kutner. 1977. Comparison of coccidioidin and spherulin in complement fixation tests for coccidioidomycosis. J. Clin. Microbiol. **6:**33–41.

Huppert, M., E. T. Peterson, S. H. Sun, P. A. Chitjian, and W. J. Derrevere. 1968. Evaluation of a latex particle agglutination test for coccidioidomycosis. Am. J. Clin. Pathol. **49:**96–102.

Huppert, M., N. S. Spratt, K. R. Vukovich, S. H. Sun, and E. H. Rice. 1978. Antigenic analysis of coccidioidin and spherulin determined by two-dimensional electrophoresis. Infect. Immun. **20:**541–551.

Huppert, M., S. H. Sun, and J. W. Bailey. 1967b. Natural variability in *Coccidioides immitis*, p. 323–328. *In* L. Ajello (ed.), Coccidioidomycosis. University of Arizona Press, Tucson.

Jacobs, P. H., and B. Russell. 1973. Dermatophyte test medium for systemic fungi (Letter). J. Am. Med. Assoc. **224:**1649.

Johnson, J. E., B. Jeffrey, and M. Huppert. 1985. Detection of *Coccidioides immitis* antibody by five commercially available immunodiffusion kits, p. 264–272. *In* H. E. Einstein and A. Catanzaro (ed.), Coccidioidomycosis. National Foundation for Infectious Diseases, Washington, D.C.

Kaplan, W., and M. K. Clifford. 1964. Production of fluorescent antibody reagents specific for the tissue form of *Coccidioides immitis*. Am. Rev. Respir. Dis. **89:**651–658.

Kaplan, W., and Kraft, D. 1969. Demonstration of pathogenic fungi in formalin-fixed tissues by immunofluorescence. Am. J. Clin. Pathol. **52:**420–432.

Kaufman, L., and M. J. Clark. 1974. Value of the concomitant use of complement fixation and immunodiffusion tests in the diagnosis of coccidioidomycosis. Appl. Microbiol. **28**:641–643.

Kaufman, L., P. G. Standard, M. Huppert, and D. Pappagianis. 1985. Comparison and diagnostic value of the coccidioidin heat stable (HS and tube precipitin) antigens in immunodiffusion. J. Clin. Microbiol. **22**:515–518.

Klotz, S. A., D. J. Drutz, M. Huppert, S. H. Sung, and P. L. DeMarsh. 1984. The critical role of CO_2 in the morphogenesis of *Coccidioides immitis* in cell-free subcutaneous chambers. J. Infect. Dis. **150**:127–134.

Kyriakos, M. 1977. Myospherulosis of the paranasal sinuses, nose and middle ear. A possible iatrogenic disease. Am. J. Clin. Pathol. **67**:118–130.

Kozub, W. R., N. Leas, R. J. Otto, and C. K. Dutt. 1977. Application of counterimmunoelectrophoresis in coccidioidomycosis, p. 445–452. *In* L. Ajello (ed.), Coccidioidomycosis: current clinical and diagnostic status. Symposia Specialists, Miami, FL.

Land, G., G. L. Dorn, J. M. Hill. 1977. The isolation of disseminated *Coccidioides immitis* by an improved blood culture technique, p. 19–30. *In* L. Ajello (ed.), Coccidioidomycosis. Symposia Specialists, Miami, Florida.

Larsen, R. A., J. A. Jacobson, A. H. Morris, and B. A. Benowitz. 1985. Acute respiratory failure caused by primary pulmonary coccidioidomycosis. Am. Rev. Respir. Dis. **131**:797–799.

Levine, H. B., J. M. Cobb, and G. M. Scalarone. 1969. Spherule coccidioides in delayed dermal sensitivity reactions of experimental animals. Sabouraudia **7**:10–32.

Levine, H. B., J. M. Cobb, and C. E. Smith. 1960. Immunity to coccidioidomycosis induced in mice by purified spherule, arthrospore and mycelial vaccines. Trans. N.Y. Acad. Sci. **22**:436–449.

Levine, H. B., A. Gonzalez-Ochoa, and D. R. Ten Eyck. 1973. Dermal sensitivity to *Coccidioides immitis*. Am. Rev. Respir. Dis. **107**:379–386.

Levine, H. B., R. L. Miller, and C. E. Smith. 1962. Influence of vaccination on respiratory coccidioidal disease in *Cynomolgous* monkeys. J. Immunol. **89**:242–251.

Lones, G. 1967. Studies of intermediary metabolism in *Coccidioides immitis*, p. 349–353. *In* L. Ajello (ed.), Coccidiodomycosis. University of Arizona Press, Tucson.

Lones, G. W., and C. L. Peacock. 1960a. Role of carbon dioxide in the dimorphism of *Coccidioides immitis*. J. Bacteriol. **79**:308–309.

Lones, G. W., and C. L. Peacock. 1960b. Studies on the growth and metabolism of *Coccidioides immitis*. Ann. N.Y. Acad. Sci. **89**:102–108.

Lonkey, S. A., A. Catanzaro, K. M. Moser, and H. Einstein. 1976. Acute coccidioidal pleural effusion. Am. Rev. Respir. Dis. **114**:681–688.

Lundergan, L. L., S. S. Kerrick, and J. N. Galgiani. 1985. Coccidioidomycosis at a university outpatient clinic: a clinical description, p. 47–54. *In:* H. E. Einstein, and A. Catanzaro (ed.) Coccidioidomycosis. National Foundation for Infectious Diseases, Washington, D.C.

Lupan, D. M., and P. Nziramasanga. 1986. Collagenolytic activity of *Coccidioides immitis*. Infect. Immun. **51**:360–361.

Meyer, P. R., A. N. Hui, and M. Biddle. 1982. *Coccidioides immitis* meningitis with arthroconidia in the cerebrospinal fluid: report of the first case and review of the arthroconidia literature. Hum. Pathol. **13**:1136–1138.

Mulder, C., S. Kiddy, and K. Lou. 1966. A rapid latex slide test for coccidioidomycosis. Bacteriol. Proc., p. 141.

Nelson, A. 1974. The surgical treatment of pulmonary coccidioidomycosis, p. 1–48. Curr. Probl. Surg., October, 1974. Year Book Medical Publishers, Inc., Chicago.

Nziramasanga, P., and D. M. Lupan. 1985. Elastase activity of *Coccidioides immitis*. J. Med. Microbiol. **19**:109–114.

Omieczynski, D. T., F. E. Swatek, and S. W. Becker. 1965. A rapid method for the isolation of *Coccidioides immitis*. Health Lab. Sci. **2**:35–39.

Pappagianis, D. 1980a. Epidemiology of coccidioidomycosis, p. 63–85. *In* D. A. Stevens, (ed.), Coccidioidomycosis. A text. Plenum Medical Book Co., New York.

Pappagianis, D. 1980b. Serology and serodiagnosis of coccidioidomycosis, p. 97–112. *In* D. A. Stevens (ed.), Coccidioidomycosis. A text. Plenum Medical Book Co., New York.

Pappagianis, D. 1981. Coccidioidomycosis, p. 681–691. *In* M. Samter (ed.), Immunological Diseases, 3rd ed, vol. I. Little, Brown, & Co., Boston.

Pappagianis, D., and R. Crane. 1977. Survival in coccidioidal meningitis since introduction of amphotericin B, p. 223–237. *In* L. Ajello (ed.), Coccidioidomycosis. Current clinical and diagnostic status. Symposia Specialists, Miami, Florida.

Pappagianis, D., and H. E. Einstein. 1978. Tempest from Tehachapi takes toll. West. J. Med. **129**:527–530.

Pappagianis, D., G. Kobayashi. 1958. Production of extracellular polysaccharide in cultures of *Coccidioides immitis*. Mycologia **50**:229–238.

Pappagianis, D. and G. S. Kobayashi. 1960. Approaches to the physiology of *Coccidioides immitis*. Ann. NY Acad. Sci. **89**:109–121.

Pappagianis, D., I. Krasnow, and S. Beall. 1976. False-positive reactions of cerebrospinal fluid and diluted sera with the coccidioidal latex agglutination test. Am. J. Clin. Pathol. **66**:916–921.

Pappagianis, D., S. Lindsay, and S. Beall. 1985a. A single immunodiffusion test for the detection of tube precipitin (TP) and complement fixation (CF) antibodies. American Society for Microbiology Annual Meeting, Abstract F-19.

Pappagianis, D., N. Lindsey, C. E. Smith, and M. T. Saito. 1965. Antibodies in human coccidioidomycosis: immunoelectrophoretic properties. Proc. Soc. Exp. Biol. Med. **118**:118–122.

Pappagianis, D., A. Ornelas, and R. Hector. 1985b. Guanine plus cytosine content of the DNA of *Coccidioides immitis*. Sabouraudi **23**:451–454.

Pappagianis, D., E. W. Putman, and G. S. Kobayashi. 1961. Polysaccharide of *Coccidioides immitis*. J. Bacteriol. **82**:714–723.

Pappagianis, D., M. T. Saito, and K. Van Hoosear. 1972. Antibody in cerebrospinal fluid in non-meningitic coccidioidomycosis. Sabouraudia **10**:173–179.

Pappagianis, D., C. E. Smith, G. S. Kobayashi. 1956. Relationship of the *in vivo* form of *Coccidioides immitis* to virulence. J. Infect. Dis. **98**:312–319.

Pappagianis, D., C. E. Smith, G. S. Kobayashi, and M. T. Saito. 1961. Studies of antigens from young mycelia of *Coccidiodes immitis*. J. Infect. Dis. **108**:35–44.

Pappagianis, D., C. E. Smith, M. T. Saito, and G. S. Kobayashi. 1957. Preparation and property of a complement-fixing antigen from mycelia of *Coccidioides immitis*. Public Health Serv. Publ. **575**:57–63.

Pappagianis, D., and the Vaccine Study Group. 1986. Evaluation of the protective efficacy of the killed *Coccidioides immitis* vaccine in man. 26th Interscience Conference, Antimicrobial Agents in Chemotherapy, Abstract 784, p. 240.

Petersen, E. A., B. A. Friedman, E. D. Crowder, and D. Rifkind. 1976. Coccidioidouria: clinical significance. Ann. Intern. Med. **85**:34–38.

Ray, J. G., Jr. 1967. The agar-gel precipitin-inhibition test used in serum titration of human and animal coccidioidomycosis, p. 215-220. *In* L. Ajello (ed.), Coccidioidomycosis. University of Arizona Press, Tucson.

Reep, B. R., and W. Kaplan. 1972. The use of N-acetyl-L-cysteine and dithiothreitol to process sputa for mycological and fluorescent antibody examinations. Health Lab. Sci. **9:**118–124.

Resnick, S., H. B. Levine, D. Pappagianis, and J. McKerrow. 1986. Isolation and characterization of an elastase and a metalloproteinase from *Coccidioides immitis*. J. Invest. Dermatol. **86:**502 (abstract).

Roberts, G. D., and N. L. Goodman. 1986. The diagnostic mycology laboratory, p. 29–32. *In* G. A. Sarosi and S. F. Davies (ed.), Fungal diseases of the lung. Grune and Stratton, Inc., Orlando, Florida.

Roberts, J. A., J. M. Counts, and H. G. Crecelius. 1970. Production in vitro of *Coccidioides immitis* spherules and endospores as a diagnostic aid. Am. Rev. Respir. Dis. **102:**811–813.

Rodenbiker, H. T., J. P. Ganley, J. N. Galgiani, and S. G. Axline. 1981. Prevalence of chorioretinal scar associated with coccidioidomycosis. Arch. Ophthalmol. **99:**71–75.

Sawaki, Y., M. Huppert, J. W. Bailey, and Y. Yagi. 1966. Patterns of human antibody reactions in coccidioidomycosis. J. Bacteriol. **91:**422–427.

Scalarone, G. M., H. B. Levine, S. D. Chaparas, and J. M. Cobb. 1973. Properties and assay of spherulins from *Coccidioides immitis* in delayed sensitivity responses of animals. Sabouraudia **11:**222–234.

Shifrine, M., D. Pappagianis, and J. Neves. 1975. Double electroimmunodiffusion: a rapid diagnostic test for canine coccidioidomycosis. Am. J. Vet. Res. **36:**819–820.

Shungu, D., M. Valiant, V. Tutlane, E. Weinberg, B. Weissberger, L. Koupal, H. Gadebusch, and E. Stapley. 1983. Gelrite as an agar substitute in bacteriological media. Appl. Environ. Microbiol. **46:**840–845.

Sippel, J. E., and H. B. Levine, 1969. Sugars and amino acids as carbon, nitrogen, or energy sources for *Coccidioides immitis* spherules and endospores. Appl. Microbiol. **18:**522–524.

Smith, C. D., and N. L. Goodman. 1975. Improved culture method for the isolation of *Histoplasma capsulatum* and *Blastomyces dermatitidis* from contaminated specimens. Am. J. Clin. Pathol. **68:**276–280.

Smith, C. E. 1943. Coccidioidomycosis. Med. Clin. North Am. **27:**790–807.

Smith, C. E. 1955. Coccidioidomycosis. Ped. Clin. North Am. **2:**109–125.

Smith, C. E., M. T. Saito, and S. A. Simons. 1956. Pattern of 39,500 serologic tests in coccidioidomycosis. J. Am. Med. Assoc. **160:**546–552.

Sobel, R. A., W. G. Ellis, S. L. Neilsen, and R. L. Davis. 1984. Central nervous system coccidioidomycosis: a clinicopathologic study of treatment with and without amphotericin B. Hum. Pathol. **15:**980–995.

Standard, P. G., and L. Kaufman. 1977. Immunological procedure for the rapid and specific identification of *Coccidioides immitis*. J. Clin. Microbiol. **8:**149–153.

Stevens, D., R. L. Stiller, P. L. Williams, and A. M. Sugar. 1983. Experience with ketoconazole in three major manifestations of progressive coccidioidomycosis. Am. J. Med. **74:**(Suppl. 1B):58–63.

Sun, S. H., M. Huppert, and K. P. Vukovich. 1976. Rapid *in vitro* conversion and identification of *Coccidioides immitis* cultures. J. Clin. Microbiol. **8:**149–153.

Thompson, D. W., W. Kaplan, and B. J. Phillips. 1977. The effect of freezing and the influence of isolation medium on the recovery of pathogenic fungi from sputum. Mycopathol. **61:**105–109.

Walch, H. A., and R. K. Walch. 1967. Studies with induced mutants of *Coccidioides immitis*, p. 339–347. *In* L. Ajello (ed.), Coccidioidomycosis. University of Arizona Press, Tucson.

Wallace, J. M., A. Catanzaro, K. M. Moser, and J. H. Harrell, II. 1981. Flexible fiberoptic bronchoscopy for diagnosing pulmonary coccidioidomycosis. Am. Rev. Respir. Dis. **123:**286–290.

Warlick, M. A., S. F. Quan, and R. E. Sobonya. 1983. Rapid diagnosis of pulmonary coccidioidomycosis. Cytologic v. potassium hydroxide preparations. Arch. Intern. Med. **143:**723–725.

Weisman, I. M., A. J. Moreno, A. L. Parker, W. C. Sippo, and W. J. Liles. 1986. Gastrointestinal dissemination in coccidioidomycosis. Am. J. Gastroenterol. **81:**589–593.

Wieden, M., J. N. Galgiani, and D. Pappagianis. 1983. Comparison of immunodiffusion techniques with standard complement fixation assay for quantitation of coccidioidal antibodies. J. Clin. Microbiol. **18:**529–534.

Willett, F. M., and A. Weiss. 1945. Coccidioidomycosis in Southern California. Report of a new endemic area with review of 100 cases. Ann. Intern. Med. **23:**349–375.

Wilson, J. W. 1968. The importance of the portal of entry in certain microbial infections. The primary cutaneous "chancriform" syndrome. Dis. Chest. (Suppl.1)**54:**299–304.

Winn, W. A. 1968. A long term study of 300 patients with cavitary-abscess lesions of the lung of coccidioidal origin. Dis. Chest. (Suppl.1)**54:**268–272.

Wolf, J. E., J. R. Little, D. Pappagianis, and G. S. Kobayashi. 1986. Disseminated coccidioidomycosis in a patient with the acquired immune deficiency syndrome. Diagn. Microbiol. Infect. Dis. **5:**331–336.

Wood, J., G. Friedly, M. Zartarian, S. Aarnaes, and L. M. de la Maza. 1982. Alternatives to the standardized laboratory branch complement fixation test for detection of antibodies to *Coccidioides immitis*. J. Clin. Microbiol. **16:**1030–1033.

Yturraspe, D. 1971. Clinical evaluation of latex particle agglutination test in the diagnosis of canine coccidioidomycosis. J. Am. Vet. Assoc. **158:**1249–1256.

Zimmer, B. L., and D. Pappagianis. 1986a. Comparison of immunoblot analysis of spherule-endospore-phase extracellular protein and mycelial phase antigen of *Coccidioides immitis*. Infect. Immun. **53:**64–70.

Zimmer, B. L., and D. Pappagianis. 1986b. Taxonomic and physiologic characteristics of *Coccidioides immitis*, p. 165–168. *In* L. Leive (ed.), Microbiology—1986. American Society of Microbiology, Washington, D.C.

CHAPTER 63

Cryptococcosis

H. J. SHADOMY and N. G. WARREN

Disease: Cryptococcosis.

Etiologic Agent: *Cryptococcus neoformans.*

Source: Airborne contaminant found worldwide, often in soil mixed with pigeon droppings, and nesting sites such as window ledges, window air conditioners, barns, etc. First isolated from peach juice.

Clinical Manifestations: *Cryptococcus neoformans* causes primary pulmonary lesions. The organism may disseminate to the central nervous system, less often to skin, bones, and other sites. Found in approximately 8% of patients with acquired immune deficiency syndrome. In 30 to 50% of cases an associated, underlying disorder exists (leukemia, lymphosarcoma, sarcoidosis, Hodgkins' disease, lupus erythematosis, tuberculosis, etc.). There is a strong association of dissemination and the immunocompromised condition, although in over half of patients with cryptococcosis, no apparent underlying disease is known.

Pathology: Little or no tissue response is elicited by this encapsulated organism, and pathology generally occurs from pressure exerted by the mechanical displacement by multiplying organisms of surrounding tissue cells.

Laboratory Diagnosis: Direct examination of cerebrospinal fluid and urine after centrifugation or other body fluids with India ink demonstrates round to oval budding and encapsulated yeasts from 3 to 6 and 5 to 10 μm in diameter. Aerobic growth on standard bacteriologic media without added cycloheximide.

Epidemiology: Worldwide, in all ages, in males more frequently than females. Associated with pigeon droppings, and pigeon breeders are known to have a significantly greater percentage of positive cryptococcin skin tests, but not more disease, than the general population. May also be found in soil apparently devoid of bird feces.

Treatment: Recent clinical studies indicate amphotericin B combined with flucytosine to be the regimen of choice for cryptococcosis.

Prevention and Control: None.

Description of Disease

Cryptococcosis is an opportunistic systemic fungal infection characterized by primary pulmonary infection which may be subclinical or overt, occasionally acute, but most often slow and insidious. Except for anecdotal instances, the disease is produced by only one species, *Cryptococcus neoformans*. The disease is seen sporadically in cats, horses, and cows—particularly bovine mastitis. Although associated with pigeons, presumably because of their high body temperature, they do not acquire the disease.

Communicability and Transmissibility

Cryptococcus neoformans produces disease in humans and animals by inhalation of the organism from an exogenous source, principally pigeon feces or soil contaminated with it. Cryptococcosis is not known to be transmissible from human to human, animal to animal, or animal to human.

Clinical Manifestations of the Disease

Cryptococcosis generally begins as a primary pulmonary infection which may spread hematogenously to the skin, bones, central nervous system (CNS), and abdominal viscera. At the time of diagnosis, up to 70% of all patients exhibit neurologic disturbances.

Pulmonary Cryptococcosis

Pulmonary cryptococcosis is often asymptomatic. Scant, blood-tinged sputum may be present. X rays often show cryptococcomas resembling a tumor, more often in the upper lobes. The disease may heal by encapsulation of the lesion, remain stable, or progress and disseminate. Pulmonary lesions do not calcify upon healing, and residual evidence of past cryptococcosis may not be found. However, benign pulmonary cryptococcosis is found occasionally at autopsy, with the presence of healed, encapsulated subpleural granulomas. Active granulomatous lesions may penetrate the bronchial wall, and large numbers of organisms may be found in the sputum.

CNS Cryptococcosis

Development of CNS cryptococcosis is usually insidious, with nonspecific symptoms such as headache which increases in frequency and severity, diplopia, vertigo, nausea, vomiting, or irritability present for weeks to months. The CNS cryptococcosis may last from months up to 20 years or more. It is not unusual for family or close friends to note impairment of memory and interrelationship with others. An insidious onset of disease may be associated with development of granulomatous lesions of the meninges; sudden and severe onset more often has been associated with rapidly spreading cerebral lesions. Papilledema occurs in up to one-third of these cases, and frequent spinal taps may be necessary to prevent irreversible optic damage. CNS cryptococcosis may be complicated by hydrocephalus, seizures (late in the course of infection), and dementia.

Osseous Cryptococcosis

Bone lesions are seen in approximately 5 to 10% of reported cases of cryptococcosis. There is a predilection for cranial bones, vertebrae, and bony prominences. They are frequently associated with pain and swelling. They are osteolytic and may extend to the skin; joints are usually spared. Granulation tissue may be seen in localized cutaneous lesions.

Cutaneous Cryptococcosis

Single or multiple cutaneous lesions occur by dissemination from a primary pulmonary site. They are seen most frequently on the face and scalp and may be papules, acneform pustules, or soft subcutaneous abscesses that ulcerate. They may resemble carcinoma or gumma. Granulation tissue may be seen in localized cutaneous lesions. Through cutaneous extension, or by hematogenous spread, mucosal lesions also may occur. They appear as nodules, granulomas, or deep or superficial ulcers. Primary inoculation into the skin or subcutaneous tissue (morticians, laboratory workers) develop chancreform lesions which heal without dissemination.

Visceral Cryptococcosis

Hematogenous spread from the lungs may produce visceral involvement resembling tuberculosis. Any organ or tissue may be involved. The more common of these rare sites of cryptococcosis are renal abscesses, prostatitis, testis, heart, and eye.

Pathology

Although found in nature in the unencapsulated form, upon entrance into the body, *C. neoformans* rapidly develops a mucopolysaccharide capsule. The capsule both gives a mucoid quality to lesions and inhibits cellular reaction to its presence by masking both heat-labile and heat-stable opsonins that bind to the cell wall of the organisms, preventing phagocytosis. In those instances when *C. neoformans* does elicit a cellular response, the lesions cannot be differentiated from other granulomatous diseases. In some tissues, macrophages and giant cells may be seen with ingested yeast cells, or with yeast cells in close proximity. Plasma cells and small lymphocytes may also be present, but granulomas are usually not seen.

C. neoformans is seen most often in the lungs, although in advanced disease, pulmonary infection may have healed leaving only a single subpleural nodule. Miliary granulomas are seen in progressive pulmonary cryptococcosis, and there may be conspicuous fibrosis.

In CNS cryptococcosis, throughout the brain, particularly in the gray matter, characteristic lesions consist of multiple organisms in cysts, with no inflammatory response. The gross appearance of brain and meninges is essentially normal. The subarachnoid space often contains a white, gelatinous or mucoid exudate.

Diagnostic Features

Cryptococcosis should be considered in patients who have complaints referable to the CNS, even if they are nonspecific or mild. Individuals who exhibit behavioral changes or impaired judgment and memory; those who have progressively severe headaches that do not subside with medication or retroorbital pain; individuals who exhibit nausea, vomiting, vertigo, diplopia, irritability, seizures, sommolence, and clumsiness, all should be examined for cryptococcal meningitis.

Patients who have underlying diseases in which any of the above symptoms are present but are uncommon, who are receiving corticosteroid therapy or are otherwise immunocompromised, or individuals with acquired immunodeficiency syndrome (AIDS) should also be examined for *C. neoformans* infection.

Prognosis

The accidental findings of healed cryptococcal lung lesions at autopsy, studies indicating a far greater than average number of positive cryptococcin skin tests in pigeon breeders, and the lack of development of pulmonary disease in the presence of increased positive skin tests among laboratorians working with the organism make it probable that pulmonary cryptococcosis is an opportunistic fungus infection that is uncommonly progressive in normal individuals.

On the other hand, CNS cryptococcosis is believed to be uniformly fatal without treatment. Even with rigorous therapeutic regimens, this form of disease is fatal in 25 to 30% of cases. In addition, many patients will die from their underlying disease even after cure of cryptococcosis, and severe complications from the disease may also lead to death even after a biological cure.

Treatment with amphotericin B is followed by a 20 to 25% relapse rate. Treatment with flucytosine alone often has led to the emergence of drug-resistant mutants. However, combined therapy with lesser doses of these drugs, along with causing fewer side affects, has lowered the relapse rate to about half of that with amphotericin B alone.

Epidemiologic Aspects

Occurrence

Cryptococcosis is worldwide in distribution. Reported cases in males outnumber those found in females almost 2 to 1, although little significant difference can be seen related to age or race. Case estimates of human cerebral meningitis in the United States approach 300 per year (Kaufman and Blumer, 1978), and the number of subclinical respiratory infections is much higher. Pigeon breeders have a higher occurrence of antibody to *C. neoformans*; however, overt infection in this group is no higher than that associated with any other occupation (Fink et al., 1968).

Cryptococcosis often may be found in patients with immunosuppressive disorders such as malignant neoplasms (Kaplan et al., 1977) and kidney dysfunctions requiring organ transplantation (Schroter et al., 1976). More recently, the opportunistic potential of cryptococcosis has been emphasized by the number of cases seen in individuals with AIDS. At one major health care facility, *C. neoformans* was detected in approximately 8% of patients treated for AIDS, making it the fourth most commonly recognized microorganism, following cytomegalovirus, *Pneumocystis carinni*, and *Mycobacterium avium* complex (Kovacs et al., 1985).

C. neoformans has been found in the environment and in animals. Ajello (1967) has stated that soil is the ultimate source of the fungus, and the organism has been recovered from milk, grass, insects, peaches, and the mesquite bush. Cats (Holzworth, 1952), dogs (Seibold et al., 1953), and cattle (Pounder et al., 1952) as well as wild animals (Barron, 1955) are susceptible to the disease and are probably infected, as are humans, through inhalation of the organism from the environment.

The pigeon plays a major role in the ecology of *C. neoformans*. Large numbers of organisms may be recovered from accumulated debris, dust, and droppings associated with pigeon roosts found in attics and other nooks of old buildings. However, the high temperature of birds allows the organism to pass through the gut but does not cause avian disease.

Transmission

Cryptococcosis is not known to be transmitted from animal to animal, human to human, or animal to human. The fungus is found in the environment, and inhalation of dust containing the organism produces pulmonary disease which then can disseminate to other body sites. In its environmental state, *C. neoformans* usually is nonencapsulated and small enough to be inhaled into the pulmonary alveolar spaces (Rippon, 1982).

Etiologic Agent

Characterization

Cryptococcus neoformans is essentially the only etiologic agent of cryptococcosis. Only rarely have other cryptococci been implicated in disease. *C. albi-*

dus has been associated with meningitis (Da Cuhna and Luisins, 1973; Melo et al., 1980) and pulmonary disease (Krumholz, 1972), and *Cryptococcus laurentii* has been found in a lung abscess (Damalam et al., 1977) and a cutaneous infection (Lynch et al., 1981).

Cryptococci are round to oval budding, encapsulated yeasts, ranging in size from 3 to 6 to 5 to 10 μm, but some may attain a size of up to 14 μm (Fig. 1). Although the round, budding form usually is observed, in rare instances a rudimentary pseudohypha can be formed (Fig. 2) (Rippon, 1982). All species assimilate inositol as a carbon source for growth. However, the ability to ferment sugars is lacking. Ascospores, teliospores, and ballistospores are not formed. The teliomorphic (sexual) state of *C. neoformans* has been identified as a basidiomycete (Shadomy, 1970) and named *Filobasidiella neoformans* (Kwon-Chung, 1975).

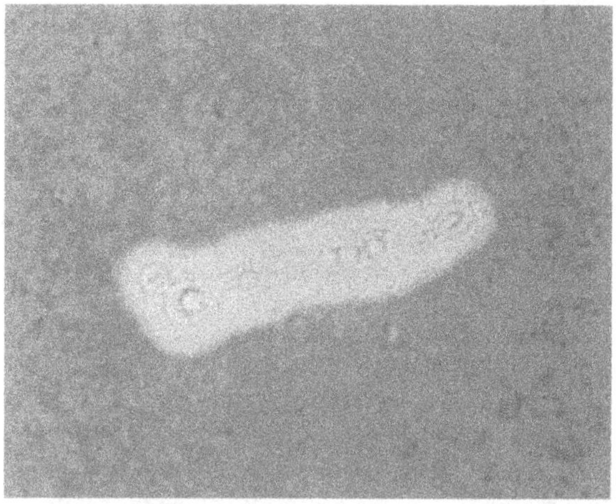

FIG. 2. Rudimentary pseudohypha of *Cryptococcus neoformans*. India ink preparation.

Specimen Collection and Transport

Cerebrospinal fluid should be submitted to the laboratory in a sterile tube. At least 2 ml (more, if available) is needed for culture and direct examination. Blood and bone marrow should be collected in heparin or sodium polyanethosulfonate and submitted in sterile tubes. Alternately, blood products may be inoculated at the bedside directly into one of a variety of blood culture media bottles furnished by the hospital microbiology laboratory.

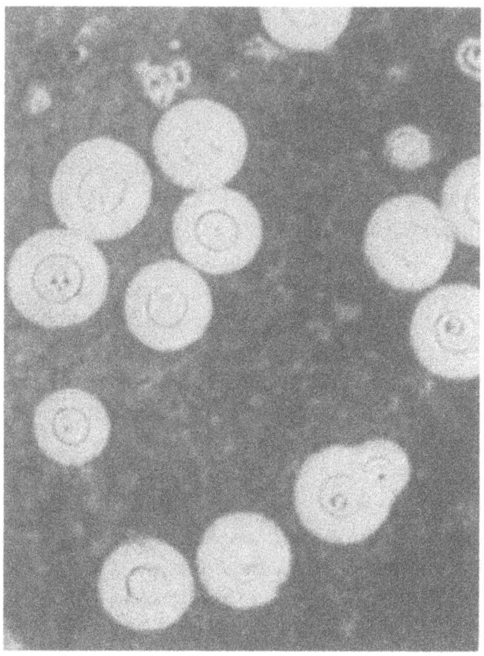

FIG. 1. Oval, budding, encapsulated yeast cells of *Cryptococcus neoformans*. Note budding cell with narrow isthmus between two cells. India ink preparation.

A series of early-morning expectorated sputa should be submitted in separate sterile containers. At least 5 ml of specimen is desirable for adequate laboratory studies.

All specimens should be transported to the laboratory as soon as possible for mycologic studies. If a delay is anticipated, then sputum and other respiratory specimens may be stored at 4°C. Every effort should be made to assure that cerebrospinal fluids are delivered to the laboratory in an expedient manner.

Isolation Procedures

Cerebrospinal fluids should be centrifuged at 2,000 × *g* for 10 min, and the resulting sediment should be used for microscopic preparation and for inoculation of media. An alternate method for processing these fluids is the use of membrane filters. The cerebrospinal fluid should be centrifuged as above, a drop of material removed from the sediment for microscopic preparation, and the sediment resuspended in the supernatant fluid by vortexing. The resuspended fluid then can be passed through a sterile, 0.22-μm membrane filter, and the entire filter can be placed on mycologic media. The membrane filter method offers the advantage of concentrating very few organisms.

Sputum specimens can be processed in two ways. The classic method involves direct inoculation of the specimen (purulent, bloody portions, if present, should be selected) onto the mycologic media. The second method (Haley and Callaway, 1978) includes a concentration step. An equal volume of digestant, such as 0.5% N-acetyl-L-cysteine in a 2.9% sodium citrate, is added to the sputum and mixed by vortexing. Enough sterile distilled water is added to bring the volume up to 50 ml, and the mixture is centrifuged at 2,000 × *g* for 15 min. After decanting the supernatant fluid, the sediment can be inoculated to mycologic media. If an overwhelming bacterial contamination problem is suspected, enough chloramphenicol to give a final concentration of 0.05 to 0.1 mg/ml may be added to the sediment before inoculation of media.

If blood cultures are directly inoculated upon collection at the bedside, 5 to 8 ml of specimen should be placed into a biphasic culture bottle. If heparinized blood is submitted to the laboratory, the specimen can be processed by standard lysis-centrifugation methods, and the sediment can be inoculated to mycologic media. Bone marrow specimens should be inoculated directly to mycologic media.

The importance of direct microscopic examination of clinical specimens, especially cerebrospinal fluids, cannot be overemphasized. Often it is the direct smear that leads to a tentative diagnosis when time may be of the utmost importance. Ideally, several smears should be prepared, and a thorough, careful search for mycotic elements should be undertaken. "Budding" fat globules and white blood cells can mimic fungal structures.

The classic method for the direct examination of clinical specimens for presence of cryptococci is the India ink method. A drop of prepared sediment is added to approximately 2 loopfuls of India ink on a glass slide. After mixing, the preparation should be brownish in color. If the mixture is too dark, 2 to 3 loopfuls of sterile distilled water may be added. After mounting with a cover slip, the preparation can be examined under low- and high-power magnification for the presence of budding yeast cells surrounded with a clear halo. The capsule of the cryptococci will repel carbon particles of India ink, giving a clear, well-demarcated zone around each cell. White blood cells also may repel carbon particles, but the halo has a fuzzy, irregular edge.

Another method of smear examination that is being used more and more frequently is the calcofluor procedure where cryptococci and other fungi can be detected by their apple green fluorescence. Equal amounts of 0.1% calcofluor and prepared specimen are mixed on a clean glass slide and mounted with a cover slip. If the specimen requires digestion (such as sputum sediment) then a drop of KOH can be added. The specimen is viewed with a fluorescent microscope having a 365-nm exciter filter and a barrier filter that will transmit light at 410 nm. This combination of filters frequently is used in the mycobacteriology laboratory for the detection of acid-fast bacilli stained with Auramine O. The preparation is viewed for typical morphology of cryptococci. Plant fibers may fluoresce a yellow color; blood and pus cells may exhibit a dull fluorescence.

Few types of media need to be used to isolate cryptococci. *The organisms are inhibited, however, by media containing cycloheximide.* Sterile specimens should be inoculated to brain-heart infusion agar; biphasic blood culture bottles should contain brain-heart infusion agar and broth. If specimens are from normally contaminated sites, brain-heart infusion agar containing chloramphenicol or gentamicin may be employed. The media are incubated at 30°C under normal atmospheric conditions and examined for 4 weeks. Visible growth of cryptococci may be detected as early as 2 days but may require a longer time. The colony will appear as a creamy, yeastlike colony that often is mucoid due to capsule production.

Culture Identification

Identification of cryptococci may be accomplished by performing tests to determine microscopic morphology and carbohydrate and nitrate assimilations. Once the purity of the isolate has been ascertained by streaking a single colony on Sabouraud dextrose agar and blood agar, morphology agar plates may be prepared. A single colony is streaked by barely scratching the surface of a cornmeal agar (without dextrose, plus 0.3% Tween 80) plate. After placing a sterile cover slip over the scratched area, the plate is incubated at 28 to 30°C for 24 to 48 h. Cryptococci appear as round to slightly oval budding yeasts without pseudohyphae. All *Cryptococcus* species are urease-positive. The finding of round, budding yeasts that generally produce no pseudohyphae but are positive for urease is suggestive of the genus *Cryptococcus*.

Species identification is determined by carbohydrate and nitrate assimilation tests. Table 1 indicates the expected reaction of the *Cryptococcus* spp. that usually are encountered in the clinical laboratory. Several kits are commercially available to aid in species identification. *C. neoformans* assimilates dextrose, maltose, sucrose, galactose, inositol, D-xylose, trehalose, and dulcitol. Most strains assimilate cellobiose and raffinose. Lactose, melibiose, and potassium nitrate are not assimilated by *C. neoformans*.

Another test for the rapid presumptive identification of *C. neoformans* is the detection of a dark brown pigment on media containing an extract of *Guizotia abyssinica* seed (Shields and Ajello, 1966). After several days, most colonies of *C. neoformans* turn dark brown when streaked onto this medium. Several other *Cryptococcus* species can turn light brown on the medium, but *C. neoformans* is the only species that yields a truly dark color.

Serotyping

Serotyping of *Cryptococcus neoformans* isolates is not a procedure usually performed in the clinical laboratory. Four serotypes (A, B, C, and D) have been recognized and are associated with two teleomorphic states. *C. neoformans* var. *neoformans* consists of serotypes A and D; serotypes B and C comprise *C. neoformans* var. *gatti*. The usual biochemical tests do not aid in separating the two varieties, nor, at the present time, is there any evidence of differences in disease symptoms or susceptibility to antifungal agents. The major difference between the two varieties is in geographic distribution. *C. neoformans* var. *gatti* is found almost exclusively in southern California whereas *C. neoformans* var. *neoformans* is associated with most infections elsewhere in the United States (Bennett et al. 1977).

Serodiagnosis of Cryptococcosis

Investigations during the past several decades have resulted in the development of serologic procedures that have diagnostic and prognostic usefulness for cryptococcosis patients who were believed previously to be essentially immunologically inactive. Procedures for determining cryptococcal antibody include tube agglutination, indirect fluorescent antibody techniques, and enzyme immunoassay. These are useful in detecting early or localized cryptococcosis. Cryptococcal antigen can be detected by latex agglutination (LA) techniques and has both diagnostic and prognostic value. The LA procedure is rapid, has proved to be successful in the specific detection of antigen in sera and cerebrospinal fluid, and is more sensitive than the India ink stain. LA titers of 1:8 or greater usually are indicative of active cryptococcosis. Titers of 1:4 or less may be found in symptomatic patients who have negative cultures. Rare false-positive reactions can occur but are usually obtained from patients with rheumatoid arthritis. Rheumatoid factor can be eliminated from sera by boiling the specimen with disodium EDTA or treating the specimen with pronase (Stockman and Roberts, 1983).

Treatment

Amphotericin B

This drug has been the backbone of treatment for cryptococcosis since 1958, primarily because there is no significant development of resistance during therapy. Early regimens of amphotericin B were overshadowed by severe renal toxicity. However, more recently, regimens with lowered doses of the drug

TABLE 1. Physiologic properties of some *Cryptococcus* species

	Carbohydrate assimilations														
Species	*Dextrose*	*Maltose*	*Sucrose*	*Lactose*	*Galactose*	*Cellobiose*	*Raffinose*	*Inositol*	*D-Xylose*	*Trehalose*	*Melibiose*	*Dulcitol*	KNO_3	*Urease*	*Growth at 37°C*
C. neoformans	+	+	+	0	+	+/−	+/−	+	+	+	0	+	0	+	+
C. albidus	+	+	+	+/−	−/+	+	+/−	+	+	+	−/+	+	+	+	+
C. laurentii	+	+	+	+	+	+	+	+	+	+	+	+	0	+	+
C. luteolus	+	+	+	−/+	+	+	+	+	+	+	+	+/−	0	+	+/−
C. terreus	+	+/−	0	+	+/−	+	0	+	+	+	0	+/−	+	+	+/−
C. uniguttulatus	+	+	+	0	+/−	−/+	+/−	+	+	−/+	0	0	0	+	0

+ = Positive; 0 = negative; +/− = most are positive, occasional isolate is negative; −/+ = most are negative, occasional isolate is positive.

have been administered concomitantly with corticosteroids. Side effects include fever, nausea, and azotemia. Intrathecal injections may occasionally be used in cryptococcal meningitis in patients who relapse or fail or respond otherwise to treatment.

Flucytosine

Flucytosine (5-fluorocytosine) is rarely used alone in the therapy of cryptococcosis because of the development of drug resistance by the organism and lowered leukocyte and/or platelet counts.

Combination Therapy

This regimen of amphotericin B and flucytosine now is considered the drug treatment of choice in cryptococcal meningitis. It has been shown to be more effective than amphotericin B alone, without the numbers of ensuing side effects, since lower doses of both drugs are used. Azotemia still may occur, however. Fewer failures and relapses occur, although numbers of deaths from the disease have not changed with the dual-drug treatment.

Literature Cited

Ajello, L. 1967. Comparative ecology of respiratory mycotic disease agents. Bacteriol. Rev. **31**:6–24.

Barron, C. N. 1955. Cryptococcosis in animals. J. Am. Vet. Med. Assoc. **127**:125–132.

Bennett, J. E., K. J. Kwon-Chung, and D. H. Howard. 1977. Epidemiologic differences among serotypes of *Cryptococcus neoformans*. Am. J. Epidemiol. **105**:582–586.

Da Cuhna, T., and J. Lusins. 1973. *Cryptococcus albidus* meningitis. South. Med. J. **66**:1230–1231.

Damalam, A., P. Yesudian, and A. S. Thambiah. 1977. Cutaneous infection by *Cryptococcus laurentii*. Br. J. Dermatol. **97**:221–223.

Fink, J. N., J. J. Barboriak, and L. Kaufman. 1968. Cryptococcal antibodies in pigeon breeder's disease. J. Allergy **41**:297–301.

Haley, L. D., and C. S. Callaway. 1978. Laboratory methods in medical mycology, 4th ed. HEW Publication No. (CDC) 78-8361, Washington, D.C.

Holzworth, J. 1952. Cryptococcosis in a cat. Cornell Vet. **42**:12–15.

Kaplan, M. H., P. P. Rosen, and D. Armstrong. 1977. Cryptococcus in a cancer hospital. Cancer **39**:2265–2274.

Kaufman, L., and S. Blumer. 1978. Cryptococcosis, the awakening giant. p. 176–184. *In* The black and white yeasts. Pan American Health Org. Sci. Pub. 356. Proceedings IV Int. Conf. Mycoses, Washington, D.C.

Kovacs, J. A., A. A. Kovacs, M. Polis, W. C. Wright, V. J. Gill, C. U. Tuazon, E. P. Gelman, H. C. Lane, R. Longfield, G. Overtuff, A. M. Macher, A. S. Fauci, J. E. Parrillo, J. E. Bennett, and H. Masur. 1985. Cryptococcosis in the acquired immunodeficiency syndrome. Ann. Intern. Med. **103**:533–538.

Krumholz, R. A. 1972. Pulmonary cryptococcosis. Am. Rev. Respir. Dis. **105**:421–423.

Kwon-Chung, K. J. 1975. A new genus, *Filobasidiella,* the perfect state of *Cryptococcus neoformans*. Mycologia **67**:1197–1200.

Lynch, J. P. III, D. R. Schaberg, D. G. Kissner, and C. A. Kauffman. 1981. *Cryptococcus laurentii* lung abscess. Am. Rev. Respir. Dis. **123**:135–138.

Melo, J. C., S. Srinivasan, M. L. Scott, and M. J. Raff. 1980. *Cryptococcus albidus* meningitis. J. Infect. Dis. **2**:79–82.

Pounden, W. P., J. M. Anderson, and R. F. Jaeger. 1952. A severe mastitis problem associated with *Cryptococcus neoformans* in a large dairy herd. Am. J. Vet. Res. **13**:121–128.

Rippon, J. W. 1982. Medical mycology. The pathogenic fungi and the pathogenic actinomycetes, 2nd ed. Saunders, Philadelphia, p. 532–558.

Schroter, G. P., D. R. Temple, B. S. Husberg, R. Weil, and T. E. Starzl. 1976. Cryptococcosis after renal transplantation: report of ten cases. Surgery **79**:268–277.

Seibold, H. R., C. S. Roberts, and E. M. Jordan. 1953. Cryptococcosis in a dog. J. Am. Vet. Med. Assoc. **122**:213–215.

Shadomy, H. J. 1970. Clamp connections in two strains of *Cryptococcus neoformans,* p. 67–72. *In* D. G. Adhern (ed.), Recent trends in yeast research. Georgia State University, Atlanta.

Shields, A. B., and L. Ajello. 1966. Medium for selective isolation of *Cryptococcus neoformans*. Science **115**:208–209.

Stockman, L., and G. D. Roberts. 1983. Specificity of the latex test for cryptococcal antigen: a rapid, simple method for eliminating interference factors. J. Clin. Microbiol. **17**:945–947.

CHAPTER 64

Dermatophytosis

IRENE WEITZMAN

Disease: Dermatophytosis (ringworm, tinea).
Etiologic Agents: Species of the anamorphic genera *Microsporum, Trichophyton,* and *Epidermophyton floccosum*.
Source: Geophilic species acquired mainly from soil; zoophilic species mainly from animals; anthropophilic species primarily from humans; direct contact or from fomites.
Clinical Manifestations: Tinea barbae (beard), tinea capitis (scalp), tinea corporis (body), tinea cruris (groin), tinea favosa (favus), tinea imbricata, tinea manuum (hands), tinea pedis (feet), tinea unguium (nails); may be acute, chronic, or subclinical; range from slight erythema and scaling to highly inflammatory with folliculitis, suppurating boggy lesions (kerion), and extensive areas of alopecia and scarring.
Pathology: Usually limited to nonliving keratinized tissue (i.e., the stratum corneum of the skin, hair, and nails).
Laboratory Diagnosis: Observation of typical fungal structures in skin, hair, and nails; isolation of the etiologic agent from scrapings or hairs.
Epidemiology: Worldwide; some species geographically limited.
Treatment: Topical (clotrimazole, econazole, miconazole and newer imidazoles, haloprogin, tolnaftate, undecylenic acid) and systemic (griseofulvin, ketoconazole).
Prevention and Control: Wood light screening of scalps and suspected animal reservoirs (i.e., cats and dogs); routine inspection of scalps of children; hairbrush technique for culture of nonfluorescing scalps; improve hygiene and discourage sharing of clothing and accessories.

Description of Disease

Dermatophytosis (tinea or ringworm) is a superficial infection or colonization of the keratinized tissue of humans and lower animals (e.g., skin, hair, nails, fur) caused by a group of closely related keratinophilic fungi collectively termed *dermatophytes*. These fungi belong to three anamorphic (imperfect or asexual) genera, *Microsporum, Trichophyton,* and *Epidermophyton,* phylum *Deuteromycota,* those capable of reproducing sexually are classified in the teleomorphic genus *Arthroderma* (Weitzman et al., 1986) of the *Ascomycota*.

Although the infections caused by dermatophytes are largely cutaneous, reactions may range from mild to severe as a consequence of the host reaction to the metabolic products of the fungus, the virulence of the infecting strain or species, the anatomic location of the infection, and local environmental factors. Dermatophytic fungi are generally unable to penetrate deeper tissue, probably because of nonspecific inhibitory factors in serum (King et al., 1975) and inhibition of fungal keratinases (Dei Cas and Vernes, 1986). Occasionally, subcutaneous tissue is invaded (e.g., in Celsus kerion or Majocchi granuloma [Wilson et al., 1954]), sometimes with extensive lesions (Baker and Para, 1984) or mycetomalike processes (Barson, 1985; West and Kwon-Chung, 1980). Rare, generalized, systemic infections have also been reported (Araviysky et al., 1975; Swart and Smit, 1979).

Dermatophytoses comprise several distinct clinical entities depending on the area of localization (e.g., tinea capitis, ringworm of the scalp; tinea corporis, ringworm of the body; tinea pedis, ringworm of the feet). A single infecting species may evoke different symptoms depending on the site of infection, and different species may produce similar reactions in the same anatomic location.

Communicability and Transmissibility

The transmission and development of a dermatophyte infection are influenced by a variety of factors, including host preference and natural habitat of the fungus, virulence of the pathogen, location of the infection, individual host susceptibility, and local environmental factors.

Dermatophytes are grouped into three categories based on host preference and natural habitat. This grouping is helpful in determining the likely source of an infection. Geophilic dermatophytes (e.g., *Microsporum gypseum*) are soil-inhabiting fungi with a global distribution, usually of low virulence, and only rarely pathogenic for humans and lower animals. However, occasional outbreaks have been reported originating from contact with infected garden or nursery soil and secondarily involving human contacts (Alsop and Prior, 1961). Transmission from soil to humans via lower animals is considered very rare.

Animals are natural hosts to zoophilic species, but infections are frequently communicable from animals to humans either directly through animal contact or by indirect transmission. These infections are characteristically inflammatory, often suppurative, and sometimes self-healing. In urban areas, human infections (e.g., tinea capitis and tinea corporis) by *Microsporum canis* are most commonly transmitted by infected cats, sometimes dogs. Outbreaks often involve contact of susceptible children with stray kittens (Lawson and McLeod, 1957). Human-to-human transmission may occur, but virulence is apparently lost after repeated transmission (Duncan, 1945). *Microsporum canis* infections, acquired from cats and dogs, are also important in rural areas. However, the main zoophilic dermatophytes in rural areas are *Trichophyton verrucosum,* usually acquired through contact with infected cattle, and *Trichophyton mentagrophytes* var. *mentagrophytes,* acquired by contact with farm animals and dogs, but more often by handling grain and straw contaminated by infected mice; feed bins in barns are probably contaminated by hairs shed by infected rodents frequenting these areas (Georg, 1960; Georg et al., 1956). *Trichophyton verrucosum* has been isolated from wooden gateposts, wooden troughs, and stalls used by infected cattle (Rippon, 1985), and infected cattle hairs carried in clothing were reported to cause infections in a family (Georg et al., 1956).

Anthropophilic species primarily infect humans; transmission is often the result of either close human contact or indirectly, by sharing clothing, towels, combs, brushes, and the like. Tinea capitis (ringworm of the scalp) is a highly contagious disease and may spread rapidly within the family, institution, or school community. *Microsporum audouinii* caused major epidemics in prepubertal children throughout the United States and Canada in the 1900s to the early 1950s; thereafter it was replaced by *Trichophyton tonsurans,* which remains the chief etiologic agent of this disease in North America (Rippon, 1985). Large-scale epidemics of tinea capitis in other parts of the world or in geographically restricted areas are caused by endemic or localized species.

Transmission of tinea corporis is most frequently associated with tinea capitis in the immediate contacts of these patients. There is little conclusive evidence of other modes of transmission. However, a high rate of infection due to *Trichophyton rubrum,* among family contacts in India, has been reported under certain conditions involving highly susceptible (e.g., malnourished) individuals, an increase in virulence resulting from human passage, and environmental factors such as warmth and humidity (Georg, 1960).

Transmission of tinea cruris (jock itch) is primarily associated with shared washing and sanitary facilities, shared towels, and clothing. However, *Epidermophyton floccosum* has also been isolated from urinal bedpans, toilets, and sheets of patients in hospital wards during an epidemic (Neves and Canova-Xavier, 1964). Tinea cruris is frequently associated with tinea pedis, and infection is likely to have originated and been transferred from this source.

The transmission of tinea pedis (athlete's foot) is controversial. Some believe that exogenous exposure to infected skin scales leads to active clinical disease. Others consider exogenous exposure relatively unimportant, as most people are exposed to dermatophytes and are often asymptomatic carriers. In such cases, clinically active infections presumably result when conditions are suitable for multiplication of the fungus, such as changes in immunologic status or local environmental conditions. There is sufficient evidence to support both points of view (Rosenthal, 1974; Silva-Hutner et al., 1981).

Tinea unguium (infected nails) is unlikely to serve as a source of exogenous infection, although it may serve as a focus for endogenous reinfection. That only one or a few nails may be infected for years while others remain uninfected indicates, at best, a very low level of contagion.

Clinical Manifestations of Disease

Clinical manifestations of dermatophytosis are variable based on a number of factors (e.g., anatomic location, immune status, age and sex, virulence of the fungus, and environmental conditions). These clinical presentations will be discussed alphabetically under the anatomic site infected.

Tinea Barbae

Ringworm of the beard and moustache affects men only, is most common in rural areas, and is usually caused by zoophilic fungi (i.e., *T. verrucosum* and *T. mentagrophytes* var. *mentagrophytes*). These infections are highly inflammatory and present as acute, deep, pustular folliculitis that may progress to suppurating, nodular, boggy lesions (kerion). Similar severe infections in endemic urban areas may be caused by the anthropophilic fungi *Trichophyton violaceum* and *Trichophyton schoenleinii*. A less severe form appearing as dry, erythematous, scaly lesions may occur. Beard hairs are usually invaded.

Tinea Capitis

Tinea capitis is a dermatophyte infection of the scalp, eyebrows, and eyelashes caused by members of the genera *Microsporum* and *Trichophyton*. The infection may range from mild, almost subclinical, with slight erythema and a few patchy areas of scaling with dull gray hair stumps to a highly inflammatory reaction with folliculitis, kerion formation, and extensive areas of scarring and alopecia, sometimes accompanied by fever, malaise, and regional lymphadenopathy. Both the skin surface and hairs are involved.

Tinea Corporis

Tinea corporis, or ringworm of the glabrous skin, involves the trunk, shoulders, limbs, and nonhairy parts of the face. The infection classically manifests itself as circular, discrete, or concrescent lesions with varying erythema and scaling and raised, well-demarcated, erythematous, vesicular margins. Central healing is common. Infections may be inflammatory or noninflammatory, extensive, and chronic.

Tinea Cruris

Tinea cruris is an infection of the groin, perineal, and perianal area usually caused by *E. floccosum* or *T. rubrum*. Lesions are erythematous to tawny brown and covered with thin, dry scales. They are usually bilateral and often asymmetric, extending down the sides of the inner thigh and exhibiting a raised, sharply marginated border that is frequently studded with small vesicles. The penis and scrotum are usually spared. Infections by *T. rubrum* are usually more extensive, spreading to the buttocks, waist, and outer thighs.

Tinea Favosa (Favus)

Favus is a severe, chronic infection involving the scalp and glabrous skin, characterized by the formation of cup-shaped or saucerlike crusts resembling honeycombs (scutula), which originate in the hair follicles and are composed of dense masses of mycelium and epithelial debris. *Trichophyton schoenleinii* is the most common etiologic agent.

Tinea Imbricata

Tinea imbricata is a geographically limited, specialized manifestation of tinea corporis. It is characterized by the formation over extensive areas of the body of concentric rings of scales with raised borders. *Trichophyton concentricum* is the sole etiologic agent.

Tinea Manuum

Tinea manuum is a dermatophytic infection of the hand, most frequently presenting as unilateral, diffuse hyperkeratosis of the palms and fingers with accentuation of the flexural creases. It is most commonly caused by *T. rubrum*.

Tinea Pedis

The feet, especially the soles and toe webs, are most frequently involved in tinea pedis. The most common clinical manifestation is the intertrigenous form, which presents with maceration, peeling, and fissuring, mainly in the spaces between the fourth and fifth toes. Another common presentation is the chronic, squamous, hyperkeratotic type, where fine silvery scales cover pinkish skin usually of the soles, heels, and sides of the foot (moccasin foot). An acute inflammatory condition, characterized by the formation of vesicles, pustules, and sometimes bullae, is most frequently caused by *T. mentagrophytes*. The more chronic agents of tinea pedis are *T. rubrum*, *T. mentagrophytes* var. *interdigitale*, and *E. floccosum*.

Tinea Unguium

Tinea unguium is an invasion of the nail by a dermatophyte, whereas onychomycosis is a nondermatophytic infection or colonization of the nails. There are two main types of nail involvement: invasive subungual and superficial white mycotic infection (leukonychia trichophytica). Invasive subungual infection begins at the lateral or distal edges of the nail, appearing as white or yellow spots that eventually spread to the entire nail plate, causing it to become discolored, brown to black, pitted or grooved, brittle, friable, and thickened, sometimes cracked, usually separating from the nail bed, and accumulating subungual debris underneath. In the superficial white infection, opaque-white, well-demarcated single or multiple spots appear on the surface of the nail, which becomes rough, soft, and crumbly.

Pathology

In the susceptible host, the dermatophytes typically colonize the nonliving keratinized tissue (i.e., the stratum corneum of the skin, hair, and nails), occasionally penetrating the upper layer of the stratum granulosum, but do not occur between living cells (Beare et al., 1972). Occasionally, living tissue may be invaded, producing an acute purulent reaction or a granulomatous reaction that is usually of the nodular type (Chandler et al., 1980). In the nodular, granulomatous perifolliculitis of the legs caused by *T. rubrum* (Majocchi granuloma), fungal hyphae and arthroconidia may be observed in the hair follicle and in the inflammatory infiltrate in the dermis (Wilson et al., 1954).

The sequence of events leading to the establishment of an infection of the scalp caused by *M. audouinii* has been described (Kligman, 1955) and probably is similar in the early stages to infections caused by other dermatophytes. Hyphae, growing centrifugally from the point of origin in the stratum corneum mass at the follicular orifices, entwine themselves around the hair and become appressed to it. As the fungus proliferates on the hair surface, it grows downward on the hair, and the hyphae break up into chains of cells (primary spore formation). After the fourth day of inoculation, the intrafollicular hyphae penetrate the hair and descend to the upper limit of the keratogenous zone to form the Adamson fringe. External branches of the extrapilary hyphae then segment into short chains of ectothrix spores (arthroconidia). Infected hairs continue to grow and carry the infected portion out of the follicle. The weakened hairs usually break off about 2 to 5 mm above the scalp surface. In contrast, in certain *Trichophyton* infections, the arthroconidia are formed within the hair shaft (endothrix infection), causing it to become so fragile that it breaks flush with the scalp. The size and arrangement of the arthroconidia are species dependent.

Dermatophytes may differ in the type of keratin they will attack; that is, some will invade skin and nails but not hair (*Epidermophyton*), whereas others may attack hair and skin but rarely nails (*Microsporum*).

Diagnostic Features

The clinical picture of dermatophytosis varies with the infecting species, the site of the infection, and the host response. Dermatophytosis may mimic other diseases; hence, definitive diagnosis requires the isolation of a dermatophyte from the skin and its appendages.

Tinea Capitis

Ringworm of the scalp is strongly indicated by patchy alopecia (baldness), scaling, and erythema, sometimes with severe inflammation, acute pustular folliculitis, and kerion formation. This description also generally applies to tinea barbae. Hair loss, scutulum formation, and crusting is diagnostic of favus.

Tinea capitis may be differentiated from seborrheic dermatitis, psoriasis, alopecia areata, lupus erythematosis, pyoderma, impetigo, and secondary syphilis. Inflammatory tinea barbae resembles pyoderma caused by staphylococci.

Tinea Corporis

Concentric, circular, scaly lesions with a raised, erythematous, sometimes vesicular border and central clearing are suggestive of ringworm of the body. Tinea corporis mimics psoriasis, pityriasis rosea, nummular eczema, seborrheic dermatitis, secondary syphilis, and the like.

Tinea Cruris

Fungal infection of the groin should be considered when sharply marginated, reddish, dry, scaly lesions extending down to the inner thighs are observed, especially in men. Differential diagnosis should consider candidiasis, psoriasis, seborrheic dermatitis, contact dermatitis, and erythrasma.

Tinea Pedis

Ringworm of the feet is indicated when a patient presents with fissuring, scaling, and maceration of the toe webs; hyperkeratosis with fine silvery scaling covering an erythematous skin; or an acute vesicular eruption of the soles. Tinea pedis may mimic erythrasma, pustular psoriasis, bacterial infection, contact dermatitis, or secondary syphilis.

Tinea Unguium

Dystrophic nails with an accumulation of subungual debris is highly suggestive of a dermatophyte infection. Dystrophic nails most closely resembles psoriasis and, along with white superficial infection, also mimics those caused by nondermatophytic fungi. The manifestation of abnormal nails must be differentiated from a variety of other conditions due to systemic disease or to external or congenital causes (Rippon, 1982).

Prognosis

Dermatophytic infections are not considered life threatening; these fungi grow on dead keratinized tissue and rarely penetrate deeper into living tissue. Some infections may even heal spontaneously, such as tinea capitis caused by *M. audouinii,* which usually resolves by early adolescence, and some inflammatory infections caused by zoophilic fungi. The prognosis of other dermatophytic infections, especially the chronic noninflammatory kind, has improved markedly since the advent of griseofulvin and other newer antifungal agents. Certain dermatophytic infections are more refractory to treatment, however, and others may require prolonged administration of oral griseofulvin. Tinea unguium of the toenail is the most resistant to treatment and commonly recurs. Relapse and reinfection are also common in tinea pedis. Rare and invasive dermatophytic infections have been reported to be successfully treated with griseofulvin, amphotericin B, or ketoconazole (Baker and Para, 1984).

Epidemiologic Aspects

Occurrence and Transmission

Dermatophytoses occur worldwide, but the true number of infections is not known because cases are not reportable. However, based on the sales of antidermatophytic medications, the prevalence of infection appears considerable (Ajello, 1970). A clinical survey of dermatologic diseases of more than 20,000 people in the United States revealed that fungal infections constituted the second largest group of skin conditions in the general population (Johnson and Roberts, 1977). Although dermatophytoses have a global distribution, the anatomic site of the infection and the infecting species may vary with the geographic location. Tinea pedis is most commonly found in temperate climates where shoes are worn, whereas it is less frequent in unshod native populations in tropical regions. In warm and humid areas, tinea corporis is encountered more frequently. Tinea imbricata, caused by *T. concentricum,* the sole etiologic agent, is geographically limited to rural areas of Southeast Asia, Indonesia, and the South Pacific and to the Indian population of Central America, the Central Amazon region of Brazil, San Blas Island, and coastal Mexico (Rippon, 1985).

Overall, *T. rubrum* is the most common dermatophyte isolated worldwide and is most often associated with tinea pedis, tinea cruris, and frequently tinea corporis (Rippon, 1985; Sinski and Flouras, 1984). However, this distribution is also subject to geographic variation.

In ringworm of the glabrous skin, where the feet serve as the reservoir of infection, the chief etiologic agents are *T. rubrum* and *T. mentagrophytes,* whereas in regions where tinea capitis is common, the causative fungi of tinea corporis are usually those infecting the scalp in the particular geographic region. According to Rippon (1985), the anthropophilic agents of tinea capitis (*M. audouinii, T. tonsurans, T. violaceum,* and *T. schoenleinii*) are being eradicated from their traditional areas with the exception of *T. tonsurans,* which has displaced *M. audouinii* to become the dominant agent of tinea capitis in North America. *Trichophyton tonsurans* remains dominant in parts of Central and South America and the Caribbean. At present, "the seat of tinea capitis" is in Africa (Rippon, 1985). *Microsporum audouinii* is still prevalent and persists in large sections of Africa; in other areas of Africa, large-scale epidemics are caused by highly endemic, localized species such as *T. soudanense, T. violaceum, T. schoenleinii,* and *M. ferrugineum.* The species *T. violaceum* and *T. schoenleinii* are still common in central and eastern Europe and the eastern Mediterranean through Iran and Iraq. *Trichophyton violaceum* also extends to China and the Phillippines. In Portugal, Italy, Malaysia, Hong Kong, Indonesia, Australia, New Zealand, and parts of South America, *M. canis* is still the dominant or the most common agent of tinea capitis and is also responsible for occasional outbreaks in western Europe (Rippon, 1985).

Some dermatophytes are associated with certain occupations: *T. mentagrophytes* has been described in farmers, sheepshearers, sheepherders, laboratory workers handling rodents, and the military in Southeast Asia; *T. verrucosum* is mostly acquired by farmers handling infected cattle; *M. nanum* in pig farmers; *M. equinum* and *T. equinum* in horse handlers; and *M. gypseum* in agricultural and nursery workers (Kane and Krajden, 1983).

Zoophilic fungi are normally transmitted to humans via animals, and geophilic fungi are usually acquired from soil; secondary human-to-human transmission may then follow. Anthropophilic fungal infections may be acquired by close human contact indirectly via infected skin scales and hair, aerosols containing infectious arthroconidia (Friedman et al., 1960), or even by dust in the home (Mackenzie, 1961; Shimmura, 1985). Apparent sources of infection for widespread epidemics of tinea capitis are shared brushes, combs, caps, common bed linen, barber's scissors and clippers, and the backs of theater seats, as dermatophytes have been isolated from all of these sources (Philpot, 1977).

Tinea cruris is usually acquired from a reservoir in the nails or feet of the host, but may be indirectly transmitted to others via shared clothing, towels, infected bed sheets, toilet seats, urinal bedpans, and benches (Lundell, 1974; Neves and Canova-Xavier, 1964).

Transmission of tinea pedis is controversial; the evidence for both points of view have been reviewed (Philpot, 1977; Rosenthal, 1974; Silva-Hutner et al., 1981). Some claim that prolonged exogenous exposure is sufficient to incite clinical disease, whereas others consider such exposure relatively unimportant (Sulzberger et al., 1942). More important than exogenous exposure is the immunologic status of the host and local factors, such as moisture, maceration, trauma, and occlusion, that may lower the normally high degree of resistance (Allen and King, 1978; Allen and Taplin, 1973; Rosenthal, 1974). Scales and hairs may serve as infecting units for lengthy periods, as they remain viable for years (Guirges, 1981; Philpot, 1977; Rippon, 1985).

Susceptibility, Resistance, and Carriage

Susceptibility or resistance to dermatophytic infections may be attributed to a variety of factors, some relating to age, heredity, predisposing local factors, and underlying medical problems. Other important mechanisms include rapidity of cell turnover in the epidermis (Berk et al., 1976), serum inhibitory factors (King et al., 1975), and the immune response of the host.

Age plays a role in tinea capitis caused by *M. audouinii,* essentally a prepuberal disease that resolves with adolescence, probably because of the change in the sebum content to more fungistatic long-chain fatty acids (Rothman and Lorinez, 1963).

Heredity may be important in determining susceptibility to tinea imbricata caused by *T. concentricum* and to chronic dermatophytosis caused by *T. rubrum*. An autosomal recessive mode of inheritance or an autosomal dominant inheritance with incomplete penetrance may determine susceptibility to tinea imbricata infections (Hay, 1985). Chronic dermatophytosis is frequently found in patients and their relatives, who are often atopic. Both conditions (i.e., susceptibility to chronic *T. rubrum* infections and atopy in families) suggest involvement of genetic factors (Svejgaard, 1986).

Local and environmental factors that cause a predisposition to tinea pedis or to tinea corporis include friction, occlusion, trauma, and warm, moist environmental surroundings (Allen and King, 1978; Allen and Taplin, 1973; Baer and Rosenthal, 1966; Ryall, 1980).

Underlying diseases such as lymphocytic leukemia and immunosuppression, particularly involving the use of steroids, may result in invasive cutaneous infection (Baker and Para, 1984), whereas chronic *T. rubrum* infections have been associated with diabetes mellitus, Cushing's disease, atopy or lymphoma (Hay, 1982), and defects in cell-mediated immunity (Svejgaard, 1986).

Increased epidermal turnover may result in shedding of the infecting fungus (Berk et al., 1976; Jones et al., 1974). A serum inhibitory factor that prevents deep penetration of dermatophytes has been identified as unsaturated transferrin, which inhibits fungal growth by binding iron, a metal essential for growth of dermatophytes present in the epidermis (King et al., 1975).

The major role played by cell-mediated immunity in host defense against dermatophytosis has been summarized in several review articles (Kaaman, 1985a; Svejgaard, 1986). Patients with acute inflammatory dermatophytic infections, which resolve spontaneously, usually manifest a delayed-type reaction to intradermal injection of the dermatophytic antigen preparation trichophytin, whereas patients with chronic *T. rubrum* infections, often associated with atopy, show a lack of or an immediate response to trichophytin, both linked to impairment of cell-mediated immunity. Some patients with impaired cell-mediated immunity resulting from kidney transplantation, systemic steroid therapy, malignant diseases, and acquired immune deficiency syndrome may develop severe, chronic dermatophytosis (Svejgaard, 1986).

The carrier state in the human-to-human transmission of dermatophytes is controversial or may play a minor role. With regard to tinea pedis, healthy individuals with no clinically apparent infection have been found to harbor dermatophytic fungi (Rosenthal, 1974). These "unsuspected" cases may be considered important in the spread of tinea pedis (Verrone et al., 1985) if one accepts the concept of exogenous exposure playing a major role in the transmission of dermatophytosis. A carrier state has been postulated to explain an asymptomatic facial lesion caused by *T. soudanense,* which is endemic in Africa, in a black woman whose family had been residents of the United States for more than 100 years (Rippon and Medenica, 1964). Subclinical infections of the scalp, especially in the early stages, may play a more important role in the transmission of tinea capitis among children. Animals are frequently asymptomatic carriers (e.g., *M. canis* in cats); typically, the animal has no clinically apparent infection, but transmits the infection to humans when handled.

Etiologic Agents

Characterization

The etiologic agents of the dermatophytoses belong to the anamorphic (imperfect or asexual) genera *Microsporum, Trichophyton,* and *Epidermophyton.* For practical purposes, the identification of the dermatophytes in the clinical laboratory is largely based on the asexual form (e.g., morphology and arrangement of the conidia and colony characteristics) rather than on the teleomorph (sexual form). Descriptions of colonies are based on growth on Sabouraud dextrose agar containing 4% glucose and neopeptone. Sometimes these criteria alone are insufficient for specific identification; hence, physiologic tests may be required in conjunction with morphology.

Most dermatophytes produce two kinds of conidia in culture: large multicellular macroconidia and smooth, usually unicellular microconidia. The three genera may be distinguished on the basis of their microscopic morphology and the presence or absence of microconidia.

The genus *Microsporum* is characterized by the production of macroconidia and microconidia, which may be numerous to rare depending on the species. Macroconidia are typically rough walled (ranging from minutely roughened, echinulate to verrucose or warty), spindle shaped (ranging from obovate to cylindrofusiform), and usually borne singly on the hyphae. Macroconidia may be thin or thick walled, have 2 to 15 cells, and range from 6 to 160 μm long by 6 to 25 μm wide. Microconidia are pyriform to clavate.

In the genus *Trichophyton,* both microconidia and macroconidia may be produced, the former usually in greater abundance. Macroconidia are smooth, typically thin walled, range in shape from clavate to cylindric, and are usually borne singly along the hyphae. They vary in size (8 to 86 μm long by 4 to 14 μm wide) and number of septa (1 to 12). Microconidia may be spherical, pyriform, or clavate and are borne singly or in clusters.

Epidermophyton species lack microconidia. Macroconidia are numerous, typically smooth walled and fairly thick, broadly clavate (20 to 40 by 6 to 12 μm), have two to five cells, and are borne singly or more often in clusters of two or three. Only two species have been described; *E. floccosum* is the only pathogen. The characteristics of the pathogenic species are described below in alphabetic order.

Microsporum

Microsporum audouinii

No teleomorph is known for *M. audouinii.* The species is anthropophilic and the etiologic agent of prepuberal tinea capitis (gray patch, ectothrix) and tinea corporis. It is prevalent in sections of Africa and in isolated pockets in the Middle East and Romania, but almost extinct in formerly prevalent areas in North America and Europe (Rippon, 1985). Colonies grow slowly and are grayish white to light tan, velvety, and flat; reverse is colorless, salmon pink to light reddish brown. Conidia are usually absent. Mycelium is typically sterile, with thick-walled terminal chlamydospores, and often apiculate. Macroconidia, when present, are large, irregular, thick walled, and smooth to echinulate; microconidia, are clavate. Growth is poor on rice grains.

Microsporum audouinii var. *langeronii* is prevalent in parts of Africa. Colonies are rose tan with radial folding. More sporulation occurs than with *M. audouinii.* The organism is pathogenic for guinea pigs.

Microsporum audouinii var. *rivalieri* is endemic in parts of Africa and Florida. Growth is rapid, producing grayish-white, folded colonies. Macroconidia are long and spindle shaped, and few microconidia are produced on soil–hair medium (Zaias et al., 1965). Hyphae are bulbous and pectinate. This strain is also pathogenic for guinea pigs.

Microsporum canis

The teleomorph of *M. canis* is known as *Arthroderma (Nannizzia) otae.* This zoophilic organism is a significant etiologic agent of tinea capitis in South America, Italy, Portugal, Malaysia, Hong Kong, the Philippines, and Indonesia (Rippon, 1985) and of tinea corporis, rarely tinea barbae. Its distribution is scattered and worldwide. Colonies grow rapidly and are white to pale buff and downy; reverse is bright yellow, later becoming orange brown. Macroconidia, fusiform with an asymmetric knobbed apex, are usually numerous and measure 18 to 125 by 5 to 25 μm; they have up to 15 septa and thick, rough walls (Fig. 1). Microconidia are rare and clavate. Good growth and sporulation is produced on rice grains. The fungus perforates hair in vitro.

Microsporum canis var. *distortum* is velvety to cottony, with radial grooves; reverse is colorless to dull yellowish tan. Macroconidia are abundant, distorted, and bizarre; microconidia are also abundant.

Microsporum cookei

The teleomorph of *M. cookei* is *Arthroderma (Nannizzia) cajetani.* This geophilic species, which has a worldwide distribution, is rarely pathogenic. Growth is rapid, and colonies are yellowish, reddish tan, or greenish buff and powdery to granular; reverse is dark purple red. Macroconidia (30 to 50 by 10 to 15 μm) are numerous and fusiform; some resemble

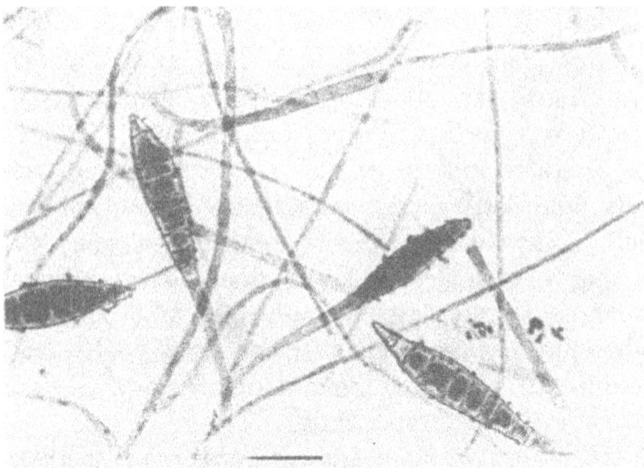

FIG. 1. Rough-walled macroconidia of *Microsporum canis*. Bar = 10 μm. (Courtesy of A. A. Padhye.)

those of *M. gypseum,* whereas others have thicker walls. Microconidia are abundant.

Microsporum equinum

No teleomorph is known for *M. equinum*. This species is zoophilic and causes ringworm of horses and humans; its geographic distribution is widely scattered. Colonies are white, pale buff to pale salmon and downy, velvety to finely powdery, some with radial grooves; reverse is buff to salmon. Macroconidia are infrequent, elliptic to broadly fusiform, thick walled, and verrucose, measure 18 to 60 by 5 to 13 μm, and consist of two to four cells. The fungus does not perforate hair in vitro; this characteristic distinguishes *M. equinum* from *M. canis,* which it closely resembles (Kane et al., 1982; Padhye et al., 1979).

Microsporum ferrugineum

No teleomorph is known for *M. ferrugineum*. This anthropophilic species is the etiologic agent of gray patch, ectothrix, tinea capitis, and tinea corporis. It is currently established in large areas of eastern Europe, was formerly predominant in Asia and Central Africa, has been largely eliminated from Japan, and is decreasing in Korea (Rippon, 1985). Colonies grow very slowly and are yellowish to rust colored, waxy, heaped, and folded, rarely white and velvety. Sporulation is absent on routine media; only sterile hyphae, sometimes with prominent septa (bamboo hyphae), are produced. Fusiform macroconidia are found on freezing agar and diluted Sabouraud agar (Vanbreuseghem et al., 1970). The organism is usually urease positive in urea broth, forms light colonies on Lowenstein–Jensen medium, and fails to grow better at 37°C compared with 25°C. The formation of flat, spreading colonies on lactose agar differentiates *M. ferrugineum* from *T. soudanense* (Weitzman and Rosenthal, 1983/84).

Microsporum gallinea

No teleomorph has been established for *M. gallinae,* a zoophilic agent of ringworm in chickens and other fowl and occasionally of tinea capitis and tinea corporis in humans. Its distribution is scattered and worldwide. Colonies, which show moderately rapid growth, are white tinged with pink, downy or satiny, and flat with radial folding; reverse shows raspberry-red pigment diffusing into the medium, macroconidia (15 to 50 by 6 to 8 μm) are moderately abundant, frequently curved and blunt tipped, and usually smooth walled, but may be echinulate at tip; 2 to 10 septa are typical. Microconidia are pyriform to clavate.

Microsporum gypseum Complex

The *M. gypseum* complex encompasses *M. gypseum* and *M. fulvum,* which are similar in colonial and microscopic morphology. Before 1963, *M. fulvum* was considered conspecific with *M. gypseum.* The discovery of the teleomorphs revealed *M. fulvum* to be a distinct species (teleomorph, *Arthroderma (Nannizzia) fulvum*) and the anamorph, *M. gypseum,* to be two species (teleomorphs, *Arthroderma (Nannizzia) incurvatum* and *Arthroderma (Nannizzia) gypseum*). Members of this complex can be clearly differentiated on the basis of mating studies, which are not performed in routine clinical laboratories; hence the term *M. gypseum* complex. These species are geophilic, with a global distribution. They occasionally cause ringworm in humans and animals. Colonies grow rapidly and are pale buff or rosy buff to light cinnamon, flat, and granular to floccose; reverse is buff to reddish brown; macroconidia (25 to 60 by 7 to 15 μm) are abundant and ellipsoid to fusiform, with up to five septa. Microconidia are clavate and may be few to moderately abundant.

Microsporum nanum

The teleomorph of *M. nanum* is *Arthroderma (Nannizzia) obtusum*. This species is geophilic or zoophilic, causing ringworm in pigs and occasionally tinea capitis or tinea corporis in humans. Distribution is sporadic and worldwide. Colonies grow moderately rapidly and are buff to light cinnamon, flat, and powdery to downy; reverse is orange tan, later becoming reddish brown, macroconidia are abundant and obovoid to elliptic, measure 10.5 to 30 by 6.5 to 13 μm, and usually consist of one to two cells. Microconidia are few to moderate and pyriform to clavate in shape.

Microsporum persicolor

The teleomorph of *M. persicolor* is *Arthroderma (Nannizzia) persicolor*. This geophilic or zoophilic organism is associated with rodents, especially bank and field voles; it is a rare agent of tinea capitis, corporis, or pedis in humans. Distribution is sporadic and worldwide. Colonies grow rapidly and are yellowish buff to peach and powdery to downy; reverse is reddish brown. Macroconidia are rare to moderate, clavate, thin walled, and smooth to finely echinulate (more evident under oil immersion and when grown on soil and hair). Microconidia are abundant, spherical to pyriform, and borne singly along the sides of the hyphae or in grapelike clusters. Spirals are numerous in fresh isolates. This fungus resembles *T. mentagrophytes,* but may be differentiated by its echinulate macroconidia (smooth in *T. mentagrophytes*) and formation of peach to wine-rose colonies on cereal agar or sugar-free peptone agar (Padhye et al., 1973a).

Microsporum praecox

No teleomorph has been established for *M. praecox.* It is geophilic, has been isolated from sites associated with horses (DeVroey et al., 1983), and occasionally causes tinea corporis in France, Belgium, and the United States (DeVroey et al., 1983; Weitzman and McMillen, 1980). Colonies grow rapidly and are cream to yellowish tan and powdery, with radial or concentric folding; reverse is yellow to orange. Macroconidia (40 to 90 by 7 to 17 μm) are abundant, thin walled, long, fusiform, and verruculose, consisting of three to nine cells; some have apical appendages. Microconidia are absent.

Microsporum racemosum

The teleomorph of *M. racemosum* is known as *Arthroderma (Nannizzia) racemosum.* This geophilic fungus has been isolated from soil in Venezuela and Romania and from lesions in two patients in Chicago (Daum and McCloud, 1976; Rippon and Andrews, 1978). Growth is rapid, producing colonies that are white to cream, flat, and powdery to granular; reverse is wine red. Macroconidia are abundant and often have a long terminal filament; they are ellipsoid to fusiform, finely echinulate, and thin walled, measure 15 to 25 by 40 to 60 μm, and consist of 5 to 10 cells. Microconidia are numerous, clavate, and arranged in racemes or grapelike clusters.

Microsporum vanbreuseghemii

The teleomorph of *M. vanbreuseghemii* is *Arthroderma (Nannizzia) grubyi.* The organism is geophilic, rarely causing ringworm in animals and humans. Colonies grow rapidly and are flat, white, cream yellow, or pink to deep rose, and granular to downy; reverse is colorless to pale yellow; macroconidia (44 to 87 by 8 to 14 μm) are numerous, cylindrofusiform, thick walled, and finely echinulate, with up to 12 septa. Microconidia are abundant and pyriform to obovate in shape.

Trichophyton

Trichophyton ajelloi

The teleomorph of *T. ajelloi* is known as *Arthroderma uncinatum.* It is geophilic, rarely causing ringworm in animals or humans. Distribution is worldwide. Colonies grow rapidly and are cream to orange tan, flat, and powdery to downy; reverse is usually blackish purple, occasionally buff. Macroconidia (20 to 65 by 5 to 10 μm) are numerous, fusiform to cylindrofusiform, and thick walled, consisting of 5 to 12 cells. This species resembles *M. vanbreuseghemii* except for its smooth walls). Microconidia are rare.

Trichophyton concentricum

No teleomorph is known for *T. concentricum,* which is anthropophilic, causing tinea imbricata in certain Pacific Islands, Southeast Asia, sections of Mexico, and Central and South America. Colonies grow very slowly and are cream colored, beige, or honey to orange brown, heaped and convoluted, and glabrous, later becoming velvety. Conidia are absent on routine media; only chlamydospores and distorted hyphae resembling rudimentary antlers of *T. schoenleinii* are produced. Thiamine stimulates growth in 50% of cultures.

Trichophyton equinum

No teleomorph is known for *T. equinum.* The organism is zoophilic, causing ringworm in horses; human infection is rare. Distribution is sporadic and worldwide; however, an autotrophic variety has been reported in Australia and New Zealand (Smith et al., 1968). Colonies grow rapidly and are white or cream-colored to tan with a yellowish periphery, velvety, and sometimes folded; reverse is yellow, becoming orange to reddish brown with age. Macroconidia are rare, clavate to cylindric, thin walled, and have three to four septa. Microconidia are abundant, globose, pyriform to elongate, sometimes stalked, and arranged singly along the hyphae. Most isolates require nicotinic acid, except var. *autotrophicum.*

Trichophyton gourvilii

No teleomorph has been established for *T. gourvilii.* The species is anthropophilic, geographically limited to parts of Africa, and chiefly causes tinea capitis (endothrix type). Growth is slow, producing lavender to purple colonies that resemble those of *T. violaceum;* they are heaped, irregularly folded, and initially waxy, becoming velvety to powdery with age.

Sporulation is poor, initially producing only irregular hyphae and arthroconidia. Macroconidia are not observed in routine culture media; microconidia are pyriform and borne singly along with hyphae.

Trichophyton megninii

No teleomorph is known for *T. megninii*. This fungus is anthropophilic, causing tinea barbae (dry) and less commonly tinea corporis and tinea capitis (ectothrix, conidia in chains). It is endemically restricted to Portugal, Sardinia, and parts of Africa, but has been reported in Canada (Kane and Fischer, 1975). Colonies grow slowly and are white, becoming pink with age, and velvety with radial folding; reverse is deep red. Macroconidia are uncommon, long, cylindric, and thin walled. Microconidia are numerous and pyriform to clavate. Colonies require L-histidine for growth and are urease positive.

Trichophyton mentagrophytes

The teleomorphs of *T. mentagrophytes* are *Arthroderma benhamiae* and *Arthroderma vanbreuseghemii*. The organism may be anthropophilic (var. *interdigitale*), or zoophilic (var. *mentagrophytes*, var. *erinacei*, var. *quinckeanum*); it is a common agent of tinea pedis, tinea corporis, tinea unguium, tinea cruris, tinea manuum, and tinea barbae and rarely causes tinea capitis (ectothrix, microides). Distribution is worldwide. Growth is rapid, with colonies varying in pigmentation (cream, tan, pale yellow, or pink) and texture (powdery, granular, velvety, or cottony); reverse is usually cream, tan, or reddish brown, but may be yellow, deep yellow to orange, pink to red, or brown with a diffusible melanoid pigment. Macroconidia range from rare to abundant and are mostly clavate (Fig. 2), sometimes cylindric, thin walled, and have two to five cells. Microconidia are usually abundant, spherical (Fig. 2), pyriform or clavate, and borne singly along the sides of the hyphae or in grapelike clusters. Spiraled hyphae or nodular organs are common in some varieties. The fungus perforates hair in vitro, is typically urease positive, and has no special nutritional requirements. *Trichophyton mentagrophytes* var. *mentagrophytes* has a powdery to granular surface, abundant microconidia, small to moderate numbers of macroconidia, and spirals. *Trichophyton mentagrophytes* var. *interdigitale* is velvety to downy, with no macroconidia and small to moderate numbers of microconidia.

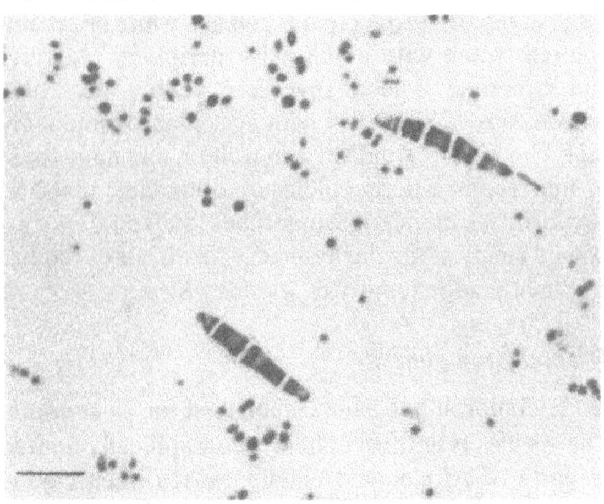

FIG. 2. Smooth-walled macroconidia and spherical microconidia of *Trichophyton mentagrophytes*. Bar = 10 μm. (Courtesy of A. A. Padhye.)

Trichophyton raubitschekii

No teleomorph has been determined for *T. raubitschekii*. This anthropophilic organism, the etiologic agent of tinea corporis and tinea cruris, has been isolated in Canada and the United States. Colonies grow rapidly and are buff, granular to velvety, and flat with a raised center, developing radial grooves with age; reverse is blood red. Macroconidia, abundant on primary isolation, are long, narrow, pencil shaped with blunt ends, and consist of five to nine cells. Microconidia are numerous, subspherical, globose to clavate, and borne laterally along the sides of the hyphae, either sessile or on a short stalk. The fungus has no special nutritional requirements and does not perforate hair in vitro. It resembles *T. rubrum* morphologically, but unlike *T. rubrum* is urease positive, lacks red pigment on cornmeal dextrose agar, and forms restricted colonies on lactose agar (Kane et al., 1981).

Trichophyton rubrum

No teleomorph is known for *T. rubrum*. This anthropophilic fungus is the most common etiologic agent of tinea pedis, tinea cruris, and tinea unguium, often causes tinea corporis, tinea barbae, tinea manuum, and rarely causes tinea capitis. It is the most frequently isolated dermatophyte worldwide (Rippon, 1985; Sinski and Flouras, 1984). Colonies exhibit slow to moderate growth and are white, cream, pink to purple, and flat to raised, often with a central umbo; they may be powdery, granular, or cottony. Reverse is usually wine red. Macroconidia may be absent, rare, or numerous; they are cylindric with rounded ends, sometimes clavate, and have three to eight cells and thin walls. Microconidia may be rare to numerous, clavate or pyriform, and are born singly along the sides of the hyphae. The organism is urease negative, has no special nutritional requirements, does not perforate hair in vitro, produces red pigment on cornmeal dextrose agar, and reacts as (−) or (a) mating type when paired with *A. simii* (+) or (A) (Young, 1972).

Seven colony types of *T. rubrum* have been described: downy, granular, melanoid, African, Rodhainii, dysgonic, and hyperpigmented (Young, 1972).

Downy Type. Colonies are white, pink to tan, or greenish pink, fluffy, and hemispheric or umbonate; reverse is wine red, sometimes orange. Number of microconidia varies; macroconidia are absent.

Granular Type. Colonies are creamy white to pink, with a powdery suede to powdery velvet texture; reverse is red tan to wine red. Macroconidia and microconidia may be rare to numerous.

Melanoid Type. The melanoid type resembles the downy type, but is characterized by a brown (melanoid) diffusible pigment on the reverse.

African Type. Colonies are cream to buff pink, compactly heaped and folded, and have a powdery center with a flat periphery; reverse is tan to red tan, with diffusible red pigment produced occasionally; macroconidia and chlamydospores are numerous; microconidia are variable.

Rodhainii Type. Colonies grow slowly and are glabrous, folded, and deep purple, becoming white and velvety to granular with age. Chlamydospores are numerous; numbers of macroconidia and microconidia may increase with aging.

Dysgonic Type. Colonies are growing, usually deep purple, and wooly to granular; peripheral growth is commonly submerged and feathery. Pyriform microconidia and chlamydospores are variable; macroconidia occasional, and hyphae often stunted.

Hyperpigmented Type. Colonies are violet to reddish violet and have a velvety surface with radial furrows; reverse is wine red, sometimes with diffusible pigment. Occasional macroconidia and many chlamydospores are produced.

Trichophyton schoenleinii

No teleomorph is known for the anthropophilic fungus *T. schoenleinii*, the etiologic agent of favus as well as tinea corporis and tinea unguium. Its distribution encompasses eastern Europe, Scandinavia, the Near East, regions of the Middle East, and Africa, with scattered foci in the United States, Canada, and Central and South America (Rippon, 1985). Colonies grow very slowly when first isolated and are white, cream, or pale yellow to shades of brown, irregularly heaped and folded, and waxy, becoming suedelike or velvety with age. Macroconidia are absent, and the clavate microconidia are absent to rare. Hyphae are irregularly branched and broadened (favic chandeliers), and chlamydospores are common. There are no special nutritional requirements, and growth is not stimulated at 37°C.

Trichophyton simii

The teleomorph of *T. simii* is *Arthroderma simii*. The organism may be geophilic (isolated from soil in India) or zoophilic (causing ringworm in small mammals, especially monkeys, and poultry); human infection is rare. Disease is endemic in India, but scattered human cases have been reported elsewhere. Colonies, which grow rapidly, are cream to buff, flat, and granular; reverse is yellow to reddish brown. Macroconidia are abundant, clavate, fusiform or cylindric, and thin walled, with 4 to 10 septa, often in small clusters. Endochlamydospores form in macroconidia in old cultures. Microconidia are abundant, pyriform to clavate, and borne singly along hyphae.

Trichophyton soudanense

No teleomorph has been found for *T. soudanense*. This anthropophilic organism, the etiologic agent of tinea capitis (large-spored endothrix) as well as tinea corporis and tinea unguium, is endemic in parts of Africa; scattered cases reported elsewhere are mostly attributable to immigrants from Africa. Colonies are slow growing, deep yellow, apricot, or violet, flat, leathery, and suedelike to velvety, with an eyelash or fringed margin; reverse is deep yellow. Macroconidia are absent on routine media. Microconidia may be absent or rare and are pyriform to clavate. Reflexive hyphae (lateral branches growing in opposite direction from main hyphae) are characteristic. Chlamydospores are common. The fungus produces light brown to black colonies on Lowenstein–Jensen medium, is urease negative; grows better at 37 than at 25°C, and usually forms restricted colonies on lactose agar (Weitzman and Rosenthal, 1983/84).

Trichophyton terrestre

The teleomorphs of *T. terrestre* are *Arthroderma quadrifidum*, *Arthroderma lenticularum*, and *Arthroderma insingulare*. This geophilic organism, rarely pathogenic for humans, is more often found as a contaminant of human skin. Distribution is worldwide. Colonies show rapid growth and are white or cream to pale yellow and powdery or granular, becoming velvety to cottony with age; reverse is pale yellow or yellow brown to rose red. Macroconidia are abundant in granular colonies, rare or absent in others; they are clavate to cylindric with rounded ends, have thin walls, and consist of two to six cells. Microconidia are numerous, clavate to pyriform, and appear singly or in clusters. Transitional forms (intermediate between microconidia and macroconidia) are characteristic. This fungus resembles *T. mentagrophytes* morphologically and physiologically, but differs by the presence of transitional conidia and by the inability to grow at 37°C.

Trichophyton tonsurans

No teleomorph is known for *T. tonsurans*. This anthropophilic organism is the dominant etiologic agent of epidemic tinea capitis in children and adults (black dot, large-spored endothrix) in North America, parts of South America, and the Caribbean; an important cause of tinea corporis in adults and children in Chicago (Bronson et al., 1983); and, less commonly, a cause of tinea unguium. It is widely distributed, with variable frequency in other parts of the world. Colonies grow slowly, and their color, texture, and topography vary with the isolate (e.g., white, cream, pale yellow, sulfur yellow, tan, or rose; powdery, suedelike, or velvety; flat, flat with a raised or depressed center and radial folding, or raised with concentric or irregular folding); reverse is yellow to mahogany red. Macroconidia are rare, clavate, usually curved at the top, and have two to three cells. Microconidia are numerous, variable in size and shape (Fig. 3), clavate to elongate, some swollen like balloons, and often attached to branched, thickened, and knobby hyphae. Growth is stimulated by thiamine. Some strains perforate hair (Matsumoto et al., 1983).

Trichophyton verrucosum

No teleomorph has been established for *T. verrucosum*. This zoophilic organism is a common agent of ringworm in cattle and dairy herds worldwide and the etiologic agent of inflammatory tinea barbae, tinea capitis (large-spored ectothrix), and tinea corporis; colonies grow very slowly and are yellow ochre, glabrous and folded (var. *ochraceum*); grayish white, glabrous, heaped, and folded, becoming velvety with age (var. *album*); and flat, grayish white, velvety, and disk shaped (var. *discoides*). Only hyphae and chlamydospores (characteristically in chains) are observed in routine culture media. Pyriform to clavate microconidia are produced in media containing thiamine. Macroconidia are rare and variable in shape. Growth is stimulated at 37°C. Some isolates require thiamine; most require thiamine and inositol.

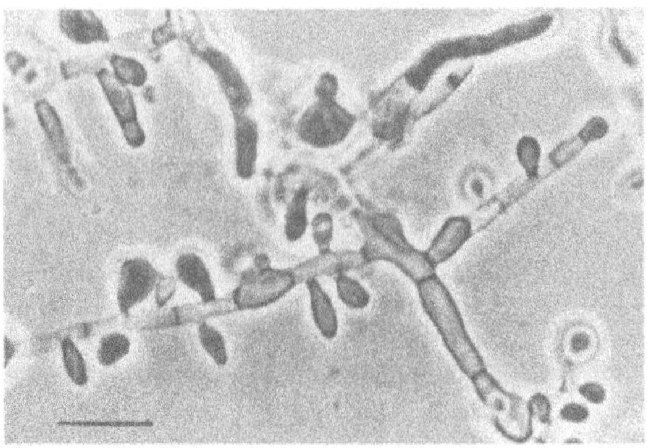

FIG. 3. Microconidia of *Trichophyton tonsurans*. Bar = 10 μm. (Courtesy of A. A. Padhye.)

Trichophyton violaceum

No teleomorph is known for *T. violaceum*. This anthropophilic organism is the dominant agent of tinea capitis (black dot, endothrix) in children and adults in eastern Europe and common in southern Europe, Africa, the Near East, and most of Asia. It is also an important etiologic agent of tinea corporis, tinea barbae, and tinea unguium in endemic areas. Colonies grow very slowly and are lavender to deep purple (occasionally nonpigmented), glabrous, waxy, heaped, and convoluted, sometimes becoming velvety and developing white sectors; reverse is purple. Microconidia and macroconidia are absent on routine culture media. Sterile hyphae and sometimes chlamydospores are observed. Growth and sporulation are stimulated by thiamine.

Trichophyton yaoundei

No teleomorph is known for *T. yaoundei*, an anthropophilic fungus that causes tinea capitis (endothrix) chiefly in equatorial Africa. Colonies grow very slowly and are cream to tan (becoming deep tan to chocolate brown with prolonged incubation), glabrous, waxy, heaped, and irregularly folded; brown pigment may diffuse into medium. Conidia are absent on routine culture media; only thick, irregular, nonsporulating hyphae and chlamydospores are found. Microconidia, produced on casein agar, are pyriform and appear singly along the sides of the hyphae. There are no special nutritional requirements (Georg et al., 1963).

Epidermophyton

Epidermophyton floccosum

No teleomorph is known for *E. floccosum*. This anthropophilic organism is an important etiologic agent of tinea cruris, tinea pedis, and tinea unguium; it does not invade hair. Distribution is global. Colonies grow slowly and are yellowish olive or olive green, flat or heaped and folded in the center (radial folding often extending towards the periphery), powdery, suedelike, or velvety, frequently with white cottony tufts on aging; reverse is orange brown. Macroconi-

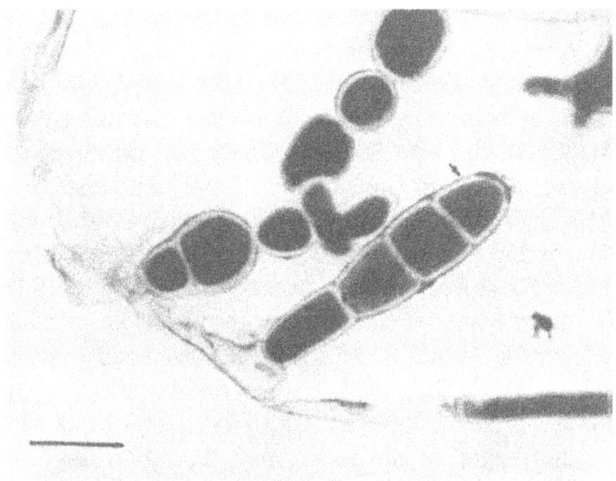

FIG. 4. Smooth-walled macroconidium (*arrow*) of *Epidermophyton floccosum*. Bar = 10 μm. (Courtesy of A. A. Padhye.)

dia (20 to 40 by 6 to 8 μm) are abundant, broadly clavate, smooth with moderately thick walls, and borne singly (Fig. 4) or more often in clusters of two to three. Microconidia are absent.

Specimen Collection and Transport

The site to be cultured is usually cleansed by wiping the area with a sterile gauze pad moistened with 70% alcohol or with soap and water. If lesions are inflamed and painful, use sterile distilled water. Avoid using cotton balls because they leave fibers, which may be mistaken for fungous hyphae on direct microscopic examination.

Glabrous Skin

Scrape epidermal scales from the active border of the lesion with a sterile scalpel (the center is usually the area of healing).

Feet

Remove and discard the macerated tissue and outer debris before scraping the interdigital spaces with a sterile scalpel (including the space between the fourth and fifth toes). If vesicles are present, as in vesicular tinea pedis, remove the roof of the vesicle with a sterile scissors for microscopic examination and culture. Do not collect blister fluid.

Hairs

Examine the hairs with a Wood lamp (filtered ultraviolet light). Hair stumps, which fluoresce a bright green (Wood light positive), are preferable for microscopic examination and for culture. However, not all invaded hairs fluoresce, depending on the infecting species. Pluck broken or fluorescent hairs with a fine forceps or eyebrow tweezers. Nonfluorescent hair stubs broken at the skin surface (black-dot endothrix infections) should be dug up from the follicle with the point of a sterile scalpel; it may be difficult to grasp the hairs with a tweezer. In favus, the scutulum at the mouth of the hair follicle is suitable for culture and microscopic examination.

Approximately 5 to 12 hair stumps should be collected. Alternately, sterile disposable brushes may be used to collect hairs. This method is especially suitable to collect hairs from animals and small children.

Nails

Examine the nail to determine if the infection is on the surface, as in white superficial mycotic infection, or is invasive subungual. For the former, scrape the chalky white areas with the point of a scalpel; for the latter, proceed as follows: if the nails are thick, bore a small hole or groove with the tip of a scalpel, discard surface material containing contaminants or nonviable fungi, and collect the waxy, yellowish subungual debris underneath for culture and examination (Silva-Hutner et al., 1981). Otherwise, remove debris underneath the nail at the tip and collect the friable material at the juncture of the nail with the nail bed.

As dermatophytes can survive for years if kept dry, they need not be cultured immediately after collection. Specimens may be collected in clean pill envelopes, in sterile petri dishes, or between surface-sterilized glass slides. If the sample is to be mailed, it is best to collect it in a paper packet. Do not place specimens in sealed glass vials, as the high humidity generated in an airtight container encourages the growth of contaminants.

Isolation and Identification Procedures

The standard medium employed in medical mycology for the isolation and identification of dermatophytes is Sabouraud dextrose agar. It is essentially a glucose (4%)–peptone medium with a pH of 5.6. The Emmons modification, with less glucose (2%) and a higher pH (between 6.8 and 7.0), is preferred for better growth, inducing more sporulation, and for maintenance of stock cultures. As Sabouraud dextrose agar will permit the growth of saprophytic fungi and bacteria, other media are recommended for isolation of dermatophytes. These media include Sabouraud dextrose agar with the addition of 500 mg/liter cycloheximide (Actidione; The Upjohn Co., Kalamazoo, Mich.) and 40 to 50 mg/liter chloramphenicol (Chloromycetin; Parke, Davis & Co., Detroit, Mich.) or dermatophyte test medium as recommended by Taplin et al. (1969). Dermatophyte test medium contains antibacterial antibiotics and cycloheximide, in addition to phenol red, a pH indicator that turns from yellow (acid) to red (alkaline) with dermatophytic growth. Although dermatophyte test medium is effective as an isolation medium, it has its limitations because certain nondermatophytic fungi will turn it red (Merz et al., 1970; Salkin, 1973). These media are all commercially available. A medium lacking cycloheximide should always be included in primary isolation, as some nondermatophytic fungi cause lesions resembling those produced by dermatophytes but are inhibited by cycloheximide (Gentles and Evans, 1970; Hay and Moore, 1984). A special medium for the early detection and identification of *T. verrucosum* has been formulated (Kane and Smitka, 1978).

Generally, cultures should be incubated aerobically at 25 to 30°C (include 37°C if *T. verrucosum* is suspected). They should be examined for growth within 1 to 2 weeks of incubation. If *T. verrucosum* or other slow-growing dermatophytes are suspected, the culture should be retained up to 4 weeks before it is discarded as negative.

If good growth is present and sporulation is indicated, prepare a teased mount and examine microscopically for the presence of characteristic conidia. Teased mounts are prepared by tearing apart, with two stiff needles, small fragments of aerial mycelium from the mature portion of a colony in a drop or two of mounting fluid (usually lactophenol cotton blue), then adding a cover slip. Identification of dermatophytes is often based solely on the morphology of the conidia and the characteristics of the colony on Sabouraud dextrose agar. However, additional procedures may be necessary, such as preparation of slide cultures to examine the arrangement of undisturbed conidia, other media (e.g., potato or cornmeal dextrose agar) may be utilized to enhance sporulation, and additional tests may be required for definitive identification, especially for *Trichophyton* species. These include in vitro hair perforation, urease activity, nutritional requirements, optimum temperature for growth, pigment formation on special media, and the like. It is best to subculture primary isolates from the periphery to minimize contamination.

Preliminary Identification

The direct microscopic examination of clinical specimens (hair, skin, and nails) and observation of fungal structures may provide the etiology of the infection, indicating the causative genus and sometimes even the species. This is especially the case with regard to the observation of the size and arrangement of the arthroconidia in the infected hair. Typically, hairs infected by *Microsporum canis, M. audouinii,* and *M. ferrugineum* appear as a mosaic sheath of arthroconidia 2 to 3 μm in diameter surrounding the hair shaft (ectothrix); the hairs fluoresce brilliant green. Arthroconidia 5 to 8 μm in diameter, appearing as sparse chains on the outside of the hair shaft, are typical of the *M. gypseum* complex; hairs are not fluorescent. Invasion by *T. verrucosum* is characterized by a sheath of large-spored arthroconidia (6 to 10 μm in diameter, megaspore ectothrix) around the hair; hairs are not fluorescent. Nonfluorescent short hair stumps filled with arthroconidia arranged in chains 5 to 8 μm in diameter (endothrix) are characteristic of *T. tonsurans, T. violaceum, T. soudanense, T. gourvilii,* and *T. yaoundii* (Rebell and Taplin, 1974). Long hairs filled with hyphal elements, empty areas suggesting air tunnels, and fat droplets are typical of *T. schoenleinii*. Hairs may show a dull fluorescence.

In skin scrapings, all three genera of the dermatophytes are similar in appearance and are observed as hyaline, septate hyphae of varying length, 2 to 4 μm in diameter, that often branch and frequently fragment into arthroconidia.

The simplest example of the direct microscopic examination, in providing a preliminary identification, is the potassium hydroxide preparation. Skin scrapings or hairs are placed in a few drops of 10% KOH (20% for nails) on a glass slide, a cover slip is added, and the preparation is heated gently over a low flame (avoid boiling) to clear. The cover slip is then gently pressed to flatten the preparation before examination. The calcofluor white technique, used alone or with KOH, may simplify the detection of fungal elements and obviate the inaccuracies associated with the interpretation of KOH preparations, especially for the novice, and is easier to prepare than the periodic acid–Schiff stain (Hageage and Harrington, 1984).

Definitive Identification and Biotyping

Most dermatophytes are identified on the basis of their microscopic morphology and gross colony characteristics on Sabouraud dextrose agar. Sometimes, the fungus does not sporulate or appears atypical and may, therefore, require additional media to stimulate sporulation or special physiologic tests in conjunction with morphology for definitive identification. Some of the more frequently used tests include growth on polished rice grains; determination of special nutritional requirements; in vitro hair perforation; urease activity; pigment production on special media; optimum temperature of growth; and mating studies (performed in special reference laboratories).

Stimulation of Sporulation

Cornmeal agar and potato dextrose agar, both available commercially, are routinely used to stimulate sporulation in pathogenic fungi. Sometimes other media, not commercially available, may be necessary for this purpose when routine media do not succeed. These special media include polished rice grains (Conant, 1936); cereal agar (Padhye et al., 1973b); and Borelli lactrimel medium and a modification of this formulation (Kaminski, 1985).

Growth on Polished Rice Grains

Microsporum audouinii, unlike most dermatophytes, grows poorly on polished rice grains and produces a brownish discoloration. This test is especially useful to differentiate *M. audouinii* from atypical isolates of *M. canis* and from other *Microsporum* spp. that grow well and sporulate profusely on this medium (Conant, 1936).

Nutritional Requirements

Nutritional tests were originally described by Georg and Camp (1957). They devised this method as a routine aid in the identification of dermatophytes, especially those species with an absolute or partial growth requirement for certain vitamins or amino acids. This test is helpful in distinguishing between certain *Trichophyton* species that resemble each other morphologically. The media are available commercially as *Trichophyton agars* nos. 1 through 7 (Difco Laboratories, Detroit, Mich.). This technique employs a casein basal agar that is vitamin free (*Trichophyton* agar no. 1) to which various vitamin solutions have been added as follows: no. 2, casein agar plus inositol; no. 3, casein agar plus thiamine and inositol; no. 4, casein agar plus thiamine; and no. 5, casein agar plus nicotinic acid. *Trichophyton* agar no. 6 is an ammonium nitrate basal medium, which is similar to casein agar except for the substitution of ammonium nitrate for casein; no. 7 contains the ammonium nitrate basal medium plus the amino acid L-histidine. As a result of these tests, Georg and Camp (1957) demonstrated that *T. tonsurans* has a partial requirement for thiamine; *T. violaceum* and 50% of isolates of *T. concentricum* are stimulated by thiamine; 16% of strains of *T. verrucosum* require thiamine and 84% require thiamine and inositol; *T. equinum* requires nicotinic acid; and *T. megninii* requires L-histidine.

It is important to inoculate these media with a small fragment of aerial mycelium (pin-head size) and avoid carrying over nutrients from the previous medium. Appropriate quality control organisms should be included each time the nutritional tests are performed to verify the proper reaction of the media.

In Vitro Hair Perforation Test

The in vitro hair perforation test, devised by Ajello and Georg (1957) to distinguish between atypical isolates of *T. mentagrophytes* and *T. rubrum*, has also been employed as an additional means to distinguish *M. canis* from *M. equinum* (Kane et al., 1982; Padhye et al., 1979). The method requires placing short segments of human hair in petri dishes and autoclaving them at 120°C for 10 min, followed by the addition of 25 ml of sterile distilled water and 2 to 3 drops of 10% sterilized yeast extract. Plates are then inoculated with several fragments from the test culture and incubated at 25°C for up to 21 days. Hair fragments showing growth are removed from the petri dish after 10 to 14 days, placed in a few drops of lactophenol cotton blue on a slide, covered with a cover slip, and heated gently and briefly to intensify staining and aid detection of the perforations. The presence of wedge-shaped perforations along the hair constitutes a positive test; *T. mentagrophytes* and *M. canis* perforate hair, whereas *T. rubrum* and *M. equinum* do not. Results of hair perforation tests with other dermatophytes are described by Ajello and Padhye (1985).

Urease Activity

Urease activity is most frequently used to differentiate *T. mentagrophytes* (typically urease positive) from *T. rubrum* (usually urease negative). It may also be used to differentiate between *T. rubrum* and *T. raubitschekii* (Kane et al., 1981) and *T. rubrum* from *T. megninii* (Rosenthal and Sokolsky, 1965). Christensen urea agar or broth may be used, but Kane and Fischer (1971) consider the broth to be more sensitive for this purpose. After the medium is inoculated, it is incubated for up to 7 days at 25 to 28°C. A positive reaction within this period is indicated by the appearance of a purple-red color in the medium; a negative reaction, indicating a lack of urease activity, is observed as no noticeable color change or a slight change to pale pink. Only pure cultures are to be employed in this test. Screw caps must be loosened for aeration. When testing unknown isolates, *T. rubrum* and *T. mentagrophytes* should be used for quality control.

Pigment Production on Special Media

Cornmeal agar with the addition of 1 to 2% glucose is helpful as an additional means of separating *T. rubrum* from *T. mentagrophytes*. The former produces a red pigment on this medium when incubated for 2 to 3 weeks at room temperature, whereas the latter does not. The formation of pink to deep rose colonies on cereal agar by *M. persicolor* may aid in distinguishing it from *T. mentagrophytes*, which lacks such colony colors (Padhye et al., 1973b).

Optimum Temperature for Growth

The ability to grow better at 37°C than room temperature, in conjunction with the nutritional tests, differentiate *T. verrucosum* from *T. schoenleinii,* which it closely resembles. The test is performed by placing a small inoculum of equal size on the surface of each of two SDA slants. One tube is incubated at 37°C, the other at 25°C. After 2 weeks, the amount of growth and colony diameter is compared. The ability to grow better at 37°C than 25°C also helps to distinguish *T. soudanense* from *M. ferrugineum* (Weitzman and Rosenthal, 1983/84). In addition, the ability or inability to grow at 37°C is useful in differentiating the similar-appearing fungi *T. terrestre* (does not grow at 37°C) from *T. mentagrophytes* (grows at 37°C).

Mating Studies

Some dermatophytes are difficult to identify by routine methods either because they are so similar (*M. gypseum* complex, *T. terrestre* complex, and *T. mentagrophytes* complex) or because they are so atypical of the species. In such cases, definitive identification requires the induction of the teleomorphic state (sexual form, e.g., gymnothecia with asci and ascospores), which in heterothallic species involves pairing the unidentified isolate on special media with tester strains of known teleomorphs comprising both the (A) or (+) or (a) (−) mating types. This method is best left to special reference laboratories.

An additional aid in the identification of atypical isolates of *T. rubrum* involves pairing the putative culture with both mating types of *Arthroderma simii. Trichophyton rubrum* characteristically reacts as (a) or (−) mating type, producing enhanced growth or abortive sexual structures when paired with *A. simii* (A) or (+) mating type (Young, 1972). This technique is also best left to reference laboratories.

Immunologic Aspects

Serodiagnostic tests for dermatophytosis are not routinely available. However, antibodies are specifically induced and are observed with a variety of newer techniques under certain clinical conditions, such as tinea capitis (especially with kerion formation), tinea complicated with dermatophytids, inflammatory tinea corporis, and chronic *T. rubrum* infections (Svejgaard, 1986). *Trichophyton* antigens are available commercially for skin testing, but these products remain crude and lack specificity (Kaaman, 1985b). A delayed-type skin reaction is interpreted as indicative of actual or previous dermatophytosis. However, atopic patients with chronic dermatophytosis manifest impaired cell-mediated immunity and may have almost total anergy of the delayed type to the *trichophytin* antigen and a hyporeactivity to tuberculin (Kaaman, 1985a). The demonstration of hypersensitivity to the intradermal injection of trichophytin appears to have limited diagnostic or prognostic value (Rippon, 1982).

Treatment, Prevention, and Control

Treatment

Treatment of dermatophytosis must take several factors into consideration: the anatomic site involved (foot, scalp, nails, skin); the extent and severity of the infection; and the character of the host response (Jones, 1982).

Topical as well as systemic medications are available for treatment of dermatophyte infections. Current topical agents include miconazole, clotrimazole, and econazole. These agents, which belong to the imidazole family of antifungal agents, are effective in treating both dermatophyte and yeast infections (Hernandez, 1985). Other topical antifungal agents used in limited dermatophyte infections include haloprogin, tolnaftate, and undecylenic acid. Systemic agents are usually reserved for tinea capitis, tinea unguium, and widespread or recalcitrant infections. Griseofulvin, which is orally administered, is the most commonly used systemic agent. Stein (1983) recommends twice-weekly shampooing with a selenium sulfide suspension, which renders the patient noninfectious more rapidly than griseofulvin, as an important adjunct for the treatment of tinea capitis with griseofulvin. Ketoconazole, another orally administered drug, has been approved by the Food and Drug Administration for the treatment of severe, recalcitrant cutaneous dermatophytosis that has not responded to conventional therapy. However, patients treated with this drug should be monitored because of the risk of liver toxicity. More detailed information on the treatment of dermatophyte infections are provided by Jones (1982) and Hernandez (1985).

Prevention and Control

Prevention and control of dermatophyte infections must take into consideration the area invaded and the etiologic agent. In fluorescent ectothrix tinea capitis caused by *M. audouinii, M. canis,* or *M. ferrugineum,* all contacts should be screened with the Wood light. Those infected should be treated promptly. In nonfluorescent infections of the scalp (e.g., *T. tonsurans, T. violaceum*) detection is more difficult, especially in minimal cases. In this situation, scalps should be checked carefully for spotty alopecia and lesions; suspicious areas should be cultured. The hair brush technique is helpful in detecting these minimal infections (MacKenzie, 1963). Good hygiene should be impressed upon those who are infected (i.e., not to share combs, brushes, and headgear).

Routine inspection of scalps of young children should be performed, when feasible, especially before school entry. All outbreaks in schools or institutions should be reported to the proper authorities.

If an infection is caused by a zoophilic fungus, it is important to locate the animal reservoir and treat or eliminate the infected animal. In infections caused by *M. canis* (tinea capitis, tinea corporis, tinea barbae), the infected cat or dog can usually be detected by the Wood light and treated systemically with griseofulvin.

It is sometimes more difficult to detect and eliminate cattle ringworm caused by *T. verrucosum*, because infective hairs do not fluoresce and hair or scales have been shown to survive for years on fomites such as wooden fences (Rippon, 1985). Systemic griseofulvin is effective but expensive for large herds of cattle. Good hygiene and sanitation, as well as fungicidal sprays and washes, have been effective in controlling infections (Rippon, 1982). Vaccines in cattle are currently being tried and appear promising (Spanoghe et al., 1985).

In tinea cruris or tinea corporis caused by anthropophilic fungi, infected clothing, towels, and bedding should be sterilized, and infected individuals should not permit others to share these items.

Prevention of tinea pedis may be enhanced by good foot hygiene, which includes regular washing of the feet and thorough drying; application of foot powder; avoidance of excessive moisture and occlusion by wearing sandals or other well-ventilated shoes; avoidance of trauma to the feet, especially blistering by ill-fitting footgear; and not sharing towels, socks, or shoes. If tinea pedis is considered contagious (i.e., transferred by infected shed skin scales), then control may be accomplished by educating infected individuals not to expose others by walking barefoot near swimming pools, public showers or baths, locker rooms, and the like and not to share towels, socks, or shoes; and encouraging frequent hosing of floors of public baths, swimming pools, and the like. However, Sulzberger et al. (1942) and Baer et al. (1956) concluded, based on experience and observation, that exogeneous measures designed to avoid contact with pathogenic fungi or to disinfect the environment were useless. According to these investigators, individuals carry pathogenic fungi in quiescent foci on nails, feet, and groin, and exacerbation occurs when trigger factors lower resistance. Measures for prevention should instead be based on the maintenance of local resistance to infection by individual care and hygiene of the feet and groin.

Literature Cited

Ajello, L. 1970. The medical mycological iceberg. Proceedings of the International Symposium on Mycoses. Pan. Am. Health Organ. Sci. Publ. **205**:3–12.

Ajello, L., and L. K. Georg. 1957. *In vitro* hair cultures for distinguishing between atypical isolates of *Trichophyton mentagrophytes* and *Trichophyton rubrum*. Mycopathologia **8**:3–17.

Ajello, L., and A. A. Padhye. 1985. Dermatophytes and the agents of superficial mycoses, p. 514–525. *In* E. H. Lennette, A. Balows, W. J. Hausler, Jr., and H. J. Shadomy (ed.), Manual of clinical microbiology, 4th ed. American Society for Microbiology, Washington, D.C.

Allen, A. M., and R. D. King. 1978. Occlusion, carbon dioxide and fungal skin infections. Lancet **1**:360–362.

Allen, A. M., and D. Taplin. 1973. Epidemic *Trichophyton mentagrophytes* infection in servicemen. J. Am. Med. Assoc. **226**:864–867.

Alsop, J., and A. P. Prior. 1961. Ringworm infection in a cucumber greenhouse. Br. Med. J. **1**:1081–1082.

Araviysky, A. N., R. A. Araviysky, and G. A. Eschkov. 1975. Deep generalized trichophytosis. Mycopathologia **56**:47–65.

Baer, R. L., and S. A. Rosenthal. 1966. The biology of fungous infections of the feet. J. Am. Med. Assoc. **197**:1017–1020.

Baer, R. L., S. A. Rosenthal, J. Z. Litt, and H. Rogachefsky. 1956. Experimental investigations on mechanism producing acute dermatophytosis of the feet. J. Am. Med. Assoc. **160**:184–190.

Baker, R. L., and M. F. Para. 1984. Successful use of ketoconazole for invasive cutaneous *Trichophyton rubrum* infection. Arch. Intern. Med. **144**:615–617.

Barson, W. J. 1985. Granuloma and pseudogranuloma of the skin due to *Microsporum canis*. Arch. Dermatol. **121**:895–897.

Beare, J. M., J. C. Gentles, and D. W. R. MacKenzie. 1972. Mycology, p. 694–805. *In* A. Rook, D. S. Wilkonson, and F. J. G. Ebling (ed.), Textbook of dermatology. Blackwell Scientific Publications, Oxford.

Berk, S. H., N. S. Penneys, and G. D. Weinstein. 1976. Epidermal activity in annular dermatophytosis. Arch. Dermatol. **112**:485–488.

Bronson, D. M., D. R. Desai, S. Barsky, and S. McMillen Foley. 1983. An epidemic of infection with *Trichophyton tonsurans* revealed in a 20-year survey of fungal infections in Chicago. J. Am. Acad. Dermatol. **8**:322–330.

Chandler, F. W., W. Kaplan, and L. Ajello. 1980. Color atlas and text of the histopathology of mycotic diseases, p. 118–119. Year Book Medical Publishers, Inc., Chicago.

Conant, N. F. 1936. Studies in the genus *Microsporum*. I. Cultural studies. Arch. Dermatol. **33**:665–683.

Daum, V., and D. J. McCloud. 1976. *Microsporum racemosum:* first isolation in the United States. Mycopathologia **59**:183–185.

Dei Cas, E., and A. Vernes. 1986. Parasitic adaptation of pathogenic fungi to mammalian hosts. Crit. Rev. Microbiol. **13**:173–218.

DeVroey, C., C. Wuytack-Raes, and F. Fossoul. 1983. Isolation of saprophytic *Microsporum praecox* Rivalier from sites associated with horses. Sabouraudia **21**:255–257.

Duncan, J. T. 1945. A survey of fungous diseases in Great Britain. Br. Med. J. **2**:715–718.

Friedman, L., V. J. Derbes, E. P. Hodges, and J. T. Sinski. 1960. The isolation of dermatophytes from the air. J. Invest. Dermatol. **35**:3–5.

Gentles, J. V., and E. G. V. Evans. 1970. Infection of the feet and nails with *Hendersonula toruloidea.* Sabouraudia **8:**72–75.
Georg, L. K. 1960. Epidemiology of the dermatophytoses: sources of infections, modes of transmission and epidemicity. Ann. N.Y. Acad. Sci. **89:**69–77.
Georg, L. K., and L. B. Camp. 1957. Routine nutritional tests for the identification of dermatophytes. J. Bacteriol. **74:**113–121.
Georg, L. K., P. Doupagne, S. R. Pattyn, and H. Neves. 1963. *Trichophyton yaoundei.* A dermatophyte indigenous to Africa. J. Invest. Dermatol. **41:**19–24.
Georg, L. K., E. A. Hand, and R. A. Menges. 1956. Observations on rural and urban ringworm. J. Invest. Dermatol. **27:**335–353.
Guirges, S. Y. 1981. Viability of *Trichophyton schoenleinii* in epilated hairs. Sabouraudia **19:**155–156.
Hageage, G. J., and B. J. Harrington. 1984. Use of calcoflour white in clinical mycology. Lab. Med. **15:**109–112.
Hay, R. J. 1982. Chronic dermatophyte infections. I. Clinical and mycological features. Br. J. Dermatol. **106:**1–7.
Hay, R. J. 1985. *Tinea imbricata.* The factors affecting persistent dermatophytosis. Int. J. Dermatol. **24:**562–564.
Hay, R. J., and M. K. Moore. 1984. Clinical features of superficial fungal infections caused by *Hendersonula toruloidea* and *Scytalidium hyalinum.* Br. J. Dermatol. **110:**677–683.
Hernandez, A. D. 1985. Superficial fungal infections of the skin, p. 642–645. *In* R. E. Rakel (ed.), Conn's current therapy. The W. B. Saunders Co., Philadelphia.
Johnson, M. L. T., and J. Roberts. 1977. Prevalence of dermatological disease among persons 1–74 years of age: United States, p. 1–6. Advance data from vital and health statics. U.S. Department of Health, Education, and Welfare no. 4. U.S. Department of Health, Education, and Welfare, Washington, D.C.
Jones, H. E. 1982. Therapy of superficial fungal infections. Med. Clin. N. Am. **66:**873–893.
Jones, H. E., J. H. Reinhardt, and M. G. Rinaldo. 1974. Acquired immunity to dermatophytes. Arch. Dermatol. **109:**840–848.
Kaaman, T. 1985a. Dermatophytic antigens and cell-mediated immunity in dermatophytosis, p. 117–134. *In* M. R. McGinnis (ed.), Current topics in medical mycology, vol. 1. Springer-Verlag, New York.
Kaaman, T. 1985b. Skin reactivity in atopic patients with dermatophytosis. Mykosen **28:**183–190.
Kaminski, G. W. 1985. The routine use of modified Borelli's lactritmeal agar (MBLA). Mycopathologia **91:**57–59.
Kane, J., and J. B. Fischer. 1971. The differentiation of *Trichophyton rubrum* and *T. mentagrophytes* by use of Christensen's urea broth. Can. J. Microbiol. **17:**911–913.
Kane, J., and J. B. Fischer. 1975. Occurrence of *Trichophyton megninii* in Ontario. Identification with a simple cultural procedure. J. Clin. Microbiol. **2:**111–114.
Kane, J., and S. Krajden. 1983. Dermatophytosis, p. 143–182. *In* A. F. DiSalvo (ed.), Occupational mycoses. Lea & Febiger, Philadelphia.
Kane, J., A. A. Padhye, and L. Ajello. 1982. *Microsporum equinum* in North America. J. Clin. Microbiol. **16:**943–947.
Kane, J., I. F. Salkin, I. Weitzman, and C. Smitka. 1981. *Trichophyton raubitschekii sp. nov.* Mycotaxon **13:**259–266.
Kane, J., and C. Smitka. 1978. Early detection and identification of *Trichophyton verrucosum.* J. Clin. Microbiol. **8:**740–747.
King, R. D., H. A. Khan, J. C. Foye, J. H. Greenberg, and H. E. Jones. 1975. Transferrin, iron and dermatophytes. 1. Serum dermatophyte inhibitory component definitively identified as unsaturated transferrin. J. Lab. Clin. Med. **86:**204–212.
Kligman, A. M. 1955. Tinea capitis due to *M. audouinii* and *M. canis.* Arch. Dermatol. **71:**313–337.
Lawson, G. T. N., and W. J. McLeod. 1957. *Microsporum canis*—an intensive outbreak. Br. Med. J. **2:**1159–1160.
Lundell, E. 1974. *Epidermophyton floccosum* endemic in the sauna of a trade school hostel. Mykosen **17:**219–220.
MacKenzie, D. W. R. 1961. The extra-human occurrence of *Trichophyton tonsurans var. sulfureum* in a residential school. Sabouraudia **1:**58–64.
MacKenzie, D. W. R. 1963. "Hairbrush diagnosis" in detection and eradication of non-fluorescent scalp ringworm. Br. Med. J. **2:**363–365.
Matsumoto, T., A. A. Padhye, and L. Ajello. 1983. *In vitro* hair perforation by a new subvariety of *Trichophyton tonsurans var. sulfureum.* Mycotaxon **18:**235–242.
Merz, W. G., C. L. Berger, and M. Silva-Hutner. 1970. Media with pH indicators for the isolation of dermatophytes. Arch. Dermatol. **102:**545–547.
Neves, H., and N. Canova-Xavier. 1964. The transmission of tinea cruris. Br. J. Dermatol. **76:**429–436.
Padhye, A. A., F. Blank, P. J. Koblenzer, S. Spatz, and L. Ajello. 1973a. *Microsporum persicolor* infection in the United States. Arch. Dermatol. **108:**561–562.
Padhye, A. A., A. S. Sekhon, and J. W. Carmichael. 1973b. Ascocarp production by *Arthroderma* and *Nannizzia* species on keratinous and non-keratinous media. Sabouraudia **11:**109–114.
Padhye, A. A., I. Weitzman, and L. Ajello. 1979. Mating behavior of *Microsporum equinum* with *Nannizzia otae.* Mycopathologia **69:**87–90.
Philpot, C. M. 1977. Some aspects of the epidemiology of tinea. Mycopathologia **62:**3–13.
Rebell, G., and D. Taplin. 1974. Dermatophytes. Their recognition and identification. University of Miami Press, Coral Gables, Fl. (Now published by University of Texas Press, Austin.)
Rippon, J. W. 1982. Medical mycology: the pathogenic fungi and the pathogenic actinomycetes, 2nd ed., p. 154–248. The W. B. Saunders Co., Philadelphia.
Rippon, J. W. 1985. The changing epidemiology and emerging patterns of dermatophyte species, p. 209–234. *In* M. R. McGinnis (ed.), Current topics in medical mycology. Springer-Verlag, New York.
Rippon, J. W., and T. W. Andrews. 1978. *Microsporum racemosum.* Second clinical isolation from the United States and the Chicago area. Mycopathologia **64:**187–190.
Rippon, J. W., and M. Medenica. 1964. Isolation of *Trichophyton soudanense* in the United States. Sabouraudia **3:**301–302.
Rosenthal, S. A. 1974. The epidemiology of tinea pedis, p. 515–526. *In* H. M. Robinson, Jr. (ed.), The diagnosis and treatment of fungal infections. Charles C. Thomas, Publisher, Springfield, Ill.
Rosenthal, S. A., and H. Sokolsky. 1965. Enzymatic studies with pathogenic fungi. Dermatol. Int. **4:**72–79.
Rothman, S., and A. L. Lorinez. 1963. Defense mechanisms of the skin. Annu. Rev. Med. **14:**215–242.
Ryall, C. 1980. The ecology of athlete's foot. New Sci. **14:**628–630.
Salkin, I. F. 1973. Dermatophyte test medium: evaluation with nondermatophytic pathogens. Appl. Microbiol. **26:**134–137.
Shimmura, Y. 1985. Isolation of dermatophytes from human cases of dermatophytosis and from house dust. Jpn. J. Med. Mycol. **26:**74–80.

Silva-Hutner, M., I. Weitzman, and S. A. Rosenthal. 1981. Cutaneous mycoses (dermatomycoses), p. 863–929. *In* A. Balows and W. J. Hausler Jr. (ed.), Diagnostic procedures for bacterial, mycotic and parasitic infections, 6th ed. American Public Health Association, Washington, D.C.

Sinski, J. T., and K. Flouras. 1984. A survey of dermatophytes isolated from human patients in the United States from 1979 to 1981 with chronological listings of worldwide incidence of five dermatophytes often isolated in the United States. Mycopathologia **85**:97–120.

Smith, J. M. B., R. D. Jolly, L. K. Georg, and M. D. Connole. 1968. *Trichophyton equinum var. autotrophicum.* Sabouraudia **6**:296–304.

Spanoghe, L., J. D. Kuiper, and A. H. Brethouwer. 1985. Een onderzoek naar de werkzaamheid van het trichofytie-vaccin LTF 130 onder praktijkomstandigheden. (Studies on the effect of ringworm vaccine LTF 130 under field conditions.) Tijdschr. Diergeneeskd. **110**:1011–1014.

Stein, D. H. 1983. Superficial fungal infections. Pediatr. Clin. N. Am. **30**:545–561.

Sulzberger, M. B., R. L. Baer, and R. Hecht. 1942. Common fungous infections of the feet and groins. Arch. Dermatol. **45**:670–675.

Svejgaard, E. 1986. Recent trends in the immunology of dermatophytosis. Microbiol. Sci. **3**:154–159.

Swart, E., and F. J. A. Smit. 1979. *Trichophyton violaceum* abscesses. Br. J. Dermatol. **101**:177–184.

Taplin, D., N. Zaias, G. Rebell, and H. Blank. 1969. Isolation and recognition of dermatophytes on a new medium (DTM). Arch. Dermatol. **99**:203–209.

Vanbreuseghem, R., C. De Vroey, and M. Takashio. 1970. Production of macroconidia by *Microsporum ferrugineum* Ota, 1922. Sabouraudia **7**:252–256.

Verrone, C., F. Tuci, G. M. Palleschi, E. M. DiFonzo, and E. Panconesi. 1985. Incidenza della tinea pedis "sconosciuta." G. Ital. Dermatol. Venerol. **120**:385–390.

Weitzman, I., M. R. McGinnis, A. A. Padhye, and L. Ajello. 1986. The genus *Arthroderma* and its later synonym *Nannizzia*. Mycotaxon **25**:505–518.

Weitzman, I., and S. McMillen. 1980. Isolation in the United States of a culture resembling *M. praecox*. Mycopathologia **70**:181–186.

Weitzman, I., and S. A. Rosenthal. 1983/84. Studies in the differentiation between *Microsporum ferrugineum* Ota and *Trichophyton soudanense* Joyeux. Mycopathologia **84**:95–101.

West, B. C., and K. J. Kwon-Chung. 1980. Mycetoma caused by *Microsporum audouinii*. Am. J. Clin. Pathol. **73**:447–454.

Wilson, J. W., O. A. Plunkett, and A. Gregersen. 1954. Nodular granulomatous perifolliculitis of the legs caused by *Trichophyton rubrum*. Arch. Dermatol. Syphilol. **69**:258–277.

Young, C. N. 1972. Range of variation among isolates of *Trichophyton rubrum*. Sabouraudia **10**:164–170.

Zaias, N., G. Rebell, and D. Taplin. 1965. *Microsporum rivalieri* Vanbreuseghem isolated from tinea capitis in Florida. Sabouraudia **4**:201–204.

CHAPTER 65

Histoplasmosis Capsulati

L. AJELLO

Disease: Histoplasmosis capsulati (histoplasmosis, classic histoplasmosis, small form histoplasmosis, Darling's disease, Tingo Maria fever, reticuloendothelial cytomycosis).

Etiologic Agent: Anamorph, *Histoplasma capsulatum* var. *capsulatum* (teleomorph, *Ajellomyces capsulatus*).

Source: Environment, predominantly habitats enriched with the feces of gregarious birds (blackbirds, starlings, pigeons, seagulls) and bats (caves, mine shafts, attics of buildings, hollow trees). Infections acquired through inhalation or rarely through traumatic implantation.

Clinical Manifestations: Basically pulmonary. Most infections asymptomatic or benign and self-limiting. Symptomatic forms with fever, night sweats, weight loss, and hemoptysis. In disseminated cases hepatomegaly and splenomegaly develop, with anemia and leukopenia. Chronic pulmonary infections associated with cough, dyspnea, chest pain, hemoptysis, and weight loss. Cavities may develop in the apex or subapical regions of the lungs.

Pathology: Epithelioid and giant cell granulomas of the lung. Invasion of cells of the reticuloendothelial system in the adrenal glands, bone marrow, gastrointestinal tract, liver, lymph nodes, and spleen. Lesions tend to become calcified.

Laboratory Diagnosis: Microscopic examination of stained smears of clinical materials or histologic sections of tissues treated with fungal stains (Gomori methenamine or periodic acid–Schiff stains). Fluorescent antibody staining of parasitic cells in tissues with specific conjugates. Isolation and identification of etiologic agent from sputum, buffy coat, or biopsied or autopsied tissue. Serologic tests for complement fixation and immunodiffusion are extremely useful for diagnosis and for monitoring the effect of therapy and the course of disease. The tissue-form cells of *H. capsulatum* var. *capsulatum* appear as small, oval yeast cells (2 to 3 by 3 to 5 μm) with or without buds. Daughter cells are attached to mother cells by a narrow attachment point.

Epidemiology: Global; especially prevalent in the United States in and around the Mississippi and Ohio River valleys, Mexico, Central and South America, and parts of Africa and Asia. Rare in Australia, Europe, and Oceania. Occupational disease among spelunkers and bat guano miners. Outbreaks associated with disturbance of avian and chiropteran habitats where deposits of guano have accumulated. In addition to humans, a wide variety of wild and domesticated animals are subject to infections. Histoplasmosis capsulati is not contagious.

Treatment: Supportive therapy is sufficient in most primary infections due to their self-limiting nature. Amphotericin B is the drug of choice in progressive disseminated histoplasmosis capsulati, meningitis, and cavitary pulmonary disease.

Prevention and Control: Wearing of protective clothing and masks before entering or working in infested habitats. Decontamination where economically and physically feasible with 3 to 5% formalin solutions.

Description of Disease

Histoplasmosis capsulati is the rational name for infections caused by *Histoplasma capsulatum* var. *capsulatum*. This nomenclatural preference stems from the occurrence of *H. capsulatum* in three recognized varieties, each of which causes a distinct clinical or histopathogenic expression (or both) of histoplasmosis. The other two forms of the disease are histoplasmosis duboisii and histoplasmosis farciminosi. Their respective etiologic agents are *H. capsulatum* var. *duboisii* and *H. capsulatum* var. *farciminosum* (Ajello, 1983; Weeks et al., 1985).

Fundamentally, histoplasmosis capsulati is a respiratory disease almost invariably acquired through inhalation of the airborne microconidia of *H. capsulatum* var. *capsulatum*. These microconidia are produced abundantly along with the strikingly ornamented tuberculate macroconidia in soil enriched with the feces of gregarious birds and bats. In addition to humans, many lower animals, both wild and domestic, are susceptible to infection by *H. capsulatum* var. *capsulatum*. Histoplasmosis capsulati is not known to be contagious. All infections have an environmental source.

Etiologic Agent

The anamorph is *Histoplasma capsulatum* var. *capsulatum* (Darling, 1906). Synonyms are *Cryptococcus capsulatus, Torulopsis capsulatus, Posadasia capsulata, Posadasia pyriforme,* and *Histoplasma pyriforme*. The teleomorph is *Ajellomyces capsulatus* (Kwon-Chung) McGinnis et Katz (1979); synonym, *Emmonsiella capsulata*.

Clinical Manifestations

Histoplasmosis capsulati has a global distribution that manifests itself in a wide variety of clinical forms. Epidemiologists have estimated that approximately 95% of all infections are asymptomatic and self-limited (Rippon, 1982). Their existence is detected through the elicitation of positive skin test reactions with histoplasmin with or without the finding of residual lung calcifications on roentogenographic examination. The symptomatic expressions of histoplasmosis capsulati can be differentiated into three broad clinical forms: acute pulmonary, disseminated, and chronic pulmonary.

Acute Pulmonary Histoplasmosis Capsulati

The victims of acute pulmonary histoplasmosis capsulati, a fulminating and frequently fatal form, are primarily children 1 year of age or less. They develop pulmonary flulike symptoms that encompass malaise, chest pains, cough, fever, chills, myalgia, anorexia, and weight loss. Hepatomegaly and splenomegaly also occur in association with such blood disorders as anemia, leukopenia, and thrombocytopenia. X-rays reveal diffuse densities in the lungs. Blood and bone marrow smears contain abnormal numbers of macrophages filled with the tissue-form yeast cells of *H. capsulatum* var. *capsulatum*. Unless specifically treated, the infections may end fatally within 1 to 28 weeks.

Disseminated Histoplasmosis Capsulati

Disseminated histoplasmosis capsulati, a rare form of the disease with estimated rates of 1 per 1,000,000 to 1 per 5,000,000, is best regarded as an opportunistic disease. Those in whom the infection progresses beyond a benign and self-limited episode are believed to have impaired immunologic systems or to suffer from some abnormal or predisposing condition. The primary pulmonary infection spreads by way of the reticuloendothelial system to the various internal organs. Frequent foci of dissemination are the adrenal glands. Caseation can be so extensive that it leads to adrenal insufficiency (Addison disease). Anemia, fever, hepatomegaly and splenomegaly, leukopenia, weight loss, and lymphadenopathy are other consequences of dissemination. Ulcerations in the oral cavity, pharynx, larynx, stomach, and large or small bowel may also develop. Unless this form of histoplasmosis capsulati is promptly diagnosed and treated, the fatality rate is high. Amphotericin B is the treatment of choice.

Chronic Cavitary Pulmonary Histoplasmosis Capsulati

Chronic cavitary pulmonary histoplasmosis capsulati occurs essentially only in adults. It may directly follow the development of an acute pulmonary infection or be the consequence of activation of a long-dormant primary infection. The symptoms most commonly associated with cavitary pulmonary histoplasmosis capsulati comprise chest pain, cough, dyspnea, fever, hemoptysis, and weight loss. The cavities may be single and large or multiple and small. Sputum may be produced abundantly in this form of the disease and contain large numbers of the tissue-form cells of *H. capsulatum* var. *capsulatum*. Without adequate therapy, the yeast cells may be disseminated to all vital organs.

Pathology

Histologically, histoplasmosis capsulati is marked by invasion of macrophages by the unicellular, budding cells of the tissue form of *H. capsulatum* var. *capsu-*

latum. These cells typically develop as round to oval hyaline yeastlike cells that measure 2 to 5 μm in diameter. They multiply by the production of single buds that are attached to the mother cell by a narrow attachment point.

Tissue response is characterized by the marked proliferation of histiocytes. Epithelioid or giant cell granulomas may also develop. These granulomas may or may not undergo central caseation. Generally, they evolve into fibrocaseous nodules that may eventually calcify (Chandler et al., 1980).

Laboratory Diagnosis

Direct Examination of Clinical Material

Because the invasive form of *H. capsulatum* var. *capsulatum* is small, special stains or procedures are required to detect its presence in such clinical specimens as sputum and impression smears of tissues. Unless examined carefully at the highest magnification, histologic sections rarely reveal the presence of the etiologic agent when stained with hematoxylin and eosin. Giemsa and Wright stains are recommended for the examination of smears. For histologic studies, Gridley, periodic acid–Schiff, and Gomori methenamine silver staining are the methods of choice. These so-called fungal stains avidly stain the cell wall of the yeast cells, which will typically appear as round to oval cells measuring 2 to 5 μm in diameter (Chandler et al., 1980). Specialized laboratories may have available the highly sensitive and specific fluorescein-labeled *H. capsulatum* var. *capsulatum* antiglobulins.

The in vivo yeast form cells of the *H. capsulatum* var. *duboisii* are significantly larger, measuring 8 to 15 μm in diameter. In contrast, the tissue form cells of the *farciminosi* variety cannot be distinguished from those of the *capsulatum* variety. Currently, the *farciminosi* variety can only be distinguished in cultures by its poor growth on mycologic media and its failure to produce microconidia and macroconidia. Special media are required to induce isolates to produce conidia (Weeks et al., 1985).

Isolation of Etiologic Agent

In general, the recovery of *H. capsulatum* var. *capsulatum* from clinical materials presents few problems. To ensure success, the use of several different media is recommended. The media of choice are Sabhi agar enriched with 5% sheep blood, yeast extract phosphate agar with ammonium hydroxide, and Sabouraud dextrose agar with chloramphenicol or gentamicin and cycloheximide. If quantities permit, several tubes of each medium should be inoculated with the clinical specimens and incubated at 25°C.

Description of *Histoplasma capsulatum* var. *capsulatum*

Growth usually becomes evident within 10 days. The colonies are initially white and downy. As conidia are produced, some isolates assume a tan to golden color.

Mounts made in lactophenol cotton blue will usually reveal the presence of two types of asexual conida: microconidia and macroconidia. The small hyaline, spherical, smooth microconidia measure 2 to 4 μm in diameter. They are borne apically on short, simple conidiophores. Generally the microconidia are few in number; in some isolates, however, they may be the dominant form.

The hyaline macroconidia are conspicuous because of their size and ornamentation. They range in diameter from 7 to 25 μm and have a thick cell wall. Although most are spherical, some are pyriform. Generally the macroconidia are heavily ornamented, with digitate tubercles of variable length (1 to 8 μm). As some isolates form thick-walled, smooth macroconidia, they present problems in identification. Nevertheless, doubts concerning the identity of such isolates can be resolved by determining whether or not they can be converted from their mycelial to their yeast form.

All three varieties of *H. capsulatum* are dimorphic. That is to say, depending on the environment in which they grow, these molds develop in a mycelial or a unicellular yeastlike form. This phenomenon can be used to verify the identity of mycelial isolates suspected of being *H. capsulatum* in vitro by growth on such media as brain–heart infusion agar enriched with blood, Kelly medium, or Kurung–Yegian medium (Kurung and Yegian, 1954). Mycelia streaked on the surface of agar slants of these media, when incubated at 37°C, will begin to form a soft, white, yeastlike growth around the edges of the inoculum.

Portions should be picked and streaked over the surface of fresh tubes of the conversion media and again incubated at 37°C. Within 3 or 4 days the new growth, on microscopic examination, will be found to be composed of unicellular, smooth, thin-walled cells measuring 1.5 to 3.5 μm in diameter. Some will be noted to be multiplying by a budding process. In form and size they will be similar to the yeast cells of the *capsulatum* variety as found in tissues from cases of histoplasmosis capsulati. Occasionally, isolates are encountered that require multiple serial transfers before conversion.

It should be noted that several saprophytic fungi produce hyaline, tuberculate conidia similar to those of *H. capsulatum* (namely, *Chrysosporium tubercu-*

latum, *Renispora flavissima*, and *Sepedonium chrysosporium*). Such species can be differentiated from *H. capsulatum* by their failure to convert to a unicellular form when grown at 37°C on nutritionally rich media and by their negative reactions to the exoantigen test for *H. capsulatum* (Gaur and Lichtwardt, 1980; Standard and Kaufman, 1976).

Immunology and Serology

Resistance to Infection

Natural resistance to overt infection by *H. capsulatum* var. *capsulatum* is high. It has been estimated that infections are asymptomatic in 95% of individuals who become infected by this geophilic mould. Overt infections occur in those who are immunologically impaired, as exemplified by victims of acquired immunodeficiency syndrome, as well as others with no discernible susceptibility factor. Immunologic evidence indicates that recovery from infection confers a high degree of resistance to reinfection. However, effective vaccines have yet to be developed to prevent histoplasmosis.

Serologic Tests

Highly useful complement-fixation, counterimmunoelectrophoresis, immunodiffusion, and latex agglutination procedures have been developed for diagnosing and monitoring the effect of therapy in cases of histoplasmosis capsulati. These tests are applicable to testing cerebrospinal fluid, peritoneal fluid, plasma, and serum.

The complement-fixation test with yeast-form antigen is more sensitive than tests utilizing histoplasmin. However, it should be noted that specificity is not as high as desired: cross-reactions may occur with sera from cases of blastomycosis and coccidioidomycosis. For this reason, it is recommended that immunodiffusion and counterimmunoelectrophoresis be performed in parallel with complement-fixation tests. As determined by Kaufman (1980), agreement between the results of immunodiffusion and counterimmunoelectrophoresis is greater than 90%.

The latex agglutination test with histoplasmin is satisfactory for the detection of acute pulmonary infections. Negative results may occur, however, in cases of chronic cavitary histoplasmosis capsulati.

Procedures for conducting these tests and preparing reagents and control sera are presented by Palmer et al. (1977). Reagents used in serologic testing for histoplasmosis capsulati may be purchased from Meridian Diagnostics, Inc., Cincinnati, Ohio and Microbiological Associates, Inc., Bethesda, Maryland.

Histoplasmin (Parke, Davis & Co., Detroit, Mich.) and Histolyn-CYL (Berkeley Biologicals, Berkeley, Calif.) for skin tests are most useful for epidemiologic and ecologic studies. They have limited value in diagnosis unless there is a history of prior tests eliciting negative reactions.

Prognosis and Therapy

As stated previously, the vast majority of infections by *H. capsulatum* var. *capsulatum* are benign and self-limiting. As such, they do not require specific treatment. The acute form of this disease, generally the result of massive exposure to the infectious conidia of the etiologic agent, may require therapy with antifungal drugs. The disseminated and chronic cavitary pulmonary forms of the disease, as well as histoplasmosis in immunocompromised individuals, also require specific treatment.

Currently, amphotericin B is the drug of choice for all forms of histoplasmosis capsulati that require treatment. Recommended dosages and duration of treatment are presented by George and Penn (1986).

Ketoconazole, an orally administered drug, may be used as an alternative to the intravenously administered amphotericin B. Its advantages and limitations, as well as recommended dosages, are also succinctly summarized by George and Penn (1986).

Literature Cited

Ajello, L. 1983. Histoplasmosis—a dual entity. Histoplasmosis capsulati and histoplasmosis duboisii. Ig. Mod. **79**:3–30.

Chandler, F. W., W. Kaplan, and L. Ajello. 1980. Color atlas and text of the histopathology of mycotic diseases. Year Book Medical Publishers, Inc., Chicago.

Gaur, P. K., and R. W. Lichtwardt. 1980. Comparative study of a new *Chrysosporium* species with *Histoplasma capsulatum*. Sabouraudia **18**:105–114.

George, R. B., and R. L. Penn. 1986. Histoplasmosis, p. 69–85. *In* G. A. Sarosi and S. F. Davies (ed.), Fungal diseases of the lung. Grune and Stratton, Orlando, Fla.

Kaufman, L., and E. Reiss. 1985. Serodiagnosis of fungal diseases, p. 924–945. In E. H. Lennette, A. Balows, W. J. Hausler and H. J. Shadomy (ed.), Manual of clinical microbiology, 4th ed. American Society for Microbiology, Washington, D.C.

Kurung, J. M., and D. Yegian. 1954. Medium for maintenance and conversion of *Histoplasma capsulatum* to yeast-like phase. Am. J. Clin. Pathol. **24**:505–508.

Palmer, D. F., L. Kaufman, W. Kaplan, and J. J. Cavallaro. 1977. Serodiagnosis of mycotic diseases. Charles C Thomas, Publisher, Springfield, Ill.

Rippon, J. W. 1982. Medical mycology, 2nd ed. The W. B. Saunders Co., Philadelphia.

Standard, P. G., and L. Kaufman. 1976. Specific immunological test for the rapid identification of members of the genus *Histoplasma*. J. Clin. Microbiol. **3**:191–199.

Weeks, R. J., A. A. Padhye, and L. Ajello. 1985. *Histoplasma capsulatum* variety *farciminosum*: a new combination for *Histoplasma farciminosum*. Mycologia **77**:964–970.

CHAPTER 66

Hyalohyphomycosis

A. A. PADHYE

Disease: Hyalohyphomycosis (beauvariomycosis, cephalosporiosis, fusariosis, paecilomycosis, scopulariopsosis).

Etiologic Agents: *Ascomycota,* genera *Anixiopsis, Microascus,* and *Pseudallescheria; Basidomycota,* genera *Coprinus* and *Schizophyllum; Deuteromycota,* genera *Acremonium, Beauvaria, Cylindrocarpon, Fusarium, Lecythophora, Myriodontium, Paecilomyces, Penicillium, Scedosporium, Scopulariopsis, Scytalidium, Tritirachium,* and *Volutella.*

Source: Environment; the etiologic agents live in soil, water, and organic debris. Infections are contracted by contamination of damaged skin, or in the cornea through trauma.

Clinical Manifestations: The etiologic agents of hyalohyphomycosis are opportunistic pathogens limited in their capability to invade tissue. The disease may be manifested as mild cutaneous infection or acute corneal keratitis. Subcutaneous infections follow traumatic implantation of the fungus in tissue of an immunocompromised host or colonization of the damaged tissue by fungi. The systemic, invasive infections manifest a wide variety of clinical symptoms such as cellulitis, bursitis, nephritis, endocarditis, invasive infections in bone marrow transplant recipients and burn victims, fungemia, and rarely involvement of the central nervous system in immunocompromised hosts.

Pathology: Tissue is invaded by hyaline, septate, branched, or unbranched hyphal elements with walls that are parallel or distorted in shape. Some hyphal elements may be moniliform or toruloid.

Laboratory Diagnosis: Hyalohyphomycosis is diagnosed when 1) a commonly occurring saprobic, hyaline fungus is isolated repeatedly, especially from sterile tissue or body fluids; 2) direct microscopic examination shows the presence of hyaline, septate hyphal elements compatible with the fungus isolated in culture repeatedly; and 3) histologic findings correlate with the findings from direct examination and the fungus isolated in culture.

Epidemiology: All of the etiologic agents known so far are ubiquitous in nature and distributed worldwide. Most of the species have a wide temperature tolerance, and as a result, these fungi can live and multiply in diverse ecologic areas. The fungi can contaminate air-conditioning systems and equipment and instruments in a hospital environment. The source of all of the opportunistic hyalohyphomycoses can be traced to the environment. Immunocompromised patients become hosts of these infections.

Treatment: Early diagnosis and prompt treatment with effective antifungal agents lead to a better chance of cure. Cutaneous infections are effectively treated with griseofulvin. Corneal infections are successfully treated with topical 5% aqueous solution of natamycin. Imidizole compounds, such as miconazole and ketoconazole, have been used with success. Amphotericin B is the drug of choice for invasive and disseminated infections.

Prevention and Control: Hyalohyphomycosis can be prevented by avoiding traumatic incidents, properly maintaining air-conditioning systems, and rigidly following sterility requirements in a hospital environment.

Introduction

In the last two decades, there has been a striking increase in the incidence of opportunistic fungal infections. Many modern treatments, such as broad-spectrum antibiotics, corticosteroids, immunosuppressive drugs for organ transplant patients, chemotherapy with cytotoxic drugs for cancer, prolonged use of indwelling catheters, and prosthetic heart valves, and diseases such as diabetes and immunodeficiency syndrome impair or destroy defense systems and permit proliferation and invasion of vital organs by opportunistic molds and yeasts. The majority of opportunistic mycoses are caused by exogenous fungi. Some yeasts, such as *Candida albicans* and *Malassezia furfur*, which are components of the normal microbial flora of the body, also become opportunistic pathogens in compromised hosts. Opportunistic fungi that are not components of the normal body flora exist in nature either as free-living saprobes or plant parasites. Until recently these fungi, when encountered in clinical samples, were considered contaminants rather than pathogens. Today, a significant number of these soil-inhabiting and plant-parasitic molds that have latent pathogenic capabilities have come into prominence as opportunistic pathogens in immunosuppressed hosts (Ajello, 1982; Ajello and McGinnis, 1984).

The saprobic molds can be divided into 1) those that have hyaline or lightly colored cell walls and 2) those whose cell walls contain melanins and thus are dark colored. An umbrella term, *phaeohyphomycosis*, was originally proposed by Ajello et al. (1974) as a substitute for the incorrect term *phaeo-sporotrichosis*, and later Ajello (1981) expanded the definition of this disease entity to include all superficial, cutaneous, subcutaneous, and systemic infections of humans and lower animals by darkly pigmented dematiaceous fungi, which in their invasive form in host tissue develop as filamentous, darkly pigmented, mycelial elements. To date, 71 species of dematiaceous fungi classified in 39 genera have been described as etiologic agents of phaeohyphomycosis (Ajello, 1986).

Description of Disease

The disease name *hyalohyphomycosis* was proposed as a complement to phaeohyphomycosis (Ajello, 1982, 1986). The group name was given to opportunistic mycotic infections, the etiologic agents of which have a basic tissue form consisting of hyaline, septate, mycelial elements that may be branched and are occasionally toruloid (Ajello, 1982; Ajello and McGinnis, 1984). The term hyalohyphomycosis does not replace well-established disease names such as aspergillosis and penicilliosis, the etiologic agents of which develop in tissues in the form of hyaline, septate, branched mycelium and thus qualify for consideration as hyalohyphomycoses; rather, it is an umbrella term for infections caused by occasionally encountered opportunistic fungi that have the appropriate tissue form. Use of this term is intended to discourage the coining and proliferation of unnecessary new disease names based on the genera of the etiologic agents and to replace names such as *beauvariomycosis, cephalosporiosis, fusariosis, paecilomycosis,* and *scopulariopsosis,* which have gained entry into some medical mycological text books. Thus, the disease names phaeohyphomycosis and hyalohyphomycosis were introduced not to disrupt nomenclature of well-established disease names, but to deal in a rational manner with the seemingly ever-growing number of genera and species of fungi that cause superficial, cutaneous, subcutaneous, and systemic opportunistic mycoses.

Etiologic Agents

To date, 36 pathogenic species belonging to 17 genera of *Ascomycota, Basidiomycota,* and *Deuteromycota* have been described in the literature under various disease names as causal agents of hyalohyphomycosis (Ajello, 1986; Ajello and McGinnis, 1984). These 36 species of etiologic agents of hyalohyphomycosis are listed in Table 1. Of these 36 species, three species of the genus *Fusarium* and two species of the genus *Paecilomyces* have been reported as opportunistic causal agents more frequently than others.

Common Causal Agents of the Genus *Fusarium*

Human infections with *Fusarium* spp. are caused in most cases by three species, namely, *F. moniliforme, F. oxysporum,* and *F. solani*. These infections are usually confined to keratitis (Austwick, 1984; Rebell, 1981; Rebell and Forster, 1980; Zapater, 1980; Zapater and Arrechea, 1975), onychomycosis (Austwick, 1984; Rush-Munro et al., 1971), and colonization of burn wounds (Wheeler et al., 1981). Infections of the deeper tissues causing endophthalmitis, leg ulcers, facial granuloma, osteomyelitis, and infection of the central nervous system have been associated with trauma, chronic granulomatous disease, use of corticosteroids, and immunosuppression (Abramosky et al., 1974; Cho et al., 1973; Kiehn et al., 1985; Matsuda and Matsumoto, 1986; Mutton et al., 1980). Disseminated infections due to *Fusarium* spp. are rare, but have occurred in immunocom-

TABLE 1. Etiologic agents of hyalohyphomycosis

Genus and species	Literature citation
Cutaneous and onychomycosis	
Anixiopsis fulvescens	Gonzalo and Campos (1984), Rippon et al. (1970), Todaro et al. (1984)
Anixiopsis stercoraria	Gueho et al. (1985), Vanbreuseghem and DeVroey (1980)
Fusarium oxysporum	Rebell (1981), Rush-Munro et al. (1971)
Microascus cinereus	Agarwal and Singh (1980)
Paecilomyces lilacinus	Takayasu et al. (1977)
Scopulariopsis brevicaulis	Frey and Muir (1981)
Scytalidium hyalinum	Campbell and Mulder (1977)
Subcutaneous	
Fusarium oxysporum	Abramosky et al. (1974), Attapattu and Anandakrishnan (1986), Benjamin et al. (1970)
Fusarium moniliforme	Austwick (1984), Wheeler et al. (1981)
Fusarium solani	Wheeler et al. (1981)
Paecilomyces lilacinus	Harris et al. (1979), Jade et al. (1986)
Paecilomyces marquandii	Harris et al. (1979)
Pseudallescheria boydii	Winston et al. (1977)
Scytalidium hyalinum	Zaatari et al. (1984)
Corneal	
Acremonium kiliense	Mendoza et al. (1985)
Acremonium potroni	Forster et al. (1975)
Beauvaria alba	McDonnel et al. (1985)
Beauvaria bassiana	Ishibashi et al. (1984), Sachs et al. (1985)
Cylindrocarpon lichenicola	Booth et al. (1985)
Cylindrocarpon tonkinense	Matsumoto et al. (1979)
Cylindrocarpon vaginae	Booth et al. (1985)
Fusarium dimerum	Zapater (1980)
Fusarium moniliforme	Zapater (1980), Zapater and Arrechea (1975)
Fusarium nivale	Rebell (1981), Zapater (1980)
Fusarium solani	Zapater (1980), Forster and Rebell (1975)
Lecythophora hoffmannii	McGinnis (1978)
Paecilomyces lilacinus	Agrawal et al. (1979), Gordon and Norton (1985), Gugnani et al. (1978), Mosier et al. (1977), O'Day (1977)
Penicillium citrinum	Gugnani et al. (1978)
Penicillium expansum	Gugnani et al. (1978)
Pseudallescheria boydii	Bakerspigel (1971), Elliott et al. (1977)
Tritirachium oryzae	Rodrigues et al. (1975)
Volutella cinerscens	Rebell and Forster (1980)
Systemic	
Acremonium roseo-griseum	Ward et al. (1961)
Corprinus cinereus	deVries et al. (1971), Speller and MacIver (1971)
Fusarium chlamydosporum	Kiehn et al. (1985)
Fusarium moniliforme	Frelier et al. (1985), Poirot et al. (1985), Young et al. (1978)
Fusarium oxysporum	Young and Meyers (1979)
Fusarium solani	Cho et al. (1973)
Lecythophora hoffmannii	Rinaldi et al. (1982)
Lecythophora mutabilis	Ahmad et al. (1985), Slifkin and Bowers (1975)
Myriodontium keratinophilum	Maran et al. (1985)
Paecilomyces fumoso-roseus	Elliott et al. (1984), Georg et al. (1962)
Paecilomyces javanicus	Allevato et al. (1984)
Paecilomyces lilacinus	Gordon (1984), Heard et al. (1986), Rockhill and Klein (1980), Sherwood and Dansky (1983)

TABLE 1. (*Continued*)

Genus and species	*Literature citation*
Paecilomyces viridis	Segretain et al. (1964)
Paecilomyces variotii	McClellen et al. (1976), Silver et al. (1971)
Penicillium chrysogenum	Upshaw (1974)
Penicillium spinulosum	Delore et al. (1955)
Pseudallescheria boydii	Arnett and Hatch (1975), Galgiani et al. (1984), Rippon and Carmichael (1976), Roberts et al. (1977), Shih and Lee (1984), Winston et al. (1977)
Scedosporium inflatum	Malloch and Salkin (1984)
Schizophyllum commune	Kern and Ucker (1986)
Scopulariopsis brevicaulis	Sekhon et al. (1974), Wheat et al. (1984)

promised patients (Benjamin et al., 1970; Gutman et al., 1975; Poirot et al., 1985; Steinberg et al., 1983; Weitzman, 1986). Mycotic keratitis is most often caused by *F. solani* and white onychomycosis by *F. oxysporum*. The most common species reported to cause systemic infections are *F. moniliforme, F. oxysporum,* and *F. solani*. Morphologic features of these three *Fusarium* spp. are summarized in Table 2.

Common Causal Agents of the Genus *Paecilomyces*

The second important group of species frequently reported as opportunistic pathogens in human and animal hyalohyphomycosis belong to the genus *Paecilomyces*. Species of *Paecilomyces* occur worldwide as saprophytes in soil, on decomposing organic matter, and as occasional insect pathogens. Of the 31 accepted species (Samson, 1974), five have been described in the literature as animal and human pathogens (Table 1). These species are rare pathogens of humans, but in immunocompromised patients they have been known to cause endocarditis following cardiac valve replacement (Allevato et al., 1984; McClellen et al., 1976; Silver et al., 1971), endophthalmitis following intraocular lens implantation (Mosier et al., 1977; O'Day, 1977), orbital cellulitis (Austwick, 1984), chronic maxillary sinusitis (Rockhill and Klein, 1980), cutaneous infections (Harris et al., 1979; Jade et al., 1986; Takayasu et al., 1977), pulmonary infection (Gordon, 1984; Heard et al.,

TABLE 2. Morphologic features of three commonly isolated *Fusarium* species[a]

Characteristic	F. moniliforme	F. oxysporum	F. solani
Growth rate on potato dextrose agar (7–10 days)	4–5 cm	4–5 cm	3–3.5 cm
Culture pigmentation	Peach, salmon, vinaceous, purple to violet	Grayish white to gray to purple to violet	Dull brown
Microconidia	Fusoid to clavate, 5–12 by 1.5–2.5 μm, produced in chains from lateral phialides	Oval, ellipsoid, cylindric, or slightly curved, 5–12 by 2–3.5 μm from simple lateral phialides	Cylindric to oval, 8–16 by 2–4 μm, from laterally branched conidiophores bearing phialides
Macroconidia	Not produced in some strains; when present equilaterally fusoid, 3–7 septate, 25–60 by 2.5–4 μm	3–5 septate, 27–60 by 3–4 μm	Inequilaterally fusoid, with widest point above center, length variable, 1–5 septate (32–55 by 4–6 μm), 5–9 septate (35–100 by 4.3–8 μm)
Chlamydospores	Generally absent	Globose, smooth, borne singly or in pairs, intercalary or on short lateral branches	Globose, smooth, borne singly or in pairs or short lateral branches or intercalary

[a] Adapted from Booth (1977).

TABLE 3. Morphologic features of *Paecilomyces lilacinus* and *Paecilomyces variotii*[a]

Characteristic	P. lilacinus	P. variotii
Growth on malt extract agar after 2 wk at 25°C	55–75 mm in diameter	70–80 mm in diameter
Temperature tolerance	Up to 40°C	Up to 50°C
Colony morphology	Velvety to floccose, white becoming vinaceous to vinaceous brown	Velvety to powdery, olive buff to olivaceous brown
Conidiophores	Originating from submerged or aerial hyphae, forming synnemata, robust, thick and rough-walled, giving rise to verticillate branches with whorls of 2–4 phialides	Dense whorls of verticillately or irregularly arranged branches, each bearing 2–7 phialides, smooth, slightly roughened or encrusted walls
Phialides	Swollen basal parts, tapering into distinct necks about 1 μm wide	In whorls or solitary, cylindric to ellipsoid at base, tapering abruptly into necks 1–2 μm wide, sometimes slightly bent away from the main axis; phialides with apical wall thickening
Conidia	In divergent chains, becoming entangled, conidia ellipsoid to fusiform, smooth or slightly rough, hyaline, purple in mass	Conidia in long chains, hyaline, subglobose to ellipsoid
Chlamydospores	Absent	In old cultures, brown to dark brown, smooth to slightly rough-walled, globose, subglobose to pyriform

[a] Adapted from Samson (1974).

1986), pyelonephritis (Sherwood and Dansky, 1983), and corneal ulcers (Gordon and Norton, 1985).

It is important that each of the clinically significant isolates be tested for its in vitro sensitivity to various antimycotics. Isolates of *P. lilacinus* and *P. marquandii* are resistant to amphotericin B and 5-fluorocytosine but sensitive to miconazole and ketoconazole. In contrast, isolates of *P. variotii* are sensitive to amphotericin B and 5-fluorocytosine (Gordon and Norton, 1985; Harris et al., 1979). The morphologic characters of *P. lilacinus* and *P. variotii* are summarized in Table 3.

Ecology and Epidemiology

The etiologic agents listed in Table 1 are ubiquitous in nature. They occur as saprophytes in soil, in water, on decaying vegetation, and as plant pathogens. They can survive and multiply in diverse environments and grow at 37°C; many can tolerate up to 40°C. The fungi can contaminate air conditioners and hospital equipment and instruments. Generally, they do not play a major role as pathogens in animal and human mycoses because of their limited ability to invade tissues. They are commonly known as colonizers of tissue damaged by trauma. In an immunocompromised host, however, these fungi can cause invasive, disseminated infections.

Clinical Manifestations

Clinically, hyalohyphomycotic infections can be divided into cutaneous, subcutaneous, corneal, and systemic infections.

Cutaneous Infections

Symptoms of cutaneous infections are similar to those of dermatophytosis. The fungi invade the keratinized layers, namely, skin, hair, and nails. Seven species of nondermatophytic fungi are known to cause infections of the skin and nails (Table 1). In none of the reported cases was the patient immunosuppressed.

Subcutaneous Infections

Invasive infections of subcutaneous tissues are rare and usually occur only in immunocompromised hosts. Subcutaneous tissue involvement, which differs in appearance from cutaneous infection, has been described as 1) a heaped-up keratotic mass with accompanying erythema; 2) violaceous nodules breaking down to black ulcers; 3) indurated, erythematous lesions becoming necrotic and black and developing scabs; 4) erythematous, nodular lesions, discrete and painful; and 5) disseminated erythematous eruptions and plaques along with fluctuant, subcutaneous nodules (Matsuda and Matsumoto, 1986; Weitzman, 1986). In reports describing such infections, the patients were leukemic, suffering from chronic granulomatous disease, or were renal transplant recipients under treatment with corticosteroids.

Corneal Infections

Corneal infections, also known as keratomycosis, are described as mycotic ulcerative keratitis. These infections often follow very minor abrasions of the cornea caused by a variety of materials such as dust, vegetable and other organic material, animal matter, and metal. The causal agents are thus introduced through traumatic implantation. Fungal keratitis is usually seen in apparently healthy persons. Infection is progressive, leading to perforation of the cornea and blindness within a few weeks. Eighteen species classified in 10 genera are known to be etiologic agents of hyalohyphomycotic keratitis (Table 1).

Systemic Infections

Systemic infections in compromised hosts show a wide variety of clinical manifestations, such as cellulitis, bursitis, nephritis in renal transplant patients and endocarditis in prosthetic valve recipients, invasive infections in bone marrow transplant patients and burn victims, fungemia, and involvement of the central nervous system.

Diagnosis

As many of the fungi listed in Table 1 are found in the environment and are frequent contaminants of clinical specimens, their isolation may pose a diagnostic dilemma. A commonly occurring saprobic fungus should be considered a pathogen when 1) it is isolated repeatedly, especially from sterile tissue or body fluids; 2) direct microscopic observations of the clinical specimen are compatible with the morphology of the fungus isolated in culture; and, more important, 3) histopathologic examination of the biopsy tissue shows tissue invasion and morphologic features of the fungus in tissue correlating with the morphologic features of the fungus isolated in culture.

The diagnosis is established by 1) direct examination, 2) isolation of the etiologic agent in pure culture, and 3) histopathology.

Direct Examination

Wet mounts of skin scrapings in 10% potassium hydroxide should be examined for the hyaline, septate hyphae and arthrospores. Smears of the corneal scrapings stained by the Gram, Giemsa, and Grocott methenamine silver methods should be examined for the presence of hyphal elements.

Isolation of Etiologic Agents

Scrapings, corneal scrapings, and biopsy tissue should be cultured on a medium routinely used in medical mycology, such as Sabouraud dextrose agar fortified with chloramphenicol (50 μg/ml) and gentamicin (5 μg/ml) or chloramphenicol and cycloheximide (500 μg/ml) and brain–heart infusion agar containing 5 to 10% sheep blood. However, many of the etiologic agents of hyalohyphomycosis, especially those causing keratitis, are sensitive to cycloheximide. When culturing corneal scrapings, the culture plates should be inoculated in a series of C-shaped cuts (Rebell and Forster, 1980) and incubated at 25 to 28°C. The plates should be examined daily for fungal growth originating from the C cuts. Growth away from the C cuts should be considered air contaminants. Once the fungal growth has appeared, it should be promptly subcultured on fresh medium for identification. Sporulation media such as potato dextrose agar, cornmeal agar, and oatmeal agar may be required for conidiation and identification of the isolate.

Histopathologic Examination

Tissue slides stained with fungal stains such as periodic acid–Schiff and Gomori methenamine silver (Chandler et al., 1980) show invasion of the tissue by hyaline, septate hyphae, which are variable in length, branched and unbranched, regular in shape with parallel walls, or distorted. The small number of hyphal elements may be moniliform or torulold.

Treatment

The diversity of fungi encountered as opportunistic pathogens in hyalohyphomycosis makes it imperative that each isolate, when determined to be clinically significant, be tested for its in vitro susceptibility by a battery of antifungal agents. Currently, cutaneous lesions caused by nondermatophytic fungi are treated effectively with griseofulvin (Rush-Munro et al., 1971; Todaro et al., 1984). Cutaneous and subcutaneous lesions caused by agents such as *P. lilacinus* and *Fusarium* spp. have been treated successfully with surgical excision and concomitant treatment with amphotericin B or a combination of amphotericin B and 5-fluorocytosine. Amphotericin B is the drug of choice for treating invasive infections caused by *Fusarium* species (Blazer et al., 1984;

Matsuda and Matsumoto, 1986; Weitzman, 1986). For keratitis, a treatment with topical antifungal drugs suitable for use in the eye is a 5% aqueous suspension of the polyene antibiotic natamycin (pimaricin). Imidizole compounds such as clotrimazole, miconazole, econazole, and ketoconazole have been used and seem to be as effective as natamycin (Austwick, 1984; Rebell, 1981; Zapater and Arrechea, 1975).

Literature Cited

Abramosky, C. R., D. Quinn, W. D. Bradfor, and N. F. Conant. 1974. Systemic infection by *Fusarium* in a burned child. J. Pediatr. **84:**561–564.

Agarwal, G. P., and S. M. Singh. 1980. *Microascus cinereus* infection of human nail. Indian J. Med. Sci. **34:**263–265.

Agrawal, P. K., B. Lal, S. Waheb, O. P. Srivastava, and S. C. Misra. 1979. Orbital paecilomycosis due to *Paecilomyces lilacinus* (Thom) Samson. Sabouraudia **17:**363–370.

Ahmad, S., R. J. Johnson, S. Hillier, W. R. Shelton, and M. G. Rinaldi. 1985. Fungal peritonitis caused by *Lecythophora mutabilis*. J. Clin. Microbiol. **22:**182–186.

Ajello, L. 1981. The gamut of human infections caused by dematiaceous fungi. Jpn. J. Med. Mycol. **22:**1–5.

Ajello, L. 1982. Hyalohyphomycosis: a disease entity whose time has come. Newsl. Med. Mycol. Soc. New York **10:**3–5.

Ajello, L. 1986. Hyalohyphomycosis and phaeohyphomycosis: two global disease entities of public health importance. Eur. J. Epidemiol. **2:**243–251.

Ajello, L., L. K. Georg, R. T. Steigbiegel, and C. J. K. Wang. 1974. A case of phaeohyphomycosis caused by a new species of *Phialophora*. Mycologia **66:**490–498.

Ajello, L., and M. R. McGinnis. 1984. Nomenclature of human pathogenic fungi, p. 363–377. *In* W. Weuffen, G. Gerenesi, D. Groschel, B. Kemter, A. Kramer, and A. P. Krasilnikow (ed.), Handbuch des Antiseptik, Band I, Grundlagen der Antiseptik. Veb. Verlag Volk und Gesundheit, Berlin.

Allevato, P. A., J. M. Ohorodnik, E. Mezger, and J. F. Eisses. 1984. *Paecilomyces javanicus* endocarditis of native and prosthetic aortic valve. Am. J. Clin. Pathol. **82:**247–252

Arnett, J. C., and H. B. Hatch. 1975. Pulmonary allescheriasis. Arch. Intern. Med. **135:**1250–1253.

Attapattu, M. C., and C. Anandakrishnan. 1986. Extensive subcutaneous hyphomycosis caused by *Fusarium oxysporum*. J. Med. Vet. Mycol. **24:**105–111.

Austwick, P. K. C. 1984. *Fusarium* infections in man and animals, p. 129–140. *In* Moss, M. and J. E. Smith (ed), The applied mycology of *Fusarium*. Cambridge University Press, London.

Bakerspigel, A. 1971. Fungi isolated from keratomycosis in Ontario, Candida I. *Monosporium apiospermum* (*Allescheria boydii*). Sabouraudia **9:**109–112.

Benjamin, R. P., J. L. Callaway, and N. F. Conant. 1970. Facial granuloma associated with *Fusarium* infection. Arch. Dermatol. **101:**598–600.

Blazer, B. R., D. D. Hurd, D. C. Stover, J. W. Alexander, and P. B. McGlave. 1984. Invasive *Fusarium* infections in bone marrow transplant recipients. Am. J. Med. **77:**645–651.

Booth, C. 1977. *Fusarium,* laboratory guide to the identification of major species. Commonwealth Mycological Institute, Kew, Surrey, England.

Booth, C., Y. M. Clayton, and M. Usherwood. 1985. *Cylindrocarpon* species associated with mycotic keratitis. Proc. Indian Acad. Sci. Plant Sci. **94:**422–436.

Campbell, C. K., and J. L. Mulder. 1977. Skin and nail infection by *Scytalidium hyalinum* sp. nov. Sabouraudia **15:**161–166.

Chandler, F. W., W. Kaplan, and L. Ajello. 1980. A colour atlas and textbook of histopathology of mycotic diseases. Wolfe Medical Publications Ltd., London.

Cho, C. T., T. S. Vats, J. T. Lowman, J. W. Brandsberg, and F. E. Tosh. 1973. *Fusarium solani* infection during treatment of acute leukemia. J. Pediatr. **83:**1028–1031.

Delore, P., J. Coudert, R. Lambert, and J. Fayalle. 1955. Un cas de mycose bronchique avec localisations musculaires septicemiques. Presse Med. **63:**1580–1582.

deVries, G. A., R. F. O. Kemp, and D. C. E. Speller. 1971. Endocarditis caused by *Coprinus delicatulus,* p. 185–186. *In* Drouhet, E. (ed), Comptes Rendus des Communications, Vth Congress, International Society for Human and Animal Mycology. ISHAM Paris, Imprimerie Louis-Jean Publications scientifiques et Litteraihes, Paris.

Elliott, I. D., C. Halde, and J. Shapiro. 1977. Keratitis and endophthalmitis caused by *Petriellidium boydii*. Am. J. Ophthalmol. **83:**16–18.

Elliott, G. S., M. S. Whitney, W. M. Reed, and J. R. Tuite. 1984. Antemortem diagnosis of paecilomycosis in a cat. J. Am. Vet. Med. Assoc. **184:**93–94.

Forster, R. K., and G. Rebell. 1975. Animal model of *Fusarium solani* keratitis. Am. J. Ophthalmol. **79:**510–515.

Forster, R. K., G. Rebell, and W. Stiles. 1975. Recurrent keratitis due to *Acremonium potroni*. Am. J. Ophthalmol. **79:**126–128.

Frelier, P. F., L. Sigler, and P. E. Nelson. 1985. Mycotic pneumonia caused by *Fusarium moniliforme* in an alligator. Sabouraudia **23:**399–402.

Frey, D. and D. B. Muir. 1981. Onychomycosis caused by *Scopulariopsis brevicaulis*. Aust. J. Dermatol. **22:**123–126.

Galgiani, J. N., D. A. Stevens, J. R. Graybill, D. L. Stevens, A. J. Tillinghast, and H. B. Levine. 1984. *Pseudallescheria boydii* infections treated with ketoconazole. Chest **86:**219–224.

Georg, L. K., W. M. Williamson, E. B. Tilden, and R. E. Getty. 1962. Mycotic pulmonary disease of captive tortoises due to *Beauvaria bassiana* and *Paecilomyces fumoso-roseus*. Sabouraudia **2:**80–86.

Gonzalo, M., and R. Campos. 1984. Dermatofitosis por *Aphanoascus fulvescens*. J. Med. Vet. Mycol. **22:**311–314.

Gordon, M. A., 1984. *Paecilomyces lilacinus* (Thom) Samson, from systemic infection in an armadillo (*Dasypus novemcinectus*). Sabouraudia **22:**109–116.

Gordon, M. A., and S. A. Norton. 1985. Corneal transplant infection by *Paecilomyces lilacinus*. Sabouraudia **23:**295–301.

Gueho, E., J. Villlard, and R. Guinet. 1985. A new human case of *Anixiopsis stercoraria* mycosis: discussion of its taxonomy and pathogenicity. Mykosen **28:**430–436.

Gugnani, H. C., S. Gupta, and R. S. Talwar. 1978. Role of opportunistic fungi in ocular infections in Nigeria. Mycopathologia **65:**155–166.

Gutman, L., S. M. Chou, and R. S. Pore. 1975. Fusariosis, myasthenic syndrome and aplastic anemia. Neurology **25:**922–926.

Harris, L. F., B. M. Dan, and A. W. Lefkowitz. 1979. *Paecilomyces* cellulitis in a renal transplant patient; successful treatment with intravenous miconazole. South. Med. J. **72:**897–898.

Heard, D. J., G. H. Cantor, E. R. Jacobson, B. Purich, L. Ajello, and A. A. Padhye. 1986. Hyalohyphomycosis caused by *Paecilomyces lilacinus* in an Aldabra tortoise. J. Am. Vet. Med. Assoc. **189:**1143–1145.

Ishibashi, Y., Y. Matsumoto, and K. Takei. 1984. The effect of intravenous miconazole on keratitis. Am. J. Ophthalmol. **98:**433–437.

Jade, K. B., M. F. Lyons, and J. W. Gnan. 1986. *Paecilomyces lilacinus* cellulitis in an immunocompromised patient. Arch. Dermatol. **122:**1169–1170.

Kern, M. E., and F. A. Uecker. 1986. Maxillary sinus infection caused by the homobasidiomycetous fungus *Schizophyllum commune*. J. Clin. Microbiol. **23:**1001–1005.

Kiehn, T. E., P. E. Nelson, E. M. Bernard, F. F. Edwards, B. Koziner, and D. Armstrong. 1985. Catheter-associated fungemia caused by *Fusarium chlamydosporum* in a patient with lymphocytic lymphoma. J. Clin. Microbiol. **21:**501–504.

Malloch, D., and I. F. Salkin. 1984. A new species of *Scedosporium* associated with osteomyelitis in humans. Mycotaxon **21:**247–255.

Maran, A. G. D., K. Kwon, J. R. Milne, and D. Lamb. 1985. Frontal sinusitis caused by *Myrodontium keratinophilum*. Br. Med. J. **290:**207.

Matsuda, T., and T. Matsumoto. 1986. Disseminated hyalohyphomycosis in a leukemic patient. Arch. Dermatol. **122:**1171–1175.

Matsumoto, T., J. Masaki, and T. Okabe. 1979. *Cylindrocarpon tonkinense*: as a cause of keratomycosis. Trans. Br. Mycol. Soc. **72:**503–504.

McClellen, J. R., J. D. Hamilton, J. A. Alexander, W. G. Wolfe, and J. B. Reed. 1976. *Paecilomyces varioti* endocarditis on a prosthetic aortic valve. J. Thorac. Cardiovasc. Surg. **71:**472–475.

McDonnel, P. J., T. P. Werblin, L. Sigler, and W. R. Green. 1985. Mycotic keratitis due to *Beauvaria alba*. Cornea **3:**213–216.

McGinnis, M. R. 1978. Human pathogenic species of *Exophiala, Phialophora,* and *Wangiella*, p. 37–59. *In* Ajello, L. (ed), The black and white yeasts. Scientific publication no. 356. Pan American Health Organization, Washington, D.C.

Mendoza, L., A. Donato, and A. Padhye. 1985. Canine mycotic keratoconjunctivitis caused by *Acremonium kiliense*. J. Med. Vet. Mycol. **23:**447–450.

Mosier, M. A., B. Lusk, T. M. Pettit, D. H. Howard, and J. Rhodes. 1977. Fungal endophthalmitis following intraocular lens implantation. Am. J. Ophthalmol. **83:**1–8.

Mutton, K. J., T. J. Lucas, and J. L. Harkness. 1980. Disseminated *Fusarium* infection. Med. J. Aust. **2:**624–625.

O'Day, D. M. 1977. Fungal endophthalmitis caused by *Paecilomyces lilacinus* after intraocular lens implantation. Am. J. Ophthalmol. **83:**130–131.

Poirot, J. L., J. P. Laporte, E. Gueho, F. Tessier, A. Verny, M. Marteau, P. Roux, and G. Duhamel. 1985. A propos d'un cas de mycose profunde a *Fusarium verticillioides*. Med. Malad. Infect. **10:**529–532.

Rebell, G. 1981. *Fusarium* infections in human and veterinary medicine, p. 210–220. *In* Nelson, P. E., T. A. Toussoun, and R. J. Cook (ed), *Fusarium*: diseases, biology, taxonomy. Pennsylvania State University Press, University Park, Pa.

Rebell, G., and R. K. Forster. 1980. Fungi of keratomycosis, p. 553–561, *In* E. H. Lennette, A. Balows, W. J. Hausler, and J. P. Traut (ed.), Manual of clinical microbiology, 3rd ed. American Society for Microbiology, Washington, D.C.

Rinaldi, M. G., E. L. McCoy, and D. P. Winn. 1982. Gluteal abscess caused by *Phialophora hoffmannii* and review of the role of this organism in human mycoses. J. Clin. Microbiol. **16:**181–185.

Rippon, J. W., and J. W. Carmichael. 1976. Petriellidiosis (allescheriosis): four unusual cases and review of the literature. Mycopathologia **58:**117–124.

Rippon, J. W., F. C. Lee, and S. McMillen. 1970. Dermatophyte infection by *Aphanoascus fulvescens*. Arch. Dermatol. **102:**552–555.

Roberts, F. J., P. Allen, and T. K. Maybee. 1977. *Petriellidium boydii* (*Allescheria boydii*) endocarditis associated with porcine valve replacement. Can. Med. Assoc. J. **117:**1250–1252.

Rockhill, R. C., and M. D. Klein. 1980. *Paecilomyces lilacinus* as the cause of chronic maxillary sinusitis. J. Clin. Microbiol. **11:**737–739.

Rodrigues, M. M., P. Laibson, and W. Kaplan. 1975. Exogenous ulcer caused by *Tritirachium roseum*. Am. J. Ophthalmol. **80:**804–806.

Rush-Munro, F. M., H. Black, and J. M. Dingley. 1971. Onychomycosis caused by *Fusarium oxysporum*. Aust. J. Dermatol. **12:**18–29.

Sachs, S. W., J. Baum, and C. Mies. 1985. *Beauvaria bassiana* keratitis. Br. J. Ophthalmol. **69:**548–550.

Samson, R. A. 1974. *Paecilomyces* and some allied hyphomycetes. Studies in mycology no. 6. Centraalbureau voor Schimmelcultures, Baarn, Netherlands.

Segretain, G., H. Fromentin, P. Destombes, E. R. Brygoo, and A. Dodin. 1964. *Paecilomyces viridis* n. sp. champignon dimorphque, agent d'une mycose generalisee de *Chameleo lateralis* Gray. C. R. Acad. Sci. (Paris) **259:**258–261.

Sekhon, A. S., D. J. Williams, and J. H. Haravey. 1974. Deep scopulariopsosis: a case report and sensitivity studies. J. Clin. Pathol. **27:**837–843.

Sherwood, J. A., and A. S. Dansky. 1983. *Paecilomyces* pyelonephritis complicating nephrolithiasis and a review of *Paecilomyces* infections. J. Urol. **130:**526–528.

Shih, L., and N. Lee. 1984. Disseminated petriellidiodis (allescheriasis) in a patient with refractory acute lymphoblastic leukemia. J. Clin. Pathol. **237:**78–82.

Silver, M. D., P. G. Tuffinell, and W. G. Bigelow. 1971. Endocarditis caused by *Paecilomyces varioti* affecting an aortic valve allograft. J. Thorac. Cardiovasc. Surg. **61:**278–281.

Slifkin, M., and H. M. Bowers. 1975. *Phialophora mutabilis* endocarditis. Am. J. Clin. Pathol. **63:**120–129.

Speller, D. C. E., and A. G. Maclver. 1971. Endocarditis caused by *Coprinus* species: a fungus of the toadstool group. J. Med. Microbiol. **4:**370–374.

Steinberg, G. K., R. H. Britt, D. R. Enzmann, J. L. Finlay, and A. M. Arvin. 1983. *Fusarium* brain abscess. J. Neurosurg. **58:**598–601.

Takayasu, S., M. Akagi, and Y. Shimizu. 1977. Cutaneous mycosis caused by *Paecilomyces lilacinus*. Arch. Dermatol. **113:**1687–1690.

Todaro, F., G. Criseo, and C. Urzi. 1984. A propos d'un cas de tinea pedis par *Anixiopsis fulvescens* (Cooke) de Vries, var. *fulvescens* (Cooke) de Vries. Bull. Soc. Fr. Mycol. Soc. **13:**239–242.

Upshaw, C. B. 1974. *Penicillium* endocarditis of aortic valve prosthesis. J. Thorac. Cardiovasc. Surg. **68:**428–431.

Vanbreuseghem, R., and C. DeVroey. 1980. Dermatophytic infection by *Anixiopsis stercoraria* in a wild boar (*Sus scrofa*). Mykosen **23:**183–187.

Ward, H. P., W. J. Martin, J. C. Ivins, and L. A. Weed. 1961. *Cephalosporium* arthritis. Proc. Staff Meet. Mayo Clin. **36:**337–343.

Weitzman, I. 1986. Saprophytic molds as agents of cutaneous and subcutaneous infection in the immunocompromised host. Arch. Dermatol. **122:**1161–1168.

Wheat, L. J., M. Bartlett, M. Ciccarell, and J. W. Smith. 1984. Opportunistic *Scopulariopsis* pneumonia in an immunocompromised host. South. Med. J. **77:**1608–1609.

Wheeler, M. S., M. R. McGinnis, W. A. Schell, and D. H. Walker. 1981. *Fusarium* infection in burned patients. Am. J. Clin. Pathol. **75:**304–311.

Winston, D. J., M. C. Jordon, and J. Rhodes. 1977. *Allescheria boyddii* infections in the immunosuppressed host. Am. J. Med. **63:**830–835.

Young, C. A., and A. M. Meyers. 1979. Opportunistic fungal infection by *Fusarium oxysporum* in a renal transplant patient. Sabouraudia **17:**219–223.

Young, N. A., K. J. Kwon-Chung, T. T. Kubota, A. E. Jennings, and R. I. Fisher. 1978. Disseminated infection by *Fusarium moniliforme* during treatment of malignant lymphoma. J. Clin. Microbiol. **7:**589–594.

Zaatari, G. S., R. Reed, and R. Morewessel. 1984. Subcutaneous hyphomycosis caused by *Scytalidium hyalinum*. Am. J. Clin. Pathol. **82:**252–256.

Zapater, R. C. 1980. The genus *Fusarium* as agent of keratomycosis in Argentina, p. 35–38. *In* Kuttin, E. S. and G. L. Baum (ed), Proceedings of the VIIth Congress, International Society for Human and Animal Mycology, Jerusalem. Excerpta Medica, Amsterdam.

Zapater, R. C., and A. Arrechea. 1975. Mycotic keratitis by *Fusarium*. Ophthalmologica **170:**1–12.

CHAPTER 67

Mycetoma

EL SHEIKH MAHGOUB

Disease: Mycetoma (Madura foot)

Etiologic Agents: Eumycetes, *Madurella mycetomatis, M. grisea, Petrilidium boydii, Leptosphaeria senegalensis, Cephalosporium* spp., *Aspergillus nidulans,* other fungi. Aerobic actinomycetes, *Streptomyces somaliensis, Actinomadura madurae, A. Pelletierii, Nocardia brasiliensis.*

Source: Agricultural soil.

Clinical Manifestations: Chronic progressive swelling with sinuses discharging grains and pus, commonly on the lower limbs.

Pathology: Subcutaneous granuloma spreading through facial planes and commonly affecting bone.

Laboratory Diagnosis: Isolation of organisms by culture, identification in histopathologic sections, and serologic diagnosis by counterimmunoelectrophoresis.

Epidemiology: A disease of tropical and subtropical countries, affecting males more than females, particularly farmers and animal herders.

Treatment: Actinomycetoma by streptomycin plus any one of cotrimoxazole, dapsone, rifampicin, or sulphadoxine pyrimethamine. Eumycetoma by one of the imidazoles, namely ketoconazole or itraconazole.

Prevention and Control: Difficult.

Description of Disease

Mycetoma (Madura foot) is primarily a disease of humans, but occasionally animals (goats, dogs, and horses) may be affected (Gumaa et al., 1978). It is caused by bacteria (actinomycetes) or fungi (eumycetes). Mycetoma is therefore subdivided into actinomycetoma and eumycetoma. This is a basic differentiation that shows up in all aspects, particularly treatment.

Actinomycetoma is caused by *Actinomadura madurae, A. pelletierii, Nocardia brasiliensis,* and *Streptomyces somaliensis. Eumycetoma* is caused by *Madurella mycetomatis, M. grisea, Petrilidium boydii, Leptosphaeria senegalensis, Aspergillus spp., Cephalosporium spp.,* and other less frequent fungi (Mahgoub and Murray 1973).

Communicability

Mycetoma is neither transmitted from human to human nor from animals to humans. The causative organisms are saprobic (Ajello et al., 1979), and humans get infected by entry of organisms from soil into subcutaneous tissue through a breach of skin most commonly caused by a thorn prick. The incubation period is between 10 to 12 weeks.

Clinical Manifestations of the Disease

Mycetoma appears as a small painless nodule, the size of a peanut that progressively and slowly increases in size. The swelling may remain intact or sinuses open discharging pus and grains. The color, size, and consistency of these grains may lead to provisional diagnosis of the causative organism (Murray, 1966). Because of the relative painless nature of the disease, patients do not seek early medical advice. About 70% of cases involve the foot (Fig. 1), although other parts of the body may be affected (Fig. 2).

Radiographs of the lesion show dense granulomata in the soft tissue and cavities in the underlying bone. Cavities are usually multiple and small in actinomycetoma, except that caused by *A. madurae*, and few and large in eumycetoma (Fig. 3; Mahgoub and Murray, 1973). However, x-ray appearance of cranial mycetoma shows mainly dense bone and very occasional cavities (Gumaa et al., 1986).

On examination the skin is stretched and smooth because it loses its normal creases. It may be hypo- or hyperpigmented at times and does not move freely over the underlying swelling, which is generally firm and sometimes nodular. The regional lymph nodes are usually small and shotty due to secondary bacterial infection, and in about 1% of cases they may be large and actually form a new focus of infection due to lymphatic spread (El Hassan and Mahgoub, 1972). However, the usual method of spread is by direct contiguity.

Pathology

After implantation of the causative organism in the skin, it proliferates and spreads into the adjacent subcutaneous tissue. Next, bone is affected because mycetoma agents seem to be osteophilic. Actinomycetes, particularly *A. pelletierii* and *N. brasiliensis*, tend to excite a quicker and wider pathologic reaction. Spread tends to follow easy routes through facial planes, but sometimes there is a genuine lymphatic spread. This is commoner in the case of the actinomycetes because of their fine filaments and is aided by repeated surgical intervention. Hematologic spread has not been recorded so far. In the case of cranial mycetoma, the swelling spreads from the scalp through bone into the cranial cavity. This causes relevant symptoms and signs of the central nervous system resulting from pressure rather than involvement of the brain cells themselves.

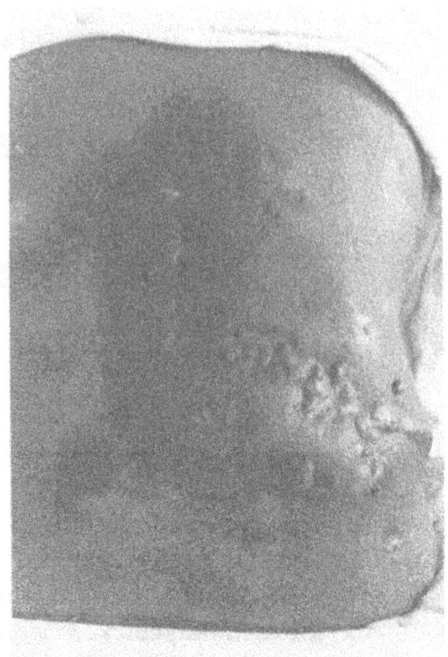

FIG. 2. Mycetoma back due to *M. mycetomatis*.

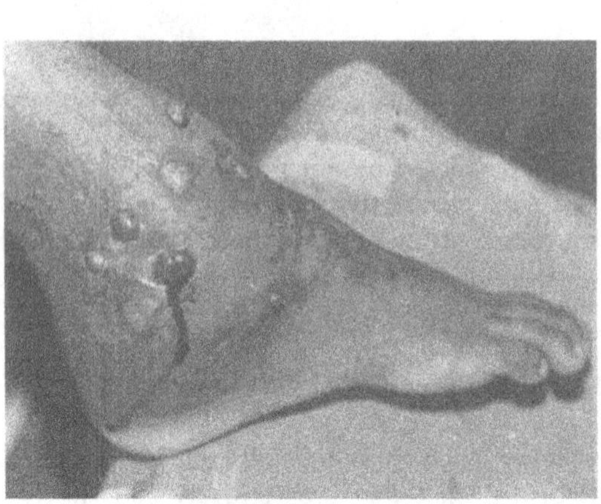

FIG. 1. Mycetoma pedis caused by *Madurella mycetomatis*.

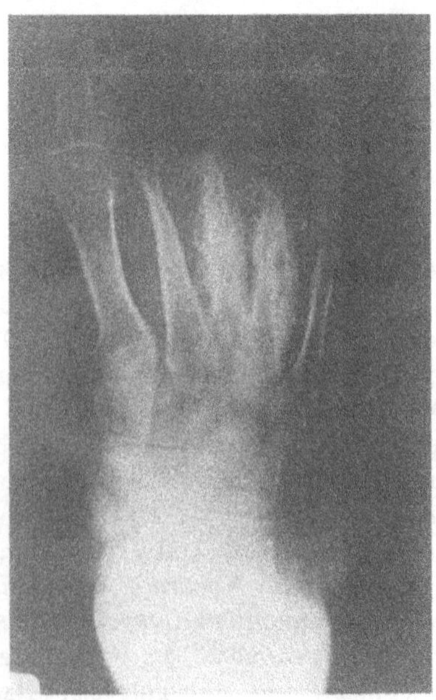

FIG. 3. Radiograph of mycetoma due to *S. somaliensis*, note cavities and bone erosion.

Like all fungal infections the reaction to mycetoma agents is mainly granulomatous, composed of a central pyogenic zone of polymorphonuclear leukocytes (PMNL) surrounding actinomycetic or eumycetic grains. This is surrounded in turn by a zone of lymphocytes, plasma cells, and multinucleated giant cells (MNGC). Although the MNGC are more frequently seen in eumycetoma, they are also indicative of good host resistance in both types of mycetoma. The grain is the most essential structure in the histopathologic field. Depending on the causative agent each grain exhibits characteristic features that make diagnosis easy. When stained with hematoxylin and eosin (H/E) *M. mycetomatis* shows a brown pigment through the whole grain, *S. somaliensis* is eosinophilic, and both *A. madurae* and *A. pelletierii* are basophilic. Pale grain fungi are eosinophilic, but broad fungal hyphae are seen. Grains are further distinguished by their size and shape (Figs. 4 and 5). For detailed histopathology of mycetoma see Mahgoub and Murray (1973).

Diagnostic Features

It may not be difficult to diagnose a well-established mycetoma on the foot with characteristic swelling and sinuses discharging grains, but it is the early lesion that should be sought and treated.

In endemic areas like those in Africa, Asia, or South and Central America any swelling, particularly in the lower limbs, that is relatively painless should be considered a mycetoma till proven otherwise. The swelling should be differentiated from tumors like lipoma, fibrolipoma hard neurofibroma or thorn granuloma, or from cystic lesions like a cold abscess or implantation dermoid. It is also to be differentiated from malignant tumors like melanoma. Osseous mycetoma is to be differentiated from chronic osteomyelitis, tuberculosis, or Kaposi's sarcoma (Davies, 1958). In endemic areas, mycetoma should be suspected in unusual sites like the scalp, (Gumaa et al., 1986), the jaw (Gumaa et al., 1977), and maxillary sinus (Mahgoub, unpublished data).

Prognosis

There is no record of self-cure in mycetoma: once a swelling appears it progressively increases. However, the severity of the lesion depends on the causative organism and the patient's immunologic competence. On the whole, actinomycetoma is more aggressive, resulting in ill-defined margins, more bone destruction, and secondary bacterial infection. This picture is more pronounced in the case of *A. pelleteirii* and *N. brasillensis* mycetoma. On the other hand, euymycetoma is well circumscribed and the lesion is smaller if compared with an actinomycetoma for the same length of time. However, very severe and painful eumycetoma due to *M. mycetomatis* were seen. Apart from cranial mycetoma,

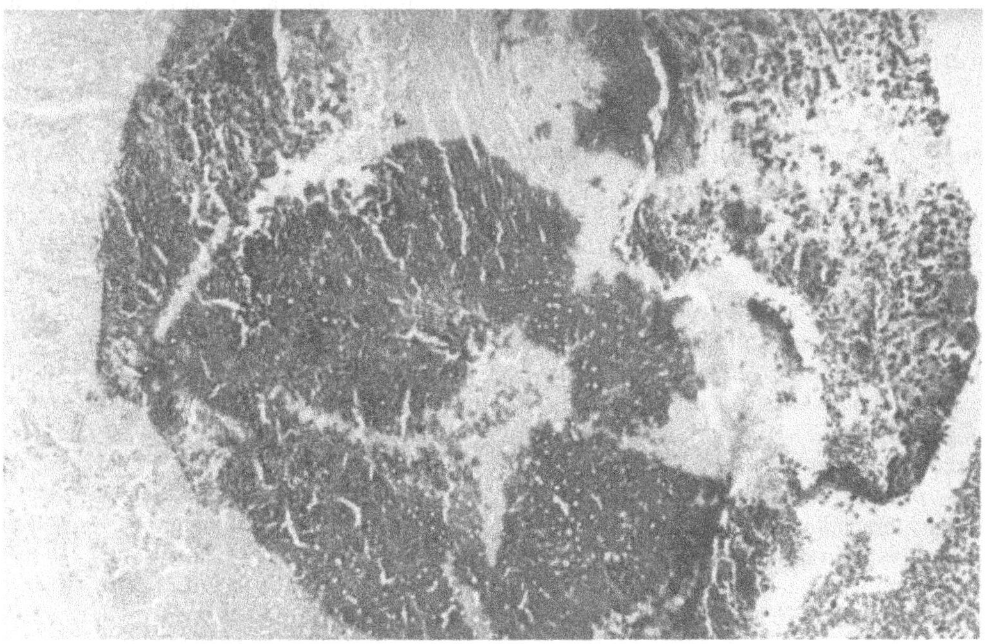

FIG. 4. Histopathologic section showing *M. mycetomatis* grain; hematoxylin-eosin ×400.

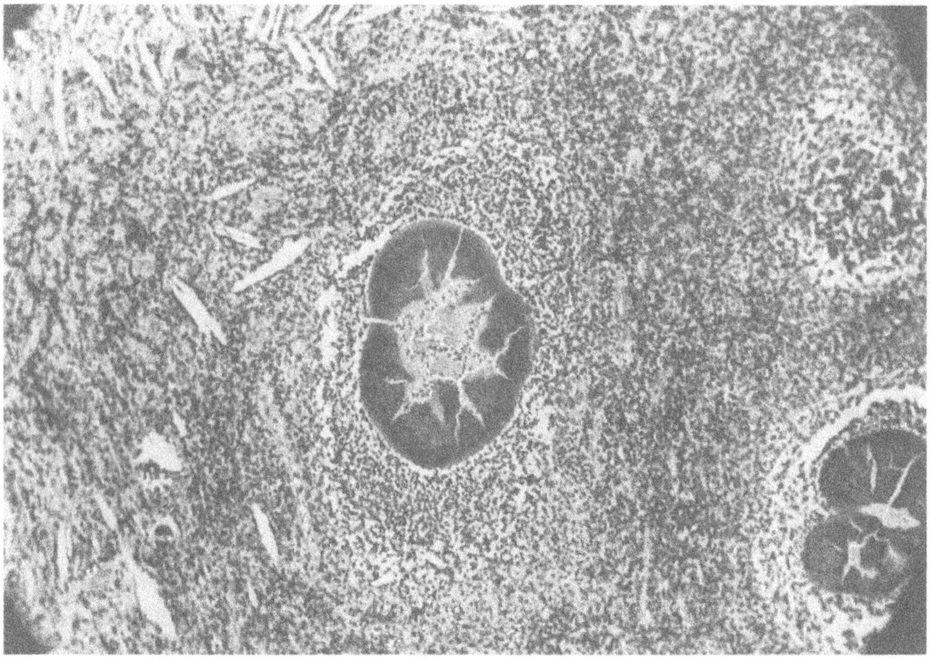

FIG. 5. Histologic section showing *S. somaliensis* grains; hematoxylin-eosin ×400.

which has a bad prognosis and can ultimately kill the patient, mycetoma, nowadays, has a good prognosis. This to a great extent depends on the right diagnosis and selection of the correct treatment. Relapse occurs and sometimes with resistance to the original drug if treatment is not carried to the end when precipitating antibodies disappear from the blood.

Epidemiologic Aspects

Occurrence and Transmission

Mycetoma is a disease of tropical and subtropical parts of the world extending between the Tropics of Cancer and Capricorn. However, within this zone more cases have been reported from countries like Mexico, Venezuela, Senegal, the Sudan, Somal, and India (Mahgoub and Murray, 1973; Mariat, 1963).

In the Sudan about 400 new cases are seen every year (Mahgoub, 1977). The amount of rain seems to influence the prevalence of species. Whereas *N. brasiliensis, A. pelletierii,* and *P. boydii* are found in areas with high precipitation, *S. somaliensis* and *M. mycetomatis* prevail in dryer regions.

The existence of the causative organisms in the soil makes persons involved in cultivation and animal herding more prone than others. In endemic areas minor injuries may help transmit the organism from soil into subcutaneous tissue. Hence, persons with other professions like traders and pupils are afflicted with mycetoma.

Susceptibility

While in tissue mycetoma agents always exist in the grain form, their saprophytic existence is in the mycelial form. Grains remained viable in soil for as long as 120 days (Abbott, 1956). Many persons are exposed to injury and implantation of nycetoma agents in their tissues. However, only few develop the disease. These seem to be deficient at least partially in their cell-mediated immunity (CMI) (Mahgoub et al., 1977). The majority of patients were anergic to tuberculin and dinitrochlorobenzine (DNCB), and only about 30% of their lymphocytes transformed to lymphoblasts, whereas in normal subjects 70 to 80% of lymphocytes transformed in the presence of phytohemagglutinin. That susceptibility to mycetoma was dependent on deficiency in CMI was proven by animal experiments. Trials have failed to produce mycetoma in any laboratory animal except in the naturally athymic nude mice (Mahgoub, 1978). In these, well-developed mycetoma grains were seen after 24 days from injection of suspension of *M. mycetomatis* mycelial hyphae in saline. Grains of *A. pelletierii* also developed in goats treated with hydrocortisone and injected subcutaneously in their legs (Gumaa, unpublished data).

The legitimate question is whether there are many persons with deficient CMI in the tropics. This may be true and can be explained by malnutrition, multiparasitemia, and viral diseases in childhood.

Etiologic Agents

Characterization

The causative agents of mycetoma are either slow-growing aerobic actinomycetes or true fungi. To a great extent they are identified according to their morphologic characteristics, although in some cases biochemical activity is assessed for confirmation of diagnosis (Mahgoub and Murray, 1973). On primary isolation the colonies of the actionmycetes are well developed in about 10 days. Those of *S. somaliensis*, *A. pelletierii*, and *N. brasiliensis* are heaped up, granular, and dry in appearance. *S. somaliensis* is dirty yellow in color and its center may turn black as it ages (Fig. 6). *A. pelletierrii* starts as white in color but soon aquires a coral-red color, *N. brasiliensis* can have shades of color between beige and orange. *A. madurae* is white in color, waxy in appearance, and rather flat. When stained by the modified Zhiel Neelsen, all actinomycetes fail to take the stain except some strains of *N. brasiliensis*.

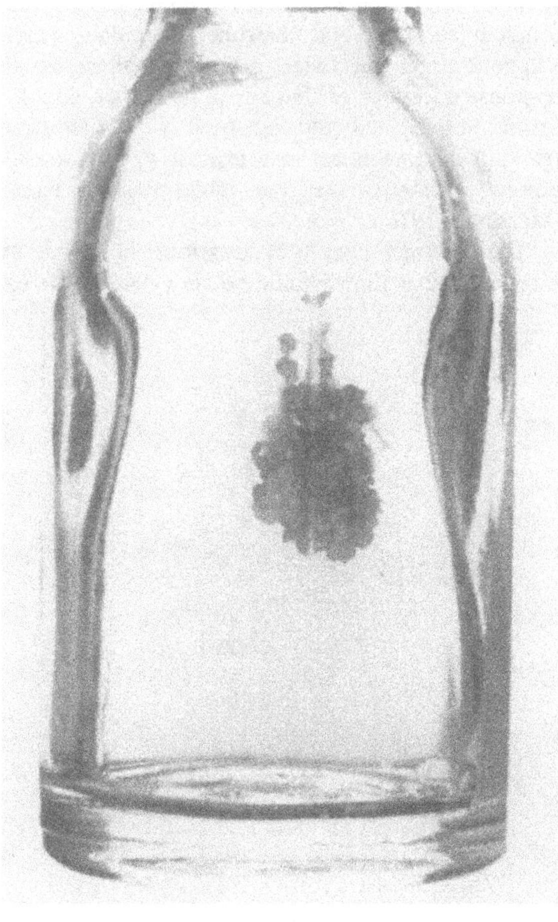

FIG. 6. Ten-day-old culture of *S. somaliensis* on Sabouraud agar.

The pale grain fungi, namely *P. boydii, Cephalosporium* spp., and *Aspergillus nidulans* are fast growers. Colonies are developed within 3 to 7 days. For their particular characteristics, which are varied, see Mahgoub and Murray (1973). The dematiaceous mycetoma agents, *M. mycetomatis, M. grisea,* and *Leptosphaeria senegalensis,* are slow growers. Colonies are well developed after 10 days. They are velvety in texture and heaped up with groves. *M. mycetomatis* is characterized by the secretion of a brown pigment that diffuses in the medium. *L. senegalensis* produces no pigment, but it sends brown-black filaments into the agar surrounding the aerial mycelium. Similarly, *M. grisea* produces no pigment and is differentiated from the rest of the fungi and actinomycetes in that it grows better at 30°C, whereas the optimum temperature for others is 37°C.

Specimen Collection and Transport

The basic specimen for identification of a mycetoma agent is the grain. Apart from the color, which has been described, the consistency is variable. Those of *S. somaliensis* and *M. mycetomatis* are hard because they contain a hard cement material; all other grains are soft. For easy collection it may be tempting to collect the grains that are discharged through sinuses. However, this should be avoided as much as possible because often such grains are dead and contaminated with bacteria. The specimen is placed in a sterile dry gauze and delivered immediately to the laboratory. One-half of the specimen is used for culture, and the other half is placed in 10% formol saline for histologic study.

Five milliliters of blood is collected in a sterile container, serum is separated, and the product is kept at −20°C for determination of antibodies.

Isolation and Identification

Using a dissection microscope or a magnifying lens, grains are fished out of the biopsy specimen and placed separately on the medium. They can be washed in sterile normal saline to free them from contaminating debris and quickly rinsed in 70% alcohol to kill contaminant bacteria.

For the actinomycetes, both Löwenstein Jensen medium and 2% glucose peptone (Sabourauds) agar are used. True fungal grains are cultured onto ordinary bacteriologic blood agar and subcultured on Sabouraud and Czapek dox agar for identification. Cultures are inspected daily, and when there is visible growth they should be subcultured immediately on fresh media for fear of contamination. All actinomycetes are incubated at 37°C, whereas fungi are grown at temperatures of 37°C and 26°C.

Identification is carried out on the morphologic picture mentioned. For detailed biochemical identification see Mahgoub and Murray (1973).

Serologic Identification

Determination of antibodies in the sera of patients with mycetoma is well established (Gordon and Mahgoub, 1976; Mahgoub, 1974). Counterimmunoelectrophoresis (CIE) is the test most used (Gumaa and Mahgoub, 1975), and to a lesser degree the immunodiffusion (ID) test is used. In both tests specific precipitation bands develop between the antibody and homologous antigen. Cytoplasmic antigens are prepared from young cultures of organisms, whether on solid media or stir-cultures in liquid media. After development of precipitation lines, slides are washed in buffer, dried at 37°C, and stained by naphthalene black (Fig. 7). This makes faint lines clearer and provides a permanent preparation. The test is not only useful in diagnosis but also in follow-up of treatment as intensity and number of precipitation bands diminish with progress of successful treatment. However, there is some cross-reactivity between *A. pelletierii* and *A. mudurae*. For preparation of antigens, particularly eumycetic antigens, several strains are to be tried to obtain a good one. The enzyme-linked immunosorbent assay (ELISA) has been tried (McLaren et al., 1978), but it was so sensitive that there was some cross-reactivity between the actinomycetes and true fungi.

Treatment

Actinomycetoma

The treatment of mycetoma depends first and foremost on identification of the causative organism or at least the decision whether it is an actinomycetoma or eumycetoma. So far no effective drug against both types is available.

For *A. Pelletierii,* and most strains of *S. somaliensis* and *N. brasiliensis,* streptomycin sulphate (1 g daily for 1 month, and on alternate days afterwards) is given together with cotrimoxazole (two tablets, each of 480 mg, morning and evening). Some strains of *S. somaliensis* and *N. brasiliensis,* as well as *A. madurae,* respond better to a combination of streptomycin sulphate and dapsone, diaminodiphenylsulphone (100 mg twice a day). If there is no response to either of these regimens, then streptomycin sulphate should be tried with rifampicin (two 300-mg tablets before breakfast) or with sulphadoxine pyrimethamine (one tablet twice a week) (Mahgoub, 1976).

The average length of treatment is about 10 months. Instructions should be very clear to the pa-

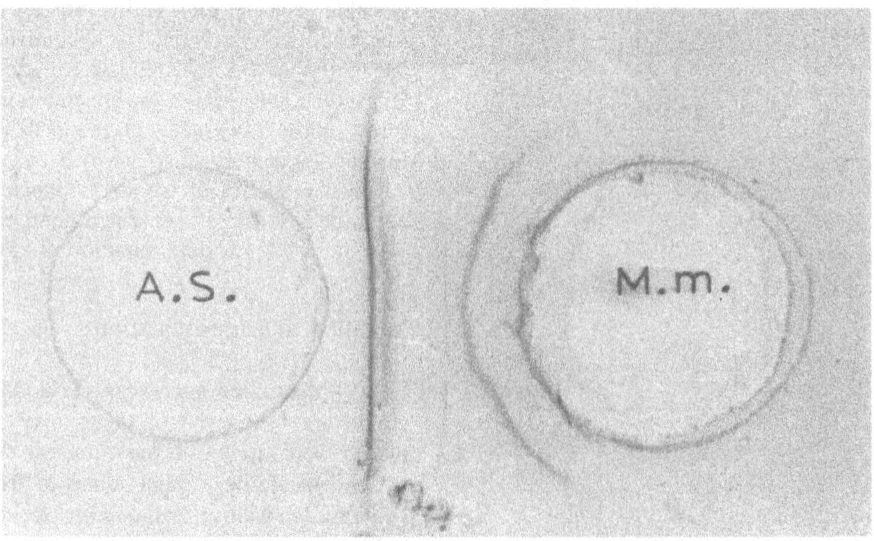

FIG. 7. Counterimmunoelectrophoresis test; note precipitation lines between *M. mycetomatis* (M.m.) antigen and homologons antiserum (A.S.).

tient that he should not interrupt treatment for fear of development of resistance and, worse, recurrence. Patients should be followed-up carefully for fear of development of leukopenia and anemia as a result of prolonged treatment with sulphonomides or dapsone. In such cases the patient should be taken off treatment until laboratory tests return to normal.

Whenever possible, bulk reduction surgery is performed and followed by the appropriate medical treatment.

Eumycetoma

Perhaps one may confidently say that successful medical treatment in eumycetoma has so far only been reported in cases due to *M. mycetomatis* (Mahgoub and Gumaa, 1984). This was obtained by ketoconazole (200 mg twice daily; Fig. 8). Treatment with this drug in pale grain mycetoma was not so successful (Rod Hay, personal communication). Although hepatotoxic effect due to ketoconazole has been reported in a few cases in Europe (Heiberg and Svejgard, 1981), no such effect was noticed in Saudi Arabia or the Sudan. Similarly, no hepatotoxicity was noticed in the treatment of patients with mucocutaneous leishmaniasia with ketoconazole (El Hassan, personal communication). However, the liver function test should be assessed before starting treatment. Like actinomycetoma, treatment is continued for about 10 months.

Clinical improvement is assessed by noting closure of sinuses, appearance of skin, reduction of swelling, and radiologic changes. Serologic follow-up is done by means of CIE. The number and intensity of precipitation lines decrease with good response to treatment and ultimately disappear. Only then should treatment be stopped.

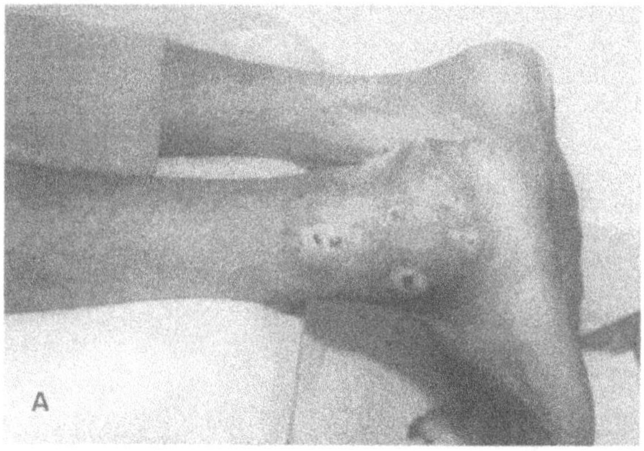

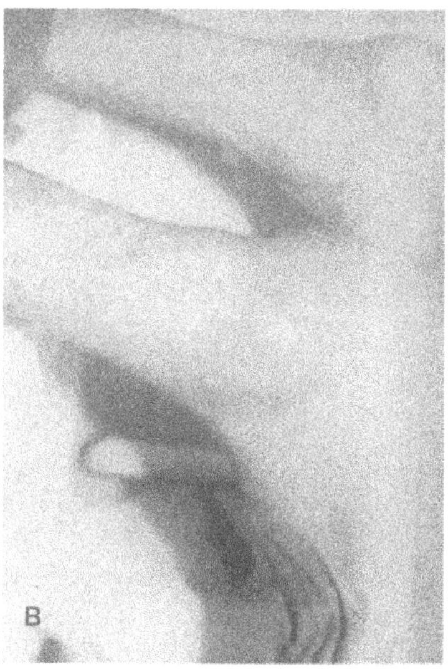

FIG. 8. Mycetoma pedis caused by *M. mycetomatis* before (A) and after (B) treatment with ketoconazole.

Prevention

That the majority of mycetoma cases are on the feet led some researchers to postulate that perhaps its morbidity can be controlled simply by wearing shoes and minimizing barefootedness. This is not true because it is the host's preparedness through the standard of CMI that determines the susceptibility to mycetoma (Mahgoub et al., 1978).

Certainly there is more mycetoma among the lower socioeconomic section of the population, who are certainly malnourished and stricken with viral, bacterial, and parasitic infections.

Literature Cited

Abbott, P. G. 1956. Mycetoma in the Sudan. Trans. R. Soc. Trop. Med. Hyg. **50:**11–24.

Ajello, G., J. Brown, E. S. Mahgoub, and L. Ajello. 1979. A note on the isolation of pathogenic aerobic actinomycetes from Sudanese soils. Curr. Microbiol. **2:**25–26.

Davies, A. G. M. 1958. The bone changes of Madura foot observations on Uganda Africans. Radiology **70:**841–847.

El Hassan, A. M., and E. S. Mahgoub. 1972. Lymphnode involvement in mycetoma. Trans. R. Soc. Trop. Med. Hyg. **66:**1865–1869.

Gordon, M. A., and E. S. Mahgoub. 1976. Immune response to aerobic pathogenic actinomycetaceae, chapter 16. *In* Rose and Friedman (eds.), Manual of clinical immunology. American Society. for Microbiology, Washington D.C.

Gumaa, S. A., and E. S. Mahgoub. 1975. Counterimmunoelectrophoresis in the diagnosis of mycetoma and its sensitivity as compared to immunodiffustion. Sabouraudia **13:**309–315.

Gumaa, S. A., A. A. Satir, A. H. Shihata, and E. S. Mahgoub. 1975. Tumor of the mandible caused by *Madurella mycetomii*. Am. J. Trop. Med. Hyg. **24**:471–474.

Gumaa, S. A., F. H. A. Mohamed, E. S. Mahgoub, S. E. Adam, A. M. El Hassan, and S. E. Imbabi. 1978. Mycetoma in goats. Sabouraudia **16**:217–223.

Gumaa, S. A., E. S. Mahgoub, and M. A. El Seed. 1986. Mycetoma of the head and neck. Am. J. Trop. Med. Hyg. **35**:594–600.

Heiberg, J. K., and E. Svejgard. 1981. Toxic hepatitis during ketoconazole treatment. Br. Med. J. **283**:825–826.

McLaren, M., E. S. Mahgoub, and E. Georgakopoulos. 1978. Preliminary investigation on use of ELISA in serodiagnosis of mycetoma. Sabouraudia **16**:225.

Mahgoub, E. S. 1975. Serological diagnosis of mycetoma. Proceedings of the 3rd. International Conference on the Mycoses of the Pan American Health Organization, Sao Paulo, Brazil, August 1974, PAHO Scientific Publication No. 304, Washington D.C.

Mahgoub, E. S. 1976. Medical management of mycetoma. W.H.O. Bull. **54**:303–310.

Mahgoub, E. S. 1977. Mycoses of the Sudan. Symposium on tropical mycoses. Trans. R. Soc. Trop. Med. Hyg. **71**:184–188.

Mahgoub, E. S. 1978. Experimental infection of athymic nude New Zealand mice "nu nu strain" with mycetoma agents. Sabouraudia **16**:211–216.

Mahgoub, E. S. 1985. Mycetoma. Semin. Dermatol. **4**:230–239.

Mahgoub, E. S., I. G. Murray. 1973. Mycetoma. William Heinemann Medical Book Limited, London.

Mahgoub, E. S., S. A. Gumaa, and A. M. El Hassan. 1977. Immunological status of mycetoma patients. Bull. Soc. Pathol. Exot. **70**:48–54.

Mahgoub, E. S., and S. A. Gumaa. 1984. Ketoconazole in the treatment of eumycetoma due to *Madurella mycetomii*. Trans. R. Soc. Trop. Med. Hyg. **78**:376–379.

Mariat, F. 1963. Sur la distribution geographique et la repartition des agents des mycetomes. Bull. Soc. Pathol. Exot. **56**:35–45.

Murray, I. G. 1966. Some aspects of medical mycology in the tropics. Trans. R. Soc. Trop. Med. Hyg. **60**:544–561.

CHAPTER 68

Nocardiosis

H. J. SHADOMY and N. G. WARREN

Disease: Nocardiosis.

Etiologic Agents: *Nocardia asteroides, Nocardia brasiliensis, Nocardia caviae* (*Nocardia otitidis-caviarum*). The latter two species are most often associated with mycetoma, which is covered in Chapter 67. They do, however, produce occasional nocardiosis.

Source: Airborne contaminant found in soil, probably throughout the world. Unlike actinomycosis, *Nocardia* is not a normal inhabitant of humans or animals.

Clinical Manifestations: *Nocardia asteroides* causes primary pulmonary lesions with extensive abscess formation, and with hematogenous spread to various organs, including kidney. There is a definite predilection for the central nervous system. There is a strong association of nocardial infections with disorders of the cellular immune system, long-term corticosteroid or other immunosuppressive treatment, lymphoma, and leukemia.

Pathology: Consolidation of one or more lobes of the lung is common, often with extension to the pleural cavity, and formation of sinus tracts throughout the abdominal cavity. Sulfur granules are not associated with these tracts. Pyogenic lesions of the brain often are produced.

Laboratory Diagnosis: Staining of pus, sputum, etc. by Gram, Giemsa, or Kinyoun's modification of the acid-fast stain; exhibit branching, "beaded," hyphae approximately 0.5 to 1 μ in diameter. Growth aerobic on standard bacteriologic media.

Epidemiology: Worldwide, in all ages, in males approximately twice as often as females. Seem to be especially prone to produce disease in patients on long-term immunosuppressive drug therapy.

Treatment: Sulfadiazine, sulfoxazole, minocycline. With pleuritis, trimethoprim-sulfamethoxazole appears to be most useful. Surgical intervention may be indicated at times.

Description of Disease

Nocardiosis is a generalized infectious disease of humans, characterized by primary pulmonary lesions that may be subclinical or overt, acute or chronic. Cavitation is not rare; the disease may spread hematogenously to other organs with a predilection for the central nervous system. In various organs it often produces pyogenic abscesses. Patients often are debilitated by other diseases or medication. Nocardiosis also may be seen in animals, most frequently in dogs and cattle, particularly dairy cattle.

The species found in nocardiosis most frequently is *N. asteroides*. Occasional cases are reported as caused by *N. brasiliensis,* and even fewer by *N. caviae*. Other species may be found anecdotally in the literature.

Communicability and Transmissibility

Nocardia asteroides is a free-living soil organism. It may become airborne where it is inhaled, leading to a primary lung infection. It probably is not transmissible from human to human, animal to animal, or animal to human, although there is some question of this in the literature.

Clinical Manifestations

Lesions and symptoms vary so widely that identification of the causal agent in nocardiosis may be difficult. It must be differentiated from miliary actinomycosis or tuberculosis, fungus diseases, and neoplasms. Pulmonary nocardiosis is characterized by anorexia, malaise, weight loss, fever (38 to 41°C), night sweats, and pleurisy. The pleura may become greatly thickened. Sinus tracts may traverse the abdominal cavity or occasionally penetrate the chest wall.

At times *Nocardia asteroides* may be introduced into subcutaneous tissue traumatically, where it may initiate a localized, chronic infection. At times these lesions appear to heal, only to metastasize to the brain.

Primary Pulmonary Nocardiosis

The usual type of disease produced is primary pulmonary infection caused by inhalation of the organism, with no site in the lung being favored. There may be a solitary abscess, scattered infiltrates, acute necrotizing pneumonia, or progressive fibrosis and extension to the pleura and chest wall.

Coughing usually is nonproductive in early disease, but becomes mucopurulent and blood tinged as the disease progresses.

Cavitation may lead to hemoptysis; consolidation of one or more lobes may occur.

Metastasis to the Brain

The most common site for development of secondary lesions with nocardiosis is the brain. Up to 15 to 27% dissemination has been reported. Symptoms include headache, lethargy, convulsions, peripheral numbness, nuchal rigidity, aphasia, tremor, and paresis. Metastasis may occur from a primary pulmonary site or, rarely, it has been reported from a primary subcutaneous infection.

Cutaneous Nocardiosis

Dissemination to the skin is not seen. Cutaneous nocardiosis typically is seen after local trauma and contact with soil. In those autoinoculated individuals with a normal immunologic response, the infection remains localized and may resolve spontaneously or with a short course of appropriate antimicrobials. However, in immunosuppressed individuals, as discussed earlier, dissemination may occur.

In the skin several types of lesions may be seen, ranging from pyogenic to granulomatous responses. Cellulitis, pustules, pyoderma, and, occasionally, subcutaneous abscesses occur. Lymphocutaneous lesions resembling sporotrichosis also occur at times.

Pathology

The pathologic picture of nocardiosis resembles that seen with infections caused by pyogenic, necrotizing bacteria. Involved organs exhibit multiple abscesses with central necrosis. Organized granulomas with giant cell infiltration are seen only rarely. There is less host response, fibrosis, scarring, and sinus tract formation with nocardiosis than seen in actinomycosis. Sulfur granules are not seen in the former, but they are seen typical in the latter disease.

Lung changes may be a combination of consolidation, abscesses, miliary and nodular lesions, and cavities with adhesions. Other organs also demonstrate suppurative abscesses, which are of the general pyogenic cellular type.

In tissue, Nocardia forms short and long branching filaments 0.5 to 1 μm in diameter, which are not stained by hematoxylin and eosin. However, the hyphal strands may be visualized with modified Gram stain (e.g., Brown-Brenn), Giemsa, and Kinyoun's modified acid-fast stain. The delicate hyphae give a typical "beaded" appearance with these stains. The methenamine silver stain also adequately demonstrates the organism, but the beaded appearance helpful in identification is lost. Special fungus stains are not used for Nocardia.

Diagnostic Features

Pulmonary nocardiosis rarely is seen in the early phases, and because of the insidious nature of the disease, it usually is well developed before medical care is sought.

Nocardiosis should be considered in patients on corticosteroid or other immunosuppressive therapy who exhibit lung abscesses or consolidation or signs and symptoms characteristic of brain abscesses.

Prognosis

The prognosis of nocardiosis, particularly after metastisis, is not good. Patients generally are debilitated with a poorly functioning immune response, which, even with chemotherapy, is unable to eliminate the disease. Involvement of the central nervous system is fatal in more than 85% of cases (Beaman et al., 1976).

Epidemiologic Aspects

Occurrence

The nocardiae are common inhabitants of the environment; *Nocardia* species have been recovered from various soils of the world, including those in the United States (Kumar and Mohapatra, 1968; Rippon, 1977). Samples of room dust, air, and tap water also have been shown to harbor these microorganisms (Houang et al., 1980; Stevens et al., 1981).

Because norcardial infections are not reported routinely to public health officials, the true incidence of disease is unknown. It has been estimated that at least 500 to 1000 new cases of nocardiosis are detected each year. While persons of any age may be susceptible to nocardial disease, frequently the patients are in the age range of 21 to 50 years. Infections in males usually outnumber those found in females by a ratio of at least 2 : 1 (Beaman et al., 1976). With the exception of mycetomas, about 85% of nocardial infections are pulmonary in origin, and nearly 45% of these will disseminate to involve other organ systems. *Nocardia asteroides* is most often the etiologic agent, followed by *Nocardia brasiliensis* and *Nocardia caviae* (*Nocardia otitidis-caviarum*). Primary cutaneous infections are usually due to *N. brasiliensis* (Stevens, 1983).

Isolation of the nocardiae can occur in a wide range of clinical situations. Nocardiae have been recovered as seemingly casual isolates or colonizers, without evidence of true infection (Frazier et al., 1975; Hosty et al., 1961). Opportunistic nocardial disease, usually caused by *N. asteroides*, has been associated with patients who were immunosuppressed because of corticosteroids, irradiation, and/or chemotherapy (Young et al., 1971). Conversely, *N. brasiliensis* often has caused disease in apparently immunocompetent individuals (Smego and Gallis, 1984). Additionally, the nocardiae have been associated with nosocomially acquired infections (Rosett and Hodges, 1978).

Transmission

Traumatic insult typically precedes primary cutaneous nocardiosis. Because these organisms are found in the soil, many opportunities exist for environmental inoculation. Injection of the organism into tissue has occurred via thorn and rusty nail puncture, cat scratch, burns (Satterwhite and Wallace, 1979), bone fractures (Smego and Gallis, 1984), and tooth extraction (Harris, 1980).

The respiratory tract offers another portal of entry for *Nocardia* species. Inhalation of the microorganisms from the environment is the most commonly suggested mode of infection. Dissemination of the disease can occur through bacteremia, pleural extension, or lymphatic spread (Curry, 1980).

Although human-to-human or animal-to-human transmission has not been conclusively documented, several reports have alluded to these modes of infection. Infections caused by *N. brasiliensis* have been found in animals (Gezuele, 1972), and it has been postulated that cats may serve as vectors for the transmission of this organism to humans (Ajello et al., 1961). Likewise, the report of an outbreak of *N. asteroides* infections in a hospital renal unit (Houang et al., 1980) and the subsequent serologic evaluation of the episode (Stevens et al., 1981) have offered the strongest suggestion of transmission to date.

Etiologic Agents

Characterization

Nocardiae are gram-positive, acid-fast variable, lysozyme-resistant, nonmotile, obligate aerobes. Depending on the age of the culture, the microscopic appearance can vary greatly, from coccoid to bacillary forms, to abundant formation of branched filaments.

Colonies are usually dry, crusty, and roughened in appearance but may have a leathery texture. Aerial mycelia are often evident, and carotenoid pigments may be produced.

Usually the genus *Nocardia* can be assigned by observation of characteristic microscopic and macroscopic morphology; however, difficulty in separating the genera *Nocardia*, *Mycobacterium*, and *Corynebacterium* can occur. Mycolic acid determinations may be useful in these situations. The corynebacteria contain about 32 carbon atoms, the nocardia about 50 carbon atoms, and the mycobacteria about 80 carbon atoms.

Specimen Collection and Transport

Early morning expectorated sputa, collected on 3 separate days, should be submitted in separate sterile containers. Specimens that grossly or microscopically resemble saliva should not be accepted unless specifically requested by the physician. At least 5 ml of specimen is desirable for adequate laboratory studies. If the patient experiences difficulty in expectorating, then sputum induction by hypertonic saline aerosolization should be considered. Naturally produced sputum after aerosol induction also should be cultured.

Pus, exudate, or drainage should be submitted in a sterile tube or syringe. When the amount is small,

0.5 to 1.0 ml of sterile physiologic saline can be added to the specimen to maintain moist conditions. Collection of these specimens by using a sterile swab, although not unsatisfactory, should be discouraged, as fibers can make smear interpretation difficult.

Surgically removed tissue such as lymph nodes and biopsy specimens should be submitted in sterile containers with 0.5 to 1.0 ml of sterile physiologic saline or between moistened gauze pads if transport is not prolonged.

Sterile fluids, such as cerebrospinal, pleural, or joint should be submitted in sterile tubes. Blood or bone marrow should be collected in heparin or sodium polyanethosulfonate and submitted in sterile tubes.

All specimens should be transported to the laboratory for processing as soon as possible. Although specimens do not require special temperatures during transport, if a delay is anticipated, then the specimens may be stored at 4°C.

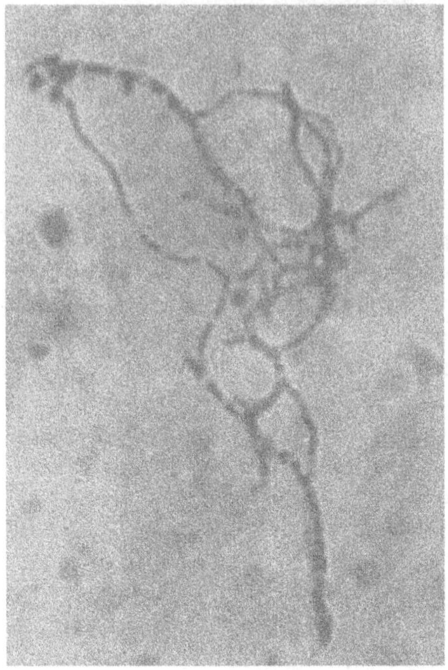

FIG. 1. *Nocardia asteroides* in sputum showing typical beading of stained hyphae. Modified Kinyoun acid-fast stain.

Isolation Procedures

Specimens should be examined grossly for the presence of granules or purulent areas and those portions should be selected for further processing. Sterile tissues should be minced with scissors or homogenized briefly with a tissue grinder. Sterile body fluids usually do not need to undergo any preparation unless the volume is large, in which case, they should be centrifuged first. Cerebrospinal fluids may be passed through a membrane filter and the filter placed directly onto the isolation medium. Sputa should be digested with dithiothreitol or N-acetyl-L-cysteine (without sodium hydroxide) before inoculation to media (Haley and Callaway, 1978).

Gram-stained smears of the clinical material should be prepared and examined microscopically. Nocardiae are gram-positive organisms that frequently exhibit extensively branching filaments as well as beaded bacillary forms (Fig. 1). Fuchsin-stained smears also may be prepared; however, the nocardiae are very susceptible to harsh decolorants and can be rendered nonacid-fast if the standard Ziehl-Neelsen or Kinyun procedures are used. For that reason, when acid-fast studies are requested for aerobic actinomycetes, a modified Kinyoun stain using 1% H_2SO_4 as a decolorizer should be used. Both *N. asteroides* and *N. brasilensis* are acid fast; many other actinomycetes are not.

Nocardia species grow readily on laboratory media without antibiotics. Specimens should be inoculated to blood agar, brain-heart infusion agar, and Middlebrook and Cohn 7H-10 or 7H-11 agar. Cultures will grow equally well on the Lowenstein-Jensen medium used for the isolation of mycobacteria. Cultures should be incubated at 35 to 37°C and held for 4 weeks. Growth may be enhanced by the addition of 5 to 10% CO_2 into the atmosphere. Growth can be detected in 10 days to 2 weeks and will appear as folded, waxy colonies, at first. Older cultures may develop areas of aerial mycelia giving a dry, powdery appearance. Often there is a musty, damp soil-like odor.

Culture Identification

Most aerobic actinomycetes from clinical specimens can be identified by the use of a few physiologic tests (Haley and Callaway, 1978). The flow chart in Table 1 indicates an approach to the identification of these organisms. When obvious growth is observed, the culture should be examined using a dissecting microscope to detect the presence of aerial mycelia. If aerial mycelia are observed, then it can be assumed that the isolate is an aerobic actinomycete, and additional tests can be performed. If no aerial mycelia are seen, then slide cultures using brain-heart infusion agar should be prepared. Within 1 to 4 weeks *Nocardia, Streptomyces,* and *Actinomadura* species will produce extensive tertiary branching or anastomosing filaments. *Mycobacterium* and *Rhodococcus* species will not develop aerial mycelia but may exhibit some primary branching. To further rule out the my-

TABLE 1. Flow chart for the identification of some aerobic actinomycetes

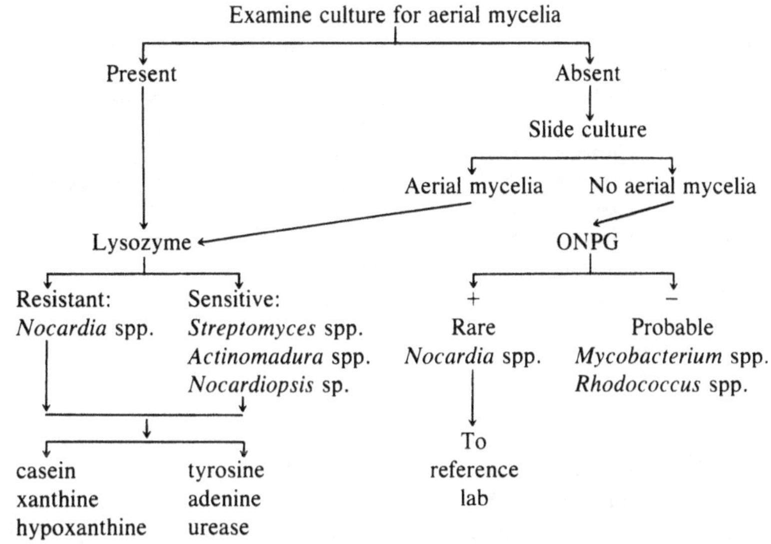

cobacteria and rhodococci, the ONPG disk test usually used for the identification of *Enterbacteriaceae* may be done. *Nocardia, Streptomyces,* and *Nocardiopsis* species are positive within 24 hours, whereas *Actinomadura, Mycobacterium,* and *Rhodococcus* species are negative.

Procedures to determine lysozyme resistance, urease broth activity, and ability to digest casein, xanthine, hypoxanthine, tyrosine, and adenine are used routinely to identify aerobic actinomycetes. Most *Nocardia* species will grow in the presence of lysozyme, whereas growth of *Streptomyces* and *Actinomadura* species will be inhibited. *Nocardia asteroides, N. caviae,* and *N. brasiliensis* are urease positive, and *N. asteroides* will not digest the various test substrates. Table 2 indicates the expected reactions of some aerobic actinomycetes to these physiologic tests, and interpretation of the reactions can be found in Table 3.

Additional Identification Procedures

Although the most common nocardiae can be identified routinely by standard techniques, an occasional variant or a less common species may prove to be more difficult. For the most part, the following supplemental tests are best performed by a reference laboratory that has the resources and technical expertise to identify uncommon isolates.

The detection of acid production on various carbohydrate-containing substrates may be helpful in separating members of the different aerobic actinomycete species. Although reference laboratories may study oxidative reactions of as many as 23 carbohydrates, the one that can be most easily incorporated into a clinical laboratory setting is the test for acid production on lactose-containing media. Cultures grown in trypticase soy broth can be inoculated to a basal medium containing lactose (Gordon et al.,

TABLE 2. Physiologic differentiation of some aerobic actinomycetes*

	Decomposition utilization of							
Organism	*Casein*	*Xanthine*	*Hypoxanthine*	*Tyrosine*	*Adenine*	*ONPG*	*Lysozyme*	*Urease*
Nocardia asteroides	−	−	−	−	−	+	R	+
Nocardia brasiliensis	+	−	V	+	−	+	R	+
Nocardia caviae	−	+	+	−	−	+	R	+
Streptomyces griseus	+	+	+	+	+	+	S	V
Actinomadura madurae	+	−	−	+	−	−	S	−
Nocardiopsis dassonvillei	+	V	+	+	+	+	S	V

* Temperature = 28 to 30°C. V = variable; R = resistant; S = sensitive; + = positive; − = negative.

TABLE 3. Interpretation of some aerobic actinomycete physiologic tests[a]

Test	*Incubation time*	*Appearance of positive reaction*	*Appearance of negative reaction*	*Positive control organism*	*Negative control organism*
ONPG	4 to 24 h	Yellow	Colorless	*N. asteroides*	*A. madurae*
Lysozyme	1 week	Growth-resistant	No growth-sensitive	*N. asteroides*	*S. griseus*
Casein	2 weeks	Clearing of media	No clearing	*S. griseus*	*N. asteroides*
Xanthine	3 weeks	Clearing of media	No clearing	*S. griseus*	*N. asteroides*
Hypoxanthine	3 weeks	Clearing of media	No clearing	*S. griseus*	*N. asteroides*
Tyrosine	4 weeks	Clearing of media	No clearing	*S. griseus*	*N. asteroides*
Adenine	4 weeks	Clearing of media	No clearing	*S. griseus*	*N. asteroides*
Urease	4 weeks	Pink to red	Straw-colored	*N. asteroides*	*A. madurae*

[a] Temperature = 28 to 30°C.

1974) and observed for the yellow acid color of the indicator. *Nocardiopsis dassonvillei,* an aerobic actinomycete that has been shown to be an opportunistic pathogen (Philip and Roberts, 1984), and *Streptomyces griseus* have very similar physiologic test patterns. Acid production from the oxidation of lactose will separate these two organisms; *N. dassonvillei* gives a negative reaction, whereas *S. griseus* is positive (Berd, 1973).

Another procedure used by reference laboratories when the standard techniques fail to reveal the identity of an aerobic actinomycetes is the study of whole cell constituents. Use of thin-layer chromatography or standard paper chromatography (Haley and Callaway, 1978) will detect the type of 2,6-diaminopimelic acid (DAP) and diagnostic sugars present in the cell. *Streptomyces* species contain L-DAP, whereas *Nocardia* species, *Actinomadura* species, and *N. dassonvillei* contain meso-DAP. Arabinose and galactose are found as part of the whole cell constituents in *Nocardia* species, and madurose will be detected in *Actinomadura* species.

Culture Serotyping

A serotyping scheme for cultures of *N. asteroides* has been developed that can aid in epidemiologic investigations of outbreaks (Pier and Fichtner, 1971). Four antigenic types (I, II, III, and IV) have been detected by immunodiffusion tests of soluble extracellular antigens of *N. asteroides.* Precipitinogens of types I, II, and III appear to be unique to *N. asteroides,* whereas type IV may be found in other *Nocardia* species as well as several other aerobic actinomycetes. Serotyping of a group of *N. asteroides* recovered from humans indicated that type IV was the most common (27.3%), followed closely by type I (25%), then a mixture of types III and IV (13.9%), type III (13.6%), type II (6.8%) and a mixture of types II and IV (4.5%) (Pier and Fichtner, 1981).

Serology

Diagnosis of nocardiosis by serologic methods has met with limited success. A complement-fixation test (Shainhouse et al., 1978) and a microimmunodiffusion test (Blumer and Kaufman, 1979) have been developed that suggest that infected individuals could be identified. Specificity of these tests is hindered by a high degree of cross-reactivity with other disease entities such as tuberculosis, actinomycosis, and histoplasmosis.

To date, the most promising serologic tool for the diagnosis of nocardiosis appears to be western blot analysis. Although this technique is not readily incorporated into the average-sized clinical laboratory, it does offer hope that serologic detection of human nocardiosis is achievable and that more conventional methods can be developed in the future. In an initial study of patients with proven nocardiosis, all sera contained IgG antibody that reacted with one or two extracellular antigens present in culture filtrates of *N. asteroides.* No sera from patients with tuberculosis cross-reacted with *N. asteroides* antigen; however, sera from 12% of the normal control group demonstrated antibody (Sugar et al., 1985).

Treatment

Chemotherapy generally is prolonged (12 to 18 months), and relapse may occur many months after it is completed.

Sulfadiazine is the treatment of choice because of its penetration into cerebrospinal and other fluids. Surgical drainage of empyema or abscesses (e.g., subcutaneous or cerebral) often is used, but must be followed by prolonged treatment as above. Metastatic lesions can occur even during or after an apparently effective course of treatment.

Sulfisoxazole may be used initially when fluid loading and urine alkalinization are difficult, as urine alkalinization and copious amounts of fluid are required during sulfadiazine treatment. Sulfisoxazole also is used after patients show continued progressive improvement with sulfadiazine.

In recent years sulfisoxazole combined with trimethoprim has been considered superior to other regimens; however, both treatment failures and late relapses have been recorded (Stamm et al., 1983).

Before the advent of sulfonamides, therapy primarily was supportive. Even today supportive care cannot be underestimated. Even with the best of care and adequate regimens of sulfonamides, as indicated by antimicrobial susceptibility testing, reports indicate survival rates of 92% in pulmonary disease, but only 52% when brain abscesses are present.

Literature Cited

Ajello, L., W. W. Walker, D. L. Dungworth, G. L. Brumfield. 1961. Isolation of *Nocardia brasiliensis* from a cat, with a review of its prevalence and geographic distribution. J. Am. Vet. Med. Assoc. **138**:370–376.

Beaman, B. L., J. Burnside, B. Edwards, and W. Causey. 1976. Nocardial infections in the United States, 1972–1974. J. Infect Dis. **134**:286–289.

Berd, D. 1973. Laboratory identification of clinically important aerobic actinomycetes. Appl. Microbiol. **25**:665–681.

Blumer, S. O., and L. Kaufman. 1979. Microimmunodiffusion test for nocardiosis. J. Clin. Microbiol. **10**:308–312.

Curry, W. A. 1980. Human nocardiosis. A clinical review with selected case reports. Arch. Intern. Med. **140**:818–826.

Frazier, A. R., E. C. Rosenow III, and G. D. Roberts. 1975. Nocardiosis, a review of 25 cases occurring during 24 months. Mayo Clin. Proc. **50**:657–663.

Gezuele, E. 1972. Fatal infection by *Nocardia brasiliensis* in an armadillo. Sabouraudia. **10**:63–65.

Gordon, R. E., B. A. Barnett, J. E. Handerman, and C. H. Pang. 1974. *Nocardia coeliaca, Nocardia autotrophica* and the nocardin strain. Int. J. Syst. Bacteriol. **24**:54–63.

Haley, L. D., and C. S. Callaway. 1978. Laboratory methods in medical mycology, 4th ed. U.S. Department of Health, Education and Welfare. HEW Publication No. (CDC) 78-8361. Washington, D.C.

Harris, G. K. 1980. Nocardiosis: a literature review and a case report of *Nocardia asteroides* infection. Am. J. Med. Tech. **46**:44–48.

Hosty, T. S., C. McDurmont, L. Ajello, L. K. Georg, G. L. Brumfield, and A. A. Calix. 1961. Prevalence of *Nocardia asteroides* in sputa examined by a tuberculosis diagnostic laboratory. J. Lab. Clin. Med. **58**:107–114.

Houang, E. T., I. S. Lovett, F. D. Thompson, A. R. Harrison, A. M. Joekes, and M. Goodfellow. 1980. *Nocardia asteroides* infection—a transmissible disease. J. Hosp. Infect. **1**:31–40.

Kumar, E. T., and L. N. Mohapatra. 1968. Studies on aerobic actinomycetes isolated from soil. I. Isolation and identification of strains. Sabouraudia **6**:140–146.

Philip, A., and G. D. Roberts. 1984. *Nocardiopsis dassonvillei* cellulitis of the arm. Clin. Microbiol. Newsl. **6**:14–15.

Pier, A. C., and R. E. Fichtner. 1971. Serologic typing of *Nocardia asteroides* by immunodiffusion. Am. Rev. Respir. Dis. **103**:698–707.

Pier, A. C., and R. E. Fichtner. 1981. Distribution of serotypes of *Nocardia asteroides* from animal, human, and environmental sources. J. Clin. Microbiol. **13**:548–553.

Rippon, J. W. 1977. *Nocardia:* a geographic prevalence [letter]. Arch. Dermatol. **113**:237.

Rosett, W., and G. R. Hodges. 1978. Recent experiences with nocardial infections. Am J. Med. Sci. **276**:279–284.

Satterwhite, T. K., and R. J. Wallace, Jr. 1979. Primary cutaneous nocardiosis. J. Am. Med. Assoc. **242**:333–336.

Shainhouse, J. A., A. C. Pier, and D. A. Stevens. 1978. Complement fixation antibody test for human nocardiosis. J. Clin Microbiol. **8**:516–519.

Smego, R. A., Jr., and H. A. Gallis. 1984. The clinical spectrum of *Nocardia brasiliensis* infection in the United States. Rev. Infect. Dis. **6**:164–180.

Stamm, A. M., D. W. McFall, and W. E. Dismukes. 1983. Failure of sulfonamides and trimethoprim in the treatment of nocardiosis. Arch. Intern. Med. **143**:383.

Stevens, D. A., A. C. Pier, B. L. Beaman, P. A. Morozumi, I. S. Lovett, and E. T. Houang. 1981. Laboratory evaluation of an outbreak of nocardiosis in immunocompromised hosts. Am. J. Med. **71**:928–934.

Stevens, D. A. 1983. Clinical and clinical laboratory aspects of nocardial infections. J. Hyg. Camb. **91**:377–384.

Sugar, A. M., G. K. Schoolnik, and D. A. Stevens. 1985. Antibody response in human nocardiosis: identification of two immunodominant culture-filtrate antigens derived from *Nocardia asteroides*. J. Infect. Dis. **151**:895–901.

Young, L. S., D. Armstrong, A. Blevins, and P. Lieberman. 1971. *Nocardia asteroides* infection complementing neoplastic disease. Am. J. Med. **50**:657–663.

CHAPTER 69

Paracoccidioidomycosis

RICARDO NEGRONI

Disease: Paracoccidioidomycosis (South American blastomycosis, paracoccidioidal granuloma, Lutz's mycosis).

Etiologic Agent: *Paracoccidioides brasiliensis.*

Source: Probably soil, rivers, or lakesides in subtropical regions of Latin America.

Clinical Manifestations: It is a systemic mycosis of subacute or chronic evolution. In children and adolescents, there is lymph node involvement and also involvement of other organs and the reticuloendothelial system. The disease shows a more chronic course in adults, when there is involvement of the lungs, skin, mucoses, adrenal glands, and lymph nodes.

Pathology: It is a granulomatous and suppurative disease. The causative agent is found free in the suppurative areas or inside giant cells. It can also be found in the skin, mucous membranes, lymph nodes, lungs, liver, spleen, adrenal glands, brain, bones, etc.

Laboratory Diagnosis: *Paracoccidioides brasiliensis* can be seen in biopsy specimens of the affected tissue, pus, or bronchial washings. It may be isolated from cultures where both phases of the microorganism are shown. Reproduction of the disease may be obtained by inoculation in the guinea pig or hamster. A positive immunodiffusion test in agar gel is particularly important, especially when it produces identity bands with a positive control serum.

Epidemiology: This mycosis is endemic in Latin America from southern Mexico to northern Argentina and Uruguay. It has been observed in areas with humid subtropical climates, near large rivers. *P. brasiliensis* has been isolated from soil samples in endemic areas. Spontaneous infection in animals has not been described. Man-to-man or animal-to-man transmission is unknown. Apparently, the infection is acquired by inhalation of the fungal spores. Progressive disease is more frequent in men 30 to 60 years of age, mainly among rural workers.

Treatment: Sulfonamides, amphotericin B, ketoconazole, and itraconazole have been used. Surgical procedures (tracheostomy, removal of cerebral granulomas, etc.) are sometimes necessary.

Prevention and Control: No specific control measures are known, and no useful vaccine has been produced for prevention of this disease.

Description of Disease

Paracoccidioidomycosis (South American blastomycosis, paracoccidioidal granuloma, Lutz's mycosis) is a systemic, granulomatous, and suppurative mycosis with endemic areas known to be the humid subtropical regions of Latin America. The infection is probably acquired by inhalation of the spores of its etiologic agent, *Paracoccidioides brasiliensis,* in its filamentous phase. Asymptomatic infection is frequent in the endemic areas. There is an increase in prevalence up to the age of 30 years, and there is no difference between sexes (Albornoz, 1982; Londero, 1984). Symptomatic manifestations are rare in childhood and are more frequent between 30 and 60 years, predominantly in men; the ratio of men to women is 15 : 1 (Del Negro et al., 1982; Lacaz et al., 1984). Rural populations have higher exposure to this mycosis than city inhabitants. Paracoccidioidomycosis has not been definitely confirmed in lower animals.

Infection

This disease has neither man-to-man nor animal-to-man transmission. *P. brasiliensis* may be found saprophytically in the soil or along river banks in its filamentous phase, and the infection is most frequently acquired by the inhalation of spores. Cutaneous primary infection is less common, even though it has occasionally been described (Albornoz, 1982; Martins Castro et al., 1975). The incubation period may range from 2 weeks to several years (Negroni, 1966, 1968).

Clinical Manifestations

The clinical manifestations of paracoccidioidomycosis vary according to age of the patient and immune status. Among children and adolescents, the disease shows a subacute and severe course, particularly with involvement of lymph nodes and organs of the reticuloendothelial system. Among adults over 30 years of age, paracoccidioidomycosis is more chronic and involves the lungs, skin, mucous membranes, nodes, and adrenal glands (Lacaz et al., 1984; Rubinstein and Negroni, 1981).

The following clinical forms of paracoccidioidomycosis have recently been accepted: paracoccidioidomycosis infection, and paracoccidioidomycosis disease, with a subacute form (youth type) and a chronic form (adulthood type).

Paracoccidioidomycosis Infection

This term includes all apparently healthy individuals of the endemic areas who show delayed-hypersensitivity skin reactions against *P. brasiliensis* antigens that are not related to cross-reactions with other fungal antigens, particularly *Histoplasma capsulatum*. It also includes those cases with confirmation of pulmonary or nodular granulomas due to *P. brasiliensis* that do not have clinical manifestations. It would also include the asymptomatic latency period between infection and occurrence of clinical manifestations which may last years, as has been proved in some cases of paracoccidioidomycosis disease outside of endemic areas (Del Negro et al., 1982; Negroni, 1966).

Symptomatic Primary Infection

Symptomatic primary infection is rarely diagnosed, as its clinical picture is not different from that of other acute pulmonary infections of bacterial or viral origin. An acute inflammatory reaction of lung tissue is produced with lymphangitis and hilar lymph node involvement. There is no typical radiologic manifestation. There may be early hematogenous dissemination, generally asymptomatic with metastasis in other organs. Infection is generally self-limited; there is regression of lesions which remain latent. Less frequently, there is calcification of the lungs, nodes, or other organs. Reactivation of latent sites of infection results in progressive paracoccidioidomycosis of adulthood and explains those cases occurring outside endemic areas. Primary infection of the skin is still observed less frequently. Appearance is that of a typical chancriform syndrome, similar to sporotrichosis, and it usually cures spontaneously (Albornoz, 1982; Martins Castro et al., 1975).

Paracoccidioidomycosis Disease

All clinical cases with progressive evolution that do not show a tendency to spontaneous regression are included under this classification.

Acute Form (Youth Type). This form of the disease is seen in children and adolescents of both sexes. It probably results from the progressive hematogenous dissemination of a pulmonary primary infection. It has a subacute course, with rapid impairment of the overall condition, septic fever, asthenia, anorexia, loss of body weight, leukocytosis, and anemia. Nodular hyperthrophy of multiple localization and hepatosplenomegaly are the most characteristic signs (Fig. 1). Lesions are rarely seen in lungs and mucous membranes. Generally, there are ulcerated acneiform papules in the skin, or even scrofuloderma; osteomyelitic localization and gastrointestinal impairment are also frequent. In moderate cases, the overall condition is more stable; lesions occur within a nodular group, and their evolution is slower. Pulmonary lesions of the miliary type have rarely been observed. The clinical picture resembles a septicemia, lymphoma, or leukemia.

Chronic Form (Adulthood Type). This occurs in individuals more than 30 years of age, mainly men. Overall condition is gradually impaired, with loss of body weight, asthenia, and exertional dyspnea. Lesions may be uni- or multifocal. In 20 to 30% of cases, there are only pulmonary changes. Initially they are rather atypical, appearing as nodular or infiltrative radiologic images located in one of the lungs, and the patient is usually asymptomatic. After some months or years, the radiologic picture shows bilateral and symmetric lesions, generally parahilarly dis-

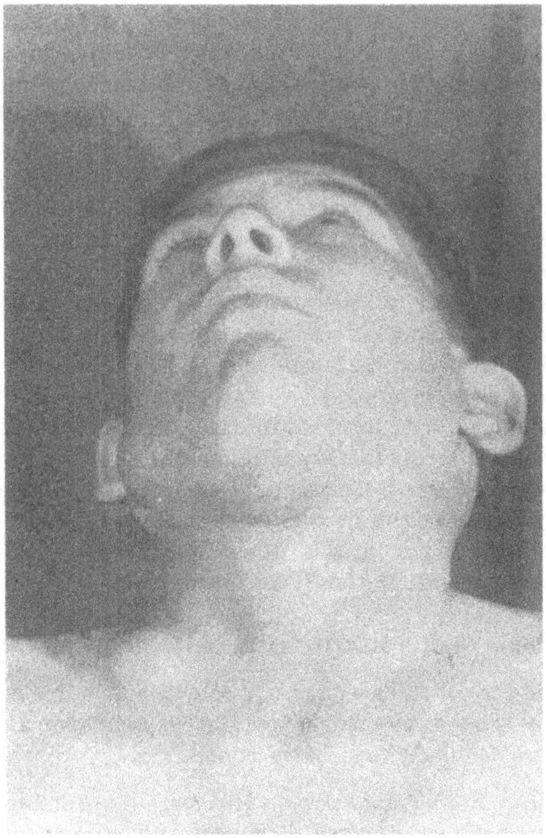

FIG. 1. Ganglionary involvement of the neck in an acute disseminated form of paracoccidioidomycosis in a young patient.

tributed, resembling the wings of a butterfly and formed by micronodules, infiltrates, and lineal shadows (Fig. 2). They are now known to be caused by retrograde progression through the lymph vessels of the pulmonary interstice, starting at the hilar nodes (Tuber et al., 1985). Patients lose body weight and show asthenia, thoracic pain, cough with mucopurulent expectoration (rarely hemoptic), mild fever, and exertional dyspnea. Most patients follow a markedly chronic evolution and die of cachexia or respiratory impairment. Less than 10% of the cases show secondary mucosal, nodular, or skin lesions.

Multifocal presentation is confirmed in 70 to 80% of patients, starting with pulmonary or mucocutaneous lesions. Lymphohematic dissemination of *P. brasiliensis* causes lesions in skin, oropharyngeal mucosa, larynx, lymph nodes, lungs, adrenal glands, abdominal organs, and central nervous system. The lesions start slowly and produce gradual asthenia and loss of body weight.

Oropharyngeal mucous membrane lesions are shallow ulcers, with granulomatous fundi, microabscesses, and punctate hemorrhages; they have an in-

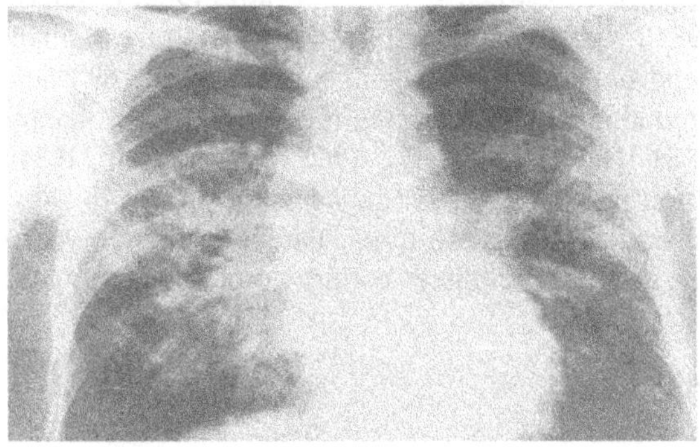

FIG. 2. Micronodular lesion of the lungs in a chronic disseminated form of the disease.

filtrated hard basis (mulberrylike stomatitis) (Fig. 3). A particularly hard, violet-red edema (thromboidlike lip) is observed in the lips and jugal mucosa. Cervical and supraclavicular nodes are generally hypertrophic and hard, and usually there are abscesses or even caseous necrosis, opening toward the skin as sinus tracts or ulcerations, which is clinically similar to tuberculous scrofuloderma.

Laryngeal attack causes dysphonia, dysphagia, and obstructive dyspnea. Skin lesions are more frequently seen as a consequence of microtraumatisms, and they are papular or papulopustular. They ulcerate, and their fundus turns vegetative or papillomatous. Adrenal gland involvement is seen in over 40% of cases; there may be minimal adrenal impairment (which can only be shown by gland stimulation tests) or even a typical Addison syndrome. As a consequence of CNS involvement, there have been reports of abscessed granulomas, generally located at the posterior fossa and also, to a lesser extent, subacute meningoencephalitis of clear cerebrospinal fluid or meningomieloradiculitis. Other probable localizations are arteries, testicles, epididymis, prostate, bones, joints, liver, and spleen (Angulo-Ortega and

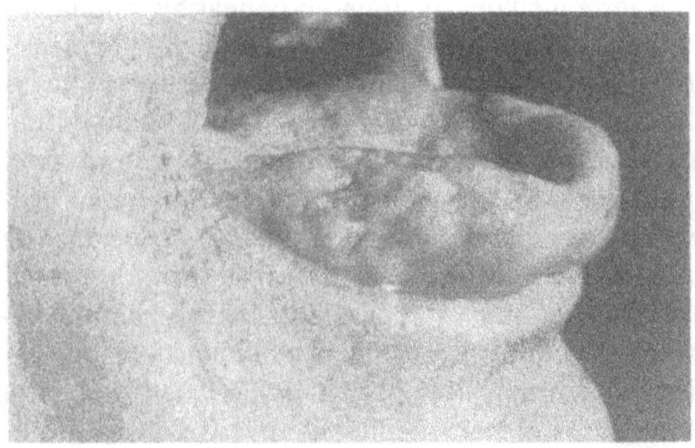

FIG. 3. Ulcerative lesion in the tongue of an adult patient.

Pollack, 1971; Franco and Montenegro, 1982; Negroni, 1966). Spontaneous evolution is always terminal after some years.

Posttherapeutic Latency

Clinical cure of paracoccidioidomycosis is achieved by effective therapy; there is remission of lesions and they remain latent with occasional relapses. Regression of paracoccidioidomycosis results in intense fibrosis, which may give rise to tracheal or laryngeal stenosis, buccal atresia, and cardiorespiratory impairment due to pulmonary fibrosis (paradoxal cure) (Negroni, 1968; Padilha Goncalves and Bardy, 1946).

Pathology

Infection is most commonly acquired by inhalation. The infectious structures are probably the aleuroconidia of the filamentous phase of *P. brasiliensis*. At the pulmonary alveolus they turn into yeastlike, multiple budding structures, and, initially, they cause an acute nonspecific inflammatory response of polymorphonuclear leukocytes and lymphocytes. They rapidly enter the hilar and mediastinal lymph nodes and subsequently invade the bloodstream and colonize different organ systems. Specific immune response, particularly cell-mediated immunity, appears after an evolution period of 10 to 20 days.

The inflammatory reaction changes, and giant cells and follicular and compact granulomas occur, which generally limit the infection and, in most cases, cause a complete remission. Acute disseminated infection of the youth is seen when cellular immune response fails during the primary infection; a rapidly progressive paracoccidioidomycosis subsequently occurs with involvement of many organs, particularly those with monocytic-histiocytic system. *P. brasiliensis* remains in some sites. These lesions, either pulmonary or extrapulmonary, are reactivated when there is a failure in the defensive mechanisms, particularly cellular immunity.

The most important latent sites of infection are generally hilar and mediastinal nodes, from which there is a new invasion of the pulmonary parenchyma by retrograde progression, through the lymph vessels of the interstice (Angulo-Ortega and Pollak, 1971; Tuber et al., 1985).

P. brasiliensis may invade all organs; histopathologic characteristics of this disease are related to the host defensive capacity. In serious cases, there are extensive suppurative areas and lax granulomas with great amounts of *P. brasiliensis;* mild and moderate forms show compact granulomas with few organisms (Franco et al. 1982, Negroni 1968).

Diagnostic Features

Acute paracoccidioidomycosis of youth must be considered for the differential diagnosis of serious febrile syndromes accompanied by adenopathies and hepatosplenomegaly. Softening and suppuration of a node, as well as residency in the endemic areas, are suggestive of paracoccidioidomycosis.

The disease may resemble a septicemia, lymphoma, leukemia or acute miliary tuberculosis. In these cases, *P. brasiliensis* may be seen in the purulent secretion of the nodes, and cultures are usually positive. Specific serologic reactions, immunodiffusion, counterimmunoelectrophoresis, and complement fixation are positive, and titers are high.

Chronic paracoccidioidomycosis of adulthood should be suspected among men over 30 years of age who have undertaken rural activities within the endemic areas and who have mucosal, skin, or pulmonary lesions of chronic evolution. Patients should be carefully interviewed, because there is sometimes a period of 20 to 30 years between the infection and the incidence of symptomatology. Skin and mucosal lesions are rather typical and allow correct diagnosis. It is possible, though, to confuse it with tuberculosis, cancer, sarcoidosis, histoplasmosis or X-histiocytosis. *P. brasiliensis* is generally observed in sputum, bronchial washings, and biopsies of mucosal, nodular or pulmonary lesions, both in wet preparations and in tissue sections stained with the PAS or Grocott technique. Culture isolation is usually more difficult for this clinical presentation. A greater percentage of isolations is achieved by intratesticular inoculation of the guinea pig or hamster. Serologic studies are positive and specific in about 90% of cases.

Prognosis

Paracoccidioidomycosis disease at one time was always terminal, but this situation changed in 1940 with the introduction of sulfonamides. At present, most patients survive but may continue with serious sequelae, particularly cardiorespiratory insufficiency. Relapses are also frequent. Manifestations of the acute presentation in youth or CNS or adrenal gland involvement are signs of poor prognosis.

Epidemiologic Aspects

Paracoccidioidomycosis is a systemic mycosis of limited geographic distribution. It is seen in Latin America from southern Mexico at 23° north latitude to 34.5° south in northeastern Argentina and north-western Uruguay. There have been no actoctonous cases in El Salvador, Nicaragua, the Caribbean Islands, Surinam, or Chile. The disease is more frequent in Brazil, Colombia, Venezuela, and Argentina. It does not have homogeneous distribution; the endemic areas are of humid subtropical climate with a great deal of rivers and gallery woods. It is not found in the semiarid steppes or in equatorial forests with mean annual temperatures above 25°C. It is found between sea level and 2,000 m altitude in regions of intense agricultural activity (Mackinnon, 1972; Restrepo et al., 1972).

Prevalence of paracoccidioidomycosis is not very well known. As this disease is not obligatorily reported, the number of diagnosed cases in the different countries is unknown. Based on personal reports, through 1978 there had been 6,000 cases of this mycosis in Latin America, of which 4,000 were reported in Brazil. Prevalence and incidence of paracoccidioidomycosis infection have been evaluated through survey reports of intradermal tests with paracoccidioidin. These studies have presented various methodologic problems including reagents of variable specificity and small study populations. In any event, infection has been confirmed more frequently in rural areas than in cities. The number of positive intradermal tests rises from early childhood to age 30 with no difference among sexes in frequency of infection. The percentage of positive skin tests varies between endemic areas (Conti-Díaz, 1972; Londero, 1984).

Most cases of paracoccidioidomycosis disease occur in those between the ages of 30 and 60 years. It is more frequent in men than in women, with a mean ratio of 15 : 1. The high ratio has been attributed to work conditions, even though one should consider that women might be less sensitive for hormonal reasons and have a better phagocytic system. Paracoccidioidomycosis of childhood is infrequent, and there is no difference between sexes. Most patients are rural workers of low socio-economic condition. There are no confirmed data on racial susceptibility to this mycosis.

P. brasiliensis lives in nature, probably in the soil, in humid places with little climatologic changes and great amounts of organic decayed substances. Occasionally, it has been isolated from soil samples in Brazil, Argentina, and Venezuela (Albornoz, 1972; Negroni, 1966). In its filamentous phase, this microorganism may develop in sterile soil, vegetal residues, water, and agar with soil and water extracts as the only nutrients (Lacaz et al., 1984; Restrepo, 1985). It has therefore been suggested that it may live by rivers or lakes (Restrepo, 1985). The infection is most commonly acquired by inhalation, though there may be cases of primary cutaneous or mucosal infection.

Paracoccidioidomycosis has no man-to-man transmission, and the family cases observed are only indicative of a greater genetic susceptibility of certain individuals.

Even though positive skin tests to paracoccidioidin have been confirmed in several animal species, there is no certainty about progressive paracoccidioidomycosis among lower animals (Conti-Díaz, 1972; Del Negro et al., 1982).

Etiologic Agent

Characteristics

Paracoccidioidomycosis is caused by a single species of a diphasic fungus, *Paracoccidioides brasiliensis*. This microorganism shows a very typical aspect in its tissue phase, in the form of spherical cells, of a diameter about 5 to 25 μm, with a thick and refracting wall, of double lining and several peripheral buds. Single buds may occasionally occur, as well as chains of 3 to 4 structures; sometimes it has 2 or 3 buds, and cryptosporulation, or accelerated multiple budding, is eventually observed. In this case the mother cell is absolutely surrounded by small blastospores, resembling a pilot wheel. It is similar to *Blastomyces dermatitidis,* differing from it as its buds are produced through a narrow pedicle out of the mother cell through a canaliculus of the cell wall. Wet preparations show numerous and highly refracting lipidic inclusions at the cytoplasma of these cells (Fig. 4).

Typical multiple-budding cells are found in suppurative lesions, particularly if they are localized in the lymph nodes. Other structures may be found free, in supurative sites, or within giant cells in the granulomas. Sometimes atypical forms with broken or collapsed cell walls are shown. Short filaments have been found less frequently in tissue among immune-

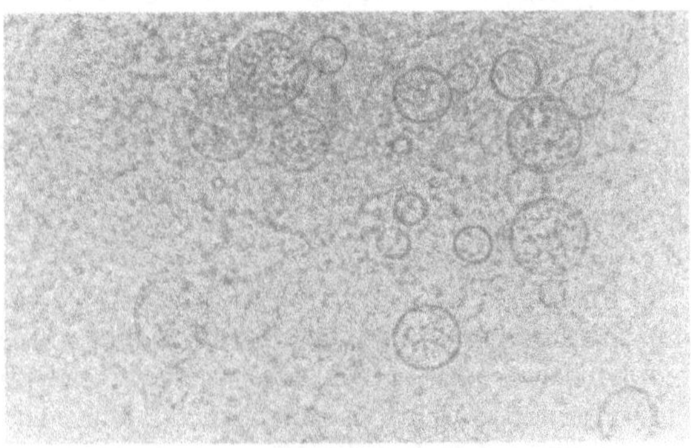

FIG. 4. Yeast phase of *Paracoccidioides brasiliensis* obtained from human lesions in a fresh preparation.

depressed hosts. Small forms, 5 μm in diameter, may be seen in patients with active hematogenous dissemination, which may be mistaken for *Histoplasma capsulatum* (Lacaz et al., 1984; Londero, 1984; Rubinstein and Negroni, 1981). Routine staining of tissue sections with hematoxylin-eosin shows few details of this microorganism. The cell wall takes a reddish color with the PAS or Gridley technique, and with Gomori or Grocott staining it turns dark brown (Fig. 5). The multinucleate nature of *P. brasiliensis* may be confirmed with ferrum hematoxylin, as well as the fact that these nucleoli multiply and move toward the periphery before budding (Lacaz et al., 1984; Negroni, 1966).

An external electrodense layer with numerous connected microfibrils has been shown in ultrastructural studies of the parasitic phase, where there is prevalence of alpha-glucan, as well as a second, more internal layer, less dense and homogeneous, formed by beta-glucan and chitin. The remaining structures are typical of a eukaroytic cell with plasmic membrane, endoplasmic reticulum, mitochondrias, several nuclei with nucleoli, ribosomes, lipid corpuscles, and glucogen (Carbonell, 1972; Del Negro et al., 1982).

Culture media containing carbohydrates and incubated at 28°C for 20 to 30 days yield cottonlike and whitish colonies similar to mouse hair. Microscopic examination shows branched, septated, and hyaline mycelia of 2 to 3 μm in diameter and numerous intercalated and terminal clamidospores which are spherical and vary in size. Fructification is best observed in nutritionally deficient media, such as distilled water or soil extract agar, where there are pearform aleuriconidia of 3 to 7 μm with smooth or slightly rugose walls, as well as arthroconidia and arthroaleurioconidia (Del Negro et al., 1982; Restrepo et al., 1986).

In culture media with serum or blood, after 8 to 10 days of incubation at 37°C, it is possible to see small, wrinkled, high, cerebriform colonies that are pasty

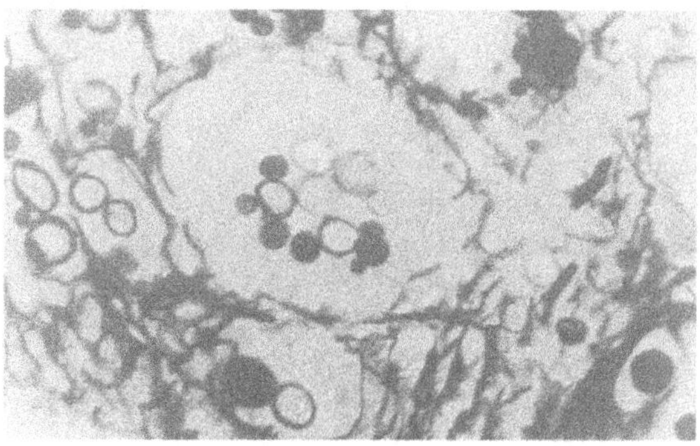

FIG. 5. Histopathologic examination of an experimental paracoccidioidomycosis stained by Grocott's method.

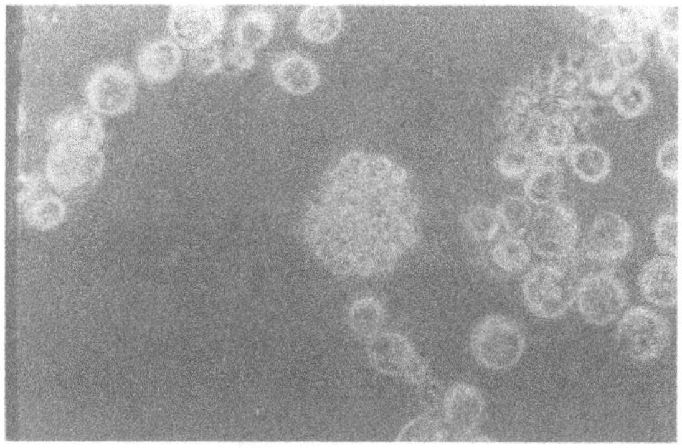

FIG. 6. Yeast phase of *Paracoccidioides brasiliensis* from culture at 37°C.

and have a whitish to yellowish color. Microscopic examination of these colonies will show budding structures similar to those observed in lesions (Fig. 6).

P. brasiliensis is a diphasic and dithermic fungus that turns from its filamentous to its yeast phase or vice versa by mere changes in the incubation temperature. When it changes from its filamentous to its yeast phase, some interseptal areas widen their diameter and turn spherical. After a short period they separate from the filament and start budding. Inverse variation is characterized by the production of germ tubes which subsequently give rise to filaments.

In the ultrastructure of the filamentous phase, a cell wall thinner than that of the yeast phase is confirmed, where the external layer is fibrilar and is oriented in the direction of the main axis of the daughter cell; there is a predominance of an amorphous substance formed by chitin and beta glucans (Carbonell, 1972; Del Negro et al., 1982). Septa show a single pore as those seen in *Ascomycotina* and have a highly electrodense Woromin corpuscle. Interseptal segments are multinucleated; they show lomosomes and mesosomes as every organella of a eukaryotic cell.

Biochemical activity of *P. brasiliensis* is of no taxonomic interest. All isolates possess urease and need thiamine for their growth (Lacaz et al. 1984).

This microorganism is pathogenic for several laboratory animals including guinea pig, hamster, mouse, rat, and also the chorioallantoic membrane of the chicken embryo.

Collection and Transport of Clinical Materials

There are several materials useful in the diagnosis of this disease: sputum, bronchial or bronchoalveolar washings, pus obtained by puncture of abscesses or soft adenopathies, and biopsies of cutaneomucosal, nodular, and pulmonary lesions. Materials should be collected in sterile containers, and biopsies should be kept in vials with sterile isotonic saline solution. A second biopsy or part of it should be fixed in 10% Formol for histopathologic examination. Nonfixed materials should be delivered promptly to the mycology laboratory. Should other forms of transportation be utilized, effort must be made to maintain temperature at or slightly below 4°C.

A fasting blood sample without anticoagulant should be obtained for serologic testing. Serum should be separated, and merthiolate-sodium borate should be added to a final concentration of 1/10,000 and kept frozen at −20°C until processed.

Microscopic Examination, Culture, and Inoculation

Microscopic examination of culture by placing a portion between cover glass and glass slide with a drop of sterile saline solution has an important diagnostic value when studying nonfixed materials, as it usually allows the observation of *P. brasiliensis* in its typical morphology. Smear staining, usually employed in bacteriology, is not useful.

Examination of tissue sections is also a useful diagnostic procedure, as *P. brasiliensis* is generally found in great amounts in lesions and can be identified quite early owing to its relatively large size. Special staining methods such as PAS, Gridley, Gomori, and Grocott allow better visualization of this microorganism, particularly when there are few parasites.

Even though direct immunofluorescence is not a routine technique it may also be performed, with a specific conjugate, for identification of *P. brasiliensis* in tissue sections or smears of clinical materials (Kaplan, 1972).

Cultures are not routinely obtained from lesions in man for this pathology; however, they should be tried on different media such as Sabouraud dextrose agar or lactrimel, with the addition of chloramphenicol, 50 μg/ml, and cycloheximide, 400 μg/ml, and incubated at 28°C or blood agar, Kelly hemoglobin agar, or brain-heart infusion agar with blood with the addition of chloramphenicol and incubated at 32° to 37°C. The former media allow the development of the filamentous phase in 20 to 40 days, while the latter are used for demonstration of the yeast phase in 8 to 10 days.

A greater proportion of positive cultures has been observed in contaminated materials such as sputum, skin, or mucosal biopsies, using culture media containing yeast extract, without dextrose but with antimicrobial agents and cycloheximide. Useful culture medium are those of Smith and Goodman (1975) with yeast extract, phosphate and ammonium hydroxide and that of Restrepo, et al. (1972) with yeast extract–penicillin–streptomycin and cycloheximide. These culture media are seeded in Petri dishes and incubated at 28°C for 1 month. Prior to inoculation, sputa or bronchial washings should be digested with 4% sodium hydroxide, trypsin-pancreatin, or N-acetylcystene (Del Negro et al., 1982; Pollak, 1972).

Recognition of the filamentous phase is sometimes difficult because of the lack of a typical fructification. Useful procedures are exoantigen extraction with distilled water followed by immunodiffusion with a control rabbit serum (Standard and Kaufman technique, 1977) or conversion to yeast phase in cultures incubated at 37°C or by inoculation into susceptible laboratory animals.

Both guinea pigs and hamsters may be intratesticularly inoculated, either with clinical material in sterile physiologic solution and antibiotics or with cultures. These animals develop intense orchitis in 15 to 20 days, and, in the case of the hamster, disseminated infection occurs. Cultures from testicular material are regularly obtained.

P. brasiliensis is not frequently isolated from blood cultures of patients with acute disseminated infections. Blood samples of 10 to 20 ml should be obtained in heparin or sodium citrate, and they should be seeded in dextrose broth or liquid Sabouraud medium and incubated at 28°C.

Immunologic Aspects

Over 90% of patients with active paracoccidioidomycosis show antibodies that can be confirmed through any of the serologic tests routinely used. Several antigens may be employed and are prepared according to two main procedures, nitrogen polysaccharide obtained from the yeast phase by autoclavation (Fava Netto, 1961), and metabolic antigens obtained by filtration of the yeast phase of *P. brasiliensis* in liquid media (Restrepo, 1966; Negroni, 1968). The first procedure has been used for tube precipitation tests and complement fixation. The metabolic antigens are used in agar gel immunodiffusion, counterimmunoelectrophoresis (Fig. 7), complement fixation, and ELISA. The antibodies detected by tube precipitation are those appearing earlier, and they tend to disappear after an evolution period of some months. Antibodies demonstrated by complement fixation and by agar gel immunodiffusion appear later and persist during active disease. Titers are higher in the most seriously affected patients and generally turn negative once clinical and mycological cure has been achieved. There is occasional persistence of positive serologic reactions with low titers (Negroni, 1972). The ELISA technique allows a better discrimination in patients with progressive and remitting lesions.

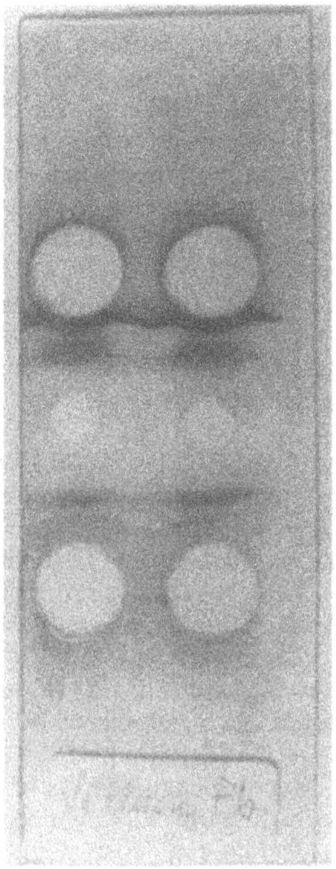

FIG. 7. Counterimmunoelectrophoresis with secondary immunodiffusion of a positive serum sample.

Agar gel immunodiffusion and agarose counterimmunoelectrophoresis are more specific than complement fixation, as the latter shows crossed reactions with histoplasmin in over 25% of cases (Negroni et al., 1976).

Delayed hypersensitivity skin tests with antigens of *Paracoccidioides brasiliensis* may be positive both in patients with paracoccidioides disease and in apparently healthy individuals in endemic areas (paracoccidioides infection).

Over 50% of patients with severe forms of this mycosis show negative specific skin tests. They generally turn positive once clinical cure has been obtained with therapy (Negroni and Robles, 1974). This behavior of intradermoreaction with paracoccidioidin may be due to a depression in cell-mediated immunity in seriously affected patients. Depression in T-lymphocyte activity has been observed with all study techniques, both in vivo and in vitro, which are aimed at estimating this aspect of immunity. This depression may be due to the existence of inhibitor plasmic substances such as alpha globulin and complexes of antigen-antibody complement and also to an increase in repressor T cells. This problem generally disappears with antifungal therapy, although recovery is not always absolute.

Treatment, Prevention, and Control

Sulfonamide therapy was introduced in 1940 by Oliveira Ribeiro, and, when used adequately, success is over 60%. Early cessation of therapy or interruption frequently results in sulfonamide resistance. Currently used sulfonamides are of slow clearance (sulfadimethoxine or similar), administered at doses of 1 g/day or 20 mg/kg/day, or cotrimoxazole, four tablets per day. Treatment duration should be of at least 3 years.

Amphotericin B is very active for paracoccidioidomycosis; it causes rapid regression of all lesions but does not avoid relapses. Therefore, its use has been suggested for very severe cases, and once clinical cure is achieved, patients are changed to sulfonamides. Amphotericin B is administered intravenously, at an initial dose of 0.2 mg/kg/day increasing up to 0.8 mg/kg/day until reaching 1,500 mg for each treatment series.

Toxicity of this drug should be controlled for periodically with the following tests: hemogram, blood potassium, creatinine and urea, and full urine analysis (Negroni, 1972; Rubinstein and Negroni, 1981). Excellent results have been reported with ketoconazole, with about 95% clinical cures after 1 year of therapy. It is administered orally at doses of 200 to 400 mg/day. Therapy may be interrupted after 1 year when cutaneous tests for paracoccidioidin are positive and serology is negative or titers are low. Reports of relapses with this drug have been less than those of other therapies (Del Negro et al., 1982; Negroni, 1980).

During the past 3 years, a triazolic compound, itraconazole, has been under evaluation. Its activity is similar, and it is administered at 50 mg/day for 6 to 12 months. Side effects have not been significant.

Sometimes, surgery is necessary to treat fibrous stenosis of the mouth, larynx, and trachea or for the removal of cerebral granulomas.

Since paracoccidioidomycosis is transmitted neither man to man nor animal to man, and since there is little information about the ecology of its causing agent, there are no control measures applicable to the prevention of this disease. No useful vaccine has been developed, and the disease is not reported obligatorily.

Literature Cited

Albornoz, M. B. 1972. Isolation of *Paracoccidioides brasiliensis* from rural soil in Venezuela. Proceedings of First Pan American Symposium on Paracoccidioidomycosis. Pan Am. Health Org. Sci. Pub. **254**:71–75.

Albornoz, M. B. 1982. Paracoccidioidomicosis-infection. p. 91–96. *In* G. Del Negro, C. da S. Lacaz, and A. M.

Fiorillo (ed.), Paracoccidioidomicose (blastomicose sulamericana). Editora da Universidade de Sao Paulo, Sarvier, Brazil.
Angulo-Ortega, A., and L. Pollak. 1971. Paracoccidioidomycosis. p. 507–576. *In* R. D. Baker (ed.), The pathologic anatomy of the mycoses. Springer-Verlag, New York.
Bustamante-Simon, B., J. C. McEwen, A. M. Tabares, M. Arango, and A. Restrepo. 1985. Characteristics of conidia produced by the mycelial form of *Paracoccidioides brasiliensis*. J. Med. Vet. Mycol. **23:**407–414.
Carbonell, L. M. 1972. Ultrastructure of *Paracoccidioides brasiliensis* in culture. Proceedings of First Pan American Symposium on Paracoccidioidomycosis. Pan Am. Health Org. Sci. Publ. **254:**21–28.
Conti-Díaz, I. A. 1972. Skin tests with paracoccidioidin and their importance. Proceedings of First Pan American Symposium on Paracoccidioidomycosis. Pan Am. Health Org. Sci. Pub. **254:**197–202.
Del Negro, G., C. da S. Lacaz, and A. M. Fiorillo. 1982. Paracoccidioidomicose (blastomicose sulamericana). Editora da Universidade de Sao Paulo, Sarvier, Brazil.
Fava Netto, C. 1961. Contribuição para o estudo immunológico da blastomicose de Lutz (blastomicose sulamericana). Rev. Inst. Lutz **21:**99–194.
Franco, M. F., and M. R. Montenegro. 1982. Anatomía patológica. p. 97–117. *In* G. Del Negro, C. da S. Lacaz, and A. M. Fiorillo (ed.), Paracoccidioidomicose (blastomicose sulamericana). Editora da Universidade de Sao Paulo, Sarvier, Brazil.
Kaplan, W. 1972. Application of immunofluorescence to the diagnosis of paracoccidioidomycosis. Proceedings of First Pan American Symposium on Paracoccidioidomycosis. Pan Am. Health Org. Sci. Pub. **254:**224–226.
Lacaz, C. da S., E. Porto, and E. Costa Martins. 1984. Micología médica. Fungos, actinomycetos e algas de interesse médico. Sarvier, Editorial Médica Panamericana, Buenos Aires.
Londero, T. A. 1984. Paracoccidioidomicosis (*Blastomicosis sudamericana*), p. 209–217. *In* Braude, A. (ed.). Enfermedades infecciosas. Editorial Panamericana, Buenos Aires.
López, R., and A. Restrepo. 1983. Spontaneous regression of pulmonary paracoccidioidomycosis. Mycopathologia **83:**187–189.
Mackinnon, J. E. 1972. Geographical distribution and prevalence of paracoccidioidomycosis. Proceedings of First Pan American Symposium on Paracoccidioidomycosis. Pan Am. Health Org. Sci. Pub. **254:**45–51.
Martins Castro, R., L. C. Cucé, and C. Fava Netto. 1975. Paracoccidioidomicose. Inoculação accidental "in anima novile." Relato de um caso. Med. Cut. Ibero-Latin-Am. **3:**289–292.
Negroni, P. 1966. Las blastomicosis y coccidioidomicosis. Micosis profundas, Vol. III. Comisión de Investigaciones Científicas de la Provincia de Buenos Aires, La Plata, Argentina.
Negroni, R. 1968. Observaciones personales sobre la micosis de Lutz. Blastomicosis sudamericana, en la Argentina. Estudio clínico e inmunológico. Tesis de Doctorado. Universidad de Buenos Aires, Buenos Aires, Argentina.
Negroni, R. 1968. Nuevos estudios sobre antígenos para pruebas serológicas en la blastomicosis sudamericana. Rev. Dermatol. Ibero Latinoam. **10:**409–416.
Negroni, P. 1972. Prolonged therapy for paracoccidioidomycosis: approaches, complications and risks. Proceedings of the First Pan American Symposium on Paracoccidioidomycosis. Pan Am. Health Org. Sci. Pub. **254:**147–155.
Negroni, R. 1980. Ketoconazole in the treatment of paracoccidioidomycosis and histoplasmosis. Proceedings, Fifth International Conference on the Mycoses, Caracas, Venezuela. Pan Am. Health Org. Sci. Pub. **396:**361–365.
Negroni, R., and A. M. Robles. 1974. El valor pronóstico dela prueba cutánea en paracoccidioidomicosis. Med. Cut. Ibero-Latin Am. **2:**453–457.
Negroni, R., Am. M. Robles, and C. Iovannitti de Flores. 1976. Estudio de las reacciones serológicas cruzadas entre antígenos del *P. brasiliensis* e *H. capsulatum*. Rev. Assoc. Arg. Microbiol. **8:**68–73.
Padilha Goncalves, A., and C. Bardy. 1946. Aspectos clínicos e radiológicos da blastomicose brasileira pulmonar. Hospital (Río de J.) **30:**1021–1041.
Pollak, L. 1972. Mycological diagnosis of paracoccidioidomycosis. Proceedings of First Pan American Symposium on Paracoccidioidomycosis. Pan. Am. Health Org. Sci. Pub. **254:**193–196.
Restrepo, A. 1966. La prueba de inmunodifusión en el diagnóstico de la paracoccidioidomicosis. Sabouraudia **4:**223–230.
Restrepo, A. 1985. The ecology of *Paracoccidioides brasiliensis:* a puzzle still unsolved. J. Med. Vet. Mycol. **23:**323–334.
Restrepo, A., D. L. Greer, and L. H. Moncada. 1972. Relationship between environment and paracoccidioidomycosis. First Pan American Symposium on Paracoccidioidomycosis. Pan Am. Health Org. Sci. Pub. **254:**84–92.
Restrepo, B. I., J. C. McEwen, M. A. Salazar, and A. Restrepo. 1986. Morphological development of the conidia produced by *Paracoccidioides brasiliensis* mycelial form. J. Med. Vet. Mycol. **24:**337–339.
Robledo, M. V. 1972. Disseminated paracoccidioidomycosis with arteritis. Proceedings of First Pan American Symposium on Paracoccidioidomycosis. Pan Am. Health Org. Sci. Pub. **254:**139–141.
Rubinstein, P., and R. Negroni. 1981. Micosis broncopulmonares del adulto y del niño, 2d. Ed. Editorial Beta, Buenos Aires, Argentina.
Standard, P. G., and L. Kaufman. 1977. Immunologic procedure for the rapid and specific identification of *Coccidioides immitis* cultures. J. Clin. Microbiol. **5:**149–153.
Tuber, R. M., R. El Ibralum, C. E. Godoy, and T. De Britos. 1985. Pathology of the human pulmonary paracoccidioidomycosis with arteritis. Mycopathologia **92:**179–188.

CHAPTER 70

Phaeohyphomycosis

MICHAEL R. McGINNIS and ANTHONY E. HILGER

Disease: Phaeohyphomycosis.
Etiologic Agents: Heterogeneous group of dematiaceous fungi classified in various taxonomic groups. Representative taxa include *Exophiala jeanselmei, Scytalidium* synanamorph of *Hendersonula toruloidea, Phaeoannellomyces werneckii,* and *Xylohypha bantiana.*
Source: Soil and woody plant material.
Clinical Manifestations: Superficial, cutaneous, subcutaneous, or systemic infections with varied clinical presentations.
Pathology: Variable depending on organ system involved. Dematiaceous yeast cells or hyphae, or both, in tissue is diagnostic for phaeohyphomycosis.
Laboratory Diagnosis: Isolation and identification of etiologic agent from biopsy or surgical specimens, exudates, aspirates, hair, nail, and skin. Mycoserology not available.
Epidemiology: Worldwide, especially for some forms in tropical and subtropical regions of the world. Subcutaneous and systemic infections sporadic.
Treatment: Medical, surgical, or a combination of both management approaches is typically required. Amphotericin B, 5-fluorocytosines and azoles like miconazole, ketoconazole, and clotrimazole have been effective.
Prevention and Control: Difficult because of the varied spectrum of infections and diverse taxa of etiologic agents.

Description of Disease

Definition

Phaeohyphomycosis encompasses a heterogeneous group of fungal infections caused by a variety of fungi that grow in tissue, either as dematiaceous yeasts, pseudohyphae, hyphae of various shapes and sizes, or a combination of all of these forms. Clinically, phaeohyphomycosis is classified on the basis of tissues involved and the pathology produced (McGinnis, 1983). The disease may be superficial, cutaneous, subcutaneous, or systemic. The infection has an impact on public health, both in humans and animals, such as birds, cats, cattle, fish, and horses.

When mycetomas caused by dark-colored fungi are excluded, mycoses caused by dematiaceous fungi fall into two categories: phaeohyphomycosis or chromoblastomycosis. Because of the unique characteristics of tumefaction, draining sinuses, and granules that typify mycetoma, this disease is not confused with phaeohyphomycosis or chromoblastomycosis. These latter two diseases are distinguished from each other by differences in their clinical manifestations, tissue forms, and, to a degree, by the taxonomic classification of the etiologic agents. In particular, chromoblastomycosis is delimited by the presence of clusters of muriform sclerotic cells. Some pathogens like *Phialophora verrucosa* and *Fonsecaea pedrosoi* may cause both diseases. Phaeohyphomycosis is distinguished from hyalohyphomycosis (Ajello, 1986; McGinnis, 1983; McGinnis et al. 1985a), a similar group of heterogeneous mycoses, the latter having hyaline hyphae in tissue.

Clinical Manifestations (Table 1)

Superficial Phaeohyphomycosis

Infections involving the superficial boundaries of the body in which there is little or no tissue damage are known as superficial phaeohyphomycosis. Examples of this form of the disease are *Piedraia hortae,* which causes black piedra, and *Phaeoannellomyces werneckii* (tinea nigra). Infections caused by *Piedraia hortae* (Fig. 1) are characterized by hard, black, carbonaceous nodules that are firmly attached to the hair. The nodules are ascostromata, and contain locules within which ascospores are formed. Tinea nigra is an infection involving the stratum corneum, in which the fungus produces yeast cells, pseudohyphae, and hyphae. The dematiaceous nature of the fungus contributes to the dark color of the lesions.

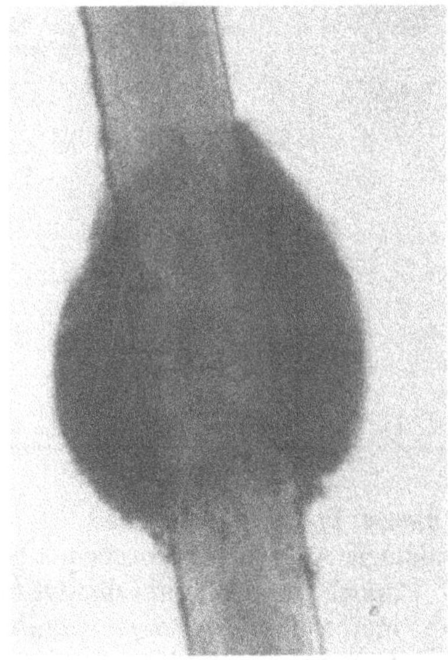

Fig. 1. Nodule of *Piedraia hortae* on hair. Original magnification ×175; reproduced at 45%.

Cutaneous Phaeohyphomycosis

Cutaneous and corneal infections involving keratinized tissues of a living host are known as cutaneous phaeohyphomycosis. Nonliving tissues are typically involved, and frequently there is extensive tissue destruction. The agents of this disease grow as dematiaceous hyphae in tissue. Clinically, infections may resemble those of chromoblastomycosis. Dermatomycoses, or infections of the skin, are often caused by fungi like the *Scytalidium* synanamorph associated with the pycnidial fungus *Hendersonula toruloidea*. *Curvularia geniculata, C. lunata,* and *C. senegalensis* are often diagnosed as the cause of mycotic keratitis. *Botryodiplodia theobromae* and *Scytalidium* have been occasionally associated with onychomycosis. In contrast, dermatophytosis is caused by pathogenic members of the genera *Epidermophyton, Microsporum,* and *Trichophyton.* In tissue, these fungi produce hyaline hyphae and, occasionally, arthroconidia.

Table 1. Clinical classification of phaeohyphomycosis

Disease	*Representative fungi*
Superficial	*Phaeoannellomyces werneckii* (syn. *Exophiala werneckii*)
	Piedraia hortae
Cutaneous	*Alternaria* sp.
	Curvularia lunata
	Exserohilum rostratum (syn. *Drechslera rostrata*)
	Scytalidium synanamorph of *Hendersonula toruloidea*
Subcutaneous	*Alternaria alternata*
	Bipolaris spicifera (syn. *Drechslera spicifera*)
	Exophiala jeanselmei
	E. spinifera
	Phialophora parasitica
	P. richardsiae
	Wangiella dermatitidis
Systemic	*Bipolaris hawaiiensis* (syn. *Drechslera hawaiiensis*)
	B. spicifera
	Exserohilum rostratum
	Wangiella dermatitidis
	Xylohypha bantiana (syn. *Cladosporium bantianum*) *C. trichoides*)

Subcutaneous Phaeohyphomycosis

Subcutaneous infection following trauma is one of the more common forms of phaeohyphomycosis. After implantation of the fungus, localized lesions develop, consisting of abscesses that are defined by fibrosis. Fungal morphology exhibited in histologic sections of infected tissues is not pathognomonic for any specific etiologic agent. Species identification must be done from isolates cultivated in the laboratory. Of the various agents of this form of phaeohyphomycosis, *Exophiala jeanselmei* is the most commonly reported. Other fungi such as *Phialophora richardsiae* and *P. parasitica* are occasionally recovered from subcutaneous disease.

SYSTEMIC PHAEOHYPHOMYCOSIS

The most serious infections caused by dematiaceous fungi are systemic diseases. The infections are often initiated in the lungs, with subsequent dissemination by hematogenous means to other organs such as the liver, pancreas, or brain. Systemic phaeohyphomycosis may have a grave prognosis, depending on the status of the host and the organ involved. For example, cerebral phaeohyphomycosis caused by *Xylohypha bantiana* is often fatal. Species of *Bipolaris* and *Exserohilum*, such as *B. hawaiiensis*, *B. spicifera* and *E. rostratum*, are being recovered with increasing frequency in systemic phaeohyphomycosis.

Pathology

Because subcutaneous and systemic infections result in more morbidity and mortality than do superficial and cutaneous diseases, only the pathology of the former two clinical forms will be discussed.

Subcutaneous phaeohyphomycosis may involve any portion of the body. The lesions tend to develop slowly and may become chronic. Abscesses, which are typically tender, fluctuant, and erythematous, occur following traumatic implantation of the etiologic agent. Verrucose lesions resembling chromoblastomycosis may occur. Lesions caused by fungi like *Exophiala jeanselmei* typically are solitary and encapsulated by dense collagenous connective tissue. A wide area of granulation tissue often surrounds the lesions, the central portion of which often contains purulent exudate. Only rarely do these cyst-like lesions rupture through the skin. The lesions almost always remain localized with no regional lymph node involvement. There are no systemic signs or symptoms. Pale-brown fungal elements consisting of septate hyphae, pseudohyphae, and yeast cells are abundant in the central purulent exudate and the peripheral granulomatous zone (Chandler et al., 1980).

Systemic phaeohyphomycosis, which generally originates from the respiratory tract, involves vital internal organs such as the liver, pancreas, and the central nervous system. *Xylohypha bantiana* causes very serious problems because it is highly neurotropic. Brain lesions caused by *X. bantiana* appear as encapsulated abscesses or generalized inflammatory infiltrates, resulting in headache, nausea, vomiting, fever, nuchal pain, and rigidity (Bennett et al., 1973).

In the brain, there is a mixed purulent and granulomatous inflammatory tissue reaction with abscess (single or multiple) formation. The majority of lesions are located in the white matter of the frontal lobe of the cerebrum. Abscesses consist of a central zone of neutrophils and necrotic debris, and may or may not be sharply defined. *Xylohypha bantiana* occurs as pale-brown hyphae with associated round cells in tissue (Fig. 2). The hyphae tend to be in clusters, often with swollen cells in the central portion.

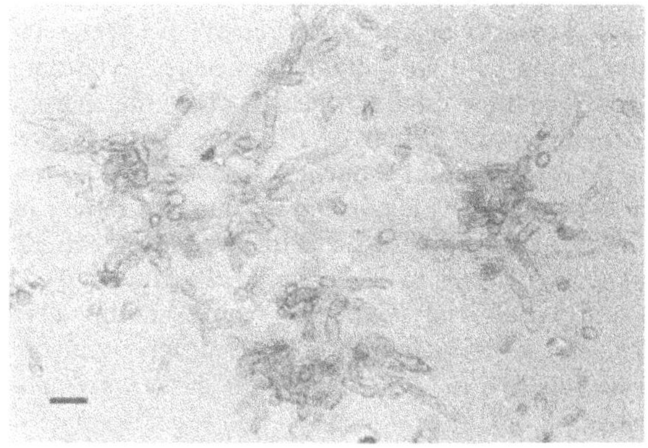

FIG. 2. *Xylohypha bantiana* producing dematiaceous hyphae in brain tissue. Unstained; bar = 10 μm.

Diagnostic Features

The diagnostic feature of phaeohyphomycosis in tissue is pale-brown hyphae that are often irregular in size and shape. Some fungi such as *Wangiella dermatitidis* frequently produce yeast cells, some of which have septa in one plane. These cells are different from the sclerotic bodies of chromoblastomycosis because they are neither muriform, large, nor thick-walled. *Phaeosclera dematioides* (McGinnis et al., 1985b) forms occasional muriform cells in tissue that are hyaline to lightly pigmented. These cells differ from sclerotic bodies in that they lack the dark chestnut-brown color, have thinner cell walls, and are produced both in vitro and in vivo.

When dematiaceous fungi are suspected in tissue, but the brown color is not distinct, unstained sections should be examined. If dematiaceous fungal elements are present, they appear pale to yellow-brown in color. Because the brown color of the fungus is due to the presence of melanin, either the Fontana-Mason silver stain or Lillie's ferrous iron stain can be used to stain the tissue sections to more clearly visualize their dematiaceous nature. There is a good correlation between the accentuation of the melanin stain and the amount of melanin in the cell wall (Wood and Russell-Bell, 1983).

Prognosis

Phaeohyphomycosis has an excellent prognosis, especially when considering that a number of patients have some underlying condition such as diabetes, tuberculosis, leprosy, or leukemia. To date, not enough cases of phaeohyphomycosis have been seen in patients with autoimmune deficiency syndrome (AIDS) to draw valid conclusions about the success of treatment. Bennett et al. (1973) discussed 35 cases of cerebral phaeohyphomycosis caused by *Fonsecaea pedrosoi, W. dermatitidis,* and *X. bantiana.* In the 34 published cases, and their case of chronic meningitis caused by *X. bantiana,* 29 patients died. Three patients survived more than 2 months after their lesions were drained or excised; three patients were reported to be alive when the cases were published.

Epidemiologic Aspects

Occurrence

The frequency of occurrence of phaeohyphomycosis is unknown because mycoses are not reportable diseases. On the basis of a few large studies and individual published case reports, phaeohyphomycosis is not a rare disease, especially in the tropics. In a study of Indians living in southeastern Venezuela, Volcan et al. (1980) found that 54% of 355 people they studied had black piedra. Molero et al. (1980) studied 2,587 other patients seen in a dermatology clinic and found that 2.3% of these patients had black piedra. The incidence of this mycosis decreases and becomes virtually nonexistent in nontropical regions of the world. The high incidence of *Piedraia hortae* colonization of hair has been associated, at various times, with completing puberty, wearing long hair, genetic factors, and predisposition due to other infections.

The *Scytalidium* synanamorph associated with *Hendersonula toruloidea* is occasionally reported as an agent of cutaneous phaeohyphomycosis. Clinically, lesions caused by *Scytalidium* resemble those caused by *Trichophyton rubrum.* In skin and nail tissue, *Scytalidium* occurs as dematiaceous hyphae. Even though a number of patients manifesting this disease are residents of Great Britain when diagnosed, it has been assumed that their initial infections were acquired in their countries of origin. All of the 57 patients studied by Moore (1978) were immigrants from tropical and subtropical regions.

The most common cause of subcutaneous phaeohyphomycosis is *Exophiala jeanselmei.* Other fungi such as *E. spinifera* (Fig. 3), *W. dermatitidis, Phialophora richardsiae, Bipolaris spicifera* (Fig. 4), and *Exserohilum rostratum* (Fig. 5) are being reported with increased frequency. These and similar pathogens are normally associated with wood and soil. A recent study by McGinnis et al. (1986) regarding the taxonomy of medically important species of *Bipolaris* and *Exserohilum* listed 66 isolates comprising six species isolated from proven and suspected cases of subcutaneous and systemic phaeohyphomycosis. A similar trend is being observed with species of *Curvularia* (Fig. 6.).

Systemic phaeohyphomycosis, especially involving the brain, is unusual. *Xylohypha bantiana* (Fig. 7) is the most frequently reported agent. Of the 35 cases summarized by Bennett et al. (1973), the patients lived in 19 countries that had various climatic conditions. They found that the male to female ratio was 24 : 11, and the third decade of life was the most

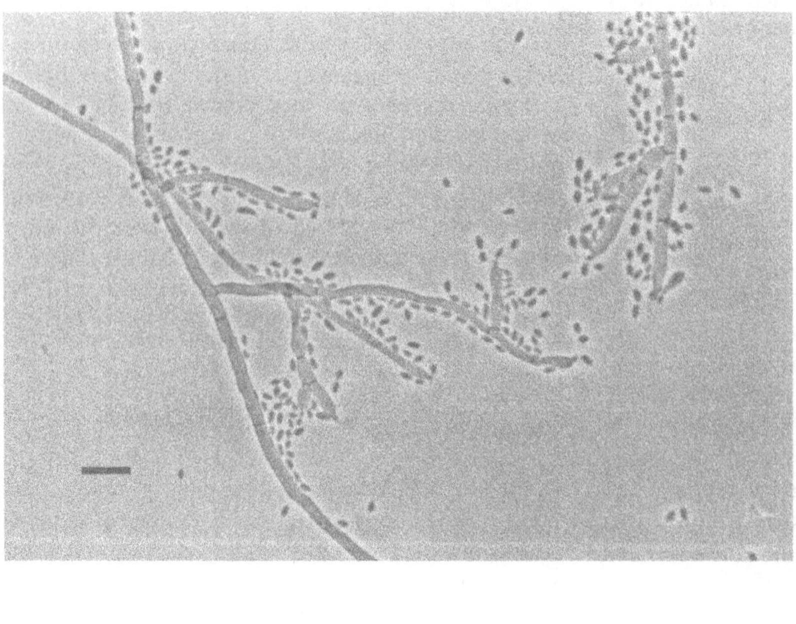

FIG. 3. Annellides on spinelike conidiophores of *Exophiala spinifera.* Bar is 10 μm.

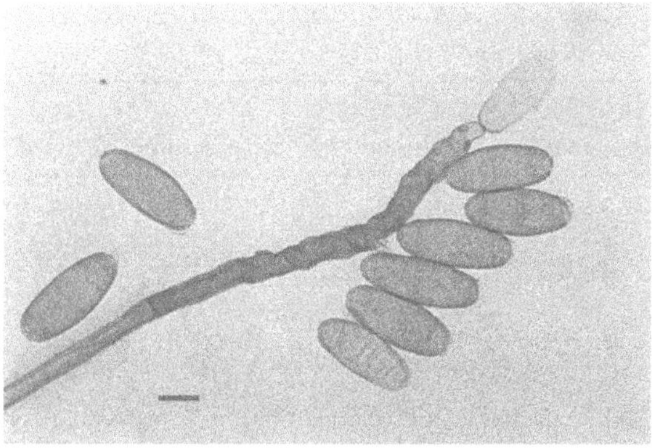

FIG. 4. Conidia of *Bipolaris spicifera* forming from a sympodial conidiophore. Bar is 10 μm.

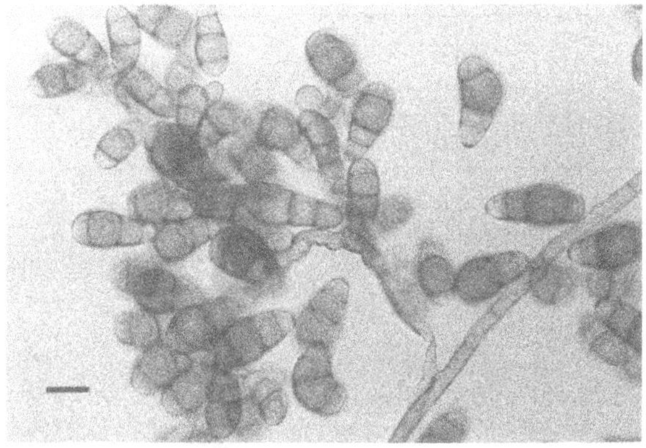

FIG. 6. Curved conidia of *Curvularia verruculosa* forming from a sympodial conidiophore. Bar is 10 μm.

common age of onset. The authors noted neither unusual incidence of preexisting illness nor any occupational predisposition. Since the review by Bennett et al. (1973), additional cases of central nervous system infections have been reported, and the spectrum of etiologic agents has been expanded to include fungi like *B. spicifera* and *B. hawaiiensis*.

As mentioned previously, patients with diseases such as diabetes and leukemia and individuals receiving antimicrobic agents, immunosuppressant drugs, and steroids may be predisposed to the opportunistic pathogens capable of causing phaeohyphomycosis. Unfortunately, in many instances, information regarding underlying conditions is not included in case reports.

Susceptibility and Resistance to Antifungal Agents

Because there is no standardized in vitro method of susceptibility testing of antifungal drugs, it is difficult to correlate susceptibility data to the clinical outcome of phaeohyphomycosis. Table 2 contains a summary of published in vitro susceptibility data for several representative etiologic agents of phaeohyphomycosis. In many instances, either a single or only a few isolates were used for the summarized values. The in vitro response to various antifungal

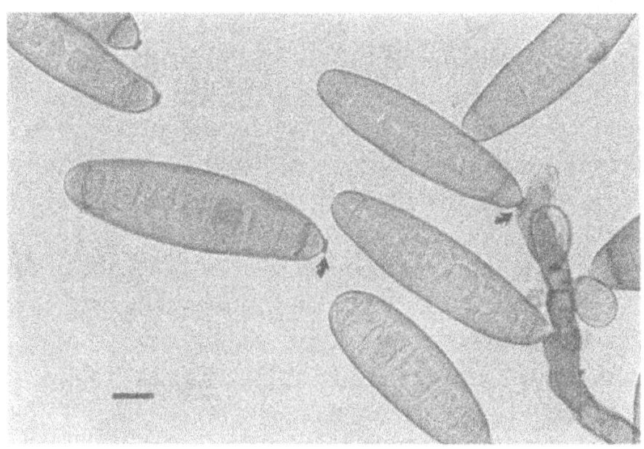

FIG. 5. Conidia of *Exserohilum rostratum* forming from a sympodial conidiophore. Note presence of strongly protruding hila (*arrows*). Bar is 10 μm.

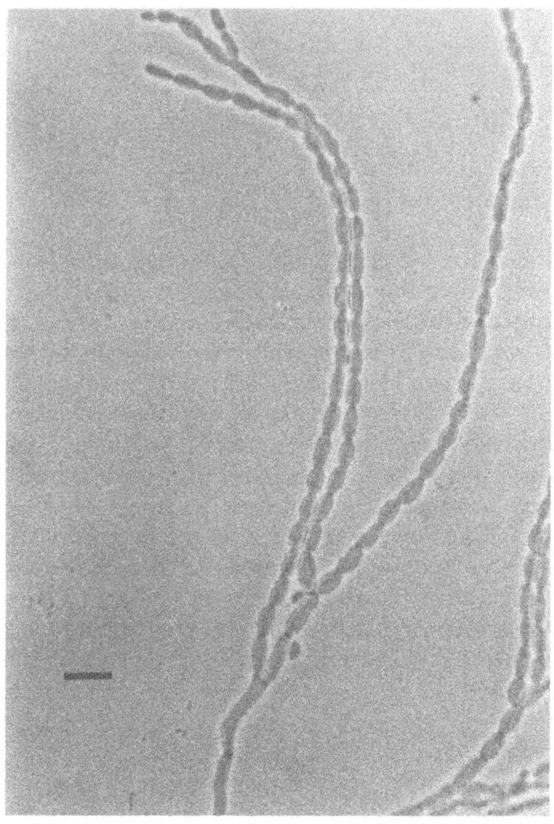

FIG. 7. Long, sparsely branching chains of blastoconidia produced by *Xylohypha bantiana*. Bar is 10 μm.

TABLE 2. Minimal inhibitory concentrations for some major antifungal agents[a]

	Antifungal agent (μg/ml)					
Fungus	*Amphotericin B*	*5-Fluorocytosine*	*Itraconazole*	*Ketoconazole*	*Miconazole*	*Pimarcin*
Aureobasidium pullulans	—[b]	—	0.1	1	—	—
Curvularia lunata	1.56	—	0.1	—	3.12	—
Exophiala jeanselmei	6–>32	0.03–>100	0.0001	8–100	0.35–>100	1.5
Exserohilum longirostratum	—	—	—	0.78	—	—
E. rostratum	0.35	—	—	—	0.35	0.75
Phaeoannellomyces werneckii	>32	>100	0.1	1–8	>100	—
Phialophora richardsiae	2–>100	>100	—	2–4	>100	—
Scedosporium apiospermum	1.56–>100	>100	0.01	0.25–10	>0.016–4	—
Wangiella dermatitidis	1–>32	0.78–>100	—	12.5–16	>100	—
Xylohypha bantiana	4–>32	1–>100	—	1–16	256	—

[a] Modified from references McGinnis and Rinaldi (1985) and Van Cutsem et al. (1987).
[b] Published data unavailable.

drugs is variable and, in many instances, minimal inhibitory concentration values are greater than levels achievable with normal therapy. Some fungi may develop resistance to some of the major antifungal drugs during therapy (McGinnis and Rinaldi, 1985). Resistance emerging during therapy has not been reported for any of the agents of phaeohyphomycosis. For susceptibility testing methods, the reader is referred to the review by McGinnis and Rinaldi (1985).

Etiologic Agents

Characterization

The agents of phaeohyphomycosis comprise a diverse spectrum of taxonomic groups. They vary from ascomycetes like *Pseudallescheria boydii* and its anamorph *Scedosporium apiospermum* to hyphomycetes such as *E. jeanselmei* and *X. bantiana*. A number of the hyphomycetous opportunistic pathogens have synanamorphs, that is, additional anatomic forms that can be classified in separate genera. Dematiaceous blastomycete genera such as *Phaeoannellomyces* and *Phaeococcomyces* are often seen in association with other genera. The term "black yeast" is usually applied to fungi that have dematiaceous blastomycete synanamorphs, the most frequently isolated examples being *E. jeanselmei* and *W. dermatitidis*. The most stable and unique form produced (anamorph) should be used to derive the name of the fungus.

Because the spectrum of pathogens is so great, reference works dealing with dematiaceous fungi should be consulted when attempting to identify an unusual etiologic agent of phaeohyphomycosis (Carmichael et al., 1980; Domsch et al., 1980; Ellis, 1971; McGinnis, 1980). Criteria for the recognition of representative opportunistic pathogens are included in Table 3.

Specimen Collection and Examination

Clinical specimens useful for the diagnosis of phaeohyphomycosis include biopsy and surgical specimens, aspirates, exudates, hair, pus, and keratinized tissue such as stratum corneum and nails. The clinical presentation will determine which specimen should be collected. Aseptic technique should be applied during collection whenever possible. Very small biopsy and surgical specimens are frequently prone to dehydration. Such specimens can be protected from dehydration by placing them in a tube containing a small amount of sterile saline solution or water. Tissue specimens should be homogenized to free the fungal elements prior to planting them onto isolation media. Animal inoculation to recover dematiaceous fungi from clinical specimens is inappropriate.

Clinical specimens should be examined microscopically for the presence of fungal structures. When black piedra is suspected, the gross appearance of hard, black, firmly attached nodules along the hair is diagnostic. Ascospores can be seen at 100× magnification in preparations of nodules. Potassium hydroxide (KOH) in a concentration of 15 to 20% is used as a mounting medium for examining clinical specimens. The clearing process can be accelerated by gently warming the preparation. Cellufluor has limited value for the detection of dematiaceous fungi that are darkly pigmented in clinical specimens because melanin interferes with binding of this fluorescent compound. The agents of

TABLE 3. Characteristics of representative dematiaceous opportunistic agents of phaeohyphomycosis

Fungus	*Characteristics*
Alternaria alternata	Branching chains of obclavate muriform conidia having a distal beak arising from a distinct conidiophore, which may proliferate in a sympodial manner.
Bipolaris spicifera syn. *Drechslera spicifera*	Conidiophores geniculate and sympodial in development. Conidia oblong to cylindrical, consistently with 3 septa, rounded at both ends, basal cells truncate with slightly protruding hila.
Curvularia verruculosa	Conidiophores geniculate and sympodial in development. Conidia curved, central cell unevenly swollen, with 3 septa, end cells lighter in color than central cells, cell wall roughened.
Exophiala jeanselmei	One-celled annelloconidia, accumulating in balls at the apices of cylindrical to lageniform annellides. Some annellides cylindrical and intercalary. *Phaeoannellomyces* synanamorph typically present. Growth at 37°C.
Exophiala spinifera	One-celled annelloconidia, accumulating in balls at the apices of robust cylindrical annellides borne on spinelike conidiophores. Some annellides cylindrical and intercalary. *Phaeoannellomyces* synanamorph having polysaccharide capsule typically present.
Exserohilum rostratum syn. *Drechslera rostrata*	Conidiophores geniculate and sympodial in development. Conidia predominantely ellipsoidal, 7 to 9 septa, with strongly protruding truncate hila. Other conidia fusiform, rostrate, up to 14 septa, with strongly protruding truncate hila.
Phaeoannellomyces werneckii syn. *Exophiala werneckii*	Two-celled annellidic yeast typically having its conidiogenous locus at one end of the cell. As the conidiogenous zone elongates, the septum appears to be displaced toward the opposite end of the yeast cell.
Phaeococcomyces exophialae syn. *Phaeococcus exophialae*	One-celled yeast producing multiple blastoconidia.
Phialophora parasitica	One-celled phialoconidia accumulating in balls at the apices of long, tapering, cylindrical phialides having slightly flaring collarettes. Crusty material often on cell wall of phialides and hyphae.
Phialophora richardsiae	One-celled phialoconidia that are either globose or ellipsoidal, accumulating at the apices of flask-shaped to lageniform phialides having saucer-shaped or slightly flaring collarettes.
Scedosporium apiospermum	One-celled, clavate to cylindrical, truncate annelloconidia, accumulating in groups at the apices of cylindrical annellides, which typically have a slight swelling immediately below the apex. Some conidia solitary on denticles along the hyphae. *Pseudallescheria boydii* teleomorph often present.
Scytalidium synanamorph of *Hendersonula toruloidea*	One-celled arthroconidia in chains forming from hyphalike conidiophores. Pycnidia of *H. toruloidea* easily formed on banana slices.
Wangiella dermatitidis	Polymorphic fungus producing one-celled phialoconidia, accumulating in balls at the apices of phialides without visible collarettes. Conidia tend to slime down and along the hyphae. Some conidiogenous cells annellides, others developing sympodially. *Phaeococcomyces* synanamorph usually present. Growth at 40°C.
Xylohypha bantiana syn. *Cladosporium bantianum*, *C. trichoides*	One-celled blastoconidia occurring in extremely long, sparsely branching chains arising from hyphalike conidiophores. Growth at 42°C.

phaeohyphomycosis are readily recognized by their pale-brown color.

In KOH preparations, a varied morphologic gamut of fungal forms can often be seen. The tissue form may consist of yeast cells, pseudohyphae, toruloid hyphae, hyphae that are short or long, or a combination of any of these forms. These forms should not be confused with the black granules of *Madurella mycetomatis* and similar fungi that cause mycetoma, as well as the muriform sclerotic bodies typical of chromoblastomycosis. It should be noted that some fungi (*Fonsecaea pedrosoi* and *Phialophora verrucosa*) may cause either phaeohyphomycosis or chromoblastomycosis. In tissue, they grow in the form characteristic of the disease.

Isolation and Identification Procedures

Once the clinical specimens have been collected, processed, and examined microscopically, they are ready to be planted onto the isolation media and incubated at 25 to 30°C for 4 weeks. Colonies of dematiaceous fungi are often visible within a few days.

For even the slowest growing of the dematiaceous fungi, 4 weeks is adequate for their recovery.

Routine mycological media such as mold inhibitory agar and Sabouraud glucose agar with and without cycloheximide and chloramphenicol are used for the recovery of dematiaceous pathogens. Other media, including potato glucose agar, V-8 juice agar, and cornmeal agar, are used for the identification of these opportunistic pathogens because they promote production of conidia. Isolation media are too enriched and tend to favor vegetative growth rather than conidiation. Conidial morphology and conidiogenesis are important characteristics used in the identification process.

Pure cultures are absolutely necessary for the identification of dematiaceous fungi. Upon primary isolation, all isolates should be considered potentially contaminated. The difficulty is that many dematiaceous fungi are polymorphic, and if more than one anatomic form or anamorph is present, the laboratorian must be certain that the isolate is a pure culture. The presence of different anamorphs in a single pure culture is valuable information for the identification process. In addition to microscopic characteristics, colonial qualities such as texture, color, growth rate, and reverse color are important. Generally, sexual fruiting bodies, or teleomorphs, are absent unless the isolate is homothallic. For example, the cleistothecia of *Pseudallescheria boydii* often form in isolates first identified as its anamorph *Scedosporium apiospermum*. Some dematiaceous fungi such as species of *Phoma* produce pycnidia that must be differentiated by the presence of conidia from teleomorphs producing asci that contain ascospores. Methods for ensuring purity of isolates have been summarized by McGinnis (1980).

To study conidiogenesis, slide culture preparations should be used. It is also important that the colony be viewed with the aid of a dissecting microscope to correlate and understand more clearly where and what is occurring during conidiogenesis. If a phase-contrast microscope is used, a mounting medium like lactophenol should be used. Cotton blue may be added to the lactophenol when a light microscope without phase-contrast optics is used. The identification process will require observation of the culture at different stages of development and several microscopic fields of view. It is strongly recommended that unusual pathogens be deposited in a major culture collection to ensure their availability for future study.

Immunologic Aspects

Mycoserology is not used to diagnose phaeohyphomycosis at this time, and little is known about the host's immune response to the agents of this disease. At selected institutions, exoantigens are available that can be used to serologically identify in vitro isolates of several fungi, including *E. jeanselmei, E. spinifera, W. dermatitidis,* and *X. bantiana.*

Treatment

Worldwide, the majority of cases of phaeohyphomycosis are superficial, cutaneous and corneal, and subcutaneous. Systemic infections are unusual and often have a poor prognosis regardless of therapy, especially when lesions occur in the brain.

Superficial infection of hair by *Piedraia hortae* is controlled by removing the infected hair. Antifungal agents are generally unnecessary. Infections caused by *Phaeoannellomyces werneckii* can be managed with Whitfield's ointment or topical antifungal agents such as miconazole. Although *Scytalidium* causes dematiaceous dermatomycotic infections, it is not responsive to griseofulvin. Whitfield's ointment and clotrimazole have been used successfully. Subcutaneous phaeohyphomycosis caused by *Exophiala jeanselmei* can be resolved by excision of the lesion. Some clinicians prefer to use either ketoconazole or 5-fluorocytosine in conjunction with surgical management. Systemic infections involving the brain require medical management (amphotericin B) and surgery.

Literature Cited

Ajello, L. 1986. Chromblastomycosis and phaeohyphomycosis: two global disease entities of public health importance. Eur. J. Epidemiol. **2**:243–251.

Bennett, J. E., H. Bonner, A. E. Jennings, and R. I. Lopez. 1973. Chronic meningitis caused by *Cladosporium trichoides*. Am. J. Clin. Pathol. **59**:398–407.

Carmichael, J. W., W. B. Kendrick, I. L. Conners, and L. Sigler. 1980. Genera of hyphomycetes. University of Alberta Press, Edmonton.

Chandler, F. W., W. Kaplan, and L. Ajello. 1980. A colour atlas and textbook of the histopathology of mycotic diseases. Wolfe Medical Publications, Ltd., London.

Domsch, K. H., W. Gams, and T. Anderson. 1980. Compendium of soil fungi. Academic Press, London.

Ellis, M. B. 1971. Dematiaceous hyphomycetes. Commonwealth Mycological Institute, Kew.

McGinnis, M. R. 1980. Laboratory handbook of medical mycology. Academic Press, New York.

McGinnis, M. R. 1983. Chromoblastomycosis and phaeohyphomycosis: new concepts, diagnosis, and mycology. J. Am. Acad. Dermatol. **8**:1–16.

McGinnis, M. R., L. Ajello, and W. A. Schell. 1985a. Mycotic diseases: a proposed nomenclature. Int. J. Dermatol. **24**:9–15.

McGinnis, M. R., R. A. McKenzie, and M. E. Connole. 1985b. *Phaeosclera dematioides,* a new etiologic agent of phaeohyphomycosis in cattle. Sabouraudia **23**:133–135.

McGinnis, M. R. and M. G. Rinaldi. 1985. Antifungal drugs: mechanisms of action, drug resistance, susceptibility testing, and assays of activity in biological fluids. *In* V. Lorian (ed.). Antibiotics in laboratory medicine. 2nd. Ed., Williams and Wilkins, Baltimore.

McGinnis, M. R., M. G. Rinaldi, and R. E. Winn. 1986. Emerging agents of phaeohyphomycosis: pathogenic species of *Bipolaris* and *Exserohilum*. J. Clin. Microbiol. **24:**250–259.

Molero, B., G. S. Volcan, and C. E. Medrano. 1980. Differences in the prevalence of piedra between the mixed population of a city in Bolivar State, Venezuela and the Indian population of that region. PAHO Sci. Pub. **396:**77–81.

Moore, M. K. 1978. Skin and nail infections by non-dermatophytic filamentous fungi. Mykosen **(suppl. 1):**128–132.

Van Cutsem, J., F. Van Gerven, and P. A. J. Janssen. 1987. Activity of orally, topically, and parenterally administered itraconazole in the treatment of superficial and deep mycoses: animal models. Rev. Infect. Dis. **9(suppl. 2):**S15–S32.

Volcan, G. S., C. E. Medrano, and B. Molero. 1980. Notes on the *Piedraia hortae* cycle. PAHO Sci. Publ. **396:**63–69.

Wood, C. and B. Russel-Bell. 1983. Characterization of pigmented fungi by melanin staining. Am. J. Dermatopathol. **5:**77–81.

CHAPTER 71

Rhinosporidiosis

H. S. RANDHAWA and GAURI BAZAZ-MALIK

Disease: Rhinosporidiosis.
Etiologic Agent: *Rhinosporidium seeberi* (Wernicke) Seeber, 1912.
Source: Probably stagnant water of pools, ponds, tanks, and rivers; or soil, dust, etc.
Clinical Manifestations: Polypoid or warty lesions of mucous membranes of the nasopharynx, larynx, trachea, bronchus, urogenital and anal canals, conjunctiva of the eye and its adnexa, etc. Rarely, generalized or disseminated disease involving the skin and visceral organs. Symptoms include itching, mucoid discharge, epistaxis, dysphagia, or dyspnea due to obstruction. Mucosal lesions appear as pink to purplish-red polyps with dotted surface owing to presence of sporangia. Ocular lesions mostly on palpebral conjunctiva, sometimes on the conjunctiva of bulb, limbus, caruncle, or canthus, causing lacrimation, photophobia, or eversion of lid. Fatal disseminated infections rare.
Pathology: Lesions characterized by stroma rich in collagen containing sporangia and spores in various stages of development. Tissue reaction of chronic inflammatory type with predominance of lymphocytes, plasma cells, and lesser number of neutrophils. Microabscesses occasionally present. Foreign-body giant-cell reaction also seen.
Laboratory Diagnosis: By demonstration of typical sporangia with spores in KOH wet mounts and suitably stained tissue sections. Carminophilic feature characteristic of *R. seeberi*. Cultivation of *R. seeberi* recently achieved on monolayers of an epitheloid human rectal tumor (HRT) cell line in humidified air with 5% CO_2. Pathogenicity test in laboratory animals of no diagnostic value; no serologic tests available.
Epidemiology: Worldwide distribution, highly endemic in peninsular India and Sri Lanka. Preponderance of disease in males, no racial predilection, more frequent occurrence in children and young adults. Acquired presumably through intimate exposure to stagnant water of tanks, ponds, pools, and rivers or soil/dust, but *R. seeberi* not yet isolated from environmental sources. Person-to-person transmission not documented.
Treatment: Radical surgical excision of lesions followed by electrocoagulation of the base. No satisfactory treatment available for disseminated visceral infection or extensive mutilating disease.
Prevention and Control: No control measures other than avoiding exposure to suspected natural reservoirs of *R. seeberi*.

Description of Disease

Rhinosporidiosis is a chronic granulomatous submucosal or subcutaneous mycotic disease of humans and animals caused by *Rhinosporidium seeberi* (Wernicke) Seeber, 1912. It is generally characterized by the formation of polyps, papillomas, or warty lesions. The lesions are hyperplastic, elastic, or friable and highly vascular, sessile, or pedunculated with the nose as most frequently and the eye as the second most frequently involved organs. Rarely, the disease may involve the pharynx, larynx, urethra, vagina, anus, skin, and visceral organs. For a detailed account of rhinosporidiosis, reference may be made to the comprehensive monograph of Karunaratne (1964) and to the excellent medical mycology textbooks by Emmons et al. (1977) and Rippon (1982).

History

It is interesting to note that rhinosporidiosis and coccidioidomycosis, caused by morphologically allied etiologic agents, were discovered in Buenos Aires (Argentina) by Guillermo Seeber (1900) and Posadas (1892), respectively, both of whom were contemporary students of Professor R. Wernicke. Seeber described the first case of rhinosporidiosis in his thesis; the patient was a 19-year-old farmer whose large nasal polyp impeded breathing. It was stated in the thesis that Professor Malbran, also from Argentina, had seen a similar case in 1892. A microscopic section of rhinosporidial nasal polyp in an Indian patient (native of Bihar) was exhibited by Major O'Kinealy (1903) at a meeting of the Laryngological Society of London; the condition was described as "localized sporospermosis." Based on a study of material of the same case, Minchin and Fanthum (1905) concluded that the etiologic agent was a sporozoan related to *Neosporidia* and *Haplosporidia,* and they named it *Rhinosporidium kinealyi.* A complete account of the life cycle of the organism was provided by Ashworth (1923), who concluded that it was a fungus and emended the name to *R. seeberi.* Subsequent investigators largely contributed to the knowledge of clinical spectrum and geographic distribution of rhinosporidiosis. The latest historical landmark in studies on rhinosporidiosis is provided by Levy et al. (1986), who have achieved the in vitro cultivation of *R. seeberi* on monolayers of an epitheloid human rectal tumor (HRT) cell line maintained in humidified air with 5% CO_2.

Clinical Manifestations

Rhinosporidiosis may be clinically classified as nasopharyngeal, ocular, cutaneous, or systemic (disseminated).

Nasopharyngeal Rhinosporidiosis

The primary infection in rhinosporidiosis most often occurs on the nasal mucosa, the initial symptom being the sensation of foreign body in the nose. The infection may be accompanied by coryza and mild to intense pruritis with mucoid and frequently blood-stained nasal discharge or epistaxis. Fully developed lesions may be sessile or pedunculated polyps, warty, or of some bizarre shape. The polyp in the nares may extend into the pharynx or externally project over the lip. The surface of the lesion may be mucoid or ulcerated, and the mass may be papillate or highly lobulated, giving it a cauliflowerlike appearance. The lesions vary from red to purplish-red or variegated, which is due to their profuse vascularization, congestion, and necrosis. In some cases, the polyps may lead to obstruction, causing dyspnea or dysphagia. The lesions are generally painless, with epistaxis as one of the predominent symptoms (Allen and Dave, 1936; Purandare and Deoras, 1953).

Ocular Rhinosporidiosis

The lesions in ocular rhinosporidiosis are frequently located on the conjunctiva between the lid and the eyeball. They are pink to red, initially flat and small; they later become pedunculated and freely movable with little discomfort to the patient. Almost 90% of the ocular rhinosporidial cases involve the palpebral conjunctiva; in the remaining cases, the lesions occur on the conjunctiva covering the bulb, limbus, caruncle, or canthi. Large lesions may incite lacrimation, photophobia, and eversion of the lid. In patients with lacrymal sac disease, swollen eyelids may be the only presenting symptom (Beiske et al., 1982; Karunaratne, 1964; Owor and Wamukota, 1978; Peters and DeBelly, 1969; Suseela and Subramaniam, 1975).

Some investigators have used the term oculosporidiosis for rhinosporidiosis of the eyes and its adnexa (Kuriakose, 1963; Peters and DeBelly, 1969). This neologism seems to be entirely misplaced, because it contravenes the very basis of mycological nomenclature according to which the names of mycoses are derived from the generic names of their respective etiologic agents and not from the names of organs involved in the infection.

Cutaneous Disease

Cutaneous rhinosporidiosis has been infrequently reported; it mostly develops adjacent to a mucocutaneous site of infection. Rarely, it may develop away from the mucosal lesion by scratching and autoinoculation. Cutaneous rhinosporidiosis unrelated to any mucosal lesion is very rare (Dhayagude, 1941; Forsyth, 1924; Madhavan et al., 1978). Skin lesions begin as tiny papules that become wartlike with further growth, but they mostly remain unpedunculated. They are relatively painless unless located on the sole of the foot.

Involvement of Other Organs and Systemic Disease

Rhinosporidial lesions rarely occur on the mucosa of palate, epiglottis, larynx, trachea, bronchus, urogenital and anal canals, and parotid gland. These may represent a primary infection or an extension from a preexisting lesion elsewhere in the body (Dhayagude, 1941; Kutty and Unni, 1969; Palaniswamy and Bhandari, 1982; Purandare and Deoras, 1953; Reddi, 1954; Subramanyam and Ramana Rao, 1960; Thomas et al., 1956). In general, the sporangia and spores (endospores) of *R. seeberi* are confined to the rhinosporidial lesions, and there is no involvement of regional lymph nodes and adjacent tissues. However, a few fatal cases of systemic infection including involvement of the liver, spleen, lung, and brain have been reported (Agrawal et al., 1959; Rajam et al., 1955). In rare cases, involvement of the bone with osteolytic lesions may also occur (Chatterjee et al., 1977).

Pathology

On gross examination, the specimen from a typical case of rhinosporidiosis would be indistinguishable from an inflammatory or neoplastic polyp but for the fact that it may reveal minute, dotlike opaque bodies (sporangia) within and/or underneath the covering epithelium. The diagnosis of rhinosporidiosis is readily established by examining hematoxylin and eosin-stained sections from the material, the characteristic feature being the presence in the collagen-rich stroma or covering epithelium of globular bodies representing sporangia in various stages of development (Figs. 1, 2). The sporangial wall is chitinous and well developed, with a thickness of 5 μm in mature sporangia. Some of the collapsed sporangia may assume a semilunar shape. As the covering mucosa in rhinosporidiosis undergoes squamous metaplasia, mucinous cysts that generally characterize atopic nasal polyps are usually lacking. The initial tissue reaction around the sporangia is polymorphonuclear, which later develops into a chronic granulomatous inflammation with lymphocytes and plasma cells predominating, but neutrophils may also be seen (Fig. 3). Frank pyogenic-type inflammation occasionally associated with microabscesses occurs in ulcerated lesions. Eosinophils are inconspicuous in contrast to the reaction seen in atopic nasal polyps. A foreign-body giant-cell reaction may also occur (Fig. 4), especially around sporangia that have ruptured or degenerated. Sometimes the inflammatory reaction may compress the blood supply, leading to local ischemic necrosis (Fig. 5).

The sporangial development of *R. seeberi* in tissues begins with a spore (endospore) about the size of a red blood cell and that has a chitinous wall, clear protoplasm, nucleus as well as nucleolus, and a karyosomelike body. Nuclear division commences

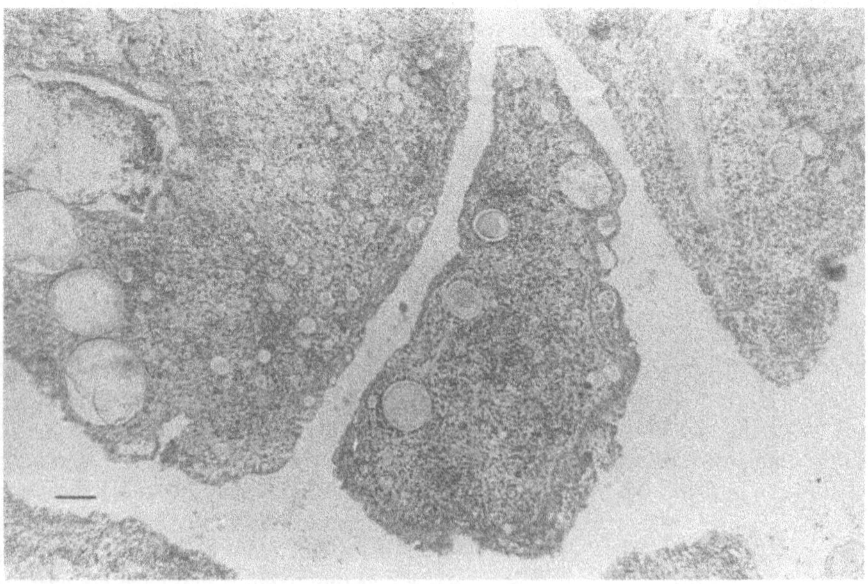

FIG. 1. Section through a rhinosporidial nasal polyp (human case), depicting its papillate character. The covering epithelium and the underlying stroma show the developing sporangia of *Rhinosporidium seeberi*. Also seen is a diffuse mononuclear infiltrate of variable density. Hematoxylin and eosin; bar = 100 μm.

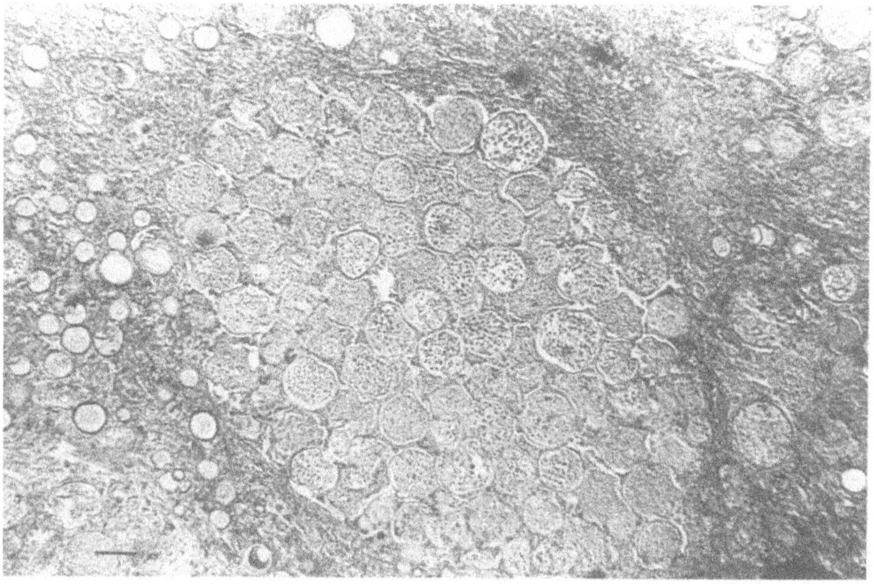

FIG. 2. Section through rhinosporidial polyp from another human case, focusing on a cluster of synchronously maturing sporangia, apparently derived from the spores (endospores) of their parent sporangium, with a rim of infiltrate. Hematoxylin and eosin; bar = 100 μm.

when the "trophic stage" or developing sporangium attains a size of about 50 μm. With further growth, nuclear divisions, and differentiation, a sporangium develops about 16,000 to 20,000 young spores, about two-thirds of which mature. A mature sporangium measures approximately 250 to 350 μm, and it releases the spores through a pore in its wall. Mature sporangia may rupture through the epithelial covering, in which case the freed spores, 6 to 7 μm, would be demonstrable microscopically in the exudates

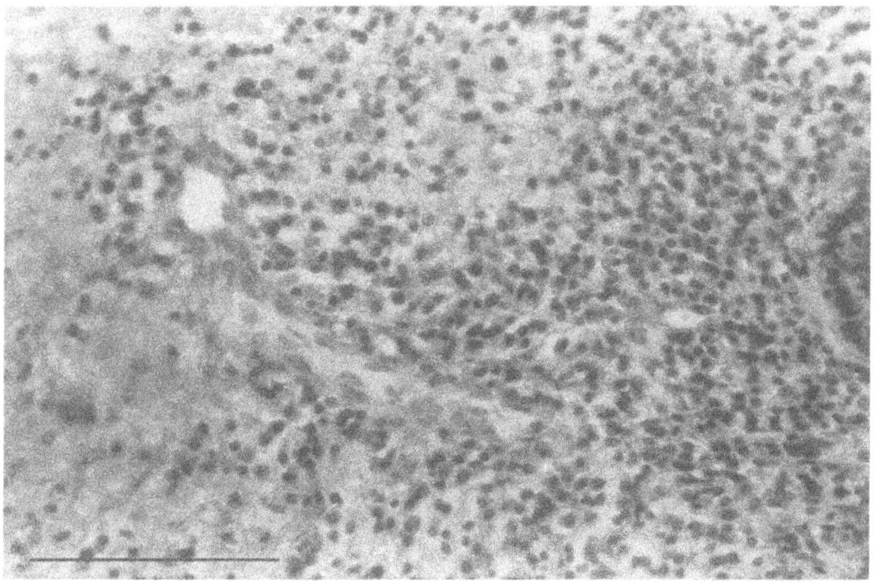

FIG. 3. Another field of section shown in Fig. 2 illustrating the type of infiltrate, comprising mainly lymphocytes and plasma cells; an arteriole is seen traversing the field transversely. Hematoxylin and eosin; bar = 100 μm.

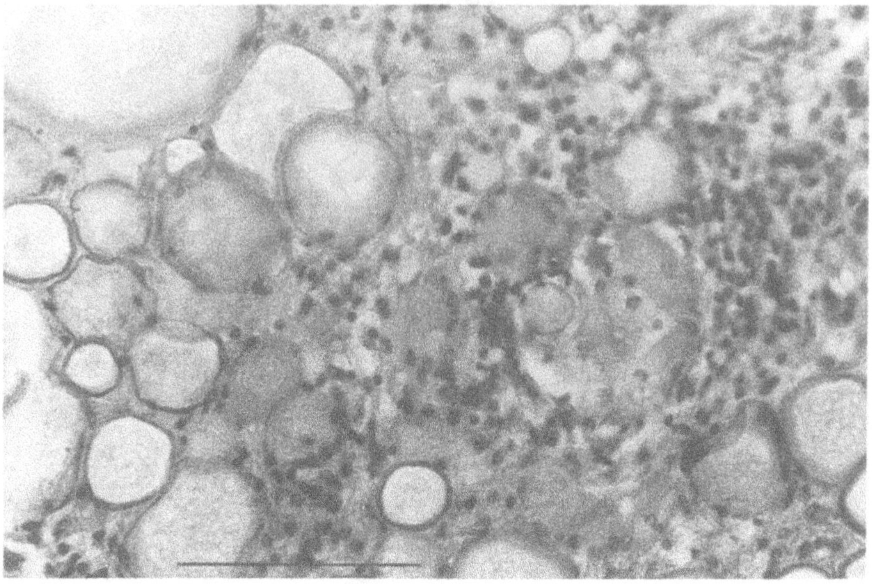

FIG. 4. A view of the same section focusing on a developing foreign body granuloma around a sporangium. Hematoxylin and eosin; bar = 100 μm.

from the lesions. A mucoid envelope surrounds each developing spore which is further strengthened by an overlying layer of chitin.

Diagnostic Features

Rhinosporidiosis should be considered in the differential diagnosis of any lesion of mucosal membranes presenting as oval, rounded, lobulated, pedunculated, or sessile polyp or bizarre growth. Typical lesions are pink to purplish-red and friable, and they have barely visible white sporangia. Sessile growths often assume the form of small nodules or elevated leaflike structures with rounded or dentate raised margins. Atypical lesions or those occurring in unusual anatomical sites should be differentiated from

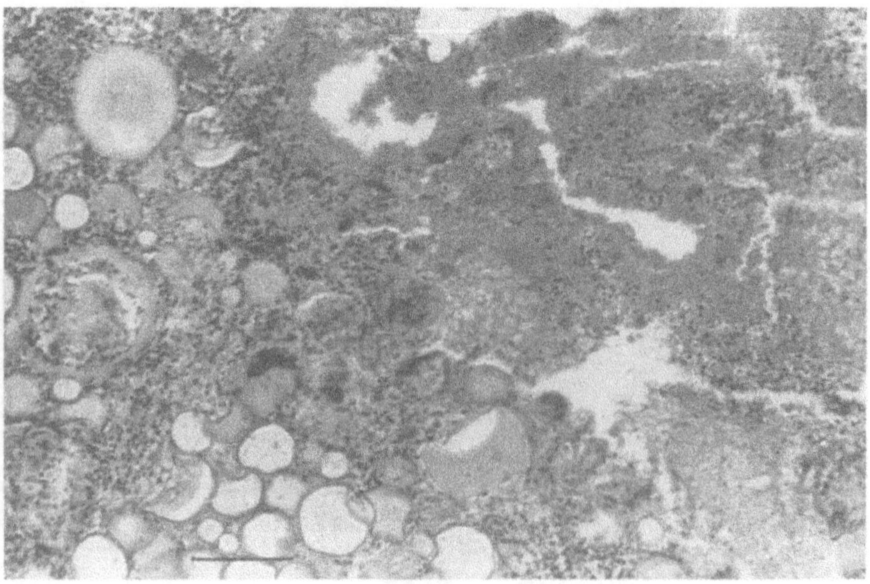

FIG. 5. An area of ischemic necrosis from section shown in Fig. 2, demarcated from viable tissue by a pallisade of foreign body giant cells. Hematoxylin and eosin; bar = 100 μm.

warts, neoplasias, hemangiomas, condylomata, mucoceles, or hemorrhoids. Rhinosporidiosis may also occur in association with syphilis, leprosy, leishmaniasis, squamous carcinoma, rodent ulcer, and molluscum contagiosum. The red color of lesions is in part due to the vascularity but more so to the hemorrhagic exudate on the surface. On removing the blood, the growth presents a grayish-pink appearance.

Prognosis

Rhinosporidiosis is a chronic infection that may go on for years with little pain or discomfort to the patient. In fact, long duration of the disease is a marked feature of rhinosporidiosis. Allen and Dave (1936) have recorded cases of rhinosporidiosis of 30 and 35 years' duration. The disease usually manifests as a solitary polyp that causes no difficulty unless it is large enough to obstruct a passage or exert pressure on neural bundles, causing pain, or on the vasculature, resulting in ischemic necrosis which may lead to spontaneous regression or total disappearance of nasal rhinosporidial lesions in certain cases. In one instance, a nasal mass of 15 years' duration was reported to have disappeared spontaneously without leaving a trace (Forsyth, 1924). Recurrence of rhinosporidiosis, especially of the nose, is a common feature. Apparently, there is no immunity to reinfection. Hematogenous dissemination of the infection may occur, but it is rare. Disseminated infections may terminate fatally, as has been reported by Rajam et al. (1955) and Agrawal et al. (1959); in the former case, *R. seeberi* was histologically demonstrable in the spleen, kidney, liver, lung, and other viscera, apart from its presence in blood, urine, and ascitic fluid. A single fatal case of tracheobronchial rhinosporidiosis is also on record (Subramanyam and Ramana Rao, 1960).

Epidemiologic Features

Occurrence and Transmission

Rhinosporidiosis has a worldwide geographic distribution, although it has been most frequently reported from India and Sri Lanka (Alonso et al., 1987; Hoff and Hall, 1986; Jimenez et al., 1984; Naik and Shukla, 1980; Raju and Jamalabadi, 1983; Roberts, 1982; Wahab et al., 1976; Wright et al., 1979). In India, the disease is especially prevalent in the southern, central, and eastern states, but more recently it has been increasingly reported from other parts of the country (Dey et al., 1973; Grover, 1975; Gupta et al., 1976, 1984; Naik et al., 1979; Paul et al., 1978; Verma et al., 1970).

It is widely believed that rhinosporidiosis is acquired from exogenous sources such as contaminated water and soil or dust. The possibility of person-to-person transmission of the disease has been suggested in certain instances of intrafamilial infection, but it has not been documented. Such cases may have actually acquired the disease through their independent exposure to a natural reservoir of *R. seeberi* rather than through direct contact with each other. Rhinosporidiosis is believed to have an insidious onset, and virtually nothing is known about the incubation period except for the observations of Jimenez et al. (1984) stating that the appearance of polypoid growth in their two patients with conjunctival infection required only 6 to 16 days following accidental injury to the eye by contaminated soil and/or dust. These two cases may be taken as illustrative of the incubation time required for manifestation of acute rhinosporidial lesions.

The frequent history of extended exposure of the patients to the stagnant water of ponds, pools, and rivers and the clustering of rhinosporidiosis cases in the subjects so exposed suggests that *R. seeberi* is an inhabitant of such natural sites that serve as reservoirs of the infection. It has been postulated that fish and aquatic insects might be hosts of *R. seeberi*, but so far the disease has been reported only from horses, mules, cattle, canines, and certain aquatic birds (Ainsworth and Austwick, 1973; Jungerman and Schwartzman, 1972). The reports of Davidson and Nettles (1977) on the occurrence of a rhinosporidial nasal polyp in a migratory wood duck (*Aix sponsa*) and of Fain and Herin (1957) on nasal rhinosporidiosis in a goose (*Alopochen aegyptiacus*) and a wild duck (*Dendrocygna viduata*) provide additional support to the hypothesis of an aquatic habitat for *R. seeberi*. Furthermore, the infected migratory aquatic birds indicate a possible means of transmission of the etiologic agent over great distances.

Susceptibility, Resistance, and Carrier States

Rhinosporidiosis may occur at any age, but it is more frequently reported in children and young adults. Its greater occurrence in males as reported in numerous studies may be related more to their greater exposure to the etiologic agent through outdoor activities than to any physiological or hormonal differences between the two sexes. It may be pointed out, however, that the ratio of male to female involvement also varies to a certain extent according to the anatomical site of lesions, age of subjects, and geographic region. For instance, in a recent study by Chitravel et al. (1980), females outnumbered the males by a ratio of 1.6 : 1 in a series of 34 cases of rhinosporidiosis detected in a population sample of 730 inhabitants of Agasteeswaram village, Kanyakumari District, Tamil Nadu, India.

The importance of trauma as a predisposing factor in primary infection and in recurrence of human rhinosporidiosis has been stressed by a number of investigators, although it may not be an essential prerequisite (Allen and Dave, 1936; Jimenez et al., 1984; Karunaratne, 1964; Kaye, 1938; Orlandi, 1926). In an attempt to explain the high endemicity of rhinosporidiosis in the south Indian states of Tamil Nadu and Kerala, Amesur (1958) speculated that the inhalation of sand containing rare minerals such as thorium and uranium found in these areas might, by inciting irritation and scratching of the nasal mucosa, predispose the native population to *R. seeberi* infection. Based on his veterinary experience, Rao (1938) was of the opinion that trauma was an important predisposing factor. He observed that in the majority of the cases of rhinosporidiosis in cattle studied by him, the growth was on the margin of the perforation made in the septum for insertion of the nose string, or close to the lesions of nasal schistosomiasis caused by *Schistosoma nasalis*. In a discussion of the mode of infection in a series of 48 cases of nasal rhinosporidiosis investigated from the Poona area, Mandlik (1937) emphasized that the disease in some localities was waterborne. He also suggested the importance of increased water pressure encountered during diving for anchoring or proper implantation of the spores of *R. seeberi* in tissues of the host.

No racial difference in susceptibility to infection has been recognized. Nothing is known about the carrier states of the etiologic agent.

Etiologic Agent

Rhinosporidium seeberi (Wernicke) Seeber, 1912, is the sole etiologic agent of rhinosporidiosis in humans and animals. Its synonyms include *Coccidium seeberi* Wernicke, 1900; *Rhinosporidium kinealyi* Minchin and Fanthum, 1905; *Rhinosporidium equi* Zschokke, 1913; and *Rhinosporidium ayyari* Allen and Dave, 1936. As with the etiologic agents of several systemic mycoses, *R. seeberi* was described as a protozoan by the early investigators. In his classical researches, Ashworth (1923) brought forward evidence to show that the etiologic agent was a fungus, and this conclusion has been generally accepted by subsequent investigators. Ashworth's evidence for classifying *R. seeberi* as a fungus included the repeated synchronous mitosis, the late division of the cytoplasm for spore formation, absence of residual cytoplasm, presence of a mucoid matrix, and the formation of cellulose in the wall of the sporangium. Recent studies on the histochemical and ultrastructural features of *R. seeberi* (Easley et al., 1986; Savino and Margo, 1983) also support the concept that it is a fungus.

The systematic position of *R. seeberi* remains uncertain, but it has been tentatively classified in the class *Phycomycetes*, order *Chytridiales*, family *Olpidiaceae*. It has been related to the genus *Synchytrium* which includes plant pathogenic fungi, inducing gall formation on the tissues of the host plant. From analogy with *Coccidioides immitis*, the etiologic agent of coccidioidomycosis, which has a saprobic mold form but multiplies by the formation of sporangia (spherules) with sporangiospores (endospores) in the host, it has been speculated that *R. seeberi* may be a dimorphic pathogen with a free-living saprobic existence in nature. So far, however, the fungus has been always encountered in the lesions of humans and animals, and it has not been isolated from nature.

Characterization

The characterization of *R. seeberi* is based entirely on its in vivo morphology as seen in the lesions occurring in infected humans and animals. This is done by examination of KOH-digested wet mounts or stained smears of material taken directly from lesions or by study of suitably stained histological sections of biopsy specimens.

Specimen Collection and Transport

Discharge or biopsy from the lesion is collected in a sterile, tightly capped or rubber-stoppered, thick-walled glass vial or bottle. A part of this material should be sent to a diagnostic mycology/microbiology laboratory for routine direct microscopic examination and culture, and the rest should be preserved in 10% formol saline for histopathologic study.

Isolation and Identification Procedures

Ever since Seeber (1900) described the first case of rhinosporidiosis in a thesis from Buenos Aires, all attempts to cultivate the etiologic agent in vitro have been unsuccessful. The credit for a breakthrough in this knotty problem goes to Levy et al. (1986) from the School of Veterinary Medicine, North Carolina State University, Raleigh. Levy et al. achieved the successful cultivation of *R. seeberi* from a canine nasal polyp, employing monolayers of an epitheloid human rectal tumor (HRT) cell line maintained in 25-cm^2 plastic tissue culture flasks incubated at 34°C in humidified air with 5% CO_2. The fungus continued to multiply for over 7 months in three subcultures, and interestingly it stimulated proliferation of the epithelial cells leading to the formation of polyplike structures. It was observed that the spores of *R. seeberi* required a period of aging or development, or both, before reinitiating the growth cycle. The spore to mature sporangium developmental cycle required 3 to 35 days, and the endospores were released over several days through a pore in the sporangial wall. In cultures at 27°C, it was reported that *R. seeberi* failed to induce proliferation of HRT cells, which was probably due to the quiescent nature of these cells at this temperature. However, fungal growth was not markedly inhibited and was independent of HRT proliferation. Furthermore, it was inferred that the ability of *R. seeberi* to induce cellular proliferation appeared to require close contact with actively growing organisms and associated mammalian cells.

The identification procedures for *R. seeberi* are quite simple. They include microscopic examination of stained smears and wet mounts/squash preparations in 10 to 15% KOH of material taken from rhinosporidial polyps or other lesions as well as a study of suitably stained tissue sections.

Preliminary Identification

Gross examination of the polyp would indicate the presence of dull white subsurface sporangia, the identity of which can be tentatively determined by direct microscopy of wet KOH mounts of macerated material taken from the lesion. Sometimes, direct microscopy of discharge or exudate from the infected area may also reveal the spores and/or sporangia of *R. seeberi*.

Definitive Identification

The tentative identification of *R. seeberi* made by direct microscopy can be confirmed by histologic examination of stained sections of infected tissue. *R. seeberi* is readily identifiable in H&E-stained preparations by the typical sporangia and spores, and it does not require any special stain. Of the various fungal stains, the results are usually good with the Gomori's silver methenamine technique (Fig. 6), whereas they may be variable or negative with the Gridley or PAS (periodic acid-Schiff) techniques. Besides, the inner walls of sporangia and the outer walls of spores are well stained by Mayer's mucicarmine technique. *Cryptococcus neoformans* is the only other carminophilic fungal pathogen, and that can be differentiated without any difficulty because of its strikingly different morphology. Pathogenicity test in laboratory animals is of no diagnostic value for want of a suitable experimental model of rhinosporidiosis.

Immunologic Aspects

Studies on the immunology of rhinosporidiosis are lacking except for some exploratory investigations of Chitravel et al. (1981, 1982) which need to be extended. In a series of 37 patients with rhinosporidiosis, leukocyte migration inhibition (LMI) was found

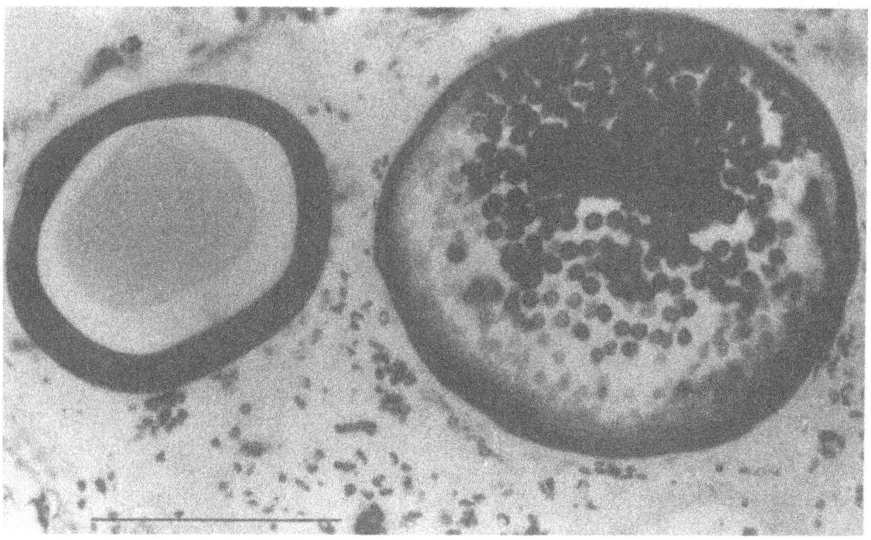

FIG. 6. Same section focusing on two sporangia of *R. seeberi*—one thick-walled and without spores, and the second larger, relatively thin-walled and with numerous mature spores. Gomori's silver methenamine; bar = 100 μm.

to be higher than in the healthy volunteers, and T-cell count was significantly low in patients with infection of 1 to 2 years' duration whereas it was normal in patients with infection of 3 years' or longer duration. Besides, presumptive evidence in support of circulating rhinosporidial antigen was presented.

Treatment, Prevention, and Control

Treatment

Surgical excision of the lesions is the most successful treatment available so far. Excision should be done as radically as possible because of the known tendency of rhinosporidiosis for recurrence. Electrocoagulation of the base of the lesions has been recommended in order to avoid autoinfection (Beiske et al., 1982). The erroneous practice of using a snare for excision of lesions, despite its criticism more than five decades ago (Allen and Dave, 1936), is still recommended in some textbooks; this is unfortunate in view of the risk of recurrence of the disease due to incomplete removal of the lesions. In a postoperative clinical trial with dapsone (diaminodiphenylsulfone) given orally at a daily dose of 100 mg, Nair (1979) reported that 71.4% of the treated group comprising 32 patients with nasal and nasopharyngeal rhinosporidiosis had no recurrence of the disease during a follow-up period of 3 years, whereas 93% of the control group required surgery for recurrent infection. In patients with disseminated or mutilating rhinosporidiosis, no medical treatment is currently available. The report of Peralta and Steiner (1976) on the cure of rhinosporidiosis in cattle by treatment with calcium nicotibine and 60% sodium iodide remains to be substantiated.

Prevention and Control

The etiologic agent *R. seeberi* has defied all attempts at its isolation from environmental sources, and its natural habitat therefore remains undefined. Furthermore, the pathogenesis of rhinosporidiosis is poorly understood for want of a satisfactory animal model of the disease. Until advancement is made in these directions, no tangible control measures other than avoidance of exposure to the natural reservoirs of *R. seeberi* can be suggested.

Acknowledgments. The authors thank Dr. Zia U. Khan, Department of Medical Mycology of this institute, for discussions; Mr. T. Kowshik for invaluable bibliographic assistance; Mr. S. Mazumdar for photomicrography; and Mr. Vijay Ahuja for preparation of the typescript.

Literature Cited

Agarwal, S., K. D. Sharma, and J. B. Shrivastava. 1959. Generalized rhinosporidiosis with visceral involvement: report of a case. AMA Arch. Dermatol. **80:**22–26.

Ainsworth, G. C., and P. K. C. Austwick. 1973. Fungal diseases of animals, 2d Ed., p. 117–119. Commonwealth Agricultural Bureaux, Farnham Royal, Slough, U.K.

Allen, F. R. W. K., and M. L. Dave. 1936. The treatment of rhinosporidiosis in man based on study of 60 cases. Indian Med. Gaz. **71:**376–394.

Alonso, M. J., R. G. De Tejada, P. Diaz, J. Ruiz, and J. Prieto. 1987. Description of a case of rhinosporidiosis in Spain. Mycopathologia **97:**9–16.

Amesur, C. A. 1958. Rhinosporidiosis—its prevalence in south India, p. 377–378. *In* H. M. Desai, D. V. Doshi, and N. H. Wadia (ed.), Indian year-book of medical sciences. Current Technical Literature Co., Bombay. Cited in Karunaratne, 1964.

Ashworth, J. H. 1923. On *Rhinosporidium seeberi* (Wernicke, 1903) with special reference to its sporulation and affinities. Trans. R. Soc. Edinb. **53:**301–342.

Beiske, K., B. Lindtjorn, and J. Olafsson. 1982. Rhinosporidiosis affecting the palpebral conjunctiva: a case report. Ethiopian Med. J. **20:**135–139.

Chatterjee, P. K., C. R. Khatua, S. N. Chatterjee, and N. Dastidar. 1977. Recurrent multiple rhinosporidiosis with osteolytic lesions in hand and foot. A case report. J. Larygol. Otol. **91:**729–734.

Chitravel, V., S. Subramanian, B. M. Sundaram, and K. Thiruneelakantan. 1980. Rhinosporidiosis in a south Indian village. Sabouraudia **18:**241–247.

Chitravel, V., T. Sundararaj, S. Subramanian, M. Kumaresan, and S. Kunjithapadam. 1981. Cell mediated immune response in human cases of rhinosporidiosis. Sabouraudia **19:**135–142.

Chitravel, V., T. Sundraraj, S. Subramanian, M. Kumaresan, and S. Kunjithapadam. 1982. Detection of circulating antigen in patients with rhinosporidiosis. Sabouraudia **20:**185–191.

Davidson, W. R., and V. F. Nettles. 1977. Rhinosporidiosis in a wood duck. J. Am. Vet. Med. Assoc. **171:**989–990.

Dey, N. C., H. L. F. Grueber, and T. K. Dey. 1973. Medical mycology, p. 186–191. Allied Agency, Calcutta.

Dhayagude, R. G. 1941. Unusual rhinosporidial infection in man. Indian Med. Gaz. **76:**513–515.

Easley, J. R., D. J. Meuten, M. G. Levy, M. J. Dykstra, E. B. Breitschwerdt, E. A. Holzinger, and R. C. Cattley. 1986. Nasal rhinosporidiosis in the dog. Vet. Pathol. **23:**50–56.

Emmons, C. W., C. H. Binford, J. P. Utz, and K. J. Kwon-Chung. 1977. Medical mycology, 3d Ed., p. 464–470. Lea and Febiger, Philadelphia.

Fain, A., and V. Herin. 1957. Deux cas de rhinosporidiose nasale chez une oie et un canard sauvages a Astruda (Ruanda-urundi). Mycopathol. Mycol. Appl. **8:**54–61.

Forsyth, W. L. 1924. *Rhinosporidium kinealyi.* Lancet **1:**951–952.

Grover, S. 1975. Rhinosporidiosis (epidemiological and clinical study). J. Indian Med. Assoc. **64:**93–95.

Gupta, K., A. K. Gupta, and S. B. Jain. 1984. Rhinosporidiosis of conjunctiva (report of a case from Patiala). J. Postgrad. Med. **30:**119–120.

Gupta, R. L., B. S. Darbari, M. P. Dwivedi, Q. P. Billore, and M. M. Arora. 1976. An epidemiological study of rhinosporidiosis in and around Raipur. Indian J. Med. Res. **64:**1293–1299.

Hoff, B., and D. A. Hall. 1986. Rhinosporidiosis in a dog. Can. Vet. J. **27:**231–232.

Jimenez, J. F., D. E. Young, and A. J. Hough, Jr. 1984. Rhinosporidiosis. A report of two cases from Arkansas. Am. J. Clin. Pathol. **82:**611–615.

Jungerman, P. F., and R. M. Schwartzman. 1972. Veterinary medical mycology, p. 40–47. Lea and Febiger, Philadelphia.

Karunaratne, W. A. E. 1964. Rhinosporidiosis in man. Athlone Press, London.

Kaye, H. 1938. A case of rhinosporidiosis on the eye. Br. J. Ophthalmol. **22:**449–455.

Kuriakose, E. T. 1963. Oculosporidiosis: rhinosporidiosis of the eye. Br. J. Ophthalmol. **47:**346–349.

Kutty, M. K., and P. N. Unni. 1969. Rhinosporidiosis of the urethra. A case report. Trop. Geogr. Med. **21:**338–340.

Levy, M. G., D. J. Meuten, and E. B. Breitschwerdt. 1986. Cultivation of *Rhinosporidium seeberi* in vitro: interaction with epithelial cells. Science **234:**474–476.

Madhavan, M., C. Ratnakar, and K. S. Mehdiratta. 1978. Rhinosporidial infection of the forehead. (Report of a case.) J. Postgrad. Med. **24:**235–236.

Mandlik, G. S. 1937. A record of rhinosporidial polypi with some observations on the mode of infection. Indian Med. Gaz. **72:**143–147.

Naik, K. G., and S. M. Shukla. 1980. Rhinosporidiosis in Zambia. Med. J. Zambia **14:**78–80.

Naik, R. S., R. S. Siddiqui, and V. Naik. 1979. Urethronasal rhinosporidiosis. J. Indian Med. Assoc. **72:**238–239.

Nair, K. K. 1979. Clinical trial of diaminodiphenylsulfone (DDS) in nasal and nasopharyngeal rhinosporidiosis. Larynogoscope **89:**291–295.

Orlandi, N. 1926. Granuloma der conjunctiva durch *Rhinospordium seeberi.* Virchows Arch. Pathol. Anat. **262:**314–327.

Owor, R., and W. M. Wamukota. 1978. Rhinosporidiosis in Uganda: a review of 51 cases. East Afr. Med. J. **55:**582–586.

Palaniswamy, R., and M. Bhandari. 1982. Rhinosporidiosis of male terminal urethra. J. Urol. **129:**598–599.

Paul, J., A. R. Khan, and J. P. Gupta. 1978. Rhinosporidiosis in Jammu. Indian J. Pathol. Microbiol. **27:**73–77.

Peralta, C., and R. Steiner. 1976. Rhinosporidiosis en bovinos. Gac. Vet. **38:**113–119; abstract 1557, Rev. Med. Vet. Mycol. **12:**287, 1977.

Peters, H. J., and C. G. DeBelly. 1969. Conjunctival polyp caused by *Rhinosporidium seeberi.* Am. J. Clin. Pathol. **51:**256–259.

Purandare, N. M., and S. M. Deoras. 1953. Rhinosporidiosis in Bombay. Indian J. Med. Sci. **7:**603–610.

Rajam, R. V., G. S. Viswanathan, A. R. Rao, P. N. Rangiah, and V. C. Anguli. 1955. Rhinosporidiosis—a study with report of a fatal case of systemic dissemination. Indian J. Surg. **17:**269–298.

Raju, G. C., and M. H. Jamalabadi. 1983. Rhinosporidiosis in Trinidad. Trop. Geogr. Med. **35:**257–258.

Rao, M. A. N. 1938. Rhinosporidiosis in bovines in the Madras presidency, with a discussion on the possible modes of infection. Indian J. Vet. Sci. Anim. Husbandry **8:**187–198.

Reddi, D. B. 1954. Rhinosporidiosis (a review based on the study of 156 biopsies). Antiseptic **51:**702–711.

Rippon, J. W. 1982. Medical mycology, p. 325–334. W. B. Saunders, Philadelphia.

Roberts, W. B. O. 1982. Rhinosporidiosis in the nasal membranes: a case report. Ethiopian Med. J. **20:**141–143.

Savino, D. F., and C. E. Margo. 1983. Conjunctival rhinosporidiosis. Light and electron microscopic study. Ophthalmology **90:**1482–1489.

Seeber, G. R. 1900. Un nuevo esporozoario parasito del hombre. Dos casos encontrados en polypos nasales. Tesis, Univ. Nac. de Buenos Aires. (Cited in Karunaratne, 1964.)

Subramanyam, C. S. V., and A. V. Ramana Rao. 1960. A fatal case of tracheobronchial rhinosporidiosis. Br. J. Surg. **47:**411–413.

Suseela, V., and K. S. Subramaniam. 1975. Rhinosporidiosis and the eyes. Indian J. Ophthalmol. **23:**1–4.

Thomas, T., N. Gopinath, and R. H. Betts. 1956. Rhinosporidiosis of the bronchus. Br. J. Surg. **44:**316–319.

Verma, P. L., and G. S. Kohli. 1970. Rhinosporidiosis in Haryana. Indian J. Otolaryngol. **22:**220–222.

Wahab, S. M. A., and T. A. Talaat. 1976. Rhinosporidiosis in Sudan. Trop. Geogr. Med. **28:**60–62.

Wright, I. G., T. D. Carter, M. M. M. Hayes, and M. M. Friedman. 1979. Rhinosporidiosis in Rhodesia—a report of three cases. Central Afr. J. Med. **25:**55–56.

CHAPTER 72

Sporotrichosis

M. MIYAJI and H. YAMAGUCHI

Disease: Sporotrichosis.
Etiologic Agents: *Sporothrix schenckii, Sporothrix schenckii* var. *luriei.*
Source: Soil and living or nonliving plants by accidental inoculation through exposed skin or inhalation.
Clinical Manifestations: Indolent nodules and ulcers, mostly limited to the skin and subcutaneous tissues. Rarely dissemination to the bones and internal organs. Subacute or chronic.
Pathology: A mixed purulent and granulomatous inflammatory reaction.
Laboratory Diagnosis: Mycologic isolation of etiologic agent from pus or biopsy materials; sporotrichin skin test; asteroid body in tissue sections.
Epidemiology: Worldwide, especially in Mexico, Brazil, France, and Japan. Occupational disease of agricultural workers. Mostly sporadic cases, but sometimes outbreaks from direct contact with contaminated plants.
Treatment: Potassium iodide, amphotericin B, miconazole and ketoconazole, and thermotherapy.
Prevention and Control: Disinfection of epidemic sources, protective clothing and equipment, and education.

Description of Disease

Sporotrichosis is a subacute or chronic disease of humans and lower animals caused by the hyphomycetous soil fungus *Sporothrix schenckii.* The disease is worldwide in distribution. Infection is usually limited to the skin, subcutaneous tissue, and lymphatic glands. In most cases lesions are indolent nodules that often suppurate, ulcerate, and drain. Rarely, the infection may hematogenously disseminate to the mucous membranes, bones, joints, and internal organs, including the central nervous system. Pulmonary infection occurs occasionally from inhaled conidia.

Communicability and Transmissibility

Sporotrichosis is not communicable from person to person or from animal to human. *Sporothrix schenckii* exists in nature as a saprophyte. It is found in soil and on living plants or various plant materials. Infections are usually contracted by accidental inoculation of the fungus into the skin and subcutaneous tissue. The incubation period varies from 1 to 3 weeks or more, but in some cases extends to 6 months after inoculation. Rarely, the lung is primarily infected by inhalation of *S. schenckii* conidia.

Clinical Manifestations of Disease

In 1912 de Beurmann and Gougerot classified sporotrichosis as consisting of six different forms: 1) lymphatic; 2) disseminated; 3) epidermal; 4) mucosal; 5) skeletal; and 6) visceral. Since then various classifications of the disease have been proposed by many medical mycologists, who fall generally into two groups. One group regards the cutaneous lesions as the key manifestation and classifies the disease from this standpoint. The other places the cutaneous lesions on the same rank with the extracutaneous (mucosal, skeletal, visceral, and the like). For example, Emmons et al. (1977) classified sporotrichosis as 1) cutaneous lymphatic, 2) cutaneous nonlymphatic, and 3) disseminated. On the other hand, Conant et al. (1971) classified the disease as 1) cutaneous lymphatic, 2) localized cutaneous, 3) disseminated, 4) mucosal, 5) skeletal, and 6) visceral. Clinically, as almost all of the lesions of sporotrichosis are limited to the skin, subcutaneous tissue, and lymphatic glands, the authors concur with the former group.

In this chapter, clinical forms of sporotrichosis are classified as 1) cutaneous and 2) extracutaneous. Cutaneous sporotrichosis is further differentiated as 1) the cutaneous lymphatic form; 2) the localized cutaneous form; and 3) the disseminated cutaneous form.

Cutaneous Sporotrichosis

Cutaneous Lymphatic Form

More than two-thirds of patients with sporotrichosis usually manifest the cutaneous lymphatic form of the disease. The primary lesion usually develops through accidental inoculation of the fungus. In most patients one of the extremities is affected (Fig. 1), usually the hand or finger. The lesion initially appears as a hard, movable, indolent nodule at the site of fungal inoculation. Later, it becomes adherent to the overlying skin.

At first, the surface of the primary skin lesion is pink, becoming purple to black as it progresses to form a necrotic nodule (sporotrichotic chancre). The nodules eventually break down, ulcerate, and discharge pus. At later stages of the infection, several nodules appear along the course of the lymphatic glands, draining the area of the primary lesion (Fig. 2). The intervening lymphatic vessels can be palpated as if they were hard cords. Usually, the larger lymph nodes of the axilla or groin are not affected because the infection subsides before these lymph nodes are involved. When the large lymph nodes are involved, some may become suppurative. In the course of the infection, the primary skin lesion often heals; the secondary lesions, however, persist for months or years if untreated, without pain or fever.

Localized Cutaneous Form

The localized cutaneous form of the disease occurs in about one-third of patients with sporotrichosis. The lesion is confined to the area of inoculation, such as the face, neck, trunk, or arm (Fig. 3), without involving the lymphatic glands. In this form of infection the lesions may be ulcerative as well as verrucous, papillomatous, acneiform, or erythematoid in appearance.

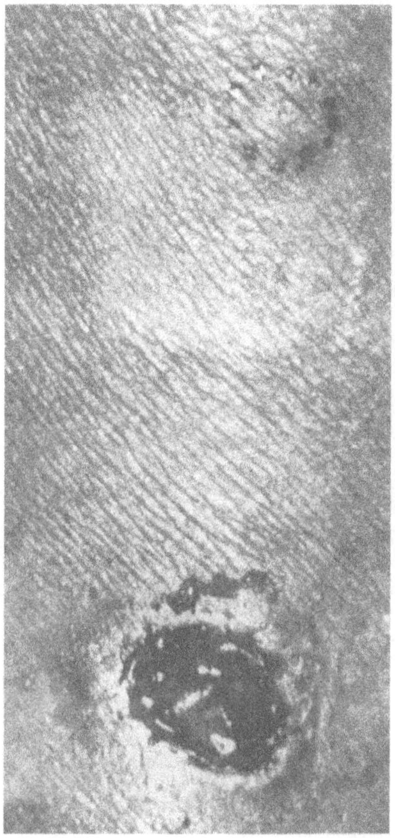

FIG. 1. Cutaneous lymphatic sporotrichosis in a 52-year-old woman. An ulcer and a subcutaneous nodule encircled by black dots are visible on the right forearm.

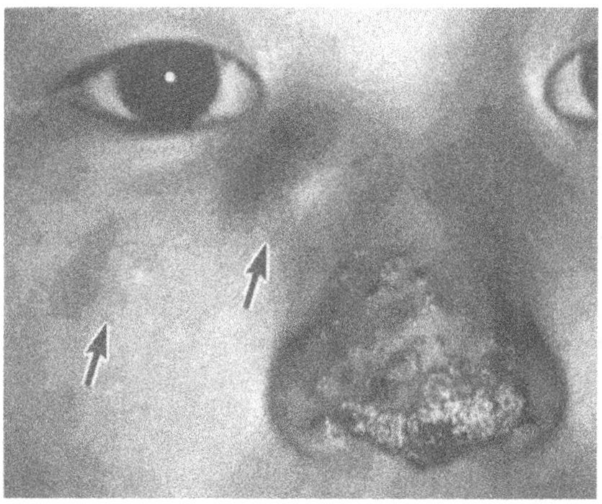

FIG. 2. Cutaneous lymphatic sporotrichosis in a 10-year-old girl, showing multiple small lesions with crusts on the nose and two subcutaneous nodules (*arrows*).

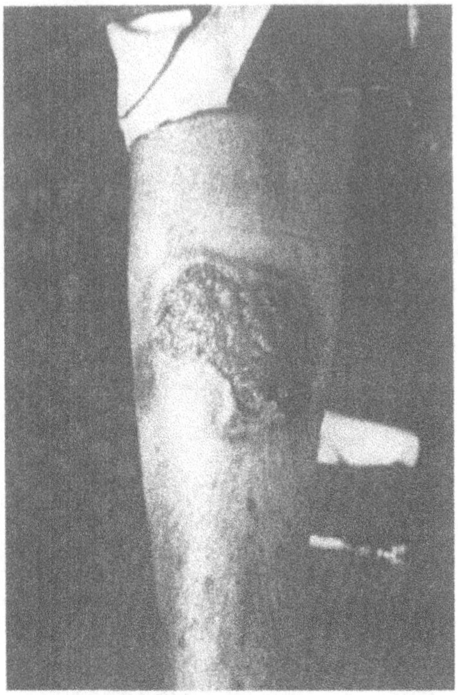

FIG. 3. Localized cutaneous sporotrichosis in an 84-year-old woman. A large ulcerative lesion is visible on the right forearm.

Concerning the etiologic basis of this form of the disease, two interesting factors are believed to be involved. One is that some patients with this form of sporotrichosis have been sensitized, without clinical manifestations, to *S. schenckii* by repeated exposure to the fungus (González-Ochoa, 1965). The other factor, proposed by Kwon-Chung (1975), is that some strains of *S. schenckii* isolated from patients with the localized cutaneous form of the disease cannot grow at temperatures above 35°C on conventional culture media, whereas most of the fungal strains isolated from those with the cutaneous lymphatic form readily grow in vitro at 38°C.

Clinicians sometimes encounter patients with clinical manifestations that are not typical of either the cutaneous lymphatic form or the localized cutaneous form of sporotrichosis. To accommodate these clinical manifestations Kariya (1975) has advocated recognition of an "intermediate" form of the disease, which is defined as follows: 1) A primary localized cutaneous lesion enlarges to the size of a pigeon egg, with subsequent spread to the lymphatic glands; 2) in patients with the cutaneous lymphatic form of the disease, secondary lesions of the lymphatic glands rapidly proceed to spontaneous healing, but the primary skin lesion persists; and 3) multiple lesions occur principally on the face without any lymphatogeous dissemination. Of 100 cases of the disease studied at Chiba University, Japan (Kariya and Iwatsu, 1979), 52 were of the cutaneous lymphatic form, 34 of the localized cutaneous form, and 14 of the intermediate form.

Disseminated Cutaneous Form

The disseminated cutaneous form of sporotrichosis is rare, probably beginning with one nodule manifested as the cutaneous lymphatic form of the disease. The causative agent may spread from this nodule via the bloodstream, resulting in the development of multiple nodules on the trunk, face, scalp, arms, and legs. Subsequently, these nodules often develop into ulcers with crusts. Compromised patients with disorders such as Hodgkin's disease, multiple myeloma, leukemia, diabetes, nephrosis, and especially alcoholism may be predisposed to develop the disseminated form of sporotrichosis.

Extracutaneous Sporotrichosis

Sporotrichosis of the mucous membranes may occur primarily by accidental fungal inoculation or follow the skin lesions of the disseminated cutaneous form. Usually, the mucous membrane of the nasopharynx, mouth, or larynx is affected. The disease starts as erythematous lesions, which gradually become ulcerative, suppurative, vegetative, or papillomatous and are accompanied by swelling of the regional lymph nodes.

The eye is also affected by the fungus. The conjunctiva is chiefly involved, with infection usually resulting from accidental inoculation of the fungus through injury by contaminated thorns.

Skeletal sporotrichosis is rare, but may occur without preceding cutaneous lesions. The bones, joints, tendon sheaths, or muscles are affected.

Pulmonary sporotrichosis, a prominent form of visceral sporotrichosis, may occur by inhalation of *S. schenckii* conidia. Because the tracheobronchial lymph nodes are involved, the early stage of the disease may be misdiagnosed as primary tuberculosis, sarcoidosis, histoplasmosis, or coccidioidomycosis. Although rare, other visceral forms of sporotrichosis, including pyelonephritis, orchitis, and mastitis, have been reported. Sporotrichosis of the central nervous system occurs much less frequently.

Histopathology

The histopathologic characteristics of sporotrichosis, regardless of clinical form, consist of a mixed purulent and granulomatous inflammatory reaction (Fig. 4). This reaction, however, is not specific for the disease; such inflammatory tissue responses are found in certain other deep-seated mycoses and in bacterial infections such as tuberculosis, anthrax, glanders, syphilis, and tularemia.

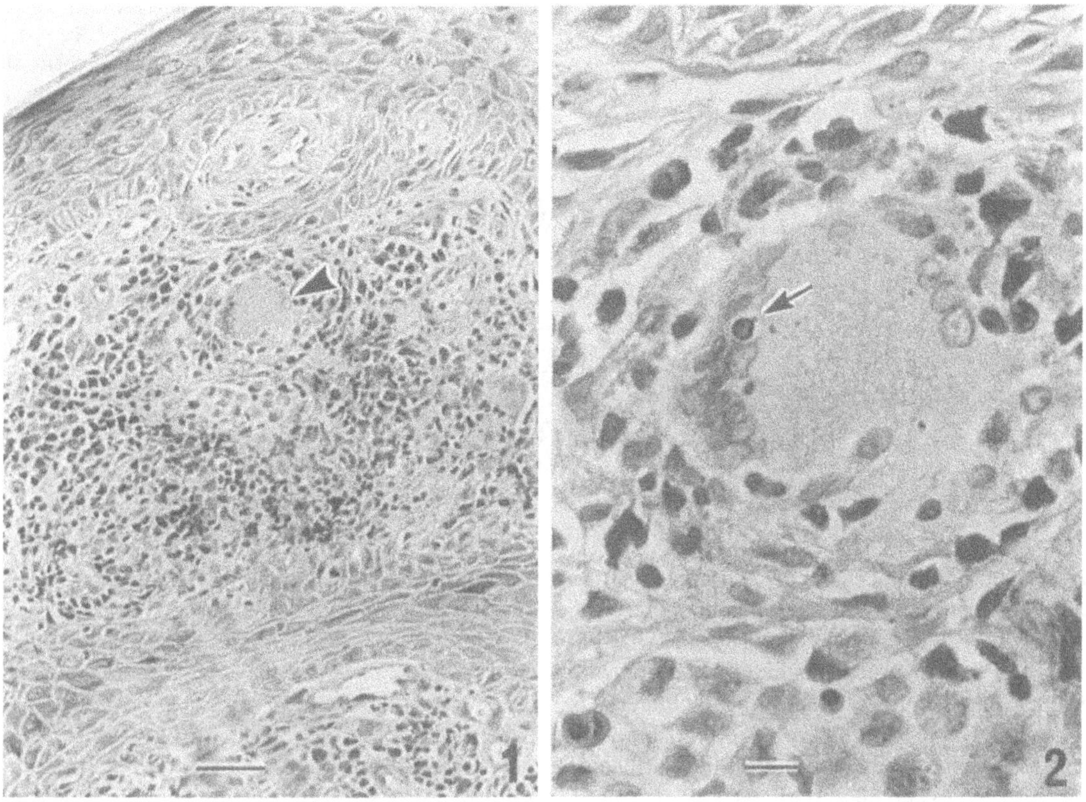

FIG. 4. (1) Granulomatous inflammatory reaction in the dermis of a patient with sporotrichosis. *Arrowhead* indicates a giant cell. Periodic acid–Schiff stain. Bar = 50 μm. (2) Higher magnification of the giant cell shown in (1). A budding yeast cell is seen within the giant cell (*arrow*). Periodic acid–Schiff stain. Bar = 10 μm.

On clinical examination, primary skin lesions are often mistaken for a neoplasm, particularly when they are verruciform. Microscopically, the tissue responses involved in these lesions are a combination of purulent and granulomatous reactions. The granulomas consist chiefly of histiocytes, some of which develop into epithelioid cells. Microabscesses are usually seen in the central area of the granulomatous tissue.

It is somewhat difficult to detect the fungal element in tissue sections, even with special fungal stains. If the primary epidermal lesion is papillomatous, it is characterized by hyperkeratosis, parakeratosis, pseudoepitheliomatous hyperplasia, and intraepidermal abscesses. These histopathologic findings are also obtained with other chronic deep mycoses such as blastomycosis, chromomycosis, coccidioidomycosis, and paracoccidioidomycosis. Because of this pseudoepitheliomatous hyperplasia, this form of sporotrichosis may be mistaken for skin cancer.

In the secondary lesions of sporotrichosis, which are produced by metastases through the lymphatic vessels draining the primary lesion, subcutaneous nodules become more prominent in appearance than in the primary lesion. Histopathologically, such nodules consist of three distinct zones. In the inner zone, the central part of the nodule, there is a chronic suppurative lesion characterized by plentiful infiltration of neutrophils. The intermediate zone is chiefly composed of epithelioid cells containing Langhans giant cells. The outer zone primarily consists of plasma cells, lymphocytes, and fibroblasts. In many cases, however, this typical three-zone arrangement cannot be distinguished so clearly. It is usually difficult to demonstrate yeast-form cells of *S. schenckii* in tissue sections prepared from secondary lesions, as well as from the primary lesion.

Although diagnosis of sporotrichosis based solely on histopathology is difficult, valid information can be obtained from histopathologic examinations. One or more yeast-form cells surrounded by a stellate, radial corona known as Splendore–Hoeppli material (the asteroid body), which is brightly eosinophilic on hematoxylin and eosin staining, may be demonstrated in tissues with a mixed inflammatory response, particularly within microabscesses (Fig. 5). The asteroid body, first described by Splendore in

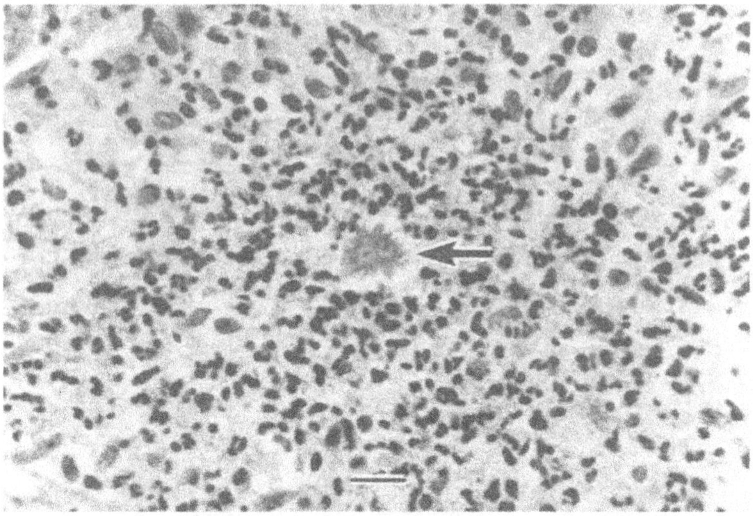

FIG. 5. Asteroid body in a microabscess. Two yeast-form cells are located at the center of the asteroid body. Hematoxylin and eosin stain. Bar = 20 μm.

Brazil in 1908 (Emmons et al., 1977), is usually 10 to 20 μm, but sometimes up to 100 μm in diameter. The centrally located yeast-form cells are somewhat basophilic and 3 to 5 μm in diameter. According to Lurie and Still (1960), Splendore–Hoeppli material is a deposit of glycoprotein nature resulting from a localized antigen–antibody reaction on the surface of the yeast in the hypersensitized host.

The asteroid body has been demonstrated frequently in tissue sections from patients with sporotrichosis in South America, South Africa, and Japan, but, strangely, with much less frequency in the United States. In a survey of 200 cases of sporotrichosis at the dermatologic clinic of Chiba University Hospital in Japan from May 1965 to August 1983, the asteroid body was detected in 54 cases (29.2%) (Ito et al., 1986).

However, the asteroid body is not pathognomonic for sporotrichosis. Such eosinophilic material may be deposited around colonies of some bacteria, foreign bodies, and other species of fungi, particularly *Coccidioides immitis, Aspergillus* spp., *Candida* spp., and certain fungal species that cause mycetoma.

As mentioned previously, it is rather difficult to demonstrate yeast-form cells of *S. schenckii* in infected tissue from patients with sporotrichosis. However, when a dozen or more tissue sections stained by the periodic acid–Schiff method or with Grocott methenamine silver are carefully examined, the yeast-form cells can be detected at the rate of more than 90%. They exist sporadically in the tissues, some engulfed in giant cells (Fig. 4). The yeast-form cells that constitute the asteroid body are single or budding, and round, oval, or elongate (cigar shaped). The round or oval cells generally measure 2 to 6 μm or more in diameter, the elongate cells 2 by 3 to 3 by 10 μm.

Interestingly, whereas histopathologic demonstration of yeast-form cells in the infected tissues is difficult, many fungal colonies can be readily recovered from biopsy materials. Nishimura and Miyaji (1975) obtained samples from the crusts and scales adherent to the edges of ulcerative lesions in 13 patients with the cutaneous form of sporotrichosis, stained these sections by the periodic acid–Schiff method, and subjected them to histopathologic examination. As a result, they succeeded in demonstrating numerous, variously shaped fungal elements (yeast and filamentous forms) of *S. schenckii* in almost all specimens (Fig. 6). Furthermore, perithecium-like organs were observed in sections prepared from a patient with sporotrichosis.

In 1969, Ajello and Kaplan described a new variety of *S. schenckii* termed *S. schenckii* var. *luriei,* which was isolated from a patient with the nondisseminated type of sporotrichosis in South Africa. In culture, the mycelium-form cells of *S. schenckii* var. *luriei* are indistinguishable from those of typical *S. schenckii.* In tissue, however, it forms not only small budding cells but large, spheroid, thick-walled cells. These huge cells are 16 to 18 μm in diameter and produce daughter cells by fission, budding, or both.

Epidemiology

Sporotrichosis have been reported in patients as young as 87 days and as old as 87 years. The disease tends to occur more frequently in male than in female patients. According to Emmons et al. (1977), this tendency is probably attributable to the increased likelihood of men being exposed to *S. schenckii* while working outdoors. However, in the 200 cases of the diseases studied at Chiba University, the male : female ratio was 1 : 2 (Ito et al., 1986). There is no

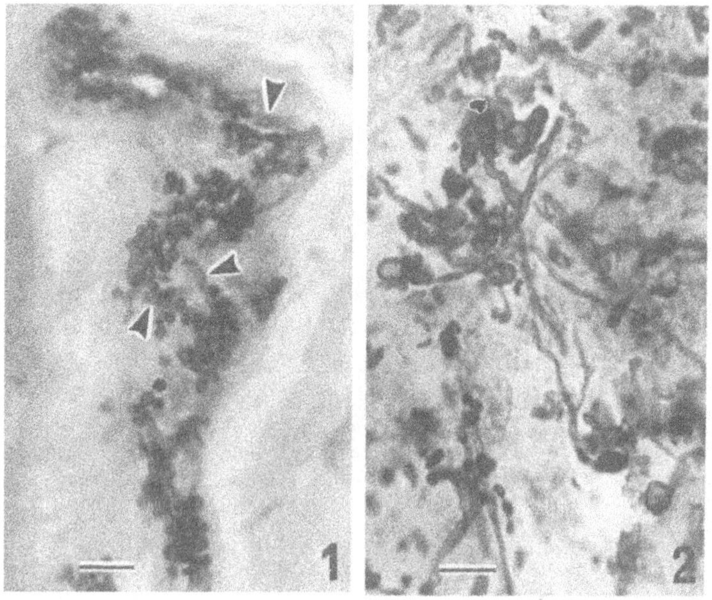

FIG. 6. (1) Parasitic form of *Sporothrix schenckii* in scales. Numerous conidia are produced along hyphae (*arrowheads*). Periodic acid–Schiff stain. Bar = 10 μm. (2) Hyphae and yeast-form cells of *S. schenckii* in crusts. Periodic acid–Schiff stain. Bar = 10 μm.

significant difference among races in susceptibility to the fungus.

Sporothrix schenckii is a normal inhabitant in nature. The fungus has been isolated from soil, plants, plant debris, and other sources. Therefore, sporotrichosis is found mainly in persons who are in active contact with soil or plants. It is an occupational disease of farmers, potters, florists, and gardeners.

The disease occurs worldwide, but is found mainly in temperate and tropical zones. Cases occur constantly, but sporadically, in the north-central part of the United States. Many cases have been reported in Mexico and Brazil. In South Africa, a great epidemic occurred in the gold mines of the Witwatersrand; about 3,000 natives working in the mines were infected with *S. schenckii* that had grown on mine timbers (Du Toit, 1942). In Japan, more than 2500 cases of sporotrichosis were reported between 1946 and 1982 (Fukushiro, 1984).

Etiologic Agents

Sporothrix schenckii (Hektoen and Perkins, 1900), synonym *Sporotrichum schenckii* (Hektoen and Perkins [Matruchot, 1910]). *Sporothrix schenckii* is dimorphic. At room temperature its growth on conventional media is mycelial, but in infected tissues it takes the form of tiny, cigar-shaped yeast cells. The yeast-form cells also develop in cultures grown in a CO_2 atmosphere at 37°C on blood agar containing cysteine or on brain heart infusion agar (Difco).

The fungus grows rapidly on Sabouraud dextrose agar at 27°C (Fig. 7). Colonies of mycelium-form cells are smooth and moist and vary in color from white to gray, dark, or almost black. Their membranous surface is wrinkled or folded. In some strains fine, grayish, velvety aerial hyphae grow upward at the periphery of the colonies.

Microscopically, hyphae are thin (1 to 2 μm in width), hyaline, branching, and septate. Conidiophores are hyaline, septate, elongate, and tapering, and their apical portion is often swollen. Two types of conidia are produced. The first type is one celled, hyaline, globose to clavate, and truncate, measures 2 to 3 by 3 to 5 μm in diameter, and is produced sympodially from distinct short denticles on the swollen apical portions of the conidiophores or from short

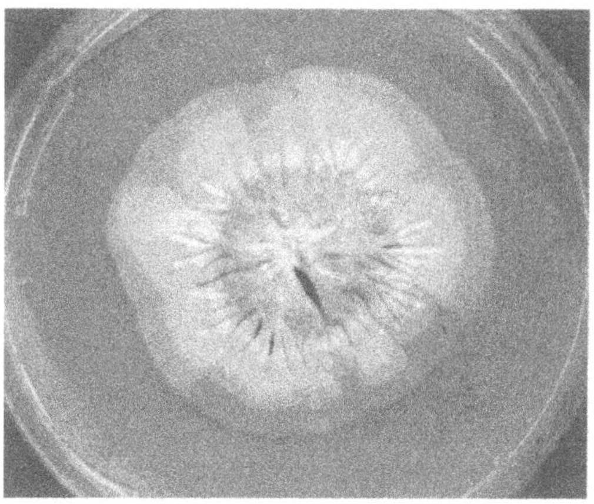

FIG. 7. Giant colony of *Sporothrix schenckii* grown on Sabouraud dextrose (2%) agar at 27°C for 45 days.

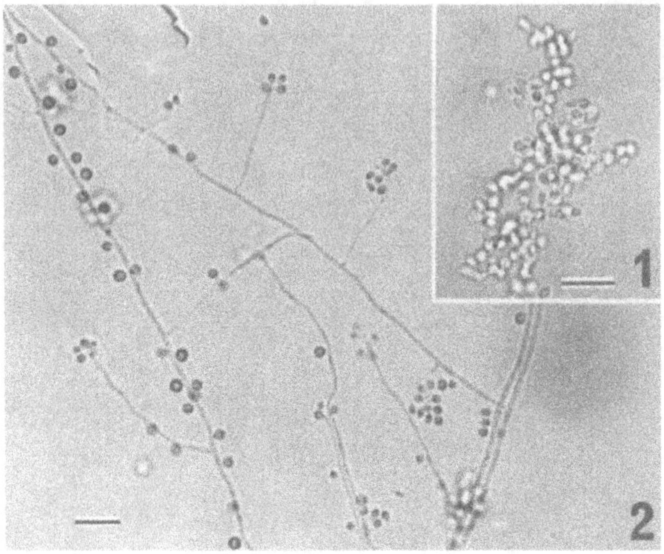

FIG. 8. (1) Yeast-form cells of *S. schenckii* grown on brain–heart infusion dextrose (1%) agar at 37°C and mounted with lactophenol solution. Bar = 10 μm. (2) Mycelium-form cells of *S. schenckii* grown on dextrose (2%) agar at 27°C and mounted with lactophenol cotton blue solution. Bar = 10 μm.

denticles along the conidiophores. The second type is one celled, globose to oblong, has a dark, thick wall, measures 2 to 3 by 3 to 6 μm in diameter, and is produced from short denticles along hyphae.

Yeast-like colonies grown on blood agar are of a creamy consistency. The surface is moist, smooth, and white. The yeast cells are ovoid to globose or elongated (cigar shaped). The ovoid cells are 2 to 6 μm or more in diameter, whereas the cigar-shaped cells measure 2 to 3 by 3 to 10 μm. These yeast-like cells obtained from in vitro cultures appear similar to those seen in vivo (Fig. 8). The morphology of *S. schenckii* var. *luriei* on culture media is indistinguishable from that of *S. schenckii*.

The teleomorph of *S. schenckii* has not been established. On the basis of ecologic, morphologic, and physiologic properties, Mariat (1977) has suggested a possible phylogenetic relationship between *S. schenckii* and the ascomycetous fungi of the genus *Ceratocystis*, especially *C. stenoceras*. In terms of immunologic properties, *S. schenckii* is also closely related to *C. stenoceras* (Andrieu et al., 1971).

Diagnostic Features

Laboratory Diagnosis

It is almost impossible to diagnose sporotrichosis by direct examination of smears of pus from lesions, because the parasitic yeast-form cells of *S. schenckii* are observed very rarely or only with difficulty on stained clinical specimens.

Histopathologic diagnosis using stained histologic sections prepared from biopsy materials is also difficult, because only few yeast-form cells appear in the infected tissues and their histopathologic features are not specific for sporotrichosis. In numerous cases, especially in chronic sporotrichosis, the asteroid body may be observed diagnostically. It is readily seen on staining with hematoxylin and eosin. For definitive diagnosis, it is necessary to recover the causative agent, *S. schenckii*, from pus aspirated from a fluctuant lesion or from biopsy materials.

Sabouraud dextrose agar supplemented with both chloramphenicol (50 mg/liter) and cycloheximide (500 mg/liter) is typically used for isolation. After inoculation, all cultures are grown at room temperature. Within 3 to 4 days, gray to brown mycelial colonies appear on the agar. Animal inoculation is not necessary for laboratory diagnosis of sporotrichosis. If such tests must be performed, however, male rats and mice are recommended. In animals inoculated intraperitoneally with yeast-form cells of *S. schenckii*, a typical orchitis develops in the lesions, in which numerous yeast cells grow.

Immunologic Diagnosis

Most patients with sporotrichosis develop delayed-type hypersensitivity and specific antibodies against the causative agent. Therefore, both a delayed-type skin reaction (sporotrichin reaction) and serologic reactions are applied toward diagnosis of the disease. Hypersensitivity can be determined by intradermal skin tests with the specific antigen sporotrichin. Two kinds of antigens are used; one is a 0.1% heat or chemically killed yeast-form cell suspension, and the other is a 1 ~ 10% culture fluid of mycelium-phase *S. schenckii*. Chemically, the latter antigen is chiefly composed of rhamnomannan, which is strongly reactive and highly specific. The sporotrichin skin test has significant diagnostic value. The test is conducted as follows: One-tenth milliliter of the antigen is inoculated intradermally. Forty-eight hours later, the diameters of the consequent erythema and induration are measured in two directions at right angles. Induration of 5 mm or more in diameter is considered positive. A positive skin reaction to the antigen occurs in persons who have had immunosensitizing contact with *S. schenckii,* including almost all patients with sporotrichosis. At Chiba University, a positive sporotrichin test has been seen in 132 of 137 cases.

Serologic tests can be used to advantage for the diagnosis of sporotrichosis. A wide variety of serodiagnostic tests have been developed for this mycosis. The tube agglutination and latex particle agglutination methods, considered by many workers to be the most reliable and specific tests for sporotrichosis, are of particular value for the diagnosis of extracutaneous infections. According to Conant et al. (1971), agglutinin titers greater than 1 : 40 are diagnostic.

Differential Diagnosis

It is not difficult for experienced dermatologists to diagnose clinically the cutaneous lymphatic form of sporotrichosis. As many infected patients display a wide variety of clinical appearances, however, it is necessary to differentiate sporotrichosis from tuberculosis, tularemia, syphilis, leprosy, glanders, pyogenic infections, other deep mycoses such as chromomycosis, coccidioidomycosis, blastomycosis, and paracoccidioidomycosis, and various kinds of drug eruptions. Histopathologic and cultural studies are mandatory for an indisputable diagnosis of sporotrichosis.

Prognosis

The prognosis is good in patients with the cutaneous lymphatic or localized cutaneous form of sporotrichosis when they are adequately treated. Although little is known about spontaneous resolution, it may occur more frequently particularly in the localized cutaneous form, than is generally thought. However, the disseminated cutaneous form, as well as the visceral, mucosal, and skeletal forms, are serious and often life-threatening because these patients usually are immunocompromised and respond poorly to the treatment with potassium iodide.

Treatment, Prevention, and Control

Treatment

Potassium Iodide

Potassium iodide is usually dissolved in water or milk. The resultant saturated solution (1 g/ml) should be given orally in starting doses of 10 drops (1 ml) three times daily after meals and gradually increased until maintenance doses (3 to 4 ml) are reached. Usually, the treatment should be continued 3 to 4 weeks after lesions disappear. If unfavorable side effects occur, administration should be stopped, then resumed after a few days with an initial dose.

The mode of administration of potassium iodide at Chiba University is as follows: In adults, 0.6 g per day is given initially, and the daily dose is gradually increased to 1.8 to 2.1 g. In children, 0.1 g is administered initially, and the dose is gradually increased to 0.3 to 0.6 g according to age. The average treatment period is 3 months. According to Kariya and Iwatsu (1979), significant side effects were noted in only 1 of 100 cases. The disease recurred in 2 of the 100 cases.

Antifungal Chemotherapy

The extracutaneous and systemic forms of sporotrichosis are resistant to iodides and thermotherapy but often respond well to treatment with amphotericin B, given intravenously in a total dose of 1.5 to 2.0 g over 6 to 8 weeks. There is some evidence that oral ketoconazole and intravenous miconazole may be of value in patients who do not respond to amphotericin B because of drug resistance, clinical resistance, or drug intolerance.

Thermotherapy

Sporothrix schenckii is temperature sensitive both in vivo and in vitro (Mackinnon and Conti-Diaz, 1962). On the basis of this finding, heat treatment (thermotherapy) has been developed and advocated for skin lesions, especially in patients who cannot tolerate iodides or amphotericin B. An attempt is made to kill the fungus in the cutaneous lesions by raising the temperature of the infected area above 39°C for more than 1 hour four times daily. This temperature can be attained with hot packs or an electric heating pad, as well as with *kairo* (Japanese traditional pocket heater).

Prevention and Control

Sporotrichosis can be prevented when epidemic sources are identified. For example, the massive epidemic in the South African gold mines was controlled by disinfecting the timber on which *S. schenckii* grew as a saprophyte. Laboratory infections can be prevented by reasonable precautions among technicians, who should be made aware of the risk of autoinoculation, especially of the eyelids. For the most part, however, the simple precautions needed to prevent infection in agricultural workers are difficult to enforce because they cannot afford the clothing needed to protect the face, hands, legs, and feet.

Literature Cited

Ajello, L., and W. Kaplan. 1969. A new variant of *Sporothrix schenckii*. Mykosen **12**:633–644.

Andrieu, S., J. Biquet, and S. Massamba. 1971. Étude immunologique comparée de *Sporothrix schenckii* et des souches saprophytes voisines. Sabouraudia **9**:206–209.

Conant, N. F., S. T. Smith, R. D. Baker, and J. L. Callaway (ed.). 1971. Manual of clinical mycology, p. 451. The W. B. Saunders Co., Philadelphia.

de Beurmann, L., and H. Gougerot. 1912. Les sporotrichoses, p. 225. Felix Alcan, Paris.

Du Toit, C. J. 1942. Sporotrichosis on the Witwatersrand. Proc. Mine Med. Offic. Assoc. **22**:111–127.

Emmons, C. W., C. H. Binford, J. P. Utz, and K. J. Kwon-Chung (ed.). 1977. Medical mycology, p. 406–424. Lea & Febiger, Philadelphia.

Fukushiro, R. 1984. Epidemiology and ecology of sporotrichosis in Japan. Zbl. Bakt. Hyg. A. **257**:228–233.

González-Ochoa, A. 1965. Contribuciones recientes al conociemiento de la esporotrichosis. Gac. Med. Mex. **95**:463–474.

Ito, M., S. Okamoto, and H. Kariya. 1986. Survey of 200 cases of sporotrichosis. Dermatologica **172**:209–213.

Kariya, H. 1975. A classification of sporotrichosis. Jpn. J. Med. Mycol. **16**:87–91.

Kariya, H., and T. Iwatsu. 1979. Statistical survey of 100 cases of sporotrichosis. J. Dermatol. **6**:211–217.

Kwon-Chung, K. J. 1975. The relationship between temperature *in vitro* and type of disease produced in human by strains of *Sporothrix schenckii*. Abstr. Annu. Meet. Am. Soc. Microbiol. **75**:88.

Lurie, H. I., and W. J. S. Still. 1969. The "capsule" of *Sporotrichum schenckii* and the evolution of the asteroid body. Sabouraudia **7**:64–70.

Mackinnon, J. E., and I. A. Conti-Diaz. 1962. The effect of temperature on sporotrichosis. Sabouraudia **2**:56–59.

Mariat, F. 1977. Taxonomic problems related to the fungal complex *Sporothrix schenckii/Ceratocystis* spp., p. 265–270. *In* K. Iwata (ed.), Recent advances in medical and veterinary mycology. University of Tokyo Press, Tokyo.

Nishimura, K., and M. Miyaji. 1975. Studies on the parasitic forms of *Sporothrix schenckii* in scales and crusts (III). Clinical study of thirteen cases. Jpn. J. Med. Mycol. **16**:57–65.

CHAPTER 73

Zygomycosis

L. AJELLO

Disease: Zygomycosis (phycomycosis, mucormycosis, subcutaneous phycomycosis, entomophthoromycosis).

Etiologic Agents: Mucorales—*Absidia corymbifera, Apophysomyces elegans, Cunninghamella bertholletiae, Mucor ramosissimus, Rhizomucor pusillus, Rhizopus arrhizus, Rhizopus microsporus, Rhizopus rhizopodiformis, Saksenaea vasiformis;* Entomophthorales—*Basidiobolus ranarum, Conidiobolus coronatus,Conidiobolus incongruus.*

Source: Environment. The etiologic agents flourish in soil and organic debris. Infections are contracted by inhalation or through traumatic incidents.

Clinical Manifestations: Protean—depend either on the site invaded or on the immunologic and physiologic status of the host, or both. Zygomycosis may be acute and fulminating, especially in its rhinocerebral form, which occurs primarily in acidotic diabetics; primary cutaneous infections are rare, usually nosocomial. Subcutaneous infections following traumatic implantation in exposed parts of the body produce swelling of the involved tissues. Chronic nasal invasion leads to dramatically disfiguring tumefactions of nasal, labial, and facial tissues. Pulmonary infections resemble those caused by bacteria and viruses; X-rays reveal densities; pain and bloody sputum may be the first symptoms.

Pathology: Tissue is invaded by broad, generally aseptate, hyaline, randomly branched mycelium. Tissue reaction is typically a severe necrosis. Angioinvasion accompanied by infarction is common.

Laboratory Diagnosis: Zygomycosis is diagnosed by demonstrating broad, rarely septate, hyaline mycelial elements in clinical specimens or histologic preparations. Isolating and identifying a zygomycete compatible with clinical signs as well as direct and histologic findings confirm the diagnosis.

Epidemiology: All of the etiologic agents are distributed worldwide. Infections with *Basidiobolus ranarum* and *Conidiobolus coronatus* are most prevalent in the tropical regions of Africa and Asia. Infections with *Rhizopus arrhizus* are most prevalent in persons afflicted with diabetes mellitus. Nosocomial infections have been traced to contaminated air-conditioning systems and bandages. The source of all these infections can ultimately be traced to the environment; they are not contagious.

Treatment: Amphotericin B is used for pulmonary and disseminated cases. Acidosis should be controlled in diabetics with rhinocerebral zygomycosis, coupled with amphotericin B treatment and surgical debridement. Oral iodides are effective in subcutaneous zygomycosis.

Prevention and Control: Zygomycosis can be prevented by avoiding traumatic incidents, by wearing protective clothing, and properly maintaining air-conditioning systems in medical centers to avoid dissemination of zygomycete spores, especially near immunocompromised persons.

Description of Disease

Zygomycosis is the preferred name for a polymorphic disease caused by many fungi that has been variously called phycomycosis, subcutaneous phycomycosis, mucormycosis, entomophthoromycosis, hyphomycosis destruens, and others. The term phycomycosis became obsolete when the old two-kingdom system (*Plantae* and *Animalia*) for classifying all living things was replaced by the currently widely accepted five-kingdom system developed and espoused by Whittaker and Margulis (1978): *Monera, Proctoctista, Fungi, Plantae,* and *Animalia.*

In abandoning the two-kingdom system, the division *Thallophyta,* one of the major divisions of the old plant kingdom, together with the class *Phycomycetes* of the subdivision *Fungi,* became nomenclatural relics. The fungi, which had formerly been components of the class *Phycomycetes,* were reclassified in 12 phyla and dispersed in the kingdoms *Proctoctista* (*Labyrinthulamycota, Acrasiomycota, Myxomycota, Plasmodiophoromycota, Hyphochytridiomycota, Chytridiomycota, Oomycota*) and *Fungi* (*Zygomycota, Ascomycota, Basidiomycota, Deuteromycota, Mycophycophyta*) (Margulis and Schwartz, 1982). On the basis of this fundamental change, the term phycomycosis was replaced by zygomycosis (Ajello et al., 1976). Objection was also raised to using the names mucormycosis and entomophthoromycosis for diseases caused by the members of the orders *Mucorales* and *Entomophthorales.* They are all zygomycetes, and using the all-encompassing term zygomycosis simplifies disease nomenclature.

The synonym "hyphomycosis destruens" merits special attention, since it had been given to an equine disease caused by *Hyphomyces destruens,* a supposed phycomycete (Bridges and Emmons, 1961). However, *H. destruens* was discovered to be a *Pythium* species by Austwick and Copland (1974) and finally described and validly named *P. insidiosum* by de Cock et al. in 1987. The *Pythium* species are members of the order *Peronosporales,* belonging to the phylum *Oomycota* of the kingdom *Proctoctista* and thus are not members of the kingdom *Fungi.*

Zygomycosis in most cases is acquired by inhaling the airborne spores of the causative agents that are produced in soil or decaying organic matter. Less commonly, direct inoculation through the skin by some traumatic incident is the mode of infection. Other animals, including insects, are also victims of the pathogenic zygomycete. The disease is not contagious. The environment is the common source of all infections.

Etiologic Agents

The class *Zygomycetes* encompasses two orders that are of medical importance: *Mucorales* and *Entomophthorales.* These are separated from each other primarily on the basis of their asexual reproduction. Members of the *Mucorales* produce sporangiospores in closed sporangia, whereas members of the *Entomophthorales* produce sporangiola (unicellular sporangia) that are forcibly discharged. Members of the genus *Pilobolus,* classified in the family *Pilobolaceae,* are the only *Mucorales* that produce sporangia that are shot from their sporangiophores. The genera and species of molds that have been incriminated as agents of zygomycosis in humans are listed in Table 1. The many synonyms that some of them have accumulated are listed in Table 2.

Clinical Manifestations

Zygomycosis in many instances is an opportunistic disease; developing in persons suffering from some underlying condition or disease. But in some instances, the predisposing conditions are inapparent and remain unknown.

Because of the broad spectrum of clinical expression, it is best to describe zygomycosis by its four basic types: rhinocerebral, cutaneous, subcutaneous, and pulmonary and disseminated.

Rhinocerebral Zygomycosis

This dramatic form of the disease is most frequently encountered in persons with cases of acute diabetes. In patients in a metabolic acidotic state, the disease may be fulminating, causing death within a few days. Infections are initiated in the nasal turbinates and paranasal sinuses. They spread rapidly to the orbital tissues, progressing to the forepart of the brain. Clinical signs include a thick, dark nasal discharge and swelling of orbital tissue. When the eyes become in-

TABLE 1. Etiologic agents of zygomycosis

Order: *Mucorales*	Order: *Entomophthorales*
Absidia corymbifera	*Basidiobolus ranarum*
Apophysomyces elegans	*Conidiobolus*
Cunninghamella bertholletiae	*C. coronatus*
Mucor ramosissimus	*C. incongruus*
Rhizomucor pusillus	
Rhizopus	
R. arrhizus	
R. microsporus	
R. rhizopodiformis	
Saksenaea vasiformis	

TABLE 2. Synonyms of the pathogenic zygomycetes

Mucorales	
Absidia corymbifera	*Rhizopus arrhizus*
= *A. cornealis*	(Cont.)
= *A. italiana*	= *R. batatas*
= *A lichtheimii*	= *R. cambodia*
= *A ramosa*	= *R. delemar*
= *A. regnieri*	= *R. formosaensis*
= *A truchisii*	= *R. japonicus*
= *Lichtheimia cornealis*	= *R. kasanensis*
= *L. corymbifera*	= *R. maydis*
= *L. italiana*	= *R. nodosus*
= *L. italica*	= *R. oryzae*
= *L. ramosa*	= *R. ramosis*
= *Mucor cornealis*	= *R. rouxii*
= *M. corymbifer*	= *R. thermosus*
= *M. lichtheimi*	= *R. tritici*
= *M. ramosus*	= *R. trubinii*
= *Pseudoabsidia vulgaris*	= *R. usamii*
Rhizomucor pusillus	*Rhizopus microsporus*
= *Mucor pusillus*	= *R. bovinus*
= *Rhizomucor parasiticus*	*Rhizopus rhizopodiformis*
Rhizopus arrhizus	= *Mucor rhizopodiformis*
= *Mucor arrhizus*	
= *Mucor norvegicus*	= *R. chinensis*
= *Mucor oryzae*	= *R. cohnii*
	= *R. equinus*
Entomophthorales	
Conidiobolus coronatus	
= *Boudierella coronata*	
= *Conidiobolus villosus*	
= *Delacroixia coronata*	
= *Entomophthora coronata*	
Conidiobolus incongruus	
= *C. gonimodes*	

volved, ophthalmoplegia, protosis, and proptosis ensue. Retroorbital headaches are not uncommon, but fever is rare. Involvement of the hard palate with perforation may also occur.

The prime agent of rhinocerebral zygomycosis, *Rhizopus arrhizus*, has a marked propensity to invade blood vessels. In the brain, this results in necrosis of cell walls and thromboses with softening of the frontal lobes, leading to lethargy, coma, and death within a matter of days.

Cutaneous Zygomycosis

Primary infections are the rarest form of the disease. Hesseltine and Ellis (1964) described a chronic case that was finally correctly diagnosed 24 years after its inception. It had begun as a flat, erythematous lesion on the left cheek with undefined boundaries and a soft consistency. In time, the infection spread to involve both lips, the nasal septum, and chin. Definitive diagnosis of this unique case was based on the demonstration of hyaline, aseptate, broad hyphal elements in tissue sections, and isolation of *Mucor ramosissimus*.

Nosocomial cutaneous zygomycosis has been recorded in Latin America and the United States. The infections were secondary to underlying primary problems. Notable for their epidemiologic workup were the cases wherein gauze-dressed wounds were covered with adhesive tape that proved to be contaminated with *Rhizopus rhizopodiformis* (Bottone et al., 1979). The resulting infections caused the development of erythematous papular lesions with tissue necrosis. Histologic examination revealed the cardinal sign of zygomycosis—the presence of broad, aseptate hyaline mycelium in the biopsied tissue.

Rhizopus rhizopodiformis has also been found to cause skin infections around indwelling catheters in renal-transplant patients (Feuilhade et al., 1982). *Rhizopus arrhizus*, in addition to rhinocerebral infections, can infect the skin of acidotic diabetic patients. In 19 patients with breaks in their skin, *R. arrhizus* and *R. rhizopodiformis* were the etiologic agents encountered most frequently. *Rhizomucor pusillus* proved to be the most common agent, infecting eight patients with intact skin. However, all had severe, immunocompromising conditions (Winn et al., 1987).

Thus, primary cutaneous infections may occur through normal skin but most commonly occur via skin breaks resulting from surgery, insertion of indwelling catheters, or other trauma. The predisposing factors for cutaneous zygomycosis are complex. They may be any of a variety of underlying diseases, trauma, or the physiologic state of a given individual. Treatment with amphotericin B with or without attendant surgery has been found to be effective.

Subcutaneous Zygomycosis

Basidiobolus ranarum and *Conidiobolus coronatus*, two species of zygomycetes, both classified in the order *Entomophthorales*, cause this form of zygomycosis.

B. ranarum infections in humans, with few exceptions, have occurred in Africa and Asia. The infections most frequently involve the trunk and limbs, beginning as subcutaneous, painless, indurated nodules that slowly increase in size. Within a few months, a large, firm mass forms that is sharply limited from the adjacent tissue. It is believed that most infections are acquired directly through the skin by direct contact with contaminated soil. However, it has been speculated that insect bites may play a role in acquisition of infections (Martinson, 1982). Inhalation of the infectious spores of *B. ranarum* may have led to invasion of the maxillary sinus and palate of a patient in the United States (Dworzack et al., 1978).

Patients with chronic lesions caused by *Conidiobolus coronatus* usually have a history of nasal obstruction with slow but progressive involvement of the tissues of the nose, cheeks, and upper lip. The tumefactions begin in the nasal turbinates and spread to the paranasal sinuses and eventually to the subcutaneous tissues of the face. The resulting tumefactions are frequently dramatically disfiguring. Although the tumefactions cause discomfort, the patients do not complain of pain (Fromentin and Ravisse, 1977).

Pulmonary and Disseminated Zygomycosis

Primary pulmonary infections usually occur in compromised hosts—persons with an underlying disease or other predisposing factor. Diabetics and patients with leukemia and lymphoma are among the prime victims (Marchevsky et al., 1980). The infections result from inhaling the airborne spores of the pathogenic zygomycetes. Well-documented cases have been caused by *Absidia corymbifera, Conidiobolus incongruus, Cunninghamella bertholletiae, Rhizomucor pusillus, Rhizopus arrhizus, R. microsporus,* and *R. rhizopodiformis.*

This form of zygomycosis is progressive and usually results in death. The clinical signs include bronchitis and pneumonia. X-rays reveal nodular and nonspecific densities and infarctions. Bloody sputum, pain, and a plural friction rub are frequently experienced. Unless controlled, the primary pulmonary infection spreads, involving such other vital organs as the spleen, stomach, and heart.

Pathology

Agents of zygomycosis incite inflammatory tissue reactions with the formation of abscesses and suppurative necrosis. Blood vessel invasion is frequent, with resulting necrosis of blood vessel walls and thrombi. Mycelium, composed of hyaline and broad, rarely septate hyphae, is generally abundant in tissues. It is randomly branched and has a highly irregular diameter. Some hyphae may assume bizarre shapes with cell walls of varying thickness.

In most instances, the hyphae of the zygomycetes stain well with hematoxylin and eosin. However, they stain poorly with the periodic acid-Schiff and Gomori methenamine silver procedures. In many cases of zygomycosis, a diagnosis can be established, without cultures, solely on the basis of the distinctive mycelium of the zygomycetes. Occasionally, sporangia with well-delineated columellas and filled with spores will be encountered in tissue sections. These are apt to be found primarily in nasal and pulmonary tissues that are well aerated (Chandler et al., 1980).

Laboratory Diagnosis

Direct Examination of Clinical Material

This procedure is the simplest and fastest way to diagnose zygomycosis. Wet mounts in 10% potassium hydroxide of sputum, bronchial washings, cerebrospinal fluid, scrapings from nasal sinuses, oral mucosa, and portions of biopsied tissues should be made and examined under the microscope. The presence of broad—10 to 15 μm in diameter—rarely septate, hyaline mycelial elements is usually a sufficient basis to establish a diagnosis of zygomycosis. However, the etiology of the invading parasite can only be determined by its isolation and identification, because all of the genera and species of pathogenic zygomycetes have similar tissue forms.

Isolation of Etiologic Agents

Zygomycetes are readily isolated from clinical materials and biopsied tissue. They are not fastidious and grow well on the media widely used in medical mycology. It is wise to add antibacterial antibiotics such as chloramphenicol, gentamycin, or polymyxin B to the isolation medium to prevent overgrowth by contaminating bacteria. However, cycloheximide should not be used because of the sensitivity of the zygomycetes to this antifungal antibiotic, which is commonly used to inhibit the growth of saprophytic fungi in isolation media.

Multiple tubes should be inoculated per clinical specimen to increase the chances of isolation. Cultures should be incubated at 25°C and 37°C. The etiologic agents grow rapidly in contrast to the growth rates of pathogenic ascomycetes and deuteromycetes. The *Mucorales* quickly fill the isolation tubes with loose, flocose, grayish-white mycelium. In contrast, the colonies of the *Entomophthorales— Basidiobolus ranarum* and *Conidiobolus* species are waxy and leathery with a thin, flat growth. The obverse is smooth or grooved with radial furrows. Its color ranges from gray to pale yellow. At maturity, when the ballistic conidia are produced, the surface of the colonies becomes covered with a thin, cottony film composed of spores. In petri dish colonies, the inner surface of the dish's cover is peppered with the spores that have been hydrostatically shot from their sporangiophores.

Description of the Pathogenic *Mucorales*

Many of the *Mucorales* produce asexual, nonmotile sporangiospores in closed structures known as sporangia. Other members of the order form their spores in subdivisions of the sporangia. When elongated, they are referred to as merosporangia (viz. *Syncephalastrum racemosum*). When they are spherical and contain small groups of spores or single sporangiospores, they are referred to as sporangiola. Single-spored sporangiola in some families are incorrectly referred to as conidia when they are discharged as a whole. Through the fusion of homothallic or heterothallic gametangia, zygospores are produced. On the basis of their ornamentation, color, suspensor size, and ornamental projections, zygospores are of taxonomic value in determining families and genera. However, zygospores are seldom if ever used to identify the pathogenic *Mucorales,* since they are rarely formed in routine culture procedures.

Absidia corymbifera

This fungus is fast-growing, producing a thick mass of grayish mycelium. As sporangia are produced, the colony assumes a darker color. Branched runners or stolons are produced that bear rhizoids at their nodes and clusters of sporangiophores at the arched internodes. The pyriform sporangia range from 10 to 120 μm in diameter. A spherical or ovoidal columella arises from the apophysis formed by the swelling of the tip of the sporangiophore. Numerous sporangiospores are produced that are smooth and pale yellow to greenish. They are globose to ellipsoidal with sizes ranging from 2 to 3 to 3 to 5 μm. Since this species is heterothallic, zygospores are seldom encountered. This species is strongly thermotolerant, growing at temperatures as high as 52°C.

Apophysomyces elegans

At 25°C, plates of Sabouraud dextrose agar are overgrown within 7 days by the creamy white floccose aerial mycelium. Colonies grown on this medium and other isolation media are sterile. A procedure described by Ellis and Ajello (1982) is required to induce sporulation and permit identification of this species. The isolate should be grown on plates of a cornmeal-glucose-sucrose-yeast extract agar (cornmeal agar [Difco Laboratories, Detroit] 17 g; glucose, 2 g; sucrose, 3 g; yeast extract [Difco], 1 g; distilled water, 1 liter) for 7 days at 30°C. A 3-mm^2 block of agar with its adherent fungal growth should then be cut out of the culture plate and placed on previously sterilized and solidified 1% agar in tap water. This preparation should be incubated at 20°C for 5 to 7 days, by which time sporulation should have occurred.

Sporangiophores arise at the ends of stolonlike hyphae. They are not always opposite the underlying rhizoids. The dark-colored sporangiophores with their prominent apophyses range up to 300 μm long and 3 to 5 μm wide near their conspicuous funnel-shaped apophyses. The sporangiophores characteristically become thick-walled, with narrow lumens below the apophyses. When mature, the pyriform columellate sporangia are multispored with thin, hyaline, deliquescent cell walls. The smooth sporangiospores are oblong or rarely subglobose and subhyaline. They measure 5 to 8 μm by 4 to 5 μm. *A. elegans* is a thermotolerant zygomycete, growing well at incubation temperatures as high as 42°C.

Cunninghamella bertholletiae

Colonies are light gray and floccose and attain diameters of 6 cm in 3 days at 25°C on Sabouroud dextrose agar. Mycelium is hyaline, aseptate (septation infrequently developed), and broad (3 to 6 μm in diameter). Sporangiophores are erect, branched verticillately with terminal vesicles. Vesicles are globose to subglobose (30 to 65 μm wide), covered with spine-like denticles that bear spherical, oval to ellipsoidal sporangiola, 6 to 18 μm by 6 to 15 μm. Sporangiolar cell walls are smooth, rough, or finely echinulate. The sporangiospores have irregular and verticillate branches. This species is temperature-tolerant; most isolates grow well at 42° to 45°C.

Isolates of *C. bertholletiae* must be distinguished from those of the saprophytic *C. elegans*. The two species cannot be distinguished from each other on the basis of their morphologic characteristics. However, *C. elegans* does not produce zygospores when mated with positive and negative tester strains of *C. bertholettiae* nor does it grow at 40°C. Weitzman and Crist (1980a) identified *C. bertholettiae* isolates by their ability to grow at $\geq$ 40°C and their production of predominantly smooth sporangiospores along with rough and, more rarely, echinulate ones.

Mucor ramosissimus

Colonies of these species are floccose, cinnamon buff to grayish-olive. Sporangiophores are sympodially branched, with shorter branches, and occasionally circinate. Sporangia are globose, 15 to 17 μm wide; columella are subglobose to elongate. Sporangiospores are ellipsoidal, 7 to 8 μm long. Maximum temperature for growing this species is 36°C.

Rhizomucor pusillus

Colonies are floccose, gray to brownish. Monopodially branched sporangiophores have short rhizoids and small, spherical sporangia 60 to 70 μm in diameter, with dark walls. Columellae are subglobose to pyriform. Sporangiospores are 3 to 4 μm, globose to subglobose. Isolates are thermotolerant to 54° to 58°C. Zygospores, when produced, are like those of *R. miehei*.

Rhizopus arrhizus

Colonies are floccose, grayish-yellow to grayish-brown. Colonies produce abundant, conspicuous sporangiophores 1 mm or more long. Rhizoids have 4 to 8 branches, are hyaline to dark, and develop at the base of the sporangiophores. The globose, dark sporangia range from 50 to 200 μm in diameter; some are as small as 35 μm and some are as large as 250 μm. Columellae are globose. Sporangiospores are 6 to 8 μm wide, rhomboidal, and striate. Isolates are tolerant to 42° to 44°C.

Rhizopus microsporus

Colonies resemble those of *R. arrhizus;* they are floccose, and grayish-yellow to grayish-brown. Sporangiophores are 300 to 800 μm long, subtended by rhizoids 80 to 100 μm long. Sporangia are 40 to 150 μm in diameter, and columellae are subglobose. Sporangiospores are subglobose, rhomboidal, smooth, or slightly striated, 5 to 6 μm. Colonies will grow at temperatures as high as 46° to 48°C.

Rhizopus rhizopodiformis

Colonies are floccose, creamy white to dark gray. Sporangia are generally inconspicuous, 200 to 1,000 μm long, subtended by heavily branched rhizoids. Sporangia are 40 to 130 μm wide; columellae are elongated to pyriform with distinct apophyses. Sporangiospores are subglobose to somewhat rhomboidal, 4 to 6 μm wide, smooth to slightly striated. This species is also thermotolerant; it will grow at 50° to 52°C.

Saksenaea vasiformis

Growth on Sabouraud dextrose agar is floccose and white; it rapidly covers the surface of the growth medium. Hyphae are hyaline, broad, and basically aseptate. With rare exceptions, colonies are sterile: sporangiophores, sporangia, and sporangiospores are not produced. To identify this species, sporulation may be induced by growing suspected isolates of this species on plates of hay infusion agar (Ellis and Ajello, 1982). Growth is rapid but still sterile. A block of the sterile growth, however, when placed in a petri dish of sterile distilled water, will form the distinctive, flask-shaped sporangia with subtending dematiaceous rhizoids characteristic of this species. These are the diagnostic hallmarks of *S. vasiformis.*

Alternatively, a block of the sterile mycelial growth that develops on corn meal–sucrose–yeast extract agar can be placed on the surface of a plate of 1% water agar and incubated at 32°C for 5 to 7 days. The growth surrounding the block should then be examined for the presence of flask-shaped sporangia. Isolates can be grown in temperatures up to 44°C.

Description of the Pathogenic *Entomophthorales*

As its name implies, most of the fungi classified in this order are insect pathogens. However, several species of the genera *Basidiobolus* and *Conidiobolus* are capable of causing disease in lower animals and humans. A basic characteristic of the species classified in these two genera is their production of forcibly discharged unicellular sporangia. In contrast, none of the fungi in the *Mucorales* with the exception of the *Pilobolus* spp. produce ballistic spores.

Basidiobolus haptosporus

Isolates grow rapidly on Sabouraud dextrose agar at 25°C. Colonies are glabrous, flat, and waxy with radial folds, pale gray to yellowish on the surface, with a chamois reverse. Aerial mycelium develops as the colonies age. When grown in petri dishes, the inner surface of the lids becomes covered with an opaque film composed of the mold's forcibly discharged globose, single-celled sporangiola.

The cultures are composed of broad (5 to 15 μm wide), irregularly branched, aseptate to rarely septate mycelia. Single-celled sporangiola (erroneously referred to as conidia by some investigators) are produced at the tips of erect sporangiophores (30 to 300 μm long and 5 to 10 μm wide). At maturity, owing to the accumulation of cytoplasm, the tip of the sporangiophore (40 to 50 μm) becomes swollen, attaining a diameter of 25 to 30 μm. This subsporangial vesicle serves as the propulsive organ for the apical sporangiola. The spherical sporangiola range in diameter from 16 to 45 μm. They are projected for distances of 1 to 2 cm by the rupture and contraction of the subsporangiolal vesicle.

The sporangiophores of *B. haptosporus* are phototropic, and their formation is induced by exposure of the cultures to light.

The ballistic sporangiola on a favorable medium germinate to form (1) a mycelial filament, (2) another ballistic sporangiolum borne at the tip of an elongated germ tube, or (3) an adhesive sporangiolum.

Adhesive sporangiola are usually produced at the tips of erect sporangiophores (2 to 3 μm wide and 70 to 400 μm long). They are ellipsoid or pyriform and contain a single nucleus. They range from 17 to 85 μm in length. The sporangiola per se, without their appendages, measure 13.65 μm by 10 to 20 μm. They bear a slender apical appendage (4 to 10 μm by 2 to 3 μm) terminated by a knoblike adhesive tip. These adhesive spores are not forcibly discharged and are not phototropic. Once detached, they may germinate to form mycelium or another adhesive spore.

Sexual reproduction in *B. haptosporus* is homothallic. Zygospores are produced when adjacent hyphae that function as gametangia join. Distinctive paired beaks develop that persist on the resulting zygospores. The zygospores are thick-walled and spherical and range from 30 to 50 μm in diameter.

In old cultures of *B. haptosporus* thick-walled chlamydoconidia are produced. They vary considerably in shape and size, some measuring 20 by 40 μm.

Taxonomists disagree on the number of valid species classified in the genus *Basidiobolus*, since the criteria used to define and identify the species are varied and transitory. Some have gone so far as to advise that in view of this variability, it is best to consider *B. haptosporus* to be a synonym of *B. ranarum* along with *B. heterosporus*, *B. lacertae*, *B. magnus*, *B. meristosporus*, *B. myxophilus*, and *B. philippinensis*. The only other valid species in the genus would then be the nonpathogenic *B. microsporus* (McGinnis, 1980).

Following a study of the antigenic relationships among six species of *Basidiobolus* (*B. haptosporus*, *B. heterosporus*, *B. magnus*, *B. meristosporus*, *B. microsporus*, and *B. ranarum*), Polonelli and Morace (1984) came to a different conclusion. These investigators found that isolates of *Basidiobolus* from human cases of subcutaneous zygomycosis were identifiable as *B. haptosporus* and *B. meristosporus*. However, they made no conclusions regarding the taxonomic validity of the other species of *Basidiobolus*.

Conidiobolus coronatus

Isolates grow forming glabrous colonies that become covered with short, white aerial mycelium that bears sporangiophores. These produce primary single sporangiola that are forcibly discharged by a ballistic mechanism. The ballistic spores are 10 to 20 μm in diameter. They have a conspicuous papilla and have a tendency to produce a ring, or corona (hence the name of the species), of secondary sporangiospores. Under unfavorable growing conditions, the spores develop a surface coating of short, hairlike appendages.

Conidiobolus incongruus

This species closely resembles *C. coronatus* in its gross and microscopic development. However, it is separated from that species by its failure to produce villose sporangiospores and by the production of numerous thick-walled, spherical zygospores.

Immunology and Serology

In view of the ubiquity of potentially pathogenic zygomycetes in the environment and the relative rarity of zygomycosis cases, a high level of inherent immunity presumably exists in the human population. Benign, asymptomatic infections undoubtedly occur, but the absence of skin test reagents makes it impossible to test this hypothesis. Analogy with coccidioidomycosis and histoplasmosis, however, gives support to such a possibility.

Clinical cases of zygomycosis occur predominantly in immunologically impaired or compromised persons. Traumatic wounds also permit the entry and invasion of tissues by the infectious air- and soilborne spores of the zygomycetes. The severity of the infection depends on the etiologic agent involved and the underlying physiologic and immunologic status of the victim.

Although results of preliminary studies on the development of serologic diagnostic tests for zygomycosis have been reported, none have yet reached the specificity and sensitivity necessary for routine practical use (Marchevsky et al., 1980; Pierce et al., 1982).

Prognosis and Therapy

The prognosis in cases of zygomycosis greatly depends on the etiologic agent involved and the immunologic and physiologic status of the patient. In general, infections caused by *Basidiobolus haptosporus* and *Conidiobolus coronatus* are not life-threatening (Edington, 1964). Some subcutaneous infections, in fact, resolve spontaneously. However, most cases by the other zygomycetes require treatment. Iodides are the treatment of choice (KI, 30 mg/kg of body weight daily, for 1 month or more). Such antifungal agents as amphotericin B and nystatin have had little or no effect. However, ketoconazole has been reported effective in some cases of subcutaneous zygomycosis (Drouhet and Dupont, 1983). Of the two reported infections caused by *C. incongruus*, the first (in a 15-month-old boy) was cured with amphotericin B (Eckert et al., 1973; Gilbert et al., 1970). The second (in a 20-year-old Thai woman) rapidly disseminated and resulted in death, despite treatment with clotrimazole (Busapakum et al., 1983).

Zygomycosis, caused by members of the *Mucorales*, usually develops as a systemic infection in compromised persons. In 32 patients with zygomycosis, of whom one had an infection caused by *Absidia corymbifera*, four by *Rhizopus arrhizus*, and six by *R. rhizopodiformis*, their predisposing or associated factors were diabetes mellitus (10 patients), acute leukemia (10), carcinomas (3), and chronic lymphocytic leukemia, Cushing syndrome, diverticulitis, immaturity, macroglobulinemia, and ulcerative colitis (1 each) (Marchevsky et al., 1980). Underlying diseases are generally, although not exclusively, associated with specific types of infection—for example, acidotic diabetes mellitus with rhinocerebral zygomycosis, malnutrition with gastrointestinal zygomycosis.

Amphotericin B is usually the treatment of choice for zygomycosis caused by the members of the *Mucorales*. Surgery, in conjunction with chemotherapy, is frequently required to effect a cure. Response to the azole antifungal drugs has not been encouraging. Until more effective drugs become available, prognosis will remain poor.

Literature Cited

Ajello, L., D. F. Dean, and R. S. Irwin. 1976. The zygomycete *Saksenaea vasiformis* as a pathogen of humans with a critical review of the etiology of zygomycosis. Mycologia **69**:52–62.

Austwick, F. K. C., and J. W. Copland. 1974. Swamp cancer. Nature **250**:84.

Bottone, E. J., I. Weitzman, and B. A. Hanna. 1979. *Rhizopus rhizopodiformis:* emerging etiological agent of mucormycosis. J. Clin. Microbiol. **9**:530–537.

Bridges, C. H., and C. W. Emmons. 1961. A phycomycosis of horses caused by *Hyphomyces destruens*. J. Am. Vet. Med. Assoc. **138**:579–589.

Busapakum, R., U. Youngchaiyud, S. Sriumpal, G. Segretain, and H. Fromentin. 1983. Disseminated infection with *Conidiobolus incongruus*. Sabouraudia **21**:323–330.

Chandler, F. W., W. Kaplan, and L. Ajello. 1980. Color atlas and text of the histopathology of mycotic diseases. Year Book Medical Publishers, Chicago.

de Cock, A. W. A. M., L. Mendoza, A. A. Padhye, L. Ajello, and L. Kaufman. 1987. *Pythium insidiosum* sp. nov., the etiologic agent of pythiosis. J. Clin. Microbiol. **25**:344–349.

Drouhet, E., and B. Dupont. 1983. Laboratory and clinical assessment of ketoconazole in deep seated mycoses. Am. J. Med **74**(1B):30–47.

Dworzack, D. L., A. S. Follack, G. R. Hodges, W. G. Barnes, L. Ajello, and A. Padhye. 1978. Zygomycosis of the maxillary sinus and palate caused by *Basidiobolus haptosporus*. Arch. Intern. Med. **138**:1274–1276.

Eckert, H. L., G. H. Khoury, R. S. Pore, E. F. Gilbert, and J. R. Gaskell. 1972. Deep *Entomophthora* phycomycotic infection reported for the first time in the United States. Chest **61**:392–394.

Edington, G. M. 1964. Phycomycosis in Ibadan, West Nigeria. Trans. R. Soc. Med. Hyg. **58**:242–245.

Ellis, J. J., and L. Ajello. 1982. An unusual source for *Apophysomyces elegans* and a method for stimulating sporulation of *Saksenaea vasiformis*. Mycologia **74**:144–145.

Feuilhade, M., J. Revuz, and R. Touraine. 1982. Mucormycose cutanee primitive chez un transplante renal. Ann. Dermatol. Vener. **109**:765–766.

Fromentin, H., and P. Ravisse. 1977. Les entomophthoromycoses tropicales. Acta Trop. **34**:375–394.

Gilbert, E. F., G. H. Khoury, and R. S. Pore. 1970. Histological identification of *Entomophthora* phycomycosis. Arch. Pathol. **90**:583–587.

. Hesseltine, C. W., and J. J. Ellis. 1964. An interesting species of *Mucor, M. ramosissimus*. Sabouraudia **3**:151–154.

Marchevsky, A. M., E. J. Bottone, S. A. Geller, and D. K. Giger. 1980. The changing spectrum of disease, etiology, and diagnosis of mucormycosis. Hum. Pathol. **11**:457–464.

Margulis, L., and S. K. V. Schwartz. 1982. Five kingdoms. An illustrated guide to the phyla of life on earth. W. H. Freeman, San Francisco.

Martinson, F. D. 1972. Clinical, epidemiological and therapeutic aspects of entomophthoromycosis. Ann. Soc. Belg. Med. Trop. **52**:329–342.

McGinnis, M. R. 1980. Recent taxonomic developments and changes in medical mycology. p. 109–135. *In* M. Starr, J. L. Ingraham, and A. Balows (ed.) Annual review of microbiology. Annual Reviews Inc., Palo Alto, CA.

Pierce, P. F., S. L. Solomon, L. Kaufman, V. F. Garagusi, R. H. Parker, and L. Ajello. 1982. Zygomycete brain abscesses in narcotic addicts with serologicsl diagnosis. J. Am. Med. Assoc. **248**:2881–2882.

Polonelli, L., and G. Morace. 1984. Antigenic relationships of *Basidiobolus* species. Igiene Moderna **81**:1331–1343.

Weitzman, I., and M. Y. Crist. 1980a. Studies with clinical isolates of *Cunninghamella*. I. Mating behavior. Mycologia **71**:1024–1033.

Weitzman, I., and M. Y. Crist. 1980b. Studies with clinical isolates of *Cunninghamella*. II. Physiological and morphological studies. Mycologia **72**:661–669.

Whittaker, R. H., and L. Margulis. 1978. Protist classification and the kingdoms of organisms. Biosystems **10**:3–18.

Winn, R. E., P. D. Ramsey, and B. D. Adams. 1987. Cutaneous zygomycosis: report of two cases and review of literature. Rev. Infec. Dis. (in press).

SECTION IV

Parasitic Infections

CHAPTER 74

Acanthamoebiasis and Naegleriosis

G. S. VISVESVARA

Disease: Acanthamoebiasis (granulomatous amebic encephalitis, keratitis, sinusitis, bone infection); naegleriosis (primary amebic meningoencephalitis).

Etiologic Agents: *Naegleria fowleri; Acanthamoeba astronyxis, Acanthamoeba castellanii, Acanthamoeba culbertsoni, Acanthamoeba hatchetti, Acanthamoeba palestinensis, Acanthamoeba polyphaga, Acanthamoeba rhysodes.*

Source: Soil and fresh water.

Clinical Manifestations: Primary amebic meningoencephalitis: sudden onset, headache, high fever, nuchal rigidity, nausea, coma, and death. Granulomatous amebic encephalitis: insidious onset, usually chronic, confusion, apathy, dizziness, hallucinations, diplopia, seizures, hemiparesis, skin ulcers, coma, and death. Keratitis: inflammation of cornea, severe ocular pain, 360° or partial paracentral stromal ring infiltrate, recurrent breakdown of corneal epithelium, corneal lesion refractory to medication. Sinusitis: progressive nasal congestion, epistaxis, nasal tenderness, frontal headache. Bone infection: inflammation of tissue surrounding the bone, feculent smell.

Pathology: Primary amebic meningoencephalitis: acute hemorrhagic necrotizing meningoencephalitis, olfactory lobes severely affected. Granulomatous amebic encephalitis: subacute or chronic encephalitis with necrotizing granulomatous foci in the cerebellum, midbrain and brain stem, and posterior aspect of the brain.

Laboratory Diagnosis: Isolation of etiologic agents from the cerebrospinal fluid, brain tissue, lung tissue, corneal scrappings or corneal biopsy, or skin ulcers.

Epidemiology: Worldwide. Swimming, diving, and water sports for naegleriosis; immunosuppression and debilitation because of diabetes, alcoholism, or radiation therapy for *Acanthamoeba* encephalitis; use of contaminated contact lenses, contact lens-care solutions, or trauma to cornea for *Acanthamoeba* keratitis.

Treatment: Amphotericin B with miconazole and rifampin for naegleriosis; no information available for *Acanthamoeba* encephalitis; oral ketoconazole together with topical miconazole, or a combination of propamidine isethionate (Brolene) and neomycin have helped some patients, but corneal transplant seems to be necessary for most patients with *Acanthamoeba* keratitis.

Prevention and Control: Avoiding swimming, diving, and water sports in polluted ponds and lakes, proper maintenance and adequate chlorination of swimming pools, hot tubs, Jacuzzis, and the like, meticulous care of contact lenses and lens-care solutions, and avoiding use of contact lenses during swimming, diving, and water sports.

Description of Disease

Naegleriosis (primary amebic meningoencephalitis) is caused by a small free-living pathogenic ameboflagellate, *Naegleria fowleri*. Naegleriosis is an acute, fulminating, rapidly fatal disease involving principally the brain and meninges. Butt (1966) coined the term *primary amebic meningoencephalitis* to distinguish this disease from secondary brain abscesses caused by *Entamoeba histolytica* as a result of dissemination from a primary site such as the intestine.

In contrast, acanthamebiasis is caused by pathogenic free-living amebae belonging to the genus *Acanthamoeba*. The disease may manifest as (1) granulomatous amebic encephalitis, a chronic or subacute encephalitis with focal or multifocal necrosis of the brain tissue and formation of multinucleated giant cells and occasionally associated with papula or chronic ulceration of the skin; (2) keratitis, an indolent, chronic inflammation of the cornea leading to loss of visual acuity and blindness; (3) granulomatous sinusitis, a chronic inflammation of the mucosal lining of the nasal and paranasal sinus; and (4) bone infection.

Clinical Manifestations

Primary Amebic Meningoencephalitis

The symptoms of primary amebic meningoencephalitis consist of high fever, headache, rhinitis, nuchal rigidity, nausea, vomiting, drowsiness, coma and finally, death. Diplopia and convulsions may also occur. Death ensues within 5 to 10 days after the onset of symptoms. Primary amebic meningoencephalitis usually occurs in healthy children and young adults who have a history of swimming, diving, or water sports (Griffin, 1978; John, 1982; Martinez, 1985).

Granulomatous Amebic Encephalitis

Symptoms of granulomatous amebic encephalitis may not be apparent and develop gradually over time. They include confusion, apathy, dizziness, hallucinations, diplopia, seizures, hemiparesis, coma, and death. Some patients may also develop nodules, papules, or nonhealing skin ulcers. The disease typically occurs in immunocompromised, chronically ill, diabetic, or otherwise debilitated individuals. Patients with granulomatous amebic encephalitis usually have no history of contact with water before symptoms develop (Martinez, 1985).

Acanthamoeba Keratitis

Acanthamoeba keratitis is characterized by inflammation of the cornea, severe ocular pain, a characteristic 360° or partial paracentral stromal ring infiltrate, recurrent breakdown of the corneal epithelium, and a corneal lesion refractory to usual medications. Some degree of ocular trauma, exposure to contaminated water, or contact lens wear usually precedes development of the disease (Centers for Disease Control, 1986).

Acanthamoeba Sinusitis

Symptoms of *Acanthamoeba* sinusitis include progressive nasal congestion, epistaxis, nasal tenderness, and frontal headaches (Gonzalez et al., 1986).

Acanthamoeba Bone Infection

Acanthamoeba bone infection is characterized by an acute inflammation of the connective tissue surrounding the bone that fails to resolve, developing an external fistula that produces a feculent smell (Borochovitz et al., 1981).

Laboratory Diagnosis

Direct Examination

Samples suspected to contain *Naegleria* or *Acanthamoeba* should be examined directly as a wet mount with a microscope. This approach will permit rapid diagnosis and intervention with appropriate therapy. Cerebrospinal fluid (CSF) of patients with suspected infections should be centrifuged at 250 × *g* for 10 min to concentrate the organisms before making the wet mount. Although not necessary, the slide may be warmed to 37°C because heating will stimulate the amebae, especially if they are *Naegleria*, to move. In previous years only *N. fowleri* was detected antemortem in the CSF of patients with primary amebic meningoencephalitis. Recently, however, *Acanthamoeba* also has been detected in the CSF of a few patients (Visvesvara, 1987).

If amebae are seen on wet mounts, smears of the samples should be fixed in appropriate fixative (e.g., Schaudinn) and stained with trichrome or iron hematoxylin. Wright or Giemsa staining is less satisfactory. In the iron hemtoxylin-stained smears, the nuclei of the amebae are black, whereas the cytoplasm is grayish or bluish. In contrast, in trichrome-stained preparations, the nuclei stain red and the cytoplasm stains bluish-green. In Wright- or Giemsa-stained smears the nucleus is a delicate pink and the cytoplasm appears sky blue (Martinez, 1985).

The CSF from patients with primary amebic meningoencephalitis is usually grayish to yellow-white, but may be tinged with pink or even red due to the presence of red blood cells. White blood cell counts range from as low as 300/mm^3 to as high as 26,000/

mm^3, with a predominance of polymorphonuclear neutrophils (PMN). Glucose levels are reduced, whereas protein levels are high. Although these findings are similar to those seen in purulent bacterial meningoencephalitis, the two conditions can be differentiated by the absence of bacteria in primary amebic meningoencephalitis (Martinez, 1985).

In granulomatous amebic encephalitis due to *Acanthamoeba*, the CSF has characteristic pleocytosis with lymphocytes predominating. Glucose levels are usually normal, but protein levels are increased (Martinez, 1985).

Histologic Diagnosis

If present, amebae can be identified in fixed and stained brain sections by their characteristic nucleus, which possesses a large, darkly staining nucleolus. The nucleolus is usually surrounded by a clear space, giving it the appearance of a "target" or "bull's eye" (Martinez, 1985).

Primary amebic meningoencephalitis may be differentiated from granulomatous amebic encephalitis by the sizes of amebae, their nuclei, and their nucleoli. The characteristic feature, apart from the obvious difference in size, is the nucleus-to-nucleolus ratio, which is 2.7 in *Acanthamoeba* and 1.9 in *Naegleria*. In contrast to *Acanthamoeba*, cysts of *N. fowleri* are generally not seen in the brain tissue of patients with primary amebic meningoencephalitis (Visvesvara, 1987).

It may not be possible to identify amebae because of poor fixation and staining. In such cases immunofluorescent or immunoperoxidase staining of tissue sections can be carried out to identify the species using rabbit antisera prepared against various species of *Naegeria* and *Acanthamoeba* (Visvesvara, 1987).

Etiologic Agents

Only one species of *Naegleria*, *N. fowleri*, is known to cause primary amebic meningoencephalitis. In contrast, several species of *Acanthamoeba*, including *A. astronyxis*, *A. castellanii*, *A. culbertsoni*, *A. hatchetti*, *A. palestinensis*, *A. polyphaga*, and *A. rhysodes*, are known to cause acanthamoebiasis.

Morphology

The trophozoites of *Naegleria* are smaller than those of *Acanthamoeba*, measuring from 10 to 15 μm. They exhibit eruptive locomotion by producing smooth hemispherical bulges anteriorly and a sticky uroid with several trailing filaments posteriorly. The

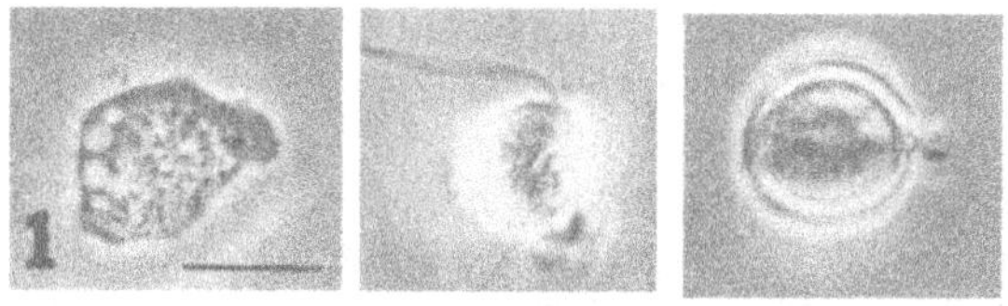

FIG. 1. Tropohozoite, flagellate, and cyst stages of *Naegleria fowleri*. Bar = 10 μm.

trophozoites may transform into a transient, pear-shaped biflagellate stage due to certain changes in the environment. The trophozoite differentiates into a spherical, smooth-walled cyst measuring 7 to 15 μm when environmental conditions are unfavorable. The trophozoite, the flagellate, and the cyst forms are uninucleate and the nucleus is characterized by the presence of a large, centrally located, densely staining nucleolus (Fig. 1).

The trophozoites of *Acanthamoeba* measures 15 to 45 μm and have fine, tapering hyaline projections on the body surface called *acanthopodia*. They move slowly by producing broad hemispherical bulges in the directions of the movement. The acanthopodia are constantly formed and resorbed. When environmental conditions become adverse, the trophozoites differentiate into double-walled cysts measuring 10 to 25 μm. The cysts are usually characterized by an outer wrinkled ectocyst and an inner stellate, polygonal, round, oval or triangular endocyst. Both trophozoites and cysts are uninucleate, and the nucleus is characterized by the presence of a large, centrally located, densely staining nucleolus (Fig. 2).

Collecting, Handling, and Storing of Specimens

Personnel responsible for collecting, handling, storing, and processing specimens should be proficient in aseptic techniques and must take appropriate precautions, such as wearing surgical masks and gloves. Specimens such as CSF, brain tissue, lungs, skin lesions, and cornea should be processed promptly.

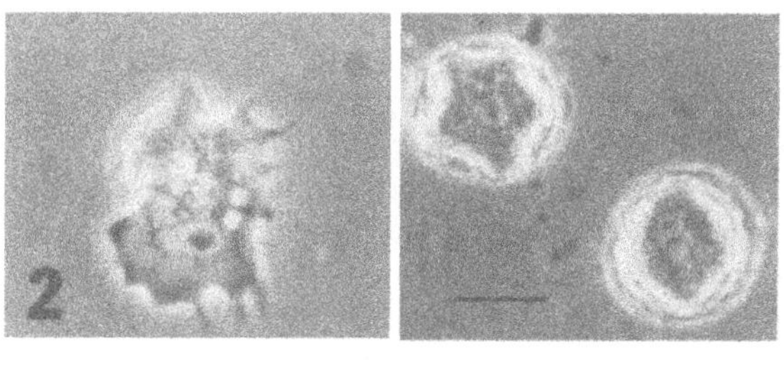

FIG. 2. Trophozoite and cyst stages of *Acanthamoeba castellanii*. Bar = 10 μm.

During the interim, however, they should either be kept at room temperature (24°C) or refrigerated, but not frozen. They should be handled in a biologic safety cabinet or a laminar flow hood. Unused CSF and biopsy or tissue materials and serum samples should be stored at −20°C for the detection of antigen and/or antibody. Portions of unused tissue should be preserved in 10% neutral buffered formalin for future histologic studies.

Isolation and Culture

Both *Acanthamoeba* and *Naegleria* can be easily cultured from biologic samples (e.g., CSF, corneal smears, brain, or corneal biopsy specimen) as well as from the environment (e.g., pond, lake, or river water and soil samples). Liquid specimens should be centrifuged at 250 × *g* for 10 min to concentrate the amebae/flagellates/cysts, the supernatant removed by suction, and the sediment mixed with about 0.5 ml of ameba saline (Page, 1967). The resuspended sediment can then be pipetted onto the center of a nonnutrient agar plate that has been covered with a suspension of *Escherichia coli*. The nonnutrient agar plate is prepared by dissolving 1.5 g Difco agar in ameba saline and autoclaving the mixture at 15 psi for 15 min. When the agar has cooled to about 50°C, 10 to 15 ml is poured aseptically into a 100-mm petri dish and kept at room temperature till the agar gels. Agar plates prepared by these procedures can be stored at 4°C for up to 3 months. Tissue or biopsy specimens should be triturated with about 0.5 ml of ameba saline and a few drops of the triturated material deposited on the center of the plate. The plates should be incubated at 37°C if the sample originates from the CSF or brain and at 30°C if the sample originates from the cornea, contact lens-care solutions, or the environment. Plates should be examined daily for 10 days with a microscope using a low-power (10×) objective. If amebae are seen, the area should be marked with a wax pencil. The agar should be cut with a small wire spatula and transferred face down onto another nonnutrient agar plate coated with bacteria and incubated as described previously. Both *Acanthamoeba* and *Naegleria* can be cultivated and maintained indefinitely in this fashion. After several days of incubation, the amebae start differentiating into cysts; both trophozoites and cysts can usually be seen after 4 to 5 days of incubation.

Identification of the amebae to the generic level is based on their morphology and motility, the characteristic features of the cyst, and an enflagellation experiment. This experiment is performed by mixing sedimented CSF, water containing the amebae, or a loopful of amebae from the agar plates with 0.5 to 2 ml of sterile distilled water. After incubating the tube at the appropriate temperature (37° or 30°C) for 2 to 3 h, several drops are removed and placed in the center of a slide, a coverslip is applied, and the material is examined microscopically for flagellates. If the sample contains *Naegleria,* it is not uncommon to observe free-swimming flagellates within 2 to 3 h of incubation.

Interactions with Cell Cultures

Clinical material may also be inoculated onto monolayers of cultured cells such as monkey kidney (Vero), MRC-human embryonic lung cells, and Hela cells. If present, amebae will grow at the expense of the cell culture and produce cytopathic effects (CPEs) such as vacuolation of the cells, disruption of the cell cytoplasm, nuclear pyknosis, and finally total disruption of the cell sheet (Visvesvara and Balamuth, 1975; Visvesvara and Callaway, 1974).

Animal Pathogenicity Tests

Mice, guinea pigs, rabbits, dogs, sheep, and monkeys have been used as experimental animals for determining the infectivity, pathogenicity, and virulence of ameba isolates (Griffin, 1978; John, 1982; Martinez, 1985). The most common and widespread method of infection is by the intranasal route, which is also the natural route for the human *Naegleria* infections. Mice are anesthetized with ether, and 0.01 to 0.02 ml of the ameba suspension containing 1,000 to 10,000 amebae is instilled into the nostril. Mice develop typical symptoms of meningoencephalitis, such as bristling of fur, aimless circling, partial paralysis, and coma, and die within 4 to 8 days if *Naegleria* amebae were present in the inoculum. If *Acanthamoeba* were used for the intranasal instillation, the mice die of encephalitis or pneumonitis within 5 days to several months. At necropsy, the brain is swollen, edematous, and necrotic, and numerous amebae are apparent. Once inside the brain, the amebae may secrete an enzyme or toxin that breaks down the surrounding tissue, or ingest bits of host cytoplasm and erode the host cell, or both. Unlike *Acanthamoeba, Naegleria* has not been known to produce cysts in the brain.

Antigenic and Isoenzyme Analysis

Acanthamoeba and *Naegleria* are antigenically distinct; however, species within the genera share certain common antigens. Although *N. fowleri* shares certain common antigens with other species of *Naegleria,* it is significantly different from other species. These differences have clearly been demonstrated by the gel diffusion and immunoelectrophoresis reactions (Visvesvara and Healy, 1975; Willaert, 1976). With these tests, one can differentiate the pathogenic *Naegleria fowleri* from the other *Naegleria* spp. Antigenic differences also exist among *Acanthamoeba* spp. (Visvesvara and Balamuth, 1975; Willaert, 1976).

Electrophoretic patterns of isoenzymes in cell-free extracts of *N. fowleri* reveal that irrespective of their geographic origin, all strains of *N. fowleri* are basically homogeneous. *N. fowleri* can be easily distinguished from the other species of *Naegleria* by comparing the patterns of leucine aminopeptidase, propionyl esterase, malate dehydrogenase, acid phosphatase, and glucose phosphate isomerase (DeJonckheere, 1982; Pernin, 1985).

Isoenzymes of *Acanthamoeba* have also been analyzed (Costas and Griffiths, 1984; Daggett et al., 1982), but the results are not as clear-cut as for *Naegleria*.

Pathology and Pathogenesis

Naegleria fowleri

Naegleria fowleri causes an acute hemorrhagic necrotizing meningoencephalitis, generally leading to death within a week. At necropsy, the brain is inflamed and edematous. The olfactory lobes are usually severely affected, with many hemorrhagic and necrotic areas filled with numerous amebae and neutrophils (Fig. 3). The reaction in the deeper regions of the cortex is somewhat less pronounced. Amebae are usually found in pockets surrounded by a clear space with few or no neutrophils. Inflammatory response appears not to have caught up with the invading parasites. Amebae can be easily recognized by their characteristic nuclear morphology and can usually be visualized by direct or indirect immunofluorescence (IIF) or immunoperoxidase (IIP) techniques.

Acanthamoeba spp.

Acanthamoeba spp. produce a necrotizing granulomatous, subacute or chronic encephalitis, which results in death within 8 days to several months (Martinez, 1985). The lesions of *Acanthamoeba* infections are most prominent in the cerebellum, midbrain and brainstem, and posterior aspect of the basal ganglion. The olfactory lobes and the anterior region of the cerebral hemispheres are affected to a lesser extent. Amebae are usually present in the affected tissue and are characterized by their larger size and clearly staining nucleus, with a prominent, densely staining nucleolus (Fig. 4). One of the characteristic features of granulomatous amebic encephalitis is the presence of amebic cysts in the brain tissue (Fig. 4). The species of ameba is identified by immunohistochemical procedures.

Serology

Serologic tests are not useful for the diagnosis of *Naegleria* infections because most patients die too soon to produce antibody. In one patient who recovered from primary amebic meningoencaphalitis,

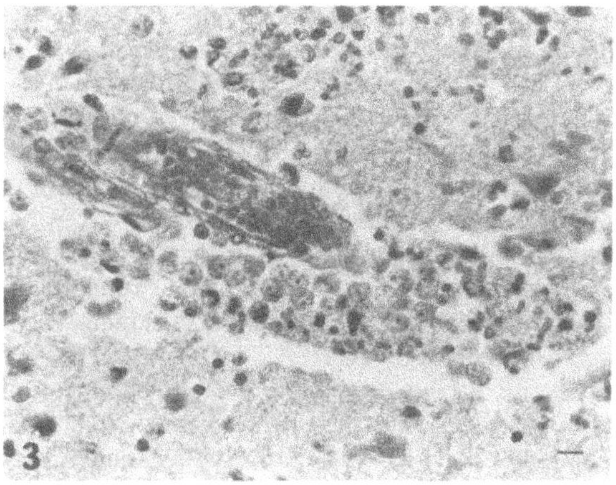

FIG. 3. Section of the brain of a patient with primary amebic meningoencephalitis due to *Naegleria fowleri*. Hematoxylin and eosin. Bar = 10 μm.

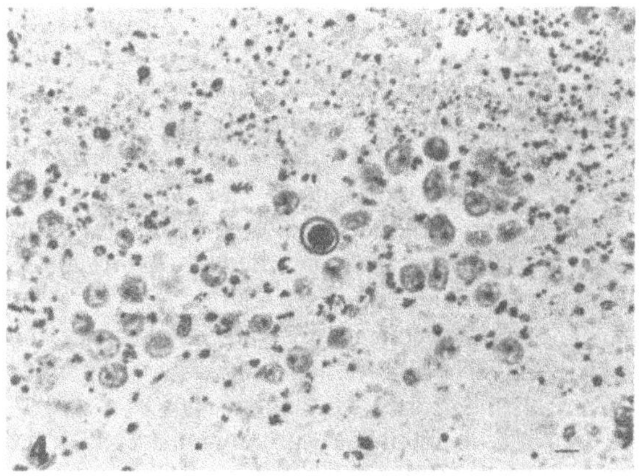

FIG. 4. Section of the brain of a patient with granulomatous amebic encephalitis due to *Acanthamoeba culbertsoni*. Note the presence of double-walled cyst. Hematoxylin and eosin. Bar = 10 μm.

however, an IIF titer of 4,096 against *N. fowleri* was demonstrated in the serum 7, 10, and 42 days after admission to the hospital (Seidel et al., 1982). Presence of antibodies to *Acanthamoeba* in patients suffering from upper respiratory tract distress (Eldridge and Tobin, 1967) indicates that many inapparent infections due to *Acanthamoeba* may exist in nature. Kenney (1971) detected CF antibodies to *Acanthamoeba culbertsoni* strain A-1 in 2 of 1,000 sera from randomly selected patients. One of the patients who had antibody titer to *Acanthamoeba* died of granulomatous amebic encephalitis. Recently, an increase in the ameba immobilization antibody titer over a 16-month period was demonstrated in the serum of a Nigerian patient who partially recovered from an *A. rhysodes*-induced disease of the central nervous system (Visvesvara, 1987). IIF and precipitin antibody has also been demonstrated in the sera of patients suffering from *Acanthamoeba* keratitis (Jones et. al., 1975; Theodore et. al., 1985).

Epidemiology

These small, free-living amebae have a worldwide distribution. They have been isolated from soil, fresh water, sea water, meltwater from Antarctica, the latex of *Euphorbia* plants, swimming pools, physiotherapy pools, hot tubs, dialysis units, cooling towers of electric and nuclear power plants, and recently from contact lens-care solutions. *Acanthamoeba* has also often been found as a contaminant of bacterial and fungal cultures as well as mammalian cell cultures. The contaminated mammalian cell cultures exhibited cytopathic changes similar to those produced by certain viruses; hence, these amebae were mistaken for transformed cell types presumably containing viruses and were erroneously designated as lipovirus and Ryan virus. *Acanthamoeba* and *Naegleria* have also been isolated from the nose and throat of humans with respiratory ailments as well as from apparently healthy individuals, which suggests that the cysts of these amebae are airborne. These observations indicate the tremendous public health importance of these organisms.

The concept that the ubiquitous, small, free-living amebae could have the potential to become human pathogens was developed by Culbertson (1961). In 1958, he and his colleagues injected into mice and monkeys a tissue culture fluid believed to contain an unknown simian virus. The animals died of meningoencephalitis, and amebae were demonstrated in the brain sections. Culbertson and his colleagues also isolated amebae from an aliquot of the tissue culture medium that had been stored frozen. They identified this isolate as *Acanthamoeba* sp. strain Lilly A-1, which is now designated as *Acanthamoeba culbertsoni*. They also isolated from soil and fresh water many strains of *Acanthamoeba* possessing a wide range of virulence, from none to very high virulence for mice. On the basis of his experiments, Culbertson suggested that human infections due to these amebae might exist in nature. Fowler and Carter (1965) and Butt (1966) reported cases of human amebic meningoencephalitis in Australia and the United States, respectively. Although at that time the organism was suspected to be *Acanthamoeba*, it is now generally considered to be *N. fowleri*, on the basis of epidemiology and clinical picture.

Almost all cases of *Naegleria* meningoencephalitis have occurred in children or young adults in good health who had a history of swimming or contact with fresh water before their illness. It is now fairly well established that *N. fowleri* amebae present in the water enter via the nasal passages of the host during swimming, diving, or other water sports. The amebae quickly cross the cribriform plate and invade the olfactory lobes of the brain. In contrast, most patients with *Acanthamoeba* infections are immunosuppressed, diabetic, alcoholic, or receiving radiation therapy. It is also possible that some individuals, especially those who are immunologically compromised, may inhale *Acanthamoeba* cysts that are present in airborne dust. Acanthamoebae have been isolated from dust in the air and external environment (Martinez, 1985).

In addition to causing granulomatous amebic encephalitis, *Acanthamoeba* causes diseases of the eye. The first case of *Acanthamoeba* keratitis in the United States was reported in a rancher from Texas in 1973 (Jones et al., 1975), but only 31 patients were diagnosed with this condition from 1973 to 1985. However, 24 cases occurred in 9 months from October 1985 to June 1986, which, according to The Centers for Disease Control, represents a striking increase over those reported in previous years. Of these patients, 20 (83%) used contact lenses, and 10 (50%) used homemade saline to clean their lenses. Apparently a similar increase has been observed in the use of contact lenses, particularly during the past 5 years, from 14.5 million in 1980 to 23.1 million in 1985 (Centers for Disease Control, 1986). It is believed that trauma to the eyes or the use of contact lenses or contaminated lens-cleaning/soaking solution may predispose the wearer to developing *Acanthamoeba* keratitis.

Several strains (two of which killed mice) of *Acanthamoeba* have also been isolated from stool samples of patients with diarrhea. It is quite likely that, under certain conditions, some strains of *Acanthamoeba* may cause transitory gastrointestinal infections.

Treatment, Prevention, and Control

Treatment

Pathogenic *N. fowleri* is extremely sensitive to amphotericin B, an antifungal agent. Rapid identification of the etiologic agent along with aggressive therapy appears to be the key to success. At least three patients with *Naegleria* have survived primary amebic meningoencephalitis after receiving intrathecal and intravenous amphotericin B therapy alone or in combination with miconazole and rifampin (Seidel et. al., 1982). Very little information is available on the efficacy of any drugs in granulomatous amebic encephalitis. A Nigerian patient apparently made a partial recovery after he was treated with sulfamethazine (Visvesvara, 1987). *Acanthamoeba* spp. are resistant to most of the commonly used antibacterial, antifungal, antiviral, and antiprotozoal agents. According to recent studies however, combinations of oral ketoconazole and topical miconazole or of topical propamidine isethionate (Brolene) and neomycin have helped some individuals with *Acanthamoeba* keratitis, but surgical intervention seems to be necessary for most patients (Centers for Disease Control, 1986; Theodore et al., 1985).

Prevention and Control

Naegleria trophozoites and cysts are susceptible to chlorine at 0.5 mg/liter, and 99% of the organisms are killed provided the contact time is at least 30 min. Acanthamoebae are resistant to killing by freezing, dessiccation, a variety of antimicrobial agents, and levels of chlorine that are routinely used to disinfect municipal drinking water, swimming pools, and hot tubs. Other disinfectants, such as iodine, ozone, and Deciquam 222, are not active against these amebae at concentrations routinely used to kill bacteria (Martinez, 1985). Recent studies indicate that thermal disinfection systems for contact lenses are superior to cold chemical disinfection in preventing the growth of *Acanthamoeba* (Centers for Disease Control, 1986).

Infections due to these amebae, especially *N. fowleri*, can be minimized by avoiding swimming, diving, and water sports in organically and thermally polluted lakes and ponds. If one must swim, the head should not be held under water while swimming and nose clips should be worn while diving or splashing in water. Contact lenses should not be worn while swimming or using hot tubs. Those using swimming pools, hot tubs, and the like should take a shower and wash the feet before entering these recreational facilities. Swimming pools, whirlpool baths, and Jacuzzis must be properly maintained, adequately chlorinated, and periodically checked for any possible cracks or crevices that may adversely affect the functioning of the pool. Contact lenses and contact lens-care solutions should be kept meticulously clean, and contact lens wearers should rigidly follow the directions of the manufacturers and/or their eye care physicians.

Literature Cited

Borochovitz, D., A. J. Martinez, and G. Patterson, 1981. Osteomyelitis of a bone graft of the mandible with *Acanthamoeba castellanii* infection. Hum. Pathol. **12**:573–576.

Butt, C. G. 1966. Primary amebic meningoencephalitis. N. Engl. J. Med. **274**:1473–1476.

Centers for Disease Control. 1986. *Acanthamoeba* keratitis associated with contact lenses—United States. Morbid. Mortal. Weekly Rep. **35**:405–408.

Costas, M., and A. J. Griffiths. 1984. The esterases and acid-phosphatases of *Acanthamoeba* (Amoebida, Acanthamoebidae). Protistologica **20**:33–41.

Culbertson, C. G. 1961. Pathogenic *Acanthamoeba* (*Hartmannella*). Am. J. Clin. Pathol. **35**:195–202.

Daggett, P. M., T. K. Sawyer, and T. A. Nerad. 1982. Distribution and possible interrelationships of pathogenic and nonpathogenic *Acanthamoeba* from aquatic environments. Microb. Ecol. **8**:371–386.

DeJonckheere, J. 1982. Isoenzyme patterns of pathogenic and non-pathogenic *Naegleria* spp. using agarose isoelectric focusing. Ann. Microbiol. (Inst. Pasteur) **133**:319–342.

Eldridge, A. E., and J. O. H. Tobin. 1967. Ryan virus. Br. Med. J. **1**:299.

Fowler, M., and R. F. Carter. 1965. Acute pyogenic meningitis probably due to *Acanthamoeba* sp.: a preliminary report. Br. Med. J. **2**:740–742.

Gonzalez, M. M., E. Gould, G. Dickinson, A. J. Martinez, G. S. Visvesvara, T. J. Cleary, and G. Henseley. 1986. Acquired immunodeficiency syndrome associated with *Acanthamoeba* infection and other opportunistic organisms. Arch. Pathol. Lab. Med. **110**:749–751.

Griffin, J. L. 1978. Pathogenic free-living amoebae, p. 507–549. *In* J. P. Kreier (ed)., Parasitic protozoa, vol. 2. Academic Press, New York.

John, D. 1982. Primary amebic meningoencephalitis and the biology of *Naegleria fowleri*. Annu. Rev. Microbiol. **36**:101–123.

Jones, D. B., G. S. Visvesvara, and N. M. Robinson. 1975. *Acanthamoeba polyphaga* keratitis and *Acanthamoeba* uveitis associated with fatal meningoencephalitis. Trans. Ophthalmol. Soc. U.K. **95**:221–232.

Kenney, M. 1971. The micro-Kolmer complement fixation test in routine screening of soil amoeba infection. Health Lab. Sci. **8**:5–10.

Martinez, A. J. 1985. Free-living amebas: natural history, prevention. diagnosis, pathology, and treatment of disease, p. 156. CRC Press, Inc., Boca Raton, Florida.

Page, F. C. 1967. Taxonomic criteria for limax amoebae with descriptions of 3 new species of *Hartmanella* and 3 of *Vahlkampfia*. J. Protozool. **14**:499–521.

Pernin, P., M. Cariou, and A. Jacquier. 1985. Biochemical identification and phylogenetic relationships in free-living amoebas of the genus *Naegleria*. J. Protozool. **32**:592–603.

Seidel, J. S., P. Harmatz, G. S. Visvesvara, A. Cohen, J. Edwards, and J. Turner. 1982. Successful treatment of primary amebic meningoencephalitis. N. Engl. J. Med. **306:**346–348.

Theodore, F. H., F. A. Jakobiec, K. B. Juechter, P. Ma, R. Troutman, P. M. Pang, and T. Iwamoto. 1985. The diagnostic value of a ring infiltrate in acanthamebic keratitis. Ophthalmology **92:**1471–1479.

Visvesvara, G. S. 1987. Laboratory diagnosis, p. 193–215. *In* I. Rondanelli (ed.), Amphizoic amoebae. Piccin Nuova Libraria, Pavia, Italy.

Visvesvara, G. S., and W. Balamuth. 1975. Comparative studies on related free-living and pathogenic amebae with special reference to *Acanthamoeba*. J. Protozool. **22:**245–256.

Visvesvara, G. S., and C. S. Callaway. 1974. Light and electron microscopic observations on the pathogenesis of *Naegleria fowleri* in mouse brain and tissue culture. J. Protozool. **21:**239–250.

Visvesvara, G. S., and G. R. Healy. 1975. Comparative antigenic analysis of pathogenic and free-living *Naegleria* species by the gel diffusion and immunoelectrophoresis techniques. Infect. Immun. **11:**95–108.

Willaert, E. 1976. Etude immuno-taxonomique des genres *Naegleria* et *Acanthamoeba* (Protozoa: Amoebida). Acta Zool. Pathol. **65:**1–239.

CHAPTER 75

African Trypanosomiasis

E. G. LONG

Disease: Gambian trypanosomiasis or West African sleeping sickness, Rhodesian trypanosomiasis or East African sleeping sickness.
Etiologic Agents: *Trypanosoma brucei gambiense, T.b. rhodesiense.*
Source: Humans, game animals, domestic animals.
Clinical Manifestations: Chancre at site of inoculation, intermittent fever, lymphadenopathy, headache, arthralgia, emaciation, pruritus, irritability, insomnia, severe neurologic disturbances.
Pathology: Lymphadenitis, anemia, immune-complex disease, hypocomplementemia, pancarditis, vasculitis, meningoencephalitis.
Laboratory Diagnosis: Demonstration of trypanosomes in blood, lymph node aspirate, or cerebrospinal fluid; cultivation in an artificial medium, animal inoculation, serodiagnosis.
Epidemiology: *Trypanosoma b. gambiense*—West Africa from Senegal south to Angola and east to Lake Victoria. *Trypanosoma b. rhodesiense*—East Africa from Sudan south to Mozambique and west to Lake Victoria. The disease is transmitted by male and female flies of the genus *Glossina* (tsetse flies) or congenitally.
Treatment: For early disease, suramin or alternatively pentamidine isethionate (for *T.b.* gambiense); for late disease with central nervous system involvement, melarsoprol.
Prevention and Control: Clearing of vegetation that provides habitats for the flies, application of residual insecticide to resting sites of the flies, insect traps, active surveillance, and prompt diagnosis and treatment.

Description of the Disease

African trypanosomiasis, or African sleeping sickness, is a spectrum of systemic disease, ranging in severity from fulminant to mild and chronic, caused by two species of hemoflagellates, *Trypanosoma brucei rhodesiense* and *T.b. gambiense*. Parasites are transmitted during the bite of a blood-seeking fly of the genus *Glossina*, better known as tsetse fly; both sexes are hematophagous. Introduced trypanosomes multiply at the site of the bite and may cause an inflammatory nodule known as a trypanosomal chancre. In a few days, the parasites are disseminated by the lymphatic system and enter the bloodstream, causing intermittent fever, headache, malaise, fatigue, night sweats, pruritus, and facial edema; later the parasites invade the lymph nodes and cause generalized lymphadenopathy. In advanced cases, the disease progresses to the central nervous system, and the patient exhibits irritability, epileptiform attacks, somnolence, and abnormal behavior. Coma is common in late stages of the disease, hence the term "sleeping sickness." Death may result from congestive heart failure following pancarditis, meningitis, asthenia, or, more often, a secondary infection, usually pneumonia, due to the immunosuppresive effects of trypanosomiasis.

Communicability and Transmissibility

The principal vectors of *T.b. rhodesiense* belong to the *Glossina morsitans* group of flies, and the vectors of *T.b. gambiense* belong to the *G. palpalis* and *G. fusca* groups. Others among the 30 species of *Glossina* in Africa may be capable of transmitting the disease (Molyneux and Ashford, 1983). Adult flies, especially gravid females, need a blood meal every 4 to 5 days, and when biting introduce parasites that are present in their saliva after the maturating process. Mechanical transmission may also occur when infective trypanosomes from a previous blood meal are present in the mouth parts and are deposited in a host during a second blood meal. Flies become infective for humans 2 to 5 weeks after an infected blood meal and remain infective for their life-span of up to 3 months.

Infections may also be transmitted congenitally (Burke et al., 1974). With *T.b. gambiense* infections, chronic disease may persist for as long as 23 years (Molyneux and Ashford, 1983).

Clinical Manifestations of the Disease

The clinical features of Gambian and Rhodesian sleeping sickness are similar except that *T.b. gambiense* generally causes a milder, prolonged disease, whereas *T.b. rhodesiense* is more virulent and usually causes an acute, fulminating disease, fatal within weeks or a few months; some patients, however, may remain asymptomatic for several months.

African trypanosomiasis has two major sequential presentations. Stage I, the hemolymphatic stage, is infectious without central nervous system (CNS) involvement. Stage II, the meningoencephalitic stage, occurs when the parasites invade the CNS.

Hemolymphatic Stage

Stage I involves two phases. First, an indurated chancre induced by multiplying trypanosomes develops at the site of a bite by an infected tsetse fly. Next, parasites are disseminated by the lymphatics and enter the lymph nodes and bloodstream. This systemic infection results in intermittent fever, lymphadenopathy, splenomegaly, anemia, abnormal electrocardiogram (Greenwood and Whittle, 1980), headache, arthralgia, cramps, leg weakness, exquisite sensitivity to pain (Kerandel's sign), edema of the limbs and face, cardiac pain, and a pulse of 100 to 140 (Apted, 1970; Beaver et al., 1984). Patients with Gambian trypanosomiasis frequently show enlargement of the postcervical lymph nodes (Winterbottom's sign). Additional symptoms include jaundice, icterohemorrhagic manifestations (Bourgeade et al., 1985), and a transient skin rash (Cochran and Rosen, 1983).

Meningoencephalitic Stage

Stage 11 of the disease follows invasion of the CNS and causes deterioration of mental and locomotor functions. Hallmarks of this usually fatal stage are severe headaches, pruritus, muscle spasms, trembling of hands and feet, athetosis, neck pain and stiffness, spastic or flaccid paralysis, and uncontrollable sleepiness. The patient frequently lapses into coma, becomes apathetic, anorexic, and emaciated, and may die from meningitis, heart failure, or, more likely, secondary bacterial or parasitic infections. Bacterial pneumonia is a common cause of death in advanced stages of the disease (Apted, 1970; Beaver et al., 1984; Greenwood and Whittle, 1980).

Rhodesian trypanosomiasis progresses more rapidly than Gambian trypanosomiasis, and the disease can be fatal during stage I, before CNS involvement.

Pathology

The pathologic events consequent to the disease appear to be immune mediated and are exemplified by intense B lymphocyte proliferation (Greenwood and Whittle, 1980). What activates B lymphocytes remains unknown; Seed (1969) identified from *T.b.* gambiense a protein that increased vascular permeability in rabbits, but the substance might have induced an immunopathologic response in the walls of the blood vessels (Greenwood and Whittle, 1980). Whatever the stimulus for B lymphocyte proliferation, the presence of trypanosomes in tissues is followed by infiltration of lymphocytes and plasma cells. The histologic appearance of tissue from the trypanosome chancre illustrates a vasculitis, interstitial multiplication of trypanosomes, and mononuclear cell infiltration of the interstitium with local inflammatory foci (Poltera, 1985).

Organs and tissues involved in trypanosomiasis are the skin, lymph nodes, erythrocytes, connective tissue, heart, serous membranes, respiratory system, endocrine system, CNS, and peripheral nervous system (Ormerod, 1970; Poltera, 1985). Changes in infected lymph nodes include follicular hyperplasia, sinus histiocytosis, and perivascular and mononuclear cell infiltrations; in advanced cases, the lymph nodes become fibrous, shrunken, and exhausted of their lymphocytes (Poltera, 1985). Anemia may be caused by complement-mediated lysis of erythrocytes coated with immune complexes or from a direct toxic effect on the erythrocytes by a trypanosome-derived product (Wery et al., 1982). The heart is especially vulnerable in Rhodesian sleeping sickness (Ormerod, 1970) and is affected by a pancarditis caused by lymphocytic infiltration and heavy deposition of immunoglobulins IgG and IgM and complement-C_3 (Poltera et al., 1980).

In advanced trypanosomiasis the brain shows vasculitis with perivascular cuffing by lymphocytes, macrophages, and plasma cells, leading to a nonspecific lymphoplasmacytic meningoencephalitis (Greenwood and Whittle, 1980; Hume Adams 1986; Poltera, 1985; Wery et al., 1982). The pathogenesis of African sleeping sickness, a matter of speculation and debate, may result from lymphokines produced by proliferating B lymphocytes, or tissue damage caused by autoantibodies or immune-complex deposits (Greenwood and Whittle, 1980).

Diagnostic Features

African trypanosomiasis is endemic in an environment where it must be differentiated from malaria, tuberculosis, heart failure, glomerulonephritis, mental illness, and meningitis (Buyst, 1973). Nonspecific laboratory findings associated with the disease include an increased erythrocyte sedimentation rate, a positive Coombs test, thrombocytopenia, excessively high levels of cerebrospinal fluid (CSF) and serum IgM with a moderate increase of IgG, and the presence of morular cells of Mott in the CSF—these are plasma cells with surface agglomerations of IgM, giving the cells a mulberry appearance in stained preparations, and are considered pathognomonic of trypanosomiasis.

Rhodesian trypanosomiasis is by far the more virulent disease, and parasites are more easily detected in blood during its early, febrile stages. Parasitemia fluctuates and may be absent during afebrile periods; thus repeated examinations of blood films may be necessary. Trypanosomes of *T.b. gambiense* are difficult to find in blood or lymph node aspirates, and serologic diagnosis may be necessary to confirm suspected disease.

Sleeping sickness can be rapidly fatal and must be treated as a medical emergency. Persons from nonendemic areas have a more rapid course of disease than do local inhabitants (Molyneux and Ashford, 1983).

Prognosis

Recovery is likely if the disease is treated in its early stages. Prognosis is poor after the CNS becomes involved, and the outcome of therapy is complicated by the toxicity of the drug used to treat CNS disease. Rhodesian trypanosomiasis can be fatal in less than 2 weeks, before CNS involvement, and both diseases will be fatal if left untreated.

Epidemiologic Aspects

Occurrence and Transmission

Rhodesian sleeping sickness occurs throughout most of East Africa from Ethiopia south through Sudan, Kenya, Tanzania, Uganda, Rwanda, Burundi, Malawi, Zambia, and Zimbabawe to Botswana. *Trypanosoma b. gambiense* ranges in a broad swath from Senegal in the West, through all of Central Africa, to the Lake Victoria region, where it overlaps with *T.b. rhodesiense* in Uganda and occurs as far south as northern Angola. The most active foci of *T.b. gambiense* are in Cameroon, Congo, Côte d'Ivoire, southern Sudan, Uganda, and Zaire (Molyneux and Ashford, 1983).

The principal vectors of *T.b. rhodesiense—Glossina morsitans, G. pallidipes,*and *G. swynnertoni*—are savanna-dwelling flies. *Trypanosoma b. gambiense* has as its principal vectors *G. palpalis, G. fuscipes*, and *G. tachinoides*, which are riverine dwellers, and *G. fusca*, which prefers humid forested areas (Molyneux and Ashford, 1983). *Glossina* species show peaks of activity in the early morning and evening and rest when temperatures exceed 13°C to 17°C (Molyneux and Ashford, 1983). The flies locate their hosts by means of visual and olfactory cues and are attracted to moving objects.

Reservoir hosts play an important role in the transmission of *T.b. rhodesiense*, and game animals, especially bushbuck, are preferred hosts for the *morsitans* group of flies, such as *G. pallidipes* (Allsopp, 1972, Wijers, 1974). Humans do not appear to be good reservoir hosts for *T.b. rhodesiense*. It is assumed that humans are both victims and reservoirs of T.b. *gambiense*, although domestic animals, such as pigs and dogs, can be infected with the parasite (Mehlitz et al., 1981).

Factors influencing the maintenance and spread of trypanosomiasis in man and reservoir hosts are extremely complex and include wind dispersal of *Glossina* (Molyneux et al., 1979), hunting, fishing (Wijers, 1974), farming (Davies, 1982), and disruption of ecosystem complexes (Sindiga, 1984). Rainfall is an important influence, as both its plenitude or dearth may increase the prevalence of the disease. Observations in Kenya have shown that when *Glossina* populations increase after heavy rains, the incidence of trypanosomiasis in lower animals also increases (Griffin and Allonby, 1979); conversely, drought conditions in Botswana correlate with increased incidence of disease when populations move closer to the water sources in the *Glossina*-infested "fly belt" (Davies, 1982).

Imported Trypanosomiasis

Since 1968, 14 cases of African trypanosomiasis have been diagnosed in the United States (Centers for Disease Control, Parasitic Disease Drug Service). Persons at high risk are visitors to game parks (tourists and hunters), missionaries who often work in remote locations, and residents of trypanosomiasis-endemic areas. Since no vectors for sleeping sickness exist outside Africa and blood from persons who have lived in Africa is not accepted for transfusions, imported cases pose no special hazard in spreading the disease.

Susceptibility, Resistance, and Immunity

Sex, age, and ethnicity do not appear to affect a person's susceptibility to trypanosomiasis. The higher prevalence of infection in males is probably related to the greater likelihood of their entering *Glossina*-infested areas. Binz and Watson (1972) suggest that the disease in Europeans is more severe than in Africans because the latter have antibodies induced by protective repeated inoculations of nonpathogenic animal trypanosomes.

Untreated trypanosomiasis is a fatal disease despite an intense antibody response to the infecting organisms, because the parasites can produce populations of antigenic variants. Although the host may destroy one specific variant, it is immunologically exhausted before all potential variants can be eliminated (Mansfield, 1981).

A single antigenic type or clone in a trypanosome infection can produce hundreds of variable antigenic types (VATs). The VAT repertoire is genetically determined (Steinert and Pays, 1985), and as each VAT is destroyed by specific antibodies, a new antigenically distinct population expands to perpetuate the infection. Host antibodies do not stimulate the generation of new VATs, but only eliminate the previous parasitemia, thus the host remains one step behind the parasite. Succeeding waves of VATs are responsible for chronicity in the disease (Sendashonga and Black, 1982; Vickerman, 1985).

The antigens that characterize each VAT are specific glycoproteins that form a surface coat on trypanosomes in the mammalian but not the *Glossina* hosts. In the mammalian host, surface antigens called variant-specific glycoproteins (VSGs) are found only on slender, rapidly dividing trypomastigotes; VSGs are not found on accompanying short, stumpy forms which can continue their cycle only inside the fly (Cross, 1975). Vickerman (1985) describes the VSG as a protective wrapping for the parasites, shielding their nonvariable antigens from the immune response. Trypanosomal antigenic variation diminishes the likelihood of a vaccine for African trypanosomiasis because protective immunity is VAT specific, and an effective vaccine would need to stimulate humoral responses to several hundred antigenically different populations of *T. brucei* trypanosomes.

Variant-specific antigens (VSAs) stimulate strong B-lymphocyte responses, and although both IgM and IgG antibodies are raised, IgM levels are raised earlier, and this antibody is more efficient in antitrypanosomal activity (Mansfield, 1981). Paradoxically, humoral responses to nontrypanosomal antigens may be suppressed while infected hosts continue to respond immunologically to a succession of VSAs (Mansfield, 1981). The exact mode of trypanosome-induced immunosuppression is unclear, but there is experimental evidence from animal studies that both B-lymphocyte (Goodwin et al., 1972) and T-lymphocyte (Mansfield and Wallace, 1974) functions are suppressed. Possible explanations of these observations include polyclonal B-cell activation leading to exhaustion of B-cells (Greenwood, 1974), simultaneous activation of antigen nonspecific helper and suppressor T-cells (Jayawardena et al., 1978), and uncontrolled clonal expansion and IgM synthesis by VSA-stimulated B-cells (Mansfield, 1981).

Etiologic Agents

Characterization

The two etiologic agents of African trypanosomiasis are elongate, spindle-shaped flagellates in the family *Trypanosomatidae*, order *Kinetoplastida*, class *Zoomastigophora*. Members of the genus *Trypanosoma* are all obligate parasites.

Trypanosomes recovered from humans are pleomorphic and range from slender forms, about 30 μm long with a free flagellum, to stumpy forms, about 15 μm long without a free flagellum. In Giemsa-stained preparations, the cytoplasm appears pale blue, vacuolated, with a scattering of dark blue volutin granules. The centrally located nucleus stains dark red, as does the kinetoplast, a spherical body at the posterior end. At the ultrastructural level, the kinetoplast is continuous with a long mitochondrion and is closely associated with a basal body. The flagellum originates from the basal body and proceeds anteriad along the edge of an undulating membrane, to project freely at the anterior end of the trypomastigote.

In the tsetse fly, trypomastigotes undergo several variations, including an elongate form called an epimastigote, with the kinetoplast lying just anterior to the nucleus.

Life Cycle (Fig. 1)

The trypanosomes that *Glossina* ingests with a blood meal may contain a mixture of slender and stumpy trypomastigotes. It is believed that the stumpy forms initiate the cycle in the fly, although it has been shown that slender forms can also infect *Glossina* (Molyneux and Ashford, 1983) and transform into slender, procyclic trypomastigotes. These migrate through the peritrophic membrane into the ectoperitrophic space, where they reproduce by binary fission and multiply enormously. The parasites then migrate forward through the esophagus to the salivary ducts, which they follow to the salivary glands. There they transform to epimastigotes, which attach to the epithelial cells of the salivary glands by the flagellum. The attached epimastigotes undergo a final transformation into slender, metacyclic trypomastigotes that are infective for humans. Development in the fly takes 2 to 5 weeks, depending on temperature.

Mature metacyclic forms introduced into a human with the saliva of a biting fly reside briefly in the interstitial spaces of the subcutaneous tissue at the site of the bite; after an incubation period of about 1 to 2 weeks, they are disseminated in the hemolymphatic system. In mammalian hosts the doubling time for dividing trypomastigotes is about 6 h.

The glycoprotein coat of VSAs is not present on the stumpy forms that are infective for *Glossina*, but the coat is present on metacyclic trypomastigotes transmitted by the biting fly.

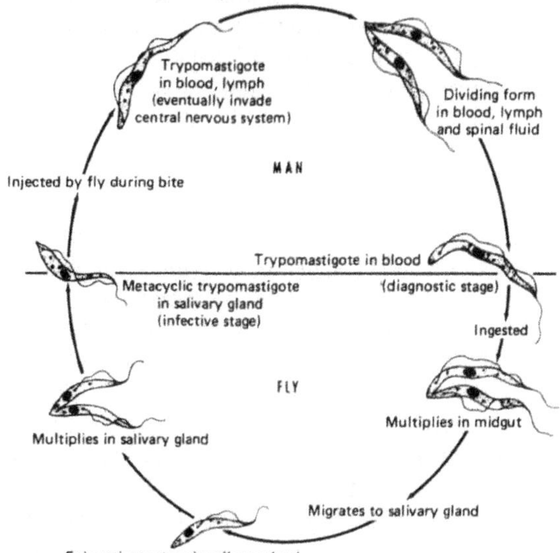

FIG. 1. Life cycle of *Trypanosoma brucei gambiense/rhodesiense*. (Reprinted with permission from the Centers for Disease Control Training and Laboratory Program Office, Atlanta, Georgia.)

Specimen Collection and Preparation

Definitive parasitologic diagnosis depends on demonstrating trypanosomes in an aspirate from the chancre, in blood, lymph node aspirate, cerebrospinal fluid, or bone marrow. Unfortunately, these specimens are often inadequate for diagnosis, and additional evidence must be sought in the patient's symptoms, travel history, exposure to the disease in the laboratory or in an endemic area, physiologic abnormalities, and serodiagnosis.

Blood Smears

Specimens should be drawn at the beginning of a febrile attack, and thick blood smears should be prepared and stained as described for malaria. A careful search of the entire smear should be made. Trypanosomes of *T.b. gambiense* are difficult to detect in blood films because of scant parasitemia and several films should be examined. Examination of a drop of fresh blood under a coverslip at a magnification of × 400 may reveal the parasites by their motility, but this technique is not recommended for definitive diagnosis because the parasites may be confused with microfilariae or spirochetes.

Blood Concentration Techniques

These techniques are useful in patients with suspected trypanosomiasis when direct blood smears are negative. Volumes of about 10 ml of blood are required.

I. *Fractional centrifugation technique*
  1. Mix 9 ml of blood and 1 ml of 5% sodium citrate in a 15-ml conical centrifuge tube.
  2. Centrifuge 100 × *g* for 10 min.
  3. Transfer the supernatant fluid to a second tube and centrifuge at 250 × *g* for 10 min.
  4. Transfer supernatant fluid again and centrifuge at 750 × *g* for 10 min.
  5. Examine sediment directly for motile organisms, and stain dried and fixed smears with Giemsa.

II. *Modified fractional centrifugation*
  1. Rehydrate phytohemagglutinin P (Difco) according to manufacturer's directions.
  2. Add solution to citrated blood (see I–1) to give a concentration of 0.1 mg of phytohemagglutin/ml of blood; use a 15-ml conical centrifuge tube.
  3. Shake the suspension and allow it to stand for 5 min.
  4. Centrifuge at 150 × *g* for 10 min.
  5. Transfer the supernatant fluid and centrifuge at 700 × *g* for 10 min.

6. Examine the fresh and stained smears of the sediment.

III. Buffy coat concentration technique (Ash and Orihel, 1987)

1. Centrifuge 5 ml of oxalated or citrated blood in a Wintrobe tube at 700 × *g* for 30 min.
2. Use a fine capillary tube to remove buffy coat cells between plasma and packed erythrocytes.
3. Examine fresh and stained smears of the white cell suspension.

IV. Microhematocrit tube centrifugation technique (Woo and Rogers, 1974)

1. Fill 3 "Red Tips" heparinized capillary tubes with blood from a finger stick, to a height of 55 mm.
2. Seal the end of the tube with plasticine.
3. Spin at 12,000 rpm for 4 min in a microcapillary centrifuge.
4. Mount the capillary tube on a microscope slide with wax or plasticine.
5. Examine the plasma-buffy cell interface with the oil-immersion lens for motile trypanosomes.
6. Cut the capillary tube at the blood–buffy coat interface, and express the buffy coat for staining and examination on glass slides.

V. DEAE–cellulose separation technique

These techniques employ a column of DEAE–cellulose—an anion exchanger—to separate negatively charged blood components from less negatively charged trypanosomes. In principle, heparinized blood is diluted in a buffer or balanced salt solution and passed through a DEAE–cellulose column. Blood components are retained, the eluate is centrifuged, and the sediment is examined for trypanosomes (Lanham and Godfrey, 1970). This technique has been modified for field use (Lumsden et al., 1981) and for isolation of trypanosomes from tissues, such as bone marrow, brain, spleen, liver, lymph nodes, kidney, and testis (Ekejindu et al., 1986).

These methodologies are complex and must be tailored to an investigator's specific requirements; refer to the original papers for details.

Aspirates and Cerebrospinal Fluid

Fluid can be obtained from a chancre by puncturing the edge of the lesion with a syringe needle and using gentle pressure on the nodule to express exudate. The fluid should be smeared and stained and can be examined wet for motile trypanosomes. Lymph node fluid can be obtained and processed in a similar manner, but it is less useful for Rhodesian sleeping sickness, in which trypanosomes are more likely to be present in the blood. Bone-marrow examination is useful when parasites cannot be found in other specimens, but it should be used as a last resort. A specimen should be obtained from the iliac crest or greater trochanter rather than by sternal puncture (Apted, 1970), as this is less traumatic. Cerebrospinal fluid is essential for determining the stage of the disease, because trypanosomes are seldom present before CNS invasion; the fluid should be centrifuged, and the sediment should be stained with Giemsa. Invasion of the CNS will increase the number of mononuclear cells and the protein in the CSF, and trypanosomes may be observed.

Cultivation of Trypanosomes

The usefulness of in vitro cultivation of African trypanosomes for diagnosis is limited by the necessity to prepare and store media and by the delay of a week or more before detectable numbers of trypanosomes are present. Nevertheless, this avenue can be useful in cases of chronic Gambian disease without apparent parasitemia, as well as for the production of trypanosomes for investigational purposes. A large number of suitable media are given by Taylor and Baker (1978).

Animal Inoculation of Trypanosomes

Rat or mouse infectivity can be a sensitive test for Rhodesian but not for Gambian sleeping sickness. Animals should be injected intraperitoneally with 150 μl of heparinized blood and be monitored daily by examination of blood from tail snips (Dukes et al., 1984).

Serodiagnosis

Serodiagnotic tests such as enzyme immunoassay (ELISA), indirect immunofluorescence test, radial gel immunodiffusion, immunoprecipitation, and indirect agglutination are useful epidemiologic tools even though they are unreliable for individual diagnoses. These tests are hampered by a lack of standardization and the necessity to detect a wide spectrum of antibodies induced by the parasites' VSAs. Antitrypanosomal antibodies can be detected 2 weeks after acquisition of infection, and overall tests show a sensitivity of 90% and a false positivity of 5%; infections such as visceral leishmaniasis and perhaps nonhuman trypanosomiasis result in cross-reacting antibodies (Van Meirvenne and Le Ray, 1985).

In recent years, card agglutination tests such as the TESTRYP$_R$ CATT strip (Smith Kline) have proved to be sensitive, rapid, and well suited for field application (Croft, 1985). The test is similar to the Rapid Plasma Reagin (RPR) test used for syphilis and requires only a drop of a patient's blood to be mixed with one drop of a suspension of trypomastigotes fixed in formalin and stained with Coomassie blue. The specimen and antigen are mixed on a plastic-coated card, and a positive result is indicated by the agglutination of blue particles within 10 min.

New technologies such as isoenzyme analysis, DNA analysis, and monoclonal antibodies are being developed to identify parasites from individual patients, determine a possible relationship between clinical disease and parasite variation, identify reservoir hosts, and study taxonomic relationships (Gibson and Miles, 1985).

Treatment, Prevention, and Control

Treatment and Prophylaxis

The drugs used for the treatment of African trypanosomiasis are toxic and, in the United States, must be obtained from the Parasitic Disease Drug Service of the Centers for Disease Control, Atlanta, Georgia.

In the early stages of the disease, pentamidine isethionate is more effective against Gambian trypanosomiasis than against Rhodesian trypanosomiasis. It is administered by intramuscular injection at a dosage of 4 mg/kg daily for up to 10 days. The drug has been used prophylactically in West Africa, but reports indicate resistance and fears that the drug may suppress an infection below patency, permitting CNS invasion (Molyneux and Ashford, 1983).

Suramin is the drug of choice against both forms of the disease before CNS involvement, but it has a wide range of serious side effects. This drug may cause severe idiosyncratic hypotensive reactions, and a test dose of 100–200 mg should be given. Suramin is given by intravenous injection of 1 dose of 1.0 g on days 1, 3, 7, 14, and 21 (total of 5 g). An alternative to suramin is berenil (Abaru et al., 1984).

Melarsoprol is effective in the meningoencephalitic stage of sleeping sickness but is toxic. The drug is administered intravenously at a dosage of 3.6 mg/kg daily for 3 days; this regimen is repeated once or twice on a 7- to 10-day schedule.

The only promising development in chemotherapy for African trypanosomiasis comes in a report by Van Nieuwenhove et al. (1985) of a successful trial of DL-alpha-difluoromethylornithine (DFMO) in the Sudan for early and late *T.b. gambiense* infections. Side effects were not severe, and the drug could be administered orally.

Prevention and Control

A strategy detailed by Molyneux (1983) involves an active search for cases to treat. This is necessary because patients with Rhodesian sleeping sickness may be too ill to seek treatment, and those with Gambian sleeping sickness may not consider their relatively mild symptoms enough cause to travel to a medical center.

The other arm of attack, chemoprophylaxis, is limited by the ineffectiveness of pentamidine against *T.b. rhodesiense*, resistance by some strains of *T.b. gambiense*, and the requirement that it be administered by intramuscular injection every 6 months.

Vector control involves clearing of vegetation around areas of human habitation, applying of insecticides to vegetation where the flies rest and to the ground where juvenile stages of *Glossina* develop, and using traps that contain insecticide-impregnated strips.

Literature Cited

Abaru, D. E., D. A. Liwo, D. Isakina, and E. E. Okori. 1984. Retrospective long-term study of effects of berenil by follow-up of patients treated since 1965. Trop. Med. Parasitol. **35:**148–150.

Allsop, R. 1972. The role of game animals in the maintenance of endemic and enzootic trypanosomiases in the Lambine Valley, South Nyanza District, Kenya. Bull. W.H.O. **47:**735–746.

Apted, F. I. C. 1970. Clinical manifestations and diagnosis of sleeping sickness, chap. 35. *In* H. W. Mulligan, and W. H. Potts, (ed.) The African trypanosomes. John Wiley & Sons, Inc., New York.

Ash, L. R., and T. H. Orihel. 1987. Parasites: a guide to laboratory procedures and identification. ASCP Press, Chicago, chap. 8.

Beaver, P. C., R. C. Jung, and E. W. Cupp. 1984. The blood and tissue parasites, chap. 10. *In* Clinical parasitlogy, 9th ed. Lea & Febiger, Philadelphia.

Binz, G., and H. J. C. Watson. 1972. Observations on the patterns of blood IgM levels in the populations of endemic and non-endemic sleeping sickness areas in Kenya. Bull. W.H.O. **47:**757–767.

Bourgeade, A., Y. Nosny, B. Faugere, and P. Pene. 1985. African trypanosomiasis of icterohemorrhagic form. Bull. Soc. Pathol. Exot. Filiales **75:**908–913.

Burke, J., K. Bengosi, and N. L. Diantete. 1974. Un cas de trypanosomiase africaine (*T. gambiense*) congenitale. Ann. Soc. Belg. Med. Trop. **54:**1–4.

Buyst, H. 1973. The diagnosis of sleeping sickness. Trop. Doc. **3:**110–112.

Cochran, R., and T. Rosen. 1983. African trypanosomiasis in the United States. Arch. Dermatol. **119:**670–674.

Croft, S. L. 1985. Serodiagnostic tests for West African trypanosomiasis. Parasitol. Today **1:**115.

Cross, G. A. M. 1975. Identification, purification and properties of clone-specific glycoprotein antigens constituting the surface coat of *Trypanosoma brucei*. Parasitology **71**:393–417.

Davies, J. E. 1982. Sleeping sickness and the factors affecting it in Botswana. J. Trop. Med. Hyg. **85**:67–71.

Dukes, P., L. R. Rickman, R. Killick-Kendrick, I. Kakoma, F. K. Wurada, P. de Raadt, and P. Morrow. 1984. A field comparison of seven diagnostic techniques for human trypanosomiasis in the Luangwa Valley, Zambia. Tropenmed. Parasitol. **35**:141–147.

Ekejindu, G. O. C., D. B. Maikaje, and I. M. Ekejindu. 1986. The combined procedures of Ficol-Paque centrifugation and avian exchange separation for the recovery of trypanosomes from blood and other tissues. Trans. R. Soc. Trop. Med. Hyg. **80**:879–882.

Gibson, W. C., and M. A. Miles. 1985. Application of new technologies to epidemiology. Br. Med. Bull. **41**:115–121.

Goodwin, L. G., D. G. Green, M. W. Guy, and A. Voller. 1972. Immunosuppression during trypanosomiasis. Br. J. Exp. Pathol. **53**:40–43.

Greenwood, B. M. 1974. Possible role of a B-cell mitogen in hypergammaglobulinaemia in malaria and trypanosomiasis. Lancet **i**:435–436.

Greenwood, B. M., and H. C. Whittle, 1980. The pathogenesis of sleeping sickness. Trans. R. Soc. Trop. Med. Hyg. **74**:716–725.

Griffin, L., and E. W. Allonby. 1979. Studies on the epidemiology of trypanosomiasis in sheep and goats in Kenya. Trop. Anim. Health Prod. **11**:133–142.

Hume Adams, J., L. Haller, F. Y. Boa, F. Douas, A. Dago, and K. Konian. 1986. Human African trypanosomiasis (*T. b. gambiense*): a study of 16 fatal cases of sleeping sickness with some observations on acute reactive arsenical encephalopathy. Neuropathol. Appl. Neurobiol. **12**:81–94.

Jayawardena, A. N., B. H. Waksman, and D. D. Eardley. 1978. Activation of distinct helper and suppressor T cells in experimental trypanosomiasis. J. Immunol. **121**:622–628.

Lanham, S. M., and D. G. Godfrey. 1970. Isolation of Salivarian trypanosomes from man and other mammals using DEAE–cellulose. Exp. Parasitol. **28**:521–534.

Lumsden, W. H. R., C. D. Kimber, P. Dukes, L. Haller, A. Stanghellini, and G. Duvallet. 1981. Field diagnosis of sleeping sickness in the Ivory Coast. I. Comparison of the miniature avion-exchange/centrifugation technique with other protozoological methods. Trans. R. Soc. Trop. Med. Hyg. **75**:242–250.

Mansfield, J. M. 1981. Chap. 4. Immunology and immunopathology of African trypanosomiasis, In Parasitic Diseases, vol. I, Marcel Dekker Inc., New York.

Mansfield, J. M., and Wallace, J. H. 1974. Suppression of cell-mediated immunity in experimental African trypanosomiasis. Infect. Immun. **10**:335–339.

Mehlitz, D., U. Brinkmann, and L. Haller. 1981. Epidemiological studies on the animal reservoir of gambiense sleeping sickness. Part 1. Review of literature and description of the study area. Tropenmed. Parasitol. **32**:129–133.

Molyneux, D. H. 1983. Selective primary health care: strategies for control of disease in the developing world. VIII. African trypanosomiasis. Rev. Infect. Dis. **5**:945–956.

Molyneux, D. H., and Ashford, R. W. 1983. Trypanosoma and Leishmania. International Publications Service, Taylor and Francis Inc., New York.

Molyneux, D. H., D. A. Baldry, and C. Fairhurst. 1979. Tsetse movement in wind fields: possible epidemiological and entomological implications for trypanosomiasis and its contol. Acta. Trop. (Basel) **36**:53–65.

Ormerod, W. E. 1970. Pathogenesis and pathology of trypanosomiasis in man, chap. 31. *In* H. W. Mulligan and W. H. Potts (ed.), The African trypanosomes. John Wiley & Sons, Inc., New York.

Poltera, A. A. 1985. Pathology of human African trypanosomiasis with reference to experimental African trypanosomiasis and infections of the central nervous system. Br. Med. Bull. **41**:169–174.

Poltera, A. A., A. Hockmann, and P. H. Lambert. 1980. A model for cardiopathy induced by *Trypanosoma brucei brucei* in mice. A histologic and immunopathologic study. Am. J. Pathol. **99**:325–351.

Seed, J. R. 1969. *Trypanosoma gambiense* and *T. lewisi:* increased vascular permeability and skin lesions in rabbits. Exp. Parasitol. **26**:214–223.

Sendashonga, C., and S. J. Black. 1982. Humoral responses against *Trypanosoma brucei* variable antigen are induced by degenerating parasites. Parasitol. Immunol. **4**:245–257.

Sindiga, I. 1984. Sleeping sickness in Kenya. Maasailand. Soc. Sci. Med. **18**:183–187.

Steinert, M., and E. Pays. 1985. Genetic control of antigenic variation in trypanosomes. Br. Med. Bull. **41**:149–155.

Taylor, A. E. R., and J. R. Baker (ed.). 1978. Methods of cultivating parasites *in vitro*. Academic Press, London.

Van Meirvenne, N., and D. Le Ray. 1985. Diagnosis of African and American trypanosomiasis. Br. Med. Bull. **41**:156–161.

Van Nieuwenhove, S., P. J. Schecter, J. Declerq, G. Bone, J. Burke, and A. Sjoerdsma. 1985. Treatment of *Gambiense* sleeping sickness in the Sudan with oral DFMO. Trans. R. Soc. Trop. Med. Hyg. **79**:692–698.

Vickerman, K. 1985. Developmental cycles and biology of pathogenic trypanosomes. Br. Med. Bull. **41**:105–114.

Wery, M., P. M. Mulumba, P. H. Lambert, and L. Kazyumba. 1982. Hematologic manifestations, diagnosis, and immunopathology of African trypanosomiasis. Semin. Hematol. **19**:83–92.

Wijers, D. J. 1974. The complex epidemiology of Rhodesian sleeping sickness in Kenya and Uganda. III. The epidemiology in the endemic areas along the lakeshore between the Nile and the Yala Swamp. Trop. Geogr. Med. **26**:307–318.

Woo, P. T. K., and D. J. Rogers. 1974. A statistical study of the sensitivity of the hematocrit centrifuge technique in the detection of trypanosomes in blood. Trans. R. Soc. Trop. Med. Hyg. **68**:319–326.

CHAPTER 76

Amebiasis

DONALD J. KROGSTAD

Disease: Amebiasis (amebic dysentery, amebic colitis, amebic liver abscess).
Etiologic Agent: *Entamoeba histolytica.*
Source: Oral ingestion of material contaminated with human stool containing cysts of *E. histolytica* (fecal-oral transmission).
Clinical Manifestations: Intestinal infection is most often benign (commensal) with few or no symptoms. Among symptomatic persons, diarrhea may be bloody and contain pus. Extraintestinal infection is most often in the liver, but may occur at any site with fever, malaise, and clinical findings consistent with a mass lesion.
Pathology: In the intestine, flask-shaped ulcers are typical and are wider at the base than at the epithelial surface. In the liver (and at other sites of metastatic infection), the center of the abscess contains primarily necrotic debris. Amebae may be found at the edge of the abscess—in the area of transition between the abscess and the surrounding normal tissue.
Epidemiology: Amebiasis is present throughout the world. It is more prevalent in areas where sanitation is poor and fecal-oral transmission is common.
Treatment: Metronidazole is the treatment of choice for active (invasive) intestinal infection and for extraintestinal (metastatic) infection. Treatments used for persons passing *E. histolytica* cysts (commensal or asymptomatic infection) include diloxanide furoate, paromomycin, diiodohydroxyquin, and furazolidone.
Prevention and Control: The two most important steps are: 1) reduction of fecal-oral contamination, and 2) identification and treatment of persons with asymptomatic intestinal infection, who pose the greatest risk of transmission because they excrete the infectious cyst form of the parasite.

Description of Disease

Amebiasis (infection due to *Entamoeba histolytica*) occurs only among humans and selected primates. Its most common manifestations are diarrhea and liver abscess. Its public health significance results primarily from the risks posed by asymptomatic carriers—to the general population (if they contaminate food or the water supply) and to their close contacts.

Communicability and Transmissibility

Amebiasis is transmitted by the oral ingestion of material contaminated with stool that contains cysts of *E. histolytica*. Stool that contains trophozoites is not infectious because trophozoites (unlike cysts) are killed by dessication in the external environment and by gastric acid after oral ingestion (Krogstad et al., 1978a). Person to person transmission associated with food handlers who are passing cysts is the most common mode of transmission (Krogstad et al., 1978b). Massive outbreaks may occur if plumbing cross-connections permit mixing of sewage containing *E. histolytica* cysts with drinking water (Davis and Ritchie, 1948; LeMaistre et al., 1956).

Clinical Manifestations of Disease

Both intestinal and extraintestinal disease typically have an incubation period of several weeks to months. Amebiasis rarely progresses from asymptomatic or initial infection to serious illness within less than 7 to 10 days.

Asymptomatic (Commensal) Intestinal Infection

Many persons with intestinal infection have few or no symptoms and excrete primarily the cyst form of the parasite. The reasons for this lack of pathogenicity are incompletely defined and remain one of the most important unknowns about this disease (Krogstad, 1986; Ravdin, 1986). (See the discussion of Sargeaunt's and Mirelman's studies, below.)

Active Intestinal Disease (Amebic Colitis, Amebic Dysentery)

A minority of persons with intestinal infection develop diarrheal disease, often with red cells and pus in the stool (Adams and MacLeod, 1977a). Typical symptoms and signs include fever, abdominal pain, and tenesmus. In its most severe form, amebic dysentery may produce complications similar to those of ulcerative colitis, including bacterial peritonitis secondary to intestinal perforation and toxic megacolon.

Extraintestinal (Metastatic) Amebic Infection

This complication results from hematogenous dissemination of the parasite from the gastrointestinal tract (Adams and MacLeod, 1977b). However, many patients with extraintestinal infection cannot recall an episode of diarrhea consistent with amebic colitis, and most have no evidence of *E. histolytica* in their stool when they present with extraintestinal infection. Signs and symptoms of amebic liver abscess include fever, pain and a mass in the right upper quadrant, an elevated right hemidiaphragm, and right lower lobe atelectasis (Krogstad et al., 1978b; Ravdin, 1986). Signs and symptoms of extraintestinal amebic abscesses at other sites include fever and the effects of a slowly enlarging mass lesion.

Localized Cutaneous Disease

Although it is quite rare, this form of amebiasis is important because it may be confused with squamous cell carcinoma and treated surgically rather than with antiamebic agents. When they occur, these lesions are most frequently in the perianal or genital regions.

Pathology

In active intestinal disease, the parasite produces flask-shaped ulcers in which the base is substantially wider than the neck at the epithelial surface. Organisms are typically found at the edge of the ulcer and are readily identifiable with either hematoxylin and eosin or Giemsa staining of infected tissue sections. Because of the parasite's relatively small size (12 to 50 μm), tissue sections must be examined with either high dry (440 ×) or oil immersion (1,000 ×) magnification (Melvin and Healy, 1985; Wheatley, 1951).

In extraintestinal (metastatic) infection, the center of the abscess typically contains only necrotic material without identifiable amebae. When amebae can be identified, they are typically at the edge of the abscess—in the zone of transition between the necrotic debris of the abscess and the normal surrounding tissue. Staining should be performed as for intestinal disease.

The pathologic features of both intestinal and extraintestinal disease are those of tissue destruction. They are consistent with the parasite's species name, which was given by Schaudinn in 1903 to describe its lytic effects on infected tissue (Kean et al., 1978; Stilwell, 1955).

Diagnostic Features

Amebiasis should be considered in patients with dysentery (diarrhea with blood and pus) or hepatic mass lesions (Adams and MacLeod, 1977a, 1977b). Amebiasis is most common in areas of the Third World with inadequate sanitation. However, it is also endemic in the United States and may be acquired by life-long residents of the United States who have never traveled abroad (Krogstad et al., 1978a, 1978b; Ravdin, 1986). Eosinophilia is not a feature of amebiasis because *E. histolytica* is a protozoan parasite, and only helminthic parasites typically produce eosinophilia.

Diagnosis of intestinal amebic infection requires identification of the parasite in the stool. Diagnosis of extraintestinal (metastatic) amebic infection requires evidence of a mass lesion in the liver or elsewhere (including the central nervous system) *and* either a positive serologic test or identification of the organism in tissue sections.

Differential Diagnosis

Although intestinal amebic infection may be confused clinically with many other invasive pathogens, the most important distinction is between amebic infection and inflammatory bowel disease (Healy and Kraft, 1972; Juniper, 1969; Krogstad et al., 1978a, 1978b). This is because the pathology of inflammatory bowel disease is often similar to that of amebiasis (toxic megacolon, pseudopolyps, loss of normal haustral markings) and because the immunosupressive drugs used to treat inflammatory bowel disease exacerbate amebic infection and may cause fatal dissemination.

Prognosis

Death is unusual in amebiasis, but may result from rupture of an extraintestinal abscess in the liver or elsewhere or from intestinal perforation with secondary bacterial peritonitis and septicemia. On occasion, death may result from exsanguination in patients with invasive intestinal disease.

Epidemiologic Aspects

Occurrence and Transmission

Amebiasis occurs throughout the world. Its prevalence is greatest in areas with inadequate sanitation and frequent fecal-oral transmission. In the United States, the actual prevalence of amebic infection is unknown. The low prevalence of seropositivity in the U.S. population (<1%) indicates that infection is rare since infected persons typically remain seropositive for years after their initial infection. In the developing world, the prevalence of intestinal infection varies greatly, but is often ≥10%.

Foci of amebic infection remain in the United States. They are most often found in rural areas, among recent immigrants from the developing world, or in institutions for the mentally retarded (especially among incontinent patients). Outbreaks (case clusters) have often been associated with food handlers who have asymptomatic intestinal infection and contaminate food with infectious cysts from their stool.

The low number of amebiasis cases reported in the United States does not reveal a seasonal pattern. However, in developing countries, the incidence of symptomatic infection often peaks in the dry season. This observation is consistent with transmission by the fecal-oral route. The reduced availability of water in the dry season increases fecal-oral transmission because the limited water available is used for both washing and drinking.

Pathogenesis—Susceptibility, Resistance, and Commensal Versus Invasive Infection

Human susceptibility to amebic infection is widespread and may be universal. The pathogenesis of amebic disease involves at least three distinct steps: 1) adherence via an N-acetylgalactosamine-inhibitable lectin, 2) cytolysis which requires calcium, phospholipase A activity, and an intact acid vesicle system, and 3) phagocytosis of the dead mammalian cell (Ravdin, 1986; Ravdin and Guerrant, 1982; Ravdin et al., 1986). The most critical question in the pathogenesis of amebiasis is the nature of the factor(s) responsible for the commensal (asymptomatic) carrier state versus invasive intestinal infection.

Work by Sargeaunt and by others has shown that the isoenzyme patterns (zymodemes) of less virulent (nonpathogenic) amebae are different from those of pathogenic isolates (Allason-Jones et al., 1986; Goldmeier et al., 1986; Mathews et al., 1986). They have suggested that patients with isolates belonging to nonpathogenic zymodermes may be at little or no risk of amebic disease and may therefore not require treatment (Allason-Jones et al., 1986). However, recent studies by Mirelman et al. (1986) have shown that isolates with nonpathogenic isoenzyme patterns can be converted to pathogenic isoenzyme patterns in vitro, and vice versa. Coincident with these changes, these isolates acquire (or lose) in vitro pathogenicity as defined by the destruction of monolayers of mammalian fibroblasts or Chinese hamster ovary (CHO) cells (Mirelman, 1987; Mirelman et al., 1986). These results suggest that isoenzyme patterns (zymodemes) may be phenotypic rather than genotypic traits (Krogstad, 1986). It is not known whether similar transitions occur in vivo in the gastrointestinal tract.

The cyst form of the parasite is infectious because its double cyst membrane permits it to withstand both dessication in the external environment and gastric acid on oral ingestion. In contrast, trophozoites are killed by dessication or gastric acid.

Etiologic Agent

Identification

The morphologic identification of *E. histolytica* is one of the most difficult tasks in clinical microbiology. *E. histolytica* is distinguished from most other amebae by its characteristic morphology (Melvin and Healy, 1985). Its nucleus has a central karyosome, and its peripheral chromatin is distributed evenly. The central karyosome of *E. coli* is larger and more eccentric; its peripheral chromatin is more clumped and irregular. Furthermore, the chromatoid bars of *E. histolytica* are smooth; those of *E. coli* are ragged. *E. histolytica* is distinguished from *E. hartmanni* (cyst 5 to 8 μm, trophozoite 6 to 10 μm) only by size (cyst 10 to 20 μm, trophozoite 12 to 50 μm). This distinction requires a carefully calibrated ocular micrometer and oil immersion magnification. It is important clinically because *E. hartmanni* is not pathogenic.

E. histolytica may be distinguished from white cells in the stool by using supravital or permanent stains to demonstrate nuclear morphology: methylene blue with fresh specimens, and Wheatley's trichrome for material preserved in polyvinyl alcohol (PVA) fixative. This distinction is important clinically because patients with inflammatory bowel disease (and other conditions that produce white cells in the stool) may otherwise be misdiagnosed as having amebic colitis and inappropriately treated with potentially toxic antiamebic agents including emetine and dehydroemetine (Krogstad et al., 1978b).

As noted, Sargeaunt et al. categorized amebae by their isoenzyme patterns and found that isolates from asymptomatic and symptomatic patients fall into different subgroups (zymodemes). Neither this technique nor culture (Diamond et al., 1978) is generally available for clinical use.

Specimen Collection and Transport

Amebae deteriorate steadily in stool specimens held prior to examination. This process is more rapid at 37° than at 4°C. Therefore, aliquots of a stool specimen should be preserved in formalin (for cysts) and in PVA (for trophozoites) if more than 1 to 2 h will elapse before the specimen is examined.

The patient should not have received soaps, antimicrobials, barium, or other agents which interfere with stool examination (Table 1) for at least 7 to 10 days before stool examination (Juniper, 1969). Specimens obtained by sigmoidoscopy should be procurred with a glass or metal instrument because amebae adhere to cotton swabs. Biopsy of a rectal valve often yields positive results in patients with amebic colitis who have had consistently negative stool examinations.

Serum for antibody testing should be examined by indirect hemagglutination (IHA), enzyme-linked immunosorbent assay (ELISA), counterelectrophoresis, or gel diffusion (Healy and Kraft, 1972; Milgram et al., 1966; Walls, 1985). These techniques yield similar results in most series, although there has been more experience with the IHA. Because amebic infections are often subacute or chronic, paired (acute and convalescent) specimens demonstrate titer increments less frequently in amebiasis than in bacterial or viral infections.

TABLE 1. Substances that interfere with stool examination for parasites[a]

Antibiotics	Enemas
Tetracyclines	Hypertonic salt
Sulfonamides	Soap
Antiparasitic drugs	Tap water
Antiprotozoal agents	Antidiarrheal preparations
Laxatives, antacids	Bismuth
Castor oil	Kaolin compounds
Magnesium hydroxide	
Radiologic contrast medium	
Barium sulfate	

[a] Adapted from Juniper (1969) with the permission of the publisher.

Culture

Infected stool is suspended in buffered saline solution and added to Diamond's TYS GSM-9 culture medium with gastric mucin. Within 2 to 3 days, amebae can be seen by microscopy adhering to the surface of the glass or plastic tissue culture flask.

Immunologic Aspects

Most patients with invasive intestinal disease (≥80%) and virtually all of those with extraintestinal infection (96 to 100%) have serum antibodies to *E. histolytica*. The IHA test is widely available and has been used by most investigators. Titers ≥1:256 are considered positive (Walls, 1985).

Studies by Pillai and Mohimen (1982) indicate that patients severely ill with intestinal or extraintestinal amebiasis often have circulating immune complexes (containing antibodies to *E. histolytica* and parasite antigens). They suggested that these immune complexes, which fall rapidly with treatment, contribute significantly to the myalgias, arthralgias, and other systemic symptoms seen in these patients.

Recent studies have demonstrated a cell-mediated immune response to *E. histolytica* among patients with liver abscess (Salata et al., 1986). The profound effect of steroids on the course of amebic infection suggests that cell-mediated immunity plays an important role in the normal response to *E. histolytica*. As judged by the prevalence of antibody to *E. histolytica* among persons with extraintestinal infection, antibodies alone are not protective.

Treatment, Prevention, and Control

Treatment

Patients with extraintestinal or invasive intestinal disease should generally be treated with metronidazole. Alternatives include emetine or dehydroemetine alone and dehydroemetine (or emetine) plus chloroquine combinations. Patients with asymptomatic intestinal infection should be treated with diloxanide furoate, diiodohydroxyquin, paromomycin, or furazolidone.

Prevention and Control

Prevention of amebiasis depends on: 1) the identification and treatment of asymptomatic persons passing cysts, and 2) water supplies with adequate sedimentation and filtration and without plumbing cross-connections.

Persons passing cysts are difficult to identify because they are usually asymptomatic. However, they must be identified and treated to prevent additional infections. This is usually accomplished by follow-up stool examinations for well persons living with persons who have symptomatic amebiasis.

Filtration and sedimentation are the most important steps in preventing waterborne transmission of amebiasis because amebic cysts (like those of *Giardia lamblia*) are resistant to chlorine concentrations that are effective against bacteria (1 to 2 ppm). The concentrations necessary to kill amebic cysts and viruses ($\geq$10 ppm) are unpalatable and are therefore not used in public water systems.

Literature Cited

Adams, E. B., and I. N. MacLeod. 1977a. Invasive amebiasis: I. Amebic dysentery and its complications. Medicine **56:**315–323.

Adams, E. B., and I. N. MacLeod. 1977b. Invasive amebiasis: II. Amebic liver abscess and its complications. Medicine **56:**325–334.

Allason-Jones, E., A. Mindel, P. Sargeaunt, and P. Williams. 1986. *Entamoeba histolytica* as a commensal intestinal parasite in homosexual men. N. Engl. J. Med. **315:**353–356.

Davis, C., and L. S. Ritchie. 1948. Clinical manifestations and treatment of epidemic amebiasis occurring in occupants of the Mantetsu apartment building, Tokyo, Japan. Am. J. Trop. Med. **28:**817–823.

Diamond, L. S. 1982. A new liquid medium for the xenic cultivation of *Entamoeba histolytica* and other lumen-dwelling protozoa. J. Parasitol. **68:**958–959.

Diamond, L. S., D. R. Harlow, and C. C. Cunick. 1978. A new medium for the axenic cultivation of *Entamoeba histolytica* and other *Entamoeba*. Trans. R. Soc. Trop. Med. Hyg. **72:**431–432.

Goldmeier, D., P. G. Sargeaunt, A. B. Price, et al. 1986. Is *Entamoeba histolytica* in homosexual men a pathogen? Lancet **i:**641–644.

Healy, G. R., and S. C. Kraft. 1972. The indirect hemagglutination test for amebiasis in patients with inflammatory bowel disease. Am. J. Dig. Dis. **17:**97–103.

Juniper, K. 1969. Parasitic diseases of the intestinal tract, p 472–555. *In:* M. Paulson (ed). Gastroenterologic medicine, Lea & Febiger, Philadelphia.

Kean, B. H., K. E. Mott, and A. J. Russell (ed). 1978. Tropical medicine and parasitology: classical investigations. Cornell University Press, Ithaca, NY, p. 71–168.

Krogstad, D. J. 1986. Isoenzyme patterns and pathogenicity in amebic infection. N. Engl. J. Med. **315:**390–391.

Krogstad, D. J., H. C. Spencer, Jr., and G. R. Healy. 1978a. Amebiasis. N. Engl. J. Med. **298:**262–265.

Krogstad, D. J., H. C. Spencer, Jr., G. R. Healy, N. N. Gleason, D. J. Sexton, and C. A. Herron. 1978b. Amebiasis: epidemiologic studies in the United States, 1971–1974. Ann. Intern. Med. **88:**89–97.

Le Maistre, C. A., R. Sappenfield, C. Culbertson, F. R. N. Carter, A. Offutt, H. Blacke, and M. M. Brooke. 1956. Studies of a waterborne outbreak of amebiasis, South Bend, Indiana: I. Epidemiological aspects Am. J. Trop. Med. Hyg. **64:**30–45.

Mathews, H. M., D. M. Moss, G. R. Healy, and D. Mildvan. 1986. Isoenzyme analysis of *Entamoeba histolytica* isolated from homosexual men. J. Infect. Dis. **153:** 793–795.

Melvin, D. M., and G. R. Healy. 1985. Intestinal and urogenital protozoa, p. 631–650. *In:* E. H. Lennette, A. Balows, W. J. Hausler, Jr., and H. J. Shadomy (ed). American Society for Microbiology, Washington, D.C.

Milgram, E. A., G. R. Healy, and I. G. Kagan. 1966. Studies on the use of the indirect hemagglutination test in the diagnosis of amebiasis. Gastroenterology **50:**645–649.

Mirelman, D. 1987. Effect of culture condition and bacterial associates on the zymodemes of *Entamoeba histolytica*. Parasitol. Today **3:**37–40.

Mirelman, D., R. Bracha, R. A. Chayen, A. Aust-Kettis, and L. S. Diamond. 1986. *Entamoeba histolytica:* effect of growth conditions and bacterial associates on isoenzyme patterns and virulence. Exp. Parasitol **62:**142–148.

Pillai, S., and A. Mohimen. 1982. A solid-phase sandwich radioimmunoassay for *Entamoeba histolytica* proteins and the detection of circulating antigens in amebiasis. Gastroenterology **83:**1210–1216.

Ravdin, J. I. 1986. Pathogenesis of disease caused by *Entamoeba histolytica:* studies of adherence, secreted toxins and contact dependent cytolysis. Rev. Infect. Dis. **8:**247–260.

Ravdin, J. I., and R. L. Guerrant. 1982. A review of the parasite cellular mechanisms involved in the pathogenesis of amebiasis. Rev. Infect. Dis. **4:**1185–1207.

Ravdin, J. I., P. H. Schlesinger, C. F. Murphy, I. Y. Gluzman, and D. J. Krogstad. 1986. Acid intracellular vesicles and the cytolysis of mammalian target cells by *Entamoeba histolytica* trophozoites. J. Protozool. **33:**478–486.

Salata, R. A., A. Martinez-Palamo, H. W. Murray, L. Conales, N. Trevino, E. Segovia, C. F. Murphy, and J. I. Ravdin. 1986. Patients treated for amebic liver abscess develop cell-mediated immune responses effective *in vitro* against *Entamoeba histolytica*. J. Immunol. **136:**2633–2639.

Stilwell, G. G. 1955. Amebiasis: its early history. Gastroenterology **28:**606–622.

Walls, K. W. 1985. Serodiagnostic tests for parasitic diseases, p. 945–948. *In:* E. H. Lennette, A. Balows, W. J. Hausler, Jr., and H. J. Shadomy (ed). Manual of clinical microbiology. American Society for Microbiology, Washington, D.C.

Wheatley, W. B. 1951. A rapid staining procedure for intestinal amoebae and flagellates. Am. J. Clin. Pathol. **21:**990–991.

CHAPTER 77

American Trypanosomiasis (Chagas' Disease)

MARIO E. CAMARGO

Disease: American trypanosomiasis (Chagas' disease).

Etiologic Agent: *Trypanosoma cruzi.*

Source: A zoonosis of wild animals reaching the human population through domiciliation of insect vectors of the family *Reduviidae*. Transmission occurs by contact with feces of the vector, by transfusion of blood or organ transplant from infected donors, or by transplacental infection.

Clinical Manifestations: An acute infection, from mild or asymptomatic to severe, is present in the early stages of the disease, subsiding after few weeks and giving place to a chronic infection that lasts for life. The infection remains unnoticed or subclinical in 30 to 70% of patients, otherwise showing cardiac disturbances with arrhythmias, myocardiopathy and sudden death, and digestive manifestations such as megaesophagus or megacolon.

Pathology: With a parasite-mediated pathogenesis in the acute phase, the disease seems to be immunologically maintained in the chronic phase. As a result, changes in the electrical conductivity and dilation of heart chambers may occur; megaviscera are a consequence of denervation of ganglion cells in the parasympathetic myenteric plexus.

Laboratory Diagnosis: Parasitologic diagnosis through detection of trypanosomes by direct microscopic examination of blood samples or buffy coat is positive in 90% or more in the acute phase of infection. In the chronic phase, enrichment techniques such as xenodiagnosis or hemoculture are necessary, and are positive in about 50% of cases. For serologic diagnosis, a number of tests are available, including direct agglutination, passive hemagglutination, indirect immunofluorescence, and immunoenzymatic and complement fixation tests. With the usual crude *T. cruzi* antigens, a sensitivity of 98% or more is seen in the chronic phase. Specificity is high, cross-reactions being observed with leishmaniases. In acute and congenital forms of the disease, anti-*T. cruzi* IgM antibodies can be detected in serum.

Epidemiology: With only occasional cases in the southern regions of the United States, the infection is most often seen in Mexico and Central America. It is highly prevalent in South America, with an estimated 10 million or more infected patients. Although mostly rural, a high rate of infection can be found in urban areas among the underprivileged, as a consequence of extensive migrations.

Treatment: Therapy with trypanosomicide products can result in parasitologic cure in the acute phase of infection by drug-susceptible *T. cruzi* strains. No apparent benefits are seen in the chronic phase, when therapy is limited to support measures according to symptoms and eventual correction of megaviscera.

Prevention and Control: Control of domestic infestation with triatomine insects is essential, through insecticides and by improving housing and living conditions. Education of the population is of paramount importance for success. Serologic screening of blood donors or adding trypanosomicide chemicals such as gentian violet to blood are obligatory measures to prevent infection by blood transfusion.

Description of Disease

American trypanosomiasis (Chagas' disease) is a widespread endemic disease, with more than 10 million of infected people in Latin America. The mortality and decreased life expectancy it causes renders it one of the major public health problems in many areas in the continent. Attacking young persons and causing permanent tissue lesions that frequently lead to serious cardiac and digestive disturbances, Chagas' disease constitutes a heavy social burden.

Sources and Reservoirs

A zoonosis originally found in many orders of wild animals, such as *Marsupialia, Chiroptera, Carnivora, Rodentia, Primata,* and others, trypanosomiasis reaches the human population through domiciliation of insect vectors of several genera of the family *Reduviidae*. This spread usually results from human invasion and destruction of their natural ecotopes. The wide proliferation of vectors in or close to poor living conditions, the high susceptibility to the infection of animals found in or around houses (e.g., dogs, cats, goats, opossums, and rats) and of humans, and the long-standing parasitemia in infection contribute to the widespread diffusion of the disease in the human population.

After describing the causative hemoflagellate, *Trypanosoma cruzi,* in the blood of reservoirs and identifying its vectors, Carlos Chagas, a Brazilian physician, then identified the infection in humans and described the disease (Chagas, 1909).

Etiologic Agent

In its life cycle the parasite undergoes different stages of development with corresponding changes in morphology. In culture media and in the intestinal tract of the invertebrate host it develops as epimastigotes, a flagellar form with an anterior kinetoplast. Trypomastigotes, the nonmultiplying flagellar form with a posteriorly located kinetoplast, and amastigotes, intracellular multiplying forms with no flagella, are found in the vertebrate hosts and may develop in cells cultures.

Trypanosoma cruzi is placed in the section *Stercoraria* of the family *Trypanosomatidae* (Hoare, 1964), which includes trypanosomes that develop throughout the digestive tube of the vector and are transmitted by direct contact with triatomine feces. Another section, *Salivaria,* includes trypanosomes such as *T. brucei* and *T. congolense,* which develop anteriorly in the insect and are transmitted by inoculation of parasites located in salivary gland of the vectors. Populations of *T. cruzi* have a high degree of heterogeneity, the research of which is of fundamental importance for studies on the epidemiology, clinical polymorphism, diagnosis, prognosis, treatment, and pathology of the disease (Morel et al., 1984). At the phenotype level, *T. cruzi* strains may be subdivided according to zymodemes, or profiles of isoenzymes, which have revealed three principal groups of strains, all infective for humans. They show different geographic distributions and have been suggested to be linked to different forms of chronic disease (Miles, 1983). At the genotype level, *T. cruzi* strains may be subdivided according to schizodemes, profiles of restriction endonuclease digests of kinetoplast DNA.

Vectors and Modes of Transmission

Vectors and Natural Transmission

Vectors of *T. cruzi* are hematophagous insects of the subfamily Triatominae, family *Reduvidae,* order *Hemiptera,* with about 98 species belonging to 15 genera occurring only in neoartic and neotropic regions. They can be found in the American continents from as far north as Salt Lake City in the United States down to Patagonia, Argentina. Most species are sylvatic and, although frequently infected with *T. cruzi,* have no epidemiologic importance. The main vectors in different areas of Latin America are *Triatoma infestans, Panstrongylus megistus, Rhodnius prolixus, Triatoma dimidiata,* and *Triatoma brasiliensis,* which after migration from their natural ecotopes become adapted to houses, thus maintaining a close proximity to humans and domestic animals.

After an almost unnoticeable sting, the flat insect fills with blood and becomes rounded, which causes simultaneous elimination of infected feces. Parasites penetrate the pierced skin or are carried by the patient, after scratching, to mucosae such as in the eyes.

Transmission by Blood Transfusion

In addition to triatomines, blood transfusion is assuming great importance in the transmission of the disease, especially in urban areas. A consequence of extensive migrations of residents from endemic areas to cities, posttransfusion infection is attributable to the subpatent parasitemia in chronic infection, which in a large percentage is clinically inapparent, and to selling of blood, which frequently exploits the underprivileged.

Congenital Infection

Transplacental transmission of *T. cruzi* has been observed and can result in congenital infection. However, because the incidence varies in different populations, being virtually absent in some, congenital infection is probably associated with other contributing factors.

Accidental Infection

The risk of accidental infection in the laboratory is corroborated by a number of cases. Strict rules must be followed for protection of researchers and technicians.

Clinical Manifestations and Pathogenesis

In the acute phase, infection is frequently a clinically mild disease; in others, however, it is expressed as an infectious state with fever, malaise, and, as a result of myocarditis, tachycardia and hypotension. Severe cases of heart failure or meningoencephalitis also occur. At the inoculation point a local reaction may develop and be observed for 1 to 2 months, either as a *chagoma,* a hard, erythematous cutaneous swelling, or as the *Romaña signal,* a unilateral edema of both eyelids with enlargement of surrounding lymph nodes. After 1 to 4 months the infection subsides, with parasitemia decreasing as the immune response increases, and enters the chronic phase. This phase may be a quiescent stage in which the disease apparently does not progress: the indeterminate, sometimes lifelong, form of Chagas' disease. In a significant percentage of patients, however, that varies with geographic area but reaches 50% or more and increases with age, cardiac disturbances characteristic of Chagas' disease can be found. These disturbances are mainly dilatation of heart chambers, change in electrical conductivity and right bundle branch block, apical left ventricular aneurysm, and a high prevalence of sudden death. Megaviscera in the digestive tract, which affects 8 to 10% of patients, results from denervation of ganglion cells in the parasympathetic myenteric plexus, causing megaesophagus and megacolon.

Whereas the acute phase depends on a parasite-mediated pathogenesis, the chronic phase is probably a disease started by the parasite but immunologically maintained. Usually developing many years after initial infection, the serious disturbances found in the chronic phase have no relationship with intensity of parasitism, seeming to be parasite independent. There is evidence that the pathogenesis of the disease is related to an autoimmune mechanism presumably induced by host cells and parasite cross-reacting antigens from the parasite (Kierszenbaum and Hudson, 1985).

Laboratory Diagnosis of Chagas' Disease

Parasitologic Diagnosis

Fresh Blood Examination and Concentration Techniques

In the acute, invasive phase of the infection, actively moving trypanosomes can be found in most cases either by direct microscopic examination of fresh blood or after simple concentration techniques. The sediment of centrifuged serum delicately separated from clotted blood and cleared of remaining red cells by a previous slow centrifugation (Strout, 1962) or the buffy coat of hematocrit tubes are suitable materials for this purpose. The microhematocrit method, with a 10-min observation of each buffy coat spread between slide and coverglass from six 50-μl capillary tubes (Feilij et al., 1983), is most useful for testing infants or children in the diagnosis of congenital infection. This technique detects parasitemias as low as 1,000 trypanosomes per milliliter of blood.

Xenodiagnosis and Hemoculture

In the subpatent parasitemia phase of infection enrichment procedures based on parasite multiplication are used. The most common is xenodiagnosis but hemoculture may also be employed as reported by Chiari and Dias (1975). For xenodiagnosis, 20 to 40 third- or fourth-instar laboratory-reared vector triatomines made to fast for 3 or 4 weeks are allowed to feed on the patient. In addition to being unpleasant, direct feeding of triatomines on patients may lead to hypersensitive skin reactions; thus, an artificial technique was devised (Cedillos et al., 1982; Freitas et al., 1955). With this method, the insects feed on the patient's heparinized, warmed venous blood through a cow-gut membrane bag that is placed in a small pot containing the insects. Preferably the species used are the local vectors, such as *T. infestans* or *R. prolixus,* among others. A giant triatomine species found in New Mexico, *Dipetalogaster maximus,* which has the advantage of ingesting about six times more blood, has also been successfully used. After feeding, the insects are kept at 27° to 30°C with relative humidity around 85%, and feces are examined periodically for flagellates for up to 90 days. In the acute phase of infection positivity is 86 to 100%, but in chronic cases only about 50% (Cerisola et al., 1974; Schenone et al., 1968, 1974).

Serologic Diagnosis

Importance and Limitations

Most patients are seen in the chronic stage of infection; they are frequently asymptomatic or show signs and symptoms that are, at best, only suggestive of the disease. At this time parasitologic diagnosis is of limited application, as sensitivity is low and requires weeks to produce a result. The diagnostic gap is filled, however, by serodiagnosis. Serologic tests have several applications, among them to confirm a suspected infection and to aid in the selection of workers for jobs in which the physical stress required may aggravate a nonevident chagasic heart impairment, or where sudden death may represent a serious risk. Millions of tests are done yearly to screen for *T. cruzi*-infected blood donors in Latin America. These tests are also useful for evaluating antiparasite chemotherapy and for seroepidemiologic surveys.

Although their sensitivity is high and their specificity is satisfactory for most purposes, serologic tests provide only a probable diagnosis, as they depend on the variable characteristics of patients' immune responses. Therefore, their results must be evaluated according to other factors, such as prevalence of the disease (Galen and Gambino, 1975) as indicated by clinical, epidemiologic, and other data.

Serologic Tests

Several tests have been developed, differing in the parasite antigens employed and techniques used, that reveal the basic antigen–antibody reaction as a perceptible signal. This signal can be the lysis of red cells, as in the complement fixation tests (Freitas and Almeida, 1949; Guerreiro and Machado, 1913; Maekelt, 1960; Neva and Gam, 1977); the agglutination of trypanosomes or antigen-coated erythrocytes, as in the direct agglutination (Vattuone and Yanovsky, 1971) and hemagglutination (Camargo et al., 1973; Romaña, 1961) tests; fluorescence or development of color, as in immunofluorescence tests (Camargo, 1965; Camargo and Amato Neto, 1974), the microscope immunoperoxidase test (Ferreira et al., 1975), or the enzyme-linked immunosorbent assays (Camargo et al., 1986a,b,c; Lemesre et al., 1986; Voller et al., 1975); radionuclide emission, as in radioimmunoassay (Tarleton et al., 1984; Voller et al., 1977), and trypanosomal lysis, as in the complement-mediated lysis of living trypomastigotes (Krettli et al., 1982).

Antigens

Antigens may vary from parasite cells or crude extracts to partially purified fractions or immunochemically well-defined components of the parasite. Limitations in the immunologic specificity of the tests derive mainly from cross-reactions frequently observed with sera from cases of visceral or mucocutaneous leishmaniasis (Camargo and Rebonato, 1969) and possibly of *Trypanosoma rangeli* infections. A very high specificity can be obtained by substituting highly purified *T. cruzi* components, which are characteristic of the parasite, for the less purified extracts usually employed as antigens in these tests (Camargo et al., 1986b; Schechter et al., 1983). Figure 1 depicts the results of immunoenzymatic assay of serum samples, as described previously (Camargo et al., 1986b), from (A) patients with Chagas' disease, (B) residents in a tropical area with a high probability of no Chagas' disease or leishmaniasis, and (C) patients with kala-azar. With antigen E, a partially purified *T. cruzi* extract, differentiation between chagasic and nonchagasic sera was excellent; however, most sera from patients with kala-azar were also reactive. GP-25, an immunochemically well-defined glycoprotein antigen derived from *T. cruzi*-specific membrane components, produced much better specificity, with nonreactivity for kala-azar sera. Reactivity was lower for chagasic sera, however, and discrimination from nonchagasic sera was poorer. The expense involved should limit tests with such specific antigens to selected applications. A more practical approach for a specific test is based on blocking an enzyme-labeled monoclonal antibody from binding to a parasite extract on a solid phase. The high specificity of the assay derives from the monoclonal being directed to a *T. cruzi* characteristic antigen (Lemesre et al., 1986).

Standardization of Serodiagnosis

In these tests, the signal elicited must be proportional to the total amount of reacting antibodies; signal intensity may vary according to a series of factors, however, and may be disturbed by interfering "noises." Significant differences in diagnosis may be observed among laboratories, from diversities in type, batches, and source of antigens, and in the quality of the signal. Differences in criteria to evaluate test results contribute also for such disagreement. To cope with the urgent need for standardization of Chagas' disease serology, a collaborative study is being pursued in the American continents under the auspices of the Special Programme for Research and Training in Tropical Diseases (TDR) (World Health Organization/World Bank/United Nations Development Program) (Camargo et al., 1986c). In a progressively expanding network of laboratories, the ability of tests to differentiate between chagasic and nonchagasic serum samples has been evaluated with the help of selected panels of coded sera. In patients with Chagas' disease, serum titers

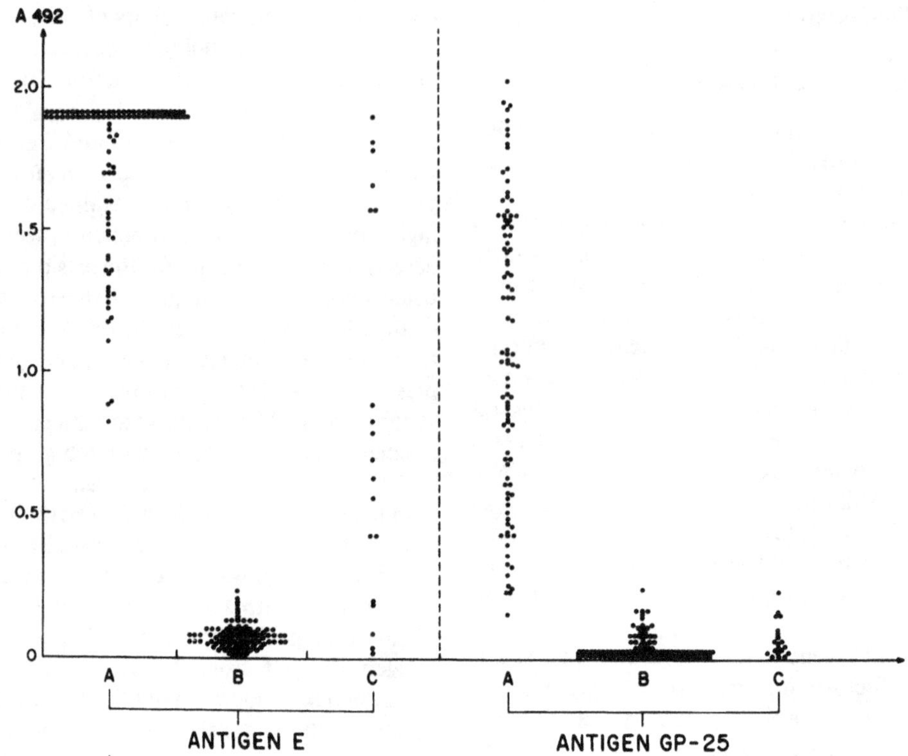

FIG. 1. Immunoenzymatic assay for Chagas' disease with a *Trypanosoma cruzi* crude antigen (E) and a purified glycoprotein antigen (GP-25). Sera were obtained from patients with Chagas' disease (A), without Chagas' disease or leishmaniasis (B), and with kala-azar (C) (Camargo et al., 1986b).

are shown to distribute around a median value in a Gaussianlike curve. Some activity is also observed in nonchagasic individuals, although with much lower median values; this finding is attributable to "natural" antibodies not related to the infection, but reacting with epitopes found in *T. cruzi* antigens and to signal-eliciting nonimmune impulses. These activity curves can overlap, and a cutoff must be chosen that best discriminates between these populations with a minimum of false-positive and -negative results. The same test can show wide variation in different laboratories, as depicted in Fig. 2 for the immunofluorescence test. The reproducibility of patterns of such titer frequency curves, as obtained in different circumstances and with different serum panels (Fig. 3), indicates the validity of the procedure. However, such evaluation should be greatly facilitated by the development of a rigorously standardized, sensitive, and specific reference test, with well-defined antigen, reagents, and technique.

SERODIAGNOSIS IN THE ACUTE AND CHRONIC PHASES

Antibodies to *T. cruzi* antigens can be detected in serum a few weeks after the patient is infected and could be found in 17 of 20 cases of posttransfusional Chagas' disease at the time the parasitologic diagnosis was made (Camargo et al., 1986a). Immunoglobulin (Ig)M antibodies, which appear early in the infection (Camargo and Amato Neto, 1974), are followed shortly by IgG antibodies; IgM antibodies are usually present for only 1 to 3 months, however, whereas IgG antibodies remain for life. Nonetheless, serologic negativity can result from an effective antiparasite chemotherapy instituted in the acute phase of infection. In chronic cases antiparasite therapy is infrequently followed by decreasing test titers, except for eventual negativity of the trypomastigote lysis test (Krettli et al., 1982), which should indicate a parasitologic cure. IgM antibodies can also be found in congenital cases of Chagas' disease (Stagno and Hurtado, 1970) for a few weeks after birth. In chronic cases, "natural" IgM antibodies can produce nonspecific positive results. For tests using whole parasites as antigens, such natural IgM antibodies should be inactivated previously, as in the direct agglutination test (Storni et al., 1975), or simply dismissed, as in the immunofluorescence test, by employing a fluorescent conjugate specific for IgM antibodies.

In chronic cases, test sensitivity is 96 to 98% or higher. No correlation is found between test titers and clinical forms or gravity of the infection.

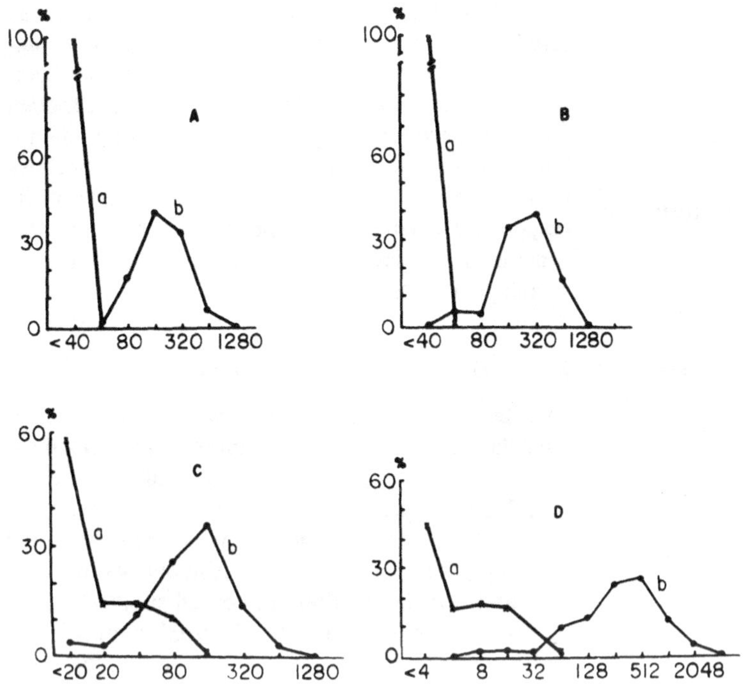

Fig. 2. Titer frequency curves from immunofluorescence test for Chagas' disease, for chagasic (B) and nonchagasic (A) sera, in different laboratories (Camargo et al., 1986c). Published with permission of the Pan American Health Organization.

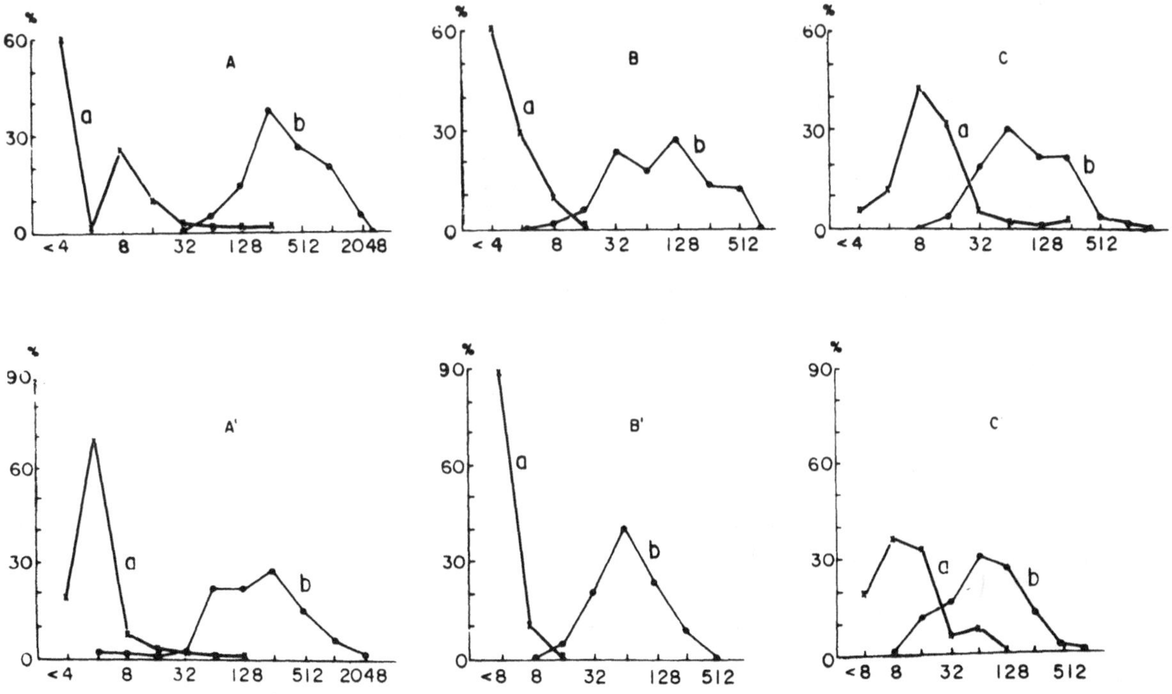

Fig. 3. Similar titer curve patterns for the same laboratories and tests, obtained months apart and with different serum panels. (a) Nonchagasic sera; (b) chagasic sera. (A, A') Direct agglutination test in laboratory A; (B, B') hemagglutination test in laboratory B; (C, C') direct agglutination test in laboratory C (Camargo et al., 1986c). Published with permission of the Pan American Health Organization.

Tests for Screening of Blood Donors and for Seroepidemiology

Because of the risk of infection through blood transfusion, two independent tests should be conducted to screen infected donors, using a low cutoff to ensure high sensitivity.

For epidemiologic purposes serologic tests are useful for determining not only the prevalence rates of the infection but also, through the study of low-age population layers, the presence of active transmission and the effect of sanitary measures to stop it. An example is the serologic survey of rural areas in Brazil, performed between 1975 and 1980, in which 1.3 million fingertip blood samples were assayed by the immunofluorescence test; this amount corresponds to a 1 to 2% sample of residents in every Brazilian county. In this fashion serologic prevalence could be mapped out (Camargo et al., 1984) as exemplified in Figs. 4 and 5.

Tests for Routine Serodiagnosis

For routine diagnosis of Chagas' disease inexpensive tests are necessary, as the disease is not only widespread in underdeveloped and developing countries but afflicts the most underprivileged in particular. The hemagglutination test is therefore highly recommended, as it is easy to perform and relies on inexpensive, stable, and standardized reagents. Although a fluorescence microscope is required, the immunofluorescence test is adequate for routine use because of its simple standardization and the low cost of its reagents. Moreover, with the development of halogen bulbs and interference filters, common microscopes can be adapted for fluorescence.

Heterophile and Anti-Tissue Antibodies

Circulating antibodies that react with the endothelial, vascular, and interstitial structures (EVI antibodies) of mouse, guinea pig, and bovine heart muscle are found in many patients infected with *T. cruzi* (Cossio et al., 1974). These antibodies, which may be found in other pathologies such as leishmaniasis and malaria and even in low titers in normal subjects, could be absorbed from serum by *T. cruzi* epimastigotes and by guinea pig erythrocytes. A heterophilic nature was suggested, as the EVI antibodies did not react with human tissue in indirect immunofluorescence tests (Khoury et al., 1983); at present, how-

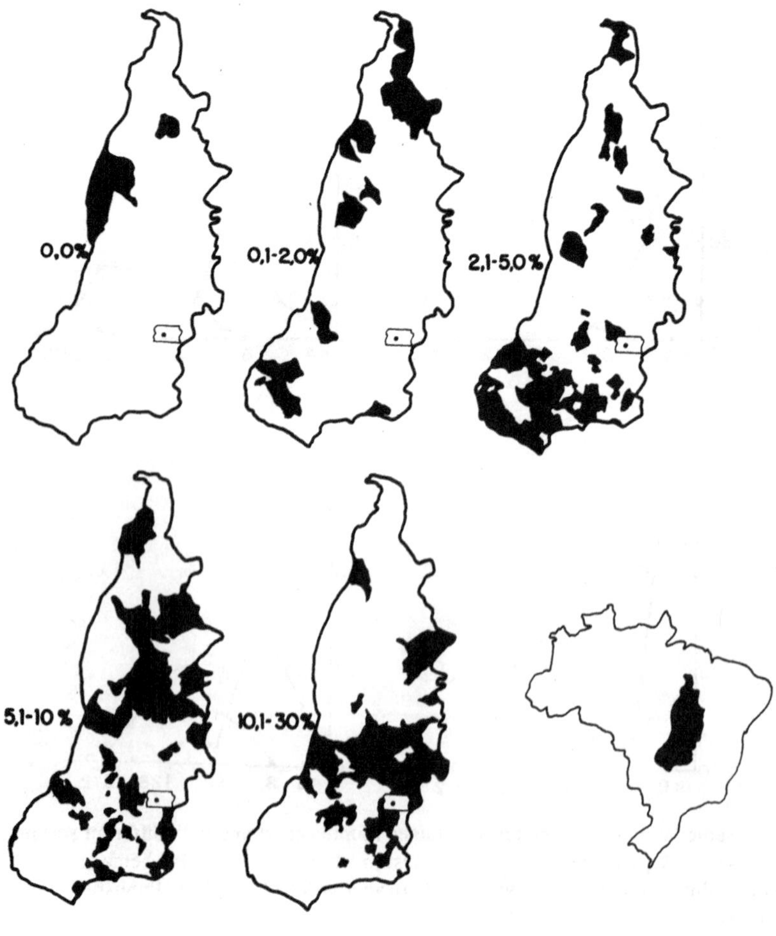

FIG. 4. Serologic prevalence of Chagas' disease in Brazilian populations: State of Goiás (Camargo et al., 1984). Published with permission of the Instituto de Medicina Tropical de São Paulo.

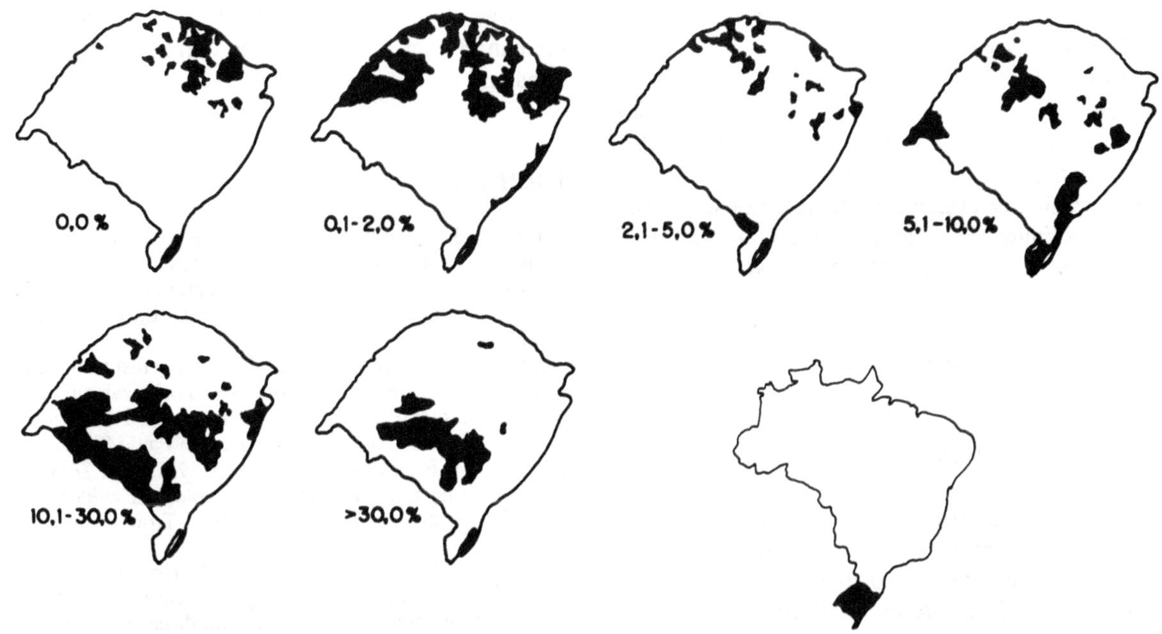

FIG. 5. Serologic prevalence of Chagas' disease in Brazilian populations: State of Rio Grande de Sul (Camargo et al., 1984). Published with permission of the Instituto de Medicina Tropical de São Paulo.

ever, they are regarded as a complex system represented by several antibodies. Pure EVI staining was found to correspond to reaction with laminin, a glycoprotein of connective tissue structures (Szarfman et al., 1982). Other antibodies demonstrated in chagasic sera reacted with β-adrenoceptors of the myocardium (Borda et al., 1984) and with the Schwann cell sheath or nerve body (Ribeiro dos Santos et al., 1979). Also, a monoclonal antibody raised against *T. cruzi* epimastigotes was found to react to a wide range of mammalian neural elements (Snary et al., 1983); another, raised against rat dorsal root ganglia, was found to react with *T. cruzi* (Wood et al., 1982). Presence or titers of EVI antibodies could not be correlated with clinical symptoms or signs occurring in chronic Chagas' disease or with the severity of the disease (Peralta et al., 1982). However, deposits of IgG, IgA, IgM, and C_3 could be demonstrated in the interstitial area, in the capillaries, and on the surfaces of endocardial or myocardial fibers in patients with chagasic cardiomyopathy when EVI antibodies were present in sera (Molina et al., 1984). In acute Chagas' disease anti-sheep erythrocyte agglutinin titers of 112 to 1,972 or higher are usually seen, but fall to 56 or less within 1 to 6 months. In contrast to infectious mononucleosis or Forssman antibodies, such agglutinins are readily absorbed with guinea pig kidney or ox cell antigens. They have been useful in laboratory detection of nonsuspected acute Chagas' disease in endemic areas or posttransfusional infections through the occasional observation of a strong sheep cell agglutination when complement fixation or heterophile antibody tests for serodiagnosis of other diseases are performed (Camargo et al., 1986a).

Epidemiology

In the Northern Hemisphere, Chagas' disease occurs in Mexico and Central America, with rare cases registered in the United States. In the Southern Hemisphere it is found in all South American countries. In Brazil, several surveys indicate a median serologic prevalence of 5.3%, with an estimated total of 4 to 5 million chagasic patients. About 100,000 new cases occur each year, of which 10,000 to 20,000 are caused by blood transfusion (Coura, 1985; Dias, 1985). With approximately 10 million infected patients in the continent, *T. cruzi* represents a serious risk of disease for more than 40 million people residing in endemic areas and in urban areas, where infected people are continuously migrating.

Treatment

Drugs such as Nifurtimox and Benznidazole show varying chemotherapeutic activity against different strains of *T. cruzi*. In the acute phase of infection, parasitologic cure is attainable through chemotherapy when the infecting strain is susceptible to the drug. Permanent serologic negativity is then observed. New improved drugs are necessary, for which pharmaceutical research is, however, very limited or lacking.

In the chronic phase, the clinical effects of chemotherapy continue to be unimpressive and controversial. Organic lesions are already established or no modification can be seen in aggravation of progressive organic impairment, probably because the parasite is of limited importance in the immunologically dependent chronic stage of the disease.

Prevention and Control

Control of Vectors

To interrupt transmission of *T. cruzi* infection to humans, vectors must be controlled. Spraying houses with insecticides such as benzene hexachloride will reduce domestic infestations with triatomine insects. Such campaigns demand an attack phase directed at every house in an infested area, followed by selective spraying of reinfested houses. Its success depends on making local people aware of the problem so that they take an active part in the campaign. Improving living conditions by replacing adobe huts with better houses and by smoothing walls with kaolin to prevent cracks and subsequent domiciling by triatomine insects represent other important measures in the campaign. Control should be based, however, on sanitary education and general improvement of living conditions.

Vaccine

Development of a vaccine seems limited by the involvement of the host's immune response in the pathogenesis of the disease. Experimental evidence suggests that protective immunity may be sometimes strain specific.

Prevention of Blood Transfusion Infections

The high percentage of infected candidates who donate blood, in endemic zones and in urban areas, make routine careful serologic screening necessary. Another important measure is adding gentian violet to blood, which inactivates trypomastigotes in a few hours; this measure should be obligatory in endemic zones. However, the resulting strong blood coloration and batch variations in activity make gentian violet less than ideal for preventing transmission of *T. cruzi* by blood transfusion. Current investigations indicate amphiphilic amines as possible alternative blood additives (Hammond et al., 1985).

Acknowledgments. I am most grateful to Drs. Zigman Brener and A. Walter Ferreira for suggestions and revision of the text.

Literature Cited

Borda, E., J. Pascual, P. Cossio, M. de la Vega, R. Arana, and L. Sterin-Borda. 1984. A circulating IgG in Chagas' disease which binds to β-adrenoceptors of myocardium and modulates their activity. Clin. Exp. Immunol. **57**:679–686.

Camargo, M. E. 1965. Fluorescent antibody test for diagnosis of American trypanosomiasis. Technical modification employing culture forms of *Trypanosoma cruzi* in a slide test. Rev. Inst. Med. Trop. São Paulo **8**:227–234.

Camargo, M. E., and V. Amato Neto. 1974. Anti-*Trypanosoma cruzi* IgM antibodies as serological evidence of recent infection. Rev. Inst. Med. Trop. São Paulo **16**:200–202.

Camargo, M. E., A. W. Ferreira, and S. Hoshino-Shimizu. 1986a. Serodiagnosis of Chagas' disease. *In* K. Walls (ed.), Immunoserology of parasitic diseases. Academic Press, Inc., New York (in press).

Camargo, M. E., A. W. Ferreira, B. A. Peres, L. Mendonça-Previato, and J. Scharfstein. 1986b. *Trypanosoma cruzi* antibodies, p. 368–382. *In* H. U. Bergmeyer (ed.), Methods of enzymatic analysis, 3rd ed., vol. 11. Verlagsgesellschaft, Weinheim.

Camargo, M. E., S. Hoshino-Shimizu, and G. R. Siqueira. 1973. Hemagglutination with preserved sensitized cells, a practical test for routine serologic diagnosis of American trypanosomiasis. Rev. Inst. Med. Trop. São Paulo **15**:81–85.

Camargo, M. E., and C. Rebonato. 1969. Cross-reactivity in fluorescence tests for *Trypanosoma* and *Leishmania* antibodies. A simple inhibition procedure to ensure specific results. Am. J. Trop. Med. Hyg. **18**:500–505.

Camargo, M. E., E. L. Segura, I. G. Kagan, J. M. P. Souza, J. R. Carvalheiro, and M. C. S. Guimarães. 1986c. Three years of collaboration on the standardization of Chagas' disease serodiagnosis on the American continent. An appraisal. Bull. Pan Am. Health Organ. **20**:233–244.

Camargo, M. E., G. R. Silva, E. A. Castilho, and A. C. Silveira. 1984. Inquérito sorológico da prevalência da infecção chagásica no Brasil, 1975/1980. Rev. Inst. Med. Trop. São Paulo **26**:192–204.

Cedillos, R. A., J. W. Torrealba, R. J. Tonn, W. Mosca, and A. Ortega. 1982. El xenodiagnóstico artificial en la enfermedad de Chagas. Bol. Of. Sanit. Panam. **93**:240–249.

Cerisola, J. A., R. Rohwedder, E. L. Segura, C. E. Del Prado, M. Alvarez, and G. J. W. De Martini. 1974. El xenodiagnóstico—Instituto Nacional de Diagnóstico y Investigacion de la Enfermedad de Chagas "Dr. Mario Fatala Chaben," Buenos Aires.

Chagas, C. 1909. Nova trypanosomiase humana. Estudos sobre a morphologia e cyclo evolutivo do *Schizotrypanum cruzi* n. gen., n. sp., agente etiológico de nova entidade mórbida do homem. Mem. Inst. Oswaldo Cruz Rio de J. **1**:159–218.

Chiari, E., and J. C. P. Dias. 1975. Nota sobre uma nova técnica de hemocultura na doença de Chagas na sua fase crônicas. Rev. Soc. Bras. Med. Trop. **9**:133–136.

Cossio, P. M., C. Diez, A. Szarfman, E. Kreutzer, B. Candiolo, and R. M. Arana. 1974. Chagasic cardiopathy. Demonstration of a serum gamma-globulin factor which reacts with endocardium and vascular structures. Circulation **49:**13–21.

Coura, J. R. 1985. Doença de Chagas como endemia urbana, p. 356–361. *In* J. R. Cançado and M. Chuster (ed.), Cardiopatia chagásica. Fundação Carlos Chagas, Belo Horizonte.

Dias, J. C. P. 1985. História natural, p. 99–113. *In* J. R. Cançado and M. Chuster (ed.), Cardiopatia chagásica. Fundação Carlos Chagas, Belo Horizonte.

Feilij, H., H. Muller, and S. M. Gonzalez Cappa. 1983. Direct micromethod for diagnosis of acute and congenital Chagas' disease. J. Clin. Microbiol. **18:**327–330.

Ferreira, A. W., M. E. Camargo, and O. S. Nakahara. 1975. *Trypanosoma cruzi:* immunoperoxidase antibody test for serologic diagnosis. Exp. Parasitol. **37:**131–137.

Freitas, J. L. P., and J. O. Almeida. 1949. Nova técnica de fixação do complemento para moléstia de Chagas. Hospital (R. de Janerio) **35:**787–800.

Freitas, J. L. P., V. Nussenzweig, V. Amato Neto, and R. Sontag. 1955. Estudo comparativo entre xenodiagnóstico praticado "in vivo" e "in vitro" em formas crônicas da moléstia de Chagas. Hospital (Rio de J.) **47:**181–188.

Galen, R. S., and R. S. Gambino. 1975. Beyond normality: the predictive value and efficiency of medical diagnosis. John Wiley & Sons, Inc., New York.

Guerreiro, C., and A. Machado. 1913. Da reação de Bordet e Gengou na moléstia de Carlos Chagas como elemento diagnóstico. Bras. Med. **27:**225–226.

Hammond, D. J., J. Hogg, and W. E. Gutteridge. 1985. *Trypanosoma cruzi:* possible control of parasite transmission by blood transfusion using amphiphilic cationic drugs. Exp. Parasitol. **60:**32–42.

Hoare, C. A. 1964. Morphological and taxonomic studies on mammalian trypanosomes. X. Revision of the systematics. J. Protozool. **11:**200–207.

Khoury, E. L., C. Diez, P. M. Cossio, and R. M. Arana. 1983. Heterophil nature of EVI antibody in *Trypanosoma cruzi* infection. Clin. Immunol. Immunopathol. **27:**283–288.

Kierszenbaum, F., and L. Hudson. 1985. Autoimmunity in Chagas disease: cause or symptom? Parasitol. Today **1:**4–9.

Krettli, A. U., J. R. Cançado, and Z. Brener. 1982. Effect of specific treatment on the levels of lytic antibodies in human Chagas' disease. Trans. R. Soc. Trop. Med. Hyg. **76:**334–340.

Lemesre, J. L., D. Afchain, O. Orozco, M. Loyens, F. S. Breniere, P. Desjeux, Y. Carlier, U. Martin, J. A. Nogueira-Queiroz, D. Le Ray, and A. Capron. 1986. Specific and sensitive immunological diagnosis of Chagas' disease by competitive antibody enzyme immunoassay using a *Trypanosoma cruzi*-specific monoclonal antibody. Am. J. Trop. Med. Hyg. **35:**86–93.

Maekelt, G. A. 1960. Die Komplementbindungsreaktion der Chagaskrankheit. Z. Tropenmed. Parasitol. **11:**152–186.

Miles, M. A. 1983. The epidemiology of South American trypanosomiasis—biochemical and immunological approaches and their relevance to control. Trans. R. Soc. Trop. Med. Hyg. **77:**5–23.

Molina, H. A., J. Milei, and R. Storino. 1984. Chronic Chagas' myocardiopathy. Demonstration of in vivo bound immunoglobulins in heart structures by the immunoperoxidase technique. Cardiology **71:**297–306.

Morel, C. M., A. M. Gonçalves, M. P. Deane, E. Chiari, M. Carneiro, and A. J. Romanha. 1984. Schizodeme characterization of natural and artificial populations of *Trypanosoma cruzi* as a tool in the study of Chagas' disease, p. 253–275. *In* B. N. Newton and F. Michal (ed.), New approaches to the identification of parasites and their vectors. Schwabe, Basel.

Neva, F. A., and A. A. Gam. 1977. A complement fixation antigen from *Trypanosoma cruzi* grown in cell cultures. Am. J. Trop. Med. Hyg. **26:**37–36.

Peralta, J. M., D. A. Manigot, E. O. A. Muscelli, T. C. R. Magalhães, E. A. Almeida, and A. Bastos. 1982. Anticorpos EVI e NP na infecção chagásica crônica. Estudo em pacientes com diferentes formas clínicas. Rev. Inst. Med. Trop. São Paulo **24:**6–10.

Ribeiro dos Santos, R., J. O. Marquez, C. C. Von Gal Furtado, J. C. Ramos, A. R. Martini, and F. Köberle. 1979. Antibodies against neurons in chronic Chagas' disease. Tropenmed. Parasitol. **30:**19–23.

Romana, C. 1961. Aplication del metodo de hemaglutinacion al diagnostico de la enfermedad de Chagas. Rev. Soc. Argent. Biol. **37:**73–76.

Schechter, M., J. E. Flint, A. Voller, F. Guhl, C. J. Marinkelle, and M. A. Miles. 1983. Purified *Trypanosoma cruzi* specific glycoprotein for discriminative serological diagnosis of South American trypanosomiasis (Chagas' disease). Lancet **2:**939–941.

Schenone, H., E. Alfaro, H. Reyes, and E. Taucher. 1968. Valor del xenodiagnóstico en la infeccion chagásica cronica. Bol. Chil. Parasitol. **23:**149–154.

Schenone, H., E. Alfaro, and A. Rojas. 1974. Bases y rendimiento del xenodiagnóstico en la infeccion chagasica humana. Bol. Chil. Parasitol. **29:**24–26.

Snary, D., J. E. Flint, J. N. Wood, M. T. Scott, M. D. Schapman, J. Dodd, T. M. Jessell, and M. A. Miles. 1983. A monoclonal antibody with specificity to *Trypanosoma cruzi,* central and peripheral neurons and glia. Clin. Exp. Immunol. **54:**617–624.

Stagno, S., and R. Hurtado. 1970. *Trypanosoma cruzi.* Una nueva causa de infeccion congenita capaz de inducir la formaccion de anticuerpos IgM. Bol. Chil. Parasitol. **25:**90–91.

Storni, R., F. L. Bolsi, and J. F. Yanovsky. 1975. Reaccion de agglutinacion direta para diagnostico de la enfermedad de Chagas. Utilizacion sistematica del 2-mercaptoethanol para la eliminacion de las aglutininas inespecificas. Medicina (Buenos Aires) **35:**67–72.

Strout, R. G. 1962. A method for concentrating hemoflagellates. J. Parasitol. **48:**100.

Szarfman, A., V. P. Terranova, S. I. Rennard, J. M. Foidart, M. F. Lima, J. I. Scheinman, and G. R. Martin. 1982. Antibodies to laminin in Chagas' disease. J. Exp. Med. **155:**1161–1171.

Tarleton, R. L., C. L. Schultz, M. Grögl, and R. E. Kuhn. 1984. Diagnosis of Chagas' disease in humans using a biotin-3H-avidin radioimmunoassay. Am. J. Trop. Med. Hyg. **33:**34–40.

Vattuone, N. H., and J. F. Yanovsky. 1971. *Trypanosoma cruzi:* agglutination activity of enzyme-treated epimastigotes. Exp. Parasitol. **30:**349–355.

Voller, A., D. E. Bidwell, A. Barlett, and R. Edwards. 1977. A comparison of isotopic and enzyme immunoassays for tropical parasitic diseases. Trans. R. Soc. Trop. Med. Hyg. **71:**431–437.

Voller, A., O. C. Draper, D. E. Bidwell, and A. Barlett. 1975. Microplate enzyme-linked immunosorbent assay for Chagas' disease. Lancet **1:**426–429.

Wood, J. N., L. Hudson, T. M. Jessel, and M. Yamamoto. 1982. A monoclonal antibody defining antigenic determinants on subpopulations of mammalian neurones and *Trypanosoma cruzi* parasites. Nature (London) **296:**34–38.

CHAPTER 78

Neurologic Angiostrongyliasis: Parasitic Eosinophilic Menigoencephalitis

MICHAEL M. KLIKS, WILLIAM K. K. LAU, and NICHOLAS E. PALUMBO

Disease: *Angiostrongyliasis cantonensis** (parasitic eosinophilic meningitis, meningoencephalitis, or radiculomyeloencephalitis; Alicata's disease). Intraabdominal infection by a related species, *Angiostrongylus costaricensis,* occurs in Central America and perhaps elsewhere.

Etiologic Agents: *Angiostrongylus cantonensis** (*Nematoda: Strongylata*). Possibly *A. malayensis* and *A. mackerrasae* where they occur in Peninsular Southeast Asia and Australia, respectively.

Sources of Infection: Accidental or purposeful ingestion (as food) of third-stage larvae, usually with infected or contaminated intermediate hosts (snails, slugs) or transport hosts (freshwater prawns, crabs, or shrimp), planaria, fish, or inanimate objects (fomites) such as lettuce and other vegetables and dishware and water that were in contact with infected intermediate or transport hosts.

Clinical Manifestations: The illness usually presents as an acute, mild meningeal irritation with headache and nuchal rigidity and intracranial pressure resulting in nausea, vomiting, and decreased sensorium; low-grade fever may be present. More severe infections also give rise to symptoms of meningoencephalitis and radiculoencephalitis with paresthesias and/or cranial and other nerve paresis. Ocular infection may cause diplopia and other visual symptoms. The first signs of infection are dry cough with sneezing and rhinorrhea caused by transit of larvae through the lungs during the first minutes of their migration to the central nervous system. Rarely, a chronic form of the disease may have a delayed onset with mild symptoms of meningitis.

Associated Patient History: There is no marked age- or sex-related distribution pattern; however, outbreaks and clusters of cases do occur among family members and persons sharing meals of raw infected mollusks and crustaceans. In areas where ingestion of these foods raw is not common practice, patients usually give no history of known association with potential intermediate hosts. Infections in infants and young children are often associated with observed ingestion of small snails and slugs. Infections in adolescent and young adult males have been associated with ingestion of mollusks persuant to dares by peers.

Laboratory Diagnosis: Examination of stained of cerebrospinal fluid smears reveals the characteristic eosinophilic pleocytosis. During the first week of symptomatic illness, living larvae may be recovered by centrifugation of cerebrospinal fluid. Ocular infection may be revealed by slit-lamp examination. Serologic confirmation of presumptive

continued

* Ubeleker (1986) has reviewed the genus *Angiostrongylus* and transfered *A. cantonensis, A. malaysiensis,* and *A. mackerrasae* to *Parastrongylus.*

continued

physical diagnosis is possible using a microenzyme-linked immunosorbent assay for parasite-specific antibodies in serum or cerebrospinal fluid; titers may not rise to detectable levels until the second week after infection.

Epidemiology: Human infections have been most commonly acquired in Taiwan, Japan, Hong Kong, throughout mainland and peninsular Southeast Asia, Micronesia, Melanesia, Polynesia, and Hawaii. More recently, human cases have been diagnosed in Madagascar and the Ivory Coast of Africa and in Cuba. During the last several years, infections in rodents and mollusks have been recorded in Puerto Rico and the New Orleans area. Spread of the disease among rodents has been associated with dispersal of the African land snail (*Achatina fulica*). Many human cases have been related to ingestion as food of transport hosts such as freshwater prawns or by the accidental ingestion of small mollusks.

Treatment: There is no specific anthelmintic drug of established safety and efficacy. Symptomatic treatment is directed at controlling central nervous system inflammation and hypertension and mediating pain and fever.

Prevention and Control: Information regarding avoidance of contact with, or ingestion of, potential sources of infection must be disseminated by public health authorities and educators in the enzootic areas. Rodent and snail control programs are expensive, require continuous expenditures, and are seldom fully effective. Appropriate food hygiene measures should be adopted at vegetable farms, at points of sale, and by the consumer to avoid accidental ingestion of infected slugs or planaria on leafy vegetables.

Description of Disease

Angiostrongyliasis refers to infection of humans with the bursate nematode of rats, *Angiostrongylus cantonensis,* which most commonly presents as an acute eosinophilic meningitis or meningoencephalitis. However, ocular and pulmonary infections with or without neurologic symptoms also occur. It is important to note that this disease may follow a chronic course with a 2- to 4-month indistinct prepatent period (Reid and Wallis, 1984). Given the rapidity and volume of interstate and international travel and the geographic expansion of angiostrongyliasis into prime regions of tourism, it is very likely that eosinophilic meningitis will be seen by practitioners outside of the current endemic areas. The recent diagnosis of neurologic angiostrongyliasis in a 16-year-old in Utah (Fisher, 1983) who had lived in rural Kauai, Hawaii exemplifies the need to consider infection by this parasite in cases of meningitis with eosinophilic pleocytosis of the cerebrospinal fluid (CSF).

History and Distribution

The adult males and females of the parasite were first noted in the pulmonary arteries of naturally infected rats in China in 1935. The life cycle (Fig. 1) in rodents was later shown to involve migration through the central nervous system (CNS) by the developing worms after ingestion of infective larvae contained within, or released from, infected molluskan intermediate hosts (Mackerras and Sandars, 1955).

As a human disease, worms were first noted in the CSF of a patient in Taiwan as early as 1944 (Nomura and Lin, 1945). This wartime finding was not widely known until after 1961, when the work of J. E. Alicata, L. R. Ash, and their associates at the University of Hawaii and Leon Rosen of the National Institutes of Health showed that *A. cantonensis* was the cause of the typical symptoms of eosinophilic meningitis throughout the Pacific Basin and Southeast Asia (Alicata, 1962; Horio and Alicata, 1961; Rosen et al., 1962).

An outbreak of eosinophilic meningitis of unknown etiology involving more than 200 cases was described on Ponape, Eastern Carolines (Federated States of Micronesia) in 1946 through 1948 (Bailey, 1948). During the next two decades infection in rodents and human cases with symptoms characteristic of the disease appeared in the Philippines (1946), Sumatra and New Caledonia (1951), Thailand (1957), Tahiti (1958), Vietnam and Malaysia (1960), and the islands of the far western Pacific, dispersed by the activities and social stresses of the wars in Asia. Alicata and Jindrak (1970) have summarized the fascinating history of this zoonosis up to 1969.

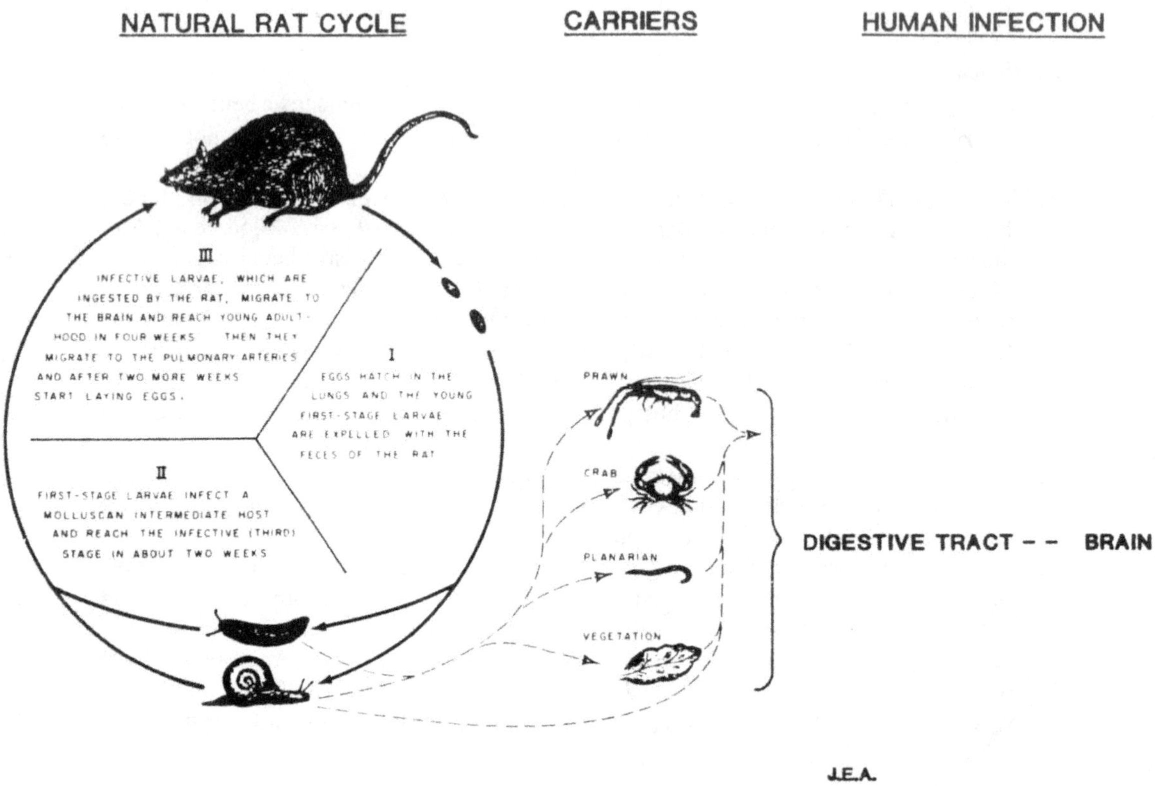

FIG. 1. Life cycle of rat lungworm, *Angiostrongylus cantonensis,* and possible avenues of human infection. (Adapted from Alicata, 1964. *Parasitic Infections of Man and Animals in Hawaii,* Technical Bulletin No. 61, Hawaii Agricultural Experimental Station. Courtesy of the College of Tropical Agriculture and Human Resources, University of Hawaii at Manoa.)

Postwar commerce and travel and the efficient vectorial capacity of the giant African land snail, *Achatina fulica,* facilitated the spread of larval worms in mollusks and adult worms in rats. The robust hosts and well-adapted parasites traveled as supercargo eastward and southward through Micronesia, onto Guam and the Marianas, throughout Melanesia, southward to the Cook Islands, Samoa, and northeastward to Hawaii, where presumptive human infections were noted during the late 1950s.

Subsequently the parasite was introduced into Africa, where human cases occurred on the islands of Madagascar (1964), Mauritius (1965), and Reunion (1968). The parasite was detected in rats in Egypt in 1977, and a presumptive human case was diagnosed in the Ivory Coast in 1979 (Fig. 2). The introduction of the parasite onto the African continent, where *A. fulica* is native and abundant, is particularly ominous.

More recently, isolated cases or sporadic outbreaks have occurred in Hong Kong (Ko et al., 1984), Queensland, Australia, Tahiti, New Hebrides (Bowden, 1981), New Britain (Scrimqeour, 1984), Okinawa, Java, Western Samoa (Reid and Wallis, 1984), and American Samoa (Beck et al., 1980; Enzenauer et al., 1983; Kliks et al., 1982). During the last 5 years, occasional presumptive cases have been acquired on the Hawaiian islands of Oahu, Molokai, Hawaii (Kliks, Lau, and Gee, unpublished observations; Koo et al., in review) and Kauai (Fisher, 1983).

The parasitocenose continues to spread into new areas. In 1973 the first presumptive human case occurred in Cuba (Pasqual et al., 1981). Infected rodents were detected for the first time during routine surveys in Puerto Rico in 1984 (Anderson et al., 1986) and near the port area of New Orleans in 1984 (B. G. Campbell and M. D. Little, personal communication). Infection in rats and mollusks in Fiji was first noted in 1981 (Uchikawa et al., 1984). Two human cases were detected in Bombay, India in 1980 by Sharma et al. (1981).

It has been amply demonstrated that this host-parasite system is readily dispersed by human activities and can rapidly become entrenched in new environments. *Angiostrongylus cantonensis* will no doubt continue to expand its range wherever biologic and social conditions permit the establishment of large populations of susceptible mollusks. The estab-

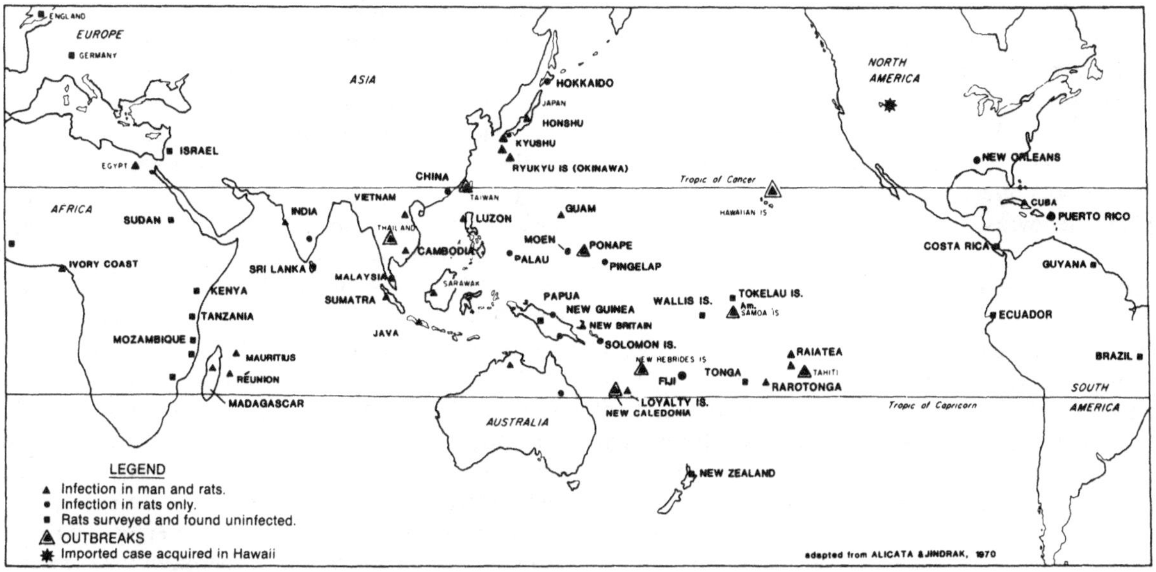

FIG. 2. Known distribution of *Angiostrongylus cantonensis* in humans and rodents through 1987. (Adapted from Alicata and Jindrak, 1970. *Angiostrongyliasis in the Pacific and Southeast Asia*. Courtesy of Charles C Thomas, Publisher, Springfield, Ill.)

lishment of this disease in India, tropical Africa, temperate North America, and the neighboring Caribbean Islands indicates that many more human cases will occur in these areas in the near future.

Clinical Manifestations

Based on our review of data on more than 600 clinically characterized cases reported in the literature (Cross, 1978; Kliks et al., 1982; Kuberski and Wallace, 1979; Punyagupta et al., 1975; Yii, 1976) and on our personal experience with 15 presumptive cases seen in Hawaii during the past 5 years (Gee et al., 1984), the following provides a general description of the clinical aspects of angiostrongyliasis cantonensis.

The incubation period after ingestion of larvae ranges from 1 to more than 30 days. The severity and spectrum of signs and symptoms are related to the dose of larvae ingested. Most human infections are probably subclinical; however, clinically apparent neurologic angiostrongyliasis typically presents as a meningitis characterized by headache, nausea, vomiting, nuchal rigidity, and paresthesias, sometimes accompanied by low-grade fever. Severe, intermittent, occipital and bitemporal headache is the most common symptom. Symptoms related to meningeal irritation and increased intracranial pressure are also common. Blurred vision, diplopia, photophobia, and retroorbital pain may occur secondary to ocular involvement. Mild, transient pulmonary symptoms of congestion and rales have been noted very early in the infection concurrent with the migration of larvae through the lung parenchyma.

In more severe cases of radiculomyeloencephalitis, exquisitely painful paresthesias of the trunk and extremities, usually unilateral, may involve peripheral as well as cranial nerves. Signs of sensorium impairment including lethargy, general motor weakness, and cranial nerve paralysis may occur. Typically affected are nerves VII and VI, with facial paralysis and lateral rectus paralysis, respectively. Diminished to absent deep tendon reflexes and weakness of limbs are common in heavy, severe infections. Occasionally frank psychosis may be seen.

In contrast to other meningitides, low fever may be present in only about one-third of cases and is usually more common in children than adults. Most infections probably do not come to the attention of a physician, and the majority of those that do are self-limited and benign; only the heaviest infections cause permanent sequelae or death.

For example, in a recent outbreak among a group of Korean fishermen visiting American Samoa, 12 men became ill after a shared meal of raw or partially cooked giant African land snails. Symptoms of radiculomyeloencephalitis appeared during the first 24 h to 6 days (Kliks et al., 1982). In this group, headache, nuchal rigidity, and fever were not common and vomiting did not occur; most patients experienced excruciating pain in the limbs and trunk, three became comatose, one died after removal from a respirator, and two suffered long-term neurologic deficits.

The diagnosis of cerebral angiostrongyliasis is based on clinical signs and symptoms along with a history of residence or travel in an enzootic area and possible association with mollusks. When warranted by the presence of appropriate symptoms and indicative history, a spinal tap should be performed. Examination of CSF preparations treated with Giemsa–May–Gruenwald stain from patients with *A. cantonensis* infection reveals a characteristic eosinophilic pleocytosis as discussed below.

Pathogenesis and Pathology

When infective third-stage larvae are ingested by humans or other appropriate mammalian hosts, they penetrate the capillary beds of the gastrointestinal tract, enter the venules, and pass through the right side of the heart to reach the lungs via the pulmonary artery within a few hours after infection. Transient or intermittent dry cough with sneezing and rhinorrhea may be associated with this phase (Sharma et al., 1980; Yii, 1976). Larvae leave the lungs by the pulmonary vein, reach the left side of the heart, and are pumped into the general circulation. In experimentally infected animals, many larvae continue to circulate; some are delayed in the major organs and some enter the skeletal muscle fibers and other tissues. A few larvae reach the brain within 4 h, but most arrive 24 to 48 h after infection in rats (Jindrak, 1968) and somewhat later, presumably, in humans. In experimental rodent hosts worms may remain alive in the CNS for up to 3 weeks (John and Martinez, 1975).

Clinical symptoms are related to damage caused by the developing larvae and young adult worms in the tissues and spaces of the CNS and associated nerve roots and trunks. The neuropathology in humans has been well characterized (Jindrak and Alicata, 1965; Kliks et al., 1982; Nye et al., 1970; Sonakul, 1978) and compared with other, similar parasitic neuropathies (Bia and Barry, 1986; Brown and Voge, 1982). Third-stage larvae accumulate in the gray matter of the brain, spinal cord, and nerve roots, where they molt. They then migrate into the subarachnoid space, where they have a final molt to the adult stage and undergo a brief period of growth. In susceptible rodent hosts, and rarely in humans (Yii et al., 1968), the young adult worms leave the brain via the cerebral veins, pass through the heart, and enter the pulmonary artery, where they grow rapidly, mate, and produce eggs about 5 weeks after entering the host.

At autopsy the brain may appear grossly swollen and congested, with engorged meningeal vessels and thickened leptomeninges. Diffuse subarachnoid hemorrhage may be evident over both brain and spi-

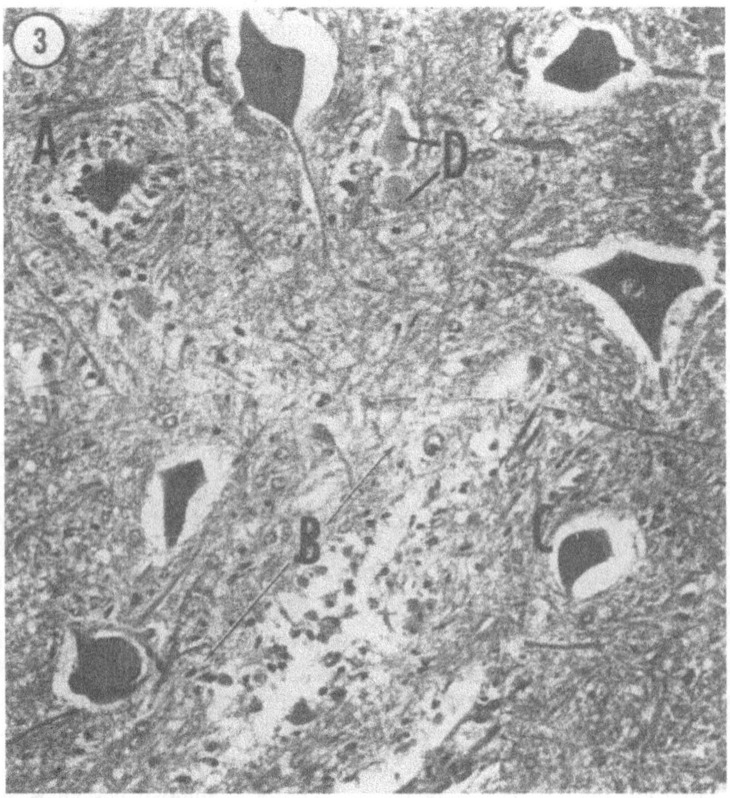

FIG. 3. Composite of typical lesions of gray matter of the spinal cord in a patient with radiculomyeloencephalitis. (A) Neuronophagia of motor neuron. (B) Empty migratory track with mononuclear cells (*arrows*). (C) Central chromatolysis. (D) Reactive axonal swelling of motor neuron. Hematoxylin and eosin. Magnification ×160; reduced by 30%. (From Kliks et al., 1982. Am. J. Trop. Med. Hyg. **31:**1114–1122; with permission.)

nal cord. Living worms are often visualized on the surface of the brain and spinal cord or in the subarachnoid spaces and can be recovered from the saline solution in which these tissues have been placed.

In fatal cases microscopic histopathologic findings are important for postmortem confirmation of infection, because living and dead larvae are frequently demonstrable in brain and spinal cord tissues. Even when larvae are not seen, the presence of characteristic cellular changes and worm tracks is pathognomonic.

Characteristic tracks in both gray and white matter are usually surrounded by a mononuclear inflammatory response, extravasated red blood cells, and injured neurons. Focal granulomatous reactions with eosinophils and foreign-body giant cells may be found around amorphous remnants of worms. Damaged neurons may display neuronophagia, central chromatolysis, and reactive axonal swelling (Fig. 3).

All central nervous systems tissues should be saved for preparation of additional blocks in case initial samples are negative. Definitive parasitologic diagnosis of worms found in CNS tissue sections is based on microanatomic criteria discussed under Recognition of Worms in Tissue Sections (below).

Diagnosis

Laboratory Findings

Peripheral blood leukocytosis is observed in about one-half and eosinophilia of greater than 10% is present in about three-fourths of confirmed cases. In cases where infection with other helminths and allergies can be ruled out, peripheral eosinophilia may be a useful indicator.

CSF is usually opalescent or turbid in appearance and in all cases shows an eosinophilic pleocytosis. Use of Giemsa–May–Gruenwald stain will facilitate the assessment of eosinophilia. After the first week of infection leukocyte counts will range from 200 to 5,000/mm^3 of which 10 to more than 90% are eosinophils. Both leukocytosis and eosinophilia usually diminish rapidly, in mild to moderate cases about 40 days after the onset of symptoms. The characteristics of nonhelminthic eosinophilic pleocytosis of the CSF have been reviewed by Busch and Oehmichen (1978).

Elevated CSF pressure of 200 to 500 mm is observed in about one-half of clinically apparent cases. A protein content of greater than 50 mg/dl and mild hypoglycorrhachia is usually present. Tungkanak et al. (1972) found increased immunoglobulins and serum proteins in both serum and CSF, but no specific antibody was detected by passive hemagglutination before the eighth day after symptoms appeared. Computed tomographic and computed axial tomographic scans and electroencephalograms are generally within normal limits (Reid and Wallis, 1984), but characteristic changes have been reported in severe cases (Ko et al., 1984).

Living young adult worms have been recovered from the spinal fluid in about 10% of cases, most often in children and usually during the second and third week of infection. Intact worms recovered may be assigned to the genus *Angiostrongylus* on the basis of readily visible morphometric characteristics (see Parasitologic Diagnosis). Practitioners are urged to submit all tentatively identified worms to experts for species identification, particularly those that may have originated in Southeast Asia, Taiwan, and Australia, where multiple and mixed infections with other *Angiostrongylus* species are common in rodents.

Presumptive Diagnosis Based on Physical and Laboratory Data

As discussed by Punyagupta (1979), a diagnosis of parasitic meningitis caused by *A. cantonensis* should be considered when a patient who lives in or has visited a known (or potential) endemic area presents with severe intermittent headache with low or moderate fever and any of the following clinical features or historic antecedents:

1. symptoms and signs of meningitis, meningoencephalitis, or radiculomyeloencephalitis
2. evidence of cranial nerve involvement, especially nerve VI (lateral rectus paralysis) and nerve VII (facial palsy)
3. evidence of spinal cord and nerve root involvement (with or without headache) such as motor–sensory abnormalities, exquisite paresthesias in the lower extremities, and bowel and bladder dysfunction
4. symptoms of ocular involvement such as blurred vision, photophobia, and diplopia
5. sensorium impairment or frankly psychotic behavior
6. history of recent ingestion of, or possible contact with, snails, slugs, freshwater prawns, planaria, or foods such as leafy green vegetables potentially contaminated with larvae

and, if laboratory tests are indicated:

7. presence of an eosinophilic pleocytosis in the CSF
8. presence of moderate to marked hypoproteinemia and/or hypoglycorrhchia of the CSF.

Differential Diagnosis from Other Eosinophilic Meningitides

The essential finding of eosinophils in the CSF contributes to the exclusion of most of the potentially confounding conditions such as migraine headache, brain tumor, psychoneurosis, purulent meningitis, tuberculosis, or viral, bacterial, and most other parasitic meningites and encephalites. The distinction of neurologic angiostrongyliasis cantonensis from other helminthic infections that occasionally present with CSF eosinophilia has been discussed by several authors (Alicata, 1970; Bia and Barry, 1986; Brown and Voge, 1982; Kuberski, 1981). The residence or recent travel history of the patient will provide important evidence for the differentiation of some of these illnesses.

In patients from Thailand and perhaps other parts of Southeast Asia, a very rare parasitic eosinophilic meningitis (often with severe ocular involvement) is caused by infection with the larvae of the spiruroid nematode *Gnathostoma*. The general symptoms of neurologic gnathostomiasis would be difficult to distinguish from angiostrongyliasis by the above criteria. In general larval gnathostome infections are associated with a much more serious, often fatal, myeloencephalitis or radiculomyeloencephalitis (Punyagupta et al., 1975). These patients usually present with paresthesias, ascending paralyses of the extremities, and coma. Ocular involvement frequently leads to enucleation. However, similar complaints and signs of severe infection were noted among the group of Korean fishermen discussed previously who ate raw snails and presumably all became infected with *A. cantonensis* (Kliks et al., 1982).

Absence of a history of ingesting raw or incompletely cooked snails or freshwater prawns and a finding of bloody or xanthocromic CSF would suggest a presumptive diagnosis of gnathostomiasis rather than angiostrongyliasis. However, definitive diagnosis relies on parasitologic findings of worms isolated from the CSF, visualized in the eye, or at necropsy in the CNS tissues at gross or microscopic examination.

Another larval helminth, the cysticercus stage of the pork tapeworm, *Taenia solium,* causes cerebral infection concurrently with *A. cantonensis* in peninsular Southeast Asia and the islands of Indonesia. This parasite is rare in Hawaii and most of the Pacific Basin and North America. However, cerebral cysticercosis is one of the most prevalent neurologic disorders in Mexico. Neurologic angiostrongyliasis is, as yet, unknown in Latin America, but will almost certainly spread to the moist coastal areas from foci in the Caribbean and the Gulf Coast of the United States. *Neither peripheral nor CSF eosinophilia is a common finding in cerebral cysticercosis.* Onset of symptoms is slow and insidious, and paresthesias suggestive of nerve root or spinal cord involvement have not been reported. The cysts are evident on computed axial tomograms as space-occupying lesions in the cerebral hemispheres and in the vesicles.

Cerebral and ocular visceral larval migrans in children caused by infection by the larvae of the cosmopolitan canine roundworm, *Toxocara canis,* may also present as a relatively mild eosinophilic meningitis. Only occasionally is an eosinophilic pleocytosis seen in the CSF of patients with toxocariasis (Gould et al., 1985). A history of association with puppies or pica is frequently present in such cases. However, the clinical features of a recently described unique case of accidental infection with the ascarid roundworm of raccoons, *Bayliascaris procyonis,* closely resembled those of typical neurologic angiostrongyliasis (Fox et al., 1985).

Serologic Diagnosis

The presumptive clinical diagnosis of neurologic angiostrongyliasis may be enhanced by serologic assay for parasite-specific antibodies or antigens in the peripheral blood or CSF as reviewed by Tharavanij (1979). A confirmed diagnosis, however, can only be made by the demonstration of adult or larval *A. cantonensis*. Unfortunately, when most cases are first seen the clinicopathologic expression of the infection is already under way, and serodiagnosis will contribute little to the determination of appropriate supportive therapy. When a suitable anthelmintic drug becomes available, definitive serodiagnostic techniques will be essential for determining specific chemotherapy and for monitoring the course of treatment.

Utility of Serodiagnosis

The existing serodiagnostic assays have not been rigorously evaluated for sensitivity, specificity, and predictive value because very few series of parasitologically confirmed, clinically defined cases of neurologic angiostrongyliasis have been studied. Cross-reactions are known to occur among the broad range of shared helminth antigenic determinants. Where such infections are common, serodiagnosis may be problematic. Furthermore, apart from a few studies in experimental animals, as reviewed by Welch et al. (1980), serodynamics of early appearance and waning of immunoglobulin isotypes in relation to the time of infection and the course of neurologic disease in humans are unknown. In experimentally infected monkeys, antibody was not demonstrable by hemagglutination until the second week of infection; however, parasite antigens were detected in the CSF after about 5 days (Chen et al., 1973).

Early diagnosis without effective and specific therapy is of little value to the patient. Given the rapidity with which the larvae reach the brain and initiate pathologic changes, the utility of antibody-detecting assays for therapeutic management after the first week is minimal. However, it is probable that a monoclonal antibody-based assay for detection of *specific antigens* in sera, sputum, or CSF may provide evidence of infection during the first days of illness. Such an assay has not been developed, but would be useful in evaluating the efficacy of recently developed, highly effective nematocidal chemotherapeutic agents, such as ivermectin, milbemycin, and albendazole (see Treatment). As mentioned previously, antigen has been detected by the relatively insensitive hemagglutination method in CSF from experimentally infected monkeys about 5 days after infection.

Nevertheless, demonstration of parasite-specific antibody can be of confirmatory value for the physician and public health epidemiologist in enzootic areas. Whenever possible, it is recommended that serial sera be collected and a portion frozen at −70°C for later study. Collections of sera from wild rats trapped in areas where transmission of angiostrongyliasis is suspected would also be useful in determining the degree of enzooticity and the parasite's range in its natural host. The authors will be pleased to provide information, advice, and serologic evaluation of sera and CSF to any physician or public health agency.

There is a great potential for the rapid dispersal of this disease in the continental United States, in Mexico, and throughout the popular vacation areas of the Caribbean, Africa, and the Pacific Basin. Hence, it is important that all suspected cases be reported to state health authorities or the Centers for Disease Control.

Assays Available

Several relatively simple, inexpensive, and sensitive assays have been developed, the most useful of which are countercurrent immunoelectrophoresis (Gee et al., 1984) and several enzyme-linked immunosorbent assay (ELISA) formats (Centers for Disease Control, 1983; Cross and Chi, 1982; Jaroonvesana et al., 1985). A novel indirect fluorescent-antibody test using a soluble affinity-purified antigen coupled to cyanogen–bromide-activated Sepharose 4B beads (Welch et al., 1980) shows considerable promise as a routine laboratory test. However, the requirement of standardized immunoreagents and a specialized fluorescent microscope may limit the clinical use of this method.

The ELISA can be performed with equipment available in most clinics and hospitals. If visually read, as described below, an ELISA could be performed in most physicians' offices without any specialized equipment. The ELISA for immunoglobulin (Ig)G and IgM antibodies was previously performed by the Centers for Disease Control, Laboratory of Parasitic Diseases, but this service is no longer available. At the present time, in addition to our own facility at the School of Medicine University of Hawaii, the ELISA for angiostrongyliasis can be performed at some clinical laboratories in Bangkok, Hong Kong, and Taiwan.

Micro-ELISA for Parasite-Specific Antibody Detection

The standard micro-ELISA (Centers for Disease Control, 1983; Voller et al., 1974), using disposable 96-well microtiter plates coated with adult whole-worm phosphate-buffered saline (PBS) extract diluted to 5 to 10 μg/ml in 0.06 M carbonate buffer (pH 9.5), has been used in our laboratory with satisfactory results over the past 6 years. Levels of IgG and IgM antibodies can be rapidly determined in both suspected human cases and in captured wild rodents parasitologically confirmed to be infected with *A. cantonensis* and one or more other helminths.

Test sera are routinely assayed (usually in triplicate) along with appropriate controls for IgG antibodies using a series of 4- or 10-fold dilutions. All reactions are carried out at 37°C, using 30-min incubations and three 1-min washes with PBS–Tween 20 (0.3%) between reagents. Affinity-purified antihuman IgG–peroxidase conjugate and ABTS substrate were used according to the manufacturers recommendations (Kirkegaard and Perry Laboratories, Gaithersberg, Md.).

Using this micro-ELISA, single sera from 10 patients known to have ingested infected snails during a 2-day period were tested 17 to 21 days after infection (13 to 18 days after onset of symptoms) (Kliks et al., 1982). Larvae were demonstrated in the brain and spinal cord of the single fatal case, but not in the CSF examined from 4 of the 10 subjects on whom one or more spinal taps were performed 20 to 40 days after infection. Titers* of sera ranged from 1 : 64 to 1 : 512 (median, 1 : 128). The titer of a postmortem serum sample (17th day postinfection) from the parasitologically confirmed case was 1 : 64.

* Criteria for determining endpoint dilution (titer) were 1) for photometric determination, that dilution at which mean absorbence of triplicate test serum samples exceeds the mean (±2 standard deviations) absorbence of triplicate, pooled, normal-control serum samples; and 2) for visual determination, that dilution at which a distinct color difference can be detected in triplicate test versus control samples.

We have subsequently assayed sera or CSF collected from 15 additional presumptive cases in Hawaii that presented with signs and symptoms typical of neurologic angiostrongyliasis. Sera and CSF samples were collected between 3 and 35 days after onset of symptoms; time from actual infection could not be determined. All subjects gave histories of ingestion of, or probable contact with, snails or slugs. No patient was severely affected. The CSF was examined in seven of these cases, but no larvae were detected. Serum titers ranged from 1 : 64 to 1 : 4096, the highest dilution tested (mean titer, 1 : 1024); the three CSF samples had titers of less than 1 : 4, 1 : 64, and 1 : 500. Sera from four normal individuals were nonreactive in this assay (less than 1 : 4), whereas sera and CSF from four patients suffering from other infectious diseases ranged from less than 1 : 4 to 1 : 64 for both serum and CSF.

Interpretation of ELISA Results

Any titered serum or CSF samples that consistently have a mean endpoint dilution of 1 : 64 or greater (when tested in duplicate) may be considered positive for antibody against *A. cantonensis*. When tested (in triplicate) at a single dilution (preferably between 1 : 100 and 1 : 500), any sample with an absorbance greater than the mean plus 2 standard deviations of a pool of at least five normal subjects may be considered positive. A rising titer during the symptomatic stages would be strongly indicative of active infection.

During the first week of symptoms, a low titer cannot be interpreted as negative; however, a mean titer of less than 1 : 64 after the second week may be considered negative. Given that parasitologic confirmation is not often achieved in nonfatal infections, the "predictive value" of this assay is not known.

Both acute and convalescent sera and CSF samples should be tested and aliquots stored frozen for further study. Residual CSF should be retained for parasitologic examination (see Parasitologic Diagnosis). Routine stool and blood examinations should be conducted to determine the possible presence of intestinal helminths or filarial infections, which may give rise to confounding cross-reactions.

Parasitologic Diagnosis

Larval and young adult worms may be recovered alive and intact from CSF, from the anterior chamber of the eye, from the surface of the brain and spinal cord at autopsy, or from the containers in which these have been held. Worms can occasionally be collected by digestion of brain or spinal cord samples, from remnants of food items, and from suspected intermediate or transport hosts if required for corroborative evidence of enzooticity.

Examination of CSF and Other Body Fluids

Recovery of larvae at spinal tap is enhanced by the use of an 18-gauge or larger aspirating needle and the application of slight negative pressure. Fourth-stage larvae and young adults may be concentrated from CSF by centrifugation at about 1000 × *g* for 3 min. The supernate is carefully poured off and saved for serodiagnosis and other laboratory tests. The sediment is examined drop by drop on a slide under low and medium power with a compound microscope. Recently molted fourth-stage larvae average about 550 μm long by 30 μm wide and have a small buccal capsule, a smooth cuticle, and a blunt conical tail; about 10 days after infection, they will have grown to 1,200 to 1,400 μm in length (Alicata and Jindrak, 1970). Living larvae have also been visualized and collected from the anterior chamber of the eye.

During the second week, most worms will have molted to become young adults and range from 2 to 5 mm in length. Males have a well-defined copulatory bursa on the oventral surface of the tail.

Recognition of Worms in Tissue Sections

Definitive parasitologic diagnosis may be made by recognizing the characteristic microanatomy of worms as seen in stained tissue sections, most frequently in the spinal cord (Fig. 4) and cerebellum. Serial transverse sections average 80 to 110 μm, but may range up to 180 μm in diameter. The cuticle is thin and finely striated longitudinally. Lateral cords, with two to four nuclei per section, are prominent. The body wall musculature is divided into four groups by lateral, dorsal, and ventral cords containing many cells per quadrant with cytoplasm not extending far into body cavity. Intestinal cells have vacuolar cytoplasm and multiple densly stained nuclei per section. The genital organs of the female are tubular, with many dense hyperchromatic nuclei per section; in the male, spicules and copulatory bursa may be visible in posterior sections (Jindrak and Alicata, 1965; Kliks et al., 1982; Mackerras and Sandars, 1955; Tangchai, et al., 1967).

Recovery of Worms by Digestion

Worms may be readily demonstrated in infected human tissue samples, the flesh of potential intermediate hosts, and remnants of suspected foodstuffs using 1% acid pepsin (10 g of pepsin, 1 liter of warm water, and 7 ml of concentrated HCl). Samples are minced, then briefly homogenized in 10 to 15 ml of

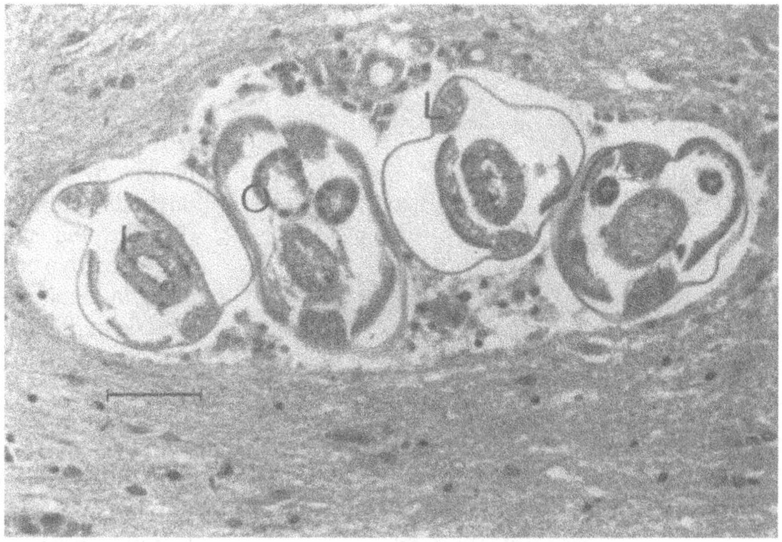

FIG. 4. Microanatomy of young adult female *Angiostrongylus cantonensis* in spinal cord from fatal human case. L = lateral cord; I = intestine; O = oviduct. Hematoxylin and eosin. Bar = 50 μm. (Original photomicrograph by John M. Hardman, Department of Pathology, School of Medicine, University of Hawaii at Manoa.)

the digest solution per gram of sample and, stirred at room temperature for 1 to 3 h, and poured through fine gauze to remove large particles. Worms may be concentrated either by use of a Baerman apparatus or by repeated centrifugation 1,000 × *g*, 3 min) and resuspension in saline until the supernate is clear. The sample may be examined on glass slides as above for CSF residues.

Second- and third-stage larvae recovered from intermediate hosts, fomites, or water are about 500 μm long and 30 to 50 μm wide and have a finely striated cuticle. Third-stage larvae have two chitenous rods at the anterior extremity, which are characteristic of this stage.

Prognosis

The course and severity of disease are dose related. When initial symptoms are mild or when symptoms are delayed for several days after the presumed infective incident, the prognosis for uneventful recovery is good. When signs and symptoms of severe cerebral and spinal nerve involvement develop within 24 to 48 h after the patient first experiences discomfort, the prognosis may be grave and appropriate supportive therapy, including artificial respiration, must be available.

Epidemiologic Aspects

It is probable that most cases of angiostrongyliasis are asymptomatic or present with such minor symptoms that they do not come to the attention of physicians. Most of these infections are acquired through the accidental ingestion of infective larvae contained within, or released from, lightly infected transport hosts. Most of the mild to moderately severe cases reported are not associated with a definite history of contact with intermediate hosts. However, because of the immediacy and intensity of symptoms in persons receiving large infective doses the source of infection can often be verified by parasitologic study of the presumptive intermediate host or fomite. The prevalence and intensity of infection in peridomestic rats and mollusks, the overall population density of susceptible hosts, and their proximity to human habitation all play important roles in the epidemiologic pattern of human cases.

Intermediate and Transport Hosts

Intermediate hosts (in which the larvae develop) and transport hosts (in which only dispersal and maintenance take place) may be purposely ingested as food, as medicine, or pursuant to a dare or practical joke.

Among terrestrial mollusks, ingestion of the giant African land snail, *A. fulica* (Fig. 5C), has been implicated as a source of human infection in Tahiti, the Ivory Coast, Ponape, American Samoa, and the Hawaiian islands. The introduction of this large, attractive snail as an ornament in formal gardens or as a curiosity has been an important factor in the dispersal of angiostrongyliasis cantonensis. Amphibious snails of the genus *Pila*, which are eaten lightly scalded and pickled in Thailand, Malaysia, and elsewhere in Southeast Asia are also an important source of infection.

A wide variety of other snails and slugs has been found naturally infected or can be experimentally in-

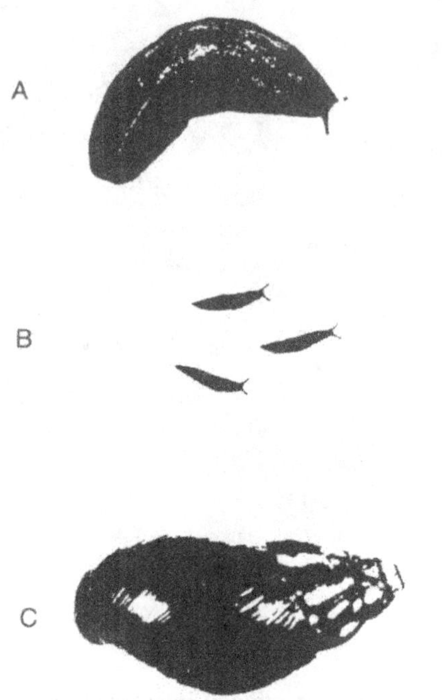

FIG. 5. Principal intermediate hosts of *Angiostrongylus cantonensis* in most of its range. (A) *Veronicella alte*. (B) *Deroceras laeve*. (C) *Achatina fulica*. Approximately two-thirds actual size. (From Alicata, 1962. Rat lungworms: a cause of parasitic meningitis of man in the Pacific. Hawaii Farm Bureau Journal, with permission.)

fected (Alicata and Jindrak, 1970). It is probable that several species of small veronicellid and derocerid slugs (Fig. 5A,B) are involved in many of the infections in infants and in infections acquired through the ingestion of vegetables contaminated with these inconspicuous creatures. Slugs are consumed as food in only a few areas of Southeast Asia, but are commonly believed to have certain medicinal properties in Asian cultures (Sharma et al., 1981).

Cases of angiostrongyliasis resulting from medicinal ingestion have been recorded among tribal peoples in Taiwan and among Japanese in Hawaii and Japan (Horio and Alicata, 1961). In American Samoa, 25 Korean fishermen were given infected giant African land snails to eat as a practical joke (Kliks et al., 1982) with tragic consequences. In Hawaii, a 16-year-old male visitor (Fisher, 1983) and a 19-year-old U.S. Marine trainee (Kliks, unpublished data) acquired serious infections by eating raw snails in response to peer pressure.

A wide variety of crustaceans can also act as transport hosts. In Tahiti, New Caledonia, American Samoa, Guam, and very recently on Hawaii (Kliks, Brown, and Palumbo, unpublished data), small outbreaks among family members have been traced to the ingestion of raw freshwater prawns (*Macrobrachium* spp.) that were crushed and sauted briefly in lemon juice. The widely cultivated Malaysian prawn, *Macrobrachium lar*, can be experimentally infected by exposure to solutions containing infective larvae. However, extensive surveys of commercially raised prawns by ourselves and others have failed to demonstrate naturally acquired infections. Land crabs (*Cardiosoma* spp.), coconut crabs (*Birgus latro*), and mangrove crabs (*Scylla seratta*) have been implicated in human infections in Southeast Asia, Tahiti, American Samoa, Ponape, and elsewhere in the Pacific islands, where they are highly regarded as food items.

More frequent than ingestion as food or folly, however, is accidental ingestion of larvae within small mollusks or terrestrial planaria (*Geoplana* spp.) commonly found in leaf litter in tropical areas. Infection rates of 80% or greater may be found in predacious planaria, and a single individual may harbor several hundred infective larvae. Both slugs and planaria may become inconspicuous when they dry up in the folds of leafy green vegetables. Foods left exposed in rural areas or campsites may rapidly become contaminated with mollusks or planaria. There is evidence that larvae may be released into slime trails or into drinking water sources and survive for some time. It has also been shown in experimental animals that infective larvae can penetrate directly through wounds or abraded skin; however, this route of infection must be very uncommon in humans.

Infection in the Definitive Rodent Hosts

As this parasitocenose spread throughout the tropical and subtropical regions of the globe, infections in humans have almost always been preceded by detection of worms in their normal rodent hosts, usually one of several peridomestic species of the genus *Rattus*. The geographic epidemiology of this disease in humans and its dispersal by rodents and mollusks has been described by Alicata (1967) and updated by Kliks (in review).

Treatment

Supportive and Symptomatic Treatment

Eosinophilic meningitis is usually a self-limiting disease, with resolution of signs and symptoms within a few days to weeks. Therapy is mainly supportive and symptomatic to relieve pain, nausea, and intracranial hypertension with aspirin, acetaminophen, and other analgesics. In more severe cases corticosteroids and osmotic brain-dehydrating agents such as mannitol have been used with some success. It has been noted that abrupt removal from dexamethasone therapy may result in a dramatic reappearance of symptoms related to intracranial pressure; these symptoms subside when the original dose is restored and withdrawal accomplished by a gradually tapered regimen (Koo et al., in review). Release of intracranial pressure by surgical procedures may also be required (Enzenhauer and Yamaoka, 1982).

Specific Anthelmintic Therapy

No specific anthelmintic agent has been shown to be more effective than analgesic and anti-inflammatory therapy alone. By the time most cases have been clinically diagnosed, the majority of larvae present will have already entered the CNS and begun to cause pathologic changes. In experimental hosts worms do not migrate synchronously to and from the CNS. Hence, administration of any presumptively effective anhelminthic drug may need to be continued into the third and fourth weeks after symptoms begin in order to affect persistent or late-arriving worms. In mild to moderate infections, it has been noted that clinical improvement coincides with the time when most young adult worms will have just migrated out of the CNS following mating.

It has been suggested that drugs effective in killing worms in the CNS may lead to a detrimental exacerbation of the inflammatory response. No such effect was noted in several cases treated with thiabendazole during the first week of infection (Kliks et al., 1982). Marked improvement in motor power and subsidence of headache was noted in one patient after administration of thiabendazole for 5 days at a dose of 1,000 mg twice daily (F. D. Pien, M.D., personal communication). This patient did experience a sudden periorbital swelling 4 h after the first dose. The reaction responded well to dexamethasone treatment.

Animal trials with newer antiparasitic drugs such as ivermectin and milbemycin D show promise in killing migrating larvae before and during their passage through the CNS (Ishii et al., 1985; Terada et al., 1986). Early serodiagnosis by detection of parasite-specific antigen may permit the timely utilization of such drugs and the monitoring of their efficacy in vivo.

Prevention and Control

Preventive measures such as dissemination of public health information and education of elementary school children will have a greater impact and be more cost effective than any control measures wherever this host–parasite system has become established. Given the ubiquitous presence and great reproductive capacities of both the rodent and molluskan hosts, and their potential for establishing both peridomestic and sylvatic habitats, it is unlikely that the continuous dispersal of this parasitocenose can be prevented short of a massive concerted campaign. The geographic pattern and rate of its spread inland and along the Gulf Coast from the recently discovered focus in the New Orleans area of the continental United States will be illustrative of the ability of *Angiostronglus cantonensis* to invade and persist in virgin territory.

Personal Prophylaxis in Enzootic Areas

Travelers and residents in the known endemic regions, including those of the southeastern United States (see Fig. 2), should avoid all contact with terrestrial mollusks and instruct children to do the same. Infants and toddlers should not be permitted to play on bare soil, among leaf litter, or in weedy areas where small slugs, snails, and planaria may lurk. At outdoor meals, food must not be left in contact with the ground in damp, shady areas. Fresh leafy vegetables and fallen, ripe fruits or ground-growing fruits such as strawberries and tomatoes should be inspected for evidence of penetration and infestation by slugs and planaria. All potential intermediate and transport hosts, particularly freshwater prawns, should be thoroughly cooked.

Control of Rodent Hosts

Local and regional authorities have maintained both focal and area-wide rodent control programs in many enzootic regions for decades without obvious effects on the incidence or spread of angiostrongyliasis cantonensis in rodents. It is probable, however, that such efforts have had some impact on the population of peridomestic rats and transmission of the disease to humans in urban and suburban settings.

Control of Intermediate and Transport Hosts

Surveys of the principal intermediate and transport hosts from many enzootic areas indicate that the prevalence of infection is such that virtual eradication of all susceptible molluskan species would be required to eliminate transmission effectively. The standard methods of control by application of molluscicides would have no direct effect on planaria, freshwater prawns, and other transport hosts. Area-wide applications would have some detrimental effects on nontarget species and would be a recurring, costly venture. However, focal control around dwellings, gardens, and vegetable farms would be appropriate in enzootic areas.

The Role of Medical Education

Providing the at-risk public with the essential preventive information is the most effective and least expensive means of minimizing human exposure. Dissemination of basic public health information on avoidance of the intermediate and transport hosts, food storage and preparation, and management of the peridomestic environment to minimize harborage for mollusks and rodents should be standard procedure in all enzootic areas. Educational information should be provided to elementary school children and to mothers, as a large portion of severe infections occur in the 1- to 15-year-old age group. Unfortunately, because of budgetary constraints, bureaucratic indifference, and politicoeconomic considerations (e.g., in prime tourist locations), these simple, effective measures are not carried out, even in the "developed" countries.

Literature Cited

Alicata, J. E. 1962. *Angiostrongylus cantonesis* (Nematoda: Metastrongylidae) as a causative agent of eosinophilic meningoencephalitis in man in Hawaii and Tahiti. Can. J. Zool. **40**:5–8.

Alicata, J. E. 1966. The presence of *Angiostrongylus cantonensis* in islands of the Indian Ocean and probable role of the giant African snail, *Achatina fulica,* in dispersal of the parasite to the Pacific islands. Can. J. Zool. **44**:1041–1049.

Alicata, J. E., and K. Jindrak. 1970. Angiostrongylosis in the Pacific and Southeast Asia. Charles C Thomas, Publisher, Springfield, Ill.

Anderson E., D. J. Gubler, K. Sorensen, J. Beddard and L. R. Ash. 1986. First report of *Angiostrongylus cantonensis* in Puerto Rico. Am. J. Trop. Med. Hyg. **35**:319–322.

Bailey C. A. 1948. An epidemic of eosinophilic meningitis, a previously undescribed disease, occurring on Ponape, Eastern Carolines. Project NM005.007, report no. 7. Naval Medical Research Institute, Bethesda, Md.

Beck, M. J., T. M. Carding, and J. E. Alicata. 1980. Eosinophilic meningitis due to angiostrongyliasis cantonensis in American Samoa. Hawaii Med. J. **39**:254–57.

Bia, F. J., and M. Barry. 1986. Parasitic infections of the central nervous system. Neurol. Clin. **4**:171–206.

Bosch, I., and M. Oehmichen. 1978. Eosinophilic granulocytes in cerebrospinal fluid: analysis of 94 cerebrospinal fluid specimens and review of the literature. J. Neurol. **219**:93–105.

Bowden, D. K. 1981. Eosinophilic meningitis in the New Hebrides: two outbreaks and two deaths. Am. J. Trop. Med. Hyg. **30**:1141–1143.

Brown, W. J., and M. Voge. 1982. Neuropathology of parasitic infections. Oxford University Press, Oxford.

Campbell, B. G., and M. D. Little. 1986. The finding of *Angiostrongylus cantonensis* in rats in New Orleans; the first report in North America. Abstract no. 137. Annual Meeting of American Society of Tropical Medicine and Hygiene, Denver, Colorado, December 7–11, 1986.

Centers for Disease Control. 1983. EIA microtitration test. Protozoal Diseases Branch. Centers for Disease Control, Atlanta.

Chen, S. N., T. Suzuki, and K. H. Lin. 1973. Studies on immunodiagnosis of angiostrongyliasis. 1. Detection of antigen and antibody in serum and cerebrospinal fluid. J. Formosan Med. Assoc. **72**:161–166.

Cross, J. H. 1978. Clinical manifestations and laboratory diagnosis of eosinophilic meningitis syndrome associated with angiostrongyliasis. Southeast Asian J. Trop. Med. Public Health **9**:161–70.

Cross, J. H., and J. C. H. Chi. 1982. ELISA for the detection of *Angiostrongylus cantonensis* antibodies in patients with eosinophilic meningitis. Southeast Asian J. Trop. Med. Public Health **13**:73–76.

Enzenauer, R. W., and R. M. Yamaoka. 1982. Eosinophilic meningitis and hydrocephalus in an infant. Arch. Neurol. **39**:380–381.

Fischer, P. R. 1983. Eosinophilic meningitis. West. J. Med. **139**:372–373.

Fox, A. S., K. R. Kazacos, N. S. Gould et. al. 1985. Fatal eosinophilic meningoencephalitis and visceral larva migrans caused by the raccoon ascarid *Baylisascaris procyonis*. N. Engl. J. Med. **312**:1619–1623.

Gee, G., W. K. Lau and M. M. Kliks. 1984. Clinical spectrum of eosinophilic meningitis due to *Angiostrongylus cantonensis* with confirmation by counter immunoelectrophoresis. J. Clin. Res. **32**:43A.

Gould, I. M., S. Newell, S. H. Green, and R. H. George. 1985. Toxocariasis and eosinophilic meningitis. Br. Med. J. **291**:1239–1240.

Horio, S. R., and J. E. Alicata. 1961. Parasitic meningoencephalitis in Hawaii. A new parasitic disease of man. Hawaii Med. J. **21**:139–140.

Ishii, A., I. Terada, and M. Sano. 1985. Studies on chemotherapy of parasitic helminths (XXIII). Effects of ivermectin on *Angiostrongylus cantonensis* in rats. Jpn. J. Parasitol. **34**:411–417.

Jaroonvesama, N., K. Chareonlarp, P. Buranasin, G. G. Zaraspe, and J. H. Cross. 1985. ELISA testing of clinical angiostrongyliasis in Thailand. Southeast Asian J. Trop. Med. Public Health **16**:111–116.

Jindrak, K. 1968. Early migration and pathogenicity of *Angiostrongylus cantonensis* in laboratory rats. Ann. Trop. Med. Parasitol. **62**:506–517.

Jindrak, K., and J. E. Alicata. 1965. A case of parasitic eosinophilic meningo-encephalitis in Vietnam probably caused by *Angiostrongylus cantonensis*. Ann. Trop. Med. Parasitol. **59**:294–300.

John, D. T., and A. J. Martinez. 1975. Animal models of human disease. Central nervous system infection with the nematode *Angiostrongylus cantonensis*. Animal model: eosinophilic meningoencephalitis in mice infected with *Angiostrongylus cantonensis*. Am. J. Pathol. **80**:345–348.

Kliks, M. M. In review. Eosinophilic meningoencephalitis: the panpacific spread of a peridomestic enzoonosis caused by *Parastrongylus cantonensis* (Chen, 1935) Ubeleker 1986, the lungworm of rats. Submitted to Social Science and Medicine.

Kliks, M. M., K. Kroenke, and J. M. Hardman. 1982. Eosinophilic radiculomyeloencephalitis: an angiostrongyliasis outbreak in American Samoa related to ingestion of *Achatina fulica* snails. Am. J. Trop. Med. Hyg. **31**:1114–1122.

Ko, R. C., M. C. Chiu, W. Kum, and S. H. Chan. 1984. First report of human angiostrongyliasis in Hong Kong diagnosed by computerized axial tomography (CAT) and ELISA. Trans. R. Soc. Trop. Med. Hyg. **78**:354.

Koo, J., F. Pien, and M. M. Kliks. In review. *Angiostrongylus* (*Pavastrongylus*) eosinophilic meningitis. Rev. Infect. Dis.

Kuberski, T. 1981. Eosinophils in cerebrospinal fluid: criteria for eosinophilic meningitis. Hawaii Med. J. **40**:97–98.

Kuberski, T., and G. D. Wallace. 1979. Clinical manifestations of eosinophilic meningitis due to *Angiostrongylus cantonensis*. Neurology **29**:1566–1570.

Mackerras, M. J., and D. F. Sandars. 1955. The life history of the rat lung-worm, *Angiostrongylus cantonensis* (Chen) (Nematoda: Metastrongylidae). Aust. J. Zool. **3**:1–21.

Nomura, S., and P. H. Lin. 1945. First case report of human infection with *Haemostrongylus ratti* Yokogawa. (in Japanese) Taiwan No. Ikai **3**:589–597.

Nye, S. W., P. Tangchai, S. Sundarakiti, and S. Punyagupta. 1970. Lesion of the brain in eosinophilic meningitis. Arch. Pathol. **89**:9–18.

Pascual, J. E., R. P. Bouli, and P. H. Aguiar. 1981. Eosinophilic meningoencephalitis in Cuba, caused by *Angiostrongylus cantonensis*. Am. J. Trop. Med. Hyg. **30**:960–962.

Punyagupta, S. 1979. Angiostrongyliasis: clinical features and human pathology, p. 138–150. *In* J. H. Cross (ed.), Studies on Angiostrongyliasis in Eastern Asia and Australia, special publication no. 44. U.S. Naval Medical Research Unit 2, Taipei, Taiwan.

Punyagupta, S., P. Juttijudata, and T. Bunnag. 1975. Eosinophilic meningitis in Thailand. Clinical studies of 484 typical cases probably caused by *Angiostrongylus cantonensis*. Am. J. Trop. Med. Hyg. **25**:233–249.

Reid, I. R., and W. E. Wallis. 1984. The chronic and severe forms of eosinophilic meningitis. Aust. N. Z. J. Med. **14**:163–165.

Rosen, L., R. Chappell, G. L. Laguer, G. D. Wallace, and P. P. Weinstein, 1962. Eosinophilic meningoencephalitis caused by a metastrongylic lung-worm of rats. J. Am. Med. Assoc. **179**:620–624.

Scrimgeour, E. M. 1984. Distribution of *Angiostrongylus cantonensis* in Papua, New Guinea. Trans. R. Soc. Trop. Med. Hyg. **78**:776–779.

Sharma, K. D., D. M. Renapurkar, and M. K. Bhopale. 1981. Study of a focus of *Angiostrongylus cantonensis* in Greater Bombay. Bull. Haffkine Inst. **9**:38–46.

Sonakul, D. 1978. Pathological findings in four cases of human angiostrongyliasis. Southeast Asian J. Trop. Med. Public Health **9**:220–227.

Tangchai, P., S. W. Nye, and P. C. Beaver. 1967. Eosinophilic meningo-encephalitis caused by angiostrongyliasis in Thailand. Am. J. Trop. Med. Hyg. **16**:454–460.

Terada, M., A. M. Dharejo, and M. Sano. 1986. Effects of milbemycin D on the motility of *Angiostrongylus cantonensis* and *A. costaricensis*. Jpn. J. Parasitol. **35**:497–504.

Tharavanij, S. 1979. Immunology of angiostrongyliasis, p. 151–175. *In* J. H. Cross (ed.), Studies on Angiostrongyliasis in Eastern Asia and Australia, special publication no. 44. U.S. Naval Medical Research Unit 2, Taipei, Taiwan.

Tungkanak, R., S. Sirisinha, and S. Punyagupta. 1972. Serum and cerebrospinal fluid in eosinophilic meningoencephalitis: immunoglobulins and antibody to *Angiostrongylus cantonensis* in Thailand. Am. J Trop. Med. Hyg. **21**:415–420.

Ubelaker, J. E. 1986. Systematics of species referred to the genus *Angiostrongylus*. J. Parasit. **72**:237–244.

Uchikawa, R., M. Takagi, S. Matayoshi and A. Sato. 1984. The presence of *Angiostrongylus cantonensis* in Viti Levu, Fiji. J. Helminthol. **58**:231–234.

Voller, A., D. E. Bidwell, G. Huldt, and E. A. Engvau. 1974. A microplate method of ELISA and its application to malaria. Bull. W.H.O. **51**:209–218.

Welch, J. S., C. Dobson, and G. R. Campbell. 1980. Immunodiagnosis and seroepidermiology of *Angiostrongylus cantonensis* zoonosis in man. Trans. R. Soc. Trop. Med. Hyg. **74**:614–623.

Yii, C. Y. 1976. Clinical observations on eosinophilic meningitis and meningoencephalitis caused by *Angiostrongylus cantonensis* on Taiwan. Am. J. Trop. Med. Hyg. **25**:233–249.

Yii, C. Y., C. Y. Chen, J. W. Fresh, J. Chen, and J. H. Cross. 1968. Human angiostrongyliasis involving the lungs. Chin. J. Microbiol. **1**:148–150.

CHAPTER 79

Anisakiasis

JEFFREY W. BIER

Disease: Anisakiasis (phocanemiasis, herring worm disease, cod worm disease, seal worm disease, contracaeciasis, hysterothylaciasis).
Etiologic Agents: *Anisakis simplex, Anisakis typica, Pseudoterranova* (*Phocanema, Terranova, Porrocaecum*) *decipiens, Contracaecum* sp., *Hysterothylacium* (*Thynnascaris*) sp.
Source: Infected fish or shellfish.
Clinical Manifestations: Variable abdominal pain may be accompanied by nausea, diarrhea, or vomiting. Mild to moderate circulating eosinophilia. Seroconversion within 10 days.
Pathology: Eosinophilic granuloma surrounding attachment site(s); edema more pronounced around nematode.
Laboratory Diagnosis: Confirmed diagnosis from recovered anisakid nematode. May be presumptively diagnosed by food history, eosinophil count, and serology.
Epidemiology: Worldwide, especially where raw seafoods are commonly consumed. May be transmitted in undercooked seafood that has not reached an internal temperature of 65°C for 30 s.
Treatment: Removal of worm by endoscope or surgical resection of lesion and nematode.
Prevention and Control: Abstinence from raw and undercooked seafoods.

Description of Disease

Transmission

The transmission of anisakid nematodes through their marine environment to their definitive hosts or humans involves intermediate hosts, which may be required or optional (paratenic). Transmission of the disease in Europe and North America has been attributed to fin fish (Deardorff et al., 1986; Van Thiel et al., 1960), although in Japan squid and octopus also have been implicated (Oshima, 1987). The life cycle of all members of the family *Anisakidae* fit the following pattern: Viable eggs in feces pass into the aquatic environment, where they embryonate, deposit at least a second-stage cuticle, hatch, penetrate the gut of an invertebrate intermediate host, grow, and develop. The predatory second intermediate host consumes the infected first host, and development of the parasite continues. It may then pass to additional intermediate hosts or mature in the intestinal tract of the definitive host, if mating and production of eggs occurs.

Clinical Manifestations

Anisakiasis (phocanemiasis, herring worm disease, seal worm disease, contracaeciasis) is a disease resulting from infection with any member of the family *Anisakidae*. Both mild and severe forms of the disease are known. The mild form predominates among cases reported in the United States, whereas the severe form dominates in Japan. Of the 27 genera described in the family, only 4 have been reported to infect humans. *Anisakis* (Van Thiel et al., 1960), *Pseudoterranova* (Little and MacPhail, 1972), *Con-*

tracaecum (Schaum and Miller, 1967), and *Hysterothylacium* (Petter, 1969), *Anisakis simplex,* and *Pseudoterranova decipiens* are responsible for all but two of the recorded cases. The first documented cases of anisakiasis with tissue involvement occurred in Holland (Van Thiel et al., 1960); Jackson (1975) provides a more complete historic account. The recovered larva was found to be similar to that in herring, and all patients had consumed green herring, a lightly marinated Dutch dish. The first identification was *Eustoma rotundata,* then *Stomachus contortus, Anisakis marina,* and finally *A. simplex* (Van Thiel, 1976). A retrospective study in Japan of cases attributed to *Ascaris lumbricoides* revealed that anisakids were involved (Yokogawa and Yoshimura, 1967).

Gastric Anisakiasis

Patients usually report suffering epigastric pain, frequently with nausea and occasionally with vomiting. The onset of symptoms may be as early as 1 h after ingestion of raw fish (Oshima, 1972) or require 12 h or longer (Deardorff et al., 1986). Occult blood is frequently present in the gastric juice or feces (Yokogawa and Yoshimura, 1967). Induced increases or decreases in gastric secretion enhanced mucosal penetration of guinea pigs by anisakid larvae (Asami and Inoshita, 1967). Circulating eosinophils (4 to 41%) without pronounced leukocytosis are present in most cases (Oshima, 1972). Exploratory laparotomy, first used to diagnose anisakiasis, led to resection of the lesion. The endoscope was introduced for the diagnosis and treatment of stomach anisakiasis in Japan (Namika et al., 1970). The biopsy forceps on the fiberoptic device allowed removal of the larvae and confirmatory morphologic identification. As Oshima (1972) predicted, the endoscope increased the frequency of diagnosis of gastric anisakiasis in Japan. In 1984, 3,141 cases of gastric anisakiasis were reported at a meeting of the Japanese Society of Digestive Tract Endoscopy (Oshima, 1987). Untreated stomach anisakiasis becomes chronic. The symptoms are milder, and diagnosis is more difficult. The worm dies and is absorbed over time, leaving an apparently self-perpetuating granulomatous lesion. Analysis of roentgenographic findings in 130 cases of acute gastric anisakiasis revealed a threadlike gastric filling defect in 47.7%, a rigid margin in 78.5%, and edema in 80.5% (Sugimachi et al., 1985).

Intestinal Anisakiasis

Penetration by anisakid nematodes in the intestine is characterized by either constant or intermittent abdominal pain. Symptoms frequently do not appear for 7 to 14 days after the suspect meal. Pain is generally associated with the lower right quadrant, the abdomen is not distended, and rebound tenderness is not experienced. Diarrhea, fever, vomiting, and nausea occur, and occult blood may be present (Oshima, 1972; Van Thiel et al., 1960). Duodenal and rectal anisakiasis may be diagnosed and treated with an endoscope (Cocheton et al., 1984).

Other Locations

Anisakid nematodes penetrate the gut of experimental animals (Myers, 1963) and may be found wandering in the abdominal cavity. Cellular reaction to these wandering nematodes is noted, as is the presence of serous fluid in the abdominal cavity. In humans, disease from this condition is not documented. A phlegmonous tumor was incidentally noted and removed from a patient being treated for an iliac aneurysm (Little and MacPhail, 1972). Sections of this tumor revealed a large nematode that resembled *Pseudoterranova.* The immediate danger to the patient is from perforative peritonitis. Adhesions and possible stenosis of areas of the intestine could result, depending on the amount of cellular reaction to the worm. Pharyngeal anisakiasis has been noted in cases from Japan (Morishita and Nishimura, 1965) and Korea (Kim et al., 1971); symptoms were those of a sore throat. These cases may be related to those in the United States of "tickle throat" anisakiasis (Desowitz, 1986). Such patients recall either few or no symptoms, but report having consumed raw or inadequately cooked fish within the past few hours to as long as 2 weeks before either coughing up a worm or feeling a tickling sensation in the back of the throat and extracting a worm. A case of pleural effusion was diagnosed as probable anisakiasis for the following reasons (Kobayashi et al., 1968): 1.5 h after consuming fish (salmon) and shellfish sushi, the patient experienced urticaria, nausea, and diarrhea of 1 week's duration, then unproductive cough accompanied by upper left thoracic pain. A roentgenogram of the chest revealed pleural effusion and atelectasis of the lower left lobe. The white blood cell count was $9{,}700/mm^3$ with 31% eosinophils. Thoracentesis yielded fluid with numerous eosinophils. No ova or parasites were found on fecal examination. Biopsy and a complement fixation test were negative for *Trichinella.* Pleural effusion and eosinophils in straw-colored fluid were removed at biopsy over a 4-month period, and serum immunoelectrophoresis 5 months after onset formed a specific band with *Anisakis* antigen, but not with *Toxicara* antigen. Mild eosinophilia persisted after 1 year. In France, an *Anisakis* larva was found and removed from a patient on two occasions within 17 days (Fabresse et al., 1984). The physicians associated the infections with polyarthritis that the patient experienced concurrent with the second exposure. Similar associations have not been noted elsewhere.

Diagnosis

Laparotomy, the earliest method for diagnosis of anisakiasis, was always reported in conjunction with resection of the lesion, usually with a nematode. Soon after their introduction in Japan, the endoscope and biopsy forceps replaced laparotomy and resection as the method of choice for the treatment of stomach anisakiasis (Namiki et al., 1970). Where the incidence of anisakiasis is lower, reports of adaptation of the technique take longer to appear. Nineteen years passed before the first case of anisakiasis was diagnosed by endoscopy in the United States (Deardorff et al., 1986). This technique was first applied to a case of duodenal anisakiasis in France (Cocheton et al., 1984). Other techniques such as the test for occult blood and differential blood counts are part of routine clinical procedures. Serodiagnosis of anisakiasis requires reagents that are not routinely available commercially. The following reports briefly summarize new developments in this field (Smith and Wooten [1978] have reviewed the subject extensively): A complement fixation test (Daniels, 1962) produced negative results in surgically proven cases of anisakiasis and positive results in 2.5% of sera tested from people who had little likelihood of prior exposure (Pollak and Kamplemacher, 1966). An immunofluorescence test was both more sensitive and more specific than complement fixation, but cross-reacted with the sera of patients with toxicariasis (Ruitenberg, 1970). Immunoelectrophoresis on starch demonstrated cross-reactivity to antigens from both *Toxicara* sp. and *Ascaris* sp. (Suzuki, 1968) in 19 of the 20 proteins resolved. An indirect fluorescent-antibody test to *A. simplex* hemoglobin was specific (Suzuki et al., 1974), but required complex biochemical purification of *Anisakis* hemoglobin. An immunoglobulin E response in anisakiasis was demonstrated by passive cutaneous anaphylaxis (Nagase, 1973) and the Shultz–Dale technique (Tozuka, 1978). A radioallergosorbent test for anisakiasis (Desowitz et al., 1985) did not cross-react with sera from patients with ascariasis or toxicariasis. The test requires a cell-free extract of *A. simplex* larvae. This reagent is not commercially available; the viscera of Pacific rockfish and Atlantic herring are suggested sources of larvae. A saline cell-free extract is coupled to cyanogen-bromide-activated microbeads. Test serum is incubated with the microbeads, then with radioiodated anti-human immunoglobulin E, washed, and counted on a gamma or beta counter. The *A. simplex* antigen is stable for several years frozen at −20°C or lower.

Pathology

The localized pathology (edema, hemorrhage, inflammation) surrounding a penetrating anisakid larvae is easily attributed to the nematode. When the nematode is not readily apparent, diagnosis is more complicated, as the parasites living and metabolizing or decaying in the tissues are exposing the host to their products. These three subjects are discussed below.

Lesion Histology

The response to anisakid nematodes in the tissues is varied. In the most benign form the invasion site is characterized by infiltration and proliferation of neutrophils, associated with a few eosinophils and giant cells. Little or no evidence of edema, fibrosis, hemorrhage, or vascular damage is observed (Deardorff et al., 1983). The eosinophilic phlegmonous reaction consists of edematous thickening of the mucosa and submucosa, extensive infiltration of eosinophils, vessels with inflammatory cells and the exuding of fibrin, lymphocytes, monocytes, neutrophils, and plasma cells in the submucosa. Larvae are surrounded by a thin layer of eosinophils. Normally this reaction is observed in the intestine within the first week of infection. The abscess type of reaction is marked by eosinophils, histiocytes, and lymphocytes around the worm and extending into the submucosa, surrounded by a granulomatous zone. Necrosis, hemorrhage, and eosinophilic infiltration surround the granuloma. The cuticle of the larva is distorted, and inner structures lose definition. This type of lesion is typical in chronic cases of stomach or intestinal anisakiasis. The abscess–granulomatous type of reaction is characterized by a reduced central abscess; generally the lymphocytes are increased and may dominate the eosinophils in the granuloma. Tissue collagenization may be in progress. Foreign-body giant cells surround larval remains, which may or may not be identifiable, and many eosinophils infiltrate the debris. This lesion is characteristic of anisakiasis of more than 6 months' duration. Finally, the granulomatous lesion consists of granulation tissue with eosinophilic infiltration and no recognizable worm, with occasional debris. This form is found rarely in old cases of stomach or intestinal anisakiasis (Oshima, 1972).

Identification of Larvae in Sections

The few species of larval anisakids that are found in cross section are representatives of the genera associated with human illness. *Anisakis* larvae are Y shaped, with lateral cords having a thin stem and divergent branches in the pseudocoelom. *Pseudoterranova* and *Contracaecum* larvae have more compact lateral cords with a broad stem. *Pseudoterranova* has an intestinal cecum in addition to the glandular ventriculus that it shares with the rest of the family. *Contracaecum* has a ventricular appendix in addition to the intestinal cecum. The preceding three genera have a ventral glandular excretory organ associated with the left lateral cord. *Hysterothylacium* has the same digestive appendages as *Contracaecum;* it differs in having a well-developed tubular excretory system in the left lateral cord.

Parasite Products

Soluble extracts of *A. simplex* larvae are selectively chemotactic for eosinophils over a range of 0.5 to 50 μg of protein. Tanaka and Torisu (1978) observed a dose–response curve and presented evidence that the response was not related to the degranulation of mast cells. A similar response was observed with culture supernatants, indicating that a similar phenomenon may be expected in vitro.

Some information is available on the nature of these excretory–secretory products, although much additional effort is needed before we understand the mechanisms leading to the pathology observed in anisakiasis. Oxidative enzymes, phosphatases, leucine aminopeptidase, and phosphorylase were demonstrated in the excretory gland (Ruitenberg, 1970). Leucine aminopeptidase is released during molting (Davey and Kan, 1967). The esophageal glands and the feces are other sources of excretory–secretory products. Proteases were demonstrated in the excretory–secretory products of *A. simplex* (Mathews, 1982). The excretory–secretory products of *A. simplex* and *Pseudoterranova* sp. type hemagglutinin were immunosuppressive to mitogen-induced lymphoblast transformation (Raybourne et al., 1983). At a protein concentration of 1 μg/ml, the excretory–secretory fraction with a molecular weight greater than 10,000 inhibited concanavilin-A-induced rat spleen lymphocyte transformation. It was demonstrated that these products are formed in vivo, as serum from a patient with anisakiasis reacted with excretory–secretory products collected in vitro, electrophoresed on sodium dodecyl sulfate–polyacrylamide gel, and electrotransferred to nitrocellulose (Raybourne et al., 1986). Dose-dependent uncoupling of oxidative phosphorylation in the muscle mitochondria of fish infected with *A. simplex* has been observed (Boczon and Bier, 1986). Excretory–secretory products and the cell-free cytoplasmic fraction of *A. simplex* larvae also increased the Mg^{2+}-stimulated adenosine triphosphatase activity in coupled rat muscle mitochondria. The foregoing data indicate significant biologic activity in the excretory–secretory products.

Prognosis

The prognosis for patients with anisakiasis is good. The only two reported deaths associated with anisakiasis were attributed to perforation peritonitis and staphylococcal enteritis after intestinal resection. Of the first 11 reported cases, 2 patients were later pensioned because of asthma; however, a similar connection has not yet been noted in Japan, where significant numbers of cases occur.

Epidemiologic Aspects

Occurrence and Transmission

Anisakiasis is transmitted to the definitive host in seafoods. The male and female larvae molt in the intestinal tract and become mature adults. Unembryonated eggs exit the host with the feces. The eggs embryonate and molt before hatching under favorable conditions. The hatched larvae may attach to the substrate (*P. decipiens*) (Myers 1960) or be free-swimming like the larvae of *A. simplex*. These larvae can infect a variety of copepods (McClelland 1983) or other invertebrates. After a period of growth and development in the first invertebrate hosts, the larvae become infective for larger invertebrates or fish which can transmit the larvae to the definitive host. Larval anisakids may be present in either the flesh or the viscera of fish and shellfish. The larvae in the flesh are of primary concern. The effect on anisakid nematodes of pickling fish viscera is unknown. However, personal observation indicates that pickled viscera products are not readily available in the United States. The risk of acquiring anisakiasis from raw and undercooked fish can be reduced by avoiding species with a high prevalence, such as cod, sole, flounder, fluke, salmon, mackerel, and herring. The larvae are present in other species, but less regularly; however, the danger of infection remains.

Susceptibility

Evidence of variation in susceptibility to anisakid infections is lacking. The United States has a preponderance of cases with which little or no discomfort or known pathology is associated. Of 20 cases, 17 were classified as lacking clinical significance (Desowitz, 1986). In Japan, a large majority of cases are considered clinically significant and characterized by pain and suffering (Oshima, 1987). It has been postulated that the difference could lie in a different strain or strains of parasite; however, *Pseudoterranova decepiens* and *A. simplex* larvae from Japan cannot be distinguished morphologically from those from the

United States. It also seems unlikely that the virulent strain of both species would be found in Japan, but not the United States. Chen (1971) proposed that anisakiasis is not found in Taiwan because the Taiwanese eat raw fish at the end of the meal rather than the beginning. In the United States, raw fish may be consumed more frequently in conjunction with a full meal. An increased prevalence of infection in men, noted in both Japan and Holland, was attributed to the increased tendency among men to consume raw fish with alcoholic beverages. No experimental data are available on these conditions.

Etiologic Agents

Characterization

The anisakid nematodes are the aquatic relatives of the large human roundworm *A. lumbricoides* and its terrestrial relatives. There have been 27 genera described in this family and hundreds of species placed in these genera. At present, reported cases of human anisakiasis have involved only four genera: *Anisakis, Pseudoterranova, Contracaecum,* and *Hysterothylacium*. The following descriptions characterize the larva with a boring tooth and describe the changes that occur at the molt, which is triggered by an increase in host body temperature in forms that mature in homoiotherms.

Anisakis

Larvae are slender, pink, and measure about 25 mm long by 1.5 mm in diameter. The anterior excretory pore is ventral to the boring tooth; the ventral excretory gland and large excretory gland nucleus are not usually visible. The lips are vaguely outlined beneath the cuticle; the muscular esophagus (properly the pharynx) is followed by the glandular ventriculus, then the intestine. The ventricular intestinal junction is angled. The rectal glands are usually visible, as is the fine caudal cuticular spine at the end of the conical tail. The cuticle has fine striae, which are visible with high contrast. The lips become distinct and the cuticular striae moderately coarse after the molt, which occurs between 2 and 4 days at mammalian body temperatures.

Pseudoterranova

Larvae are stout, tan, and measure 20 to 35 mm long by 1.5 mm in diameter. The anterior excretory gland is ventral to the stout boring tooth. The excretory gland and nucleus, found at the level of the anterior intestine, are usually not visible. The lips are vaguely outlined beneath the cuticle. The muscular esophagus is followed by the cylindric glandular ventriculus. The junction with the intestine is approximately perpendicular. The intestinal cecum extends anteriorly dorsal to the ventriculus and ends in a ligament, which attaches to the left lateral cord. The rectal glands are usually visible, as is the caudal spine. The cuticle has moderate striae, which are visible with average contrast. The lips become distinct and the cuticular striae coarse after the molt, which occurs after 2 to 4 days at mammalian body temperatures.

Contracaecum

Larvae are delicate, white, and measure 10 to 20 mm long by about 0.8 mm in diameter. The anterior excretory pore is ventral to the stout boring tooth. The excretory gland nucleus is prominent under the anterior intestine, and the excretory gland is intermittently distinct to the midbody. The lips are not discernible. The muscular esophagus is followed by the subcylindric ventriculus. The intestinal cecum is dorsal to the ventriculus and extends almost to the level of the nerve ring, where it attaches by a ligament to the left lateral cord. The ventricular appendix extends posteriorly almost to the end of the first one-fourth of body length. The rectal glands are distinct, as is the caudal spine. The lips become distinct, with defined interlabia, and the excretory gland and nucleus become obscure after the molt, which occurs after 5 to 6 days at mammalian body temperatures.

Hysterothylacium

Larvae are delicate to stout, measuring 6 to 30 mm long by 0.5 to 1.2 mm in diameter. The excretory pore is ventral, near the level of the nerve ring, and the tubular excretory gland is not visible. The boring tooth is usually distinct. The lips are sometimes discernible beneath the cuticle. The muscular esophagus is followed by the subcylindric glandular ventriculus. The intestinal cecum is dorsal to the ventriculus and muscular esophagus; its length varies depending on the species. The ventricular appendix is ventral to the intestine and may extend to near midbody depending on the species. Genital primordia are often visible as tubular structures in the pseudocoelom. Lateral alae may be apparent over the entire length of the body. The rectal glands are distinct, as is the caudal spine; the tail may be covered with spines. The stimuli to molt, which results in the formation of distinct lips and interlabia, are unknown.

Specimen Preservation

Identification is easier when complete specimens are present. Although identification from sections is possible, the orientation of the specimen or the area available is normally less than ideal. Obtaining complete specimens can prove difficult due to the condition of the parasite in situ or damage inflicted in processing. Rapid fixation of specimens is recommended; the author prefers specimens fixed in glacial acetic acid. Formalin or alcoholic fixation is adequate. Specimens must be cleared for internal organs to be visible; clearing agents include crystal phenol liquefied with ethanol, concentrated lactic acid, and glycerol.

Treatment, Prevention, and Control

Treatment

Surgical resection and endoscopic treatment of anisakiasis have been discussed previously. Diagnosis and treatment with an endoscope is the preferred treatment for gastric and duodenal anisakiasis in Japan. Conservative treatment of diagnosed intestinal anisakiasis by administration of antibiotics and isotonic glucose and resection of lesions in chronic intestinal and gastric anisakiasis is recommended (Oshima, 1972, 1987). Captive marine mammals have been treated successfully with thiabendazole for anisakid infections. A dose of 45 mg/kg body weight repeated after 3 months, then at 6-month intervals, is recommended (Ridgeway, 1972). Levamisole was also effective, and piperazine less so. Anthelmintic treatment has not been evaluated in humans.

Prevention and Control

Cooking fish and seafood to an internal temperature of 65°C for at least 30 s could eliminate anisakiasis (Bier, 1976). In practice, attaining this temperature is difficult to ensure. Consumers have complained of receiving broiled fish that appeared thoroughly cooked, yet contained a live anisakid.

Freezing at −20°C for 96 h kills all anisakid nematodes in fish (Deardorff et al., 1985). The temperature is critical, as living parasites have recovered from fish frozen at least 1 month at −17°C. Consumer complaints have been received about recovering living parasites from commercially frozen flounder, salmon, monkfish, and fish fillet. The effects of using cryopreservatives to enhance the appearance, flavor, and texture of frozen fish products have not been evaluated in terms of parasite survival. Visual inspection of fish and fish fillets is not an effective means of finding and removing nematodes from fish. Candling under controlled conditions has been studied (Power, 1957). Commercial fish fillets are candled to remove parasites (anisakids) in Canada, Iceland, Greenland, Paraguay, Uruguay, and Chile. The Codex Committee on Fish and Fisheries Products of the Codex Alimentarius Commission has drafted standards requiring candling of "all fisheries products in which parasites can be found" (Valdimarsson et al., 1985). Candling is not effective in detecting parasites in fish with pigmented flesh. Sonography may prove a good method for detecting parasites in such fish. Attempts to use this method in Nova Scotia, however, have not been successful (McClelland et al., 1983).

Control of definitive hosts has been used to reduce the incidence of the parasite in inshore fisheries. In Nova Scotia, harvesting gray seal pups and placing a bounty on adults has not resulted in reducing the parasite burden in fishery stocks or controlling the expansion of the seal population (McClelland et al., 1983). Selective fishing of unparasitized stock places undue pressure on the fishery and tends to deplete the stock (Rae, 1972). Treatment of the definitive hosts with anthelmintics or by immunization has been attempted (Odense, 1978). Anthelmintics require repeated administration. Immunity is not developed through repeated exposure to natural infections (McClelland, 1980). Presentation of the antigen in a different manner has proved effective in inducing immunity. The field of immunizing wild animal stocks is just beginning to be explored.

Literature Cited

Asami, K., and Y. Inoshita. 1967. Experimental anisakiasis in guinea pigs: factors influencing infection of larvae in host. Jpn. J. Parasitol. **18:**415–422.

Bier, J. W. 1976. Experimental anisakiasis: cultivation and temperature tolerance determinations. J. Milk Food Technol. **39:**132–137.

Boczon, K., and J. W. Bier. 1986. *Anisakis simplex:* uncoupling of oxidative phosphorylation in the muscle mitochondria of infected fish. Exp. Parasitol. **61:**270–279.

Chen, E. R. 1971. Species of marine fishes found with *Anisakis* larvae in Taiwan, p. 301. *In* J. H. Cross (ed.), Proceedings of the 7th Southeast Asian Seminar on Tropical Medicine and Public Health: Infectious Diseases of the Gastrointestinal System in Southeast Asia and the Far East. Taipei, Taiwan. SEAMEO, Bangkok.

Cocheton, J. J., T. Gombeaud, and J. P. Ferroir. 1984. Anisakiase duodenale aigue extraction endoscopique. Presse Med. **13:**744.

Daniels, J. J. H. M. 1962. De eosinofile flegmone van het maagdarmkanaal veroorzaakt door de haringworm. Ned. Tijdschr. Geneeskd. **106:**131–132.

Davey, K. G., and S. P. Kan. 1967. An endocrine basis for ecdysis in a parasitic nematode. Nature (London) **214:**737–738.

Deardorff, T. L., T. Fukumura, and R. B. Raybourne. 1986. Invasive anisakiasis. A case report from Hawaii. Gastroenterology **90:**1047–1050.

Deardorff, T. L., M. M. Klicks, and R. S. Desowitz. 1983. Histopathology induced by larval *Terranova* (HA) (Nematoda: Anisakidae) in experimentally infected rats. J. Parasitol. **69:**191–195.

Deardorff, T. L., R. B. Raybourne, and R. S. Desowitz. 1985. Behavior and viability of third-stage larvae of *Terranova* (HA) and *Anisakis simplex* (type 1) under coolant conditions. J. Food Prot. **47**:49–52.

Desowitz, R. S. 1986. Human and experimental anisakiasis in the United States. Hokkaido J. Med. Sci. **5**:358–371.

Desowitz, R. S., R. B. Raybourne, H. Ishikura, and M. M. Klicks. 1985. Radioallergosorbent test (RAST) for the serodiagnosis of human anisakiasis. Trans. R. Soc. Trop. Med. Hyg. **79**:609–612.

Fabresse, F. X., H. Essioux, M. Meyran, P. Larroque, and H. Chelton. 1984. Polyarthritis ague anisakiasis. J. Med. Paris **13**:1004.

Jackson, G. J. 1975. The "new disease" status of human anisakiasis and North American cases: a review. J. Milk Food Technol. **38**:769–773.

Kim, C. H., B. S. Chung, Y. I. Moon, and S. H. Chun. 1971. A case report of human infection with *Anisakis* in Korea. Korean J. Parasitol. **9**:39–43.

Kobayashi, A., T. Koyama, M. Kumada, T. Suguro, and K. Koito. 1968. Skin test with somatic antigen and E-S (excretions and secretions) antigens from *Anisakis* larvae. Jpn. J. Parasitol. **17**:407–418.

Little, M. D., and J. C. MacPhail. 1972. Large nematode from the body cavity of a man in Massachusetts. Am. J. Trop. Med. Hyg. **21**:948–950.

Mathews, B. E. 1982. Behaviour and enzyme release by *Anisakis* sp. larvae. J. Helminthol. **56**:177–183.

McClelland, G. 1980. *Phocanema decipiens:* moulting in seals. Exp. Parasitol. **49**:128–136.

McClelland, G., R. K. Misra, and D. J. Marcogliese. 1983. Variations in the abundance of larval anisakines, sealworm (*Phocanema decipiens*) and related species in cod and flatfish from southern Gulf of St. Lawrence and the Breton Shelf. Can. Tech. Rep. Fish. Aquat. Sci. **1201**:1–51.

Morishita, K., and T. Nishimura. 1965. Studies on the *Anisakis* larva: *Anisakis* larva from human buccal mucosae. Jpn. J. Parasitol. **14**:54.

Myers, B. J. 1960. On the morphology and life history of *Phocanema decipiens*. Can. J. Zool. **38**:381–384.

Myers, B. J. 1963. The migration of *Anisakis*-type larvae in experimental animals. Can. J. Zool. **41**:147–148.

Nagase, K. 1973. Studies on *Anisakis*. Acta Sch. Med. Univ. Gifu **21**:85–108.

Namiki, M., T. Morooka, H. Kochi, N. Ueda, T. Sekiya, K. Nakagawa, T. Furuta, T. Hutoguro, and H. Kamata. 1970. Diagnosis of acute stomach anisakiasis. I To Cho **5**:1437–1440.

Odense, P. H. 1978. Some aspects of the codworm problem. Fish. Environ. Can. Fish. Mar. Serv. Ind. **106**:1–12.

Oshima, T. 1972. *Anisakis* and anisakiasis in Japan and adjacent areas. Prog. Med. Parasitol. Jpn. **4**:301–393.

Oshima, T. 1987. Anisakiasis—is the sushi bar guilty? Parasitol. Today **3**:44–48.

Petter, A. J. 1969. Enquete sur le nematodes de poissons de la region nantaise. Ann. Parasitol. Hum. Comp. **44**:25–35.

Pollak, M. F., and E. H. Kamplemacher. 1966. Haringwormsiekte in 1965 en voorgaande jaren. Versl. Volksgezondn. **12**:344–355.

Power, H. E. 1957. The effect of various lighting conditions on the efficiency of candling cod fillets for detection of parasites. J. Fish. Res. Board Can. **15**:537–542.

Rae, B. B. 1972. A review of the cod-worm problem in the North Sea and in western Scottish waters 1958–1970. Mar. Res. Dep. Agric. Fish. Scotl. **2**:12–24.

Raybourne, R., R. S. Desowitz, M. M. Klicks, and T. L. Deardorff. 1983. *Anisakis simplex* and *Terranova* sp: inhibition by larval excretory–secretory products of mitogen-induced rodent lymphoblast proliferation. Exp. Parasitol. **55**:289–298.

Raybourne, R. B., T. L. Deardorff, and J. W. Bier. 1986. *Anisakis simplex:* production dynamics of larval excretory–secretory proteins and their cytostatic action in mammalian cell cultures. Exp. Parasitol. **62**:92–97.

Ridgeway, S. H. 1972. Homeostasis in the aquatic environment, p. 590–747. *In* S. H. Ridgeway (ed.), Mammals of the sea: biology and medicine. Charles C Thomas, Publisher, Springfield, Ill.

Ruitenberg, E. J. 1970. Anisakiasis: pathogenesis, serodiagnosis and prevention. Thesis, University of Utrecht, Netherlands.

Schaum, E., and W. Miller. 1967. Die Heterocheilidiasis eine Infection des Menschen mit Larven von Fish Ascariden. Dtsch. Med. Wochenschr. **92**:2230–2233.

Smith, J. W., and R. Wooten. 1978. Anisakis and anisakiasis. Adv. Parasitol. **16**:93–148.

Sugimachi, K., K. Inokuchi, T. Ooiwa, and Y. Ishii. 1985. Acute gastric anisakiasis. J. Am. Med. Assoc. **253**:1012–1013.

Suzuki, T. 1968. Studies on the immunological diagnosis of anisakiasis. Jpn. J. Parasitol. **17**:213–220.

Suzuki, T., Y. Sato, T. Yamashita, H. Sekisawa, and M. Otsuru. 1974. Anisakiasis: preparation of stable antigen for indirect fluorescent antibody test. Exp. Parasitol. **35**:418–424.

Tanaka, J., and M. Torisu. 1978. *Anisakis* and eosinophil. J. Immunol. **122**:302–320.

Tozuka, M. 1978. Allergic reaction of guinea pig intestine sensitized with *Anisakis* larvae. Jpn. Smooth Muscle Soc. J. **14**:308–315.

Valdimarsson, G., H. Einarsson, and F. King. 1985. Detection of parasites in fish muscle by candling technique. J. Assoc. Off. Anal. Chem. **68**:549–551.

Van Thiel, P. H. 1976. The present state of anisakiasis and its causative agent. Trop. Geogr. Med. **28**:75–85.

Van Thiel, P. H., F. C. Kuipers, and T. H. Roskam. 1960. A nematode parasitic to herring, causing acute abdominal syndrome in man. Trop. Geogr. Med. **12**:97–113.

Yokogawa, M., and H. Yoshimura. 1967. Clinicopathologic studies on larval anisakiasis in Japan. Am. J. Trop. Med. Hyg. **16**:723–728.

CHAPTER 80

Ascariasis

JERRY E. JONES

Disease: Ascariasis (roundworm infection, ascaridiasis).
Etiologic Agent: *Ascaris lumbricoides.*
Source: Human.
Clinical Manifestations: Cough, dyspnea, fever, wheezing, rales during larval migration through lungs. Adult phase may generate symptoms of obstruction, and bolus of worms produces mechanical small bowel obstruction.
Pathology: Larva—eosinophilic infiltration and granuloma formation around and in paths of migrating larvae. Adult—little change in intestinal mucosa. Mechanical obstruction if migration is to a small orifice or if large numbers aggregate in lumen.
Laboratory Diagnosis: Identification of characteristic eggs or adult worms. Direct fecal smears, sedimentation techniques, and rarely sputum or gastric aspiration for larvae.
Epidemiology: Worldwide distribution; highest rates in children. High endemic areas exist where prevalence often exceeds 50%. Transmitted by ingestion of eggs.
Treatment: Mebendazole, pyrantel pamoate (embonate), piperazine citrate.
Prevention and Control: Combination of approaches, including education, sanitation, and chemotherapy.

Description of Disease

The hookworm and the malarial parasite have been the objects of international campaigns of far-reaching scope while the ascarid has received only causal attention. And yet this parasite is of very great importance because it is so widespread and because the pathological effect on the individual may be so very serious.

Cram (1926)

Ascariasis is the infection caused by the largest nematode to infect humans, *Ascaris lumbricoides*. It has been known to physicians since the dawn of history and, with the possible exception of the pinworm (*Enterobius vermicularis*), it is the most widely prevalent of all human roundworms (Pawlowski and Arfaa, 1984). Although most frequent in tropical countries, this infection is also common in many temperate areas. The mortality and morbidity due to ascariasis are both relatively low, often leads to the assumption that ascariasis is a fairly benign intestinal parasite. However, the absolute number of fatal cases is high, primarily because of intestinal complications. Therefore, because of its global presence and potential to cause severe symptoms, *A. lumbricoides* must be considered a major intestinal parasite.

Characterization and Life Cycle

As an adult, *A. lumbricoides* is elongated, cylindric, and tapers at both ends. The mature male worm measures 12 to 31 cm in length and 2 to 4 mm in diameter. The mature female measures 20 to 35 cm in length and 3 to 6 mm in diameter. Specimens measuring up to 45 cm have been reported. The adults live in the lumen of the small intestine and maintain their position by strong, muscular motions of their bodies. Each adult may live for 12 months, with some living for up to 2 years. (Beaver et al., 1984). Sexual reproduction occurs in the lumen, producing fertilized eggs that are passed into the fecal stream.

At the time of deposit, the fertilized egg is broadly ovoid, measuring 67 to 75 μm long and 35 to 50 μm wide. Microscopically, it is easily identifiable by its distinct morphologic characteristics. A coarsely mammillated, albuminoid shell serves as a protective layer. A coarsely granular, spherical egg cell, which matures to the first larval stage, is enclosed. Female worms without males produce infertile eggs that are larger and more elliptic. These infertile eggs have a very globular cellular material, often appearing as mass of disorganized granules and globules that completely fill the shell. Both fertilized and unfertilized eggs, which are biled stained by the time they reach the environment, may be passed.

The eggs of *A. lumbricoides* are extremely resistant to environmental conditions. They survive putrefaction and can withstand considerable desiccation and cold. The eggs must remain in the soil approximately 2 to 3 weeks before they become infective. Once infective, the eggs may remain viable in the soil for a year or longer.

The eggs are ingested when an individual comes in contact with the fecally contaminated soil. The infective larvae hatch in the duodenum and penetrate the intestinal wall. Here the larvae enter the lymphatic channels or the mesenteric vessels and pass into the right side of the heart. The larvae are then trapped by the pulmonary capillaries, where they perforate into the alveoli. This migration from intestine to pulmonary capillary bed takes 3 to 4 days. Once in the lung, the larvae double in length and molt into third-stage larvae. At this point, they begin migration up the trachea to the intestine, where they mature in 2 to 3 months to become sexually active worms. Figure 1 depicts the life cycle of *A. lumbricoides*. The entire cycle can be completed in 60 to 90 days.

Clinical Manifestations

Understanding the clinical spectrum of *A. lumbricoides* requires recognition of several key factors pertaining to this parasite. First, because of its large size, it can produce symptoms by mechanical obstruction. In addition, because of its mating habits, it tends to aggregate with other worms, thereby in-

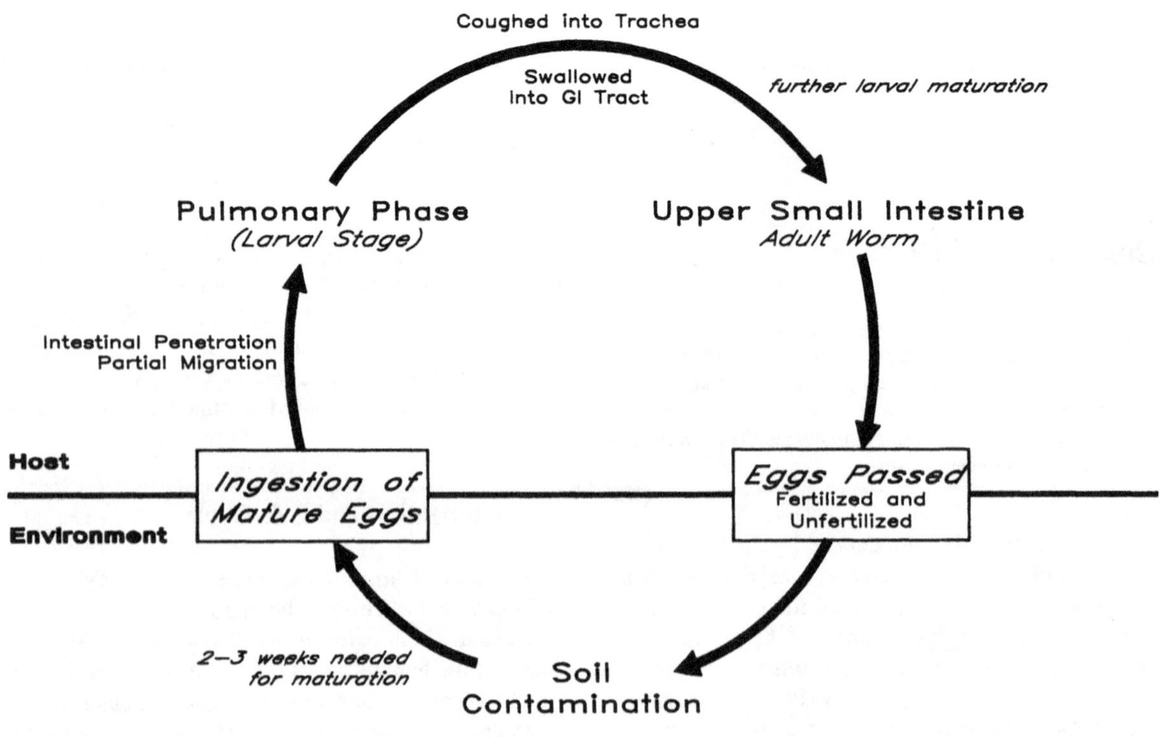

FIG. 1. Life cycle of *Ascaris lumbricoides*. Cycle is completed 60 to 90+ days after adult worm passes fertilized eggs.

creasing its obstructive potential. As a single adult, this parasite also migrates using a spiral motion, which may lead it into small orifices and cause obstruction. Therefore, either as a single adult or as a bolus of worms, the potential for life-threatening manifestations exist. Second, the *Ascaris* allergen is one of the most potent allergens of parasitic origin. Allergic symptoms can be generated that may invoke systemic hypersensitivity reactions. Third, other factors such as poor living conditions, inadequate nutrition, and coexistent or past parasitic or infectious diseases will all have roles in the clinical signs and symptoms of ascariasis. Fourth, recognizing the life-cycle phase involved at the time of clinical presentation will help place the often disparate signs and symptoms into an understandable sequence. The often complex signs and symptoms may be easily overlooked when viewed out of the context of a single parasitic infection.

Larval Migration Phase

The initial abdominal migration of *Ascaris* larvae is usually asymptomatic. On exposure, the larvae migrate through the liver, lymphatic channels, and bloodstream to the pulmonary capillary bed without exciting much host reactivity. On repeated infection the host may react violently, with a hypersensitivity reaction occurring along the path of larval migration. This reactivity has produced generalized symptoms of urticaria, angioneurotic edema, and bronchial asthma. Once in the pulmonary capillary bed, the larvae break into the alveolar air sac, causing hemorrhage and bronchial damage. A cough is produced, which may be productive because the alveolar sac fills with serous exudate. Bronchial spasm is often produced, accompanied by wheezing and rhonchi that may be heard on auscultation of the lung fields. The cough continues for 4 to 5 days. With heavy infection or in an allergic individual, severe pulmonary symptoms may be generated. Severe asthmatic reactions, associated with wheezing, dyspnea, cough, and pulmonary edema, have been reported. At this point in the clinical course, there is moderate fever and eosinophilia. Roentgenographic findings of shifting, mottled lung infiltrations are suggestive of viral pneumonia. This infiltration pattern is transient, varying from day to day with spontaneous clearing in 3 to 14 days. This lung phase, which is associated with peripheral eosinophilia, has been called *Löffler syndrome*. Examination of sputum may also show eosinophils and Charcot–Leyden crystals. Occasionally larvae are seen in the sputum and may be found in gastric washings. In general, the host reaction is proportional to the number of larvae involved during migration and to the allergic status of the host (Koino, 1923; Pawlowski, 1982; Takata, 1951).

Adult Intestinal Phase

Once coughed and swallowed into the intestinal tract, the larvae mature into adult worms. Roentgenographic studies show that 87% of the worms reside in the jejunum (Makidono, 1956). These worms remain stationary by their strong muscular action and appear little affected by normal intestinal peristaltic action. Intermittent collicky cramps, loss of appetite, and fretfulness are general symptoms. Migration of adult worms may lead to biliary and pancreatic duct obstruction (Choi and Wong, 1984; Houang et al., 1980). Therefore, single worms may cause significant clinical problems. Migration of worms to the liver, forming liver abscesses, has been reported (Beaver et al., 1984; Rossi and Bisson, 1983). Further migration of adult worms has led to their passage through the mouth and nose and even the eustachian tubes (Berkowitz et al., 1980). In rare cases, worms have been aspirated and listed as causing death through asphyxia. Migration of single adults through the other end of the intestinal tract has lead to perforation of the appendix and acute appendicitis. A bolus of adult worms can also form a mechanical small-bowel obstruction (Katz et al., 1985; Wynne and Ellman, 1983), most commonly at the ilioceoleal valve. The complications of such obstruction have included perforation and gangrene. Blumethal and Schultz (1975) found that most cases of intestinal obstruction occurred in children between the ages of 1 and 5 years, with an incidence of approximately 2 per 1,000 infected children per year. The fatality from such complications has been estimated to be as low as 6 deaths per 100,000 cases. However, because of the many infected individuals, several hundred deaths would be expected among the millions infected by this parasite. Figure 2 shows the clinical spectrum of this infection, highlighting the clinical manifestations by life-cycle phase involved.

Pathology

The pathogenic effects of *A. lumbricoides* depend on three basic mechanisms: (1) The immune reaction of the host, (2) the mechanical effects of the adult worms, and (3) the effects of adult worms on host nutrition. The immune reaction of the host involves the allergic potential of the parasite as well as the immune status of the host. This process is triggered primarily as a reaction to the larval phase of the parasite (Gelpi and Mustafa, 1967). Unless large numbers of larvae are migrating simultaneously, the first passage from the intestine through the liver and lungs provokes no marked pathologic changes. However, subsequent migrations may cause intense tissue reactions in the liver and lungs even when relatively few larvae are migrating (Baird et al., 1986;

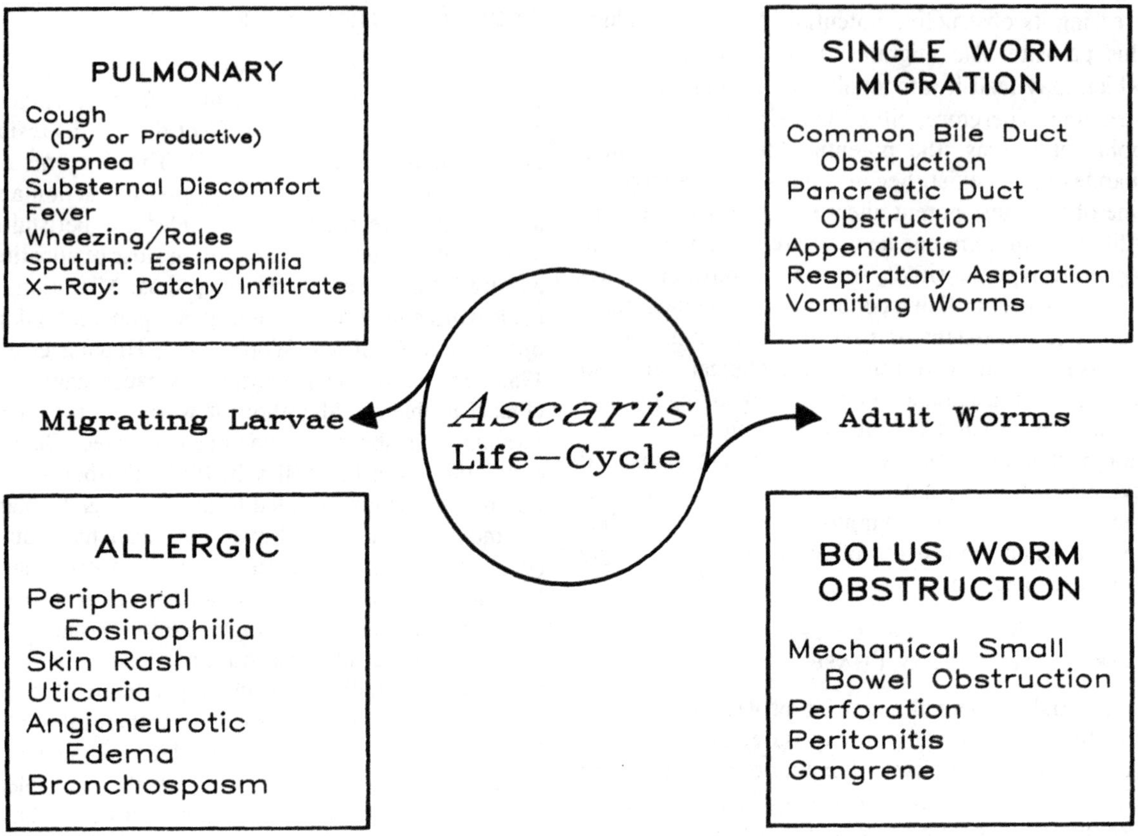

FIG. 2. Clinical manifestations of *Ascaris lumbricoides*.

Pawlowski, 1978, 1982). The tissue reaction involves focal eosinophilic infiltration and granuloma formation around and in the paths of migrating larvae. In later stages there may be fibrosis of the periportal and interlobular spaces as a result of the intense local cellular reaction around the larvae in the tissues. As this process occurs in the lungs, sloughing of bronchial epithelium may occur. Intense interstitial reactions produce the classic pulmonary pathology of *Ascaris* pneumonitis (Spillman, 1975). Histologically, there is infiltration of eosinophils, epithelioid cells, and macrophages. During this phase, the alveolar sacs fill with serous exudate and sputum may show larvae. As noted previously, the exudate will demonstrate many eosinophils and Charcot-Leyden crystals (Beaver et al., 1984).

The adult worms are believed to cause few pathologic changes as a result of their presence in the intestinal lumen, although they may cause serious damage as a result of their migratory and obstructive activities. In the animal model (*Ascaris suum*), significant hypertrophy of the intestinal muscle layers accompanied by a corregated appearance of the mucosa, a shortening of the crypt depth, and a diminished amount of mucus has been observed (Stephenson et al., 1980b). Whether this pattern of change occurs in the human host is unclear, although a disordered pattern in the small bowel described as a coarsening of the mucosal folds is reported (Lagundoye, 1972).

The local intestinal hypersensitivity pattern produced by the metabolites of the adult worm has not been studied extensively. However, some believe that ascariasis itself is responsible for decreases in growth, absorption of various host nutrients, and impaired carbohydrate absorption (Cerf et al., 1981; Layrisse and Vargas, 1975; Stephenson, 1980).

In the animal model, strong inhibitors of pepsin, cathepsin E, trypsin, and chymotrypsin have been reported (Lee and Atkinson, 1976). However, the impact of these findings in humans is uncertain (Pawlowski and Arfaa, 1984). The nutritional implications of ascariasis have been reviewed by Schultz (1982), who raises several issues regarding past studies and the conclusions drawn from them. He notes that studies claiming a positive correlation between ascariasis and protein energy malnutrition have not found a significant difference in weight between the infected and the uninfected. Also, he comments that several studies have shown no significant improvement in nutritional status after treatment intervention. Thus, much remains uncertain regarding the nutritional impact of ascariasis (Calloway, 1982).

Diagnostic Features

The diagnosis of ascariasis is made by identifying the characteristic eggs or adults of this species. During the pulmonary migration phase, a diagnosis can be made by identifying the larval stage in sputum or gastric washings; however, this etiology is often not considered at this point.

The adult female worm passes approximately 200,000 eggs per 24 h. This number has been found to vary depending on worm load and the age of the worm (Sinniah, 1982). The Beaver direct smear method was found to give higher egg counts than the other methods tested (the Stoll technique and the Katz technique). However, the Katz technique was found to be more reliable in estimating worm load (Sinniah, 1982).

In the office setting, simple stool examination using the wet preparation can be utilized for diagnosis (Jones, 1980). The characteristic eggs are easily seen under the low-power (10×) objective and can be readily distinguished at a magnification of 45×. However, care must be taken in identification, as false-positive results have been reported when eggs are confused with fecal debris. A direct smear of 2 mg of fecal material (approximately the amount of stool on the tip of a wooden applicator stick) normally has two to six eggs per female worm (Pawlowski and Arfaa, 1984). The digital rectal examination can be used in the office setting to collect small samples of stool for slide preparation (Jones, 1981a, 1984). Stool kits are available for the office setting (Meridian Diagnostic, Inc., Cincinnati, Ohio).

Reference laboratory procedures for demonstrating the presence of *Ascaris* eggs, including centrifugation and sedimentation, are available. Formalin–ether sedimentation (after Ritchie), formalin–ether with preserved specimens (after Brooke), acid–sodium sulfate–triton NE–ether concentration (after Hunter), and the Kato thick smear have all been recommended (Hunter et al., 1976).

Frequently, the adult worm may be brought to the office after being passed through the rectum or oral cavity. This passage may produce a dramatic reaction in the patient or the patient's family (Jones, 1981b, 1982). An explanation of the life cycle of the worm, treatment, and the course of the infection may be necessary to help the patient and family to deal with the psychologic impact of "passing a worm."

The adult worm can be readily identified by its

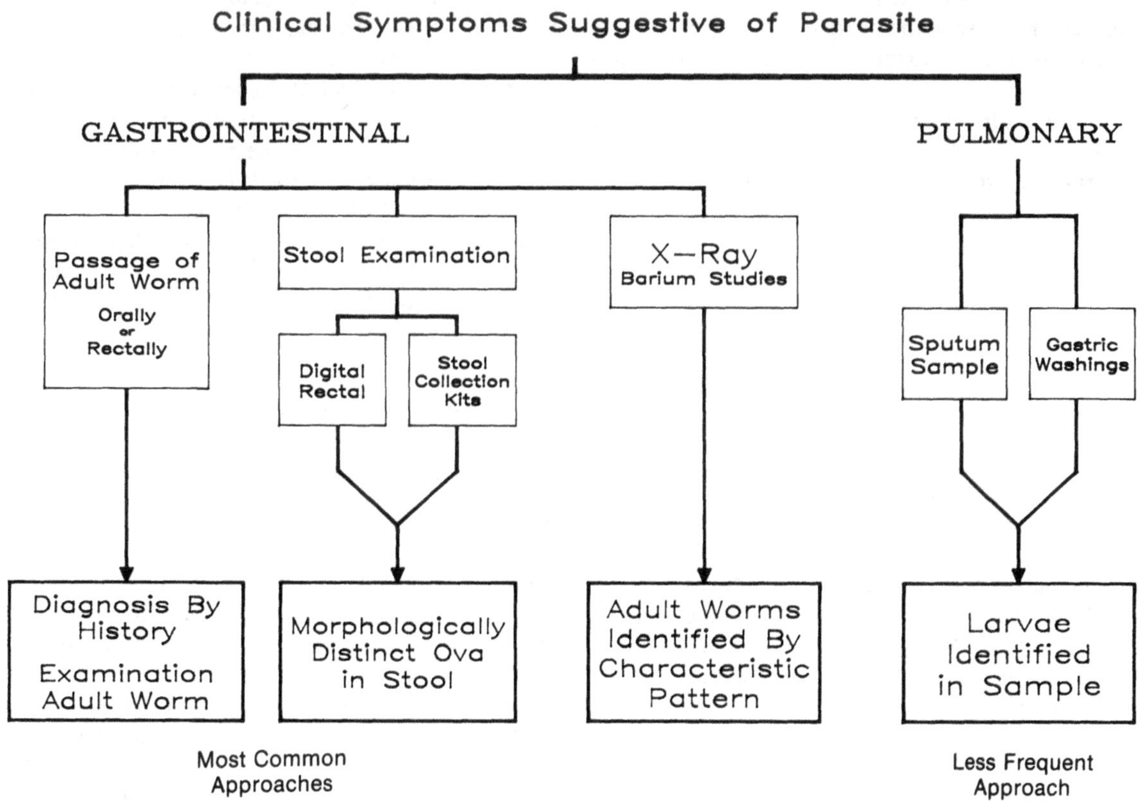

FIG. 3. Diagnostic approach to *Ascaris lumbricoides*.

size and earthwormlike appearance. Males are smaller and have curved tails. In many cases only one worm will be passed. However, it is important to warn the patient and the patient's family that other worms may be passed following treatment.

Occasionally, ascariasis will be diagnosed by radiographic examination. Barium will often show distinct worms in the bowel lumen. The worms may be seen as curved, linear, negative shadows surrounded by the contrast material. If the worms have ingested the barium sulfate, they may be outlined by the finer linear shadows of their own intestines. A group of worms lying parallel to each other and filling the host's intestinal lumen produces an X-ray shadow of parallel lines that have been called the *tramway sign* because of their resemblance to the parallel tracks of trolley car lines (Beaver et al., 1984).

The larval phase passed in the sputum and found in gastric washings consists of third-stage larvae. They are 1.2 mm in length and have poorly developed lips. Differentiation of *Ascaris* larvae from hookworm and *Strongyloides* larvae is accomplished by noting their size and mouth development. Hookworm and *Strongyloides* larvae are smaller and lack the distinct lip development of *Ascaris* larvae.

Indirect methods of diagnosing ascariasis have included skin testing, the radioallergosorbent test, and measurement of serum immunoglobulin E antibodies (Pawlowski and Arfaa, 1984; Shephard et al., 1982; Turner et al., 1980). The interpretation of these tests is difficult because of the complex antigenic structure of *Ascaris* and its extensive antigenic cross-reactivity.

Figure 3 summarizes the diagnostic approach to *A. lumbricoides*. The approach varies depending on the primary symptom complex presented to the physician.

Epidemiologic Aspects

Occurrence and Transmission

A. lumbricoides has been estimated to infect up to one-fourth of the world's population (Pawlowski and Arfaa, 1984; Peters, 1978; Walsh and Warren, 1979). Marked regional differences in prevalence rates occur, with areas of high endemicity scattered throughout the world. As a species, *Ascaris* is most prevalent in crowded, rural areas where sanitary facilities are lacking or not used secondary to behavioral factors. Ascariasis has a wide geographic distribution, however, and has been identified in urban as well as rural populations. The species flourishes in warm, moist climates or in moist, temperate regions where fecally contaminated soil is common. The frequency of ascariasis increases sharply during the first 2 to 3 years of life and usually remains maximal between the ages of 4 and 14 years. Depending on the geographic area, the prevalence then either persists at slightly lower levels or declines with advancing age. Therefore, the intensity of worm burden is usually maximal in children, who are also at greatest risk for severe complications from this parasite. It must also be recognized that ascariasis is a "family infection" and will frequently aggregate in family units. Pawlowski (1982) estimates that the global external environment is contaminated *daily* by 9×10^{14} eggs of *A. lumbricoides*.

Transmission of *A. lumbricoides* is by oral ingestion of the fully embryonated eggs. The eggs are highly resistant to desiccation and to thermal changes under 45°C (113°F) (Hunter et al., 1976). Studies by Beaver (1952, 1975) have demonstrated the unique relationships between the particulate elements of soil and the eggs of *A. lumbricoides*. Eggs in clay or heavy loam soils are protected against desiccation and freezing. Also, rainfall tends to improve the distribution of eggs both horizontally and vertically. This distribution among the particulate elements helps to mix the eggs and to provide them with protective layers against environmental hazards. Eggs have been reported to survive as long as 10 to 15 years (Brudastov et al., 1970; Krasnonos, 1978). Under less than ideal conditions, however, eggs probably survive an average of 10 to 12 months (Beaver et al., 1984). The eggs are transmitted by direct eating of soil (pica) and by carriage on feet or footwear, the fur and feet of domesticated animals, flies and other invertebrates, fruits and vegetables, water, and even dust (i.e., airborne transmission) (Beaver, 1975; Ghadirian et al., 1979; Glickman et al., 1981; Rukmono et al., 1977). Therefore, the occurrence and transmission of *A. lumbricoides* depend on the density of the human population, levels of sanitation and agricultural development, and regional environmental conditions.

Susceptibility, Resistance, and Carriage

Susceptibility

All ages and both sexes are susceptible to infection with *A. lumbricoides*. Children are at greater risk for both intensity and complications of infection. Exposure to ascariasis is related to many factors, including climate and seasonal variations, soil conditions, level of sanitary practices, use of night soil for fertilizer, density of the human population, nutritional status of the host, particular cultural habits, and immune status of the host.

Although not clearly defined, the immune reaction of the host invokes both cellular and humoral responses. The resistance appears to be primarily focused on the initial larvae migration phases and the preadult or fourth-stage larvae. In a nonimmune host, cellular responses involve eosinophils and neutrophils and are not solely T cell mediated (Johnstone et al., 1981). In addition to the cell-mediated responses, a nonspecific response is postulated. This response has been believed to involve the eosinophil chemotactic factor and the neutrophil chemotactic factor (Pawlowski, 1982).

On a humoral level, antibodies of the immunoglobulin M class are detected around the sixth day of infection. Antibodies of the immunoglobulin G class follow and appear at maximum levels some 3 to 4 weeks into the infection. The circulating antibodies continue during the first 2 to 3 months.

Analysis of the protein–polysaccharide antigens in the animal model reveal the presence of multiple determinants. These complex antigens may be immunogenic or allergenic. Studies reveal a potentiation of immunoglobulin E by *Ascaris* infection, but only a small portion of these globulins has specific reactions to *Ascaris* (Strejan, 1975; Turner et al., 1979).

Carrier states are a reflection of the chronic endemicity of ascariasis. In a hyperendemic area, a reinfection rate of 55% was found 7 months after mass treatment (Hayashi, 1980). Fluctuations in *Ascaris* infections were documented by Shield et al., (1984), but after treatment, levels of infection returned to pretreatment numbers. Therefore, constant reinfections keep ascariasis at relatively steady levels in many parts of the world. Reports from Iran, Colombia, Kenya, and South Korea demonstrate overall prevalence rates of 25 to 86% (Botero, 1975; Ghadirian et al., 1979; Soh, 1973; Stephenson et al., 1980a). This endemic nature reflects complex social, cultural, and environmental factors in many parts of the world. In addition, many infected individuals with no symptoms are responsible for passing large numbers of eggs into the environment.

Treatment, Prevention, and Control

Treatment

Drugs are available to treat the adult intestinal phase of ascariasis effectively. Table 1 displays these drugs and their mode of action against the adult worm. Mebendazole (Vermox) has been shown effective in 84 to 100% of cases. A chewable tablet, its ease of administration works well for patients more than 2 years of age. As it has been shown to be embryotoxic in some rodents, it is not recommended during pregnancy. Pyrantel pamoate (Antiminth) works well as a single-dose agent, with up to 90% efficacy reported. Doses for children are fairly well tolerated. In large mass treatment programs, a lower dosage has been used with good results (Soe, 1981).

Piperazine citrate (Antepar), the oldest and perhaps most widely used anthelmintic, has 70 to 80% efficacy in single doses, which increases to 90% if used for 2 days. Sporadic hypersensitivity reactions and neurotoxic reactions have been reported (Pawlowski and Arfaa, 1984).

The pulmonary phase of *A. lumbricoides* infection is without treatment. There is no conclusive evidence that any anthelmintic drugs currently used are

TABLE 1. Drugs for the treatment of *Ascaris lumbricoides*[a]

Drug (trade name)	*Dose*	*Action*	*Reported efficacy*	*Concerns*
Mebendazole (Vermox)	100 mg twice per day for 3 days	Inhibits phosphorylation in mitochondria	84–100%	Embryotoxic in some rodents Contraindicated in pregnancy
Pyrantel pamoate (Antiminth)	10 mg/kg (maximum, 1 g)	Neuromuscular transmission blockade Spastic ACh-like paralysis	90%	May be used as single dose Mass treatment programs: lower dose (2.5 mg/kg) has been used (Soe, 1981)
Piperazine citrate (Antepan)	75 mg/kg (maximum, 3.5 g) for 2 days (also used single dose)	Neuromuscular blockade Anticholinergic action at myoneural junction	90% 70–80% (single dose)	Sporadic hypersensitivity reactions Neurotoxic reactions reported

[a] Adapted from *The Medical Letter* (1986), Johnston (1981), Pawlowski and Arfaa (1984), and Soe (1981).

effective against the larval phase as they migrate through the liver, lungs, or intestinal mucosa. In severe hypersensitivity reactions, corticosteroid therapy may be of benefit (Phills et al., 1972).

Signs and symptoms suggesting ilieus after the passage of an adult worm may signal an obstruction from a bolus of adult worms. Initially, a conservative approach to treatment is indicated unless volvulus, perforation, or peritonitis is suspected. Complete intestinal obstruction also constitutes a reason for surgical intervention (Blumenthal and Schultz, 1975; Darby and Westphal, 1972; Waller and Othersen, 1970).

Conservative treatment includes nasogastric tube placement and decompression, rehydration, fluid and electrolyte balance, and treatment through the nasogastric tube with an anthelmintic. Piperazine citrate has been used in a dosage of 75 mg/kg up to the maximum dose. This dose is flushed with saline, and the nasogastric tube is clamped for 1 to 2 h. The treatment can be repeated every 8 h. It may require 2 to 4 days for the worms to be passed. However, the symptoms of obstruction usually improve within the first 24 h. (Darby and Westphal, 1972). Laparotomy is indicated if the patient fails to respond to this conservative approach.

Prevention and Control

Ascariasis is sustained by fecal contamination of the soil. This contamination is augmented in many areas by the indiscriminate defecation of preschool children, particularly around their immediate home environment. Ideally, providing adequate facilities for proper disposal of feces and prevention of soil contamination in areas immediately adjacent to houses would help eliminate this parasitic problem. However, control and ultimate eradication of this infection has proved difficult.

In many developed countries, the prevalence of ascariasis has decreased. This reduction has been attributed solely to improved levels of sanitation, without any specific control interventions (Pawloski and Arfaa, 1984). Other countries have reported success in periodic mass treatment programs (Cabrera et al., 1975; Morishita, 1972). After periodic treatment programs, however, high reinfection rates have been reported. In addition, no practical means of killing *Ascaris* eggs has been discovered. As noted previously, no parasitic stage is more resistant to chemical and physical harm than that of *Ascaris* eggs (Beaver et al., 1984). Therefore, a combined approach, including chemotherapy, sanitation, and education, is necessary to control ascariasis. Educational approaches would need to include instruction of all people in the use of toilet facilities, with special focus on educating children in hygienic practices such as handwashing and proper food handling (Benenson, 1985).

Finally, prevention and control in edemic areas will take several years of application before maximum benefit will be obtained. Therefore, patience and persistence must be maintained before long-term outcomes can be measured.

Literature Cited

Baird, J. K., M. Mistrey, M. Pimsler, and D. H. Connor. 1986. Fatal human ascariasis following secondary massive infection. Am. J. Trop. Med. Hyg. **35**:314–318.

Beaver, P. C. 1952. Observations on the epidemiology of ascariasis in a region of high hookworm endemicity. J. Parasitol. **38**:445–453.

Beaver, P. C. 1975. Biology of soil-transmitted helminths: the massive infection. Health Lab. Sci. **12**:116–125.

Beaver, P. C., R. C. Jung, and E. W. Cupp. 1984. Oxyuroidea and Ascaridoidea, p. 302–334. *In* Clinical parasitology, 9th ed. Lea & Febiger, Philadelphia.

Benenson, A. S. (ed.). 1985. Control of communicable diseases in man, 14th ed. American Public Health Association, Washington, D.C.

Berkowitz, F. E., E. Sochet, and G. Packirisamy. 1980. *Ascaris* in the middle ear. S. Afr. Med. J. **58**:630.

Blumenthal, D. S., and M. G. Schultz. 1975. Incidence of intestinal obstruction in child infected with *Ascaris lumbricoides*. Am. J. Trop. Med. Hyg. **24**:801–805.

Botero, D. 1975. Epidemiology and public health importance of intestinal nematode infections in Latin America. Prog. Drug. Res. **19**:28–43.

Brudastov, A. N., L. N. Krasnonos, V. R. Lemelev, and S. K. Kholmukhamedov. 1970. Infectivity of *Ascaris lumbricoides* eggs to man and guinea-pigs after 10 years in the soil. Med. Parasitol. Parasit. Dis. (Moscow) **39**:447–451.

Cabrera, B. D., P. V. Arambulo, and G. P. Portillo. 1975. Ascariasis control and/or eradication in a rural community in the Philippines. Southeast Asian J. Trop. Med. Public Health **6**:510–518.

Calloway, D. H. 1982. Nutritional requirements in parasitic diseases. Rev. Infect. Dis. **4**:891–895.

Cerf, B. J., J. E. Rohde, and T. Soesanto. 1981. *Ascaris* and malnutrition in a Balinese village: a conditional relationship. Trop. Geogr. Med. **33**:367–373.

Choi, T. K., and J. Wong. 1984. Severe acute pancreatitis caused by parasites in the common bile duct. J. Trop. Med. Hyg. **87**:211–214.

Cram, E. B. 1926. Ascariasis in preventive medicine. Am. J. Trop. Med. **6**:91–114.

Darby, C. P., and M. Westphal. 1972. The morbidity of human ascariasis. J. S.C. Med. Assoc. **68**:104–108.

Gelpi, A. P., and A. Mustafa. 1967. Seasonal pneumonitis with eosinophilia. A study of larval ascariasis in Saudi Arabia. Am. J. Trop. Med. Hyg. **16**:646–657.

Ghadirian, E., N. A. Croll, and T. W. Gyorkos. 1979. Socio-agricultural factors and parasitic infections in the Caspian littoral region of Iran. Trop. Geogr. Med. **31**:485–491.

Glickman, I. T., G. U. Chaudry, J. Costantino, F. B. Clask, R. H. Cypess, and L. Winslow. 1981. Pica patterns, toxocariasis, and elevated blood lead in children. Am. J. Trop. Med. Hyg. **30**:77–80.

Hayashi, S. A. 1980. A model for the evaluation and assessment of the effect of control of the soil-transmitted helminthiasis, p. 265–273. *In* Collected papers on the control of soil-transmitted helminthiasis, vol. 1. Asian Parasite Control Organization, Tokyo.

Houang, M., A. Franklin, and X. Arozena. 1980. *Ascaris lumbricoides* detected in common bile duct following routine cholecystectomy and common bile duct sphincterotomy. Br. J. Radiol. **53**:804–805.

Hunter, G. W., J. C. Swartzwelder, and D. F. Clyde. 1976. Tropical medicine, 5th ed. The W. B. Saunders Co., Philadelphia.

Johnston, T. S. 1981. Diagnosis and treatment of five parasites. Drug Intell. Clin. Pharm. **15**:103–110.

Johnstone, C. R., Lerenthal, and E. L. Soulsby. 1981. *Ascaris suum:* T-cell responses of C57BL/6J mice in vitro and in vivo. Exp. Parasitol. **51**:243–256.

Jones, J. E. 1980. Office parasitology. Am. Fam. Physician **22**:86–90.

Jones, J. E. 1981a. Identification of intestinal nematodes using the digital rectal examination. J. Fam. Pract. **12**:563–565.

Jones, J. E. 1981b. The royal roundworm: *Ascaris lumbricoides*. J. Fam. Pract. **13**:271–276.

Jones, J. E. 1982. Psychosocial impact of common parasitic diseases. J. Fam. Pract. **14**:1139–1143.

Jones, J. E. 1984. Identification of intestinal parasites in an office setting: a comparison with reference laboratory results. Fam. Pract. Res. J. **3**:148–154.

Katz, Y., D. Varsano, B. Siegal, and A. Bar-Yochai. 1985. Intestinal obstruction due to *Ascaris lumbricoides* mimicking intussusception. Dis. Colon Rectum **28**:267–269.

Koino, S. 1923. Infection experiments of ascariasis in the human body. II. Communication on the young larvae expectorated in *Ascaris* pneumonia. Jpn. Med. World **4**:46–47.

Krasnonos, L. 1978. Many-year viability of ascarid eggs (*Ascaris lumbricoides* L. 1758) in soil of Samarkand. Med. Parasitol. Parasit. Dis. (Moscow) **47**:103–105.

Lagundoye, S. B. 1972. Disordered small bowel pattern in ascariasis. Trop. Geogr. Med. **24**:226–231.

Layrisse, M., and A. Vargas. 1975. Nutrition and intestinal parasitic infections. Prog. Food Nutr. Sci. **1**:645–667.

Lee, D. L., and H. J. Atkinson. 1976. Physiology of Nematodes, 2nd ed. Macmillan, London.

Makidono, J. 1956. Observations on *Ascaris* during fluoroscopy. Am. J. Trop. Med. Hyg. **5**:699–702.

Morishita, K. 1972. Studies on epidemiological aspects of ascariasis in Japan and basic knowledge concerning its control, p. 3–153. *In* K. Morishita, Y. Komiya, and H. Matsubayashi (ed.), Progress of medical parasitology in Japan, vol. 4. Meguro Parasitological Museum, Tokyo.

Pawlowski, Z. S. 1978. Ascariasis. Clin. Gastroenterol. **7**:157–178.

Pawlowski, Z. S. 1982. Ascariasis: host–pathogen biology. Rev. Infect. Dis. **4**:806–814.

Pawlowski, Z. S., and F. Arfaa. 1984. Ascariasis. *In* K. S. Warren and A. F. Mahmoud (ed.), Tropical and geographic medicine. McGraw–Hill Book Co., New York.

Peters, W. 1978. Medical aspects—comments and discussion II. The relevance of parasitology to human welfare today, p. 25–41. *In* E. R. A. Taylor and R. Muller (ed.), Symposia of the British Society for Parasitology, vol. 16. Blackwell Scientific Publications, Oxford.

Phills, J. A., A. J. Harrold, G. V. Whiteman, and L. Perelmutter. 1972. Pulmonary infiltrates, asthma and eosinophilia due to *Ascaris suum* infection in man. N. Engl. J. Med. **286**:965–970.

Rossi, M. A., and F. W. Bisson. 1983. Fatal case of multiple liver abscesses caused by adult *Ascaris lumbricoides*. Am. J. Trop. Med. Hyg. **32**:523–525.

Rukmono, B., I. S. Ismid, R. Indriyono, N. Suzuki, and T. Hara. 1977. Soil pollution with *A. lumbricoides* in Sawahlunto and Serpong, p. 229–233. *In* Proceedings of the 4th Conference of the Asian Parasite Control Organization. Tokyo.

Schultz, M. G. 1982. Ascariasis: nutritional implications. Rev. Infect. Dis. **4**:815–819.

Shephard, E. G., C. M. Macfarlane, and J. R. Joubert. 1982. A radioallergosorbent test (RAST) for *Ascaris lumbricoides*. Allergy **37**:231–239.

Shield, J., G. Anian, R. Ostwald, and R. Arnhold. 1984. Reinfection with intestinal helminths after treatment with mebendazole and fluctuations in individual *Ascaris lumbricoides* infections with time. Papua New Guinea Med. J. **27**:89–94.

Sinniah, B. 1982. Daily egg production of *Ascaris lumbricoides:* the distribution of eggs in the faeces and the variability of egg counts. Parasitology **84**:167–175.

Soe, B. S. 1981. Ascariasis and its control problems in Korea. Seoul J. Med. **22**:323–341.

Soh, C. T. 1973. Control of soil-transmitted helminths in Korea. Yonsei Rep. Trop. Med. **4**:102–125.

Spillman, R. K. 1975. Pulmonary ascariasis in tropical communities. Am. J. Trop. Med. Hyg. **24**:791–800.

Stephenson, L. S. 1980. The contribution of *Ascaris lumbricoides* to malnutrition in children. Parasitology **81**:221–233.

Stephenson, L. S., M. C. Latham, and M. L. Oduori. 1980a. Costs, prevalence and approaches for control of *Ascaris* infection in Kenya. J. Trop. Pediatr. **26**:246–263.

Stephenson, L. S., W. G. Pond, M. C. Nesheim, L. P. Krook, and D. W. T. Crompton. 1980b. *Ascaris suum:* nutrient absorption, growth and intestinal pathology in young pigs experimentally infected with 15-day-old larvae. Exp. Parasitol. **49**:15–25.

Strejan, G. H. 1975. Physico-chemical and biologic properties of the *Ascaris* allergen. Dev. Biol. Stand. **29**:79–89.

Takata, I. 1951. Experimental infection of man with *Ascaris* of man and the pig. Kitasato Arch. Exp. Med. **23**:49–59.

The Medical Letter. 1986. Drugs for parasitic infections, (issue 706). **28**:9–16.

Turner, K. J., L. Feddema, and E. H. Quinn. 1979. Non-specific potentiation of IgE by parasitic infections in man. Int. Arch. Allergy Appl. Immunol. **58**:232–236.

Turner, K. J., E. H. Fisher, and A. S. McWilliam. 1980. Homology between roundworm (*Ascaris*) and hookworm (*N. americanus*) antigens detected by human IgE antibodies. Aust. J. Exp. Biol. Med. Sci. **58**:249–257.

Waller, C. E., and H. B. Othersen. 1970. Ascariasis: surgical complications in children. Am. J. Surg. **120**:50–54.

Walsh, J. A., and K. S. Warren. 1979. Selective primary health care: an interim strategy for disease control in developing countries. N. Engl. J. Med. **301**:967–974.

Wynne, J. M., and B. A. Ellman. 1983. Bolus obstruction by *Ascaris lumbricoides*. S. Afr. Med. J. **63**:644–646.

CHAPTER 81

Babesiosis

TRENTON K. RUEBUSH, II

Disease: Babesiosis (piroplasmosis).
Etiologic Agents: *Babesia microti, B. divergens* (probably identical to *B. bovis*), and other species of *Babesia.*
Source: Wild and domestic animals by the bite of an infected tick; from other humans by blood transfusion.
Clinical Manifestations: Fever, chills, myalgia, fatigue, anemia, jaundice, and hemoglobinuria.
Pathology: Intraerythrocytic protozoan parasites, hemolytic anemia, and renal insufficiency or failure secondary to hemolysis.
Laboratory Diagnosis: Identification of organisms on thick or thin blood smears stained with Giemsa stain. Isolation of organisms in small laboratory animals. Indirect fluorescent antibody test for *B. microti.*
Epidemiology: Worldwide in wild and domestic animals. Human infections have only been reported from Europe and North America.
Treatment: No completely effective drug treatment known. Combination of clindamycin and quinine appears to be effective in *B. microti* infections. Patients with severe illness may require exchange transfusion.
Prevention and Control: Avoidance of areas infested by ticks. Prompt removal of attached ticks.

Description of Disease

Babesiosis is a tickborne hemoprotozoan disease of wild and domestic animals that is occasionally transmitted to humans. Although more than 90 different species of *Babesia* have been described in animals, only 2 species are known to cause infections in humans: *Babesia microti,* a parasite of rodents, and *B. divergens,* a parasite of cattle that is probably identical to *B. bovis.*

Communicability and Transmissability

Babesiosis is normally transmitted from animals to humans by various species of hard-bodied or *Ixodid* ticks. The incubation period varies from 7 to 28 days. Although humans probably play no role in the transmission or maintenance of *Babesia* infections in nature, occasional cases of blood transfusion-induced babesiosis have been recorded.

Clinical Manifestations of Disease

The clinical features of human babesiosis depend on the species of *Babesia* causing the infection, the presence or absence of the spleen, and the age of the patient.

Patients infected with *B. microti* usually have a gradual onset of illness with chills, fever, diaphoresis, myalgia, and fatigue. Fever and hepatosplenomegaly are found on physical examination. In most patients the acute illness lasts 1 to 4 weeks, but may be followed by persistent malaise and fatigue for several months. *Babesia microti* infections and Lyme disease are transmitted by the same tick species, *Ixodes dammini,* and may occur simultaneously. In patients who have undergone splenectomy, *B. microti* infections tend to be more protracted and severe; however, fatalities are rare.

Babesia divergens infections in humans have a more rapid onset, with high fever, chills, jaundice, and dark urine. Fever, hypotension, and jaundice are evident on physical examination. The course of illness is fulminant and results in death in most patients. All patients with *B. divergens* infections who have been reported thus far had undergone splenectomy before becoming infected for reasons that included trauma, surgical accidents, and Hodgkin's disease.

Pathology

Following the bite of an infected tick, *Babesia* organisms enter red blood cells directly without a preliminary exoerythrocytic stage such as occurs in malaria. Within the red blood cell the parasite divides by budding, usually forming two to four daughter cells. When the infected cell ruptures, other erythrocytes are invaded and the cycle is repeated. The hemolytic anemia, jaundice, and hemoglobinuria that occur in severe *Babesia* infections are apparently due to both the rupture of infected erythrocytes and an autoimmune reaction directed against the red blood cell membrane.

The presence of the spleen is important in resistance to *Babesia* infections. Patients with splenectomies, regardless of the species of *Babesia* causing the infection, tend to have higher levels of parasitemia and more severe illnesses than patients with intact spleens. Splenectomy in a patient with a subclinical *Babesia* parasitemia may lead to a recurrence of parasitemia and symptoms.

Diagnostic Features

Babesiosis should be considered in any patient with fever and a history of a tick bite or exposure to ticks. If the patient is asplenic the probability of a *Babesia* infection is increased. Anemia is usually present, but may not be severe in *B. microti* infections. Although human babesiosis has been reported only from Europe and North America, *Babesia* infections in wild and domestic animals are worldwide in distribution, and the diagnosis must be considered in patients from other areas.

A definitive diagnosis of babesiosis requires identification of characteristic organisms on blood smears or the isolation of the parasite in animals inoculated with the patient's blood. As serologic cross-reactions between *Babesia* and *Plasmodium* may occur, such tests cannot be relied on alone for diagnosis.

Prognosis

Human *B. microti* infections are usually self-limited, although malaise, fatigue, and parasitemia may persist for several months. Even in splenectomized patients *B. microti* infections are rarely fatal and resolve without permanent sequelae. In contrast, most patients infected with *B. divergens* have died despite aggressive supportive therapy, including blood transfusions, renal dialysis, and various antiprotozoal drugs.

Epidemiologic Aspects

Occurrence and Transmission

Although *Babesia* infections in animals occur worldwide, the first human case of babesiosis was not reported until 1957. Most of the cases recorded since that time have been caused by *B. microti,* a parasite of rodents, and were acquired along the coast of New York and Massachusetts. A human case has also been reported from Wisconsin. *Ixodes dammini,* the tick responsible for transmission of the parasite among rodents, is thought to be the vector to humans.

Babesia divergens infections in humans have been reported from Yugoslavia, Ireland, France, Scotland, and Russia. This organism normally parasitizes cattle. *Ixodes ricinus* is the presumed vector.

Several cases of blood transfusion-induced *B. microti* infection have been reported in recent years. These cases are probably attributable to the tendency of *B. microti* to cause prolonged asymptomatic parasitemia.

Susceptibility, Resistance, and Carrier States

Babesia enter the bloodstream through the bite of an infected tick. Although splenectomized patients are probably no more susceptible to infection than patients with functioning spleens, they do tend to have more serious illnesses. The patient's age also seems to influence the severity of illness caused by *B. microti.* Older patients generally have more severe illnesses, whereas young patients experience only mild or asymptomatic infections.

Babesia microti infections may persist for several months after the acute illness, probably because the organism can alter its surface antigens in response to specific antibody (antigenic variation). Organisms may also survive in banked blood for several weeks.

Etiologic Agents

Characterization

Babesia spp. are small, round, pyriform or ring-shaped intraerythrocytic protozoan parasites belonging to the phylum *Apicomplexa*. Although more than 90 different species of *Babesia* have been named, organisms that are apparently identical morphologically often carry several names due to the practice of creating a new species name each time the organism is identified in a different vertebrate host.

Specimen Collection and Transport

When the diagnosis of babesiosis is suspected, thick and thin blood smears should be obtained and stained with Giemsa stain. In patients with low levels of parasitemia, thick films may be necessary to detect organisms; thin films aid in distinguishing *Babesia* spp. from malaria parasites. An acute-phase serum specimen should be obtained as soon as possible after the onset of illness followed by convalescent-phase serum 14 to 21 days later.

Isolation and Identification Procedures

Isolation of *B. microti* from infected patients may be achieved by intraperitoneal inoculation of the patient's blood into small laboratory animals, such as hamsters or gerbils. Parasitemia becomes patent in inoculated animals within 2 to 4 weeks. Although *B. divergens* appears to have a narrower host range than *B. microti,* this species has also been isolated in gerbils.

Babesia parasites are often mistaken for *Plasmodium* spp., particularly *B. microti,* which typically appears as a small ring form similar to *P. falciparum*. *Babesia* can be distinguished from malaria parasites by the lack of pigment in erythrocytes infected with *Babesia* spp. The species identification of these organisms is based on their morphology and the epidemiologic features of the case. No biochemical tests are available for species identification.

Immunologic Aspects

Serologic tests should not be considered a substitute for parasitologic diagnosis based on the identification of organisms on blood smears. Several serologic tests for *Babesia* have been developed and have been used for clinical and epidemiologic studies. The most commonly used test for *B. microti* infections is the indirect fluorescent antibody test. Elevated titers appear within a few weeks after the onset of illness, reach a peak of 1 : 1,024 or above within 4 to 6 weeks, and then gradually decline over the next 12 months. Serologic cross-reactions occur between *B. microti, Plasmodium* spp., and other species of *Babesia,* but titers are generally highest to the infecting organism.

Treatment, Prevention, and Control

Treatment

No completely effective drugs are available for treating *Babesia* infections in humans. As *B. microti* infections are usually self-limited, symptomatic treatment is sufficient in most cases. In severe cases a combination of clindamycin and quinine has been of value. Exchange transfusion has been used in several patients when more conservative therapy failed. Although Pentamidine, a drug that is effective against *Babesia* infections in animals, is probably the most appropriate therapy for human *B. divergens* infections, blood transfusion and renal dialysis may also be required.

Prevention and Control

The only completely effective means of preventing *Babesia* infections is by avoiding areas infested with ticks. Insect repellants may be of some benefit. Prompt removal of attached ticks may also help to prevent infection, as ticks appear to transmit *Babesia* organisms only after they have been feeding for several hours.

Literature Cited

Gombert, M. E., E. J. C. Goldstein, J. L. Benach, M. J. Tenenbaum, E. Grunwaldt, M. H. Kaplan, and L. K. Eveland. 1982. Human babesiosis: clinical and therapeutic considerations. J. Am. Med. Assoc. **248:**3005–3007.

Healy, G. R., and T. K. Ruebush II. 1980. Morphology of *Babesia microti* in human blood smears. Am. J. Clin. Pathol. **73:**107–109.

Rosner, F., M. H., Zarrabi, J. L. Benach, and G. S. Habicht. 1984. Babesiosis in splenectomized adults: review of 22 reported cases. Am. J. Med. **76:**696–701.

Ruebush, T. K., II. 1980. Human babesiosis in North America. Trans. R. Soc. Trop. Med. Hyg. **74:**149–152.

Ruebush, T. K., II, E. S. Chisholm, A. Sulzer, and G. R. Healy. 1981. Development and persistence of antibody in persons infected with *Babesia microti*. Am. J. Trop. Med. Hyg. **30:**291–292.

Spielman, A., M. L. Wilson, J. F. Levine, and J. Piesman. 1985. Ecology of *Ixodes dammini*-borne human babesiosis and Lyme disease. Am. Rev. Entomol. **30:**439–460.

Wittner, M., K. S. Rowin, H. B. Tanowitz, J. F. Hobbs, S. Saltzman, B. Wenz, R. Hirsch, E. Chisholm, and G. R. Healy. 1982. Successful chemotherapy of transfusion babesiosis. Ann. Intern. Med. **96:**601–604.

CHAPTER 82

Balantidiasis

T. KOYAMA

Disease: Balantidiasis (balantidial dysentery).

Etiologic Agent: *Balantidium coli.*

Source: Domestic livestock, especially pigs, and some wild mammals by ingesting the trophozoites or cysts in their stools.

Clinical Manifestations: Diarrhea, dysentery, sometimes bloody dysentery, nausea, vomiting, epigastric pain, headache, and loss of appetite. May be asymptomatic, acute, or chronic.

Pathology: Ulcers in the large intestine, sometimes extending down to the muscularis mucosae and accompanied by lymphocytic infiltration and occasionally coagulation necrosis and hemorrhage.

Laboratory Diagnosis: Detection of etiologic agent from fecal specimens in the form of trophozoite or cyst. Direct smear, concentration methods, cultivation, and permanently stained slides useful in diagnosis.

Epidemiology: Worldwide; highest prevalence rates in tropical and subtropical regions. Occupational disease of livestock producers, veterinarians, and abattoir workers. Sporadic cases or outbreaks from consumption of water or food contaminated with trophozoites or cysts.

Treatment: Carbarsone, oxyquinoline, paromomycin, metronidazole, and tetracyclines.

Prevention and Control: Avoid contact with infected animals, especially pigs, and prevent ingestion of food or drink contaminated with the feces of infected animals.

Description of Disease

Human balantidiasis (balantidial dysentery) is an infectious disease of the cecum and colon in humans. It has also been found in chimpanzees, orangutans, rhesus monkeys, cynomolgus monkeys, pigs, peccaries, and other animals; and occasionally in dogs, rats, sheep, and cattle. The most common of these carriers is the pig, which is probably the main natural host. A single species, *Balantidium coli,* the causative agent of this disease, occurs in humans and domestic animals. The disease has been reported worldwide, although most cases occur in the warmer regions.

Communicability and Transmissibility

The pig seems to be the natural host of *B. coli,* and humans usually become infected by swallowing either cysts or trophozoites passed out in the pig's feces. Consequently, most human infections are in country dwellers and farm workers.

Clinical Manifestations

The ciliates not only inhabit the lumen of the large intestine but invade the intestinal wall, penetrating the mucosa and submucosa and producing undermining lesions similar to those caused by invasive *Ent-*

amoeba histolytica. The infection therefore may be asymptomatic in many individuals, but sometimes produces acute- or chronic-type disease. The irritation produced by the parasites leads to diarrhea or dysentery and, as the lesions progress, blood vessels become eroded and bloody dysentery results. In such cases, the external characteristics of the stool are similar to those seen in amoebic dysentery.

The patient may suffer from concurrent nausea, epigastric pain, vomiting, tenesmus, and intestinal colic, in addition to diarrhea. Loss of appetite, headache, insomnia, loss of weight, and muscular weakness have often been observed in these infections.

In most patients recovery occurs even without treatment, but in debilitated individuals infection with *B. coli* may be serious and even fatal (Faust et al., 1970; Zaman, 1978).

Dorfman et al. (1984) reported a case of fatal balantidial infection with appendicular and pulmonary involvement in a Venezuelan farmer. They believed that this patient died of peritonitis caused by a perforation of the appendix produced by the destructive action of the ciliates.

Diagnosis depends on demonstration of *B. coli* in the feces of the patient. In acute dysenteric stools, motile trophozoites are seen in abundance; in semiformed and formed stools, cysts are seen.

Pathology

The ciliate is normally a nonpathogenic commensal inhabitant of the lumen of the large intestine. Apparently it cannot penetrate the intact intestinal mucosa by itself; if the latter is damaged by infection with some other organisms, however, *Balantidium* may be a secondary invader and found deep in the ulcer. The ulcers sometimes extend down to the muscularis mucosae and are accompanied by lymphocytic infiltration and occasionally coagulation necrosis and hemorrhage. The protozoa often occur in groups in the tissues (Fig. 1) or in the capillaries, lymph ducts, and regional lymph nodes (Faust et al., 1976; Flynn, 1973; Zaman, 1978). Only rarely do the parasites perforate the bowel by penetrating through the muscle layers and cause peritonitis (Dorfman et al., 1984). The general pathology produced by *B. coli* resembles that of *E. histolytica*. However, unlike *E. histolytica*, the parasites do not invade extraintestinal tissues. The lesions in pigs and other animals are similar to those in humans.

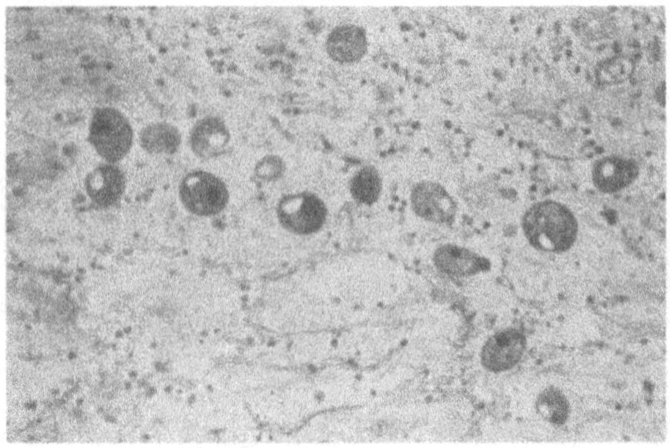

FIG. 1. Many *Balantidium coli* trophozoites in tissue from the large intestine of a pig. Darkly stained macronuclei are found. Hematoxylin and eosin stain. Magnification × 100; reduced by 37%. A color plate of this figure appears in Color Plate 4 at the beginning of this volume.

Epidemiologic Aspects

Occurrence and Transmission

Human infections are generally rare, the infection rate being about 1% or less (Bano, 1984; Chavalittamrong and Jirapinyo, 1984; Levine, 1973). Nuti et al. (1979) divided 80 cases of *B. coli* infection by symptom and reported that 60 patients had mild, nonspecific symptoms, 17 were asymptomatic, and only 3 had classic dysenteric symptoms.

The infection rate is usually higher in pigs than in humans. Diverse rates have been reported by many research workers: 21 to 100% (Levine, 1973), 77 to 92% (Letonja et al., 1975), 16.6% (Indermühle, 1978), 2.75% (Akinboade et al., 1983), and 10% in suckling piglets and 28.6% in almost-grown pigs (Marti and Hale, 1986).

The parasite has been reported in many parts of the world. However, the highest prevalence rates are seen in tropical and subtropical regions.

The pig is the usual source of human infection. Levine (1973) reported that contact with swine has been noted in more than half of the human cases documented. An outbreak of balantidiasis on Truk Island involved 110 persons who drank water that had become contaminated by pig feces during a typhoon (Walzer et al., 1972). Indeed, Hoare (1962) regards balantidiasis as an anthropozoonosis, as the vast majority of infected humans showed a history of contact with pigs.

Resistance

The cysts are more resistant to environmental conditions than the trophozoites. Most strains from pigs survive heating to 47°C for more than 15 min, but survive at room temperature for less than 3 days (Levine, 1973). It is also reported that the cysts remain viable at room temperature for at least 2 weeks (Zaman, 1978).

Etiologic Agent

Characteristics

B. coli is the largest protozoon and the shape of the trophozoite is ovoid (Fig. 2), measuring about 60 to 70 by 40 to 60 μm. The living trophozoites have a rapid rotary, boring motion. The entire surface is covered by cilia arranged in longitudinal, slightly spiral rows. At the anterior end is a funnel-shaped depression known as the peristome, lined with slightly longer cilia, which leads to the cytostome and the cytopharynx. The animal feeds by drawing a current of water down the peristome by means of the cilia that line it.

Any food particles, such as starch grains, bacteria, red cells, and fat droplets, carried in this current are ingested through the cytostome in small food vacuoles, which then circulate along a defined path through the body of the ciliate while their contents are being digested. Eventually any undigested remnants are expelled through a permanent pore or cytopyge near the hind end of the organism. The cytoplasm contains two contractile vacuoles, a number of food vacuoles, a reniform macronucleus, and a small dotlike micronucleus, which is often difficult to see. *B. coli* reproduces asexually by transverse binary fission and sexually by conjugation followed by binary fission.

The parasite encysts in the lumen of the host's large intestine, and the cysts, passed out in the feces, are the transmissive stage. The cysts are large and spherical (Fig. 3), measuring about 50 to 60 μm in diameter. The cyst wall is thick and transparent. The cyst does not stain well, as it frequently does not take enough stain throughout the wall. The macronucleus generally can be distinguished. No nuclear or cytoplasmic division occurs in the cyst. Thus, the number of nuclei in the cyst remains the same as that in the trophozoite. The cysts can survive outside the host for some time. After being swallowed by a new host they hatch, presumably in the large intestine.

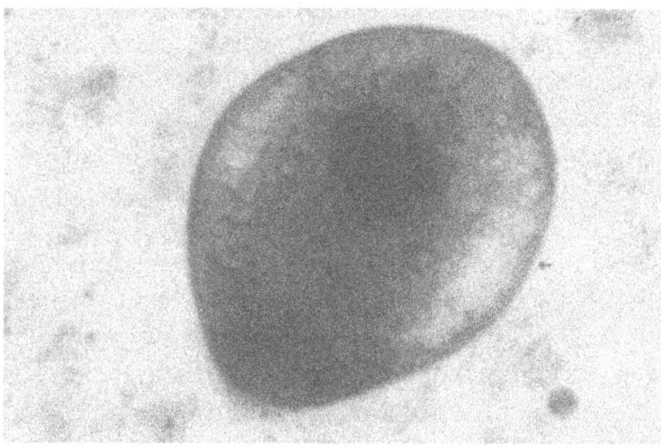

FIG. 2. *Balantidium coli* trophozoite from the feces of a chimpanzee. Hematoxylin and fast green FCF stain. Magnification × 650; reduced by 37%. A color plate appears in Color Plate 4 at the beginning of this volume.

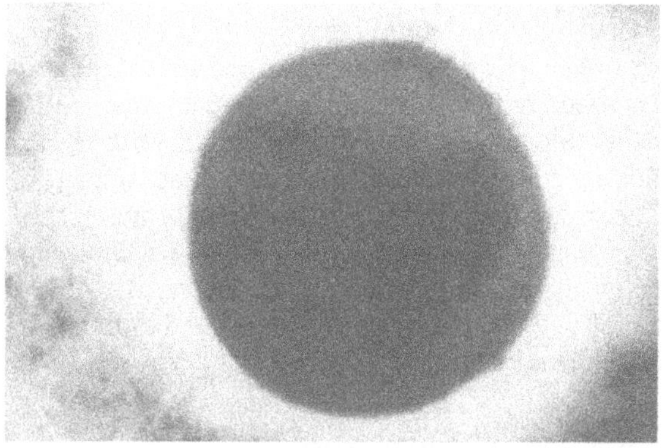

FIG. 3. *Balantidium coli* cyst from the feces of a chimpanzee. Hematoxylin and fast green FCF stain. Magnification × 700; reduced by 37%. A color plate of this figure appears in Color Plate 4 at the beginning of this volume.

Isolation and Identification Procedures

Balantidium infection can be diagnosed easily by finding the ciliates, in the form of trophozoites or cysts, in fecal specimens. Trophozoites are more likely to be found in loose stools and cysts in formed or semiformed stools. The parasites will grow in the ordinary culture media for intestinal amebae. In addition to direct examination, methods of concentrating the parasites and multiplying them in culture media may be very useful for isolation and identification.

DIRECT SMEAR

The direct smear is prepared by spreading a small amount of fecal material (approximately 2 mg), either alone or mixed with a drop of physiologic saline, on a slide. If too much material is used for the direct smear, the film on the slide will be too thick for an accurate examination; a smear of less than 2 mg will result in a film too thin for examination, thus decreasing the chances of finding any organisms. The direct wet mount is also useful for quickly detecting the trophozoite stage of a protozoon by its motility.

When the direct smear is stained with iodine solution, some cellular organs of the protozoan cyst are more apparent (e.g., yellow–gold cytoplasm, brown glycogen material, paler refractile nuclei), and protozoan parasites may be identified more precisely. This iodine solution consists of distilled water (100.0 ml), potassium iodide (1.0 g), and powdered iodine crystals (1.5 g). After a drop of iodine solution has been placed on the direct smear, the smear can be covered with a cover slip and observed under the microscope.

In general, protozoan organisms should never be identified on the basis of a wet mount alone; permanent stained smears should be examined to confirm the specific identification of suspected organisms. The protozoan cysts may also be seen more easily after fecal concentration procedures, which permits many cysts to be collected; moreover, the trophozoites will become more numerous and more visible after the cultivation.

Concentration Methods

Fecal concentration procedures allow the detection of small numbers of organisms that may be missed by using only a direct smear.

There are two types of concentration procedures: flotation and sedimentation techniques. These methods are designed to separate protozoan organisms from excess fecal debris through differences in specific gravity. The formalin–ether technique, one of the sedimentation techniques, is most useful for collecting cysts of *B. coli:*

1. Place 0.5 g of fecal sample with 2 to 3 ml of saline solution in a small test tube and comminute and stir thoroughly.
2. Filter the fecal suspension through gauze into a centrifuge tube after increasing its volume to 15 ml with saline solution.
3. Centrifuge the filtrate for 2 min at approximately 1,200 xg.
4. Decant the supernatant, add 7 to 10 ml of 10% formalin, and fix the sediment for 10 min.
5. Add 3 ml of ether, stopper, and shake for 30 s.
6. Centrifuge for 1 to 2 min at 2,000 xg after removal of the stopper from the tube.
7. Decant the supernatant with the fecal residue peeled from the tube wall by means of an applicator stick.
8. Place some of the sediment on a slide and examine microscopically.

Cultivation

Tempelis and Lysenko (1957) used an agar slant of Difco *Entamoeba histolytica* medium (Difco Laboratories, Detroit, Mich.) overlaid with Balamuth egg yolk infusion plus rice starch, 500 U/ml of streptomycin, and 250 U/ml of penicillin. This medium was used successfully to establish clone cultures from single microorganism. Zaman (1964) used Balamuth medium (Balamuth, 1946) without any additive and Jones medium (Jones, 1946), which consists of 5% horse serum in phosphate-buffered saline with yeast extract.

In general, all media that support the growth of *E. histolytica* also support the growth of *B. coli.* Addition of starch is as essential for the growth of *B. coli* as for the growth of *Entamoeba.*

The temperature range at which the parasites multiply is between 20 and 43°C (Cox, 1961). The parasites are not known to encyst in cultures. Kim et al. (1979) obtained good results using Sabouraud glucose agar medium to isolate *B. coli* from a plantar ulcer in a patient with leprosy. They reported that the ciliate also multiplied in liquid dextrose–peptone medium (pH 7.2).

Permanently Stained Slides

Correct identification of the protozoa depends on examination of a permanently stained smear. For this purpose, the Heidenhain iron alum–hematoxylin staining method and the trichrome staining method are very effective, providing well-stained slides of protozoa (Adam et al., 1971; Burrows, 1965; Garcia and Ash, 1979). Both techniques, however, are very complicated and time consuming. Most laboratories prefer to use one of the shorter and simpler procedures, such as modified Kohn one-step staining technique, for fixing and staining protozoan parasites in fresh fecal specimens (Gleason and Healy, 1965). Fixation and staining are accomplished in a single solution with no need for subsequent destaining or differentiation. The technique requires a minimum of working time and reagents. Stained fecal smears have shown no fading or deterioration over a period of 2 years. The quality of modified Kohn solution itself remains stable for a long time. Moreover, the technique is very effective in producing well-stained slides of almost any kind of intestinal protozoa containing *B. coli.* Thus, this fixative stain offers many advantages for hospital and clinical laboratories, where time, space, personnel, and equipment for routine parasitologic examinations are often limited.

Modified Kohn solution consists of chlorazol black E (5 g) and a basic solution of 90% ethyl alcohol (170 ml), methyl alcohol (160 ml), glacial acetic acid (20 ml), liquid phenol (20 ml), 1% phosphotungstic acid (12 ml), and distilled water (618 ml). To prepare the fixative stain, 5 g of chlorazol black E is put in a mortar and ground while the basic solution is added until all of the dye appears to be in solution. The solution should be left for 4 to 6 weeks to ripen, then filtered before use. To stain protozoan parasites with this solution, fresh feces containing protozoa are smeared on slides; the slides are then immersed immediately in the solution. Ordinarily, the protozoa will stain well if kept at room temperature for 1 to 4 h. Stock fixative stain may be diluted with basic solution, according to the preference of the investigator. After immersion in the stain, the slides are drained briefly, washed for 10 to 15 s in 95% ethyl alcohol, immersed twice for 5 min each in absolute alcohol, immersed in xylene for 5 min, and mounted in balsam. The protozoa stain dark green or black, and each of cellular structures is well differentiated.

Identification

In sections of infected intestine, parasites are easily recognized by their large size and large macronuclei; however, to determine their exact morphologic features and to identify the parasites as *B. coli*, the parasites in permanently stained smears should be investigated thoroughly.

The body length, the morphologic features of such cellular organs as the cilia, and the presence of a peristome, two kinds of nuclei, contractile vacuoles, food vacuoles, and others in each of the two forms of parasites, trophozoite and cyst, are very important for the identification of this species (see Characteristics).

At present, serologic tests for detecting the infection in humans and animals are under investigation, but are not yet available for practical use (Zaman, 1978).

Treatment, Prevention, and Control

Treatment

Many antiamebic drugs are also active against *B. coli*. Paromomycin, carbarsone, and oxyquinoline have been used for treatment of balantidiasis (Zaman, 1978). Beasley and Walzer (1972) found metronidazole partially successful when they used it in five patients at the dose recommended for amebiasis. Recently, however, many investigators have reported that metronidazole is very effective against balantidiasis. Garcia-Laverde and De Bonilla (1975) reported that metronidazole was given in 20 cases of balantidiasis at total doses of 2.5 g (over 5 days) or 7.5 g (over 10 days) to children and 5 g (over 5 days) or 12.5 g (over 10 days) to adults. In all 20 patients, the parasites could not be isolated from the feces after the fourth day of treatment. Moreover, these investigators reported that frequent posttreatment stool examinations yielded negative results, and there were no side effects. Nicholson (1978) used metronidazole (800 mg three times per day for 5 days) to treat successfully a man with *B. coli* infection. Nuti et al. (1979) gave 600 to 1,200 mg of metronidazole daily for 10 days and eliminated the parasites from the feces in 94.2% of 80 cases. Other clinical trials with imidazole compounds have shown them to be very effective against the parasite (Biagi, 1973; Botero, 1973). Tetracyclines have also been reported as effective against *B. coli* (Aréan and Koppisch, 1956; Zaman, 1978).

Prevention and Control

Sanitary measures designed to prevent ingestion of cysts or feces, especially pig feces, should prevent balantidial infections. Preventive methods are similar to those employed for *E. histolytica* infections; *B. coli*-infected pigs are probably the main danger, however, as human infection depends exclusively on contact with these animals or contamination of food or water with their feces.

Literature Cited

Adam, K. M. G., J. Paul, and V. Zaman. 1971. Medical and veterinary protozoology, 1st ed. Churchill Livingstone, Ltd., Edinburgh.

Akinboade, O. A., N. A. Sadiq, and O. O. Dipeolu. 1983. The incidence of *Balantidium coli* among local and exotic pigs in Ibadan, Nigeria. Rev. Elev. Med. Vet. Pays Trop. **36**:133–136.

Areán, V. M. and E. Koppisch. 1956. Balantidiasis. A review and report of cases. Am. J. Pathol. **32**:1089–1115.

Balamuth, W. 1946. Improved egg yolk infusion for cultivation of *Entamoeba histolytica* and other intestinal protozoa. Am. J. Clin. Pathol. **16**:380–384.

Bano, L. 1984. A preliminary report on prevalence of balantidiasis and diphyllobothriasis in Afghan refugees in Peshawar. Pakistan Vet. J. **4**:229.

Beasley, J. W., and Walzer, P. D. 1972. The effectiveness of metronidazole treatment of *Balantidium coli* infection. Trans. R. Soc. Trop. Med. Hyg. **66**:519.

Biagi, F. 1973. Metronidazole in the treatment of balantidiasis. Trans. R. Soc. Trop. Med. Hyg. **67**:143.

Botero, D. 1973. Effectiveness of nitrimidazine in the treatment of *Balantidium coli* infection. Trans. R. Soc. Trop. Med. Hyg. **67**:145.

Burrows, R. B. 1965. Microscopic diagnosis of the parasites of man, 1st ed. Yale University Press, New Haven, Conn.

Chavalittamrong, B., and P. Jirapinyo. 1984. Intestinal parasites in pediatric patients with diarrhoeal diseases in Bangkok. South. Asian J. Trop. Med. Public Health **15**:385–388.

Cox, F. E. G. 1961. The cultivation of *Balantidium coli* throughout its viable temperature range. Ann. Trop. Med. Parasitol. **55**:305–308.

Dorfman, S., O. Rangel, and L. G. Bravo. 1984. Balantidiasis: report of a fatal case with appendicular and pulmonary involvement. Trans. R. Soc. Trop. Med. Hyg. **78**:833–834.

Faust, E. C., P. F. Russell, and R. C. Jung. 1970. Craig and Faust's clinical parasitology, 8th ed. Lea & Febiger, Philadelphia.

Flynn, R. J. 1973. Parasites of laboratory animals, 1st ed. Iowa State University Press, Ames, Iowa.

Garcia, L. S., and L. R. Ash. 1979. Diagnostic parasitology—clinical laboratory manual, 2nd ed. C. V. Mosby Co., St. Louis.

Garcia-Laverde, A., and L. De Bonilla. 1975. Clinical trials with metronidazole in human balantidiasis. Am. J. Trop. Med. Hyg. **24:**781–783.

Gleason, N. N., and G. R. Healy. 1965. Modification and evaluation of Kohn's one-step staining technic for intestinal protozoa in feces or tissue. Am. J. Clin. Pathol. **43:**494–496.

Hoare, C. A. 1962. Reservoir hosts and natural foci of human protozoal infections. Acta Trop. **19:**281–317.

Indermühle, N. A. 1978. Endoparasitenbefall beim Schwein. Schweiz. Arch. Tierheilkd. **120:**513–525.

Jones, W. R. 1946. The experimental infection of rats with *Entamoeba histolytica*. Ann. Trop. Med. Parasitol. **40:**130–140.

Kim, Y. P., J. S. Lee, J. B. Kahng, and S. D. Park. 1979. Studies of new culture and staining methods for Ciliata, *Balantidium coli,* found parasitized in a plantar ulcer of a leprosy patient. Korean J. Dermatol. **17:**319–327.

Letonja, T., A. Henriquez, G. Reyes, and L. Zapata. 1975. Prevalencia de infecciones por *Balantidium coli* en cerdos de Santiago, Chile. Bol. Chil. Parasitol. **30:**88–89.

Levine, N. D. 1973. Protozoan parasites of domestic animals and of man, 2nd ed. Burgess Publishing Co., Minneapolis.

Marti, O. G., and O. M. Hale. 1986. Parasite transmission in confined hogs. Vet. Parasitol. **19:**301–314.

Nicholson, N. W. 1978. Case report of *Balantidium coli* infection. East Afr. Med. J. **55:**133.

Nuti, M., S. Sanguigni, and C. De Bac. 1979. Osservazioni su di un focolaio endemico di balantidiasi (studio di 80 casi). Ann. Med. Nav. **84:**641–646.

Tempelis, C. H., and M. G. Lysenko. 1957. The production of hyaluronidase by *Balantidium coli.* Exp. Parasitol. **6:**31–36.

Walzer, P. D., F. N. Judson, K. B. Murphy, G. R. Healy, D. K. English, and M. G. Schultz. 1972. Balantidiasis outbreak in Truk. Am. J. Trop. Med. Hyg. **22:**33–41.

Zaman, V. 1964. Studies on the immobilization reaction in the genus *Balantidium*. Trans. R. Soc. Trop. Med. Hyg. **58:**255–259.

Zaman, V. 1978. *Balantidium coli,* p. 633–653. *In* J. P. Kreier (ed.), Parasitic protozoa, 1st ed., vol. II. Academic Press, Inc., New York.

CHAPTER 83

Capillariasis

JOHN H. CROSS

Disease: Intestinal capillariasis, capillariasis philippinensis.

Etiologic Agent: *Capillaria philippinensis*.

Source: Fish-eating birds are suspected reservoir hosts, and freshwater and brackish-water fish the intermediate hosts and source of infection.

Clinical Manifestations: Borborygmus, abdominal pain, and intermittent diarrhea in the acute or initial stage followed by intractable voluminous diarrhea, vomiting, anorexia, weight loss, body and muscle weakness, extreme emaciation, pallor, anasarca, and dehydration in later or chronic disease. There is a protein-losing enteropathy, low levels of serum potassium, carotene, calcium, and protein, and malabsorption of sugars and fats.

Pathology: The parasite is located in the jejunum, causing sloughing of the mucosa, denuded flattened villi, atrophied crypts of Lieberkuhn, and infiltration of plasma cells, lymphocytes, macrophages, neutrophils, and eosinophils in the lamina propia. Autoinfection results in the production of large numbers of parasites. Pathologic changes in other organs are due to malnutrition and hypokalemia. Death results from untreated infections.

Laboratory Diagnosis: Stool examination for the presence of eggs, larvae, or adults of *C. philippinensis*. Several stool specimens should be examined if parasites are not found in persons with symptoms.

Epidemiology: The disease is reported primarily from the Philippine Islands and Thailand, with single case reports from Japan and Iran. Freshwater fish serve as intermediate hosts and transmit the infection to people who eat raw freshwater or brackish-water fish. Fish-eating birds are believed to be natural or reservoir hosts for the parasite. Nearly 2,000 cases and more than 100 deaths have been documented from rural areas of Northern Luzon in the Philippines.

Treatment: Mebendazole, 400 mg/day in divided doses for 20 days, 30 days for relapses. Albendazole is also effective at the same dosages. Thiabendazole is also effective, but causes side effects, and incomplete treatment may occur. Inadequate treatment results in relapse.

Prevention and Control: Educate populations in endemic areas to avoid eating raw freshwater or brackish-water fish. Treat all infected persons and promote sanitary disposal of their feces.

Description of Disease

Intestinal capillariasis (capillariasis philippinensis) is a parasitic disease of the small intestines of humans caused by a tiny nematode, *Capillaria philippinensis*. Humans acquire the disease by eating uncooked small freshwater or brackish-water fish. It is a zoonotic parasitosis, with fish-eating birds considered the natural definitive hosts and fish the intermediate hosts. The disease is reported from humans in the Philippine Islands and Thailand (Cross and Bhaibulaya, 1983) and with single case reports from Japan (Mukai et al., 1983) and Iran (Hooghooghi-Rad et al., 1986).

Communicability and Transmissibility

Intestinal capillariasis is acquired by people who eat raw freshwater or brackish-water fish infected with larval stages of the parasite located within the intestines of the fish. The larvae develop into adult worms in the small intestines, and the female worms produce larvae that may develop into adults ("autoinfection") or pass in the feces. Female worms also produce eggs that pass in the feces. The prepatent period in experimental animals, the period between infection and the presence of eggs in the feces, is approximately 25 days. The incubation period in humans, the period between eating of raw fish and symptoms, is estimated to be 1 to 2 months.

Clinical Manifestations of Disease

Asymptomatic Intestinal Capillariasis

Symptoms do not develop until the parasite is well established in the intestinal tract, an estimated 4 to 8 weeks after ingestion of larvae from fish. Asymptomatic infections have been detected during stool surveys in endemic areas; however, all persons found passing eggs and other forms of the parasite eventually developed characteristic symptoms.

Acute Intestinal Capillariasis

Initial symptoms are vague, with what is described by Filipinos as a "gurgling stomach" (borborygmus). Abdominal pain and intermittent diarrhea follow.

Chronic Intestinal Capillariasis

Diarrhea and other symptoms become more severe with time and, if untreated, patients experience voluminous diarrhea, vomiting, body weakness, weight loss, malaise, edema, anorexia, muscle wasting, and cachexia. Anasarca, dehydration, pallor, and extreme emaciation follow (Watten et al. 1972; Whalen et al. 1969).

Relapses of Intestinal Capillariasis

Death usually occurs in 4 to 6 months in untreated patients. Relapses may occur if treatment is not complete (Alcantara et al., 1985; Singson, 1974). Relapses occur from a few months to as long as 11 months after inadequate treatment. Relapses are believed to result from the inability of the drugs to affect the larval stages of the parasite. The adult worms are eliminated, but if medication is not continued long enough, the larvae develop into adults and reestablish the infection with reappearance of symptoms.

Immunity

There is no evidence of acquired immunity. Infected persons may experience repeated relapses (Alcantara et al., 1985; Singson, 1974). Antibodies may be produced in low levels, but are not protective. Rosenberg et al. (1970) studied immunoglobulin (Ig) in a number of patients and found high levels of IgE and low levels of IgG, IgM, and IgA. Several months after treatment, the immunoglobulin levels returned to normal.

Pathology

Adult and larval *C. philippinensis* are found in very large numbers in the small intestine, especially in the jejunum. Autoinfection is responsible for the hyperparasitism; as many as 200,000 worms per liter of bowel fluid have been found at autopsy (Cross and Bhaibulaya, 1983). Histologic examination of the intestinal tissue (Fig. 1) shows sloughing of the mucosa and denuded and flattened villi; the crypts of Lieberkuhn are atrophic and the mucosal glands depleted; and the lamina propia is infiltrated with plasma cells, lymphocytes, macrophages, neutrophils, and eosinophils (Fresh et al., 1972). In one autopsy, sections of *Capillaria* worms werë found in the liver, but not in other organs (Fresh et al. 1972), and any pathologic changes in these organs are attributed to malnutrition and hypokalemia. Ultrastructural studies of human tissue from jejunal biopsy and experimental infections in Mongolian gerbils showed a loss of epithelial cells, microulceration of the epithelium, compressive degeneration of the cells, mechanical separation, and homogenous material at the oral tip of the parasite. These changes are considered responsible for fluid, protein, and electrolyte loss (Sun et al., 1974). Death is probably caused by irreversible effects of electrolyte loss resulting in heart failure and secondary bacterial infection resulting in septicemia (Whalen et al., 1969).

Diagnostic Features

In endemic areas, patients complaining of borborygmus, abdominal pain, and diarrhea usually have eggs, larvae, or adult *C. philippinensis* in their stools. These can be detected by direct microscopic examination of the feces or examination of the Formalin–ether or ethyl acetate concentration. One stool ex-

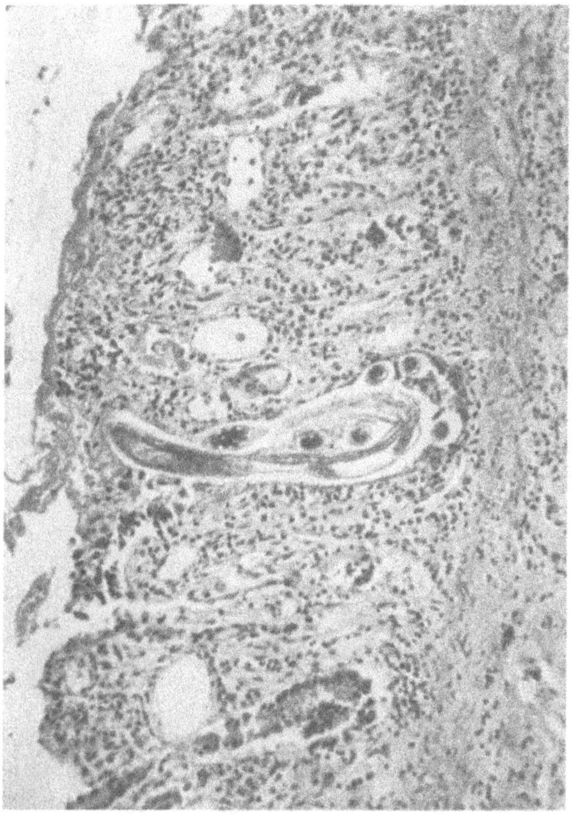

FIG. 1. Intestinal capillariasis; section of jejunum with adult worm in gland, denuded villi, sloughing mucosa, and cellular infiltration. (From Cross, J. H. and Bhaibulaya, M., 1983. Intestinal capillariasis in the Philippines and Thailand. *In:* Croll, N. and Cross, J. H. (ed.). Human ecology and infectious diseases. Used with permission of Academic Press, N.Y.)

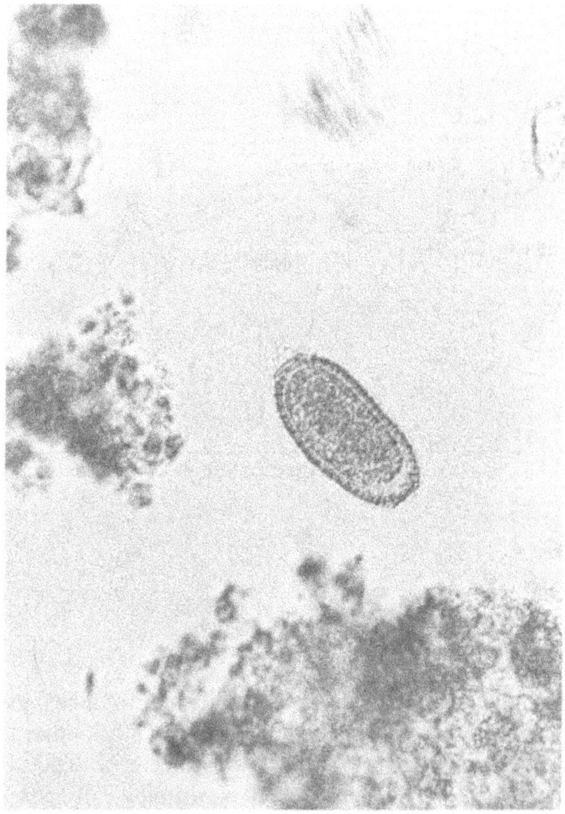

FIG. 2. Egg of *Capillaria philippinensis* from human feces. Note flattened opercular plugs at each end and striated shell; measures 36 to 45 μm by 20 to 22 μm.

amination is usually sufficient; sometimes, however, two or three stool samples must be examined before the parasite can be demonstrated. In some cases the parasites are found only after one or two treatments with an anthelmintic. The parasites can also be found in jejunal aspirates or biopsies. Serum levels of potassium, carotene, and calcium, as well as levels of total protein, are low; malabsorption of fats and sugars and a protein-losing enteropathy are usually evident (Watten et al., 1972; Whalen et all., 1969); Malabsorption may be often demonstrated by X-ray examination of the small intestine. Laboratory technicians not familiar with the eggs of *C. philippinensis* (Fig. 2) often incorrectly identify them as those of *Trichuris trichiura*.

Serologic testing has no diagnostic value. Immunologic tests using antigens prepared from another species of *Capillaria* as well as C. *philippinensis* were inconclusive (Banzon et al., 1975; Cross and Chi, 1978).

Prognosis

Mortality was appreciable during the epidemic of the disease in the Philippines in 1966 to 1967, with 77 (7.4%) deaths occurring in 1,037 confirmed infections. Since specific treatment regimes have been developed, death is rare. Once irreversible pathologic lesions develop, death is common. Incomplete treatment or inadequate dosages of anthelmintics lead to relapse with reappearance of symptoms and eggs and worms in the feces.

Epidemiologic Aspects

Capillariasis philippinensis was unknown until 1964 (Chitwood et al., 1964) and not fully appreciated as an important parasitosis until an epidemic of "diarrhea" occurred in Northern Luzon in the Philippines in the mid-1960s. The geographic distribution of reported cases of intestinal capillariasis now extends from the Far East to the Middle East (Fig. 3).

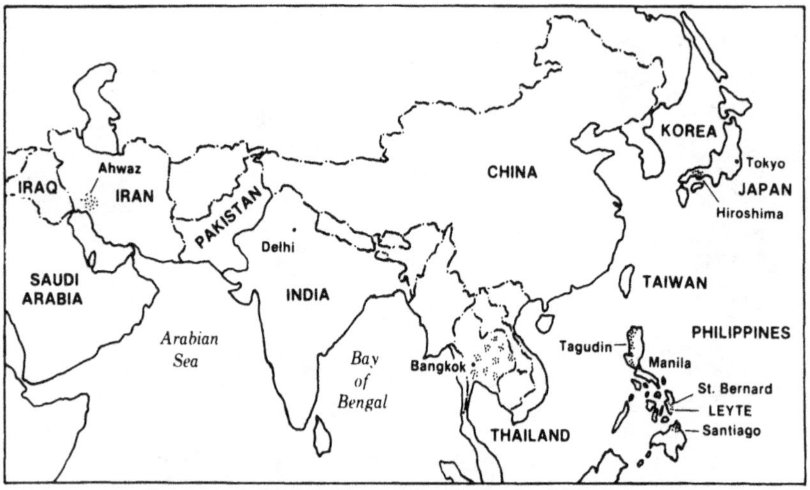

Geographic Distribution of Capillariasis philippinensis

FIG. 3. Geographic distribution of intestinal capillariasis, indicating endemic areas in the Philippines and Thailand and single case reports from Japan and Iran.

Occurrence

PHILIPPINES

Intestinal capillariasis is endemic to provinces in Northern Luzon and more recently in Southern Leyte in the Central Philippines. Since case reporting began in 1967, nearly 1,900 parasitologically confirmed cases have been seen in Northern Luzon. Twice as many men as women have acquired the disease, and most cases occur in adults 20 to 40 years of age. Infection rates in Southern Leyte are not known, but it is believed that several hundred persons have been infected, with a number of deaths. In 1983, a stool survey in the area showed 14% of 362 stools examined to be positive for *C. philippinensis* eggs, larvae, or adult worms. A few cases have also been reported from Northeastern Mindanao (Carney et al., 1987), but these infections may have been acquired in Southern Leyte. A case possibly acquired on the island of Bohol has also been reported; however, this individual had traveled to several areas of the Philippines (Salazar et al., 1973).

THAILAND

The disease was first recognized in Thailand in 1973 (Pradatsundarasar et al., 1973), and since then there have been additional reports from different areas of the country (Cross and Bhaibulaya, 1983). An epidemic of more than 100 cases was also reported from Thailand, and sporadic cases continue to be seen. A number of deaths have also been reported from this country.

JAPAN

One man from Hiroshima was found to be infected with *C. philippinensis* and treated. He had symptoms typical of the disease, but the source of infection was not determined (Mukai et al., 1983).

IRAN

An Iranian farmer living 20 miles southwest of Awaz, Koozestan Province, Iran, has also been found infected. He ate small fish uncooked and demonstrated characteristic symptoms of intestinal capillariasis. The parasite was identified in his feces. He was successfully treated with mebendazole (Hooghooghi-Rad et al., 1986).

Transmission

In all areas where intestinal capillariasis has been documented, people frequently eat raw freshwater or brackish-water fish. In the Philippines and Thailand, several species of small fish have been experimentally infected with the parasite, and infective-stage larvae recovered from the intestines of these fish were fed to monkeys and Mongolian gerbils. Both species of animals developed infections, but the monkeys did not suffer ill effects of the infection, whereas all of the gerbils died in 5 to 6 weeks. Fish-eating birds exposed to infective larvae from fish also became infected, and some died. Fish (*Hypseleotris bipartita*) recovered from lagoons from endemic areas in Northern Luzon were fed to gerbils, three of which developed patent infections and died. This species of fish is commonly eaten uncooked by Ilocano populations in Luzon, and these experiments proved that fish serve as a natural source of infection (Cross and Bhaibulaya, 1983). Studies have also shown that gerbils fed only two larvae from fish can develop fatal infections; 2,000 to 5,000 worms have been recovered from these animals.

Susceptibility, Resistance, and Carriage

Although not experimentally documented, it is believed that all humans exposed to infective-stage larvae from fish will develop patent infections. The infection rate among the many people who eat raw fish in the endemic areas of the Philippines is low because very few fish are infected with larval stages of the parasite. Although many wild and laboratory animals have been exposed to infection, only Mongolian gerbils, monkeys, multimammate rats, some wild fish-eating birds, and a few wild rats become infected. A large number of animals, including humans, have been fed embryonated eggs of *C. philippinensis* (Cross et al., 1970), but eggs hatch and develop only in freshwater and brackish-water fish in the Philippines and Thailand (Bhaibulaya et al. 1979; Cross et al., 1972). Fish-eating birds are the only animals indigenous to endemic areas in which experimental infections developed (Bhaibulaya and India-Ngarm, 1979; Cross and Basaca-Sevilla, 1983).

Human immunity to infection does not develop. Those left untreated usually die from the infection. Furthermore, relapses continue to occur in patients who are not treated with adequate doses of an anthelmintic. Inadequate treatment leads to relapse (Singson, 1974). Alcantara et al. (1985) reported a patient who experienced 16 relapses from 1966 to 1979. This patient denied eating raw fish following his first episode of the disease; he was finally cured after receiving continuous anthelmintic therapy for 10 months.

Current theory holds that capillariasis philippinensis is a zoonosis, with fish-eating birds the natural hosts and freshwater and brackish-water fish the intermediate hosts. The parasite has a bird–fish life cycle, and humans acquire the disease when the fish are eaten. The carrier and definitive host are birds, and fish are intermediate hosts. One fish-eating bird, the yellow bittern (*Ixobrychus* sp.), has been found naturally infected with a male *C. philippinensis* (Velasquez, personal communication).

Life Cycle

The life cycle of *C. philippinensis* is unique; it has been achieved experimentally and natural transmission through fish established. The adult worms reside primarily in the jejunum of both humans and experimentally infected animals. Eggs pass in the feces and, in a moist environment, embryonate in 5 to 10 days at ambient temperature. The eggs must reach water where they are ingested by fish. The eggs hatch in the intestine of the fish, and larvae grow from 130 to 150 μm to 250 to 300 μm in 3 weeks. In the Philippines, experimentally infected fish species include *Hypeleotris bipartita, Ambassis miops, Eleotris melanosoma, Sicyopterus* sp., and *Poicelia reticulata* (an introduced species) (Cross et al., 1972). In Thailand, six species of fish have been experimentally infected: *Cyprinus carpio, Puntius gonionotus, Aplocheilus panchox, Gambusia holbrooki, Rasbora borapetenis,* and *Trichopsis vittalus* (Bhaibulaya et al., 1979). Mammals can be infected by feeding larvae from fish or whole fish. Patent infections developed in monkeys in 22 to 96 days, whereas in Mongolian gerbils adult male and female worms were found at necropsy as early as 11 days after infection, with eggs appearing after 25 days. The first generation of female worms in gerbils produced larvae, which, in turn, matured into adult worms in 10 to 12 days. The second generation of female worms produced eggs that could be detected in the feces of the animals as early as 25 days after infection. Most female worms will produce eggs, but a small number will produce living larvae. Autoinfection is an integral part of the life cycle and maintains the infection (Cross et al., 1978). In gerbils, female worms can change from producing eggs to producing larvae; some simultaneously produce both larvae and eggs. Although this life cycle has not been confirmed in humans, it is probably similar to that found experimentally in animals.

Etiologic Agent

The parasite, *Capillaria philippinensis,* is a small roundworm or nematode of the superfamily Trichuroidea and closely related to *Trichuris trichiura* and *Trichinella spiralis*. The species may lie between the latter two, as it is capable of producing eggs like *T. trichiura* and larvae like *T. spiralis*.

Characterization of *Capillaria philippinensis*

Adult Worms

The parasite was first described by Chitwood et al. (1968). Female worms measure 2.3 to 5.3 mm and males 1.5 to 3.9 mm in length (Fig. 4). The worms are slender, measuring 3 to 8 μm at the anterior end and widening posteriorly to 23 to 47 μm. The male spicule is posterior, measuring 230 to 300 μm with a sheath and about 440 μm in length, without spines. The anus is subterminal, and the tail has two pairs of papillae. The female vulva is salient, immediately posterior to the esophagus, and without a flap. The uterus may contain thick-shelled unembryonated eggs, thick- and thin-shelled embryonated eggs, or larvae. The anus is subterminal. The esophagus, which extends from the mouth to the area of the

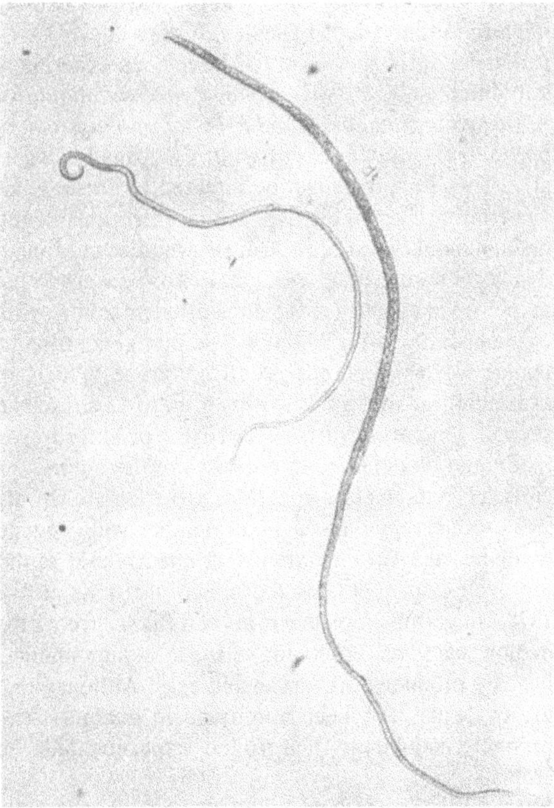

FIG. 4. Adult male (*left*) and female (*right*) *Capillaria philippinensis*. The anterior half contains the esophagus, surrounded by stichocytes forming a stichosome. The posterior ends are slightly wider, containing the digestive tract and reproductive organs. Males range in length from 1.5 to 3.9 mm and females from 2.3 to 5.3 mm. The widest areas range in size from 23 to 47 μm. (From Cross, J. H. and Bhaibulaya, M. 1983. Intestinal capillariasis in the Philippines and Thailand. *In:* Croll, N. and Cross, J. H. (ed.). Human ecology and infectious diseases. Used with permission of Academic Press, N.Y.)

vulva in female worms and to midbody in males, is surrounded by stichocytes, which form the stichosome characteristic of the superfamily Trichuroidea.

EGGS

The eggs passed in the feces are peanut-shaped, with striated shells and flattened bipolar plugs or opercula, and measure 36 to 45 μm by 20 to 22 μm (see Fig. 2). The eggs are usually unembryonated when passed, but on rare occasions may contain embryos or larvae. Thin-shelled eggs are also found.

LARVAL STAGES

It is not unusual to find all larval stages of the parasite in the feces. The stichocytes are present in all stages, with the esophagus occupying four-fifths of the body length in the first stage to half of the total body length in the fourth and fifth stages. The genital organs begin to appear in the third stage and are developed in the fourth and young adult stages. Figure 5 depicts a second-stage larva as it appears in the feces.

Specimen Collection and Transport

Fecal specimens are collected, sent to the laboratory, and examined microscopically by direct or after Formalin–ether or ethyl acetate concentration. Repeated stool examinations may be necessary to confirm the diagnosis. Intestinal aspirates can also be obtained and sent to the laboratory for microscopic examination. Specimens can be preserved in 10% Formalin until examined. Eggs, larvae, and adults may all be found in persons with severe disease.

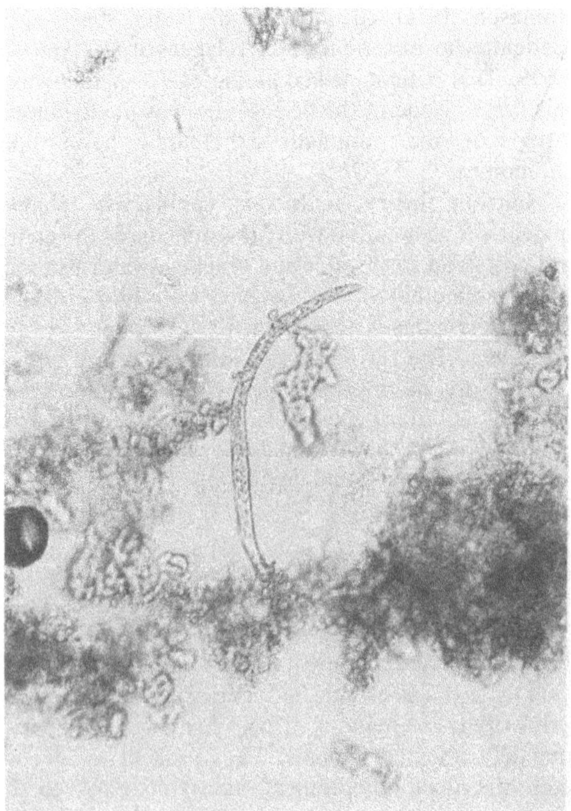

FIG. 5. *Capillaria philippinensis* larva found in the feces of a human with intestinal capillariasis. All stages of the parasite may be found in the stool of patients with intractable diarrhea.

Isolation and Identification Procedures

Portions of the specimen should be placed onto a microscope slide, a drop of saline added, coverslipped, and the slide examined under 10× magnification using a compound microscope. Once an egg is identified, higher magnification should be used. Aspirates or watery stools can be placed in a petri dish, saline added, and the specimen examined under a stereoscopic microscope. Larvae and adult worms can be collected with a capillary pipette and placed onto a slide for further high-power microscopic examination. Worms can be fixed and preserved in 10% Formalin for further study.

Preliminary and Definitive Identification

It is important that the microscopist be trained to identify the eggs of the parasite and be able to differentiate those of *C. philippinensis* from those of *T. trichiura*. It is also important to be able to identify the capillarid larval stages. The microscope should be equipped with an ocular micrometer to permit measurement of the eggs.

In endemic areas, a presumptive diagnosis can be made based on symptoms of borborygmus, abdominal pain, and diarrhea, particularly in those with intractable diarrhea, weight loss, muscle wasting, and cachexia. The definitive diagnosis is made by demonstrating the egg, larval, or adult form of the parasite in the bowel or stool of the patient. Serologic tests are of no value.

In some suspected cases when eggs or the parasite were not identified, anthelmintics were given and the parasite was subsequently found in the feces. Microscopic stool examinations, therefore, should be done on all persons in endemic countries presenting with gastroenteritis with borborygmus, abdominal pain, and diarrhea. One fatal case seen at a Manila hospital had several previous admissions for the same complaint, but was examined only for enteric bacterial infection. The diagnosis could have been made on the first admission had the stools been examined microscopically for ova and parasites.

Treatment, Prevention, and Control

Treatment

Early in the 1967 epidemic in the Philippines, thiabendazole, in dosages of 25 mg/kg of body weight in two divided oral doses, or 1 g per day for 30 days, was given. A number of relapses and side effects occurred, but the drug continued to be widely used until mebendazole was found to be effective (Singson et al., 1975). The recommended dosage for mebendazole is 400 mg/day in divided oral doses, for 20 days for new infections and 30 days for relapses. Attempts to lower the dosage or shorten the treatment time result in relapses (Basaca-Sevilla and Cross, 1985). Subsequently, albendazole (Zentel) was found as effective as mebendazole at 400 mg/day in divided oral doses for 20 days. More recently, albendazole has been found effective at similar dosages for only 10 days. Mebendazole, while affecting adult worms, may have no effect on the larval stages. With further trials, the time of treatment with albendazole may be further reduced. Very few persons experience side effects attributable to either mebendazole or albendazole. Symptoms of diarrhea and borborygmus disappear as early as 4 days and abdominal pain diminishes within a week following initiation of treatment.

Persons with chronic infections and with severe manifestations of the disease should also be given fluid and electrolyte replacement (especially potassium), an antidiarrhetic, and a high protein diet (Whalen et al., 1968). Most, given adequate treatment, survive. Death occurs when treatment is greatly delayed.

Prevention and Control

Capillariasis philippinensis can be prevented avoiding the consumption of raw freshwater or brackish-water fish. Educational programs should help, but it is extremely difficult to change eating habits that have existed for generations. Treatment of all infected persons and sanitary disposal of feces is important in control. In the first community to experience capillariasis in the Philippines in 1967, the first infected persons seeded the lagoon with *C. philippinensis* eggs by washing their fecally contaminated bed linens in the lagoon waters. This amply exposed fish to the eggs, and more than 200 cases of intestinal capillariasis were documented. Anthelmintic treatment of all infected persons is important in control of the parasitosis. The widespread use of thiabendazole is credited with the control of the epidemic.

Literature Cited

Alcantara, A. K., C. Uylangco, and J. H. Cross. 1985. An obstinate case of intestinal capillariasis. Southeast Asian J. Trop. Med. Public Health **16:**410–413.

Banzon, T. C., Lewert, R. M., and M. G. Yagore. 1975. Serology of *Capillaria philippinensis* infection: reactivity of human sera to antigens prepared form *Capillaria obsignata* and other helminths. Am. J. Trop. Med. Hyg. **24:**256–263.

Basaca-Sevilla, V., and J. H. Cross. 1985. Attempts to modify treatment of intestinal capillariasis. Southeast Asian J. Trop. Med. Public Health **16:**546–548.

Bhaibulaya, M., and S. Indra-Ngarm. 1979. *Amaurornis phoenicurus* and *Ardeola bacchus* as experimental definitive hosts for *Capillaria philippinensis* in Thailand. Int. J. Parasitol. **9**:321–322.

Bhaibulaya, M., S. Indra-Ngarm, and M. Anathapruit. 1979. Freshwater fishes of Thailand as experimental intermediate host for *Capillaria philippinensis*. Int. J. Parasitol. **9**:105–109.

Carney, W. P., R. Nocun, A. D. Vergel, I. G. Pagaran, A. S. Mercado, and J. H. Cross. 1987. Intestinal parasites of man in Agusan del Norte, Mindanao, Philippines, with emphasis on schistosomiasis and capillariasis. Philipp. J. Microbiol. Infect. Dis. **16**:5–9.

Chitwood, M. B., C. Velasquez, and N. G. Salazar. 1964. Physiological changes in a species of *Capillaria* (Trichuroidea) causing a fatal case of human intestinal capillariasis, p. 797. *In* Proceedings of the First International Congress on Parasitology, Rome, vol. 2.

Chitwood, M. B., C. Velasquez, and N. G. Salazar. 1968. *Capillaria philippinensis sp. n.* (Nematoda: Trichuroidea) from intestine of man in the Philippines. J. Parasitol. **54**:368–371.

Cross, J. H., T. C. Banzon, M. D. Clarke, V. Basaca-Sevilla, R. H. Watten, and J. J. Dizon. 1972. Studies on the experimental transmission of *Capillaria philippinensis* in monkeys. Trans. R. Soc. Trop. Med. Hyg. **66**:819–827.

Cross, J. H., T. C. Banzon, K. D. Murrell, R. H. Watten, and J. J. Dizon. 1970. A new epidemic diarrheal disease caused by the nematode, *Capillaria philippinensis*. Int. Trop. Public Health **7**:124–131.

Cross, J. H., T. C. Banzon, and C. N. Singson. 1978. Further studies on *Capillaria philippinensis:* development of the parasite in the Mongolian gerbil. J. Parasitol. **64**:208–213.

Cross, J. H., and V. Basaca-Sevilla. 1983. Experimental transmission of *Capillaria philippinensis* to birds. Trans. R. Soc. Trop. Med. Hyg. **77**:511–514.

Cross, J. H., and M. Bhaibulaya. 1983. Intestinal capillariasis in the Philippines and Thailand, p. 103–136. *In* N. Croll and J. H. Cross (ed.), Human ecology and infectious diseases. Academic Press, Inc., New York.

Cross, J. H., and J. C. H. Chi. 1978. The Elisa test in the detection of antibodies to some parasitic diseases in Asia, p. 178–182. *In* Proceedings of the 18th SEAMEO–TROPMED Seminar, Kuala Lumpur, Malaysia.

Fresh, J. W., J. H. Cross, V. Reyes, G. E. Whalen, C. V. Uylangco, and J. J. Dizon. 1972. Necropsy findings in intestinal capillariasis. Am. J. Trop. Med. Hyg. **21**:169–173.

Hooghooghi-Rad, N., S. Maraghi, and A. Narenj-Zadeh. 1988. *Capillaria philippiensis* infection in Khoozestan Province, Iran: case report. Am. J. Trop. Med. Hyg. **37**:135–137.

Mukai, T., S. Shimizu, M. Yamamoto, I. Horiuchi, S. Kishimoto, G. Kajiyama, H. Kawamura, M. Tsuji, and A. Miyoshi. 1983. A case of intestinal capillariasis. Jpn. Arch. Intern. Med. **30**:163–169.

Pradatsundarasar, A., K. Pecharanond, C. Chintanawongs, and P. Ungthavor. 1973. The first case of intestinal capillariasis in Thailand. Southeast Asian J. Trop. Med. Public Health **4**:131–134.

Rosenberg, E. B., G. E. Whalen, H. Bennich, and S. G. O. Johansson. 1970. Increased circulating IgE in a new parasitic disease—human intestinal capillariasis. N. Engl. J. Med. **283**:1148–1149.

Salazar, N. G., F. Sy, and M. Elayda. 1973. *Capillaria philippinensis:* a case of intractable diarrhea. Phillip. J. Microbiol. Infect. Dis. **7**:17–22.

Singson, C. N. 1974. Recurrences in human intestinal capillariasis. Philipp. J. Microbiol. Infect. Dis. **3**:7–13.

Singson, C. N., T. C. Banzon, and J. H. Cross. 1975. Mebendazole in the treatment of intestinal capillariasis. Am. J. Trop. Med. Hyg. **24**:932–934.

Sun, S. C., J. H. Cross, H. S. Berg, S. L. Kau, C. N. Singson, T. C. Banzon, and R. H. Watten. 1974. Ultrastructural studies of intestinal capillariasis—*Capillaria philippinensis* in human and gerbil hosts. Southeast Asian J. Trop. Med. Public Health **5**:524–533.

Watten, R. H., W. M. Becker, J. H. Cross, J-J. Gunning, and J. Jarimillio. 1972. Clinical studies on capillariasis philippinensis. Trans. R. Soc. Trop. Med. Hyg. **66**:828–834.

Whalen, G. E., G. T. Strickland, J. H. Cross, C. V. Uylangco, E. B. Rosenberg, R. A. Gutman, R. H. Watten, and J. J. Dizon. 1969. Intestinal capillariasis—a new disease in man. Lancet **1**:13–16.

CHAPTER 84

Clonorchiasis

HAN-JONG RIM

Disease: Clonorchiasis (Chinese or oriental liver fluke infection).
Etiologic Agent: *Clonorchis sinensis.*
Source: The ingestion of raw, inadequately cooked, or dried, salted, or pickled flesh of freshwater fish infected with *Clonorchis* metacercariae.
Clinical Manifestations: Indigestion, fullness of abdomen, loss of appetite, epigastric distress, weakness, discomfort in right upper quadrant, diarrhea, edema, and some hepatomegaly.
Pathology: Obstructive disturbances of the biliary tract, as well as bile retention and hyperplasia of the biliary system with glandular proliferation of papillomatous and adenomatous types.
Laboratory Diagnosis: Finding the characteristic eggs in the feces by stool examinations.
Epidemiology: Common in the Far East, especially in southern Korea, China, Taiwan, Japan, and northern Vietnam, where the eating of raw fish is widespread.
Treatment: Praziquantel is the drug of choice.
Prevention and Control: Reduction of the sources of infection by treatment, sterilization of feces, protection of fish ponds from contamination with night soil, control of snail hosts, and health education to avoid eating raw fish.

Description of Disease

Clonorchiasis is an infectious disease of the biliary passages caused by the Chinese or oriental liver fluke, *Clonorchis sinensis,* the most important liver fluke of humans. This infection is common in the Far East, especially in southern Korea, China, Taiwan, Japan, and northern Vietnam, where the eating of raw fish containing infectious metacercariae is commonplace. Clonorchiasis occurs in all parts of the world where there are Asian immigrants from endemic areas. Dogs, hogs, cats, and rats are reservoir hosts. The clinical manifestations of the infection range from asymptomatic in light infections to severe symptoms, including marked gastrointestinal disturbances with a syndrome associated with biliary obstruction, in heavy infections.

Communicability and Transmissibility

The human and animal reservoir hosts acquire the infection from the ingestion of raw, inadequately cooked, or even dried, salted, or pickled flesh of infected freshwater fish. The flukes reach maturity within 16 to 25 days after encysted larvae are ingested. The infected individuals may pass viable eggs for as long as 30 years. It is not directly transmitted from person to person.

Clinical Manifestations

Clinical manifestations result from the number of worms, infection period, condition of the worms, complications, and the chances of heavy infection due to the long infection period. In the early stage of infection or when the number of worms is small, both subjective and objective symptoms are almost nil. However, the symptoms are aggravated in accordance with the intensity and the duration of infection. The clinical manifestations of clonorchiasis may be conveniently classified as follows.

Invasion by Immature Worms

The earliest indications of illness are usually general malaise, abdominal discomfort, and occasional diarrhea. These symptoms are caused by the mechanical and toxic irritation of biliary tract for 3 to 4 weeks during invasion of biliary capillaries by immature worms. This stage is characterized by significant digestive disturbances and general symptoms associated with marked eosinophilia. Heavier early infections produce sudden effects with chills and fever up to 40°C, followed by liver pain and enlargement and jaundice. The jaundice is caused by obstruction of the intrahepatic biliary trees by worms or inflammatory changes. However, in light infections, an acute stage cannot usually be observed.

Infection by Mature Worms

In light and early infections, infected persons usually show no symptoms. In progressive cases, symptoms include indigestion, fullness of the abdomen, loss of appetite, epigastric distress unrelated to meals, and weakness. In moderate infections, the clinical symptoms include loss of appetite, indigestion, fullness of the abdomen, epigastric distress unrelated to meals, discomfort in the right upper quadrant, diarrhea, edema, and some hepatomegaly. In heavy infections, the symptoms are weakness and lassitude, epigastric discomfort, paresthesia, loss of weight, palpitation of the heart and tachycardia, diarrhea, vertigo, tetanic cramps and tremors, and toxemia from liver impairment. In the later stage of heavy infection or severe cases, the symptoms include marked gastrointestinal disturbances, with a syndrome associated with portal hypertension, splenomegaly, ascites, and edema.

Complications

As for the complications of clonorchiasis, formation of calculi in the intrahepatic biliary passages is one of the most characteristic pathologic features. It is sometimes accompanied by suppurative cholangitis, cholecystitis, and biliary abscesses or so-called cholangiohepatitis and ultimately can cause primary liver cancer, especially cholangiocarcinoma.

Pathology

The classic histopathologic response in human infection with *Clonorchis sinensis* was described by Hou (1955). The degree of pathologic change depends on the intensity as well as the duration of infection.

In the early stages of invasion by immature worms of *C. sinensis,* the mucosa of the biliary tree is frequently found to be edematous and the biliary epithelium desquamated by the mechanical and toxic irritation of the worms for 3 to 4 weeks during the invasion of the biliary capillaries. In addition, some proliferation and inflammatory reactions are initiated in the biliary epithelium, soon followed by thickening of the ducts and formation of crypts. Subsequent events include metaplasia of the biliary epithelial cells into goblet cells (early in the course of infection), which may proliferate to produce many small glandlike structures in the mucosa, giving the bile a persistent and excessively high mucus content (Flavell, 1981). In cases of long-term heavy infection, this continuous hyperplasia eventually results in formation of adenomatous tissue.

Pathogenesis

The following factors are considered the pathogenesis of *C. sinensis:* 1) Mechanical action in which bile ducts are blocked, caused by numerous *C. sinensis* living in the bile ducts in heavily infected cases; 2) destructive action in which partial damage and desquamation of mucosa of the bile ducts occurs, caused by adult parasites sucking the blood of the host through the mucous membrane of the bile duct; 3) chemical action caused by the parasite's metabolites; 4) inflammatory lesions of the bile ducts caused by irritation of worms as well as bacterial infections; and 5) reaction of the host, affected by such factors as age, nutrition, resistance, and the existence of concomitant diseases.

The main causes of dysfunction are a mechanical obstruction of the biliary tract by the worms, congestion of bile, and the effects of soluble substances (metabolites) released from the flukes into the ducts and surrounding tissues. The severity of the dysfunction depends on the number of worms present and the length of infection.

Diagnostic Features

Clinical diagnosis is suggestive in patients with an enlarged liver and symptoms of hepatitis in endemic areas where uncooked fish is eaten. Advanced infections require differentiation from malignancy, cirrhosis of the liver, and other causes of hepatic enlargement. The parasitologic diagnosis is based on finding the characteristic eggs in the feces or duodenal bile drainage. The eggs require differentiation from those of other *Opisthorchis* spp. and heterophid flukes (Lee et al., 1984). The direct smear method of recovering eggs from the feces is possible in cases of heavy infection, but in light infections eggs may escape recognition by this method. The Formalin–ether method or other sedimentation techniques are more reliable for detecting the eggs in feces.

To correlate the *Clonorchis* infection with clinical significance, a quantitative examination may be helpful. The egg count in terms of number per gram of feces (EPG) is important in estimating the worm load and applicable in quantitative analysis of *Clonorchis* infection, as well as in assessing the efficacy of treatment. A number of attempts have been made to use immune reactions for the diagnosis of infection with *C. sinensis*. However, immunodiagnosis is still regarded as a supplementary diagnostic method. The prognosis is usually good in light infections and less satisfactory in heavy or chronic infections.

Epidemiologic Aspects

Occurrence and Geographic Distribution

The infection of *C. sinensis* is largely confined to the Orient, from Vietnam to Japan. The endemic areas include Japan, southern Korea, essentially all of the mainland of China except the Northwest, Taiwan, and northern Vietnam. The incidence of clonorchiasis varies considerably from one district to another, even within small regions. In other parts of the world imported cases may occur in immigrants from Asia.

The heavy endemic areas of clonorchiasis in Korea are scattered throughout the country, with the most extensive and intensive endemic regions found mainly along the Nakdong River and the lower reaches of other rivers. The southern part of China, particularly Kwangtung Province, is the most seriously affected. The prevalence rate varies between 0.08 and 57% in different parts of China, but in many areas are 5% or more (Chung et al., 1979). In Taiwan, there are three highly endemic areas: Meinung, Kaohsiung Hsien in southern Taiwan; the Sun-Moon Lake in the central area; and Miao-Li in the middle of Taiwan. The Red River Delta in northern Vietnam is a highly endemic area. The heaviest infections occurred among the inhabitants in and around Haiphong and Hanoi, and for the entire region the infection rate averaged more than 40% in adults and 8% in children (Komiya, 1966).

In Japan, *C. sinensis* infection was widely distributed except in Hokkaido, the northern part of Japan on a century ago. At present, however, the infection rate is markedly low, which is attributed to the decreased density of the snail population in the endemic area, due to water pollution from factories, insecticides, land reclamation, and the like (Komiya and Suzuki, 1964).

Definitive and Reservoir Hosts

Clonorchis sinensis inhabits the bile ducts in the liver and occasionally the pancreas of humans. The natural definitive hosts other than humans are the dog, hog, cat, wild cat, marten, badger, mink, weasel, and rat. Experimentally susceptible animals (cats, dogs, rabbits, and guinea pigs) were demonstrated.

Intermediate Hosts

First (Snail) Intermediate Hosts

The first intermediate snail hosts are the species of *Parafossarulus* and *Bithynia*. The important species in China, Korea, and Japan is *P. manchouricus* (i.e., *P. striatulus* and *P. striatulus japonicus*) (Fig. 1). The other susceptible snails in China are reported as *Bithynia fuchsiana*, *Bithynia* (*Alocinma*) *longicornis*, *Assiminea lutea*, and *Melanoides tuberculata* (Gibson and Sun, 1971).

Second Intermediate Hosts

Numerous species of freshwater fishes serve as the second intermediate hosts of *C. sinensis*. The majority of fish intermediate hosts are confined to *Cyprinidae*. There are 9 families involving at least 80 species of fishes that act as the second intermediate host of *C. sinensis*. Among them 71 species belong to the family *Cyprinidae*, two to the family *Eleotridae*, and one to each of the families *Bagridae*, *Cyprinodontidae*, *Clupeidae*, *Osmeridae*, *Cichlidae*, *Ophiocephalidae*, and *Gobiidae*. In addition to these fishes, certain freshwater shrimps (*Caridinia nilotica gracilipes*, *Macrobrachium superbum*, and *Palemonetes sinensis*) have been incriminated as a source of infection in children in Fukien Province, China (Tang et al., 1963).

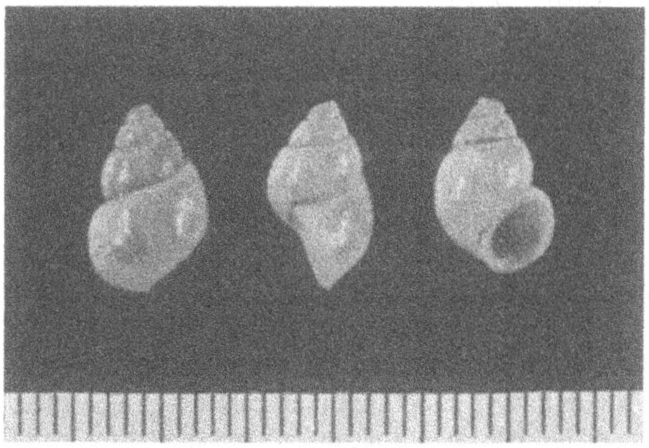

FIG. 1. *Parafossarulus manchouricus*, the snail intermediate host of *Clonorchis sinensis*.

More than 43 species from 5 families of freshwater fish were found to harbor *Clonorchis* metacercariae in China. The most heavily infected species, *Pseudorasbora parva*, *Rhodeus oceliatus*, and *Cultriculus kneri*, are widely distributed and numerous. The larger fish, such as those of the genera *Ctenopharyngodon*, *Hypophthalmichthys*, *Carassius*, and *Cyprinus*, generally have a low rate of metacercariae and a light degree of infection, with the exception of the genus *Mylopharyngodon*. However, *Ctenopharyngodon idellus*, *Mylopharyngodon aethiops*, and *Culter alburnus* are commonly eaten raw by Chinese people.

Fourteen species of freshwater fishes belonging to the families *Cyprinidae*, *Ophiocephalidae*, and *Cichlidae* are known as intermediate hosts for the parasite in Taiwan. *Pseudorasbora parva* and *Cultriculus kneri* are commonly infected with large numbers of metacercariae, but are seldom eaten. Species such as *Tilapia mossambica* and *Ophiocephalus maculatus* are commonly consumed raw, but their rates of metacercarial infection are low. More than 22 species of fishes belonging to the families *Cyprinidae*, *Osmeridae*, and *Eleotridae* have been reported to be naturally infected in Japan. Of these fishes, *Clonorchis* metacercariae are found most frequently in *Pseudorasbora parva*, *Sarcocheilichthys variegatus*, and *Tribolodon hakonensis*.

In Korea, more than 36 species of freshwater fishes belonging to the families *Cyprinidae*, *Bagridae*, and *Clupeidae* were found to serve as intermediate hosts. Among these fishes, infection with *Clonorchis* metacercariae is found most frequent in *Pseudorasbora parva*, *Sarcocheilichthys sinensis*, *Hemibarbus labeo*, *Acanthorhodeus gracilis*, *Acanthorhodeus taenianalis*, *Puntungia herzi*, *Pseudogobio esosinus*, *Gnathopogon* species, and *Acheilognathus limbata*. Of these fishes, *P. parva* shows an extremely high degree of infection with *Clonorchis* metacercariae (Fig. 2). Occasionally small fishes are eaten raw by people who live near the water, but normally they are eaten only after cooking. On the other hand, large fishes such as *Cyprinus carpio* and *Carassius carassius*, which have an extremely low rate of infestation, are frequently eaten by the inhabitants of endemic areas.

FIG. 2. *Pseudorasbora parva*, the most heavily infected freshwater fishes with *Clonorchis* metacercariae.

Mode of Human Infection

Clonorchis infection is acquired by eating uncooked fish containing infectious metacercariae. The intensity of human infection depends on the eating habits of the population.

In Hong Kong and South China, people with a marked preference for raw fish in their diet are notably infected with *C. sinensis*. Fish is served raw in thin slices either with vegetables and condiments or with hot rice gruel. In the endemic area of South Fukien Province, the residents eat raw shrimp, which are considered to be the only source of infection.

It is well known that most Japanese and Korean people like to eat raw fish. In endemic areas, heavily infected small fish such as *Pseudorasbora parva* are not generally eaten raw, but may be eaten undercooked and thus transmit infection. The large fish *Cyprinus carpio* and *Carassius carassius*, which are frequently eaten raw, have a low rate of infestation. However, the repeated consumption of these raw fish finally leads to heavy infection of the individual and a high incidence in the population.

Some workers have investigated whether humans can be infected by drinking water from a stream contaminated by the infected flesh of dead fish. Infection may occasionally be transmitted in this fashion, but has not been proved to occur under field conditions. Another possible mode of human infection is the accidental transference of encysted metacercariae to the mouth during the handling of fish.

Etiologic Agent

Characterization

ADULT WORM

The adult worm of *C. sinensis* is a flat, elongated, aspinous, flabby worm tapering anteriorly and somewhat rounded posteriorly. The living worm is transparent and slightly pink or brownish. The size of the worm varies according to its age, the species of the host, the number of parasites in one host, and its location in large or small bile ducts, its minimum and maximum lengths measuring 4 and 20 mm, respectively. Generally speaking, *C. sinensis* measures 8 to 15 mm long by 1.5 to 4.0 mm wide and about 1.0 mm thick. The oral sucker is directed anteriorly and slightly larger than the ventral sucker, which is situated about one-fourth the length of the body from the anterior end. The digestive organ begins at the oral sucker and consists of a small globose pharynx, a short esophagus, and a bifurcated intestine that terminates blindly near the posterior end of the worm. The deeply lobed testes lie in tandem at the posterior of the body. The anterior testis has four main branches and the posterior five. These main branches are divided into irregular branchlets. The seminal vesicle ascends to the genital atrium in front of the ventral sucker, where it opens through a poorly developed ejaculatory duct without a cirral pouch, or cirrus. The small, slightly lobate ovary lies on the median line anterior to the anterior testis. The minutely follicular vitellaria lie lateral to the intestinal ceca, ranging from the level of the anterior margin of the ventral sucker to the level of the ovary. Slightly behind the equator, a much larger seminal receptacle lies behind the ovary. A vitelline duct joins the oviduct between the common opening of the Laurer canal and the seminal receptacle with the ootype surrounded by the Mehlis gland. A loosely coiled and convoluted uterus, arising from the ootype, lies in the middle of the body and terminates in the genital atrium. A sacculate bladder with a terminal excretory pore, lying in the posterior part of the body, receives collecting tubules from a branching network of capillaries on each side of the body (Figs. 3 and 4).

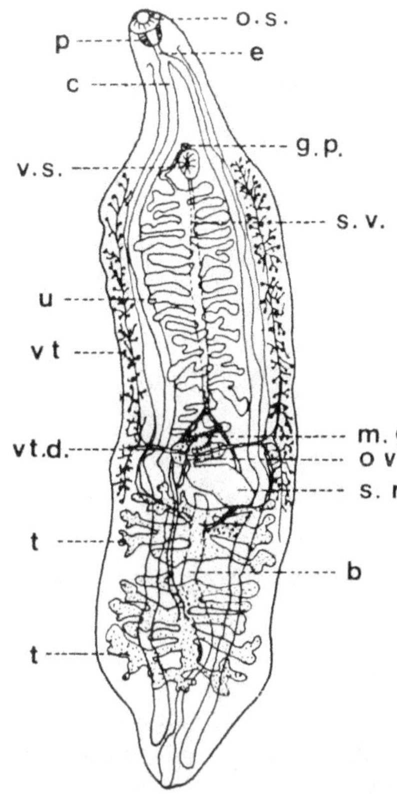

FIG. 4. Internal structure of the adult *C. sinensis*. *b* = Bladder; *c* = ceca; *e* = esophagus; *g.p.* = genital pore; *m.g.* = Mehlis' gland; *o.s.* = oral sucker; *ov* = ovary; *p* = pharynx; *s.r.* = seminal receptacle; *s.v.* = seminal vesicle; *t* = testes; *u* = uterus; *v.s.* = ventral sucker; *vt* = vitellaria; *vt.d.* = vitelline duct.

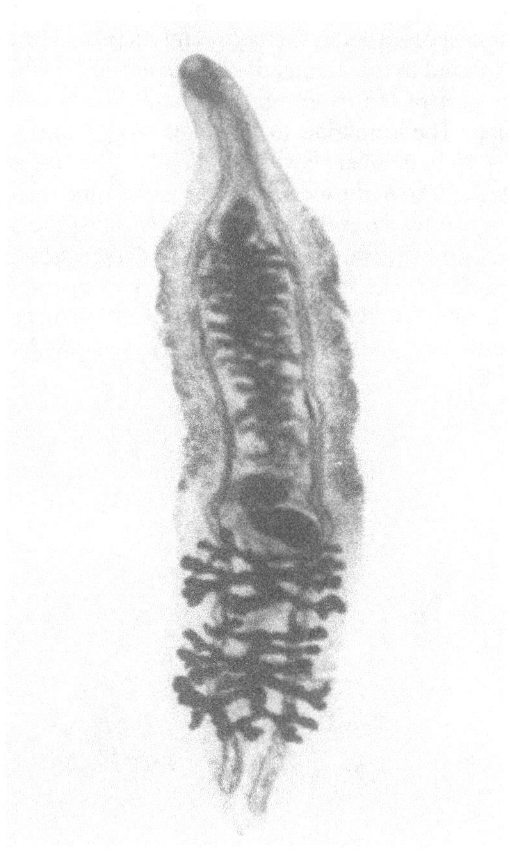

FIG. 3. Adult *Clonorchis sinensis*. Acetocarmine stain. ×20.

EGG

The yellow to light-brown egg measures 26 to 30 by 15 to 17 μm and is ovoid, narrowing a little posteriorly into the operculum. The operculum takes the shape of a watchglass with a prominent shoulder and is found on the narrow anterior end of the shell (Figs. 5 and 6). At the thickened posterior end is a small median protuberance, usually comma shaped. The egg surface is covered with a membranous projection measuring 0.05 to 0.25 μm as shown by scanning

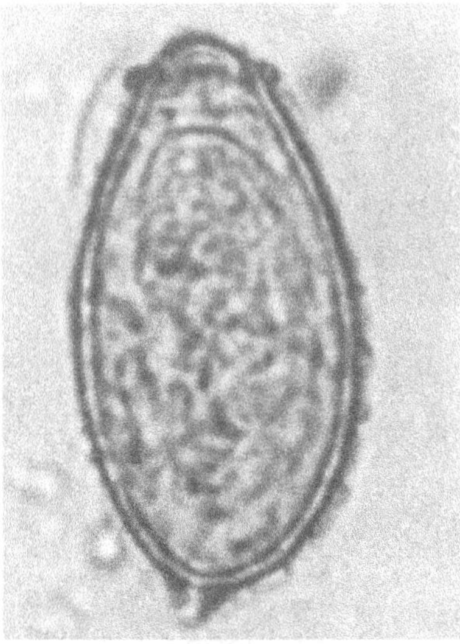

FIG. 5. Ovum of *C. sinensis*. Magnification ×2,300.

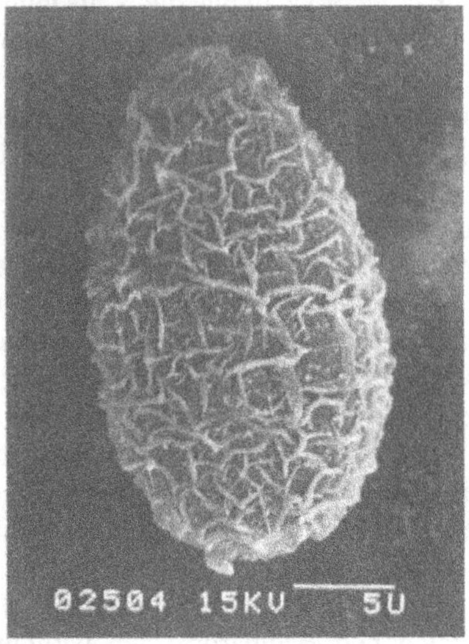

FIG. 6. Scanning electron photomicrograph of the ovum of *C. sinensis*. Magnification ×2,200.

electron microscopy. There are two groups of abnormal eggs, namely, "abnormal" and "incomplete." Both have the same shell structure as normal eggs. The former, however, has no operculum, although the etching of the arabesque polygonal pattern can be found in the surface. Its content is degenerated, and a droplike substance is recognized. The latter holds no miracidium, although it has an operculum (Fig. 7).

LARVAL STAGES

The eggs are deposited in the biliary passages of the definitive host, pass into the intestinal tract, and leave the body in the feces. Although the eggs contain fully developed miracidia, they do not hatch upon reaching water. The miracidium hatches only after ingestion by suitable species of operculated snails (i.e., *Parafossarulus, Bithynia*). The miracidia hatch in either the intestine or the rectum of the snail, then penetrate the wall of these organs and become sporocysts in the perirectal regions. The redia is found in the connective tissue around the posterior part of the esophagus of snails 14 days after feeding. The cercariae in the redia vary from 3 to nearly 50 in number, corresponding to their size and maturity. The mature cercaria has a trunk measuring 130 to 170 by 60 to 80 μm; the tail is 330 to 380 μm long, with a membranous keel on its dorsal and ventral surface. The cercaria possesses two eye spots and a number of penetration glands; its cuticle is spinous and carries a number of slender hairs (Fig. 8).

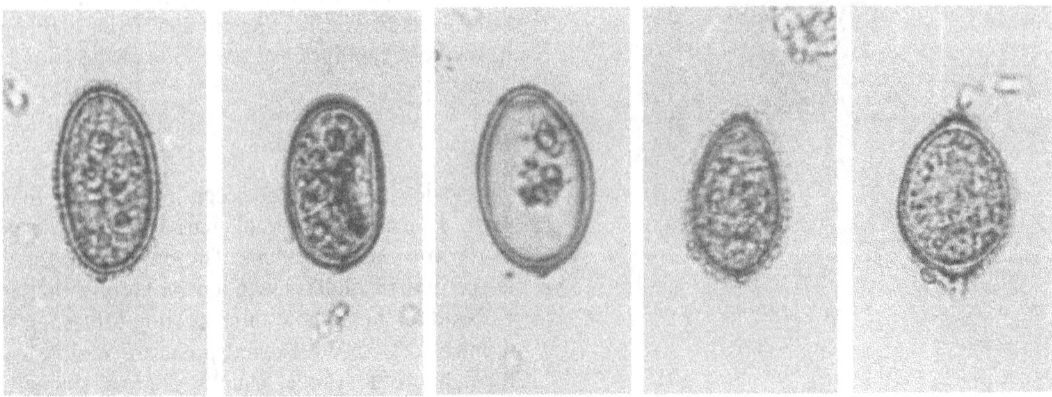

FIG. 7. Abnormal eggs of *C. sinensis*. Magnification ×1,000.

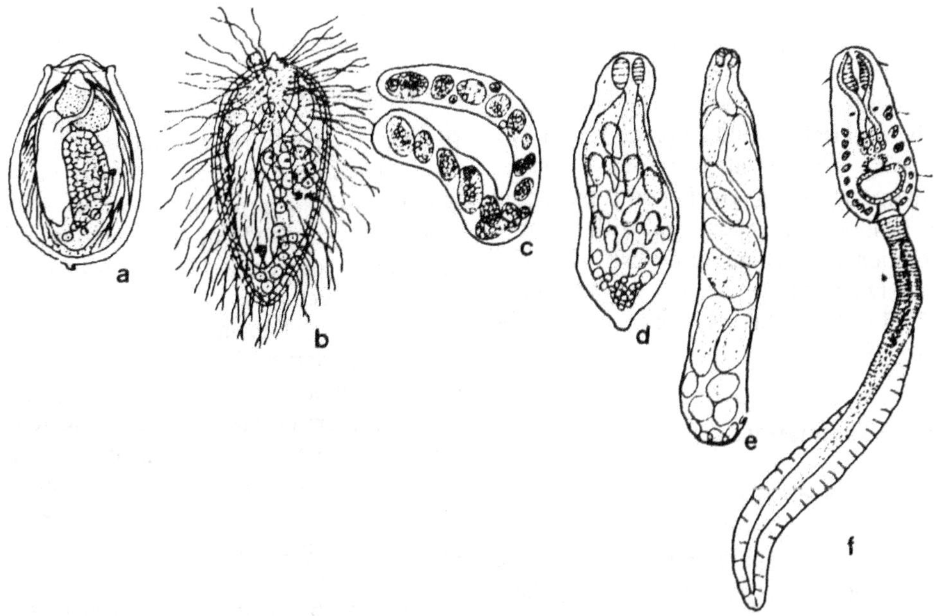

FIG. 8. Larval stages of *Clonorchis sinensis*. (a) Mature larva (miracidium) within egg shell; (b) miracidium; (c) matured sporocyst; (d) redia; (e) redia with developing cercariae; (f) cercaria. (Modified from Faust and Khaw [1927] with permission.)

The free-swimming cercaria perishes unless it encounters a fish within 24 to 48 h. The cercaria penetrates beneath the scales and, less often, on the scales, fins, and gills of the freshwater fish serving as the second intermediate host. In about 35 days cercariae develop into encysted metacercariae that have lost the larval organs. The metacercaria, measuring 138 by 115 μm, has an outer and an inner hyaline wall secreted by the parasite and a surrounding capsule formed by the tissues of the fish (Fig. 9).

Development on Final Host

When the definitive host, human or dog, ingests living metacercariae by consuming raw fish, the cyst is freed from the flesh of the fish by the action of gastric juices in the stomach. In the duodenum the outer layer of the true cyst is dissolved by trypsin, whereas the inner layer is ruptured by the activity of the larva. In 4 to 7 h the excystation of metacercariae, the majority of the young worms have migrated from the duodenum through the common duct directly by way of ampula of Vater to the distal biliary ducts, where they become mature worms in about 1 month. The usual habitat of the fluke in the host is the intrahepatic ducts, occasionally the gallbladder and the bile duct. In cases having fairly numerous flukes, most are found in the peripheral fine hepatic ducts. In heavy infections, flukes may be found in the gallbladder and the pancreatic duct.

Specimen Collection and Transport

The materials for stool examination should be collected in a clean container without contamination by urine or other extrinsic substances. The amount of feces requested may range from 5 to 10 g for qualitative and quantitative examinations. If possible, information regarding the consistency of the specimen should be included on the label with other pertinent data. When large numbers of fecal specimens are

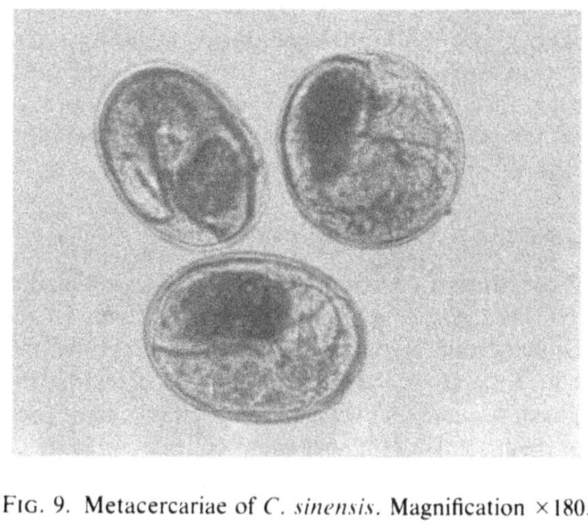

FIG. 9. Metacercariae of *C. sinensis*. Magnification ×180.

being collected in the field, where it is impractical to establish a routine of prompt examination, fecal specimens can be preserved and examined later at an institutional or hospital laboratory. For shipment of specimens the materials should be preserved in 10% Formalin solution, which is advantageous to prepare a concentration method. A satisfactorily preserved specimen is 1 ml of feces mixed well with 14 ml of Formalin solution in a 20-ml screw-cap vial. The minimum recommended specimen is a 1-ml sample of feces in 7 ml of Formalin solution in a 10-ml vial (Beaver et al., 1984).

Parasitologic Examinations

The parasitologic diagnosis is based on finding the characteristic eggs in the feces or duodenal bile drainage. The eggs require differentiation from those of *Opisthorchis* and heterophyid flukes. The following methods of stool examination have been used to detect the eggs of *Clonorchis sinensis*.

Direct Smear

The direct smear method is the most simple test for detection of helminth eggs. In the preparation of a wet film, small amounts of material from several parts of the stool specimen (usually 2 to 3 mg) are mixed with a drop of normal saline or tap water on a clean slide, and a cover slip is placed over the preparation. In clonorchiasis, the direct smear method of recovering eggs from feces is possible in cases of heavy infection; in light infections, however, eggs may escape recognition by this method.

Cellophane Thick Smear

Since the introduction of the cellophane thick smear technique (Kato thick smear method) by Kato and Miura in 1954, it has been recognized as one of the most effective methods of mass stool examination. This technique is very simple, economical, and highly efficient for the detection and identification of the eggs of *Ascaris, Trichuris,* hookworms, and other helminths, including *Clonorchis*. It was improved and adapted for quantitative as well as qualitative diagnosis by Martin and Beaver (1968).

Concentration Method

Concentration can be accomplished by sedimentation, flotation, or a combination of the two. Of the sedimentation methods, Formalin–ether centrifugal sedimentation is the most useful for detecting the eggs of *Clonorchis* in feces. This method can be used for fresh stools or for specimens collected previously and preserved in Formalin solution. Flotation methods, however, are less satisfactory for detecting the eggs of *Clonorchis* or other trematodes.

Quantitative Method

In surveys of infected populations, it is sometimes desirable to determine the intensity of infection by obtaining the approximate number of eggs. The quantitative methods of egg counting in fecal samples are of four types: direct smear (Beaver direct egg count technique), dilution (Stoll egg counting technique), cellophane thick smear (Kato–Katz technique, Martin–Beaver technique), and concentration method. Of these procedures, the Stoll dilution egg counting method and the Kato–Katz technique are most suitable for field work. There is no exact correlation between egg count and number of worms harbored by the host; however, the counts provide an approximation of the relative intensity of *Clonorchis* infection. By the levels of egg counts, patients can be divided into four groups: light infection, EPG between 1 and 999; moderate infection, EPG between 1,000 and 9,999; heavy infection, EPG between 10,000 and 29,000; and very heavy infection, EPG 30,000 or more (Rim, 1986). In practice, however, this method has some disadvantages in the qualitative stool examination. Errors can be made because the EPG count may appear to be negative during the initial phase of infection or in cases of light infestation.

Other Methods

There are some techniques for detecting liver fluke eggs in other than stool examination. One is an examination of duodenal fluid collected by duodenoscopy in which the aim is to collect eggs as near as possible to where they were laid and before they become diluted in the feces. The other method is the string test (Enterotest, HEDECO Co., Palo Alto, California), which uses a cotton thread fixed to a very light weight, enclosed in a capsule, except for the free end of the thread. The patient swallows the capsule and the end of the thread is held outside the mouth. After digestion of the capsule, the weight unrolls the encapsulated part of the thread and descends to the terminal part of the duodenum. The weight and thread are withdrawn after about 3 h, and all material sticking to the thread is collected by scraping (Ambroise-Thomas and Goullier, 1984). However, all of these techniques present certain inconveniences to the patient and examiner alike.

Immunologic Aspects

Several attempts have been made to use immune reaction for diagnosing *C. sinensis* infection. However, immunodiagnosis is still regarded as a supplementary diagnostic method. Since the 1930s, various serologic tests have been employed to detect *Clonorchis* infection; including the intradermal test, complement fixation (CF), indirect hemagglutination (IHA), gel-double diffusion (DD), immunoelectrophoresis (IEP), indirect fluorescent antibody (IFA), and enzyme-linked immunosorbent assay (ELISA).

The intradermal test has been widely used in Korea for the last three decades for the purpose of epidemiologic study of clonorchiasis. The adult worm antigen of *C. sinensis* have been used with Veronal-buffered saline extract (VBS antigen, 1 : 10,000 dilution). Although the intradermal test is a rapid, sensitive, and useful epidemiologic tool for studying the prevalence of *Clonorchis* infection, cross-reactions with other helminth infections and moderate false-negative reactions are the main disadvantages in its practical application. The CF, IHA, DD, and IEP tests are fairly specific, but their clinical application is difficult because of low sensitivity and complicated procedures. The IFA test has been evaluated utilizing frozen sections of adult worms of *Clonorchis* as antigen; this test has a great practical advantage, because the antigen is easy to prepare and provides consistent readings (Kwon et al., 1984). Recently Hahm et al. (1984) made a comparison of the sensitivity of three immunodiagnostic tests (i.e., ELISA, IFA and CF) on 55 human cases of clonorchiasis. The results showed that ELISA was the most sensitive and specific of the three in the diagnosis of clonorchiasis. Some investigators believe the ELISA technique to be the immunodiagnostic tool for clonorchiasis and other parasitic diseases. However, further investigations are needed before this assay can be applied for practical use.

Treatment, Prevention, and Control

Treatment

Praziquantel is the treatment of choice for clonorchiasis. The most effective regimen is 25 mg/kg three times daily (total dose, 75 mg/kg) administered orally at 5- to 6-h intervals over a single day. For large-scale treatment, a single dose of 40 mg/kg is recommended. In cases of heavy infection, however, a dosage of 25 mg/kg three times daily for 2 consecutive days is needed for a complete cure. Praziquantel appears to well tolerated, and side effects (headache, dizziness, and abdominal distress) are mild and transient (Rim et al., 1981).

Prevention and Control

The control of clonorchiasis is fundamentally an enterprise directed at reducing or eliminating transmission of the disease, so that no new infections, reinfections, or superinfections occur. There are several methods of control in endemic areas: 1) reduction of the sources of infection by treatment, 2) sterilization of feces, 3) protection of fish ponds from contamination with night soil, 4) control of snail hosts, and 5) health education.

The control of clonorchiasis is theoretically very simple, because the infection can only be contracted by way of encysted metacercariae, which are introduced when the intermediate host (fish) is eaten. Measures to prevent stools containing viable eggs from reaching bodies of water containing the snail intermediate hosts would apply to the human population, as water pollution by reservoir animal hosts cannot be controlled. This method alone would not be sufficient to produce tangible benefits. The most practical method of preventing human infection is to avoid eating raw, freshly pickled, or undercooked freshwater fish. However, it is exceedingly difficult to carry out such simple measures in the face of centuries-old traditions, to which the population clings with great tenacity. Therefore, in the districts where it is customary to eat raw fish, educational propaganda stressing the importance of thoroughly cooking all freshwater fish appears to be the most effective means of preventing clonorchiasis (Rim, 1986).

Literature Cited

Ambroise-Thomas, P., and A. Goullier. 1984. Parasitological examinations and immunodiagnostic advances in fluke infections. Arz. Forsch. Drug Res. **34**:9b, 1129–1132.

Beaver, P. C., R. C. Jung, and E. W. Cupp. 1984. Clinical parasitology. 9th ed. Lea & Febiger, Philadelphia.

Chung, H. L., W. C. Ts'ao, L. Y. Ho, and C. P. Hsü. 1979. Progress in studies on clonorchiasis sinensis in New China. Report from the Peking Tropical Medicine Research Institute, Peking, People's Republic of China.

Faust, F. C., and O. K. Khaw. 1927. Studies on clonorchiasis. Am. J. Hyg. Monograph. No. 8, 1–288.

Flavell, D. J. 1981. Liver-fluke infection as an aetiological factor in bile-duct carcinoma of man. Trans. R. Trop. Med. Hyg. **75**:814–824.

Gibson, J. B., and T. Sun. 1971. Clonorchiasis, p. 546–566. *In* R. A. Marcial-Rojas (ed.) Pathology of protozoal and helminthic diseases with clinical correlation. Williams & Wilkins, Baltimore.

Hahm, J. H., J. S. Lee, and H. J. Rim. 1984. Comparative study on the indirect immunofluorescent antibody test, complement fixation test and ELISA in diagnosis of human clonorchiasis. Korea Univ. Med. J. **21**:177–184 (In Korean.)

Hou, P. C. 1955. The pathology of *Clonorchis sinensis* infestation of the liver. J. Pathol. Bacteriol. **70**:53–64.
Kato, K., and M. Miura. 1954. Comparative examinations. Jpn. J. Parasitol. **3**:35.
Komiya, Y. 1966. *Clonorchis* and clonorchiasis, p. 53–106. *In* B. Dawes (ed.), Advances in parasitology, vol. 4. Academic Press, Inc., New York.
Komiya, Y., and N. Suzuki. 1964. The distribution of *Clonorchis* infection in Japan with remarks on its epidemiology, p. 601–660. *In* K. Morishita, Y. Komiya, and H. Matsubayashi (ed.) Progress of medical parasitology in Japan, vol. 1. Meguro Parasitological Museum, Tokyo.
Kwon, K. H., J. S. Lee, and H. J. Rim. 1984. The use of IFAT in the diagnosis of human clonorchiasis. Korea Univ. Med. J. **21**:91–100. (In Korean.)
Lee, S. H., S. W. Hwang, J. Y. Chai, and B. S. Seo. 1984. Comparative morphology of eggs of heterophids and *Clonorchis sinensis* causing human infections in Korea. Korean J. Parasitol. **22**:171–180.
Martin, L. K., and P. C. Beaver. 1968. Evaluation of Kato thick smear technique for quantitative diagnosis of helminth infections. Am. J. Trop. Med. Hyg. **17**:382–391.
Rim, H. J. 1986. The current pathobiology and chemotherapy of clonorchiasis, vol. 24. Korean J. Parasitol. Suppl. Monogr. Ser. No. 3, 1–141.
Rim, H. J., K. S. Lyu, J. S. Lee, and K. H. Joo. 1981. Clinical evaluation of the therapeutic efficacy of praziquantel (Embay 8440) against *Clonorchis sinensis* infection in man. Ann. Trop. Med. Parasitol. **75**:27–33.
Tang, C. C., Y. K. Lin, P. C. Wang, P. H. Chen, C. T. Tang, T. S. Chen, C. T. Lin, C. K. Huang, C. F. Chen, and S. H. Chen. 1963. Clonorchiasis in South Fukien with special reference to the discovery of crayfishes as second intermediate hosts. Chinese Med. J. **82**:545–562.

CHAPTER 85

Cryptosporidiosis

WILLIAM L. CURRENT and LYNNE S. GARCIA

Disease: Cryptosporidiosis.

Etiologic Agents: *Cryptosporidium* spp. The species believed to be responsible for cryptosporidiosis of humans and domesticated mammals is *Cryptosporidium parvum*. Although taxonomically imprecise, *C. parvum* will be referred to as *Cryptosporidium*, because the generic portion of the name is used in the medical literature to convey the concept of this protozoan parasite and the diseases it may cause.

Source: Humans become infected by ingestion of environmentally resistant oocysts that are released in the feces. Infection can be direct by fecal–oral transmission from person to person or from animal to human or indirect by ingestion of food or water contaminated with oocysts. *Cryptosporidium* may cause traveler's diarrhea and outbreaks of diarrhea among children attending day-care centers.

Clinical Manifestations: Acute onset of profuse, watery diarrhea is often accompanied by abdominal cramping, nausea with or without vomiting, headache, anorexia, and low-grade fever. In immunodeficient persons, an intestinal infection often develops into a prolonged, life-threatening, choleralike illness. Extraintestinal infections may cause respiratory and hepatic disease in immunodeficient patients.

Pathology: Histologic findings associated with heavily infected regions of the intestine include villous atrophy, an increase in crypt length, and mild to moderate mononuclear cell infiltration of the lamina propria. Extraintestinal infections may result in bronchiolitis, pneumonitis, and acute and gangrenous cholecystitis.

Laboratory Diagnosis: Definitive diagnosis requires identification of the small (4 to 6 μm) oocysts in stool or sputum samples. The most common technique for demonstrating oocysts is acid-fast staining. Negative staining and concentration procedures may be useful, but are more difficult to interpret in light infections. Diagnosis is also confirmed by identification of the small (2 to 6 μm) endogenous stages in the microvillous region of tissues obtained by biopsy and processed for histology or transmission electron microscopy.

Epidemiology: Worldwide. More prevalent in developing countries where food and water hygiene practices are inadequate. Waterborne transmission has been reported, foodborne transmission is highly probable, and direct fecal transmission from person to person or from animal to human is well documented.

Treatment: Presently, no drug has been shown to be effective in the treatment of cryptosporidiosis. Oral and parenteral rehydration therapy is recommended if fluid loss is extensive. Antidiarrheal drugs may lessen the severity of the clinical manifestations.

Prevention and Control: Observe good personal hygiene and the general precautions used to prevent exposure to waterborne and foodborne enteropathogens. Risk factors include ingestion of water and food in areas where hygiene practices are inadequate and contact with mammals that have diarrhea, especially calves, rodents, kittens, and puppies. Persons with severe immune deficiencies are at risk of developing a life-threatening infection if exposed to *Cryptosporidium* oocysts.

Description of Disease

Cryptosporidiosis is an infectious disease, primarily of the gastrointestinal tract, of mammals, birds, and reptiles that is caused by small protozoan parasites (*Cryptosporidium* spp.) that are similar taxonomically to the other coccidia, *Isospora belli, Sarcocystis* spp., and *Toxoplasma gondii,* that infect humans. Infections are acquired by ingestion of environmentally resistant oocysts that are passed sporulated (infective) in the feces (Current, 1986; Current and Reese, 1986).

In immunocompetent humans, at least one species (*C. parvum*) may produce mild to severe diarrhea lasting from several days to more than a month (Current et al., 1983; Koch et al., 1985). In immunodeficient persons, especially those with the acquired immune deficiency syndrome (AIDS), cryptosporidiosis often presents as a prolonged, life-threatening, choleralike illness (Navin and Juranek, 1984). Cryptosporidial diarrhea in many immunodeficient persons becomes irreversible, and fluid loss may be extensive (Centers for Disease Control, 1982a). Respiratory and biliary cryptosporidiosis demonstrate that the parasite is not confined to the gastrointestinal tract of immunodeficient persons (Forgacs et al., 1983; Guarda et al., 1983). As no effective therapy has been found, the finding of cryptosporidiosis in the immunodeficient host often carries an ominous prognosis.

Communicability and Transmissibility

Cryptosporidiosis can be transmitted from person to person or from animal to human via the fecal–oral route. Oocysts, the environmentally resistant form of the parasite that transmits infection from one host to another, are passed in the feces of infected hosts. Oocysts that are infective at the time they are passed in the feces can become disseminated and contaminate food, drinking water, and environmental surfaces. Although the infectious dose is not known, an accidental infection of a researcher suggests that ingestion of only a few oocysts can result in clinical cryptosporidiosis (Blagburn and Current, 1983). Oocysts of *Cryptosporidium* are very resistant to chemicals and disinfectants and remain viable for months if kept moist and cool. We routinely store oocysts in 2.5% (wt/vol) aqueous solution of potassium dichromate and use solutions of 25% (vol/vol) commercial bleach to sterilize these resistant forms before in vitro excystation and inoculation of sporozoites into cell cultures. Oocysts stored at 4°C may remain viable and infective for more than 1 year (Current, 1984). Thus, it is apparent that oocysts of *Cryptosporidium* can remain viable for extended periods in aqueous environments and that routine chlorination of drinking water should have little effect on their infectivity.

Clinical Manifestations

Much of the published literature presents gastrointestinal cryptosporidiosis as a short-term (3 to 12 days), often choleralike illness in immunocompetent persons or as a prolonged, life-threatening, choleralike disease in immunodeficient persons. Clinicians should, however, be aware that cryptosporidiosis does not always fit within these two divergent categories. Immunocompetent persons as well as those with the clinical and laboratory features of AIDS have been shown to carry asymptomatic infections (Current et al., 1983; Zar et al., 1985), immunocompetent persons have been shown to harbor infections for more than a month (Koch et al., 1985), and there are several unpublished cases in which AIDS patients have apparently cleared an infection in the absence of a demonstrable improvement is their immune status.

Cryptosporidiosis in Immunocompetent Persons

Most of the 18 cases of cryptosporidiosis in immunocompetent humans reported before 1983 and the numerous cases reported since (reviewed by Current [1986], Fayer and Ungar [1986], and Navin and Juranek [1984]) describe a self-limited, choleralike illness or a short-term, flulike gastrointestinal illness. The most common reported symptoms include profuse, watery diarrhea (choleralike), abdominal cramping, nausea, vomiting, low-grade fever, anorexia, and headache (flulike). An accidental laboratory infection of a researcher indicates that the incubation period following oral exposure to oocysts of *Cryptosporidium* is 4 to 5 days (Blagburn and Current, 1983). This incubation period is similar to that reported for laboratory and farm mammals following oral inoculation of *Cryptosporidium* oocysts (Current et al., 1983). In most healthy, well-nourished persons, diarrheal illness attributed to cryptosporidiosis usually lasts 3 to 12 days. Oral or parenteral fluid replacement therapy is sometimes required for previously healthy patients that develop severe cryptosporidial diarrhea. In poorly nourished children, especially those less than 2 years old, infection with diarrhea lasting more than a month is more common than in older, healthier persons, and fluid replacement therapy is often recommended for these younger patients (Current, 1986).

Cryptosporidiosis in Immunodeficient Persons

In many immunodeficient persons, diarrheal illness due to gastrointestinal cryptosporidiosis becomes progressively worse and may be a major factor leading to death. Such life-threatening infections have been documented in individuals on immunosuppressive chemotherapy (Lewis et al., 1985; Meisel et al., 1976; Miller et al., 1983), persons with congenital immunodeficiencies such as hypogammaglobulinemia (Current et al., 1983; Lasser et al., 1979; Sloper et al., 1982), and patients with AIDS (Koch et al., 1983; Malbranche et al., 1983; Navin and Juranek, 1984; Pitlik et al., 1983b). Fluid loss by patients with AIDS and gastrointestinal cryptosporidiosis is often extensive; 3 to 6 liters of stool per day is common, and as much as 17 liters of watery feces per day has been reported (Centers for Disease Control, 1982). Reports of respiratory (Forgacs et al., 1983; Kocoshis et al., 1984; Ma et al., 1984) and hepatic (Blumberg et al., 1984; Guarda et al., 1983; Pitlik et al., 1983b) cryptosporidiosis demonstrates that *Cryptosporidium* is not always confined to the gastrointestinal tract of immunodeficient persons. The importance of *Cryptosporidium* as a cause of respiratory and hepatic disease in immunodeficient and perhaps in immunocompetent human beings has not been determined.

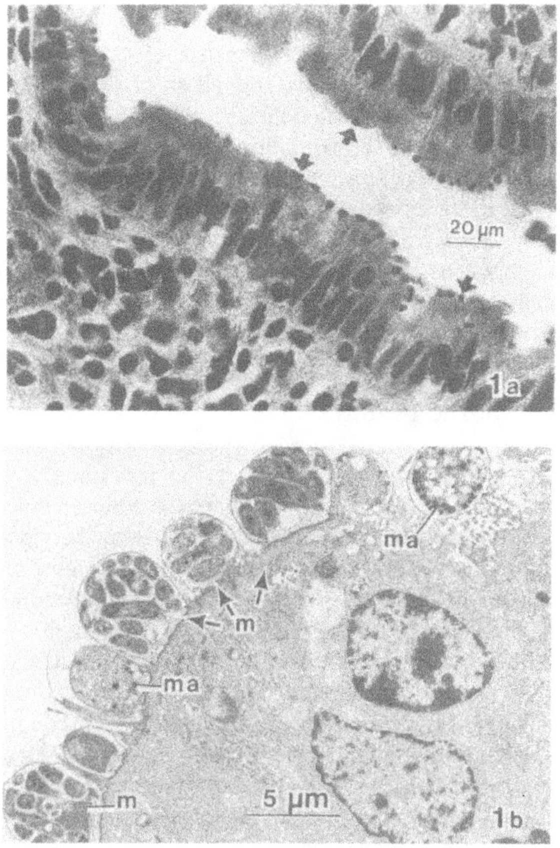

FIG. 1. (a) Bright-field photomicrograph of a histologic section of a small bowel biopsy specimen from a 15-year-old boy with hypogammaglobulinemia and prolonged cryptosporidiosis. Note the large number of mononuclear cells (mostly plasma cells) in the lamina propria and the numerous developmental stages of *Cryptosporidium* (*arrows*) within the brush border of the gut epithelium. Hematoxylin and eosin. (b) Transmission electron micrograph of the mucosal surface of the ileum of a mouse infected experimentally with *Cryptosporidium*. The meronts (m) and macrogametes (ma) shown here are each within a parasitophorous vacuole confined to the microvillous region of the enterocytes.

Pathology

Much of our knowledge of the pathology of human cryptosporidiosis has been derived from histologic studies of tissues obtained from naturally infected, immunodeficient persons with the disease. In immunodeficient persons with gastrointestinal cryptosporidiosis, developmental stages of *Cryptosporidium* have been found in the microvillous region of epithelial cells of the pharynx, esophagus, stomach, duodenum, jejunum, ileum, appendix, colon, and rectum (Fig. 1). The jejunum and ileum are usually the most heavily infected regions of the gut (reviewed by Navin and Juranek, 1984). Histologic lesions in heavily infected areas of the small intestine include villous atrophy, an increase in crypt length, and mild to moderate infiltration of the lamina propria with mononuclear cells (Fig. 1). Similar microscopic lesions occur in the small intestine of experimentally infected mice (Heine et al., 1984a), pigs (Moon and Bemrick, 1981), and lambs (Angus et al., 1982) and in the small and large intestines of germ-free calves monoinfected with *Cryptosporidium* (Heine et al., 1984b; Pholenz et al., 1984). From studies of experimental infections in germ-free calves, it appears that malabsorption and impaired digestion of the small intestine coupled with malabsorption in the large intestine are major factors responsible for the watery diarrhea associated with cryptosporidiosis. Malabsorption and impaired digestion has also been reported in immunosuppressed patients with cryptosporidiosis (Sloper et al., 1982; Navin and Juranik, 1984). However, the secretory (often described as choleralike) diarrhea common to many immunodeficient persons with cryptosporidiosis suggests hypersecretion into the gut. Slow marker intestinal profusion tests of a patient with AIDS and gastrointestinal cryptosporidiosis suggested profuse fluid secretion into the duodenum (Andrean et al., 1983). Carefully controlled studies are needed to determine the actual pathophysiologic mechanism(s) responsible for diarrhea in animals infected with *Cryptosporidium*.

Extraintestinal infections with *Cryptosporidium* have been reported in patients with immune deficiencies, especially those with AIDS (Blumberg et al., 1984; Forgacs et al., 1983; Guarda et al., 1983; Kocoshis et al., 1984). The presence of *Cryptosporidium* in the microvillous epithelium of the respiratory tract of immunodeficient patients has been associated with bronchiolitis and pneumonitis and the respiratory distress normally associated with these disease conditions. Acute and gangrenous cholecystitis have been associated with the presence of endogenous stages of *Cryptosporidium* in epithelial cells of the gallbladder and biliary tree of immunodeficient persons.

Diagnostic Features

Cryptosporidiosis should be considered along with the other known causes of diarrheal illness, especially if the patient's stool is watery. Acute cryptosporidiosis in immunocompetent persons is most often reported as a self-limited flulike or choleralike illness characterized by profuse watery diarrhea, abdominal cramping, nausea, anorexia, and headache. Mild or asymptomatic infections have been reported and may be common; therefore, physicians may consider the use of diagnostic procedures for the detection of *Cryptosporidium* infection before placing a patient on immunosuppressive chemotherapy. Delaying immunosuppressive chemotherapy until after a mild or asymptomatic *Cryptosporidium* infection is cleared may avoid a life-threatening illness. Prolonged gastrointestinal infections in immunodeficient patients are usually associated with extensive fluid loss, marked weight loss, and depression. *Cryptosporidium* should also be considered as a possible cause of respiratory and hepatic disease in immunodeficient patients.

Definitive diagnosis of cryptosporidiosis requires demonstration of oocysts in stool or sputum samples or the demonstration of developmental stages of the parasite in epithelial cells of tissues obtained by biopsy. Although a number of techniques have been reported to be useful for demonstrating oocysts in stool and sputum samples, the acid-fast staining (Fig. 2a) and to a lesser extent the negative staining (Fig. 2b) and concentration procedures (Fig. 2c) discussed below are the most widely used. These techniques are rapid, inexpensive, and sensitive; therefore, the use of expensive, invasive biopsy procedures is rarely needed for the diagnosis of cryptosporidiosis.

Prognosis

Gastrointestinal cryptosporidiosis is a self-limited diarrheal illness in immunocompetent persons, with recovery the usual outcome. Fluid loss may be extensive and the patient may require rehydration therapy. Antidiarrheal compounds may be of value in reducing the severity of the illness. Development of acquired immunity appears to be responsible for the clearance of developmental stages of *Cryptosporidium* from the infected mucosae of immunocompetent persons. The extent and duration of protective immunity to reinfection are not known.

The finding of cryptosporidiosis in immunodeficient persons usually carries an ominous prognosis,

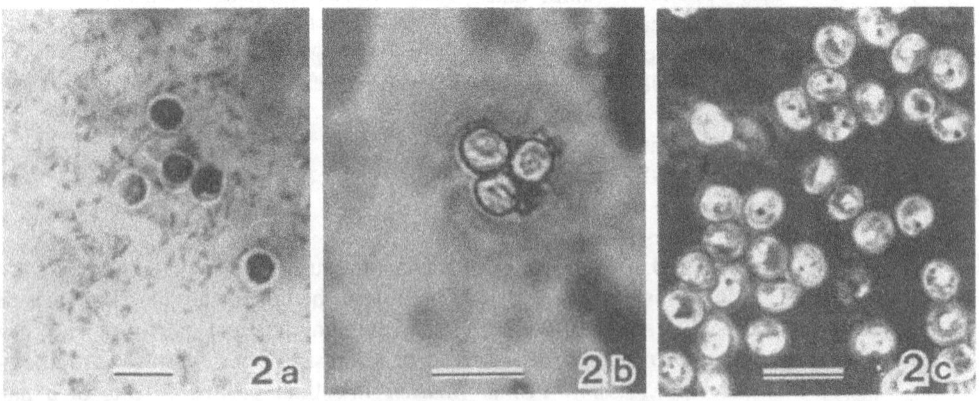

FIG. 2. Oocysts of *Cryptosporidium* in smears (a and b) and in a concentration (c) of feces from persons with clinical cryptosporidiosis. (a) In acid-fast stains (Garcia et al., 1983), the oocysts stain bright red, whereas yeast do not. (b) Kinyoun carbol fuchsin negative stain (Current, 1983). Yeast, bacteria, and other fecal debris stain darkly; however, the oocysts do not stain. When placed in an oil medium, the water-containing oocysts are refractile. (c) Sheather's sugar flotation (Current et al., 1983) viewed with phase-contrast microscopy. The oocysts appear as bright, refractile bodies with one to six dark particles inside. Examination of oocysts with each of these techniques should reveal the presence of four sporozoites. Bars = 10 μm.

because no effective drug has been identified and because these individuals are not capable of mounting an immune response of sufficient magnitude to stop proliferation of *Cryptosporidium* within the epithelium of the affected organ(s). Discontinuation of immunosuppressive chemotherapy should be considered if *Cryptosporidium* is diagnosed as the cause of diarrheal illness in cancer patients or in organ transplant recipients.

Epidemiologic Aspects

Transmission and Sources of Infection

Studies of experimental infections in farm and laboratory animals have established that cryptosporidiosis is transmitted by environmentally resistant oocysts that are fully sporulated and infective at the time they are passed in the feces (Current, 1986; Current et al., 1983). Several early reports clearly demonstrate that calves are a source of human infection (Anderson et al., 1982; Current et al., 1983; Reese et al., 1982), and it has been suggested that companion animals such as rodents, puppies, and kittens are also potential reservoir hosts (Current et al., 1983; Koch et al., 1985). These findings, coupled with a large list of animals (see Current, 1986 for some potential reservoir hosts) reported to be infected by a parasite that appears to readily cross host species barriers, resulted in the view that zoonotic transmission accounts for most human infections. This view may be correct for persons living and working in environments where they are exposed to fecal contamination from potential reservoir hosts; however, it is probably not applicable to most urban dwellers, whose exposure to animal feces is minimal. More recent evidence indicates that person-to-person transmission is common. Outbreaks of cryptosporidiosis have been reported among children attending day-care centers (Alpert et al., 1984; Centers for Disease Control, 1984), several hospital-acquired infections have been investigated (Koch et al., 1985; Baxby et al., 1983), at least one waterborne outbreak has been documented (D'Antonio et al., 1986), and this protozoan parasite is now recognized as a cause of traveler's diarrhea (Jokipii et al., 1983; Soave et al., 1985).

Prevalence and Potential Impact on Human Health

The importance of *Cryptosporidium* as a widespread cause of human diarrheal illness is just beginning to be recognized. Studies conducted in selected areas of Australia, Costa Rica, the United Kingdom, the United States, Brazil, Peru, South Africa, Central Africa, Bangladesh, Spain, Denmark, Liberia, and Haiti (see Current [1986] for references) have provided data on the importance of *Cryptosporidium* as a cause of human diarrheal illness. In the populations studied, the incidence of *Cryptosporidium* oocysts in stools of persons with diarrhea ranged from 2 to 14%. In most of these surveys *Cryptosporidium* was the most common parasite found and, in several, this protozoan was considered the most significant of the enteropathogens known to cause diarrheal illness. Other findings common to most of these studies were that children had a significantly higher prevalence of cryptosporidiosis than did adults and that infections were often seasonal, with a higher prevalence during the warmer, wetter months. Another important finding from the standpoint of infection control was that small numbers of oocysts may be present in the feces for up to 2 weeks following resolution of diarrhea.

Diarrheal illness is a major cause of morbidity and mortality, especially in young children living in developing countries. Estimates by Walsh and Warren (1979) suggest that in Asia, Africa, and Latin America alone, there are as many as 5 billion episodes of diarrhea and 5 to 10 million diarrhea-associated deaths annually. Although the surveys mentioned above demonstrate that cryptosporidiosis is a major cause of diarrheal illness worldwide, it is not clear to what extent this parasitic disease contributes to morbidity and mortality. Additional, detailed studies employing proper diagnostic tools should define more clearly the impact of cryptosporidiosis on human health.

Etiologic Agents

Taxonomic Relationships, Host Specificity, and Life Cycle

Organisms of the genus *Cryptosporidium* are small protozoan parasites assigned to the family *Cryptosporidiidae,* suborder *Eimeriorina,* order *Eucoccidiorida,* subclass *Coccidiasina,* phylum *Apicomplexa* (Levine, 1984). Other true coccidia (*Eimeriorina*) that infect humans include *Toxoplasma gondii, Isospora belli,* and *Sarcocystis* spp. The type species, *C. muris,* was described by Tyzzer in 1907, and a second species, *C. parvum,* infecting the same host was described 5 years later (Tyzzer, 1912). Oocysts indistinguishable from those of *C. muris* and *C. parvum* have been recovered from calves, and, on the basis of oocyst morphology, it is the latter species (not *C. muris*) that is associated with all of the previously well-documented cases of cryptosporidiosis in mammals (Upton and Current, 1985). A more thorough discussion of the taxonomy and validity of the 19 named species of *Cryptosporidium* can be found elsewhere (Current, 1986).

The life cycle of *Cryptosporidium* (Fig. 3) can be divided into six major developmental events: excystation, the release of infective sporozoites from oocysts; merogony, the asexual reproductive phase; gametogony, the formation of macro- and microgametes; fertilization, the fusion of macro- and microgamete nuclei to form the zygote; oocyst wall formation, the process resulting in environmentally resistant oocysts; and sorogony, the formation of infective sporozoites within the oocyst wall. The most important aspects of the life cycle and morphology of *Cryptosporidium*, discussed in detail by Current and Reese (1986), Current and Long (1983), Current and Haynes (1984), Goebel and Brandler (1982), and Reduker et al. (1985), are summarized below.

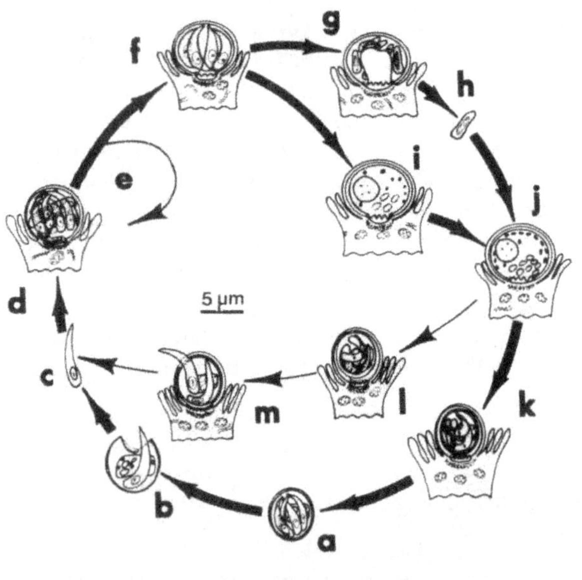

FIG. 3. Life cycle of human and calf isolates of *Cryptosporidium* in experimentally infected mice. (a) Sporulated oocyst in feces; (b) excystation in intestine; (c) free sporozoite in intestine before penetration into the microvillous region of an ileal enterocyte; (d) type I meront (six or eight merozoites); (e) recycling of type I merozoite; (f) type II meront (four merozoites); (g) microgametocyte, with approximately 16 microgametes. Microgamete (h) fertilizes macrogamete (i) to form zygote (j); approximately 80% of the zygotes form thick-walled oocysts (k), which sporulate within the host cell. Almost all of the thick-walled oocysts pass unaltered in the feces and are the resistant forms that transmit the infection to another host. About 20% of the zygotes do not form an oocyst wall, and their sporozoites are surrounded only by a unit membrane (l). Sporozoites within autoinfective, thin-walled oocysts (l) are released into the intestinal lumen (m) and reinitiate the endogenous cycle (at c).

The life cycle of *Cryptosporidium* (Fig. 3) differs somewhat from that of other single-host (monoxenous) coccidia (e.g., *Eimeria* and *Isospora* spp.) that infect humans and domesticated animals. The intracellular stages of *Cryptosporidium* occur within a parasitophorous vacuole confined to the microvillous region of the host cell (Fig. 1b), whereas comparable forms of *Eimeria* and *Isospora* spp. usually occupy a vacuole deep within the host cell. *Cryptosporidium* can complete its life cycle within cells of the mucosal epithelium of a variety of organs, whereas *Eimeria* and *Isospora* spp. are usually confined to the gut mucosa. The oocysts of *Cryptosporidium* sporulate within the host cells and are infective when passed in the feces, whereas those of *Eimeria* and *Isospora* spp. do not sporulate until they are passed from the host and exposed to oxygen and temperatures below that of the mammalian body. Approximately 20% of the oocysts of *Cryptosporidium* within host cells do not produce the environmentally resistant bilayered oocyst wall, their sporozoites are surrounded only by a single unit membrane. Soon after a thin-walled oocyst is released from the host cell, the unit membrane surrounding the four sporozoites ruptures, and these invasive forms enter adjacent host cells and reinitiate the developmental cycle. These thin-walled, autoinfective oocysts are not passed in the feces. The majority (80%) of the oocysts of *Cryptosporidium* developing in the host cells are similar to those of other coccidia infecting mammals in that they produce multilayered, environmentally resistant walls and are passed in the feces. It is the thick-walled form that transmits an infection to a suitable host. The presence of autoinfective, thin-walled oocysts and type I meronts that can recycle may explain why a small number of ingested oocysts can cause an overwhelming infection and why immunodeficient persons may have persistent, life-threatening disease in the absence of repeated exposure to environmentally resistant, thick-walled oocysts.

Specimen Collection and Transport

Prior to 1980, most human cases were diagnosed after examination of small- or large-bowel biopsy material, often using both light and electron microscopy. However, because biopsy specimens may be taken from uninfected areas of the mucosa, and because accurate stool diagnostic techniques have been developed within the past 4 years, cases have more recently been diagnosed by identification of oocysts in fecal material.

Biopsy specimens should be submitted for routine histologic processing and examination. Fecal specimens can be submitted as fresh material, in 10% Formalin, or in sodium acetate–acetic acid–formalin (SAF) preservatives. Most of the recommended stains for demonstrating *Cryptosporidium* oocysts can not be performed from stool preserved in polyvinyl alcohol fixative. As many laboratories are presently using a single collection vial (polyvinyl alcohol) from which routine examinations for parasites can be performed, they may have to incorporate a vial of 10% Formalin into the system to accommodate requests for *Cryptosporidium*.

The number of organisms will vary from specimen to specimen; consequently, multiple stool samples should be submitted, particularly in the case of a light infection or an individual in the "carrier" state. In general, the more loose and watery the stool, the more oocysts will be present. In formed stools, the small number of oocysts present may be very difficult to find.

As respiratory cryptosporidiosis has been reported in AIDS patients (Forgacs et al., 1983; Kocoshis et al., 1984), sputum samples from immunodeficient patients with undiagnosed respiratory illness should be placed in 10% Formalin and examined for *Cryptosporidium* oocysts using stool examination procedures described below. Infection of the gallbladder and biliary tree should result in oocysts being passed in the stool.

Preliminary Identification

Oocysts seen in direct smears or concentration preparations are generally round and measure from 4 to 6 μm. Unless the numbers are very heavy and the viewer has some experience identifying *Cryptosporidium*, it will be very easy to confuse oocysts with yeast cells. Identification based on unconcentrated and concentrated wet preparations are often considered tentative and must be confirmed with one of the staining techniques. The four sporozoites are often not visible, particularly when the specimen is very watery (severe diarrhea).

Definitive Identification

Acid-Fast Staining

The routine stains (trichrome, iron hematoxylin) used for stool parasitology are not acceptable for the identification of *Cryptosporidium* oocysts. Most laboratories are using one of the acid-fast methods (Fig. 2a), many of which are very satisfactory in demonstrating the organisms in stool material (Baxby et al., 1984; Garcia et al., 1983; Ma and Soave, 1983). The oocysts do not all stain uniformly, some being more difficult to see. Although the cold acid-fast method is excellent, some workers believe that better stain penetration of the oocyst may be obtained using a hot acid-fast method (Garcia et al., 1983). Whatever stain the laboratory selects should be acceptable if oocysts in positive control smears stain well. Subtle differences in staining characteristics of oocysts are probably not clinically relevant.

Key points concerning laboratory diagnosis are as follows: 1) If the specimen is the watery diarrhea typical of gastrointestinal cryptosporidiosis, numerous oocysts should be caught up in the flecks of mucus. 2) The more normal the stool (semiformed, formed), the fewer the organisms and the more artifact material will be present. 3) Several stool concentrates and at least five or six acid-fast smears should be examined before a patient is considered negative, particularly if the patient is being monitored for efficacy of experimental therapy. 4) The normal parasitologic stains used for stool material do not adequately stain *Cryptosporidium* oocysts. 5) Screening techniques using auramine–rhodamine (or other similar stains) may reveal the organisms; however, in formed stool specimens containing more artifact material, these identifications are considered tentative and should be confirmed using other staining methods.

Other Oocyst Identification Procedures

Several non-acid-fast techniques are used commonly in research laboratories, and occasionally in some clinical laboratories for identification of *Cryptosporidium* oocysts in stool or sputum samples. These procedures include negative staining (Current, 1983; Heine, 1982) (Fig. 2b) and Sheather sugar flotation–phase contrast microscopy (Current et al., 1983) (Fig. 2c). As noted previously for the acid-fast techniques, considerable experience is required with these latter two procedures to obtain an accurate diagnosis, particularly if the stool specimen contains few oocysts. The use of polyclonal or monoclonal antibodies should also prove very helpful in providing a more sensitive method of detecting organisms in clinical specimens (Garcia et al., 1987; Sterling and Arrowood, 1985).

Serologic Diagnosis

Preliminary results from serologic methods for the diagnosis of *Cryptosporidium* look promising; however, additional patients must be studied before the actual diagnostic potential of these methods can be evaluated (Campbell and Current, 1983; Tzipori and Campbell, 1981; Ungar et al., 1986). Serologic tests should be very useful in estimating the prevalence of this infection in different parts of the world.

Treatment, Prevention, and Control

Treatment

Treatment of cryptosporidiosis, especially in immunodeficient persons, has been frustrating for the physician and has been unsuccessful in most cases. To date, there have been no controlled studies and all therapeutic information is based on isolated reports. The list of unsuccessful attempts to treat cryptosporidiosis in immunodeficient patients is growing rapidly. Many of these attempts have been reviewed (Fayer and Ungar, 1986; Hart and Baxby, 1985; Navin and Juranek, 1984; Soave, 1984). None of the more than 65 antiparasitic or antimicrobial drugs evaluated for the control of cryptosporidiosis in humans and animals (Fayer and Ungar, 1986) has proved particularly effective. Of the more than 40 antimicrobial and antiparasitic drugs administered to immunodeficient patients with intestinal cryptosporidiosis, spiramycin is the only one that appears to have some efficacy. Portnoy et al. (1984) reported that oral administration of 3 g per day of this macrolide antibiotic resulted in improvement of diarrhea in nine patients with AIDS and cryptosporidiosis. However, three of these patients continued to shed *Cryptosporidium* oocysts in their feces. Numerous other physicians have reported to us that spiramycin is not effective in the treatment of cryptosporidiosis in immunodeficient persons. Carefully controlled studies are needed to determine if spiramycin therapy is of any value in the treatment of cryptosporidiosis. Until an effective drug is found to treat cryptosporidiosis, supportive therapy is the only avenue available to most physicians.

Antidiarrheal compounds may provide a little symptomatic relief in some cases of cryptosporidiosis in immunodeficient persons; however, they may be of no value in others. Antidiarrheal compounds usually lessen the severity of gastrointestinal cryptosporidiosis in most immunocompetent persons. Another form of supportive therapy that may be of value in lessening the impact of gastrointestinal cryptosporidiosis is oral and parenteral rehydration therapy. Such rehydration therapy is often required for children with severe cryptosporidial diarrhea. Parenteral nutrition may help sustain the nutritional status of some immunodeficient patients with cryptosporidiosis; however, this procedure has not been successful in sustaining some persons with AIDS and cryptosporidiosis.

The major factor determining the severity and duration of a *Cryptosporidium* infection appears to be the immune status of the host. Discontinuation of immunosuppressive chemotherapy, allowing restoration of immune function, has resulted in clearance of intestinal cryptosporidiosis from several patients (Lewis et al., 1985; Miller et al., 1983). These findings suggest that immunomodulation therapy may be of value in the treatment of cryptosporidiosis in immune deficient patients. Transfer factor prepared from lymphoid cells obtained from calves immune to *Cryptosporidium* was reported to have some effect on the course of cryptosporidiosis in several AIDS patients (Louie et al., 1985). Oral administration of bovine transfer factor resulted in resolution of diarrhea in six of seven patients with AIDS and gastrointestinal cryptosporidiosis. Four of these patients stopped shedding oocysts in their feces; the other three continued to shed small numbers of the organisms. One patient showed no symptomatic or parasitologic improvement during or following treatment. More extensive studies are needed to confirm the results obtained from this small number of AIDS patients and to determine if bovine transfer factor therapy or other forms of immunomodulation are of any value in the treatment of cryptosporidiosis in persons with severe immune deficiencies.

Prevention and Control

It is now established that cryptosporidiosis is transmitted by environmentally resistant oocysts via the fecal–oral route common to other enteric pathogens. Physicians, veterinarians, and others involved in educating the public of risk factors should broaden their educational role to include cryptosporidiosis. This educational role should be approached aggressively because the disease appears to be common among most populations studied, because many potential reservoir hosts have been identified, and because persons with compromised immune systems are at risk of developing a life-threatening infection following oral exposure to oocysts. Recognizing when the infection is present in humans and domesticated animals is important so that persons at risk of developing serious illness can be warned. It is also important that these high-risk individuals are aware of standard precautions to avoid exposure to waterborne and food-borne enteropathogens.

Literature Cited

Alpert, G., L., M. Bell, C. E. Kirkpatrick, L. D. Budnick, J. M. Campos, H. M. Friedman, and S. A. Poltkin. 1984. Cryptosporidiosis in a day care center. N. Engl. J. Med. **311**:860–861.

Anderson, B. C., T. Donndelinger, R. M. Wilkins, and J. Smith. 1982. Cryptosporidiosis in a veterinary student. J. Am. Vet. Med. Assoc. **180**:408–409.

Andrean, T., R. Modigliani, Y. Charpentier, A. Calian, J. C. Brouet, M. Liance, J. R. Lachance, B. Messing, and B. Vernisse. 1983. Acquired immunodeficiency with intestinal cryptosporidiosis: possible transmission by Haitian whole blood. Lancet **1:**1187–1191.

Angus, K. W., S. Tzipori, and E. W. Gray. 1982. Intestinal lesions in specific pathogen-free lambs associated with a *Cryptosporidium* from calves with diarrhea. Vet. Pathol. **19:**587–592.

Baxby, D., N. Blundell, and C. A. Hart. 1984. The development and performance of a simple, sensitive method for detection of *Cryptosporidium* oocysts in faeces. J. Hyg. **93:**317–324.

Baxby, D., C. A. Hart, and C. Taylor. 1983. Human cryptosporidiosis: a possible case of hospital cross infection. Br. Med. J. **287:**1760–1761.

Blagburn, B. L., and W. L. Current. 1983. Accidental infection of a researcher with human *Cryptosporidium*. J. Infect. Dis. **148:**772–773.

Blumberg, R. S., P. Kelsey, T. Perrone, R. Dickersin, M. Laquaglia, and J. Ferruci. 1984. Cytomegalovirus- and *Cryptosporidium*-associated acalculous gangrenous cholecystitis. Am. J. Med. **76:**1118–1123.

Campbell, P. N., and W. L. Current. 1983. Demonstration of serum antibodies to *Cryptosporidium* sp. in normal and immunodeficient humans with confirmed infections. J. Clin. Microbiol. **18:**165–169.

Centers for Disease Control. 1982a. Human cryptosporidiosis—Alabama. Morbid. Mortal. Weekly Rep. **31:**252–254.

Centers for Disease Control. 1982b. Cryptosporidiosis: assessment of chemotherapy of males with acquired immune deficiency syndrome (AIDS). Morbid. Mortal. Weekly Rep. **31:**589–592.

Centers for Disease Control. 1984. Cryptosporidiosis among children attending day care centers—Georgia, Pennsylvania, Michigan, California, New Mexico. Morbid. Mortal. Weekly Rep. **33:**599–601.

Current, W. L. 1983. Human cryptosporidiosis. N. Engl. J. Med. 309:1326–1327.

Current, W. L. 1984. *Cryptosporidium* and cryptosporidiosis, p. 355–373. *In* M. S. Gottlib, and J. E. Groopman (ed.), Acquired immune deficiency syndrome. UCLA Symposia on Molecular and Cellular Biology, New Series, vol. 16. Alan R. Liss, Inc., New York.

Current, W. L. 1986. Cryptosporidium: its biology and potential for environmental transmission. Crit. Rev. Environ. Control **17:**21–51.

Current, W. L., and T. B. Haynes. 1984. Complete development of *Cryptosporidium* in cell culture. Science **224:**603–605.

Current, W. L., and P. L. Long. 1983. Development of human and calf *Cryptosporidium* in chicken embryos. J. Infect. Dis. **148:**1108–1113.

Current, W. L., and N. C. Reese. 1986. A comparison of endogenous development of three isolates of *Cryptosporidium* in suckling mice. J. Protozool. **33:**98–108.

Current, W. L., N. C. Reese, J. V. Ernst, W. S. Bailey, M. B. Heyman, and M. D. Weinstein. 1983. Human cryptosporidiosis in immunocompetent and immunodeficient persons: studies of an outbreak and experimental transmission. N. Engl. J. Med. **308:**1252–1257.

D'Antonio, R. G., R. E. Winn, J. P. Taylor, T. L. Gustafson, W. L. Current, M. M. Rhodes, G. W. Gary, and R. A. Zajac. 1986. A waterborne outbreak of cryptosporidiosis in normal hosts. Ann. Intern. Med. **103:**886–888.

Fayer, R., and B. L. P. Ungar. 1986. *Cryptosporidium* spp. and cryptosporidiosis. Microbiol. Rev. **50:**458–483.

Forgacs, P., A. Tarshis, P. Ma, M. Federman, L. Mele, M. L. Silverman, and J. A. Shea. 1983. Intestinal and bronchial cryptosporidiosis in an immunodeficient homosexual man. Ann. Intern. Med. **99:**793–794.

Garcia, L. S., T. C. Brewer, and D. A. Brackner. 1987. Fluorescent detection of *Cryptosporidium* oocysts in human fecal specimens by using monoclonal antibodies. J. Clin. Microbiol. **25:**119–121.

Garcia, L. S., D. A. Bruckner, T. C. Brewer, and R. Y. Shimzu. 1983. Techniques for the recovery and identification of *Cryptosporidium* oocysts from stool specimens. J. Clin. Microbiol. **18:**185–190.

Goebel, E., and U. Brandler. 1982. Ultrastructure of microgametogenesis, microgametes, and gametogony of *Cryptosporidium* sp. in the small intestine of mice. Protistologica **18:**331–334.

Guarda, L. A., S. A. Stein, K. A. Cleary, and N. G. Ordonez. 1983. Human cryptosporidiosis in the acquired immune deficiency syndrome. Arch. Pathol. Lab. Med. **107:**562–566.

Hart, A., and D. Baxby. 1985. Management of cryptosporidiosis. J. Antimicrob. Chemother. **15:**3–4.

Heine, J. 1982. Eine einfache Nachweismethode fur Kryptosporidien in Kot. Zentralbl. Veterinaermed. Reihe B **29:**324–327.

Heine, J., H. W. Moon, and D. B. Woodmansee. 1984a. Persistent *Cryptosporidium* infection in congenitally athymic (nude) mice. Infect. Immun. **43:**856–859.

Heine, J., J. F. L. Pholenz, H. W. Moon, and G. N. Woode. 1984b. Enteric lesions and diarrhea in gnotobiotic calves monoinfected with *Cryptosporidium* species. J. Infect. Dis. **150:**768–775.

Jokipii, L., S. Pohjola, and A.M.M. Jokipii. 1983. Cryptosporidium: a frequent finding in patients with gastrointestinal symptoms. Lancet **1:**358–361.

Koch, K. L., D. J. Phillips, R. C. Aber, and W. L. Current. 1985. Cryptosporidiosis in hospital personnel: evidence for person-to-person transmission. Ann. Intern. Med. **102:**593–596.

Koch, K. L., T. V. Shankey, G. S. Weinstein, R. E. Dye, A. B. Abt, W. L. Current, and M. E. Eyster. 1983. Cryptosporidiosis in a patient with hemophilia, common variable hypogammaglobulinemia, and the acquired immunodeficiency syndrome. Ann. Intern. Med. **99:**337–340.

Kocoshis, S. A., M. L. Cibull, and T. E. Davis. 1984. Intestinal and pulmonary cryptosporidiosis in an infant with severe combined immune deficiency. J. Pediatr. Gastroenterol Nutr. **3:**149–157.

Lasser, K. H., K. J. Lewin, and F. W. Ryning. 1979. Cryptosporidial enteritis in a patient with congential hypogammaglobulinemia. Hum. Pathol. **10:**234–240.

Levine, N. D. 1984. Taxonomy and review of the coccidian genus *Cryptosporidium*(Protozoa:Apicomplexa). J. Protozool. **31:**94–98.

Lewis, I. J., C. A. Hart, and D. Baxby. 1985. Diarrhoea due to *Cryptosporidium* in acute lymphoblastic leukaemia. Arch. Dis. Child. **60:**60–62.

Louie, E., W. Borkowsky, P. H. Klesius, T. B. Haynes, S. Gordon, and H. S. Lawrence. 1985. Treatment of cryptosporidiosis with oral bovine transfer factor. Interscience Conference on Antimicrobial Agents and Chemotherapy, abstract 817, Minneapolis, Minnesota, September 29–October 2. American Society For Microbiology, Washington D.C.

Ma, P., and R. Soave. 1983. Three step stool examination for cryptosporidiosis in ten homosexual men with protracted watery diarrhea. J. Infect. Dis. **147:**824–828.

Ma, P., T. G. Villanueva, D. Kaufman, and J. F. Gillooley.

1984. Respiratory cryptosporidiosis in the acquired immune deficiency syndrome. J. Am. Med. Assoc. **252**:1298–1301.

Malebranche, R., E. Arnoux, J. M. Guerin, G. D. Piere, A. C. Laroche, C. P. Guichard, R. Elie, P. H. Morisset, T. Spira, R. Mandeville, P. Drotman, T. Seemayer, and J. P. Dupuy. 1983. Acquired immunodeficiency syndrome with severe gastrointestinal manifestations in Haiti. Lancet **2**:873–878.

Meisel, J. L., D. R. Perera, C. Meligro, and C. E. Rubin. 1976. Overwhelming watery diarrhea associated with *Cryptosporidium* in an immunosuppressed patient. Gastroenterology **70**:1156–1160.

Miller, R. A., R. E. Holmberg, and C. R. Clausen. 1983. Lifethreatening diarrhea caused by *Cryptosporidium* in a child undergoing therapy for acute lymphocytic leukemia. J. Pediatr. **103**:256–259.

Moon, H. W., and W. J. Bemrick. 1981. Fecal transmission of calf cryptosporidia between calves and pigs. Vet. Pathol. **18**: 248–255.

Navin, T. R., and D. D. Juranek. 1984. Cryptosporidiosis: clinical, epidemologic, and parasitologic review. Rev. Infect. Dis. **6**:313–327.

Pholenz, J. F. L., G. N. Woode, and D. Woodmansee. 1984. Typhlitis and colitis in gnotobiotic calves experimentally infected with 10^8–10^9 chemically purified *Cryptosporidium* oocysts. Conference of Research Workers in Animal Disease, Chicago, November 12–13, abstract no. 273. Conference of Research Workers in Animal Diseases, Chicago, IL.

Pitlika, S. D., V. Fainstein, D. Garza, L. Guarda, R. Boliver, A. Rios, R. L. Hopfer, and P. A. Mansell. 1983a. Human cryptosporidiosis: spectrum of disease. Report of six cases and reviewes of the literature. Arch. Intern. Med. **143**:2269–2273.

Pitlik, S., V. Fainstein, A. Rios, L. Guarda, P. W. A. Mansell, and E. M. Hersh. 1983b. Cryptosporidial cholecystitis. N. Engl. J. Med. **308**:976.

Portnoy, D. M. E. Whiteside, E. Buckley, and C. L. MacLeod. 1984. Treatment of intestinal cryptosporidiosis with spiramycin. Ann. Intern. Med. **101**:202–204.

Reduker, D. W., C. A. Speer, and J. A. Blixt. 1985. Ultrastructure of *Cryptosporidium* oocysts and excysting sporozoites as revealed by high resolution scanning electron microscopy. J. Protozool. **32**:708–711.

Reese, N. C., W. L. Current, J. V. Ernst, and W. S. Bailey. 1982. Cryptosporidiosis of man and calf: a case report and results of experimental infections in mice and rats. Am. J. Trop. Med. Hyg. **31**:226–229.

Sloper, K. S., R. R. Dourmashkin, R. B. Bird, G. Slavin, and A. B. D. Webster. 1982. Chronic malabsorption due to cryptosporidiosis in a child with immunoglobulin deficiency. Gut **23**:80–82.

Soave, R. 1984. Therapy and prevention of coccidiosis, p. 232–236. *In* L. Leive and D. Schlessenger, (ed.), Microbiology—1984. American Society for Microbiology, Washington, D.C.

Soave, R., and P. Ma. 1985. Cryptosporidiosis travelers diarrhea in 2 families. Arch. Intern. Med. **145**:70–72.

Sterling, C. R., and M. J. Arrowood. 1986. Detection of *Cryptosporidium sp.* infections using a direct immunofluorescent assay. Pediatr. Infect. Dis. **5**. S139-S142.

Tyzzer, E. E. 1907. A sporozoan found in the peptic glands of the common mouse. Proc. Soc. Exp. Biol. Med. **5**:12–13.

Tyzzer, E. E. 1912. *Cryptosporidium parvum* (sp. nov.), a coccidian found in the small intestine of the common mouse. Arch. Protistenkd. **26**:394–418.

Tzipori, S., and I. Campbell. 1981. Prevalence of *Cryptosporidium* antibodies in 10 animal species. J. Clin. Microbiol. **14**:455–456.

Ungar, B. L. P., R. Soave, R. Fayer, and T. E. Nash. 1986. Enzyme immunoassay detection of immunoglobulin M and G antibodies to *Cryptosporidium* in immunocompetent and immunocompromised persons. J. Infect. Dis. **153**:570–578.

Upton, S. J., and W. L. Current. 1985. The species of *Cryptosporidium* (Apicomplexa:Cryptosporiidae) infecting mammals. J. Parasitol. **71**:625–629.

Walsh, J. A., and K. S. Warren. 1979. Selective primary care. An interim strategy for disease control in developing countries. N. Engl. J. Med. **301**:967–974.

Zar, F., P. J. Geiseler, and V. A. Brown. 1985. Asymptomatic carriage of *Cryptosporidium* in the stool of a patient with acquired immune deficiency syndrome. J. Infect. Dis. **151**:195.

CHAPTER 86

Diphyllobothriasis

H. KAMO

Disease: Diphyllobothriasis (broad tapeworm infection, fish tapeworm infection).

Etiologic Agents: *Diphyllobothrium latum* (worldwide), *Diphyllobothrium dalliae* (western Alaska), *Diphyllobothrium nihonkaiense* (Japan), and *Diphyllobothrium pacificum* (Peru, Chile).

Source: Fish as second intermediate or paratenic hosts of each *Diphyllobothrium* species by eating raw or undercooked flesh, hard roe, or liver.

Clinical Manifestations: Asymptomatic or subclinical in most cases. Some patients experience fatigue, weakness, diarrhea, numbness of extremities, a "sensation of hunger," and a "craving for salt." Occasional vomiting of worms, intestinal obstruction, abdominal pain suspicious of acute abdomen. Most characteristic, but rare, is *D. latum* pernicious anemia.

Pathology: Intestinal malabsorption caused by direct mechanical irritation of intestinal mucosa or by competition for available nutrients. *Diphyllobothrium latum* prevents vitamin B_{12} from combining with intrinsic factor, thus blocking absorption of vitamin B_{12} and resulting in deficiency.

Laboratory Diagnosis: Recovery and identification of characteristic eggs or proglottids in fecal matter. A few months of negative-phase during egg-discharging periodicity is possible.

Epidemiology: Prevalent in subarctic and temperate regions of the Northern Hemisphere, especially in areas from Siberian rivers Yenisei and Ob to Baltic Sea of Eurasian continent. Some transferred centers in North and South America. Sporadic, imported cases on other islands and continents. Some species other than *D. latum* found in Alaska, Canada, Peru, Chile, and Japan. Consumption of dishes containing raw fish or raw fish products is always responsible for infection.

Treatment: Niclosamide (Yomesan) or praziquantel (Biltricide); bithionol (Bithin) plus purgative.

Prevention and Control: Education of population to avoid eating raw or undercooked fish and hard roe or liver; mass treatment of cestode carriers; purification of waste water and sewage disposal.

Description of Disease

Diphyllobothriasis means literally an invasion of the human intestines by any species of tapeworm belonging to the genus *Diphyllobothrium* (Cestoda: Diphyllobothriidae). Recently several species other than *Diphyllobothrium latum* have been isolated from humans in different parts of the world. However, most of them are known in somewhat restricted areas, and information regarding their hygienic, pathogenic effects or biologic aspects is not necessarily sufficient.

Usually the term diphyllobothriasis (broad tapeworm infection, fish tapeworm infection) applies to infection with *D. latum,* the best known and most important species, which has been experienced by humans ever since the Stone Age, at least in Northern Europe.

Communicability and Transmissibility

The host becomes infected by eating raw or undercooked fish of certain species, which play the role of second intermediate or paratenic host in each area. Spread of the disease is favored by the popularity of eating raw fish and by rapid, remote transportation of human and fresh foods. On the other hand, suitable ecologic conditions for the parasite and susceptible intermediate hosts exist over large areas. The risk of infection or new contamination in countries where the worms are not endemic should not be neglected.

Clinical Manifestations

In cases of voluntary infection with *D. latum,* general fatigue is a common complaint during its initial course, followed by epigastric pain, mild fever, and diarrhea as the worm reaches maturity (about 20 days postinfection). In contrast, in cases of natural infection the parasitosis is frequently asymptomatic. In many cases *D. latum* may be harbored for decades without causing ill effects, whereas in rare cases carriers suffer from tapeworm pernicious anemia.

Symptoms and Signs in Nonanemic Carriers

Some carriers present a variety of clinical manifestations, although most experience very little discomfort. Statistical analysis (Saarni et al., 1963) showed that worm carriers had a significantly higher frequency of fatigue, weakness, diarrhea, numbness of the extremities, and "sensation of hunger" than controls. A "craving for salt" was also more frequent among worm carriers. These findings do not exclude the possibility of severe abdominal symptoms in carriers of *D. latum,* although no difference between worm carriers and controls was noticed with regard to the frequency of abdominal pain. Vomiting of worms occurs, sometimes in connection with intestinal obstruction, occasionally during pregnancy, but often without any demonstrable cause (von Bonsdorff, 1947). In several cases abdominal pain and nausea, with or without vomiting of worms, have aroused suspicion of perforating peptic ulcer, gallstone colic, appendicitis, and the like, occasionally even resulting in surgical intervention. Only a few cases of obstruction have been known. Patients under 30 years of age experienced more symptoms than older patients.

Optic neuropathy with central scotoma was found in four nonanemic worm carriers with low serum levels of vitamin B_{12} (Björkenheim, 1966). The signs of optic neuropathy disappeared when serum levels of vitamin B_{12} normalized after expulsion of the worm. Asthenopic symptoms were significantly more common among tapeworm carriers with subnormal serum levels of vitamin B_{12} than among those with normal levels.

In cases of infection with other species, such as *Diphyllobothrium pacificum* and *Diphyllobothrium nihonkaiense,* fatigue, weakness, diarrhea, and other minor disorders of the digestive tract were commonly observed, but "sensation of hunger," "craving for salt," and optic neuropathy have never been observed.

Tapeworm Pernicious Anemia

The most serious complication of diphyllobothriasis is tapeworm anemia, which is only known in connection with *D. latum* infection and occurs in less than 2% (overall, 0.01 to 0.02%) of worm carriers. The clinical picture of tapeworm anemia is similar to that of genuine pernicious anemia, although the gastric juice in tapeworm pernicious anemia is usually normal. The patient frequently shows slight jaundice, with "anemic fever" and glossitis in the form of atrophic changes. The tongue appears smooth and shiny from loss of the papillae, with occasional small ulcers. Edema and hemorrhage are also common features. The liver and spleen are seldom enlarged. Nervous system manifestations are also common to genuine pernicious anemia. Paresthesia, disturbances of motility and coordination, and impairment of deep sensation are the most common neurologic manifestations. They are relatively mild, but are sometimes (nearly 50%) severe enough to reduce the capacity to work. Absence of neurologic symptoms and signs is more common in younger patients than in those of advanced age. The hematologic profile is characterized by macrocytic, megaloblastic anemia with leukopenia, thrombocytopenia, and increased hemolysis. Nucleated erythrocytes and normoblasts are occasionally seen in the peripheral blood. The red blood cell count is reduced to a much greater degree than is the hemoglobin. The mean corpuscular hemoglobin of the erythrocytes is increased. Hypersegmented neutrophils with six nuclear lobes or more are common. The leukopenia is mainly due to the decrease in neutrophils; the actual number of eosinophils is unchanged. Levels of vitamin B_{12} in the blood serum decrease markedly to less than 100 pg/ml. The bone marrow in this disease exhibits the same picture as in genuine pernicious anemia, particularly with regard to the occurrence of megaloblasts and promegaloblasts. An increased eosinophil count is occasionally noted in the marrow. Giant metamyelocytes predominate, and sometimes increase before the appearance of megaloblasts. Hemolytic jaundice of moderate degree is involved: the urine contains urobilin, but no bilirubin, the feces contain an abundance of urobilin, hemosiderosis is observed in the liver, and the blood serum contains unconjugated bilirubin (von Bonsdorff, 1977).

Pathology

Diphyllobothrium latum is most frequently attached to the wall of the ileum, less commonly to that of the jejunum, seldom at other levels, and rarely wandering in the gallbladder. A patient may be burdened a single or multiple strobilae.

The predominance of diffuse general symptoms without special characteristics indicates a nonspecific general action on the nutritional state of the host, assuming the existence of some kind of intestinal malabsorption. This malabsorption may be caused by direct mechanical irritation of the intestinal mucosa or by competition for available nutrients. Mild disturbances in folic acid absorption and partial vitamin B_{12} deficiency result from this competition.

The same endogenous factors that lead to the development of genuine pernicious anemia promote the development of vitamin B_{12} deficiency and tapeworm pernicious anemia in carriers of *D. latum*. The tapeworm plays the role of an exogenous factor (i.e., the living *D. latum* in the intestine takes up vitamin B_{12} in competition with vitamin B_{12} receptors of the host). The following factors are conducive to the development of vitamin B_{12} deficiency: (1) an inadequate supply of vitamin B_{12} in the diet; (2) a relative intrinsic factor deficiency, due to endogenous or exogenous damage to the gastric mucosa; (3) location of the parasite proximally in the intestine or the presence of many worms; and 4) an increased requirement for vitamin B_{12}. Moreover, malabsorption of folates caused by the worm and decrease in vitamin B_{12} absorption due to changes in the intestinal bacterial flora may be considered possible contributory factors (von Bonsdorff, 1977).

Diagnostic Features

As a rule, symptoms are vague and difficult to differentiate from those of some other diseases. Specific diagnosis is usually based on recovery and identification of the characteristic eggs or proglottids in fecal matter.

Eggs are usually numerous and readily demonstrated in a direct fecal smear. However, sedimentation with Formalin–ether for concentrating eggs must be recommended, because few eggs are produced at the beginning of the patent period in the egg-discharging periodicity (Kamo et al., 1986).

Prognosis

In patients having no predisposition to pernicious anemia, the prognosis is good. The tapeworm is easily expelled with the aid of adequate vermifuges. This expulsion can usually be accomplished without danger even in patients with fairly low blood values indicative of anemia; thereafter, the anemia must be treated by modern methods.

Epidemiologic Aspects

Geographic Distribution

Human infection with *D. latum* occurs in various parts of the world. However, it is most prevalent in the subarctic and temperate regions of the Northern Hemisphere, especially in the area from the Siberian rivers Yenisei and Ob to the Baltic Sea of the Eurasian continent. A high prevalence of infection is reported in northern parts of the European Soviet Union, the Volga Basin, and Finland. In ancient times diphyllobothriasis was more abundant in other countries of northern Europe; autochthonous foci of infection existed previously, and to a certain extent persist, in the lake districts of northern Italy and western Switzerland as well as in areas surrounding the river Danube and its tributaries. After transfer by immigrants from infested areas, the parasite established new endemic centers, especially in North and South America. Sporadic, apparently imported cases have been recorded in many other countries, including the British Isles, Australia, the Federal Republic of Madagascar, Argentina, and Venezuela (von Bonsdorff and Bylund, 1982). Infection with *Diphyllobothrium dalliae* is common in western Alaska (Rausch and Hilliard, 1970). *D. pacificum* is epidemic in Peru and Chile. Human infection with *D. nihonkaiense* is fairly common in Japan, especially along the coast of the Sea of Japan.

Source of Infection and Mode of Transmission

Human infection results from eating raw fish or raw fish products such as roe or liver. In different geographic regions the infection is attributable to different dishes and dietary habits, which are usually local or regional. In the Finnish lake district slightly salted filets of various freshwater fishes are customarily eaten without further preparation. Slightly salted hard roe of pike and burbot is also regarded as a delicacy at home as well as in restaurants in Finland. These dietary habits probably also apply to endemic regions of the Soviet Union. In some regions of Northern Europe liver of burbot is often eaten, especially by the fishing population (von Bonsdorff and Bylund, 1982). In the Jewish population of both Europe and the United States, gefilte fish constitutes a unique mode of transmission. Housewives and children usually become infected from tasting the dish before cooking.

Numerous species of freshwater fish may serve as second intermediate hosts of *D. latum*, and the plerocercoids in small intermediate hosts usually accumulate in large predatory fish as paratenic hosts. The most important transmitters of infection to humans are pike (*Essox lucius*), perch (*Perca fluviatilis*), burbot (*Lota lota*), and ruff (*Acerina cernua*) in the Eurasian continent; northern pike (*Essox lucius*), walleyed or blue pike (*Stizostedion vitreum*), sand pike or sauger (*Stizostedion canadense*), burbot, and yellow perch (*Perca flavescens*) in North America; and rainbow trout (*Salmo gairdneri*) in southern Chile. Blackfish (*Dallia pectoralis*) plays a role in transmission of *D. dalliae* to Alaskan Eskimos.

Oncorhynchus species and anadromous fishes are also important in transmission to humans, for example, masu salmon (*Oncorhynchus masou*) and pink salmon (*Oncorhynchus gorbusha*) for *D. nihonkaiense* in Japan. Their roe and liver are not dangerous. Some marine fishes are suspected of transmitting several species, especially *D. pacificum*, in Peru and Chile.

Etiologic Agents

Characterization

The genus *Diphyllobothrium* is related to medically important genera, *Spirometra* and *Diplogonoporus*, in *Diphyllobothriidae*. *Diphyllobothrium* is characterized by having one set of reproductive organs in each proglottid, an external seminal vesicle, and parallel or rosette-shaped uterine loops. By contrast, *Diplogonoporus* has two sets of reproductive organs in each proglottid, and *Spirometra* has a partly embedded seminal vesicle in the wall of a cirrus pouch and a simple spiral form of uterus.

Species Parasitic in Humans

Numerous tapeworm species belonging to the genus *Diphyllobothrium* have been reported to infect humans; many of these species, however, are regarded as identical to *D. latum* or *D. dendriticum*. The following species reported in humans seem to be generally accepted as valid species other than *D. latum*: *D. alascense*, *D. dalliae*, and *D. lanceolatum* in Alaska; *D. dendriticum* in Alaska, Ireland, and Siberia; *D. cordatum* in Greenland; *D. ursi* in Alaska and Canada; *D. pacificum* in Peru, Chile, and Japan; and *D. cameroni*, *D. nihonkaiense*, and *D. yonagoense* in Japan. *Diphyllobothrium latum* is the most important and best-known species, whereas *D. dalliae* in western Alaska (Rausch and Hilliard, 1970), *D. nihonkaiense* in Japan (Yamane et al., 1986), and *D. pacificum* in Peru and Chile (Baer, 1969; Cattan et al., 1977) are regionally important.

Morphology and Life Cycle of *D. latum*

The morphology and life cycle of *D. latum*, one of the largest tapeworms in humans, is shown in Fig. 1. Its strobila is creamy white in color, with a length of about 2 to 15 m and a maximal width of 20 mm, and consists of 3,000 proglottids or more. The lanceolate scolex (lateral view), measuring 2.0 by 1.0 mm on average, has dorsal and ventral depressions called *bothria*, which are quite agile and serve as holdfast and locomotor organs. The slender neck ranges in length from about 4 to 30 mm. The thin body increases slowly in width, reaching its maximum at the end of the middle part of strobila. Each gravid proglottid contains both male and female sex organs, the uterus of which can be seen even in unstained mounts as a brown, central rosette due to the accumulation of numerous ova. The uterine pore opens posteriorly apart from the genital atrium on the midline of the ventral surface, through which eggs are expelled. The ovoid, thin-shelled, operculated egg, measuring 65 by 45 μm on average, is yellowish brown in color. Eggs are periodically discharged in enormous numbers with human feces (about 200,000 eggs per gram of feces). The embryo develops within the shell in 10 to 12 days under favorable conditions (O_2 concentration of 8 to 12 mg/liter at 18 to 20°C). The ciliated oncosphere, the coracidium, hatches when it is exposed to adequate illumination for 30 to 60 s at 50 to 100 lux, in which two wavelengths, 500 to 600 nm (yellow) and 250 nm (ultraviolet), are optimal (Grabiec et al., 1963). The coracidium released from the eggshell is spherical, having a diameter of about 50 μm, and swells gradually to double this size in the water. The free-swimming coracidium swallowed by the first intermediate host, a copepod, within half a day to 5 days or more can continue to develop. Different copepod species are of epidemiologic significance in different countries and continents at different seasons depending on geographic and ecologic factors. After being ingested by the copepod, the coracidium discards its ciliated envelope, penetrates the intestinal wall of the copepod, and enters its body cavity, where a procercoid infectious

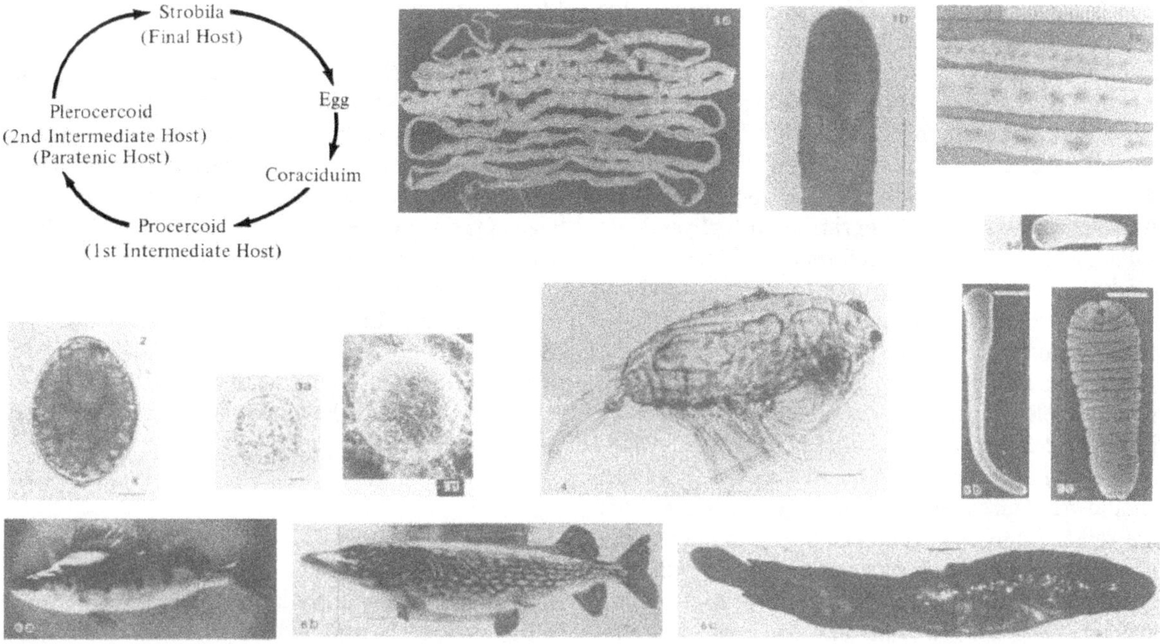

FIG. 1. Life cycle of *Diphyllobothrium latum*. (1) Strobila: (a) Whole body; (b) scolex and bothrium (bar = 1 mm); (c, *top to bottom*) mature, gravid, and terminal gravid proglottids. (2) Egg (bar = 10 μm). (3) Coracidium: (a) Oncosphere (bar = 10 μm); (b) ciliated envelope (scanning electron microscopy). (4) Procercoid in copepod (bar = 10 μm). (5) Plerocercoid (a,b) killed in Formol–saline and (c) examined by scanning electron microscopy (bar = 1 mm). (6) Source of infection: (a) Perch; (b) pike; (c) burbot.

to fish develops within 16 to 18 days at room temperature. The procercoid, including its hooked cercomer, measures 500 to 600 μm. A copepod measuring only 1 to 2 mm may harbor several procercoids. When the infected copepod is consumed by a plankton-eating fish the procercoid sheds the cercoma, penetrates the ventricle or intestinal wall, and migrates to the muscle or visceral organs of this second intermediate host, where it develops into the plerocercoid infectious to the final host.

The glistening, gray-white plerocercoid may reach a maximum length of 4 to 5 cm (average, 6 to 10 mm). The plerocercoids usually lie unencysted in the host tissue, but also occur enclosed in connective tissue cysts. Large predatory fish can be infected with plerocercoids by feeding on smaller fish. In this way they gradually accumulate in large predatory fish, the carriers or paratenic hosts.

On eating the raw or insufficiently cooked flesh of infected fishes, humans, the most suitable final hosts, become infected. During the third week or later, eggs begin to appear in the feces; thereafter eggs are discharged periodically, alternately presenting negative phases and patent periods at irregular intervals (Kamo et al., 1986).

Some domestic animals such as dogs, cats, and pigs have been recorded as natural or experimental hosts. Other species of mammals have been recorded by different authors as hosts of *D. latum*, including wolves, foxes, bears, and seals. In several cases the authors base their reports on worm species of doubtful identity.

Diphyllobothrium nihonkaiense is closely related to *D. latum*, especially in the morphologic characteristic of strobilia. It is clearly distinguished from *D. latum* by various characteristics such as morphology of embryonal hooks and plerocercoids and electrophoretic aspects. Its second intermediate host is quite different; accordingly their geographic distribution is separated. *Diphyllobothrium dalliae* differs from *D. latum* in having smaller strobilae and measures as much as 522 mm long by 9 mm wide. The plerocercoid occurs in a high proportion of blackfish in western Alaska. *Diphyllobothrium pacificum* has distinct morphologic characteristics, which readily distinguish it from other species. The second intermediate hosts are suspected to be some marine fishes commonly used as food by humans and Otariidae.

Specimen Collection and Transport

Eggs of *D. latum* are viable after storage for several months or even a year under refrigeration. If the fecal specimen cannot be examined within this period, small portions can be preserved by emulsification in 10% Formalin, a widely used fixative. This solution is most suitable for concentration techniques to enhance recovery of ova. Adult worms (strobilae) or portions (proglottids) expelled spontaneously or after treatment should be adequately preserved for further examination. The following procedure may be recommended: The living worm is transferred to tap water after rinsing with physiologic saline, a few drops of chloroform are added, and the worm is placed in a refrigerator (temperature, about 4°C) for 2 to 3 h. When removed from the refrigerator, the worm is dead and relaxed. Before fixing the worm, the following morphometric data should be noted: scolex length and dorsoventral height, neck length, total length of worm, and maximum width and length of segments. A solution of 5% Formalin should be poured gently over large specimens that are relaxed and extended. The strobilae are usually stained in Semichon acetic carmine or Ehrlich acid hematoxylin, and sagittal and transverse sections are stained with hematoxylin and eosin or with trichrome staining solution.

Identification

The species of *Diphyllobothrium* are difficult to distinguish, as the strobilae show a wide range of morphologic variation. Accurate identification should be based not only on the morphologic characteristics of strobilae from natural hosts but also on those of strobilae of the same age from experimental hosts. Moreover, the morphologic characteristics of plerocercoids, embryonal hooks, and other structures are very important. Electrophoretic methods (e.g., isoelectric focusing, isozyme characterization) and serologic methods also give some promising results. In general laboratories the following identification standards can be applied practically.

Eggs

Among yellowish-brown, ovoid, operculated eggs from human feces, fairly large (approximately 60 to 70 μm by 40 to 50 μm) and thin-shelled (about 1 μm) eggs with small, shallow, almost invisible (and some-

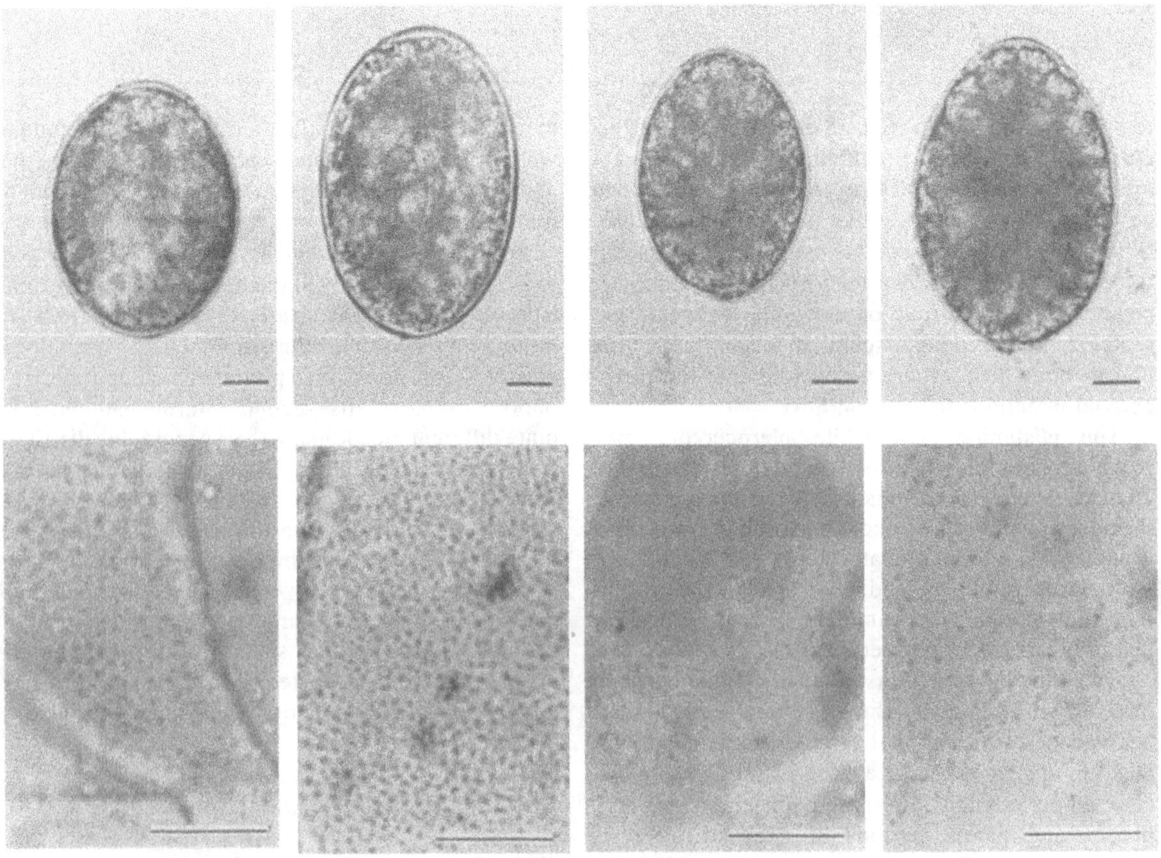

Fig. 2. Eggs showing pits on eggshell surface of (*left to right*) *Diphyllobothrium pacificum*, *Diphyllobothrium dalliae*, *Diphyllobothrium nihonkaiense*, and *Diphyllobothrium latum* (bars = 10 μm).

times lacking) pits on their surface can be identified as *D. latum* in the Eurasian and American continents, whereas they should be identified as *D. nihonkaiense* in Japan. It must be kept in mind that eggs of *D. ursi* and *D. dendriticum* are found occasionally in Alaska and Canada. Relatively small (about 50 to 65 μm by 35 to 45 μm) but relatively thick-shelled (about 1.5 to 2.5 μm) eggs, which show recognizable pits on their surface, can be identified as *D. dalliae* in Alaska and as *D. pacificum* in Peru and Chile. Pits on eggshell surfaces can be observed on fragments of broken eggshells through the immersion lens of a light microscope (Maejima et al., 1983) (Fig. 2).

Strobilae

The standard morphologic characteristics of *D. latum* are as follows: The scolex is lanceolate (rarely ovate or even cordate), with a frontal pit measuring about 2.0 by 1.0 mm (lateral view) on average. The neck ranges in length from about 4 to 30 mm (average, 15 mm). The thin body increases slowly in width, reaching its maximum at the end of the central part of the strobila. In whole-mount preparations of mature proglottids the seminal vesicle is seen from the ventral side, which is situated dorsally and caudally to the cirrus sac. The anterior part of the uterus, containing full-developed eggs, forms four to eight large loops to each side of the midline. Vitellaria and testes are arranged in two distinct lateral fields, which with few exceptions are not connected with each other anterior to the cirrus sac. The ovary varies somewhat in shape, but has no anterior horns. In sagittal sections of mature proglottids the cirrus sac is situated horizontally, with the seminal vesicle situated dorsally and caudally to it. There is a constriction between the segments and often an area without testes and vitellaria. The uterine pore opens far from the genital atrium (Andersen, 1975) (Figs. 3 to 7).

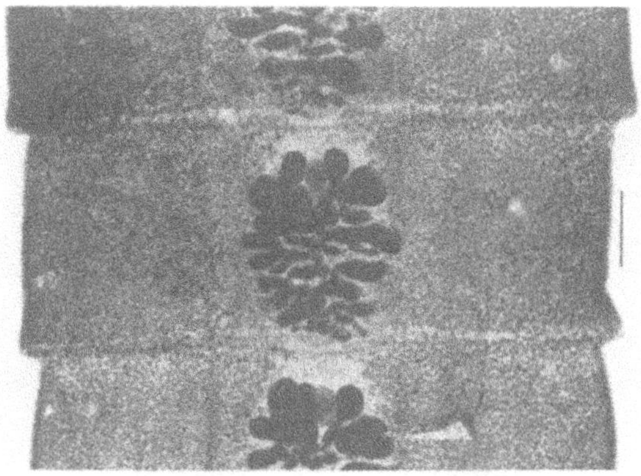

FIG. 4. Gravid proglottids at broadest level (bar = 1.5 mm).

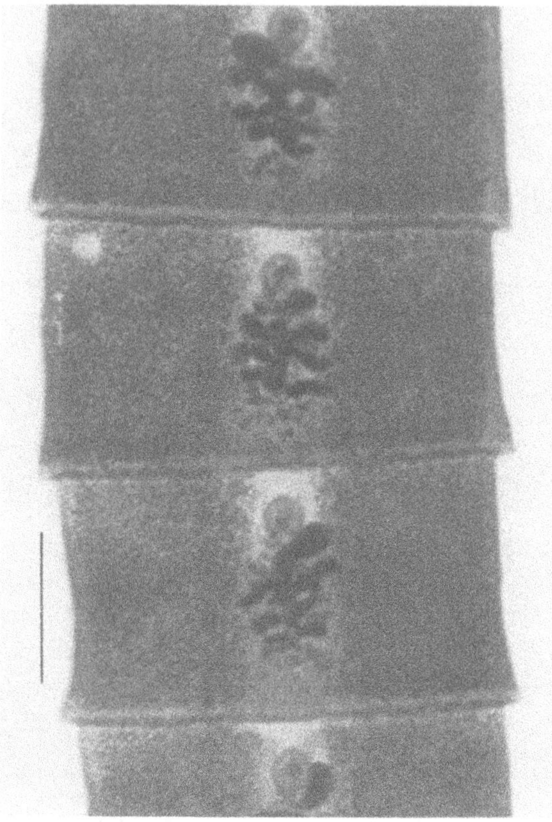

FIG. 3. Early mature proglottids of *Diphyllobothrium latum* (bar = 1.5 mm).

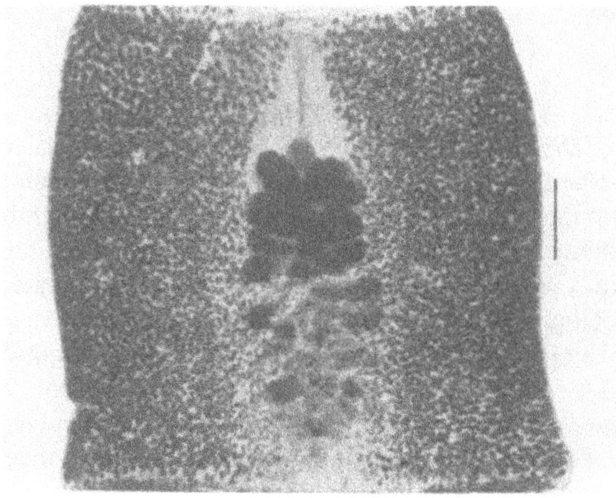

FIG. 5. Terminal gravid proglottid, broad type (bar = 1.5 mm).

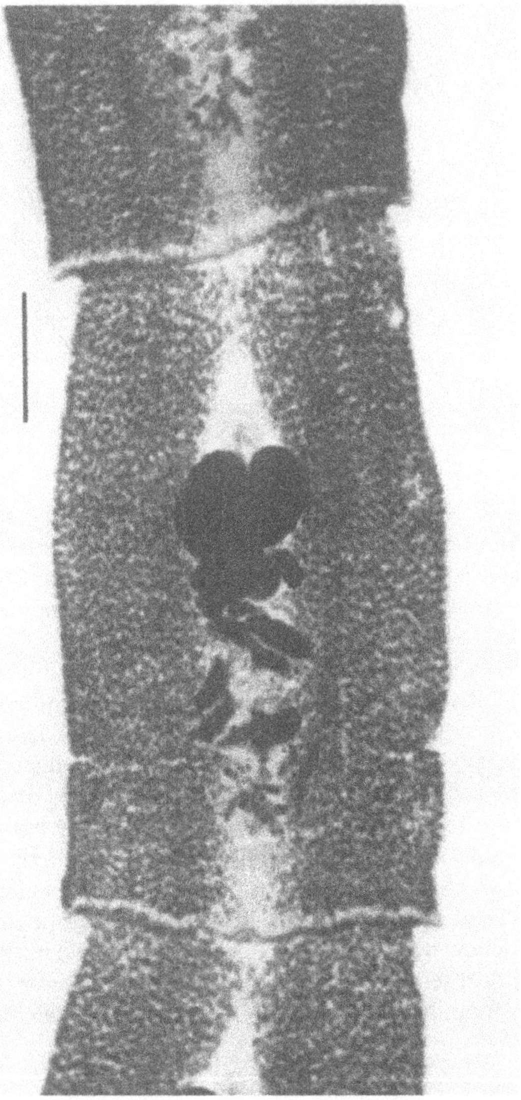

FIG. 6. Terminal gravid proglottid, slender type (bar = 1.5 mm).

Diphyllobothrium nihonkaiense is closely related to and very difficult to distinguish from *D. latum* only on the basis of strobilar morphology. It has a slightly thinner body and a larger scolex. In sagittal sections of mature proglottids the cirrus sac is situated obliquely, and the seminal vesicle is larger relative to the cirrus sac (Yamane et al., 1986) (Fig. 8). *Diphyllobothrium dalliae* differs macroscopically from *D. latum* in having a much less attenuated strobila with relatively few pregravid segments and a wide, lanceolate scolex. In whole-mount preparations, the testes and vitellaria are confluent across as much as the anterior one-quarter of the segment; the ovarian lobes are rounded to reniform and are often confluent posteriorly. In sagittal sections the vagina is strongly reflexed ventrad just ventral to the seminal vesicle, which is posterior to the proximal end of the relatively large cirrus sac (Rausch and Hilliard, 1970). *Diphyllobothrium pacificum* has relatively small strobilae (up to 2 m long by 10 mm wide). It is distinguishable from other species by its leaf-shaped scolex and the longitudinal protuberance with transverse grooves along the midline between the anterior border and the genital opening, which lies almost in the middle of the mature proglottid.

PLEROCERCOIDS

If plerocercoids are obtained from a suspected source of infection, they are quite helpful in identifying the species. When killed in Formol–saline (4% formaldehyde in 1% NaCl), plerocercoids became contracted and wrinkled, and each species presents uniform, characteristic features. Plerocercoids of *D.*

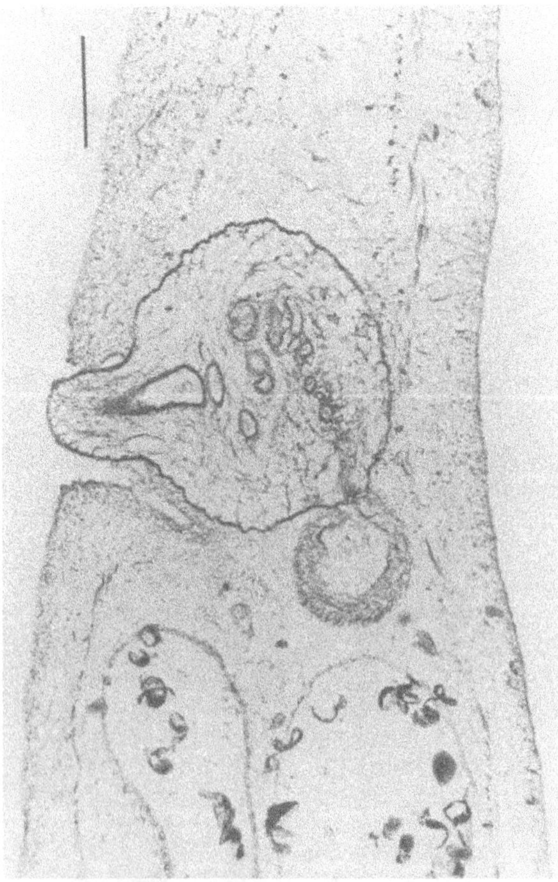

FIG. 7. Sagittal section of mature proglottid of *Diphyllobothrium latum*, showing cirrus sac and seminal vesicle (bar = 0.2 mm).

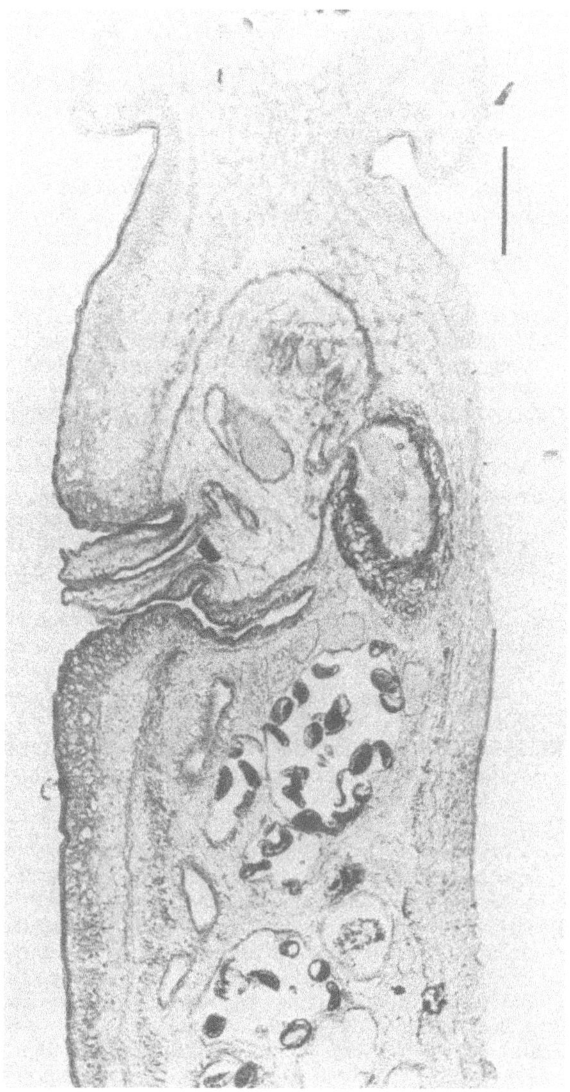

FIG. 8. Saggital section of mature proglottid of *Diphyllobothrium nihonkaiense*, showing cirrus sac and seminal vesicle (bar = 0.2 mm).

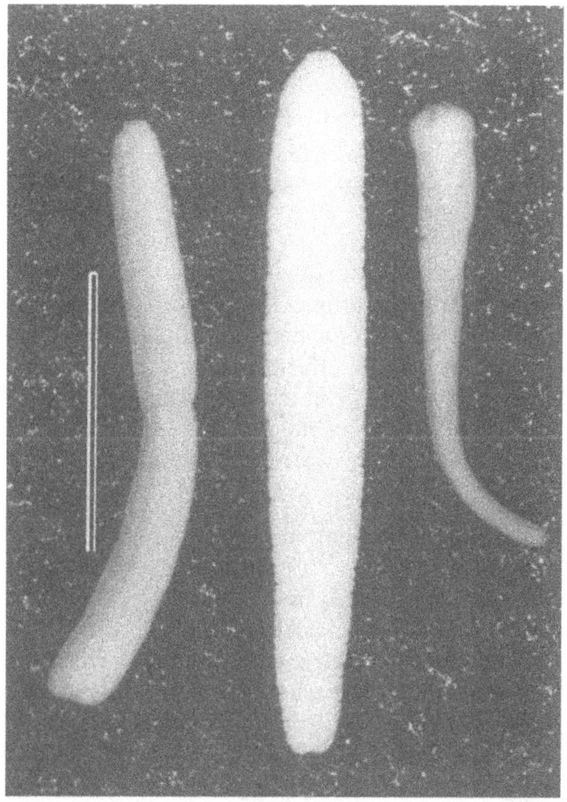

FIG. 9. Plerocercoids of (*left to right*) *Diphyllobothrium dendriticum*, *Diphyllobothrium nihonkaiense*, and *Diphyllobothrium latum* killed in Formol–saline (bar = 5 mm).

latum, for example, present a clubate form, withdrawing their scoleces into the broadest anterior part of the larvae and tapering rapidly toward the posterior end. With those of *D. dendriticum*, in contrast, the scoleces are not fully withdrawn and the body gradually becomes broader toward the broadest posterior end of the larvae. The plerocercoids of *D. nihonkaiense* are somewhat like those of *D. latum*, but the broadest part of the larvae is far behind the scolex, and the body becomes slender more gradually toward the posterior end of the larvae (Halvorsen, 1970; Kuhlow, 1953) (Fig. 9).

Treatment, Prevention, and Control

Treatment

The broad tapeworm is easily expelled with the aid of various vermifuges. To effect a cure, it is necessary that the scolex be expelled. If the scolex is not recovered or seen in the feces after treatment, it is necessary to wait 3 months to ascertain that the patient is no longer passing proglottids or eggs.

It must be borne in mind that the phenomenon termed "egg-discharging periodicity" was found in a *D. latum* carrier, and no eggs were seen in feces for 30 to 140 days or more during the negative phase (Kamo et al., 1986). Niclosamide (Yomesan; Phenasal in the Soviet Union) is given orally in a single dose of 2 g (four tablets) chewed thoroughly after a light meal. No purgatives are required. A few patients experience nausea and abdominal pain. Praziquantel (Biltricide) is as effective as niclosamide against broad tapeworm infection in a single dose of

5 to 10 mg per kilogram of body weight. Unfortunately, use of niclosamide or praziquantel produces a macerated and digested worm, so it is usually impossible to identify the scolex in the material passed after treatment.

In Japan, bithionol (Bithin) is commonly prescribed (30 to 40 mg/kg in two doses at 30-min intervals). It is given after a light breakfast, and a purgative (sodium surfate 30 g with 300 ml of water for adults) is administered 2 h after the second dose. A complete strobila with the scolex is usually expelled. There may be occasional side effects such as mild headache, nausea, or abdominal pain. Paromomycin (Aminocidin) is also effective administered in a single dose of 50 mg/kg in the same way as bithionol, but it is often impossible to identify the scolex because of maceration or partial digestion of the worm.

Prevention and Control

Although any point in the life cycle of the parasite can be theoretically attacked for prevention and control, practically such measures should include the following: (1) against adult worms in the reservoir hosts, treatment of cestode carriers; (2) against viable eggs contaminating the biotope, purification of waste water and sewage disposal; and (3) against transmission of infective larvae from fish to humans, changing dietary habits to avoid eating raw or undercooked fish. Education of the population exposed to infection is important, and the best results have been obtained with visits to people in their homes (Fedorova et al., 1970). By this method, in combination with mass expulsion of the worm, a considerable decrease in the prevalence of tapeworm infection has been attained in many regions in both the Soviet Union and Finland. It must be borne in mind that fish is a valuable and important food, and even fish or hard roe that possibly harbors infective larvae is harmless if it has been cooked, fried, or smoked at a high temperature or deep-frozen for 1 or 2 days.

Literature Cited

Andersen, K. 1975. Studies of the helminth fauna of Norway XXXIV: the morphological stability of *Diphyllobothrium* Cobbold. A comparison of adult *D. dendriticum* (Nitzsch), *D. latum* (L.) and *D. ditremum* (Creplin) developed in different hosts. Norw. J. Zool. **23**:45–53.

Baer, J. G. 1969. *Diphyllobothrium pacificum*, a tapeworm from sea lions endemic in man along the coastal area of Peru. J. Fish. Res. Board Can. **26**:717–723.

Björnkenheim, B. 1966. Optic neuropathy caused by vitamin-B_{12} deficiency in carriers of the fish tapeworm, *Diphyllobothrium latum*. Lancet **1**:688.

Cattan, P. E., A. Atias, B. B. Babero, and D. Torres. 1977. Helmintofauna de Chile V. Primer hallazgo de *Diphyllobothrium pacificum* (Nybelin, 1931) Margolis, 1956, en lobos marinos de la costa Chilena. Rev. Iber. Parasitol. **37**:285–290.

Fedorova, S. P., V. I. Khodakova, and E. P. Razumova. 1970. Investigation of the effectiveness of some methods of sanitary education in foci of diphyllobothriasis in the Volgograd region (in Russian). Med. Parasitol. **39**:453–457.

Grabiec, S., A. Guttowa, and W. Michajłow. 1963. Effect of light stimulus on hatching of coracidia of *Diphyllobothrium latum* (L.). Acta Parasitol. Pol. **11**:229–238.

Halvorsen, O. 1970. Studies of the helminth fauna of Norway XV: on the taxonomy and biology of plerocercoids of *Diphyllobothrium* Cobbold, 1858 (Cestoda, Pseudophyllidea) from north-western Europe. Nytt Mag. Zool. (Oslo) **18**:113–174.

Kamo, H., S. Yazaki, S. Fukumoto, J. Maejima, and H. Kawasaki. 1986. Evidence of egg-discharging periodicity in experimental human infection with *Diphyllobothrium latum* (Linné, 1758) (in Japanese). Jpn. J. Parasitol. **35**:53–57.

Kuhlow, F. 1953. Bau und Differential-diagnose heimischer *Diphyllobothrium* Plerocercoide. Z. Tropenmed. Parasitol. **4**:186–202.

Maejima, J., S. Yazaki, S. Fukumoto, and H. Kamo. 1983. Morphological comparison of eggs between marine species and fresh-water species in Diphyllobothriid cestodes (in Japanese). Jpn. J. Parasitol. **32**:27–42.

Rausch, R. L., and D. K. Hilliard. 1970. Studies on the helminth fauna of Alaska. XLIX. The occurrence of *Diphyllobothrium latum* (Linnaeus, 1758) (Cestoda: Diphyllobothriidae) in Alaska, with notes on other species. Can. J. Zool. **48**:1201–1219.

Saarni, M., W. Nyberg, R. Gräsbeck, and B. von Bonsdorff. 1963. Symptoms in carriers of *Diphyllobothrium latum* and in non-infected controls. Acta. Med. Scand. **173**:147–154.

von Bonsdorff, B. 1947. In which part of the intestinal canal is the fish tapeworm found? Acta Med. Scand. **129**:142–155.

von Bonsdorff, B. 1977. Diphyllobothriasis in man. Academic Press, Inc., New York.

von Bonsdorff, B., and G. Bylund. 1982. The ecology of *Diphyllobothrium latum*. Ecol. Dis. **1**:21–26.

Yamane, Y., H. Kamo, G. Bylund, and B.-J. Wikgren. 1986. *Diphyllobothrium nihonkaiense* sp. nov. (Cestoda: Diphyllobothriidae)—revised identification of Japanese broad tapeworm. Shimane J. Med. Sci. **10**:29–48.

CHAPTER 87

Dracunculiasis

DONALD R. HOPKINS

Disease: Dracunculiasis (guinea worm disease, dracontiasis, dracunculosis).
Etiologic Agent: *Dracunculus medinensis.*
Source: Drinking water obtained from stagnant freshwater ponds or step wells containing infected copepods.
Clinical Manifestations: A painful blister ruptures to reveal the long white worm, which emerges through the skin over several days or weeks. Arthritis, arthralgias, abscesses, cellulitis, and tetanus may be associated with the worm's emergence or secondary infection of the local ulcer it causes.
Pathology: Local secondary infection. Adult worms that do not emerge may become calcified in subcutaneous tissues or muscle. Aberrant worms may damage the central nervous system, genitourinary system, or other organs.
Laboratory Diagnosis: Usually not necessary. First-stage larvae may be demonstrated microscopically in fluid ejected by the adult worm.
Epidemiology: Remaining endemic areas include mainly India and Pakistan and a band of sub-Saharan African countries extending from Senegal to Ethiopia. Sporadic, rural distribution. Working-age adults and schoolchildren most affected. Seasonal, often coinciding with planting or harvest periods of peak demand for agricultural labor.
Treatment: Traditionally, by rolling the emerging worm slowly around a small stick. Aspirin and tetanus toxoid. Mebendazole or niridazole may reduce symptoms or aid emergence of the worm marginally.
Prevention and Control: Provision of clean drinking water, education of affected populations to filter or boil their water and not to enter drinking water sources while the worm is emerging, and vector control by chemically treating water sources with temephos are effective. A campaign is under way to eradicate dracunculiasis by exploiting the International Drinking Water Supply and Sanitation Decade (1981–1990).

Description of Disease

Dracunculiasis (guinea worm disease, dracontiasis, dracunculosis) is a painful condition caused by infection with the filarial parasite *Dracunculus medinensis*. Humans are infected by drinking contaminated water, usually from stagnant sources such as ponds or stepwells. There is no animal reservoir of *D. medinensis,* although in North America particularly a closely related parasite that does not infect humans, *D. insignis,* is commonly found in skunks, foxes, raccoons, and other wild animals (Muller, 1971). Because of its seasonality, high prevalence, and predilection for older children and adults and the duration of its painful manifestations, dracunculiasis has severe adverse social and economic effects on the health, agricultural production, and education of affected populations.

History

Dracunculiasis is an ancient disease, thought by some to have been the "fiery serpent" that plagued the Israelites on the shores of the Red Sea. In addition to ancient Egypt, the disease was known in classic Greece and Rome. It was also described clearly by the medieval Arabic physician Avicenna. The age-old practice of rolling the slowly emerging worm around a small stick may well be the origin of the staff of Aesculapius, symbol of the healing arts.

The description of the role of the cyclops intermediate host for this worm in the 1870s by a Russian scientist, Aleksei Fedchenko, was the first time such an intermediate host was recognized to be necessary for a parasite of humans.

Life Cycle

Dracunculiasis is the only infectious disease that is *only* transmitted to humans by drinking contaminated water. When stagnant water containing infected cyclops or copepod intermediate vectors is drunk by humans, gastric secretions kill the microscopic crustaceans, allowing the third-stage larvae to escape and penetrate the stomach or small intestine. The male and female worms apparently mate in the thorax or abdomen after 3 or 4 months. Thereafter, the male worms die and are resorbed, but the females continue to grow and migrate towards the lower extremities (Muller, 1971).

About a year after the larvae are ingested, the mature female worms, which are only 1 to 2 mm wide but up to 1 m (2 to 3 feet) long, are packed with first-stage larvae and ready to emerge. The worm produces a painful blister on the skin at the chosen site, which ruptures when that part of the body comes in contact with water. When immersed in water, the worm's uterus contracts to expel hundreds of thousands of first-stage larvae. Larvae must be ingested by an appropriate copepod intermediate vector, usually a species of *Cyclops,* for development after almost 2 weeks into third-stage larvae, which are infective for humans. The larvae are too small to be seen with the naked eye, but copepods are visible as tiny darting specks if water containing them is held up to a light.

Clinical and Pathologic Manifestations

The infection remains entirely asymptomatic in humans until the worm is about to emerge. General signs and symptoms that may occur around that time include generalized urticaria, pruritis, nausea, and occasionally dyspnea (Gilles, 1985). The appearance of such manifestations and their severity, if present, are highly variable. Eosinophilia is not usually seen in dracunculiasis, but may appear during the early stages of infection. Sometimes the adult worm presents as a serpiginous cord that can be seen and felt just beneath the skin. Other worms provoke formation of a large abscess on the leg, arm, or elsewhere and are discovered at its center if the abscess is incised and drained.

By far the most common mode of appearance begins with the raising of a painful blister on the skin around the head of the worm that is about to emerge. This blister is the result of a local inflammatory reaction caused by an irritant released by the worm. When the affected part of the body is immersed or doused in water, the blister ruptures, and the worm ejects a milky fluid containing thousands of larvae. This dramatic debut is accompanied by relief of the pain occasioned by the blister, and by formation of a clean shallow ulcer at the base of the former blister, through the center of which the white body of the worm protrudes. Larvae are released in diminishing numbers each time the protruding worm comes in contact with water over the next several days. The body of the worm itself may require up to 2 or 3 weeks to fully emerge, as usually only 1 to 2 cm comes out each day.

Histologic sections through tissues surrounding an emerging worm reveal a localized inflammatory infiltrate of neutrophils, monocytes, and eosinophils. It is probably this reaction that creates resistance to forcible rapid removal of the worms once they have broken the skin. Intact adult worms that are palpated beneath the skin are surrounded by no such reaction, and can be removed easily through a small incision. Adult worms that do not emerge and that are not removed from the body may become encysted and sometimes calcify. The latter occurrence is manifest by bizarre but characteristic radiologic findings (Reddy et al., 1968).

The worms usually emerge from the foot, ankle, or lower leg, and most affected patients have only one worm emerge in a "guinea worm season." However, worms have been known to emerge from other parts of the body, such as the head, neck, chest, scrotum, vulva, arm, hip, and the like (Muller, 1971). They have also caused paraplegia by invading the spinal cord (Gilles, 1985). Rare reports of adult worms emerging from very young infants suggest that some may even penetrate the uterus. Although most patients experience only one worm emerging at a time, 2 to 5 worms emerging at the same time is not uncommon, and up to 24 or more may emerge simultaneously in severe infections.

No effective immunity develops, so persons in endemic communities may be infected repeatedly, year after year, if they are reexposed. However, each year's ingested larvae either emerge or die in the body during the following year; the worms do not survive more than 1 year in humans.

The shallow ulcer surrounding an emerging worm often heals spontaneously about 4 to 6 weeks after the worm appears. Frequently, however, such ulcers become secondarily infected, resulting in local inflammation, cellulitis, deep abscesses, or worse. Tetanus is a not infrequent secondary complication, and is the main reason for occasional fatalities from this disease (Hopkins, 1983). With or without secondary infection, inflammation and pain of nearby major joints is a common consequence of dracunculiasis infections, and may persist for weeks or months (McLaughlin et al., 1984; El Garf, 1985). In less than 1% of cases, an affected joint, which may or may not have been directly invaded by one of the worms, may become permanently frozen.

Diagnosis

At the stage of infection when the adult worm is emerging through the skin, this disease is not likely to be mistaken for any other. In endemic areas, experienced villagers even recognize the prodromal signs as the worm prepares to emerge; they also know that plunging the painful blister into water will cause it to rupture and thereby relieve the pain.

After the worm has emerged, the residual cutaneous ulcer can be distinguished from other tropical ulcers only by history. While the worm is still emerging, fluid obtained from the worm can be inspected microscopically for signs of the numerous larvae, if such confirmation is deemed necessary. The radiologic picture of one or more calcified worms in tissues is also highly specific evidence of prior infection. Color photographs of a typical lesion with the emerging worm, so-called "recognition cards," are helpful aids to diagnosis for inexperienced persons.

A past history of worms emerging through the skin, residence in an endemic area, or x-ray evidence of infection may be the only clue to the true origin of a rare patient's neurologic ailment, the diagnosis of which can only be presumptive until the worm is seen at surgery.

Epidemiology

The sources of infection for humans are stagnant pools of surface water such as are found in ponds, step wells, or residual depressions in river beds during the dry season. The copepod intermediate vectors do not thrive in flowing rivers or springs, which thus are not associated with the disease.

Geographic Distribution

Formerly, dracunculiasis was widespread throughout much of Africa, the Middle East, and southern Asia, including parts of the Soviet Union, Persia (Iran), and the Indian subcontinent. It was imported into the Western Hemisphere with African slaves, but died out. It was eliminated from the southern Soviet Union in the 1930s by a vigorous control program. Provision of clean drinking water has apparently gradually eliminated the infection from most of the Middle East, except for a small focus reported in Yemen and an adjacent area of Saudi Arabia (World Health Organization, 1982). Iran reportedly eliminated the infection in the 1970s. The major remaining endemic foci are six states in Western India, parts of Pakistan, and a band of 19 countries above the equator in sub-Saharan Africa: Benin, Burkina Faso, Cameroon, Central African Republic, Chad, Côte d'Ivoire, Ethiopia, Gambia, Ghana, Guinea, Kenya, Mali, Mauritania, Niger, Nigeria, Senegal, Sudan, Togo, and Uganda (World Health Organization, 1986).

Within affected countries, transmission of the disease is entirely rural, as most urban populations have water that is not grossly contaminated. Distribution of the disease in the endemic countryside is decidedly focal, with heavily infected villages and unaffected villages sometimes adjoining one another. Where the disease is reported to medical authorities at all, there are usually at least 20 to 100 actual cases for each one that is officially recorded.

Populations Affected

The most heavily affected age group is usually 15- to 50-year-olds, the most physically active part of the population, apparently because they tend to drink more water. Infants and older persons are rarely affected, but schoolchildren suffer considerably (Kale, 1977; Nwosu et al., 1982). Depending on the area studied, men are often found to be more heavily parasitized than women, but the reverse is also commonly true. Farmers in endemic areas tend to be particularly affected, perhaps because their work causes them to drink heavily from surface sources of water (Belcher et al., 1975).

Seasonality

In drier areas such as the Sahelian zone of Africa, dracunculiasis is transmitted, and hence obvious disease is most prevalent during the short rainy season when surface sources of drinking water are available. At other times, populations in such areas must necessarily rely on deep wells, rivers, or some other water source that cannot transmit the infections, either because it is physically inaccessible to contamination or because it will not sustain the requisite intermediate vector (Hopkins, 1983).

In areas that experience an abundant rainy season and a distinct dry season, such as countries nearer the equator in Africa, the disease is transmitted during the dry season, when surface water sources are still present but much more polluted than during the rainy season. Because the incubation period in humans averages exactly 12 months, the parasites' cycle remains perfectly synchronized with the local ecology. Sometimes the disease is transmitted year-round.

Social and Economic Impact

The duration of incapacity from guinea worm disease, the often high prevalence rates of infection in older children and adults, and its appearance at the time of year when demand for labor is greatest in agricultural communities combine to magnify the impact of this infection. In one study, the average duration of incapacitation from dracunculiasis was 5 weeks; in another, it was 100 days (Belcher et al, 1975; Kale, 1977). A recent review of school absenteeism in an area of southwestern Nigeria found that whereas primary school children who were absent for all other reasons were not in attendance for an average of 1 week during the school year, pupils who were absent because of guinea worm disease were out of school for an average of 7 to 8 weeks (Ilegbodu et al., 1986).

In India, annual incidence rates in affected villages tend to range up to about 20%, rarely more. Affected villages in Africa, on the other hand, often manifest an incidence of more than 30%, with some areas having many villages with incidence rates greater than 50% (Hopkins, 1984). Nwosu and colleagues (1982) found that in an area of eastern Nigeria, rates of school absenteeism rose from the usual rate of about 13% to a peak of more than 60% during the "guinea worm season." For farmers, such an incapacitating illness would be unwelcome at any time but, depending on the local ecology, its peak prevalence occurs during the planting season, at the beginning of the rains, or at harvest time just after the rains end.

Prevention and Control

Treatment

The ancient practice of winding the emerging worm around a small stick is still widely practiced, as modern medicine can offer little better treatment once a person is infected. Aspirin can help relieve local pain, antibiotics can cure secondary infections, and immunization with tetanus toxoid can prevent a potentially fatal complication. Mebendazole and niridazole have been reported to ease symptoms and speed emergence of the worm marginally (Gilles, 1985). No drug has been found to be suitable for mass treatment of victims. Preventing the infection is a far more effective control measure.

Water Supply

As the infection is only transmitted by drinking water, and the parasite does not persist beyond 1 year in humans if exposure ceases, it can be prevented altogether by providing clean drinking water. Usually this means substituting a piped water supply, draw well, or drilled borehole well as the source of drinking water and educating affected populations to use them instead of the former sources, which may still be preferred for taste, convenience, or other reasons. In Nigeria, providing piped drinking water to a town of 30,000 persons in the 1960s reduced the incidence of dracunculiasis from more than 60% to zero in 2 years (Muller, 1971). Providing safe drinking water obviously yields other benefits in addition to preventing dracunculiasis.

Health Education

Where safe sources of drinking water cannot be provided immediately, villagers can be taught to prevent the infection by boiling drinking water beforehand if sufficient fuel is available, by filtering water through a cloth or a recently developed monofilament nylon material, and by not entering any source of drinking water while they have a prodromal blister or emerging guinea worm on any part of their body. Health education alone has been shown to reduce the incidence of dracunculiasis by more than 80% in 1 year in several areas (Hopkins, 1984).

Vector Control

Another effective temporary control measure is to treat all drinking water sources with a chemical pesticide, temephos (Abate), to provide a concentration of 1 ppm at 6-week intervals, beginning just before and extending through the transmission season. Temephos is safe, colorless, odorless and tasteless in the recommended concentration, and very effective. It has been used to reduce the incidence of guinea worm infection by more than 85% in 1 year (Hopkins, 1984). It too is suitable for reducing or eliminating dracunculiasis in affected areas even before permanent sources of safe drinking water can be provided.

Eradication

The advent of the International Drinking Water Supply and Sanitation Decade (1981 to 1990) provides an unprecedented opportunity to eradicate dracunculiasis, as one of the main goals of the Decade is to provide safe drinking water to all unserved populations. Combined with the other control measures just described, and active surveillance to detect virtually all cases of the disease, India began a national Guinea Worm Eradication program in 1980. Since then, Pakistan and at least eight African countries have begun or planned national programs against the disease. In May 1986, the World Health Assembly adopted a resolution calling for the elimination of dracunculiasis, the first such resolution since the successful Smallpox Eradication Program.

Literature Cited

Belcher, B. W., F. K. Wurapa, W. B. Ward, and I. M. Lourie. 1975. Guinea worm in Southern Ghana: its epidemiology and impact on agricultural productivity. Am. J. Trop. Med. Hyg. **24**:243–249.

El Garf, A. 1985. Parasitic rheumatism: rheumatic manifestations associated with calcified guinea worm. J. Rheumatol. **12**:976–979.

Gilles, H. M. 1985. Clinical and therapeutic aspects of *Dracunculus medinensis*, p. 19–22. *In* Contributed papers to workshop on opportunities for control of dracunculiasis. National Academy Press, Washington, D.C.

Hopkins, D. R. 1983. Dracunculiasis: an eradicable scourge. Epidemiol. Rev. **5**:208–219.

Hopkins, D. R. 1984. Eradication of dracunculiasis, p. 93–114. *In* P. G. Bourne (ed.), Water and sanitation. Academic Press, Inc., New York.

Ilegbodu, V. A., O. O. Kale, R. A. Wise, J. H. Christensen, B. L. Steele, Jr., and L. A. Chambers. 1986. Impact of guinea worm disease on children in Nigeria. Am. J. Trop. Med. Hyg. **35**:962–964.

Kale, O. O. 1977. The clinico-epidemiological profile of guinea worm in the Ibadan district of Nigeria. Am. J. Trop. Med. Hyg. **26**:208–214.

McLaughlin, G. E., P. D. Utsinger, W. F. Trakat, D. Resnick, and R. A. Moidel. 1984. Rheumatic syndromes secondary to guinea worm infestation. Arthritis Rheum. **27**:694–697.

Muller, R. 1971. *Dracunculus* and dracunculiasis, p. 73–151. *In* B. Dawes (ed.), Advances in parasitology, vol. 9. Academic Press, Inc., New York.

Nwosu, A. B. C., E. O. Ifezulike, and A. O. Anya. 1982. Endemic dracontiasis in Anambra State of Nigeria: geographical distribution, clinical features, epidemiology and socioeconomic impact of the disease. Ann. Trop. Med. Parasitol. **76**:187–200.

Reddy, C. R. R. M., M. D. Sivaprasad, G. Parvathi, and P. S. Chari, 1968. Calcified guinea worm: clinical, radiological and pathological study. Ann. Trop. Med. Parasitol. **62**:399–406.

World Health Organization. 1982. Dracunculiasis surveillance. Weekly Epidemiol. Rec. **57**:65–67.

World Health Organization. 1986. Dracunculiasis: global surveillance summary—1985. Weekly Epidemiol. Rec. **61**:29–32.

CHAPTER 88

Echinococcosis (Hydatid Disease)

PETER M. SCHANTZ

Diseases: Hydatid diseases caused by the larval stages of *Echinococcus* spp. Cystic hydatid disease (*Echinococcus granulosus*); alveolar hydatid disease (*E. multilocularis*); polycystic hydatid disease (*E. vogeli*).

Etiologic Agents: *Echinococcus granulosus; Echinococcus multilocularis; Echinococcus vogeli.*

Source: Ingestion of *Echinococcus* eggs by direct handling of dogs or in foods and fomites contaminated with feces of infected dogs or other canid or felid definitive hosts.

Clinical Manifestations: Signs and symptoms are produced by slow-growing cyst(s) and vary according to the size, number, and location of the cysts.

Pathology: Cysts of *E. granulosus* are unilocular, have a laminated, noncellular wall, and are usually surrounded by a fibrous pericyst of host origin. Larvae of *E. multilocularis* are not circumscribed by a cyst wall; instead, they grow by exogenous proliferation of germinative membrane, causing microvesicles that may metastasize from the liver to the brain or lungs. Cysts of *E. vogeli* proliferate externally to form new cysts and internally to form septae that divide the cavity into numerous microcysts.

Laboratory Diagnosis: Demonstration of specific antibodies in patient serum; demonstration of parasitic elements (protoscolices, laminated membrane, hooklets) in material obtained at surgery or biopsy.

Epidemiology: Worldwide. *Echinococcus granulosus* occurs in dog–sheep (cattle, horse, camel, pig) cycles. Highest prevalence in Mediterranean countries, southern South America, Australia, and Central Asia. *Echinococcus multilocularis* occurs in fox–rodent cycles in the Northern Hemisphere; cats and dogs also susceptible. *Echinococcus vogeli* occurs in wild canid (dog)–rodent cycles in Central and South America.

Treatment: Surgical removal of larval cysts; inoperable cases may respond to benzimidazole drugs.

Prevention and Control: Prevent dogs from becoming infected by controlling access to viscera of sheep and other livestock; strict controls in slaughter of herbivorous animals; periodic treatment of high-risk dogs with praziquantel; health education; strict hygienic precautions when handling foxes or their furs.

Description of Disease

The hydatid diseases are caused by the cystic larval (metacestode) stages of cestodes of the genus *Echinococcus*. These parasites have life cycles involving two hosts. The adult tapeworms are found in the intestines of carnivores, the larval forms in the viscera of a variety of mammalian intermediate hosts. Transmission is often accomplished by the predator–prey relationship existing between the definitive and intermediate hosts. Humans and intermediate hosts become infected by ingesting eggs passed in the feces of definitive hosts. Four species are recognized: *E. granulosus* (Batsch, 1786); *E. multilocularis* Leuck-

TABLE 1. Animal hosts and diseases caused by *Echinococcus* species

Characteristic	E. granulosus	E. multilocularis	E. vogeli	E. oligarthrus
Definitive hosts	Dogs and other canids	Foxes, dogs, other canids, and domestic cats	Wild and domestic canids	Wild felids
Intermediate hosts	Sheep, cattle, horses, pigs, and other domestic and wild ungulates	Rodents	Rodents	Rodents
Human disease	Cystic hydatid disease	Alveolar hydatid disease	Polycystic hydatid disease	Unknown

art, 1862; *E. oligarthrus* (Diesing, 1862), and *E. vogeli* Rausch and Bernstein, 1972. The definitive and intermediate hosts as well as the characteristics of the disease produced in humans vary for each species (Table 1). *Echinococcus granulosus*, the cause of cystic hydatid disease, occurs worldwide and is of considerable public health and economic importance in some countries; *E. multilocularis* and *E. vogeli* cause the alveolar and polycystic forms of hydatid disease, respectively. Limited to life cycles involving wild animals, the geographic distributions of the latter species are more limited than that of *E. granulosus*. *Echinococcus oligarthrus* occurs in wild felids and rodents in Central and South America and has not yet been reported to cause disease in humans.

Adult-stage *Echinococcus* species are small tapeworms measuring 2 to 11 mm in length (Rausch and Bernstein, 1972; Schantz, 1982). The strobila consists of a scolex (with four suckers and a double row of rostellar hooks) and two to five proglottids (Fig. 1). Morphologic characteristics that, in combination, can be used for taxonomic discrimination within the genus *Echinococcus* include the form of the strobila, position of the genital pore in the mature and gravid proglottids, size of rostellar hooks, number and distribution of testes, and form of the gravid uterus (Table 2). Ovoid eggs (diameter, 30 to 36 μm) containing single, fully differentiated embryos (oncospheres) are shed with the feces of infected definitive hosts (Fig. 2). When ingested by a suitable intermediate host, digestive processes and other factors in the host gut result in hatching and release of the oncospheres. The activated oncospheres penetrate the intestinal mucosa, enter venous and lymphatic vessels, and are distributed to other anatomic sites. Most larvae develop in the liver, but some may reach the lungs or other organs. Development at the elective site includes degeneration of the oncospheral tissue, vesiculation, and production of protoscolices. Mechanisms of growth and proliferation and other characteristics of the metacestodes are unique for each species of *Echinococcus* and will be described.

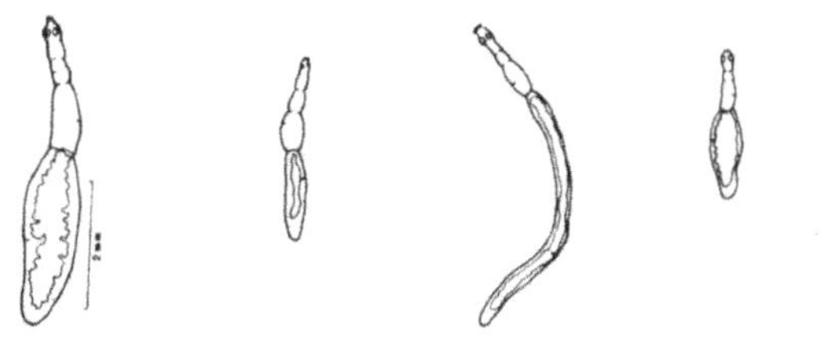

FIG. 1. Comparative morphology of adult-stage *Echinococcus* species. (Adapted from Rausch and Bernstein [1972].)

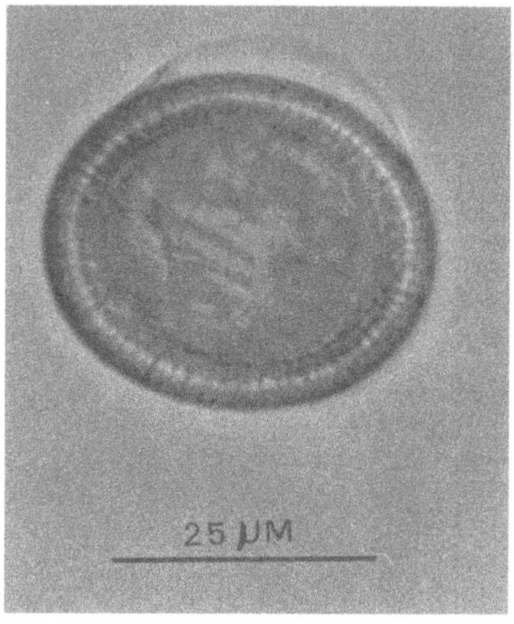

FIG. 2. Taeniid tapeworm (*Echinococcus* spp.) egg freshly shed from proglottid with part of outer egg membrane intact. Several of the three pairs of hooklets of the oncosphere are evident.

TABLE 2. Comparative morphologic characteristics of adult *Echinococcus* species[a]

Taxonomic criteria	E. granulosus	E. vogeli	E. multilocularis	E. oligarthus
Length of strobila (mm)	2.0–7.0	4.0–11.0	1.2–3.7	1.9–2.9
Length of rostellar hooks (μm)				
Large (average)	31–49 (37–42)	49–57 (53)	28–34 (31)	43–60 (52)
Small (average)	22–39 (29–34)	30–47 (43)	23–31 (27)	28–45 (39)
Number of proglottids (range)	3–6	2–3	2–6	2–3
Number of testes (average)	25–80 (32–68)	50–67 (56)	16–35 (18–26)	15–46 (29)
Distribution of testes, anterior/posterior genital pore	Equally anterior/posterior	Majority anterior	Majority posterior	Majority posterior
Position of genital pore relative to middle of segment	Near posterior	Posterior	Anterior	Anterior
Mature proglottid	Penultimate	Penultimate	Antepenultimate	Antepenultimate
Form of uterus sacculations	Lateral tubular	Long	Saclike	Saclike
Ratio of anterior part of strobila to gravid segment	1:0.9–1.3	1:1.9–3.0	1:0.3–0.8	1:0.9–1.1

[a] Adapted from Rausch and Bernstein (1972) and Eckert et al. (1981).

Cystic Hydatid Disease (*Echinococcus granulosus*)

Life Cycle and Geographic Distribution

The life cycle of *E. granulosus* involves dogs and other canids as definitive hosts and domestic and wild ungulates as intermediate hosts (Fig. 3). In quantitative terms, the sheep is the most important intermediate host; however, swine, cattle, goats, horses, camels, and other species assume dominant importance in certain regions. The adaptability of *E. granulosus* to a wide variety of host species has made possible the present broad geographic distribution from as far north as the Arctic circle to as far south as Argentina and New Zealand (Fig. 4). The occurrence of the parasite in a particular host assem-

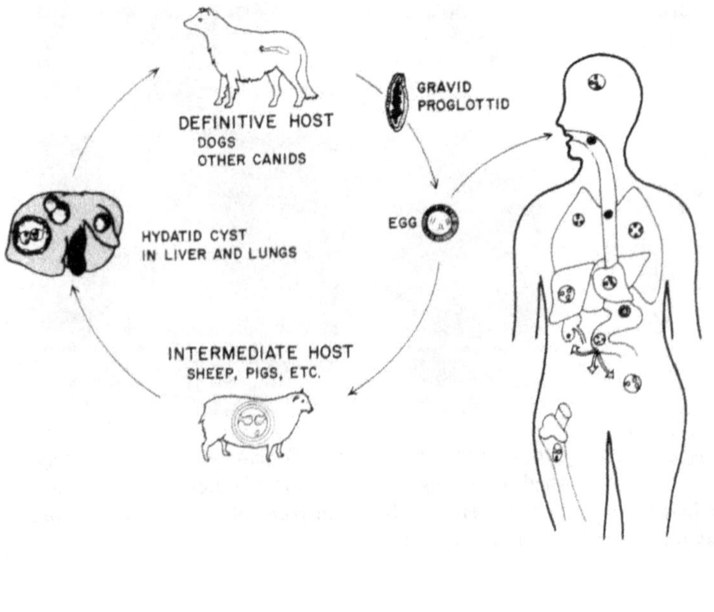

FIG. 3. *Echinococcus granulosus* life-cycle diagram and mode of infection for human beings.

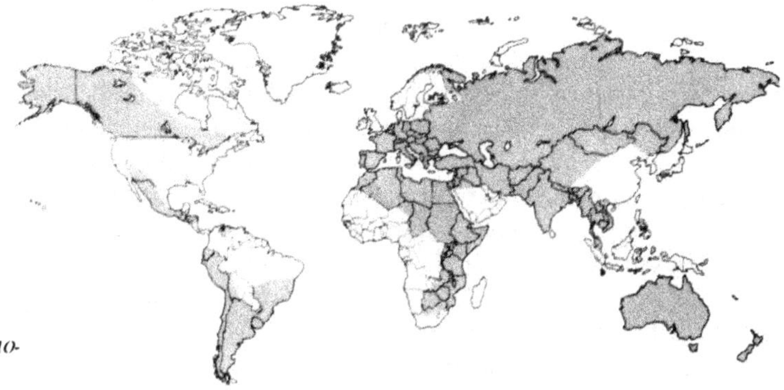

FIG. 4. Geographic distribution of *Echinococcus granulosus* infection.

blage (e.g., wolf–moose; dog–sheep; dog–horse) reflects a degree of host-parasite adaptation that may vary from one cestode population to another. At higher latitudes of North America and Eurasia, the northern sylvatic strain of *E. granulosus* occurs in the wolf, moose, and caribou. Although the confinement of this strain to sylvatic animal intermediate hosts limits the risk of human exposure, high infection rates are reported among some indigenous human populations that hunt or herd these wild or semidomesticated hosts (Wilson et al., 1968). The greatest public health problem is associated with the parasite in cycles that involve dogs and domesticated animals.

Populations of *E. granulosus* in domesticated animals are commonly termed *pastoral strains,* but it has become apparent that important differences exist among populations (Kumaratilake and Thompson, 1982). Strains of *E. granulosus* recovered from sheep in Europe are different from those recovered from horses, and both are distinct from those recovered from cattle (Kumaratilake and Thompson, 1984). Three different strains exist in Australia and are distinguished by developmental, morphologic, and biochemical characteristics (Thompson et al., 1984). Two strains are maintained primarily in dog–sheep cycles, one on the Australian mainland and the other on the island of Tasmania. The third strain is perpetuated in a dingo–macropod marsupial cycle. As yet, no agreement has been reached on the formal taxonomic status of these infraspecific variants, and the mechanisms by which these parasites, often existing sympatrically, are able to maintain their genetic distinctiveness remains unclear.

The greatest prevalence of cystic hydatid disease in humans is found in countries of the temperate zones, including southern South America, the entire Mediterranean coast, the southern and central Soviet Union, central Asia, southern Australia, and parts of Africa (Schantz, 1982). In the United States and many western European countries, most infections are diagnosed in immigrants from endemic countries such as Italy, Greece, the Middle East, and South America.

Epidemiology

The close association of humans with dogs and sheep and the nearly universal habit of feeding dogs with viscera of home-butchered animals create optimal conditions for perpetuation of the tapeworm and transmission to humans. Under such conditions canine infection occurs repeatedly, producing high prevalence rates and environmental contamination with cestode eggs. Infected dogs may harbor thousands of tapeworms in the intestine, with each worm shedding approximately 800 eggs every 2 weeks. Humans become infected through hand-to-mouth contact, particularly in the course of playful and intimate contact between children and dogs.

Infective eggs adhere to hairs around the anus of infected dogs and are also found on the muzzle and paws. Indirect transfer of eggs, either through contaminated water and food or through flies and other arthropod vectors, can also result in human infection (Lawson and Gemmell, 1983). Taeniid eggs are extraordinarily resistant to environmental factors and can survive many months under most climatic conditions.

Pathogenesis and Pathology

After their ingestion, cestode eggs hatch to release embryos in the small intestine. Penetration through the mucosa leads to blood-borne distribution to the liver and other sites, in which development of the hydatid cyst begins. Although most primary infections in humans consist of a single cyst, 20 to 40% of patients have multiple cysts or multiple organ involvement. The liver is the most common site for hydatid cysts (50 to 70%), followed by the lungs (25%) and, less frequently, the spleen, kidneys, heart, bones, central nervous system, and elsewhere. The fully developed hydatid cyst is filled with fluid and typically unilocular; however, multilocular or chambered cysts are not uncommon. Structurally, the cyst consists of an inner germinative layer of cells supported by an acellular laminated membrane of mucopolysaccharide material (endocyst). Small secondary cysts, called *brood capsules,* develop internally from the germinative layer and produce multiple protoscolices by asexual budding. A protoscolex is a scolex in which the rostellum and suckers are deeply withdrawn into the post-sucker region (Fig. 5). Surrounding the hydatid cyst is a connective tissue adventitial reaction of variable intensity (ectocyst).

Growth is by concentric enlargement. Rates of growth of hydatid cysts are highly variable and range from 1 to 5 cm in diameter per year. The slowly growing hydatid cyst is often well tolerated until it causes dysfunction because of its size. In humans the cyst may attain a volume of many liters and contain many thousands of protoscolices. Rupture or leakage of cysts may precipitate allergic reactions as a consequence of sensitization to echinococcal antigens. At the same time, seeding of the peritoneum or pleura with protoscolices may result in extensive secondary cyst formation.

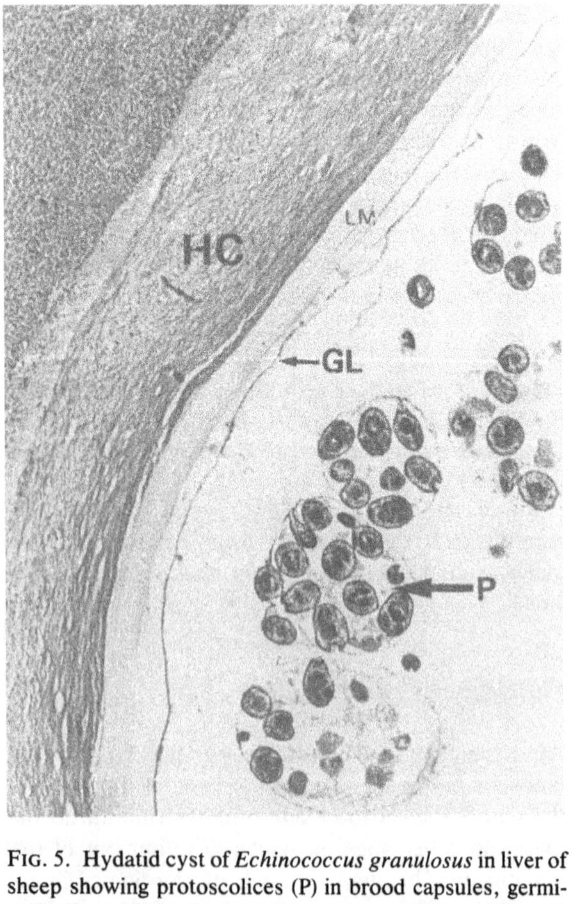

FIG. 5. Hydatid cyst of *Echinococcus granulosus* in liver of sheep showing protoscolices (P) in brood capsules, germinative layer (GL), laminated membrane (LM), and fibrous host capsule (HC). Hematoxylin and eosin stain. Magnification approximately ×350.

Clinical Manifestations

Clinical manifestations of cystic hydatid disease may be absent or result from the mass effect and anatomic position of the slowly growing cyst(s). The interval between first infection and clinical manifestations is variable and often prolonged for many years. Most infections are diagnosed in patients between 10 and 50 years of age. The signs and symptoms of hepatic hydatid disease may include hepatic enlargement (with or without a palpable mass in the right upper quadrant), right epigastric pain, nausea, and vomiting. Rupture or leakage usually results in acute or intermittent allergic manifestations. Complications that may exist at the time of initial presentation include traumatic or spontaneous rupture, thoracobilia, and biliary fistula (Amir-Jahed et al., 1975; Little, 1976). Twenty-five percent of patients with hepatic cysts also have cysts in their lungs.

Leakage or rupture of hydatid cysts in the lungs causes chest pain, coughing, dyspnea, and hemoptysis. Hydatid membranes may be expectorated, sometimes resulting in spontaneous cure. Emergency complications that may exist at the time of presentation include cyst rupture and secondary bacterial infection. Nearly 40% of patients with pulmonary hydatidosis have liver involvement as well.

The first symptom of cerebral cysts may be raised intracranial pressure or focal epilepsy, whereas kidney cysts may be manifested by loin pain or hematuria. Bone cysts are often asymptomatic until pathologic fractures occur, and, because of their resemblance, they are often misdiagnosed as tuberculous lesions.

Diagnosis

The diversity of clinical manifestations and the difficulty of demonstrating the deeply located cysts often make diagnosis very difficult. Hydatid disease usually comes to the attention of the clinician for four major reasons: when a large cyst has some mechanical effect on body function; when allergic phenomena or other symptoms such as eosinophilia occur; following the accidental traumatic rupture of a cyst with consequent acute allergic reactions; and on the incidental finding of cysts during roentgenography, body scanning, or surgery for other clinical reasons. A history of possible exposure to *E. granulosus* is very helpful and should be sought by probing the patient's residence and travel history. If past exposure is likely, further diagnostic workup includes a variety of clinical and laboratory procedures.

Nonimmunologic Diagnostic Methods

Sensitive methods are now available for identifying and characterizing mass lesions in internal organs. Plain-film roentgenography permits the detection of hydatid cysts in the lungs, but usually only calcified cysts can be demonstrated in other sites (Balikian and Mudarris, 1974). Radioisotopic and ultrasonic scanning (Kourias et al., 1970), computerized tomography (Ozgen et al., 1979; Scherer et al., 1978), and angiography (Garti and Deutsche 1971; Gutgemann et al., 1976) are useful for visualizing the avascular cyst(s) in many organs. Of these techniques, computerized axial tomography is the most specific in that it allows a more detailed characterization of the nature of space-occupying lesions than the "cold spot/no cold spot" diagnosis typically provided by other scanning methods (Grabbe et al., 1981; Irnberger, 1980; Scherer et al., 1978; Schulze et al., 1980) (Fig. 6). Nevertheless, some cysts, particularly those in extrahêpatic abdominal locations, may not be demonstrated by computed axial tomographic scans (Guarnera and Varela-Diaz, 1984). Ultrasound imaging techniques are almost as useful as computed axial tomography and have the added advantages of avoiding radiation exposure and relatively low cost (Grabbe et al., 1981; Schulze et al., 1980). The availability of portable, generator-powered ultrasound devices permits their use in field studies. Hydatid cysts must be differentiated from cavitary tuberculosis, abscess, and benign or malignant neoplasms.

Closed aspiration of hydatid cysts should not be attempted because accidental spilling of the contents could cause secondary spread or anaphylaxis. Cyst membranes, protoscolices, or hooklets may sometimes be demonstrated in the sputum or in bronchial washings (Allen and Fullmer, 1972); identification is facilitated by staining the hooklets (Ishak, 1972) (Fig. 7). An intravenous pyelogram is helpful in visualizing renal cysts (Silber and Moyad, 1972). Eosinophilia is present in one-third or fewer of cyst carriers (Little, 1976).

Hydatid cysts removed at surgery are frequently misidentified by inexperienced persons, particularly when protoscolices are lacking or the larva is otherwise degenerated or anomalous. Nevertheless, most lesions can be identified and differentiated from other larval cestodes (i.e., cysticerci or coenuri) and nonparasitic cysts by the characteristic appearance of the laminated membranes (Slais, 1973).

Immunologic Diagnostic Methods

Serodiagnostic tests, including enzyme-linked immunosorbent assay, indirect hemagglutination, and gel diffusion methods, can be useful if factors that can influence the results are understood (Rickard & Lightowlers, 1986; Schantz and Gottstein, 1986). First, negative test results do not rule out a diagnosis of cystic hydatid disease, because some cyst carriers do not have detectable antibodies. The presence of a detectable immune response has been associated with the location, integrity, and vitality of the larval cysts. Cysts in the liver are more likely to elicit antibody than cysts in the lungs and, regardless of locali-

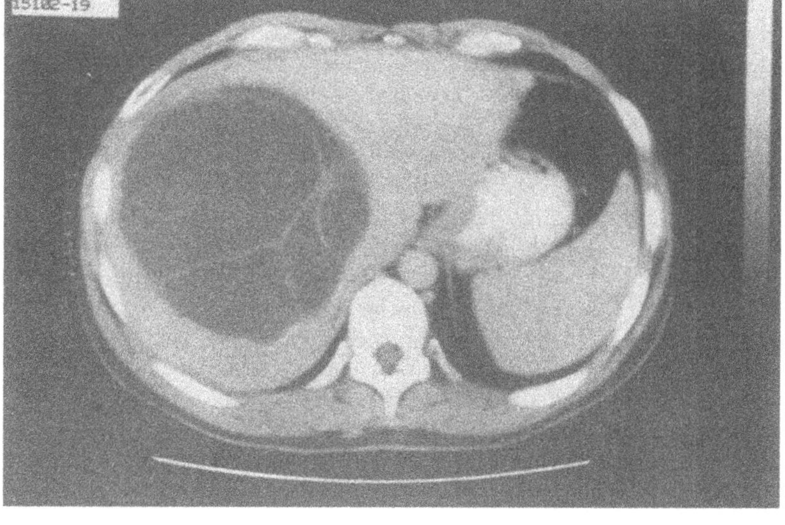

FIG. 6. Abdominal computed tomographic scan showing large primary hydatid cyst (*Echinococcus granulosus*) in the right lobe of the liver. Note multiple internal septations indicating secondary (daughter) cyst formation (Courtesy Dr. D. S. Rubin.)

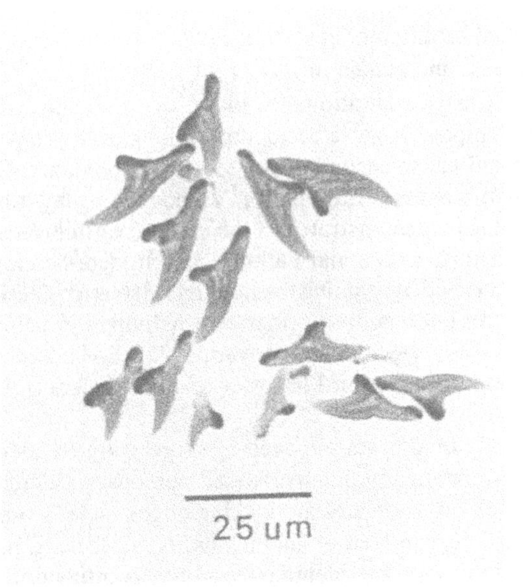

FIG. 7. *Echinococcus granulosus* protoscolex hooklets stained with acid fuchsin.

zation, tests are least sensitive for diagnosing intact healthy cysts. Such healthy metacestodes produce a low level of antigenic stimulation, and nearly 50% of carriers of this kind of cyst in the lungs may be serologically negative. Fissuration or rupture of the cyst is followed by abrupt stimulation of antibodies. Senescent or dead cysts apparently cease to stimulate the host, and carriers of such cysts may be seronegative.

Indirect immunofluorescence, enzyme-linked immunosorbent assay, and indirect hemagglutination tests using *E. granulosus* antigens are sensitive screening procedures, test positivity varying from approximately 60 to 90% depending on the characteristics of the cases. Tests using whole *E. granulosus* antigens are not specific, however, and false-positive reactions should be anticipated in sera from persons with other helminthic infections, liver cirrhosis, or cancer (Schantz and Gottstein, 1986). The specificity of serologic reactions can be confirmed by the demonstration of antibody to *Echinococcus* antigen 5. Only larval *Taenia solium* (cysticercosis) has been shown to induce cross-reacting antibody to antigen 5 in human sera. The clinical manifestations of cysticercosis, however, rarely mimic hydatid disease and should not hamper accurate clinical diagnosis. Tests useful for demonstrating arc 5 include immunoelectrophoresis, double diffusion, and counterimmunoelectrophoresis, the latter two tests being more sensitive (Coltorti and Varela-Diaz 1978).

A major limitation of basing test positivity on the arc 5 criterion is that, preoperatively, only about two-thirds of patients with cystic hydatid disease are arc 5 positive (Rickard, 1984; Varela-Diaz and Coltorti, 1984; Varela-Diaz et al., 1975a,b,c), and, in patients with hyaline-type cysts in the lungs, the arc 5 positivity rate may be less than 50% (Varela-Diaz and Coltorti, 1984; Yarzabal et al., 1974). Serologic confirmation of a presumptive clinical diagnosis of cystic hydatid disease cannot be obtained in the patient who is arc 5 negative. The probability that such a patient actually has the disease, however, increases with the intensity of the serologic response measured by other criteria. Thus, with immunoelectrophoresis or double diffusion, demonstration of three or more noncharacterized bands is a highly reliable serologic indicator of infection (Rickard, 1984; Varela-Diaz and Coltorti, 1984; Varela-Diaz et al., 1975a). A cutoff titer of 1:512 in the indirect hemagglutination assay is associated with a minimum of false-positive reactions (Picardo and Guisantes, 1981; Varela-Diaz et al., 1975b). Lower indirect hemagglutination titers (1:128 to 1:256) are measured in the sera of approximately 20% of those with hydatid disease (Kagan and Norman, 1979; Varela-Diaz et al., 1975b). However, nonspecific reactions in the sera of persons without hydatid disease are not infrequent in this low range of titers.

Treatment

Although the disease may not progress or may resolve spontaneously, the risks of anaphylactic reactions, dissemination following rupture, and relentless growth, which renders subsequent therapy more difficult or impossible, provides a strong argument for early therapy.

Surgical extirpation of the hydatid cyst remains the most effective treatment. Extirpation is usually accomplished by using generous surgical exposure and aspiration of part of the cyst contents, followed by injection of a protoscolicidal agent (hypertonic saline or silver nitrate) and complete removal of the endocyst (Sayek et al., 1980). The residual cavity is then treated by suture plication, marsupialization, closure after filling with saline, or omentoplasty. An alternative method of removal is en bloc excision, a procedure particularly applicable to splenic cysts and, to a lesser extent, to hepatic lobectomy. Diagnosis and surgical removal of intact cysts before the onset of complications result in high cure rates with low surgical mortality (1 to 4%). In contrast, secondary or recurrent disease often requires repeated surgical intervention, with surgical mortality reaching 20% (Amir-Jahed et al., 1975).

Until recently, there has been little to offer patients with anatomically irresectable lesions (e.g., those in whom the spine is involved or in whom spontaneous rupture, operative spillage, or widespread metastatic cyst formation has occurred). Under such conditions, treatment with mebendazole or albendazole may be attempted, although the results are upredictable and adverse reactions have been reported (Davis et al., 1986; Morris et al., 1985; Schantz, 1982). Pulmonary and liver cysts of *E. granulosus* appear to respond relatively well, but cysts in other locations, particularly in the brain, bone, and eye, respond poorly if at all. Mebendazole (50 mg to 150 mg/kg of body weight per day, as a single oral dose for 3 months) or albendazole (10 to 15 mg/kg per day, in divided oral doses for 1 month) are recommended; many patients require repeated courses. Albendazole is also recommended for prevention of recurrent disease secondary to spontaneous or accidental cyst rupture with spillage of contents. Reported side effects have included neutropenia, liver toxicity, and hair loss. As long-term prognosis in individual patients is difficult to predict, prolonged follow-up is required to determine the eventual outcome of treated cases.

Prevention and Control

For individuals in endemic areas, preventive measures include careful personal hygiene, strict dietary regulation of pet dogs to preclude ingestion of sheep offal, and avoidance of dogs that are not so regulated. Prophylactic treatment of dogs with praziquantel at intervals just short of the prepatent period (e.g., 45 to 50 days) prevents egg excretion in dogs so treated.

The principal objective of control of cystic hydatid disease is reducing parasite prevalence in definitive and intermediate hosts to levels below those necessary for continued transmission (Eckert et al., 1981). Measures that have been used in various circumstances include health education, control of livestock slaughtering in abattoirs and on farms, safe disposal of infected viscera, dog control, and periodic diagnostic testing and mass treatment of dogs with praziquantel. The results achieved by control programs in New Zealand, Tasmania, Cyprus, Falkland Islands, Argentina, and Chile have been spectacular, given the considerable difficulties involved (Gemmell and Lawson, 1986). Marked reductions in prevalence of cysts in sheep following the initiation of control methods, as well as in the numbers of new diagnosed human cases, indicate that the fundamental requirements for effective control are now tried and proved.

Alveolar Hydatid Disease (*Echinococcus multilocularis*)

Life Cycle and Geographic Distribution

The definitive and intermediate hosts of *E. multilocularis* are foxes and small cricetid rodents, respectively (Schantz, 1982) (Fig. 8). In the comparatively simple biotic conditions of the arctic tundra zone, the cestode is maintained in cycles involving the arctic fox, *Alopex,* and several kinds of voles of the genera

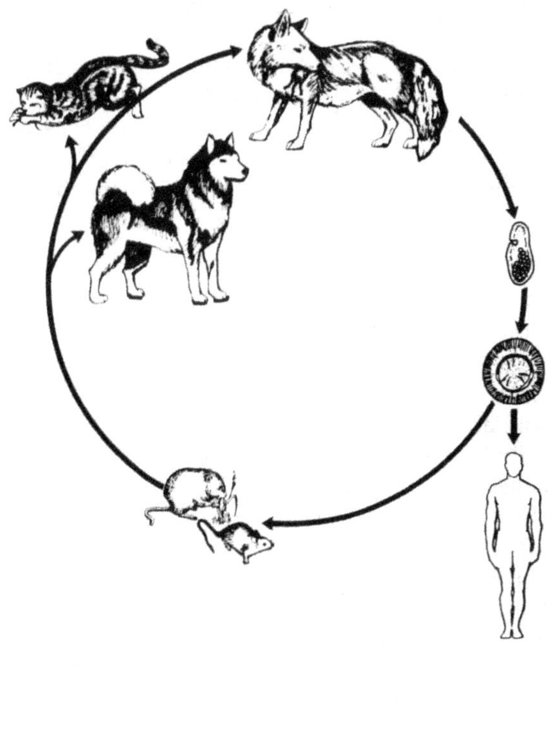

FIG. 8. Life cycle diagram of *Echinococcus multilocularis*.

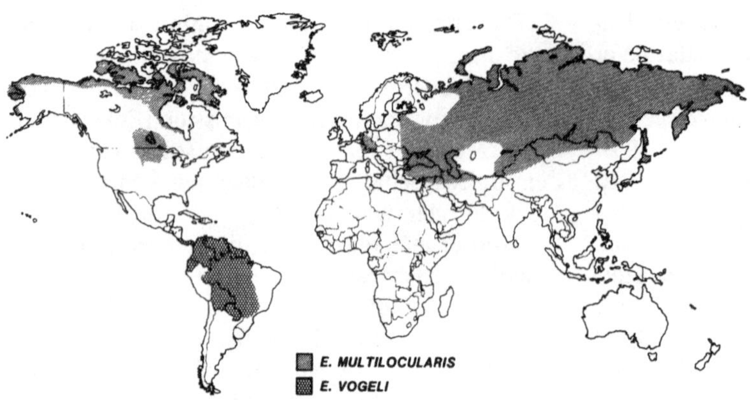

FIG. 9. Geographic distribution of *Echinococcus multilocularis* and *Echinococcus vogeli* infections.

Microtus and *Clethrionomys*. In more complex biomes a greater variety of host species are involved. The red fox, *Vulpes vulpes*, is the most important definitive host outside the Arctic circle. Other carnivores that have been found naturally infected are domestic dogs and cats, coyotes, and wolves.

Echinococcus multilocularis has an extensive geographic range in the Northern Hemisphere (Fig. 9), including an endemic region in central Europe (central and southern Germany, western Austria, Switzerland, and southern France) and most of northern Eurasia from Bulgaria and Turkey through the Soviet Union, extending eastward to several of the Japanese islands. Recent reports of this species from China, India, Iran, and central North America have extended our knowledge of the southernmost limits of its range. Human cases are reported regularly from endemic regions in central Europe, Siberia, the Japanese island of Hokkaido, and North America. Among exposed Eskimo populations in Alaska, approximately one new case per year has been diagnosed since the infection was first discovered there in 1950 (Wilson and Rausch, 1980). Thus far, only a single human case has been diagnosed in the contiguous United States (Minnesota) in association with an apparently expanding focus of enzootic transmission in the north–central United States (Gamble et al., 1979).

Epidemiology

Echinococcus multilocularis is largely restricted to wild animal hosts and is thereby separated ecologically from humans. Human exposure is often determined by occupational and avocational pursuits. Hunters, trappers, and persons who work with fox fur in endemic areas are frequently exposed. Dogs that become infected by capturing and eating infected voles are also important as a source of human infection. When infected voles exist in villages as commensal rodents (as, for example, in some Eskimo villages of the tundra zone in North America), the cycle of the cestode is readily completed and a hyperendemic focus is produced (Wilson and Rausch, 1980).

Pathogenesis and Pathology

Compared with *E. granulosus*, larval *E. multilocularis* is more progressive and damaging to the host. Human disease caused by *E. multilocularis* has many of the characteristics of malignant neoplasia. Growth in humans is different from that in natural rodent hosts (Figs. 10 and 13) in that the larval mass is inhibited from completing its development and remains in the proliferative stage indefinitely. It then continues to invade and destroy the hepatic parenchyma, and retrogressive stages within the mass result in necrosis of the central portion. Macroscopically the hepatic lesion usually appears as one or more firm-to-solid, whitish, rounded masses slightly elevated above the surrounding tissue of the surface of the liver. When transected, it appears to consist of a central cavity surrounded by dense, pale tissue that lacks a clearly defined border with respect to the adjacent hepatic tissue (Fig. 11). In advanced cases the cavity usually contains turbid yellowish-to-brown fluid with particles or fragments of necrotic tissue. Microscopically, cysts of various sizes are scattered or closely aggregated in a matrix of connective tissue. Only rarely are brood capsules and calcareous corpuscles present. Metastasis by a hematogenous route is not uncommon, and secondary foci usually occur in the brain and lungs.

Clinical Manifestations

The insidious and progressive nature of alveolar hydatid disease usually results in delayed onset of

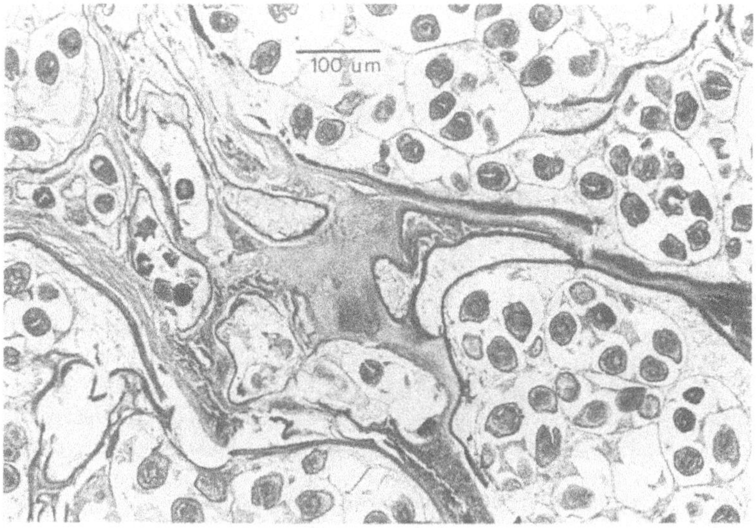

FIG. 10. Larval *Echinococcus multilocularis* in liver of infected rodent, showing characteristic alveolarlike microvesicles with protoscolices.

symptoms and diagnosis, by which time the lesions have invaded extensively or metastasized and are inoperable. The clinical signs of the disease usually do not become evident until middle age. Initial symptoms are generally vague. Mild upper-quadrant and epigastric pain with hepatomegaly may progress to obstructive jaundice. Occasionally the initial manifestations are caused by metastases in the lungs or brain. Patients eventually succumb to hepatic failure or invasion of contiguous structures. The proportion of surgically resectable cases has varied from 20 to 60% (Kasai et al., 1980; Rausch and Wilson, 1980). Of the nonresectable cases, 90% of patients died within 10 years (Mosimann, 1980).

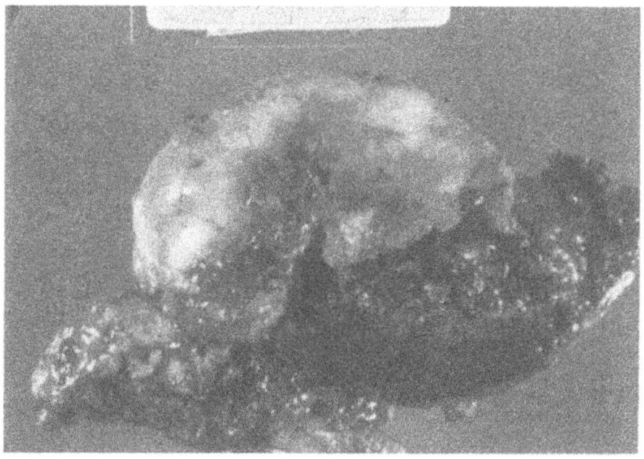

FIG. 11. Alveolar hydatid lesion involving left lobe of the liver after surgical resection. Note the bulging inflammatory mass, which measured 12 cm in maximum dimension.

Diagnosis

Nonimmunologic Diagnostic Methods

Alveolar hydatid disease caused by *E. multilocularis* is typically observed in elderly persons and mimics hepatic carcinoma and cirrhosis, with which it is commonly confused. Plain-film roentgenography shows hepatomegaly and characteristic scattered radiolucencies outlined by calcific rings measuring 2 to 4 mm (Thompson et al., 1972).

In Germany and other areas where both cystic and alveolar hydatid disease occur, the two can be distinguished by selective angiography (Gutgemann et al., 1976; Rudofsky et al., 1975) and by computerized axial tomography (Grabbe et al., 1981; Irnberger, 1980; Schulze et al., 1980). The characteristic computerized tomographic image in cystic hydatid disease shows sharply contoured cyst(s) (sometimes with internal daughter cysts) and marginal calcifications. In contrast, the computerized tomographic image of alveolar hydatid disease reveals indistinct solid tumor with a central necrotic area and perinecrotic plaquelike calcifications (Didier et al., 1985; Grabbe et al., 1981; Irnberger, 1980; Schulze et al., 1980) (Fig. 12).

Exploratory laparotomy is often performed in order to diagnose and delineate the extent to which the lesion has invaded. Needle biopsy confirms the diagnosis if larval elements can be demonstrated. Even in the absence of protoscolices, the larval membranes can be readily demonstrated using periodic acid–Schiff staining (Fig. 13), or the sterile larva can be reared to the adult stage after secondary passage in susceptible rodents (Rausch and Wilson, 1973).

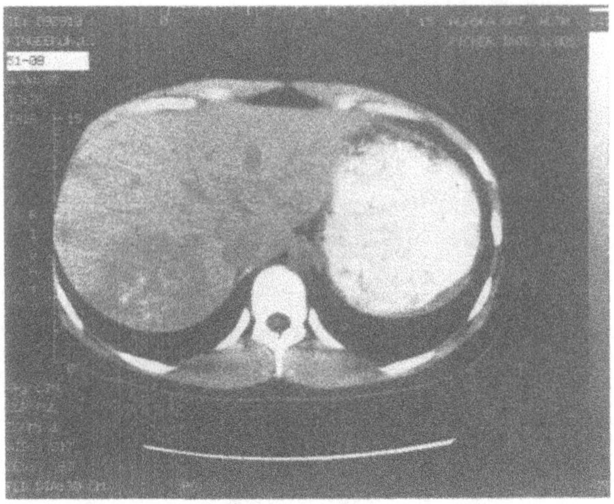

FIG. 12. Abdominal computed tomographic scan of patient with *Echinococcus multilocularis* lesion in right lobe of the liver. Note irregular low-density zones alternating with scattered areas of calcification.

Immunologic Diagnostic Methods

Problems of sensitivity appear less important in serodiagnosis of *E. multilocularis* infections than in that of *E. granulosus* infections. Most infected persons appear to have a detectable immune response. Common antigens are shared by *E. granulosus* and *E. multilocularis*, and cross-reactions occur with most tests. When crude *Echinococcus* antigens are used, however, nonspecific reactions create the same difficulties as described previously for diagnosis of cystic hydatid disease. Fortunately, recent research appears to have largely solved this problem for diagnosis of alveolar hydatid disease. The purification of a highly specific antigen from *E. multilocularis* antigens in enzyme-linked immunosorbent assays yielded a sensitivity of 94% for diagnosis of 78 patients with alveolar hydatid disease and completely eliminated nonspecific reactions with sera of patients with filariasis, schistosomiasis, cysticercosis, and other helminthic infections (Gottstein, 1985). Comparison of titers produced against both crude and purified *E. multilocularis* antigens permitted the differentiation of patients with alveolar from those with cystic hydatid disease in 95% of cases (Gottstein et al., 1983, 1986). Some asymptomatic persons with *E. multilocularis* infections who are negative in serologic tests using *E. granulosus* antigens can be diagnosed serologically using the purified *E. multilocularis* antigen (Gottstein et al., 1983; Lanier et al., 1987; Rausch et al., 1987).

Treatment

Resection of the entire larval mass is the aim of surgery for this otherwise progressive and fatal disease (Kasai al., 1980; Wilson and Rausch, 1980). Usually the entire affected lobe(s) must be removed. When involvement is more extensive, wedge resections of

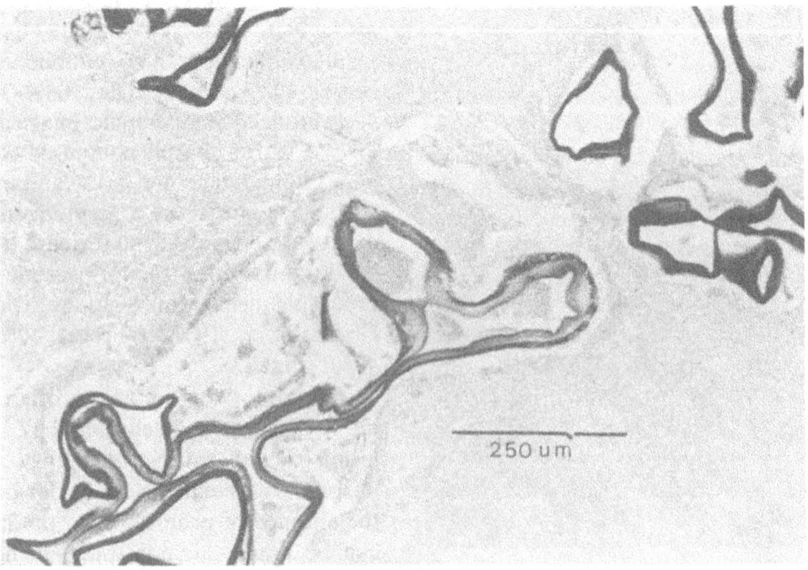

FIG. 13. Histologic appearance of alveolar hydatid disease in a section of human liver, showing absence of protoscolices and calcareous corpuscles; the laminated membrane is darkly stained with periodic acid–Schiff stain.

the lesions may be attempted. Because alveolar hydatid disease is often not diagnosed until parasitic invasion is well advanced, the lesion is frequently not resectable; thus, partial resections and biliodigestive and hepatodigestive anastomoses are performed as palliative measures, predominantly to ensure bile passage. For treatment of nonresectable cases, continuous mebendazole therapy (40 mg/kg of body weight per day) results in clinical improvement and prolonged survival time; however, the effect is inhibitory rather than lethal to the parasite. Recent reports on therapy with albendazole suggest similar effectiveness (Wilson et al., 1987).

Prevention and Control

Although *E. multilocularis* was eradicated from Rebun Island in Japan by eliminating foxes, control of *E. multilocularis* infections in sylvatic hosts has not been pursued elsewhere because of its obvious impracticalities. Personal preventive measures include avoidance of contact with foxes and other potentially infected final hosts. Control of pet dogs and cats must be practiced to prevent them from preying on rodents. Patent infections in dogs and cats that are liable to eat infected rodents can be prevented by monthly prophylactic treatments with praziquantel (5 mg/kg of body weight). Potentially exposed human populations should be educated about the dangers of the disease to promote better personal hygiene and sanitation and to motivate them to take effective measures to prevent their pets from eating rodents.

Polycystic Hydatid Disease (*Echinococcus vogeli*)

Life Cycle and Geographic Distribution

Relatively little is known about the biology of this parasite *E. vogeli*. In regions investigated, the natural life cycle of *E. vogeli* involves the bush dog (*Speothos venaticus*) and the paca (*Cuniculus paca*) as definitive and intermediate hosts, respectively (D'Allessandro et al., 1981; Rausch and Bernstein, 1972). Larvae of *E. vogeli* have also been found in agoutis (*Dasyprocta* spp.) and spiny rats (*Proechimys* spp.). Domestic dogs are also suitable definitive hosts.

At present, *E. vogeli* is believed to occur within the range of its requisite hosts in neotropical America (Fig. 9). Fifteen cases of polycystic hydatid disease have been reported from Costa Rica, Panama, Colombia, Equador, and Venezuela (D'Allesandro et al., 1979).

Epidemiology

Echinococcus vogeli appears to be maintained in cycles involving bush dogs and pacas and, to a lesser extent, agoutis and spiny rats. In endemic areas dogs are commonly fed viscera of paca, and human infections are probably acquired from the feces of infected hunting dogs. Bush dogs are rare and avoid human beings; therefore, they probably play little role in direct exposure of humans (Rausch and Bernstein, 1972).

Pathogenesis and Pathology

Echinococcus vogeli has been identified as the cause of a polycystic form of human hydatid disease with characteristics intermediate between cystic and alveolar hydatid disease (D'Allesandro et al., 1979). The primary localization is in the liver, but cysts may spread to contiguous sites. These infections produce multiple fluid-filled vesicles (5 to 80 mm in diameter) with numerous brood capsules and protoscolices. Proliferation of vesicles sometimes leads to destruction of much of the liver and involvement of adjacent structures.

Clinical Manifestations

Among the 15 reported cases, age at onset of symptoms leading to diagnosis ranged from 13 to 74 years. Hepatomegaly or the presence of tumorlike masses in the liver were typical findings. Most of the reported cases were lost to follow-up; however, their prognosis must be poor due to the extensive nature of the disease.

Diagnosis

The diagnosis of hydatid disease was not suspected preoperatively in most of the cases of polycystic hydatid disease thus far reported. Techniques that can demonstrate cystic or alveolar hydatid disease should also be useful for diagnosing larval *E. vogeli* infections. Limited experience has shown that most patients have high titers of antibodies to *E. granulosus* antigens including, arc 5 (Varela-Diaz et al., 1978). The length and relative proportions of the rostellar hooks differ between species of *Echinococcus* and permit confirmation of the species identification in cases where doubt exists (Rausch et al., 1978).

Treatment

Little has been reported regarding treatment, but management criteria similar to those for cystic and alveolar hydatid disease should apply to polycystic disease.

Literature Cited

Allen, A. R., and C. D. Fullmer. 1972. Primary diagnosis of pulmonary echinococcosis by the cytologic technique. Acta Cytol. **16:**212.

Amir-Jahed, A. K., R. Fardin, A. Farzad, and K. Bakshandeh. 1975. Clinical echinococcosis. Ann. Surg. **182:**541–546.

Balikian, J. P., and F. F. Mudarris. 1974. Hydatid disease of the lungs, a roentgenologic study of 50 cases. Am. J. Roentgenol. **122:**692.

Coltorti, E. A., and V. M. Varela-Diaz. 1978. Detection of antibodies against *Echinococcus granulosus* arc 5 antigens by double diffusion test. Trans. R. Soc. Trop. Med. Hyg. **72:**226–229.

D'Alessandro, A., R. L. Rausch, C. Cuello, and N. Aristizabal. 1979. First observation of *Echinococcus vogeli* in man, with a review of human cases of polycystic hydatid disease in Colombia and neighboring countries. Am. J. Trop. Med. Hyg. **28:**303–317.

D'Alessandro, A., R. L. Rausch, G. A. Morales, S. Collet, and D. Angel. 1981. *Echinococcus* infection in Colombian animals. Am. J. Trop. Med. Parasitol. **30:**1263–1276.

Davis, A., Z. S. Pawlowski, and H. Dixon. 1986. Multicentre clinical trials of benzimidazolecarbamates in human echinococcosis. Bull. W.H.O. **64:**383–388.

Didier, D., S. Weiler, P. Rohmer, A. Lasseque, J. P. Deschamps, D. Vuitton, J. P. Miquet, and F. Weill. 1985. Hepatic alveolar echinococcosis. Correlative US and CT study. Radiology **154:**179–186.

Eckert, J., M. A. Gemmell, and E. J. L. Soulsby (ed.). 1981. FAO/UNEP/WHO Guidelines for Surveillance, Prevention and Control of Echinococcosis/Hydatidosis, p. 147. VPH/81.28. World Health Organization, Geneva.

Gamble, W. G., M. Segal, P. M. Schantz, and R. L. Rausch. 1979. Alveolar hydatid disease in Minnesota: first human case acquired in contiguous United States. J. Am. Med. Assoc. **241:**904–908.

Garti, I., and V. Deutsch. 1971. The angiographic diagnosis of echinococcosis of the liver and spleen. Clin. Radiol. **22:**466–469.

Gemmell, M. A., and J. R. Lawson. 1986. Epidemiology and control of echinococcosis/hydatidosis, p. 189–216. *In* R. C. A. Thompson (ed.), The biology of echinococcus and hydatid disease. Allen & Unwin, London.

Gottstein, B. 1985. Purification and characterization of a specific antigen from *Echinococcus multilocularis*. Parasite Immunol. **70:**385–389.

Gottstein, B., J. Eckert, and H. Fey. 1983. Serological differentiation between *Echinococcus granulosus* and *E. multilocularis* infections in man. Z. Parasitenkd. **69:**347–356.

Gottstein, B., P. M. Schantz, T. Todorov, A. G. Salmot, and P. Jacquier, 1986. An international study on the serological differential diagnosis of human cystic and alveolar echinococcosis. Bull. W.H.O. **64:**101–105.

Grabbe, E., P. Kern, and M. Heller. 1981. Human echinococcosis: diagnostic value of computed tomography. Tropenmed. Parasitol. **32:**35–39.

Guarnera, E. A., and V. M. Varela-Diaz. 1984. Immunodiagnosis of abdominal hydatid disease. Difficulties in the radiological and scintillographic localization of cysts. Med. J. Aust. **140:**493–495.

Gutgemann, A., C. Kaufer, C. H. Prange, E. Raschke, E. Bucheler, and H. J. Biersack. 1976. Diagnostik und Chirurgie der Leberechinococcen. Langenbecks Arch. Chir. **340:**285–290.

Irnberger, V. T. 1980. Radiographische, sonographische, und computertomographische erscheinungsbilder des leberechinokokkus. Wien. Med. Wochenschr. **22:**718–724.

Ishak, K. G. 1972. Acid-fast staining of hooklets of *Taenia echinococcus*. Lancet **1:**556.

Kagan, I. G., and L. Norman. 1979. Evaluation of *Echinococcus* antigens for the diagnosis of hydatid disease. Rev. Iber. Parasitol. **39:**153–164.

Kasai, Y., I. Koshino, and N. Kawanishi. 1980. Alveolar echinococcosis of the liver. Ann. Surg. **191:**145–152.

Kourias, B., E. Gyftaki, P. Peveretos, and D. Binopoulos. 1970. The value of pre- and post-operative scanning in liver echinococcosis. Br. J. Surg. **57:**178–182.

Kumaratilake, L. M., and R. C. A. Thompson. 1982. A review of the taxonomy and speciation of the genus *Echinococcus,* Rudolphi 1801. Z. Parasitenkd. **68:**121–146.

Kumaratilake, L. M., and R. C. A. Thompson. 1984. Morphological characterization of Australian strains of *Echinococcus granulosus*. Int. J. Parasitol. **14:**467–477.

Lanier, A. P., D. E. Trujillo, P. M. Schantz, J. F. Wilson, and B. Gottstein. 1987. Sensitivity, specificity and clinical correlations of purified *Echinococcus multilocularis* and *E. granulosus* antigens for serodiagnosis of alveolar hydatid disease. Am. J. Trop. Med. Hyg. (in press).

Lawson, J. R., and M. A. Gemmell. 1983. Hydatidosis and cysticercosis: the dynamics of transmission. Adv. Parasitol. **22:**261–308.

Little, J. M. 1976. Hydatid disease at Royal Prince Alfred Hospital, 1964 to 1974. Med. J. Aust. **1:**903–908.

Morris, D. L., P. W. Dykes, S. Marriner, J. Bogan, F. Burrows, H. Skeene-Smith, and M. J. Clarkson. 1985. Albendazole—objective evidence of response in human hydatid disease. J. Am. Med. Assoc. **253:**2053–2057.

Mosimann, F. 1980. Is alveolar hydatid disease of the liver incurable? Ann. Surg. **192:**118–122.

Ozgen, T., A. Erbengi, V. Bertan, S. Saglam, O. Gurcay, and T. Pirnar. 1979. The use of computerized tomography in the diagnosis of cerebral hydatid cysts. J. Neurosurg. **50:**339–342.

Picardo, N. G. A., and J. A. Guisantes. 1981. Comparison of three immunological tests for seroepidemiological purposes in human echinococcosis. Parasite Immunol. **3:**191–199.

Rausch, R. L., and J. J. Bernstein. 1972. *Echinococcus vogeli* spp. n. (Cestoda: Taeniidae) from the bush dog, *Speothos venaticus* (Lund). Tropenmed. Parasitol. **23:**25–34.

Rausch, R. L., V. R. Rausch, and A. D'Alessandro. 1978. Discrimination of the larval stages of *Echinococcus aligarthrus* (Diesing, 1863) and *E. vogeli* Rausch and Bernstein, 1972 (Cestoda:Taeniidae). Am. J. Trop. Med. Hyg. **27:**1195–1202.

Rausch, R. L., and J. R. Wilson. 1973. Rearing of the adult *Echinococcus multilocularis* Leuckart. 1863, from sterile larvae from man. Am. J. Trop. Med. Hyg. **22:**357–358.

Rausch, R. L., J. F. Wilson, P. M. Schantz, and B. J. McMahon. 1987. Spontaneous death of *Echinococcus multilocularis:* cases diagnosed by EmZ ELISA, and clinical significance. Am. J. Trop. Med. Hyg. **36:**576–585.

Rickard, M. D. 1984. Serological diagnosis and post-operative surveillance of human hydatid disease. Pathology **16:**207–210.

Rickard, M. D., and M. W. Lightowlers. 1986. Immunodiagnosis of hydatid disease, p. 217–249. *In* R. C. A. Thompson, (ed.), The biology of echinococcus and hydatid disease. Allen & Unwin, London.

Rudofsky, G., W. Wolfert, and R. M. Rau. 1975. Angiographische darstellung der Leberarterien beim *Echinococcus alveolaris*. Med. Klin. **70:**1641–1644.

Sayek, I., R. Yalin, and Y. Sanoc. 1980. Surgical treatment of hydatid disease of the liver. Arch. Surg. **115:**847–850.

Schantz, P. M. 1982. Echinococcosis, p. 231–277. *In* J. Steele and P. Arambulo (ed.), Handbook of zoonoses, vol. III. CRC Press, Boca Raton, Fla.

Schantz, P. M., and B. Gottstein. 1986. Echinococcus (hydatidosis), p. 69–107. *In* K. F. Walls and P. M. Schantz (ed.), Immunodiagnosis of parasitic diseases, vol. I. Academic Press, Inc., Boca Raton, Fla.

Scherer, U., U. Bull, R. Rothe, J. Eisenburg, F. W. Schildberg, P. Meister, and J. Lissner. 1978. Computerized tomography and nuclear imaging of the liver: a comparative study in 83 cases. Eur. J. Nucl. Med. **3:**71–80.

Schulze, V. K., K.-H. Hubener, K. Klott, H. Jenss, and R. Bahr. 1980. Computertomographische und sonographische diagnostik der echinokokkose. Fortschr. Rontgenstr. **132:**514–521.

Silber, S. J., and R. A. Moyad. 1972. Renal echincoccosis. J. Urol. **108:**669–672.

Slais, J. 1973. Functional morphology of cestode larva. Adv. Parasitol. **11:**395.

Thompson, R. C. A., L. M. Kumaratilake, and J. Eckert. 1984. Observations on *Echinococcus granulosus* of cattle origin in Switzerland. Int. J. Parasitol. **14:**283–291.

Thompson, W. M., D. P. Chisholm, and R. Tank. 1972. Plain film roentgenographic findings in alveolar hydatid disease—*Echinococcus multilocularis*. Am. J. Roentgenol. **116:**345–351.

Varela-Diaz, V. M., and E. A. Coltorti. 1984. Immunodiagnostic confirmation of hydatid disease in patients with a presumptive diagnosis of infection. Rev. Inst. Med. Trop. Sao Paulo **26:**87–96.

Varela-Diaz, V. M., E. A. Coltorti, and A. D'Alessandro. 1978. Immunoelectrophoresis tests showing *Echinococcus granulosus* arc 5 in human cases of *Echinococcus vogeli* and cysticercosis-multiple myeloma. Am. J. Trop. Med. Hyg. **27:**554–557.

Varela-Diaz, V. M., E. A. Coltorti, U. Prezioso, M. H. Lopez-Lemes, J. A. Guisantes, and L. A. Yarzabal. 1975a. Evaluation of three immunodiagnostic tests for human hydatid disease. Am. J. Trop. Med. Hyg. **24:**312–319.

Varela-Diaz, V. M., J. A. Guisantes, M. I. Ricardes, L. A. Yarzabal, and E. A. Coltorti. 1975b. Evaluation of whole and purified hydatid fluid antigens in the diagnosis of human hydatidosis by the immunoelectrophoresis test. Am. J. Trop. Med. Hyg. **24:**298–303.

Varela-Diaz, V. M., M. H. Lopez-Lemes, U. Prezioso, E. A. Coltorti, and L. A. Yarzabal. 1975c. Evaluation of four variant of the indirect hemagglutination test for human hydatidosis. Am. J. Trop. Med. Hyg. **24:**304–311.

Wilson, J. F. A. C. Diddams, and R. L. Rausch. 1968. Cystic hydatid disease in Alaska. A review of 101 autochthonous cases of *Echinococcus granulosus* infection. Am. Rev. Respir. Dis. **98:**1–12.

Wilson, J. F. and R. L. Rausch. 1980. Alveolar hydatid disease. A review of clinical features of 33 indigenous cases of *Echinococcus multilocularis* infection in Alaskan Eskimos. Am. J. Trop. Med. Hyg. **29:**1340–1349.

Wilson, J. F., R. L. Rausch, B. J. McMahon, P. M. Schantz, D. E. Trujillo, and M. O'Gorman. 1987. Albendazole therapy in alveolar hydatid disease: a report of favorable results in two patients after short term therapy. Am. J. Trop. Med. Hyg. **37:**162–168.

Yarzabal, L. A., J. Leiton, and M. H. Lopez-Lemes. 1974. The diagnosis of human pulmonary hydatidosis by the immunoelectrophoresis test. Am. J. Trop. Med. Hyg. **23:**662–666.

CHAPTER 89

Enterobiasis

E. G. LONG

Disease: Enterobiasis, oxyuriasis.
Etiologic Agents: *Enterobius vermicularis* (*Oxyuris vermicularis*), pinworm, threadworm, oxyurid.
Source: Humans.
Clinical Manifestations: Pruritus ani, perianal eczematous dermatitis, insomnia, enuresis, gastrointestinal disturbances, and genitourinary infections (usually in women).
Pathology: Mild inflammation and ulceration of the colonic mucosa, granulomatous lesions in ectopic sites, intense irritation of perianal skin.
Laboratory Diagnosis: Demonstration of eggs and sometimes female worms on perianal skin.
Epidemiology: Cosmopolitan, but more frequent in temperate than in tropical climates; primarily a disease of children.
Treatment: Pyrantel pamoate or mebendazole.
Prevention and Control: Individual and group treatment and improved personal hygiene.

Description of Disease

Enterobiasis, or oxyuriasis, is a cosmopolitan, helminthic disease, but occurs more frequently in cold and temperate climates than in the tropics. The disease is usually seen in children; in adults, it is more common in women than in men. Humans are the only known hosts for the agent, *Enterobius vermicularis*. Clinical symptoms arise when female worms migrate out of the anus to oviposit and cause intense irritation of the perianal region, when they invade the appendix and cause inflammation, or when they migrate to organs within the female genital or urinary system to provoke granulomatous reactions.

Communicability and Transmissibility

Adult worms emerge from the anus at night to oviposit, and the tickling sensation they cause, as well as the pruritus that follows, results in an unbearable urge to scratch the perianal area. Eggs that lodge under the nails and contaminate the fingertips can be transferred directly to the mouth and cause reinfection or serve as a source of infection to others. Infection can also be acquired when soiled night clothes and bed linens are shaken and airborne eggs are inhaled. Other sources of infection are contaminated food, dust from the rooms of infected persons, and toys and furniture touched by contaminated fingers. Egg laying and the symptoms of enterobiasis begin about 1 month after infection (Beaver et al., 1984).

Clinical Manifestations of Disease

Intestinal Enterobiasis

Intestinal symptoms are absent in most infections, but heavy parasite loads may cause diarrhea and mucoid, blood-streaked stools (Symmers, 1950). Characteristic symptoms of infection are abdominal pain, pruritis ani, and excoriations, which result from scratching the perianal skin. Associated symptoms related to the irritation and consequent insomnia include nocturnal tooth-grinding, irritability, behavioral problems, lassitude, anorexia, and enuresis. Sachdev and Howards (1975) attribute enuresis to reflex stimulation of the bladder in infants, who may be too young to localize the site of irritation. Symptomatic infections are sometimes accompanied by a slight eosinophilia, but this finding is not consistent and is thus of little diagnostic value (Mayers and Purvis, 1970). The relationship between *E. vermicularis* and appendicitis remains undetermined, but the parasites have been found in the tissues of inflamed appendices (Morgensen et al., 1985). Bouree and Dubourdieu (1984), in a literature review, reported that *E. vermicularis* was recovered from 55% of diseased appendices in which parasites were present.

Ectopic Enterobiasis

Complications can ensue when female worms invade organs outside their normal habitat in the large intestine. This spread occurs most frequently in women, because the vagina is well within range of the worm's migratory capabilities. Frequently, worms remain unnoticed within the vagina until their eggs are seen on Papanicolaou-stained cervicovaginal smears (Avram et al., 1984; Garud et al., 1980; Wong and Becker, 1982). More serious complications arise when the worms invade organs of the genital tract or traverse the fallopian tubes to the peritoneum. *E. vermicularis* has been implicated in salpingitis, omentitis, and periappendicitis (Nutting et al., 1980), oophoritis (Beckman and Holland, 1981; Kogan et al., 1983), peritonitis (Kahn et al., 1981), and cervicitis (Bhaskaran et al., 1980; Campbell et al., 1981). Wandering worms and accompanying enteric bacteria can also enter the urethra and cause recurrent urinary tract infections in children (Kropp et al., 1978) and perianal abscesses when the worms invade anal glands (Mortensen and Thomson, 1984). Unusual infections reported in men are postcoital urethritis (Ogungi, 1983) and suppurative balanoposthitis (Symmers, 1950).

Pathology

Pinworms lack a buccal apparatus suited to tissue penetration and may gain entrance to extraintestinal sites by utilizing small abrasions in the gut mucosa; investigations by Chandrasoma and Mendis (1977), however, suggest that the worms may be capable of penetrating the entire width of the intestine and that both males and females may be involved in anorectal lesions, whereas smaller males may be more frequently involved in tissue invasion. Hematogenous transport may carry pinworms to distant organs, as they have been recovered from the liver (Daly and Baker, 1984), spleen, lungs, and breast (Sachdev and Howards, 1975). Secondary bacterial infection is a common consequence of the pinworm's extraintestinal activities, and abscess formation may result from enteric bacteria that accompany worms or enter lesions caused by the worms.

Intestinal lesions caused by *E. vermicularis* are usually small, mucosal ulcerations or submucosal abscesses that result from secondary bacterial infection (Neafie et al., 1976). In extraintestinal organs, lesions show eosinophilic and lymphocytic infiltration, ulceration, necrosis, and granuloma formation (Sun, 1982).

Diagnostic Features

Enterobiasis should be suspected in children with behavioral problems accompanied by symptoms of insomnia, scratching of the buttocks, and lesions in the perianal area. The infection must also be considered in solitary abscesses or granulomatous lesions in the female genital tract. Stool examination is not adequate for diagnosis; repeated adhesive-tape preparations from the perianal skin or preferably presumptive treatment, may be necessary.

Prognosis

Intestinal enterobiasis is not life threatening, but can be extremely irritating. The infection is easily eliminated but may recur unless all close contacts, especially household members, are treated. Potentially fatal complications arise from ectopic infections with secondary bacterial infection.

Epidemiologic Aspects

Oxyuriasis is cosmopolitan but has a higher prevalence in cold climates than in the tropics, a difference that Beaver et al. (1984) attribute to the greater amount of clothing worn in temperate countries. Rates of infection in children range from 69 to 89% in Belgium, 30% in England, 9 to 12% in Mexico, and 1 to 12% in Surinam (Muller, 1975). Crowded conditions in institutions such as asylums and orphanages are conducive to rates of infection as high as 100%. The infection occurs more commonly in children than in adults and more often in women than in men. Whites have higher rates of infection than blacks, possibly because of racial immunity (Wagner and Eby, 1983). Infected children spread the infection by scratching the irritated perianal area, then transferring eggs to food and household items. Eggs on contaminated bedclothes are released into the atmosphere when these fomites are disturbed; the infection can then be acquired by inhalation of airborne eggs. Enterobiasis is not considered a sexually transmitted infection, but it is occasionally seen in homosexual men who practice oroanal sex (Baker and Peppercorn, 1982; McMillan, 1978).

Enterobiasis occurs frequently with infections of the flagellate *Dientamoeba fragilis;* there is some evidence, although inconclusive, that *D. fragilis* may be transmitted by the eggs of *E. vermicularis* (Burrows and Swerdlow, 1956; Yang and Scholten, 1977).

Etiologic Agents

Characterization

E. vermicularis belongs to the superfamily *Oxyuroidea* within the order *Oxyurida*. These parasites are short, stout nematodes with a distinct esophageal bulb and a single spicule in males; the females produce eggs that are flattened on one side.

Pinworms are white and resemble short lengths of cotton thread. Males measure 2 to 5 mm long and 0.1 to 0.2 mm wide and have a single copulatory spicule. The females measure 8 to 13 mm long and 0.3 to 0.5 mm wide; they have paired uteri that occupy most of the body cavity and open at a vulva situated at the anterior third of the body. A gravid female may produce about 11,000 eggs and may burst from distention of the uteri after emergence from the uterus; males die after copulation and are passed with the feces. Both sexes have a distinct esophageal bulb, and the cuticle is extended at the anterior end as small flaps called *cervical alae* and down the body as two small ridges. In cross section, the lateral alae resemble two small spurs and help to identify the worm in tissue sections. Eggs are 50 to 60 μm long and 20 to 30 μm wide; they are characteristically flattened on one side and have a thick, transparent, colorless shell. The shell consists of two layers and is coated on the outside by an albuminous gel that enables the egg to adhere to skin and clothes. The embryo is surrounded by a thin membrane and becomes infective after exposure to atmospheric oxygen for 6 h.

Life Cycle

Worms mate in the intestine and the gravid female emerges from the anus at night, lays her eggs on the perianal skin, and dies shortly thereafter; occasion-

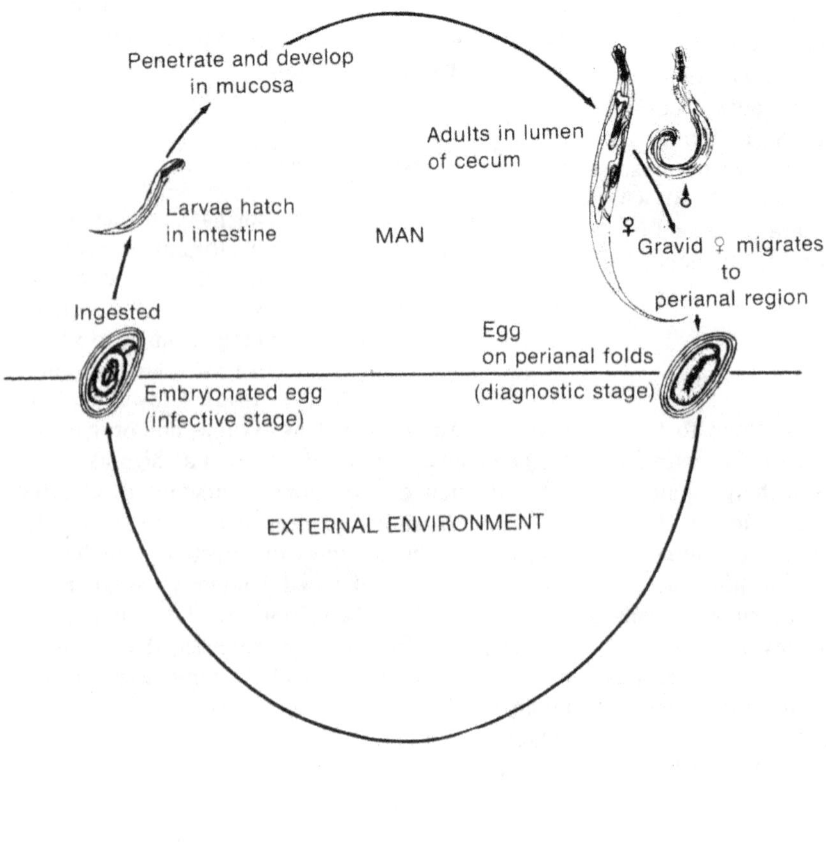

FIG. 1. Life cycle of *Enterobius vermicularis*. (Reprinted with permission from the Centers for Disease Control Training and Laboratory Program Office, Atlanta, Georgia.)

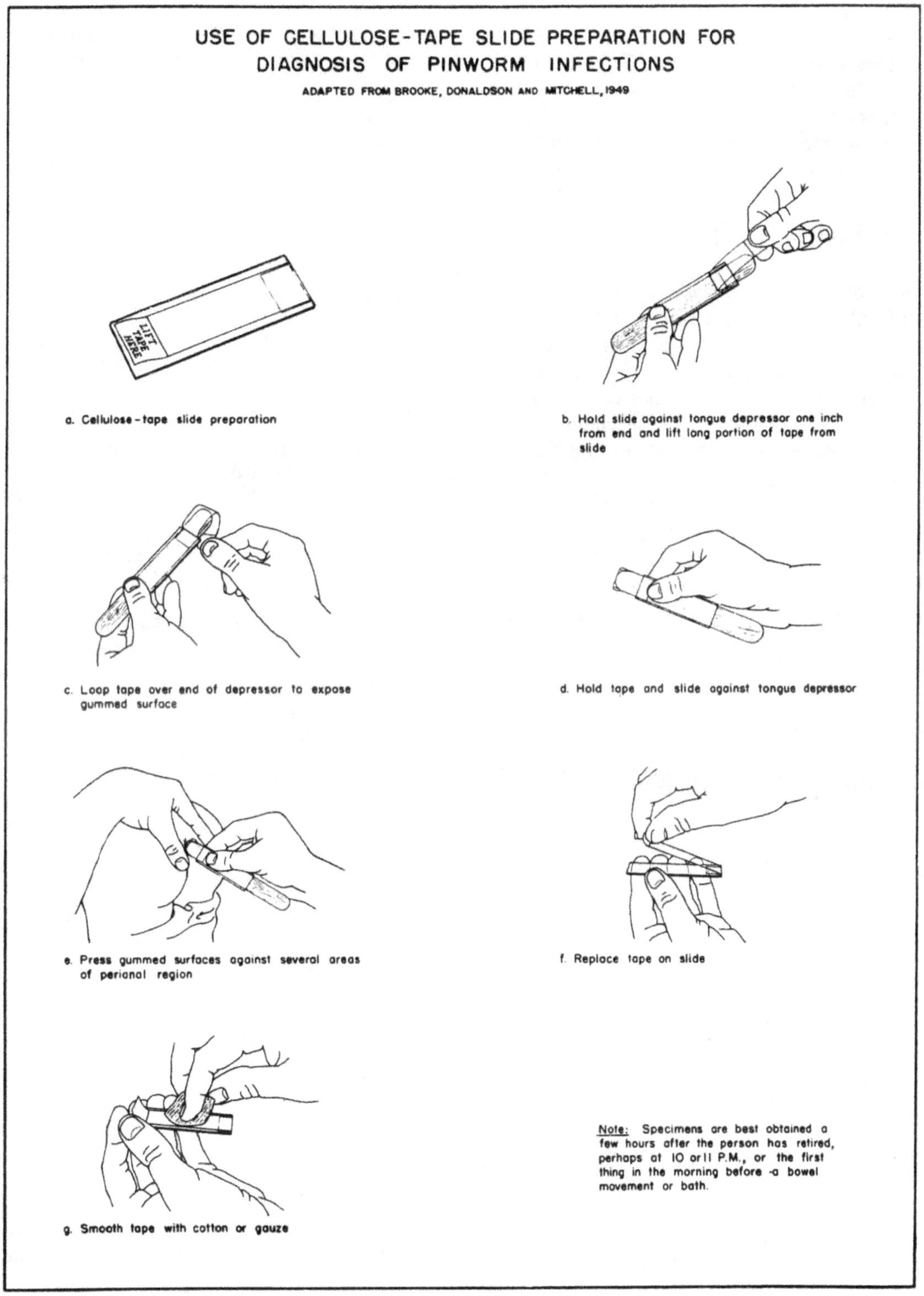

FIG. 2. Use of the cellulose-tape slide preparation for diagnosis of pinworm infections. (Reprinted with permission from the Centers for Disease Control Training and Laboratory Program Office, Atlanta, Georgia.)

ally, female worms may migrate to the vagina. Eggs become infective after 6 h in air and will remain infective for up to 2 weeks under normal household conditions. Ingested eggs hatch in the duodenum, and juvenile worms develop within the crypts of the small intestine. Adults move to the cecum and large intestine, where they attach to the mucosa (Fig. 1). Eggs develop into adults in 15 to 43 days, and adult worms live for a maximum of 53 days (Muller, 1975).

Specimen Collection and Preparation

Eggs and adults of *E. vermicularis* are occasionally seen in fecal specimens, but stool examinations are not recommended for detection of enterobiasis. The diagnostic method of choice is the adhesive-tape slide preparation (Fig. 2). The adhesive tape should be transparent, not frosted. The specimen can be better visualized and air bubbles eliminated if a drop of toluene is placed on the slide before smoothing on the tape. In heavy infections eggs can be seen as tightly packed masses, but solitary eggs can be missed, as they are transparent and colorless and can be mistaken for air bubbles or vice versa. A magnification of 400× is satisfactory for viewing the J-shaped embryo within the egg. Specimens should be taken for 7 days before a patient can be considered uninfected.

Treatment, Prevention, and Control

Treatment

Pyrantel pamoate, the current drug of choice, is effective against adult worms when given as a single dose of 11 mg/kg and repeated after 2 weeks. Mebendazole, an alternative drug that eliminates both juvenile and adult parasites, should be given as a single dose of 100 mg and repeated in 2 weeks.

Prevention and Control

All members of a family or of an institution should be treated when an infestation of enterobiasis is detected, and treatment should be repeated after 2 weeks to eliminate reinfections from eggs in the environment. Children should have their fingernails kept short to protect them from abrading the perianal skin and acquiring secondary bacterial infections and to prevent the transport of contaminated material under the nails. Good personal hygiene with frequent handwashing is necessary, especially in persons who care for children and infants. Fumigation will not destroy eggs in the environment.

Literature Cited

Avram, E., M. Yakovlevitz, and A. Schacter. 1984. Cytologic detection of *Enterobius vermicularis* and *Strongyloides stercoralis* in routine cervicovaginal smears and cytograms. Acta Cytol. **28**:468–470.

Baker, R. W., and M. A. Peppercorn. 1982. Enteric diseases of homosexual men. Pharmacotherapy **2**:32–42.

Beaver, P. C., R. C. Jung, and E. W. Cupp. 1984. Oxyuroidea and Ascaridoidea, p. 302–306. *In* Clinical parasitology, 9th ed. Lea & Febiger, Philadelphia.

Beckman, E. N., and J. B. Holland. 1981. Ovarian enterobiasis—a proposed pathogenesis. Am. J. Trop. Med. Hyg. **30**:74–76.

Bhaskaran, C. S., M. M. Reddy, and P. V. Ramalakshmi. 1980. Unusual parasitic infestation of cervix detected by cytology. Indian J. Pathol. Microbiol. **23**:51–52.

Bouree, P., and M. Dubourdieu. 1984. Parasitic appendicitis. A propos of four cases of acute appendicitis. Bull. Soc. Pathol. Exot. Filial. **77**:81–89.

Burrows, R. B., and M. A. Swerdlow. 1956. *Enterobius vermicularis* as a probable vector of *Dientamoeba fragilis*. Am. J. Trop. Med. Hyg. **5**:258–265.

Campbell, J. S., S. Mandavia, W. Threlfall, P. McCarthy, and C. N. Loverys. 1981. Pinworm granuloma of cervix uteri—incidental observation following IUD use and cone biopsy. J. Trop. Med. Hyg. **84**:215–217.

Chandrasoma, P. T., and K. N. Mendis. 1977. *Enterobius vermicularis* in ectopic sites. Am. J. Trop. Med. Hyg. **26**:644–649.

Daly, J. J., and G. F. Baker. 1984. Pinworm granuloma of the liver. Am. J. Trop. Med. Hyg. **33**:62–64.

Garud, M. A., U. Saraiya, M. Paraskar, and J. Khokhawalla. 1980. Vaginal parasitosis. Acta Cytol. **24**:34–35.

Khan, J. S., R. J. Steele, and J. Stewart. 1981. *Enterobius vermicularis* infestation of the female genital tract causing generalised peritonitis. Case report. Br. J. Obstet. Gynaecol. **88**:681–683.

Kogan, J., M. Alter, and H. Price. 1983. Bilateral *Enterobius vermicularis* salpingo-oophoritis. Postgrad. Med. **73**:309–310.

Kropp, K. A., G. A. Cichocki, and N. K. Bansal. 1978. *Enterobius vermicularis* (pinworms), introital bacteriology and recurrent urinary tract infections in children. J. Urol. **120**:480–482.

Mayers, C. P., and R. J. Purvis. 1970. Manifestations of pinworms. Can. Med. Assoc. J. **103**:489–493.

McMillan, A. 1978. Threadworms in homosexual males. Br. Med. J. **1**:367.

Morgensen, K., E. Pahle, and K. Kowalski. 1985. *Enterobius vermicularis* and acute appendicitis. Acta Chir. Scand. **151**:705–707.

Mortensen, N. J., and P. Thomson. 1984. Perianal abscess due to *Enterobius vermicularis*. Report of a case. Dis. Colon Rectum **27**:677–678.

Muller, R. 1975. *Enterobius*, p. 89–91. *In* Worms and disease. William Heinemann Medical Books Ltd., London.

Neafie, R. C., D. H. Connor, and W. M. Meyers. 1976. Enterobiasis, p. 455–459. *In* Pathology of tropical and extraordinary diseases. Armed Forces Institute of Pathology, Washington, D.C.

Nutting, S. A., R. Murphy, and F. G. Inglis. 1980. Abdominal pain due to *Enterobius vermicularis*. Can. J. Surg. **23**:286–287.

Ogungi, F. O. 1983. Postcoital pinworm infection. J. Hyg. Epidemiol. Microbiol. Immunol. **27**:103–105.

Sachdev, Y. V., and S. S. Howards. 1975. *Enterobius vermicularis* infestation and secondary enuresis. J. Urol. **113**:143–144.

Sun, T. 1982. Enterobiasis (oxyuriasis), p. 145–149. *In* Pathology and clinical features of parasitic diseases. Masson Publishing USA, New York.

Symmers, W. S. C. 1950. Pathology of oxyuriasis. Arch. Pathol. **50:**475–516.

Wagner, E. D., and W. C. Eby. 1983. Pinworm prevalence in California elementary school children, and diagnostic methods. Am. J. Trop. Med. Hyg. **32:**998–1001.

Wong, J. Y., and S. N. Becker. 1982. *Enterobius vermicularis* ova in routine cervicovaginal smears. Light and scanning electron microscope observations. Acta Cytol. **26:**484–487.

Yang, J., and T. H. Scholten. 1977. *Dientamoeba fragilis:* a review with notes on its epidemiology, pathogenicity, mode of transmission, and diagnosis. Am. J. Trop. Med. Hyg. **26:**16–22.

CHAPTER 90

Fascioliasis and Fasciolopsiasis

GEORGE V. HILLYER

Disease: Fascioliasis, fasciolopsiasis.

Etiologic Agents: *Fasciola hepatica, Fasciola gigantica, Fasciolopsis buski.*

Source: Ingestion of encysted metacercaria, generally on aquatic vegetation consumed raw, including watercress (*F. hepatica, F. gigantica*), water chestnut, or water caltrop (*F. buski*).

Clinical Manifestations: Acute fascioliasis generally occurs before the worm reaches maturity (6 to 8 weeks) and can cause a febrile fasciolitic eosinophilic syndrome involving painful enlargement of the liver, which together with fever and high eosinophilia forms a triad of symptoms of diagnostic importance. Chronic fascioliasis is generally of an obstructive biliary or inflammatory nature, as the adult worms lodge in the bile ducts. Many lightly infected individuals may be asymptomatic.

Fasciolopsiasis can cause severe, gripping pain and hunger preceding meals and intermittent diarrhea. Adult worms attaching to the mucosa of the duodenum or jejunum produce ulceration and prominent inflammatory changes. As in fascioliasis, many lightly infected individuals may be asymptomatic.

Pathology: In fascioliasis, alterations of the liver forming burrows during the migration of the young parasite correspond to the acute phase; those due to lodging of the parasite in the biliary channels, including hyperplasia of the ductal epithelium and periportal fibrosis, constitute the chronic phase of the parasite.

Laboratory Diagnosis: Definitive diagnosis of *Fasciola* infection is usually achieved by finding the eggs in feces, but only after 6 to 8+ weeks of infection; however, "false" fascioliasis due to the ingestion of eggs in infected livers must be ruled out. Generally, the number of eggs detected in feces is small. In acute fascioliasis a triad of symptoms of an individual in an endemic area with a history of eating watercress is suggestive, as are the alterations in liver enzymes (e.g., alkaline phosphatase, serum glutamic oxalacetic transaminase); detection of specific antibodies by serology is a powerful adjunct.

For fasciolopsiasis, detection of eggs in the feces is diagnostic, as their appearance in large numbers in feces precedes symptoms.

Fasciola and *Fasciolopsis* eggs are indistinguishable; thus, patient history, including dietary considerations and travel history, are important adjuncts for diagnosis.

Epidemiology: Human *Fasciola* infection is found in areas where sheep, cattle, and other large herbivorous animals are raised, including North and South America, the Caribbean, Europe, the Middle East, and Australia, and where watercress is eaten raw. Fasciolopsiasis is found primarily in the People's Republic of China and also in India, Bangladesh, Formosa, Southeast Asia, and Indonesia, where water caltrops and other water plants are consumed raw by humans.

Treatment: For fascioliasis, bithionol and triclabendazole are effective. For fasciolopsiasis, praziquantel is the drug of choice.

Prevention and Control: Education of the public on the dangers of eating uncooked, contaminated vegetables; prevention of fecal contamination of both humans and herbivorous animals to prevent infection of the snail intermediate hosts.

Description of Disease

Fascioliasis is a disease caused by digenetic trematodes of the genus *Fasciola; F. hepatica* and *F. gigantica* are the two species causing human disease. *Fasciolopsis buski,* also a digenetic trematode, causes fasciolopsiasis.

Etiologic Agents

Fasciola hepatica is a flattened, leaf-shaped, hermaphroditic trematode possessing both ovaries and testes, which are highly branched and occupy a large part of the body. The adult parasite measures 18 to 32 mm in length by 7 to 14 mm in width. It is grayish brown in color when fresh, with a flattened body that usually shows the dark outline of the blood-filled ceca. The anterior end of the parasite forms a conical projection that broadens at the shoulders, then gradually narrows towards the posterior end. At the narrow tip of the conical projection is the muscular oral sucker, which surrounds the mouth of the parasite. On the ventral surface near the base of the cone there is an acetabulum, or ventral sucker. The alimentary canal consists of two main branches of the ceca, with an elaborate system of secondary and tertiary lateral branches. The habitat of the worm is the intrahepatic biliary tree. The eggs are large, ovoid, operculate, and yellowish brown, measuring 130 to 148 by 60 to 90 μm. *Fasciola gigantica* is larger and more lanceolate, but the widening of the body is not as distinct as in *F. hepatica* (Boray, 1982). It is the common liver fluke in widespread areas of Africa and Asia, overlapping with *F. hepatica* in some parts of the world. The life cycle, pathology, and symptomatology parallel those of *F. hepatica,* but *F. gigantica* employs different snails as intermediate hosts (e.g., *Lymnaea auricularis, Fossaria ollula*), development of the various stages is slower, and metacercariae survive longer at higher temperatures and are more susceptible to dessication (Boray and Enigk, 1964; Sinclair, 1967).

Fasciolopsis buski is one of the largest flukes. The adult fluke is a flat, large, fleshy worm 2 to 7 cm long, 2 cm wide, and 0.5 to 3 mm thick. The intestine includes a prepharynx, a pharynx, an esophagus, and an unbranched ceca. The markedly branched testes are placed one behind the other in the posterior half of the worm. The branched ovary lies to the right of the midline. The highly branched uterus occupies the anterior half of the worm. The adult fluke inhabits the adult intestine, particularly the duodenum, but sometimes may be found in the stomach or in the large intestine of man (Rim, 1982).

In Japan, a *Fasciola* sp. has been reported that resembles *F. gigantica* morphologically, but resembles *F. hepatica* ecologically (Watanabe, 1965).

Humans typically are infected with *F. hepatica* when they ingest the metacercariae encysted on watercress. The wall of the metacercaria is digested in the stomach, and the encysted larva burrows through the wall of the gut into the abdominal cavity. Within several weeks the larva has penetrated into the liver, tunneling its way through and ingesting the liver parenchyma. By 5 to 6 weeks postinfection the worm settles in the bile ducts, and 2 to 3 weeks later it has become an egg-laying mature worm. At this time (6 to 8 weeks postinfection) eggs can be seen in the feces. Although the adult worm is hermaphroditic, cross-fertilization between two worms is thought to be the most common form of sexual reproduction. The embryos within the eggs undergo cleavage before leaving the interior of the worm. However, the final formation of the ciliated miracidium in fresh water is temperature dependent. Under optimal conditions in Puerto Rico, a mature ciliated miracidium is formed within 2 to 3 weeks after leaving the mammalian host (Hillyer, 1981). The egg has an operculum through which the miracidium emerges upon exposure to daylight. The miracidium swims in fresh water and must find and penetrate a suitable snail intermediate host of the family *Lymnaeidae*. Once inside the snail, the miracidium sheds its cilia and is transformed into a saclike sporocyst, usually in the pulmonary cavity of the snail. As the sporocyst grows, new larval forms called *mother* or *primary rediae* develop. These rediae break free from the sporocyst and migrate to the digestive gland (i.e., hepatopancreas) of the snail. As the mother rediae grow, new daughter rediae develop and eventually form cercariae. The miracidium-to-cercaria life cycle varies with temperature and with the species of the snail intermediate host. In Puerto Rico, this cycle is completed in 28 to 30 days in the case of *Lymnaea* (=*Fossaria* =*Bakerilymnaea*) *cubensis* and in 57 to 60 days when *L.* (=*Pseudosuccinea*) *columella* is the host (Hillyer, 1981). Other snail intermediate hosts are listed by Pantelouris (1965). The cercariae emerge from the snail during daylight, encyst on vegetation, and form the metacercaria, a process that occurs within 24 h of attachment of the cercaria to the vegetation; thereafter, the metacercaria is infective. Under appropriate environmental conditions the metacercaria can survive for 1 year. The life cycle continues when the metacercaria is ingested by the mammalian host. In humans, the adult worms can live for at least 6 years. In this documented case, infection was acquired when the patient ingested wild watercress (*Nasturtium officinale*) containing infectious cysts while camping (Jones et al., 1977).

Regarding *F. buski,* when the immature egg is discharged in feces and deposited in fresh water it requires 3 to 7 weeks at 27 to 32°C to complete its development, after which the miracidium hatches. The miracidium penetrates the soft tissues of planorbid snails (e.g., *Segmentina* spp., *Hippeutis* spp.) and over the next month transforms into the sporocyst, first- and second-generation rediae, and finally cercariae. The free-swimming cercariae, almost identical to those of *F. hepatica,* usually encyst on the seed pods of water caltrop, on the bulbs of the water chestnut, and on the roots of other aquatic vegetation and mature in 3 to 4 weeks. Humans are infected by ingesting the raw aquatic plants; the metacercariae excyst in the duodenum, become attached to the nearby intestinal wall, and develop into adult worms in 1 to 3 months (Rim, 1982; Barlow, 1925). Each adult worm produces numerous eggs daily: Barlow (1925) estimated 10,000 to 15,000 eggs per worm per day; Stoll and Kwei (1927) estimated 25,000 eggs per worm per day; and Sadun and Maiphoom (1953) found that the average egg count of infected persons in central Thailand was more than 5,000 eggs per gram of feces.

Source

Humans are infected with *F. hepatica* primarily by ingesting the infectious metacercariae in watercress, and probably accidentally while sucking or chewing on grass blades containing the cysts.

The most common source of infection with *F. buski* is ingestion of uncooked aquatic vegetables or nuts, most commonly water caltrop as well as water chestnut, water lotus or hyacinth stems, and water bamboo, containing the infectious metacercariae. In the case of water caltrop nuts or water chestnut tubers, infection occurs when the hull or skin is peeled off with the teeth and lips.

Clinical Manifestations

The acute phase of human fascioliasis usually presents with a transitory period of dyspepsia, followed by an abrupt onset of high fever and abdominal pain of variable intensity localized to the right hypochondrium or epigastrium. Other symptoms such as prostration, anorexia, sweating, myalgias, joint and bone pains, violent headache, nausea, and vomiting may also develop. Acute fascioliasis causes what is known as a febrile fasciolitic eosinophilic syndrome, involving a painful enlargement of the liver that, together with fever and marked (80 to 90%) eosinophilia, forms a triad of symptoms of diagnostic importance when seen in an endemic area (Naquira-Vildoso and Marcial-Rojas, 1971).

The chronic phase of fascioliasis occurs when the flukes gain access to the bile ducts, mature, and lay eggs. In this phase symptoms are very few and are usually of an obstructive biliary or inflammatory nature. Surgical complications of the extrahepatic bile ducts caused by fascioliasis have been described, the clinical picture being similar to that of choledocholithiasis. The unsuspected presence of the disease may be found upon exploration of the common bile duct or its branches. Most of the cases observed in this phase of the disease have been discovered in this fashion (Naquira-Vildoso and Marcial-Rojas, 1971). A liver scan may reveal multiple focal defects, which disappear after successful chemotherapy (Hillyer et al., 1984).

In Spain, surgeons extracted four adult *F. hepatica* worms from the principal biliary tract of a patient with gallstones (Pina Arroyo et al., 1982). One suspects that the number of individuals operated on for liver tumors and found to have *F. hepatica* worms instead is much larger than that indicated by the literature. Two cases of suspected hydatid cyst or hepatoma were found to be *F. hepatica* after surgery (Noya et al., 1982).

Ectopic fascioliasis has been observed; for example, worms have been discovered in a furunculoid or cystic mass excised from the subcutaneous or intermuscular tissues of the abdominal or chest wall or in a subcutaneous swelling on the sternum (see Beaver et al. [1984]). Park et al. (1984) reported a case of ectopic human *Fasciola* spp. infection in the cecal wall of a Korean woman. She presented with nausea, vomiting, and epigastric tenderness, and 1 week later a palpable mass was discovered in the right iliac fossa, resulting in a clinical diagnosis of carcinoma of the colon. The patient underwent cecal resection, and the removed mass proved to be an inflammatory reaction containing numerous tracts made by the migrating fluke.

It should be pointed out that many infected individuals are asymptomatic, and infection has been detected incidentally, both in Peru and in Puerto Rico, on stool or serologic examination (Bendezú et al., 1982; Naquira-Vidoso and Marcial-Rojas, 1971). In the Puerto Rican study a mountain region was selected because bovine fascioliasis was hyperendemic in farms upstream from farms where watercress was grown locally for human consumption and the appropriate snail intermediate host was present. Of 110 human fecal samples examined, 12 had small numbers of *F. hepatica* eggs (1 to 2 eggs per gram); 11 of the 12 patients had a history of eating watercress, and all were asymptomatic (Bendezú et al., 1982).

Some textbooks report that a clinical condition called *halzoun* (known as *marra syndrome* in the Sudan) develops after the ingestion of infected sheep livers containing *F. hepatica* worms, causing a pharyngeal fascioliasis in which the adult worms lodge on the pharyngeal mucosa. In Lebanon the disease is linked with the eating of raw or undercooked sheep or goat liver or lymph nodes, and in the Sudan with the ingestion of various raw visceral organs of sheep, goats, cattle, or camels. Several minutes to half an hour or more after eating, there is discomfort and a prickling sensation deep in the throat; pain may later extend to the ears. Edematous congestion of the fauces, tonsils, larynx, Eustachian tubes, nasal passages, conjunctivae, and lips is sometimes marked. Nasal and lachrymal discharges, episodic sneezing and coughing, dyspnea, dysphagia, dysphonia, and frontal headache are common. Schacher et al. (1969), however, believe that this condition, originally described in 1905, is probably caused by nymphs of *Linguatula serrata* encapsulated in the liver and lymph nodes of domestic herbivores, which serve as the natural intermediate hosts of this parasite (see also Beaver et al. [1984]).

Experimenting with fasciolopsiasis, Barlow (1925) swallowed 132 *F. buski* cysts and found eggs in his stools 1 month later, unaccompanied by noticeable symptoms. Two weeks later, mild symptoms in the form of epigastric pain and hunger preceding meals appeared and continued for several days, subsided, and then renewed with increasing severity. The major subsequent symptoms were intermittent diarrhea accompanied by severe gripping pain, and "the gas passed was of an offensive nature and generated in large quantities." At the conclusion of this 3-month experiment, 124 flukes were recovered. The adult worms attach to the mucosa of the duodenum or jejunum and produce ulceration and prominent inflammatory changes. The traumatic effect may produce an increase in secretion of mucus and minimal bleeding. In severe and late infection the large number of worms present causes more extensive ulceration, producing intestinal stasis and occasionally obstruction, edema, and eosinophilia. The symptoms usually develop in 30 to 60 days after exposure, especially in massive infections; however, infected individuals do not always show noticeable symptoms. In children the abdomen is frequently distended, and the skin is harsh and dry. There is extreme prostration, and death results from intoxication and exhaustion (Rim, 1982).

Pathology

In fascioliasis, alterations of the liver during the migration of the young parasite correspond to the acute phase; changes caused by the lodging of the parasite in the biliary channels constitute the chronic phase of the disease. The migration of excysted metacercariae through the wall of the intestine causes no significant lesions in humans. About 2 weeks after infection the developing flukes penetrate the Glisson capsule and burrow through the parenchyma of the liver; after approximately 6 to 8 weeks they invade the bile ducts. These burrows are rarely identified in a recently infected individual. Following passage of the worm, the burrows heal and only scar tissue remains. Subcapsular hematomas of the liver and even intraabdominal hemorrhages have been described in patients who have undergone laparotomy for acute abdominal pain and shock. In the acute phase, liver function tests are usually abnormal and leukocytosis and eosinophilia are present. When the flukes have invaded the bile ducts, the liver parenchymal lesions are usually healed, and symptoms are related to obstruction of the bile ducts. There is hyperplasia of the ductal epithelium and periportal fibrosis (Meyers and Neafie, 1976). Laparoscopic findings include yellowish-white nodules 1 to 4 mm in diameter on the liver surface. Close to these nodules, ribbed or veriform formations of a color and consistency similar to that of the nodules are also observed in some patients. It is thought that these formations correspond to microabscesses caused by contact between the parasite and the intrahepatic bile ducts (Uribarrena et al., 1985).

In fasciolopsiasis, the worms cause ulceration, inflammation, and hypersecretion of the intestinal mucosa around their site of attachment. There may be minimal bleeding, and ileus and abscesses within the mucosa usually accompany heavy infections. In light infections, worms inhabit the duodenum and jejunum; they may also occupy the stomach, ileum, and colon in heavy infections. Large numbers of worms may cause intestinal obstruction, and the absorption of toxic or allergic metabolites causes generalized edema, most markedly in the face. Leukocytosis and absolute eosinophilia are common (Connor and Neafie, 1976). Sometimes the patient will vomit adult worms, as in one fatal case reported by Sadun and Maiphoom (1953) and in a child reported by Hadidjaja et al. (1982).

Laboratory Diagnosis

Definitive diagnosis of *Fasciola* infection is usually achieved by finding the eggs in feces. However, "false" fascioliasis can occur when *F. hepatica* eggs are found in the stool following ingestion of infected liver. In such cases, diagnosis requires placing the patient on a liver-free diet and performing repeated follow-up stool examinations. For "true" fascioliasis, repeated stool examinations are often needed to demonstrate the eggs; nonetheless, in many cases eggs are never recovered. *Fasciola hepatica* worms mature and lay eggs after 8 weeks of infection. As symptoms of acute fascioliasis occur earlier in the infection, eggs are virtually never found in the stools; thus, indirect methods such as serodiagnosis constitute a useful adjunct for identifying infection (see below). Serum enzymes to detect liver function (e.g., alkaline phosphatase, serum glutamic oxalacetic transaminase, serum glutamic pyruvic transaminase, sorbitol dehydrogenase) increase during acute fascioliasis (Schiappacasse et al., 1985; Sinclair, 1967).

Because of the possibility of false fascioliasis and the frequent absence of *F. hepatica* eggs in acutely ill patients, serodiagnosis is clearly a viable and excellent alternative for detecting infections. The methods and antigens used have been reviewed recently (Hillyer, 1986). With crude antigens, immunoelectrophoresis is the method of choice to detect immunoprecipitin "arc 2" (Biguet et al., 1962; Capron et al., 1971). Alternatively, counterelectrophoresis and Ouchterlony immunodiffusion are likewise effective. Counterelectrophoresis, which provides a reading in less than 1 h, is particularly useful when a rapid diagnosis is required (Hillyer, 1975; Hillyer and Capron, 1976). When using these last two methods, a requirement for specificity is to use in adjacent wells serum from individuals with fascioliasis whose primary precipitin is to arc 2. Thus, by observing the linking of lines with the unknown patient serum, one can diagnose the patient as having fascioliasis. The additional advantage of the precipitation tests is that they become seronegative several months after successful therapy (Hillyer and del Llano de Diaz, 1976; Hillyer et al., 1984). Enzyme-linked immunosorbent assay is an extremely sensitive method, becoming seropositive by 2 weeks postinfection (Levine et al., 1980) and seronegative after therapy (Hillyer et al., 1984). Cross-reactivity of crude extracts with other parasitic infections have been resolved by gel filtration in the case of *F. hepatica* (Santiago de Weil et al., 1984) and *F. gigantica* (Mansour et al., 1983), and "arc 2" antigen can be isolated by immunoaffinity chromatography (Santiago de Weil et al., 1984). Other candidate antigens have been isolated by electroelution (Santiago and Hillyer, 1986). When only crude *F. hepatica* antigens are available for use in the enzyme-linked immunosorbent assay and schistosomiasis is endemic, a circumoval precipitin test with *Schistosoma mansoni* eggs and patient serum is also done. Positive serology with *F. hepatica* antigen in conjunction with a negative circumoval precipitin test is strongly suggestive of fascioliasis, as the test does not cross-react with serum from *Fasciola* infection (Hillyer and Sagramoso de Ateca, 1980).

Western blot analysis of sera from 13 patients with fascioliasis reacted with *F. hepatica* excretory-secretory antigens shows that antibodies recognize two antigenic polypeptides of 63 and 17 kDa. A marked decrease in intensity of recognition of the 63 kDa antigen was observed in patients treated and cured of *F. hepatica* for 2 years. Antibody reactivity to the 17 kDa antigen disappeared within 1 year after cure. These two antigenic polypeptides are not recognized by the serum of patients with trichinosis or with schistosomiasis mansoni. Thus, the 63 and 17 kDa *F. hepatica* excretory-secretory antigens are excellent candidates for the immunodiagnosis of human fascioliasis and for the prediction of cure (Hillyer and Soler de Galanes, unpublished).

For fasciolopsiasis, diagnosis is normally accomplished by detecting eggs in feces. As these eggs usually appear in the feces before symptoms develop, and in sizable numbers compared with those of *F. hepatica*, egg detection is the method of choice. Confirmation can be achieved by sifting feces after chemotherapy with praziquantel and finding evacuated worms with blisters (sic) on the tegument (Harinasuta et al., 1984).

Fasciola and *Fasciolopsis* eggs are morphologically indistinguishable and also can be confused with those of *Echinostoma*. Thus, the dietary history of the patient and the region of infection are important considerations.

Epidemiology

Human infection with *F. hepatica* is found in sheep- and cattle-raising areas in North and South America, the Caribbean, Europe, the Middle East, and Australia. Cattle, water buffalo, and other large herbivorous animals harbor *F. gigantica*, and the excreted eggs infect the appropriate snail intermediate host. Eventually cercariae are released, which in turn encyst in aquatic vegetation.

Several human infections with a Japanese strain of *Fasciola* sp. have been recorded, but its speciation is unclear (Watanabe, 1965).

Fasciolopsis buski is a common parasite of humans and pigs where people commonly eat uncooked metacercariae-laden plants; it is found primarily in China, but is also endemic in India, the Dacca area of Bangladesh, Formosa, Southeast Asia, and Indonesia. The parasite was probably introduced into Bangladesh from India in 1948, when large-scale migrations of people from Bihar and Assam, both endemic for *F. buski*, took place (Gilman et al., 1982). Sadun and Maiphoom (1953) examined the epidemiology of fasciolopsiasis in central Thailand and found that human infection was endemic in areas where water caltrops (*Trapa bicornis*) were cultivated. The prevalence and intensity of infection were highest among children 10 to 14 years of age, presumably because this age group was responsible for picking up the water caltrops on their way to and from school; however, infection was prevalent at all ages and in both sexes. Cross (1969) provided a literature review (and excellent bibliography) on human fasciolopsiasis in Southeast Asia and the Far East. The disease was first reported in a child in Indonesia who vomited seven worms (Hadidjaja et al., 1982). There is, however, a paucity of information on the current prevalence of fasciolopsiasis (see Rim, 1982).

Treatment, Prevention, and Control

Treatment

For infection with *Fasciola* sp., bithionol (30 to 50 mg/kg orally, on alternate days for 2 or 3 weeks) and hexachloroparaxylene (25 to 35 mg/kg orally, twice daily after food) are probably effective (Campbell and Garcia, 1986; Hillyer et al., 1984). Praziquantel (75 mg/kg orally, in three divided doses over 1 day) was shown effective in an infected Yemenite (Schiappacasse et al., 1985), although in a more extensive study by Knobloch et al. (1985) only 21% of 34 Peruvian patients with fascioliasis stopped excreting *F. hepatica* eggs even though half of the patients had been treated repeatedly. Dugernier et al. (1986) found that high doses of mebendazole (4 g/day for 3 weeks) cured a patient with invasive *F. hepatica*. Triclabendazole is a novel anthelminthic that has been found to be highly effective against adult and immature stages of *F. hepatica* (Bennett and Köhler, 1987; Eckert et al., 1984).

For infection with *F. buski*, the drug of choice is praziquantel. In a trial conducted in schoolchildren in Thailand, this drug was 100% effective when given as a single dose of 15 mg/kg. Side effects occurred in 21% of the children, but were mild or transient; sleepiness was the most common side effect (Harinasuta et al., 1984). Hexylresorcinol, bephenium, dichlorophen, and tetrachloroethylene were previously used for this infection (Campbell and Garcia, 1986). Of the latter group, tetrachloroethylene is the most effective; however, because of its potential for causing anaphylactic reactions, especially when worm burdens are heavy, the drug should be administered under clinical supervision and preferably with prior antihistamine treatment (Rabbani et al., 1985).

Prevention and Control

An important preventive measure is to educate the public regarding the mode of transmission of the parasite and the dangers of eating uncooked, contaminated vegetables. Prevention of fecal contamination of waters used for plant growing where infected sheep and cattle (*F. hepatica*) or swine and humans (*F. buski*) are present is also essential.

Acknowledgments. The investigations of the author have been supported by USPHS–NIH Grant No. AI22906 and the Parasitic Diseases Research Program of Puerto Rico through USPHS–NIH Grant No. RR-2657.

Literature Cited

Barlow, C. H. 1925. The life cycle of the human intestinal fluke *Fasciolopsis buski* (Lankester). Am. J. Hyg. mongr. series. no. 4.

Beaver, P. C., R. C. Jung, and E. W. Cupp. 1984. Clinical parasitology, 9th ed. Lea & Febiger, Philadelphia.

Bendezú, P., A. Frame, and G. V. Hillyer. 1982. Human fascioliasis in Corozal, Puerto Rico. J. Parasitol. **68**:297–299.

Bennett, J. L. and P. Köhler. 1987. *Fasciola hepatica:* action *in vitro* of triclabendazole on immature and adult stages. Exp. Parasit. **63**:49–57.

Biguet, J., A. Capron, and P. Tran Van Ky. 1962. Les antigenes de *Fasciola hepatica*. Etude electrophoretique et immunoeléctrophoretique. Identification des fractions et comparaison avec les antigènes correspondant a sept autres helminthes. Ann. Parasitol. Hum. Comp. **37**:221–231.

Boray, J. C. 1982. Fascioliasis, p. 71–88. *In* J. H. Steele, G. V. Hillyer, and C. E. Hopla (ed.), CRC handbook series in zoonoses. Section C: parasitic zoonoses, vol. III. CRC Press, Inc., Boca Raton, Fla.

Boray, J. C., and K. Enigk. 1964. Laboratory studies on the survival and infectivity of *Fasciola hepatica*- and *F. gigantica*-metacercariae. Z. Tropenmed. Parasitol. **15**:324–331.

Campbell, W. C., and E. G. Garcia. 1986. Trematode infections in man, p. 385–399. *In* W. C. Campbell and R. S. Rew (ed.), Chemotherapy of parasitic diseases. Plenum Publishing Corp., New York.

Capron, A., A. Vernes, and J. Fruit. 1971. Le diagnostic immunologique de la distomatose hépatica à *F. hepatica*. Bilan personnel à propos de 430 observations. Etude comparée de trois techniques diagnostiques (immunoelectrophorese, reaction de fixation de complement et immunofluorescence). Rev. Pathol. Comp. **70**:31–36.

Connor, D. H., and R. C. Neafie. 1976. Fasciolopsiasis, p. 528–529. *In* C. H. Binford and D. H. Connor (ed.), Pathology of tropical and extraordinary diseases, vol. 2, Armed Forces Institute of Pathology, Washington, D.C.

Cross, J. H. 1969. Fasciolopsiasis in southeast Asia and the Far East: a review, p. 177–196. *In* C. Harinasuta (ed.), Proceedings of the Fourth Southeast Asian Seminar on Parasitology and Tropical Medicine, Schistosomiasis and other Snail-Transmitted Helminthiasis. Watana Panich Press, Bangkok.

Dugernier, T., A. Geubel, G. Bigaignon, J.-Y. Cesbron, and E. Coche. 1986. Human fascioliasis: cure by mebendazole? A case report. Gastroenterol. Clin. Biol. **10**:513–516.

Eckert, J., G. Schneiter, and K. Wolff. 1984. FASINEX (Triclabendazole)—ein neues fasziolizid. Berliner und Münchner Tierärztliche Wochenschrift **91**:349–356.

Gilman, R. H., G. Mondal, M. Maksud, K. Alam, E. Rutherford, J. B. Gilman, and M. U. Khan. 1982. Endemic focus of *Fasciolpsis buski* infection in Bangladesh. Am. J. Trop. Med. Hyg. **31**:796–802.

Hadidjaja, P., H. M. Dahri, R. Roesin, S. S. Margono, J. Djalins, and M. Hanafiah. 1982. First autochtonous case of *Fasciolopsis buski* infection in Indonesia. Am. J. Trop. Med. Hyg. **31**:1065.

Harinasuta, T., D. Bunnag, and P. Radomyos. 1984. Efficacy of praziquantel on fasciolopsiasis. Arzneimittelforsch **34**:1214–1215.

Hillyer, G. V. 1975. Use of counterelectrophoresis to detect infections of *Fasciola hepatica*. J. Parasitol. **61**:557–559.

Hillyer, G. V. 1981. Fascioliasis in Puerto Rico—a review. Bol. Asoc. Med. P.R. **73**:94–101.

Hillyer, G. V. 1986. Fascioliasis, paragonimiasis, clonorchiasis, and opisthorchiasis, p. 39–68. *In* K. W. Walls and P. M. Schantz (ed.), Immunodiagnosis of parasitic diseases, vol. 1. Academic Press, Inc., New York.

Hillyer, G. V., R. H. Bermudez, and G. Ramírez de Arellano. 1984. Use of immunologic techniques to predict success of therapy in human fascioliasis: a case report. Bol. Asoc. Med. P.R. **76**:116–119.

Hillyer, G. V., and A. Capron. 1976. Immuodiagnosis of human fascioliasis by counterelectrophoresis. J. Parasitol. **62**:1011–1013.

Hillyer, G. V., and A. del Llano de Diaz. 1976. Use of immunologic techniques to detect chemotherapeutic success in infection with *Fasciola hepatica*. I. Rabbit infections. Am. J. Trop. Med. Hyg. **25**:307–311.

Hillyer, G. V., and L. Sagramoso de Ateca. 1980. Antibody responses in murine schistosomiasis and fascioliasis. Am. J. Trop. Med. Hyg. **29**:598–601.

Jones, E. A., J. M. Kay, H. P. Milligan, and D. Owens. 1977. Massive infection with *Fasciola hepatica* in man. Am. J. Med. **63**:836–842.

Knobloch, J., A. Delgado, A. Alvarez, U. Reymann, and R. Bialek. 1985. Human fascioliasis in Cajamarca/Peru. I. Diagnostic methods and treatment with praziquantel. Trop. Med. Parasit. **36**:88–90.

Levine, D. M., G. V. Hillyer, and S. I. Flores. 1980. Comparison of counterelectrophoresis, the enzyme-linked immunosorbent assay, and Kato fecal examination for the diagnossis of fascioliasis in infected mice and rabbits. Am. J. Trop. Med. Hyg. **29**:602–608.

Mansour, N. S., F. G. Youssef, E. M. Mikhail, and F. N. Boctor. 1983. Use of partially purified *Fasciola gigantica* worm antigen in the serological diagnosis of human fascioliasis in Egypt. Am. J. Trop. Med. Hyg. **32**:550–554.

Meyers, W. M., and R. C. Neafie. 1976. Fascioliasis (p. 524–527). *In* C. H. Binford and D. H. Connor (ed.), Pathology of tropical and extraordinary diseases, vol. 2, Armed Forces Institute of Pathology, Washington, D.C.

Naquira-Vildoso, F., and R. Marcial-Rojas. 1971. Fascioliasis, p. 477–489. *In* R. A. Marcial-Rojas (ed.), Pathology of protozoal and helminthic diseases. The Williams & Wilkins Co., Baltimore.

Noya, M., M. Lema, J. Castillo, L. Rodriguez-Cameselle, and W. M. Ali Al-Shaban. 1982. Parasitación por *Fasciola hepatica*. Claves diagnósticas. Rev. Clin. Esp. **166**:177–179.

Pantelouris, E. M. 1965. The common liver fluke. Pergamon Press, Inc., Elmsford, N.Y.

Park, C. I., J. Y. Ro, H. Kim, and Y. Gutierrez. 1984. Human ectopic fascioliasis in the cecum. Am. J. Surg. Pathol. **8**:73–77.

Pina Arroyo, J., J. García García, J. L. Villar Galan, and A. Gomez Alonso. 1982. *Fasciola hepatica* en la vía biliar principal. Rev. Esp. Enferm. Apar. Dig. **62**:496–501.

Rabbani, G. H., R. H. Gilman, I. Kabir, and G. Mondel. 1985. The treatment of *Fasciolopsis buski* infection in children: a comparison of thiabendazole, mebendazole, levamisole, pyrantel pamoate, hexylresorcinol and tetrachloroethylene. Trans. R. Soc. Trop. Med. Hyg. **79**:513–515.

Rim, H.-J. 1982. Fasciolopsiasis, p. 89–97. *In* J. H. Steele, G. V. Hillyer, and C. E. Hopla (ed.), CRC handbook series in zoonoses. Section C: parasitic zoonoses, vol. III. CRC Press, Inc., Boca Raton, Fla.

Sadun, E. H., and C. Maiphoom. 1953. Studies on the epidemiology of the human intestinal fluke, *Fasciolopsis buski* (Lankester) in central Thailand. Am. J. Trop. Med. Hyg. **2**:1070–1084.

Santiago, N., and G. V. Hillyer. 1986. Isolation of potential serodiagnostic *Fasciola hepatica* antigens by electroelution from polyacrylamide gels. Am. J. Trop. Med. Hyg. **35**:1210–1217.

Santiago de Weil, N., G. V. Hillyer, and E. Pacheco. 1984. Isolation of *Fasciola hepatica* genus-specific antigens. Int. J. Parasitol. **14**:197–206.

Schacher, J. F., S. Saab, R. Germanos, and N. Boustany. 1969. The aetiology of Halzoun in Lebanon: recovery of *Linguatula serrata* nymphs from two patients. Trans. R. Soc. Trop. Med. Hyg. **63**:854–858.

Schiappacasse, R. H., D. Mohammadi, and A. J. Christie. 1985. Successful treatment of severe infection with *Fasciola hepatica* with praziquantel. J. Infect. Dis. **152**:1339–1340.

Sinclair, K. B. 1967. Pathogenesis of *Fasciola* and other liver-flukes. Helminthol. Abstr. **36**:115–134.

Stoll, N. R., and W. S. Kwei. 1927. Egg-worm correlations in cases of *Fasciolopsis buski*. J. Parasitol. **13**:166–172.

Uribarrena, R., F. Borda, M. Muñoz, and A. Rivero-Puente. 1985. Laparoscopic findings in eight cases of liver fascioliasis. Endoscopy **17**:137–138.

Watanabe, S. 1965. A revision of the genus *Fasciola* in Japan, with particular reference to *F. hepatica* and *F. gigantica*. p. 359–381. *In* K. Morishita, Y. Komiya, and H. Matsubayashi (ed.), Progress of medical parasitology in Japan, vol. II. Meguro Parasitological Museum, Tokyo.

CHAPTER 91

Filariasis

D. O. FREEDMAN and T. B. NUTMAN

Disease: Filariasis (bancroftian filariasis, brugian filariasis, onchocerciasis, loiasis, streptocerciasis, mansonellosis).
Etiologic Agents: *Wuchereria bancrofti, Brugia malayi, Brugia timori, Onchocerca volvulus, Mansonella perstans, Mansonella ozzardi,* and *Mansonella streptocerca.*
Source: Human-to-human transmission via arthropod vector.
Clinical Manifestations: Lymphangitis, lymphedema, elephantiasis, lymphadenopathy, pruritus, angioedema, dermatitis, keratitis, subcutaneous nodules. May be asymptomatic.
Pathology: Immune-mediated inflammatory reaction around both adult and larval stages of the parasite in the lymphatic system, skin, eyes, and subcutaneous tissues.
Laboratory Diagnosis: Identification of microfilariae in blood or skin snips. Identification of adult parasites in the lymphatic system or tissue. Serology helpful in nonendemic persons, but is not species-specific.
Epidemiology: Worldwide; mostly tropical. Lymphatic forms predominantly in India, China, Indonesia, and South Pacific. Onchocerciasis in Africa and Central America.
Treatment: Diethylcarbamazine and ivermectin as microfilaricides. Diethylcarbamazine and suramin as macrofilaricides, dependent on species.
Prevention and Control: Mass treatment with diethylcarbamazine and vector control.

Description of Disease

The filariases are vectorborne human parasitic diseases with a complex life cycle; they are caused by nematodes, which, in their mature stage, reside in the lymphatic system or in the connective tissue. Although the adult parasite itself may provoke impressive and chronic inflammatory reactions in tissues, it can live for years or decades continuously producing huge numbers of larvae. These larvae, called microfilariae, can themselves cause significant, debilitating, and long-term chronic pathology.

There are eight filarial species in which humans are the definitive hosts: *Wuchereria bancrofti, Brugia malayi, Brugia timori* (all causing lymphatic filariasis), *Onchocerca volvulus* (causing onchocerciasis or river blindness), *Loa loa* (eyeworm), *Mansonella perstans, Mansonella ozzardi,* and *Mansonella streptocerca*. However, only three, *W. bancrofti, B. malayi,* and *O. volvulus,* are responsible for the majority of human disease.

The World Health Organization (1984) estimates that over 300 million people are constantly exposed to the vectors of lymphatic filariasis and approximately 30 million to that of onchocerciasis, indicating the magnitude of the public health problem. Of importance to clinicians are the striking differences in the clinical manifestations of disease in those individuals native to the endemic areas and in those newly arrived or resident in the area for a relatively short period (Wartman and King, 1947).

Life Cycle and Mode of Transmission

The rather complex life cycle of the filariae is schematized in Fig. 1, and Table 1 provides more explicit details for individual species. Maturation of the in-

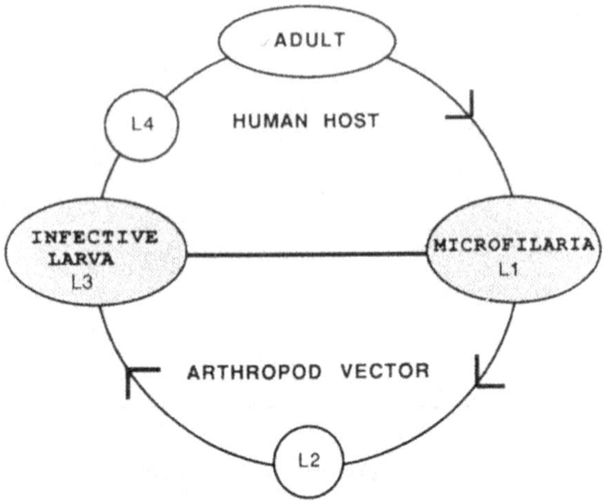

FIG. 1. Schematized life cycle of the filariae. The horizontal line represents the exchange of larvae between the host and the vector during a blood meal. L2 = the larval development stage in the vector; L4 = the immature adult larval stage in humans.

fective (third-stage) larvae in the arthropod vector takes 1 to 3 weeks, whereas it is thought to take a minimum of 4 to 6 months for the development of a mature adult female in the human host after infection. Clinical manifestations usually take longer to become apparent, and thus the disease is rare in the short-term traveler to endemic areas. Exposure to infective larvae must be intense and prolonged for infection to occur.

Clinical Manifestations

Disease expression varies with the species of filaria. Newly exposed individuals characteristically have manifestations of typical acute symptomatology, with more rapid progression to chronic or irreversible pathology, than do natives of the endemic area. Profound eosinophilia and high serum levels of immunoglobulin E are common.

Lymphatic Filariasis

Asymptomatic Disease

Asymptomatic microfilaremia is a common manifestation of bancroftian or brugian filariasis, and in all endemic areas a proportion of the infected individuals never become symptomatic. In contrast, many of the chronically infected persons are amicrofilaremic.

Filarial Fevers

Acute attacks of retrograde adenolymphangitis accompanied by fever, chills, and malaise (lasting 3 to 15 days each) are typical initial manifestations of disease and can occur several times per year. Lympha-

TABLE 1. Characteristics of the filariae

Organism	*Periodicity*	*Distribution*	*Vector*	*Location of adult*	*Location of microfilariae*
Wuchereria bancrofti	Nocturnal	Cosmopolitan Mainly India China, Indonesia	*Culex* *Anopheles* *Aedes*	Lymphatic	Blood
	Subperiodic	Eastern Pacific	*Aedes*	Lymphatic	Blood
Brugia malayi	Nocturnal	Southeast Asia Indonesia, India	*Mansonia* *Anopheles*	Lymphatic	Blood
	Subperiodic	Indonesia Southeast Asia	*Coquilletidia* *Mansonia*	Lymphatic	Blood
Loa loa	Diurnal	West and Central Africa	*Chrysops* (deerfly)	Subcutaneous	Blood
Mansonella ozzardi	None	South and Central America Caribbean	*Culicoides* *Simulium*	Undetermined	Blood
Mansonella perstans	None	South and Central America, Africa	*Culicoides*	Body cavities, mesentery, perirenal	Blood
Onchocerca volvulus	None	South and Central America, Africa	*Simulium*	Subcutaneous	Skin
Mansonella streptocerca	None	West and Central Africa	*Culicoides*	Subcutaneous	Skin

denitis most often affects the groin, the axilla, and, in men infected with *W. bancrofti,* the lymphatic vessels of the genitalia, leading to funiculitis, orchitis, and epidydymitis. In some parts of the world infected nodes may suppurate and form sterile abscesses. In nonendemic persons the initial attacks can be milder and unaccompanied by fever.

Chronic Disease

After 10 to 15 years (or as little as 6 months in nonendemic persons) of filarial fevers, episodes of chronic obstructive disease due to damaged lymphatic vessels begin to occur. Reversible lymphedema of the legs, scrotum, arms, penis, or breasts progresses to elephantiasis in the distribution of the affected lymphatic vessels over a period of years. Chyluria, a rare event, occurs in bancroftian filariasis when increased pressure in the renal lymphatic vessels causes rupture into the urinary tract. In many areas the most common chronic manifestation is hydrocele, which probably evolves from recurrent orchitis.

Tropical Pulmonary Eosinophilia

Tropical pulmonary eosinophilia, occurring in less than 1% of infected individuals, is believed to reflect a state of immunologic hyperresponsiveness to the parasite in which rapid clearance of microfilariae occurs in the lung, leading to local release of inflammatory mediators. It is most often seen in young men and is characterized clinically by nocturnal cough and wheezing, low-grade fever, adenopathy, and high-grade eosinophilia. Chest radiographs show diffuse interstitial markings, and a restrictive pattern is most often found on pulmonary function testing. In the absence of effective treatment, chronic pulmonary fibrosis may develop.

Onchocerciasis

Dermatitis, caused by migration of microfilariae through the skin, is manifested by pruritus, localized edema, erythema, and pigment changes. In more chronic cases lichenification can occur. Subcutaneous nodules, present in many but not all patients, contain masses of adult worms and are characteristically found over the lower trunk in Africa and over the head, neck, and upper body in Central America.

Lymphadenopathy is frequently found in inguinal and femoral areas. When it occurs in stretched-out abdominal skin, the so-called "hanging groin" can result.

Ocular onchocerciasis in lightly infected individuals may cause a punctate keratitis or manifest as free microfilariae visible in the anterior chamber or vitreous humor. Heavier infection leading to eventual blindness can cause sclerosing keratitis, uveitis, glaucoma, or chorioretinal lesions.

Loiasis

Endemic individuals usually have an asymptomatic microfilaremia with occasional subconjunctival migration of an adult worm (eyeworm) or infrequent episodes of localized angioedema known as *Calabar swellings*. Nonendemic persons, who are usually amicrofilaremic, have severe allergic symptoms with frequent and incapacitating Calabar swellings, pruritus, and urticaria (Nutman et al., 1986). Subcutaneous nodules containing dying adult worms often develop with treatment. Nephropathy, peripheral neuropathy, and encephalopathy can occur with *L. loa* and can be worsened with drug therapy.

Mansonella Ozzardi

Mansonella ozzardi is usually considered nonpathogenic for humans, with the diagnosis made only during mass blood surveys. Some poorly controlled studies have attributed arthralgia, headache, pruritus, and pulmonary symptoms to this parasite.

Mansonella perstans

The symptomatology attributable to *M. perstans* is poorly defined owing to the high incidence of coinfection with other filarial parasites. Transient localized angioedema, pruritus, neurologic symptoms, fever, and headache have all been ascribed to this species (Adolph et al., 1962). Pericarditis and hepatitis have also been reported.

Mansonella streptocerca

The skin of the upper trunk and the inguinal lymph nodes are predominantly affected with pruritus, rash, and large areas of hypopigmentation as the dominant symptoms of *M. streptocerca* infection. Many infected individuals are completely asymptomatic. Inguinal adenopathy is almost always present in microfilaremic individuals (Meyers et al., 1972).

Nonhuman Filariae

Subperiodic *Brugia malayi* infection is a zoonosis that can cause natural, patent human infection. Several animal filariae, namely, *Dirofilaria immitis* and *Dirofilaria repens* in dogs, as well as *Dirofilaria tenuis* in racoons, can infect humans rarely. The parasites die in the larval stages before reaching maturity and cause few symptoms. Localization is to the lungs in *D. immitis* infection (appearing as coin lesions) and to the subcutaneous tissues and lymph nodes in the other dirofilaria infections (Dissanaike, 1979).

Pathophysiology

After inoculation into a susceptible human host, the third-stage larvae of bancroftian and brugian filariasis mature over a period of months to become adult parasites, dwelling in the afferent lymph vessels or lymph nodes located in virtually any part of the body. It is not known if any pathology is associated with the immature parasites. The adult lymphatic-dwelling parasite is thought to induce a local inflammatory reaction, with cell-mediated and humoral components leading to vessel dilatation and hypertrophy of the vessel walls. The worm itself does not seem to cause blockage of the vessel, but its death leads to necrosis, fibrosis, and obstruction of lymph flow within the lumen. Collateralization of vessels takes place over time as elephantiasis develops.

Onchocercal dermatitis is recognized microscopically by increased numbers of fibroblasts through the dermal layer and a moderate infiltrate of lymphocytes, histiocytes, plasma cells, and eosinophils; polymorphonuclear leukocytes are generally absent (Buck, 1974). Microfilariae are relatively difficult to find on biopsy specimens, although granulomas may form around dead larvae. Lymph nodes are scarred and fibrotic. Ocular pathology is not well characterized, with most of the tissue reactions seen caused by infiltrates of lymphocytes, eosinophils and mast cells.

The other filarial species cause generalized nonspecific inflammatory responses around both adults and microfilariae. This tissue reaction is often intensified by treatment.

Diagnostic Features

Clinical diagnosis is often necessary in suspected filarial infection, as patients are often amicrofilaremic. Bacterial infection, thrombophlebitis, or trauma may be mistaken for acute filarial adenolymphangitis. Filarial lymphangitis is retrograde, which helps to differentiate it from bacterial lymphangitis. The long incubation period and prolonged exposure necessary to acquire filarial disease should be kept in mind. Chronic lymphedema may be caused by malignancy, postoperative changes, and congenital malformations, as well as renal or cardiac failure. Furthermore, it is thought that a foreign body reaction to silica dust introduced into traumatized legs accounts for elephantiasis in some parts of the world. Patients with filarial lymphedema are frequently amicrofilaremic; thus, diagnosis depends on the clinical history as well as the physical examination and may be supported by a positive serology, if present. In cases of orchitis and epidydymitis, the sexually transmitted diseases must be considered.

Tropical pulmonary eosinophilia can be confused with Loeffler's syndrome, allergic bronchopulmonary aspergillosis, eosinophilic pneumonia, drug reactions, or other helminthic infections (hookworm, *Ascaris, Strongyloides*). A typical clinical history, exposure in appropriate areas, and response to diethylcarbamazine help with the diagnosis.

O. volvulus, L. loa, M. perstans, M. ozzardi, and *M. streptocerca* all cause overlapping syndromes and must be separated by a complete search for microfilariae in blood and skin, in addition to a slit-lamp examination in patients from areas endemic for onchocerciasis. The patient should be examined with a slit lamp after sitting for 10 min with the head forward and bowed to allow microfilariae located posteriorly and inferiorly to become visible. Scabies and contact dermatitis enter the differential diagnosis. Leprosy must be considered if there are chronic skin changes.

Natural History

Little mortality is associated with filarial disease. Untreated individuals remaining in endemic areas may develop elephantiasis in the case of lymphatic filariasis or blindness in the case of onchocerciasis; others, however, remain either uninfected or as asymptomatic microfilaremic carriers of the parasite. The determinants of this variable host response to continued exposure to the infectious agent are unclear. The other filarial species have no known consistent chronic sequelae, but complications associated with profound hypereosinophilia, such as endomyocardial fibrosis, have been associated with loiasis.

Epidemiologic Aspects

Table 1 indicates the approximate worldwide distribution of the parasite. Reasonable prevalence data exist only for lymphatic filariasis and onchocerciasis (Sasa, 1976). It is estimated that 300 million people, mostly in India and Southeast Asia, live in areas endemic for lymphatic filariasis, with approximately 100 million actually infected. Thirty million people are constantly exposed to onchocerciasis, with many fertile areas of Africa uninhabitable because of the disease.

Nocturnally periodic microfilariae circulate in the peripheral circulation between 10:00 p.m. and 2:00 a.m. and are usually undetectable at other times, whereas diurnal or subperiodic microfilariae circulate at all times during the day but peak in intensity late in the afternoon. This periodicity is apparently an evolutionary adaptation to the feeding habits of the vector in a particular area. The skin-dwelling microfilariae do not have a periodicity. Except for subperiodic *B. malayi* infection in Indonesia, there is little evidence for any clinically important nonhuman reservoir.

Laboratory Diagnosis

Characterization of the Parasite

Filariae are nematodes of the superfamily *Filarioidea*, of which at least 100 species have been described to date. The adult worms are threadlike, with smooth cuticles, and taper toward each end. Adult males measure 15 to 40 mm by 0.1 mm, females 50 to 100 mm by 0.3 mm. The microfilariae, the usual stage used for laboratory diagnosis, are unique in that they are much more undifferentiated than the first-stage larva of other nematodes. They measure 180 to 310 μm long by 3 to 8 μm wide, are bluntly rounded anteriorly, and have pointed tail ends. Some species of microfilariae retain their egg shell or sheath after release from the adult female. The intermediate stages between microfilaria and adult in both the vector and the host are rarely seen and are unimportant in diagnosis.

Isolation and Identification of Microfilariae

A knowledge of where the patient has been is critical in deciding which specimens to obtain and at what time of day to obtain them. Figure 2 shows schematically the morphologic criteria used in differentiating the microfilariae. The presence or absence of a sheath, the configuration of the caudal nuclei, and whether the microfilaria is found in blood or in skin is sufficient information to make a diagnosis in most cases.

Blood

Although a thick blood smear (Fig. 3) performed as for the diagnosis of malaria may be sufficient for the diagnosis of a heavily infected individual, this technique is not sensitive enough to detect microfilaremias of less than 100 mf/ml of blood. Blood smears may be all that is available in the field; however, one of three commonly used concentration procedures is mandatory to rule out microfilaremia or to detect low-level infection.

Knott's Concentration

1. Place 1 ml of anticoagulated (ethylenediaminetetraacetate or heparin) blood in a conical centrifuge tube with 9 ml of 2% Formalin.
2. Centrifuge at 1,500 rpm for 1 min or let stand overnight.
3. Decant supernatant, spread the sediment on a slide, and air-dry.
4. Fix with methanol.
5. Stain with Giemsa (1:40) or hematoxylin (Knott, 1939).

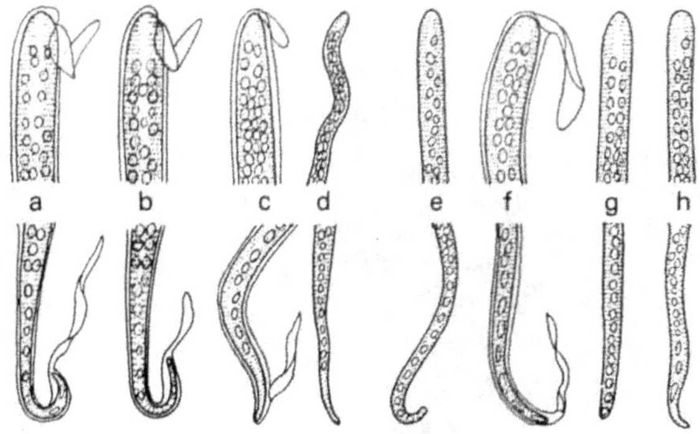

FIG. 2. Differential characterization of the microfilariae of (a) *Brugia malayi;* (b) *Brugia timori;* (c) *Wuchereria bancrofti;* (d) *Onchocerca volvulus;* (e) *Mansonella streptocerca;* (f) *Loa loa;* (g) *Mansonella perstans;* and (h) *Mansonella ozzardi*. Note the presence or absence of a sheath and the relation of the terminal nuclei to the end of the tail.

Membrane Filtration

1. Collect 1 to 10 ml of anticoagulated (ethylenediaminetetraacetate or heparin) blood.
2. Place a 5-μm Nuclepore membrane (3 μm if *M. perstans* or *M. ozzardi* is possible) in a Swinney adaptor (Fig. 4).
3. With a syringe, pass blood through apparatus.
4. Pass 50 ml of water through to lyse red cells.

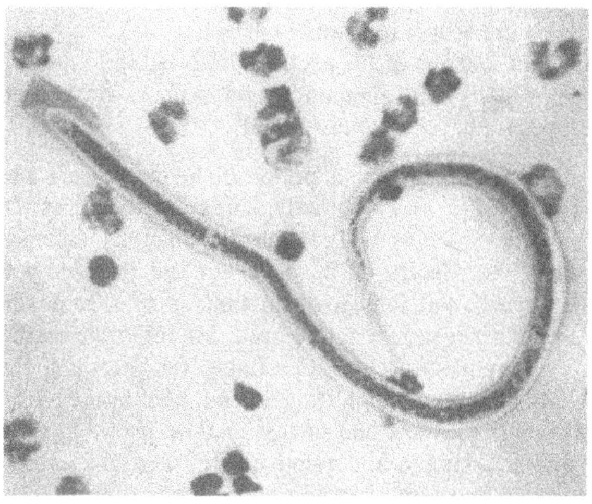

FIG. 3. Hematoxylin-stained thick film showing a microfilaria of *Wuchereria bancrofti*.

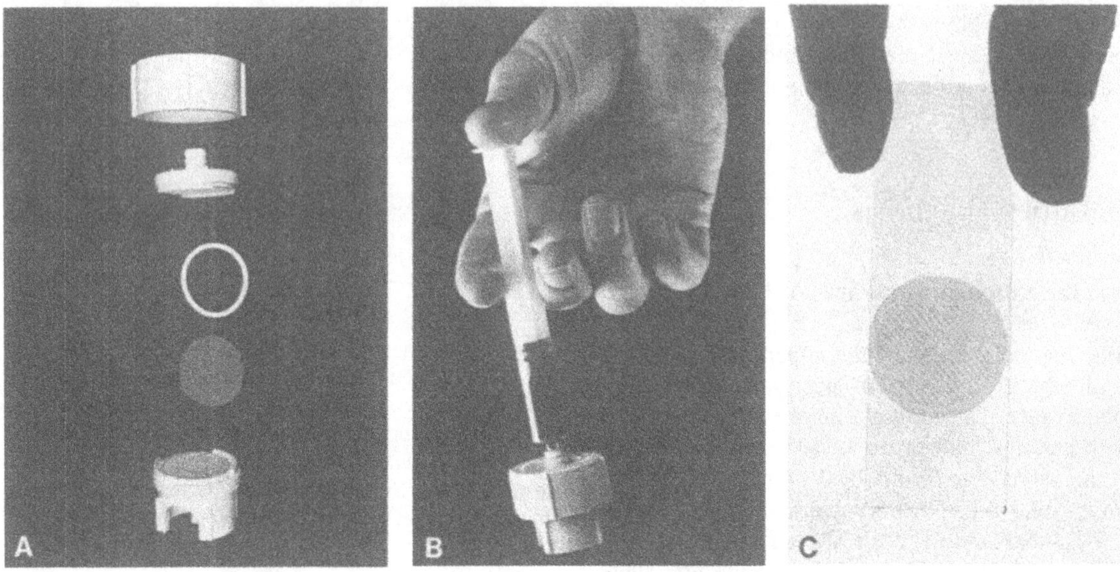

FIG. 4. Nuclepore filtration procedure. (A) A 3- or 5-μm membrane is placed in a Swinney adaptor. (B) Blood is forced through the membrane. (C) The stained membrane is ready for microscopy.

5. Remove the membrane and place it on a microscope slide to be examined wet, or
6. Air-dry the membrane, fix with methanol, and stain with Giemsa or hematoxylin (Abaru and Denham, 1976).

Saponin Lysis Concentration

1. Take 10 ml of anticoagulated (ethylenediaminetetraacetate or citrate) blood, centrifuge it at 1,500 rpm for 10 min, and discard the plasma.
2. Mix the cells with 50 ml of 0.5% Saponin in saline (used by all automated hematology laboratories).
3. Leave for 10 min to lyse the red cells, then centrifuge for 10 min.
4. Spread the sediment on a slide.
5. Examine wet for motile microfilariae, or
6. Add 2 drops of 1% acetic acid to kill the microfilaria, fix with methanol, and stain as described previously. (McQuay, 1970).

Some parasitologists prefer the hematoxylin stain (Hira, 1977), as it distinctly stains the sheath of all sheathed species a light purple, whereas Giemsa stains the sheaths of *W. bancrofti* and *B. malayi* a faint pink, but fails to stain that of *L. loa* at all. Wright stain should not be used. Microfilariae on filters may not always be suitable for morphologic study no matter which stain is used. Membrane filtration is the easiest and fastest method if one has on hand the simple and relatively inexpensive equipment. Otherwise, Saponin lysis allows one to concentrate a large volume of blood. A minimum of two daytime and nighttime blood samples each are necessary in areas where nocturnally periodic filariae coexist with nonperiodic or diurnally periodic forms. The microfilariae remain alive in anticoagulated blood kept at room temperature for at least 2 days; thus, immediate examination is not necessary. Of note is that microfilariae can acquire jet lag as easily as their hosts and may take several weeks to readjust their periodicity to local time. A single dose of diethylcarbamazine can in many cases provoke daytime microfilaremia in nocturnally periodic disease (McMahon, 1982); this practice is discouraged outside of mass surveys, however, as it is less sensitive and can result in drug reactions in heavily microfilaremic persons.

SKIN

A skin snip of such thickness that the outer parts of the dermal papillae are sampled is the optimal method for diagnosis of onchocerciasis and streptocerciasis. Although a needle and blade can be used (Fig. 5), a corneoscleral biopsy punch (Fig. 6) makes it possible to obtain snips of appropriate size and depth in a fast, painless, and bloodless manner. Blood contaminating a skin snip may result in one of the blood-borne microfilariae escaping into the specimen, with *M. ozzardi* most frequently incriminated (Nathan, 1979).

In the diagnostic workup, all patients should have skin snips from each scapula, from each iliac crest, and from the lateral aspect of each calf, although in Africa the highest yield will be from the iliac crest and in Central America from over the scapula. *Mansonella streptocerca* is most often found in snips taken over the upper trunk. Each snip should be transferred to a drop of normal saline in the well of a microtiter plate or on a microscope slide. In heavier infection, motile microfilariae will be visible emerging from the skin snip under low-power microscopy in 10 to 60 min; however, negative snips should be covered, to prevent evaporation, and checked periodically for at least 24 h. If the patient has been in an area endemic for streptocerciasis, it is then necessary to fix the skin and to stain the microfilariae for identification.

Deep skin biopsy is not necessary and multiple skin snips will have a higher yield than one random, traumatic deep biopsy. A 50-mg dose of diethylcarbamazine (Mazzotti test) can be given to a patient suspected of harboring *O. volvulus* but in whom no microfilariae can be detected after a rigorous search of blood, of skin snips, and after a careful ophthalmologic examination. Within 15 min to 24 h the infected patient will develop pruritus and erythema over involved areas as microfilariae die in the skin. Severe reactions in both the skin and eye may occur in more heavily infected individuals, emphasizing the importance of reserving this maneuver for very lightly infected persons.

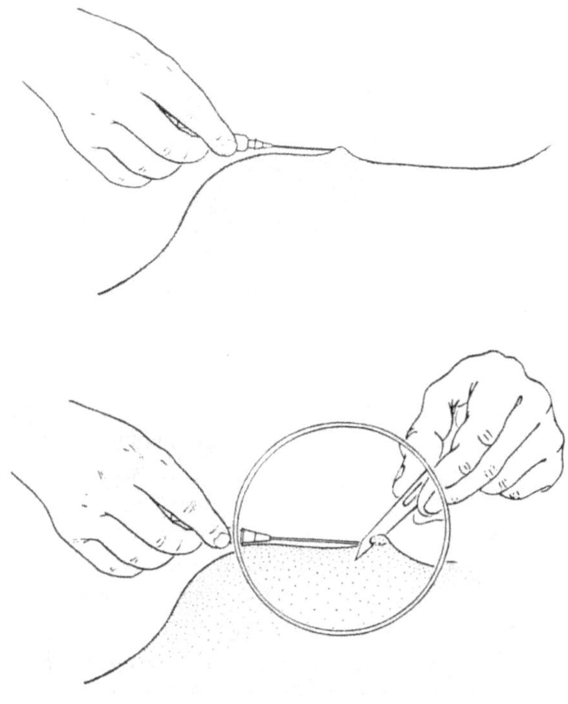

FIG. 5. Skin snip technique, using a blade to slice a thin piece of skin that has been tented up with a needle.

URINE AND OTHER FLUIDS

Microfilariae of *O. volvulus* have been found in the urine of some infected persons during mass epidemiologic studies in hyperendemic regions, with microfilarial counts being well correlated to skin counts (Buck, 1973). Examination of the urine, however, is no substitute for properly performed skin snips in the diagnostic workup of a single patient.

Microfilariae of *W. bancrofti* have also been found in the urine of the occasional asymptomatic, amicrofilaremic patient, but are usually not present in chyluric patients. Chyluric urine can be diagnosed with xylene or ether, which dissolves the top floating, creamy layer after the specimen is shaken. Incidental findings of microfilariae of *W. bancrofti* have been made in cytologic preparations of cervix, vagina, and gastric brushings This finding presumably represents blood contaminating the specimen.

Microfilariae of *L. loa* and *O. volvulus* have been described in the cerebrospinal fluid of severe cases; diethylcarbamazine treatment may provoke encephalopathy in these cases.

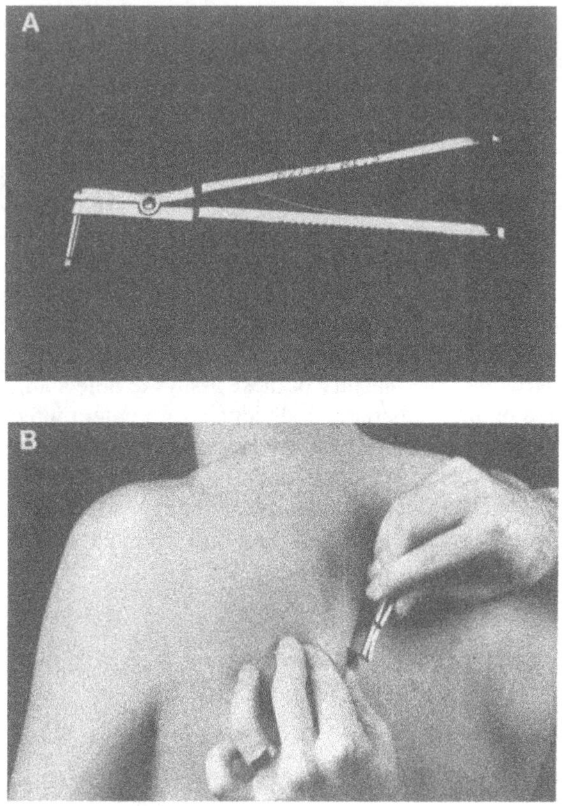

FIG. 6. (A) Corneoscleral biopsy punch; (B) technique for taking a skin snip from over the scapula.

Identification of the Adult Parasite

An excised subcutaneous nodule containing a coil of threadlike white worms 8 to 10 cm in length in the appropriate clinical setting is diagnostic of *O. volvulus* on visual inspection. Occasionally an adult *Loa loa* parasite is excised while it is crawling across the conjunctiva or under the skin. In the absence of such observations an experienced pathologist is needed to identify the sections of adult worms, which can be seen in specimens of diverse human body tissues.

Serologic and Immunologic Aspects

Serologic and other immunodiagnostic procedures (including skin testing and in vitro tests of parasite-specific cell-mediated immunity) have been in use for filariasis and onchocerciasis since 1916. However, none has been particularly useful as a diagnostic tool because most individuals from endemic regions will have developed antibodies against the filarial parasite.

Although techniques such as complement fixation, gel diffusion, and indirect immunofluorescence assays have been used in the past, the currently available serologic tests for the diagnosis of filarial diseases are immunoassays (either indirect hemagglutination assays or the highly sensitive enzyme-linked immunosorbent assays) using as their basis crude antigens derived from either adult worms of *B. malayi* or the dog filarid *D. immitis* (Ambroise-Thomas and Peyron, 1986). None of these assays can either distinguish among the eight filarial parasites that commonly infect humans or distinguish among actively infected patients, those merely exposed (but not infected), and those successfully treated. Thus, at present, serologic evaluation in filarial diseases remains useful in only two situations: 1) in those individuals from nonendemic areas (and therefore initially seronegative) who are exposed to or infected with filarial parasites; and 2) as a mechanism to evaluate response to therapy on a longitudinal basis, as antifilarial antibodies tend to diminish after treatment (Nutman et al., 1986).

More recently, and with more promise, assays have been developed in both lymphatic filariasis and onchocerciasis that can detect circulating parasite antigen either in the blood (reviewed by Hamilton [1985]) or, in some cases, in the urine of infected individuals. The sensitivity of these assays and, in particular, the inability of these assays to detect antigen in amicrofilaremic individuals with patent infections awaits further improvement. However, these assays should overcome both the problem of species specificity and the inability of antibody assays to distinguish between infected and uninfected individuals.

Treatment, Prevention, and Control

Treatment

Diethylcarbamazine (DEC) is microfilaricidal to all species of human filariae except *M. ozzardi* and *M. perstans*. Dosing regimens (Hawking, 1979) depend on species. Fever, chills, arthralgias, and hypotension can occur during the treatment of any species, but reactions are most commonly seen in onchocerciasis and in microfilaremic loiasis. This pathology probably represents a hypersensitivity reaction to parasite antigens released from dying worms. The ocular lesions of onchocerciasis can be exacerbated during treatment, and concurrent steroid therapy is advisable. The importance of ruling out onchocerciasis before initiating therapy for any other species cannot be overemphasized because of the potentially disastrous consequences to the eye. Suppression of microfilaremia may vary from weeks to months.

The adult stage of *L. loa* is quite sensitive to DEC, and patients should receive repeated courses of treatment until eosinophilia and serology normalizes. There is good indirect and some direct evidence (Ottesen, 1985) that DEC is macrofilaricidal to *W. bancrofti*; thus, repeated and aggressive therapy may be indicated even in patients with advanced disease. Early removal of infected nonendemic persons from the area tends to hasten clinical recovery.

Suramin is the only agent macrofilaricidal to *O. volvulus* and is extremely toxic, requiring hospitalization of the patient for prolonged periods. It is impractical to use in developing countries and otherwise should only be given in consultation with an expert.

Ivermectin is a new broad-spectrum anthelmintic agent that shows great promise (Greene et al., 1985). A single oral dose can suppress microfilaremia in onchocerciasis for many months with what are probably tolerable side effects. Trials in other forms of filariasis are ongoing. It has not so far been shown to be macrofilaricidal to any species.

Prevention and Control

Mass treatment programs with DEC have been attempted, with mixed results (Sasa, 1976). In Tahiti, regular large-scale mass DEC treatment over 25 years reduced microfilaria rates to negligible levels; when the program was stopped, however, there was an annual 1% increase in the prevalence. At present, a single dose of DEC is given on a mass basis once a year, and rates of microfilaremia are again falling. In India, programs involving either mass treatment or selective treatment of infected persons on a large-scale basis have failed, probably due to drug side effects, poor compliance by individuals who feel well, or insufficient availability of the drug. Continuous use of low-dose DEC in medicated salt seems to have been highly efficacious in China, but is a difficult strategy to implement in most areas. Vector control is difficult in the mosquito-borne filariaes.

The Onchocerciasis Control Program in West Africa has been ongoing for 10 years, employing a strategy of aggressive vector control using aerial spraying with insecticide. This expensive project has reduced disease prevalence and intensity in many areas; however, whether this effect will be sustained once spraying ceases is difficult to predict.

Literature Cited

Abaru, D. E., and D. A. Denham. 1976. A comparison of the efficacy of the Nuclepore and Millipore filtration systems for detecting microfilariae. Southeast Asian J. Trop. Med. Public Health **7:**367–369.

Adolph, P. E., I. G. Kagan, and R. M. McQuay. 1962. Diagnosis and treatment of *Acanthocheilonema perstans* filariasis. Am. J. Trop. Med. Hyg. **11:**76–88.

Ambroise-Thomas, P., and F. Peyron. 1986. Filariasis, p. 233–256. *In* K. W. Walls and P. M. Shantz (ed.), Immunodiagnosis of parasitic diseases, vol. 1. Academic Press, Inc., New York.

Buck, A. A. 1973. Microfilariae in onchocerciasis in Africans. Z. Tropenmed. Parasitol. **24:**336–338.

Buck, A. A. (ed.). 1974. Onchocerciasis, symptomatology, pathology, and diagnosis, p. 9–14. World Health Organization, Geneva.

Dissanaike, A. S. 1979. Zoonotic aspects of filarial infection in man. Bull. W.H.O. **57:**349–357.

Greene, B. M., H. R. Taylor, E. W. Cupp, R. P. Murphy, A. T. White, M. A. Aziz, et al. 1985. Comparison of ivermectin and diethylcarbamazine in the treatment of onchocerciasis. N. Engl. J. Med. **313:**133–138.

Hamilton, P. G. 1985. Application of immunoassay methods in the serodiagnosis of human filariasis. Rev. Infect. Dis. **7:**837–843.

Hawking, F. 1979. Diethylcarbamazine and new compounds for the treatment of filariasis. Adv. Pharmacol. Chemother. **16:**129–194.

Hira, P. R. 1977. *Wuchereria bancrofti:* the staining of the microfilarial sheath in Giemsa and haematoxylin for diagnosis. Med. J. Zambia **11:**93–96.

Knott, J. I. 1939. A method for making microfilarial surveys in day blood. Trans. R. Soc. Trop. Med. Hyg. **33:**191–196.

McMahon, J. E. 1982. The examination time/dose interval in the provocation of nocturnally periodic microfilariae of *Wuchereria bancrofti* with diethylcarbamazine and practical uses of the test. Tropenmed. Parasitol. **33:**28–30.

McQuay, R. M. 1970. Citrate saponin acid method for the recovery of microfilariae from blood. Am. J. Clin. Pathol. **54:**743–746.

Meyers, W. M., D. H. Connor, L. E. Harman, K. Fleshman, R. Moris, and R. C. Neafie. 1972. Human streptocerciasis. A clinicopathologic study of 40 Africans including identification of the adult filaria. Am. J. Trop. Med. Hyg. **21:**528–545.

Nathan, M. B. 1979. A comparison of *Mansonella ozzardi* microfilaria densities in the blood and in skin snips from three areas of the body. Trans. R. Soc. Trop. Med. Hyg. **73:**338–340.

Nutman, T. B., K. D. Miller, M. Mulligan, and E. A. Ottesen. 1986. *Loa loa* infection in temporary residents of endemic regions: recognition of a hyperresponsive syndrome with characteristic clinical manifestations J. Infect. Dis. **154:**10–18.

Ottesen, E. A. 1985. Efficacy of diethylcarbamazine in eradication of infection with the lymphatic dwelling filariae of humans. Rev. Infect. Dis. **7:**341–356.

Sasa, M. 1976. Human filariasis: a global survey of epidemiology and control. University Park Press, Baltimore.

Wartman, W. B., and B. G. King. 1947. Filariasis in American armed forces in WWII. Medicine (Baltimore) **26:**334–394.

W.H.O. Fourth Expert Committee on Lymphatic Filariasis. 1984. World Health Organization technical report series no. 702. World Health Organization, Geneva.

CHAPTER 92

Giardiasis

JERRY E. JONES

Disease: Giardiasis (giardia enteritis, lambliasis).

Entiologic Agent: *Giardia lamblia, Giardia intestinalis* (Br.).

Source: Humans; probably beaver and other wild or domestic animals (cases reported in beavers, dogs, and gerbils).

Clinical Manifestations: Acute stages: diarrhea, often explosive; epigastric discomfort; flatulence; belching (usually with offensive odor); upper abdominal distention and tenderness; nausea; anorexia. Chronic or subacute stage: loose to mushy stools alternating with constipation; intermittent diarrhea; weight loss; malnutrition.

Pathology: No consistently defined morphologic changes of the intestinal mucosa. Organisms feed on mucous secretions. Trophozoite stage interferes with the integrity of the microvillous border of the upper small intestine. Exact mechanism of attachment is not clearly delineated.

Laboratory Diagnosis: Direct fecal smear; concentration methods; string test; duodenal aspiration and biopsy; immunodiagnosis (enzyme-linked immunosorbent assay, a research tool, is under development).

Epidemiology: Endemic worldwide; prevalence highest in the tropics and subtropics; isolated outbreaks reported from day-care environments, residential facilities, contaminated surface water supplies. A low dose of the infectious agent (10 to 25 cysts) reported to produce human infection.

Treatment: Quinacrine, metronidazole, furazolidone, and tinidazole (the latter unavailable in the United States).

Prevention and Control: Sanitary disposal of feces. Education of patients, families, and institutional personnel and inmates, especially workers at day-care centers, regarding personal hygiene. Examination of household members and other suspected contacts. Boiling of water is best method to kill cysts. Protection of public water supplies against contamination with human or animal feces. If epidemic outbreak, determine source of infection (e.g., day-care center, surface water source, foodborne) and institute appropriate preventive or control measures.

Description of Disease

It is a matter of regret that seemingly so little consideration has been given to the problem of the carriers of lamblia cysts, who may be a source of considerable trouble in the future.

Porter (1916)

Giardiasis is the human infection caused by the flagellated protozoan *Giardia lamblia*. This organism is distributed worldwide among most cultural groups. Once colonized in humans, its clinical expression may vary from no symptoms to those of steatorrhea and malnutrition. Its cosmopolitan distribution adds this parasite to the list of parasitic burdens already existing in many developing countries. In this setting, the protozoan is often present simultaneously with other parasites or infectious agents. In developed nations, giardiasis has become an important problem, particularly among certain subsets of the population or during epidemic outbreaks. The common thread that maintains this parasite as an active pathogen is fecal contamination. Fecal contamination with the cysts of *G. lamblia* is needed for its transmission to a new host. It is this factor that must be considered in the treatment, prevention, and control of giardiasis.

Characterization

Giardia lamblia is a flagellated protozoan belonging to the phylum *Sarcomastigophora*. It has two forms in its life cycle, the trophozoite and the cyst. The vegetative (actively feeding) trophozoite stage is responsible for the colonization of this species in humans. In this stage, the organism resembles a longitudinally cut pear, measuring 10 to 20 μm in length (average 14 μm), 6 to 15 μm in width, and 1 to 3 μm in thickness. The pear shape is bowed ventrally, with an ovoid adhesive (sucking) disk occupying the anterior ventral surface. Thus, the trophozoite form is convex dorsally and concave ventrally. The dorsal surface provides an area for diffusion of nutrients. The ovoid sucking disk appears to be a rigid structure connected to the exterior by a canal containing flagella. It is hypothesized that the beating of the flagella produces a negative pressure in the concavity for its attachment to the microvilli (Owen et al., 1979). Eight flagella are present in four symmetric pairs, and there are two distinct nuclei. This morphologic shape easily distinguishes the parasite from other protozoan species, giving it a "monkey-face" appearance on light microscopy. *Giardia lamblia* multiplies by binary fission. Therefore, as long as host nutrition and immediate environmental conditions are satisfactory, the parasite can maintain itself indefinitely. However, when environmental factors threaten the parasite, it is capable of forming a more resistant cyst stage. The cysts are oval, 8 to 14 μm in length (average 12 μm) and 6 to 10 μm in width. They contain four nuclei, usually positioned near one end. The cysts are passed through the remaining intestine and become part of the fecal stream. Figure 1 demonstrates the life cycle in humans. The trophozoite is rarely passed in feces and rapidly dies on exposure.

Once passed in the fecal stream, the cysts survive best in wet, cool conditions. They can withstand standard concentrations of chlorine used routinely in water purification systems. The cysts are then reintroduced into humans by the oral route through water, food, hands, or other bodily contact. Once ingested, the cysts apparently need the acidic environment of the stomach to signal excystation (Bingham and Meyer, 1979). Excystation is accomplished in the duodenum, and the new trophozoites make their home among the intestinal villi. Each cyst divides to form two trophozoites, which obtain nourishment by diffusion through the cell wall. Once the trophozoite has established itself, it may multiply by binary fission approximately every 5 h, thus rapidly establishing its presence in great numbers.

Communicability and Transmissibility

Giardia lamblia is transmitted by viable cysts that are swallowed. Rendtorff (1954) was the first to dem-

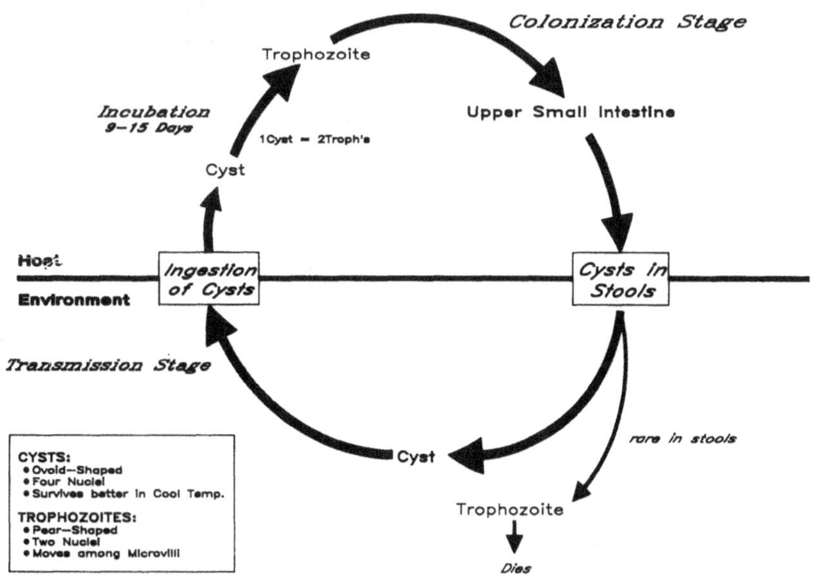

FIG. 1. Life cycle of *Giardia lamblia*.

onstrate this transmission clearly in human studies. He showed that cysts of *Giardia* could be transmitted by a water source (Rendtorff and Holt, 1954). These cysts have also been shown to remain viable in water for up to 3 months (Raizman, 1976). As few as 10 cysts can effectively establish an infection in human volunteers. The results of these studies clearly demonstrate the high degree of infectivity of *Giardia*.

Rendtorff (1954) was able to clarify the natural course of the infection, showing an average of 9 days (range, 6 to 15 days) from the point of cyst ingestion to the point of changes in stool. Cysts are then passed in the feces for the next 5 to 41 days (average 18 days). In two cases, cysts were passed for 129 and 132 days, respectively. Rendtorff noted that 12 of 14 cases studied cleared spontaneously. In comparison, Walzer et al. (1971) found the incubation period to be 12 to 15 days, and Jokipii and Jokipii (1977) found it to be 12 to 19 days.

It has become clear that *Giardia* is spread by the fecal–oral route. Epidemiologically, factors involved in its transmission are contaminated water and food; clustering of individuals in day-care and institutional facilities; homosexual oral–genital sex; and person-to-person spread, especially between close family contacts. Children are especially at risk. It has generally been found that giardiasis has an inverse correlation with the level of sanitary practices.

TABLE 1. Clinical manifestations attributed to *Giardia lamblia*

Acute	*Chronic*
Diarrhea, often foul smelling	Recurrent diarrhea
Weight loss	Mushy stools
Abdominal cramps	Anorexia
Greasy stools	Nausea
Flatulence	Fatigue
Vomiting	Periodic constipation
Belching	Abdominal distention
Fever, usually low grade	Upper epigastric distention
Anorexia	Substernal burning
Nausea	Belching, often sulfuric
Headache	Intolerance to meat
Chills	Urticaria
Mucus in stools	Erythema nodosum
	Steatorrhea
	Overt malabsorption: abnormalities in lactose, xylose, vitamin A, vitamin B_{12}, fat

Clinical Manifestations

Giardiasis may present to the physician with a variety of clinical manifestations. Rendtorff (1954) clearly showed that 12 of 14 cases of proven *Giardia* infection cleared spontaneously without overt signs or symptoms. He recognized that all individuals infected showed a change in the consistency of stools passed, but not enough subjectively for the infected individuals to report it as a recognizable problem. Therefore, it would appear that the majority of healthy individuals infected with cysts of *G. lamblia* will probably develop very few subjective changes recognizable during the infection. With some individuals, however, the manifestations will proceed to more clearly recognizable intestinal symptoms. Table 1 lists the clinical manifestations reported to occur with giardiasis.

For the clinician, the most characteristic early phase of giardiasis is an acute diarrheal illness of brief duration. Usually lasting 2 to 3 days, this period may be associated with low-grade fever, anorexia, nausea, and abdominal cramping. Many individuals will not seek health care at this point and will seem to recover with very few residual effects. Within 2 to 3 months, however, intermittent or even protracted bouts of diarrhea may occur. During this time, cramps, abdominal pain, bloating, and flatulence usually occur. Individuals will often complain of foul-smelling breath or flatulence. Brodsky and coworkers (1974) found that 96% of travelers developing giardiasis experienced diarrhea during the acute phase. Likewise, weight loss, abdominal cramps, and greasy stools occurred in more than one-half of the individuals infected. Flatulence, vomiting, belching, and fever occurred in less than one-third of cases.

A chronic phase of giardiasis may then develop. This phase may involve 2 to 3 years of recurrent bouts of diarrhea. Lassitude, headache, and myalgia have been associated with this phase of the disease. Urticaria also has been reported in both adults and children (Hamrick and Moore, 1983; Weisman, 1979). During this period, the individual will experience continued weight loss, anorexia, and a pattern of malabsorption. In children this phase may present as failure to thrive. Abnormalities in absorption of lactose, xylose, vitamin A, vitamin B_{12}, and fat have been identified (Cordingley and Crawford, 1986; Gillon, 1985; Meyer and Radulescu, 1979).

Therefore, it is important to recognize that *Giardia lamblia* will present at various phases of infection to a variety of health care providers. Its potential for acute and chronic infection may cloud the clinical picture and confuse the clinician when investigating oubreaks in families, day-care centers, or institutions. Figure 2 outlines the possible spectrum of *Giardia* infections. This outline shows that a number of host responses are possible and that during the course of the infection, one individual may move from one clinical spectrum to another. Likewise, one

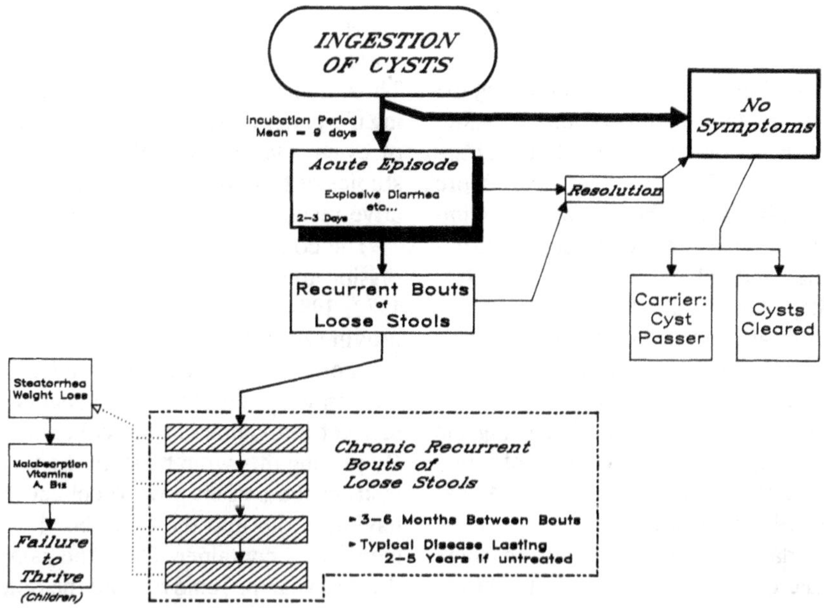

FIG. 2. Possible course of *Giardia lamblia* in humans.

family may show several different clinical presentations. It is suspected that this variation in clinical symptoms may be related to differences in the virulence of the parasite and to host immunologic responses.

Pathology

It is generally believed that the trophozoite stage of *G. lamblia* interferes with the integrity of the brush border of the upper small intestine. The exact mechanism has not been clearly delineated, but several investigations have suggested the following: First, *Giardia* trophozoites literally "line" the intestine, producing mechanical obstruction of the mucosa. This obstruction blocks absorptive activities, thus decreasing the absorption of fats and fat-soluble vitamins (Meyer and Radulescu, 1979; Sealy and Schuman, 1981). Some investigators believe that there is a direct microvillous attachment, with mucosal and villous irritation (Meyer and Radulescu, 1979; Owen et al., 1979; Wolfe, 1979). This irritation is believed by some to produce an increased lymphocytic reaction (Meyer and Radulescu, 1979; Wright and Tomkins, 1977). A correlation between intraepithelial lymphocytic infiltration and malabsorption has been postulated (Meyer and Radulescu, 1979; Wright and Tomkins, 1977). However, other investigators have found no correlation between the degree of mucosal injury and the severity of malabsorption (Dupont and Sullivan, 1986). Some believe that this irritation is marked, and direct tissue invasion has been reported (Blankinsopp et al., 1978; Saha and Ghosh, 1977). This irritation has been believed to stimulate a more rapid mucosal turnover, thus decreasing the digestive functions of the microvilli (Meyer and Radulescu, 1979).

Some investigators have suggested that the degree of infectivity is related to associated factors such as fungal or bacterial overgrowth (Gillon, 1984; Tandon et al., 1977). Others have postulated that certain strains of *G. lamblia* exist with differing abilities to produce symptoms, whereas some strains are postulated to produce an enterotoxin responsible for the biochemical interference (Korman et al., 1986; Sealy and Schuman, 1981).

Other studies have focused on host defense mechanisms. It has been recognized that individuals with certain hypogammaglobulinemic disorders have an increased prevalence of giardiasis (Ament and Robin, 1972). In addition, hypochlorhydria and achlorhydria have been shown to facilitate infection with *Giardia* (Giannella et al., 1973; Wolfe, 1979). However, some studies have demonstrated no difference in immunoglobulin levels between those proved to be actively infected and healthy controls (Jones and Brown, 1974).

In truth, it is likely that a combination of factors is involved in the pathogenesis of giardiasis. Clearly, absorption of fats and certain biochemical processes are impeded at the brush border of the small intestine (Welsh et al., 1984). The ability of the host to respond to this interference is apparently initiated through a complex immune response involving both thymus-dependent and thymus-independent mechanisms. Therefore, those with a defect in their immune response would be at higher risk for this and other infections.

Diagnostic Features and Approaches

The diagnosis of giardiasis is made by identifying the characteristic cyst or trophozoite of the parasite. This identification is accomplished by obtaining appropriate samples of stool or duodenal fluid. More recently, great emphasis has been placed on immunodiagnosis using fecal samples or serum specimens from infected individuals (Goka et al., 1986; Green et al., 1985; Smith et al., 1981; Ungar et al., 1984; Visvesvara et al., 1980). However, the clinician must deal with individual cases and recognize that the pattern or phase of giardiasis will affect cyst passage. Thus, the best approach to diagnosing the infection may vary depending on the phase of infection. For example, it has been demonstrated that cysts will be passed during the acute phase of the infection 5 to 41 days (Rendtorff, 1954). In some cases, this passage of cysts may be longer. During this phase, moderate to high numbers of parasites are reported to be consistently present in stools. On the other hand, studies of individuals with chronic disease have tended to show a more cyclic pattern of cyst passage (Porter, 1916; Tsuchiya, 1931). Numerous cysts are passed at 2-week intervals. Although cysts are found daily in smaller numbers, they are more difficult to find during the intercyclic periods. It has also been reported that in individuals with chronic infections, cysts are more likely to be found in loose or diarrheal stools than in formed stools (Porter, 1916). Therefore, if a patient has a history more consistent with a chronic, recurrent infection, collecting stools during the loose or diarrheal period could aid the diagnosis; however, this theory has not been proved conclusively. Nonetheless, one study of chronic giardiasis in children found that the number of cysts passed fell into three patterns: high, low, and mixed. Increased numbers in diarrheal stools were not reported (Danciger and Lopez, 1975). Recognition of the phase of infection and proper timing of fecal collection is important in the diagnosis of giardiasis. In acute infection, cysts are more apt to be passed initially and continuously for several days to weeks. During this phase, collecting stool samples every other day would be satisfactory. However, the cyclic pattern found in chronic infections would suggest that collecting stools 4 to 7 days apart would be better. This approach would provide an overlap such that a 10- to 14-day period would pass and one complete cycle could be evaluated. Most of the patterns of cyst passage identified in adults and children would thus be included.

It must also be remembered that certain medications and radiologic examinations will interfere with the diagnosis of giardiasis. Therefore, stool must be collected before upper gastrointestinal and barium enema studies are ordered. Also, antibiotics, antidiarrheal agents, and even antacids have been reported to interfere with cyst passage (Jones, 1980). Thus, a medication history is important in determining if a patient could be blocking appropriate passage of cysts. One must "think" parasites before other studies or treatments are initiated to avoid false-negative results on stool examinations.

The collection of fecal specimens is often embarrassing and frequently difficult to accomplish (Jones, 1982, 1984). Stool collection kits are helpful in improving compliance as well as giving clear instructions on how to collect the sample. The use of plastic kitchen wrap under the toilet seat is a helpful way to instruct individuals in the collection of a specimen. By placing this wrap beneath the toilet seat, one can instruct the patient how to collect the stool without having to dip into the toilet bowl or pass the specimen into a container. Once the sample is collected, the wrap can be removed and the remaining stool can be dropped into the toilet bowl. The stool specimen is placed in a prepared kit containing two vials: polyvinyl alcohol (PVA) and buffered Formalin. Once mixed, the specimen can be returned to the laboratory or office. Collection kits are available commercially (Meridian Diagnostics, Inc., Cincinnati, Ohio).

When dealing with individuals in the office setting, a simple technique for specimen collection is the digital rectal examination (Jones, 1981, 1984). A small specimen from the digital rectal examination is placed on a slide prepared with 1 to 2 drops of normal saline. This technique avoids transport and additional cost, but requires a clinician with interest in examining a microscopic slide. With a little experience, however, the cysts of *G. lamblia* are easily seen in wet preparations. Trophozoites are rarely seen in stool specimens.

Morphologic characteristics that distinguish the cyst are its shape, size, and intracytoplasmic structures. Approximately the size of a white cell, the cyst has a distinctive oval shape. Often the cyst will show retraction of the cystoplasm from the cyst wall. On wet preparation, a refractile body called the *axostyle* will be seen obliquely within the cytoplasm. One to two of the four nuclei are usually seen in any one field of focus. These structures can be seen at a magnification of 45× and will allow the diagnosis to be made in the office setting with minimal equipment. Dilute iodine dropped on the edge of the cover slip will cause the cysts to take up a yellowish color that outlines the nuclei.

Trophozoites are easily recognized on wet preparations by their irregular motion and activity. The trophozoite is pear shaped, and on special stains the two large nuclei give it its characteristic "monkey-face" appearance. The flagella are rarely visible on routine stains, but can be outlined using special flagellar stains.

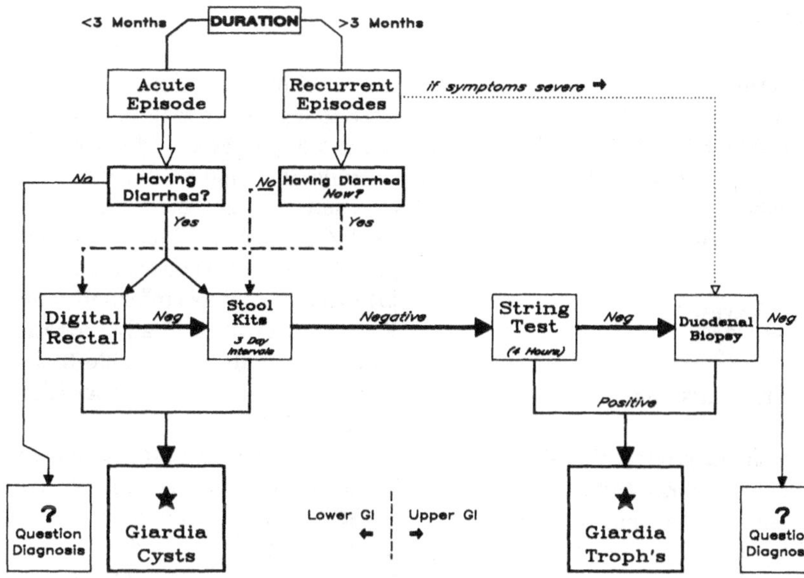

FIG. 3. Alogram for diagnosis of *Giardia lamblia*.

Figure 3 displays a diagnostic approach to individuals with symptoms suggesting giardiasis. This approach is intended to be applied to individual patients from a clinician's perspective. Certain adjustments would be indicated if one is undertaking investigation of an epidemic.

The first fact to be established using this approach is whether the infection is acute or chronic. If symptoms have lasted less than 3 months or if this episode is the first reported, one has a good chance of demonstrating the cysts in fecal specimens if diarrhea is present. Along with the history of the patient and a physical examination, a digital rectal sample is obtained to establish the presence of blood and to prepare a wet preparation for microscopic examination. If the stool is negative for blood and the wet preparation is negative for pus, cysts should be carefully sought under the 10× and 45× objectives. If blood or pus is seen, additional or other causes of symptoms should be explored. If no cysts are identified, the individual should be asked to collect three separate stool samples, each from a different bowel movement and preferably on a different day. These samples are returned to the office, and a repeat wet preparation is made from the PVA and Formalin vials. If no cysts are identified, the samples are sent to the laboratory for concentration and trichrome staining. If they are negative for cysts, the individual should return to the office after an overnight fast. The string test is then used to obtain duodenal fluid (Jones, 1986). Wet preparations are made from this sample and screened for the presence of trophozoites. If results continue to be negative, the individual can be examined by endoscopy and samples, including those from brush biopsies, can be obtained. Examination and preparation of tissue specimens have been described previously (Sun, 1980).

For an individual with a history of chronic or recurrent bouts or a history of recurrent symptoms longer than 3 months, one should alter the previous approach slightly. If there is a present or recent history of diarrheal stools, one can proceed along the same diagnostic course. However, if the individual is between bouts of diarrhea or has constipation, the fecal specimens should be collected on a 5- to 7-day cycle instead of alternating days. Also, one may wish to utilize the string test earlier, as cyst passage may be diminished. Likewise, if the chronic symptoms and malabsorption have become severe, one may want to go directly to duodenal intubation and biopsy.

On a more research-oriented and epidemiologic level, several newer approaches for immunodiagnosis are under investigation. It is clear that giardiasis produces host responses at both the humoral and cellular levels. Serum and fecal antibodies have been demonstrated (Jones and Brown, 1974). These antibodies have been used to separate infected individuals from noninfected groups. Most recently, a specific serum anti-*G. lamblia* (Goka et al., 1986) immunoglobulin M has been described that appears short-lived compared with the serum anti-*G. lamblia* immunoglobulin G. This difference may prove helpful in separating cases of acute or recent giardiasis from those exposed and treated previously. Likewise, more attempts are being made to understand the various antigenic properties of the trophozoite stage. Several investigators believe that various strains of *G. lamblia* may be responsible for the variation in pathogenicity (Korman et al., 1986). One investigator has described a specific giardial antigen that shows promise of becoming a specific marker for the immunologic response to infection (Rosoff and Stibbs, 1986). At present, these procedures are limited as to clinical application and by the difficult production of antigen antibodies.

Epidemiologic Aspects

Occurrence and Transmission

Giardiasis occurs in all age groups. Worldwide studies report its prevalence to range from 2 to 25% (DeGirolami et al., 1985; Dupont and Sullivan, 1986; Mahmoud and Warren, 1975; Meyer and Jarroll, 1980). The incidence appears to be higher in children, with some population groups showing rates of 100% by 2 years of age (Dancescu and Tintareanu, 1964). In highly endemic areas infection is acquired early, usually within the first year of life. In one study group, more than 40% of the children less than 7 years of age acquired the infection (Farthing et al., 1986; Gilman et al., 1985). This finding is readily understandable, given the high degree of infectivity (10 cysts) of *G. lamblia* and the usually poor hygienic practices of children. Likewise, infected children are more apt to transmit giardiasis in settings where crowding or increased exposure is likely (i.e., day-care centers, institutions, and family units). Clearly, day-care centers have been found to offer many opportunities for transmission of *G. lamblia* (Pickering et al., 1984; Schuman, 1983).

Adult infections are more apt to result from sporadic epidemic outbreaks of food or water contamination. Such outbreaks have been reported in travelers (Isaac-Renton et al., 1986; Jokipii et al., 1985; Stiles and Kretchman, 1974; Walzer et al., 1971), skiers (Moore et al., 1969), and campers (Barbour et al., 1976; Meyer, 1973; Meyer and Radulescu, 1979). Likewise, certain adult groups are at risk from sexual practices, including anal–oral sex (Meyers et al., 1977).

Endemic forms result from direct person-to-person spread via contaminated hands, toys, food, or bodily parts (Blondell and Dedman, 1984; Harter et al., 1984; Meyers et al., 1977). Epidemic forms have resulted from contaminated surface water, sewer contamination, and occasionally food supply contamination (Craun, 1979; Dykes et al., 1980; Hopkins et al., 1985; Osterholm et al., 1981). Certainly, endemic and epidemic forms can share the same means of transmission. It is established, however, that fecal contamination with the cysts of *G. lamblia* is needed for transmission.

Susceptibility, Carriage, and Resistance

The level of susceptibility has been shown to be related to a series of factors. As discussed previously, age is a major factor. Children are more susceptible to active infection because of their increased risk of exposure and lack of immune development (Gillon, 1984; Gilman et al., 1985). Also, individuals with impaired immune responses are at higher risk for showing symptoms of this infection at any age (Meyer and Radulescu, 1979). Finally, reports have suggested that individuals with achlorhydria or hypochlorhydria have increased susceptibility to giardiasis (Wolfe, 1978).

It has been demonstrated in the animal model that T-cell function is necessary for the development of resistance to reinfection with *G. lamblia* (Roberts-Thomson et al., 1976; Stevens et al., 1978). It has also been shown in humans that a certain degree of resistance develops after epidemic outbreaks (Istre et al., 1984). Therefore, some form of resistance to repeated exposure to *Giardia* cysts is possible. However, repeat infections have occurred in the same individual. With the present epidemiologic evidence, this resistance must develop during childhood exposure. Thus, by adulthood, a degree of resistance to this infection is available. In one study, investigators found 14% of a group of asymptomatic, healthy adults to have serum antibodies to *Giardia* (Smith et al., 1981).

The carrier state has become an increasing concern to health care providers. Recognizing the many individuals who will be asymptomatic but transmit cysts, there will be a high rate of potential transmitters of this organism. Various investigators disagree on the treatment of persons with asymptomatic passage of cysts (Davidson, 1984; Pancorbo et al., 1985; Pickering et al., 1984; Vogt, 1985).

Treatment, Prevention, and Control

Treatment

Treatment for giardiasis has been available since the introduction of quinacrine in the late 1930s (Galli-Valerio, 1937). This drug has remained the treatment of choice for adults because of its low cost and effectiveness.

Metronidazole has gained common use for the treatment of giardiasis and been shown as effective as quinacrine (Davidson, 1984; Karsousi, 1979). In some cases, it may be better tolerated in children because quinacrine has a bitter taste. As its potential teratogenicity has not been resolved, the application of metronidazole is limited. Liquid suspensions of

TABLE 2. Comparative cure rates of available drugs and standard treatment regimens for giardiasis

	Cure rate (%)			
Study (year)	*Quinacrine*	*Metronidazole*	*Furazolidone*	*Tinidazole*[a]
Bassily et al. (1970)	100	95	80	—
Garg (1972)	—	95	94	—
Levi et al. (1977)	—	87	71	96
Nair et al. (1979)	—	84	80	—
Karsousi (1979)	75	100	—	—
Sabcharreon et al. (1980)	100	60	—	—
Craft et al. (1981)	70	—	92	—
Davidson (1984)[b]	93	92	84	—

[a] Unavailable in the United States.
[b] Grouped data.

metronidazole are not available within the United States, but may be used in children where available.

Furazolidone, although somewhat less effective than quinacrine and metronidazole, is available in both tablet and suspension form (Davidson, 1984). It may be used in young children. However, some investigators report frequent nausea and vomiting with this drug.

Outside the United States, tinidazole has been used successfully for the treatment of giardiasis. Its advantage is its administration in a single 2-g dose, which improves compliance and the ability to ensure effective treatment (Speelman, 1985).

Table 2 compares the rate of treatment success in studies evaluating the effectiveness of these drugs (Bassily et al., 1970; Craft et al., 1981; Davidson, 1984; Garg, 1972; Karsousi, 1979; Levi et al., 1977; Nair et al., 1979; Sabcharreon et al., 1980). Over the past 10 years, one can see the variation reported using the same drug. Davidson (1984) groups these studies and suggests that quinacrine and metronidazole are equally effective.

TABLE 3. Adult and pediatric treatment regimens for *Giardia lamblia*

Drug	*Regimen*[a]
Adults	
Quinacrine	100 mg p.o. t.i.d. × 5 days
Metronidazole	250 mg p.o. t.i.d. × 5 days
Furazolidone	100 mg p.o. t.i.d. × 7 days
Tinidazole[b]	150 mg p.o. t.i.d. × 7 days (2 g single dose)
Pediatric[c]	
Quinacrine	<8 yr: 2 mg/kg p.o. t.i.d. × 5–7 days
	>8 yr: 100 mg p.o. t.i.d. × 5–7 days
Metronidazole	<30 lb (<13.6 kg): 125 mg p.o. b.i.d. × 10 days
	31–40 lb (14.1–18.1 kg): 125 mg p.o. t.i.d. × 10 days
	41–60 lb (18.6–27.2 kg): 250 mg p.o. b.i.d. × 7–10 days
	>60 lb (>27.2 kg): 250 mg p.o. t.i.d. × 7–10 days
Furazolidone	1 mo–1 yr: ½ to 1 tsp q.i.d. × 7 days
	1–4 yr: 1 to 1½ q.i.d. × 7 days
	>5 yr: 1 tbsp q.i.d. × 7 days
Tinidazole[b]	50 mg/kg (maximum 2 g) single dose

[a] p.o. = per oral; t.i.d. = three times a day; q.i.d. = four times a day; b.i.d. = twice a day; tsp = teaspoon; tbsp = tablespoon.
[b] Unavailable in the United States.
[c] Data from Wolfe (1979).

The accepted standard drug regimens are shown in Table 3. Age-related differences are demonstrated (Gorski, 1985; Medical Letter, 1986; Wolfe, 1979). Pediatric doses are adjusted by age for quinacrine and furazolidone and by weight for metronidazole and tinidazole (Wolfe, 1979). A study by McIntyre (1986) using *Giardia* isolates found considerable variation in their susceptibility to the drugs listed. He suggested that this variation represents different strains of *Giardia* and may, in part, account for treatment failures.

Derivatives of nitroimidazoles are also being tested for the treatment of giardiasis (Rossignol et al., 1984). The newer derivatives, such as ornidazole and secnidazole, may offer alternative drug choices in the future.

Prevention and Control

Strategies for the control of giardiasis must center on the effective elimination of fecal contamination and on adequate recognition and treatment of cases and carriers; some, however, do not treat carriers (Vogt, 1985). Approaches must include education at the individual patient level, the community level to include workers in day-care centers and institutions, the public health care level to include public works and water systems, and the regional level to educate environmental workers and planners. Health care workers also need education regarding the transmission of and treatment projects available for giardiasis. Physicians are especially important in education, as they must first recognize the signs and symptoms, carry out appropriate diagnostic steps, and then provide the best treatment options for each case.

Serodiagnosis offers epidemiologic approaches to study the distribution of giardiasis, monitor its spread, and evaluate its natural course. With improved techniques for detecting antigenic determinants, the possibility of vaccine development has been suggested (Stevens, 1985). Until vaccines becomes possible or practical, however, the prevention and control of giardiasis will require a concerted effort at individual education directed toward personal hygiene; control of fecal wastes; understanding water supply treatments; and education of families and adult personnel of day-care centers.

For epidemic outbreaks, emergency water supplies are best boiled (Benenson, 1985). If boiling is not possible, the water supply can be treated with hypochlorite or iodine, using 0.1 to 0.2 ml (2 to 4 drops) of household bleach or 0.5 ml of 2% tincture of iodine per liter for 70 min. Longer treatment, preferably overnight, may be required if the water is cold. Attempts must be made to identify the source of the outbreak. The community water source can be filtered and public water supplies protected against contamination with human or animal feces.

Literature Cited

Ament, M. E., and C. E. Robin. 1972. Relation of giardiasis to abnormal intestinal structure and function in gastrointestinal immunodeficiency syndromes. Gastroenterology **62:**216–226.

Barbour, A. G., C. R. Nichols, and T. Furushima. 1976. An outbreak of giardiasis in a group of campers. Am. J. Trop. Med. Hyg. **25:**384–389.

Bassily, S., A. Farid, J. Mikhail, D. Kent, and J. Lehman. 1970. The treatment of *Giardia lamblia* infection with mepacrine, metronidazole and furazolidone. J. Trop. Med. Hyg. **73:**15–18.

Benenson, A. S. (ed.). 1985. Control of communicable diseases in man, 14th ed. American Public Health Association, Washington, D.C.

Bingham, A. K., and E. A. Meyer. 1979. Giardia excystation can be induced in vitro in acidic solutions. Nature (London) **277:**301–302.

Blankinsopp, W. K., J. A. Gibson, and G. P. Haffenden. 1978. Giardiasis and severe jejunal abnormality. Lancet **1:**994.

Blondell, R. D., and E. B. Dedman. 1984. An extended family with giardiasis. J. Fam. Pract. **19:**338–392.

Brodsky, R. E., H. C. Spencer, and M. G. Schultz. 1974. Giardiasis in American travelers to the Soviet Union. J. Infect. Dis. **130:**319.

Cordingley, F. T., and G. P. Crawford. 1986. Giardia infection causes vitamin B_{12} deficiency. Aust. N.Z. J. Med. **16:**78–79.

Craft, J. C., T. Murphy, and J. D. Nelson. 1981. Furazolidone and quinacrine: comparative study of therapy for giardiasis in children. Am. J. Dis. Child. **135:**164–166.

Craun, G. F. 1979. Waterborne giardiasis in the United States: a review. Am. J. Public Health **69:**817–819.

Dancescu, P., and J. Tintareanu. 1964. Investigation concerning the spread of lambliasis in a children's collectivity. Microbiol. Parazitol. Epidemiol. **9:**343–350.

Danciger, M., and M. Lopez. 1975. Numbers of *Giardia* in the feces of infected children. Am. J. Trop. Med. Hyg. **24:**237–242.

Davidson, R. A. 1984. Issues in clinical parasitology: the treatment of giardiasis. Am. J. Gastroenterol. **79:**256.

DeGirolami, P. C., J. C. Dunn, and M. Federman. 1985. Infections caused by intestinal protozoa. Pathol. Annu. **20:**463–505.

Dupont, H. L., and P. S. Sullivan. 1986. Giardiasis: the clinical spectrum, diagnosis and therapy. Pediatr. Infect. Dis. **5:**S131–S138.

Dykes, A. C., D. Juranek, R. Lorenz, S. Sinclair, W. Jakubowski, and R. Davies. 1980. Municipal water borne giardiasis: an epidemiologic investigation—beavers implicated as a possible reservoir. Ann. Intern. Med. **92:**165–170.

Farthing, M. J. G., L. Mata, J. Urrutia, and R. Kronmal. 1986. Natural history of *Giardia* infection in infants and children in rural Guatamala and its impact on physical growth. Am. J. Clin. Nutr. **43:**395–405.

Galli-Valerio, B. 1937. La lambliase et son traitement par l'atebrine. Schweiz. Med. Wochenschr. **67:**1181–1182.

Garg, B. J. 1972. Furazolidone and metronidazole in the treatment of giardiasis. Indian J. Pediatr. **39**:264–266.

Giannella, R. A., S. A. Broitman, and N. C. Zamcheck. 1973. Influence of gastric acidity on bacterial and parasite enteric infections: a perspective. Ann. Intern. Med. **78**:271–276.

Gillon, J. 1984. Giardiasis: review of epidemiology, pathogenetic mechanisms and host responses. Q. J. Med. **209**:29–39.

Gillon, J. 1985. Clinical studies in adults presenting with giardiasis to a gastro-intestinal unit. Scott. Med. J. **30**:89–95.

Gilman, R. H., K. Brown, G. Visvesvara, G. Monoal, R. Greenberg, B. Sack, F. Brandt, and K. Cohn. 1985. Epidemiology and serology of *Giardia lamblia* in a developing country: Bangladesh. Trans. R. Soc. Trop. Med. Hyg. **79**:469–473.

Goka, A. K. J., D. Rolston, V. Mathan, and M. Farthing. 1986. Diagnosis of giardiasis by specific IgM antibody enzyme-linked immunosorbent assay. Lancet **2**:184–186.

Gorski, E. D. 1985. Management of giardiasis. Am. Fam. Physician **32**:157–164.

Green, E. L., M. A. Miles, and D. C. Warhurst. 1985. Immunodiagnostic detection of giardia antigen in faeces by a rapid visual enzyme-linked immunosorbent assay. Lancet **2**:691–693.

Hamrick, H. J., and G. W. Moore. 1983. Giardiasis causing urticaria in a child. Am. J. Dis. Child. **137**:761–763.

Harter, L., F. Frost, G. Grunenfelder, K. Perkins-Jones, and J. Libby. 1984. Giardiasis in an infant and toddler swim class. Am. J. Public Health **74**:155–156.

Hopkins, R. S., P. Shillam, B. Gasparo, L. Eisnach, and R. Karlin. 1985. Waterborne disease in Colorado: three years' surveillance and 18 outbreaks. Am. J. Public Health **75**:254–262.

Isaac-Renton, J. L., W. Black, R. Mathias, E. Proctor, and C. Sherlock. 1986. Giardiasis in a group of travelers—attempted use of a serological test. Can. J. Public Health **77**:86–88.

Istre, G. R., T. Dunlop, B. Gasparo, and R. Hopkins. 1984. Waterborne giardiasis at a mountain resort: evidence for acquired immunity. Am. J. Public Health **74**:602–604.

Jokipii, A. M. M., M. Hemila, and L. Jokipii. 1985. Prospective study of acquisition of cryptosporidium, *Giardia lamblia*, and gastrointestinal illness. Lancet **2**:487–489.

Jokipii, A. M. M., and L. Jokipii. 1977. Prepatency of giardiasis. Lancet 1095–1097.

Jones, E. G., and W. R. Brown. 1974. Serum and intestinal fluid immunoglobulins in patients with giardiasis. Dig. Dis. **19**:791–796.

Jones, J. E. 1980. Office parasitology. Am. Fam. Physician **22**:86–90.

Jones, J. E. 1981. The identification of intestinal nematodes using the digital rectal examination. J. Fam. Pract. **12**:563–656.

Jones, J. E. 1982. The psychosocial impact of common parasitic infections. J. Fam. Pract. **14**:1139–1143.

Jones, J. E. 1984. Identification of intestinal parasites in an office setting: a comparison with reference laboratory results. Fam. Pract. Res. J. **3**:148–154.

Jones, J. E. 1986. String test for diagnosing giardiasis. Am. Fam. Physician **34**:123–126.

Karsousi, S. 1979. Giardiasis in infancy and childhood: a prospective study of 160 cases with comparison of quinacrine (Atabrine) and metronidazole (Flagyl). Am. J. Trop. Med. Hyg. **28**:19–23.

Korman, S. H., S. LeBlancq, D. Spira, J. Elon, R. Reifen, and R. Deckelbaum. 1986. *Giardia lamblia:* identification of different strains from man. Z. Parasitenkd. **72**:173–180.

Levi, G. C., C. A. deAviula, and V. A. Neto. 1977. Efficacy of various drugs for treatment of giardiasis. Am. J. Trop. Med. Hyg. **26**:564–565.

Mahmoud, A. A. F., and K. S. Warren. 1975. Algorithms in the diagnosis and management of exotic diseases. II. Giardiasis. J. Infect. Dis. **131**:621–624.

McIntyre, P. 1986. Chemotherapy in giardiasis: clinical responses and in vitro drug sensitivity of human isolates in axenic culture. J. Pediatr. **108**:1005–1010.

Medical Letter. 1986. Drugs for parasitic infection. **28**(issue 706):9–16.

Meyer, E. A., and E. L. Jarroll. 1980. Giardiasis. Am. J. Epidemiol. **111**:1–12.

Meyer, E. A., and S. Radulescu. 1979. Giardia and giardiasis. Adv. Parasitol. **17**:1.

Meyer, W. T. 1973. Epidemic giardiasis. A continued elusive entity. Rocky Mt. Med. J. **70**:48–49.

Meyers, J. D., H. A. Kuharic, and K. K. Holmes. 1977. *Giardia lamblia* infection in homosexual man. Br. J. Vener. Dis. **53**:54–55.

Moore, G. T., W. Cross, D. McGuire, C. Mollohan, N. Gleason, C. Healy, and L. Newton. 1969. Epidemic giardiasis at a ski resort. N. Engl. J. Med. **281**:402–407.

Nair, K. V., M. Sharma, S. Mithal, and B. Tandon. 1979. Success of metronidazole and furazolidone in the treatment of giardiasis. J. Indian Med. Assoc. **72**:162–165.

Osterholm, M. T., J. Forfang, T. Ristinen, A. Dean, J. Washburn, J. Godes, R. Rude, and J. McCullough. 1981. An outbreak of foodborne giardiasis. N. Engl. J. Med. **304**:24–28.

Owen, R. L., P. Nemanic, and D. Stevens. 1979. Ultrastructural observations in giardiasis in a murine model. I. Intestinal distribution, attachment and relationship to the immune system of *Giardia muris*. Gastroenterology **76**:757–769.

Pancorbo, J. M. C., M. T. G. Munoz, and J. L. Badin. 1985. Giardiasis: treatment of carriers. Lancet **2**:951.

Pickering, L. K., W. Woodward, H. DuPont, and P. Sullivan. 1984. Occurrence of *Giardia lamblia* in children in day care center. J. Pediatr. **104**:522–526.

Porter, A. 1916. The cysts of *Giardia (lamblia) intestinalis* in human dysenteric faeces. Lancet **1**:1166–1169.

Raizman, R. E. 1976. Giardiasis: an overview for the clinician. Am. J. Dig. Dis. **21**:1070–1074.

Rendtorff, R. C. 1954. The experimental transmission of human intestinal protozoan parasites. II. *Giardia lamblia* cysts given in capsules. Am. J. Hyg. **59**:209–220.

Rendtorff, R. C., and C. J. Holt. 1954. The experimental transmission of human intestinal protozoan parasites. IV. Attempts to transmit *Entamoeba coli* and *Giardia lamblia* by water. Am. J. Hyg. **60**:327–338.

Roberts-Thomson, I. C., D. P. Stevens, A. A. F. Mahmoud, and K. S. Warren. 1976. Acquired resistance to infection in an animal model of giardiasis. *J. Immunol.* **117**:2036–2037.

Rosoff, J. D., and H. N. Stibbs. 1986. Isolation and identification of a *Giardia lamblia*-specific stool antigen (GSA 65) useful in coprodiagnosis of giardiasis. J. Clin. Microbiol. **23**:905–910.

Rossignol, J. F., H. Maisonneuve, and Y. W. Cho. 1984. Nitroimidazoles in the treatment of trichomoniasis, giardiasis, and amebiasis. Int. J. Clin. Pharmacol. Ther. Toxicol. **22**:63–72.

Sabcharreon, A., T. Chongsuphajaisiddhi, and P. Attanath. 1980. Treatment of giardiasis in children with quinacrine, metronidazole, tinidazole and midazole. Southeast Asian J. Trop. Med. Public Health **11**:280–284.

Saha, T. K., and T. K. Ghosh. 1977. Invasion of small intestinal mucosa by *Giardia lamblia* in man. Gastroenterology **72**:402–405.

Schuman, S. H. 1983. Clues for suspecting giardiasis in a day care center. J. Fam. Pract. **17**:23–24.

Sealy, D. P., and S. H. Schuman. 1981. Giardiasis: a common and underrecognized enteric pathogen. J. Fam. Pract. **12**:47–54.

Smith, P.D., F. Gilun, W. Brown, and T. Nash. 1981. IgG antibody to *Giardia lamblia* detected by enzyme-linked immunosorbent assay. Gastroenterology **80**:1476–1480.

Speelman, P. 1985. Single-dose tinidazole for the treatment of giardiasis. Antimicrob. Agents Chemother. **27**:227–229.

Stevens, D. P. 1985. Selective primary health care: strategies for control of disease in the developing world. XIX. Giardiasis. Rev. Infect. Dis. **7**:530–535.

Stevens, D. P., D. M. Frank, and A. A. F. Mahmoud. 1978. Thymus dependency of host resistance to *Giardia muris* infection studies in nude mice. J. Immunol. **120**:680–682.

Stiles, M. A., and L. A. Kretchman. 1974. Giardiasis and the adult traveler. Postgrad. Med. **56**:134–138.

Sun, T. 1980. The diagnosis of giardiasis. Am. J. Surg. Pathol. **4**:265–271.

Tandon, B. N., R. Tandon, B. Satpathy, and S. Shriniwas. 1977. Mechanism of malabsorption in giardiasis: a study of bacterial flora and bile salt deconjugation in upper jejunum. Gut **18**:176–181.

Tsuchiya, J. 1931. A study on variabilities in dimensions and numbers of discharged cysts of *Giardia lamblia* (Stiles 1915) from day to day under normal conditions. Am. J. Hyg. **13**:544–567.

Ungar, B. L. P., R. Yolken, T. Nash, and T. Quinn. 1984. Enzyme-linked immunosorbent assay for the detection of *Giardia lamblia* in fecal specimens. J. Infect. Dis. **149**:90–97.

Visvesvara, G. S., P. Smith, G. Healy, and W. Brown. 1980. An immunofluorescence test to detect serum antibodies to *Giardia lamblia*. Ann. Intern. Med. **93**:802–805.

Vogt, R. L. 1985. Giardiasis in day care centers. J. Pediatr. **106**:857–858.

Walzer, P. D., M. S. Wolfe, and M. G. Schultz. 1971. Giardiasis in travelers. J. Infect. Dis. **124**:235–237.

Weisman, B. L. 1979. Urticaria and *Giardia lamblia* infections. Ann. Allergy **42**:91.

Welsh, J. D., R. Poley, J. Hensley, and M. Bhatia. 1984. Intestinal disaccharidase and alkaline phosphatase activity in giardiasis. J. Pediatr. Gastroenterol. Nutr. **3**:37–40.

Wolfe, M. S. 1978. Medical intelligence, current concepts in parasitology giardiasis. N. Engl. J. Med. **298**:319.

Wolfe, M. S. 1979. Giardiasis. Pediatr. Clin. N. Am. **26**:295–303.

Wright, S. G., and A. M. Tomkins. 1977. Quantification of the lymphocyte infiltrate in jejunal epithelium in giardiasis. Clin. Exp. Immunol. **29**:408–412.

CHAPTER 93

Hookworm

JERRY E. JONES

Disease: Hookworm (ancylostomiasis, necatoriasis, uncinariasis).

Etiologic Agents: *Ancylostoma duodenale, Necator americanus, Ancylostoma ceylanicum* (rarely).

Source: Humans; cats and dogs for *A. ceylanicum*.

Clinical Manifestations: Classic: hypochromic microcytic anemia of iron deficiency. Acute infection: maculopapular rash, pruritis (skin), cough, rales, eosinophilia (lungs), epigastric pain, nausea, vomiting, diarrhea (bowels). Chronic infection: epigastric pain, pica, hypoproteinemia, edema, fatigue, dyspnea.

Pathology: Focal hemorrhages (lung); direct mechanical attachment to intestinal mucosa with cytolysis of mucosa and ingestion of red blood cells.

Laboratory Diagnosis: Direct fecal smears; concentration methods; dilution egg count techniques for worm burdens.

Epidemiology: Endemic in tropical and subtropic countries where temperature, rainfall, and deposit of feces are favorable for larval development. Infection is transmitted by skin penetration of filariform larvae, rarely by oral transmission of *A. duodenale*.

Treatment: Mebandazole, pyrantel pamoate (embonate), bephenium hydroxynaphthoate, tetrachlorethylene.

Prevention and Control: Identification and treatment of heavily infected individuals with simultaneous approaches to controlling the disposal of human feces and protecting the skin from contact with larvae.

Description of Disease

> One must wonder what stage of advancement and civilization man would have reached by this date had not this veritable monster interfered with the mental and physical development of every generation during the developing and learning periods of life. And even now man is pouring into the ever empty stomachs of this blood-sucking parasite a veritable river of his life blood.
>
> Bass (1914)

Hookworms represent the human infection with the intestinal nematodes *Ancylostoma duodenale, Necator americanus,* and rarely *Ancylostoma ceylanicum* (Gilman, 1982). Once considered a major scourge of mankind, the infections are now less dramatic in most parts of the world than those previously encountered. However, it has been estimated that hookworms parasitize more than 900 million people (Schad and Banwell, 1984). This number suggests that nearly one-fourth of the world's population remains at risk for this intestinal parasitic infection. Therefore, hookworms remain an important parasitic problem worldwide, with many population groups still infected by this blood-sucking parasite.

Hookworm disease is so named because the two major human species pass morphologically identical eggs in the feces. Thus, the two species cannot be separated by stool examination alone. The separation of species has little clinical significance, and the treatment approaches are the same. Therefore, "hookworm disease" has come to represent the diagnosis routinely made in the clinical and laboratory settings. Historically, this diagnosis has referred primarily to the intestinal phase of the adult worms.

Characterization

Hookworm infection is a disease transmitted in warm, moist climates that have appropriate soil suited for its larval-phase transmission. The continued survival of this species depends on chance contact with a human host while it is in its infective or filariform larval stage. Individuals come in contact with this stage of the parasite when they provide exposed skin to the larvae. Breaks in the skin surface such as scales, cracks, fissures, and hair follicles are used to initiate skin penetration.

Once within the new host, the parasites are carried to the lungs. Here they break into the alveolar spaces and move by cough and migration through the respiratory tree to the pharynx. The larvae are swallowed and pass into the small intestine. They mature and attach to the mucosa of the small intestine, living from 1 to 5 years depending on the species. Morphologically, the adult worms differ from one another by specialized mouth parts and reproductive organs. The adults are pink or creamy white; the female (10 to 13 mm) is slightly longer than the male (8 to 11 mm). Adults frequently migrate from site to site, creating multiple sites of attachment. Sexual reproduction takes place in the small intestine, where each female worm can pass 10,000 to 25,000 eggs per day depending on the species (Hill, 1926). Once passed into the fecal stream, the eggs undergo a number of cell divisions before passage out of the host. The eggs then embryonate in the feces and hatch in the fecally contaminated soil as a rhabditiform (first-stage) larva. This stage feeds on microbes and organic particles while developing into the infective or filariform larva. The filariform stage then migrates to the soil surface and seeks contact with a human host. In tropical environments, the filariform larvae will survive 3 to 4 weeks, and under optimal conditions they may survive for 3 to 4 months. If contact is made, the larvae seeks a port of entry through breaks in the integrity of the skin. Figure 1 depicts this parasitic life cycle. Penetration of the skin is facilitated by enzymes secreted by the larvae, their boring movements, and a specialized "oral spear" at the mouth parts (Beaver et al., 1984).

Clinical Manifestations

Over time, hookworm disease has come to refer specifically to the microcytic, hypochromic anemia caused by the intestinal, adult phase of the parasites. Hookworms may produce symptoms localizing at the site of infection or along its course of development from skin contact to the pulmonary system or the gastrointestinal tract. The host may also show indirect effects such as fatigue, dyspnea, and edema as a result of anemia and protein loss. A cellular immune response involving eosinophils may occur during the larval or the adult phase. The eosinophilia may result directly from the worms or indirectly from their products or wastes. Therefore, signs and symptoms of hookworm infection may reflect multisystemic involvement. The severity of these symptoms relate to the number of worms and the duration of infection, with the nutritional and immune status of the host playing an important role. Recognizing the direct relationship between signs and symptoms

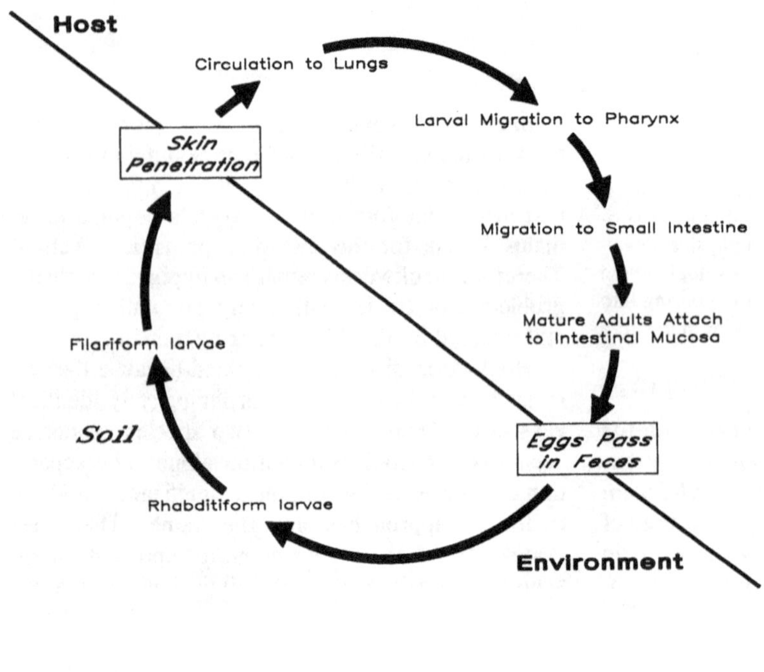

FIG. 1. Life cycle of the hookworm.

and worm burden, it has been helpful to classify hookworm disease by the severity of infection using fecal egg counts. It has been noted that the clinical severity of the infection is light with individuals showing less than 5 eggs per milligram of feces, whereas individuals with more than 20 eggs per milligram have significant anemia. Infections are considered massive when stools show egg counts exceeding 50 per milligram of feces (Beaver et al., 1984).

Acute infection in a previously unexposed host may involve allergic as well as direct-contact reactions. Cutaneous lesions develop at the sites of skin penetration, producing itching. Erythematous, papular eruptions develop and may last 7 to 10 days. With heavy infections, unrelieved pruritus may be produced. The rash occurs within several hours of larval exposure. The severity depends on the degree of allergic response.

In an unexposed host, the larvae reach the lung within the first 24 h. They break into the alveoli, causing minute focal hemorrhages. Heavy infections are generally required to produce pneumonitis severe enough to reach clinical significance. In Japan a clinical syndrome called *Wakana disease*, attributable to *A. duodenale,* has been described. In this syndrome, the larvae of this species appear to pass directly into the lungs after ingestion by the host of contaminated vegetables. The initial symptoms reported after ingestion are nausea, vomiting, salivation, itching of the pharynx, and hoarseness. In several days severe pulmonary symptoms develop, including cough, dyspnea, and hypereosinophilia. Patchy infiltration may be seen on chest X ray (Harada, 1962).

Abdominal symptoms may be severe in a previously unexposed host. Initially, volunteers who were infected repeatedly had prominent intestinal symptoms, including intense nausea, pain, weight loss, and severe diarrhea. These symptoms decreased with repeated infections (Ogilvie et al., 1978). Acute massive intestinal bleeding has been reported rarely especially in children (de la Riva et al., 1981).

Chronic hookworm infection usually produces gradual blood and protein loss (Tandon et al., 1969). It is estimated that *A. duodenale* causes approximately 0.15 ml (range, 0.05 to 0.30 ml) of blood loss per day, whereas *N. americanus* will average 0.03 ml (range, 0.01 to 0.04 ml) of blood loss per day (Variyam and Banwell, 1982). In addition, focal hemorrhages at the sites of hookworm attachment add to the total blood and protein loss. Depletion of iron stores will occur more rapidly in menstruating women and in growing children, who have increased iron requirements (Roche and Layrisse, 1966). As anemia and hypoproteinemia develop, additional symptoms include fatigue, weakness, dyspnea on exertion, cardiac palpitation, and pallor. In heavy infections of long duration, mental and physical retardation have been described (Hunter et al., 1976). In addition, edema and high output failure may ensue as the severity of anemia and protein loss increases.

The peripheral blood picture will vary depending on the stage and severity of infection. In unexposed or allergic individuals, the eosinophilia may be high. In established disease, a mild eosinophilia (7 to 14%) may remain. The reticulocyte count is usually normal. Table 1 summarizes the clinical manifestations of hookworm infection.

Pathology

The pathophysiology of hookworm disease relates directly to the stage involved and the number of invading organisms. Penetration of the skin by larvae excites a local tissue reaction at the point and the course of invasion. This reaction may vary from erythematous papules to vesiculation and edema. The bacteria carried by the larvae are believed to be principally responsible for the tissue reactions, as axenic larvae of *N. americanus* fail to produce a reaction when they penetrate the skin (Miller, 1979). Secondary bacterial infection tends to develop if the lesions are scratched continuously.

TABLE 1. Clinical manifestations of hookworm disease

Skin ("ground itch")
Local erythema[a]
Maculopapular rash[a]
Intense pruritus[a]
Lung (pneumonitis/bronchitis)
Cough
Wheezing, rales
Low-grade temperature
Eosinophilia (marked in hyperacute infection)
Wakana disease
Gastrointestinal
Acute
Epigastric pain
Nausea, vomiting
Tenderness
Diarrhea, anorexia
Hemorrhage
Chronic
Epigastric pain
Pica
Hemorrhage
Other systems
Blood: Iron deficiency anemia, hypoproteinemia/edema
Heart: High output failure, murmur, fatigue, dyspnea, edema
Bone marrow: Depletion of iron stores

[a] May be more prominent on reinfection.

Circulatory transfer to the lung occurs rapidly for the maturing larvae. Trapped by their size at the alveolar–vascular bed, they penetrate into the air sacs, producing minute focal hemorrhages. Interestingly, they tend to produce less sensitization than other roundworm species and, except in cases of heavy infection or Wakana disease, excite little pulmonary reaction. In experimental models using the rat, it has been suggested that the changes produced in the lungs are due to active movement of the larvae through the arterioles (Croll and Ma, 1978).

Gastrointestinal pathology relates to the direct mechanical attachment to the intestinal mucosa. Portions of the villi are sucked into the mouth parts. Studies using the dog hookworm revealed that a worm could attach, ingest, and digest a bolus of tissue within 10 min (Kalkofen, 1974). In humans, the adult worms have been shown to secrete an anticoagulant and an acetylcholineasterase (Ogilvie et al., 1978). It is unclear whether or not cytolytic enzymes are secreted into the villi at the site of attachment (Gilman, 1982). The lesions generated are characterized by focal hemorrhage, tissue cytolysis, and a vigorous neutrophilic response. In the dog model, these lesions repair rapidly without sequelae when treated. Adult worms have a strong sucking action and have been shown to ingest blood, plasma, saline, and formulin (Roche and Layrisse, 1966). The worms may burrow through the lamina propria and proceed to the muscularis mucosa. However, the muscularis mucosa appears to prevent further penetration (Gilman, 1982). The adult worms tend to abandon sites of attachment frequently, particularly when mating and when conditions are crowded. Thus, blood loss is believed to occur via two mechanisms: by direct ingestion of red blood cells and by the tissue trauma that results when the worm attaches and feeds. Long-term infections and heavy worm burdens produce the classic iron deficiency anemia of hookworm disease.

Hypoproteinemia with hypoalbuminemia may also occur in individuals with iron deficiency anemia due to hookworm infection (Tandon et al., 1969). The degree of protein loss correlates with the degree of anemia. In highly endemic areas, however, other nutritional and host factors may play as important role in this protein loss (Mayoral et al., 1967; Variyam and Banwell, 1982).

The immune response in humans is unclear. Ogilvie reported a rise in specific immunoglobulin IgE and IgG antibody to the secretory products of the worms. These antibodies appeared to play no protective role, however, because the infection could be initiated and the cycle completed in their presence (Gilman 1982; Ogilvie et al., 1978).

Other systemic manifestations reflect the indirect changes caused by the presence of adult worms or their secretions. The eosinophilic response varies from mild to moderate. Acquired platelet dysfunction has also been described (Kueh et al., 1983).

Diagnostic Features

The diagnosis of hookworm disease is made by identifying the characteristic eggs in fecal specimens. With each female passing 10,000 to 25,000 eggs per day, the diagnosis can be made readily in clinically significant infections. One study revealed a cyclic production of eggs depending on worm burden and species. This study also reported that the egg output declined as parasite burdens increased (Anderson and Schad, 1985). Morphologically, the eggs are oval and measure 60 by 40 μm. The shell is thin, and the fertilized eggs are usually in early cleavage stages.

Stools can be examined by direct smears and, if negative, by a concentration method such as zinc sulfate. To calculate worm burden, the Stoll dilution egg count technique or Beaver direct egg count technique can be used (Beaver, 1950; Stoll, 1961).

Occasionally, it may be necessary to differentiate the rhabditiform larvae of the hookworm from that of *Strongyloides*. Rhabditiform larvae of the hookworm may be found in fresh stool samples that are held too long before examination. Careful microscopy of the mouth parts will show a short buccal cavity in *Strongyloides* as compared with a long buccal cavity in hookworm larvae. Stool kits are helpful in storage and transport of samples if immediate microscopy is unavailable.

Epidemiologic Aspects

Occurrence and Transmission

The larval phase needed for transmission of the hookworm limits this parasite to warm, moist areas of the world. Transmission appears to be particularly successful in sandy loam soils and in areas populated by dung beetles (Miller, 1954). Therefore, hookworm disease occurs in nearly all subtropical and tropical countries. Species of hookworm overlap, and both major species are present in many countries.

Necator americanus is predominant in the southern United States, Central and South America, and the West Indies. It is also prevalent in southern Asia, Indonesia, the Philippines, Polynesia, and Central and South Africa.

Ancylostoma duodenale is the predominant hookworm species in North Africa, southern Europe, northern India, China, and Japan. This species is also present in southern India, Indonesia, Burma, the Philippines, the southern and central Pacific Islands, West Africa, Australia, and Paraguay.

Recent prevalence studies continue to show focal areas of high endemicity. Schoolchildren show the highest rates, with one study finding 86 to 90% infected (Michaelsen, 1985). Age-related rates have also been found in studies from Ethiopia and Nigeria. In Ethiopia children 8 to 11 years old were found to have the highest rates for hookworm (Tedla and Jemaneh, 1985), whereas in Nigeria young adults 15 to 24 years old had the highest rates (Udonsi, 1985).

Rainfall patterns have an important role in endemicity, as moisture and dampness are necessary for the larva to migrate to the soil surface. Hookworm disease is endemic only in those tropical and subtropical areas where the rainfall averages 50 or more inches per year (Hunter et al., 1976). The larva appear to need the water film to migrate vertically to the soil surface. Periods of dryness will kill the organisms (Schad and Banwell, 1984). Longitudinal studies have shown a rise in egg counts 2 months after a short rainy season. The intensity of infection peaked 7 months after the rainy season (Knight and Merrett, 1981).

Temperature has also been shown to be important for survival of the eggs and larval phases of both species. Temperatures between 26.7 and 32.2°C (80 and 90°F) are optimal. *Necator* species are more tolerant of high temperature than are *Ancylostoma* species (Beaver et al., 1984), which seem to tolerate lower temperatures. This finding may parallel those of Tedla and Jemaneh (1985), who reported transmission to be inversely associated with altitude (i.e., prevalence generally decreased with increasing altitudes). Afework and colleagues (1985) also reported higher rates of *Ancylostoma* than *Necator* in a cooler gorge area. This finding also supports the role of temperature in the epidemiology of hookworm disease.

It has been noted that *A. duodenale* has the ability to arrest its larval phase of development. This ability appears to provide the species with a means of adapting to seasonally unfavorable conditions (Schad et al., 1973).

Susceptibility, Resistance, and Carriage

As with many other parasitic infections, various cultural, social, and nutritional factors relate to the susceptibility of hookworm infection. Because of the silent bloodletting caused by the hookworm, an already nutritionally deficient host can become more depleted of iron. A study in Guam found that malnourished children developed more serious consequences from acute anemia caused by hookworms than did those children regarded as better nourished (Zimmerman, 1946). However, it has also been noted that when poorly nourished individuals heavily infected with hookworm are nutritionally repleted, the total worm burden and daily loss of blood remains essentially unchanged (Tripathy et al., 1971). Some investigators believe that improving nutrition may protect the individual by limiting the number of worms that complete development (Gilman, 1982). Therefore, preexisting malnutrition and avitaminosis would appear to lower host ability to sustain a subclinical infection, as a smaller inoculum of worms would be more likely to produce clinical symptoms.

Schad and Anderson (1985) reported quantitative data suggesting that after effective treatment of heavily infected individuals, this group was more likely to become heavily reinfected. They postulated that this pattern reflects an underlying susceptibility to hookworm infection.

Although certain animal species have demonstrated a distinct immune response to hookworm invasion, no clear evidence is available that humans can develop a functional protective immunity (Gilman, 1982).

Some authors have suggested, based on epidemiologic evidence, that immunity may develop. They note that in some population groups, a static worm population develops during the second and fifth decades. As worms are acquired steadily during early life, one would expect a steady increase in worm burden. Therefore, with the finding of this static level, a population with acquired immunity is postulated (Ogilvie and Worms, 1976). In human studies, repeated infections with small inocular of *Necator* did not reduce overall worm load. However, a rise in specific IgG and IgE antibodies was reported (Ogilvie and Worms, 1976; Ogilvie et al. 1978). Studies in a single volunteer showed that repeated larval challenges produced a greater degree of skin migration, suggesting that local immune factors in the skin may have a role in some immune responses (Ball and Bartlett, 1969). No conclusive studies on intestinal secretions of IgA have been reported.

Carriage refers to individuals with few worms and no clinical evidence of hookworm disease. In these situations, a distinction between hookworm infection and hookworm disease is made. Hookworm infection implies that eggs of hookworms are identified in stool examinations. It is important to recognize that routine stool examination merely demonstrates the presence of the parasite, but provides no exact evidence of the magnitude of the hookworm burden. As discussed previously, other methods involving egg count techniques provide data on the intensity of the infection. Clinical cases present signs and symptoms when the intensity of the infection is greater (i.e., usually more than 5 eggs per milligram of feces).

Clinical cases are severe only when numerous parasites are present.

Breastfeeding does not seem to offer any protection to infants against hookworm infection. One study reported no significant difference in the presence of hookworm ova between a group that was breastfed only and another that was breastfed and received supplements (Khan et al., 1983).

Treatment, Prevention, and Control

Treatment

Treatment of hookworm disease must be considered from a variety of standpoints, both for the individuals infected and for the communities involved. For individuals in a setting where reinfection is unlikely and where "having worms" has psychologic and social complications out of proportion to their pathogenicity (Jones, 1982), all infections are justifiably treated. In contrast, in highly endemic areas where reinfection will occur, treatment decisions will be based on worm burden or severity of illness. This approach would include individuals severely ill from other causes that may be aggravated by hookworm infection (Gilles, 1985). In addition to direct drug treatment for species of hookworms, supplemental therapy includes appropriate iron replacement. Iron supplements for infected individuals in endemic areas have been effective in controlling anemia (Basta et al., 1979).

Drugs widely used in the treatment of hookworm disease are shown in Table 2. Pyrantel pamoate (embonate), mebendazole, bephenium hydroxynaphthoate, and tetrachlorethylene are used most extensively. A comprehensive review of the chemotherapy of hookworm infections is available (Sharma and Charles, 1982). On an individual basis, mebendazole seems to be equally effective against both species of hookworm. Other commonly used drugs tend to favor one species or the other in their effectiveness. The toxic effects of tetrachlorethylene have limited its usefulness in the private clinic setting; however, because of its availability and lower cost it is easily applied to mass treatment programs (Schad and Banwell, 1984). Multiple intestinal helminthic infections must also be considered in mass treatment programs. In these cases, drugs that are effective against more than one helminth may be useful. Mebendazole, pryantel pamoate, and thiabendazole are good examples of broad-spectrum anthelmintics that have been used for both individual and community treatment.

More recently, clinical trials of albendazole have been evaluated using multiple- and single-dose regimens (Bassily et al., 1984; Chitchang et al., 1983; Misna et al., 1985; Ramalingam et al., 1983; Vivaran et al., 1982). Success has varied depending on the intensity of infection, the dosage used, and the age of the patient at treatment. Single-dose therapy has also produced varying results, and additional studies are needed before single-dose therapy with albendazole can be widely accepted.

Prevention and Control

Prevention and control of hookworm disease depend on the application of techniques that will have a direct impact on the factors responsible for its endemicity. These factors may vary in importance from one geographic region to another, but the basic principles are similar. First, it is necessary for a filariform larva to come in contact with the skin; entry is usually through the feet. Methods of control and prevention that protect individuals from direct skin contact with the soil will decrease exposure. Historically, control has been achieved in some countries

TABLE 2. Widely used drugs for treatment of hookworms

Drug (trade name)	*Dosage*	*Side effects*	*Broad spectrum*
Mebendazole (Vermox)	100 mg p.o. b.i.d. for 3 days (chewable tablet)	Mild, occasional diarrhea; abdominal pain; vomiting	Yes
Pyrantel pamoate (embonate) Antiminth	10–12 mg/kg single dose (maximum, 1 g)	Same as mebendazole	Yes
Bephenium hydroxynaphthoate	*Ancyclosma*, 5 g single dose; *Necator*, 5 g t.i.d. for 1 day	Nausea, vomiting, diarrhea	Yes
Tetrachloroethylene[a]	0.1 mg/kg (maximum, 5-ml dose)	Burning epigastric pain; vomiting; drowsiness	No

[a] To be taken on an empty stomach with no eating for 3 h thereafter; must treat for Ascaris before using.

by patient education and simply by the wearing of shoes. Second, defecation habits that expose work or other areas to human feces must be addressed. Installation of simple latrines or careful waste control in certain areas has been successful. Additionally, community education is necessary to ensure understanding of the reasons for introducing new or different habits to a rural or semirural environment. Education regarding the use of night soil as fertilizer is also important. Finally, an acceptable identification and treatment program must be utilized to treat heavily infected individuals and to minimize the continuous presence of infection within the population group. These treatment approaches will vary depending on local, social, behavioral, and economic conditions.

Yearly mass or selective chemotherapy has not been successful without application of the approaches outlined previously. A combination of improved environmental sanitation, health education, and chemotherapy offers the best approach to prevention and control (Block, 1985; Gilles, 1985; Udonsi, 1985; Variyam and Banwell, 1982; Warren, 1981).

Literature Cited

Afework, M., T. Dejene, L. Jemaneh, and S. Tedla. 1985. *Ancylostoma duodenale* and *Necator americanus* in the Abay (Blue Nile) gorge. Ethiop. Med. J. **23:**135–136.

Anderson, R. M., and G. A. Schad. 1985. Hookworm burdens and faecal egg counts: an analysis of the biological basis of variation. Trans. R. Soc. Trop. Med. Hyg. **79:**812–825.

Ball, P. A. J., and A. Bartlett. 1979. Serological reactions to infection with *Necator americanus*. Trans. R. Soc. Trop. Med. Hyg. **63:**362–369.

Bass, C. C. 1914. Hookworm disease: a world disease of long duration. Am. J. Trop. Dis. Preventive Med. **1:**669–671.

Bassily, S., N. El-Masry, B. Trabolsi, and Z. Farid. 1984. Treatment of ancylostomiasis and ascariasis with albendazole. Ann. Trop. Med. Parasitol. **78:**81–82.

Basta, S. S., K. Soerkirman, and N. Scrimshaw. 1979. Iron deficiency anemia and the productivity of adult males in Indonesia. Am. J. Clin. Nutr. **32:**916–925.

Beaver, P. C. 1950. The standardization of fecal smears for estimating egg production and worm burden. J. Parasitol. **36:**451–456.

Beaver, P. C., R. C. Jung, and E. W. Cupp. 1984. Clinical parasitology. Lea & Febiger, Philadelphia.

Block, M. 1986. Hookworm infection. Rev. Infect. Dis. **8:**306.

Chitchang, S., S. Leelayoova, and T. Piamjinda. 1983. Albendazole in the treatment of hookworm infestation in Thailand. J. Med. Assoc. Thai. **66:**45–48.

Croll, N. A., and K. Ma. 1978. The location of parasites within their hosts: the passage of *Nippostrongylus brasiliensis* through the lungs of the laboratory rat. Int. J. Parasitol. **8:**289–295.

de la Riva, H., D. G. Escamida, and A. Frati. 1981. Acute massive intestinal bleeding caused by hookworm. J. Am. Med. Assoc. **246:**68.

Gilles, H. M. 1985. Selective primary health care: strategies for control of disease in the developing world. XVII. Hookworm infection and anemia. Rev. Infect. Dis. **7:**111–118.

Gilman, R. H. 1982. Hookworm disease: host–pathogen biology. Rev. Infect. Dis. **4:**824–829.

Harada, Y. 1962. Wakana disease and hookworm allergy. Yonago Acta Med. **6:**109–118.

Hill, R. B. 1926. The estimation of the numbers of hookworms harbored by the use of the dilution egg count method. Am. J. Hyg. (Suppl. 2) **6:**19–41.

Hunter, G. W., J. C. Swartzwelder, and D. F. Clyde. 1976. Tropical medicine, 5th ed. The W. B. Saunders Co., Philadelphia.

Jones, J. E. 1982. Psychosocial impact of common parasitic diseases. J. Fam. Pract. **14:**1139–1143.

Kalkofen, V. P. 1974. Intestinal trauma resulting from feeding activities of *Ancylostoma caninum*. Am. J. Trop. Med. Hyg. **23:**1046–1053.

Khan, M. V., M. M. Shahidullah, and T. Begum. 1983. Role of breast feeding in preventing acquisition of roundworm and hookworm in Dhaka slum children. Indian J. Pediatr. **50:**493–495.

Knight, R., and T. G. Merrett. 1981. Hookworm infection in rural Gambia. Seasonal changes, morbidity and total IgE levels. Ann. Trop. Med. Parasitol. **75:**299–314.

Kueh, Y. K., L. Chan, B. Lim, and H. Wong. 1983. Transient non-thrombocytopenic purpura in hookworm infection. Scand. J. Haematol. **30:**174–176.

Mayoral, L. G., K. Tripathy, F. T. Garcia, S. Klake, D. Bolanos, and J. Ghitis. 1967. Malabsorption in the tropics: a second look. I. The role of protein malnutrition. Am. J. Clin. Nutr. **20:**866–883.

Michaelsen, K. F. 1985. Hookworm infection in Kweneng District, Botswana. A prevalence survey and a controlled treatment trial. Trans. R. Soc. Trop. Med. Hyg. **79:**848–851.

Miller, A. 1954. Dung beetles (Coleoptera, Scarabaeidae) and other insects in relation to human feces in a hookworm area of southern Georgia. Am. J. Trop. Med. Hyg. **3:**372–387.

Miller, T. A. 1979. Hookworm infection in man. Adv. Parasitol. **17:**315–384.

Misna, P. K., N. K. Pande, and S. C. Jagota. 1985. Albendazole in the treatment of internal helminthiasis in children. Curr. Med. Res. Opin. **9:**516–519.

Ogilvie, B. M., A. Bartlett, R. Godfrey, J. Turton, M. Worms, and R. Yeates. 1978. Antibody response in self-infections with *Necator americanus*. Trans. R. Soc. Trop. Med. Hyg. **72:**66–71.

Ogilvie, B. M., and M. J. Worms. 1976. Immunity of nematode parasites of man with special preference to *Ascaris* hookworm and filariae, p. 380–407. *In* S. Cotter and E. Sadum. (ed.), Immunology of parasitic infections. Blackwell Scientific Publications, Ltd., Oxford.

Ramalingam, S., B. Sinniah, and V. Krishnan. 1983. Albedazole, an effective single dose, broad spectrum anthelmintic drug. Am. J. Trop. Med. Hyg. **32:**984–989.

Roche, M., and M. Layrisse. 1966. The nature and causes of hookworm anemia." Am. J. Trop. Med. Hyg. **15:**1029–1100.

Schad, G. A., and R. M. Anderson. 1985. Predisposition to hookworm infection in humans. Science **228:**1537–1540.

Schad, G. A., and J. G. Banwell. 1984. Hookworms, p. 359–372. *In* Warren and Mahmoud (ed.), Tropical and geographical medicine. McGraw-Hill Book Co., New York.

Schad, G. A., A. Chowdhury, C. Dean, V. Kochar, T. Nawalinski, J. Thomas, and J. Tonascia. 1973. Arrested development in human hookworm infections: an adaption to a seasonally unfavorable external environment. Science **180**:502–504.

Sharma, S., and E. S. Charles. 1982. Chemotherapy of hookworm infections. Prog. Drug. Res. **26**:9–54.

Stoll, N. R. 1961. Dilution egg counting for hookworm. *Ascaris Trichuris* et. W. H. O. mimeograph report. World Health Organization, Geneva.

Tandon, B. N., R. Kohli, A. Saraya, K. Ramachandram, and O. Frakash. 1969. Relationship of anemia and hypoproteinaemia to the functional and structural changes in the small bowel in hookworm disease. Gut **10**:360–365.

Tedla, S., and L. Jemaneh. 1985. Distribution of *Ancylostoma duodenale* and *Necator americanus* in Ethiopia. Ethiop. Med. J. **23**:149–158.

Tripathy, K., F. T. Garcia, and H. Lotero. 1971. Effect of nutritional repletion on human hookworm infection. Am. J. Trop. Med. Hyg. **20**:219–223.

Udonsi, J. K. 1985. Effectiveness of seasonal community-based mass-expulsion chemotherapy in the control of human hookworm infections in endemic communities. Public Health **99**:295–301.

Variyam, E. P., and J. G. Banwell. 1982. Hookworm disease: nutritional implications. Rev. Infect. Dis. **4**:830–835.

Viravan, C., S. Migasena, D. Bunnag, and T. Harinasuta. 1982. Clinical trials of albendazole in hookworm infections. Southeast Asian J. Trop. Med. Public Health **13**:654–657.

Warren, K. S. 1981. The control of helminths: nonreplicating infectious agents of man. Annu. Rev. Public Health **2**:101–115.

Zimmerman, H. M. 1946. Fatal hookworm disease in infancy and childhood on Guam. Am. J. Pathol. **22**:1081–1100.

CHAPTER 94

Hymenolepiasis

E. G. LONG

Disease: Hymenolepiasis, dwarf tapeworm infection, rat tapeworm infection.
Etiologic Agents: *Hymenolepis nana, Hymenolepis nana fraterna, Hymenolepis diminuta,* dwarf tapeworms.
Source: Humans, rats and mice, insects.
Clinical Manifestations: Abdominal pain, diarrhea, anorexia, headache, pallor, keratoconjunctivitis, irritability, eosinophilia.
Pathology: Allergic reactions, impairment of protein digestion, minimal tissue pathology.
Laboratory Diagnosis: Demonstration of eggs in feces.
Epidemiology: More common in tropical than in temperate countries, adapted to transmission in arid climates, higher rates of infection in children than in adults.
Treatment: Praziquantel, niclosamide, paromomycin.
Prevention and Control: Individual and mass chemotherapy, improved personal hygiene.

Description of Disease

Hymenolepiasis (dwarf tapeworm infection) is a cosmopolitan cestode infection encountered more frequently in the tropics than in colder climates. The disease is acquired by ingestion of eggs of *Hymenolepis nana* or insects containing larval stages of *Hymenolepis diminuta,* the rat tapeworm. Most infections are seen in children and are usually caused by *H. nana,* which can bypass an intermediate host and is easily acquired by fecal–oral transmission. Worms attach to the mucosa of the small intestine, where they cause little tissue damage but provoke gastrointestinal disturbances and allergic reactions. Autoinfection by *H. nana* can maintain infections for many years in children, and the worm can disseminate and cause ectopic infections in immunosuppressed hosts. *Hymenolepis nana fraterna* is a murine cestode, morphologically identical to *H. nana,* that only occasionally infects humans.

Communicability and Transmissibility

Hymenolepis diminuta is quite rare, and *H. nana* is responsible for most infections. Transmission of the latter occurs directly between humans on ingestion of eggs in fecally contaminated food or water. Unlike *H. diminuta, H. nana* does not require a insect intermediate host; its eggs are immediately infective upon passage, and it is well suited for spread among family members or institutionalized children.

Clinical Manifestations of Disease

Intestinal Infection

Light worm burdens are largely asymptomatic, whereas patients with heavy infections and those with infections over 5 years old are likely to have symptomatic diseases; excretion of more than 15,000 eggs per gram of feces is almost always symptomatic

(Pawlowski, 1984). The most consistent manifestations of disease are anorexia, headache, abdominal pain, diarrhea, and generalized weakness; other associated symptoms are epileptoid convulsions, irritability, eosinophilia up to 15%, and keratoconjunctivitis. *Hymenolepis nana* is responsible for most symptomatic infections; *H. diminuta* may cause diarrhea, but is relatively innocuous.

Disseminated Infection

Manipulation of the immune system of infected mice has shown that *H. nana* can disseminate to extraintestinal sites (Lucas et al., 1979). One report of disseminated cysts in blood vessels and all deep organs in an immunosuppressed patient with Hodgkin's disease (Connor, 1976; Lucas et al., 1979) suggests that *H. nana* can cause invasive disease in such patients.

Pathology

Worms attached to the intestinal mucosa cause insignificant tissue damage, although rodent infections result in extensive villous atrophy and fusion (Martin and Holland, 1984). The symptomatology of hymenolepiasis is attributed to host reactions to worm allergens and metabolites (Sun, 1982). These products are responsible for clinical manifestations at sites distant from the intestine, such as the brain and eye.

A strong association between *H. nana* and phlyctenular keratoconjunctivitis was reported from Egypt by Al-Hussaini et al. (1979). These investigators suggest that worm antigens provoke the necrotic ulcerations of the conjunctiva and cornea seen in 73.4% of children with *H. nana* infections.

Diagnostic Features

The eggs of *H. nana* are infective immediately upon passage and, as the parasite can readily establish itself in families, it should be suspected in outbreaks of enteritis in these situations. Hymenolepiasis should also be considered in patients from tropical countries with keratoconjunctivitis or epileptoid fits. The disease bears similarities to enterobiasis in transmission and clinical signs, and stool examinations should accompany cellophane-tape preparations in children suspected of having enterobiasis. Hymenolepiasis should always be considered in the diagnosis of enteritides in patients exposed to household infestations by rats and mice.

Prognosis

Hymenolepiasis is not a fatal illness, but persistent infection may indicate an immunodeficient state. *Hymenolepis nana* has shown a potential for causing disseminated infection, at least in animal models, and thus should be considered a hazard in infected persons undergoing therapeutic immunosuppression or suffering from an immunodeficiency disease. Children tend to lose their infections as they grow older and become resistant to reinfection.

Epidemiologic Aspects

Occurrence and Transmission

Hymenolepiasis occurs worldwide, but is more prevalent in warm than in cold climates. It infects children more readily than adults and is readily spread among children in institutions or refugee camps. Rates of infection around the world range from 26% in Eastern Europe and 20% in Chile to 95% in the Punjap area of India (Muller, 1975). Infection with *H. nana* spreads from person to person by fecal–oral transmission, and thus is especially prevalent in children because of their careless personal hygiene and handwashing practices. Transmission of *H. nana* can occur efficiently even under adverse environmental conditions, as the parasite requires neither an intermediate or paratemic host nor an incubation period for maturation of the embryo. In hot arid areas of the world, such as Iran and Western Australia, hymenolepiasis is a common intestinal infection (Pawlowski, 1984). *Hymenolepis diminuta* is a common parasite of rats, mice, and related rodents, but can be transmitted to humans via accidental ingestion of an insect intermediate host. Thus, household rodents and their accompanying fleas can serve as sources of human infection (Akinboade et al., 1981).

Susceptibility, Resistance, and Infection

It is known that adults are more resistant than children to infection by *H. nana* and that infected persons lose their infections in time; however, the mechanisms involved have not been well studied. Pawlowski (1984) reports that in a study of hymenolepiasis in 93 Punjabi children, 26% were constantly positive, 34% were consistently negative, and 39% had intermittent infections during 22 weeks of observation. Studies in rats and mice show that a primary infection will induce strong resistance to reinfection by *H. nana* and *H. diminuta* (Bhopale and Johnri,

1981; Hopkins, 1982). This resistance is thymus dependent (Hindsbo et al., 1982; Lucas et al., 1979) and can be adoptively transferred by peritoneal exudate cells from sensitized donors (Gupta et al., 1980). Resistance to reinfection may also work to the parasite's advantage by reducing overcrowding, and appears to involve secretory "crowding factors" that inhibit the development of newcomers (Roberts and Insler, 1982). All age groups appear to be susceptible to *H. diminuta* infection (Beaver et al., 1984).

Etiologic Agents

Characterization

Hymenolepis nana adults achieve lengths of 15 to 40 mm and are only 0.5 to 1.0 mm wide; because of crowding effects, their size is inversely proportional to their number in the intestine (Beaver et al., 1984). The scolex is 0.3 mm wide, bears four suckers, and has a retractable rostellum arm with a ring of 20 to 30 backward-pointing hooks. A thin neck leads to a strobila of about 200 proglottids. Mature proglottids at the posterior portions of the strobila are wide and short, measuring 0.22 by 0.85 mm. Each proglottid is hermaphroditic with a central, lobed ovary and three rounded testes. Gravid proglottids are almost completely filled by a uterus laden with up to 200 eggs. The eggs are ovoid or circular and measure 30 to 47 mm in diameter. The complex eggs are composed of an outer transparent shell and an embryonic oncosphere centrally suspended in a gelatinous layer. The oncosphere has the three parts of hooks characteristic of all larval cestodes, and its membrane carries two polar knobs bearing 4 to 8 filaments each.

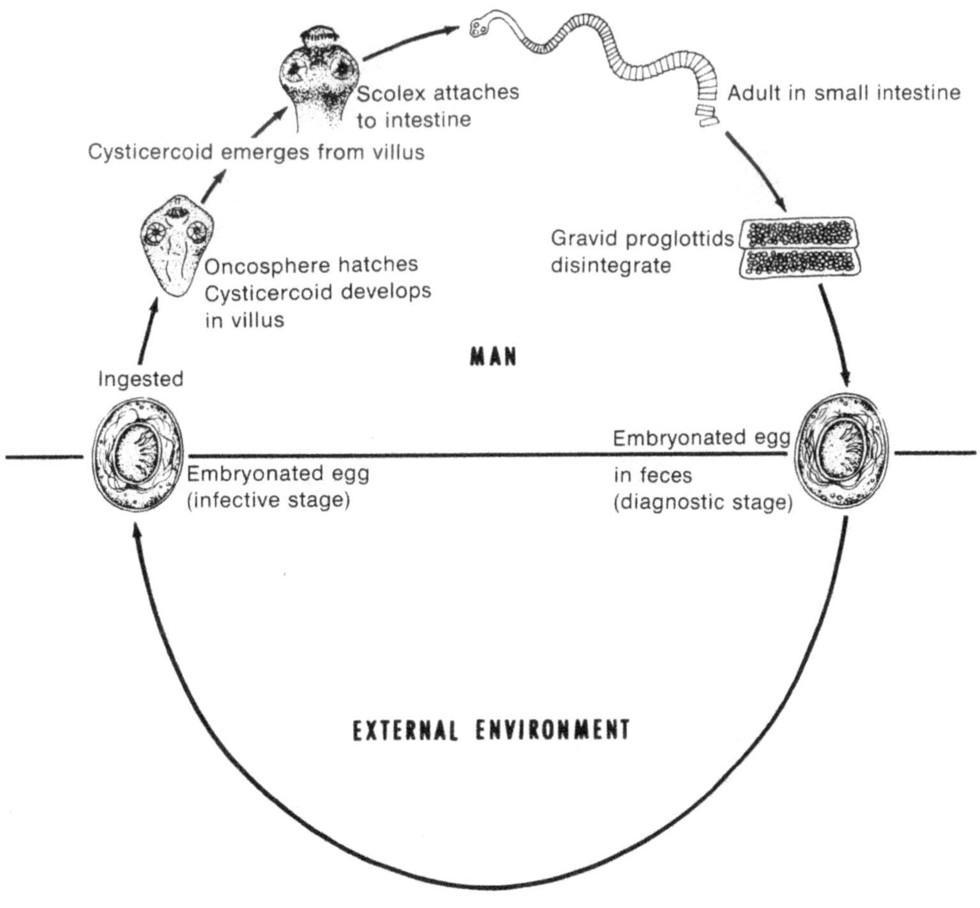

FIG. 1. Life cycle of *Hymenolepis nana*. (Reprinted with permission from the Centers for Disease Control, Parasitology Section, Training and Laboratory Program Office, Atlanta, Georgia.)

Adults of *H. diminuta* are larger worms that measure 300 to 600 mm long and 4.0 mm wide and have up to 1,000 proglottids. The scolex is about 0.4 mm wide, with four suckers and a retractable but unarmed rostellum. Mature proglottids have an internal structure similar to that of *H. nana*. Eggs are 60 to 80 mm in diameter, have a thick yellow outer shell, and an oncosphere with no polar filaments.

Life Cycle

The oncosphere of *H. nana* is released from the egg after digestion of the outer shell by trypsin. It penetrates a villus of the upper lumen and after 4 days becomes a cysticercoid. Four days later, the cysticercoid reenters the intestinal lumen and over an additional 10 to 12 days develops into a mature tapeworm. Egg production begins about 30 days after infection, and eggs are released when gravid terminal proglottids disintegrate within the intestinal lumen (Fig. 1). The worms live up to 4 months, but infections may persist for years by internal autoinfections, when eggs hatch within the intestine, and by external autoinfections, when infected persons reinfect themselves by fecal containment. Eggs can survive in the environment for up to 11 days.

The eggs of *H. diminuta* are not infective to humans and must first be ingested by an arthropod intermediate host. Suitable hosts include fleas, flour beetles, earwigs, and millipedes. Fleas acquire the infection when their larvae ingest the eggs and the cysticercoid persists through the insect's metamorphosis. Within the insect the liberated oncosphere penetrates the intestinal wall and remains in the hemocoelom as a tailed cysticercoid (Fig. 2). The egg of

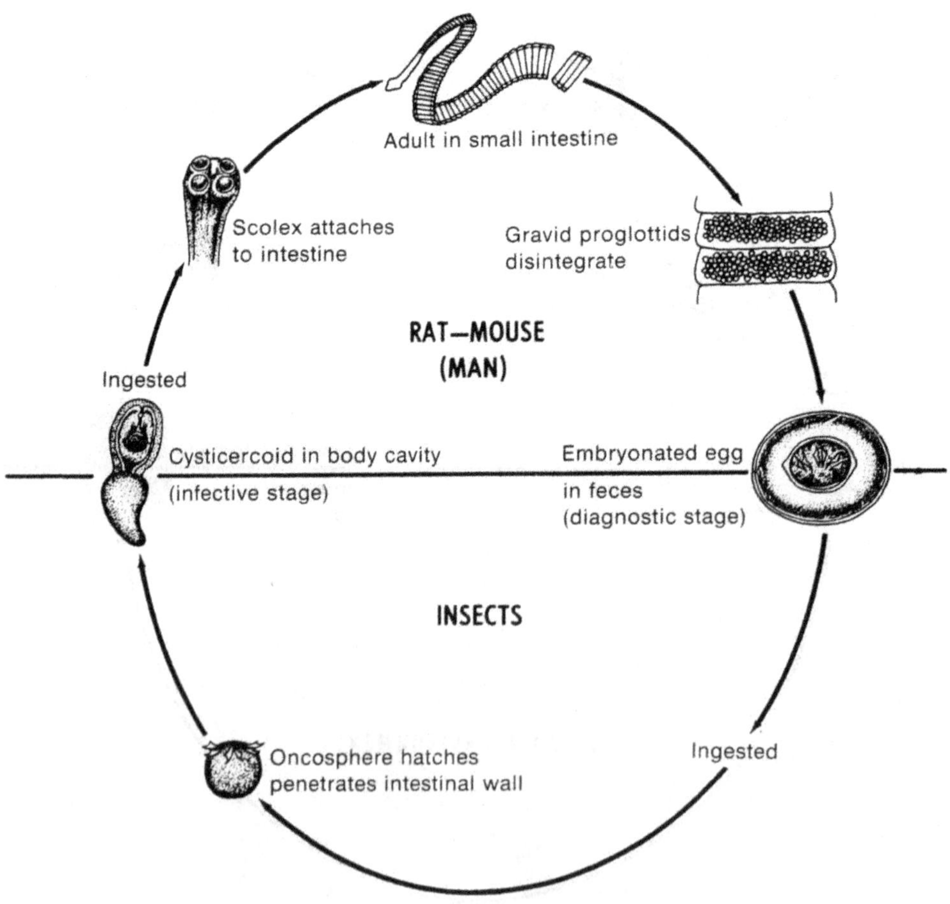

FIG. 2. Life cycle of *Hymenolepis diminuta*. (Reprinted with permission from the Centers for Disease Control, Parasitology Section, Training and Laboratory Program Office, Atlanta, Georgia.)

H. diminuta is not affected by proteolytic enzymes in the human gut (Salem et al., 1980) and appears to require mechanical breakage of the outer shell (Holmes and Fairweather, 1982). This action may be provided by the feeding and digestive processes of an arthropod: Humans acquire *H. diminuta* infection when an infected intermediate host is accidentally swallowed and digested. The released cysticercoid attaches to the intestinal lumen and develops into a tapeworm.

Specimen Collection and Processing

Fecal specimens should be examined soon after collection or preserved in 5 or 10% buffered Formalin for delayed examination. In old preserved specimens, the polar filaments of *H. nana* eggs may be difficult to see. Eggs can be concentrated by the zinc–sulfate flotation or Formalin–ethyl acetate concentration techniques, although some may be lost with the latter technique.

Identification

The characteristics of adults and eggs of the hymenolepids are described under Characterization and illustrated in Figures 1 and 2. It may help to remember that the species name *diminuta,* somewhat illogically, refers to the larger worm and egg.

Treatment, Prevention, and Control

Treatment

The new schistosomicide praziquantel is excellent for treatment (Schenone, 1980) and can be given as a single dose of 10 to 25 mg/kg to adults or children. An alternative choice is niclosamide, to be given as a single dose of 2 g for 7 days to adults; 1.0 g on the first day and daily doses of 0.5 g for 6 days to children weighing 11 to 34 kg; and 1.5 g on the first day and daily doses of 1.0 g for 6 days to children weighing 34 kg (Medical Letter, 1986).

Prevention and Control

Personal hygiene, especially handwashing and proper disposal of feces, are the best methods of preventing transmission. Treatment of infected cases should extend to close personal contacts such as household members. Insect and rodent control will reduce the chances of acquiring *H. diminuta* infections and also *H. nana* infections, as the latter has retained the option of using an insect intermediate host.

Literature Cited

Akinboade, O. A., O. O. Dipeolu, F. O. Ogunji, and G. O. Adegoke. 1981. The parasites obtained and the bacteria isolated from house rats (*Rattus rattus* Linnaeus, 1758) caught in human habitations in Ibadan, Nigeria. Int. J. Zoonoses **8:**26–32.

Al-Hussaini, M. K., R. Khalifa, A. T. Al-Ansary, G. H. Hussain, and K. M. Moustafa. 1979. Phlyctenular eye disease in association with *Hymenolepsis nana* in Egypt. Br. J. Ophthalmol. **63:**627–631.

Beaver, P. C., R. C. Jung, and E. W. Cupp. 1984. Cyclophyllidean tapeworms, p. 509–512. *In* Clinical parasitology, 9th ed. Lea & Febiger, Philadelphia.

Bhopale, K. K., and G. N. Johri. 1981. The stimulation of immunity to *Hymenolepis nana* in mice exposed to low-level infections. J. Hyg. Epidemiol. Microbiol. Immunol. **25:**1–5.

Connor, D. H., A. K. Sparks, A. J. Strano, R. C. Neafie, and B. Juvelier. 1976. Disseminated parasitosis in an immunosuppressed patient. Arch. Pathol. Lab. Med. **100:**65–68.

Groll, E. 1980. Praziquantel for cestode infections in man. Acta Trop. **37:**293–296.

Gupta, R. K., S. M. Elizabeth, S. L. Kaushik, and G. N. Johri. 1980. *Hymenolepis nana:* transfer of acquired immunity in mice through sensitized peritoneal exudate cells. Experientia **36:**128–129.

Hindsbo, O., J. Andreassen, and J. Ruitenberg. 1982. Immunological and histopathological reactions of the rat against the tapeworm *Hymenolepis diminuta* and the effects of antithymocyte serum. Parasite Immunol. **4:**59–76.

Holmes, S. D., and I. Fairweather. 1982. *Hymenolepsis diminuta:* the mechanism of egg hatching. J. Parasitol. **85:**237–250.

Hopkins, C. A. 1982. Immunological memory in mice to adult *Hymenolepsis diminuta* (Cestoda). J. Parasitol. **68:**32–38.

Lucas, S. B., O. A. Hassounah, M. Doenhoff, and R. Muller. 1979. Aberrant form of *Hymenolepsis nana:* possible opportunistic infection in immunosuppressed patients. Lancet. **2:**1372–1373.

Martin, J., and C. Holland. 1984. Scanning electron microscope studies of the mucosa of rats infected with *Hymenolepsis diminuta* (Cestoda). J. Helminthol. **58:**93–99.

Medical Letter. 1986. Drugs for parasitic infections. **28:**9–18.

Muller, R. 1975. *Hymenolepsis nana,* p. 60–62. *In* Worms and disease. William Heinemann Medical Books, Ltd., London.

Pawlowski, Z. S. 1984. Cestodiasis: taeniasis, diphyllobothriasis, hymenolepiasis and others, p. 482–485. *In* K. S. Warren and A. A. F. Mahmound (ed.), Tropical and geographical medicine. McGraw-Hill Book Co., New York.

Roberts, L. S., and G. D. Insler. 1982. Developmental physiology of cestodes. XVII. Some biological properties of putative "crowding factors" in *Hymenolepsis diminuta*. J. Parasitol. **68**:263–269.

Salem, S. A., H. A. Sidky, and L. Abdel-Rehim. 1980. A study of the direct mode of transmission of the intestinal tapeworm *Hymenolepsis nana* (Rudolphi, 1819). J. Egypt Soc. Parasitol. **10**:95–99.

Schenone, H. 1980. Praziquantel in the treatment of *Hymenolepis nana* infections in children. Am. J. Trop. Med. Hyg. **29**:320–321.

Sun, T. 1982. Diphyllobothriasis, hymenolepiasis and dipylidiasis, *In* Pathology and clinical features of parasitic diseases. Masson Publishing USA, New York.

CHAPTER 95

Isosporiasis

LYNNE S. GARCIA, ROBERT L. OWEN, and WILLIAM L. CURRENT

Disease: Isosporiasis.
Etiologic Agent: *Isospora belli.*
Source: Food, water, or environmental surfaces, including skin, contaminated with human fecal material containing mature (sporulated) *I. belli* oocysts.
Clinical Manifestations: May range from mild gastrointestinal distress to severe diarrhea or dysentery. May include headaches, malaise, anorexia, nausea, abdominal discomfort, flatulence, steatorrhea, diarrhea, and weight loss. Immunosuppressed patients often present with profuse diarrhea associated with weakness, anorexia, and weight loss. Usually self-limited in the immunocompetent patient.
Pathology: Architectural changes in intestinal mucosa range from mild broadening of the villus tips to a flat mucosal surface with marked crypt elongation. There may be infiltrates of lymphocytes, plasma cells, and eosinophils in the lamina propria.
Laboratory Diagnosis: Routine histology may reveal developing stages in the intestinal epithelial cells. Diagnosis most often based on the recovery of oocysts from human stool using the fecal concentration method (either sedimentation or flotation). Fresh specimens or those fixed in 10% formalin are preferred. Oocysts not easily seen in concentrates from fecal specimens preserved in polyvinyl alcohol (PVA).
Epidemiology: Worldwide; may be more common in the tropics than in temperate regions. More frequent diagnosis in the tropics may only reflect greater awareness of and facility in diagnosis of stool parasites.
Treatment: Drug of choice is trimethoprim–sulfamethoxazole (investigational drug for treatment of this infection).
Prevention and Control: Good personal hygiene and proper disposal of human fecal material.

Description of Disease

Communicability and Transmissibility

Isospora belli is thought to be the only species of *Isospora* that infects humans, and no other reservoir hosts are recognized for this infection. Transmission is through ingestion of mature, sporulated (infective) oocysts contaminating food, water, or environmental surfaces. Sexual transmission by direct oral contact with the anus or perineum has also been postulated (Forthal and Guest, 1984), although this mode of transmission is probably much less common. The oocysts, thought to be very resistant to environmental conditions and commonly used disinfectants, may remain viable for months if kept cool and moist.

The oocysts are usually released in the stool in the immature, unsporulated state; thus, several days may be required before they are infective (Long, 1982). Sporulated, mature oocysts have also been observed in the lumen in intestinal aspirates (Trier et al., 1974), which may be important in direct person-to-person transmission. Patients infected with *Isos-*

pora need not be isolated; however, enteric precautions should be maintained so that stools are properly handled.

Clinical Manifestations

Clinical symptoms may range from mild gastrointestinal distress to severe diarrhea or dysentery (Beaver et al., 1984; Brandborg et al., 1970). Approximately 1 week after ingestion of infective oocysts, the patient may develop headaches, malaise, anorexia, nausea, abdominal discomfort, flatulence, steatorrhea, diarrhea, and weight loss. Bowel movements (usually 6 to 10 per day) are watery to soft, foamy, and offensive, suggesting a malabsorption process. Patients who are immunosuppressed, particularly those with the acquired immune deficiency syndrome (AIDS), often present with profuse diarrhea associated with weakness, anorexia, and weight loss (Kobayashi et al., 1985). Malabsorption, characterized by a low 5-h D-xylose test and positive fecal fat, has been observed in these patients (Whiteside et al., 1984). In the immunocompetent patient, the disease may be self-limiting, with spontaneous resolution after several weeks, or symptoms may last for months or years (Trier et al., 1974).

Pathology

Examination of the mucosal specimens described by Brandborg et al. (1970) revealed damage and probable destruction of parasitized epithelial cells, infiltrates of lymphocytes, plasma cells, and eosinophils in the lamina propria, capillary dilatation, and excess collagen. Intestinal mucosal changes ranged from mild broadening of the villus tips to a flat mucosal surface with associated crypt elongation. In a patient described by French et al. (1964), a jejunal biopsy showed no parasites, but the mucosa was flat and increased numbers of lymphocytes were seen in the epithelial cell layer and lamina propria. Forthal and Guest (1984) reported on *Isospora* enteritis in three homosexual men, one of whom had chronic inflammation of the small bowel on endoscopic examination. Charcot–Leyden crystals, formed from the breakdown products of eosinophils, are often seen in the stool of patients with *I. belli* infection.

Diagnostic Features

Isosporiasis should be considered in patients who have a history of chronic diarrhea and weight loss. Severe and chronic enteritis has been attributed to *Isospora* infections, particularly in the debilitated patient. Although this infection is not new, it has become more widely recognized as a potential pathogen in those with AIDS and others who may have immune deficiencies for a variety of reasons. In some of these patients, additional organisms may also be present, including *Cryptosporidium* (Bruckner and Garcia, 1985). As *Isospora* is thought to occur worldwide, the patient will not necessarily have a significant travel history.

Prognosis

In the immunocompetent patient, the infection and symptoms will probably be mild and of short duration. In the immunosuppressed or immunodeficient patient, however, the diarrhea and symptoms can be very severe and prolonged. Information on therapy is based on isolated anecdotal reports, but drug failures have been reported (Current, 1985). Also, relapses (or reinfection) have been reported after therapy (DeHovitz et al., 1986; Whiteside et al., 1984). Although deaths have been reported (Brandborg et al., 1970), the prognosis is usually good with early diagnosis and therapy.

Epidemiologic Aspects

Occurrence and Transmission

Although this infection has been found in various parts of the world, certain tropical areas in the Western Hemisphere appear to contain some well-defined endemic locations (Faust et al., 1961). Human infections are generally more common in the tropics than in temperate regions. The infection is acquired by ingestion of infective oocysts, usually through food or drink. Faust et al. (1961) reported on 33 cases of *I. belli* infection from Columbia and tabulated many case reports from the Western Hemisphere, except Canada. Eaton (1966) later reported the presence of *I. belli* infection from Canada.

Susceptibility, Resistance, and Carriage

Reports indicate that the immunocompetent host can acquire the infection, as evidenced by laboratory infections (Connal, 1922; Cook, 1961). In these and in experimental infections, however, the illness was usually self-limiting within less than 2 weeks (Matsubayashi and Nozawa, 1948; McCracken, 1972). Many of the infections have occurred in patients with AIDS, most of whom may have had other agents that also caused diarrhea and weight loss. In one study by DeHovitz et al. (1986), diarrhea in AIDS patients in Haiti stopped within 2 days of the initiation of therapy with oral trimethoprim–sulfamethoxazole; however, recurrent symptomatic isosporiasis developed in 47% of the patients. With appropriate therapy the infection seems to be eradicated easily. Unfortunately, however, many AIDS patients develop drug reactions to trimethoprim. A true carrier state may be difficult to detect, particularly if the patient is asymptomatic and the numbers of oocysts shed in the stool are very low.

Etiologic Agent

Characterization

Organisms of the genus *Isospora* have environmentally resistant life-cycle stages called oocysts. The oocyst initially contains one sporont, which matures to form two sporocysts, each of which will contain four sporozoites. Both the asexual (schizogonic) and sexual (gametogenic) stages in the life cycle (Fig. 1) have been described in studies of intestinal mucosal biopsy specimens (Brandborg et al., 1970; Trier et al., 1974). Intracellular development occurs primarily in the distal duodenum and proximal jejunum (Fig. 2). During their life cycle the trophozoites mature and through schizogony produce merozoites (Figs. 3 and 4), which, when released from the mature schizonts, penetrate other epithelial cells and reinitiate the asexual cycle. After approximately 1 week, both macro- and microgametocytes (Figs. 5 and 6) are produced, fertilization occurs, and the zygote produces a thick wall and becomes an oocyst

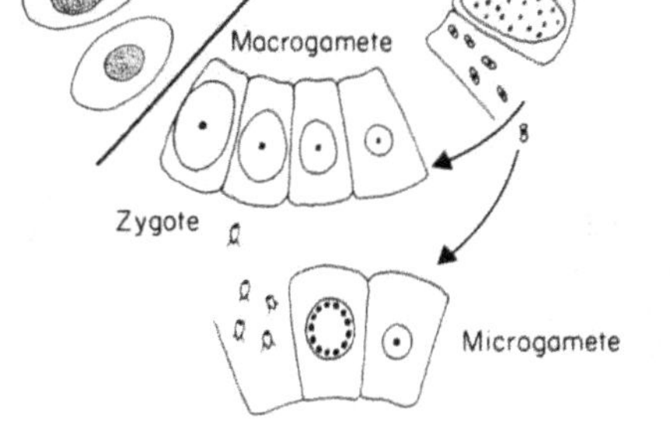

FIG. 1. Representative life cycle of *Isospora belli*. (Reproduced with permission from Garcia and Bruckner, 1988.)

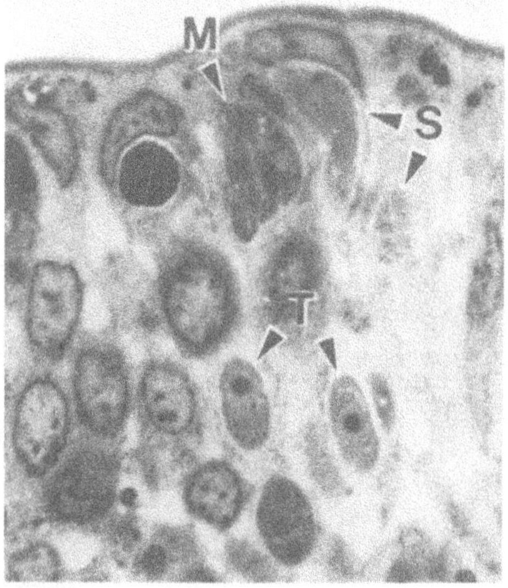

FIG. 2. Jejunal biopsy from a patient with *Isospora* and severe diarrhea. Trophozoites (T) divide asexually within enterocytes by schizogony (S) to produce merozoites (M). Magnification ×1100. (Reproduced courtesy of Dr. Robert Owen.)

(Fig. 7), which is passed in the stool. Oocysts are elongate to ovoid and measure 20 to 33 by 10 to 19 μm. There is some variation, with some oocysts being more fusiform or ovoid. Although sporulation has been reported to occur in the gut lumen (Trier et al., 1974), usually the oocyst passed in fresh stool will contain a single immature sporont. Further development outside of the host results in the production of two mature sporocysts, each containing four sporozoites (Fig. 7). It is this stage containing the sporozoites that, if ingested, is infective.

Specimen Collection and Transport

Fecal specimens submitted for the examination for parasites should be collected fresh or placed in 10% formalin or sodium acetate–acetic acid–formalin (SAF) fixative. The oocysts are very difficult to recognize in specimens that have been preserved in polyvinyl alcohol (PVA). Oocysts are not shed continuously in the stool during infections, so the recommended number of specimens is three, collected over a period of no more than 10 days. This recommendation is standard for the examination of stool for ova or other parasites. Patients suspected of having *Isospora* infections but whose stools are negative after

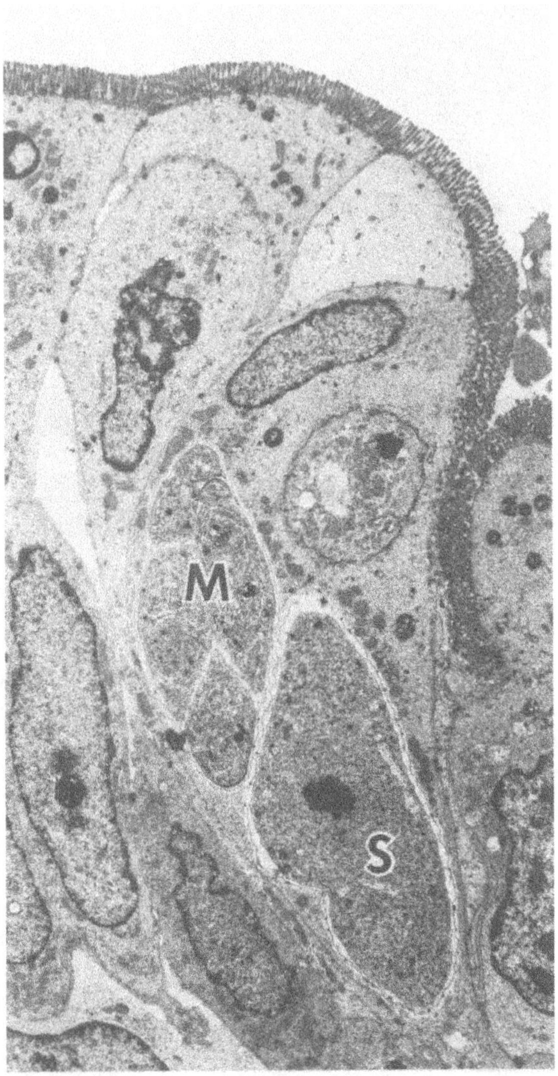

FIG. 3. Short microvilli and numerous apical phagolysosomes reflect damage to enterocytes in a patient with severe *Isospora* infection. A schizont (S) that has not yet divided and a group of merozoites (M) still clustered together in their parasitophorous vacuole distend and distort a single enterocyte. Transmission electron micrograph. Magnification ×3230. (Reproduced courtesy of Dr. Robert Owen.)

multiple stool examinations may be diagnosed using other means, such as aspiration of intestinal contents or biopsy of the jejunal–duodenal mucosa. Multiple specimens should be obtained. Using routine histologic stains, both asexual and sexual stages can be detected within the epithelial cells. In many patients, however, only the asexual stages are seen (Liebman et al., 1980; Strickland, 1984), which may explain why oocysts were not detected by routine fecal examinations. Oocysts have also been detected in duodenal aspirates (Long, 1982).

If a collection or storage system is specifically requested for a specimen known to be positive for *Isospora,* a 2% potassium dichromate solution can be used and will allow sporulation to occur; however, this collection solution is neither routinely used nor recommended and does not act as a fixative.

Identification

Oocysts found in stool material can be immature (a single cell or sporont) or can contain two sporoblasts or sporocysts. The oocysts measure approximately 12 by 27 μm and are ovoid and somewhat more narrow at one end (Fig. 8.A). In freshly passed stools, particularly if the patient has diarrhea, the entire oocyst may be filled by a single sporont that is not well developed. As the oocysts mature, two well-formed

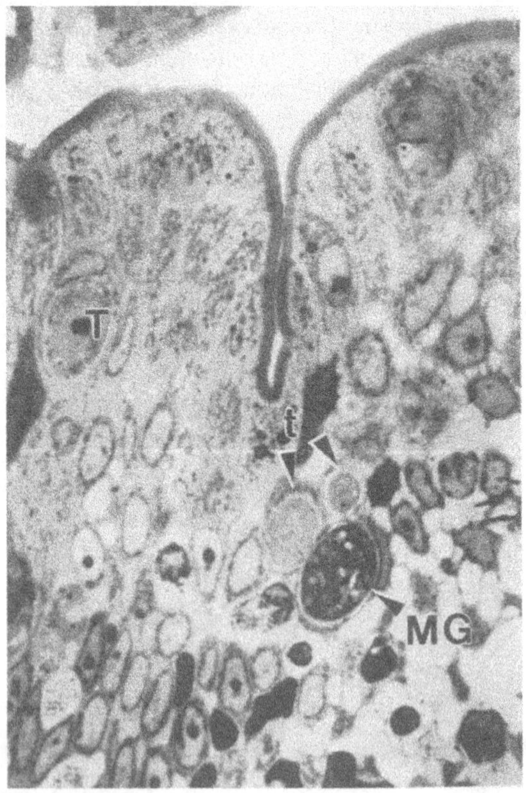

FIG. 4. Large trophozoite (T) and small trophozoites (t) are distinguished from banana-shaped merozoites (Figs. 2 and 3) by their rounded contour. Trophozoites either become schizonts or develop into microgametes or macrogametocytes (MG); the latter is characterized by varicolored wall-forming bodies, granules, and vacuoles. Semithin plastic-embedded section stained with toluidine blue. Magnification ×1100. (Reproduced courtesy of Dr. Robert Owen.)

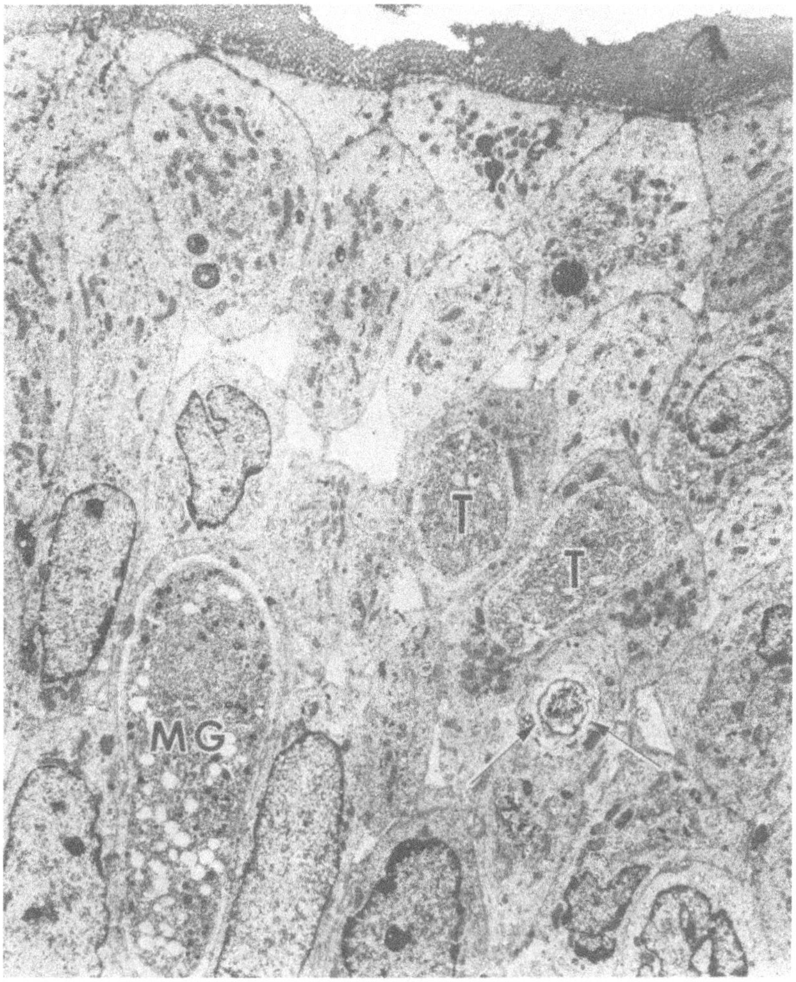

FIG. 5. In a tangential section, jejunal enterocytes are parasitized by *Isospora* trophozoites (T) and a female macrogametocyte (MG), which if fertilized by a migrating male microgamete (not shown) would develop into an oocyst. The apical end of a merozoite is filled with characteristic dark rod-like micronemes (*arrows*). Transmission electron micrograph. Magnification ×3300. (Reproduced courtesy of Dr. Robert Owen.)

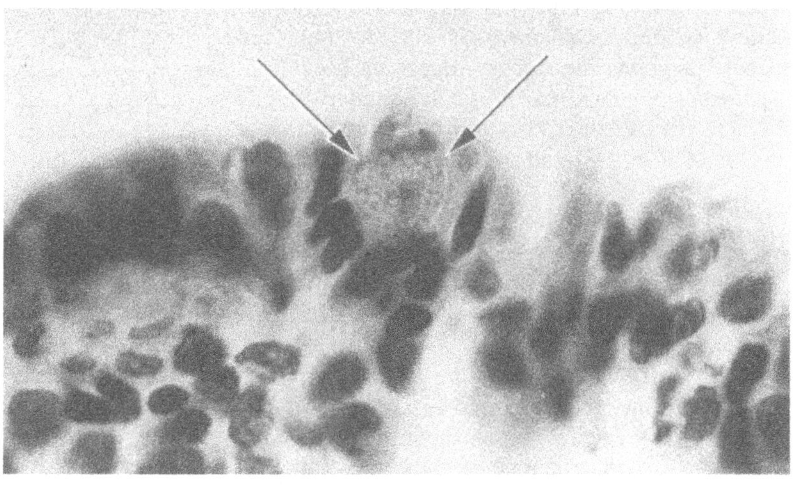

FIG. 6. Oocyst (*arrows*) being extruded into the jejunal lumen. Paraffin-embedded, Giemsa-stained section. Magnification ×1200. (Reproduced courtesy of Dr. Robert Owen.)

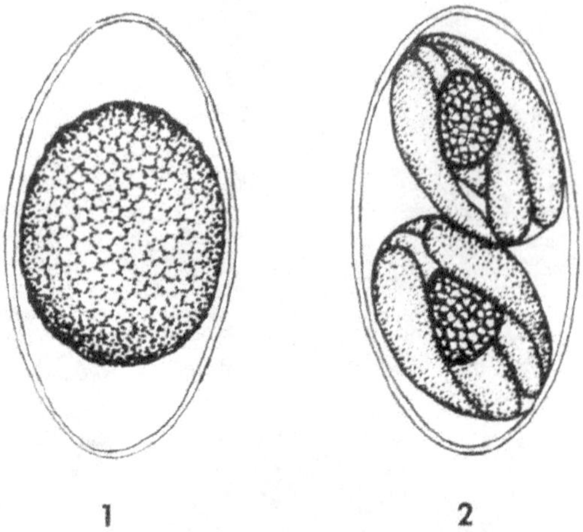

FIG. 7. *Isospora belli* oocysts. (1) Immature (unsporulated); (2) mature (sporulated). (Illustration by N. Kitamura.)

sporoblasts become visible (Fig. 8.B). When mature, the oocyst will contain two sporocysts, each of which contains four sporozoites. The sporozoites, which are long, slender, crescent-shaped bodies clumped together within the sporocyst, may be difficult to see using light microscopy.

The diagnosis of *Isospora* infection is usually based on finding oocysts in stool concentrates or, less frequently, by direct wet-mount examination during routine examinations for ova and parasites. The morphology of the oocyst is poor in specimens preserved in polyvinyl alcohol fixative (PVA). Oocysts can be stained with auramine–rhodamine, acid-fast procedures, and Giemsa stains. When stained with auramine–rhodamine, the oocyst wall and the sporoblasts or sporocysts will fluoresce bright yellow (Forthal and Guest, 1984). Using a modified acid-fast stain, the oocyst will appear pink with bright red–purple sporocysts (Ng et al., 1984). Giemsa stains both the oocyst and the sporocyst pale blue. The trichrome stain, which is routinely used for the permanent stained fecal smear, is not recommended because it stains the oocyst poorly or not at all. *Cryptosporidium* and *Isospora* have similar staining characteristics, but *Cryptosporidium* oocysts are smaller (4 to 6 μm) and round compared with the oblong (12 to 15 by 23 to 33 μm) *Isospora* oocysts (Ma and Kaufman, 1984).

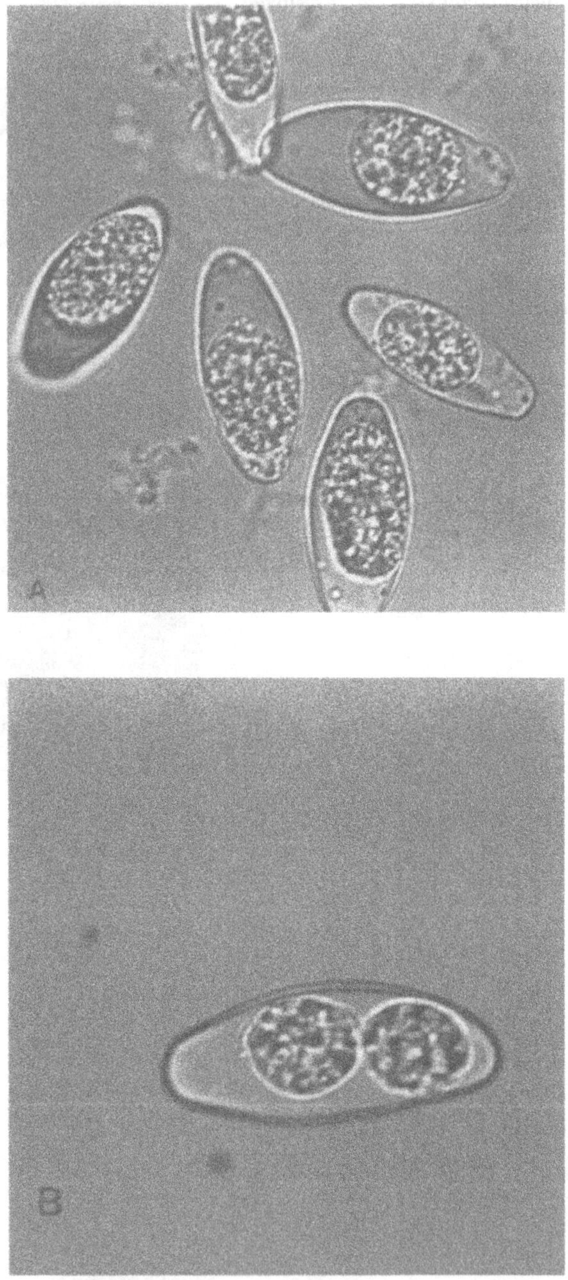

FIG. 8. Representative oocysts of *Isospora belli* as seen in a wet-smear preparation of stool. (1) Oocysts containing single sporonts; (2) oocyst containing two sporoblasts.

Treatment, Prevention, and Control

Although *Isospora* infections are usually self-limiting, therapy has been reported as successful when patients were treated with metronidazole, with combinations of pyrimethamine–sulfadiazine, and with trimethoprim–sulfamethoxazole over extended periods. The drug of choice is trimethoprim–sulfamethoxazole (DeHovitz et al., 1986), which is classified as an investigational drug for treatment of this infection (Medical Letter, 1986).

As the infection is acquired through ingestion of contaminated food or water containing viable oocysts from fecal material, prevention would include good personal hygiene and proper disposal of fecal material. These measures are effective against other fecal-borne infections and should be successful in eliminating transmission of *I. belli* infections.

Literature Cited

Beaver, P. C., R. C. Jung, and E. W. Cupp. 1984. Clinical parasitology. Lea & Febiger, Philadelphia.

Brandborg, L. L., S. B. Goldberg, and W. C. Breidenbach. 1970. Human coccidiosis—a possible cause of malabsorption. N. Engl. J. Med. **283**:1306–1313.

Bruckner, D. A., and L. S. Garcia. 1985. *Isospora belli*, ASCP Microbiology MB 85-5 (MB-146), ASCP Check Sample (Continuing Ed. Program).

Connal, A. 1922. Observations on the pathogenicity of *Isospora hominis*, Rivolta, emend. Dobell, based on a second case of human coccidiosis in Nigeria; with remarks on the significance of Charcot-Leyden crystals in the faeces. Trans. R. Soc. Trop. Med. Hyg. **16**:223–245.

Cook, E. B. M. 1961. Safety in the public health laboratory. Public Health Rep. **76**:51–56.

Current, W. L. 1985. Human enteric coccidia. II. *Isospora belli* and *Sarcocystis* spp. Clin. Microbiol. Newsl. **7**:175–178.

DeHovitz, J. A., J. W. Pape, M. Boncy, and W. D. Johnson. 1986. Clinical manifestations and therapy of *Isospora belli* infection in patients with the acquired immunodeficiency syndrome. N. Engl. J. Med. **315**:87–90.

Eaton, R. D. P. 1966. Human coccidiosis in Saskatchewan. Can. Med. Assoc. J. **95**:1384–1385.

Faust, E. C., L. E. Giraldo, G. Caicedo, and R. Bonfante. 1961. Human isosporiasis in the Western Hemisphere. Am. J. Trop. Med. Hyg. **10**:343–349.

Forthal, D. N., and S. S. Guest. 1984. *Isospora belli* enteritis in three homosexual men. Am. J. Trop. Med. Hyg. **33**:1060–1064.

French, J. M., J. L. Whitby, and G. W. Whitfield. 1964. Steatorrhea in a man infected with coccidiosis (*Isospora belli*). Gastroenterology **47**:642–648.

Garcia, L. S., and D. A. Bruckner. 1988. Diagnostic medical parasitology. Elsevier Science Publishing Co., New York.

Kobayashi, L. M., M. P. Kort, O. G. Berlin, and D. A. Bruckner. 1985. *Isospora* infection in a homosexual man. Diagn. Microbiol. Infect. Dis. **3**:363–366.

Liebman, W. M., M. M. Thaler, A. Delorimier, L. L. Brandborg, and J. Goodman. 1980. Intractable diarrhea of infancy due to intestinal coccidiosis. Gastroenterology **78**:579–584.

Long, P. L. 1982. The biology of the coccidia. University Park Press, Baltimore.

Ma, P., and D. Kaufman. 1984. *Isospora belli* diarrheal infection in homosexual men. A.I.D.S. Res. **1**:327–328.

Matsubayashi, H., and T. Nozawa. 1948. Experimental infection of *Isospora hominis* in man. Am. J. Trop. Med. Hyg. **28**:633–637.

McCracken, A. W. 1972. Natural and laboratory-acquired infection by *Isospora belli*. South. Med. J. **65**:800–818.

Medical Letter. 1986. Drugs for parasitic infections. **28**:9–18.

Ng, E., E. K. Markell, R. L. Fleming, and M. Fried. 1984. Demonstration of *Isospora belli* by acid-fast stain in a patient with acquired immune deficiency syndrome. J. Clin. Microbiol. **20**:384–386.

Strickland, G. T. 1984. Hunter's tropical medicine, 6th ed. The W. B. Saunders Co., Philadelphia.

Trier, J. S., P. C. Moxey, E. M. Schimmel, and E. Robles. 1974. Chronic intestinal coccidiosis in man: intestinal morphology and response to treatment. Gastroenterology **66**:923–935.

Whiteside, M. E., J. S. Barkin, R. G. May, S. D. Weiss, M. A. Fischi, and C. L. MacLeod. 1984. Enteric coccidiosis among patients with the acquired immunodeficiency syndrome. Am. J. Trop. Med. Hyg. **33**:1065–1072.

CHAPTER 96

Leishmaniasis

THOMAS R. NAVIN

Disease: Leishmaniasis (visceral leishmaniasis or kala-azar, cutaneous leishmaniasis, mucocutaneous leishmaniasis).

Etiologic Agents: *Leishmania donovani, Leishmania major, Leishmania tropica, Leishmania aethiopica, Leishmania mexicana, Leishmania braziliensis,* and *Leishmania peruviana*.

Source: Transmitted to humans by phlebotomine sandflies from a wide variety of reservoir hosts.

Clinical Manifestations: Visceral leishmaniasis is associated with fever, weight loss, anorexia, splenomegaly, and pancytopenia. Cutaneous leishmaniasis is associated with chronic skin ulcers. Mucocutaneous leishmaniasis is caused by metastatic spread of primary cutaneous lesions to the mouth, nose, and pharynx, where destruction of mucosa occurs.

Pathology: In humans, parasites occur as intracellular amastigotes that infect macrophages and other phagocytes. In cutaneous and mucocutaneous disease, parasites are found at the site of the cutaneous lesions. In visceral leishmaniasis, parasites are most commonly found in the spleen, liver, and bone marrow.

Laboratory Diagnosis: Identification of amastigotes in or culture of promastigotes from infected tissue.

Treatment: Pentavalent antimonial agents (Pentostam or Glucantime).

Prevention and Control: Prevention and control strategies vary with different geographic areas.

Description of Disease

Leishmaniasis is an infectious disease caused by the protozoan parasite *Leishmania*. The three common clinical manifestations of the disease are cutaneous leishmaniasis, which is associated with chronic ulcers of the skin; mucocutaneous leishmaniasis, which is caused by the metastatic spread of *Leishmania* from a primary cutaneous source to the mucous membranes of the nose, mouth, or pharynx, where it causes destructive lesions; and visceral leishmaniasis or kala-azar, which is associated with fever, chronic weight loss, anorexia, marked splenomegaly, and pancytopenia.

Although the species that cause each of these syndromes vary somewhat, usually visceral leishmaniasis is caused by *Leishmania donovani* and mucocutaneous leishmaniasis is caused by *Leishmania braziliensis*. Cutaneous leishmaniasis can be caused by wide range of species, including *Leishmania major, Leishmania tropica, Leishmania aethiopica, Leishmania mexicana, Leishmania braziliensis,* and *Leishmania peruviana*.

Communicability and Transmissibility

Leishmaniasis is a vector-borne disease, being transmitted from a vertebrate reservoir host to humans by the bite of an infected phlebotomine sandfly. Laboratory-acquired infections have occurred rarely. Direct person-to-person transmission from an open lesion to a skin cut is theoretically possible, but routine precautions in examining patients and obtaining specimens should eliminate this risk.

Clinical Manifestations

The three major types of leishmaniasis are visceral, cutaneous, and mucocutaneous disease. A number of rarer clinical presentations will be discussed.

Visceral Leishmaniasis

Visceral leishmaniasis is almost always caused by *L. donovani*. The onset of symptoms is usually gradual, and fever, anorexia, and weight loss are common. The spleen and liver both enlarge, and in time the spleen can become enormous. Decreased hemoglobin, white blood cell count, and platelet count are very common and worsen as the disease progresses. Although asymptomatic visceral leishmaniasis occurs, after symptoms appear the patient's condition usually worsens until the patient is treated or dies.

Cutaneous Leishmaniasis

The clinical manifestations of cutaneous leishmaniasis differ somewhat according to the species of *Leishmania* causing the infection. In all types, however, the cutaneous lesion usually first appears as a papule at the site of inoculation by the infected sandfly. This papule enlarges slowly, often leading to ulceration of the central area. In diffuse cutaneous leishmaniasis, an uncommon form of the disease, no ulceration occurs. The natural history of untreated cutaneous leishmaniasis is not well described and probably varies considerably with the species involved. Lesions caused by *L. peruviana* are thought to heal within about 4 months, lesions caused by *L. mexicana, L. tropica,* and *L. major* heal within about 1 year, and cutaneous lesions caused by *L. aethiopica* and *L. braziliensis* heal very slowly or not at all.

Mucocutaneous Leishmaniasis

After initial inoculation into the skin, certain species of *Leishmania* can metastasize to the mucous membranes of the nose, mouth, or pharynx. Metastatic spread can occur while the primary lesion is present or up to many years later; sometimes there is no evidence of a primary lesion. Slowly progressive ulceration of the mucosa and underlying cartilage causes severe mutilation. By far the most common species to cause mucocutaneous diseases is *Leishmania braziliensis braziliensis*, but *Leishmania braziliensis panamensis* and *Leishmania braziliensis guyanensis* msy also at times be associated with mucocutaneous disease. A few cases of mucocutaneous leishmaniasis have been reported with infections due to *L. donovani* and *L. aethiopica*.

Rarer Forms of Leishmaniasis

Leishmaniasis recidivans is associated with chronic tuberculoid lesions, usually of the face, that progress slowly and respond poorly to treatment. Post-kala-azar dermal leishmaniasis is seen in the Indian subcontinent and less commonly in Africa one to several years after the patient has recovered from visceral leishmaniasis. The disease is manifested by chronic cutaneous nodules or hypopigmented macules. As mentioned, diffuse cutaneous leishmaniasis is characterized by slowly progressing plaques, papules, or nodules that do not ulcerate.

Pathology

The promastigote stage of *Leishmania* is injected into the skin of the vertebrate host at the time the female phlebotomine sandfly takes a blood meal. Promastigotes are phagocytized at the site of inoculation and convert into the intracellular amastigote stage. In the visceral form of the disease, amastigotes are presumably transported to the spleen, liver, bone marrow, and other organs within these phagocytes. The histologic picture seen in the tissues involved is one of accumulation of mononuclear phagocytic cells.

Diagnostic Features

As it is rare to see cutaneous, mucocutaneous, or visceral leishmaniasis in the United States, these infections are often confused with other diseases. Probably the most helpful feature that will suggest the diagnosis is a history of travel to an endemic area. The differential diagnosis of cutaneous leishmaniasis includes chronic cutaneous tuberculosis, fungal and spirochetal infections, and lupus vulgaris, as well as basal or squamous cell carcinomas. Mucocutaneous leishmaniasis may be mistaken for paracoccidioidomycosis, carcinomas, midline granuloma, leprosy, and other destructive lesions of the nasopharynx. Visceral leishmaniasis can be confused with other chronic systemic infections, such as brucellosis, histoplasmosis, and malaria, as well as with lymphoma.

A definitive diagnosis of leishmaniasis requires examining infected tissue for amastigotes (either by direct microscopic examination or by testing for homologous DNA) or culturing samples for promastigotes. In the case of visceral leishmaniasis, a positive serologic test is suggestive.

Prognosis

Cutaneous leishmaniasis caused by *L. mexicana, L. major, L. tropica,* and *L. peruviana* is usually a self-limited infection, although the infection can last for more than a year. Cutaneous leishmaniasis caused by *L. braziliensis* also can resolve without treatment, but secondary lesions along the lymphatic glands or in the nose, mouth, or pharynx can occur. Therapy of cutaneous leishmaniasis with adequate doses of pentavalent antimonial agents usually cures the infection. Mucocutaneous leishmaniasis typically is slowly progressive, does not heal without therapy, and often responds poorly even to appropriate therapy. Patients with asymptomatic visceral leishmaniasis have been described, but symptomatic infections are usually fatal unless treated. Treatment of visceral leishmaniasis with pentavalent antimonial agents is usually rapidly effective unless the patient is severely malnourished or suffers from a concurrent infection.

Epidemiologic Aspects

Occurrence and Transmission

Leishmaniasis is usually a disease of rural areas in developing countries. In some areas the disease is found predominantly in hot, humid regions; in others it is found in hot, dry regions. Because leishmaniasis usually occurs in rural areas that lack adequate health care facilities, its worldwide occurrence is difficult to estimate.

Although very rare in the United States, cases of indigenous cutaneous leishmaniasis have been reported from Texas (Gustafson et al., 1985). There have been no reports of indigenous visceral or mucocutaneous leishmaniasis in the United States. Approximately 20 cases of cutaneous leishmaniasis in travelers are reported to the Centers for Disease Control each year. Although less common, cases of visceral and mucocutaneous leishmaniasis are also reported (Centers for Disease Control, unpublished data).

Cutaneous leishmaniasis is endemic in Latin America from the Yucatan peninsula in the north, along the Atlantic coast of Central America, and most of South America, with the exception of the southernmost areas of Chile, Uruguay, and Argentina. It is especially prevalent in low-lying tropical forests. In the Old World, cutaneous leishmaniasis is found in all countries surrounding the Mediterranean Sea, as well as in the Middle East, Afghanistan, southern Russia, and the countries bordering the Sahara Desert to the south.

Mucocutaneous leishmaniasis is most common in Brazil, south of the Amazon, but cases have been reported from Colombia, Central America, and rarely from the Sudan.

Visceral leishmaniasis is most common in central and southern Brazil, Kenya and surrounding countries, and eastern India. Areas where the disease is less common include the Pacific coast of Central America, Colombia, and countries bordering the Mediterranean Sea (World Health Organization, 1984).

Susceptibility, Resistance, and Carriage

Little is known about factors that might affect the outcome of an infection with *Leishmania* in a particular person (Pearson, 1983). Whether one develops cutaneous, mucocutaneous, or visceral disease seems to depend predominantly on the species of the infecting parasite, although even here there are cases that suggest that host factors play an important role in determining the extent of disease. In inbred mice susceptibility to leishmaniasis is under genetic control, but it is not known whether this is true for humans as well.

Leishmaniasis is predominantly a sylvatic disease, and humans are usually infected when they live in or travel to rural areas. Although asymptomatic carriers of visceral leishmaniasis are known to occur, their importance in the maintenance and transmission of the disease is probably small.

Etiologic Agents

Characterization

Leishmania exist in two forms, the promastigote, which is found in the sandfly and in cultures, and the amastigote, which is found within macrophages of the host. Both promastigotes and amastigotes have similar staining characteristics; with Giemsa stain their nuclei stain purple, their cytoplasms stain a light blue, and their kinetoplasts stain a dark blue. Promastigotes are long and slender and have an anterior flagellum that pulls them forward; both the bodies and the flagellum measure about 20 μm in length. Amastigotes are round or oval and measure from 2 to 5 μm in diameter.

A number of media, including semidefined media, have been described to cultivate promastigotes. Tests comparing the sensitivity of different media have shown divergent results, and it is not possible to select one medium that is best for all isolation requirements. The reader should consult technical papers available in the literature for descriptions of different media (Evans et al., 1984; Hockmeyer et al., 1981; World Health Organization, 1984). Growth is best between 23 and 26°C. Amastigotes can be grown in macrophages, fibroblasts, and other cell lines.

Cutaneous Leishmaniasis

Tissue containing amastigotes can be collected by scraping lesions and smearing the material obtained on a microscope slide. Although there may be more sensitive methods, direct observation of amastigotes from lesion scrapings is the simplest method to diagnose cutaneous leishmaniasis. Barbed dental nerve broaches (available from dental supply houses), scalpels, capillary tubes, and other instruments with sharp blades are suitable for obtaining material for smears. Work in our laboratory in Guatemala, where lesions are caused by both *L. mexicana mexicana* and *L. braziliensis braziliensis,* has shown that the listed instruments are roughly equal in their usefulness in preparing slides. In our experience, more important than the instrument used is the care with which a smear is prepared. The keys to preparing an adequate specimen are to obtain as much tissue as possible, avoid making a smear with excessive red blood cells, and make as thin a smear as possible. Many authorities state that the best location from which to obtain scrapings is the upper dermis at the border of a lesion just beyond the ulcer. A small incision is made with a scalpel blade through intact epidermis at the edge of an ulcer, and the upper dermis beneath intact skin is scraped. In Guatemala, we have compared scrapings made from such cuts with scrapings made both at the center of ulcers and at the border of ulcerated lesions; we have found no differences between the three areas. At least for leishmaniasis in Guatemala, if one takes care to thoroughly debride the lesion, it would appear that amastigotes can be found wherever there is inflammation.

Superficial scrapings can also be used to inoculate culture medium. The problem with this approach is the risk of contamination of the cultures with bacteria and fungi found on the surface of the skin. The risk of contamination appears to be lessened for samples taken from beneath intact epidermis, but, again, in our experience, more important than the location of the sample is the care with which ulcers are debrided and washed and the attention paid to detecting and counteracting contamination in cultures. We remove scabs and devitalized tissue with a scalpel. We then scrub the lesion thoroughly, first with a surgical scrub brush and mild soap and then with a surgical scrub brush soaked in alcohol. We place a sterile field around the lesion and work with sterile surgical gloves. With these precautions we have had equal success culturing material obtained from the base of the ulcer and material from cuts made through intact skin just beyond the ulcer.

In addition to scrapings, aspirates of lesions are useful. Although aspirates sample only small amounts of tissue, and for this reason microscope slides made from aspirates are not particularly advantageous, they are helpful for inoculating culture medium since the risk of contamination can be lessened. While excellent results have been obtained with 26-gauge needles, a larger gauge needle allows more material to be obtained. The needle is attached to a 3- or 5-ml syringe, and a small amount of sterile culture medium or sterile saline without preservative is drawn into the syringe. The needle is inserted into intact skin adjacent to the edge of the lesion. Suction is applied, and the syringe is rotated and moved forward to obtain a small amount of tissue from the edge of the lesion. The aspirated material is expelled into a tube containing culture medium.

Biopsy specimens are also helpful in collecting material for impression smears, cultures, DNA analysis, or animal inoculation. The advantages of a biopsy specimen include the large amount of tissue sampled and the ability to analyze the material histologically for *Leishmania* and other agents.

Visceral Leishmaniasis

Material for diagnosing visceral leishmaniasis can be obtained from spleen, bone marrow, liver, and lymph nodes. Splenic aspirates are probably the most sensitive method to diagnose the infection; for a patient with an adequate prothrombin time and platelet count the procedure, if performed by an experienced person, is relatively safe (Chulay and Bryceson, 1983). A 1.25-inch, 21-gauge needle is attached to a 5-ml syringe. The needle is inserted just beneath the skin midway between the edges of the spleen, 2 to 4 cm below the costal margin. The needle is aimed cranially at a 45° angle to the abdominal wall. The plunger of the syringe is pulled back to the 1-ml mark, and with a quick in-and-out movement the needle is pushed into the spleen to the full needle depth and then immediately withdrawn completely, maintaining suction throughout. The aspirate is expelled into culture media, and the material remaining in the syringe is used to prepare a smear on a glass slide.

Bone marrow aspirates are also useful and have the advantage that more people are familiar with their use. Bone marrow aspirates are handled in the same way as splenic aspirates.

Finally, biopsies or aspirates of the liver or lymph nodes can be made, but are not as useful as splenic or bone marrow aspirates.

Isolation and Identification Procedures

Direct Examination

Slides made from superficial scrapings, aspirates, impression smears, or histologic sections can be used for direct microscopic examination. Slides are fixed in methanol and stained with Giemsa. Positive identification of *Leishmania* amastigotes should be based on observing all of the following: a nucleus, a kinetoplast, a cell wall, and cytoplasm. Histologic sections and smears made from spleen and bone marrow usually show intracellular amastigotes, but other smears often contain only isolated, extracellular amastigotes.

Culture Techniques

As stated previously, several culture media suitable for isolating *Leishmania* promastigotes have been described (Evans et al., 1984; Hockmeyer et al., 1981). Studies comparing these media have produced conflicting results, and at present it is not possible to select one medium that is superior to others for isolating all species. One easily prepared and well-tested medium is NNN, which consists of defibrinated rabbit blood and blood agar (World Health Organization, 1984).

The most important aspect of culturing is avoiding bacterial and fungal contamination. A number of techniques have been described to reduce contamination. Antibiotics are usually added to culture media used for primary isolation of parasites from humans or animals. *Leishmania* tolerate high doses (up to 200 μg/ml) of penicillin and gentamicin and common recommendations are to add 100 μg of each drug per ml of culture media. Some species tolerate the antifungal agent 5-fluorocytosine, but other species do not. One strategy is to use one tube of culture medium with 100 μg/ml 5-fluorocytosine and one tube without.

Cultures should be examined at approximately 1-week intervals for 4 weeks before they are discarded as negative. Positive cultures should be subcultured to new medium as soon as they are identified.

For laboratories without experience in isolating *Leishmania*, another approach to avoiding contamination is to inject patient material into a hamster (*Mesocricetus auratus*) or a Balb/c mouse. Material from patients with visceral leishmaniasis should be injected intraperitoneally, and material from cutaneous or mucocutaneous lesions should be injected into the hind footpad, the base of the tail, or the skin of the nose.

Isoenzyme Analysis

Although many methods for characterizing *Leishmania* are now available (see below), isoenzyme analysis has become the standard (Kreutzer et al., 1983). The method involves culturing large numbers of promastigotes, which are then disrupted by freeze-thawing. The soluble extract of the lysed promastigotes is analyzed by thin-layer electrophoresis. Electrophoretic patterns of a dozen or more enzymes from the strain in question are compared with patterns for reference strains. Isoenzyme analysis is available from World Health Organization diagnostic reference centers.

Other Identification Techniques

Methods to analyze parasite DNA have recently been described that offer rapid identification of species (Wirth and Pratt, 1982). Infected tissue, such as that from a punch biopsy, is first pressed onto nitrocellulose paper. After preliminary processing to fix the DNA to the nitrocellulose, samples are incubated with kinetoplast DNA isolated from known reference strains and labeled with radioactive phosphorus. Samples with homologous kinetoplast DNA will bind to the radiolabeled reference DNA and will show up as dark spots when developed on radiographic paper. Although the procedure is available only as a research tool, it offers several advantages over traditional identification techniques. First, the method does not depend on the cultivation of parasites, so problems related to contamination of cultures or the need for specialized culture media do not arise. This advantage makes the method particularly suited to laboratories that do not have experience culturing *Leishmania*. Second, the samples are stable for months at room temperature, so laboratories can send samples to reference centers without risk of degradation. Sample stability and the ability to analyze a number of samples at the same time makes the technique particularly suited for field studies in which a number of samples are collected in remote areas and brought back to a central laboratory for processing.

Monoclonal antibodies have been produced recently that can distinguish between subspecies of *Leishmania* (Pratt and David, 1982). The most common way to employ these antibodies in the identification of *Leishmania* is to examine cultured promastigotes by indirect immunofluorescence. After being dried and fixed, promastigotes are incubated with the reference monoclonal antibody. The promastigotes are then incubated with fluorescein-labeled anti-mouse immunoglobulin G and examined with a fluorescence microscope. Recently, monoclonal antibodies have been used to examine histologic sections directly, making identification possible even when cultures are not available (Lynch et al., 1986). Monoclonal antibodies are not commercially available, and this test is also available only at a few centers as a research tool.

Other identification procedures with more specialized applications include radiorespirometry, analysis of restriction endonuclease fragments, and excreted factor serotyping.

Immunologic Aspects

Although serologic tests for leishmaniasis are poorly standardized, in certain instances they can be an important adjunct to direct examination of tissue and culture. This is especially true in the case of visceral leishmaniasis, because antibody levels are often high by the time the patient seeks medical attention. Serology is of marginal utility for mucocutaneous leishmaniasis and is of almost no help in diagnosing an individual case of cutaneous leishmaniasis.

The most widely used serologic test is indirect immunofluorescence, which is available in the United States from the Centers for Disease Control. A titer of 1:16 is a borderline positive reaction, with titers of 1:32 and above very suggestive of leishmaniasis. Enzyme-linked immunosorbent assays have also been used successfully. Serologic tests cannot distinguish between different species of *Leishmania,* and cross-reactions occur with *Trypanosoma cruzi,* making differentiation between Chagas' disease and visceral leishmaniasis difficult on serologic grounds alone.

Treatment, Prevention, and Control

Treatment

The drug of choice in the treatment of cutaneous, mucocutaneous, and visceral leishmaniasis is either Pentostam or Glucantime (World Health Organization, 1984). Both drugs are pentavalent antimony compounds and appear to be similar in terms of toxicity and efficacy. Pentostam is available in the United States as an investigational drug from the Centers for Disease Control. Pentostam comes in 100-ml bottles containing 100 mg of antimony per ml. Although the undiluted drug can be given intramuscularly, newer regimens that call for larger doses require that the drug be given intravenously, either as undiluted solution or diluted in 5% dextrose. Glucantime is not available in the United States; it comes in 5-ml vials containing 85 mg of antimony per ml. Dosage regimens for pentavalent antimonial drugs are undergoing revision, and experts in this field should be consulted for current recommendations. In general it has been found that 20 mg of antimony per kg of body weight once a day for 20 days is reasonably well tolerated and results in a better cure rate than lower doses.

Amphotericin B is useful as an alternative therapy for patients who have failed to respond to appropriate doses of Pentostam or Glucantime. The drug is especially useful for patients with mucocutaneous leishmaniasis, which is often unresponsive to traditional therapy.

Prevention and Control

Cutaneous and mucocutaneous leishmaniasis are zoonoses that are usually maintained in sylvatic reservoirs. Elimination of these reservoirs and the sandflies that transmit the disease is usually impractical (Marsden, 1984). Control strategies in this setting, therefore, usually focus on preventing contact between humans and sandflies. Heavy clothing that covers arms and legs and insect repellent are helpful. These approaches, however, are often impractical for the subsistence farmer working in a tropical forest.

Visceral leishmaniasis is often transmitted in or around the house, and insecticide spraying in this setting may be helpful in controlling sandflies. In several areas where insecticide spraying to control malaria has been stopped, rates of visceral leishmaniasis have risen. Control of infected dogs in areas where they are reservoirs has also been helpful.

Literature Cited

Chulay, J. D., and A. D. Bryceson. 1983. Quantitation of amastigotes of *Leishmania donovani* in smears of splenic aspirates from patients with visceral leishmaniasis. Am. J. Trop. Med. Hyg. **32**:475–479.

Evans, D. A., S. M. Lanham, C. I. Baldwin, and W. Peters. 1984. The isolation and isoenzyme characterization of *Leishmania braziliensis* subsp. from patients with cutaneous leishmaniasis acquired in Belize. Trans. R. Soc. Trop. Med. Hyg. **78**:35–42.

Gustafson, T. L., C. M. Reed, P. B. McGreevy, M. G. Pappas, J. C. Fox, and P. G. Lawyer. 1985. Human cutaneous leishmaniasis acquired in Texas. Am. J. Trop. Med. Hyg. **34**:58–63.

Hockmeyer, W. T., P. A. Kager, P. H. Rees, and L. D. Hendricks. 1981. The culture of *Leishmania donovani* in Schneider's insect medium: its value in the diagnosis and management of patients with visceral leishmaniasis. Trans. R. Soc. Trop. Med. Hyg. **75**:861–863.

Kreutzer, R. D., M. E. Semko, L. D. Hendricks, and N. Wright. 1983. Identification of *Leishmania* spp. by multiple isozyme analysis. Am. J. Trop. Med. Hyg. **32**:703–715.

Lynch, N. R., C. Malave, R. B. Ifante, R. L. Modlin, and J. Convit. 1986. In situ detection of amastigotes in American cutaneous leishmaniasis, using monoclonal antibodies. Trans. R. Soc. Trop. Med. Hyg. **80**:6–9.

Marsden, P. D. 1984. Selective primary health care: strategies for control of disease in the developing world. XIV. Leishmaniasis. Rev. Infect. Dis. **6**:736–744.

Pearson, R. D., D. A. Wheeler, L. H. Harrison, and H. D.

Kay. 1983. The immunobiology of leishmaniasis. Rev. Infect. Dis. **5**:907–927.

Pratt, D. M., and J. R. David. 1981. Monoclonal antibodies that distinguish between New World species of *Leishmania*. Nature (London) **291**:581–583.

Wirth, D. F., and D. M. Pratt. 1982. Rapid identification of *Leishmania* species by specific hybridization of kinetoplast DNA in cutaneous lesions. Proc. Natl. Acad. Sci. USA **79**:6999–7003.

World Health Organization. 1984. The leishmaniases: Report of a W.H.O. expert committee. Technical report series no. 701. World Health Organization, Geneva.

CHAPTER 97

Malaria

E. G. LONG

Disease: Malaria (English), paludismo (Spanish), paludisme (French).

Etiologic Agents: *Plasmodium (Laverania) falciparum, Plasmodium (Plasmodium) vivax, Plasmodium (Plasmodium) ovale,* and *Plasmodium (Plasmodium) malariae.*

Source: Humans.

Clinical Manifestations: Headache, malaise, diarrhea, myalgia, moderate fever progressing to paroxysms every 48 or 72 h or intense chills (rigors) followed by high fevers and ending with profuse sweating.

Pathology: Hemolysis, anemia, occlusion of cerebral capillaries and venules, hepatosplenomegaly, renal dysfunction, pulmonary edema, circulatory collapse, cardiac and gastrointestinal lesions.

Laboratory Diagnosis: Demonstration of parasitic forms in blood films by Giemsa or Wright staining. Immunodiagnostic methods.

Epidemiology: Sub-Saharan Africa, Saharan oases, the Americas from Mexico to Argentina, Turkey, Saudi Arabia, Asia from Iran to the Phillipines, New Guinea, Oceania. Transmission is from person to person by infected female *Anopheles* mosquitoes and occasionally by blood transfusions or common use of hypodermic syringes; can also be transmitted congenitally.

Treatment: Complicated because of increasing resistance to antimalarial drugs. Chloroquine phosphate, quinine dihydrochloride, or quinidine hydrochloride. For chloroquine-resistant *P. falciparum* malaria, quinine sulfate combined with pyrimethamine–sulfadoxine or with tetracycline. Quinine dihydrochloride or quinidine gluconate. For prevention of *P. vivax* or *P. ovale* relapses, primaquine phosphate.

Prevention and Control: Education, elimination of *Anopheles* mosquitoes by insecticides or destruction of breeding sites, treatment of infected persons, prophylaxis, and protection such as bed nets and insect repellents.

Description of Disease

Malaria gets its name from the Italian *mal aria,* meaning bad air. The disease is transmitted from person to person by infected female *Anopheles* mosquitoes, which require a blood meal for egg production. Parasites first infect the liver, where they multiply, and in some cases remain dormant, within hepatocytes. Later, large numbers of organisms leave ruptured hepatocytes and infect erythrocytes, in which they grow, multiply, and undergo sexual differentiation. This erythrocytic stage of the disease is responsible for the pathology of malaria. Large numbers of erythrocytes are destroyed by rupture and by host immunologica mechanisms, and hemolysis is responsible for the anemia associated with malaria. The most frequent cause of death is cerebral edema caused by blockage of cerebral capillaries by infected erythrocytes in patients with *Plasmodium falciparum* malaria. Other important complications are renal failure; pulmonary edema; algid malaria, which resembles septic shock; and a nephrotic syndrome seen in quartan malaria caused by *Plasmodium malariae*.

The most important species of human malaria parasite is *P. falciparum,* which is responsible for most of the morbidity and mortality of the disease. *Plasmodium vivax, Plasmodium ovale,* and *P. malariae* seldom cause fatal disease, although symptoms may be severe. Infections with *P. malariae* sometimes result in a nephrotic syndrome. Malaria remains the most important infectious disease of the tropics and a constant threat to temperate countries that harbor suitable *Anopheles* vectors.

Communicability and Transmissibility

Malaria is usually transmitted from person to person by blood-sucking infected *Anopheles* mosquitoes. Parasites appear in the bloodstream of an infected patient 3 to 4 days after infection with *P. vivax* or *P. ovale,* 10 days with *P. falciparum,* and 30 to 40 days with *P. malariae;* and sexual forms, which are infective for mosquitoes, appear about 84 h after the parasites begin dividing within erythrocytes. Mosquitoes can transmit infective sporozoites 10 to 12 days after becoming infected themselves, and continue to do so for a lifetime of about 1 month.

Other means of infection are blood transfusions, multiple use of contaminated hypodermic needles by drug addicts, and congenital or connatal transmission.

Clinical Manifestations of Disease

The periodic paroxysms of malaria are commonly believed to be pathognomonic of the disease, but malaria can mimic almost any febrile illness. Symptoms may begin after an incubation period of 8 to 14 days for *P. falciparum,* 12 to 17 days for *P. vivax,* 11 to 16 days for *P. ovale,* and 30 to 40 days for *P. malariae*.

In a primary attack of falciparum malaria, the initial signs and symptoms may be deceptively mild, with influenzalike symptoms (Manson-Bahr and Apted, 1982). The patient first experiences headache, myalgia, chills, irritability, nausea, vomiting, and diarrhea. Later, symptoms include fevers with temperatures as high as 40°C accompanied by convulsions. The initial symptoms increase in severity, and the triad of chills, fever, and sweating recur at intervals of 48 h after the release of parasites and metabolites from ruptured erythrocytes. In many infections, paroxysms are not periodic, and symptoms may occur at random intervals, perhaps because swarms of erythrocytes are entering the bloodstream at different times.

Other common manifestations are hepatosplenomegaly, jaundice, anemia, leukopenia, albuminuria with hyaline or granular casts (Bruce-Chwatt, 1985), and labial herpes (Miller, 1984).

Nonimmune children have more severe disease and have higher parasitemias and higher parasitemia rates. In endemic areas, infected infants with passively acquired maternal antibodies show milder symptoms and have lower parasitemias until their immunity wanes.

Clinically, malaria can be asymptomatic, severe and complicated, recrudescent, or relapsing.

Asymptomatic Malaria

In malaria-endemic areas, repeated infections lead to a state of partial protection in which symptoms may be mild or absent. Untreated patients with *P. vivax* or *P. ovale* infection may recover after a week of illness, but the disease may relapse in 8 to 40 weeks. Falciparum malaria may remain latent for 1 year, and patients with *P. malariae* can remain asymptomatic for life. Subcurative therapy may also lead to the disappearance of symptoms in patients with patent infections.

Falciparum Malaria and Severe Complicated Malaria

Plasmodium falciparum infections are responsible for most of the mortality due to malaria. The pathogenicity of this protozoan is related to its reproductive potential, its capacity to infect reticulocytes as well as erythrocytes, and its tendency to block capillaries and venules. Bray (1963) estimated that sporozoite inocula of equal density for the four plasmodia would result in an initial parasitemia of *P. falciparum* 20 times as great as *P. malariae,* 4 times as great as *P. vivax,* and 3 times as great as *P. ovale*. Multiplication within an erythrocyte results in an average of 20 to 24 progeny in contrast to 16 for *P. vivax* and 8 for *P. ovale* and *P. malariae*. In *P. falciparum* infections, 50% of erythrocytes may be invaded, although the host usually succumbs before this level is reached (Beaver et al., 1984). Recurrences of *P. falciparum* malaria in an untreated patient may be due to recrudescences of the disease from organisms sequestered in the bone marrow or viscera. Unlike infection with *P. vivax* or *P. ovale,* dormant hepatic organisms occur. Most of the severe symptoms and pathologic effects of malaria are caused by *P. falciparum*.

The World Health Organization has defined severe, complicated malaria as the demonstration of asexual forms of *P. falciparum* in a patient with a potentially fatal manifestation or complication of malaria in whom other diagnoses have been excluded. The signs of severe malaria are as follows: hyperparasitemia, with a blood smear showing asexual forms in more than 5% of erythrocytes; cerebral malaria, manifested by unarousable coma in the absence of other encephalopathies; and other signs of cerebral dysfunction such as confusion, stupor, convulsions, or psychoses, which can be caused by fevers of nonmalarious etiologies. Other clinical findings are as follows: severe anemia, with a hematocrit of less than 20%; hemoglobin level of less than 4.4 mmol/liter or 7.1 g/dl; jaundice, with a serum bilirubin of more than 50 μmol/liter; fluid or electrolyte disturbances requiring intravenous therapy; renal failure, with a urine output of less than 400 ml in 24 h; a serum creatinine of more than 265 μmol/liter or more than 30 mg/dl and refractory to rehydration; hyperthermia with a rectal temperature of more than 30°C; and hypoglycemia, with a blood glucose concentration of less than 2.2 mmol/liter or less than 40 mg/dl. Complications include aspiration bronchopneumonia, septicemia, pulmonary edema, shock, bleeding and clotting disturbances, retinal hemorrhages, vomiting of orally administered drugs, and hemoglobinuria (World Health Organization, 1986). Malarial hemoglobinuria results from the rapid and massive hemolysis of both parasitized and normal erythrocytes and is characterized by the passage of black urine. The typical patient is a nonimmune resident in an endemic area who has had repeated attacks of malaria and has taken irregular courses of treatment or prophylaxis (World Health Organization, 1986).

Vivax Malaria

The symptoms of vivax malaria can be severely debilitating, although the disease is seldom fatal. The main pathologic event is hemolysis leading to anemia. Splenic rupture may result from trauma or palpation. The incubation period is 12 to 17 days for *P. vivax*. Fever and other symptoms of a primary attack are irregular at first, and paroxysms attain a 48-h periodicity after 2 days of illness. The term "benign tertian" malaria for vivax malaria refers to recurrence of fever and chills every third day. Untreated or partially treated vivax malaria is notorious for causing relapsing disease: this may be a short-term relapse, 8 to 10 weeks after a primary attack, or a long-term relapse, 30 to 40 weeks later (Bruce-Chwatt, 1985). Relapses are due to maturation of dormant liver forms, hypnozoites, which develop at different rates before invading the bloodstream (Fig. 1).

Ovale Malaria

The life cycle, morphology, and disease of *P. ovale* are similar to those of *P. vivax*, but *P. ovale* does not cause relapses as often as the latter.

Quartan Malaria

Quartan malaria was formerly known as benign quartan malaria, because paroxysms sometimes recurred every fourth day. The limited pathogenicity of *P. malariae* is probably due to a longer incubation period, a lower rate of reproduction and longer erythrocytic cycle than the other malarias, and its preference for older erythrocytes. This disease may recrudesce decades after a primary infection. The most important complication of this disease is a nephrotic syndrome, which may occur in children in endemic areas (Hendrickse et al., 1972).

Congenital Malaria

During pregnancy, parasite densities may be greater in placental blood than in the peripheral circulation, and parasites may be found in the cord blood of newborn infants from infected mothers (McGregor, 1984). Despite this finding, true congenital infections are uncommon even in endemic areas, probably because of protective antibodies acquired transplacentally (McGregor, 1984). Congenital malaria is encountered more often in babies born of mothers infected for the first time (Bruce-Chwatt, 1985), and most of these infections may have been acquired connatally rather than transplacentally (Madecki and Kretschmar, 1966).

Transfusion Malaria

Transmission of malaria through transfusion occurs most frequently when blood from an infected donor is used. Transfusion malaria has also been reported among intravenous drug abusers using contaminated syringes (Most, 1968). As in congenital malaria, malaria parasites are transferred directly into a recipient's bloodstream via infected erythrocytes. For this reason, the American Association of Blood Banks has set the following rules for blood donations from persons who may have been exposed to malaria.

1. Travelers returning from endemic areas should delay blood donations for 6 months, even if they have remained asymptomatic or have not taken antimalarial drugs.
2. Persons who have had malaria or have taken chemoprophylaxis should not donate blood for 3 years after becoming asymptomatic or completing therapy or chemoprophylaxis.
3. Immigrants or visitors from endemic areas should not donate blood for 3 years after departure, even if they have remained asymptomatic.
4. Carriers of malaria or persons who have had *P. malariae* malaria should never donate blood.

Malaria should be considered in a person who has an unexplained fever within 3 months after receiving

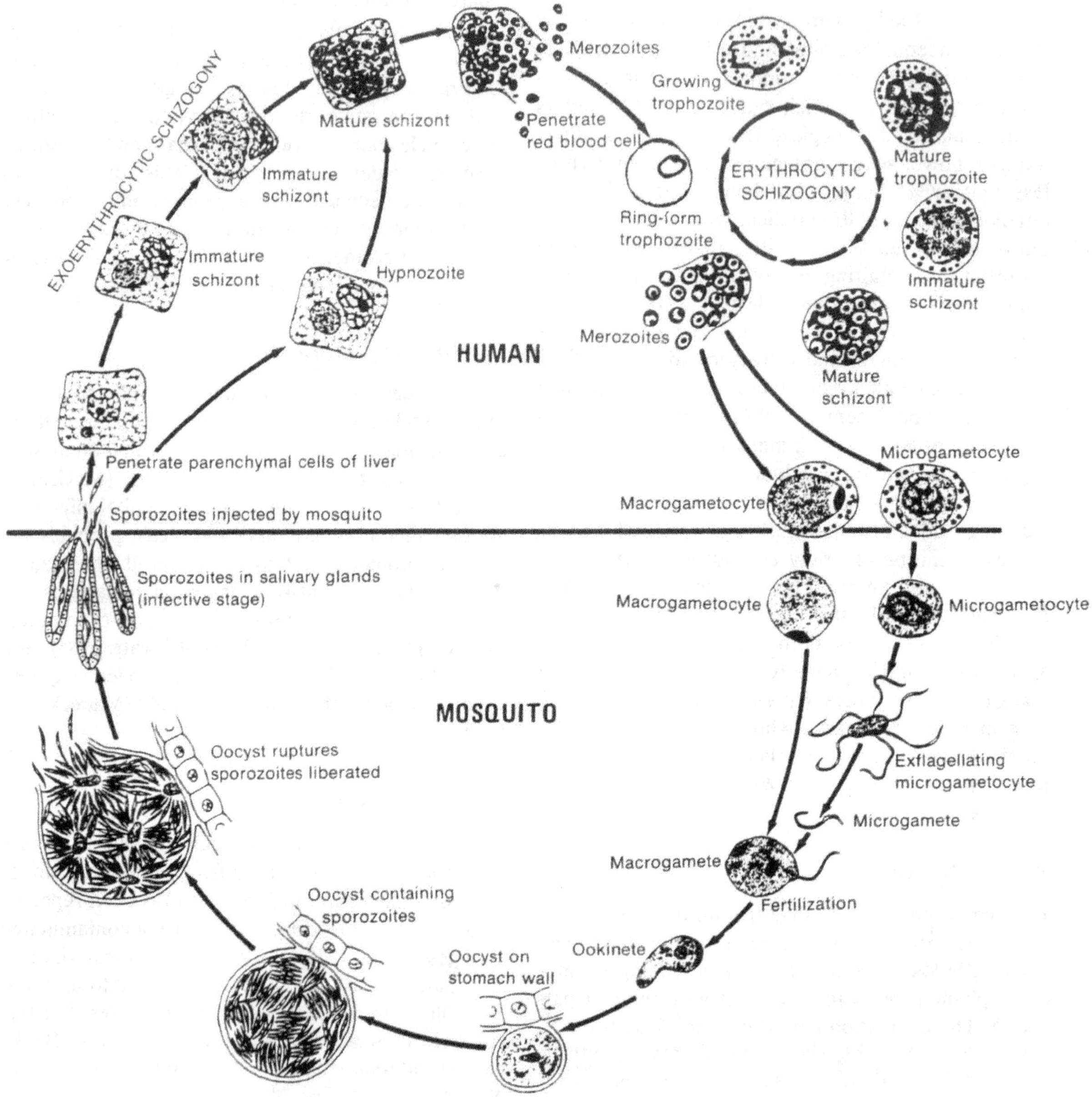

FIG. 1. Life cycle of malarial parasites based on that of *Plasmodium vivax*. (Reprinted with permission from the Centers for Disease Control, Parasitology Section, Training and Laboratory Program Office, Atlanta, Georgia.)

a blood transfusion (Guerrero et al., 1983), the indirect fluorescent antibody test is useful in identifying the donor.

RECRUDESCENCE

The reappearance of *P. falciparum* and *P. malariae* after a period of latency is due to the success of the parasites in overcoming factors that kept them at undetectable levels in the circulation. The return of a parasite and concomitant disease does not constitute true relapse because the parasitemia originates from blood stages. During the life cycle of these two parasites, pre-erythrocytic schizonts in the liver complete their development to merozoites, which invade erythrocytes and do not remain dormant in the liver.

Relapse

Plasmodium vivax and *P. ovale* may cause recurrent infections after total clearance from the blood. These recurrences are true relapses because dormant liver schizonts, or hypnozoites, activate to complete their development to produce new populations of erythrocytic merozoites.

Pathology

The principal pathologic consequence of malaria that is common to the four types of disease is a normochromic, normocytic anemia. The mechanisms of anemia include hemolysis of infected erythrocytes by escaping parasites; depression of erythropoiesis; increased phagocytosis of erythrocytes; and hemolysis of both parasitized and normal erythrocytes by immunologic mechanisms (Perrin et al., 1982; Weatherall and Abdalla, 1982). The other pathologic feature, which is responsible for most deaths associated with malaria, is the characteristic tendency of *P. falciparum*-infected erythrocytes to stick to capillary endothelium and thus block the blood circulation in vital organs. In cerebral malaria, brain tissue is damaged when infarction following blocked capillaries results in necrosis of surrounding parenchyma, demyelination of neurons, and extravasation of erythrocytes to form "ring hemorrhages" (Connor et al., 1976). The liver becomes slightly enlarged and slate gray from the malaria pigment accumulated within the Kupffer cells (Connor et al., 1976). The spleen enlarges and may weigh about 500 g in acute cases and 1,000 g in chronic malaria (Connor et al., 1976) because of massive proliferation of splenic macrophages (Perrin et al., 1982). Kidneys may appear grayish and be slightly enlarged, the capillaries contain parasitized erythrocytes, and glomeruli contain deposits of malaria pigment (Connor et al., 1976). Tubular necrosis with hemoglobin casts in the renal tubules is seen in cases with malarial hemoglobinuria (Sun, 1982). The lungs become congested and edematous, with thickened alveolar walls (Connor et al., 1976). Portions of the gastrointestinal tract may become edematous and congested with focal or diffuse hemorrhages; massive involvement of the gastrointestinal tract may be related to algid malaria (Connor et al., 1976). The sinusoids of the placenta appear to provide a sequestered site for infected erythrocytes, and occasional overwhelming parasitism can lead to maternal or fetal death (Connor et al., 1976). Lymph nodes are pigmented and enlarged from histiocytic hyperplasia, and the bone marrow shows some pigmentation and hyperplasia of erythrocytic and leucocytic precursors (Connor et al., 1976). Finally, stasis of parasitized erythrocytes within ocular capillaries may cause retinal hemorrhages in patients with cerebral malaria (Looareesuwan et al., 1983).

Diagnostic Features

Malaria should be considered in the diagnosis of any febrile illness of unknown etiology in persons who have traveled to malarious areas, those who have recently received whole-blood transfusions, drug addicts, refugees, undocumented aliens, persons who live or work near airports, and employees at research facilities where malaria-infected mosquitoes are maintained. The familiar tertian or quartan paroxysms are not always present, especially in the early stages, and the presenting symptoms of falciparum malaria are notoriously variable.

Acute malaria has been mistaken for tuberculosis, cholecystitis, yellow fever, leptospirosis, appendicitis, influenza, urinary tract infections, and other unrelated illnesses. A definitive diagnosis must be based on the observation of parasite forms in infected erythrocytes; it is important that blood smears be read by someone who is sufficiently competent to recognize the different species, especially in mixed infections, because the identity of the species is an important consideration in the selecting specific chemotherapy.

Thick and thin smears should be prepared from the patient's blood immediately if malaria is suspected. Subsequent specimens should be taken at intervals between peaks of fever and at 6-h intervals until malaria is excluded from the diagnosis. The preparation and examination of blood films are detailed under Specimen Collection and Preparation. Serologic tests are better suited to epidemiologic studies than to individual diagnoses.

Prognosis

Malaria should be treated as a medical emergency because *P. falciparum* can cause a rapidly fatal illness and cerebral malaria has a mortality of 20% even in patients who receive proper care and therapy. The other malarias are seldom fatal, although their signs and symptoms can be severely debilitating. The nephrotic symdrome that may follow *P. malariae* infections does not respond to antimalarial therapy and may cause death.

Epidemiologic Aspects

Occurrence and Transmission

Before malaria control measures began in the 1950s, the disease claimed about 2.5 million lives each year (Bruce-Chwatt, 1985). This number had been decreasing rapidly until the 1970s, when the disease began a steady resurgence because of increasing costs of control, resistance of vectors to insecticides, and resistance of the plasmodia to antimalarial drugs. The situation is especially serious in tropical Africa, where about 200 million cases are estimated to occur annually (Bruce-Chwatt, 1985).

The present classification of malaria endemicity shown below reflects the impact of economic, social, and control measures (World Health Organization, 1986).

1. Areas where malaria never existed or disappeared following social and economic development (1,292 million persons).
2. Previously malarious areas where malaria has been eradicated by successful eradication efforts and by improved social and economic conditions (800 million persons).
3. Areas where low levels of transmission are maintained by continued antimalarial measures (2,117 million persons).
4. Areas where no control measures exist and where transmission levels are high (365 million persons, mainly in Africa).

Other means of transmission have been classified according to the following scheme (World Health Organization, 1963): autochthonous–indigenous, malaria acquired through the bite of a mosquito in an endemic area; autochthonous–introduced, malaria acquired through the bite of a mosquito in an area where malaria is not endemic; imported, malaria acquired outside a specific area; induced, malaria acquired by artificial means, such as blood transfusions; relapsing, recurrence of malaria after an interval greater than the normal periodicity of the paroxysms; and cryptic, an isolated case of malaria not associated with secondary cases.

Factors that favor the transmission of malaria are high ambient temperatures and humidity, biting preference of *Anopheles* vectors for humans, and suitable breeding sites near human habitations.

Imported Malaria

The affordability of air travel, the resurgence of malaria in many tropical countries, the movement of refugees from endemic areas, and the decreased efficacy of chloroquine prophylaxis (Lobel et al., 1985) have played a major part in the increasing numbers of cases of malaria diagnosed in the United States and Europe. In 1983, 803 cases in the United States were reported to the Centers for Disease Control; in 1984, the number increased to 1,016 cases. Most of the cases (632) in 1984 were in foreign civilians, and only 1 (a congenitally acquired infection) was acquired in the United States (Centers for Disease Control, 1985). *Plasmodium vivax* was identified in 56.1% of the cases, *P. falciparum* in 29.5%, *P. malariae* in 4.5%, *P. ovale* in 2.5%, mixed infections in 0.5%, and unidentified plasmodia in 6.8%.

Imported malaria poses an additional hazard to the community, as many areas free of malaria still have anopheline vectors. In 1974, a small outbreak of vivax malaria occurred in California's Sacramento Valley when *Anopheles freeborni* mosquitoes spread the disease from infected Punjabi immigrants to three Valley residents (Singal et al., 1977). Similar outbreaks originating from infected travelers or transported, infected mosquitoes have occurred around airports in Europe (Holvoet et al., 1982; Leger et al., 1981).

Susceptibility, Resistance, and Immunity

The plasmodia infecting humans have no respect for the age, sex, or race of their victims. Children born to mothers in endemic areas are probably protected during the first year of life by passively acquired antibodies, and adults living in holoendemic areas have slowly acquired a degree of immunity that protects them from disease but maintains parasitemias at very low levels.

Certain genetically determined physiologic states provide innate resistance to malaria; the most important of these is the sickle-cell trait prevalent in African populations. The homozygous expression of sickle-cell disease (HbS/S) is fatal early in life, but protection against falciparum malaria is afforded to heterozygotes with sickle-cell trait (HbS/A). The mechanism of this protection seems to be inhibition of parasite growth in erythrocytes containing hemoglobin S under conditions of low oxygen tension (Friedman, 1978) and the possible production of a toxic heme (Orjih et al., 1985). The high frequency of β-thalassemia in formerly malarious areas of southern Europe suggests that, as with sickle-cell trait, falciparum malaria may have been selective for β-thalassemia heterozygotes. Another form of protection may be a deficiency of the enzyme glucose 6-phosphate dehydrogenase; this deficiency is controlled by a sex-linked gene and is advantageous to heterozygote females, but not to homozygous males (Bienzle et al., 1979). The mechanism of protection is unclear, but glucose 6-phosphate dehydrogenase may be essential for the respiratory processes of the parasite. Finally, epidemiologic evidence is available that the absence of the Duffy blood group protects against vivax malaria (Matthews and Armstrong, 1981; Miller et al., 1976); the deficiency may deprive parasites of a surface receptor on the erythrocyte surface (Miller et al., 1976).

In areas where human populations are continuously exposed to malaria, children are protected at birth by transplacental transfer of immunoglobin (Ig)G antibodies. This passively acquired immunity is effective for the first few weeks of life, then begins to wane rapidly (Udezue, 1985), leaving children vulnerable to infection. After the first year of life, repeated infections provide the antigenic challenge to maintain active immunity, and this corresponds to rising titers of antibody that peak at 5 or 6 years of age (McGregor et al., 1965). Immunity to malaria affords protection against disease, but not infection; low levels of parasitemia are necessary for maintenance of that immunity (Collins et al., 1977) This consideration is important for previously immune persons who must live in nonmalarial countries for long periods and thus lose their immunity; students from tropical Africa frequently succumb to severe malaria on returning home after a year or two of study abroad. Hoffman et al. (1986) have shown that antibodies against *P. falciparum* sporozoites are detectable only after 10 years of exposure in an endemic area. Protective immunity against malaria includes the production of IgM, IgG, and IgA (McGregor, 1983); cell-mediated responses against extraerythrocytic schizonts (Jensen et al., 1983); and γ-interferon against extraerythrocytic liver schizonts (Nussenzweig and Nussenzweig, 1986).

The spleen plays an essential role in acquisition of immunity to malaria, but the mechanisms involved remain to be elucidated; splenectomy in an infected patient invariably aggravates the disease (Playfair, 1982).

The arsenal of host responses to malaria does not provide sterile immunity, and the mechanisms employed by the parasite to persist in the host are not known but may include antigenic variation (Brown, 1973), sequestration in a protected site such as the liver (McGregor, 1983), or suppression of host immune responses (Ho et al., 1986).

Etiologic Agents

Characterization

The agents of human malaria are protozoans in the genus *Plasmodium,* class *Sporozoa,* the members of which reproduce asexually and sexually to form oocysts, which in turn produce infective sporozoites. The *Sporozoa* belong to the phylum *Apicomplexa,* which is characterized by the possession of an apical complex consisting of organelles of attachment and penetration (Beaver et. al., 1984). It has been suggested that *P. falciparum* be placed in the subgenus *Laverania* because of its virulence, sequestration of mature erythrocytic forms in the visceral circulation, absence of persistent liver forms, and production of crescentic gametocytes, in contrast to the spherical gametocytes of the other *Plasmodium* species. The four *Plasmodium* species infecting humans show strict specificity for the anopheline mosquito as their definitive host and vector, and for humans as their intermediate hosts.

Life Cycle

The life cycles of the four species are essentially similar; on the life cycle of *P. vivax* is illustrated in Fig 1. The sporogonic or sexual cycle in the mosquito begins with the ingestion of microgametocytes and macrogametocytes (Greek *gamete,* wife; *gametes,* husband), with a blood meal. Within the stomach of the insect, the microgametocyte divides to form eight flagellated microgametes, and the macrogametocyte escapes its erythrocyte envelope to become a macrogamete. The microgamete attaches to a surface papilla on the macrogamete and fertilizes it to form a zygote (Greek *zygotos,* joined together). After a few hours, this zygote becomes a motile ookinete (Greek *oon,* egg; *kinesis,* motion), which penetrates the epithelium of the stomach to rest below the elastic membrane. It now becomes a sessile oocyst (Greek *kystis,* bladder) and grows and divides to produce approximately 1,000 sporozoites (Greek *sporos,* seed; *zoon,* animal). The oocyst ruptures and the sporozoites move through the hemocele to the salivary glands of the mosquito.

When the mosquito bites a human, the sporozoites are introduced with anticogulant saliva into the cutaneous capillaries. In about 30 min, sporozoites penetrate liver parenchyma cells and disappear from the bloodstream; there they develop to extraerythrocytic schizonts (Greek *schizein,* to divide). Division within the liver may result in the production of 10,000 merozoites (Greek *miros,* part), which are infective for erythrocytes. *Plasmodium vivax* and *P. ovale* have extraerythrocytic schizonts, which can remain dormant for varying periods within the liver and contribute to relapses; these forms have been named hypnozoites (Greek *hypnos,* sleep) by Krotoski (1985). *Plasmodium falciparum* and *P. malariae* do not have dormant liver stages, and all liver schizonts mature to form merozoites; thus, these two species do not cause relapsing infections, but may recrudesce from parasitemias persisting at undetectably low levels.

Merozoites are released into the bloodstream after rupture of parasitized hepatocytes, use their apical-complex organelles to attach to erythrocyte membranes, and enter the cells via invagination of the host cell membrane (Sherman, 1985; Wilson, 1982). Within the erythrocyte, merozoites assume a ringlike form and enter a phase of rapid growth as trophozoites (Greek *trophe,* food).

In 36 h, growth stops and the nuclear division begins; the parasite is now called a schizont. After the formation of about 16 nuclei, cytoplasm accumulates around each nucleus and a new crop of merozoites is born. The merozoites burst out of the erythrocytes and penetrate new cells to continue a cycle of erythrocytic schizogony. Some merozoites are influenced by unknown factors to develop into gametocytes in their host cells; these remain in the circulation until they are ingested by a feeding mosquito or are eventually eliminated by the immune system.

Infected erythrocytes typically contain golden-brown or black accumulations of pigment, which is the breakdown of product of hemoglobin on which the parasites feed. The pigment is properly called malaria pigment and not hemozoin (World Health Organization, 1963).

Specimen Collection and Preparation

A reliable diagnosis of malaria and identification of the causative species is obtainable only if blood smears are properly made and stained. Glass slides for preparing blood films should be absolutely free of grease, lint, fingerprints, or scratches. Even "precleaned" glass slides should be washed in soapy water, rinsed, and rewashed in 70 or 95% ethanol.

Blood should be obtained by finger puncture with a sterile disposable lancet. The palmar surface of the last digit of the middle or ring finger should be cleaned thoroughly with a gauze pad soaked with 70% isopropyl or ethyl alcohol, and the site should be dried with a clean, dry, gauze pad before sticking the finger. Blood collected at venipunctures can also be used, but is usually unsatisfactory because anticoagulants prevent proper adhesion of the blood film and may interfere with staining. Residual blood in the needle used for the venipuncture is satisfactory if it has not been contaminated with anticoagulant.

Blood films should be collected when a patient who has symptoms suggestive of malaria is admitted to the hospital, but the chances of detecting the infection are best when specimens are collected midway between fever spikes. If preliminary films are negative, subsequent examinations should be made at 6- or 12-h intervals for 36 h. Ideally, three films should be made: a thin film to be stained as soon as it dries; a thick film (made thinner than that recommended below under Thick Blood Films) to be stained after drying for 1 h; and a combination film to be stained after drying overnight. Specimens other than blood films that sometimes reveal infection when blood films are negative are bone marrow aspirates, intradermal fluid (Hoffman, 1986), and buffy-coat smears.

Identification

Thin Blood Films

Thin films are useful for identifying *Plasmodium* species, but lack the sensitivity of thick films. Films should be prepared as illustrated in Fig. 2. The thin, feathered edge of the film should be one cell thick and is the area to be examined for infected erythrocytes. Slides should be dried at room temperature before fixing and staining.

Thick Blood Films

Thick blood films are more useful for detecting malaria infections because the quantity of blood is much greater than in the thin film and the parasites are not obscured by erythrocytes. Identifying species may be more difficult than in thin films, and *P. ovale* is indistinguishable from *P. vivax*. The blood film should be 1.5 cm in diameter and sufficiently thin to read newsprint through (Fig. 2). The film can be made by touching the slide to a welling drop of blood and rotating the slide, without touching the skin, to form a circular smear of the required size and density.

Alternatively, two or three drops of blood can be applied to the slide in the same manner, but the smear can be made with an applicator stick or the corner of a clean slide (Fig. 2). Slides should be dried flat at room temperature for 8 to 12 h. In emergencies, slides can be stained after 1 h, but the central

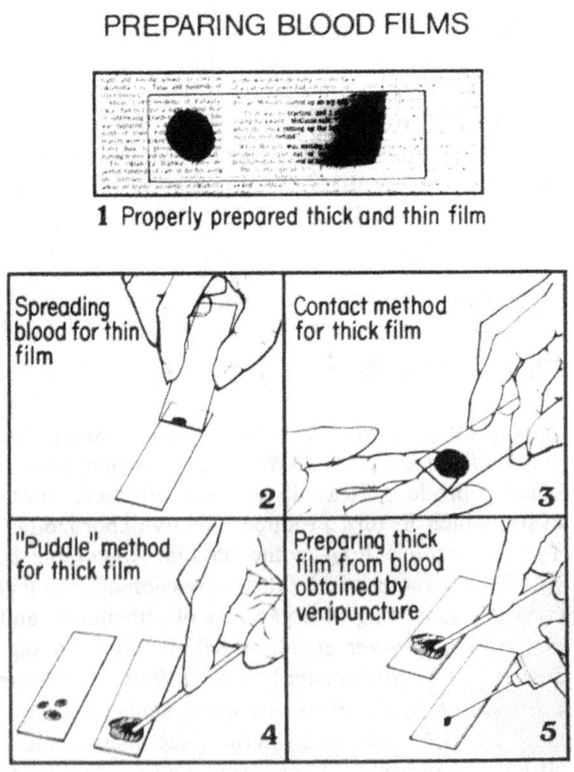

FIG. 2. Preparation of thin and thick blood films. (Reprinted with permission from the Centers for Disease Control, Parasitology Section, Training and Laboratory Program Office, Atlanta, Georgia.)

portion of the smear may fall off. Thick films should not usually be heat since, erythrocytes will be fixed and will not lyse during staining, but gentle warming of about 60° in a dry incubator or on a heat block for 1 min can produce adequate smears. Thick films that have been quick-dried will dehemoglobinize, but erythrocyte membranes may remain relatively intact. For this reason, the drying of thick films should not be hurried except in emergencies.

Combination films are convenient their but preparation requires some care. Preferably, the thin film should be made first, and the thick film should be applied in the remaining space away from the feathered end of the thin smear. Allow sufficient space between the two specimens so that the thick film is not affected by the methyl alcohol used for fixing the thin film.

Staining Blood Films

The following procedure is that used by the Centers for Disease Control.

Staining Solutions

1. Staining solutions
   Giemsa stock stain (available commercially)
2. Stock alkaline buffer—disodium hydrogen phosphate, 0.067 M solution

Na_2HPO_4 (anhydrous)	9.5 g
Distilled water	1,000 ml

   Dissolve Na_2HPO_4 in a small volume of water and add remaining water to make 1 liter.
3. Stock acid buffer—Sodium dihydrogen phosphase, 0.067 M solution

NaH_2PO_4 H_2O	9.2 g

   Prepare as for the alkaline buffer. These stock solutions should be kept in separate glass-stoppered bottles and will last for several months.
4. Buffered water

Alkaline buffer	61 ml
Acid buffer	39 ml
Distilled water	900 ml
10% aqueous solution of Triton X-100	1 ml

Triton X-100 is a nonionic surfactant that enhances the appearance of Schüffner dots and gives cleaner and brighter preparations; its use is optional. The pH of the solution should be 7.0 to 7.2; if it is outside this range, it can be adjusted with the appropriate acid or alkaline buffer, but it may be better to make a new solution. Slides that have been stained in a buffer of incorrect pH will appear too blue or red and should be restained in a buffer of neutral pH.

Staining Procedure

1. Thin films
   (a) Fix blood film by dipping briefly into absolute methanol or by pouring methanol over the smear. Be careful not to fix the thick film in a combination smear.
   (b) Allow smear to drip dry.
   (c) Stain in a 1:20 dilution of Giemsa in neutral buffered water for 20 min.
   (d) Wash briefly by dipping in neutral buffered water.
   (e) Allow to drip dry in a vertical position.
2. Thick films
   (a) Allow the film to dry for at least 8 h, if possible.
   (b) Stain in a 1:50 dilution of Giemsa in neutral buffered water for 45 to 50 min. The thick smear must *not* be fixed in methanol.
   (c) Wash in buffered water for 3 to 5 min.
   (d) Allow to drip dry in a vertical position.

When staining combination films, care must be taken to avoid fixation of the thick film by fumes or droplets of methanol. The combination should be stained in a 1 : 50 dilution of Giemsa for 45 to 50 min. A Coplin jar is satisfactory for use in staining small numbers of slides. A single slide can be stained by placing it in an inverted position in a petri dish or small tray with a thin glass rod or applicator stick to keep the surface off the bottom of the container. Stain can then be gently squirted under the slide with a syringe or pipette.

The blood films should be allowed to dry thoroughly before they are examined directly with the 100× oil-immersion lens, or they can be mounted under a glass cover slip or coated with a medium such a Flo-Texx, which does not require a cover slip; plastic cover slips are of irregular thickness and should not be used.

Slides should be examined at 100× with an oil-immersion lens. At least 100 fields of the thick smear should be examined, or it should be examined in its entirety if time permits. The appearance of the thick film is confusing to novices, but it is about 20 times as sensitive as the thin film and can be examined more rapidly; it is also essential for estimating parasite counts. The thin film should be examined for 30 min, and scanning should be confined to the feathered end of the smear where the erythrocytes are separated, in a layer one cell thick and not overlapping (Wilcox, 1960).

The principal causes of confusion in a stained preparation are granules of stain, which can resemble chromatin dots or nuclei of parasites; fibrin strands, which resemble cytoplasm of parasites; and platelets, which can be mistaken for trophozoites. Other causes of confusion are fungal spores and hy-

TABLE 1. Comparison of four species of human malaria[a]

Species	Length of asexual cycle	Stages found in circulating blood	Appearance of red blood cells: Size	Appearance of red blood cells: Stippling	Appearance of parasites: Cytoplasm	Appearance of parasites: Pigment	Appearance of parasites: No. of merozoites
Plasmodium vivax	41–47 h, usually 43–45 h; for practical purposes, cycle is considered to be 48 h	All (trophozoites, schizonts, gametocytes); wide range on given smear	Enlarged; maximum size may be 1.5–2 times normal size	+ (Schüffner dots)	Ameboid trophozoites, light blue; has irregular, "spread out" appearance in trophozoite stage	Golden brown, inconspicuous	12–24 (average 16)
Plasmodium malariae	72 h	All (trophozoites, schizonts, few gametocytes); few rings usually seen; does not stay in ring stage very long	Normal	(Ziemann dots rarely seen)	Rounded, compact trophozoites; dark blue, dense cytoplasm; band-form trophozoite occasionally seen	Dark brown, coarse, conspicuous	6–12 (average 8); "rosette" occasionally
Plasmodium ovale	49–50 h	All (trophozoites, schizonts, gametocytes)	Enlarged; maximum size may be 1.25–1.5 times normal size	+ (Schüffner dots)	Rounded, compact trophozoites; dark to medium blue, usually dense; chromatin, large	Dark brown, conspicuous	6–14 (average 8)
Plasmodium falciparum	Approximately 48 h	Rings, gametocytes; other stages develop in internal organs	Normal	(Maurer dots or clefts occasionally seen, but infrequent)	Double dots in rings common; rings often small and delicate	Black, coarse and conspicuous in gametocytes	6–32 (average 20–24)

[a] Adapted from the Centers for Disease Control, Parasitology Training Section, Training and Laboratory Program Office.

phae, bacteria, and protozoa. Failure to clean the skin properly before drawing blood may be the chief cause of contaminated thick films (Wilcox, 1960). In both thick and thin Giemsa-stained films, the cytoplasm of all parasitic forms stains blue, the nuclei stain dark red or purple, and pigment granules stain golden brown or black.

Estimation of parasitemia requires a thick film and is essential in determining the efficacy of chemotherapy and the severity of disease. The formula most conveniently used requires a white blood cell (WBC) count and is expressed as:

No. of parasites per mm^3 of blood =

$$\frac{\text{No. of parasites counted in the same fields as 100 WBC} \times \text{WBCs per mm}^3}{\text{WBCs counted}}$$

Parasite counts should be obtained before and after chemotherapy, and the information should be given to the patient's physician. Mixed infections present a special problem because there is a natural tendency for the microscopist to identify "*the* species" of malaria parasites seen. In mixed infections, one species usually predominates; *P. falciparum* may overwhelm *P. vivax,* and the latter may obscure *P. malariae*. Accurate diagnosis of a mixed infection must be based on the observation of characteristic forms of the species involved; for example, the crescentic gametocytes of *P. falciparum,* the large ameboid trophozoites and irregular schizonts of *P. vivax,* the oval or fimbriated appearance of erythrocytes parasitized by *P. ovale,* and the rosette-shaped schizonts and band forms of *P. malariae*. If the species of *Plasmodium* cannot be determined, the report should state: malaria, species unidentified; further specimens should be examined or help should be sought in identification. Malaria is a reportable disease; diagnoses and slides should be sent to state health laboratories for confirmation and recording.

One should remember that other parasites such as trypanosomes *Babesia* sp. microfilariae, and spirochetes of *Borrelia* sp. may be detected in blood films.

Species Identification

The following guides are a useful adjunct to the color atlases available in reference texts such as the *Manual of Clinical Microbiology* (Krogstad et al., 1985). For consistently accurate diagnosis of malaria, no visual aid can match experience and practice. Skillful technologists may actually find atlases unnecessary. A summary of characteristics for the four species of *Plasmodium* is given in Table 1. A common source of error in microscopic identification is the expectation of seeing Schüffner stippling in *P. vivax* infected erythrocytes. These dots may be absent even in properly stained films. Erythrocytes infected by *P. falciparum* occasionally have Maurer clefts, which resemble Schüffner stippling, but the former are relatively sparse and occur in erythrocytes that are not enlarged.

Chemotherapy alters the morphology of malaria parasites; trophozoites may appear ragged or fragmented with clumped pigment, the pigment of gametocytes may become clumped or extruded, and nuclei or cytoplasm may show degeneration.

Guide to Identification of Malaria in Thin Blood Films

In examining thin blood films for malaria, you must look at the infected red blood cells (RBC) and the parasites inside the cells.

I. Infected Red Blood Cells

(1) Look at the size of the infected red blood cells.
(2) Is Schüffner stippling present or not present?

- Infected red blood cells
  - Enlarged; Schüffner stippling present*
    - RBC twice normal size; most RBC round, only rare RBC oval or elongated — Suggests *P. vivax;* look at parasites
    - RBC 1.25 to 1.5 times normal size; 20% oval and/or fimbriated — Suggests *P. ovale;* look at parasites
  - Normal; Schüffner stippling not present — *P. malariae* (possibly *P. falciparum*)

* In poorly stained films, Schüffner stippling may not be visible.

II. Parasites

Rings of all four species may look alike. If you see rings, look for older stages. In *P. falciparum* malaria, only rings are usually seen; older stages are present only in severe infections. Therefore, if you see only rings and no older stages, the species is probably *P. falciparum*.

- Rings*
  - Small, delicate, one or two chromatin dots, about 1 to 1.5 μm in diameter, two or more rings in one RBC; no older stages present — *P. falciparum*
  - 1.5 to 2 μm in diameter, 1 or 2 chromatin dots — Look for other stages

* The containing rings are usually not enlarged, and Schüffner stippling may not be seen. Look for older stages to determine species.

- Growing Trophozoites
  - Ameboid (spread-out, irregular outline); pigment not distinct; RBC enlarged; Schüffner stippling present* — *P. vivax*
  - Compact (outline regular), pigment dark brown
    - RBC enlarged, many oval and/or fimbriated; Schüffner stippling present — *P. ovale*
    - RBC normal in size; no Schüffner stippling present — *P. malariae*

* In poorly stained films, Schüffner stippling may not be visible.

- Mature Trophozoites
  - Rounded and compact, almost fills up RBC; pigment golden brown, may not be seen; RBC enlarged (may be 1.5 to 2 times normal size); Schüffner stippling present — *P. vivax*
  - Compact; pigment dark brown; almost fills up RBC; RBC normal in size; no Schüffner stippling present — *P. malariae*
  - Compact; pigment dark brown; may not fill up RBC; RBC enlarged; many oval and/or fimbriated; Schüffner stippling present — *P. ovale*

- Immature Schizonts
  - Compact, with 2 to 24 chromatin masses; pigment golden brown (may not be seen); infected RBC enlarged; Schüffner stippling present — *P. vivax*
  - Compact with 2 to 12 chromatin masses; pigment dark brown
    - Infected RBC oval and/or fimbriated; Schüffner stippling present (may have 2 to 14 chromatin masses) — *P. ovale*
    - Infected RBC normal size; no Schüffner stippling present — *P. malariae*

- Mature Schizonts
  - 12 to 24 merozoites inside enlarged RBC: Schüffner stippling present — *P. vivax*
  - 6 to 14 merozoites inside enlarged RBC; Schüffner stippling present; RBC may be oval and/or fimbriated — *P. ovale*
  - 6 to 12 merozoites inside normal size RBC; no Schüffner stippling present — *P. malariae*

- Gametocytes
  - Large, rounded parasite, fills up RBC; RBC enlarged; Schüffner stippling present
    - RBC rounded; pigment in parasite golden brown — *P. vivax*
    - RBC oval and/or fimbriated; pigment dark brown — *P. ovale*
  - Parasite fills up RBC; pigment dark brown; RBC normal in size; no Schüffner stippling present — *P. malariae*
  - Parasite crescent or elongate — *P. falciparum*

Serodiagnosis

Serologic tests that have been investigated or applied for the diagnosis of malaria by antibody detection include the indirect fluorescent-antibody test (IFA) test, indirect hemagglutination (IHA) test, enzyme-linked immunosorbent assay (ELISA), agar–gel diffusion test, and radioimmunoasssy (RIA). Tests for detection of malaria antigens include the RIA, the ELISA, and investigational DNA probes.

No serologic test can yet match the sensitivity and specificity of direct examination of stained slides in subjects with patent parasitemias. The Centers for Disease Control place strict guidelines on use of the IFA as a diagnostic tool in individual patients: serum specimens are tested only 1) if a patient with a febrile illness is strongly suspected of having malaria despite blood films repeatedly negative for parasites 2) if they are from the donor of a blood transfusion that preceded malaria in a patient; or 3) if they are to be used for research projects and for standardizing control sera for IFA protocols.

The test of choice is the IFA, since the IHA and ELISA are less sensitive in detecting early infections and in diagnosing infections in children less than 5 years old. The IFA detects antibodies produced simultaneously with parasitemia, but the test may remain positive in patients with *P. vivax* or *P. ovale* infections for years after cure, even as long as 15 years after an infection with P. *malariae* (Ambroise-Thomas, 1974). In *P. falciparum* infections, antibody titers fall rapidly 6 months to a year after cure; this finding makes the test useful in detecting asymptomatic infections or recrudescences (Voller and Draper, 1982). The IFA has also been valuable in diagnosing malaria in symptomatic patients who do not have a parasitemia, and in asymptomatic source donors in transfusion malaria.

The IFA test performed by the Malaria Branch of the Centers for Disease Control uses slides coated with lysed thick blood films of the four human plasmodia. The preparations must contain schizonts, as these are the most reactive stages of *P. falciparum* obtained from culture, whereas *P. vivax, P. ovale,* and *P. malariae* are obtained from infected chimpanzees, *Aotus* and *Ateles* monkeys, and infected human patients (Dr. Alex Sulzer, Centers for Disease Control; personal communication). Dilutions of a test serum are tested against all four species, and a fluorescein-conjugated second antibody specific for human immunoglobulin G is added. A positive specimen is one determined to react at a titer of 1:64 or greater. Antibodies among the various humans malarias cross-react, but higher titers are obtained between homologous species and antisera.

Treatment, Prevention, and Control

Treatment and Prophylaxis

Chemotherapy of malaria is complicated and evolving because the human malarias are becoming increasingly resistant to drugs. Guidelines for treatment are detailed in the *Medical Letter* (1986), in a World Health Organization supplement (1986), in a comprehensive review by Hoffman (1986), and in publications from the Centers for Disease Control, Atlanta, Georgia. Factors that will determine the choice of antimalarial agents are the clinical condition of the patient, the expected drug sensitivity of the parasite, the tolerance of the patient to specific drugs, and the immune status of the patient to malaria (Hoffman, 1986). Table 2 outlines of the drugs of choice for various presentations of malaria and the stage of the malaria parasites affected by the drug.

As with antimalarial chemotherapy, prophylaxis must be tailored to meet host and environmental factors such as the risk of exposure, sensitivity of the disease to chloroquine, pregnancy, age, availability of medical care, and length of stay. The presence of malaria in a country should not be the sole criterion for prophylaxis, as the disease may pose a risk only in rural areas or during certain seasons. Guidelines for chemoprophylaxis have been provided by Lobel and Campbell (1986) and in the publication *Health Information for International Travel 1986.* (Centers for Disease Control, 1986). Additional information can be obtained from the Malaria Branch of the Centers for Disease Control.

Prophylaxis against malaria should not rely exclusively on chemotherapy, which in fact does not guarantee safety, but should include some of the measures mentioned below (under Prevention and Control), such as the use of insect repellents, insecticide treatment of residences, and protective clothing. Nonimmune travelers who may be exposed to chloroquine-sensitive strains should use chloroquine for prophylaxis.

The treatment of severe complicated falciparum malaria may require measures in addition to chemotherapy; exchange blood transfusions may be necessary, especially if parasitemia is greater than 10% (Peters and Hall, 1985).

Pregnancy places a woman and her fetus at a high risk of severe or fatal disease from malaria. Prompt treatment with chloroquine, quinine, and sulfadoxine–pyrimethamine should be given to patients with uncomplicated disease (Hoffman, 1986). Patients with complicated malaria should be treated with quinine; the risk of inducing abortions with quinine is not as great as formerly believed (Looareesuwan et

TABLE 2. Antimalarial drugs

Disease	Drug of choice	Stage affected	Alternative
Severe, complicated *P. falciparum* infection	Quinine dihydrochloride (i.v. infusion)[a]	Erythrocytic state except *P. falciparum;* asexual stages; gametocytes	Quinidine gluconate (i.v.), chloroquine (i.v.)
Uncomplicated chloroquine-sensitive *P. falciparum* infection	Chloroquine (oral)	Asexual stages	
Uncomplicated chloroquine-resistant *P. falciparum* infection	Sulfadoxine–pyrimethamine (Fansidar, oral)	Asexual stages	Amodiaquine, quinine sulfate, mefloquine, tetracycline, doxycycline, clindamycin
Severe *P. vivax, P. ovale,* and *P. malariae* infections	Same as for *P. falciparum* + primaquine (oral)[b]	Liver stages, trophozoites of *P. vivax,* gametocytes	
Uncomplicated *P. vivax, P. ovale,* and *P. malariae* infections	Chloroquine + primaquine (oral)[b]		Amodiaquine, quinine, mefloquine, Fansidar

[a] i.v. = Intravenous.
[b] Primaquine is essential for eradication of dormant liver stages of *P. vivax* and *P. ovale.*

al., 1985). Primaquine and tetracyclines are potentially harmful to the fetus and should not be used to treat pregnant women. If the malarial infections involve *P. vivax* or *P. ovale,* primaquine should be administered to the mother after delivery, for elimination of hypnozoites.

Prevention and Control

The goal of eradicating malaria worldwide has been abandoned, and the present hope is to reduce transmission to acceptable levels (Davidson, 1982). Classic methods of control encompass measures to prevent mosquitoes from feeding on humans; to destroy adult mosquitoes and mosquito larvae; to alter or eliminate breeding places; and to treat cases of malaria (Russell, 1952). The aim of these measures is to reduce the capacity of the vectors to transmit malaria and to reduce the reservoir of infection in the human population. A comprehensive list of the principles involved has been presented by Bruce-Chwatt (1985): For individual protection, prevention and control measures include the use of mosquito repellents, bed nets, house screening, house location, pyrethrum house spray, and antimosquito fumigants. For vector control, measures include environmental modification, chemical and biologic larvicides, insecticide space spraying, and residual insecticide spraying. Antiplasmodial measures include treatment of acute cases of malaria; prophylaxis and suppression of infection; treatment of relapses; and, it is hoped, vaccination against malaria.

These measures do not constitute an inflexible formula to be used in all situations in which malaria occurs; factors such as rainfall, geography, culture, and economics are some of the influences that must decide the choice of control measures.

One hope for control of malaria now lies in the promise of a vaccine against *P. falciparum;* attention is presently focused on an antisporozoite vaccine (Nussenzweig and Nussenzweig, 1986). This vaccine elicits the production of antibodies against a protein in the sporozoite surface membrane, the circumsporozoite protein. The gene responsible for the production has been identified and cloned, and the structure of the circumsporozoite protein has been determined. Another target for vaccination is the ring-infected erythrocyte surface antigen of *P. falciparum.* This antigen seems to aid penetration of the merozoite into the erythrocyte, and antibodies against the antigen inhibit parasite growth in vitro (Collins et al., 1986). The major problem foreseen for an estimate vaccine is antigenic variation within species and geographic strains and the large diversity of merozoite antigens (Ravetch et al., 1985).

Monitoring of drug resistance should be a component of malaria control because a successful antimalarial drug policy requires information on the development and progression of drug-resistant falciparum malaria (Spencer, 1985). Resistance can be measured in terms of patient response to therapy and can be expressed as RI, in which the disease recrudesces shortly after the disappearance of parasitemia following treatment; RII, in which the parasitemia is reduced but not cleared; and RIII, in which no reduction of parasitemia is achieved. Drug resistance can also be measured by an in vitro test using blood from an infected patient (Rieckman et al., 1978).

Literature Cited

Ambroise-Thomas, P. 1974. La reaction d'immunofluorescence dans l'etude sero-immunologique du paludisme. Bull. W.H.O. **50:**267–276.

Beaver, P. C., R. C. Jung, and E. W. Cupp. 1984. Malaria parasites and piroplasms, p 174–205. *In* Clinical parasitology, 9th ed. Lea & Febiger, Philadelphia.

Bienzle, U., I. Guggenmoos-Holzmann, and L. Luzzatto. 1979. Malaria and erythrocyte glucose-6-phosphate dehydrogenase variants in West Africa. Am. J. Trop. Med. Hyg. **28:**619–621.

Bray, R. S. 1963. The exoerythrocytic phase of malaria parasites. Int. Rev. Trop. Med. **2:**41–74.

Brown, K. N. 1973. Antibody-induced variation in malaria parasites. Nature (London) **242:**49–50.

Bruce-Chwatt, L. J. 1985. Essential malariology, p. 55–266. John Wiley & Sons, Inc., New York.

Centers for Disease Control. 1985. Malaria surveillance annual summary, 1984. Centers for Disease Control, Atlanta.

Centers for Disease Control. 1986. Health information for international travel 1986. U.S. Department of Health and Human Services publication no. (CDC) 86-8280. Centers for Disease Control, Atlanta.

Collins, W. E., R. F. Anders, M. Pappaioanou, G. H. Campbell, G. V. Brown, D. J. Kemp, R. L. Coppel, J. C. Skinner, P. M. Andrysiak, J. M. Favaloro, L. M. Corcoran, and J. R. Broderson. 1986. Immunization of *Aotus* monkeys with recombinant proteins of an erythrocyte surface antigen of *PLasmodium falciparum*. Nature (London) **323:**259–262.

Collins, W. E., R. A. Cedillos, and M. Warren. 1977. Seroepidemiology of malaria in middle America. Am. J. Trop. Med. Hyg. **26:**1105–1107.

Connor, D. H., R. C. Neafie, and W. T. Hockmeyer. 1976. Malaria, p. 273–283. *In* C. H. Binford and D. H. Connor (ed.), Pathology of tropical and extraordinary diseases. Armed Forces Institute of Pathology, Washington, D.C.

Davidson, G. 1982. Developments in malaria vector control. Br. Med. Bull. **38:**201–206.

Friedman, M. J. 1978. Erythrocyte mechanism of sickle cell resistance to malaria. Proc. Natl. Acad. Sci. USA **75:**1994–1997. Guerrero, I. C., B. C. Weniger, and M. G. Schultz. 1983. Transfusion malaria in the United States, 1972–1981. Ann. Intern. Med. **99:**221–226.

Hendrickse, R. G., A. Adeniyi, and G. M. Edington. 1972. Quartan malarial nephrotic syndrome. Collaborative clinicopathological study in Nigerian children. Lancet **1:**1143–1148.

Ho, M., H. K. Webster, S. Looareesuwan, W. Supanaranond, R. E. Phillips, P. Chanthavanich, and D. A. Warrell. 1986. Antigen-specific immunosuppression in human malaria to *Plasmodium falciparum*. J. Infect. Dis. **153:**763–771.

Hoffman, S. L. 1986. Treatment of malaria. Clin. Trop. Med. Commun. Dis. **1:**171–224.

Hoffman, S. L., R. Wistar, Jr., W. R. Ballou, M. R. Hollingdale, R. A. Wirtz, I. Schneider, H. A. Marwoto, and W. T. Hockmeyer. 1986. Immunity to malaria and naturally acquired antibodies to the circumsporozoite protein of *Plasmodium falciparum*. N. Engl. J. Med. **315:**601–606.

Holvoet, G., P. Michielsen, and J. Vandepitte. 1982. Airport malaria in Belgium. Lancet **2:**881–882.

Jensen, J. B., M. T. Boland, J. S. Allan, J. M. Carlin, J. A. Vande Waa, A. A. Divo, and M. A. S. Akood. 1983. Association between human serum-induced crisis forms in cultured *Plasmodium falciparum* and clinical immunity to malaria in Sudan. Infect. Immun. **41:**1302–1311.

Krogstad, D. J., G. S. Visvesvara, K. W. Walls, and J. W. Smith. 1985. Blood and tissue protozoa, *In* E. H. Lenette, A. Balows, W. J. Hausler, and H. J. Shadomy (ed.), Manual of clinical microbiology, 4th ed. American Society for Microbiology, Washington, D.C. 612–630.

Krotoski, W. A. 1985. Discovery of the hypnozoite and a new theory of malarial relapse. R. Soc. Trop. Med. Hyg. **79:**1–11.

Leger, N., B. Pesson, M. Bruaire, R. Cassaigne, G. Ferrand, R. Van Damme, and J. Timbal. 1981. Le paludisme des aeroports. Med. Trop. **41:**431–441.

Lobel, H. O., and C. C. Campbell. 1986. Malaria prophylaxis and distribution of drug resistance. Clin. Trop. Med. Commun. Dis. **1:**171–224.

Lobel, H. O., C. C. Campbell, I. K. Schwartz, and J. M. Roberts. 1985. Recent trends in the importation of malaria caused by *Plasmodium falciparum* into the United States from Africa. J. Infect. Dis. **152:**613–617.

Looareesuwan, S., D. A. Warrell, N. J. White, P. Chanthavanich, M. J. Warrell, S. Chantaratherakitti, S. Changswek, L. Chongmankongcheea, and C. Kanchanaranya. 1983. Retinal hemorrhage, a common sign of prognostic significance in cerebral malaria. Am. J. Trop. Med. Hyg. **32:**911–915.

Madecki, O., and W. Kretschmar. 1966. Frequency and causes of malaria in new-born babies in West Africa. Z. Tropenmed. Parasitol. **17:**195–210.

Manson-Bahr, P. E. C., and F. I. C. Apted. 1982. Malaria and babesiosis, p. 38–71. *In* Manson's tropical diseases. Balliere-Tindall, London.

Matthews, H. M., and J. C. Armstrong. 1981. Duffy blood types and vivax malaria in Ethiopia. Am. J. Trop. Med. Hyg. **30:**299–303.

McGregor, I. A., K. Williams, A. Voller, and W. Z. Billewicz. 1965. Immunofluorescence and the measurement of immune response to hyperendemic malaria. Trans. Roy. Soc. Trop. Med. Hyg. **59:**395–414.

McGregor, I. A. 1983. Current concepts concerning man's resistance to infection with malaria. Bull. Soc. Pathol. Exot. **76:**433–445.

McGregor, I. A. 1984. Epidemiology, malaria and pregnancy. Am. J. Trop. Med. Hyg. **33:**517–525.

Medical Letter. 1986. Drugs for parasitic infections. **28:**9–18.

Miller, L. H. 1984. Malaria, p. 223–239. *In* K. S. Warren and A. A. F. Mahmoud (ed.), Tropical and geographic medicine. McGraw-Hill Book Co., New York.

Miller, L. H., S. J. Mason, D. F. Clyde, and M. H. McGinniss. 1976. The resistance factor to *Plasmodium vivax* in blacks. The Duffy-blood-group genotype Fy Fy. N. Engl. J. Med. **295:**302–304.

Most, H. 1968. Manhattan: "a tropic isle?" Am. J. Trop. Med. Hyg. **17:**333–354.

Nussenzweig, V., and R. S. Nussenzweig. 1986. Development of a sporozoite malaria vaccine. Am. J. Trop. Med. Hyg. **35:**678–688.

Orjih, A. U., R. Chevli, and C. D. Fitch. 1985. Toxic heme in sickle cells: an explanation for death of malaria parasites. Am. J. Trop. Med. Hyg. **34:**223–227.

Perrin, L. H., L. J. Mackey, and P. A. Miescher. 1982. Hematology of malaria in man. Semin. Hematol. **19:**70–82.

Peters, W., and A. P. Hall. 1985. Treatment of severe falciparum malaria. Br. Med. J. **291**:1146–1147.

Playfair, J. H. L. 1982. Immunity to malaria. Br. Med. Bull. **38**:153–159.

Ravetch, J. V., J. Young, and G. Poste. 1985. Molecular genetic strategies for the development of antimalarial vaccines. Bio. Technol. **3**:729–740.

Rieckman, K. H., G. H. Campbell, L. J. Sax, and J. E. Mrena. 1978. Drug sensitivity of *P. falciparum:* an *in vitro* technique. Lancet **1**:22–23.

Russell, P. F. 1952. Malaria. Basic principles briefly stated, p. 123. Blackwell Scientific Publications, Ltd., Oxford.

Sherman, I. W. 1985. Membrane structure and function of malaria parasites and the infected erythrocyte. Parasitology **91**:609–645.

Singal, M., P. K. Shaw, R. C. Lindsay, and R. R. Roberto. 1977. An outbreak of introduced malaria in California possibly involving secondary transmission. Am. J. Trop. Med. Hyg. **26**:1–9.

Spencer, H. C. 1985. Drug-resistant malaria—changing patterns mean difficult decisions. Trans. R. Soc. Trop. Med. Hyg. **79**:748–758.

Sun, T. 1982. Malaria, p. 29–39. *In* Pathology and clinical features of parasitic diseases. Masson Publishing USA, New York.

Udezue, E. O. 1985. Persistence of malarial antibody in Nigerian children born in the United Kingdom and its clinical relevance. Trans. R. Soc. Trop. Med. Hyg. **79**:427–429.

Voller, A., C. C. Draper. 1982. Immunodiagnosis and seroepidemiology of malaria. Br. Med. Bull. **38**:173–177.

Weatherall, D. J., and S. Abdalla. 1982. The anaemia of *Plasmodium falciparum* malaria. Br. Med. Bull. **38**:147–151.

Wilcox, A. 1960. Manual for the microscopical diagnosis of malaria in man. Public Health Service, U.S. Department of Health, Education, and Welfare, Washington, D.C.

Wilson, R. J. 1982. How the malaria parasite enters the red blood cell. Nature (London) **295**:368–369.

World Health Organization. 1963. Terminology of malaria and malaria eradication. World Health Organization, Geneva.

World Health Organization. 1986. Severe and complicated malaria. Trans. R. Soc. Trop. Med. Hyg. (Suppl.) **80**:1–50.

CHAPTER 98

Microsporidiosis

ANN CALI and ROBERT L. OWEN

Disease: Microsporidiosis (Nosematosis and Encephalitozoonosis).
Etiologic Agents: *Nosema connori, Enterocytozoon bieneusi, Nosema* or nosema-like organisms, *Pleistophora* sp., and *Encephalitozoon cuniculi.*
Source: Unknown.
Clinical Manifestations: *Nosema connori* and *Enterocytozoon bieneusi* have been associated with diarrhea and malabsorption. *Nosema* or nosema-like organisms in the eye have caused pain and blindness. *Pleistophora* in the skeletal muscles caused muscle weakness and contractures, and *Encephalitozoon cuniculi* in the liver caused hepatitis.
Pathology: Intracellular protozoans cause hypertrophy or rupture of host cells, with granulomatous foci in muscle, liver, and eye. In intestine, results in malabsorption, diminution of villi, and elongation of crypts; inflammatory cells in lamina propria increase.
Laboratory Diagnosis: Light microscopic demonstration of spore with PAS+ granule in anterior end or electron microscopic demonstration of polar tubule inside spore.
Epidemiology: Probably worldwide, North America, France, Africa, and Ceylon. Associated with immunocompromised tissues or individuals.
Treatment: Unknown.
Prevention and Control: Unknown.

Description of Disease

Microsporidiosis is an infection with an obligate intracellular protozoan parasite of the phylum *Microspora*. Because there are hundreds of *Microsporida*, the nature of the disease is variable. The host range is extremely wide, extending from invertebrates, especially insects, to vertebrates, particularly fish and mammals. In humans, four genera of *Microsporida* have been demonstrated: *Nosema,* generally found in insects; *Pleistophora,* a pathogen of fish and insects; *Encephalitozoon,* diagnosed in many mammals; and *Enterocytozoon,* identified only from the intestinal enterocytes of humans.

Those organisms infecting the intestinal tract (*Nosema connori* and *Enterocytozoon bieneusi*) have been associated with diarrhea and malabsorption. *Nosema* or nosema-like organisms in the eye have caused pain and blindness. *Pleistophora* in the skeletal muscles caused muscle weakness and contractures, and *Encephalitozoon cuniculi* in the liver caused hepatitis.

Overview of Disease

Clinically apparent infections with the *Microsporida* have been identified thus far only in immunocompromised patients or in locations, such as the central nervous system and eye, that are somewhat separated from immune effector mechanisms by diffusion barriers. Although it is possible that microsporidian infections are only opportunistic in humans, it would be remarkable if this were so considering the ubiquitous occurrence of the *Microsporida* in immunologically normal animals throughout the animal kingdom. It seems more likely that only overwhelming and sustained infections in immunocompromised individuals are being identified because of the insensitivity of our current tools for identifying microsporidian infections. With an increased awareness of the *Microsporida* and development of serologic, microscopic, and tissue culture methodologies for their ready identification, acute self-limited infection and chronic subclinical infection with *Microsporida* will no doubt be diagnosed with increasing frequency. In this regard, infection with the *Microsporida* appears to parallel medical experience with cryptosporidia, which is also an obligate intracellular protozoan parasite and, until 1980, also was identified only in immunocompromised humans. With the development of widespread facilities for laboratory identification, cryptosporidia is now recognized as a cause of acute enteric disease in humans, paralleling the range of clinical conditions observed with cryptosporidia in veterinary practice (Current and Owen, in press). Because of this similarity in spectrum and severity of human cryptosporidia infection to that observed in veterinary infections, it is anticipated that review of experience with microsporidian infection in various animal species may be of similar use in guiding investigation of the variety of clinical syndromes associated with human microsporidian infections.

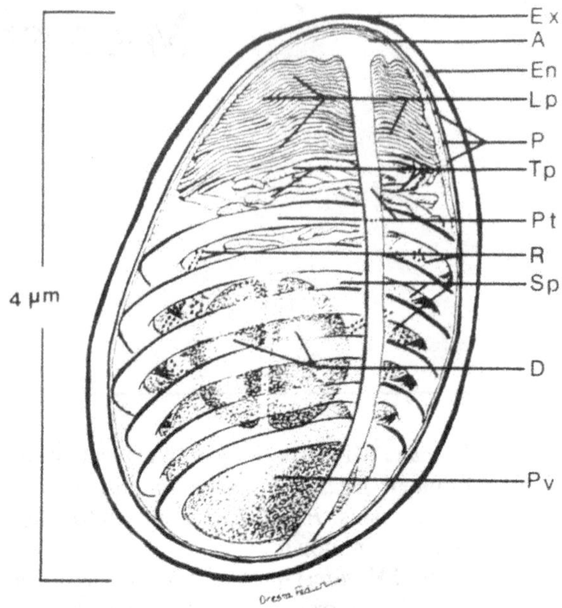

FIG. 1. Diagram of the internal structure of a microsporidian spore. The spore coat has an outer electron-dense region called the exospore (Ex) and an inner thicker electron-luscent region, endospore (En). A unit membrane (P) separates the spore coat from the spore contents. The extrusion apparatus [anchoring disc (A), polar tubule (Pt), lamellar polaroplast (lp), and tubular polaroplast (Tp)] dominates the spore contents and is diagnostic for microsporidian identification. The posterior vacuole (Pv) is a membrane-bound vesicle that sometimes contains a "membrane whirl" or "glomerular-like" structure or flocculent material or some combination of these structures. The spore cytoplasm is dense and contains ribosomes (R) in tightly coiled helical array. The nucleation may consist of a single nucleus or a pair of abutted nuclei, diplokaryon (D). The size of the spore depends on the particular species and can vary from less than 1 μm to more than 10 μm. The number of polar tubule coils is also variable from a few to 30 or more, again depending on the species observed.

General Characteristics

Classification

In 1980, protozoologists revised the phylum protozoa; creating six new phyla, one of which was the phylum *Microspora* (Levine et al., 1980). Within this phylum is the order *Microsporida,* the group that contains the four genera of parasites so far associated with human infections. The term *microsporidia* is often used as a common name for this group, and microsporidiosis is used for a disease caused by any of the *Microsporida.*

Nosematosis is a name sometimes used to describe a disease caused by *Microsporida* from the genus *Nosema,* and the term encephalitozoonosis is used for disease caused by infections from *Encephalitozoon;* however, the general term microsporidiosis is preferable until there is sufficient clinical experience to indicate whether differing clinical syndromes indeed are specific to various genera.

Morphology and Life Cycle of the *Microsporida*

The *Microsporida* are characterized by small oval spores that vary in size from less than 1 μm to about 10 μm in length. The spore is the stage usually observed when one detects the presence of a microsporidian infection. Because it is the spores that are diagnostic, a detailed ultrastructural description will follow (Fig. 1).

The spore coat is divided into three parts. The outermost layer, the exospore, is electron dense and originates from material secreted at the beginning of sporogony. This layer is two to four times as thick as a membrane and may be ornamented with ridges, filaments, or tubular structures. The next region is called the endospore; it is electron lucent, contains chitin, and is variable in its thickness but is usually many times thicker than the exospore. This layer is always thinnest in the region of the attachment of the polar tubule anchoring disc. The innermost layer is a membrane surrounding the complex of structures called the extrusion apparatus, a posterior vacuole, and the sporoplasm (the nucleus and cytoplasm of the spore).

The extrusion apparatus consists of a long polar tubule, also known as a polar filament, attached inside the anterior of the spore by a mushroom-shaped

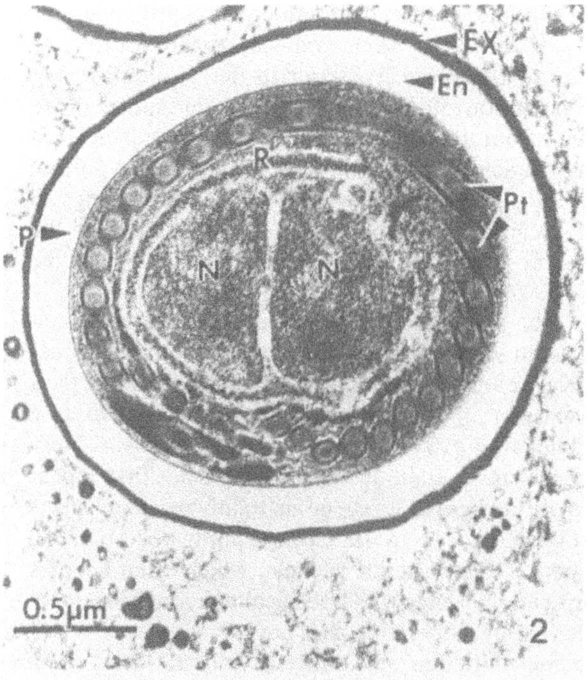

FIG. 2. Electron micrograph of *Nosema connori* spore in the adrenal gland from a child. Note the arrangement of paired abutted nuclei (N) typical of diplokaryotic genera. An electron-dense exospore wall (Ex) surrounds the electron-lucent endospore (En). A well-defined plasmalemma (P) limits the sporoplasm, which contains numerous ribosomes (R) and several cross-sections of the coiled polar tube (Pt) AFIP # 75-10043. (From Strano et al., 1976.)

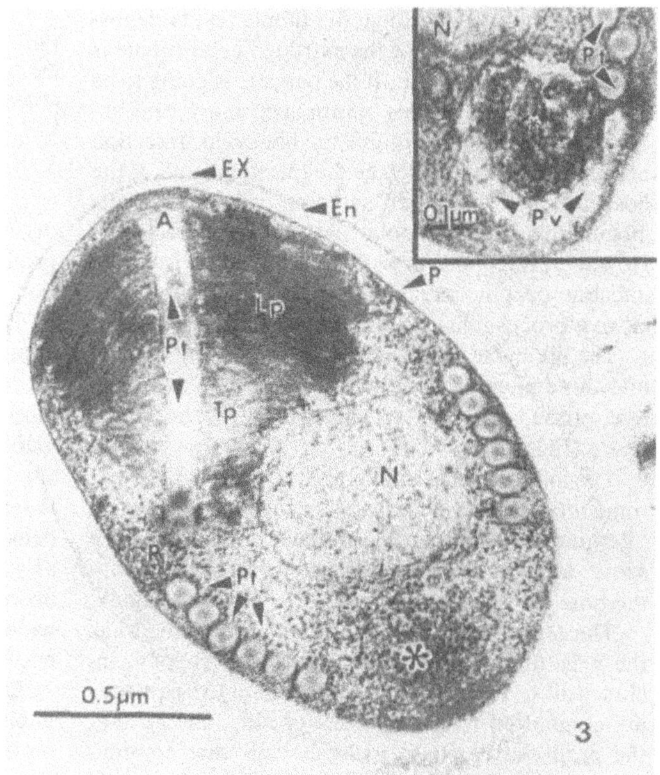

FIG. 3. Electron micrograph of *Glugea americanus* spore from the angler fish, *Lophius americanus,* depicting the structure of a typical uninucleate spore. The anterior end is filled with the anchoring disc (A), polar tube (Pt), and the various lamellar (Lp) and tubular (Tp) polaroplast membranes of the extrusion apparatus. Ribosomes (R), polar tube cross-sections (Pt), and a single nucleus (N) occupy the mid-portion of the spore. The posterior end (*) contains a vacuole that is not evident in this oblique section. *Insert:* The posterior vacuole (Pv). It often contains a "dense body" composed of tubular material that may appear "glomerular" (*). The posterior vacuole is often abutted to the coils of the polar tube (Pt). (From Takvorian and Cali, 1986.)

anchoring disc complex and an elaborate lamellar membrane system, called the polaroplast, that surrounds the polar tubule and fills the anterior third to half of the spore. Posterior to the polaroplast is a tubular convoluted membrane system that seems to emit from the lamellar system, traversing beneath it and attached at several points to the straight portion of the polar tubule. The polar tubule is quite long and coils around the contents of the remainder of the spore, the sporoplasm, and posterior vacuole (Fig. 1).

The sporoplasm may be binucleated or mononucleated depending on the genus of *Microsporida* (Figs. 2 and 3). The binucleated organisms have their nuclei in an abutted arrangement, referred to as a diplokaryon. The surrounding cytoplasm is very dense and contains numerous ribosomes that appear to be in helical array or in tightly packed sheets (Figs. 2 and 4). In the most posterior region of the spore is the posterior vacuole, which often appears irregular in shape with flocculent material inside. Occasionally, structures that are glomerular or membrane whirled in appearance have been observed inside the vacuole (Fig. 3,insert).

In the presence of the proper environmental stimuli, the anterior polar complex of the spore is "triggered," the polar tubule everts, and extrusion occurs. This usually takes place in the intestine of the host where pH and ionic conditions initiate extrusion. The polar tubule everts with a great deal of force, enabling it to pierce adjacent host cells. The sporoplasm travels through the tubule and is deposited wherever the end of the extruded polar tubule is embedded in host tissue. If the parasite is going to be successful, the infective sporoplasm must gain access to host cells to continue the life cycle. Infection of mucosal surfaces such as the intestinal wall of the host are easy to explain as being caused by direct inoculation from the polar tubule to the host cell. However, infections in tissues far removed from the intestine become infected by a less direct and more elusive process that is often not well understood.

The life cycle of the *Microsporida* can be divided into three phases: the infective phase, the proliferative phase, and the sporogony or spore-forming phase (Fig. 5).

The infective phase includes free spores, the environmental factors necessary to cause the spore to "germinate," the sporoplasm that emerges from the spore, and any forms that transfer the infection into the host cells, where the proliferative phase begins.

The proliferative phase, sometimes referred to as the vegetative phase, schizogony or merogony, includes all cell growth and division until the parasites are committed to spore formation. In some genera, the proliferative phase includes cells that continuously divide, producing hundreds or thousands of

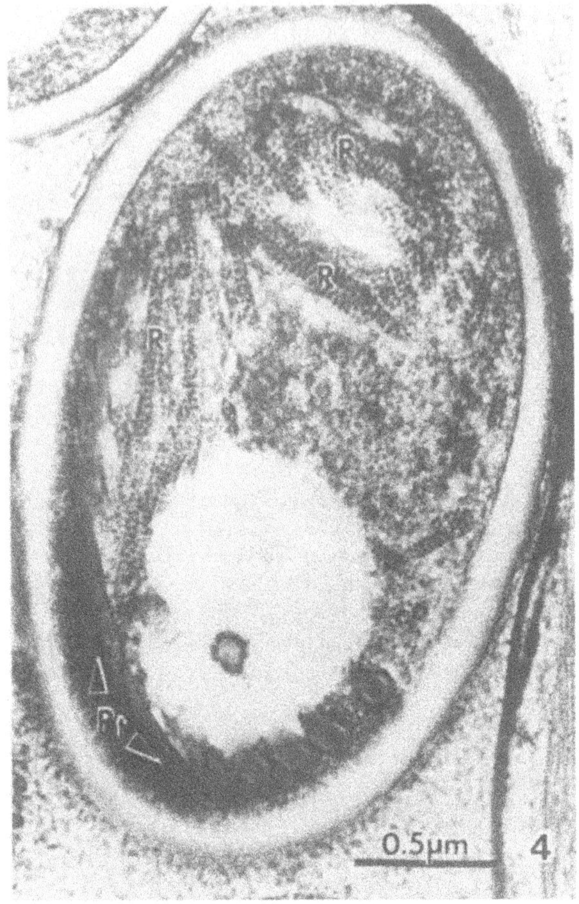

FIG. 4. Electron micrograph of *Pleistophora* sp. spore from the muscle of a patient with AIDS. The ribosomes (R) are arranged in helical array or in tightly packed sheets. (From Cali et al., in preparation.)

organisms. In other genera, relatively few cells are produced in this phase, most of the parasite replication occurring later in sporogony.

The sporogony phase includes all forms of the organism from commitment to spore formation as sporonts until the development of mature spores. A morphologic means of identifying this phase of development is the change in appearance of the parasite plasmalemma (outer membrane). Material is secreted by the parasite, which forms an electron-dense addition to the plasmalemma often referred to as a "thickened membrane." This was initially demonstrated in the genus *Nosema,* and it has been widely reported in other genera since that time; no exceptions are known.

Once this "thickened membrane" is formed, the number of cell divisions that follow vary depending on the genus. In some genera, such as *Nosema* and *Encephalitozoon,* each sporont cell divides only one

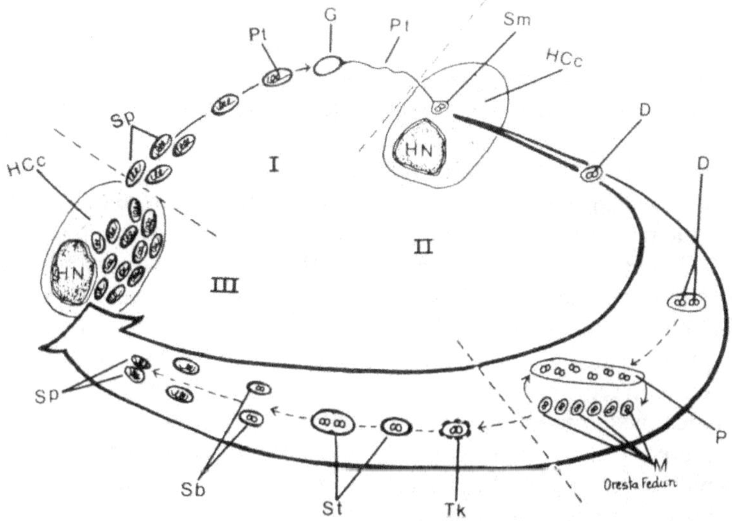

FIG. 5. The life cycle of *Nosema apis* in intestinal epithelial cells of the honeybee. Phase one (I) includes mature spores (Sp.) shed into the environment (e.g., via fecal material). Exposure to the proper environmental conditions is necessary to trigger germination (G) of the spore (this is often in the gut of a new host). This results in the extrusion of the polar tubule (Pt), at the end of which emerges the sporoplasm (Sm). It is generally believed that the polar tubule must pierce the intestinal lining or the sporoplasm will be destroyed by the host digestive enzymes. Phase II begins here and includes all proliferative growth and multiplication to the point of commitment to spore formation. *Nosema* species characteristically grow in direct contact with the host cell cytoplasm (HCc); there is no phagosome or parasitophorous vacuole isolating them. Diplokaryotic nuclei (D) divide, producing more pairs of nuclei within a single cell; this occurs many times without cytokinesis, resulting in the production of multinucleate plasmodia (P). These plasmodia divide producing many diplokariotic cells, which may repeat this process resulting in the formation of many parasite cells, sometimes called meronts (M). When the plasmalemma of these meronts begins to form electron-dense deposits on its surface, the cell is committed to spore formation and has therefore entered the third phase (III) of the cycle, sporogony. These electron-dense deposits become confluent around the cell surface producing a "thick membrane" (Tk); the sporont (St) grows, and one more nuclear division occurs producing a sporont with two diplokarya. The sporont divides producing two sporoblasts (Sb), which each develop into a mature spore. This entire process takes place in the cytoplasm of the parasitized host cells. When this cell dies the spores are freed into the surrounding environment. In the intestine, the spores may pass out with fecal material and thus enter Phase I again.

time to produce two sporoblasts, each of which undergoes a metamorphosis to produce a spore. In the genus *Thelohania*, eight spores are produced and in other polysporous genera, such as *Pleistophora*, the majority of cell divisions occur during sporogony. Additionally, some *Microsporida* are pansporoblastic, that is, they produce an envelope that surrounds the developing parasites inside the host cell cytoplasm. This envelope may be a membrane or a thick wall of secreted material, but in either case it is present through sporoblast and spore formation.

The parasite cells from the proliferative and sporogonic phases of the parasite, develop within the cytoplasm of the infected host cells. Although parasite development is variable depending on the genus of *Microsporida*, all these intracellular cycles produce sporonts, and sporoblasts, and terminate in the production of more spores (Fig. 5).

Spores are the resistant stage of the parasite that may pass into the environment. They are able to resist desiccation, treatment with disinfectants, and antibiotics. They can remain viable at room temperature for weeks to months and may remain viable for months or years in water just above freezing, 3 to 5°C.

Distribution of Hosts

Microsporida have been studied for many years in insects, primarily due to their economic importance. Of these the first studied, and probably the most economically important, is *Nosema bombycis* (Nageli, 1857) from the silk worm. Since that time, the *Microsporida* have been reported in every major group of animals including the protozoa, the invertebrates, and all classes of the vertebrates. The vast majority of research on the *Microsporida* has been conducted on those infecting insects and fish; however, one species, *Encephalitozoon cuniculi*, infecting many mammals and humans, has probably been the single most studied species.

Pathology—Host-Parasite Interaction

Although there are many genera of *Microsporida* and a vast assemblage of hosts, there are relatively few types of intracellular host-parasite interfaces during their developmental cycle. These relationships can be divided into four categories (Cali, 1986). They are direct contact, parasite-induced indirect contact, host-induced indirect contact, and host-parasite-induced indirect contact. Genera representing the first three of these relationships have been found in human infections and will be highlighted in the following relationship descriptions. Thus far, no examples of host-parasite-induced indirect contact with *Microsporida* have been identified in humans.

Microsporida considered to have direct contact are those that maintain direct contact between their plasmalemma and the host cell cytoplasm throughout their development. The only change occurs at the commencement of sporogony, when the parasites' plasmalemma thickens due to the deposition of material on its surface. This type of relationship occurs in such genera as *Nosema* (Cali, 1971), *Unikaryon* (Canning et al., 1974), and *Enterocytozoon* (Desportes et al., 1985). These *Microsporida* avoid the host mechanism of forming phagosomes to surround them and instead are found among the cytoplasmic organelles, as though they were a natural part of this environment. Host distribution is variable depending on the genus. *Nosema* is a genus that seems to be ubiquitous. It has been reported from numerous host groups including invertebrates and vertebrates. *Unikaryon* is found in trematode worms (Canning et al., 1974) and insects (Togubaye and Marchand, 1984). *Enterocytozoon* is a new genus created for the *Microsporida* found in the intestinal lining cells (enterocytes) of humans. It is placed in this group because of its host-parasite contact and because it is similar to *Unikaryon*.

Microsporida in the parasite-induced indirect contact group begin development in direct contact with the host cell cytoplasm but develop a parasite vesicle that isolates them from direct contact with the host cell cytoplasm. These genera are known as "pansporoblastic." Many genera comprise the "pansporoblastic group," in which a progression of host-parasite contacts occurs. This progression ranges from minimal to extreme isolation from the host cytoplasm. Minimal isolation is present in several genera, including *Thelohania*. In these genera, the pansporoblastic vesicles form during sporogony, are very thin, and may or may not persist. The only member of this group found in mammals is *Thelohania apodemi*, reported from the brain of a field mouse (Doby et al., 1963). Pansporoblastic microsporidians with maximal isolation from the host cytoplasm include the genus *Pleistophora*. In this genus, the parasite plasmalemma is surrounded by secreted material during early proliferative development. This amorphous material persists, becomes modified by further secretions to form a "sporophorous vesicle" during sporogony, and eventually houses the mature spores. *Pleistophora* is most commonly found in the muscle tissue of fish and amphibians. However, it has a broad host range that includes reptiles, insects, decapods, snails, and humans.

Microsporida with host-induced indirect contact are separated from host cytoplasm from the earliest proliferative forms to the final spore forms. All stages are enclosed in a host-produced, phagosome-like, limiting vesicle. The genus *Encephalitozoon* is an example of this host-parasite relationship. Although it is typical for parasitized cells to surround any foreign material that comes into their cytoplasm by a phagosome membrane, most of the other genera of *Microsporida* seem to have eluded this mechanism; *Encephalitozoon* is an exception (Cali, 1971; Pakes et al., 1975; Sprague and Vernick, 1971). The genus *Encephalitozoon* is found primarily in mammals and, on rare occasion, in birds, reptiles, and humans.

Diagnostic Characteristics

Microscopic Techniques

Microsporida are intracellular organisms that have an elusive nature, even to the trained eye of the pathologist. They are small, difficult to visualize at the light microscopic level, and are relatively poorly stained by the usual histologic stains. Routine hematoxylin and eosin staining is usually of little help. Special stains such as Grocott's fungi stain (GMS) (Fig. 6). Ziehl-Neelsen's acid-fast stain (Fig. 7) Giemsa (Fig. 8), and Heidenhain's hematoxylin (Fig. 9) help to demonstrate the presence of some microsporida in light microscopic preparations (Luna, 1968; Gray et al., 1969). The presence of a periodic acid-Schiff (PAS) positive granule in the anterior end of mature spores is diagnostic for *Microsporida*, at the light microscope level (Fig. 10). Fresh tissue squash preparations made from biopsy specimens may be viewed directly by phase-contrast microscopy. The spores are quite refractile and may be birefringent (Fig. 11).

Sections of plastic-embedded tissue, .5 to 1-μm thick, mounted on glass slides, and stained with toluidine blue with or without basic fuchsin (Bennett and Radimska, 1966; Grimley, 1964) can often reveal the presence of microsporidian spores and/or developing stages. These are not as easily seen in 5 to 7-μm thick paraffin-embedded tissues (Fig. 12).

The best tool for the identification of Microsporida is the electron microscope. The presence of a

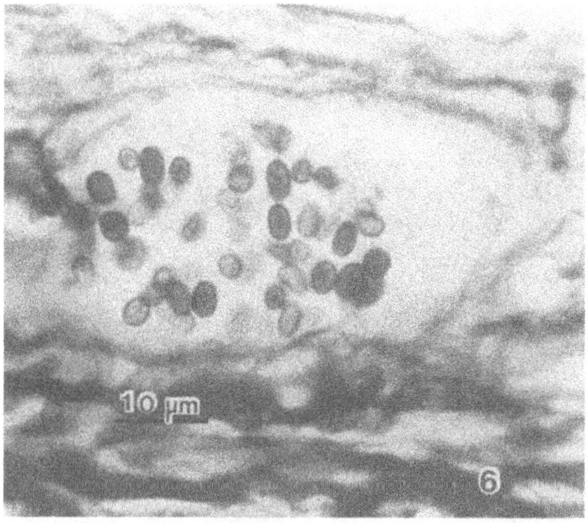

FIG. 6. *Nosema connori* spores in the muscularis of the jejunum. The spore coats are well illustrated by Grocott's methenamine silver (GMS) stain AFIP # 71-5887. (From Strano et al., 1976.)

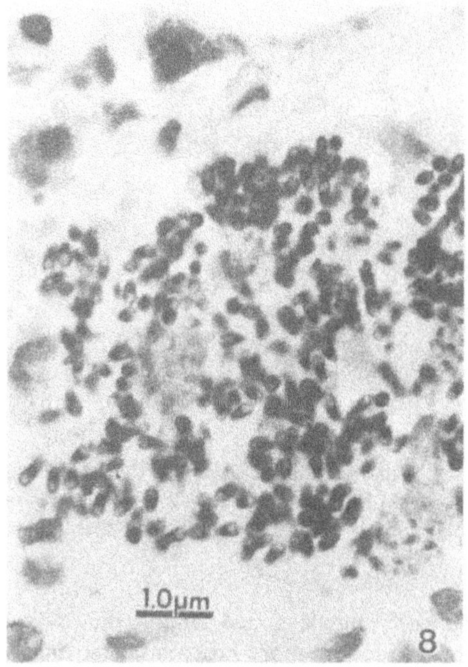

FIG. 8. *Pleistophora* sp. spores from the muscle of a patient with AIDS. The spores are very obvious in this Geimsa-stained preparation, AFIP # 84-6479. (From Cali et al., in preparation.)

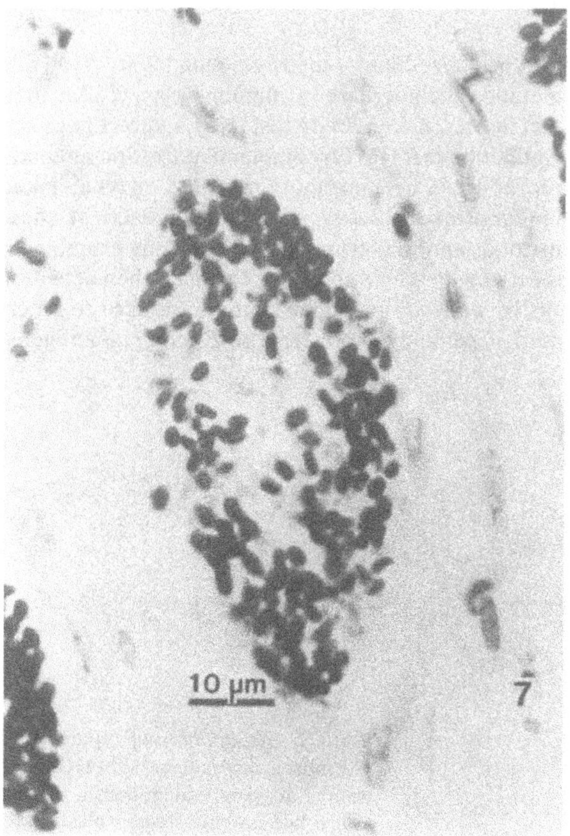

FIG. 7. *Nosema connori* spores in the heart. The spores are acid-fast when stained with Ziehl-Neelsen stain AFIP # 71-864. (From Strano et al., 1976.)

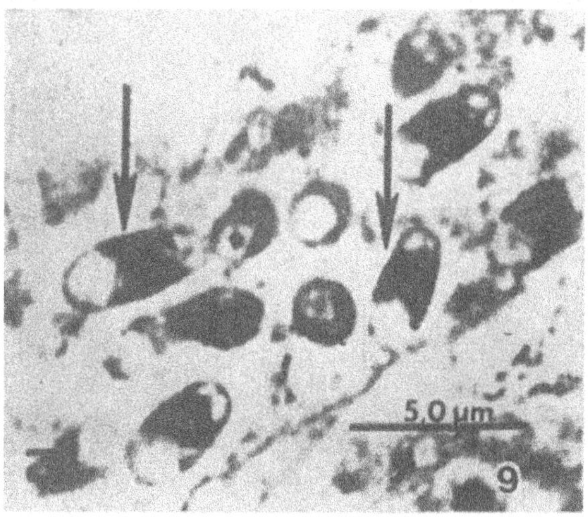

FIG. 9. *Nosema apis* spores from the midgut of the honey bee. The spores stain very intensively with Heidenhain's hematoxylin. (From Gray et al., 1969.)

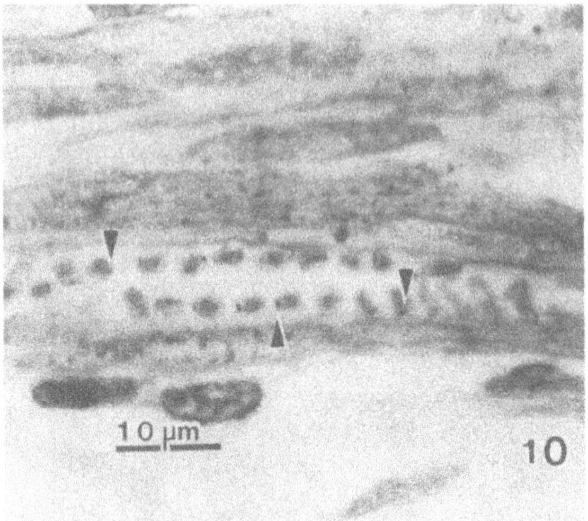

FIG. 10. *Nosema connori* spores in the muscularis. The anterior end of the spores contain a PAS positive (*arrows*) granule, AFIP # 71-5883. (From Strano et al., 1976.)

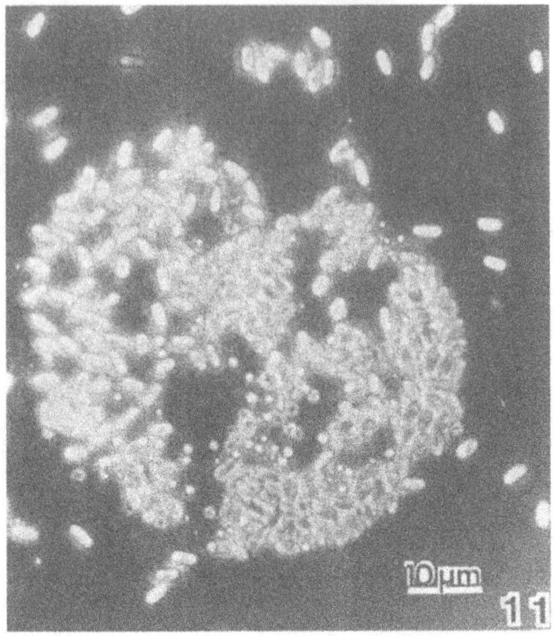

FIG. 11. *Nosema apis* spores in the intestinal epithelium of the honey bee. The spores appear highly refractile when observed in a fresh squash preparation by phase contrast microscopy. (From Cali, 1970.)

single polar tubule coiled around the spore cytoplasm, inside the spore coat, is diagnostic for Microsporida at the electron microscope level (Fig. 3).

IMMUNODIAGNOSTIC TECHNIQUES

Serologic approaches have been developed for identification of infection with *E. cuniculi* in animals. These methods share the problem of assuring that the antigen used is specific for this organism. Development of tissue culture techniques for production of large and consistent batches of *E. cuniculi* spores made such tests feasible and reproducible. Individual tests using either whole spores or processed extracts vary in complexity, sensitivity, specificity, and need for specialized and expensive equipment. Available methods include carbon immunoassay (CIA), indirect fluorescent antibody test (FA), indirect microagglutination test (IMT), complement fixation antibody procedure (CF), and, more recently, enzyme-linked immunosorbent assay (ELISA). In each of these methods antibody from the subject being examined is reacted with whole spores or extracts, then detection systems of varying sophistication are used to determine if antibody is adherent to *E. cuniculi* antigens.

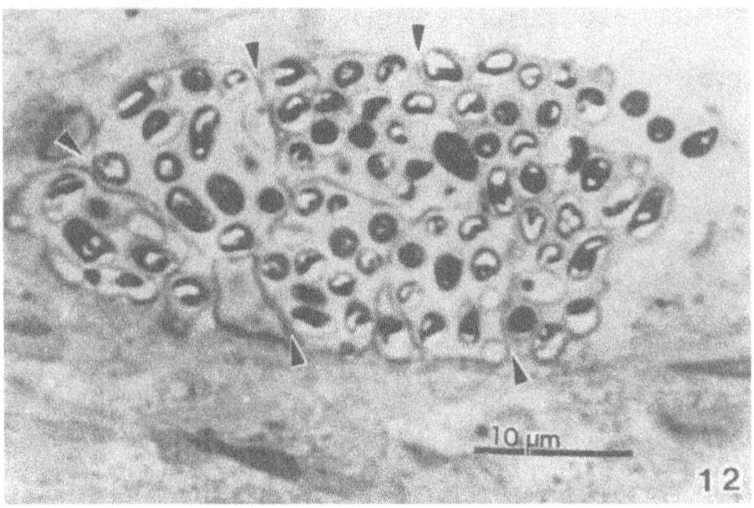

FIG. 12. *Pleistophora* sp. spores from the muscle of a patient with AIDS. The spores are easily identified in this 0.5 µm toluidine blue-stained plastic section. Clusters of dark staining spores are contained in membrane limited vacuoles (*arrows*). (From Cali et al., in preparation.)

In the carbon immunoassay, also known as the India ink immunoreaction, spores are incubated with serum. When this is rabbit serum, it has been empirically determined that added microscopic carbon particles adsorb to heavy chains of IgG, staining the spores. An advantage of this test is that it does not require equipment more complex than a light microscope; however, carbon particles from different batches of India ink vary in affinity for rabbit IgG. Also, usefulness of this test for detection of early infections is limited by failure of carbon particles to adhere to IgM, which is initially generated in infections. Carbon particles do not adhere to Ig in sera from mice, rats, and humans, but either IgM or IgG can be indirectly detected by applying antispecies rabbit serum which then attracts the carbon (Kellett and Bywater, 1980). This indirect method was used to detect clinical *E. cuniculi,* infection in a child with unexplained seizures (WHO, 1983). When using this indirect technique, however, care must be taken that rabbit antisera do not contain antibodies to *E. cuniculi,* which Bywater et al. (1980) found in 24 of 29 commercially available rabbit antisera samples.

The indirect fluorescent antibody test (FA or IFAT), also sometimes called indirect immunofluorescence test (IIF), uses fluorescence-labeled antispecies serum to detect immunoglobulin in test samples reacting with *E. cuniculi*. This method may be more sensitive than the indirect carbon method described but requires a fluorescence microscope. Fluorescent antisera can be used from species other than rabbits, but *E. cuniculi* is found in multiple mammals and caution is still necessary to avoid false-positive results due to *E. cuniculi* antibodies in animal-derived test reagents.

Using the carbon immunoassay, Bergquist et al. (1984a) first suspected *E. cuniculi* in a boy born in Columbia and hospitalized at age 2 with generalized seizures. Initial serologic studies with CIA were confirmed by FA against *E. cuniculi*. Because animals infected with *E. cuniculi* excrete the organism in urine, this boy's urine was screened and 1.5×2.5 μm organisms were recovered, which reacted with a fluorescent rabbit anti-*E. cuniculi* conjugate. Intraperitoneal injection of urine into mice known to be free of *E. cuniculi* produced ascites containing organisms consistent with *E. cuniculi,* which were subsequently cultured in dog kidney cells. The specificity of the boy's antibodies were investigated by reacting his serum with a crude *E. cuniculi* antigen preparation after it had been treated with sodium dodecyl sulphate, separated electrophoretically into bands on a polyacrylamide gel, and applied to a nitrocellulose sheet. The boy's serum reacted with antigens in at least 12 bands, supporting the conclusion that his high titer positive serology was indeed a response to *E. cuniculi* (WHO, 1983).

The inherent trade-off between sensitivity and specificity in serologic methods limits their use in diagnosis of individual patients, but such methods are quite useful for epidemiologic purposes. The FA method, using antigen from *E. cuniculi* spores from tissue culture, was used in initial epidemiologic studies of the prevalence of *Microsporida* in humans. Whether this test will pick up past infection with other genera and species of microsporidians and whether antibodies to other protozoa will cross-react with *E. cuniculi* antigens is not yet known. Within these limitations, some tantalizing data have emerged. Singh et al. (1982) using banked sera from England found low positive titers in 6 of 69 healthy adults, 3 of 20 children suspected of toxocariasis, and 1 of 18 with toxoplasmosis. Surprisingly, 33 of 92 Ghanians with malaria, 38 of 89 Nigerians with tuberculosis, and 13 of 70 Malayasians with filariasis had titers up to 1:256 against *E. cuniculi*. In Sweden antibodies to *E. cuniculi* were detected in 14 of 115 travelers returning from the tropics but in none of 48 nontravelers (WHO, 1983). Serology was positive in 8 of 54 patients with Chagas' disease, 2 of 23 with filariasis, and 38 of 100 with malaria. The high apparent prevalence of positive serology to *E. cuniculi* in patients with parasitic or mycobacterial diseases from developing countries may reflect environmental exposure due to sanitary conditions, dietary content, or patterns of living but may also reflect increased susceptibility to infection. In a Swedish screening clinic for homosexual men with acquired immunodeficiency syndrome (AIDS) or related symptoms of prolonged fever, weight loss, and persistent adenopathy, 10 of 30 had antibodies to *E. cuniculi* (Bergquist et al., 1984b). All patients with positive titers had traveled to the tropics. This 33% prevalence of positive *E. cuniculi* serology compared with 12% of nonselected Swedish travelers may reflect decreased resistance to infection or direct person-to-person transmission associated with sexual habits.

Skin hypersensitivity testing, indirect microagglutination tests, and complement fixation tests are also reported to be useful in detection of *E. cuniculi* in rabbits (Pakes et al., 1984). Enzyme-linked immunosorbent assays for *E. cuniculi* have been described that hold promise for standardization and quantification of antibodies to *E. cuniculi* (Cox et al., 1981). For a further discussion of immunodiagnosis see Canning and Lom (1986).

Specimen Collection, Transport, and/or Storage

If microsporidiosis is suspected, biopsy samples of affected tissue should be taken and fixed in formalin for light microscopy and (2 to 4%) gluteraldehyde (buffered pH 7.4) for electron microscopy. For best results, tissues should be fixed immediately. Currently, the only means of identification is by microscopic observation. If tissues are to be processed elsewhere, it is best to change the tissues to fresh fixative after a few hours of fixation and keep them refrigerated until transported. Fixed tissues do not need to be kept cold during transport if they are express mailed. Tissues may be held in fixative for weeks or longer (do not store tissues in buffer).

Unfixed biopsy material containing microsporidia can be stored and remain viable for months to years, if frozen in the same manner used for the storage of tissue culture cells. Spores in fresh tissue or isolated and concentrated by centrifugation may be stored at 3 to 5°C in saline or water for months. Additionally, blood sera can be frozen and stored for later immune testing (although diagnostic techniques are not yet available, it would be desirable to have the sera for comparative purposes).

Microsporida in Mammals

The Genus *Encephalitozoon*

The first report of *Microsporida* in a mammal was in 1923 when Levaditi et al. (1923a) reported the occurrence of an organism causing encephalitis in a rabbit. Although they named it *Encephalitozoon cuniculi,* published several papers on it (Levaditi et al., 1923a,b,c, 1924a,b,c), and suggested that it might be a microsporidian, it was not confirmed to be a member of the *Microsporida* until several years later. In 1964, Weiser and Lainson et al. simultaneously studied this material and placed it into *Nosema,* the genus that it best fit into at the time. In 1970, however, the genus *Nosema* was redefined on the basis of characteristics that could only be resolved and demonstrated with the electron microscope (Cali, 1971). These differences include paired abutted nuclei (diplokarya) in the developmental stages and in the spores of *Nosema* (Figs. 2 and 13). In *E. cunicili* (or *N. cuniculi* as it was transiently called), it was demonstrated that the nuclei were not in the diplokaryon arrangement in any stage of development and that there was a parasitophorous vesicle separating the parasite from direct contact with the host cell cytoplasm (Figs. 14 and 15). On the basis of these characteristics, *Encephalitozoon* was made a genus in the *Microsporida* and *E. cuniculi* the type for the new genus (Cali, 1971). Thus, *Nosema* and *Encephalitozoon* are not synonymous but, instead, are two valid genera of *Microsporida* and both have been identified as causing infections in humans.

Encephalitozoon Species

The original species of *Encephalitozoon* (the type) is *cuniculi* isolated from a rabbit (Levaditi et al., 1923a). Its ultrastructure was described in three publications. Cali (1971) established the genus characteristics with this species, Sprague and Vernick (1971) compared it with other microsporidian genera, and Pakes et al. (1975) compared it with the *Microsporida* found in mice and in hamsters. *Encephalitozoon cuniculi* was the only species identified in these three mammals. Subsequently, it has been reported in many other hosts (see below).

The characteristics of *Encephalitozoon cuniculi* are: proliferative cells contain one to several isolated nuclei (no diplokarya); each sporont cell produces

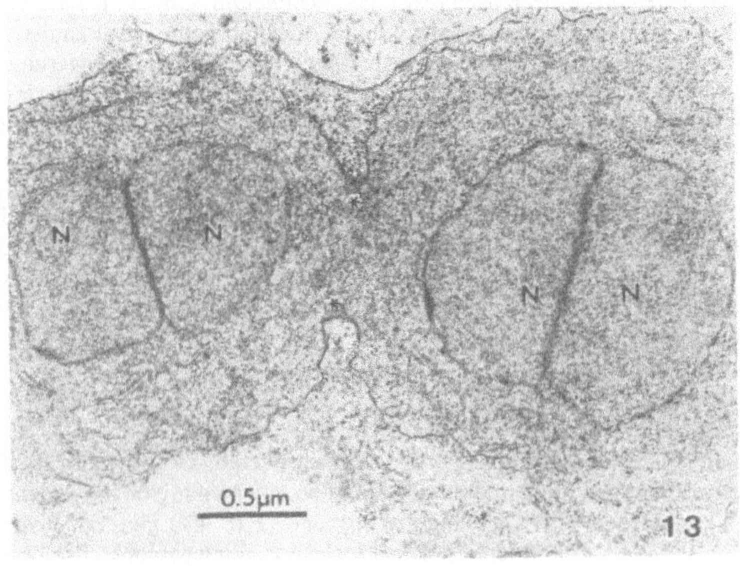

FIG. 13. Electron micrograph of *Nosema bombycis* proliferative stage undergoing division (*). The nuclei (N) are in the diplokaryon arrangement (paired and abutted) typical of this genus. Note the direct contact between the organism and the host cytoplasm. (From Cali, 1971.)

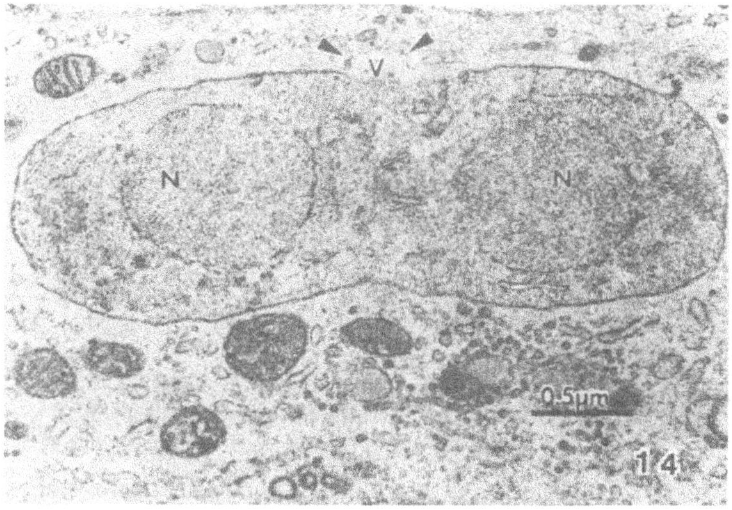

FIG. 14. Electron micrograph of *Encephalitozoon cuniculi* proliferative stage about to divide. The nuclei are isolated (N). Note the vacuole (V) in which the organism is developing; it isolates the protozoan from direct contact with the host cytoplasm. (From Cali, 1971.)

two sporoblasts (disporoblastic sporont); each sporoblast produces one spore; and each spore contains a single nucleus, 5 to 6 polar tubule coils, and measures 1.5 to 2 μm in length. All of these developmental stages occur in the cytoplasm of the host cell inside a phagosome or a host-formed parasitophorous vacuole, which is not typically found in other microsporidian infections (see Pathology above) (Figs. 14, 15, and 16). Tissues infected included brain, liver, kidney, and spleen.

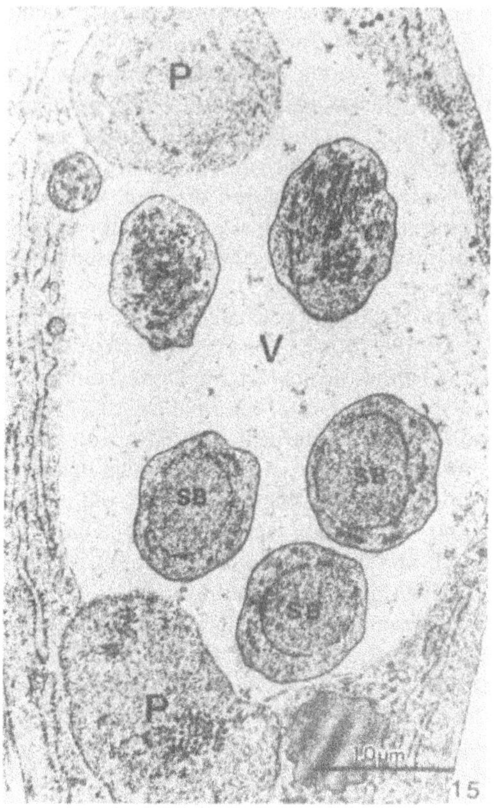

FIG. 15. Electron micrograph of *Encephalitozoon cuniculi* sporoblasts (SB), sporonts (S), and proliferative forms (P) in a parasitophorous vacuole (V). (From Pakes et al., 1975.)

Three species of *Encephalitozoon* were named as causitive agents for human infections in articles before the nature of this parasite was known. Because *Encephalitozoon cuniculi* is the organism for which this genus was created, it is called the "type" species for the genus. The importance of this is that any subsequent species added to the genus must conform to the type characteristics. If the type is further described at a later date the additional species added to it must be checked for conformity to these new or additional characteristics.

Encephalitozoon chagasi was described from a Brazilian infant who had convulsions and died 2 days after birth (Torres, 1927). Spores were described only by size, 3.5 × 1.5 μm. In 1964, when Weiser placed *E. cuniculi* into the *Microsporida,* he stated that the pathology (type) material for *E. chagasi* had been lost so no identification could be made (Weiser, 1964). Consequently, in a subsequent review of the *Microsporida,* this organism has not been accepted as a microsporidian (Sprague and Vavra, 1977).

Encephalitozoon brumpti was described from a 17-year-old Corsican who had high fever and meningitis, became comatose, and died (Coulon, 1929). The spores were described as variable in size, 1.25 to 4 × 2 to 6 μm. However, in 1974, Barker studied the type material and concluded that the organism was a yeast (Barker, 1974).

Encephalitozoon hominis was described from a 1-month-old American infant who was admitted to a hospital because of convulsive seizures and died 10 days later (Wolf and Cowen, 1937). Organisms 1.5 × 3 μm were found in the brain and spinal cord. The authors compared their material to *E. cuniculi*. How-

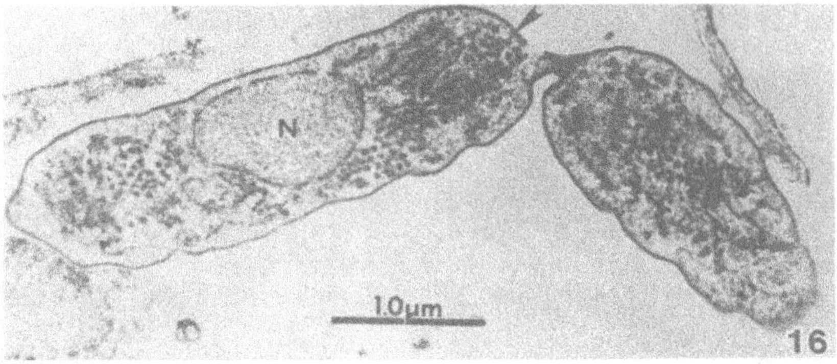

FIG. 16. Electron micrograph of *Encephalitozoon cuniculi* sporont undergoing cytokinesis. Note the thickened wall (*arrow*) and single nuclei (N). It is the occurrence of cell division at this stage of development that makes this parasite disporus. (From Pakes et al., 1975.)

ever, 2 years later, they changed their diagnosis to *Toxoplasma* (Wolf et al., 1939).

Unspecified possible microsporidian infections in humans have included an organism associated with a tumor (Marcus et al., 1973). Although this case was never proven, Petri (1969) did demonstrate *E. cuniculi* in transplantable ascities tumors in mammals. Additionally, a report of "encephalitozoon-like" organisms observed from human spinal fluid cultured in mouse tissue culture (Matsubayashi et al., 1959) contained a microsporidian; however, as they were not using a cell line, it is possible that the microsporidian came from the mouse cells.

Subsequent to these questionable species and reports, *E. cuniculi* has been demonstrated in many mammals, a few birds, and humans. Mammalian infections in addition to rabbits, mice, and hamsters include rats (Attwood and Sutton, 1965, Cowdry and Nicholson, 1924), guinea pigs (Cowdry and Nicholson, 1924; Perrin, 1943; Sebek, 1969), shrews (Nelson et al., 1969), goats (Khanna and Iyer, 1971), monkeys (Zeman and Baskin, 1985; Seibold and Fussell, 1973), dogs (Plowright, 1952; van Dellen et al., 1978), foxhounds (Cole et al., 1982), blue foxes (Vavra et al., 1971; Mohn et al., 1981; Mohn, 1983), cats, including a suricat, leopards (Vavra et al., 1971), a mountain lion, Siberian polecats (Novilla et al., 1980), and domestic cats (van Rensburg and du Plessis, 1971), and several other mammals by experimental infection or immunodiagnostic detection (Pang and Shadduck, 1985; Montrey et al., 1973; Wosu et al., 1977a, 1977b). Bird infections include three species of lovebirds (Kemp and Kluge, 1975; Novilla and Kwapien, 1978; Lowenstein and Petrak, 1980) and a parrot (Poonacha et al., 1985). Recently, it has been demonstrated in the liver of a human (Gordon et al., 1986; Terada et al., 1987). Additionally, there are reports of *E. cuniculi* in humans, as demonstrated by serodiagnostic methods (Bergquist et al., 1984a, 1984b).

Encephalitozoon cuniculi in Humans

The first morphologic demonstration of this organism in humans was published without identification of the organism (Gordon et al., 1986). Since that time, one of us (A. C.) has studied that material and identified it as *E. cuniculi* (Terada et al., 1987).

The patient was a 35-year-old homosexual man with AIDS. He was healthy until 4 months before his last hospitalization and death. His illness was characterized by general fatigue, weight loss, and occasional bouts of dark-colored diarrhea without nausea, vomiting, or abdominal pain. Small purplish-red spots on his face and chest were diagnosed as Kaposi's sarcoma.

Laboratory data consistent with hepatitis included an AST of 265 U/liter (normal 10 to 30 U/liter) ALT 158 U/liter (normal 6 to 30 U/liter), and alkaline phosphatase 384 U/liter (normal 36 to 92 U/liter). Hepatitis B surface antigen, anti-HBs, and anti-HBc were all negative. Hepatitis A IgG was positive but IgM was negative, indicating a remote but not active infection. Antibodies against Epstein-Barr virus, herpes II, and cytomegalovirus were present in low titers. Liver biopsy showed sinusoidal congestion and microgranulomas, but special stains were negative for fungi and acid-fast bacilli. After 1 month the patient developed jaundice, fever, and worsening of diarrhea. Laboratory tests confirming hepatic deterioration included bilirubin 8.4 mg/dl (normal 0.2 to 1.0 mg/dl), and further elevations in AST to 1,460 U/liter, ALT to 183 U/liter, and alkaline phosphatase to 1,377 U/liter. A repeat liver biopsy showed focal granulomatous and nonsuppurative necrosis on hematoxylin and eosin stain and gram-positive cocci-like organisms on PAS stain. Hepatocellular necrosis progressed rapidly and the patient died (Terada et al., 1987).

Electron microscopy of the liver biopsy revealed the presence of *Microsporida*. They were found in vacuoles that were located in the cytoplasm of liver cells. The earliest stage seen was a sporont dividing to form two cells, thus it was disporoblastic. A late sporoblast stage contained one nucleus, condensed cytoplasm, and five coils of the polar tube (Fig. 17). The spores were oval and 2×1 μm in size. The size, nucleation, disporous nature, and location (inside a parasitophorous vacuole) are all characteristics of *E. cuniculi* (Terada et al., 1987). Additionally, immunodeficient and immunocompetent mice exposed to *E. cuniculi,* by oral inoculation, developed the infection in the liver first (Gannon, 1980).

The Genus *Nosema* and Infections in Humans

The genus *Nosema* (Nageli, 1857) was the first microsporidian genus described and it contains hundreds of species identified in hosts ranging from invertebrates, especially insects, to humans. However, the only infections so far identified in humans have been opportunistic in nature.

NOSEMA CONNORI

The first reported case of a *Nosema* in a human was in 1973 (Margileth et al., 1973). A thymus-deficient infant from the Washington, D. C. area was hospitalized shortly after birth with intermittent diarrhea, vomiting, and acute respiratory distress. Laboratory tests demonstrated leukopenia and severe hypogammaglobulinemia. Diarrhea, malabsorption syndrome, and respiratory problems continued until his death at 4 months of age.

Autopsy tissues were processed for light and electron microscopy. *Pneumocystis carinii* was found in the lungs and disseminated microsporidiosis was found involving all the organs examined, including the lungs, stomach, small and large bowel, liver, kidney, adrenal gland, heart, smooth muscle of the intestine and blood vessel walls, and peripheral nerve fibers. The central nervous system was not examined.

The parasite spores were oval, 4×2 μm, and contained a PAS+ polar granule (Fig. 10). Some of the spores were acid fast with Ziehl-Neelsen stain (Fig. 7), and the spore coats stained brown with GMS (Fig. 6). Hematoxylin eosin did not stain the organisms well.

Electron microscopy studies revealed 10 or more coils of the polar tubule, which demonstrated that the parasite was definitely a microsporidian (Margileth et al., 1973). In 1974, Sprague named the organism *Nosema connori* because he saw two nuclei in a Heidenhain's stained light microscopic preparation (Sprague, 1974). This was confirmed in 1976 (Strano et al., 1976) when two nuclei, in diplokaryon arrangement, were electron microscopically demonstrated

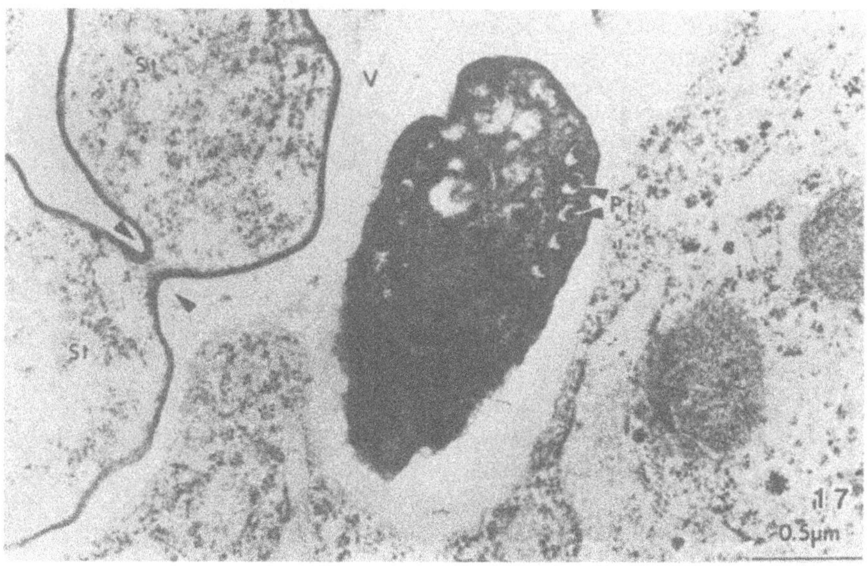

FIG. 17. Electron micrograph of *Encephalitozoon cuniculi* sporoblast developing within a vacuole (V) in the liver of a patient with AIDS. The polar tube coils (Pt) are visible. Note the dividing sporont (St) and the electron-dense wall (*arrows*) at the furrow. (From Gordon et al., 1986; Terada et al., 1987.)

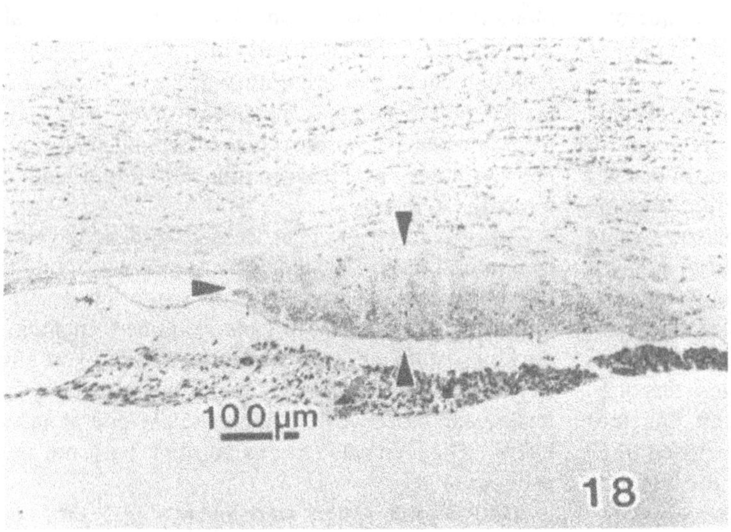

FIG. 18. Corneal disc from an 11-year-old boy from Ceylon. *Arrows* point to refractile spores just above Descement's membrane, AFIP # 73-10783. (From Strano et al., 1976.)

in the spore stage (Fig. 2). Additionally, they demonstrated spores, containing 10 to 12 polar tubule coils, directly in the host cell cytoplasm. These characteristics were again demonstrated by Shadduck et al. in 1979, when they contrasted it to a microsporidian contaminant of a nonhuman primate cell culture (Shadduck et al., 1979). The microsporidian in question was thus not *E. cuniculi,* which has approximately 5 to 6 polar tubule coils and a smaller spore, nor was it a new species of *Encephalitozoon* because the spores were in direct contact with the host cell cytoplasm and contained paired nuclei. Finally, direct contact of the parasite with the host cell cytoplasm, in combination with the abutted nuclei, supports a *Nosema* classification.

This finding was somewhat of a surprise because *Nosema* is commonly an infection in insects and all the mammalian infections previously identified were from the genus *Encephalitozoon.* It becomes less surprising when one considers that in the environment, a chance encounter with a microsporidian in the genus *Nosema* is much more likely than with *Encephalitozoon.* There are hundreds of *Nosema* species and their occurrence is quite common, whereas, there is only one *Encephalitozoon* species and it is relatively less common.

Nosema Infections of the Eye

An 11-year-old boy in Ceylon had been gored by a goat in the right eye and his right upper lid had been sutured. Six years later, he developed keratitis (Ashton and Wirasinha, 1973). A biopsy of the cornea revealed a necrotic central stroma surrounded by acute inflammatory cells. Immediately above Descemet's membrane were many oval bodies that were free and in macrophages and highly refractile when viewed by light microscopy. They were 3.5×1.5 μm and were identified as *Microsporida* (Figs. 18 and 19). These organisms were too large to be *Encephalitozoon cuniculi* and, because they were free in the host tissue, they could not be a new *Encephalitozoon* species or any pansporoblastic genus. *Nosema* is a reasonable designation but, because the nucleation

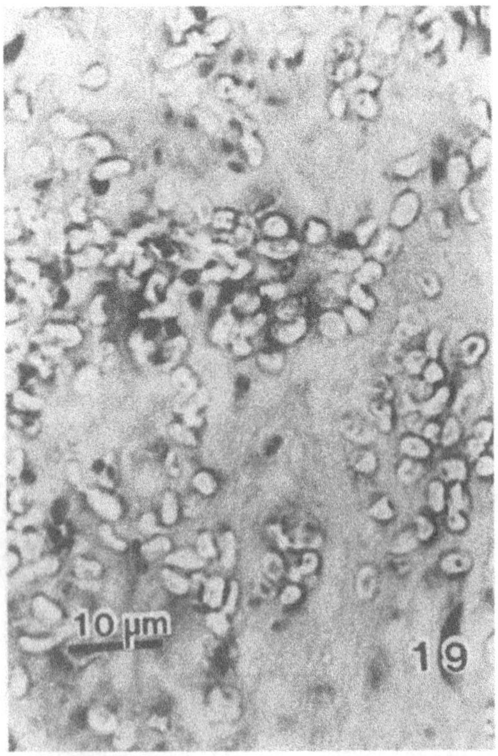

FIG. 19. Higher magnification of the spores described in Fig. 18. Note the number of spores and their refractility, AFIP # 73-10784. (From Strano et al., 1976.)

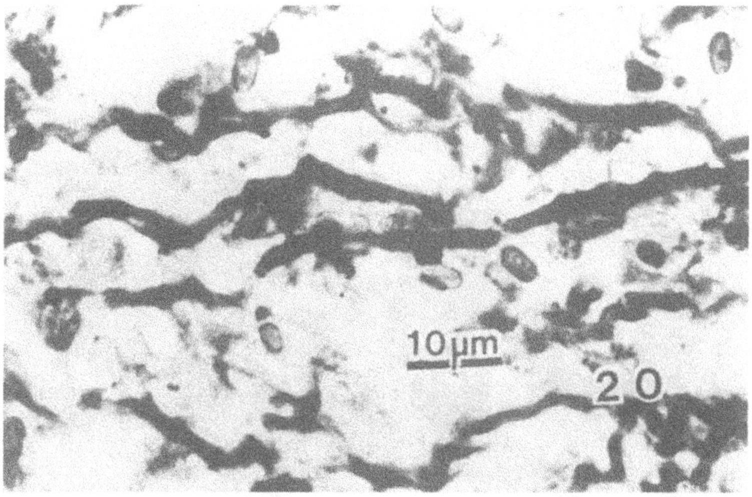

FIG. 20. Microsporidiosis of the cornea. Note the presence of numerous spores, AFIP # 79-11866. (From Pinnolis et al., 1981.)

of the organism was not observed, the genus designation is not proven. Although Sprague and Vavra (1977) and Current and Owen (in press) listed this organism as *Nosema* sp. in their reviews, Canning and Lom recently transferred this parasite into the genus *Microsporidium* (Canning and Lom, 1986), a genus created for organisms whose generic position cannot be determined. Ashton and Wirasinha (1973) had suggested that, because the boy was gored by a goat, *N. helminthorum*, a hyperparasite in the goat, was a possible diagnosis. Canning and Lom, however, pointed out that the spores were of different sizes so it could not be *N. helminthorum*. Instead, they made it a new species, *N. ceylonensis*.

A 26-year-old woman from Botswana, Africa, complained of pain in the left eye (Pinnolis et al., 1981). There was no light perception in the eye, the conjunctiva was inflamed, and a perforated corneal ulcer with marked keratouveitis and a hyphema were present. After 2 weeks of treatment with topical choramphenicol, hydrocortisone, and atropine, the eye remained painful so it was enucleated. Histopathologic examination revealed a perforated central corneal ulcer with incarceration of the iris. There was much necrosis adjacent to the area of perforation. The corneal lamellae were infiltrated with neutrophils, mononuclear cells, epithelioid granulomas, and a myriad of elongated oval organisms. These organisms were just barely visible with hematoxylin eosin staining, but were clearly visible when examined with phase contrast microscopy or stained with Grocott's methenamine silver (GMS)(Fig. 20) or the acid-fast technique. The spores were 4.5 to 5 × 2.5 to 3 μm and were found in clusters in histiocytes and free among the necrotic corneal lamellae. Electron microscopic observations were made on material that was first processed for light microscopy and then reprocessed for the electron microscope. The organisms had a thick electron-lucent spore wall and contained a polar tube that made 11 to 13 coils (Fig. 21). No vacuole was observed surrounding the organisms in the histiocytes. Because the spores were free in the host cell, the parasite was not a new *Encephalitozoon* or any pansporoblastic genus; thus, the authors classified it as a *Nosema* species. Unfor-

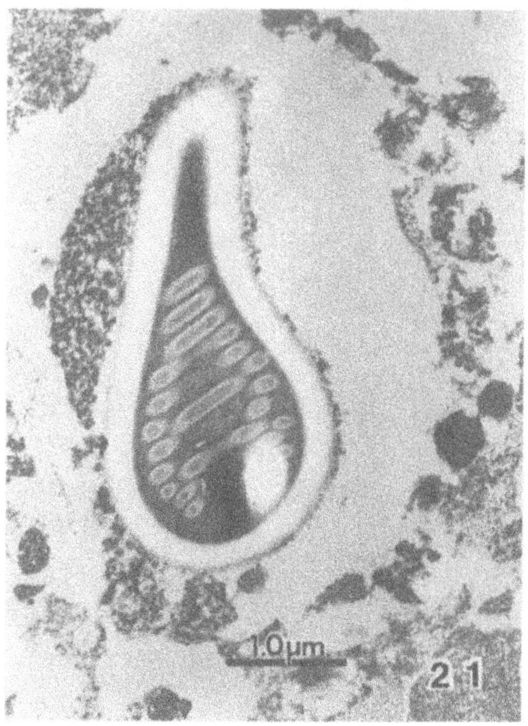

FIG. 21. Electron micrograph of a spore from the material described in Figure 20. The spore is approximately 4-μm long and contains 11 to 13 polar tube coils. (From Pinnolis et al., 1981.)

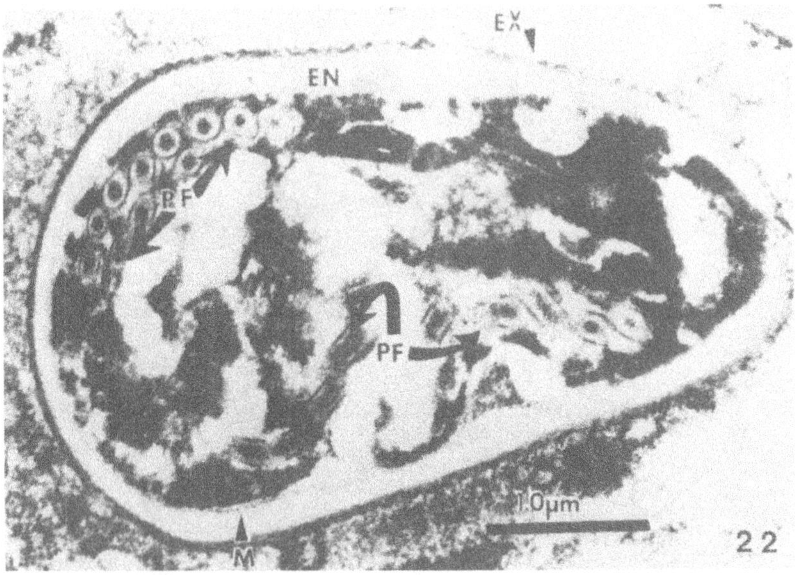

FIG. 22. Electron micrograph of *Pleistophora* sp. spore from the muscle of a patient with AIDS. The spore contains 11 polar tube coil cross-sections (Pf). (From Ledford et al., 1985.)

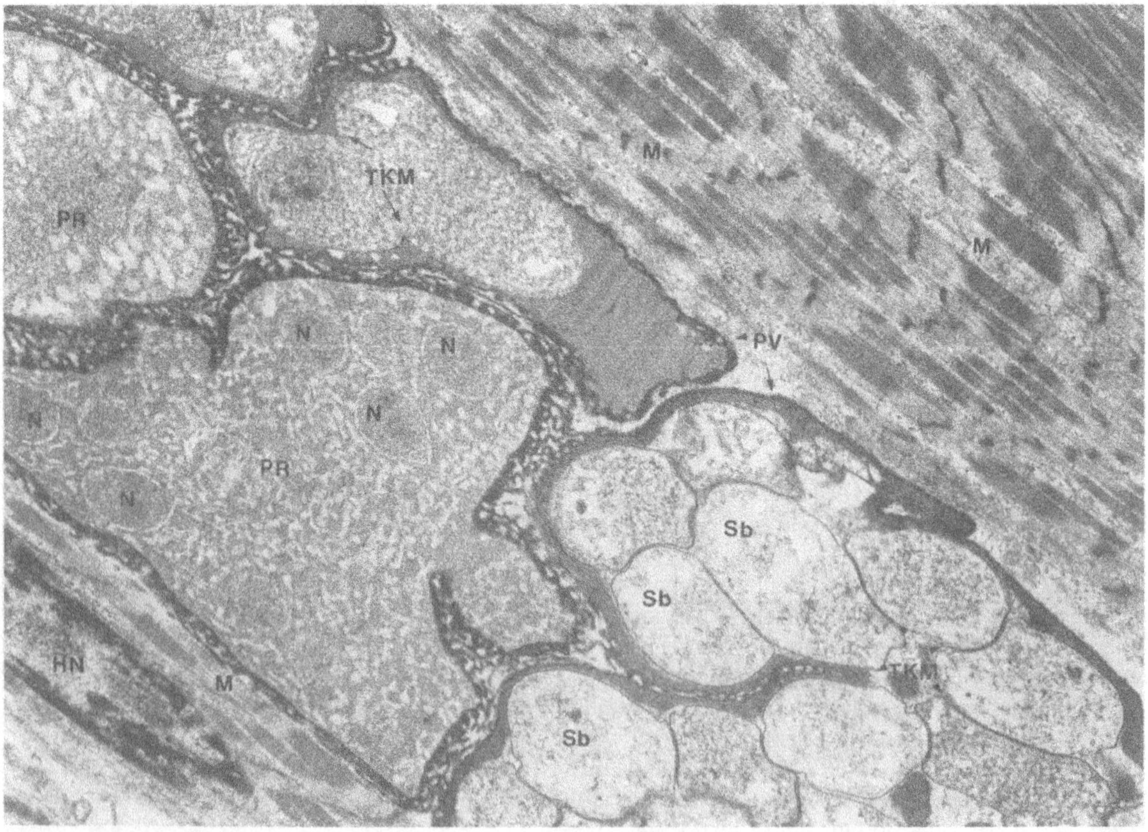

FIG. 23. Electron micrograph of *Pleistophora* sp. developing in the skeletal muscle of a patient with AIDS. The sporoblasts (Sb) and proliferative stages (PR) are enclosed in thick-walled (TKM) parasitophorous vacuoles (PV). The proliferative forms are multinucleated, but they are not in the diplokaryon arrangement. (From Cali et al., in preparation.)

tunately, they could not determine nuclear arrangement as is necessary to prove the genus designation. This organism also has been transferred to the genus *Microsporidium* and a new species name given to it, *M. africanum* (Canning and Lom, 1986).

Additional *Microsporida* Associated with Acquired Immunodeficiency Syndrome

Simultaneously, two reports of microsporidiosis in humans with acquired immunodeficiency syndrome (AIDS) were published (Ledford et al., 1985; Modigliani et al., 1985). They represent two *Microsporida* distinctly different from each other and from human microsporidian infections previously reported.

PLEISTOPHORA SP.

Ledford et al. (1985) reported the presence of a *Pleistophora* species in the skeletal muscle of a 20-year-old black male from Florida. *Pleistophora* is often found in skeletal muscles of fish (Sprague and Vavra, 1977). This genus is defined by the following three characteristics. First, the genus is pansporoblastic, that is, a thick-walled, parasite-formed, parasitophorous vesicle is formed during the life cycle that isolates the developing parasite cells from direct contact with the host cell cytoplasm. Second, all cells that are produced during the parasite life cycle lack diplokarya. Third, the sporont cells undergo multiple cell divisions resulting in the production of a variable number of spores that exceeds eight and may exceed 100. This case represents the first report of a *Pleistophora* species in a mammal.

The patient developed fever, generalized lymphadenopathy, and progressive generalized muscle weakness which was followed by proximal and distal muscle contractures during a 7-month period and by an 18-kg weight loss. Muscles were firm but not inflamed, and muscle enzyme levels were normal. Electromyogram and nerve conduction findings were consistent with diffuse myopathy. Muscle biopsy specimens from his left quadriceps and left deltoid and 5 months later from his right deltoid showed microsporidia (see below). After treatment with trimethoprim/sulfisoxazole (20 mg and 100 mg/kg body weight per day, respectively), he became afebrile and his weight stabilized. After his second muscle biopsy specimen still showed the presence of microsporidians, his antibiotics were changed to sulfadiazine (75 mg/kg per day; Ledford et al., 1985).

Light microscopy of histologic sections from above biopsies revealed an intense inflammatory reaction of plasma cells, lymphocytes, and histiocytes. Scarring and fibrosis were common findings along with atrophic and degenerating muscle fibers infiltrated with spores in clusters of 12 or more enclosed in membranes. In phagocytic cells, single spores or groups of 2 or 3 were encountered. The spores stained acid fast with Ziehl-Neelsen stain, variable with Brown-Hopps stain, and contained a PAS+ granule in one end when stained with periodic acid-Schiff reagent.

Electron microscopic examination revealed 11 coils of the polar tubule inside a typical microsporidian spore (Fig. 22). The spores were oval, measuring 2.8 × 3.2 to 3.4 μm, mononucleated, and occurred in thick-walled parasitophorous vacuoles (Fig. 12).

In subsequent electron microscopic studies of infected muscle cells, developmental stages including meronts, sporonts, sporoblasts, and spores were identified (Cali et al., in preparation). The parasite meronts and sporonts are surrounded by secreted

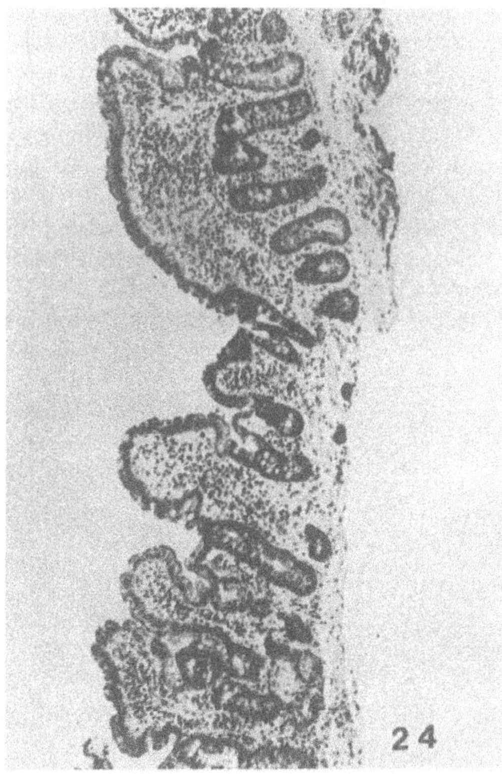

FIG. 24. Endoscopic jejunal biopsy specimen from a man with severe diarrhea and intestinal microsporida infection. Villi are shortened and crypts elongated from the normal ratio of 3 : 1 to 1 : 1. Goblet cells are depleted from the exposed surface, villi at the right are fused, and inflammatory cells in the lamina propria are increased. Microsporida were not detected with any light microscopy stain. This morphologic pattern is consistent with tropical sprue but is nondiagnostic (magnification ×70). (From Owen et al., in preparation.)

material that forms a honeycomb-like meshwork that becomes the parasitophorous vesicle wall (Fig. 23). The parasite cells may be multinucleate, but the nuclei do not occur in diplokaryon arrangement. The proliferative phase (merogony) ends when the plasmalemma thickens. The sporogonic phase includes much cell division: more than a dozen mononucleated sporoblasts are formed from each sporont and each sporoblast produces a spore (Fig. 12).

Enterocytozoon bieneusi

The second microsporidian infection found associated with AIDS is *Enterocytozoon bieneusi* (Desportes et al., 1985; Modigliani et al., 1985). *E. bieneusi* is a new genus and species of *Microsporida* created specifically for this case in the above publication by Desportes et al. Consequently, its description and this case description are one and the same.

This parasite possibly occurs more frequently than the other microsporidians in humans. One of us (A. C.) previously studied this parasite in an AIDS patient from Texas in 1983 (unpublished). Based on these findings a second patient (with AIDS) also from Texas was diagnosed. Additionally, we are currently studying still another case of this parasite from a man with AIDS in California (Owen et al., in preparation; Figs. 24, 25, and 26, and it occurs in a publication by Dobbins and Weinstein, 1985).

The published reports (Desportes et al., 1985; Modigliani et al., 1985) describe the infection from a 29-year-old Haitian who developed AIDS after living in France for 4 years. He suffered from diarrhea and malabsorption, a condition reported to be especially frequent in Haitian patients with AIDS (Malebranche et al., 1983). He had a 5-month history of diarrhea, a 17-kg weight loss, diffuse adenopathy, fever, and abdominal pain. He produced 0.3 to 1.4 liter of solid-to-watery nonbloody stools per day in one to three bowel movements. Malabsorption was documented by elevated output of fecal fat of 60.5 (normal < 21.6 mmol/day), and 120 min after ingestion of 25 g of D-xylose his serum level was only 0.53 (normal > 1.66 mmol/liter), indicating an absorptive defect in his bowel wall. A Schilling test with intrinsic factor for evaluation of B_{12} was 3 (normal $> 10\%$), indicating that his absorptive defect may have extended down as far as the terminal ileum where B_{12} is absorbed. Four upper small bowel and two lower small bowel biopsy specimens were taken that showed *Microsporida* but no viral inclusions. Stool examination revealed only *Giardia lamblia,* which cleared after treatment with pyrimethamine (50 mg daily for 5 days), sulphadiazine (2 g daily for 15 days), and metroniadazole (1.5 g daily for 7 days) but neither diarrhea nor intestinal function test improved. Eventually, he died with multiple extraintestinal opportunistic infections and also was found to have Kaposi's sarcoma at necropsy.

Histologic examination of the endoscopic biopsy specimens taken from the duodenojejunal and ileal mucosa revealed partial villus atrophy, mild inflammatory infiltrate, and round to oval basophilic bodies in approximately 50% of the villus enterocytes (Figs. 24 and 25). Electron microscopic observations of spores containing the coiled polar tubule confirm a

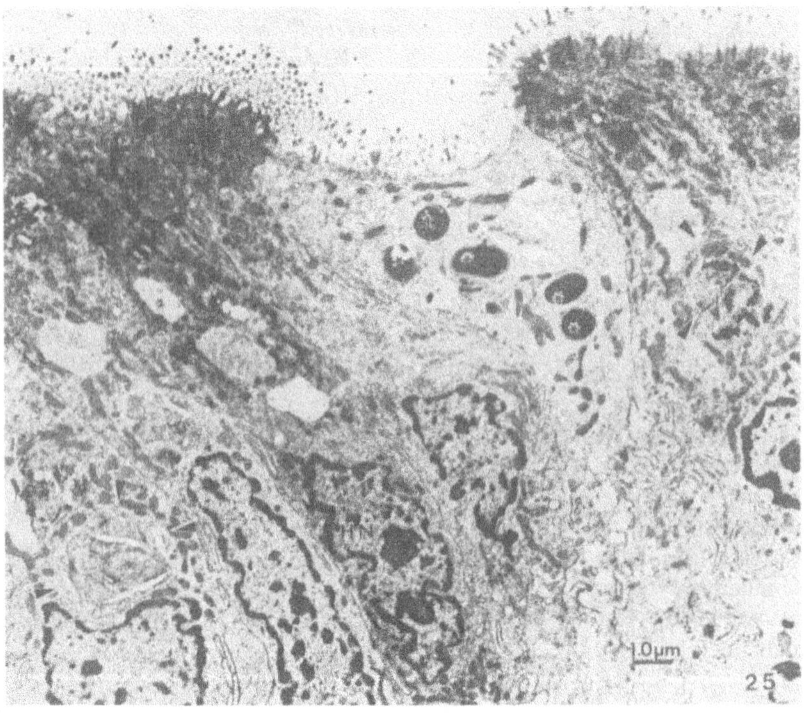

FIG. 25. Mature oval condensed microsporida spores (*) in the cytoplasm of jejunal enterocytes near the villus tip. Spores can be differentiated from an adjacent multivesicular lysosomal body (L) by their spore coats and oval shape. Dividing sporogonial forms containing spiral polar tubules are present in adjacent enterocytes (*arrows*). (From Owen et al., in preparation.)

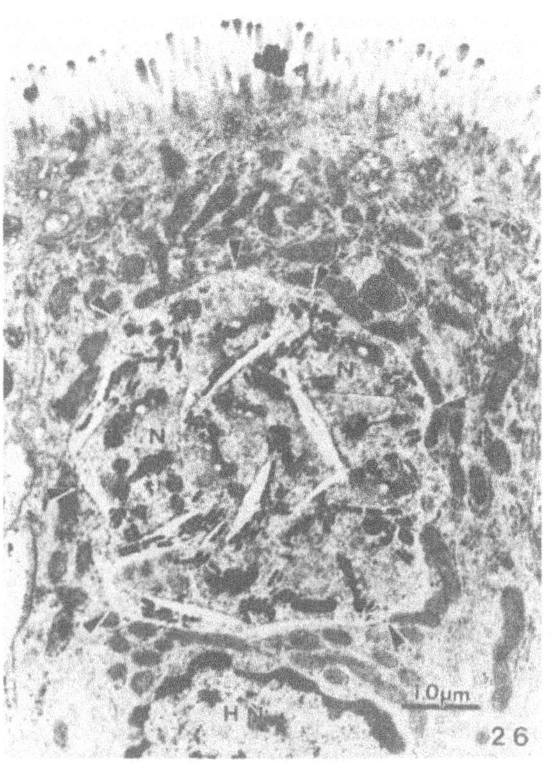

FIG. 26. Microsporida undergoing sporogonial division between the luminal microvilli and the nucleus (HN) of a jejunal enterocyte from a man with severe diarrhea. At least 11 nuclei (N) with associated vacuoles and polar tubes (*) are seen within the parasite outer membrane (*arrows*). (From Owen, in press.)

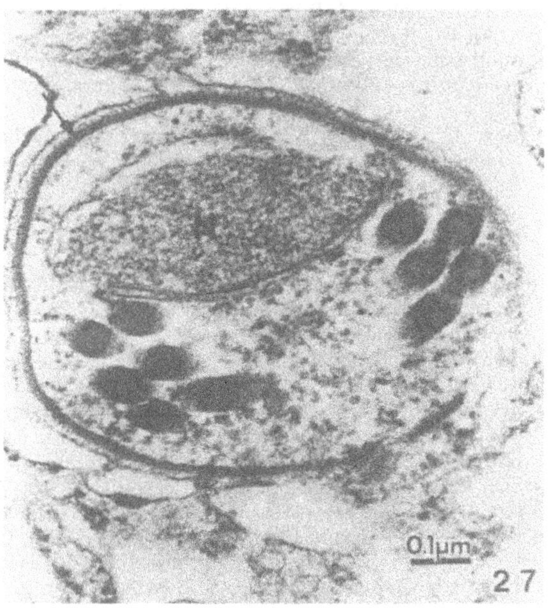

FIG. 27. Sporoblast after the fission of the sporogonial plasmodium. This stage is characterized by the occurrence of the spore wall (*arrow*). Note the presence of a nucleus (N) and the polar tube. (From Desportes et al., 1985.)

microsporidian diagnosis (Fig. 25). All stages of the parasite are found in direct contact with the host cell cytoplasm. The early proliferative cells contain nuclei in diplokaryotic arrangement, but the maturing sporogonial plasmodia contain multiple nuclei in the uninucleate form surrounded by polar tubule coils (Fig. 26). Modigliani et al. (1985) described multinucleated sporonts as being 1 to 5 μm and Desportes et al. (1985) suggested that this stage probably produces eight sporoblasts by multiple fission. Multiple spore extrusion apparati are found in the sporogonial plasmodium before it divides (Fig. 26). The spores lack a thick endospore coat and contain a single nucleus and about five coils of the polar tubule (Fig. 27). The mature condensed spore size is 1.5×0.5 μm in tissue sections (Desportes et al., 1985).

Summary

Microsporida are obligate intracellular protozoa, which are ubiquitous throughout the animal kingdom and characterized by spores containing coiled polar tubules that evert and penetrate host cells into which spore nucleus and cytoplasm are extruded. The infection is usually acquired from ingestion of or mucous membrane exposure to spores from feces, urine, or uncooked tissues of infected hosts. Because of their small size and poor staining properties, microsporidians frequently escape recognition. Only four genera, out of hundreds of species of *Microsporida,* that have been described, have thus far been associated with disease in humans. These microsporidians are *Pleistophora,* which has been associated with myositis, *Nosema* with diarrhea and eye infections, *Encephalitozoon* with hepatitis, and *Enterocytozoon* with diarrhea and intestinal mucosal colonization. In most instances, clinical infections with microsporidians have only been identified in patients with immunosuppression or immunodeficiency. In such patients, level of infection is high and persists long enough for diagnosis. Serologic studies suggest that microsporidian infections in humans may be much more common than medical experience would indicate. In patients with normal immune mechanisms, acute self-limited infections may give rise to syndromes currently ascribed by exclusion to viruses. Heightened awareness of the *Microsporida* as infectious agents and better diagnostic approaches can be expected to reveal a greater incidence of human microsporidian infection and a much broader spectrum of clinical responses as has been observed in various animal species.

Acknowledgments. We wish to thank Oresta Fedun for drawing the illustrations, Peter Takvorian for assistance with the figures, Edward Hanrahan for assistance with the manuscript, and the Armed Forces Institute of Pathology for reuse of figures.

Literature Cited

Ashton, N., and P. A. Wirasinha. 1973. Encephalitozoonosis of the cornea. Br. J. Ophthalmol. **57:**669–674.

Attwood, H. D., and R. D. Sutton. 1965. *Encephalitozoon* granulomata in rats. J. Pathol. Bacteriol. **89:**735–738.

Barker, R. J. 1974. The nature of *Encephalitozoon brumpti* Coulon, 1924. J. Parasitol. **60:**542–544.

Bennett, D., and O. Radimska. 1966. Flotation-fluid staining; toluidine blue applied to Maraglas section. Stain Technol. **41:**349.

Bergquist, N. R., G. Stintzing, L. Smedman, T. Waller, and T. Andersson. 1984a. Diagnosis of encephalitozoonosis in man by serological tests. Br. Med. J. **288:**902.

Bergquist, N. R., L. Morfeldt-Monnson, P. O. Pehrson, B. Petrini, and J. Wasserman. 1984b. Antibody against *Encephalitozoon cuniculi* in Swedish homosexual men. Scand. J. Infect. Dis. **16:**389–391.

Blazek, K., I. Koczkova, M. Lavicka, J. Vavra, and M. Stehlik. 1972. Nosematosis (Encephalitozoonosis) bei Karnivoren. Verhandlungsberichte. XIV Internat. Symp. Enkrankungen der Zootiere, Acad. Verlag DDR, 369–371.

Bywater, J. E. C., B. S. Kellett, and T. Waller. 1980. *Encephalitozoon cuniculi* antibodies in commercially-available rabbit antisera and serum reagents. Lab. Animal **14:**87–89.

Cali, A. 1970. A fine structural study of *Nosema apis* (Microsporida: Cnidospora: Protozoa). The Ohio State Univ. Dissertation. Univ. Microfilm #70-26,258.

Cali, A. 1971. Morphogenesis in the genus *Nosema, p. 431–438.* In Proc. IVth Int. Coll. Insect Pathol. Maryland, 1970.

Cali, A. 1986. Comparison of the biology and pathology of *microsporida* from different host groups. *In* Samson, R. A., Vlak, J. M. and Peters, D. (ed.), Fundamental and applied aspects of invertebrate pathology. Foundation for the 4th International Colloquim on Invertebrate Pathology, Wageningen, The Netherlands.

Canning, E. U., L. P. Foon, and L. K. Joe. 1974. Microsporidian parasites of trematode larvae from aquatic snails in West Malasia. J. Protozool. **21:**19–25.

Canning, E. U., and J. Lom. 1986. The *microsporidia* of vertebrates, p. 289. Academic Press Inc., New York.

Cole, J. R., L. T. Sangster, C. R. Sulzer, A. R. Pursell, H. C. Ellinghausen. 1982. Infections with *Encephalitozoon cuniculi* and *Leptospira interrogans,* serovars *grippotyphosa* and *ballum,* in a kennel of foxhounds. J. Am. Vet. Med. Assoc. **180:**435–437.

Coulon, G. 1929. Presence d'un nouvel *Encephalitozoon* (*Encephalitozoon brumpti* n. sp.) dans le liquide cephalo-rachidien d'un sujet atteint de meningte suraigue. Ann. Parasitol. Hum. Comp. **7:**449–452.

Cowdry, E. V., and F. M. Nicholson. 1924. Meningo-encephalitic lesions and protozoan-like parasites in brains of apparently normal laboratory animals commonly employed for experimentation. J. Am. Med. Assoc. **82:**545.

Cox, J. C., R. Horsburgh, and D. Pye. 1981. Simple diagnostic test for antibodies to *Encephalitozoon cuniculi* based on enzyme immunoassay. Lab. Animal **15:**41–43.

Current, W. L., and R. L. Owen. 1988. Cryptosporidiosis and Microsporidiosis. *In* M. J. G. Farthing and G. T. Keusch (ed.), Enteric infection: mechanisms, manifestations, and management. Chapman and Hall, Ltd., London.

Desportes, I., Y. Le Charpentier, A. Galian, F. Bernard, B. Cochand-Priollet, A. Lavergne, P. Ravisse, and R. Modigliani. 1985. Occurrence of a new microsporidian: *Enterocytozoon bieneusi* n.g., n. sp. in the enterocytes of a human patient with AIDS. J. Protozool. **32:**250–254.

Dobbins, W. O., and W. N. Weinstein. 1985. Electron microscopy of the intestine and rectum in acquired immunodeficiency syndrome. Gastroenterology **88:**738–749.

Doby, J. M., A. Jeannes, and B. Rault. 1963. *Thelohania apodemi* n. sp., premiere microsporidie du genre *Thelohania* observee chez un Mammifere. C. R. Acad. Sci. Paris **257:**248–251.

Gannon, J. 1980. The course of infection of Encephalitozoon cuniculi in immunodeficient and immunocompetent mice. Lab. Animal **14:**189–192.

Gordon, S. C., K. R. Reddy, E. E. Gould, R. McFadden, C. O'Brien, M. De Medina, L. J. Jeffers, and E. R. Schiff. 1986. The spectrum of liver disease in the acquired immunodeficiency syndrome. J. Hepatology **2:**475–484.

Gray, H., A. Cali, and J. D. Briggs. 1969. Intracellular stages in the life cycle of the microsporidan *Nosema apis*. J. Invert. Pathol. **14:**391–394.

Grimley, P. M. 1964. A tribasic stain for thin sections of plastic-embedded, OsO4-fixed tissues. Stain Technol. **39:**229.

Kellett, B. S., and J. E. C. Bywater. 1980. The indirect india-ink immunoreaction for detection of antibodies to *Encephalitozoon cuniculi* in rat and mouse serum. Lab. Animal **14:**83–86.

Kemp, R. L., and J. P. Kluge. 1975. *Encephalitozoon* sp. in the blue-masked lovebird, *Agapornis personata* (Reichenow): first confirmed report of microsporidian infection in birds. J. Protozool. **22:**489–491.

Khanna, R. S., and P. K. Iyer. 1971. A case of *Nosema cuniculi* infection in a goat. Ind. J. Med. Res. **59:**993–995.

Lainson, R., P. C. C. Garnham, R. Killick-Kendrick, and R. G. Bird. 1964. Nosematosis, a microsporidial infection of rodents and other animals, including man. Br. Med. J. **22:**470–472.

Ledford, D. K., M. D. Overman, A. Gonzalvo, A. Cali, S. W. Mester, and R. F. Lockey. 1985. Microsporidiosis myositis in a patient with the acquired immunodeficiency syndrome. Ann. Intern. Med. **102:**628–630.

Levaditi, C., S. Nicolau, and R. Schoen. 1923a. L'etiologie de l'encephalite. C. R. Acad. Sci. Paris. **177:**985–988.

Levaditi, C., S. Nicolau, and R. Schoen. 1923b. L'agent etiologique de l'encephalite epizootique du lapin (*Encephalitozoon cuniculi*). C. R. Soc. Biol. Paris **89:**984–986.

Levaditi, C., S. Nicolau, and R. Schoen. 1923c. Nouvelles donnees sur l'*Encephalitozoon cuniculi*. C. R. Soc. Biol. Paris **89:**1157–1162.

Levaditi, C., S. Nicolau, and R. Schoen. 1924a. Nouvelles recherches sur l'*Encephalitozoon cuniculi*. C. R. Soc. Biol. Paris **40:**662–666.

Levaditi, C., S. Nicolau, and R. Schoen. 1924b. Nouvelles recherches sur l'*Encephalitozoon cuniculi*. C. R. Soc. Biol. Paris **40:**662–666.

Levaditi, C., S. Nicolau, and R. Schoen. 1924c. L'etiologie de l'encephalite epizootique du lapin, dans ses rapports avec l'etude experimentale de l'encephalite lethargique *Encephalitozoon cuniculi* (nov. spec.). Ann. Inst. Pasteur. Paris **38:**651–711.

Levine, N. D., J. O. Corliss, F. E. G. Cox, G. Deroux, J. Grain, B. M. Honigberg, G. F. Leedale, A. R. Loeblich, III, J. Lom, D. Lynn, E. G. Merinfeld, F. C. Page, G. Poljansky, V. Sprague, J. Vavra, and F. G. Wallace. 1980. A newly revised classification of the *Protozoa*. J. Protozool. **27:**37–58.

Lowenstein, L. J., and M. L. Petrak. 1980. Microsporidiosis in two peach-faced lovebirds, p. 365–368. *In* R. J. Montali and G. J. Migaki (ed.), The comparative pathology of zoo animals, Proc. Symp. Nat. Zool. Park,

Smithsonian Institution, 1978. Smithsonian Institution Press, Washington, D. C.
Luna, L. G. 1968. Manual of histologic staining methods of the Armed Forces Institute of Pathology (3rd ed.), p. 258. McGraw-Hill Book Company, New York.
Majeed, S. K., and A. J. Zubaidy. 1982. Histopathological lesions associated with *Encephalitozoon cuniculi* (nosematosis) infection in a colony of Wistar rats. Lab. Animal **16:**244–247.
Malebranche, R., E. Arnoux, J. M. Guerin, G. D. Pierre, A. C. Laroche, C. Pean-Guichard, R. Elie, P. H. Maurisset, T. Spira, R. Mandeville, P. Brotman, T. Seemayer, and J. M. Dupuis. 1983. Acquired immunodeficiency syndrome with severe gastrointestinal manifestations in Haiti. Lancet **2:**873–878.
Marcus, P. B., J. J. van der Walt, and P. J. Burger. 1973. Human tumor microsporidiosis. Arch. Pathol. **95:**341–343.
Margileth, A. M., A. J. Strano, R. Chadra, R. Neafie, M. Blum, and R. M. McCully. 1973. Disseminated nosematosis in an immunologically compromised infant. Arch. Pathol. **95:**145–150.
Matsubayashi, H., T. Koike, I. Mikata, H. Takei, and S. Hagiwara. 1959. A case of *Encephalitozoon*-like body infection in man. Arch. Pathol. **67:**181–187.
Modigliani, R., C. Bories, Y. LeCharpentier, M. Salmeron, B. Messing, A. Galian, J. C. Rambaud, and I. Desportes. 1985. Diarrhoea and malabsorption in acquired immune deficiency syndrome: a study of four cases with special emphasis on opportunistic protozoan infestations. Gut **26:**179–187.
Mohn, S. F., T. Landsverk, and K. Nordstoga. 1981. Encephalitozoonosis in the blue fox-morphological identification of the parasite. Acta Pathol. Microbiol. Scand. **89:**117–122.
Mohn, S. F. 1983. Encephalitozoonosis in the blue fox (Alopex lagopus) transmission, diagnosis and control (thesis). National Veterinary Institute, Oslo, Norway.
Montrey, R. D., J. A. Shadduck, and S. P. Pakes. 1973. In vitro study of host range of three isolates of *Encephalitozoon* (*Nosema*). J. Infect. Dis. **127:**450–454.
Nageli. 1857. *Nosema bombycis* Nageli. Bot. Zeit. **15:**760.
Nelson, L. R., O. B. Mock, and R. E. Flatt. 1969. Nosematosis in the shrew. 20th Ann. Sci. Session, Am. Ass. Lab. Anim. Sciences.
Novilla, M. N., J. W. Carpenter, and R. P. Kwapien. 1980. Dual infection of Siberian polecats with *Encephalitozoon cuniculi* and *Hepatozoon mustelis* n.sp, p. 353–363. *In* R. J. Montali and G. Migaki (ed.), Comparative pathology of zoo animals. Proc. Symp. Nat. Zool. Park, Smithsonian Institution, Washington, D.C., 1978.
Novilla, M. N., and R. P. Kwapien. 1978. Microsporidian infection in the pied peach-faced lovebird (*Agapornis roseicollis*). Avian Dis. **22:**198–204.
Owen, R. L. 1988. Parasitic disease Ch. 62. *In* M. H. Sleisenger and J. S. Fordtran (ed.), Gastrointestinal disease. (4th ed.). W. B. Saunders Co., Philadelphia, PA.
Pakes, S. P., J. A. Shadduck, and A. Cali. 1975. Fine structure of *Encephalitozoon cuniculi* from rabbits, mice and hamsters. J. Protozool. **22:**481–488.
Pakes S. P., J. A. Shadduck, D. B. Feldman, and J. A. Moore. 1984. Comparison of tests for the diagnosis of spontaneous encephalitozoonosis in rabbits. Lab. Animal Sci. **34:**356–359.
Pang, V. F., and J. A. Shadduck. 1985. Susceptibility of cats, sheep, and swine to a rabbit isolate of *Encephalitozoon cuniculi*. Am. J. Vet. Res. **46:**1071–1077.
Perrin, T. L. 1943. Spontaneous and experimental *Encephalitozoon* infection in laboratory animals. Arch. Pathol. **36:**559–567.
Petri, M. 1969. Studies on *Nosema cuniculi* found in transplantable ascites tumours with a survey of microsporidosis in mammals. Acta Pathol. Microbiol. Scand. **204:**1–91.
Plowright, W. 1952. An encephalitis-nephritis syndrome in the dog, probably due to congenital *Encephalitozoon* infection. J. Comp. Pathol. **62:**83–92.
Pinnolis, M., P. R. Egbert, R. L. Font, and F. C. Winter. 1981. Nosematosis of the cornea. Case report, including electron microscopic studies. Arch. Ophthalmol. **99:**1044–1047.
Poonacha, K. B., P. D. William, and R. D. Stamper. 1985. Encephalitozoonosis in a parrot. J. Am. Vet. Med. Assoc. **186:**700–702.
Sebek, Z. 1969. The finding of the genus *Nosema* Nageli, 1857 (*Microsporidia*, Nosematidae) in the kidney of a guinea pig. Folia Parasitol. **16:**165–169.
Seibold, H. R., and E. N. Fussell. 1973. Intestinal microsporidiosis in *Callicebus moloch*. Lab. Anim. Sci. **23:**115–118.
Shadduck, J. A., W. T. Watson, S. P. Pakes, and A. Cali. 1979. Animal infectivity of *Encephalitozoon cuniculi*. J. Parasitol. **65:**123–129.
Shadduck, J. A., G. Kelsoe, and J. Helmke. 1979. A microsporidan contaminant of a nonhuman primate cell culture: ultrastructural comparison with *Nosema connori*. J. Parasitol. **65:**185–188.
Singh, M., G. J. Kane, L. Mackinley, I. Quaki, Y. E. Hian, H. B. Chuan, H. L. Chuen, and L. K. Chew. 1982. Detection of antibodies to *Nosema cuniculi* (Protozoa: *Microsporidia*) in human and animal sera by the indirect fluorescent antibody technique. Southeast Asia J. Trop. Med. Public Health **13:**110–113.
Sprague, V., and S. H. Vernick. 1971. The ultrastructure of *Encephalitozoon cuniculi* (*Microsporida*, Nosematidae) and its taxonomic significance. J. Protozool. **18:**560–569.
Sprague, V. 1974. *Nosema connori* n. sp., a microsporidian parasite of man. Trans. Am. Microsc. Soc. **93:**400–403.
Sprague, V., and J. Vavra. 1977. Systematics of the *microsporidia*, p. 510. *In* L. A. Bulla and T. C. Cheng (ed.), Comparative pathobiology, Vol. 2. Plenum Press, New York.
Strano, A. J., A. Cali, and R. Neafie. 1976. Microsporidiosis, p. 336–339. *In* C. H. Binford and D. H. Connor (ed.), Pathology of tropical and extraordinary diseases, Vol. 1. Armed Forces Institute of Pathology Press.
Takvorian, P., and A. Cali. 1986. The ultrastructure of spores (Protozoa: *Microsporida*) from *Lophius americanus*, the angler fish. J. Protozool. **33:**570–575.
Terada, S., R. Reddy, L. J. Jeffers, A. Cali, and E. R. Schiff. 1987. Microsporidan hepatitis in a patient with the acquired immunodeficiency syndrome. Ann. Intern. Med. **107:**66–67.
Togubaye, B. S., and B. Marchand. 1984. Etude ultrastructurale de *Unikaryon matteii* n. sp. (*Microsporida*, Unikaryonidae) parasite de *Nisotra* sp. (Coleoptera, Chrysomelidae) et remarques sur la validite de certaines *Nosema* d'insects. J. Protozool. **31:**339–346.
Torres, C. M. 1927. Morphologie d'un nouveau parasite l'homme, *Encephalitozoon chagasi*, n. sp. observe dans un cas de meningoencephalomyelite congenitale avec myosite et myocardite. C. R. Soc. Biol. Paris **97:**1781–1790.
Vavra, J., K. Blazek, N. Lavicka, I. Koczkova, S. Kalafa, and M. Stehlik. 1971. Nosematosis in carnivores. J. Parasitol. **57:**923–924.
van Dellen, A. F., W. S. Botha, J. Boomker, and W. E. J. Warnes. 1978. Light and electron microscopical studies on canine encephalitozoonosis: cerebral vasculitis. Onderstepoort J. Vet. Res. **45:**165–186.

van Rensburg, I. B. J., and J. L. du Plessis. 1971. Nosematosis in a cat: a case report. J. South Afr. Vet. Med. Assoc. **42**:327–331.

Weiser, J. 1964. On the taxonomic position of the genus *Encephalitozoon* Levaditi, Nicolau and Schoen, 1923 (Protozoa, Microsporida). Parasitology **54**:749–751.

Wosu, N. J., R. Olsen, J. A. Shadduck, A. Koestner, and S. P. Pakes. 1977a. Diagnosis of experimental encephalitozoonosis in rabbits by complement fixation. J. Infect. Dis. **135**:944–948.

Wosu, N. J., J. A. Shadduck, S. P. Pakes, J. K. Frenkel, K. S. Todd, and J. D. Conroy. 1977b. Diagnosis of encephalitozoonosis in experimentally infected rabbits by intradermal and immunofluorescence tests. Lab. Anim. Sci. **27**:210–216.

WHO Parasitic Diseases Surveillance. 1983. Antibody to *Encephalitozoon cuniculi* in man. WHO Weekly Epidem. Rec. **58**:30–32.

Wolf, A., and D. Cowen. 1937. Granulomatous encephalomyelitis due to an Encephalitozoon (Encephalitozoic Encephalomyelitis) a new protozoan disease of man. Bull. Neurol. Inst. New York. **6**:306–371.

Wolf, A., D. Cowen, and B. H. Page. 1939. Toxoplasmic encephalomyelitis Part III. A new case of granulomatous due to a protozoan. Am. J. Pathol **15**:657–694.

Zeman, D. H., and G. B. Baskin. 1985. Encephalitozoonosis in squirrel monkeys (*Saimiri sciureus*). Vet. Pathol. **22**:24–31.

CHAPTER 99

Paragonimiasis

JOHN H. CROSS

Disease: Paragonimiasis (oriental lung fluke disease, pulmonary distomiasis or trematodiasis, parasitic hemoptysis, endemic hemoptysis).

Etiologic Agents: *Paragonimus westermani*; *Paragonimus skrjabini* = *Paragonimus szechuanensis*; *Paragonimus miyazakii*; *Paragonimus heterotremus* = *Paragonimus tuanshanensis*; *Paragonimus africanus*; *Paragonimus uterobilateralis*; *Paragonimus mexicanus* = *Paragonimus peruvianus* = *Paragonimus ecuadoriensis*. Other species reported that may or may not infect humans.

Source: Metacercariae from crustaceans, especially raw or poorly cooked or wine-soaked crabs or crayfish. Juice containing metacercariae squeezed from crustaceans used for seasoning and medicinal purposes. Immature stages in wild animal meat eaten raw.

Clinical Manifestations: Acute symptoms during early stages of worm migration; diarrhea and abdominal pain, fever, chills, chest pain, and cough. Chronic infections cause cough, chest pain, eosinophilia, hemoptysis, and tenacious bloody sputum. Central nervous system, especially the brain, abdominal cavity and viscera, and skin may be invaded by the parasites and cause symptoms.

Pathology: A fibrous cyst surrounds the worms and fills with parasite eggs, necrotic material, and Charcot-Leyden crystals. Eggs in lung tissue produce granulomas, and abscesses form in the abdominal wall, liver, genitals, skin, and other organs. Adults produce granulomas in brain tissue.

Laboratory Diagnosis: Detection of *Paragonimus* sp. eggs in blood-streaked sputum, stools, or pulmonary exudates. Skin test reaction to adult worm antigen is usually positive and remains positive for years. The complement fixation test with similar antigen is a test for cure; the ELISA is the most sensitive and specific test available.

Epidemiology: *Paragonimus* species are reported worldwide, but most human disease is reported from Asian populations who eat raw or partially cooked crabs and crayfish. Paragonimiasis in Africa, Central and South America, and parts of China involves species that invade cutaneous and subcutaneous tissues. Freshwater snails are first intermediate hosts, crustaceans the second intermediate hosts, and carnivorous animals are reservoir hosts. Many imported cases of paragonimiasis from Southeast Asian immigrants are being reported in the United States and other countries accepting these refugees.

Treatment: Praziquantel, 25 mg/kg body weight three times a day by mouth for 3 days, is the drug of choice. Bithionol, 30 to 50 mg/kg body weight in divided oral doses two times a day every other day for 10 to 15 days, and Niclofolan, 25 mg/kg body weight by mouth three times a day for 3 days, are effective. Patients must be followed for 6 months and retreated if necessary. These drugs are also effective in treating extrapulmonary paragonimiasis, but retreatment may be required. Surgical intervention may be necessary in some patients with extrapulmonary infections.

Prevention and Control: Change eating habits, education of the population, destruction of intermediate and reservoir hosts, and treatment of infected persons or entire populations may be necessary.

The opinions or assertions contained herein are the private ones of the author and are not to be construed as official or reflecting the views of the Department of Defense or the Uniformed Services University of the Health Sciences.

Description of Disease

Paragonimiasis is a chronic disease of the lungs and other organs caused by trematodes or flukes of the genus *Paragonimus*. It is a zoonotic disease with infection occurring in wild and domestic carnivorous animals as well as in people who ingest raw or partially cooked crustaceans. Infection with *Paragonimus* species have been reported worldwide, but the major endemic areas are located in Asia where populations relish eating uncooked crabs and crayfish.

There are 48 species of *Paragonimus* reported (Yokogawa, 1986), and, although many are possible pathogens of humans, most are parasites of lower animals and not all are accepted as valid species. Seven species are considered important causes of paragonimiasis in humans: *Paragonimus westermani* reported from China, Japan, Korea, Eastern Russia, Taiwan, Philippines, Indonesia, Malaysia, Thailand, Laos, India, and Sri Lanka; *P. skrjabini* or *P. szechuanensis* in China; *P. miyazakii* in Japan; *P. heterotremus* or *P. tuanshanensis* in China, Thailand, and Laos; *P. africanus* in Cameroon and Nigeria; *P. uterobilateralis* in Cameroon, Nigeria, Liberia, and Guinea; and *P. mexicanus* or *P. peruvianus* or *P. ecuadoriensis* in Central and South America. In the United States and Canada, *P. kellicotti* has been reported on rare occasions and *P. westermani* in immigrants from Asia.

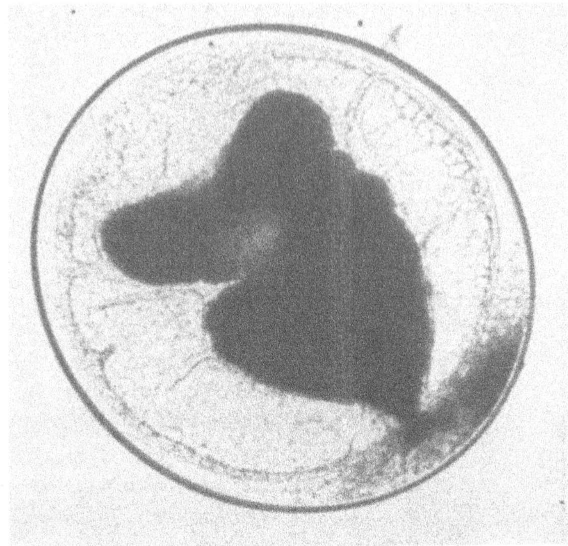

FIG. 1. Metacercaria from a crab; *Sundathelphusa* sp. from Southern Luzon, Philippines.

Communicability and Transmissibility

Paragonimiasis is acquired by humans who eat raw or improperly cooked freshwater crabs and crayfish. The infective stage of the parasite, a metacercaria (Fig. 1), is encysted in the gills, muscles, and viscera. When the crustacean host is eaten raw or juices containing the metacercariae are squeezed from the animal and ingested by the definitive host, the parasites migrate from the intestinal tract to the host's abdominal cavity. *P. westermani* infections also have been acquired by eating raw boar meat containing immature worms. The incubation period in pulmonary paragonimiasis is highly variable with symptoms in humans reported as early as 2 to 30 days to as long as several months. The prepatent period, or first appearance of eggs in the sputum, however, is 8 to 10 weeks. Some worms have been known to live for as long as 20 years.

Clinical Manifestations

Clinical manifestations are associated with the organ that is invaded by the parasite. Most clinical manifestations are pulmonic as the lungs are most commonly invaded. The parasites can also produce lesions in the central nervous system (CNS), abdominal cavity, skin, and at many other locations.

Parasite Migration

Early symptoms are associated with larval migration in which patients experience diarrhea and abdominal pain resulting from the parasite's invasion of the digestive tract and penetration into the abdominal cavity.

Pulmonary Paragonimiasis

The early symptoms are followed by fever, chills, chest pain, and cough as the parasite migrates into the lung. Once the parasite becomes established in the tissues, there may be general malaise, sweats, cough, chest pain, and eosinophilia. Pleurisy with hemoptysis, pleural effusion, persistent rales, pneumothorax, and clubbing of the fingers also occur. In chronic cases, usually only cough and blood-streaked sputum are reported.

Cerebral Paragonimiasis

The brain and spinal cord are more frequently invaded by *P. westermani* than other extrapulmonary sites. Lesions in the brain are usually caused by the migration of adult worms from the lungs; therefore, pulmonary paragonimiasis usually precedes cerebral paragonimiasis. Cerebral symptoms begin about 10 months after the appearance of pulmonary symptoms. The clinical manifestations are suggestive of an abscess, brain tumor, or Jacksonian epilepsy. Headache, vomiting, seizures, fever, paralysis, and visual disturbances may result. Eggs of the parasite are usually found within the abscess. In chronic cases, the abscess may become calcified. Cerebral paragonimiasis may be caused by species other than *P. westermani*.

Abdominal and Visceral Paragonimiasis

P. westermani and other species of *Paragonimus* may make an erratic migration in the abdominal cavity. Diarrhea, abdominal pain, and abscess formation in the liver and spleen are reported. Hepatic damage with eosinophilic abscess and abnormal liver function is occasionally seen with *P. skrjabini* in China (Hu et al., 1982).

Cutaneous Paragonimiasis

Adult and larval *Paragonimus* may migrate to the skin and cause permanent or transient subcutaneous eosinophilic nodules. This has been reported occasionally with *P. westermani* and more often with *P. skrjabini* in China and *P. mexicanus* in Latin America.

Paragonimiasis in Other Organs

Paragonimus species have been reported in many other locations in the body. Both adult and larval worms have been recovered from the eye, pericardium, myocardium, lymph nodes, male and female genital organs, face, bone marrow, urinary bladder, neck, and gluteal region.

Pathology

Pathologic changes are associated with the location of the worms.

Pulmonary Paragonimiasis

The young worms passing through the intestinal wall, abdominal cavity, diaphragm, and into the lung are assisted in their migration by the secretion of lytic substances that digest tissue. The worms settle in areas close to bronchi and an inflammatory reaction develops around the worms with infiltration of neutrophils and eosinophils. A fibrous cyst (Fig. 2) eventually forms containing one, two, or more worms; cellular debris; brownish pus; Charcot-Leyden crystals; and eggs of the parasite. Eggs in the lung tissue provoke granulomas with the infiltration of eosinophils and lymphocytes and, eventually, giant cells and fibroblasts.

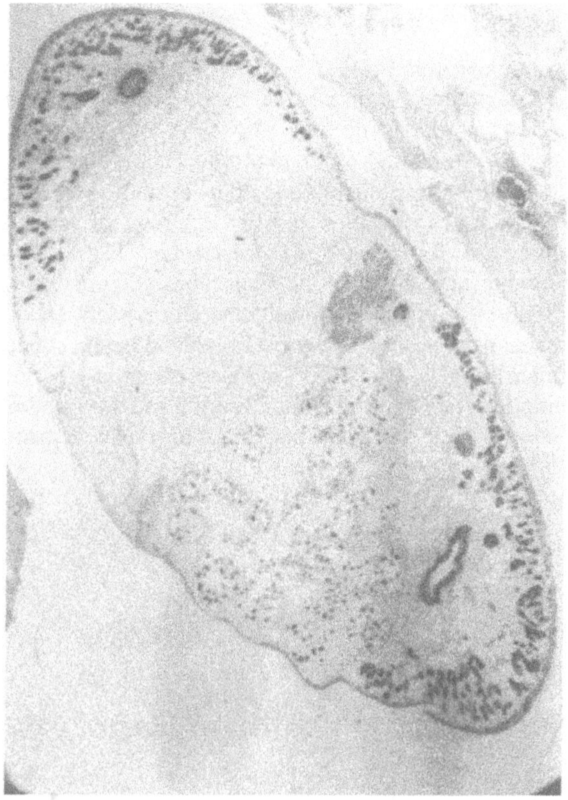

FIG. 2. Lung tissue with section of *Paragonimus westermani*.

Cerebral Paragonimiasis

Adult worms that first develop in the lung and migrate to the brain are thought to be the cause of cerebral paragonimiasis. The parasite lies in a cavity, is surrounded by a membrane, and eventually surrounded by a cyst which becomes necrotic and fills with eggs, cellular debris, and Charcot-Leyden crystals. Small granulomas may form around eggs that enter surrounding tissue.

Abdominal Paragonimiasis

Erratic worms may become encysted in any organ or tissue of the body. The parasite becomes surrounded by an inflammatory reaction forming an abscess.

Cutaneous Paragonimiasis

Several reports describe *Paragonimus* species in the skin and subcutaneous tissue. Worms found at cutaneous sites are usually immature and are difficult to identify specifically. Migrating subcutaneous nodules are not unusual. When cysts form, they contain the parasite, an eosinophilic infiltrate, and may be abscessed.

Diagnostic Features

Egg Detection

Pulmonary paragonimiasis westermani should be considered in Asians who eat raw or partially cooked crabs and crayfish and who have symptoms of cough, fever, and produce a tenacious bloody sputum. Sputum specimens should be examined directly or after centrifugation in a 2% solution of sodium hydroxide for the eggs of *P. westermani* (Fig. 3). The sputum may also contain Charcot-Leyden crystals. Sputum is often swallowed; therefore, stool specimens from suspected cases also should be examined microscopically by direct and after formalin-ether or ethyl acetate concentration. Sputum and stool specimens should be examined simultaneously at least three times in patients suspected of having paragonimiasis. Aspirates from pleural effusions may also contain *Paragonimus* eggs.

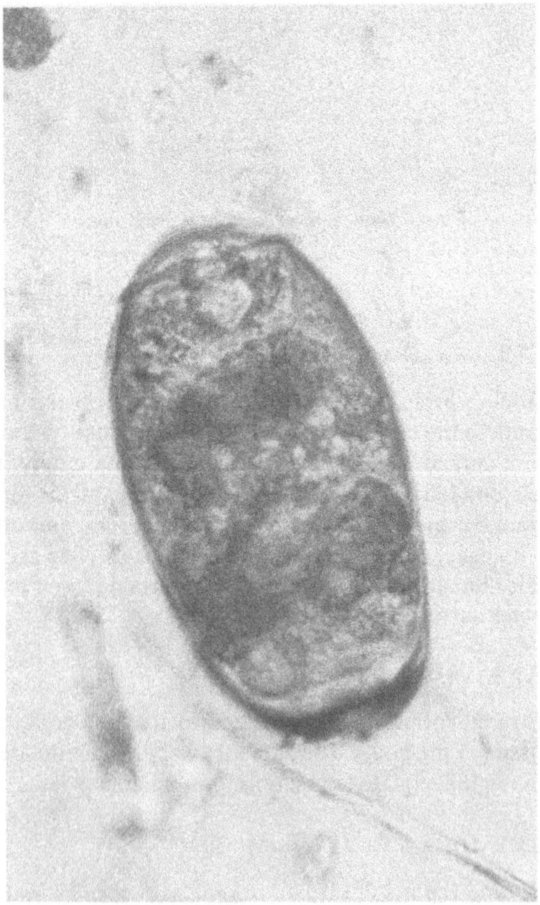

FIG. 3. Egg of *Paragonimus westermani*. Eggs measure 80 to 120 × 50 to 60 μm, are yellowish-brown in color, and have a large flattened operculum and pronounced operculum villi.

Radiographic Examination

Chest x-rays may show abnormalities, but it is not possible to differentiate pulmonary paragonimiasis from tuberculosis. Many patients are treated for tuberculosis with little success until it is realized that they have paragonimiasis. Cerebral angiography or computer tomographic scan may be of value in cerebral and hepatic paragonimiasis (Hara et al., 1985).

Immunodiagnosis

Although a number of immunodiagnostic tests are available for paragonimiasis, a positive serologic reaction is only suggestive of infection. A confirmed diagnosis can be made only after demonstration of the parasite eggs in sputum, stool, or biopsy specimen. Serologic tests may be helpful, especially for extrapulmonary paragonimiasis and in establishing treatment failure or success. An intradermal test with a veronal buffered saline extract (VBS) of adult *P. westermani* is usually positive in persons with the parasitosis. The test is of value in epidemiologic studies, but reactions remain positive for a number of years.

A complement fixation test (CFT) using the VBS antigen correlates well with active infections. The CFT with *P. westermani* antigen may cross-react with other species of *Paragonimus*, however. Precipitin tests, such as those used in gel diffusion, immunoelectrophoresis, and counterimmunoelectrophoresis, have been of diagnostic value as well as in species identification (Knobloch, 1984). The enzyme-linked immunosorbent assay (ELISA) gives excellent results with high sensitivity and specificity. Serologic tests are of most value in providing a presumptive diagnosis of cerebral and other extrapulmonary *Paragonimus* infections.

Prognosis

Most cases of pulmonary paragonimiasis are self-limiting, lasting 5 to 10 years without reinfection. Pulmonary infections involving large numbers of worms cause more disease and, consequently, a poor prognosis. Death is rare in pulmonary infection, but lesions in ectopic locations caused by parasites in the central nervous system, pericardium, or visceral organs may cause severe disease. Specific treatment improves the prognosis. Secondary bacterial infections, on the other hand, may lead to complications.

Epidemiologic Aspects

Paragonimiasis occurs throughout the world where there are susceptible snails and crustaceans and people who eat crustaceans uncooked.

Occurrence and Transmission

The reported 48 species are from various areas of the world. Human infections are documented from China, Japan, Korea, Russia, Taiwan, Indonesia, Malaysia, Thailand, Laos, India, Nigeria, Cameroon, Liberia, Guinea, Mexico, Guatemala, Panama, Columbia, Peru, Ecuador, Costa Rica, Honduras, Venezuela, El Salvador, Nicaragua, Canada, and the United States.

Most human infections by *P. westermani* are reported from the Far East, but few prevalence studies have been done. Surveys for paragonimiasis are conducted in endemic areas by first skin testing the population and then asking those with positive reactions to provide sputum and stool specimens which are examined microscopically for *Paragonimus* eggs. There are only sporadic reports of the disease from Africa and South America. *P. kellicotti* has been occasionally reported in humans in the United States (Pachucki et al., 1984); in recent years there are increasing reports of infections of *P. westermani* or possibly *P. heterotremus* in Southeast Asians immigrants (Johnson et al., 1982).

Paragonimiasis is widespread in China where there are 21 species of crustacean second intermediate hosts (Zhong et al., 1981). *Potamon* species of crabs may be eaten poorly cooked or after soaking in a weak rice wine—"drunken crabs." Raw shrimp are also eaten in China; raw crab sauce, crab curd, and raw crayfish also have been reported as sources of infection. Uncooked *Eriocheir japonicus*, or the hairy crab, was eaten in the past on Taiwan, but the parasite and its intermediate hosts have nearly disappeared from former endemic areas due to educational programs and to changes in ecology from pollution related to industrialization of the island (Cross, 1984).

Transmission in Japan is attributed to eating uncooked *Potamon* species of crabs and to contamination of utensils used in chopping crabs in the preparation of crab soup. The chopping blocks and knives become contaminated with the *Paragonimus* metacercaria and become dislodged and mixed with vegetables chopped with the contaminated equipment. Juice squeezed from crabs is used for medicinal purposes in Japan and in Korea and Koreans eat raw crab after immersion in soy sauce. Crayfish juice is also used to treat measles in Korea. Filipinos in endemic areas eat crabs that are roasted but not cooked long enough to destroy the infective stages of the parasite. Chopped crabs are also pressed and squeezed through cloth to collect the juice to season meat. *Paragonimus* infections in Japan have followed eating small uncooked slices of wild boar meat (Miyazaki and Hirose, 1976). The meat contained immature forms of the parasite. In Africa, where some women eat raw crab to increase fertility, others eat crabs because of the shortage of other foods.

Life Cycle of *Paragonimus* Species

P. westermani and other species usually live in cysts singly or in pairs near the bronchioles or bronchi of the lung. Eggs are deposited in the lungs, are coughed up in the sputum, or are swallowed and passed in the feces. The eggs must reach water where they mature in 3 weeks; the developed miracidium hatches from the egg, swims in the water, and enters a snail first intermediate host. There are several families of snails throughout the world that serve as hosts, but the most important species for *P. westermani* is *Semisulcospira libertina*. Species of *Thiara* and *Brotia* are also important in Southeast Asia. The parasite develops in the snails and reproduces by polyembryony, producing cercariae which escape from the snails and search for a crustacean second intermediate host. The latter includes crab species of *Eriocheir*, *Potamon*, and *Sundathelphusa*, and crayfish of the species *Cambaroides*. The parasite encysts in the muscles, gills, and viscera of the crustaceans and matures into a metacercaria (Fig. 1) in several weeks. When the crustacean is eaten uncooked, the metacercaria excysts in the digestive tract, penetrates the intestinal tissue, and enters the abdominal cavity and then the abdominal wall. It emerges from the wall, reenters the abdominal cavity, and penetrates the diaphragm, enters the thoracic cavity and penetrates the lung. The parasite reaches sexual maturity in several weeks and produces eggs in 8 to 10 weeks. The parasite may live as long as 20 years.

Etiologic Agents

Characteristics of Paragonimiasis

Adult Worms

P. westermani, the most important species infecting humans, is a reddish-brown plumb, oval, or coffee bean-shaped fluke measuring approximately 7 to 16 mm long, 4 to 8 mm wide, and 3 to 5 mm thick. The tegument is covered with spines. There is an oral sucker and a midbody ventral sucker. When flattened, fixed, and stained, internal structures can be seen. There is a short esophagus, which divides into two blind ceca that extend posteriorly. Genital organs consist of a large multilobed ovary behind the ventral sucker, an egg-filled uterus extends laterally, and the paired irregularly lobed testes are situated behind the ovary and uterus. The branched follicles of the vitellaria are in the lateral fields the entire length of the body (Fig. 4).

Eggs

The immature yellowish-brown eggs measure 80 to 120 × 50 to 60 μm and are irregularly shaped. The operculum is large and flattened with a pronounced operculum rim. The egg shell has varying thickness, with the thickest portion at the aboperculated end (Fig. 3).

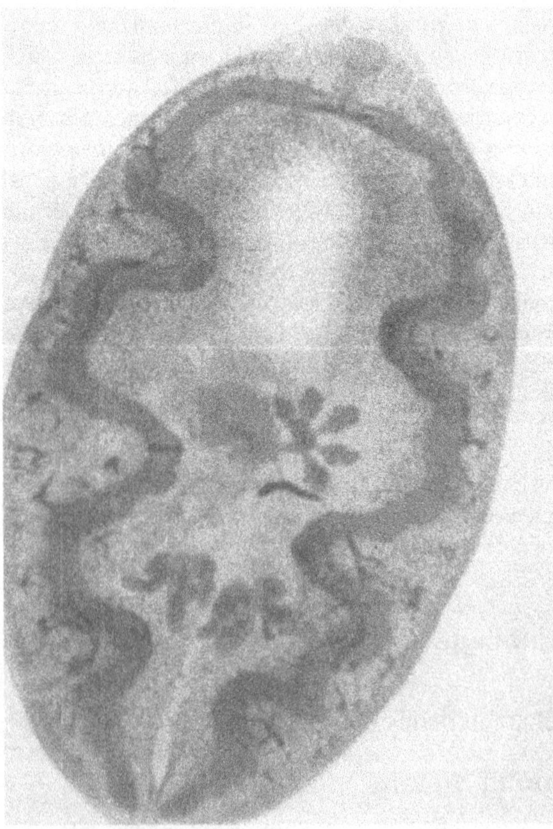

FIG. 4. Adult *Paragonimus westermani,* flattened and stained to see internal organs. The adults measure 7 to 16 mm long, 4 to 8 mm wide, and 3 to 5 mm thick.

Specimen Collection and Transport

Sputum specimens can be collected and examined directly microscopically for *Paragonimus* eggs; however, it is advisable to collect the blood-streaked and/or tenacious sputum and place it in 2% sodium hydroxide for centrifuge concentration. Stool specimens can be examined microscopically for eggs or placed into 10% formalin and examined after ether or ethyl acetate concentration for eggs. The specimens are to be sent to the laboratory in suitable containers for processing and microscopic examination.

Sputum and stool specimens should be examined simultaneously from suspected cases of pulmonary paragonimiasis. These specimens should be examined at least three times in suspected cases with initial negative results. Aspirates collected aseptically from pleural effusion should be sent to the laboratory for concentration and microscopic examination, as well as culturing for bacteria.

Isolation and Identification Procedures

Sputum and stool specimens should be placed onto a microscope slide, a cover glass placed on top of the specimen, and the slide screened for *Paragonimus* eggs under low magnification. Suspected eggs should be examined further under higher magnification.

In extrapulmonary paragonimiasis, biopsied tissue should be sectioned, stained with hematoxylin and eosin, and examined microscopically for worms and *Paragonimus* eggs. It is difficult to identify sections of worms sufficiently to species.

Preliminary Identification

Preliminary identification of the parasitosis is based on finding *Paragonimus* eggs, but species identification is made only with the recovery of adult worms. Positive serologic tests also provide preliminary evidence of infection.

Definitive Identification

Definitive identification is made by recovering worms by surgery or on rare occasions when worms are expectorated (Vanijanonta et al., 1981). The specimen is fixed, stained, and mounted onto a glass slide and examined. Worms recovered from cutaneous or subcutaneous lesions can be examined similarly and identified. Careful study of the eggs and measurements may be helpful in species identification in cases living where more than one species of

Paragonimus is endemic. Serologic tests also may aid in species-specific diagnosis (Knobloch, 1984).

Immunologic Aspects

Antibodies are produced to *Paragonimus* antigen, which can be detected by various serologic tests (Cross, 1973). Antigens for serologic tests are prepared from adult worms. The use of these tests are particularly useful in epidemiologic surveys and in the presumptive diagnosis of pulmonary and extrapulmonary paragonimiasis. Antibodies are detectable in serum and in cerebral spinal fluid. Some of the tests may remain positive even after successful treatment but others such as the complement fixation test become negative after several months (Tsuji, 1984). The ELISA seems to be the most sensitive and specific.

Immunoglobulin E antibody is high in patients with paragonimiasis and can also be detected in pleural exudates (Yokogawa et al., 1976).

Treatment, Prevention, and Control

Treatment

There are several drugs available today for the specific treatment of paragonimiasis. Emetine hydrochloride and chloroquine were used in the past, but these have been replaced by bithionol, niclofolan, and praziquantel.

Bithionol is available in the United States from the Parasitic Disease Drug Service at the Centers for Disease Control, Atlanta, Georgia. It is given orally in dosages of 30 to 50 mg/kg body weight in divided doses in the morning and at night every other day for 10 to 15 days. The drug is also useful in the treatment of extrapulmonary paragonimiasis, but two or more courses of treatment may be required for cerebral paragonimiasis. There may be transient side effects with the drug, such as diarrhea, skin rash, and urticaria.

Niclofolan has been used in Korea with good results in oral dosages of 25 mg/kg body weight three times a day for 3 days (Rim and Chang, 1980). This drug is not available in the United States.

Praziquantel is currently the drug of choice and is available in the United States. It is a broad-spectrum trematodicidal drug given orally in dosages of 25 mg/kg body weight three times a day for 3 days. The drug is effective against extrapulmonary as well as pulmonary infections. Side effects, although transient, may occur, such as headache, rash, dizziness, low-grade fever, and gastrointestinal disturbances (Johnson et al., 1982; Rim et al., 1981; Yokogawa, 1986).

Surgery may be required in the treatment of extrapulmonary paragonimiasis, especially if cerebral, abdominal, and cutaneous lesions are present.

Patients receiving chemotherapy should be observed and reexamined for parasite eggs at least 6 months after treatment. Retreatment may be required, especially in cases of extrapulmonary paragonimiasis.

Prevention and Control

Paragonimiasis can easily be prevented by changing the eating habits of populations in endemic areas. If crustacean intermediate hosts are well cooked before consumption, infection can be eliminated. Treatment of individual cases or mass treatment of the population with praziquantel should be part of a control program. Destruction of the snail and crustacean intermediate hosts also could be part of a control program, but it is difficult. Educating the people on the hazards of eating raw, poorly cooked, or "drunken crab," as well as discouraging the use of juice from crushed crustaceans for seasoning and medicinal purposes should be a major part of any control program. The implementation of control programs is difficult and expensive, and it is nearly impossible to change habits that have been practiced for generations in the endemic populations.

Literature Cited

Cross, J. H. 1973. Immunodiagnosis of some parasitic diseases of the respiratory system. J. Formosan Med. Assoc. **72**:407–412.

Cross, J. H. 1984. Changing patterns of some trematode infections in Asia. Arzneimittelforsch. **34**:1224–1226.

Hu, X., R. Feng, Z. Zheng, J. Liang, H. Wang, and T. Lu. 1982. Hepatic damage in experimental and clinical paragonimiasis. Am. J. Trop. Med. Hyg. **31**:1148–1155.

Johnson, R. J., E. C. Jong, S. B. Dunning, W. L. Carberry, and B. H. Minshew. 1985. Paragonimiasis: diagnosis and the use of praziquantel treatment. Rev. Infect. Dis. **7**:200–206.

Knobloch, J. 1984. Application of different *Paragonimus* antigens to immunodiagnosis of human lung fluke infections. Trematode infections and chemotherapy in Southeast and East Asia. Arzneimittelforsch **34**:1208–1210.

Miyazaki, I., and H. Hirose. 1976. Immature lung fluke first found in the muscle of wild boar in Japan. J. Parasitol. **62**:835–837.

Pachucki, C. T., R. A. Levandowski, V. A. Brown, B. H. Sonnenkalf, and M. J. Vruno. 1984. American paragonimiasis treated with praziquantel. N. Engl. J. Med. **311**:582–583.

Rim, H. J., and Y. S. Chang. 1980. Chemotherapeutic effects of niclofolan and praziquantel in the treatment of paragonimiasis. Korea Univ. Med. J. **17**:113–128.

Rim, H. J., Y. S. Chang, J. S. Lee, K. H. Joo, W. H. Suh,

and M. Tsuji. 1981. Clinical evaluation of praziquantel (Embay 8440: Biltricide) in the treatment of *Paragonimus westermani*. Korean J. Parasit. **19**:27–37.

Tsuji, M. 1984. Pre- and posttreatment serodiagnosis for paragonimiasis. Arzneimittelforsch. **34**:1204–1207.

Yokogawa, M., S. Kojima, K. Araki, H. Tomioka, and S. Yoshida. 1976. Immunoglobulin E: raised levels in sera and pleural exudate of patients with paragonimiasis. Am. J. Trop. Med. Hyg. **25**:581–586.

Yokogawa, M. 1986. Paragonimiasis: current concepts, laboratory diagnosis and chemotherapy. *In* Parasites and parasitic infections, p. 79–92. Asian J. Clin. Sci. Mono. 7.

Vanijanonta, S., P. Radomyos, D. Bunnag, and T. Harinasuta. 1981. Pulmonary paragonimiasis with expectoration of worms: a case report. Southeast Asian J. Trop. Med. Public Health **12**:104–106.

Zhong, H. L., L. Y. He, Z. B. Xu, and W. J. Cao. 1981. Recent programs in studies of *Paragonimus* and paragonimiasis control in China. China Med. J. **94**:483–494.

CHAPTER 100

Pneumocystosis

JAMES W. SMITH and MARILYN S. BARTLETT

Disease: Pneumocystosis (*Pneumocystis* pneumonia).
Etiologic Agent: *Pneumocystis carinii* (may be multiple species).
Source: Various mammals. Humans likely acquire infection by inhalation of infectious material from other humans.
Clinical Manifestations: Most subclinical infections. Organisms remain latent in lungs. Immunocompromised hosts and malnourished infants develop interstitial pneumonia.
Pathology: Foamy intraalveolar exudate with slightly increased inflammatory cells in immunocompromised. Malnourished children show interstitial plasma cell pneumonia.
Laboratory Diagnosis: Demonstration of *Pneumocystis* cysts and/or trophozoites in pulmonary material.
Epidemiology: Worldwide. Clinical disease in 60 to 80% of patients with acquired immunodeficiency syndrome in North America; most arise from activation of latent infection.
Treatment: Trimethoprim/sulfamethoxazole or pentamidine. Reactions to drugs are common. Some do not respond to therapy.
Prevention and Control: Prophylaxis with drugs in susceptible persons.

Description of Disease

Pneumocystosis is a diffuse interstitial pneumonia caused by *Pneumocystis carinii*, which occurs in most patients with acquired immune deficiency syndrome (AIDS), and in some persons who are immune compromised by congenital diseases, by therapy of neoplastic diseases, or by immunosuppressive therapy for transplantation or autoimmune diseases. The infection is probably acquired by inhalation and is a very common latent infection that becomes manifest when the individual becomes immunocompromised (Young, 1984).

Pneumocystis are probably protozoan parasites that infect a number of animal species (Juranek, 1982). There are differences among the *Pneumocystis* strains infecting humans and those infecting various animal species (Frenkel 1974; Graves et al., 1986; Walzer and Rutlege, 1980).

Communicability and Transmissibility

Apparently the infection is acquired by inhalation of cysts and is widespread, as more than 90% of children have developed antibody to *Pneumocystis* by the age of 4 years (Meuwissen et al., 1977; Pifer et al., 1978). In most patients, *Pneumocystis* pneumonia appears to arise from reactivation of latent infection (Mills, 1986), although in occasional patients, it may be a primary infection in a previously unexposed individual or a reinfection. Additional studies are needed to determine if the respiratory route is the only means of transmission.

Clinical Manifestations

The usual presenting symptoms are increasing shortness of breath with fever, and a chest radiograph that shows a diffuse interstitial pattern. The depression of the pO_2 is often greater than would be predicted based on the severity of changes in the chest radiograph. Radiologic changes generally are diffuse, but in some instances may be localized to certain areas of the lung. Gallium scans show increased pulmonary uptake. The disease varies greatly in the rapidity of progression. Some patients may be relatively well and develop respiratory distress over a period of several days, whereas in others, the onset is gradual over a period of weeks. The latter is characteristic of the infection in patients with AIDS (Kovacs et al., 1984).

Pathogenesis and Pathology

The *Pneumocystis* organisms are often attached to and appear to damage type I pneumocytes leading to damage to the alveolar walls. The organisms are present in large numbers in the alveolar spaces, and are embedded in a matrix that appears to be derived from products of the organism. There is some round-cell infiltration in the alveolar septa, although this is usually minimal in immunocompromised patients and greatest in malnourished infants, in whom the disease was first described as plasma cell interstitial pneumonitis. In hematoxylin and eosin stained sections the alveolar exudate often has dot-like areas in the foamy material, which, in special stains, can be shown to be *Pneumocystis* organisms. Other organs only rarely contain organisms, and then only in patients with severe pulmonary disease.

Disseminated *Pneumocystis* infections have been reported with increasing frequency in the last year. In patients with AIDS, *Pneumocystis* has been reported from liver, spleen, GI tract, bone marrow cutaneous nodules, eye, and lymph nodes (Coulman et al., 1987; Grimes et al., 1987; Hegmon and Rasmassen, 1987; Macher et al., 1987; Pelor et al., 1987). The role of *P. carinii* as a possible agent in extrapulmonary infections must be considered in patients with AIDS.

Diagnostic Features

Pneumocystosis should be suspected in an immunocompromised patient who develops pulmonary disease. *Pneumocystis* pneumonia is the presenting manifestation of AIDS in 60% of patients; and more than 80% of AIDS patients will develop *Pneumocystis* pneumonia at some time during their clinical course (Mills, 1986). The infection is diagnosed by demonstrating *Pneumocystis* organisms in pulmonary material. Traditionally, material obtained by invasive transbronchial or transthoracic procedures has been required (Blumenfeld and Wagner, 1984; Coleman et al., 1983; Hartman et al., 1985; Ognibene et al., 1984); however, recently two groups have shown that *Pneumocystis* can often be detected in induced sputum, saving an invasive procedure for more than half the patients (Bigby et al., 1986; Pitchenik et al., 1986). In any specimen, demonstration of organisms requires good specimens, good preparations, and good stains which are examined by competent individuals.

Prognosis

The prognosis for a first episode of *Pneumocystis* pneumonia is good if the disease is diagnosed early and the patient is promptly placed on therapy. If the disease is not diagnosed until late, it may prove fatal before there is adequate opportunity to respond to therapy. The most common therapy with trimethoprim/sulfamethoxazole usually leads to a response within 4 to 5 days. Many AIDS patients develop allergic reactions to trimethoprim/sulfamethoxazole and some infections fail to respond to this drug combination (Gordin et al., 1984). Therapy with pentamidine isethionate may be required, but this drug is also toxic and treatment failures have been noted (Mills, 1986). In AIDS patients, the initial episode of *Pneumocystis carinii* may respond readily to chemotherapy, whereas subsequent relapses become increasingly difficult to treat. Relapse occurs in a significant proportion of patients.

Epidemiologic Aspects

Pneumocystis infects a wide variety of animal species, including humans (Juranek, 1982), but there appear to be differences in *Pneumocystis* strains affecting various species (Graves et al., 1986; Walzer and Rutlege, 1980). These taxonomic relationships have not been well defined, and the number of species is not certain. Although a separate species infecting humans (*Pneumocystis jiroveci*) has been proposed (Frenkel 1974), it has not been widely accepted. The life cycle is not completely understood. Studies of transmission have demonstrated that the infection can be spread by the respiratory route, either by exposing animals in a colony (Hughes et al., 1983) or by intranasal installation of *Pneumocystis*-infected material (Furuta et al., 1985). Clinically manifest infection usually is not recognized in normal humans; it is compromised individuals who develop clinical disease.

Both T-cell and B-cell immunity appear to be important, as people with defects in either system have developed *Pneumocystis* pneumonia; however, most patients have had T-cell deficiency. Latent infection occurs as is reflected by organisms demonstrable in lungs of persons who reach autopsy for other reasons (Ruskin, 1976). The importance of the latent state in transmission is not clear. Latent infections are believed to be responsible for most infections that develop when patients are immune compromised (Mills, 1986).

Susceptibility

The organism is usually susceptible to trimethoprim/sulfamethoxazole or pentamidine isethionate and, more recently, has been found to be susceptible to dapsone. Several additional drugs currently are being evaluated in laboratory models or in limited human studies. Certain strains infecting humans appear to be resistant to trimethoprim/sulfamethoxazole and also possibly to pentamidine isethionate, but because the human organisms cannot be cultured reproducibly in vitro, the nature of this resistance has not been defined.

Etiologic Agent

Characterization and Taxonomy

The taxonomy of *Pneumcystis carinii* is not clear, and there is still some debate about whether it is a fungus or a protozoan. It is generally believed to be a protozoan parasite in the phylum *Apicomplexa*, with an uncertain relationship to other parasites in this phylum. Organisms apparently proliferate both within the cyst stage and during the free trophozoite stage. Synaptonemal complexes have recently been described (Matsumoto et al., 1984), which implies that reduction division takes place. It is possible that additional forms of the parasite have yet to be discovered. Most work with *Pneumocystis* has used the rat model described by Frenkel et al. (1966). Ordinary laboratory rats have latent infection with *Pneumocystis* and develop *Pneumocystis* pneumonia after immunosuppression by adrenal corticosteroids for 6 to 8 weeks. Morphologically, rat infection resembles the infection of humans.

Pneumocystis organisms from rats can be grown in culture for short periods, but continuous culture has not been achieved (Smith and Bartlett, 1984). *Pneumocystis* from humans have not been cultured reproducibly. Culture has no role in diagnosis, although it has been useful for experimental studies.

There are differences among *Pneumocystis* organisms infecting various species. It has been shown that sera that react well with *Pneumocystis* from rats may react poorly or not at all with *Pneumocystis* from humans and vice versa (Graves et al., 1986; Walzer and Rutledge, 1980). Differences in ability to culture human and rat *Pneumocystis* are noted above.

Specimen Collection

Specimens for diagnosis of *Pneumocystis* pneumonia are pulmonary specimens from patients suspected to have the infection. *Pneumocystis* material in the alveoli is quite tenaceous, and does not readily reach the tracheobronchial tree and the sputum. Examination of ordinary expectorated sputum is not useful for diagnosis of pneumocystosis; however, recent work (Bigby et al., 1986; Pitchenik et al., 1986) suggests that in the hands of experienced persons, induced sputum may be useful for diagnosing the infection in AIDS patients but not other immunocompromised patients. If this examination is negative, there should be prompt progression to an invasive procedure.

Specimens obtained by invasive procedures, either transbronchial or transthoracic, are most commonly examined. Transbronchial procedures used include bronchial washing, transbronchial brushing (Hartman et al., 1985), transbronchial biopsy (Blumenfeld et al., 1984; Broaddus et al., 1985; Feldman, 1975; Jenkins et al., 1979), and bronchoalveolar lavage (Broaddus et al., 1985; Ognibene et al., 1984). Bronchoalveolar lavage and transbronchial biopsy have been effective in the hands of most who have used them. Although Hartman et al. (1985) have reported a good yield for bronchial brushings, our laboratory has found this procedure to be of little value in diagnosing *Pneumocystis*. Bronchial washing is not as good as bronchoalveolar lavage because it is less effective at getting material from the alveoli. Transthoracic needle aspirates may be useful, but transthoracic cutting needle biopsies are discouraged because of frequent serious complications.

Open-lung biopsy is sometimes used as the initial procedure or after the above are negative when it is imperative to establish the diagnosis, when the conditions considered in the differential diagnosis can only be diagnosed by open-lung biopsy, or when establishing the diagnosis rapidly is important because the patient is deterioriating rapidly (Prober et al., 1984). Open-lung biopsy is generally not required in AIDS patients, but may be helpful in other immunosuppressed patients with pulmonary disease.

Before collecting induced sputum or transbronchial specimens, it is important that the patient brush his teeth and clean the oropharynx by gargling and rinsing well. Immunocompromised patients often have oral thrush or large numbers of bacteria that make microscopic examination difficult.

There are a variety of infectious and noninfectious processes that can cause pulmonary infiltrates in immunocompromised patients. The diagnostic tests selected are in part determined by the abilities of various diagnostic procedures to diagnose the conditions that might be affecting the specific patient. In addition, the type or types of specimens will depend on the competence of the persons obtaining the specimens, and the competence of the persons examining the specimens. Therefore, methods selected must be tailored to each institution and each patient.

Laboratory Diagnosis

Laboratory diagnosis is based on morphologic demonstration of *Pneumocystis* organisms in various types of pulmonary material. In microbiology laboratories, this is generally in smears of pulmonary material. In some institutions, *Pneumocystis* is usually diagnosed by the surgical pathologist in histologic sections.

Specimen Handling

It is important that specimens be properly handled and promptly transported to the laboratory. If smears are to be made of biopsy material, the specimen must not be placed in fixative, and should be submitted in a moist sponge in a sterile container and not floating in saline.

Bronchial washings, bronchoalvealor lavage material, and induced sputum specimens are processed by centrifugation with subsequent smearing of sediment onto clean glass slides. Slides are then allowed to air dry, are fixed in absolute methanol or acetone for 2 to 3 min, air dried, and stained using appropriate procedures. Bloody transbronchial specimens represent a particular problem because the numerous erythrocytes dilute the pulmonary cells. If a specimen comes in several containers, some of which are bloody, the bloody material should be processed separately from the nonbloody or less bloody material. Six to eight slides should be made. Multiple samples can be on different locations of the same slides, that is, biopsy and wash.

Processing of Tissue

Open-lung biopsy or transbronchial biopsy tissue for impression smears must be blotted on sterile towel or other absorbent material to remove excess fluid from the surface, because if the surface is very wet, the tenacious *Pneumocystis*-containing material from the alveoli will not be pulled from the tissue to the slide. It is best that the tissue stick slightly when removed from the slide. The impressions should be placed close together so that there is little empty space in the area to be screened. Again six to eight similar slides should be prepared. For transbronchial biopsies multiple impression smears are made in an area approximately 0.5 cm in diameter. The area may be outlined with a diamond marking stylus to aid in screening. Transbronchial tissue used to prepare impression smears cannot be used for permanent sections, so if permanent sections are desired, separate specimens must be obtained. For open-lung biopsies, similar procedures are used. If areas of lung show different gross appearances, impression smears can be made of these in different areas of the slide, generally in rows. Gay et al. (1985) have found that grinding of tissue and staining with a cyst wall stain improves sensitivity, but we have not had experience with this method. It may be helpful to perform frozen sections of a portion of the lung, which can also be stained with cyst wall stains; however, it may be best not to do frozen sections on tissue from AIDS patients because of contamination of the cryostat.

Stains

Stains commonly used for *Pneumocystis* are of two types, cyst wall stains and trophozoite stains (Smith and Bartlett, 1982). The former stain only the walls of the *Pneumocystis* cysts, whereas the latter stain free *Pneumocystis* organisms and intracystic contents, but do not stain the cyst walls. In addition, immunologic stains, both immunofluorescent and immunoperoxidase, have been described (Levin et al., 1983; Lim et al., 1974), but are not used diagnostically as reagents are not commercially available. Each laboratory should be proficient in performing and examining one cyst wall stain and one organism stain.

Cyst Wall Stains

A variety of cyst wall stains have been described, and all appear to work reasonably well. The most widely used are toluidine blue 0 (Chalvardjian and Grawe, 1963; Gosey et al., 1985) and methenamine silver nitrate, although some use cresyl echt violet (Bowling et al., 1973) or Gram-Weigert (Rosen, 1974). Initially, we used the Gram-Weigert routinely, but have since adopted a rapid methenamine silver nitrate stain. Each laboratory should select a cyst wall stain that works well in its hands. It is important to run a control with the cyst wall stain to be certain that the stain is working properly. The best control is material that is positive for *Pneumocystis*, although if this is not available, fungus-containing material can be used. The procedure for rapid methenamine silver is given.

Methenamine silver nitrate stain (Grocott, 1955)

has long been used to stain cyst walls of *Pneumocystis*. It usually has not been used for staining smears because of the long staining time (4 hours). More recently, staining techniques that heat the stain to speed the staining process allow a staining time of less than 30 min and are convenient to use on smears. A number of rapid methenamine silver nitrate staining methods have been described, which often use a microwave oven but a review of these is beyond the scope of this chapter. We use slight modifications of the method of Brinn (1983). The modifications are that we perform the chromic acid step at room temperature rather than in the microwave because we believe this gives more even staining of the organisms, and we counterstain with fast green rather than fast red. The staining procedure is as outlined below. It can be performed on tissue sections that have been brought to water or on fixed smears.

Reagents: Methenamine solution, 3% (store at room temperature); aqueous silver nitrate 5% (store refrigerated at 4°C); borax solution 15% (store at room temperature); chromic acid 10% (store at room temperature), chromic acid (chromium tetroxide) 10 g dissolved in distilled water 100 ml (store at room temperature); 1% sodium metabisulfite (store at room temperature); 1% gold chloride (store in refrigerator); and 0.2% fast green counterstain (store at room temperature).

Staining procedure: Specimen slides and control are stained simultaneously. The following instructions are for staining two or three slides. Except for the microwave silver staining, the procedure is performed on a staining rack, although for larger numbers of slides, coplin jars could be used.

1. Make up methenamine silver nitrate working solution at time of use; 3% methenamine solution, 20 ml; 5% aqueous silver nitrate, 1.0 ml; 5% borox solution, 1.5 ml; and distilled water, 17.0 ml. Place working solution in a 100-mm glass Petri dish. This will be adequate to stain one or two specimen slides' plus a control slide.
2. Flood or immerse slides in 10% chromic for 10 minutes.
3. Rinse with tap water.
4. Clear slides by covering with 1% sodium metabisulfite for 1 min.
5. Rinse with distilled water.
6. Place slides in a Petri dish with prepared working methenamine silver nitrate solution (1 above).
7. Place slides in microwave oven, cover with paper towel to prevent spatter. Microwave at 50% power for 135 s (time and power setting will need to be worked out for the specific microwave oven used).
8. Let stand 2 min. (There should be visible grey-brown silver deposit evident on the Petri dish and slide).
9. Rinse in distilled water.
10. Tone by flooding with gold chloride for 10 s.
11. Flood with 5% sodium thiosulfate for 1 min.
12. Rinse in tap water.
13. Flood with fast green counterstain, 1 min (other stains may be used for counterstain if desired including Giemsa).

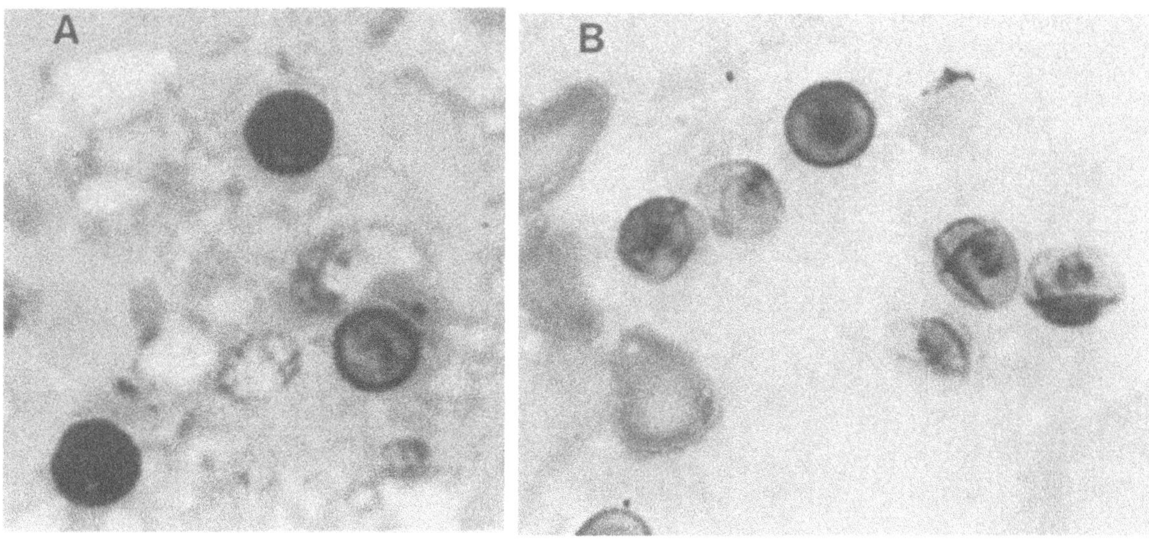

FIG. 1. *Pneumocystis carinii* cysts, methenamine silver nitrate stain, magnification × 1,000. Note that the dark-staining areas are more evident in the cysts in (B).

14. Rinse with water.
15. Air dry and examine with oil immersion or dehydrate, clear in xylene, and mount with coverslip.

If more slides are to be stained, double the amount of working methenamine solution prepared and use a 150-mm glass Petri dish or coplin jar. Microwave for 90 s, turn 180°, and microwave an additional 90 s. We have had coplin jars break from the heat.

Pneumocystis stain black (Fig. 1). The intensity of staining may vary slightly. Organisms may appear collapsed or may have a more intensely staining dot-like area. There is no budding.

The toluidine blue 0 stain is widely used for staining *Pneumocystis* organisms (Chalvardjian and Grawe, 1963; Gosey et al., 1985). It requires the use of concentrated acid to prepare the sulfation reagent, and the original method used ether. The modification developed by Gosey et al., uses the sulfation reagent described for the cresyl echt violet stain (Bowling, et al., 1973), which requires glacial acetic acid and concentrated sulfuric acid but not ether and is stable for 1 week. Disposal of concentrated acids can present a problem in some laboratories.

Toluidine blue 0 stain can vary from lot to lot and even solutions made from the same lot may vary. Much of the background is digested by the strong acids of the sulfation reagent resulting in a relatively clean background. *Pneumocystis* cysts, yeasts, and other fungal elements stain with this stain. The cysts stain lavender and have an appearance similar to that in methenamine silver (Fig. 2), there is little background as this is digested during sulfation.

Cresyl echt violet stain provides similar morphology, but it also requires use of a sulfation reagent of glacial acetic acid and concentrated sulfuric acid (Bowling et al., 1973). Results are similar to those with toluidine blue 0 (Fig. 2).

The Gram-Weigert stain stains the cysts dark blue and has a pale red background (Rosen, 1984). We used the Gram-Weigert stain for a number of years, but found that it was less reproducible than the rapid methenamine silver stain. The key step is decolorization by aniline-xylene. Appearance of *Pneumocystis* cysts is similar to the other cyst wall stains (Fig. 3).

Organism Stains

Giemsa stain should be made fresh (within 1 h). We prefer to use azure B Giemsa stock stain diluted 1:20, with phosphate buffered water, pH 7.2 to which Triton X-100 (final concentration 0.01%) has been added. This provides good staining of cells and of bacteria and most fungi, as well as of *Pneumocystis* organisms. After fixation in methanol or acetone, the slides are stained in Giemsa for approximately 30 min, although the time may vary with different lots of stain. A Wright-Giemsa stain can be performed, but gives less definitive staining.

The Diff-Quick (American Scientific Products) or similar rapid hematologic stain can be performed in approximately 30 s; the first step is fixation (10 s); the second step is addition of the red dye (5 s); and the

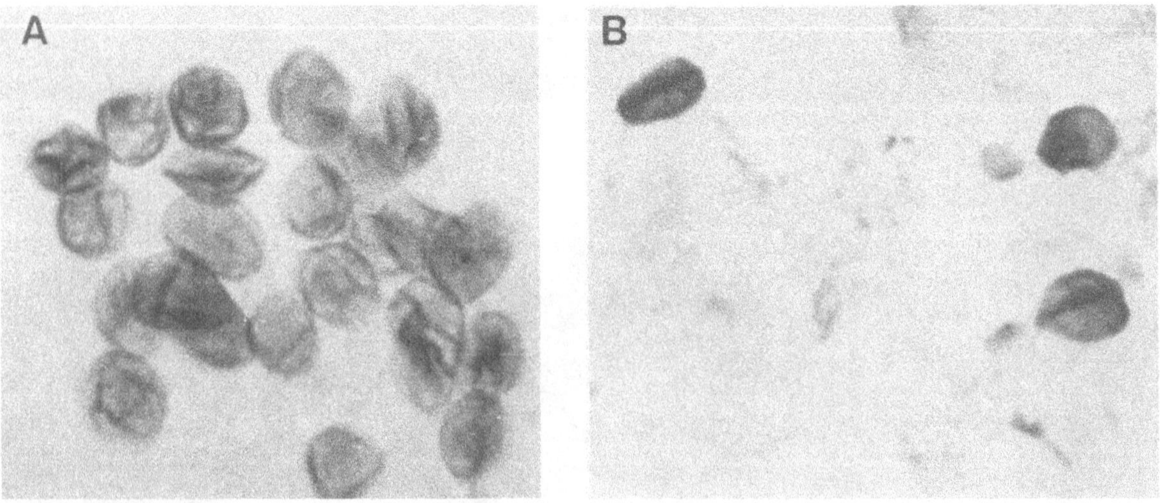

FIG. 2. *Pneumocystis carinii* cysts, toluidine blue 0 (A) and Cresyl echt violet (B), magnification × 1,000. The cysts appear similar to cysts stained with methenamine silver nitrate. In severe infections, there may be large clumps of cysts such as those in (A).

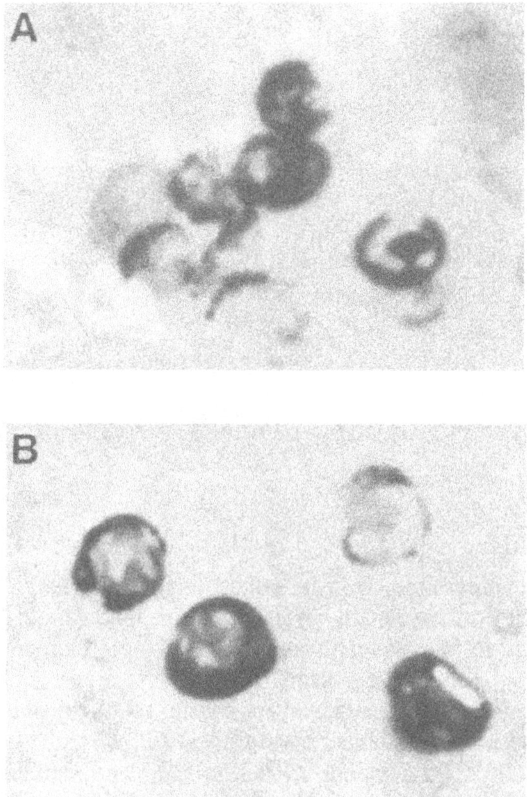

FIG. 3. *Pneumocystis carinii* cysts, Gram-Weigert stain, magnification × 1,000. There is some irregularity of staining of the cysts.

third step is the addition of blue dye (10 s). The slide is then allowed to drip dry or can be gently blotted on blotting paper. The Diff-Quick provides staining similar to that of Giemsa stain, and has the advantage that the laboratory can examine these slides while the other stains are being performed.

Giemsa or Diff-Quick stain *Pneumocystis* trophozoites and intracystic organisms. Nuclei stain red and cytoplasm stains pale blue in an ideal stain (Fig. 4). It should be noted that all bacteria will stain blue in these stains, whether they are gram-positive or gram-negative.

Diff-Quick does not provide as delicate a stain as Giemsa, so we generally perform both a Diff-Quick and a Giemsa. If the Diff-Quick is negative, the Giemsa is carefully examined. Bacteria may contaminate Diff-Quick reagents; therefore, we prefer to use dropper bottles rather than coplin jars for the stain as this decreases contamination.

It is not necessary to have separate control slides with the Giemsa or Diff-Quick stains, as the staining of the cellular background shows whether or not the stain is proper.

Immunologic Stains

Both immunofluorescent (Lim et al., 1974) and immunoperoxidase (Levin et al., 1983) stains for *Pneumocystis* have been described. However, reagents are not commercially available. There are no large series in which the usefulness of immunologic stains has been compared with that of conventional stains (see Addendum).

Use of Stains

Specimens for *Pneumocystis* examination should be stained with both a cyst wall stain and with an organism stain. The cyst wall stain stains *Pneumocystis* cysts, as well as fungi and some bacteria, whereas the organism stains stain the free trophozoite stages, bacteria, fungi, and actinomycetes. Although some laboratories use only a cyst wall stain or only an organism stain, it is recommended that both be used. This is especially important for inexperienced examiners. In some patients, there may be large clumps of trophozoite stages and very few cysts, and the diagnosis might be missed if an organism stain were not performed. Conversely, more cysts are evident in cyst wall stains than in organism stains.

Identification of Pneumocystis

Cysts measure about 5 μ in diameter and trophozoites vary from 2 to 6 μ.

In cyst wall stains, many of the cysts contain a dot-like area that stains more deeply than the rest of the cyst (Fig. 1, 2, and 3). This dot-like area appears to be related to the thickness of the cyst wall (Watts, 1985). The cysts do not show budding, and some of them are collapsed and cup shaped. If only occasional organisms are present, fungal elements may be confused with *Pneumocystis* cysts (Young et al 1976). Glove starch also has been confused with cysts (Hubbard, 1984). Differentiation of *Pneumocystis* from fungi is particularly important in AIDS patients, who often have oral candidiasis.

In organism stains, mature cysts contain eight intracystic bodies, often arranged in a clock-like fashion. Intracystic organisms and free-living organisms have red nuclei and pale blue cytoplasm. Cyst walls do not stain, but in some instances, they can be recognized by negative staining around the group of intracystic organisms. In other instances, cysts are not recognized, either because they are empty, or because the negative staining is not evident. Many more cysts will be recognized with cyst wall stains than with organism stains. *Pneumocystis* organisms often occur in clumps, and are recognized by the multiple red nuclei in a mass of indistinct blue cytoplasm. Often, some of the organisms can be recognized to be cysts with others being trophozoites.

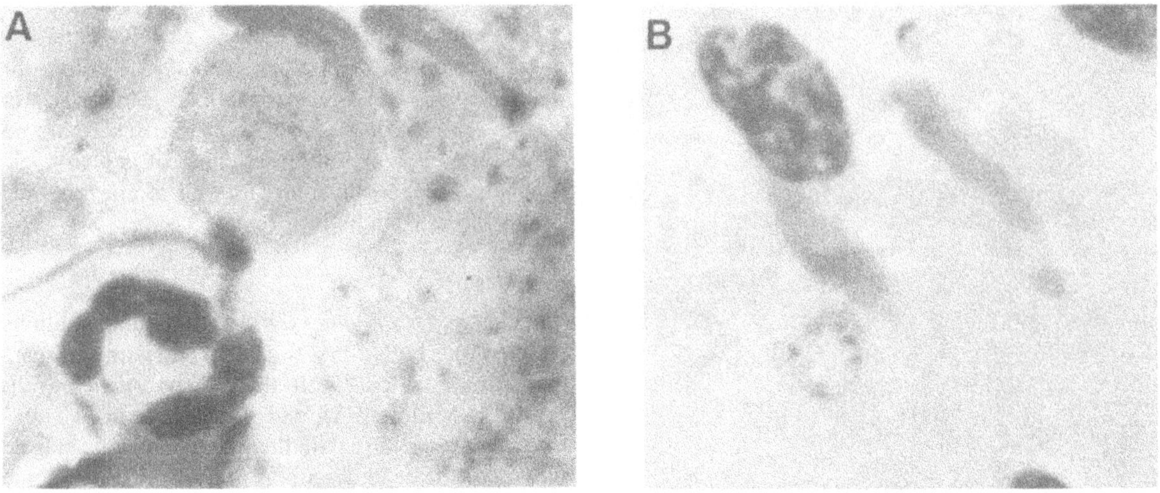

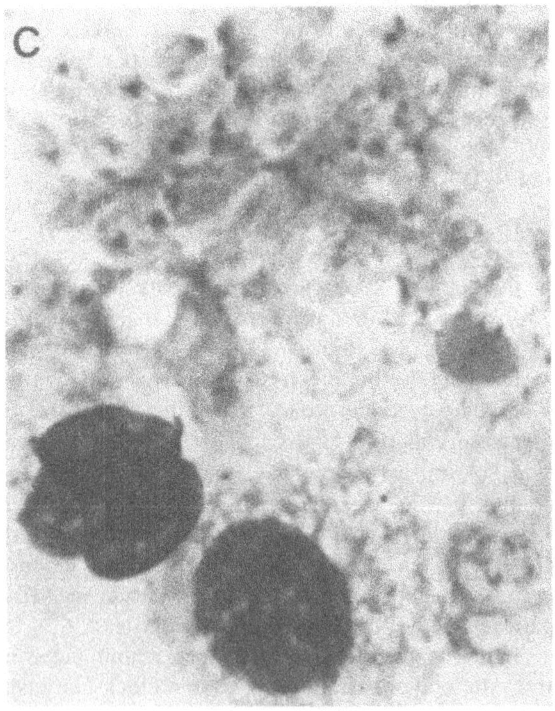

FIG. 4. *Pneumocystis carinii* cysts and trophozoites, Giemsa stain, magnification × 1,000. A clump of trophozoites (A) shows dark nuclei and cytoplasm with indistinct borders. A cyst with six intracystic organisms is shown in (B). (C) shows a large mass of *Pneumocystis* trophozoites and cysts. Many of the cysts are immature. The unstained cyst walls appear as clear areas.

Some cysts are immature and contain only one or a few intracystic bodies. It is important to differentiate cell fragments from *Pneumocystis* organisms. Cell fragments containing granules in some instances may be impossible to differentiate from *Pneumocystis*; however, in *Pneumocystis* infection, one should be able to find typical cysts or typical large clumps of *Pneumocystis* organisms. Finding suspicious organisms should stimulate a very careful search of slides, and perhaps staining of additional slides.

In AIDS patients, *Pneumocystis* cysts may continue to be evident in large numbers for 3 weeks or longer despite good response to therapy (DeLorenzo et al., 1985).

Immunodiagnosis

A number of immunodiagnostic procedures have been described and used in diagnosing *Pneumocystis* infection, although none are ideal and none are recommended for routine use.

Serology

Serologic tests for *Pneumocystis* have been used for a number of years and are based on detection of antibody to *Pneumocystis* organisms. Early serologic studies showed that most people developed *Pneumocystis* infection as children, and therefore, the presence of antibody does not necessarily imply recent infection (Meuwissen et al., 1977; Pifer et al., 1978). In addition, immunocompromised people often do not develop antibody, and may have severe *Pneumocystis* with no detectable antibody. Thus, the test does not have sufficient sensitivity or specificity to aid in clinical decision making.

Antigen Detection

Antigen detection tests are quite controversial (Hughes, 1985; Pifer, 1985). Methods using counterimmunoelectrophoresis and latex agglutination have been described, with the latter appearing somewhat more sensitive but of undetermined specificity. They have been positive in some patients with *Pneumocystis*; however, positive tests also have been found in a number of other immunocompromised patients who did not have clinical *Pneumocystis*, and a number of patients with proven *Pneumocystis* have had negative tests (Pifer, 1985). Antigen detection requires further study to better define its sensitivity and specificity, and at this time, its use in clinical decision making cannot be recommended (Hughes, 1985).

Differential Diagnosis in the Laboratory

The differential diagnosis of *Pneumocystis* in the laboratory includes recognition of other infections that may cause interstitial pneumonitis in the immunosuppressed patient population. Radiographic and clinical findings similar to those of pneumocystosis may develop in patients with histoplasmosis, nocardiosis, mycobacteriosis, cytomegalovirus pneumonia, and other infectious processes. It also should be noted that AIDS patients may have pulmonary infection by more than one agent. Thus, a variety of agents should be looked for systematically when screening for *Pneumocystis*. We routinely perform a modified acid-fast stain which would detect nocardial or mycobacterial infections, and often perform immunofluorescent stains for *Legionella*, especially if pleomorphic rods are seen in Giemsa-stained smears.

When there are numerous organisms, there is little difficulty in diagnosing *Pneumocystis*. Some patients, especially non-AIDS patients, may have rather severe pneumonia caused by *Pneumocystis*, with few organisms detectable in smears, whereas others may have minimal clinical symptoms with numerous organisms. The reasons for this are not clear.

Pulmonary infiltrates also may be caused by noninfectious processes in both AIDS and other patients. Several antineoplastic agents can cause severe pulmonary damage.

Treatment, Prevention, and Control

Treatment

Current recommended treatment for *Pneumocystis* is trimethoprim/sulfamethoxazole, and patients who do not respond to this therapy are usually changed to pentamidine isethionate (Wharton et al., 1986). Unfortunately, many patients, particularly AIDS patients, develop allergic reactions to the trimethoprim/sulfamethoxazole therapy, and it may need to be discontinued because of these reactions (Gordin et al., 1984). Pentamidine isethionate is toxic, and therefore, cannot be used for long-term therapy or prophylaxis. More recently, dapsone, generally in combination with trimethoprim, has been recommended as an efficacious treatment for *Pneumocystis* (Hughes and Smith, 1984; Leoung et al., 1986), and pyrimethamine has been used experimentally. A number of other drugs have shown promise in research situations, but clinical studies are not complete.

Prevention

Prevention of infection with *Pneumocystis* is probably not possible because it is so widespread. Development of clinical disease may be prevented by giving prophylactic therapy with trimethoprim/sulfamethoxazole or, more recently, with dapsone plus trimethoprim. In addition, intermittent chemoprophylaxis may provide the best prophylactic regimen for patients with AIDS (Hughes et al., 1987). With prophylaxis, the incidence of infection has dropped greatly; however, because of allergic reactions, prophylaxis has not been widely used in AIDS patients, although it usually has been satisfactory in cancer chemotherapy patients. Prophylaxis has generally not been used in renal transplant patients because of the potential renal toxicity.

Infection Control

For infection control, patients with *Pneumocystis* should be placed in respiratory isolation, as they may be shedding large numbers of organisms.

Addendum

Use of monoclonal antibodies for immunofluorescent staining of *Pneumocystis* has been described (Gill et al., 1987; Linder et al., 1987; Kovacs et al., 1988) and is being developed commercially. In one study, the method showed a sensitivity of 98.4% and a specificity of 100% in 126 specimens, of which 45 were positive by toluidine blue 0. In another study of induced sputum (Kovacs et al., 1988), it showed a greater sensitivity (92%) than that with Diff-Quik (76%) or toluidine blue (80%) and a specificity of 100%.

Literature Cited

Bigby, T. D., D. Margolskee, J. L. Curtis, P. F. Michael, D. Sheppard, W. K. Hadley, and P. C. Hopewell. 1986. Usefulness of induced sputum in diagnosis of *Pneumo-*

cystis carinii pneumonia in patients with the acquired immunodeficiency syndrome. Am. Rev. Respir. Dis. **133**:515–518.

Blumenfeld, W., E. Wagner, and W. K. Hadley. 1984. use of the transbronchial biopsy for diagnosis of opportunistic pulmonary infections in acquired immunodeficiency syndrome (AIDS). Am. J. Clin. Pathol. **81**:1–5.

Bowling, M. C., I. M. Smith, and S. L. Wescott. 1973. Rapid staining procedures for *Pneumocystis carinii*. Am. J. Med. Technol. **39**:267–268.

Brinn, N. T. 1983. Rapid metallic histological staining. J. Histotechnology **6**:125–129.

Broaddus, C., M. D. Dake, M. S. Stulbarg, W. Blumenfield, W. K. Hadley, J. A. Golden, and P. C. Hopewell. 1985. Bronchoalveolar lavage and transbronchial biopsy for the diagnosis of pulmonary infections in the acquired immunodeficiency syndrome. Ann. Intern. Med. **102**:747–752.

Chalvardjian, A. M., and L. A. Grawe. 1963. A new procedure for the identification of *Pneumocystis carinii* cysts in tissue sections and smears. J. Clin. Pathol. **16**:383–384.

Coleman, D. L., P. M. Dodek, J. M. Luce, J. A. Golden, M. W. Gold, and J. F. Murray. 1983. Diagnostic utility of fiberoptic bronchoscopy in patients with *Pneumocystis carinii* pneumonia and the acquired immunodeficiency syndrome. Am. Rev. Respir. Dis. **128**:795–799.

Coulmon, C. A., I. Greene, and W. R. Archibald. 1987. Cutaneous Pneumocystosis. Ann. Intern. Med. **106**:396–398.

DeLorenzo, L. J., G. P. Maguire, G. P. Wormser, M. M. Davidian, and D. J. Stone. 1985. Persistence of *Pneumocystis carinii* pneumonia in the acquired immunodeficiency syndrome. Chest **88**:79–83.

Feldman, N. T. 1975. An assessment of transbronchial lung biopsy. N. Engl. J. Med. **293**:299–300.

Frenkel, J. K., J. T. Good, and J. A. Schultz. 1966. Latent Pneumocystis infection of rats, relapse and chemotherapy. Lab. Invest. **15**:1559–1577.

Frenkel, J. K. 1974. *Pneumocystis jiroveci* n. sp. from man. Morphology, physiology and immunology in relation to pathology. Natl. Cancer Inst. Monogr. **43**:13–30.

Furuta, A., K. Veda, K. Fujiwara, and K. Yamanouchi. 1985. Immune response of mice infected with *Pneumocystis carinii*. Infect. Immun. **47**:544–548.

Gay, J. D., T. S. Smith, and D. M. Ilstrup. 1985. Comparison of processing techniques for detection of *Pneumocystis carinii* cysts in open lung biopsy specimens. J. Clin. Microbiol. **21**:150–151.

Gill, V. J., G. Evans, F. Stock, J. E. Parillo, H. Masur, and J. A. Kovacs. 1987. Detection of *Pneumocystis carinii* by fluorescent-antibody stain using a combination of three monoclonal antibodies. J. Clin. Microbiol. **25**:1837–1840.

Gordin, F. M., G. L. Simon, C. B. Wofsy, and J. Mills. 1984. Adverse reactions to trimethoprim-sulfamethoxazole in patients with the acquired immunodeficiency syndrome. Ann. Intern. Med. **100**:495–499.

Gosey, L. L., R. M. Howard, F. G. Witebsky, F. P. Ognibene, T. C. Wu, V. J. Gill, and J. D. MacLowry. 1985. Advantages of modified toluidine blue O stain and bronchoalveolar lavage for diagnosis of *Pneumocystis carinii* pneumonia. J. Clin. Microbiol. **22**:803–807.

Graves, D. C., S. J. McNabb, M. A. Worley, T. D. Dawns, and M. H. Ivey. 1986. Analyses of rat *Pneumocystis carinii* antigens recognized by human and rat antibodies by using western immunoblotting. Infect. Immun. **54**:96–103.

Grimes, M. M., J. D. LaPook, M. H. Bar, H. S. Wasserman, and A. Dwork. 1987. Disseminated *Pneumocystis carinii* infection in a patient with acquired immunodeficiency syndrome. Human Pathol. **18**:307–308.

Grocott, R. G. 1955. A stain for fungi in tissue sections and smears using Gomori's methenamine silver nitrate technique. Am. J. Clin. Pathol. **25**:975–979.

Hartman, B., M. Koss, A. Hui, W. Baumann, L. Athos, and T. Boylen. 1985. *Pneumocystis carinii* pneumonia in the acquired immunodeficiency syndrome (AIDS). Diagnosis with bronchial brushings, biopsy and bronchoalveolar lavage. Chest **87**:603–607.

Hegmon, M. R., and P. Rasmussen. 1987. *Pneumocystis carinii* involvement of the bone marrow in acquired immunodeficiency syndrome. Am. J. Clin. Pathol. **87**:780–783.

Hubbard, J. D. 1984. Examination of *Pneumocystis* cysts (letter). J. Pediatr. **105**:171–172.

Hughes, W. T., D. L. Bartley, and B. M. Smith. 1983. A natural source of infection due to *Pneumocystis carinii*. J. Infect. Dis. **147**:595.

Hughes, W. T., G. K. Rivera, M. J. Schell, D. Thornton, and L. Lott. 1987. Successful intermittent chemoprophylaxis for *Pneumocystis carinii* pneumonitis. N. Engl. J. Med. **316**:1627–1632.

Hughes W. T., and B. L. Smith. 1984. Efficacy of diaminodiphenylsulfone and other drugs in murine *Pneumocystis carinii* pneumonitis. Antimicrob. Agents Chemother. **26**:436–440.

Hughes, W. T. 1985. Serodiagnosis of *Pneumocystis carinii* (letter). Chest **87**:700.

Jenkins, R., R. L. Myerowitz, T. Kavic, and S. Slasky. 1979. Diagnostic yield of transbronchoscopic biopsies. Am. J. Clin. Pathol. **72**:926–930.

Juranek, D. D. 1982. Pneumocystosis, p. 197–205. *In* L. Jacobs and P. Arambulo (ed.), Section C parasitic zoonoses. *In* J. H. Steele, (ed.), CRC handbook series in zoonoses. CRC Press, Inc., Boca Raton, FL.

Kovacs, J. A., J. W. Hiemenz, A. M. Macher, D. Stover, H. W. Murray, J. Shelhammer, H. C. Lane, C. Urmacher, C. Honig, D. L. Longo, M. M. Parker, C. Nathanson, J. B. Parrillo, A. S. Fauci, A. Pizzo, and H. Masur. 1984. *Pneumocystis carinii* pneumonia. A comparison between patients with the acquired immunodeficiency syndrome and patients with other immunodeficiencies. Ann. Intern. Med. **100**:663–671.

Kovacs, J. A., V. L. Ng, H. Masur, G. Leoung, W. K. Hadley, G. Evans, H. C. Lane, F. P. Ognibene, J. Shelhamer, J. E. Parrillo, and V. J. Gill. 1988. Diagnosis of *Pneumocystis carinii* pneumonia: improved detection in sputum with use of monoclonal antibodies. N. Engl. J. Med. **318**:589–593.

Leoung, G. S., J. Mills, P. Hopewell, W. Hughes, and C. Wofsy. 1986. Treatment of first episode *Pneumocystis carinii* pneumonia (PCP) in AIDS patients with dapsone and trimethoprim. Ann. Intern. Med. **105**:45–48.

Levin, M., R. McLeod, Q. Young, C. Abrahams, P. D. Walzer, and S. A. Kabins. 1983. *Pneumocystis* pneumonia: importance of gallium scan for early diagnosis and description of a new immunoperoxidase technique to demonstrate *Pneumocystis carinii*. Am. Rev. Respir. Dis. **128**:1182–1185.

Lim, S. K., W. C. Eveland, and R. J. Porter. 1974. Direct fluorescent-antibody method for the diagnosis of *Pneumocystis carinii* pneumonitis from sputa or tracheal aspirates from humans. Appl. Microbiol. **27**:144–149.

Linder, E., L. Lundin, and H. Vorma. 1987. Detection of

Pneumocystis carinii in lung-derived samples using monoclonal antibodies to an 82 kDa parasite component. J. Immunol. Methods **98:**57–62.

Macher, A. M., D. S. Bardenstein, L. E. Zimmerman, C. K. Steigman, L. Pastore, D. M. Poretz, and L. J. Eron. 1987. *Pneumocystis carinii* choroiditis in a male homosexual with AIDS and disseminated pulmonary and extrapulmonary *P. carinii* infection. N. Engl. J. Med. **316:**1092.

Matsumoto, Y., and Y. Yoshida. 1984. Sporogony in *Pneumocystis carinii.* Synaptonemal complexes and meiotic nuclear divisions observed in precysts. J. Protozool. **31:**420–428.

Meuwissen, J. H. E. Th., I. Tauber, A. D. E. M. Leeuwenberg, P. J. A. Beckers, and M. Sieben. 1977. Parasitologic and serologic observations in infection with pneumocystis in humans. J. Infect. Dis. **136:**43–49.

Mills, J. 1986. *Pneumocystis carinii* and *Toxoplasma gondii* infections in patients with AIDS. Rev. Infect. Dis. **8:**1001–1011.

Ognibene, F. P., J. Shelhamer, V. Gill, A. M. Macher, D. Loew, M. M. Parker, E. Gelmann, A. S. Fauci, J. E. Parrillo, and H. Masur. 1984. The diagnosis of *Pneumocystis carinii* pneumonia in patients with the acquired immunodeficiency syndrome using subsegmental bronchoalveolar lavage. Am. Rev. Respir. Dis. **129:**929–932.

Pelon, V. A., R. M. Ehols, J. S. Cello, and S. L. Elmendorf. 1987. Disseminated *Pneumocystis carinii* infection in AIDS. N. Engl. J. Med. **316:**1410–1411.

Pifer, L. L., W. T. Hughes, S. Stango, and D. Woods. 1978. *Pneumocystis carinii* infection: evidence for high prevalence in normal and immunosuppressed children. Pediatrics **61:**35–41.

Pifer, L. L. 1985. Serodiagnosis of *Pneumocystis carinii* (letter). Chest **87:**698–700.

Pitchenik, A. E., P. Ganjei, A. Torres, and H. Baier. 1986. Sputum examination for the diagnosis of *Pneumocystis carinii* pneumonia in the acquired immunodeficiency syndrome. Am. Rev. Respir. Dis. **133:**226–229.

Prober, C. G., H. Whyte, and C. R. Smith. 1984. Open lung biopsy in immunocompromised children with pulmonary infiltrates. AJDC **138:**60–63.

Rosen, P. 1974. Rapid identification of *Pneumocystis carinii* (letter). Am. J. Clin. Pathol. **61:**578.

Ruskin, J., 1976. *Pneumocystis carinii,* p. 691–746. *In* J. S. Remington and J. O. Klein (ed.), Infectious diseases of the fetus and newborn infant. W. B. Saunders, Philadelphia.

Smith, J. W., and M. Bartlett. 1982. Laboratory diagnosis of *Pneumocystis carinii* infection. Clin. Lab. Med. **2:**393–405.

Smith, J. W., and M. Bartlett. 1984. In vitro cultivation of *Pneumocystis,* p. 107–137. *In* L. Young (ed.), *Pneumocystis carinii* pneumonia, Vol. 22. *In* Claude Lenfant (exec. ed.), Lung biology in health and disease. Marcel Dekker, Inc., New York.

Walzer, P. D., and M. E. Rutledge. 1980. Comparison of rat, mouse, and human *Pneumocystis carinii* by immunofluorescence. J. Infect. Dis. **142:**449.

Watts, J. C. 1985. *Pneumocystis carinii* pneumonitis. The nature and diagnostic significance of the methenamine silver-positive "intracystic bodies." Am. J. Surg. Pathol. **9:**744–751.

Wharton, B. M., D. L. Coleman, C. B. Wofsy, J. M. Luce, W. Blumenfeld, W. K. Hadley, L. Ingram-Drake, P. A. Volberding, and P. C. Hopewell. 1986. Prospective randomized trial of trimethoprim-sulfamethoxazole versus pentamidine for *Pneumocystis carinii* pneumonia in the acquired immunodeficiency syndrome. Ann. Intern. Med. **105:**37–44.

Young, L. S. (ed.) 1984. *Pneumocystis carinii,* p. 243. Pneumonia, pathogenesis, diagnosis, treatment, Vol. 22. *In* C. Lenfant (exec. ed.), Lung biology in health and disease. Marcel Dekker, Inc., New York.

Young, R. C., J. E. Bernett, and E. W. Chu. 1976. Organisms mimicking *Pneumocystis carinii.* Lancet **2:**1082–1083.

CHAPTER 101

Schistosomiasis

S. E. MADDISON and F. O. RICHARDS, Jr

Disease: Schistosomiasis, bilharziasis.

Etiologic Agents: *Schistosoma mansoni, S. haematobium, S. japonicum, S. intercalatum,* and *S. mekongi.*

Source: Water harboring schistosome cercariae which penetrate unbroken skin of humans. Cercariae are shed into water by infected snails.

Clinical Manifestations: Acute itching may occur when the cercariae penetrate the skin. The acute phase of illness, with fever and blood eosinophilia as principal features, is self-limited, but may be life-threatening. Mild to severe illness can occur in the chronic phase of infection. *S. haematobium* infections characteristically result in hematuria and dysuria, and occasionally upper urinary tract disease. Portal hypertension with hepatosplenomegaly results from the schistosome species that cause intestinal schistosomiasis. Rarely, the lung, heart, and other organs are involved.

Pathology: Granulomatous infiltration and fibrosis around schistosome eggs deposited in liver, intestine, bladder, and other tissues. Fibrosis around portal veins and ureter or ureterovesical junction cause portal hypertension and obstructive uropathy, respectively.

Laboratory Diagnosis: Definitive diagnosis is made when schistosome eggs are demonstrated in stool, urine, or tissue specimens. Cryptic or ectopic schistosomiasis may require serologic evidence of infection.

Epidemiology: *S. mansoni*—throughout most of Africa, in some areas of the Middle East and Caribbean Islands, and in some parts of South America, notably Brazil, Suriname, and Venezuela. *S. japonicum*—some parts of People's Republic of China, the Philippines, and Indonesia. *S. haematobium*—throughout most of Africa and the Middle East. *S. intercalatum*—parts of West Africa, notably Cameroon, Zaire, Gabon, and Central African Republic. *S. mekongi*—parts of Southeast Asia, notably along the Mekong River and parts of Laos, Thailand, and Kampuchea. Humans become infected during the course of work (crop irrigation, canal cleaning, fishing), play, and other exposure to water harboring infected snails.

Treatment: Numerous safe drugs have recently been introduced for treating human schistosomiasis. The most important of these are oxamniquine (for the treatment of *S. mansoni*), metrifonate (*S. haematobium*), and praziquantel (effective against all schistosome species).

Prevention and Control: Measures designed to prevent infection include mollusciciding, water and sanitation engineering, and health education. The availability of safe, single-dose, oral medications has made population-based chemotherapy the most popular and cost-effective control measure for schistosomiasis. Research on vaccine development is in progress.

Description of Disease

Life Cycle of the Parasite

When an individual is exposed to water harboring snails infected with a *Schistosoma* spp., the larval cercariae shed by the snails penetrate the unbroken skin of the individual and migrate through the lungs (2 to 3 weeks after exposure) to the liver. Here the parasites develop into mature male or female adults. The *Schistosoma* spp. that cause intestinal disease migrate to the mesenteric veins, where they copulate and remain in pairs for the rest of their lives (5 to 10 or more years). The females deposit eggs in the venules of the intestinal wall. Some of the eggs remain trapped in the intestinal wall, but others make their way into the lumen of the intestine and are passed out in the feces. On coming in contact with water the eggs hatch, and the released larval miracidia may establish infection in the appropriate species of snail. Multiplication of the larvae occurs in the snails, which may shed cercariae for several weeks. The life cycle of *S. haematobium* differs only in that the adults migrate to the vesicle veins and deposit their eggs in the wall of the bladder and, in some patients, the ureters. Eggs traverse the bladder wall and are passed into water in urine.

Clinical Manifestations

Acute Disease

Persons with schistosomal infection may be asymptomatic or may manifest a spectrum of disease conditions. The presence and nature of illness associated with schistosomal infection have been correlated with numerous variables. From a clinical perspective, the most important of these are as follows: (1) the species of the infecting schistosome; (2) duration of infection; and (3) worm burden harbored by the patient (often termed the intensity of infection).

Acute schistosomiasis (Katayama fever, Katayama syndrome) is a febrile condition thought to be a hypersensitivity reaction (similar to serum sickness) to schistosomal antigens. The syndrome occurs 18 to 58 days after exposure to contaminated water and is most commonly observed among infected expatriates and travelers (Harries et al., 1986; Hiatt et al., 1979). Patients complain of a flu-like illness, with fatigue, headache, arthralgia, and night sweats. They may have a history of weight loss, cough, and diarrhea that is sometimes bloody. Hepatosplenomegaly is frequently present. Acute schistosomiasis is a self-limited condition, and most symptoms resolve without treatment in 4 to 6 weeks. The severity of Katayama fever appears to increase with worm burden of the patient, although this relationship is less clear than in chronic forms of schistosomiasis (Clark et al., 1970; Hiatt et al., 1979).

Chronic Disease

Infections with *Schistosoma japonicum, S. mansoni, S. intercalatum,* or *S. mekongi* can cause chronic intestinal schistosomiasis. This condition is often asymptomatic. Symptoms, when they occur, are of insidious onset, and clinical disease is usually related to the intensity of the individual's infection. Intensity of the infection is estimated by the number of eggs excreted in feces, which correlates closely with worm burden. The higher the fecal egg count (usually expressed in terms of eggs per gram), the more likely the patient will suffer signs and symptoms of chronic schistosomiasis.

Inflammation of the intestinal wall and mucosa may cause crampy abdominal pain, dysentery, polyposis, or obstruction. However, the most consistent clinical finding in intestinal schistosomiasis is hepatosplenomegaly (Hadidjaja et al., 1985). Liver function is usually not compromised, but obstruction to portal blood flow can cause ascites and splenomegaly.

The most important sequellae of chronic intestinal schistosomiasis are hematemesis and/or cor pulmonale. Portal hypertension stimulates the development of portasystemic shunts, and bleeding from esophageal varices may result in uncontrollable hematemesis. Shunted blood from the portal system may carry eggs into the pulmonary arterial bed, causing arteritis, pulmonary vascular obstruction, and right heart failure (cor pulmonale).

Patients with schistosomiasis haematobium are more often symptomatic than those patients with intestinal schistosomiasis, but as with the intestinal disease, the severity of symptoms is related to the intensity of infection. Granulomas in the bladder wall cause bleeding, mucosal dysplasias, and organ dysfunction, particularly in the trigone area. Symptoms include complete or terminal hematuria, dysuria, frequency, and urgency. Recent ultrasonographic studies of patients with urinary schistosomiasis have demonstrated that obstructive upper urinary tract disease, such as hydroureter and hydronephrosis, may occur in as many as 24% of cases (Doehring et al., 1985a). *S. haematobium* infections may be complicated by bacteruria, lower urinary tract infection, chronic pyelonephritis, and cor pulmonale. Squamous cell carcinoma, an unusual form of bladder cancer, has been associated with urinary schistosomiasis (Schwartz, 1981).

Ectopic schistosomiasis occurs when eggs reach uncommon locations outside of the portal venous system, lungs, or urinary tracts. Eggs are deposited in these tissues by aberrantly migrating worms or abnormal blood shunts. The incidence of ectopic schistosomiasis is not known. Granulomatous nodules, which often are found under the skin over the perineum, also have been observed on the uterine cervix (Schwartz, 1984). The most dramatic ectopic lesions are those found in the central nervous system (CNS). *S. japonicum* is the most common species to invade nervous tissue, resulting in cerebral granulomatous masses that incite personality change and seizures (Reyes et al., 1964). *S. mansoni* and *S. haematobium* are more likely to localize in the spinal column, where they can result in paraplegia (Scrimgeour and Gajdusek, 1985). Transverse myelitis, probably caused by granulomas around *S. mansoni* eggs in the spinal cord, has been recently reported in United States expartriates (Centers for Disease Control, 1984).

Communicability and Transmissibility

Schistosomiasis is not directly communicable from human to human or from animals to humans. The prerequisite for acquiring infection is the exposure of humans to water harboring cercariae. Although some experimental laboratory animals are susceptible to the larval cercariae, animal reservoirs do not play a major role in maintaining the life cycle. A possible exception is *S. japonicum*, which is infective for water buffalo, dogs, rats, and some other mammals. In endemic areas in the tropics, snails become infected when infected humans contaminate water (canals, lakes, streams) by defecating and/or urinating into it. Humans begin to pass viable schistosome eggs within 6 to 8 weeks of exposure, and continue to do so as long as they harbor adult female schistosome worms.

Pathology

The pathology of *S. mansoni* and *S. haematobium* has been summarized by Cheever (1982) and Laughlin (1984). The intensity of the infection influences the pathology in the early as well as the chronic stages of the disease. The early lesions from the penetration of cercariae (schistosomal dermatitis) are characterized by edema and eosinophilic/mononuclear infiltrates of both the dermis and epidermis. The reactions are most severe after exposure to avian schistosomes and occur most rarely with *S. japonicum*. They resolve in 7 to 10 days.

Early stages of infection are frequently asymptomatic and remain undetected. The pneumonitis that may occur when large numbers of larvae migrate through the lungs has not been characterized histologically. Severe Katayama fever is thought to resemble serum sickness, but the pathology has not been described.

In contrast to its acute stages, the pathology of chronic schistosomiasis caused by *S. mansoni*, *S. japonicum*, and *S. haematobium* has been widely studied. In most patients chronic schistosomiasis is not associated with clinical disease, and one must use caution in attributing disease to a proved *Schistosoma* infection. When disease does result, the lesions are usually distinctive.

In chronic infections of *S. mansoni* and *S. japonicum*, schistosomal eggs incite granulomatous reactions in the liver and intestine. These granulomata are characteristic of a florid cellular host immune response. The granulomas have a marked eosinophilic component, with monocytes and some neutrophils. In older granulomas, a fibrotic response, with monocytes, fibroblasts and giant cells, is characteristic. Hepatosplenic disease is associated with portal hypertension due to a characteristic Symmers' clay stem fibrosis around the hepatic portal veins. The liver parenchyma between the fibrotic areas and egg granuloma is normal. The chronic granulomas of *S. japonicum* are characterized by necrosis and calcification, compared with those seen with *S. mansoni* infections. In both *S. mansoni* and *S. japonicum* infections, granulomatous lesions throughout the small intestine may appear grossly as "sandy patches."

Two types of intestinal complication may occur and are most frequently seen in Egyptian patients infected with *S. mansoni*. The first is colonic polyposis, in which single or numerous polyps occur. These granulomatous lesions are friable, and severe intestinal bleeding is common. The second pathologic intestinal complication is a localized intense inflammation and fibrosis that develops into a mass of lesions known as bilharzomas. Splenomegaly results from a combination of increased intrasplenic venous pressure (which relates from the hepatic portal hypertension) and splenic hyperplasia. Early in the course of infection, both the red pulp and the germinal centers of the lymphoid follicles show cellular proliferation. In chronic disease, distended venous sinuses and dense splenic cords are observed.

In severe chronic *S. haematobium* infection large numbers of eggs in the bladder and, in some patients, the ureters, are surrounded by the granulomatous immune response. Eggs in the vesical veins do not pass to the liver but can pass directly to the lungs. Thus, in severe disease hepatosplenomegaly is not observed, but cardiopulmonary schistosomiasis may occasionally occur. In the bladder and ureters calcification of eggs may occur. Grossly the lesions with calcified eggs appear as characteristic "sandy patches" in the bladder wall. Radiologically the calcification appears as a linear lesion in the submucosa.

Ectopic schistosomiasis may be due to any of the species of *Schistosoma* that infect humans. The lesions are the result of the host's granulomatous response to eggs deposited in tissues unassociated with the usual habitat of the adult worms. The sites include the skin of the genitalia and lower abdomen, the spinal cord, and the brain.

Diagnostic Features

The clinician should consider this infection in foreign nationals, travelers, and expatriates of endemic countries, particularly those who spent extended periods in rural areas. Patients should be questioned about swimming or bathing in natural bodies of water. A history of itch or rash after such exposure should be sought. A transient papular eruption occasionally is observed within hours of skin penetration by larval cercariae (cercarial dermatitis).

Clinical findings will vary. In acute schistosomiasis, the most important physical findings include elevation of body temperature and hepatosplenomegaly (Hiatt et al., 1979). Bronchospasm and/or urticarial rash may also be observed. Blood eosinophilia is an important finding, and often the first indication that the patient has a parasitic disease. Other considerations in the differential diagnosis should include strongyloidiasis, migrating hookworm or ascaris, trichinosis, and filariasis. Diagnosis should be attempted by multiple urine and stool examinations. Serologic testing or rectal biopsy may be required in early (4 to 6 weeks postexposure) infections (Harries et al., 1986). In the course of the evaluation, the clinician should bear in mind potentially lethal febrile conditions commonly acquired in developing countries, such as malaria, typhoid fever, and hepatitis B.

Hepatosplenomegaly is the most common physical finding in chronic intestinal infections. Clinicians should examine patients with this condition for ascites, dermal evidence of portal hypertension (caput medusae), and chronic heart and lung disease. Patients with urinary schistosomiasis often have painful and grossly bloody micturition; clear urines are almost invariably positive when examined for blood. Routine blood studies, including liver function tests, coagulation, urea nitrogen, and creatinine, are generally normal in all forms of chronic schistosomiasis. Blood eosinophilia is uncommon. Anemia may occur secondary to hypersplenism, hematemesis, or severe polyposis. Diagnosis is established by demonstration of eggs in stool or urine. Rectal, liver, or bladder biopsies are not necessary to establish the diagnosis, but radiologic, ultrasound, and other studies may be indicated to delimit the extent of organ dysfunction or the presence of sequelae. Serologic testing does not play an important role in diagnosing chronic schistosomiasis in endemic areas.

Prognosis

The prognosis is excellent: even without treatment, virtually all persons with light infections remain asymptomatic. Modern drugs can rapidly cure the infection, or worm burdens can be drastically reduced to the point of minimizing the risk for disease.

Prognosis must be guarded when sequellae of chronic and heavy infections are present. Incapacitation or death may occur as a result of hematemesis, ascites, intestinal, or CNS complications and renal, cardiac or respiratory compromise. Treatment with antischistosomal drugs can reverse some organ damage (Doehring et al., 1985b; Hadidjaja et al., 1985).

Epidemiologic Aspects

Occurrence

S. mansoni is found throughout Africa, in some areas of the Middle East and Caribbean islands, and in some parts of South America, notably Brazil, Suriname, and Venezuela. African slaves brought *S. mansoni* to South America and the Caribbean, where susceptible snails were already present. *S. haematobium,* also carried by the slaves, was not established because of the absence of appropriate snail hosts.

S. haematobium occurs throughout Africa and the Middle East.

S. japonicum is a parasite of the Far East. It also occurs in some islands of the Philippines and Indonesia. Due to control efforts, transmission of *S. japonicum* has been markedly reduced in most of the People's Republic of China (PRC).

The two *Schistosoma* spp. of lesser importance to humans are much more focal. *S. intercalatum* occurs in small areas in West Africa, notably Cameroon Zaire, Gabon, and Central African Republic. *S. mekongi* is limited to parts of Southeast Asia, notably along the Mekong River and parts of Laos, Thailand, and Kampuchea.

In the United States schistosomiasis is seen in Puerto Ricans, expatriates and tourists, and, more recently, in refugees from the Far East. The disease also may be found in African nationals. Transmission cannot occur in the continental United States, however, because snail species susceptible to those schistosomes that infect humans do not exist in the country.

Immunity in Humans

Concomitant immunity has been clearly demonstrated in experimentally infected animals. It is conferred by persistence of a small number of egg-producing adult worms from an initial infection long after resistance to a challenge exposure is demonstrable (Smithers, 1976). Reports of probable immunity developing in humans as a result of infection has, until recently, been anecdotal, based on a very small number of uncontrolled studies (Warren, 1973b). However, the ongoing study by Butterworth et al. (1985) in an area highly endemic for *S. mansoni* in Kenya indicates that some children do acquire a degree of immunity. Baseline parasitologic and serologic data as well as information on water contact were obtained, after which the children were treated with oxamniquine. The fecal egg counts of the immune group remained low for the year of observation after chemotherapy despite frequent contact with water harboring infected snails. In contrast, the egg counts of the susceptible group increased during the year of postchemotherapy follow-up. The antibody responses to adult worm antigen increased in both groups 5 weeks after treatment, then reverted to pretreatment levels. The serologic response of the immune group to egg antigens decreased progressively after treatment, whereas the antiegg response of the nonimmune group showed no change throughout 1 year's follow-up. Butterworth et al. (1985) discussed the possibility that the immunity observed may be directly related to duration of infection, a concept that originated in an earlier study by Kloetzel and da Silva (1967) of adult immigrant farmers in an endemic area.

Etiologic Agents

Characterization

The *Schistosoma* are digenetic trematodes, the adults of which are approximately 0.6 to 2.5 cm in length. Adult worms living for as long as 20 years have been documented, but more likely the average life span is 5 to 10 years. The eggs vary in size from 60 × 90 μ for *S. japonicum* to 114 to 175 × 48 to 68 μ for *S. mansoni*. The appearance of the eggs is quite characteristic for each of the three main species infecting humans, and can be identified without special stains. *S. mansoni* belongs to the group of *Schistosoma* with eggs that have a lateral spine. *S. haematobium* and *S. intercalatum* each have terminal spines. The *S. japonicum* and *S. mekongi* eggs are more rounded and have a minute lateral hook that is difficult to see.

Specimen Collection

Fecal Specimens

If feces are to be examined on the same day as collected, at least 2 g of stool should be placed in a clean container. If blood and/or mucus are present in the feces, they should be included. When fecal specimens must be stored or transported before examination, 10% formalin has traditionally been used as a preservative. Recently Bundy et al. (1985) have reported that sodium azide is a more convenient preservative. Five milligrams of sodium azide powder is sprinkled over 2 g of feces and thoroughly mixed with a stick. To obtain an accurate egg count, exactly 1 g of feces is added (by displacement of 1 ml of fluid) to 7 ml of the formol-saline in a calibrated 15 ml centrifuge tube. It is essential that the feces be well mixed with the formol-saline solution (Knight et al., 1976). Stool examinations should be negative on 3 successive days before the patient is declared uninfected.

Urine Specimens

Thirty minutes before the specimen is collected, patients should be given a fluid load of 250 ml of water or carbonated beverage. An entire midday urine specimen is collected into a clean plastic or glass container. If preservation is required a final concentration of 0.1% sodium azide can be used. Make up a 10% solution in water and add the appropriate amount to the urine. Alternatively, formaldehyde at a final concentration of 10% in the urine (volume by volume) will be satisfactory.

Serum Specimens

In late acute and chronic phases of schistosomiasis the antibody levels remain comparatively stable. Therefore, the testing of a single serum specimen will suffice. If the level of reactivity is very low, however, a second, confirmatory serum should be obtained. If the patient is suspected of being in an early acute phase of infection (4 to 6 weeks after possible exposure), a second specimen obtained 2 weeks after the first may show an increase in specific antibody level. The presence of antibodies does not constitute absolute proof of current parasitic infection especially in persons originating from endemic areas. Antibodies in travelers to endemic areas, however, is highly suggestive of recent contact with the parasite.

Definitive Diagnosis

Demonstration of the Parasite

Stool examinations: If the patient has a heavy infection (400 eggs/g or greater) with one of the intestinal schistosomes, eggs will be detected in a simple saline smear of feces. However, a sensitive modified formol-ether concentration technique is the method of choice (Knight et al., 1978). This method allows detection and quantitation of all other helminthic eggs, and diagnosis is possible even in very light (less than

20 eggs/g) infections. Ethyl acetate may be used to replace the ether (Young et al., 1979). The cellophane thick smear Kato technique is also used for egg counts (Arap Siongok et al., 1976) but is not as sensitive for low egg counts as the modified formol-ether technique, and hookworm eggs are often not visualized. In addition, 24 h are required for clearing of the fecal preparation to observe the eggs. Modifications of the Kato technique, using metal or plastic templates, have been developed to allow rapid and quantitative testing in field and control studies (Peters et al., 1980). Egg counts of 0 to 100 eggs/g indicate low intensity infections, 100 to 400 eggs/g medium, and greater than 400 are high intensity infections.

Urine examinations: Unpreserved urine specimens should be processed as soon as possible. A simple method for examining small numbers of urine is to allow the specimens to stand for 15 to 30 min, preferably in a conical urine sedimenting flask. The supernatant fluid is decanted and the sediment centrifuged. The entire deposit is examined microscopically. Counts can be determined by measuring the volume of the initial specimen and counting all the eggs in the centrifugation deposit. The presence of red and white blood cells should be noted. Urine also can be examined by passing 10 ml through one of the following types of filters: (1) polycarbonate filter 13 mm in diameter and pore size 8 to 12 μ (N800 CPR 1300) supported in a 13-mm Poptop* filter holder no. FH013 P1 001–10, both of which are available from Nuclepore Corp., Pleasanton, CA (Warren et al., 1978; Richards et al., 1984) and filter kits with polycarbonate membranes are obtainable from PATH, Seattle, WA; (2) Whatman no. 1 filter paper; and (3) Nytrel TI 20 HD 20-μ pore size monofilament polyamide filter (L'Union Gazes a Bluter, 42360 Panissieres, France). Both the paper and the polyamide filters are supported in a 25-mm Swinnex filter support (Millipore Corp., Bedford, MA; Mott et al., 1982). The polycarbonate and polyamide filters are examined microscopically without staining. The paper filter is sprayed with ninhydrin, to stain the eggs, and allowed to dry. The eggs are counted and reported as eggs per 10 ml of urine. If urines are negative or only have 1 or 2 eggs per 10 ml, the entire specimen should be filtered to confirm the diagnosis. An egg count of greater than 50 eggs per 10 ml urine is indicative of high intensity infection which is often associated with gross hematuria and lower urinary tract dysfunction.

Urines, particularly on standing before examination, may cause clogging of the filters. In such instances 10 ml of urine should be centrifuged and the entire deposit examined microscopically.

Tissue diagnosis: Schistosomiasis seldom can be confirmed by needle biopsy of the liver, but this procedure may confirm other possibilities, such as cirrhosis, considered in the diagnosis. Wedge biopsy usually allows definitive diagnosis of egg granulomas and Symmers' fibrosis but can only be justified if surgery is required for other reasons (Cheever, 1982). In light infections or partially treated cases eggs may not be demonstrable on repeated stool examinations (Laughlin, 1984). In some such patients, eggs may be found in a rectal biopsy specimen, but this procedure is not recommended for routine diagnosis. Occasionally in a long-standing, chronic urinary infection eggs cannot be detected in urine, but biopsy specimens of sandy, granulomatous lesions in the bladder wall will contain eggs. Biopsies of ectopic schistosomal skin lesions on the genitalia or skin of the lower abdomen may be diagnostic (El-Mofty and Cahill, 1964). Biopsy of suspected schistosomal spinal cord lesions has been reviewed by Adeloye et al. (1982). Prompt chemotherapy, including steroids, is indicated in such patients.

DNA Probes

Although studies are limited, early reports indicate that various species of *Schistosoma* can be differentiated by means of DNA probes (McCutchan et al., 1984; Walker et al., 1986). Because the parasite's eggs can usually be readily demonstrated in schistosomiasis, however, it is unlikely that DNA probes will play a role in diagnosis.

Presumptive Diagnosis

Antibody Detection

Indirect immunofluorescence assay (IIF): The most frequently used assay in the United States for antibodies in schistosomiasis uses adult worms embedded in Tissue-Tek OCT. Cryostat sections are prepared and stored at −20°C (Wilson et al., 1974). The test is positive in 85% parasitologically proved cases (Walls and Wilson, 1983). This assay cannot differentiate between early active infection and chronic infection of long duration. Cross-reactivity occurs between the various species of schistosomes that infect humans. Essentially the only nonspecificity reported with other helminthic infections occurs at a low level with sera of some trichinosis patients. However, the spotty patterns of this nonspecific fluorescence is readily distinguished from the specific gut-associated or diffuse fluorescence characteristic of schistosome infections.

Use of Rossman's fixed sections of adult worms

* Use of trade names is for identification only and does not imply endorsement by the Public Health Service or by the U.S. Department of Health and Human Services.

in IIF allows only detection of antibodies directed against the gut-associated antigens of the worms. This reactivity is stronger in acute than chronic cases (Nash, 1982). A radioimmunoassay (RIA) is used by Nash (1982) with a purified phenol-soluble proteoglycan antigen (PSAP) to distinguish acute from chronic infections. The acute infection antibody responses in this assay are predominantly of the IgM class. In addition, the antibody levels in the PSAP/RIA correlate with intensity of infection as determined by fecal egg counts.

Enzyme-linked immunosorbent assay (ELISA): Although ELISA has been widely reported in diagnosing schistosomiasis, its use as a routine diagnostic procedure is limited. This is largely due to the fact that crude schistosome antigens in ELISA are not specific, and specific purified antigens are not available commercially. A standardized kinetic-based ELISA (Hancock and Tsang, 1986), using a purified microsomal antigen of *S. mansoni* (MAMA) (Tsang et al., 1983), was recently introduced into the Centers for Disease Control reference diagnostic laboratory. An internally developed standard reference serum pool arbitrarily designated as having 100 activity units per μl, is used to prepare a standard curve for each ELISA plate. The absorbance values of the unknowns are converted to activity units per μl. The sensitivity and specificity of this assay are both 99% for infection with *S. mansoni*. Because the antigen is species specific, the sera of patients suspected of having *S. japonicum* or *S. mekongi* infection are checked by immunoblot with JAMA (Tsang et al., 1984). Until HAMA is available from *S. haematobium* adult worms, suspected schistosomiasis haematobia patients' sera are tested by immunoblot with a crude preparation of *S. haematobium*.

Serologic tests are usually negative in cases of "swimmer's itch" or cercarial dermatitis caused by the cercariae of avian schistosomes (Moore et al., 1968; Wilson et al., 1976).

Treatment, Prevention, and Control

Treatment

Excellent drugs have been introduced during the past decade that have revolutionized the approach to treating schistosomiasis. The newer agents are safe, orally administered, and provide cure in 60 to 90% of patients; in patients not cured by a course of treatment, egg burdens are often reduced to 5 to 10% of pretreatment levels. Pregnancy and (in the case of oxamniquine) epilepsy are the only contraindications to the use of these new drugs. Several stool and urine examinations should be routinely obtained 4 to 6 months after treatment to evaluate the success of therapy, and another course administered if eggs are still being passed.

Praziquantel (Biltricide, Miles) is considered the most important development in the field of antischistosomal therapy. The drug is nontoxic and effective against all species of human schistosomes. Praziquantel acts on the tegument of the parasites, resulting in loss of osmotic regulation and death. A single dose of 40 mg/kg is effective against *S. mansoni, S. intercalatum,* and *S. haematobium* infections. *S. japonicum* and *S. mekongi* infections should be treated with 3 doses of 20 mg/kg, given in 1 day. Side effects include mild to moderate abdominal pain, occasional nausea, and dizziness.

Oxamniquine (Vansil, Pfizer) is effective treatment against *S. mansoni,* particularly the New World strains. It is thought to affect nucleic acid synthesis of the parasite. Doses vary according to the age of the patient and the geographic origin of the parasite. For infections acquired in the Caribbean and South America, 15 mg/kg in a single dose is effective therapy for adults; children should receive 20 mg/kg. Infections in children and adults acquired in Africa should be treated with 30 mg/kg per day in divided doses twice daily for 2 to 3 days. Side effects include drowsiness, hallucination, and gastrointestinal discomfort. Oxamniquine may aggravate epilepsy, and this drug should not be used to treat patients with a history of seizure disorder.

Metrifonate (Bilarcil, Bayer) is active against *S. haematobium*. The drug is metabolized to its active form, dichlorvos, which is a cholinesterase inhibitor. Disruption of synaptic neurotransmission causes paralysis of the schistosomes. Metrifonate is usually given in three doses of 7.5 to 10 mg/kg, each dose separated by a 2-week rest interval. Mild nausea and occasional vomiting may accompany the use of this drug.

Niradazole (Ambilhar, Ciba-Geigy) is an antiquated drug when compared with praziquantel, oxamniquine, and metrifonate. However, it is still widely used in some endemic countries. It is active against *S. mansoni, S. haematobium,* and *S. japonicum*. The drug affects energy metabolism by inhibiting glucose uptake. Niradazole is administered in a dose of 25 mg/kg per day divided twice daily for 7 to 10 days. Almost all persons treated complain of muscle weakness, abdominal pain, vomiting, and insomnia. Unpleasant ideation, hallucinations, and even psychosis may occur. The drug should not be administered to patients with severe hepatosplenic disease, who appear to be at heightened risk for severe CNS side effects.

Patients with the major complications of chronic schistosomiasis should be managed conservatively. Portasystemic shunting, splenectomy, urologic procedures, and other surgical interventions should be avoided unless there is recurrent hematemesis, complications of hypersplenism, or hydronephrosis. The reader is referred to the chapter by Laughlin (1984) for detailed discussion of the management of these patients.

Prevention

Travelers to endemic areas who plan to swim, wade, or wash in rural fresh water bodies should be warned of the risk of schistosomiasis. There is no practical way for these individuals to distinguish snail-infested from noninfested water. Although some medical and travel literature suggests that rapidly flowing waters are safer than lakes and stagnant pools, travelers have been infected after exposure to river water (Istre et al., 1984). Therefore, all fresh water contact in rural areas of endemic countries should be avoided. For campers in relatively primitive settings, bathing water can be made safe by heating to 50°C (122°F) for 5 min or with iodine or chlorine treatment (concentration of 1 mg/liter). Water left to stand for 3 days is safe because cercariae die within 48 hours of being shed from the snail. If initiated immediately after water contact, the application of rubbing alcohol or vigorous towel drying over the exposed areas also may reduce the risk or intensity of subsequent infection (Istre et al., 1984). No effective chemical (oral or topical) prophylaxis is available for preventing schistosomiasis.

Strategies to prevent schistosomiasis among indigenous populations include chemical snail control (mollusciciding), water and sanitation engineering, programs in health education (directed toward changing of defecation and urination habits), and vaccination. The feasibility of a vaccine for schistosomiasis has recently improved as a result of better understanding of host immunity and immunogen production. Two major problems exist with the experimental vaccines so far developed: only partial immunity to infection has been achieved, and the most effective immunogens appear to be living, attenuated (irradiated) schistosomulae, which are difficult to produce and administer on a large scale (von Lichtenberg, 1985). Several laboratories are using recombinant DNA technology to identify and produce potent synthetic immunogens that will obviate these problems.

Control

Given the tools and financial resources currently available, interruption of schistosomiasis transmission in endemic communities is not feasible. Accordingly, the goal of modern control programs in endemic countries is to reduce disease morbidity by maintaining low infection prevalence and intensity. These efforts require indefinite expenditures to prevent resurgence of disease. Control programs are evaluated by epidemiologic studies that use quantitative parasitologic techniques to measure the reduction in infections of heavy worm burden.

The cornerstone of modern schistosomiasis control strategies is population-based chemotherapy programs. Target groups for chemotherapy vary (Warren, 1982). One of the most popular strategies aims treatment at school children because they are (1) usually available to health workers; (2) highly infected; and (3) important in maintaining intense transmission of the infection. The more comprehensive control efforts also incorporate focal snail control, environmental sanitation engineering, and health education programs.

Literature Cited

Adeloye, A., A. O. Folami, and P. U. Aghadiung. 1982. *Schistosoma haematobium* granuloma of the spinal cord in a Nigerian boy. Neurochirugia **25:**100–102.

Arap Siongok, T. K., A. A. F. Mahmoud, J. H. Ouma, K. S. Warren, A. S. Muller, A. K. Handa, and H. B. Houser. 1976. Morbidity in schistosomiasis mansoni in relation to intensity of infections: study of a community in Machakos, Kenya. Am. J. Trop. Med. Hyg. **15:**273–284.

Bundy, D. A., J. D. Foreman, and M. H. Golden. 1985. Sodium azide preservation of fecal specimens for Kato analysis. Parasitology **90**(pt. 3)**:**464–469.

Butterworth, A. E., M. Capron, J. S. Cordingley, P. R. Dalton, D. W. Dunne, H. C. Kariuki, G. Kimani, D. Koech, M. Mugabi, J. H. Ouma, M. A. Prentice, B. A. Richardson, T. K. Arap Siongok, R. F. Sturrock, and D. W. Taylor. 1985. Immunity after treatment of human schistosomiasis mansoni. II. Identification of resistant individuals, and analysis of their immune responses. Trans. R. Soc. Trop. Med. Hyg. **79:**393–408.

Centers for Disease Control. 1984. Acute schistosomiasis with transverse angelitis in American students returning from Kenya. Morbid. Mortal. Weekly Rep. **31:**445–447.

Cheever, A. W. 1982. Pathology and pathogenesis. p. 742–743. *In* T. E. Nash (mod.), Schistosome infection in humans: perspectives and recent findings. Ann. Intern. Med. **97:**740–754.

Clark, W. D., P. M. Cox, L. H. Ratner, and R. Correa-Coronas. 1970. Acute schistosomiasis mansoni in 10 boys: an outbreak in Caguas, Puerto Rico. Ann. Intern. Med. **73:**379–385.

Doehring, E., J. H. H. Ehrich, F. Reider, M. Dittrich, G. Schmidt-Ehry, and J. Brodehl. 1985a. Morbidity in urinary schistosomiasis: relation between sonographical lesions and pathological urine findings. Trop. Med. Parasitol. **36:**145–149.

Doehring, E., F. Reider, G. Schmidt-Ehry, J. H. H. Ehrich. 1985b. Reduction of pathological findings in urine and bladder lesions in infection with *Schistosoma haematobium* after treatment with praziquantel. J. Infect. Dis. **152:**809–810.

Domingo, E. O., E. Tiu, and P. A. Peters. 1980. Morbidity in schistosomiasis japonica in relation to intensity of infection: study of a community in Leyte, Philippines. Am. J. Trop. Med. Hyg. **29:**858–867.

El-Mofty, A. M. and Cahill, K. M. 1964. Cutaneous manifestations of schistosomiasis. Dermatologia Tropica et Ecologia Geographica **3:**157–161.

Hadidjaja, P., N. Syamsuddin, and J. S. Ismid. 1985. The impact of schistosomiasis mass treatment on hepatosplenomegaly in Mapu Valley, Central Sulawesi, Indonesia. Southeast Asian J. Trop. Med. Public Health **16:**401–404.

Hancock, K., and V. C. W. Tsang. 1986. Development and optimization of the FAST-ELISA for detecting antibodies to *Schistosoma mansoni.* J. Immunol. Methods **92:**167–176.

Harries, A. D., R. Fryatt, J. Walker, P. L. Chiodini, and A. D. M. Bryceson. 1986. Schistosomiasis in expatriates returning to Bastain from the tropics: a controlled study. Lancet **1:**86–89.

Hiatt, R. A., Z. R. Sotomayor, G. Sanchez, M. Zambrama, and W. B. Knight. 1979. Factors in the pathogenesis of acute schistosomiasis mansoni. J. Infect. Dis. **139:**659–666.

Istre, G. R., R. E. Fontaine, J. Tarr, and R. S. Hopkins. 1984. Acute schistosomiasis among Americans rafting the Omo River, Ethiopia. J. Am. Med. Assoc. **251:**508–510.

Kloetzel, K., and J. A. da Silva. 1967. *Schistosomiasis mansoni* acquired in adulthood: behaviour of egg counts and intradermal test. Am. J. Trop. Med. Hyg. **16:**167–169.

Knight, W. B., R. A. Hiatt, B. L. Cline, and L. S. Ritchie. 1976. A modification of the formol-ether concentration technique for increased sensitivity in detecting *Schistosoma mansoni* eggs. Am. J. Trop. Med. Hyg. **25:**818–823.

Laughlin, L. W. 1984. Schistosomiasis, p. 708–710. *In* G. T. Strickland (ed.), Hunter's tropical medicine (6th ed.), W. B. Saunders, Philadelphia.

McCutchan, T. F., A. J. G. Simpson, J. A. Mullins, A. Sher, T. E. Nash, F. Lewis, and C. Richards. 1984. Differentiation of schistosomes by species, strain and sex using cloned DNA markers. Proc. Natl. Acad. Sci. USA **81:**889–893.

Moore, G. T., R. L. Kaiser, R. S. Lawrence, S. M. Patham, and I. G. Kagan. 1968. Intradermal and serologic reactions to antigens from *Schistosoma mansoni* in schistosome dermatitis. Am. J. Trop. Med. Hyg. **17:**86–91.

Mott, K. E., R. Baltes, J. Bambagha, and B. Baldassini. 1982. Field studies of a reusable polyamide filter for detection of *Schistosoma haematobium* eggs by urine filtration. Trop. Med. Parasitol. **33:**227–228.

Nash, T. E. 1982. Diagnostic serologic responses, p. 749–750. *In* T. E. Nash (mod.), Schistosome infection in humans: perspectives and recent findings. Ann. Intern. Med. **97:740–754.**

Reyes, V. A., M. G. Yogare, and L. P. Pardo. 1964. Studies on cerebral schistosomiasis. J. Philippine Med. Assoc. **40:**87–100.

Richards, F. O., F. Hassan, B. L. Cline, and M. A. El Alamy. 1984. Evaluation of quantitative techniques for *Schistosoma haematobium* eggs in urine preserved with carbolfuchsin. Am. J. Trop. Med. Hyg. **33:**857–861.

Schwartz, D. A. 1981. Helminths in the induction of cancer. II. *Schistosoma haematobium* and bladder cancer. Trop. Geogr. Med. **33:**1–7.

Schwartz, D. A. 1984. The uterine cervix and schistosomiasis in West Africa. Gynecol. Oncol. **19:**365–370.

Scrimgeour, E. M., and D. C. Gajdusek. 1985. Involvement of the central nervous system in *Schistosoma mansoni* and *S. haematobium* infection: a review. Brain **108:**1023–1038.

Peters, P. A., M. El Alamy, K. S. Warren, and A. A. F. Mahmoud. 1980. Quick Kato smear for field quantification of *Schistosoma mansoni* eggs. Am. J. Trop. Med. Hyg. **29:**217–219.

Smithers, S. R. 1976. Immunity to trematode infections, p. 296–332. S. Cohen and E. H. Sadun (ed.). Immunology of parasitic infections. Blackwell Scientific Publications, Oxford.

Tsang, V. C. W., K. Hancock, M. A. Kelley, B. C. Wilson, and S. E. Maddison. 1983. *Schistosoma mansoni* adult microsomal antigens a serologic reagent. II. Specificity of antibody responses to the *S. mansoni* microsomal antigen (MAMA). J. Immunol. **130:**1366–1370.

Tsang, V. C. W., K. Hancock, S. E. Maddison, A. L. Beatty, and D. M. Moss. 1984. Demonstration of species-specific and cross-reactive components of the adult microsomal antigens for *Schistosoma mansoni* and *S. japonicum* (MAMA and JAMA). J. Immunol. **132:**2607–2613.

von Lichtenberg, F. 1985. Conference on contended issues of immunity to schistosomes. Am. J. Trop. Med. Hyg. **34:**78–85.

Walker, T. K., D. Rollinson, and A. J. G. Simpson. 1986. Differentiation of *Schistosoma haematobium* from related species using cloned ribosomal RNA gene probes. Mol. Biochem. Parasitol. **20:**123–131.

Walls, K. W. and M. Wilson. 1983. Immunoserology in parasitic infections, p. 191–241. *In* R. M. Aloisis, and J. Hyun (ed.), Immunodiagnostics, Alan R. Liss, Inc., New York.

Warren, K. S. 1973a. Schistosomiasis, p. 25–27. The evolution of a medical literature. M.I.T. Press, Cambridge, MA.

Warren, K. S. 1973b. Regulation of the prevalence and intensity of schistosomiasis in man: immunology or ecology? J. Infect. Dis. **127:**595–609.

Warren, K. S. 1982. Selective primary health care: strategies for control of disease in the developing world. I. Schistosomias. Rev. Infect. Dis. **4:**715–726.

Warren, K. S., T. K. Arap Siongok, H. B. Houser, J. H. Ouma, and P. A. Peters. 1978. Quantification of infection with *Schistosoma haematobium* in relation to epidemiology and selective chemotherapy. I. Minimal number of daily egg counts in urine necessary to establish intensity of infection. J. Infect. Dis. **138:**849–855.

Wilson, M., A. J. Sulzer, and K. W. Walls. 1974. Modified antigens in the indirect immunofluorescence test for schistosomiasis. Am. J. Trop. Med. Hyg. **23:**1072–1076.

Young, K. H., S. L. Bullock, D. M. Melvin, and C. L. Spruill. 1979. Ethyl acetate as a substitute for diethyl ether in the formalin-ether sedimentation technique. J. Clin. Microbiol. **10:**852–853.

CHAPTER 102

Strongyloidiasis

ROBERT M. GENTA

Disease: Strongyloidiasis, Anguillulosis.
Etiologic Agents: *Strongyloides stercoralis, Strongyloides fülleborni.*
Source: Soil contaminated with feces of infected humans, monkeys, or dogs.
Clinical Manifestations: Often asymptomatic. May cause mild to severe abdominal symptoms. Opportunistic behavior in immunocompromised hosts, with dissemination to extraintestinal organs, pulmonary hemorrhages, sepsis, and death.
Pathology: Adult females in small intestinal mucosa may cause mild to severe inflammation. Migrating larvae may cause inflammation, granulomas, or hemorrhages in any organ.
Laboratory Diagnosis: Demonstration of larvae (rarely eggs) in stools, duodenal fluid, sputum, other body fluids. Fecal culture and/or baermanization may be helpful. Positive serology (IgG and IgE) is highly suggestive, but not diagnostic.
Epidemiology: Worldwide, especially Southeast Asia, South America, Africa. Endemic foci present in United States and Europe.
Treatment: Thiabendazole.
Prevention and Control: Avoid contact with contaminated soil and infected animals. Screening of candidates for immunosuppression is important to prevent hyperinfection and dissemination.

Description of Disease

Strongyloidiasis (known in some Francophone countries as Anguillulosis) is an intestinal parasitosis caused by the nematode *Strongyloides stercoralis*. In some parts of the world, notably Africa and New Guinea, human infections caused by *S. fülleborni* have been reported (Hira and Patel, 1980; Kelly et al., 1976).

The growing importance of strongyloidiasis is related to the unique ability of this nematode to replicate within its host and behave as a potentially fatal opportunistic pathogen in immunocompromised patients.

Life Cycle

The infection is acquired when third-stage filariform larvae, usually found in contaminated soil, penetrate the human skin, enter a blood or a lymphatic vessel, and proceed to the lungs. Here they break into the alveolar spaces, migrate through bronchi, trachea, esophagus, and stomach, and reach the duodenum, where female worms complete their maturation. Males are apparently nonparasitic, and are passed with the stools after migration is completed. The adult females establish themselves just under the intestinal mucosa, where they lay a small number of eggs (probably no more than 20 to 30 per day) in a rather irregular fashion. As the eggs hatch, they liberate small, sluggishly moving rhabditiform larvae, which reach the intestinal lumen and, for the most part, are passed out with the feces. In a favorable external environment, the rhabditiform larvae moult into the long, slender, and swift filariform larvae which are the skin-penetrating infective forms of the parasite. While still in the intestine, some rhabditiform larvae become filariform, and repenetrate the colonic mucosa or the perianal skin, thus starting a new parasitic generation within the same host. This endogenous cycle, known as autoinfection, allows the virtually indefinite persistence of the parasite, and appears to be regulated by the host's immune responses. In the presence of immunosuppression, the internal replication of parasites may increase dramatically (hyperinfection), and large numbers of filariform larvae, as well as adults, may migrate to extraintestinal sites (dissemination), eventually causing the host's death.

If the external conditions are appropriate, the rhabditiform larvae may initiate a free-living cycle. This ability to temporarily relinquish parasitism gives *Strongyloides* the versatility to survive for several generations in the absence of adequate hosts. Each free-living cycle, however, includes the infective filariform stage, the return ticket to the parasitic way of life.

Epidemiology and Transmission

Strongyloidiasis is present in most tropical and temperate regions of the world. Highly endemic areas have been identified in Southeast Asia and South America, particularly rural Brazil, where prevalence rates as high as 60% have been reported. Endemic foci are present in the southeastern United States, especially in the Appalachian region (Walzer et al., 1982). In Europe, areas of high prevalence are found in Portugal, Italy, the Balkanic peninsula, and the Soviet Union (Scaglia et al., 1984). A large portion of the cases diagnosed in industrialized countries are imported. Because of the long persistence of *S. stercoralis* in its host, even individuals who have left an endemic area many years before should be considered at risk. In North America, Great Britain, and Australia, particularly high infection rates have been found in (1) veterans of war with a history of captivity in tropical regions, (2) Southeast Asian refugees who have spent time in relief camps, and (3) immigrants from the Caribbean islands (Gill and Bell, 1979; Vermund et al., 1986, Genta et al., 1987).

S. stercoralis is transmitted by fecal contamination of soil. There is evidence, however, that direct interhuman transmission may occur, particularly in institutions for the mentally handicapped. A few infections apparently acquired from infected animals (dogs and monkeys) have been reported. Institutionalized persons and animal caretakers should therefore be considered at risk.

Clinical Manifestations

Strongyloidiasis may be considered as a spectral disease, with a wide range of manifestations modulated by the complex relationship between host responses and parasite.

Chronic Strongyloidiasis

The overwhelming majority of immunocompetent individuals with chronic *S. stercoralis* infection appear to develop a relatively harmonious relationship with their parasite and have no signs or symptoms that allow to differentiate them from a comparable noninfected population. When symptoms are present, they tend to be intermittent and nonspecific. Vague abdominal pain, postprandial fullness, nausea, diarrhea, and constipation may appear during the course of an infection (Milder et al., 1981). Weight loss and malabsorption have been reported in some cases, but a causal relationship with the parasite has not been demonstrated. In rare instances, *S. stercoralis* infection has been associated with severe gastritis, colitis, and acute appendicitis. A characteristic sign of strongyloidiasis is "larva currens," a highly pruritic dermatitis caused by the intradermal migration of filariform larvae on the perianal, gluteal, and abdominal skin. This sign, virtually unknown in South America, has been most frequently found in Caucasian patients who had acquired the infection in Southeast Asia.

Disseminated Strongyloidiasis

When a patient harboring *S. stercoralis* develops a condition that impairs the immune system, the host-parasite balance may be disrupted. The mechanisms that normally control the rate of autoinfection may fail to operate, and large numbers of parasites may develop and disseminate to extraintestinal organs. This event, known as dissemination, has been most frequently reported in association with the use of immunosuppressive regimens of corticosteroids, but it has occurred in patients who received other immunosuppressive drugs, chemotherapy, as well as in untreated individuals with lymphomas, leukemias, leprosy, and extreme malnutrition (Igra-Siegman et al., 1981). Recently, a few cases of dissemination associated with the acquired immunodeficiency syndrome have been described. Although it is clear that the common denominator in all these patients is a derangement of the immune responses, the precise nature of the mechanisms involved in the parasite's escape from its host controls remains to be defined.

The dramatic increase in the number of parasites that accompanies dissemination is always associated with severe symptoms. Typically, a patient under steroid therapy develops severe abdominal pain and diarrhea, sometimes bloody. Shortly thereafter, a patchy pneumonia may appear, often associated with blood-stained sputum or hematemesis, which may cause the patient's death. Most commonly, these patients develop gram-negative sepsis, possibly because of the large numbers of minute colonic lacerations produced by the penetrating larvae. If untreated, the majority of these patients die of septic shock, whereas some develop meningitis or brain abscesses.

Immune Responses

Patients with chronic, uncomplicated strongyloidiasis develop parasite-specific lymphoproliferative responses in vitro. These responses are partially masked by the presence of yet unidentified serum blocking factors that appear to specifically inhibit the responses directed against *Strongyloides* antigens (Genta et al., 1983). Because most patients with disseminated disease are pharmacologically immunosuppressed, the lack of lymphoproliferative activity found in these individuals is difficult to interpret.

More than 90% of all patients with strongyloidiasis, including those with severe disseminated disease, have detectable levels of serum IgG antibodies specifically directed against *S. stercoralis* filariform larval antigens. Larval-specific circulating IgA antibodies also have been demonstrated in more than 80% of infected persons (Genta, Frei and Linke, 1987). No association could be found between the magnitude of these responses and the clinical manifestations of disease or the presence of immunosuppression.

IgE antibodies specific for filariform antigens were demonstrated by a radioallergosorbant test in the serum of approximately 85% of infected individuals. Significantly lower titers of specific IgE were found in patients with disseminated disease, indicating the possibility of a protective role of this class of immunoglobulins. However, the number of patients studied were small, and these data must be interpreted cautiously (Genta et al., 1986).

Pathology

In chronic uncomplicated strongyloidiasis the pathologic findings are usually confined to the duodenal and jejunal mucosa. In light infections the mucosal architecture is preserved, and a modest inflammatory response, consisting of eosinophils and mononuclear cells, may be seen in the lamina propria around the female worms. In more severe cases, the inflammation may extend into the mucosa and cause superficial ulcerations, and submucosal edema may be prominent. In rare cases of long-standing severe infections, fibrosis of the small intestinal wall has been described.

In disseminated strongyloidiasis, most of the small intestine may be affected by severe inflammation, edema, petechial hemorrhages, mucosal sloughing, and patchy ulcerations. Female worms, eggs, and larvae can often be seen in both the mucosa and the exudate adherent to the ulcerated areas. The colon may be grossly edematous, and microscopically large numbers of filariform larvae can be seen penetrating its walls. Larvae can be seen in the mesenteric lymph nodes, sometimes surrounded by granulomas with giant cells. The pulmonary findings usually consist of patchy bronchopneumonia and intraalveolar hemorrhages, in some cases very extensive. The identification of larvae in the alveoli usually requires the careful observation of numerous sections taken from various areas of the lungs.

When meningitis is present, the inflammatory exudate consists mostly of neutrophils and mononuclear cells. Eosinophils are usually not prominent, and only rarely larvae have been identified in the cerebrospinal fluid. In cases of dissemination, filariform larvae, with or without associated inflammatory changes, have been seen in tissue sections obtained from virtually every organ, including liver, pancreas, kidneys, and skin (Caymmi-Gomes, 1980).

Diagnosis and Laboratory Findings

The possibility of strongyloidiasis should be considered in all candidates for immunosuppressive therapy with a suggestive geographic history, irrespective of whether they are symptomatic or not. Because of their vagueness in chronic, uncomplicated infections, the clinical symptoms are of little help of establishing, or even suggesting, the diagnosis of strongyloidiasis. One exception is "larva currens," the almost pathognomonic dermatitis that should alert the clinician about the possibility of *S. stercoralis* infection.

The sudden development of abdominal pain, diarrhea, and pneumonia in an immunocompromised patient with a consistent geographic history should suggest disseminated strongyloidiasis, and every effort should be made to identify the parasite.

Nonspecific laboratory tests are of limited value in establishing the diagnosis. Moderate eosinophilia (10 to 25%) is found in 60 to 80% of patients with chronic uncomplicated strongyloidiasis. However, in many patients, and particularly in severe disseminated infections, the peripheral eosinophil count may be normal, or even low. Total serum IgE is elevated (above 250 IU/ml) in slightly more than half of the patients with *S. stercoralis* infection. Therefore, strongyloidiasis should be considered when investigating a patient with an elevation of either peripheral eosinophilia or total serum IgE. However, the absence of these laboratory abnormalities does not rule out a parasitic infection.

Parasitologic Diagnosis

The definitive diagnosis of strongyloidiasis rests on the positive identification of *S. stercoralis* (or *S. fülleborni*) parasitic forms (larvae, adults, or eggs) in feces or other specimens.

Stool Examination

Because of the female worms' scanty and irregular egg output, fecal examination methods used for the diagnosis of other intestinal parasites lack sensitivity in strongyloidiasis. Therefore, alternative techniques are frequently necessary to demonstrate the parasite.

Specimen Collection

Whenever strongyloidiasis is suspected, fresh stools should be examined. Fixatives, such as formalin and PVA used in some commercial stool collection kits, kill the larvae. This makes them less readily visible in a direct preparation, and precludes the application of more sensitive special techniques. If they cannot be examined, immediately fresh stools should be stored in a refrigerator at temperatures between 4 and 8°C, to maintain the larvae alive while preventing their molting.

Biohazard Precautions

The majority of larvae in the fresh stools of a patient with strongyloidiasis are rhabditiform (noninfective). However, filariform larvae (the tissue-penetrating, infective stage) also may be present; their number increases with time, as the rhabditiform larvae molt when stools are left at room temperature. Therefore, all unfixed specimens from a patient with suspected strongyloidiasis must be considered infective, and gloves must always be worn while handling such specimens. An aspirator bottle containing 70% ethanol should always be available in the parasitology area. If accidental skin contact occurs, immediate washing with alcohol will kill the larvae and effectively prevent their penetration.

All materials used to process and analyze such specimens should be disinfected with alcohol, clorox, or autoclaved after use.

Direct Wet Preparation

A small quantity of feces is placed in a drop of saline or iodine and covered with a cover glass. The preparation is then examined microscopically.

Formalin-Ether Concentration

Approximately 2 ml of feces are added to a tube containing saline, filtered through gauze, and centrifuged. The pellet is resuspended in 10% formalin and ether (or ethyl acetate) is added. The preparation is centrifuged, the supernatant is removed, and a small amount of the pellet of feces is placed on a glass slide with saline and examined microscopically. Because of the concentration of parasite ova and larvae achieved, this technique is more sensitive than the direct wet preparation.

Baermann's Method

This technique, also known as baermanization, exploits the tendency of *Strongyloides* larvae to migrate from a colder to a warmer area. For its performance, the following equipment is needed: a glass funnel measuring at least 10 cm in diameter, with a short rubber hose with a screw-clamp fitted onto the stem; a round metal sieve inserted into the funnel. An appropriate amount of fresh feces is placed on a cheese cloth, which is placed on the sieve; the funnel is filled with tap water at 42°C to the level sieve, and the preparation is left to remain in a warm environment for at least 30 min (Fig. 1). Then, the screw-clamp is opened to allow 1 to 2 ml of water to escape into a Petri dish. A drop of Lugol's solution is then mixed with the water, and the preparation is examined for larvae under a dissecting microscope. This is the most sensitive parasitologic technique for the diagnosis of strongyloidiasis (Jones and Abadie 1954; Pereira-Lima and Delgado, 1961).

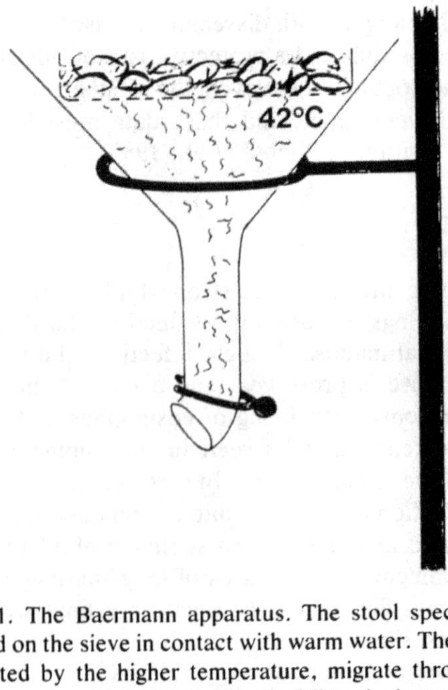

FIG. 1. The Baermann apparatus. The stool specimen is placed on the sieve in contact with warm water. The larvae, attracted by the higher temperature, migrate through the filter into the water and eventually fall to the bottom of the funnel where they can be recovered and counted.

Fecal Culture

A small amount of stools is thoroughly mixed with an approximately equal weight of sterile peat moss and tap water. The mixture, which should be wet but not soaked, is placed into a Petri dish and incubated at room temperature for 5 to 7 days. If the strain of *S. stercoralis* present in the stools is able to go through the free-living cycle, large numbers of filariform larvae may then be seen crawling on the lid and the surface of the preparations. This technique is very sensitive, because the free-living cycle amplifies the number of worms and therefore allows the detection of even light infections. Its drawback is represented by the fact that not all strains of *S. stercoralis* grow in culture.

Duodenal Aspiration

Duodenal fluid for parasitologic examination may be obtained by aspiration by means of the Enterotest®, a gelatin capsule containing 140 cm of coiled thread. The patient swallows the capsule with several sips of water, and the thread is carried by the gastrointestinal peristalsis into the first portions of the jejunum. After 4 h, the thread is pulled out, the fluid squeezed into a Petri dish, and examined for larvae under a dissecting microscope. Some studies report a high sensitivity of this test (Beal et al., 1970).

Examination of Other Specimens

In disseminated strongyloidiasis, larvae (and occasionally adult worms and eggs) have been recovered from sputum, urine, pleural, ascitic, and cerebrospinal fluid. In all cases, the best method is the direct observation under a dissecting microscope of the sediment of centrifuged specimens. In the case of thick, mucous sputum, prior fluidification with KOH may be necessary to visualize the larvae.

Morphology

Rhabditiform Larvae

This is the form most commonly encountered in specimens of intestinal origin. They measure between 250 and 550 μm in length and 15 to 25 μm in diameter. They have a bulbar esophagus extending for less than one-third their length. Their movements are sinusoidal, but slow and nondirectional (Fig. 2A).

Filariform Larvae

They can be found in stools and intestinal fluids, but they are most commonly encountered in sputum, body fluids, and urine in disseminated disease. Filariform larvae are the form seen in fecal cultures. They measure between 400 and 640 μm in length and 16 μm in diameter; their slender (filariform) esophagus occupies the cephalic 40% of their body length. Their sinusoidal movements are extremely rapid and clearly directional. They are easily distinguished from other nematode larvae because of their characteristically notched (trifid) tail (Fig. 2B).

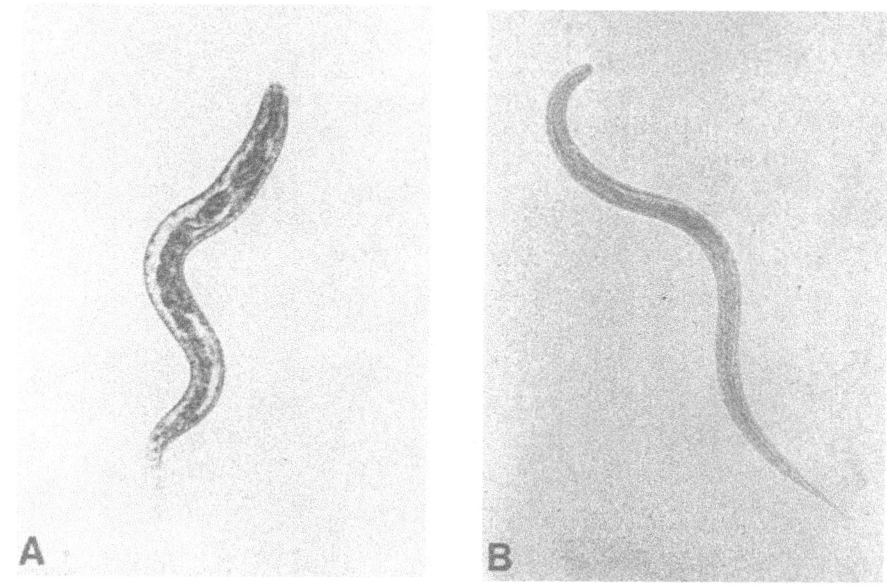

FIG. 2. Rhaditiform larvae (A) are the stage most commonly found in the stools of infected patients. They can be identified by their proprominent esophagus. Filariform larvae (B) are found in cultures, and occasionally in specimens from patients with disseminated strongyloidiasis. They are slender, fast moving, and have a fine tubular esophagus.

Adult Females

Parasitic adult females are rarely recovered from specimens of intestinal origin or sputum. They measure 1.5 to 2.5 mm in length and 40 to 50 μm in diameter, have a translucent whitish color, and contain 8 to 15 eggs in the uterus, which occupies more than the caudal half of their body. They move actively, usually in an apparent effort to liberate themselves from the mucous material in which they are often entrapped.

Free-Living Adults

These forms are seen in fecal cultures or recovered from contaminated soil. The adult males are 650 to 1000 μm long and 40 to 50 μm wide, with a clearly visible gubernaculum. Females are 0.9 to 1.7 mm long, 50 to 85 μm wide, with thin-shelled embryonated eggs in their uterus.

Eggs

S. stercoralis eggs are almost never found in the stools, and their presence in these specimens should suggest *S. fülleborni* infection. They measure 55 to 60 μm × 28 to 32 μm and have a thin shell; inside, a coiled embryo may be visible (Levine, 1980).

Identification of *S. stercoralis* in Tissue Sections

All parasitic forms of *S. stercoralis* may be found in biopsy or autopsy tissue specimens. In the small intestine, the simultaneous presence in the submucosa of eggs, usually embryonated, cross-sections of adult worms (20 to 50 μm in diameter), and sections of small larvae (8 to 16 μm in diameter) containing fine basophilic granules is unique to strongyloidiasis. With very few exceptions, the only form seen in other organs is the filariform larva. Cross-sections are usually basophilic on hemotoxylin and eosin staining, and contain fine darker basophilic granules. In good cross-sections, the lateral alae can often be seen. In transverse sections, the fine granules, the small intestinal tube, and the size of the larvae help in the identification of the parasites (Fig. 3).

Immunoserology

A number of sensitive and specific serologic tests have been proposed recently as useful tools in the diagnosis of strongyloidiasis. Most of them are still in an experimental phase, their main limitation being a readily available source of standardized antigens. However, the results are extremely promising, and it is possible that these tests may soon become available to more laboratories.

Antigens

For the performance of these immunologic tests, a steady source of antigens is provided by animals experimentally infected with a human strain of *S. stercoralis*. Parasite-free, immunosuppressed monkeys or dogs can be maintained in the laboratory as larval donors (Genta, 1986). Their feces are collected daily, cultured, and then baermanized to separate filariform

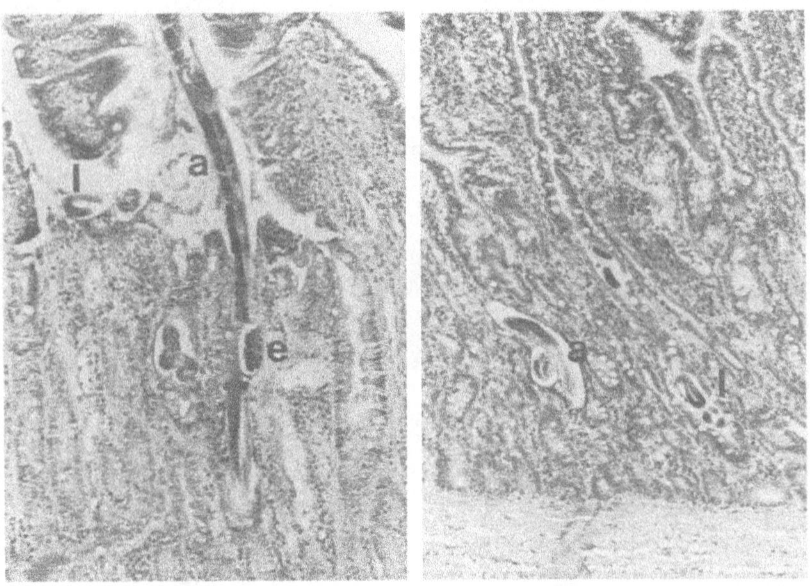

FIG. 3. Adult worms (a), larvae (l), and eggs (e) of *S. stercoralis* in the small intestinal mucosa of an infected host.

larvae. The larvae are then washed, and either fixed in 1% formalin or used for the preparation of soluble antigens.

Some authors have used antigens prepared from *S. ratti,* maintained by repeated passage in laboratory rats. Although the sensitivity of these antigens appears to be slightly lower than that of *S. stercoralis,* their specificity has not been tested. If proved satisfactory by further testing, *S. ratti* antigens might eventually become the standard products for strongyloidiasis serology (Neva et al., 1981).

Indirect Immunofluorescence Test

Formalin-fixed filariform larvae are used as antigens. After repeated washings in a buffer solution, approximately 100 larvae are incubated for 30 min at 37°C with 100 μl of test serum at appropriate dilutions. After three washings, fluorescein-conjugated antihuman IgG is added at the appropriate concentration. After 30 min incubation at 37°C, the fluorescence on the larval surface is evaluated using a standard fluorescent microscope. The sensitivity of this test is between 85 and 90%, and its specificity is highly satisfactory; rare weak positive results have been found in some patients with tropical eosinophilia, a condition so rare in most parts of the world, that problems of interpretation are usually negligible (Grove and Blair, 1981; Genta and Weil, 1982).

Enzyme-Linked Immunosorbant Assay (ELISA)

A saline extract from sonicated filariform larvae is used as antigen to sensitize standard polyvinyl ELISA plates. Test sera are then added to each well in appropriate dilutions, incubated for 30 min at 37°C, and alkaline phosphatase-conjugated antihuman IgG is then added, with *p*-nitrophenylphosphate used as the substrate to complete the test. Results are expressed as the net absorbance, which is calculated by subtracting the absorbance of a nonsensitized well with test serum from the absorbance of the sensitized well with serum. Standardized negative, low positive, and high positive serum pools are tested on each plate to monitor intra- and interassay variability. In our laboratory, the sensitivity of this test is between 85 and 95%. No false-positive results have been found in nonparasitized persons. Borderline results have been observed in occasional patients with filariasis; it is important to consider, however, that in these patients *S. stercoralis* infection had been excluded on the basis of one stool examination, leaving the possibility of undetected occult infections (Genta, 1986).

Radioallergosorbant Test (RAST)

We recently have developed a solid-phase radioimmunosorbant test for the detection of parasite-specific serum IgE antibodies. Paper disks are activated and sensitized with soluble *S. stercoralis* antigens, then incubated with the test sera. Specific IgE are then measured using an antihuman IgE conjugated with ^{125}I (Phadebas, Pharmacia) and the results are expressed as the percent specific binding, a value linearly correlated to the amount of specific IgE present in the serum.

The sensitivity of the test is around 85%. Borderline results have been obtained in patients with *Ascaris lumbricoides* infections and in some allergic individuals with total IgE levels higher than 2500 IU/ml (McRury et al., 1986).

Use and Significance of Serology in Strongyloidiasis

Because of the lack of sensitivity of stool examination, serology is more important in strongyloidiasis than in other intestinal helminthiasis. The results of the tests described are extremely encouraging, and standardized antigens for their widespread use should become generally available. The interpretation of serology results may be different in highly endemic areas and in the industrialized world. In the latter nosologic context, a positive test should be an indication for an aggressive parasitologic search for *S. stercoralis,* particularly if the patient is a candidate for immunosuppression or is already receiving immunosuppressive treatment. In endemic areas, the significance of positive serology is less clear. High antibody levels may reflect previous exposure rather than current infection. Also, in some parts of the world antihelminthic self-treatment is quite common, and elevated specific IgG and IgE titers may persist long after treatment. Until these important questions are resolved, we suggest that positive serology should be used as an indication of possible infection, but it should not substitute the parasitologic demonstration of the parasite. The only exception is an immunosuppressed patient with signs and symptoms suggestive of disseminated infection: in such a case, positive serology should be an indication for immediate treatment, even if parasites cannot be readily demonstrated (Genta and Walzer, 1988).

Treatment, Prevention, and Control

Treatment

The drug of choice for strongyloidiasis is thiabendazole. The recommended regimen is 25 mg/kg per day for 2 days. This dose appears to be effective in 80 to 90% of patients with chronic, uncomplicated infections. Multiple courses may be necessary in certain cases.

Immunocompromised patients do not appear to respond as well to this treatment, and higher doses (up to 50 mg/kg per day) for longer periods have been recommended.

Thiabendazole is not without side effects, and nausea, vomiting, foul-smelling urine, dizziness, and other complaints are reported in a high percentage of patients (Grove, 1982). Of the several other benzimidazoles that have been tried for the treatment of strongyloidiasis, only cambendazole has been reported to have very high cure rates. Recently, however, the production of this drug has been discontinued. A new potent antiparasitic agent, Ivermectin, is currently being evaluated for its effectiveness against intestinal helminths (Campbell et al., 1983), and it appears to be very promising for the treatment of strongyloidiasis. To date, however, no reports of extensive clinical trials are available.

Prevention and Control

The only known way to prevent strongyloidiasis is to avoid contact with contaminated soil or infected animals. Specimens from infected patients should be treated as potentially infective, but there is no need for isolation.

The control of the infection will have to hinge on improving sanitary conditions in endemic areas, with particular emphasis on proper disposal of human excreta.

Acknowledgments. This work was supported in part by the Medical Research Service of the Veterans Administration.

Literature Cited

Beal, C. B., P. Viens, R. G. Grant, and J. M. Huges. 1970. A new technique for sampling duodenal contents. Am. J. Trop. Med. Hyg. **19**:349–352.

Campbell, W. C., M. H. Fisher, and E. O. Stapley. 1983. Ivermectin. A new potent antiparasitic agent. Science **221**:823–825.

Caymmi-Gomes, M. 1980. Mecanismos patologicos relacionados a autoinfecçaõ na estrongiloidose humana fatal. Revist. Patol. Trop. **9**:165–261.

Genta, R. M. 1986. Strongyloidiasis, p. 183–199. *In* K. Walls and P. Schantz (ed.), Immunodiagnosis of parasitic diseases, vol. I. Academic Press, New York.

Genta, R. M., and P. D. Walzer. 1988. Strongyloidiasis. *In* P. D. Walzer and R. M. Genta (ed.), Parasitic infections in the immunocompromised host, Marcel Dekker, New York.

Genta, R. M., and G. J. Weil. 1982. Antibodies to *Strongyloideds stercoralis* larval surface antigens in chronic strongyloidiasis. Lab. Invest. **47**:87–90.

Genta, R. M., R. W. Douce, and P. D. Walzer. 1986. Diagnostic implications of parasite-specific immune responses in immunocompromisd patients with strongyloidiasis. J. Clin. Microbiol. **23**:1099–1103.

Genta, R. M., E. A. Ottesen, A. A. Gam, F. A. Neva, M. Wittner, H. B. Tanowitz, and P. D. Walzer. 1983. Cellular responses in human strongyloidiasis. Am. J. Trop. Med. Hyg. **32**:990–994.

Genta, R. M., R. Weesner, R. W. Douce, T. Huitger-O'Connor, and P. D. Walzer. 1987. Strongyloidiasis in United States veterans of the Vietnam and other wars. J. Am. Med. Assoc. **258**:49–52.

Genta, R. M., D. F. Frei, and M. J. Linke. 1987. Demonstration and partial characterization of parasite-specific IgA responses in human strongyloidiasis. J. Clin. Microbiol. **25**:1505–1510.

Gill, G. V., and D. R. Bell. 1979. *Strongyloides stercoralis* infection in former Far East prisoners of war. Br. Med. J. **2**:572–574.

Grove, D. I. 1982. Treatment of strongyloidiasis with thiabendazole: an analysis of toxicity and effectiveness. Trans. R. Soc. Trop. Med. Hyg. **76**:114–118.

Grove, D. I., and J. Blair. 1981. Diagnosis of human strongyloidiasis by immunofluorescence using *Strongyloides ratti* and *S. stercoralis* larvae. Am. J. Trop. Med. Hyg. **30**:344–349.

Hira, P. R., and B. G. Patel. 1980. Human strongyloidiasis due to the primate species *Strongyloides fülleborni*. Trop. Geo. Med. **32**:23–29.

Igra-Siegman, Y., R. Kapila, P. Sen, Z. C. Kaminski, and D. B. Louria. 1981. Syndrome of hyperinfection with *Strongyloides stercoralis*. Rev. Infect. Dis. **3**:397–407.

Jones, C. A., and S. H. Abadie. 1954. Studies in human strongyloidiasis. II. A comparison of the efficiency of diagnosis by examination of feces and duodenal fluid. Am. J. Clin. Pathol. **24**:1154–1158.

Kelly, A., M. D. Little, and M. Voge. 1976. *Strongyloides fülleborni*-like infections in man in Papua New Guinea. Am. J. Trop. Med. Hyg. **25**:694–699.

Levine, N. D. 1980. Nematode parasites of domestic animals and of man. Burgess Publishing Co., Minneapolis, MN.

McRury, J., I. T. deMessias, P. D. Walzer, T. Huitger, and R. M. Genta. 1986. Specific IgE responses in human strongyloidiasis. Clin. Exp. Immunol. **65**:631–638.

Milder, J. E., P. D. Walzer, G. Kilgore, I. Rutherford, and M. Klein. 1981. Clinical features of *Strongyloides stercoralis* infection in an endemic area of the United States. Gastroenterology **80**:1481–1488.

Neva, F. A., A. A. Gam, and J. Burke. 1981. Comparison of larval antigens in an enzyme-linked immunosorbant assay for strongyloidiasis in humans. J. Infect. Dis. **144**:427–432.

Pereira-Lima, J., and P. G. Delgado. 1961. Diagnosis of strongyloidiasis: importance of Baermann's method. Am. J. Digest. Dis. **6**:899–904.

Scaglia, M., R. Brustia, S. Gatti, A. M. Bernuzzi, M. Strosselli, A. Malfitano, and D. Cappelli. 1984. Autochthonous strongyloidiasis in Italy: an epidemiological and clinical review of 150 cases. Bull. Soc. Pathol. Exp. **77**:328–332.

Vermund, S. H., F. Lafleur, and S. McLeod. 1986. Parasitic infections in a New York City Hospital: trends from 1971 to 1984. Am. J. Public Health **76**:1024–1026.

Walzer, P. D., J. E. Milder, J. G. Banwell, G. Kilgore, M. Klein, and R. Parker. 1982. Epidemiologic features of *Strongyloides stercoralis* infection in an endemic area of the United States. Am. J. Trop. Med. Hyg. **31**:313–319.

CHAPTER 103

Taeniasis and Cysticercosis

MARCIA L. RHOADS and K. DARWIN MURRELL

Diseases: Taeniasis: beef tapeworm, pork tapeworm. Cysticercosis: beef, pork, and human, beef measles, human neurocysticercosis, or cerebral cysticercosis.

Etiologic Agents: Adult (tapeworm) and larval (cysticercus) stages of the cestode parasites *Taenia solium* (human to pig life cycle or accidentally human to human) and *Taenia saginata* (human to cattle life cycle).

Source: Taeniasis: adult tapeworm in definitive host (human) derived from ingestion of raw or undercooked pork or beef containing cysticerci. Cysticercosis: larval stage developing in intermediate host (cattle, pigs, humans) after ingestion of eggs passed in stool of tapeworm carriers.

Clinical Manifestations: Taeniasis: generally asymptomatic. Human cysticercosis: varied, can mimic brain tumor, including seizures, hydrocephalus, meningitis, arachnoiditis, headache, nausea, vomiting, impaired vision, confusion, dizziness, and ataxia.

Pathology: Taeniasis: slight inflammatory response of intestine to tapeworms. Cysticercosis: viable cysticerci well tolerated. Inflammatory response to dead or dying cysticerci consists of infiltrate of eosinophils and lymphoid cells; calcification follows death of cysticerci. In humans, location and number of cysts and/or inflammatory response to dead cysts may have severe consequences when strategic areas of the central nervous system are involved.

Diagnosis: Taeniasis: detection of egg, proglottid, or scolex in stool sample or anal swab. Proglottid but not egg can be used for species differentiation. Human cysticercosis: Computed tomography is the most useful procedure for diagnosis of neurocysticercosis, especially when supported by biopsy or serology procedures.

Epidemiology: Taeniasis and cysticercosis transmitted via two life cycle stages of cestodes; the cyst embedded in bovine or porcine muscle tissue and the egg shed in human feces. Humans acquire taeniasis by ingestion of raw or undercooked beef or pork containing cysts. Cysticercosis infection in animals and humans results from egg contamination of feed or water sources via infected human feces; autoinfection can occur in humans. *T. saginata* taeniasis-cysticercosis found in cattle-rearing countries worldwide. Frequently occurs in the United States and Europe, with an increasing prevalence in a number of developed countries. *T. solium* taeniasis-cysticercosis occurs only rarely in highly developed countries, being most prevalent in poor underdeveloped countries. Human cysticercosis infections are on the rise in the United States due to an increase in immigrant populations from endemic areas.

Treatment: Niclosamide or praziquantel for taeniasis. For human cysticercosis, supportive therapy (antiseizure medication, decompression of hydrocephalus with ventricular shunts). Also surgical removal of cysts and chemotherapy with praziquantel.

Prevention and Control: Eradication in beef and pork through mandatory and improved surveillance procedures at slaughter. Eradication in humans through screening and drug therapy. Ultimately, by improvements in sanitary and economic conditions and in public health education.

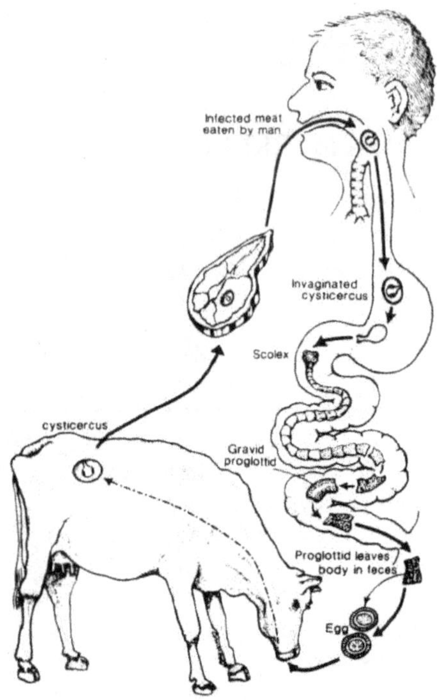

FIG. 1. Life cycle and transmission pattern of the beef tapeworm *Taenia saginata*. Reprinted with permission from Murrell et al., 1986.

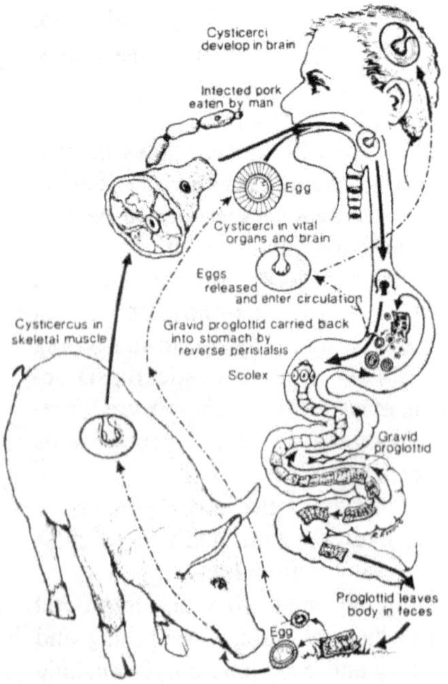

FIG. 2. Life cycle and transmission patterns of the pork tapeworm *Taenia solium*. Reprinted with permission from Murrell et al., 1986.

Description of Diseases

Taeniasis and cysticercosis are diseases caused by the adult and larval stages of the cestode or tapeworm parasites *Taenia saginata* and *Taenia solium* in their definitive host (humans) and intermediate hosts (cattle, pigs, humans). Both species are meat-borne parasites that localize as adults in the intestines of the human host. These intestinal infections, termed taeniasis, normally produce only mild symptoms. Eggs passed in the feces of human carriers can cause further disease if ingested by cattle, pigs, or humans. In these intermediate hosts, the egg develops to the larval (cysticercus) stage, and the disease is termed cysticercosis. The larval stage *of T. saginata* infects cattle, whereas *T. solium* larvae can infect both pigs and humans. Although larvae invade mainly skeletal muscles, *T. solium* larvae frequently invade the central nervous system of humans, and is, consequently, a serious public health problem. In livestock, the potential for infection necessitates continuous meat surveillance procedures and, when detected, requires either additional processing of cattle and pig carcasses at slaughter or results in their condemnation, causing a significant economic burden. Disruption of the infection at either life cycle stage effectively would eliminate the disease. The most direct control approach is improvement in sanitation and hygiene for both humans and livestock.

Etiologic Agents

Life Cycle

The life cycle of *T. saginata* and *T. solium* is illustrated in Figs. 1 and 2, respectively.

Development in Definitive Host

The adult stages of *T. saginata* and *T. solium* are flat tapeworms consisting of a scolex (head), and strobila (chain) of proglottids (segments). *Taenia saginata* can contain 1,000 to 2,000 proglottids and can vary in length from 4 to 12 m; *T. solium* usually contains fewer than 1,000 proglottids and is usually less than 3 m in length. Mature gravid proglottids, containing a uterus filled with as many as 80,000 eggs, detach from the strobila either singly or in multiples and pass out of the intestine through the anus.

Development in Intermediate Hosts

Ingestion of the *T. saginata* egg by cattle or the *T. solium* egg by pigs or humans results in the hatching of the larval stage (onchosphere). Activated by host gastric juices, the onchosphere penetrates the host's intestinal mucosa and enters the blood or lymphatic system. The primary site in cattle for development of the *T. saginata* onchosphere to the infective cysticercus stage is the dilated lymphatic spaces in skeletal muscles; some also reach full development in the heart. In pigs and humans, the cysticerci of *T. solium* develop in skeletal muscle, brain, myocardium, and the eye. In humans, localization in the central nervous system is frequent. The cysticercus is composed of a scolex invaginated within a fluid-filled vesicle or bladder. *Taenia saginata* cysticerci are 7 to 10 × 4 to 6 mm, and *T. solium* cysticerci measure 5 to 20 mm in diameter. The cysticerci in cattle and pigs become infective for humans in 10 to 12 weeks after ingestion of eggs, and the life cycle is completed when these cysticerci are ingested via raw or inadequately cooked beef or pork. The life span of the cysticerci in cattle and pigs is reported to be as short as a few months to as long as 3 years.

Clinical Symptoms

Taeniasis

Infections of adult *T. saginata* and *T. solium* in humans produce only mild symptoms. These might include discharge of proglottid, abdominal pain, nausea, loss of weight, change in appetite, and headache (Pawlowski and Schultz, 1972).

Human Cysticercosis

Taenia solium cysticerci that develop in muscle tissue present few symptoms, although pain results if muscle cysts encroach on nerve endings. The death and calcification of muscle cysts is a more important cause of discomfort as is eye involvement. The most common locations of cysts in the eye are vitreous humor and subretina, and the inflammatory response to the parasite can lead to retinal detachment or atrophy, iridocyclitis, and chorioiditis. Myocardial cysticercosis frequently occurs in massive infections, yet clinical signs rarely result (Pawlowski, 1983). Involvement of the central nervous system can cause serious clinical disease resulting from either an inflammatory response (to dead or dying cysticerci) or an obstruction (involving the foramina and ventricular system of the brain). Symptoms include seizures, hydrocephalus, headaches, dizziness, arterial thrombosis, loss of vision, and nausea (Brown and Voge, 1985; Grisolia and Wiederholt, 1982; Nash and Neva, 1984).

Immunologic Considerations

In taeniasis, only slight inflammatory responses are induced in the intestinal tract despite the large size of these tapeworms. Blood levels of eosinophils and IgE often are elevated.

In human cysticercosis, viable cysts appear to be well tolerated by the host. In the absence of massive infection, generally a latent period of 4 to 5 years occurs, which might correspond to the life span of cysts (Nash and Neva, 1984). Eventually, dead and dying cysts invoke an inflammatory cellular infiltrate consisting of eosinophils and lymphocytes. Surprisingly, the humoral antibody response of humans to cysticerci, especially to those localized in the central nervous system, is weak and variable (Flisser et al., 1979, 1980). This is a major reason for the current lack of a serodiagnostic test for cysticercosis.

Diagnosis

Taeniasis

Eggs of *T. saginata* and *T. solium* can be detected in fecal smears or anal swabs. Eggs are not distributed uniformly throughout the stool, and fecal smears are accurate for diagnosis in only about 77% of cases involving a single examination; the anal swab method is about 85% accurate. Tapeworm species can be determined by examination of proglottid morphology (if a proglottid is recovered) but not by eggs alone. The proglottid is pressed between glass slides and the number of uterine branches counted. For *T. saginata* the number of branches is >18; for *T. solium*, it is <12. Specific diagnosis also can be done by an experienced observer through examination of a scolex, which is usually voided after drug treatment.

Human Cysticercosis

Biopsy

Definitive diagnosis can be accomplished by surgical removal and biopsy of nodules located on the surface of skin or in subcutaneous tissues and muscles.

Radiology

Calcified intramuscular cysts can be detected by x-ray analysis. Ocular cysts can be detected by opthalmoscopy. Computed axial tomography (CAT scan) can detect cysticerci in the central nervous system, but this procedure is not totally specific. Although the detection of multiple cystic or calcified lesions is considered highly suggestive of cysticercosis, the detection of single lesions or nonspecific findings such as hydrocephalus are less diagnostically certain. Also, cysts located in certain regions of the brain (subarachnoids or intraventricles) often are not detectible by this procedure (Nash and Neva, 1984).

Serology

Immunodiagnostic tests include complement fixation, immunofluorescence, indirect hemagglutination assay, and enzyme-linked immunosorbent assay (ELISA). These tests exhibit varying degrees of nonspecificity or lack of sensitivity. Currently, the ELISA shows the highest accuracy. Using either a crude extract of pig cysticerci as antigen or a purified fraction thereof (antigen B), antibodies were detected by ELISA in sera or cerebrospinal fluid (CSF) in 70 to 80% of clinically diagnosed cases (Diwan et al., 1982; Espinoza et al., 1982). However, nonspecificity has been demonstrated (Coker-Vann et al., 1984). False-positive reactions were eliminated by combining ELISA with an immunoblotting technique (Gottstein et al., 1986; Gottstein et al., 1987). The complementary use of an ELISA designed for the detection of larval *T. solium* antigens and one designed for the detection of anti-*T. solium* antibodies in CSF was effective in detecting 69% of patients with a clinical diagnosis of cysticercosis and 100% of patients for whom cysticercosis was confirmed surgically or histologically; no false-positive reactions were found (Estrada and Kuhn, 1985). Recently, an ELISA using any one of three purified proteins isolated from the scolex of *T. solium* metacestodes by monoclonal antibody-immunoaffinity chromatography was able to detect 100% of patients with cysticercosis, with no false-positive reactions (Nascimento et al., 1987). At present, no immunodiagnostic test with the necessary specificity and sensitivity has been developed for cysticercosis; therefore, these procedures are of greatest value when complemented and correlated with other diagnostic procedures.

Beef and Pork Cysticercosis

Diagnosis in livestock is made by direct meat inspection by organoleptic means using prescribed procedures designed to optimize detection of lesions (Snyder and Murrell, 1986). However, these postmortem examinations for cysts are extremely insensitive and have an error rate estimated at 25% or higher (Dewhirst et al., 1967). As with human cysticercosis, no serologic diagnostic tests are yet available, although several research efforts have yielded encouraging results (Geerts et al., 1981; Harrison and Sewell, 1980; Rhoads et al., 1985).

Epidemiology

Transmission

Animal to Human

The cysticerci of *T. saginata* and *T. solium* are transmitted to humans through the ingestion of raw or inadequately cooked beef or pork.

Human to Animal

Humans harboring tapeworms can contaminate the environment with large numbers of eggs (250,000 to >500,000 eggs per day per individual). These eggs can survive for long periods and are resistant to moderate dessication, disinfectants, and low temperature (Froyd, 1962). Tapeworm eggs can be transmitted directly to pigs and cattle by the handling and feeding of these animals by farm workers with egg-contaminated hands. Infected farm workers also can cause transmission by contaminating hayfields or hay, silage, and other feeds, as well as irrigation water used on field crops, through indiscriminate defecation. In many areas of Latin America, it is common practice to let pigs roam free to clean up human wastes. A source of indirect contamination of the environment with eggs is the effluent from sewage treatment plants used for the fertilization of agricultural lands. Taeniid eggs can survive many sewage treatment processes (Silverman and Griffiths, 1955), a problem exacerbated by the introduction in recent years of detergents and chemicals in households and industry, which reduce the efficiency of these processes. Transmission can also occur by birds, earthworms, or beetles that have fed on raw sewage sludge because eggs can survive passage through their digestive tracts and can then contaminate pastures. Eggs also can be dispersed by wind.

Human to Human

Autoinfection of humans can occur with *T. solium* eggs transmitted directly from anus to mouth, or transmission can occur between individuals through contaminated hands or food. The development of human cysticercosis by internal autoinfection (whereby gravid proglottids enter the stomach via vomiting or reverse peristalsis; Fig. 2) has been cited repeatedly. However, the frequency of this occurrence is questioned (Pawlowski, 1983; Webbe, 1967), and it would appear that humans most commonly become infected like pigs through the ingestion of *T. solium* eggs contaminating the environment (i.e., drinking water contaminated by raw or inadequately treated sewage or eating vegetables fertilized with raw sewage).

Distribution

T. saginata Cysticercosis and Taeniasis

High infection rates of *T. saginata* cysticercosis in cattle are reported for central and East African countries (as high as 20% in slaughtered cattle in Kenya and even higher in Ethiopia), Caucasian and South Central Asian USSR republics, Near East countries, and certain regions of Yugoslavia. Moderate infection rates are reported for Europe, Southeast Asia, and South America; low infection rates are reported for the United States, Canada, and Australia. Some developed countries have had an increase in *T. saginata* infections in cattle since World War II. In the United States, a recent estimate of infection in slaughtered cattle was 0.0002% (Snyder and Murrell, 1986). Incidence rates, however, vary greatly among the states with 80% or more of infected cattle originating from California, Arizona, New Mexico, and Texas, states which border on Mexico (Hird and Pullen, 1979; Schultz, 1971). It should be noted also that infection rates in cattle almost certainly are underreported because inspection at slaughter is, at present, relatively insensitive. In humans, the prevalence of *T. saginata* taeniasis may reach 50% or more in certain regions in Africa, East Europe, and in Caucasian and South Central Asian USSR republics. Although human tapeworm infections are not reportable to health authorities in the United States, data collected in 1981 by the Centers for Disease Control indicate that 443 patients in the United States were treated for *T. saginata*.

T. solium Taeniasis and Cysticercosis

Areas endemic for *T. solium* are Central and South America, Central and South Africa, Southeast Asia, and South and East Europe. Infection rates in pigs in the USSR and East Europe are less than 0.01%. In the United States, 6 of about 88 million hogs slaughtered in federally inspected plants were detected during 1981 (FSIS, 1981). However, human cysticercosis infections have been on the rise in the United States in recent years (Brown and Voge, 1985; Miller et al., 1984; Richards et al., 1985), probably due to the influx of large numbers of immigrants from Latin America, particularly Mexico.

Treatment, Prevention, and Control

Treatment

Taeniasis

Niclosamide (reported efficacy between 85 and 95%; Pawlowski and Schultz, 1972; Pawlowski, 1970) and praziquantel (95% efficacy with single dose of 10 mg per kg body weight; Rim et al., 1979) are effective for human taeniasis. Praziquantel has the advantage in that it effectively removes the scolex and causes destruction of the strobila without release of gravid proglottids and it has minimal side effects (Groll, 1980).

Cysticercosis

Praziquantel has been shown also to be effective in humans for *T. solium* cysticerci (Rim et al., 1980). Previously, the only treatment available was the surgical removal of cysts. In a series of studies with praziquantel (Groll, 1982), side effects in patients were common but could be alleviated with corticosteroids. The majority of patients showed clinical improvement. Although determination of the overall benefits of praziquantel for patients with neurocysticercosis needs further study, initial results are optimistic. Traditional supportive therapy includes antiseizure and steroid medication and decompression for hydrocephalus. Praziquantel is effective also against cysticerci in cattle and pigs (Liu, 1981; Walther and Koske, 1979).

Prevention and Control

In highly endemic areas where improvements in sanitary and economic conditions would not be expected to occur rapidly, one approach for control is the mass chemotherapy of humans, which has had some success in East Europe and the USSR (Pawlowski, 1983). Another measure considered for endemic areas is animal vaccination. Also, the institution and enforcement of veterinary meat inspection would be of value in reducing transmission. It is here that the development of an accurate and rapid immunodiagnostic test could make a substantial contribution to control. A serologic test would also be of great value in the United States where efforts to have employees of cattle feedlot operations submit to examination and treatment by public health agencies have not been successful. The elimination of the disease therefore must instead rely on improved meat inspection procedures. Research presently is directed toward the development of an enzyme-linked immunosorbent assay which could be automated for large-scale use in packing houses. Some research has been carried out also on the inactivation of cysticerci in beef and pork by γ-irradiation, but the data are not yet complete (Verster et al., 1977).

Literature Cited

Brown, W. J., and M. Voge. 1985. Cysticercosis. A modern day plague. Pediatr. Clin. North Am. **32**:953–968.

Coker-Vann, M., P. Brown, and D. C. Gajdusek. 1984. Serodiagnosis of human cysticercosis using a chromatofocused antigenic preparation of *Taenia solium* cysticerci in an enzyme-linked immunosorbent assay (ELISA). Trans. Roy. Soc. Med. Hyg. **78**:492–496.

Dewhirst, L. W., J. D. Cramer, and J. J. Sheldon. 1967. An analysis of current inspection procedures for detecting bovine cysticercosis. J. Am. Vet. Med. Assoc. **150**:412–417.

Diwan, A. R., M. Coker-Vann, P. Brown, D. B. Subianto, R. Yolken, R. Desowitz, A. Escobar, C. J. Gibbs, Jr., and D. C. Gajdusek. 1982. Enzyme-linked immunosorbent assay (ELISA) for the detection of antibody to cystercerci of *Taenia solium*. Am. J. Trop. Med. Hyg. **31**:364–369.

Espinoza, B., A. Flisser, A. Plancarte, and C. Larralde. 1982. Immunodiagnosis of human cysticercosis: ELISA and immunoelectrophoresis, p. 163–170. *In* A. Flisser, K. Willms, J. P. Laclette, C. Larralde, C. Ridaura, and F. Beltran (ed.), Cysticercosis. Present state of knowledge and perspectives, Academic Press, Inc., New York.

Estrada, J. J., and R. E. Kuhn. 1985. Immunochemical detection of antigens of larval *Taenia solium* and anti-larval antibodies in the cerebrospinal fluid of patients with neurocysticercosis. J. Neurol. Sci. **71**:39–48.

Flisser, A., R. Perez-Montford, and C. Larralde. 1979. The immunology of human and animal cysticercosis: a review. Bull. W. H. O. **57**:839–856.

Flisser, A., E. Woodhouse, and C. Larralde. 1980. Human cysticercosis: antigens, antibodies and non-responders. Clin. Exp. Immunol. **39**:27–37.

Froyd, G. 1962. Longevity of *Taenia saginata* eggs. J. Parsitol. **48**:279.

FSIS. 1981. Statistical summary: federal meat and poultry inspection for fiscal year 1981. USDA, Washington, D. C.

Geerts, S., T. Vervoot, V. Kumar, and F. Ceulemans. 1981. Isolation of fraction no. 10 from *Taenia saginata* and evaluation of its specificity for the diagnosis of bovine cysticercosis. Zeitschrift fur Parasitenkunde **66**:201–206.

Gottstein, B., V. C. W. Tsang, P. M. Schantz. 1986. Demonstration of species-specific and cross-reactive components of *Taenia solium* metacestode antigens. Am. J. Trop. Med. Hyg. **35**:308–313.

Gottstein, B., D. Zini, and P. M. Schantz. 1987. Species-specific immunodiagnosis of *Taenia solium* cysticercosis by ELISA and immunoblotting. Am. J. Trop. Med. Hyg. (submitted)

Grisolia, J. S., and W. C. Wiederholt. 1982. CNS cysticercosis. Arch. Neurol. **39**:541–544.

Groll, E. 1980. Praziquantel for cestode infections in man. Acta Trop. **37**:293–296.

Groll, E. W. 1982. Chemotherapy of human cysticercosis with praziquantel, p. 207–218. *In* A. Flisser, K. Willms, J. P. Laclette, C. Larralde, C. Ridaura and F. Beltran (ed.), Cysticercosis. Present state of knowledge and perspectives, Academic Press, Inc., New York.

Harrison, L. J. S., and M. M. H. Sewell. 1980. The antigenic activity of chromatographic fractions of a saline extract of *Taenia saginata* (Goeze, 1782) proglottides. Vet. Immunol. Immunopathol. **1**:361–369.

Hird, D. W., and M. M. Pullen. 1979. Tapeworms, meat and man: a brief review and update of cysticercosis caused by *Taenia saginata* and *Taenia solium*. J. Food Prot. **42**:58–64.

Liu, X. G. 1981. Treatment of *Cysticercosis cellulosae* with praziquantel. Lianoning Anim. Husb. Vet. Med. **4**:32.

Murrell, K. D., R. Fayer, and J. P. Dubey. 1986. Parasitic organisms, p. 311–377. *In* A. M. Pearson and T. R. Dutson (ed.), Advances in meat research, vol. 2., AVI Publishing, Westport, CT.

Nascimento E., C. A. Tavares, and J. D. Lopes. 1987. Immunodiagnosis of human cysticercosis (*Taenia solium*) with antigens purified by monoclonal antibodies. J. Clin. Micro. **25**:1181–1185.

Nash, T. E., and F. A. Neva. 1984. Recent advances in the diagnosis and treatment of cerebral cysticercosis. N. Engl. J. Med. **311**:1492–1496.

Pawlowski, Z. S. 1970. Inorganic tin compounds as the alternative drug in human taeniarhynchosis. J. Parasitol. **56**:261.

Pawlowski, Z. S. 1983. Taeniasis and cysticercosis, p. 313–348. *In* J. H. Steele (ed.), Parasitic zoonosis, vol. 1, CRC handbook series in zoonosis. CRC Press, Boca Raton, FL.

Pawlowski, Z. S., and M. G. Schultz. 1972. Taeniasis and cysticercosis (*Taenia saginata*), p. 269–343. *In* B. Dawes (ed.), Advances in parasitology, vol. 10. Academic Press, Inc., New York.

Rhoads, M. L., K. D. Murrell, G. W. Dilling, M. M. Wong, and N. F. Baker. 1985. A potential diagnostic reagent for bovine cysticercosis. J. Parasit. **71**:779–787.

Richards, Jr., F., P. M. Schantz, E. Ruiz-Tiben, and F. Sorvillo. 1985. Cysticercosis in Los Angeles County. J. Am. Med. Assoc. **254**:3444–3448.

Rim, H. J., S. B. Park, J. S. Lee, and K. S. Joo. 1979. Therapeutic effects of praziquantel (Embay 8440) against *Taenia solium* infection. Korean J. Parasitol. **17**:67–72.

Rim, H. J., C. R. Wan, and J. W. Chu. 1980. Studies on the human cysticercosis and its therapeutic trial with praziquantel (Embay 8440). Korea U. Med. J. **17**:459–474.

Schultz, M. G. 1971. Parasitic diseases along the Mexican–USA frontier. Salud Publica Mex. **13**:377–380.

Silverman, P., and R. Griffiths. 1955. A review of methods of sewage disposal in Great Britain, with special reference to the epizootiology of *Cysticercus bovis*. Ann. Trop. Med. Parasitol. **49**:436–450.

Snyder, G. R., and K. D. Murrell. 1986. Bovine cysticercosis, p. 161–170. *In* G. Woods (ed.), Practices in veterinary public health and preventive medicine. Iowa State Press, Ames.

Verster, A., T. A. du Plessis, and L. W. van den Heever. 1977. The eradication of tapeworms in pork and beef carcasses by irradiation. Radiat. Phys. Chem. **9**:769–773.

Walther, M., and J. K. Koske. 1979. The efficacy of praziquantel against *Taenia saginata* cysticercosis in naturally infected calves. Z. Tropenmed. Parasitol. **30**:401–403.

Webbe, G. 1967. The hatching and activation of taeniid ova in relation to the development of cysticercosis in man. Z. Tropenmed. Parasitol. **18**:354–369.

CHAPTER 104

Toxocariasis

S. L. JOSEPHSON

Disease: Toxocariasis (visceral larva migrans, ocular larva migrans).
Etiologic Agents: *Toxocara canis,* rarely *Toxocara cati.*
Source: Soil, hands, or fomites contaminated with canine or cat feces containing embryonated eggs.
Clinical Manifestations: Often asymptomatic with eosinophilia. Systemic migration of larvae (visceral larva migrans) into somatic tissues resulting in eosinophilia, abdominal pain, respiratory symptoms, hepatomegaly, and fever. Ocular infections presenting as uveitis, strabismus, neuroretinitis, optic neuritis, and endophthalmitis.
Pathology: Eosinophilic abscesses and eosinophilic granulomas in affected tissues that can include lung, liver, kidney, myocardium, brain, and eye. Also intraocular disorganization and retinal detachment in the eye.
Laboratory Diagnosis: Identification of larvae in affected tissues. Enzyme-linked immunosorbent assay titers $\geq 1:32$ in visceral larva migrans and $\geq 1:8$ in ocular larva migrans are diagnostic for *T. canis*.
Epidemiology: Worldwide in temperate and tropical climates. Most infections in children with a history of geophagia and contact with puppies.
Treatment: Diethylcarbamazine, thiabendazole, mebendazole, and corticosteroids. Vitrectomy and retinal detachment surgery to prevent complications. Laser photocoagulation to kill larvae in the eye.
Prevention and Control: Discourage geophagia in children and worm puppies.

Description of Disease

Toxocariasis (visceral larva migrans, ocular larva migrans) in humans is a systemic parasitic infestation by migrating larvae of *Toxocara canis* and occasionally *Toxocara cati,* the common intestinal ascarids (nematode roundworms) of canids and felids. Infection can be asymptomatic, subacute, or acute with visceral and ocular manifestations. Although toxocariasis is infrequently diagnosed, serologic evidence suggests that a significant portion of the world's population living in temperate and tropical climates have been exposed to the parasite. Toxocariasis characteristically occurs in infants and children having a history of geophagia and exposure to puppies.

Communicability and Transmissibility

Toxocara is transmitted from the definitive host, a dog or cat, to humans by way of soil, hands, or fomites contaminated with eggs originally passed in the animal's feces. Human toxocariasis results from the ingestion of embryonated eggs containing second- or third-stage larvae. Because eggs passed in feces are unembryonated and become infective (embryonated) only after 2 to 5 weeks in soil, direct human contact with fresh dog or cat feces will not lead to infection. Human-to-human transmission does not occur owing to the inability of the parasite to complete its life cycle in humans. As a result, infected humans do not shed eggs.

Most of the reported cases of toxocariasis have been attributed to *T. canis* (Glickman and Schantz, 1981). Infected puppies are the primary source of infective eggs; this reflects a high rate of prenatal infection in dogs. Young children are most likely to become infected because of their lack of personal hygiene and propensity for playing in soil. Those children who regularly ingest soil are at a particularly high risk of becoming infected.

Clinical Manifestations

Human infections with Toxocara are predominantly asymptomatic or subclinical, usually going undetected. Toxocara, however, is capable of producing two clinical syndromes, visceral larva migrans and ocular larva migrans (ocular toxocariasis). Toxocariasis characteristically involves the dissemination of infective larvae from the small bowel to various somatic tissues via the circulation, followed by active migration of larvae through somatic tissues.

Asymptomatic Toxocariasis

Asymptomatic infections have been demonstrated experimentally in humans after ingestion of Toxocara eggs, with the only sign of infection being eosinophilia (Chaudhuri and Saha, 1959).

Visceral Larva Migrans

Visceral larva migrans is usually a disease of young children (average age, 2 years) with a history of geophagia (Brown, 1970). Somatic migration of larvae may involve the lungs, heart, kidney, muscles, and central nervous system. Signs and symptoms resulting from the larval migration and the immunologic response to the invading parasite include fever, cough, wheezing, rash, hepatomegaly, anorexia, pallor, petit mal seizures, and abdominal discomfort (Glickman and Schantz, 1981; Shields, 1984). Additional signs include hypergammaglobulinemia, leukocytosis with marked and persistent eosinophilia, liver enzyme changes, pulmonary infiltrates, and isoagglutinin titers to A and B blood group antigens (Glickman and Schantz, 1981). The latter finding is due to the presence of A and B blood group-like substances on the surface of Toxocara larvae (Smith et al., 1983).

Central nervous system infections are rare. Both severe toxocaral meningoencephalitis (Shochet, 1967) and benign, self-limited meningitis (Gould et al., 1985) have been described. Toxocara infections also have been associated with behavioral changes (Dolinsky et al., 1985) and epilepsy (Woodruff et al., 1966). Although Toxocara has not been proven to be a cause of epilepsy in children, epileptics serologically do demonstrate a higher than normal rate of infection with the parasite (Glickman et al., 1979; Woodruff et al., 1966).

Toxocara can also cause poliomyelitis and myocarditis in humans (Friedman and Hervada, 1960; Woodruff et al., 1966). With the exception of myocardial and central nervous system involvement, the visceral form of toxocariasis is not life threatening.

Ocular Larva Migrans (Ocular Toxocariasis)

Toxocara larvae occasionally migrate to the eye where they produce a variety of intraocular manifestations broadly classified as ocular larva migrans or ocular toxocariasis. These manifestations include uveitis, poor vision or strabismus, diffuse lateral subacute neuroretinitis, optic neuritis, granulomatous retinal lesions, and endophthalmitis, which is often misidentified as retinoblastoma (Brown, 1970; Molk, 1982; Oppenheim et al., 1985). Ocular infection is typically unilateral. Unlike visceral larva migrans, ocular toxocariasis occurs more often in older children and adults (average age, 7.5 years) who are not geophagic (Brown, 1970). Furthermore, eosinophilia and hepatosplenomegaly usually are absent or mild in ocular infections (Brown, 1970).

Ocular toxocariasis occurs concurrently with or subsequent to visceral larva migrans in approximately 2% of cases (Brown, 1970). The reason for this and other differences between ocular and systemic disease has been attributed to inoculum size (Glickman and Schantz, 1981). The relatively low dose of infective eggs ingested by older children or adults appears to be insufficient to elicit a significant eosinophilic and antibody response. Therefore, a small number of larvae are able to migrate unhindered. In the absence of a significant immune response, manifestations of visceral larva migrans do not occur. On those rare occasions when a larva penetrates a blood vessel wall in the eye and migrates into the subretinal space or vitreous cavity, ocular toxocariasis occurs (Kielar, 1983).

Pathology

Both the migration of larvae and the ensuing immune inflammatory response to infection produce characteristic pathologic manifestations. Circulating larvae bore through blood vessel walls and migrate through the surrounding tissue, leaving tracks of hemorrhage and necrosis (Glickman and Schantz, 1981). Eosinophilic abscesses and allergic-type eosinophilic granulomas appear in affected tissues (Glickman and Schantz, 1981). Larvae lodged in blood vessel walls can result in focal granulomatous vasculitis (Neafie and Connor, 1976). Many histologic sections are usually needed to detect larvae in affected tissues.

Although the pathology of visceral larva migrans is similar in most respects to that of ocular larva migrans, some differences do exist, mainly related to the unique structure of the eye (Shields, 1984). For example, patients with endophthalmitis will demonstrate intraocular disorganization and retinal detachment with focal thickening in the periphery. Cyclitic membrane may be present. Other ocular manifestations will result in retinal granuloma without pronounced ophthalmitis.

Diagnosis

Diagnosis of toxocariasis is based on patient history, the clinical presentation, immunodiagnostic results, and, occasionally, histologic examination of affected tissues. Signs and symptoms associated with toxocariasis are not pathognomonic. Consequently, additional steps are needed to confirm a clinical diagnosis of toxocariasis. Histologic identification of larvae in tissues is the only certain way to confirm a diagnosis. However, this is usually impractical in view of the large number of sections required to detect any larvae and the uncertainty of finding intact, identifiable larvae (Shields, 1984). Immunodiagnosis has evolved into a reliable alternative with the development of an enzyme-linked immunosorbent assay (ELISA) for detecting antibody to *T. canis*. ELISA has a sensitivity and specificity of 78 and 92%, respectively, at a titer of 1 : 32, and 90 and 91%, respectively, at a titer of 1 : 8 (Glickman et al., 1978, 1985).

Visceral larva migrans should be suspected in any child with a history of geophagia and exposure to puppies and with fever, eosinophilia, and hepatomegaly. Percutaneous liver biopsy specimens can be examined to detect larvae but often reveal only granulomatous type reactions (Beaver et al., 1952). An antibody titer of 1 : 32 or greater in the ELISA test, along with appropriate signs and symptoms of visceral larva migrans, is considered to be indicative of toxocariasis. Furthermore, the ELISA can be used to detect antibody in cerebrospinal fluid.

Diagnosis of ocular toxocariasis is based on patient history, physical examination, slit-lamp examination, indirect ophthalmoscopy, and ELISA testing. Ultrasonography, computed tomography, and cytology also can be helpful in developing an accurate diagnosis (Shields, 1984; Edwards and Pordell, 1985). Signs and symptoms of visceral larva migrans, a history of geophagia, and contact with puppies are all significant but are often absent in cases of ocular toxocariasis. Slit-lamp findings consistent with ocular toxocariasis include keratic precipitates, cyclitic membrane, aqueous flare and cells, and secondary cataract (Shields, 1984). Ophthalmoscopy can be used to detect posterior pole and peripheral fundus granulomas, early retinal detachment, and larvae (Oppenheim et al., 1985). A serum antibody titer of 1 : 8 or higher in patients with appropriate ocular signs and symptoms is consistent with a diagnosis of ocular toxocariasis. ELISA also has been used to detect antibody in intraocular fluid (Felberg et al., 1981). Finally, the presence of eosinophils in vitreous fluid obtained by needle aspiration also supports a diagnosis of ocular toxocariasis (Shields, 1984).

Prognosis

Most cases of visceral larva migrans are mild with symptoms resolving within several weeks or months. However, infestation of the myocardium or central nervous system or an exaggerated immunologic response to infection can be fatal.

Ocular toxocariasis, although not life-threatening, often results in blindness in the affected eye.

Epidemiologic Aspects

Source of Infection

T. canis is found in tropical and temperate regions all over the world. The overall prevalence of *T. canis* has been estimated to be 15% (range 0 to 93%) in dogs and approaches 100% in puppies (Glickman and Schantz, 1981; Ugochukwu and Ejimadu, 1985). *T. cati* infections in cats also occur worldwide with a prevalence rate of 24 to 67% in North America (Glickman and Schantz, 1981).

The high prevalence of Toxocara in dogs and cats, together with an expanding pet population, has resulted in a high degree of environmental contamination. In North America and Europe Toxocara eggs have been recovered from up to 24.4% of soil samples taken from public and private lands (Borg and Woodruff, 1973; Childs, 1985).

Human Infection

Seroepidemiologic studies have demonstrated that human infection with Toxocara is quite common despite only sporadic reports of human toxocariasis. Studies in Britain (De Savigny et al., 1979), Japan (Matsumura and Endo, 1983), Chile (Herskovic and Astorga, 1985), and North America (Hermann et al., 1985) show that 2.6 to 8.8% of the population have been exposed to the parasite at some time in their lives. Moreover, seropositivity among certain defined groups is even higher. For instance, prevalence of infection was 23.1% in kindergarten children from

North Carolina (Worley et al., 1984) and 15.7% in dog breeders in Britain (Woodruff et al., 1978). Factors such as climatic conditions, lack of hygiene, age, contact with dogs, and geophagia determine the risk of infection.

Etiologic Agents

T. canis is the primary agent of toxocariasis. Although human infection with *T. cati* can occur, its overall significance has yet to be determined.

Characterization

Infective larvae of *T. canis* and *T. cati* average 400 μm in length and 15 to 21 μm in diameter and have single, small, lateral alae (Neafie and Connor, 1976). Adult female *T. canis* are 6 to 18 cm in length, whereas *T. cati* females measure 4 to 12 cm. Adult male *T. canis* are 4 to 10 cm long compared with a length of 3 to 6 cm for adult male *T. cati* (Glickman and Schantz, 1981).

Life Cycle in the Definitive Host

Dogs are infected with *T. canis* transplacentally or by ingesting eggs containing infective larvae or by ingesting larvae in the milk of lactating bitches. Larvae penetrate the intestinal mucosa of the dog and travel in the circulation to the liver, lungs, and heart. Larvae are then carried throughout the body, migrate into somatic tissues, and become encysted. These larvae often survive for years.

Alternatively, larvae can complete their life cycle by following a tracheal route from the lungs. This occurs frequently in young puppies. The larvae migrate up the trachea to the pharynx and are swallowed. Once in the intestine, the larvae develop into adults. Adult female Toxocara then produce eggs, which are passed in the feces.

The large number of prenatal infections appear to be due to hormonal changes in the pregnant bitch that permit encysted larvae to migrate again. Larvae can then cross the placenta and infect the unborn pup.

The life cycle of *T. cati* is similar to that of *T. canis* with the exception of transplacental transmission, which does not occur in cats.

Life Cycle in Humans

Toxocara eggs become embryonated in warm, moist soil within 2 to 5 weeks. Humans are infected by ingesting embryonated eggs, which hatch in the small intestine. The hatched larvae then penetrate the mucosa, enter the portal circulation, and travel to the liver and lungs, where they enter the systemic circulation. Larvae can then spread to various organs including liver, lung, eyes, and the central nervous system. After penetrating blood vessels, the larvae migrate extensively in somatic tissues and are ultimately encapsulated. The tracheal migration that can occur in the definitive host does not occur in humans. Therefore, larvae do not develop into adults, and eggs are not produced.

Identification

Toxocara larvae in tissues and eggs in soil or animal feces can be identified on the basis of morphology. Toxocara in human infections can be identified serologically by ELISA with *T. canis* embryonated egg antigen or excretory–secretory antigen (exoantigen) to detect toxocaral antibody (Glickman et al., 1985; Speiser and Gottstein, 1984). The excretory–secretory antigen appears to be the best antigen for discriminating between ocular toxocariasis and retinoblastoma (Glickman et al., 1985).

Treatment, Prevention, and Control

Treatment

Diethylcarbamazine and thiabendazole have been used to treat visceral larva migrans with some success (Elliot et al., 1985). These drugs have produced clinical improvement, although eosinophilia has persisted. Somewhat better results have been obtained with mebendazole, with clinical improvement and disappearance of eosinophilia (Bekhti, 1984). Corticosteroids can be used to decrease inflammation, especially in ocular toxocariasis (Molk, 1982). Furthermore, vitrectomy and retinal detachment surgery, in some circumstances, can prevent ocular complications (Shields, 1984). Finally, laser photocoagulation can be used to kill larvae that have been detected ophthalmoscopically (Fitzgerald and Rubin, 1974).

Prevention and Control

Several steps can be taken to prevent human infection with Toxocara. Worming of puppies will help reduce the level of environmental contamination. Piperazine salts are effective in eliminating adult toxocara in dogs. However, new drugs are needed to kill encysted larvae as well as the adult parasite. Cleaning up cat and dog feces also can be helpful in preventing infection. Finally, an effort should be made to discourage geophagia.

Literature Cited

Beaver, P. C., C. H. Snyder, G. M. Carrera, J. H. Dent, and J. W. Lafferty. 1952. Chronic eosinophilia due to visceral larva migrans. Pediatrics **9:**7–19.

Bekhti, A. 1984. Mebendazole in toxocariasis. Ann. Intern. Med. **100:**463.

Borg, O. A., and A. W. Woodruff. 1973. Prevalence of infective ova of Toxocara species in public places. Br. Med. J. **4:**470–472.

Brown, D. H. 1970. Ocular toxocara canis: a clinical review. J. Pediatr. Ophthalmol. **7:**182–191.

Chaudhuri, R. N., and T. K. Saha. 1959. Tropical eosinophilia: experiments with *Toxocara canis*. Lancet **2:**493–494.

Childs, J. E. 1985. The prevalence of Toxocara species in backyards and gardens of Baltimore, Maryland. Am. J. Pub. Health **75:**1092–1093.

De Savigny, D. H., A. Voller, and A. W. Woodruff. 1979. Toxocariasis: serological diagnosis by enzyme immunoassay. J. Clin. Pathol. **32:**284–288.

Dolinsky, Z. S., C. A. Hardy, R. G. Burright, and P. J. Donovick. 1985. The progression of behavioral and pathological effects of the parasite *Toxocara canis* in the mouse. Physiol. Behav. **35:**33–42.

Edwards, M. G., and G. R. Pordell. 1985. Ocular toxocariasis studied by CT Scanning. Radiology **157:**685–686.

Elliot, D. L., S. W. Tolle, L. Goldberg, and J. B. Miller. 1985. Pet associated illness. N. Engl. J. Med. **313:**985–995.

Felberg, N., J. A. Shields, and J. L. Federman. 1981. Antibody to *Toxocara canis* in the aqueous humor. Arch. Ophthalmol., **99:**1563–1564.

Fitzgerald, C. R., and M. L. Rubin. 1974. Intraocular parasite destroyed by photocoagulation. Arch. Ophthalmol. **91:**162–164.

Friedman, S., and A. R. Hervada. 1960. Severe myocarditis with recovery in a child with visceral larva migrans. J. Pediatr. **56:**91–96.

Glickman, L. T., R. H. Cypress, P. K. Crumrine, and D. A. Gitlin. 1979. Toxocara infection and epilepsy in children. J. Pediatr. **94:**75–78.

Glickman, L. T., R. B. Grieve, S. S. Lauria, and D. L. Jones. 1985. Serodiagnosis of ocular toxocariasis: a comparison of two antigens. J. Clin. Pathol. **38:**103–107.

Glickman, L. T., and P. M. Schantz. 1981. Epidemiology and pathogenesis of zoonotic toxocariasis. Epidemiol. Rev. **3:**230–250.

Glickman, L., P. Schantz, R. Dombrowske, and R. Cypess. 1978. Evaluation of serodiagnostic tests for visceral larva migrans. Am. J. Trop. Med. Hyg. **27:**492–498.

Gould, I. M., S. Newell, S. H. Green, and R. H. George. 1985. Toxocariasis and eosinophilic meningitis. Br. Med. J. **291:**1239–1240.

Herrmann, N., L. T. Glickman, P. M. Shantz, M. G. Weston, and L. M. Domanski. 1985. Seroprevalence of zoonotic toxocariasis in the United States: 1971–1973. Am. J. Epidemiol. **122:**890–896.

Herskovic, P., and B. Astorga. 1985. Human toxocara infection in Chile. Rev. Med. Chile. **113:**18–21.

Kielar, R. A. 1983. *Toxocara canis* endophthalmitis with low ELISA titer. Ann. Ophthalmol. **15:**447–449.

Matsumura, K., and R. Endo. 1983. Seroepidemiological study on toxocaral infection in man by enzyme-linked immunosorbent assay. J. Hyg. Camb. **90:**61–65.

Molk, R. 1982. Treatment of toxocaral optic neuritis. Clin. Neuroophthalmol. **2:**109–112.

Neafie, R. C., and D. H. Connor. 1976. Visceral larva migrans, p. 433–436. *In* C. H. Binford and D. H. Connor (ed.), Pathology of tropical and extraordinary diseases, vol. 2. Armed Forces Institute of Pathology, Washington, D. C.

Oppenheim, S., G. Rogell, and R. Peyser. 1985. Diffuse unilateral subacute neuroretinitis. Ann. Ophthalmol. **17:**336–338.

Shields, J. A. 1984. Ocular toxocariasis. A review. 1984. Surv. Ophthalmol. **28:**361–381.

Shochet, S. S. 1967. Human *Toxocara canis* encepalopathy in a case of visceral larva migrans. Neurology **17:**227–229.

Smith, H. V., J. R. Kusel, and R. W. A. Girdwood. 1983. The production of human A and B blood group like substances by in vitro maintained second stage *Toxocara canis* larvae: their presence on the outer larval surfaces and in their excretions/secretions. Clin. Exp. Immunol. **54:**625–633.

Speiser, F., and B. Gottstein. 1984. A collaborative study on larval excretory/secretory antigens of *Toxocara canis* for the immunodiagnosis of human toxocarasis with ELISA. Acta Tropica **41:**361–372.

Ugochukwu, E. I., and K. N. Ejimadu. 1985. Studies of the prevalence of gastrointestinal helminths of dogs in Calabar, Nigeria. Int. J. Zoon. **12:**214–218.

Woodruff, A. W., B. Bisseru, and J. C. Bowe. 1966. Infection with animal helminths as a factor in causing poliomyelitis and epilepsy. Br. Med. J. **1:**1576–1579.

Woodruff, A. W., A. De Savigny, and D. E. Jacobs. 1978. Study of toxocaral infection in dog breeders. Br. Med. J. **2:**1747–1748.

Worley, G., J. A. Green, T. E. Frothingham, R. A. Sturner, K. W. Walls, V. A. Pakalnis, and G. S. Ellis. 1984. *Toxocara canis* infection: clinical and epidemiological associations with seropositivity in kindergarten children. J. Infect. Dis. **149:**591–597.

CHAPTER 105

Toxoplasmosis

KENNETH W. WALLS

Disease: Toxoplasmosis.

Etiologic Agent: *Toxoplasma gondii.*

Source: Ingestion of raw or undercooked meat, congenital transfer, blood transfusion or tissue transplant, ingestion of oocysts from the environment contaminated with feces from infected cats.

Clinical Manifestations: Asymptomatic or subclinical in most cases. Adults usually complain of an "infectious mononucleosis-like" syndrome. When acquired during pregnancy may cause congenital malformations and retardation or fetal death. One of the most common causes of posterior chorioretinitis. In immunoincompetent or immunosuppressed patients, it frequently results in a fatal disseminated disease with heavy cerebral involvement.

Pathology: Acute disease is usually accompanied by lymphadenopathy with reticular cell hyperplasia. Pulmonary necrosis, myocarditis, and hepatitis caused by tissue necrosis are common. In congenital disease, hepatitis and pneumonia are followed by central nervous system involvement resulting in hydrocephalus, retinochoroiditis, and cerebral calcifications. Retinitis is associated with intense inflammation and necrosis due to the proliferation of the parasite within the tissue cells. In addition to all of these characteristics, immunocompromised patients demonstrate a unique encephalitis with necrotizing lesions accompanied by inflammatory infiltrates.

Laboratory Diagnosis: Serologic tests are the principal method of diagnosis. Indirect immunofluorescence and enzyme immunoassays for both IgM and IgG antibodies are used. Isolation and histologic demonstration of the parasite are difficult and the results are obscured by the chronic nature of the infection, resulting in positive demonstration of the parasite in asymptomatic individuals.

Epidemiology: Ubiquitous, toxoplasmosis is found worldwide in all warm-blooded animals. There is higher prevalence in warm humid climates and at lower altitudes. Although transmitted by cats, cat ownership is not significant, probably because transmission results primarily from the feral cat. Although the modes of transmission are known, no definite epidemiologic patterns have been determined.

Treatment: A combination of pyrimethamine (Daraprim), tri-sulfa, and leucovorin (folinic acid) is the only successful therapy. Side effects occur and treatment must be monitored closely.

Prevention and Control: Education to minimize exposure. Avoid eating raw meat, wash hands after working in the soil, keep the cat litterbox clean daily, and keep personal cat indoors if possible and feed it commercial food.

Description of Disease

Toxoplasmosis is perhaps the most underrated parasitic infection. Affecting populations throughout the world with prevalence rates from less than 1% to greater than 90%, it is insidious in nature, usually oligosymptomatic, and frequently undiagnosed; in certain patient types, however, the disease is highly debilitating and occasionally fatal. It is a major cause of congenital and perinatal disease, one of the leading causes of posterior chorioretinitis, and has recently become one of the leading causes of death among patients with acquired immunodeficiency syndrome (AIDS).

Although the parasite *Toxoplasma gondii* has been recognized for most of the 20th century, progress in its diagnosis and treatment has occurred mostly in the last decade. During this time, better knowledge has been gained concerning transmission, prevention, detection, and serologic diagnosis. Despite these advances, there remain tremendous needs to understand the pathogenesis, epidemiology, and immunology of the disease. In addition to the many basic questions that must be answered, there are four major areas of obvious need: 1) Is a vaccine feasible for humans? 2) How should pregnant patients be managed? 3) Should asymptomatic congenital cases be treated? and 4) What is the optimum diagnostic approach and treatment of cerebral toxoplasmosis in patients with AIDS? It is hoped that the next decade will see the resolution of some of these important problems.

Communicability and Transmissibility

Although transmission in the laboratory has been shown to occur readily through the ingestion of either oocysts in cat feces or tissue cysts in meat from chronically infected animals and by transplacental transfer when toxoplasmosis is acquired during pregnancy, in nature these modes of transmission have been established primarily by epidemiologic inference. Except congenitally, transmission from person to person has not been documented, and no animals other than felines have been incriminated in fecal transmission. Cases have been documented in which transmission occurred through blood transfusion or tissue transplantation (Remington and Cavanaugh, 1965; Reynolds et al., 1966; Seigel et al., 1971). Natural outbreaks have indicated that under the proper circumstances ingestion is a highly efficient method of transmission.

Clinical Manifestations of Disease

It is important to recognize that toxoplasmosis is characterized by both disease and infection. Fortunately, most cases fall into the latter category, and acquisition may go unnoticed because of the mild symptomatology. The infection subsequently becomes "chronic," with persistent tissue cysts and no symptoms. In the usual situation there are no further complications, and the patient becomes latently infected for life. A small percentage of cases will develop into acute disease, with any or all of the symptoms of lymphadenopathy, fever, rash, malaise, and dyspnea. Further progression can lead to central nervous system (CNS) involvement and occasionally death.

Four basic clinical manifestations can be described for the disease: acute acquired, congenitally acquired, ocular, and immunosuppressed reactivation.

Acute Acquired Disease

Most cases of acute acquired disease go undiagnosed or are misdiagnosed as "fever of unknown origin" or "infectious mononucleosis" because of the nondescript and benign characteristics of the signs and symptoms. Accidental laboratory infections and natural outbreaks in which the epidemiologic data are sufficient indicate that symptoms occur 10 to 14 days after exposure and persist from 3 weeks to several months. Characteristically the patient exhibits fever, malaise, fatigue, and dyspnea. Though specific therapy is available, most adult patients are treated symptomatically and spontaneously recover in a relatively short period. If symptoms persist or become more severe, therapy with pyrimethamine and sulfadiazine is presently recommended (see Treatment, Prevention, and Control). Infants and young children, because of a less active immune defense, should be given specific therapy at the earliest recognition of the disease.

Congenitally Acquired Disease

Congenital transmission occurs only when the mother acquires her *primary* infection during pregnancy. Although exceptions have been reported, data are clear that previous infections elicit antibody production, which provides protection to the fetus. The manifestations of an initial infection during pregnancy range in severity from none, due to failure to transmit the infection to the fetus, to asymptomatic infection, to severe congenital malformations, or to stillbirth. It has been shown that fetal acquisition and fetal outcome depend on the trimester in which the infection was acquired (Desmonts and Couvreur, 1974). Maternal infections occurring during the first trimester result in relatively few fetal infections, but most are fatal, whereas those acquired during the third trimester result in a high percentage of infected infants, but most survive with few or no symptoms. The "classic triad" to describe the congenitally infected newborn includes cerebral calcifications, chorioretinitis, and hydrocephalus; it is apparent from the preceding discussion, however, that there is, in fact, no pathognomonic symptomatology to describe the disease. Congenital toxoplasmosis is frequently confused with other congenital infections, such as rubella, cytomegalovirus infection, herpes infection, and syphilis. In the absence of the triad of symptoms, chorioretinitis and mental or physical retardation are the most commonly recognized symptoms. Indications and therapeutic regimes for treatment during pregnancy to prevent transmission will be discussed later. Treatment of the acutely affected infant should begin as soon after delivery as practical. As in acute acquired disease, pyrimethamine and sulfonamides are presently recommended. More perplexing, however, is the disposition of asymptomatically infected infants. Studies have shown that many, if not all, of these children develop complications, such as chorioretinitis or mental or physical retardation, which do not become apparent for several years (Remington and Desmonts, 1976). Longitudinal studies are necessary to evaluate the effect of therapy on these infants, but it would appear that treatment should be instituted at the earliest possible moment to attempt to prevent subsequent damage. As will be discussed in a subsequent section, diagnosis of these individuals depends almost exclusively on serologic data from the mother during pregnancy and from the infant and mother at delivery.

Ocular Disease

Toxoplasmosis is one of the most common causes of posterior chorioretinitis. Appearing as either a bilateral or a unilateral infection, toxoplasmic retinitis is one of the late-appearing manifestations of the disease. The frequent occurrence of active necrotic lesions accompanied by scarred healed lesions suggests the possibility that most of these cases are, in fact, the result of congenital disease, with a long delay of sufficiently damaging lesions to result in visual impairment and clinical recognition.

Diagnosis of these cases is made primarily by the recognition of typical lesions after other possible causes are ruled out. In the absence of involvement of other tissues in the patient, serologic titers are usually very weak and are of questionable assistance in diagnosis. As will be discussed under Clinical Diagnosis, this failure to detect a strong response may indicate either that the diagnosis is commonly made in the late stages of the disease or that the ocular lesion is immunologically protected with poor antibody synthesis. Untreated, these infections frequently lead to blindness or severe visual impairment in one or both eyes. Specific therapy is effective in alleviating symptoms.

Steroid therapy should be avoided in ocular toxoplasmosis whenever possible. The immediate response due to the reduction of inflammation suggests positive results, but in fact reflects symptomatic relief only. Relapse occurs readily on cessation of steroid administration and leads to rapid expansion of the lesion. When steroid therapy is necessitated by the clinical situation, specific therapy should be instituted simultaneously.

Immunosuppressed Reactivation

Two types of patients are involved in the immunosuppressed reactivated form of disease: therapeutically immunosuppressed patients in whom the immunosuppression is reversible; and patients with AIDS, in whom it appears the immunosuppression is permanent. In the case of patients with AIDS, antibody studies frequently indicate the presence of preexisting infection that has remained inactive until the loss of immunity, at which time the parasite is again capable of disseminating and causing fulminating, frequently fatal systemic disease. In fact, toxoplasmosis is presently one of the leading causes of death among patients with AIDS. In addition, chemotherapeutically suppressed patients who are serologically negative are at risk of acquiring the infection from a positive tissue or blood donor (Seigel et al., 1971). A unique feature of the immunosuppressed cases is the development of encephalitis and cerebral involvement in *adults*. Although this pathology is a frequent symptom in newborn congenital cases, CNS involvement in adults is unusual. Accompanied by the pathology of acute acquired disease, the CNS involvement is the primary concern in patient management. Specific therapy in therapeutically suppressed individuals is usually sufficient to inhibit the parasite until the patient's own immune system is restored and the disease is controlled by natural mechanisms.

Pathology

Upon entry into the host, *Toxoplasma* is first concentrated in the regional lymph nodes, from which it is disseminated by an active parasitemia. Biopsy and necropsy data have demonstrated *Toxoplasma* in virtually every tissue. The extent of damage in each is dependent on the intensity of infection in that area and the length of time necessary for immunity to develop. Although lymph node, ocular, and CNS involvement are clinically most obvious, there does not appear to be any real predilection to these tissues. Each form of the disease has its associated pathology, which may or may not be shared with the other types.

Acute Acquired Disease

In acute acquired disease, which is characterized by a febrile lymphadenopathic illness, dissemination can be extensive. Lymph node involvement is usually local, with generalized lymphadenopathy being rare. Affected nodes show reticular cell hyperplasia with an absence of necrosis or fibrosis. The nodes are enlarged and may or may not be tender. This enlargement may persist for extended periods.

A cutaneous rash is commonly described in acute cases. In accidental laboratory infections this rash has been limited in some cases to the region affected, whereas in naturally acquired infections the rash may develop over the entire body. It is usually of relatively short duration.

Myocarditis has been reported and may be more common than is suggested by the data (Hooper, 1957; Kass et al., 1952; Sexton et al., 1953). Pathology results from the development of intracellular cysts and their subsequent rupture in the musculature of the heart wall. The extent of pathology appears to depend on the number of parasites involved and the degree of immunity maintained by the patient. Chronic myocarditis has been described, but does not seem to be a common feature of the disease.

Although clinically less frequently recognized, involvement of both the liver and lungs is common. Only a mild transient hepatitis has been reported (Vischer et al., 1967), but cysts are frequently found on biopsy or autopsy of liver tissue from suspected cases; however, little or no tissue damage is seen. Similarly, clinically significant pulmonary involvement is unusual, but the common finding of cysts and, in the case of acute disease, tachyzoites in biopsy or autopsy material indicates the lungs are probably involved in nearly all cases of disseminated disease. Only minor pathology results even when large numbers of organisms are present.

Congenitally Acquired Disease

Although most infected neonates are born oligo- or asymptomatic, a great variety of complications occur in those with generalized disease. The severity of complications depends on when during gestation the fetus became infected; the earlier in the pregnancy that the mother acquires the infection, the more likely are severe consequences. When apparent, signs and symptoms include a wide range of generalized signs such as fever, jaundice, rash, lymphadenopathy, pneumonitis, myocarditis, and hepatosplenomegaly; ocular involvement including microophthalmia, chorioretinitis, and strabismus; central nervous system involvement including microcephalus, hydrocephalus, and cerebral calcifications; and a variety of hematologic abnormalities. The lack of immunity allows extensive proliferation of the parasite and its dissemination to the CNS with subsequent involvement of the brain. Internal hydrocephalus results from the aqueductal blockage occurring when lesions form in the aqueduct wall. The pathogenesis of this process is discussed thoroughly by Frenkel (1972), who describes the accompanying periaqueductal and periventricular vasculitis and necrosis as unique to toxoplasmosis.

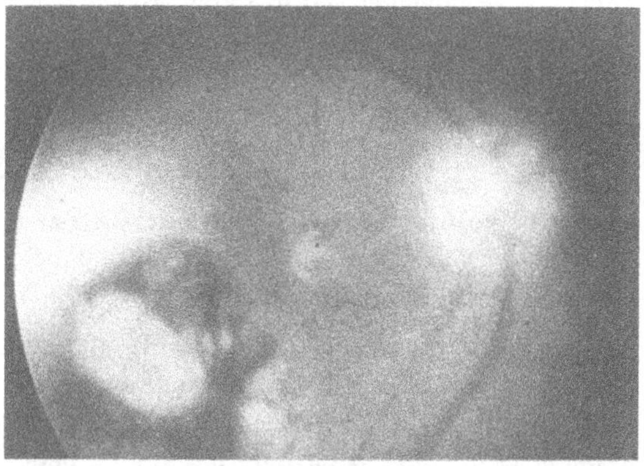

FIG. 1. Toxoplasmic chorioretinitis. Note the dark healed lesion along with the highly inflamed active lesion.

As with the CNS involvement, the lack of immunity in newborn infants permits extensive dissemination of the parasite into the eye where, even with a fully competent immune system, antibody production is minimal. Probably as a result of this poor immune response, the lesion continues to progress slowly and may not be recognized until much later in life. Although the disease process in these cases is not well understood, it is apparent there are two types of lesions (Fig. 1): a localized self-limiting lesion, which usually appears as a scarred area; and a fulminating necrotic lesion, which spreads rapidly to extensively destroy the retinal tissue. Still not understood is the reason for the frequency with which only one eye is affected; if the first eye is left untreated, however, the second eye will subsequently become involved.

Autopsy of stillborn or aborted fetuses has demonstrated the organism in virtually every tissue without the appearance of a predilection for any one area. It is of interest that other than those problems associated with brain and eye damage, toxoplasmosis does not result in fetal malformations.

Ocular Disease

Particularly in young children, ocular disease may go unnoticed for many years as it remains relatively inactive (Remington and Desmonts, 1976). When it becomes sufficiently active to cause visual damage, it is characterized by large yellowish-white lesions, usually accompanied by one or more healed scarred lesions. Although not pathognomonic, these lesions are sufficiently characteristic to alert the ophthalmologist to the possible diagnosis. There is diffuse retinal involvement with a granulomatous inflammatory reaction and the production of a vitreous exudate.

Relapses are common and in some cases appear to occur even after specific therapy (Crawford, 1966; Vietzke et al., 1968). Because of the frequent occurrence of chorioretinitis in congenital toxoplasmosis and the presence of healed lesions in most active cases, it is assumed by many that this form of the disease is most frequently of congenital origin.

Immunosuppressed Reactivation

As many as 70 to 80% of patients with AIDS have been diagnosed as having toxoplasmic encephalitis (Luft et al., 1984; Pitchenik et al., 1983). Similarly, frequent cases have been reported among patients suffering from various malignant diseases who acquire cerebral toxoplasmosis after therapeutic immunosuppression, blood transfusion, tissue transplant, or lymphocyte replacement (Ruskin and Remington, 1976). Although all of the clinical conditions described for the acute acquired form of toxoplasmosis have been described in immunosuppressed patients, central nervous system involvement is by far the most striking and may ultimately result in a fatal outcome. The atypical serologic results seen in immunocompromised patients with toxoplasmic encephalitis, along with the common finding of cysts in brain tissue of chronic naturally acquired human infections and in experimental animal infections, suggest that most of these immunocompromised cases are caused by reactivation of an existing infection. Probably as a result of the proliferation of tachyzoites, a diffuse encephalitis is seen with one or more mass lesions distinguished by computer-assisted tomographic scanning (Wong et al., 1984). These may be multiple or single lesions, which may be homogeneous or show ring enhancement on contrast. Mononuclear pleocytosis, with a mildly increased protein level, is apparent in the cerebrospinal fluid. Histologic examination of the lesions produced in the brain revealed massive necrotizing but limited lesions with some inflammatory infiltrate. Relapse may occur in the event inadequate or inappropriate medication has been administered (Emerson et al., 1981). In such cases some regression of symptoms may occur initially, but relapse with an increase in symptoms and lesions identifiable by computer-assisted tomographic scanning may occur. Resolution of lesions and a marked improvement in symptoms have been reported in patients in whom diagnosis was made early and specific therapy instituted promptly (Townsend et al., 1975).

Diagnostic Features

Although one might describe a "typical" case of toxoplasmosis, there is, in fact, no clinical presentation to alert the clinician specifically to a diagnosis of toxoplasmosis. Particularly in adults, the disease mimics influenza or infectious mononucleosis, with spontaneous cure normally occurring in 3 to 6 weeks (although much longer periods have been reported). Particularly women who, during pregnancy, show unusual fever, lymphadenopathy, and malaise, should be suspect and appropriate laboratory methods used for diagnosis. Because specific therapy can relieve symptoms, all adults in whom these symptoms persist or in whom tests for mononucleosis are negative should receive further diagnostic follow-up.

In infants, toxoplasmosis can easily be confused with many of the congenital diseases. It is frequently confused with, but not limited to, the TORCH complex (*TO*xoplasma, *R*ubella, *C*ytomegalovirus, and *H*erpesvirus). Determining when and if the mother acquired active toxoplasmosis is extremely difficult, and even if she has acquired the infection while pregnant, transplacental transfer may not have occurred (Desmonts and Couvreur, 1974); therefore, all infants with compatible symptoms must be assumed infected and a thorough diagnostic workup initiated immediately. A combination of chorioretinitis, hydrocephaly, and cerebral calcifications is virtually diagnostic, but only occurs in a very small percentage of infected neonates. Diagnosis in the infant frequently depends on clinical evaluation of both the mother and the child.

Toxoplasmosis should be immediately suspected in any immunocompromised patient, whether therapeutically immunosuppressed or naturally deficient, in whom there are signs or symptoms of brain and CNS involvement. Early diagnosis is vital to the survival of these patients.

Prognosis

Morbidity and mortality in toxoplasmosis appear to be very low. Serologic data indicate infection rates vary from as low as 1% in some populations to as high as 85% in others (Feldman and Miller, 1956; Gibson and Coleman, 1958), yet clinical toxoplasmosis occurs in less than 1% of infections. Fatalities occur almost exclusively in untreated immunodeficient individuals. Abortion, stillbirth, and neonatal death occur in congenitally acquired infections, but can be minimized by proper therapeutic management. Therapy does not appear to cure the disease, but prevents multiplication of the parasite until the natural immune mechanisms are capable of suppressing the disease. The parasite probably remains encysted in the host tissue indefinitely, but causes no further complications unless immunosuppression occurs. Infected neonates, even if asymptomatic at birth, appear to be at risk of subsequently developing mental or physical retardation (Wilson et al., 1980). The long-term effect of therapy on this group has not been determined, but limited data indicate a positive response.

Epidemiologic Aspects

Occurrence and Transmission

Toxoplasmosis is ubiquitous and is found in all warm-blooded animals throughout the world. Most cases are undiagnosed or misdiagnosed due to the nondescript symptomatology, and as a consequence the actual incidence of the disease can only be estimated from serologic prevalence data. In the United States, serology indicates rates of 20 to 30% along the sea coasts, with rates of less than 1% in the Rocky Mountains and the desert Southwest (Walls et al., 1967). Similarly, in Central America a direct correlation between prevalence and altitude of residence has been reported (Walton, 1966a). In France, rates vary from an estimated 85% in Paris to rates in other areas of the country comparable to those seen in the United States. Similar variations are reported for other ethnic and geographic populations.

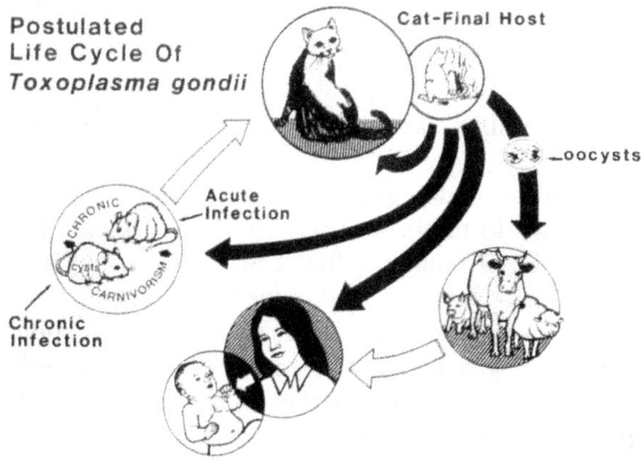

FIG. 2. The epidemiologic cycle of toxoplasmosis. Cats are the central figure, disseminating the parasite as oocysts in feces. These oocysts are then ingested through contaminated foods and the like to food animals and to humans.

Figure 2 illustrates the three recognized modes of transmission: ingestion of undercooked contaminated meat containing cysts present in the tissues of chronically infected animals; ingestion of oocysts from cat feces in the environment; and transplacental transfer during pregnancy. Pork and lamb are the primary meats that have been identified as common sources of infection. Only a few reports have implicated beef, whereas at least one study has shown that chickens may also harbor the parasite (McCulloch et al., 1966). Epidemiologic evidence has shown that undercooked hamburger accounted for a mild epidemic among students at Cornell University (Kean et al., 1969). As beef has rarely been incriminated as a source of infection, it is assumed that the hamburgers were most likely contaminated with pork. None of the original meat was available for study.

Domestic cats (and all members of Felidae) experience a form of toxoplasmosis similar to that seen in humans and also experience an intestinal infection in which the sexual forms of *Toxoplasma gondii* develop and produce oocysts that are passed in the feces. These oocysts are very resistant to adverse climatic conditions and are highly infectious (Ruiz et al., 1973). Data from two natural outbreaks suggest that the infection acquired by oocyst ingestion is usually mild, resembling infectious mononucleosis, but with limited morbidity (Bennenson et al., 1982; Teutsch et al., 1979). In one of these outbreaks, however, one patient was pregnant and transmission to the fetus resulted in spontaneous abortion with recovery of the parasite from fetal tissues. In both outbreaks, oocysts were implicated as the source of infection. One outbreak occurred in servicemen on jungle military maneuvers who refilled their canteens for drinking from an open pond later proved to be contaminated, probably from wild cats defecating in the vicinity of the pond. The other outbreak occurred at a riding stable, where it was shown epidemiologically that a cat passing oocysts had defecated in the riding ring. Apparently oocysts were dispersed in the dust created by the horses, thus exposing the riders.

There is no evidence of person-to-person transmission except as noted previously (i.e., blood transfusion or tissue transplantation). Experimentally, tachyzoites have been demonstrated in body fluids other than blood, but have not been associated with transmission. In humans parasitemia occurs, especially in acute stages of the disease, but the parasite has not been demonstrated in other fluids.

It has been shown that both flies and cockroaches are capable of disseminating the parasite from contaminated fecal material, but direct transmission to humans has not been noted (Wallace, 1971, 1972).

In nature, small rodents appear to be the reservoir, with feral cats becoming infected by ingesting them. Although it is clear that cats play a major role in the epidemiology of the disease, there is no correlation between cat ownership and infection, probably because domestic cats more commonly eat prepared food and defecate in known locations, decreasing their chances of acquiring and transmitting the infection to the owner. Feral cats, on the other hand, are much more likely to become infected and defecate in unknown locations (such as under shrubs and bushes where gardeners are likely to work in the soil). If this suggestion is true, then exposure is as likely for nonowners as for owners of cats. Likewise, although definitive data are not available, it is assumed that cats, both domestic and feral, which are common inhabitants of farms, serve as the source of infection for cattle, pigs, and sheep.

Susceptibility, Resistance, and Carrier States

Both innate and acquired immunity are involved in resistance to toxoplasmosis. The role of innate immunity is, of course, difficult to assess in humans, but is easily demonstrated in the laboratory by comparing the susceptibility of rats and mice to infection. When using a highly virulent strain of *T. gondii,* fewer than 10 organisms will cause a rapidly fatal infection in mice, whereas inoculation with as many as 10^6 parasites of the same strain causes no apparent disease in rats but produces numerous tissue cysts, from which parasites may be recovered months or even years later.

Acquired immunity appears to provide lasting protection and is of major importance in preventing congenital transmission. Relapses in acute cases improperly or insufficiently treated have been shown (Emerson et al., 1981; Vietzke et al., 1968), but reinfection, if it occurs, appears only to cause changes in antibody levels with little or no clinical response. In congenital infections, the data are clear that transmission rarely, if ever, occurs if the mother has existing antibody before conception. For the fetus to become infected the mother must acquire her *primary* infection during pregnancy. The cellular nature of this immunity involves both macrophages and lymphocytes in the resistance to toxoplasmosis. Experimentally, immune animals can be made susceptible by corticoid treatment (Frenkel, 1966a). Similarly, as has been discussed previously, human patients who are immunocompromised are highly susceptible. That this resistance is not associated with antibody is apparent from the fact that patients with AIDS frequently exhibit reasonable levels of immunoglobulin (Ig)G antibody while illiciting a fulminating and fatal infection. Other humoral factors such as opsonins and interferon have been shown to activate the cellular elements and perhaps play an important role (Ruskin et al., 1969).

Although "carrier state" as described for other infectious diseases, does not apply to toxoplasmosis, such a condition does exist but is of selective consequences. Serologic evidence and the findings that *Toxoplasma* can be isolated from nonclinical cases of toxoplasmosis indicate that even after therapy viable cysts of *Toxoplasma* persist probably indefinitely in the tissues of the host (Remington and Cavanaugh, 1965; Walls et al., 1963). These cysts play no role in person-to-person transmission but, as described previously, serve as the source of reactivation in the event of an immunologic failure. In addition, the existence of these cysts in the tissues of food animals is a common source of human infection.

Etiologic Agent

Characterization

Although isolates of varying antigenic makeup and virulence have been described, only one species, *Toxoplasma gondii,* is recognized. The life cycle of this parasite is illustrated in Fig. 3. Infection occurs with the ingestion of cysts or oocysts. Tachyzoites penetrate the gut wall and are disseminated by lymph and blood throughout the body. In the tissues, the organisms invade host cells, where multiplication takes place by endodyogeny, a unique process in which the mother cell develops two internal daughter cells that are released to continue the multiplication. In the absence of immunity, this multiplication continues until the host cell disintegrates and the tachyzoites invade other cells. In the presence of immunity, the tachyzoites develop a cyst wall within the host cell and continue to develop slowly. In felines, in addition to this cycle, the ingested sporozoites invade the epithelial cells of the intestinal mucosa and develop gametes. These gametes then unite to form oocysts containing up to eight cystizoites, which after being passed in the feces mature to become infectious sporozoites.

Toxoplasma is an obligate intracellular parasite related to the coccidia and malaria. The reproductive form, known as the tachyzoite, is an ellipsoid organism measuring about 3 by 6 μm, with a distinct subterminal nucleus and an anterior structure called the conoid (Fig. 4A). Although parasitemia has been reported histologically, the small size and relatively few organisms present make finding free tachyzoites unusual except with special stains and in special tissue sections. Frequently, a characteristic rosette of multiplying tachyzoites will be seen within cells. These organisms are very labile and degenerate rapidly outside the host.

Cysts, which originally develop within a host cell, eventually usurp the entire cell and become an intercellular structure (Fig. 4B). Bradyzoites develop slowly within the cyst by endodyogeny and may reach numbers of hundreds or more per cyst. A mature cyst may be as large as 100 μm and tends to retain the shape of the cell in which it develops; thus, in the brain it may be very spherical, whereas in striated muscle it may be elongated. Much more resistant than the tachyzoites, cysts remain viable in the tissues of animals even after slaughter. Most of the cysts are killed by freezing, but this treatment cannot be assumed to "sterilize" the meat. The cysts are destroyed by cooking.

Oocysts of *T. gondii* are distinguishable from most other coccidian oocysts by their small size, but are difficult to differentiate from the small *Isospora bigemina* cysts. As many as 10^6 immature oocysts

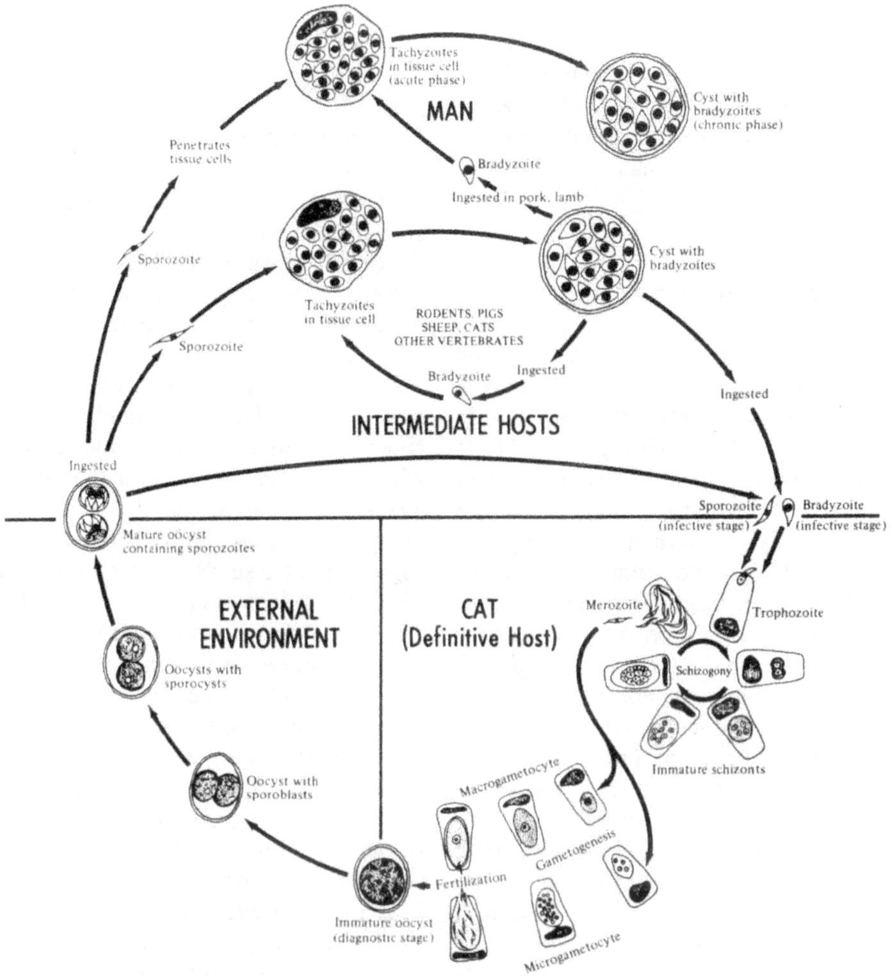

FIG. 3. The life cycle of *Toxoplasma gondii*. The asexual stages occur in all warm-blooded animals and the sexual stages only in the *Felidae*.

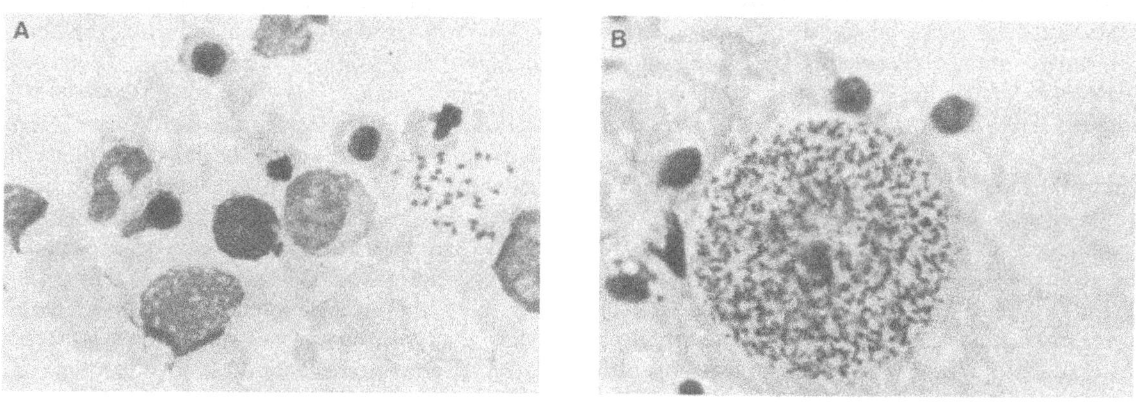

FIG. 4. (A) Tachyzoites of *Toxoplasma gondii* seen intracellularly in bone marrow cells. (B) Tissue cyst of *Toxoplasma gondii* in the brain of a mouse. Note the many bradyzoites enclosed in the cyst.

are passed each day starting 2 weeks after the initial infection of the cat and will continue for 3 to 4 weeks. These oocysts have been shown to be resistant to environmental conditions and may remain viable for months and perhaps years (Ruiz et al., 1973). Accidental laboratory exposure has suggested these oocysts are highly infectious, but usually result in mild disease.

Specimen Collection and Transport

Because of the poor viability of *Toxoplasma* outside the host and its somewhat fragile nature, it is important that specimens for culture or histology be obtained and transported to the laboratory as quickly as possible. Tissue specimens should be neither frozen nor allowed to dessicate. If the specimen is to be shipped or must be held overnight or longer, it should be suspended in tissue culture medium or whole blood with anticoagulant. The patient's own blood is acceptable. Refrigeration is not necessary for shipping, but is suggested if the specimen is to be held in the laboratory overnight. For histologic purposes an enlarged lymph node, when available, is adequate; however, a biopsy specimen from an affected organ is desirable. When possible, the specimen should be submitted fresh rather than embedded, as paraffin embedding interferes with some histologic techniques. For attempting to culture the organism, these same specimens can be used and, during the acute phases of the disease, whole blood or cerebrospinal fluid containing anticoagulant is often productive. Expedience cannot be overemphasized, but with proper handling a delay of 1 or 2 days does not seem to jeopardize the possibility of culture and allows for overnight shipment to an appropriate laboratory for processing.

For antibody determinations, the blood should be drawn and the serum collected as aseptically as possible. Blood should be allowed to clot and the serum clarified by centrifugation to remove cellular elements, although minor hemolysis does not interfere with most serologic procedures for toxoplasmosis. Serum samples should be refrigerated until tested; although IgM studies may be anticipated, however, freezing is not required. Mailed specimens should be sent by the most expeditious method, but need not be sent refrigerated.

Although antibody determinations are less frequently made using cerebrospinal fluid, this procedure is recommended in cases of suspected CNS involvement (Luft et al., 1984). In such cases, processing of the sample is similar to that of serum. The sample should be handled as aseptically as possible, refrigerated if storage is necessary, and delivered to the laboratory as soon as practical. Successive specimens should be obtained if possible to assess the progression of the disease or monitor the effects of therapy.

Isolation and Identification Procedures

Because *T. gondii* is an obligate intracellular parasite, isolation requires inoculation into living cells. Successful isolation has been reported by the inoculation of suspected tissue into cell cultures, but data on the use of this procedure are limited (Shepp, 1985). As a consequence, tissue culture has not been accepted as the procedure of choice. Mice are highly susceptible and have routinely been the animal of choice for this purpose. Tissue samples are ground with physiologic saline in a mortar and pestle and inoculated intraperitoneally. Whole blood or cerebrospinal fluid can be inoculated, but some reports suggest greater success if the buffy coat of blood or the cellular elements of cerebrospinal fluid are used.

Cell cultures can be monitored regularly to watch for cytopathology and the organisms observed directly if infection occurs. Due to unknown factors, probably the virulence of the organism, mice may die in 3 to 7 days or may survive with no overt symptoms of infection. Cultures isolated from humans usually are not lethal to mice. Mice surviving more than 4 weeks should be tested for antibody and, if positive, should undergo necropsy to demonstrate the organisms in the tissues. Tissue sections or impression smears of brain, liver, and spleen should reveal the parasite when treated with Wright–Giemsa stain. If stained tissue does not reveal any organisms and further confirmation is desired, suspensions of brain, liver, or spleen should be inoculated into clean mice and these mice evaluated 4 to 6 weeks later. If this approach is unsuccessful, a second subculture would not be productive.

In cell culture, tachyzoites of typical morphology should appear in rosettes and clusters within cells. As the infection continues, cytopathology becomes evident with the destruction of cells and the liberation of free tachyzoites. Depending on the number of organisms introduced, close observation may be necessary to detect infection in the early stages. Occasional agitation of the culture flasks will facilitate the dissemination of tachyzoites breaking out of isolated infected cells and help detection of the cytopathic effect. Characteristic cysts cannot be expected in these cell cultures.

In the tissues of mice infected more than a month earlier, the typical tissue cyst should be seen. These cysts are normally easily identified by their large size and the numbers of morphologically identifiable parasites they contain. Depending on the numbers of organisms in the inoculum, cysts may be rare and extensive microscopy may be necessary. Particularly in brain tissue, free tachyzoites may be seen, but because of their small size may be difficult to locate. Pathologic tissue contains few organisms; therefore, even under ideal conditions the number of cysts that develop is limited, and it is not unusual to fail to find cysts in serologically positive animals. For this reason, subinoculation is recommended if confirmation is necessary.

Specific Antigen Detection

Techniques for detecting specific toxoplasma antigen in tissue and body fluids are important in determining acute infection. A peroxidase–antiperoxidase technique has been used to identify antigen and whole tachyzoites in tissue when conventional histologic techniques have proved inadequate (Conley, 1981). The detection of cysts in tissue only confirms infection, whereas the demonstration of free tachyzoites or antigen is highly indicative of current active disease.

Similarly, the detection of circulating antigen in blood and other body fluids is a major criterion for establishing the activity of the disease. An enzyme-linked immunosorbent assay procedure has been reported as both sensitive and specific for antigenemia (Araujo and Remington, 1980).

Immunodiagnosis

Both histologic results and isolation of the parasite can be misleading, as both of these procedures may be positive in "normal" individuals with chronic toxoplasmosis. As has been discussed previously, the nondescript symptomatology normally exhibited in toxoplasmosis makes clinical diagnosis extremely difficult. Therefore, immunologic procedures are the primary methods for diagnosing toxoplasmosis.

Skin Test

A delayed-type hypersensitivity is demonstrable with toxoplasma antigens (Frenkel, 1971). Unfortunately, reactivity does not develop for at least several months after infection and consequently only serves to determine immune status. The simplicity and accuracy of the skin test should make it ideal for the evaluation of immunity in pregnant women. However, its usefulness has been limited because the antigens have not been standardized, and as with other delayed-type skin tests, a second clinic visit by the patient for evaluation is required. As a consequence, skin tests have not been used except for population evaluations.

Lymphocyte Transformation

Several reports have suggested the use of antigen-stimulated lymphocyte transformation to measure immunity (Krahenbuhl et al., 1971; Maddison et al., 1979). Although these reports have been promising, the requirements for performing the test and the difficulty of the procedure have prevented its use other than in research situations.

Serum Antibody Tests

The fact that *T. gondii* appears to be highly immunogenic is both an advantage and a disadvantage. It has made possible the development of a variety of serologic procedures, but at the same time confuses the diagnosis because humoral antibody apparently remains indefinitely after infection due to the chronic persistence of the parasite in the tissues. The prolonged presence of antibody has resulted in the development of two types of tests: those that are qualitative or semiquantitative and intended to measure the presence of antibody, which implies immunity; and those that are quantitative and may or may not differentiate between IgG and IgM antibody to determine the likelihood of active infection. Of the many tests reported, only a few are available outside specialty laboratories and are applicable to diagnosis at the local clinical level.

Qualitative Tests

Designed only to detect the presence or absence of antibody, qualitative tests are used for screening and should not be interpreted as indicative of active disease. In general, these tests are simple, rapid, inexpensive, and available either in kit form or at local public health or clinical laboratories. Some of these tests are described as follows.

Direct Agglutination

Not presently available in the United States, direct agglutination is a very promising procedure in which a suspension of *Toxoplasma* organisms is reacted with a patient's serum, resulting in agglutination if antibody is present (Desmonts and Remington, 1980). Some false-positive results are reported, but its simplicity and low cost make direct agglutination a prime candidate for screening tests.

Latex Agglutination

In the latex agglutination test, latex particles are coated with an extract of *T. gondii* and reacted with a patient's serum (Payne et al., 1984). Preliminary results are available in 90 min, with final results after overnight incubation. As with the direct agglutination, simplicity and economy are positive factors. The test has been reported to give some false-positive reactions in sera containing rheumatoid factor and has failed to detect some congenital cases, but appears to be very effective in screening adults (Walls and Remington, 1983).

Enzyme Immunoassays (EIA)

Three types of EIA are available. Solid-phase EIA tests are performed on plates or beads on which *Toxoplasma* antigen has been affixed. A final color reaction indicates positivity. These tests are more expensive and time consuming than the agglutination procedures unless numerous tests are performed simultaneously. Results are usually recorded automatically on an instrument that makes reading less subjective and faster. All of these qualities are advantageous to the larger laboratories. In general, these procedures are more accurate, and some are semiquantitative. Positive results can be used to alert the clinician to potential toxoplasmosis infection, but extreme caution should be exercised in extending the interpretation beyond that of a screening test.

Other variations are slight modifications of the procedure: In one, latex particles are coated with antigen, reacted with a patient's serum, and then entrapped on a filter, where the EIA test is completed; and in the other, the antigen is attached to the filter, where the entire test is performed. Speed of performance is the major advantage of these tests over all others. These extremely simple tests are performed in special filter containers and can be completed in less than 15 min. These procedures are strictly screen tests, but can be performed in the clinic while the patient is present. The preliminary results of these tests enable additional specimen collection and, if necessary, immediate referral for quantitative testing. Clinical data indicate these tests to be highly sensitive, specific, and ideal for screening, but not applicable to quantitative testing.

Within the limits described, all of the above screening tests can be used as "rule-out" procedures when active disease is suspected, but should never be used for the positive diagnosis of toxoplasmosis.

Quantitative Tests

The diagnosis of toxoplasmosis depends to a large extent on detecting an increase in antibody titers, the persistence of unusually high antibody titers, or the presence of IgM antibody. The quantitative procedures are more time consuming, more difficult to perform, and more expensive, but are designed to measure titer levels and changes to aid in establishing a diagnosis. It is critical for these procedures to measure accurately and reproducibly the amount of antibody present and in the appropriate tests to clearly identify IgM antibody. There appears to be no advantage in specifically measuring IgG antibody, although some procedures are designed to do so and it does not detract from the results. In general, quantitative tests do not lend themselves to the clinician's office or small laboratory because of the expense and usual requirement for special equipment or training.

Sabin–Feldman Dye Test

The first successful quantitative serologic procedure for toxoplasmosis, the Sabin–Feldman dye test (Sabin and Feldman, 1948) is now performed only in specialized laboratories and has for the most part been superceded by newer methods. This test is highly specific and sensitive and has been thoroughly studied for its clinical significance. The dye test primarily measures IgG antibody and for many years has remained the standard against which other tests are measured. Its acceptance has been inhibited by the need for specialized training and special reagents, including live *Toxoplasma* organisms.

Indirect Hemagglutination

The indirect hemagglutination (IHA) test (Lewis and Kessel, 1961) was accepted for some time as a substitute for the Sabin–Feldman dye test. It is simple to perform, inexpensive, and reasonably rapid. It has now been shown (Remington and McCloud, 1986), however, that IHA becomes positive much later than other serologic procedures and as a consequence may be negative in documented cases, especially congenitally acquired cases. The test has been used successfully for population studies and could be used as a screening test to determine immunity in pregnant women. It should not be used as a diagnostic procedure or where recent infection is suspected. The test is quantitative and reproducible and can be used in conjunction with other serologic tests to indicate recency of infection. In sequential serum specimens, recent infection would be indicated when early specimens are positive by the dye test and negative by IHA; subsequently, however, specimens convert to positive in both tests.

Indirect Immunofluorescence Test

The indirect immunofluorescence (IIF) test (Kelen et al., 1961; Sulzer and Hall, 1967) is the most frequently used procedure in the clinical laboratory for the serologic diagnosis of toxoplasmosis. Studies have shown a direct positive correlation between the results of IIF and those of the dye test (Walton et al., 1966b). Indirect immunofluorescence has three major advantages over the dye test: It uses a more or less permanent antigen of Formalin-killed *Toxoplasma* organisms fixed on a slide; all special reagents are commercially available; and both IgG and IgM antibodies can be measured independently. Its only major disadvantage compared with the dye test is the requirement for a microscope with an ultraviolet light source. As with the dye test, IIF is also labor intensive and requires special training; however, it is less demanding than the dye test.

Perhaps the most important advantage of IIF is its ability to measure IgM antibodies specifically. The IgM antibodies rise somewhat before the IgG antibodies and quickly decline to very low levels or disappear by 3 to 6 months. "High" levels (1:256 or greater) persist for even shorter periods. As a result, IgM antibodies are used to imply recent or current disease activity. Unfortunately, this concept is clouded by the fact that rheumatoid factor (Camargo et al., 1972) and antinuclear antibodies (Araujo et al., 1971) cause false-positive reactions in a small percentage of sera and must always be considered when interpreting these data. In addition, IgG antibodies occasionally block the detection of IgM, resulting in false-negative reactions (Ordonez et al., 1982). Newer modifications of the procedure introduce column separation methods for separating IgG and IgM before testing, which avoids this problem. Despite these limitations with the test, IgM and IgG indirect immunofluorescence tests are considered to be sensitive, accurate, and reproducible.

Enzyme Immunoassays

Technically, enzyme immunoassay procedures (Walls et al., 1977) are similar to IIF with an enzyme substituted for the fluorocein label and the reaction read by color development rather than microscopic fluorescence. The major advantages of EIA are that soluble antigens can be used and that the tests tend to be more sensitive than IIF because reactivity is measured by a multiplicative enzyme reaction. Enzyme immunoassay correlates well with IIF, but is technically much more difficult to perform and should be monitored by a more stringent quality control system. Most of the procedures described have differed primarily in the solid phase, to which a soluble antigen has been attached. These solid phases include polystyrene plates, metal and polystyrene beads, and membranes. Each of the procedures appears to give satisfactory results. Although reactivity can be read visually, more accurate and reproducible results are obtained by using one of the many special colorimeters available.

Most of the problems associated with the detection of IgM antibody by IIF have been overcome by the "capture" EIA methodology (Duermeyer et al., 1979). In this procedure, the solid phase is coated with highly specific anti-μ antibodies so that only the IgM from the patient's serum is retained and can then be reacted with *Toxoplasma* antigen. Blocking of the IgM by IgG is avoided by this removal of the IgG before reaction with the antigen. False-positive rheumatoid reactions are eliminated because the labeled detection measures the presence of antigen and not the presence of IgM. Two modifications of this procedure have been described: a double-sandwich immunoglobulin-M enzyme-linked immunosorbent assay (DS-IgM-ELISA) (Duermeyer et al., 1980; Naot and Remington, 1980), in which a labeled anti-*Toxoplasma* antibody is used as the detecting system; and an enzyme-labeled antigen (Schmitz et al., 1980) or reversed ELISA (Franco et al., 1981), in which a labeled antigen is the detector. The latter systems eliminate the need for a labeled anti-*Toxoplasma* antiserum, thus avoiding one procedural step as well as the possibility of interaction between the capture antibody and the detecting antibody in the sandwich technique.

The sensitivity and specificity of both of these methods have been documented to be superior to those of IIF. In one study, IIF-IgM failed to detect IgM in 9 of 122 sera from cases of recently acquired toxoplasmosis that were positive by reverse ELISA (Franco et al., 1983). In another study, DS-IgM-ELISA was 88% positive, whereas IIF-IgM was 77% positive in 26 patients 1 month after the onset of toxoplasmic lymphadenopathy. In addition, studies have shown IIF-IgM to detect as few as 25% of congenital cases, whereas DS-IgM-ELISA detected 77%.

The specificity of the EIA methods has been well documented (Franco et al., 1981). There are no false-positive reactions due to rheumatoid factor, antinuclear antibody, or macrogammaglobulin. No false-positive results were detected in "normal" individuals or in patients suspected of having toxoplasmosis but in whom the disease had been ruled out (Naot et al., 1981). In addition, reactivity from previously *Toxoplasma*-positive sera was removed by destruction or specific elimination of IgM.

Early results suggested that IgM levels decreased rather rapidly and disappeared within 3 to 6 months. However, using the new techniques that separate IgM from IgG, it is apparent that IgM antibodies are masked by the IgG. Studies have shown that in some cases IgM antibodies persist as long as 1 to 2 years (Brooks et al., 1987). High levels of IgM are significant and with appropriate clinical symptomatology are diagnostic, but low levels—even with symptomatology—must be confirmed by more extensive clinical and serologic follow-up.

The capture EIA methods, then, measure specific anti-*Toxoplasma* IgM antibodies without the interferences seen in the IIF-IgM test. They are more sensitive and specific than any previous procedures and are an invaluable adjunct to the noninvasive criteria necessary for diagnosis of active toxoplasmosis.

Other Tests

Complement fixation. Once very popular, the complement fixation test is used in few laboratories today. Complement-fixing antibodies appear somewhat later than those measured in other tests and disappear much sooner. For these reasons, complement fixation was used mainly as a measure of current activity. The newer tests for measuring IgM antibody have replaced complement fixation in this role. The lack of standardization of antigens and the relative poor sensitivity of complement fixation have caused this test to fall into disfavor.

Immunoglobulin M immunosorbent agglutination assay. The IgM immunosorbent agglutination assay, a promising new test, combines the capture technology with a direct agglutination test (Desmonts et al., 1981). Anti-μ antibody is coated on the solid phase, the patient's serum introduced and, after washing, a suspension of *Toxoplasma* organisms added and allowed to agglutinate. Too new for clinical acceptance, the data presented indicate this test to be simple, rapid, and inexpensive, with good sensitivity and specificity in screening for IgM antibody.

Screening Tests

Screening tests are perhaps as much abused as used. The temptation to draw quantitative conclusions and, thus, establish a diagnosis is great, but totally invalid. Screening tests are invaluable for determining the presence or absence of antibody, from which important conclusions may be drawn; however, clinical decisions may be reached only with the recommended quantitative procedures. The primary use of these tests is to evaluate the immune status of women during pregnancy and of their children at birth, a negative test indicating no infection or susceptibility, a positive test indicating immunity or infection to be confirmed by quantitative testing. Various procedures, each with advantages and disadvantages, have been designed for screening. Basically each is intended to be simple, rapid, inexpensive, and sensitive. Some of the tests may be read as "weak," "moderate," or "strong," but these interpretations are only guidelines and in general have no real meaning in quantitation.

Diagnosis

Clinical, pathologic, and histologic data are required for the definitive diagnosis of toxoplasmosis, but each has problems that limit its usefulness. Histology requires invasive procedures, which in some cases are hazardous and in some the patient's condition makes even simple surgery undesirable. Serologic methods usually provide the primary diagnostic information. As with any disease, diagnosis is a compilation of clinical, histologic, and serologic information.

Three basic criteria are required for diagnosis of acute toxoplasmosis: Isolation of *Toxoplasma* or identification of the organism or its antigens from tissue or body fluids; characteristic histology, especially in lymph nodes; and the presence of both IgM and IgG antibodies at diagnostic levels. As will be discussed, because rarely are all of these criteria met, presumptive diagnosis can be made with varying degrees of confidence with each. Definitive diagnosis requires fulfillment of all three criteria.

Clinical Diagnosis

As mentioned previously, no syndrome in itself defines toxoplasmosis. In the immunologically normal adult, infection may result in no specific symptoms and will usually go unnoticed. It is important to recognize this possibility, because pregnant women may experience an infection without recognizable symptoms and may transmit the disease to the fetus with disastrous effects. Unexplained prolonged lymphadenopathy, fever, or malaise should suggest toxoplasmosis. More severe symptoms rarely occur in the immunocompetent adult.

In immunocompromised individuals one should be particularly alert to the mild symptoms described above as precursors to more serious complications. In one study, more than one-half of immunoincompetent individuals with toxoplasmosis exhibited neurologic problems (Ruskin and Remington, 1976; Wong et al., 1984). In such individuals toxoplasmosis should be suspected any time the combination of symptoms of the generalized infection and neurologic involvement are seen. Neurologic involvement may include disorientation, seizures, and loss of consciousness.

Although the lesions are not pathognomonic, ocular infections can frequently be recognized by the experienced ophthalmologist by the typical lesions, which are rarely seen in other diseases (O'Connor, 1974). Usually, these are old latent cases that have become active only recently or in which the lesion has become sufficiently severe to cause clinical problems and recognition. The patient may show no other signs or symptoms of the disease. A history of an earlier diagnosis of active or suspected toxoplasmosis is helpful in reaching a decision.

Clinically, congenital toxoplasmosis is virtually impossible to diagnose. Chorioretinitis, perhaps the one constant finding, is not present in 100% of cases and may be due to other causes as well. Any infant showing marked abnormalities should be suspect and further workup pursued immediately. Hydrocephalus, cerebral calcifications, and mental retardation are other symptoms that are highly suggestive of toxoplasmosis but, again, may be due to a variety of other causes and diseases.

Congenital infection clearly can occur without apparent clinical involvement of the mother during pregnancy or the neonate at delivery (Wilson et al., 1980). However, these "normal" infants are at high risk of developing adverse sequelae later in life. Toxoplasmosis is one of the most prevalent congenital infections, yet many cases go undiagnosed because of the frequent lack of symptoms at birth. Screening may help to alleviate this problem. Estimates have shown the economic feasibility of screening compared to the cost of maintaining the abnormal children resulting from undiagnosed cases. In one study (Wilson and Remington, 1980), the clinical costs for maintaining all children abnormal as a result of toxoplasmosis were estimated to be up to $222 million per year, whereas screening of all pregnancies was estimated to cost $110 million—one-half as much.

Absolute minimal testing should include a test as early as possible during pregnancy and another on both mother and child at delivery. Obviously, the mother would be retested if symptoms became apparent, and ideally she would be tested at least twice during pregnancy even without symptoms. It should be noted that the cost estimates cited for the care of abnormal children are not based on current markedly inflated medical costs. Furthermore, since these estimates were determined the costs for serologic screening tests have diminished.

Histology

Although the histologic changes in some tissues, especially lymph nodes, are characteristic, the prime objective in histology is to demonstrate the presence of tachyzoites or free antigen within the tissues. It has been shown that the histology of lymph nodes is sufficiently characteristic for the diagnosis to be established without demonstrating the parasite (Dorfman and Remington, 1973); however, the lymph node is the only tissue in which histology is so reliable.

The tachyzoite or antigen is difficult to demonstrate in tissue, as the tachyzoite does not stain well with hematoxylin and free antigen is not visualized by staining. Immunofluorescent staining techniques have facilitated success in visualizing tachyzoites or free toxoplasma antigen in tissue, but these procedures present problems with autofluorescence in formalinized tissue and are relatively insensitive. A very successful and highly sensitive peroxidase–antiperoxidase system has been reported (Conley et al., 1981) for detecting parasites and antigen in tissue. Cysts are readily demonstrated in tissues by conventional histologic procedures; active disease cannot be diagnosed in this way, however, because cysts have been identified in tissue at least 2 years after infection.

Isolation of *Toxoplasma*

Isolation of *Toxoplasma* from blood or other body fluids clearly indicates active proliferation of the organism and is a strong indicator of active disease and thus a major criterion for diagnosis. However, as mentioned previously, isolation or identification of the organism in tissue can be very misleading, because *T. gondii* has been isolated from tissues of patients with totally unrelated infections and from autopsy specimens from patients dying from causes other than toxoplasmosis (Remington and Cavanaugh, 1965; Walls et al., 1963). Therefore, one must be aware that *Toxoplasma* is frequently found in association with other infections, and identification of active *Toxoplasma* may not be a diagnosis of the *clinical* problem. *Toxoplasma* is difficult to culture, and even in active cases isolation is frequently negative. Therefore, a negative isolation does not rule out active disease.

Serology

Although serology is no diagnostic panacea, it is the most reliable diagnostic criterion. Because of the large number of positive titers in normal individuals, results must be carefully interpreted. Screening tests should be used cautiously, and quantitative tests must be properly interpreted. Rising titers are diagnostic, but because of the insidious nature of toxoplasmosis, diagnosis is frequently not attempted until the disease is well established and beyond the time when rising titers must be expected. The level of IgG antibody titers is not necessarily indicative of the intensity of infection, but there is a correlation between titer level and activity. An unusually high single titer (> 1:1,024) is highly suggestive, but does not serve as the definitive diagnostic criterion. IgG antibodies appear to persist for life, often at levels high enough to be considered clinically important. Serologic diagnosis can only be reached reliably by finding elevated titers in both the IgG and IgM antibodies. These interpretations are based on titers obtained with the dye test or IIF techniques, but other procedures are generally in agreement. Despite exceptions, one can assume that the simultaneous presence of an IgG titer of greater than 1 : 1,024 and IgM of greater than 1 : 64 is diagnostic.

The time interval between the appearance of IgM and IgG antibodies is very short, so that *clinically* one does not see IgM-positive sera that are IgG negative. In a few documented cases sera were obtained that were IgM positive and IgG negative, but these were special cases. A second specimen should be obtained from patients showing this phenomenon to ensure the specificity of the reaction. A negative IgM titer may occur after a few weeks even in active cases due to blocking by IgG, even though the level of IgG may not be unusually high (Pyndiah et al., 1979). In cases where active toxoplasmosis is suspected and the IgM titer is negative, one of the IgM-specific EIA procedures or one of the column separation procedures must be used.

Serodiagnosis during pregnancy is of special concern. Toxoplasmosis is transmitted to the fetus only during primary infection of the mother, when immunity is absent. With but few exceptions, infection of the fetus will not occur when there is preexisting antibody (Desmonts and Couvreur, 1974). It is important, then, that serologic testing be done before pregnancy if possible or at the earliest obstetric visit. This baseline is invaluable if in midterm there is a suspected exposure and antibody determinations are of borderline levels. Patient management during pregnancy depends on when infection occurs. By pretesting, a temporal relationship can be established. If the mother's pretest is positive *at any level,* it can be assumed that she is immune and no further testing is necessary. On the other hand, if she is negative she is susceptible and should be retested periodically during pregnancy. Both mother and child, *even when asymptomatic,* should also be tested at delivery. Studies have shown that children born with congenital toxoplasmosis but without symptoms may develop symptoms years later (Wilson et al., 1980). It is important to rule out maternal chronic toxoplasmosis at birth. IgG antibodies do not reach their maximum levels until 1 to 2 months after infection, whereas IgM antibodies begin to decline shortly thereafter. If the mother is not tested until her first obstetric visit, which is usually in the second or third month, and the IgG titer is stable, infection probably occurred before pregnancy. If the IgG titer is low, infection probably occurred years ago. Current infection is assumed if simultaneous testing of 3-week sequential sera shows rising antibody levels and the presence of IgM antibodies by EIA.

Maternal IgM is transmitted to the fetus only in cases of placental damage, but the IgG passes quite freely. At delivery, both mother and child should be tested simultaneously to determine the significance of infant antibody. As antibody production begins in utero, the infant will express its own IgM and both its own and maternal IgG. Placental leakage can be detected by a variety of procedures, but these methods are usually not available in the diagnostic laboratory. The consequences of congenital toxoplasmosis suggest that in the presence of compatible symptomatology, it is advisable to assume that *any* IgM detected in the infant is diagnostic, and therefore appropriate steps must be taken. A second specimen taken 2 weeks later will clarify this point and allow reevaluation of patient management. Transferred IgM resulting from placental leakage should begin to decline rapidly, but may not disappear completely by the second specimen if the original levels were high. If infant IgG or IgM titers increase or remain at high levels, diagnosis is virtually assured.

Immunoincompetent individuals rarely show IgM antibody levels even during acute episodes (Remington and McCloud, 1986). IgG antibodies are frequently at low levels but are usually present. IgG antibodies may reflect an abnormal immune system or suggest the preexistence of IgG antibody and a reactivated infection. All tissue transplant patients should be pretested to determine susceptibility and should be followed throughout immunosuppression to detect possible activity. These patients are particularly at risk when receiving a transplant from a seropositive donor. Serology is less valuable in patients with AIDS because of their unusual antibody activity. A positive serologic test before the onset of symptoms is an indication of a preexisting infection, and these patients should be monitored closely. The existence of a titer or even a rising titer in AIDS cases is sometimes confusing, as a variety of causes may result in serologic changes in these individuals and diagnosis may depend on biopsy and acute clinical diagnosis.

In ocular infections, serum antibody levels may persist at very low levels and IgM antibody is usually negative (Desmonts, 1966). Ophthalmologists thoroughly familiar with toxoplasmic chorioretinitis may use titers of 1 : 2 or 1 : 4 as supportive data for diagnosis; this diagnosis must be done with a great deal of caution, however, because a very high percentage of normal individuals exhibit titers at this level. We have seen titer levels change with the activity of the lesion (unpublished observations), but there is no correlation between titer level and degree of involvement. As a solution to the low-titer dilemma, a method has been described to determine the "antibody load" in the eye (Desmonts, 1966). The ratio of antibody to total gamma globulin in the aqueous humor is compared to the ratio of antibody to gamma globulin in the serum. Because of a lower total protein in the eye, if the ratio is higher for the eye, it is assumed there is local antibody production and the lesion is toxoplasmic. The low antibody levels probably reflect the immunologic isolation of the eye. In addition, it is widely presumed that most cases of ocular toxoplasmosis are of congenital origin and thus exhibit stable, old antibody levels.

Treatment, Prevention, and Control

Treatment

Acute toxoplasmosis responds rather dramatically to appropriate specific therapy, which makes early diagnosis important. The effect of therapy on long-term sequelae in infants has not been evaluated; however, adults show an immediate response to therapy with remission of symptomatology. These results suggest that if the child develops normal immunity, treatment should prevent further progression of the disease. As will be discussed, complications are associated with the recommended therapeutic regimen; these complications are well recognized, however, and can be avoided or alleviated to allow therapy under most clinical situations. Nonetheless, because of these hazards, therapy must be instituted judiciously.

General Considerations

To date, only pyrimethamine in combination with sulfadiazine or trisulfapyrimidines can be recommended. Other sulfa analogues have not proved to be as effective. As seen in Fig. 5, pyrimethamine and the sulfas interrupt the folic acid pathway, which is critical for the parasite. The primary side effect of pyrimethamine treatment is a decrease in bone marrow activity. If folinic acid as calcium leucovorin is included, most patients will overcome this inhibition without decreasing the effect of therapy on the *Toxoplasma* organisms. Because this inhibition is temporary, an immediate response will normally be elicited in patients reacting strongly to pyrimethamine or in whom the leucovorin is not sufficient to reverse the problem by cessation of therapy and vitamin supplementation. A thorough discussion of pyrimethamine, its chemistry, and its metabolism has been presented elsewhere (Frenkel, 1966b). Duration and dosage depend on the individual patient and the severity of disease, and variations will be discussed briefly. A suggested regimen (McCabe and Remington, 1983) for most acute cases consists of pyrimethamine (loading dose, 1 mg/kg up to 50 mg twice a day for the first 2 days; maintenance dose, 1 mg/kg up to 25 mg every other day), sulfadiazine (loading dose, 75 mg/kg up to 4 g for the first day; maintenance dose, 100 mg/kg/day up to 6 g/day in two divided doses), or leucovorin (10 mg injectable or oral given simultaneously with pyrimethamine). Patients should undergo peripheral blood cell and platelet counts twice each week during therapy.

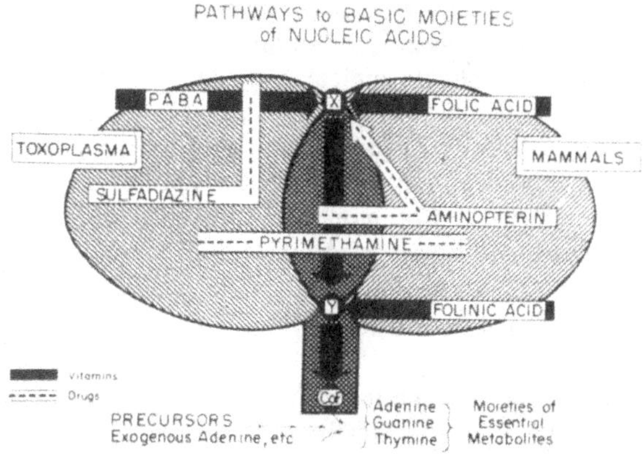

FIG. 5. The effects of pyrimethamine and sulfonamides on the metabolism of *Toxoplasma* and a mammalian host. Note that both drugs interrupt both metabolic systems, but only the mammalian host can utilize folinic acid to restore the pathway. (From Frenkel [1966], with permission.)

Acute Acquired Disease

Although there is some controversy, it is generally believed that in immunologically normal individuals therapy is not necessary unless symptoms become severe or persist. If initiated, therapy should be continued for a minimum of 21 days. We have seen cases in which therapy of shorter duration resulted in relapse.

Infection Acquired During Pregnancy

Therapeutic recommendations are perhaps most controversial and difficult in cases of infection during pregnancy. Pyrimethamine has been shown to be teratogenic (Krahe, 1965) and therefore should not be used during the first trimester. As an alternative, spiramycin, which is not available in the United States, has been used with limited success in Europe. The side effects of spiramycin are minimal and it is not teratogenic, but its efficacy seems to be less than that of pyrimethamine and sulfadiazine. In the United States there is no useful alternative therapy for first-trimester infections. If spiramycin is available, it has been suggested that therapy be initiated immediately on diagnosis and continued until term. Side effects are mild and may be minimized by giving therapy intermittently as 3 weeks of spiramycin (2 to 3 g/day) and 2 weeks of rest (Remington and McCloud, 1986).

Therapeutic abortion, an extremely controversial alternative to therapy, should not be considered without a full understanding of the risks of first-trimester toxoplasmosis. As discussed previously, only a small percentage of infected mothers will transmit the infection to the fetus, and only a portion of these will develop symptoms. On the other hand, data suggest that if transmission does occur during the first trimester, the consequences are frequently tragic.

Therapy with pyrimethamine and sulfadiazine initiated later in pregnancy has been shown effective in reducing fetal infections by one-half to two-thirds, but does not eliminate them entirely (Remington and Desmonts, 1983). As with spiramycin, therapy with daraprim and sulfadiazine should be given for the duration of the pregnancy in intermittent periods of 2 weeks' treatment and 4 weeks' rest.

Congenital Infection

Treatment of infants with congenital toxoplasmosis consists of dosages considerably different from those stated previously (for specific recommendations, see Remington and Desmonts [1983]). As stated earlier, disease sequelae may occur years after birth; consequently, immediate therapy at the earliest recognition of infection, even in the absence of symptoms, is important. Although longitudinal studies are still needed to determine the efficacy of these treatments, at least two studies have suggested the possible effectiveness of therapy in both symptomatic and asymptomatic congenital cases (Stagno et al., 1977; Wilson et al., 1980). As stated in the specific recommendations, in infants diagnosed as infected either by clinical signs or by serology, treatment should continue for at least 21 days and should include folinic acid. This treatment should be repeated 3 or 4 times during the first year. A healthy child not shown to be infected but born to a mother diagnosed as acquiring toxoplasmosis during pregnancy should receive a single course of treatment.

Ocular Disease

In most instances of ocular toxoplasmosis, an active lesion is recognized at diagnosis. Treatment is primarily intended to prevent further spread of the infection. Therapy, as outlined above, is effective in halting the proliferation of the tachyzoites. A favorable clinical response has been reported in up to 80% of treated cases (O'Connor, 1974). Chronic relapsing cases that have persisted for years have resolved after a single course of treatment (Wong et al., 1984). In some cases, severe inflammation and necrosis dictates the use of corticosteroid therapy. When such therapy is warranted, it must be realized that the immune mechanisms are paralyzed and specific therapy must be initiated simultaneously and continued for at least 2 weeks beyond the steroid program. The usual course of pyrimethamine–sulfadiazine–folinic acid therapy is 4 to 8 weeks.

Immunosuppressed Disease

Therapy in immunosuppressed patients, particularly those with AIDS, is complicated by a number of factors. Definitive diagnosis is difficult to establish in cerebral infections and in many cases is made by inference, as invasive procedures in the brain are obviously withheld unless considered unavoidable; the acuteness of the disease makes immediate therapy essential, even if only empiric; variable patient responses require a tailored therapeutic program; and many patients with AIDS are hypersensitive to sulfadiazine. Despite these problems, therapy for CNS toxoplasmosis is reported to be very successful. In one study, of six patients with AIDS and CNS toxoplasmosis who were treated with pyrimethamine and sulfadiazine, one entered therapy very late in the disease and died, but the other five showed remission of signs and symptoms (Wong, 1984). Two relapses occurred, but responded to a second course of therapy. The dosage of pyrimethamine was relatively constant, but that of sulfadiazine varied from 4 to 6 g and from 13 days to 9 months. In another study of 81 immunoincompetent patients without AIDS, 80% showed marked improvement with treatment (Ruskin and Remington, 1976). To date, there is no resolution to the dilemma of those patients who are hypersensitive to sulfa drugs. Patients with AIDS who have received sulfonamides for other infections frequently have developed a sensitivity to them and cannot tolerate the usual therapeutic program recommended for toxoplasmosis. Anecdotal reports suggest exceedingly high doses of pyrimethamine alone have been successful, but the number of cases and the documentation are too weak to accept this treatment as an alternative. As yet, there is no other drug that can be substituted. Clindamycin and trimethoprim–sulfamethoxazole, which are used for other AIDS-related infections, have proved ineffective. Because as many as 12% of patients with AIDS will develop toxoplasmosis, therapy should be instituted even empirically on clinical suggestion (Wong, 1984). Clinical resolution may serve as a diagnostic aid. Therapy should continue for 4 to 6 weeks beyond complete resolution of signs and symptoms. Patients with AIDS must be monitored carefully for relapse and retreated as necessary.

Treatment and Control

There is no vaccine for toxoplasmosis, and prophylactic treatment is unrealistic. Simple sanitary avoidance is by far the best preventive method. Epidemiologically, it is well established that humans become infected almost exclusively by ingestion of the cyst or oocyst of the parasite. Populations at risk, namely, seronegative pregnant women and patients

TABLE 1. Methods for preventing transmission of toxoplasmosis

1. Cook all meat at least medium well done to an internal temperature of 60°C (150°F).
2. Wash hands thoroughly after handling raw meat.
3. Wear gloves when gardening and wash hands thoroughly immediately upon returning to the house.
4. An indoor litterbox should be emptied daily. A pregnant woman should leave this chore for another person.
5. A "house" cat should be kept indoors as much as possible and fed commercially prepared food.

with AIDS, should avoid contact with cat feces and undercooked meat. Table 1 indicates basic procedures to help ensure avoidance of the infective stages of the parasite. Perhaps the least obvious, but most important precaution is to wear gloves when working in the garden and to wash hands thoroughly afterward. A domestic cat usually defecates in a litterbox, which can be appropriately cleaned, and usually is fed well, which decreases the possibility of the cat catching and eating wild rodents. As was indicated previously, the oocyst in cat feces requires 2 to 4 days after excretion to mature; consequently, if the litterbox in the house is emptied daily, any contaminated feces is discarded before it becomes infective. The feral cat is a greater hazard because its food source is the wild rodent and it defecates indiscriminately, often in gardens or under shrubs. Although freezing decreases the number of viable organisms found in meat, it does not sterilize; therefore, cooking is necessary. To ensure safety, meat should be cooked to an internal temperature of 60°C (150°F), or "medium well done." Until a vaccine is developed, prevention and control are best accomplished by obstetricians and other physicians advising their patients of the risks and methods of avoiding contamination.

Literature Cited

Araujo F. G., E. V. Barnett, L. O. Gentry, and J. S. Remington. 1971. False-positive anti-toxoplasma fluorescent-antibody tests in patients with antinuclear antibodies. Appl. Microbiol. **22**:270–275.

Araujo, F. G., and J. S. Remington. 1980. Antigenemia in recently acquired acute toxoplasmosis. J. Infect. Dis. **141**:144–150.

Bennenson, M. W., E. T. Takafuji, S. M. Lemon, R. L. Greenup, and A. J. Sulzer. 1982. Oocyst-transmitted toxoplasmosis associated with ingestion of contaminated water. N. Engl. J. Med. **307**:666–669.

Brooks, R. G., R. E. McCabe, and J. S. Remington. 1987. Role of serology in the diagnosis of toxoplasmic lymphadenopathy. Rev. Infect. Dis. **9**:775–782.

Camarago, M. E., P. G. Leser, and A. Rocca. 1972. Rheumatoid factors as a cause for false positive IgM anti-toxoplasma fluorescent tests. A technique for specific results. Rev. Inst. Med. Trop. Sao Paulo **14**:310–313.

Conley, F. K., K. A. Jenkins, and J. S. Remington. 1981. *Toxoplasma gondii* infection of the central nervous system: use of the peroxidase–antiperoxidase method to demonstrate *Toxoplasma* in formalin-fixed paraffin embedded tissue sections. Hum. Pathol. **12**:690–698.

Crawford, J. B. 1966. Toxoplasma retinochoroiditis. Arch. Ophthalmol. **76**:829–832.

Desmonts, G. 1966. Definitive serologic diagnosis of ocular toxoplasmosis. Arch. Ophthalmol. **76**:839–851.

Desmonts, G., and J. Couvreur. 1974. Congenital toxoplasmosis. A prospective study of 378 pregnancies. N. Engl. J. Med. **290**:1110–1116.

Desmonts, G., Y. Naot, and J. S. Remington. 1981. Immunoglobulin M-immunosorbent agglutination assay for diagnosis of infectious diseases: diagnosis of acute congenital and acquired *Toxoplasma* infections. J. Clin. Microbiol. **14**:486–491.

Desmonts, G., and J. S. Remington. 1980. Direct agglutination test for diagnosis of *Toxoplasma* infection: method for increasing sensitivity and specificity. J. Clin. Microbiol. **11**:562–568.

Dorfman, R. F., and J. S. Remington. 1973. Value of lymph node biopsy in the diagnosis of acute acquired toxoplasmosis. N. Engl. J. Med. **289**:878–881.

Duermeyer, W., F. Wielaard, and J. van der Veen. 1979. A new principle for the detection of specific IgM antibodies applied in an ELISA for hepatitis A. J. Med. Virol. **4**:25–32.

Duermeyer, W., F. Wielaard, H. van Gruijthuijsen, and J. Swinkels. 1980. Enzyme-linked immunosorbent assay for detection of immunoglobulin M antibodies against *Toxoplasma gondii*. J. Clin. Microbiol. **12**:805–806.

Emerson, R. G., D. S. Jardine, E. S. Milvenan, B. J. D'Souza, G. J. Elfenbein, G. W. Santos, and R. Saral. 1981. Toxoplasmosis: a treatable neurologic disease in the immunologically compromised patient. Pediatrics **67**:653–655.

Feldman, H., and L. T. Miller. 1956. Serological study of toxoplasmosis prevalence. Am. J. Hyg. **64**:320–335.

Franco, E. L., K. W. Walls, and A. J. Sulzer. 1981. Reverse enzyme immunoassay for detection of specific anti-*Toxoplasma* immunoglobulin M antibodies. J. Clin. Microbiol. **13**:859–864.

Franco, E. L., K. W. Walls, A. J. Sulzer, and J. C. Soto. 1983. Diagnosis of acute acquired toxoplasmosis with the enzyme-labelled antigen reversed immunoassay for immunoglobulin M antibodies. J. Immunoassay **4**:373–393.

Frenkel, J. K. 1948. Dermal hypersensitivity to toxoplasma antigens (toxoplasmins). Proc. Soc. Exp. Biol. **68**:634–639.

Frenkel, J. K. 1966a. Effects of cortisone, total body radiation and nitrogen mustard on chronic latent toxoplasmosis. Am. J. Pathol. **33**:618–619.

Frenkel, J. K. 1966b. Toxoplasmosis, p. 296–321. *In* K. Benirschke (ed.), Comparative aspects of reproductive failure. Springer-Verlag, New York.

Frenkel, J. K. 1972. Toxoplasmosis. Mechanisms of infection, laboratory diagnosis and management. Curr. Top. Pathol. **54**:28–75.

Gibson, C. L., and N. Coleman. 1958. The prevalence of toxoplasma antibodies in Guatemala and Costa Rica. Am. J. Trop. Med. **7**:334–338.

Hooper, A. D. 1957. Acquired toxoplasmosis. Arch. Pathol. **64**:1–9.

Kaas, E. H., S. B. Andrus, R. D. Adams, F. C. Turner, and H. A. Feldman. 1952. Toxoplasmosis in the human adult. Arch. Intern. Med. **89:**759–782.

Kean, B. H., A. C. Kimball, and W. N. Christenson. 1969. An epidemic of acute toxoplasmosis. J. Am. Med. Assoc. **208:**1002–1004.

Kelen, A. E., L. Ayllon-Leindl, and N. A. Labzoffsky. 1961. Indirect fluorescent antibody method in serodiagnosis of toxoplasmosis. Can. J. Microbiol. **8:**545–554.

Krahe, M. 1965. Untersuchungen uber die teratogene wirkung von medikamenten zur behandlung der toxoplasmose wahrend der schwangerschaft. Arch. Gynakol. **202:**104–109.

Krahenbuhl, J. L., J. D. Gaines, and J. S. Remington. 1971. Lymphocyte transformation in human toxoplasmosis. J. Infect. Dis. **125:**283–288.

Lewis, W. P., and J. F. Kessel. 1961. Hemagglutination in the diagnosis of toxoplasmosis and amebiasis. Am. Med. Assoc. Arch. Ophthalmol. **66:**471–476.

Luft, B. J., R. G. Brooks, F. K. Conley, R. E. McCabe, and J. S. Remington. 1984. Toxoplasmic encephalitis in patients with acquired immune deficiency syndrome. J. Am. Med. Assoc. **252:**913–917.

Maddison, S. E., S. B. Slemenda, S. M. Teutsch, K. W. Walls, I. G. Kagan, W. R. Mason, F. Bell, and J. Smith. 1979. Lymphocyte proliferative response in 31 patients after an outbreak of toxoplasmosis. Am. J. Trop. Med. Hyg. **28:**955–961.

McCabe, R. E., and J. S. Remington. 1983. The diagnosis and treatment of toxoplasmosis. Eur. J. Clin. Microbiol. **2:**95–104.

McCulloch, W. F., B. G. Foster, J. L. Braun, and F. H. Top. 1966. Toxoplasmosis in Iowa. *In* Proceedings of the 11th Pacific Science Congress, Tokyo, Japan. **6:**4.

Naot, Y., G. Desmonts, and J. S. Remington. 1981. IgM enzyme-linked immunosorbent assay test for the diagnosis of congenital *Toxoplasma* infection. J. Pediatr. **98:**32–36.

Naot, Y., and J. S. Remington. 1980. An enzyme-linked immunosorbent assay for detection of IgM antibodies to *Toxoplasma gondii:* use for diagnosis of acute acquired toxoplasmosis. J. Infect. Dis. **142:**757–766.

O'Connor, G. R. 1974. Manifestations and management of ocular toxoplasmosis. Bull. N.Y. Acad. Med. **50:**192–210.

Ordonez, G. A., J. T. Newman, and M. J. Stone. 1982. Serological diagnosis of *Toxoplasma gondii* infections by rapid separation of serum immunoglobulins M and G with CM Bio-Gel A. J. Clin. Microbiol. **16:**751–753.

Payne, R. A., J. M. Francis, and W. Kwantes. 1984. Comparison of a latex agglutination test with other serological tests for the measurement of antibodies to *Toxoplasma gondii*. J. Clin. Pathol. **37:**1293–1297.

Pitchenik, A. E., M. A. Fischl, and K. W. Walls. 1983. Evaluation of cerebral mass lesions in acquired immunodeficiency syndrome, letter. N. Engl. J. Med. **308:**1099.

Pyndiah, N., U. Krech, P. Price, and J. Wilhelm. 1979. Simplified chromatographic separation of immunoglobulin M from G and its application to *Toxoplasma* indirect immunofluorescence. J. Clin. Microbiol. **9:**170–174.

Remington, J. S., and E. N. Cavanaugh. 1965. Isolation of the encysted form of *Toxoplasma gondii* from human skeletal muscle and brain. N. Engl. J. Med. **273:**1308–1310.

Remington, J. S., and G. Desmonts. 1976. Toxoplasmosis, p. 191–332. *In* J. S. Remington and J. O. Klein (ed.), Infectious diseases of the fetus and newborn infant. The W. B. Saunders Co., Philadelphia.

Remington, J. S., and G. Desmonts. 1983. Toxoplasmosis, p. 143. *In* J. S. Remington and J. O. Klein (ed.), Infectious diseases of the fetus and newborn infant. The W. B. Saunders Co., Philadelphia.

Remington, J. S., and R. McCloud. 1986. Toxoplasmosis, p. 1521–1535. *In* A. I. Braude, C. E. Davis, and J. Fierer (ed.), Infectious diseases and medical microbiology, 2nd ed. The W. B. Saunders Co., Philadelphia.

Reynolds, E. S., K. W. Walls, and R. I. Pfeiffer. 1966. Generalized toxoplasmosis following renal transplantation. Arch. Intern. Med. **118:**401–405.

Ruiz, A., J. K. Frenkel, and L. Cerdas. 1973. Isolation of *Toxoplasma* from soil. J. Parasitol. **59:**204–206.

Ruskin, J., J. McIntosh, and J. S. Remington. 1969. Studies on the mechanisms of resistance to phylogenetically diverse intracellular organism. J. Immunol. **103:**252–259.

Ruskin, J., and J. S. Remington. 1976. Toxoplasmosis in the compromised host. Ann. Intern. Med. **84:**193–199.

Sabin, A. B., and H. A. Feldman. 1948. Dyes as microchemical indicators of a new immunity phenomenon affecting a protozoan parasite (*Toxoplasma*). Science **108:**660–663.

Schmitz, H., U. von Deimling, and B. Flehmig. 1980. Detection of IgM antibodies to cytomegalovirus (CMV) using an enzyme-labelled antigen (ELA). J. Gen. Virol. **50:**59–68.

Sexton, R. C., D. E. Eyles, and R. E. Dillman. 1953. Adult toxoplasmosis. Am. J. Med. **14:**366–377.

Shepp, D. H., R. C. Hackman, F. K. Conley, J. B. Anderson, and J. D. Meyers. 1985. *Toxoplasma gondii* reactivation identified by detection of parasitemia in tissue culture. Ann. Intern. Med. **103:**218–221.

Siegel, S. E., M. N. Lunde, A. H. Gelderman, R. H. Halterman, J. A. Brown, A. S. Levine, and R. G. Graw, Jr. 1971. Transmission of toxoplasmosis by leucocyte transfusion. Blood **37:**388–394.

Stagno, S., D. W. Reynolds, C. S. Amos, A. J. Dahle, F. P. McCollister, I. Mohindra, R. Ermocilla, and C. A. Alford. 1977. Auditory and visual defects resulting from symptomatic and subclinical congenital cytomegaloviral and *Toxoplasma* infections. Pediatrics **59:**669–678.

Sulzer, A. J., and E. C. Hall. 1967. Indirect fluorescent antibody tests for parasitic diseases. IV. Statistical study of the variation in the indirect fluorescent antibody (IFA) test for toxoplasmosis. Am. J. Epidemiol. **86:**401–407.

Teutsch, S. M., D. D. Juranek, A. Sulzer, J. P. Dubey, and R. K. Sikes. 1979. Epidemic toxoplasmosis associated with infected cats. N. Engl. J. Med. **300:**695–699.

Townsend, J. J., J. S. Wolinsky, J. R. Baringer, and P. C. Johnson. 1975. Acquired toxoplasmosis: a neglected cause of treatable nervous system disease. Arch. Neurol. **32:**332–343.

Vietzke, W. M., A. H. Gelderman, P. M. Grimley, and M. P. Valsamis. 1968. Toxoplasmosis complicating malignancy. Cancer **21:**816–827.

Vischer, T. L., C. Bernheim, and E. Englebrecht. 1967. Two cases of hepatitis due to *Toxoplasma gondii*. Lancet **2:**919–920.

Wallace, G. D. 1971. Experimental transmission of *Toxoplasma gondii* by filth-flies. Am. J. Trop. Med. Hyg. **20:**411–413.

Wallace, G. D. 1972. Experimental transmission of *Toxoplasma gondii* by cockroaches. J. Infect. Dis. **126:**545–547.

Walls, K. W., S. L. Bullock, and D. K. English. 1977. Use of the enzyme-linked immunosorbent assay (ELISA) and its microadaptation for the serodiagnosis of toxoplasmosis. J. Clin. Microbiol. **5:**273–277.

Walls, K. W., I. G. Kagan, and A. Turner. 1967. Studies on the prevalence of antibodies to *Toxoplasma gondii*. 1. U.S. military recruits. Am. J. Epidemiol. **85**:87–92.

Walls, K. W., and J. S. Remington. 1983. Evaluation of a commercial latex agglutination method for toxoplasmosis. Diagn. Microbiol. Infect. Dis. **1**:265–271.

Walls, K. W., J. J. Taraska, and M. Goldman. 1963. Isolation of *Toxoplasma gondii* from cysts in human brain. J. Parasitol. **49**:930–931.

Walton, B. C., I. Arjona, and B. M. Benchoff. 1966a. Relationship of toxoplasma antibodies to altitude. Am. J. Trop. Med. Hyg. **15**:492–495.

Walton, B. C., B. M. Benchoff, and W. H. Brooks. 1966b. Comparison of the indirect fluorescent antibody test and methylene blue dye test for detection of antibodies to *Toxoplasma gondii*. Am. J. Trop. Med. Hyg. **15**:149–152.

Wilson, C., and J. S. Remington. 1980. What can be done to prevent congenital toxoplasmosis? Am. J. Obstet. Gynecol. **138**:357–363.

Wilson, C. B., J. S. Remington, S. Stagno, and D. W. Reynolds. 1980. Development of adverse sequelae in children born with subclinical congenital toxoplasma infection. Pediatrics **66**:767–774.

Wong, B., J. W. M. Gold, A. E. Brown, M. Lange, R. Fried, M. Grieco, D. Mildvan, J. Giron, M. L. Tapper, C. W. Lerner, and D. Armstrong. 1984. Central-nervous-system toxoplasmosis in homosexual men and parenteral drug abusers. Ann. Intern. Med. **100**:36–42.

CHAPTER 106

Trichinellosis

H. R. GAMBLE and K. DARWIN MURRELL

Disease: Trichinellosis (trichinosis; trichiniasis).
Etiologic Agent: *Trichinella spiralis.*
Source: Domestic swine, feral swine, bear, horse, walrus, other wild animals by ingestion of raw or undercooked meat harboring infective *T. spiralis* larvae.
Clinical Manifestations: Abdominal pain, diarrhea (intestinal phase); fever, myalgia, malaise, periorbital edema (muscular phase). Severity from mild to acute is dependent on number of larvae ingested.
Pathology: Inflammatory enteritis, vasculitis, and hemorrhage in eyes, brain, pericardium, endocardium, lungs, muscle fiber necrosis, eosinophilia.
Laboratory Diagnosis: Presumptive diagnosis based on history of exposure and clinical signs; confirmation by serology, either bentonite flocculation or ELISA test; definitive diagnosis by muscle biopsy.
Epidemiology: Trichinellosis is distributed worldwide, particularly in temperate and arctic regions. Where prevalence in swine is low, it is maintained in wild animal reservoirs. Human infection most often occurs in groups of people exposed to a common source, often by consumption of ethnic dishes requiring minimal cooking or curing.
Treatment: Removal of intestinal worms by mebendazole, thiabendazole, levamisole, or pyrantel. Therapy for muscle phase using mebendazole coupled with corticosteroids to reduce inflammation. Specific treatment for complications.
Prevention and Control: Eradication in domestic animals through surveillance and improved management. Prevention of disease through commercial meat processing or proper cooking by consumer.

Description of Disease

Trichinellosis (trichinosis, trichiniasis) is caused by the parasitic nematode *Trichinella spiralis,* which completes its life cycle in a single host (Fig. 1). Infective larvae develop to adult worms in the intestinal epithelium of the host and produce the disseminating newborn larval stage which migrates to skeletal muscle fibers and matures to the infective first stage. Infection with *Trichinella* results exclusively from ingestion of muscle tissue containing infective larvae. Virtually all mammals including humans are susceptible to infection, although not all to the same degree.

The distribution of trichinellosis is worldwide, although it is most frequently reported in temperate and arctic regions. In countries where the prevalence of infection is low in domestic swine, human disease results primarily from ingestion of infected game meats.

Four subspecies of *Trichinella* have been identified (Lichtenfels et al., 1983). *Trichinella spiralis spiralis,* the domestic pig type; *T. s. nativa,* an arctic type; *T. s. nelsoni,* a tropical type; and *T.* s. *pseudospiralis,* a non-cyst-forming type. These *Trichinella* types differ in infectivity for various hosts; their relative pathogenicity in human infection is not known.

Clinical Manifestations

Trichinellosis can be divided into three phases: the intestinal, the muscular, and the convalescent. The time of onset of symptoms of each phase of trichinellosis following exposure is variable and depends largely on the number of larvae ingested.

In cases of exposure to large numbers of larvae (1,000 or more), abdominal symptoms can develop as early as 2 to 7 days after ingestion (Gould, 1970a). Abdominal disease is characterized by pain, nausea, vomiting, loss of appetite, mild fever, and diarrhea caused by inflammation of the intestinal mucosa where worms have penetrated and established. The duration of intestinal disease depends on the persistence of adult *T. spiralis*; without treatment, worms may survive for 6 to 8 weeks in the intestine.

In most cases where moderate or low numbers of parasites are ingested, the abdominal disease is mild or absent, but the general trichinosis syndrome or muscular phase develops. Symptoms resulting from a strong inflammatory response associated with muscle invasion include fever, myalgia, periorbital edema, malaise, and eosinophilia. Muscles most severely affected include the extraocular muscles; masseters; muscles of the larnyx, tongue, diaphragm, and neck; intercostals; and muscular attachments to joints and tendons (Murrell and Nelson, 1984). Other symptoms include facial edema, headache, fainting, urticaria, splinter hemorrhages in the nailbeds, conjunctivitis, loss of appetite, hoarseness, dysphagia, dyspnea, and edema of the legs. Symptoms may develop from 1 week to several weeks following exposure. Generally, the earlier the development of symptoms, the more severe the disease.

Complications occurring during the muscular phase result from ectopically migrating larvae or allergic reactions to parasite products. Pulmonary symptoms result from involvement of respiratory muscles. Neurologic complications caused by larval migration through the brain include foci of intracerebral hemorrhage and meningitis. Other neurologic symptoms include dizziness, ataxia, hysteria, and psychotic disturbances (Murrell and Nelson, 1984). Myocarditis is the most frequent complication of trichinosis and the most common cause of death. Larval migration through the myocardium results in dysrhythmia, tachycardia, hypotension, elevated venous pressure, and peripheral edema (Murrell and Nelson, 1984). Damage to the myocardium may result in chronic cardiac dysfunction.

The convalescent phase of trichinellosis is associated with a gradual cessation of symptoms. However, muscular symptoms may persist for several years owing to the presence of live larvae in muscular cysts. Rehabilitation during the convalescent phase includes physical therapy as well as mental therapy to overcome the feelings associated with the knowledge that the parasites remain in the patient's tissue.

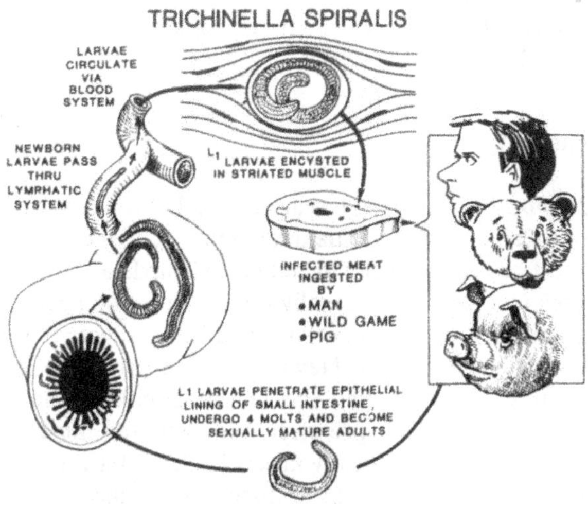

FIG. 1. Life cycle of *Trichinella spiralis*.

The severity of clinical trichinellosis has been classified as asymptomatic, mild, moderate, or severe by Kassur and Januszkiewicz (1968).

Asymptomatic Trichinellosis

Exposure to the disease is confirmed. Serologic tests are positive, and there may be transient eosinophilia. No additional symptoms are present.

Mild Trichinellosis

History of exposure, positive serology, and eosinophilia. A complex of symptoms are present but mildly expressed; fever < 38°C for less than 1 week. Recovery is rapid, less than 3 weeks, and hospitalization is not usually required.

Moderate Trichinellosis

Symptoms, fever, myalgia, and periorbital edema are well expressed with temperature > 38°C for up to 2 weeks. Hypoalbuminemia may occur. Recovery in 5 to 7 weeks with complications rarely occurring. Hospitalization is more likely.

Severe Trichinellosis

A full complex of intensely expressed symptoms are displayed with fever > 39°C for more than 2 weeks. Hypoalbuminemia is more pronounced (< 2.5 g). Allergic vasculitis results in edema and hemorrhage, particularly in the eyes. Recovery is slow, and neurological, cardiovascular, or pulmonary complications are common. Larval densities in muscle tissue in severe trichinellosis range from several hundred to several thousand per gram. Hospitalization is required.

Pathology

T. spiralis developing larvae and adult worms cause an inflammatory enteritis in the mucosa of the small intestine, occasionally resulting in diarrhea. Gross intestinal pathology, usually restricted to the ileum, is limited to small ulcerations and hemorrhagic erosions (Gould, 1970b). Newborn larvae produced in the intestine move into the general circulation and are carried to their preferred site—striated muscle tissue. Penetration of myofibers is characterized by damage due to mechanical disruption of the sarcolemma (Despommier, 1976). Muscle damage is initially related to the intensity of infection. Electromyography may demonstrate acute myositis or prolonged diffuse myopathic dysfunction (Pawlowski, 1983). Pathological mechanisms associated with fever and myalgia are not well characterized. The onset of myalgia and fever is often accompanied by allergic vasculitis resulting in fluid leakage and hemorrhage (Pawlowski, 1983). Fluid leakage is most generally seen as periorbital edema, and hemorrhages are often seen in the bulbar conjunctiva and fingernail beds.

Laboratory tests generally show marked eosinophila (15 to 89% of WBCs) and leukocytosis (Gould, 1970c). Serum may show hypoalbuminemia, hypokalemia, and elevated levels of creatine phosphokinase and 1,6-diphosphofructoaldolase (Pawlowski, 1983).

Severe trichinellosis often is accompanied by complications due to ectopically migrating larvae or hypersensitivity reactions. Complications include ocular, neurological, cardiovascular, and pulmonary manifestation. Myocarditis is the major cause of death due to trichinellosis (Andy et al., 1977).

Diagnostic Features

Trichinellosis is a difficult disease to diagnose owing to generally vague symptoms. Differential diagnosis requires consideration of a variety of diseases with characteristics similar to the various stages of trichinellosis, the abdominal stage, the general trichinellosis syndrome, allergic manifestations, and an array of complications. Trichinellosis should be considered when symptoms of the general trichinellosis syndrome develop (fever, myalgia, periorbital edema), when there is a history of exposure, or when a high eosinophilia develops. Diagnosis is based on (1) history of exposure, (2) presence of symptoms, (3) laboratory testing including serology, and (4) direct demonstration of parasites (Pawlowski, 1983). Definitive diagnosis can only be accomplished by demonstration of the parasite in muscle, a procedure requiring biopsy of 0.5 to 1.0 g of muscle, usually deltoid or gastrocnemius tissue. Muscle biopsy will be positive as early as 2 weeks following exposure to *T. spiralis*. Serology by bentonite flocculation and ELISA has proved useful for confirmation but generally becomes positive only several weeks following exposure.

Prognosis

The severity of trichinellosis is dependent on several factors including the number of larvae ingested and the age and physical health of the individual. Death occurs in a small percentage of cases generally from myocardial, neurological, or pulmonary involvement (Andy et al., 1977). In mild and moderate cases, recovery is complete within a few months of onset. In moderate to severe cases, muscular pain, weakness, and fatigue may persist for several years following infection (Gould, 1970a). Myocardial involvement requires particular attention, because unsuccessful attempts by newborn larvae to penetrate the heart muscle may result in complications if the patient becomes too active before complete healing has occurred (Murrell and Nelson, 1984).

Epidemiologic Aspects

Occurrence and Transmission

Trichinellosis occurs worldwide. Prevalence of human disease varies considerably among countries according to prevalence rates in pigs and wildlife, meat inspection requirements, and eating habits. In Denmark, no human trichinellosis has been reported in 75 years, but in Poland the number of annual cases has exceeded 1,000 (Kim, 1983). In the United States, the number of cases of human trichinellosis reported to the Centers for Disease Control has declined from approximately 400 per year in the 1940s to approximately 100 per year in the past decade (Schantz, 1983; Schantz et al., 1977). However, considerable evidence indicates that recorded clinical cases represent only a small portion of human infections. Cadaver surveys in the United States have demonstrated infection rates of 16.1% of the population in 1943–44 (Wright et al., 1943, 1944) and 4.2% in the period 1966–70 (Zimmerman et al., 1973). Similar surveys in other countries indicate similar disproportionate ratios of actual infection rates versus clinical cases (Kim, 1983). These disparities suggest that most cases of human trichinellosis are asymptomatic or undiagnosed. The most recent national survey of prevalence of trichinellosis in swine indicates a rate of 0.1% in the United States (Zimmerman, 1975). Regional surveys have demonstrated rates as high as 0.73% (Schad et al., 1985a,b).

Human trichinellosis is acquired exclusively by ingestion of raw or undercooked unprocessed meat containing infective trichinae. The major source of human infection in the United States is pork; however, bear meat, feral swine, and walrus meat have also been implicated in outbreaks (Schantz, 1983). In some countries, where pork is specifically tested for *Trichinella,* most human cases now result from eating wild mammals. In general, any carnivore is a potential source of trichinellosis, so all meat from carnivorous game requires thorough cooking. Several large outbreaks of trichinellosis in France and Italy have been traced to horsemeat; however, the sources of these infections were not determined (Ancelle et al., 1985).

Human trichinellosis does not follow prevalence patterns based on age, sex, or seasonal distribution (Schantz, 1983). However, there is a pattern of group outbreaks in which a number of individuals are exposed to a common source of infection. Common-source outbreaks are often linked to speciality foods associated with holidays or celebrations. Ethnic dishes frequently serve as sources of infection, especially sausage in its many variations. Outbreaks also occur from ingestion of sausage or hamburger prepared from game but "spiked" with pork; in these instances, the source may be domestic pork or wild animal meat.

Susceptibility and Resistance

Trichinellosis is acquired exclusively by the ingestion of trichinous meat. No general patterns of susceptibility based on sex or age are apparent in the population, although children appear less prone to complications and recover more quickly (Gould, 1970a; Pawlowski, 1983). Trichinellosis appears milder during pregnancy but more severe in lactating women; infection appears to be more detrimental to the fetus than to the mother (Pawlowski, 1983). Racial differences in susceptibility may exist, as suggested by high worm burdens found in Kenyans without clinical signs (Murrell and Nelson, 1984).

Etiologic Agent

Characterization

Trichinella spiralis is a parasitic nematode comprising the single genus of the family *Trichinellidae,* Ward 1907. *T. spiralis* has been differentiated into several subspecies—*T. s. spiralis, T. s. nativa, T. s. nelsoni,* and *T. s. pseudospiralis* (Lichtenfels et al., 1983)—based on distribution (Dick, 1983) and biological (Dick, 1983) and biochemical properties (Curran et al., 1985; Dick et al., 1985; Flockhart et al., 1982; Mydzynski and Dick, 1985). *T. s. spiralis* is considered the subspecies associated with the domestic pig and is most generally associated with human disease. The role of the other subspecies in human disease is unclear; many isolates from wild animals are poorly infective for domestic swine (Murrell et al., 1985) and their infectivity for humans cannot be tested experimentally.

Trichinella spiralis has a direct life cycle completing its development in a single host (Fig. 1). Following ingestion of trichinous meat, the infective larvae are digested free of the muscle cysts and invade epithelial tissue in the upper portion of the small intestine. The parasite develops through four molts to reach the adult stage in about 4 to 5 days following infection, at which time the male and female adult worms mate, and the females produce live larvae. Newborn larvae migrate via the circulatory system to skeletal muscle tissue, where they penetrate individual muscle cells. Newborn larvae develop to the mature infective first-stage larvae within the muscle capsule by 17 to 21 days after initial exposure. Infective larvae can remain viable for several years encapsulated in muscle tissue in some hosts, although calcification appears to occur much earlier in rodents.

Diagnosis

The differential diagnosis of human trichinellosis is based on (1) history of exposure, (2) physical symptoms, (3) laboratory testing including serology, and (4) direct demonstration of the parasite (Pawlowski, 1983).

History of Exposure

A history of exposure to infected meat is an important diagnostic tool in identifying trichinellosis. Ascertaining exposure is most difficult in isolated cases or in cases where other individuals in a common-source outbreak have not been identified. When symptoms suggest trichinellosis is possible, an effort should be made to determine potential exposure to any raw or undercooked meat, including pork, bear, or other game meats. The time of exposure should be correlated with the phase of disease; for example, symptoms indicative of latter stages of the disease would indicate that a longer period of time had elapsed since exposure. If suspected meat can be recovered, large samples should be examined by the digestion technique (Zimmerman, 1983) for the presence of *T. spiralis* larvae. An early identification of the source of an infectious meal may prevent additional exposures.

Physical Symptoms

Classical symptoms of trichinellosis including fever, myalgia, and periorbital edema are present in almost all infected individuals. A variety of other symptoms may be present (Gould, 1970a); however, none are pathognomonic for trichinellosis.

Laboratory Tests

The most characteristic feature of trichinellosis in laboratory tests is a marked eosinophilia (Gould, 1970c). Eosinophilia (500 cells/mm^3 or 5% of peripheral blood leukocytes) occurs 2 to 3 weeks following exposure and persists throughout the course of the disease. In some cases of fatal trichinellosis, eosinophilia has been absent in the latter stages of disease (Gould, 1970c).

Other laboratory findings with less value in diagnosis include leukocytosis, elevated levels of creatine phosphokinase, lactic dehydrogenase, 1,6-diphosphofructoaldolase, and hypoalbuminemia (Gould, 1970c; Pawlowski, 1983).

Serologic Testing

Specific serologic tests can be extremely valuable in providing confirmatory diagnosis of trichinellosis. A variety of tests have been used for the diagnosis of trichinellosis summarized by Kagan and Norman (1970) including indirect immunofluorescence (Jackson, 1959; Sadun et al., 1962), passive hemagglutination (Kagan and Bargai, 1956), bentonite flocculation (Sadun and Norman, 1955), and ELISA (Ljungstrom, 1974). The most widely used tests in the United States are the bentonite flocculation (BF) and the enzyme-linked immunosorbent assay (ELISA).

The BF test is based on the aggregation of *T. spiralis* antigen-coated bentonite particles in the presence of specific antibodies (Kagan and Norman, 1970). The bentonite particles are typically coated with an acid-soluble extract of *T. spiralis* muscle larvae, called Melcher's antigen (Melcher, 1943). Serial dilutions of serum are mixed with coated bentonite particles on a slide and allowed to react for 15 min. Greater than 50% agglutination of particles is considered a positive result. The results of the BF test are expressed as endpoint titer of flocculation; a titer of 1:5 or higher is considered diagnostic for trichinellosis.

The ELISA test is based on the detection of specific antibodies binding to *T. spiralis* antigens coated onto the plastic wells of microtiter plates. Antigens used in the ELISA include Melcher's antigen, a partially purified stichosome antigen, or stichosome products collected from in vitro-cultured *T. spiralis* muscle larvae (Gamble et al., 1983; Seawright et al., 1983; Richards et al., 1985). ELISA values are compared to known negative and positive controls, and results are expressed as endpoint titers.

Both the BF test and ELISA are rapid and simple to perform. The specificity of these tests is excellent, using the BF test with Melcher's antigen and the ELISA with culture-derived secretory antigens (Richards et al., 1985). Some reports suggest that ELISA is slightly more sensitive than BF, detecting positive sera in the second week of disease, as compared to the third week for BF (Pawlowski, 1983); however, other studies suggest no difference in the sensitivity of the two tests (Richards et al., 1985). In general, the temporal appearance of detectable antibody titers is directly related to the intensity of infection but not necessarily the clinical expression of disease. No differences have been observed in the temporal appearance of IgM and IgG antibodies (Feldmeier et al., 1986). Antibody titers in the BF and ELISA may persist for a year or longer after infection. One drawback to serologic testing is that detectable antibodies are not produced in the early stages ($<$ 2 weeks) of infection.

Parasite Detection

Muscle biopsy is the only means for direct verification of the presence of *T. spiralis*. Larvae may be detected in tissue as early as 2 weeks after infection (Gabryel et al., 1974; Gerwel et al., 1970). A biopsy of 0.5 to 1.0 g of muscle from the deltoid or gastrocnemius should be examined by several procedures: (1) direct microscopic inspection of compressed tissue pieces; (2) histologic sectioning; and (3) digestion of tissue to recover larvae. Digestion of the muscle in artificial gastric fluid (1% pepsin–1% HCl at 37°C for 1 to 4 h) will release any larvae present and is the method of choice. Larvae in the early stages of infection will not resist digestion, so direct inspection is also required. Digestion will give quantitative results as to infection level and will aid in determining a course of treatment.

Treatment, Prevention, and Control

Treatment

Treatment for trichinellosis will vary depending on the severity of the disease and related complications. In all cases, treatment should be applied to remove adult worms from the intestine. In moderate and severe cases, additional treatment is necessary to suppress tissue inflammatory responses and kill larvae in muscle tissue.

Several drugs are effective against adult worms in the intestine (Medical Letter, 1986; WHO, 1986) including mebendazole (5 mg/kg/day, 200 mg minimum/day, for 5 days [WHO]; 200–400 mg t.i.d. for 3 days, then 400–500 mg t.i.d. for 10 days [Medical Letter]), thiabendazole (25 mg/kg b.i.d., maximum of 3 g/day, for 5 days), levamisole (2.5 mg/kg/day, maximum 150 mg/day, for 3 days), and pyrantel (10 mg/kg/day for 3–5 days). Levamisole and pyrantel will only remove adult worms for the intestine and are not effective against the muscle stages (Campbell and Denham, 1983). The benzimidazole compounds mebendazole and thiabendazole are active against *Trichinella* larvae, and their use may provoke hypersensitive reactions (Campbell and Denham, 1983); therefore, these compounds should be used in conjunction with corticosteroids. Because adult worms remain in the intestine for 6 weeks or longer, treatment to remove these worms should be applied in all cases regardless of the severity of the disease.

For treatment of acute, severe trichinellosis, the WHO Guidelines (1986) recommend corticosteroids (prednisolone, 40–60 mg/day) and mebendazole (5 mg/kg/day, in 2 doses, 400 mg maximum/day) until fever and allergic symptoms disappear. Although treatment of muscle stage larvae is less effective than treatment for intestinal worms, mebendazole therapy has shown considerable promise in ridding muscle tissue of larvae (Ozeretskovskaya et al., 1981; Pawlowski, 1983).

In mild or moderate trichinellosis, corticosteroids (prednisolone, 20–40 mg/day) are used only in patients with fever, allergic symptoms, high leukocytosis, or eosinophilia. In the absence of these symptoms, antipyretic or analgesic treatment is recommended (WHO, 1986). Corticosteroids and benzimidazoles are generally not recommended for pregnant women.

Prevention and Control

Several strategies exist for the prevention of human exposure to trichinellosis. Some countries, including those in the European Economic Community, inspect slaughtered hogs for trichinae. The United States has no mandatory program for the inspection of swine for trichinellosis. The U.S. Department of Agriculture regards all pork as potentially infected with trichinae and provides regulations governing the preparation of commercial "ready-to-eat" pork products (curing, cooking, freezing) and guidelines to the consumer for cooking and freezing fresh pork to inactivate the parasite. The consumer should cook all pork to at least 170°F (76.6°C) to allow for a safety margin, because the thermal death point for *T. spiralis* in pork is 135°F (57.2°C) if the meat is kept at that temperature for about 4 min. Pork less than 6 inches thick can also be rendered safe if frozen to 5°F (−15°C) for 20 days, −10°F (−23°C) for 10 days, or −20°F (−29°C) for 6 days. These guidelines should extend to cover all game meats that potentially harbor trichinae. Because many outbreaks are associated with ethnic dishes, intensified education efforts should be targeted at these high-risk groups (Stehr-Green and Schantz, 1986).

Prevention of human trichinellosis is most directly related to control of the disease in swine. Because *T. spiralis* has such a broad host range and is well established in wild-animal reservoirs, it is not possible to totally eradicate the parasite. However, specific programs can aid in drastically reducing the incidence in domestic swine, the major source of human infection; eradication from domestic swine is possible, as demonstrated in Denmark and England. Prevention of transmission of trichinellosis to swine can be substantially effected by application of prudent management practices. The dissemination of information to producers concerning modes of transmission, including sylvatic sources, should aid in overall control. Management practices aimed at reducing potential exposure of swine to trichinellosis include (1) thorough cooking of garbage, including household scraps, fed to hogs, (2) stringent rodent control, and (3) avoidance of exposure of live hogs to dead hog or wild-animal carcasses.

In addition to improved management, surveillance programs can be valuable in the identification of foci of swine trichinellosis. Serologic antemortem or slaughterhouse surveys coupled with farm eradication programs can help reduce the incidence of trichinellosis in hogs reaching the consumer.

Literature Cited

Ancelle, T., J. Dupouy-Camet, F. Heyer, C. Faurant, and J. Lapierre. 1985. Outbreak of trichinosis due to horsemeat in the Paris area. Lancet **ii:**660.

Andy, J. J., J. P. O'Connell, R. C. Daddario, and W. C. Roberts. 1977. Trichinosis causing extensive ventricular mural endocarditis with superimposed thrombosis. Am. J. Med. **63:**824–829.

Campbell, W. C., and D. A. Denham. 1983. Chemotherapy. p. 335–366. *In* W. C. Campbell (ed.), *Trichinella* and trichinosis. Plenum, New York.

Curran, J., D. L. Baillie, and J. M. Webster. 1985. Use of genomic DNA restriction fragment length differences to identify nematode species. Parasitology **90:**137–144.

Despommier, D. D. 1976. Musculature. p. 269–285. *In* C. R. Kennedy (ed.), Ecological aspects of parasitology. Elsevier North-Holland, Amsterdam.

Dick, T. A. 1983. Species and intraspecific variation. p. 31–73. *In* W. C. Campbell (ed.), *Trichinella* and trichinosis. Plenum, New York.

Dick, T. A., J. Curran, and G. Klassen. 1985. Genetics and molecular biology of *Trichinella*. p. 118–128. *In* C. W. Kim (ed.), Trichinellosis. State University of New York Press, Albany.

Feldmeier, H., H. Fischer, and G. Blaumeiser. 1987. Kinetics of humoral response during the acute and convalescent phase of human trichinosis. Zbl. Bakt. Hyg. A **264:**221–234.

Flockhart, H. A., S. E. Harrison, A. R. Dobinson, and E. R. James. 1982. Enzyme polymorphism in *Trichinella*. Trans. R. Soc. Trop. Med. Hyg. **76:**541–544.

Gabryel, P., C. Gerwel, L. Gustowska, W. Kociecka, and A. Pawlowski. 1974. Muscle biopsy in human trichinellosis. p. 388–393. *In* Proc. VIth Int. Cong. Infect. Parasit. Dis., Warsaw, Vol. II.

Gamble, H. R., W. R. Anderson, C. E. Graham, and K. D. Murrell. 1983. Diagnosis of swine trichinosis by enzyme-linked immunosorbent assay (ELISA) using an excretory-secretory antigen. Vet. Parasitol. **13:**349–361.

Gerwel, C., W. Kociecka, and Z. Pawlowski. 1970. Parasitologic examination of muscles several years after trichinosis. Epidemiol. Rev. **24:**262–269.

Gould, S. E. 1970a. Clinical manifestations: symptomatology. p. 269–306. *In* S. E. Gould (ed.), Trichinosis in man and animals. Charles C. Thomas, Springfield, IL.

Gould, S. E. 1970b. Anatomic pathology. p. 147–189. *In* S. E. Gould (ed.), Trichinosis in man and animals. Charles C. Thomas, Springfield, IL.

Gould, S. E. 1970c. Clinical pathology: diagnostic laboratory procedures. p. 190–221. *In* S. E. Gould (ed.), Trichinosis in man and animals. Charles C. Thomas, Springfield, IL.

Jackson, G. J. 1959. Fluorescent antibody studies of *Trichinella spiralis* infection. J. Infect. Dis. **105:**97–117.

Kagan, I. G., and U. Bargai. 1956. Studies on the serology of trichinosis with hemagglutination, agar diffusion tests, and precipitin ring tests. J. Parasitol. **42:**237–245.

Kagan, I. G., and L. G. Norman. 1970. The serology of trichinosis. p. 222–268. *In* S. E. Gould (ed.), Trichinosis in man and animals. Charles C. Thomas, Springfield, IL.

Kassur, B., and J. Januszkiewicz. 1968. Clinical classification of trichinosis. Epidemiol. Rev. **22:**134–139.

Kim, C. W. 1983. Epidemiology. II. Geographic distribution and prevalence. p. 445–500. *In* W. C. Campbell (ed.), *Trichinella* and trichinosis. Plenum, New York.

Lichtenfels, J. R., K. D. Murrell, and P. A. Pilitt. 1983. Comparison of three subspecies of *Trichinella spiralis* by scanning electron microscopy. J. Parasitol. **69:**1131–1140.

Ljungstrom, I. 1974. Antibody response to *Trichinella spiralis*. p. 449–459. *In* C. W. Kim (ed.), Trichinellosis. Intext Education Publishers, New York.

Medical Letter. 1986. Drugs for parasitic infections. Med. Lett. **28:**15.

Melcher, L. 1943. An antigenic analysis of *Trichinella spiralis*. J. Infect. Dis. **73:**31–40.

Murrell, K. D., and G. S. Nelson. 1984. Trichinosis. p. 689–695. *In* G. T. Strickland (ed.), Hunter's tropical medicine. Saunders, New York.

Murrell, K. D., D. A. Leiby, C. Duffy, and G. A. Schad. 1985. Susceptibility of domestic swine to wild animal isolates of *Trichinella spiralis*. p. 301–305. *In* C. W. Kim (ed.), Trichinellosis. State University of New York Press, Albany.

Mydzynski, L. J., and T. A. Dick. 1985. Use of enzyme polymorphisms to identify genetic differences in the genus *Trichinella*. J. Parasitol. **71:**671–673.

Ozeretskovskaya, N. N., T. M. Morenets, E. V. Pereverzeva, A. M. Bronstein, N. L. Veretennikova, M. O. Kolosova, A. M. Poverenny, V. K. Podgorodnichenko, and D. A. Kagorodin. 1981. Therapeutical properties of benzimidazoles in trichinellosis and the side effects of the treatment. p. 287–290. *In* C. W. Kim, E. J. Ruitenberg, and J. S. Teppema (ed.), Trichinellosis. Reedbooks, Chertsey, U.K.

Pawlowski, Z. S. 1983. Clinical aspects in man. p. 367–401. *In* W. C. Campbell (ed.), *Trichinella* and trichinellosis. Plenum, New York.

Richards, F., J. Fried, P. M. Schantz, D. Despommier, D. Murrell, and H. R. Gamble. 1985. The performance of four serological tests in the diagnosis of human trichinosis. Proc. Am. Soc. Trop. Med. Hyg. **34:**148.

Sadun, E. H., and L. Norman. 1955. The use of an acid soluble protein fraction in the flocculation test for the diagnosis of trichinosis. J. Parasitol. **41:**476–482.

Sadun, E. H., R. I. Anderson, and J. S. Williams. 1962. Fluorescent antibody test for the serological diagnosis of trichinosis. Exp. Parasitol. **12:**424–433.

Schad, G. A., M. Kelly, D. A. Leiby, K. Blumrick, C. Duffy, and K. D. Murrell. 1985a. Swine trichinosis in mid-Atlantic slaughterhouses: possible relationship to hog marketing systems. Prev. Vet. Med. **3:**391–399.

Schad, G. A., D. A. Leiby, C. H. Duffy, and K. D. Murrell. 1985b. Swine trichinosis in a New England slaughterhouse. Am. J. Vet. Res. **46:**2008–2010.

Schantz, P. M. 1983. Trichinosis in the United States 1947–1981. Food Technol. **37:**83–86.

Schantz, P. M., D. D. Juranek, and M. G. Schultz. 1977. Trichinosis in the United States, 1975: increase in cases attributed to numerous common source outbreaks. J. Infect. Dis. **136:**712–715.

Seawright, G. L., D. D. Despommier, W. Zimmerman, and R. S. Isenstein. 1983. Enzyme immunoassay for swine trichinellosis using antigens purified by immunoaffinity chromatography. Am. J. Trop. Med. Hyg. **32:**1275–1284.

Stehr-Green, J. K., and P. M. Schantz. 1986. Trichinosis in southeast Asian refugees in the United States. Am. J. Public Health **76:**1238–1239.

WHO. 1986. Guidelines on surveillance prevention, and control of trichinellosis. W. C. Campbell, R. B. Griffiths, A. Mantovani, Z. Matyas, and Z. S. Pawlowski (ed.). Document WHO/VPH/86.1. WHO, Geneva.

Wright, W. H., K. B. Kerr, and L. Jacobs. 1943. Studies on trichinosis. XV. Summary of the findings on *Trichinella spiralis* in a random sampling and other samplings of the population of the United States. Public Health Rep. **58:**1293–1313.

Wright, W. H., L. Jacobs, and A. C. Walton. 1944. Studies on trichinosis. XVI. Epidemiological considerations based on the examination for trichinae of 5,313 diaphragms from 189 hospitals and in 37 states and the District of Columbia. Public Health Rep. **59:**669–681.

Zimmerman, W. J. 1975. Trichinosis, p. 187–204. *In* J. A. Libby (ed.), Meat hygiene. Lea and Febiger, Philadelphia.

Zimmerman, W. J. 1983. Surveillance in swine and other animals by muscle-examination, p. 515–528. *In* W. C. Campbell (ed.), *Trichinella* and trichinosis. Plenum Press, New York.

Zimmerman, W. J., J. H. Steele, and I. G. Kagan. 1973. Trichiniasis in the U.S. population, 1966–1970: prevalence, epidemiologic factors. Health Serv. Rep. **88:**606–623.

CHAPTER 107

Trichomoniasis

HENRY M. MATHEWS

Disease: Trichomoniasis, trichomonal vaginitis.
Etiologic Agent: *Trichomonas vaginalis.*
Source: Direct contact with secretions containing the organism.
Clinical Manifestations: In women: vaginal itching and/or discharge which may be copious. In men: prostatitis or urethritis. Many infections (in men and women) are asymptomatic.
Pathology: Edema and inflammation of the cervical epithelium.
Laboratory Diagnosis: Demonstration of the organism in fresh or stained secretions. Cultivation in one of several culture media.
Epidemiology: Worldwide. Highest prevalence in women of childbearing age. Prevalence in men is poorly understood because of difficulties in diagnosis.
Treatment: Metronidazole.
Prevention and Control: Education about the disease and its transmission. Treatment of all infected persons.

Description of Disease

Trichomoniasis is an infection of the genitourinary tract caused by the flagellated protozoan *Trichomonas vaginalis*. Infection in women may be asymptomatic or characterized by vaginal itching and discomfort with or without an accompanying discharge (Fouts and Kraus, 1980). In men, infections are most often transient and asymptomatic, but occasionally persistent infection with prostatitis, urethritis, or epididymitis may occur (Krieger, 1981). The disease derives its public health significance from its worldwide distribution, the economic burden for treatment, and the social and psychological impact on patients with persistent infections.

Communicability and Transmissibility

Humans are the only natural hosts for *Trichomonas vaginalis,* and person-to-person spread is primarily through sexual contact, although infection may be acquired by contact with articles contaminated with infected vaginal discharges. Infections occur worldwide in all races, and prevalence is usually highest in the age group 16 to 35.

Clinical Manifestations

Trussell (1947) summarized the clinical presentation of vaginal trichomoniasis as reported in the literature up to that time and suggested that 20 to 65% of infected persons had no symptoms and less than 5% had severe symptoms. More recently, Krieger (1981) estimated that up to 25% of culture-proven infections occurred in asymptomatic women.

Typical symptoms usually associated with trichomoniasis in women include vaginal discharge, which may be copious, frothy, and malodorous; a "strawberry cervix"; and severe itching of the vulva and inner thighs (Honigberg, 1978; Hume, 1978). These symptoms may occur singly or in combination. Although most men are asymptomatic, some may complain of minor itching or penile discharge.

Atypical Infections

Trichomonas vaginalis has been isolated from atypical sites on several occasions. McLaren and coworkers (1983) described two cases of pulmonary infection with *T. vaginalis* in infants born to mothers with histories of vaginal infections with *Trichomonas*. Both infants developed significant respiratory symptoms that resolved (in 1 case after treatment with metronidazole) in time and after no trichomonads could be detected. Littlewood and Kohler (1966) described a *T. vaginalis* infection of the urinary tract in a newborn girl who had no evidence of vaginal infection. A case of pulmonary trichomoniasis was described by Rebhun (1964) in a man with chronic bronchitis, "hay fever," and a fever of unknown origin. Although the flagellate was not identified as to species, the patient reported frequent oral-vaginal contact. His symptoms did not recur after he altered his sexual practices and presumably was no longer exposed.

Pathology

The pathological effects associated with *T. vaginalis* infections result from a complex interrelationship between the strain of parasite and several host factors. A recent study correlating parasite pathogenicity and pathological changes in the cervical epithelium showed that approximately half of the cases had no detectable damage (Honigberg et al., 1984). Typical pathological changes associated with vaginal infections include "strawberry cervix," characterized by increased vascularity, edema, and inflammation of the cervical epithelium. Basal hyperactivity and squamous metaplasia of the endocervical mucosa are also seen (Honigberg, 1978). The sometimes marked cellular changes found in chronic trichomoniasis may complicate the interpretation of exfoliative cytology examinations. Although the prevalence of trichomoniasis is higher in women with cervical carcinoma than in the general population, the cause-and-effect relationship remains unclear (Krieger, 1981).

The pathology of *T. vaginalis* infections in males remains controversial. Most infected males are asymptomatic, but workers in Europe tend to consider *T. vaginalis* infection in the male a more significant disease than do their American counterparts (Krieger, 1981). In the United States, Gardner et al. (1986) used an immunoperoxidase technique to clearly identify *T. vaginalis* organisms in the prostate gland of five patients. The inflammatory changes were consistent with acute or chronic infections, and in one patient, organisms were seen mixed with inflammatory exudate near microabscesses.

Diagnostic Features

Trichomonas infection should be suspected in women presenting with vaginal discharge and/or itching. Many cases, however, are diagnosed incidentally during examination for cervical cytology or for other sexually transmitted diseases (STDs). Most men have no symptoms, but trichomoniasis should be considered in those who have urethritis, prostatitis, or epididymitis. Sexual partners of infected women should be considered infected.

Definitive diagnosis is made by demonstrating the motile organism in fresh secretions or by cultivating in one of several culture media. Although several serodiagnostic methods have been proposed, none has found wide acceptance. Diagnosis based solely on clinical criteria is unreliable owing to low sensitivity and specificity (Fouts and Kraus, 1980; McLellan et al., 1982).

Epidemiologic Aspects

Trichomonas vaginalis has a cosmopolitan distribution and ranks as one of the most frequently occurring sexually transmitted diseases; yet data on incidence and prevalence are often confusing (Krieger, 1981). Jirovec and Petru (1968) summarized prevalence findings in surveys done in Europe and reported rates as high as 37% in healthy female populations and 65% in clinic populations. Krieger's literature review (1981) yields similar data for the United States.

Prevalence in females is highest in the sexually active age groups, particularly among persons with multiple partners. Because most epidemiologic studies have been conducted without standardized sampling and diagnostic techniques, the true prevalence of *T. vaginalis* infections remains unclear. The difficulty of making the diagnosis in males makes estimates of true prevalence in men almost impossible.

Occurrence and Transmission

The distribution of *Trichomonas vaginalis* is worldwide, and no population group appears to have any degree of innate resistance to infection. Under normal circumstances, humans are the only natural host.

Transmission is almost always by sexual contact. The absence of a cystic stage leaves the organism susceptible to drying. Several studies (reviewed by

Honigberg, 1978; Neinstein et al., 1984) have demonstrated that *T. vaginalis* can survive brief exposures to tap water, mineral water, urine, and rather wide temperature changes, but only if protected from drying. Thus it has been possible to culture organisms from toilet seats, bath tubs, and damp clothing contaminated with trichomonads. It is, therefore, theoretically possible for infection to be transmitted from one female to another in this manner. It is less likely that infections in men would be acquired by these means.

Although nonsexual transmission has been suggested to explain cases of *T. vaginalis* in young girls, the possibility of sexual abuse (Jones et al., 1985; Neinstein et al., 1984) must be considered in many cases. Sexual abuse should be considered when young people (not yet sexually active) are diagnosed with *T. vaginalis* or any other sexually transmitted disease. A full workup for trichomoniasis (and other STDs) should be done in all cases of sexual abuse.

Etiologic Agent

Characterization

Trichomonas vaginalis is a pyriform flagellate approximately 10 to 30 μm long, with four anterior flagella and a trailing flagellum attached to an undulating membrane. A slender axostyle projects from the posterior. The motile organism "swims" with a characteristic jerky motion. When not motile, they often attach themselves to debris or to other organisms, forming clumps.

Trichomonads are aerotolerant anaerobes with a fermentative metabolism. They have cytoplasmic granules, termed hydrogenosomes, which function in the metabolism of pyruvate and in H_2 formation (Muller, 1980). The hydrogenosome is an organelle limited to a few eukaryotic species that lack mitochondria. The evolutionary implications of this organelle are interesting but fall outside the scope of this presentation.

Strain Differences

Pathogenicity

Isolates of *T. vaginalis* vary widely in their capacity to cause disease in the human host. As yet, we have no well-defined marker to measure differences in virulence. A number of assays have been developed or proposed to measure the pathogenic potential of an isolate. Honigberg (1961) developed a subcutaneous assay using the mouse to measure the intrinsic pathogenicity of *T. vaginalis* isolates. Using this assay, Honigberg and co-workers (1984) noted a strong relationship between subcutaneous abscess volume in mice caused by *T. vaginalis* and pathological changes in the cervical epithelium of the patient from whom the isolate was obtained.

The effects of *T. vaginalis* on cell cultures have been studied by several investigators (Alderete and Perlman, 1984; Honigberg, 1978; Krieger et al., 1985b). Cell destruction appears to be contact-dependent and related to in vivo measures of pathology. Krieger et al. (1983) have correlated beta-hemolytic activity of *Trichomonas* isolates to virulence. Warton and Honigberg (1983) measured lectin receptors of the surface of organisms from symptomatic and asymptomatic patients and showed that receptors for soybean agglutinin (SBA) and garden pea agglutinin (GPA) were more abundant on parasites with higher pathogenicity levels.

The clinical significance of differences in pathogenicity or of the in vitro correlates of pathogenicity are not clear. It is not known, for example, if an isolate maintains the same level of pathogenicity when transmitted to a different host. No studies have yet attempted long-term follow-up of patients from whom organisms with measured pathogenicity were isolated, nor is it known whether asymptomatic infections may become symptomatic because of changes in the host's physiology or in the parasite.

Antigenic Differences

Antigenic differences among isolates (strains) of *T. vaginalis* have been recognized for many years, first by European workers (for pertinent references see Honigberg, 1970, 1978). Recently Alderete (1983) used a highly sensitive radioimmunoprecipitation (RIP) assay combined with polyacrylamide gel electrophoresis (PAGE) to demonstrate antigenic differences among several isolates. Su-Lin and Honigberg (1983) used a quantitative fluorescent antibody method to measure the antigenic interrelationships among clones of five *T. vaginalis* isolates. They found that each isolate had unique as well as shared antigens, but they were not able to associate antigenic characteristics with either geographic distribution or pathogenicity assays. Torian et al. (1984) used a panel of monoclonal antibodies and demonstrated differences in reactivity among isolates with immunofluorescent (IF) and enzyme-linked immunosorbent assays (ELISAs). Krieger and co-workers (1985a), using their panel of nine monoclonal antibodies, showed different patterns of reactivity to a battery of 88 *Trichomonas* isolates from different parts of North America. Their results suggested that some of the antigenic differences were associated with the geographic origin of the isolate. Although the antigenic diversity of *T. vaginalis* has been clearly demonstrated by several workers with an array of different immunologic methods, no clear understanding has emerged of the role(s) of this diversity in the host-parasite relationship.

Isoenzyme Analysis

Isoenzyme analysis of cloned and uncloned isolates of *T. vaginalis* has shown that a number of zymodemes can be identified (Gradus and Matthews, 1985; Soliman et al., 1982). Intraspecific differences were noted in lactate dehydrogenase, glucose phosphate isomerase, phosphoglucomutase, malate dehydrogenase, hexokinase, and glucose-6-phosphate dehydrogenase in cloned cultures (Soliman et al., 1982). Working with uncloned populations, Gradus and Matthews (1985) found differences in the isoenzymes of esterase, superoxide dismutase, and acid phosphatase in four isolates. As yet, these isoenzyme differences have not been associated with virulence or other biological phenomena or with geographic distribution. The role, if any, of the various isoenzyme types remains to be defined.

Diagnosis

The two most commonly used diagnostic methods for *Trichomonas vaginalis* infection—clinical presentation and wet-mount examination of vaginal secretions—are very unreliable (Fouts and Kraus, 1980; Krieger, 1981; McLellan et al., 1982). The symptoms commonly associated with trichomoniasis, discharge and pruritis, are nonspecific, and reliance on these alone lead to overdiagnosis. The wet-mount method of examining fresh secretions from vaginal or urethral swabs is rarely false-positive, because it relies on observing the characteristic motility of the organism, but one may miss 50% of infected patients when using this technique.

Wet Mounts and Staining

The wet-mount preparation has been used as a primary diagnostic tool despite its lack of sensitivity. It remains an effective preliminary screening test but should not be used to assess cure when parasite numbers are low. Philip et al. (1987) compared the wet-mount method to a quantitative culture method and found the wet mount was positive only when parasite counts exceeded 10^5/ml of vaginal secretions.

A variety of staining methods have been used to enhance detection of trichomonads in fixed smears of secretions or of cytologic or histologic preparations. The acridine orange (AO) method (Fripp et al., 1976; Levett, 1980; Mason et al., 1976) appeared to be more sensitive than other staining methods including Giemsa, Leishman, and the Diff-Quick stains. The improved sensitivity of the AO method is due to the differential uptake of the dye by trichomonads. The Papanicolaou (Pap) stain is frequently used, but it may lose specificity owing to loss of morphologic detail (Perl, 1972). Unequivocal identification of *T. vaginalis* in cytologic or histologic preparations is difficult, because the commonly used stains do not distinguish the trichomonads from the surrounding cells. To overcome this difficulty, Bennett et al., (1980) and O'Hara et al. (1980) have used an immunoperoxidase technique to identify trichomonads in fixed material. This method identifies the organism by the presence of specific antigens on the surface, but it is too time-consuming and expensive for use in the routine diagnostic laboratory.

Chang et al. (1986) have developed and evaluated a pool of monoclonal antibodies to *T. vaginalis* in a direct fluorescence test to detect organisms in vaginal smears and found it to be more sensitive than the wet mount alone. This pool of antibodies has become the basis for a commercial test kit (Integrated Diagnostics, Inc.). The assay is reported to detect trichomonads at concentrations of 300 organisms or more per milliliter of vaginal or urethral discharge (package insert, Feb. 1986).

Culture Methods

Cultivation of the organism remains the "gold standard" for diagnosing trichomoniasis. Studies comparing the efficiency of culture methods to wet mounts have shown that culture detects about twice as many cases as does the wet mount alone (Fouts and Krause, 1980; Philip et al., 1987).

Various media are available for primary isolation and continued cultivation of *Trichomonas* (Bromke, 1986; Diamond, 1957;, Feinberg and Whittington, 1957; Linstead, 1981; Smith, 1983; Smith et al., 1982). Most cultures will show growth in 3 or 4 days, but incubation for a full 7 days is recommended to obtain the highest rate of true positives. Smith (1986) found that a blind passage of cultures judged negative at 7 days would slightly increase the positivity rate, but this is not a practical step for most laboratories. The same study confirmed that culture, although more expensive, detected twice the number of infections as wet mounts alone.

Serologic Methods

The application of serologic methods to the diagnosis of *T. vaginalis* infection began in the 1940s and included a variety of assays (see Honigberg, 1978). More recently, several serologic assays have been described, including indirect hemagglutination (Kuberski, 1978; Mathews and Healy, 1983), IF (Mason, 1979), and several ELISAs (Alderete, 1984; Cogne et al., 1985; Street et al., 1982). The sensitivity and specificity of these tests vary, partly because of the parasite strain used for antigen. None of these tests has received acceptance as a diagnostic tool perhaps because of the perceived accuracy and simplicity of the wet-mount examination. The role of serologic tests may lie in the epidemiologic assessment of trichomoniasis in selected populations. Serologic screening of apparently healthy female employees produced a seropositivity rate of 30%, suggesting a high level of trichomoniasis in the general population (Mathews and Healy, 1983).

Ackers and co-workers introduced a radioimmunoassay for detecting IgA antibody in vaginal secretions (1975) and semen samples (1978). The presence or absence of antibody in vaginal secretions could not be correlated with clinical states, but the authors noted a "slight suggestion" that low parasite counts were associated with the presence of specific antibody. ELISA tests have also been used to measure antibody in vaginal secretions (Alderete, 1984; Street et al., 1982; Su, 1982). Although IgG and IgA antibodies are detectable in vaginal washings, they appear to have no diagnostic significance.

Watt et al., (1986) described an antigen capture assay that, although not a serologic assay in the strict sense, is designed to detect *Trichomonas vaginalis* antigens in vaginal fluids. The authors claim that the assay is sensitive (able to detect as few as 100 organisms/ml) and specific and requires only 1 h to complete. This type of assay holds great promise as a diagnostic test that will gain wide acceptance.

Treatment, Prevention, and Control

Treatment

Metronidazole was introduced in the United States in 1963 and remains the only effective specific therapy for trichomoniasis (Lossick, 1982). For most cases, 2.0 g of metronidazole in a single dose (or 250 mg 3 times daily for 7 days) will eradicate the infection. Sexual partners should be treated simultaneously with the same dose.

Some infections are not cured with the standard dose of metronidazole but may be cured with a second course of therapy. Although investigators have shown that some isolates of *T. vaginalis* are relatively resistant to metronidazole (Krajden et al., 1986; Muller et al., 1980); most such infections can be eradicated with large doses of metronidazole (Lossick et al., 1986).

Treatment is recommended for all cases of trichomoniasis, in both men and women. Asymptomatic sex partners of infected persons should also be treated. Treatment with metronidazole should be withheld during the first trimester of pregnancy and during the first 6 weeks of life (Lossick, 1982).

Prevention and Control

Prevention and control measures for *T. vaginalis* infection include educating the public about the disease and its mode of transmission and encouraging infected persons and their partners to seek treatment and to avoid sexual contact. Patients infected with flagellates that do not respond to the standard regimen of metronidazole should be retreated with higher doses if necessary to eradicate the parasite (Lossick, 1982).

Literature Cited

Ackers, J. P., W. H. R. Lumsden, R. D. Catterall, and R. Coyle. 1975. Antitrichomonal antibody in the vaginal secretions of women infected with *T. vaginalis*. Br. J. Vener. Dis. **51**:319–323.

Ackers, J. P., R. D. Cattarall ,W. H. R. Lumsden, and A. McMillan. 1978. Absence of detectable local antibody in genitourinary tract secretions of male contacts of women infected with *Trichomonas vaginalis*. Br. J. Vener. Dis. **54**:168–171.

Alderete, J. F. 1983. Antigen analysis of several pathogenic strains of *Trichomonas vaginalis*. Infec. Immun. **39**:1041–1047.

Alderete, J. F. 1984. Enzyme-linked immunosorbent assay for detecting antibody to *Trichomonas vaginalis:* use of whole cells and aqueous extract as antigen. Br. J. Vener. Dis. **60**:164–170.

Alderete, J. F., and E. Pearlman. 1984. Pathogenic *Trichomonas vaginalis* cytotoxicity to cell culture monolayers. Br. J. Vener. Dis. **60**:99–105.

Bennett, B. D., J. Bailey and W. A. Gardner, Jr. 1980. Immunocytochemical identification of trichomonads. Arch. Pathol. Lab. Med. **104**:247–249.

Bromke, B. J. 1986. A serum-free, lipid-supplemented medium for the growth of *Trichomonas vaginalis*. J. Microbiol. Methods **6**:55–59.

Chang, T. H., S. Y. Tsing, and S. Tzeng. 1986. Monoclonal antibodies against *Trichomonas vaginalis*. Hybridoma **5**:43–51.

Cogne, M., P. Brasseur, and J. J. Ballet. 1985. Detection and characterization of serum antitrichomonal antibodies in urogenital trichomoniasis. J. Clin. Microbiol. **21**:588–592.

Diamond, L. S. 1957. The establishment of various trichomonads of animals and man in axenic cultures. J. Parasitol. **43**:488–490.

Feinberg, J. G., and M. J. Whittington. 1957. A culture medium for *Trichomonas vaginalis* Donne and species of *Candida*. J. Clin. Pathol. **10**:327–329.

Fouts, A. C., and S. J. Kraus. 1980. *Trichomonas vaginalis:* reevaluation of its clinical presentation and laboratory diagnosis. J. Infect. Dis. **141**:137–143.

Fripp, P. J., P. R. Mason, and H. Super. 1975. A method for the diagnosis of *Trichomonas vaginalis* using acridine orange. J. Parasitol. **61**:966–967.

Gardner, W. A., Jr., D. E. Culberson, and B. D. Bennett. 1986. *Trichomonas vaginalis* in the prostate gland. Arch. Pathol. Lab. Med. **110**:430–432.

Gradus, M. S., and H. M. Matthews. 1985. Electrophoretic analysis of soluble proteins and esterase, superoxide dismutase and acid phosphotase isoenzymes of members of the protozoan family *Trichomonadidae*. Comp. Biochem. Physiol. **81B**:229–233.

Honigberg, B. M. 1961. Comparative pathogenicity of *Trichomonas vaginalis* and *Trichomonas gallinae* to mice. I. Gross pathology, quantitative evaluation of virulence, and some factors affecting pathogenicity. J. Parasitol. **47**:545–571.

Honigberg, B. M. 1970. Trichomonads. *In* G. J. Jackson, R. Herman, and L. Singer (ed.), Immunity to parasitic animals, Vol. 2. Appleton, New York.

Honigberg, B. M. 1978. Trichomonads of importance in human medicine. p. 275–454. *In* J. P. Kreier (ed.), Parasitic protozoa, Vol. 2. Academic Press, New York.

Honigberg, B. M., P. K. Gupta, M. R. Spence, J. K. Frost, K. Kuczynska, L. Choromanski, and A. Warton, 1984. Pathogenicity of *Trichomonas vaginalis:* cytopathologic and histopathologic changes of the cervical epithelium. Obstet. Gynecol. **64**:179–184.

Hume, J. C. 1978. Trichomoniasis—eight reasons why you should take it seriously. Med. Times **106**:59–63.

Jirovec, O., and M. Petru. 1968. *Trichomonas vaginalis* and trichomoniasis. Adv. Parasitol. **6**:117–188.

Jones, J. G., T. Yamauchi, and B. Lambert. 1985. *Trichomonas vaginalis* infestation in sexually abused girls. Am. J. Dis. Child. **139**:846–847.

Krajden, S., J. G. Lossick, E. Wilk, J. Yang, J. S. Keystone, and K. Elliot. 1986. Persistent *Trichomonas vaginalis* infection due to a metronidazole-resistant strain. Can. Med. Assoc. J. **134**:1373–1374.

Krieger, J. N. 1981. Urologic aspects of trichomoniasis. Invest. Urol. **18**:411–417.

Krieger, J. N., M. A. Poisson, and M. F. Rein. 1983. Beta-hemolytic activity of *Trichomonas vaginalis* correlates with virulence. Infect. Immun. **41**:1291–1295.

Krieger, J. N., K. K. Holmes, M. R. Spence, M. F. Rein, W. M. McCormack, and W. R. Tam. 1985a. Geographic variation among isolates of *Trichomonas vaginalis:* demonstration of antigenic heterogeneity by using monoclonal antibodies and the indirect immunofluorescence technique. J. Infect. Dis. **152**:979–984.

Krieger, J. N., J. I. Ravdin, and M. F. Rein. 1985b. Contact-dependent cytopathogenic mechanisms of *Trichomonas vaginalis*. Infect. Immun. **50**:778–786.

Kuberski, T. 1978. Evaluation of the indirect hemagglutination technique for study of *Trichomonas vaginalis* infections, particularly in men. Sex. Transm. Dis. **5**:97–102.

Levett, P. N. 1980. A comparison of five methods for the detection of *Trichomonas vaginalis* in clinical specimens. Med. Lab. Sci. **37**:85–88.

Linstead, D. 1981. New defined and semi-defined media for the cultivation of the flagellate *Trichomonas vaginalis*. Parasitology **83**:125–137.

Littlewood, J. M., and H. G. Kohler. 1966. Urinary tract infection by *Trichomonas vaginalis* in a newborn baby. Arch. Dis. Child. **41**:693–695.

Lossick, J. G. 1982. Treatment of *Trichomonas vaginalis* infections. Rev. Infect. Dis. (Suppl.)**4**:S801–S818.

Lossick, J. G., M. Muller, and T. E. Gorrell. 1986. In vitro drug susceptibility and doses of metronidazole required for cure in cases of refractory vaginal trichomoniasis. J. Infect. Dis. **153**:948–955.

Mason, P. R. 1979. Serodiagnosis of *Trichomonas vaginalis* infection by the indirect fluorescent antibody test. J. Clin. Pathol. **32**:1211–1215.

Mason, P. R., H. Super, and P. J. Fripp. 1976. Comparison of four techniques for the routine diagnosis of *Trichomonas vaginalis* infection. J. Clin. Pathol. **29**:154–157.

Mathews, H. M., and G. R. Healy. 1983. Evaluation of two serologic tests for *Trichomonas vaginalis* infection. J. Clin. Microbiol. **17**:840–843.

McLaren, L. C., L. E. Davis, G. R. Healy, and C. G. James. 1983. Isolation of *Trichomonas vaginalis* from the respiratory tract of infants with respiratory disease. Pediatrics **71**:888–890.

McLellan, R., M. R. Spence, M. Brockman, L. Raffel, and J. L. Smith. 1982. The clinical diagnosis of trichomoniasis. Obstet. Gynecol. **60**:30–34.

Muller, M. 1980. The hydrogenosome. *In* G. W. Gooday, D. Lloyd, and A. P. J. Trinci (ed.), The eukaryotic microbial cell. Cambridge University Press, Cambridge, U.K.

Muller, M., J. G. Meingasser, W. A. Miller, and W. J. Ledger. 1980. Three metronidazole-resistant strains of *Trichomonas vaginalis* from the United States. Am. J. Obstet. Gynecol. **138**:808–812.

Neinstein, L. S., J. Goldenring, and S. Carpenter. 1984. Nonsexual transmission of sexually transmitted diseases: an infrequent occurrence. Pediatrics **74**:67–76.

O'Hara, C. M., W. A. Gardener, Jr., and B. D. Bennett. 1980. Immunoperoxidase staining of *Trichomonas vaginalis* in cytologic material. Acta Cytol. **24**:448–451.

Perl, G. 1972. Errors in the diagnosis of *Trichomonas vaginalis* infection as observed among 1199 patients. Obstet. Gynecol. **39**:7–9.

Philip, A., P. Carter-Scott, and C. Rogers. 1987. An agar culture technique to quantitate *Trichomonas vaginalis* from women. J. Infect. Dis. **155**:304–308.

Rebhun, J. 1964. Pulmonary trichomoniasis associated with a fever of unknown origin. Calif. Med. **100**:443–444.

Smith, R. F. 1983. Viability of *Trichomonas vaginalis* in vitro at four temperatures. J. Clin. Microbiol. **18**:834–836.

Smith, R. F. 1986. Incubation time, second blind passage, and cost considerations in the isolation of *Trichomonas vaginalis*. J. Clin. Microbiol. **24**:139–140.

Smith, R. F., R. Welch, R. Stickney, S. Venerable, S. Fredrickson, K. Shima, and P. Hilton. 1982. Serum-free modified Trichosel medium for the isolation of *Trichomonas vaginalis*. Curr. Microbiol. **7**:153–156.

Soliman, M. A., J. P. Ackers, and R. D. Catterall. 1982. Isoenzyme characterisation of *Trichomonas vaginalis*. Br. J. Vener. Dis. **58**:250–256.

Street, D. A., D. Taylor-Robinson, J. P. Ackers, N. F. Hanna, and A. McMillan. 1982. Evaluation of an enzyme-linked immunosorbent assay for the detection of antibody to *Trichomonas vaginalis* in sera and vaginal secretions. Br. J. Vener. Dis. **58**:330–333.

Su, K. E. 1982. Antibody to *Trichomonas vaginalis* in human cervicovaginal secretions. Infect. Immun. **37**:852–857.

Su-Lin, K. E., and B. M. Honigberg. 1983. Antigenic analysis of *Trichomonas vaginalis* strains by quantitative fluorescent antibody methods. Z. Parasitenkd. **69**:161–181.

Torian, B. E., R. J. Connelly, R. S. Stephens, and H. H. Stibbs. 1984. Specific and common antigens of *Trichomonas vaginalis* detected by monoclonal antibodies. Infect. Immun. **43**:270–275.

Trussell, R. E. 1947. *Trichomonas vaginalis* and trichomoniasis. Charles C. Thomas, Springfield, IL.

Warton, A., and B. M. Honigberg. 1983. Analysis of surface saccharides in *Trichomonas vaginalis* strains with various pathogenicity levels by fluorescein-conjugated plant lectins. Z. Parasitenkd. **69**:149–159.

Watt, R. M., A. Philip, S. M. Wos, and G. J. Sam. 1986. Rapid assay for immunological detection of *Trichomonas vaginalis*. J. Clin. Microbiol. **24**:551–555.

Index to Volume I

Note: Page numbers of chapter titles appear in **boldface** type.

H

J

K

L

Index to Volume II

Note: Virus *families* (suffix *-viridae*) refer to English plural vernacular forms, e.g., *Poxviridae, see also* Poxviruses.

Most virus *genera* (suffix *-virus*) are listed in plural vernacular form, e.g., *Cytomegalovirus* is listed as Cytomegaloviruses.

Virus *species* are listed in singular vernacular form, e.g., Orf virus.

Most *infections* are also listed, e.g., Orf.

Page numbers of chapter titles appear in **boldface** type.

G

N

Q

R

S